DICTIONARY OF CHEMISTRY AND CHEMICAL TECHNOLOGY

German–English

DICTIONARY OF CHEMISTRY AND CHEMICAL TECHNOLOGY

German
English

Editor in Chief

HELMUT GROSS

Department of Applied Linguistics, Technische Universität Dresden,

German Democratic Republic

Compiled by

William Athenstaedt

Wolfgang Borsdorf

Helmut Gross

Helmut Hildebrand

Joachim Knepper

Fridrun Pfeifer

ELSEVIER

Amsterdam – Oxford – New York – Tokyo

1984

PUBLISHED IN COEDITION WITH VEB VERLAG TECHNIK, BERLIN 1984

This book is exclusively distributed in all non-socialist countries with the exception of the Federal Republic of Germany, West-Berlin, Austria and Switzerland by
Elsevier Science Publishers B.V.
Molenwerf 1
P.O. Box 211, 1000 AE Amsterdam, The Netherlands

Distributors for the United States and Canada
Elsevier Science Publishing Company, Inc.
52 Vanderbilt Avenue
New York, NY 10017

Library of Congress Cataloging in Publication Data
Main entry under title:

Dictionary of chemistry and chemical technology.

Includes index.
1. Chemistry–Dictionaries–German. 2. Chemistry, Technical–Dictionaries–German. 3. German language–Dictionaries–English. I. Gross, Helmut, Dipl.-Sprachlehrer. II. Athenstaedt, William.
QD5.D54 1984 540'.3'31 84-1561

ISBN 0-444-99617-6 (U.S.)

Registered trade marks, designs and patents are not explicitly marked in this dictionary.

Printed in the German Democratic Republic

PREFACE

This dictionary has been compiled by a team of scientific co-workers of the Department of Applied Linguistics at the Technical University of Dresden. It comprises the special terms of an important branch of the natural sciences. It is intended for all those working in the field of chemistry and chemical technology as an aid in their study of the literature pertinent to these disciplines.

About 56,000 terms have been gathered from all branches of chemistry and chemical technology as well as from related fields of science.

The dictionary is based on the comprehensive material of the "Dictionary of Chemistry and Chemical Technology", English–German, prepared in close cooperation with numerous scientists of the Technical University of Dresden under the direction of the same Editor. This material has been revised from both scientific and lexicographic standpoints. Many important and new words have been added and obsolete words eliminated.

As in the English–German volume, the authors have provided additional information for many of the entries by indicating the scientific branch concerned and adding short definitions or examples.

The help of our chief scientific adviser, Dozent Dr. *H. J. Bär* (Department of Chemistry of the Technical University of Dresden) and that of several English guest lecturers in our Department, is gratefully acknowledged.

All suggestions for improving this dictionary are welcome, and these should be directed to the publisher.

Helmut Gross

DIRECTIONS FOR USE · HINWEISE FÜR DIE BENUTZUNG

1. Examples of the alphabetical arrangement
Beispiele für die alphabetische Ordnung

oberflächenhärten
  durch Diffusion o.
Oberflächenhärten
Oechsle-Waage
Ofen
  O. mit Sohlebeheizung
  außenbeheizter O.
  Kasseler O.
Öfen / gekoppelte
Ofenalterung
OH-Radikal
Oktadekan

Öl
Olbrücke
ölen
Ölindustrie
Öl-in-Wasser-Emulsion
Olivin
ölreaktiv
Orbital
  sp-hybridisiertes O.
  molekulares O.
π-Orbital
Orbitaltheorie

Dimethylarsinsäure
bis-Dimethylarsyl
Dimethyläther
1,1-Dimethyläthylen
asymm.-Dimethyläthylen
symm.-Dimethyläthylen
Dimethyläthylkarbinol
Dinitrokresol
4,6-Dinitro-*o*-kresol
Diphenylphenylen
N,N'-Diphenyl-*p*-phenylendi-
  amin

Italicalized symbols denoting position or configuration like *asymm.-, symm.-, prim.-, sek.-, tert.-, cis-, trans-, m-, o-, p-, N-, D-, L-,* α-, β-, γ- are disregarded in alphabetization.

Kursiv gesetzte Stellungs- und Konfigurationssymbole wie *asymm.-, symm.-, prim.-, sek.-, tert.-, cis-, trans-; m-, o-, p-, N-,* α-, β-, γ-, *D-, L-* bleiben bei der alphabetischen Einordnung unberücksichtigt.

2. Signs used in the dictionary
Bedeutung der Zeichen

/ geeicht / auf Auslauf = auf Auslauf geeicht

( ) aktives (wirksames) Chlor = aktives Chlor *oder* wirksames Chlor
fume chamber (cupboard, closet) = fume chamber *or* fume cupboard *or* fume closet

[ ] Polymer[es] = Polymer *oder* Polymeres
Drehungswinkel [/ optischer] = Drehungswinkel *oder* optischer Drehungswinkel
Dickstoff-Abwärts[bleich]turm = Dickstoff-Abwärtsturm *oder* Dickstoff-Abwärtsbleichturm
vibration[al] band = vibration band *or* vibrational band
safety[-relief] valve = safety valve *or* safety-relief valve
high-density [bleaching] stage = high-density stage *or* high-density bleaching stage

( ) These brackets contain explanations
Diese Klammern enthalten Erklärungen

ABBREVIATIONS · ABKÜRZUNGEN

abbr — abbreviation for / Abkürzung für
agric — agricultural and silvicultural chemistry / land- und forstwirtschaftliche Chemie
Am — American English / amerikanisches Englisch
bioch — biochemistry / Biochemie
biol — biology / Biologie
bot — botany / Botanik
build — chemistry and technology of building materials / Chemie und Technologie der Baustoffe
ceram — ceramics / Keramik
chem — chemistry / Chemie
coal — coal chemistry and technology / Chemie und Technologie der Kohle
coat — chemistry and technology of organic coating materials / Chemie und Technologie der Anstrichstoffe
coll — colloid chemistry / Kolloidchemie
cosmet — chemistry and technology of cosmetics and perfumes / Chemie und Technologie der Kosmetika und Riechstoffe
cryst — crystallography and crystal chemistry / Kristallografie und Kristallchemie
distil — distillation apparatus and practice / Destillationstechnik
dye — chemistry and technology of organic dyes / Chemie und Technologie der organischen Farbstoffe
e. g. — exempli gratia, for example / zum Beispiel
el chem — electrochemistry / Elektrochemie
esp — especially / besonders
f — feminine noun / Femininum
ferm — chemistry and technology of fermentation / Gärungschemie und -technologie
fert — fertilizer technology / Technologie der Düngemittelherstellung
filtr — filtration equipment and practice / Filtervorrichtungen und Filtertechnik
food — food chemistry and technology / Lebensmittelchemie und -technologie
geoch — geochemistry / Geochemie
geol — geology / Geologie
glass — glass technology / Glasherstellung
lab — laboratory technique and apparatus / Laboratoriumstechnik und -geräte
m — masculine noun / Maskulinum
med — medicine / Medizin
met — metallurgy and science of metals / Metallurgie und Metallkunde
min — mineralogy and petrography / Mineralogie und Petrografie
min tech — mineral technology / Aufbereitungstechnik
mine — mining / Bergbau
n — neuter noun / Neutrum
nomencl — chemical nomenclature / chemische Nomenklatur
nucl — nuclear science and technology / Kernphysik und Kerntechnik
org chem — organic chemistry / organische Chemie

pap	pulp and paper chemistry and technology / Zellstoff- und Papierchemie und -technologie
petrol	petroleum chemistry and technology / Erdölchemie und -technologie
pharm	pharmaceutical chemistry / pharmazeutische Chemie
phot	photographic chemistry / fotografische Chemie
phys chem	physical chemistry / physikalische Chemie
pl	plural / Plural
plast	plastics chemistry and technology / Chemie und Technologie der Plaste
rubber	rubber technology / Technologie der Gummiherstellung
s.	see / siehe
s. a.	see also / siehe auch
soil	soil chemistry and soil science / Bodenchemie und Bodenkunde
specif	specifically / im engeren Sinne
sugar	sugar industry / Zuckerindustrie
tann	chemistry and technology of leather manufacture / Gerbereichemie und -technologie
tech	technology / Technik und Technologie
text	textile chemistry / Textilchemie
tox	toxicology / Toxikologie

A

a *s.* absolut
A *s.* Anode
Ä. *s.* Äther
Aasappretur *f (tann)* flesh finish
Aasschmiere *f (tann)* dubbin[g], stuffing mixture
Aasseite *f (tann)* flesh layer (side)
Abaka *m* 1. abaca, Musa textilis Née; 2. abaca [fibre], Manil[l]a fibre (hemp) *(leaf fibres from 1.)*
Abakafaser *f s.* Abaka 2.
Abart *f* variety, modification, species
Abbau *m* 1. degradation, decomposition, disintegration, breaking-down, breakdown, fragmentation; *(bioch)* katabolism, catabolism; 2. working, exploitation *(of a mine)*; mining, extracting, winning *(as of coal)*
A. mit Peptisiermitteln *(rubber)* peptization
A. nach Weerman Weerman degradation
Barbier-Wielandscher A. Barbier-Wieland degradation
Braunscher A. Braun degradation *(of tertiary amines)*
Curtiusscher A. Curtius method (reaction, rearrangement) *(the decomposition of acid azides to give isocyanates and nitrogen)*
fotochemischer A. photochemical destruction
Hofmannscher A. Hofmann degradation *(of quaternary ammonium hydroxides)*
Lossenscher A. Lossen rearrangement *(of aromatic hydroxamic acids or their derivatives into isocyanates)*
mechanischer A. *(rubber)* mechanical (mill) breakdown
oxydativer A. oxidative degradation (breakdown)
Ruffscher A. Ruff degradation *(of sugars)*
thermischer A. thermal degradation
von-Braunscher A. *s.* Braunscher A.
abbaubar degradable, decomposable
biologisch a. biodegradable
Abbaubarkeit *f* degradability, decomposability
biologische A. biodegradability
abbauen 1. to degrade, to decompose, to disintegrate, to break down, to fragment; 2. to work, to exploit *(a mine)*; to mine, to extract, to win *(e. g. coal)*
mit Peptisiermitteln a. *(rubber)* to peptize
sich a. to degrade
zu Dextrin a. to dextrinate, to dextrinize
Abbaugeschwindigkeit *f* decomposition rate
Abbaugrad *m* degree of degradation
Abbaukurve *f (tox)* decline (disappearence, loss, degradation, decay) curve, [residue-]-persistence curve *(of pesticides)*
Abbaumittel *n* decomposing (disintegrating) agent, disintegrant; *(rubber)* plasticizing (peptizing) agent, plasticizer, peptizer
Abbauprodukt *n* degradation (decomposition, breakdown) product
Abbaureaktion *f* degradative (decomposition) reaction
Abbaustufe *f* stage in degradation (decomposition)
Abbauwirkung *f (rubber)* peptizing effect
abbeizen to pickle, to scour *(seeds, metal)*; to remove *(paint, varnish)*
Abbeizen *n* pickling, scouring *(of seeds, metal)*; removal *(of paint, varnish)*
Abbeizmittel *n* 1. pickling (scouring) agent; 2. paint remover; varnish remover
Abbe-Refraktometer *n* Abbe refractometer
Abbe-Spektrometer *n* Abbe spectrometer
Abbindebeschleuniger *m (build)* setting (cementing) accelerator
Abbindedauer *f (build)* setting (set-up) time
abbinden *(build)* to set, to harden; *(tann)* to settle down
Abbinden *n (build)* set[ting], hardening
Abbinderegler *m (build)* quick-setting (rapid-cementing) agent, quick hardener
Abbindeverhalten *n (build)* setting behaviour
Abbindeverzögerer *m (build)* [setting] retarder, retarding admix[ture]
Abbindeverzögerung *f (build)* retarding of the set
Abbindezeit *f s.* Abbindedauer
Abblasedruck *m* relieving (blow-off) pressure
abblasen 1. to blow off *(e. g. steam)*; 2. *(pap)* to blow [off] *(a digester)*; 3. *(ceram)* to dust *(a ware before firing)*
Abblasen *n* 1. blow-off, blowing-off *(as of steam)*; 2. *(pap)* blow[ing] *(of a digester)*, digester blow; *(filtr)* blow discharge; 3. *(ceram)* dusting *(of a ware before firing)*
Abblasleitung *f* blow pipe, blowpipe
Abblasprodukt *n (pap)* blow-off
gasförmiges A. blow-pit gas, gaseous blow-off
abblättern to peel [off], to shell [off], to flake [off], to exfoliate
Abbot-Cox-Färbeverfahren *n* Abbot-Cox process
Abbrand *m* calcine *(product of roasting)*; burn-off; *(nucl)* burn-up
Abbrandauslauf *m* calcine-discharge outlet
Abbrandkühler *m* calcine cooler
Abbrandmittel *n (agric)* desiccant
Abbrandsammelbehälter *m* calcine-collecting tank
abbrausen *(min tech)* to spray *(classified material)*
Abbrausen *n* **mit Wasser** water spraying
abbrechen to terminate, to shortstop, to arrest *(e. g. a reaction)*; to terminate, to break *(a chain)*
abbremsen to slow down, to decelerate, to retard; *(nucl)* to moderate
Abbremsung *f* slowing-down, deceleration, retardation; *(nucl)* moderation
abbrennen 1. to burn off; *(text)* to singe; 2. to deflagrate *(as of explosives)*
Abbrennen *n* 1. burning-off; *(text)* singeing; 2. deflagration *(as of explosives)*
explosionsfreies A. deflagration
Abbrennlöffel *m* deflagrating (deflagration) spoon
Abbrennmittel *n (agric)* deflagrating agent
abbröckeln to exfoliate, to peel [off], to crumble away

Abbröckeln *n* exfoliation, peeling
Abbruch *m* termination *(as of a reaction)*; termination, breaking *(of a chain)*
Abbruchreaktion *f* termination reaction; chain-terminating (chain-breaking, chain-stopping) reaction
Abbruchstadium *n* termination stage
abbuffen *(tann)* to buff
Abbuffen *n (tann)* grain buffing
Abdampf *m* exhaust (waste, dead) steam
abdampfen to evaporate; to bleed *(gas chromatography)*
Abdampfen *n* evaporation; bleeding *(gas chromatography)*
Abdampfgefäß *n* evaporating vessel
Abdampfkolben *m* evaporating flask
Abdampfleitung *f* waste-steam line
Abdampfpfanne *f* evaporating pan
Abdampfrückstand *m* dry matter (residue, substance)
Abdampfschale *f* evaporating dish, capsule
Abdampfschalenhalter *m* dish tongue
Abdampftemperatur *f* evaporation temperature
Abdampftiegel *m s.* Abdampfschale
abdarren *(ferm)* to kiln-dry
Abdarrtemperatur *f (ferm)* kiln temperature
abdecken to cover [over] *(e.g. a vessel)*; to coat, to cover *(leather)*; to stop off *(to protect desired areas from chemical action)*
Abdecken *n* **von verschossenen Farbpartien** *(text)* fade covering
Abdeckpappe *f* carpet felt, felt brown
Abdeckplatte *f* cover [plate]
abdestillieren 1. to distil off, to strip [off, out], to remove by distillation; 2. *(of a liquid)* to distil off
abdichten to make tight, to seal, to pack; to ca[u]lk; to waterproof; to gasproof
Abdrift *f (agric)* blow-off, drift *(of pesticides)*
A. des Sprühmittels spray drift
Abdruck *m* impression, mark, stamp; print; *(plast)* replica
abdrucken / sich to set off *(of printing ink)*
Abdrucken *n* offset *(of printing ink)*
Abdrücken *n* **von oben** *(plast)* top ejection
Abel-Apparat *m s.* Abel-Flammpunktprüfer
Abel-Flammpunkt[s]prüfer *m* Abel flash-point apparatus, Abel apparatus (tester)
Abel-Gerät *n s.* Abel-Flammpunktprüfer
Abel-Pensky-Flammpunkt[s]prüfer *m* Abel-Pensky apparatus (tester)
Abel-Test *m* Abel test
Abernathyit *m (min)* abernathyite
Aberration *f* aberration
chromatische A. chromatic aberration
sphärische A. spheric[al] aberration
Abfall *m* 1. waste [material, product], refuse, scrap; rubbish, *(Am)* garbage; *(plast, glass)* cull; 2. fall, decrease, decline, reduction, drop *(as of measuring values)*
aufgearbeiteter A. *(plast)* reground material
flüssiger radioaktiver A. radioactive effluent
heißer A. *(nucl)* hot (high-activity, high-level) waste
hochaktiver A. *s.* heißer A.
radioaktiver A. radioactive (atomic) waste
Abfallauge *f* waste (spent) lye, waste (spent) liquor
Abfallbeseitigung *f* waste disposal
Abfälle *mpl s.* Abfall 1.
abfallen 1. to go to waste, to be left over; 2. to fall, to decrease, to decline, to reduce, to drop *(as of measuring values)*
im Farbton a. to be off-shade
Abfallen *n* fall, decrease, decline, reduction, drop *(as of measuring values)*
Abfallgummi *m* rubber scrap (waste), scrap (waste) rubber
Abfallgut *n* waste [material, product]
Abfallholz *n (pap)* waste wood
Abfallkalk *m (agric)* waste (by-product) lime
Abfallkübel *m* waste jar; waste (disposal) crock *(in laboratories)*
Abfallpapier *n* waste (old) paper
Abfallprodukt *n* waste [product], by-product
Abfallsäure *f* waste (residuary) acid
Abfallsäurebetrieb *m (dye)* waste-acid plant
Abfallschlicker *m (ceram)* casting scrap
Abfallstoffe *mpl s.* Abfall 1.
Abfallsubstanz *f* waste [material, product]
Abfallvernichtung *f* waste disposal
Abfallverwertung *f* waste utilization
abfangen to catch, to trap; *(cryst)* to capture *(e.g. a trace element)*
Abfangen *n* **eines Spurenelements** *(cryst)* capture of a trace element
Abfangreaktion *f* trapping reaction
abfärben to lose colour, to bleed, to stain, to mark off; *(tann)* to colour
abfehmen *(glass)* to skim
Abfehmer *m (glass)* skimmer
abfeimen *s.* abfehmen
abfiltern *s.* abfiltrieren
abfiltrieren to filter off
in Filterpressen a. to filter-press
abflammen *(text)* to singe
abflecken *(text)* to stain, to mark off
abfließen to flow [off], to run [off], to drain [off]
a. lassen to drain [off], to run [off], to discharge
vorsichtig a. lassen to decant
Abfließen *n* flowing-off, draining-off, drainage, discharging, discharge, effluence, efflux, outflow
Abfluß *m* 1. s. Abfließen; 2. discharge [point], outlet, outflow, drain; s. Abflußrohr; 3. *(substance)* runoff, effluent, effluence
Abflußbecken *n* [bench] sink
Abflußhahn *m* discharge cock
Abflußkanal *m* effluent channel
Abflußleitung *f* discharge (drain, drainage, runoff, outlet) line; waste line
Abflußmenge *f* outflow, runoff, amount of discharge

Abflußöffnung *f* discharge opening (outlet, door, aperture, port)
Abflußregler *m* effluent (outlet) weir
Abflußrohr *n* discharge (drainage, runoff, effluent, delivery, outlet) tube, discharge (drainage, runoff, effluent, delivery, outlet) pipe; waste tube (pipe)
Abflußstutzen *m s.* Abflußrohr
Abflußventil *n* discharge valve
abführen 1. to lead off (away), to carry off, to drain off (away), to draw off, to withdraw *(liquids or gases)*; to exhaust *(gases)*; to dissipate, to eliminate *(heat)*; 2. *(pharm)* to purge
Abführmittel *n* cathartic, purgative
Abfüllabteilung *f* filling plant (room), *(esp)* bottling plant (department, room), bottle house, bottlery
Abfüllautomat *m* filling machine, *(esp)* bottling (bottle-filling) machine, bottle filler, bottler
A. für Flüssigkeiten liquid filler (filling machine)
abfüllen 1. to barrel, to rack; to bottle [in, up]; 2. to decant
auf Fässer a. to barrel, to rack
auf Flaschen a. to bottle [in, up]
Abfüllerei *f s.* Abfüllabteilung
Abfüllhahn *m* racking cock
Abfüllmaschine *f s.* Abfüllautomat
abfüllreif bottle-ripe
Abfüllschlauch *m* racking hose
Abgabe *f* delivery, liberation, release *(as of gases)*; release, donation, transfer *(of electrons, ions)*
Abgabeschieber *m* outlet valve
Abgang *m* 1. loss; 2. waste [material, product]; 3. *s.* Abgänge
Abgänge *mpl (min tech)* [waste] tailings, refuse
Abgangsseite *f* discharge end
Abgas *n* waste (exit) gas, off-gas; excess (tail) gas; flue (stack) gas; exhaust gas; combustion gas; *(pap)* digester relief gas
abgasbeständig *(text)* fast to burnt gas fumes, fast to gas fading
Abgasbeständigkeit *f (text)* gas-fume fastness, fastness to burnt gas fumes, fastness to gas fading
abgasecht *s.* abgasbeständig
Abgasechtheit *f s.* Abgasbeständigkeit
Abgasempfindlichkeit *f s.* Ausbleichen in Abgasatmosphäre
abgasen *(pap)* to relieve *(a digester)*
Abgasen *n (pap)* relief [of gases]
A. am Kocherdeckel *(pap)* digester top relief
Abgaskanal *m* waste-gas flue
Abgasleitung *f* waste-gas line; *(pap)* relief line
Abgassammelkanal *m* waste-gas flue
Abgasseite *f* waste-gas side
Abgasturm *m (pap)* tail-gas tower
Abgasventil *n (pap)* relief (blow-through) valve
Abgasvorwärmer *m* economizer
abgautschen *(pap)* to couch
abgeben to deliver, to liberate, to release, to give off, to evolve *(e. g. gas, heat)*; to yield; to release, to donate, to transfer *(e. g. electrons, ions)*
Abgeber *m* material (substance) to be extracted
abgedeckt / mit Metallgewebe (Siebgewebe) *(filtr)* screen-covered
abgehen / durch den Schornstein to be lost at the stack
abgelagert mature *(wine)*
Abgeschäumtes *n* skimmings
Abgeschöpftes *n* skimmings
abgestanden flat, stale, insipid *(beverages)*
abgießen to pour off, to drain [off], to run [off]
vorsichtig a. to decant
Abgleichmethode *f* balance method
abgraten *(plast)* to deburr, to deflash
Abgraten *n (plast)* deburring, deflashing, flash removal
Abgratmaschine *f (plast)* deflashing machine
Abhängigkeit *f* dependence
Abhängigkeitsverbundwirkung *f (tox)* dependent joint action
Abhäsion *f* abhesion *(loss of adhesion)*
abhäsiv abhesive
abhaspeln to reel off, to unreel
Abhebeformmaschine *f* pattern-draw[ing] machine
abheben *(ceram)* to strip *(to remove the dried ware from the plaster moulds)*
sich a. *(ceram)* to peel [off], to shell [off] *(of a glaze)*
Abhebeprüfung *f* peeling test *(for testing the strength of an adhesive-bonded joint)*
abhebern to siphon [off]
Abhitze *f* waste heat
Abhitze[dampf]kessel *m* waste-heat boiler
Abhitze[koks]ofen *m* waste-heat oven
Abhitzerückgewinnung *f* waste-heat recovery
Abietat *n* abietate *(salt or ester of abietic acid)*
Abietin *n s.* Koniferin
Abietinsäure *f* abietic acid *(a resin acid)*
abimpfen *(bioch)* to subculture
Abimpfung *f (bioch)* subculture
a-Bindung *f* a-bond, axial bond
abklappbar retractable
abklären to clarify, to refine
sich a. to clarify, to refine
Abklärflasche *f (lab)* aspirator
Abklärung *f* clarification, refining
abklingen to die down (away), to quieten down *(of a reaction)*; to fade away, to subside *(of vibrations)*
Abklingen *n* **des latenten Bildes** *(phot)* latent-image regression
abkochen to boil [down]; to boil off (out); to decoct; *(text)* to bowk, to buck, to kier-boil
Abkochen *n* boiling; boiling-off, boil-off, boiling-out, boil-out; decoction; *(text)* bowking, bucking, kier-boiling, kiering
A. in breitem Zustand *(text)* open-width boil-out
Abkochung *f* decoction
Abkömmling *m* derivative
abkratzen to scrape off, to scratch off
abkreiden *(coat)* to chalk, to become chalky
abkühlen to cool [down]; to chill; to quench
a. lassen to allow to cool
rasch a. to chill; to quench

sich a. to cool [down]; to chill
Abkühlmittel *n* refrigerant, refrigerating medium
Abkühlung *f* cooling; chilling; quenching
Abkühlungsgesetz *n* / **Newtonsches** Newton law for cooling (heat loss)
Abkühlungskurve *f* cooling curve
Abkühlungsvorrichtung *f (plast)* shrink (cooling) fixture, shrinkage (cooling) jig *(for mouldings)*
Abkühlverfahren *n* **mit Kühltrommeln** dry [drum-] cooling method, chill-roll method *(margarine making)*
Abkühlverlust *m* cooling loss
Abkühlzone *f* cooling zone (compartment, section)
ablagern to sediment, to deposit *(of liquids)*; *(food)* to age, to mature; to season *(timber)*
im Tank a. to tank *(e. g. pigments)*
in Borke a. *(tann)* to age
sich a. to sediment, to settle [out], to deposit, to subside
sich wieder a. to redeposit
Ablagerung *f* 1. *(substance)* sediment, settling[s], bottom sediment (settlings), bottoms, foots, lees, deposit, scale, *(esp of biological matter)* fouling; 2. *(process)* sedimentation, deposition; *(food)* ageing, maturing; seasoning *(of timber)*
Ablagerungsgeschwindigkeit *f* rate of sedimentation (deposition)
Ablaß *m* 1. discharge, outlet, drain; vent, relief *(for lessening excessive pressure)*; 2. *s.* Ablaßöffnung
ablassen to let off (out), to discharge, to exhaust, *(liquid also)* to run [off], to drain [off], to sewer, to tap, *(gas also)* to blow off; *(vessel)* to discharge, to drain, to exhaust, to tap, to empty
Ablaßhahn *m* discharge (delivery, drain, runoff) cock, dispensing spigot
Ablaßkanal *m* sewer, drainage channel
Ablaßleitung *f* discharge (delivery, drain, drainage, runoff, outlet) line; waste line
Ablaßöffnung *f* discharge opening (outlet, door, aperture, port); vent, relief *(for lessening excessive pressure)*
Ablaßrohr *n* discharge (drainage, runoff, effluent, delivery, outlet) tube, discharge (drainage, runoff, effluent, delivery, outlet) pipe; waste tube (pipe)
Ablaßventil *n* discharge (drain, outlet) valve
Ablaßvorrichtung *f* discharging apparatus (device), discharger
Ablauf *m* 1. discharge [point], outlet, outflow, drain; 2. *(substance)* runoff, effluent, effluence; *(sugar)* runoff syrup; 3. course *(as of a reaction or process)* ✦ **auf A. geeicht (geteilt)** calibrated (graduated) to deliver, calibrated for delivery
Ablauf I *m (sugar)* high green syrup
Ablaufbrett *n* drain[ing] board, drying (apparatus) rack
aufhängbares A. wall-mounting draining board
ablaufen 1. *(of liquids)* to drain [off], to run [off], to flow [off]; *(of surface coatings)* to curtain, to flow, to sag; 2. to proceed, to take its course
a. lassen to drain [off], to run [off], to siphon
Ablaufende *n* discharge end
Ablaufkanal *m* sewer, drainage channel
Ablaufleitung *f* discharge (delivery, drain, drainage, runoff, outlet) line; waste line
Ablaufrinne *f* effluent channel
Ablaufrohr *n* discharge (drainage, runoff, effluent, delivery, outlet) tube, discharge (drainage, runoff, effluent, delivery, outlet) pipe; waste tube (pipe); *(distil)* downpipe, downspout, downtake, downcomer, rundown pipe
Ablaufschurre *f* discharge chute
Ablaufsirup *m (sugar)* centrifugal syrup
Ablaufstutzen *m s.* Ablaufrohr
Ablaufwehr *n* outlet (effluent, overflow) weir, outlet (effluent, overflow) lip; *(distil)* exit (overflow) weir
Ablauge *f* waste (spent) lye, waste (spent) liquor; *(pap)* sulphite spent (waste) liquor, red liquor
Ablaugengewinnung *f (pap)* waste-liquor (spent-liquor) recovery
Ablaugenregeneration *f s.* Ablaugengewinnung
Ablaugetrockensubstanz *f (pap)* waste-liquor (spent-liquor) solids
abläutern *(ferm)* to lauter; *(plast)* to refine
ablegen to set off *(of printing ink)*
ableiten 1. to lead away (off), to carry off, to drain away (off), to draw off, to withdraw *(liquids or gases)*; to dissipate, to eliminate *(heat)*; 2. to derive *(one chemical compound from another)*; 3. *s.* ablenken
Ableitung *f* 1. leading-away, leading-off, carrying-off, drainage, withdrawal *(of liquids or gases)*; dissipation, elimination *(of heat)*; 2. derivation *(of one chemical compound from another)*; 3. *s.* Ablenkung
Ableitungselektrode *f* reference junction *(in pH measurement)*
Ableitungsrohr *n s.* Ablaufrohr
Ablenkblech *n* deflector [plate]
ablenken to deflect, to turn off
Ablenkkammer *f* dust trap
Ablenkkeil *m (petrol)* whipstock
Ablenkplatte *f* deflector [plate]
Ablenkrinne *f (glass)* deflector
Ablenkrolle *f* snub pulley
Ablenktrommel *f* snub pulley
Ablenkung *f* deflection
Ablenkwand *f* deflector [plate]
Ableseblatt *n (lab)* burette [meniscus] reader, anti-parallax card
Ablesefehler *m* reading error
parallaktischer A. parallax error
Ableselupe *f* reading lens, burette meniscus magnifier
Ablesemikroskop *n* reading microscope
ablesen to read off, to take a reading
Ablesevorrichtung *f* reading device
Ablesung *f* reading
ablöschen to quench *(e. g. coke)*; to slake *(esp lime)*
in (mit) Öl a. to oil-quench
in (mit) Wasser a. to water-quench

Ablöschmittel *n* quenching medium (agent)
Ablöschung *f* quenching *(as of coke)*; slaking *(esp of lime)*
Ablösearbeit *f (phys chem)* [electronic] work function, work of separation
Ablösefestigkeit *f (plast)* peel strength
ablösen to strip, to detach, to peel *(e.g. a foil)*; to release *(as from a mould)*; to dissolve *(e.g. a soluble substance from a vessel wall)*
sich a. to peel [off], to scale [off], to flake [off], to shell [off]; *(phys chem)* to separate *(as of electrons)*; *(cryst)* to part, to cleave; *(phot)* to frill *(of an emulsion)*
Ablösungsebene *f (cryst)* cleavage (parting) plane
Ablösungseffekt *m*, **Ablösungswirkung** *f* release action (effect) *(as of silicon oil in a mould)*
Abluft *f* outlet (exit, leaving) air
Ablufttemperatur *f* outlet air temperature
abmessen to measure [off]; to meter
Abmessung *f* measurement; size, dimension
abnabeln *(glass)* to shear *(a drop on the feeder)*
Abnabeln *n (glass)* shear[ing] *(of a drop on the feeder)*
Abnahme *f* 1. *(apparatus)* take-off, *(plast also)* haul-off; 2. *(action)* take-off, withdrawal, discharge *(of liquids or gases)*; pick-up *(of the paper web from the wire)*; 3. *(process)* decrease, loss, drop, fall, reduction *(as of measuring values)*
Abnahmefilz *m (pap)* pick-up (lick-up) felt
oberer A. top felt, overfelt
unterer A. bottom (lower) felt
Abnahmegeschwindigkeit *f* take-off rate; *(plast)* haul-off rate; *(pap)* pick-up rate
Abnahmeschaber *m* [roll] doctor
Abnahmetisch *m (ceram)* delivery table
Abnahmevorrichtung *f* take-off equipment
Abnahmewalze *f* take-off roll, discharge roll; *(pap)* pick-up roll
abnehmen 1. to take off, *(plast also)* to haul off; to withdraw, to discharge *(liquids or gases)*; to pick up *(the paper web from the wire)*; 2. to decrease, to drop, to fall, to reduce *(as of measuring values)*
die Bahn a. *(pap)* to take (pick) up the web
mittels Schaber a. to doctor off
abnorm anomalous, abnormal, irregular
abnutschen to filter [off] by suction, to filter under (with) suction
Abnutschen *n* suction filtration
abnutzen / sich to wear [out]; *(by friction)* to abrade
Abnutzung *f* wear
A. durch Abrieb abrasion, abrasive, wear, attrition
abnutzungsbeständig wear-resistant; *(to friction)* abrasion-resistant
Abnutzungsbeständigkeit *f* wear resistance; *(to friction)* abrasion resistance
Abnutzungsfaktor *m* wear factor
Abnutzungsprüfer *m*, **Abnutzungsprüfmaschine** *f* abrasion-testing machine, abrasion tester (machine), abrader
Abnutzungsprüfung *f* abrasion test
Abnutzungswiderstand *m* wear resistance; *(to friction)* abrasion resistance
abölen *(tann)* to oil, to stuff
Abperleffekt *m* water repellency (repellent effect)
abpipettieren to withdraw by pipette, to pipette off
abplatzen to exfoliate, to flake [off], to scale [off], to chip [off], to peel [off], to spall
abpressen to press [off], to squeeze [off], to force out, to expel, to express
in Filterpressen a. to filter-press
Abpreßwalze *f* squeezing (squeeze) roll; *(pap)* lumpbreaker roll
Abprodukt *n* waste [product], by-product
abpuffern to buffer [off]
abpumpen to pump off, to evacuate
abquetschen to squeeze [off]
Abquetschfläche *f (plast)* land [area]; *(plast)* cut-off *(of a press mould)*
Abquetschform *f (plast)* flash mould
Abquetschrand *m (plast)* cut-off *(of a press mould)*
Abquetsch- und Füllform *f* (**Füllpreßwerkzeug** *n*) *(plast)* semipositive mould
Abquetschwalze *f* squeezing (squeeze) roll
Abquetschwerkzeug *n (plast)* flash mould
abrahmen to cream [off], to skim *(milk)*
Abrasion *f* abrasion
Abraumsalz *n* abraum salt
Abraumsalze *npl* / **Staßfurter** Stassfurt [salt] deposits
abreiben to abrade, to scour
abreibend abrasive
abreißen to tear off, to pull off; to break *(as of paper webs)*
Abreißen *n* **der Papierbahn** *(pap)* break in the web, sheet break
Abreißmethode *f (phys chem)* detachment method
Abrieb *m* 1. *(process)* abrasion, abrasive wear, attrition; 2. *(substance)* rubbings, dust
abriebbeständig *s.* abriebfest
Abriebbeständigkeit *f s.* Abriebfestigkeit
abriebfest abrasion-resistant
Abriebfestigkeit *f* abrasion resistance; scuff resistance *(in sliding or rolling friction)*
Abriebprüfmaschine *f* abrasion-testing machine, abrasion tester (machine), abrader
Abriebprüfung *f* abrasion test
A. in der Trommel nach Cochrane Cochrane [abrasion] test
Abriebsabnutzung *f s.* Abnutzung durch Abrieb
Abriebverlust *m* abrasion loss
Abriebversuch *m* abrasion test
Abriebwiderstand *m* abrasion resistance
Abriebzahl *f* abrasion index
abriegeln to block
Abrin *n* abrine, *N*-methyltryptophan
Abriß *m* **der Papierbahn** *(pap)* break in the web, sheet break
abrollen *(pap)* to unreel, to reel off, to unwind

sich a. *(ceram)* to crawl *(of glazing material)*
Abrollgestell *n (pap)* reel-off (reeling-off, reel, back) stand, unwind[ing] stand
abrösten to roast, to burn *(e.g. ores)*
abs. *s.* absolut
ABS *s.* Akrylnitril-Butadien-Styrol
absacken 1. to bag [up] *(bulk material)*; 2. to sag *(of coatings)*
absättigen to saturate *(valencies)*
Absättigung *f* saturation *(of valencies)*
Absatzgestein *n* sedimentary rock
absatzweise batch[wise] *(operation)*
a. arbeitend batch *(apparatus)*
absäuern *(text)* to acidify
Absäuerung *f (text)* acidification; *(pap)* acid souring (stage) *(a bleaching stage)*
absaugen to suck off, to siphon [off], to withdraw, to exhaust; *(filtr)* to filter [off] by suction
Absaugen *n* suction, siphoning, withdrawal, exhaustion
Absauger *m* 1. aspirator, exhauster; 2. *s.* Absaugvorrichtung
Absaugkasten *m (plast)* plenum chamber
Absaugpumpe *f* suction (withdrawal) pump
Absaugvorrichtung *f* suction (exhausting) device; *(text)* suction extractor
Abschabemesser *n* scraper (doctor) knife, scraper (doctor) blade, scraper
abschaben to scrape off, to doctor off
abschälen to peel [off] *(e.g. a foil)*; to skim *(as in the knife-discharge centrifuge)*
sich a. *(ceram)* to peel [off], to shell [off], to exfoliate, to flake [off] *(of a glaze)*
Abschaum *m* scum
abschäumen to skim [off], to scum
Abschäumer *m* skimmer
Abschäumlöffel *m* skimmer, skimming ladle
abscheidbar separable, precipitable
Abscheidbarkeit *f* separability, precipitability
Abscheidegrad *m* separation (collection) efficiency
Abscheidekammer *f* separating (settling, settlement, fall-out) chamber, drop-out box
abscheiden 1. *(of human agent or apparatus)* to separate, to eliminate, to segregate, to remove; *(when bottoms are being formed)* to sediment[ate], to precipitate; *(min tech)* to discard *(tailings)*; 2. *(of a liquid)* to deposit, to sediment; to evolve, to set free, to liberate *(gases)*
elektrochemisch (elektrolytisch) a. to electrodeposit
sich a. to separate [out], *(of precipitates also)* to settle [down, out], to sediment, to set, to precipitate, to deposit, to subside
sich baumartig a. *(cryst)* to tree
sich elektrochemisch (elektrolytisch) a. to plate out
Abscheiden *n* 1. separation, elimination, segregation, removal; *(when bottoms are being formed)* sedimentation, precipitation; *(min tech)* discard *(of tailings)*; 2. deposition, sedimentation *(of bottoms by a liquid)*; evolution, liberation *(of gases)*
akustisches A. [ultra]sonic precipitation, [ultra]sonic agglomeration
elektrisches A. electrostatic (electrical) precipitation
elektrochemisches (elektrolytisches) A. electrodeposition
elektrostatisches A. *s.* elektrisches A.
Abscheider *m* separator, settler, precipitator, catcher, trap; *(min tech)* collector; *(nucl)* trap
A. für mitgerissene Flüssigkeit entrainment separator
Abscheideraum *m s.* Abscheidekammer
Abscheidezyklon *m* centrifugal [cyclone] collector, centrifugal [cyclone] separator, cyclone
Abscheidung *f* 1. *s.* Abscheiden; 2. *(substance)* deposit, sediment, settling[s], bottom sediments (settlings), B.S., bottoms ✦ **zur A. gelangen** *s.* abscheiden / sich
A. von Flüssigkeit weeping *(as of a gel)*
elektrochemisch (elektrolytisch) aufgebrachte A. electrodeposit
galvanisch aufgebrachte A. electrodeposit
Abscheidungskonstante *f* distribution (partition, segregation) coefficient *(in zone melting)*
Abscheidungsmittel *n* precipitating agent, precipitant, precipitator
Abscheidungspotential *n* deposition potential
abscheren to shear [off]
Abscheren *n* shear[ing]
Abscherwirkung *f* shearing effect (action)
Abschirmbeton *m* shielding (radiation) concrete, concrete for [atomic] radiation shielding
abschirmen to shield, to screen [off], to blanket
Abschirmung *f* / **diamagnetische** diamagnetic screening
Abschirmungskonstante *f (nucl)* screening constant
Abschirmungszahl *f (nucl)* screening constant
Abschlamm *m* sludge
abschlämmen to elutriate
Abschlämmen *n* elutriation
abschleifen to grind [off], to abrade; to [sand-]paper *(to treat with abrasive paper)*
auf der Fleischseite a. *(tann)* to fluff
den Narben a. *(tann)* to buff
abschleudern to centrifuge [off], to spin; *(text)* to hydroextract *(water)*
Abschleudern *n* centrifugation, spinning; *(text)* hydroextraction *(of water)*
abschließen to terminate *(a process)*; to seal *(e.g. a vessel)*
glatt a. to be flush
wasserdicht a. to waterproof
Abschluß *m* termination *(of a process)*; seal *(as of a vessel)*; exclusion *(as of air during a reaction)*
trockener A. dry seal
Abschlußdüse *f (plast)* shut-off nozzle
Abschlußwasserschüssel *f* water seal
abschmelzen *(met)* to liquate, to melt down, to smelt; *(glass)* to seal [off]

abschmieren 1. to lubricate; 2. to set off *(of printing ink)*
Abschminkpapier *n* cleansing tissue [paper], facial tissue [paper]
abschmutzen *(dye, coat, tann)* to bleed [off]; to set off *(of printing ink)*
Abschmutzen *n (dye, coat, tann)* bleeding; offset *(of printing ink)*
Abschmutzmakulatur *f s.* Abschmutzpapier
Abschmutzpapier *n* set-off (tympan) paper
abschnappen to be starved of liquid *(of a pump)*
abschneiden to cut off (away), to shear [off]
Abschneider *m s.* Abschneidetisch
Abschneidetisch *m (ceram)* cutting-off table, cutter
Abschnitt *m* 1. cutting, chip *(of a substance)*; 2. section, part *(of equipment)*; 3. stage *(of a process)*; 4. period *(of time)*; 5. *(glass)* cutoff *(of the gather)*
A. abnehmender Trocknungsgeschwindigkeit falling-rate drying period
A. konstanter Trocknungsgeschwindigkeit constant-rate drying period
Abschnitte *mpl (pap)* shavings
abschöpfen to skim [off]; *(met)* to rabble
Rahm a. *(food)* to cream
Abschrägung *f* bevelling, sloping, tapering; *(on a device)* bevel, slope, taper
Abschreckalterung *f (met)* quench ag[e]ing
Abschreckbad *n* quenching bath
abschrecken 1. to quench, to cool rapidly, *(met also)* to chill; 2. to deter *(animals)*
direkt a. to direct-quench
in Öl a. to oil-quench
in Wasser a. to water-quench
abschreckend repellent *(odour, smell)*
abschreckhärten to quench-harden
Abschreckhärtung *f* quench hardening
Abschreckmittel *n* 1. quenching medium (agent); 2. repellent, deterrent *(against animals)*
A. gegen Blattläuse greenfly repellent
A. gegen Nagetiere rodent repellent
Abschrecköl *n* quench[ing] oil
Abschreckstoff *m* repellent, deterrent *(against animals)*
Abschreckung *f* 1. quenching, rapid cooling, *(met also)* chill[ing]; 2. repelling, deterring *(of animals)*
A. in Öl oil quenching
A. in Wasser water quenching
direkte A. direct quenching
Abschreckungsmittel *n s.* Abschreckmittel
Abschreckwirkung *f* repellency, deterrent action *(on animals)*
abschuppen / sich to scale [off], to flake [off], to chip
abschwächen to weaken, to thin *(e.g. a solution)*; to attenuate *(an action)*; to tone down *(a colour)*; *(phot)* to reduce
Abschwächer *m* thinner, thinning agent; *(phot)* reducer
Farmerscher A. *(phot)* Farmer's reducer
proportionaler (proportional wirkender) A. *(phot)* proportional reducer
subtraktiver (subtraktiv wirkender) A. *(phot)* subtractive (cutting, subproportional) reducer
superproportionaler A. *(phot)* superproportional reducer
überproportionaler A. *s.* superproportionaler A.
Abschwächung *f* weakening, thinning *(as of a solution)*; attenuation *(of an action)*; *(phot)* reduction
örtliche (partielle) A. *(phot)* local reduction
Abschwächungsmittel *n s.* Abschwächer
abschwemmen to float off, to elutriate
abschwenkbar retractable
abseifen to soap
abseigern *(met)* to liquate, to melt down, to smelt
abseihen to strain
Abseihen *n* straining
absenken to lower, to sink
Absetzabstand *m* settling (sedimentation) distance
Absetzanlage *f* settling (sedimentation, precipitation) unit, settling (sedimentation, precipitation) plant, settler, precipitator, *(waste-water treatment also)* clarification unit (plant)
A. mit horizontaler Durchströmung horizontal-flow [clarification] unit *(waste-water treatment)*
A. mit vertikaler Durchströmung vertical-flow (up-flow) clarification unit *(waste-water treatment)*
Absetzapparat *m* settling (sedimentation, precipitation) apparatus, settler, precipitator
absetzbar settleable *(e.g. contaminants)*
Absetzbecken *n* settling (sedimentation, precipitation) basin, settling (sedimentation, precipitation) tank, settler, precipitator, *(waste-water treatment also)* quiescent tank
Absetzbehälter *m s.* Absetzgefäß
Absetzbottich *m s.* Absetzgefäß
Absetzbütte *f (pap)* draining tank (chest), drainer
absetzen to sediment, to deposit *(of a liquid)*
sich a. to settle [down, out], to sediment, to set, to precipitate, to deposit, to subside
sich wieder a. to redeposit
Absetzen *n* settling, sedimentation, setting, precipitation, deposition, subsiding
A. im Schwerefeld gravitational (gravity) sedimentation
behindertes A. *(min tech)* hindered settling
freies A. *(min tech)* free settling
gestörtes A. *s.* behindertes A.
hartes A. *(coat)* hard settling
Absetzer *m* settler, precipitator
absetzfähig settleable *(e.g. contaminants)*
Absetzfläche *f* settling (sedimentation) area
Absetzgefäß *n* settling (sedimentation, precipitation) tank, settling (sedimentation, precipitation) basin, settler, precipitator, separator tank; settling chamber *(in the pebble-heater process)*
Absetzgeschwindigkeit *f* settling (sedimentation, precipitation) rate, settling (sedimentation, precipitation) velocity

Absetzglas *n* **nach Imhoff** Imhoff [sediment] cone *(waste-water investigation)*
Absetzgrube *f* settling (sedimentation, precipitation) pit
Absetzkammer *f* settling (settlement) chamber
Absetzkonus *m* settling cone
Absetzkurve *f* settling curve
Absetzleistung *f* settling (sedimentation) performance
Absetzplatte *f (glass)* dead plate
Absetzraum *m* settling (sedimentation) chamber
Absetzschleuder *f* sedimentation centrifuge
Absetzstoffänger *m (pap)* sedimentation (gravity) save-all
Absetztank *m s.* Absetzgefäß
Absetzverhinderungsmittel *n* antisettling agent, sedimentation inhibitor
Absetzverhütungsmittel *n s.* Absetzverhinderungsmittel
Absetzweg *m* settling (sedimentation) distance
Absetzzeit *f* time of settling (sedimentation)
Absetzzentrifuge *f* sedimentation centrifuge
absieben to screen [out], to sieve [out]; to scalp [out] *(to separate out coarser grades)*
absieden to decoct
Absieden *n* decoction
absinken 1. to sediment, to settle [out], to sink, to subside *(of particles)*; 2. to decrease, to fall, to reduce, to drop *(as of measuring values)*
Absinken *n* 1. sedimentation, settling *(of particles)*; 2. decrease, fall, reduction, drop *(as of measuring values)*
Absinkgeschwindigkeit *f* settling (sedimentation) velocity, settling (sedimentation) rate
Absinth *m* 1. absinth[e], Artemisia absinthium L.; 2. *s.* Absinthbranntwein
absinthartig absinthine
Absinthbranntwein *m* absinth[e]
Absinthiin *n* absinthiin *(a bitter glycoside of Artemisia absinthium L.)*
Absinthin *n* absinthin *(a bitter principle of Artemisia absinthium L.)*
Absinthlikör *m* absinth[e]
Absitzbecken *n s.* Absetzbecken
Absitzbehälter *m s.* Absetzgefäß
absitzen to settle [down, out], to sediment, to set, to precipitate, to deposit, to subside
  a. lassen to allow to deposit, to sediment, to set *(from liquids)*
Absitzen *n* settling, sedimentation, setting, precipitation, deposition, subsiding
Absitzenlassen *n* sedimentation, setting
Absitzgefäß *n*, **Absitztank** *m s.* Absetzgefäß
Absolu [de concret] *n (cosmet)* absolute essence, absolute from concrete
Absolu d'enfleurage *n (cosmet)* absolute of enfleurage, enfleurage absolute
absolut absolute
  a. trocken absolutely dry, bone-dry, B. D., *(pap also)* oven-dry, oven-dried, O. D.
Absolutbetrag *m* absolute value
Absolutmethode *f* absolute method
Absolutwert *m* absolute value
absondern 1. to separate, to eliminate, to isolate, to segregate; 2. *(biol)* to secrete, to exude *(e. h. resin)*
  sich a. 1. to separate, to segregate *(from a mixture)*; 2. *(biol)* to exude
Absonderung *f* 1. separation, elimination, isolation, segregation *(act or process)*; *(biol)* secretion *(process or product)*
Absonderungsstoff *m (biol)* secretion, *(of useless substances also)* excretum
Absorbat *n* absorbate, absorbed material (substance)
Absorbend *m* material to be absorbed
Absorbens *n* absorbent, absorber, absorbing agent (material, substance)
Absorber *m* absorber, absorption (absorbing) apparatus, absorption unit
absorbierbar absorbable
Absorbierbarkeit *f* absorbability
absorbieren to absorb, to imbibe
  wieder a. to re-absorb
absorbierend absorbing, absorbefacient, absorbent, absorptive
  schwach a. poorly absorbing
Absorbierkolonne *f* absorbing (absorption) column, absorption tower (stack)
Absorbiersystem *n* absorption system (unit, plant)
Absorbierung *f* absorption
Absorptiometer *n* absorptiometer
Absorptiometrie *f* absorptiometry
absorptiometrisch absorptiometric
Absorption *f* absorption
  A. der Röntgenstrahlen (Röntgenstrahlung) X-ray absorption
  A. durch die Haut skin absorption
  A. in Öl oil absorption
Absorptionsanalyse *f* absorptiometry
Absorptionsanlage *f* absorption plant (unit, system)
  A. für Rauchgase fume scrubber
Absorptionsapparat *m* absorbing (absorption) apparatus, absorption unit, absorber
Absorptionsbande *f* absorption band
Absorptionsbasis *f (pharm)* absorption base
Absorptionsbatterie *f* absorption train *(for gases)*
Absorptionsbereich *m* absorption region
absorptionsbeschleunigend absorbefacient
Absorptionsbodenkolonne *f* plate-type absorption tower
absorptionsfähig absorptive, absorbent
Absorptionsfähigkeit *f s.* Absorptionsvermögen
Absorptionsflasche *f* absorption bottle
Absorptionsflüssigkeit *f* absorption (absorbing) liquid
absorptionsfördernd absorbefacient
Absorptionsgefäß *n* absorption bulb; absorption vessel
Absorptionsgeschwindigkeit *f* absorption rate (velocity)
Absorptionsgesetz *n* law of absorption
  Bouguer-Lambertsches A. Bouguer-Lambert law

[of absorption], Lambert's absorption law
Henrysches A. Henry's law [of absorption]
Lambertsches A. *s.* Bouguer-Lambertsches A.
Absorptionsgrad *m* absorptive power, absorptivity
Absorptionsgrundlage *f (pharm)* absorption base
Absorptionskälteanlage *f* absorption refrigerating (refrigeration) system
Absorptionskältemaschine *f* absorption refrigerating (refrigeration) machine
Absorptionskoeffizient *m* absorption coefficient
atomarer A. atomic absorption coefficient
Bunsenscher A. Bunsen [absorption] coefficient
Absorptionskolben *m* absorption flask
Absorptionskolonne *f* absorption (absorbing) column, absorption tower (stack)
Absorptionskurve *f* absorption curve
Absorptionsküvette *f* absorption cell
Absorptionslinie *f* absorption line
Absorptionsmaschine *f* absorption machine
Absorptionsmaximum *n* absorption maximum
Absorptionsmittel *n* absorbing agent (material, substance), absorbent, absorber
Absorptionsöl *n* absorption (absorbent) oil
mageres A. lean oil *(for the absorption column)*
Absorptionspipette *f* absorption [gas] pipet
Absorptionsprozeß *m* absorption process
Absorptionsquerschnitt *m (nucl)* capture cross section
Absorptionsrohr *n* absorption tube
Absorptionsröhrchen *n* absorption tube
Absorptionsröhre *f* absorption tube
Absorptionssäule *f* absorption (absorbing) column, absorption tower (stack)
Absorptionspektrometer *n* absorption spectrometer
Absorptionsspektroskopie *f* absorption spectroscopy
Absorptionsspektrum *n* absorption spectrum
Absorptionssystem *n* absorption system (unit, plant)
Absorptionsturm *m* absorption (absorbing) tower
Absorptionsverfahren *n* absorption process
Absorptionsvermögen *n* absorbency, absorbancy, absorptivity, absorbing (absorption, absorptive) capacity
Absorptionsvorgang *m* absorption process
Absorptionszahl *f* absorptivity, absorptive power
absorptiv absorptive, absorbent
Absorptiv *n* absorbate, absorbed material (substance)
abspalten to split off, to cleave off, to separate, to eliminate *(atomic groups from molecules)*
Brom a. to debrominate
Halogenwasserstoff a. to dehydrohalogenate
Sauerstoff a. to deoxidize, to deoxidate, to deoxygenate
Schichten a. to flake
sich a. to split [off], to cleave [off]
Wasser a. to dehydrate
Wasserstoff a. to dehydrogenate
Xanthatgruppen a. to dexanthate
Abspaltung *f* splitting[-off], cleavage; separation, elimination; spalling *(of ore)*; removal *(as of water)*; *(nucl)* spallation
Abspaltungsreaktion *f* elimination reaction
Abspannen *n (plast)* stripping *(of a mould)*
absperren to shut off; to seal *(painting technology)*
Absperrgröße *f* degree of retentivity *(of a filter)*
Absperrhahn *m* stopcock
Absperrmittel *n* sealer, sealing paint *(painting technology)*
Absperrschieber *m* gate valve
Absperrstein *m* **des Speisers** *(glass)* feeder gate (plug)
Absperrventil *n* block (cut-off, shut-off, stop) valve
Absperrvorrichtung *f* cut-off (shut-off, stopping) device
absplittern to splinter off, to come off in splinters, to chip [off]
Absplittern *n (ceram)* chipping *(as of glaze)*
absprengen *(glass)* to burn off
Absprengen *n (glass)* burn-off, flame cutoff
Absprengkappe *f (glass)* moil
Absprengperlen *fpl* glass filler rings
abspringen to scale [off], to flake [off], to exfoliate, to peel [off], to shell [off]
abspulen to unreel, to reel off, to unwind
abspülen to rinse, to swill; *(min tech)* to spray
Abstand *m* distance, space, spacing, range; separation *(e. g. in a crystal lattice)*; interval
abstauben *(ceram)* to dust *(a ware before firing)*
Abstaubmaschine *f (ceram)* dusting machine, duster
abstechen to tap [out], to draw off *(liquid metals)*; to dig out *(e. g. a filter cake)*
Abstehzeit *f (rubber)* drop period
abstellen to turn off, to shut off *(water, gas)*; to shut [down], to stop *(a machine)*
absterben to fade away, to die down (away), to decay, to subside
Abstich *m* 1. tapping *(of liquid metals)*; 2. *s.* Abstichloch
Abstichgaserzeuger *m* slagging [ash] producer
A. von Ebelmen Ebelmen producer
Abstichgenerator *m s.* Abstichgaserzeuger
Abstichloch *n (met)* taphole, tapping hole
Abstichmaschine *f* tapping device *(calcium carbide manufacture)*
Abstichöffnung *f s.* Abstichloch
Abstichpfanne *f (met)* tapping receiver
Abstichrinne *f (met)* tapping channel, runner
Abstichschirm *m* tapping screen *(of a carbide furnace)*
Abstichschnauze *f s.* Abstichrinne
abstoppen to stop [off, up]; *(rubber)* to shortstop *(polymerization)*
Abstoppmittel *n (rubber)* shortstopping agent, shortstop, stopper
Abstoßeffekt *m* repulsive effect
abstoßen to repel; to discard, to reject; to repulse

Narben a. *(tann)* to degrain
abstoßend repellent; repulsive
Abstoßung *f* repulsion
elektrostatische A. electrostatic repulsion
Abstoßungskraft *f* repulsive force, force of repulsion
Abstoßungsmittel *n (text)* repellent
abstrahlen to radiate, to emit rays
Abstrahlung *f* radiation, emission [of radiation]
Abstrahlungsverlust *m* radiation loss, loss by radiation
abstreichen to skim [off]
Abstreicher *m* skimmer; plough *(on conveyors)*
abstreifen to scrape off, to doctor off; to wipe; to skim [off]; *(distil)* to strip [off, out]; *(plast)* to eject
Abstreifen *n* **vom Stempel** *(plast)* top ejection
Abstreifer *m* scraper (doctor) knife, scraper (doctor) blade, scraper; wiper; skimmer; *(distil)* stripper
Abstreiferkolonne *f (distil)* stripper [column], strip action still, stripping column (still), side stripper
Abstreifermesser *n* scraper (doctor) knife, scraper (doctor) blade, scraper
Abstreiferöl *n* stripping oil
Abstreiferteil *m s.* Abstreiferzone
Abstreiferzone *f* stripping (exhausting) section *(of a fractionating column)*
Abstreifmeißel *m*, **Abstreifmesser** *n s.* Abstreifermesser
Abstreifplatte *f* stripper plate
Abstrippzone *f s.* Abstreiferzone
abströmen to flow off (away), to run off *(of liquids)*; to escape *(of gases)*
Abstufung *f (phot)* gradation
abstumpfen to deaden, to dull *(e.g. paint)*; *(tann)* to raise the basicity *(of chrome liquor)*
Absud *m* decoction, decoctum
absüßen to sweeten, to dulcify, to edulcorate
Absüßen *n* sweetening, dulcification, edulcoration
Absüßpumpe *f (sugar)* sweet-water pump
Absüßwasser *n (sugar)* sweet water
Abszisin *n (bioch)* abscisin
abtauchen to dip
Abtauchen *n* dip[ping]
abtauen to defrost, to thaw [off]
abtönen to shade, to tint, to tone
Abtöner *m*, **Abtönmittel** *n* tinting agent
Abtönung *f* 1. shading, tinting; 2. shade; *(phot)* gradation
abtoppen to skim *(petroleum)*
Abtränkbrühe *f (tann)* duster
abtreiben to distil off, to strip [off, out]; *(met)* to cupel
Abtreiben *n* simple distillation, stripping; *(met)* cupellation
Abtreiber *m s.* Abtreiberkolonne
Abtreiberkolonne *f (distil)* stripper [column], stripping column (still), side stripper, strip action still
Abtreibeteil *m* stripping (exhausting) section *(of a fractionating column)*
Abtreibkolonne *f s.* Abtreiberkolonne
abtrennbar separable
Abtrennbarkeit *f* separability
abtrennen to separate [off], to isolate *(something from a mixture)*; to remove *(e.g. an acid)*; to abstract, to separate by distillation
Äthan a. to de-ethanize
Eiweiß a. to deproteinize
Paraffin (Wachs) a. to dewax
Abtrennung *f* separation, isolation *(from a mixture)*; removal *(as of an acid)*; abstraction, separation by distillation
Abtrennungsarbeit *f* work of separation, work function [of electrons], electronic work function
Abtrieb *m* simple distillation, stripping
Abtriebsgerade *f (distil)* stripping operating line
Abtriebskolonne *f (distil)* stripper [column], strip action still, stripping column (still), side stripper
Abtriebssäule *f s.* Abtriebskolonne
Abtriebsteil *m* stripping (exhausting) section *(of a fractionating column)*
Abtrift *f* blow-off, drift *(of pesticides)*
abtrocknen to dry [off, up]
Abtropfbrett *n* drying (apparatus) rack, drain[ing] board
aufhängbares A. wall-mounting draining board
abtropfen to drain [off], to drip (trickle) off
a. lassen to drain [off]
Abtropfen *n* drainage
Abtropfenlassen *n* drainage
Abtropfgestell *n* drying (apparatus) rack
Abtropfkanal *m* draining pan *(in flow coating)*
abtupfen / mit Fließpapier to blot
Abukumalit *m (min)* abukumalite
abwandern to migrate out *(as of ions)*
Abwärme *f* waste heat
Abwärmerückgewinnung *f* waste-heat recovery
Abwärts-Dickstoffturm *m (pap)* down-flow high density tower
Abwärtsgasen *n* down-run[ning], down-steaming *(in water-gas production)*
A. mit Dampfüberhitzung back run
Abwärtsstrom *m* down[ward] flow
Abwaschbecken *n* sink
abwaschen to wash [up], to rinse
Abwasser *n* waste water, waste (discharge) water, waste, effluent [water], *(esp municipally)* sewage; foul water; *(pap)* white water
A. der Registerpartie *(pap)* tray water, water from the tray
faser- und füllstoffreiches A. *(pap)* rich white water
geklärtes A. *(pap)* filtered water
gewerbliches A. *s.* industrielles A.
häusliches A. household waste water, domestic sewage
industrielles A. industrial waste water, factory (works, trade) effluent
kommunales A. [municipal] sewage
phenolhaltiges A. phenolic waste
rohes A. raw (crude) waste, raw (crude) sewage
städtisches A. town sewage

Abwasserablauf *m* waste-water flow, effluent discharge
Abwasseraufbereitung *f s.* Abwasserbehandlung
Abwasserbehälter *m (pap)* white-water chest
Abwasserbehandlung *f* waste-water treatment, waste (sewage) treatment
Abwasserbehandlungsanlage *f* waste-treatment (sewage-treatment) plant
Abwasserbeseitigung *f* waste-water disposal, sewage (effluent) disposal
Abwasserbeseitigungsanlage *f* waste-water [disposal] plant, sewage [disposal] plant, waste-water works, sewage works
Abwasserkanal *m* sewer
Abwasserlast *f* pollution (contaminant, waste) load
Abwasserleitung *f* waste line; drain line
Abwasserpilz *m* sewage fungus
Abwasserpumpe *f* waste-water pump, sewage pump; *(pap)* backwater (white-water) pump
Abwasserreinigung *f* waste-water purification, sewage purification
biologische A. biological purification of sewage, biological elimination
Abwasserreinigungsanlage *f* waste-water purification (treatment) plant, sewage (effluent) purification plant, sewage (effluent) treatment plant
Abwasserrohr *n* waste pipe; sewage (sewer) pipe
Abwassersammelbehälter *m (pap)* white-water chest
Abwasserschlamm *m* sewage sludge
Abwasserstrom *m* waste-water flow
Abwasserteich *m* waste-water pond
Abwasserüberwachung *f* waste-water control
Abwasserversenkung *f* deep-well disposal
Abwasserverwertung *f* utilization of waste water, utilization of sewage
Abwehrferment *n* defensive enzyme
Abwehrmittel *n*, **Abwehrstoff** *m (agric)* repellent
abweichen to deviate, to vary, to be different, to differ, to deflect; to depart *(from a norm, a standard or a regular value)*
Abweichung *f* deflection, deviation, departure, variation
A. vom Reziprozitätsgesetz *(phot)* reciprocity [law] failure
mittlere quadratische A. standard deviation *(statistics)*
zulässige A. tolerance, allowance
abweisen to repel
abweisend repellent
a. gegen klebende (klebrige) Stoffe antistick
a. gegen Öl oil-repellent
a. gegen Wasser water-repellent, hydrophobic, hydrophobe
Abweisendausrüstung *f* repellent finish
Abweisungsvermögen *n* repellency
abwelken *(tann)* to sam[my] *(to dry partially)*
Abwelkpresse *f (tann)* samm[y]ing machine
abwerfen to discharge, to dump *(as from a conveyor)*

abwickeln *(pap)* to unwind, to unreel, to reel off
Abwickelvorrichtung *f (rubber)* let-off arrangement
abwiegen to weigh [out, up]
Abwurf *m* 1. discharge, discharging; 2. discharge [point]
Abwurfende *n* discharge end *(as of a conveyor)*
Abwurföffnung *f* discharge opening (outlet, door) *(as of a conveyor)*
Abwurframpe *f* coke wharf
Abwurfstelle *f* discharge point *(as of a conveyor)*
Abwurfwagen *m* tripper
abzentrifugieren to centrifuge off
Abziehbild *n (ceram)* decalcomania, decal, litho, transfer
Abziehbilderpapier *n* transfer (decalcomania) paper
Abziehbilderrohpapier *n* transfer base paper
abziehen 1. to draw off, to withdraw, to drain off *(liquids)*; to barrel, to rack *(e. g. wine)*; to bottle [in, up] *(e. g. wine)*; to tap *(slag)*; to strip *(a dye)*; to run off *(e. g. a layer)*; *(phot)* to print; *(pap)* to proof; to transfer *(designs)*; 2. to escape, to issue *(as of vapours)*
auf Fässer a. to barrel, to rack
auf Flaschen a. to bottle [in, up]
Farbe a. to decolour[ize], to strip, to remove a dye *(from yarn or fabric)*
Narben a. *(tann)* to degrain
Abziehflotte *f* stripping bath
Abziehgeschwindigkeit *f (plast)* haul-off rate
Abziehhahn *m* racking cock
Abziehhalle *f* racking room
Abziehhilfsmittel *n* decolourizing (stripping) agent (assistant), decolourizer
Abziehlack *m* decorators' size
Abziehmittel *n s.* Abziehhilfsmittel
Abziehpapier *n s.* Abzugspapier 1.
Abzug *m* 1. *(lab)* [exhaust] hood, fume hood (chamber, closet); discharge [point], outlet, drain; *(phot)* [contact] print; *(pap)* proof [sheet]; 2. drawing-off, withdrawal *(of liquids or gases)*; escape, issue *(as of vapours)*
Abzug ... *s.* Abzugs ...
Abzugsdampf *m* exhaust (waste) steam
Abzugsende *n* discharge end
Abzugsgeschwindigkeit *f (plast)* haul-off rate
Abzugshaube *f* [fume] hood
Abzugskanal *m* discharge duct
Abzugskasten *m* local exhaust hood
Abzugsleitung *f* discharge (drain, drainage, outlet) line; waste line
Abzugsöffnung *f* discharge opening (outlet)
Abzugspapier *n* 1. [galley] proof paper, proofing (stripping) paper; 2. *(phot)* printing (print) paper
Abzugspumpe *f* withdrawal pump
Abzugsrohr *n* discharge (drainage, outlet) pipe, discharge (drainage, outlet) tube; waste pipe (tube); offtake [pipe]; effluent pipe; vent pipe; fume pipe; *(lab)* vapour tube, fume chamber (duct, exhaust manifold) *(of the Kjeldahl digestion apparatus)*

seitliches A. side tube *(of a fractionating flask)*
Abzugsschleuse *f* sink
Abzugsschrank *m (lab)* [exhaust] hood, fume hood (chamber, closet)
Abzugsventil *n* outlet valve
Abzugswalze *f (plast)* haul-off roll
Abzweig *m* branch
abzweigen to branch
Abzweigstück *n* multiway union
Abzweigung *f* branching; branch
Acajou *n s.* Acajugummi
Acajugummi *n* cashew (cashawa) gum *(from Anacardium occidentale L.)*
Acetarsol *n s.* Azetarsol
Aceton *n s.* Azeton
Achat *m (min)* agate
Achatablaufdüse *f s.* Achatausflußrohr
Achatausflußrohr *n* agate tube (jet) *(of a Redwood viscosimeter)*
Achatmörser *m* agate mortar
Achatplanlager *n* agate plane *(of a precision balance)*
Achatpolierstein *m* agate burnisher
Achatsteinglätte *f s.* Achatsteinglätteinrichtung
Achatsteinglätteinrichtung *f (pap)* flint-glazing machine, flint glazer, stone burnisher
Achondrit *m (min)* achondrite *(a meteorite)*
Achse *f (cryst)* axis
dreizählige A. axis of threefold symmetry, threefold axis of symmetry
kristallografische A. crystal[lographic] axis
optische A. optic[al] axis
sechszählige A. axis of sixfold symmetry, sixfold axis of symmetry
vierzählige A. axis of fourfold symmetry, fourfold axis of symmetry
zweizählige A. axis of twofold symmetry, twofold axis of symmetry
Achsenabschnitt *m (cryst)* intercept, parameter
rationaler A. rational intercept
Achsensymmetrie *f (cryst)* axial symmetry
achsensymmetrisch *(cryst)* axially symmetric[al], axisymmetric
Achsenverhältnis *n (cryst)* [crystallographic] axial ratio, ratio of the intercepts
Achsenwinkel *m (cryst)* axial angle
achtatomig octatomic
Achtergruppe *f* octet
Achtergruppierung *f* octet
Achterring *m s.* Achtring
Achterschale *f* octet
achtflächig *(cryst)* octahedral, oct.
Achtflächner *m (cryst)* octahedron
Achtring *m* eight-membered ring
achtwertig octavalent
Achtwertigkeit *f* octavalency
aci-Form *f* aci form *(of nitro compounds)*
Ackerkrume *f* topsoil
ACTH *s.* Hormon / adrenokortikotropes
Adamin *m (min)* adamite, adamine *(basic zinc arsenate)*
Adamsit *m (min)* adamsite *(a variety of muscovite)*
Adamsit *n* $NH(C_6H_4)_2AsCl$ adamsite, 10-chloro-5,10-dihydrophenarsazine
Adams-Katalysator *m* Adams' catalyst *(platinum oxide)*
Addend *m* addend
addieren to add
sich a. to add
Addition *f* addition
A. im Anti-Markownikow-Sinn *s.* durch Peroxide ausgelöste A.
durch Peroxide ausgelöste (initiierte) A. peroxide-initiated addition, anti-Markovnikov addition
elektrophile A. electrophilic addition
ionoide A. ionic addition
kationoide A. *s.* elektrophile A.
1,2-Addition *f* 1,2-addition
1,4-Addition *f* 1,4-addition
Additionseffekt *m* additive effect
Additions-Eliminierungs-Mechanismus *m* addition-elimination mechanism
additionsfähig capable of addition
Additionsfähigkeit *f* additive capacity, capacity for addition, addition reactivity
Additionsmischkristall *m* addition solid solution
Additionsname *m* additive name
Additionspolymer[es] *n* addition polymer
Additionspolymerisation *f* addition polymerization
Additionsprodukt *n* addition product
Additionsreaktion *f* addition reaction
Additionsrollenpapier *n* tabulating paper, paper for calculating machines
Additionsverbindung *f* addition (additive) compound
Additionsvermögen *n* additive capacity, capacity for addition, addition reactivity
additiv additive
Additiv *n* additive *(a substance added to another in small amounts; specif a substance added to mineral-oil products in a quantity from 1 to 10%)*
rostverhinderndes (rostverhütendes) A. rust-preventing additive
Additivität *f* additive capacity, capacity for addition, addition reactivity
Additivname *m* additive name
Addukt *n* adduct
Adduktkautschuk *m* adduct rubber
Adenase *f* adenase
Adenin *n* adenine, 6-aminopurine
Adenosindiphosphat *n s.* Adenosindiphosphorsäure
Adenosindiphosphorsäure *f* adenosine diphosphoric acid, adenosine diphosphate, ADP
Adenosinmonophosphat *n s.* Adenosinmonophosphorsäure
Adenosin-5'-monophosphat *n s.* Adenosin-5'-phosphorsäure
Adenosinmonophosphorsäure *f* adenosine monophosphoric acid, adenosine monophosphate, AMP, adenylic acid, AA
Adenosinphosphat *n s.* Adenosinphosphorsäure

Adenosinphosphorsäure *f* adenosinephosphoric acid, adenosine phosphate, *(specif)* s. Adenosinmonophosphorsäure
Adenosin-5'-phosphorsäure *f* adenosine 5'-phosphoric acid, adenosine 5'-monophosphate, AMP, muscle adenylic acid
Adenosinpyrophosphat *n* s. Adenosindiphosphorsäure
Adenosinpyrophosphorsäure *f* s. Adenosindiphosphorsäure
Adenosintriphosphat *n* s. Adenosintriphosphorsäure
Adenosintriphosphatase *f* adenosine triphosphatase
Adenosintriphosphorsäure *f* adenosine triphosphoric acid, adenosine triphosphate, ATP
Adenylpyrophosphat *n* s. Adenosintriphosphorsäure
Adenylpyrophosphorsäure *f* s. Adenosintriphosphorsäure
Adenylsäure *f* s. Adenosinmonophosphorsäure
Ader *f (geol)* lode, vein
anstehende A. outcrop
Adergneis *m* veined gneiss
Adermin *n* adermin[e], 3-hydroxy-4,5-di-(hydroxymethyl)-2-methylpyridine *(vitamin B_6)*
Aderpapier *n* batik paper; creased paper *(a variety of batik paper)*
Adhärend *m* adherend *(a body to be attached to another by an adhesive)*
Adhärens *n* adhesive (adhesion) agent, adhesive (adhesion) substance, adhesive
adhärieren to adhere
adhärierend adherent, adhesive
Adhäsion *f* adhesion, adherence
Adhäsionsarbeit *f* adhesional work, work of adhesion
Adhäsionsbeschleuniger *m* adhesion promoter
Adhäsionsenergie *f* adhesion energy
adhäsionsfähig adherent, adhesive
Adhäsionsfähigkeit *f* adhesiveness, adhesion (adhesive) capacity, adhesion (adhesive) power, adherence
adhäsionsfeindlich abhesive
Adhäsionsfestigkeit *f* adhesive (adhesion) strength
Adhäsionskraft *f* s. Adhäsionsvermögen
Adhäsionsprüfer *m* adhesion tester
Adhäsionsspannung *f* adhesion (adhesive) tension, adhesion (adhesive) stress
Adhäsionsvermögen *n* adhesiveness, adhesion (adhesive) capacity, adhesion (adhesive) power, adherence
adhäsiv adhesive, adherent
Adiabate *f* adiabat[ic], adiabatic curve (line)
adiabatisch adiabatic
adiatherman atherm[an]ous
Adion *n* adion *(an ion adsorbed on a surface)*
Adipat *n* adipate
Adipinsäure *f* $HOOC[CH_2]_4COOH$ adipic acid, hexanedioic acid
Adjuvans *n (pharm)* adjuvant
Adkins-Katalysator *m* Adkins' catalyst
Admiralitätskohle *f* Admiralty steam coal
ADP s. Adenosindiphosphat
Adrenalin *n* adrenaline, 1-(3,4-dihydroxyphenyl)-2-methylaminoethanol
Adrenalon *n* adrenalone *(an acetophenone derivative)*
Adrenokortikotropin *n* adrenocorticotropic hormone, ACTH, corticotropin
Adsorbat *n* adsorbate, adsorptive, adsorbed material (substance); adsorption complex, adsorbent-adsorbate complex (system), adsorbate
Adsorbend *m* material to be adsorbed
Adsorbens *n* adsorbent [material], adsorbing agent (material, substance)
Adsorbensschicht *f* adsorbent bed
Adsorber *m* adsorber *(apparatus)*; s. Adsorbens
adsorbierbar adsorbable
Adsorbierbarkeit *f* adsorbability
adsorbieren to adsorb
chemisch a. to chemisorb
adsorbierend adsorbing, adsorbent, adsorptive
Adsorbierung *f* adsorption
Adsorption *f* adsorption
aktivierte A. activated adsorption
apolare A. apolar (non-polar) adsorption
bevorzugte A. s. selektive A.
chemische A. chemical (activated) adsorption, chemisorption, chemosorption
negative A. negative adsorption
physikalische A. physical (reversible, van der Waals) adsorption
polare A. polar adsorption
positive A. positive adsorption
reversible A. s. physikalische A.
selektive A. selective (preferential) adsorption
van-der-Waalssche A. s. physikalische A.
Adsorptionsanalyse *f* adsorption analysis
chromatografische A. chromatographic adsorption analysis, adsorption chromatography
radiometrische A. radiometric adsorption analysis
Adsorptionsapparat *m* adsorber
Adsorptionschromatografie *f* adsorption chromatographie, chromatographic adsorption [analysis]
Adsorptionsenergie *f* adsorption energy (heat)
Adsorptionserscheinung *f* adsorption phenomenon
adsorptionsfähig adsorbent, adsorptive, adsorbing
Adsorptionsfähigkeit *f* adsorptive capacity (power), adsorptiveness
Adsorptionsfilm *m* adsorbed film
Adsorptionsgesetz *n* adsorption law, law of adsorption
Gibbsches A. Gibbs adsorption law
Adsorptionsgleichgewicht *n* adsorption equilibrium
Adsorptionsgleichung *f* adsorption equation
Gibbsche A. Gibbs adsorption equation
Adsorptionsindikator *m* adsorption indicator
Adsorptionsisostere *f* adsorption isostere
Adsorptionsisotherme *f* adsorption isotherm

Freundlichsche A. Freundlich [adsorption] isotherm
Langmuirsche A. Langmuir [adsorption] isotherm
Adsorptionsklärmittel *n* sweetener *(for dry cleaning)*
Adsorptionskolonne *f* adsorption (adsorbing, adsorbent) column
Adsorptionskraft *f* adsorptive capacity (power), adsorptiveness
Adsorptionskurve *f* adsorption curve
Adsorptionsmedium *n* adsorption medium
Adsorptionsmittel *n* adsorbent [material], adsorbing agent [material, substance); chromatographic packing *(in adsorption chromatography)*; sweetener *(for dry cleaning)*
Absorptionsmittelschicht *f* adsorbent bed
Adsorptionspotential *n* adsorption potential
Adsorptionsprozeß *m* adsorption process
Adsorptionsquellung *f* adsorption swelling
Adsorptionssäule *f* adsorption (adsorbing, adsorbent) column
Adsorptionsschicht *f* adsorbed layer
monomolekulare A. unimolecular [adsorbed] layer, monomolecular layer, monolayer
multimolekulare A. multimolecular [adsorbed] layer, multilayer
Adsorptionsturm *m* adsorption tower
Adsorptionsverbindung *f* adsorption compound
Adsorptionsverdrängung *f* adsorption displacement
Adsorptionsverfahren *n* adsorption process
Adsorptionsvermögen *n* adsorptive capacity (power), adsorptiveness
Adsorptionsvorgang *m* adsorption process
Adsorptionswaage *f* adsorption balance
Adsorptionswärme *f* adsorption heat (energy)
differentiale (differentielle) A. differential heat of adsorption
integrale A. integral heat of adsorption
Adsorptionszentrum *n* adsorption centre (site)
Adsorptionszone *f* adsorption zone
adsorptiv adsorptive, adsorbent, adsorbing
Adsorptiv *n* adsorptive, adsorbate, adsorbed material (substance)
Adstringens *n* astringent, styptic
adstringent *s.* adstringierend
Adstringenz *f* astringency, stypticity
adstringierend astringent, styptic
ÄDTE *s.* Äthylendiamintetraessigsäure
Adular *m (min)* adularia *(a variety of orthoclase)*
Advektion *f* advection
Advektionsnebel *m* advection fog
AeDTE *s.* Äthylendiamintetraessigsäure
aerob *(biol)* aerobiotic, aerobic, oxybiotic
Aerobier *m*, **Aerobiont** *m (biol)* aerobe
Aerobiose *f (biol)* aerobiosis, oxybiosis
Aerofall-Mühle *f* Aerofall mill *(an autogenous mill)*
Aerogel *n* aerogel
aerolisieren *s.* aerosolieren
Aerolith *m (geol)* aerolite, aerolith, stone (stony) meteorite, meteoric stone
Aeromultizyklon *m* multitube cyclone separator
aerophil air-avid
Aerosol *n* aerosol
Aerosolbombe *f (agric)* aerosol bomb (projector)
Aerosoldose *f s.* Aerosolsprühdose
Aerosolfarbe *f* aerosol paint
aerosolieren to aerosolize, to nebulize
Aerosolisierung *f* aerosolization
Aerosolrasiercreme *f* aerosol shave (shaving cream)
Aerosolspray *m* aerosol spray
Aerosolsprühdose *f* aerosol spray can, aerosol dispenser (bomb)
Aerosolsprühgerät *n (agric)* aerosol bomb (projector)
Aerosoltreibmittel *n* aerosol propellant
Aerosolzerstäuber *m* aerosol bomb (projector); *s.* Aerosolsprühdose
Aerozyklon *m* cyclone air separator
Aeschynit *m* aeschynite *(a mineral containing cerium, titanium, and thorium)*
Affenschaukel *f (glass)* birdcage, bird swing *(a glass thread spanning the inside of a bottle)*
Affichenpapier *n* poster paper
Affinade *f* affinated (affination) sugar, washed raw sugar
Affination *f* affination *(treatment of sucrose crystals to free them from residual molasses)*
Affinität *f* affinity
A. zu Schwefel sulphur affinity, affinity for sulphur
A. zu Wasser water affinity, affinity for water
A. zur Faser affinity for the fibre
chemische A. chemical affinity
Affinitätskonstante *f* affinity (dissociation, ionization) constant
Affinitätskurve *f* affinity curve
Afterkristall *m* pseudomorph
A-Füllmasse *f (sugar)* first fillmass, white (high-grade) massecuite
Agalmatolith *m (min)* agalmatolite *(aluminium dihydroxide tetrasilicate)*
Agar[-Agar] *m(n)* agar[-agar], agar gel, Japan agar (isinglass), Chinese gelatin
Agarizin *n s.* Agarizinsäure
Agarizinsäure *f* agaric acid
Agarnährboden *m* agar medium, nutrient agar
Agar[nähr]platte *f* nutrient agar plate
Agarröhrchenmethode *f* agar-tube method
Agathendisäure *f* agathene dicarboxylic acid, agathic acid
Agathsäure *f s.* Agathendisäure
Agave *f* agave, Agave L.
Agavefaser *f* agave
Agens *n* agent
aktives A. active agent
angreifendes A. attacking agent
chemisches (chemisch wirksames) A. chemical agent
mutationsauslösendes A. mutagen, mutagenic agent

nitrierendes A. nitrating agent
nitrosierendes A. nitrosating agent
sulfonierendes A. sulphonating agent
wirksames A. active agent
Agglomerat *n* agglomerate
Agglomeration *f* agglomeration
Agglomerationsneigung *f* tendency to agglomerate
agglomerieren to agglomerate
Agglomerierung *f* agglomeration
Agglugen *n s.* Agglutinogen
Agglutinating-Index *m (coal)* agglutinating value
Agglutination *f* agglutination
agglutinieren to agglutinate
agglutinierend agglutinant
Agglutinogen *n* agglutinogen
Aggregat *n* aggregate
körniges A. granular aggregate
Aggregatbildner *m* aggregating agent
Aggregatbildung *f* aggregate formation
Aggregation *f* aggregation
A. des Bodens soil aggregation
Aggregationsvorgang *m* aggregation process
Aggregatstabilität *f* aggregate stability
Aggregatzustand *m* state of aggregation (matter)
nematischer A. nematic state *(of liquid crystals)*
smektischer A. smectic state *(of liquid crystals)*
Aggregatzustandsänderung *f* change in (of) state
aggregieren to aggregate
sich a. to aggregate
Aggregierung aggregation
aggressiv aggressive, offensive *(chemicals)*
Ägirin *m (min)* aegirine *(iron(III) potassium disilicate)*
A-Glas *n* glass A, alkali glass
Aglukon *n* aglucone *(non-sugar portion of a glucoside)*
Aglykon *n* aglycone *(non-sugar portion of a glycoside)*
Agmatin *n* $NH{=}C(NH_2)NH[CH_2]_4NH_2$ agmatine, 4-guanidinobutylamine
Agon *n* agon, prosthetic (active) group, coenzyme
Agrarchemie *f s.* Agrikulturchemie
Agricolit *m (min)* agricolite, eulytite, eulytine *(bismuth orthosilicate)*
Agrikulturchemie *f*, **Agrochemie** *f* agricultural chemistry, agrochemistry
Agrochemikalie *f* agricultural chemical, agrochemical
Aguilarit *m (min)* aguilarite *(silver sulphide)*
A-Harz *n* A-stage (one-stage) resin, resol
Ähnlichkeitsverbundwirkung *f (tox)* similar joint action
A-Horizont *m (soil)* A-horizon, eluvial horizon
Ahornsaft *m* maple sap
Ahornsirup *m* maple syrup
Ahornzucker *m* maple sugar
Ahrens-Verfahren *n* Ahrens process *(gas production)*
AH-Salz *n* 6,6 salt, hexamethylenediamine salt of adipic acid, hexamethylenediamine adipate
Aikinit *m (min)* aikinite *(a complex sulphide of lead, copper, and bismuth)*
Airlift *m* 1. air lift, mammoth (air-lift) pump; 2. air lift, air-lift pump *(catalyst-handling system in catalytic cracking)*
Airliftförderung *f* air lifting
Airliftkracken *n s.* Airliftkrackverfahren
Airliftkrackverfahren *n* air-lift process (method), riser cracking
Airliftsystem *n* air-lift system
Air-slip-Verfahren *n (plast)* air-slip forming
Ajax-Northrup-Ofen *m* Ajax-Northrup [coreless induction] furnace, Ajax-Northrup high-frequency induction furnace
Ajax-Wyatt-Induktionsofen *m s.* Ajax-Wyatt-Ofen
Ajax-Wyatt-Niederfrequenzinduktionsofen *m s.* Ajax-Wyatt-Ofen
Ajax-Wyatt-Ofen *m* Ajax-Wyatt furnace
Ajmalin *n* ajmaline *(a rauwolfia alkaloid)*
Ajowanöl *n* ajowan oil *(essential oil from Carum copticum (L.)Benth. et Hook.)*
Akanthit *m (min)* acanthite *(silver sulphide)*
akarizid acaricidal, miticidal
Akarizid *n* acaricide, miticide
Akaroidgummi *n s.* Akaroidharz
Akaroidharz *n* acaroid (accroides, xanthorrhoea, yakka, grass-tree) gum, acaroid resin, Botany Bay gum, Resina Acaroidis *(from Xanthorrhoea specc.)*
Gelbes A. yellow grass-tree gum, Botany Bay gum *(from Xanthorrhoea hastilis R.Br.)*
Rotes A. red acaroid resin, red gum *(from Xanthorrhoea specc.)*
Akazetin *n* acacetin, 5,7-dihydroxy-4′-methoxyflavone
Akaziengummi *n* acacia (Arabic) gum, gum arabic *(from Acacia specc., esp from Acacia senegal (L.) Willd.)*
Akazienrinde *f (tann)* stick (wattle) bark
Akkommodationskoeffizient *m* accommodation coefficient *(chemisorption)*
Akku *m s.* Akkumulator
Akkumulation *f* accumulation
Akkumulator *m* accumulator, [storage] battery
hydraulischer A. hydraulic accumulator
Akkumulator[en]säure *f* [storage] battery acid
Akkumulatorzelle *f* storage cell (battery)
Akku-Säure *f s.* Akkumulatorensäure
Akmit *m (min)* acmite *(iron(III) sodium metasilicate)*
Akonitin *n* aconitine *(alkaloid)*
Akonitsäure *f* $HOOC{-}CH{=}C(COOH)CH_2{-}COOH$ aconitic acid, propene-1,2,3-tricarboxylic acid
Akridan *n* acridan, 9,10-dihydroacridine
Akridin *n* acridine
Akridinfarbstoff *m* acridine dye
Akridonfarbstoff *m* acridone dye
Akridonringschluß *m* acridonation
Akriflavin *n* acriflavine, *(pharm also)* flavine
Akrolein *n* $CH_2{=}CHCHO$ acrolein, acraldehyde, propenal
Akrylaldehyd *m s.* Akrolein
Akrylamid *n* $CH_2{=}CHCONH_2$ acrylamide, propenamide

Akrylat *n* acrylate
Akrylatkautschuk *m* acrylate-butadiene rubber, acrylate (acrylic, polyacrylate) rubber
Akryl-Butadien-Kautschuk *m s.* Akrylatkautschuk
Akrylelastomer[es] *n* acrylate (acrylic, polyacrylate) elastomer
Akrylfaser *f* acrylic (polyacrylonitrile) fibre
Akrylfaserstoff *m* acrylic (polyacrylonitrile) fibre
Akrylharz *n* acrylic (acrylate, acrylic-acid) resin
Akryllack *m* acrylic lacquer
Akrylnitril *n* $CH_2=CHCN$ acrylonitrile, cyanoethylene, vinyl cyanide, propene nitrile
Akrylnitril-Butadien-Kautschuk *m* acrylonitrile-butadiene rubber
Akrylnitril-Butadien-Styrol *n* acrylonitrile-butadiene styrene
Akrylnitril-Butadien-Styrol-Harz *n* acrylonitrile-butadiene-styrene resin
Akrylnitril-Butadien-Styrol-Kopolymer[es] *n* acrylonitrile-butadiene-styrene copolymer
Akrylnitril-Butadien-Styrol-Kunststoff *m* acrylonitrile-butadiene-styrene plastic
Akrylnitril-Butadien-Styrol-Mischpolymer[es] *n* acrylonitrile-butadiene-styrene copolymer
Akrylnitril-Butadien-Styrol-Plast *m* acrylonitrile-butadiene-styrene plastic
Akrylnitril-Butadien-Styrol-Polymer[es] *n* acrylonitrile-butadiene-styrene polymer
Akrylonitril *n s.* Akrylnitril
Akryloylchlorid *n* $CH_2=CHCOCl$ acryloyl chloride
Akrylsäure *f* $CH_2=CHCOOH$ acrylic acid, propenoic acid
Akrylsäureamid *n s.* Akrylamid
Akrylsäureester *m* acrylic (acrylic-acid) ester
Akrylsäuremethylester *m* $CH_2=CHCOOCH_3$ methyl acrylate
Akrylsäurenitril *n s.* Akrylnitril
Aktinid[enelement] *n s.* Aktinoidenelement
aktinisch actinic
Aktinium *n* Ac actinium
Aktinium-Emanation *f s.* Radon-219
Aktiniumfluorid *n* AcF_3 actinium fluoride
Aktiniumhydroxid *n* $Ac(OH)_3$ actinium hydroxide
Aktiniumphosphat *n* $AcPO_4$ actinium phosphate
Aktiniumreihe *f s.* 1. Aktiniumzerfallsreihe; 2. Aktinoidenreihe
Aktiniumsulfid *n* Ac_2S_3 actinium sulphide
Aktiniumzerfallsreihe *f* actinium [decay] series
Aktinoid *n s.* Aktinoidenelement
Aktinoidenelement *n* actinoid [element]
Aktinoidengruppe *f s.* Aktinoidenreihe
Aktinoidenkontraktion *f* actinoid contraction
Aktinoidenreihe *f* actinoid group (series)
Aktinolith *m (min)* actinolite *(an inosilicate)*
Aktinometer *n* actinometer
Aktinometrie *f* actinometry
Aktinomyzin *n* actinomycin *(antibiotic)*
Aktinon *n s.* Radon-219
Aktinorhodin *n* actinorhodine *(a quinonoid pigment from Actinomycetes)*
Aktinouran *n* AcU actinouranium *(the uranium isotope of mass 235)*
Aktithiazinsäure *f* actithiazic acid, 2-(5-carboxypentyl)-4-thiazolidone
aktiv active
  lichtelektrisch a. photoactive
  nicht a. *s.* aktivitätsfrei
  optisch a. optically active
Aktivanode *f* sacrificial anode
Aktivation *f* activation
Aktivator *m* activator, activating agent (substance); promoter, promoting agent; *(rubber)* [polymerization] initiator, initiating agent, reaction catalyst; *(met)* energizer *(additive to a carburizer)*
  A. für den Beschleuniger *(rubber)* activator of cure (vulcanization), accelerator activator
Aktivatoreffekt *m (tox)* activation
Aktivatorterm *m* activator term (level)
Aktivchlor *n* available (active) chlorine
Aktivchlorbedarf *m* available chlorine demand
Aktivchlorgehalt *m* available chlorine content
Aktiverde *f* active (activated) earth
aktivieren to activate; *(radiochemistry)* to radioactivate; *(rubber)* to boost
Aktivierung *f* activation; radioactivation *(radiochemistry)*
Aktivierungsanalyse *f* activation (radioactivation) analysis
Aktivierungsenergie *f* activation energy
Aktivierungsentropie *f* activation entropy
Aktivierungsmethode *f* activation method
Aktivierungsmittel *n* activating agent (substance), activator; *(met)* energizer *(additive to a carburizer)*
Aktivierungsprozeß *m* activation process
Aktivierungsquerschnitt *m* activation cross section
Aktivierungsstadium *n* initiation stage
Aktivierungswärme *f* heat of activation
Aktivierungszusatz *m (met)* energizer *(additive to a carburizer)*
Aktivität *f* activity
  herbizide A. herbicidal activity
  katalytische A. catalytic activity
  lichtelektrische A. photoactivity
  optische A. optical activity (rotation, rotatory power)
Aktivitätsanalyse *f* activation (radioactivation) analysis
Aktivitätsfaktor *m* activity factor (coefficient)
aktivitätsfrei *(nucl)* cold *(allowing little or no possibility of contact with radioactivity)*
Aktivitätskoeffizient *m* activity coefficient (factor)
  individueller A. individual activity coefficient
  kinetischer A. kinetic activity coefficient (factor)
  mittlerer A. mean activity coefficient (factor)
Aktivkohle *f* active (activated) charcoal, active (activated) carbon
Aktivkohleanlage *f (petrol)* charcoal plant
Aktivruß *m* active (reinforcing) black
Aktivsauerstoff *m* available oxygen
Aktivstoff *m*, **Aktivsubstanz** *f* active substance (agent, ingredient, principle, material)
Aktivtonerde *f* activated alumina

Akzelerator *m (plast)* accelerator
Akzent *m (nomencl)* prime *(with locants)*
Akzeptor *m* acceptor; electron acceptor
  A. des Elektronenpaares electron-pair acceptor
Akzeptoreigenschaften *fpl* acceptor ability (power)
Akzeptorniveau *n* acceptor level
Akzeptorstörstelle *f* acceptor impurity
Akzeptorterm *m* acceptor level
Akzeptorverhalten *n* acceptor behaviour
Akzeptorverunreinigung *f* acceptor impurity
Akzessibilität *f* accessibility
Akzessorien *npl* accessory minerals (components, constituents
Akzidenzdruck *m* job printing
Akzidenzfarbe *f* job-press ink
AL *s.* Alginatfaserstoff
Alabamin *n s.* Astat
Alabandin *m (min)* alabandite, manganblende *(manganese(II) sulphide)*
Alabaster *m (min)* alabaster *(calcium sulphate-2-water)*
Alabasterglas *n* alabaster glass
Alabasterkarton *m* alabaster board (cardboard)
Alan *n* $(AlH_3)_x$ alane, aluminium hydride
Alanat *n* $M^I[AlH_4]$ tetrahydridoaluminate, alanate *(a complex hydride)*
Alanin *n* $CH_3CH(NH_2)COOH$ alanine, 2-aminopropionic acid
Alantkampfer *m* alant camphor, alantolactone, helenin
Alarmapparat *m* alarm
Alarmgrenze *f* danger point *(waste-water treatment)*
Alarmvorrichtung *f* alarm
Alaun *m* $M^IM^{III}(SO_4)_2 \cdot 12H_2O$ alum, *(specif)* $KAl(SO_4)_2 \cdot 12H_2O$ potassium (potash) alum
  gebrannter A. dried (burnt, exsiccated) alum
alaungar alum-tanned, alum-dressed, alumed
alaungerben to taw
Alaungerbung *f* tawing, alum tannage
alaunhaltig aluminous, aluminiferous
Alaunleder *n* alum leather
Alaunlösung *f (pap)* alum liquor *(aluminium sulphate solution)*
Alaunmehl *n* alum flour (meal) *(fine crystalline potash alum)*
Alaunschiefer *m* alum schist (shale, slate)
Alaunspat *m s.* Alunit
Albany-Schlicker *m (ceram)* Albany slip *(produced of Albany clay)*
Albany-Ton *m (ceram)* Albany clay
Albert-Effekt *m* Albert reversal
Albertit *m* albertite *(a bituminous mineral)*
Albit *m (min)* albite *(sodium aluminosilicate)*
Albumin *n* albumin, *(sometimes)* albumen *(one of a class of simple proteins)*
albuminartig albuminous
Albuminleim *m* blood glue
Albuminoid *n* albuminoid
Albuminsol *n* albumin sol
Albuminverfahren *n (phot)* albumin process
Albumose *f* albumose *(a protein derivative)*
Albumosesilber *n (pharm)* silver protein
$AlCl_3$-KW-Komplex *m* / **inaktiver** *s.* Aluminiumchlorid-Kohlenwasserstoff-Komplex / inaktiver
Aldanit *m (min)* aldanite
Aldehyd *m* aldehyde
  aliphatischer A. aliphatic aldehyde, alkanal
Aldehyd C12[L] *m* dodecanal, aldehyde C-12 [lauric]
Aldehydaminbeschleuniger *m (rubber)* aldehyde-amine accelerator
Aldehyddehydrogenase *f* aldehyde dehydrogenase
Aldehyddimerisation *f s.* Aldolkondensation
aldehydfrei aldehyde-free
Aldehydgerbung *f* aldehyde tannage
Aldehydgruppe *f* -CHO aldehyde (aldehydic) group
Aldehydharz *n* aldehyde resin
Aldehydigkeit *f s.* Aldehydranzigkeit
Aldehydkarbonsäure *f s.* Aldehydsäure
Aldehydkondensation *f* aldehyde condensation
Aldehydranzigkeit *f* rancidity with formation of aldehydes
Aldehydreagens *n* / **Feders** Feder solution for aldehydes
Aldehydsäure *f* aldehyde (aldehydic) acid
Aldehydsynthese *f* / **Gattermannsche** Gattermann aldehyde synthesis
Aldimin *n* aldimine
Aldohexonsäure *f* aldonic (glyconic) acid *(any of several acids derived from an aldose; general formula* $HOCH_2[CHOH]_4COOH$*)*
Aldohexose *f* aldohexose *(a hexose containing an aldehyde group)*
Aldoketen *n* $RCH{=}C{=}O$ aldoketene
Aldol *n* aldol *(any of a class of 3-hydroxyaldehydes), (specif)* $CH_3CH(OH)CH_2CHO$ acetaldol, aldol, 3-hydroxybutyraldehyde
Aldoladdition *f s.* Aldolkondensation
Aldolase *f* aldolase
Aldolisation *f*, **Aldolisierung** *f s.* Aldolkondensation
Aldolkondensation *f* aldol condensation (addition), aldolization
  gekreuzte A. crossed aldol condensation
Aldonsäure *f* $HOCH_2[CHOH]_nCOOH$ aldonic acid *(any of a class of acids derived from aldoses)*
Aldose *f* aldose *(any of a class of sugars containing one -CHO group per molecule)*
Aldosteron *n* aldosterone *(a steroid hormone)*
Aldoxim *n* $R{-}CH{=}NOH$ aldoxime
Aleppogallen *fpl (tann)* Aleppo (Smyrna, Turkish) galls *(from Quercus infectoria Oliv.)*
Aleppokiefer *f* Aleppo pine, Pinus halepensis Mill.
Alepponuß *f* pistachio nut *(from Pistacia vera L.)*
Aleuritinsäure *f* aleuritic acid, 9, 10, 16-trihydroxyhexadecanoic acid
Aleuron *n* aleurone *(reserve protein material)*
Aleuronkörner *npl* aleurone grains
Aleuronschicht *f* aleurone layer
Alexandrit *m (min)* alexandrite (beryllium aluminate)
Alfa-Butter *f* Alfa butter *(made by the Alfa process)*

Alfa-Butterung *f s.* Alfa-Verfahren
Alfapapier *n* esparto paper
Alfa-Verfahren *n* Alfa process *(a buttermaking process)*
Alfin-Katalysator *m (rubber)* Alfin catalyst
Alfin-Kautschuk *m* Alfin [catalyzed] polymer
Alfin-Polymerisation *f (rubber)* Alfin [catalyzed] polymerization
Algarobilla *f (tann)* algarob[ill]a *(seeds of Caesalpinia brevifolia (Baill.)Clos.)*
Algarotpulver *n (pharm)* algaroth [powder] *(consisting principally of antimony chloride oxide)*
Algenbekämpfungsmittel *n* algicide
Algensäure *f s.* Alginsäure
Alginat *n* alginate *(salt or ester of alginic acid)*
Alginatcreme *f (pharm)* alginate cream
Alginatfaden *m (text)* alginate thread
Alginatfaser *f (text)* alginate fibre
Alginatfaserstoff *m (text)* alginate fibre
Alginatseide *f (text)* alginate yarn
Alginit *m* alginite *(a kind of coal)*
Alginsäure *f* alginic acid *(a polymer of D-mannuronic acid)*
Alginsäureester *m* alginic ester, alginate
Aliphaten *pl* aliphatics, aliphatic compounds
Aliphatenchemie *f* aliphatic chemistry
aliphatisch aliphatic
Alit *m* alite *(a constituent of portland-cement clinker)*
alitieren to alitize *(to make metals oxidation-resistant by heating them with powdered aluminium)*
Alizarin *n* alizarin
Alizarinblau *n* alizarin blue
Alizarinbordeaux *n* alizarin bordeaux *(mostly equalling quinalizarin, mordant violet 26)*
Alizarinbrillantgrün *n* alizarin cyanine green, acid green 25
Alizarinfarbstoff *m* alizarin dye *(important mordant dye)*
Alizarinfluorblau *n* alizarin fluorine blue
Alizaringelb *n* alizarin yellow
Alizarinindig[o]blau *n* alizarin indigo blue
Alizarinorange *n* alizarin orange
Alizarinprimverosid *n* alizarin primveroside, ruberythric acid
Alizarinprobe *f (food)* alizarin test
Alizarinrot *n* alizarin red
Alizarinsaphirol B *n* alizarin saphirol B, acid blue 45
Alizarinviolett *n* alizarin violet
Alizarinzyanin *n* alizarin cyanine
Alizarinzyaningrün *n* alizarin cyanine green
Alizarolprobe *f (food)* alizarol (alizarin-alcohol) test
Alizyklen *pl* alicyclics, alicyclic compounds
alizyklisch alicyclic, cycloaliphatic
Alk. *s.* Alkohol
alkal. *s.* alkalisch
Alkali *n* alkali
  aktives (effektives) A. *(pap)* active (effective) alkali (chemical) *($NaOH + Na_2S$, expressed as Na_2O)*
  eingestelltes (normales, standardisiertes) A. standard alkali
  wirksames A. *s.* aktives A.
alkaliähnlich alkali-like
Alkalialkoholat *n* alkali alkoxide (alcoholate)
alkaliarm poor in alkali
Alkaliaufwand *m (pap)* amount of alkali required
Alkaliazetylid *n* alkali acetylide (carbide)
Alkalibedarf *m (pap)* amount of alkali required
Alkalibehandlung *f (pap)* alkali [extraction] stage, caustic [extraction] stage, alkaline-washing stage
  zweite A. bei hoher Stoffdichte high-density second caustic extraction
alkalibeständig alkali-stable, alkali resistant, alkali-resisting, alkali-fast, alkali-proof
  nicht a. alkali-unstable, alkali-labile
Alkalibeständigkeit *f* alkali resistance, resistance (stability) to alkali[es]
alkalibildend alkaligenous
Alkalibindemittel *n* alkali-binding agent
Alkaliblau *n* alkali blue, Nicholson blue
Alkaliboden *m* alkali soil, solonetz
Alkalichelat *n* alkali chelate
alkaliecht *(dye)* alkali-fast, fast to alkali[es]
alkaliempfindlich alkali-sensitive, sensitive to alkalies (alkaline reagents)
Alkaliempfindlichkeit *f* alkali sensitivity, sensitivity to alkalies (alkaline reagents)
Alkalien *npl* alkalies, alkalis
Alkalien ... *s.* Alkali ...
Alkaliextraktion *f s.* Alkalibehandlung
Alkalifehler *m* alkali error *(of a glass electrode)*
Alkalifeldspat *m* alkali feldspar
alkalifest *s.* alkalibeständig
Alkalifestigkeit *f s.* Alkalibeständigkeit
Alkaligehalt *m* alkali content
Alkaligestein *n* alkali rock
Alkaliglas *n* alkali glass, glass A
Alkaliglasur *f (ceram)* alkaline glaze
Alkaligranit *m* alkali granite
Alkalihalogenid *n* alkali halogenide (halide)
alkalihaltig alkali-containing
Alkalihydroxid *n* alkali hydroxide
Alkalikalkgestein *n* calc-alkali[c] rock
Alkalikarbid *n* alkali metal carbide (acetylide)
Alkalikarbonat *n* alkali carbonate
Alkalikochung *f (pap)* alkaline cook
alkalilabil *s.* alkaliunbeständig
Alkalilauge *f* lye
Alkalilignin *n (pap)* alkali lignin
alkalilöslich alkali-soluble, soluble in alkali[es]
Alkalimenge *f (pap)* quantity of caustic
Alkalimetall *n* alkali[ne] metal
Alkalimetallhalogenid *n* alkali halogenide (halide)
Alkalimetallhydroxid *n* alkali hydroxide
Alkalimetallkarbonat *n* alkali carbonate
Alkalimetalloxid *n* alkali oxide
Alkalimetallpolymer[es] *n*, **Alkalimetallpolymerisat** *n (rubber)* alkali metal polymer

Alkalimetallpolymerisation *f (rubber)* alkali metal polymerization, alkali-metal catalyzed polymerization
Alkalimetallsalz *n* alkali [metal] salt
Alkalimeter *n* alkalimeter *(for measuring the proportion of alkali in a solution)*
Alkalimetrie *f* alkalimetry
alkalimetrisch alkalimetric
Alkalinität *f* alkalinity *(referring to hydrogen carbonates dissolved in water)*
Alkaliregenerat *n (rubber)* alkali reclaim, alkaline type of reclaim
Alkalireserve *f (med)* alkali reserve
Alkalirückgewinnung *f (pap)* alkali (soda) recovery
Alkalisalz *n* alkali [metal] salt
alkalisch alkaline, basic ✦ **a. machen** to make alkaline, to alkalify, to alkal[in]ize ✦ **a. reagieren** to react alkaline ✦ **a. stellen** *(dye)* to make alkaline, to alkal[in]ize, to basify
 a. aufgeschlossen (gekocht) *(pap)* alkaline-cooked
 schwach a. alkalescent, subalkaline
 stark a. superalkaline
Alkalischmachen *n* alkal[in]ization
Alkalischmelze *f* alkali[ne] fusion, alkali melt
alkalisierbar alkalifiable, alkalizable
alkalisieren to alkal[in]ize, to alkalify, to make alkaline
Alkalisierung *f* alkal[in]ization; *(pap)* alkaline purification, alkali refining
Alkalisierungsturm *m (pap)* caustic (alkaline extraction) tower *(multistage bleaching)*
Alkalisilikat *n* alkali silicate
Alkalispaltung *f* alkali cleavage
Alkalität *f* alkalinity, basicity, basic strength
Alkaliturm *m* 1. *(pap)* reaction tower *(pulping with chlorine)*; 2. *s.* Alkalisierungsturm
alkaliunbeständig alkali-unstable, alkali-labile
alkaliunlöslich alkali-insoluble, insoluble in alkali[es]
Alkali-Veredlungslauge *f (pap)* alkali refining liquor
Alkaliverfahren *n (rubber)* alkali [reclaiming] process
Alkaliverhältnis *n (pap)* alkali[-to-wood] ratio, chemical[-to-wood] ratio, ratio of chemical to wood
Alkaliverlust *m (pap)* loss of chemical ✦ **den A. decken** to make up for the loss of chemical
Alkaliwäsche *f* 1. *s.* Alkalibehandlung; 2. *(petrol)* caustic[-soda] wash
Alkalizellulose *f* alkali cellulose, *(Am)* soda cellulose
 zerfaserte A. crumbs
Alkalizusatz *m (pap)* alkali make-up, make-up chemical
Alkalizyanid *n* alkali cyanide
Alkaloid *n* alkaloid
 A. mit Chinolinring quinoline-type alkaloid
 A. mit Isochinolinring isoquinoline-type alkaloid
alkaloidartig alkaloid-like
Alkaloidbase *f* alkaloid base
alkaloidisch alkaloidal
Alkaloidlösung *f* alkaloidal solution
Alkaloidvergiftung *f* alkaloid poisoning
Alkalose *f (med)* alcalosis
Alkamin *n* alkamine, amino alcohol
Alkan *n* alkane, paraffin [hydrocarbon], saturated hydrocarbon
 geradkettiges (unverzweigtes) A. normal alkane
Alkanal *n* alkanal
Alkanhalogenid *n* alkyl halide, haloalkane
Alkanna *f* 1. alkanna, alkanet, Alkanna tinctoria (L.)Tausch.; 2. alkanna, alkanet *(root from 1. containing the colouring matter alkannin)*
Alkannafarbstoff *m s.* Alkannin
Alkannarot *n s.* Alkannin
Alkannin *n* alkannin *(a red crystalline colouring matter)*
Alkanol *n* alkanol
Alkanolamin *n* alkanolamine
Alkanon *n* alkanone
Alkanphosphonigsäure *f* $RP(OH)_2$ alkylphosphonous (alkylphosphinic) acid, *(Am also)* alkanephosphonous acid
Alkanreihe *f* alkane family
Alkarsin *n* $[As(CH_3)_2]_2O$ cacodyl oxide, alkarsin
Alkazidanlage *f* alkazid plant *(gas purification)*
Alken *n* alkene, olefin [hydrocarbon], ethylenic hydrocarbon
 fluoriertes A. fluoroalkene, fluoro-olefin
Alkenderivat *n* alkene derivative
Alkenimin *n* alkeneimine
Alkenreihe *f* alkene series (family), olefin series
Alkensäure *f* alkenoic acid *(any of the monounsaturated fatty acids)*
Alkenylbenzol *n* alkenylbenzene, aromatic-alkene compound
Alkermes *m* grains of kermes *(the dried bodies of various female scales of the genus Kermes)*
Alkiminchelat *n* alkimine chelate
Alkin *n* alkyne, acetylenic hydrocarbon
Alkinsäure *f* alkynoic acid
Alkinylbenzol *n* alkynylbenzene
Alkohol *m* alcohol, *(specif)* C_2H_5OH ethyl alcohol
 absoluter A. absolute (dehydrated, anhydrous) alcohol
 denaturierter A. denatured alcohol, methylated spirit
 dreiwertiger A. trihydric alcohol
 einwertiger A. monohydric alcohol
 gewöhnlicher A. fermentation alcohol *(ethanol)*
 höherer A. higher alcohol
 mehrwertiger A. polyhydric (polyhydroxy, polyfunctional) alcohol, polyalcohol, polyol
 primärer A. primary alcohol
 reiner A. *s.* absoluter A.
 sekundärer A. secondary alcohol
 technischer A. commercial (industrial) alcohol (spirit)
 tertiärer A. tertiary alcohol
 vergällter A. *s.* denaturierter A.
 vierwertiger A. tetrahydric alcohol

wasserfreier A. *s.* absoluter A.
zweiwertiger A. dihydric alcohol, diol
Alkohol C 11 *m* $CH_3[CH_2]_{10}CH_2OH$ 1-undecanol, alcohol C-11
Alkohol C 12 *m* $CH_3[CH_2]_{10}CH_2OH$ 1-dodecanol, alcohol C-12
alkoholarm light *(beverage)*
alkoholartig alcohol-like
Alkoholat *n* alkoxide, alcoholate
Alkoholauszug *m* alcoholic extract
Alkoholdampf *m* alcohol vapour
Alkoholdehydrogenase *f* alcohol dehydrogenase
Alkoholentwöhnungsmittel *n* alcohol deterrent
Alkoholextrakt *m s.* Alkoholauszug
alkoholfrei alcohol-free, non-alcoholic, soft
Alkoholgehalt *m* alcohol[ic] concentration, alcohol[ic] content ✦ **von geringem A.** light *(beverage)*
alkoholhaltig alcoholic, spirituous
alkoholisch alcoholic, spirituous
schwach a. light *(beverage)*
alkoholisierbar alcoholizable
alkoholisieren to alcoholize
Alkoholisierung *f* alcoholization
Alkoholkraftstoff *m* alcohol fuel
alkohollöslich alcohol-soluble, spirit-soluble, soluble in alcohol
Alkoholmesser *m s.* Alkoholometer
Alkohol[o]meter *n* alcohol[o]meter
Alkoholometrie *f* alcoholometry
alkoholometrisch alcoholometric
Alkoholprobe *f* alcohol test
A. zur Milchuntersuchung alcohol coagulation test
Alkoholtest *m* alcohol test
alkoholunlöslich insoluble in alcohol
Alkoholyse *f* alcoholysis ✦ **einer A. unterwerfen** to alcoholyze
Alkosol *n* alcosol *(a colloidal system in which the liquid is alcohol)*
Alkoxygruppe *f s.* Alkoxylgruppe
Alkoxylgruppe *f* $C_nH_{2n+1}O$- alkoxyl (alkoxy) group, alkoxyl (alkoxy) residue
Alkyd *n* alkyd [resin]
Alkydharz *n* alkyd [resin]
kurzöliges A. short-oil alkyd
langöliges A. long-oil alkyd
mittelöliges A. medium-oil alkyd
styrolisiertes A. styrenated alkyd
Alkydharzlack *m* alkyd varnish
Alkyl *n* alkyl
Alkylamin *n* alkylamine
Alkylans *n* alkylating agent
Alkylarylsilikon *n s.* Alkylarylsiloxan
Alkylarylsiloxan *n* alkyl aryl siloxane
Alkylarylsulfonat *n* alkylarenesulphonate *(any of a class of detergents)*
Alkylat *n* alkylate
Alkylation *f s.* Alkylierung
Alkylbenzol *n* alkylbenzene
Alkylbrücke *f* alkyl bridge
Alkylderivat *n* alkyl derivative
Alkyldihalogenid *n* alkyl dihalide
Alkylen *n s.* Alken
Alkylgruppe *f* alkyl group (residue)
Alkylhalogenid *n* alkyl halide, haloalkane
alkylieren to alkylate
Alkylierung *f* alkylation, alkanation
A. mit Fluorwasserstoff *s.* A. mit Fluorwasserstoffsäure
A. mit Fluorwasserstoffsäure (Flußsäure) hydrofluoric-acid alkylation, HF alkylation
A. mit Schwefelsäure sulphuric-acid alkylation
thermische A. thermal alkylation
Alkylierungsmittel *n* alkylating agent
Alkylierungsverfahren *n* alkylation process
Alkyljodid *n* alkiodide
1,2-Alkylkarboniumumlagerung *f* 1,2-shift of alkyl, alkyl shift
Alkylmonohalogenid *n* alkyl monohalide
Alkylperoxid *n* alkyl peroxide
Alkylphenolharz *n* alkylphenol resin
Alkylphenolnovolak *m* alkyl phenol novolak
Alkylphosphonsäure *f* $RP=O(OH)_2$ alkylphosphonic acid
Alkylquecksilberverbindung *f* alkylmercury [compound], *(incorrectly)* mercury alkyl
Alkylradikal *n* alkyl radical, *(specif)* free alkyl radical
Alkylrest *m* alkyl residue (group)
Alkylschwefelsäure *f* alkylsulphuric acid
Alkylsilikon *n* alkyl silicone
höheres A. higher alkyl silicone
Alkylsulfid *n* alkyl sulphide, thioether
Alkylsulfonat *n* alkyl sulphonate
Alkylsulfonsäure *f* alkylsulphonic acid
Alkylverbindung *f* alkyl compound
Alkylwanderung *f* migration of an alkyl group
Alkylzyanid *n* alkyl cyanide
Allanit *m (min)* allanite, orthite *(a sorosilicate)*
Allantoin *n* allantoin, glyoxylic diureide
Allantoinsäure *f* $(NH_2-CO-NH)_2CHCOOH$ allantoic acid, diureidoacetic acid
all-cis-Isomer[es] *n* all-cis isomer
Allelopathie *f (biol)* allelopathy
Allelopathikum *n (biol)* allelopathic
allelopathisch *(biol)* allelopathic
allelotrop allelotropic
Allelotropie *f* allelotropism
Allemontit *m* allemontite *(an arsenic antimony mineral)*
Allen *n* allene, *(specif)* $CH_2=C=CH_2$ allene, propadiene
Allenisomerie *f* allene isomerism
Allen-Kegel *m*, **Allen-Konus** *m* Allen cone classifier
Allen-Moore-Zelle *f* Allen-Moore cell
Allergen *n (med)* allergen
Alleskleber *m* all-purpose adhesive
Allesreiniger *m* all-purpose cleaner
Allgemeinempfindlichkeit *f (phot)* overall sensitivity, emulsion speed
Allgemeinformel *f* general formula
Allgemeinschleier *m (phot)* overall fog

Allgewürz *n* allspice *(from Pimenta dioica (L.) Merr.)*
Allheilmittel *n* panacea, cure-all
Allihn-Kühler *m* Allihn (bulb) condenser
allochthon *(geol)* allochthonous
Allokatalyse *f* allocatalysis
Allokrotonsäure *f s.* Isokrotonsäure
Allopalladium *n (min)* allopalladium
Allophan *m (min)* allophane *(a phyllosilicate)*
Allophanamid *n* $H_2N-CO-NHCONH_2$ allophanamide, carbamyl urea, biuret
Allophansäureamid *n s.* Allophanamid
Allose *f* allose *(an aldohexose)*
allotriomorph *(cryst)* allotriomorphic, anhedral, xenomorphic
allotrop allotropic
Allotropie *f* allotropism, allotropy *(existence of an element in two or more different modifications)*
Allozimtsäure *f* $C_6H_5CH=CHCOOH$ allocinnamic acid, *cis*-cinnamic acid, *cis*-3-phenylacrylic acid
All-sliming-Verfahren *n* all-sliming process *(gold recovery)*
All-trans-Isomer[es] *n* all-trans isomer
Allyl *n* allyl
Allylaldehyd *m s.* Akrolein
Allylalkohol *m* $CH_2=CHCH_2OH$ allyl alcohol, AA, 2-propen-1-ol
p-**Allylanisol** *n* $CH_2=CHCH_2C_6H_4OCH_3$ *p*-allylanisole, 1-allyl-4-methoxybenzene
Allylbromid *n* $CH_2=CHCH_2Br$ 3-bromopropene, allyl bromide
Allylbromierung *f* allylic bromination
Allylchlorid *n* $CH_2=CHCH_2Cl$ 3-chloropropene, allyl chloride
Allylen *n s.* Propin
Allylester-Kunststoff *m* allyl plastic
Allylgruppe *f* allyl group (residue)
Allylharz *n* allyl[ic] resin
Allylharz-Kunststoff *m* allyl plastic
Allylierung *f* allylation
Allylradikal *n* allyl radical, *(specif)* free allyl radical
Allylrest *m* allyl residue (group)
Allylsenföl *n* allyl mustard oil
allylständig allyl
Allylumlagerung *f* allyl rearrangement
Allylwanderung *f* allyl migration
Allzweckkautschuk *m* general-purpose rubber
Allzwecksynthesekautschuk *m* general-purpose synthetic rubber
Almandin *m (min)* almandine, almandite *(aluminium iron(II) orthosilicate)*
Almén-Nylander-Probe *f* Almen-Nylander test *(of sugar)*
Aloin *n* aloin *(a cathartic obtained from aloe)*
Aloinprobe *f* aloin test *(of blood)*
Alphaspektrum *n* alpha-particle spectrum
Alphastrahlen *mpl* alpha rays
Alphastrahlenquelle *f* alpha-radiation source
Alphastrahler *m* alpha emitter
Alphastrahlung *f* alpha radiation
Alphateilchen *n* alpha particle
Alphazellulose *f* alpha cellulose
Alphazerfall *m* alpha decay (disintegration)
Alphazerfallsenergie *f* alpha disintegration energy
Alphinateilchen *n (nucl)* alphina particle
ALS *s.* Alginatseide
Alstonit *m (min)* alstonite
Altait *m (min)* altaite *(lead telluride)*
Altblei *n* scrap lead
Alteisen *n* scrap iron
altern 1. to age *(e.g. a rubber product)*; 2. to undergo ageing, to age, *(of precipitates also)* to digest
Alternativparameter *m* alternative parameter
Altersbestimmung *f* age determination
  A. durch Radionuklide *s.* radioaktive A.
  absolute (physikalische, physikalisch-chemische) A. *s.* radioaktive A.
  radioaktive A. radioactive (chemical) dating
Alterung *f* 1. ag[e]ing; 2. digestion *(of a precipitate)*
  A. durch Licht light ageing
  A. im Geer-Ofen *(rubber)* Geer oven ageing
  A. im Wärmeschrank *(rubber)* [air] oven ageing
  A. im Zellenofen *(rubber)* test-tube ageing
  A. in der Sauerstoffbombe *(rubber)* oxygen bomb ageing
  beschleunigte A. *s.* künstliche A.
  künstliche A. artificial (accelerated) ageing
  natürliche A. natural ageing
  thermische A. heat ageing, thermosenescence
alterungsbeständig resistant to ageing, age-resistant, non-ageing
Alterungsbeständigkeit *f* ageing resistance, resistance to ageing
Alterungseigenschaften *fpl* ageing characteristics (performance, properties)
Alterungsgeschwindigkeit *f* rate of ageing
Alterungsprüfung *f* ageing test
  A. nach Geer *(rubber)* Geer oven test
  beschleunigte A. *s.* künstliche A.
  künstliche A. artificial (accelerated) ageing test
Alterungsschutzmittel *n* antiageing agent, antiager, age-resistor
  A. gegen Oxydation und Biegerisse *(rubber)* anti-flex-cracking antioxidant
Alterungstest *m* ageing test
Alterungsverhalten *n s.* Alterungseigenschaften
Alterungsversuch *m* ageing test
Alterungswiderstand *m s.* Alterungsbeständigkeit
Altgeschmack *m (food)* stale flavour (taste)
Altgummi *m* rubber scrap (waste), scrap (waste) rubber
Altgummibrecher *m* breaking mill for scrap rubber
Althopfen *m (ferm)* old hop
Altkatalysator *m* used catalyst
Altmaterial *n* waste [material]
Altmetall *n* scrap [metal]
Altöl *n* used oil
Altpapier *n* waste (old) paper
  aufbereitetes (regeneriertes) A. recovered (repulped) stock, recovered (repulped) waste paper, old-paper stock
Altpapieraufbereitung *f*, **Altpapierregeneration** *f* recovery of waste paper

Altpapierstoff *m s.* Altpapier / aufbereitetes
Altpapierstoffärbung *f* waste-paper colouring
Altrose *f* altrose *(an aldohexose)*
Altstoff *m* waste [material]
Alttuberkulin *n (pharm)* old tuberculin
Altwurzel *f* elecampane *(dried root of Inula helenium L.)*
Alu *n s.* Aluminium
Aluminat *n* aluminate
Aluminatlauge *f* aluminate liquor (solution), sodium aluminate solution *(Bayer process)*
Aluminatlösung *f s.* Aluminatlauge
Aluminid *n* aluminide
aluminieren to aluminize
Aluminit *m (min)* aluminite *(hydrous aluminium sulphate)*
Aluminium *n* Al aluminium, *(Am)* aluminum
Aluminiumalkyl *n (incorrectly for)* Trialkylaluminium
Aluminiumarsenat(V) *n* $AlAsO_4$ aluminium arsenate
Aluminiumazetat *n* $Al(CH_3COO)_3$ aluminium acetate
basisches A. $Al(OH)(CH_3COO)_2$ aluminium acetate hydroxide
Aluminiumazetylazetonat *n* $Al[CH(CO-CH_3)_2]_3$ aluminium acetylacetonate
Aluminiumblech *n* sheet aluminium
Aluminiumbromid *n* $AlBr_3$ aluminium bromide
Aluminiumbronze *f* aluminium bronze
Aluminiumbronzepulver *n* aluminium bronze powder
Aluminiumchlorat *n* $Al(ClO_3)_3$ aluminium chlorate
Aluminiumchlorid *n* $AlCl_3$ aluminium chloride
Aluminiumchlorid-Kohlenwasserstoff-Komplex *m* / **inaktiver** *(petrol)* complex out *(in liquid-phase isomerization)*
Aluminiumerz *n* aluminium ore
Aluminiumfluorid *n* AlF_3 aluminium fluoride
wasserfreies A. anhydrous aluminium fluoride
Aluminiumfluorid-Hydrat *n* aluminium fluoride hydrate
Aluminiumfolie *f* aluminium foil
Aluminiumfolien-Kaschierpapier *n* aluminium foil backing paper
Aluminium(I)-halogenid *n* aluminium(I) halogenide, aluminium monohalogenide (monohalide)
Aluminium(III)-halogenid *n* aluminium(III) halogenide, aluminium trihalogenide (trihalide)
aluminiumhaltig aluminous, aluminiferous
Aluminiumhexafluorosilikat *n* $Al_2[SiF_6]_3$ aluminium hexafluorosilicate, aluminium fluorosilicate
Aluminiumhütte *f* aluminium works
Aluminiumhydroxid *n* $Al(OH)_3$ aluminium hydroxide, aluminium trihydroxide
Aluminiumhydroxiddiazetat *n* $Al(OH)(CH_3COO)_2$ aluminium diacetate hydroxide
Aluminiumhydroxiddiformiat *n* $Al(OH)(HCOO)_2$ aluminium diformate hydroxide
Aluminiumjodid *n* AlI_3 aluminium iodide
Aluminiumkarbid *n* Al_4C_3 aluminium carbide
Aluminiumlack *m* aluminium lake
Aluminiumlegierung *f* aluminium alloy
Aluminiummonohalogenid *n s.* Aluminium(I)-halogenid
Aluminiumnaphthenat *n* aluminium naphthenate
Aluminiumnitrat *n* $Al(NO_3)_3$ aluminium nitrate
Aluminiumnitrid *n* AlN aluminium nitride
Aluminiumorthohydroxid *n s.* Aluminiumhydroxid
Aluminiumoxid *n* Al_2O_3 aluminium oxide
Aluminiumoxidhydrat *n* hydrous aluminium oxide
Aluminiumpapier *n* aluminium (silver) paper
Aluminiumpulver *n* aluminium powder
gesintertes A. sintered aluminium powder, S.A.P.
Aluminium-Raffinationselektrolyse *f* **nach Hoopes** Hoopes [electrolytic-refining] process
Aluminiumrhodanid *n s.* Aluminiumthiozyanat
Aluminiumseife *f* aluminium soap
Aluminiumsilikatglas *n* aluminosilicate glass
Aluminiumstaub *m* aluminium powder
Aluminiumstearat *n* $Al(C_{17}H_{35}COO)_3$ aluminium stearate, aluminium tristearate
Aluminiumsulfat *n* $Al_2(SO_4)_3$ aluminium sulphate
Aluminiumsulfid *n* Al_2S_3 aluminium sulphide
Aluminiumthiozyanat *n* $Al(SCN)_3$ aluminium thiocyanate, aluminium rhodanide
Aluminiumtrialkyl *n (incorrectly for)* Trialkylaluminium
Aluminiumtriäthyl *n (incorrectly for)* Triäthylaluminium
Aluminiumtriformiat *n* $(HCOO)_3Al$ aluminium triformate
Aluminiumtrihydroxid *n s.* Aluminiumhydroxid
Aluminiumtrimethyl *n (incorrectly for)* Trimethylaluminium
Aluminiumwasserstoff *m* $(AlH_3)_x$ aluminium hydride, alane
Aluminogel *n* alumina gel, gelatinous aluminium hydroxide
Aluminothermie *f* aluminothermics, aluminothermy, Goldschmidt's process
aluminothermisch aluminothermic
Alumogel *m(n) (min)* alumogel, kliachite *(gel of aluminium hydroxide)*
Alumosilikat *n* aluminosilicate
Alunit *m (min)* alunite, alumite, alumstone *(hydrous aluminium potassium sulphate)*
Alunitisation *f*, **Alunitisierung** *f* alunitization
Alunogen *m (min)* alunogene, feather alum, hair salt *(aluminium sulphate-18-water)*
Amagat *n s.* Amagat-Einheit
Amagat-Einheit *f* Amagat unit *(molar volume of a gas at 0°C and 1 atmosphere)*
Amalgam *m (min)* amalgam
Amalgam *n* amalgam *(an alloy of mercury)*
Amalgamation *f* amalgamation
europäische A. barrel amalgamation
Amalgamations ... *s.* Amalgamier ...
Amalgamator *m s.* Amalgamierapparat
Amalgamelektrode *f* amalgam electrode
Amalgamfaß *n* amalgamating barrel
Amalgamierapparat *m* amalgamation apparatus, amalgamator

amalgamieren to amalgam[ate], to amalgamize
Amalgamieren *n* amalgamation
Amalgamierfaß *n* amalgamating barrel
Amalgamierherd *m* amalgamating table
Amalgamierpfanne *f* amalgamating (amalgamation) pan
Amalgamierplatte *f* amalgamating (amalgamation) plate
Amalgamiertisch *m* amalgamating table
Amalgamierung *f* amalgamation
Amalgamierungs ... *s.* Amalgamier ...
Amalgamierverfahren *n* amalgamation process
Amalgamplatte *f* amalgamated plate
Amalgamverfahren *n* mercury-cell process *(electrolysis)*
Amalgamzelle *f* mercury cell *(electrolysis)*
Amalgamzersetzer *m* amalgam decomposer *(electrolysis)*
Amarant *n* amaranth *(a red acid azo dye)*
Amarogentin *n* amarogentin *(bitter substance of gentian and swertia)*
Amazonenstein *m*, **Amazonit** *m (min)* amazonite, amazonstone *(potassium aluminosilicate)*
Amberbaum *m* / **Amerikanischer** [American] sweet gum, Liquidambar styraciflua L.
  Orientalischer A. oriental sweet gum, Liquidambar orientalis Mill.
Amberglimmer *m* amber mica
Amberholz *n* sweet gum *(heartwood of Liquidambar styraciflua L.)*
ambient ambient, surrounding
Amblygonit *m (min)* amblygonite
Amboinokino *n* East India kino, Malabar kino *(kino gum from Pterocarpus marsupium Roxb.)*
Ambra *f* **[/ graue, natürliche]** *(cosmet)* ambergris, ambergrease
Ambrettemoschus *m* musk ambrette, 2,6-dinitro-3-methoxy-4-*tert.*-butyltoluene
Ameisenbekämpfungsmittel *n* ant poison
Ameisenfreßlack *m* ant syrup
Ameisenöl *n* / **synthetisches** artificial ant oil, furfural
Ameisensäure *f* $HCOOH$ formic acid, methanoic acid
Ameisensäureamid *n* $HCONH_2$ formamide, methanamide
Ameisensäureäthylester *m* $HCOOC_2H_5$ ethyl formate
Ameisensäuremethylester *m* $HCOOCH_3$ methyl formate
Ameisensäurenachweis *m* formic-acid test
Ameisenvertilgungsmittel *n* ant poison
Amerizium *n* Am americium
Amethyst *m (min)* amethyst *(a variety of quartz)*
Amianth *m (min)* amiant[h]us *(fine silky asbestos)*
Amid *n* $RCONH_2$ amide; M^INH_2 [metal] amide
  polymeres A. polyamide
Amidase *f* amidase
Amidbindung *f* amide linkage
amidieren to amidate
Amidierung *f* amidation
Amidin *n* amidine
N-**Amidinoglyzin** *n* *N*-guanylglycine, guanidinoacetic acid
Amidinogruppe *f* $-C(=NH)NH_2$ amidine group
Amidinokarbonylgruppe *f s.* Amidinogruppe
Amidogruppe *f* $-CONH_2$ amido group
Amidokohlensäure *f* NH_2COOH carbamic acid
Amidol-Entwickler *m (phot)* amidol developer
Amidophosphorsäure *f* $NH_2P(=O)(OH)_2$ amidophosphoric acid, phosphamic acid
Amidoquecksilber(II)-chlorid *n* $[Hg(NH_2)]Cl$ amidomercury(II)chloride, infusible white precipitate
Amidoschwefelsäure *f* NH_2SO_3H amidosulphuric acid
Amidosulfat *n* $NH_2SO_3M^I$ amidosulphate
Amidosulfonsäure *f s.* Amidoschwefelsäure
Amidstickstoff *m* amide nitrogen
Amikron *n* amicron, subsubmicron *(a disperse particle invisible under the microscope)*
amikroskopisch amicroscopic, submicroscopic
Amin *n* amine
  biogenes A. biogenic amine
  primäres A. RNH_2 primary amine
  sekundäres A. R_1R_2NH secondary amine
  tertiäres A. $R_1R_2R_3N$ tertiary amine
aminartig amine-like
Aminäscher *m (tann)* hair loosening by amines
Aminase *f* aminase
Aminbeschleuniger *m (rubber)* amino accelerator
aminieren to aminate
Aminierung *f* amination
  reduktive A. reductive amination
Aminoalkohol *m* amino alcohol, alkamine
Aminoanteil *m* amino moiety
Aminoanthrachinon *n* aminoanthraquinone
Aminoäthan *n s.* Äthylamin
1-Aminoäthanol *n* $CH_3CH(NH_2)OH$ 1-aminoethanol
2-Aminoäthanol-(1) *n* $HOCH_2CH_2NH_2$ 2-aminoethanol, 2-aminoethyl alcohol, monoethanolamine, MEA
Aminoäthansäure *f s.* Aminoessigsäure
2-Aminoäthansulfonsäure *f* $NH_2CH_2CH_2SO_3H$ 2-aminoethanesulphonic acid, taurine
1-Aminoäthylalkohol *m s.* 1-Aminoäthanol
2-Aminoäthylalkohol *m s.* 2-Aminoäthanol-(1)
Aminoäthylbenzol *n* aminoethylbenzene, phenylethylamine
Aminoazetal *n* $NH_2CH_2CH(OC_2H_5)_2$ aminoacetal
Aminoazetanilid *n* $NH_2C_6H_4NHCOCH_3$ aminoacetanilide
p-**Aminoazobenzol** *n* $NH_2C_6H_4N=NC_6H_5$ *p*-aminoazobenzene, aniline yellow
Aminoazobenzolsulfonsäure *f* $NH_2C_6H_4N=NC_6H_4SO_3H$ aminoazobenzene sulphonic acid
Aminoazoverbindung *f* aminoazo compound
Aminobenzoesäure *f* $NH_2C_6H_4COOH$ aminobenzoic acid
o-**Aminobenzoesäure** *f* $C_6H_4(NH_2)COOH$ *o*-aminobenzoic acid, anthranilic acid
p-**Aminobenzoesäure** *f* $NH_2C_6H_4COOH$ *p*-aminobenzoic acid, PABA
Aminobenzoesäureäthylester *m* $NH_2C_6H_4COOC_2H_5$ ethyl aminobenzoate

Aminobenzoesäurebutylester *m* $NH_2C_6H_4CO$-$O[CH_2]_3CH_3$ butyl aminobenzoate
Aminobenzol *n* aminobenzene, phenylamine, aniline
p-**Aminobenzolsulfonamid** *n* $NH_2C_6H_4SO_2NH_2$ *p*-aminobenzenesulphonamide, sulphanilamide
Aminobenzolsulfonsäure *f* $NH_2C_6H_4SO_3H$ aminobenzenesulphonic acid, anilinesulphonic acid
Aminobenzoyl-I-Säure *f s.* Aminobenzoyl-J-Säure
Aminobenzoyl-J-Säure *f* aminobenzoyl J acid
Aminobernsteinsäure *f* $HOOCCH(NH_2)CH_2COOH$ aminosuccinic acid, aspartic (asparaginic) acid, aminobutanedioic acid
2-Aminobornan *n* bornylamine, 2-aminobornane
Aminochinolin *n* aminoquinoline, quinolylamine
Aminodikarbonsäure *f* amino dicarboxylic acid
Aminodimethylbenzol *n* $NH_2C_6H_3(CH_3)_2$ aminodimethylbenzene
Aminodinitrophenol *n* $NH_2C_6H_2(NO_2)_2OH$ aminodinitrophenol
1-Aminododekan *n* $CH_3[CH_2]_{11}NH_2$ dodecylamine, 1-aminododecane
Aminoessigsäure *f* NH_2CH_2COOH aminoacetic acid, glycine
4-Aminofolsäure *f* 4-aminofolic acid, aminopterin
Aminoformiat *n* NH_2COOM^I aminoformate
Aminogruppe *f* $-NH_2$ amino group (residue)
Amino-G-Salz *n* amino G salt
Amino-G-Säure *f* $NH_2C_{10}H_5(SO_3H)_2$ amino-G acid, 2-naphthylamine-6,8-disulphonic acid
2-Aminokamphan *n s.* 2-Aminobornan
Aminokarbonsäure *f s.* Aminosäure
Aminokomponente *f* amino moiety
Aminolyse *f* aminolysis
Aminomethan *n* CH_3NH_2 methylamine
aminomethylieren to aminomethylate
Aminomethylierung *f* aminomethylation
Aminonaphthalin *n* naphthylamine
Aminonaphthalindisulfonsäure *f* $NH_2C_{10}H_5(SO_3H)_2$ naphthylaminedisulphonic acid, aminoaphthalenedisulphonic acid
Aminonaphthalinsulfonsäure *f* $NH_2C_{10}H_6SO_3H$ naphthylaminesulphonic acid, aminonaphthalenesulphonic acid
Aminonaphtholdisulfonsäure *f* $NH_2C_{10}H_4(OH)$-$(SO_3H)_2$ aminonaphtholdisulphonic acid
Aminonaphtholsulfonsäure *f* $NH_2C_{10}H_5(OH)SO_3H$ aminonaphtholsulphonic acid
Aminooxydase *f* amino oxidase
6-Aminopenizillansäure *f* 6-aminopenicillanic acid, 6-APA
Aminophenetol *n* $NH_2C_6H_4OC_2H_5$ aminophenetol, phenetidine, aminophenol ethyl ether
Aminophenol *n* $NH_2C_6H_4OH$ aminophenol, hydroxyaniline
Aminophenoläthyläther *m s.* Aminophenetol
p-**Aminophenol-Entwickler** *m (phot)* para-aminophenol developer, paraminophenol developer
p-**Aminophenylarsonsäure** *f* $C_6H_4NH_2AsO(OH)_2$ *p*-aminophenylarsonic acid, arsanilic acid
Aminophenylessigsäure *f* aminophenylacetic acid
p-**Aminophenylsulfonamid** *n s. p*-Aminobenzolsulfonamid
Aminoplast *m* aminoplastic
Aminoplastharz *n* amino resin
Aminopolykarbonsäure *f* amino polycarboxylic acid
2-Aminopropionsäure *f* $CH_3CH(NH_2)COOH$ 2-aminopropionic acid, alanine
γ-**Aminopropyltriäthoxysilan** *n* γ-aminopropyl triethoxysilane, γ-APT
Amino-R-Salz *n* $NH_2C_{10}H_5(SO_3H)SO_3Na$ amino-R salt
Amino-R-Säure *f* $NH_2C_{10}H_5(SO_3H)_2$ amino-R acid, 2-naphthylamine-3,6-disulphonic acid
Aminosalizylsäure *f* $HOC_6H_3(NH_2)COOH$ aminosalicylic acid, amino-2-hydroxybenzoic acid
Aminosäure *f* amino acid
  essentielle (unentbehrliche) A. essential amino acid
Amino-ε-Säure *f* $NH_2C_{10}H_5(SO_3H)_2$ epsilon acid, ε-acid, 1-naphthylamine-3,8-disulphonic acid
Aminosäureanion *n* amino-acid anion
Aminosäurechelat *n* amino-acid chelate
Aminosäureeinheit *f* amino-acid unit
Aminosäurefolge *f* amino-acid sequence, sequence of amino-acid residues
Aminosäureoxydase *f* amino-acid oxidase
Aminosäurereihenfolge *f s.* Aminosäurefolge
Aminosäurerest *m* amino-acid residue
Aminosäuresequenz *f s.* Aminosäurefolge
Amino-S-Säure *f* $NH_2C_{10}H_5(SO_3H)_2$ amino-S acid, 1-naphthylamine-4,8-disulphonic acid
Aminoteil *m* amino moiety
Aminotoluol *n* $CH_3C_6H_4NH_2$ aminotoluene, toluidine
Aminotransferase *f* transaminase
Aminoverbindung *f* amino compound
Aminoxid *n* amine oxide
Aminoxylol *n* $NH_2C_6H_3(CH_3)_2$ aminoxylene, xylidine, aminodimethylbenzene
Aminozucker *m* amino sugar
Aminsalz *n* amine salt
Aminsynthese *f* / **Gabrielsche** Gabriel phthalimide synthesis, Gabriel synthesis of primary amines
Aminzahl *f* amine value
Amin-Zucker-Bräunung *f (food)* amino-sugar browning, Maillard-Type browning
Ammeter *n s.* Amperemeter
Ammin *n* ammine *(coordination chemistry)*
Amminkomplex *m* ammine complex
Ammonalaun *m s.* Ammoniakalaun
Ammoniak *n* NH_3 ammonia
  A. in Gasform *s.* gasförmiges A.
  flüssiges A. liquid ammonia
  gasförmiges A. gaseous ammonia, ammonia gas
  synthetisches A. synthetic ammonia
  verflüssigtes A. liquefied (liquid) ammonia
  wasserfreies A. anhydrous ammonia
  wäßriges A. *s.* Ammoniakwasser 2.
Ammoniakabwasser *n* ammonia waste, spent ammonia (gas) liquor *(of a gas-works)*
Ammoniakalaun *m* $NH_4Al(SO_4)_2 \cdot 12H_2O$ ammonia

(ammonium) alum
ammoniakalisch ammoniac[al]
Ammoniakanlage *f* ammonia plant
Ammoniakäscher *m (tann)* hair loosening by ammonia
Ammoniakat *n* ammoniate, ammonate
Ammoniakbegasung *f (agric)* ammonia fumigation
Ammoniakbildner *mpl* ammonifyers, ammonifying bacteria
Ammoniakdampf *m* ammonia fume (vapour)
Ammoniakdestillationsapparat *m* ammonia-distillation apparatus
Ammoniakentwicklung *f* evolution of ammonia
Ammoniakflüssigkeit *f s.* Ammoniakwasser 2.
Ammoniakgas *n* ammonia gas, gaseous ammonia
Ammoniakgummi *n* [gum] ammoniac, *(specif)* Persian ammoniac *(from Dorema ammoniacum Don and related specc.)*
Afrikanisches A. Moroccan ammoniac *(from Ferula tingitana L. and Ferula communis L. var. brevifolia Marcz)*
Ammoniak-Gummiharz *n s.* Ammoniakgummi
ammoniakhaltig ammoniac[al]
Ammoniakhydrat *n* ammonia hydrate
Ammoniakkältemaschine *f* ammonia refrigerating machine
Ammoniakkondensator *m* ammonia condenser
Ammoniaklösung *f* **[/ wäßrige]** *s.* Ammoniakwasser 2.
Ammoniak-Luft-Gemisch *n* ammonia-air mixture
Ammoniakseife *f* ammonium soap
Ammoniak-Soda-Verfahren *n* [Solvay's] ammonia soda process, Solvay process
Ammoniakstickstoff *m* ammonia (ammoniacal) nitrogen
Ammoniaksynthese *f* ammonia synthesis
Fausersche A. Fauser [ammonia] process
Ammoniaksyntheseofen *m* ammonia converter
Ammoniakverbrennung *f* ammonia oxidation
Ammoniakverdampfer *m* ammonia vaporizer
Ammoniakverdampfung *f* ammonia evaporation (vaporization)
Ammoniakverflüssiger *m* ammonia condenser
Ammoniakwäsche *f* ammonia scrubbing
Ammoniakwäscher *m* ammonia scrubber
Ammoniakwasser *n* 1. *(tech)* ammonia[cal] liquor, crude ammonia liquor, ammonia water, gas liquor; 2. household [aqua] ammonia, ammonia water (solution, spirit)
konzentriertes (verdichtetes) A. concentrated ammoniacal liquor
Ammonifikation *f*, **Ammonifizierung** *f (agric)* ammonification
Ammonio-Gruppe *f* $-N^+H_3$ ammonio group
Ammonisator *m*, **Ammonisierapparat** *m (fert)* ammoniator
ammonisieren to ammoniate *(fertilizers)*
Ammonisiergranulator *m (fert)* ammoniator-granulator
Ammonisier-Granuliertrommel *f (fert)* ammoniator-granulator drum, rotary ammoniator-granulator, rotary-drum ammoniator, reaction-granulation drum

Ammonisiertrommel *f (fert)* ammoniation drum
Ammonisierung *f* 1. ammoniation *(of fertilizers)*; 2. *s.* Ammonifizierung
Ammonium *n* ammonium
Ammoniumalaun *m s.* Ammoniakalaun
Ammoniumaluminiumchlorid *n* $NH_4Cl \cdot AlCl_3$ aluminium ammonium chloride
Ammoniumaluminiumsulfat *n* $NH_4Al(SO_4)_2$ aluminium ammonium sulphate
Ammoniumamalgam *n* ammonium amalgam
Ammoniumamidokarbonat *n s.* Ammoniumkarbamat
Ammoniumazetat *n* CH_3COONH_4 ammonium acetate
Ammoniumazid *n* NH_4N_3 ammonium azide
Ammoniumbase *f* / **quartäre (quaternäre)** quaternary ammonium base
Ammoniumbenzoat *n* $C_6H_5COONH_4$ ammonium benzoate
Ammoniumbisulfitkochsäure *f (pap)* ammonia-base [sulphite] acid, ammonia-base [sulphite] liquor
Ammoniumborfluorid *n s.* Ammoniumfluoroborat
Ammoniumbromid *n* NH_4Br ammonium bromide
Ammoniumchlorat *n* NH_4ClO_3 ammonium chlorate
Ammoniumchlorid *n* NH_4Cl ammonium chloride, salmiac
Ammoniumchromalaun *m* $NH_4Cr(SO_4)_2 \cdot 12H_2O$ chrome ammonium alum, chrome alum ammonium
Ammoniumdichromat *n* $(NH_4)_2Cr_2O_7$ ammonium dichromate
Ammoniumdihydrogenorthophosphat *n* $NH_4H_2PO_4$ ammonium dihydrogenorthophosphate
Ammoniumdithionat *n* $(NH_4)_2S_2O_6$ ammonium dithionate
Ammoniumeisenalaun *m* $NH_4Fe(SO_4)_2 \cdot 12H_2O$ ammonium iron alum, ferric ammonium alum
Ammoniumeisen(III)-oxalat *n* $(NH_4)_3[Fe(C_2O_4)_3]$ ammonium iron(III) oxalate, ferric ammonium oxalate
Ammoniumeisen(II)-zyanid *n s.* Ammoniumhexazyanoferrat(II)
Ammoniumfluorid *n* NH_4F ammonium fluoride
Ammoniumfluoroborat *n* $[NH_4][BF_4]$ ammonium fluoroborate, ammonium tetrafluoroborate
Ammoniumformiat *n* $HCOONH_4$ ammonium formate
Ammoniumgallium(III)-sulfat *n* $Ga_2(SO_4)_3 \cdot (NH_4)_2SO_4$ ammonium gallium(III) sulphate
Ammoniumgruppe *f* ammonium group (residue)
Ammoniumheptamolybdat-4-Wasser *n* $(NH_4)_6[Mo_7O_{24}] \cdot 4H_2O$ ammonium heptamolybdate-4-water
Ammoniumhexachloroplatinat(IV) *n* $(NH_4)_2[PtCl_6]$ ammonium hexachloroplatinate(IV), ammonium chloroplatinate(IV)
Ammoniumhexachlorostannat(IV) *n* $(NH_4)_2[SnCl_6]$ ammonium hexachlorostannate(IV), pink salt
Ammoniumhexafluorosilikat *n* $(NH_4)_2[SiF_6]$ ammonium hexafluorosilicate, ammonium fluorosilicate

Ammoniumhexafluorozirkonat *n* $(NH_4)_2[ZrF_6]$ ammonium hexafluorozirconate

Ammoniumhexazyanoferrat(II) *n* $(NH_4)_4[Fe(CN)_6]$ ammonium hexacyanoferrate(II), ammonium cyanoferrate(II)

Ammoniumhydrogenfluorid *n* NH_4HF_2 ammonium hydrogenfluoride

Ammoniumhydrogenkarbonat *n* NH_4HCO_3 ammonium hydrogencarbonate

Ammoniumhydrogenorthophosphat *n* $(NH_4)_2HPO_4$ ammonium hydrogenorthophosphate

Ammoniumhydrogensulfat *n* NH_4HSO_4 ammonium hydrogensulphate

Ammoniumhydrogensulfid *n* NH_4HS ammonium hydrogensulphide

Ammoniumhydrogensulfit *n* NH_4HSO_3 ammonium hydrogensulphite

Ammoniumhydrogentartrat *n* $NH_4OOC[CHOH]_2$-COOH ammonium hydrogentartrate

Ammoniumhydroxid *n* NH_4OH ammonium hydroxide

quartäres (quaternäres) A. quaternary ammonium hydroxide

Ammoniumhypophosphit *n* $NH_4PH_2O_2$ ammonium hypophosphite, *(better)* ammonium phosphonate

Ammoniumjodat *n* NH_4IO_3 ammonium iodate

Ammoniumjodid *n* NH_4I ammonium iodide

Ammoniumkalziumarsenat *n* NH_4CaAsO_4 ammonium calcium arsenate

Ammoniumkalziumphosphat *n* NH_4CaPO_4 ammonium calcium phosphate

Ammoniumkarbamat *n* $[NH_4][CO_2NH_2]$ ammonium carbamate (aminoformate)

Ammoniumkarbaminat *n s.* Ammoniumkarbamat

Ammoniumkarbonat *n* $(NH_4)_2CO_3$ ammonium carbonate

handelsübliches A. commercial ammonium carbonate, salt of hartshorn, sal volatile *(a mixture consisting of ammonium hydrogencarbonate and ammonium carbamate)*

Ammoniumkupfer(II)-sulfat *n* $(NH_4)_2SO_4 \cdot CuSO_4$ ammonium copper(II) sulphate, cupric ammonium sulphate

Ammoniumlaktat *n* $CH_3CH(OH)COONH_4$ ammonium lactate

Ammoniummagnesiumkarbonat *n* $(NH_4)_2CO_3 \cdot MgCO_3$ ammonium magnesium carbonate

Ammoniummagnesiumsulfat *n* $(NH_4)_2Mg(SO_4)_2$ ammonium magnesium sulphate

Ammoniummanganat(VII) *n s.* Ammoniumpermanganat

Ammoniummangan(II)-phosphat *n* NH_4MnPO_4 ammonium manganese(II) phosphate

Ammoniummangan(II)-sulfat *n* $(NH_4)_2SO_4 \cdot MnSo_4$ ammonium manganese(II) sulphate

Ammoniummetaantimonat(V) *n* NH_4SbO_3 ammonium metaantimonate(V)

Ammoniummetaarsenat(III) *n* NH_4AsO_2 ammonium metaarsenite

Ammoniummetaperjodat *n* NH_4IO_4 ammonium metaperiodate, ammonium periodate, ammonium tetraoxoiodate(VII)

Ammoniummetavanadat *n* NH_4VO_3 ammonium metavanadate, ammonium trioxovanadate(V)

Ammoniummolybdat *n* $(NH_4)_2MoO_4$ ammonium molybdate

handelsübliches A. commercial ammonium molybdate, ammonium paramolybdate

normales A. *s.* Ammoniummolybdat

Ammoniumnitrat *n* NH_4NO_3 ammonium nitrate

Ammoniumnitrit *n* NH_4NO_2 ammonium nitrite

Ammoniumorthophosphat *n* $(NH_4)_3PO_4$ ammonium orthophosphate

Ammoniumoxalat *n* $NH_4OOC{-}COONH_4$ ammonium oxalate

Ammoniumparamolybdat *n s.* Ammoniummolybdat / handelsübliches

Ammoniumpentasulfid *n* $(NH_4)_2S_5$ ammonium pentasulphide

Ammoniumperchlorat *n* NH_4ClO_4 ammonium perchlorate

Ammoniumpermanganat *n* NH_4MnO_4 ammonium permanganate

Ammoniumperoxochromat *n* $(NH_4)_3CrO_8$ ammonium peroxochromate

Ammoniumperoxodisulfat *n* $(NH_4)_2S_2O_8$ ammonium peroxodisulphate

Ammoniumpersulfat *n (incorrectly for)* Ammoniumperoxodisulfat

Ammoniumphosphat *n* ammonium phosphate, *(specif) s.* Ammoniumorthophosphat

Ammoniumpikrat *n* $NH_4OC_6H_2(NO_2)_3$ ammonium picrate

Ammoniumpraseodymsulfat *n* $(NH_4)_2SO_4 \cdot Pr_2(SO_4)_3$ ammonium praseodymium sulphate

Ammoniumradikal *n* ammonium radical, *(specif)* free ammonium radical

Ammoniumrest *m* ammonium residue (group)

Ammoniumrhodanid *n s.* Ammoniumthiozyanat

Ammoniumsalizylat *n* $C_6H_4(OH)COONH_4$ ammonium salicylate

Ammoniumsalz *n* ammonium salt

Ammoniumseife *f* ammonium soap

Ammoniumselenat *n* $(NH_4)_2SeO_4$ ammonium selenate

Ammoniumstearat *n* $CH_3[CH_2]_{16}COONH_4$ ammonium stearate

Ammoniumstickstoff *m s.* Ammoniakstickstoff

Ammoniumsulfamat *n s.* Ammoniumsulfamidat

Ammoniumsulfamidat *n* $NH_4OSO_2NH_2$ ammonium sulphamate, AMS

Ammoniumsulfaminat *n (incorrectly for)* Ammoniumsulfamidat

Ammoniumsulfat *n* $(NH_4)_2SO_4$ ammonium sulphate

Ammoniumsulfid *n* $(NH_4)_2S$ ammonium sulphide

Ammoniumsulfit *n* $(NH_4)_2SO_3$ ammonium sulphite

Ammoniumtartrat *n* $NH_4OOC[CHOH]_2COONH_4$ ammonium tartrate

Ammoniumtellurat *n* $(NH_4)_2TeO_4$ ammonium tellurate

Ammoniumtetrachloroplatinat(II) *n* $(NH_4)_2[PtCl_4]$ ammonium tetrachloroplatinate(II), ammonium chloroplatinate(II)
Ammoniumtetrazyanoplatinat(II) *n* $(NH_4)_2[Pt(CN)_4]$ ammonium tetracyanoplatinate(II), ammonium cyanoplatinate(II)
Ammoniumtetroxojodat *n s.* Ammoniummetaperjodat
Ammoniumthiosulfat *n* $(NH_4)_2S_2O_3$ ammonium thiosulphate
Ammoniumthiozyanat *n* NH_4SCN ammonium thiocyanate (rhodanide)
Ammoniumtrithiokarbonat *n* $(NH_4)_2CS_3$ ammonium trithiocarbonate
Ammoniumvalerat *n* $CH_3[CH_2]_3COONH_4$ ammonium valerate (valerianate)
Ammoniumverbindung *f* ammonium compound
quartäre (quaternäre) A. quaternary ammonium compound
Ammoniumwolframat *n* $(NH_4)_2WO_4$ ammonium wolframate (tungstate)
Ammoniumzyanat *n* NH_4OCN ammonium cyanate
Ammoniumzyanid *n* NH_4CN ammonium cyanide
Ammonolyse *f* ammonolysis
ammonolytisch ammonolytic
Ammonpulver *n* ammonium powder *(explosive)*
Ammonsalpetersprengstoff *m* ammonium nitrate explosive
Ammonsalz *n s.* Ammoniumsalz
Ammonsalzküpe *f (text)* ammonia vat
Ammonsulfatsalpeter *m (agric)* ammonium nitrate sulphate
amöbizid amoebicidal
Amöbizid *n* amoebicide
amorph amorphous, amor.
AMP *s.* Adenosinmonophosphat
Ampere *n* ampere
Amperemeter *n* amperemeter, ammeter
Amperometrie *f* amperometry
amperometrisch amperometric
Amphetamin *n* amphetamine, 1-phenyl-2-propylamine
Amphetaminsulfat *n* amphetamine sulphate
Amphibol *m (min)* amphibole, hornblende *(an inosilicate)*
Amphibolasbest *m (min)* amphibole asbestos
Amphibolfamilie *f*, **Amphibolgruppe** *f (min)* amphibole group
Amphibolit *m* amphibolite
amphi-Stellung *f* amphi-position
Ampho-Ion *n* amphion, amphoteric (ampholyte, dipolar, dual, hybrid) ion, zwitterion
Ampholyt *m* ampholyte, amphoteric electrolyte, amphiprotic substance
Ampholytoid *n* ampholytoid *(an amphoteric soil colloid)*
amphoter amphoteric, amphiprotic
Ampulle *f* ampoule, ampul[e]
amu *s.* Masseeinheit / atomare
Amygdalin *n* amygdalin *(a glycoside obtained from bitter almonds)*
Amygdaloid *m (geol)* amygdaloid
***n*-Amylaldehyd** *m s.*Pentanal
Amylalkohol *m* $C_5H_{11}OH$ amyl alcohol, *(specif)* $CH_3[CH_2]_3$ CH_2OH *n*-amyl alcohol, 1-pentanol
Amylase *f* amylase
dextrinogene A. dextrinogenic amylase
Amylaseaktivität *f* amylolytic (diastatic) activity
Amylasewirkung *f* amylolytic (diastatic) action
***n*-Amyläthin** *n s.* Heptin-(1)
Amylazetat *n* $CH_3COOC_5H_{11}$ amyl acetate
***n*-Amylazetylen** *n s.* Heptin-(1)
***n*-Amylchlorid** *n* $CH_3[CH_2]_3CH_2Cl$ 1-chloropentane, *n*-amyl chloride
***n*-Amylen-(1)** *n s.* Penten-(1)
Amylenhydrat *n s.* 2-Methylbutanol-(2)
Amylgruppe *f* $CH_3[CH_2]_3CH_2$- amyl group (residue)
Amylhalogenid *n* amyl halide
Amylkarbinol *n s.* Hexanol-(1)
Amylnitrit *n* $C_5H_{11}ONO$ amyl nitrite, *(specif)* $(CH_3)_2CHCH_2CH_2ONO$ [ordinary] amyl nitrite, 3-methyl-1-butyl nitrite
Amylodextrin *n* amylodextrin
amyloid amyloid[al]
Amyloid *n (pap)* amyloid *(cellulose treated with concentrated sulphuric acid)*; *(med)* amyloid *(antibody globulin abnormally deposited in animal tissues)*
Amylopektin *n* amylopectin
Amylose *f* amylose
Amyloverfahren *n* amylo fermentation process *(for obtaining alcohol from starchy materials without the use of malt)*
Amylrest *m s.* Amylgruppe
Anabolismus *m* anabolism
Anacardiumgummi *n* cashew (cashawa) gum *(from Anacardium occidentale L.)*
anaerob anaerobic
Anaerobier *m* anaerobe
fakultativer A. facultative anaerobe
Anaerobiont *m s.* Anaerobier
Anaerobiose *f* anaerobiosis *(life in the absence of oxygen)*
Analeptikum *n (pharm)* analeptic, central nervous system stimulant
Analgetikum *n (pharm)* analgesic, pain-reliever, pain-killer
analgetisch analgesic
Analog[es] *n* analogue
Analysator *m* analyser
Analyse *f* analysis ✦ **zur A.** *s.* analysenrein
A. mit radioaktiven Reagenzien radiometric analysis
chemische A. chemical analysis
coulometrische A. coulometric analysis
elektrochemische A. electroanalysis
fotometrische A. photometric analysis
gravimetrische A. gravimetric analysis
kolorimetrische A. colorimetric analysis
polarografische A. polarographic analysis
potentiometrische A. potentiometric analysis
praktische A. commercial analysis
qualitative A. qualitative analysis
quantitative A. quantitative analysis

radiometrische A. radiometric analysis
röntgenchemische A. X-ray chemical analysis
röntgenspektrochemische A. X-ray spectrochemical analysis
röntgenspektroskopische A. X-ray spectroscopic analysis
sensorische A. sensory estimation (evaluation, rating, test), organoleptic estimation (evaluation, rating, test) *(of food)*
technische A. commercial analysis
thermische A. thermal analysis
thermogravimetrische A. thermogravimetric analysis, TGA
volumetrische A. volumetric (titrimetric, mensuration) analysis
Analysenbefund *m* analytical result
Analysenbereich *m* analysing range
Analysenfehler *m* analytical error
Analysengang *m* course of analysis
Analysengeschwindigkeit *f* analysing speed
Analysenprobe *f* analytical sample
analysenrein of reagent purity, analytical-reagent-quality, reagent-grade
Analysenwaage *f* analytical balance
A. mit Dämpfung damped balance
Analysenwert *m* analytical value
analysierbar analysable
analysieren to analyse
Analytik *f* analytical chemistry
Analytiker *m* analyst, analyser, analytical chemist
analytisch analytic[al]
Analzim *m (min)* analcime, analcite *(a zeolite)*
a-Name *m s.* Aza-Benennung
Anaphorese *f (phys chem)* anaphoresis
Anaphrodisiakum *n (pharm)* antiaphrodisiac
anästhesieren to anaesthetize
anästhesierend anaesthetic
Anästhesierungsmittel *n*, **Anästhetikum** *n* anaesthetic
anästhetisieren *s.* anästhesieren
Anatas *m (min)* anatase *(titanium dioxide)*
Anatoxin *n* anatoxin *(detoxicated toxin)*
Anatto *n(m)* annatto, annotta, arnatto, arnotta *(a colouring matter from Bixa orellana L.)*
Anattofarbstoff *m s.* Anatto
anätzen to etch; *(med)* to cauterize *(by chemicals)*
anbacken to bake on, to stick on
anbläuen to blue
anbluten *(dye, coat, tann)* to bleed [off, through]
Anbrenncharakteristik *f (rubber)* scorch characteristic
anbrennen 1. to light, to ignite *(e.g. a burner)*; 2. *(rubber)* to scorch, to burn, to set up, to fire up, to cure up, to prevulcanize, to precure *(to undergo unintended vulcanization)*; 3. to burn on *(as of moulding sand on a casting)*
Anbrennen *n* 1. *(rubber)* scorch[ing], burning, set[-ting]-up, firing-up, curing-up, prevulcanization, precure, precuring; *(rubber)* pile (bin) curing *(during storage)*; 2. burning-on *(as of moulding sand on castings)*
Anbrennkurve *f (rubber)* scorch curve
Anbrennperiode *f (rubber)* scorch period
Anbrennpunkt *m (rubber)* scorch point
Anbrenntendenz *f (rubber)* scorch[ing] tendency, scorchiness
Anbrennzeit *f (rubber)* scorch time
anbuttern *(food)* to prechurn
Andalusit *m (min)* andalusite, andaluzite *(aluminium oxide silicate)*
Andersonit *m (min)* andersonite
Anderthalbfachbindung *f* one-and-a-half bond, one-and-one-half bond, three-halves bond
Änderung *f* change, alteration
Andesin *m (min)* andesine
Andesit *m* andesite *(volcanic rock)*
Andradit *m (min)* andradite
Androgen *n* androgen *(any of a class of sex hormones)*
andrucken *(pap)* to proof
andrücken to press, to contact
Andruckpapier *n* galley proof paper, proof[ing] paper
Andruckwalze *f* nip roll
aneinanderhaften to adhere
Aneinanderhaften *n* adherence, adhesion
aneinanderhaftend adherent, adhesive
Anelektrolyt *m* non-electrolyte
anellieren *(org chem)* to anellate, to fuse
Anellierung *f (org chem)* anellation, fusion
angulare A. angular anellation
lineare A. linear anellation
Anellierungsname *m* fusion name
Anellierungspräfix *n* fusion prefix
Anellierungsseite *f*, **Anellierungsstelle** *f* side (point, position) of fusion
anerkennen to approve, to schedule, to register *(e.g. a pesticide)*
Anerkennung *f* / **amtliche** [government] approval, registration, *(Am also)* label clearance *(as of a pesticide)*
Anethol *n* $CH_3OC_6H_4CH{=}CHCH_3$ anethole
anfahren to start up *(e.g. a reactor)*
Anfang *m* **der Trockenpartie** *(pap)* wet end of the dryer section
Anfangeisen *n (glass)* gathering (punty) iron
Anfänger *m (glass)* gatherer
Anfangsbelag *m* initial deposit *(of a pesticide)*
Anfangsfeuchte[beladung] *f* initial moisture content, IMC
Anfangsgeschwindigkeit *f* initial rate (velocity)
A. des Färbens *(text)* strike
Anfangsglied *n* initial member; *(nucl)* parent element *(of a decay series)*
Anfangskonzentration *f* initial concentration
Anfangskriechen *n* primary creep
Anfangslöslichkeit *f* initial solubility
Anfangsmenge *f* initial amount
Anfangspunkt *m* initial point
Anfangsretention *f* / **maximale** maximum initial retention, MIR *(crop protection)*
Anfangssiedepunkt *m* initial boiling point, I. B. P.
Anfangsspreitungskoeffizient *m* initial spreading coefficient

Anfangstemperatur *f* initial temperature
Anfangswert *m* initial value
Anfangszustand *m* initial (original) state, primary condition
anfärbbar dyeable; paintable; *(biol)* stainable
Anfärbbarkeit *f* dyeability, dye receptivity; paintability; *(biol)* stainability
anfärben to tint, to tinge, *(with soluble colouring matter)* to dye, *(chiefly Am)* to colo[u]r, *(with suspended colouring matter)* to paint; *(biol)* to stain *(for microscopical investigation)*; *(tann)* to pretan, to colour *(to tan weakly)*
direkt a. to dye directly
anfärbend / nicht non-dyeing
normal a. regular-dyeing
stark a. deep-dyeing
Anfärbevermögen *n* dyeability, dyeing (tinctorial) power
anfeuchten to moisten, to wet, to damp[en], to humidify, *(text also)* to dew
in Sägespänen a. *(tann)* to sawdust
wieder a. to rewet
anflanschen to flange
anflecken to stain
anfressen to corrode, to eat, to attack
punktförmig a. to pit
angarnieren *(ceram)* to stick up
Angärung *f* primary fermentation, pre-fermentation
angefault putrid
Angelikasäure *f* $CH_3CH=C(CH_3)COOH$ angelic acid, 2-methylisocrotonic acid, *cis*-2-methyl-2-butenoic acid
angerben to pretan, to colour
Anger-Mühle *f*, **Anger-Prallmühle** *f* Anger mill
Angle-Methode *f* *(rubber)* angle method *(for determining the tear propagation strength)*
Angle-Probe *f* *(rubber)* angle test *(for determining the tear propagation strength)*; angle test piece
Anglesit *m* *(min)* anglesite *(lead sulphate)*
angreifbar attackable; vulnerable *(as to attack by reagents)*; *(physically)* affectable, affectible; *(chemically)* corrodible
Angreifbarkeit *f* attackability; vulnerability *(as to attack by reagents)*; *(physically)* affectability, affectibility; *(chemically)* corrodibility
angreifen to attack; *(physically)* to affect; *(chemically)* to corrode, to eat
Angriff *m* attack
A. von der Rückseite [her] back-side attack, rear attack *(as in Walden inversion)*
bakterieller A. bacterial attack
elektrophiler A. electrophilic attack
nukleophiler A. nucleophilic attack
angular angular
Anguß *m* *(plast)* gate, sprue
direkter A. direct gate
ringförmiger A. ring gate
Angußausdrückstift *m* *(plast)* sprue ejector
Angußbuchse *f* *(plast)* sprue bush[ing], feed bush[ing]
Angußfarbe *f* *(ceram)* engobe
Angußkegel *m* *(plast)* gate, sprue
Angußmasse *f* *(ceram)* engobe
Angußsteg *m* *(plast)* inlet
Angußverteiler *m* *(plast)* runner; *(plast)* spreader *(of an injection mould)*
beheizter A. hot runner
Angußzieher *m* *(plast)* sprue puller
anhaften to adhere
Anhaften *n* adherence, adhesion
anhaftend adherent, adhesive
Anhänger[etiketten]karton *m* tag [card]board
Anhängestäubegerät *n* *(agric)* traction duster
anharmonisch anharmonic
Anharmonizität *f* anharmonicity
Anharmonizitätskonstante *f* anharmonicity (anharmonic) constant
anhäufen to accumulate
sich a. to accumulate, to collect
Anhäufung *f* accumulation
anheben to elevate, to raise *(e.g. temperature, boiling point)*; to energize *(electrons into an excited state)*
Anhebung *f* elevation, rise *(as of temperature, boiling point)*; energization *(of electrons into an excited state)*
anheizen to heat up
Anheizzeit *f* heat-up period (time), heating-up period (time), coming-up time
anhenkeln *(ceram)* to handle
Anhydrid *n* anhydride
inneres (intramolekulares) A. inner (internal, intramolecular) anhydride
anhydrisieren to anhydr[id]ize
Anhydrisierung *f* anhydr[id]ization
Anhydrit *m* *(min)* anhydrite *(anhydrous calcium sulphate)*
Anhydritbinder *m* anhydrite binder
Anhydrobase *f* anhydrobase
Anhydroglukose *f* anhydroglucose
Anhydrozucker *m* anhydrosugar
Anilid *n* C_6H_5NHCOR anilide *(an N-acyl derivative of aniline)*
Anilidoessigsäure *f* *s.* Anilinoessigsäure
Anilin *n* $C_6H_5NH_2$ aniline, aminobenzene
salzsaures A. *s.* Anilinhydrochlorid
technisches A. aniline oil
Anilinblau *n* aniline blue; *(specif)* *s.* spritlösliches A.
spritlösliches A. *(biol)* aniline blue [2 B], spirit blue, acid blue 20 *(triphenylrosaniline hydrochloride)*
Anilinchlorhydrat *n* *s.* Anilinhydrochlorid
Anilindruck *m* aniline (flexographic) printing, flexography
Anilindruckfarbe *f* aniline (flexographic) ink
Anilinfarbe *f* aniline dye
Anilinfarbstoff *m* aniline dye
Anilinformaldehydharz *n* aniline-formaldehyde resin
Anilingelb *n* $C_6H_5N=NC_6H_4NH_2$ aniline yellow, *p*-aminoazobenzene, solvent yellow 1
Anilingummidruck *m* *s.* Anilindruck

Anilinharz *n* aniline[-formaldehyde] resin
Anilinhydrochlorid *n* $C_6H_5NH_2 \cdot HCl$ aniline hydrochloride, aniline salt
Anilinoderivat *n* anilino derivative
Anilinoessigsäure *f* $C_6H_5NHCH_2COOH$ anilinoacetic acid, *N*-phenylglycine
Anilinöl *n* aniline oil *(commercial grade of aniline)*
Anilinpunkt *m* aniline [cloud] point
Anilinpurpur *m* aniline purple, mauveine
Anilinsalz *n s.* Anilinhydrochlorid
Anilinschwarz *n* aniline black, pigment black 1
Anilin-*m*-sulfonsäure *f* $C_6H_4(NH_2)SO_3H$ aniline-*m*-sulphonic acid, metanilic acid
Anilin-*o*-sulfonsäure *f* $C_6H_4(NH_2)SO_3H$ aniline-*o*-sulphonic acid, orthanilic acid
Anilin-*p*-sulfonsäure *f* $NH_2C_6H_4SO_3H$ aniline-*p*-sulphonic acid, sulphanilic acid
Anilintrübungspunkt *m* aniline [cloud] point
Anilinvergiftung *f* aniline poisoning
Anilinwasser *n* aniline water
animalisch animal
animalisieren *(text)* to animalize
Anime-Kopal *m* Anime copal *(from Hymenaea courbaril L. or Trachylobium hornemannianum Hayne), (specif)* Brazil (Colombia) copal *(from Hymenaea courbaril L.)*
Anion *n* anion
  komplexes A. complex anion
anionaktiv *s.* anionenaktiv
Anionaustausch[er]harz *n s.* Anionenaustauschharz
Anionbase *f* anion base
Anioncharakter *m* anionic nature
Anionelektrode *f* anion electrode
anionenaktiv anion-active, anionic
Anionenaustausch *m* anion exchange
Anionenaustauscher *m* anion exchanger, anionite
  A. auf Kunstharzbasis anion exchange resin
  schwach basischer A. weak-base anion exchanger, weak-base deionizer (resin)
  stark basischer A. strong-base anion exchanger, strong-base deionizer (resin)
Anionenaustauschfähigkeit *f* anion exchangeability
Anionenaustauschharz *n* anion exchange resin
Anionenfehlstelle *f* anion vacancy
Anionenkomplex *m* anion complex
Anionenleerstelle *f*, **Anionenlücke** *f* anion vacancy
Anionenstrom *m* anionic current
anionisch anionic, anion-active
anionoid anionoid
Aniontensid *n* anionic tenside
Anis *m* 1. anise, Pimpinella anisum L.; 2. ani[se]seed *(fruit from 1.)*
Anisidin *n* $CH_3OC_6H_4NH_2$ anisidine, aminophenol methyl ether
Aniskampfer *m s.* Anethol
anisodesmisch *(cryst)* anisodesmic
anisodimensional anisodimensional
Anisol *n* $CH_3OC_6H_5$ anisole, methoxybenzene, methyl phenyl ether
Anisöl *n* aniseed oil *(from Pimpinella anisum L. and Illicium verum L.)*
anisotrop anisotropic, aeolotropic
Anisotropie *f* anisotropy, aeolotropism
Anisotropiefaktor *m* anisotropy (dissymmetry) factor
Anisotropieglied *n* anisotropy term
Anissäure *f* $CH_3OC_6H_4COOH$ anisic acid, *p*-methoxybenzoic acid
Ankerit *(min)* ankerite *(a variety of dolomite)*
Ankermischer *m*, **Ankerrührer** *m* anchor agitator (mixer), horseshoe mixer
Ankersäule *f* buckstay *(of a furnace)*
ankleben 1. to stick [on]; to paste on; to glue on; 2. to adhere, to stick
anklebend adherent, adhesive
Anklemmrührer *m* portable mixer
Ankochperiode *f (pap)* impregnation (penetration) period
ankohlen to char
ankondensieren to fuse to *(of rings)*
ankuppeln *(dye)* to couple
Ankylit *m (min)* ancylite
Anlage *f* 1. plant; unit; 2. arrangement *(as of a trial)*; layout *(of a plant)*
  A. zum Süßen *(petrol)* sweetening plant
  A. zur Doktorbehandlung *(petrol)* doctor-treating unit
  A. zur Entschwefelung desulphurization plant
  A. zur Hydrodesulfurierung *(petrol, coal)* hydrodesulphurization plant
  A. zur Lösungsmittelentparaffinierung *(petrol)* solvent-dewaxing plant
  A. zur Propanentasphaltierung *(petrol)* propane-deasphalting plant
  A. zur Propanentparaffinierung *(petrol)* propane-dewaxing plant
  chemische A. chemical plant
Anlagentechnik *f* plant technology
anlagern 1. to attach, to add; 2. to gain
  koordinativ a. to coordinate *(e.g. molecules)*
  sich a. to become attached, to attach [oneself], to undergo addition
  sich koordinativ a. to coordinate
Anlagerung *f* attachment, *(esp chemical-bond theory)* addition
Anlagerungskomplex *m* addition complex
Anlagerungsprodukt *n* addition product
Anlagerungsreaktion *f* addition reaction
Anlagerungsrichtung *f* direction of addition
Anlagerungsverbindung *f* addition (additive) compound
anlassen to temper *(metals)*; to start [up] *(e.g. a machine)*
Anlaßfarbe *f* temper colour
Anlaßöl *n* tempering oil
anlaufen to tarnish, *(esp of lacquers)* to blush, *(esp of oil varnish)* to bloom
  weiß a. to blush *(esp of nitrocellulose lacquer)*
Anlauffarbe *f* temper colour
Anlegegoniometer *n (cryst)* contact goniometer
Anlegeöl *n* gold size *(for attaching gold leaf to surfaces)*

Anlockmittel *n*, **Anlocktoff** *m* attractant
Anlösung *f* partial solution
anmachen to temper, to mix *(e. g. mortar, concrete)*
Anmachwasser *n* tempering (mixing) water *(manufacture of concrete)*; *(ceram)* water of plasticity
Anmeldung *f* **eines Schädlingsbekämpfungsmittels** pesticide-chemical petition
Annabergit *m (min)* annabergite, nickel bloom *(nickel orthoarsenate)*
Annahme *f* 1. assumption, hypothesis; 2. acceptance; 3. acquisition *(as of a configuration)*
Goudsmit-Uhlenbecksche A. Goudsmit and Uhlenbeck assumption *(of rotating electrons)*
annässen to moisten
Annatto *n(m) s.* Anatto
annehmen 1. to assume *(a theory)*; 2. to accept; 3. to acquire *(a configuration)*; 4. to take on *(a colour)*
Farbstoff a. *(text)* to take the dye
Annihilation *f (phys chem)* annihilation [radiation]
Annihilationsspektrum *n* annihilation spectrum
Anode *f* anode, anelectrode, positive electrode
Anodeneffekt *m* anode effect
Anodenlaufzeit *f* anode life
Anodenpotential *n* anode potential
Anodenraum *m* anode compartment
Anodenschlamm *m* anode slime (sludge, mud)
Anodenspannung *f* anode voltage
Anodenstrahl *m* anode ray
Anodenstrom *m* anode current
Anodenverfahren *n* anode process
anodisch anodic
anodisieren *(met)* to anodize
Anodisierung *f (met)* anodizing, anodic oxidation (coating)
Anolyt *m* anolyte *(electrolyte in the anode compartment)*
anomal anomalous, abnormal
Anomalie *f* anomaly
optische A. optical anomaly
α-Anomalie *f (rubber)* vitrification, glass (second-order) transition
a-Nomenklatur *f* „a" nomenclature
anomer anomeric
Anomer[es] *n* anomer
Anomit *m (min)* anomite
anordnen to arrange, to align, to aline *(e. g. parts of an apparatus)*
sich a. to align, to aline *(as of chain molecules)*
Anordnung *f* arrangement, alignment, alinement
natürliche A. *s.* periodische A.
periodische A. periodic arrangement *(of the elements)*
räumliche A. spatial arrangement (characteristics)
Anorganiker *m* inorganic chemist
anorganisch inorganic
anorganisch-chemisch inorganic-chemical
Anorthit *m (min)* anorthite
Anorthoklas *m (min)* anorthoclase *(an aluminosilicate)*
Anoxybiose *f* anaerobiosis *(life in the absence of oxygen)*
Anpaßstück *n* adapter
anpasten to make into a paste, to paste *(e. g. coal for hydrogenation)*
Anpolymerisation *f* grafting
anpolymerisieren to graft
Anprall *m* impact, impingement
anprallen to impact, to impinge
Anpreßdruck *m (pap)* plug pressure *(in perfecting engines)*
anrauhen to roughen
anregbar excitable
Anregbarkeit *f* excitability
anregen to excite *(e. g. an atom)*; to activate, to initiate, to start *(e. g. a reaction)*
die Kristallisation a. to induce crystallization
Anregung *f* excitation *(as of an atom)*; activation, initiation *(as of a reaction)*
A. von Atomen atomic excitation
thermische A. thermal excitation
Anregungsenergie *f* excitation energy
Anregungsfunktion *f* excitation function
Anregungsmittel *n (pharm)* stimulant, stimulus, stimulatory drug
Anregungsniveau *n* excitation (excited) level
Anregungspotential *n* excitation potential
Anregungsspannung *f* excitation voltage
Anregungswahrscheinlichkeit *f* excitation probability
Anregungszustand *m* excited state
Anreibbarren *m* pressure bar *(of a roller mill)*
anreiben *(coat)* to grind; to paste *(e. g. coal for hydrogenation)*
Anreibung *f (coat)* grinding; pasting *(e. g. in coal hydrogenation)*
Anreicher ... *s.* Anreicherungs ...
anreichern to enrich, to fortify *(e. g. food by vitamins)*; to concentrate, to beneficiate *(ore)*; *(pap)* to fortify *(the cooking acid)*
mit Benzol a. to benzolize
mit Enzymen (Fermenten) a. to enzymize
mit Kohlenstoff a. to carbonize
mit Ozon a. to ozonize, to ozonify
mit Sauerstoff a. to oxygenate, to oxygenize, to enrich by oxygen
mit Vitaminen a. to vitaminize, to enrich (fortify) by vitamins
sich a. to accumulate
Anreicherung *f* 1. enrichment, fortification *(as of food by adding vitamins)*; concentration, beneficiation *(of ore)*; *(pap)* fortification *(of the cooking acid)*; 2. accumulation
Anreicherungsanlage *f (min tech)* concentration plant, concentrator
Anreicherungsfaktor *m* enrichment factor
Anreicherungsgerät *n (min tech)* concentrating machine, concentrator
Anreicherungsgrad *m* degree of enrichment, enrichment factor
Anreicherungsherd *m (min tech)* concentrating (concentrator, concentrate) table

Anreicherungshorizont *m (soil)* enrichment (illuvial) horizon, accumulate layer
Anreicherungsschicht *f (phys chem)* enriched layer
Anreicherungsverfahren *n* concentration process (method)
Anreißen *n (text)* first break, *(Am)* first perceptible step *(just perceptible alteration of colour)*
anrühren to temper, to mix *(e.g. mortar, concrete, paint)*; to paste *(e.g. coal for hydrogenation)*
ansammeln / sich to accumulate, to collect
Ansammlung *f* accumulation, collection
Ansatz *m* 1. batch, [initial] charge, charge (charging) stock, feed[stock]; *(plast)* formulation; 2. crust *(of undesired solid matter)*; 3. *s.* Ansatzstück
Ansatzbuch *n* batch book
Ansatzrohr *n* attached tube; stem *(of a gas burette)*
Ansatzstück *n* extension limb; lateral
ansäuern to acidify, to make acidic, to acidulate; *(food)* to sour; to leaven *(dough)*
erneut a. to reacidify
Ansäuern *n* acidification, acidulation; *(food)* souring; leavening *(of dough)*
Ansäuerungsmittel *n* acidulant
Ansaugdruck *m* suction pressure
ansaugen to draw [in], to suck, to aspirate
Ansaugen *n* drawing, suction, aspiration
Ansaugleistung *f* suction capacity
Ansaugleitung *f* suction line
Ansaugrohr *n* suction pipe (tube)
Ansaverbindung *f (org chem)* ansa-compound
anschärfen *(tann)* to sharpen, to strengthen, to mend *(the lime liquor)*
Anschärf[ungs]mittel *n (tann)* sharpener, sharpening agent
Anschlagpapier *n* poster paper
anschließen to connect, to join; to attach
Anschliff *m* ground (polished) face, ground (polished) surface, ground (polished) section *(in direct-light microscopy)*
Anschliffläche *f* / **polierte** *s.* Anschliff
Anschlifftechnik *f (min)* polished-face (polished-surface, polished-specimen) technique
Anschluß *m* 1. connection, joining; 2. *s.* Anschlußstelle; 3. *s.* Anschlußstück
Anschlußleitung *f* connecting (connection) pipe, connection
Anschlußrohr *n* connecting tube
Anschlußstelle *f* joint, junction
Anschlußstück *n* connecting piece, connection, joint
anschmutzen to stain; *(text)* to soil
künstlich a. *(text)* to soil artificially
Anschmutzung *f* staining; *(text)* soiling, *(if artificially also)* soiling operation (procedure)
Anschnitt *m (plast)* feed (fill) orifice
fächerförmiger A. *(plast)* fan gate
Anschütz-Aufsatz *m (distil)* Anschütz head
anschwänzen *(ferm)* to sparge
Anschwänzvorrichtung *f (ferm)* sparger
Anschwänzwasser *n (ferm)* sparge water
anschwellen to swell [up]
Anschwemmfilter *n* precoat[-type] filter, precoated filter
Anschwemmgut *n (filtr)* precoat
Anschwemmklärfilter *n* precoat clarifier
Anschwemmschicht *f (filtr)* precoat layer (bed)
anschwöden *(tann)* to flood, to paint *(the flesh side of hides with lime)*
ansetzen 1. to prepare *(e.g. a reaction mixture)*; 2. to charge *(a furnace)*; 3. to scale, to become encrusted *(with a hard deposit)*; to become covered *(e.g. with rust)*; 4. *(tech)* to attach, to fit
Grünspan a. to become covered with verdigris
Kesselstein a. to scale, to fur
Rost a. to become covered with rust, to gather rust, to rust
sich a. to deposit
ansprechen / auf Düngung *(agric)* to respond to fertilizing
Anspringcharakteristik *f (rubber)* scorch characteristic
anspringen *(rubber)* to scorch, to burn, to set up, to fire up, to cure up, to prevulcanize, to precure *(to undergo unintended vulcanization)*
Anspringen *n (rubber)* scorch[ing], burning, set[ting]-up, firing-up, curing-up, prevulcanization, precure, precuring; *(rubber)* pile (bin) curing *(during storage)*
Anspringkurve *f (rubber)* scorch curve
Anspringperiode *f (rubber)* scorch period
Anspringpunkt *m (rubber)* scorch point
Anspringtendenz *f (rubber)* scorch[ing] tendency, scorchiness
Anspringzeit *f (rubber)* scorch time
anspruchsvoll *(agric)* exacting *(with reference to nutrition)*
anstechen to tap *(e.g. a cask, a furnace, a glass tank)*
anstehen *(geol)* to crop out, to outcrop
Anstehen *n (geol)* outcrop[ping]
Anstehendes *n (geol)* bedrock
ansteigen to increase, to rise, to grow
Ansteigen *n s.* Anstieg
Anstellbottich *m (ferm)* pitching vessel
anstellen 1. to turn on *(an apparatus)*; to start, to set going (in motion) *(a machine)*; 2. to carry out, to run *(an experiment)*; 3. to pitch *(yeast)*
Versuche a. to experiment[alize], to run (carry out) experiments
Anstellhefe *f (ferm)* pitching (inoculating) yeast
Anstelltemperatur *f (ferm)* pitching (fermenter set) temperature
Anstieg *m* increase, rise, growth
Anstoß *m* impact, impingement
Anstoßatom *n* knocked-on atom
Anstoßelektron *n* impact electron
anstoßen to impact, to impinge
anstreichen to paint, to brush, to coat
Anstrich *m* paint, coat [of paint]
Anstrichaufbau *m s.* Anstrichsystem
Anstrichbindemittel *n* binder, coating vehicle
Anstrichfarbe *f* paint
A. für Unteranstrich undercoat material, undercoater

chemisch trocknende A. chemical-reaction paint
ofentrocknende A. stoving paint
plastische A. plastic paint
rostschützende A. antirust (rust-protective, rust-resisting) paint
seewasserbeständige A. marine paint
thixotrope A. thixotropic (gel) paint
Anstrichfehler *m* paint-film defect (fault)
Anstrichfilm *m* paint film
Anstrichmasse *f (ceram)* coating slip
Anstrichmittel *n s.* Anstrichstoff
Anstrichmittelwanne *f* sump *(in the flow-coating process)*
Anstrichschaden *m s.* Anstrichfehler
Anstrichstoff *m* surface coating, *(if pigmented)* paint
wäßriger A. water[-base] paint
Anstrichsystem *n* paint (coating) system
Anstrichtechnik *f* painting technology
Anstrichtrocknung *f* paint drying
ANT *s.* α-Naphthylthioharnstoff
Antagonismus *m* antagonism
Antagonist *m* antagonist
Antaphrodisiakum *n* antiaphrodisiac
Antazidum *n (pharm)* [gastric] antacid *(an agent that counteracts superacidity)*
anteigen to make into a paste, to paste, *(rubber also)* to make into a dough
Anteigmittel *n* pasting agent
Anteil *m* moiety, portion, proportion, constituent [part], component
disperser A. disperse[d] phae
elastischer A. *(rubber)* rebound, recovery
hochsiedender (höhersiedender) A. *(distil)* less volatile component, high-boiling component
kristalliner A. crystalline fraction
leichterflüchtiger (leichtersiedender) A. *(distil)* more volatile component, M.V.C., low-boiling component, light[er] component
prozentualer A. percentage
saurer A. acid component
schwererflüchtiger (schwerersiedender) A. *s.* hochsiedender A.
schwerflüchtiger (schwersiedender) A. *s.* hochsiedender A.
unverseifbarer A. unsaponifiable matter (residue)
Anthanthronfarbstoff *m* anthanthrone dye
Anthelminthikum *n (pharm)* anthelmint[h]ic, vermifuge, helminthagogue
anthelminthisch *(pharm)* anthelmint[h]ic
Anthophyllit *m (min)* anthophyllite *(an inosilicate)*
Anthophyllitasbest *m* anthophyllite asbestos
Anthozyan *n* anthocyan[in] *(any of a class of plant pigments)*
Anthozyanidin *n* anthocyanidin *(any of a class of plant pigments)*
Anthozyanin *n s.* Anthozyan
anthrachinoid anthraquinonoid
Anthrachinolin *n* anthraquinoline
Anthrachinolinchinon *n* anthraquinolinequinone
Anthrachinon *n* anthraquinone
Anthrachinonakridon *n* anthraquinoneacridone
Anthrachinonchinolin *n s.* Anthrachinolinchinon
Anthrachinondisulfonsäure *f* anthraquinonedisulphonic acid
Anthrachinonfarbstoff *m* anthraquinone (anthraquinonoid) dye
Anthrachinonkarbonsäure *f* anthraquinonecarboxylic acid
Anthrachinonklasse *f* anthraquinone class
Anthrachinonküpenfarbstoff *m* anthraquinone vat dye
Anthrachinonreihe *f* anthraquinone series
Anthrachinonsulfonsäure *f* anthraquinonesulphonic acid
Anthranilsäureester *m* anthranilate
Anthranol *n* anthranol, 9-anthrol, 9-hydroxyanthracene
Anthrapurpurin *n* anthrapurpurine, 1,2,7-trihydroxy-anthraquinone
Anthraxylon *n (coal)* anthraxylon
Anthrazen *n* anthracene
Anthrazenblau *n* anthracene blue
Anthrazenöl *n* anthracene oil
Anthrazenreihe *f* anthracene series
Anthrazit *m* anthracite
Anthrazitgruppe *f* anthracite group
Anthrimid *n (dye)* anthrimid
Anthron *n* anthrone, 9*H*-anthracen-9-one
Antiabsetzmittel *n* antisettling agent, sedimentation inhibitor
Antialkoholikum *n* alcohol deterrent
Antiaphrodisiakum *n* antiaphrodisiac
Antiarthritikum *n* antiarthritic
Antibackmittel *n (fert)* anticaking agent
antibakteriell antibacterial
Antibase *f* antibase
Antibiose *f* antibiosis
Antibiotikum *n* antibiotic [agent, substance]
A. gegen Rickettsien antirickettsial antibiotic
antibiotisch antibiotic
Antichlor *n (text, pap)* antichlor *(for removing chlorine after bleaching)*
Antidepressivum *n* antidepressant
Antidiabetikum *n* antidiabetic
Antidiarrhoikum *n* antidiarrhoeal, styptic
Antidiazotat *n* $RN{=}NOM^{I}$ antidiazo compound
antidiuretisch antidiuretic
Antidot *n* antidote
Antiemetikum *n* antiemetic
Antienzym *n s.* Antiferment
Antiepileptikum *n* antiepileptic
Antifebrilium *n s.* Antifiebermittel
Antiferment *n* antiferment, antienzyme
Antiferromagnetismus *m* antiferromagnetism
Antifertilitätspräparat *n (pharm)* antifertility agent
Antifiebermittel *n* antipyretic, febrifuge
Antifilzausrüstung *f (text)* antifelting treatment
Antifouling[anstrichmittel] *n*, **Antifoulingfarbe** *f* antifouling paint *(for underwater structures)*
antifungal antifungal, fungicidal
Antigen *n (bioch)* antigen
agglutinierbares A. agglutinogen

Antigenität *f s.* Antigenwirkung
Antigenwirkung *f (bioch)* antigenicity
Antigorit *m (min)* antigorite *(a variety of serpentine)*
Antihaftmittel *n* antiblocking agent *(for plastics and paper)*
Antihaftvermögen *n* antistick properties
Antihaut[bildungs]mittel *n (coat)* antiskinning (anticreaming) agent
Antihidrotikum *n* antihidrotic, antiperspirant, perspiration check
Antihistaminikum *n*, **Antihistaminpräparat** *n* antihistamine, antihistaminic drug
Antihormon *n* antihormone
antiinflammatorisch anti-inflammatory
anti-Isomer[es] *n* anti isomer
Antikatalysator *m* anticatalyst, negative catalyst, inhibitor, inhibiting substance, retarder
Antikatalyse *f* anticatalysis, negative catalysis, inhibition
antikatalytisch anticatalytic
Antikatode *f* anticathode
Antikglas *n* antique glass
Antiklebemittel *n*, **Antikleber** *m* antisticking agent
Antiklebvermögen *n* antistick properties
Antiklinalfalle *f (geol)* anticlinal trap
Antiklopfeigenschaften *fpl* antiknock properties
Antiklopfmittel *n* antiknock [additive, agent, compound, dope, material, reagent, substance], knock inhibitor (reducer, suppressor), octane improver
Antiklopfwirkung *f* antiknock effect
Antikoagulans *n*, **Antikoaguliermittel** *n* anticoagulant [drug]
Antikoinzidenz *f (nucl)* anticoincidence
Antikoinzidenzzähler *m (nucl)* anticoincidence counter
anti-Konformation *f*, **anti-Konstellation** *f* anti[periplanar] conformation
Antikonvulsans *n*, **Antikonvulsivum** *n* anticonvulsive
antikonzeptionell contraceptive
Antikonzeptionsmittel *n*, **Antikonzipiens** *n* contraceptive
Antikörper *m* antibody, immune body
Antikörperprotein *n* antibody (immune) protein
antikorrosiv anticorrosive
antileukozytär antileucocytic
Antimaculepapier *n* tympan (set-off) paper
antimagnetisch antimagnetic
Antimalariamittel *n*, **Antimalariapräparat** *n* antimalarial [drug]
Antimalariawirksamkeit *f* antimalarial activity
Anti-Markownikow-Addition *f* anti-Markovnikov addition, peroxide-initiated addition
Antimaterie *f* antimatter
Antimetabolit *m* antimetabolite
antimikrobiell, antimikrobisch antimicrobial, microbicidal
Antimon *n* Sb antimony
graues (metallisches) A. gray (metallic, gamma) antimony
schwarzes A. black (beta) antimony
Antimonat *n* antimonate
Antimonblüte *f (min)* antimony bloom, flowers of antimony, valentinite *(antimony(III) oxide)*
Antimon(III)-bromid *n* $SbBr_3$ antimony(III) bromide, antimony tribromide
Antimonbutter *f* antimony butter, butter of antimony, mineral butter *(antimony(III) chloride)*
Antimon(III)-chlorid *n* $SbCl_3$ antimony(III) chloride, antimony trichloride
Antimon(V)-chlorid *n* $SbCl_5$ antimony(V) chloride, antimony pentachloride
Antimonelektrode *f* antimony electrode
Antimonfahlerz *n (min)* grey copper [ore], copper grey, tetrahedrite *(antimony copper sulphide, often containing iron and silver)*
Antimon(III)-fluorid *n* SbF_3 antimony(III) fluoride, antimony trifluoride
Antimon(V)-fluorid *n* SbF_5 antimony(V) fluoride, antimony pentafluoride
Antimongelb *n* antimony yellow, Naples yellow *(lead antimonate)*
Antimonglanz *m s.* Antimonit
Antimonhalogenid *n* antimony halogenide (halide)
antimonhaltig antimoniferous, antimonial, stibial
Antimon(III)-hydrid *n s.* Antimonwasserstoff
Antimonid *n* antimonide
Antimonit *m (min)* antimonite, stibnite, antimony glance, grey antimony *(antimony(III) sulphide)*
Antimon(III)-jodid *n* SbI_3 antimony(III) iodide, antimony triiodide
Antimon(V)-jodid *n* SbI_5 antimony(V) iodide, antimony pentaiodide
Antimonkarmin *n (coat)* antimony cinnabar (vermilion), antimonial cinnabar
Antimonocker *m (min)* antimony ochre
Antimon(III)-oxid *n* Sb_2O_3 antimony(III) oxide, antimony trioxide
Antimon(III, V)-oxid *n* Sb_2O_4 antimony(III, V) oxide, antimony tetraoxide
Antimon(V)-oxid *n* Sb_2O_5 antimony(V) oxide, antimony pentaoxide
Antimon(III)-oxidchlorid *n* SbOCl antimony(III) chloride oxide
Antimon(V)-oxidchlorid *n* $SbOCl_3$ antimony(V) trichloride oxide
Antimon(III)-oxidsulfat *n* $(SbO)_2SO_4$ antimony(III) oxide sulphate
Antimonpent[a] ... *s.* Antimon(V)- ...
Antimonregulus *m* antimony regulus, regulus of antimony
Antimonsäure *f* antimonic acid
Antimon(III)-selenid *n* Sb_2Se_3 antimony(III) selenide, antimony triselenide
Antimonspeise *f (met)* antimonial speiss
Antimon(III)-sulfat *n* $Sb_2(SO_4)_3$ antimony(III) sulphate, antimony trisulphate
Antimon(III)-sulfid *n* Sb_2S_3 antimony(III) sulphide, antimony trisulphide
Antimon(V)-sulfid *n* Sb_2S_5 antimony(V) sulphide, antimony pentasulphide

Antimon(III)-tellurid *n* Sb_2Te_3 antimony(III) telluride, antimony tritelluride
Antimontetroxid *n s.* Antimon(III, V)-oxid
Antimontri ... *s.* Antimon(III)- ...
Antimonwasserstoff *m* SbH_3 antimony hydride, antimony trihydride, stibine
Antimonweiß *n* antimony white *(antimony(III) oxide)*
Antimonylkaliumtartrat *n (incorrectly for)* Kaliumantimonotartrat
Antimonzinnober *m s.* Antimonkarmin
Antimutagen *n* antimutagen
Antimykotikum *n* antimycotic, antifungal drug, fungicide
antimykotisch antimycotic, antifungal, fungicidal
a. wirken to have antifungal activity
Antimyzin *n* antimycin *(antibiotic)*
Antimyzinlakton *n* antimycin lactone
Antineuralgikum *n* antineuralgic
Antineutrino *n* antineutrino
Antineutron *n* antineutron
Antinukleon *n* antinucleon
Antioxydans *n s.* Antioxydationsmittel
Antioxydationsmittel *n* antioxidant [agent], antioxidizer, antioxidizing agent, antioxygen
Antioxygen *n s.* Antioxydationsmittel
Antiozonant *m*, **Antiozonisator** *m* antiozonant, antiozidant, sunproofing agent
Antipartikel *f* antiparticle
Antipellagravitamin *n* pellagra-preventive factor, pp factor *(either nicotinic acid or nicotinamide)*
±**anti-periplanar** non-eclipsed, staggered *(stereochemistry)*
Antiperspirant *n* antiperspirant, antihidrotic, perspiration check
antiperspirierend antiperspirant
Antiperthit *m (min)* antiperthite *(a tectosilicate)*
Antiphlogistikum *n* antiphlogistic
antiphlogistisch antiphlogistic
antiplasmodisch antiplasmodial
Antipode *m* **[/ optischer]** antipode, optical antipode (opposite), mirror-image isomer
Antipodenpaar *n* pair of antipodes
optisches A. racemic pair
antiproteolytisch antiproteolytic
Antiprothrombin *n* antiprothrombin
Antiproton *n* antiproton, negative proton
Antipyretikum *n* antipyretic, febrifuge
antipyretisch antipyretic
antirachitisch antirachitic
Antireflexbelag *m* antireflection (antiflare) coating
Antireflexschicht *f* antireflection (antiflare) layer
Antirheumatikum *n* antirheumatic
Antirostadditiv *n* rust-inhibiting (rust-preventing) additive
Antischaummittel *n* antifoaming (antifrothing, defoaming) agent, antifoam [agent, compound], antifoamer, defoamer, foam breaker (destroyer, inhibitor, killer), froth-preventing agent, froth suppressor
Antischleiermittel *n (phot)* antifog[ging] agent, antifoggant
Antischweißlotion *f s.* Antitranspirationslotion
Antischweißmittel *n s.* Antitranspirationsmittel
Antiscorcher *m (rubber)* antiscorcher, antiscorching agent, retarder
Antiseptikum *n* antiseptic [agent]
A. gegen Schimmel antimildew agent
antiseptisch antiseptic
Antiskabiosum *n* scabicide, scabieticide
Antispasmodikum *n* antispasmodic
Antispritzmittel *n (food)* antispatterer
Antistatikmittel *n*, **Antistatikum** *n* antistatic [additive, agent], static eliminator
antistatisch antistatic
anti-Stellung *f* anti position ✦ **in anti-Stellung stehen** to be anti
Antisterilitätsvitamin *n* antisterility vitamin
Antistreßmineral *n (geol)* antistress mineral
antisymmetrisch antisymmetric
anti-syn-Isomerie *f* anti-syn isomerism
Antiteilchen *n* antiparticle
Antithrombin *n* antithrombin
Antitoxin *n* antitoxin
antitoxisch antitoxic
Antitranspirationslotion *f* antiperspirant lotion
Antitranspirationsmittel *n* antiperspirant, antihidrotic, perspiration check
Antitrypsin *n* antitrypsin
Antitussivum *n* antitussive
Antiurease *f* antiurease
Antivergrauungsmittel *n (text)* antiredeposition agent
Antivitamin *n* antivitamin
Antiweinsäure *f* $HOOC[CHOH]_2COOH$ mesotartaric acid
Antlerit *m (min)* antlerite *(copper(II) tetrahydroxide sulphate)*
antreiben to drive; to operate
Antrieb *m* drive
Antriebsrolle *f* driving pulley
Antriebstrommel *f* driving pulley
Antriebswelle *f* drive shaft
Antu *s.* α-Naphthylthioharnstoff
Anvulkanisation *f (rubber)* scorch[ing], set[ting]-up, firing-up, burning, curing-up, precure, precuring, prevulcanization; *(rubber)* pile (bin) curing *(during storage)*
Anvulkanisationscharakteristik *f (rubber)* scorch characteristic
Anvulkanisationsgeschwindigkeit *f (rubber)* cure rate
Anvulkanisationskurve *f (rubber)* scorch curve
Anvulkanisationsneigung *f (rubber)* scorch[ing] tendency, scorchiness
Anvulkanisationsperiode *f (rubber)* scorch period
Anvulkanisationspunkt *m (rubber)* scorch point
Anvulkanisationstendenz *f s.* Anvulkanisationsneigung
Anvulkanisationszeit *f (rubber)* scorch time
anvulkanisieren *(rubber)* to scorch, to set up, to fire up, to burn, to cure up, to precure, to prevulcanize *(to undergo unintended vulcanization)*
anwachsen to increase, to rise

Anwachsen *n* increase, rise
anwärmen to preheat
Anwärmloch *n (glass)* glory hole
Anwärmsektion *f* heating section
Anwärmzeit *f* heat-up period (time), heating-up period (time), coming-up time
Anwärmzone *f* preheating zone (compartment)
anweichen *(tann)* to presoak
anwendbar applicable
Anwendbarkeit *f* applicability
anwenden to use, to apply
lokal a. *(pharm)* to use topically (in topical applications)
Anwendung *f* application, use
Anwendungsbereich *m*, **Anwendungsgebiet** *n* field of application
Anwendungsweise *f* mode (method) of application
Anzahl *f* number, quantity
A. der Freiheiten (Freiheitsgrade) *(phys chem)* variance, number of degrees of freedom
A. der Zonendurchgänge number of zone passes *(in zone melting)*
A. je Zeiteinheit rate
anzapfen to tap
Anzeige *f* indication, reading *(of a measuring instrument)*
Anzeigebereich *m* indicating range, scale span *(of a measuring instrument)*
Anzeigegerät *n*, **Anzeigeinstrument** *n* indicating instrument, indicator
Anzeigelampe *f* indicating (pilot) lamp
anzeigen to indicate
Anzeiger *m* indicator *(in indicating instruments)*
anziehen to attract; to draw [in] *(e. g. a liquid)*
Feuchtigkeit a. to gain moisture
Anziehung *f* attraction
elektrische A. electrical attraction
intermolekulare A. molecular attraction
magnetische A. magnetic attraction
van-der-Waalssche A. van der Waals attraction
zwischenmolekulare A. *s.* intermolekulare A.
Anziehungskraft *f* attractive force (power)
Coulombsche A. Coulomb (coulombic) force of attraction
Anziehungskräfte *fpl* / **van-der-Waalssche** van der Waals forces [of attraction], van der Waals attractive forces
anzünden to light, to kindle; to ignite; to inflame
AO *s.* Atomorbital
AP *s.* Anilinpunkt
Apatit *m (min)* apatite *(calcium fluorophosphate)*
Apertur *f* aperture
numerische A. numerical aperture
Apex *m* apex *(of a liquid cyclone)*
Apexdüse *f* apex opening
Apfeläther *m* apple oil *(main constituent isoamyl valerianate)*
Apfelessenz *f* apple essence, apple oil *(esp in alcohol solution)*
Apfelmost *m* [sweet] cider
Apfelöl *n s.* Apfelessenz
Apfelsaft *m* apple juice
Apfelsäure *f s.* Äpfelsäure
Äpfelsäure *f* HOOCCH(OH)CH_2COOH malic acid, hydroxysuccinic acid, hydroxybutanedioic acid
natürliche Ä. *s.* *L*(-)-Äpfelsäure
***l*-Äpfelsäure** *f s.* *L*(-)-Äpfelsäure
***L*(-)-Äpfelsäure** *f* (-)-malic acid, ordinary (common) malic acid
Äpfelsäuredehydr[ogen]ase *f* malic acid dehydrogenase
Äpfelsäure-Milchsäure-Gärung *f* malo-lactic fermentation
Apfelsineneinwickelpapier *n* orange wrapper
Apfelsinenschale *f* 1. orange peel *(from Citrus sinensis (L.) Osbeck)*; 2. *s.* Apfelsinenschaleneffekt
Apfelsinenschaleneffekt *m (coat)* orange peel, bad flow *(a surface defect)*
Apfelsinenschalenhaut *f s.* Apfelsinenschaleneffekt
Apfelsinenschalenöl *n* [sweet] orange-peel oil *(from Citrus sinensis (L.) Osbeck)*
Apfelwein *m* cider
Aphizid *n* aphicide
Aphrodisiakum *n* aphrodisiac
Aphthitalit *m (min)* aphthitalite *(potassium sodium sulphate)*
API-Dichte *f (petrol)* API gravity
API-Grad *m (petrol)* API degree
Apiol *n* apiol, 2,5-dimethoxy-3,4-methylenedioxy-1-allylbenzene
API-Skala *f (petrol)* API scale
AP-Kautschuk *m* EP-rubber, ethylene-propylene rubber, EPR
Aplit *m (geol)* aplite
Aplom *m (min)* aplome *(a garnet)*
Apoenzym *n s.* Apoferment
Apoferment *n* apoferment, apoenzyme, colloid carrier, protector
Apokrensäure *f (soil)* apocrenic acid *(a fulvic acid)*
apolar apolar, non-polar
Apomorphin *n* apomorphine *(alkaloid)*
Apophyllit *m (min)* apophyllite *(a phyllosilicate)*
Apophyse *f (geol)* apophysis
Aporphin *n* aporphine
Aporphinalkaloid *n* aporphine alkaloid
Apostroph *m (nomencl)* prime *(with locants)*
Apotheker *m* pharmacist, pharmaceutic[al] chemist
Apothekerwaage *f* hand balance
Apo-Umlagerung *f* apo rearrangement
Apparat *m* apparatus
A. für kontinuierliche Extraktion continuous-extraction apparatus, continuous extractor
A. nach Schopper-Riegler *(pap)* Schopper-Riegler apparatus
A. zur Kohlensäurebestimmung carbon dioxide apparatus
Brühlscher A. Brühl receiver
Kippscher A. Kipp [gas] generator, Kipp's apparatus
Marshscher A. Marsh's testing apparatus

Apparatebau *m* apparatus construction
Apparateeinheit *f* unit
Apparategruppe *f* set
Apparateklemme *f* apparatus clamp
Apparatekonstante *f* apparatus constant
Apparatur *f* apparatus, equipment
A. zum tiegelfreien (tiegellosen) Zonenschmelzen floating-zone apparatus (unit)
A. zum Zonenschmelzen zone-melting apparatus (unit)
A. zur Zonenreinigung zone-refining apparatus (unit), zone refiner
Beckmannsche A. Beckmann apparatus
Appearance-Potential *n* appearance potential *(mass spectroscopy)*
Applikation *f (agric)* application, placement *(of fertillizers or pesticides); (pharm)* application, administration, dosage *(of a medicine)*
A. aus der Luft aerial (air-to-ground) application
A. unter Abschirmung directed application
A. vom Flugzeug aus aeroplane application
A. vor dem Auspflanzen preplanting application
Applikationsart *f s.* Applikationsmethode
Applikationsgerät *n (agric)* [mechanical] applicator, application apparatus
Applikationsmethode *f (agric)* placement method; *(pharm)* mode of administration
applizieren *(agric)* to apply, to place *(fertilizers or pesticides)*; *(pharm)* to apply, to administer *(a medicine)*
Appret *n s.* Appreturmittel
Appretbrechmaschine *f s.* Appreturbrechmaschine
Appreteur *m (text)* finisher
appretieren *(text, tann)* to finish, to dress, to season
Appretur *f* 1. finish; 2. finishing, dressing, seasoning; 3. *s.* Appreturmittel
antistatische A. *(text)* antistatic finish
glatttrocknende A. *(text)* smooth-drying finish
griffgebende A. *(text)* stiffening finish
knirschende A. *(text)* rustling finish
schrumpffreie A. *(text)* unshrinkable (shrink-resist) finish
stärkehaltige A. *(text)* starchy finish
wasserabstoßende (wasserabweisende) A. *(text)* water-repellent finish
wasserdichte A. *(text)* waterproof finish
Appreturbrechmaschine *f (text)* finish-breaking machine
Appreturfoulard *m (text)* finishing padder
Appreturmittel *n (text, tann)* finish, dressing [agent], sizing material (agent, chemical, substance), size
Appreturöl *n* textile oil
aprotisch aprotic
APS *s.* Apurinsäure
6-APS *s.* 6-Aminopenizillansäure
APT-Kautschuk *m* ethylene-propylene terpolymer, EPT
Apurinsäure *f* apurinic acid *(a derivative of ribonucleic acid)*
Aquamarin *m (min)* aquamarine *(a variety of beryl)*
Aquametrie *f* aquametry
Aquarellfarbe *f* water colour
Aquarellpapier *n* water-colour [drawing] paper, aquarel[le] paper
aquatisch *(geol)* aqueous, aq.
äquatorial equatorial *(stereochemistry)* ✦ **ä. stehen** to be equatorial
äquilibrieren to equilibrate
Äquilibrierung *f* equilibration
Äquilibrierungsverfahren *n* equilibration process
äquimolar equimolar, equimolecular
Äquipartitionsprinzip *n,* **Äquipartitionstheorem** *n* equipartition principle, law of equipartition of energy
Äquipotentialfläche *f* equipotential surface, potential energy surface
äquivalent equivalent
Äquivalent *n* equivalent
fotochemisches Ä. photochemical equivalent
Äquivalentgewicht *n s.* Äquivalentmasse
Äquivalentladung *f* equivalent charge
Äquivalentleitfähigkeit *f* equivalent conductance
Ä. bei unendlicher Verdünnung equivalent conductance at infinite dilution
Ä. der Ionen equivalent ion[ic] conductance
Äquivalentmasse *f* equivalent (combining) weight
Äquivalenz *f* equivalence, equivalency
Ä. von Energie und Masse mass-energy equivalence
Äquivalenzgesetz *n* / **fotochemisches** law of the photochemical equivalent, Stark-Einstein law
Stark-Einsteinsches Ä. *s.* fotochemisches Ä.
Äquivalenzleitfähigkeit *f s.* Äquivalentleitfähigkeit
Äquivalenzpunkt *m* equivalent (equivalence, equivalency) point, point of neutrality *(in titration)*
Aquoion *n* aquo-ion, hydrated ion
Aquopentamminkobalt(III)-chlorid *n* $[Co(NH_3)_5H_2O]Cl_3$ aquapentaamminecobalt(III) chloride
aquotisieren to aquate, to subject to aquation *(as in the formation of coordination complexes)*
Aquotisierung *f* aquotization, aquation, aquatization, hydrolysis *(the replacement by water molecules of a coordinated atom or group in a coordination complex)*
Aquoverbindung *f* aquo compound
Araban *n* $(C_5H_{10}O_5)_n$ araban *(a pentosan)*
Arabinose *f* arabinose, pectinose
Arabit *m* $CH_2OH(CHOH)_3CH_2OH$ arabitol *(a sugar alcohol)*
Arachidonsäure *f* arachidonic acid
Arachinsäure *f* $CH_3[CH_2]_{18}COOH$ arachidic acid, eicosanoic acid
Arachisöl *n* arachis (peanut) oil
Arachnolysin *n* arachnolysine *(a haemolysin secreted by certain spiders)*
Aragonit *m (min)* aragonite *(calcium carbonate)*
Aralkylsilikon *n s.* Alkylarylsiloxan
Aralkylsiloxan *n s.* Alkylarylsiloxan
Aralkylsulfonat *n (incorrectly for)* Alkylarylsulfonat
Aräometer *n* areometer, hydrometer, densimeter
Aräometerskale *f* areometer (hydrometer) scale

Aräometrie *f* areometry, hydrometry
aräometrisch areometric[al], hydrometric[al]
Ararobapulver *n (pharm)* araroba, Goa powder *(from Andira araroba Aguiar)*
Arbeit *f* work *(physics)*
äußere A. external work
geleistete A. work done
gesamte A. total work
gewonnene A. *s.* geleistete A.
maximale A. [der Reaktion] maximum useful work, work of reaction
umgesetzte A. *s.* geleistete A.
arbeiten / mit dem Lötrohr to blowpipe
Arbeitsaufwand *m* expenditure of work
arbeitsaufwendig / wenig low-labour
Arbeitsbühne *f* operating floor (platform); drilling floor *(of a rotary-drilling installation)*
Arbeitsbütte *f (pap)* service (machine, pulp, supply, stuff) chest
Arbeitsdruck *m* working (operating) pressure
Arbeitselektrode *f* working electrode
Arbeitsgang *m* operation
Arbeitsgeschwindigkeit *f* operating speed
Arbeitsherd *m* hearth *(of an air furnace)*
Arbeitsinhalt *m* operating hold-up
A. einer Kolonne liquid operating hold-up of a column
Arbeitsleistung *f* performance
Arbeitslinie *f (distil)* working (operating) line
Arbeitsloch *n s.* Arbeitsöffnung
Arbeitslösung *f* working solution
Arbeitsmethode *f* working (operating) method
Arbeitsöffnung *f (glass)* gathering hole (opening)
Arbeitsrichtung *f (pap)* machine (making, long, grain) direction
Arbeitsschablone *f* working template *(chromatography)*
Arbeitsschutz *m* industrial safety
Arbeitsschutzanordnung *f* safety regulation
Arbeitsschutzsalbe *f* barrier cream
Arbeitsspannung *f* working voltage
Arbeitsstoff *m* working substance
Arbeitsstrom *m* working current
Arbeitsstufe *f* stage of operation
Arbeitstemperatur *f* working (operating) temperature
Arbeitstisch *m* laboratory desk
Arbeitsverlust *m (rubber)* hysteresis
Arbeitsvermögen *n* energy *(of a physical system)*
Arbeitswanne *f (glass)* working (refining) chamber, working (refining) end, refiner, nose
Arbeitsweise *f* operation, processing *(of an apparatus)*; working (operating) method, procedure, *(esp lab)* technique
A. mit totalem Rücklauf total-reflux operation
diskontinuierliche A. batch processing (operation)
periodische A. *s.* diskontinuierliche A.
Arborizid *n* brushkiller, silvicide
Arbutin *n* $HOC_6H_4OC_6H_{11}O_5$ arbutin *(glycoside)*
Arcatom-Schweißverfahren *n* atomic hydrogen [arc] welding
Archil *n* archil, orchil *(a lichen dye)*
Arekaidin *n* arecaidine *(alkaloid)*
Arekanuß *f* areca (betel) nut *(from Areca cathechu L.)*
Arekolidin *n* arecolidine *(alkaloid)*
Arekolin *n* arecoline *(methyl ester of arecaidine)*
Arfvedsonit *m (min)* arfvedsonite *(an inosilicate)*
Arg *s.* Arginin
Argentin *m (min)* argentine *(a pearly variety of calcite)*
Argentit *m (min)* argentite *(silver sulphide)*
Argentometrie *f* argentometry
argentometrisch argentometric
Argentopyrit *m (min)* argentopyrite *(iron silver sulphide)*
Arge-Synthese *f* Arge synthesis
Argillit *m (geol)* argillite
Arginin *n* arginine, 2-amino-5-guanidinovaleric acid
Argon *n* Ar argon
Argonatmosphäre *f* argon atmosphere
Argonklathrat *n* argon clathrate
Argyrodit *m (min)* argyrodite *(germanium silver sulphide)*
Aridye-Verfahren *n* Aridye process *(pigment printing process)*
Arin *n* aryne *(any of a class of transient dehydrogenated derivatives of aromatic compounds), (specif)* benzyne
Aristolochiagelb *n s.* Aristolochiasäure
Aristolochiasäure *f* aristolochic acid, aristolochin, aristolochia yellow
Aristolochin *n s.* Aristolochiasäure
Arkanit *m (min)* arcanite *(potassium sulphate)*
Arkansit *m (min)* arkansite *(titanium dioxide)*
Arkose *f (geol)* arkose
arm poor, with a low content; lean *(gas)*; low-grade *(ores)*; barren, weak *(solution)*; thin *(soil)*
Arm *m* arm, blade *(as of a mixing machine)*
Armcoeisen *n* Armco iron *(with less than 1% impurities)*
Armgas *n* lean gas
armieren to reinforce
Armierung *f* reinforcement
Armlauge *f* barren solution *(in cyaniding)*
Armstrong-Verfahren *n* Armstrong method *(for producing 1-naphthol-2-sulphonic acid)*
Arndt-Eistert-Synthese *f* Arndt-Eistert synthesis *(converting a carboxylic acid to its next higher homologue)*
Arnika[blüten]öl *n* arnica flowers oil
Arnold-Probe *f* Arnold's test *(for detecting acetoacetic acid in urine)*
Aroma *n* 1. aroma, flavour; 2. *s.* Aromastoff
butterähnliches A. buttery flavour
Aromabakterien *pl*, **Aromabildner** *mpl* aroma[-forming] bacteria, aroma organisms (producers)
Aromabildung *f* aroma development
Aromadendrin *n* aromadendrin, 3,4′,5,7-tetrahydroxyflavanone
Aromafülle *f (food)* ful[l]ness

Aromakomposition *f s.* Aromaträger
Aromastoff *m* aroma (aromatic) substance (body, ingredient), aromatizing product, flavouring material (matter, substance)
Aromaten *pl* aromatic hydrocarbons (compounds), aromatics
  mehrkernige A. polynuclear aromatics
aromatenfrei free from aromatic hydrocarbons
aromatisch *(org chem)* aromatic *(containing the benzene ring)*; *(food)* aromatic, spiced, spicy
aromatisieren *(org chem)* to aromatize *(to convert into aromatic hydrocarbons)*; *(food)* to aromatize, to flavour
Aromatisierung *f (org chem)* aromatization *(conversion into aromatic hydrocarbons)*; *(food)* aromatization
Aromaträger *m* aroma compound
Aromaverlust *m* aroma loss
Aroniumion *n* aronium ion
Aroylbenzoesäure *f* aroylbenzoic acid
Aroylgruppe *f* aroyl group (residue)
Arrak *m* arrack
arretieren to arrest, to lock, to stop
Arretierung *f* 1. arrestment, arresting; 2. arresting mechanism
Arretiervorrichtung *f* arresting mechanism
Arrojadit *m (min)* arrojadite *(a phosphate of sodium, iron, and manganese)*
Arsanilsäure *f* $NH_2C_6H_4AsO(OH)_2$ arsanilic acid, *p*-aminophenylarsonic acid
Arsen *n* As arsenic
  gelbes A. yellow arsenic, α-arsenic
  graues (metallisches) a. gray (metallic) arsenic, γ-arsenic
  schwarzes A. black arsenic, β-arsenic
Arsenat(III) *n* arsenite
Arsenat(V) *n* arsenate
Arsen(III)-bromid *n* $AsBr_3$ arsenic(III) bromide, arsenic tribromide
Arsenbutter *f* butter of arsenic *(arsenic(III) chloride)*
Arsen(III)-chlorid *n* $AsCl_3$ arsenic(III) chloride, arsenic trichloride
Arsen(V)-chlorid *n* $AsCl_5$ arsenic(V) chloride, arsenic pentachloride
Arsendampf *m* arsenic vapour
Arseneisensinter *m (min)* pitticite
Arsenerz *n* arsenic ore
Arsen(III)-fluorid *n* AsF_3 arsenic(III) fluoride, arsenic trifluoride
Arsen(V)-fluorid *n* AsF_5 arsenic(V) fluoride, arsenic pentafluoride
arsenhaltig arsenical, arsenian
Arsen(III)-hydrid *n s.* Arsin 2.
Arsenid *n* arsenide
Arsenik *n* As_2O_3 [white] arsenic, arsenic(III) oxide
  gelbes A. yellow arsenic [sulphide], arsenic yellow, king's yellow (gold), royal yellow, orpiment [yellow] *(technically pure arsenic(III) sulphide)*
  rotes A. *(tann)* red arsenic *(a mixture of arsenic sulphides)*
  weißes A. *s.* Arsenik
Arsenikalie *f* arsenical
Arsenikbrocken *mpl (glass)* glassy (dense) arsenic
Arsenikschwöde *f (tann)* arsenic paint
Arsenikvergiftung *f* arsenic poisoning
Arseninsektizid *n* arsenical insecticide
Arsen(III)-jodid *n* AsI_3 arsenic(III) iodide, arsenic triiodide
Arsen(V)-jodid *n* AsI_5 arsenic(V) iodide, arsenic pentaiodide
Arsenkies *m s.* Arsenopyrit
Arsenkobalt *m (min)* arsenical cobalt *(cobalt arsenide)*
Arsenkobaltsulfid *n* CoAsS cobalt arsenosulphide
Arsenobenzol *n* $C_6H_5As{=}AsC_6H_5$ arsenobenzene
Arsenoferrit *m (min)* arsenoferrite *(iron arsenide)*
Arsenolith *m (min)* arsenolite *(arsenic(III) oxide)*
Arsenopyrit *m (min)* arsenopyrite, arsenic[al] iron, arsenical pyrite, mispickel *(iron sulpharsenide)*
Arsenosulfid *n* $M^{II}AsS$ arsenosulphide, sulpharsenide
Arsen(III)-oxid *n* As_2O_3 arsenic(III) oxide, arsenic trioxide, arsenic
Arsen(V)-oxid *n* As_2O_5 arsenic(V) oxide, arsenic pentaoxide
Arsen(III)-oxidchlorid *n* AsOCl arsenic(III) chloride oxide
Arsenpent[a] ... *s.* Arsen(V)- ...
Arsen(III)-phosphid *n* AsP arsenic(III) phosphide
Arsenprobe *f* arsenic test
  Bettendorfsche A. Bettendorf's test [for arsenic]
  Gutzeitsche A. Gutzeit's test [for arsenic]
  Marshsche A. Marsh's test [for arsenic]
Arsen(III)-säure *f* H_3AsO_3 arsenious acid
Arsen(V)-säure *f* H_3AsO_4 arsenic acid
Arsensäureanhydrid *n s.* Arsen(V)-oxid
Arsen(III)-selenid *n* As_2Se_3 arsenic(III) selenide, arsenic triselenide
Arsenspeise *f (met)* arsenical speiss
Arsenspiegel *m* arsenic mirror, stain of arsenic
Arsensulfid *n (tann)* arsenic sulphide *(a mixture mainly consisting of tetraarsenic tetrasulphide and arsenic trisulphide)*
Arsen(III)-sulfid *n* As_2S_3 arsenic(III) sulphide, arsenic trisulphide
Arsen(V)-sulfid *n* As_2S_5 arsenic(V) sulphide, arsenic pentasulphide
Arsentri ... *s.* Arsen(III)- ...
Arsentrisulfidsol *n* arsenious sulphide sol
Arsenvergiftung *f* arsenic poisoning
Arsenwasserstoff *m s.* Arsin 2.
Arsin *n* 1. AsR_3 arsine *(an organic compound)*; 2. AsH_3 arsine, arsenic(III) hydride, arsenic trihydride
Arsoniumgruppe *f*, **Arsoniumrest** *m* arsonium group (residue)
Arterit *m* arterite, arteritic migmatite *(veined gneiss)*
Arylalkylsilikon *n s.* Alkylarylsiloxan
Aryläther *m* aryl (aromatic) ether
Arylazogruppe *f*, **Arylazorest** *m* $ArN{=}N-$ arylazo group (residue)

Arylen *n* arylene *(any of a class of bivalent radicals derived from an aromatic hydrocarbon)*
Arylesterase *f* arylesterase *(a ferment of human blood serum)*
Arylgruppe *f* aryl group (residue)
Arylhalogenid *n* aryl halide
arylieren to arylate
Arylierung *f* arylation
Arylphenolumlagerung *f* aryl-phenol rearrangement
Arylradikal *n* aryl radical, *(specif)* free aryl radical
Arylrest *m s.* Arylgruppe
Arylsilikon *n s.* Arylsiloxan
Arylsiloxan *n* aryl siloxane
Arylsulfonsäure *f* arylsulphonic acid
Arylverknüpfung *f* aryl coupling
Aryn *n s.* Arin
Arzneibuch *n* pharmacop[o]eia ✦ **den Anforderungen des Arzneibuchs entsprechen** to be of pharmacopoeial quality
arzneilich medicinal
Arzneimittel *n* pharmaceutic[al], pharmaceutical preparation, medicinal drug, medicament
A. auf Sulfonamidbasis sulpha drug
Arzneimittelforschung *f* drug research
Arzneimittelindustrie *f* pharmaceutic[al] industry
Arzneimittelvergiftung *f* drug intoxication
Arzneimittelverordnung *f* medication
Arzneipflanze *f* medicinal (officinal) plant
Arzneistoff *m* medicinal substance
Arzneiverordnung *f s.* Arzneimittelverordnung
Asa foetida *f* asafetida, devil's dung, food of the gods *(a gum resin from Ferula specc.)*
Asant *m s.* Asa foetida
A-Säure *f* A acid, 6-amino-1-naphthol-5-sulphonic acid
Asbest *m* asbestos, asbestus
Asbestaufschlämmung *f* asbestos milk (suspension)
Asbestband *n* asbestos tape
Asbestdiaphragma *n* asbestos diaphragm
Asbestdichtung *f* asbestos gasket (packing); asbestos joint
Asbestdrahtnetz *n* asbestos gauze
Asbestfaser *f* asbestos fibre
Asbestfausthandschuh *m* asbestos mitt[en]
Asbestfilter *n* asbestos filter
Asbestfilz *m* asbestos felt
Asbestfingerling *m* asbestos finger cot
Asbestgarn *n* asbestos yarn
Asbestgewebe *n* asbestos cloth, woven asbestos
Asbesthandschuh *m* asbestos glove
Asbestine *f (pap)* asbestine
Asbestkleidung *f* asbestos clothing
Asbestmasseplatte *f (filtr)* asbestos-pulp disk
Asbestmehl *n* asbestos powder
Asbestmembran *f s.* Asbestdiaphragma
Asbestmörtel *m* asbestos mortar
Asbestpackung *f* asbestos packing (gasket)
Asbestpapier *n* asbestos paper
Asbestpappe *f* asbestos board
Asbestplatte *f* asbestos plate (mat, board)
Asbestpulver *n* asbestos powder
Asbestrohr *n* asbestos pipe
Asbestscheibe *f* asbestos disk
Asbestschiefer *m* asbestos slate
Asbestschnur *f* asbestos cord
Asbeststoff *m*, **Asbesttuch** *n s.* Asbestgewebe
Asbestunterlage *f s.* Asbestplatte
Asbestwolle *f* asbestos wool
Asbestzement *m* asbestos cement
Asbolan *m (min)* asbolan[e], asbolite, black cobalt *(earthy manganese dioxide containing cobalt oxide)*
Asche *f* ash[es]
äußere A. *(coal)* extraneous ash
innere A. *(coal)* inherent ash
vulkanische A. volcanic ash
Asche ... *s.* Aschen ...
aschenarm low-ash
Aschenaustrag *m* ash removal
Aschenaustragsschleuse *f* ash-discharging vessel
Aschenaustragung *f* ash removal
Aschenbestimmung *f* ash determination
Aschenentfernung *f* ash removal
Aschenerweichungspunkt *m* ash fluid temperature
Aschenfall *m* ash pit
aschenfrei ash-free, ashless, free from ash
Aschenfreiheit *f* freedom from ash
Aschengehalt *m* ash content ✦ **mit hohem A.** highash ✦ **mit niedrigem A.** low-ash
Aschengehaltskurve *f* ash curve
Aschengrube *f* ash pit
aschenhaltig ashen, ashy
schwach a. low-ash
stark a. high-ash
Aschenkasten *m* ash pan
Aschenkeller *m* ash pit
Aschenraum *m* ash pit
aschenreich high-ash
Aschenschleuse *f s.* Aschenaustragsschleuse
Aschenschüssel *f* ash pan
Aschensinter *m* sintered fly ash
Aschentrichter *m* ash hopper
Aschentuff *m (geol)* ash tuff
Aschenwaage *f* ash scale
Aschenzone *f* ash zone
Äscher *m* 1. *(tann)* lime; 2. *s.* Äschergrube; 3. *s.* Äschern
angeschärfter Ä. sharpened lime *(lime milk treated with sodium sulphide)*
fauler Ä. rotten (dead) lime
frischer Ä. fresh (head) lime
milder Ä. mellow lime liquor
toter Ä. *s.* fauler Ä.
Äscherbrühe *f (tann)* lime liquor
Äscherfaß *n (tann)* liming tub
Äscherflüssigkeit *f s.* Äscherbrühe
Äschergang *m (tann)* round of lime
Äschergrube *f (tann)* lime pit
äschern *(tann)* to lime
Äschern *n (tann)* liming
aschist *(geol)* aschistic
Asebotoxin *n* asebotoxin

Asepsis *f* asepsis
aseptisch aseptic
Asiderit *m (geol)* asiderite
Askaridol *n* ascaridole, 1,4-peroxido-*p*-menthene-2
Askorbigen *n* ascorbigen
Askorbinsäure *f* ascorbic acid
Askorbinsäureoxydase *f* ascorbic [acid] oxidase
Äskulin *n* aesculin *(a coumarin derivative)*
Asp *s.* Asparaginsäure
Asparagin *n* asparagine, *(specif)* $HOOCCH(NH_2)CH_2CONH_2$ β-asparagine
Asparaginase *f* asparaginase *(an enzyme that hydrolyzes asparagine)*
Asparaginsäure *f* $HOOCCH(NH_2)CH_2COOH$ aspartic acid, asparaginic acid, aminosuccinic acid
Aspergillsäure *f* aspergillic acid
Asphalt *m* 1. *(min)* [natural, native] asphalt[e], asphaltum, mineral (earth) pitch; 2. *(build)* [artificial] asphalt
natürlicher A. *s.* Asphalt 1.
Asphaltbasis *f* asphalt base *(of crude petroleum)*
Asphaltbasisöl *n* asphalt-base petroleum (crude), asphaltic petroleum, asphaltic-base crude oil
Asphaltbeton *m* asphalt[ic] concrete
Asphaltbinder *m* asphaltic binder
Asphaltbitumen *n* asphaltic bitumen
Asphalten *n* asphaltene
Asphaltfarbe *f* asphalt paint
Asphaltfotografie *f* asphalt (bitumen) process
Asphaltgestein *n* asphalt rock
asphalthaltig asphaltic
Asphalthartpappe *f* bitumen board
asphaltisch asphaltic
Asphaltit *m* asphaltite *(natural asphalt)*
Asphaltlack *m* asphalt varnish
Asphaltmakadam *m(n)* asphalt macadam
Asphaltmastix *m* asphalt mastic
Asphaltogensäure *f* asphaltogenic acid
Asphaltöl *n s.* Asphaltbasisöl
Asphaltpapier *n* asphalt (tar, pitch) paper, tarred [brown] paper
Asphaltsand *m* asphaltic (tar) sand
Asphaltsee *m* asphalt (pitch) lake
Asphaltverfahren *n s.* Asphaltfotografie
Asphaltzement *m* asphaltic cement
Aspirationspsychrometer *n* aspiration psychrometer
Aspirator *m* aspirator
Asplund-Defibrator-Verfahren *n (pap)* Asplund process
Asp-NH₂ *s.* Asparagin
Assamkautschuk *m* Assam (Indian) rubber *(from Ficus elastica Roxb.)*
Assimilat *n* assimilate, photosynthate
Assimilation *f* assimilation
magmatische A. *(geol)* magmatic digestion
assimilierbar assimilable
Assimilierbarkeit *f* assimilability
assimilieren to assimilate
Assoziation *f (chem)* [molecular] association; *(min)* association
Assoziationsflüssigkeit *f* associated liquid
Assoziationsgrad *m* degree of association
Assoziationskolloid *n* micellar colloid
Assoziationskonstante *f* association constant
Assoziationswärme *f* heat of association
assoziieren to associate
sich a. to associate
Ast *m s.* Astknoten
Astat *n* At astatine
Astat-Emanation *f s.* Radon-218
astatisch astatic
Ästefänger *m s.* Astfang
Asterismus *m (cryst)* asterism
Astfang *m*, **Astfänger** *m (pap)* knot[ter] screen, knotter, jag-knotter
Asthenosphäre *f (geol)* asthenosphere
Astknoten *m (pap)* knot *(in wood)*
ASTM-Destillation *f* ASTM distillation
ASTM-Spezifikation *f* ASTM specification
ASTM-Standardmethode *f* standard ASTM method
ASTM-Verfahren *n* ASTM method
Aston-Massenspektrograf *m* Aston mass spectrograph
Astra-Druckschneckenkühler *m* Astra pressure cooler *(margarine making)*
Astrakanit *m (min)* astrak[h]anite, blödite *(magnesium sodium sulphate)*
Astro-Dye-Verfahren *n (text)* astro dyeing *(local cheese dyeing)*
Astrom-Entrindungsmaschine *f (pap)* Astrom chain barker
Astrophyllit *m (min)* astrophyllite
asym. *s.* asymmetrisch
Asymmetrie *f* asymmetry, dissymmetry
Asymmetriepotential *n* asymmetry potential
Asymmetriezentrum *n* asymmetric centre, centre of asymmetry
asymmetrisch asymmetric[al], dissymmetric, unsymmetrical
AT *s.* Alttuberkulin
Atakamit *m (min)* atacamite *(basic copper chloride)*
Ataraktikum *n (pharm)* tranquil[l]izer, tranquillizing drug
Ataxit *m* ataxite *(an iron meteorite)*
Atemgift *n* respiratory poison
Atemschutzgerät *n* respiratory protection apparatus
Atemschutzmaske *f* respirator
a-Term *m (nomencl)* „a" term
Äthal *n s.* Hexadekanol-(1)
Äthan *n* CH_3CH_3 ethane
Äthanal *n* CH_3CHO ethanal, acetaldehyde
Äthanalsäure *f s.* Glyoxylsäure
Äthanamid *n* CH_3CONH_2 acetamide, ethanamide
Äthandial *n* OHC–CHO ethanedial, oxaldehyde, glyoxal
Äthandiamid *n* $NH_2COCONH_2$ ocamide, oxalic acid diamide
1,2-Äthandiamin *n s.* Äthylendiamin
Äthandikarbonsäure-(1,2) *f* $HOOCCH_2CH_2COOH$ ethane-1,2-dicarboxylic acid, butanedioic acid, succinic acid

Äthandiol-(1,2) *n* $HOCH_2CH_2OH$ 1,2-ethanediol, 1,2-dihydroxyethane, ethylene glycol
Äthandisäure *f* HOOC–COOH ethanedioic acid, oxalic acid
Äthanol *n* CH_3CH_2OH ethanol, ethyl alcohol
Äthanolal *n s.* Glykolaldehyd
Äthanolamin *n* $HOCH_2CH_2NH_2$ ethanolamine, 2-aminoethanol, colamine
Äthanolgehalt *m* ethanol content (strength)
äthanolisch ethanolic
äthanollöslich ethanol-soluble, soluble in ethanol
Äthanolsäure *f s.* Glykolsäure
äthanolunlöslich ethanol-insoluble, insoluble in ethanol
Äthanolyse *f* ethanolysis
Äthanoyl *n* acetyl, ethanoyl (*for compounds s.* Azetyl ...)
Äthansäure *f* CH_3COOH ethanoic acid, acetic acid
Äthansäureanhydrid *n* $(CH_3CO)_2O$ ethanoic anhydride, acetic anhydride
Äthansulfonsäure *f* $CH_3CH_2SO_3H$ ethanesulphonic acid
Äthanthiol *n* CH_3CH_2SH ethanethiol, ethyl hydrosulphide
Äthen *n* $H_2C=CH_2$ ethene, ethylene
Äthenkarbonsäure *f* $CH_2=CHCOOH$ ethylenecarboxylic acid, propenoic acid, acrylic acid
Äthenkohlenwasserstoff *m s.* Äthylenkohlenwasserstoff
Äthenol *n* $CH_2=CHOH$ ethenol, vinyl alcohol
Äthenylgruppe *f s.* Vinylgruppe
Äther *m* R–O–R' ether, aether, *(specif)* $C_2H_5OC_2H_5$ diethyl ether, ordinary ether
  absoluter Ä. *s.* wasserfreier Ä.
  aromatischer Ä. aromatic (aryl) ether
  einfacher Ä. R–O–R symmetrical ether
  gemischter Ä. R'–O–R" mixed (unsymmetrical) ether
  symmetrischer Ä. *s.* einfacher Ä.
  unsymmetrischer Ä. *s.* gemischter Ä.
  wasserfreier Ä. absolute ether
  zyklischer Ä. cyclic ether
Ätherat *n* etherate
Ätherbildung *f* ether formation
Ätherbindung *f* ether bond (link, linkage) *(state)*
Ätherdampf *m* ether vapour
Ätherextraktion *f* ether extraction
ätherhaltig containing ether, ethereal
ätherisch ethereal, etherial; essential, volatile, ethereal *(oils)*
ätherlöslich ether-soluble, soluble in ether
atherman atherm[an]ous
Ätherspaltung *f* ether cleavage
ätherunlöslich ether-insoluble, insoluble in ether
Äthin *n* $HC\equiv CH$ acetylene, ethyne
Äthindikarbonsäure *f* $HOOCC\equiv CCOOH$ acetylenedicarboxylic acid, butynedioic acid
Äthinkohlenwasserstoff *m* acetylenic (acetylene) hydrocarbon, alkyne
Äthinylbenzol *n* $C_6H_5C\equiv CH$ ethynylbenzene, acetylenylbenzene, phenylacetylene
Äthinylgruppe *f* $-C\equiv CH$ ethynyl (acetylenyl) group, ethynyl (acetylenyl) residue
äthinylieren to ethynylate
Äthinylierung *f* ethynylation
Äthinylkarbinol *n s.* Propin-(2)-ol-(1)
Äthinylrest *m s.* Äthinylgruppe
Äthisteron *n* ethisterone, 17α-hydroxy-4-pregnen-20-yn-3-one
Äthoxyanilin *n* $C_6H_4(NH_2)OC_2H_5$ ethoxyaniline, phenetidine, aminophenol ethyl ether
Äthoxyäthan *n s.* Diäthyläther
2-Äthoxyäthanol *n* $C_2H_5OCH_2CH_2OH$ 2-ethoxyethanol, 2-ethoxyethyl alcohol
2-Äthoxyäthylalkohol *m s.* 2-Äthoxyäthanol
Äthoxybenzoesäure *f* $C_2H_5OC_6H_4COOH$ ethoxybenzoic acid
Äthoxybenzol *n* $C_6H_5OC_2H_5$ ethoxybenzene, ethyl phenyl ether, phenetole
Äthoxylinharz *n* ethoxylene resin
m-**Äthoxyphenol** *n* $C_2H_5OC_6H_4OH$ *m*-ethoxyphenol, resorcinol monoethyl ether
o-**Äthoxyphenol** *n* $C_2H_5OC_6H_4OH$ *o*-ethoxyphenol, catechol monoethyl ether, guaethol
p-**Äthoxyphenol** *n* $C_2H_5OC_6H_4OH$ *p*-ethoxyphenol, quinol monoethyl ether
p-**Äthoxyphenylharnstoff** *m* *p*-ethoxyphenylurea, dulcin
Äthyl *n s.* Äthylgruppe
Äthylal *n* $CH_2(OC_2H_5)_2$ ethylal, diethoxymethane, formaldehyde diethyl acetal
Äthylalkohol *m* CH_3CH_2OH ethyl alcohol, ethanol
  wasserfreier A. absolute [ethyl] alcohol
Äthylaluminium *n* $(C_2H_5)_3Al$ triethylaluminium, ATE
Äthylamin *n* $C_2H_5NH_2$ ethylamine
Äthylaminobenzoat *n* $NH_2C_6H_4COOC_2H_5$ ethyl aminobenzoate
Äthylat *n* $C_2H_5OM^I$ ethylate, ethoxide
Äthyläthen *n s.* Äthyläthylen
Äthyläther *m (incorrectly for)* Diäthyläther
Äthyläthylen *n* $CH_2=CHCH_2CH_3$ 1-butene, ethylethylene
Äthylazetat *n* $CH_3COOC_2H_5$ ethyl acetate, acetic ester
Äthylazetoazetat *n* $CH_3COCH_2COOC_2H_5$ ethyl acetoacetate
Äthylazetylen *n* $HC\equiv CCH_2CH_3$ ethyl acetylene, 1-butyne
Äthylbenzin *n* ethyl gasoline
Äthylbenzoat *n* $C_6H_5COOC_2H_5$ ethyl benzoate
Äthylbenzol *n* $C_2H_5C_6H_5$ ethylbenzene, phenylethane
Äthylbromid *n* CH_3CH_2Br bromoethane, ethyl bromide
Äthylbutyrat *n* $CH_3CH_2CH_2COOC_2H_5$ ethyl butyrate
Äthylchlorid *n* CH_3CH_2Cl chloroethane, ethyl chloride
Äthyldimethylmethan *n s.* 2-Methylbutan
Äthyldipropylthiokarbamat *n* $(C_3H_7)_2NCOSCH_2CH_3$ ethyl-*N,N*-dipropylthiocarbamate, EPTC *(a herbicide)*
Äthyldisulfid *n s.* Diäthyldisulfid

Äthyldithioäthan *n s.* Diäthyldisulfid
Äthylen *n* $H_2C=CH_2$ ethylene, ethene; *(broadly) s.* Äthylenkohlenwasserstoff
Äthylen-Äthylakrylat-Kopolymerisat *n* ethylene ethyl acrylate copolymer, EEA
Äthylenbromhydrin *n* $BrCH_2CH_2OH$ 2-bromoethanol, ethylene bromohydrin
Äthylenbromid *n s.* Äthylendibromid
Äthylenbrücke *f* ethylene bridge
Äthylenchlorhydrin *n* $ClCH_2CH_2OH$ 2-chloroethanol, ethylene chlorohydrin
Äthylenchlorid *n s.* Äthylendichlorid
Äthylendiamin *n* $NH_2CH_2CH_2NH_2$ ethylenediamine, 1,2-ethanediamine
Äthylendiamintetraazetat *n* ethylenediamine tetraacetate
Äthylendiamintetraessigsäure *f* ethylenediamine tetraacetic acid, EDTA
Äthylendibromid *n* $BrCH_2CH_2Br$ 1,2-dibromoethane, ethylene dibromide, EDB
Äthylendichlorid *n* $ClCH_2CH_2Cl$ 1,2-dichloroethane, ethylene dichloride
Äthylen-1,2-dikarbonsäure *f* $HOOCCH=CHCOOH$ ethylene-1,2-dicarboxylic acid, butenedioic acid
Äthylenglykol *n* $HOCH_2CH_2OH$ ethylene glycol, 1,2-hydroxyethane, 1,2-ethanediol
Äthylenglykolmonoäthyläther *m* $C_2H_5OCH_2CH_2OH$ ethylene glycol ethyl ether, 2-ethoxyethanol
Äthylenharnstoff *m* 2-imidazolidone, ethyleneurea
Äthylenimin *n* ethyleneimine, aziridine
Äthylenisomerie *f* ethylene isomerism
Äthylenkohlenwasserstoff *m* ethylenic hydrocarbon, alkene, olefin[e]
Äthylenmilchsäure *f* $HOCH_2CH_2COOH$ ethylenelactic acid, 3-hydroxypropionic acid
Äthylenoxid *n* ethylene oxide, oxiran, epoxyethane
äthylenoxidisch ethylene-oxidic
Äthylen-Propylen-Kautschuk *m* ethylene-propylene rubber, EP-rubber, EPR
Äthylen-Propylen-Kopolymerisat *n* / **fluoriertes** fluorinated ethylene-propylene resin, FEP
Äthylen-Propylen-Terpolymerisat *n* ethylene-propylene terpolymer, EPT
Äthylenreihe *f* ethylene series
Äthylen-Vinylazetat-Kopolymerisat *n* ethylene-vinylacetate copolymer, EVA
Äthylessigsäure *f* $CH_3CH_2CH_2COOH$ ethylacetic acid, butyric acid
Äthylester *m* ethyl ester
Äthylfluid *n* ethyl fluid *(an antiknock additive, mainly consisting of tetraethyl lead)*
Äthylformiat *n* $HCOOC_2H_5$ ethyl formate
Äthylglykol *n s.* Äthoxyäthanol
Äthylgruppe *f* C_2H_5— ethyl group (residue)
Äthylhalogenid *n* ethyl halide
Äthylhexanoat *n* $CH_3[CH_2]_4COOC_2H_5$ ethyl hexanoate, ethyl caproate
Äthylhydrogensulfat *n* $C_2H_5HSO_4$ ethyl hydrogensulphate, ethylsulphuric acid
Äthylhydrosulfid *n* C_2H_5SH ethyl hydrosulphide, ethanethiol
Äthylidenchlorid *n* CH_3CHCl_2 1,1-dichloroethane, ethylidene chloride
Äthylidenmilchsäure *f* $CH_3CH(OH)COOH$ lactic acid, ethylidenelactic acid, 2-hydroxypropionic acid
äthylieren to ethylate
Äthylierung *f* ethylation
Äthylierungsmittel *n* ethylating reagent
Äthylisopropylkarbinol *n s.* 2-Methylpentanol-(3)
Äthyljodid *n* CH_3CH_2I iodoethane, ethyl iodide
Äthyljodidverbindung *f* ethiodide
Äthylkaprinat *n* $CH_3[CH_2]_8COOC_2H_5$ ethyl decanoate, ethyl caprate
Äthylkapronat *n* $CH_3[CH_2]_4COOC_2H_5$ ethyl hexanoate, ethyl caproate
Äthylkarbamat *n* $NH_2COOS_2H_5$ ethyl aminoformate, ethyl carbamate
Äthylkrotonat *n* $CH_3CH=CHCOOC_2H_5$ ethyl crotonate
2-Äthyl-*cis*-krotonylharnstoff *m* 2-ethyl-*cis*-crotonylurea, ectylurea
Äthyllaktat *n* $CH_3CH(OH)COOC_2H_5$ ethyl lactate
Äthyllaurat *n* $CH_3[CH_2]_{10}COOC_2H_5$ ethyl laurate
Äthylmerkaptan *n* CH_3CH_2SH ethanethiol, ethyl hydrosulphide, ethyl mercaptan
Äthylmethylazetylen *n* $CH_3C\equiv CCH_2CH_3$ 2-pentyne, ethyl methyl acetylene
Äthylmethylkarbinol *n s.* Butanol-(2)
Äthylmethylsulfid *n* $C_2H_5SCH_3$ ethyl methyl sulphide
Äthylnitrat *n* $C_2H_5NO_3$ ethyl nitrate
Äthylnitrit *n* C_2H_5ONO ethyl nitrite
Äthylphenyläther *m* $C_2H_5OC_6H_5$ ethyl phenyl ether, phenetole
Äthylphenylazetat *n* $C_6H_5CH_2COOC_2H_5$ ethyl phenyl acetate
Äthylphenylketon *n* $C_6H_5COCH_2CH_3$ ethyl phenyl ketone, propiophenone
Äthylphthalat *n* $C_6H_4(COOCH_2CH_3)_2$ diethyl phthalate
Äthylpropionat *n* $CH_3CH_2COOC_2H_5$ ethyl propionate
Äthylradikal *n* ethyl radical, *(specif)* free ethyl radical
Äthylrest *m* C_2H_5– ethyl residue (group)
Äthylrot *n* ethyl red
Äthylschwefelsäure *f* $C_2H_5HSO_4$ ethylsulphuric acid, ethyl hydrogensulphate
Äthylsilikat-Gießverfahren *n (ceram)* ethyl silicate casting process
Äthylsilikon *n* ethyl silicone
Äthylstearat *n* $CH_3[CH_2]_{16}COOC_2H_5$ ethyl stearate
Äthylsulfat *n s.* Diäthylsulfat
Äthylsulfid *n s.* Diäthylsulfid
Äthylsulfonsäure *f (incorrectly for)* Äthansulfonsäure
Äthylsulfursäure *f. s.* Äthylschwefelsäure
Äthylthioalkohol *m s.* Äthanthiol
Äthylthioäthan *n* $(C_2H_5)_2S$ ethylthioethane, diethyl sulphide
Äthylurethan *n* $NH_2COOS_2H_5$ ethyl aminoformate, ethyl carbamate

Äthylvalerat *n* $CH_3[CH_2]_3COOC_2H_5$ ethyl valerate
Äthylvalerianat *n s.* Äthylvalerat
Äthylxanthogensäure *f* C_2H_5OCSSH ethylxanthogenic acid, xanthogenic acid, dithiocarbonic *O*-ethyl ester, ethoxydithioformic acid
Äthylzellulose *f* ethylcellulose
Äthylzinnamat *n* $C_6H_5CH{=}CHCOOC_2H_5$ ethyl cinnamate
Äthylzyanid *n* CH_3CH_2CN ethyl cyanide, propionitrile
Äthylzyklohexan *n* ethylcyclohexane
Atisin *n (pharm)* aticine, atisine, atesine
A. T. Koch *s.* Alttuberkulin
Atmolyse *f* atmolysis
Atmometer *n* atmometer, atmidometer
atmophil atmophile, atmophilic
Atmosphäre *f* atmosphere
  kontrollierte A. controlled atmosphere *(of a furnace)*
  oxydierende A. oxidizing atmosphere
  physikalische A. physical atmosphere, atm *(1 atm = 0.101 MPa)*
  reduzierende A. reducing atmosphere
  technische A. technical atmosphere, at *(1 at = 0.0981 MPa)*
Atmosphärendruck *m* atmospheric (air) pressure
atmosphärisch atmospheric[al]
Atmung *f* breathing, respiration
  anaerobe A. anaerobic respiration
  luftfreie (sauerstofffreie) A. *s.* anaerobe A.
Atmungsfähigkeit *f* breathability *(of a coating)*
Atmungsferment *n* respiratory enzyme
  Warburgsches A. cytochrome oxidase
  Warburgsches gelbes A. Warburg's [yellow] enzyme, old yellow enzyme
Atmungsinhibitor *m* respiration inhibitor
Atmungskatalysator *m* respiratory catalyst
Atmungskoeffizient *m*, **Atmungsquotient** *m* respiratory quotient (ratio)
Atom *n* atom
  A. auf Zwischengitterplatz interstitial [atom]
  adsorbiertes A. adatom, adsorbed atom
  angeregtes A. excited (activated) atom
  angestoßenes A. knocked-on atom
  endständiges A. terminal (end) atom
  heißes (hoch angeregtes) A. *s.* hochenergiereiches A.
  hochenergiereiches A. hot atom
  hochionisiertes A. stripped atom
  ionisiertes A. ionized atom, atomic ion
  markiertes A. labelled (tagged) atom, label
  neutrales A. neutral atom
  normales A. normal atom *(which has all its electrons in their normal orbitals)*
  stabiles A. stable atom
Atom-% *s.* Atomprozent
Atomabfall *m s.* Atommüll
Atomabstand *m* atomic (interatomic) distance, atomic (interatomic) spacing
Atomaggregat *n* atomic aggregate, aggregate of atoms
Atomanordnung *f s.* Atomkonfiguration
atomar atomic[al]
Atomart *f* atomic species
Atom-Atom-Konformation *f* eclipsed conformation
Atomaufbau *m* atomic structure
Atomaufspaltung *f* atom splitting
Atombau *m* atomic structure
Atombindigkeit *f s.* Atombindungszahl
Atombindung *f* atomic (covalent, homopolar, nonpolar) bond, atomic link (linkage), [shared-]electron-pair bond *(state)*
Atombindungszahl *f* covalency, covalence
Atomdrehung *f* atomic rotation
Atomdurchmesser *m* atomic diameter
Atomelektron *n* atomic electron
Atomenergie *f s.* Kernenergie
Atomfaktor *m s.* Atomformfaktor
Atomformfaktor *m (cryst)* atomic form (structure, scattering) factor
Atomforschung *f* atomic research
Atomfrequenz *f* atomic frequency
Atomgerüst *n* atomic framework
Atomgewicht *n s.* Atommasse / relative
  absolutes A. *s.* Atommasse / absolute
  relatives A. *s.* Atommasse / relative
Atomgewichtsbestimmung *f s.* Bestimmung der relativen Atommasse
Atomgitter *n* atom[ic] lattice
Atomgramm *n* gram atom, gram-atomic weight
Atomgröße *f* atomic size
Atomgruppe *f* group of atoms
  dreiwertige A. triad
  einwertige A. monad
  fünfwertige A. pentad
  mehrwertige A. polyad
  vierwertige A. tetrad
  zweiwertige A. diad, dyad
Atomhülle *f* atomic electron shell, atomic envelope (sheath)
Atomigkeit *f s.* Wertigkeit
Atomion *n* atomic ion, ionized atom
Atomisator *m* atomizer
Atomiseur *m (agric)* low-volume mist blower, airblast sprayer, fog generator (appliance), nebulizer
atomisieren to atomize
Atomisierung *f* atomization
atomistisch atomistic
Atomizität *f s.* Wertigkeit
Atomkern *m* [atomic] nucleus
Atomkernenergie *f s.* Kernenergie
Atomkernspaltung *f s.* Kernspaltung
Atomkernstabilität *f* nuclear stability
Atomkette *f* atomic chain, chain of atoms
Atomkonfiguration *f* atom (atomic) arrangement, [atom, atomic] configuration
Atomkristall *m* covalent crystal
Atomladung *f* atomic charge
Atomlehre *f* atomic theory
Atom-Lücke-Konformation *f* staggered conformation
Atommasse *f s.* absolute A.

absolute A. atomic mass
relative A. [chemical] atomic weight
Atommassenskala *f* mass scale
Atommeiler *m* [atomic] pile
Atommodell *n* atom[ic] model
Bohr-Rutherfordsches A. Bohr atom [model], Bohr-Rutherford atom [model]
Bohrsches A. *s.* Bohr-Rutherfordsches A.
Bohr-Sommerfeldsches A. *s.* Bohr-Rutherfordsches A.
Rutherfordsches A. Rutherford atom [model], nuclear atom [model], nuclear model of the atom
Atommolekül *n* atomic molecule
Atommüll *m* atomic (radioactive) waste
Atomniveau *n* atomic [energy] level
Atomnummer *f* atomic (ordinal, nuclear charge) number
effektive A. effective atomic number, E. A. N.
ungerade A. odd atomic number
Atomorbital *n* atomic orbital, AO
Atomparachor *m* [gram-]atomic parachor
Atomphysik *f* atomic physics
Atompolarisation *f* atom[ic] polarization
Atomprozent *n* atomic percentage, atom %
Atomradius *m* atomic radius
kovalenter A. covalent radius [of atoms]
Atomreaktor *m s.* Kernreaktor
Atomrefraktion *f* atomic refraction
Atomrest *m s.* Atomrumpf
Atomrotation *f* atomic rotation
Atomrumpf *m* atomic core (kernel, trunk, torso, residue)
Atomschale *f s.* Atomhülle
Atomschwingung *f* atomic vibration
Atomspaltung *f* atom splitting
Atomspektroskopie *f* atomic spectroscopy
Atomspektrum *n* atomic (line) spectrum
Atomsprengstoff *m s.* Kernsprengstoff
Atomstrahl *m* atomic beam (ray)
Atomstrahlapparat *m* atomic-beam apparatus
Atomstrahlmethode *f* atomic-beam method
Atomstrahlspektroskopie *f* atomic-beam spectroscopy
Atomstrahlung *f* atomic radiation
Atomstruktur *f* atomic structure
Atomsuszeptibilität *f* atomic susceptibility, susceptibility per gram atom
Atomtheorie *f* atomic theory
Atomumlagerung *f* atomic shift
Atomumwandlung *f* atomic transmutation
Atomverband *m* atomic union
Atomverbindung *f* atomic compound
Atomverhältnis *n*, **Atomverhältniszahl** *f* atomic proportion, atomic ratio [value], proportion by atoms
Atomverschiebung *f* atomic shift
Atomvolumen *n* atomic volume
Atomwärme *f* atomic heat
Atomwertigkeit *f s.* Atombindungszahl
Atomzerfall *m* atomic (radioactive) decay (disintegration)
Atomzertrümmerung *f* atom splitting
Atomzustand *m* atomic state
hybridisierter A. hybrid[ization] atomic state
atoxisch non-toxic, atoxic
ATP *s.* Adenosintriphosphat
ATPase *f s.* Adenosintriphosphatase
atro absolutely dry, bone-dry, B. D., *(pap also)* oven-dry, oven-dried, OD
Atrolaktinsäure *f* $CH_3C(C_6H_5)(OH)COOH$ atrolactinic acid, 2-hydroxy-2-phenylpropionic acid
Atropasäure *f* $CH_2{=}C(C_6H_5)COOH$ atropic acid, 2-phenylacrylic acid
Atropin *n* atropine, *DL*-hyoscyamine *(alkaloid)*
Atropinisomer[es] *n* atropo-isomer
ATS *s.* Ablaugetrockensubstanz
Attacke *f* / **elektrophile** electrophilic attack
nukleophile A. nucleophilic attack
Attapulgit *m* attapulgite, palygorskite *(a phyllosilicate)*
Attraktion *f* attraction
Attraktionskraft *f* attractive force (power)
Attraktivstoff *m* attractant
ätzbar corrodible; *(text)* dischargeable
Ätzbarkeit *f* corrodibility; *(text)* dischargeability
Ätzbaryt *m* caustic baryta *(barium hydroxide)*
Ätzdruck *m (text)* discharge print[ing]
Ätzdruckpaste *f (text)* discharge-printing paste
ätzen to etch; to corrode; *(text)* to discharge; *(med)* to cauterize *(e. g. by chemicals)*
makroskopisch ä. to macroetch
Ätzen *n* etching; corrosion; *(text)* discharge, discharging; *(med)* cauterization *(e. g. by chemicals)*
Ä. mit Oxydationsmitteln *(text)* oxidation discharge
elektrolytisches Ä. electrolytic etch[ing], electroengraving
ätzend corrosive; caustic; *(med)* cauterant, cauterizing
Ätzfigur *f (cryst)* etch (corrosion) figure
Ätzflüssigkeit *f* etching acid (fluid); engraver's acid *(usually nitric acid)*
Ätzgift *n* corrosive poison
Ätzkali *n* caustic potash *(potassium hydroxide)*
Ätzkalk *m* 1. caustic (burnt) lime, quicklime *(calcium oxide)*; 2. slaked (hydrated) lime, slacklime *(calcium hydroxide)*
Ätzlösung *f* etching solution
Ätzmittel *n* etching reagent, etchant; engraver's acid *(usually nitric acid)*; *(text)* discharging agent; *(med)* caustic [agent], cauterant, cautery
Ä. zur Makroätzung macroetching reagent
Frysches Ä. Fry reagent
Steadsches Ä. Stead's reagent
Ätzmittelemulsion *f (agric)* contact emulsion *(for weed control)*
Ätznatron *n* caustic [soda] *(sodium hydroxide)*
Ätznatronschmelze *f* caustic-soda fusion
Ätzpaste *f s.* Ätzdruckpaste
Ätzreserve *f (text)* discharge resist
Ätzsublimat *n* corrosive mercuric (mercury) chloride, sublimate *(mercur(II) chloride)*

Ätzung *f* etching; corrosion; *(text)* discharge, discharging; chemigraph, chemitype *(an engraving made by chemigraphy)*
makroskopische Ä. macroetching
Ätzweiß *n (text)* white discharge
Ätzzeit *f* etching time
Audibert-Arnu-Dilatometer *n* Audibert-Arnu dilatometer
Audibert-Arnu-Dilatometerverfahren *n* Audibert-Arnu method
Auerbachit *m (min)* auerbachite *(zirconium orthosilicate)*
aufarbeiten to process *(e. g. ores)*; to reprocess, to rework *(used or discarded material)*
Topprückstände a. *(petrol)* to run resid
zu Crepe a. *(rubber)* to crêpe
Aufbau *m* 1. *s.* Aufbauen; 2. structure, constitution *(of a molecule)*; composition *(of a chemical compound or mixture)*; set-up, structure, build *(of an apparatus)*; *(tech)* build, construction
A. nach Rezept formulation
chemischer A. chemical structure
streifiger A. *(geol)* banded structure
zonaler A. *(geol)* zonal (zonary) structure, zoning
Aufbaubestandteil *m s.* Aufbauelement 1.
Aufbauelement *n* 1. structural element (entity, unit); 2. *(coal)* constituent, mazeral
aufbauen to build up *(a chemical compound)*, to synthesize; to set *(an apparatus)*
ein Mischungsrezept a. to compound
gezielt a. to make to measure, to tailor[-make] *(e. g. polymers)*
nach Maß a. *s.* gezielt a.
nach Rezept a. to formulate
Aufbauen *n* building-up, build-up *(of a chemical compound)*, synthesis; setting *(of an apparatus)*
Aufbaugranulieren *n* pelletizing
Aufbauprinzip *n (phys chem)* building[-up] principle, aufbau principle
Aufbaustoff *m* [detergency] builder *(added to synthetic detergents to increase their cleansing action)*
aufbereiten to prepare; to recover; to treat, to condition *(water)*; to recondition *(waste oil)*; to dress, to beneficiate *(ore)*; *(pap)* to break [in, up] *(rags)*
auf nassem Wege a. *s.* naß a.
auf trockenem Wege a. *s.* trocken a.
in der Setzmaschine a. *(min tech)* to jig
naß a. *(min tech)* to wet-clean
pneumatisch a. *s.* trocken a.
trocken a. *(min tech)* to dry-clean
zu Crepe a. *(rubber)* to crêpe
Aufbereitung *f* preparation; recovery; dressing, beneficiation *(of ore)*; *(pap)* breaking *(of rags)*
A. auf nassem Wege *s.* nasse A.
A. auf trockenem Wege *s.* trockene A.
A. durch Flotation *s.* flotative A.
A. zu Crepe *(rubber)* crêpeing, creping
bergbauliche A. mineral dressing, minerals beneficiation
flotative A. floatation beneficiation
magnetische A. magnetic concentration
nasse A. *(min tech)* wet cleaning (washing)
pneumatische A. *s.* trockene A.
trockene A. *(min tech)* dry (pneumatic) cleaning
Aufbereitungsanlage *f* preparation plant; *(min tech)* ore dressing plant
Aufbereitungsherd *m (min tech)* concentrator (concentrating, concentrate) table
Aufbereitungskonzentrat *n (min tech)* concentrate
Aufbereitungstechnik *f* mineral technology
Aufbereitungsteil *m (glass)* conditioning section (zone) *(of the feeder channel)*
aufbewahren to store, to keep
im Brutschrank a. to incubate
kühl a. to store in a cool place
lichtgeschützt a. to keep screened from the light
Aufbewahrung *f* storing, storage, keeping
A. in der Brutkammer incubation
Aufbewahrungstemperatur *f* storage (holding) temperature
Aufblähung *f* **der Kohle** intumescence (swelling) of coal
Aufblähungsmittel *n (build)* expanding agent; *(rubber, plast)* blowing (inflating, sponging) agent
Aufblasekonverter *m* [top-blown] basic oxygen converter, [top-blown] basic oxygen furnace
Aufblaseverfahren *n* top-blown oxygen converter process, basic oxygen converter (furnace, steel) process, oxygen process of steelmaking, oxygen lance process
Aufblaskonverter *m s.* Aufblasekonverter
Aufblasverfahren *n s.* Aufblaseverfahren
Aufblasverhältnis *n (plast)* blow-up ratio
aufblättern to exfoliate, to delaminate, to cleave *(of laminated material)*
Aufblättern *n* exfoliation, delamination, cleavage *(of laminated material)*
aufbrausen to effervesce
Aufbrausen *n* effervescence, effervescency
aufbrausend effervescent, effervescing
aufbrennen to fire on *(on-glaze decorations)*
Aufbrenntemperatur *f* firing-on temperature
aufbringen to apply *(organic coatings)*; to superimpose *(reagents)*
Wasserzeichen a. *(pap)* to watermark
Aufbringen *n* application *(of organic coatings)*; superimposing *(of reagents)*
A. von Wasserzeichen *(pap)* watermarking
Aufdampfen *n* evaporation coating, coating by evaporation
A. im Vakuum vacuum coating by evaporation
Aufdockvorrichtung *f (text)* batching equipment
aufdringlich objectionable *(smell)*
aufeinanderfolgend consecutive
Aufenthaltswahrscheinlichkeit *f* probability to find a particle at a given place *(quantum mechanics)*
Aufenthaltszeit *f* retention (hold-up, holding) time,

retention (hold-up, holding) period
Auffangbehälter *m* receiving tank, receiver
A. für Katalysator catalyst surge hopper
Auffangelektrode *f* collector (collecting) electrode, collector
auffangen to collect, to catch, to trap, to receive
Auffänger *m* receiver, catcher; electron acceptor; target *(as of an X-ray tube)*
Auffanggefäß *n* receiver, catcher, catch pot, trap; *(met)* tapping receiver
Auffangrinne *f (tech)* pickup; *(glass)* receiver
Auffangvorrichtung *f* collecting (catching, trapping, receiving) device
auffärben *(text)* to redye
aufflammen to flash
Aufflammen *n* flash; *(text)* afterflaming
Auffflußspalt *m (pap)* gate, slot, slice *(of the headbox)*
auffrischen *(text)* to revive
auffüllen to fill up; to refill
Aufgabe *f* charging, feed[ing], filling, loading, furnishing
Aufgabeapparat *m* charging (feeding, feed) mechanism
Aufgabebecherwerk *n* directly fed bucket elevator
Aufgabebehälter *m* feed tank
Aufgabeboden *m (distil)* feed tray (plate)
Aufgabeende *n* feed end
Aufgabegut *n* feed[stock], charge [stock], load, batch; feed slurry (pulp) *(as in centrifugation)*; prefilt [feed, slurry] *(as in filtration)*
Aufgabekasten *m* feed box
Aufgabekohle *f* feed[stock] coal
Aufgabeöffnung *f* charging (filling, feed) hole, charging (filling, feed) door, feed inlet (opening)
Aufgaberinne *f*, **Aufgaberutsche** *f* charging (feed) chute, feed launder
Aufgabeschieber *m* feed gate
Aufgabeschurre *f s.* Aufgaberinne
Aufgabeseite *f* feed end
Aufgabetrichter *m* feed (feeding, charge, charging, input, loading) hopper, feed (feeding, charge, charging, input, loading) funnel
Aufgabetrog *m* feed tank
Aufgabevorrichtung *f* feeder, loader
Aufgabewalze *f* feed[ing] roll
aufgasen *(pap)* to fortify *(the cooking acid)*
Aufgasen *n (pap)* fortification *(of the cooking acid)*
aufgeben to charge, to feed, to fill, to load, to furnish *(filling material)*
Aufgeben *n s.* Aufgabe
Aufgeber *m* feeder, loader
aufgehen to rise *(of yeast, dough)*; *(tann)* to plump *(of pelts)*
a. lassen to raise, to leaven *(dough)*
aufgeschliffen *(lab)* ground-in
aufgießen *(ceram)* to coat
Aufglasur *f (ceram)* on-glaze, overglaze
Aufglasurdekor *n (ceram)* on-glaze decoration, overglaze decoration
Aufglasurfarbe *f (ceram)* on-glaze colour, overglaze (enamel) colour
Aufglasurmalerei *f (ceram)* on-glaze painting, overglaze painting
Aufguß *m* infusion, *(by boiling also)* decoction
Aufgußverfahren *n (ferm)* infusion mashing (method, process)
aufhäufen to accumulate
sich a. to accumulate
Aufhäufung *f* accumulation
Aufheizabschnitt *m s.* Aufheizzeit
aufheizen to heat up
Aufheizgeschwindigkeit *f* heating rate
Aufheizperiode *f s.* Aufheizzeit
Aufheizsektion *f* heating section
Aufheizung *f* heating[-up]
Aufheizzeit *f* heat[ing]-up period, heat[ing]-up time, coming-up time
Aufhelleffekt *m s.* Aufhellungseffekt
aufhellen to brighten; to raise *(a colour)*; to bleach, to clear
Aufheller *m* brightener, brightening agent
optischer A. optical (fluorescent) brightener, optical (brightening, bleaching, whitening) agent
Aufhellung *f* / **optische** optical bringhtening (bleaching)
Aufhellungseffekt *m* brightening effect
Aufhellungsmittel *n s.* Aufheller
aufkalken *(agric)* to lime *(to a higher pH value)*
aufklappen to dismantle *(e. g. a Sweetland filter)*
aufkleben *(tann)* to paste *(damp hides on boards or metal plates)*
Aufklebepapier *n* pasting (lining) paper
aufklotzen *(text)* to pad
aufkochen to boil up
a. lassen to boil up, to bring to the boil
wieder a. to reboil
Aufkochen *n* boiling-up; *(of a liquid)* boiling-up, ebullition
aufkochend ebullient
Aufkocher *m s.* Aufkochofen
Aufkochofen *m (petrol)* reboiler [furnace]
Aufkohl... *s.* Aufkohlungs...
aufkohlen *(met)* to carburize
im Salzbad a. to bath-carburize, to liquid-carburize
in der Randschicht (Randzone) a. to case-carburize
in festen Kohlungsmitteln a. to pack-carburize
in flüssigen Mitteln a. *s.* im Salzbad a.
in gasförmigen Mitteln a. to gas-carburize
Aufkohlen *n (met)* carburizing, carburization
A. im Salzbad bath (liquid, liquid-salt, molten-salt) carburizing
A. in der Randschicht (Randzone) case carburizing
A. in festen Kohlungsmitteln solid (pack, solid-pack) carburizing
A. in flüssigen Mitteln *s.* A. im Salzbad
A. in gasförmigen Mitteln gas carburizing
Aufkohlung *f (met)* carburizing, carburization

flüssige A. *s.* Aufkohlen im Salzbad
Aufkohlungsbad *n (met)* carburizing bath
Aufkohlungsgas *n (met)* carburizing gas
Aufkohlungsgemisch *n (met)* carburizing mixture
Aufkohlungsgeschwindigkeit *f (met)* carburizing rate
Aufkohlungshitze *f (met)* carburizing heat
Aufkohlungsmittel *n (met)* carburizing material (medium, agent, compound), [case-hardening] carburizer
Aufkohlungsofen *m (met)* carburizing furnace (oven)
Aufkohlungspulver *n (met)* carburizing powder
Aufkohlungssalz *n (met)* carburizing salt
Aufkohlungsschicht *f (met* [carburized] case
Aufkohlungstiefe *f (met)* carburizing (carburization, case) depth
Aufkohlungsverfahren *n (met)* carburizing process
Aufkohlungswirkung *f (met)* carburizing action
Aufkohlungszone *f (met)* [carburized] case
aufkonzentrieren to concentrate *(an acid)*; *(pap)* to fortify *(the cooking acid)*
Aufkonzentrieren *n* concentration *(of an acid)*; *(pap)* fortification *(of the cooking acid)*
aufkrausen *(tann)* to pommel
Aufladung *f* charge, charging, electrification
[elektro]statische A. electrostatic charge, static electrification
Auflagehumus *m* raw humus, mor
Auflageplatte *f* bed plate
auflaufen / auf das Sieb *(pap)* to enter onto the wire
Auflaufkasten *m (pap)* flow (stuff, breast) box, headbox
Auflaufleder *n (pap)* apron
Auflaufrahmen *m (pap)* deckle
Auflichtelektronenmikroskop *n* direct-light electron microscope
Auflichtmikroskopie *f* direct-light microscopy
auflockern to loosen; to fluff; to agitate
auflösbar dissoluble, dissolvable
Auflösbarkeit *f* dissolubility, dissolvability
Auflöseholländer *m (pap)* breaker (broke) beater
auflösen 1. to put into solution, to dissolve; 2. to dissolve, to disintegrate, to decompose, to break up *(something into its constituent elements)*; to break [in, up], to repulp *(waste paper)*
sich a. 1. to pass into solution, to dissolve; 2. to dissolve, to disintegrate, to decompose, to break up *(into its constituent elements)*
Auflösung *f* 1. dissolution; 2. dissolution, disintegration, decomposition, breakup *(into constituent elements)*; breaking *(of waste paper)*; 3. *s.* Auflösungsvermögen
A. im Gestein *(geol)* intrastratal solution
Auflösungsgeschwindigkeit *f* dissolution rate
Auflösungsprozeß *m* dissolving process
Auflösungsvermögen *n (phot)* resolving power, resolution
Aufmachungseinheit *f (text)* package
Aufnahme *f* 1. acceptance, acquisition *(as of electrons)*; uptake, take-up *(of substances)*; absorption *f (of gases)*; pickup *(as of glue by paper)*; 2. *(phot)* taking; photograph
autoradiografische A. autoradiograph, radioautograph
makrofotografische A. photomacrograph
mikrofotografische A. photomicrograph
Aufnahmeeisen *n (glass)* gathering iron
aufnahmefähig absorptive, absorbent
Aufnahmefähigkeit *f s.* Aufnahmevermögen 2.
Aufnahmemasse *f* loading *(wood preservation)*
Aufnahmematerial *n (phot)* negative material
Aufnahmespule *f (text)* winding bobbin
Aufnahmetisch *m (glass)* casting table
Aufnahmevermögen *n* 1. capacity; 2. absorbency, absorbancy, absorptivity, absorbing (absorption, absorptive) capacity
aufnehmbar absorbable; *(agric)* available *(of nutrients)*
Aufnehmbarkeit *f* absorbability; *(agric)* availability *(of nutrients)*
aufnehmen 1. to accept, to acquire *(e. g. electrons)*; to take up *(a substance)*; to absorb *(gases)*; to pick up *(e. g. liquids)*; 2. to take a photograph
Farbe a. *(text)* to take the dye
Glas aus der Schmelze a. to gather glass
Aufnehmer *m* absorbent, absorber, absorbing agent (material, substance)
aufpfropfen to graft *(polymers)*
Aufpolymerisation *f* grafting
aufpolymerisieren *s.* aufpfropfen
Aufprall *m* impingement, impact
aufprallen to impinge, to impact
aufpressen *(pap, text, tann)* to emboss
aufquellen to swell [up]
Aufquellung *f* swelling *(act or state)*
aufrahmen *(rubber, plast, food)* to cream
Aufrahmung *f (rubber, plast, food)* creaming, rising of cream
Aufrahmungsfähigkeit *f* creamability, creaming ability (potential, power)
Aufrahmungsmittel *n* creaming agent
Aufrahmungspotential *n*, **Aufrahmungsvermögen** *n s.* Aufrahmungsfähigkeit
Aufrahmungsvorgang *m* creaming process
aufrauhen to roughen; *(text)* to raise [a nap], to nap
aufrechterhalten to maintain, to keep up
Aufrechterhaltung *f* maintenance, upkeep
A. des Gleichgewichts equilibration
Aufrollapparat *m (pap)* reeling machine, reel[er], winder
aufrollen *(pap)* to make into a roll, to work up into a reel, to wind [up], to reel [up]; to roll, to upend, to fold back *(a rubber stock)*
Aufrollen *n (pap)* reeling, winding; rolling, folding *(of a rubber stock)*; *(ceram)* crawling *(a defect during glazing)*
dichtes A. *(pap)* tight winding
hartes (klanghartes) A. *s.* dichtes A.
Aufrollstange *f (pap)* winder (rewind) shaft
Aufrolltrommel *f (pap)* reel-up (reeling-up) drum

(cylinder), reel (reeling) drum (cylinder)
Aufrollvorrichtung *f* winding (wind-up) arrangement, winding (wind-up) equipment
aufrühren to agitate; to repulp, to reslurry
aufsättigen to resaturate, to reconcentrate
aufsaugbar absorbable
Aufsaugbarkeit *f* absorbability
aufsaugen to absorb, to suck (soak) up, to imbibe
  wieder a. to re-absorb
Aufsaugen *n* absorption, suction, imbibition
aufsaugend absorbing, absorptive, absorbent
Aufsauger *m* absorber, absorbing (absorption) apparatus
aufsaugfähig absorptive, absorbent
Aufsaugfähigkeit *f* absorptivity, absorbency, absorbing (absorption, absorptive) capacity, absorbing power
Aufsaugung *f* absorption, suction, imbibition
aufschäumbar *(plast)* expandable, foamable
Aufschäumbarkeit *f (plast)* expandability, foamability
aufschäumen 1. to foam, to froth *(a substance)*; *(plast)* to expand, to foam; *(glass)* to reboil; 2. *(of a substance)* to foam [up], to froth [up], to effervesce
aufschäumend effervescent, effervescing
Aufschlag *m* impact, impingement
aufschlagen *(pap)* to refine, to clear, to brush out, to break down, to potch, to poach; *(tann)* to handle, to haul *(hides out of the tanning liquor)*
Aufschläger *m (pap)* refiner, refining (perfecting) engine, refining (perfecting) machine
aufschlämmen to suspend
Aufschlämmung *f* suspension
aufschließbar digestible
aufschließen to digest, to decompose *(by heat or solvents)*; *(min)* to develop, to open up; *(biol)* to macerate; *(pap)* to cook, to pulp, to reduce to pulp, to make into pulp; *(pap)* to repulp *(waste paper)*
  intensiv a. to cook soft *(cellulose)*
  mit Säure a. *(fert)* to acidulate
  unvollständig a. to cook raw *(cellulose)*
Aufschließgestell *n* digestion stand *(of a Kjeldahl apparatus)*
Aufschließung *f s.* Aufschluß
Aufschluß *m* digestion, decomposition *(by heat or solvents)*; *(min)* development; *(biol)* maceration; *(pap)* cooking, pulping
  A. im Bombenrohr (Einschmelzrohr, Schießrohr) sealed-tube decomposition
  A. mit Säure *(pap)* acid pulping; *(fert)* acidulation
  alkalischer A. *(pap)* alkaline pulping
  chemischer A. *(pap)* [full] chemical pulping
  halbchemischer A. *(pap)* semichemical pulping
  mechanischer A. des Holzes *(pap)* mechanical (groundwood) pulping
  saurer A. *(pap)* acid pulping
Aufschlußbohrung *f (petrol)* 1. exploration drilling; 2. exploration (exploratory) well, wildcat
  erfolglose A. unproductive (dry) well, duster
Aufschlußchemikalie *f s.* Aufschlußmittel
Aufschlußgrad *m* degree of digestion (decomposition); *(pap)* degree of cooking
Aufschlußlauge *f s.* Aufschlußlösung
Aufschlußlösung *f (pap)* pulping (cooking, digestion) liquor
Aufschlußmittel *n* digesting (decomposing) agent; *(pap)* pulping (cooking) agent (chemical)
Aufschlußmittelgemisch *n* digestion mix
Aufschlußverfahren *n* decomposition process; *(pap)* pulping process
  alkalisches A. *(pap)* alkaline process
aufschmelzen to flux on, to melt on, to fuse on
aufschmieren to smear *(e. g. a lubricant)*
Aufschrumpfen *n* **unter Vakuum** *(plast)* vacuum snap-back forming
aufschwemmen to suspend, *(min tech)* to pulp
Aufschwemmung *f* suspension
aufschwimmen to float
aufspalten 1. *(mechanically)* to split [up], to cleave, to delaminate; 2. *(chemically)* to cleave, to break [down], to crack [up]; to resolve *(racemic mixtures)*; 3. *s.* sich a.
  durch Solvolyse a. to solvolyze
  in Fibrillen (Teilfäserchen) a. to fibrillate *(fibres)*
  sich a. 1. *(mechanically)* to split, to delaminate, to cleave; 2. *(chemically)* to cleave, to break up, to crack; to dissociate *(into ions)*
Aufspaltung *f* 1. splitting[-up], cleavage, delamination; 2. cleavage, fission, breakdown *(of a chemical compound)*, breaking *(of a chemical bond)*; resolution *(of racemic mixtures)*
Aufspaltungsbild *n (phys chem)* splitting pattern
Aufspannplatte *f (plast)* clamping (mounting) plate
aufspeichern to store, to accumulate
  sich a. to accumulate
Aufspeicherung *f* storage, accumulation
aufsprengen *(org chem)* to rupture *(a ring)*
Aufsprengung *f (org chem)* rupture *(of a ring)*
aufspritzen to splash on *(e. g. wet material onto a roller dryer)*; to sputter *(a metallic film)*
aufsprühen to spray on *(e. g. wet material onto a roller dryer)*
aufspüren to prospect *(e. g. ore deposits)*
aufstärken to fortify, *(distil)* to dephlegmate
Aufstärkung *f* fortification, *(distil)* dephlegmation
aufsteigen to rise, to ascend, to pass up[wards]
  in Blasen a. to bubble
aufstellen to install, to erect, to put up, to set [up]
Aufstellung *f* arrangement, setting *(as of an apparatus)*
aufsticken *(met)* to nitride
Aufstickung *f (met)* nitride hardening, nitrogen [case-]hardening, nitriding, nitridation
aufstreichen to brush on, to spread on, to smear
Aufstrich *m* coat[ing]
Aufstrom *m* ascending (upward) current, rising stream
aufströmen to entrain *(coal dust in gasification)*
Aufströmen *n* entrainment *(of coal dust in gasification)*

Aufstromklassieren *n* hydraulic classification
Aufstromklassierer *m* hydraulic classifier
auftauen to thaw, to defrost
Auftrag *m* 1. coat[ing]; 2. *s.* Auftragen
auftragen to apply *(e. g. organic coatings); (tann)* to swab *(lime paint)*
Wasserzeichen a. *(pap)* to watermark
Auftragen *n* application *(as of organic coating material)*; *(tann)* swabbing *(of lime paint)*
A. von Wasserzeichen *(pap)* watermarking
galvanisches A. electrodeposition
auftragend sein *(pap)* to bulk high
Auftragewalze *f* application (applicator, feed, feeding) roll
Auftragewerk *n (text)* coating system
Auftragmaschine *f* coating machine, coater
Auftragschweißen *n* hard [sur]facing
Auftragwalze *f s.* Auftragewalze
auftreffen to impinge
Auftreffplatte *f* target *(of an X-ray tube)*
Auftreiber *m (lab)* reamer
auftreten to occur, to appear
Auftreten *n* occurrence, appearance
A. von Anlauffarben tarnish decolo[u]rization
Auftrieb *m* buoyancy
Auftriebskraft *f* buoyancy (buoyant) force
Auftriebsmethode *f* buoyancy method *(for measuring gas density)*
aufwallen to boil [up], to bubble [up]
Aufwallen *n* boiling, bubbling, ebullience, ebullition
aufwallend ebullient
Aufwandmenge *f* amount of application
aufwärmen *(glass)* to warm in
Aufwärmloch *n (glass)* glory hole
Aufwärts-Dickstoffturm *m (pap)* upflow high-density tower
Aufwärtsgasen *n* uprun[ning], upsteaming
Aufwärtskläranlage *f* vertical-flow unit, up-flow unit *(waste-water treatment)*
Aufwärtsstrom *m* up[ward] flow
Aufwärtsziehmaschine *f (glass)* updraw machine
Aufwärtsziehverfahren *n (glass)* updraw (Schuller) process
aufweichen to soak; to soften
aufweisen / einen konstanten Wert to show a constand reading
Aufweitverbindung *f* expanded joint *(of pipes)*
aufwerfen *(ceram)* to warp
Aufwerfen *n (ceram)* warpage, warping
aufwickeln *(pap, text)* to reel [up], to wind
Aufwickeln *n (pap, text)* reeling[-up], winding
dichtes A. *(pap)* tight winding
hartes (klanghartes) A. *s.* dichtes A.
Aufwickelspule *f (text)* winding bobbin
Aufwickeltrommel *f (pap)* reel-up (reeling-up) drum (cylinder), reel (reeling) drum (cylinder)
Aufwickelvorrichtung *f* winding (wind-up) arrangement (equipment)
aufwinden *(text)* to wind
Aufwirkmaschine *f (food)* dough forming (moulding) machine
Aufzehrung *f (geol)* assimilation, magmatic digestion
Aufzieheigenschaft *f s.* Aufziehvermögen
aufziehen to attach, to absorb *(dyes)*; *(tann)* to handle, to haul *(hides out of the tanning liquor)*
Aufziehen *n (dye)* absorption, strike; *(tann)* handling, hauling
langsames A. *(dye)* slow strike
mäßiges A. *(dye)* moderate strike
schnelles A. *(dye)* rapid strike
Aufziehgeschwindigkeit *f (dye)* rate (speed) of absorption
Aufziehkarton *m* mounting board
Aufziehvermögen *n (dye)* absorptive (absorbing) capacity, absorptive (absorbing) power, *(Am)* pile-on property
Augenbrauenstift *m* eyebrow (eye-make-up) pencil
augenreizend lachrymatory
Augenreizstoff *m* lachrymator
Augensalbe *f* eye ointment
Augenschatten *m*, **Augenschattenschminke** *f s.* Lidschatten
Augentropfen *pl* eye drops
Augenwasser *n* eyewash
Auger-Ausbeute *f (phys chem)* Auger yield
Auger-Effekt *m (phys chem)* Auger effect
Auger-Elektron *n* Auger electron
Auger-Elektronen-Ausbeute *f* Auger yield
Augit *m (min)* augite *(an inosilicate)*
Auramin *n* $[(CH_3)_2NC_6H_4]_2C{=}NH$ auramine, 4,4'-*bis*-dimethylaminobenzophenoneimide
Aurat *n* aurate
Aureolin *n* aureolin, cobalt (Indian) yellow *(potassium hexanitrocobaltate)*
Aurichalzit *m (min)* aurichalcite *(a basic copper zinc carbonate)*
Aurid *n* auride
Aurin *n* aurine dyestuff, *(specif)* $(HOC_6H_4)_2$-$C{=}C_6H_4{=}O$ aurine
Auripigment *n (min)* auripigment *(arsenic(III) sulphide)*; orpiment [yellow], yellow arsenic [sulphide], arsenic yellow, king's yellow (gold), royal yellow *(arsenic(III) sulphide)*
Auron *n* aurone *(any of several flavonoids)*
ausäthern to extract (shake out) with ether, *(broadly)* to extract, to shake out
Ausäthern *n* extraction with ether, *(broadly)* extraction, shaking-out
ausbalancieren to equilibrate, to counterbalance
Ausbalancieren *n* equilibration
Ausbesserungslack *m* touch-up paint
Ausbeute *f* yield, recovery, gain ration ✦ **mit hoher A.** high-yield
brauchbare A. fair yield
Ausbeutung *f* exploitation *(of mineral resources)*
sekundäre A. *(petrol)* secondary recovery
ausbilden / sich to form
Ausbildung *f* formation
A. von Bindungen bond formation

Ausbiß *m (mine)* outcrop; *(petrol)* [surface] seepage
Ausblasbehälter *m (pap)* blow pit (tank, vat), receiving (wash) tank
ausblasen to blow out; *(pap)* to blow [off] *(a digester)*
Ausblasen *n (pap)* blowing, blow *(of the digester)*
A. mit Druckluft air blowing
Ausblasgas *n (pap)* blow-pit gas, gaseous blow-off
Ausblasleitung *f* blow[-out] line
Ausblasrohr *n* blow pipe, blowpipe
Ausblasschieber *m*, **Ausblasventil** *n* blow[-off] valve, blow-off
ausbleichen to bleach out *(something)*; to fade
Ausbleichen *n* **in Abgasatmosphäre** *(text)* gas fading
ausbleichend fading
nicht a. fadeless
ausbleien to lead-line, to lead-clad
Ausbleiung *f* lead lining (cladding)
ausblenden to stop out (down) *(a region of the spectrum)*
ausblühen to bloom [out], to effloresce *(of crystals)*
Ausblühen *n* blooming, efflorescence *(of crystals)*
A. von Schwefel *(rubber)* sulphur blooming, sulphuring-up
ausblühend / nicht *(rubber)* non-blooming
Ausblühung *f (cryst, min)* bloom, efflorescence; *(ceram)* scumming
ausbluten *(dye, coat, tann)* to bleed [off, through], to mark off
Ausbrand *m* burn-out
ausbreiten to spread [out], to diffuse, to propagate
sich a. to spread [out], to diffuse, to propagate
Ausbreitung *f* spreading, diffusion, propagation
Ausbreitungskoeffizient *m* spreading coefficient, SC
Ausbrennartikel *m (text)* burnt-out fabric
ausbrennen *(text)* to burn out; *(coal, petrol)* to burn out (off)
Ausbrenner *m (text)* burnt-out fabric
ausbringen 1. to discharge; 2. to yield, to produce; 3. *(agric)* to apply, to place *(fertilizers or pesticides)*
breitwürfig (flächenhaft) a. *(agric)* to apply (place) broadcast
Ausbringen *n* 1. discharge, discharging; 2. yield, recovery; 3. *(agric)* application, placement, placing *(of fertilizers or pesticides)*
A. aus der Luft air-to-ground application
A. in flüssiger Form spraying
A. unter Abschirmung directed application
A. vom Flugzeug aus aeroplane application
A. vom Tragschrauber aus autogiro application
A. vor dem Auspflanzen preplanting application
ganzflächiges A. overall application, *(in presence of crops also)* overhead application
Ausbringungsgerät *n (agric)* [mechanical] applicator, application apparatus
Ausbringungsmethode *f*, **Ausbringungsweise** *f* *(agric)* application (distribution, placement) method
Ausbringungszeit *f (agric)* time of application
Ausbruch *m (petrol)* blow-out
Ausbruchpreventer *m (petrol)* blow-out preventer
ausdampfen to evaporate
Ausdampfen *n* evaporation
ausdämpfen to steam; *(distil)* to strip [off, out]
Ausdämpfer *m (distil)* [side] stripper
Ausdämpf[er]kolonne *f (distil)* stripper [column], stripping column (still), steam-stripping still
Ausdämpfsektion *f* stripping section *(of a fractionating column)*
Ausdämpfungsteil *m s.* Ausdämpfsektion
ausdehnbar expansible, expandable, extensible, extensile, extendible, extendable
Ausdehnbarkeit *f* expansibility, expandability, extensibility, extendibility
ausdehnen to expand, to extend
sich a. to expand *(with heat)*; to spread *(over an area)*
Ausdehnung *f* expansion, extension, dilatation
kubische A. cubical expansion
lineare A. linear expansion
prozentuale A. expansion percentage
räumliche A. cubical expansion
thermische A. thermal expansion
ausdehnungsfähig expansive
Ausdehnungsfähigkeit *f* expansiveness
Ausdehnungskoeffizient *m* expansion coefficient
kubischer A. *s.* räumlicher A.
linearer A. coefficient of linear expansion
räumlicher A. coefficient of cubic[al] expansion
thermischer A. coefficient of thermal expansion
Ausdehnungskondenswasserableiter *m* expansion trap
Ausdehnungsmesser *m* extensometer
Ausdehnungsthermometer *n* expansion thermometer
Ausdehnungsvermögen *n* expansiveness
Ausdehnungszahl *f* expansion coefficient
ausdestillieren to distil out
Ausdrückbolzenfeder *f (plast)* return spring
ausdrücken 1. press out, to squeeze [out], to express; 2. to push out, to blow out, to discharge *(solid material)*; *(plast)* to eject
Ausdrücken *n* 1. squeeze, expression; 2. discharge *(of solid material)*; *(plast)* ejection
A. von Hand *(plast)* hand ejection
A. von unten *(plast)* bottom ejection
automatisches A. *(plast)* automatic ejection
Ausdrücker *m (plast)* ejector, knockout
Ausdrückhilfsvorrichtung *f (plast)* extractor
Ausdrückkolben *m (plast)* ejection ram
Ausdrückleitung *f (pap)* blow[-out] line
Ausdrückmaschine *f (coal)* pusher machine
Ausdrückplatte *f (plast)* ejector (ejection, knockout) plate

Ausdrückrahmen *m (plast)* ejector frame
Ausdrückstange *f (plast)* ejector (knockout) bar, pull rod
Ausdrückstempel *m (plast)* ejection pad
Ausdrückstift *m (plast)* ejector (knockout) pin
mit Federkraft betätigter A. *(plast)* spring ejector
Ausdrücktraverse *f (plast)* ejection connecting bar
ausfahren *(ceram)* to draw *(the kiln)*; *(filtr)* to slide out *(e. g. a Kelly filter)*
Ausfall *m* 1. loss *(of material)*; failure, breakdown, outage *(of an apparatus)*; falling-off *(in production)*; 2. discharge [point]
A. durch Ermüdung fatigue failure
ausfällbar precipitable
Ausfällbarkeit *f* precipitability
ausfallen 1. *(from a solution)* to come down, to sediment, to settle [down, out], to set, to precipitate, to deposit, to subside, to separate out; 2. *(of an apparatus)* to fail, to break down
ausfällen to precipitate, to sediment[ate], to throw down; to cement *(a metal by a more active one)*
ausfällend precipitative
Ausfallklappe *f* discharge door
Ausfallkonus *m* discharge cone
Ausfallöffnung *f* discharge opening (outlet, door, aperture, port)
Ausfallseite *f* discharge end
Ausfällung *f* precipitation, sedimentation; cementation *(of a metal by a more active one)*
Ausfällungsanlage *f* precipitation plant, precipitator
Ausfällungsmittel *n* precipitant, precipitating agent, precipitator
ausflammen *(glass)* to sting out
Ausflammverlust *m (glass)* sting-out loss
ausfließen to flow out, to discharge, to issue, to effuse, to emanate
Ausfließen *n* outflow, discharge, effusion, effluence, emanation
freies A. *(petrol)* natural flow
ausfließend effluent
ausflocken to flocculate, to coagulate, to clot, to curd[le]
Ausflockung *f* flocculation, coagulation, clotting, curdling
Ausflockungsmittel *n* flocculating (coagulating) agent, flocculant, coagulant, coagulator
Ausfluß *m* 1. *(substance)* effluent, efflux, outflow; 2. *(process)* effluence, efflux, outflow; 3. *(apparatus)* drain, outlet
Ausflußbürette *f* gravity-flow burette
Ausflußdauer *f* efflux time, time of outflow
Ausflußdüse *f (pap)* slice nozzle
Ausflußöffnung *f* discharge opening (outlet, door, aperture, port)
Ausflußplastometer *n* **nach Marzetti** Marzetti plastometer
Ausflußrohr *n* effluent (outflow, discharge) pipe, effluent (outflow, discharge) tube
Ausflußschlitz *m*, **Ausflußspalt** *m (pap)* gate, slice, slot *(of the headbox)*
Ausflußzeit *f s.* Ausflußdauer

ausformen to shape out, to perfect
Ausformen *n* **des Reifens** *(rubber)* tyre shaping
ausfressen *(met)* to scour *(the furnace lining)*
Ausfressung *f (met)* scouring *(in the furnace lining)*
Ausfressungen *fpl (rubber)* backrinding *(of mould-parting lines)*
ausfrieren to freeze out; to demarg[ar]inate, to destearinate, to destearinize, to winterize *(oils)*
Ausfrieren *n* freezing-out; demargarination, destearinization, winterization *(of oils)*
Ausfriertasche *f* cold trap
ausführen / einen Versuch to run (carry out) an experiment
ausfüttern *(met)* to line
Ausgangselement *n* parent element
Ausgangsflüssigkeit *f (distil)* feed liquor
Ausgangsgestein *n* parent (mother, source) rock
Ausgangsgut *n* starting material
Ausgangskonzentration *f* initial concentration
A. der Beimengung (Verunreinigung) initial solute concentration *(in zone melting theory)*
A. des gelösten Stoffes *s.* A. der Beimengung
Ausgangslinie *f* **im Massenspektrum** parent [mass] peak
Ausgangslösung *f* initial (parent) solution
Ausgangsmaterial *n* starting (parent, raw) material, initial (parent) substance, stock, source
A. für Krackverfahren cracking feed (feedstock, stock)
Ausgangsöl *n* charge oil
Ausgangsprodukt *n* initial product
Ausgangspunkt *m* starting point; zero [point] *(of a scale)*
Ausgangsquerschnitt *m* original cross section
Ausgangsspalt *m* exit slit
Ausgangsstellung *f* original position
Ausgangsstoff *m*, **Ausgangssubstanz** *f s.* Ausgangsmaterial
Ausgangsverbindung *f* parent compound
Ausgangszustand *m* original (initial) state
ausgebraucht spent *(solution)*
ausgehärtet / nicht *(plast)* undercured
ausgeheizt *(rubber)* fully cured
ausgehen *(geol)* to outcrop, to crop out
Ausgehendes *n (geol)* outcrop
ausgekleidet / basisch *(met)* basic-lined
sauer a. *(met)* acid-lined
ausgemauert / feuerfest firebrick-lined
ausgereift mature
ausgerüstet / flammfest (flammsicher) flame-proofed
knitterarm (knitterbeständig, knitterecht, knitterfest) a. anticreased, crease-proofed
Ausgiebigkeitsfaktor *m*, **Ausgiebigkeitswert** *m (coat)* yield value
ausgießen to pour out, to diffuse
den Hafen a. *(glass)* to teem
Ausgießverfahren *n (plast)* slush moulding *(for hollow bodies consisting of PVC paste)*
Ausgleich *m* compensation; counterbalance *(esp of a force)*; makeup *(esp for material lost or used up)*

Ausgleichbecken *n* balancing tank; equalization tank *(waste-water treatment)*
Ausgleichbehälter *m* balancing tank
Ausgleichbunker *m* surge hopper
ausgleichen to compensate; to [counter]balance; to make up *(esp losses of material)*; to equilibrate, to level [out] *(esp weight)*
Ausgleicher *m* expansion joint
Ausgleichgefäß *n (tech)* expansion tank (vessel); *(lab)* levelling bottle
ausgleichglühen *(met)* to soak
Ausgleichinstrument *n* null-balance instrument
Ausgleichkolben *m* levelling bulb
Ausgleichmeßinstrument *n* null-balance instrument
Ausgleichs ... *s. a.* Ausgleich ...
Ausgleichsdüngung *f* compensation fertilization
Ausgleichsentwickler *m (phot)* compensating developer
ausglühen to glow [thoroughly], to heat [thoroughly]; *(met)* to anneal
Ausglühen *n* **/ vollständiges** *(met)* true (full) annealing
Ausguß *m* drain, sink; spout *(as of a beaker)*; *(petrol)* mud outlet *(of a rotary-drilling installation)*
Ausgußbecken *n* [bench] sink
Ausgußleitung *f* waste line
Aushängen *n* **an der Luft** *(text)* exposure to the air
aushärten 1. *(met)* to precipitation-harden *(relating to steel)*; to age[-harden] *(relating to light metal)*; 2. *(coat, plast)* to cure; 3. to harden *(fats)*; 4. to set *(of concrete)*
Aushärtung *f* 1. *(met)* precipitation hardening *(of steel)*; ageing, age-hardening *(of light metal)*; 2. *(coat, plast)* curing, cure; 3. hardening *(of fats)*; 4. setting, set *(of concrete)*
  vorzeitige A. *(plast)* premature curing *(moulding defect)*
Aushärtungsgeschwindigkeit *f (coat, plast)* speed of cure
Aushärtungszeit *f (coat, plast)* cure (curing) time
aushebern to siphon
Ausheizung *f (rubber)* full cure (vulcanization), complete cure (vulcanization)
Aushöhlung *f* cavity
  formgebende A. mould cavity
auskalken to lime out *(in manufacturing organic intermediates)*
auskippen to dump
ausklauben *(min tech)* to remove, to pick [off]
  von Hand a. to remove by hand, to pick [off] by hand, to [hand-]pick
auskleiden to line, to coat *(e. g. an oven)*
  mit Blei a. to lead-line, to lead-clad
  mit einem Gitterwerk a. to honeycomb
  mit Filz a. to felt
  mit Graphit a. to line with graphite, to graphitize, *(Am also)* to graphite
Auskleidung *f (material)* lining; *(act)* lining, cladding
  A. aus feuerfesten Steinen firebrick lining
  A. aus Nickelstahl nickel-steel lining
  A. mit Blei lead lining (cladding)
  basische A. *(met)* basic lining
  feuerfeste A. *(met)* refractory lining
  saure A. *(met)* acid lining
Auskleidungswerkstoff *m* lining (cladding) material
auskochen to decoct; *(text)* to boil off (out)
Auskochen *n* decoction; *(text)* boiling-off, boiling-out
auskohlen to carbonize *(wool)*
Auskohlen *n* carbonization, carbonizing *(of wool)*
auskondensieren to condense out
Auskopierpapier *n (phot)* print[ing]-out paper, P. O. P.
auskratzen to scratch out, to scrape out, to rake [out]
auskreiden *(coat)* to chalk, to become chalky
auskristallisieren to crystallize [out]
  wieder a. to recrystallize
Auslaß *m* 1. *(act)* discharge; 2. discharge [point], outlet, exit, drain, outflow, runoff
auslassen to discharge, to let out, to drain; *(food)* to render *(for gaining fat)*
Auslaßrohr *n* discharge (outlet, outflow) pipe, discharge (outlet, outflow) tube
Auslaßschleuse *f* outlet sluice
Auslaßventil *n* delivery (discharge) valve
Auslauf *m* discharge [point], outlet, exit, drain, outflow, runoff ✦ **auf A. geeicht (geteilt, graduiert, justiert)** calibrated (graduated) to deliver, calibrated for delivery
Auslaufdüse *f (pap)* slice nozzle
auslaufen 1. *(of liquids)* to run out, to flow out, *(unintentionally)* to leak out; 2. *(of vessels)* to discharge, *(unintentionally)* to leak; 3. *(of paint, writing ink on paper)* to spread, to feather, to run
Auslaufende *n* discharge end
Auslaufgeschwindigkeit *f* discharge velocity; *(pap)* spouting (stock) velocity, speed of the stock
Auslauföffnung *f* discharge opening (outlet, door, aperture, port)
Auslaufrohr *n* discharge (outlet, outflow) pipe, discharge (outlet, outflow) tube
Auslaufrutsche *f*, **Auslaufschurre** *f* discharge chute
Auslaufventil *n* tap
  gekrümmtes A. bib tap
Auslaufverlust *m* exit loss
auslaugbar leachable, extractable, extractible
Auslaugbarkeit *f* leachability, extractability
Auslaugbehälter *m*, **Auslaugbottich** *m* leach tank, leaching tank (vat, vessel)
auslaugen to leach [out], to lixiviate, to extract; *(soil)* to eluviate, to dilute *(nutrients from the eluvial horizon)*
Auslaugung *f* leach[ing], lixiviation, extraction; *(soil)* eluviation, chemical denudation
Auslaugungsverfahren *n (ferm)* infusion mashing (process)
auslenken to deflect
Auslenkung *f* deflection

Auslenk[ungs]winkel *m* angle of deflection
Auslese *f* selection
auslesen to select, to pick out; *(min tech)* to remove, to pick [off]
von Hand a. to remove by hand, to pick [off] by hand, to [hand-]pick
Auslesen *n* selection; *(min tech)* removal, picking
A. von Hand hand picking (cleaning, sorting)
auslöschen to extinguish; to quench *(an electric arc)*
auslösen 1. to initiate, to bring about *(a reaction)*; 2. to dissolve out *(a substance by solvents)*
Auslösezähler *m (nucl)* self-quenched (self-quenching) counter
Auslösezeit *f (nucl)* time of liberation
Auslösung *f* 1. initiation *(of a reaction)*; 2. dissolving-out *(of substances by solvents)*
Auslösungsgeschwindigkeit *f* **des Lignins** *(pap)* rate of delignification
ausmahlen to grind [thoroughly]
Ausmahlung *f* grinding
feine A. fine grinding
grobe A. coarse grinding
Ausmahlungsgrad *m* degree of fineness
ausmauern to brick-line
Ausmauerung *f* brick lining *(act or material)*
ausmustern *(text)* to cast
ausnutzen to utilize; to exhaust *(e. g. chemicals)*
Ausnutzung *f* utilization; exhaustion *(as of chemicals)*
Ausnutzungsgrad *m* efficiency; *(agric)* recovery *(of fertilizer by a crop)*
auspressen 1. to press out, to express, to force out, to squeeze [out]; 2. *(plast)* to extrude
Auspressen *n* 1. pressing, expression, squeeze; 2. *(plast)* extrusion, extruding
Auspuffgas *n* exhaust gas
auspumpen to evacuate, to exhaust, to pump out
Auspumpen *n* evacuation, exhaustion
ausquetschen to squeeze [out], to press out, to express, to force out
Ausquetschen *n* squeeze, pressing, expression
ausräuchern to fumigate, to smoke [out]
Ausräuchern *n* fumigation, smoking
Ausräumer *m* scraper, wiper
ausrecken *(tann)* to strike (set) out
ausreifen to ripen, to mature
Ausreifung *f* ripening, maturation
ausrichten to align
sich a. to align; *(of molecules)* to orient
Ausrichtung *f* alignment, directional distribution; orientation *(of molecules)*
ausrühren / mit Mononitrotoluol to detoluate *(to remove nitro compounds in TNT production)*
Ausrühren *n* **mit Mononitrotoluol** detoluation *(removal of nitro compounds in TNT production)*
ausrüsten 1. to equip; 2. *(pap, text)* to finish
knirschend a. *(text)* to scroop
Ausrüster *m (text)* finisher
Ausrüstung *f* 1. equipment; 2. *(pap, text)* finish[ing]
antistatische A. antistatic finish
chemische A. chemical proofing
glatttrocknende A. smooth-drying finish
hydrophobe A. water-repellent finish
knitterarme (knitterbeständige, knitterechte, knitterfeste) A. crease-resistant (non-crease, anticrease) finish (treatment), *(Am)* crush proofing
schmutzabstoßende (schmutzabweisende) A. dirt-repellent treatment
schmutzfreigebende A. soil release, SR
schrumpffreie A. unshrinkable (shrink-resist) finish
Ausrüstungsabteilung *f (pap)* finishing department (plant)
Ausrüstungsgegenstände *mpl* equipment
Ausrüstungsmittel *n* / **verrottungshemmendes** *(text)* antimildew agent
Ausrüstungssaal *m (pap)* finishing room
Aussalzchromatografie *f* salting-out chromatography
Aussalzeffekt *m* salting-out effect
aussalzen to salt out, *(esp soap)* to grain, to cut
Aussalzung *f* salting out, *(esp of soap)* graining, cutting
fraktionierte A. fractionation with salts
aussaugen to suck out, to exhaust
ausschaben to rake
ausschaufeln to scoop [out], to shovel out
Ausschaufeln *n* **von Hand** hand scooping
ausscheiden 1. *(of human agent or apparatus)* to separate, to eliminate; (min tech) to discard *(tailings)*; *(classifying)* to screen out; 2. *(of a liquid)* to sediment, to deposit; to evolve, to set free, to liberate *(gases)*; 3. *(biol)* to excrete, *(esp resins)* to exude
sich a. to separate [out]; to sediment, to deposit, to precipitate, to subside, to set, to settle [out] *(from a liquid as a sediment)*; *(biol)* to exude *(esp of resins)*
Ausscheiden *n* 1. separation, elimination; *(min tech)* discard; 2. sedimentation, deposition, precipitation *(of sediment by a liquid)*; evolution, liberation *(of gases)*; 3. *(biol)* excretion, *(esp of resins)* exudation
Ausscheidung *f* 1. *s.* Ausscheiden; 2. *(substance)* sediment, settling[s], bottom sediment (settlings), B. S., bottoms, deposit, foots, lees; *(biol)* excretion product, excretum, *(esp of resins)* exudate ✦ **zur A. gelangen** *s.* sich ausscheiden
Ausscheidungsgrad *m* separation efficiency
Ausscheidungshärten *n* precipitation hardening
Ausscheidungsprodukt *n s.* Ausscheidung 2.
Ausscheidungsverfahren *n* / **Steffensches** Steffen [separation] process *(for recovering sugar from beet molasses)*
ausschlacken to slag
Ausschlag *m* deflection *(of an indicator)*
ausschlagen to deflect
Ausschlagen *n* deflection *(of an indicator)*
Auschlag[meß]instrument *n* deflection instrument
ausschlämmen to elutriate

Ausschlämmen *n* elutriation
ausschleudern to jet; to centrifuge; *(nucl)* to emit, to eject, to expel *(particles)*
Ausschließungsprinzip *n* / **Paulisches** [Pauli] exclusion principle
Ausschluß *m* exclusion
ausschmelzen *(met)* to fuse, to melt, to smelt; *(food)* to render *(for gaining fat)*
Ausschmelzmodell *n* investment (fusible alloy) pattern *(in investment casting)*
ausschneiden *(plast)* to blank
ausschöpfen *(glass)* to ladle
Ausschuß *m* refuse, reject, scrap, broke, waste [material]; *(plast)* cull; *(pap) s.* Ausschußpapier
Ausschußpapier *n* mill (machine) broke, broken paper (material), brokes, waste paper (stuff)
ausschütteln to shake out, to extract by shaking
mit Äther a. to extract (shake out) with ether
Ausschütteln *n* shake, shaking[-out], extraction by shaking, solvent partition
A. mit Äther extraction with ether
ausschütten to dump [out], to pour out
ausschwefeln to sulphur *(e.g. a vat)*; *(rubber)* to sulphur up
Ausschwefeln *n* sulphuration *(as of vats)*; *(rubber)* sulphur blooming, sulphuring-up
ausschwefelnd / nicht *(rubber)* non-blooming
ausschwimmen *(coat)* to float off *(to segregate horizontally)*; to flood *(to segregate vertically)*
Ausschwimmen *n* **in horizontaler Richtung** *(coat)* floating *(separation of pigments)*
A. in vertikaler Richtung *(coat)* flooding *(separation of pigments)*
ausschwitzen *(bot, text)* to exude *(e.g. resin or lubricant)*; *(petrol)* to sweat
Ausschwitzen *n (bot, text)* exuding, exudation *(of resin or lubricant)*
A. des Schmälzmittels *(text)* lubricant exudation
Ausschwitzungsprodukt *n* exudate, exudation
Aussehen *n* appearance, look
blumenkohlähnliches A. cauliflower appearance *(of a coke button)*
glänzendes A. gloss
mattes A. dullness
Außenanstrichfarbe *f* exterior paint
außenbeheizt externally heated
Außenbeheizung *f* external heating
Außenbeständigkeit *f* outdoor (exterior) durability
Außenbewetterung *f (rubber)* outdoor weathering
aussenden to emanate, to emit, to issue, to expel, to eject *(e.g. radioactive particles)*
Außendruck *m* external pressure
Aussendung *f* emanation, emission, expulsion, ejection *(as of radioactive particles)*
Außendurchmesser *m* outer (outside) diameter, O.D.
Außenelektron *n* outer (outside, external, valence) electron
Außenlack *m* exterior paint, *(if transparent)* exterior varnish
Außenluft *f* external (outdoor) air
Außenorbitalbindung *f* outer-orbital bond *(chemical-bond theory)*
Außenorbitalkomplex *m* outer-orbital complex *(chemical-bond theory)*
Außenrohrschlange *f* external coil
Außenrüttler *m* external vibrator
Außenschale *f* external (outer, outermost) shell
Außenschleifen *n (ceram)* external grinding
außenseitig external
Außenzahnradpumpe *f* external-gear pump
Außenzylinder *m* outer cylinder
Außerbetriebsetzung *f* shut[down]
außermittig off-centre
aussetzen 1. to expose *(to radiation)*; 2. to fail *(of an engine)*
dem Rauch a. to fumigate
einer Röntgenstrahlung a. to expose to X-rays
Aussetzen *n* 1. exposure *(to radiation)*; 2. failure *(of an engine)*
aussieben to screen [out]
aussommern *(ceram)* to weather
aussondern to separate, to eliminate; *(min tech)* to discard *(tailings)*
Aussondern *n* separation, elimination; *(min tech)* discard
aussortieren to cull, to eliminate; *(min tech)* to discard *(tailings)*
Aussortieren *n* culling, elimination; *(min tech)* discard
Ausspritzapparat *m* rinser
Ausspritzer *m (ceram)* spit-out *(a defect)*
ausspülen to rinse [out], to scour, to flush [out], to swill out
ausstanzen *(plast)* to blank
Ausstattung *f* equipment
Ausstattungskarton *m* fancy board
Ausstattungspapier *n* fancy (letter, note) paper, *(Am)* correspondence (decorated) paper
aussteinen to brick-line
Ausstellungsglas *n* museum jar
Ausstoß *m* 1. [production] output, make; 2. discharge *(of solid material)*; 3. *(plast)* ejection; lift, set of mouldings *(produced in one pressing operation)*
automatischer A. *(plast)* automatic ejection
ausstoßen 1. to eject, to emit, to emanate, to issue, to expel *(e.g. radioactive particles)*; 2. to push out, to discharge *(solid material)*; *(rubber)* to dump; *(plast)* to eject; 3. *(tann)* to set out, to strike out *(hides)*
Ausstoßen *n* 1. ejection, emission, emanation, expulsion *(as of radioactive particles)*; 2. pushing-out, discharge *(of solid material)*; *(rubber)* dumping; *(plast)* ejection; 3. *(tann)* setting-out, striking-out
Ausstoßer *m (plast)* ejector
Ausstoßmaschine *f (coal)* pusher machine
Ausstoßrate *f* output rate
Ausstoßtemperatur *f (rubber)* dump temperature
ausstrahlen to emit, to emanate, to eject, to issue, to expel *(e.g. radioactive particles)*; to [ir]radiate *(light)*

Ausstrahlung *f* emission, emanation, ejection, expulsion *(as of radioactive particles)*; [ir]radiation *(of light)*
spezifische A. emittance
Ausstrahlungsverlust *m* radiation loss
ausstreichen *(geol)* to outcrop, to crop out
Ausstreichen *n (geol)* outcrop[ping]
ausströmen 1. *(of liquids)* to flow out, to discharge, to issue, to effuse, *(unintentionally)* to leak out; *(of gases, steam)* to escape; 2. to emanate, to emit *(heat)*
a. lassen to discharge, to issue, to run
Ausströmen *n* 1. *(of liquids)* outflow, discharge, effluence, effusion, *(unintentionally)* leak[age]; *(of gases, steam)* escape; 2. emanation, emission *(of heat)*
Ausströmgeschwindigkeit *f s.* Ausströmungsgeschwindigkeit
Ausströmöffnung *f* discharge opening (outlet, door, aperture, port)
Ausströmungsgeschwindigkeit *f* discharge velocity; *(pap)* spouting (stock) velocity, speed of the stock
Ausströmverlust *m* exit loss
aussüßen *(pap)* to [re]causticize *(to convert soda or potash into NaOH or KOH)*
austarieren to tare, to [counter]balance
Austausch *m* exchange, interchange, substitution, replacement
nukleophiler A. nucleophilic substitution
Austauschabteilung *f* / **regenerative** regeneration section *(of a plate pasteurizer)*
Austauschadsorption *f* exchange adsorption
austauschaktiv substitutable
Austauschaktivität *f* substitutability
Austauschazidität *f (soil)* exchange acidity
austauschbar exchangeable, interchangeable, substitutable, replaceable
Austauschbarkeit *f* exchangeability, interchangeability, substitutability, replaceability
Austauschboden *m (distil)* [exchange] plate, tray
gelochter A. perforated plate (tray), sieve plate (tray)
Austauschchromatografie *f* [ion-]exchange chromatography
Austauschdüngung *f* exchange fertilization
austauschen to exchange, to interchange, to substitute, to replace
Austauschentartung *f (phys chem)* exchange degeneracy
Austauscher *m* exchanger
indirekter A. surface exchanger
Austauscherbett *n* ion-exchange bed
Austauscherharz *n* ion-exchange resin
chelatbildendes A. chelate [ion-exchange] resin
Austauscherkorn *n* ion-exchange bead
Austauschermasse *f* [ion-]exchange medium
austauschfähig substitutable
Austauschfähigkeit *f* substitutability, exchange capacity (property)
Austauschfeuchtemenge *f* / **spezifische** drying rate
Austauschgerbstoff *m* / **synthetischer** exchange (replacement) syntan
Austauschgeschwindigkeit *f* exchange rate
Austauschglied *n s.* Austauschintegral
Austauschglocke *f (distil)* bubble cap, dome
Austauschharz *n s.* Austauscherharz
Austauschintegral *n* exchange integral (term) *(quantum chemistry)*
Austauschkalk *m (soil)* exchangeable calcium
Austauschkapazität *f* exchange capacity (property), substitutability
Austauschkolonne *f* rectifying (rectification) column
Austauschkomplex *m s.* Austauschmaterial
Austauschkraft *f* exchange (interchange) force
Austauschmaterial *n (soil)* exchange material, ion-exchange compound, [base-]exchange complex
Austauschmöglichkeit *f* exchange possibility
Austauschname *m* replacement name
Austauschreaktion *f* exchange (replacement, displacement, substitution) reaction
Austauschsäule *f s.* Austauschkolonne
Austauschstoff *m* substitute
Austauschtrennung *f* exchange separation
Austauschvermögen *n s.* Austauschkapazität
Austenit *m (met)* austenite *(a solid solution of carbon in gamma iron)*
austenitisch *(met)* austenitic
austenitisieren *(met)* to austenitize
Austenitisierung *f (met)* austenitizing
Austenitisierungstemperatur *f (met)* austenitizing temperature
Austrag *m* 1. *(act)* discharge; *(min tech)* discard; 2. discharge [point], outlet, exit; 3. *(el chem)* drag-out
Austragapparat *m* discharge apparatus (device), discharger
Austragdüse *f* discharge nozzle; skimming nozzle *(of a centrifuge)*
Austrageinrichtung *f s.* Austragapparat
austragen to discharge; *(min tech)* to discard *(tailings)*
Austragen *n* discharge; *(min tech)* discard
Austragende *n* discharge (outlet) end
Austragklappe *f* discharge door
Austragkonus *m* discharge cone
Austragmesser *n* discharge knife
Austragöffnung *f* discharge opening (outlet, door, aperture, port)
Austragpflug *m* plough, *(Am)* plow *(on conveyors)*
Austragrinne *f* discharge chute
Austragrohr *n* discharge (outlet, offtake) pipe, discharge (outlet, offtake) tube
Austragrost *m* discharge grate (grating)
Austragrutsche *f* discharge chute
Austrags ... *s.* Austrag ...
Austragschieber *m* discharge gate
Austragschleuse *f* outlet sluice, exit lock
Austragschlitzschieber *m* discharge gate
Austragschnecke *f* discharge scroll

Austragschurre *f* discharge chute
Austragseite *f* discharge (outlet) end
Austragstelle *f* discharge [point], outlet, exit
Austragventil *n* discharge (outlet) valve
Austragvorrichtung *f* discharging apparatus (device), discharger
Austragwalze *f* discharge roll
Australit *m* australite *(a tektite)*
Austreibekolonne *f (distil)* stripper [column], stripping column (still), side stripper, strip action still
austreiben to expel, to dispel, to drive off (out), to sweep out *(gases from liquids)*; *(distil)* to strip [off, out], to distil off
CO_2 a. to decarbonate
Fluor a. to defluorinate
Lösungsmittel a. to desolventize
Austreiben *n* expulsion *(of gases from liquids)*; *(distil)* stripping; stripping, desorption *(of sorbed gases from sorbent material)*
A. mit Luft air stripping
A. von CO_2 decarbonation
Austreiber *m* generator *(of an absorption refrigeration system)*
austreten to escape *(of gases, steam)*; to exude *(esp of resins)*
Austreten *n* escape *(of gases, steam)*; exudation *(esp of resins)*
Austrieb *m (plast)* fan, fin, flash; *(rubber)* excess stock (rubber), overflow, spew, spue
Austriebnut *f (plast)* groove spew
Austritt *m* 1. *s.* Austreten; emission *(as of heat, light, or electrons)*; 2. discharge [point], exit, outlet, outflow
A. an der Oberfläche *(petrol)* surface seepage
Austrittsarbeit *f* work of separation, electronic work funktion, work function [of electrons]
Austrittsende *n* discharge end
Austrittsgeschwindigkeit *f* discharge velocity
Austrittsöffnung *f* discharge opening (outlet, door, aperture, port)
Austrittsrohr *n* discharge (outlet) pipe (tube), offtake [pipe]; outflow (effluent) pipe
Austrittsschwelle *f* exit threshold
Austrittsspalt *m* exit slit; *(pap)* slice, slot, gate *(of the headbox)*
A. des Extruderkopfes *(plast)* die lips
Austrittsstelle *f* discharge [point], exit, outlet, outflow
Austrittsstrahl *m* exit beam
Austrittstemperatur *f* exit temperature
A. des Gases exit gas temperature
Austrittsverlust *m* exit loss
austrocknen to dry [out, up], to exsiccate
Austrockner *m (tann)* drying oven
Austrocknung *f* drying, exsiccation
Austrocknungsgrad *m* degree of drying
Ausvulkanisation *f* full cure (vulcanization), complete cure (vulcanization)
Ausvulkanisationszeit *f* vulcanization (vulcanizing) time
ausvulkanisiert *(rubber)* fully cured
Auswahl *f* selection, picking
auswählen to select, to pick
Auswahlregel *f (phys chem)* selection principle (rule)
auswalzen to mill *(soap chips)*; *(plast)* to calender
Platten a. *(rubber)* to sheet [out]
zu einem Fell a. *s.* Platten a.
auswandern to migrate out *(as of ions)*
auswaschen to wash [out], to rinse [out], to flush [out], to scour *(e.g. a vessel)*; to scrub *(gases)*; to wash *(a filter cake)*; to lixiviate, to leach [out], to wash out *(soluble substances)*; to elute *(chromatography)*; *(soil)* to eluviate, to dilute, to leach [out] *(nutrients from the eluvial horizon)*
gründlich a. to wash thoroughly
Auswaschen *n* **mit Säure** acid wash[ing]
A. mit Waschöl oil wash[ing]
A. mit Wasser water wash[ing]
Auswaschkolonne *f (petrol)* scrubber column
Auswaschung *f (soil)* eluviation, dilution, leach[ing] *(of nutrients from the eluvial horizon)*
Auswaschungshorizont *m (soil)* eluvial horizon, A-horizon
Auswaschungsverlust *m (soil)* leaching loss
Auswaschverlust *m* washing-out loss
Auswässerungsgrad *m (phot)* washing rate
auswechselbar exchangeable, interchangeable, substitutable, replaceable
Auswechselbarkeit *f* exchangeability, interchangeability, substitutability, replaceability
auswechseln to exchange, to interchange, to substitute, to replace
Ausweichreaktion *f* evasion (dodge) reaction
auswerfen to push out, to discharge *(solid material)*; *(rubber)* to dump; *(plast)* to eject
Auswerfen *n* pushing-out, discharge *(of solid material)*; *(rubber)* dumping; *(plast)* ejection
Auswerfer *m (plast)* ejector, knockout
Auswerferstift *m (plast)* ejector (knockout) pin
Auswerfstempel *m (plast)* ejection pad
Auswintern *n (ceram)* wintering, weathering
Auswurfvorrichtung *f (plast)* ejector, knockout
auszementieren to cement *(a metal by a more active one)*
ausziehbar extractable, extractible, digestible
ausziehen 1. to extract, to digest; to lixiviate, to wash out, to leach [out] *(soluble substances)*; to decoct *(to extract by boiling)*; to exhaust *(e.g. dye liquor)*; 2. to pull out *(glass into a capillary)*
Platten a. *(rubber)* to sheet [out] *(on the calender)*
zu einem Fell a. *s.* Platten a.
Ausziehen *n* extraction, digestion; lixiviation, washing-out, leach[ing] *(of soluble substances)*; decoction *(extraction by boiling)*; exhaustion *(as of dye liquor)*
A. des Heizschlauchs *(rubber)* de-bagging [operation]
A. von Platten *(rubber)* sheet calendering, sheeting[-out]
A. zu Fellen *s.* A. von Platten
Ausziehtusche *f* India ink

Auszug *m* extract, extractive [material, matter, substance], essence; decoction; lixivium, leachate
alkoholischer A. alcoholic extract
wäßriger A. water extract (leachate)
authigen *(geol)* authigene, authigen[et]ic, authigenous
Authigenese *f (geol)* authigenesis
autochthon *(geol)* autochthonous, in place, in situ
autogen autogenous
Autogenbrennschneiden *n s.* Autogenschneiden
Autogenmühle *f* autogenous mill
Autogenschmelzen *n* autogenous smelting
Autogenschneiden *n* autogenous (oxygen-acetylene, oxyacetylene, oxy-gas) cutting, autogenous gas cutting
Autogenschweißen *n* oxyacetylene (oxygen-acetylene, autogenous gas) welding, oxywelding
Autohydratation *f (geol)* autohydration
Autoionisation *f* autoionization, preionization
Autokatalyse *f* autocatalysis, self-catalysis
autokatalytisch autocatalytic
Autoklav *m* 1. autoclave, [large-]pressure cooker; 2. *s.* Autoklavheizpresse ✦ **im Autoklaven behandeln (kochen)** to autoclave
Autoklavenbehandlung *f* autoclaving
Autoklavenpresse *f s.* Autoklavheizpresse
Autoklavheizpresse *f (rubber)* autoclave press, vulcanizer autoclave, pot heater [vulcanizer]
autoklavieren to autoclave
Autoklavieren *n* autoclaving
Autokollimationsspektrograf *m* autocollimating spectrograph
Autolack *m* automobile lacquer, automotive coating (finish)
Autolith *m (geol)* autolith, cognate inclusion
Autolysat *n (biol)* autolysate
Autolyse *f (biol)* autolysis, autolytic decomposition
autolytisch *(biol)* autolytic
Autometamorphose *f (geol)* autometamorphism
Autometasomatose *f (geol)* autometasomatism
automorph *(geol)* automorphic, idiomorphic, idiomorphous, euhedral
Autopneumatolyse *f (geol)* autopneumatolysis
Autoprotolyse *f* autoprotolysis *(a reaction mechanism)*
Autoradiochromatografie *f* autoradiochromatography
Autoradiografie *f* autoradiography, radioautography
autoradiografisch autoradiographic, radioautographic
Autoradiogramm *n* autoradiograph, radioautograph
Autorazemisation *f*, **Autorazemisierung** *f* autoracemization
autotroph *(biol)* autotrophic
Autotypie[druck]papier *n* autotype (half-tone) paper
Autovakzine *f* autovaccine, autogenous vaccine
Autoxydation *f* autoxidation, autooxidation, induced (spontaneous) oxidation
durch Kautschukgifte beschleunigte A. metallic poisoning
Autoxydationsreaktion *f* autoxidation reaction
Autoxydator *m* autoxidator
Autrometer *n* autrometer *(an automatic analyser)*
Autunit *m (min)* autunite *(calcium uranyl phosphate)*
Auxin *n* auxin *(a growth regulator in plants)*
auxochrom auxochromic
Auxochrom *n* auxochrome, auxochromic group
negatives A. negative (acidic) auxochromic group
positives A. positive (basic) auxochromic group
auxotroph *(biol)* auxotrophic
Avaram-Rinde *f (tann)* Avaram bark *(from Cassia auriculata L.)*
Aventurinfeldspat *m (min)* aventurine feldspar
Aventuringlas *n* aventurine glass
Aventuringlasur *f (ceram)* aventurine glaze
Aventurinquarz *m (min)* aventurine quartz *(silicon(IV)oxide)*
Avitaminose *f* avitaminosis, vitamin deficiency
Avivage *f (text)* reviving
avivieren *(text)* to revive
Aviviermittel *n (text)* reviving agent
Avocadoöl *n* avocado oil *(from Persea americana Mill.)*
Awapfefferwurzel *f (pharm)* kava, cava *(from Piper methysticum G. Forst.)*
axial axial
Axialkolbenpumpe *f* axial-piston pump
Axialkompressor *m* axial[-flow] compressor
Axiallager *n* thrust bearing
Axiallüfter *m* propeller (axial-flow) fan
A. mit Leiträdern vaneaxial fan
Axialpumpe *f* propeller (axial-flow) pump
axialsymmetrisch axially symmetric[al], axisymmetric
Axialventilator *m* propeller (axial-flow) fan
Axialverdichter *m* axial[-flow] compressor
Axinit *m (min)* axinite *(a cyclosilicate)*
Axonometrie *f (cryst)* axonometry
AZ *s.* 1. Azetatfaserstoff; 2. Azetylzahl
Aza-Benennung *f*, **Aza-Name** *m (nomencl)* „a"[-term] name
Aza-Nomenklatur *f* „a" nomenclature
Azelainsäure *f* $HOOC[CH_2]_7COOH$ azelaic acid, nonanedioic acid
azeotrop azeotropic
Azeotrop *n* azeotrope, azeotropic mixture
Azeotropbildner *m (distil)* azeotrope (azeotroping, entraining) agent, azeotrope (azeotroping, entraining) former, azeotropic[-distillation] agent, entrainer
Azeotropdestillation *f* azeotropic distillation
Azeotropie *f* azeotropy
azeotropisch azeotropic
Azetal *n* $R–CH(OR')_2$ acetal, *(specif)* $CH_3CH(OC_2H_5)_2$ acetal, 1,1-diethoxyethane
Azetaldehyd *m* CH_3CHO acetaldehyde, ethanal

Azetaldehydzyanhydrin *n* $CH_3CH(OH)CN$ acetaldehyde cyanohydrin, lactonitrile, 2-hydroxypropane nitrile
Azetaldol *n* $CH_3CH(OH)CH_2CHO$ acetaldol, aldol, c-hydroxybutyraldehyde
Azetalharz *n* acetal resin
azetalisieren to acetalate
Azetalisierung *f* acetalation
Azetamid *n* CH_3CONH_2 acetamide
p-**Azetamidobenzolsulfochlorid** *n* *p*-acetamidobenzenesulphonyl chloride, *p*-acetylaminobenzenesulphonyl chloride
Azetamidogruppe *f* $-NHCOCH_3$ acetamido group, acetylamino group
Azetanhydrid *n* $(CH_3CO)_2O$ acetic anhydride
Azetanilid *n* $C_6H_5NHCOCH_3$ acetanilide, *N*-phenylacetamide
Azetarsol *n* acetarsol, 3-acetylamino-4-hydroxybenzenearsonic acid
Azetat *n* acetate
Azetatelementarfaden *m* acetate filament
verseifter A. saponified acetate filament
Azetatfaser *f* acetate fibre
Azetatfaserstoff *m* acetate fibre
Azetatfilm *m* acetate film
Azetat-Ion *n* acetate ion
Azetatokomplex *m* acetate complex
Azetatpuffer *m* acetate buffer
Azetatrayon *m(n)* *s.* Azetatseide
Azetatseide *f* cellulose acetate rayon, acetate filament yarn
polyfile A. acetate multifilament yarn
Azetatstapelfaser *f s.* Azetatfaser
Azetessigester *m s.* Azetessigsäureäthylester
Azetessigsäure *f* CH_3COCH_2COOH acetoacetic acid, 3-oxobutyric acid
Azetessigsäureäthylester *m* $CH_3COCH_2COOC_2H_5$ ethyl acetoacetate, acetoacetic ester
Azetessigsäuredekarboxylase *f* acetoacetic decarboxylase
Azetessigsäurekarboxylase *f* acetoacetic carboxylase
Azetimeter *n* acetimeter
Azetoin *n* $CH_3CH(OH)COCH_3$ acetoin, dimethylketol, 3-hydroxy-2-butanone
Azetolyse *f* acetolysis
Azetometer *n s.* Azetimeter
Azeton *n* CH_3COCH_3 acetone, propanone, dimethylketone
Azeton-Benzol-Verfahren *n* benzol-acetone process *(dewaxing process)*
Azetonchloroform *n* $(CH_3)_2C(OH)CCl_3$ acetone chloroform, chloretone, 1,1,1-trichloro-2-hydroxy-2-methylpropane
Azetondikarbonsäure *f* $O{=}C(CH_2COOH)_2$ acetonedicarboxylic acid, ADA, 3-oxoglutaric acid, 3-oxopentanedioic acid
Azetonharz *n* acetone resin
Azetonkörper *m* acetone body
azetonlöslich acetone-soluble
Azetophenon *n* $C_6H_5COCH_3$ acetophenone, acetylbenzene, methyl phenyl ketone
Azetotartrat *n* acetotartrate
Azetotoluid[id] *n s.* Azettoluidid
Azetoxybenzoesäure *f* $CH_3COOC_6H_4COOH$ *o*-acetoxybenzoic acid, *O*-acetylsalicylic acid
Azetsäure *f s.* Äthansäure
Azettoluidid *n* $CH_3CONHC_6H_4CH_3$ acet-toluidide, acetyl-toluidine, acetamidotoluene
Azetylaminoanthrachinon *n* acetylaminoanthraquinone
p-**Azethylaminobenzolsulfonsäurechlorid** *n s.* *p*-Azetamidobenzolsulfochlorid
Azetylaminogruppe *f* $-NHCOCH_3$ acetylamino group, acetamido group
Azetylation *f* acetyl[iz]ation
Azetylazeton *n* $CH_3COCH_2COCH_3$ acetylacetone, 2,4-pentanedione
Azetylbenzol *n* $C_6H_5COCH_3$ acetylbenzene, methyl phenyl ketone, acetophenone
Azetylbutyrylzellulose *f* cellulose acetate butyrate, CAB
Azetylchlorid *n* CH_3COCl acetyl chloride
Azetylen *n* $HC{\equiv}CH$ acetylene, ethyne, ethine; *(broadly) s.* Azetylenkohlenwasserstoff
Azetylenbrenner *m* oxyacetylene blowpipe (torch)
Azetylendikarbonsäure *f* $HOOCC{\equiv}CCOOH$ acetylenedicarboxylic acid, butynedioic acid
Azetylenid *n s.* Azetylid
Azetylenkarbonsäure *f* acetylenic acid *(any carboxylic acid having a triple bond)*
Azetylenkohlenwasserstoff *m* acetylenic hydrocarbon, alkyne
Azetylenruß *m* acetylene black
Azetylen-Sauerstoff-Brennschneiden *n* oxyacetylene (oxygen-acetylene, autogenous gas) cutting
Azetylen-Sauerstoff-Schweißen *n* oxyacetylene (oxygen-acetylene, autogenous gas) welding, oxywelding
Azetylenschwarz *n* acetylene black
Azetylensilber *n* Ag_2C_2 silver acetylide (carbide)
Azetylentetrachlorid *n* $CHCl_2CHCl_2$ 1,1,2,2-tetrachloroethane, acetylene tetrachloride
Azetylenylbenzol *n* $C_6H_5C{\equiv}CH$ acetylenylbenzene, ethynylbenzene, phenylacetylene
Azetylenylgruppe *f* $-C{\equiv}CH$ acetylenyl group (residue), ethynyl group (residue)
Azetylenylkarbinol *n s.* Propin-(2)-ol-(1)
Azetylessigsäure *f s.* Azetessigsäure
Azetylesterase *f* acetyl esterase
Azetylformaldehyd *m s.* Brenztraubensäurealdehyd
Azetylgruppe *f* CH_3CO- acetyl group (residue)
Azetylharnstoff *n* $CH_3CONHCONH_2$ acetylurea, *N*-monoacetylurea
Azetylid *n* $M^I_2C_2$ acetylide
azetylierbar acetyl[at]able
azetylieren to acetylate
Azetylierung *f* acetyl[iz]ation
Azetylierungskatalysator *m* acetylation catalyst
Azetylierungsmittel *n* acetylation (acetylating) agent

Azetylkarbonsäure *f s.* Brenztraubensäure
Azetyl-Koenzym A *n*, **Azetyl-Koferment A** *n* acetyl coenzyme A
Azetylmethylkarbinol *n s.* Azetoin
Azetylpapier *n* acetylated paper
***N*-Azetyl-*N*-phenylglyzin** *n* $C_6H_5N(COCH_3)CH_2COOH$ *N*-acetylphenylglycine, *N*-phenylaceturic acid
Azetylrest *m s.* Azetylgruppe
Azetylsalizylat *n* acetylsalicylate
Azetylsalizylsäure *f* $CH_3COOC_6H_4COOH$ *O*-acetylsalicylic acid, *o*-acetoxybenzoic acid
***N*-Azetyltoluidin** *n s.* Azettoluidid
Azetylureid *n s.* Azetylharnstoff
Azetylzahl *f* acetyl number (value)
Azetylzellulose *f* $[C_6H_7O_5(COCH_3)_3]_x$ acetylated cellulose, cellulose acetate, CA
azid acidic *(having the character of an acid)*
Azid *n* M^IN_3 azide, hydrazoate
Azidifikation *f* acidification
azidifizieren to acidify
Azidifizierung *f* acidification
Azidimeter *n* acidimeter, acidometer
Azidimetrie *f* acidimetry
azidimetrisch acidimetric
Azidität *f* acidity, acid strength
  aktuelle A. active acidity
  hydrolytische A. hydrolytic acidity
  potentielle A. potential (total, reserve) acidity
Aziditätsbestimmung *f* determination of acidity
Aziditätsgrad *m* degree of acidity
Aziditätskonstante *f* acidity constant
Azidivinylphosphinoxid *n* azidivinylphosphine oxide
Azidoid *n (soil)* acidoid
Azidokomplex *m* acido complex
Azidolyse *f* acidolysis
azidolytisch acidolytic
azidophil acidophilic, acidophilous, oxyphil[e], oxyphilic, oxyphilous
Azidose *f (med)* acidosis
Azimethylen *n* CH_2N_2 azimethylene, diazomethane
Aziminobenzol *n* aziminobenzene, 1,2,3-benzotriazole
Azimutalquantenzahl *f* azimuthal (orbital, secondary) quantum number
Azin *n* azine *(any of two classes of organic compounds containing two or more N atoms)*
Azlakton-Kondensation *f* / **Erlenmeyer-Plöchlsche** Erlenmeyer-Plöchl azlactone synthesis *(to yield α-amino acids)*
A-Z-Lösung *f (biol)* Hoagland solution *(a nutrient solution containing microelements)*
Azoapparatur *f* azo[-dye] unit
Azobenzol *n* $C_6H_5N{=}NC_6H_5$ azobenzene
Azobindung *f* azo link
Azobrücke *f* azo link
Azodikarbonamid *n* $NH_2CON{=}NCONH_2$ azodicarbonamide, azoformamide
Azodispersionsfarbstoff *m* disperse azo dye
Azofarbstoff *m* azo dye
  auf der Faser erzeugter A. *s.* unlöslicher A.
  unlöslicher A. azoic [dye]
Azogelb *n* azo yellow
Azogruppe *f* $-N{=}N-$ azo group
Azoimid *n* HN_3 azoimide, hydrogen azide, hydrazoic acid
Azokomponente *f* azo component
Azokörper *m* $R-N{=}R'$ azo compound
Azokupplung *f* azo coupling
Azol *n* azole *(any of a class of heterocyclic compounds containing N)*
Azomethan *n* $CH_3N{=}NCH_3$ azomethane
Azomethin *n s.* Azomethinverbindung
Azomethinfarbstoff *m* azomethine dye
Azomethinverbindung *f* $Ar-N{=}CR_2$ azomethine [compound], *N*-arylimide, Schiff base
Azophenylen *n* azophenylene, phenazine
Azoreihe *f* azo series
Azostickstoff *m* nitrogen of the diazonium group
Azotometer *n* azotometer, nitrometer
Azoverbindung *f* $R-N{=}N-R'$ azo compound
Azoxybenzol *n* azoxybenzene
Azoxygruppe *f* azoxy group
Azoxyverbindung *f* azoxy compound
Azurit *m (min)* azurite, blue copper ore *(copper dihydroxide dicarbonate)*
A-Zustand *m (plast)* A stage
azyklisch acyclic, non-cyclic[al]
Azylgruppe *f* acyl group (residue)
Azylhalogenid *n* acyl halide *(acid halide of a carboxylic acid)*
azylierbar acylable
azylieren to acylate
Azylierung *f* acylation
Azylierungsmittel *n* acylating agent
Azylium-Ion *n* acylium ion
Azyloin *n* acyloin *(a keto alcohol of the general formula $R-CO{=}CH-R'OH$)*
Azyloinkondensation *f* acyloin condensation (synthesis)
Azyloinmethode *f* acyloin method
Azyloinsynthese *f* acyloin synthesis (condensation)
Azylradikal *n* acyl radical, *(specif)* free acyl radical
Azylrest *m s.* Azylgruppe
Azylverschiebung *f*, **Azylwanderung** *f* acyl migration

B

Babassufett *n*, **Babassuöl** *n* babassu (babussu) oil *(palm oil from Orbignya speciosa Berk.)*
Babbit-Metall *n* Babbit metal
Babcock-Kugelringmühle *f* Babcock & Wilcox pulverizer (ball-and-ring mill), Babcock mill
Babingtonit *m (min)* babingtonite *(an inosilicate)*
Babo-Blech *n* Babo boiling plate *(of the Babo funnel)*
Babo-Trichter *m* Babo funnel
Babulrinde *f (tann)* Babul bark *(from Acacia nilotica (L.) Del.)*
Babypresse *f (pap)* baby (pony) press

Bachbildung *f* channelling *(as in column packings)*
Backe *f (tech)* jaw *(of a jaw breaker)*; *(lab)* jaw *(of a clamp)*
  feste B. anvil jaw
Backeigenschaft *f* 1. *(food)* baking property (quality, value, characteristics); 2. *(coal)* caking property
backen *(food)* to bake; *(chem)* to bake *(amines for the purpose of sulphonation)*; *(coal)* to cake
Backenbrecher *m* jaw breaker (crusher)
Backenwerkzeug *n (plast)* bar mould
Bäckerhefe *f* bakery (baker's) yeast, barm
Bäckermargarine *f* bakery (confectionary, cake, pastry) margarine
Backerrohr *n* tubular heater
Backfähigkeit *f* 1. *(coal)* caking power; 2. *s.* Backeigenschaft 1.
Backfähigkeitsverlust *m (coal)* loss of caking power
Backfähigkeitsverminderung *f (coal)* reduction of caking power
Backfähigkeitszahl *f (coal)* index of caking power, caking index
  B. nach Campredon Campredon index
  B. nach Roga Roga index
Backfett *n* pastry fat
Backhefe *f s.* Bäckerhefe
Backkohle *f* caking coal
Backmargarine *f. s.* Bäckermargarine
Backofen *m* baking oven
Backpulver *n* baking powder
Backqualität *f s.* Backeigenschaft 1.
Backsteintee *m* brick tea
Backverfahren *n* baking process *(for sulphonating amines)*
Backverhalten *n (coal)* caking properties
Backvermögen *n s.* Backfähigkeit 1.
Backzahl *f s.* Backfähigkeitszahl
Bacon-Hochdruckzelle *f* Bacon high-pressure hydrogen cell
Bad *n* bath, *(tech also)* dip, steep; *(text)* bath, liquor
  alkalisches B. alkaline bath
  altes (stehendes) B. *(text)* standing bath
Badan *m (tann)* badan *(roots from Bergenia specc.)*
badaufkohlen *(met)* to liquid-carburize, to bath-carburize
Badaufkohlen *n (met)* liquid (liquid-salt, molten-salt, bath) carburizing
Baddeleyit *m (min)* baddeleyite *(zirconium dioxide)*
badeinsetzen *s.* badaufkohlen
Bademittel *n (agric)* [animal] dip *(as for controlling vermin)*
Badeöl *n* bath oil
Badepräparat *n* bath preparation
  brausendes (sprudelndes) B. bubble bath
Badepulver *n* bath powder
Badesalz *n* bath salt
Badezusatz *m* bath preparation
Badflüssigkeit *f* bath fluid
Badian *m* badian, star aniseed *(a condiment from Illicium verum Hook. fil.)*
badzementieren *s.* badaufkohlen
Baekeland-Verfahren *n* Baekeland process *(condensation of phenols and formaldehyde to yield resins)*
Baeyer-Spannung *f* Baeyer's angle strain
Baeyer-Villiger-Reaktion *f* Baeyer-Villiger oxidation *(of ketones into esters)*
Bagasse *f* [sugar cane] bagasse, begass[e], megass[e] *(remains of sugar cane)*
Bagassefeuerung *f* bagasse furnace
Bahn *f* 1. orbit *(of electrons surrounding a nucleus)*; path *(of particles leaving an atom)*; 2. *(pap, plast, text)* web
  Bohrsche B. Bohr orbit
  endlose B. *(plast)* web
  stabile (stationäre) B. stable orbit *(of an electron)*
Bahnabnahme *f* **[mit Oberfilz]** *(pap)* lick-up
Bahnabriß *m s.* Bahnriß
Bahndrehimpuls *m* moment of momentum
Bahnentrockner *m (pap, text)* sheeting (web) dryer
Bahnriß *m (pap)* break in the web, sheet break
  B. in der Naßpartie wet[-end] break
  B. in der Rollenschneidmaschine slitter break
  B. in der Trockenpartie dry break
Bahnspannung *f (pap)* sheet (web, paper) tension, pull (tension) of the web
Baikalit *m (min)* baikalite *(a variety of salite)*
Bainit *m (met)* bainite
  oberer B. upper bainite
  unterer B. lower bainite
Baisalz *n* bay (solar) salt
Bajonettkupplung *f* bayonet coupling (joint)
Baker-Nathan-Effekt *m s.* Hyperkonjugation
bakteriell bacterial
Bakterien *pl* / **aerobe** aerobic bacteria
  anaerobe B. anaerobic bacteria
  aromabildende B. aroma[-forming] bacteria
  denitrifizierende B. denitrifying bacteria, denitrifiers
  desulfurierende B. sulphur-reducing bacteria, sulphate reducers
  nitrifizierende B. nitrifying bacteria, nitrobacteria *(collectively for nitrite and nitrate bacteria)*
  säurebildende B. acid-forming (acid-producing) bacteria, acid-formers, acid-producers
  sporenbildende B. spore-forming (spore-producing) bacteria, spore formers
  stäbchenförmige B. rod-shaped bacteria
  stickstoffbindende B. nitrogen-fixing bacteria
  thermophile (wärmeliebende) B. thermophilic bacteria
bakteriendicht bacteria-tight
Bakteriendünger *m* bacterial manure
Bakterieneinwirkung *f* bacterial attack
Bakterienfarbstoff *m* bacterial pigment
Bakterienfärbung *f* bacteria staining
Bakterienfilter *n* bacteriological filter; germ-tight (germ-proofing) filter

Bakterienkultur *f* bacterial culture
Bakterienpräparat *n* bacteria preparation
Bakterientätigkeit *f* bacterial activity
bakterientötend bacteri[o]cidal
Bakterienwachstum *n* bacterial growth
bakterienwachstumshemmend bacteriostatic
Bakteriologie *f* bacteriology
bakteriologisch bacteriological
Bakteriolyse *f* bacteriolysis
Bakteriolysin *n* bacteriolysin
Bakteriophag[e] *m* [bacterio]phage
Bakteriostase *f* bacteriostasis
Bakteriostatikum *n* bacteriostat[ic]
bakteriostatisch bacteriostatic
bakterizid bacteri[o]cidal
Bakterizid *n* bactericide
äußerlich anwendbares B. external bactericide
Balasrubin *m (min)* balas ruby
Balata *f* balata *(rubber-like raw material from Mimusops balata Crueger)*
Baldrianöl *n* valerian oil *(from Valeriana officinalis L.)*
Baldriansäure *f* $CH_3[CH_2]_3COOH$ *n*-valeric acid, pentanoic acid
Baldriansäureäthylester *m* $CH_3[CH_2]_3COOC_2H_5$ ethyl *n*-valerate
Balg *m (rubber)* diaphragm, bladder
Balkenrührer *m* straight-arm paddle agitator
Balkenwaage *f* beam balance
gleicharmige B. equal-arm balance
ungleicharmige B. unequal-arm balance
Ballast[stoff] *m* impurity, diluent *(as of fuel); (food)* bulk
Ballastventilboden *m* ballast tray
ballen to ball [up]
sich b. to ball [up]
Ballen *m (rubber, text)* bale; pack
Ballenlisseuse *f (text)* bale backwashing unit
Ballenöffner *m (text)* bale breaker
Ballenspalter *m*, **Ballenspaltmaschine** *f* bale cutter (splitting machine, splitter)
Ballenzerteiler *m* agglomerate breaker, lump-breaker *(as in mixers)*
Ballformpresse *f (rubber)* ball moulding press
ballig bearbeiten to crown, to camber *(e. g. rolls, profiles)*
Balligkeit *f* crown *(of rolls and profiles)*
Balligwerden *n (text)* caking
Balling-Grad *m (sugar)* [degree] Balling
Ballon *m* balloon flask; demijohn; *(esp for acids)* carboy; *(rubber)* buld
Ballonausgießer *m* carboy pourer
Ballonentleerer *m* carboy emptier
Ballonkipper *m* carboy tipper
Ballpresse *f (rubber)* ball moulding press
Bally-Scholl-Reaktion *f* Bally-Scholl reaction
Balmer-Formel *f* Balmer formula
Balmer-Serie *f* Balmer series *(of the hydrogen spectrum)*
Balmer-Terme *mpl* Balmer terms
Balsam *m* balsam, balm
Indischer B. *s.* Peruanischer B.
Peruanischer B. Peru (black) balsam *(from Myroxylon balsamum (L.) Harms var. pereirae)*
Schwarzer B. *s.* Peruanischer B.
Balsamharz *n*, **Balsamkolophonium** *n* gum (common) rosin
Baly-Gefäß *n*, **Baly-Rohr** *n* Baly cell (tube) *(for measuring absorption)*
Bambus *m* bamboo, Bambusa Schreb.
Bananenbindung *f* bent bond *(C–C double bond)*
Banbury-Innenmischer *m* Banbury [mixer], intensive mixer
Banbury-Kneter *m s.* Banbury-Innenmischer
Banbury-Lancaster-Verfahren *n (rubber)* hot Banbury process, thermodynamic process *(a reclaiming process)*
Banbury-Mischer *m s.* Banbury-Innenmischer
Band *n* 1. band, tape; apron, belt, band, strand *(of a conveyor)*; 2. band *(spectroscopy)*; 3. strip chart *(of a recorder)*
erlaubtes B. *(phys chem)* allowed band
verbotenes B. *(phys chem)* forbidden band, energy gap
Bandabwurfwagen *m* tripper
bandagieren *(rubber)* to wrap
Bandantrieb *m* chart drive *(of a strip-chart recorder)*
Banddurchhang *m* belt sag
Bande *f* band *(of a spectrum)*
Bandeindruck *m* chain mark *(a defect in glass)*
Bandenfolge *f* band sequence *(of a spectrum)*
Bandengruppe *f* band group (system), series of bands *(of a spectrum)*
Bandenkante *f*, **Bandenkopf** *m* band edge (head) *(of a spectrum)*
Bandenmitte *f* origin (centre) of the band *(of a spectrum)*
Bandenreihe *f s.* Bandenfolge
Bandenspektrum *n* band (banded, molecular) spectrum
Bandensystem *n s.* Bandengruppe
Bändertheorie *f* **der Festkörper** band theory of solids
Bänderung *f (geol)* banding
Bandfilter *n* band (belt, linear belt) filter
Bandförderer *m* band (belt) conveyor
Bandheizkörper *m* band (strip) heater, heater band
Bandheizung *f* band (strip) heating
Bandklassierer *m* drag classifier
Bandmesser *n* band knife
Bandmesserspaltmaschine *f (tann)* band knife splitting machine
Bandmischer *m* ribbon mixer (blender)
Bandrührer *m* ribbon-blade agitator
Bandschleifenwagen *m* tripper
Bandschliff *m (glass)* banding
Bandschnecke *f* ribbon flight *(a screw conveyor)*
Bandschneckenmischer *m* ribbon mixer (blender)
Bandschreiber *m* strip-chart recorder
Bandspannung *f* belt tension

Bandtheorie *f* **der Festkörper** band theory of solids
Bandtrockner *m* band (belt, belt tunnel, apron, conveyor) dryer
Bandwaage *f* conveyor scale, [feed-]belt weigher, weigh[ing] belt
Bandwerkstoff *m* belting
Bandzellenfilter *n* travelling-pan filter, TP filter
Bank *f (geol)* bed, stratum; *(glass)* bench, siege *(of a pot furnace)*
Bankbürette *f* microburet[te]
Banknotenpapier *n* banknote (money) paper, *(Am)* currency paper
Bankpostpapier *n* bank paper (post), bond [paper]
Baptifolin *n* baptifoline *(alkaloid)*
Baptitoxin *n* cytisine, baptitoxine *(a lupine alkaloid)*
Baratte *f (text)* baratte, xanthator, [xanthating] churn
Barbados-Aloe *f (pharm)* Barbados (Curacao) aloe *(from Aloe vera L.)*
Barbier-Wieland-Reaktion *f* Barbier-Wieland degradation
Barbitalnatrium *n* barbital sodium
Barbiturat *n* barbiturate
Barbitursäure *f* barbituric acid, pyrimidinetrione
Barbotage *f* barbotage
Barcol-Härte *f* Barcol hardness
Bari-Sol-Verfahren *n* Barisol (Bari-Sol) process *(for deparaffinization)*
Baritflintglas *n* barium flint [glass]
Baritkronglas *n* barium crown [glass]
Barium *n* Ba barium
Bariumarsenat *n* $Ba_3(AsO_4)_2$ barium arsenate
Bariumarsenid *n* Ba_3As_2 barium arsenide
Bariumazetat *n* $Ba(CH_3COO)_2$ barium acetate
Bariumazetylid *n s.* Bariumkarbid
Bariumazid *n* $Ba(N_3)_2$ barium azide, barium hexanitride
Bariumborid *n* BaB_6 barium boride
Bariumbromat *n* $Ba(BrO_3)_2$ barium bromate
Bariumbromid *n* $BaBr_2$ barium bromide
Bariumchlorat *n* $Ba(ClO_3)_2$ barium chlorate
Bariumchlorid *n* $BaCl_2$ barium chloride
Bariumchromat *n* $BaCrO_4$ barium chromate
Bariumdichromat *n* $BaCr_2O_7$ barium dichromate
Bariumdiphosphat *n* $Ba_2P_2O_7$ barium diphosphate, barium pyrophosphate
Bariumdithionat *n* BaS_2O_6 barium dithionate
Bariumdivanadat(V) *n* $Ba_2V_2O_7$ barium divanadate(V)
Bariumflintglas *n* barium flint [glass]
Bariumformiat *n* $Ba(HCOO)_2$ barium formate
Bariumgetter *m* barium getter
bariumhaltig barytic
Bariumheptoxodivanadat(V) *n s.* Bariumdivanadat(V)
Bariumhexachloroplatinat(IV) *n* $Ba[PtCl_6]$ barium hexachloroplatinate(IV), barium chloroplatinate(IV)
Bariumhexafluorosilikat *n* $Ba[SiF_6]$ barium hexafluorosilicate, barium fluorosilicate
Bariumhexazyanoferrat(II) *n* $Ba_2(Fe(CN)_6]$ barium hexacyanoferrate(II), barium cyanoferrate(II)
Bariumhydrid *n* BaH_2 barium hydride
Bariumhydrogenarsenat *n* $BaHASO_4$ barium hydrogenarsenate
Bariumhydrogenorthophosphat *n* $BaHPO_4$ barium hydrogenorthophosphate, barium hydrogenphosphate
Bariumhydrogenphosphat *n s.* Bariumhydrogenorthophosphat
Bariumhydrogensulfid *n* $Ba(HS)_2$ barium hydrogensulphide
Bariumhydroxid *n* $Ba(OH)_2$ barium hydroxide
Bariumhypochlorit *n* $Ba(ClO)_2$ barium hypochlorite
Bariumhypophosphit *n* $Ba(PH_2O_2)_2$ barium hypophosphite, *(better)* barium phosphinate
Bariumjodat *n* $Ba(IO_3)_2$ barium iodate
Bariumjodid *n* BaI_2 barium iodide
Bariumkalzit *m (min)* bariumcalcite
Bariumkarbid *n* BaC_2 barium carbide
Bariumkarbonat *n* $BaCO_3$ barium carbonate
Bariumkronglas *n* barium crown [glass]
Bariumlack *m* barium lake
Bariummanganat(VI) *n* $BaMnO_4$ barium manganate
Bariummanganat(VII) *n s.* Bariumpermanganat
Bariummetasilikat *n* $BaSiO_3$ barium metasilicate, barium trioxosilicate
Bariummolybdat *n* $BaMoO_4$ barium molybdate
Bariumnitrat *n* $Ba(NO_3)_2$ barium nitrate
Bariumnitrid *n* Ba_3N_2 barium nitride
Bariumnitrit *n* $Ba(NO_2)_2$ barium nitrite
Bariumorthosilikat *n* Ba_2SiO_4 barium orthosilicate, barium tetraoxosilicate
Bariumoxid *n* BaO barium oxide
Bariumperchlorat *n* $Ba(ClO_4)_2$ barium perchlorate
Bariumpermanganat *n* $Ba(MnO_4)_4$ barium permanganate
Bariumperoxid *n* BaO_2 barium peroxide
Bariumperoxodisulfat *n* BaS_2O_8 barium peroxodisulphate
Bariumpyrophosphat *n s.* Bariumdiphosphat
Bariumpyrovanadat(V) *n s.* Bariumdivanadat(V)
Bariumrhodanid *n s.* Bariumthiozyanat
Bariumselenat *n* $BaSeO_4$ barium selenate
Bariumsulfat *n* $BaSO_4$ barium sulphate
  gefälltes B. precipitated barium sulphate
Bariumsulfid *n* BaS barium sulphide, barium monosulphide
Bariumsulfit *n* $BaSO_3$ barium sulphite
Bariumtellurat *n* $BaTeO_4$ barium tellurate
Bariumtetrachloroplatinat(II) *n* $Ba[PtCl_4]$ barium tetrachloroplatinate(II), barium chloroplatinate(II)
Bariumtetrasulfid *n* BaS_4 barium tetrasulphide
Bariumtetrazyanoplatinat(II) *n* $Ba[Pt(CN)_4]$ barium tetracyanoplatinate(II), barium cyanoplatinate(II)
Bariumtetroxosilikat *n s.* Bariumorthosilikat
Bariumthiosulfat *n* BaS_2O_3 barium thiosulphate

Bariumthiozyanat *n* ($Ba(SCN)_2$) barium thiocyanate, barium rhodanide
Bariumtrioxosilikat *n s.* Bariummetasilikat
Bariumtrisulfid *n* BaS_3 barium trisulphide
Bariumwolframat *n* $BaWO_4$ barium wolframate, barium tungstate
Bariumzyanid *n* $Ba(CN)_2$ barium cyanide
Barker-Turm *m (pap)* Barker tower *(milk-of-lime system)*
Barkometer *n (tann)* barkometer, barktrometer
Barkometerwert *m (tann)* Bk figure *(indicating the density of a solution)*
Barn *n (nucl)* barn *(a unit of area for measuring cross section)*
Barograf *m* barograph
Barometer *n* barometer
Barometerdruck *m* barometric pressure
Barometerformel *f* barometer formula
Barometerrohr *n* barometer tube
Barometrie *f* barometry
barometrisch barometric
Baroskampfer *m* Baros (Borneo, Sumatra) camphor *(from Dryobalanops aromatica Gaertn. f.)*
barotrop barotropic
Barré-Effekt *m (text)* barré effect *(a defect in weaving, knitting, or printing processes)*
Barren *m* [pressure] bar *(of a roller mill)*; *(met)* ingot, billet
Barrenguß *m* ingot casting
Barrenlänge *f* ingot length *(in zone melting)*
Barriere *f* barrier
Bart *m (tech)* burr *(as on castings)*
Bartgrasöl *n* citronella oil *(from Cymbopogon nardus (L.) Rendle and C. winterianus Jowitt)*
Bartlett-Kraft *f (nucl)* Bartlett force
Bart-Reaktion *f* Bart reaction *(for preparing aromatic arsonic acids)*
Barvoys-Verfahren *n* Barvoys process *(for cleaning coal)*
Barylith *m (min)* barylite *(barium beryllium disilicate)*
Baryon *n (nucl)* baryon, barion
Barysphäre *f (geol)* barysphere, siderosphere, earth's core (nucleus)
Baryt *m (min)* barite, baryte[s], heavy spar *(barium sulphate)*
Barytageschicht *f s.* Barytschicht
Barytbeton *m* baryte concrete
Baryterde *f s.* Bariumoxid
Barytgelb *n* baryta (ultramarine, Steinbühl) yellow, yellow ultramarine, lemon chrome, gelbin *(barium chromate)*
Barytokalzit *m (min)* barytocalcite *(barium calcium carbonate)*
Barytpapier *n* barite paper, baryta [coated] paper
Barytsalpeter *m (min)* nitrobarite *(barium nitrate)*
Barytschicht *f (phot)* baryta coating (layer)
Barytverfahren *n (sugar)* baryta (barium saccharate) process
Barytwasser *n* baryta water
Barytweiß *n* fixed (permanent) white, blanc fixe *(precipitated barium sulphate)*
Barytzinkweiß *n* zinc baryta white, Orr's white
Basalt *m* basalt
basalthaltig basaltic
basaltisch basaltic
Basaltware *f (ceram)* basalt ware
Basaltwolle *f* basalt wool
Base *f* base
  B. für Sulfitkochsäure *(pap)* bisulphite liquor base
  Fischersche B. Fischer base, 1,3,3-trimethyl-2-methyleneindoline
  korrespondierende B. conjugate base
  Millonsche B. $[Hg_2N]OH \cdot 2H_2O$ Millon's base
  organische B. organic base
  Schiffsche B. $ArN{=}CR_2$ Schiff base, *N*-arylimide, azomethine [compound]
  schwache B. weak base
  starke B. strong base
  stickstoffhaltige B. nitrogenous base
  Trögersche B. Tröger's base *(a doubly tertiary amine)*
Basenanhydrid *n* basic anhydride
Basenaustausch *m* base exchange
Basenaustauscher *m* base exchanger
Basenaustauscherenthärtung *f* base-exchange water softening
Basenaustauschfähigkeit *f* base-exchange capacity
Basenaustauschkapazität *f* base-exchange capacity
Basenbindungsvermögen *n* base-binding capacity (power)
Basengehalt *m* basicity *(of a solution)*; *(soil)* alkalinity
basengesättigt *(soil)* base-saturated
Basenkatalyse *f* base (basic) catalysis
  allgemeine B. general base catalysis
basenkatalysiert base-catalyzed
Basenmineralindex *m* base-mineral index
Basensättigung *f (soil)* base saturation
Basensättigungsgrad *m (soil)* degree of base saturation, base status, base-saturation percentage
Basenstärke *f* basic strength
Basenumtausch *m (soil)* base exchange
basenungesättigt *(soil)* base-unsaturated
Baseose *f (med)* alcalosis
Basilikumöl *n* basil oil *(from Ocimum basilicum L.)*
basiphil *(bot)* basophilous *(growing preferably in alkaline soils)*
Basis *f* base, basis; *(phys chem)* base region
  B. des Roherdöls base of crude petroleum
  asphaltische B. *(petrol)* asphalt base
  gemischte B. *(petrol)* mixed (intermediate) base
  naphthenische B. *(petrol)* naphthene base
  paraffinische B. *(petrol)* paraffin base
basisch basic, alkaline; *(met)* basic ✦ **b. stellen** to basify, to make alkaline, to alkal[in]ize
  b. ausgekleidet (zugestellt) *(met)* basic-lined
  schwach b. feebly basic, low-alkalinity
  stark b. strongly basic, high-alkalinity

Basischstellen *n* basification
Basisgebiet *n (phys chem)* base region
Basisgewicht *n s.* Masse je Flächeneinheit
Basisraum *m*, **Basisschicht** *f*, **Basiszone** *f (phys chem)* base region
Basit *m* basite, basic (subsilicic) rock
Basizität *f* alkalinity, basic strength, basicity *(of a solution)*; basic capacity *(of an acid)* ✦ **die B. schwächend** base-weakening
  B. nach Schorlemmer *(tann)* Schorlemmer basicity *(of chrome liquors)*
  aktuelle B. active alkalinity
  Freiberger B. *(tann)* Freiberg value for basicity *(of chrome liquors)*
Basizitätsbestimmung *f (tann)* precipitation figure test *(applied to chrome liquors)*
Basoid *n (soil)* basoid *(colloidal substance saturated with OH ions)*
basophil basophile, basophilic, basophilous *(having an affinity for basic dyes)*; *(bot)* basophilous *(growing preferably in alkaline soils)*
Basophiler *m (med)* basophil[e]
Bassin *n* basin, tank
Bassoragummi *n* gum bassora *(any of various low-grade kinds of gum tragacanth or other similar gums)*
Bassorin *n* bassorin *(a pectin-like substance obtained from certain gums)*
bastardisieren to hybridize *(chemical-bond theory)*
Bastardisierung *f* hybridization *(chemical-bond theory)*
  tetraedrische B. tetrahedral hybridization
  trigonale B. trigonal hybridization, sp^2 hybridization
sp^2-Bastardisierung *f s.* Bastardisierung / trigonale
Bastard-Kardamom *m(n)* bastard Siamese cardamom *(from Amomum xanthioides Wall.)*
Bastardorbital *n* hybrid [bond] orbital
Bastardstruktur *f* hybrid structure
Bastfaser *f* bast fibre
Bastfaserbündel *n* bast fibre bundle
Bastit *m (min)* bastite, schillerspar *(an inosilicate)*
Bastnäsit *m (min)* bastnaesite, bastnäsite *(cerium carbonate fluoride)*
Bastseide *f* raw (gum) silk, grege, greige
Batch *m (rubber)* batch
Batch-off-Vorrichtung *f (rubber)* batch-off equipment
Bathmetall *n* bath metal *(a copper zinc alloy)*
bathochrom bathochromic
Bathochromie *f* bathochromic shift (displacement)
Batholith *m (geol)* batholith, bathylith
  zusammengesetzter B. composite batholith
bathyal *(geol)* bathyal
Batikdruck *m* batik printing
Batikfärberei *f* batik dyeing
Batikpapier *n* batik paper
Batsch[ing]öl *n* batching oil *(for steeping jute fibres)*

Batterie *f* battery *(a group of uniform devices)*; bench *(a group of retorts in a coke oven)*; *(el chem)* [storage] battery
  galvanische B. voltaic battery
batteriegespeist battery-operated
Batteriekohle *f* battery carbon
Batteuse *f (cosmet)* batteuse *(agitator kettle for extractions)*
Batylalkohol *m* batyl alcohol, glycerol 1-octadecyl ether
Batzen *m* [large] lump; *(ceram)* blank, clot; *(pap)* knot
Bau *m* 1. structure, constitution *(of a molecule)*; set-up, structure, build *(of an apparatus)*; 2. *(act)* building, build[ing]-up, construction, structure, setting
Bauart *f* build, make, design
Baubestandteil *m s.* Bauelement
Bauchetikett *n* body label
Bauchstäuber *m* chest-type hand duster *(for pesticides)*
Baueinheit *f s.* Bauelement
Bauelement *n* structural element (entity, unit) *(as of a chemical compound)*
Bauer-Mühle *f* Bauer double-disk refiner
Baufehler *m (cryst)* structural defect
Bauformel *f* constitutional formula *(of molecules)*
Bauglas *n* structural glass
Baukalk *m* building lime
Baukastenprinzip *n* modular (building brick) principle
Baukeramik *f* building (structural) ceramics
Baumé-Grad *m* degree Baumé
Baumé-Skale *f* Baumé scale
Baumfärbeapparat *m (text)* beam dyeing machine
Baumfärbeautoklav *m (text)* beam autoclave
Baumfärberei *f (text)* beam dyeing
Baumkristall *m* dendrite
Baumwachs *n (agric)* grafting wax
Baumwollaffinität *f (dye)* affinity for cotton
Baumwollblau *n* cotton blue
Baumwolldichtung *f* cotton packing
Baumwolle *f* cotton
  egrenierte B. cotton lint
  tote B. dead cotton
  unreife B. dead cotton
  zyanäthylierte B. cyanoethylated cotton
Baumwollegreniermaschine *f* cotton gin
Baumwollentkörnungsmaschine *f* cotton gin
Baumwollfaden *m* cotton thread
Baumwollfarbstoff *m* cotton dye, *(Am also)* cotton color
Baumwollfaser *f* cotton fibre
Baumwollfaserdichtung *f* cotton fibre gasket
Baumwollfeinheit *f* cotton count
Baumwollfilz *m (pap)* cotton felt
  B. mit Asbestzusatz asbestos felt
Baumwollgewebe *n* cotton fabric (cloth)
Baumwollhadern *pl (pap)* cotton rags
Baumwollhalbstoff *m (pap)* cotton [rag] pulp
Baumwollinters *pl* [cotton] linters

Baumwollkalanderwalze *f (pap)* cotton bowl (roll) *(of a calender)*
Baumwollkapsel *f* cotton boll
Baumwollkernöl *n* cotton[seed] oil
Baumwollkord *m (rubber)* cotton cord
Baumwollkurzhaar *n* cotton fuzz
Baumwollpackung *f* cotton packing
Baumwollpapier *n* cotton paper
Baumwollsaathartfett *n* hydrogenated cotton[-seed] oil
Baumwollsaatkuchen *m* cotton cake
Baumwollsaatlezithin *n* cottonseed lecithin
Baumwollsaatöl *n*, **Baumwollsamenöl** *n* cotton[seed] oil
Baumwollstoff *m* cotton fabric (cloth)
Baumwolltrockenfilz *m (pap)* cotton dry[er] felt
Baumwollumpen *pl (pap)* cotton rags
Baumwollwachs *n* cotton wax
Baumwollwalze *f (pap)* cotton bowl (roll) *(of a calender)*
Baupappe *f* building [paper] board
Bauplatte *f* building (structural) board
Bauprinzip *n* building principle
Baur-Moschus *m* Baur musk *(synthetic musk)*
Bausand *m* builder's sand
bauschig *(text, pap)* bulky
  b. sein to bulk high
Bauschigkeit *f (text, pap)* bulkiness
Baustein *m* building block (stone), structural element (entity, unit) *(as of a chemical compound)*
Baustoff *m* building (construction) material
  keramischer B. ceramic building material
Bauterrakotta *f (ceram)* architectural terra-cotta
Bauxit *m* bauxite, beauxite
bauxitisch bauxitic
Bauzement *m* building cement
Bauziegel *m* building brick
Bayberrytalg *m* myrtle (bayberry, myrica) tallow (wax) *(from Myrica specc.)*
Bayer-Verfahren *n* Bayer process *(digestion of bauxite in sodium hydroxide solution)*
Bayöl *n* bay (myrcia) oil *(from Pimenta racemosa (Mill.) I. W. Moore)*
Bdellium *n* / **Indisches** Indian bdellium *(balsamic resin from Commiphora mukul Engl.)*
Beanspruchung *f* stress, strain, load
  dynamische B. dynamic stress
  schwingende B. vibrating stress
  sinusförmig schwingende B. waved stress
  statische B. static stress
Beanspruchungs-Dehnungs-Diagramm *n* stress-strain diagram
Beanspruchungsgeschwindigkeit *f* / **mittlere** mean rate of stressing
bearbeitbar workable
Bearbeitbarkeit *f* workability
bearbeiten to work, to treat, to process, to machine
  ballig b. to crown, to camber *(e.g. rolls, profiles)*
Bearbeitung *f* working, treatment, processing, machining
Bearbeitungsspannung *f* fabrication stress
Beattie-Bridgeman-Gleichung *f* Beattie and Bridgeman equation
Bebeerin *n*, **Bebirin** *n* curine, bebeerine *(alkaloid)*
bebrausen *(min tech)* to spray *(classified material)*
Bebrausen *n* **mit Wasser** water spraying
Bebrausungsdüse *f* spray [nozzle]
Becher *m* cup; *(lab)* beaker; *(tech)* bucket, scoop *(as of an elevator)*
  rotierender B. rotary (rotating) cup
  schnellrotierender B. spinning cup
Becheraufzug *m*, **Becherelevator** *m s.* Becherwerk
Becherfließzahl *f (plast)* moulding index, cup flow figure
Becherglas *n* beaker
Becherschließzeit *f s.* Becherfließzahl
Becherteilung *f* pitch of buckets
Becherversprüher *m* spinning-cup atomizer (sprayer), rotary-cup atomizer (sprayer)
Becherwerk *n* bucket elevator
  B. mit Becherstrang continuous-bucket elevator
  B. mit Einzelbechern spaced-bucket elevator
Becherwerkextrakteur *m* basket band extractor
  kombinierter B. rectangular basket extractor
  stehender B. vertical basket extractor
Becherwerkextraktor *m s.* Becherwerkextrakteur
Becherzeit *f s.* Becherfließzahl
Becherzerstäuber *m s.* Becherversprüher
Becken *n* basin, tank, pond, pool
Becker-Ofen *m* Becker oven
Beckmann-Thermometer *n* Beckmann thermometer
Bedeckungsgrad *m (phys chem)* degree of coverage
bedienen to operate, to run, to handle *(e.g. an apparatus)*
Bedienung *f* operation, running, handling *(as of an apparatus)*
Bedienungsbühne *f* operating floor (platform), bench
  koksseitige B. coke-side bench *(of a coke oven)*
  maschinenseitige B. pusher-side bench *(of a coke oven)*
bedruckbar printable
Bedruckbarkeit *f* printability, printing properties
bedrucken to print
Bedrucken *n* printing
  beidseitiges B. *(text)* double-face printing
Bedruckstoff *m* stock *(in graphic industry)*
beeinflussen to influence, to affect, *(esp negatively)* to interfere with
  störend b. to interfere with
Beeinflussung *f* influence, affection, *(esp negatively)* interference
beeinträchtigen to interfere with
beenden to finish, to complete; to terminate, to break *(the growth of a chain molecule)*
Beendigung *f* finishing, completion; termination *(of*

chain growth)
beeteln *(text)* to beetle *(to produce a mangle effect)*
Beetle-Maschine *f (text)* beetling machine, beetler
befeuchten to moisten, to wet, to damp[en], to humidify, to water, to dew; *(pap)* to wet out (up)
Befeuchter *m* moistener, humidifier *(specif for gummed surfaces)*; *(pap)* wetting machine
Befeuchtung *f* moistening, wetting, damp[en]ing, humidification, watering, dewing; *(pap)* wetting-up, wetting-out
Befeuchtungsanlage *f* humidifier
Befeuchtungsapparat *m* humidifier
Befeuchtungsdüse *f* humidifying nozzle
Befeuchtungsmaschine *f (text)* damping machine
befeuern to fire, to heat
  direkt b. to direct-fire
beflecken to stain
beflocken *(plast)* to flock
Beflocken *n (plast)* flocking, flock spraying
befreien to liberate *(from contaminants)*
  von flüchtigen Bestandteilen b. to devolatilize
  von Kationen b. to decationize
Befreiung *f* liberation *(from contaminants)*
Befund *m* result
  röntgenografischer B. X-ray result
begasen *(agric, food)* to fumigate, to gas
Begasse *f s.* Bagasse
Begasung *f (agric, food)* fumigation, gassing, exposure to gas
Begasungsfilter *n* air stone; *(if tubular)* gas dispersion (distribution) tube
Begasungsmittel *n (agric, food)* fumigant
Begasungsröhrchen *n* gas dispersion (distribution) tube
begichten to burden, to charge *(a blast furnace)*
Begichtung *f* charging *(of a blast furnace)*
Begichtungsöffnung *f* throat
begießen to water, to irrigate; *(ceram)* to engobe
Beginn-Kochpunkt *m*, **Beginn-Siedepunkt** *m* initial boiling point, I. B. P.
Begleitalkaloid *n* companion alkaloid
Begleiter *m s.* Begleitstoff
Begleitreaktion *f* concurrent reaction
Begleitstoff *m*, **Begleitsubstanz** *f* companion [substance], accompanying substance, admixture
Beguß *m*, **Begußmasse** *f (ceram)* engobe
Begußton *m (ceram)* slip clay
Behälter *m* vessel, receptacle, holder, tank, bin, silo, basin, reservoir, *(funnel-shaped)* hopper, *(esp for shipping goods)* container
  unterirdischer B. underground reservoir (storage tank)
Behälterpappe *f* container board
behandeln to treat, to process
  anodisch b. *(met)* to anodize
  chemisch b. to treat with chemicals, to chemicalize
  die Oberfläche b. to surface, to finish
  im Autoklaven b. to autoclave
  im Faß b. *(tann)* to drum
  in der Küpe b. *(dye)* to vat
  in Kleienbeize (Schrotbeize) b. *(tann)* to drench
  mit Arsen b. to arsenicate
  mit Bleicherde b. to clay
  mit Bromat b. to bromate
  mit Chlor b. to chlorinate, to chlorinize
  mit Chlorid b. to chloridize
  mit Chlorwasserstoff b. to hydrochlorinate
  mit Dampf b. to steam
  mit dem Dephlegmator b. to dephlegmate
  mit Formaldehyd b. to formolize, to formalinize
  mit Gips b. to plaster *(wine)*; to burtonize, to gypsum *(brewing water)*
  mit Hitze b. to heat-treat; to bake
  mit Kalk b. to lime out *(manufacture of organic intermediates)*
  mit Kopfdünger b. *(agric)* to topdress
  mit Lake b. to brine
  mit Lehm b. to clay
  mit Ozon b. to ozonize, to ozonify
  mit Salzlake (Salzlösung) b. *(food, tann)* to brine
  mit Säure b. to treat with acid; *(tann)* to drench
  mit Schwefel b. to sulphur
  mit schwefliger Säure b. *(sugar)* to sulphite
  mit Silikonen b. to silicone-treat, to siliconize
  nochmals (wiederholt) b. to re-treat
Behandlung *f* treatment, processing
  B. im Faß *(tann)* drumming
  B. mit Bleicherde clay treatment
  B. mit Chlorwasserstoff hydrochlorination
  B. mit Doktorlauge (Doktorlösung) *(petrol)* doctor treatment (sweetening)
  B. mit Lake (Salzlake, Salzlösung) *(food, tann)* brining
  B. mit Säure acid treatment
  B. mit schwefliger Säure *(sugar)* sulphitation
  B. mit Silikonen siliconization, siliconizing, silicone treatment
  B. nach dem Anstreichverfahren brush treatment *(wood preservation)*
  B. nach dem Furnos-Verfahren Furnos treatment *(wood preservation)*
  B. nach dem Sprühverfahren spray treatment *(wood preservation)*
  B. ohne Druckanwendung *s.* drucklose B.
  B. vor dem Auspflanzen preplanting treatment *(of a soil with pesticides)*
  B. während der Winterruhe dormant treatment *(of plants with pesticides)*
  anodische B. *(met)* anodizing, anodization, anodic treatment (oxidation, coating)
  chemische B. treatment with chemicals, chemicalization
  drucklose B. non-pressure treatment *(wood preservation)*
  enzymatische B. *s.* fermentative B.
  fermentative B. enzyme treatment
  nochmalige B. re-treatment
  schmutzabstoßende (schmutzabweisende) B. *(text)* dirt-repellent treatment
  thermische B. heat treatment; baking

thermische B. mit Promotor *(rubber)* promoted heat treatment
wiederholte B. re-treatment
Behandlungsgefäß *n* treater
Behandlungskammer *f (plast)* plenum chamber
Behandlungsverfahren *n* treating process
Beharrungsvermögen *n* inertia
Beharrungszustand *m* steady (stationary) state
beharzen *(plast)* to resin
beheizbar heatable
beheizen to heat, to fire, to warm [up]
beheizt/direkt direct-fired
indirekt b. indirect-fired
mit Dampf b. steam-heated
mit Öl b. oil-fired, oil-heated
Beheizung *f* heating, firing, warming[-up]
B. durch Bogenentladung electrical-discharge heating
B. durch Elektronenbeschuß electron-bombardment heating
B. mit Gas gas-fired heating
B. mit Kohle coal-fired heating
B. mit Öl oil-fired heating
dielektrische B. dielectric (electronic, radio-frequency) heating
direkte B. direct heating
elektrische B. electric heating
indirekte B. indirect (external) heating
induktive B. induction heating
Beheizungsapparat *m* heater
Beheizungsgas *n* fuel (heating) gas
Behen-Öl *n* ben (behen) oil, oil of ben *(from Moringa aptera Gaertn., less frequently from M. oleifera Lam.)*
Behenolsäure *f* $CH_3[CH_2]_7C{\equiv}C[CH_2]_{11}COOH$ behenolic acid, 13-docosynoic acid
Behensäure *f* $CH_3[CH_2]_{20}COOH$ docosanoic acid, behenic acid
Behenylalkohol *m* $CH_3[CH_2]_{20}CH_2OH$ docosyl alcohol, behenyl alcohol
Behinderung *f* hindrance, inhibition
B. der freien Drehbarkeit einer Bindung bond hindrance (inhibition) *(chemical-bond theory)*
sterische B. steric hindrance (inhibition) *(chemical-bond theory)*
beidseitigglatt *(pap)* glazed on both sides
Beigeschmack *m* foreign flavour (taste)
Beilstein-Probe *f* Beilstein's test
beimengen to admix
Beimengen *n* admixture
Beimengung *f (act)* admixture; *(substance)* admixture, impurity; solute *(zone melting)*
gesundheitsschädliche B. deleterious impurity
beimischen to admix
Beimischung *f (act)* admixture; *(substance)* admixture, impurity
beimpfen to inoculate *(e. g. a fermentation tank)*
Beimpfung *f* inoculation *(as of a fermentation tank)*
Beinschwarz *n* bone (animal) black
Beiprodukt *n* by-product, coproduct
beißend acrid, pungent *(taste, smell)*
Beistoff *m* inert ingredient; corrective *(in building up active-substance mixtures)*
Beiwert *m*, **Beizahl** *f* index; factor; coefficient
Beizbad *n (met)* scouring bath
Beizbehandlung *f* mit Säure *(tann)* drenching
Beize *f* 1. mordant *(for treating textiles or microscopic preparations)*; stain *(for treating rubber, glass, or wood)*; *(tann)* bate; *(agric)* pickle; *(met)* scouring agent, pickle; 2. *s.* Beizen
basische B. *(text)* metallic mordant
beizen to mordant *(textiles or microscopic preparations)*; to stain *(rubber, glass, or wood)*; *(tann)* to bate; *(agric)* to pickle; *(met)* to pickle, to scour *(for removal of scale)*
mit Schwefelsäure b. *(met)* to vitriol
Beizen *n* mordanting *(of textiles or microscopic preparations)*; staining *(of rubber, glass, or wood)*; *(tann)* bating; *(agric)* pickling; *(met)* pickling, scouring *(for removal of scale)*
Beizenfärberei *f (text)* mordant dyeing
Beizenfarbstoff *m (text)* mordant dye[stuff], adjective dye
Beizenverfahren *n (text)* chromate process, chromate [dyeing] method
Beizereiabwasser *n (met)* pickling wastes
Beizmittel *n*, **Beizstoff** *m s.* Beize 1.
Beizung *f s.* Beizen
Bekämpfung *f* / **biologische** biological control
Bekämpfungsmaßnahme *f (agric)* control measure
bekleben *(pap)* to line, to laminate, to paste, to paper
Beklebepapier *n* liners, lining paper, pasting [paper]
beklebt / einseitig *(pap)* single-lined
zweiseitig b. double-lined
Bekleidungsleder *n* garment leather
beklopfen to rap
beladen to charge, to feed, to load
mit Benzol b. to benzolize
beladen / mit Kohlenstoff fouled with carbon *(catalyst)*
mit Koks b. coke-contaminated *(catalyst)*
Beladeöffnung *f* charging (filling) hole (door), feed hole (inlet)
Beladung *f* 1. charging, feeding, loading; 2. load, content; *(of liquids)* concentration
Belag *m* coat[ing], cover[ing], layer, *(if thin)* film; *(esp of alien substance)* scale, encrustation; overlay *(for wood)*; *(food)* bloom
deckender B. coverage *(as of insecticides)*
reflexmindernder B. antiflare (antireflection) coating
sekundärer B. secondary deposit *(of pesticides)*
Belastbarkeit *f (tech)* load-bearing capacity; *(distil)* loading capacity
B. eines Lagers bearing capacity (strength)
belasten 1. to load, to weight; *(text, pap)* to weight *(with fillers or sizing material)*; 2. to stress; *(distil)* to load *(a column)*
Belastungsschaumzahl *f (text)* lather value in presence of dirt

Belastungsverhalten *n* loading behaviour
Belastungswiderstand *m* load resistance
Beleben *n* activating *(floatation)*
Beleber *m* activator *(floatation)*
Belebtschlamm *m* activated (active, biological) sludge *(waste-water treatment)*
Belebtschlammanlage *f* activated-sludge plant
Belebtschlammbecken *n* activated-sludge tank (chamber)
Belebtschlammverfahren *n* activated-sludge process (method)
Belebungs ... *s.* Belebtschlamm ...
belegen to cover, to coat; *(rubber)* to skim[coat] *(frictioned tissue)*
beidseitig (zweiseitig) b. to double-coat
Belegen *n* covering, coating; *(rubber)* skim coating
Beleuchtungsmittel *n* illuminant
Beleuchtungsstärke *f* illuminance
belichten *(phot)* to expose [to light], to irradiate
Belichtung *f (phot)* exposure [to light], irradiation
✦ **bei B.** on exposure to light, under illumination
lange B. *(phot)* prolonged exposure
Belichtungsbereich *m (phot)* range of exposure
Belichtungsbreite *f (phot)* exposure latitude (range)
Belichtungsdauer *f (phot)* exposure time
Belichtungsschleier *m (phot)* optical fog
Belichtungsskale *f (phot)* exposure scale
Belichtungsspielraum *m*, **Belichtungsumfang** *m s.* Belichtungsbreite
Belichtungszeit *f (phot)* exposure time, duration of exposure
Belit *m* belite *(a crystal type in portland cement clinker)*
Belladonnaalkaloid *n* belladonna alkaloid
Belleek-Porzellan *n* Belleek china
Bell-Verfahren *n (met)* Bell process *(removal of P and Si by iron oxide)*
belüften to aerate
Belüftung *f* aeration
Belüftungsanlage *f s.* Belüftungssystem
Belüftungsbecken *n* aeration (aerated) pond *(waste-water treatment)*
Belüftungshahn *m* aeration cock
Belüftungskolben *m* aeration flask
Belüftungsleitung *f* aeration line
Belüftungsmittel *n (build)* air-entraining additive (admixture, compound, agent)
Belüftungssystem *n* aeration (ventilation) system; aeration (air-diffusion) system *(waste-water treatment)*
Belüftungstank *m* aeration tank *(waste-water treatment)*
Bemalbarkeit *f* paintability
bemessert *(pap)* equipped (fitted) with bars (knives)
Bemesserung *f (pap)* filling, tackle
benachbart adjacent, neighbouring, vicinal *(substituents)*
Bence-Jones-Eiweißkörper *m* Bence-Jones protein
Bender-Prozeß *m (petrol)* Bender (lead-sulphide) process *(for sweetening distillates)*
Benedict-Metall *n* Benedict metal *(a copper-nickel alloy)*
Benedict-Nickel *n* Benedict nickel *(an alloy consisting of Zn, Ni, Pb, and Sn)*
benennen *(nomencl)* to name
Benennung *f (nomencl)* naming, denomination; name, term
nach den IUPAC-Regeln gebildete B. IUPAC name
nach der Genfer Nomenklatur gebildete B. Geneva name
systematische B. systematic name
unsystematische B. unsystematic name
Benennungssystem *n (nomencl)* naming system
benetzbar wettable, hydrophilic, hydrophile
leicht b. easily wetted
nicht b. non-wettable, hydrophobic, hydrophobe
Benetzbarkeit *f* wettability
benetzen to perfuse, *(with water also)* to wet, to moisten, to humidify, to dew, to water, to damp[en]; *(of a liquid)* to suffuse
Benetzung *f* perfusion, *(with water also)* wetting, moistening, humidification, dewing, watering, damp[en]ing
vollkommene B. complete wetting
Benetzungsfähigkeit *f* wetting power (ability)
Benetzungskoeffizient *m* spreading coeffient, SC
Benetzungsmittel *n* wetting agent (aid), wetter
Benetzungsspannung *f* wetting tension
Benetzungsverfahren *n (agric)* steeping method *(for seed protection)*
Benetzungsvermögen *n* wetting power (ability)
Benetzungswärme *f* heat of wetting
Bengalkino *n* Bengal (butea) gum *(from Butea superba Roxb.)*
Bengough-Stuart-Verfahren *n* Bengough-Stuart process, chromic-acid [anodizing] process
Benitoit *m (min)* benitoite *(a cyclosilicate)*
Ben-Öl *n* oil of ben *(from Moringa aptera Gaertn., less frequently from M. oleifera Lam.)*
Benoxinat *n* benoxinate
Bentonit[ton] *m* bentonite [clay]
Benzalazeton *n* $C_6H_5CH{=}CHCOCH_3$ benzalacetone, 4-phenyl-3-buten-2-one
Benzalchlorid *n* $C_6H_5CHCl_2$ benzal chloride, αα-dichlorotoluene, benzylidene chloride
Benzaldehyd *m* C_6H_5CHO benzaldehyde
Benzalgrün *n* malachite (benzal) green
Benzamid *n* $C_6H_5CONH_2$ benzamide
Benzaminsäure *f* $C_6H_4(NH_2)COOH$ *m*-aminobenzoic acid
2,3-Benzanthrazen *n* 2,3-benzanthracene, naphthacene
Benzanthron *n* benzanthrone
Benzanthronchinolin *n* benzanthronequinoline
Benzanthronfarbstoff *m* benzanthrone dye
Benzanthronreihe *f* benzanthrone series
Benzen *n s.* Benzol
Benzeniumion *n* phenonium ion, benzene carbonium ion

Benzidin *n* $NH_2C_6H_4C_6H_4NH_2$ benzidine, 4,4'-diaminobiphenyl
Benzidinbase *f* benzidine base
Benzidinprobe *f* benzidine test
Benzidinumlagerung *f* benzidine conversion (rearrangement, transformation)
  halbe (halbseitige) B. semidine rearrangement (transformation)
Benzil *n* $C_6H_5COCOC_6H_5$ benzil, bibenzoyl
Benzil-2,2'-dikarbonsäure *f* benzil-2,2'-dicarboxylic acid
Benzilsäure *f* $(C_6H_5)_2C(OH)COOH$ benzilic acid
Benzilsäureumlagerung *f* benzilic-acid rearrangement
Bis-**Benzimidazolaufheller** *m* *bis*-benzimidazole brightener
Benz-in *n* C_6H_4 benzyne
Benzin *n* *(chem)* benzin[e]; *(as a motor fuel)* gasoline, petrol, [motor] spirit; *(esp for technical purposes or as a reformer feedstock)* [petroleum] naphtha
  butanfreies B. debutanized gasoline
  direkt herausdestilliertes B. straight-run gasoline, distillate gasoline, straight-run benzine, S. R. B.
  gebleites B. leaded (ethyl) gasoline
  gesüßtes B. sweet gasoline
  hochklopffestes B. high-octane gasoline
  hochoktaniges (hochoktanzahliges) B. high-octane gasoline
  instabiles B. unstabilized (unstable) gasoline, *(Am also)* wild gasoline
  klopffestes B. antiknock gasoline
  leichtes B. gasoline *(with boiling range 30 to 100 °C)*, light gasoline (benzine, spirit, naphtha)
  mit Tetraäthylblei versetztes B. *s.* gebleites B.
  reformiertes B. reformed gasoline
  saures B. sour gasoline
  schweres B. heavy gasoline *(boiling range 150 to 210 °C)*
  stabiles (stabilisiertes) B. stabilized (stable) gasoline
  süßes B. sweet gasoline
  unstabiles (unstabilisiertes) B. *s.* instabiles B.
  verbleites B. *s.* gebleites B.
  wildes B. *s.* instabiles B.
Benzinadditiv *n* gasoline additive
Benzinbereich *m* gasoline range
Benzindampf *m* gasoline vapour
Benzingewinnungsanlage *f* gasoline plant
Benzinraffination *f* gasoline refining
Benzinrückgewinnung *f* *(petrol)* naphtha recovery
Benzinsiedebereich *m* gasoline range
Benzinwäscher *m* *(petrol)* naphtha wash tower
Benzoat *n* benzoate
Benzoazurin *n* benzoazurin[e]
5,6-Benzochinolin *n* benzo*[f]*quinoline, 5,6-benzoquinoline
7,8-Benzochinolin *n* benzo*[h]*quinoline, 7,8-benzoquinoline
p-**Benzochinon** *n* $O{=}C_6H_4{=}O$ *p*-benzoquinone, quinone *(proper)*, 1,4-cyclohexadienedione
Benzodiazin *n* benzodiazine
Benzoe *f*, **Benzoeharz** *n* benzoin, benzoin (Benjamin) gum (resin) *(from Styrax specc.)*
Benzoesäure *f* C_6H_5COOH benzoic acid, benzene carboxylic acid
Benzoesäureanhydrid *n* $(C_6H_5CO)_2O$ benzoic anhydride
Benzoesäureäthylester *m* $C_6H_5COOC_2H_5$ ethyl benzoate
Benzoesäurebenzylester *m* $C_6H_5COOCH_2C_6H_5$ benzyl benzoate
Benzoesäuremethylester *m* $C_6H_5COOCH_3$ methyl benzoate
Benzoesäurephenylester *m* $C_6H_5COOC_6H_5$ phenyl benzoate
o-**Benzoesäuresulfimid** *n* *o*-sulphobenzoic imide, saccarin
Benzofuran *n* benzofuran, coumarone, cumarone
benzoid benzenoid
Benzoin *n* $C_6H_5{-}CH(OH){-}COC_6H_5$ benzoin, α-hydroxybenzyl phenyl ketone
Benzoinkondensation *f* benzoin condensation
Benzol *n* C_6H_6 benzene, *(esp commercially)* benzole, benzol ✦ **mit B. anreichern (beladen, sättigen)** to benzolize
  anorganisches B. $B_3N_3H_6$ inorganic benzene, borazole, triborine triamine
  technisches B. commercial benzole
  vordestilliertes B. once-run benzole
90er Benzol *n* 90's benzole
Benzolabkömmling *m* benzene derivative
Benzolabscheider *m* benzole separator
Benzolabtreiber *m* benzole still
benzolähnlich benzene-like
Benzolanlage *f* benzole plant
Benzolazimid *n* benzeneazimide, benzotriazole, aziminobenzene
Benzolboronsäure *f* $C_6H_5B(OH)_2$ benzeneboronic acid
Benzoldampf *m* benzene vapour
Benzolderivat *n* benzene derivative
Benzoldestillieranlage *f* benzole still
Benzoldiazoanilid *n* $C_6H_5N{=}NNHC_6H_5$ benzenediazoanilide, diazoaminobenzene, 1,3-diphenyltriazen
Benzoldiazoniumchlorid *n* $[C_6H_5N{\equiv}N]Cl$ benzenediazonium chloride
Benzol-*m*-dikarbonsäure *f* $C_6H_4(COOH)_2$ benzene-*m*-dicarboxylic acid, *m*-phthalic acid, isophthalic acid
Benzol-*o*-dikarbonsäure *f* $C_6H_4(COOH)_2$ benzene-*o*-dicarboxylic acid, *o*-phthalic acid, phthalic acid
Benzol-*p*-dikarbonsäure *f* $HOOCC_6H_4COOH$ benzene-*p*-dicarboxylic acid, *p*-phthalic acid, terephthalic acid
Benzoldruckextraktion *f* benzene-pressure extraction
benzolgesättigt benzolized
Benzolgewinnung *f* benzene (benzole) recovery
Benzolhexachlorid *n* $C_6H_6Cl_6$ benzene hexachloride, BHC, hexachlorocyclohexane

Benzolhexakarbonsäure *f* $C_6(COOH)_6$ benzenehexacarboxylic acid, mellitic acid
Benzolkarbonsäure *f* benzene carboxylic acid *(specif)* C_6H_5COOH benzoic acid
Benzolkern *m* benzene ring (nucleus)
Benzolkohlenwasserstoffe *mpl* benzene (aromatic) hydrocarbons, aromatics
Benzolkondensator *m* benzene (benzole) condenser
benzollöslich soluble in benzene
Benzolmonosulfonsäure *f s.* Benzolsulfonsäure
Benzolphosphonsäure *f* $C_6H_5PO(OH)_2$ benzene phosphonic acid
Benzolpumpe *f* benzene (benzole) pump
Benzolreihe *f* benzene series
Benzolring *m* benzene ring (nucleus)
Benzol-Schwefeldioxid-Verfahren *n* sulphur dioxide-benzole process *(for dewaxing petroleum)*
Benzolsulfinsäure *f* $C_6H_5SO_2H$ benzenesulphinic acid
Benzolsulfonamid *n* $C_6H_5SO_2NH_2$ benzene sulphonamide
Benzolsulfonsäure *f* $C_6H_5SO_3H$ benzenesulphonic acid
Benzolsulfonsäureamid *n s.* Benzolsulfonamid
Benzoltetrakarbonsäure *f* $C_6H_2(COOH)_4$ bezenetetracarboxylic acid
Benzolthermometer *n* benzole thermometer
 B. nach Casella Casella benzole thermometer
Benzoltrikarbonsäure *f* $C_6H_3(COOH)_3$ benzenetricarboxylic acid
benzolunlöslich insoluble in benzene
Benzolvorerzeugnis *n s.* Benzolvorprodukt
Benzolvorlauf *m* benzole forerunnings
Benzolvorprodukt *n* once-run benzole
Benzolwäscher *m* benzole washer (scrubber)
Benzolwaschöl *n* benzole wash (absorbing) oil
Benzonitril *n* C_6H_5CN benzonitrile, cyanobenzene
Benzoorange R *n* benzoorange R, direct orange 8
Benzopersäure *f* $C_6H_5CO-O-OH$ perbenzoic acid
Benzophenanthren *n* benzophenanthrene
Benzophenon *n* $C_6H_5COC_6H_5$ benzophenone, benzoyl benzene, diphenyl ketone
Benzopyrazin *n* benzpyrazine, quinoxaline, 1,4-benzodiazine
2,3-Benzopyridin *n* 2,3-benzpyridine, quinoline, 1-benzazine
3,4-Benzopyridin *n* 3,4-benzpyridine, isoquinoline, 2-benzazine
Benzopyrimidin *n* benzpyrimidine
Benzopyron *n* benzopyrone
2,3-Benzopyrrol *n* 2,3-benzpyrrole, indole
Benzotriazol *n* benzotriazole
Benzotrichlorid *n* $C_6H_5CCl_3$ benzotrichloride, α,α,α-trichlorotoluene
Benzoylameisensäure *f* $C_6H_5COCOOH$ benzoylformic acid, phenylglyoxylic acid
Benzoylaminoessigsäure *f s.* Benzoylglyzin
Benzoylbenzol *n s.* Benzophenon
Benzoylglykokoll *n s.* Benzoylglyzin
Benzoylglyzin *n* $C_6H_5CONHCH_2COOH$ benzoylglycine, benzoylaminoacetic acid, hippuric acid
Benzoylgrün *n s.* Benzalgrün
Benzoylgruppe *f* C_6H_5CO-benzoyl group (residue)
Benzoylhydroperoxid *n s.* Benzopersäure
benzoylieren to benzoylate
Benzoylierung *f* benzoylation
 zweifache B. dibenzoylation
Benzoyl-I-Säure *f*, **Benzoyl-J-Säure** *f* benzoyl J acid
Benzoylnaphthalin *n* $C_{10}H_7COC_6H_5$ benzoylnaphthalene, naphthyl phenyl ketone
Benzoylperoxid *n* $C_6H_5-CO-O-O-CO-C_6H_5$ benzoyl peroxide, dibenzoyl peroxide
Benzoylphenylkarbinol *n s.* Benzoin
Benzoylrest *m s.* Benzoylgruppe
Benzphenanthren *n* benzophenanthrene
Benzpyren *n* benzopyrene
Benzylalkohol *m* $C_6H_5CH_2OH$ benzyl alcohol
Benzylazetat *n* $CH_3COOCH_2C_6H_5$ benzyl acetate
Benzylbenzoat *n* $C_6H_5CO-OCH_2C_6H_5$ benzyl benzoate
Benzylbenzol *n* $CH_2(C_6H_5)_2$ benzylbenzene, diphenylmethane
Benzylbutyrat *n* $CH_3CH_2CH_2CO-OCH_2C_6H_5$ benzyl butyrate
Benzylchlorid *n* $C_6H_5CH_2Cl$ α-chlorotoluene, benzyl chloride
Benzylessigsäure *f* $C_6H_5CH_2CH_2COOH$ benzylacetic acid, hydrocinnamic acid, 3-phenylpropionic acid
Benzylglyoxylsäure *f s.* Phenylbrenztraubensäure
Benzylgruppe *f* $C_6H_5CH_2-$ benzyl group (residue)
Benzylidenchlorid *n* $C_6H_5CHCl_2$ $\alpha\alpha$-dichlorotoluene, benzylidene chloride, benzal chloride
Benzylisochinolin *n* benzylisoquinoline
Benzylkarbinol *n s.* 2-Phenyläthanol
Benzylpenaldinsäure *f* benzylpenaldic acid, penaldic-G acid
Benzylpenillosäure *f* benzylpenilloic acid, penilloic-G acid
Benzylpenillsäure *f* benzylpenillic acid, penillic-G acid
Benzylpenizillin *n* benzylpenicillin
Benzylpenizilloinsäure *f* benzylpenicilloic acid, penicilloic-G acid
Benzylphenol *n* $C_6H_5CH_2-C_6H_4OH$ benzyl phenol
Benzylphenylkarbinol *n s.* 1,2-Diphenyläthanol
Benzylpropionat *n* $C_2H_5CO-OCH_2C_6H_5$ benzyl propionate
Benzylradikal *n* benzyl radical, *(specif)* free benzyl radical
Benzylrest *m s.* Benzylgruppe
Benzylsalizylat *n* $HOC_6H_4CO-OCH_2C_6H_5$ benzyl salicylate
Benzylzellulose *f* benzyl cellulose
Benzylzinnamat *n* $C_6H_5CH=CHCO-OCH_2C_6H_5$ benzyl cinnamate, cinnamein
Benzylzyanid *n* $C_6H_5CH_2CN$ benzyl cyanide, ω-cyanotoluene, phenylacetonitrile
Benzyn *n* C_6H_4 benzyne
beobachten to observe, to watch, to study
Beobachtung *f* observation
Beobachtungsfehler *m* observational error

Beobachtungsrohr *n (lab)* observation tube
Berberin *n* berberine *(an isoquinoline alkaloid)*
Berechnung *f* **von Trennstufe zu Trennstufe** *(distil)* tray-to-tray calculation (procedure)
Beregnungsdüngung *f* dressing by spray irrigation
Beregnungsprüfung *f*, **Beregnungsversuch** *m (text)* rain test
Bereich *m* 1. region, range *(as of measurement or state)*; 2. *(cryst)* domain *(in ferromagnetic substances)*; 3. sphere *(as of a science)*
kristalliner B. crystalline region
plastischer B. plastic range
bereiten to prepare, to make [ready, up]
bereitet / frisch freshly prepared
Bereitung *f* preparation, making
Berg *m* peak *(as of a chromatogram)*
Bergamottöl *n* bergamot oil *(from Citrus aurantium L. ssp. bergamia)*
Bergbausprengstoff *m* mining explosive (powder)
Bergblau *n* verditer blue *(a basic copper carbonate)*
Bergbreite *f* peak width *(as in a chromatogram)*
Berge *pl (min tech)* tail[ings], waste tailing, refuse
Bergeaustrag *m (min tech)* tailings (refuse) discharge (extraction)
Bergeaustragsöffnung *f (min tech)* tailings-discharge (refuse-discharge) port
Berggrün *n* malachite green *(ground malachite or similar pigment made synthetically)*
Bergius-Hochdruckverfahren *n*, **Bergius-Hydrierverfahren** *n* Bergius [hydrogenation] process
Bergkork *m (min)* mountain cork *(an asbestos)*
Bergkristall *m (min)* rock crystal *(a variety of quartz)*
Bergkupfer *n* native copper
Bergleder *n (min)* mountain leather *(an asbestos)*
Bergmann-Serie *f* Bergmann series *(spectroscopy)*
Bergmilch *f (min)* rock milk, agaric mineral *(calcium carbonate)*
Bergtalg *m*, **Bergwachs** *n (min)* ozokerite, earth (ader) wax, native paraffin
berichtigen to correct
berieseln to sprinkle, to spray; to scrub *(gases)*
Berieselung *f* **zur Staubbindung** *(coal)* dust proofing
Berieselungsgefrierverfahren *n (food)* spray freezing process
Berieselungskondensator *m* atmospheric condenser
Berieselungskühler *m* spray cooler
Berieselungsverflüssiger *m* atmospheric condenser
Berkefeld-Filter *n* Berkefeld filter
Berkelium *n* Bk berkelium
Berl-Sattel[körper] *m* Berl saddle *(a filling body)*
Bernstein *m* amber, succinite
Bernsteinlack *m* amber varnish
Bernsteinöl *n* amber oil
Bernsteinsäure *f* $HOOC-CH_2CH_2-COOH$ succinic acid, butanedioic acid
Bernsteinsäuredehydrogenase *f* succinodehydrogenase, succinic [acid] dehydrogenase
Bernsteinsäuredialdehyd *m* $OHC-CH_2CH_2-CHO$ succindialdehyde, 1,4-butanedial
Bernsteinsäurediäthylester *m* diethyl succinate
Bernsteinsäuredibenzylester *m* dibenzyl succinate
Bernsteinsäuredichlorid *n* $ClCO-CH_2CH_2-COCl$ succinyl chloride
Bernsteinsäureimid *n* succinimide, 2,5-dioxopyrrolidine
Bernsteinsäuremonoamid *n* $NH_2OC-CH_2CH_2-COOH$ succinic acid monoamide, succinamic acid
Bernsteinsäureoxydase *f* succinoxidase, succinic [acid] oxidase
Berstdruckfestigkeit *f* bursting strength
bersten to crack, to break, *(esp of surfaces)* to burst
Berstfestigkeit *f* bursting strength
Berstscheibe *f* rupture (bursting) disk *(in pressure relief devices)*
Berthelot-Bombe *f* Berthelot bomb
Berthelot-Gleichung *f* Berthelot equation
Berthelot-Kalorimeter *n* Berthelot calorimeter
Berthelot-Mahler-Bombenkalorimeter *n* Berthelot-Mahler bomb calorimeter
Berthelot-Mahler-Kröcker-Bombe *f* Mahler (Kröcker) bomb
Berthelot-Prinzip *n* Thomson-Berthelot principle
Berthollide *npl* berthollid[e]s, berthollide (non-Daltonian, non-daltonide, non-stoichiometric) compounds
Berthollidverbindungen *fpl s.* Berthollide
Bertrandit *m (min)* bertrandite
beruhigen to kill *(a smelt)*; to deoxidize, to deoxidate *(steel)*
Beruhigungsbecken *n* stabilization basin *(waste-water treatment)*
Beruhigungskammer *f* settling chamber *(in the pebble-heater process)*
Beruhigungsmittel *n (pharm)* sedative; *(met)* killing agent
berühren to contact
sich b. to contact
Berührungsfläche *f* surface (area) of contact
Berührungsgift *n* [direct] contact poison, [direct] contact toxicant
Berührungsgrenze *f* contact boundary
Berührungslinie *f* **der Walzen** roll nip *(e. g. between calender rolls)*
Berührungsmetamorphose *f (geoch)* contact metamorphism (metamorphosis)
Berührungstrocknen *n* contact (conduction, indirect) drying
Berührungswinkel *m* contact angle *(in testing surface-active substances)*
Berührungszeit *f* time of contact
Berührungszone *f* area (surface) of contact
Beryll *m (min)* beryl *(beryllium aluminium silicate)*
Beryllat *n* beryllate

Beryllerde *f s.* Berylliumoxid
Beryllid *n* beryllide
Beryllium *n* Be beryllium
Berylliumbromid *n* $BeBr_2$ beryllium bromide
Berylliumchlorid *n* $BeCl_2$ beryllium chloride
Berylliumfluorid *n* BeF_2 beryllium fluoride
Berylliumhalogenid *n* beryllium halogenide (halide)
Berylliumhydrid *n* BeH_2 beryllium hydride
Berylliumhydroxid *n* $Be(OH)_2$ beryllium hydroxide
Berylliumjodid *n* BeI_2 beryllium iodide
Berylliumkarbid *n* Be_2C beryllium carbide
Berylliumkarbonat *n* $BeCO_3$ beryllium carbonate
Berylliumnitrat *n* $Be(NO_3)_2$ beryllium nitrate
Berylliumorthosilikat *n* Be_2SiO_4 beryllium orthosilicate, beryllium tetraoxosilicate
Berylliumoxid *n* BeO beryllium oxide
Berylliumsulfat *n* $BeSO_4$ beryllium sulphate
Berylliumsulfid *n* BeS beryllium sulphide
Berylliumtarget *n* beryllium target
Berylliumtetroxosilikat *n s.* Berylliumorthosilikat
Beryllonit *m (min)* beryllonite
Berzelianit *m (min)* berzelianite *(copper(I) selenide)*
besanden *(ceram)* to sand *(a mould)*
Besandung *f (ceram)* sanding *(of a mould)*
Besatz *m (ceram)* setting
Besatzdichte *f (ceram)* density of setting
Besatzfläche *f (ceram)* setting space
Besatzhöhe *f (ceram)* setting height
Besatzraum *m (ceram)* setting space
besäumen *(plast)* to trim
Besäummaschine *f (plast)* trimming machine, trimmer
Beschaffenheit *f* quality, constitution, nature, *(of man-made products also)* make
  brekzienartige B. brecciation
  grießartige B. grittiness
  klumpige B. lumpiness
  körnige B. graininess, grain
  mehlige B. mealiness
  stückige B. lumpiness
beschichten to coat, to overlay, to laminate, *(esp with metal)* to plate
  durch Tauchen b. to dip-coat
Beschichten *n* coating, overlaying, laminating, *(esp with metal)* plating
  B. aus Lösungen solution coating
  B. durch Streichen spread coating
  B. durch Tauchen dip coating
  B. endloser Bahnen web coating
  B. mit Rakel knife coating
  B. mittels Extruders extrusion coating
  B. über Schneckenpresse extrusion coating
  einseitiges B. *(pap)* one-sided (single-sided) coating
  zweiseitiges B. *(pap)* double coating
beschichtet / mit Gummi rubber-coated, rubber-covered
  mit Schaumstoff b. foam-backed
Beschichtung *f* 1. coat; 2. *s.* Beschichten
beschicken to feed, to charge, to fill, to load, to furnish; *(nucl)* to fuel *(a reactor)*; *(met)* to burden
  zwangsläufig b. to force-feed
Beschicker *m* stoker *(a mechanical device for feeding solid fuel)*
Beschickertrog *m* feeding trough
Beschickung *f* 1. feed[ing], charging, filling, loading, furnishing; *(nucl)* fuelling; *(met)* burdening; 2. *s.* Beschickungsmaterial
  automatische B. automatic feed
  ruhende B. static charge *(of an intermittent gas-making retort)*
  selbsttätige B. automatic feed
Beschickungsautomat *m* automatic feeder
Beschickungsbehälter *m* feed tank
Beschickungsbühne *f* feeding (charging) platform
Beschickungsbunker *m* feeding (charging) bin, *(if funnel-shaped)* feeding (charging) hopper (funnel)
Beschickungseinrichtung *f s.* Beschickungsvorrichtung
Beschickungshöhe *f (met)* stock level
Beschickungsmaterial *n* feed[stock], charge [stock], load, batch; *(met)* burden
Beschickungsmulde *f* charging box *(for an open-hearth furnace)*
Beschickungsoberfläche *f (met)* stock line
Beschickungsoberkante *f s.* Beschickungsoberfläche
Beschickungsöffnung *f* feed inlet (opening, hole, door), charging (filling) hole (door)
Beschickungsrinne *f* feed (charging) chute
Beschickungsrohr *n* feed pipe (tube)
Beschickungsrutsche *f* feed (charging) chute
Beschickungssäule *f (met)* stock column
Beschickungsschleuse *f* inlet sluice, entry lock
Beschickungsschurre *f* feed (charging) chute
Beschickungsseite *f* feed end
Beschickungstrichter *m* feed (feeding, charging, charge, input, loading) hopper, feed (feeding, charging, charge, input, loading) funnel
Beschickungstür *f* feed (feeding, charging, filling) door
Beschickungsvorrichtung *f* feeder, loader, *(for solid fuel also)* stoker
Beschickungszone *f (plast)* feed zone (section)
beschießen *(nucl)* to bombard
Beschießen *n (nucl)* bombardment
Beschlag *m* bloom; *(glass)* tarnish *(defect)*
beschlagen to bloom
beschleunigen to accelerate, to promote, to speed [up]
Beschleuniger *m* accelerator, promoter, promoting agent ✦ **ohne B.** *(rubber)* unaccelerated, non-accelerated
  B. mit verzögertem Vulkanisationseinsatz delayed-action accelerator
  anorganischer B. *(rubber)* inorganic accelerator
  basischer B. *(rubber)* basic accelerator

langsamer B. *(rubber)* slow (slow-acting, slow-action) accelerator
linearer B. *(nucl)* linear accelerator
mittelschneller (mittelstarker) B. *(rubber)* moderate (medium, medium-speed) accelerator
organischer B. *(rubber)* organic accelerator
saurer B. *(rubber)* acidic accelerator
schneller (schnellwirkender) B. *(rubber)* fast (fast-curing, rapid) accelerator
schwacher (schwachwirkender) B. *s.* langsamer B.
starker (starkwirkender) B. *s.* schneller B.
Beschleunigeraktivator *m (rubber)* accelerator activator, activator of cure (vulcanization)
Beschleunigerbatch *m (rubber)* accelerator masterbatch
Beschleunigerdosierung *f (rubber)* accelerator level
beschleunigerfrei *(rubber)* unaccelerated, non-accelerated
beschleunigerhaltig *(rubber)* accelerated
Beschleunigersystem *n (rubber)* accelerating system
Beschleunigervormischung *f (rubber)* accelerator masterbatch
Beschleunigerwirkung *f (rubber)* accelerating activity
beschleunigt *(rubber)* accelerated
nicht b. *(rubber)* unaccelerated, non-accelerated
Beschleunigung *f* acceleration, promotion
negative B. deceleration
Beschleunigungsdruckhöhe *f* acceleration head *(in a pump)*
Beschleunigungseffekt *m* accelerating effect
Beschleunigungshöhe *f* acceleration head *(in a pump)*
Beschleunigungskammer *f (nucl)* accelerating chamber
Beschleunigungsspannung *f* accelerating potential
Beschleunigungsverhältnis *n* relative centrifugal force *(centrifuging)*
Beschleunigungsvermögen *n* accelerating ability
Beschleunigungszone *f* accelerating zone
Beschleusung *f* sewerage
beschmieren to smear, to stick
beschmutzen to stain, to smudge, to smear, to soil, to pollute
Beschneidemaschine *f* trimmer, trimming machine
beschneiden to cut, to trim
Beschuß *m (nucl)* bombardment
beschweren *(pap, text)* to weight, to load *(with fillers or sizing material)*
Beschwerungsmaterial *n* weighting (loading) material (agent), load[ing]
Beschwerungsstoff *m* high-gravity solid *(in dense-media separation)*
beseitigen to eliminate, to remove
den Geruch b. to deodorize
Beseitigung *f* elimination, removal
besetzen *(cryst, nucl)* to occupy; *(phys chem)* to fill, to populate *(an energy band with electrons)*; *(tech)* to charge *(e. g. a furnace)*; to fit *(e. g. with knives)*
Besetzung *f (cryst, nucl)* occupation; *(phys chem)* filling, population *(of an energy band with electrons)*
Besetzungsgrad *m (phys chem)* degree of filling (population) *(of energy bands)*
Besetzungsinversion *f (phys chem)* population inversion *(of the energy level of atoms)*
Besetzungszahl *f* occupation number *(number of electrons in a shell)*
besprengen to sprinkle, to water
bespritzen to spray, to sprinkle
besprühen to spray, to dew
Besprühen *n* **aus der Luft** *(agric)* aerial spraying
B. mit Wasser water spraying
B. vom Flugzeug aus *(agric)* aeroplane spraying
B. zur Staubbindung dust proofing *(as of coal)*
elektrostatisches B. electrostatic spraying
Bessemer-Birne *f* Bessemer converter
Bessemer-Kleinbirne *f*, **Bessemer-Kleinkonverter** *m* baby Bessemer converter
Bessemer-Konverter *m* Bessemer converter
Bessemer-Konverterstahl *m* Bessemer steel
Bessemer-Konverterverfahren *n s.* Bessemer-Verfahren
bessemern *(met)* to bessemerize, to convert
Bessemern *n (met)* bessemerizing, converting
Bessemer-Roheisen *n* Bessemer pig (iron)
Bessemer-Schlacke *f* Bessemer (acid) slag
Bessemer-Stahl *m* Bessemer (acid) steel
Bessemer-Verfahren *n* Bessemer (converter) process, *(specif)* acid [Bessemer] process
basisches B. basic Bessemer (converter) process, basic process, Thomas[-Gilchrist] process
saures B. acid Bessemer (converter) process, acid process
beständig stable, resistant, resisting, persistent, durable, fast, proof
b. gegen Alkalien alkali-stable, alkali-resistant, alkali-resisting, alkali-fast, alkali-proof
b. gegen oxydative Einflüsse stable (resistant) to oxidation, oxidatively stable
b. gegen Säuren acid-stable, acid-resistant, acid-resisting, acid-fast
b. gegen Wasser stable to water
an der Luft b. stable in air
chemisch b. chemically resistant (stable), resistant (stable) to chemical attack
gut b. sein to last well
thermisch b. *(relating to decomposition)* thermally stable; *(relating to deformation)* heat-resistant
Beständigkeit *f* stability, resistance, persistence, durability, fastness, proofness
B. des Schaumes *(coll)* stability (persistence, lifetime) of the foam; *(ferm)* head retention, firmness of the head
B. gegen Alkalien stability (resistance) to al-

kali[es], alkali resistance
B. gegen chemische Einwirkungen *s.* chemische B.
B. gegen den Koronaeffekt resistance to corona [discharge], corona resistance
B. gegen hartes Wasser resistance to hard water
B. gegen hohe Temperaturen resistance to high temperature[s], high-temperature stability (durability, resistance)
B. gegen Lösungsmittel solvent resistance
B. gegen oxydative Einflüsse resistance to oxidation, oxidation (oxidative) resistance
B. gegen Säuren stability (resistance) to acids, acid resistance
chemische B. chemical resistance (stability), resistance (stability) to chemical attack
thermische B. *(relating to decomposition)* thermal stability (resistance, endurance), thermostability; *(relating to deformation)* heat resistance
Beständigmachen *n* proofing
Bestandteil *m* constituent, component, ingredient
aktiver B. *s.* wirksamer B.
azetonlöslicher B. *(plast)* acetone-soluble matter
dispergierender B. dispersion (dispersive, disperse) medium
disperser B. disperse[d] phase, internal phase
färbender B. colouring principle
flüchtiger B. volatile (fugitive) constituent
gasförmiger B. *s.* flüchtiger B.
giftiger B. toxic principle
integranter (integrierender) B. integral constituent
leichtflüchtiger B. *s.* flüchtiger B.
makropetrografischer B. macrocomponent, macroconstituent
mikropetrografischer B. microcomponent, microconstituent
nichtzuckerartiger B. non-sugar
wesentlicher B. integral (essential) constituent
wirksamer B. active ingredient (principle)
zuckerfreier (zuckerfremder) B. non-sugar, *(esp)* aglycon *(of a glycoside)*
Bestandteile *mpl* / **flüchtige** volatile matter, volatiles, v. m., VM
nichtflüchtige B. non-volatile matter
α-Bestandteile *mpl (coal)* alpha fraction
β-Bestandteile *mpl (coal)* beta fraction
γ-Bestandteile *mpl (coal)* gamma fraction
bestäuben to dust, to powder
mit Talkum b. *(rubber)* to soapstone
bestimmen *(quantitatively)* to determine, to estimate; *(qualitatively)* to identify
Bestimmung *f (quantitatively)* determination, estimation; *(qualitatively)* identification
B. der Methoxylgruppen nach Zeisel Zeisel methoxyl determination
B. der relativen Atommasse atomic-weight determination
B. der relativen Molekülmasse molecular-weight determination
B. der relativen Molekülmasse nach der Mikromethode von Rast Rast microprocedure, micro Rast
B. der relativen Molekülmasse nach Rast Rast's molecular-weight determination, Rast method
B. der Wirksamkeit test for potency
Barfoedsche B. Barfoed's test *(for monosaccharides)*
blinde B. *s.* Blindbestimmung
kolorimetrische B. colorimetric determination
nochmalige B. redetermination
bestrahlen to [ir]radiate; *(nucl)* to bombard
mit Röntgenstrahlen b. to X-ray
Bestrahlen *n* [ir]radiation; *(nucl)* bombardment
B. mit energiereicher (harter) Strahlung high-energy irradiation
Bestrahlung *f* [ir]radiation; *(nucl)* bombardment
Betadickenmesser *m* beta[-absorption] gauge
Betafit *m (min)* betafite
Betain *n* betaine, *(specif)* $(CH_3)_3N^+CH_2COO^-$ betaine, lycine, trimethylglycine, oxyneurine
Betaspektrum *n* beta-ray spectrum
Betastrahlen *mpl* beta rays
Betastrahlen-Dickenmesser *m* beta[-absorption] gauge
Betastrahlenquelle *f* beta-radiation source
Betastrahler *m* beta-emitter, β-emitter
Betastrahlung *f* beta radiation
Betateilchen *n* beta particle
betätigt / hydraulisch hydraulic-operated
Betätigungsorgan *n* actuator
betatop *(nucl)* betatopic
Betatron *n* betatron
betäubend anaesthetic
Betaumwandlung *f s.* Betazerfall
Betazellulose *f* beta cellulose
Betazerfall *m* beta decay (disintegration), beta-ray decay (disintegration)
Betazerfallsenergie *f* beta decay (disintegration) energy
Betelnuß *f* betel nut *(from Areca cathechu L.)*
BET-Gleichung *f (phys chem)* BET equation, Brunauer-Emmett-Teller relationship
Bethabarraholz *n (dye)* bethabar[r]a wood *(from Tabebuia specc.)*
Bethellisieren *n* Bethell treatment *(wood preservation)*
Betol *n* betol, 2-naphthyl salicylate
Beton *m* concrete
B. mit Haufwerksporosität single-sized concrete
armierter B. reinforced concrete
belüfteter B. air-entrained (air-entraining) concrete
bewehrter B. reinforced concrete
entfeinter B. no-fines concrete
erdfeuchter B. earth-moist concrete
fetter B. rich (good) concrete
feuerfester B. refractory concrete
grüner (junger) B. green concrete

magerer B. lean (lean-mixed, poor) concrete
plastischer B. plastic concrete
steifer B. earth-moist concrete
vorgefertigter B. precast concrete
vorgepackter B. prepacked (grouted) concrete
vorgespannter B. prestressed concrete
weicher B. plastic concrete
Betonbelüfter *m* air-entraining additive (admixture, agent, compound)
Betonblock *m* concrete block
Betonfestigkeit *f* concrete strength
Betonmasse *f* concrete mass
Betonmauerstein *m* concrete brick
Betonpumpe *f* concrete pump
Betonstein *m* concrete brick
Betonturm *m (pap)* concrete acid tower
Betonverdichtung *f* compaction of concrete
Betonzuschlag[stoff] *m* concrete aggregate
Betrag *m* amount, quantum, value
betreiben to operate, to drive, to run
Betrieb *m* 1. plant, works, factory; 2. operation ✦ **außer B. [befindlich]** out of operation, idle ✦ **außer B. setzen** to put out of operation, to cut out of service, to stop *(a machine)*; to shut [down], to close down *(a factory)* ✦ **in B. [befindlich]** in operation, at (in) work, working, on-stream ✦ **in B. nehmen** *s.* in B. setzen ✦ **in B. sein** to be in operation, to run, to be running (working, operating) *(of a machine)* ✦ **in B. setzen** to put (set) in operation, to set in action, to start [up], to prime *(a machine)*
chemischer B. chemical plant (works)
diskontinuierlicher B. batch operation
ganzjähriger B. year-round operation
kontinuierlicher (stetiger) B. continuous operation
betrieben / diskontinuierlich batch
hydraulisch b. hydraulic-operated
kontinuierlich b. continuous
mit Atomkraft b. atomic-powered
mit Gas b. gas-fired, gas-heated
mit Kernenergie b. atomic-powered
mit Luft b. air-driven
stetig (ununterbrochen) b. continuous
Betriebsanlage *f* plant, works
betriebsbereit operable, ready for operation (use)
Betriebsbereitschaft *f* operability
Betriebschemiker *m* industrial (works) chemist
Betriebsdampf *m* operating steam
Betriebsdestillation *f* works distillation
Betriebsdrehzahl *f* operating speed
Betriebsdruck *m* operating (working) pressure
betriebsfertig *s.* betriebsbereit
Betriebsinhalt *m (distil)* [column] hold-up, [liquid] operating hold-up
Betriebslinie *f (distil)* operating line
Betriebssäurewecker *m (food)* bulk starter
Betriebsstillegung *f* plant shut-down
Betriebsstoffwechsel *m (bioch)* energy metabolism
Betriebsstörung *f* upset, stoppage, breakdown
Betriebstemperatur *f* temperature of operation, operating (working) temperature
Betriebsunterbrechung *f* downtime, down period
Betriebsverhalten *n* operating characteristics
Betriebswasser *n* industrial (service, process) water
Betriebszustand *m* operating state
stationärer B. operating steady state
Bett *n (tech, geol)* bed
ruhendes (statisches) B. *(tech)* fixed (static) bed
Betthöhe *f (tech)* bed depth
Betts-Verfahren *n* Betts process *(for refining lead)*
Betulin *n* betulin, betula (birch) camphor *(a triterpenoid alcohol)*
Beuche *f (text)* 1. kier boiling, kiering, bowking, bucking; 2. *s.* Beuchflotte
Beuchechtheit *f (text)* fastness to kier-boiling
beuchen *(text)* to kier-boil, to bowk, to buck
Beuchen *n s.* Beuche 1.
Beuchfaß *n (text)* kier, bowking (bucking) tub
Beuchflotte *f (text)* kier (bowking, bucking) liquor, kier (bowking, bucking) lye
Beuchhilfsmittel *n (text)* kier[-boiling] assistant
Beuchkessel *m (text)* [bowking] kier
Beugung *f* diffraction
B. der Röntgenstrahlen X-ray diffraction
Beugungsbild *n* diffraction pattern
Beugungserscheinung *f* diffraction phenomenon
Beugungsgitter *n* diffraction grating
Beugungsring *m* diffraction ring
Beugungsspektrum *n* diffraction (normal) spectrum
Beugungswinkel *m* diffraction angle
Beutel *m* bag
Beutelfilter *n* bag filter
beuteln to bolt, to sift *(e. g. flour)*
Beutelpapier *n* bag paper
Beutelschließmaschine *f*, **Beutelschweißmaschine** *f* bag sealing machine
bewässern to water, to irrigate
Bewässerung *f* watering, irrigation
bewegen to move; to agitate *(e. g. a reaction mixture)*; *(rapidly up and down or to and fro)* to jig
beweglich mobile, movable
Beweglichkeit *f* mobility
B. der Nährstoffe *(biol)* nutrient mobility
elektrophoretische B. electrophoretic mobility
Bewegtbett *n* moving bed
Bewegtbettverfahren *n* moving-bed process
Bewegung *f* motion; agitation *(as of a reaction mixture)*
Brownsche B. Brownian motion (movement)
drehende B. rotary (rotational) motion
fortschreitende B. translational motion
laminare B. laminar (streamlined) flow
pulsierende B. pulsation
rotierende B. rotary (rotational) motion
Bewegungsenergie *f* kinetic energy
Bewegungsgesetze *npl* / **Newtonsche** Newton laws of motion
Bewegungsgröße *f* momentum
Bewegungsrichtung *f* direction of motion
bewehren to reinforce

Bewehrung *f* reinforcement
Beweis *m* proof
beweisen to proof
bewettern to weather *(to expose to the open air)*
Bewetterung *f* weathering *(exposure to the open air)*
Bewetterungsechtheit *f* weathering fastness
Bewetterungsprüfung *f* weathering testing
Bewetterungsversuch *m* weathering test
bewiesen / experimentell experimentally true
bewirken to cause, to effect, to bring about, to produce, to give rise to
bewittern *s.* bewettern
Bewitterung *f s.* Bewetterung
Bewoid-Leim *m (pap)* Bewoid size
Bewoid-Verfahren *n (pap)* Bewoid process
bezeichnen *(nomencl)* to notate, to designate, to name
  mit Buchstaben b. to letter
Bezeichnung *f (nomencl)* notation, designation, name
  freie B. non-proprietary name
  geschützte B. proprietary name
  nichtgeschützte B. non-proprietary name
  Stocksche B. Stock notation
Bezeichnungssystem *n (nomencl)* system of notation (naming), notation system
Bezeichnungsweise *f (nomencl)* notation, manner (method) of notation
  Stocksche B. Stock notation (scheme)
Beziehung *f* relation
  Debyesche B. *(phys chem)* Debye relation
  Duprésche B. *(phys chem)* Dupré equation
  gegenseitige B. correlation
beziffern to number, to index, to indicate, to label
Bezifferung *f* numbering, indexing, indication, labelling
  B. im Uhrzeigersinn clockwise numbering
Bezifferungssystem *n* numbering system
Bezirk *m* region, range *(as of measurement or state)*; *(cryst)* domain *(in ferromagnetic substances)*
  kristalliner B. crystalline region
  Weißscher B. *(cryst)* Weiss [molecular magnetic] field
Bezugsbasis *f* basis
  B. Masse der handelsüblich trockenen Substanz commercial dry basis, CDW
  B. Masse des feuchten Stoffs wet[-weight] basis, WWB
  B. Trockenmasse bone-dry-weight basis, BDWB, dry[-weight] basis
  B. Trockenstoffmasse (Trockensubstanzmasse) *s.* B. Trockenmasse
Bezugseinheit *f* reference unit
Bezugselektrode *f* reference (comparison) electrode
Bezugskraftstoff *m* reference fuel
Bezugspapier *n* facing paper
Bezugsspannung *f* reference voltage
Bezugsstandard *m* reference standard
Bezugssubstanz *f* reference (standard) substance
Bezugssystem *n* reference system
Bezugstreibstoff *m* reference fuel
Bezugszelle *f* reference cell
B-Füllmasse *f (sugar)* intermediate massecuite, second fillmass
B-Harz *n* B-stage resin, resitol
B-Horizont *m (soil)* B-horizon, illuvial horizon
Biallyl *n (incorrectly for)* Hexadien-(1,5)
Biaryl *n* biaryl
biaxial *(cryst)* biaxial
Biazetyl *n* $CH_3COCOCH_3$ biacetyl, butanedione
Bibeldruckpapier *n* bible paper, [Oxford] India paper
Bibenzoyl *n* $C_6H_5COCOC_6H_5$ bibenzoyl, benzil
Bibenzyl *n* $C_6H_5CH_2CH_2C_6H_5$ bibenzyl, *sym.*-diphenylethane
Bibergeil *n (pharm)* castor
Bicheroux-Verfahren *n* Bicheroux process *(flat-glass manufacture)*
Bieberit *m (min)* bieberite, cobalt vitriol *(cobalt(II) sulphate-7-water)*
Biegeeigenschaft *f* flexural property
Biegeermüdung *f s.* Biegerißbildung
Biegefestigkeit *f* bending (transverse) strength, *(Am)* flexural strength
Biegemodul *m* flexural modulus
biegen to bend, to flex
Biegen *n* bend[ing], flex[ing], flexion
Biegeofen *m (glass)* bending furnace
Biegeprobe *f*, **Biegeprüfung** *f* bend[ing] test
  B. in der Kälte cold-bend test
Biegerißbildung *f* flex[ural] cracking
Biegerißfestigkeit *f*, **Biegerißwiderstand** *m* flex (flex-cracking) resistance, resistance to flex cracking
Biegeschwingung *f* bending vibration *(spectroscopy)*
Biegesteifigkeit *f* flexural rigidity, stiffness in bend (flexure)
Biegewalze *f (glass)* bending roll *(Colburn sheet process)*
biegsam flexible, pliable, pliant
Biegsamkeit *f* flexibility, pliability
  B. bei niedriger Temperatur cold flex
Biegung *f* bend, flexure
Biegungs ... *s.* Biege ...
Bienengift *n* bee poison
Bienenharz *n* bee glue, propolis, balm
Bienenkorbkoks *m* beehive[-oven] coke
Bienenkorbofen *m* beehive [coke, coking] oven
Bienenkorbofenkoks *m* beehive[-oven] coke
Bienenvorwachs *n s.* Bienenharz
Bienenwachs *n* beeswax
  gebleichtes B. bleached beeswax, white wax
Bier *n* beer; *(if top-fermented and strongly hopped)* ale
  dunkles B. dark beer
  grünes B. new beer
  helles B. pale (light) beer
  leichtes B. mild beer
  obergäriges B. top-fermented (top-fermentation) beer

schwach gehopftes B. mildly hopped beer
stark gehopftes B. strongly hopped beer
untergäriges B. bottom-fermented (low-fermentation) beer
bierartig beery
Bierbeschauglas *n* beer inspection glass
Bierbrauen *n* brewing
Bierbrauer *m* brewer
Bierbrauerei *f* brewery
Bierdeckelpappe *f* coaster board
Bierer-Davis-Bombe *f* Bierer-Davis oxygen bomb
Bieressig *m* beer vinegar
Bierfilzpappe *f s.* Bierdeckelpappe
Bierhefe *f* beer yeast, brewing (brewer's, brewery) yeast
Bierstein *m* beer stone (scale) *(on the inside surfaces of brewing apparatus)*
Biertreber *pl* brewer's grains, malt spent grains
Bierwürze *f* [beer] wort
Biestmilch *f* colostrum, beestings, first milk
Bifaser *f s.* Bikomponentenfaser
bifunktionell bifunctional, difunctional
Biguanid *n* $NH_2C(=NH)NHC(=NH)NH_2$ biguanide, diguanide
Biharnstoff *m* $NH_2CONH-NHCONH_2$ biurea, dicarbamylhydrazine
Bihexyl *n s. n*-Dodekan
Bikoloreffekt *m (text)* bicolour effect
Bikolorin *n s.* Äskulin
Bikomponentenfaser *f (text)* bicomponent (conjugate) fibre
Bilanz *f* balance
Bild *n (phot)* image
äußeres latentes B. surface latent image
latentes B. latent image
negatives B. negative image
oberflächliches latentes B. *s.* äußeres latentes B.
positives B. positive image
Bildband *n s.* Bildstreifen
bilden to form; to produce
Blasen b. to bubble *(of gas or water)*; to blister *(of metal or paint)*; to vesicate, to blister *(of the skin)*
Chelate b. to chelate
ein Sol b. to solate
eine Bindung b. to make a bond
eine Kruste b. to encrust, to incrust
einen Bleibaum b. to tree
einen Bodenkörper (Bodensatz) b. *s.* einen Niederschlag b.
einen Komplex b. to complex
einen Niederschlag b. to precipitate, to sediment, to set, to settle [down, out], to deposit, to subside; *(of a liquid)* to sediment, to deposit; *(of vapour)* to condense
Klumpen b. to clot, to clog
Kristalle b. to crystallize [out]
Luppen b. *(met)* to ball [up]
Runzeln b. to wrinkle *(painting technology)*
Schlacke b. to slag; *(coal)* to clinker
sich b. to form
bildsam plastic; ductile
wenig b. *(ceram)* short *(of a clay body)*
Bildsamkeit *f* plasticity; ductility
Bildstein *m (min)* pencil stone, agalmatolite *(a variety of pyrophyllite)*
Bildstreifen *m (phot)* film strip
Bildung *f* formation; production
B. eines Bodenkörpers (Bodensatzes, Niederschlags) sedimentation, deposition
B. von Einschlußverbindungen clathrate formation
Bildungsenthalpie *f* enthalpy of formation, heat of formation at constant pressure
Bildungsfunktion *f* formation function
Bildungsgeschwindigkeit *f* rate (velocity) of formation
Bildungskonstante *f* formation constant
Bildungsreaktion *f* formation reaction
Bildungswärme *f* heat of formation
atomare B. atomic heat of formation
molare B. heat of formation per mole
Bilirubin *n (bioch)* bilirubin
Biliverdin *n (bioch)* biliverdin
Billetkarton *m* ticket board
Billetpapier *n* letter paper
Billiter-Zelle *f* Billiter cell *(electrolysis)*
Billitonit *m* billitonite *(a tektite)*
Bilsenkraut *n* henbane, Hyoscyamus L., *(specif) s.* Schwarzes B.; *(pharm)* hyoscyamus herb *(from Hyoscyamus niger L.)*
Schwarzes B. black henbane, Hyoscyamus niger L.
Bimetall *n* bimetal
bimetallisch bimetallic
Bimetallthermometer *n* bimetallic thermometer
bimolekular bimolecular
Bimsbeton *m* pumice concrete
Bimsen *n (tann)* buffing
Bimskiesbeton *m s.* Bimsbeton
Bimsmaschine *f (tann)* buffing machine
Bimsstaub *m* pumice powder
Bimsstein *m* pumice [stone]
bimssteinartig pumiceous
Bimssteinpulver *n* pounce
Bimssteintuff *m* pumice tuff
binär binary
Binde ... *s. a.* Bindungs ...
Bindeglied *n* binding link; *(tech)* link
Bindekörper *m (coat)* binder
Bindemittel *n* binding (bonding) agent (material), binder, bonding, cementing agent (material), cement, agglutinant; adhesive agent (substance), adhesive; *(coat)* vehicle, medium; *(geol)* cement, cementing agent, agglutinant
binden 1. to bond, to bind, to link *(atoms)*; to agglutinate *(particles)*; 2. *(build)* to set, to harden
komplex b. to complex
koordinativ b. to coordinate *(e.g. molecules)*
Binder *m (build)* binder, binding agent (material)
hydraulischer B. hydraulic binder
Bindestrich *m* bonding dash *(in structural formulae)*

Bindeton *m* bond[ing] clay
bindig *(soil)* tenaceous
Bindigkeit *f* 1. covalence *(chemical-bond theory)*; 2. *(soil)* tenacity
maximale B. maximum covalence
Bindung *f* 1. binding; agglutination; 2. *(chemical-bond theory)* bonding, linking, binding *(process)*; bond, link[age] *(state)*; 3. *(agric)* fixation *(of atmospheric nitrogen)* ✦ **die B. eingehen** to bond ✦ **eine B. bilden (herstellen)** to make a bond
anderthalbfache B. one-and-a-half bond, one-and-one-half bond, three-halves bond
äquatoriale B. equatorial bond, e-bond
axiale B. axial bond, a-bond
chemische B. chemical bonding *(process)*; chemical bond *(state)*
dative B. *s.* koordinative B.
delokalisierte B. delocalized bond
doppelte B. double bond (link)
dreifache B. triple bond (link)
einpolare B. *s.* homöopolare B.
elektrostatische B. *s.* heteropolare B.
elektrovalente B. *s.* heteropolare B.
gebogene B. bent bond *(C–C double bond)*
glykosidische B. glycosidic bond
halbpolare B. *s.* koordinative B.
heteropolare B. heteropolar bond (link, linkage), polar bond (link, linkage), ionic bond (link, relationship), electrostatic (electrovalent) bond, electrovalence, elestrovalency *(state)*
homöopolare B. homopolar (non-polar, covalent, atomic) bond, homopolar link[age], [shared-] electron-pair bond, unitarian bond *(state)*
ionogene B. *s.* heteropolare B.
komplexe B. complex bond
koordinative B. coordinate (coordinative, coordination, dative, semipolar, half-polar, donor-acceptor) bond, coordinate (dative) covalency, semipolar double bond *(state)*
kovalente B. *s.* homöopolare B.
lokalisierte B. localized bond
mehrfache B. multiple bond
metallische B. metal[lic] bond
nichtlokalisierte B. *s.* delokalisierte B.
peptidartige B. peptide bond
polare B. *s.* heteropolare B.
semipolare B. *s.* koordinative B.
semizyklische B. semicyclic bond
silikatische B. silicate bond
symbiotische B. *(agric)* symbiotic fixation *(of atmospheric nitrogen)*
unitarische B. *s.* homöopolare B.
unpolare B. *s.* homöopolare B.
van-der-Waalssche B. van der Waals bond
zwischenmolekulare B. *s.* van-der-Waalssche B.
π-Bindung *f* π bond, pi bond
σ-Bindung *f* σ bond, sigma bond
Bindungsachse *f* bond axis
Bindungsart *f* bond type, kind of bond (link, linkage)
Bindungsbildung *f* bond formation
Bindungsbruch *m* bond breaking
Bindungscharakter *m* bond character
Bindungsdipol *m* bond dipole
Bindungsdipolmoment *n* bond dipole moment, dipole moment of linkage
Bindungsdissoziationsenergie *f* bond-dissociation energy
Bindungsdublett *n s.* Bindungselektronenpaar
Bindungselektron *n* bonding electron, linkage (valency, outermost, optical) electron
Bindungselektronenpaar *n* bonding electron pair, bonding pair of electrons
Bindungsenergie *f* bond (bonding, binding) energy
bindungsfähig bondable
Bindungsfähigkeit *f* bondability, bonding (binding, combining) power, bonding ability (capacity)
Bindungsfestigkeit *f* bond (bonding, binding) strength
Bindungsgrad *m*, **Bindungsgradzahl** *f* bond number (order)
Bindungskraft *f* 1. bond (bonding, binding, combining, linkage) force; 2. *s.* Bindungsfähigkeit
Bindungslänge *f* bond length (distance)
bindungslos non-bonded
Bindungsmoment *n* bond moment
Bindungsorbital *n* bond orbital
Bindungsordnung *f s.* Bindungsgrad
Bindungsparachor *m* structural parachor *(chemical-bond theory)*
Bindungsrefraktion *f* bond refraction
Bindungsrichtung *f* bond direction (orientation)
Bindungsspaltung *f* bond fission (cleavage)
Bindungssphäre *f* boundary (bounding) surface *(chemical-bond theory)*
Bindungsstärke *f s.* Bindungsfestigkeit
Bindungsstrich *m* bonding dash *(in structural formulae)*
Bindungssystem *n* bond (linkage) system
Bindungstheorie *f* chemical-bond theory
B. der Elektronenpaarbindungen (Valenzstrukturen) electron-pair (valence-bond) theory, VB theory, Heitler-London-Slater-Pauling theory, HLSP theory
Bindungstyp *m s.* Bindungsart
Bindungsvermögen *n s.* Bindungsfähigkeit
Bindungsverschiebung *f* bond shift[ing], bond migration
Bindungswertigkeit *f s.* Bindigkeit
Bindungswinkel *m* bond angle
Bindungszahl *f* bond number
Bindungszustand *m* bonding state
Bingham-Körper *m (coll)* Bingham body
Binnendruck *m* cohesion (internal) pressure
Biochemie *f* biochemistry, biological chemistry
Biochemiker *m* biochemist
biochemisch biochemical
Biofilter *n* biofilter, biological filter
Bioflavonoid *n* bioflavonoid
Biogas *n* fermentation gas
biogen biogenic, biogenous
Bioingenieurwesen *n* bioengineering

Biokatalysator *m* biocatalyst, biochemical catalyst, ergone
bioklastisch *(geol)* bioclastic
Biokristall *m* biocrystal
Biolith *m* biolith, biogenic (organic) rock
biologisch biological
Biolumineszenz *f* bioluminescence
Bios I *n* inositol
Bios II *n* biotin
Biose *f* biose *(monosaccharide containing two carbon atoms)*
Biosid *n* bioside
Biosynthese *f* biosynthesis
Biotest *m (tox)* bioassay [test], biological assay
biotisch biotic
Biotit *m (min)* biotite, black (dark) mica
Biozid *n* biocide *(a pesticide for controlling microbes)*
Biphenyl *n* $C_6H_5C_6H_5$ biphenyl, phenylbenzene
Biphenyldikarbonsäure *f* $HOOC-C_6H_4-C_6H_4-COOH$ biphenyl-2,2′-dicarboxylic acid, diphenic acid
Biphthalidensäure *f s.* Benzil-2,2′-dikarbonsäure
Biquarz *m* biquartz
Biradikal *n* biradical
Birch-Hückel-Reaktion *f* Birch reduction *(of organic compounds by metallic sodium dissolved in liquid ammonia)*
Birkeland-Eyde-Verfahren *n* Birkeland-Eyde process *(for manufacturing nitric acid)*
Birkenhaarwasser *n* birch hair lotion
Birkenkampfer *m* birch (betula) camphor, betulinol *(a triterpenoid alcohol)*
Birkenknospenöl *n* oil of birch buds
Birkenöl *n* [sweet-]birch oil
Birkenrindenöl *n* birch bark oil
Birkenteer *m* birch tar
Birkenteeröl *n* birch tar oil
Birkenwasser *n* birch water
Birkenwein *m s.* Birkenwasser
Birnenäther *m* pear essence *(alcoholic solution of amyl acetate)*
Birnenöl *n* pear (banana) oil *(amyl acetate)*
Bisabolen *n* bisabolene *(a monocyclic sesquiterpene)*
bis-axial bis-axial
Bisbeeit *m (min)* bisbeeite *(a basic aluminium copper sulphate)*
Bischler-Napieralski-Reaktion *f* Bischler-Napieralski reaction *(for the synthesis of isoquinoline)*
Bis-β-chloräthyläther *m s.* 2,2′-Dichlordiäthyläther
Bischofit *m (min)* bischofite *(magnesium chloride)*
Bis-harnstoff *m s.* Biharnstoff
Biskuitbrand *m (ceram)* biscuit firing, biscuitting
Biskuitbrandware *f s.* Biskuitware
Biskuitporzellan *n (ceram)* biscuit [porcelain], bisque
Biskuitware *f (ceram)* biscuit, bisque, bisquitted (biscuit-fired) ware
Bismarckbraun *n* Bismarck (vesuvine) brown, vesuvin
Bismit *m (min)* bismite *(bismuth(III) oxide-3-water)*
Bismutan *n s.* Bismutin
Bismutat *n* bismuthate
Bismuthin[it] *m (min)* bismuthin[it]e, bismuth glance *(bismuth(III) sulphide)*
Bismutin *n* BiH_3 bismuthine, bismuth hydride
Bismutit *m (min)* bismutite *(bismuth carbonate oxide)*
Bisphenol A *n* $HOC_6H_4C[CH_3]_2C_6H_4OH$ bisphenol A, 2,2-di-*p*-hydroxyphenylpropane
Bister *m(n)* manganese brown
Bisulfit *n (incorrectly for)* Hydrogensulfit
Bisulfitzellstoff *m* sulphite pulp
bitter bitter
 b. machen to embitter, to imbitter *(e. g. beer)*
Bittererde *f s.* Magnesiumoxid
Bitterholz *n s.* Quassiaholz
Bitterkleesalz *n s.* Kleesalz
Bittermandelessenz *f s.* Bittermandelöl / künstliches
Bittermandelöl *n* bitter almond oil, amygdala amara oil
 künstliches B. 1. artificial (synthetic) essential oil of almonds *(chemically benzaldehyde)*; 2. *s.* unechtes B.
 unechtes B. *(cosmet)* mirbane (myrbane) oil, essence of mirbane *(chemically nitrobenzene)*
Bittermandelölgrün *n* malachite green, basic green 4, green verditer *(a triphenylmethane dye)*
Bittermandelölkampfer *m* $C_6H_5CH(OH)COC_6H_5$ bitter almond oil camphor, benzoin, α-hydroxybenzyl phenyl ketone
Bitterquelle *f* bitter spring
Bittersalz *n* $MgSO_4 \cdot 7H_2O$ bitter salt, Epsom salt[s], magnesium sulphate-7-water; *(min)* epsomite *(magnesium sulphate-7-water)*
Bittersäure *f* bitter acid
Bitterspat *m (min)* magnesite, bitter spar *(magnesium carbonate)*
Bitterstoff *m* bitter principle (substance), *(ferm also)* bittern
 nichtglykosidischer B. amaroid
Bitterstoffwert *m (ferm)* bitterness value *(of hops)*
Bitterwasser *n* bitter water
Bitterwert *m s.* Bitterstoffwert
Bitterwurzel *f* gentian [root]
Bitumen *n* bitumen, *(specif)* asphaltic bitumen
 geblasenes B. *s.* Blasbitumen
Bitumenanstrich *m* bituminous coating
Bitumendachpappe *f* asphaltic felt
Bitumenemulsion *f* bituminous emulsion, emulsified bitumen
Bitumenfarbe *f* bituminous paint
bitumenhaltig bituminous
Bitumenlack *m* bituminous varnish
Bitumenpapier *n* asphalt (tar, pitch) paper, tarred [brown] paper
Bitumenpappe *f* bitumen board
Bitumenpreßmasse *f* bituminous plastic, bituminous moulding composition

Bitumensand *m* bituminous sand
Bitumenschiefer *m* bituminous (oil) shale
Bituminit *m (min)* bitumenite
bituminös, bitumisch bituminous
Biuret *n* $NH_2CONHCONH_2$ biuret, ureidoformamide
Biuretprobe *f* biuret test
Biuretreaktion *f* biuret reaction
bivalent bivalent, divalent
Bivalenz *f* bivalence, divalence
bivariant bivariant, divariant
Bivinyl *n* $CH_2{=}CHCH{=}CH_2$ 1,3-butadiene, bivinyl
Bixin *n* bixin *(chief colouring matter of annatto)*
bizyklisch bicyclic
Bizyklus *m* bicyclic compound
Bladder *m (rubber)* bladder
Blähen *n* intumescence, swelling *(of coal)*
Blähgrad *m*, **Blähindex** *m s.* Blähungsgrad
Blähmittel *n (rubber, plast)* blowing (inflating, sponging) agent; *(build)* expanding agent
Blähprobe *f (coal)* swelling test
Blähschiefer *m* expanded shale
Blähschlamm *m* bulking sludge
Blähton *m* expanded (foamed) clay, lightweight expanded clay [aggregate], foamclay
Blähungsgrad *m (coal)* swelling index (number)
  B. ohne Belastung der Kohle *s.* freier B.
  freier B. free-swelling index
Blähvermögen *n* swelling power *(of coal)*
Blähzahl *f s.* Blähungsgrad
Blanc fixe *n* blanc fixe, permanent white *(precipitated barium sulphate)*
Blankfilter *n* polishing filter
Blankfiltration *f* polishing [filtration]
blankfiltrieren to polish
Blankfix *n s.* Blanc fixe
blankglühen to bright-anneal
Blankglühen *n* bright anneal[ing]
Blankglühofen *m* bright-annealing furnace
Blankkochen *n (sugar)* blank boiling
blankstoßen *(tann)* to glaze
Blasanlage *f (plast)* blow moulder, blow-moulding machine
Blasbitumen *n* blown bitumen, [air-]blown asphalt, mineral rubber
Blas-Blas-Verfahren *n (glass)* blow-and-blow process
Bläschen *n* bubble
Bläschenbeton *n* air-entrained (air-entraining) concrete
Bläschenbildung *f* bubbling
Blasdruck *m* blowing pressure
Blase *f* 1. bubble; 2. distillation boiler, [re]boiler, [still] pot; 3. blister *(a defect in material), (met also)* blow-hole, gas cavity, *(plast also)* void, *(glass also)* cat eye, *(pap)* bell; 4. vesication, blister *(on the skin)* ✦ **Blasen bilden** to bubble *(of gas or water)*; to blister *(of metal or paint)*; to vesicate, to blister *(of the skin)* ✦ **in Blasen aufsteigen** to bubble ✦ **in Blasen aufsteigen lassen** to bubble *(a gas through a liquid)*
  äußere B. *(met)* subcutaneous blow-hole
  offene B. *(plast)* open bubble *(a moulding defect)*
Blasebalg *m* bellows
  B. mit Fußbetrieb foot bellows
Blasegas *n* blow gas
blasen to blow
  mit Druckluft b. to air-blow
Blasen *n* blow[ing]
  B. mit Bodenwind bottom blowing
  B. mit Luft air blowing
  B. von schlauchförmigen Vorformlingen *(plast)* blowing of tubular parisons
  seitliches B. *(met)* side blowing
  Blasenbildung *f* bubbling, formation of bubbles; *(tech)* blistering, blister formation; *(med)* vesication, blistering ✦ **zur B. reizen** *(med)* to vesicate, to blister
Blasendestillationsanlage *f* / **kontinuierliche** continuous shell still
Blasendestillierapparat *m* pot still
Blasendruck *m* bubble pressure
  maximaler B. maximum bubble pressure
Blasendruckmethode *f* bubble method, maximum bubble pressure method
Blasenflüssigkeit *f (distil)* reboiler liquid
blasenfrei *(plast)* free from voids; *(met)* free from blow-holes
Blasengärung *f* bubble fermentation
Blasenkammer *f (nucl)* bubble chamber
Blasenkupfer *n* blister copper
Blasenlassen *n (glass)* blocking
Blasenmethode *f s.* Blasendruckmethode
Blasenrückstand *m* still residue
Blasenzähler *m* bubble counter
Blasenziehen *n* blistering, blister formation
blasenziehend *(med)* vesicant, vesicatory, blistering
Blaseperiode *f* blow period
Blasfolie *f (plast)* blown film
Blasform *f (met)* [air-blast] tuyère, twyer
Blasformebene *f (met)* tuyère level
Blasformen *n (plast)* blow moulding
  B. von Folienhalbzeug sheet blow moulding
Blasformteil *n (plast)* blow moulding
Blashochofen *m (met)* blast furnace
Blaskopf *m* blow head
Blaslanze *f (met)* [oxygen] lance
Blasluft *f* blow air; *(glass)* puff
Blasmaschine *f* blow moulder, blow-moulding machine
Blasöl *n* blown oil
Blasrohr *n* blow pipe, blowpipe
Blasstahl *m* basic oxygen [furnace] steel
Blasstahlkonverter *m* basic oxygen converter (furnace), top-blown basic oxygen converter (furnace)
Blasstahlverfahren *n* basic oxygen [converter, furnace, steel] process, top-blown oxygen converter process, oxygen process of steelmaking, oxygen-lance process
Blasstahlwerk *n* [basic] oxygen steel plant
Blastank *m (pap)* blow (wash, receiving) tank, blow pit (vat)
blastogranitisch *(geol)* blastogranitic

blastophitisch *(geol)* blastophitic
blastoporphyrisch *(geol)* blastoporphyritic
Blasverfahren *n* blowing process, *(met also)* pneumatic process
Blaswerkzeug *n (plast)* blow[ing] mould
Blatt *n* 1. *(tech)* leaf *(of a filter)*; paddle, blade, shovel *(as of an agitator)*; 2. *(pap)* sheet; *(cryst)* folium; *(ceram)* bat *(for producing flat ware)*
Blattapplikation *f (agric)* foliage application *(as of pesticides)*
blattbildend *(pap)* sheet-forming
Blattbildung *f (pap)* sheet formation
blättchenförmig lamellar
Blattdüngung *f (agric)* leaf dressing
Blatter *f (glass)* blister *(a defect)*
Blätter *npl* shavings *(in soap manufacture)*
Blättergelatine *f s.* Blattgelatine
blätterig *s.* blättrig
Blätterkohle *f* cannel (candle, jet) coal
Blätterserpentin *m (min)* antigorite *(a variety of serpentine)*
Blattfarbstoff *m* leaf pigment
Blattfaser *f* leaf fibre
Blattfilter *n* leaf filter
Blattformermaschine *f (ceram)* bat-making machine, batting-out machine
blattförmig lamellar
Blattgelatine *f* sheet gelatin
Blattgold *n* leaf gold, gold leaf (foil)
Blattgrün *n* leaf green, chlorophyll
Blattlänge *f (pap)* length of the sheet
Blattlausbekämpfungsmittel *n* aphicide
blättrig lamellar, foliated; *(min)* spathic, spathose
Blattrührer *m* vane stirrer, paddle (leaf) agitator
Blattscheibe *f s.* Blattformermaschine
Blattverbrennungen *fpl (agric)* foliage burn *(as by pesticides)*
Blattzinn *n* tin foil
Blau *n* blue *(sensation or substance)*
Berliner B. Prussian (Berlin) blue *(a complex iron cyanide)*
Braunschweiger B. *s.* Bremer B.
Bremer B. Bremen blue, blue verditer, copper blue *(a basic copper carbonate)*
Meldolas B. Meldola's blue *(an oxazine dye)*
Neuwieder B. *s.* Bremer B.
Pariser B. *s.* Berliner B.
Preußisch B. *s.* Berliner B.
Thénards B. Thénard blue, cobalt blue (ultramarine), king's blue *(cobalt aluminate)*
Turnbulls B. Turnbull's blue *(a complex iron cyanide)*
Blauanlaufen *n* blueing *(of tools)*; blooming *(of oil varnishes)*
Blaudruck *m* 1. blue print, blueprint, blue-printing; 2. *s.* Blaupause
Blaudruckverfahren *n* ferroprussiate process
blauempfindlich *s.* blausensibilisiert
bläuen to blue
Blaufarbenglas *n* smalt, powder blue *(cobalt(II) potassium silicate)*
Blaugas *n s.* Kokswassergas
Blau-Gas *n* Blau gas *(an oil gas)*
Blauglas *n* blue glass
Blauholz *n* Campeachy (Campechy) wood, logwood *(from Haematoxylum campechianum L.)*
Blaumasse *f (text)* cuprammonium cellulose
Blaupackpapier *n* mill wrapper (wrapping)
Blaupause *f* blue print, cyanotype ✦ **eine B. herstellen** to blueprint
Blaupauspapier *n* blue-print paper, cyano paper
Blausäure *f* HCN hydrogen cyanide, hydrocyanic acid
Blausäureglykosid *n* cyanogenetic (cyanophoric) glycoside
Blauschlamm *m*, **Blauschlick** *m* blue mud
Blauschönung *f (ferm)* blue fining
blausensibilisiert blue-sensitive, blue-sensitized
Blaustich *m (text)* blue cast
blaustichig bluish
Blauton *m* blue shade
Bläuung *f* blueing
Bläuungsmittel *n* blueing [agent] *(for improving the degree of whiteness)*
Blauwassergas *n s.* Kokswassergas
Blauwert *m* blue value
Blaw-Knox-Mühle *f* Blaw-Knox mill *(a jet mill)*
Blech *n (met)* sheet, plate
gelochtes B. perforated plate
Blechbüchse *f*, **Blechdose** *f* tin can, tin
Blei *n* Pb lead ✦ **mit B. auskleiden** to lead-line, to lead-clad ✦ **mit B. überziehen** to lead-coat
raffiniertes B. refined lead
Bleiabfälle *mpl* scrap lead
Bleiakkumulator *m* lead (lead-acid) accumulator (battery)
Bleialkalisilikatglas *n* lead-alkali silicate glass
Bleialkyle *npl* / **gemischte** mixed lead alkyls, MLA *(an antiknock agent)*
Blei-Antimon-Legierung *f* antimonial lead
Blei(II)-arsenat(V) *n s.* Blei(II)-orthoarsenat(V)
Bleiauskleidung *f* lead lining, internal lead cladding
Blei(II)-azetat *n* $Pb(CH_3COO)_2$ lead(II) acetate, lead diacetate
Blei(IV)-azetat *n* $Pb(CH_3COO)_4$ lead(IV) acetate, lead tetraacetate
Bleiazetatpapier *n* lead[-acetate] paper
Bleiazid *n* $Pb(N_3)_2$ lead azide, lead azoimide
Bleibaum *m* lead tree ✦ **einen B. bilden** to tree
Bleibaustein *m* lead brick *(for screening radiation)*
Bleibenzin *n* leaded (ethyl) gasoline
Bleibenzoat *n* $Pb(C_6H_5COO)_2$ lead benzoate
Bleiblock *m* lead block
B. nach Trauzl Trauzl lead block *(for testing explosives)*
Bleiblockausbauchung *f* lead-block expansion *(in testing explosives)*
Bleiblockprobe *f* lead-block expansion test *(for evaluating explosives)*
B. nach Trauzl Trauzl [lead-block] test
Bleibromat *n* $Pb(BrO_3)_2$ lead bromate
Blei(II)-bromid *n* $PbBr_2$ lead(II) bromide, lead dibromide
Bleibronze *f* leaded bronze

Bleichanlage *f* bleach[ing] plant, bleachery
Bleichapparat *m* bleaching apparatus, bleacher
Bleichbad *n s.* Bleichflotte
bleichbar bleachable
Bleichbarkeit *f* bleachability
Bleichbedarf *m s.* Bleichmittelbedarf
Bleichbeginn *m* beginning of the bleaching period
Bleichbottich *m* bleaching vat (chest)
Bleichchlor *n* / **aktives (wirksames)** *(pap)* available chlorine
Bleichdauer *f* bleaching time
Bleiche *f* bleach[ing]
kalte B. *(pap)* cold bleach[ing]
natürliche B. *(text)* natural bleach[ing], grass bleach[ing], grassing
optische B. optical bleach[ing]
warme B. *(pap)* warm bleach[ing]
bleichecht fast to bleach[ing], resistant to bleaching
Bleichechtheit *f* fastness to bleach[ing], resistance to bleaching, bleach-fastness
Bleicheffekt *m* bleaching efficiency
bleichen 1. *(of human agent)* to bleach, to whiten, to brighten, to decolour[ize], to discolour; 2. *(of substances)* to bleach, to discolour
auf eine höhere Weiße b. *(pap)* to whiten, to brighten
mit Schwefeldioxid b. *(text)* to stove
unvollständig b. *(pap)* to underbleach
Bleichen *n* bleach[ing], whitening, brightening, decolourization, discolouration
B. mit Schwefeldioxid *(text)* stoving
elektrolytisches B. *(pap)* electrolytic bleach[ing]
unvollständiges B. *(pap)* underbleaching
Bleichende *n* end of the bleaching period, end of bleaching
Bleicher *m* 1. *s.* Bleichapparat; 2. bleacher *(profession)*
Bleicherde *f* 1. bleaching (decolourizing, discolouring) earth, bleaching (decolourizing, discolouring) clay; 2. *s.* Bleicherdeboden
aktivierte (künstlich aktive) B. activated clay *(petroleum refining)*
naturaktive (natürliche) B. naturally occurring clay, natural (non-activated) clay *(petroleum refining)*
säureaktivierte B. acid clay *(petroleum refining)*
Bleicherdebehandlung *f* clay treating (treatment) *(of oils)*
B. nach dem Kontaktverfahren *(petrol)* clay contacting
Bleicherdeboden *m* podzol[ic] soil, podzol
Bleicherdekontakt *m* clay catalyst (contact)
Bleicherderaffination *f s.* Bleicherdebehandlung
Bleicherei[anlage] *f s.* Bleichanlage
bleichfähig bleachable
Bleichfähigkeit *f* bleachability
Bleichflotte *f*, **Bleichflüssigkeit** *f* bleaching bath (liquor, liquid, lye, solution), bleach
Bleichgrad *m* degree of bleaching
Bleichhilfsmittel *n* bleaching assistant
Bleichholländer *m* *(pap)* [bleaching] potcher, bleaching (potching, poaching) engine, po[a]cher
Bleichkalk *m* chlorinated lime, chloride of lime, bleaching powder
Bleichkammer *f* bleaching chamber
Bleichkessel *m* bleaching vat (chest)
Bleichkufe *f* bleaching vat (chest)
Bleichlauge *f* bleaching liquor (solution)
Javellesche B. Javel[le] water *(aqueous solution of potassium hypochlorite)*
Bleichlaugenbehälter *m* bleach-liquor tank
Bleichlorat *n* $Pb(ClO_3)_2$ lead chlorate
Blei(II)-chlorid *n* $PbCl_2$ lead(II) chloride, lead dichloride
Blei(IV)-chlorid *n* $PbCl_4$ lead(IV) chloride, lead tetrachloride
Bleichlorit *n* $Pb(ClO_2)_2$ lead chlorite
Bleichlösung *f* bleaching liquor (solution)
Bleichmittel *n* bleaching agent (material), bleach; decolourizing agent, decolourizer, decolourant
optisches B. optical bleaching (brightening, whitening) agent, optical bleach (brightener), fluorescent brightening (whitening) agent, fluorescent brightener (white dye)
oxydativ wirkendes B. oxidizing bleaching agent
Bleichmittelaufwand *m* *(pap)* bleach requirements (demand)
Bleichmittelbedarf *m* *(pap)* bleach requirements (demand)
Bleichmittelverbrauch *m* *(pap)* bleach consumption
Bleichmoostorf *m* sphagnum peat
Bleichpulver *n s.* Bleichkalk
Blei(II)-chromat *n* $PbCrO_4$ lead(II) chromate
Bleichsand *m* bleached sand
Bleichschlamm *m* *(pap)* bleach sludge
Bleichsoda *f* bleaching soda
Bleichstufe *f* bleaching stage
Bleichstiefel *m* *(text)* J box
Bleichton *m* bleaching (discolouring) clay
Bleichtrommel *f* *(pap)* tumbler
Bleichturm *m* *(pap)* bleaching tower, bleacher
Bleichung *f s.* Bleichen
Bleichverfahren *n* bleaching process; *(phot)* bleach-out process
Bleichverhältnis *n* *(pap)* bleach ratio
Bleichwirkung *f* bleaching effect (action); bleaching efficiency
Bleichzeit *f* bleaching time
Bleidi... *s.a.* Blei(II)-...
Bleidiantimonat(V) *n* $Pb_2Sb_2O_7$ lead diantimonate(V)
Bleidiarsenat(V) *n* $Pb_2As_2O_7$ lead diarsenate
Bleichdichromat *n* $PbCr_2O_7$ lead dichromate
Bleidichtung *f* lead packing; lead gasket *(for parts without relative motion)*
Bleidioxid *n s.* Blei(IV)-oxid
Bleidiphosphat *n* $Pb_2P_2O_7$ lead diphosphate, lead pyrophosphate
Bleidithionat *n* PbS_2O_6 lead dithionate
Bleidraht *m* lead wire
Bleielektrode *f* lead electrode
Bleiempfindlichkeit *f* lead susceptibility *(of fuels)*

bleien to lead *(fuels)*
Bleien *n* leading *(of fuels)*
Bleiessig *m*, **Bleiextrakt** *m* vinegar of lead, Goulard's extract *(aqueous solution of basic lead acetates)*
Bleifarbe *f* lead paint
Bleiferrat(III) *n* $Pb(FeO_2)_2$ lead ferrite(III)
Blei(II)-fluorid *n* PbF_2 lead(II) fluoride, lead difluoride
Blei(II)-formiat *n* $Pb(HCOO)_2$ lead(II) formate
bleifrei lead-free, leadless
Bleifritte *f (ceram)* lead frit
Bleigehalt *m* lead content ✦ **mit hohem B.** rich in lead; high-leaded *(fuel)*
Bleiglanz *m (min)* lead glance, galena, galenite *(lead(II) sulphide)*
Bleiglas *n* lead glass
bleiglasiert *(ceram)* lead-glazed
Bleiglasur *f (ceram)* lead glaze
Bleiglätte *f* litharge, yellow lead oxide *(lead(II) oxide)*
bleihaltig lead-containing, plumbiferous
Blei(II)-hexafluorosilikat *n* $Pb[SiF_6]$ lead(II) hexafluorosilicate, lead fluorosilicate
Blei(II)-hexazyanoferrat(II) *n* $Pb_2[Fe(CN)_6]$ lead(II) hexacyanoferrate(II)
Blei(II)-hexazyanoferrat(III) *n* $Pb_3[Fe(CN)_6]_2$ lead(II) hexacyanoferrate(III)
Blei(II)-hydrogenarsenat(V) *n* $PbHAsO_4$ lead(II) hydrogenarsenate
Bleihydroxid *n* $Pb(OH)_2$ lead hydroxide
Bleijodat *n* $Pb(IO_3)_2$ lead iodate
Blei(II)-jodid *n* PbI_2 lead(II) iodide, lead diiodide
Bleikammer *f* lead chamber
Bleikammerkristalle *mpl* chamber crystals *(nitrosylsulphuric acid)*
Bleikammerverfahren *n* [lead-]chamber process *(for obtaining sulphuric acid)*
Bleikaprinat *n* $Pb(C_9H_{19}COO)_2$ lead decanoate, lead caprate
Bleikapronat *n* $Pb(C_5H_{11}COO)_2$ lead hexanoate, lead caproate
Bleikaprylat *n* $Pb(C_7H_{15}COO)_2$ lead octanoate, lead caprylate
Bleikarbonat *n* $PbCO_3$ lead carbonate
Bleikrankheit *f s.* Bleivergiftung
Bleikristallglas *n* lead crystal glass
Bleilässigkeit *f (ceram)* lead solubility
Bleilegierung *f* lead alloy
Bleilöslichkeit *f (ceram)* lead solubility
Bleimantelverfahren *n (rubber)* lead press technique
Bleimantelvulkanisation *f (rubber)* lead press cure
Bleimennige *f* red lead [oxide], minium *(lead(II) orthoplumbate)*
Bleimetaarsenat(III) *n* $Pb(AsO_2)_2$ lead metaarsenite
Bleimetaarsenat(V) *n* $Pb(AsO_3)_2$ lead metaarsenate
Bleimetaborat *n* $Pb(BO_2)_2$ lead metaborate
Bleimetaphosphat *n* $Pb(PO_3)_2$ lead metaphosphate
Blei(II)-metaplumbat *n* $Pb[PbO_3]$ lead(II) metaplumbate
Bleimetasilikat *n* $PbSiO_3$ lead metasilicate
Bleimetatitanat *n* $PbTiO_3$ lead metatitanate
Bleimetavanadat(V) *n* $Pb(VO_3)_2$ lead metavanadate
Bleimolybdat(VI) *n* $PbMoO_4$ lead molybdate(VI)
Bleimonoxid *n s.* Blei(II)-oxid
Blei(II)-nitrat *n* $Pb(NO_3)_2$ lead(II) nitrate
Blei(II)-nitrit *n* $Pb(NO_2)_2$ lead(II) nitrite
Bleiofen *m* lead blast furnace
Blei(II)-orthoarsenat(V) *n* $Pb_3(AsO_4)_2$ lead(II) orthoarsenate, lead(II) arsenate
Blei(II)-orthophosphat *n* $Pb_3(PO_4)_2$ lead(II) orthophosphate, lead(II) phosphate
Blei(II)-orthophosphit *n s.* Bleiphosphit
Blei(II)-orthoplumbat *n* $Pb_2[PbO_4]$ lead(II) orthoplumbate
Blei(II)-oxalat *n* $Pb(C_2O_4)$ lead(II) oxalate
Bleioxid *n* / **rotes** *s.* Bleimennige
Blei(II)-oxid *n* PbO lead(II) oxide, lead monooxide
Blei(IV)-oxid *n* PbO_2 lead(IV) oxide, lead dioxide
Blei(II,IV)-oxid *n s.* 1. Blei(II)-orthoplumbat; 2. Blei(II)-metaplumbat
Blei(II)-oxidazetat *n* $Pb_2O(CH_3COO)_2$ lead(II) acetate oxide
Bleioxidrot *n s.* Bleimennige
Bleipackung *f* lead packing
Bleipapier *n* lead[-acetate] paper
Blei(II)-perchlorat *n* $Pb(ClO_4)_2$ lead(II) perchlorate
Blei(II)-peroxodisulfat *n* PbS_2O_8 lead(II) peroxodisulphate
Bleipfanne *f* lead pan
Blei(II)-phosphat *n* lead(II) phosphate; *(specif) s.* Blei(II)-orthophosphat
Bleiphosphit *n* $PbPHO_3$ lead phosphite, *(better)* lead phosphonate
Bleiphtalozyanin *n* lead phthalocyanine
Bleipikrat *n* $Pb[OC_6H_2(NO_2)_3]_2$ lead picrate
Bleiplatte *f* lead plate
Bleipyro... *s.* Bleidi...
Bleiraffination *f* lead refining
Blei(II)-rhodanid *n s.* Blei(II)-thiozyanat
Bleirohr *n* lead pipe
Bleirohrschlange *f* lead coil
Bleisalz *n* lead salt
Bleisalzlösung *f* solution of lead salts
Bleisammler *m s.* Bleiakkumulator
Bleischachtofen *m* lead blast furnace
Bleischlamm *m* lead sludge
Bleischlange *f s.* Bleirohrschlange
Bleischmelzofen *m* lead blast furnace
Bleiseife *f* lead soap
Blei(II)-selenat *n* $PbSeO_4$ lead(II) selenate
Blei(II)-selenid *n* PbSe lead(II) selenide
Bleistearat *n* $Pb(C_{17}H_{35}COO)_2$ lead stearate
Bleisuboxid *n* lead suboxide *(a mixture of lead and lead(II) oxide)*
Blei(II)-sulfat *n* $PbSO_4$ lead(II) sulphate
Blei(II)-sulfid *n* PbS lead(II) sulphide
Bleisulfidbehandlung *f s.* Bleisulfidsüßen
Bleisulfidsüßen *n (petrol)* lead-sulphide sweetening (treating), Bender sweetening
Bleisulfidverfahren *n (petrol)* Bender [lead-sulphide] process
Blei(II)-sulfit *n* $PbSO_3$ lead(II) sulphite

Blei(II)-tellurid *n* PbTe lead(II) telluride
Bleitetraäthyl *n (incorrectly for)* Tetraäthylblei
Bleitetraazetat *n s.* Blei(IV)-azetat
Bleitetrachlorid *n s.* Blei(IV)-chlorid
Bleitetramethyl *n (incorrectly for)* Tetramethylblei
Bleithiosulfat *n* PbS_2O_3 lead thiosulphate
Blei(II)-thiozyanat *n* $Pb(SCN)_2$ lead(II) thiocyanate
Bleiüberzug *m* lead coating
Bleiventilator *m* lead fan
Bleivergiftung *f* lead poisoning, saturnism, plumbism
Bleivitriol *n s.* Blei(II)-sulfat
Bleiwasserstoff *m* PbH_4 lead hydride, plumbane
Bleiweiß *n* white lead, ceruse *(lead carbonate hydroxide)*
feines B. flake white
Blei(II)-wolframat *n* $PbWO_4$ lead(II) wolframate, lead(II) tungstate
Blei-Zinn-Lot *n* lead-tin solder
Bleizucker *m* $Pb(CH_3COO)_2$ sugar of lead, salt of Saturn, lead(II) acetate
Bleizyanat *n* $Pb(OCN)_2$ lead cyanate
Bleizyanid *n* $Pb(CN)_2$ lead cyanide
Bleizylinderprobe *f s.* Bleiblockprobe
Blende *f* 1. orifice; screen; diaphragm; 2. *(min)* blende
blenden *(petrol)* to blend
Blenden[einlauf]kante *f* orifice edge
Blendenmischer *m* orifice mixer
Bleu de Lyon *n* Lyons blue, bleu de Lyon *(chlorine salt of triphenylrosaniline)*
blind dull, tarnished *(metal)*
b. werden to tarnish *(of metals)*
Blindbestimmung *f* blank determination
Blindboden *m* false bottom
Blindflansch *m* blind flange, blank
Blindprobe *f s.* Blindversuch
Blindrohr *n* dummy tube
Blindscheibe *f* blind
Blindtitration *f* blank titration
Blindversuch *m* blank, blank experiment (test, trial), negative control ✦ **einen B. anstellen (machen)** to run a blank
Blindwerden *n* tarnish[ing] *(of metals)*
Blindwert *m* blank reading; *(tox)* predosage level
Blisterkupfer *n* blister copper
Blitzdämpfen *n (text)* flash ageing
Blitzdämpfer *m (text)* flash ager
Blitzfotolyse *f s.* Blitzlichtfotolyse
Blitzlicht *n (phot)* flash-light
Blitzlichtfotolyse *f* flash photolysis *(an analytical method)*
Blitzlichtpulver *n (phot)* flashing powder, flash-light
Blitzpasteurisierapparat *m (food)* flash pasteurizer
Blitzpulver *n* lycopodium powder *(from club mosses)*
Blitzröstofen *m (met)* flash roaster
B. nach Nichols-Freeman Nichols-Freeman flash roaster
Blitzröstung *f (met)* flash (shower) roasting
Bloch-Wand *f (cryst)* Bloch wall
Block *m* 1. block; 2. set; 3. *(met)* ingot *(of iron)*; pig *(of non-ferrous metals)*
Blockbild *n* block diagram
Blockblei *n* pig lead
Blockeis *n* block (can, cake) ice
Blocken *n (plast)* blocking *(undesired adhesion between two sheets)*
Blockform *f* ingot mould
Blockgießen *n*, **Blockguß** *m* ingot casting
blockieren to block
Blockierung *f* blocking
blockig blocky
Blockkokille *f* ingot mould
Blockkopolymer[es] *n*, **Blockkopolymerisat** *n* block copolymer
Blockkopolymerisation *f* block copolymerization
Blocklehm *m* boulder clay
Blockmischpolymerisat *n s.* Blockkopolymer
Blockmischpolymerisation *f s.* Blockkopolymerisation
Blockpackung *f* bulk package
Blockpolymerisation *f s.* 1. Blockkopolymerisation; 2. Substanzpolymerisation
Blockpresse *f* block press
Blockschälchen *n (lab)* clearing (staining) well *(as for microscopy)*
Blockschaltbild *n* block diagram
Blockung *f s.* Blockierung
Blockzelle *f* cavity cell *(spectrometry)*
Blockzitwer *m* Cassumunar ginger, Zingiber cassumunar Roxb.
Blödit *m (min)* blödite, astrak[h]anite *(magnesium sodium sulphate)*
Blondiermittel *n*, **Blondierpräparat** *n* hair bleach (bleaching agent), blondizing agent
Blöße *f (tann)* pelt
Blubber *m (food)* [whale] blubber
Blume *f* bouquet, aroma, flavour *(as of wine)*; head *(on beer)*; *(tann)* exudation, bloom
Blumendünger *m* flower fertilizer
Blumenkohlende *n*, **Blumenkohlkopf** *m* cauliflower end *(of a piece of oven coke)*
Blumenkohlstruktur *f* cauliflower appearance *(of a coke button)*
Blumenseidenpapier *n* flower tissue
Blutalbumin *n s.* Blutserumalbumin
Blutalbuminleim *m* blood[-albumin] glue
blutdrucksenkend *(pharm)* hypotensive, antihypertensive
blutdrucksteigernd *(pharm)* pressor
Blutduftstoff *m* blood scent
bluten *(dye, coat, tann)* to bleed [off, through]
Bluten *n (dye, coat, tann)* bleeding
Blütenöl *n (cosmet)* flower oil
absolutes B. absolute flower oil, flower absolute
konkretes B. floral concrete
Blutfarbstoff *m* blood pigment
Blutfleck *m* blood stain
blutgefäßerweiternd vasodilating
blutgefäßverenge[r]nd vasoconstrictive
Blutgerinnung *f* blood coagulation (clotting)
Blutholz *n s.* Blauholz
Blutkohle *f* blood char[coal]

Blutlack *m* button lac
Blutlaugensalz *n* / **gelbes** $K_4[Fe(CN)_6]$ yellow prussiate of potash, yellow potassium prussiate, potassium hexacyanoferrate(II)
rotes B. $K_3[Fe(CN)_6]$ red prussiate of potash, red potassium prussiate, potassium hexacyanoferrate(III)
Blutmehl *n* blood meal, dried blood *(a fertilizer)*
Blutplasma *n* blood plasma
Blutserum *n* blood serum
Blutserumalbumin *n* blood [serum] albumin, serum albumin, seralbumin
Blutstein *m (min)* blood-stone, red iron ore *(iron(III) oxide)*
blutstillend styptic
Blutstillstift *m* styptic pencil
Blutstillungsmittel *n* styptic
Blutverdünnung *f* blood dilution
Blutwasseralbumin *n s.* Blutserumalbumin
Blutzucker *m* blood sugar
Bobine *f (text)* bobbin
Bock *m (tann)* horse
Boden *m* 1. *(agric, soil, geol)* soil; 2. *(tech)* bottom, base *(as of a vessel)*; *(distil)* plate, tray; deck *(as of flat screens)*; 3. floor ✦ **mit halbkugeligem (rundem) B.** round-bottomed
B. zur Neuverteilung redistributor *(in packed columns)*
idealer B. *s.* theoretischer B.
neutraler B. sweet soil
praktischer B. *(distil)* actual (practice) plate
theoretischer B. *(distil)* theoretical (ideal, perfect) plate
versetzter B. *(glass)* offset punt (base) *(a defect)*
Bodenaggregat *n* soil aggregate
Bodenaggregation *f* soil aggregation
Bodenanalyse *f* soil analysis
Bodenatmung *f* soil respiration
Bodenazidität *f* soil acidity
Bodenbegasung *f* soil fumigation
Bodenbegasungsmittel *n* soil fumigant
Bodenbegiftung *f* soil poisoning
Bodenbegiftungsmittel *n* soil poison
Bodenbelag *m* floor covering, flooring
Bodenbildung *f (soil)* pedogenesis
Bodenblasen *n (met)* bottom blowing
Bodenchemie *f* soil chemistry
Bodendesinfektion *f* soil sterilization, *(specif)* soil fumigation
Bodendesinfektionsmittel *n* soil sterilant, *(specif)* soil fumigant
Bodendispergierung *f (agric)* autodisintegration *(with sodium ions in excess)*
bodeneigen *(geol)* autochthonous, in place, in situ
Bodenentleerung *f* bottom discharge
Bodenentseuchung *f s.* Bodendesinfektion
Bodenentwicklung *f s.* Bodenbildung
Bodenerhärtungsmittel *n s.* Bodenstabilisator
Bodenfeuchte *f*, **Bodenfeuchtigkeit** *f* soil moisture
Bodenfliese *f* floor tile
Bodenflüssigkeit *f s.* Bodenwasser
Bodenfraktion *f (distil)* bottom fraction
bodenfremd *(geol)* allochthonous
Bodenfruchtbarkeit *f* soil fertility
Bodenfungizid *n* soil fungicide
Bodengare *f* tilth
Bodengefüge *n s.* Bodenstruktur
Bodenhorizont *m* soil horizon
Bodenkolloid *n* soil colloid
Bodenkolonne *f (distil)* plate (tray) column (tower)
Bodenkomplex *m* / **adsorbierender** *(soil)* base-exchange complex, ion-exchange compound, exchange material
Bodenkörper *m* precipitate, sediment, deposit, settling[s], subsidence, bottom settlings (product), bottom sediment[s], bottoms, lees, foots, dregs ✦ **einen B. bilden** to sediment, to deposit *(of a liquid)*
Bodenkörpermenge *f* amount (quantity) of precipitate
Bodenkörperregel *f (coll)* solid-phase (disperse-phase) rule
Bodenkorrosion *f* soil corrosion
Bodenkunde *f* soil science, pedology
Bodenkundler *m* soil scientist, pedologist
bodenkundlich pedologic[al]
Bodenlanze *f* soil injector
Bodenlösung *f* soil solution
Bodenluft *f* soil air
Bodenmelioration *f* soil conditioning (amendment)
Bodenmikrobiologie *f* soil microbiology
Bodenmüdigkeit *f* soil exhaustion (sickness)
Bodennährstoff *m* soil nutrient
Bodenplatte *f* 1. bottom plate; 2. *s.* Bodenfliese
Bodenprobe *f* sample of soil
Bodenprodukt *n (distil)* bottom product, bottoms
Bodenprofil *n* soil profile
Bodenreaktion *f* soil reaction
Bodensatz *m s.* Bodenkörper
Bodensatzwäscher *m (pap)* dregs washer
Bodensäule *f s.* Bodenkolonne
Bodenschicht *f* bottom layer
Bodenschlange *f* tank-bottom coil
Bodenstabilisator *m* soil stabilizer (stabilizing agent)
bodenständig *s.* bodeneigen
Bodenstein *m* bottom (lower, fixed, under) millstone, bed (base) stone, bedder *(of an edge-runner mill)*
Bodensteine *mpl* bottom brickwork
Bodenstruktur *f* soil structure
Bodenstrukturverbesserung *f s.* Bodenmelioration
Bodenventil *n* bottom valve
Bodenverbesserungsmittel *n* soil conditioner (ameliorant), [soil] amendment
Bodenverdichtung *f* soil compaction
Bodenversauerung *f* soil souring; soil sourness
Bodenwasser *n* soil water (solution)
Bodenwirkungsgrad *m (distil)* plate efficiency [factor], tray efficiency [factor]
B. nach Murphree Murphree [plate] efficiency
Bodenzahl *f (distil)* plate number
praktische (tatsächliche) B. actual plate number, number of actual plates (trays)

theoretische B. ideal plate number, number of theoretical plates (trays)
wirkliche B. *s.* praktische B.
Bogen *m* 1. sheet *(of paper or pulp)*; 2. [electric] arc
Bogendruckmaschine *f (pap)* sheet-fed press
Bogenentladung *f* arc (spark) discharge
bogengeglättet *(pap)* sheet-calendered
Bogengewicht *n s.* Masse je Bogen
Bogenhalbstoff *m (pap)* lap[ped] pulp, laps (sheets) of pulp, pulp (solid pulp) board
Bogenkalander *m (pap)* sheet calender
Bogenlampe *f* arc lamp
Bogenlampenkohle *f* arc carbon
Bogenlänge *f (pap)* length of sheet
Bogenpapier *n* sheet (ream) paper, sheeted paper, paper in sheets
Bogensatinage *f (pap)* sheet calendering
bogensatiniert *(pap)* sheet-calendered
Bogenschneiden *n (pap)* sheeting
Bogensortierung *f (pap)* sheet sorting
Bogenspektrum *n* arc spectrum
Bogenstreichmaschine *f (pap)* sheet coater
Bogenzähler *m*, **Bogenzählgerät** *n (pap)* sheet counter, sheet-counting device
Boghead-Kännel-Kohle *f* boghead cannel [coal]
Bogheadkohle *f* boghead [coal], bitumenite
Böhmit *m (min)* boehmite *(a crystalline form of aluminium oxide and hydroxide)*
Bohnerwachs *n* floor wax (polish)
Bohnerz *n* pea ore
Bohranlage *f* drilling rig
bohrbar drillable
Bohrbarkeit *f* drillability
Bohrdruckmesser *m* drillometer
bohren / nach Erdöl to bore (drill, prospect) for oil
Bohren *n (petrol)* drilling
Bohrer *m (petrol)* drill; auger *(as for samples of soil)*
Bohrflüssigkeit *f* drilling fluid
schlammartige B. *s.* Bohrschlamm
Bohrgarnitur *f* drilling string
Bohrgerät *n* drilling rig
Bohrgestänge *n* drill pipe
Bohrkern *m (geol)* core
Bohrklein *n* drill cuttings
Bohrloch *n* borehole, *(esp petrol)* [oil] well
Bohrlochabsperrvorrichtung *f (petrol)* blow-out preventer
Bohrlochkopf *m* well head
Bohrlochkopfgas *n* casing-head gas
Bohrlochverfahren *n* borehole producer method *(in underground gasification)*
Bohrlochwand[ung] *f* borehole wall
Bohrmast *m* drilling mast
Bohrmeißel *m* drilling bit
Bohrplattform *f* drilling platform
Bohrrohr *n* drill (casing) pipe
Bohrschlamm *m* drilling mud
Bohrschmand *m* drill cuttings
Bohrseil *n* drilling cable
Bohrspülung *f* drilling fluid
Bohrstrang *m* drilling string
Bohrturm *m* [drilling] derrick
stationärer B. fixed derrick
Bohrung *f* 1. *s.* Bohren; 2. boring, bore, [bore]hole; perforation, orifice *(of a screen)*; 3. *(petrol)* [oil] well
B. auf Neuland exploration drilling; exploration (exploratory) well, wildcat
freifließende B. flowing well
fündige B. discovery well
gerichtete B. directional drilling
trockene B. unproductive (dry) well, duster
Bohrwerkzeug *n* drilling tool
Bol *m s.* Bolus
Bollmann-Extrakteur *m* Bollmann extractor
Bolometer *n* bolometer *(a resistance thermometer)*
bolometrisch bolometric
Boltzmann-Faktor *m* Boltzmann factor
Boltzmann-Konstante *f* Boltzmann constant
Boltzmann-Statistik *f* [Maxwell-]Boltzmann statistics
Boltzmann-Theorem *n* Boltzmann distribution law
Boltzmann-Verteilung *f* Boltzmann distribution
Bolus *m (min)* bole
B. alba *s.* weißer B.
roter B. red bole
weißer B. white bole, bolus alba, china (porcelain) clay, kaolin[e]
Bombage *f* crown[ing], camber *(of rolls or profiles)*
bombardieren *(nucl)* to bombard
Bombardierung *f (nucl)* bombardment
Bombay-Katechu *n* Pegu catechu (cutch), black (dark) catechu (cutch) *(from Acacia catechu Willd.)*
Bombay-Macis *m* Bombay mace *(from Myristica malabarica Lam.)*
Bombe *f* bomb; [pressure] cylinder
Berthelotsche B. *s.* kalorimetrische B. nach Berthelot
kalorimetrische B. calorimeter (calorimetric, explosion) bomb, bomb calorimeter
kalorimetrische B. nach Berthelot Berthelot (Kröcker, Mahler) bomb
Reidsche B. Reid apparatus
vulkanische B. *(geol)* bomb
Bombenkalorimeter *n s.* Bombe / kalorimetrische
Bombenmethode *f* bomb method *(for determining sulphur)*
Bombenofen *m (lab)* tube (tubular) furnace, Carius furnace
Bombenrohr *n* tube for sealing, sealing (sealed) tube
B. nach Carius Carius tube
Bombensauerstoff *m* cylinder oxygen
bombieren to crown, to camber *(rolls or profiles)*
Bombierung *f s.* Bombage
Boms *m (ceram)* case *(a piece of kiln furniture)*
Bonbonsirup *m* starch syrup
Boot *n (lab)* boat
Bootform *f* boat conformation (form) *(stereochemistry)*
Bor *n* B boron
Boran *n* borane, boron hydride, *(specif)* BH_3 borane(3)

Boranat *n* $M^I[BH_4]$ hydridoborate, tetrahydridoborate
Borat *n* borate, *(specif)* $M_3^IBO_3$ orthoborate, trioxoborate
Boratglas *n* borate glass
Boräthan *n s.* Diboran(6)
Boräthyl *n s.* Bortriäthyl
Boratperoxyhydrat *n* peroxyborate *(an addition compound)*
Boratphosphor *m* borate phosphor
Boratpuffer *m* borate buffer
Borax *m* $Na_2B_4O_7 \cdot 10H_2O$ borax, sodium tetraborate-10-water, *(min also)* tincal
B. usta *s.* gebrannter B.
gebrannter (kalzinierter) B. $Na_2B_4O_7$ calcined (burnt, anhydrous, dehydrated) borax, borax usta, sodium tetraborate
oktaedrischer B. $Na_2B_4O_7 \cdot 5H_2O$ octahedral borax, jeweller's borax, sodium tetraborate-5-water
Boraxglas *n* borax glass
Boraxperle *f* borax bead
Boraxsee *m* borax lake
Borazin *n* borazene, borazine, *(specif)* $B_3N_3H_6$ borazene, borazole
Borazit *m (min)* boracite *(a tectoborate)*
Borazol *n* $B_3N_3H_6$ borazene, borazole
Bor(III)-bromid *n* BBr_3 boron(III) bromide, boron tribromide
Borbutan *n s.* Tetraboran
Bor(III)-chlorid *n* BCl_3 boron(III) chloride, boron trichloride
Bordeauxbrühe *f* Bordeaux mixture *(a fungicide)*
Bordeaux-Terpentin *n (m)* Bordeaux turpentine, French oil turpentine *(from Pinus pinaster Ait.)*
Bördelflansch *m* lap-joint flange
Bördelrand *m* bead
Bördelverbindung *f* flared-fitting joint *(of tubing)*
Bordüngemittel *n* boron fertilizer
Bor(III)-fluorid *n* BF_3 boron(III) fluoride, boron trifluoride
Borfluorwasserstoffsäure *f s.* Tetrafluoroborsäure
Borhydrid *n* boron hydride, borane
Borid *n* boride
Borin *n* borine *(any of a class of compounds B_nH_{n+2})*
Borinkarbonyl *n* BH_3CO borine carbonyl
Bor(III)-jodid *n* BI_3 boron(III) iodide, boron triiodide
Borkarbid *n* B_4C boron carbide
Borkarbonylhydrid *n s.* Borinkarbonyl
Borkezustand *m (tann)* crust condition *(of leather)*
Bormangel *m* boron deficiency
Bornan *n (org chem)* bornane
Borneokampfer *m* Borneo (Baros, Sumatra) camphor, Malay[an] camphor *(from Dryobalanops aromatica Gaertn. f.)*
Borneol *n* borneol, bornyl alcohol, 2-hydroxybornane
Borneolazetat *n s.* Bornylazetat
Bornesit *m* $C_6H_{11}O_5$-OCH_3 bornesitol *(1-O-methyl ether of myo-inositol)*
Born-Haber-Kreisprozeß *m* Born-Haber [thermochemical] cycle
Bornit *m (min)* bornite, peacock ore, horse-flesh ore, purple copper [ore] *(copper(II) iron(II) sulphide)*
Bornitrid *n* BN boron nitride
Bornylalkohol *m s.* Borneol
Bornylamin *n* bornylamine, 2-aminobornane
Bornylan *n s.* Bornan
Bornylazetat *n* bornyl acetate, borneol acetate
Bornylchlorid *n* 2-chlorobornane, bornyl chloride
Bornylen *n* bornylene, 2-bornene
Borobutan *n s.* Tetraboran
Borosilikat *n* borosilicate
Borosilikatglas *n* borosilicate (hard) glass
Borosilikatkronglas *n* borosilicate crown [glass]
Boroskampfer *m s.* Borneokampfer
Borowolframat *n (incorrectly for)* Wolframatoborat
Bor(III)-oxid *n* B_2O_3 boron(III) oxide, boron trioxide, boric oxide
Borphosphat *n (incorrectly for)* Borphosphoroxid
Borphosphid *n* BP boron phosphide
Borphosphoroxid *n* BPO_4 boron phosphorus oxide *(a double oxide of B_2O_3 and P_2O_5)*
Borsäure *f* boric acid, *(specif)* $B(OH)_3$ orthoboric acid, trioxoboric acid
Borsäureanhydrid *n s.* Bor(III)-oxid
Borsäureester *m* boric-acid ester, borate ester
Borstahl *m* boron steel
Bort *m* boort, boart, bort *(a diamond of inferior quality)*
Bort *n* boort, boart, bort *(abrasive diamond powder)*
Borte *f* bulb edge *(of window glass)*
Bortri... *s. a.* Bor(III)-...
Bortriäthyl *n* $B(CH_2CH_3)_3$ triethylborine, boron triethyl
Borverbindung *f* boron compound ✦ **mit Borverbindungen versetzen** to boronate *(fertilizers)*
Borwasserstoff *m* boron hydride, borane
Bose-Einstein-Gas *n* Bose-Einstein gas
Bose-Einstein-Statistik *f* Bose-Einstein statistics
Bose-Teilchen *n* boson
Boson *n* boson
bossieren *(ceram)* to emboss
Boswellinsäure *f* boswellic acid *(a mixture of two isomeric hydroxytriterpene acids)*
Botany-Bay-Harz *n* Botany Bay gum *(from Xanthorrhoea hastilis R. Br.)*
Botany-Bay-Kino *n* Botany Bay kino *(from Eucalyptus resinifera Sm.)*
Boten-Ribonukleinsäure *f*, **Boten-RNS** *f* messenger ribonucleic acid, messenger RNA
Bottich *m* tub, vat, trough, beck
oberster B. *(dye)* top tub (vat)
unterster B. *(dye)* bottom tub (vat)
Boucherie-Verfahren *n* Boucherie process *(wood preservation)*
Boudouard-Gleichgewicht *n* producer-gas equilibrium
Boudouard-Reaktion *f* air-carbon reaction
Bouillon *f (pharm)* broth
Bouillonkultur *f (pharm)* broth culture
Bouillonverdünnungsmethode *f (pharm)* broth di-

lution method
Boulangerit *m (min)* boulangerite *(antimony(III) lead(II) sulphide)*
Bourbonal *n* $C_6H_3(OH)(OC_2H_5)CHO$ bourbonal, 3-ethoxy-4-hydroxybenzaldehyde
Bourdon-Röhre *f* Bourdon [pressure] gauge
Bournonit *m (min)* bournonite *(sulphide of antimony, lead, and copper)*
Bouveault-Blanc-Reduktion *f* Bouveault-Blanc reduction *(of esters to alcohols)*
Boyle-Kurve *f* Boyle curve
Boyle-Punkt *m*, **Boyle-Temperatur** *f* Boyle point (temperature)
BP *s.* Brennpunkt
Brackelsberg-Ofen *m (met)* Brackelsberg furnace
Brackett-Serie *f* Brackett series *(spectroscopy)*
brackig brackish
Brackwasser *n* brackish water
Bradley-Mühle *f* Bradley mill *(a pendulum roller mill)*
  B. mit drei Pendeln (Pendelrollen) Bradley three-roll[er] mill
Bragg-Methode *f* Bragg method [of crystal analysis], Bragg rotating crystal method
Bragg-Spektrometer *n* Bragg spectrometer
Bramme *f (met)* slab
Brand *m* 1. burning; *(ceram)* firing, baking; 2. fire
Brandausbreitung *f* spread of a fire
Brandbekämpfung *f* fire fighting
Brandfleck *m* scorch, burn, burned spot
Brandgel *n* incendiary gel
Brandprobe *f (met)* fire assay
Brandriß *m (glass, ceram)* fire (firing) crack
Brandschiefer *m* carbonaceous shale
Brandschutz *m* fire protection (prevention)
brandsicher fireproof
  b. machen to fireproof
Brandsichermachen *n* fireproofing
Branntkalk *m* quicklime, burned (burnt, caustic, lump, stone, unslaked, anhydrous) lime, calcined limestone, calx, calcium oxide
Branntwein *m* alcohol, spirit[s]
  denaturierter (vergällter) B. denatured alcohol
Branntweinbrennerei *f* alcohol plant, distillery
Branntweinhefe *f* distillery (distillers') yeast
Brasilholz *n (dye)* Brazil[ian] redwood, brazilwood *(from Caesalpinia specc.)*
  gelbes B. yellow Brazil wood, dyer's mulberry *(from Chlorophora tinctoria Gaudich.)*
Brasilin *n* brazilin, brasilin *(a natural dye)*
Brasilnuß *f* Brazil (para, cream) nut *(seeds from Bertholletia excelsa Humb. et Bonpl.)*
Brasilsäure *f* brazilic (brasilic) acid
Brassidinsäure *f*, **Brassinsäure** *f* brassidic acid, *trans*-13-docosenoic acid
Brassylsäure *f* $HOOC[CH_2]_{11}COOH$ brassylic acid, tridecanedioic acid
Brauchwasser *n* industrial (process, service) water
Brauchwasseraufbereitung *f* industrial-water treatment
brauen to brew
Brauer *m* brewer
Brauerei *f* 1. brewery, brewhouse; 2. brewing
Brauereiabwasser *n* brewery waste water
Brauereichemie *f* brewing chemistry
Brauereiindustrie *f* brewing industry
Brauereimalz *n* brewer's malt, malt for brewing
Brauereitechnologie *f* brewing technology
Brauereiwasser *n* brew[ing] water, brewing liquor
Braugerste *f* brewing (brewer's, malting) barley
Brauhaus *n s.* Brauerei 1.
Brauhopfen *m* hop for brewing
Brauindustrie *f s.* Brauereiindustrie
Braukessel *m* brew kettle, [wort] copper
Braumalz *n s.* Brauereimalz
Braun *n* brown *(sensation or substance)*
  Florentiner B. Florence brown, Vandyke red *(copper(II) hexacyanoferrate(II))*
  Kasseler B. Cassel brown (earth), ulmin brown *(bituminous earthy brown coal)*
Braunbleierz *n (min)* brown (green) lead ore, pyromorphite
Brauneisenerz *n (min)* brown iron ore (stone), limonite
Braunerde *f (soil)* brown earth
Braunfärbung *f* brown colo[u]ration
Braunglas *n* amber glass
Braunglasur *f (ceram)* brown glaze
Braunholzpappe *f s.* Braunschliffpappe
Braunhuminsäure *f (soil)* brown humic acid
Braunit *m (min)* braunite *(a manganese silicate)*
Braunkohle *f* lignite, brown coal, *(according to the ASTM coal classification)* lignitic coal
  bituminöse B. bituminous lignite, lignitous (subbituminous) coal
  braune B. brown lignite
  erdige B. earthy brown lignite
  erhärtete B. *s.* verfestigte B.
  faserige B. fibre brown coal
  holzartige B. *s.* xylitische B.
  lignitische B. *s.* xylitische B.
  lockere (nichterhärtete, nichtverfestigte) B. *s.* unverfestigte B.
  schwarze B. black lignite
  steinkohlenähnliche B. *s.* bituminöse B.
  unverfestigte B. brown coal
  verfestigte B. lignite *(according to the ASTM coal classification)*
  xylitische B. xylite, woody lignite (brown coal)
Braunkohlenbrennstaub *m s.* Braunkohlenstaub
Braunkohlenbrikett *n* brown-coal briquette
Braunkohlengaserzeuger *m* brown-coal generator
Braunkohlenholz *n s.* Braunkohlenxylit
Braunkohlenkoks *m* lignite (brown-coal) coke
Braunkohlenlignit *m s.* Braunkohlenxylit
Braunkohlenschwelbrikett *n* carbonized brown-coal briquette
Braunkohlenstadium *n* lignite stage
Braunkohlenstaub *m* pulverized (powdered) brown coal
Braunkohlenteer *m* lignite (brown-coal) tar
Braunkohlenxylit *m (n)* xylite, woody lignite (brown coal)
Braunlehm *m* brown loam

Braunschliff *m (pap)* brown mechanical pulp
Braunschliffpappe *f* brown mechanical pulp board, leather board
Braunstein *m* MnO_2 manganese dioxide
Braunton *m* brown shade
Brauntonung *f (phot)* sepia toning
Bräunung *f (food)* browning
B. vom Maillard-Typus Maillard-type browning, amino-sugar browning
Bräunungsgrad *m (food)* degree of browning
Bräunungsmittel *n (food)* browning aid (ingredient, material)
Bräunungsreaktion *f (food)* browning reaction
nichtenzymatische B. *s.* Bräunung vom Maillard-Typus
Bräunungszusatz *m s.* Bräunungsmittel
Braunwerden *n (food)* 1. browning; 2. darkening *(of wine)*
Braupfanne *f* brew kettle, [wort] copper
Brauqualität *f* brewing quality (value) *(as of barley)*
Braureis *m* brewer's rice
Brause *f* 1. spray[er], sprinkler; 2. *s.* Brauselimonade
Brauselimonade *f* effervescent soft drink, effervescent lemonade
brausen 1. to spray, to sprinkle; 2. to effervesce
Brausen *n* 1. spraying, sprinkling; 2. effervescence, effervescency
brausend effervescent, effervescing
Brausepulver *n* effervescent powder (salt)
Brauwasser *n* brew[ing] water, brewing liquor
Brauwert *m s.* Brauqualität
Brauzucker *m* brewing (brewer's) sugar
Bravais-Gitter *n (cryst)* Bravais lattice
Bravaisit *m (min)* bravaisite *(aluminium dihydrogentetrasilicate)*
Breakermischung *f (rubber)* breaker stock (compound)
Brechbacke *f* crusher jaw
feststehende B. fixed jaw
schwingende B. moving jaw, swing[ing] jaw
Brecheisen *n (rubber)* [mould] breaker, mould-breaking (mould-clearing) jack, mould cracker
brechen 1. to break; to mill, to crush *(rock, ore)*; to chop *(bark)*; to roll *(flax)*; *(phys chem)* to refract, to break *(light)*; 2. to break; *(coll)* to break, to crack *(of emulsions)*
eine Emulsion b. to break (crack) an emulsion, to demulsify, to desemulsify, to de-emulsify
Brechen *n* breaking, breakage; milling, crush[ing] *(of rock, ore)*; chopping *(of bark)*; rolling *(of flax)*; *(phys chem)* refraction, breaking *(of light)*
B. einer Emulsion 1. breaking (cracking) of an emulsion, demulsification, desemulsification, de-emulsification; 2. emulsion breakdown
Brecher *m* crusher
Brecherrahmen *m* fixed seat *(of a jaw crusher)*
Brecherwalzwerk *n* breaker, breaking (breakdown) mill; *(rubber)* cracker (cracking) mill
Brechgut *n* material being *(or* to be) crushed
Brechkegel *m* crushing cone (head)
Brechkoks *m* broken coke
Brechmaul *n*, **Brechmaulöffnung** *f* feed opening *(of a crusher)*
Brechmittel *n (pharm)* emetic
Brechnuß *f* 1. nux vomica, poison nut, Strychnos nux-vomica L.; 2. *(pharm)* strychnos seed, nux vomica, poison nut *(from 1.)*
Brechnußpulver *n (pharm)* powdered strychnos seed *(from Strychnos nux-vomica L.)*
Brechnußsamen *mpl s.* Brechnuß 2.
Brechplatte *f* crushing (breaker) plate
Brechpunkt *m* breaking point
B. nach Fraass Fraass breaking point
Brechschwinge *f s.* Brechbacke / schwingende
Brechung *f (phys chem)* refraction
Brechungsdispersion *f* refractive dispersion
Brechungsindex *m*, **Brechungskoeffizient** *m* refractive (refraction) index
Brechungsmesser *m* refractometer
Brechungsquotient *m s.* Brechungsindex
Brechungsvermögen *n* refractivity
Brechungswinkel *m* angle of refraction
Brechungszahl *f s.* Brechungsindex
Brechwalze *f* crushing (crusher) roll
Brechwalzwerk *n* crushing (crusher) rolls
Brechweinstein *m* $K[C_4H_2O_6Sb(OH)_2] \cdot 0.5H_2O$ tartar emetic, potassium antimonotartrate
Brechwurz *f* ipecac[uanha], Cephaëlis ipecacuanha (Brot.) A. Rich. and C. acuminata Karsten
Brasilianische B. Brazilian ipecac[uanha], Cephaelis ipecacuanha (Brot.) A. Rich.
Brechwurzel *f (pharm)* ipecac[uanha] root *(from Cephaëlis ipecacuanha (Brot.) A. Rich. and C. acuminata Karsten)*
Brasilianische B. Rio ipecac[uanha] *(from Cephaëlis ipecacuanha (Brot) A. Rich.)*
Brechzahl *f s.* Brechungsindex
Brechzahlmesser *m s.* Brechungsmesser
Bredigit *m (min)* bredigite *(calcium orthosilicate)*
Brei *m* pulp, paste, magma
dünner B. slurry
brei[art]ig pulpy
Breitbandantibiotikum *n* broad-spectrum antibiotic
Breitbleiche *f (text)* open-width bleaching
Breitbrenner *m* flat-flame burner
Breitbrenneraufsatz *m* flat burner head, burner wing top (tip), [burner] flame spreader
Breitenwirksamkeit *f* broad-spectrum effectiveness *(as of pesticides)*
Breitfärbemaschine *f (text)* padding machine (mangle), pad[der]
Breithalter *m (text)* expander
Breithauptit *m (min)* breithauptite *(nickel antimonide)*
Breitschlitzdüse *f (plast)* slot die *(for sheet forming)*
Breitspektrumantibiotikum *n s.* Breitbandantibiotikum
Breitstrahldüse *f* slot (flat-spray) nozzle
Breitwaschmaschine *f (text)* open-width washing machine
breitwürfig *(agric)* broadcast *(application of chemicals)*
Brekzie *f* breccia *(fragmental rock)*

brekzienartig brecciated
Brekzienbildung *f* brecciation
brekziös *s.* brekzienartig
bremsen to slow down, to decelerate, to retard; to inhibit *(a reaction)*; *(nucl)* to moderate
Bremsen *n* slowing down, deceleration, retardation, retarding; inhibition *(of a reaction)*; *(nucl)* moderation
Bremsenöl *n (agric)* anti-gad-fly oil
Bremsmittel *n* retarding agent (material), retarder
Bremsstoff *m (nucl)* moderator
Bremsstrahlung *f (nucl)* bremsstrahlung
Bremssubstanz *f (nucl)* moderator
Bremsung *f (nucl)* moderation
Bremsvermögen *n* stopping power
Bremsvorrichtung *f* **der Zentrifuge** centrifuge brake kit
Bremswirkung *f* drag effect *(Debye-Hückel theory of strong electrolytes)*
Brennapparat *m* s. Brenngerät
brennbar combustible, burnable, *(esp relating to liquids)* [in]flammable
  nicht b. incombustible, non-combustible, *(esp relating to liquids)* non-flammable
Brennbares *n* combustible
Brennbarkeit *f* combustibility, *(esp relating to liquids)* [in]flammability
Brennbereich *m (ceram)* firing range
Brenndauer *f* burning time; *(ceram)* firing time
Brenneigenschaften *fpl (ceram)* firing properties
Brennelement *n s.* Brennstoffelement 1.
brennen 1. to burn; 2. *(ceram)* to fire, to bake; *(text)* to crab; to distil *(alcohol)*; to calcine *(e.g. limestone)*
  weiß b. *(ceram)* to fire to a white colour
  zu Porzellan b. to porcelainize
Brennen *n* 1. burning; 2. *(ceram)* firing, baking; *(text)* crabbing; distillation *(of alcohol)*; calcining, calcination *(as of limestone)*
Brenner *m* 1. *(lab)* burner; 2. *(tech)* burner; [welding] gun, torch; 3. distiller *(profession)*
  B. für flüssigen Brennstoff liquid-fuel burner
  B. zum Anheizen (Anlassen) starting burner *(of a furnace)*
Brennerei *f* distillery, alcohol plant
Brennereibetrieb *m* 1. *s.* Brennerei; 2. distillation (distilling, distillery) operation
Brennereihefe *f* distillery (distillers') yeast
Brennereiindustrie *f* distilling industry
Brennereimaische *f* distillery mash
Brennereischlempe *f* distillery slop (vinasse), distillers' wash, spent mash (wash)
Brennerflamme *f* burner flame
Brennerhals *m (glass)* port neck
Brennerhaus *n* burner (hot) house *(carbon-black production)*
Brennermaul *n (glass)* port mouth (opening)
Brennermund *m* burner throat
Brennermundstück *n* burner tip
Brennermündung *f* s. Brennermaul
Brennerrinne *f* burner channel
Brennerrohr *n* burner tube, barrel *(as of a Bunsen burner)*
Brennerstein *m* burner tile
Brennerzunge *f (glass)* tongue [tile], midfeather, mid-wall
Brennfarbe *f (ceram)* fired colour
Brennfehler *m (ceram)* firing defect (fault)
Brenngas *n* combustible gas
Brenngerät *n* distillation apparatus *(in alcohol distillation)*
Brenngeschwindigkeit *f* burning rate, rate of burning; rate of combustion
brennhärten to flame-harden
Brennhärtung *f* flame hardening
Brennhilfsmittel *n (ceram)* piece (item) of kiln furniture
Brennintervall *n (ceram)* firing range
Brennkammer *f* combustion chamber (space); *(ceram)* firebox, firing box (chamber)
Brennkanal *m* 1. *(ceram)* firing channel; 2. *s.* Brennschacht
Brennkapsel *f (ceram)* fireclay box, saggar, sagger
Brennkegel *m* **nach Seger** Seger cone
Brennkurve *f (ceram)* firing curve
Brennmalz *n* distillery malt
Brennmaterial *n* fuel
Brennofen *m (ceram)* kiln, stove
  B. mit direktem (offenem) Feuer open-flame kiln
  intermittierender (periodisch arbeitender, periodischer) B. periodic kiln
Brennöl *n* burning oil
Brennplatte *f (ceram)* bat
Brennprobe *f* burning test
Brennpunkt *m* burning (fire) point
Brennraum *m* furnace chamber; combustion chamber (space)
Brennreife *f (ceram)* maturity
Brennriß *m (glass, ceram)* fire (firing) crack
Brennrohr *n s.* Brennerrohr
Brennschacht *m* combustion chamber *(of a hot-blast stove)*
Brennschneiden *n* oxygen cutting
  autogenes B. oxygen-acetylene (oxyacetylene) cutting
Brennschwindung *f (ceram)* fire (firing) shrinkage, firing contraction
Brennspiritus *m* mineralized methylated spirit
Brennstaub *m* pulverized (powdered) fuel
brennstaubgefeuert pulverized-fuel-fired
Brennstoff *m* [combustion] fuel
  fester B. solid fuel
  fossiler B. fossil fuel
  gasförmiger B. gaseous fuel
  künstlicher B. prepared fuel
  mineralischer B. *s.* fossiler B.
  rauchfreier (rauchloser) B. smokeless fuel
  rußfreier B. *s.* rauchfreier B.
  synthetischer B. prepared fuel
  veredelter B. prepared fuel
Brennstoffbett *n* fuel bed
  festes B. fixed fuel bed
  fluidisiertes (kochendes) B. *s.* wirbelndes B.
  ruhendes B. *s.* festes B.

wirbelndes B. fluidized fuel bed
Brennstoffchemie *f* fuel chemistry
Brennstoffchemiker *m* fuel chemist
Brennstoffeinsparung *f* economy in fuel
Brennstoffelement *n* 1. *(nucl)* fuel element; 2. *(el chem)* fuel cell
Brennstoffofen *m* fuel-heated furnace
Brennstofforschung *f* fuel research
Brennstoffschicht *f*, **Brennstoffschüttung** *f s.* Brennstoffbett
Brennstoffstaub *m s.* Brennstaub
Brennstofftechnologie *f* fuel technology
Brennstoffverwertung *f* fuel utilization
Brennstoffzelle *f (el chem)* fuel cell
Brennstütze *f (ceram)* post, upright, prop
Brenntemperatur *f (ceram)* firing temperature
maximale B. peak firing temperature
Brennunterlage *f (ceram)* bat
Brennverhalten *n (ceram)* firing behaviour
Brennversuch *m* burning test
Brennzeit *f* burning time; *(ceram)* firing time
Brennzone *f* combustion zone *(of a blast furnace)*; calcining zone (compartment)
Brenzkatechin *n* $C_6H_4(OH)_2$ catechol, pyrocatechol, *o*-dihydroxybenzene
Brenzkatechindimethyläther *m* $C_6H_4(OCH_3)_2$ catechol dimethyl ether, veratrol, 1,2-dimethoxybenzene
Brenzkatechinmonoäthyläther *m* $C_2H_5OC_6H_4OH$ catechol monoethyl ether, *o*-ethoxyphenol, guäthol
Brenzkatechinmonomethyläther *m* $CH_3OC_6H_4OH$ catechol methyl ether, guaiacol, *o*-methoxyphenol
brenzlig empyreumatic *(smell)*
Brenzschleimsäure *f* pyromucic acid, furan-2-carboxylic acid
Brenztraubensäure *f* $CH_3COCOOH$ pyruvic acid, 2-oxopropionic acid
Brenztraubensäurealdehyd *m* CH_3COCHO pyruvic aldehyde, methylglyoxal, 2-oxopropionaldehyde
Brenzweinsäure *f* $HOOCCH(CH_3)CH_2COOH$ pyrotartaric acid, methylsuccinic acid, 2-methyl-1,4-butanedioic acid
Brevifolin *n (tann)* brevifolin
Brevifolinkarbonsäure *f (tann)* brevifolincarboxylic acid
Brevilagin *n (tann)* brevilagin
Brewsterit *m (min)* brewsterite *(a tectosilicate)*
Briefhüllenpapier *n s.* Briefumschlagpapier
Briefmarkenpapier *n* [postage] stamp paper
Briefpapier *n* letter (note) paper, *(Am)* correspondence paper
Briefumschlagfutterseide *f* envelope lining [tissue]
Briefumschlagpapier *n* envelope paper
Briefumschlagseide *f s.* Briefumschlagfutterseide
Brightstock *m*, **Brightstock-Öl** *n (petrol)* bright stock
Brikett *n* briquet[te]
pechgebundenes B. pitch-bound briquette
Brikettfabrik *f* briquetting (briquette) plant
Brikettfestigkeit *f* briquette strength
Brikettieranlage *f* briquetting (briquette) plant (installation)
brikettierbar briquettable
Brikettierbarkeit *f* briquettability, briquetting properties (qualities)
Brikettierdruck *m s.* Brikettierpreßdruck
Brikettiereigenschaften *fpl s.* Brikettierbarkeit
brikettieren to briquette
brikettierfähig *s.* brikettierbar
Brikettierfähigkeit *f s.* Brikettierbarkeit
Brikettierkohle *f* briquetting (briquette) coal
Brikettiermaschine *f* briquetting (briquette) machine
Brikettierpech *n* briquetting pitch
Brikettierpreßdruck *m* briquetting pressure
Brikettierpresse *f* briquetting (briquette) press
Brikettiersteinkohle *f* briquetting (briquette) coal
Brikettierung *f* briquetting
Brikettierungs... *s.* Brikettier...
Brikettierverfahren *n* briquetting method (technique)
Brikettierwalzen *fpl* briquetting rolls
Brikettierwerk *n s.* 1. Brikettfabrik; 2. Brikettiermaschine
Brikettkohle *f* briquetting (briquette) coal
Brikettkoks *m* briquette coke
Brikettpech *n* briquetting pitch
Brikettpresse *f s.* Brikettierpresse
Brikettwalzenpresse *f* roll-type briquetting (briquette) machine, Belgium roll machine
brillant brilliant
Brillantalizarinblau *n* brilliant alizarin blue
Brillantfarbstoff *m* brilliant dye
Brillantgelb *n* brilliant yellow
Brillantgrün *n* brilliant (emerald) green *(a basic triphenylmethane dye)*; emerald green, chrome green, Mittler's green, Guignet's green *(hydrated chromium oxide)*
Brillantine *f* brilliantine
Brillantrosa *n* brilliant pink
Brillanz *f* brilliance, brilliancy
Brillouin-Polyeder *n (cryst)* Brillouin polyhedron
Brillouin-Zone *f (cryst)* Brillouin zone
Brinellhärte *f* Brinell hardness
Brinellprobe *f* Brinell test
Brinellzahl *f* Brinell number
bringen:
auf Typ b. *(dye)* to bring to standard strength
in Kontakt b. to contact
in Lösung b. to bring (put) into solution
ins Gleichgewicht b. to bring into equilibrium, to equilibrate
zum Erlöschen b. to extinguish
zum Kochen b. to bring (raise) to the boil
zum Schäumen b. to foam, to froth
zum Stehen b. to arrest *(a reaction)*
zur Ausflockung b. to coagulate, to curdle
zur Deckung b. to make coincide
zur Explosion b. to explode
zur Kristallisation b. to crystallize out
zur Reaktion b. to react
Brisanz *f* brisance, shattering power

Brisanzwert *m* brisance value
Bristolkarton *m* Bristol board
Britannia-Metall *n* Britannia metal *(an alloy containing Sn, Sb, Cu, and Bi)*
Brochantit *m (min)* brochantite *(a basic copper sulphate)*
Brocken *m* lump *(as of coal, sugar)*; bat *(as of clay or plaster)*
bröcklig friable, crumbly
Bröckligkeit *f* friability, crumbliness
Brom *n* Br bromine
Bromabspaltung *f* debromination
Bromal *n* Br_3CCHO bromal, tribromoacetaldehyde
Bromalid *n (org chem)* bromalide
Bromalkan *n* bromoalkane
Bromallylalkohol *m* bromoallyl alcohol
Bromaminsäure *f* bromamine acid, 1-amino-4-bromoanthraquinone-2-sulphonic acid
Bromanil *n* bromanil, tetrabromo-*p*-benzoquinone
Bromanilsäure *f* bromanilic acid, 3,6-dibromo-2,5-dihydroxy-*p*-benzoquinone
Bromargyrit *m (min)* bromargyrite *(silver bromide)*
Bromat *n* M^IBrO_3 bromate ✦ **mit B. behandeln** to bromate
Bromäthan *n* CH_3CH_2Br bromoethane
Bromäthanal *n* $BrCH_2CHO$ bromoethanal, bromoacetaldehyde
2-Bromäthanol *n* $BrCH_2CH_2OH$ 2-bromoethanol, 2-bromoethyl alcohol
Bromäthansäure *f s.* Bromessigsäure
Bromäthen *n* $CH_2{=}CHBr$ bromoethylene, vinyl bromide
Bromäthyl *n s.* Bromäthan
2-Bromäthylalkohol *m s.* 2-Bromäthanol
Bromäthylen *n s.* Bromäthen
Bromatometrie *f* bromatometry
bromatometrisch bromatometric
Bromazetaldehyd *m* $BrCH_2CHO$ bromoacetaldehyde
Bromazetol *n* $CH_3CBr_2CH_3$ bromacetol, 2,2-dibromopropane
Bromazeton *n* CH_3COCH_2Br bromoacetone
Brombenzoesäure *f* BrC_6H_4COOH bromobenzoic acid
Brombenzol *n* C_6H_5Br bromobenzene
***N*-Brombernsteinsäureimid** *n* *N*-bromosuccinimide, NBS
1-Brombutan *n* $CH_3CH_2CH_2CH_2Br$ 1-bromobutane
2-Brombutan *n* $CH_3CH_2CHBrCH_3$ 2-bromobutane
Brombutansäure *f s.* Brombuttersäure
Brombuttersäure *f* bromobutyric acid
Brombutylkautschuk *m* bromobutyl (brominated butyl) rubber
Bromchlorargyrit *m (min)* bromchlorargyrite, embolite
Bromdampf *m* bromine vapour
Bromdekahydrat *n s.* Brom-10-Wasser
Bromdesulfonierung *f* bromodesulphonation
Bromelin *n* bromelin *(an enzyme from Ananas comosus (L.) Merr.)*
Bromellit *m (min)* bromellite *(beryllium oxide)*
Bromessigsäure *f* $CH_2BrCOOH$ bromoacetic acid
Brometon *n* $(CH_3)_2C(OH)CBr_3$ brometone, 1,1,1-tribromo-2-methyl-2-propanol
Brom(III)-fluorid *n* bromine(III) fluoride, bromine trifluoride
Brom(V)-fluorid *n* bromine(V) fluoride, bromine pentafluoride
Bromgelatineplatte *f (phot)* gelatin bromide plate
Bromgoldsäure *f s.* Tetrabromogold(III)-säure
bromhaltig bromine-containing
Bromhydrin *n* bromohydrin
Bromid *n* M^IBr bromide
Bromidpapier *n (phot)* bromide (bromic-silver) paper
bromieren to brominate
  zweifach b. to dibrominate
Bromierung *f* bromination
  zweifache B. dibromination
Bromierungsgeschwindigkeit *f* rate of bromination
Bromierungsmittel *n* brominating agent
Bromierungsreaktion *f* bromination reaction
Bromismus *m (tox)* bromism
2-Bromisobuttersäure *f* $(CH_3)_2C(Br)COOH$ 2-bromoisobutyric acid, 2-bromo-2-methylpropionic acid
Bromitentschlichtung *f (text)* sodium-bromite desizing
Bromkampfer *m* bromocamphor, bromated (brominated, monobrominated) camphor
Bromkresol *n* $CH_3C_6H_3(Br)OH$ bromocresol
Bromkresolgrün *n* bromocresol green *(a pH indicator)*
Bromkresolpurpur *m* bromocresol purple *(a pH indicator)*
Brommethan *n* CH_3Br bromomethane
Brommethyl *n s.* Brommethan
2-Brom-2-methylpropan *n* $(CH_3)_3CBr$ 2-bromo-2-methylpropane
Brommonofluorid *n* BrF bromine monofluoride
Brommonosilan *n* SiH_3Br bromomonosilane, bromosilane
Bromoantimonat *n* bromoantimonate
Bromoaurat(I) *n* $M^I[AuBr_2]$ bromoaurate(I), dibromoaurate(I)
Bromoaurat(III) *n* $M^I[AuBr_4]$ bromoaurate(III), tetrabromoaurate(III)
Bromoform *n* $CHBr_3$ bromoform, tribromomethane
Bromoiridat(III) *n* $M_3^I[IrBr_6]$ bromoiridate(III), hexabromoiridate(III)
Bromoiridat(IV) *n* $M_2^I[IrBr_6)$ bromoiridate(IV), hexabromoiridate(IV)
Bromojodat(I) *n* $M^I[IBr_2]$ bromoiodate(I)
Bromokuprat *n* bromocuprate
Bromometrie *f* bromometry
bromometrisch bromometric
Bromonium *n* bromonium
Bromoniumion *n* bromonium ion
Bromoosmat(III) *n* $M_3^I[OsBr_6]$ bromoosmate(III), hexabromoosmate(III)
Bromoosmat(IV) *n* $M_2^I[OsBr_6]$ bromoosmate(IV), hexabromoosmate(IV)
Bromopalladat(II) *n* $M_2^I[PdBr_4]$ bromopalladate(II), tetrabromopalladate(II)

Bromopalladat(IV) *n* $M_2^I[PdBr_6]$ bromopalladate(IV), hexabromopalladate(IV)
Bromoplatinat(II) *n* $M_2^I[PtBr_4]$ bromoplatinate(II), tetrabromoplatinate(II)
Bromoplatinat(IV) *n* $M_2^I[PtBr_6]$ bromoplatinate(IV), hexabromoplatinate(IV)
Bromoplatin(II)-säure *f* $H_2[PtBr_4]$ bromoplatinic(II) acid, tetrabromoplatinic(II) acid
Bromoplatin(IV)-säure *f* $H_2[PtBr_6]$ bromoplatinic(IV) acid, hexabromoplatinic(IV) acid
Bromorhenat(IV) *n* $M_2^I[ReBr_6]$ bromorhenate (IV), hexabromorhenate(IV)
Bromorhodat(III) *n* $M_3^I[RhBr_6]$ bromorhodate(III), hexabromorhodate(III)
Bromoselenat(IV) *n* $M_2^I[SeBr_6]$ bromoselenate(IV), hexabromoselenate(IV)
Bromostannat *n* bromostannate
Bromotellurat *n* $M_2^I[TeBr_6]$ bromotellurate, hexabromotellurate
Bromotitanat *n* $M_2^I[TiBr_6]$ bromotitanate, hexabromotitanate
Bromozinkat *n* bromozincate
Brompentafluorid *n* BrF_5 bromine pentafluoride
Bromphenol *n* BrC_6H_4OH bromophenol
Bromphenolblau *n* bromophenol blue, tetrabromophenolsulphonphthalein *(a pH indicator)*
Bromphosgen *n* $COBr_2$ bromophosgene, carbonyl bromide, carbon dibromide oxide
Brompropansäure *f s.* Brompropionsäure
1-Brompropen *n* $CH_3CH{=}CHBr$ 1-bromopropene
3-Brompropen *n* $CH_2{=}CHCH_2Br$ 3-bromopropene, allyl bromide
Brompropionsäure *f* bromopropionic acid
Bromsäure *f* $HBrO_3$ bromic acid
Bromsilan *n* bromosilane, *(specif)* SiH_3Br bromomonosilane, bromosilane
Bromsilber *n (phot)* silver bromide
Bromsilberdruck *m* bromide print
Bromsilberkristalle *mpl (phot)* silver-bromide crystals
Bromsilberpapier *n (phot)* bromide (bromic-silver) paper
***N*-Bromsukzinimid** *n N*-bromosuccinimide
Bromthymol *n* bromothymol, 4-bromo-2-isopropyl-5-methylphenol
Bromthymolblau *n* bromothymol blue, dibromothymolsulphonphthalein *(a pH indicator)*
Bromtoluol *n* bromotoluene
Bromtrichlorsilan *n (incorrectly for)* Trichlorbromsilan
Bromtrifluorid *n* BrF_3 bromine trifluoride
Bromwasser *n* bromine water
Brom-10-Wasser *n* $Br_2 \cdot 10H_2O$ bromine-10-water, bromine decahydrate
Bromwasserstoff *m* HBr hydrogen bromide
Bromwasserstoffabspaltung *f* dehydrobromination
Bromwasserstoffsäure *f* HBr hydrobromic acid
α-Bromxylol *n* $CH_3C_6H_4CH_2Br$ α-bromoxylene
Bromzahl *f* bromine number
Brönner-Säure *f (dye)* Brönner's acid *(2-naphthylamine-6-sulphonic acid)*
Bronze *f* bronze, *(specif)* tin bronze
Bronzefleck *m* bronze speck *(a defect in paper)*
Bronzelack *m* bronzing lacquer
Bronzepapier *n* bronze paper
Bronzepulver *n* bronze powder
Bronzetinktur *f* bronzing liquid (fluid)
bronzieren to bronze
Bronzit *m (min)* bronzite *(an inosilicate)*
Brookit *m (min)* brookite *(titanium dioxide)*
Brotfruchtbaum *m* breadfruit tree, Artocarpus communis I.R. et G. Forst.
Brotmehl *n* bread flour
Brotteig *m* bread dough
Brotweizen *m* common wheat, Triticum aestivum L.
Brownmillerit *m (min)* brownmillerite *(an aluminium calcium iron oxide)*
Bruch *m* 1. *(process)* breaking, breakage, fracturing, fracture, rupture; 2. *(result)* break, fracture, rupture; *(min)* fracture *(texture of a broken surface)*; *(min)* cleavage *(the particular manner in which a mineral may be cleft or split)*; scrap *(discarded metal collected for melting down)*; cullet *(broken glass for re-melting)*; curd *(the coagulated part of milk used for cheese making)*; casse *(a disorder in wine)*
  B. der C-C-Bindung C-C bond rupture, C-C rupture
  B. durch Stoß breakage by impact
  glasiger B. vitreous fracture
  hakiger B. hackly fracture
  interkristalliner B. intercrystalline fracture
  intrakristalliner B. transcrystalline fracture
  weißer B. white casse *(a disorder in wine)*
  würfelförmiger B. *(glass)* dice
Bruchbearbeitung *f (food)* curd treatment
Bruchbildung *f (food)* curd formation
Bruchdehnung *f* strain (elongation, extension) at break, breaking elongation (extension); *(pap)* tensile stretch
Bruchfestigkeit *f* breaking strength (resistance); bursting strength *(the capacity of a material to resist pressure)*; *(ceram)* crushing strength; *(food)* curd firmness (strength)
Bruchfläche *f* fracture surface
Bruchglas *n* cullet
brüchig brittle
  b. machen to embrittle
  b. werden to embrittle
Brüchigkeit *f* brittleness
Bruchkorn *n (food)* curd grain (particle)
Bruchlast *f (pap)* breaking strain
Bruchmesser *n (food)* curd knife
Bruchmodul *m* modulus of rupture
Bruchpapier *n* waste stuff
Bruchpunkt *m* breaking point, point of rupture
bruchsicher shatterproof
Bruchstück *n* fragment; *(nucl)* fission fragment
Bruchteil *m* fraction
Bruchzeit *f (rubber)* flex-life time
Brucit *m (min)* brucite *(magnesium hydroxide)*
Brücke *f* bridge *(chemical-bond theory)*; *(plast, rubber)* cross-link[age]; *(glass)* bridge [wall]

aliphatische B. aliphatic bridge
Brückenatom *n* bridge atom
Brückenbildung *f* bridging; *(pap)* arching *(of chips in the silo)*
Brückenbindung *f* bridge (bridging) bond
polysulfidische B. polysulphidic bridge (cross-link, link)
Brückenglied *n* *(org chem)* binding link; *(plast, rubber)* bridge-type cross-link
Brücken-Ion *n* bridged ion
Brücken-Kohlenwasserstoff *m* bridged hydrocarbon
Brückenkopf *m* bridgehead *(in bridge-ring systems)*
Brückenkopfatom *n* bridgehead atom
Brückenmethode *f s.* Brückenverfahren
Brückenname *m*, **Brückenpräfix** *n (nomencl)* bridge name
Brückenring *m* bridge[d] ring
Brückenringsystem *n* bridge-ring system
Brückenringverbindung *f* bridge-ring compound
Brückensauerstoff *m* bridging oxygen
Brückenschaltung *f* bridge [circuit]
Wheatstonesche B. Wheatstone bridge circuit
Brückenverbindung *f* bridge-type bond
Brückenverfahren *n* bridge method
Brückenwand *f (glass)* bridge [wall]
Brückenwannenofen *m (glass)* bridge-type furnace
Brüden *m* vapour
Brüdenabscheider *m* vapour condenser, demister
Brüdenaustritt *m* vapour outlet
Brüdendampf *m* vapour
Brüdenhaube *f* vapour hood, air dome
Brüdenkondensat *n* vapour condensate
Brüdenkondensator *m s.* Brüdenabscheider
Brüdenraum *m* vapour chamber (space); flash chamber *(of a flash evaporator)*; body *(of an evaporator)*
Brüdenschlottrockner *m* cascade (tower) dryer
Brüdenverdichtung *f* vapour compression
Brühe *f* 1. broth *(a culture medium)*; 2. *(tann)* liquor; 3. *(agric)* wash *(pest control)*
Bordelaiser B. Bordeaux mixture *(fungicide)*
Burgunder B. Burgundy mixture, soda bordeaux *(fungicide)*
maskierte B. *(tann)* masked liquor (solution)
unmaskierte B. *(tann)* straight liquor (solution)
Brühebehälter *m (agric)* spray tank
brühen *(text)* to bowk, to kier-boil
Brühen *n (text)* bowking, bocking, kier boiling
Brühenmesser *m (tann)* bark[tr]ometer
brünieren to brown *(metal surfaces)*
Brunnen *m* 1. well; *(natural fountain)* spring, *(specif)* mineral spring; *(petrol)* [oil] well; 2. mineral waters
fließender B. *(petrol)* flowing well
Brustkalander *m (rubber)* inverted L type of calender
Brustwalze *f (pap)* breast roll
brütbar *(nucl)* fertile
Brütbarkeit *f (nucl)* fertility
Brüten *n (nucl)* fertilization
Brüter *m s.* Brutreaktor
brutfähig *s.* brütbar
Brutfähigkeit *f s.* Brütbarkeit
Brutmaterial *n (nucl)* fertile material
Brutreaktor *m (nucl)* breeder reactor
schneller B. fast breeder reactor
Brutschrank *m* incubator ✦ **im B. aufbewahren** to incubate
B. mit natürlicher Luftumwälzung gravity convection incubator
Brutstoff *m s.* Brutmaterial
Bruttoformel *f* empirical formula, *(specif)* empirical molecular formula
einfachste B. simplest [possible] formula, stoichiometric formula
wahre B. true (empirical molecular) formula
Bruttogleichung *f* over-all equation *(of a reaction)*
Brutvorgang *m (nucl)* fertilization
Bruun-Kolonne *f (distil)* Bruun column
Bruzin *n* brucine, 2,3-dimethoxystrychnine *(alkaloid)*
Bruzinpapier *n* brucine paper *(for detecting HNO_2)*
BSB *s.* Sauerstoffbedarf / biochemischer
BSB_5-Wert *m s.* Sauerstoffbedarf / fünftägiger biochemischer
BTÄ *(abbr)* Bleitetraäthyl, *s.* Tetraäthylblei
Buchbinderpappe *f* [book]binder's board
Buchbinderschneidemaschine *f (pap)* trimming machine, trimmer
Buchdruckfarbe *f* letterpress ink
Buchdruckpapier *n* book[-printing] paper, *(Am)* text paper
Bucheckernöl *n* beechnut oil
Buchenholz *n* beechwood
Buchenholzteerkreosot *n* beechwood creosote
Bucherer[-Lepetit]-Reaktion *f* Bucherer reaction *(the conversion of a naphthylamine to a naphthol or vice versa)*
Bücherpapier *n s.* 1. Buchdruckpapier; 2. Bücherschreibpapier
Bücherpappe *f* [book]binder's board
Bücherschreibpapier *n* ledger [paper], account book paper
Büchner-Nutsche *f*, **Büchner-Trichter** *m* Büchner filter (funnel)
Buchöl *n s.* Bucheckernöl
Büchse *f* can, preserve can (tin)
Büchsenbier *n* canned beer
Büchsenmilch *f* canned milk
Buchstabenbuna *m* lettered buna rubber, letter grade of buna
Buchweizenhonig *m* buckwheat honey
Buchweizenmehl *n* buckwheat flour
Budde-Effekt *m (phot)* Budde effect
buffieren *(tann)* to buff
Bufotenin *n* bufotenine *(a toad poison)*
Bügelarmausrüstung *f s.* Bügelfreiausrüstung
bügelecht fast to ironing
Bügelechtheit *f* fastness to ironing
Bügelfreiausrüstung *f (text)* no-iron finish
Bügelmethode *f (phys chem)* detachment method
bügeln *(glass)* to flatten; *(text)* to press, to iron

Bugspriet *m* bowsprit *(stereochemistry)*
Bühne *f s.* Bedienungsbühne
Bukett *n* bouquet *(as of wine)*
Bukkokampfer *m* buchucamphor, buccocamphor, diosphenol, 2-hydroxy-6-isopropyl-1-methyl-2-cyclohexen-1-one
Bullers-Ring *m (ceram)* Bullers ring *(for measuring temperatures in a kiln)*
Bülwern *n (glass)* blocking
Buna *m (n) s.* Bunakautschuk
Bunakautschuk *m* buna [rubber]
Bunakrümel *mpl (npl)* crumbs of buna synthetic rubber
Bunalatex *m* buna latex
bündeln to concentrate *(e.g. rays)*
Bündelung *f* concentration *(as of rays)*
Bunker *m* bunker, bin, silo, [storage] tank; *(if funnel-shaped)* hopper
Bunker-C-Öl *n s.* Bunkeröl C
Bunkerkohle *f* bunker coal
Bunkeröl *n* bunker fuel [oil]
Bunkeröl C *n* bunker C fuel [oil]
Bunsenbrenner *m* Bunsen burner
Bunsenflamme *f* Bunsen flame
Bunsentrichter *m* Bunsen (long-stemmed) funnel
Bunsenventil *n* Bunsen valve
Buntätzen *n (text)* coloured discharge
Bunte-Bürette *f* Bunte gas burette
Buntglas *n* coloured glass
Buntglaspapier *n* diaphanic paper
Buntkupferkies *m (min)* peacock (horse-flesh, purple copper) ore, bornite *(copper(II) iron(II) sulphide)*
Buntpapier *n* coloured paper
Buntpappe *f* tinted cardboard
Buntstift *m* crayon
Buntwaschmittel *n* detergent for the coloured wash
Bürette *f* burette
 B. mit automatischer Nullpunkteinstellung automatic zero burette
 B. mit Schellbachstreifen Schellbach burette
 B. nach Squibb Squibb burette
 automatische B. automatic burette
 Buntesche B. Bunte gas burette
Bürettenbürste *f* burette brush
Bürettenhahn *m* burette stopcock (valve)
Bürettenhalter *m s.* Bürettenklemme
Bürettenklemme *f* burette clamp (holder)
 zweiarmige B. double-beam burette holder
Bürettenquetschhahn *m* burette pinchcock
Bürettensperrhahn *m s.* Bürettenhahn
Bürettenspitze *f* burette jet (tip, outlet, tube) *(a replacement part)*
Bürettentrichter *m* burette filler (funnel)
Burgers-Versetzung *f (cryst)* screw dislocation
Burgunderharz *n*, **Burgunderpech** *n* Burgundy pitch
Bürste *f* brush
bürsten to brush
Bürstenentstauber *m* brush sifter
Bürstenfeuchter *m (pap)* brush damper
Bürstenglättung *f (pap)* brush polishing
Bürstenschaber *m (pap)* brush doctor
Bürstenstreichmaschine *f* brush coater (spreader, spreading machine)
Bürstenstrich *m* brush coating
Bürstfärberei *f*, **Bürstfärbung** *f* brush dyeing
Bürstmaschine *f* brush[ing] machine
Burt-Filter *n* Burt filter *(a leaf filter with rotating filter drum)*
Burton-Clark-Spaltverfahren *n (petrol)* Burton-Clark [cracking] process
burtonisieren to burtonize, to gypsum *(brewing water)*
Burton-Spaltverfahren *n (petrol)* Burton [cracking] process
Butadien *n* butadiene; *(specif) s.* Butadien-(1,3)
Butadien-(1,3) *n* $\overset{1}{C}H_2{=}CHCH{=}CH_2$ 1,3-butadiene, butadiene
Butadien-Akrylnitril-Kautschuk *m* butadiene-acrylonitrile rubber, nitrile[-butadiene] rubber, NBR
Butadien-Akrylnitril-Mischpolymerisat *n* butadiene-acrylonitrile copolymer
Butadienanlage *f* butadiene plant
Butadienkautschuk *m* butadiene rubber, BR
Butadienmischpolymerisat *n* butadiene copolymer
Butadienpolymer[es] *n*, **Butadienpolymerisat** *n* butadiene polymer
Butadien-Styrol-Kautschuk *m* butadiene-styrene rubber, styrene-butadiene rubber, SBR
 ölgestreckter B. oil-extended styrene-butadiene rubber, OE-SBR, oil-extended (oil-masterbatched) polymer, OEP
 ölhaltiger (ölplastizierter) B. *s.* ölgestreckter B.
Butadien-Styrol-Latex *m* butadiene-styrene latex
Butadien-Styrol-Mischpolymerisat *n* butadiene-styrene copolymer
Butadien-Vinylpyridin-Mischpolymerisat *n* butadiene-vinylpyridine copolymer
Butadiin *n* $HC{\equiv}CC{\equiv}CH$ butadiyne, biacetylene
Butan *n* C_4H_{10} butane; *(specif) s. n*-Butan
***i*-Butan** *n* $(CH_3)_3CH$ isobutane, 2-methylpropane
***n*-Butan** *n* $CH_3CH_2CH_2CH_3$ *n*-butane
Butanabtrennung *f* debutanization
***n*-Butanal** *n* $CH_3CH_2CH_2CHO$ *n*-butanal, *n*-butyric aldehyde
Butandial-(1,4) *n* $OHCCH_2CH_2CHO$ 1,4-butanedial, succindialdehyde
Butandikarbonsäure-(1,4) *f* $HOOC[CH_2]_4COOH$ butane-1,4-dicarboxylic acid, hexanedioic acid, adipic acid
Butandiol-(2,3) *n* $CH_3CH(OH)CH(OH)CH_3$ 2,3-butanediol, 2,3-dihydroxybutane
Butandion-(2,3) *n* $CH_3COCOCH_3$ 2,3-butanedione, biacetyl
Butandisäure *f* $HOOCCH_2CH_2COOH$ butanedioic acid, ethane-1,2-dicarboxylic acid, succinic acid
Butandisäuremonoamid *n* $NH_2OCCH_2CH_2COOH$ succinic acid monoamide, succinamic acid
butanfrei butane-free
butanhaltig butane-containing
Butanol-(1) *n* $CH_3[CH_2]_2CH_2OH$ 1-butanol, *n*-butyl alcohol
Butanol-(2) *n* $CH_3CH_2CH(OH)CH_3$ 2-butanol, SBA
sek.-**Butanol** *n s.* Butanol-(2)

tert.-**Butanol** *n* $(CH_3)_3COH$ 2-methyl-2-propanol, *tert*-butyl alcohol
Butanol-(3)-al-(1) *n* $CH_3CH(OH)CH_2CHO$ 3-hydroxybutanal, 3-hydroxybutyraldehyde, acetaldol, aldol
Butanol-Azeton-Gärung *f* butylic fermentation
Butanon(-2) *n* $CH_3COC_2H_5$ 2-butanone, 2-oxobutane
Butansäure *f s.* Buttersäure
Butantetrol *n* $CH_2OH[CHOH]_2CH_2OH$ butanetetrol, tetrahydroxybutane
Butantrennkolonne *f* debutanizer
Buteakino *n* butea gum, gum butea, Bengal kino *(from Butea superba Roxb.)*
Butein *n* butein *(a chalcone derivative)*
Buten-(1) *n* $CH_2{=}CHCH_2CH_3$ 1-butene
Buten-(2) *n* $CH_3CH{=}CHCH_3$ 2-butene
Buten-(2)-al-(1) *n* $CH_3CH{=}CHCHO$ 2-butenal, crotonic aldehyde
Butendisäure *f* $HOOCCH{=}CHCOOH$ butenedioic acid, ethylene-1,2-dicarboxylic acid
cis-**Butendisäureanhydrid** *n* *cis*-butenedioic anhydride, maleic anhydride, 2,5-furandione
Butenpolymer[es] *n* butylene polymer
cis-**Buten-(2)-säure** *f* $CH_3CH{=}CHCOOH$ *cis*-2-butenoic acid, isocrotonic acid
trans-**Buten-(2)-säure** *f* $CH_3CH{=}CHCOOH$ *trans*-2-butenoic acid, crotonic acid
Butin *n* 1. butin, 7,3',4'-trihydroxyflavanone; 2. butyne *(either of two isomeric alkynes)*
Butin-(1) *n* $HC{\equiv}CCH_2CH_3$ 1-butyne, ethyl acetylene
Butin-(2) *n* $CH_3C{\equiv}CCH_3$ 2-butyne
Butindisäure *f* $HOOCC{\equiv}CCOOH$ butynedioic acid, acetylene-dicarboxylic acid
Bütte *f* vat, tub
Büttenersatzpapier *n* imitation hand-made paper
Büttenpapier *n* vat (hand-made, genuine) paper
B. mit imitiertem Büttenrand *s.* imitiertes B.
imitiertes B. imitation hand-made paper
Büttenpapierfabrik *f* vat mill, hand-made paper mill
Büttenrand *m (pap)* deckle [edge], deckel ✦ **mit B.** deckled ✦ **mit zweiseitigem B.** double-deckled
echter B. deckle [edge], deckel
zweiseitiger B. double deckle
Butteraroma *n* butter aroma, buttery flavour
butterartig buttery
Butterbereitung *f s.* Butterherstellung
Butterbrotpapier *n* greaseproof (grease-resistant) paper
Butterei *f* butter factory, butter-making plant
Buttererzeugung *f s.* Butterherstellung
Butterfarbe *f* butter colour
Butterfaß *n*, **Butterfertiger** *m* [butter] churn
Butterfett *n* butterfat
Butterformmaschine *f* butter-moulding (butter-pat) machine
Buttergelb *n* $C_6H_5N{=}NC_6H_4N(CH_3)_2$ butter yellow, *p*-dimethylaminoazobenzene
Butterherstellung *f* churning, butter-making
Butterkorn *n* butter grain (granule)
Buttermilch *f* buttermilk
Buttermilchpulver *n* powdered (dried) buttermilk
Buttermischmaschine *f* butter-blending machine
buttern to churn [to butter]
Buttern *n* churning, butter-making
Butterpapier *n* butter [parchment] paper
Buttersalz *n* butter salt
Buttersäure *f* C_3H_7COOH butyric acid, butanoic acid; *(specif) s. n*-Buttersäure
i-**Buttersäure** *f* $(CH_3)_2CHCOOH$ isobutyric acid, 2-methylpropanoic acid
n-**Buttersäure** *f* $CH_3CH_2CH_2COOH$ *n*-butyric acid, *n*-butanoic acid
Buttersäureäthylester *m* $CH_3CH_2CH_2COOC_2H_5$ ethyl butyrate
Buttersäurebakterien *pl* butyric acid bacteria
Buttersäurebenzylester *m* $CH_3CH_2CH_2COOCH_2C_6H_5$ benzyl butyrate
Buttersäuregärung *f* butyric fermentation
Butterschmalz *n* rendered butter
Butterserum *n* butter serum
Butterung *f* churning, butter-making
Butterungsanlage *f s.* Butterungsmaschine
Butterungsdauer *f* churning time (period)
butterungsfähig churnable
Butterungsfähigkeit *f* churnability
Butterungsmaschine *f* [butter] churn, butter-making machine
Butterungstemperatur *f* churning temperature
Butterungsverfahren *n* churning (butter-making) process
Butterungszeit *f* churning time (period)
Büttgeselle *m (pap)* dipper
n-**Butylalkohol** *m s.* Butanol-(1)
sek.-**Butylalkohol** *m s.* Butanol-(2)
tert.-**Butylalkohol** *m s. tert.*-Butanol
Butylaminobenzoat *n* butyl aminobenzoate
n-**Butylazetat** *n* $CH_3COO[CH_2]_3CH_3$ *n*-butylacetate
Butylazetylen *n* $HC{\equiv}C[CH_2]_3CH_3$ 1-hexyne, *n*-butylacetylene
n-**Butylbromid** *n* $CH_3CH_2CH_2CH_2Br$ 1-bromobutane, *n*-butyl bromide
prim.-n-**Butylbromid** *n s. n*-Butylbromid
sek.-**Butylbromid** *n* 2-bromobutane, *sec*-butyl bromide
tert.-**Butylbromid** *n* $(CH_3)_3CBr$ 2-bromo-2-methylpropane, *tert*-butyl bromide
Butylchlorid *n* butyl chloride, *(specif)* $CH_3CH_2CH_2CH_2Cl$ 1-chlorobutane, *n*-butyl chloride
1-Butylen *n s.* Buten-(1)
2-Butylen *n s.* Buten-(2)
2,3-Butylenglykol *n* $CH_3[CHOH]_2CH_3$ 2,3-butylene glycol, 2,3-butanediol
Butylenpolymer[es] *n* butylene polymer
Butylgruppe *f* C_4H_9- butyl group (residue)
Butylhalogenid *n* butyl halide
Butylkarbinol *n s.* Pentanol-(1)
Butylkautschuk *m* butyl rubber; BR
Butylkautschukmischung *f* butyl [rubber] compound
Butylkautschukvulkanisat *n* butyl [rubber] vulcanizate
Butyllatex *m* butyl latex
Butyllösung *f (rubber)* butyl cement

Butylmischung *f s.* Butylkautschukmischung
Butylperbenzoat *n* $C_6H_5CO–O_2–C_4H_9$ butyl perbenzoate
Butylradikal *n* butyl radical, *(specif)* free butyl radical
Butylregenerat *n (rubber)* butyl reclaim
Butylreifen *m* butyl tyre
Butylrest *m s.* Butylgruppe
Butylschlauch *m* butyl [inner] tube
Butylstearat *n* $C_{17}H_{35}COOC_4H_9$ butyl stearate
n-**Butyraldehyd** *m* $CH_3CH_2CH_2CHO$ *n*-butyric aldehyde, *n*-butanal
Butyrat *n* butyrate
Butyrometer *n* butyrometer
Butyron *n* $CH_3CH_2CH_2COCH_2CH_2CH_3$ 4-heptanone, butyrone
Bytownit *m (min)* bytownite *(a tectosilicate)*
BZ *s.* Backzahl
bz-Phase *f (coll)* smectic phase
B-Zustand *m (plast)* B stage

C

C *s.* Chemiefaserstoff
CA 1. *s.* Zelluloseazetat; 2. *(abbr)* anorganische Chemiefaserstoffe
CAB 1. *s.* Zelluloseazetatbutyrat; 2. *(abbr)* critical air blast, *s.* Luftmenge / kritische
Cabalglas *n* cabal glass
Cabanholz *n s.* Camholz
CAB-Test *m (coal)* critical air blast test
CAB-Wert *m (coal)* critical air blast value
Cairngormstone *m (min)* cairngorm [stone], smoky quartz
Calaverit *m (min)* calaverite *(silver-containing gold(III) telluride)*
Caliche *m* caliche, natural Chilean saltpetre
C-Alkaloid *n* C-alkaloid, calabash-curare alkaloid
Camba[l]holz *n s.* Camholz
Camholz *n* camwood *(from Baphia nitida Afz.)*
Campecheholz *n* Campeachy (Campechy) wood, logwood *(from Haematoxylum campechianum L.)*
Camps-Reaktion *f* Camps reaction *(formation of hydroxyquinolines by ring closure)*
Canaigre *n (tann)* canaigre *(roots of Rumex hymenosepalus Torr.)*
Cancrinit *m (min)* cancrinite *(a tectosilicate)*
Candelkohle *f*, **Cannelkohle** *f* cannel (candle, jet) coal
Canneloidkohle *f* canneloid coal
Cannizzaro-Reaktion *f* Cannizzaro reaction *(aldehyde dismutation)*
CAP *s.* Zelluloseazetatpropionat
Carbon-Test *m (petrol)* carbon-residue test
Carbon-Wert *m (petrol)* carbon-residue value
Carius-Aufschluß *m* Carius method *(for determining halogens, sulphur, and phosphorus in organic compounds)*
Carius-Rohr *n* Carius tube
Carnot-Prozeß *m* Carnot cycle
Carrageen *n* carrag[h]een, chondrus *(from marine algae Chondrus crispus and Gigartina mamillosa)*
Carthagenakautschuk *m* tuno gum *(from Castilloa elastica Cerv.)*
Carthamusöl *n* carthamus (safflower) oil *(from the seeds of Carthamus tinctorius L.)*
Casale-Verfahren *n* Casale process *(ammonia synthesis)*
Cascara-sagrada-Rinde *f (pharm)* cascara sagrada [bark] *(from Rhamnus purshianus DC.)*
Casinghead-Benzin *n* casing-head gasoline
Casinghead-Gas *n* casing-head gas
Cassiopeium *n s.* Lutetium
Castner-Zelle *f (el chem)* Castner cell
Cat-Benzin *n* cat-cracked gasoline
Catcracken *n* catalytic (catalyst, cat) cracking
Catcracker *m* catalytic (catalyst, cat) cracker
Catcrack-Verfahren *n* catalytic-cracking process
Catenan *n s.* Catena-Verbindung
Catena-Verbindung *f* catenation compound
Catformen *n (petrol)* catforming
Catforming-Verfahren *n (petrol)* catforming process
Cay-Cay-Butter *f* cay-cay fat *(from Irvingia oliveri Pierre)*
Cayennepfeffer *m* Cayenne (hot) pepper, chilli
CaZ *s.* Zetanzahl
C-C-Abstand *m s.* C-C-Bindungslänge
C-C-Bindung *f* **[/ einfache]** *s.* C-C-Einfachbindung
C=C-Bindung *f s.* C-C-Doppelbindung
C≡C-Bindung *f s.* C-C-Dreifachbindung
C-C-Bindungslänge *f* C-C bond length
C-C-Doppelbindung *f* carbon[-carbon] double bond, C=C bond
C-C-Dreifachbindung *f* carbon[-carbon] triple bond, C≡C bond
C-C-Einfachbindung *f* carbon[-carbon] single bond, carbon-carbon bond, C-C bond
C-C-Kernabstand *m s.* C-C-Bindungslänge
CC-Ruß *m* conducting (conductive) channel black
CCSC-Verfahren *n (pap)* cold [caustic] soda process, cold caustic semichemical process
CCT *s.* Conradson-Carbon-Test
C-C-Verknüpfung *f s.* C-C-Vernetzung
C-C-Vernetzung *f* carbon-carbon cross-linking
C-C-Vernetzungsstelle *f* carbon[-to]-carbon cross-link, C-C cross-link
CDAA *s.* 2-Chlor-*N,N*-diallylazetamid
CDEC *(abbr)* 2-Chlor-allyl-*N,N*-diäthyldithiokarbamat *(a herbicide)*
CE *s.* Eiweißchemiefaserstoff
Ceara-Kautschuk *m* Ceara rubber *(from Manihot glaziovii Muell. Arg.)*
CED *(abbr)* cohesive energy density, *s.* Energiedichte / kohäsive
Celdecor-Pomilio-Verfahren *n (pap)* Celdecor-Pomilio process
Celsius-Skale *f*, **Celsius-Thermometerskale** *f* centigrade (Celsius) scale
C-endständig C-terminal *(amino acids in proteins)*
Cerini-Dialysator *m* Cerini dialyzer
Cermet *n* cer[a]met, ceramal, ceramel

Cervantit *m (min)* cervantite *(antimony(III,V) oxide)*
Ceylanit *m*, **Ceylonit** *m (min)* ceylonite, ceylanite, pleonaste *(a variety of spinel)*
Ceylon-Kardamom *m (n)* Ceylon cardamom *(from Elettaria major Sm.)*
Ceylon-Zimt *m* Ceylon cinnamon *(from Cinnamomum zeylanicum Bl.)*
CFR-Motormethode *f* CFR motor method *(for measuring the antiknock qualities of fuels)*
CFR-Prüfmotor *m* CFR [test] engine, Cooperative Fuel Research engine *(for measuring the antiknock qualities of fuels)*
C-Füllmasse *f (sugar)* low-grade massecuite, aftermassecuite
C-Futter *n* carbon lining *(of a blast furnace)*
Chabasit *m (min)* chabasite, chabazite *(a tectosilicate)*
Chagrinleder *n* shagreen
Chalkanthit *m (min)* chalcanthite, blue (copper) vitriol *(copper(II) sulphate-5-water)*
Chalkogen *n* chalcogen
Chalkogenid *n* chalcogenide *(a binary compound of a chalcogen)*
Chalkomenit *m (min)* chalcomenite *(copper(II) selenite)*
Chalkon *n* chalcone, chalkone, *(specif)* $C_6H_5COCH{=}CHC_6H_5$ chalcone, chalkone, benzalacetophenone
Chalkophanit *m (min)* chalcophanite *(hydrous manganese and zinc oxide)*
Chalkophyllit *m (min)* chalcophyllite, chalkophyllite *(a basic arsenate and sulphate of aluminium and copper)*
Chalkopyrit *m (min)* chalcopyrite, chalkopyrite, copper pyrites *(copper(II) iron(II) sulphide)*
Chalkosin *m* chalcocite, chalcosine *(copper(I) sulphide)*, *(specif)* low-chalcocite *(the variety stable below 105 °C)*
Chalkosin(-H) *m (min)* high-chalcocite *(the variety of copper(I) sulphide stable above 105 °C)*
Chalkosphäre *f (geol)* chalcosphere
Chalkotrichit *m (min)* chalcotrichite, capillary cuprite, plush copper *(a variety of cuprite)*
Chalzedon *m (min)* chalcedony *(a variety of quartz)*
Chamaenol *n* chamenol, 1-methoxy-2-hydroxy-4-isopropylbenzene
Chamaezin *n* chamecin *(a tropolone derivative)*
Chamäleon *n* / **mineralisches** chameleon mineral *(potassium manganate)*
Chaminsäure *f* chaminic acid *(a monoterpene derivative)*
Chamoispapier *n (phot)* cream paper
Chamosit *m (min)* chamosite *(a phyllosilicate)*
Champacablütenöl *n (cosmet)* Champaca oil *(from Michelia longifolia Blume and M. champaca L.)*
Chamsäure *f* chamic acid *(a monoterpene derivative)*
Chance-Kegel *m (min tech)* Chance cone
Chance-Sandflotationsverfahren *n*, **Chance-Sandschwimmverfahren** *n* Chance [sand-floatation]
Channel[-Black]-Anlage *f* channel black plant
Channel-Black-Verfahren *n* channel process
Channel-Ruß *m* channel (impingement) black *(a gas black)*
Charakter *m* **des Roh[erd]öls** base of crude petroleum
  abweisender C. repellency
  anionischer C. anionic nature
  asphaltischer C. asphalt base *(of crude petroleum)*
  kationischer C. cationic nature
  ölabweisender C. oil repellency
  wasserabweisender C. water repellency
Charakteristik *f* characteristics; characteristic curve
Chardonnet-Seide *f* chardonnet silk
Charge *f* charge [stock], batch, feed[stock], load; batch *(product)*
Chargenbetrieb *m* batch processing (operation)
Chargendestillation *f* batch distillation
Chargenmasse *f* batch weight
Chargenmischer *m* batch mixer
Chargennitrierung *f* batch nitration
Chargennummer *f* maker's serial number
Chargentrockner *m* batch dryer
Chargenwaage *f* batch scale
chargenweise batchwise
chargieren 1. to charge, to load, to feed, to fill, to furnish; 2. *(text)* to weight *(to add sizing materials)*
Chargiertür *f* charging (filling, feeding, feed) door
Charpy-Prüfung *f* Charpy test *(for measuring the breaking strength of materials under impact)*
C-Harz *n* C-stage resin, resite
Chassis *n* / **absolutes** *(cosmet)* absolute chassis
Chaulmoograöl *n (pharm)* chaulmoogra (hydnocarpus, gynocardia) oil *(from Hydnocarpus specc.)*
Chaulmoograsäure *f* chaulmoogric acid, 13-2'-cyclopentenyltridecanoic acid
Chaulmugraöl *n s.* Chaulmoograöl
Chavikolmethyläther *m* $CH_2{=}CHCH_2C_6H_4OCH_3$ chavicol methyl ether, esdragol, 1-allyl-4-methoxybenzene
C-H-Bindung *f* carbon-hydrogen bond
Chebulagsäure *f* chebulagic acid *(an ellagitannin)*
Chebulinsäure *f* chebulinic acid *(a gallotannin)*
Chebulsäure *f* chebulic acid *(a gallotannin)*
Chedakristall *m* chadacryst
Chelat *n s.* Chelatverbindung
chelatartig chelate-like
Chelataustauscher *m s.* Chelatharz
chelatbildend chelate-forming, chelating
Chelatbildner *m* chelating agent
Chelatbildung *f* chelation, chelate formation
Chelatbildungskonstante *f* chelate formation constant
Chelatbindung *f* chelate linkage
Chelatdonatorgruppe *f* chelate donor group
Chelateffekt *m* effect of chelation, chelate effect
Chelatfarbstoff *m* chelate pigment
chelatgebunden chelated
Chelatgruppe *f* chelating group
Chelatharz *n* chelate [ion-exchange] resin, chelating resin

Chelation *f s.* Chelatbildung
Chelatkatalyse *f* chelate catalysis
Chelatkomplex *m s.* Chelatverbindung
Chelatometrie *f* chelatometry, chelatometric titration
chelatometrisch chel[at]ometric
Chelator *m s.* Chelatbildner
Chelatring *m* chelate ring
Chelatringbildung *f* formation of chelate rings
Chelatstabilität *f* chelate stability
Chelatstabilitätskonstante *f* chelate stability constant
Chelatstruktur *f* chelate structure
Chelatverbindung *f* chelate [compound, complex], scissor compound, crab's-claw complex
Chelidonsäure *f* chelidonic acid, γ-pyrone-2,6-dicarboxylic acid
Chemie *f* chemistry
  C. der Heterozyklen (heterozyklischen Verbindungen) heterocyclic chemistry
  C. der Hochpolymeren polymer (high polymeric) chemistry
  C. der Kohlenstoffverbindungen chemistry of the carbon compounds
  C. der Koordinationsverbindungen chemistry of coordination compounds, coordination chemistry
  C. der Milch [und Milchprodukte] dairy chemistry
  C. der siliziumorganischen Verbindungen organosilicon chemistry
  C. des Farbensehens colour vision chemistry
  aliphatische C. aliphatic chemistry
  allgemeine C. general chemistry
  analytische C. analytical chemistry
  angewandte C. applied chemistry
  anorganische C. inorganic chemistry
  biologische C. biological chemistry, biochemistry
  forensische C. forensic (legal) chemistry
  fotografische C. photographic chemistry, chemistry of photography
  geologische C. geological chemistry, geochemistry
  gerichtliche C. *s.* forensische C.
  heiße C. hot-atom chemistry
  industrielle C. industrial (technical) chemistry
  klinische C. clinical chemistry
  kosmetische C. cosmetic chemistry
  landwirtschaftliche C. agricultural chemistry, agrochemistry
  makromolekulare C. polymer (high polymeric) chemistry
  metallurgische C. metallurgical chemistry
  milchwirtschaftliche C. dairy chemistry
  mineralogische C. mineral[ogical] chemistry
  organische C. organic chemistry
  pharmazeutische C. pharmaceutic[al] chemistry
  physikalische C. physical chemistry
  physiologische C. physiologic[al] chemistry
  präparative C. preparative chemistry
  reine C. pure chemistry
  siliziumorganische C. organosilicon chemistry
  synthetische C. synthetic chemistry
  technische C. technical (industrial) chemistry
  theoretische C. theoretical chemistry
  toxikologische C. toxicological chemistry
Chemieanlage *f* chemical plant
Chemieausrüstung *f* chemical equipment
Chemiebetrieb *m* chemical works (plant)
Chemiefaser *f (text)* man-made fibre, staple [fibre] *(a fibre of relatively short length cut from continuous filaments)*
Chemiefaserindustrie *f* man-made-fibre industry
Chemiefaserstoff *m (text)* man-made fibre
  C. aus natürlichen Polymeren natural polymer fibre
  C. aus synthetischen Polymeren synthetic [polymer] fibre
Chemieholz *n (pap)* pulpwood
Chemieindustrie *f* chemical [processing] industry
Chemieingenieur *m* chemical engineer
Chemieingenieurtechnik *f* chemical engineering technology
Chemieingenieurwesen *n* chemical engineering
Chemielaborant *m* laboratory assistant
Chemiepumpe *f* chemical (process) pump
Chemieschliff *m (pap)* chemigroundwood
Chemieseide *f* man-made continuous filament yarn
Chemietechnik *f* chemical technology
Chemietechnologe *m* chemical technologist
Chemiewerk *n* chemical works (plant)
Chemiezellstoff *m* dissolving (rayon) pulp
Chemikal *n s.* Chemikalie
Chemikalie *f* chemical
  fotografische C. photographic chemical
  pharmazeutische C. pharmaceutic[al] chemical
  technische C. industrial chemical
Chemikalienbeständigkeit *f* 1. chemical resistance (stability); 2. resistance to chemicals (chemical attack)
Chemikalienfestigkeit *f s.* Chemikalienbeständigkeit 2.
Chemikalienkosten *pl* chemical costs
Chemikalienrückführung *f*, **Chemikalienrückgewinnung** *f* chemical recovery, recovery (conservation) of chemicals
Chemikalienverbrauch *m* chemical consumption
Chemikalienverhältnis *n (pap)* chemical[-to-wood] ratio
Chemikalienverlust *m (pap)* loss of chemical
Chemikalienzulauf *m* chemical feed inlet
Chemikalienzuteilvorrichtung *f* chemical feeder
Chemiker *m* chemist
  verantwortlicher C. chemist in charge
Chemilumineszenz *f* chemiluminescence
chemisch chemical
Chemischreiniger *m (text)* dry cleaner
Chemischreinigung *f (text)* dry cleaning
Chemisierung *f* chemicalization *(as of agriculture)*
Chemismus *m* chemism
chemisorbieren to chemisorb, to chemosorb
Chemisorption *f* chemisorption, chemosorption, chemical adsorption
Chemolumineszenz *f s.* Chemilumineszenz
chemosorbieren *s.* chemisorbieren

Chemosorption *f s.* Chemisorption
Chemosteril[is]ans *n*, **Chemosterilisierungsmittel** *n* chemosterilant
chemotaktisch *(biol)* chemotactic *(moving in relation to chemical agents)*
Chemotaxis *f (biol)* chemotaxis, chemotaxy *(movement in relation to chemical agents)*
Chemotaxonomie *f* chemotaxonomy, chemical taxonomy *(biochemical systematics)*
Chemotechniker *m* chemical technician
Chemotherapeutikum *n* chemotherapeutant, chemotherapeutic agent
chemotherapeutisch chemotherapeutic[al]
Chemotherapie *f* chemotherapy, chemotherapeutics
chemotropisch *(biol)* chemotropic
Chemotropismus *m (biol)* chemotropism *(orientation in relation to chemicals)*
Chemurgie *f* chemurgy *(industrial utilization of organic raw materials)*
Chenopodiumöl *n (pharm)* chenopodium oil *(from Chenopodium ambrosioides L. var. anthelminthicum Gray)*
Cheshunt-Mischung *f* Cheshunt compound *(a pesticide consisting of $CuSO_4$ and $(NH_4)_2CO_3$)*
Chiastolith *m* chiastolite *(a variety of andalusite)*
Chibouharz *n* tacamahaca [gum], West Indian elemi *(from Bursera gummifera L.)*
Chicagoblau *n* Chicago blue
Chicagosäure *f* $HOC_{10}H_4(NH_2)(SO_3H)_2$ Chicago acid, 8-amino-1-naphthol-5,7-disulphonic acid, 2S-acid
Chicle *m*, **Chiclegummi** *n* chicle (zapota) gum *(from Achras sapota L.)*
chiffrieren *(nomencl)* to cipher
Chiffriersystem *n (nomencl)* ciphering system
Chiffrierung *f (nomencl)* ciphering, cipher notation
Chilenit *m* chilenite *(a natural alloy of silver and bismuth)*
Chilesalpeter *m* Chile saltpetre (nitre, nitrate), Chilean (Chilian) nitrate, soda nitre *(sodium nitrate)*
  roher C. caliche, natural Chilean saltpetre
Chimonanthin *n* chimonanthine *(alkaloid)*
Chimylalkohol *m* $C_{16}H_{33}OCH_2CH(OH)CH_2OH$ chimyl alcohol, 2,3-dihydroxypropyl hexadecyl ether
Chinaalkaloid *n s.* Chinarindenalkaloid
Chinagelb *n* yellow arsenic [sulphide], arsenic (king's) yellow (gold), royal yellow, orpiment [yellow] *(technically pure arsenic(III) sulphide)*
Chinagras *n* China (Chinese) grass *(Boehmeria nivea (L.) Gaudich.)*
Chinagrün *n* Chinese green, locao, locaonic acid *(natural dye from Rhamnus specc.)*
Chinaholzöl *n* tung (China wood) oil *(from the seeds of Aleurites fordii Hemsl.)*
Chinaldin *n* quinaldine, 2-methylquinoline
Chinaldinalkyljodid *n* quinaldine alkiodide
Chinaldinchelat *n* quinaldine chelate
Chinalizarin *n* quinalizarin, 1,2,5,8-tetrahydroxyanthraquinone
Chinapapier *n* China (Chinese, Indian) paper, India [proof] paper
Chinarinde *f* cinchona [bark], red bark *(from Cinchona specc.)*
Chinarindenalkaloid *n* cinchona (quinoline-type) alkaloid
Chinasäure *f* $(HO)_4C_6H_7COOH$ quinic acid, 1,3,4,5-tetrahydroxycyclohexane-1-carboxylic acid
Chinasilber *n* China (Chinese) silver *(an alloy of copper, nickel, tin, and silver)*
Chinatalg *m* Chinese vegetable tallow *(from Sapium sebiferum (L.) Roxb.)*
Chinäthylin *n* quinethyline, dihydroquinine
Chinawachs *n* Chinese [tree] wax, insect wax, vegetable spermaceti *(secreted by scales)*
Chinaweiß *n* Chinese white *(zinc oxide)*
Chinazolin *n* quinazoline, 5,6-benzpyrimidine
Chinesischweiß *n s.* Chinaweiß
Chinhydron *n (org chem)* quinhydrone
Chinhydronelektrode *f* quinhydrone electrode
Chinidin *n* quinidine *(a cinchona alkaloid)*
Chinidinsulfat *n* quinidine sulphate
Chinin *n* quinine *(a cinchona alkaloid)*
Chininhydrochlorid *n* quinine hydrochloride
Chininsulfat *n* quinine sulphate
Chiniofon *n* chiniofon, 8-hydroxy-7-iodoquinoline-5-sulphonic acid
Chinizarin *n* quinizarin, 1,4-dihydroxyanthraquinone
Chinizarinkondensation *f* quinizarin condensation
chinoid quin[on]oid
Chinolat *n* quinolate *(salt or ester of quinolinic acid)*
Chinolin *n* quinoline, chinoline, 2,3-benzpyridine
Chinolinäthyljodid *n* quinoline ethiodide
Chinolinblau *n* cyanine blue, pigment blue 15
Chinolinfarbstoff *m* quinoline dye
Chinolin-4-karbonsäure *f* quinoline-4-carboxylic acid, cinchoninic acid
Chinolinsäure *f* quinolinic acid, pyridine-2,3-dicarboxylic acid
Chinolinsynthese *f* quinoline synthesis
  Combessche C. Combes quinoline synthesis
  Friedländersche C. Friedländer quinoline synthesis
  Skraupsche C. Skraup quinoline synthesis
Chinolizidin *n* quinolizidine
Chinolizidinring *m* quinolizidine ring
Chinolumlagerung *f* quinol rearrangement
Chinolylamin *n* quinolylamine, aminoquinoline
Chinon *n* quinone, chinone, *(specif) p*-benzoquinone, 1,4-cyclohexadienedione
***p*-Chinon** *n p*-benzoquinone, 1,4-cyclohexadienedione
Chinondiimin *n* quinonediimine
Chinondioximvernetzung *f*, **Chinondioximvulkanisation** *f (rubber)* quinoid cure
Chinonfarbstoff *m* quinonoid dye
Chinoniminfarbstoff *m* quinone imine dye
Chinonmethid *n* quinone methide
Chinonmonoxim *n* $C_6H_4O(NOH)$ quinone monoxime
Chinonring *m* quinone ring

Chinoxalin *n* quinoxaline, 1,4-benzodiazine
Chinuklidin *n* quinuclidine, 1,4-ethylenepiperidine
Chiralität *f* chirality *(stereochemistry)*
Chitin *n* chitin *(a polysaccharide)*
chitinig, chitinös chitinous
Chitosamin *n* chitosamine, glucosamine
Chlathrat *n*, **Chlathratverbindung** *f* clathrate [inclusion compound], cage compound
Chloanthit *m (min)* chloanthite, white nickel, nickel-skutterudite *(nickel arsenide)*
Chlor *n* Cl chlorine ✦ **mit C. behandeln** to chlorinate, to chlorinize
aktives (wirksames) C. active (available) chlorine
Chloral *n* CCl_3CHO chloral, trichloroacetaldehyde
Chloralhydrat *n* $Cl_3CCH(OH)_2$ chloral hydrate, trichloroacetaldehyde hydrate
Chloralkalielektrolyse *f* electrolysis of alkali-metal chlorides
Chloralose *f* chloralose *(a hypnotic)*
Chlorameisensäure *f* ClCOOH chloroformic acid
Chlorameisensäureester *m* chloroformic acid ester, chloroformate, chlorocarbonate
Chlorameisensäuretrichlormethylester *m* $ClCOOCCl_3$ trichloromethyl chloroformate, diphosgene
Chloramin *n* chloramine, *(specif)* NH_2Cl chloramine
Chloramphenikol *n* chloramphenicol
Chloranil *n* chloranil, tetrachloro-*p*-benzoquinone
Chloranilin *n* $C_6H_4(Cl)NH_2$ chloroaniline
Chloranthrachinon *n* chloroanthraquinone
Chlorapatit *m (min)* chlorapatite *(calcium phosphate chloride)*
Chlorargyrit *m (min)* chlorargyrite, cerargyrite, horn silver *(silver chloride)*
Chlorat *n* M^ICIO_3 chlorate
Chloräthan *n* CH_3CH_2Cl chloroethane
2-Chloräthanol *n* $ClCH_2CH_2OH$ 2-chloroethanol, 2-chloroethyl alcohol
Chloräthansäure *f* $ClCH_2COOH$ chloroethanoic acid, chloroacetic acid
Chloräthen *n* $CH_2{=}CHCl$ chloroethylene, chloroethene, vinyl chloride
Chloräthyl *n s.* 1. Chloräthan; 2. Chloräthylgruppe
2-Chloräthylalkohol *m s.* 2-Chloräthanol
Chloräthylen *n s.* Chloräthen
Chloräthylgruppe *f* $ClCH_2CH_2$-chloroethyl group
Chloratit *n* chloratite *(a chlorate explosive)*
Chlorator *m s.* Chlorierungskessel
Chloratsprengstoff *m* chlorate explosive
Chloraufschluß *m (pap)* pulping with chlorine
C. nach Pomilio-Celdecor Celdecor-Pomilio process
Chlorazetal *n* $ClCH_2CH(OC_2H_5)_2$ chloroacetal, diethyl chloroacetal
Chlorazetaldehyd *m* $ClCH_2CHO$ chloroacetaldehyde, chloroethanal
Chlorazeton *n* CH_3COCH_2Cl chloroacetone, acetonyl chloride, chloropropanone
Chlorazetophenon *n* chloroacetophenone; *(specif)* *s.* α-Chlorazetophenon
α-Chlorazetophenon *n* $C_6H_5COCH_2Cl$ α-chloroacetophenone, phenacyl chloride
ω-Chlorazetophenon *n s.* α-Chlorazetophenon
Chlorbenzilat *n* chlorobenzilate, ethyl 4,4′-dichlorobenzilate
Chlorbenzoesäure *f* ClC_6H_4COOH chlorobenzoic acid
Chlorbenzol *n* C_6H_5Cl chlorobenzene
Chlorbenzolkarbonsäure *f s.* Chlorbenzoesäure
Chlorbernsteinsäure *f* $HOOCCH_2CH(Cl)COOH$ chlorosuccinic acid
Chlorbleiche *f* chlorine bleaching; *(text)* chemic[k]
Chlorbromsilberpapier *n*, *(phot)* chlorobromide paper
1-Chlorbutan *n* $CH_3[CH_2]_2CH_2Cl$ 1-chlorobutane
Chlorbutylkautschuk *m* chlorobutyl rubber, chlorinated butyl rubber
2-Chlor-*N,N*-diallylazetamid *n* 2-chloro-*N,N*-diallylacetamide, CDAA *(a herbicide)*
Chlordimethylarsin *n* $(CH_3)_2AsCl$ chlorodimethylarsine, cacodyl chloride
Chlordinitrobenzol *n* $(NO_2)_2C_6H_3Cl$ chlorodinitrobenzene
Chlordioxid *n* ClO_2 chlorine dioxide
Chlordioxidbleiche *f (pap)* chlorine dioxide bleaching
Chlordioxid-Bleichlauge *f (pap)* chlorine dioxide bleaching liquor
Chlordioxid-Bleichstufe *f (pap)* chlorine dioxide bleaching stage
chlorecht fast to chlorine
Chlorechtheit *f* chlorine fastness, fastness to chlorine
Chlorelektrode *f* chlorine electrode
Chlorellagsäure *f* chlorellagic acid
chloren to chlorinate, to chlorinize, to chlore; *(text)* to chemick *(to treat with calcium hypochlorite)*
Chloressigsäure *f* $CH_2ClCOOH$ chloroacetic acid
Chloreton *n* $(CH_3)_2C(OH)CCl_3$ chloretone, acetone chloroform, 1,1,1-trichloro-2-hydroxy-2-methylpropane
Chlorgas *n* chlorine gas
chlorhaltig chlorine-containing
Chlorheptoxid *n s.* Chlor(VII)-oxid
1-Chlorhexan *n* $CH_3[CH_2]_4CH_2Cl$ 1-chlorohexane
Chlorhydrin *n* chlorohydrin
Chlorhydrochinon *n* $C_6H_3(OH)_2Cl$ chlorohydroquinone, chloroquinol
Chlorid *n* M^ICl chloride
chloridfrei free from chloride
Chloridpapier *n (phot)* [silver-]chloride paper
Chloridschmelze *f* chloride melt
Chloridsole *f* chloride brine
chlorieren to chlorinate *(to introduce chlorine)*; *(min tech)* to chloridize, to chloridate *(to treat with chlorine or with a chloride)*
Chlorierer *m s.* Chlorierungskessel
Chlorierung *f* chlorination; *(min tech)* chloridization, chloridation
C. bei niedriger Stoffdichte *(pap)* low-density chlorination
C. der Seitenkette side-chain chlorination
C. in saurem Medium *(pap)* acidic chlorination
fotochemische C. photochemical chlorination

Chlorierungsbehälter *m*, **Chlorierungsgefäß** *n s.* Chlorierungskessel
Chlorierungskessel *m* chlorinating vessel, chlorinator
Chlorierungsmittel *n* chlorinating agent
Chlorierungsstufe *f* chlorination (chlorine) stage
Chlorierungsturm *m* chlorination tower, chlorinator; *(pap)* reaction tower *(pulp bleaching)*
Chlorierungsverfahren *n* method of chlorination
Chlor-IPC *n* $ClC_6H_4NHCOOCH(CH_3)_2$ chloro-IPC, C-IPC, chloroisopropyl-*N*-phenylcarbamate *(a herbicide)*
Chlorit *m (min)* chlorite *(any of a series of phyllosilicates)*
Chlorit *n* M^IClO_2 chlorite
Chloritbleiche *f (text)* chlorite bleaching
Chloritbleichechtheit *f (text)* chlorite bleaching fastness
Chloritisation *f*, **Chloritisierung** *f (geol)* chloritization
Chloritoid *m (min)* chloritoid *(a neso-subsilicate)*
Chloritschiefer *m (geol)* chlorite schist
Chlorkalk *m* chlorinated lime, chloride of lime
Chlorkalziumröhrchen *n (lab)* calcium chloride tube
Chlorkalziumzylinder *m* [gas] drying jar
Chlorkautschuk *m* chlorinated rubber
Chlorkautschuk[anstrich]farbe *f* chlorinated-rubber paint
Chlorkautschuklack *m* chlorinated-rubber lacquer
Chlorknallgas *n* chlorine detonating gas
Chlorknallgaskette *f* hydrogen-hydrochloric acid cell
Chlorkohlensäure *f s.* Chlorameisensäure
Chlorkohlenwasserstoff *m* chlorinated hydrocarbon
Chlorkresol *n* $CH_3C_6H_3(OH)Cl$ chlorocresol
Chlorlignin *n (pap)* chlorolignin, chlorinated lignin
Chlormethan *n* CH_3Cl chloromethane, methyl chloride
Chlormethin *n* chlormethine, mustine, *N*-di-(2-chloroethyl)methylamine hydrochloride
Chlormethyl *n s.* Chlormethan
***o*-Chlormethylbenzol** *n* 2-chloro-1-methylbenzene, *o*-chlorotoluene
Chlormethylierung *f* chloromethylation
Chlormonosilan *n* SiH_3Cl chloromonosilane, chlorosilane
Chlormonoxid *n s.* Chlor(I)-oxid
Chlornaphthalin *n* chloronaphthalene
Chlornitrobenzol *n* $C_6H_4Cl(NO_2)$ chloronitrobenzene
Chlornitroparaffin *n* chloronitroparaffin
Chloroantimonat *n* chloroantimonate
Chloroargentat *n* chloroargentate
Chloroaurat(I) *n* $M^I[AuCl_2]$ chloroaurate(I), dichloroaurate(I)
Chloroaurat(III) *n* $M^I[AuCl_4]$ chloroaurate(III), tetrachloroaurate(III)
Chlorobromat *n* chlorobromate
Chlorochromat *n* $M^I[CrO_3Cl]$ chlorochromate
Chloroform *n* $CHCl_3$ chloroform, trichloromethane
chloroformlöslich soluble in chloroform
Chloroformlöslichkeit *f* solubility in chloroform
Chlorogensäure *f* chlorogenic acid, 3-[3,4-dihydroxycinnamoyl]quinic acid
Chlorogold(III)-säure *f* $H[AuCl_4]$ chloroauric(III) acid, tetrachloroauric(III) acid
Chloroguanid *n* chloroguanide, 1-(*p*-chlorophenyl)-5-isopropylbiguanide
Chloroiridat(III) *n* $M_3^I[IrCl_6]$ chloroiridate(III), hexachloroiridate(III)
Chlorojodat *n* chloroiodate
Chlorokadmat *n* chlorocadmate
Chlorokomplex *m* chlorocomplex
Chloroktahydrat *n s.* Chlor-8-Wasser
Chlorokuprat *n* chlorocuprate
Chloromagnesit *m (min)* chloromagnesite *(magnesium chloride)*
Chloromanganat *n* chloromanganate
Chloromerkurat *n* chloromercurate
Chloromolybdat *n* chloromolybdate
Chloroniobat *n* chloroniobate
Chloroosmat(III) *n* $M_3^I[OsCl_6]$ chloroosmate(III), hexachloroosmate(III)
Chloropalladat(II) *n* $M_2^I[PdCl_4]$ chloropalladate(II), tetrachloropalladate(II)
Chloropentamminkobalt(III)-chlorid *n* $[Co(NH_3)_5Cl]Cl_2$ chloropentaamminecobalt(III) chloride
Chloropentamminplatin(IV)-chlorid *n* $[Pt(NH_3)_5Cl]Cl_3$ chloropentaammineplatinum(IV) chloride
Chlorophosphat *n* chlorophosphate
Chlorophyll *n* chlorophyll, leaf green
Chlorophyllase *f* chlorophyllase
Chlorophyllid *n* chlorophyllide
Chlorophyllin *n* chlorophyllin
Chlorophyllkorn *n*, **Chloroplast** *m* chloroplast
Chloroplatinat(II) *n* $M_2^I[PtCl_4]$ chloroplatinate(II), tetrachloroplatinate(II)
Chloroplatinat(IV) *n* $M_2^I[PtCl_6]$ chloroplatinate(IV), hexachloroplatinate(IV)
Chloroplatin(IV)-säure *f* $H_2[PtCl_6]$ chloroplatinic(IV) acid, hexachloroplatinic(IV) acid
Chloroplumbat *n* chloroplumbate
Chloropren *n* $CH_2{=}CClCH{=}CH_2$ chloroprene, 2-chloro-1,3-butadiene
Chloroprenkautschuk *m* chloroprene rubber, CR
Chlororhenat *n* chlororhenate
Chlororhodat(III) *n* $M_3^I[RhCl_6]$ chlororhodate(III), hexachlororhodate(III)
Chlororuthenat *n* chlororuthenate
Chlorosäure *f* chloro acid
Chloroschwefelsäure *f* $SO_2(OH)Cl$ chlorosulphuric acid
Chlorose *f (bot)* chlorosis
Chlorostannat *n* chlorostannate
Chlorosulfat *n* chlorosulphate
chlorosulfonieren *s.* chlorsulfonieren
Chlorotellurat(IV) *n* $M_2^I[TeCl_6]$ chlorotellurate(IV), hexachlorotellurate(IV)
Chlorothallat *n* chlorothallate
Chlorothionit *m (min)* chlorothionite *(a copper potassium chloride sulphate)*
Chlorotitanat(IV) *n* $M_2^I[TiCl_6]$ chlorotitanate(IV), hexachlorotitanate(IV)

Chlorowolframat *n* chlorowolframate, chlorotungstate
Chlor(I)-oxid *n* Cl_2O chlorine(I) oxide, dichlorine monooxide
Chlor(IV)-oxid *n* ClO_2 chlorine(IV) oxide, chlorine dioxide
Chlor(VII)-oxid *n* Cl_2O_7 chlorine(VII) oxide, dichlorine heptaoxide
Chloroxybenzol *n s.* Chlorphenol
Chlorozinkat *n* chlorozincate
Chlorozirkonat(IV) *n* $M_2^I[ZrCl_6]$ chlorozirconate(IV), hexachlorozirconate(IV)
Chlorparaffin *n* chlorinated paraffin
1-Chlorpentan *n* $CH_3[CH_2]_3CH_2Cl$ 1-chloropentane
Chlorphenol *n* HOC_6H_4Cl chlorophenol
Chlorphenylendiamin *n* $ClC_6H_3(NH_2)_2$ chlorophenylenediamine
Chlorpikrin *n* CCl_3NO_2 chloropicrin, trichloronitromethane
1-Chlorpropan *n* $CH_3CH_2CH_2Cl$ 1-chloropropane
2-Chlorpropan *n* $CH_3CHClCH_3$ 2-chloropropane
3-Chlor-1,2-propandiol *n* $HOCH_2CH(OH)CH_2Cl$ 3-chloro-1,2-propanediol, [glycerol] α-monochlorohydrin
Chlorpropanon *n* CH_3COCH_2Cl chloropropanone, chloroacetone, acetonyl chloride
3-Chlorpropen-(1) *n* $CH_2{=}CHCH_2Cl$ 3-chloropropene
Chlorpropham *n s.* Chlor-IPC
2-Chlorpropionsäure *f* $CH_3CHClCOOH$ 2-chloropropionic acid
3-Chlorpropionsäure *f* CH_2ClCH_2COOH 3-chloropropionic acid
Chlorpropyl *n s.* 1-Chlorpropan
3-Chlorpropylen *n s.* 3-Chlorpropen-(1)
Chlorpropylenoxid *n* 3-chloropropylene oxide, chloromethyloxiran, α-epichlorhydrin
Chlorretention *f*, **Chlorrückhaltevermögen** *n* chlorine retention
Chlorsäure *f* $HClO_3$ chloric acid
Chlorschiefer *m (geol)* chlorite schist
Chlorsilan *n* chlorosilane, *(specif)* SiH_3Cl chloromonosilane
  organisches C. organochlorosilane
Chlorsilber *n (phot)* silver chloride
Chlorsilberpapier *n (phot)* [silver-]chloride paper
chlorsulfonieren to chlorosulphonate
Chlorsulfonierung *f* chlorosulphonation
Chlortetrazyklin *n* chlorotetracycline, CTC, aureomycin *(antibiotic)*
Chlortetroxid *n* ClO_4 chlorine tetraoxide
o-**Chlortoluol** *n* $CH_3C_6H_4Cl$ *o*-chlorotoluene
α-**Chlortoluol** *n* $C_6H_5CH_2Cl$ α-chlorotoluene
Chlortrifluoräthylen *n* $ClFC{=}CF_2$ chlorotrifluoroethylene, CFE
Chlortrocknung *f* chlorine drying
Chlorturm *m s.* Chlorierungsturm
Chlorung *f* chlorination
  C. über den Durchbruchspunkt hinaus breakpoint chlorination *(water treatment)*
Chlorungsmittel *n* chlorinating agent
Chlorungsverfahren *n* method of chlorination
Chlorverbindung *f* / **organische** organochlorine compound
Chlorverbrauch *m* chlorine consumption
Chlorverbrauchszahl *f (pap)* chlorine number
Chlorverflüssigung *f* chlorine liquefaction
Chlorwasser *n* chlorine water, chlorine-water solution
Chlor-8-Wasser *n* $Cl_2 \cdot 8H_2O$ chlorine-8-water, chlorine octahydrate
Chlorwasserstoff *m* HCl hydrogen chloride
  trockener (wasserfreier) C. anhydrous hydrogen chloride
Chlorwasserstoffanlagerung *f* hydrochlorination
Chlorwasserstoffgas *n* hydrochloric-acid gas
Chlorwasserstoffsäure *f* HCl hydrochloric acid
α-**Chlorxylol** *n* $CH_3C_6H_4CH_2Cl$ α-chloroxylene
Chlorzahl *f (pap)* chlorine number
Chlorzelle *f* chlorine cell *(electrolysis)*
Chlorzinklauge *f* zinc-chloride solution
Chlorzurückhaltung *f s.* Chlorretention
Chlorzyan *n* CNCl cyanogen chloride, chlorine cyanide
Chlorzyanhydrin *n* $CCl_3CH(OH)CN$ chlorocyanohydrin
Cholesterin *n* cholesterol, cholesterin
cholesterisch cholesteric
Cholsäure *f* cholic acid, 3,7,12-trihydroxy-5β-cholan-24-oic acid
Chondrit *m* chondrite *(a meteoric stone)*
Chondrodit *m (min)* chondrodite *(a neso-subsilicate)*
Chorismasäure *f*, **Chorisminsäure** *f* chorismic acid
C-Horizont *m (soil)* C-horizon
Christbaum *m (petrol)* Christmas tree
Chrom *n* Cr chromium
Chromalaun *m* $M^ICr(SO_4)_2 \cdot 12H_2O$ chrome alum, *(specif)* $KCr(SO_4)_2 \cdot 12H_2O$ potassium (potash) chrome alum, chrome potash alum
Chromammin *n* chromammine, chromium ammine
Chromat *n* $M_2^ICrO_4$ chromate
Chromatgelatine *f* chrome (chromatic) gelatin
chromat[is]ieren to chromatize
Chromatit *m (min)* chromatite *(calcium chromate)*
Chromatografie *f* chromatography
  C. mit normaler Substanzmenge full-scale chromatography
  C. mit Phasenumkehr (umgekehrter Phase) reversed-phase (rear-phase) chromatography
Chromatografiegefäß *n* chromatography tank
  C. für die aufsteigende Methode ascending chromatography tank
Chromatografiekammer *f* chromatography chamber, chromatographic cabinet
Chromatografiepapier *n* chromatographic paper
chromatografieren to chromatograph
Chromatografierohr *n* chromatographic tube
Chromatografiesäule *f* chromatographic column
chromatografisch chromatographic
Chromatogramm *n* chromatogram
  äußeres C. external (liquid) chromatogram
  fließendes (flüssiges) C. *s.* äußeres C.
Chromatometrie *f* chromatometry, chromatomet-

ric method, dichromate titration
chromatometrisch chromatometric
Chromatopackverfahren *n* chromatopack method *(variant of paper chromatography)*
Chromatophor *n (biol)* chromatophore
Chromatopileverfahren *n* chromatopile method *(variant of paper chromatography)*
Chromatopsie *f* colour vision
Chromatverfahren *n (text)* chromate [dyeing] method
Chromaventurin *m*, **Chromaventuringlas** *n* chrome (green) aventurine
Chrom(II)-azetat *n* $Cr(CH_3COO)_2$ chromium(II) acetate, chromous acetate
Chrom(III)-azetat *n* $Cr(CH_3COO)_3$ chromium(III) acetate, chromic acetate
Chrombeize *f (text)* chrome (chromium) mordant
Chrombeizverfahren *n (text)* chrome mordant process
Chrom(II)-bromid *n* $CrBr_2$ chromium(II) bromide, chromium dibromide
Chrom(III)-bromid *n* $CrBr_3$ chromium(III) bromide, chromium tribromide
Chrombrühe *f (tann)* chrome liquor
Chrom(II)-chlorid *n* $CrCl_2$ chromium(II) chloride, chromium dichloride
Chrom(III)-chlorid *n* $CrCl_3$ chromium(III) chloride, chromium trichloride
Chromdi... *s. a.* Chrom(II)-...
chromdiffundieren to chromize
Chromdioxid *n s.* Chrom(IV)-oxid
Chromeisenerz *n s.* Chromit
Chromentwicklungsfarbstoff *m* afterchrome (chrome-developed) dye
Chromerz *n* chromium ore
Chromfarbstoff *m s.* Chromierungsfarbstoff
chromfeucht *(tann)* blue wet
Chrom(II)-fluorid *n* CrF_2 chromium(II) fluoride, chromium difluoride
Chrom(III)-fluorid *n* CrF_3 chromium(III) fluoride, chromium trifluoride
Chromformiat *n* $Cr(HCOO)_3$ chromium formate
chromgar, chromgegerbt chrome tanned
Chromgelatine *f* chrome (chromatic) gelatin
Chromgelb *n* chrome yellow, lemon chrome *(lead chromate)*
Chromgerbbrühe *f (tann)* chrome liquor
chromgerben to chrome
Chromgerbung *f* chrome (chromium) tannage, chroming
Chromglimmer *m (min)* chrome mica
Chromgrubengerbung chrome pit tannage
Chrom(II)-hydroxid *n* $Cr(OH)_2$ chromium(II) hydroxide
Chrom(III)-hydroxid *n* $Cr(OH)_3$ chromium(III) hydroxide
chromieren *(dye)* to chrome; to chromize *(metals)*
Chromierfarbstoff *m s.* Chromierungsfarbstoff
Chromierung *f (dye)* chroming; chromizing *(of metals)*
 C. aus der flüssigen Phase salt-bath chromizing *(of metals)*
 C. aus der Gasphase gas chromizing *(of metals)*
Chromierungsfarbstoff *m* chrome dye
Chromit *m (min)* chromite, chrome iron ore, chromic iron *(chromium(III) iron(II) oxide)*
Chrom(II)-jodid *n* CrI_2 chromium(II) iodide, chromium diiodide
Chrom(II)-karbonat *n* $CrCO_3$ chromium(II) carbonate
Chromkarbonyl *n* $Cr(CO)_6$ chromium carbonyl, chromium hexacarbonyl
Chromkomplex *m* chromium complex
Chromleder *n* chrome leather
 frisch gegerbtes C. blue chrome leather
Chromleim *m* chrome glue
Chrommolybdänstahl *m* chrome molybdenum steel
Chrommonosulfid *n s.* Chrom(II)-sulfid
Chrommonoxid *n s.* Chrom(II)-oxid
Chrom-Muskovit *m (min)* fuchsite *(a phyllosilicate)*
Chromnickelstahl *m* chrome nickel steel
Chrom(III)-nitrat *n* $Cr(NO_3)_3$ chromium(III) nitrate
Chromnitrid *n* CrN chromium nitride
Chromocker *m (min)* chrome ochre *(a phyllosilicate)*
Chromodruck *m* [multi]colour printing
Chromoersatzkarton *m* imitation chromo board
Chromogen *n* chromogen *(a compound containing a chromophore)*
Chromoisomer[es] *n* chromoisomer
Chromoisomerie *f* chromoisomerism
Chromokarton *m* chromo board
Chromon *n* chromone, benzopyrone
chromophor chromophoric
Chromophor *m* chromophore, chromophoric group
Chromoproteid *n* chromoprotein
Chromorange *n* chrome orange *(a basic lead chromate)*
Chromorohpapier *n* chromo base (body) paper
Chrom(III)-orthophosphat *n* $CrPO_4$ chromium(III) orthophosphate, chromium(III) phosphate
Chromosmiumessigsäurelösung *f* / **Flemmings** *(biol)* Flemming solution *(a fixative)*
Chromosphäre *f* [solar] chromosphere
Chromotropsäure *f* chromotropic acid, 4,5-dihydroxynaphthalene-2,7-disulphonic acid
Chrom(II)-oxalat *n* CrC_2O_4 chromium(II) oxalate, chromous oxalate
Chrom(II)-oxid *n* CrO chromium(II) oxide, chromium monooxide
Chrom(III)-oxid *n* Cr_2O_3 chromium(III) oxide, dichromium trioxide
Chrom(IV)-oxid *n* CrO_2 chromium(IV) oxide, chromium dioxide
Chrom(VI)-oxid *n* CrO_3 chromium(VI) oxide, chromium trioxide
Chrom(VI)-oxidchlorid *n s.* Chromylchlorid
Chromoxidgrün *n* chrome [oxide] green, green cinnabar, oil green *(chromium(III) oxide)*
Chromoxidgrün feurig *n s.* Chromoxidhydratgrün
Chromoxidhydratgrün *n* chrome (emerald) green, Mittler's (Guignet's) green, transparent chromium oxide *(hydrated chromium oxide)*

Chrompapier *n* chromo paper
Chromphosphid *n* CrP chromium phosphide
Chromrot *n* chromate (chrome, Persian) red, American vermilion *(a basic lead chromate)*
Chromsäure *f* H_2CrO_4 chromic acid
Chromsäureoxydation *f* chromic-acid oxidation
Chromsäureverfahren *n* chromic-acid process, Bengough-Stuart process *(anodic oxidation)*
Chromstahl *m* chromium (chrome) steel
Chrom(II)-sulfat *n* $CrSO_4$ chromium(II) sulphate
Chrom(III)-sulfat *n* $Cr_2(SO_4)_3$ chromium(III) sulphate
Chrom(II)-sulfid *n* CrS chromium(II) sulphide, chromium monosulphide
Chrom(II,III)-sulfid *n* Cr_3S_4 chromium(II,III) sulphide, trichromium tetrasulphide
Chrom(III)-sulfid *n* Cr_2S_3 chromium(III) sulphide, dichromium trisulphide
Chromtetrasulfid *n s.* Chrom(II,III)-sulfid
Chromtri... *s.a.* Chrom(III)-...
Chromtrioxid *n s.* Chrom(VI)-oxid
Chromverstärker *m (phot)* chromium intensifier
Chromylchlorid *n* CrO_2Cl_2 chromyl chloride, chromium(VI) dichloride dioxide
Chromzinnober *m s.* Chromrot
Chrysamin *n* chrysamine *(a coal-tar dye)*
Chrysanthemumsäure *f* chrysanthemumic acid *(a cyclopropane derivative)*
Chrysatropasäure *f* chrysatropic acid, scopoletin, 7-hydroxy-6-methoxycoumarin
Chrysen *n* chrysene, 1,2-benzophenanthrene
Chrysergonsäure *f* chrysergonic acid *(a fungal pigment from Claviceps purpurea (Fr.) Tul.)*
Chrysin *n* chrysin, 5,7-dihydroxyflavone
Chrysoberyll *m (min)* chrysoberyl, gold beryl *(beryllium aluminate)*
Chrysoidin *n* $(NH_2)_2C_6H_3N{=}NC_6H_5$ chrysoidine, 2,4-diaminoazobenzene
Chrysokoll *m (min)* chrysocolla *(copper(II) metasilicate)*
Chrysolith *m (min)* chrysolite *(a nesosilicate)*
Chrysophanol *n s.* Chrysophansäure
Chrysophansäure *f* chrysophanic acid, chrysophanol, 1,8-dihydroxy-3-methylanthraquinone
Chrysopras *m (min)* chrysoprase *(a variety of quartz)*
Chrysotil *m (min)* chrysotile, Canadian asbestos
Chrysotilasbest *m (min)* serpentine asbestos
C/H-Verhältnis *n* C/H ratio
Chymase *f*, **Chymosin** *n* chymosin, chymase, rennin
CI *(abbr)* Colour Index
Cinchonaalkaloid *n s.* Chinarindenalkaloid
CI-Nummer *f s.* Colour-Index-Nummer
CIPC *s.* Chlor-IPC
cis-Addition *f* cis addition
cis-cis-Isomer[es] *n* cis-cis isomer
cis-Form *f* cis form
cis-Isomer[es] *n* cis isomer
cis-Lage *f* cis position
cisoid cisoid
cis-orientiert cis-oriented
cis-ständig cis ✦ **c. [angeordnet] sein** to be cis
cis-Stellung *f* cis position
cis-trans-Gemisch *n* mixture of cis and trans isomers
cis-trans-Isomer[es] *n* cis-trans isomer
cis-trans-Isomerie *f* cis-trans isomerism, geometric isomerism
C-Kette *f* chain of carbon atoms
C_3-Kette *f* three-carbon chain
C_6-Kette *f* six-carbon chain
Cladinose *f* cladinose *(a monosaccharide)*
Claisen-Kolben *m* Claisen [distilling] flask
Claisen-Kondensation *f* Claisen condensation *(between esters or between esters and ketones)*
Claisen-Schmidt-Kondensation *f* Claisen-Schmidt condensation *(for preparing unsaturated aldehydes or ketones)*
Claisen-Umlagerung *f* Claisen rearrangement *(of allyl ethers)*
Clarain *m* clarain *(constituent of bright coal)*
Clarkeit *m* clarkeite *(a uranium mineral)*
Clarkes *npl*, **Clarke-Zahlen** *fpl (geoch)* Clarke numbers, clarkes
Claudetit *m (min)* claudetite *(arsenic(III) oxide)*
Claude-Verfahren *n* Claude process *(synthesis of ammonia from nitrogen and hydrogen)*
Clausius-Mosotti-Formel *f*, **Clausius-Mosotti-Gleichung** *f* Clausius-Mosotti equation
Clausthalit *m (min)* clausthalite *(lead selenide)*
Clayden-Effekt *m (phot)* Clayden effect
Clayton-Gas *n* Clayton gas *(mixture of SO_2 and N_2)*
Cleavelandit *m (min)* cleavelandite *(sodium aluminosilicate)*
Clemmensen-Reduktion *f* Clemmensen reduction *(of aldehydes or ketones to hydrocarbons)*
Clerici-Lösung *f* Clerici's solution *(of thallium malonate and thallium formate)*
Cleveit *m* cleveite *(a rare-earth mineral containing thorium)*
Cleveland-Flammpunkt[s]prüfer *m*, **Cleveland-Gerät** *n* Cleveland open tester (cup), Cleveland apparatus
Cleve-Säure *f* Cleve's acid *(1-naphthol 5-sulphonic acid or any of several 1-naphthylamine sulphonic acids)*
Cleve-Säure-1,6 *f* $NH_2C_{10}H_6SO_3H$ Cleve's acid-1,6, Cleve's 1,6 acid, Cleve's β acid, 1-naphthylamine-6-sulphonic acid
Cleve-Säure-1,7 *f* $NH_2C_{10}H_6SO_3H$ Cleve's acid-1,7, Cleve's 1,7 acid, 1-naphthylamine-7-sulphonic acid
Cleve-Säure-6 *f s.* Cleve-Säure-1,6
Clitocybin *n* clitocybine *(antibiotic)*
Clusius-Dickel-Verfahren *n* Clusius-Dickel method, thermal-diffusion method *(for separating isotopes)*
Clusius-Trennrohr *n* Clusius column *(for separating isotopes)*
CN *s.* Chemiefaserstoff aus natürlichen Polymeren
C/N-Verhältnis *n (soil)* C/N ratio
CoA *s.* Koferment A
Cochrane-Trommelprüfung *f* Cochrane [abrasion] test
C-O-Doppelbindung *f* carbonyl double bond

CO_2-Erstarrungsverfahren *n (met)* CO_2 process *(a mouldmaking process)*
Coerulein *n (dye)* coerulein
C=O-Gruppe *f* =C=O group, carbonyl group
CO-Hämoglobin *n* carboxyhaemoglobin
Cohuneöl *n* cohune (corozo-nut) oil *(seed oil from Orbignya cohune (Mart.) Dahlgr.)*
Coin-Technik *f (text)* coin technique
Co-Kontakt *m* cobalt catalyst
Colburn-Verfahren *n (glass)* Colburn [sheet] process, Libbey-Owens process, LOF-Colburn process
Coldcreme *f* cold cream
Coldrubber *m* cold [polymerized] rubber, low-temperature polymer (rubber), LTP
Colemanit *m (min)* colemanite *(a hydrous calcium borate)*
Collin-Ofen *m* Collin oven *(a coke oven)*
CO_2-Löscher *m* carbon dioxide fire extinguisher
Colour-Index-Nummer *f* Colour Index number, CI No.
compoundieren to compound *(oils)*
Compoundöl *n* compounded oil *(a lubricant)*
Compreg *n* compreg, compressed resin-impregnated wood
Compton-Effekt *m* Compton effect, Compton scattering
Compton-Elektron *n* Compton electron
Compton-Rückstoßteilchen *n* Compton recoil particle
Compton-Streuung *f s.* Compton-Effekt
Compton-Verschiebung *f* Compton shift
Compton-Wellenlänge *f* Compton wavelength *(of an electron)*
Conducting-Channel-Ruß *m* conducting (conductive) channel black
Conradson-Carbon *n* Conradson coke (carbon) residue
Conradson-Carbon-Test *m* Conradson [carbon] test
Conradson-Carbon-Wert *m s.* Conradson-Verkokungswert
Conradson-Methode *f* Conradson [coking] method, Conradson carbon residue method
Conradson-Test *m s.* Conradson-Carbon-Test
Conradson-Verkokungswert *m*, **Conradson-Verkokungszahl** *f* Conradson coke number (value), Conradson value
Containerpappe *f* container board
COOH-Gruppe *f* -COOH carboxyl group
Cooperative-Fuel-Research-Motor *m s.* CFR-Prüfmotor
Cope-Eliminierung *f* Cope elimination *(pyrolysis of amine oxides)*
Copiapit *m (min)* yellow copperas *(a basic iron magnesium sulphate)*
Coppée-Flammofen *m* Coppée oven *(a coke oven)*
Coquimbit *m (min)* coquimbite *(iron(III) sulphate-9-water)*
Cordit *m* cordite *(an explosive)*
Corning-Band-Maschine *f (glass)* corning ribbon machine
Corning-Glas *n* corning glass *(for electrodes)*
Cornu-Prisma *n* Cornu prism *(for spectral analysis)*
Cornwallit *m (min)* cornwallite *(copper(II) tetrahydroxide orthoarsenate)*
Corpus-luteum-Präparat *n (pharm)* luteoid
Cotton-Effekt *m (phys chem)* Cotton effect *(anomalous optical rotation near absorption bands)*
Cottonhartfett *n* hydrogenated cotton[seed] oil
Cottonöl *n* cotton[seed] oil
Cottrell-Abscheider *m* Cottrell precipitator
Cottrell-Entstaubungsverfahren *n* Cottrell [electric precipitation] process
Cottrell-Filter *n*, **Cottrell-Staubfilter** *n s.* Cottrell-Abscheider
Couepinsäure *f* couepic acid, licanic acid *(either of two isomeric oxoalkenoic acids)*
Couette-Apparat *m s.* Couette-Viskosimeter
Couette-Viskosimeter *n* Couette viscometer, rotating-cylinder viscometer of Couette
Coulomb-Feld *n* Coulomb field
Coulomb-Gesetz *n* Coulomb law
Coulomb-Glied *n*, **Coulomb-integral** *n* Coulomb term (integral)
Coulomb-Potential *n* Coulomb potential
Coulometer *n* coulo[mb]meter, voltameter
  coulometrisches C. coulometric coulometer
  kolorimetrisches C. colorimetric coulometer
Coulometrie *f* coulometry
  C. bei konstantem Potential *s.* potentiostatische C.
  C. bei konstanter Stromstärke *s.* amperostatische C.
  C. mit kontinuierlich geändertem Potential potential scanning coulometry
  amperostatische C. amperostatic (galvanostatic) coulometry, coulometry at constant current
  potentiostatische C. potentiostatic coulometry, coulometry at constant potential
coulometrisch coulometric
Coulteria-Rotholz *n (dye)* Lima wood *(from Caesalpinia tinctoria (H.B.K.) Benth.)*
Coupage *f (food)* blending
CO_2-Verfahren *n (met)* CO_2 process *(a mouldmaking process)*
Cowrikopal *m s.* Kaurikopal
CP *s.* Zellulosepropionat
Craqueléeglas *n s.* Krakeleeglas
Craqueléeglasur *f s.* Krackglasur
Crazing-Effekt *m (rubber)* crazing
Creep-Test *m (rubber)* creep test
Creme *f* cream
  enthaarende C. depilatory cream
  fettfreie C. *s.* nichtfettende C.
  hautnährende C. nourishing cream, skin food
  nichtfettende C. greaseless cream
cremeartig, cremig creamy
Crêpe *m s.* Crepekautschuk
Crepekautschuk *m* crêpe rubber, crêpe, crepe
Crescent-Methode *f (rubber)* crescent tear test, crescent method *(for determining tearing strength)*
Crescent-Probe *f (rubber)* crescent test-piece

Criegee-Reaktion *f* Criegee reaction *(for splitting glycol compounds)*
Crinis veneris *(min)* cupid's darts, flèche d'amour, love arrows *(a fibrous variety of rutile)*
Cristobalit *m (min)* cristobalite *(either of two modifications of silicon dioxide)*
Croning-Formmaske *f* shell mould *(foundry)*
Croning-Formmaskenverfahren *n* Croning process, C process, shell-moulding process *(foundry)*
Cronstedtit *m (min)* cronstedtite *(a phyllosilicate)*
Crookes-Glas *n* Crookes glass *(absorbing ultraviolet light)*
Crookesit *m (min)* crookesite *(selenide of copper, silver, and thallium)*
Cross-Verfahren *n (petrol)* Cross process
Cross-Zellulose *f* Cross cellulose
Croupon *m (tann)* butt
crouponieren *(tann)* to butt
Crude *n s.* 1. Crudeasbest; 2. Roherdöl
Crudeasbest *m* crude asbestos
CS *s.* Synthesefaserstoff
C-Säure *f* $NH_2C_{10}H_5(SO_3H)_2$ C acid, 2-naphthylamine-4,8-disulphonic acid
C-Stahl *m* carbon steel
CSV *s.* Sauerstoffverbrauch / chemischer
C-terminal C-terminal *(of amino acids in proteins)*
$CuCl_2$-Verfahren *n (petrol)* copper chloride [sweetening] process
Cudbear *m* cudbear, persio, persis *(dried paste of archil, a lichen dye)*
Cuen *n s.* Kupferäthylendiamin
Cuite *f* cuit, bright silk *(completely degummed silk)*
Curaçao-Aloe *f (pharm)* Curacao (Barbados) aloe *(from Aloë vera L.)*
Curare *n* curare, curara *(an arrow poison from several menispermaceae and loganiaceae)*
Curie-Punkt *m*, **Curie-Temperatur** *f* Curie point (temperature), magnetic transition point
Curium *n* Cm curium
curlatieren *(pap)* to curlate
Curlatieren *n (pap)* curlation
Curlator *m (pap)* curlator
Curometer *n (rubber)* curometer *(used in determining the vulcanization curve)*
Cutback-Bitumen *n* cutback [bitumen], bitumen cutback
CuZ *s.* Kupferzahl
CV-Anlage *f (rubber)* CV unit
Cycloversion-Verfahren *n* cycloversion process *(of catalytic reforming)*
CZ *s.* Zellulosechemiefaserstoff
Czako-Hahn *m (lab)* T-shape 120° bore stopcock
Czapek-Dox-Medium *n*, **Czapek-Dox-Nährboden** *m (biol)* Czapek-Dox medium
C-Zustand *m (plast)* C stage

D

***d*-** *s.* dextrogyr
D *s.* Dichte
2,4-D *s.* 2,4-Dichlorphenoxyessigsäure
Dakin-Reaktion *f* Dakin reaction *(oxidation of phenolic aldehydes to polyphenols)*
Dalapon *n* CH_3CCl_2COONa dalapon, sodium 2,2-dichloropropionate *(a herbicide)*
Daltonide *npl* daltonides, daltonide compounds
Dammar[harz] *n* dammar [resin], gum dammar *(esp from several species of the family Dipterocarpaceae)*
  Schwarzes D. black dammar resin *(from Canarium specc.)*
Dämmbeton *m* insulation concrete
Dampf *m* vapour; *(water)* steam; fume ✦ **in D. überführen** to vaporize ✦ **mit D. behandeln** to treat with steam, to steam ✦ **mit D. beheizt** steam-heated
  D. konstanten Drucks constant-pressure steam
  direkter D. live (prime, direct, open) steam
  gesättigter D. saturated vapour; saturated steam
  gespannter D. *s.* direkter D.
  indirekter D. exhaust steam
  perlender D. sparge steam
  trockengesättigter D. dry saturated vapour
  über Kopf abgehender D. overhead vapour
  überhitzter D. superheated steam
Dampfabstreifer *m*, **Dampfabstreiferkolonne** *f (distil)* stripper [column], stripping column (still), strip action still
Dampfanschluß *m* steam joint
Dampfantrieb *m* steam drive
Dampfaufbereitung *f (ceram)* hot preparation, steam tempering
Dampfautoklav *m* steam autoclave
Dampfbad *n* steam bath
Dampfbedarf *m* steam requirement (demand)
dampfbehandelt steamed; steam-cured *(concrete)*
Dampfbehandlung *f* steaming; steam curing *(of concrete)*
dampfbeheizt steam-heated
Dampfbeheizung *f* steam heating
Dampfblase *f* vapour bubble; steam bubble
Dampfdekatur *f (text)* steam blowing
Dampfdestillation *f* steam distillation
dampfdicht vapour-tight; steam-tight
Dampfdichte *f* vapour density
Dampfdichtebestimmung *f* vapour-density determination (measurement)
Dampfdichtebestimmungsmethode *f* vapour-density method
Dampfdichtemessung *f s.* Dampfdichtebestimmung
Dampfdruck *m* vapour pressure (tension)
  D. nach Reid Reid vapour pressure, R.V.P.
Dampfdruckerniedrigung *f* vapour-pressure lowering
  relative D. relative lowering of vapour pressure
Dampfdruckgefälle *n* vapour-pressure gradient
Dampfdruckkurve *f* vapour-pressure curve
Dampfdrucksterilisator *m (med)* autoclave
Dampf-Druckstrahlpumpe *f* steam-jet pump (injector)
Dampfdurchdringtiefe *f (distil)* [static] submergence
Dampfdurchflußmesser *m* steam flowmeter
Dampfdurchtrittsschlitz *m (distil)* slot

Dampfdüsenblasverfahren *n (glass)* steam-blowing process
Dampfeinlaß *m* steam inlet (entrance)
Dampfeinlaßkopf *m (pap)* steamfit, steam joint
Dampfeintritt *m s.* Dampfeinlaß
dampfen to steam
dämpfen 1. to steam; *(pap)* to presteam *(chips before cooking)*; *(text)* to age *(to fix dyeings and prints)*; 2. to damp[en], to attenuate *(e.g. the violence of a reaction)*
Dämpfen *n* 1. steaming; *(pap)* presteaming *(of chips before cooking)*; *(text)* ageing *(fixing of dyeings and prints)*; *(food)* steam deodorization *(of fats)*; 2. damp[en]ing, attenuation *(as of a violent reaction)*
Dampfentfettung *f* vapour degreasing
Dampfentwickler *m* steam generator, boiler
Dämpfer *m* steamer; *(text)* [steam] ager; *(food)* deodorizer *(for fats)*
dampferhärtet steam-cured *(concrete)*
Dampferhärtung *f* steam curing *(of concrete)*
Dämpferpassage *f (coat)* steaming
Dampferzeuger *m* steam generator, boiler
Dampferzeugung *f* steam generation (raising)
Dampferzeugungsanlage *f* steam generating (raising) plant, boiler plant
Dämpfestutzen *m s.* Dampfhals
dampfflüchtig steam-volatile
Dampf-Flüssigkeits-Gemisch *n* vapour-liquid mixture
dampfförmig vaporous
Dampffüllapparat *m (pap)* steam chip distributor
Dampfgefäß *n* steam autoclave
Dampfgeschwindigkeit *f* vapour rate (velocity)
dampfgetrieben steam-driven
dampfgetrocknet steam-dried
Dampfgummi *n* $(C_6H_{10}O_5)_x$ dextrin[e], British (starch) gum
Dampfhals *m (distil)* riser [tube], chimney *(of a tray)*
Dampfhärten *n* steam curing *(of concrete)*
Dampfheizschlange *f* steam coil
Dampfheizung *f* steam heating
Dampfheizungsrohr *n* steam pipe (tube)
Dampfkalorimeter *n* steam calorimeter
Dampfkamin *m s.* Dampfhals
Dampfkammer *f* steam chamber (chest)
Dampfkanal *m (plast)* steam channel
Dampfkanne *f (lab)* steam can
Dampfkessel *m* [steam] boiler
Dampfkesselkohle *f* steam[-raising] coal
Dampfknetwerk *n* pug
Dampfkohle *f* steam[-raising] coal
Dampfkondensat *n* steam condensate
Dampfkopf *m (pap)* steamfit, steam joint
Dampfleitung *f* steam line
Dampfleitungsnetz *n* steam main
Dampfleitungsrohr *n* steam pipe (tube)
Dampf-Luft-Gemisch *n* vapour-air mixture
Dampfmachen *n (tann)* tempering
Dampfmantel *m* steam jacket ✦ **mit D.** steam-jacketed
Dampfmenge *f* / **verbrauchte** amount of steam used
Dampfmesser *m* steam meter (gauge)
Dampfphase *f* vapour phase
Dampfphasekracken *n (petrol)* vapour-phase cracking
Dampfphasenchromatografie *f* vapour-phase chromatography
Dampfphaseninhibitor *m* vapour-phase inhibitor, V.P.I.
Dampfphasenisomerisierung *f* vapour-phase isomerization
Dampfphasennitrierung *f* vapour-phase nitration
Dampfphasenoxydation *f* vapour-phase oxidation
Dampfphaseverfahren *n* vapour-phase process
Dampfpumpe *f* steam pump
Dampfpunkt *m* steam point, boiling point of water
Dampfraum *m* vapour chamber (space); steam chamber (chest)
Dampfregister *n* steam battery
Dampfrohr *n* steam pipe (tube)
Dampfsammler *m* steam collector (accumulator)
Dampfschlange *f* steam coil
Dampfschmalz *n (food)* steam lard
Dampfspannung *f s.* Dampfdruck
Dampfspannungsthermometer *n* vapour-pressure thermometer
Dampfspeicher *m* steam collector (accumulator)
Dampfstauer *m* expansion trap
Dampfsterilisation *f* steam sterilization
Dampfstoßverfahren *n (plast)* steam-moulding process
Dampfstrahl *m* steam jet
Dampfstrahlapparat *m* steam-jet apparatus
Dampfstrahlejektor *m* steam-jet (steam-motivated, steam-operated) ejector
Dampfstrahlinjektor *m* steam-jet injector (pump)
Dampfstrahlkältemaschine *f*, **Dampfstrahlkühlanlage** *f* steam-jet refrigerating machine
Dampfstrahlkühlung *f* steam-jet refrigeration
Dampfstrahlmaschine *f s.* Dampfstrahlkältemaschine
Dampfstrahlpumpe *f* steam-jet pump (injector)
Dampfstrahlsauger *m s.* Dampfstrahlejektor
Dampfsublimation *f* sublimation in steam
Dampftrichter *m* steam-heated funnel
Dampftrockenapparat *m* steam dryer (drying apparatus)
Dampftrockenschrank *m* steam drying oven
Dampftrockner *m* vapour dryer; steam dryer
Dampfturbinenöl *n* steam-turbine oil
Dampfturbogebläse *n* steam-driven turboblower
Dampfüberhitzer *m* steam superheater
dampfundurchlässig vapour-tight; steam-tight
Dampfung *f* steaming *(in manufacturing water gas)*
Dämpfung *f* 1. steaming; *(pap)* presteaming *(of chips before cooking)*; *(rubber)* hysteresis; 2. damp[en]ing, attenuation *(as of a violent reaction)*
Dämpfungseinrichtung *f* damping device *(as of chemical balances)*
Dämpfungsfaktor *m* damping factor
Dämpfungsflüssigkeit *f* damping fluid

Dämpfungsgerät *n* **nach Roelig** *(rubber)* Roelig hysteresis apparatus
Dämpfungsmittel *n* damping medium
Dämpfungsschleife *f (rubber)* tensile hysteresis loop
Dämpfungsverhalten *n* damping properties
Dämpfungswaage *f* damped balance
Dämpfungszylinder *m* dash pot
Dampfventil *n* steam valve
Dampfverbrauch *m* steam consumption
Dampfversprühung *f* steam atomization
Dampfvulkanisation *f (rubber)* steam vulcanization (curing, cure)
Dampfzerstäubung *f* steam atomization
Dampfzufuhr *f* steam supply
Dampfzustand *m* vapour state
Dampfzylinder *m* steam cylinder
Dampfzylinderöl *n* steam-cylinder [lubricating] oil, steam-cylinder stock
Danburit *m (min)* danburite *(calcium borosilicate)*
Daniell-Element *n*, **Daniell-Kette** *f* Daniell cell
Danner-Verfahren *n* Danner process *(for manufacturing glass tubing)*
DAP *s.* Diallylphthalatharz
Darmfett *n* gut fat
Darmöl *n (pharm)* intestinal lubricant
Darre *f* 1. [drying] kiln, kiln dryer; *(ferm)* malt [drying] kiln, oast; 2. *s.* Darrhaus
darren to kiln-dry, to kiln, to desiccate *(e. g. malt)*
Darren *n* kiln-drying, kilning, desiccating *(as of malt)*
Darrhaus *n* oast-house
Darrhorde *f* [kiln] floor
Darrmalz *n* kiln (kiln-dried, cured) malt
Darrmasse *f (pap)* dry wood weight, moisture-free weight
Darrofen *m* [drying] kiln, kiln dryer
Darrtemperatur *f* kiln temperature
darstellbar / rein isolable *(natural product)*
darstellen to prepare, to make, to recover; to isolate *(natural products)*
  rein d. to isolate *(natural products)*
Darstellung *f* preparation, making, recovery; isolation *(of natural products)*
  D. mit Dreieckskoordinatensystem triad grouping
  grafische D. graph
Darstellungsbedingung *f* condition of preparation
Darstellungsmethode *f* method of (for) preparation, preparative method
Darzens-Erlenmeyer-Claisen-Kondensation *f s.* Darzens-Reaktion
Darzens-Reaktion *f* Darzens [glycidic ester] condensation
Dasymeter *n* dasymeter
Daten *pl* data
  kritische D. *(phys chem)* critical constants (data)
  röntgenografische D. X-ray data
Datolith *m (min)* datolite *(calcium hydroxide borosilicate)*
Dauerbeanspruchung *f* repeated stress
Dauerbehandlung *f* long-term treatment
Dauerbetrieb *m* continuous operation (working)
Dauerbiegebeanspruchung *f* repeated flexural stress
Dauerbiegefestigkeit *f* repeated flexural strength, flex[ing] life; *(tann)* bending endurance
Dauerbiegespannung *f* repeated flexural stress
Dauerelektrode *f* [Söderberg] continuous electrode, self-baking electrode
Dauererhitzung *f (food)* vat (holding, holder) pasteurization
Dauerfestigkeit *f (plast)* endurance (fatigue) limit
Dauerfixierung *f (text)* permanent set
Dauerform *f* permanent mould *(foundry)*
  metallische D. permanent metal mould, gravity die
Dauerformgießverfahren *n s.* Dauerformgußverfahren
Dauerformguß *m* permanent-mould casting, gravity die casting
Dauerformgußstück *n* gravity die casting, diecasting
Dauerformgußverfahren *n* permanent-mould casting process, gravity die-casting process
Dauergießform *f s.* Dauerform
dauerhaft durable, permanent, stable
Dauerhaftigkeit *f* durability, permanence, stability, strength, life
Dauerhaftigkeitsprüfung *f* durability test
Dauerhumus *m* stable humus
Dauerknickversuch *m* flex-cracking test
Dauermilch *f* preserved milk
Dauermilchwaren *fpl* milk preserves
Dauermilchwerk *n* preserved-milk factory
Dauerpasteurisation *f (food)* vat (holding, holder) pasteurization
Dauerschlagwerk *n* impact fatigue testing machine
Dauerschwingfestigkeit *f* endurance (fatigue) limit
Dauerstandfestigkeit *f* creep resistance
Dauerwanne *f (glass)* continuous tank
Dauerwärmebeständigkeit *f* continuous heat resistance
Dauerwellflüssigkeit *f (cosmet)* permanent-waving solution
Dauerwellpräparat *n (cosmet)* permanent-wave preparation
Daumenbrecher *m* sawtooth crusher
Daunendruckpapier *n* featherweight paper, *(Am)* bulking paper
Dawsonit *m (min)* dawsonite *(aluminium sodium carbonate dihydroxide)*
Dazit *m (geol)* dacite *(an extrusive rock)*
2,4-DB *s.* 4(2',4'-Dichlorphenoxy)buttersäure
DBPC *s.* 2,6-Di-*tert.*-butyl-*p*-kresol
DDNP *s.* Diazodinitrophenol
DDT *s.* Dichlordiphenyltrichloräthan
DE *s.* Defometerelastizität
Dead-stop-Titration *f* dead-stop titration
Dead-stop-Titrationskurve *f* dead-stop titration curve
dealkylieren to dealkylate
Dealkylierung *f* dealkylation
Deäthanisator *m* de-ethanizer
deäthanisieren to de-ethanize

Deäthanisierung *f* de-ethanization
de-Broglie-Beziehung *f* de Broglie relationship
de-Broglie-Gleichung *f* de Broglie equation
de-Broglie-Wellen *fpl* de Broglie waves
Debutanisator *m* debutanizer
debutanisieren to debutanize
Debutanisierung *f* debutanization
Debutanisierungskolonne *f* debutanizer
Debye *n s.* Debye-Einheit
Debye-Einheit *f* Debye [unit], D
Debye-Falkenhagen-Effekt *m* Debye-Falkenhagen effect *(dispersion of conductance)*
Debye-Funktion *f* Debye function
Debye-Hückel-Theorie *f* Debye-Hückel theory
Debye-Länge *f* Debye length
Debye-Scherrer-Aufnahme *f*, **Debye-Scherrer-Diagramm** *n (cryst)* Debye-Scherrer diagram (pattern, photograph), X-ray powder diagram
Debye-Scherrer-Methode *f (cryst)* Debye-Scherrer[-Hull] method, [X ray] powder method
Debye-Temperatur *f* [Debye] characteristic temperature
Deck *n* deck *(of a concentrating table)*
Deckablauf *m (sugar)* wash syrup
Deckanstrich *m (coat)* finishing (top) coat, finish *(result)*; finish[ing] *(act)*
Deckappretur *f*, **Deckauftrag** *m* coating (top) finish
Deckbogen *m (plast)* surfacing sheet
Deckdruck *m (text)* blotch printing
Decke *f* 1. cover; 2. *s.* Deckgebirge
Deckel *m* lid, cap, top, cover, hood ✦ **mit einem D. verschlossen (versehen)** lidded
D. der Petrischale Petri dish top
Deckelleitrad *n (pap)* deckle (strap) pulley
Deckelrahmen *m (pap)* deckle frame
Deckelriemen *m (pap)* deckle (boundary) strap
oberer D. upper run of the deckle strap
unterer D. lower run of the deckle strap
Deckelriemenführungsrad *n*, **Deckelriemenrolle** *f (pap)* deckle (strap) pulley
decken to cover, to top *(with a finish or another dye)*; to cover *(as of a pigment)*
die Chemikalienverluste d. *(pap)* to make up for the loss of chemical
einander (sich) d. to coincide *(stereochemistry)*
deckend / einander coincident *(stereochemistry)*
Deckerdruck *m (text)* blotch printing
Deckfähigkeit *f s.* Deckkraft
Deckfarbe *f* final (top) coat
Deckgebirge *n (mine)* overburden, roof rock, rock cover
Deckglas *n* cover glass; slide cover glass *(microscopy)*
Deckglaskultur *f* [hanging-]drop culture
Deckgrün *n* chrome green *(a mixture of iron blue and chrome yellow)*
Deckgummi *m* rubber cover *(of a conveyor)*
Deckkraft *f* coverage, covering (opacifying, obliterating, hiding) power
Decklack *m* coating lacquer, finish
Decklauge *f (fert)* covering lye
Deckmasse *f* cover coat
Deckmittel *n* covering agent (material, medium)
Deckplatte *f* cover [plate]
Deckschicht *f* 1. covering layer, cover, coating, overlay; 2. *s.* Deckgebirge
oxidische D. oxide coating *(on metals)*
Deckung *f* coincidence *(in space or time)*; *(cryst)* self-coincidence; *(phot)* extinction, optical density ✦ **zur D. bringen** to make coincide
D. der Chemiekalienverluste *(pap)* make-up of chemical loss
deckungsgleich superimposable
nicht d. non-superimposable
Deckvermögen *n s.* Deckkraft
Dedolomitisierung *f (geol)* dedolomitization
Dees *pl s.* Duanten
Defäkation *f (sugar)* defecation, liming
defekt defective, faulty
Defekt *m* defect, fault, flaw; *(cryst)* defect
Defektelektron *n* defect electron, [electron] hole
Defektelektronenleitung *f* hole conduction
Defektgitter *n* defect lattice
Defektleiter *m (phys chem)* defect conductor
Defektleitung *f (phys chem)* hole conduction
Defibrator *m (pap)* pulpwood grinder
defibrieren *(pap)* to defibre, to defibrate, to reduce to fibres, *(Am also)* to [de]fiberize
Defibrierung *f (pap)* defib[e]ring, defibration, *(Am also)* [de]fiberization
defibrillieren *(pap)* to fibrillate
Defibrillierung *f (pap)* fibrillation
defibrinieren to defibrinate *(blood)*
Defibrinierung *f* defibrination *(of blood)*
Deflagration *f* deflagration
deflagrieren to deflagrate
de-Florez-Krackprozeß *m* de Florez process
Defo... *s. a.* Defometer...
Defoliationsmittel *n (agric)* defoliant
Defomeßgerät *n*, **Defometer** *n (rubber)* Defo plastometer
Defometerelastizität *f (rubber)* Defo elasticity
Defometerhärte *f (rubber)* Defo hardness
Defometerwert *m (rubber)* Defo value
Defometerzahl *f (rubber)* Defo number
Deformation *f* deformation, strain
bleibende D. residual (plastic, permanent) deformation, residual set
elastische D. elastic deformation
irreversible (plastische) D. *s.* bleibende D.
postkristalline D. *(geol)* postcrystalline deformation
präkristalline D. *(geol)* precrystalline deformation
Deformationsgeschwindigkeit *f* rate of deformation
Deformationsschwingung *f* deformation vibration *(of molecules)*
deformieren to deform, to strain
Degorgement *n (ferm)* disgorgement *(for removing lees)*
degorgieren *(ferm)* to disgorge *(for removing lees)*
Degradation *f (soil)* degradation
degradieren *(soil)* to degrade
Degras *m(n) (tann)* degras, sod oil, moellon

degummieren *(text)* to boil off (out), to degum
Degummieren *n (text)* boil[ing]-off, degumming
dehalogenieren to dehalogenate
Dehalogenierung *f* dehalogenation
dehnbar extensible, expansible, expandable, expandible, *(esp of metal)* ductile
Dehnbarkeit *f* extensibility, expansibility, expandability, *(esp of metal)* ductility
dehnen to extend, to expand, to stretch
Dehnfuge *f* expansion joint
Dehngrenze *f* yield strength *(of a metal)*
0,2%-Dehngrenze *f* yield strength 0.2% offset
Dehnung *f* extension, expansion, stretch, elongation
  D. je Längeneinheit *s.* relative D.
  bleibende D. plastic elongation, offset
  elastische D. reversible elongation, *(text also)* stretch
  irreversible D. *s.* bleibende D.
  relative D. unit elongation
  reversible D. *s.* elastische D.
Dehnungsausgleicher *m* expansion joint
dehnungsfähig *s.* dehnbar
Dehnungsfähigkeit *f s.* Dehnbarkeit
Dehnungsfuge *f* expansion joint
Dehnungsmesser *m*, **Dehnungsmeßgerät** *n* extensometer
Dehnungsrest *m (rubber)* elongation (tensile) set, residual elongation, permanent [set at] elongation
Dehnungs-Spannungs-Kurve *f* stress-strain curve *(expressed as tons or pounds per squ. in.)*; load-elongation (load-extension) curve *(expressed as inches per inch)*
Dehydrase *f (incorrectly for)* Dehydrogenase
Dehydratation *f* dehydration
Dehydratationsmittel *n* dehydrating agent, dehydrator
Dehydration *f s.* Dehydrierung
dehydratisieren to dehydrate
Dehydratisierung *f s.* Dehydratation
Dehydrazetsäure *f* dehydracetic (dehydroacetic) acid, DHA, 3-acetyl-2-hydroxy-6-methylpyran-4-one
Dehydrier... *s.* Dehydrierungs...
dehydrieren to dehydrogenate, to dehydrogenize
dehydrierend dehydrogenative
Dehydrierung *f* dehydrogen[iz]ation
Dehydrierungskatalysator *m* dehydrogenation (dehydrogenating) catalyst
Dehydrierungsmittel *n* dehydrogenation (dehydrogenating) agent
Dehydroazetsäure *f s.* Dehydrazetsäure
Dehydrobase *f* dehydro base
Dehydrobenzol *n* C_6H_4 benzyne
Dehydrobilirubin *n* dehydrobilirubin, biliverdin
Dehydrochinasäure *f* dehydroquinic acid, 1,3,4-trihydroxy-5-oxocyclohexanecarboxylic acid
Dehydroessigsäure *f s.* Dehydrazetsäure
Dehydrogenase *f* dehydrogenase
Dehydrogeraniumsäure *f* dehydrogeranic acid *(an alkenoic acid)*
dehydrohalogenieren to dehydrohalogenate
Dehydrohalogenierung *f* dehydrohalogenation
Dehydroisomerisierung *f* dehydroisomerization
Dehydrozyklisierung *f* dehydrocyclization
D-Einheit *f* difunctional (bifunctional) unit, D unit *(structural element of macromolecules)*
deinken to deink *(waste paper)*
Deinking-Anlage *f (pap)* deinking plant
Deionisation *f* deionization
Deionisationspotential *n* deionization potential
Deionisationszeit *f* deionization time
deionisieren to deionize
Deionisierung *f* deionization
Deionisierungsmittel *n* deionizer
Deisobutanisator *m* deisobutanizer
deisobutanisieren to deisobutanize
Deisobutanisierung *f* deisobutanization
Dekaboran *n* $B_{10}H_{14}$ decaborane
Dekadien-(2,4)-säure *f* $CH_3[CH_2]_4[CH{=}CH]_2COOH$ 2,4-decadienoic acid
Dekahydrat *n* decahydrate
Dekahydronaphthalin *n* decahydronaphthalene
Dekameter *n* dekameter, DK-meter, dielectrometer
Dekametrie *f* dielectrometry *(measurement of dielectric constants)*
Dekan *n* $C_{10}H_{22}$ decane
Dekanal *n* $CH_3[CH_2]_8CHO$ decanal
Dekanamid *n* $CH_3[CH_2]_8CONH_2$ decanamide
Dekan-1,10-dikarbonsäure *f* $HOOC[CH_2]_{10}COOH$ decane-1,10-dicarboxylic acid, dodecanedioic acid
Dekandisäure *f* $HOOC[CH_2]_8COOH$ decanedioic acid, sebacic acid
Dekansäure *f* $CH_3[CH_2]_8COOH$ decanoic acid
Dekansäureanhydrid *n* $(C_9H_{19}CO)_2O$ decanoic anhydride
Dekansäureäthylester *m* $CH_3[CH_2]_8CO{-}OC_2H_5$ ethyl decanoate
Dekansäuremethylester *m* $CH_3[CH_2]_8CO{-}OCH_3$ methyl decanoate
Dekantation *f* decantation
Dekanteur *m s.* Dekantiergefäß
dekantieren to decant, to pour off
Dekantiergefäß *n*, **Dekantiertopf** *m* decanter, decanting jar
Dekantierung *f* decantation, decanting, pouring-off
Dekapeptid *n* decapeptide
  zyklisches D. cyclodecapeptide
dekarbonisieren *(petrol)* to decarbonize, to decoke
Dekarbonisierung *f (petrol)* decarbonization, decoking
Dekarbonylierung *f* decarbonylation
Dekarboxylase *f* decarboxylase
dekarboxylieren to decarboxylate
Dekarboxylierung *f* decarboxylation
Dekatierechtheit *f (text)* fastness to decatizing
dekatieren *(text)* to decatize, to decate, to hot-press
Dekatieren *n s.* Dekatur
Dekatur *f (text)* decatizing, decating, hot-pressing
Dekaturechtheit *f (text)* fastness to decatizing
Dekokt *n* decoctum, decoction
Dekoktionsverfahren *n (ferm)* decoction process

(mashing)
Dekontaminationsindex *m (nucl)* decontamination index
dekontaminieren *(nucl)* to decontaminate
Dekontaminierung *f (nucl)* decontamination
Dekontaminierungsmittel *n (nucl)* decontaminating agent (chemical, substance)
Dekorationsfolie *f s.* Dekorfolie
Dekorationspapier *n* decorating paper
Dekorationsschichtstoff *m* decorative laminate
Dekorationsseidenpapier *n* decorating tissue paper, decoration tissue
Dekorbrand *m (ceram)* decorating (decoration, enamel) firing
Dekorbrandofen *m (ceram)* decorating kiln
Dekorfolie *f* decorating sheeting, decorative foil (sheet); overlay [paper] *(for particle board)*
Dekorofen *m s.* Dekorbrandofen
Dekrepitation *f (cryst)* decrepitation
dekrepitieren *(cryst)* to decrepitate
Dekulator *m (pap)* deculator, stock deaerator
Delessit *m (min)* delessite *(a phyllosilicate)*
Delftware *f (ceram)* delftware, delf[t], delph[ware]
delignifizieren *(pap)* to delignify
Delignifizierung *f (pap)* delignification, lignin removal, dissolution of lignin
Delignifizierungsmittel *n (pap)* delignifying agent
Delikateßmargarine *f* high-class table margarine
delokalisieren *(phys chem)* to delocalize
Delokalisierung *f (phys chem)* delocalization
Delokalisierungseffekt *m* delocalization effect
Delokalisierungsenergie *f* delocalization (resonance, mesomeric) energy
Delphinin *n* delphinine *(alkaloid)*
Delphinsäure *f* $(CH_3)_2CHCH_2COOH$ isovaleric acid, 3-methylbutyric acid, delphinic acid
Delphintran *m* dolphin oil
Deltaelektron *n* delta electron
Deltastrahl *m* delta ray
demargarinieren to demargarinate, to destearinate, to destearinize, to winterize *(oils)*
Demargarinieren *n*, **Demargarinisation** *f* demargarination, destearinization, winterization *(of oils)*
De-Mattia-Biegeprüfmaschine *f*, **De-Mattia-Knickermüdungsprüfer** *m* De Mattia [flexing] machine
Demethanisator *m* demethanizer
demethanisieren to demethanize
Demethanisierung *f* demethanization
demethylieren to demethylate
Demethylierung *f* demethylation
Demijohn *m* demijohn *(a large glass vessel enclosed in wicker-work)*
Demineralisation *f* demineralization
demineralisieren to demineralize
Demjanow-Umlagerung *f* Demjanov rearrangement *(of primary cycloaliphatic amines)*
Demulgator *m* demulsifier
demulgieren to demulsify, to de-emulsify, to break, to crack *(an emulsion)*
Demulgieren *n*, **Demulgierung** *f* demulsification, de-emulsification, breaking, cracking *(of an emulsion)*
Denaturation *f s.* Denaturierung
denaturieren 1. to denature, to denaturize *(e.g. common salt, native proteins)*; to methylate, to denature *(ethanol)*; 2. to denature, to denaturize *(of native proteins)*
Denaturierung *f* 1. denaturation *(as of common salt, native proteins)*; methylation, denaturation *(of ethanol)*; 2. denaturation *(change in the molecular structure of native proteins)*
Denaturierungsmittel *n* denaturant, denaturing agent
Dendrit *m (cryst)* dendrite
dendritisch *(cryst)* dendritic[al]
Denitration *f* denitration
Denitrator *m s.* Denitrierturm
Denitrierapparat *m* denitrator
denitrieren to denitrate
Denitrierturm *m* denitration tower, denitrator [tower]
Denitrierung *f* denitration
Denitrierungs... *s.* Denitrier...
Dentrifikanten *mpl s.* Denitrifikationsbakterien
Denitrifikation *f* denitrification *(reduction of nitrates brought about by denitrifying bacteria)*
Denitrifikationsbakterien *pl* denitrifying bacteria, denitrifiers
Denitrifikatoren *mpl s.* Denitrifikationsbakterien
denitrifizieren to denitrify
Denitrifizierung *f s.* Denitrifikation
de-Nora-Zelle *f* de Nora cell *(a mercury cell)*
Densimeter *n* densimeter, densitometer, hydrometer; *(phot)* densitometer
Densimetrie *f* densimetry
Densitometer *n (phot)* densitometer
Densitometrie *f (phot)* densitometry
Densograf *m s.* Densitometer
Densometer *n s.* Densitometer
Dentalporzellan *n* dental porcelain
Dentin *n* dentin[e]
Depentanisator *m* depentanizer
depentanisieren to depentanize
Depentanisierung *f* depentanization
Dephlegmation *f s.* Dephlegmierung
Dephlegmator *m (distil)* dephlegmator, [countercurrent] partial condenser, partial-condensation head
dephlegmieren *(distil)* to dephlegmate
Dephlegmierung *f (distil)* dephlegmation, partial condensation
dephosphorylieren to dephosphorylate
Dephosphorylierung *f* dephosphorylation
Depilation *f* depilation
Depilatorium *n* depilator, depilatory [agent], depilitant, hair remover
Depiliercreme *f* depilatory cream
depilieren to depilate, to remove hair
Depilierung *f* depilation
Depolarisation *f* depolarization
Depolarisationsgrad *m* degree of depolarization
Depolarisator *m* depolarizer
depolarisieren to depolarize

Depolymerisation *f* depolymerization
depolymerisieren to depolymerize
Depolymerisierung *f* depolymerization
Deponie *f* disposal
Depot *n* 1. storehouse, storage, store; 2. *(geol)* deposit
Depotfett *n* depot fat
Depression *f* depression
Depropanisator *m* depropanizer
depropanisieren to depropanize
Depropanisierung *f* depropanization
Depropanisierungskolonne *f* depropanizer
deproteinisieren to deproteinize
Deproteinisierung *f* deproteinization
Deprotonierung *f* deprotonation
Deprotonierungs-Protonierungs-Reaktion *f* deprotonation-protonation reaction
Depsid *n (org chem)* depside
Depsidon *n (org chem)* depsidone
Derbylith *m (min)* derbylite *(an iron antimonate titanate)*
Derbyrot *n s.* Chromrot
Derivat *n* derivative
organisches D. organoderivative
desaktivieren to inactivate, to deactivate, to block
Desaktivierung *f* inactivation, deactivation, blocking
Desaktivierungszeit *f* inactivation (deactivation) period
Desamidase *f* deamidase, desamidase
desamidieren to deamidate, to desamidate
Desamidierung *f* deamidation, desamidation
Desaminase *f* deaminase, desaminase
desaminieren to deaminate, to desaminate
Desaminierung *f* deamination, desamination
oxydative D. oxidative deamination
Descloizit *m (min)* descloizite *(a basic lead zinc copper vanadate)*
Desensibilisator *m (phot)* desensitizer
desensibilisieren *(phot)* to desensitize
Desensibilisierung *f (phot)* desensitization
Deserpidin *n* deserpidine, 11-demethoxyreserpine *(a rauwolfia alkaloid)*
deshalogenieren *s.* dehalogenieren
desilifizieren *(geoch)* to desilicate
Desilifizierung *f (geoch)* desilication
Desinfektion *f* disinfection
Desinfektionsmittel *n*, **Desinfiziens** *n* disinfectant
desinfizieren to disinfect
desinfizierend disinfectant
Desinfizierung *f* disinfection
Desintegrationstheorie *f* theory of radioactive disintegration
Desintegrator *m s.* 1. Desintegratorgaswäscher; 2. Desintegratormühle
Desintegratorgaswäscher *m* disintegrator [washer]
D. nach Theisen Theisen disintegrator
D. nach Zschocke Zschocke disintegrator
Desintegratormühle *f* disintegrator, *(specif)* cage (squirrel-cage) disintegrator (mill), bar mill; *(pap)* chip crusher, chipbreaker, rechipper
Desintegratorwäscher *m* disintegrator [washer]
desintegrieren to disintegrate
Desmin *m (min)* desmine, stilbite *(a tectosilicate)*
Desmoenzym *n*, **Desmoferment** *n* desmo-enzyme, extra-cellular enzyme
Desmolase *f* desmolase
Desmolyse *f* desmolysis
desmotrop desmotropic
Desmotropie *f* desmotropy, desmotropism, dynamic isomerism
Desodorans *n* deodorant, deodorizer
Desodorant-Lotion *f* deodorant lotion
Desodoration *f* deodorization
Desodoreur *m* deodorizer *(an apparatus for deodorization of fats and oils)*
desodorieren to deodorize
desodorierend deodorant
Desodorierer *m s.* Desodoreur
Desodorierung *f* deodorization
Desodorierungsmittel *n* deodorant, deodorizer
Desodorisation *f s.* Desodoration
desodorisieren *s.* desodorieren
desorbierbar desorbable
Desorbierbarkeit *f* desorbability
desorbieren to desorb, to strip [off, out]
mit Wasserdampf d. to steam
Desorption *f* desorption, stripping
Desorptionskurve *f* desorption curve
Desosamin *n* desosamine *(a xylohexose derivative)*
Desoxycholsäure *f* deoxycholic acid *(a bile acid)*
Desoxydation *f* deoxid[iz]ation, deoxygenation
Desoxydationsmittel *n* deoxidant, deoxidizer, deoxidizing agent, *(met also)* scavenger
desoxydieren to deoxidize, to deoxidate, to deoxygenate, *(met also)* to scavenge
Desoxydieren *n s.* Desoxydation
Desoxykortikosteron *n* deoxycorticosterone
Desoxykorton *n s.* Desoxykortikosteron
Desoxypentose *f* deoxypentose
Desoxypentosenukleinsäure *f* deoxypentose nucleic acid
Desoxyribonuklease *f* deoxyribonuclease
Desoxyribonukleinsäure *f* deoxyribonucleic acid, DNA
Desoxyribonukleoprotein *n* deoxyribonucleoprotein
Desoxyribose *f* deoxyribose *(a monosaccharide)*
desozon[is]ieren to deozonize
destabilisieren to destabilize, to make unstable
Destillans *n* material being distilled
Destillat *n* distillate
leichtes D. light distillate
mittleres D. middle distillate
Destillatabnahme *f* product take-off, distillate drain
Destillatbenzin *n* straight-run gasoline, distillate gasoline, straight-run benzine, S.R.B.
Destillateur *m* distiller
Destillatfangrinne *f* distillate [collection] gutter
Destillatfraktion *f* distillate fraction
Destillatheizöl *n* distillate fuel oil
Destillation *f* distillation
D. eines Mehrkomponentensystems (Mehrstoffgemischs) multicomponent distillation

D. eines Zweikomponentensystems (Zweistoffgemischs) binary distillation
D. im Vakuum vacuum distillation, distillation under vacuum (reduced pressure)
D. mit fallendem Film falling-film distillation
D. mit Zusatzstoff[en] codistillation
D. nach ASTM ASTM distillation
D. unter vermindertem Druck *s.* D. im Vakuum
abbauende D. *s.* trockene D.
absteigende D. distillation by descent
aufsteigende D. distillation by ascent
azeotrop[isch]e D. azeotropic distillation
destruktive D. *s.* trockene D.
differentielle D. differential distillation
direkte D. straight (straight-run, simple) distillation
diskontinuierliche D. batch distillation
einfache D. *s.* direkte D.
einfache kontinuierliche D. simple continuous distillation
erneute D. redistillation, rerun[ning]
erste D. primary distillation
extrahierende (extraktive) D. extractive distillation
fraktionierende (fraktionierte) D. fractional (fractionating) distillation, fractionation
geschlossene D. equilibrium distillation
gewöhnliche D. simple (straight) distillation
halbkontinuierliche D. semicontinuous distillation
integrale D. equilibrium distillation
isobare D. isobaric distillation, distillation at constant pressure
isotherme D. isothermal distillation, distillation at constant temperature
katalytische D. catalytic distillation
kontinuierliche D. continuous distillation
kontinuierliche D. von Zweistoffgemischen continuous binary distillation
mehrmalige einfache D. simple batch distillation
nochmalige D. redistillation, rerun[ning]
offene D. differential distillation
primäre D. primary distillation
schonende D. gentle distillation
stetige D. continuous distillation
stetige D. von Zweistoffgemischen continuous binary distillation
trockene D. dry (destructive, pyrogenic) distillation
zersetzende D. *s.* trockene D.
Destillations... *s. a.* Destillier...
Destillationsanlage *f* distillation (distilling) plant (unit), distillery, still
diskontinuierlich arbeitende D. batch-distillation plant (unit)
kontinuierlich arbeitende D. continuous-distillation plant (unit)
Destillationsapparat *m* distillation apparatus, distiller, still, boiler
D. für Benzolvorprodukt once-run[ning] still
D. mit fallendem Film falling-film still
D. mit rotierender Verdampferfläche rotary still
D. mit Verteilerbürsten wiped-film still
D. nach Savalle Savalle's still
Destillationsapparatur *f* distillation assembly
Destillationsbenzin *n s.* Destillatbenzin
Destillationsbereich *m* distillation range
Destillationsdruck *m* distillation pressure
Destillationseinheit *f* distillation (distilling) unit
Destillationserzeugnis *n* distillation product, running
Destillationsfraktion *f* distillate fraction
Destillationsgas *n* distillation gas
Destillationsgeschwindigkeit *f* distillation rate
Destillationsgut *n* distilland, material to be distilled; material being distilled
Destillationskurve *f*, **Destillationslinie** *f* distillation curve
Destillationsmaterial *n s.* Destillationsgut
Destillationsprobe *f* distillation test
Destillationsprodukt *n* distillation product, running
Destillationsretorte *f* distillation (distilling) retort
Destillationsrückstand *m* distillation (still) residue
kurzer D. short residue (residuum)
langer D. long residue (residuum)
Destillationsstufe *f* distillation stage
Destillationstemperatur *f* distillation (distilling) temperature
Destillationsturm *m* distillation tower
Destillationsverlust *m* distillation loss
Destillationsvorgang *m* distillation process
Destillationsvorlage *f* distillate (distillation, still) receiver
Destillationswasser *n* distillation water
destillativ by [means of] distillation
Destillatkühler *m* distillate cooler
Destillatöl *n* distillate oil
Destillatsammelrinne *f* distillate [collection] gutter
Destillatsammler *m s.* Destillationsvorlage
Destillatschmieröl *n* distillate lubricating oil
Destillatstock *m (petrol)* distillate stock
Destillatvorlage *f s.* Destillationsvorlage
Destillatzusammensetzung *f* distillate composition
Destillier... *s. a.* Destillations...
Destillierarbeit *f* distillation (distilling) operation
Destillieraufsatz *m* distillation head, stillhead, distillation connecting tube
D. nach Claisen Claisen stillhead
destillierbar distillable
mit Dampf (Trägerdampf, Wasserdampf) d. steam-distillable
Destillierbarkeit *f* distillability
Destillierbetrieb *m* distillation (distilling) plant, distillery; distillation (distilling) operation
Destillierblase *f* still, pot, reboiler, [distillation] boiler
D. mit direkter Beheizung direct-fired reboiler
Destilliereinrichtung *f* distillation equipment
destillieren to distil
erneut d. to redistil, to rerun
fraktioniert d. to fractionate, to fraction
mit Dampf (Trägerdampf, Wasserdampf) d. to steam-distil
nochmals d. *s.* erneut d.
stufenweise d. to fractionate, to fraction

wiederholt d. to redistil, to rerun
Destillieren *n s.* Destillation
Destilliergefäß *n* distilling vessel, still pot, reboiler, [distillation] boiler
Destillierhaus *n* still house
Destillierkolben *m* distillation (distilling) flask
Destillierkolonne *f* distillation (distilling) column
D. mit Glockenböden bubble-cap (bubble-tray, bubble-plate) column
Destillierkopf *m s.* Destillieraufsatz
Destillierofen *m* distillation (distilling) furnace, retort furnace (oven)
Destillierrohr *n* distillation tube
Destilliersäule *f* distillation (distilling) column
destilliert / doppelt twice-distilled
dreifach d. triple-distilled
unter Vakuum d. vacuum-distilled, distilled in vacuo
zweifach d. twice-distilled
Destilliervorlage *f* distillate (distillation, still) receiver
Destilliervorstoß *m* adapter
Destruktion *f* destruction
destruktiv destructive
desulfurieren to desulphurize, to desulphur
Desulfurierung *f* desulphur[iz]ation
Desylchlorid *n* $C_6H_5COCHClC_6H_5$ desyl chloride, α-chloro-α-phenylacetophenone
Detachiermittel *n (text)* stain remover, spotting agent
Detachur *f (text)* stain removal
Detailzeichenpapier *n* detail paper
Detektor *m* detector *(of a gas chromatograph)*
Detergens *n* [synthetic] detergent, syndet, soapless soap
Detergent *n* 1. detergent *(for holding in suspension insoluble matter)*; 2. *s.* Detergens
Detergentzusatz *m* detergent additive
Detonation *f* detonation
Detonationsgeschwindigkeit *f* detonation rate
Detonationsübertragung *f* transmission of detonation
detonieren to detonate
detosylieren to detosylate
Detosylierung *f* detosylation
Detritus *m* 1. *(geol)* detritus, detrital material; 2. detritus, tripton *(suspended non-living debris in water)*
Detroit-Lichtbogenschaukelofen *m* Detroit rocking [arc] furnace
deuterieren to deuterate, to deuterize
Deuterierung *f* deuteration
Deuterium *n* ^2_1H, D deuterium, heavy hydrogen
schweres D. ^3_1H, T tritium
Deuteriumoxid *n* D_2O deuterium oxide, heavy water
Deuteron *n (nucl)* deuteron
Devitrit *m (glass)* devitrite *(a product of devitrification)*
Devulkanisation *f* devulcanization
devulkanisieren to devulcanize
Dewar-Gefäß *n (lab)* Dewar flask (jar), Dewar; Dewar [vessel] *(for shipping liquid gases)*
Deweylith *m (min)* deweylite *(a serpentine)*
Dexanthogenierung *f* dexanthation
Dextran *n* dextran[e] *(a polysaccharide)*
Dextranase *f* dextranase
Dextrin *n* $(C_6H_{10}O_5)_x$ dextrin[e], British (starch) gum
Dextrinbildung *f* dextrinization
Dextrinleim *m* dextrin adhesive (glue)
dextrinogen dextrinogenic
Dextrinogenamylase *f* dextrinogenic amylase
Dextrinstärke *f* soluble starch
dextrogyr dextrogyrate, dextrogyre, dextrogyratory, dextrogyrous, dextrorotatory, right-rotating
Dextronsäure *f s. D*-Glukonsäure
Dextropimarsäure *f* (+)−pimaric acid, dextropimaric acid
Dextrose *f* dextrose, *D*-glucose *(a monosaccharide)*
Dezen-(1) *n* $CH_2{=}CH[CH_2]_7CH_3$ 1-decene
Dezimalwaage *f* decimal balance
Dezin-(1) *n* $HC{\equiv}C[CH_2]_7CH_3$ 1-decyne
Dezin-(5) *n* $CH_3[CH_2]_3C{\equiv}C[CH_2]_3CH_3$ 5-decyne
***n*-Dezylalkohol** *m* $CH_3[CH_2]_8CH_2OH$ 1-decanol, *n*-decyl alcohol
***n*-Dezylen** *n s.* Dezen-(1)
***n*-Dezylsäure** *f s.* Dekansäure
DFB *s.* Druckfeuerbeständigkeit
DH *s.* Defometerhärte
D.I. *s.* Diesel-Index
diablastisch *(cryst)* diablastic
Diagenese *f (geol)* diagenesis
diagenetisch *(geol)* diagenetic
Diagonalbeziehung *f* diagonal relationship *(in the periodic system)*
Diagonalschneidemaschine *f (pap)* angle cutting machine, angle cutter
Diagonalschnittpapier *n* angle-cut (cater-cornered) paper, angle (angular) paper
Diagramm *n* graph, diagram, chart
Diagrammband *n s.* Diagrammstreifen
Diagrammpapier *n* graph paper, recorder chart
Diagrammpapierantrieb *m* chart drive *(of a strip-chart recorder)*
Diagrammstreifen *m* strip chart *(of a recorder)*
Dialdehyd *m* dialdehyde
Dialkylalkoxyphosphin *n* $R_2P(OR)$ alkyl dialkylphosphinite
Dialkyläther *m* dialkyl ether
Dialkylbenzol *n* dialkylbenzene
Dialkylboran *n* dialkylborane
Dialkylchlorphosphin *n* R_2PCl dialkylphosphinous chloride, *(Kosolapoff's nomenclature)* dialkylchlorophosphine
Dialkyl-dialkylamino-phosphin *n* $R_2P(NR_2)$ *NN*-dialkyl-dialkylphosphinous amide, dialkyl-dialkylamino-phosphine, *(Kosolapoff's nomenclature)* dialkyl-dialkylphosphine amide
Dialkylhydroxyphosphin *n* $R_2P(OH)$ dialkylphosphinous acid
dialkylieren to dialkylate
Dialkylierung *f* dialkylation
Dialkylmalon[säure]ester *m* dialkylmalonic ester
Dialkylphosphorigsäurechlorid *n* $(RO)_2PCl$ dialkoxychlorophosphine, dialkylphosphorochlori-

dite, *(Am)* dialkyl chlorophosphite
Dialkylsulfid *n* alkyl sulphide, thioether, thiaalkane
Dialkylzink *n* dialkylzinc
Diallylphthalat *n* $C_6H_4(COOCH_2CH=CH_2)_2$ diallyl phthalate
Diallylphthalatharz *n* diallyl phthalate resin
Dialogit *m (min)* dialogite, rhodochrosite *(manganese(II) carbonate)*
Dialursäure *f* dialuric acid, 5-hydroxybarbituric acid
Dialysat *n* dialyzate
Dialysator *m* dialyzér, dialytic cell
Dialyse *f* dialysis
Dialysenpresse *f* filter-press dialyzer
dialysierbar dialyzable
dialysieren to dialyze
Dialysierfläche *f* dialyzing area
Dialysiergut *n* material to be dialyzed; material being dialyzed
Dialysierhülse *f* dialysis tubing
Dialysiermembran *f* dialyzing membrane
Dialysierzelle *f* dialyzer, dialytic cell
Diamagnetikum *n* diamagnet, diamagnetic [substance]
diamagnetisch diamagnetic
Diamagnetismus *m* diamagnetism
Diamant *m* diamond
schwarzer D. carbonado, black (carbon) diamond
diamanten adamantine
Diamantfarbstoff *m* diamond dye
Diamantgitter *n (cryst)* diamond lattice
Diamantglanz *m* brilliant lustre
Diamantgrün *n* emerald (brilliant) green *(a basic triphenylmethane dye)*
diamanthart adamantine
Diamantmörser *m* diamond (crushing, percussion) mortar
Diamantpackung *f (cryst)* diamond packing
Diamantschneider *m* diamond cutter
Diamantschwarz *n* diamond black
Diamantstruktur *(cryst)* diamond structure
Diamanttinte *f* diamond ink *(for etching glassware)*
Diamid *n* H_2N-NH_2 diamide, hydrazine
Diamin *n* diamine
Diaminchelat *n* diamine chelate
Diaminoanthrachinon *n* diaminoanthraquinone
2,4-Diaminoazobenzol *n* $(NH_2)_2C_6H_3N=NC_6H_5$ 2,4-diaminoazobenzene, chrysoidine
Diaminobenzol *n* $C_6H_4(NH_2)_2$ diaminobenzene, phenylenediamine
4,4'-Diaminobiphenyl *n* $NH_2C_6H_4-C_6H_4NH_2$ 4,4'-diaminobiphenyl, benzidine
1,6-Diaminohexan *n* 1,6-diaminohexane, hexamethylene diamine
2,6-Diaminohexansäure *f* $NH_2[CH_2]_4CH(NH_2)COOH$ 2,6-diaminohexanoic acid, lysine
2,4-Diaminophenol *n* $C_6H_3(OH)(NH_2)_2$ 2,4-diaminophenol, 4-hydroxy-*m*-phenylenediamine
1,2-Diaminopropan *n s.* 1,2-Propandiamin
Diaminostilben *n* $NH_2C_6H_4CH=CHC_6H_4NH_2$ diaminostilbene, stilbenediamine, 1,2-diphenylethylenediamine
Diaminotoluol *n* $CH_3C_6H_3(NH_2)_2$ diaminotoluene, toluylenediamine
diaminvernetzt *(rubber)* diamine-cross-linked
Diamminquecksilber(II)-chlorid *n* $[Hg(NH_3)_2]Cl_2$ diamminemercury(II) chloride, diammine mercuric chloride, fusible white precipitate
Diammoniumhydrogenphosphat *n* $(NH_4)_2HPO_4$ diammonium hydrogenphosphate
Diamorphin *n* diamorphine, diacetylmorphine, heroin *(a narcotic)*
Diamylphthalat *n* $CH_4(COOC_5H_{11})_2$ diamyl phthalate
Dian *n* $HOC_6H_4-C(CH_3)_2-C_6H_4OH$ bisphenol A, 2,2-di-*p*-hydroxyphenylpropane
Dianisidin *n* dianisidine, diaminodimethoxybiphenyl
Dianthrachinonindigo *m(n)* dianthraquinoneindigo
Dianthrachinonylamin *n s.* Dianthrimid
Dianthrimid *n* dianthrimide, dianthraquinonylamine
Diantimonat(V) *n* diantimonate(V)
Diantimonpentoxid *n* Sb_2O_5 diantimony pentaoxide, antimony(V) oxide
Diantimon(V)-säure *f* $H_4Sb_2O_7$ diantimonic(V) acid
Diaphaniepapier *n* diaphanic paper
Diaphoretikum *n* diaphoretic, sudorific
diaphoretisch diaphoretic, sudorific
Diaphorit *m (min)* diaphorite *(an antimony lead silver sulphide)*
Diaphragma *n* diaphragm, membrane
Diaphragmaelement *n s.* Diaphragmazelle
Diaphragmasack *m* membrane bag *(in dialysis)*
Diaphragmaverfahren *n* diaphragm [cell] process *(electrolysis)*
Diaphragmazelle *f* diaphragm cell *(electrolysis)*
Diaphragmen ... *s.* Diaphragma ...
Diarsenat(III) *n* diarsenite
Diarsenat(V) *n* diarsenate
Diarsendisulfid *n s.* Tetrarsentetrasulfid
Diarsenpentasulfid *n* As_2S_5 diarsenic pentasulphide, arsenic(V) sulphide
Diarsenpentoxid *n* As_2O_5 diarsenic pentaoxide, arsenic(V) oxide
Diarsen(V)-säure *f* $H_4As_2O_7$ diarsenic acid
Diarsentrioxid *n* As_2O_3 diarsenic trioxide, arsenic(III) oxide, arsenic
Diarsentriselenid *n* As_2Se_3 diarsenic triselenide, arsenic(III) selenide
Diarsentrisulfid *n* As_2S_3 diarsenic trisulphide, arsenic(III) sulphide
Diaspor *m (min)* diaspore *(α-aluminium hydroxide oxide)*
Diasporton *m* diaspore clay
Diastase *f s.* Amylase
diastereomer diastereo[iso]meric
Diastereomer[es] *n* diastereo[iso]mer, epimer[ide]
Diäthen *n s.* Butadien-(1,3)
Diätherat *n* dietherate
diatherm[an] diathermanous, diathermic
Diathermansie *f* diatherma[n]cy
Diäthin *n s.* Diazetylen
1,1-Diäthoxyäthan *n* 1,1-diethoxyethane, acetal
Diäthoxymethan *n* $CH_2(OC_2H_5)_2$ diethoxymethane,

diethylformal, formaldehyde diethyl acetal
Diäthylamin *n* $(C_2H_5)_2NH$ diethylamine
Diäthyläther *m* $C_2H_5OC_2H_5$ diethyl ether
Diäthylazetylen *n* $CH_3CH_2C{\equiv}CCH_2CH_3$ 3-hexyne, diethylacetylene
Diäthyldisulfid *n* $(C_2H_5)_2S_2$ diethyl disulphide
Diäthyldithiokarbaminsäure *f* $(C_2H_5)_2{-}NCS{-}SH$ diethyldithiocarbamic acid
Diäthylen *n s.* Butadien-(1,3)
Diäthylendiamin *n* diethylenediamine, piperazine
Diäthylenglykol *n* $HOCH_2CH_2OCH_2CH_2OH$ diethylene glycol, di(-2-hydroxyethyl) ether
Diäthylenglykoldimethyläther *m* diethylene glycol dimethyl ether, diglyme
***N,N*-Diäthylenharnstoff** *m* bisethyleneurea
Diäthylenoxid *n s.* Tetrahydrofuran
Diäthylformal *n s.* Diäthoxymethan
Diäthylmagnesium *n* $Mg(C_2H_5)_2$ diethylmagnesium
Diäthylmalonat *n* $CH_2(COOC_2H_5)_2$ diethyl malonate, malonic ester
Diäthylrhodamin *n* diethylrhodamine
Diäthylsebazat *n* $C_2H_5OCO[CH_2]_8COOC_2H_5$ diethyl sebacate
Diäthylstilböstrol *n* diethylstilboestrol
Diäthylsukzinat *n* $C_2H_5OCO[CH_2]_2COOC_2H_5$ diethyl succinate
Diäthylsulfat *n* $(C_2H_5)_2SO_4$ diethyl sulphate
Diäthylsulfid *n* $(C_2H_5)_2S$ diethyl sulphide, ethylthioethane
Diäthylzink *n* $Zn(C_2H_5)_2$ diethylzinc
Diatomeenerde *f* diatomaceous (infusorial, diatom) earth, kieselguhr
Diatomeenschlamm *m* diatom ooze
diaxial diaxial
1,4-Diazanaphthalin *n* 1,4-diazanaphthalene, 1,4-benzodiazine, quinoxaline
Diazazyanin *n (dye)* diazacyanine
Diazetonalkohol *m* $CH_3COCH_2C(CH_3)_2OH$ diacetone alcohol, 4-hydroxy-4-methyl-2-pentanone
Diazetyl *n s.* Biazetyl
Diazetylen *n* $HC{\equiv}C{-}C{\equiv}CH$ biacetylene, butadiyne
Diazetylmorphin *n* diacetylmorphine, diamorphine, heroin *(a narcotic)*
1,2-Diazin *n* 1,2-diazine, pyridazine
1,3-Diazin *n* 1,3-diazine, pyrimidine
1,4-Diazin *n* 1,4-diazine, pyrazine
Diazoamidobenzol *n s.* Diazoaminobenzol
Diazoaminobenzol *n* $C_6H_5N{=}NNHC_6H_5$ diazoaminobenzene, 1,3-diphenyltriazen
Diazoaminoverbindung *f* $RN{=}NNHR$ diazoamino compound
Diazoanhydrid *n* diazo anhydride
Diazodinitrophenol *n* $(NO_2)_2C_6H_2N_2O$ diazodinitrophenol, D. D. N. P.
Diazofarbstoff *m* diazo dye
Diazokomponente *f* diazo (diazonium) component
Diazokupplung *f* diazo coupling
1,2-Diazol *n* 1,2-diazole, pyrazole
Diazolösung *f* diazo solution
Diazomethan *n* CH_2N_2 diazomethane
Diazomethanreaktion *f* diazomethane reaction
Diazoniumgruppe *f* $[ArN{\equiv}N]^+$ diazonium group
Diazonium-Ion *n* diazonium ion
Diazoniumsalz *n* diazonium salt
Diazoniumsalzlösung *f* diazonium solution
Diazopapier *n* diazotype paper
Diazoreaktion *f* diazo reaction
  Paulysche D. Pauly [protein] reaction
Diazotat *n* $R{-}N{=}N{-}OM$ diazoate, diazotate
diazotierbar diazotizable
diazotieren to diazotize
  erneut d. to rediazotize
Diazotierung *f* diazotization
Diazotierungskomponente *f* diazo (diazonium) component, *(dye also)* primary component
Diazotypie *f* diazotype, diazotypy, diazo (dyeline) print, whiteprint, diazotype (dyeline) process *(a blue-printing process)*
Diazotypiepapier *n* diazotype paper
Diazoverbindung *f* diazo compound
Diazoxid *n* diazo oxide
Diazylperoxid *n* diacyl peroxide
Dibenzanthrachinon *n* dibenzanthraquinone
Dibenzanthrazen *n* dibenzanthracene
Dibenzopyran *n* dibenzopyran, dibenzo[a,e]pyran, xanthene
Dibenzopyrenchinon *n* dibenzopyrenequinone
Dibenzo-γ-pyron *n* dibenzopyrone, xanthone
Dibenzopyrrol *n* dibenzopyrrole, carbazole
Dibenzoyl *n s.* Bibenzoyl
dibenzoylieren to dibenzoylate
Dibenzoylierung *f* dibenzoylation
Dibenzoylperoxid *n* $C_6H_5OCOOCOC_6H_5$ dibenzoyl peroxide, benzoyl peroxide
Dibenzyl *n s.* Bibenzyl
Dibenzyläther *m* $(C_6H_5CH_2)_2O$ dibenzyl ether
Dibenzylsukzinat *n* $(CH_2COOCH_2C_6H_5)_2$ dibenzyl succinate
Diboran *n* diborane
Diboran(4) *n* B_2H_4 diborane(4)
Diboran(6) *n* B_2H_6 diborane(6), diborane *(proper)*
Diboranid *n* diboranide
Diborid *n* diboride
Dibortetrachlorid *n* B_2Cl_4 diboron tetrachloride
Dibromanthrachinon *n* dibromoanthraquinone
1,2-Dibromäthan *n* $BrCH_2CH_2Br$ 1,2-dibromoethane
Dibrombenzol *n* $C_6H_4Br_2$ dibromobenzene
Dibromdichlorsilan *n* $SiBr_2Cl_2$ dibromodichlorosilane
Dibromid *n* dibromide
dibromieren to dibrominate
Dibromierung *f* dibromination
Dibromindigo *m(n)* dibromoindigo
Dibrommethan *n* CH_2Br_2 dibromomethane
Dibrompropan *n* dibromopropane
Dibromthymolsulfophthalein *n* dibromothymolsulphonphthalein, bromothymol blue *(a pH indicator)*
Dibromverbindung *f* dibromo compound
2,6-Di-*tert*.-butyl-*p*-kresol *n* 2,6-di-*tert*-butyl-*p*-cresol, 4-methyl-2,6-di-*tert*-butylphenol *(an antioxidant)*
Dibutylphthalat *n* $C_6H_4(COOC_4H_9)_2$ dibutyl phthalate, DBP

Dibutylsebazat *n* $C_4H_9OCO[CH_2]_8COOC_4H_9$ dibutyl sebacate
Dichlon *n* dichlone, 2,3-dichloro-1,4-naphthoquinone *(a herbicide)*
Dichloranilin *n* $Cl_2C_6H_3NH_2$ dichloroaniline
Dichloranthrachinon *n* dichloroanthraquinone
1,1-Dichloräthan *n* CH_3CHCl_2 1,1-dichloroethane
1,2-Dichloräthan *n* $ClCH_2CH_2Cl$ 1,2-dichloroethane
Dichloräthansäure *f* $Cl_2CHCOOH$ dichloroethanoic acid, dichloroacetic acid
Dichloräther *m s.* 1,2-Dichlordiäthyläther
1,1-Dichloräthylen *n* $CH_2{=}CCl_2$ 1,1-dichloroethylene, vinylidene chloride
2,2-Dichlorazetamid *n* $CHCl_2CONH_2$ 2,2-dichloroacetamide
Dichlorbenzol *n* $C_6H_4Cl_2$ dichlorobenzene
p-**Dichlorbenzol** *n* $C_6H_4Cl_2$ *p*-dichlorobenzene, paradichlorobenzene
Dichlorderivat *n* dichloro derivative
1,2-Dichlordiäthyläther *m* $CH_3CH_2OCHClCH_2Cl$ 1,2-dichloroethyl ethyl ether, 1,2-dichlorodiethyl ether
2,2′-Dichlordiäthyläther *m* $ClCH_2CH_2OCH_2CH_2Cl$ di-2-chloroethyl ether, 2,2′-dichlorodiethyl ether
Dichlordiäthylsulfid *n* $(ClCH_2CH_2)_2S$ dichlorodiethyl sulphide, di-chloroethyl sulphide
Dichlordifluormethan *n* Cl_2CF_2 dichlorodifluoromethane
Dichlordiphenyltrichloräthan *n* dichlorodiphenyltrichloroethane, DDT *(a contact insecticide)*
Dichloressigsäure *f* $Cl_2CHCOOH$ dichloroacetic acid
Dichlorfluormethan *n* Cl_2CHF dichloromonofluoromethane
Dichlorheptoxid *n* Cl_2O_7 dichlorine heptaoxide, chlorine(VII) oxide
Dichlorid *n* dichloride
Dichlormethan *n* CH_2Cl_2 dichloromethane
Dichlornaphthalin *n* $C_{10}H_6Cl_2$ dichloronaphthalene
Dichlorotetramminplatin(IV)-chlorid *n* $[Pt(NH_3)_4Cl_2]Cl_2$ dichlorotetraamineplatinum(IV) chloride
Dichloroxid *n* Cl_2O dichlorine monooxide, chlorine(I) oxide
tris-**(2,4-Dichlorphenoxyäthyl)phosphit** *n* *tris*-(2,4-dichlorophenoxyethyl) phosphite, 2,4-DEP *(a herbicide)*
4-(2′,4′-Dichlorphenoxy)buttersäure *f* 4,(2′,4′-dichlorophenoxy)butyric acid, 2,4-DB *(a herbicide)*
2,4-Dichlorphenoxyessigsäure *f* 2,4-dichlorophenoxyacetic acid, 2,4-D *(a herbicide)*
Dichlorprop *n* $Cl_2C_6H_3OCH(CH_3)COOH$ dichlorprop, 2,4-DP, 2-(2′,4′-dichlorophenoxy)propionic acid *(a herbicide)*
Dichlorpropan *n* $C_3H_6Cl_2$ dichloropropane
Dichlortriazin *n* dichlorotriazine
Dichroismus *m (cryst, coll)* dichroism
dichroitisch *(cryst, coll)* dichroic
Dichromat *n* $M_2^ICr_2O_7$ dichromate
dichromatisch dichromatic, dichroic
Dichromtrioxid *n* Cr_2O_3 dichromium trioxide, chromium(III) oxide
dicht tight, proof, impermeable, impenetrable, impervious; leaktight, leakproof; close-grained, fine-grained, compact *(structure)*
  d. [ab]schließend tight-fitting
dichtbrennend *(ceram)* dense-burning
Dichtdruck *m* sealing pressure
Dichte *f* 1. density, *(specif)* mass density, D *(mass of a substance per unit volume)*; strength *(of a solution)*; 2. *s.* optische D.
  D. in API-Graden *(petrol)* API gravity
  D. nach dem Brand *(ceram)* fired density
  optische D. optical density
  scheinbare D. apparent density
  wahre (wirkliche) D. true density
Dichteanreicherung *f* gravity concentration
Dichtebestimmung *f* densimetry
  D. von Flüssigkeiten hydrometry
Dichtekurve *f (phot)* characteristic curve, H and D curve
Dichtemesser *m* densimeter
Dichtemessung *f* densimetry
dichten to make tight, to seal, to pack, to ca[u]lk
Dichtesortierung *f* density separation (cut), gravity concentration
Dichteströmung *f* density current
Dichteverhältnis *n* specific gravity, S.G. *(where temperatures of substance and standard are equal)*
Dichtewaage *f* density balance
Dichtezahl *f* specific gravity, S.G. *(where temperatures of substance and standard are different)*
Dichtfläche *f* seal face, sealing [sur]face, gasketing area
dichtgepackt *(cryst)* close-packed
Dichtheit *f*, **Dichtigkeit** *f* tightness, proofness
Dichtmaterial *n*, **Dichtmittel** *n* sealant
dichtpolen to pole down *(copper for eliminating sulphur)*
Dichtpolen *n* poling-down *(of copper for eliminating sulphur)*
Dichtring *m* packing (sealing) ring
Dichtsatz *m* packing set
Dichtschnur *f* packing (sealing) strip
dichtschweißen to seal-weld
Dichtstoff *m s.* Dichtmaterial
Dichtung *f* packing, seal, *(for parts without relative motion also)* static seal, gasket, *(for moving parts also)* dynamic seal
  druckausgeglichene (druckentlastete) D. balanced seal
  eingefaßte D. envelope gasket
  halbmetallische D. semimetallic packing
  umhüllte D. envelope gasket
Dichtungsdruck *m* sealing pressure
Dichtungsfläche *f* seal face, sealing [sur]face, gasketing area
Dichtungsflüssigkeit *f* sealing liquid (fluid)
Dichtungskarton *m* fitting cardboard
Dichtungsmanschette *f* gasket
Dichtungsmasse *f* lute, luting, ca[u]lking compound
Dichtungsmaterial *n* packing (sealing, gasketing) material

D. für Rohrgewindeverbindungen pipe dope
Dichtungspappe *f* fitting board
Dichtungsring *m* packing (sealing) ring
Dichtungssatz *m* packing set
Dichtungsschnur *f* packing (sealing) strip
Dichtungswerkstoff *m s.* Dichtungsmaterial
Dickablauge *f (pap)* concentrated (evaporated, thick) black liquor
Dickdruckpapier *n* featherweight paper, *(Am)* bulking paper
Dicke *f* thickness; *(text)* size *(of fibrous material)*; *(glass)* substance, strength *(of flat glass)*; *(pap)* caliper
Dickenmesser *m* thickness gauge (tester), caliper
dickflüssig viscid, viscous, viscose, turbid, thick, semiliquid, syruplike, syrupy, ropy
d. werden to thicken, to inspissate
Dickflüssigkeit *f* viscosity, thickness, ropiness
Dickglas *n* thick [sheet] glass
dickgriffig sein *(pap)* to bulk high
Dickinsonit *m (min)* dickinsonite *(manganese(II) sodium hydrogenorthophosphate)*
Dickit *m (min)* dickite *(aluminium hydroxide silicate)*
Dicklauge *f* concentrated liquor; *(pap)* concentrated (evaporated, thick) black liquor
Dicklaugenaustritt *m* concentrated liquor outlet
Dicklegung *f* souring *(of milk)*
Dickmaische *f* thick (heavy) mash
Dickmilch *f* fermented (cultured, set) milk
Dicksaft *m* condensed (inspissated) juice; *(sugar)* thick juice
Dicksaftfilter *n (sugar)* thick-juice filter
Dicksaftfiltration *f (sugar)* thick-juice filtration
Dicksaftpumpe *f (sugar)* thick-juice pump
Dicksaftsaturation *f (sugar)* thick-juice saturation
Dicksaft[vor]wärmer *m (sugar)* thick-juice heater
Dickschlamm *m* thickened liquor (sludge), thick slime, *(as discharge of a concentrator also)* thickener pulp
Dickschlammaustrag *m* thickened-liquor outlet, thick-slime discharge
Dickschlammzone *f* sludge zone
Dickstoff *m (pap)* slush (high-density) pulp, slush (thick) stock
Dickstoff-Abwärts[bleich]turm *m (pap)* downflow high-density tower
Dickstoff-Aufwärts[bleich]turm *m (pap)* upflow high-density tower
Dickstoffbehälter *m (pap)* slush pulp storage
Dickstoffbleiche *f (pap)* high-density bleaching, bleaching at high consistency
Dickstoffbleichstufe *f (pap)* high-density [bleaching] stage
Dickstoffbleichturm *m (pap)* high-density bleacher
Dickstoffendbleiche *f (pap)* final bleaching at high consistency
Dickstoffmahlung *f (pap)* high-consistency refining, HCR
Dickstoffpumpe *f* sludge pump; *(pap)* thick-stock pump
Dickstoffvorratsbehälter *m (pap)* slush pulp storage
dickwandig thick-walled, heavy-wall
Dickwerden *n* thickening *(as of milk)*
Diderivat *n* di-derivative
Didezylphthalat *n* $C_6H_4(COOC_{10}H_{21})_2$ didecyl phthalate, DDP
Didier-Bubiag-Verfahren *n* Didier-Bubiag process *(of coal gasification)*
Didym *n* 1. didymium *(a mixture containing neodymium and praseodymium)*; 2. *s.* Didymmetall
Didymerde *f s.* Didymoxid
Didymmetall *n* didymium *(an alloy consisting of neodymium and praseodymium)*
Didymoxid *n* didymium oxide
Dieckmann-Reaktion *f* Dieckmann reaction *(intramolecular condensation of esters)*
Dieisentrioxid *n* Fe_2O_3 diiron trioxide, iron(III) oxide, ferric oxide
Dieisentrisulfid *n* Fe_2S_3 diiron trisulphide, iron(III) sulphide, ferric sulphide
Dielektrikum *n* dielectric [material], non-conductor
dielektrisch dielectric, non-conducting
Dielektrizitätskonstante *f* dielectric constant, permittivity
Dielektrizitätskonstante-Messer *m* dielectrometer, dekameter, DK-meter
Dielektro[be]heizung *f* dielectric (radio-frequency) heating
Dielektrometrie *f*, **Dielkometrie** *f s.* Dekametrie
Diels-Alder-Reaktion *f*, **Diels-Alder-Synthese** *f* Diels-Alder [diene] reaction, Diels-Alder [diene] synthesis
Dien *n* diene *(a compound with two esp conjugated C-C double bonds)*
Dienkautschuk *m* diene rubber
dienophil dienophilic
Dienophil *n* dienophile
Dienpolymer[es] *n*, **Dienpolymerisat** *n* diene[-based] polymer
Diensynthese *f* diene synthesis
Diels-Aldersche D. *s.* Diels-Alder-Reaktion
Diesel-Index *m* diesel index
Dieselklopfen *n* diesel knock
Dieselkraftstoff *m* diesel fuel (oil)
Dieselöl *n s.* Dieselkraftstoff
Dieselschmieröl *n* diesel engine oil
Dieseltreibstoff *m* diesel fuel (oil)
Diester *m* diester
Differentialdestillation *f* differential distillation
Differentialdetektor *m* differential detector *(gas chromatography)*
Differentialflotation *f* differential (selective) floation
Differentialfotometrie *f* differential photometry
Differentialkalorimeter *n* differential calorimeter
Differentialmanometer *n* differential manometer
Differentialthermoanalyse *f* differential thermal analysis, DTA
Differential-Thermogravimetrie *f* derivative (differential) thermogravimetry
Differentialtitration *f* differential titration
Differentiation *f* / **gravitative** *(geoch)* gravitative differentiation
Differenzdruck *m* differential pressure (head)

Differenzdruckmesser *m*, **Differenzmanometer** *n* differential-pressure meter, differential manometer
Diffraktion *f* diffraction
Diffraktionserscheinung *f* diffraction phenomenon
diffundieren to diffuse
Diffusat *n* diffusate
Diffusatzelle *f* diffusate cell, *(with water also)* water cell
Diffuseur *m* diffuser, diffusor, diffusing tank, diffusion cell
Diffusion *f* diffusion
D. von Festkörpern solid diffusion
Diffusionsabwasser *n (sugar)* diffusion [waste] water, diffusion pulp water, [beet] pulp water
Diffusionsapparat *m s.* Diffuseur
Diffusionsbatterie *f* diffusion battery
diffusionsfähig diffusible
Diffusionsfähigkeit *f* diffusibility
Diffusionsgeschwindigkeit *f* diffusion rate (velocity)
diffusionsglühen to homogenize *(alloys)*
Diffusionsglühen *n* homogenization, homogenizing *(of alloys)*
Diffusionsgrenzstrom *m (phys chem)* limiting (maximum) diffusion current
Diffusionskoeffizient *m*, **Diffusionskonstante** *f* diffusion coefficient
Diffusionslänge *f* diffusion length
Diffusionsmethode *f* diffusion method
Diffusionsmischen *n* diffusive mixing
Diffusionsnebelkammer *f* diffusion cloud chamber
Diffusionspotential *n* diffusion (liquid-liquid, liquid junction) potential
Diffusionspumpe *f* [vapour] diffusion pump
D. mit Quecksilberfüllung mercury-vapour pump
Diffusionspumpenöl *n* diffusion-pump oil
Diffusionssaft *m (sugar)* diffusion (raw) juice
Diffusionsschicht *f* diffusion layer
Diffusionsschnitzel *npl (sugar)* wet [beet] pulp, diffusion cossettes
Diffusionsstrom *m (phys chem)* diffusion current
maximaler D. maximum (limiting) diffusion current
Diffusionsstromkonstante *f* diffusion-current constant
Diffusionsverchromung *f* chromizing, chromium impregnation
Diffusionsverlust *m (sugar)* diffusion loss
Diffusionsvermischen *n* diffusive mixing
Diffusionsvermögen *n* diffusivity
diffusorisch diffusional
Difluordichlormethan *n s.* Dichlordifluormethan
Difluorid *n* difluoride
Diformiat *n* diformate
difunktionell difunctional, D, bifunctional
Digalliumtrioxid *n* Ga_2O_3 digallium trioxide, gallium(III) oxide
m-**Digallussäure** *f* digallic acid, *m*-digallic acid, gallic acid 3-monogallate, 5,6-dihydroxy-3-carboxyphenyl ester of gallic acid
Digenit *m (min)* digenite *(a copper sulphide)*
digerieren to digest *(by heat or solvents)*
Digerieren *n* digestion *(by heat or solvents)*
Digerierkolben *m* digestion flask, digester
Digerman *n* Ge_2H_6 digermane, germanium hexahydride
Digermanat *n* $M_2^IGe_2O_5$ digermanate
Digestion *f (bioch, pharm)* digestion
Digestionskolben *m s.* Digerierkolben
Digestivum *n (pharm)* digestive, digester
Digestor *m s.* Digerierkolben
Digestorium *n (lab)* [fume] hood, fume chamber (cupboard, closet)
Digitoninfällung *f* digitonin precipitation *(for detecting vegetable fat)*
Digitoxin *n* digitoxin *(alkaloid)*
Diglykolsäure *f* $O(CH_2COOH)_2$ diglycolic (diglycollic) acid, dicarboxymethyl ether
Diglyzerid *n* diglyceride
Digoxin *n* digoxin *(alkaloid)*
Dihalogenalkan *n* alkyl dihalide
Dihalogenid *n* dihalogenide, dihalide
Diharnstoff *m s. p*-Urazin
Diheptadezylketon *n* $CH_3[CH_2]_{16}-CO-[CH_2]_{16}CH_3$ diheptadecyl ketone, stearone, 18-pentatriacontanone
Dihexyl *n (incorrectly for) n*-Dodekan
Dihexylphthalat *n* $C_6H_4(COOC_6H_{13})_2$ dihexyl phthalate, DHP
Dihydrat *n* dihydrate
Dihydrit *m (min)* dihydrite, pseudomalachite *(a hydrous basic copper phosphate)*
Dihydrobenzol *n* dihydrobenzene, cyclohexadiene
Dihydrochinin *n* dihydroquinine, hydroquinine, quinethyline
Dihydrogenarsenat(V) *n* $M^IH_2AsO_4$ dihydrogenarsenate, dihydrogenorthoarsenate
Dihydrogendodekawolframat(VI) *n* $M_6^I[H_2W_{12}O_{40}]$ dihydrogendodecawolframate(VI), dihydrogendodecatungstate(VI)
Dihydrogendodekawolframsäure *f* $H_6[H_2W_{12}O_{40}]$ dihydrogendodecawolframic acid, dihydrogendodecatungstic acid
Dihydrogenmonophosphat *n s.* Dihydrogenorthophosphat
Dihydrogenorthophosphat *n* $M^IH_2PO_4$ dihydrogenorthophosphate, dihydrogenphosphate
Dihydrogenphosphat *n s.* Dihydrogenorthophosphat
Dihydrogensalz *n* dihydrogen salt *(of a tribasic acid)*
Dihydrostufe *f* dihydric stage
Dihydroxid *n* dihydroxide
2,5-Dihydroxybenzoesäure *f* $(HO)_2C_6H_3COOH$ 2,5-dihydroxybenzoic acid, gentisic acid
Dihydroxybernsteinsäure *f* $HOOC[CHOH]_2COOH$ dihydroxysuccinic acid, tartaric acid, 2,3-dihydroxybutanedioic acid
1,5-Dihydroxynaphthalin-3,7-disulfonsäure *f* 1,5-dihydroxynaphthalene-3,7-disulphonic acid, red acid
Dihydroxyphenylalanin *n* dihydroxyphenylalanine, dopa, DOPA
2,3-Dihydroxypropanal *n* $CH_2(OH)CH(OH)CHO$ 2,3-dihydroxypropanal, glyceraldehyde

2,3-Dihydroxypropansäure *f* $CH_2(OH)CH(OH)COOH$ 2,3-dihydroxypropionic acid, glyceric acid
Dihydroxyverbindung *f* dihydroxy compound
Dihydroxyxylol *n* $(CH_3)_2C_6H_2(OH)_2$ dihydroxyxylene, xylorcinol
3,4-Dihydroxyzimtsäure *f* $(HO)_2C_6H_3CH{=}CHCOOH$ 3,4-dihydroxycinnamic acid, caffeic acid
Diisodezylphthalat *n* $C_6H_4(COOC_{10}H_{21})_2$ diisodecyl phthalate, DIDP
Diisooktylphthalat *n* $C_6H_4(COOC_8H_{17})_2$ diisooctyl phthalate, DIOP
Diisopropyläther *m* $(CH_3)_2CHOCH(CH_3)_2$ diisopropyl ether
Diisopropylmethan *n* $(CH_3)_2CHCH_2CH(CH_3)_2$ diisopropylmethane, 2,4-dimethylpentane
Diisozyanat *n* diisocyanate *(a compound with two -N=C=O groups)*
Dijoddisulfid *n* I_2S_2 diiodine disulphide
Dijodid *n* diiodide
Dijodmethan *n* CH_2I_2 diiodomethane
Dijodpentoxid *n* I_2O_5 diiodine pentaoxide, iodine(V) oxide
Dijodtetroxid *n* I_2O_4 diiodine tetraoxide
3,5-Dijodtyrosin *n* $HOC_6H_2(I_2)CH_2CH(NH_2)COOH$ 3,5-diiodotyrosine, iodogorgoic acid
Dikabutter *f*, **Dikafett** *n* Di[k]ka butter *(from Irvingia gabonensis Baill.)*
Dikaliumdisulfat *n* $K_2S_2O_7$ dipotassium disulphate, potassium disulphate
Dikaliumhydrogenphosphat *n* K_2HPO_4 dipotassium hydrogenphosphate, potassium hydrogenorthophosphate
Dikalziumdiphosphat *n* $Ca_2P_2O_7$ dicalcium diphosphate, calcium pyrophosphate
Dikarbid *n* dicarbide
Dikarbonsäure *f* dicarboxylic acid
Dikatechin *n* dicatechin *(a derivative of hydroxyflavan)*
Diketoester *m* diketo ester
Diketon *n* diketone
Dikobalttrioxid *n* Co_2O_3 dicobalt trioxide, cobalt(III) oxide
Dikobalttrisulfid *n* Co_2S_3 dicobalt trisulphide, cobalt(III) sulphide
Dikrotalin *n* dicrotaline *(alkaloid)*
Dikrotalinsäure *f s.* Dikrotolsäure
Dikrotolsäure *f* $HOOCCH_2C(CH_3)(OH)CH_2COOH$ dicrotalic acid, 3-hydroxy-3-methylglutaric acid
dilatant *(coll)* dilatant
Dilatanz *f (coll)* dilatancy, inverse plasticity
Dilatation *f* dilatation
Dilatometer *n* dilatometer
Dilatometertest *m* dilatometer test
  D. nach Audibert-Arnu Audibert-Arnu dilatometer test
Dilatometrie *f* dilatometry
dilatometrisch dilatometric
Dilitursäure *f* dilituric acid, nitromalonylurea, 5-nitrobarbituric acid
Dillöl *n* dill seed oil *(from Anethum graveolens L.)*
Dimangansilizid *n* Mn_2Si dimanganese silicide, manganese(II) silicide
Dimangantrioxid *n* Mn_2O_3 dimanganese trioxide, manganese(III) oxide
Dimensionsstabilisierung *f* dimensional stabilization
Dimensionsstabilität *f* dimensional stability
dimer dimeric
Dimer[es] *n* dimer
Dimerisierung *f* dimerization
1,2-Dimethoxybenzol *n* $C_6H_4(OCH_3)_2$ 1,2-dimethoxybenzene
Dimethoxymethan *n* $H_2C(OCH_3)_2$ dimethoxymethane, methylal, formaldehyde dimethyl acetal
Dimethylamin *n* $(CH_3)_2NH$ dimethylamine
p-**Dimethylaminoazobenzol** *n* $C_6H_5N{=}NC_6H_4N(CH_3)_2$ *p*-dimethylaminoazobenzene, butter (oil, methyl) yellow
Dimethylanilin *n* $(CH_3)_2C_6H_3NH_2$ dimethylaniline, aminoxylene, aminodimethylbenzene
Dimethylarsinchlorid *n s.* Dimethylchlorarsin
Dimethylarsinsäure *f* $(CH_3)_2AsOOH$ dimethylarsinic acid, cacodylic acid
bis-**Dimethylarsyl** *n s.* Kakodyl
Dimethyläther *m* CH_3OCH_3 dimethyl ether
1,1-Dimethyläthylen *n* $CH_2{=}CH(CH_3)_2$ 1,1-dimethylethylene, 2-methylpropene
asymm.-**Dimethyläthylen** *n s.* 1,1-Dimethyläthylen
symm.-**Dimethyläthylen** *n s.* Buten-(2)
Dimethyläthylkarbinol *n s.* 2-Methylbutanol-(2)
Dimethylazetylen *n* $CH_3C{\equiv}CCH_3$ dimethylacetylene, 2-butyne
2,5-Dimethylbenzoesäure *f* $(CH_3)_2C_6H_3COOH$ 2,5-dimethylbenzoic acid, *p*-xylylic acid
Dimethylbenzol *n* $C_6H_4(CH_3)_2$ dimethylbenzene, xylene
Dimethylchinon *n* dimethylbenzoquinone, xyloquinone
Dimethylchlorarsin *n* $(CH_3)_2AsCl$ chlorodimethylarsine, cacodyl chloride
Dimethyldichlorsilan *n* $(CH_3)_2SiCl_2$ dimethyldichlorosilane
Dimethylenimin *n* aziridine, dimethyleneimine
Dimethylformal *n* $H_2C(OCH_3)_2$ formaldehyde dimethyl acetal, dimethoxymethane, methylal
Dimethylformamid *n* $(CH_3)_2NCHO$ dimethyl formamide, DMF
Dimethylglyoxal *n* $CH_3COCOCH_3$ dimethylglyoxal, biacetyl, 2,3-butanedione
Dimethylkarbinol *n s.* Propanol-(2)
Dimethylketol *n* $CH_3CH(OH)COCH_3$ dimethylketol, acetoin, 3-hydroxy-2-butanone
Dimethylketon *n* CH_3COCH_3 dimethyl ketone, acetone, propanone
2,3-Dimethylnaphthalin *n* 2,3-dimethylnaphthalene, guaiene
Dimethylolharnstoff *m* $HOCH_2NH{-}CO{-}NHCH_2OH$ dimethylolurea, 1,3-bishydroxymethylurea
Dimethylphenol *n* $(CH_3)_2C_6H_3OH$ dimethylphenol, xylenol, hydroxyxylene
Dimethylphthalat *n* $C_6H_4(COOCH_3)_2$ dimethyl phthalate, DMP
2,2-Dimethylpropansäure *f s.* 2,2-Dimethylpropionsäure
2,2-Dimethylpropionsäure *f* $(CH_3)_3CCOOH$ 2,2-dimethylpropionic acid, pivalic acid

Dimethylsulfat *n* $(CH_3)_2SO_4$ dimethyl sulphate
Dimethylsulfoxid *n* $(CH_3)_2SO$ dimethyl sulphoxide, DMSO
Dimethylterephthalat *n* $C_6H_4(COOCH_3)_2$ dimethyl terephthalate, DMT
Dimethylzink *n* $Zn(CH_3)_2$ dimethyl zinc
dimolekular bimolecular
Dimolybdäntrioxid *n* Mo_2O_3 dimolybdenum trioxide, molybdenum(III) oxide
Dimolybdat *n* $M_2{}^IMo_2O_7$ dimolybdate
dimorph dimorphic, dimorphous
Dimroth-Kühler *m* Dimroth condenser
Dinatriumhydrogenarsenat(III) *n s.* Dinatriumhydrogenorthoarsenat(III)
Dinatriumhydrogenarsenat(V) *n s.* Dinatriumhydrogenorthoarsenat(V)
Dinatriumhydrogenorthoarsenat(III) *n* Na_2HAsO_3 disodium hydrogenorthoarsenite, sodium hydrogenarsenite
Dinatriumhydrogenorthoarsenat(V) *n* Na_2HAsO_4 disodium hydrogenorthoarsenate, sodium hydrogenarsenate
Dinatriumhydrogenorthophosphat *n*, **Dinatriumhydrogenphosphat** *n* Na_2HPO_4 disodium hydrogenorthophosphate, sodium hydrogenphosphate
Dinatriummethylarsonat *n* $CH_3As(=O)(ONa)_2$ disodium methyl arsonate, DMA
Dinatriumpentazyanonitrosylferrat(II) *n* $Na_2[Fe(CN)_5(NO)]$ disodium pentacyanonitrosylferrate(II), sodium nitroprusside, sodium nitroprussiate
Dinatriumsalz *n* disodium salt
Dinickeltrioxid *n* Ni_2O_3 dinickel trioxide, nickel(III) oxide
Dinitranilin *n s.* Dinitroanilin
Dinitrid *n* dinitride
dinitrieren to dinitrate
Dinitrierung *f* dinitration
Dinitril *n* dinitrile
Dinitrilfaser *f* dinitrile fibre
Dinitrilfaserstoff *m* dinitrile fibre
Dinitroaminophenol *n (incorrectly for)* Aminodinitrophenol
Dinitroanilin *n* $(NO_2)_2C_6H_3NH_2$ dinitroaniline
Dinitroanthrachinon *n* dinitroanthraquinone
Dinitrobenzoesäure *f* $C_6H_3(NO_2)_2COOH$ dinitrobenzoic acid
Dinitrobiphenyl *n* $NO_2C_6H_4-C_6H_4NO_2$ dinitrobiphenyl
Dinitrobiphenyldikarbonsäure *f* dinitrobiphenyldicarboxylic acid, dinitrodiphenic acid
Dinitrochlorbenzol *n s.* Chlordinitrobenzol
Dinitrodiphensäure *f* dinitrodiphenic acid, dinitrobiphenyl-dicarboxylic acid
Dinitrodiphenyl *n (incorrectly for)* Dinitrobiphenyl
2,4-Dinitrofluorbenzol *n s.* Fluor-2,4-dinitrobenzol
Dinitrokörper *m* dinitro body
Dinitrokresol *n* $CH_3C_6H_2(NO_2)_2OH$ dinitrocresol
4,6-Dinitro-*o*-kresol *n* 4,6-dinitro-*o*-cresol, DNOC, DNC *(a pesticide)*
Dinitromischsäure *f* dinitro mixed acid
Dinitrophenol *n* $HOC_6H_3(NO_2)_2$ dinitrophenol
Dinitrophenolat *n* dinitrophenate
Dinitrophenylderivat *n* dinitrophenyl derivative, DNP derivative
Dinitrophenyldisulfid *n* $NO_2C_6H_4S-SC_6H_4NO_2$ dinitrophenyl disulphide
Dinitroresorzin *n* $C_6H_2(OH)_2(NO_2)_2$ dinitroresorcinol
Dinitrotoluol *n* $C_6H_3(CH_3)(NO_2)_2$ dinitrotoluene
Dinitrotoluylsäure *f* $C_6H_2(CH_3)(NO_2)_2COOH$ dinitrotoluic acid
Dinitroverbindung *f* dinitro compound
2,4-Dinitro-6-zyklohexylphenol *n* $C_6H_{11}C_6H_2(NO_2)OH$ 2,4-dinitro-6-cyclohexylphenol, DNOCHP
Dinoseb *n* Dinoseb, DNBP, 2-*sec*-butyl-4,6-dinitrophenol *(a pesticide)*
Dioktadezylamin *n* $CH_3[CH_2]_{17}NH[CH_2]_{17}CH_3$ dioctadecylamine
Dioktylketon *n* $CH_3[CH_2]_7CO[CH_2]_7CH_3$ 9-heptadecanone, di-octyl ketone, nonylone
Dioktylphthalat *n* $C_6H_4(COOC_8H_{17})_2$ dioctyl phthalate
Diol *n* $C_nH_{2n}(OH)_2$ diol, dihydric alcohol
Diolefin *n* diolefin *(a hydrocarbon with two double bonds)*
diolefinisch diolefinic
Diopsid *m (min)* diopside *(calcium magnesium silicate)*
Dioptas *m (min)* dioptase, emerald copper (malachite) *(a copper silicate)*
Diorit *m* diorite *(an igneous rock)*
Diorsellinsäure *f s.* Lekanorsäure
Diosphenol *n* diosphenol, buchucamphor, buccocamphor, 2-hydroxy-6-isopropyl-1-methyl-2-cyclohexen-1-one
Dioxan *n* dioxan
Dioxid *n* dioxide
Dioxindol *n* dioxindole, 2,3-dihydro-3-hydroxy-2-oxoindole
Dioxoborat *n* M^IBO_2 dioxoborate, metaborate
Dioxoborsäure *f* HBO_2 dioxoboric acid, metaboric acid
Dioxodisiloxan *n* $[Si_2O_3H_2]_x$ dioxodisiloxane
1,3-Dioxophthalan *n s.* Phthalsäureanhydrid
Dipenten *n* dipentene, (±)-4-isopropenyl-1-methylcyclohexene
Dipeptid *n* dipeptide
Dipeptidase *f* dipeptidase
Diperjodat *n* $M_4{}^II_2O_9$ dimesoperiodate, enneaoxodiiodate(VII)
Diphensäure *f* $HOOCC_6H_4-C_6H_4COOH$ diphenic acid, *o,o'*-bibenzoic acid, biphenyl-2,2-dicarboxylic acid
Diphenyl *n (incorrectly for)* Biphenyl
Diphenylamin *n* $(C_6H_5)_2NH$ diphenylamine, DPA, phenylaniline
Diphenylaminblau *n* $(C_6H_5NHC_6H_4)_3COH$ diphenylamine blue, *tris*-(4-anilinophenyl)methanol
Diphenylaminchlorarsin *n* 10-chloro-5,10-dihydrophenarsazine, diphenylamine chloroarsine
Diphenylaminorange *n s.* Tropäolin 00
1,2-Diphenyläthan *n* $C_6H_5CH_2CH_2C_6H_5$ 1,2-diphenylethane, bibenzyl

1,2-Diphenyläthanol *n* $C_6H_5CH(OH)CH_2C_6H_5$ 1,2-diphenylethanol
Diphenyläthen *n* $C_6H_5CH{=}CHC_6H_5$ diphenylethylene, stilbene
Diphenyläther *m* $C_6H_5OC_6H_5$ diphenyl ether
Diphenyläthin *n s.* Diphenylazetylen
Diphenylazetylen *n* $C_6H_5C{\equiv}CC_6H_5$ diphenylacetylene, diphenylethyne, tolane
Diphenylbenzol *n* $(C_6H_5)_2C_6H_4$ diphenylbenzene
Diphenyldiimid *n* $C_6H_5N{=}NC_6H_5$ diphenyldiimide, azobenzene
Diphenyldikarbonsäure-(2,2′) *f s.* Diphensäure
Diphenyldiketon *n* $C_6H_5COCOC_6H_5$ diphenyl diketone, diphenylglyoxal, benzil, bibenzoyl
Diphenyldisulfid *n* $C_6H_5SSC_6H_5$ diphenyl disulphide
Diphenylenimid *n s.* Diphenylenimin
Diphenylenimin *n* $C_6H_4NHC_6H_4$ diphenyleneimine, carbazole, dibenzopyrrole
Diphenylessigsäure *f* $(C_6H_5)_2CHCOOH$ diphenylacetic acid
Diphenylfarbstoff *m* diphenyl dye
Diphenylglykolsäure *f* $(C_6H_5)_2C(OH)COOH$ diphenylglycollic acid, benzilic acid
Diphenylglyoxal *n s.* Diphenyldiketon
Diphenylimid *n s.* Diphenylenimin
Diphenylin *n* $NH_2C_6H_4C_6H_4NH_2$ diphenyline, 2,4′-diaminobiphenyl
Diphenylisomerie *f* diphenyl isomerism
Diphenylketon *n* $C_6H_5COC_6H_5$ diphenyl ketone, benzophenone, benzoyl benzene
Diphenylmethanfarbstoff *m* diphenylmethane dye
Diphenyloxid *n s.* Diphenyläther
Diphenylphenylen *n s.* Diphenylbenzol
***N,N′*-Diphenyl-*p*-phenylendiamin** *n* $(C_6H_5NH)_2C_6H_4$ *N,N′*-diphenyl-*p*-phenylenediamine, DPPD
Diphenylthiokarbazon *n* diphenylthiocarbazone, dithizone
1,3-Diphenyltriazen *n* $C_6H_5N{=}NNHC_6H_5$ diazoaminobenzene, 1,3-diphenyltriazene
Diphosgen *n* $ClCOOCCl_3$ diphosgene, trichloromethyl chloroformate
Diphosphan *n* P_2H_4 diphosphane
Diphosphat *n* $M_4^IP_2O_7$ diphosphate, pyrophosphate
Diphosphit *n* $M_2^IP_2H_2O_5$ diphosphite, *(better)* diphosphonate
Diphosphoglyzerinsäure *f* diphosphoglyceric acid
Diphosphopyridinnukleotid *n s.* Nikotinamid-adenin-dinukleotid
Diphosphorsäure *f* $H_4P_2O_7$ diphosphoric acid, pyrophosphoric acid
Diphosphortetrajodid *n* P_2I_4 diphosphorus tetraiodide
Diphosphortriselenid *n* P_2Se_3 diphosphorus triselenide
Diphosphortrisulfid *n* P_2S_3 diphosphorus trisulphide, phosphorus(III) sulphide, phosphorous sulphide
Diphthalyl *n* biphthalyl, diphthalyl
Diphthalylsäure *f* diphthalylic acid, benzil-2,2′-dicarboxylic acid
Diphtherietoxin *n* diphtheria toxin
Dipikolinsäure *f* dipicolinic acid, pyridine-2,6-dicarboxylic acid
Dipikrylamin *n* $(NO_2)_3C_6H_2NHC_6H_2(NO_2)_3$ dipicrylamine, di-2,4,6-trinitrophenylamine, hexite
Dipol *m* dipole
Dipolachse *f* dipole axis
dipolar dipolar
Dipolassoziation *f* dipole association
Dipol-Dipol-Wechselwirkung *f* dipole-dipole interaction
dipolfrei non-dipolar
Dipolglied *n* dipole term
Dipolion *n* dipolar (dual) ion
Dipolkraft *f* dipole force
Dipol-Ladungs-Wechselwirkung *f* charge-dipole interaction
Dipolmessung *f* dipole measurement
Dipolmolekül *n* dipolar (dipole) molecule
Dipolmoment *n* dipole moment
Dipolstrahlung *f* dipole radiation
Dipropylazetylen *n* $CH_3[CH_2]_2C{\equiv}C[CH_2]_2CH_3$ 4-octyne, dipropylacetylene
Dipropylketon *n* $CH_3[CH_2]_2CO[CH_2]_2CH_3$ 4-heptanone, di-*n*-propyl ketone
Dip-Verfahren *n (rubber)* dip [reclaiming] process
Dipyr *m (min)* dipyre *(a variety of scapolite)*
Diradikal *n* biradical, diradical
Direktabschreckung *f* direct quenching *(of metals)*
Direktapplikation *f* / **dosierte** *(tox)* topical application *(for testing the efficiency of an insecticide)*
Direktdampf *m* live (open, prime, direct) steam
Direktdruck *m* direct printing
Direktexpansionskühler *m* direct-expansion chiller
Direktfarbstoff *m* direct (substantive) dye (dyestuff)
Direktmethode *f (tox)* direct-feeding test *(for detecting pesticide residues by contacting animals with the substance to be checked)*
Direkt-Positiv-Prozeß *m (phot)* reversal process
Direktspinnverfahren *n (text)* direct spinning system
Direktsynthese *f* direct synthesis *(as for obtaining chlorosilanes)*
Direktverfahren *n (dye)* flushing process
direktziehend *(dye)* direct, substantive
Dirheniumheptoxid *n* Re_2O_7 dirhenium heptaoxide, rhenium(VII) oxide
Dirheniumtrioxid *n* Re_2O_3 dirhenium trioxide, rhenium(III) oxide
Dirhodiumtrioxid *n* Rh_2O_3 dirhodium trioxide, rhodium(III) oxide
dirigieren to direct *(a substituent into a position)*
dirigierend / nach der meta-Stellung meta-directing
nach der ortho- und para-Stellung d. ortho-para-directing
Disaccharid *n* $C_{12}H_{22}O_{11}$ disaccharide
Disauerstoff *m* O_2 dioxygen
Disazofarbstoff *m* bisazo dye
Dischwefeldibromid *n* S_2Br_2 disulphur dibromide
Dischwefeldichlorid *n* S_2Cl_2 disulphur dichloride
Dischwefeldifluorid *n* S_2F_2 disulphur difluoride
Dischwefeldijodid *n* S_2I_2 disulphur diiodide

Dischwefelheptoxid *n* S_2O_7 disulphur heptaoxide
Dischwefelpentoxiddichlorid *n s.* Disulfurylchlorid
Dischwefelsäure *f* $H_2S_2O_7$ disulphuric acid
Dischwefeltrioxid *n* S_2O_3 disulphur trioxide, sulphur(III) oxide
Dischwefelwasserstoff *m* H_2S_2 hydrogen disulphide, disulphane
Disco-Schwelverfahren *n (coal)* Disco process
disekundär disecondary
Diselendibromid *n* Se_2Br_2 diselenium dibromide
Diselendichlorid *n* Se_2Cl_2 diselenium dichloride
Diselendijodid *n* Se_2I_2 diselenium diiodide
Disilan *n* Si_2H_6 disilane
Disilberfluorid *n* Ag_2F disilver fluoride
Disilberhydrogenphosphat *n* Ag_2HPO_4 disilver hydrogenphosphate, silver hydrogenorthophosphate
Disilberpentazyanonitrosylferrat(II) *n* $Ag_2[Fe(CN)_5NO]$ disilver pentacyanonitrosylferrate(II), silver nitroprusside, silver nitroprussiate
Disilikan *n s.* Disilan
Disilikat *n* disilicate
Disilikoäthan *n s.* Disilan
Disilizid *n* disilicide
Disiliziumhexabromid *n* Si_2Br_6 hexabromodisilane, disilicon hexabromide
Disiliziumhexachlorid *n* Si_2Cl_6 hexachlorodisilane, disilicon hexachloride
Disiloxan *n* $(SiH_3)_2O$ disiloxane
diskontinuierlich discontinuous, batch[wise] *(operation)*
d. arbeitend batch *(apparatus)*
Dislokation *f (cryst)* dislocation
Dislokationsmetamorphose *f (geol)* dynamometamorphism, dynamic metamorphism
Dismembrator *m* pin-disk (pinned-disk) disintegrator (mill), pin mill
dismulgieren to demulsify, to desemulsify, to de-emulsify, to break, to crack *(an emulsion)*
Dismulgieren *n* demulsification, desemulsification, de-emulsification, breaking, cracking *(of an emulsion)*
Dismutation *f* dismutation
dismutativ dismutative
dismutieren to dismutate
Dismutierung *f s.* Dismutation
Dispensation *f (pharm)* dispensation
Dispensatorium *n (pharm)* dispensatory
dispensieren *(pharm)* to dispense
Dispergens *n s.* Dispersionsmittel 1.
dispergierbar dispersible
Dispergierbarkeit *f* dispersibility
dispergieren to disperse, to deflocculate
dispergierend dispersive
Dispergier[hilfs]mittel *n s.* Dispersionsmittel 1.
Dispergierung *f* dispersion, dispersal, deflocculation
Dispergierungs... *s.* Dispergier...
Dispergiervermögen *n* dispersive (dispersing, dispersant) power, dispersive property
Dispergierwirkung *f* dispersive (dispersing, dispersant) action, dispersing effect
dispers disperse
Dispersant *n s.* Dispersionsmittel 1.
Dispersantadditiv *n s.* Dispersionszusatz
Dispersantwirkung *f s.* Dispergierwirkung
Dispersion *f* 1. dispersion, dispersal *(of particles)*; 2. dispersion *(of waves), (as to quantity also)* dispersivity, differential refractivity
grobe D. coarse dispersion *(of particles)*
paramagnetische D. paramagnetic dispersion *(of waves)*
reziproke relative D. reciprocal relative dispersion constringence, Abbe value (number), ν-value, v-value *(optics)*
spezifische D. specific dispersivity *(of waves)*
Dispersionsanalyse *f* dispersion analysis
Dispersionsazofarbstoff *m* disperse azo dye
Dispersionseffekt *m* dispersion effect
D. der Leitfähigkeit *(el chem)* dispersion of conductance, Debye-Falkenhagen effect
Dispersionsfarbstoff *m* disperse[d] dye
Dispersionsformel *f* dispersion formula
Dispersionsgrad *m* degree of dispersion; *(coll)* dispersity
Dispersionskolloid *n* dispersion colloid
Dispersionskonstante *f* dispersive constant
Dispersionskraft *f s.* 1. Dispersionsvermögen; 2. Dispersionskräfte
Disperionskräfte *fpl* dispersion (London) forces *(part of the intermolecular forces)*
Dispersionskurve *f* dispersion curve
Dispersionsmedium *n* dispersion (dispersive, disperse) medium
Dispersionsmittel *n* 1. dispersing (dispersion, deflocculating) agent, dispersant, deflocculant, deflocculator; 2. *s.* Dispersionsphase
Dispersionsmühle *f* dispersing mill
Dispersionsphase *f* continuous phase, disperse (dispersive, dispersion) medium *(of a disperse system)*
Dispersionsverfahren *n* dispersion method
Dispersionsvermögen *n* dispersive (dispersing) power (property)
Dispersionszusatz *m* dispersing (dispersant) additive
Dispersität *f* dispersity
Dispersitätsgrad *m s.* Dispersionsgrad
Dispersoid *n* dispersoid
Dispersoidanalyse *f* dispersoid analysis
Dispersoidologie *f* dispersoidology, *(seldom used for)* colloid chemistry
Dispersum *n (phys chem)* internal (discontinuous, disperse, dispersed) phase
Dispiro-Verbindung *f* dispiro compound (hydrocarbon)
disproportionieren to disproportionate
Disproportionierung *f* disproportionation
Disproportionierungsreaktion *f* disproportionation reaction
Dissimilation *f (bioch)* dissimilation
Dissipation *f* dissipation

dissipativ dissipative
Dissoziation *f* dissociation, ionization, splitting-up
  elektrolytische D. electrolytic dissociation
  thermische D. thermal dissociation
Dissoziationsdruck *m* dissociation pressure
Dissoziationsenergie *f* dissociation energy
Dissoziationsgeschwindigkeit *f* velocity of dissociation
Dissoziationsgleichgewicht *n* dissociation equilibrium
Dissoziationsgrad *m* degree (fraction) of dissociation
  thermischer D. degree of thermal dissociation
Dissoziationsgrenze *f* dissociation limit
Dissoziationskonstante *f* dissociation (ionization) constant, *(with acids and bases also)* affinity constant; instability constant *(of complex ions)*
Dissoziationskontinuum *n* dissociation continuum
Dissoziationswärme *f* heat of dissociation
dissoziieren to dissociate, to split up
Distanzstück *n* spacer
Distex-Verfahren *n* Distex process, extractive distillation
Disthen *m (min)* disthene, cyanite, kyanite *(aluminium oxide orthosilicate)*
Distickstoffmonoxid *n* N_2O dinitrogen monooxide, nitrogen(I) oxide
Distickstoffpentasulfid *n* N_2S_5 dinitrogen pentasulphide, nitrogen(V) sulphide
Distickstoffpentoxid *n* N_2O_5 dinitrogen pentaoxide, nitrogen(V) oxide
Distickstofftetroxid *n* N_2O_4 dinitrogen tetraoxide
Distickstofftrioxid *n* N_2O_3 dinitrogen trioxide, nitrogen(III) oxide
disubstituiert disubstituted
Disulfan *n* H_2S_2 disulphane, hydrogendisulphide
Disulfat *n* $M_2^IS_2O_7$ disulphate
Disulfid *n* disulphide
Disulfidbindung *f s.* Disulfidbrücke
Disulfidbrücke *f* disulphide bridge (crosslink, link, bond)
Disulfiram *n* disulphiram, tetraethylthiouram disulphide
Disulfit *n* $M^I_2S_2O_5$ disulphite
Disulfitomerkurat(II) *n* $M^I_2[Hg(SO_3)_2]$ disulphitomercurate(II), sulphitomercurate(II)
Disulfonat *n* disulphonate
disulfonieren to disulphonate
Disulfonierung *f* disulphonation
Disulfonsäure *f*, **Disulfosäure** *f* disulphonic acid
Disulfurylchlorid *n* $S_2O_5Cl_2$ disulphuryl chloride
Ditantalat *n* $M^I_4Ta_2O_7$ ditantalate
Ditellurat *n* $M^I_2Te_2O_7$ ditellurate
Diterpen *n* diterpene
ditertiär ditertiary
Dithalliummonoxid *n* Tl_2O dithallium monooxide, thallium(I) oxide
Dithiazin *n* dithiazine, 3,3′-diethylthiadicarbocyanine
Dithiazolanthrachinonfarbstoff *m* dithiazolanthraquinone dye
Dithioarsenat(III) *n* M^IAsS_2 dithioarsenite, metathioarsenite
Dithiokarbamat *n* NH_2CSSM^I dithiocarbamate
Dithiokarbamatbeschleuniger *m (rubber)* dithiocarbamate accelerator
Dithiokarbamidsäure *f* NH_2CSSH dithiocarbamic acid
Dithiokarbaminsäure *f s.* Dithiokarbamidsäure
Dithiokohlensäure *f* HOCSSH dithiocarbonic acid, xanthic acid
Dithiokohlensäure-*O*-äthylester *m* $C_2H_5O-CSSH$ dithiocarbonic *O*-ethyl ester, xanthogenic acid, ethylxanthogenic acid
Dithionat *n* $M^I_2S_2O_6$ dithionate
Dithionit *n* $M^I_2S_2O_4$ dithionite; *(dye)* $Na_2S_2O_4$ [sodium] dithionite, [sodium] hydrosulphite
Dithionitbleiche *f (pap)* sodium hydrosulphite bleaching
Dithionit-Natronlauge *f (dye)* caustic hydrosulphite solution
Dithionsäure *f* $H_2S_2O_6$ dithionic acid
Dithiooxalsäure *f* HS−CO−CO−SH *or* HO−CS−CS−OH dithio-oxalic acid
Dithiooxalsäurediamid *n s.* Dithiooxamid
Dithiooxamid *n* $H_2NSCCSNH_2$ dithiooxamide, rubeanic acid
Dithiophosphat *n* $M^I_2PS_2O_2$ dithiophosphate
Dithiophosphorsäure *f* $H_3PS_2O_2$ dithiophosphoric acid
Dithiosalizylsäure *f* 1. $C_6H_4(OH)CSSH$ dithiosalicylic acid, 2-hydroxybenzenethionothiolic acid
Dithiosäure *f* C_nH_{2n+1} CSSH dithio acid
Dithioverbindung *f* dithio compound
Dithizon *n* $C_6H_5N{=}N{-}CS{-}NHNHC_6H_5$ dithizone, diphenylthiocarbazone
Dititantrioxid *n* Ti_2O_3 dititanium trioxide, titanium(III) oxide
Dititantrisulfid *n* Ti_2S_3 dititanium trisulphide, titanium(III) sulphide
ditrigonal-skalenoedrisch *(cryst)* ditrigonal-scalenohedral
Diuranat *n* $M^I_2U_2O_7$ diuranate
Diurantrisulfid *n* U_2S_3 diuranium trisulphide, uranium(III) sulphide
Diuretikum *n (pharm)* diuretic
diuretisch *(pharm)* diuretic
Divanadat *n* $M^I_4V_2O_7$ divanadate
Divanadinpentoxid *n* V_2O_5 divanadium pentaoxide, vanadium(V) oxide
Divanadintrioxid *n* V_2O_3 divanadium trioxide, vanadium(III) oxide
Divanadintrisulfid *n* V_2S_3 divanadium trisulphide, vanadium(III) sulphide
divariant bivariant, divariant
Divarikatinsäure *f* divaricatinic acid, 2-hydroxy-4-methoxy-6-propylbenzoic acid
Divarikatsäure *f* divaricatic acid *(a depside of divaric acid)*
Divarsäure *f* divaric acid, 2,4-dihydroxy-6-propylbenzoic acid
Dividivi *pl (tann)* divi-divi, libi-dibi *(husks of Caesalpinia coriaria (Jacq. Willd.)*
Divinyl *n s.* Butadien-(1,3)
Divinyläther *m* $CH_2{=}CHOCH{=}CH_2$ divinyl ether

Divinylensulfid *n s.* Thiophen
Diwasserstoff *m* H_2 dihydrogen
Diwismuttriselenid *n* Bi_2Se_3 dibismuth triselenide, bismuth(III) selenide
Diwolframat *n* diwolframate, ditungstate
Dizyan *n* N≡C−C≡N dicyanogen, cyanogen [gas], oxalonitrile
1,1-Dizyanäthen *n* $CH_2=C(CN)_2$ 1,1-dicyanoethylene
Dizyandiamid *n* $NH_2C(=NH)NHCN$ dicyandiamide
Dizyanoaurat(I) *n* $M^I[Au(CH)_2]$ dicyanoaurate(I), cyanoaurate(I)
Djelutung *m*, **Djelutungharz** *n* jelutong *(a copal from Dyera specc. and Parthenium argentatum A. Gray)*
Djenkolsäure *f* $CH_2[SCH_2CH(NH_2)COOH]_2$ djenkolic acid
DK *s.* 1. Dielektrizitätskonstante; 2. Dieselkraftstoff
DK-Meter *n s.* Dekameter
DK-Metrie *f s.* Dekametrie
D-Konfiguration *f* D configuration
DMF *s.* Dimethylformamid
DNBP *s.* Dinoseb
DNC *s.* 4,6-Dinitro-*o*-kresol
DNOC *s.* 4,6-Dinitro-*o*-kresol
DNOCHP *s.* 2,4-Dinitro-6-zyklohexylphenol
DNOK *s.* 4,6-Dinitro-*o*-kresol
DNP-Derivat *n s.* Dinitrophenylderivat
DNS *s.* Desoxyribonukleinsäure
Döbereiner-Triaden *fpl* triads of Döbereiner
Docht *m* wick
 D. für Spirituslampen burner wick
Dochtlampe *f* wick[-fed] lamp
Docke *f (text)* skein
Dodekaeder *n (cryst)* dodecahedron
Dodekahydrat *n* dodecahydrate
Dodekamolybdatophosphat *n* $M^I_3[PMo_{12}O_{40}]$ dodecamolybdophosphate
Dodekan *n* $C_{12}H_{26}$ dodecane; *(specif) s. n*-Dodekan
***n*-Dodekan** *n* $CH_3[CH_2]_{10}CH_3$ *n*-dodecane
Dodekanal *n* $CH_3[CH_2]_{10}CHO$ dodecanal, dodecylaldehyde
Dodekandisäure *f* $HOOC[CH_2]_{10}COOH$ dodecanedioic acid
Dodekan-1-karbonsäure *f* $CH_3[CH_2]_{11}COOH$ tridecanoic acid, dodecane-1-carboxylic acid
Dodekanol-(1) *n* $CH_3[CH_2]_{11}OH$ 1-dodecanol, *n*-dodecyl alcohol, alcohol C−12
Dodekansäure *f* $CH_3[CH_2]_{10}COOH$ dodecanoic acid, lauric acid
Dodekansäureäthylester *m* $CH_3[CH_2]_{10}COOC_2H_5$ dodecanoic acid ethyl ester, ethyl laurate
1-Dodekanthiol *n* $CH_3[CH_2]_{11}SH$ dodecanethiol
Dodekawolframatophosphat *n* $M^I_3[PW_{12}O_{40}]$ dodecawolframophosphate, dodecatungstophosphate
Dodezen-(1) *n* $CH_3[CH_2]_9CH=CH_2$ 1-dodecene
Dodezen-(2)-disäure *f* $HOOC[CH_2]_8CH=CHCOOH$ 2-dodecenedioic acid, traumatic acid
Dodezensäure *f* $C_{11}H_{21}COOH$ dodecenoic acid
Dodezin-(2) *n* $CH_3[CH_2]_8C≡CCH_3$ 2-dodecyne
Dodezylaldehyd *m s.* Dodekanal
***n*-Dodezylalkohol** *m s.* Dodekanol-(1)
Dodezylamin *n* $CH_3[CH_2]_{10}CH_2NH_2$ dodecylamine
α-Dodezylen *n s.* Dodezen-(1)
Dodezylgruppe *f* $CH_3[CH_2]_{10}CH_2-$ dodecyl group
***n*-Dodezylmerkaptan** *n s.* 1-Dodekanthiol
Dodge-Backenbrecher *m* Dodge [jaw] crusher
Doebner-Miller-Reaktion *f* Doebner-Miller reaction (synthesis) *(formation of quinoline and its derivatives)*
Doebner-Synthese *f* Doebner synthesis *(formation of substituted cinchoninic acids)*
Dokosan *n* $C_{22}H_{46}$ docosane, *(specif)* $CH_3[CH_2]_{20}CH_3$ *n*-docosane
Dokosanol-(1) *n* $CH_3[CH_2]_{20}CH_2OH$ 1-docosanol
Dokosansäure *f* $C_{21}H_{43}COOH$ docosanoic acid
Dokosensäure *f* $C_{21}H_{41}COOH$ docosenoic acid
Dokosen-(11)-säure *f* $CH_3[CH_2]_9CH=CH[CH_2]_9COOH$ 11-docosenoic acid, cetoleic acid
Dokosen-(13)-säure *f* 13-docosenoic acid
Dokosin-(13)-säure *f* $CH_3[CH_2]_7C≡C[CH_2]_{11}COOH$ 13-docosynoic acid
prim.-n-**Dokosylalkohol** *m s.* Dokosanol-(1)
Doktorbehandlung *f (petrol)* doctor treatment (treating, sweetening)
Doktorlauge *f*, **Doktorlösung** *f (petrol)* doctor solution
doktor-negativ *(petrol)* sweet[ened], doctor-sweet
doktor-positiv *(petrol)* sour
Doktorsüßen *n s.* Doktorbehandlung
Doktorsüßungsverfahren *n (petrol)* doctor process
Doktortest *m (petrol)* doctor test
Doktorverfahren *n s.* Doktorsüßungsverfahren
Dokumentenpapier *n* document paper, *(Am)* deed paper
Dolerophanit *m (min)* dolerophanite, dolerophane *(copper(II) oxide sulphate)*
dollieren *(tann)* to buff
Dolomit *m* 1. *(min)* dolomite *(calcium magnesium carbonate)*; 2. *s.* Dolomitgestein
Dolomitgestein *n* dolomite [rock]
dolomitisch dolimitic
dolomitisieren *(geol)* to dolomitize
Dolomitisierung *f (geol)* dolomitization
Dolomitkalk *m* dolomitic (dolomite) lime *(calcium magnesium oxide)*
Dolomitkalkstein *m* dolomitic limestone
Dolomitknolle *f* coal ball *(a petrifaction in coal)*
Dolomitstein *m* dolomite brick
Dom *m* stillhead
Domeykit *m (min)* domeykite *(copper arsenide)*
Donator *m* donor, donator
 D. des Elektronenpaares electron-pair donor
Donator-Akzeptor-Bindung *f* donor-acceptor bond, semipolar (half-polar, coordinate, dative) bond, dative covalence
Donator-Akzeptor-Komplex *m* donor-acceptor complex, charge transfer complex
Donator-Akzeptor-Wechselwirkung *f* donor-acceptor interaction
Donatoratom *n* donor atom

Donatorgruppe *f* donor group
Donatormolekül *n* donor molecule
Donatorniveau *n* donor level
Donatorsolvens *n* donor solvent
Donatorstärke *f* donor power
Donatorstörstelle *f s.* Donatorverunreinigung
Donatorsubstanz *f* donor reagent
Donatorterm *m s.* Donatorniveau
Donatorverunreinigung *f* donor impurity
Donnan-Effekt *m* Donnan effect
Donnan-Gleichgewicht *n* Donnan [membrane] equilibrium
Donnan-Potential *n* Donnan potential
Donnelly-Krackverfahren *n* Donnelly process
Donnerwurzel *f* elecampane *(dried root of Inula helenium L.)*
Donor *m s.* Donator
Dopa *n s.* Dihydroxyphenylalanin
Dope *m(n),* **Dope-Mittel** *n* dope *(a substance added to mineral-oil products in a quantity below 1%)*
dopen to dope *(to treat mineral-oil products with a dope)*
Dope-Stoff *m s.* Dope
Dopingmittel *n (pharm)* dope
Doppelarmkneter *m* double-arm mixer
Doppelbandpolieren *n (glass)* twin polishing
Doppelbandschleifen *n (glass)* twin grinding
Doppelbild *n (phot)* ghost image
Doppelbindung *f* double bond (link, linkage)
  gehäufte D. *s.* kumulierte D.
  gekreuzte D. crossed double bond
  halbpolare D. *s.* semipolare D.
  isolierte D. isolated double bond
  konjugierte D. conjugated double bond
  kumulierte D. cumulated double bond
  mesomeriefähige D. resonating double bond
  nichtkonjugierte D. unconjugated double bond
  semipolare D. semipolar (half-polar, dative, coordinate) bond, dative covalence
  semizyklische D. semicyclic [double] bond
Doppelbindungscharakter *m* double-bond character
Doppelbindungselektron *n* double-bond electron
Doppelbindungssystem *n* double-bond system
doppelbrechend double-refracting, doubly refracting (refractive), birefringent
Doppelbrechung *f* double refraction, birefringence
  elektrische D. electrical double refraction
  positive D. positive double refraction
Doppeleinzelheizer *m (rubber)* twin curing unit, double (twin) press, twin heater
Doppelfalzungen *fpl (pap)* double folds
doppelfarbig *(cryst, coll)* dichroic
Doppelfarbigkeit *f (cryst, coll)* dichroism
Doppelflügelwäscher *m (min tech)* logwasher
Doppelfokussierung *f* double focusing
Doppelgaserzeuger *m,* **Doppelgasgenerator** *m* predistillation [gas] producer
Doppelgebläse *n* double bulb blower
Doppelheizer *m s.* Doppeleinzelheizer
Doppelkegelbindung *f* sandwich bond *(coordination chemistry)*
Doppelkegelstruktur *f* sandwich structure *(coordination chemistry)*
Doppelkegel-Trommelmischer *m s.* Doppelkonusmischer
Doppelkolbenpresse *f* double-ram press
Doppelkolonne *f* double column
Doppelkonusmischer *m* double-cone blender (mixer)
Doppelkreppapier *n* double-crêpe paper
Doppelkrümmer *m* return bend
Doppellaktatverfahren *n (soil)* double-lactate method
Doppelleerstelle *f* divacancy *(a crystal defect)*
Doppellinie *f* doublet *(spectroscopy)*
Doppelmolekül *n* double[d] molecule
Doppelmuffe *f (lab)* clamp holder, bosshead
Doppelpaddelmischer *m* double-arm mixer
Doppelpechpapier *n* tarred [brown] paper, tar (pitch, asphalt) paper
Doppelpfeil *m* double-headed arrow
Doppelpistole *f s.* Doppelspritzpistole
Doppelpresse *f* dual press
Doppelquarz *m* biquartz
Doppel-Reifeneinzelheizer *m (rubber)* double (twin) tyre press
Doppelrohraustauscher *m s.* Doppelrohrwärmeaustauscher
Doppelrohrkondensator *m* double-pipe condenser
Doppelrohrkristallisator *m* double-pipe crystallizer
Doppelrohrverflüssiger *m s.* Doppelrohrkondensator
Doppelrohrwärmeaustauscher *m,* **Doppelrohrwärmeübertrager** *m* double-pipe heat exchanger
Doppelrührwerk *n* double agitator
Doppelsalz *n* double salt
  sulfatisches D. double sulphate
Doppelsäule *f* double column
Doppelscharlach *m (dye)* Biebrich [scarlet] red, scarlet red
Doppelscheibenmühle *f (pap)* double-disk refiner
  D. System Bauer Bauer double-disk refiner
Doppelscheibenrefiner *m s.* Doppelscheibenmühle
Doppelschicht *f* double layer
  diffuse D. *(phys chem)* diffuse double layer
  elektrische (elektrochemische) D. electrical double layer
Doppelschichtfilm *m (phot)* double-coated film
Doppelschnecke *f* twin screw (worm)
Doppelschneckenextruder *m* twin-screw extruder
Doppelsitzventil *n* double-seat[ed] valve
Doppelspalt *m* double slit
Doppelspat *m* / **Isländischer** Iceland spar *(a transparent variety of calc-spar)*
Doppelspritzpistole *f (coat)* two-nozzle [spray] gun
Doppelstrahlinstrument *n* double-beam instrument *(spectroscopy)*
Doppelsulfat *n* double sulphate

Doppelsuperphosphat *n* double (concentrated, triple, treble) superphosphate *(a fertilizer)*
doppeltwirkend double-acting
Doppelwalzenpresse *f* double-roll press
Doppelwalzentrockner *m* twin-drum dryer
doppelwandig double-walled
Doppelwellendampfmischer *m (pap)* double-shaft pulp (steam) mixer
Doppelwellenmischer *m* double-shaft[ed] mixer
Doppelwellenmuldenmischer *m* twin-rotor mixer
Doppler-Breite *f* Doppler half-width *(of a spectral line)*
Doppler-Effekt *m* Doppler effect
Doppler-Verbreiterung *f* Doppler broadening *(of a spectral line)*
Doppler-Verschiebung *f* Doppler shift *(spectroscopy)*
Dorn *m* 1. *(tech)* pin; 2. [extruder] core, mandrel *(extrusion moulding)*
Dorr-Eindicker *m* Dorr thickener
dörren 1. to dry, to desiccate *(fruit)*; to dry-cure *(meat)*; 2. *s.* darren
Dörren *n* 1. drying, desiccation *(of fruit)*; dry curing *(of meat)*; 2. *s.* Darren
Dörrgemüse *n* dried vegetables
Dorr-Klassierer *m* Dorr [rake] classifier
  D. mit Schüssel Dorr bowl classifier
Dörrobst *n* dry (dried) fruit
Dorr-Rechenklassierer *m s.* Dorr-Klassierer
Dorschleberöl *n*, **Dorschlebertran** *m* cod-liver oil
Dosenkonserven *fpl* canned food
Dosenmilch *f* canned milk
Dosieranlage *f* dosing plant
Dosierapparat *m* dosing (metering, proportioning) apparatus
Dosierbandwaage *f* weighing (balanced-weigh) belt
Dosiereinrichtung *f* dosing (metering, proportioning) mechanism
dosieren to dose, to meter, to proportion
Dosiergerät *n s.* Dosierapparat
Dosierlöffel *m (lab)* measuring spoon
Dosiermaschine *f* dosing machine
Dosierpumpe *f* 1. dosing (metering, proportioning, controlled-volume) pump; 2. *(plast)* spinning pump
Dosierschraube *f* proportioning screw
Dosiertank *m* weigh tank
Dosierung *f* dosage, dosing, metering, proportioning
Dosierungs ... *s.* Dosier ...
Dosierventil *n* control valve
Dosimeter *n* dosimeter, dosemeter, dosage meter
Dosimetrie *f* dosimetry, dosage measurement
Dosis *f* dose, dosage
  kleinste wirksame D. minimum effective dose
  letale D. lethal (fatal) dose
  mittlere effektive D. median effective dose
  mittlere letale D. median lethal dose
  subletale D. sublethal dose
  therapeutische D. therapeutic dose
  tödliche D. *s.* **letale D.**
  toxische D. toxic dose
Dosiseffekt *m s.* Dosiswirkung
Dosiskonstante *f* dosage constant
Dosisleistung *f* dose (dosage) rate
Dosisleistungsmesser *m* dose rate meter
Dosismesser *m*, **Dosismeßgerät** *n s.* Dosimeter
Dosismessung *f s.* Dosimetrie
Dosisprotrahierung *f* dose protraction
Dosisrate *f s.* Dosisleistung
Dosiswirkung *f* dose (dosage) response (effect)
Dosiswirkungskurve *f* dose-response (dose-effect) curve
Dotter *m(n)* [egg] yolk, vitellus
Dotteröl *n* cameline (dodder) oil *(from the seeds of Camelina sativa Crantz)*
Doublettspektrum *n* doublet spectrum
Doublettstruktur *f* doublet structure
Doverit *m (min)* doverite *(a calcium yttrium carbonate fluoride)*
d. P. *s.* Packung / dichteste
2,4-DP *s.* Dichlorprop
DPN *(abbr)* Diphosphopyridinnukleotid, *s.* Nikotinamid-adenin-dinukleotid
DPPD *s.* *N,N'*-Diphenyl-*p*-phenylendiamin
Drachenblutharz *n* dragon's blood [resin] *(from Daemonorops draco Blume or from Dracaena specc.)*
Dragée *n* dragée
Dragierkessel *m* coating pan
Draht *m* 1. wire; 2. *(text)* twist
  falscher D. *(text)* false twist
Drahtbarren *m* wire bar
Drahtemaille *f*, **Drahtemaillelack** *m* wire enamel
Drahterteilung *f s.* Drahtgebung
Drahtgebung *f (text)* twist
Drahtgewebe *n* wire cloth (gauze), metal fabric
Drahtglas *n* wire[d] glass
Drahtkorb *m* **für Exsikkatoren** desiccator cage (guard)
Drahtnetz D. wire gauze (net)
Drahtnetzspirale *f* gauze plug *(in a combustion tube)*
Drahtrohrmodell *n* wire model *(as of a pipe system)*
Drahtsieb *n* wire screen (sieve), metal screen
Drahtspirale *f* wire spiral
drahtumwickelt wire-wound
Draht- und Schnurfang *m (pap)* string catcher
Drainage *f s.* Dränage
Drakorubin[harz] *n* dracorubin
Drakorubinpapier *n* dragon's blood resin paper
Drall *m* 1. spin, angular momentum, moment of momentum; 2. *(text)* twist
Dralldüse *f* swirl[-plate] nozzle, hollow-cone nozzle
Drän *m s.* Dränrohr
Dränage *f* drainage
Dränageleitung *f* drain line
dränieren to drain
Dränkammer *f* draining pan *(in flow coating)*
Dränrohr *n* drainage pipe (tube), drain

Dräntunnel *m s.* Dränkammer
Dravit *m (min)* dravite *(a magnesium-containing tourmaline)*
Drechsel-Waschflasche *f (lab)* Drechsel bottle
Drehachse *f* axis of rotation; *(cryst)* axis of symmetry
dreizählige D. *(cryst)* threefold (triad) axis of symmetry
sechszählige D. *(cryst)* sixfold (hexad) axis of symmetry
vierzählige D. *(cryst)* fourfold (tetrad) axis of symmetry
zweizählige D. *(cryst)* twofold (diad) axis of symmetry
Drehbandkolonne *f (distil)* spinning band column, rotating-strip column
drehbar rotatable
Drehbarkeit *f* rotatability
freie D. free rotation
Drehbewegung *f* rotary (rotational) motion, rotation
Drehbohren *n* rotary drilling
Drehdiagramm *n* X-ray rotation photograph
Drehdurchführung *f* rotary joint *(as for fluids in piping)*
Dreheinschlagpapier *n* twisting paper
(+)-drehend dextro[rotatory], dextrorotary, dextrorotating, dextrogyrate, dextrogyratory, dextrogyre, dextrogyrous
(–)-drehend laevo[rotatory], laevorotary, laevorotating, laevogyrate, laevogyratory, laevogyre, laevogyrous
Drehfilter *n* rotary (revolving) filter
Drehhalter *m s.* Drehkreuz
Drehimpuls *m* angular momentum, moment of momentum, spin
innerer D. intrinsic angular momentum *(of elementary particles)*
Drehimpulsentwicklung *f* angular momentum expansion
Drehimpulsquantenzahl *f* rotation[al] quantum number
Drehknotenfänger *m (pap)* rotary (rotating, revolving) strainer, [revolving] drum strainer
Drehkocher *m (pap)* rotary (revolving) boiler (digester)
Drehkolbenpumpe *f* lobe pump
Drehkreuz *n* spider
Drehkristall *m* rotating (rotation) crystal
Drehkristallaufnahme *f*, **Drehkristalldiagramm** *n* rotating-crystal photograph (diagram), X-ray rotation photograph
Drehkristallkamera *f* rotating-crystall camera
Drehkristallverfahren *n* rotating-crystal method
D. von Bragg Bragg method [of crystal analysis], Bragg treatment
Drehkülbelform *f (glass)* paste mould
Drehmaschine *f (ceram)* jiggering machine, jigger
Drehmoment *n* torque
Drehofen *m* rotary kiln, *(esp met)* rotary furnace
Drehofenmantel *m* rotary-kiln shell
Drehrichtung *f* direction of rotation
Drehrohr *n* revolving tube, rotating cylinder
Drehrohrofen *m* rotary kiln
Drehrost *m* rotating (revolving) grate
Drehscheibe *f* 1. rotary (rotating) disk *(as of an extractor)*; 2. *(ceram)* potter's wheel
Drehscheibenextrakteur *m*, **Drehscheibenextraktor** *m* rotary-disk contactor (extractor, tower)
Drehscheibenkolonne *f s.* Drehscheibenextrakteur
Drehschieberölpumpe *f* rotary slide-valve oil pump
Drehschieberverdichter *m* [sliding-]vane compressor
Drehschwingung *f* twisting vibration
Drehsieb *n* rotary screen
Drehspäne *mpl* turnings
Drehspulgalvanometer *n* moving-coil galvanometer
D. von D'Arsonval D'Arsonval galvanometer
Drehtank *m (glass)* revolving pot
Drehtisch *m (petrol)* rotary table (machine)
Drehtischantrieb *m (petrol)* rotary-machine drive
Drehtrommel *f* rotary (rotating) drum (cylinder), tumbler, tumbling barrel
Drehtrommeltrockner *m* rotary (rotatory) dryer
Drehung *f* 1. rotation; 2. *(text)* twist[ing] ✦ **ohne D.** *(text)* twistless
D. der Polarisationsebene *s.* optische D.
magnetische D. [der Polarisationsebene] magnetic rotation, Faraday effect
molare D. molar rotation
molekulare D. molecular rotation
optische D. optical rotation
spezifische D. specific rotation
drehungsfixiert *(text)* twist-set
drehungsfrei *(text)* twistless
Drehungsvermögen *n* rotatory power
magnetisches D. magnetic rotatory power
molares D. molar rotatory power
optisches D. optical activity (rotatory power)
spezifisches D. specific rotatory power
Drehungswinkel *m* **[/ optischer]** *s.* Drehwert
Drehvermögen *n s.* Drehungsvermögen
Drehverteiler *m* rotating (revolving) distributor
Drehwaage *f* torsion balance
Drehwanne *f (glass)* revolving pot
Drehwert *m* amount (angle) of rotation
Drehwertanteil *m* rotatory contribution
Drehwinkel *m* **[/ optischer]** *s.* Drehwert
Drehzahl *f* number of revolutions, rate of rotation, speed
kritische D. critical speed
spezifische D. specific speed
Drehzerstäuber *m* rotary (rotary-cup, spinning-cup) atomizer *(of an oil burner)*
Dreiäschersystem *n s.* Dreigrubenäschersystem
dreiatomig triatomic
Dreibandenspektrum *n* three-banded spectrum
dreibasig tribasic *(acid)*
Dreieckskoordinatensystem *n* triangular diagram
Drei[er]elektronenbindung *f* three-electron bond (link, linkage)

Dreierreaktion *f* termolecular reaction
Dreierring *m s.* Dreiring
Dreierstoß *m* ternary (triple, threefold, three-body) collision
Dreifachaufsatz *m* triple adapter, three-neck[ed] adapter
Dreifachbindung *f* triple bond (link, linkage)
dreifachnegativ [geladen] trinegative
dreifachpositiv [geladen] tripositive
Dreifarbendruckverfahren *n* three-colour process
Dreifarbigkeit *f (cryst)* trichroism
Dreifingerklemme *f (lab)* burette clamp
dreifunktionell trifunctional
Dreifuß *m* 1. *(lab)* tripod; 2. *(ceram)* [wedge] stilt *(a piece of kiln furniture)*
Dreifußstativ *n (lab)* tripod retort stand
dreiglied[e]rig three-membered
Dreigrubenäschersystem *n (tann)* three-pit [liming] system
Dreigutapparat *m s.* Dreiproduktapparat
Dreigutscheider *m s.* Dreiproduktscheider
dreihalsig three-neck[ed]
Dreihalskolben *m* three-neck bottle (flask)
Dreihalsrundkolben *m* round-bottom three-neck bottle (flask)
Dreikant *m (ceram)* saddle *(a piece of kiln furniture)*
Drei-Kohlenstoff-Tautomerie *f* three-carbon[-atom]-tautomerism
Dreikolbendruckpumpe *f* three-piston pump
Dreikomponentensystem *n* ternary (tertiary, three-component) system
Dreikörperverdampfer *m* triple-effect evaporator (evaporating unit)
Dreimaischverfahren *n* three-mash method
Dreipressenschleifer *m (pap)* three-pocket grinder
Dreiprodukt[en]apparat *m* three-product unit
Dreiprodukt[en]scheider *m* three-product separator
Dreiring *m* three-membered [carbon] ring
Dreiringhypothese *f* three-membered ring hypothesis
dreisäurig triacid *(base)*
Dreischichtplatte *f* three-layer board
Dreistoffgemisch *n* ternary (three-component) mixture
Dreistofflegierung *f* ternary (three-component) alloy
Dreistoffsystem *n* ternary (three-component) system
Dreistufenbleiche *f (pap)* three-stage bleaching
Dreistufen-Gegenstromwäsche *f* three-stage countercurrent washing
Dreistufenverdampfer *m s.* Dreikörperverdampfer
Dreistufenwäsche *f* three-stage washing
Dreiwalzenkalander *m* three-bowl (three-roll) calender
Dreiwalzenmühle *f*, **Dreiwalzenstuhl** *m* three-roll mill
Dreiweg[e]hahn *m (lab)* three-way cock, 3-way cock; *s.* Dreiwegventil
  D. mit Bohrung senkrecht zur Achse T-bore stopcock
  D. nach Czako T-shape 120° bore stopcock
Dreiwegventil *n* three-way valve
dreiwertig trivalent, tervalent
Dreiwertigkeit *f* trivalency, tervalency
dreizählig 1. *(cryst)* triad, threefold; 2. *s.* dreizähnig
dreizähnig tridentate *(ligand)*
Dreizellenapparat *m* **zur Elektrodialyse** three-compartment electrodialyzing device, three-chamber cell
Dreizentrenbindung *f* three-centre bond
Dreizentrenorbital *n* three-centre orbital
Dressler-Muffel *f (ceram)* Dressler muffle
Dressler-Ofen *m (ceram)* Dressler kiln
Drewboy-Scheider *m (coal)* Drewboy separator
Drift *f* drift *(as of zero point)*
drillen *(agric)* to drill *(e.g. fertilizers)*
drillfähig *(agric)* drillable *(as of fertilizers)*
Drillfähigkeit *f (agric)* drillability *(as of fertilizers)*
Drillingspumpe *f* triplex pump
Drillometer *n* drillometer
Drittluft *f* tertiary air
Droge *f* drug
  pflanzliche D. vegetable drug
  tierische D. animal drug
Drogenkunde *f* pharmacognosy, pharmacogn[os]ia
Drogenkundler *m* pharmacognosist
drogenkundlich pharmacognostic
Droseron *n* droserone, 3,5-dihydroxy-2-methyl-1,4-naphthoquinone
Drosselklappe *f*, **Drosselklappenventil** *n* butterfly valve
drosseln to slow down, to choke, to throttle
Drosselpfropfen *m* porous plug
Drosselscheibe *f* orifice plate *(flow measurement)*
Drosselstelle *f* constriction, restriction, throttle
Drosselventil *n* butterfly valve
Druck *m* 1. pressure; 2. printing *(process)*; print *(printed material)*
  D. am Umfang peripheral pressure
  D. auf Kammzug *(text)* top printing
  D. der expandierenden Gaskappe *(petrol)* gas cap drive
  D. in der Düse *(plast)* die pressure
  atmosphärischer D. atmospheric (air) pressure
  barometrischer D. barometric pressure
  flexografischer D. flexographic printing, flexography
  innerer D. intrinsic (cohesion) pressure
  kritischer D. critical pressure
  lithografischer D. lithographic printing, lithography
  osmotischer D. osmotic pressure, OP
  reduzierter D. *s.* verminderter D.
  reiner (scharfer) D. clean print
  verminderter D. reduced pressure
Druckabfall *m* pressure drop
  D. infolge Reibung friction drop
Druckanschluß *m* pressure port
Druckanstieg *m* pressure increase, increase in pressure

Druckaufnahmefläche *f (plast)* pressure pad
Druckausgleich *m* pressure compensation
Druckausgleichsgefäß *n* pressure-compensating vessel
Druckball *m* pressure bulb
Druckbegrenzungsventil *n* pressure relief valve
Druckbehälter *m* 1. pressure vessel (tank); 2. *(pap)* pressure container (accumulator)
Druckbirne *f s.* Druckfaß
Druckbombe *f* bomb
Druckdecke *f s.* Drucktuch
Druckdeckenwäscher *m s.* Drucktuchwäscher
Druckdestillat *n* pressure distillate, P. D.
Druckdestillation *f* pressure distillation
druckdicht pressure-tight
Druckdifferential *n* pressure differential
Druckdifferenz *f* pressure difference
Druckdurchtränkung *f (pap)* pressure impregnation, penetration under pressure, forced penetration
Druckdüse *f* pressure nozzle
Druckeigenschaften *fpl (pap)* printing properties
Druckeintritt *m* pressure port
drucken to print
drücken 1. to press, to push *(e. g. a lever, a button)*; 2. to push out, to discharge *(e. g. coke from a coke oven)*; 3. to blow *(a liquid into a reservoir)*; 4. to squeeze, to extract under pressure; 5. *(min tech)* to depress *(to cause to sink)*
Drucken *n* printing
Druckenergie *f* pressure energy
Druckentnahmestelle *f* pressure tap
Drücker *m (min tech)* depressant
Druckereihilfsmittel *n* printing additive
Druckerhöhung *f* pressure increase, increase in pressure
Druckerschwärze *f* printing (printer's) ink
Druckextraktion *f* extraction under pressure
  D. mit Benzol benzene-pressure extraction
druckfähig *(pap)* printable
Druckfähigkeit *f (pap)* printability
Druckfarbe *f* printing ink
  D. für Flachdruck planographic [printing] ink
  D. für Heliogravüre photogravure ink
  D. für Hochdruck typographic [printing] ink
  D. für Kupferstich copperplate engraving ink
  D. für Offsetdruck offset [printing] ink
  D. für Rotationstiefdruck rotogravure ink
  D. für Siebdruck screen-process ink
  D. für Silk-Screen-Druck silk-screen ink
  D. für Stahlstich steel-plate [engraving] ink
  D. für Steindruck lithographic [printing] ink, litho ink
  kurze D. short ink
  lange D. long ink
Druckfärben *n* pressure dyeing
Druckfarbenaufnahmevermögen *n (pap)* ink receptivity
Druckfarbenbindemittel *n (pap)* ink binder
Druckfarbenentfernung *f* deinking *(of waste paper)*
Druckfaß *n* blowcase, acid egg
Druckfestigkeit *f* compressive (compression) strength, strength in compression
Druckfeuerbeständigkeit *f* refractoriness under load
Druckfilter *n* pressure filter, pressure-filtration funnel
Druckfilternutsche *f* pressure nutsche
Druckfiltration *f*, **Druckfiltrieren** *n* pressure filtration
Druckflasche *f* pressure cylinder
Druckflüssigkeitsspeicher *m* hydraulic accumulator
Druckgaserzeuger *m* pressurized-gas producer
Druckgasflasche *f s.* Druckflasche
Druckgasgenerator *m s.* Druckgaserzeuger
Druckgefälle *n* pressure difference
druckgießen to [pressure-]diecast
Druckgießen *n* [pressure] diecasting, pressure casting
Druckgießform *f* diecasting (pressure-casting) die
Druckgießmaschine *f* [pressure-]diecasting machine
  D. für Kaltkammerverfahren cold-chamber [diecasting] machine
  D. für Warmkammerverfahren hot-chamber [diecasting] machine
Druckgießverfahren *n* diecasting process
Druckgrün *n* chrome green *(a mixture of iron blue and chrome yellow)*
Druckgrund *m* stock *(for printing)*
Druckguß *m* [pressure] diecasting, pressure casting ✦ **im D. herstellen** *s.* druckgießen
Druckgußform *f s.* Druckgießform
Druckgußlegierung *f* diecasting alloy
Druckgußmaschine *f s.* Druckgießmaschine
Druckgußstück *n*, **Druckgußteil** *n* [pressure] diecasting
Druckgußverfahren *n s.* Druckgießverfahren
Druckhärte *f* indentation hardness
Druckhöhe *f* pressure head
Druckholz *n* compressed (compression) wood
Druckhub *m* delivery (discharge) stroke *(of a pump)*
Druckhydrierung *f* hydrogenation under pressure
  spaltende D. *(petrol)* hydrocracking
Druckimprägnierung *f (pap)* pressure impregnation, penetration under pressure, forced penetration
Druckkammer *f* pressure chamber
Druckkessel *m* 1. pressure vessel (tank); 2. pressure pot *(paint spraying technique)*
Druckknopf *m* push button
Druckknopfschalter *m* push-button switch
Druckkochen *n* pressure boil[ing], boiling under pressure
Druckkörper *m (filtr)* pressure case (cylinder)
Druckkristallisation *f* piezocrystallization
Druckkühler *m* pressure cooler
Druckleiste *f (plast)* pressure pad
Druckleitung *f* pressure line; delivery line *(of a pump)*

drucklos without [the use of] pressure, pressureless
Druckluft *f* compression (compressed) air
druckluftbetätigt air-operated, air-driven
Druckluftdüse *f* air-atomizing (gas-atomizing, two-fluid) nozzle
Drucklufteintritt *m* blow port *(for blowing-off the filter cake)*
Druckluft-Flotationsapparat *m* dissolved-air floatation machine
Druckluftförderung *f* air lifting
Druckluftförderverfahren *n* air-lift process
Druckluftformen *n* **mit Vorstreckung** *(plast)* plug-assist pressure forming
Druckluftgießmaschine *f* air-operated diecasting machine
Drucklufttheber *m* air lift, mammoth (air-lift) pump
Drucklufthebersystem *n* air-lift system
Druckluftleitung *f* compressed-air line
Druckluftpistole *f* blow gun
Drucklufttrüttler *m* pneumatic (air-driven) vibrator
Druckluft-Schwefelverbrennungsofen *m* *(pap)* spray-type sulphur burner
Druckluftspritzen *n* *(coat)* compressed-air spraying
Druckluftventil *n* pneumatic (air) valve
Druckluftvernebler *m* *(agric)* power[-operated] sprayer
Druckluftversprüher *m* pneumatic (auxiliary-fluid) atomizer; *(pap)* spray[ing] gun *(for producing sulphur dioxide)*
Druckluftversprühung *f* pneumatic [nozzle] atomization, auxiliary-fluid atomization
Druckluftvibrator *m* pneumatic (air-driven) vibrator
Druckluftzelle *f* dissolved-air floatation machine
Druckluftzerstäuber *m* *s.* Druckluftversprüher
Druckluftzerstäubung *f* *s.* Druckluftversprühung
Druckmesser *m* pressure gauge, manometer
  piezoelektrischer D. piezometer
Druckminderer *m* *s.* Druckminderungsventil
Druckminderung *f* pressure drop
Druckminder[ungs]ventil *n* [pressure-]reducing valve
Druckmischer *m* bubbler
Drucknutsche *f* pressure nutsche
Druckpapier *n* print[ing] paper
Druckpaste *f* print[ing] paste
Druckpumpe *f* pressure pump
Druckreibungshöhe *f* discharge friction head
Druckrohr *n* pressure pipe (tube); delivery pipe *(of a pump)*
Drucksack *m* *(plast)* pressure bag
Drucksackmethode *f* *(plast)* pressure-bag moulding
Drucksäurebehälter *m* *(pap)* pressure container (accumulator)
Druckschärfe *f* *(text)* sharpness in print outline
Druckscheibe *f* *(plast)* pressure pad
Druckschlauch *m* pressure tubing
Druckschwingungsdämpfer *m* pulsation damper, [pulsation] snubber
Druckseite *f* delivery side *(of a pump)*
Druckseparator *m* *(pap)* selectifier [screen]
Drucksintern *n* sintering under pressure, hot pressing
Druckspannung *f* compressive stress
Druckspeicher *m* pressure tank
  hydraulischer D. hydraulic accumulator
druckspülen to jet
Druckstange *f* coke pusher ram *(for discharging coke)*
Drucksteigerung *f* pressure increase, increase in pressure
Drucksterilisator *m*, **Drucksterilisierapparat** *m* autoclave sterilizer
Druckstoß *m* pressure surge; water hammer, hydraulic shock *(as in an evaporator)*
Druckstrahlpumpe *f* injector
Druckströmung *f* pressure flow
Drucktaste *f* push button
Druckträger *m* stock *(for printing)*
Drucktuch *n* *(text)* blanket
Drucktuchwäscher *m* *(text)* blanket washer
Druckturm *m* pressure tower
Druckventil *n* pressure [control] valve; delivery valve *(of a pump)*
Druckverdampfer *m* pressure evaporator
Druckverdickungsmittel *n* *(text)* print[ing] thickener
Druckverfahren *n* printing process
Druckverformung *f* 1. *(phys)* [permanent] compression set; 2. *(plast, rubber)* compression moulding
  bleibende D. *s.* Druckverformung 1.
Druckverformungsrest *m* *s.* Druckverformung 1.
Druckvergaser *m* pressurized gas producer
Druckvergasung *f* elevated-pressure gasification
Druckverhältnis *n* compression ratio
Druckverlust *m* pressure drop
  trockener D. dry pressure drop
  trockener D. eines Kolonnenbodens dry-plate pressure drop
Druckversprüher *m* pressure atomizer
Druckversprühung *f* pressure atomization
Druckwalze *f* 1. printing roll; 2. *(tann)* grip roll; 3. pressure (compression) roll
Druckwasserreaktor *m* pressurized-water reactor, PWR
Druckwasserstoffraffination *f* hydrorefining
Druckwelle *f* blast
Druckzerstäuber *m* *s.* Druckversprüher
Druckzerstäubung *f* *s.* Druckversprühung
Druckzylinder *m* *(plast)* pressure cylinder
Drummond-Kalklicht *n* Drummond's limelight
Druse *f* *(geol)* druse, geode, vug[g], vugh
Drusenraum *m* / **miarolitischer** *(geol)* miarolitic cavity
Dry-blend-Strangpressen *n* *(plast)* dry-blend extrusion
D's *s.* Duanten

DTA *s.* Differentialthermoanalyse
Dualzerfall *m* branched (multiple) disintegration (decay), branching
Duanten *mpl (nucl)* dees *(D-shaped electrodes in a cyclotron)*
Dubbs-Krackprozeß *m* Dubbs process
Dublett *n* doublet *(a spectrum line having two close components)*; duplet *(the structure in which two atoms share a pair of electrons)*
Dublettaufspaltung *f* doublet splitting
Dublettsystem *n* doublet system
Dublett-Term *m* doublet term
Dublettzustand *m* doublet state
Duff-Reaktion *f* Duff reaction *(formylation of phenol)*
Dufour-Effekt *m (phys chem)* Dufour effect
Dufrenoysit *m (min)* dufrenoysite *(a natural sulpharsenide of lead)*
Duft *m* [pleasant] smell, odour, fragrance, aroma, scent
duftend sweet-smelling, sweet-scented, fragrant, aromatic
Duftlockstoff *m* scent attractant
Duftträger *m* perfume carrier
Dükerzulauf *m* siphon feed *(of a clarifier)*
duktil ductile
Duktilität *f* ductility
Dulong-Petit-Regel *f (phys chem)* Dulong and Petit's law
Dulzit *m* $CH_2OH[CHOH]_4CH_2OH$ dulcitol, dulcite *(a sugar alcohol)*
Dumortierit *m (min)* dumortierite *(a neso-subsilicate)*
Dung *m* farm manure, dung
Düngelanze *f* soil injector
Düngemaschine *f* fertilizing machine
Düngemischkalk *m* compound lime fertilizer
Düngemittel *n* fertilizer, *(esp of animal excreta also)* [farm] manure
Düngemittelbedarf *m* fertilizer needs (requirements)
Düngemittelindustrie *f* fertilizer industry
Düngemitteltechnologie *f* fertilizer technology
Düngemittelverbrauch *m* fertilizer consumption
düngen to fertilize, to dress; to manure, to dung
mit Kalk d. to lime, to fertilize with lime
Dünger *m* fertilizer, *(esp of animal excreta also)* [farm] manure
anorganischer (mineralischer) D. mineral fertilizer
Düngerbedarf *m s.* Düngemittelbedarf
Düngermühle *f* fertilizer mill
Düngernährstoff *m* fertilizer nutrient
Düngerphosphor *m* fertilizer phosphorus
Düngerstickstoff *m* fertilizer nitrogen
Düngerwalze *f* fertilizer roll
Düngerwert *m* manurial value
Düngesalz *n* fertilizing (manure) salt
Düngung *f* fertilization, [fertilizer] dressing, *(esp using animal excreta)* manuring ✦ **auf D. ansprechen (reagieren)** to respond to fertilizing
D. durch Flugzeuge aeroplane fertilization
aviotechnische D. *s.* D. durch Flugzeuge
Düngungsempfehlung *f* fertilization recommendation
Düngungspflug *m* fertilizing plough
Dunit *m* dunite *(a granitoid igneous rock)*
Dunkelfärbung *f* darkening
Dunkelfeldbeleuchtung *f* dark-field (dark-ground) illumination
Dunkelfeldmikroskop *n* dark-field microscope
Dunkelkammer *f* dark-room
Dunkelkammerbeleuchtung *f* dark-room illumination
Dunkelmalz *n* dark malt
Dunkelöl *n* black oil
Dunkelraum *m* dark space
Astonscher D. Aston dark space
Crookesscher (Hittorfscher) D. Crookes (Hittorf, cathode) dark space
innerer D. *s.* Crookesscher D.
Dunkelreaktion *f* dark reaction
Dunkelrotglut *f* dull redness (red heat)
Dunkelstrom *m* dark current
Dunkelwerden *n* darkening
Dunlop-Pendel *n (rubber)* Dunlop pendulum
Dunlop-Tripsometer *n (rubber)* Dunlop tripsometer
Dunlop-Verfahren *n (rubber)* Dunlop process
Dünnablauge *f (pap)* dilute (weak) black liquor
Dunnachie-Ofen *m (ceram)* Dunnachie kiln
Dünndruckpapier *n* bible paper, [Oxford] India paper
Dünnfilm *m* thin film
dünnflüssig thin, highly liquid (fluid); watery; thin-bodied *(e. g. oil)*
Dünnglas *n* thin [sheet] glass; micro-glass *(for use in microscopy)*
dünngriffig sein *(pap)* to bulk low
Dünnlauge *f s.* Dünnablauge
Dünnlaugeneintritt *m* feed liquor inlet *(as on an evaporator)*
Dünnmaische *f (ferm)* thin (lauter) mash
Dünnpergamin[papier] *n* thin glassine
Dünnpostpapier *n* thin post (letter) paper
Dünnsaft *m (sugar)* thin juice
Dünnsaftfilter *n (sugar)* thin-juice filter
Dünnsaftfiltration *f (sugar)* thin-juice filtration
Dünnsaftvorwärmer *m (sugar)* thin-juice heater
Dünnschicht *f* thin layer, [thin] film
Dünnschichtabsorber *m* wetted-wall absorber
Dünnschichtchromatografie *f* thin-layer chromatography, TLC
Dünnschichtchromatogramm *n* thin-layer chromatogram
Dünnschichtdestillation *f* thin-layer distillation
Dünnschichtdestillator *m* thin-layer distillator, film still
Dünnschichtelektrophorese *f* thin-layer electrophoresis
Dünnschichtfilm *m* thin emulsion film
Dünnschichttrockner *m* film dryer
Dünnschichtverdampfer *m* thin-layer (film-type) evaporator

D. mit rotierenden Wischern agitated-film evaporator
Dünnschichtverfahren *n (distil)* thin-layer method
Dünnschliff *m (min)* thin section
polierter D. polished thin section
Dünnschlifftechnik *f* thin-section technique
Dünnschliffverfahren *n* thin-section method
Dünnschnitt *m* thin section *(microscopy)*
Dünnsole *f* weak brine
Dünnstoffbleiche *f (pap)* low-density bleaching
Dünnstoff-Turmbleiche *f (pap)* low-density tower bleaching
dünnwandig thin-walled
Dunst *m* 1. *(of droplets)* damp, haze; 2. *(of solid particles)* fume, smoke; 3. *(unpleasant odour)* smell
Dunsthaube *f* hood, air dome
Duosolanlage *f* Duo-sol solvent extraction plant
Duosolextraktion *f* Duo-sol extraction
Duosolverfahren *n* Duo-sol [solvent extraction] process, two-solvent process
Duplexdruck *m (text)* double-face printing
Duplexkarton *m* duplex cardboard
Duplexmischer *m* duplex blender
Duplexpapier *n* duplex paper; duplex transfer paper
Duplexpappe *f* duplex (two-ply) board
Duplexpumpe *f* duplex (two-throw) pump
Duplexschmelzverfahren *n* duplex process *(for steel-making)*
Duplexstahl *m* duplex steel
Duplexverfahren *n s.* Duplexschmelzverfahren
Du-Pont-Biegeprüfmaschine *f*, **Du-Pont-Ermüdungsmaschine** *f (rubber)* Du Pont machine
Du-Pont-Grasselli-Abriebmaschine *f (rubber)* Du Pont-Grasselli-Williams machine
Du-Pont-Kettenermüdungsmaschine *f s.* Du-Pont-Biegeprüfmaschine
Du-Pont-Verfahren *n* Du Pont process *(ammonia oxidation)*
Durain *m* durain *(a constituent of coal)*
Durangit *m (min)* durangite *(aluminium sodium arsenate fluoride)*
durcharbeiten 1. to work (knead) thoroughly *(e.g. a dough)*; 2. to homogenize *(an emulsion)*; 3. *(text)* to pole; 4. *(tann)* to pummel *(hides)*
im Faß d. *(tann)* to drum
durchbeißen *(tann)* to penetrate *(of a tan)*
Durchbelüftung *f* through [air] circulation
durchbeuteln to bolt, to sift *(e.g. flour)*
durchbiegen / sich to sag
Durchbiegung *f* sag[ging]
durchblasen to blow through
durchbluten *(dye, coat, tann)* to bleed [through], to strike through
Durchbluten *n (dye, coat, tann)* bleeding, strike-through
durchbohren to bore [through]
Durchbohrung *f* boring
Durchbruch *m* break-through *(in sorption processes)*
Durchbruchskurve *f* break-through curve *(of sorption processes)*
Durchbruchspunkt *m* break point *(ion exchange)*
durchdringbar penetrable, permeable
Durchdringbarkeit *f* penetrability, permeability
durchdringen to penetrate, to permeate
ganz (vollständig) d. to impenetrate
durchdringend penetrating *(e.g. an odour)*
Durchdringung *f* penetration, permeation
Durchdringungsfähigkeit *f s.* Durchdringungsvermögen
Durchdringungskomplex *m* penetration complex
Durchdringungskraft *f s.* Durchdringungsvermögen
Durchdringungsmittel *n (text)* penetrating agent
Durchdringungsvermögen *n* penetrating power, permeativity
Durchdringungszwillinge *mpl (cryst)* penetration twins
Durchdringwahrscheinlichkeit *f (nucl)* penetration probability
durchdrücken to force through
Durchfalloch *n*, **Durchfallöffnung** *f* drop hole
Durchfärbbarkeit *f (text)* penetrability
Durchfärbemittel *n (text)* penetrating agent
durchfärben 1. *(of human agent)* to dye thoroughly (completely); 2. *(of a dyestuff)* to penetrate
Durchfärbung *f* penetration dyeing
durchfeuchten to moisten thoroughly, to humidify; to macerate *(e.g. flax)*
durchfließen to flow (pass, run) through
Durchfluß *m* 1. flow, flowing-through, passage; 2. *(glass)* throat, flow hole; 3. *s.* Durchflußstrom
Durchflußbeiwert *m s.* Durchflußkoeffizient
Durchflußgeschwindigkeit *f* flow rate
Durchflußkoeffizient *m* flow (discharge) coefficient
Durchflußkurve *f* flow curve
Durchflußmengenmesser *m* flowmeter, rate meter
D. für Flüssigkeiten liquid (stream) meter
elektromagnetischer D. magnetic meter
Durchflußmengenmeßgerät *n s.* Durchflußmengenmesser
Durchflußmengenmessung *f* flow measurement
Durchflußpasteurisation *f (food)* continuous pasteurization
Durchflußstrom *m* throughput, flow rate
Durchflußstrom-Stellglied *n* flow controller
Durchflußwiderstand *m (filtr)* resistance to flow
Durchflußzahl *f* flow (discharge) coefficient
durchfressen to eat through, to corrode
durchführen to carry out, to run, to perform *(an experiment)*
Durchgang *m* passage
Durchgangsofen *m (food)* continuous bake oven
Durchgangszeit *f* retention (hold-up, holding) time, holding period
durchgasen *(agric, food)* to fumigate
Durchgasung *f (agric, food)* fumigation
Durchgasungsmittel *n (agric, food)* fumigant, fumigator
durchgerben to tan thoroughly

Durchgerbung *f (tann)* leathering
Durchgerbungszahl *f* tanning index
Durchhang *m* 1. sag[ging]; 2. *(phot)* region of underexposure, toe, foot *(of the characteristic curve)*
durchhängen to sag
Durchhängen *n* sag[ging]
durchkneten to knead [thoroughly]
durchkochen to cook thoroughly
Durchlaß *m* 1. passage; 2. *(glass)* throat, flow hole
bodengleicher (normaler) D. *(glass)* straight throat
tiefer (tiefliegender) D. *(glass)* sump (drop, submarine, submerged) throat
versenkter D. *s.* tiefer D.
Durchlaß-Abdeckstein *m (glass)* throat cover
Durchlaßgrad *m* transmission ratio, transmittance, transmittancy *(optics)*
durchlässig permeable, porous, porose
einseitig d. semipermeable
für Wärmestrahlen d. diathermanous, diathermic
Durchlässigkeit *f* permeability
optische D. transmittance
Durchlaßquerschnitt *m* flow area
Durchlaß-Seitenstein *m (glass)* throat cheek, dice (sleeper) block
Durchlauf *m* passage
durchlaufen to pass [through], to strain, to percolate, to filter
d. lassen to pass [through], to strain, to percolate, to filter
Durchlaufen *n* percolation *(passage of a liquid through a filtering medium)*
Durchlaufentwicklung *f* overrun development *(paper chromatography)*; continuous development *(thin-layer chromatography)*
Durchlaufgeschwindigkeit *f* flow rate
Durchlaufglühofen *m (met)* continuous-annealing furnace
Durchlaufglühung *f (met)* continuous annealing
Durchlaufkühlofen *m (glass)* continuous-annealing lehr
Durchlaufkühlung *f (glass)* continuous annealing
Durchlaufmahlung *f* open-circuit grinding
Durchlaufofen *m* continuous furnace
Durchlauftechnik *f s.* Durchlaufentwicklung
Durchlaufverdampfer *m* single-pass (one-pass) evaporator
Durchlaufzeit *f* retention (hold-up, holding) time, holding period
durchleiten to pass [through]
durchleuchten [/ mit Röntgenstrahlen] to X-ray
Durchlicht *n* transmitted light
Durchlichtmethode *f* transmitted-light technique
durchlüften to aerate
Durchlüftung *f* aeration; through [air] circulation *(in a dryer)*
Durchmessereffekt *m (phot)* Eberhard effect
durchmischen to intermix, to mix (blend) together, to interfuse; to mix thoroughly
Durchmischung *f* intermixture, interfusion; thorough mixing
durchnumerieren *(nomencl)* to number
Durchnumerierung *f (nomencl)* numbering
Durchnumerierungssystem *n (nomencl)* numbering system
durchpressen to squeeze through
durchräuchern 1. *(food)* to smoke thoroughly; 2. to fumigate *(a chamber)*
Durchreißen *n (pap)* further tearing
Durchreißfestigkeit *f (pap)* tear (tearing) resistance (strength), tear propagation strenght, resistance to [further] tearing
Durchreißfestigkeitsprüfung *f (pap)* tear[ing] test, tear propagation test
Durchreißprüfer *m* **nach Elmendorf** *(pap)* Elmendorf tester
durchrühren to stir, to agitate
Durchrühren *n* stirring, agitation
durchsacken to sag
Durchsacken *n* sag[ging]
Durchsatz *m* throughput, *(relating to a liquid also)* flow rate
Durchsatzstrom *m s.* Durchsatz
Durchscheinbarkeit *f* translucence, translucency
durchscheinen to show through *(of printing ink)*
Durchscheinen *n* show-through *(of printing ink)*
durchscheinend translucent, translucid
nicht d. opaque
Durchschlag *m* 1. colander, cullender; 2. breakdown *(of a dielectric)*
durchschlagen 1. *(dye, coat, tann)* to bleed [through], to strike through; 2. to puncture *(a dielectric)*
Durchschlagen *n* 1. *(dye, coat, tann)* bleeding, strike-through; 2. puncture *(of a dielectric)*
Durchschlagfestigkeit *f* **/ dielektrische (elektrische)** dielectric (breakdown, puncture) strength, electric strength
Durchschlagpapier *n* carbon copy[ing] paper, copy[ing] paper, copyings, *(Am)* manifold paper
Durchschlag[s]spannung *f* breakdown voltage
durchschlämmen *(soil)* to percolate
Durchschlämmung *f (soil)* percolation
Durchschnittsausbeute *f* average yield
Durchschnittsbetrag *m* average amount
Durchschnittsfettgehalt *m* average fat content
Durchschnittsprobe *f* average sample
Durchschnittswert *m* average (mean) value, mean
Durchschreib[e]papier *n* carbon (carbonic, carbonized) paper
Durchschubofen *m (ceram)* sliding-bat (pushed-bat) kiln
durchschütteln to shake
Durchschütteln *n* shake, shaking
durchseihen to strain, to percolate
Durchseihen *n* straining, [per]colation
durchsetzen to put through, to handle *(a definite quantity of material)*
durchsetzt / mit Streifen *(min, geol)* interbanded

Durchsicht *f* 1. examination, inspection; 2. *(pap)* look-through
durchsichtig transparent
Durchsichtigkeit *f* transparency
durchsickern 1. *(filtr)* to trickle through, to seep through, to percolate, to strain; 2. to leak
d. lassen to pass [through], to percolate, to strain *(a liquid)*
Durchsickern *n* 1. percolation, seepage; 2. leak
durchsieben to screen *(e. g. coal, gravel)*; to sieve, to sift *(e. g. flour)*
durchspülen to wash thoroughly (through)
Durchspülung *f* thorough (through) washing
Durchstrahlungsaufnahme *f (cryst)* front reflection pattern
Durchströmquerschnitt *m* flow area
durchtränken to soak, to impregnate, to imbibe, to penetrate, to saturate
Durchtränkung *f* soaking, impregnation, imbibition, penetration, saturation
Durchtränkungsgeschwindigkeit *f (pap)* rate of penetration *(of the chips)*
Durchtränkungsgrad *m (pap)* extent of penetration *(of the chips)*
Durchtränkungszeit *f (pap)* penetration time (period)
Durchtritt *m* passage
Durchwachsung *f (cryst)* intergrowth
durchwärmen *(glass)* to reheat *(the parison)*
Durchwärmen *n (glass)* reheat *(of the parison)*
durchwaschen to wash through
Durchwaschen *n* thorough (through) washing
durchweichen to soak, to wet [through]
Durchweichzone *f* soaking zone *(of an annealing furnace)*
Durchzeichenpapier *n* tracing paper
Durchzeichnung *f* **der Schatten** *(phot)* shadow detail
Durit *m* durain *(a constituent of coal)*
Durol *n* $C_6H_2(CH_3)_4$ durene, durol, 1,2,4,5-tetramethylbenzene
Duromer[es] *n* duromer
Durometer *n* durometer, hardness tester (meter)
Durometerhärte *f* durometer (Shore) hardness
Duroplast *m* thermosetting plastic (resin), thermoset [resin]
Durville-Gießverfahren *n* Durville casting process *(foundry)*
Durylsäure *f* durylic acid, 2,4,5-trimethylbenzoic acid
Duschrinne *f* runoff gutter for wet cooling *(in margarine making)*
Duschverfahren *n* ice-water (wet-cooling) method *(in margarine making)*
Düse *f* 1. nozzle, jet; *(plast)* die; 2. orifice *(in a steam trap)*
D. zum Strangpressen von Folien *(plast)* flat die
rotierende D. rotating (rotary, spinning) nozzle
Düsenbeiwert *m* nozzle coefficient
Düsenblasverfahren *n (glass)* jet process, *(specif)* air-blowing process *or* steam-blowing process
Düsenboden *m s.* Düsenlochboden
Düsenbohrung *f (text)* spinneret hole
Düsenebene *f (met)* tuyère level
Düsenfärbemaschine *f (text)* jet dyeing machine
Düsenfärbung *f (text)* dope (spin) dyeing
düsengefärbt *(text)* dope-dyed, spin-dyed, spun-dyed, mass-dyed, solution-dyed
Düsenhals *m* nozzle throat
Düsenhalter *m (plast)* die adapter
Düsenkeller *m* underjet cellar *(of an underjet coke oven)*
Düsenkondens[at]ableiter *m* orifice trap
Düsenkörper *m (plast)* die body, *(Am)* die base
Düsenkraftstoff *m s.* Düsentreibstoff
Düsenleitungen *fpl* underjet piping *(of an underjet coke oven)*
Düsenlochboden *m* base of the bushing *(glass-fibre manufacture)*
Düsenmischer *m* nozzle mixer
Düsenöffnung *f (met)* tuyère opening
Düsenpaßstück *n (plast)* die adapter
Düsenplanrahmen *m (text)* single-layer jet stenter
Düsensicherheitsventil *n* nozzle-type relief valve
Düsenstock *m (met)* tuyère stock
düsentexturiert *(text)* air-bulked
Düsentreibstoff *m* fuel for jet planes, [turbo]jet fuel
Düsentrockner *m* jet dryer
Düsentrocknung *f* jet drying
Düsenverengung *f* nozzle throat
Düsenversprüher *m* nozzle atomizer
Düsenversprühung *f* nozzle atomization
Düsenzerstäuber *m* nozzle atomizer
Düsenzerstäubung *f* nozzle atomization
Düsenzone *f (met)* tuyère zone
Dutch-Flüssigkeit *f* Dutch liquid *(1,2-dichloroethane)*
Dwight-Lloyd-Sintermaschine *f* Dwight-Lloyd sintering machine
Dynamik *f* dynamics
Dynamit *n* dynamite
Dynamometamorphose *f (geol)* dynamometamorphism, dynamic metamorphism
Dypnon *n* $C_6H_5C(CH_3){=}CHCOC_6H_5$ dypnone, α-methylchalcone, 1,3-diphenyl-2-buten-1-one
Dyskrasit *m (min)* dyscrasite *(silver antimonide)*
Dyson-Notationssystem *n (nomencl)* Dyson [notation] system
Dysprosium *n* Dy dysprosium
Dysprosiumazetat *n* $Dy(CH_3COO)_3$ dysprosium acetate
Dysprosiumbromat *n* $Dy(BrO_3)_3$ dysprosium bromate
Dysprosiumchlorid *n* $DyCl_3$ dysprosium chloride
Dysprosiumchromat *n* $Dy_2(CrO_4)_3$ dysprosium chromate
Dysprosiumkarbonat *n* $Dy_2(CO_3)_3$ dysprosium carbonate
Dysprosiumnitrat *n* $Dy(NO_3)_3$ dysprosium nitrate
Dysprosiumorthophosphat *n* $DyPO_4$ dysprosium orthophosphate, dysprosium phosphate
Dysprosiumoxid *n* Dy_2O_3 dysprosium oxide

Dysprosiumselenat *n* $Dy_2(SeO_4)_3$ dysprosium selenate
Dysprosiumsulfat *n* $Dy_2(SO_4)_3$ dysprosium sulphate
Dysprotid *n* proton don[at]or, protonic acid

E

Eagle-Mühle *f* Eagle mill *(a fluid-energy mill)*
Easy-care-Ausrüstung *f (text)* easy-care finish
Easy-Processing-Channel-Ruß *m* easy processing channel black, EPC black
Eau de Cologne *n(f)* eau de cologne, cologne [water]
Eau de Javelle *n(f)* eau de Javel[le], Javel[le] water *(aqueous solution of sodium or potassium hypochlorite)*
Eau de Labarraque *n(f)* eau de Labarraque *(aqueous solution of sodium hypochlorite)*
Ebene *f* plane
schiefe E. slope
Eberhard-Effekt *m (phot)* Eberhard effect
e-Bindung *f* e-bond, equatorial bond
Ebullioskop *n* ebullioscope, ebulliometer
Ebullioskopie *f* ebullioscopy
ebullioskopisch ebullioscopic
EC *s.* Äthylzellulose
Echelettegitter *n* echelette grating *(a diffraction grating)*
Echelongitter *n* echelon grating *(a diffraction grating)*
Echimidinsäure *f* echimidinic acid *(a C_7 trihydroxy acid)*
Echinochrom A *n* echinochrome A, 2-ethyl-3,5,6,7,8-pentahydroxy-1,4-naphthoquinone
echt fast *(dye)*; genuine *(noble metal, gem)*; real, pure *(silk)*; genuine *(leather)*
äußerst e. *(dye)* exceedingly fast
mäßig e. *(dye)* moderately fast
Echtbase *f s.* Echtfarbbase
Echt-Bütten[papier] *n* genuine handmade paper, vat paper
Echt-Büttenpapier-Herstellung *f* papermaking by hand, handmade paper making
Echtdrahtverfahren *n (text)* conventional twisting
Echtfarbbase *f*, **Echtfärbebase** *f* fast [colour] base
Echtfärben *n*, **Echtfärberei** *f* fast dyeing
Echtfärbesalz *n* fast [colour] salt
echtfarbig fast-dyed
Echtfarbstoff *m* fast dye
echtgefärbt fast-dyed
Echtheit *f* fastness *(of dyes)*; genuineness *(of noble metals or gems)*
Echtheitsprüfung *f (dye)* fastness test
Echtlichtgelb *n* hydrazine yellow, tartrazine *(a pyrazole derivative)*
Echtneublau 3 R *n* fast blue 3 R, Meldola blue, basic blue 6
Echtorange *n* fast orange
Echtorangebase *f* fast orange base
Echtpergamentpapier *n* parchment paper, vegetable parchment
Echtrosa *n* fast pink
Echtrot *n* fast red
Echtrotbase *f* fast red base
Echtrot-GL-Base *f* fast red GL base
Echtsalz *n s.* Echtfärbesalz
Echtscharlach *m* fast scarlet
Echtscharlachbase *f* fast scarlet base
Echtscharlach-G-Base *f* fast scarlet G base
Echtwollgelb *n s.* Echtlichtgelb
Eckenfeuerung *f* tangential firing
Eckventil *n* angle valve
Ecruseide *f* ecru silk *(a partially degummed silk)*
E-Cu *s.* E-Kupfer
ED_{50} *s.* Dosis / mittlere effektive
Edeleanu-Extrakt *m* Edeleanu extract *(solvent extraction)*
Edeleanu-Verfahren *n* Edeleanu process *(solvent extraction)*
Edelerde *f* activated (active) earth
Edelgas *n* inert (noble, rare) gas
Edelgaskonfiguration *f* inert-gas [electron] configuration, inert-gas [electronic] structure, octet structure
Edelgasoktett *n* inert-gas octet
Edelgasrumpf *m* inert-gas core
Edelgasschale *f* inert-gas shell
Edelmetall *n* noble (precious) metal
Edelrost *m* patina
Edelstein *m* precious stone, gem [stone]
Edelzellstoff *m (pap)* [high] alpha pulp, processed (purified wood) pulp
Edingtonit *m (min)* edingtonite *(hydrous aluminium barium silicate)*
Edison-Akkumulator *m (el chem)* Edison accumulator (cell)
Edwards-Ofen *m (met)* Edwards roaster
E-Eisen *n* electrolytic iron
Effekt *m* effect
bathochromer E. *(dye)* bathochromic effect
dielektrischer E. dielectric effect
elektromerer E. *s.* mesomerer E.
elektrophoretischer E. electrophoretic effect
elektroviskoser E. electroviscous effect
fotoelektrischer E. photoelectric effect
hypsochromer E. *(dye)* hypsochromic effect
induktiver E. *(phys chem)* inductive (induction) effect, I effect
kataphoretischer E. *s.* elektrophoretischer E.
klopfhemmender (klopfhindernder) E. antiknock effect
lichtelektrischer E. *s.* fotoelektrischer E.
longitudinaler E. *s.* elektrophoretischer E.
magnetokalorischer E. magnetocaloric effect
mechanokalorischer E. mechanocaloric effect
mesomerer E. mesomeric (electromeric, resonance) effect
piezoelektrischer E. piezo[electric] effect
sterischer E. steric effect
synergistischer E. synergistic effect, synergism
thermoelektrischer E. thermoelectric effect
thermomagnetischer E. thermomagnetic effect

Effektgarn *n* effect yarn
effektiv effective
Effektlack *m* effect varnish
Effektpapier *n* fancy stained paper
Effloreszenz *f (min)* efflorescence
effloreszieren *(min)* to effloresce
Effusion *f* effusion
Effusionsgeschwindigkeit *f* rate of effusion
effusiv *(geol)* effusive
Effusivgestein *n* effusive rock
egal *(dye)* level
egalfärben *(text)* to level, to dye level
Egalfärben *n (text)* levelling, level dyeing
egalisieren *s.* egalfärben
Egalisierer *m*, **Egalisier[hilfs]mittel** *n (text)* levelling agent, level dyeing assistant
Egalisierung *f s.* 1. Egalfärben; 2. Egalität
Egalität *f (dye)* levelness
Eggonit *m (min)* eggonite, sterrettite *(scandium phosphate)*
E-Glas *n* E-glass, glass E *(a low-alkali borosilicate glass)*
Egoutteur *m (pap)* watermarking dandy [roll], dandy [roll]
Egoutteur[wasser]zeichen *n (pap)* dandy roll watermark
Egrenieren *n (text)* ginning
Egreniermaschine *f (text)* [cotton] gin
Ehlit *m (min)* ehlite, pseudomalachite, dihydrite *(a hydrous basic copper phosphate)*
eiabtötend ovicidal
Eialbumin *n* egg albumin
eichen to calibrate *(e. g. measuring apparatus)*; to adjust *(balance weights or measures)*
Eichengalle *f* oak gall
Eichenrinde *f (tann)* oak bark
Eichkraftstoff *m* reference fuel
Eichkurve *f* calibration curve, calibrating plot
Eichlösung *f* calibrating solution
Eichmarke *f s.* Eichstrich
Eichspannung *f* reference voltage
Eichstrich *m* calibration mark
Eichtreibstoff *m* reference fuel
Eichung *f* calibration *(as of measuring apparatus)*; adjustment *(of balance weights or measures)*
Eidotter *m(n) s.* Eigelb
Eieralbumin *n* egg albumin
Eierbrikett *n* ovoid
Eiergift *n* ovicide *(a kind of pesticide)*
Eieröl *n* egg-yolk oil
Eierschalenporzellan *n (ceram)* egg-shell porcelain
Eierschaligkeit *f (ceram)* egg-shelling *(of the glaze)*
Eigelb *n* [egg] yolk, vitellus
  flüssiges E. liquid egg yolk
Eigelbnachgare *f (tann)* egging
Eigelböl *n* egg-yolk oil
Eigenabsorption *f* self-absorption
Eigendiffusion *f* self-diffusion
Eigendrehimpuls *m* intrinsic angular momentum, spin *(of elementary particles)*
Eigenfarbe *f* self-colour
Eigenfrequenz *f* natural frequency *(of vibration)*
Eigenfunktion *f (phys chem)* eigenfunction
Eigengewicht *n* own weight
Eigenhalbleiter *m* intrinsic semiconductor
Eigenhalbleitfähigkeit *f* intrinsic semiconductivity, intrinsic electrical conductivity
Eigenhalbleitung *f* intrinsic semiconduction
eigenhärtend *(plast)* self-curing
Eigenionisation *f* autoionization
Eigenleitfähigkeit *f s.* Eigenhalbleitfähigkeit
Eigenleitung *f s.* Eigenhalbleitung
Eigenmasse *f* own weight
Eigenparität *f (nucl)* intrinsic parity
Eigenpolymerisation *f* homopolymerization
Eigenpotential *n* self-potential
Eigenpotentialkurve *f* self-potential curve
Eigenschaft *f* property
  abhäsive E. abhesiveness
  abweisende E. repellency
  additive E. additive property
  backtechnische E. *(food)* baking property (quality, value, characteristics)
  extensive E. extensive property
  intensive E. intensive property
  konstitutive E. constitutive property
  ölabweisende E. oil repellency
  periodische E. periodic property *(of the elements)*
  physikalische E. physical property
  thermische E. thermal property
  wasserabweisende E. water repellency
Eigenschwingung *f* natural vibration *(of molecules)*
Eigenwert *m* eigenvalue
Eigenzustand *m* eigenstate, characteristic (proper) state
Eiglobulin *n* egg globulin
Eiklar *n* [egg] albumen, glair[e]
Eikosan *n* $C_{20}H_{42}$ eicosane
Eikosansäure *f* $CH_3[CH_2]_{18}COOH$ eicosanoic acid, arachidic acid
Eikosantetraensäure *f* $C_{19}H_{31}COOH$ eicosane-tetraenoic acid
Eikosen-(9)-säure *f* 9-eicosenoic acid, gadoleic acid
Eimer *m* bucket *(as of an elevator)*
einachsig [/ optisch] [optically] uniaxial
einarbeiten to work in, to intermingle
  Wasserzeichen e. to watermark
einatomig [mono]atomic
Einatomigkeit *f* monoatomicity
Ein-Aus-Regelung *f* automatic-start-and-stop control
Einbadchrombrühe *f (tann)* one-bath chrome liquor
Einbadchromgerbung *f* one-bath [chrome] tannage
Einbadchrom[ier]verfahren *n (text)* one-bath chroming method
Einbadgerbung *f s.* Einbadchromgerbung
Einbadverfahren *n (text)* one-bath (single-bath) method

Einbandtrockner *m* single-conveyor dryer
einbasig monobasic *(acid)*
Einbau *m* 1. pack, fill *(as in a cooling tower)*; 2. insertion *(as of atoms into interstitial lattice sites)*; 3. incorporation *(as of nutrients into organic substances)*
E. von Deuterium (schwerem Wasserstoff) deuteration
einbauen to insert *(e. g. atoms into interstitial lattice sites)*; to incorporate *(e. g. nutrients into organic substances)*
Einbaugenerator *m* built-in producer
Einbauhöhe *f* **[/ lichte]** *(plast, rubber)* mould opening, daylight [opening]
Einbauteil *n* fill member *(as of a cooling tower)*
Einbauten *mpl* pack, fill *(as in a cooling tower)*; baffles *(as for directing a fluid stream)*
einbetten to embed; *(met)* to pack *(as with a carburizing powder)*; *(plast)* to pot, *(Am)* to encapsulate
Einbettungsmasse *f (ceram)* ground-mass, matrix; *(met)* packing material
Einbettungsmittel *n (met)* packing material; embedding medium *(microscopy)*
Einbettungswerkstoff *m (met)* packing material
einblasen to blow in
Einblasen *n* **/ seitliches** *(met)* side blowing
Einbrand *m* 1. *(ceram)* burning-in, firing-on, maturing, stoving; 2. *s.* Einmalbrand
Einbrennemaillack *m* stoving (baking) enamel
Einbrennemaille *f (incorrectly for)* Einbrennemaillack
Einbrennemaillelack *m*, **Einbrennemaillelackfarbe** *f s.* Einbrennemaillack
einbrennen to burn in, *(ceram also)* to fire on, to mature *(a coating)*; *(coat)* to stove, to bake
Einbrennen *n* burning-in
Einbrennfarbe *f* stoving paint
Einbrennlack *m* stoving varnish, *(Am)* baking varnish; *(if pigmented)* stoving lacquer
Einbrennlackierung *f* stove enamelling
Einbrennofen *m (coat)* stove
Einbrennverfahren *n (coat)* stoving process; *(ceram)* firing-on process
einbringen to introduce, to place
Beton e. to pour (place) concrete
Hackschnitzel e. *(pap)* to pack the chips
Einbringen *n* introduction, placing
E. der Hackschnitzel *(pap)* chip packing (filling), packing of the chips
nesterweises E. *(agric)* spot application
Einbringtiefe *f (agric)* depth of application
Eindampfapparat *m* evaporator
eindampfen to evaporate, to concentrate by evaporation, to boil down (away), to inspissate
zur Trockne e. to evaporate to dryness
Eindampfen *n* evaporation, concentration by evaporation, boildown, boiling-down, inspissation
Eindampfer *m* evaporator
Eindampfpfanne *f* evaporating (boiling-down) pan
Eindeckersiebmaschine *f* single-deck screen
Eindickapparat *m* thickener, concentrator
Eindickbütte *f (pap)* draining tank (chest), drainer
eindicken to thicken, to concentrate, to boil down, to inspissate, *(esp food)* to condense; *(pap)* to decker; *(coat)* to body
durch Erhitzen (Hitzebehandlung) e. *s.* thermisch e.
thermisch e. to heat-thicken; *(coat)* to heat-body; to durmolize, to calorize *(e. g. linseed oil)*
Eindicker *m* 1. thickener, concentrator; *(pap)* decker; 2. *s.* Eindickmittel
E. mit Mittelsäule centre-column[-supported] thickener
E. mit Randantrieb traction thickener
Eindickfilter *n* filter thickener
Eindickmaschine *f s.* Eindicker 1.
Eindickmittel *n* thickening agent, thickener
Eindickung *f* thickening, concentration, boildown, inspissation, *(esp food)* condensation; *(pap)* deckering; *(coat)* bodying
Eindickungsgrad *m* degree of thickening
Eindick[ungs]verhinderungsmittel *n (coat)* antilivering agent
eindimensional one-dimensional
eindosen *(food)* to can, to tin
eindrehen *(ceram)* to jolley
eindringen to enter [into], to penetrate, to permeate; to diffuse *(of liquids or gases)*; *(geol)* to intrude
Eindringen *n* entering, penetrating, penetration, permeation; diffusion *(of liquids or gases)*; *(geol)* intrusion
Eindringfähigkeit *f* permeativity
Eindringhärte *f s.* Eindruckhärte
Eindringtiefe *f* [depth of] penetration
Eindringungsmittel *n (text)* penetrating agent
Eindringvermögen *n* permeativity
eindrücken to blow in *(e. g. a gas into a vessel)*
Eindruckhärte *f* indentation hardness *(materials testing)*
eindunsten to evaporate
Eindunsten *n* evaporation
Eindunstung *f* **/ solare** solar evaporation
einebnen to level
Einebnungsstange *f* coal leveller bar, levelling bar
Einebnungsvorrichtung *f* levelling device
ein-ein-wertig uniunivalent
Einelektronbindung *f s.* Einelektronenbindung
Einelektronenatom *n* one-electron atom
Einelektronenaustauschreaktion *f* one-electron transfer process *(in radical reactions)*
Einelektronenbindung *f* one-electron (single-electron) bond, singlet link[age]
Einelektronenorbital *n* one-electron orbital
Einelektronenreduktion *f* one-electron reduction
Einelektronenzustand *m* one-electron state
einengen *s.* eindampfen
Einetagenpresse *f (rubber)* single-daylight (one-daylight) press
Einfachbindung *f* single bond
Einfachbindungsorbital *n* single-bond orbital

Einfachfilter *n* single-medium filter
Einfachform *f s.* Einfachwerkzeug
Einfachgarn *n* single yarn
einfachnegativ uninegative
einfachpositiv unipositive
Einfachsalz *n* simple (single) salt
Einfachstreuung *f (phys chem)* single scattering
Einfachsubstitution *f* monosubstitution
Einfachwalzwerk *n* single-roll mill
Einfachwerkzeug *n (plast)* single-impression mould, single-cavity mould (tool)
einfachwirkend single-acting *(e. g. pump)*
Einfachzucker *m* simple sugar, monosaccharide
Einfahrhub *m* return stroke
einfallend incident *(rays)*
Einfallstelle *f (plast)* sink mark, sunk spot *(a moulding defect)*
Einfall[s]winkel *m* angle of incidence
Einfang *m (nucl)* capture
einfangen *(nucl)* to capture
Einfangquerschnitt *m (nucl)* capture cross section
einfarbig *(phot)* monochromatic, monochrome
einfetten to grease, to lubricate
einfließen to flow in
  e. lassen to infuse, to run in
Einfließen *n* inflow
Einfließenlassen *n* infusion, running-in
Einfluß *m* 1. influence; 2. *s.* Einfließen
  dirigierender E. directive influence
Einformen *n* **von Hand** *(ceram)* hand modelling (moulding)
  maschinelles E. machine moulding
einfrieren 1. *(food)* to freeze; 2. *(glass, plast, rubber)* to exhibit transition
Einfrieren *n* 1. *(food)* freezing; 2. *(glass, plast, rubber)* transition
  langsames E. *(food)* slow freezing
  schnelles E. *(food)* quick (fast, rapid, sharp) freezing
Einfriergebiet *n (plast)* transition interval
Einfrierpunkt *m s.* Einfriertemperatur
Einfriertemperatur *f (glass, plast, rubber)* [glass] transition temperature, Tg point, T_g
einfügen to insert *(e. g. atoms into interstitial lattice sites)*
Einfügen *n* insertion *(as of atoms into interstitial lattice sites)*
einführen to introduce
Einführen *n* introduction
Einführungsseil *n (pap)* leading-through tape, rope carrier
einfüllen to fill [in], to feed [in], to charge, to load
Einfülltrichter *m (lab)* chemical funnel; *(tech)* feed (feeding) funnel (hopper), charge (charging) hopper
Eingabe *f* feed[ing]
Eingangstemperatur *f* inlet temperature
eingebaut built-in
eingeben to feed, to charge
eingehen *(text)* to shrink, to contract
  eine Bindung e. to bond
  eine [chemische] Verbindung e. to enter into chemical combination, to combine
eingelagert intercalary
eingeschliffen ground-in
eingesprengt *(geol)* disseminated
eingetauscht *(soil)* exchange-adsorbed
Eingrabtest *m (text)* soil burial test
eingravieren to engrave
Eingrubenäschersystem *n (tann)* one-pit liming system
eingruppieren to classify
Eingruppierung *f* classification
Einguß *m* pouring-in ✦ **auf E. geeicht (graduiert, justiert)** calibrated to contain, calibrated for content
Einhängegestell *n (el chem)* plating rack
Einhängekühler *m* suspended condenser, cold finger [condenser]
Einheit *f* 1. unit *(amount)*; 2. *(tech)* unit, set
  E. der Röntgendosis X-ray unit
  E. der Stromstärke unit of current
  difunktionelle E. difunctional (bifunctional) unit, D unit *(a structural element of macromolecules)*
  elektrostatische E. electrostatic unit, e. s. unit
  internationale E. international unit, I. U. *(of biochemically active substances)*
  monofunktionelle E. monofunctional unit, M unit *(a structural element of macromolecules)*
  tetrafunktionelle E. tetrafunctional unit, Q unit *(a structural element of macromolecules)*
  trifunktionelle E. trifunctional unit, T unit *(a structural element of macromolecules)*
einheitlich uniform; *(relating to composition)* homogeneous
Einheitlichkeit *f* uniformity; *(relating to composition)* homogeneity
Einheitsfläche *f* unit area
Einheitswert *m* unit value
Einheitszelle *f (cryst)* unit cell
Einhordendarre *f* one-floor[ed] kiln
Einhorn-Reaktion *f (org chem)* Einhorn (haloform) reaction
Einkammereindicker *m* single-compartment thickener, unit thickener
einkernig mononuclear
einklemmen to clamp
einkochen [lassen] *(lab)* to boil away (down)
Einkochen *n (lab)* boildown, boiling-down
Einkomponentensystem *n* one-component system; *(dye)* one-pack system
Einkornbeton *m* single-sized concrete
Einkörperverdampfer *m* single-effect (once-through) evaporator
Einkristall *m* single crystal, monocrystal
Einkristallfaden *m* single-crystal fibre, whisker
einlaben *(food)* to rennet
Einlage *f* pad; *(pap)* filler [board], middle, centre, core *(of triplex board)*; *(plast)* insert
einlagern 1. to store; 2. to insert, to intercalate, to include *(e. g. atoms into a crystal lattice)*
  schichtförmig e. to interleave

Einlagerung *f* 1. storing, storage; 2. insertion, intercalation, inclusion *(as of atoms into a crystal lattice)*
Einlagerungsatom *n* interstitial [atom]
Einlagerungsfremdatom *n* impurity interstitial, interstitial impurity atom
Einlagerungshydrid *n* interstitial hydride
Einlagerungskarbid *n* interstitial carbide
Einlagerungslegierung *f* interstitial alloy
Einlagerungsmischkristall *m* interstitial [solid] solution
Einlagerungsphase *f (cryst)* interstitial phase
Einlagerungsstruktur *f* interstitial structure
Einlagerungsverbindung *f* interstitial compound
schichtförmig ausgebildete E. lamellar compound
Einlaß *m* inlet, intake
Einlaßkanal *m* inlet duct
Einlaßrohr *n* inlet (influent) pipe (tube)
Einlaßventil *n* inlet valve
Einlauf *m* 1. inlet, point of entry; 2. influent *(material)*
einlaufen 1. to flow in; 2. *(text)* to shrink, to contract
e. lassen to run in, to infuse
Einlaufen *n* 1. inflow; 2. *(text)* shrinkage, shrinking, contraction
einlaufend / nicht *(text)* non-shrinking, unshrinking, unshrinkable
Einlauföffnung *f* inlet port
Einlaufseite *f* intake
Einlaufstelle *f* point of entry, inlet
Einlaufverlust *m* entrance (entry) loss
Einlaufverteilerkasten *m* feed-splitter box
Einlegemaschine *f (glass)* batch charger (feeder)
Einlegevorbau *m (glass)* doghouse
Einlegevorrichtung *f s.* Einlegemaschine
Einlegewand *f (glass)* end (back, gable) wall
einleiten 1. to introduce, to pass in *(material)*; 2. to initiate, to start *(a reaction)*
Einleitung *f* 1. introduction, passing-in *(of material)*; 2. initiation, start *(of a reaction)*
Einleitungsrohr *n* inlet (delivery) tube (pipe)
Einling *m s.* Einkristall
Ein-Lösungsmittel-Verfahren *n (petrol)* single-solvent process
einmaischen *(ferm)* to mash, to dough [in]
Einmaischverfahren *n (ferm)* single-mash process
Einmalbrand *m (ceram)* single fire (firing)
einmischen *(rubber)* to incorporate
Einmischung *f (rubber)* incorporation
einmitten to centre
Einmitten *n* centr[e]ing
Einmuldenunterschubrost *m* single-retort [underfeed] stoker
Einnährstoffdüngemittel *n* single-nutrient fertilizer
einnehmen 1. *(pharm)* to take; 2. to occupy *(e. g. the interstices in a lattice)*; to cover *(an area)*
einölen to oil; *(tann)* to anoint
einpacken to wrap, to pack[age]; *(met)* to pack *(with a carburizing powder)*
Einpackmittel *n (met)* packing material
Einpackpapier *n* wrapping (packing) paper, *(Am)* package (packaging) paper
einpegeln / sich to level off *(as of pH value)*
Einpendelmühle *f* single-roll mill
Einphasensystem *n* one-phase system, homogeneous system
einpökeln to cure
Einpökeln *n* curing, cure
einpolar non-polar, homopolar, covalent *(bond)*
einpressen to inject (e. g. gas); *(pap, text, tann)* to emboss, to goffer; *(plast)* to mould in
Einpressen *n* injection *(as of gas)*; *(pap, text, tann)* embossing, goffering; *(plast)* moulding-in
E. von Wasser *(petrol)* water flooding
Einpreßteil *n (plast)* insert
einpudern to powder, to dust
Einrad-Karrenstäuber *m (agric)* wheelbarrow-type duster
Einreißfestigkeit *f*, **Einreißwiderstand** *m* tear initiation strength
Einrichtung *f* arrangement; equipment
Einriß *m* tear
einrühren to stir in
Einsackstelle *f s.* Einfallstelle
einsalben *(pharm)* to smear
einsalzen *(food)* to salt away (down), to rouse
trocken e. to dry-salt, to dry-cure
Einsatz *m* 1. application; *(tech)* charging, feeding; 2. *(met)* [carburized] case; 3. *s.* Einsatzgut
Einsatzbad *n (met)* carburizing bath
Einsatzbecken *n (lab)* bench sink
Einsatzbehälter *m (ceram)* setter *(a piece of kiln furniture)*
Einsatzgemisch *n (met)* carburizing mixture
Einsatzgut *n* feed[stock], charge [stock], charging stock
E. für Krackverfahren cracking feed (stock, feedstock)
Einsatzhärtbarkeit *f (met)* case hardenability
Einsatzhärte *f (met)* case hardness
Einsatzhärtebad *n (met)* case-hardening bath
Einsatzhärtekasten *m (met)* case-hardening box
einsatzhärten *(met)* to case-harden
Einsatzhärten *n (met)* case-hardening
E. mit festen Mitteln pack hardening
Einsatzhärteofen *m (met)* case-hardening furnace
Einsatzhärtung *f (met)* case-hardening
Einsatzhärtungstiefe *f (met)* case depth, depth of case
Einsatzheizkörper *m* cartridge heater
Einsatzkasten *m s.* Einsatzhärtekasten
Einsatzkohle *f* feed[-stock] coal
Einsatzmaterial *n s.* Einsatzgut
Einsatzmenge *f* amount (quantity) required, dosage
Einsatzmittel *n (met)* case-hardening material (compound)
Einsatzofen *m (petrol)* charge heater
Einsatzöl *n (petrol)* charge oil
Einsatzplatte *f* chamber plate *(chromatography)*
Einsatzpulver *n (met)* carburizing (cementing) powder

Einsatzschicht *f* **[/ gehärtete]** *(met)* [hardened] case
Einsatzschichtdicke *f (met)* case thickness
Einsatzstahl *m* case-hardening steel; case-hardened steel
legierter E. alloy case-hardening steel
Einsatzstoff *m s.* Einsatzgut
Einsatztiefe *f (met)* case depth
Einsatztopf *m (met)* case-hardening pot
Einsatztrichter *m* insertion funnel
Einsatztulpe *f (lab)* crucible adapter
Einsatzverchromung *f* chromizing
Einsatzverzögerung *f (rubber)* delayed action
Einsatzzone *f s.* Einsatzschicht
einsaugen to suck [in, up], to imbibe, to absorb
Einsaugen *n* suction, imbibition, absorption
einsäurig monoacid[ic] *(base)*
Einschalenanalysenwaage *f* single-pan analytical balance
einschalten to switch on, to turn on
Einscheibenrefiner *m (pap)* single-disk refiner
Einscheibensicherheitsglas *n* heat-treated (heat-strengthened, heat-toughened, tempered, toughened, case-hardened) glass
Einschichtensicherheitsglas *n s.* Einscheibensicherheitsglas
einschieben to insert
Einschiebung *f* insertion
Einschießbogen *m (pap)* set-off sheet, *(Am)* slip sheet
Einschlagpapier *n* wrapping (packing) paper, wrapper
E. für Papierrollen (Rollenpapier) mill wrapper (wrapping)
Einschlagseidenpapier *n* tissue wrapper (wrapping), wrapping (packing, commercial) tissue
einschleifen to grind in
einschließbar enclosable
einschließen to enclose, to include, to occlude; to [en]trap *(gases)*
Einschluß *m* enclosure, inclusion, occlusion
enallogener E. *s.* exogener E.
endogener E. *(geol)* cognate inclusion, autolith
exogener E. *(geol)* exogenous enclosure, xenolith
fremder E. *s.* exogener E.
homöogener E. *s.* endogener E.
Einschlußverbindung *f* inclusion (enclosure) compound
einschmelzen 1. to fuse (seal) in *(e.g. into a glass tube)*; 2. *(met)* to melt [down]; to smelt *(scrap metal)*
Einschmelzen *n* 1. fusing-in, sealing in *(e.g. into a glass tube)*; 2. *(met)* melting[-down], meltdown; smelting *(of scrap metal)*
Einschmelzrohr *n* tube for sealing, sealing (sealed) tube
E. nach Carius Carius tube
Einschmelzung *f (geol)* magmatic digestion, assimilation
einschmieren to smear; *(tann)* to dub
Einschneckenextruder *m* single-screw extruder (extruding machine)
einschneiden to cut *(a rubber stock)*
Einschnürung[sstelle] *f* constriction, weist, throat *(as of a tube)*, *(esp in flow measurement)* vena contracta
einschrumpfen to contract
Einschrumpfung *f* contraction
Einschubheizkörper *m* cartridge heater
einschütten to fill in; *(tech)* to charge, to load, to feed [in]
Einschütttrichter *m* [loading, input, feed] hopper
Einschwemmungshorizont *m (soil)* illuvial horizon
einseitig one-sided
einseitigglatt *(pap)* glazed on one side, machine-glazed
einsetzbar applicable
Einsetzbarkeit *f* applicability
einsetzen 1. to apply *(e.g. a certain chemical)*; to charge, to feed *(a certain quantity)*; 2. to insert *(mechanically)*; *(ceram)* to set; 3. *(met)* to carburize; 4. to start *(of a reaction)*
im Salzbad e. *s.* in flüssigen Mitteln e.
in festen Mitteln e. *(met)* to pack-carburize
in flüssigen Mitteln e. *(met)* to liquid-carburize, to bath-carburize
in gasförmigen Mitteln e. *(met)* to gas-carburize
in Zementationskästen e. *(met)* to box-carburize
Einsetzen *n* 1. application *(as of a certain chemical)*; charging, feeding *(of a certain quantity)*; 2. insertion *(mechanically)*; *(ceram)* setting; 3. *(met)* carburizing, carburization; 4. start *(of a reaction)*
E. im Salzbad *s.* E. in flüssigen Mitteln
E. in festen Mitteln *(met)* solid (pack, solid-pack) carburizing
E. in flüssigen Mitteln *(met)* liquid (liquid-salt, bath) carburizing
E. in gasförmigen Mitteln *(met)* gas carburizing
E. in Zementationskästen *(met)* box carburizing
Einsetzmulde *f* charging box
einsickern to seep in, to trickle in, to soak in, to infiltrate
Einsickern *n* seepage, trickling, soaking, infiltration
einsinken to sink in
Einsitzventil *n* single-seat[ed] valve
einspänen *(tann)* to sawdust
einspannen to clamp, to fix, to attach
Einspannrahmen *m (plast)* clamping frame
einspeisen *(tech)* to charge, to feed [in]
Einspeisevorrichtung *f* feeder
Einspeisung *f* 1. charging, feeding; 2. charge, charging stock, feed[stock]
einspielen / sich to level off *(as of pH value)*
einsprengen to sprinkle, to spray
Einsprengling *m (geol)* inset, phenocryst
Einsprengung *f (geol)* dissemination
Einspritzdruck *m* injection pressure

einspritzen to inject, *(plast also)* to mould in
Einspritzkondensator *m* jet (wet) condenser
Einspritzöffnung *f* injection port
Einspritzrate *f* injection rate
Einspritzteil *n (plast)* insert
Einspritzung *f* injection
einstäuben to dust, to powder
Einstäuben *n* dusting, powdering
Einsteigluke *f*, **Einsteigöffnung** *f* manhole, manway
Einstein-Gleichung *f* Einstein equation
Einsteinium *n* Es einsteinium
einstellen to adjust, to rectify *(e. g. instruments)*; to standardize *(chemicals)*
eine Lösung auf n e. to set a solution to N
sauer e. to acidify
sich e. to establish *(as of equilibrium)*
Einstellthermometer *n* adjustable-zero thermometer
Einstellung *f* positioning; adjustment, rectification *(as of instruments)*; standardization *(of chemicals)*; establishment *(of chemical equilibrium)*
Einstoffkraftstoff *m* monofuel
Einstoffmasse *f*, **Einstoffscherben** *m (ceram)* single-component (single-material) body
Einstoffsystem *n* one-component system
Einstofftreibstoff *m* monofuel
Einstoffversprühung *f* single-fluid atomization
Einstoffzerstäubung *f s.* Einstoffversprühung
Einstrahlinstrument *n* single-beam instrument *(spectroscopy)*
einsträngig single-strand[ed]
Einstrangkette *f* single-strand chain *(conveying)*
einströmen to flow in
Einströmen *n* inflow
Einströmgeschwindigkeit *f* rate of inflow
Einströmrohr *n* influent pipe
Einstufenhomogenisierung *f* single-stage homogenization
Einstufenreaktion *f* single-step (one-step) reaction
Einstufenverdampfer *m* single-effect (once-through) evaporator
Einstufenverfahren *n* single-step (one-step) process; *(plast)* one-shot process
einstufig single-stage
einsumpfen *(ceram)* to wet
Eintauchelektrode *f* immersion electrode
eintauchen to dip, to plunge, to immerse, to immerge
Eintauchen *n* dip[ping], plunging, immersion
Eintaucher *m* dipper *(a worker)*
Eintauchgefrierverfahren *n (food)* immersion liquid) freezing
Eintauchkolorimeter *n* immersion (dipping) colorimeter
Eintauchnutsche *f* immersion filter tube
Eintauchpyrometer *n* dipping pyrometer
Eintauchrefraktometer *n* immersion (dipping) refractometer
Eintauchrohr *n* immersion pipe (tube)
Eintauchtiefe *f* [depth of] submergence; *(distil)* static submergence
Eintauchverhältnis *n* submergence ratio
Eintauchwalze *f* immersion roll
Eintauchwalzentrockner *m* dip-feed drum dryer
Eintauschstärke *f (soil)* replacing power
einteigen to dough [in]
einteilen to classify; to divide *(e. g. a scale)*
in Abstände e. to space
in Grade e. to graduate
nach Korn[größen]klassen e. to size
Einteilung *f* classification; division *(as of a scale)* ✦
mit [genauer] E. versehen to graduate, to scale
E. nach dem Inkohlungsgrad rank classification *(of coals)*
Einteilungssystem *n* system of classification
Eintrag *m* 1. charge, charging stock, feed[stock], load; *(pap)* furnish; *(electroplating)* drag in; 2. charging, feeding, loading, introduction *(of material into a reactor)*, *(esp lab)* placing *(in a vessel)*
eintragen 1. to charge, to feed [in], to load, to introduce *(material into a reactor)*, *(esp lab)* to place *(in a vessel)*; *(pap)* to furnish; 2. to plot *(in a coordinate system)*; to register *(a trademark)*
Eintragkasten *m* feed box
Eintragmenge *f* **in der Zeiteinheit** rate of feeding, feed rate
Eintragöffnung *f* charging (filling) door (hole), feed inlet (hole)
Eintragrohr *n* feed pipe (tube)
Eintragseite *f* feed end
Eintragverteilerkasten *m* feed-splitter box
Eintragvorrichtung *f* feeding device, feeder
Eintragzelle *f* entry lock, inlet sluice
Eintragzylinder *m* feed well *(of a continuous thickener)*
eintreten 1. *(of material being charged)* to enter; 2. *(of an event)* to occur
Eintritt *m* 1. entry, entrance *(of material)*; 2. *s.* Eintrittsstelle; 3. occurrence *(of an event)*
Eintrittsöffnung *f* inlet [port], intake
Eintrittsspalt *m* entrance slit *(spectroscopy)*
Eintrittsstelle *f* point of entry, inlet, intake
Eintrittstemperatur *f* inlet temperature
Eintrittsverlust *m* entry (entrance) loss
eintrocknen to dry [up]
Einwaage *f* weighed object (portion)
Einwalzenbrecher *m* single-roll crusher
Einwalzenmühle *f*, **Einwalzenstuhl** *m* single-roll mill; *(coat)* uniroll mill
Einwalzentrockner *m* single-drum dryer
einwandern to migrate in *(as of ions)*
einwässern to steep, to soak
Einwässern *n* steeping, steep[age], soaking
Einweghahn *m* single-bore stopcock
einweichen to steep, to soak
Einweichen *n* steeping, steep[age], soaking
übermäßiges E. oversteeping *(of malt)*
Einweichflüssigkeit *f* steeping liquor, *(text also)* steep
Einweichkufe *f (text)* steeping pan
Einweichsektion *f (petrol)* soaking section *(of a pipe furnace)*

Einweichtrog *m (text)* steeping pan
Einweichwasser *n* steep[ing] water
einwerfen *(tech)* to load, to charge, to feed
einwertig monovalent, univalent; monohydric *(alcohol, phenol)*
Einwertigkeit *f* monovalence, monovalency, univalence, univalency
Einwickelmaschine *f* wrapping device (machine), wrapper
Einwickelmechanismus *m* wrapping mechanism
einwickeln to wrap
Einwickelpapier *n s.* Einschlagpapier
Einwickelseidenpapier *n s.* Einschlagseidenpapier
einwirken to act
aufeinander e. to react
aufeinander e. lassen to react
störend e. auf to interfere with, to affect
Einwirkung *f* action
Einwirkungsdauer *f*, **Einwirkungszeit** *f* exposure time, duration of exposure
Einwohnergleichwert *m* population equivalent *(water chemistry)*
Einwurföffnung *f* charging (filling) hole (door), feed hole (inlet)
einzählig *s.* einzähnig
einzähnig monodentate, unidentate *(coordination chemistry)*
Einzeldosenbehälter *m (pharm)* single-dose container
Einzeldosis *f (pharm)* single dose
Einzeldünger *m* single fertilizer
Einzeldüngung *f* straight fertilization
Einzelfaser *f* elementary fibre ✦ **in Einzelfasern zerlegen** *(pap)* to defibre, to defibrate, to reduce to fibres, to shred, *(Am also)* to [de]fiberize
Einzelglied *n* member
Einzelheizer *m (rubber)* unit vulcanizer (press), individual (watch-case) vulcanizer, individual curing unit (press)
Einzelkorngefüge *n*, **Einzelkornstruktur** *f (soil)* single-grained structure
Einzelkristall *m* single crystal, monocrystal
Einzellinie *f* single line *(spectroscopy)*
Einzelnährstoffdüngemittel *n* straight fertilizer
Einzelofen *m (ceram)* individual kiln
Einzelpfannensaturation *f (sugar)* batch carbonation
Einzelpotential *n* single[-electrode] potential
Einzelreaktion *f* single-step reaction
Einzelring *m (nomencl)* single (individual) ring
Einzelzelle *f (text)* single cell; *(min tech)* floatation unit
einziehen *(pap)* to put on *(a wire)*
in sich e. to absorb *(a fluid)*
Einzugswinkel *m* angle of nip *(of a roll crusher)*
einzwängen to squeeze *(e.g. atoms into interstices)*
ein-zwei-wertig unibivalent
Einzylinderpumpe *f* single-cylinder pump
Einzylinderschermaschine *f (plast)* single-shearing machine

Eirich-Mischer *m (ceram)* Eirich mixer *(a wet pan)*
Eis *n* ice; *(food)* ice cream ✦ **auf E. geben** to run on to ice
gestoßenes E. chopped (crushed) ice
kleinstückiges E. small ice
zerkleinertes E. *s.* gestoßenes E.
Eisbad *n* ice bath
Eisblock *m* ice cake
Eisblumenbildung *f (coat)* gas checking; *(plast)* frosting *(a defect)*
Eisblumenglas *n* frosted glass
Eisbordeaux *n (dye)* ice bordeaux
Eisen *n* Fe iron
dreiwertiges E. trivalent (ferric) iron
zweiwertiges E. bivalent (divalent, ferrous) iron
α-Eisen *n* alpha iron
β-Eisen *n* beta iron
γ-Eisen *n* gamma iron
δ-Eisen *n* delta iron
Eisenablauf *m* iron runoff
Eisenabscheider *m* tramp-iron magnet (magnetic separator)
Eisenabstich *m* 1. iron tapping; 2. *s.* Eisenabstichloch
Eisenabstichloch *n* iron taphole (tapping hole, notch)
Eisenabstichrinne *f* iron runner
Eisenalaun *m* iron alum *(any of several salts $M^{I}Fe(SO_4)_2 \cdot 12H_2O$)*
Eisenammoniakalaun *m s.* Ammoniumeisenalaun
Eisen(III)-ammoniumoxalat *n s.* Ammoniumeisen(III)-oxalat
eisenarm poor in iron, low-iron
Eisenarsenid *n* FeAs iron arsenide
Eisenauslauf *m s.* Eisenablauf
Eisenausscheider *m s.* Eisenabscheider
Eisen(II)-azetat *n* $Fe(CH_3COO)_2$ iron(II) acetate, ferrous acetate
Eisenbahnkesselwagen *m*, **Eisenbahntankwagen** *m* rail tank [car], tank car (wagon)
Eisenbakterien *pl* iron bacteria *(Leptothrix, Crenothrix, and Gallionella specc.)*
Eisenbeize *f (dye)* iron (black) liquor (mordant) *(an iron acetate solution)*
Eisenblaudruck *m* cyanotype, blueprint *(proper)*
negativer E. [negative] cyanotype
positiver E. positive cyanotype
Eisenblech *n* sheet iron, iron plate
verzinntes E. tinned sheet iron, tin plate
Eisenbohrspäne *mpl* iron borings
Eisenborid *n* iron boride
Eisen(II)-bromid *n* $FeBr_2$ iron(II) bromide, iron dibromide, ferrous bromide
Eisen(III)-bromid *n* $FeBr_3$ iron(III) bromide, iron tribromide, ferric bromide
Eisen(II)-chelat *n* iron(II) chelate, ferrous chelate
Eisen(III)-chelat *n* iron(III) chelate, ferric chelate
Eisen(II)-chlorid *n* $FeCl_2$ iron(II) chloride, iron dichloride, ferrous chloride
Eisen(II,III)-chlorid *n* $FeCl_2 \cdot 2FeCl_3$ iron(II, III) chloride, ferrosoferric chloride

Eisen(III)-chlorid *n* $FeCl_3$ iron(III) chloride, iron trichloride, ferric chloride
Eisenchlorose *f (agric)* iron chlorosis *(a plant disease caused by iron deficiency)*
Eisen(III)-chromat *n* $Fe_2(CrO_4)_3$ iron(III) chromate, ferric chromate
Eisendi . . . *s.a.* Eisen(II)- . . .
Eisen(III)-dichromat *n* $Fe_2(Cr_2O_7)_3$ iron(III) dichromate, ferric dichromate
Eisen(III)-diphosphat *n* $Fe_4(P_2O_7)_3$ iron(III) diphosphate, ferric pyrophosphate
Eisen(II)-disulfid *n* FeS_2 iron(II) disulphide
Eisenerz *n* iron ore
phosphorarmes E. Bessemer ore *(containing less than 0.09% phosphorus)*
Eisenfeilspäne *mpl* iron filings
Eisenfleck *m* iron speck *(a paper defect)*; iron stain *(in wood)*
Eisen(II)-fluorid *n* FeF_2 iron(II) fluoride, iron difluoride, ferrous fluoride
Eisen (III)-fluorid *n* FeF_3 iron(III) fluoride, iron trifluoride, ferric fluoride
Eisen(II)-formiat *n* $Fe(HCOO)_2$ iron(II) formate, ferrous formate
Eisen(III)-formiat *n* $Fe(HCOO)_3$ iron(III) formate, ferric formate
eisenfrei iron-free, non-ferrous
eisenführend *s.* eisenhaltig
Eisengehalt *m* iron content ✦ **mit hohem E.** high-iron
Eisengelb *n s.* Eisenoxidgelb
Eisenglanz *m (min)* specular iron [ore], specularite *(a variety of haematite)*
Eisenglimmer *m* micaceous iron ore
Eisengruppe *f* iron group
eisenhaltig iron-containing, ferrugin[e]ous
eisen(II)-haltig ferroan
eisen(III)-haltig ferrian
Eisen(II)-häm *n* ferrohaem, ferroheme
Eisen(III)-häm *n* ferrihaem, ferriheme
Eisen(III)-hämochromogen *n* ferrihaemochromogen, ferrihaemochrome
Eisen(II)-hämoglobin *n* ferrohaemoglobin
Eisen(III)-hämoglobin *n* ferrihaemoglobin
Eisen(II)-hexachloroplatinat(IV) *n* $Fe[PtCl_6]$ iron(II) hexachloroplatinate(IV), iron(II) chloroplatinate(IV)
Eisen(II)-hexazyanoferrat(II) *n* $Fe_2[Fe(CN)_6]$ iron(II) hexacyanoferrate(II), iron(II) cyanoferrate(II)
Eisen(II)-hexazyanoferrat(III) *n* $Fe_3[Fe(CN)_6]_2$ iron(II) hexacyanoferrate(III), iron(II) cyanoferrate(III)
Eisen(II,III)-hexazyanoferrat(III) *n* $Fe^{III}{}_4Fe^{II}{}_3[Fe(CN)_6]_6$ iron(II,III) hexacyanoferrate(III)
Eisen(III)-hexazyanoferrat(II) *n* $Fe_4[Fe(CN)_6]_3$ iron(III) hexacyanoferrate(II), iron(III) cyanoferrate(II)
Eisenhochofen *m* iron blast furnace
Eisenhüttenwesen *n s.* Eisenmetallurgie
Eisen(II)-hydroxid *n* $Fe(OH)_2$ iron(II) hydroxide, ferrous hydroxide
Eisen(III)-hydroxid *n* $Fe(OH)_3$ iron(III) hydroxide, ferric hydroxide
Eisen(III)-hydroxidsol *n* iron(III) hydroxide sol, ferric hydroxide sol
Eisen(III)-hypophosphit *n* $Fe(PH_2O_2)_3$ iron(III) hypophosphite, ferric hypophosphite, *(better)* iron(III) phosphinate
Eisenindigochelat *n* iron chelate of indigo
Eisen(II)-jodid *n* FeI_2 iron(II) iodide, iron diiodide, ferrous iodide
Eisenkarbid *n* Fe_3C iron carbide; *(met)* Fe_3C cementite, cemented carbide, iron carbide
Eisen(II)-karbonat *n* $FeCO_3$ iron(II) carbonate, ferrous carbonate
Eisenkatalysator *m* iron catalyst
Eisenkegel *m* cone *(of a blast furnace)*
Eisenkies *m (min)* pyrite, iron pyrite[s], mundic *(iron(II) disulphide)*
Eisenkitt *m* iron (rust, iron-rust) cement
Eisenklinker *m* blue brick
Eisenkontakt *m s.* Eisenkatalysator
Eisen(II)-laktat *n* iron(II) lactate, ferrous lactate
Eisenlegierung *f* iron alloy
Eisenmennige *f* stone red, red ochre (rudd)
Eisenmetall *n* ferrous metal
Eisenmetallurgie *f* ferrous (iron) metallurgy
Eisenmeteorit *m* iron meteorite, meteoric iron, [holo]siderite
Eisenmonosulfid *n s.* Eisen(II)-sulfid
Eisenmonoxid *n s.* Eisen(II)-oxid
Eisen-Nickel-Kern *m (geol)* iron-nickel core
Eisennickelkies *m (min)* pentlandite *(an iron nickel sulphide)*
Eisen(II)-nitrat *n* $Fe(NO_3)_2$ iron(II) nitrate, ferrous nitrate
Eisen(III)-nitrat *n* $Fe(NO_3)_3$ iron(III) nitrate, ferric nitrate
Eisennitrid *n* iron nitride
Eisen(II)-orthoarsenat(V) *n* $Fe_3(AsO_4)_2$ iron(II) orthoarsenate, ferrous orthoarsenate, ferrous arsenate
Eisen(III)-orthoarsenat(V) *n* $FeAsO_4$ iron(III) orthoarsenate, ferric orthoarsenate, ferric arsenate
Eisen(II)-orthophosphat *n* $Fe_3(PO_4)_2$ iron(II) orthophosphate, ferrous orthophosphate, ferrous phosphate
Eisen(III)-orthophosphat *n* $FePO_4$ iron(III) orthophosphate, ferric orthophosphate, ferric phosphate
Eisen(II)-oxalat *n* $Fe(C_2O_4)$ iron(II) oxalate, ferrous oxalate
Eisenoxid *n* iron oxide; *(specif) s.* Eisen(II)-oxid
Eisen(II)-oxid *n* FeO iron(II) oxide, iron monooxide, ferrous oxide
Eisen(II,III)-oxid *n* Fe_3O_4 iron(II,III) oxide, triiron tetraoxide, ferrosoferric oxide
Eisen(III)-oxid *n* Fe_2O_3 iron(III) oxide, diiron trioxide, ferric oxide, ferric trioxide
Eisen(III)-oxidgel *n* iron(III) oxide gel, ferric oxide gel
Eisenoxidgelb *n* ferrite yellow
Eisen(III)-oxidhydrat *n* $Fe_2O_3 \cdot nH_2O$ hydrated iron(III) oxide, hydrated ferric oxide
Eisenoxidrot *n* iron oxide red, red oxide, chemical red

Eisenoxidschwarz *n* black rouge *(iron(II,III) oxide)*
Eisenoxygenase *f s.* Zytochromoxydase
Eisenpentakarbonyl *n* $Fe(CO)_5$ iron pentacarbonyl
Eisen(II)-perchlorat *n* $Fe(ClO_4)_2$ iron(II) perchlorate, ferrous perchlorate
Eisenpfanne *f* iron tray
Eisenphosphattrübung *f* ferric phosphate haze *(of beer)*
Eisenphosphid *n* iron phosphide
Eisen(II)-phthalozyanin *n* iron(II) phthalocyanine, ferrous phthalocyanine
Eisenporphyrin *n (bioch)* iron porphyrin
Eisenporphyrinprotein *n* iron porphyrin protein
Eisenpulver *n* iron powder
Eisen(III)-pyrophosphat *n s.* Eisen(III)-diphosphat
Eisenquelle *f (pharm)* chalybeate spring
eisenreich rich in iron, high-iron
Eisen(III)-resinat *n* iron(III) resinate, ferric resinate
Eisen(II)-rhodanid *n s.* Eisen(II)-thiozyanat
Eisen(III)-rhodanid *n s.* Eisen(III)-thiozyanat
Eisenrost *m* iron rust
Eisenrot *n* red bole *(iron(III) oxide)*
Eisen(II)-salz *n* iron(II) salt, ferrous salt
Eisen(III)-salz *n* iron(III) salt, ferric salt
Eisenschlamm *m* iron sludge
Eisenschmelzklinker *m* blue brick
Eisenschrott *m* scrap iron
Eisenschwamm *m* iron sponge, sponge iron
Eisenschwarz *n s.* Eisenoxidschwarz
Eisen-Silber-Verfahren *n* silver-iron process, Vandyke (sepia negative) process, brownprint *(reprography)*
Eisen(II)-silikat *n* $FeSiO_3$ iron(II) silicate, ferrous silicate
Eisensilizid *n* iron silicide
Eisenspäne *mpl* iron borings
Eisenspat *m (min)* spathic iron [ore], siderite *(iron(II) carbonate)*
Eisenstein *m* ironstone *(a sedimentary rock rich in iron)*
Eisenstich *m s.* Eisenabstich
Eisen(II)-sulfat *n* $FeSO_4$ iron(II) sulphate, ferrous sulphate
Eisen(III)-sulfat *n* $Fe_2(SO_4)_3$ iron(III) sulphate, ferric sulphate
Eisen(II)-sulfid *n* FeS iron(II) sulphide, iron monosulphide, ferrous sulphide
Eisen(III)-sulfid *n* Fe_2S_3 iron(III) sulphide, diiron trisulphide, ferric sulphide
Eisen(II)-sulfit *n* $FeSO_3$ iron(II) sulphite, ferrous sulphite
Eisentetrakarbonyl *n* $Fe(CO)_4$ iron tetracarbonyl
Eisen(II)-thiosulfat *n* FeS_2O_3 iron(II) thiosulphate, ferrous thiosulphate
Eisen(II)-thiozyanat *n* $Fe(SCN)_2$ iron(II) thiocyanate, iron(II) rhodanide, ferrous thiocyanate
Eisen(III)-thiozyanat *n* $Fe(SCN)_3$ iron(III) thiocyanate, iron(III) rhodanide, ferric thiocyanate
Eisentiegel *m* iron crucible
Eisentri ... *s.* Eisen(III)- ...
Eisen(II)-verbindung *f* iron(II) compound, ferrous compound
Eisen(III)-verbindung *f* iron(III) compound, ferric compound
Eisenvitriol *m (min)* iron vitriol, copperas, melanterite *(iron(II) sulphate-7-water)*
Eisenvitriol *n* $FeSO_4 \cdot 7H_2O$ iron vitriol, [green] copperas, green vitriol, iron(II) sulphate-7-water
Eisenwasser *n (pharm)* ferrugin[e]ous water, chalybeate water
Eisen(III)-zyanid *n* $Fe(CN)_3$ iron(III) cyanide, iron tricyanide, ferric cyanide
Eisen-*bis*-zyklopentadienyl *n* $Fe(C_5H_5)_2$ dicyclopentadienyliron, ferrocene
Eiserzeugungsanlage *f* ice plant; ice[-making] machine
Eisessig *m* glacial acetic acid
Eisfabrik *f* ice plant
Eisfarbbase *f* ice-colour base
Eisfarbe *f*, **Eisfarbstoff** *m* ice dye (colour)
eisgekühlt ice-cooled
Eisglas *n* frosted glass
Eishydrat *n* gas hydrate, gas clathrate compound
Eiskalorimeter *n* ice calorimeter
  E. nach Bunsen Bunsen ice calorimeter
eiskalt ice-cold
Eiskrem *f* ice cream
Eiskristall *m* ice crystal
Eiskühlung *f* ice cooling (refrigeration)
Eismaschine *f (food)* ice-cream freezer, ice[-cream] freezing machine
Eismühle *f* ice crusher
Eispapier *n* ice paper
Eisschrank *m* icebox
Eiswasser *n* ice[d] water, frozen water
Eiweiß *n* 1. *(org chem)* protein; 2. *(food)* egg white, albumen, glair
Eiweißabbau *m* protein degradation (breakdown)
eiweißabbauend proteolytic, proteoclastic, protein-digesting
Eiweißabtrennung *f* deproteinization
Eiweißappretur *f (tann)* albumen finish
eiweißartig proteinaceous
Eiweißchemie *f* protein chemistry
Eiweißchemiefaser *f (text)* protein [man-made] fibre
Eiweißchemiefaserstoff *m (text)* protein [man-made] fibre
Eiweißchemiker *m* protein chemist
Eiweißeinschluß *m* protein inclusion
Eiweißfaser *f s.* Eiweißchemiefaser
Eiweißfaserstoff *m s.* Eiweißchemiefaserstoff
eiweißfrei protein-free
Eiweißgerbstofftrübung *f* protein-tannin turbidity
Eiweißkomponente *f* protein component
Eiweißkörper *m* proteic substance, protein [material, matter]
  einfacher E. simple protein
  fibrillärer E. fibrous protein
  globulärer E. globular protein
  konjugierter E. conjugated protein, proteid[e]
  regenerierter natürlicher E. regenerated naturally occurring protein

zusammengesetzter E. *s.* konjugierter E.
Eiweißleim *m* protein adhesive, glair
Eiweißmangel *m* protein deficiency
eiweißspaltend proteolytic, proteoclastic, protein-digesting
Eiweißspalter *m (text)* digester *(for protein stain removal)*
Eiweißspaltung *f* proteolysis
Eiweißstickstoff *m* protein nitrogen
Eiweißstoff *m s.* Eiweißkörper
Eiweißstoffwechsel *m* protein metabolism
Eiweißsynthese *f* protein synthesis
Eiweißtrübung *f* protein haze (turbidity) *(as of beer)*
eiweißverdauend *s.* eiweißspaltend
Eiweißverdauer *m s.* Eiweißspalter
Eiweißzersetzer *mpl* proteolytic bacteria
Eiweißzucker *m* glycoprotein, glycoproteid
Ejektor *m* ejector, eductor
Ekgonin *n* ecgonine *(alkaloid)*
ekliptisch eclipsed, opposed *(stereochemistry)*
Eklogit *m* eclogite *(a metamorphic rock)*
Eklogithülle *f*, **Eklogitschale** *f (geol)* eclogite shell
Eko *m s.* Ekonomiser
Ekonomiser *m* economizer, boiler feed preheater
Ektohormon *n* ectohormone, pheromone
Ektotoxin *n* ectotoxin
Ektylharnstoff *m* ectylurea, 2-ethyl-*cis*-crotonylurea
E-Kupfer *n* electrolytic copper
Elaidin[is]ierung *f* elaidinization *(conversion of oleic acid into its trans isomer)*
Elaidinprobe *f* elaidin test
Elaidinsäure *f* $CH_3[CH_2]_7CH{=}CH[CH_2]_7COOH$ elaidic acid, *trans*-9-octadecenoic acid
Elainsäure *f* $CH_3[CH_2]_7CH{=}CH[CH_2]_7COOH$ oleic acid, *cis*-9-octadecenoic acid
Eläolith *m (min)* elaeolite *(a tectosilicate)*
Eläostearinsäure *f* elaeostearic acid *(name of two stereoisomeres of 9,11,13-octadecatrienoic acid)*
Elast *m s.* Elastomer
Elastikator *m* elasticator *(a plasticizing agent)*
Elastin *n* elastin *(a scleroprotein)*
elastisch elastic
Elastizität *f* elasticity; *(tann)* run
Elastizitätsachse *f (cryst)* elastic axis
Elastizitätsgrenze *f* elastic limit
Elastizitätskonstante *f* elastic constant
Elastizitätsmodul *m* modulus of elasticity, Young's modulus [of elasticity], elastic modulus
Elastizitätsprüfung *f* elasticity test
elastomer elastomer[ic]
Elastomer[es] *n* elastomer
Elastomerfaden *m* elastomeric yarn
Elastomerfaser *f* elastomeric fibre, *(Am)* snap-back fiber
Elastomerfaserstoff *m* elastomeric fibre, *(Am)* snap-back fiber
Elaterit *m (min)* elaterite, elastic bitumen, mineral caoutchouc
Elbs-Reaktion *f* Elbs reaction *(formation of anthracene derivatives)*
Elefantenhautbildung *f (rubber)* crazing
elektrisch electric[al]
e. neutral uncharged
Elektrizität *f* electricity
Elektrizitätsleiter *m* conductor of electricity
Elektroabscheider *m s.* Elektrofilter
Elektroabscheidung *f s.* Elektrofiltration
Elektroaffinität *f* electroaffinity, electron affinity
Elektroanalyse *f* electroanalysis
E. bei kontrolliertem Katodenpotential controlled-potential electrodeposition
elektroanalytisch electroanalytical
Elektroätzen *n* electroengraving, electrolytic etch[ing]
Elektrobeheizung *f* electric heating
Elektrobrenner *m* electric burner
Elektrochemie *f* electrochemistry
technische E. industrial electrochemistry
Elektrochemiker *m* electrochemist
elektrochemisch electrochemical
Elektrochromatografie *f* electrochromatography
Elektrode *f* electrode
E. nach Söderberg *s.* selbstbrennende E.
massive E. solid electrode
mehrfache E. multiple electrode, polyelectrode
reversible E. reversible electrode
ringförmige E. annular electrode
selbstbackende E. *s.* selbstbrennende E.
selbst[ein]brennende E. Soderberg [continuous] electrode, [Soderberg] self-baking electrode
umkehrbare E. *s.* reversible E.
vorgebackene (vorgebrannte) E. prebaked electrode
Elektrodekantation *f s.* Elektrodekantierung
elektrodekantiert electrodecanted
Elektrodekantierung *f* electrodecantation, electric decantation, electrocremage
Elektrodenabstand *m* interelectrode distance
Elektrodenkammer *f* electrode chamber (compartment) *(of an electrodialyzer)*
Elektrodenkohle *f* electrode carbon
elektrodenlos electrodeless
Elektrodenmasse *f* electrode material, paste
E. für Söderberg-Elektroden Soderberg paste
grüne (rohe) E. green paste
Elektrodenmaterial *n* electrode material
Elektrodenpotential *n* electrode potential
Elektrodenreaktion *f* electrode reaction
Elektrodenspannung *f* electrode voltage
Elektrodenstampfmasse *f s.* Elektrodenmasse
Elektrodenwerkstoff *m s.* Elektrodenmaterial
Elektrodialysator *m* electrodialyzer
E. nach Pauli Pauli electrodialyzer
Elektrodialyse *f* electrodialysis
Elektrodispersion *f* electrodispersion, electrical dispersion
Elektroendosmose *f* electro[end]osmosis
Elektrofilter *n* electrical (electrostatic) precipitator, electrostatic filter
E. in Einzonenanordnung single-stage electrical precipitator

E. in Zweizonenanordnung two-stage electrical precipitator
Elektrofilterschlot *m* vertical-flow electrical precipitator
Elektrofiltration *f* electrical (electrostatic) precipitation, electrofiltration
Elektrographit *m* electrographite
Elektrogravimetrie *f* electrogravimetric (electrolytic deposition) analysis
Elektroheizung *f* electric heating
Elektroisolieröl *n* electrical insulating oil
Elektrokapillarität *f* electrocapillarity
Elektrokapillarkurve *f* electrocapillary curve
Elektrokeramik *f* electroceramics
elektrokinetisch electrokinetic
elektrokratisch *(coll)* electrocratic *(stabilized by electric charge)*
Elektrolichtbogenofen *m* electric-arc furnace
Elektrolumineszenz *f* electroluminescence
Elektrolyse *f* electrolysis
E. mit Quecksilberkatode mercury-cathode electrolysis
Elektrolyseraum *m s.* Elektrolysezelle
Elektrolysezelle *f* electrolysis (electrolytic) cell
E. mit Söderberg-Elektrode Soderberg cell
elektrolysieren to electrolyze
Elektrolysierzelle *f s.* Elektrolysezelle
Elektrolyt *m* electrolyte
amphoterer E. amphoteric electrolyte, ampholyte
ein-ein-wertiger E. uniunivalent electrolyte
ein-zwei-wertiger E. unibivalent electrolyte
kolloidaler E. colloidal electrolyte
schwacher E. weak electrolyte
starker E. strong electrolyte
Elektrolytbleiche *f (pap)* electrolytic bleach
Elektrolytbrücke *f* salt bridge
Elektrolyteisen *n* electrolytic iron
Elektrolytfällung *f* precipitation by electrolytes
Elektrolytgleichrichter *m* electrolytic rectifier
elektrolytisch electrolytic
Elektrolytkoagulation *f (coll)* flocculation by electrolytes
Elektrolytkondensator *m* electrolytic capacitor
Elektrolytkupfer *n* electrolytic copper
Elektrolytlösung *f* electrolytic solution, solution of electrolytes
Elektrolytnickel *n* electrolytic nickel
Elektrolytpulver *n* electrolytic powder
Elektrolytschlüssel *m s.* Elektrolytbrücke
Elektrolytsilber *n* electrolytic silver
Elektrolyttheorie *f* theory of electrolytes
Elektrolytverfahren *n* electrolytic method (process, technique)
Elektrolytvorlaufverfahren *n* ion-exclusion process *(chromatography)*
Elektrolytzink *n* electrolytic zinc
elektromagnetisch electromagnetic
Elektromagnetrolle *f* electromagnetic pulley *(in belt conveyors)*
elektromer electromeric
Elektron *n* electron ✦ **Elektronen abgeben** to release (lose) electrons ✦ **Elektronen aufnehmen** to accept (gain, acquire) electrons
anteiliges (aufgeteiltes) E. *s.* gemeinsames E.
äußeres E. outer (outside, external) electron
einsames E. unshared (unpaired, odd, nonbonding) electron
freies E. free electron
gemeinsames E. shared (sharing) electron
gepaartes E. paired electron
gestreutes E. scattered electron
nichtanteiliges (nichtbindendes) E. *s.* einsames E.
optisches E. optical (valence, outermost) electron
positives E. positive electron, posit[r]on, antielectron
primäres E. primary (initiating) electron
schnelles E. high-speed electron
sekundäres E. secondary electron
supraleitendes E. superconducting electron
ungepaartes (unpaariges) E. *s.* einsames E.
vagabundierendes E. stray electron
π-Elektron *n* π electron, pi electron, unsaturation electron
σ-Elektron *n* σ electron, sigma electron
elektronegativ electronegative
Elektronegativität *f* electronegativity
E. nach Pauling Pauling's electronegativity
Elektronegativitätsunterschied *m* electronegativity difference
Elektronenabgabe *f* release (donation) of electrons
Elektronenabgabevermögen *n* electron-releasing potency
elektronenabgebend electron-releasing, electron-donating
elektronenablösend *s.* elektronenabgebend
Elektronenablösung *f s.* Elektronenabgabe
elektronenabstoßend electron-repelling
elektronenaffin electron-affinitive
Elektronenaffinität *f* electron affinity, electroaffinity
Elektronenakzeptor *m* electron acceptor
Elektronenakzeptorstärke *f* electron acceptor strength
Elektronenanordnung *f s.* Elektronenkonfiguration
Elektronenanregung *f* electron excitation
elektronenanziehend electron-attracting
elektronenarm electron-deficient
Elektronenaufbau *m s.* Elektronenstruktur
Elektronenaufnahme *f* acceptance of electrons
elektronenaufnehmend electron-accepting
Elektronenauslösung *f s.* Elektronenabgabe
Elektronenaustausch *m* electron exchange
Elektronenaustauscher *m* electron exchanger
Elektronenaustauscherharz *n* electron-exchange resin
Elektronenbahn *f* electron[ic] orbit
Elektronenbelegung *f* electron density
Elektronenbeschleunigung *f* electron acceleration

Elektronenbeschuß *m* electron bombardment
Elektronenbeugung *f* electron diffraction
Elektronenbeugungsanalyse *f* electron diffraction analysis
Elektronenbeugungsbild *n*, **Elektronenbeugungsdiagramm** *n* electron diffraction pattern
Elektronenbeugungsversuch *m* electron diffraction experiment
Elektronen-Defektelektronen-Paar *n* electron-hole (hole-electron) pair
Elektronendichte *f* electron density
Elektronendon[at]or *m* electron donor
Elektronendrall *m* electron spin
Elektronendublett *n* doublet, duplet
Elektroneneinfang *m* electron capture
Elektroneneinfangdetektor *m* electron capture detector
Elektronenemission *f* electronic emission
  thermische E. thermionic emission
Elektronenenergie *f* electron[ic] energy
Elektronenentzug *m* electron removal
Elektronenfalle *f*, **Elektronenfänger** *m* electron trap
Elektronenformel *f* electron[ic] formula, dot formula
Elektronengas *n* electron gas
Elektronengitter *n* electron lattice
Elektronengruppe *f* electron group
Elektronenhaftstelle *f* electron trap
Elektronenhalbleiter *m* electronic semiconductor
Elektronenhülle *f* electron sheath
Elektronenkonfiguration *f* electron[ic] configuration, electron[ic] arrangement, orbital electron arrangement
  unstabile E. unstable electron configuration
Elektronenkonzentration *f* electron density
Elektronenkorrelation *f* electron correlation
Elektronenkreisbahn *f* electron[ic] orbit
Elektronenladung *f* electron[ic] charge
Elektronenlawine *f* [electron] avalanche, Townsend avalanche
Elektronenleiter *m* electronic conductor
Elektronenleitfähigkeit *f* electronic conductivity
Elektronenleitung *f* electron conduction
elektronenliefernd electron-donating, electron-releasing
Elektronenloch *n* electron hole
Elektronenloslösung *f s.* Elektronenabgabe
Elektronenlücke *f* electron gap
Elektronenmangel *m* electron deficiency, shortage of electrons
Elektronenmangelhydrid *n* electron-deficient hydride
Elektronenmangelverbindung *f* electron-deficient compound
Elektronenmasse *f* electron mass
Elektronenmikroskop *n* electron microscope
Elektronenmikroskopie *f* electron microscopy
Elektronenniveau *n* electronic (energy) level, term
Elektronenoktett *n* electron octet
Elektronenoktett-Anordnung *f* octet (eight-electron, inert-gas) structure, octet configuration
Elektronenorbital *n* electron orbital
Elektronenpaar *n* electron pair
  bindendes E. bonding pair of electrons
  einsames (freies) E. unshared (lone, free) electron pair
  gemeinsames E. pair of shared electrons
Elektronenpaarbindung *f* 1. *(state)* electron-pair bond (link, linkage), covalent (atomic, homopolar, non-polar, unitarian) bond; 2. covalent bonding *(process)*
π-Elektronenpaarbindung *f* π bond, pi pond
σ-Elektronenpaarbindung *f* σ bond, sigma bond
Elektronenpaarmethode *f* electron-pair (spin-state, valence-bond, Heitler-London-Slater-Pauling, VB, HLSP) method, method of valence-bond structures
Elektronenpolarisation *f* electron polarization
Elektronenquelle *f* electron source
Elektronenreichweite *f* range of electrons
Elektronenruh[e]masse *f* electron rest mass
Elektronenschale *f* electron shell
Elektronenschleuder *f* betatron
Elektronensextett *n* electron sextet
elektronenspendend *s.* elektronenliefernd
Elektronenspender *m s.* Elektronendonator
Elektronenspin *m* electron spin
Elektronenspinresonanz *f* electron paramagnetic (spin) resonance, EPR, ESR, paramagnetic [electronic] resonance, PMR
Elektronenspinresonanzspektroskopie *f* electron paramagnetic (spin) resonance spectroscopy, EPR spectroscopy, ESR spectroscopy
Elektronensprung *m* electron jump
Elektronensprungspektrum *n* electron jump spectrum
Elektronenstoß *m* electron impact
Elektronenstoßmethode *f* electron impact method
Elektronenstrahl *m (if bundled)* electron beam; *(if single)* electron ray
Elektronenstrahlschmelzen *n* electron-beam melting
Elektronenstruktur *f* electronic structure
Elektronentheorie *f* electronic theory
  E. der Metalle free-electron theory of metals
  E. der Valenz electronic theory of valency
Elektronenträger *m* electron carrier
Elektronenübergang *m* electronic transition
Elektronenüberschuß *m* excess of electrons
Elektronenüberschußhalbleiter *m* *n*-type semiconductor
Elektronenüberträger *m* electron carrier
Elektronenübertragung *f* electron transfer[ence]
Elektronenverschiebung *f* electron shift (displacement)
Elektronenverteilung *f* distribution of electrons
Elektronenvolt *n* electron-volt, eV, E.V.
Elektronenwanderung *f* electron migration
Elektronenwolke *f* electron cloud
elektronenziehend electron-attracting, electron-withdrawing, electrophilic
Elektronenzug *m* electron-attracting effect

Elektronenzusammenstoß *m* electron collision
Elektronenzustand *m* electronic state
elektroneutral electrically neutral
Elektroneutralität *f* electroneutrality
elektronisch electronic
Elektron-Positron-Paar *n* electron-positron pair
Elektroofen *m* electric furnace
Elektroosmose *f* electro[end]osmosis
Elektropherogramm *n* electropherogram
elektrophil electrophilic, electron-attracting, electron-withdrawing
Elektrophil *n* electrophile
Elektrophorese *f* electrophoresis
Elektrophoresegerät *n* electrophoresis (electrophoretic) apparatus
Elektrophoresekammer *f* electrophoresis cabinet, migration chamber
Elektrophoresetrog *m* electrophoresis tank
elektrophoretisch electrophoretic
elektroplattieren to [electro]plate, to electrodeposit metals
Elektroplattierung *f* [electro]plating, electrodeposition of metals
Elektroporzellan *n* electrical porcelain
elektropositiv electropositive
Elektroraffination *f* electrorefining, electrolytic refining
Elektroreinigen *n* electrical (electrostatic) precipitation
Elektrorüttler *m* electric (electrically driven) vibrator
Elektroscheiden *n* electrostatic separation
  E. mit Ionisation high-tension separation
Elektroschmelze *f* electrofusion
Elektroschmelzverfahren *n (met)* electric[-furnace] process
Elektrosortieren *n s.* Elektroscheiden
Elektrostahl *m* electrosteel, electric[-furnace] steel
Elektrostahlverfahren *n* electric[-furnace] process
elektrostatisch electrostatic
Elektrostriktion *f* electrostriction
Elektro-Teerfilter *n*, **Elektro-Teerscheider** *m* electrostatic tar filter
Elektrothermie *f* electrothermics
elektrothermisch electrothermal, electrothermic
Elektrotunnelofen *m* electric tunnel kiln
Elektro-Ultrafiltration *f* electroultrafiltration
elektrovalent electrovalent
Elektrovalenz *f* electrovalency, electrovalent (electrostatic, ionic, polar, heteropolar) bond, ionic relationship
  positive E. positive (electrochemical) valency
Elektrovibrator *m s.* Elektrorüttler
Elektrowalzenscheider *m* rotor separator
Elektrum *n (min)* electrum *(a natural alloy of gold and silver)*
Elektuarium *n (pharm)* electuary
Element *n* 1. element; 2. *(el chem)* [chemical] cell
  E. der Aktiniumreihe actinoid [element]
  E. der Lanthanreihe lanthanoid [element]
  atmophiles E. atmophile element
  biophiles E. biophile element
  chalkophiles E. chalcophile element
  chemisches E. chemical element
  dreiwertiges E. trivalent element
  einwertiges E. monovalent (univalent) element
  elektrochemisches E. [chemical] cell, voltaic (galvanic) cell (element)
  fünfwertiges E. pentavalent element
  galvanisches E. *s.* elektrochemisches E.
  lithophiles E. *(geol)* lithophile element
  mehrwertiges E. polyvalent (multivalent) element
  radioaktives E. radioactive element, radioelement
  reversibles E. *(el chem)* reversible cell (element)
  siderophiles E. *(geoch)* siderophile element
  umkehrbares E. *s.* reversibles E.
  vierwertiges E. tetravalent (quadrivalent) element
  zweiwertiges E. divalent (bivalent) element
elementar *(chem)* elemental
Elementaranalyse *f* ultimate (elementary) analysis
Elementarbestandteil *m* elementary constituent
Elementarfaden *m (text)* filament, continuous fibre (filament) *(a natural or man-made fibre of great or indefinite length)*
  schmelzgesponnener E. melt-spun filament
Elementarfadenbildung *f (text)* filament forming
Elementarfadenbündel *n (text)* strand
Elementarfadenkabel *n (text)* tow
Elementargitter *n (cryst)* translational lattice, translation grating
Elementarkörper *m (cryst)* unit cell
Elementarladung *f* **[/elektrische]** *s.* Elementarquantum / elektrisches
Elementarquantum *n* **/ elektrisches** elementary quantum (charge), unit [electric] charge
Elementarteilchen *n* elementary (fundamental, subnuclear) particle
Elementarwürfel *m (cryst)* cube
  einfacher E. simple cube
  flächenzentrierter E. face-centred cube
Elementarzelle *f (cryst)* unit cell
Elementsymbol *n* chemical sign (symbol)
Elementumwandlung *f* transmutation
Elemi[harz] *n* elemi *(from several specc. of Burseraceae, Rutaceae, and Humiriaceae)*
Elemiöl *n* elemi oil *(from Canarium luzonicum Miquel)*
Elevator *m* elevator, *(of a rotary-drilling installation also)* drill pipe elevator
Elfenbeinkarton *m* ivory cardboard
eliminieren to eliminate
Eliminierung *f* elimination
  ionische E. ionic elimination
Eliminierungs-Additions-Mechanismus *m* elimination-addition mechanism
Eliminierungsreaktion *f* elimination reaction
Elixier *n* elixir
Ellagengerbstoff *m* ellagitannin
Ellagsäure *f* ellagic acid *(a phenolic dilactone)*
Ellipsenbahn *f* elliptical orbit
Elmendorf-Prüfgerät *n (pap)* Elmendorf tester

Elmo-Pumpe *f* [Nash] Hytor pump
Elpasolith *m (min)* elpasolite *(potassium sodium hexafluoroaluminate)*
Elpidit *m (min)* elpidite *(a hydrated sodium zirconium silicate)*
Elsholtziaöl *n (cosmet)* Elsholtzia oil *(from Elsholtzia ciliata (Thunb.) Hyl.)*
Eluat *n* eluate *(liquid obtained by washing out adsorbed substances)*
Eluent *m* eluent, eluant, elutant, eluting agent (solvent)
eluieren to elute
Elution *f* elution
Elutionsanalyse *f* elution analysis
Elutionschromatografie *f* elution chromatography
Elutionsgeschwindigkeit *f* rate of elution
Elutionsmittel *n s.* Eluent
Elutriator *m* elutriator *(used for separating fine catalyst particles in moving-bed cracking)*
Eluvialhorizont *m (soil)* eluvial horizon, A-horizon
Email *n* [vitreous] enamel, *(Am)* porcelain enamel
Emailfarbe *f (ceram)* enamel (overglaze, vitrifiable) colour
Emaillack *m* enamel [varnish]
ofentrocknender E. stoving enamel
Emaillackfarbe *f* enamel paint
Emaille *f s.* Email
emaillieren to enamel
Emaillierofen *m* enamelling kiln
Emailwaren *fpl* enamel ware
Emanation *f (nucl)* [radioactive] emanation
emanieren to emanate
Emballage *f* packing (packaging) material, packing, package
Embolit *m (min)* embolite, bromchlorargyrite *(a silver halogenide)*
Emde-Abbau *m* Emde degradation *(of quaternary ammonium salts)*
Emeraldgrün *n* emerald green *(Guignet green or a mixture of Schweinfurth green and coal-tar colours)*
Emetikum *n (pharm)* emetic
Emetin *n* emetine *(alkaloid)*
Emission *f* emission, issue, *(nucl also)* ejection, expulsion
kalte E. field emission
lichtelektrische E. photoemission
thermische E. thermionic emission
Emissionselektrode *f* emitter electrode
Emissionskoeffizient *m* emission coefficient
Emissionslinie *f* emission line
Emissionsspektrometer *n* emission spectrometer
Emissionsspektroskopie *f* emission spectroscopy
Emissionsspektrum *n* emission spectrum
Emissionsvermögen *n* emissive power, emission capability *(energy emitted per unit time from each unit area of a surface)*; emissivity, emittance *(the ratio of the radiation emitted by a surface to the radiation emitted by a black body at the same temperature)*
Emitter *m* emitter
Emitterbereich *m s.* Emitterzone
Emitterelektrode *f* emitter [electrode]
Emitterübergang *m*, **Emitterübergangsschicht** *f* emitter junction
Emitterzone *f* emitter region
emittieren to emit, to issue, *(nucl also)* to eject, to expel
emittierend emissive
EMK *s.* Kraft / elektromotorische
EMK-Normal *n* standard of emf, standard of electromotive force
Emmonsit *m (min)* emmonsite *(a hydrous oxide of iron and tellurium)*
E-Modul *m* elastic modulus, Young's modulus [of elasticity], modulus of elasticity
empfängnisverhütend contraceptive
empfindlich sensitive *(instrument, material)*; labile, sensitive *(chemicals)*; delicate *(e. g. test)*
e. gegen Schwefel *(agric)* sulphur-shy
Empfindlichkeit *f* sensitivity, sensitiveness *(of instruments or material)*, *(phot also)* speed; lability *(of chemicals)*
E. gegen Reibung sensitiveness to friction
Empfindlichkeitsbereich *m (phot)* range of sensitivity
Empfindlichkeitsmesser *m (phot)* sensitometer
Empfindlichmachen *n (phot)* sensitization, sensitizing
empirisch empirical, through trial and error
Emplektit *m (min)* emplectite *(bismuth(III) copper(I) sulphide)*
emporheben to lift; to buoy up *(in a liquid)*; to energize *(electrons into an excited state)*
emporsteigen to rise, to ascend, to pass up[wards]
emporziehen *(glass)* to pull upward[s]
Emprotid *n* proton acceptor
empyreumatisch empyreumatic *(smell)*
Emscherbrunnen *m* Imhoff tank *(water purification)*
Emulgator *m* 1. emulsification machine, emulsifier; 2. emulsifier, emulsifying agent
nichtionogener E. non-ionic emulsifier
Emulgens *n s.* Emulgator 2.
emulgierbar emulsifiable, emulsible
Emulgierbarkeit *f* emulsifiability, emulsibility
emulgieren to emulsify
Emulgierfähigkeit *f* emulsifying power
Emulgiermaschine *f s.* Emulgator 1.
Emulgiermittel *n s.* Emulgator 2.
Emulgiermühle *f* emulsifying mill
Emulgierung *f* emulsification
Emulgiervermögen *n* emulsifying power
Emulsin *n* emulsin *(a mixture of enzymes that are active on β-glycosides)*
Emulsion *f* emulsion
E. niedriger Empfindlichkeit *s.* geringempfindliche E.
beständige E. tight emulsion
feinkörnige E. *(phot)* fine-grain[ed] emulsion
feste E. lisoloid *(a colloidal system consisting of a liquid surrounded by a solid phase)*
fotografische E. photographic (sensitive) emulsion

gehärtete E. *(phot)* hardened emulsion
geringempfindliche E. *(phot)* slow emulsion
grobkörnige E. *(phot)* coarse-grain[ed] emulsion
hart arbeitende E. *s.* kontrastreich arbeitende E.
hochempfindliche E. *(phot)* fast (high-speed) emulsion
kontrastarm arbeitende E. *(phot)* low-contrast emulsion
kontrastreich arbeitende E. *(phot)* high-contrast emulsion
lichtempfindliche E. *s.* fotografische E.
panchromatische E. *(phot)* panchromatic emulsion
pharmazeutische E. pharmaceutic[al] emulsion
schnellbrechende E. quick-breaking emulsion
wäßrige E. aqueous emulsion

emulsionieren to emulsify
emulsionsartig emulsive
Emulsionsbeständigkeit *f* emulsion stability
Emulsionsbildner *m* emulsifying agent, emulsifier
Emulsionsbildung *f* emulsification
Emulsionsbrecher *m*, **Emulsionsentmischer** *m s.* Emulsionsspalter
Emulsionsentmischung *f s.* Emulsionsspaltung
Emulsionskolloid *n s.* Emulsoid
Emulsionsmischpolymerisation *f* emulsion copolymerization
Emulsionsöl *n* emulsion oil
Emulsionspolymer[es] *n*, **Emulsionspolymerisat** *n* emulsion polymer
Emulsionspolymerisation *f* emulsion polymerization
Emulsions-Polyvinylchlorid *n* emulsion polyvinylchloride
Emulsionsschicht *f (phot)* emulsion layer
Emulsionsschleier *m (phot)* emulsion fog
Emulsionsspalter *m* demulsifier
Emulsionsspaltung *f* breaking (cracking) of emulsions, de-emulsification, demulsification
Emulsionsspinnverfahren *n (text)* emulsion spinning
Emulsionsspülung *f (petrol)* emulsion-type mud
Emulsionsstabilisator *m* emulsion stabilizer
Emulsionsstabilität *f* emulsion stability
Emulsionsträger *m (phot)* emulsion support
Emulsionstyp *m* emulsion type
Emulsionsunterlage *f s.* Emulsionsträger
Emulsionsverdichtung *f*, **Emulsionsverdickung** *f* emulsion thickening, creaming
Emulsionsvermittler *m* emulsifying agent, emulsifier
Emulsionswäsche *f* emulsion scouring
Emulsoid *n* emulsoid [colloid]
Emulsor *m s.* Emulgator 1.
Enamin *n (org chem)* enamine
enantiomer *s.* enantiomorph
Enantiomer[es] *n* enantiomorph, enantiomer, enantiomorphous form (isomer), optical isomer (antipode, opposite), antimer, mirror-image isomer
enantiomorph enantiomorphous, enantiomorphic, enantiomeric
Enantiomorphie *f* enantiomorphism, optical isomerism, mirror-image isomerism (relationship)
enantiotrop enantiotropic
Enantiotropie *f* enantiotropy
Enargit *m (min)* enargite *(arsenic(III) copper(I,II) sulphide)*
Endatom *n* terminal (end) atom
Endbleiche *f* final bleaching
Endbleichstufe *f (pap)* whitening stage *(in which the pulp reaches its maximum whiteness)*
Enddruck *m* final pressure, *(vacuum technology also)* ultimate (maximum) vacuum
Ende *n* end; finish
E. der Trockenpartie *(pap)* dry end of the dryer section
E. der Umwandlung finish of transformation
verjüngtes E. constricted end *(of a pipe)*

Enderzeugnis *n s.* Endprodukt
Endfallgeschwindigkeit *f* terminal [falling] velocity
Endfestigkeit *f* final strength
Endfeuchte[beladung] *f* final moisture content, FMC
Endfläche *f* base *(of a crystal)*
Endgas *n* end gas
Endgruppe *f* terminal (end) group
C-terminale E. C-terminal group (residue) *(in proteins)*

Endgruppenanalyse *f* end group analysis
Endgruppenbestimmung *f* end group assay
Endhypochloritbleiche *f (pap)* last hypochlorite treatment, final hypochlorite stage
Endkochpunkt *m* final boiling point, F. B. P.
Endkomponente *f* final (end) component
Endkonzentration *f* final concentration (strength)
Endlauge *f* final lye *(as in fertilizer manufacture)*
Endlosfaser *f s.* Elementarfaden
Endlosgarn *n (text)* continuous-filament yarn
Endlöslichkeit *f* final solubility
Endlossieb *n (pap)* endless wire
Endmelasse *f (sugar)* final molasses
Endoenzym *n* endoenzyme, endocellular (intracellular) enzyme (ferment)
endogen endogenous
endokrin endocrine
Endomethylenbrücke *f* endomethylene bridge ✦ mit E. endomethylene-bridged
Endomorphose *f (geol)* endomorphism
Endopeptidase *f* endopeptidase
endo-Produkt *n* endo product
Endosmose *f* endosmosis
endosmotisch endosmotic
endotherm endothermic
Endotoxin *n* endotoxin
endozyklisch endocyclic
Endphase *f* end phase
Endprodukt *n* final (finished, consumer) product; *(phys chem)* final (end) product *(as of a radioactive transformation)*
Endprodukthemmung *f (bioch)* feedback [inhibition]

Endpunkt *m* end (final) point *(as of a titration)*
Endpunktsbestimmung *f* end-point determination
  E. nach Gay-Lussac Gay-Lussac method
  E. nach Mohr Mohr method
  E. nach Volhard Volhard method
Endpunktserkennung *f* end-point detection
Endschneide *f* end (terminal) knife edge *(of a balance)*
Endschneidenhaltung *f* stirrup *(of a balance)*
Endsiedepunkt *m* final boiling point, F. B. P.
Endspreitungskoeffizient *m (coll)* final spreading coefficient
endständig terminal
Endtemperatur *f* final (end) temperature
Endtrocknung *f* final drying
Endung *f (nomencl)* ending, termination
Endvakuum *n* ultimate (maximum) vacuum
Endvergärung *f* end fermentation
Endwäsche *f* final washing
Endweiße *f (pap)* final brightness
Endzustand *m* final state (condition)
Energie *f* energy ✦ **von geringer E.** low-energy
  freie E. free energy
  freie E. der Grenzfläche interfacial free energy
  innere E. intrinsic (internal) energy
  kinetische E. kinetic energy
  molare freie E. molar free energy
  nukleare E. nuclear energy
  potentielle E. potential energy
  thermische E. thermal energy
Energieabgabe *f* energy output, release of energy
energieabgebend exoergic
Energieabnahme *f* decrease in energy
energiearm low-energy, poor in energy
Energieaufwand *m* expenditure of energy
Energieaustausch *m* interchange of energy
Energieband *n s.* Energiebereich
Energiebarriere *f* energy barrier
Energiebeitrag *m* energy contribution
Energiebereich *m* energy range (band), band
  erlaubter E. allowed band
  leerer E. empty band
  nicht besetzter E. *s.* leerer E.
  nicht zugelassener E. *s.* verbotener E.
  unbesetzter E. empty band
  verbotener E. forbidden band, energy gap
  zugelassener E. allowed band
Energieberg *m* energy barrier
Energiebetrag *m* amount of energy
Energiebilanz *f* energy balance
Energiebrutreaktor *m (nucl)* power breeder reactor
Energiedegradation *f* degradation of energy
Energiediagramm *n* energy diagram
Energiedichte *f* energy density
  kohäsive E. cohesive energy density, C. E. D.
Energiedifferenz *f* energy difference
Energiedissipation *f* dissipation of energy
Energieeinheit *f* energy unit
Energieerhaltung *f* conservation of energy
Energieerhaltungssatz *m* energy principle, law of conservation of energy
energiefreigebend exoergic
Energiegewinn *m* gain in energy
Energiegleichgewicht *n* energy balance
Energiehaushalt *m* energy balance
Energieinhalt *m* energy content
Energiekette *f* energy chain
Energielücke *f* energy gap, forbidden band
Energieniveau *n* energy level, term
  E. des Atoms atomic [energy] level
Energieniveaudiagramm *n* energy diagram
Energieprinzip *n* energy principle, law of conservation of energy
Energiequant[um] *n* energy quantum
Energiequelle *f* source of energy
  starke E. high-energy source
Energiereaktor *m* power reactor
energiereich high-energy, rich in energy
Energie-Reichweite-Beziehung *f* range-energy relation
Energieschranke *f* energy barrier
Energieschwelle *f* energy barrier
Energiesenkung *f* energy reduction
Energiestufe *f*, **Energieterm** *m* energy level, term
Energieüberführung *f* energy transfer
Energieübergang *m* energy transfer
Energieüberschuß *m* excess energy
Energieübertragung *f* energy transfer
Energieumformung *f s.* Energieumwandlung
Energieumsatz *m (bioch)* energy turnover
Energieumwandlung *f* transformation of energy, energy change
Energieunterschied *m* energy difference
Energieverbrauch *m* energy consumption
Energieverlust *m* loss of energy
Energieverteilung *f* energy distribution
Energieverteilungsgesetz *n* / **Boltzmannsches** Boltzmann distribution law
Energiewandlung *f s.* Energieumwandlung
Energiezufuhr *f* energy input
Energiezustand *m* energy state
  E. des Atoms atomic state
energisch vigorous, drastic *(treatment)*
Enfleurage *(cosmet)* enfleurage *(method for obtaining odoriferous substances by absorption with fats)*
Enfleurageöl *n (cosmet)* absolute of enfleurage, enfleurage absolute
Engel-Verfahren *n (plast)* Engel process *(a powder sintering process)*
Enghalsflasche *f* narrow-neck[ed] bottle, narrow-mouth bottle
Enghalskolben *m* narrow-neck[ed] flask
engklassiert closely graded
Engler-Kolben *m* Engler flask
Englischrot *n* polishing rouge *(iron(III) oxide)*
engmaschig close-meshed, narrow-meshed, fine-meshed
Engobe *f (ceram)* engobe
Engobeton *m (ceram)* slip clay
engobieren *(ceram)* to engobe

engporig fine-pored, finely pored (porous)
E-Nickel *n* electrolytic nickel
Enneoxodijodat(VII) *n* $M_4^IlI_2O_9$ enneaoxodiiodate(VII), dimesoperiodate
Enol *n (org chem)* enol
Enolase *f* enolase
Enolat *n (org chem)* enolate
Enolatmesomerie *f*, **Enolatresonanz** *f* enolate resonance
Enolform *f* enol[ic] form (structure)
enolisch enolic
enolisierbar enolizable
  nicht e. non-enolizable
enolisieren to enolize
Enolisierung *f* enolization
Enolisierungstendenz *f* enolization tendency
Enolkonstante *f* keto-enol constant
Enometrie *f* enometry *(determination of double bonds in fats by addition of halogens)*
Enstatit *m (min)* enstatite *(magnesium metasilicate)*
entaktivieren to deactivate; *(nucl)* to decontaminate
Entaktivierung *f* deactivation; *(nucl)* decontamination
entalkylieren to dealkylate
Entalkylierung *f* dealkylation
entamidieren to deamidate
Entamidierung *f* deamidation
entarretieren to unlock
entarten to deteriorate; *(phys chem)* to degenerate
Entartung *f* deterioration; *(phys chem)* degeneracy
  zufällige E. *(phys chem)* accidental degeneracy
Entartungsgrad *m (phys chem)* degree of degeneracy
Entartungstemperatur *m (phys chem)* degeneracy temperature
entaschen to deash
Entaschung *f* deashing
entasphaltieren to deasphalt
Entasphaltierung *f* deasphalting
  E. mit Propan propane deasphalting
Entäthaner *m* de-ethanizer
entäthanisieren to de-ethanize
Entäthanisierung *f* de-ethanization
entazetylieren to deacetylate
Entazetylierung *f* deacetylation
entbasen to dealkalize
entbasten to decorticate *(vegetable fibres)*; to degum, to scour, to boil off (out) *(silk)*
Entbasten *n* decortication *(of vegetable fibres)*; degumming, scouring, boil[ing]-off *(of silk)*
Entbastungsbad *n (text)* degumming (scouring, boiling-off) liquor, degumming bath
  gebrochenes E. broken degumming liquor
Entbastungsflotte *f s.* Entbastungsbad
Entbastungsmittel *n (text)* degumming (scouring) agent
Entbasung *f* dealkalization
entbenzol[ier]en to debenzolize
Entbenzol[ier]ung *f* debenzolization
entbinden to release, to set free, to liberate *(e. g. heat)*
Entbindung *f* release, setting-free, liberation *(as of heat)*
entbittern to debitter[ize]
Entblätterungsmittel *n (agric)* defoliant
entbromen to debrominate
Entbromung *f* debromination
Entbrühungssieb *n* drain (rinse) screen
Entbutaner *m* debutanizer
entbutanisieren to debutanize
Entbutanisierkolonne *f* debutanizer
Entbutanisierung *f* debutanization
entchloren to dechlorinate
Entchlorung *f* dechlorination
Entchlorungsmittel *n* dechlorinating agent
Entdeckungsbohrung *f (petrol)* discovery well
enteisen to de-ice, to defrost
enteisenen to deferrize *(e. g. water)*
Enteisenung *f* deferrization *(as of water)*
Enteisenungsanlage *f* iron-removal plant
Enteisung *f* de-icing, defrosting
Enteisungsanlage *f* de-icer, defroster
enteiweißen to deproteinize
Enteiweißung *f* deproteinization
Entemaillieren *n* de-enamelling
entemulsionieren to demulsify, to de-emulsify, to break, to crack *(an emulsion)*
Entemulsionieren *n* demulsification, de-emulsification, breaking, cracking *(of an emulsion)*
Enteramin *n* serotonin, enteramine, 3-(2-aminoethyl)-5-hydroxyindole
Enterokinase *f* enterokinase
entfärben to decolour[ize], to discolour, to bleach; *(pap)* to whiten, to brighten; to deink *(waste paper)*
  sich e. to decolour[ize], to discolour, to bleach out
entfärbend decolo[u]rant
Entfärber *m s.* Entfärbungsmittel
Entfärbung *f* decolo[u]rization, discolo[u]ration, bleaching; *(pap)* whitening, brightening; deinking *(of waste paper)*
  E. in Abgasatmosphäre *(text)* gas fading
Entfärbungserde *f (petrol)* decolo[u]rizing (discolouring, bleaching) earth, decolo[u]rizing clay
Entfärbungshilfsmittel *n s.* Entfärbungsmittel
Entfärbungskohle *f* decolo[u]rizing carbon (charcoal)
Entfärbungsmittel *n* decolo[u]rizing (stripping) agent (assistant), decolo[u]rizer, decolo[u]rant
entfernen to eliminate, to remove, to discharge, to abstract
  Druckfarbe e. to deink *(waste paper)*
  flüchtige Bestandteile e. to devolatilize
  Flüssigkeit e. to deliquefy
  Kesselstein e. to [de]scale
  Lignin e. to delignify *(wood)*
  Lösungsmittel e. to desolventize
Entfernen *n* elimination, removal, discharge, abstraction

Entferner *m* remover
Entfernung *f s.* Entfernen
entfetten to degrease, to defat; to scour *(wool)*
Entfettung *f* degreasing, defatting; scouring *(of wool)*
elektrolytische E. electrolytic degreasing
Entfettungsmittel *n* degreasing agent, degreaser
entfeuchten to dehydrate, to dewater, to dehumidify, to desiccate, to dry
Entfeuchtung *f* dehydration, dewatering, dehumidification, desiccation, drying
entflammbar [in]flammable
nicht e. uninflammable, non-[in]flammable, flameproof
Entflammbarkeit *f* [in]flammability
entflammen 1. to flash, to inflame, to burst into flame; 2. to inflame *(something)*
Entflammung *f* inflammation
Entflammungstemperatur *f* ignition (kindling) point (temperature)
Entflammungszeit *f* inflammation time
entflocken *(coll)* to deflocculate
sich e. to deflocculate
Entflocker *m* deflocculant, deflocculent, deflocculator, deflocculating agent
Entflockung *f* deflocculation
entfluorieren to defluorinate
Entfluorierung *f* defluorination
Entformungsmittel *n* [mould-]release agent, [mould-]release medium, mould lubricant
entfrosten to defrost, to thaw
entgasen to degas, to degasify, to outgas; *(coal)* to coke, to carbonize; to carbonize *(wood)*; *(plast)* to vent, to breathe *(the mould)*
Entgaser *m* degasser, degasifier
Entgasung *f* degassing, degasification, outgassing; *(coal)* coking, carbonization; dry distillation, carbonization *(of wood)*; *(plast)* venting, breathing *(of the mould)*
Entgasungsanlage *f* degasser, degasifier
Entgasungsextruder *m (plast)* vent[ed] extruder
Entgasungsgas *n (coal)* carbonization gas
Entgasungsgerät *n* degasser, degasifier
Entgasungsschacht *m* carbonization chamber *(of a predistillation gas producer)*
Entgasungsschneckenpresse *f (plast)* vent[ed] extruder
Entgasungsverfahren *n (coal)* carbonization process
entgegenwirken to counteract
entgerben to de-tan
entgiften to detoxicate, to detoxify; *(nucl)* to decontaminate
Entgiftung *f* detoxication, detoxification; *(nucl)* decontamination
entglasen to devitrify
Entglasung *f* devitrification
entgraten to trim, to fettle, *(plast also)* to deburr, to deflash
Enthaareisen *n (tann)* unhairing knife
enthaaren to depilate, to epilate, *(tann also)* to dehair, to unhair
enthaarend depilatory
Enthaarung *f* depilation, epilation, *(tann also)* dehairing, unhairing
Enthaarungscreme *f (cosmet)* depilatory cream
Enthaarungsmaschine *f (tann)* unhairing machine
Enthaarungsmittel *n (cosmet, tann)* depilatory [agent], depilitant, epilator, hair remover
Enthaarungspulver *n* depilatory powder
Enthaarungswachs *n* epilating wax
Enthalpie *f* enthalpy, heat content
Enthalpieänderung *f* enthalpy change
Enthalpie-Entropie-Diagramm *n* enthalpy-entropy chart (diagram), Mollier chart
enthärten to soften *(water)*
Enthärter *m* 1. softener, softening agent (material); 2. softener, softening unit
Enthärtung *f* softening *(of water)*
Enthärtungsanlage *f* softening plant (installation)
Enthärtungsmittel *n* softener, softening agent (material); alkali builder *(in soaps)*
entholzen *(text)* to decorticate
Entholzen *n (text)* decortication
Entionisation *f* deionization
entionisieren to deionize
Entionisierung *f* deionization
Entionisierungsmittel *n* deionizer
Entionisierungspotential *n* deionization potential
Entionisierungszeit *f* deionization time
Entisobutaner *m* deisobutanizer
entisobutanisieren to deisobutanize
Entisobutanisierkolonne *f* deisobutanizer
Entisobutanisierung *f* deisobutanization
entkalken to unlime, to delime, to decalcify
entkälken *(tann)* to delime
Entkalkung *f* deliming, decalcification
Entkälkung *f (tann)* deliming
Entkälkungsmittel *n (tann)* deliming agent
entkarbonisieren to decarbonate, to decarbonize *(water)*
Entkarbonisierung *f* decarbon[iz]ation *(of water)*
Entkarbonisierungsanlage *f* decarbonization plant, decarbonator
entkeimen to sterilize, to degerm, *(relating to pathogenic organisms)* to disinfect
Entkeimung *f* sterilization, degermation, *(relating to pathogenic organisms)* disinfection
Entkeimungsapparat *m* sterilizer
entkernen *(text)* to gin *(cotton fibre)*
Entkernungsmaschine *f (text)* gin
entkieseln to desilicify
Entkieselung *f* desilicification
entknäueln / sich to uncoil *(of molecules)*
Entknäuelung *f* uncoiling *(of molecules)*
entkohlen 1. *(petrol)* to decarbonize, to decoke; 2. *(met)* to decarburize; 3. *(text)* to carbonize *(raw wool)*
Entkohlung *f* 1. *(petrol)* decarbonization, decoking; 2. *(met)* decarburization; 3. *(text)* carbonization *(removal of burs from raw wool)*
entkoppeln *(bioch)* to uncouple
entkorken *(ferm)* to disgorge *(for removing lees)*

Entkorken *n (ferm)* disgorgement *(for removing lees)*
entkrusten to [de]scale
entkupfern to decopperize, to decopperate
Entkupferung *f* decopperizing
Entladeklappe *f* discharge door
entladen 1. *(tech)* to discharge, to unload; 2. *(phys)* to discharge
Entladen *n s.* Entladung
Entladevorrichtung *f* discharging apparatus (device), discharger
Entladung *f* 1. *(tech)* discharge, unloading; 2. *(phys)* discharge
Entladungselektrode *f* discharge electrode
Entladungserscheinung *f* discharge phenomenon
Entladungspotential *n* discharge potential
Entladungsröhre *f* [gas] discharge tube
Entlastungsventil *n* relief (unloading) valve
Entlaubungsmittel *n* defoliant
entleeren to discharge, to unload, to evacuate, to empty, to drain *(a vessel)*; to siphon, to drain *(a liquid)*; to dump *(bulk material)*
Entleerung *f* discharge, unloading, evacuation, emptying, drainage *(of a vessel)*; siphoning, drainage *(of a liquid)*; dumping *(of bulk material)*
Entleerungsklappe *f* discharge door
Entleerungsschieber *m* discharge gate
Entleerungsstutzen *m* discharge (runoff) pipe
Entleerungsvorrichtung *f* discharging apparatus (device), discharger
entlüften to deaerate, to degas, to bleed *(e.g. a vessel)*; to air out, to purge *(gases)*; *(ceram)* to de-air *(the clay)*; *(plast)* to breathe, to vent, to degas *(the mould)*; *(pap)* to relieve *(a digester)*
Entlüfter *m* deaerator
Entlüftung *f* 1. deaeration, degassing, bleeding *(as of a vessel)*; airing-out, purging *(of gas)*; *(ceram)* de-airing *(of clay)*; *(plast)* breathing, venting, degassing *(of the mould)*; *(pap)* relief *(of a digester)*; 2. *s.* Entlüftungseinrichtung
Entlüftungsapparat *m s.* Entlüftungseinrichtung 2.
Entlüftungseinrichtung *f* 1. air relief (vent); 2. deaerator *(as of a steam generator)*
Entlüftungskammer *f (ceram)* de-airing chamber *(of a vacuum extrusion press)*
Entlüftungskanal *m* vent channel; *(plast)* mould vent
Entlüftungsleitung *f* vent line
Entlüftungsöffnung *f* air vent (relief)
Entlüftungspause *f (plast)* dwell *(in moulding)*
Entlüftungsrohr *n* vent (ventilator, blow) pipe
Entlüftungsventil *n* vent valve
entmagnetisieren to demagnetize
Entmagnetisierung *f* demagnetization
 adiabatische E. adiabatic demagnetization
Entmagnetisierungsfaktor *m* demagnetizing factor
entmanganen to demanganize
Entmanganung *f* demanganization
Entmethaner *m* demethanizer
entmethanisieren to demethanize
Entmethanisierung *f* demethanization
entmethylieren to demethylate
Entmethylierung *f* demethylation
entmineralisieren to demineralize
Entmineralisierung *f* demineralization
entmischen to separate, to segregate; to break, to crack, to demulsify, to de-emulsify *(an emulsion)*
 sich e. to separate, to segregate; *(of emulsions)* to break, to crack, to deteriorate; *(of fertilizers)* to disintegrate
Entmischung *f* separation, segregation; *(relating to emulsions)* breaking, cracking, demulsification; disintegration *(of fertilizers)*
 liquide E. *(geol)* liquation [differentiation] *(of fused rock)*
Entnahme *f* 1. take-off, offtake, withdrawal, discharge; *(glass)* take-out; 2. discharge point; 3. *s.* Entnahmevorrichtung
 doppelphasige E. double withdrawal *(in extracting)*
 einphasige E. single withdrawal *(in extracting)*
 vollständige E. diamond separation, completion of square *(in extracting)*
 wechselphasige E. alternate withdrawal *(in extracting)*
Entnahmeende *n* discharge end
Entnahmegreifer *m (glass)* take-out tongs (jaw)
Entnahmeloch *n (glass)* gathering hole (opening)
Entnahmeöffnung *f* discharge opening (outlet, door, aperture, port)
Entnahmestelle *f* discharge point; *(glass)* gathering hole (opening)
Entnahmeteil *n (glass)* working chamber (end)
Entnahmeverhältnis *n (distil)* rate of withdrawal
Entnahmevorrichtung *f* discharging device, discharger, discharge, take-off, offtake; *(glass)* take-out [mechanism]
entnaphthal[is]ieren to denaphthalize
Entnaphthal[is]ierung *f* denaphthalization
entnehmen to take off, to withdraw, to discharge; *(glass)* to take out
 eine Probe e. to sample
entolen *(tann)* to de-olate
entölen to deoil
Entolung *f (tann)* de-olation
Entölung *f* deoiling
entorientieren to disorient
entozonisieren to deozonize
Entozonisierung *f* deozonization
entparaffinieren to deparaffin[ize], to dewax
Entparaffinierung *f* deparaffinization, dewaxing
 E. mit Lösungsmitteln solvent dewaxing
 E. mit Propan propane dewaxing
Entparaffinierungsanlage *f* dewaxing plant
Entpentaner *m* depentanizer
entpentanisieren to depentanize
Entpentanisierung *f* depentanization
entphenol[ier]en to dephenolize
Entphenolierung *f* dephenol[iz]ation
Entphenolierungsanlage *f* dephenolizing plant
Entphenolung *f s.* Entphenolierung

entphosphoren to dephosphorize
Entphosphorung *f* dephosphorization
entpickeln *(tann)* to depickle
Entpolarisierungsgrad *m* degree of depolarization
entpolymerisieren to depolymerize
Entpolymerisierung *f* depolymerization
Entpropaner *m* depropanizer
entpropanisieren to depropanize
Entpropanisierkolonne *f* depropanizer
Entpropanisierung *f* depropanization
entquellen *(tann)* to deplete
entrahmen to cream [off], to skim *(milk)*
Entrahmung *f* creaming, skimming *(of milk)*
Entrahmungsschärfe *f (food)* creaming (skimming) efficiency
Entrahmungsschleuder *f*, **Entrahmungszentrifuge** *f* milk (cream) separator, milk centrifuge, skimming machine
entrinden *(pap)* to [de]bark, to peel, *(Am also)* to ross
Entrinder *m (pap)* barking machine, barker
  hydraulischer E. hydraulic (stream) barker
entrindet *(pap)* bark-free
Entrindung *f (pap)* [de]barking, peeling, *(Am also)* rossing
  chemische E. chemical [de]barking
  hydraulische E. hydraulic [de]barking
  mechanische E. mechanical [de]barking
Entrindungsanlage *f (pap)* barking plant
Entrindungsmaschine *f (pap)* barking machine, barker
Entrindungstrommel *f (pap)* barking drum, drum barker, tumbler
Entropie *f* entropy
  E. der Nullpunktskonfiguration zero-point configurational entropy
  molare E. molar entropy
  partielle molare E. partial molar entropy
Entropieabnahme *f* entropy decrease
Entropieänderung *f* entropy change
Entropieanteil *m* entropy contribution
Entropieeffekt *m* entropy effect
Entropieeinheit *f* entropy unit
Entropieerzeugung *f* entropy production
Entropiesatz *m* law of entropy
Entropieverlust *m* loss of entropy
Entropiewert *m* entropy value
Entropiezunahme *f*, **Entropiezuwachs** *m* entropy increase, positive entropy change
Entrostungsmittel *n* rust-removing agent, rust remover
Entrußen *n* desooting
entsaften *(food)* to sap
entsalzen to desalinate, to desalt, to free from salt
Entsalzung *f* desalin[iz]ation, desalting
Entsalzungsaggregat *n* desalting (desalinating, desalination) unit, desalinator
Entsalzungsanlage *f* desalting (desalinating, desalination) plant
entsäuern to deacidify, to neutralize
Entsäuerung *f* deacidification, neutralization
  kontinuierliche E. continuous neutralization *(in margarine making)*
entschälen *(text)* to boil off (out), to degum; to scour *(silk)*
Entschälen *n (text)* boiling-off, degumming; scouring *(of silk)*
entschäumen to defoam, to skim [off]
Entschäumer *m s.* Entschäumungsmittel
Entschäumung *f* defoaming
Entschäumungsmittel *n* foam inhibitor (killer, breaker, destroyer), defoaming (antifoam, antifroth) agent, defoamer, antifoamer, froth-preventing agent, froth suppressor
entschlacken to slag, to free from slag
Entschlackung *f* slagging, *(coal also)* clinker discharge
entschlammen to desludge, to deslime
entschleimen to deslime; to degum *(oil)*
entschlichten *(text)* to desize, to free from size
Entschlichten *n (text)* desizing
  enzymatisches (fermentatives) E. enzyme desizing
  thermisches E. heat cleaning *(of glass fibre material)*
Entschlichtungsbad *n (text)* desizing bath
Entschlichtungsmittel *n (text)* desizing agent
entschwefeln to desulphur[ize]
Entschwefelung *f* desulphur[iz]ation
  hydrierende E. hydrodesulphurization, HDS
  trockene E. dry desulphurization
Entschweißbad *n (text)* scouring bath *(for wool)*
entschweißen *(text)* to scour, to degrease *(wool)*
Entschweißen *n (text)* scouring, degreasing, desuinting *(of wool)*
Entschweißungsmittel *n (text)* scouring (degreasing) agent, degreaser
entseuchen *(med)* to disinfect; *(nucl)* to decontaminate
Entseuchung *f (med)* disinfection; *(nucl)* decontamination
  radioaktive E. radioactive decontamination
Entseuchungsfaktor *m*, **Entseuchungsgrad** *m (nucl)* decontamination factor
Entseuchungsindex *m (nucl)* decontamination index
Entseuchungsmittel *n (nucl)* decontaminating agent (chemical, substance)
entsilbern to desilver[ize]
Entsilberung *f* desilverization, desilvering
entspannen to expand, to release; *(pap)* to relieve *(a digester)*
Entspanner *m* expansion valve
Entspannung *f* expansion, release, stress relief; *(pap)* relief *(of a digester)*
Entspannungsdestillation *f* / **[kontinuierliche]** flash distillation, continuous equilibrium vaporization
Entspannungsglühen *n* stress relief annealing, stress-relieving anneal
Entspannungskammer *f* flash chamber *(of a flash evaporator)*
Entspannungskühler *m* flash cooler

Entspannungsmaschine *f* expansion engine, expander [machine]
Entspannungsofen *m* stress-relieving furnace
Entspannungsventil *n* expansion valve
Entspannungsverdampfer *m* flash evaporator, flasher
Entspannungsverdampfung *f* flash (instantaneous) vaporization *(act)*; flash evaporation *(process)*
entsprechen / den Anforderungen des Arzneibuchs to be of pharmacopoeial quality
entstabilisieren to destabilize
entstauben to [de]dust
entstäuben *s.* entstauben
Entstauber *m* 1. dust catcher (collector, separator, settler); 2. elutriator *(moving-bed cracking)*
Entstaubung *f* dedusting
  elektrische (elektrostatische) E. electrical (electrostatic) precipitation
Entstäubungsapparat *m (pap)* dusting machine, duster *(for rags)*
Entstaubungsgrad *m* collection (separation) efficiency
  logarithmischer E. decontamination factor, DF
entstearin[is]ieren to destearinate, to destearinize, to demarg[ar]inate, to winterize *(oils)*
Entstearin[is]ierung *f* destearinization, demargarination, winterization *(of oils)*
entstehen to originate, to be formed, to form
Entstehung *f* origination, formation, generation, nascency; *(geoch)* genesis *(as of coal or petroleum)*
entstrahlen *(nucl)* to decontaminate
Entstrahlung *f (nucl)* decontamination
entteeren to detar
Entteerer *m* tar separator (extractor)
Entteerung *f* detarring, tar separation
enttoluolen to detoluate
Enttoluolen *n* detoluation
entwachsen to dewax
Entwachsen *n* **mit Lösungsmitteln** solvent dewaxing
entwässerbar dewaterable
Entwässerbarkeit *f* dewaterability
Entwässerer *m* dehydrator
entwässern 1. to dehydrate, to desiccate, to dewater, to dry; 2. *(pap)* to drain *(the web in the wet part)*; to decker *(to pass pulp over a wet machine)*
Entwässerung *f* 1. dehydration, desiccation, dewatering; 2. *(pap)* drainage *(of the web)*; deckering *(of pulp)*
Entwässerungsanlage *f* dehydration plant, concentrator
Entwässerungsgerät *n* dehydrator
Entwässerungsgeschwindigkeit *f (pap)* rate of drainage
Entwässerungsgrad *m (pap)* freeness [value]
Entwässerungsgradbestimmung *f (pap)* freeness test
Entwässerungsgradprüfer *m (pap)* freeness tester
Entwässerungsprüfung *f (pap)* freeness test
Entwässerungskanal *m* drainage channel
Entwässerungsleitung *f* drainage line
Entwässerungsmaschine *f (pap)* decker, thickener, concentrator, wet machine
Entwässerungsmittel *n* dewatering (dehydrating) agent, dehydrator
Entwässerungsperiode *f (pap)* drainage period
Entwässerungssieb *n* drain (rinse) screen
Entwässerungssystem *n* system of drainage, water-carriage system
Entwässerungswiderstand *m (pap)* drainage resistance
Entwässerungszeit *f (pap)* drainage period
entweichen to escape, to leak [out], to issue
Entweichen *n* escape, leak
  E. von Dämpfen outbreathing
Entwertung *f* degradation *(as of energy)*
entwesen to disinfest
Entwesung *f* disinfestation
entwickeln to evolve, to generate, to liberate, to release *(e. g. gas or heat)*; to develop, to evolve *(e. g. a method)*; to develop *(a photograph or chromatogram)*
  sich e. to evolve, to be evolved, to form, to be formed
  zur Betriebsreife e. to bring to the commercial stage
Entwickler *m (phot)* developing agent, developer; *(dye)* developing agent, developer, coupling component
  E. für Papiere *(phot)* print developer
  erschöpfter E. *(phot)* exhausted developer
  fotografischer E. photographic developer
  gerbender E. *(phot)* tanning developer
  hart (kontrastreich) arbeitender E. *(phot)* high-contrast developer
  verbrauchter E. *(phot)* exhausted developer
  weich arbeitender E. *(phot)* low-contrast (soft-working) developer
Entwicklerbad *n (phot)* developing bath
Entwicklerflecken *mpl (phot)* developer stains
Entwicklerformel *f* developer formula
Entwicklerlösung *f (phot)* developer (developing) solution
Entwicklerschale *f (phot, lab)* developing dish
Entwicklersubstanz *f (phot)* developing agent
Entwicklertank *m (phot)* developing tank
Entwicklervorschrift *f* developer formula
Entwicklerzusatz *m (phot)* developer improver
Entwicklung *f* evolution, generation, liberation, release *(as of gas or heat)*; development, evolution *(as of a method)*; development *(of a photograph or a chromatogram)*
  E. nach Sicht *(phot)* development by inspection
  E. nach Zeit *(phot)* development by time
  ausgedehnte E. *(phot)* prolonged development
  kontrollierte E. *(phot)* see-saw development
  verlängerte E. *(phot)* prolonged development
Entwicklungsbad *n (phot)* developing bath
Entwicklungsdauer *f (phot)* development time
Entwicklungsdose *f (phot)* developing tank
Entwicklungsfaktor *m (phot)* development factor, gamma value

Watkinsscher E. Watkins [development] factor
Entwicklungsfarbstoff *m* developed (ingrain) dye
Entwicklungsgerät *n* / **chromatografisches** chromatography apparatus
Entwicklungsgeschwindigkeit *f (phot)* development rate
Entwicklungskammer *f* developing (chromatography) chamber (tank)
E. für die aufsteigende Methode ascending chromatography tank
Entwicklungskoeffizient *m* / **arithmetischer** *(phot)* Watkins [development] factor
Entwicklungslabor[atorium] *n* development laboratory
Entwicklungslösung *f (phot)* developer (developing) solution
Entwicklungspapier *n (phot)* development (developing) paper
Entwicklungsschleier *m (phot)* development (chemical) fog
Entwicklungsstadium *n* development stage
Entwicklungssubstanz *f (phot)* developing agent
Entwicklungsverfahren *n (phot)* development technique
Entwicklungszeit *f (phot)* development time
Entwindungszahl *f* deconvolution count *(in mercerization of cotton)*
Entwulsten *n (rubber)* debeading, bead removal
Entwulster *m (rubber)* debeader, debeading machine, bead cutter
Entwurfsdruck *m* design pressure
Entwurfstemperatur *f* design temperature
entziehen to abstract, to extract, to remove, to withdraw
Wasser e. to dehydrate, to desiccate, to dewater, to dry
Entziehung *f* abstraction, removal
entzinken to dezinc[ify]
Entzinkung *f* dezincification
entzinnen to detin
Entzinnung *f* detinning
entzuckern to desugar[ize]
Entzuckerung *f* desugarization
Entzug *m* abstraction, extraction, removal, withdrawal, deprivation
entzündbar [in]flammable, ignitable, ignitible
leicht e. easy to ignite
nicht e. non-[in]flammable
Entzündbarkeit *f* [in]flammability, ignitability, ignitibility
entzünden to ignite, to kindle, lo light, to fire, to inflame
sich e. to ignite, to kindle, to light, to fire, to inflame
entzundern *(met)* to [de]scale, to scour
entzündlich *s.* entzündbar
entzündungshemmend *(pharm)* antiphlogistic, anti-inflammatory
Enzianviolett *n* gentian violet
Enzym *n* enzyme, ferment (*for compounds s.* Ferment)
Enzym ... *s.* Ferment ...
enzymatisch enzym[at]ic, ferment[at]ive
enzym[is]ieren to enzymize
Enzymologie *f* enzymology
Eosin *n* eosin, tetrabromofluorescein
Eosinfarbstoffsäure *f s.* Eosinsäure
eosinophil *(biol)* eosinophil[e], eosinophilic *(staining readily with eosin)*
Eosinsäure *f (dye)* bromo acid *(acid form of tetrabromofluorescein)*
Eosphorit *m (min)* eosphorite *(a hydrous aluminium manganese phosphate)*
EP *s.* 1. Epoxidharz; 2. Erstarrungspunkt
EPC-Ruß *m* EPC black, easy processing channel black
Ephedrin *n* $C_6H_5CH(OH)CH(NHCH_3)CH_3$ ephedrine, α-hydroxy-β-methylaminopropylbenzene
Ephedrinhydrochlorid *n* ephedrine hydrochloride
Epichlorhydrin *n* epichlorohydrin, α-epichlorhydrin, chloropropylene oxide, chloromethyloxiran
Epidot *m (min)* epidote *(an aluminium calcium silicate)*
Epilation *f (cosmet, tann)* epilation
Epilatorium *n s.* Epiliermittel
epilieren *(cosmet, tann)* to epilate, to depilate
Epiliermittel *n (cosmet, tann)* epilator, depilatory [agent], depilitant, hair remover
Epilierwachs *n (cosmet)* epilating wax
Epilupinin *n* epilupinine *(alkaloid)*
epimer epimeric
Epimer[es] *n* epimer[ide]
Epimerie *f* epimerism
epimerisieren to epimerize
Epimerisierung *f* epimerization
Epinephrin *n* $(HO)_2C_6H_3CH(OH)CH_2NHCH_3$ adrenaline, epinephrine, 1-(3,4-dihydroxyphenyl)-2-methylaminoethanol
Epistilbit *m (min)* epistilbite *(a tectosilicate)*
Epitaxie *f (cryst)* epitaxy *(oriented growth on a different crystalline substrate)*
Epoxid *n* epoxide
Epoxidgruppe *f s.* Epoxidring
Epoxidharz *n* epoxide (epoxy, ethoxylene) resin
Epoxidharzvernetzung *f* epoxy cure
Epoxidkleber *m* epoxide resin adhesive
Epoxidring *m* epoxide (epoxy) ring (group)
Epoxidweichmacher *m* epoxide plasticizer
Epoxyäthan *n* epoxyethane, oxiran, ethylene oxide
Epoxydation *f* epoxidation
epoxydieren to epoxidize
Epoxydierung *f s.* Epoxydation
Epoxygruppe *f s.* Epoxidring
Epoxyharz *n s.* Epoxidharz
EP-Schmiermittel *n* EP lubricant, extreme-pressure lubricant
Epsilonsäure *f* 1. $C_{10}H_5(OH)(SO_3H)_2$ epsilon acid, ε-acid, 1-naphthol-3,8-disulphonic acid; 2. *s.* Amino-ε-säure
Epsomit *m (min)* epsomite *(magnesium sulphate-7-water)*
EPTC *s.* Äthyldipropylthiokarbamat

E-PVC *s.* Emulsions-Polyvinylchlorid
Erbinerde *f s.* Erbiumoxid
Erbium *n* Er erbium
Erbiumchlorid *n* $ErCl_3$ erbium chloride
Erbiumnitrat *n* $Er(NO_3)_3$ erbium nitrate
Erbiumoxid *n* Er_2O_3 erbium oxide
Erbiumsulfat *n* $Er_2(SO_4)_3$ erbium sulphate
erbsengroß pea-size
Erbsenstein *m (min)* pisolite *(calcium carbonate)*
Erdalkali *n* alkaline earth
Erdalkalichelat *n* alkaline-earth chelate
Erdalkalien *npl* alkaline earths
Erdalkalikarbonat *n* alkaline-earth carbonate
Erdalkalimetall *n* alkaline-earth metal
Erdalkaliphosphat *n* alkaline-earth phosphate
Erdalkaliphosphor *m* alkaline-earth phosphor
erdartig earthy
Erdbecken *n* earthen basin
Erdbraun *n* umber *(a naturally occurring brown earth)*
Erdbraunkohle *f* earthy brown lignite
Erde *f* 1. soil, earth; 2. earth [globe]
  aktivierte E. activated (active) earth *(for bleaching)*
  Kasseler E. Cassel earth (brown), ulmin brown *(a natural pigment)*
  naturaktive E. natural earth
Erden *pl* / **seltene** rare earths
Erdfarbe *f (incorrectly for)* Erdpigment
Erdfaulversuch *m (text)* soil burial test
erdfeucht earth-moist
Erdgas *n* natural gas
  feuchtes (nasses) E. wet [natural] gas
  trockenes E. dry [natural] gas
  verflüssigtes E. liquefied natural gas, LNG
Erdgasabtrennung *f (petrol)* gas separation
Erdgasbenzin *n* natural gasoline
Erdgaslagerstätte *f* deposit of natural gas
Erdgaspipeline *f* natural gas pipeline
Erdgasquelle *f* gas well
Erdgasrohrleitung *f* natural gas pipeline
Erdharz *n* asphalt[e], asphaltum, mineral (earth) pitch
erdig earthy
Erdkern *m* earth's core (nucleus), siderosphere, barysphere, centrosphere
Erdkruste *f* earth's crust
Erdleitung *f* underground line, buried duct
Erdmandel *f* earth almond, Cyperus esculentus L.
Erdnuß *f* peanut, groundnut, goober, arachis, Arachis hypogaea L.
Ernußbutter *f* peanut butter
Erdnußeiweiß *n* peanut protein
Erdnußeiweißfaser *f (text)* peanut protein [staple] fibre
Erdnußfaser *f (text)* peanut fibre
Erdnußfaserstoff *m (text)* peanut fibre
Erdnußhartfett *n* hydrogenated (hardened) peanut oil
Erdnußkuchen *m* peanut cake
Erdnußlezithin *n* peanut lecithin
Erdnußöl *n* peanut oil
  gehärtetes E. *s.* Erdnußhartfett
Erdnußprotein *n s.* Erdnußeiweiß
Erdnußsäure *f* $CH_3[CH_2]_{18}COOH$ arachidic acid, eicosanoic acid
Erdöl *n* petroleum, mineral (rock) oil
  E. auf Asphaltbasis *s.* asphaltbasisches E.
  E. auf gemischter Basis *s.* gemischtbasisches E.
  E. auf Naphthenbasis *s.* naphthenbasisches E.
  E. auf Paraffinbasis *s.* paraffinbasisches E.
  asphaltbasisches (asphaltisches) E. asphalt-base petroleum (crude oil, crude), asphaltic petroleum
  gemischtbasisch asphaltisches E. intermediate asphaltic petroleum
  gemischtbasisch paraffinisches E. intermediate paraffinic petroleum
  gemischtbasisches E. mixed-base petroleum (crude oil, crude)
  naphthenbasisches E. naphthene-base petroleum (crude oil, crude), naphthenic petroleum
  naphthenisch-aromatisches E. naphthenic-aromatic petroleum
  naphthenisches E. *s.* naphthenbasisches E.
  nichtasphaltisches E. non-asphaltic petroleum
  paraffinbasisches (paraffinisches) E. paraffin-base petroleum (crude oil, crude), paraffinic petroleum
  paraffinisch-naphthenisches E. paraffinic-naphthenic petroleum
  rohes E. crude petroleum (oil), crude
Erdölanzeichen *n* oil indication (show)
Erdölasphalt *m* petroleum asphalt, asphaltic residue
Erdölbasis *f* base of crude petroleum
Erdölbildung *f* petroleum (oil) genesis
Erdölbohrloch *n* petroleum (oil) well
Erdölbohrung *f* 1. oil-well drilling; 2. *s.* Erdölbohrloch
Erdölchemie *f* petroleum chemistry, petrochemistry
Erdölchemikalie *f* petrochemical
Erdöldestillat *n* petroleum distillate
  gesüßtes E. sweet oil
  saures E. sour oil
  süßes E. sweet oil
Erdöldestillation *f* petroleum distillation
Erdöldestillationsrückstand *m* petroleum distillation residue
Erdölentstehung *f* petroleum (oil) genesis
Erdölfalle *f* oil trap
Erdölfeld *n* oil field (pool, reservoir)
  E. mit Gaskappe gas cap-drive field
  E. mit Gastrieb gas-drive field
  E. mit Wassertrieb water-drive field
  E. unter Gaskappendruck gas cap-drive field
Erdölfolgeprodukt *n* petroleum product
Erdölfraktion *f* petroleum fraction
erdölführend petroliferous, petroleum-bearing, oil-bearing
Erdölgas *n* petroleum gas
  verflüssigtes E. liquefied petroleum gas, L. P. gas, L. P. G.

Erdölgenesis *f* petroleum (oil) genesis
erdölhaltig *s.* erdölführend
Erdölharz *n* petroleum resin
Erdölindikation *f* oil indication (show)
Erdölindustrie *f* petroleum (oil) industry
Erdölkohlenwasserstoff *m* petroleum hydrocarbon
Erdölkoks *m* petroleum (still) coke
Erdöllagerstätte *f* oil deposit (occurrence)
Erdölmuttergestein *n* oil[-source] rock, mother rock
Erdölparaffin *n* petroleum wax
Erdölpech *n* petroleum pitch
Erdölprodukt *n* petroleum product
Erdölquelle *f* petroleum (oil) well
Erdölraffinerie *f* oil refinery
Erdölresiduum *n* petroleum residue
Erdölrückstand *m* petroleum residue
Erdölschwerbenzin *n* petroleum naphtha
Erdölspeichergestein *n* reservoir rock
Erdöltechnologie *f* petroleum technology
Erdölteer *m* petroleum tar
Erdölverarbeitung *f* petroleum refining
Erdölvorkommen *n* oil occurrence (deposit)
Erdölwachs *n* petroleum wax
Erdölzeresin *n* petroleum ceresin
Erdorseille *f* *(dye)* crabeye lichen, Lecanora parella Mass.
Erdpech *n* 1. *(min)* elastic bitumen, mineral caoutchouc, elaterite; 2. mineral pitch, asphalt[e], asphaltum
Erdpigment *n* earth (mineral, natural) pigment, earth colour
Erdrinde *f* earth's crust
Erdschellack *m* *(dye)* Botany Bay gum *(from Xanthorrhoea hastilis R. Br.)*
Erdschwarz *n* slate black, black chalk *(a natural pigment)*
erdverlegt buried
Erdverrottungstest *m* *(text)* soil burial test
Erdwachs *n* *(min)* earth (ader) wax, native paraffin, ozokerite
  gereinigtes E. ceresin [wax], ceresine
Erepsin *n* erepsin *(a proteolytic enzyme)*
erfassen to detect *(as in analyses)*
Erfassungsgrenze *f* detectability *(in analyses)*
Erfolgsbohrung *f* *(petrol)* discovery well
erforschen to investigate, to examine, to elucidate
Erfrischungsgetränk *n* refreshment drink
ergeben to yield
ergiebig high-yield
ergießen / sich to flush, to pour, to spill
Ergin *n* biocatalyst, biochemical catalyst, ergone
Ergobasin *n* *s.* Ergometrin
Ergometrin *n* ergometrine, ergobasine, ergonovine, ergotocine, ergostetrine *(an ergot alkaloid)*
Ergonovin *n* *s.* Ergometrin
Ergosterin *n* ergosterol
Ergosterol *n* *s.* Ergosterin
Ergostetrin *n* s. Ergometrin
Ergotamin *n* ergotamine *(an ergot alkaloid)*
Ergotismus *m* *(tox)* ergotism
Ergotozin *n* *s.* Ergometrin
Ergußgestein *n* extrusive (effusive) rock
Erhalt *m* obtaining *(as by synthesis)*
erhalten to obtain *(as by synthesis)*; to retain *(e. g. flavour)*
erhältlich / im Handel commercially available
Erhaltung *f* conservation *(as of mass or energy)*
Erhaltungsdüngung *f* *(agric)* maintenance dressing
Erhaltungssatz *m* conservation law
  E. der Energie law of conservation of energy, energy principle
  E. der Masse (Materie) law of conservation of mass (matter)
  E. der wägbaren Masse (Materie) *s.* E. der Masse
erhärten to harden, to set; *(too rapidly or unintentionally)* to set up; *(geol)* to lithify
Erhärtung *f* hardening, set[ting]; *(if too rapid or unintentional)* set-up; *(geol)* lithification, induration
erhitzen to heat
  auf (bis zur) Rotglut e. to heat to redness, to make red-hod
  auf (bis zur) Weißglut e. to incandesce
  in der Retorte e. to retort
  sich e. to heat
  unter Rückfluß e. *(distil)* to reflux
Erhitzer *m* heater
Erhitzung *f* heating
  dielektrische E. dielectric heating
  rote E. *(tann)* red heat discoloration *(of hides due to bacteria)*
Erhitzungsbehandlung *f* *(chem)* heat processing
erhitzungsbeständig resistant to heat
Erhitzungsbeständigkeit *f* thermal (heat) stability (resistance), resistance to heat
Erhitzungsgeschwindigkeit *f* heating rate
Erhitzungszyklus *m* heating cycle
erhöhen 1. to raise, to elevate *(e. g. temperature, boiling point)*; 2. to increase *(e. g. number, quantity)*
  den Weißgehalt e. *(pap)* to whiten
  sich e. 1. to rise *(as of temperature, boiling point)*; 2. to increase *(of number, quantity)*
Erhöhung *f* 1. *(act)* raising, raise, elevation; *(process)* rise, elevation *(as of temperature, boiling point)*; 2. increase *(of number, quantity)*
  E. des Weißgehalts *(pap)* whitening, brightening
erholen / sich *(tech)* to recover
Erholung *f* *(tech)* recovery
  elastische E. *(rubber)* elastic recovery, rebound
  zeitabhängige E. *(text)* recovery
Eriochalzit *m* *(min)* eriochalcite *(copper(II) chloride-2-water)*
erkalten to cool [down], to chill
  e. lassen to [allow to] chill
Erkalten *n* cooling, chilling
Erkennbarkeit *f* perceptibility, recognizability, detectability
Erkennungsmittel *n* detector substance; *(food)* indicator ingredient (substance)
Erkensator *m* *(pap)* erkensator

erkochen *(pap)* to cook
erkocht / alkalisch *(pap)* alkaline-cooked
Erkochung *f (pap)* cooking
Erlenmeyer-Kolben *m* Erlenmeyer flask
erlöschen to go out, *(gradually)* to die out
Erlöschen *n* going-out, *(gradually)* dying-out
ermitteln to elucidate, to establish, to determine, to find out
Ermittlung *f* elucidation, establishment, determination
E. der Klopffestigkeit [anti]knock rating
ermüden *(tech)* to fatigue
Ermüdung *f (tech)* fatigue
ermüdungsbeständig fatigue-resisting
Ermüdungsbeständigkeit *f* fatigue resistance (strength), resistance to fatigue
Ermüdungserscheinung *f (tech)* fatigue
Ermüdungsfestigkeit *f s.* Ermüdungsbeständigkeit
Ermüdungsgrenze *f* fatigue limit
Ermüdungsprüfung *f* fatigue test; *(rubber)* flex-cracking test
Ermüdungsschutzmittel *n* antifatigue (fatigue-preventing) agent; *(rubber)* anti-flex-cracking antioxidant
Ermüdungswiderstand *m* fatigue resistance
Ernährung *f* nourishment, diet
mineralische E. mineral nutrition
Ernährungsforschung *f* nutritional investigation
Ernährungswissenschaft *f* nutrition science
erniedrigen *s.* herabsetzen
sich e. to decrease *(as of boiling point)*
Erniedrigung *f (act)* depression, lowering; *(process)* decrease *(as of boiling point)*
Erprobung *f* **im Feldversuch** field evaluation *(as of fertilizers)*
erregbar excitable
Erregbarkeit *f* excitability
erregen to excite, to activate
Erregung *f* excitation
Erregungsenergie *f* excitation energy
Ersatz *m* 1. replacement, substitution; 2. substitute
Ersatzdüngung *f* compensation fertilization
Ersatzmittel *n*, **Ersatzstoff** *m* substitute
Erscheinung *f* phenomenon
Erscheinungspotential *n* appearance potential *(mass spectroscopy)*
erschließen 1. to elucidate; 2. *(min)* to develop
Erschließung *f* 1. elucidation; 2. *(min)* development
erschmelzen to smelt *(metals)*
erschöpfend exhaustive
Erschöpfung *f* exhaustion *(as of a dye bath)*; exhaustion, squeeze *(of natural resources)*
erschütterungsfest vibration-proof
erschütterungsfrei vibrationless
Erschütterungsvorrichtung *f* rapper *(for cleaning electrodes)*
erschweren *(text)* to weight
Erschwerungsmittel *n (text)* weighting agent, weighter
ersetzbar replaceable, displaceable, substitutable
ersetzen to replace, to displace, to substitute
Ersetzen *n* replacement, displacement, substitution
Ersetzungsname *m* replacement name
erspinnen to spin *(chemical fibres)*
Erspinnen *n* spinning *(of chemical fibres)*
E. aus der Schmelze melt spinning, [melt] extrusion
E. aus Lösungen solution (solvent) spinning
Erspinnfärbung *f* spin (dope) dyeing
erspinngefärbt spin-dyed, spun-dyed, dope-dyed, *(in a melt also)* mass-dyed, *(in solution also)* solution-dyed
Erspinnlösung *f* spinning solution, dope
erstarren to solidify, to harden, to set, to freeze, to congeal, *(esp coll)* to gel[ate]
e. lassen to solidify, to congeal, to set
wieder e. to refreeze; to regelate *(of ice)*
zu Gelee e. to jelly, to jellify, to gel[ate], to gelatinate, to gelatinize
Erstarren *n* solidification, hardening, set[ting], freezing, congealing, congelation, *(esp coll)* gelation
Erstarrung *f s.* Erstarren
Erstarrungsgestein *n s.* Eruptivgestein
Erstarrungsintervall *n* solidification range
Erstarrungskurve *f* freezing[-point] curve *(of a melt)*
Erstarrungspunkt *m*, **Erstarrungstemperatur** *f* solidification (setting) point, s. p., freezing (congealing) point (temperature)
Erstarrungswärme *f* heat of solidification
Erstbelichtung *f (phot)* first (initial) exposure
Erstdestillation *f* primary distillation
ersticken to blanket, to choke *(fire)*
Erstkomponente *f (dye)* primary (diazo, diazonium) component
Erstluft *f* primary air
Erstproduktfüllmasse *f (sugar)* first fillmass, white (high-grade) massecuite
Erstproduktzucker *m* first raw (product) sugar, high-grade sugar
Erstsubstituent *m* first substituent
Ertragsfähigkeit *f (agric)* productive capacity, crop-producing power *(of a soil)*
Ertragsgesetz *n (agric)* law of yields
Ertragswirkung *f (agric)* effect on yield *(of fertilizers)*
Erukasäure *f* $CH_3[CH_2]_7CH{=}CH[CH_2]_{11}COOH$ erucic acid, *cis*-13-docosenoic acid
eruptieren *(petrol)* to blow out
Eruption *f (petrol)* blow-out
Eruptionskopf *m s.* Eruptionskreuz
Eruptionskreuz *n (petrol)* Christmas tree
eruptiv *(geol)* eruptive, igneous
Eruptivgestein *n* eruptive (igneous) rock
basisches E. basic (subsilicic) rock, basite
intermediäres E. *s.* neutrales E.
neutrales E. neutral (intermediate) rock
saures E. acid rock
Eruptivkreuz *n s.* Eruptionskreuz
Eruptivstock *m (geol)* boss
Eruzylalkohol *m* $CH_3[CH_2]_7CH{=}CH[CH_2]_{12}OH$ erucyl alcohol, 13-docosen-1-ol
erwärmen to heat [moderately], to warm

erneut e. to reheat
mäßig e. to warm gently (moderately)
sich e. to heat, to warm [up]
Erwärmung *f* heating, warming
dielektrische E. dielectric heating
erneute E. reheat[ing]
induktive E. induction heating
Erwärmungsgeschwindigkeit *f* heating rate
Erwartungswert *m* expectancy (expectation, expected) value
erweichen to plasticize, to plasticate, to plastify, to soften; *(plast)* to flux, to soften
mit Peptisiermitteln e. *(rubber)* to peptize
thermisch e. to heat-soften
Erweichung *f* plastic[iz]ation, softening
E. mit Peptisiermitteln *(rubber)* peptization
thermische E. thermal (heat) softening (plastication)
Erweichungsbereich *m*, **Erweichungsintervall** *n* softening (plastic) range
Erweichungsmittel *n* softening agent, softener, emollient
Erweichungspunkt *m* softening point (temperature); *(glass)* Littleton [softening] point, seven-point-six temperature, 7.6 temperature *(at which the viscosity is $10^{7.6}$ poises)*
E. KS *s.* E. nach Krämer-Sarnow
E. nach Krämer-Sarnow Kraemer and Sarnow softening point (temperature) *(e.g. in investigating fats, pitch)*
E. nach Vicat *(plast)* Vicat softening point (temperature), V.S.P., Vicat needle point
E. „Ring und Kugel" ring-and-ball softening point *(e.g. in investigating fats, pitch)*
E. RuK *s.* E. „Ring und Kugel"
Erweichungstemperatur *f s.* Erweichungspunkt
Erweichungszone *f s.* Erweichungsbereich
Erweichungszustand *m* softening stage
erweitern to expand
sich e. to expand
Erweiterungsbohrung *f (petrol)* appraisal well
Erythralin *n* erythraline *(an erythrina alkaloid)*
Erythren *n* $CH_2{=}CHCH{=}CH_2$ erythrene, 1,3-butadiene
Erythrin *m (min)* erythrine, erythrite, cobalt bloom *(cobalt(II) orthoarsenate)*
Erythrinaalkaloid *n* erythrina alkaloid
Erythrit *m* $CH_2OH[CHOH]_2CH_2OH$ erythritol, 1,2,3,4-tetrahydroxybutane
Erythrodextrin *n* erythrodextrin[e]
Erythroform *f* erythro form *(stereochemistry)*
Erythrogensäure *f* erythrogenic acid, isanic acid, 17-octadecen-9,11-diynoic acid
Erythroidin *n* erythroidine *(an erythrina alkaloid)*
Erythromyzin *n* erythromycin *(an antibiotic)*
Erythrose *f* $CH_2OH[CHOH]_2CHO$ erythrose, 1,2,3-trihydroxybutyraldehyde
Erythrosin *n* erythrosine *(disodium salt of tetraiodofluorescein*
Erz *n* ore
abbauwürdiges E. pay ore
armes E. lean (low-grade) ore
bauwürdiges E. pay ore
feines E. fine ore
gemengtes E. complex ore
geringhaltiges (geringwertiges) E. lean (low-grade) ore
hochwertiges E. high-grade ore
komplexes E. complex ore
oxidisches E. oxidized ore
polymetallisches E. complex ore
primäres (protogenes) E. primary (protogenic) ore
reiches (reichhaltiges) E. high-grade ore
sulfidisches E. sulphide ore
zusammengesetztes E. complex ore
Erzanreicherung *f* concentration (enrichment) of ores
Erzaufbereitung *f* ore dressing (beneficiation)
Erzaufbereitungsanlage *f* ore-dressing (ore-beneficiation) plant
Erzaufbereitungsverfahren *n* ore-dressing (ore-beneficiation) process
Erzbett *n* ore bed
erzbildend metallogen[et]ic
Erzbrecher *m* ore crusher (breaker)
Erzbrocken *m* lump of ore
Erzbrücke *f* ore bridge
Erzbunker *m* ore bunker
erzeugen 1. *(of human agent)* to manufacture, to make, to produce *(e.g. chemicals)*; to generate *(e.g. steam)*; 2. *(of a substance)* to evolve *(e.g. fumes)*
Erzeugnis *n* product, *(esp if relating to its origin)* make; *(distil)* liquid product
feuerfestes E. refractory [product]
grobkeramisches E. heavy clay product (ware)
hochtonerdehaltiges feuerfestes E. high-alumina refractory [product]
keramisches E. ceramic article (product), ceramic
pflegeleichtes E. *(text)* wash and wear product, w & w product
schmelzgeformtes E. fusion-cast refractory [product]
Erzeugnisse *npl* / **feinkeramische** fine ceramics (ceramic ware)
oxidkeramische E. oxide-ceramic products, oxide ceramics
pyrotechnische E. pyrotechnics
Erzeugung *f* manufacture, making, make, *(esp over a specified period)* production; generation *(as of steam)*
Erzeugungsprogramm *n* production pattern
Erzfall *m* ore shoot
erzführend ore-bearing, metalliferous
Erzgangart *f* ore gangue
erzhaltig ore-bearing, metalliferous
Erzlager *n*, **Erzlagerstätte** *f* ore deposit, orebody
Erzmineral *n* ore mineral
Erzmöller *m* ore burden
Erzprobe *f (met)* 1. ore assay; 2. ore sample
erzreich rich in ore
Erztrübe *f* ore pulp

wäßrige E. aqueous pulp of ground ore
Erzverteilung *f* / **zonale (zonare)** *(geol)* zonal distribution of minerals, mineral zoning
Erzvorbereitung *f* ore preparation
Erzvorbereitungsanlage *f* ore-preparation plant
Erzvorkommen *n* ore deposit, orebody
Erzwäsche *f* 1. ore washing (cleaning); 2. ore washery
Eschka-Methode *f* Eschka method *(for determining total sulphur content)*
e. s. E. *s.* Einheit / elektrostatische
Eserin *n* eserine, physostigmine *(alkaloid)*
E-Silber *n* electrolytic silver
Esparto *m*, **Espartogras** *n* esparto (Spanish, Alfa) grass, Stipa tenacissima L.
Espartopapier *n* esparto paper
Espartowachs *n* *(pap)* esparto wax
Espartozellstoff *m* esparto pulp
Espartozellstoffabrik *f* esparto mill
Espe *f* aspen, *(Am)* European aspen, Populus tremula L.
ESR *s.* Elektronenspinresonanz
eßbar edible, eatable, esculent, comestible
Eßbarkeit *f* edibility, edibleness
Esse *f* chimney, [smoke]stack
Essence absolue [de concrète] *f* *(cosmet)* absolute from concrete
Essence absolue d'enfleurage *f* *(cosmet)* enfleurage absolute
Essence concrète *f* *(cosmet)* concrete [oil]
essentiell essential
Essenz *f* essence
Essig *m* vinegar
Essigbakterien *pl* acetic-acid bacteria
Essigester *m* *s.* Essigsäureäthylester
Essigfabrik *f* vinegar factory
Essiggeist *m* *s.* Dimethylketon
Essiggeruch *m* acetous odour
Essigherstellung *f* manufacture of vinegar, acetification
Essigmesser *m* acetimeter
Essigmutter *f* mother of vinegar *(a slimy substance consisting of microorganisms)*
Essigprüfer *m* acetimeter
Essigsäure *f* CH_3COOH acetic acid, ethanoic acid
aktivierte E. active acetate, acetyl coenzyme A
Essigsäurealdehyd *m* *s.* Azetaldehyd
Essigsäureamid *n* CH_3CONH_2 acetamide
Essigsäureamylester *m* $CH_3COOC_5H_{11}$ amyl acetate
Essigsäureanhydrid *n* $(CH_3CO)_2O$ acetic anhydride, ethanoic anhydride
Essigsäureäthylester *m* $CH_3COOC_2H_5$ ethyl acetate, acetic ester
Essigsäurebakterien *pl* acetic-acid bacteria
Essigsäurebenzylester *m* $CH_3COOCH_2C_6H_5$ benzyl acetate
Essigsäurebildung *f* acetification
Essigsäurebornylester *m* $CH_3COOC_{10}H_{17}$ borneol (bornyl) acetate
Essigsäure-*n*-butylester *m* $CH_3COO[CH_2]_3CH_3$ *n*-butyl acetate
Essigsäurechlorid *n* CH_3COCl acetyl chloride
Essigsäuregärung *f* acetic[-acid] fermentation, acetification
Essigsäuremethylester *m* CH_3COOCH_3 methyl acetate
Essigsäurephenylester *m* $CH_3COOC_6H_5$ phenyl acetate
Essigsäurepropylester *m* $CH_3COOC_3H_7$ propyl acetate
Essigsäureureid *n* $CH_3CO{-}NH{-}CO{-}NH_2$ acetylurea, *N*-monoacetylurea
Essiguntersuchung *f* acetimetry
Ester *m* ester
aktivierter E. activated ester
innerer E. intra-ester, inter-ester
zyklischer E. cyclic ester
Esterase *f* esterase
Esteraustausch *m* ester exchange
Esterbildung *f* ester formation
Esterbindung *f* ester bond, ester link[age]
Esterenolat *n* ester enolate
Esterharz *n* ester gum
Esterhydrolyse *f* ester hydrolysis
Esterkondensation *f* condensation of esters
Claisensche E. Claisen condensation
Dieckmannsche [intramolekulare] E. Dieckmann condensation (reaction)
Ester-pool *m* *(bioch)* ester pool *(a group of phosphoric-acid esters possessing a sugar component)*
Esterspaltung *f* ester cleavage
E. nach Hunsdiecker Hunsdiecker cleavage (reaction)
Esterumlagerung *f* rearrangement of esters
Esterverseifung *f* ester saponification
Esterzahl *f* ester number (value)
Estragol *n* $CH_2{=}CHCH_2C_6H_4OCH_3$ estragole, chavicol methyl ether, 1-allyl-4-methoxybenzene
Estragonöl *n* tarragon oil *(from Artemisia dracunculus L.)*
Etagenhöhe *f* daylight *(of a multiplaten press)*
Etagennutsche *f* multiplate (horizontal plate) filter
Etagenpresse *f* multidaylight (multiple-daylight, daylight, multiplaten, platen) press
Etagenvulkanisierpresse *f* *(rubber)* daylight curing press
Etagenwalzwerk *n* multiplex-roll plant
Etikett *n* label; tag
etikettieren to label; to tag
Etikettiermaschine *f* labelling machine, labeller
Etruria-Mergel *m* *(ceram)* etruria marl
Ettringit *m* *(min)* ettringite *(aluminium calcium hydroxide sulphate)*
Euchroit *m* *(min)* euchroite *(copper(II) hydroxide orthoarsenate)*
Eudalin *n* $(CH_3)_2CH{-}C_{10}H_6CH_3$ eudalene, 7-isopropyl-1-methylnaphthalene
Eudesmol *n* eudesmol *(a bicyclic sesquiterpene)*
Eudialyt *m* *(min)* eudialyte *(a cyclosilicate containing zirconium)*
Eudiometer *n* eudiometer
Eugenol *n* $C_6H_3(OH)(OCH_3)CH_2CH{=}CH_2$ eugenol, 1-allyl-4-hydroxy-3-methoxybenzene

Eukairit *m (min)* eucairite *(copper silver selenide)*
Eukalyptol *n* eucalyptol, cineole, cineol-1,8
Eukalyptuskino *n* ribbon gum kino *(from Eucalyptus specc.)*
Euklas *m (min)* euclase *(aluminium beryllium hydroxide orthosilicate)*
Eukolloid *n* eucolloid
Eukrasit *m (min)* eucrasite *(thorium silicate)*
Eukryptit *m (min)* eucryptite *(aluminium lithium silicate)*
Eulytin *m (min)* eulytine, eulytite *(bismuth(III) orthosilicate)*
Europium *n* Eu europium
Europium(II)-chlorid *n* $EuCl_2$ europium(II) chloride, europium dichloride
Europium(III)-chlorid *n* $EuCl_3$ europium(III) chloride, europium trichloride
Europiumdichlorid *n s.* Europium(II)-chlorid
Europiumoxid *n* Eu_2O_3 europium oxide
Europiumsulfat *n* $Eu_2(SO_4)_3$ europium sulphate
Europiumtrichlorid *n s.* Europium(III)-chlorid
Eutektikum *n* eutectic [mixture]
eutektisch eutectic
eutektoid eutectoid
Eutektoid *n* eutectoid
eutektoidisch eutectoid
eutroph eutrophic *(rich in dissolved plant nutrients)*
Eutrophierung *f* eutrophication
Euxenit *m* euxenite *(a rare-earth mineral)*
evakuieren to evacuate, to exhaust
Evakuierung *f* evacuation, exhaustion
evaporieren to evaporate
Evaporimeter *n* evaporimeter, evaporometer
Evaporisation *f* evaporation
Evaporometer *n s.* Evaporimeter
Evelyn-Röhrchen *n* Evelyn tube *(turbidimetry)*
Exaltation *f* **[/ optische]** [optical] exaltation *(in molar refraction)*
Exergie *f* exergy
Exhalation *f* exhalation, outgassing *(of a volcano)*
Exhaustor *m* exhauster
Exinit *m (coal)* exinite
exinitisch *(coal)* exinitic
existenzfähig capable of existence
  vorübergehend e. of fleeting existence
Exkret *n (biol)* excretum
Exkretion *f (biol)* excretion
exkretorisch *(biol)* excretory
Exoelektron *n* exoelectron
Exoenzym *n* exoenzyme
exoergonisch exoergic
exokrin exocrine
Exomorphose *f (geol)* exomorphism
Exopeptidase *f* exopeptidase
Exosmose *f* exosmosis
exotherm[isch] exothermic
Exotoxin *n* exotoxin
exozyklisch exocyclic
expandieren to expand, to dilate
expansibel expansible
Expansion *f* expansion, dilatation
  adiabatische E. adiabatic expansion
  isotherme E. isothermal expansion
Expansionsmaschine *f* expansion engine, expander [machine]
Expansionsnebelkammer *f* expansion cloud chamber
Expansionsturbine *f* turboexpander
Expansionsventil *n* expansion valve
Expansivzement *m* expanding (expansive) cement
Expektorans *n,* **Expektorantium** *n (pharm)* expectorant
Experiment *n* experiment
Experimentalchemie *f* experimental chemistry
Experimentalforschung *f* experimental research
Experimentalvorlesung *f* demonstration lecture
Experimentator *m* experimentalist, experiment[at]or
experimentell experimental
Experimentelles *n* experimental *(in treatises)*
experimentieren to experiment[alize]
Experimentieren *n* experimentation
Experimentierkunst *f s.* Experimentiertechnik
Experimentiertechnik *f* experimental technique
explodierbar *s.* explosiv
Explodierbarkeit *f s.* Explosivität
explodieren to explode
  e. lassen to explode, to blow up
Exploration *f* exploration
Explorationsbohrloch *n (petrol)* exploration (exploratory) well, wildcat
Explorationsbohrung *f* 1. *(petrol)* exploration drilling; 2. *s.* Explorationsbohrloch
explosibel explosive, explosible
  nicht e. non-explosive
Explosibilität *f* explosiveness, explosivity, explosibility
Explosion *f* explosion, blast, shot
explosionsartig explosive
Explosionsbereich *m* explosive range (limits)
Explosionsdruck *m* explosion pressure
Explosionsfähigkeit *f s.* Explosivität
Explosionsfront *f* explosion front
Explosionsgefahr *f* hazard (danger) of explosion
explosionsgefährlich explosive, explosible
explosionsgeschützt explosion-proof
Explosionsgeschwindigkeit *f* explosion velocity
Explosionsgrenze *f* explosion limit
  obere E. upper explosion limit
  untere E. lower explosion limit
Explosionskette *f* explosion chain
Explosionsmethode *f* explosion method *(for determining molar heat)*
Explosionsprodukt *n* explosion product
Explosionspunkt *m (petrol)* shot point
explosionssicher explosion-proof
Explosionstemperatur *f* explosion temperature
Explosionsverfahren *n (pap)* explosion (Masonite) process *(chemigroundwood process)*
Explosionswärme *f* heat of explosion
Explosionswirkung *f* explosive action (effect)
explosiv explosive, explosible
  nicht e. non-explosive

Explosivität *f* explosiveness, explosivity, explosibility
Explosivstoff *m* explosive
brisanter E. high explosive, H. E.
Explosivstoffchemie *f* chemistry of explosives
Exponent *m* superscript [numeral], numeric superscript
Bornscher E. Born exponent
Exponentialpapier *n* semilog[arithmic] paper
exponieren *(phot)* to expose [to light]
Exposition *f (phot)* exposure [to light]
Expositionsdauer *f*, **Expositionszeit** *f (tox)* exposure duration (time), length of exposure *(as of animals to pesticides)*
Exsikkator *m (lab)* desiccator
E. mit Einsatz filled desiccator
E. nach Scheibler Scheibler desiccator
Exsikkatordeckel *m* desiccator lid
Exsikkatorplatte *f* desiccator plate (disk)
Exsudat *n* exudate, exudation
Exsudation *f* exudation
Extender *m* extender, extending filler
Extenderweichmacheröl *n (rubber)* extending oil
Extinktion *f* extinction, optical density, *(esp colorimetry)* absorbency, absorbance
Extinktionskoeffizient *m* extinction coefficient, *(esp colorimetry)* absorbency index, absorptivity
molarer E. molar extinction coefficient, *(esp colorimetry)* molar absorptivity (absorbency index)
spezieller (spezifischer) E. absorbency index, absorptivity *(proper)*
Extinktionskonstante *f s.* Extinktionskoeffizient
Extinktionskurve *f* extinction curve
extrahierbar extractable, extractible
Extrahierbarkeit *f* extractability, extractibility
extrahieren to extract, to abstract, *(esp with water also)* to leach [out], to lixiviate
erschöpfend e. to exhaust
gemeinsam e. to coextract
Extrahieren *n* extraction, abstraction, *(esp with water also)* leach[ing], lixiviation
extrahierend extractive
Extrakt *m* extract[ive], essence
alkoholischer E. alcoholic extract
Goulards E. Goulard's extract, vinegar of lead *(aqueous solution of basic lead acetates)*
trockener E. dry extract
Extraktabnahme *f (ferm)* attenuation *(diminution of density of wort resulting from its fermentation)*
Extraktbrühe *f* extract liquor
Extraktende *n (petrol)* extract end
Extrakteur *m* extractor, extraction apparatus; contactor *(for solvent extraction)*
E. mit Förderschnecken screw-conveyor extractor
E. mit waagerechten Siebplattenförderern travelling-belt extractor
kontinuierlich arbeitender E. continuous extractor, continuous-extraction apparatus
liegender E. horizontal extractor
stehender E. vertical extractor
Extraktherstellung *f* extract manufacture
Extraktion *f* extraction, abstraction, *(esp with water also)* leach[ing], lixiviation
E. in flüssigen Systemen liquid-liquid extraction
diskontinuierliche E. discontinuous extraction
einmalige einfache E. single simple extraction
kontinuierliche E. continuous extraction
wiederholte einfache E. multiple single extraction
Extraktionsanalyse *f* extraction analysis
Extraktionsanlage *f* extraction plant
Extraktionsapparat *m* extraction apparatus, extractor; contactor *(for solvent extraction)*
E. nach Bollmann Bollmann extractor
E. nach Soxhlet Soxhlet [extractor]
Extraktionsaufsatz *m* extraction head; extractor jacket *(of an extraction apparatus)*
Extraktionsbatterie *f* extraction battery
Extraktionsbenzin *n* extraction naphtha
Extraktionsbrühe *f (tann)* leach liquor
Extraktionsgefäß *n* extraction (extracting) vessel, *(esp with water also)* leaching tank (trough, vat, vessel), leach
Extraktionsgrad *m* degree of extraction
Extraktionsgut *n* material being (*or* to be) extracted
Extraktionsharz *n* wood rosin
Extraktionshülse *f* extraction thimble
E. nach Soxhlet Soxhlet thimble
Extraktionskolben *m* extraction flask
Extraktionskolonne *f* extraction column; contactor *(for solvent extraction)*
Extraktionskolophonium *n* wood rosin
Extraktionsmaschine *f* contactor *(for solvent extraction)*
Extraktionsmittel *n* extracting agent (solvent), extractant
Extraktionspresse *f* filtration extractor
Extraktionsrückstand *m* extraction residue, marc, mark
Extraktionssäule *f* extraction column
Extraktionssystem *n* extraction system
Extraktionstemperatur *f* extraction temperature
Extraktionsturm *m* extraction tower
Extraktionswäsche *f (text)* solvent scouring
Extraktionszeit *f* **je Charge** extraction cycle
extraktiv extractive
Extraktivdestillation *f* extractive distillation
Extraktivstoff *m s.* Extraktstoff
Extraktor *m s.* Extrakteur
Extraktphase *f* extract phase
Extraktseite *f (petrol)* extract end
Extraktstoff *m* extractive material (matter, substance), extractive
Extraktstripper *m* extract stripper
Extraktverdampfer *m* extract evaporator
extranuklear extranuclear
extrazellulär extracellular
Extreme-Pressure-Schmiermittel *n* extreme-pressure lubricant, EP lubricant
Extremum *n s.* Extremwert
Extremwert *m* extreme value, extreme [point], extremum

Extrinsic-Faktor *m (bioch)* extrinsic factor
Extrinsic-Leitfähigkeit *f*, **Extrinsic-Leitung** *f* impurity electric conductivity
Extrudat *n* extrudate
Extruder *m* extruder, extruding machine, extrusion press (auger)
  schneckenloser E. screwless extruder
Extruderkopf *m* extruder (extrusion) head, *(plast also)* die head
Extrudermundstück *n* extruder (extrusion) die
extrudern *s.* extrudieren
Extrudersiegeln *n (plast)* extruded bead sealing
Extruderzylinder *m* extruder barrel
extrudieren to extrude
Extrudieren *n* extruding, extrusion
  E. mit Kühlwalzen *(plast)* chill-roll extrusion
  E. von Trockenmischung *(plast)* dry-blend extrusion
Extrusion *f (geol)* extrusion
Extrusionsblasen *n*, **Extrusionsblasformen** *n* extrusion blowing
Extrusivgestein *n* extrusive rock
Exzenterantrieb *m* eccentric drive
Exzenterpresse *f* eccentric press
Exzenterschwingsiebmaschine *f* eccentrically driven vibrating screen
exzentrisch eccentric[al], off-centre
EZ *s.* Esterzahl
E-Zink *n* electrolytic zinc

F

Fabric-Presse *f (pap)* fabric press
Fabrik *f* works, plant, factory
  chemische F. chemical works (plant)
Fabrikabwasser *n s.* Fabrikationsabwasser
Fabrikat *n* product, article; make *(as of a specific plant or country)*
Fabrikation *f* manufacture, make, making, *(esp over a specified period)* production
Fabrikationsabfälle *mpl* industrial wastes
  unvulkanisierte F. rubber scrap (waste), scrap (waste) rubber
Fabrikationsabwasser *n* industrial waste-water (sewage), factory (works, trade) effluent
Fabrikationsgang *m* manufacturing process
Fabrikationsnummer *f* reference (lot) number
Fabrikationssicherheit *f* processing safety
Fabrikationswasser *n* process (service, industrial) water, water for industrial use
Fabrikmarke *f* trademark, make
Fabrikwasser *n s.* 1. Fabrikationswasser; 2. Fabrikationsabwasser
fächeln to fan
Fächerdüse *f* fan nozzle
fächern *(pap)* to fan [down] *(sheet sorting)*
FAD *s.* Flavin-adenin-dinukleotid
fad[e] insipid, flat, stale, tasteless, dead
Faden *m* filament *(as of carbon or metal)*; thread *(as of glass, plastic, rubber, metal)*; *(text)* yarn *(for knitting or weaving fabrics)*; thread *(for sewing)*; *(glass)* string *(a defect)*; *(nucl)* pinch *(of the ion current)*
  einfacher F. *(text)* single yarn
  metallisierter F. *(text)* metallic yarn
  thermoplastischer F. *(text)* thermoplastic yarn
Fadenbildung *f s.* Fadenziehen
fadenförmig threadlike
Fadengalvanometer *n* string galvanometer
Fadengitter *n (cryst)* chain lattice
Fadenkorrektur *f* stem correction *(with liquid thermometers)*
Fadenmolekül *n* threadlike (thread, filamentary, linear) molecule
Fadenprobe *f (sugar)* string-proof test
Fadenstruktur *f* threadlike structure
Fadenthermometer *n* thread thermometer
  F. nach Mahlke Mahlke thread thermometer
Fadenziehen *n* 1. *(coat)* cobwebbing *(a defect)*; 2. *(text)* fibre drawing
fadenziehend ropy, stringy
Fadeometer *n (text)* fad[e]ometer
Fagergren-Zelle *f* Fagergren floatation machine
Fagopyrin *n* fagopyrine *(a colouring matter of buckwheat)*
fahl flat *(colour tone)*
Fahlerz *n (min)* fahlerz, fahlore
Fahrbenzin *n* **[/ normales]** motor (regular) spirit (gasoline)
Fahrdieselkraftstoff *m* automotive diesel fuel (oil)
Fahrdieselöl *n s.* Fahrdieselkraftstoff
fahren to run *(e. g. an apparatus)*
Fahrweise *f (chem)* operation, processing
  F. mit totalem Rücklauf *(distil)* total-reflux operation
  F. ohne Destillatabnahme *(distil)* total-reflux operation
  diskontinuierliche F. batch operation
  periodische F. batch operation
Fahrzeuglack *m* automobile lacquer, automotive coating (finish)
Fairfieldit *m (min)* fairfieldite *(calcium manganese(II) orthophosphate)*
Fäkalien *pl* faecal matter, faeces
Fäkalwasser *n* sanitary sewage
Faktis *m (rubber)* factice, rubber substitute
  brauner F. brown (dark) factice, brown substitute
  weißer F. white factice (substitute)
Faktor *m* factor, coefficient; volumetric (titrimetric) factor *(volumetry)*; gravimetric (analytical) factor *(gravimetry)*
  Boltzmannscher F. *(phys chem)* Boltzmann factor
  innerer F. *(bioch)* intrinsic factor *(a substance produced by stomach and intestinal mucosa)*
  sterischer F. *(phys chem)* steric (probability) factor
  van't-Hoffscher F. *(phys chem)* van't Hoff factor
Fällanlage *f* precipitator
Fällbad *n (text)* spinning (coagulating, coagulation, precipitating, precipitation) bath
fällbar precipitable
Fällbarkeit *f* precipitability

Falle *f* trap
antiklinale F. *(petrol)* anticlinal trap
chemotropische F. chemotropic trap *(for pest control)*
stratigrafische F. *(petrol)* stratigraphic trap
strukturelle (strukturgebundene) F. *(petrol)* structural trap
tektonische F. *(petrol)* structural trap
fallen to fall, to drop; to decrease, to fall, to drop, to reduce *(as of measuring values)*
fällen to precipitate, to sediment[ate], to throw down
elektrochemisch (elektrolytisch) f. to electrodeposit
Fallen *n* drop *(as of bulk material)*; decrease, fall, drop, reduction *(as of measuring values)*; fall *(as of a liquid level)*
Fällen *n s.* Fällung 1.
fällend precipitative
Fallengift *n (agric)* poison for traps
Fallfilmkolonne *f (distil)* falling-film still
Fallfilmkonzentrierer *m* falling-film concentrator *(as for sulphuric acid)*
Fallfilmverdampfer *m* falling-film evaporator, downflow evaporator
Fällflüssigkeit *f* precipitating liquid
Fallgeschwindigkeit *f* velocity (rate) of fall
Fallhärteprüfer *m* scleroscope
Fällkasten *m* precipitation box (tank), precipitator
Fällkolonne *f* precipitation column; carbonating tower *(in sodium carbonate manufacture)*
Fallkugel *f* drop weight
Fällkupfer *n* cement copper
Fallmischer *m* tumbler (tumbling) mixer, tumbler
Falloch *n* drop hole
Fallout *m* fallout *(the radioactive particles which settle from the atmosphere)*
Fallout *n* fallout *(the settling of radioactive particles from the atmosphere)*
Fallprobe *f s.* Fallprüfung
Fallprüfung *f* [drop] shatter test *(for coke)*
Fallraum *m* gravity chamber *(in dedusting)*
Fallrohr *n (distil)* downpipe, downspout, downtake, downcomer, delivery pipe
Fallrohrkondensator *m* barometric condenser
Fallstromverdampfer *m* falling-film evaporator, downflow evaporator
Fällturm *m* precipitation column; carbonating tower *(in sodium carbonate manufacture)*
Fällung *f* 1. precipitation, sedimentation; 2. precipitate
chemische F. chemical precipitation
fraktionierte F. fractional precipitation
isoelektrische F. *(coll)* isoelectric precipitation
stufenweise F. fractional precipitation
Fällungsagens *n s.* Fällungsmittel
Fällungsanalyse *f* [volumetric] precipitation analysis, precipitation titration
Fällungsanlage *f* precipitating plant, precipitator
Fällungsform *f* precipitated form *(in gravimetric analysis)*
Fällungsmaßanalyse *f s.* Fällungsanalyse
Fällungsmittel *n* precipitating agent, precipitant, precipitator
Fällungspolymerisation *f* precipitation polymerization
Fällungsreagens *n s.* Fällungsmittel
Fällungsreaktion *f* precipitation reaction
periodische (rhythmische) F. periodic reaction
Fällungstitration *f s.* Fällungsanalyse
Fällungsvermögen *n* precipitating (precipitant) power
Fällungswirkung *f* precipitating action
Falte *f (glass)* fold, lap *(a surface defect)*; *(text)* wrinkle, crease
Falten *fpl (glass)* washboard, ladder *(a surface defect)*
senkrechte F. scrub (brush) marks *(a surface defect)*
Faltenbalg *m* corrugated bellows
Faltenfilter *n* plaited (folded, fluted) filter
Faltschachtel *f* cardboard box, [paperboard] carton
Faltschachtelkarton *m* [folding] boxboard, carton
Falunit *m (min)* fahlunite
falzen *(pap)* to fold; *(tann)* to shave
im Kalkzustand f. *(tann)* to green-shave
Falzfestigkeit *f* folding endurance (resistance, strength)
Falzmaschine *f (tann)* shaving machine
Falzung *f (pap)* fold
Falzungszahl *f (pap)* number of folds
Falzwiderstand *m* folding endurance (resistance, strength)
Falzwiderstandsprüfgerät *n (pap)* fold-testing machine, folding tester
Falzwiderstandsprüfung *f (pap)* folding-endurance test
Falzzahl *f (pap)* number of folds
Famatinit *m (min)* famatinite *(antimony(III) copper(I,II) sulphide)*
Familie *f* family *(as in the periodic system)*
Fang *m s.* Fänger
Fangarbeit *f (petrol)* fishing
Fangdorn *m (petrol)* fishing tap
Fangelektrode *f* collecting electrode, collector [electrode]
Fangen *n (petrol)* fishing
Fänger *m (tech)* catcher, trap
Fangglocke *f (petrol)* overshot
Fangmagnet *m (petrol)* fishing magnet
Fangmuffe *f (petrol)* overshot
Fango *m (med)* fango *(a clay mud)*
Fangstelle *f (nucl)* trap
Fangstoff *m* 1. *(pap)* recovered stock (material); 2. getter *(vacuum technology)*
Fangstoffanlage *f (pap)* stuff catcher, pulp saver, save-all [tray]
Fangstück *n (glass)* bait
Fangtaschenelektrode *f* pocket (tulip, hollow) electrode
Fantasiepapier *n* fancy paper, *(Am)* decorated paper

Faraday-Effekt *m* Faraday effect, magnetic rotation
Faraday-Tyndall-Effekt *m* [Faraday-]Tyndall phenomenon (effect)
Farbabbeizmittel *n* paint remover
Farbablauf *m (text)* side-to-centre shading
Farbänderung *f* change in colour, alteration of colour, colour change
negative F. hypsochromic shift
positive F. bathochromic shift
Farbanstrich *m* paint coat[ing]
Farbaufhellung *f* lightening of the colour; hypsochromic shift *(dye theory)*
Farbaufnahmevermögen *n* dye receptivity
Farbbad *n* dye bath (liquor)
färbbar dyeable
Färbbarkeit *f* dyeability
Farbbase *f* dye (colour) base
Farbbeize *f* stain *(for glass, wood)*
farbbeständig colour-fast, non-discolouring, fadeless
Farbbeständigkeit *f* colour fastness (stability)
Farbbestandteil *m* colouring principle
Farbbildner *m* colour former
Farbbindemittel *n* paint binder (vehicle)
Farbbrillanz *f* brilliance, brilliancy
Farbe *f* 1. *(sensation)* colour; 2. *(substance)* colouring matter, *(Am also)* color; *(in or for suspension)* paint, pigment; *(for walls and ceilings)* distemper; *(for artists)* paint, colour; *(in or for solution)* dye; *(for typography)* [printing] ink; *(for glass)* stain ✦ **F. annehmen** to colour ✦ **F. verlieren** to discolour, to fade
F. für Anilingummidruck (Flexodruck, Flexografie) aniline (flexographic) ink
F. für Lichtdruck photogelatin ink
ausgezehrte F. *(tann)* tailing[s], tails
deckende F. body colour
fluoreszierende F. fluorescent paint
gebrauchsfertige F. *s.* streichfertige F.
glänzende F. gloss [printing] ink *(typography)*
grafische F. [printing] ink
keramische F. ceramic colour
lösungsmittelverdünnbare F. solvent-thinned paint
nachleuchtende F. phosphorescent paint
phosphoreszierende F. phosphorescent paint
plastische F. plastic paint
streichfertige F. ready-mixed (do-it-yourself) paint
wasserverdünnbare F. water-thinned paint
Färbeapparat *m* dyeing apparatus
Färbebad *n* dye bath (liquor)
Färbebase *f* dye (colour) base
Färbebeschleuniger *m (text)* dyeing accelerant, carrier
Färbebottich *m s.* Färbekufe
farbecht colour-fast, non-discolouring, fadeless
Farbechtheit *f* colour fastness (stability)
Farbechtheitsmesser *m*, **Farbechtheitsprüfer** *m* *(text)* fad[e]ometer
Färbeeigenschaft *f* dyeing property
Färbefähigkeit *f s.* Färbekraft
Färbefaß *n (tann)* dyeing drum
Färbeflotte *f (text)* dye bath (liquor)
Färbeflottenbehälter *m s.* Färbekufe
Färbegeschwindigkeit *f* rate of dyeing
Färbegestell *n* slide staining rack (tray) *(microscopy)*
Färbehilfsmittel *n* dyeing assistant (aid)
Färbehülse *f (text)* dyeing cone
Färbekasten *m* staining dish (trough) *(microscopy)*
F. nach Coplin Coplin jar
Färbekessel *m s.* Färbekufe
Färbekraft *f* colouring (tinctorial) power; *(text)* dyeing power
Färbekufe *f (text)* dye back, dye[ing] vessel, dye[ing] vat
Färbeküvette *f* staining dish (trough) *(microscopy)*
Färbelack *m* **[/ roter]** lac (Indian) lake, lac lac, lake lac
Färbemaschine *f* dyeing machine
Färbemittel *n* colouring matter (agent)
F. für die Mikroskopie microscopic stain
farbempfindlich *(phot)* colour-sensitive
Farbempfindlichkeit *f (phot)* colour sensitivity
färben to colour, *(using suspensions)* to paint, *(using solutions)* to dye, *(slightly)* to tint, to tinge, *(using dry colouring matter)* to pigment; to stain *(esp wood, glass, or tissues for microscopy)*; *(food)* to add colourants, to colour
direkt f. *(text)* to dye directly
im Faß f. *(tann)* to drum-dye
im Garn f. *(text)* to yarn-dye
im Tauchverfahren f. to dip-dye, to colour by dipping
in der Trommel f. *(text)* to drum-dye
nach Farbvorlage (Muster) f. *(text)* to match
sich f. to colour
sich dunkel (dunkler) f. to darken
unmittelbar f. *s.* direkt f.
Färben *n* colouring, *(using suspensions)* painting, *(using solutions)* dyeing, *(slight)* tinting, tinging, *(using dry colouring matter)* pigmenting; staining *(esp of wood, glass, or of tissues for microscopy)*; *(food)* colouring
F. auf stehendem Bad *(text)* standing-bath dyeing
F. im Garn *(text)* yarn dyeing
F. im Holländer *(pap)* beater dyeing (colouring)
F. im Metallbad *(text)* molten-metal dyeing
F. im Packsystem *(text)* pack[age] dyeing
F. im Stoff *s.* F. in der Masse
F. im Strang *(text)* rope (hank) dyeing
F. im Stück *(text)* piece dyeing
F. im Tauchverfahren dip dyeing, colouring by dipping
F. in der Flocke *(text)* [loose] stock dyeing
F. in der Masse *(pap)* beater dyeing (colouring), dyeing (colouring) in the pulp; *(plast)* mass dyeing
F. in der Wolle *(text)* stock dyeing
F. in Gegenwart von Lösungsmitteln *(text)* solvent dyeing
F. in Strangform *s.* F. im Strang
F. nach Farbvorlage (Muster) *(text)* matching

F. unter Druck pressure dyeing
F. unter HT-Bedingungen *(text)* high-temperature dyeing
F. von Faserstoffmischungen *(text)* union dyeing
F. von Kammzug *(text)* top dyeing
F. von Kreuzspulen *(text)* cheese dyeing
kontinuierliches F. *(text)* continuous dyeing
ungleiches (ungleichmäßiges) F. *(text)* ending

Farbenabweichung *f* chromatic aberration
Farbenatlas *m* colour chart (atlas)
Farbenbeständigkeit *f s.* Farbbeständigkeit
Farbenbindemittel *n* paint binder (vehicle)
Farbenchemie *f* 1. paint chemistry, *(Am also)* color chemistry; 2. *(incorrectly for)* Farbstoffchemie
Farbenchemiker *m* 1. paint chemist, *(Am also)* color chemist; 2. *(incorrectly for)* Farbstoffchemiker
Farbendruck *m* [multi]colour printing; colour print
Farbenfehler *m* chromatic aberration
Farbenfotografie *f s.* Farbfotografie
Farbengang *m (tann)* suspender set, round of handlers, *(Am)* rocker yard
Farbenschönheit *f* brilliance, brilliancy
Farbensehen *n* colour vision
Farbensensibilisierung *f (phot)* optical (dye) sensitization (sensitizing)
Farbentferner *m* paint remover
Farbentwickler *m (phot)* colour developer
Farbentwicklung *f (phot)* colour development
Farbenumschlag *m s.* Farbumschlag
Farbenzwischenstoff *m s.* Farbstoffzwischenprodukt
Färber *m* dyer
Färberdistel *f* safflower, safflor, Carthamus tinctorius L.
Färberei *f* 1. dyeing; 2. dye-house, dye-works
Färbereiche *f (tann)* black (dyer's) oak, Quercus velutina Lam.
Färbereihilfsmittel *n* dyeing assistant (aid)
Färberflechte *f* / **tangartige** Lima (Angola) weed, Roccella fuciformis (L.) Lam.
Färberginster *m* dyer's greenwood, Genista tinctoria L.
farberhöhend hypsochromic
Farberhöhung *f* hypsochromic shift
Färberkamille *f* dyer's (golden) chamomille, Anthemis tinctoria L.
Färberkrapp *m* madder, Rubia tinctorum L.
Färbermaulbeerbaum *m* dyer's mulberry, Chlorophora tinctoria Gaudich.
Färber-Meister *m* dyer's woodruff, Asperula tinctoria L.
Färberröte *f s.* Färberkrapp
Färberwaid *m* woad, Isatis tinctoria L.
Färbestern *m (text)* star frame
Färbeverhalten *n* dyeing properties
Färbevermögen *n* colouring (tinctorial) power
Färbezeit *f (text)* dyeing time (cycle); staining time *(microscopy)*
Farbfehler *m* 1. colour defect, off colour; 2. chromatic aberration *(optics)*
Farbfilm *m (phot)* colour film
Farbfilter *n* colour filter
Farbfleck *m* colour spot *(a defect in paper)*
Farbflotte *f (text)* dye bath (liquor)
Farbfotografie *f* colour photography
farbfrei colourless
farbgebend colour-bearing, chromophoric
Farbglas *n* stained (coloured) glass
Farbgrube *f (tann)* [suspender] pit
Farbholz *n* dyewood
Farbindikator *m* colour indicator
Farbintensität *f* colour intensity, colouring strength
Farbkomparator *m* colour comparator
Farbkraft *f* colouring (tinctorial) power; *(text)* dyeing power
farbkräftig of strong colouring (tinctorial) power; highly (intensely) coloured, brilliant
Farblack *m* colour[ed] lake
farblos colourless
Farblosigkeit *f* colourlessness
Farblösung *f s.* Farbstofflösung
Farbmalz *n* coloured (roasted, black) malt
Farbmittel *n* colouring matter (material, substance, agent)
Farbnuance *f* shade, tint, tinge, tone, hue, *(text also)* cast
Farbpigment *n* paint pigment
Farbprinzip *n* colouring principle
Farbsalz *n* dye salt
Farbschattierung *f s.* Farbnuance
Farbschönheit *f* brilliance, brilliancy
farbschwach weakly (feebly) coloured
Farbskala *f* colour range (scale, chart)
Farbspritzen *n* spray painting, paint spraying
elektrostatisches F. electrostatic paint spraying
Farbspritzpistole *f* paint spray[ing] gun, paint sprayer
elektrostatische F. electrostatic spray gun
Farbspritztechnik *f* paint spraying technique
farbstark intensely (highly) coloured, brilliant
Farbstärke *f* colouring (tinctorial) strength
Farbstoff *m* colouring matter (substance), colourant; dye[stuff] *(soluble organic compound)*; pigment *(insoluble matter of organic or inorganic origin)*; *(for wood, glass, microscopical investigation)* stain; *(biol)* pigment
F. für Holländerfärbung *(pap)* beater dye (colour)
F. für Kalanderfärbung *(pap)* calender dye (colour)
F. für Massefärbung *(pap)* beater dye (colour)
adjektiver F. adjective (mordant) dye
anthrachinoider F. anthraquinone (anthraquinonoid) dye
basischer F. basic dye; basic stain *(for microscopy)*
beizenfärbender F. *s.* adjektiver F.
blauer F. blue; *(for improving the degree of whiteness)* blueing
direktziehender F. direct (substantive) dye
dispergierter F. disperse[d] dye
echter F. fast dye

gemischter F. mixed dye
indigoider F. indigoid dye
kationischer F. cationoid dye
kombinierbarer F. compatible dye
natürlicher F. natural dye (colouring matter), biochrome
pflanzlicher F. *(for use)* plant colouring matter, vegetable dye; *(biol)* plant pigment
saurer F. acid[ic] dye
sensibilisierender F. sensitizing dye
substantiver F. substantive (direct) dye
unechter F. fugitive dye
farbstoffaffin dye-affinitive
Farbstoffaffinität *f* dye affinity ✦ **mit erhöhter F.** deep dyeing
Farbstoffaufnahme *f* dye uptake (absorption)
Farbstoffaufnahmevermögen *n* dye receptivity
Farbstoffbase *f* dye (colour) base
Farbstoffchelat *n* dye chelate
Farbstoffchemie *f* dye chemistry
Farbstoffchemiker *m* dye chemist
Farbstoffechtheit *f* dye fastness
Farbstoffixiermittel *n* dye-fixing agent, dye fixative
Farbstoffixierungsgeschwindigkeit *f* rate of dye fixation
Farbstoffklasse *f* class of dyestuffs
Farbstofflösung *f (pap)* dye solution
F. für Kalanderfärbung calender solution
Farbstoffschutzschicht *f (phot)* dye layer, backing [layer]
Farbstoffsortiment *n* range (assortment) of dyes
Farbstoffsuspension *f* pigment suspension
Farbstoffträger *m (biol)* chromatophore
Farbstoffzusatz *m (food)* 1. colouring matter, colourant; 2. addition (admixture) of colouring matter
Farbstoffzwischenprodukt *n* dye intermediate
Farbtiefe *f* depth of colour
Farbton *m* shade, tint, tone, hue, *(text also)* cast ✦ **einen F. treffen** to match a shade ✦ **im F. abfallen** to be off-shade
Farbtonänderung *f s.* Farbtonverschiebung
Farbtonbeständigkeit *f*, **Farbtonechtheit** *f* colour fastness (stability)
Farbtonumschlag *m s.* Farbtonverschiebung
Farbtönung *f s.* Farbton
Farbtonverschiebung *f* change in shade, alteration of shade
bathochrome F. bathochromic shift (displacement)
hypsochrome F. hypsochromic shift (displacement)
farbtragend chromophoric
Farbträger *m* chromophore, chromophoric group; *(coat)* substrate
Farbüberzug *m* paint coat
Farbumschlag *m* change in colour, alteration of colour, colour change; indicator transition *(titrimetry)*
Färbung *f* 1. *s.* Färben; 2. *(phenomenon)* colo[u]ration, colour (*s. a.* Farbton); *(biol)* pigmentation; 3. *(result of treatment: with suspensions)* paint, *(with solutions)* dye; *(of wood, glass, tissues for microscopy)* stain
blaue F. blueness
Gramsche F. Gram staining *(for bacteria)*
leere (matte) F. dead dyeing
stippenfreie F. *(text)* speck-free dyeing
stumpfe (tote) F. dead dyeing
Färbungsbremsmittel *n* dye retardant
Farbunterschied *m* difference in colour
Farbveränderung *f s.* Farbänderung
Farbvergleicher *m* colour comparator
Farbvergleichszylinder *m* colour-comparator tube
Farbverschiebung *f s.* Farbtonverschiebung
farbverstärkend auxochromic
farbvertiefend bathochromic
Farbvertiefung *f* bathochromic shift
Farbwanderung *f* swealing *(during the drying of textiles)*
Farbwechsel *m* change in colour, alteration of colour, colour change
Farbzahl *f (min)* colour index
Farin[zucker] *m* brown sugar, muscovado
Faser *f (biol)* fibre; *(text)* [staple] fibre *(a natural or man-made fibre of relatively short length)*
anorganische F. inorganic fibre
gemahlene F. milled fibre
gesponnene F. spun fibre
keramische F. ceramic fibre (staple)
kollagene F. *(biol)* collagen fibre *(of connective tissue)*
mineralische F. mineral fibre
native (natürliche) F. natural fibre
organische F. organic fibre
pflanzliche F. vegetable fibre
polynosische F. polynosic fibre
synthetische F. [completely, fully] synthetic fibre, synthetic polymer fibre, synthetic
tierische F. animal fibre
Faserabbau *m* fibre disintegration
Faseraffinität *f* affinity for the fibre
faserartig fibrous
Faserbanane *f* abaca, Musa textilis Née
Faserband *n (text)* sliver
faserbildend fibre-forming
Faserbildung *f* fibre formation
Faserbraunkohle *f* fibre brown coal
Faserbrei *m (pap)* fibrous pulp (mass), pulp slurry (stock), slush [of] stock
Faserbreipreßteil *n (plast)* pulp moulding
Faserbruchstücke *npl (pap)* fragments of fibres
Faserbündel *n (pap)* fibre bundle; *(text)* strand
Fäserchen *n* fibril[la]
Faserdiagramm *n* fibre diagramm (photograph)
Faserfilter *n* felt[-fabric] filter
Faserfilz *m (pap)* web of fibre[s], [paper] web, mat
Faserflug *m (text)* linters
Fasergefüge *n* fibrous structure
Fasergewebe *n* fabric [cloth]
Fasergewirre *n s.* Faservlies
Fasergut *n* / **loses** *(text)* loose stock
Faserhalbstoff *m (pap)* half-stuff, half-stock

eingetragener F. pulp (fibrous) furnish, beater charge
Faserhaut *f (text)* sheath, shell, skin
Faserholz *n (pap)* pulpwood
faserig fibrous
Faserinkrustierung *f* fibre incrustation
Faserkalk *m* fibrous (vein) chalk
Faserkohle *f* fibrous coal
Faserlänge *f* fibre length (staple), staple [length]
Faserlein *m* fibre flax *(variety of Linum usitatissimum)*
Fasermantel *m (text)* sheath, shell, skin
Fasermasse *f (pap)* fibrous pulp (mass), pulp slurry (stock), slush [of] stock
Fasermaterial *n (pap)* fibrous material
Fasermischung *f* fibre blend
Faserplatte *f* fibre board
Faserprotein *n* fibrous protein
Faserrohstoff *m* fibrous raw material, crude fibre material, raw papermaking material, raw (paper) stock
Faserrückgewinnung *f (pap)* fibre recovery
Faserrückgewinnungsanlage *f (pap)* stuff catcher, pulp saver, save-all [tray]
Faserschädigung *f* damage to fibres
Faserschutzmittel *n* fibre-protective agent
Faserserpentin *m (min)* chrysotile, Canadian asbestos
Faserstoff *m (text)* fibre, fibrous material *(a natural or man-made fibre that has a length usually many hundred or thousand times greater than its cross section)*
anorganischer F. inorganic fibre
keramischer F. ceramic fibre
natürlicher F. natural fibre
organischer F. organic fibre
pflanzlicher F. vegetable fibre
polynosischer F. polynosic fibre
synthetischer F. [completely, fully] synthetic fibre, synthetic polymer fibre, synthetic
faserstoffbildend fibre-forming
Faserstoffbildung *f* fibre formation
Faserstoffbrei *m s.* Faserbrei
Faserstoffchemie *f* chemistry of fibres
Faserstoffknoten *m (pap)* knot
Faserstoffschutzmittel *n* fibre-protective agent
Faserstoffstruktur *f (text)* fibre structure
Faserstoffsuspension *f (pap)* fibre-bearing liquid (water)
Faserstruktur *f (text)* fibre structure; *(min)* fibrous structure
Fasersuspension *f (pap)* fibre suspension, fibrous pulp, pulp slurry (stock)
Fasertorf *m* moss peat
Faserverfilzung *f (pap)* felting (matting) of the fibres
Faserverkettung *f (pap)* bonding of the fibres, interfibre bonding
Faserverlust *m (pap)* fibre (stock) loss
Faservlies *n (text)* non-woven fabric
Faservliesfilter *n* felt[-fabric] filter
F. mit Blasring reverse-jet filter
Faserwiedergewinnung *f (pap)* fibre recovery
fasrig fibrous
Faß *n* drum, vat, *(if wooden)* barrel, cask, *(if small)* keg, *(if very large)* tun ✦ **im F. behandeln (durcharbeiten)** *(tann)* to drum
Faßabfüller *m*, **Faßabfüllmaschine** *f* [cask-]racking machine, [cask] racker
Faßabfüllung *f* barrel[l]ing, racking
Faßamalgamation *f* barrel amalgamation
Faßäscher *m*, **Faßäscherung** *f (tann)* drum liming
Faßbier *n* keg (cask) beer
fassen to hold *(a certain quantity)*; to take *(a load)*
Fässeramalgamation *f* barrel amalgamation
Faßfärbung *f (tann)* drum dyeing
Faßfettung *f (tann)* drum stuffing
Faßfüller *m*, **Faßfüllmaschine** *f* [cask-]racking machine, [cask] racker
Faßgeläger *n (food)* bottoms
Faßgerbung *f* drum tannage
Faßgut *n (met)* hutch product
Fassonieren *n* profiling
Faßreinigungsmaschine *f* cask washer (washing machine)
Faßschmiere *f (tann)* drum stuffing
Fassungsvermögen *n* capacity
Fassungswinkel *m* angle of nip *(on roll crushers)*
Faßwaschmaschine *f* cask washer (washing machine)
Faßwein *m* cask (bulk) wine
Fast-Extrusion-Furnace-Ruß *m (rubber)* fast extrusion furnace black, FEF black
Fastkristall *m* liquid crystal, paracrystal
Faugeron-Ofen *m (ceram)* Faugeron kiln *(a tunnel kiln)*
Faujasit *m (min)* faujasite *(a tectosilicate)*
Fauläscher *m (tann)* dead (rotten) lime
Faulbaum *m* frangula, Rhamnus frangula L.
Faulbaumrinde *f* [alder] buckthorn bark, black dogwood bark, frangula *(from Rhamnus frangula L.)*
Amerikanische F. Persian (sacred, chittam, chittem, chittim) bark, cascara sagrada [bark] *(from Rhamnus purshianus DC.)*
Faulbecken *n*, **Faulbehälter** *m* digestion (septic) tank, digester *(waste-water treatment)*
faulen 1. *(biol)* to putrefy; 2. *(ceram)* to sour, to age
Faulen *n* 1. *(biol)* putrefaction; 2. *(ceram)* souring, ageing
faulend putrescent
faulfähig putrefiable, putrescible; digestible *(waste water)*
nicht f. unputrefiable, imputrescible; indigestible *(waste water)*
Faulfähigkeit *f* putrescibility; digestibility *(of waste water)*
Faulgas *n* digester gas *(waste-water treatment)*; *(agric)* fermentation gas
Faulgrube *f s.* Faulbecken
faulig putrid
Fäulnis *f* putrefaction
Fäulnisbakterien *pl* putrefactive bacteria
fäulnisbeständig rotproof, rot-resistant, antirot

Fäulnisbeständigkeit *f* rotproofness, rot (decay) resistance
fäulnisbewohnend *(biol)* saprophytic
Fäulnisbewohner *m* saprophyte, saprophytic organism
fäulniserregend putrefactive, putrefacient, saprogenic, saprogenous
fäulnisfähig *s.* faulfähig
Fäulnisfähigkeit *f s.* Faulfähigkeit
fäulnisfest *s.* fäulnisbeständig
Fäulnisfestigkeit *f s.* Fäulnisbeständigkeit
Fäulnisgärung *f* putrefactive fermentation
fäulnisverhindernd, fäulnisverhütend rotproofing, antirot
Fäulnisverhütungsmittel *n* rotproofing agent
Faulraum *m* [sludge-]digestion chamber, [sludge-] digestion compartment *(waste-water treatment)*
Faulschlamm *m* digested sludge *(waste-water treatment)*; sapropel *(hydrobiology)*
Faulschlammgestein *n* sapropelite
Faulschlammkohle *f* sapropelic coal
Faulung *f* digestion *(waste-water treatment)*
F. in zwei Stufen two-stage digestion
thermophile F. thermophilic digestion
Faulverfahren *n* septic-tank method *(waste-water treatment)*
Faulzeit *f* digestion time (period) *(waste-water treatment)*
Fauser-Verfahren *n* Fauser [ammonia] process
Faustregel *f* rule of thumb
Faustzahl *f* round figure
Faworski-Umlagerung *f* Faworski rearrangement *(of α-haloketones to acids or esters)*
Fayalit *m (min)* fayalite *(iron(II) orthosilicate)*
Fayence *f (ceram)* faience
Delfter F. delft[ware], delph[ware]
Fayenceware *f* faience ware
Fäzes *pl* faecal matter, faeces
Fazies *f (geol)* facies
FCC-Anlage *f (petrol)* FFC unit, fluid catalytic cracking unit (plant)
FCC-Verfahren *(petrol)* fluid catalytic [cracking] process
Febrifugin *n* febrifugine *(antimalarial alkaloid)*
Feder *f (tech)* spring
federbelastet spring-loaded
Federbildung *f (ceram)* feathering *(a glaze fault)*
Federkraftwälzmühle *f* **nach Loesche** Loesche mill
Federleichtpapier *n* featherweight paper, *(Am)* bulking paper
Federventil *n* spring[-seated] valve
Federwaage *f* spring balance
Feedermaschine *f (glass)* gob-fed machine
Feederverfahren *n (glass)* feeder (gob) process
FEF-Ruß *m (rubber)* FEF black, fast extrusion furnace black
Fehlbenennung *f* misnomer
Fehlbohrung *f (petrol)* unproductive (dry) well, duster
Fehler *m* 1. *(of material)* defect, flaw, fault; *(cryst)* defect, imperfection; 2. error *(statistics)*
methodischer F. error of method
mittlerer quadratischer F. standard deviation *(statistics)*
parallaktischer F. parallax error
persönlicher (subjektiver) F. personal error
systematischer F. systematic error
wahrer F. error of observation
zufälliger F. accidental (random) error
Fehlerausgleich *m* compensation of errors
Fehlerfunktion *f* error function, erfc
Fehlergrenze *f* limit of error
fehlerhaft defective, faulty, imperfect
Fehlerintegral *n* error function, erfc
Fehlerkompensation *f* compensation of errors
Fehlersuche *f* fault finding
Fehlfärbung *f* faulty dyeing
Fehlgärung *f* faulty fermentation
fehlgeordnet *(cryst)* disordered
Fehlgitter *n s.* Fehlstellengitter
Fehlordnung *f (cryst)* disorder
Frenkelsche F. Frenkel disorder
Fehlstelle *f (cryst)* [lattice] defect
Frenkelsche F. Frenkel defect
punktförmige F. point defect
Fehlstellengitter *n* defect lattice
Fehlstellenhalbleiter *m* extrinsic semiconductor
Fehlstellenkonzentration *f (cryst)* defect concentration
Fehlstellenpaar *n* / **Frenkelsches** *(cryst)* Frenkel (interstitial-vacancy) pair
Fehlstellenwanderung *f (cryst)* defect motion
Feilspäne *mpl* filings
Feinanteile *mpl s.* Feingut
feinausgemahlen finely ground
Feinausmahlung *f* fine grinding
Feinboden *m* fine soil *(particle size less than 2 mm)*
Feinbrechen *n* fine crushing
Feinbrecher *m* fine crusher
Feinbürette *f* microburette
Feinchemikalie *f* fine chemical
Feindestillation *f* precision distillation
feindispers finely dispersed
feinen *(met)* to refine
Feinerde *f s.* Feinboden
Feinerz *n* fine ore
Feines *n s.* Feingut
Feinfilter *n* fine filter
Feinfiltration *f* polishing [filtration]
Feinfolie *f (plast)* film *(thickness less than 0.01 inch)*
Feinfoliengießmaschine *f* film casting machine
Feingefüge *n* microscopic structure, microstructure
feingepulvert finely powdered
Feinguß *m* precision casting
Feingut *n* fine sizes, fines, fine (minus, undersize) material *(classifying)*, *(min tech also)* slime
Feingutüberlauf *m (min tech)* slime overflow
Feinheit *f* 1. [degree of] fineness; 2. *(text)* count, number
Feinheitsgrad *m* degree of fineness
Feinheitsmodul *m* fineness modulus

Feinkeramik *f* fine ceramics (ceramic ware)
Feinklassierung *f* fine-size fractionation (separation)
Feinkohle *f* fine coal, [coal] fines
Feinkorn *n* 1. *s.* Feingut; 2. *(sugar)* false grain; 3. *(geol)* grain; 4. *(phot)* fine grain
Feinkornbild *n (phot)* fine-grain image
Feinkornemulsion *f (phot)* fine-grain emulsion
Feinkornentwickler *m (phot)* fine-grain developer
Feinkornentwicklung *f (phot)* fine-grain development
feinkörnig fine-grain[ed], fine[-granular]; *(min)* close-grained *(texture)*
feinkristallin[isch] finely crystalline, fine-grained
Feinkühlen *n (glass)* fine annealing
feinmahlen to grind finely; *(pap)* to refine, to clear, to brush out
Feinmahlung *f* fine grinding; *(pap)* refining, clearing, brushing-out
Feinmanipulator *m* micromanipulator *(microscopical technique)*
feinmaschig fine-meshed, narrow-meshed, close-meshed
Feinmühle *f* fine-grinding mill, fine grinder, pulverizing mill, pulverizer
Feinpapier *n* fine paper, F.p.
Feinpapierfabrik *f* fine mill
Feinpappe *f* fine board
feinporig fine-pored
Feinpostpapier *n* bank paper (post), bond [paper]
feinpulverig finely powdered
Feinpulvriges *n (plast)* fines *(as of moulding material)*
Feinrechen *m* fine-screen unit *(waste-water treatment)*
Feinregulierung *f* delicate control
Feinsand *m* fine sand *(grain size 0.2 to 0.02 mm)*
Feinschicht-Walzentrockner *m* drum film dryer
Feinschleifen *n* fine grinding
Feinschliff *m* fine grinding
Feinseife *f* toilet soap
Feinsieb *n* fine sieve
Feinsortierer *m* fine (secondary, second) screen
Feinsortierung *f* fine (secondary) screening
Feinsteinzeug *n (ceram)* fine stoneware
Feinstmahlung *f* pulverizing
Feinstmühle *f* pulverizing mill, pulverizer
Feinstoff *m (pap)* accepted (screened) stock, accepts
Feinstoffe *mpl (plast)* fines *(as of moulding material)*
feinstreifig fine-banded, finely banded
Feinstruktur *f (phys chem)* fine structure; microscopic structure, microstructure
Feinstrukturanalyse *f (phys chem)* fine-structure analysis
Feinstrukturkonstante *f* fine-structure constant *(spectroscopy)*
  Sommerfeldsche F. Sommerfeld fine-structure constant
feinstückig small-sized
Feintrub *m (ferm)* cold sludge
feinvermahlen to grind finely
Feinvermahlen *n* fine grinding
feinverteilt finely dispersed (divided)
Feinverteilung *f* dispersion, dispersal
feinzerkleinern to comminute
Feinzerkleinerung *f* comminution
Feinzerteilung *f* dispersion, dispersal
Feinzuschlag *m (build)* fine aggregate
Feld *n (phys chem)* field
  atomares F. atomic field
  elektrostatisches F. electrostatic field
  entmagnetisierendes F. demagnetizing field,
  magnetisches F. magnetic field
  selbstkonsistentes F. self-consistent field, SCF
Feldbrennofen *m (ceram)* clamp
Felddüngungsversuch *m* field fertilization test
Feldelektronenemission *f* field emission
Feldelektronenmikroskop *n* field emission (electron) microscope
Feldemission *f* field emission
Feldmissionsmikroskop *n s.* Feldelektronenmikroskop
Feldleistung *f* field performance (efficiency) *(of pesticides)*
Feldofen *m (ceram)* clamp
Feldrichtung *f (phys chem)* field direction
Feldspat *m (min)* fel[d]spar, feldspath
  als Massebestandteil verwendeter F. *(ceram)* body spar
  erdiger F. clay-stone
  zur Glasherstellung verwendeter F. glass spar
feldspathaltig *(min)* fel[d]spathic
Feldspatoid *m*, **Feldspatvertreter** *m (min)* feldspathoid
Feldspritzrohr *n (agric)* boom sprayer, spray boom
Feldstärke *f* field strength
Feldstärkeeffekt *m* Wien effect
Feldtest *m s.* Feldversuch
Feldvalenzverbindung *f* field valency compound
Feldversuch *m (agric)* field experiment (test, trial)
Feldwirksamkeit *f* field performance (efficiency) *(of pesticides)*
Felgenband *n (rubber)* flap
Felgenbandheizer *m (rubber)* flap mould (vulcanizer)
Felit *m* felite *(a crystalline constituent of portland cement clinker)*
Fell *n (rubber)* sheet, band
Fenac *n* $Cl_3C_6H_2COOH$ fenac, 2,3,6-trichlorophenylacetic acid *(a herbicide)*
Fenchelöl *n* fennel[-seed] oil *(from Foeniculum vulgare Mill.)*
Fenchol *n s.* Fenchylalkohol
Fenchylalkohol *m* fenchyl alcohol, fenchol, 1,3,3-trimethylbicyclo[1,2,2]-heptan-2-ol
Fensterglas *n* window glass
  F. doppelter Dicke double-strength glass
  F. einfacher Dicke single-strength glass
Fensterglaszylinder *m (glass)* roller
Fensterkitt *m* [painter's, glazier's] putty
Fensterpapier *n* diaphanic paper
Fenuron *n* $C_6H_5NHCON(CH_3)_2$ fenuron, 1,1-dimethyl-3-phenylurea *(a herbicide)*

Ferberit *m (min)* ferberite *(iron(II) tungstate)*
Ferment *n* enzyme, ferment ✦ **mit Fermenten anreichern** to enzymize
eiweißabbauendes (eiweißspaltendes) F. proteolytic (protein-digesting) enzyme, protease
extrazelluläres F. extra-cellular enzyme, desmo-enzyme
fettspaltendes F. lipolytic enzyme
geformtes F. organized enzyme
Gelbes F. yellow enzyme, flavoprotein, flavin enzyme, flavoenzyme
intrazelluläres F. endocellular (intracellular) enzyme, endoenzyme
originäres F. natural enzyme
pektisches F. pectic enzyme, pectinase
pektolytisches F. pectolytic enzyme
proteinspaltendes (proteolytisches) F. *s.* eiweißabbauendes F.
stärkespaltendes F. starch-splitting (starch-reducing, starch-converting, amylolytic) enzyme, amylase
ungeformtes F. unorganized enzyme
urikolytisches F. uricolytic enzyme
Fermentaktivität *f* enzyme activity
Fermentäscher *m (tann)* enzyme unhairing
Fermentation *f* fermentation
F. im Haufen heap fermentation *(as of cocoa beans)*
Fermentationsbrühe *f* fermentation broth *(in manufacturing antibiotics)*
fermentativ enzym[at]ic, ferment[at]ive
Fermentator *m s.* Fermenter
Fermentbehandlung *f* enzyme treatment
Fermentchemie *f* enzyme chemistry
Fermentdonator *m* enzyme donor
Fermenter *m* fermenter, fermentor
fermentieren to ferment
Fermentierung *f* fermentation
fermentkatalysiert enzyme-catalyzed
Fermentreaktion *f* enzyme (enzymatic) reaction
fermentresistent enzyme-resistant
Fermentsystem *n* enzyme system
Fermenttätigkeit *f* enzyme activity
Fermentwirkung *f* enzyme action
Fermi-Dirac-Statistik *f (phys chem)* Fermi[-Dirac] statistics
Fermi-Energie *f* Fermi energy
Fermi-Fläche *f* Fermi surface
Fermi-Gas *n* Fermi-Dirac gas
Fermi-Kante *f s.* Fermi-Niveau
Fermi-Konstante *f* Fermi constant
Fermi-Niveau *n* Fermi [characteristic energy] level
Fermion *n (nucl)* fermion
Fermi-Statistik *f s.* Fermi-Dirac-Statistik
Fermi-Temperatur *f* Fermi temperature
Fermi-Theorie *f* **[des β-Zerfalls]** Fermi theory [of beta decay]
Fermium *n* Fm fermium
Fermi-Verteilung *f* Fermi-distribution
Fernambukholz *n*, **Fernambuko** *n (dye)* brazilwood *(wood of Caesalpinia specc.)*
Fernanzeige *f* remote indication
Fernbedienung *f* remote operation
Fernbetätigung *f* remote operation
Fernleitung *f* [long-distance] pipeline
Fernordnung *f (cryst)* long-range order
Fernordnungsgrad *m (cryst)* long-range order
Fernsteuerung *f* remote control
Fernthermometer *n* telethermometer, distance (recording) thermometer
Ferrat(III) *n* $M^{I}FeO_2$ ferrate(III)
Ferrichrom *n* ferrichrome
Ferrimolybdit *m (min)* ferrimolybdite, molybdic ochre *(a hydrous iron(III) molybdate)*
Ferrioxamin *n* ferrioxamine *(any of a series of metabolic products of actinomycetes)*
Ferrit *m* 1. *(met)* ferrite *(a solid solution of carbon in alpha or delta iron)*; 2. ferrite *(a magnetic material of the formula $M^{II}Fe_2O_4$)*
ferritisch ferritic
Ferrobor *n* ferroboron
Ferrochrom *n* ferrochromium
Ferroelektrikum *n* ferroelectric [material, substance]
ferroelektrisch ferroelectric
Ferroin *n* ferroin *(any of a class of complexes of tertiary heterocyclic amines)*
Ferrolegierung *f* ferro-alloy
Ferromagnetikum *n* ferromagnetic [material, substance]
ferromagnetisch ferromagnetic
Ferromagnetismus *m* ferromagnetism
Ferromangan *n* ferromanganese
Ferromolybdän *n* ferromolybdenum
Ferronickel *n* ferronickel
Ferroniob *n* ferroniobium
Ferrophosphor *m* ferrophosphorus
Ferrosilizium *n* ferrosilicon
Ferrospinell *m (min)* hercynite *(iron(II) aluminate)*
Ferrotantal *n* ferrotantalum
Ferrotellurit *m (min)* ferrotellurite
Ferrotitan *n* ferrotitanium
Ferrovanadin *n* ferrovanadium
Ferrowolfram *n* ferrotungsten
Ferroxylindikator *m* ferroxyl indicator
Ferrozen *n* $Fe(C_5H_5)_2$ ferrocene, dicyclopentadienyl iron
Ferrozenpolymer[es] *n* ferrocene polymer
Ferrozirkonium *n* ferrozirconium
Ferrozyankupfermembran *f (biol)* copper ferrocyanide membrane, *(better)* copper(II) hexacyanoferrate(II) membrane *(for demonstrating osmosis)*
fertigbearbeiten to finish
Fertigbearbeitung *f* finish[ing]
Fertigbeton *m* ready-mixed concrete
Fertigblasen *n (glass)* final blow[ing]
Fertigbleiche *f (pap)* final bleaching
Fertigbrand *m* / **eierschaliger** *(ceram)* egg-shell finish
fertigen to manufacture, to make, to produce

Fertigerzeugnis *n s.* Fertigprodukt
Fertigform *f (glass)* blow[ing] mould
Fertigformboden *m (glass)* [blow mould] bottom plate
fertiggeformt finally shaped
Fertigkochperiode *f (pap)* pulping period
fertigmahlen *(pap)* to refine, to clear, to brush out
Fertigpräparat *n (pharm)* preparation
Fertigprodukt *n* final (finished) product (stock)
Fertigsintern *n* full (final) sintering
fertigstellen to finish
Fertigstellung *f* finish[ing]
 F. in Bogen *(pap)* sheeting
Fertigteilverschäumung *f* foaming in place (situ)
Fertigung *f* processing
Fertigungslos *n (plast)* run
 kleines F. short run
Fertigungsmenge *f* production output
Fertigungsverfahren *n* manufacturing process
fertil fertile
Fertilität *f* fertility
Fertilitätsvitamin *n* antisterility vitamin
Ferulasäure *f* $CH_3OC_6H_3(OH)CH=CHCOOH$ ferulic acid, 4-hydroxy-3-methoxycinnamic acid
Feruloylchinasäure *f* feruloylquinic acid *(a depside)*
fest *(as opposed to liquid)* solid; *(mechanically)* firm, stable, compact, tight, rigid; resistant *(esp to wear)*
 f. werden to solidify, to harden, to set, to congeal, *(esp coll)* to gel[ate]
 f. werden lassen to solidify, to congeal, to set
festbacken to cake
Festbett *n (tech)* fixed (static) bed; *(petrol)* dense bed
Festbettadsorber *m* fixed-bed adsorber
Festbettkatalysator *m* fixed-bed catalyst, static catalyst
Festbettkolonne *f* fixed-bed column
Festbettkontakt *m s.* Festbettkatalysator
Festbettverfahren *n* fixed-bed process
Festbitumen *n* solid bitumen
festblasen *(glass)* to blow down
Festblasen *n (glass)* settle blow
Festbrennen *n (ceram)* firing-on, stoving *(of on-glaze decorations)*
Festbrennstoff *m* solid fuel
festdrücken *(lab)* to press down *(a precipitate in a funnel)*
Fest-Fest-Grenzfläche *f* solid-solid interface
Fest-Flüssig-Chromatografie *f* solid-liquid chromatography
Fest-Flüssig-Grenzfläche *f* solid-liquid interface
festfressen / sich to seize
Festfressen *n* seizing, seizure
Festgehalt *m* solids content (loading)
 F. der Schwarzlauge *(pap)* black liquor solids
festheften to affix
Festigkeit *f* firmness, stability, compactness, tightness, rigidity, tenacity, *(esp in materials testing)* resistance, strength
 chemische F. chemical resistance (stability), resistance (stability) to chemical attack
 dielektrische F. dielectric (breakdown) strength
 thermische F. thermal stability (resistance, endurance), heat stability (resistance, endurance), thermostability
Festigkeit-Masse-Verhältnis *n* strength-to-weight ratio
Festikeitseinbuße *f s.* Festigkeitsverlust
Festigkeitsprüfer *m*, **Festigkeitsprüfmaschine** *f* strength tester (testing machine); *(rubber)* tensile[-strength] tester, tensile[-strength] testing machine
Festigkeitsrückgang *m* loss in stability (strength)
Festigkeitsverlust *m* loss in stability (strength)
Festkautschuk *m* solid (dry) rubber
festklemmen to clamp
 sich f. to seize
Festkörper *m* solid [matter]
Festkörperdiffusion *f* solid diffusion
Festkörperlöslichkeit *f* solid solubility
Festkörperphysik *f* solid-state physics
Festkörperreaktion *f* solid-state reaction
Festkörperzustand *m* solid state
Festkraftstoff *m* solid fuel (propellant)
festlegen to fix *(nutrients in a soil)*
 biologisch f. to immobilize *(nutrients by the action of soil microorganisms)*
Festlegung *f* fixation *(of nutrients in a soil)*
 biologische F. immobilization *(of nutrients by the action of soil microorganisms)*
Festlinie *f (distil)* solidus curve (line)
Festoondämpfer *m (text)* festoon ager
Festparaffin *n* solid paraffin, paraffin wax
feststampfen to tamp
feststellbar 1. determinable, detectable, detectible; 2. fastenable
feststellen 1. to determine, to identify, to detect *(analytical chemistry)*; to establish *(e. g. the structure of molecules)*; 2. *(mechanically)* to arrest, to fasten, to fix
Feststellung *f* determination, identification, detection *(analytical chemistry)*; establishment *(as of the structure of molecules)*
Feststoff *m* solid [matter, substance]
Feststoffanteil *m* solids content (loading)
Feststoffaufnahmevermögen *n* solids-holding capacity
Feststoffbett *n* solid bed
 bewegtes F. moving bed
 ruhendes (statisches) F. fixed (static) bed
Feststoffdichte *f* true density
Feststoffdurchsatzleistung *f* solids-handling capacity
Feststoffentfernung *f* mud removal *(waste-water treatment)*
Feststoffextraktion *f* liquid-solid extraction
Feststoffgehalt *m* solids content (loading)
Feststoffgemenge *n* mixture of solids
feststoffhaltig solids-bearing
Feststoffkonzentration *f* solids concentration
Feststoffoberfläche *f* solid surface
Feststofformulierung *f (agric)* dust formulation

Feststoffrakete *f* solid (solid-fuel, solid-propellant) rocket
Feststoffsorbens *n*, **Feststoffsorptionsmittel** *n* sorbent solid
Festsubstanz *f* solid [matter]
Festtreibstoff *m* solid fuel (propellant)
Festwerden *n* solidification, hardening, setting, congelation, congealing, *(esp coll)* gelation
fett *(chem)* fatty *(e.g. oil)*
Fett *n (chem, food)* fat; *(tech)* grease
ausgelassenes F. rendered fat
ausgelassenes tierisches F. grease
festes F. solid fat
gebleichtes F. bleached fat
gehärtetes F. hardened (hydrogenated) fat
halbgehärtetes F. plastic fat
hydriertes F. *s.* gehärtetes F.
natürliches F. natural fat
pflanzliches F. vegetable (plant) fat
technisches F. commercial grease, inedible fat
tierisches F. animal fat
Fettabscheider *m* grease (fat) trap, grease remover (interceptor, separator)
Fettabweisungsvermögen *n* grease (fat) repellency
fettähnlich fatlike
Fettalkohol *m* fatty alcohol
höherer F. long-chain fatty alcohol
Fettalkoholsulfat *n s.* Fettalkylsulfat
Fettalkylsulfat *n* fatty alkyl sulphate
Fettamin *n* fatty amine *(any of a series of aliphatic amines derived from fats)*
Fettansatz *m* fat blend *(as for margarine making)*
fettartig fatlike
Fettausschlag *m (tann)* fatty [acid] spew
Fettbestimmung *f* fat determination
Fettbohne *f* soy[bean], soya[-bean], soja, Glycine max (L.)Merr.
Fettbrühe *f (tann)* fat liquor
Fettchemie *f* fat chemistry
fettdicht greaseproof, grease-resistant, fat-resistant, fat-tight
Fettdichtigkeit *f* resistance to grease (fat), grease (fat) resistance
Fetteilchen *n* grease (fat) particle
Fettemulsion *f* fat emulsion
fetten *(tech)* to grease, to lubricate; *(tann)* to oil, to stuff; to compound *(oils)*
Fetten *n (tech)* greasing, lubrication; *(tann)* oiling, stuffing
F. im Faß *(tann)* drum oiling (stuffing)
Fettfang *m*, **Fettfänger** *m s.* Fettabscheider
Fettfleck *m* grease spot, smear
Fettflecke[n] *mpl (tann)* fat spue *(a defect in leather)*
Fettfleckfotometer *n* grease-spot photometer
fettfrei non-fat[ty], fat-free
fettgar *(tann)* chamois, oil-tanned
Fettgas *n* fatty gas
Fettgehalt *m* fat content
F. der Butter butter-fat content
Fettgehaltsbestimmung *f* test for fat content
Fettgerbung *f* chamois (oil) tannage
Fettgewebe *n (biol)* adipose tissue
Fettglanz *m (min)* greasy lustre
fetthaltig fatty, adipose
Fetthärtung *f*, **Fetthydrierung** *f* fat hardening, hydrogenation of fat
Fetthydrolyse *f* fat hydrolysis, saponification of fat
fettig fatty, oily, greasy, unctuous *(consistency or substance)*
Fettigkeit *f* fattiness, oiliness, greasiness, unctuousness
Fett-in-Wasser-Emulsion *f* fat-in-water emulsion
Fettkalk *m* fat (rich) lime
Fettkäse *m* fat cheese
Fettkocher *m* fat melter (melting kettle)
Fettkohle *f* fat coal
kurzflammige F. fat short-flame coal
langflammige F. fat long-flame coal
Fettkomposition *f* fat blend *(as for margarine manufacture)*
Fettkreide *f* / **lithografische** lithographic crayon
Fettkristall *m* fat crystal
Fettkügelchen *n* fat globule
Fettkügelchenmembran *f* fat globule membrane
Fettkügelchenprotein *n* fat globule protein
Fettlicker *m (tann)* fat liquor
fettlickern *(tann)* to fat-liquor
Fettlöser *m* grease (fat) solvent
Fettlöserseife *f* fat-dissolving soap
Fettlöserwaschmittel *n* fat-dissolving washing agent
fettlöslich fat-soluble, soluble in fat, liposoluble
Fettlöslichkeit *f* solubility in fat, liposolubility
Fettlösungsmittel *n* grease (fat) solvent
Fettmischung *f* fat blend *(as for margarine manufacture)*
Fettöl *n (chem)* fat[ty] oil, fixed oil *(as opposed to volatile oil)*
Fettpech *n* fatty acid pitch
Fettphase *f* fatty phase
Fettprobe *f* fat sample
fettreich rich in fat
Fettreif *m* fat (chocolate) bloom
Fettreihe *f (org chem)* fatty (aliphatic) series
Fettsäure *f* fatty acid
F. mit drei C-Atomen three-carbon acid
F. mit einer Doppelbindung *s.* einfach ungesättigte F.
F. mit zwei C-Atomen two-carbon acid
einfach ungesättigte F. monounsaturated (monoethenoid) fatty acid
essentielle F. essential fatty acid, EFA
freie F. free fatty acid, FFA
gesättigte F. saturated fatty acid
höhere F. fat acid *(containing 12 to 24 carbon atoms)*
mehrfach ungesättigte F. polyunsaturated (polyethenoid) fatty acid
niedere F. lower fatty acid
ungesättigte F. unsaturated fatty acid
Fettsäurealkylolamid *n* fatty alkylolamide
Fettsäureamid *n* fatty [acid] amide

Fettsäureester *m* fatty acid ester
Fettsäureranzidität *f*, **Fettsäureranzigkeit** *f* lipolytic (hydrolytic) rancidity
Fettsäurerest *m* fatty acid radical
Fettschmiere *f (tann)* dubbin[g], stuffing mixture, fat liquor
fettspaltend lipolytic, fat-splitting
Fettspaltung *f* lipolysis, fat splitting, cleavage of fats
Fettstift *m* marking (wax) pencil
Fettstoff *m* fatty matter (substance)
Fettstoffwechsel *m* fat metabolism
Fettsubstanz *f* fatty matter (substance)
Fettsynthese *f* fat synthesis
Fetttröpfchen *n* fat globule
fettundurchlässig greaseproof, grease-resistant, fat-resistant, fat-tight
Fettundurchlässigkeit *f* resistance to grease (fat), grease (fat) resistance
Fettung *f (tann)* stuffing
Fettusche *f* tusche
  lithografische F. lithographic tusche
Fettverlust *m* fat loss
Fettverseifung *f* saponification of fat, fat hydrolysis
Fettzelle *f (biol)* fat (adipose) cell
feucht moist, damp, wet, *(relating to air also)* humid
  f. werden to become moist (wet), to moisten
Feuchtapparat *m (pap)* wetting machine, damper
Feuchte *f* moisture, dampness, wetness, *(relating to air also)* humidity
  F. der lufttrockenen Probe *s.* hygroskopische F.
  absolute F. absolute humidity *(of air)*
  gebundene F. bound moisture
  hygroskopische F. air-dried moisture
  kritische F. critical moisture content
  relative F. percentage (relative) humidity, R.H. *(of air)*
Feuchteanteil *m* moisture content [wet weight basis]
Feuchteaufnahme *f* moisture absorption
Feuchteaufnahmevermögen *n* moisture-carrying capacity
Feuchteausdehnung *f* moisture expansion
Feuchtebeladung *f s.* Feuchtegehalt
feuchtebeständig moisture-resistant, moistureproof
Feuchtebeständigkeit *f* moisture resistance
Feuchtebestimmung *f* estimation of moisture
Feuchtediagramm *n* humidity (psychrometric) chart
feuchtefest moisture-resistant, moistureproof
Feuchtefestigkeit *f* moisture resistance
Feuchtegefälle *n* moisture gradient
Feuchtegehalt *m* moisture content [wet weight basis]; *(relating to air)* humidity [content]
  absoluter F. moisture content dry weight basis
Feuchtegrad *m* degree of moisture
Feuchteeinrichtung *f (pap)* wetting machine, damper
Feuchtemesser *m* hygrometer; moisture meter *(for determining the percentage of moisture in a material)*
feuchten *(pap)* to wet out (up)
Feuchter *m (pap)* wetting machine, damper
Feuchtesatz *m* moisture content dry weight basis
Feuchteverlust *m* moisture loss
Feuchtglätte *f*, **Feuchtglättwerk** *n (pap)* nip rolls, intermediate rolls (calender)
Feuchtgut *n* wet (damp) product (feed) *(on drying)*
Feuchthaltemittel *n* moisturizer, humectant
Feuchtigkeit *f s.* Feuchte
Feuchtigkeits... *s. a.* Feuchte...
feuchtigkeitsabweisend moisture-repellent
feuchtigkeitsbeständig moisture-resistant, moistureproof
Feuchtigkeitsfilm *m* film of moisture
Feuchtigkeitstafel *f s.* Feuchtediagramm
Feuchtkugeltemperatur *f* wet-bulb (wet-surface) temperature
Feuchtlagerbeständigkeit *f* resistance to damp storing
Feuchtluft *f* moisture-laden air
Feuchtlufttrockner *m (ceram)* humidity dryer
Feuchtlufttrocknung *f (ceram)* humidity drying
Feuchtmaschine *f (pap)* wetting machine, damper
Feuchttrockner *m (ceram)* humidity dryer
Feuchttrocknung *f (ceram)* humidity drying
Feuchtung *f* damp[en]ing; *(pap)* wetting-out, wetting-up
Feuchtwalze *f (pap)* damping roll, damper
feuchtwarm damp warm
Feuer *n* fire
  hartes (hohes) F. *(ceram)* hard fire
Feueraluminierung *f* hot-dip aluminizing
feuerbeständig fireproof; *(materials science)* fire-resistant; *(esp ceram)* refractory
  f. machen to fireproof
Feuerbeständigkeit *f* fireproofness; *(materials science)* fire resistance; *(esp ceram)* refractoriness
Feuerbeständigmachen *n* fireproofing
Feuerbeton *m s.* Feuerfestbeton
Feuerbrücke *f* fire bridge *(of a flame furnace)*
feuerfest *s.* feuerbeständig
Feuerfestbeton *m* refractory concrete
Feuerfestigkeit *f s.* Feuerbeständigkeit
Feuerfestkeramik *f* refractory ceramics
Feuerfestmachen *n* fireproofing
Feuerfestton *m* fireclay, refractory clay
Feuerfortschritt *m (ceram)* fire travel *(in the kiln)*
feuerhemmend fire-retardant, fire-retarding
Feuerkammer *f s.* Feuerraum
Feuerlöschbrause *f* drench (safety, emergency) shower
Feuerlöschdecke *f* fire blanket
Feuerlöscher *m*, **Feuerlöschgerät** *n* [fire] extinguisher
Feuerlöschmittel *n* fire-extinguishing agent
Feuerlöschpumpe *f* fire pump
Feuerlösch-Schaummittel *n* fire-fighting foam
feuern to fire, to fuel
Feueröffnung *f* fire mouth

Feueropal *m (min)* fire opal
feuerpolieren *(glass)* to fire-polish, to fire-finish, to fire-glaze
Feuerpolitur *f (glass)* fire polishing (finishing, glazing), fire polish
Feuerraffination *f* fire refining
feuerraffinieren to fire-refine
Feuerraum *m* combustion chamber (space), furnace chamber, firebox *(of an industrial furnace)*
Feuerschutzfarbe *f* flameproofing paint
Feuerschutzmittel *n* flameproofing (fireproofing) agent, fire retardant
feuersicher *s.* feuerbeständig
Feuersichermachen *n s.* Feuerbeständigmachen
Feuerstein *m (min)* firestone, flint [stone]
Feuerstrecke *f* fire zone *(as in underground gasification)*
Feuerton *m s.* Feuerfestton
Feuerung *f* 1. firing; 2. *s.* Feuerraum; 3. *s.* Feuerungsmaterial
Feuerungsmaterial *n* fuel
Feuerveraluminierung *f* hot-dip aluminizing
Feuerverbleiung *f* hot-dip leading
Feuervergoldung *f* fire gilding
Feuerversilberung *f* fire silvering
Feuerverzinkung *f* hot[-dip] galvanizing, galvanization
Feuerverzinnung *f* hot-dip tinning, fire tinning
Feuerwerk *n* fireworks, pyrotechnics
Feuerwerker *m* pyrotechnist, pyrotechnician
Feuerwerkerei *f* pyrotechnics
Feuerwerkskörper *m* firework, pyrotechnic
Feuerwerkskunst *f s.* Feuerwerkerei
feuerwiderstandsfähig fireproof; *(materials science)* fire-resistant
Feuerwiderstandsfähigkeit *f* fireproofness; *(materials science)* fire resistance
Feuerzeug *n* lighter
  Döbereinersches F. Döbereiner's lamp
Feuerzone *f* fire zone *(as in underground gasification)*
Feuerzug *m* flue, fire tube
ff. *(abbr)* feuerfest
FF-Ruß *m (rubber)* fine furnace black, FF black
Fibrille *f (biol, text)* fibril[la]
Fibrillenbildung *f* fibrillation
fibrillieren to fibrillate
Fibrillierung *f* fibrillation
Fibrin *n (bioch)* fibrin
Fibrinogen *n (bioch)* fibrinogen
Fibroin *n* fibroin *(the insoluble protein of silk)*
fibrös fibrous
Fichte *f* spruce, Picea A. Dietr.
Fichtelit *m (min)* fichtelite *(a naturally occurring hydrocarbon)*
Fichtennadelöl *n* pine-needle oil *(from pine and fir needles)*
Fichtenöl *n* spruce turpentine, *(pap also)* sulphite turpentine
Fichtenrinde *f* spruce bark
Fiebermittel *n* antipyretic, febrifuge
Fieberrinde *f* cinchona (red) bark *(from Cinchona specc., specif from Cinchona succirubra Pavon)*
fiebersenkend antipyretic, febrifuge, antifebrile
Figuren *fpl* / **Widmannstättensche** Widmannstätten figures (lines)
Filament *n (text)* filament, continuous fibre (filament) *(a natural or man-made fibre of great or indefinite length)*
Filixextrakt *m (pharm)* male fern extract
Filixsäure *f* filixic (filicic) acid *(a mixture of homologous phloroglucine derivatives)*
Filizinsäure *f* filicinic acid, 1,1-dimethyl-2,4,6-trioxocyclohexane
Film *m* film
  dünnschichtiger F. *(phot)* thin emulsion film
  fallender F. *(distil)* falling film
  flüssig-expandierter F. *(phys chem)* liquid expanded film
  monomolekularer F. monomolecular (unimolecular) film (layer), monolayer
  panchromatischer F. *(phot)* panchromatic film, pan-film
  selbsttragender (trägerloser) F. *(plast)* self-supporting film
Filmabfall *m (phot)* waste film
Filmband *n (phot)* film strip
Filmbearbeitung *f (phot)* film processing
filmbildend film-forming
Filmbildner *m* film former, film-forming component (material, substance)
Filmbildung *f* film formation (forming)
Filmdeckung *f* film coverage *(of wetting agents)*
Filmdicke *f* film thickness
Filmgießmaschine *f (plast)* film casting machine
Filmgrundlage *f s.* Filmschichtträger
Filmschichtträger *m (phot)* film base, support
Filmstreifen *m (phot)* film strip
Filmträger *m*, **Filmunterlage** *f s.* Filmschichtträger
Filter *n(m)* filter, *(if working without pressure also)* strainer
  F. mit loser Schicht bed filter
  F. mit Obenaufgabe top-feeding filter
  F. mit Untenaufgabe bottom-feeding filter
  aschefreies F. ashless (ash-free) filter
  bakteriendichtes F. germ-tight (germ-proofing) filter, bacteriological filter
  druckloses F. *s.* hydrostatisches F.
  eingearbeitetes F. ripened filter *(water conditioning)*
  glattes F. *(lab)* plain filter
  hydrostatisches F. gravity (hydrostatic head) filter
  loses F. bed filter
  offenes F. open filter
  pulvermetallurgisches F. metal-powder filter
  zellenloses F. non-cellular filter
Filterbett *n* filter bed
Filterbeutel *m* filter bag
Filterblatt *n* filter leaf
Filterboden *m* filter plate
Filterbottich *m* filter tank (vat), lauter tub (tun)
Filtereindicker *m* filter thickener

Filtereinsatz *m* catch pot; *s.* Filterelement
Filterelement *n* filter[ing] element
Filterfilz *m (pap)* filter felt *(for recovering fibres)*
Filterfläche *f* filter area (surface)
Filterflocken *fpl* filter flakes *(a filter aid)*
Filterfotometer *n* filter photometer
Filtergeschwindigkeit *f* filtering (filtration) rate
Filtergewebe *n* filter (filtration) cloth (fabric)
Filtergleichung *f* filtration equation
Filtergut *n* prefilt [slurry, feed], material being (*or* to be) filtered
Filterhaut *f* schmutzdecke *(water treatment)*
Filterhilfe *f*, **Filterhilfsmittel** *n* filter aid, filtration accelerator
Filterhilfsschicht *f* precoat filtering medium
Filterhilfsstoff *m s.* Filterhilfe
Filterkammer *f* filter chamber
Filterkasten *m* filter tank (vat)
Filterkerze *f* filter candle (tube)
Filterkies *m* filter gravel
Filterkuchen *m* filter cake, [filter]cake
Filterkuchenleistung *f* [filter]cake capacity
Filterkuchenwäsche *f* [filter]cake washing
Filtermasse *f* filter mass (pulp)
Filtermassekuchen *m* filter pad
F. aus Baumwollfasern filtermasse
Filtermasseplatte *f* filter pad
Filtermaterial *n* filter material (medium)
angeschwemmtes F. precoating material, precoat filtering medium
Filtermatte *f* filter mat
Filtermedium *n*, **Filtermittel** *n s.* Filtermaterial
filtern to filter, to filtrate, *(without pressure also)* to strain
durch Ton f. to clay
durch Ultrafilter f. to ultrafilter
erneut f. to refilter
in Filterpressen f. to filter-press
klar f. to polish, to clarify, to filter until bright
nochmals f. to refilter
Filtern *n s.* Filtration
Filternutsche *f* nutsch[e], nutsch filter
F. nach Büchner Büchner funnel (filter)
Filterpaketchromatografie *f* chromatopack method
Filterpapier *n* filter paper
qualitatives F. qualitative filter paper
quantitatives F. quantitative filter paper
Filterpapierscheibe *f* filter paper disk
Filterpapierstreifen *m* strip of filter paper
Filterplatte *f* filter plate (disk); *(plast)* screen pack
Filterpresse *f* filter press
Filterpressenpapier *n* filter press paper
Filterpreßmasse *f* filter pad
Filterrahmen *m* filter frame (carriage)
Filterröhrchen *n (lab)* filter tube
Filterröhre *f* filter tube
Filterrückstand *m* filtration residue
Filtersack *m* filter bag
Filterscheibe *f* filter disk
F. nach Witt Witt plate
Filterschicht *f* filter bed
Filterschlauch *m* filter bag
Filterschüttschicht *f*, **Filterschüttung** *f s.* Filterschicht
Filtersieb *n* filter screen
Filterstab *m*, **Filterstäbchen** *n (lab)* filter stick
Filterstein *m* filter[ing] stone, *(for diffusing gases also)* gas [diffuser] stone, air stone, sintered bubbler
Filterstoff *m s.* Filtertuch
Filterstoffänger *m (pap)* filter save-all
Filtertiegel *m* filtering crucible
F. nach Gooch Gooch crucible (filter)
Filtertrichter *m* filter (fritted disk) funnel
Filtertrog *m* filter tank (vat)
Filtertrommel *f* filter drum
Filtertuch *n* filter (filtration) cloth (fabric)
Filterung *f* filtration
Filterungs ... *s.* Filter ...
Filterverfahren *n (coal)* percolation method *(in underground gasification)*
Filterwagen *m* filter carriage
Filterwanne *f* filter tank (vat)
Filterwasser *n (pap)* filtered water
Filterzelle *f* filter cell (unit)
Filterzentrifuge *f* [filtering] centrifugal, whizzer, wringer, extractor
Filtrat *n* filtrate
Filtratablauf *m*, **Filtratauslauf** *m*, **Filtrataustritt** *m* filtrate outlet (exit)
Filtration *f* filtration, *(without pressure also)* straining)
F. in Filterpressen filter pressing
F. mit adsorptiv wirksamen Erden earth filtration
F. mit konstanter Filtriergeschwindigkeit constant-rate filtration
F. unter Ausnutzung der Schwerkraft gravity (natural) filtration
F. unter konstantem Filtrationsdruck constant-pressure filtration
F. unter vermindertem Druck filtration under reduced pressure
F. von Flüssigkeiten liquid filtration
F. von Gasen gas filtration
Filtrationsdruck *m* filtering (filtration) pressure
Filtrationskonstante *f* filtration constant
Filtrationskurve *f* filtration curve
Filtrationszyklus *m* filtration cycle
Filtrex-Abscheider *m* louver separator
Filtrier ... *s. a.* Filter ...
filtrierbar filt[e]rable
leicht f. freely filt[e]rable, free-filtering
nicht f. unfilt[e]rable
Filtrierbarkeit *f* filterability
filtrieren to filter, to filtrate, *(without pressure also)* to strain (*for compounds s.* filtern)
Filtrieren *n s.* Filtration
Filtrierstativ *n (lab)* filter (funnel) rack (stand)
Filtrierstutzen *m* filtering jar
Filtriervorrichtung *f* filter assembly
Filz *m* felt ✦ **mit F. auskleiden** to felt
endloser F. *(pap)* endless felt
gerippter F. *(pap)* ribbed felt

Filzärmel *m (tann)* felt sleeve *(of a sammying machine)*
filzartig felt-like
Filzbildung *f* felting
Filzdichtung *f* felt packing (seal, gasket)
Filzeigenschaft *f* felting property
filzen *(text)* to felt; *(pap)* to felt together
Filzfähigkeit *f* felting power
Filzfreiausrüstung *f (text)* antifelting treatment
Filzinstandhalter *m (pap)* felt conditioner
Filzinstandhaltung *f (pap)* felt conditioning
Filzkalander *m* felt calender
Filzlauf *m (pap)* felt travel
Filzlaufdauer *f*, **Filzlaufzeit** *f (pap)* felt life
Filzleitwalze *f (pap)* felt-carrying (felt-leading) roll, felt roll
filzlos *(pap)* feltless
Filzmarke *f*, **Filzmarkierung** *f* felt mark *(a defect in paper)*
Filzpackung *f* felt packing (seal, gasket)
Filzpappe *f* felt board
Filzpolierscheibe *f* bob
Filzreinigung *f (pap)* felt cleaning
Filzsauger *m (pap)* felt suction box
Filzscheibe *f s.* Filzpolierscheibe
Filzschleife *f (pap)* endless felt
Filzschrumpfung *f (text)* felting shrinkage
Filzseite *f* felt (top) side *(of paper)*
Filzspannung *f (pap)* tension on the felt
Filzspannwalze *f (pap)* felt stretching (tightener) roll, hitch roll
Filzstrang *m (pap)* felt run
Filztrockenzylinder *m*, **Filztrockner** *m (pap)* felt dryer (drying cylinder)
Filztrum *m(n) (pap)* felt run
  rücklaufender F. return felt run
  vorlaufender F. felt run
Filztuch *n (pap)* felt [blanket]
  endloses F. endless felt
Filzvermögen *n* felting power *(of fibres)*
Filzwalze *f (pap)* felt-covered couch roll
Filzwäsche *f (pap)* 1. felt cleaning; 2. *s.* Filzwäscher
Filzwascheinrichtung *f s.* Filzwäscher
Filzwäscher *m (pap)* felt cleaner (washer)
Filzwechsel *m (pap)* felt changing
finalisieren to formulate
Finalisierung *f* formulation
Finalprodukt *n* final (finished, resulting, consumer) product
Fine-Furnace-Ruß *m (rubber)* fine furnace black, FF black
Fine-Thermal-Ruß *m (rubber)* fine thermal black, FT black *(gas black of small particle size)*
Finger *m* / **kalter** *(lab)* cold finger [condenser]
Fingerhut *m (ceram)* thimble *(a piece of kiln furniture)*
Fingerhuttinktur *f (pharm)* tincture of digitalis
Fingerling *m* fingerstall, finger cot *(protective equipment)*
Finger-print-Gebiet *n* finger-print region *(spectroscopy)*
Fingerrührer *m* finger agitator
Finkelstein-Austausch *m (org chem)* Finkelstein exchange
Firestone-Flexometer *n (rubber)* Firestone flexometer
Firestone-Plastometer *n (rubber)* Firestone[-Dillon] plastometer
Firnis *m* boiled oil
  F. von Martaban Burmese lacquer *(from Melanorrhoea usitata Wall.)*
  Japanischer F. Japanese (Chinese) lacquer *(from Rhus verniciflua Stokes)*
  lithografischer F. litho[graphic] varnish
fischartig fishy *(smell)*
Fischauge *n* fish eye *(a defect in transparent or translucent plastics)*
Fischer-Base *f (org chem)* Fischer base
Fischer-Hepp-Umlagerung *f* Fischer-Hepp rearrangement
Fischer-Tropsch-Anlage *f* Fischer-Tropsch plant
Fischer-Tropsch-Synthese *f* Fischer-Tropsch synthesis
Fischgift *n* 1. *(med)* ichthyotoxin; 2. *(agric)* fish poison *(e.g. several pesticides)*
fischig fishy *(smell)*
Fischigkeit *f* fishiness *(of milk and milk products)*
Fischkonservierung *f* fish curing
Fischleberöl *n*, **Fischlebertran** *m* fish liver oil
Fischleim *m* isinglass, ichthyocol[l], fish gelatine (glue)
Fischmehl *n* fish flour; *(as fertilizer)* fish tankage
Fischöl *n* fish oil (fat)
Fischschuppen *fpl* fish scale *(a defect in enamel)*
Fischschuppenessenz *f (cosmet)* fish scale essence
Fischschwanzbrenner *m (lab)* fish-tail (bats-wing) burner
Fischschwanzmeißel *m (petrol)* fish-tail bit
Fischsilber *n (coat, cosmet)* pearl essence
Fischtran *m* fish oil (fat)
Fischvergiftung *f* poisoning from fish, fish poisoning, ichthyism[us], ichthyotoxism
Fisetholz *n (dye)* fustet, young fustic *(from Cotinus coggygria Scop.)*
Fisetin *n* fisetin, 3,7,3′,4′-tetrahydroxyflavone
Fisetteholz *n s.* Fisetholz
Fixage *f (phot)* fixing, fixation
Fixateur *m (cosmet)* fixative *(added to a perfume)*
Fixativ *n (coat)* fixative, fixing agent
Fixierbad *n (phot)* fixing (hypo) bath, fixer
  härtendes F. hardening fixing bath
  saures F. acid fixing bath
Fixierdauer *f* fixing time
fixieren to fix
Fixiergeschwindigkeit *f (phot)* rate of fixing, fixing speed
Fixierlösung *f* fixing solution
Fixiermittel *n (phot)* fixing agent, fixer; *(text)* fixing agent
Fixiernatron *n (phot)* hyposulphite, hypo *(sodium thiosulphate)*
Fixiernatronzerstörer *m (phot)* hypo eliminator

Fixiernatronzerstörung *f (phot)* hypo elimination
Fixiersalz *n (phot)* fixing salt, fixer; *(specif) s.* Fixiernatron
saures F. acid fixer
Fixierung *f* fixation
nichtsymbiotische F. *(agric)* non-symbiotic fixation, free fixation, azofication *(of atmospheric nitrogen)*
symbiotische F. *(agric)* symbiotic fixation *(of atmospheric nitrogen)*
Fixierungsflüssigkeit *f* fixing solution
Bendasche F. *(biol)* Benda solution *(consisting of osmic acid, chromic acid and glacial acetic acid)*
Fixierungsmittel *n* fixative, fixing agent *(for fixing living tissue)*
Fixiervorgang *m* fixing process
Fixpunkt *m* fixed point
Fizin *n* ficin *(a proteinase obtained from the latex of Ficus specc.)*
F-Kalander *m (rubber)* inverted L calender
Flachband *n* flat belt
Flachdruck *m* 1. planographic (flat) printing, planography; 2. planography *(product)*
Flachdruckfarbe *f* planographic [printing] ink
Flachdruckverfahren *n* planographic process
Fläche *f* 1. surface; *(cryst)* face; 2. [surface] area *(as of an industrial plant)*; 3. plane *(esp geometry)*
offene F. open area *(as of a sieve)*
Flächenabscheider *m* envelope (screen) filter
Flächenbehandlung *f (agric)* blanket application, broadcast treatment *(as with herbicides)*
Flächendüngung *f (agric)* plain dressing
Flächenfixierung *f (text)* flat setting
Flächengebilde *n* / **textiles** textile fabric
Flächengewicht *n s.* Masse je Flächeneinheit
Flächenheizkörper *m* heating mantle
Flächenwinkel *m (cryst)* interfacial angle
flächenzentriert *(cryst)* face-centred
einseitig f. one-face centred
Flächenzentrierung *f (cryst)* face centr[e]ing
Flachglas *n* flat glass
Flachglasofen *m* flat-glass furnace
Flachglockenboden *m (distil)* low-riser plate (tray)
Flachgurt *m* flat belt
Flachkegelbrecher *m* flat-cone crusher
Flachkegelgranulator *m* short-head cone crusher
Flachriemen *m* flat belt
Flachs *m* flax, Linum usitatissimum L.
Neuseeländer F. phormium, Phormium tenax J.R. et G. Forst.
Flachsdichtung *f* flax packing
Flachsfaser *f* flax fibre
Flachsieb *n* flat sieve
Flachsortierer *m* flat screen
Flachspackung *f* flax packing
Flachsröste *f* 1. retting; 2. ret[tery] *(plant)*
Flachsrösterei *f* ret[tery]
Flachsrotte *f s.* Flachsröste
Flachsstroh *n* flax straw
Flachstrahldüse *f* slot (flat-spray) nozzle
Flachtrog *m (coal)* shallow bath
Flachware *f (glass, ceram)* flat ware
Flachwurfsieb *n* oscillating screen (strainer)
flammbeständig flame-resistant, uninflammable, non-[in]flammable, flameproof
Flammbeständigkeit *f* flame resistance (resistivity), uninflammability, non-flammability
Flamme *f* flame
freibrennende F. free (naked) flame
kleine F. small flame
leuchtende F. luminous flame
nichtleuchtende F. non-luminous flame
offene F. free (naked) flame
rauschende F. roaring flame
rußende F. smoky flame
Flammenaufprall *m* flame impingement
Flammenblasverfahren *n (glass)* flame-blowing process
Flammenbogen *m* flame (flaming) arc
Flammenfärbung *f* flame colo[u]ration
Flammenfortpflanzung *f* flame propagation
Flammenfortpflanzungsgeschwindigkeit *f s.* Flammengeschwindigkeit
Flammenfotometer *n* flame photometer
Flammenfotometrie *f* flame photometry, flame-emission (atomic-emission) spectroscopy
flammenfotometrisch flame-photometric
Flammenfront *f* flame front
Flammenführung *f* firing *(of an industrial furnace)*
aufsteigende F. up-draught firing
horizontale F. horizontal-draught firing
überschlagende F. down-draught firing
Flammengeschwindigkeit *f* flame velocity, velocity of flame propagation
flammenhärten to flame-harden
Flammenhärtung *f* flame hardening
flammenhemmend flame-retardant
Flammenionisation *f* flame ionization
Flammenionisationsdetektor *m* flame ionization detector
Flammenofen *m* reverberatory (air) furnace (kiln)
Flammenrückschlag *m* flareback, flashback
Flammenschutz ... *s.* Flammschutz ...
Flammenspektrometrie *f* flame photometry, flame-emission (atomic-emission) spectroscopy
Flammenspektroskopie *f* flame spectroscopy
Flammenspektrum *n* flame spectrum
Flammenspritzen *n* flame spraying
Flammentemperatur *f* flame temperature
Flammenverzögerungsmittel *n* flame retardant (retarder)
Flammenverzögerungsvermögen *n* flame retardancy
Flammenwächter *m* flame failure safeguard
flammfest flameproof
Flammfestausrüstung *f (text)* flameproof (flame-resistant, fireproof) finish, flameproof impregnation
Flammfestigkeit *f* flameproofness
Flammfestimprägnierung *f s.* Flammfestausrüstung
Flammfestmachen *n* flameproofing
Flammfotometrie *f s.* Flammenfotometrie
Flammfront *f* flame front

flammhärten to flame-harden
Flammhärtung *f* flame hardening
Flammkaschierung *f* flame lamination
Flammkohle *f* flame coal
Flammofen *m* reverberatory (air) furnace (kiln)
Flammpunkt *m* flash point
F. im geschlossenen Tiegel closed[-cup] flash point
F. im offenen Tiegel open[-cup] flash point
Flammpunktapparat *m*, **Flammpunktgerät** *n* *s.* Flammpunktprüfer
Flammpunktprüfer *m* flash-point apparatus (tester), flash tester
F. nach Abel-Pensky Abel-Pensky [flash-point] tester
F. nach Pensky-Martens Pensky-Martens [flash-point] tester
geschlossener F. closed[-cup] flash tester
geschlossener F. nach Pensky-Martens Pensky-Martens closed tester
geschlossener F. nach Tagliabue Tagliabue (Tag) closed tester
offener F. open[-cup] flash tester
Flammpunktprüfgerät *n* *s.* Flammpunktprüfer
Flammpunktstiegel *m* flash cup
Flammrohr *n* fire tube, flue
Flammrohrkessel *m* fire-tube boiler
Flammruß *m* *(rubber)* lampblack
Flammschutzausrüstung *f*, **Flammschutzimprägnierung** *f* *(text)* flameproof (flame-resistant, fireproof) finish, flameproof impregnation
Flammschutzmittel *n* flameproofing (fireproofing) agent
flammsicher flameproof
Flammsicherheit *f* flameproofness
Flammsichermachen *n* flameproofing
Flammspritzpistole *f* flame-spraying gun
Flammstrahlen *n* flame descaling *(for cleaning metal surfaces)*
flammwidrig flame-resistant, uninflammable, non[in]flammable, flameproof
Flammwidrigkeit *f* flame resistance (resistivity), uninflammability, non-flammability
Flansch *m* flange, socket
loser F. lap-joint (slip-on) flange
Flanschbolzen *m* flange bolt
Flanschdichtung *f* flange gasket (seal)
Flanschenrohr *n* flanged[-end] pipe
Flanschfitting *m(n)* flanged fitting
Flanschfläche *f* flange face
Flanschformstück *n* flanged fitting
Flanschschraube *f* flange bolt
Flanschstirnfläche *f* flange face
Flanschstück *n* flanged fitting
Flanschverbindung *f* flanged joint
Flasche *f* bottle, flask
Florentiner F. *(distil)* Florentine receiver
Mariottesche F. *(lab)* Mariotte bottle (flask), aspirator
Woulfesche F. *(lab)* Woulfe bottle
Flaschenabfüllerei *f* bottling plant (department, room), bottle house, bottlery, filling room
Flaschenabfüllmaschine *f* bottling (bottle-filling) machine, bottle filler, bottler
Flaschenabfüllung *f* bottle filling, bottling
Flaschenabgabe *f* *(glass)* bottle delivery
Flaschenabzug *m* bottling
Flaschenbier *n* bottle[d] beer
Flaschenbürste *f* bottle brush
Flaschenetikett *n* bottle label
Flaschenetikettiermaschine *f* bottle labeller
Flaschenfüllerei *f* *s.* Flaschenabfüllerei
Flaschenfüllmaschine *f* *s.* Flaschenabfüllmaschine
Flaschengärung *f* secondary fermentation; *s.* Flaschengärverfahren
Flaschengärverfahren *n* bottle champagnization
Flaschengas *n* bottle[d] gas
Flaschenglas *n* bottle glas
Flaschenkappe *f* bottle cap
Flaschenmilch *f* bottled milk
Flaschenofen *m* *(ceram)* bottle kiln (oven)
Flaschenreinigungsmaschine *f* *s.* Flaschenspülmaschine
Flaschenschleuder *f* bottle centrifuge
Flaschenseidenpapier *n* bottle tissue (wrapping)
Flaschenspule *f* *(text)* bottle bobbin
Flaschenspülmaschine *f* bottle-washing (bottle-cleaning, bottle-rinsing) machine, bottle washer (cleaner, rinser)
Flaschenstopfen *m* bottle stopper
Flaschenverschluß *m* bottle cap
Flaschenwaschmaschine *f* *s.* Flaschenspülmaschine
Flaschenwein *m* bottle[d] wine
Flaschenzentrifuge *f* bottle centrifuge
Flash-Destillat *n* *(petrol)* flash distillate
primäres F. primary flash distillate, P.F.D.
Flash-Kammer *f* *(distil)* flash chamber
Flash-Kurve *f* *(distil)* [single-]flash curve
Flash-Raum *m* *s.* Flash-Kammer
Flash-Röster *m* *(met)* flash roaster
Flash-Verdampfung *f* flash evaporation
mehrstufige F. multiflash (multistage flash, MSF) evaporation
Flaum *m* *(food)* bloom *(as on certain fruits or cocoa products)*
Flavan *n* flavan
Flavanoid *n* flavanoid
Flavanon *n* flavanone
Flavanthron *n* flavanthrone
Flaviansäure *f* $HOC_{10}H_6(NO_2)_2SO_3H$ flavianic acid, 2,4-dinitro-1-naphthol-7-sulphonic acid
Flavin *n* flavin[e], *(specif)* isoalloxazine
Flavin-adenin-dinukleotid *n* flavin[e] adenine dinucleotide, FAD
Flavinenzym *n*, **Flavinferment** *n* flavin[e] enzyme, flavoenzyme, flavoprotein, yellow enzyme
Flavinmononukleotid *n* flavin[e] mononucleotide, FMN, riboflavin-5′-phosphate
Flavon *n* flavone
Flavonfarbstoff *m* flavone pigment
Flavoproteid *n*, **Flavoprotein** *n* *s.* Flavinferment

Flechtenfarbstoff *m* lichen dye
Flechtensäure *f* lichen acid
Fleck *m* spot, *(if undesirable also)* speck, blotch, stain; *(chromatography)* spot
Fleckenbenzin *n* cleaner's naphtha (solvent)
Fleckenbildung *f* spotting, specking
Fleckenentferner *m* stain (spot) remover
Fleckenentfernung *f* stain (spot) removal
Fleckenentfernungsmittel *n* stain (spot) remover
Fleckenschierling *m* [poison] hemlock, Conium maculatum L.
Fleckenseife *f* scouring soap
Fleckentfernung *f s.* Fleckenentfernung
Fleckenunempfindlichkeit *f (plast)* stain resistance
Fleischaroma *n* meat flavour
Fleischbrühe *f* broth
Fleischdüngemehl *n* garbage tankage
Fleischextrakt *m* meat extract
Fleischfuttermehl *n* digester tankage, meat meal
Fleischguano *m s.* Fleischdüngemehl
Fleischmehl *n (agric)* [animal] tankage, *(as feed also)* digester tankage, meat meal, *(as fertilizer also)* garbage tankage
Fleischmilchsäure *f* $CH_2(OH)CH(OH)COOH$ sarcolactic acid, *L*-lactic acid, dextrorotatory lactic acid
Fleisch-Pepton-Agar *m(n)* meat-infusion agar *(bacteriology)*
Fleischsaft *m* meat juice
Fleischspalt *m (tann)* flesh split
Fleming-Methode *f* Fleming method *(for determining penicillin)*
Fletcher-Bleichturm *m (pap)* Fletcher bleacher
Fletton-Ziegel *m* fletton
Fletton-Ziegelton *m* Fletton brick clay
flexibel flexible
flexibilisieren *(plast)* to flexibilize
Flexibilität *f* flexibility
Flexodruck *m*, **Flexografie** *f* flexographic (aniline) printing, flexography
Flexometer *n (rubber)* flexometer
Fl. g. T. *s.* Flammpunkt im geschlossenen Tiegel
Fliegenbekämpfungsmittel *n* antifly preparation, fly poison
Fliegenfängerpapier *n* fly paper
Fliegengift *n* fly poison
Fliegenholz *n (pharm)* quassia, bitterwood, bitter ash *(from Quassia amara L. or Picrasma excelsa (Swartz) Planchon)*
Fliegenpapier *n s.* Fliegenfängerpapier
Fliegermethode *f* F3 method *(for octane rating)*
Fliehkraft *f* centrifugal force
Fliehkraftabscheider *m* centrifugal force separator (collector)
Fliehkraftabscheidung *f* centrifugal separation
Fliehkraftklassierer *m* centrifugal classifier
Fliehkraftmaschine *f* **nach Roelig** Roelig hysteresis apparatus
Fliehkraftpendelmühle *f* pendulum roller mill
Fliehkraftreiniger *m* centrifugal cleaner, *(Am)* centrifiner
Fliehkraftscheibe *f* centrifugal disk
Fliehkraftscheider *m s.* Fliehkraftabscheider
Fliehkraftsichter *m* centrifugal classifier
Fliehkraftversprüher *m* centrifugal atomizer, spinning disk [atomizer]
Fliehkraftversprühung *f* centrifugal atomization
Fliehkraftwalzenmühle *f s.* Fliehkraftpendelmühle
Fliehkraftzerstäuber *m s.* Fliehkraftversprüher
Fliehkraftzerstäubung *f s.* Fliehkraftversprühung
Fliese *f* tile, slab
  F. für Sonderzwecke special-purpose tile
  trockengepreßte F. dust-pressed tile
fließbar *s.* fließfähig
Fließbeständigkeit *f* resistance to flow
Fließbetrieb *m* continuous operation (working, processing)
Fließbett *n* fluid (fluidized, boiling) bed
Fließbettadsorption *f* fluidized (moving-bed) adsorption
Fließbettkatalysator *m* fluidized (fluid-bed, fluid) catalyst
Fließbett-Technik *f* fluid-bed (fluidized-bed, boiling-bed) technique
Fließbetttrockner *m* fluid-bed dryer
Fließbettverfahren *n* fluid (fluidized) process (operation)
Fließbettvergasung *f* fluid-bed gasification
Fließbettvulkanisation *f* fluid-bed vulcanization
Fließbild *n* flow diagram (pattern, sheet)
Fließdehnung *f* yield strain
Fließdiagramm *n s.* Fließbild
Fließdialyse *f* continuous dialysis
Fließeigenschaften *fpl s.* Fließverhalten
fließen 1. to flow, to run, to pour; *(coat)* to flow; 2. to yield *(materials testing)*
  f. lassen to run *(a liquid)*
Fließen *n* 1. flow, flux; plastic flow *(a kind of deformation)*; 2. yield *(materials testing)*
  Binghamsches F. Bingham flow, structural viscosity
  gleichmäßiges F. smooth fluidization *(fluidized-bed technique)*
  kaltes F. cold flow *(of thermoplastic adhesives)*
  Newtonsches F. *(plast)* Newtonian flow
  nicht-Newtonsches F. *(plast)* non-Newtonian flow
  quasiplastisches F. pseudoplastic flow
  schlechtes F. *(plast)* low flow
  viskoses F. viscous flow
Fließerscheinung *f* yield phenomenon *(as in metals under tension)*
fließfähig flowable
Fließfähigkeit *f* flowability, fluidity
Fließfestigkeit *f* resistance to flow
Fließformen *n (rubber)* transfer moulding
Fließgeschwindigkeit *f* flow rate (velocity)
Fließgrenze *f* yield point *(materials testing)*; *(coat)* yield value; *(met)* flow point
  praktische F. yield strength *(materials testing)*
  untere F. yield value *(materials testing)*
Fließkunde *f* rheology, science of flow

Fließkurve *f* flow curve
Fließmittel *n* mobile solvent *(chromatography)*
Fließmittelfront *f* solvent front *(chromatography)*
Fließmitteltrog *m* solvent trough *(chromatography)*
Fließpapier *n* absorbent paper
Fließpapier-Filterpresse *f* blotter press
Fließpunkt *m* melting point, m.p.; pour point *(of oils)*; *(met)* flow point
Fließpunkterniedriger *m* pour-point depressant *(for oils)*
Fließpunktprüfung *f* pour-point test *(applied to oils)*
Fließrichtung *f* direction of flow
Fließschema *n s.* Fließbild
Fließschicht *f s.* Fließbett
Fließschmelzpunkt *m* slip point
Fließspannung *f* yield stress
Fließspeiser *m (glass)* flow feeder
Fließstaubkontakt *m* fluidized (fluid-bed, fluid) catalyst
Fließtemperatur *f* flow temperature; *(met)* flow point
Fließverfahren *n* fluid (fluidized) process (operation)
Fließverfestigung *f* rheopexy
Fließverhalten *n* flow (rheological) behaviour, flow[ing] property
Fließvermögen *n* flowability, fluidity
Fließweg *m* flow path
Fließwert *m (coat)* yield value
Flint *m (min)* flint [stone], firestone *(a variety of opal)*
Flintglas *n* flint glass
Flintstein *m* 1. *(tech)* flint pebble; 2. *s.* Flint
Flitter *m (min)* spangle; *(glass)* glass frost, frost glass, tinsel
Flöckchen *n* floccule
Flockdruck *m (text)* flock print
Flocke *f* flake, *(esp in suspensions)* floc; *(text)* flock
flockegefärbt *(text)* stock-dyed
flocken to flocculate, to coagulate, to clot, to curdle
flockenartig flake-like, flocculent
Flockenbast *m (text)* cottonin, cottonized bast fibre
Flockenbildung *f* flock formation, flocculation, coagulation, clotting
flockengefärbt *(text)* stock-dyed
Flockenschlamm *m* floc *(water treatment)*
Flocker *m s.* Flockungsmittel
flockig flocculent
Flockseide *f* flock silk *(from cocoon waste)*
Flockung *f* flocculation, coagulation, clotting, curdling
  chemische F. chemical flocculation *(water treatment)*
Flockungsanlage *f* flocculation unit *(water treatment)*
Flockungsfähigkeit *f s.* Flockungsvermögen
Flockungsgeschwindigkeit *f* flocculation rate
Flockungshilfsmittel *n* flocculation aid
Flockungsklärapparat *m* [/ **kombinierter**] *s.* Flockungsreaktor
Flockungskraft *f s.* Flockungsvermögen
Flockungsmittel *n* flocculant, flocculating agent
Flockungsraum *m* flocculation (flocculating) chamber
Flockungsreaktor *m* reactor-clarifier, clari-flocculator, flocculator, flocculation tank *(water treatment)*
Flockungsverlauf *m* flocculation process
Flockungsvermögen *n* flocculating power
Flockungswert *m* flocculation value
Flockungszone *f* flocculation zone
Flokkulation *f s.* Flockung
Flokkulator *m s.* Flockungsreaktor
Floretteseide *f* floret (florette, floss) silk
Florey-Einheit *f* Florey (Oxford) unit *(an international unit of penicillin no longer used)*
Florideenstärke *f* floridean starch
Florpostpapier *n* onion skin
Fl. o.T. *s.* Flammpunkt im offenen Tiegel
Flotation *f* flo[a]tation
  differentielle F. differential (selective) floatation
  kollektive F. collective (bulk) floatation
  selektive (sortenweise) F. *s.* differentielle F.
Flotationsabgänge *mpl* floatation tailings
Flotationsanlage *f* floatation plant
Flotationsberge *pl* floatation tailings
Flotationsgerät *n*, **Flotationsmaschine** *f* floatation machine (apparatus)
Flotationsmittel *n* floatation [re]agent
  drückend wirkendes F. depressant
Flotationsöl *n* floatation oil
Flotationsprobe *f* floatation assay
Flotationsreagens *n* floatation [re]agent
Flotationsschwefel *m* floatation sulphur, gas sulphur
Flotationsstoffänger *m (pap)* floatation save-all
Flotationsverfahren *n* floatation (buoyancy) process
Flotationszelle *f* floatation cell (unit)
  pneumatische F. dissolved-air floatation machine
flotierbar floatable
Flotierbarkeit *f* floatability
flotieren to float *(e.g. ore)*
Flotieren *n* flo[a]tation
Flotte *f* liquor
Flottenaufnahme *f (text)* pick-up
Flottenkreislauf *m (text)* liquor circulation
Flottenlauf *m (text)* liquor flow
Flottenmenge *f (text)* amount of liquor
Flottenverhältnis *n (text)* liquor (bath) ratio; *(pap)* liquor[-to-wood] ratio, liquid-to-solid ratio
Flottenzirkulation *f (text)* liquor circulation
Flottenzulauf *m (text)* liquor flow
Flöz *n (mine)* stratum, layer, *(if thin also)* seam
Flözvergasung *f* underground gasification
flüchtig volatile, fugitive
  leicht f. highly (readily) volatile, high-volatile
  mit Dampf (Trägerdampf, Wasserdampf) f. steam-distillable

schwer f. difficultly volatile, slow-evaporating, heavy
schwerer f. less volatile
Flüchtiges *n* volatile matter, v.m., VM
Flüchtigkeit *f* volatility, fugacity
relative F. relative volatility
Fluellit *m (min)* fluellite *(aluminium fluoride-1-water)*
Flugasche *f* fly (flue, quick) ash
Flugaschenabscheider *m* fly-ash precipitator (collector)
Flugaschenabscheidung *f* fly-ash precipitation (collection)
Flugbahn *f (nucl)* path
Flugbenzin *n* aviation gasoline (spirit)
Flügel *m* blade, shovel, beater, vane, paddle *(as of an agitator)*
Flügelmischer *m s.* Flügelrührer
Flügelpigment *n* wing pigment *(as of butterflies)*
Flügelpumpe *f* vane pump
Flügelradanemometer *n* vane anemometer
Flügelradlüfter *m* propeller fan
Flügelradzähler *m* rotating (current, velocity) meter *(flow measurement)*
Flügelrührer *m* blade (paddle) agitator (mixer)
Flügelzellenpumpe *f* vane pump
Flugkraftstoff *m* aviation fuel
Flugmotorenbenzin *n* aviation gasoline (spirit)
Flugmotorenöl *n* aviation oil
Flugstaub *m* flue dust
Flugstaubkammer *f* dust chamber (room)
Flugstaubverfahren *n* entrained catalyst system *(a variety of the Fischer-Tropsch hydrocarbon synthesis)*
Flugstaubverlust *m* stack loss
Flugzeitspektrometer *n* time-of-flight spectrometer
Flugzeugausbringung *f* aeroplane application *(as of pesticides)*
Flugzeug-Düngerstreuen *n* aeroplane fertilization
Fluid *n* fluid
Fluidextrakt *m (pharm)* fluid (liquid) extract
Fluid-Hydroformen *n (petrol)* fluid hydroforming
Fluidisation *f* fluidization
fluidisieren to fluidize
Fluidität *f (phys chem)* fluidity *(reciprocal of viscosity)*
Fluidkrackverfahren *n (petrol)* fluid catalytic [cracking] process
Fluidsystem *n* fluid[ized] system
Fluidtechnik *f* fluid-bed (fluidized-bed, boiling-bed) technique
Fluidverfahren *n* fluid (fluidized) process (operation)
Fluo... *(in chemical compounds)* s. Fluoro...
Fluor *n* F fluorine
Fluoralkan *n* fluoroalkane, fluorinated alkane, aliphatic fluorocarbon
Fluoralken *n* fluoroalkene, fluoro-olefin
Fluoraustreibung *f* defluorination
Fluorbenzol *n* C_6H_5F fluorobenzene
Fluor-2,4-dinitrobenzol *n* fluoro-2,4-dinitrobenzene, DNFB
Fluorelastomer[es] *n* fluorinated (fluorine, fluorocarbon) polymer, fluoroelastomer
Fluoressigsäure *f* FCH_2COOH fluoroacetic acid
Fluoreszein *n* fluoresc[e]in
Fluoreszenz *f* fluorescence
sensibilisierte F. sensitized fluorescence
Stokessche F. Stokes fluorescence
Fluoreszenzanalyse *f* fluorescence analysis
Fluoreszenzfarbe *f* fluorescent paint
Fluoreszenzfarbstoff *m* fluorescent dye
Fluoreszenzindikator *m* fluorescent indicator
Fluoreszenzlicht *n* fluorescent light
Fluoreszenzmessung *f s.* Fluorometrie
Fluoreszenzmikroskopie *f* fluorescence microscopy
Fluoreszenzschirm *m* fluorescent screen
Fluoreszenzspektrum *n* fluorescence spectrum
Fluoreszenzstandard *m* fluorescent standard
Fluoreszenzstoff *m* fluorescent agent (substance)
Fluoreszenzstrahlung *f* fluorescence radiation
Fluoreszenztitration *f* fluorescence titration
fluoreszieren to fluoresce
fluoreszierend fluorescent
fluorhaltig fluorine-containing
Fluorid *n* M^IF fluoride
Fluoridglas *n* fluoride glass
fluoridieren to fluoridize, to fluoridate *(drinking water)*
Fluoridierung *f* fluoridation *(of drinking water)*
fluorieren to fluorinate
Fluorierung *f* fluorination
Fluorierungsmittel *n* fluorinating agent
Fluorimeter *n s.* Fluorometer
Fluorimetrie *f s.* Fluorometrie
fluorisieren *(incorrectly for)* fluoridieren
Fluorisierung *f (incorrectly for)* Fluoridierung
Fluorit *m (min)* fluorite, fluor-spar *(calcium fluoride)*
Fluoritstruktur *f (cryst)* fluorite structure
Fluorkarbonfaser *f* fluorocarbon fibre
Fluorkarbonfaserstoff *m* fluorocarbon fibre
Fluorkarbonkautschuk *m* fluorocarbon rubber
Fluorkarbonplast *m* fluoroplastic
Fluorkautschuk *m* fluorinated (fluorine) rubber
Fluorkohlenwasserstoff *m* fluorocarbon
Fluorkronglas *n* fluor crown glass, fluorcrown
Fluoroaluminat *n* fluoroaluminate
Fluoroantimonat *n* fluoroantimonate
Fluoroarsenat *n* fluoroarsenate
Fluoroberyllat *n* fluoroberyllate
Fluoroborat *n* fluoroborate, tetrafluoroborate
Fluoroborsäure *f* $H[BF_4]$ fluoroboric acid, tetrafluoroboric acid
Fluorochromat *n* fluorochromate
Fluoroferrat *n* fluoroferrate
Fluoroform *n* CHF_3 fluoroform, trifluoromethane
Fluorofotometer *n s.* Fluorometer
Fluorogermanat *n* fluorogermanate, hexafluorogermanate
Fluorohafnat *n* fluorohafnate
Fluorojodat *n* fluoroiodate

Fluorokieselsäure *f* $H_2[SiF_6]$ fluorosilicic acid, hexafluorosilicic acid, sand acid
Fluorolefin *n* fluoroalkene, fluoro-olefin
Fluoromanganat *n* fluoromanganate
Fluorometer *n* fluorometer, fluorimeter, fluophotometer *(for measuring fluorescence)*
Fluorometrie *f* fluorometry, fluorimetry
fluorometrisch fluorometric, fluorimetric
Fluoromolybdat *n* fluoromolybdate
Fluoroniobat *n* fluoroniobate
Fluorophor *m* fluorophore, fluorogen *(a radical which causes fluorescence)*
Fluorophosphat *n* fluorophosphate
Fluororhenat *n* fluororhenate
Fluororhodat *n* fluororhodate
Fluorosilikat *n* fluorosilicate
Fluorostannat *n* fluorostannate
Fluorotantalat *n* fluorotantalate
Fluorotellurat *n* fluorotellurate
Fluorothorat *n* fluorothorate
Fluorotitanat *n* fluorotitanate
Fluorouranat *n* fluorouranate
Fluorovanadat *n* fluorovanadate
Fluorowolframat *n* fluorowolframate, fluorotungstate
Fluoroxid *n (incorrectly for)* Sauerstoffdifluorid
Fluorozirkonat *n* fluorozirconate
Fluorsilikonkautschuk *m* fluorosilicone rubber
Fluorüberträger *m* fluorinating agent
Fluorwasserstoff *m* HF hydrogen fluoride
Fluorwasserstoffalkylierung *f* hydrofluoric-acid alkylation, HF alkylation
fluorwasserstoffsauer fluorohydric
Fluorwasserstoffsäure *f* HF hydrofluoric acid
Fluorwasserstoff[säure]verfahren *n (petrol)* hydrofluoric-acid process, HF process
FluoSolids-Kalkbrennofen *m* FluoSolids lime kiln
FluoSolids-Reaktor *m* [Dorrco] FluoSolids reactor
FluoSolids-Röstung *f* FluoSolids roasting
FluoSolids-Verfahren *n* FluoSolids process
Flur *m (ceram)* corridor *(of a dryer)*
flushen *(coat)* to flush
Flushkneter *m (coat)* flusher *(kneading machine for preparing pigment paste)*
Flushpaste *f (coat)* flushed colour
Flushverfahren *n (coat)* flushing process
Fluß *m* 1. flow, flux; 2. *s.* Flußmittel
  kalter F. *(plast)* cold flow
flüssig liquid, fluid
  f. werden to liquefy, to melt, to fuse, to flux
Flüssigbrennstoffystem *n* liquid-fuel system
Flüssig-Chromatografie *f* liquid chromatography
Flüssigdünger *m* liquid fertilizer
Flüssigextraktion *f s.* Flüssig-Flüssig-Extraktion
Flüssig-Fest-Chromatografie *f* liquid-solid chromatography, L.S.C.
Flüssig-Fest-Chromatogramm *n* liquid-solid chromatogram
Flüssig-Fest-Grenzfläche *f* liquid-solid interface
Flüssig-Flüssig-Chromatografie *f* liquid-liquid chromatography, L.L.C.
Flüssig-Flüssig-Extraktion *f* liquid-liquid extraction
Flüssig-Flüssig-Verteilung *f* liquid-liquid partition
Flüssig-Flüssig-Verteilungschromatografie *f* liquid-liquid partition chromatography
Flüssiggas *n* liquefied [petroleum] gas, L.P. gas, L.P.G., liquid gas *(liquefied hydrocarbons)*
Flüssigkeit *f* 1. *(as opposed to solid or gas)* liquid; *(tech, food, biol)* liquor; 2. *(state)* liquidity, fluidity
  anisotrope F. anisotropic (crystalline) liquid, liquid crystal, mesomorphic state
  assoziierte F. associated liquid
  Bendasche F. *(biol)* Benda solution *(a mixture of osmic, chromic, and glacial acetic acid)*
  dekantierte F. decantate
  Diverssche F. Divers' liquid *(concentrated solution of NH_4NO_3 in liquid ammonia)*
  geförderte F. liquid being pumped
  geklärte F. clarified liquid (liquor)
  kristalline F. *s.* anisotrope F.
  mitgerissene F. entrained liquid (liquor)
  Muthmannsche F. Muthmann's liquid *(1,1,2,2-tetrabromoethane)*
  Newtonsche F. Newtonian liquid (fluid)
  nicht-Newtonsche F. non-Newtonian liquid (fluid)
  normale F. normal (non-polar, non-associated) liquid
  polare F. polar liquid
  schwere F. *(min tech)* dense (heavy) medium
  überstehende F. supernatant liquid (liquor)
  Wackenrodersche F. Wackenroder's solution *(a mixture of polythionic acids)*
Flüssigkeit-Dampf-Gemisch *n* vapour-liquid mixture
Flüssigkeit-Dampf-Grenzfläche *f* liquid-vapour interface
Flüssigkeit-Flüssigkeit-Grenzfläche *f* liquid-liquid interface, dineric interface
Flüssigkeitsabgabe *f* release of liquid, *(of gels also)* weeping
Flüssigkeitsaufnahme *f* uptake of liquid, imbibition, *(text also)* pickup
Flüssigkeitsbadvulkanisation *f* liquid curing
Flüssigkeitsbarren *m* bar of solvent *(in zone melting)*
Flüssigkeitsbereich *m* liquid range
Flüssigkeitscharakter *m (phys chem)* fluidity *(reciprocal of viscosity)*
Flüssigkeits-Chromatografie *f* liquid chromatography
Flüssigkeitsdichte *f* density of a liquid
Flüssigkeitsdruckdüse *f* pressure nozzle
Flüssigkeitsdurchsatz *m* liquid throughput
Flüssigkeitseinsatz *m* batch of liquid
Flüssigkeitsfassungsvermögen *n* liquid holding volume
Flüssigkeitsfilm *m* liquid film
Flüssigkeitsfiltration *f* liquid filtration
Flüssigkeitsgemisch *n* liquid mixture

Flüssigkeits-Glasthermometer *n* liquid-in-glass thermometer
Flüssigkeitshydrat *n* liquid hydrate
Flüssigkeitshydratkristall *m* liquid hydrate crystal
Flüssigkeitsinhalt *m (distil)* column hold-up
Flüssigkeitskalorimeter *n* water calorimeter
Flüssigkeitskatode *f* pool cathode
Flüssigkeitsmischer *m* liquid mixer
kontinuierlicher F. flow (line) mixer
Flüssigkeitsniveau *n* liquid level
Flüssigkeitsphase *f* liquid phase
Flüssigkeitspotential *n* diffusion (liquid-liquid, liquid junction) potential
Flüssigkeitsringverdichter *m* liquid-piston compressor, [Nash] Hytor pump
Flüssigkeitssäule *f* column of liquid
Flüssigkeitsschicht *f* liquid layer
Flüssigkeitsspiegel *m* liquid level
Flüssigkeitsstand *m* liquid level
Flüssigkeitsstand[s]anzeiger *m* liquid-level meter; gauge glass
Flüssigkeitsstand[s]messung *f* liquid-level measurement
Flüssigkeitsstrahl *m* jet of liquid
Flüssigkeitsstrom *m* liquid (fluid) flow; *(tech)* liquor flow
Flüssigkeitsströmung *f* liquid (fluid) flow; *(tech)* liquor flow
Flüssigkeitsthermometer *n* liquid [expansion] thermometer
Flüssigkeitsverlust *m* loss of liquid
Flüssigkeitsverschluß *m* liquid seal *(as for stirrers)*
Flüssigkeitsverteiler *m* liquid distributor
Flüssigkeitsvorlage *f* liquid receiver
Flüssigkeitszerstäuber *m (incorrectly for)* Versprüher
Flüssigköder *m* wet bait *(pest control)*
Flüssiglinie *f (distil)* liquidus curve (line)
Flüssigluftsprengstoff *m* oxyliquit
Flüssigmetallbrennstoff *m (nucl)* liquid metal reactor fuel
Flüssigphase *f* liquid phase
Flüssigphaseisomerisierung *f* liquid-phase isomerization
Flüssigphasekatalysator *m* liquid-phase catalyst
Flüssigphasekracken *n* liquid-phase cracking
Flüssigphaseverfahren *n* liquid-phase process
Flüssigsauerstoff *m* liquid oxygen
Flüssigstäuber *m* liquiduster *(an apparatus for the joint spraying of pesticides in liquid and powder form)*
Flüssigstickstoff *m* liquid nitrogen
Flüssigwerden *n* liquefaction, liquefication, melting, fusing, fluxing
Flußkies *m* river gravel
Flußmesser *m* flow (rate) meter
Flußmittel *n* flux[ing agent]
flußmittelfrei flux-free
Flußsäure *f* H_2F_2 hydrofluoric acid
Flußsäurealkylierung *f* hydrofluoric-acid alkylation, HF alkylation
Flußsäureverfahren *n (petrol)* hydrofluoric-acid process, HF process
Flußspat *m (min)* fluor-spar, fluorite *(calcium fluoride)*
Flußstahl *m* ingot iron (steel)
Flußverunreinigung *f* river (stream) pollution
Flußwasser *n* river water
fluten 1. *(distil)* to flood *(e.g. a packed column)*; 2. *(petrol)* to flood *(oil sand)*
Fluten *n* 1. *(distil)* flooding *(as of a packed column)*; 2. *(petrol)* [water] flooding *(of oil sand)*; 3. *(coat)* flow coating
Flutkammer *f s.* Flutzone
Flutlackieren *n* flow coating
Fluttunnel *m s.* Flutzone
Flutung *f (distil)* flooding
Flutzone *f (coat)* flow-coating section
fluxen *(petrol)* to flux, to cut back
Fluxöl *n (petrol)* flux [oil]
fl. z. *s.* flächenzentriert
F-1-Methode *f* F1 method, research method *(for octane rating)*
F-2-Methode *f* F2 method, motor method *(for octane rating)*
FMN *s.* Flavinmononukleotid
FM-Zyklotron *n* frequency-modulated cyclotron, synchrocyclotron
Foid *m (min)* feldspathoid
fokussieren to focus
Folgereaktion *f* consecutive (consequent, successive) reaction
Folie *f (plast)* film, *(if thickness greater than 0.01 inch)* sheeting *(as a web)*, sheet *(as a piece)*; *(esp relating to metal)* foil
gegossene F. cast film (sheet)
gepreßte F. pressed sheet
geschälte F. sliced film (sheet)
geschäumte F. expanded sheet
gespritzte F. extruded film (sheet)
kalandrierte F. calendered film (sheet)
stranggepreßte F. extruded film (sheet)
Foliefaser *f* split fibre
Folienblaskopf *m (plast)* blow head
Folienblasmaschine *f (plast)* film blowing machine
Folienformung *f (plast)* film formation (forming), *(relating to products with thickness greater than 0.01 inch)* sheet formation (forming)
Foliengießmaschine *f (plast)* casting machine for film formation, solution-casting machine
Folienisolierung *f* multiple-layer insulation, superinsulation *(cooling technology)*
Folienpapier *n* foil paper
Folienschneidmaschine *f (plast)* slicing machine
Folienstrangpressen *n (plast)* film extrusion *(relating to products with thickness greater than 0.01 inch)* sheet extrusion
Folinsäure *f* folinic acid, N^5-formyltetrahydrofolic acid, leucovorin, citrovorum factor
Follikelhormon *n* follicular (oestrus-producing) hormone, oestrogen
Follikelreifungshormon *n*, **Follikelstimulierungshormon** *n* follicle-stimulating hormone, FSH

Folsäure *f* folic acid *(collectively for a series of growth factors)*
Fond *m (nucl)* background
Fondcreme *f (cosmet)* foundation cream, make-up base
Förderanlage *f* conveyor
Förderband *n* conveyor (conveying) belt, apron; *(incorrectly for)* Gurtbandförderer
Förderbandtrockner *m* conveyor (band, belt, belt tunnel, apron) dryer
Förderbandwaage *f* weighing belt
Förderbehälter *m* lift tank
Förderbohrung *f (petrol)* development (exploitation) well
Förderbraunkohle *f* raw lignite
Förderdruck *m* discharge (delivery) pressure
Fördereigenschaften *fpl* conveying characteristics
Fördereinrichtung *f* conveyor; *(for vertical transportation)* lift
Förderer *m* conveyor
  pneumatischer F. air conveyor
Fördererz *n* crude (raw, run-of-mine, as-mined) ore
Förderflüssigkeit *f* liquid being (*or* to be) pumped
Fördergefäß *n* skip [car]
Fördergerät *n* conveyor
Fördergeschwindigkeit *f* delivery rate *(as of a pump)*
Fördergurt *m* conveyor (conveying) belt
Fördergut *n* material being (*or* to be) conveyed
Förderhöhe *f* discharge (delivery) head *(as of a pump)*
Förderhub *m* discharge (delivery) stroke *(as of a pump)*
Förderkohle *f* rough (run-of-mine) coal
Förderkorb *m* skip [car]
Förderkübel *m* skip [car]
Förderlänge *f* conveyor length
Förderleistung *f* delivery, discharge
Förderleitung *f* delivery line
Förderluft *f* conveying air, *(directed upwards also)* lift air
Fördermenge *f s.* Förderleistung
fördern 1. to mine, to extract, to win *(e.g. coal)*; 2. to convey *(bulk material)*; to deliver, to discharge *(of pumps, compressors)*; 3. to promote *(e.g. a reaction)*; to stimulate *(e.g. growth)*
Förderrohr *n* delivery pipe
Förderrutsche *f* oscillating conveyor
Förderschnecke *f* screw (spiral, helix, worm, scroll) conveyor, conveying (conveyor) screw (worm)
Förderseil *n* hoisting rope
Förderseite *f* delivery side
Förderstrecke *f* conveyed length, length of travel
Förderstrom *m* rate of delivery (discharge, flow), throughput, delivery
Förderung *f* 1. mining, extracting, winning *(as of coal)*; 2. conveying *(of bulk material)*; delivery, discharge *(of pumps, compressors)*; 3. promotion *(as of a reaction)*; stimulation *(as of growth)*
  F. im Tagebau surface (open-cut, open-cast) mining, open pit method, *(esp relating to ores)* [surface] quarrying
  F. im Tiefbau (Untertagebau) deep mining, underground mining (working)
  F. mittels Airlifts air lifting
  F. mittels Druckgases gas lifting
  F. mittels Druckluft air lifting
  F. mittels Gaslifts gas lifting
  pneumatische F. air conveying
  übermäßige F. over-stimulation *(as of growth)*
Fördervolumen *n* displacement *(of a pump)*
Fördervorrichtung *f* conveyor; *(for vertical transportation)* lift
Förderwagen *m* trolley
Förderweg *m* conveyed length, length of travel
Form *f* 1. form, shape; *(isomerism)* form; 2. *(tech)* mould
  F. mit Heizkanälen *(plast)* cored mould
  blanke F. *(glass)* uncoated mould
  chinoide F. quino[no]id form
  einteilige F. *(glass)* block mould
  enantiomorphe F. enantiomorph, enantiomer, enantiomorphous form (isomer), optical antipode (opposite, isomer), antimer
  flexible F. boat form (conformation) *(stereochemistry)*
  gestaffelte F. staggered form (conformation) *(stereochemistry)*
  geteilte F. *(glass)* split mould
  getrocknete F. dry-sand mould *(foundry)*
  grüne F. green-sand mould *(foundry)*
  intermediäre F. intermediate form
  linksdrehende F. laevo[rotatory] form, (−) form *(of an optically active compound)*
  mehrteilige F. *(glass)* split mould
  metallische F. [permanent] metal mould *(foundry)*
  nasse F. green-sand mould *(foundry)*
  rechtsdrehende F. dextro[rotatory] form, (+) form *(of an optically active compound)*
  schiefe F. gauche (skew) form (conformation) *(stereochemistry)*
  starre F. chair form (conformation) *(stereochemistry)*
  syn-clinale F. *s.* schiefe F.
  tautomere F. tautomeric form, dynamic isomer
  trockene F. dry-sand mould *(foundry)*
  ungetrocknete F. green-sand mould *(foundry)*
  verdeckte F. eclipsed form (conformation) *(stereochemistry)*
  windschiefe F. *s.* schiefe F.
  zweiteilige F. *(glass)* split mould
***d*-Form** *f*, **(+)-Form** *f. s.* Form / rechtsdrehende
***l*-Form** *f*, **(−)-Form** *f. s.* Form / linksdrehende
Formal *n* formal, formaldehyde acetal *(any acetal derived from formaldehyde and an alcohol)*
Formaldehyd *m* HCHO formaldehyde, methanal
  ✦ **mit F. behandeln** to formolize, to formalinize
Formaldehydazetal *n s.* Formal
Formaldehyddiäthylazetal *n* $CH_2(OC_2H_5)_2$ formaldehyde diethyl acetal, diethoxymethane

Formaldehyddimethylazetal *n* $CH_2(OCH_3)_2$ formaldehyde dimethyl acetal, methylal, dimethoxymethane
Formaldehydgerbung *f* formaldehyde tanning
Formaldehydlösung *f* formaldehyde solution
Formaldehydnatriumsulfoxylat *n* $HOCH_2SO_2Na$ sodium formaldehydesulphoxylate
Formaldehydoxim *n s.* Formaldoxim
Formaldehydsulfoxylat *n* $HOCH_2SO_2M^I$ formaldehydesulphoxylate
Formaldehydsulfoxylsäure *f* $HOCH_2SO_2H$ formaldehydesulphoxylic acid
Formaldoxim *n* $HCH{=}NOH$ formaldoxime, formaldehyde oxime
Formalladung *f* formal charge
Formamid *n* $HCONH_2$ formamide
Formamidin *n* $HN{=}CHNH_2$ formamidine
Formänderung *f* deformation, strain
Formänderungsrest *m* residual (permanent) deformation (set)
  F. bei Dehnungsbeanspruchung residual (permanent) deformation (set) at elongation, elongation (tensile) set
  F. bei Druckbeanspruchung [permanent] compression set
Formänderungs-Spannungs-Linie *f* stress-strain curve
Formanilid *n* $HCONHC_6H_5$ formanilide, formylaniline
Formart *f* state of aggregation (matter)
Formartikel *m* moulded article (part, product), moulding
Format *n* size
Formation *f (geol)* formation
  ölführende F. *(petrol)* producing formation
Formatpapier *n* sheeted (sheet, ream) paper, paper in sheets
Formatschneider *m (pap)* guillotine cutter (cutting machine, press, trimmer)
Formatwalze *f (pap)* press roll *(of a cylinder board machine)*
Formatzylinder *m s.* Formatwalze
Formazyl *n* $-C({=}NNHC_6H_5)(N{=}NC_6H_5)$ formazyl
formbar formable, shap[e]able, ductile; mouldable
Formbarkeit *f* formability, shap[e]ability, ductility; mouldability
Formbeständigkeit *f (plast, text)* dimensional stability
  F. in der Wärme *(plast)* heat deflection (distortion) point (temperature), plastic yield [with temperature]
  F. in der Wärme nach Vicat *(plast)* Vicat softening point (temperature), V.S.P., Vicat needle point
Formboden *m (glass)* bottom plate
Formdichtung *f* moulded seal
Formeinstreichmittel *n s.* Formentrennmittel
Formel *f* formula
  allgemeine F. general formula
  angenäherte F. approximate formula
  Arrheniussche F. Arrhenius equation
  Balmersche F. Balmer formula
  Braggsche F. Bragg equation
  Clausius-Clapeyronsche F. Clapeyron-Clausius equation
  Debyesche F. Debye equation
  einfachste (empirische) F. *s.* stöchiometrische F.
  geradkettige F. straight-chain formula
  perspektivische F. perspective formula
  Ritzsche F. Ritz formula
  stöchiometrische F. stoichiometric (empirical, simplest, simplest possible) formula
  wahre F. true formula
Formelbestimmung *f* determination of formulae
Formelbild *n* graphic formula
Formelgewicht *n*, **Formelmasse** *f* formula weight
Formelregister *n* formula index
Formelsprache *f* formula (symbolic) language
Formelumsatz *m* formula conversion
formen to form, to shape; to mould
  zu Krümeln f. *(rubber)* to pelletize
  zu Kügelchen (Pellets) f. to pellet[ize], to pill
Formen *n* forming, shaping; moulding
  F. eines Rohlings *(plast)* forming of a blank
  F. in Grünsand green-sand moulding *(foundry)*
  F. in Lehm loam moulding *(foundry)*
  F. in Trocken[guß]sand dry[-sand] moulding *(foundry)*
  F. mit **Ausschmelzmodellen** investment moulding *(foundry)*
  F. mit Wachs[ausschmelz]modellen lost-wax moulding *(foundry)*
  nachträgliches F. *(plast)* postforming
Formenbau *m* mould making
Formenbrecher *m (rubber)* mould breaker (cracker), mould-breaking (mould-clearing) jack
Formeneinstreichmittel *n s.* Formentrennmittel
Formeneis *n* block (can) ice
Formengips *m* moulding plaster
Formennaht *f* parting (joint) line, match (mould) mark (seam) *(a defect in glass)*
Formenöffner *m s.* Formenbrecher
Formenrahmen *m (rubber)* dipping rack
Formenschluß *m* mould closing
Formenschmiere *f*, **Formenschmiermittel** *n (glass)* mould lubricant, dope
Formenschwindmaß *n* mould shrinkage
Formentrennmittel *n* [mould-]release agent, mould release (lubricant)
Formentrennung *f* mould release
Formfaktor *m (nucl)* form factor; *(rubber)* shape factor
Formgebung *f* forming, shaping, profiling
Formgebungsmaschine *f (ceram)* shaping machine
Formgrat *m (plast)* fan, fin
Formgußstück *n* casting
Formheizung *f (rubber)* mould cure (curing, vulcanization)
Formherstellung *f* mould making
Formhöhlung *f (plast)* mould cavity
Formhydroxamsäure *f* $HCO{-}NHOH$ or $HOCH{=}NOH$ formhydroxamic acid, formhydroximic acid
Formhydroximsäure *f s.* Formhydroxamsäure

Formiat *n* HCOOM' formate, formiate
Formkammer *f* moulding chamber
Formkoks *m* formed (shaped) coke
Formkörper *m (distil)* tile *(a piece of packing)*
Formmaschine *f* moulding machine, moulder
Formmaske *f* shell mould *(foundry)*
Formmaskenverfahren *n* **[nach Croning]** shell-moulding process, Croning process, C process *(foundry)*
Formmasse *f (plast)* moulding material (compound)
F. mit Faserstoffüllung (Faserstoffverstärkung) fibre-filled moulding material
hitzehärtbare F. *s.* wärmehärtbare F.
kittartige F. dough moulding material
wärmehärtbare F. thermosetting moulding material
Formnest *n (plast)* mould cavity
Formöffnung *f* mould (tuyère) opening *(foundry)*
formolisieren to formolize
Formolisierung *f* formolization
Formose *f* formose *(a mixture of aldoses and ketoses)*
Formplatte *f (plast)* platen; pattern plate *(foundry)*
bewegliche F. *(plast)* movable (moving) platen
feststehende F. *(plast)* stationary platen
Formpresse *f (plast)* moulding (compression) press
formpressen *(plast)* to mould
Formpressen *n (plast)* [compression] moulding
F. mit Hochfrequenzheizung (Hochfrequenzvorwärmung) high-frequency (radio-frequency) moulding
Formpreßstoff *m (plast)* compression-moulding material
Formsand *m* moulding sand
grüner F. green [moulding] sand
gut gasdurchlässiger F. open (free-venting) sand
nasser F. *s.* grüner F.
natürlicher F. natural [moulding] sand, naturally bonded sand
synthetischer F. synthetic [moulding] sand
wenig gasdurchlässiger F. poor-venting sand
Formschließeinheit *f (plast)* mould clamp
Formschließkraft *f (plast)* mould-clamping force
Formschließzeit *f (plast)* mould-closing time
Formstanze *f (plast)* punch press
Formstanzen *n (plast)* pressure forming
Formstoff *m* moulded material
Formstück *n* fitting *(for pipes and hoses)*
Formtechnik *f* moulding technique
Formteil *n* moulding, blank, shape
gegossenes F. cast moulding
nachgeformtes F. postformed moulding
nichtausgeformtes (unvollständiges) F. short moulding
Formteile *npl* moulded articles (goods, parts, products)
Formtrennmittel *n s.* Formentrennmittel
formulieren to formulate
Formulierung *f* formulation *(1. expressing with a formula; 2. compounding in accordance with a recipe)*
Formung *f s.* Formen
Formungsdruck *m (plast)* forming pressure
Formungstemperatur *f (plast)* forming temperature
Formunterteil *n (plast)* force
Formveränderung *f* deformation, strain
Formverschäumung *f (plast)* foaming in place (situ)
Formvulkanisation *f* press (mould) cure
Formylazeton *n* CH_3COCH_2CHO formylacetone, acetoacetaldehyde
Formylbenzol *n s.* Benzaldehyd
Formylessigsäure *f* formylacetic acid
Formylgruppe *f* –C(=O)H formyl group (residue)
Formylhydroperoxid *n* HCO–O–OH formyl hydroperoxide, performic acid
formylieren to formylate
Formylierung *f* formylation
Formylierungsreagens *n* formylating reagent
Formylradikal *n* formyl radical, *(specif)* free formyl radical
Formylrest *m s.* Formylgruppe
Formyltribromid *n s.* Bromoform
Formyltrichlorid *n s.* Chloroform
Formyltrijodid *n s.* Jodoform
Formzeit *f (plast)* moulding time
Forschungslabor[atorium] *n* research laboratory
Forschungsreaktor *m* research (experimental) reactor
Forsterit *m (min)* forsterite *(magnesium orthosilicate)*
Forsteriterzeugnis *n* / **feuerfestes** forsterite refractory
Forsteritporzellan *n* forsterite porcelain
Forsteritstein *m* / **feuerfester** forsterite refractory brick
Forsteritweißware *f* forsterite whiteware
fortpflanzen to propagate *(e.g. a chain molecule)*
sich f. to propagate *(as of a reaction)*
Fortpflanzung *f* propagation
Fortpflanzungsgeschwindigkeit *f* propagation velocity
Fortpflanzungsreaktion *f* propagation reaction
Fortpflanzungsrichtung *f* direction of propagation
Fortpflanzungsstadium *n* propagation stage
Fortreißfestigkeit *f (pap)* tear (tearing) resistance (strength), tear propagation strength, resistance to further tearing
Fortreißfestigkeitsprüfung *f (pap)* tear[ing] test, tear propagation test
fortschreiten to proceed, to advance
fortschwemmen to float off
fortspülen to wash (rinse, flush) away
fortwandern to migrate out *(as of ions)*
fossil *(geol)* fossil
Fossilbrennstoff *m* fossil fuel
fossilisieren to fossilize
fotoaktiv photoactive
Fotoaktivität *f* photoactivity
Fotobromierung *f* photochemical bromination, photobromination

Fotochemie *f* photochemistry
Fotochemikalie *f* photographic chemical
fotochemisch photochemical
Fotochlorierung *f* photochemical chlorination, photochlorination
Fotodissoziation *f* photodissociation
Fotoeffekt *m* photoelectric effect
fotoelektrisch photoelectric
Fotoelektrizität *f* photoelectricity
Fotoelektron *n* photoelectron
Fotoelektronenvervielfacher *m* photomultiplier [tube], multiplier phototube, secondary-emission electron multiplier
Fotoelement *n* photovoltaic (photochemical, photobarrier, photoelectrolytic, barrier-layer) cell
Fotoemission *f* photoemission
Fotoemissionszelle *f* photoelectric cell, photocell, phototube, photovalve
Fotoemulsion *f* photographic (sensitive) emulsion
Fotogravüre *f* photogravure
Fotohalogenid *n* photohalide
Fotoionisation *f* photoionization
Fotokatalysator *m* photocatalyst
Fotokatalyse *f* photocatalysis
fotokatalytisch photocatalytic, light-catalyzed
Fotokatode *f* photocathode
Fotokopie *f* [silver halide] photocopy
Fotokopieren *n* [silver halide] photocopying
Fotoldruckverfahren *n* fotol (ferrogelatin) process *(reprography)*
Fotoleitfähigkeit *f* photoconductivity
Fotoleitungseffekt *m* photoconductive effect
Fotolumineszenz *f* photoluminescence
Fotolyse *f* photolysis, photodecomposition, photodegradation
fotolytisch photolytic
Fotomaterial *n* sensitive (sensitized) material
Fotometer *n* photometer
Fotometerbank *f* bench photometer
Fotometerkopf *m* photometer head
Fotometrie *f* photometry
fotometrisch photometric
Fotoneutron *n* photoneutron
Fotooxydation *f* photooxidation
Fotopapier *n* photo[graphic] paper
Fotophorese *f* photophoresis
Fotoplatte *f* photo[graphic] plate
Fotopolymer[es] *n* photopolymer
Fotopolymerisation *f* photopolymerization
Fotoproton *n* photoproton
Fotoreaktion *f* photochemical reaction, photoreaction
Fotorohpapier *n* photographic base paper
Fotoschale *f* developing dish
Fotosekundärelektronenvervielfacher *m s.* Fotoelektronenvervielfacher
Fotosensibilisation *f* photosensitization
Fotosensibilisator *m* photosensitizer
fotosensibilisieren to photosensitize
Fotosensibilisierung *f* photosensitization
Fotostrom *m* photocurrent
Fotosynthese *f* photosynthesis
fotosynthetisch photosynthetic
Fototropie *f* phototropism, phototropy
Foto-Volta-Effekt *m* photovoltaic effect
Fotovolteffekt *m* photovoltaic effect
Fotowiderstand *m*, **Fotowiderstandszelle** *f* photoresistor, photoconductive cell
Fotozelle *f* photoelectric cell, photocell, phototube, photovalve
Foulard *m (text)* padding machine (mangle), padder, pad
Foulardbehandlung *f (text)* slop-padding
Foulardfärbung *f (text)* pad[ded] dyeing
foulardieren *(text)* to [slop-]pad
Foulardierlösung *f (text)* pad (padding) bath (liquor)
Foulard-Jigger-Verfahren *n (text)* pad-jig process
Fourcault-Verfahren *n (glass)* Fourcault [sheet-drawing] process
FP. *s.* Flammpunkt
Fragmentierung *f* fragmentation *(of hydrocarbon molecules)*
Fraktion *f* 1. *(distil)* cut, fraction; 2. size fraction *(in classifying)*
  hochsiedende (höhersiedende) F. high-boiling (higher-boiling) fraction, heavy fraction
  leichte F. *s.* niedrigsiedende F.
  mittlere F. middle fraction
  niedrigsiedende F. low-boiling fraction, light fraction
  schwere (schwerer flüchtige) F. *s.* hochsiedende F.
  schwer[er]siedende F. *s.* hochsiedende F.
  tiefsiedende F. *s.* niedrigsiedende F.
α-Fraktion *f* alpha fraction *(in the pyridine extraction of hard coal)*
β-Fraktion *f* beta fraction *(in the pyridine extraction of hard coal)*
γ-Fraktion *f* gamma fraction *(in the pyridine extraction of hard coal)*
Fraktionator *m s.* Fraktionierkolonne
Fraktionierapparat *m* fractionating apparatus
Fraktionierbürste *f (distil)* wiper
fraktionieren to fractionate, to fraction
Fraktioniergerät *n* fractionating apparatus
Fraktionierkolben *m* fractionating flask
Fraktionierkolonne *f*, **Fraktioniersäule** *f* fractionating column, fractionator
fraktioniert fractional
Fraktionierturm *m* fractionating tower
Fraktionierung *f* fractionation, *(distil also)* fractional distillation
Fraktionsabscheidegrad *m*, **Fraktionsentstaubungsgrad** *m* fractional[-weight collection] efficiency *(classifying)*
Fraktionskolben *m* fractionating flask
Fraktionssammler *m* fraction collector
Franck-Condon-Prinzip *n* Franck-Condon principle
Frangulaemodin *n* frangula-emodin, 1,3,8-trihydroxy-6-methylanthraquinone

Frangulin *n*, **Frangulosid** *n* frangulin, franguloside *(glycoside)*
Frankium *n* Fr francium
Franklinit *m (min)* franklinite *(zinc ferrite)*
Fransenfibrille *f (text)* fringed fibril[la]
Fransenmizelle *f (text)* fringed micelle
Frasch-Verfahren *n* Frasch process *(in sulphur mining)*
Fraßgift *n* stomach poison
  F. für Insekten stomach insecticide
Frästorf *m* milled peat
Frauenmilch *f* human (breast) milk
Fraunhofer-Linien *fpl* Fraunhofer lines
frei free; vacant *(orbitals)*
  f. von Feststoffen solids-free
  praktisch f. substantially free
Freialdehydigkeit *f (food)* rancidity with formation of aldehydes
Freibergit *m (min)* freibergite *(a tetrahedrite containing silver)*
freibrennend free-burning
Freidampfheizung *f*, **Freidampfvulkanisation** *f* open-steam cure (curing)
Freidrehen *n (ceram)* [free-]hand throwing
Freieslebenit *m (min)* freieslebenite *(a sulphide of antimony, lead, and silver)*
Freifallklassierer *m* non-mechanical classifier
Freifallmischer *m* tumbling mixer, tumbler
Freifallscheider *m* plate separator *(for electrostatic separation)*
Freiflußventil *n* inclined-seat valve
Freigold *n* free gold
Freihandblasen *n* off-hand glassworking
Freiharz *n (pap)* free rosin (resin)
Freiharzgehalt *m (pap)* content of free rosin
Freiharzleim *m (pap)* free-rosin size, acid size
  stabilisierter F. protected rosin size, high free protected size
Freiheit *f (phys chem)* variance, degree of freedom ✦ **in F. setzen** to liberate, to release ✦ **ohne F.** nonvariant, invariant
Freiheitsgrad *m (phys chem)* degree of freedom, variance
Freiheizung *f (rubber)* open cure (vulcanization)
Freilandversuch *m (agric)* field experiment (test, trial)
Freiluftbrenner *m* air-atomizing burner
Freilufttrocknung *f* air drying, *(ceram also)* hack drying
Freiname *m* non-proprietary name, generic name (term)
Freischwefel *m (rubber)* [true] free sulphur
freisetzen to release, to liberate, to set free
Freisetzung *f* release, liberation
Freivulkanisation *f* open cure (vulcanization)
Freiwerden *n* liberation
freiwillig spontaneous
Fremdasche *f (coal)* extraneous ash
Fremdatom *n* foreign (impurity) atom
Fremdbestandteil *m s.* Fremdstoff
Fremdeisen *n* tramp iron
Fremdgas *n* foreign gas
Fremdgeruch *m* foreign odour
Fremdgeschmack *m* foreign flavour (taste)
fremdgestaltig *(cryst)* xenomorphic, allotriomorphic, anhedral
Fremdgut *n* tramp material
Fremdhalbleiter *m* impurity semiconductor
Fremdkraftringmühle *f*, **Fremdkraftrollenmühle** *f*, **Fremdkraftwälzmühle** *f* ring (ring-roll, centrifugal attrition) mill, centrifugal grinder
Fremdleiter *m s.* Fremdstoffhalbleiter
Fremdling *m (geol)* xenocryst
Fremdmolekül *n* foreign molecule
Fremdstoff *m* admixture, impurity, foreign matter (substance, material), extraneous material (substance)
  F. in Futtermitteln feed additive
  F. in Lebensmitteln food additive
Fremdstoffhalbleiter *m* impurity (extrinsic) semiconductor
Fremdstoffkonzentration *f* solute concentration *(in zone melting)*
Fremdstromkorrosion *f* electrocorrosion
Fremdstromverfahren *n* impressed e.m.f. method *(cathodic protection for minimizing corrosion)*
Fremdsubstanz *f s.* Fremdstoff
Fremdzündung *f* external ignition
Frenkel-Defekt *m* Frenkel defect
Frenkel-Fehlordnung *f* Frenkel disorder
Frenkel-Fehlstelle *f* Frenkel defect
Frequenz *f* frequency
  F. der Seriengrenze convergence frequency *(of a spectral series)*
  charakteristische F. characteristic (proper, natural) frequency
Frequenzband *n* frequency band
Frequenzbedingung *f* / **Bohrsche** Bohr frequency condition
Frequenzfaktor *m* frequency factor
Frequenzgang *m* frequency function
Frequenzverteilung *f* frequency distribution
fressen to corrode
Friedel-Crafts-Alkylierung *f* Friedel-Crafts alkylation
Friedel-Crafts-Azylierung *f* Friedel-Crafts acylation
Friedel-Crafts-Katalysator *m* Friedel-Crafts catalyst (agent)
Friedel-Crafts-Kondensation *f* Friedel-Crafts condensation
Friedel-Crafts-Reaktion *f* Friedel-Crafts reaction (synthesis)
Friedelit *m (min)* friedelite *(a phyllosilicate)*
Friedländer-Synthese *f* Friedländer synthesis
Fries-Reaktion *f* Fries reaction (rearrangement)
Friktion *f* friction
friktionieren *(rubber)* to friction
Friktionierkalander *m s.* Friktionskalander
Friktionseffekt *m* frictional effect
  differentieller F. directional frictional effect
Friktionskalander *m* friction (frictioning, glazing) calender
Friktionsmischung *f (rubber)* friction compound

Friktionsstreifen *m (rubber)* chafer [strip]
Friktionsverhältnis *n* friction ratio
frischbereitet freshly prepared
Frischbeton *m* fresh (ready-mixed, green) concrete
Frischdampf *m* live (direct, prime, open) steam
Frischgas *n* make-up gas
frischgefällt freshly precipitated
Frischhaltepapier *n* avenized paper
Frischkatalysator *m* fresh catalyst
Frischkautschuk *m* new rubber
Frischlauge *f (pap)* white (fresh cooking) liquor
Frischluft *f* fresh air
Frischmilch *f* fresh (freshly drawn) milk
Frischsäure *f* fresh acid
Frischwasser *n* fresh water
Fritte *f* 1. *(ceram)* frit, agglomerate; 2. *(tech)* diffuser stone *(for dissolving gas in a liquid)*; 3. *(agric)* frit *(containing micronutrients)*; 4. *s.* Glasfritte
Fritteglasur *f (ceram)* fritted glaze
fritten 1. *(ceram)* to frit, to sinter, to agglomerate *(by means of heat)*; to sinter, to agglomerate *(under the influence of heat)*; 2. *(glass)* to dragladle, to dragade, to dry-gage, *(Am)* to shrend *(cullet)*
Frittenporzellan *n* fritted porcelain
Frittenwaschflasche *f (lab)* sintered-plate washbottle
Fritteofen *m (ceram)* frit kiln
Fritz-Verfahren *n* Fritz method *(a churning process)*
Frontalanalyse *f* frontal analysis *(chromatography)*
frostbeständig frost-resistant, frostproof
Frostbeständigkeit *f* frost resistance, resistance to frost (freezing)
frosten *(food)* to deep-freeze
Frosten *n (food)* deep freezing, freezing preservation
Frostschutzmittel *n* antifreeze [agent, compound, fluid], antifreezing agent (dope)
Frostschutzpapier *n* antifreeze paper
Frostschutzpappe *f* antifreeze cardboard
frostsicher frost-resistant, frostproof
Frostsicherheit *f* frost resistance, resistance to frost (freezing)
Fruchtaroma *n* fruit essence
Fruchtäther *m* fruit essence
fruchtbar fertile *(soil)*
Fruchtbarkeit *f* fertility *(of soil)*
Fruchtbrei *m* pomace, squash
Fruchtessenz *f* fruit essence
Fruchtessig *m* fruit vinegar
Fruchtfaser *f* fruit fibre
Fruchtmark *n* fruit pulp
Fruchtsaft *m* [fruit] juice
Fruchtschale *f (pharm)* cortex
Fruchtseidenpapier *n* fruit paper (tissue)
Fruchtzucker *m s.* Fruktose
Frue-Vanner *m (min tech)* Frue vanner *(a concentrating table)*
Frühzündung *f* preignition
Fruktosan *n* fructosan *(a polysaccharide)*
Fruktose *f* fructose, fruit sugar, Fru
Fruktosid *n* fructoside *(a glycoside)*
Fruktosidase *f* fructosidase
Fry-Ätzmittel *n*, **Fry-Reagens** *n* Fry reagent *($CuCl_2$ in HCl, for etching steel)*
F-Säure *f* $C_{10}H_6(OH)SO_3H$ F acid, 2-naphthol-7-sulphonic acid; $C_{10}H_6(NH_2)SO_3H$ F acid, 2-naphthylamine-7-sulphonic acid
FSH *s.* Follikelstimulierungshormon
FT-Ruß *m (rubber)* FT black, fine thermal black
Fuchs *m (met)* skimmer *(for separating the slag flowing with the molten iron)*
Fuchsin *n* fuchsine, magenta, rosaniline
Fuchsinfarbstoff *m* fuchsine (rosaniline) dye
Fuchsit *m (min)* fuchsite *(muscovite containing chromium)*
Fruchsonimin *n (dye)* fuchsonimine
Fugat *n* centrifugate
Fugazität *f* fugacity, volatility
Fühler *m* sensing device, sensor, detector
führen to conduct *(a process)*; to lead, to conduct, to pipe *(e.g. vapour in piping)*; to run *(a factory)*
im Kreislauf f. to [re]circulate, to recycle
in Rohrleitungen f. to pipe
Führungsgröße *f* reference variable (input) *(control engineering)*
Führungslager *n* guide [bearing], bearing assembly *(as of an agitator)*
Führungsrolle *f* guide roll
Führungsseil *n (pap)* leading-through tape, rope carrier
Führungstrichter *m (glass)* guide funnel
Führungswalze *f* guide roll
Füllbeton *m* poor (lean, lean-mixed) concrete
Füllbunker *m* charging bin
Fülldichte *f (pap)* chip capacity
füllen 1. to fill; *(tech)* to feed, to charge, to load, to furnish *(e.g. a furnace or reactor)*; *(distil)* to pack *(a column with packing material)*; *(lab)* to prime *(a burner with fuel)*; 2. to load, to fill *(a product with fillers)*, *(pap also)* to weight, *(rubber also)* to pigment; *(tann)* to feed, to fill *(incompletely tanned leather with additional tanning material)*
auf (in) Flaschen f. to bottle [in, up]
mit Füllkörpern f. *(distil)* to pack
wieder f. to refill
Füller *m* 1. filler [material], filling [agent, material]; 2. filler, filling machine
F. für Flüssigkeiten liquid filler (filling machine)
Fullererde *f* fuller's earth
Fuller-Lehigh-Mühle *f* Fuller-Lehigh mill
Fuller-Mühle *f* Fuller mill *(a ball-and-ring mill)*
Füllfaktor *m (plast)* bulk factor
Füllfaser *f* staple for filling
Füllform *f* positive mould
Füllgas *n* filler gas
Füllgerbung *f* plumping tannage
Füllgut *n* 1. *s.* Füllmaterial 1.; 2. *(plast)* mould charge

Füllhöhe *f* filling level, fill height
Füllhöhenmessung *f* level measurement
Füllklappe *f* charging (filling) door
Füllkörper *m* 1. filler [material], filling [agent, material]; 2. *(esp distil)* packing body, piece of packing
Füllkörper *mpl (esp distil)* packing[s], fillings
F. aus Glas glass packing
F. aus Porzellan porcelain packing
geschüttete F. dumped packing
Füllkörperhöhe *f* packed height
äquivalente F. height equivalent to a theoretical plate, HETP
Füllkörperkolonne *f*, **Füllkörpersäule** *f* packed column
Füllkörperschichthöhe *f* height of packing
Füllkörperturm *m* packed tower
Füllmaschine *f* filling machine, filler
F. für Flüssigkeiten liquid filler (filling machine)
Füllmasse *f* 1. *s.* Füllstoff 1.; 2. *s.* Füllmaterial 1.; 3. *(sugar)* massecuite, magma, fillmass
Füllmaterial *n* 1. load, batch, feed[stock], charge, charging stock *(material to be reacted)*; 2. *(distil)* packing [material]; 3. *(text)* staple for filling; 4. *s.* Füllstoff 1.
Füllmittel *n s.* Füllstoff 1.
Fülloch *n*, **Füllöffnung** *f* charging (filling) hole (door), feed inlet (hole)
Füllraum *m (plast)* loading chamber, pot
Füllraumform *f*, **Füllraumwerkzeug** *n* positive mould
Füllrohr *n* charging pipe
Füllrumpf *m s.* Fülltrichter
Füllschlitten *m (ceram)* sliding carriage
Füllstand *m* filling level, fill height
Füllstandsmessung *f* level measurement
Füllstation *f* filling plant (room), *(esp)* bottling plant (department, room), bottle house, bottlery
Füllstoff *m* 1. filler [material], filling [material, agent], loading [material, agent], load[er], *(rubber also)* pigment; 2. *s.* Füllmaterial 1.
aktiver F. *(rubber)* active (reinforcing) filler (pigment)
heller F. *(rubber)* white (non-black) filler (pigment), light-coloured filler
heller aktiver F. *(rubber)* white (non-black) reinforcing filler (pigment)
inaktiver (inerter) F. *(rubber)* inert (inactive, extending) filler, non-reinforcing filler (pigment), cheapener
mineralischer F. mineral filler
passiver F. *s.* inaktiver F.
verstärkender F. *s.* aktiver F.
Füllstoffdispergierung *f* filler dispersion
Füllstoffdosierung *f* dosage of filler, filler loading, *(rubber also)* pigment loading, pigmentation
füllstofffrei unfilled, unloaded, filler-free, *(rubber also)* non-pigmented
füllstoffhaltig filled, loaded, *(rubber also)* pigmented
Füllstoffisolierung *f* powder insulation *(cryogenics)*
Füllstoffkaolin *m(n) (pap)* filler clay
Füllstoffnester *npl (rubber)* filler specks
Füllstoffverteilung *f* filler dispersion
Füllstutzen *m* charging (filling, feeding) pipe
Fülltablett *n (plast)* charging (loading) tray *(for compression moulds)*
Fülltrichter *m* 1. [charge, feed] hopper, feed[ing] funnel; 2. cup *(of a blast furnace)*
Fülltrichter-Auslaufstutzen *m* feed throat
Füllturm *m* packing (packed) tower *(as for absorption)*
Füllung *f* 1. filling; *(tech)* feeding, charging, loading, furnishing *(as of a furnace or reactor)*; 2. loading, filling *(of a product for conditioning)*, *(rubber also)* pigmentation; 3. *s.* Füllmaterial 1. *and* 2.; 4. *s.* Füllstoff 1.
F. mit Kalkstein *(pap)* stone charging
F. mit Ruß *(rubber)* [carbon] black loading
ungenügende F. *(plast)* short shot
Füllungsgrad *m* degree of filling
Füllvorrichtung *f* **für Preßwerkzeuge** *s.* Fülltablett
Füllzylinder *m (plast)* pot
Fulminat *n* CNOM[I] fulminate
Fulminsäure *f* C≡N−OH fulminic acid, carbyloxime
Fulven *n* fulvene
Fulvenkohlenwasserstoff *m* fulvene
Fulvosäure *f* fulvic acid *(any of several water-soluble humic acids)*
Fumagillin *n* fumagillin *(antibiotic)*
Fumarase *f* fumarase
Fumarat *n* fumarate *(salt or ester of fumaric acid)*
Fumarole *f (geol)* fumarole
Fumarsäure *f* HOOC−CH=CH−COOH fumaric acid, *trans*-butenedioic acid
Fumarsäuregärung *f* fumaric-acid fermentation
Fumigazin *n* fumigacin, helvolic acid *(an antibiotic)*
Fundamentalserie *f* fundamental (Bergmann) series *(spectroscopy)*
Fundamentplatte *f* base plate
Fundbohrung *f (petrol)* discovery well
fünfbasig pentabasic *(acid)*
Fünfeck *n* pentagon *(as of a cyclic compound)*
Fünferring *m s.* Fünfring
fünfgliedrig five-membered
Fünfring *m* five-membered ring
fünfsäurig pentaacid *(base)*
Fünfstufenbleiche *f (pap)* five-stage bleaching
Fünfwalzenmühle *f*, **Fünfwalzenstuhl** *m* five-roll mill
fünfwertig pentavalent, quinquevalent; pentahydric *(alcohol)*
Fünfwertigkeit *f* pentavalence, quinquevalence
fünfzählig *s.* fünfzähnig
fünfzähnig pentadentate *(ligand)*
Fungichromin *n* fungichromin *(antibiotic)*
Fungistatikum *n* fungistat
fungistatisch fungistatic
fungitoxisch fungitoxic
fungizid fungicidal, antifungal

Fungizid *n* fungicide, *(pharm also)* antifungal drug
F. mit kurativer Wirkung *s.* direktes F.
F. zur Schorfbekämpfung scab fungicide
direktes F. direct (eradicant) fungicide
kupferhaltiges F. copper fungicide
nichtsystemisches F. non-systemic fungicide
quecksilberhaltiges F. mercurial fungicide
systemisches F. systemic fungicide
fungizidresistent fungicide-resistant
Funkenentladung *f* spark discharge
Funkenkammer *f (nucl)* spark chamber
Funkenspektrum *n* spark (flash) spectrum
Funktion *f* function
Debyesche F. Debye function
periodische F. periodic function
thermodynamische F. thermodynamic function
Funktionalität *f* functionality
Furakrylsäure *f s.* Furylakrylsäure
Fural *n s.* Furfural
2-Furaldehyd *m s.* Furfural
Furalessigsäure *f s.* Furylakrylsäure
Furan *n* furan, furfuran
Furanaldehyd *m s.* Furfural
Furandion *n* furandione
Furanharz *n* furan resin
2-Furankarbinol *n s.* Furfuryl-(2)-alkohol
2-Furankarbonal *n s.* Furfural
Furankarbonsäure *f* furancarboxylic acid
Furanose *f* furanose
Furanosid *n* furanoside
Furazan *n* furazan
Furazanring *m* furazan ring
Furchendüngung *f* furrow fertilization
Furfural *n* furfural, 2-furylaldehyde
Furfuralanlage *f* furfural extraction plant
2-Furfuraldehyd *m s.* Furfural
Furfuralextraktion *f* furfural extraction
Furfuralextraktionsanlage *f* furfural extraction plant
Furfuralharz *n* furfural resin
Furfuralkohol *m s.* Furfuryl-(2)-alkohol
Furfuralraffination *f* furfural refining
Furfuralstripper *m (distil)* furfural stripper
Furfuralverfahren *n* furfural process, furfural extraction (refining, solvent) process
Furfuralwaschturm *m* furfural treating tower
Furfuran *n* furan, furfuran
Furfurol *n (incorrectly for)* Furfural
2-Furfurylaldehyd *m s.* Furfural
Furfuryl-(2)-alkohol *m* furfuryl alcohol, 2-hydroxymethylfuran
Furfuryliden *n* 1. furfurylidene *(a bivalent group of atoms)*; 2. *s.* Furfural
Furil *n* furil, αα-furil, di-2-furylglyoxal, di-α-furyl diketone
Furnace-Ruß *m* furnace [combustion] black
Furnace-Verfahren *n (rubber)* furnace [combustion] process, continuous-furnace method *(for producing soot)*
Furnierleim *m* veneer glue
Furoin *n* furoin, αα-furoin, 1,2-difuryl-2-oxoethanol
Furol *n (incorrectly for)* Furfural
Furoylgruppe *f* furoyl group
Furylakrylsäure *f* furylacrylic acid, 3-α-furylacrylic acid, furfuralacetic acid
2-Furylaldehyd *m s.* Furfural
2-Furylkarbinol *n s.* Furfuryl-(2)-alkohol
Fusain *m* fusain *(a charcoal-like microscopic constituent of coal)*
Fusarinsäure *f* fusaric acid, 5-butylpyridine-2-carboxylic acid
Fusarsäure *f s.* Fusarinsäure
Fuselöl *n* fusel (fousel, potato, grain) oil
Fusidinsäure *f* fusidic acid *(antibiotic)*
Fusinit *m s.* Fusit
Fusion *f (nucl)* fusion
Fusionsname *m (nomencl)* fusion name
Fusit *m* fusi[ni]te *(a microscopic structure found in fusain)*
Fußventil *n* foot (suction) valve
Fustet *m s.* Fustik
Fustik *m (dye)* fustic, fustet
Alter (Echter) F. old fustic *(from Chlorophora tinctoria Gaudich.)*
Junger F. young fustic *(from Cotinus coggygria Scop.)*
Fustikholz *n s.* Fustik
Fustin *n* fustin, dihydrofisetin, 3,3′,4′,7-tetrahydroxyflavanone
Futter *n* 1. *(met)* lining, refractory; 2. *(agric)* feed[stuff], feeding stuff
basisches F. *(met)* basic lining (refractory)
saures F. *(met)* acid lining (refractory)
Futterhefe *f* feed (fodder, mineral) yeast
Futterkalk *m* feed lime
Futterkonservierung *f* feed preservation
Futterleder *n* lining leather
Futtermittel *n (agric)* feed[stuff], feeding stuff
Futtermittelzusatz *m* feed additive
füttern *(met)* to line
Futterrohre *npl (petrol)* casing
Futterrohreinbau *m (petrol)* casing, introduction of casing
Futterseidenpapier *n* envelope lining [tissue]
Futterwert *m* feed value
Futterzucker *m* feeding sugar (syrup)
Futterzusatz *m* feed supplement

G

Gabanholz *n (dye)* camwood *(from Baphia nitida Afz.)*
Gabbro *m* gabbro *(an igneous rock)*
Gabe *f (pharm)* dose
größte G. maximal dose
Gabelstapler *m* fork[lift] truck
Gabriel-Synthese *f* Gabriel phthalimide synthesis [of primary amines], Gabriel synthesis
Gadoleinsäure *f* gadoleic acid, 9-eicosenoic acid
Gadolinit *m (min)* gadolinite *(a nesosilicate containing yttrium)*
Gadolinium *n* Gd gadolinium

Gadoliniumazetat *n* $Gd(CH_3COO)_3$ gadolinium acetate
Gadoliniumbromid *n* $GdBr_3$ gadolinium bromide
Gadoliniumchlorid *n* $GdCl_3$ gadolinium chloride
Gadoliniumfluorid *n* GdF_3 gadolinium fluoride
Gadoliniumnitrat *n* $Gd(NO_3)_3$ gadolinium nitrate
Gadoliniumoxid *n* Gd_2O_3 gadolinium oxide
Gadoliniumsulfat *n* $Gd_2(SO_4)_3$ gadolinium sulphate
Gadoliniumsulfid *n* Gd_2S_3 gadolinium sulphide
Gagat *m* gagate, jet *(a mineral of the nature of coal)*
Gahnit *m (min)* gahnite, zinc spinel *(zinc aluminate)*
Gaillard-Kammer *f* Gaillard tower *(for concentrating sulphuric acid)*
Gaillard-Turbozerstäuber *m* Gaillard disperser
Galaktagogum *n (pharm)* galactagogue *(milk-ejecting agent)*
Galaktan *n* galactan *(a polysaccharide)*
Galaktarsäure *f s.* Galaktozuckersäure
Galaktometer *n* [ga]lactometer
Galaktonsäure *f* $HOCH_2[CHOH]_4COOH$ galactonic acid
Galaktose *f* galactose, Gal *(a monosaccharide)*
Galaktosidase *f* galactosidase
Galaktowaldenase *f* galactowaldenase
Galaktozuckersäure *f* $HOOC[CHOH]_4COOH$ galactosaccharic acid, mucic acid
Galakturonsäure *f* $HOC[CHOH]_4COOH$ galacturonic acid
Galambutter *f* Galam (Bambuk, shea) butter *(from Butyrospermum parkii (Don) Kotschy)*
Galbanum *n* galbanum *(a gum resin from Ferula specc.)*
Galenikum *n (pharm)* galenical
Galenit *m (min)* galena, galenite *(lead(II) sulphide)*
Galgant *m*, **Galgantwurzel** *f* galanga[l] *(from Alpinia specc.)*
Gallamid *n* $(HO)_3C_6H_2CONH_2$ gallamide
Gallanilid *n* $C_6H_5NH-CO-C_6H_2(OH)_3$ gallanilide
Gallapfel *m* nut gall, gall nut
Galle *f* 1. *(med)* bile, gall; 2. *(glass)* gall, salt water; 3. *(tann, bot)* gall
Gallein *n* gallein *(a quinonoid dye)*
Gallen *fpl* / **Chinesische (Japanische)** Chinese (Japanese) galls *(from Rhus chinensis Mill.)*
Smyrnaer (Türkische) G. Smyrna (Aleppo, Turkish, Mekka, Levant) galls *(from Quercus infectoria Oliv.)*
Gallenfarbstoff *m (med)* bile pigment
Gallenflüssigkeit *f (med)* bile, gall
Gallensäure *f* bile acid
Gallenstein *m (med)* gallstone, biliary calculus
Gallert *n* gelatin[e], jelly, gelatinous mass (substance)
gallertartig gelatinous, gelatiniform, jelly-like
Gallertbildung *f* jellification
Gallerte *f s.* Gallert
gallertig *s.* gallertartig
Gallium *n* Ga gallium
Gallium(III)-bromid *n* $GaBr_3$ gallium(III) bromide, gallium tribromide
Gallium(II)-chlorid *n* $GaCl_2$ gallium(II) chloride, gallium dichloride
Gallium(III)-chlorid *n* $GaCl_3$ gallium(III) chloride, gallium trichloride
Galliumdichlorid *n s.* Gallium(II)-chlorid
Gallium(III)-fluorid *n* GaF_3 gallium(III) fluoride, gallium trifluoride
Galliumhexazyanoferrat(II) *n* $Ga_4[Fe(CN)_6]_3$ gallium hexacyanoferrate(II), gallium cyanoferrate(II)
Gallium(III)-hydroxid *n* $Ga(OH)_3$ gallium(III) hydroxide
Gallium(III)-jodid *n* GaI_3 gallium(III) iodide, gallium triiodide
Galliummonoselenid *n s.* Gallium(II)-selenid
Galliummonosulfid *n s.* Gallium(II)-sulfid
Galliummonoxid *n s.* Gallium(II)-oxid
Gallium(III)-nitrat *n* $Ga(NO_3)_3$ gallium(III) nitrate
Gallium(I)-oxid *n* Ga_2O gallium(I) oxide, digallium oxide
Gallium(II)-oxid *n* GaO gallium(II) oxide, gallium monooxide
Gallium(III)-oxid *n* Ga_2O_3 gallium(III) oxide, digallium trioxide
Gallium(I)-selenid *n* Ga_2Se gallium(I) selenide, digallium selenide
Gallium(II)-selenid *n* GaSe gallium(II) selenide, gallium monoselenide
Gallium(III)-selenid *n* Ga_2Se_3 gallium(III) selenide, digallium triselenide
Gallium(III)-sulfat *n* $Ga_2(SO_4)_3$ gallium(III) sulphate
Gallium(I)-sulfid *n* Ga_2S gallium(I) sulphide, digallium sulphide
Gallium(II)-sulfid *n* GaS gallium(II) sulphide, gallium monosulphide
Gallium(III)-sulfid *n* Ga_2S_3 gallium(III) sulphide, digallium trisulphide
Galliumtri ... *s.* Gallium(III)- ...
Gallotannin *n s.* Gallusgerbsäure
Gallusgerbsäure *f* gallotannic acid, gallotannin, tannic acid *(proper) (glucose esterified with gallic acid or depsides of it)*
gallussauer gallic
Gallussäure *f* $C_6H_2(OH)_3COOH$ gallic acid, 3,4,5-trihydroxybenzoic acid
Gallussäureamid *n s.* Gallamid
Gallussäureanilid *n s.* Gallanilid
Gallussäure-3-monogallat *n s. m*-Digallussäure
Galmei *m* 1. *(min)* calamine, galmei, galmey; 2. *(pharm)* calamine *(zinc oxide with a small amount of ferric oxide)*
edler G. *(min)* smithsonite *(zinc carbonate)*
Galmeistein *m s.* Galmei 2.
galvanisch galvanic
Galvanisierbad *n* plating bath
Galvanisierbetrieb *m* electroplating plant
galvanisieren to electroplate, to plate
Galvanisieren *n* electroplating, electrodeposition of metals
Galvanisiergehänge *n* plating rack

Galvanismus *m* galvanism
Galvano *n* electrotype *(typography)*
Galvanolumineszenz *f* galvanoluminescence
Galvanometer *n* galvanometer
astatisches G. astatic galvanometer
ballistisches G. ballistic galvanometer
schreibendes G. galvanometer recorder
Galvanometerschreiber *m* galvanometer recorder
Galvanoplastik *f* electroforming, galvanoplastics, galvanoplasty; electrotyping *(typography)*
galvanoplastisch galvanoplastic
Galvanoskop *n* galvanoscope
Galvanostegie *f* electroplating
Galvanotechnik *f* electroplating and electroforming technology
Gambir *n* gambi[e]r, pale catechu, white cutch *(from Uncaria gambir Roxb.)*
Gamma *n s.* Gammawert
G. unendlich *s.* Gammagrenzwert
Gammagrenzwert *m (phot)* gamma infinity
Gammasäure *f* $NH_2C_{10}H_5(OH)SO_3H$ gamma acid, γ-acid, 2-amino-8-naphthol-6-sulphonic acid
Gammastrahlen *mpl* gamma rays
Gammastrahlendetektor *m* gamma-ray detector
Gammastrahlenquelle *f* gamma-ray source
Gammastrahlung *f* gamma radiation
Gammaumwandlung *f (rubber)* glass (second-order) transition, vitrification
Gammawert *m (phot)* gamma value, development factor
Gamma-Zeit-Kurve *f (phot)* time-gamma curve
Gang *m* 1. *(geol)* vein, dike, dyke, *(of metal ore also)* lode; 2. *(chem)* course *(of a reaction)*; 3. *(tech)* flight *(as of a worm shaft)* ✦ **außer G. setzen** to put out of operation, to cut out of service, to stop *(a machine)* ✦ **in G. sein** to be in operation, to be running (working, operating), to run *(of a machine)* ✦ **in G. setzen** to initiate, to start up *(a reaction)*; to put (set) in operation, to set in action, to start [up], to prime *(a machine)*
Gangart *f (mine)* gangue, gang, waste rock, matrix
Gangerz *n* vein ore
Ganggestein *n* 1. *(geol)* dike (dyke) rock, dykite; 2. *s.* Gangart
Ganglienblocker *m*, **Ganglioplegikum** *n (pharm)* ganglion blocking agent, ganglionic blockader
Gangmineral *n* gangue mineral
Gangsteigungswinkel *m* helix angle *(as of an extruder screw)*
Gangtrockner *m (ceram)* corridor dryer
Ganomatit *m (min)* ganomatite, goose dung ore *(an iron arsenate containing silver and cobalt)*
Gänsekötigerz *n s.* Ganomatit
Ganzflächenapplikation *f (agric)* overall application, *(when crops are growing also)* overhead application *(of pesticides)*
Ganzflächenbehandlung *f (agric)* overall (non-selective) treatment
Ganzflächenbesprühung *f (agric)* overall spraying
Ganzreifenregenerat *n (rubber)* whole-tyre reclaim
Ganzstoff *m (pap)* whole (finished) stuff, paper[making] stock
Ganzstoffaufbereitung *f (pap)* stock preparation
Ganzstoffmahlmaschine *f (pap)* perfecting (refining) engine (machine), refiner
Ganzstoffmahlung *f (pap)* beating of stock
Ganzstoffreinigung *f (pap)* stock cleaning (clean-up)
Ganzstoffsortierer *m (pap)* stock screen
Ganzzeug *n s.* Ganzstoff
Ganzzeugbereitung *f s.* Ganzstoffaufbereitung
Ganzzeugholländer *m (pap)* Hollander [beater, beating engine], pulp engine (grinder), stuff engine
Ganzzeugmahlung *f (pap)* beating of stock
gar unctuous *(soil)*
Gärbottich *m* fermentation cask (vat, tub, vessel), fermenter
Garbrand *m (ceram)* maturing, soaking, final firing
Garbrandbereich *m (ceram)* maturing range
Garbrandtemperatur *f (ceram)* maturing (soaking) temperature
garbrennen *(ceram)* to mature, to soak
Garbrennen *n s.* Garbrand
Gärbstahl *m* shear steel, refined steel (iron, bar), merchant bar
Gärdauer *f* fermentation time
Gardine *f (coat)* curtain *(a film fault)*
Gardinenbildung *f* curtaining, sagging *(of surface coatings)*
Gardschanbalsam *m* gurjun balsam *(from Dipterocarpus alatus Roxb.)*
Gare *f* unctuousness, [good] tilth *(of soil)*
gären to ferment, to yeast
gärfähig fermentable
Gärfähigkeit *f* fermentability
Gärflüssigkeit *f* fermentable liquid, wash
Gärführung *f* fermentation method
Gargarisma *n (pharm)* gargarism, gargle
Gärkeller *m* fermenting (fermentation) cellar
Gärkelleranlage *f* fermentation plant
Gärkellerausbeute *f* fermentation cellar output
Gärkolben *m* fermentation flask
Gärkraft *f* fermentative (fermenting) power
Garkupfer *n* refined (casting, tough-pitch) copper
Garlauge *f (fert)* refining lye
Gärlösung *f s.* Gärflüssigkeit
Garn *n (text)* spun[-staple] yarn, yarn
G. aus Rohseidenabfällen spun silk
einfädiges G. single yarn
hochvoluminöses G. high-bulk yarn
umsponnenes G. core spun yarn
Garnfärbeapparat *m* yarn-dyeing machine
Garnfärben *n* yarn dyeing
Gärniederschlag *m* cloud
garnieren *(ceram)* to handle, to stick up
Garnierer *m (ceram)* handler
Garnierit *m (min)* garnierite *(nickel silicate)*

Garniermaschine *f (ceram)* handle sticking machine
Garnkörper *m (text)* bobbin, package
Garnnummer *f* yarn count (number, size)
Gärprobe *f* fermentation test
Gärprozeß *m* fermentation process
Garpunkt *m (ceram)* maturing point
Gärraum *m* fermenting (fermentation) room
Gärreduktaseprobe *f* fermentation reductase test
Gärröhrchen *n* fermentation tube
  G. nach Durham Durham tube
Gärröhre *f s.* Gärröhrchen
Gärsalz *n* fermentation salt
Garschaum[graphit] *m* keesh, kish
Gärspund *m* fermentation bung
Gärtank *m* fermentation (fermenting) tank (vessel), fermenter
Gartemperatur *f (ceram)* maturing (soaking) temperature
Gärtemperatur *f* fermentation temperature, fermenter set temperature
Gärung *f* fermentation
  aerobe G. aerobic fermentation
  alkoholische G. alcoholic fermentation
  anaerobe G. anaerobic fermentation
  bakterielle G. bacterial fermentation
  kochende G. boiling fermentation
  milchsaure G. lactic[-acid] fermentation
  offene G. open fermentation
  oxydative G. aerobic fermentation
  schleimige G. ropy (slime) fermentation
  zellfreie G. cell-free fermentation
Gärungs ... *s.a.* Gär ...
Gärungsalkohol *m* fermentation alcohol
Gärungsamylalkohol *m* fermentation amyl alcohol
Gärungschemie *f* fermentation chemistry, zymurgy
Gärungsenzym *n* fermentation enzyme
gärungserregend fermentative, zymogenic, zymogenous
Gärungsessig *m* fermentation vinegar
Gärungsferment *n* fermentation enzyme
Gärungsgewerbe *n* fermentation industry
Gärungsgleichung *f* / **Harden-Youngsche** Harden-Young fermentation equation
gärungshemmend antifermentative
Gärungsindustrie *f* fermentation industry
Gärungsmilchsäure *f* $CH_3CH(OH)COOH$ lactic acid of fermentation, *DL*-lactic acid
Gärungsschaum *m* bloom
Gärungstechnik *f* zymotechnics, fermentation technology
gärungstechnisch zymotechnic[al]
Gärungstechnologie *f* fermentation technology
gärungsverhindernd antifermentative
Garungszeit *f (coal)* coking time
Gärverfahren *n* fermentation method
Gärverlust *m* fermentation loss
Gärvermögen *n* fermentative (fermenting) power
Gas *n* gas
  aufkohlendes G. *(met)* carburizing gas
  brennbares G. combustible gas
  gelöstes G. gas in solution
  ideales G. ideal (perfect) gas
  inertes G. inert (inactive) gas
  kohlenwasserstoffhaltiges G. hydrocarbon gas
  künstlich hergestelltes G. manufactured gas
  permanentes G. permanent gas
  reales G. real (actual) gas
  reiches G. rich gas
  staubhaltiges G. dust-laden gas
  technisches G. manufactured gas
  überschüssiges G. surplus gas
  verflüssigtes G. liquefied gas
  vollkommenes G. ideal (perfect) gas
  wirkliches G. real (actual) gas
Gasabführung *f* gas offtake
Gasabgang *m* gas outlet
Gasabgangsrohr *n* gas outlet pipe
Gasableitungsrohr *n* gas offtake pipe
Gasabscheider *m* gas separator
Gasabscheidung *f* 1. gas separation; 2. *s.* Gasentwicklung
Gasabtrennung *f* gas separation
Gasabzug *m* gas offtake
Gasabzug[s]rohr *n* gas offtake pipe; fume pipe
  fallendes G. downcomer
  steigendes G. gas uptake
Gasadsorptionschromatografie *f* gas adsorption chromatography
Gasanalyse *f* gas analysis
Gasanalysenapparat *m* **nach Orsat** Orsat gas [analysis] apparatus, Orsat analyzer
gasanalytisch gas-analytical
Gasanreicherung *f* gas enrichment
Gasanstalt *f s.* Gaswerk
Gasanzünder *m* burner lighter
gasartig gaseous
Gasaufbereitung *f* gas treating (treatment)
gasaufkohlen to gas-carburize *(steel)*
Gasaufkohlung *f* gas carburizing *(of steel)*
Gasaufnahme *f* gas absorption
Gasausbeute *f*, **Gasausbringung** *f* gas output
Gasaushauchung *f* outgassing *(of a volcano)*
Gasaustritt *m* 1. escape of gas; 2. gas outlet
Gasaustrittsöffnung *f* gas outlet
Gasaustrittstemperatur *f* exit gas temperature
Gasbehälter *m* gas tank (container), *(esp for town gas)* gasholder
gasbeheizt gas-fired, gas-heated
Gasbeheizung *f* gas[-fired] heating, gas-fuel firing
gasbeständig resistant to gases
Gasbeständigkeit *f* resistance to gases
Gasbeton *m* gas (gassy) concrete
gasbildend gas-forming
Gasbildung *f* formation of gas
Gasbläschen *n* gas bubble
Gasblase *f* gas bubble; *s.* Gaseinschluß
Gasbleiche *f (pap)* gas bleaching
Gasbohrung *f* gas well
Gasbrenner *m* gas burner
Gasbrunnen *m* gas well
Gasbürette *f* gas burette

Hempelsche G. Hempel gas burette
Gaschromatograf *m* gas chromatograph
Gaschromatografie *f* gas chromatography
G. mit Temperaturprogramm programmed temperature gas chromatography
Gaschromatogramm *n* gas chromatogram
Gascoulometer *n* gas coulometer
gasdicht gastight; gasproof
Gasdichte *f* gas density
Gasdichtemessung *f* measurement of gas density
Gasdichtewaage *f* gas [density] balance
Gasdichtigkeit *f* gastightness
Gasdichtung *f* gas seal
Gasdiffusionsverfahren *n* gaseous diffusion method *(for separating isotopes)*
Gasdispersion *f* gas dispersion
Gasdruck *m* gas pressure (drive)
Gasdurchflußmesser *m* gas flowmeter
Gasdurchgang *m* gas passage
gasdurchlässig permeable to gas
Gasdurchlässigkeit *f* permeability to gas
Gasdurchlässigkeitszahl *f (met)* permeability number *(of moulding sand)*
Gäse *f* seed *(a defect in glass)*
Gase *npl* / **nitrose** nitrous gases
gasecht *(text)* fast to burnt gas fumes, fast to gas fading
Gasechtheit *f (text)* gas-fume fastness, fastness to burnt gas fumes, fastness to gas fading
Gaseinpressen *n* gas injection
Gaseinschluß *m* blister *(a flaw in material; foundry also)* gas cavity, blow hole
gaseinsetzen *(met)* to gas-carburize
Gaseinsetzen *n (met)* gas carburizing
Gaseintritt *m* gas inlet
Gaseintrittstemperatur *f* inlet gas temperature
Gaselektrode *f* gas electrode
Gaselement *n (el chem)* gas cell
gasen 1. to gas; 2. to steam *(in producing water gas)*
abwärts g. to steam downwards
aufwärts g. to steam upwards
von oben g. *s.* abwärts g.
von unten g. *s.* aufwärts g.
Gasen *n* 1. gassing; 2. steaming, make, run *(in producing water gas)*
G. in absteigender Richtung down-steaming, down-run[ning]
G. in aufsteigender Richtung up-steaming, up-run[ning]
Gasentartung *f (phys chem)* gas degeneracy *(near absolute zero)*
Gasentladungsplasma *n* plasma of gas discharge, discharge plasma
Gasentladungsröhre *f* [gas] discharge tube
Gasentlösungsdruck *m (petrol)* dissolved-gas drive, solution gas drive
Gasentlösungslagerstätte *f (petrol)* solution, gas-drive reservoir, depletion-type reservoir
Gasentnahme *f* gas offtake
Gasentwickler *m* gas generator
Kippscher G. Kipp [gas] generator, Kipp's apparatus
Gasentwicklung *f* generation (evolution) of gas, gassing
Gaserzeuger *m* gas producer (generator)
G. mit Treppenrost step-grate producer
Gaserzeugung *f* gas manufacture (making, production)
Gaserzeugungsanlage *f* gas-making plant, gas plant
Gaserzeugungsverfahren *n* gas-making process
G. nach Didier-Bubiag Didier-Bubiag process
G. von Ahrens Ahrens process
Gas-Fest[stoff]-Chromatografie *f* gas-solid chromatography
Gasfeuerung *f* gas-fuel firing, gas heating
Gasfilter *n* gas filter, *(lab also)* gas filtering (filtration) tube
Gasfiltration *f* gas filtration
Gasflamme *f* gas flame
Gasflammkohle *f* gas flame coal
Gasflasche *f* gas cylinder
Gas-Flüssig[keits]-Chromatografie *f* gas-liquid chromatography
Gasförderung *f* gas lift[ing]
gasförmig gaseous
gasführend gas-bearing
Gas-Furnace-Verfahren *n (rubber)* gas furnace process
Gasgehalt *m* gas content
Gasgemisch *n* gas mixture
Gasgenerator *m* gas producer (generator)
Gasgesetz *n* gas law
Gasgleichgewicht *n* gas equilibrium
Gasgleichung *f* gas equation
Gasglühkörper *m*, **Gasglühlichtstrumpf** *m* gas mantle
Gashahn *m* gas [stop]cock
Gasheber *m* gas lift
Gasheizkranz *m* ring burner
Gasheizung *f* gas[-fired] heating, gas-fuel firing
Gasherstellung *f s.* Gaserzeugung
Gashydrat *n* gas hydrate *(any of a group of clathrate compounds)*
Gashydratbildung *f* formation of gas hydrates
gasieren *(text)* to gas
Gasieren *n (text)* gassing, gas singeing
Gasindustrie *f* gas industry
Gasinjektion *f* gas injection
Gasion *n* gaseous ion
Gaskalk *m (agric)* gas lime
Gaskalorimeter *n* gas calorimeter
Gaskammer *f* gas cell *(spectroscopy)*
Gaskammerofen *m (ceram)* gas chamber kiln
Gaskappe *f (petrol)* gas cap
Gaskappendruck *m (petrol)* gas cap drive
Gaskappenlagerstätte *f* gas cap-drive field (reservoir)
Gaskette *f (el chem)* gas cell
Gaskohle *f* gas[-making] coal
Gaskohlung *f* gas carburizing *(of steel)*
Gaskohlungsofen *m* gas-carburizing furnace

Gaskoks *m* gas coke
Gaskomponente *f* gaseous component
Gaskompressor *m* gas compressor
Gaskonstante *f* gas constant
allgemeine (molare) G. universal gas constant
Gaskonzentration *f* gas concentration (strength)
Gaskopf *m (petrol)* gas cap
Gaskühler *m* gas cooler
Gaskühlung *f* gas cooling
Gaslagerstätte *f* gas field (reservoir)
Gasleitung *f* gas main
Gaslift *m* gas lift
Gasliften *n*, **Gasliftförderung** *f* gas lift[ing]
Gasliftventil *n* gas-lift valve
Gasliftverfahren *n* gas-lift method
Gas-Liquidus-Chromatografie *f* gas-liquid chromatography
Gaslöslichkeit *f* gas solubility
Gasmaske *f* gas mask
Gasmesser *m* gas meter
nasser G. wet gas meter
trockener G. dry gas meter
Gasmuffelofen *m (ceram)* gas muffle kiln
Gasnitrieren *n* gas nitriding *(of steel)*
Gasöl *n* gas oil
Gasöldestillat *n* gas-oil distillate
Gas-Öl-Separator *m (petrol)* oil/gas separator, gas separator
mehrstufiger G. multistage separator
Gas-Öl-Trennung *f (petrol)* gas separation
Gas-Öl-Trennvorrichtung *f s.* Gas-Öl-Separator
Gas-Öl-Verhältnis *n* gas/oil ratio
Gasphase *f* gas phase; vapour phase *(in high-pressure hydrogenation)*
Gasphasechlorierung *f* gas-phase chlorination
Gasphasehydrierofen *m* vapour-phase converter
Gasphasehydrierung *f* vapour-phase hydrogenation
Gasphaseinhibitor *m* vapour-phase inhibitor, V.P.I.
Gasphaseisomerisierung *f* vapour-phase isomerization
Gasphasekatalysator *m* vapour-phase catalyst
Gasphasekracken *n* vapour-phase cracking
Gasphasen ... *s.* Gasphase ...
Gasphaseofen *m* vapour-phase converter
Gasphaseverfahren *n* vapour-phase process
Gaspipette *f* gas pipette
Hempelsche G. Hempel gas pipette
Gasquelle *f* gas well
Gasraum *m* 1. gas space; 2. *(pap)* top *(of a digester)*
Gasreaktion *f* gas reaction
Gasreiniger *m* gas cleaner (purifier)
Gasreinigung *f* gas cleaning (clarification, purification)
G. durch Ultraschall ultrasonic agglomeration
akustische G. sonic agglomeration
Gasreinigungsmasse *f* gas cleaning (gas-purifying) material
Gasretorte *f* gas retort
Gasretortenkoks *m* gas-retort coke
Gasretortenteer *m* gas-retort tar
Gasruß *m* gas black (soot)
Gassammelleitung *f* gas collecting main *(of a coke-oven battery)*
Gassammelröhre *f* gas collection (collecting) tube, gas sampling tube (pipette)
Gassäule *f* gas column
Gasscheider *m s.* Gasseparator
Gasschieber *m* gas valve
Gasschiefer *m* cannel (candle, jet) coal
Gas[schmelz]schweißen *n* gas (autogenous) welding
Gasschwund *m (text)* gas fading
Gassengen *n (text)* gas singeing, gassing
Gassengmaschine *f (text)* gas-singeing machine
Gasseparator *m (petrol)* oil/gas separator, gas separator
mehrstufiger G. multistage separator
Gas-Solidus-Chromatografie *f* gas-solid chromatography
Gasspüler *m* bubbler
Gasspürgerät *n* gas detector
G. für Halogenide halide leak detector
Gasstrom *m* gas stream (flow)
steigender G. stream of upward moving gas
Gasströmung *f* gas flow
Gast *m* guest *(in inclusion compounds)*
Gastechnik *f* gas technology
Gasteer *m* gas[works] tar
Gastelement *n* guest element
Gastheorie *f* / **kinetische** kinetic theory of gases
Gasthermometer *n* gas thermometer
G. konstanten Drucks constant-pressure gas thermometer
G. konstanten Volumens constant-volume gas thermometer
G. konstanter Dichte *s.* G. konstanten Volumens
Gastkomponente *f* guest component
Gastkristall *m* chadacryst
Gastrennanlage *f (petrol)* oil/gas separator, gas separator
mehrstufige G. multistage separator
Gastrennung *f* gas separation
Gastrieb *m (petrol)* gas drive
Gastrieblagerstätte *f* gas-drive field (reservoir)
gasundurchlässig impervious (impermeable) to gas, gastight; gasproof
Gasung *f s.* Gasen 2.
Gasventil *n* gas valve
Gasventilator *m* gas fan
Gasverdichter *m* gas compressor
Gasverflüssigung *f* gas liquefaction
Gasvergiftung *f* gas poisoning, gassing
Gasverteilleitung *f* distribution gas main
Gasverteilungschromatografie *f* gas-liquid partition chromatography
Gasverteilungsfritte *f* gas diffuser stone, air stone, sintered bubbler
Gasvolumeter *n* gas volumeter
Gasvolumetrie *f* gas volumetry
Gasvorlage *f* gas collecting main *(of a coke-oven battery)*

Gaswaage *f* gas [density] balance, dasymeter
Gaswäsche *f* 1. gas washing (scrubbing), wet gas cleaning; 2. gas-washing system
Gaswäscher *m* gas washer, scrubbing tower, scrubber
Gaswaschflasche *f* gas-washing bottle
Drechselsche G. Drechsel bottle
Gaswaschsystem *n* gas-washing system
Gaswaschturm *m* gas-washing tower
Gaswasser *n* ammonia water (liquor), ammoniacal (gas) liquor
Gaswechselquotient *m* gas exchange quotient
Gaswerk *n* gasworks, gas-[making] plant
Gaswerkskoks *m* gas coke
Gaswerksretorte *f* gas retort
Gaswerksteer *m* gas[works] tar
Gaszähler *m* gas meter
nasser G. wet gas meter
trockener G. dry gas meter
Gaszelle *f (el chem)* gas cell
gaszementieren to gas-carburize *(steel)*
Gaszementieren *n* gas carburizing *(of steel)*
Gaszentrifuge *f* gas centrifuge
Gaszentrifugenverfahren *n* gas centrifugation, gas centrifuge method *(for separating isotopes)*
Gaszuführung *f* gas supply (inlet) *(process)*; gas inlet *(junction)*
Gaszuführungsrohr *n* gas feed pipe (tube), gas inlet pipe, gas conductor
Gaszuleitung *f s.* Gaszuführung
Gaszustand *m* gaseous condition
gaszyanieren to gas-cyanide, to dry-cyanide, to carbonitride, to dry-nitride *(steel)*
Gaszyanieren *n* gas (dry) cyaniding (cyanization), carbonitriding, nitrocementation, ni-carbing *(of steel)*
Gatsch *m* [paraffin] slack wax
Gattermann-Koch-Synthese *f* Gattermann-Koch reaction *(for preparing phenolic aldehydes)*
Gattermann-Reaktion *f* Gattermann reaction *(for preparing halogen-substituted aromatic compounds)*
Gatterrührer *m* gate agitator (mixer), *(with horizontal paddles between stationary fingers also)* shear-bar agitator (mixer)
Gattungsname *m* generic name (term)
Gaufrage *f (pap, text, tann)* embossing, goffering
gaufrieren *(pap, text, tann)* to emboss, to goffer
Gaufrierkalander *m* embossing (goffering) calender
Gautschbrett *n (pap)* couch
Gautschbruchbütte *f (pap)* couch box (pit)
Gautsche *f (pap)* couch press
gautschen *(pap)* to couch, to line, to laminate
Gautscher *m (pap)* couchman, coucher
Gautschpresse *f (pap)* couch press
Gautschwalze *f (pap)* couch[-press] roll, couching roll
obere G. top couch-press roll
untere G. bottom couch-press roll
Gay-Lussac-Gesetz *n* Gay-Lussac law
Gay-Lussac-Turm *m* Gay-Lussac tower
Gaylussit *m (min)* gay-lussite, gaylussite *(calcium sodium carbonate)*
Gaze *f* gauze
G-Band *n s.* Grundband
geäschert / in der Grube *(tann)* pit-limed
gebändert *(geol)* banded
geben / Körper *(coat)* to body
Gebiet *n* region, range *(as of a diagram or scale)*
G. der Normalbelichtung *(phot)* region of correct (normal) exposure, straight[-line] portion, straight line *(of the characteristic curve)*
G. der Röntgenstrahlen X-ray region (range)
G. der Überbelichtung *(phot)* region of overexposure, shoulder, knee *(of the characteristic curve)*
G. der Unterbelichtung *(phot)* region of underexposure, toe, foot *(of the characteristic curve)*
geradliniges G. *s.* G. der Normalbelichtung
plastisches G. plastic range
Gebirge *n (mine)* ground
Gebläse *n* fan, [air] blower, air-blast system
Gebläsebrenner *m* blowlamp, blowtorch, bench blowpipe, blast (nozzle-mix) burner, blast (glass blower's) lamp
Gebläseentstauber *m* dust collecting fan, fan impeller collector
Gebläselampe *f s.* Gebläsebrenner
Gebläseluft *f s.* Gebläsewind
Gebläsemischen *n* air (gas) agitation
Gebläsemischer *m* air-agitated mixer
geblasen / vor der Lampe *(glass)* lamp-blown, lampworked
Gebläse[schacht]ofen *m* blast furnace *(for obtaining iron)*
G. für NE-Metalle non-ferrous blast furnace
Gebläsewind *m* [air] blast
Gebläsewindtemperatur *f* blast temperature
Gebräu *n* brew
Gebrauchsanweisung *f* direction[s] for use
Gebrauchsdosis *f (pharm)* usual dose
Gebrauchsechtheit *f (text)* fastness to wearing
Gebrauchseigenschaft *f (text)* wearing quality
gebrauchsfertig ready for use, ready-to-use
Gebrauchsleistung *f* performance
Gebrauchstemperatur *f* service temperature
Gebrauchswasser *n s.* Brauchwasser
gebraucht spent *(e.g. solution)*
gebunden combined; bonded, bound, linked, connected *(chemical-bond theory)*
chemisch g. chemically bonded
doppelt g. doubly bonded
dreifach g. triply bonded
einfach g. singly bonded
einpolar (homöopolar) g. *s.* kovalent g.
keramisch g. ceramic-bonded
koordinativ g. coordinate
kovalent g. covalently bonded
organisch g. organically bonded
unpolar g. *s.* kovalent g.
vierfach g. quadruply bonded
zweifach g. doubly bonded

Gedächtniseffekt *m (nucl)* memory effect
gedeckt / einseitig *(pap)* single-lined
zweiseitig g. double-lined
gediegen *(min, met)* virgin, elemental, native
gedoktert *(petrol)* doctor-sweet
geeicht / auf Ablauf (Auslauf) calibrated (graduated) to deliver, calibrated for delivery
auf Einguß g. calibrated to contain, calibrated for content
Geer-Alterung *f (rubber)* Geer oven ageing
Geer-Alterungsprüfung *f (rubber)* Geer oven test
Geer-Ofen *m (rubber)* Geer oven
Geer-Ofenalterung *f (rubber)* Geer oven ageing
Gefahrenklasse *f* danger class
Gefälle *n* drop, slope; *(quantitatively)* gradient
Gefällezuführung *f* gravity feed
gefärbt dyed, coloured
dunkel g. dark-coloured
im Holländer (Stoff) g. *(pap)* pulp-coloured, dyed in the beater (stuff)
in der Flocke g. *(text)* stock-dyed
in der Masse g. *s.* im Holländer g.
in der Wolle g. *(text)* stock-dyed
schwach g. feebly (weakly) coloured
Gefäß *n* vessel, receptacle, container, *(made of glass or earthenware also)* jar, *(if spherical also)* bulb, *(if large also)* vat
Geißlersches (Mohrsches) G. Geissler-Mohr absorption (potash) bulb
Weinhold-Dewarsches G. Dewar flask (jar), Dewar
gefäßerweiternd *(pharm)* vasodilating
Gefäßerweiterungsmittel *n (pharm)* vasodilator, vasodepressor
gefäßkontrahierend *s.* gefäßverengend
Gefäßofen *m (met)* vessel (closed-vessel) furnace
gefäßverengend *(pharm)* vasoconstrictive
Gefäßverengungsmittel *n (pharm)* vasoconstrictor
Gefäßversuch *m (agric)* pot experiment (study, test)
Gefluder *n*, **Gefluter** *n (min tech)* sluice
geformt / von Hand *(ceram)* hand-moulded
Gefrierapparat *m* freezing apparatus, freezer
gefrierbar freezable, congealable
Gefrier-Brandzeichen *n (tann)* freeze brand
gefrieren to freeze, to congeal
g. lassen to freeze, to congeal
schnell g. to quick-freeze
Gefrieren *n* freezing, congealing, congelation
G. in bewegter Luft blast (air-blast) freezing
langsames G. slow freezing
schnelles G. quick (rapid, fast, sharp) freezing
Gefrierfleisch *n* chilled (frozen, cold-storage) meat
Gefriergeschwindigkeit *f* freezing rate (velocity)
Gefrierkonservierung *f* freezing preservation, cold-pack method
Gefrierkurve *f* freezing[-point] curve
Gefrierlagerung *f* frozen storage
Gefriermikrotom *n* freezing microtome
Gefrierpunkt *m* freezing point (temperature)
Gefrierpunktmesser *m* cryoscope
Gefrierpunktsdepression *f s.* Gefrierpunktserniedrigung
Gefrierpunktserniedrigung *f* freezing-point depression (lowering)
molale (molare, molekulare) G. molal freezing-point[-depression] constant, molal (molar, molecular) depression constant, cryoscopic constant
Gefrierschutzmittel *n* antifreeze [agent, compound, fluid], antifreezing agent (dope)
Gefriertemperatur *f s.* Gefrierpunkt
gefriertrocknen to freeze-dry, to lyophilize, to dehydrofreeze
Gefriertrockner *m* freeze dryer (drying apparatus)
Gefriertrocknung *f* freeze drying, lyophilization, dehydrofreezing ✦ **durch G. haltbar machen** *(food)* to dehydrofreeze
Gefriertrocknungsanlage *f s.* Gefriertrockner
Gefrierverfahren *n* freezing process
Gefüge *n* 1. *(met)* [grain] structure, grain; 2. *(geol)* rock fabric, *(relating to the larger features)* structure, *(relating to the smaller features)* texture
G. der oberen Zwischenstufe *(met)* upper bainite
G. der unteren Zwischenstufe *(met)* lower bainite
G. der Zwischenstufe *(met)* bainite [structure], bainitic structure
bainitisches G. *s.* G. der Zwischenstufe
dichtes (feinkörniges) G. fine-grain structure
Gefügebestandteil *m* structural constituent, *(coal also)* maceral
Gefügekunde *f (geol)* petrofabrics
gefügelos structureless, textureless, devoid of structure (texture)
gefüllt / mit Füllstoffen *(pap, tann)* filled, loaded; *(rubber)* loaded, pigmented
mit Hackschnitzeln g. *(pap)* chip-filled *(digester)*
gegenblasen *(glass)* to blow back
Gegenblasen *n (glass)* counter blow
Gegendruck *m* back pressure
Gegendruckabfüllapparat *m*, **Gegendruckfüller** *m* counterpressure (isobarometric) filler (racker)
gegeneinanderlaufend contrarotating *(rolls)*
Gegenelektrode *f* counterelectrode
Gegen-EMK *f* counter (back) electromotive force, back e.m.f.
Gegenfluß *m s.* Gegenstrom
Gegengewicht *n s.* Gegenmasse
Gegengift *n s.* Gegenmittel
Gegenion *n* counterion, gegenion, compensating ion
gegenläufig counterrotating, contrarotating
Gegenmasse *f* counterweight, counterpoise
Gegenmittel *n (tox)* antidote
Gegenreaktion *f* reverse reaction
Gegenstrom *m* countercurrent [flow], counterflow
✦ **im G.** countercurrent[ly], in countercurrent, in counterflow
Gegenstromauswaschung *f* countercurrent washing

Gegenstromdekantation *f* countercurrent decantation
Gegenstromdestillation *f* countercurrent distillation, rectification
diskontinuierliche G. batch rectification
kontinuierliche (stetige) G. continuous rectification
unstetige G. batch rectification
Gegenströmer *m s.* Gegenstromwärmeaustauscher
Gegenstromextraktion *f* countercurrent extraction
Gegenstromextraktionsapparat *m* countercurrent contactor
G. nach Podbielniak Podbielniak centrifugal [countercurrent] contactor, Podbielniak contactor (machine)
Gegenstromführung *f* backward-feed operation *(as of multiple-effect evaporators)*
Gegenstromhydrolyse *f* countercurrent hydrolysis
Gegenstromklassieren *n* countercurrent classification
Gegenstromklassierer *m* countercurrent classifier
Gegenstromkondensator *m*, **Gegenstromkühler** *m* countercurrent condenser (cooler)
Gegenstromkühlturm *m* countercurrent cooling tower
Gegenstromprinzip *n* countercurrent principle ✦ **nach dem G.** countercurrent[ly], in countercurrent, in counterflow
Gegenstromsystem *n* countercurrent system
Gegenstromtrockner *m* countercurrent dryer
Gegenstromverdichter *m s.* Gegenstromkondensator
Gegenstromverteilung *f* countercurrent distribution
Gegenstromwärmeaustauscher *m* countercurrent heat exchanger
Gegenstromwäsche *f* countercurrent washing
Gegenuhrzeigersinn / im anticlockwise
Gegenurspannung *f* counter (back) electromotive force, back e.m.f.
Gehalt *m* content, percentage, concentration; loading *(of a gas with solid particles)* ✦ **auf G. prüfen** *(pharm, met)* to assay
G. an aktivem Bleichchlor (Chlor) *(pap)* available chlorine content
G. an Flüchtigem (flüchtigen Bestandteilen) volatile [matter] content
G. an freien Fettsäuren free-fatty-acid content
G. an Gesamttrockenmasse total-solids content
G. an Trockenmasse (Trockensubstanz) solids (solid, dry) content
G. an wirksamem Bleichchlor (Chlor) *(pap)* available chlorine content
prozentualer G. percentage
α-Gehalt *m (pap)* alpha cellulose content
gehärtet / ungenügend *(plast)* undercured
Gehäuse *n* casing, case, *(esp of pumps, motors, bearings)* housing, *(if rectangular also)* box, *(if domed, hemispherical or spherical also)* shell
Gehäusemesser *npl (pap)* shell bars, bars in the shell, bars on the casing *(of a perfecting engine)*
Geheimtinte *f* sympathetic (secret) ink
gehen:
g. lassen to raise, to leaven *(dough)*
in Lösung g. to go into solution, to dissolve
zugrunde g. *(nucl)* to decay
gehindert / sterisch sterically hindered
Gehlenit *m (min)* gehlenite *(a sorosilicate)*
Gehman-Test *m (rubber)* Gehman torsion test
Gehölzvernichtungsmittel *n* brushkiller, silvicide
Geierrinde *f (pharm)* cundurango bark *(from Marsdenia reichenbachi Triana)*
Geigenharz *n* pine rosin, colophony *(from Pinus specc.)*
Geiger-Müller-Zählrohr *n (nucl)* Geiger[-Müller] counter, Geiger[-Müller] tube, G-M counter (tube)
Geiger-Nuttall-Beziehung *f (nucl)* Geiger-Nuttall rule
Geiger-Zähler *m s.* Geiger-Müller-Zählrohr
Geißler-Röhre *f* Geissler tube
Geister *mpl* ghosts *(false images of a spectral line)*
Geisterbild *n (phot)* ghost image
Geistersalz *n s.* Hirschhornsalz
gekocht / alkalisch *(pap)* alkaline-cooked
gekörnt granular, granulate[d], grained, grainy
Gekrösefett *n* mesenteric (ruffle) fat
Gekrösestein *m (min)* tripestone *(calcium sulphate)*
Gel *n (coll)* gel
ionotropes G. ionotropic gel
resolubles (reversibles) G. reversible gel
thixotropes G. thixotrope
unelastisches G. rigid gel
geladen / einfach singly charged
einfach negativ g. uninegative
einfach positiv g. unipositive
negativ g. negatively charged, negative
positiv g. positively charged, positive
zweifach g. doubly charged
zweifach negativ g. dinegative
zweifach positiv g. dipositive
Geläger *n (ferm)* cloud
gelartig *(coll)* gel-like, gelatinous
Gelatine *f* gelatin[e]
gekörnte G. kibbled gelatin
gelatineartig gelatinous, gelatiniform
Gelatineeffekt *m (phot)* gelatin effect
Gelatineemulsion *f (phot)* gelatin emulsion
gelatinegeleimt *(pap)* gelatin-sized, glue-sized, animal-sized, animal tub-sized
gelatinehaltig gelatinous
Gelatinekultur *f* gelatin culture
Gelatineleimung *f (pap)* gelatin (glue, animal) sizing, animal tub-sizing
Gelatineschicht *f (phot)* gelatin layer
Gelatineschutzschicht *f (phot)* gelatin protective layer
Gelatinesol *n* gelatin sol

gelatinieren to gelatinize, to gelatinate, to gel[ate], to jellify, to jelly
Gelatiniermittel *n* gelatinizing (gelling) agent, gelatinizer
Gelatinierung *f* gelatin[iz]ation, gelatification, gelation, jellification
Gelatinierungstemperatur *f* gel point, gelation temperature
gelatinös gelatinous, gelatiniform
Gelb *n* yellow *(sensation or substance)*
  Kasseler G. Cassel (Turner's, Verona, mineral) yellow *(a basic lead chloride)*
  Steinbühler G. Steinbühl yellow, yellow ultramarine, lemon chrome, gelbin *(barium chromate)*
  Turners G. *s.* Kasseler G.
Gelbätze *f s.* Gelbbeize
Gelbbeere *f (dye)* buckthorn (yellow) berry *(from Rhamnus specc.)*
  Französische G. French berry, Avignon berry (grain) *(from Rhamnus infectorius L.)*
  Persische G. Persian berry *(from Rhamnus specc.)*
Gelbbeize *f (glass)* yellow (silver) stain
Gelbbleierz *n (min)* yellow lead ore, wulfenite *(lead molybdate)*
Gelbglas *n* yellow arsenic [sulphide], king's yellow (gold), arsenic (royal) yellow, orpiment [yellow] *(technically pure arsenic(III) sulphide)*
Gelbglut *f* yellow heat
Gelbguß *m* yellow brass
Gelbharz *n* Botany Bay gum *(from Xanthorrhoea hastilis R. Br.)*
Gelbholz *n* fustic
  Echtes G. old fustic *(from Chlorophora tinctoria Gaud.)*
  Kubanisches G. Cuba wood *(a sort of old fustic)*
  Ungarisches G. fustet, young fustic *(from Cotinus coggygria Scop.)*
Gelbholzextrakt *m* fustic extract
Gelbildung *f* gel formation, gelling, gelation, jellification
Gelbin *n* gelbin *(calcium chromate)*
Gelbstich *m*, **Gelbstichigkeit** *f* yellow cast, yellowish tinge
Gelbstoff *m* fulvic acid *(any of several water-soluble humic acids)*
Gelbstroh *n (pap)* straw
Gelbstrohstoff *m (pap)* coarse straw pulp, [yellow mechanical] straw pulp
Gelbton *m* yellow shade
Gelbware *f (ceram)* yellow (cane) ware
Gelbwerden *n (pap)* yellowing, discolo[u]ration
Gelbwurz[el] *f* turmeric, curcuma, Indian saffron, Curcuma longa L.
Gelchromatografie *f* gel [permeation] chromatography, GPC
Gelee *n* jelly ✦ **in G. überführen** to gelatinize, to gelatinate, to jellify, to jelly ✦ **zu G. erstarren** to jelly, to jellify, to gel[ate], to gelatinate, to gelatinize
Gelée royale *f (pharm)* royal jelly, queen-bee's nutrient jelly
geleimt *(pap)* sized
  doppelt g. double-sized
  mit Gelatine g. gelatin-sized, glue-sized, animal-sized, animal tub-sized
  mit Natronwasserglas g. silicate-sized
  mit Tierleim g. *s.* mit Gelatine g.
$^1/_2$geleimt half-sized, $^1/_2$ sized
$^1/_4$geleimt quarter-sized, $^1/_4$ sized
Gelenkschmiere *f (med)* synovial fluid
Gelfiltration *f* gel filtration *(chromatography)*
Geliereinheit *f* **des Pektins** pectin grade
gelieren to gelatinize, to gelatinate, to gel[ate], to jellify, to jelly
Geliergrad *m* **des Pektins** pectin grade
Gelierkanal *m (text)* gelatification oven
Geliermittel *n*, **Gelierstoff** *m* gelatinizing (gelling) agent, gelatinizer
Gelierung *f* gelatin[iz]ation, gelatification, gelation, jellification
Gelierungstemperatur *f* gel point, gelation temperature
Gelose *f* gelose, *(broadly)* agar [gel], agar-agar, Japan agar (isinglass), Chinese gelatin
Gelöstes *n* dissolved substance, solute
Gel-Permeations-Chromatografie *f* gel [permeation] chromatography, GPC
Gelsäule *f* gel column
Gelseminsäure *f* gelseminic acid, scopoletin, 7-hydroxy-6-methoxycoumarin
Gel-Sol-Gel-Umwandlung *f* gel-sol-gel transformation
Gel-Sol-Übergang *m*, **Gel-Sol-Umwandlung** *f* peptization, solation
Gelteilchen *n* gel particle
Gelteilchengehalt *m (plast)* gel content
Gelteilchenzählung *f (plast)* gel count
Gelzeit *f* gel[ling] time
Gelzustand *m (coll)* gel state (condition) ✦ **in den G. übergehen** to gel
Gemenge *n* 1. bulk blend, mixture, mix; 2. *(glass)* batch
  scherbenfreies G. *(glass)* raw batch
Gemengehaus *n (glass)* batch house
Gemengemischer *m (glass)* batch mixer
Gemengesatz *m (glass)* batch formula
Gemengespeiser *m (glass)* batch charger (feeder)
Gemengestein *m (glass)* batch stone *(a defect)*
Gemengteil *m (min)* constituent, component; ingredient *(of a bulk blend)*
  akzessorischer G. *(min)* accessory constituent (component)
Gemisch *n* mixture, mix
  azeotrop[isch]es G. azeotropic mixture, azeotrope
  binäres G. binary mixture
  dystektisches G. dystectic mixture *(a mixture having a maximum melting point)*
  eutektisches G. eutectic [mixture]
  Johnsons G. Johnson's mixture *(a pesticide)*
  optisch-inaktives G. optically inactive compound
  razemisches G. racemic compound

reduzierendes G. reduction mixture
thermochemisches G. thermochemical mixture
tonfreies G. *(ceram)* non-clay body

Gemischtphase *f (petrol)* mixed phase
Gemischtphasekracken *n (petrol)* mixed-phase cracking
gemittelt average
Gemüsesaatschutz *m* vegetable-seed protection
Gemüsesaft *m* vegetable juice
Genauguß *m* precision casting
Genauigkeit *f* accuracy, precision
Genauigkeitsgrad *m* degree of accuracy
Generator *m* producer, generator *(for manufacturing gas or steam)*
Generatorbetrieb *m* producer operation
Generatorgas *n* producer gas
Generatorgasgleichgewicht *n* producer-gas equilibrium
Generatorkohle *f* producer coal
Generatorschacht *m* generator *(proper, the chamber for holding the fuel)*
Generatorverfahren *n* **[/ englisches]** *(food)* generator method *(of acetification)*
generieren to generate *(e.g. gas)*
Generierung *f* generation *(as of gas)*
Genese *f*, **Genesis** *f* genesis *(as of coal or petroleum)*
genießbar edible, esculent, eatable, comestible
Genießbarkeit *f* edibility
Genin *n* genin *(non-carbohydrate portion of glycosides related to sterols)*
Gentianaviolett *n* gentian violet *(a mixture of pararosaniline derivatives)*
Gentisin *n* gentisin, 1,7-dihydroxy-3-methoxyxanthone
Gentisinsäure *f* $(HO)_2C_6H_3COOH$ gentisic acid, 2,5-dihydroxybenzoic acid
Genußreife *f* eating maturity
Geochemie *f* geochemistry
Geochemiker *m* geochemist
geochemisch geochemical
Geode *f (geol)* amygdule, amygdale, geode
Geokronit *m (min)* geocronite *(a sulphide of lead, arsenic, and antimony)*
Geosphäre *f* geosphere
gepastet *(tann)* paste-dried
gepreßt / trocken *(ceram)* dry-pressed
geprüft und anerkannt / amtlich scheduled *(e.g. pesticides)*
geradkettig straight-chain, unbranched-chain *(molecules)*
Geradsichtspektroskop *n* direct-vision spectroscope
Geranial *n* geranial, citral a, *trans*-3,7-dimethyl-2,6-octadienal
Geraniol *n* geraniol *(either of two dimethyloctadienol isomers)*
Geranium[gras]öl *n* **/ Ostindisches** Indian geranium (grass) oil, rusa oil *(from Cymbopogon martini (Roxb.) Stapf)*
Geraniumsäure *f* geranic acid *(either of two dimethylheptadienecarboxylic acid isomers)*
Gerät *n* 1. *(chem)* apparatus, *(if no specific term otherwise implicit)* device; *(esp for measuring)* instrument; 2. *(collectively)* apparatus[es], equipment
G. für den Verkokungstest apparatus for determining carbon residue
G. zur Kohlensäurebestimmung carbon dioxide apparatus
registrierendes G. recording instrument
selbstschreibendes G. [self-]recording instrument

Geräte *npl* apparatus[es], equipment
wissenschaftliche G. scientific apparatus, research equipment

Gerätefehler *m* instrumental error
Gerätekonstante *f* apparatus constant
geraten / außer Kontrolle to get out of control (hand), to run wild
Gerätschaften *pl. s.* Geräte
Gerbanlage *f* tanning plant, tanyard, tannery
Gerbbrühe *f* tanning (tan, tanner's) liquor, ooze
süße G. mellow tan liquor

Gerbeffekt *m* tanning action (effect)
gerben to tan
mit Chromsalzen g. to chrome

Gerben *n s.* Gerbung
Gerber *m* tanner
Gerberbaum *m* beam *(for mechanical treatment of hides)*
Gerberbock *m* horse
Gerberei *f* 1. tannage, tannery, tanning; 2. tanning plant, tanyard, tannery
eigentliche G. tanning proper

Gerbereiche *f* tan [bark] oak, Lithocarpus densiflora Rehd.
Gerbereichemie *f* chemistry of leather manufacture
Gerbereichemiker *m* leather chemist
Gerberlohe *f* tanbark
ausgelaugte G. spent tan

Gerberrot *n* tanner's red, phlobaphene
Gerbersumach *m* tanner's (tanning) sumac, Rhus coriaria L.
Gerberwolle *f (text)* slipe wool
Gerbextrakt *m* tanning (tannin) extract
Gerbfaß *n* tanning drum
Gerbgrube *f* tanning (tan, suspender) pit, tanning (tan, suspender) vat
Gerbholz *n* tanwood
Gerblösung *f* tanning (tannin) solution
Gerbmaterial *n s.* Gerbmittel
Gerbmethode *f* tanning method
Gerbmittel *n* tanning material, tan
pflanzliches G. vegetable tan

Gerbmittelauszug *m* tanning (tannin) extract
Gerbmittelvorrat *m* tannery stock
Gerböl *n* tanning oil
Gerbrinde *f* tan (tanner's, tanning) bark
Gerbsäure *f* tannic acid; *(specif) s.* Gallusgerbsäure
gerbsäurehaltig tanniferous

Gerbsäuremesser *m* barkometer, barktrometer, tannometer
Gerbstoff *m* 1. tanning agent, *(if of vegetable origin also)* tannin; 2. *s.* Gerbmittel
gebundener G. fixed tannin
kondensierter G. condensed (non-hydrolyzable) tannin
künstlicher G. *s.* synthetischer G.
pflanzlicher G. vegetable tannin
synthetischer G. syntan, synthetic tannin (tanning agent)
Gerbstoffauszug *m* tanning (tannin) extract
unsulfitierter G. ordinary tanning extract
Gerbstoffbrühe *f* tanning (tan, tanner's) liquor, ooze
Gerbstoffextrakt *m* tanning (tannin) extract
Gerbstoffgehalt *m* tannin content
gerbstoffhaltig tanniferous
Gerbstoffixierung *f* tannin fixation
Gerbstofflösung *f* tannin solution
Gerbstoffpflanze *f* tanniferous (tanning) plant
Gerbstoffrot *n* tanner's red, phlobaphene
Gerbtrommel *f* tanning drum
Gerbung *f* 1. tannage, tanning, tannery; 2. *(phot)* tanning
beschleunigte G. accelerated tannage
eigentliche G. tanning proper
pflanzliche G. vegetable tannage
synthetische G. syntan tannage
vegetabilische G. vegetable tannage
Gerbverfahren *n* tanning method
Gerbvermögen *n* tanning power
Gerbvorgang *m* tanning process
Gerbwert *m* tanning value
Gerbwirkung *f* tanning action (effect)
Gerhardtit *m* *(min)* gerhardtite *(copper(II) trihydroxide nitrate)*
Gerichtschemie *f* forensic (legal) chemistry
Gerichtschemiker *m* forensic (legal) chemist
geriffelt, gerillt ribbed, grooved
geringinkohlt *(coal)* low-rank
geringwertig low-grade, poor
gerinnbar coagulable, congealable
Gerinnbarkeit *f* coagulability
Gerinne *n (min tech)* sluice, trough, launder
gerinnen to coagulate, to congeal, to curd[le], to clot, *(milk also)* to sour
g. lassen to coagulate, to congeal, to curd[le]
Gerinnen *n* coagulation, congelation, curdling, clotting, *(milk also)* souring
Gerinnsel *n* clot, curd
Gerinnung *f s.* Gerinnen
Gerinnungseigenschaften *fpl* curd characteristics *(of milk)*
Gerinnungsfaktor *m* clotting factor *(any of a group of compounds involved in blood clotting)*
gerinnungshemmend anticoagulant
Gerinnungspunkt *m* coagulation (curdling) point, *(milk also)* setting point
gerippt ribbed
German *n* germane, *(specif)* GeH_4 monogermane, germane, germanium tetrahydride
Germanat *n* germanate
Germanium *n* Ge germanium
Germanium(II)-bromid *n* $GeBr_2$ germanium(II) bromide, germanium dibromide, germanous bromide
Germanium(IV)-bromid *n* $GeBr_4$ germanium(IV) bromide, germanium tetrabromide, germanic bromide
Germaniumbromoform *n* $GeHBr_3$ germanium bromoform
Germanium(II)-chlorid *n* $GeCl_2$ germanium(II) chloride, germanium dichloride, germanous chloride
Germanium(IV)-chlorid *n* $GeCl_4$ germanium(IV) chloride, germanium tetrachloride, germanic chloride
Germaniumchloroform *n* $GeHCl_3$ germanium chloroform
Germaniumdi ... *s.a.* Germanium(II)- ...
Germaniumdioxid *n s.* Germanium(IV)-oxid
Germaniumdisulfid *n s.* Germanium(IV)-sulfid
Germanium(II)-fluorid *n* GeF_2 germanium(II) fluoride, germanium difluoride, germanous fluoride
Germanium(IV)-fluorid *n* GeF_4 germanium(IV) fluoride, germanium tetrafluoride, germanic fluoride
Germaniumhexahydrid *n* Ge_2H_6 germanium hexahydride, digermane
Germaniumimid *n* $Ge(NH)_2$ germanium imide
Germanium(II)-jodid *n* GeI_2 germanium(II) iodide, germanium diiodide, germanous iodide
Germanium(IV)-jodid *n* GeI_4 germanium(IV) iodide, germanium tetraiodide, germanic iodide
Germaniummonosulfid *n s.* Germanium(II)-sulfid
Germaniummonoxid *n s.* Germanium(II)-oxid
Germanium(II)-nitrid *n* Ge_3N_2 germanium(II) nitride, trigermanium dinitride
Germanium(IV)-nitrid *n* Ge_3N_4 germanium(IV) nitride, trigermanium tetranitride
Germaniumoktahydrid *n* Ge_3H_8 germanium octahydride, trigermane
Germanium(II)-oxid *n* GeO germanium(II) oxide, germanium monooxide, germanous oxide
Germanium(IV)-oxid *n* GeO_2 germanium(IV) oxide, germanium dioxide, germanic oxide
Germaniumoxidchlorid *n* $GeOCl_2$ germanium dichloride oxide
Germanium(IV)-oxidhydrat *n* $GeO_2 \cdot xH_2O$ soluble germanium dioxide
Germaniumsäure *f* germanic acid
Germanium(II)-sulfid *n* GeS germanium(II) sulphide, germanium monosulphide, germanous sulphide
Germanium(IV)-sulfid *n* GeS_2 germanium(IV) sulphide, germanium disulphide, germanic sulphide
Germaniumtetra ... *s.a.* Germanium(IV)- ...
Germaniumtetrahydrid *n* GeH_4 germanium tetrahydride, monogermane, germane
Germizid *n* germicide

Gersdorffit *m (min)* gersdorffite *(arsenic nickel sulphide)*
Gerste *f* barley, Hordeum L.
bespelzte G. husky barley
glasige G. steely barley
milde G. mealy barley
speckige G. steely barley
Gerstenberg-Komplektor *m* Gerstenberg complector plant *(for manufacturing margarine)*
Gerstenmalz *n* barley malt, malted barley
Gerstenmehl *n* barley flour, *(if coarsely ground)* barley meal
Gerstenstärke *f* barley starch
Gerstenzucker *m* barley sugar
Geruch *m* odour, smell
beißender G. pungent smell
durchdringender G. strong smell
stechender G. pungent smell
übler G. bad smell, stench, stink
geruchlos inodorous, odourless, odour-free
g. machen to deodorize
Geruchlosigkeit *f* inodorousness
Geruchlosmachen *n* deodorization
Geruchsbelästigung *f* nasal nuisance
geruchsbeseitigend deodorant
Geruchsbeseitigung *f*, **Geruchsentfernung** *f* deodorization
geruchsfrei *s.* geruchlos
Geruchsstoff *m* odorous substance; *(food)* flavouring matter; *(cosmet)* perfume
Geruchsverbesserer *m* odour improver, deodorizer, deodorant
Geruchsverbesserung *f* odour improvement, deodorization
Geruchsverschluß *m* water seal, [siphon] trap
geruchszerstörend deodorant
Geruchszerstörung *f* deodorization
Gerüst *n* 1. skeleton *(as of a molecule)*; 2. *(tech)* framework
Gerüsteiweißstoff *m* skeletal protein, scleroprotein
Gerüstpolysaccharid *n* skeletal polysaccharide
Gerüstschwingung *f* skeletal vibration
Gerüststoff *m*, **Gerüstsubstanz** *f* [detergency] builder *(a substance added to synthetic detergents)*
Gerüstumlagerung *f* skeletal rearrangement
Gesamtablauf *m* overall course *(of a reaction)*
Gesamtalkali *n*, **Gesamtalkaligehalt** *m* total [amount of] alkali, *(pap also)* total chemical, total alkalinity *(active alkali +* Na_2CO_3*)*
Gesamtalkaloide *npl* total alkaloids
Gesamtarbeit *f (phys chem)* total work
Gesamtausbeute *f* total (overall) yield
Gesamt-Austauschkapazität *f (soil)* total exchangeable bases
Gesamtbasizität *f (tann)* overall basicity
Gesamtbelastung *f* total load *(of a balance)*
Gesamtdruck *m* total pressure
Gesamtechtheit *f (text)* all-round fastness
Gesamtemissionsvermögen *n* total emissivity (emissive power)
Gesamtenergie *f* total energy
Gesamtentropie *f* total entropy
Gesamtentstaubungsgrad *m* overall collection efficiency
Gesamtfestsubstanz *f (rubber)* total solids, T.S., solid material
Gesamtfläche *f* 1. total area; 2. *s.* Gesamtoberfläche
Gesamtflüchtigkeit *f* overall volatility
Gesamtgehalt *m* **an** SO_2 *(pap)* total sulphur dioxide
Gesamtgeschwindigkeit *f* overall rate (velocity)
Gesamthärte *f* total hardness
G. des Wassers total water hardness
Gesamtheizzeit *f (rubber)* total cure (curing, vulcanizing) time
Gesamtkontrast *m (phot)* overall contrast
Gesamtkonzentration *f* total concentration
Gesamtladung *f (el chem)* total charge
Gesamtlängenverhältnis *n (text)* stretch ratio
Gesamtleitfähigkeit *f*, **Gesamtleitvermögen** *n* total (resultant) conductance
Gesamtmasse *f* total mass
Gesamtmolarität *f* total molarity
Gesamtmolzahl *f* total number of moles
Gesamtnährstoffbedarf *m* nutrient requirements
Gesamtoberfläche *f* total surface
Gesamtorbitaldrehimpuls *m* resultant orbital angular momentum
Gesamtpolarisation *f* total polarization
Gesamtquellung *f* total swelling
Gesamtreaktion *f* total (overall) reaction
Gesamtregister *n* cumulative (collective) index
Gesamtschwefel *m* total sulphur
Gesamtschwindung *f (ceram)* total shrinkage
Gesamt-SO_2**-Gehalt** *m (pap)* total sulphur dioxide
Gesamtspindrehimpuls *m* resultant spin angular momentum
Gesamtstickstoff *m* total nitrogen
Gesamtstickstoffgehalt *m (agric)* overall nitrogen content
Gesamtstrahlungspyrometer *n* total-radiation pyrometer
Gesamtstrom *m* total current
Gesamtsymmetrie *f* total symmetry
Gesamttrockenmasse *f*, **Gesamttrockensubstanz** *f* total solids, T.S.
Gesamtübergangswahrscheinlichkeit *f (phys chem)* total transition probability
Gesamtvolumen *n* total volume
Gesamtwertigkeit *f* total valency
Gesamtwirkung *f* overall effect
Gesamtwirkungsgrad *m* overall efficiency *(as of a pump)*
thermischer G. overall thermal efficiency
Gesamtzusatzmenge *f (pap)* total make-up
Gesäß *n (glass)* siege, bench *(of a pot furnace)*
gesättigt saturated *(solution or compound)*
gesäuert pickle-cured, pickled
Geschäftsbücherpapier *n* ledger [paper], account book paper
Geschirr *n (tann)* vat
Deutsches G. *(pap)* stamping (hammer) mill, stamper, stamps, stocks

Geschirrporzellan *n* table porcelain
Geschlechtshormon *n* sex hormone
geschlossenzellig closed-cell *(foamed plastic)*
Geschmack *m* taste, flavour ✦ **von unangenehmem G.** distasteful
altöliger G. oiliness *(as of spoiled fats)*
fischiger G. fishiness
ranziger G. rancidity, rancidness
talgiger G. tallowiness
geschmacklos insipid, flavourless
Geschmacksabwertung *f* flavour reversion
geschmacksbeeinträchtigend impairing the flavour (taste)
Geschmacksbeeinträchtigung *f* impairment of the flavour (taste), flavour deterioration
Geschmacksfehler *m* flavour defect, off-flavour
Geschmacksprüfung *f* organoleptic estimation (evaluation, rating, test), sensory test
Geschmacksstoff *m* flavouring material (matter, substance)
geschmacksverändernd taste-modifying
Geschmacksveränderung *f* taste modification
Geschmacksverbesserer *m* taste improver, flavour enhancer
geschmacksverbessernd taste-improving
Geschmacksverbesserung *f* improvement in flavour (taste)
geschützt / gesetzlich proprietary
vor Licht g. protected from light
Geschwindigkeit *f* velocity, rate, *(esp of rotating parts)* speed
G. der Moleküle molecular velocity
G. der Zonenwanderung zone-travel rate, zone (zoning) speed *(in zone melting)*
G. des Aufstroms upward velocity
G. im Rohr tube velocity
durchschnittliche G. mean velocity
kritische G. critical speed *(of a rotating part)*
mittlere G. mean velocity
mittlere quadratische G. root-mean-square velocity
wahrscheinlichste G. most probable velocity *(of molecules)*
geschwindigkeitsbestimmend rate-determining, rate-controlling
Geschwindigkeitsdifferenz *f* difference of velocity
Geschwindigkeitsdruckhöhe *f* velocity head
Geschwindigkeitsfokussierung *f* velocity focus[s]ing
Geschwindigkeitsgefälle *n*, **Geschwindigkeitsgradient** *m* velocity gradient
Geschwindigkeitshöhe *f* velocity head
Geschwindigkeitskoeffizient *m* velocity coefficient
Geschwindigkeitskomponente *f* velocity component
Geschwindigkeitskonstante *f* velocity constant, specific reaction velocity
Geschwindigkeitsselektor *m* velocity selector
Geschwindigkeitsverteilung *f* distribution of velocities
Geschwindigkeitsverteilungsgesetz *n* / **Maxwellsches** Maxwell-Boltzmann velocity-distribution law
Geschwindigkeitswert *m* velocity coefficient
Gesenk *n (plast)* female form (mould), force; *(met)* die
Gesenkblock *m (plast)* cavity block
Gesenkplatte *f (plast)* [cavity] retainer plate, *(Am)* retainer
Gesetz *n* law, principle
G. der äquivalenten Proportionen law of equivalent proportions
G. der Gleichverteilung der Energie law of equipartition of energy
G. der konstanten Proportionen law of constant (definite) proportions, Proust law
G. der konstanten Wärmesummen law of constant heat summation, Hess law
G. der multiplen Proportionen law of multiple proportions
G. der Periodizität periodic law
G. der rationalen Parameterverhältnisse *(cryst)* rational index law, Haüy law
G. der unabhängigen Ionenwanderung law of independent migration of ions
G. von der Erhaltung der Energie law of conservation of energy, energy principle
G. von der Erhaltung der Masse law of conservation of mass (matter)
G. von der Rationalität der Achsenabschnitte *(cryst)* law of rational intercepts, law of rationality of intercepts
Amagatsches G. Amagat (Leduc) law *(of partial volumes in gas mixtures)*
Avogadrosches G. Avogadro law (hypothesis) *(of the number of molecules in gases)*
Beersches G. Beer law *(of light absorption)*
Bouguer-Lambertsches G. Bouguer-Lambert law [of absorption], Lambert's absorption law
Boyle-Mariottesches G. law of Boyle-Mariotte, Boyle (Mariotte) law *(of gas pressure and volume)*
Bunsen-Roscoesches G. *(phot)* Bunsen-Roscoe [reciprocity] law
Daltonsches G. Dalton law *(of partial pressures)*
Drapersches G. Draper law *(of chemically effective radiation)*
Faradaysches G. Faraday law *(electrolysis)*
Ficksches G. Fick law *(of diffusion)*
Gay-Lussacsches G. Gay-Lussac law *(of gas volume and temperature)*
Grahamsches G. der Transfusionsgeschwindigkeiten Graham law of diffusion
Grotthus[s]-Drapersches G. Grotthus[s]-Draper law
Haüysches G. *(cryst)* Haüy law, rational index law
Hesssches G. Hess law, law of constant heat summation
Kohlrauschsches G. Kohlrausch law *(of independent migration of ions)*

Lambert-Beersches G. Lambert-Beer law *(of light absorption)*
Lambertsches G. *s.* Bouguer-Lambertsches G.
Moseleysches G. Moseley law *(of the wave numbers in X-ray spectra)*
Panethsches G. Paneth rule *(radiochemistry)*
Gesichtsmaske *f* 1. face mask; 2. *(cosmet)* face pack
Gesichtspackung *f (cosmet)* face pack
Gesichtspuder *m (cosmet)* face powder
Gesichtsschutz *m* face protector
Gesichtswasser *n (cosmet)* face tonic
Gespinst *n (text)* spun[-staple] yarn
gestaffelt staggered, non-eclipsed *(stereochemistry)*
Gestalt *f* shape, form
gestalten to shape, to form
gestaltlos amorphous
Gestaltsänderung *f* deformation, strain
bleibende G. permanent deformation (set), residual deformation (set), set
Gestaltsveränderung *f s.* Gestaltsänderung
Gestänge *n (petrol)* drill pipe
Gestängeanheber *m (petrol)* [drill pipe] elevator
Gestängeverbinder *m (petrol)* tool joint *(of a rotary-drilling installation)*
Gestank *m* bad smell, stench, stink
Gestein *n* rock, *(mine also)* ground
G. der Alkalireihe alkali rock
G. der Orthoreihe ortho rock
anstehendes G. bedrock
autoklastisches G. autoclastic rock, autoclast
basisches G. basic (subsilicic) rock, basite
biogenes G. biogenic (organic) rock, biolith
bioklastisches G. bioclast
chorismatisches G. chorismite
endogenes G. endogenetic rock
exogenes G. exogenetic rock
holokristallines G. holocrystalline rock
hybrides G. hybrid rock
hydroklastisches G. hydroclastic rock
infrakrustales G. infracrustal rock
intermediäres G. *s.* neutrales G.
magmatisches G. magmatic (igneous) rock
metamorphes G. metamorphic (metamorphosed) rock, metamorphite
neutrales G. neutral (intermediate) rock
organogenes G. *s.* biogenes G.
plutonisches G. plutonic (irruptive, hypogene) rock, plutonite
polymetamorphes G. polymetamorphic rock
pyroklastisches G. pyroclastic rock
saures G. acid rock
superkrustales (suprakrustales) G. supercrustal rock
taubes G. waste rock, gangue, matrix
ultrabasisches G. ultrabasic rock
umgebendes G. enclosing rock
vulkanisches G. volcanic rock
Gesteinschemie *f* petrochemistry
gesteinschemisch petrochemical
Gesteinsgang *m* vein, dike, dyke
Gesteinsglas *n* natural glass
Gesteinsgrus *m* rock waste
Gesteinshülle *f*, **Gesteinskruste** *f* lithosphere
Gesteinskunde *f* petrography
Gesteinsmantel *m* lithosphere
Gesteinsmehl *n (agric)* crushed rocks
Gesteinsschutt *m* detritus, detrital material, rock waste
Gesteinssprengstoff *m* rock explosive
Gesteinswolle *f* rock wool
Gestell *n* 1. rack; 2. *(met)* hearth, crucible *(of a blast furnace)*
gesteuert / selbsttätig self-controlled
gestreift banded
gestrichen *(pap)* coated
einseitig g. coated on one side
in der Maschine g. machine-coated
zweiseitig g. double-coated, coated on both sides
gesundheitsschädlich dangerous (injurious) to health, harmful, deleterious
gesüßt *(petrol)* sweet[ened]
geteilt / auf Ablauf (Auslauf) calibrated (graduated) to deliver, calibrated for delivery
auf Einguß g. calibrated to contain, calibrated for content
bis zur Spitze g. calibrated to jet *(pipette)*
Getränk *n* beverage, drink, potable
alkoholfreies G. alcohol-free (non-alcoholic) beverage, soft (temperance) drink
alkoholisches G. alcoholic (spirituous) beverage, spirit
berauschendes G. intoxicating liquor, intoxicant
geistiges G. *s.* alkoholisches G.
hochprozentiges alkoholisches G. strong drink, *(Am)* hard drink
karbonisiertes (mit CO_2 imprägniertes) G. carbonated beverage
getränkt / mit Öl oil-impregnated
Getreidebegasungsmittel *n* grain fumigant
Getreidebranntwein *m* grain alcohol
Getreidekeimöl *n* cereal seed oil
Getreidemehl *n* flour, *(if coarsely ground)* [corn] meal
Getreideschlempe *f* distillers' [spent] grains
Getreidestärke *f* cereal starch
Getreidestroh *n* cereal straw
Getreidetrockner *m* grain dryer
Getreidewaschmaschine *f* grain washing machine
Getriebehebewerk *n (petrol)* drawworks
Getriebeöl *n* gear oil
getrocknet / auf dem Dachboden (Trockenboden) *(pap)* loft-dried
auf der Maschine g. *(pap)* cylinder-dried, machine-dried, steam-dried
im Sprühverfahren g. spray-dried
im Trockenschrank g. oven-dried, oven-dry, OD
im Zerstäubungsverfahren g. spray-dried
getrübt *(food)* hazy, cloudy, feculent

Getter *m(n)*, **Gettermetall** *n* getter *(vacuum technology)*
gettern to getter *(vacuum technology)*
Getterstoff *m* getter *(vacuum technology)*
Getterung *f* gettering *(vacuum technology)*
Gewässeraufsichtsorgan *n* water pollution control authority, river pollution control agency
Gewebe *n* 1. *(text)* fabric [cloth], cloth, textile [fabric], web; 2. *(biol)* tissue
baumwollenes G. cotton fabric
beschichtetes G. coated fabric
gekrepptes G. crêpe, crepe
gestrichenes G. coated fabric
gummiertes G. rubberized (rubbered, rubber-coated, proofed) fabric, proofing
kaschiertes G. coated (combined) fabric
selbstglättendes G. self-smoothing fabric
Gewebeabscheider *m* fabric (woven-fabric, woven) filter
Gewebedruck *m* textile printing
Gewebeeinlage *f (rubber)* textile insert (insertion, casing)
Gewebeextrakt *m* tissue extract
Gewebefilter *n* fabric (woven-fabric, woven) filter
Gewebeflüssigkeit *f (biol)* tissue fluid
gewebefrei fabric-free *(e.g. rubber product)*
Gewebehormon *n* tissue hormone
Gewebeinkrustierung *f (text)* loading of fabrics
Gewebekrumpfmaschine *f (text)* shrinking machine
Gewebekultur *f (biol)* tissue culture
Gewebekulturmedium *n (biol)* tissue-culture medium
Gewebekunstleder *n* coated fabric
Gewebelage *f (rubber)* carcass (casing) ply
Gewebepapier *n* reinforced paper, papyrolin *(cloth-faced or cloth-centred paper)*
Gewebeschlauch *m* woven hose
Gewebeschnitzel *npl (plast)* macerated fabric
Gewebeschnitzelpreßmasse *f (plast)* fabric-filled moulding compound (material)
Gewebezüchtung *f (biol)* tissue culture
Gewebs ... *s.* Gewebe ...
Gewerbeschutzsalbe *f* barrier cream
Gewicht *n* 1. weight *(the force by which the mass of a substance is attracted by gravity)*; 2. *(incorrectly for)* Gewichtstück; 3. *(incorrectly for)* Massestück, Wägestück
spezifisches G. specific weight
statistisches G. *(phys chem)* statistical weight
Gewichtsabweichung *f* off weight
Gewichtsanalyse *f* gravimetric analysis
Gewichtsänderung *f* weight change
Gewichtsanteile *mpl s.* Gewichtsprozent
gewichtsbetätigt weight-operated, gravity-operated
Gewichtsfunktion *f* weighting function *(control engineering)*
Gewichtskonstanz *f* constancy of weight, constant weight ✦ **bis zur G. glühen** to ignite to constant weight
gewichtsmolar molal
Gewichtsmolarität *f* molality, molal concentration
Gewichtsprozent *n* percentage (per cent) by weight
Gewichtssatz *m (incorrectly for)* Wägesatz
Gewichtsteil *m* part by weight
Gewichtstück *n* weight *(a heavy object of indefinite mass for counterbalancing)*
Gewichtsveränderung *f* variation in weight
Gewichtsverlust *m* loss in weight, weight loss
Gewichtszunahme *f* gain in weight, weight increase
Gewindefitting *m(n) s.* Gewinderohrverbindung
Gewindeformstück *n* screwed fitting
Gewinderohr *n* threaded pipe
Gewinde[rohr]verbindung *f* screwed fitting, threaded joint
gewinnen 1. to obtain, to recover *(a reaction product)*; 2. *(bioch)* to isolate *(from natural products)*; 3. *(mine)* to mine, to extract, to win
Gewinnung *f* 1. recovery *(of a reaction product)*; 2. *(bioch)* isolation *(from natural products)*; 3. *(mine)* mining, extraction, winning
G. durch Flotation floatation recovery
Gewölbe *n* crown *(of a melting furnace)*
Gewölbeschlußstein *m* bullhead
gewölbt domed, convex; dished, concave
Gewürz *n* seasoning, *(of vegetable origin)* spice, *(esp salt and pepper)* condiment
Gewürznelke *f* clove *(from Syzygium aromaticum (L.) Merr. et L.M. Perry)*
Gewürznelkenöl *n* clove (caryophyllus) oil
Gewürzpaprika *m* hot pepper, chil[l]i *(from Capsicum specc.)*
Gewürzpflanze *f* spice plant
Geyserit *m (min)* geyserite, siliceous sinter *(a deposit around hot springs and geysers)*
g-Faktor *m* **[/ Landéscher)** Landé g-factor, [Landé] splitting factor g *(gyromagnetic ratio of electrons)*
GFK *s.* Kunststoff / glasfaserverstärkter
GFP *s.* Plast / glasfaserverstärkter
GFS *s.* Glasfaserschichtstoff
Ghatti *n* gum ghatti, India gum *(from Anogeissus latifolia Wall.)*
Gibberellinsäure *f (bioch)* gibberellic acid
Gibbs-Helmholtz-Gleichung *f* Gibbs-Helmholtz equation
Gibbsit *m (min)* gibbsite, hydrargillite *(aluminium trihydroxide)*
Gibbs-Zelle *f* Gibbs cell *(electrolysis)*
Gicht *f (met)* stock, burden, charge[stock], feed; throat, top *(part of a blast furnace)*
Gichtbrücke *f (met)* hoist bridge
Gichtbühne *f (met)* charge floor
Gichtgas *n* furnace (blast-furnace) gas
Gichtglocke *f (met)* bell
Gichthöhe *f (met)* stock level
Gichtsonde *f (met)* stock level (line) indicator
Gichtstaub *m (met)* blast-furnace dust
Gichtverteiler *m (met)* distributor
drehbarer G. revolving (rotating) distributor

Gieseler-Plastometer *n* Gieseler plastometer
gießbar 1. capable of being poured, pourable; 2. castable *(molten material)*
Gießbarkeit *f* 1. pourability; 2. castability *(of molten material)*
Gießbett *n* pig bed *(foundry)*
gießen 1. to pour; 2. to cast *(molten material into moulds)*
in Kokille g. to diecast
unter Druck g. to pressure-diecast
Gießen *n* 1. pouring; 2. casting *(of molten material into moulds)*
G. in getrocknete Sandformen dry-sand casting
G. in grüne (ungetrocknete) Sandformen green-sand casting
G. mit verlorener Gußform [precision] investment casting
G. unter Druck pressure diecasting
kontinuierliches G. continuous casting
Gießer *m* founder, foundryman, caster
Gießerei *f* foundry
Gießereieisen *n* foundry [pig] iron
Gießerei[kern]harz *n* foundry resin
Gießereikoks *m* foundry coke
Gießereikupolofen *m* foundry cupola
Gießereiroheisen *n* foundry [pig] iron
Gießereischmelzkoks *m* foundry coke
Gießfähigkeit *f s.* Gießbarkeit
Gießfilm *m s.* Gießfolie
Gießfläche *f* casting area
Gießfleck *m (ceram)* casting spot (stain), flashing *(a defect)*
Gießfolie *f (plast)* cast sheet, *(if thickness less than 0.01 inch)* cast film
Gießform *f* [casting] mould
Gießharz *n* cast[ing] resin
Gießhaut *f (ceram)* casting skin
Gießkern *m (ceram, met)* core
Gießlackierung *f* curtain coating
Gießling *m* casting
Gießloch *n* casting hole
Gießlöffel *m* ladle
Gießmaschine *f* casting machine
Gießmasse *f (ceram)* casting (liquid) slip
Gießmittel *n (agric)* gravity-fed spray
Gießnaht *f* casting seam (line)
Gießpfanne *f* tilting hopper *(foundry)*
Gießrinne *f* gutter *(foundry)*
Gießschlicker *m (ceram)* casting (liquid) slip
tonfreier G. non-clay casting slip
Gießtemperatur *f* casting temperature
Gießtisch *m (glass)* casting table
Gießverfahren *n* casting process
Gift *n* poison, toxicant, toxic [substance]
ökonomisches G. *(agric)* economic poison
pflanzliches G. plant poison
protektives (protektiv wirkendes) G. *(agric)* protective toxicant
systemisches G. *(agric)* systemic poison
tierisches G. venom
vorbeugend wirkendes G. *(agric)* protective toxicant
Giftgas *n* poison[ous] gas
Giftgetreide *n* poisoned grain
gifthaltig toxiferous
Giftheber *m* siphon for poisons
giftig poisonous, toxic[al]
Giftigkeit *f* toxicity, poisonousness
G. für Pflanzen phytotoxicity
G. für Säugetiere mammalian toxicity
akute G. acute toxicity
chronische G. chronic toxicity
Giftköder *m* poison (toxic) bait
G. gegen Nagetiere rodent bait
Giftkunde *f* toxicology
Giftkundiger *m* toxicologist
Giftmehl *n (met)* white arsenic
Giftpapier *n* poisoned paper, *(esp)* insecticide paper
Giftsachverständiger *m* toxicologist
Giftschierling *m* [poison] hemlock, Conium maculatum L.
Giftstoff *m* toxicant, toxic [substance]
Giftwert *m* toxic limit *(of wood preservatives)*
Giftwirkung *f* poisoning (toxic) action (effect)
gilben to go yellow
Gingergrasöl *n* ginger-grass oil *(chiefly from Cymbopogon martini (Roxb.) Stapf var. sofia)*
Ginseng *m (pharm)* ginseng *(Panax quinquefolius L. and Panax schin-seng Nees)*
Gipfelpunkt *m* peak *(of a curve)*
Gips *m (min)* gypsum *(calcium sulphate-2-water)*
✦ **mit G. behandeln** to plaster *(wine)*; to burtonize, to gypsum *(brewing water)*
gebrannter G. calcined (anhydrous) gypsum, gypsum cement, plaster of Paris *(essentially calcium sulphate-0.5-water)*
kristalliner G. *s.* Gips
gipsen 1. to plaster *(wine)*; to burtonize, to gypsum *(brewing water)*; 2. *(agric)* to gypsum
Gipsform *f* plaster mould
gipsführend *(geoch)* gypsiferous
gipshaltig gypsiferous, gypseous
Gipsmörtel *m* gypsum mortar, plaster [mortar]
gipsreich gypseous *(brewing water)*
Girbotol-Verfahren *n* Girbotol process *(of removing hydrogen sulphide from gases)*
Girlandenrolle *f* suspended-cable idler *(conveying)*
Gisbe *f s.* Gispe
Gismondin *m (min)* gismondine, gismondite *(a hydrous calcium aluminium silicate)*
Gispe *f* seed *(a defect in glass)*
Gitter *n* 1. *(cryst)* lattice; 2. grating *(optics)* 3. *(tech)* grating, grid; checker (chequer) chamber *(of a blast-furnace stove)*
flächenzentriertes G. face-centred lattice
hexagonales G. hexagonal lattice
innenzentriertes G. body-centred lattice
kubisches G. cubic lattice
kubisch-flächenzentriertes G. face-centred cubic lattice
kubisch-innenzentriertes G. body-centred cubic lattice

metallisches G. metallic lattice
molekulares G. molecular lattice
raumzentriertes G. *s.* innenzentriertes G.
Gitterabstand *m (cryst)* lattice distance (spacing), spacing of the planes
Gitteranordnung *f (cryst)* lattice arrangement
Gitteraufbau *m (cryst)* lattice structure
Gitteraufweitung *f (cryst)* lattice expansion
Gitterbau *m (cryst)* lattice structure
Gitterbaufehler *m (cryst)* lattice defect
Gitterboden *m (distil)* [turbo]grid tray
Gitterebene *f (cryst)* lattice (atomic, net) plane
Gittereinschlußverbindung *f* lattice-enclosure compound *(chemical-bond theory)*
Gitterenergie *f (cryst)* lattice energy
Gitterexpansion *f (cryst)* lattice expansion
Gitterfehler *m (cryst)* lattice defect (imperfection)
flächenhafter G. surface defect
linienhafter G. line defect
punktförmiger G. point defect
stöchiometrischer G. stoichiometric lattice defect
Gitterfehlordnung *f (cryst)* lattice disorder
Gitterfehlstelle *f (cryst)* point defect (imperfection), lattice defect
Gitterfläche *f (cryst)* lattice face
Gitterhohlraum *m (cryst)* interstitial lattice site, interstice
Gitterkonstante *f (cryst)* lattice constant
Gitterkräfte *fpl (cryst)* lattice forces
Gitterleerstelle *f (cryst)* lattice vacancy (hole), hole [position], vacant site, vacant lattice site (position)
Gitterloch *n*, **Gitterlücke** *f s.* Gitterleerstelle
Gittermauerwerk *n* checker (chequer) brickwork, checkerwork, chequerwork
Gitterordnung *f (cryst)* lattice arrangement
Gitterorientierung *f (cryst)* lattice orientation
Gitterperiode *f s.* Gitterabstand
Gitterplatz *m (cryst)* lattice position (site, point)
unbesetzter G. *s.* Gitterleerstelle
Gitterpunkt *m (cryst)* lattice point
Gitterraum *m* checker (chequer) chamber *(of a blast-furnace stove)*
Gitterrostboden *m (distil)* [turbo]grid tray
Gitterrührer *m s.* Gatterrührer
Gitterspektrofotometer *n* grating spectrophotometer
Gitterspektrograf *m* grating spectrograph
Gitterspektroskop *n* grating spectroscope
Gitterspektrum *n* grating (diffraction, normal) spectrum
Gitterstein *m* checker (chequer) brick
Gitterstelle *f* lattice point (position, site)
Gitterstörstelle *f*, **Gitterstörung** *f s.* Gitterfehler
Gitterstruktur *f*, **Gitterverband** *m* lattice structure
Gitterverbindung *f* lattice compound *(chemical-bond theory)*
Gitterverzerrung *f (cryst)* lattice distortion
Gitterwerk *n s.* Gittermauerwerk
Gitterwerksstein *m* checker (chequer) brick
GL *s.* Glasfaserstoff
Glabratsäure *f s.* Lekanorsäure
Glacépapier *n* enamel[led] paper
gebürstetes G. brush enamel paper
Glanz *m* lustre, gloss[iness], shine
matter G. sheen *(as of powder or silk)*
metallischer G. metallic lustre
Glanzabzug *m (phot)* glossy print
Glanzappretur *f* 1. *(text)* glazed finish; 2. *s.* Glanzauftrag
Glanzauftrag *m (tann)* lustre (glossy) finish, season
Glanzausrüstung *f (text)* 1. glossing; 2. glazed finish
Glanzausrüstungsmittel *n (text)* lustring agent
Glanzbeständigkeit *f (coat)* gloss retention
Glanzbildner *m* brightener, brightening agent *(electroplating)*
Glanzbraunkohle *f* black (bituminous) lignite, lignitous (subbituminous) coal
Glanzeffekt *m* gloss, lustre, shiny effect
glänzen 1. to glaze, to season, to satine *(the leather)*; 2. to shine, to glitter, to glisten
elektrolytisch g. to plate bright *(electroplating)*
Glänzen *n* / **elektrolytisches** bright plating *(electroplating)*
glänzend lustrous, glossy, shiny
Glanzerhaltung *f* gloss retention
Glanzfarbe *f* glazing varnish; gloss [printing] ink *(typography)*
Glanzfaser *f (text)* glaze fibre
Glanzfirnis *m* gloss varnish *(typography)*
Glanzglasur *f (ceram)* bright glaze
Glanzgold *n (ceram)* bright (liquid) gold
Glanzhaltung *f* gloss retention
Glanzkohle *f* glance (bright) coal; *(specif) s.* Glanzbraunkohle
Glanzkohlenstoff *m* lustrous carbon
Glanzlack *m* gloss varnish
glanzlos lustreless, hazy, mat[t], cloudy, dull, *(of colours also)* dead
Glanzlosigkeit *f* lustrelessness, haziness, mattness, cloudiness, dullness
Glanzmesser *m (pap)* gloss meter, glossimeter, glarimeter
Glanzmetall *n* speculum metal *(an alloy chiefly consisting of copper and tin)*; *(ceram)* liquid-bright metal
Glanzpalladium *n (ceram)* bright palladium
Glanzpapier *n* flint[-glazed] paper
Glanzpappe *f* glazed board
Glanzplatin *n (ceram)* bright platinum
Glanzpressen *n (text)* glossing
Glanzsilber *n (ceram)* bright silver
Glanzstoff *m (text)* copper rayon
glanzstoßen *(tann)* to glaze, to enamel
Glanzstoßmaschine *f (tann)* glazing machine
Glanzstoßzurichtung *f (tann)* glazed finish
Glanzweiß *n* satin white (spar) *(a pigment)*
Glanzwinkel *m (cryst)* glancing (Bragg) angle
Glas *n (tech, geol)* glass; *(equipment)* glassware
G. mit Luftblasendekor bubble glass
alkaliarmes G. low-alkali glass

alkalihaltiges G. alkali glass, glass A
braunes G. amber glass
chemisch widerstandsfähiges G. chemically resistant glass
farbiges G. coloured glass
feuerfestes G. heat-resisting glass
freihandgeblasenes G. off-hand (free-blown) glass
geblasenes G. blown glass
gebogenes G. bent glass
geriffeltes (geripptes) G. fluted (ribbed) glass
geschliffenes G. cut glass
gezogenes G. drawn glass
gispiges G. seedy glass
hitzebeständiges G. heat-resisting glass
im Hafen geschmolzenes G. pot[-melted] glass
in der Wanne geschmolzenes G. tank glass
in Formen geblasenes G. mould-blown glass
kohlegelbes G. amber glass
kugelfestes (kugelsicheres) G. bullet-proof (bullet-resistant) glass
kurzes G. short glass
langes G. long glass
leicht schmelzbares G. soft [sealing] glass
leicht verarbeitbares G. sweet glass
mundgeblasenes G. handblown (hand-made) glass
opakes G. opaque (opal) glass
optisches G. optical glass
optisches G. mit hohem Brechungsindex dense glass
organisches G. organic glass
reflexfreies G. non-reflecting glass
schlieriges G. cordy glass
schußfestes G. *s.* kugelfestes G.
vorgespanntes G. prestressed (case-hardened, tempered, toughened, heat-treated, heat-strengthened) glass
vulkanisches G. *(geol)* volcanic glass
wärmeabsorbierendes G. heat-absorbing glass
weiches G. soft glass
glasähnlich glassy, glass-like, vitreous
Glasampulle *f* glass ampoule
glasarmiert glass[-fibre] reinforced
glasartig glassy, glass-like, vitreous; *(ceram)* vitrified, vitreous
nicht g. *(ceram)* non-vitrified, non-vitreous
Glasartikel *mpl* glassware
Glasätzung *f* glass etching
Glasauskleidung *f* glass lining ✦ **mit G.** glass-lined
Glasballon *m* glass balloon flask, *(if cushioned)* glass carboy, *(if enclosed in wickerwork with wicker handle)* demijohn
Glasband *n* ribbon of glass
Glasbaustein *m* glass brick (block)
Glasbecher *m* glass beaker
glasbildend glass-forming
Glasbildner *m* glass-forming substance, glass former
Glasblasen *n* glass blowing
Glasbläser *m* glass-blower
G. vor der Lampe lampworker
Glasbläserlampe *f* glass-blower's lamp
Glasbläserpfeife *f* glass-blower's pipe, blow pipe, blow[ing] iron
Glasbohrer *m* glass drill
Glasbürste *f* glass brush
Glasdeckel *m* glass lid
G. mit Schliff ground-glass lid
Glaselektrode *f* glass electrode
Glasemaille *f* glass enamel *(glass coating of enamel-like composition)*
glasemailliert glassed
Gläserbürste *f (lab)* beaker (jar) brush
Glaserdiamant *m* cutting diamond
Glaserit *m (min)* aphthitalite, glaserite *(potassium sodium sulphate)*
Glaserkitt *m* [painter's, glazier's] putty
gläsern vitreous, glassy
Glasfabrik *f* glassworks, glass factory (house)
Glasfabrikation *f s.* Glasherstellung
Glasfaden *m* glass filament
Glasfarbe *f* glass colour
Glasfaser *f* glass fibre, *(broadly)* fibre (fibrous, spun) glass
Glasfaserband *n* glass-fibre tape
Glasfasererzeugnis *n* glass-fibre (fibre-glass) product
Glasfaserfüllstoff *m* glass-fibre filler
Glasfasergarn *n* glass-fibre yarn, staple-fibre glass yarn
Glasfasergewebe *n* glass[-fibre] fabric, woven-glass-fibre cloth, glass cloth
Glasfaserlaminat *n* glass-fibre laminate
Glasfasermaterial *n* / **vorimprägniertes** prepreg
Glasfaserpapier *n* glass-fibre paper
Glasfaserpreßmasse *f* glass-fibre [reinforced] moulding compound
Glasfaserprodukt *n s.* Glasfasererzeugnis
Glasfaserschichtstoff *m* glass-fibre laminate
Glasfaserstoff *m* fibre (fibrous, spun) glass
glasfaserverstärkt glass[-fibre] reinforced
Glasfaserverstärkung *f* glass[-fibre] reinforcement
Glasfaservlies *n* glass-fibre veil, chopped (chopper) strand mat
Glasfedermanometer *n* glass-spring manometer
Glasfehler *m* glass defect
Glasfilter *n* glass filter
poröses G. porous-glass filter
Glasfilternutsche *f s.* Glasfiltertrichter
Glasfilterplatte *f* glass filter disk
Glasfiltertiegel *m* sintered-glass crucible, glass filter crucible
Glasfiltertrichter *m* sintered glass [filtering] funnel, glass filter funnel, glass suction filter [funnel]
Glasfluß *m* glass flow
Glasformer *m s.* Glasbildner
Glasformgebung *f* glass forming
Glasformstück *n* glass fitting
Glasformung *f* glass forming
Glasfritte *f* glass filter
Glasgalle *f (glass)* [glass] gall, salt water
Glasgefäß *n* glass vessel

Glasgemenge *n* glass batch
Glasgeräte *npl* glassware
Glasgespinst *n s.* Glasfaserstoff
Glasglanz *m (min)* vitreous (glassy) lustre
Glasglocke *f* glass bell, bell-jar
Glashafen *m* glass[-melting] pot
Glashafenofen *m* glass pot furnace
Glashahn *m* glass stopcock (tap)
glashart glass-hard
Glashärte *f* glass hardness
Glashaut *f (plast)* cellulose film
Glashersteller *m* glass manufacturer (maker)
Glasherstellung *f* glass manufacture (making)
Glashütte *f* glassworks, glass factory (house)
glasieren *(ceram)* to glaze
Glasiermaschine *f (ceram)* glazing machine
glasig vitreous, glassy, glass-like; *(ferm)* steely, vitreous *(malt)*; *(geol)* vitrophyric, vitreous
Glasindustrie *f* glass industry
Glaskappe *f* glass cap
Glaskeramik *f* glass ceramic, vitroceramic, vitrokeram, devitrified (neo-ceramic) glass
glasklar glass-clear
Glasklebstoff *m* glass adhesive
Glaskohle *f (el chem)* glassy carbon
Glaskolben *m* glass flask (bulb)
Glaskopf *m* **/ brauner** *(min)* brown iron ore (stone), limonite
Glaskugel *f* [glass] marble *(glass-fibre manufacture)*
Glaskugler *m (glass)* cutter
Glasküvette *f* glass cuvette
Glasmacher *m* glass-maker, [glass] blower
Glasmacherpfeife *f* glass-blower's pipe, blow pipe, blow[ing] iron
Glasmacherseife *f* glass[makers'] soap
Glasmalz *n (ferm)* steely (vitreous) malt
Glasmasse *f* glass mass
Glasmembran *f* glass membrane *(as of a glass electrode)*
Glasmesser *n* glass knife
Glas-Metall-Verschmelzung *f*, **Glas-Metall-Verschweißung** *f* glass-[to-]metal seal
Glasnageltrichter *m (lab)* glass nail funnel
Glasnutsche *f s.* Glasfiltertrichter
Glasofen *m* glass[-melting] furnace
Glaspapier *n* glass paper *(1. an abrasive paper; 2. an insulating material)*
Glasperle *f* glass bead
Glasphase *f* vitreous (glassy) phase (state)
Glasplatte *f* glass plate, *(if thin)* sheet of glass
Glasposten *m* [glass] gob, gather of glass
Glaspulver *n* glass powder
Glasrohr *n* glass pipe (tube); *(collectively)* glass piping (tubing)
Glasröhre *f* glass pipe (tube)
Glasrohrleitung *f* glass pipeline
Glasrohrmaterial *n* glass piping (tubing)
Glasrührer *m* glass stirrer
Glassatz *m (glass)* batch
Glasschale *f* glass dish
Glasscheibe *f* 1. pane [of glass] *(of a window)*; 2. *s.* Glasplatte
Glasschleifen *n* glass grinding; brilliant cut *(for decorating flat glass)*
Glasschliff *m* glass grinding
Glasschmelze *f* glass melt
Glasschmelzofen *m* glass[-melting] furnace
Glasschmelzwanne *f* [glass-]melting tank
Glasschneiden *n* glass cutting
Glasschneider *m* glass cutter *(tool or worker)*
Glasschneiderdiamant *m* cutting diamond
Glasseide *f* glass silk
Glasseidenband *n* glass silk tape
Glasseidenbeschichtung *f* glass silk coating
Glasseidengewebe *n* glass silk fabric
Glasseidenmatte *f* glass-fibre mat
Glasseidenpapier *n* transparent tissue paper, glass tissue
Glasseidenroving *m s.* Glasseidenstrang
Glasseidenspinnfaden *m* continuous glass filament
Glasseidenstrang *m* [glass-fibre] roving, glass-fibre strand
geschnittener G. chopped strand
Glasseife *f* glass[makers'] soap
Glassintertiegel *m* sintered-glass crucible
Glasspiralkolonne *f* **nach Widmer** Widmer spiral column
Glasstab *m* glass rod, cane
Glasstapelfasergarn *n* staple-fibre glass yarn, glass-fibre yarn
Glasstopfen *m* glass stopper ✦ **mit G.** glass-stoppered
eingeschliffener G. ground[-glass] stopper
Glasstopfenflasche *f* glass-stoppered bottle
Glasstruktur *f* glass structure
Glastechnologie *f* glass technology
Glastinte *f* [glass-]marking ink, glass ink
Glasträne *f* glass tear (drop)
Glastrichter *m* glass funnel
Glastropfen *m* glass drop (tear); [glass] gob *(on a glass-blower's pipe)*
Glastuff *m (geol)* vitric tuff
Glasübergang *m* glass transition
Glasübergangsbereich *m* glass transition region
Glasumwandlung *f* 1. glass transition; 2. *(rubber)* second-order transition, vitrification
Glasumwandlungspunkt *m*, **Glasumwandlungstemperatur** *f* 1. glass transition temperature, Tg [point]; 2. *(rubber)* second-order transition temperature (point)
Glasur *f (ceram)* glaze
ausgeschmolzene G. matured glaze
deckende G. opaque glaze
gebrannte G. fired glaze
gefrittete G. fritted glaze
gesprenkelte G. mottled glaze
kratzfeste G. scratch-resisting glaze
opake (trübe) G. opaque glaze
Glasuraufnahme *f (ceram)* pickup
Glasurbrand *m (ceram)* glost firing, glaze baking
Glasurbrandofen *m (ceram)* glost kiln
Glasurfehler *m (ceram)* glaze fault
Glasurlehm *m (ceram)* slip clay

Glasurofen *m (ceram)* glost kiln
Glasurschlicker *m (ceram)* glaze slip
Glasurschmelze *f (ceram)* glaze batch
Glasursitz *m (ceram)* glaze fit
Glasurspat *m (ceram)* glaze spar
Glasurton *m (ceram)* slip clay
Glasurüberzug *m (ceram)* glaze coating
Glasverarbeitungsmaschine *f* glass-forming machine
Glasverbindung *f* glass connection
Glaswanne *f (lab)* glass trough; *(glass)* glass tank
Glaswannenofen *m* glass tank furnace
Glaswaren *fpl* glassware
Glaswatte *f* glass wadding
Glaswerk *n* glassworks, glass factory (house)
Glaswolle *f* glass wool
Glaszement *m* glass cement
Glasziegel *m* glass block (brick)
Glaszustand *m* vitreous (glassy, glass-like) state
Glaszylinder *m* jar
glatt 1. smooth *(surface)*; 2. plain *(shape)*; 3. smooth *(e.g. course of reaction)*
Glattbrand *m (ceram)* glost firing
Glattbrandofen *m (ceram)* glost kiln
glattbrennen *(ceram)* to glost-fire
Glätte *f* smoothness *(of a surface)*; *(pap)* glaze
glätten to smooth, to polish; *(glass)* to flatten; *(pap)* to glaze, to smooth, to enamel, to plate, to [super]calender; *(tann)* to scud
Glätten *n* **auf der Bürstmaschine** *(pap)* brush polishing
  G. mit Achatstein *(pap)* flint glazing, flinting
  G. von Bogenpapieren sheet calendering
  G. von Rollenpapieren web calendering
Glätteprüfer *m*, **Glätteprüfgerät** *n (pap)* smoothness tester
Glättezahl *f (pap)* smoothness number
Glättfilz *m (pap)* glazing felt
Glättmaschine *f* calender [machine]
Glattofen *m (ceram)* glost kiln
Glattrohr *n* bare tube
Glättschaberstreichmaschine *f (pap)* trailing blade coater
Glattscherben *mpl (ceram)* [glost] pitchers
glattschleifen to grind smooth
glattschmelzen *(lab)* to fire-polish, to fire-finish, to fire-glaze
Glattstreicher *m* leveller
Glattwalze *f* smooth[-surfaced] roll
Glättwalze *f* spreader *(of a cylinder dryer)*; *(pap)* smoothing roll
Glattwalzenbrecher *m* smooth-roll crusher
Glätt[walzen]werk *n* calender [machine]
Glättwerkspartie *f (pap)* surfacing end
Glättwerkswalze *f* calender roll (bowl)
Glauberit *m (min)* glauberite *(calcium sodium sulphate)*
Glaubersalz *n* Glauber salt, *(min also)* mirabilite *(sodium sulphate-10-water)*
Glaukodot *m (min)* glaucodot[e] *(cobalt iron sulpharsenide)*
Glaukonit *m (min)* glauconite *(a phyllosilicate)*
Glaukophan *m (min)* glaucophane *(an inosilicate)*
Glauzin *n* glaucine *(alkaloid)*
GLC *s.* Gas-Liquidus-Chromatografie
Gleiboden *m s.* Gleyboden
gleichartig homogeneous
Gleichartigkeit *f* homogeneity
gleichgestaltig *(cryst)* isomorphic, isomorphous
Gleichgestaltigkeit *f (cryst)* isomorphism
Gleichgewicht *n* equilibrium, balance ✦ **im G. halten** to keep in equilibrium, to equilibrate ✦ **ins G. bringen (setzen)** to equilibrate, to bring into equilibrium
  bewegliches G. mobile equilibrium
  Boudouardsches G. producer-gas equilibrium
  Donnansches G. Donnan [membrane] equilibrium
  dynamisches G. dynamic equilibrium
  eingefrorenes G. retarded equilibrium
  fotochemisches G. photochemical equilibrium (stationary state), photostationary state
  heterogenes G. heterogeneous equilibrium
  homogenes G. homogeneous equilibrium
  indifferentes G. neutral equilibrium
  metastabiles G. metastable equilibrium
  radioaktives G. radioactive equilibrium
  thermisches G. thermal equilibrium
  thermodynamisches G. thermodynamic equilibrium
Gleichgewichtsapparatur *f (distil)* equilibrium still
Gleichgewichtsbedingungen *fpl* equilibrium conditions
Gleichgewichtsdampfdruck *m* equilibrium vapour pressure
Gleichgewichtsdestillation *f* equilibrium distillation
Gleichgewichtsdruck *m* equilibrium pressure
Gleichgewichtseinstellung *f* establishment of equilibrium
Gleichgewichtsfeuchte[beladung] *f* equilibrium moisture content
Gleichgewichtskasten *m* equilibrium box *(theory of gas reactions)*
  G. nach van't Hoff van't Hoff equilibrium (reaction) box
Gleichgewichtskonstante *f* equilibrium constant
Gleichgewichtskonzentration *f* equilibrium concentration
Gleichgewichtslage *f* position of equilibrium
Gleichgewichtsmethode *f* equilibrium method
Gleichgewichtsquellung *f (rubber)* equilibrium swelling
Gleichgewichtsreaktion *f* balanced reaction
Gleichgewichtssiedekurve *f* [single-]flash curve
Gleichgewichtstemperatur *f* equilibrium temperature
Gleichgewichtsverdampfung *f* equilibrium [flash] vaporization
Gleichgewichtsverhältnis *n* equilibrium ratio
Gleichgewichtsverteilung *f* equilibrium distribution

Gleichgewichtsverteilungskoeffizient *m* equilibrium distribution (segregation, partition) coefficient
Gleichgewichtswert *m* equilibrium value
Gleichgewichtszustand *m* equilibrium state
Gleichlauf *m* synchronism ✦ **mit G.** synchronized, even-speed *(e.g. rolls)*
gleichmäßig uniform, even *(also of a process)*; smooth *(e.g. course of a reaction)*; *(dye)* level
Gleichmäßigkeit *f* uniformness, evenness *(also of a process)*; smoothness *(as of a reaction)*; *(dye)* levelness
gleichschnell laufend synchronous, even-speed *(e.g. rolls)*
Gleichstrom *m* 1. cocurrent (concurrent, parallel) flow *(of two fluids)*; 2. direct current, d.c., D.C. *(of electricity)* ✦ **im G. [geführt]** cocurrent *(fluids)*
Gleichstromdestillation *f* simple (direct) distillation
  diskontinuierliche G. simple batch distillation
  kontinuierliche G. simple continuous distillation
Gleichstromlichtbogen *m* direct-current arc, d.c. arc
Gleichstrompolarograf *m* direct-current polarograph
Gleichstrompolarogramm *n* direct-current polarogram
Gleichstromverfahren *n* **nach Didier-Bubiag** Didier-Bubiag process *(of coal gasification)*
Gleichung *f* equation
  G. von Brunauer, Emmett und Teller Brunauer-Emmett-Teller equation (relationship), BET equation
  Arrheniussche G. Arrhenius equation
  Berthelotsche G. Berthelot equation
  Braggsche G. Bragg equation
  chemische G. chemical equation
  Clausius-Clapeyronsche G. Clapeyron-Clausius equation
  Debyesche G. Debye equation
  empirische G. empirical equation
  Gibbs-Duhemsche G. Gibbs-Duhem equation
  van-der-Waalssche G. van der Waals equation [of state]
  van't-Hoffsche G. van't Hoff equation
Gleichverteilung *f* equipartition, equidistribution
Gleichverteilungssatz *m* **[der Energie]** law of equipartition of energy, principle of equipartition [of energy], equipartition principle
gleichwertig equivalent
Gleichwertigkeit *f* equivalence
gleichzeitig simultaneous, coincident
Gleichzeitigkeit *f* simultaneity, simultaneousness, coincidence
Gleiswaage *f* wagon balance
Gleit[dehnungs]ausgleicher *m* slip-type expansion joint
Gleitebene *f (cryst)* glide plane
gleiten to slide, to slip
Gleiten *n* slide, slip
Gleitfähigkeit *f* slip
Gleitlager *n* plain (sliding) bearing
Gleitmittel *n* lubricant, lubricating agent, lube, slip additive (agent); *(pharm)* intestinal lubricant; *(plast)* external lubricant; mould lubricant *(foundry)*
Gleitrichtung *f (cryst)* glide direction
Gleitschutzwirkung *f* antislip effect
Gleitwinkel *m* angle of slide; angle of repose *(of bulk material)*
Glessit *m* glessite *(a fossil resin resembling amber)*
Gley[boden] *m* gley [soil]
Gliadin *n* gliadin *(a prolamin)*
Gliederbandförderer *m* apron conveyor
Gliederwalze *f* section roller
glimmen to glow [feebly]; to smoulder, to burn faintly
Glimmen *n* feeble glow; smouldering
Glimmentladung *f* glow discharge
Glimmer *m* mica
glimmerähnlich mica-like
glimmerartig micaceous
Glimmermineral *n* mica
Glimmersandstein *m* micaceous sandstone
Glimmerschiefer *m* mica schist
Glimmschicht *f* / **negative** cathode[-glow] layer *(spectroscopy)*
Gliotoxin *n* gliotoxin *(an antibiotic)*
glitschig slippery
Globin *n* globin *(protein component of haemoglobin)*
globulär globular
Globulin *n* globulin *(any of a class of simple proteins)*
Glocke *f (distil)* [bubble] cap, bubbler; crown *(of a melting furnace)*; bell, cone *(of a blast furnace)*; *(lab)* bell jar *(for protecting objects esp under vacuum)*
  G. mit Dampfkamin *(distil)* cap-and-riser assembly
  G. mit gezacktem Rand *(distil)* serrated bubbler
  Brühlsche G. *(lab)* Brühl receiver
  große G. large bell *(of a blast furnace)*
  kleine G. small bell *(of a blast furnace)*
Glockenboden *m (distil)* bubble[-cap] tray, bubble[-cap] plate
Glockenbodenkolonne *f (distil)* bubble-cap (bubble-tray, bubble-plate) column
Glockenbronze *f* bell bronze
Glockengasbehälter *m* liquid seal gasholder
Glockenkappe *f (distil)* [bubble] cap, bubbler
Glockenkolonne *f s.* Glockenbodenkolonne
Glockenmessing *n* bell brass
Glockenmetall *n* bell metal
Glockenmühle *f* cone mill, conical grinder, rotary crusher
Glockenstange *f* bell beam *(of a blast furnace)*
Glockentrichter *m (lab)* thistle funnel (tube)
  G. mit Schleife und Kugel thistle funnel with safety bulb
Glocken- und Trichter[gicht]verschluß *m* bell and hopper, cup and cone *(of a blast furnace)*

Glover *m s.* Gloverturm
Gloversäure *f* Glover [tower] acid, brown oil of vitriol, B.O.V. *(chamber process)*
technisch reine G. best brown oil of vitriol, b.b.o.v., B.B.O.V.
Gloverturm *m* Glover tower *(chamber process)*
Gloverturmsäure *f s.* Gloversäure
Glover-West-Ofensystem *n* Glover-West system
Glover-West-Retorte *f* Glover-West coking (continuous vertical) retort
Glu *s.* Glutaminsäure
Glühaufschluß *m* calcination
Glühbrand *m (ceram)* biscuit firing, biscuitting
Glühdraht *m* incandescent filament
glühelektrisch thermionic
Glühelektron *n* thermionic electron, thermoelectron
Glüh[elektronen]emission *f* thermionic emission
glühen 1. to ignite, *(esp limestone, ores)* to calcine; *(ceram)* to bake; *(met)* to anneal; 2. to glow
bis zur Gewichtskonstanz (Massekonstanz) g. to ignite to constant weight
graphitisierend g. *(met)* to graphitize
homogenisierend g. *(met)* to homogenize
in Schutzgas g. *(met)* to bright-anneal
nochmals g. to re-ignite
normalisierend g. *(met)* to normalize
stabilisierend g. *(met)* to stabilize
Glühen *n* 1. ignition, *(esp of limestone, ores)* calcination; *(ceram)* baking; *(met)* anneal[ing]; 2. glow
entspannendes G. *(met)* stress-relieving anneal, stress relief [anneal]
isothermes G. *(met)* isothermal anneal
rekristallisierendes G. *(met)* recrystallization anneal
glühend glowing; *(geol)* igneous *(magma)*; red-hot *(metal)*; incandescent *(gas)*
Glühfaden *m* incandescent filament
Glühfadenpyrometer *n* disappearing-filament (hot-filament) pyrometer
Glühfarbe *f* annealing colour
Glühkasten *m*, **Glühkiste** *f (met)* annealing box
Glühkörper *m* incandescent mantle
Glühlampe *f* incandescent [lamp]
Glühofen *m (ceram)* heating furnace, hardening-on kiln; *(met)* annealing furnace
Glühphosphat *n (fert)* thermal (fused, calcined) phosphate
Glühröhrchen *n* ignition [test] tube
Glührohrprobe *f* ignition tube test
Glührückstand *m* residue on ignition
Glühschale *f (lab)* igniting dish
rechteckige G. combustion barge
Glühschiffchen *n* combustion boat
Glühstrumpf *m* incandescent mantle
Glühtemperatur *f* annealing temperature
Glühtopf *m (met)* annealing pot (can)
Glühung *f s.* Glühen
Glühverlust *m* loss on ignition, ignition loss
Glühzone *f* incandescent zone, zone of incandescence
Glukagon *n* glucagon *(a protein produced by the pancreas)*
Glukokortikoid *n* glucocorticoid *(an adrenal cortex hormone)*
Glukomannan *n* glucomannan
Glukoneogenese *f (bioch)* gluconeogenesis *(formation of glucose from proteins)*
D-Glukonsäure *f* $CH_2OH[CHOH]_4COOH$ *D*-gluconic acid
Glukonsäurelakton *n* gluconolactone
Glukoproteid *n* glucoprotein
Glukopyranose *f* glucopyranose
Glukosamin *n* glucosamine, GlcN, chitosamine
Glukosazon *n* glucosazone
Glukose *f* glucose, *(specif) D*-glucose, dextrose
***d*-Glukose** *f s. D*-Glukose
***D*-Glukose** *f D*-glucose, dextrose
Glukose-Oxydase *f* glucose oxydase *(a flavin enzyme)*
Glukosephosphat *n* glucose phosphate
Glukoserest *m* glucose residue
Glukosesirup *m* glucose syrup
Glukosid *n* glucoside *(a glycoside that yields glucose on hydrolysis)*
Glukosidase *f* glucosidase
glukosidisch glucosidic, glucosidal
Glukozuckersäure *f* $HOOC[CHOH]_4COOH$ glucosaccharic acid, saccharic acid, glucaric acid *(one form of 2,3,4,5-tetrahydroxyhexanedioic acid)*
Glukuronid *n* glucuronide, glucuronoside
Glukuronidase *f* glucuronidase
Glukuronsäure *f* $OHC[CHOH]_4COOH$ glucuronic acid
Glukuronsäurelakton *n* glucuronolactone
Glu-NH_2 *s.* Glutamin
Glutamat *n* glutamate *(salt or ester of glutamic acid)*
Glutamin *n* $NH_2-CO-CH_2CH_2CH(NH_2)COOH$ glutamine
Glutaminase *f* glutaminase
Glutamindehydrogenase *f s.* Glutaminsäuredehydrogenase
Glutaminsäure *f* $HOOC-CH_2CH_2CH(NH_2)-COOH$ glutamic acid, 1-aminopropane-1,3-dicarboxylic acid
Glutaminsäuredehydrase *f s.* Glutaminsäuredehydrogenese
Glutaminsäuredehydrogenase *f* glutamic acid dehydrogenase
Glutaminylzysteinylglyzin *n s.* Glutaminzysteinglykokoll
Glutaminzysteinglykokoll *n* glutamylcysteinylglycine, glutathione
Glutaraldehyd *m* $OHC-[CH_2]_3-CHO$ glutaraldehyde, glutaric dialdehyde, 1,3-pentanedial
Glutardialdehyd *m s.* Glutaraldehyd
Glutarsäure *f* $HOOC-[CH_2]_3-COOH$ glutaric acid, propane-1,3-dicarboxylic acid
Glutarsäuredialdehyd *m s.* Glutaraldehyd
Glutathion *n* glutathione, glutamylcysteinylglycine

Glutbeständigkeit *f* resistance to glow heat
Glutbeständigkeitsprobe *f* glow bar test, glowing hot-body test
Glutelin *n* glutelin *(any of a class of simple proteins)*
Gluten *n (bioch)* gluten
Glutfestigkeit *f s.* Glutbeständigkeit
Gluzinerde *f s.* Berylliumoxid
Gluzinium *n s.* Beryllium
Glv. *s.* Glühverlust
Gly *s.* Glykokoll
Glykan *n* glycan, polysaccharide
Glykocholsäure *f* glycocholic acid *(a bile acid)*
Glykogen *n* glycogen, animal (liver) starch
Glykogenabbau *m* glycogenolysis
Glykogenase *f* glycogenase
Glykogenese *f* glycogenesis
Glykogenolyse *f* glycogenolysis
Glykogensäure *f s. D*-Glukonsäure
Glykokoll *n* NH_2CH_2COOH glycine, glycocoll, aminoacetic acid
Glykokollkupfer *n* glycine copper, copper glycine
Glykol *n* $C_nH_{2n}(OH)_2$ glycol, diol, dihydric alcohol, diatomic alcohol; *(specif) s.* 1,2-Glykol
1,2-Glykol *n* $HOCH_2CH_2OH$ ethylene glycol, 1,2-dihydroxyethane, 1,2-ethanediol
Glykolaldehyd *m* $HOCH_2CHO$ glycollaldehyde, hydroxyacetaldehyde
Glykolbad *n (lab)* glycol bath
Glykolchlorhydrin *n* $ClCH_2CH_2OH$ 2-chloroethanol, glycol chlorohydrin
Glykoldibromid *n s.* 1,2-Dibromäthan
Glykolharnstoff *m* glycollylurea, hydantoin, 2,4-imidazolidinedione
Glykolipid *n* glycolipid[e]
Glykolmonoäthyläther *m s.* Äthylenglykolmonoäthyläther
Glykolsäure *f* $HOCH_2COOH$ glycolic (glycollic) acid, hydroxyacetic acid
glykolspaltend glycol-splitting
Glykolyse *f* glycolysis
d-**Glykonsäure** *f s. D*-Glukonsäure
Glykoproteid *n* glycoprotein, glycopeptide
Glykose *f s.* Glukose
Glykosid *n* glycoside
  herzaktives G. cardiac glycoside
Glykosidase *f* glycosidase
Glykosidbindung *f* glycosidic bond (linkage)
glykosidisch glycosidic
Glykosurie *f (med)* glycosuria
Glykozyamin *n* $HN{=}C(NH_2)NHCH_2COOH$ glycocyamine, guanidinoacetic acid
Glyoxal *n* $OHC{-}CHO$ glyoxal, oxaldehyde, ethanedial
Glyoxalase *f* glyoxalase
Glyoxalin *n s.* Imidazol
Glyoxalsäure *f s.* Glyoxylsäure
Glyoxylsäure *f* $OHC{-}COOH$ glyoxylic acid
Glyptal[harz] *n* glyptal [resin], glycerol phthalic resin, phthalic glyceride resin
Glyzeraldehyd *m s.* Glyzerinaldehyd
Glyzerat *n* glycerate *(salt or ester of glyceric acid)*
Glyzerid *n* glyceride *(any of the esters of glycerol)*
  gemischtes (gemischtsäuriges) G. mixed (component, partial) glyceride
Glyzeridöl *n* glyceride oil
Glyzerin *n* $CH_2(OH)CH(OH)CH_2OH$ glycerol, *(less desirably)* glycerin, 1,2,3-trihydroxypropane
Glyzerinaldehyd *m* $CH_2(OH)CH(OH)CHO$ glyceraldehyde, 2,3-dihydroxypropanal
Glyzerinbad *n (lab)* glycerol bath
Glyzerin-α-chlorhydrin *n* $HOCH_2CH(OH)CH_2Cl$ [glycerol] α-chlorohydrin, 3-chloro-1,2-propanediol
Glyzerindiazetat *n* glycerol diacetate, diacetylglycerol, diacetin
Glyzeringärung *f* glycerol fermentation
Glyzerinmonoazetat *n* glycerol monoacetate, monoacetin
Glyzerinmonostearat *n* glycerol monostearate, GMS, monostearin
Glyzerinphosphorsäure *f* $C_3H_5(OH)_2OPO_3H_2$ glycerophosphoric acid
Glyzerin-Phthalsäure-Harz *n s.* Glyptalharz
Glyzerinsäure *f* $CH_2(OH)CH(OH)COOH$ glyceric acid, 2,3-dihydroxypropionic acid
Glyzerintrimyristat *n s.* Glyzerintrimyristinsäureester
Glyzerintrimyristinsäureester *m* glycerol trimyristate, trimyristin
Glyzerintrinitrat *n* $C_3H_5(ONO_2)_3$ glycerol trinitrate, glyceryl trinitrate
Glyzerintripalmitat *n s.* Glyzerintripalmitinsäureester
Glyzerintripalmitinsäureester *m* glycerol tripalmitate, tripalmitin
Glyzerintristearat *n s.* Glyzerintristearinsäureester
Glyzerintristearinsäureester *m* glycerol tristearate, tristearin
Glyzerol *n s.* Glyzerin
Glyzerophosphat *n* glycerophosphate
Glyzerophthalat *n* glycerophthalate
Glyzerylmonostearat *n s.* Glyzerinmonostearat
Glyzidester *m* glycidic ester
Glyzidylätherharz *n* glycidyl ether resin
Glyzin *n* NH_2CH_2COOH glycine, aminoacetic acid
Glyzinerde *f s.* Berylliumoxid
Gmelinit *m (min)* gmelinite *(a tectosilicate)*
Gneis *m (geol)* gneiss
Gneist *m (tann)* scud *(fat and lime soap remaining on hides or skins)*
Goapulver *n (pharm)* Goa powder, araroba *(from Andira araroba Aguiar)*
Goethit *m (min)* goethite, göthite *(iron hydroxide oxide)*
Gold *n* Au gold
  kolloidales G. gold sol
Goldamalgam *m (min)* amalgam
Goldaventurin *m* gold aventurine
Gold(I)-bromid *n* AuBr gold(I) bromide, gold monobromide, aurous bromide
Gold(III)-bromid *n* $AuBr_3$ Gold(III) bromide, gold tribromide, auric bromide

Gold(I,III)-bromid *n* Au_2Br_4 gold(I,III) bromide, digold tetrabromide, auroso-auric bromide
Goldbromsäure *f s.* Tetrabromogold(III)-säure
Goldbromwasserstoff *m s.* Tetrabromogold(III)-säure
Goldbronze *f* gold bronze
Gold(I)-chlorid *n* AuCl gold(I) chloride, gold monochloride, aurous chloride
Gold(III)-chlorid *n* $AuCl_3$ gold(III) chloride, gold trichloride, auric chloride
Gold(I,III)-chlorid *n* Au_2Cl_4 gold(I,III) chloride, digold tetrachloride, auroso-auric chloride
Goldchlorwasserstoffsäure *f s.* Tetrachlorogold(III)-säure
Golddekor *n (ceram)* gold decoration
Goldfolie *f* gold foil
goldführend auriferous, gold-bearing
Goldgehalt *m* gold content
goldhaltig gold-bearing, *(geoch also)* auriferous
Gold(III)-hydrogenbromid *n s.* Tetrabromogold-(III)-säure
Goldhydrosol *n* gold hydrosol
Gold(I)-hydroxid *n* AuOH gold(I) hydroxide, aurous hydroxide
Gold(III)-hydroxid *n* $Au(OH)_3$ gold(III) hydroxide, auric hydroxide, auric acid
Gold(I)-jodid *n* AuI gold(I) iodide, gold monoiodide, aurous iodide
Gold(III)-jodid *n* AuI_3 gold(III) iodide, gold triiodide, auric iodide
Goldklumpen *m* gold nugget
Goldmono ... *s.a.* Gold(I)- ...
Goldmonoxid *n s.* Gold(I)-oxid
Goldorange *n* methyl (gold) orange
Gold(I)-oxid *n* Au_2O gold(I) oxide, digold monooxide, aurous oxide
Gold(III)-oxid *n* Au_2O_3 gold(III) oxide, digold trioxide, auric oxide
Goldpräparat *n* gold preparation
Goldquarz *m* auriferous quartz
Goldrubinglas *n* gold ruby [glass]
Goldsand *m* auriferous sand (gravel), wash
Goldsäure *f s.* Gold(III)-hydroxid
Goldschmidt-Radius *m* Goldschmidt radius *(of ions)*
Goldschmidt-Verfahren *n* Goldschmidt process, aluminothermics, aluminothermy
Goldschwefel *m* golden antimony sulphide, antimonial saffron, antimony red *(antimony(V) sulphide)*
Goldseife *f (geoch)* gold-placer
Goldsol *n* gold sol
Goldsolreaktion *f (med)* gold sol test
Goldstaub *m* gold dust
Gold(I)-sulfid *n* Au_2S gold(I) sulphide, digold sulphide, aurous sulphide
Gold(III)-sulfid *n* Au_2S_3 gold(III) sulphide, digold trisulphide, auric sulphide
Gold(I,III)-sulfid *n* Au_2S_2 gold(I,III) sulphide, digold disulphide, auroso-auric sulphide
Goldtönung *f (phot)* gold toning
Goldtri ... *s.* Gold(III)- ...
Goldwäsche[rei] *f* gold washing
Goldzahl *f (coll)* gold number
Gold(I)-zyanid *n* AuCN gold(I) cyanide, gold monocyanide, aurous cyanide
Gold(III)-zyanid *n* $Au(CN)_3$ gold(III) cyanide, gold tricyanide, auric cyanide
Gomartharz *n* tacamahac[a] gum, West Indian elemi *(from Bursera gummifera L.)*
Goniometer *n (cryst)* goniometer
goniometrisch *(cryst)* goniometric
Gooch-Tiegel *m* Gooch crucible (filter)
Goodrich-Flexometer *n (rubber)* Goodrich flexometer
Goodyear-Winkelmaschine *f (rubber)* Goodyear angle machine
Gordon-Plastikator *m (rubber)* Gordon plasticator
Goslarit *m (min)* goslarite, zinc vitriol, white vitriol (copperas) *(zinc sulphate-7-water)*
Gottignies-Ofen *m (ceram)* Gottignies kiln *(an electric multipassage kiln)*
Gould-Jacobs-Reaktion *f* Gould-Jacobs reaction *(formation of 4-hydroxyquinolines)*
GÖV, G.Ö.V. *s.* Gas-Öl-Verhältnis
GPF-Ruß *m (rubber)* general-purpose furnace black, GPF black
Grad *m* degree; grade *(of purity of chemicals)* ✦ **in Grade [ein]teilen** to graduate
  G. API *(petrol)* degree API
  G. Baumé degree Baumé, °Bé
  G. Celsius degree centigrade (Celsius), deg C, °C
  G. Fahrenheit degree Fahrenheit, deg F, °F
  G. Sugar degree sugar solution
  G. Twaddell degree Twaddell, °Tw
Gradation *f (phot)* gradation, development factor, gamma value
Gradationskurve *f (incorrectly for)* Schwärzungskurve
Gradeinteilung *f* graduation, division
Gradient *m* gradient
Gradient-Dünnschichtchromatografie *f* gradient [thin-]layer chromatography
Gradient[en]elution *f* gradient elution
Gradientschicht *f* gradient layer
Gradientschichtchromatografie *f* gradient [thin-] layer chromatography
Gradientschichttechnik *f* gradient layer technique
gradieren to graduate *(solutions by evaporation)*
Gradieren *n* graduation *(of solutions by evaporation)*
graduieren to graduate, to calibrate
graduiert / auf Ablauf (Auslauf) graduated (calibrated) to deliver, graduated for delivery
  auf Einguß g. calibrated to contain, calibrated for content
Graduierung *f* graduation
gradweise gradual
Graftpolymer[es] *n* graft [polymer]
Graham-Salz *n* Graham's salt *(a sodium metaphosphate glass)*

grainieren *(pap)* to grain, to press
Gram-Farbstoff *m* Gram stain *(bacteriology)*
Gram-Färbung *f* Gram staining *(bacteriology)*
Graminin *n* graminin *(a polysaccharide)*
Gramizidin *n* gramicidin *(an antibiotic)*
Grammäquivalent *n* gram equivalent, g. equiv., val
Grammatitstrahlenstein *m (min)* tremolite *(an inosilicate)*
Grammatom *n* gram atom, gram-atomic weight
Grammion *n* gram ion
Grammmol[ekül] *n* gram molecule (mole), gram-molecular weight, mole
Grammsuszeptibilität *f* susceptibility per gram, specific (mass) susceptibility
gramnegativ Gram-negative *(bacteriology)*
grampositiv Gram-positive *(bacteriology)*
Granalie *f* granule, pellet, agglomerate
Granat *m (min)* garnet
Granat[schel]lack *m* garnet lac
Granit *m* granite
graniten granitic
Granitisation *f (geol)* granitization
granitisch granitic
granitisieren *(geol)* to granitize
Granitisierung *f (geol)* granitization
Granitwalze *f (pap)* granite roll
Granulat *n* granular material; *(molten droplets which have solidified)* shot
Granulatformen *n* agglomerating, agglomeration *(of powder by means of a liquid)*
Granulatformer *m s.* Granulator
Granulation *f* granulation, granulating, graining, *(solidification of molten droplets also)* shotting, *(of powder by means of a liquid also)* agglomeration
Granulatkorn *n* pellet, granule, *(formed from powder also)* agglomerate
Granulator *m* granulator, pelletizer, granulating (pelletizing) machine
Granulieranlage *f* granulation plant
Granulierapparat *m s.* Granulator
granulieren to granulate, to grain
granuliert granulate[d], grained
Granulierteller *m* pan granulator
Granuliertrommel *f* rotary-drum granulator, [rotary] granulation drum
Granulierung *f s.* Granulation
Granulierverfahren *n (ceram)* granule method
granulometrisch granulometric *(classifying)*
granulös granular, grainy, granulate
Granulum *n* granule, pellet
Graphit *m* graphite ✦ **mit G. auskleiden** to line with graphite, to graphitize, *(Am also)* to graphite ✦ **mit G. überziehen** to coat with graphite, to graphitize, *(Am also)* to graphite
primärer G. kish, keesh *(foundry)*
graphitartig graphite-like
Graphitbildung *f* graphitization
Graphitboot *n* graphite boat *(in zone melting)*
Graphitbrenner *m* graphite burner
Graphit-Einlagerungsverbindung *f* intercalation (lamellar) compound of graphite
Graphitelektrode *f* graphite electrode
graphitgebremst *(nucl)* graphite-moderated
Graphitgitter *n* graphite lattice
Graphitglühen *n* graphitizing
graphithaltig graphitic, containing graphite
Graphithydrogensulfat *n* graphite hydrogensulphate
graphitieren to graphitize, *(Am also)* to graphite
Graphitierung *f* graphitization
graphitisch graphitic
graphitisieren *s.* graphitieren
graphitmoderiert *(nucl)* graphite-moderated
Graphitpapier *n* graphite paper
Graphitreaktor *m (nucl)* graphite-moderated reactor
natriumgekühlter G. sodium graphite reactor
Graphitsalz *n* graphite (graphitic) salt
Graphitsäure *f* graphitic acid
Graphitschicht *f* graphite layer
Graphitschiffchen *n* graphite boat *(in zone melting)*
Graphitstab *m* graphite rod
Graphitstruktur *f* / **wabenförmige** *(cryst)* honeycomb graphitic structure
Graphittiegel *m* graphite (plumbago) crucible
graphitüberzogen graphite-faced
Graphitverbindung *f* graphitic compound
Graphit-Wärmeaustauscher *m*, **Graphit-Wärmeübertrager** *m* graphite heat exchanger
Grappierzement *m* grappier cement
Grasbaumharz *n* acaroid gum (resin), accroides (grass-tree, black-boy) gum *(from Xanthorrhoea specc.)*
Gräserbekämpfungsmittel *n* grass killer
Grasöl *n* / **Indisches** *(cosmet)* lemon-grass oil, East Indian verbena oil *(from Cymbopogon specc.)*
Grat *m* burr *(produced in cutting metal)*; flash, fin *(on castings)*; *(plast)* burr, flash; *(rubber)* flash, rind
Gratlinie *f*, **Gratnaht** *f (plast)* joint (spew, flash) line
Gratus-Strophantin *n* G-strophantin
Grauguß *m* grey [cast] iron; *(from a mould)* grey-iron casting
globularer (sphärolithischer) G. nodular cast iron
Grauhuminsäure *f (soil)* grey humic acid
Graukalk *m* 1. grey lime (acetate), vinegar salt *(crude calcium acetate)*; 2. *(agric)* dolomitic lime *(calcium magnesium oxide)*
Graukarton *m* grey cardboard
Graupappe *f* grey (news) board
Grauschleier *m (phot)* grey fog; *(text)* frosting effect
Grauschwefel *m* wind-blown sulphur
Grauspießglanz *m (min)* grey antimony, antimony glance, antimonite, stibnite *(antimony(III) sulphide)*
Grauwacke *f (geol)* greywacke
Gravimetrie *f* gravimetry
gravimetrisch gravimetric[al]
Gravitation *f* gravitation, gravity

Gravitationsdifferentiation *f (geol)* gravitative differentiation
Gravitationsfeld *n* gravitational field
Gravitationskonstante *f* gravitational constant
Gravitationskraft *f* force of gravitation (gravity), gravitational force (pull)
Gray-Dampfphase-Prozeß *m s.* Gray-Prozeß
Gray-King-Kokstypus *m* Gray-King [assay] coke type
Gray-King-Test *m*, **Gray-King-Verkokungstest** *m* Gray-King assay [test]
Gray-Prozeß *m* Gray process *(for treating gasoline with bleaching earth)*
Great-Northern-Schleifer *m (pap)* Great Northern grinder
Greaves-Etchells-Ofen *m* Greaves-Etchells furnace *(an arc-heated furnace)*
Greenalith *m (min)* greenalite *(a hydrous iron(II) silicate)*
Greenawalt-Pfanne *f*, **Greenawalt-Sinterpfanne** *f (met)* Greenawalt sintering machine
Greenockit *m (min)* greenockite *(cadmium sulphide)*
Greenovit *m (min)* greenoughite, greenovite *(a variety of titanite)*
Grège[seide] *f* grege, greige, gum silk
Greifer *m* scoop
Greisen *m (geol)* greisen
Greisenbildung *f (geol)* greisening, greisenization
Grenzdextrin *n* residual (limit) dextrin
Grenze *f* 1. limit; 2. boundary *(as of an atom or phase)*
0,2%-Grenze *f* yield strength 0.2 % offset *(materials testing)*
Grenzemulsion *f* marginal emulsion
Grenzenergie *f* **/ Fermische** Fermi [characteristic energy] level
Grenzfläche *f* boundary (bounding) surface, interface
G. fest-fest solid-solid interface
G. fest-flüssig solid-liquid interface
G. flüssig-flüssig liquid-liquid interface, dineric interface
G. flüssig-gasförmig liquid-vapour interface
grenzflächenaktiv surface-active
Grenzflächenaktivität *f* surface (interfacial) activity
Grenzflächenenergie *f* interfacial energy
Grenzflächenerscheinung *f* interfacial phenomenon
Grenzflächenfilm *m* interfacial film *(in emulsions)*
Grenzflächenreibungsarbeit *f* interfacial work
Grenzflächenspannung *f* interfacial tension
Grenzflächenwinkel *m* interfacial angle
Grenzformel *f* **[/ mesomere]** resonance formula
Grenzfrequenz *f* limiting frequency
Grenzkohlenwasserstoff *m* alkane, saturated hydrocarbon, paraffin [hydrocarbon]
Grenzkohlenwasserstoffreihe *f* alkane family, paraffin series
Grenzkonformation *f* full conformation *(stereochemistry)*
Grenzkonzentration *f* limiting concentration; *(tox)* maximum permissible concentration
Grenzkorngröße *f* **/ obere** upper size
untere G. lower size
Grenzleitfähigkeit *f* equivalent conductance at infinite dilution
Grenzlinie *f* interface line *(between two components)*
Grenzschicht *f* boundary layer
Grenzstrom *m (phys chem)* [limiting, maximum] diffusion current
Grenzstromtitration *f* amperometric titration
Grenzstruktur *f* limiting structure
mesomere G. resonance (resonating) structure, contributing [mesomeric] form
Grenzverhältnis *n* limiting ratio
Grenzviskosität *f* intrinsic viscosity
Grenzviskositätszahl *f* limiting viscosity number
Grenzwert *m* limiting value, limit
G. für Arsen *(tox)* arsenic limit *(expressed as As_2O_3)*
G. für Blei *(tox)* lead limit
Grenzwertbestimmung *f* limit test
Grenzzustand *m* limiting state
Grieß *m* oversize [material, product] *(air classifying)*
grießartig gritty
Grießprobe *f (glass)* powder test
Griff *m* 1. handle *(of an apparatus)*; 2. *(pap, text, tann)* feel, handle, *(pap also)* bulk, *(text also)* hand *(property of material)*
weicher G. *(text)* soft handle
Griffappretur *f (text)* stiffening
griffig of good feel (handle), *(pap also)* bulky
g. sein to have a good feel (handle), *(pap also)* to bulk
Griffigkeit *f (pap, text, tann)* feel, handle, *(pap also)* bulk, *(text also)* hand
Griffin-Mühle *f* Griffin [ring-roll] mill
Grignardierung *f (org chem)* grignardization
Grignard-Reagens *n (org chem)* Grignard reagent
Grignard-Reaktion *f*, **Grignard-Synthese** *f (org chem)* Grignard reaction (synthesis)
Grignard-Verbindung *f (org chem)* Grignard compound
Grignard-Verfahren *n (org chem)* Grignard method (process)
Grisein *n* grisein *(an antibiotic)*
grob 1. coarse *(screen)*; 2. coarse, rough *(e.g. bulk material)*; 3. rough *(method)*
grobätzen to macroetch
Grobausmahlung *f* coarse grinding
Grobbrechen *n* coarse crushing
Grobbrecher *m* coarse crusher
grobdispers coarse-disperse, coarsely dispersed
grobfaserig coarse-fibred
Grobfilter *n* roughing filter
Grobgefüge *n* macrostructure
Grobgut *n* coarse material; oversize [material, product], tailings, tails *(classifying)*; *(pap)* oversize chips

Grobgutaustrag *m* sand discharge *(of a classifier)*
Grobkeramik *f* 1. heavy ceramics *(branch)*; 2. heavy clay product (ware)
Grobkeramikindustrie *f* heavy clay industry
Grobklassieren *n* coarse sizing
Grobkorn *n* oversize [material, product], tailings, tails *(classifying)*
grobkörnig coarse-grained
Grobkörnigkeit *f* coarseness
grobkristallin coarse-crystalline
grobmahlen to crush, to grind coarsely
Grobmahlung *f* crushing, coarse grinding
grobnarbig *(tann)* coarse-grained
Grobsand *m* coarse sand, grit
Grobsortierer *m* coarse screen
Grobsortierung *f* coarse [material] screening
Grobspäne *mpl (pap)* oversize chips
Grobstoff *m (pap)* groundwood (screen) rejects, rejected stock, junk, screen[ing]s, tail[ing]s
Grobton *m* coarse clay
Grobtrub *m (ferm)* coarse sludge
Grobvakuum *n* low vacuum
Grobwaschmittel *n* heavy-duty detergent
grobzerkleinern to crush, to grind coarsely
Grobzerkleinerung *f* crushing, coarse grinding
Grobzuschlag *m* coarse aggregate
Großbetrieb *m* large-scale plant; *(as opposed to pilot plant)* full-scale plant
Größe *f (phys chem)* quantity
extensive G. extensive quantity
intensive G. intensive quantity
konstante G. constant [quantity]
kritische G. critical constant
partielle molare G. partial molar (molal) quantity
Größenbestimmung *f* size determination *(classifying)*
Größenordnung *f* order of magnitude
Größenverteilung *f* size distribution *(classifying)*
Großerzeuger *m* large producer
Großfertigung *f s.* Großproduktion
großindustriell large-scale
Großkoks *m* large coke
Großleistungszentrifuge *f* high-capacity centrifuge
Großpackung *f* bulk package
großporig large-pored
Großproduktion *f* large-scale production (manufacture, fabrication); *(as opposed to pilot-plant production)* full-scale factory production
Großproduzent *m* large producer
Großprozeß *m* large-scale process
Großraumgärverfahren *n* bulk champagnization (process), charmat process
Großraumzentrifuge *f* high-capacity centrifuge
Großringverbindung *f* macrocyclic (macroring, large-ring) compound
großstückig blocky, lumpy
großtechnisch large-scale; *(as opposed to pilot-scale)* full-scale
g. hergestellt produced on the large scale
Grossular *m (min)* grossular[ite] *(a garnet)*
Großversuch *m* large-scale test
Grübchen *n (plast)* pit *(a moulding defect)*
Grübchenbildung *f (plast)* pitting
Grube *f* 1. pit; 2. mine
Grubenäscher *m (tann)* pit liming *(process)*; pit lime *(substance)*
Grubengas *n* mine gas, filty, fire-damp
Grubengerbung *f* pit tannage
Grubenkohle *f* run-of-mine coal
Grubenwasser *n* mine water
Grudekoks *m* granular coke
Grün *n* green *(sensation or substance)*
Böttgers G. *s.* Kasseler G.
Chinesisches G. Chinese green, locao *(a natural dye from Rhamnus specc.)*
Kasseler G. Cassel (manganese) green *(barium manganate)*
Mittlers G. Mittler's (emerald, chrome) green *(hydrated chromium oxide)*
Pariser G. *s.* Schweinfurter G.
Rinmanns G. Rinmann's (cobalt) green *(consisting essentially of cobalt and zinc oxides)*
Rosenstiehls G. *s.* Kasseler G.
Scheeles G. Scheele's green *(copper arsenite)*
Schweinfurter G. Paris (emerald, Schweinfurth) green *(copper acetate arsenite)*
Spanisches G. *s.* Grünspan
Wiener G. *s.* Schweinfurter G.
Grünablauf *m (sugar)* high green syrup, first molasses
Grünbleierz *n (min)* green (brown) lead ore, pyromorphite
Grund *m (tann)* scud *(remaining fat and lime soap on hides or skins)*
Grundanstrich *m* priming *(act)*; first (priming, ground) coat, priming *(result)*
Grundanstrichfarbe *f* priming (prime) paint, priming, primer; *(Am)* priming (ground) color
Grundanstrichmittel *n* priming, primer
Grundausstattung *f* small-scale equipment
Grundband *n (phys chem)* ground band
Grundbegriff *m* basic term
Grundbestandteil *m* main (principal, chief) component, main constituent (ingredient), base, basis
Grundchemikalie *f* key chemical
Grunddünger *m* basal fertilizer
Grunddüngung *f* basal dressing
Grundeinheit *f* 1. base molecule (unit), repeating unit *(of a polymer)*; 2. base unit *(of a system of units)*
Gründeldruck *m (text)* blotch printing
Grundelektrolyt *m* supporting (basic) electrolyte
Grundfarbe *f* 1. primary colour *(theory of colours)*; 2. *s.* Grundanstrichfarbe
Grundfläche *f* basis, base
Grundformel *f* fundamental formula
stöchiometrische G. stoichiometric (simplest) formula
Grundfraktion *f (petrol)* primary fraction
Grundfrequenz *f* fundamental frequency
Grundgerüst *n* backbone, skeleton *(of a chain*

molecule)
zyklisches G. parent ring system
Grundgestein *n* bedrock
Grundieranstrich *m s.* Grundanstrich
grundieren *(coat)* to prime, to ground; *(tann, text)* to bottom
Grundierfarbe *f (dye)* bottoming dye[stuff]; *s.* Grundanstrichfarbe
Grundiermasse *f* sizing [material, substance], size
Grundiermittel *n (coat)* priming, primer
Grundierung *f* 1. *s.* Grundanstrich; 2. *(tann, dye)* bottoming
Grundkohlenwasserstoff *m* parent hydrocarbon
Grundkomponente *f s.* Grundbestandteil
Grundkörper *m* parent (mother) substance
Grundlack *m* base (bottom) lacquer; *(if transparent)* base varnish
Grundlage *f* base, basis
Parrsche G. Parr's basis *(in Seyler's coal chart)*
Grundlagencreme *f (cosmet)* make-up base, foundation cream
flüssige G. foundation lotion
Grundlagenforschung *f* fundamental research
Grundlinie *f* persistent (ultimate) line *(spectral analysis)*
Grundlösung *f* primary (basis, basic) solution
Grundluft *f* soil air
Grundmann-Synthese *f* Grundmann synthesis *(for obtaining aldehydes)*
Grundmasse *f (ceram, geol)* matrix, ground-mass
Grundmetall *n* principal (base) metal *(of an alloy)*
Grundmischung *f (rubber)* masterbatch, base stock (compound, mix), mother (blank) stock
Grundmolekül *n* basic (base, fundamental) molecule, repeating (base) unit, monomer, structural element (unit) *(of a polymer)*
bifunktionelles G. bifunctional unit
Grundniveau *n s.* Grundzustand
Grundoperation *f* unit operation
G. der chemischen Verfahrenstechnik chemical engineering unit operation
Grundplatte *f* base (bottom) plate, bed [plate]
Grundplattenpumpe *f* pedestal-mounted pump
Grundprozeß *m* unit process
Grundsatz *m* principle
Grundschlamm *m* bottom sludge *(in waste-water)*
Grundschwingungsbande *f (phys chem)* fundamental band
Grundstoff *m* parent substance, key chemical
chemischer G. chemical element
unentbehrlicher G. *(agric)* essential element
Grundstrom *m (phys chem)* residual current
Grundstruktur *f* basic structure
Grundsubstanz *f* parent (mother) substance
Grundterm *m s.* Grundzustand
Grundviskosität *f* intrinsic viscosity
Grundwasser *n* ground (subsoil, subsurface, underground, plerotic) water
Grundwasserspiegel *m* ground water level, water table
Grundwerk *n (pap)* bedplate, dead (beater) plate *(of a hollander beater)*
Grundwerkfassung *f*, **Grundwerkkasten** *m (pap)* bedplate box
Grundwerksmesser *npl (pap)* bedplate bars (knives) *(of a beater)*
Grundzustand *m (phys chem)* ground state (level, term), normal state, basic term
grünen to green *(canned vegetables)*
Grünfestigkeit *f* green strength *(1. of moulding sand; 2. of ceramic ware)*
Grünform *f* green-sand mould *(foundry)*
Grünformsand *m* green [moulding] sand *(foundry)*
Grünglas *n* green glass
Grüngußsand *m s.* Grünformsand
Grünlauge *f (pap)* green liquor
Grünlaugenbehälter *m (pap)* green liquor storage
Grünlaugenklärer *m*, **Grünlaugenklärtank** *m (pap)* green liquor clarifier
Grünlaugenklärung *f (pap)* green liquor clarification
Grünlaugenvorratstank *m (pap)* green liquor storage
Grünling *m* green compact *(in powder metallurgy)*
Grünmalz *n (ferm)* green malt
Grünmasse *f (tann)* green weight *(of hides and skins)*
Grünöl *n* green (anthracene) oil
Grünpellet *n* green (moist) pellet
Grünsand *m* green [moulding] sand *(foundry)*
Grünsandform *f* green-sand mould *(foundry)*
Grünsandformen *n* green-sand moulding *(foundry)*
Grünschlick *m* green mud
Grünschwefel *m* green sulphur *(in gas purification)*
Grünsirup *m (sugar)* high green syrup, first molasses
Grünspan *m* verdigris, aerugo *(a mixture of basic copper(II) acetates)*
gereinigter (kristallisierter) G. neutral verdigris *(copper(II) acetate hydrate)*
Grünstärke *f* raw (wet-end) starch
Grünstein *m (min)* greenstone
Grünung *f* greening *(of canned vegetables)*
Gruppe *f* group *(of atoms)*; family, group *(in the periodic system)*; set, battery, bank *(of equipment)*
G. der Eisenmetalle iron group
aktive G. active (prosthetic) group, coenzyme, agon *(enzymology)*
aktivierende G. activating group
auxochrome G. *(dye)* auxochromic group, auxochrome
bathochrome G. *(dye)* bathochromic group, bathochrome, bathychrome
brückenbildende G. bridging group
chromophore G. *(dye)* chromophoric group, chromophore
die Substitution desaktivierende G. substitution-deactivating group

endständige G. terminal group
farberhöhende G. hypsochromic group
farbgebende (farbtragende) G. *s.* chromophore G.
farbvermehrende (farbverstärkende) G. *s.* auxochrome G.
farbvertiefende G. *s.* bathochrome G.
funktionelle G. functional group
haptophore G. haptophoric group, haptophore *(portion of a toxin molecule which binds it to a body cell)*
hypsochrome G. *(dye)* hypsochromic group, hypsochrome
löslichmachende G. solubilizing group
meta-dirigierende G. meta-directing group, meta director
nukleophile G. nucleophilic (electron-releasing) group
nullte G. zero group *(in the periodic system)*
ortho-para-dirigierende G. ortho-para-directing group, ortho-para director
osmophore G. osmophore
prosthetische G. prosthetic (active) group, coenzyme, agon *(enzymology)*
reaktionsfähige (reaktive) G. reactive group
terminale G. terminal group
toxophore G. toxophoric group, toxophore
zybotaktische G. *(phys chem)* cybotactic group *(of molecules exhibiting crystal-like arrangement in certain liquids)*

Gruppenanalyse *f* group analysis
Gruppeneinteilung *f* group separation *(analytical chemistry)*
Gruppengeschwindigkeit *f* group velocity *(of waves)*
Gruppennummer *f* group number *(coal classification)*
Gruppenparameter *m* group parameter *(coal classification)*
Gruppenreagens *n* group reagent
Gruppentrennung *f* group separation *(analytical chemistry)*
Gruppenversuch *m* group experiment
Gruppenziffer *f s.* Gruppennummer
gruppieren to group
Gruppierung *f* grouping
Grus *m (soil)* grit; *(coal)* breeze
G-Salz *n* G salt, 2-naphthol-6,8-disulphonic acid dipotassium salt
G-Säure *f* $HOC_{10}H_5(SO_3H)_2$ G acid, 2-naphthol-6,8-disulphonic acid
GSC *s.* Gas-Feststoff-Chromatografie
g-Strophanthin *n* G-strophanthin
g. T. *s.* Tiegel / geschlossener
GU *s.* Gummifaserstoff
Guajakharz *n* guaiac [rosin], gum guaiac *(from Guajacum officinale L. and Guajacum sanctum L.)*
Guajakol *n* $CH_3OC_6H_4OH$ guaiacol, *o*-methoxyphenol
Guajen *n* 1. $C_{15}H_{24}$ guaiene *(a sesquiterpene)*; 2. $C_{10}H_6(CH_3)_2$ 2,3-dimethylnaphthalene, guaiene
Guajol *n* guaiol *(a sesquiterpene alcohol)*
Guanajuatit *m (min)* guanajuatite *(bismuth selenide)*
Guanidin *n* $NH{=}C(NH_2)_2$ guanidine, iminourea
Guanidinbeschleuniger *m* guanidine accelerator
Guanidinessigsäure *f s.* Guanidinoessigsäure
Guanidinoessigsäure *f* $HN{=}C(NH_2)NHCH_2COOH$ guanidinoacetic acid
Guanin *n* guanine, 6-hydroxy-2-aminopurine
Guano *m* guano *(a fertilizer esp from partly decomposed bird excrements)*
Guanosin *n* guanosine, guanine riboside *(a nucleoside)*
Guarana *n*, **Guaranapaste** *f* guarana *(a paste from the seeds of Paullinia cupana Kunth containing caffeine)*
Guaruma-Wachs *n* guaruma wax *(from Calathea lutea G.F.W. Mey.)*
Guäthol *n* $C_2H_5OC_6H_4OH$ guäthol, catechol monoethyl ether
Guayana-Elemi *n* elemi of Guiana *(from Icica viridiflora Lam.)*
Guayule *f*, **Guayule-Kautschuk** *m* guayule rubber *(from Parthenium argentatum A. Gray)*
Guggenheim-Verfahren *n* Guggenheim process *(for obtaining pure sodium nitrate from Chile saltpetre)*
Gugul *n* Indian bdellium *(balsamic resin from Commiphora mukul Engl.)*
Guignetgrün *n* Guignet's (chrome, emerald, Mittler's) green *(hydrated chromic oxide)*
Guillotinehadernschneider *m (pap)* guillotine rag cutter
Guillotineschneider *m (pap)* guillotine cutter (cutting machine, press, trimmer)
Guineakörner *npl* grains of paradise *(from Aframomum melegueta Schum.)*
Gulf-HDS-Verfahren *n (petrol)* Gulf HDS process *(for hydrogenating desulphurization)*
Gum *m (petrol)* gum
aktueller G. existent (preformed) gum
möglicher (potentieller) G. potential (ultimate) gum
vorgebildeter (vorhandener) G. *s.* aktueller G.

gumbildend *(petrol)* gum-forming
Gumbildung *f (petrol)* gum formation, gumming
Gumbildungstest *m (petrol)* gum test
Gumgehalt *m (petrol)* gum content
Gummi *m* rubber ✦ **mit G. beschichten (überziehen)** to rubber-cover, to rubber-coat
halbharter G. half-hard rubber, semiebonite
mikroporöser G. microporous rubber
poröser G. porous rubber

Gummi *n* [vegetable] gum
arabisches G. gum arabic, acacia (Arabic, Arabian) gum *(from Acacia specc., esp from Acacia senegal (L.) Willd.)*

Gummi arabicum *n s.* Gummi / arabisches
Gummi Ghatti *n* gum ghatti, India gum *(from Anogeissus latifolia Wall.)*
Gummiabfälle *mpl* rubber scrap (waste), scrap (waste) rubber

Gummiarabikum *n s.* Gummi / arabisches
gummiartig rubber-like, rubbery
Gummiartikel *m* rubber article (product)
Gummiauskleidung *f* rubber lining
Gummiband *n* rubber band
endloses G. rubber belt
Gummibelag *m* rubber coating
Gummichemie *f* rubber chemistry
Gummichemiker *m* rubber chemist
Gummideckplatte *f* rubber cover
Gummidichtung *f (on moving parts)* rubber packing (seal); *(with static application)* rubber gasket
Gummielastizität *f* rubber elasticity
Gummielementarfaden *m* rubber filament
gummieren to rubber[ize], to rubber-cover, to rubber-coat
Gummierung *f* 1. *(act)* rubberizing, proofing; 2. *(layer)* rubber coat[ing]; *(inner surfaces)* rubber lining
Gummifaden *m* rubber thread
Gummifaser *f* rubber fibre
Gummifaserstoff *m* rubber fibre
Gummifinger *m (lab)* rubber fingerstall (finger cot)
Gummigutt *n* [gum] gamboge, camboge, cambogia *(from Garcinia specc.)*
Gummihaar *n* rubberized hair
Gummiharz *n* gum resin
Gummiindustrie *f* rubber [manufacturing] industry
gummiisoliert rubber-insulated
Gummilack *m* gum lac *(crude shellac)*
Gummilösung *f* rubber solution; rubber cement
Gummilösungsmittel *n* rubber solvent
Gummimanschette *f* **für Frittentiegel** *(lab)* crucible adapter (holder), filter adapter
Gummi-Metall-Verbindung *f* 1. *(act)* rubber-to-metal bonding; 2. *(result)* rubber-to-metal bond
Gummipackung *f s.* Gummidichtung
Gummiquetscher *m* squeegee
Gummiriemen *m* rubber belt
Gummiring *m* rubber ring (annulus)
Gummirohr *n* rubber tube
Gummisack *m (plast)* rubber bag
Gummisack-Formverfahren *n s.* Gummisack-Preßverfahren
Gummisack-Preßverfahren *n (plast)* [pressure] bag moulding
Gummisackverfahren *n* / **abgewandeltes** *(plast)* autoclave moulding
Gummischeibe *f* rubber washer
Gummischlauch *m* rubber hose (tubing)
Gummischlauchmaterial *n* rubber tubing
Gummischnittfaden *m* cut rubber thread
Gummistempel *m (ceram)* rubber stamp
Gummistopfen *m* rubber stopper (bung)
doppelt durchbohrter G. two-hole rubber stopper
durchbohrter G. bored rubber stopper
Gummistöpsel *m s.* Gummistopfen
Gummit *m* gummite *(a rubber-like mixture of uranium minerals)*
Gummitaschenventil *n* pinch valve
Gummitechnologe *m* rubber technologist
Gummitechnologie *f* rubber technology
Gummitreibriemen *m* rubber belt
Gummiwalze *f* rubber-covered roll
Gummiwaren *fpl* rubber goods (products, articles)
technische G. mechanical rubber goods
Gummiwerk *n* rubber factory, rubber[-manufacturing] plant
Gummiwischer *m (lab)* rubber-tipped glass rod, [rubber] policeman, bobby
Gummizucker *m* pectin sugar, arabinose, pectinose
Gumtest *m (petrol)* gum test
Gur-Dynamit *n* guhr dynamite
Gurgelmittel *n (pharm)* gargarism, gargle
Gurjunbalsam *m* gurjun (gurjan, gardjan, gargan) balsam *(from Dipterocarpus alatus Roxb.)*
Gurjunbalsamöl *n* gurjun balsam oil, Indian wood oil
Gurt *m* belt
gemuldeter G. troughed belt
Gurtbandförderer *m* belt (band) conveyor
G. mit Gummigurt rubber belt conveyor
gemuldeter G. troughed-belt conveyor
Gurtbecherwerk *m* belt[-and-bucket] elevator
Gurtdurchhang *m* belt sag
Gurtförderer *m s.* Gurtbandförderer
Gurtreiniger *m* belt cleaner *(of a belt conveyor)*
Gurtspannung *f* belt tension
Gurtwerkstoff *m* belting
Guß *m* 1. *(act)* casting, *(of metal also)* founding, pouring; 2. *(product)* cast[ing]
fallender G. top casting
kontinuierlicher (ununterbrochener) G. continuous casting
Gußblase *f* blow hole, blister *(a material fault)*
Gußblock *m* ingot
Gußeisen *n* cast iron
siliziumlegiertes G. silicon cast iron
weißes G. white cast iron
Gußeisenarmierung *f* cast-iron armouring
Gußeisenretorte *f* cast-iron retort
gußeisern cast-iron
Gußfehler *m* casting defect, *(on the surface also)* scar
Gußform *f* [casting] mould
Gußglas *n* rolled glass
Gußhaut *f* skin *(foundry)*
Gußlegierung *f* cast[ing] alloy
Gußnaht *f* casting seam (line)
Gußrohr *n* cast pipe
Gußstahl *m* cast steel
Gußstahlfilterpresse *f* cast-steel filter press
Gußstahl-Hochdruckautoklav *m* cast-steel high-pressure autoclave
Gußstück *n*, **Gußteil** *n* casting
Gußtisch *m (glass)* casting table
Gußverfahren *n (rubber, plast)* flow casting
Gußwalze *f* forming roll *(in manufacturing sheet glass)*

Gut *n* material, stuff, *(if already treated also)* product
abgeröstetes G. *(met)* calcine
aufschwimmendes G. *(min tech)* floating material (product, fraction), floats
getrocknetes G. dry product
magnetisches G. magnetics
magnetisierbares G. magnetics
nichtmagnetisches G. non-magnetics
nichtmagnetisierbares G. non-magnetics
trocknendes G. material being dried
zu behandelndes G. material to be handled
zu trocknendes G. material to be dried
Gutabscheider *m* product collector (separator)
Gutaufgabe *f s.* Guteintrag
Gutaustrag *m*, **Gutaustritt** *m* 1. discharge *(act or process)*; 2. discharge [point]
Gutbrandbereich *m (ceram)* maturing range
Gutbrandtemperatur *f (ceram)* maturing temperature
Gutbrennen *n (ceram)* maturing
Gütegrad *m* grade
Guteintrag *m* 1. charging *(act or process)*; 2. charging point
Güteklasse *f* grade
Gütekontrolle *f* quality control
Gutentnahme *f s.* Gutaustrag
Gutfeuchte *f* product moisture
Gutstoff *m (pap)* accepted (screened) stock, accepts
Guttapercha *f(n)* gutta-percha *(from Payena and Palaquium specc.)*
Gutverlust *m* product loss
Gutverweilzeit *f* residence time
GV *s.* Glühverlust
Gynokardsäure *f* gynocardic acid *(a mixture of acids found in chaulmoogra oil)*, *(specif)* chaulmoogric acid, 13-2′-cyclopentenyltridecanoic acid
Gyro-Dampfphase[krack]verfahren *n*, **Gyro-Spaltverfahren** *n* Gyro [vapour-phase] process
Gyttja *f (soil)* gyttja

H

H *s.* Heizwert
Haar *n* hair, *(text collectively)* hair fibre
Haarbehandlungsmittel *n (cosmet)* hair preparation
Haarbleichmittel *n* hair bleach (bleaching agent)
Haareisen *n (tann)* unhairing knife
haarentfernend depilatory
Haarentfernung *f* depilation, epilation
Haarentfernungscreme *f* depilatory cream
Haarentfernungsmittel *n (cosmet, tann)* depilatory [agent], depilitant, epilator, hair remover
Haarfarbe *f s.* Haarfärbemittel
Haarfärbemittel *n* hair dye
chemisch wirkendes H. permanent hair dye
Haarkosmetikum *n* hair cosmetic
Haarkristall *m* whisker
Haarlack *m* hair lacquer
haarlockernd depilatory
Haarlockerung *f* depilation, epilation, *(tann also)* unhairing, dehairing
enzymatische (fermentative) H. *(tann)* enzyme unhairing
Haarlockerungsmittel *n s.* Haarentfernungsmittel
Haarlotion *f* hair lotion
Haarnadelrohr *n* hairpin tube
Haarnadelrohrbündel *n* hairpin coil
Haarnadelwärmeaustauscher *m*, **Haarnadelwärmeübertrager** *m* heat exchanger with hairpin tubes
Haarpflegemittel *n* hair cosmetic (preparation)
Haarreinigungsmittel *n* hair wash
Haarrißbildung *f (ceram)* crazing *(a defect in glazes)*; *(glass)* crizzling *(a defect)*
Haarrisse *mpl (ceram)* crazing *(a defect in glazes)*; *(glass)* crizzle *(a defect)*; *(ceram, glass)* crackle *(for decorative purposes)*
Haarrißglasur *f (ceram)* crackle glaze
Haarrissigwerden *n s.* Haarrißbildung
Haarröhrchen *n* capillary [tube]
Haarröhrchenwirkung *f* capillarity, capillary action
Haartöner *m* hair tint
Haarwäsche *f s.* Haarwaschmittel
Haarwaschmittel *n* hair wash
Haarwellotion *f* waving lotion
Haber-Bosch-Verfahren *n* Haber[-Bosch] process
Habitus *m* habit *(as of crystals)*
Hacke *f s.* Hackmaschine
hacken *(pap)* to chip, to chop
Hacker *m s.* Hackmaschine
Hackmaschine *f (pap)* chipper, chopper, chipping (chopping) machine
Hackmesser *n (pap)* chipper knife
Hackschnitzel *npl (pap)* [wood] chips, chippings
Hackschnitzelbehälter *m s.* Hackschnitzelsilo
Hackschnitzeldurchtränkung *f (pap)* penetration of wood (chips) *(with cooking liquor)*
Hackschnitzelfüllung *f (pap)* chip filling (packing), filling with chips
Hackschnitzelimprägnierung *f s.* Hackschnitzeldurchtränkung
Hackschnitzellagerung *f (pap)* chip storage
Hackschnitzelsilo *m (pap)* chip silo (bin, storage bin)
über dem Kocher angeordneter H. overhead bin
Hackschnitzelsortiermaschine *f (pap)* chip screen
Hackschnitzelspeicher *m (pap)* chip loft
Häckselmaschine *f (pap)* chopping machine, chopper, cutter *(for straw)*
häckseln *(pap)* to chop, to cut *(straw)*
Hackspan *m (pap)* [wood] chip
Hackspanlänge *f (pap)* chip length
Hadern *pl (pap)* rags
Hadernaufbereitung *f (pap)* pulping of rags
Hadernaufbereitungsanlage *f (pap)* rag mill
Hadernaufschluß *m (pap)* pulping of rags
Hadernbleiche *f (pap)* bleaching of rag pulp (stock)

Haderndrescher *m (pap)* rag willow (thrasher), devil
Hadernfaser *f (pap)* rag fibre
Haderngehalt *m (pap)* rag content
Hadernhalbstoff *m (pap)* all-rag furnish, rag pulp (stuff, stock), non-woody pulp
H. aus Tauen old-rope stock
Hadernhalbstoffbleiche *f (pap)* bleaching of rag pulp (stock)
Hadernhalbstoffpapier *n* [all-]rag paper
Hadernhalbzeug *n s.* Hadernhalbstoff
Hadernkocher *m (pap)* rag (bleach) boiler
Hadernkochung *f (pap)* cooking of rags
Hadernpapier *n* [all-]rag paper
Hadernpapierfabrik *f* rag mill
Hadernpapierfabrikation *f,* **Hadernpapierherstellung** *f* rag paper making
Hadernpappe *f* rag board
Hadernschneider *m (pap)* rag cutter (chopper)
Hadernseidenpapier *n* rag tissue paper
Hadernstäuber *m (pap)* rag duster
Hadernstoff *m s.* Hadernhalbstoff
Hadsel-Mühle *f,* **Hadsel-Prallmühle** *f* Hadsel mill
Haematit *m s.* Hämatit
Hafen *m (glass, ceram)* pot ✦ **den H. ausgießen** *(glass)* to teem
eingeglaster H. *(glass)* glazed pot
gedeckter H. *s.* geschlossener H.
geschlossener H. *(glass)* hooded (covered) pot; *(ceram)* closed pot
glasierter H. *(glass)* glazed pot
offener H. *(glass, ceram)* open pot
verdeckter H. *s.* geschlossener H.
Hafenbank *f (glass)* siege, bench *(of a pot furnace)*
Hafenglas *n* pot[-melted] glass
Hafenofen *m (glass)* pot furnace
Hafenschmelze *f (glass)* pot melting
Hafentemperofen *m (glass)* pot arch
Hafermalz *n* oat malt, malted oat
Haferstärke *f* oat starch
Hafnat *n* $M_2^IHfO_3$ hafnate
Hafnium *n* Hf hafnium
Hafniumdioxid *n s.* Hafnium(IV)-oxid
Hafniumkarbid *n* HfC hafnium carbide
Hafnium(IV)-oxid *n* HfO_2 hafnium(IV) oxide, hafnium dioxide
Hafniumoxidchlorid *n* $HfOCl_2$ hafnium dichloride oxide
Hafniumsulfat *n* $Hf(SO_4)_2$ hafnium sulphate
Hafniumtetrachlorid *n* $HfCl_4$ hafnium tetrachloride
HAF-Ruß *m (rubber)* high abrasion furnace black, HAF black
Haftarbeit *f (phys chem)* adhesional work, work of adhesion
haften to adhere, to stick
Haften *n* adhesion, adherence, sticking
H. am Werkzeug *(plast)* mould sticking
haftend, haftfähig adhesive, adherent
Haftfähigkeit *f* adhesiveness, adherence, adhesive (adhesion) power (capacity), sticking power (capacity)
anfängliche H. initial retention *(of pesticides)*
Haftfestigkeit *f* tenacity, adhesive strength
Haftgrundierung *f,* **Haftgrundmittel** *n (coat)* self-etch pretreatment primer, wash (self-etching) primer
Haftinhalt *m (distil)* [liquid] hold-up
Haftkleber *m* contact (pressure-sensitive) adhesive (cement)
Haftmittel *n* adhesive (adhesion, sticking) agent, adhesive, sticker; *(rubber)* bonding agent; *(plast)* coupling (anchoring) agent; *(text)* coupling agent *(for laminates)*; *(agric)* deposit builder *(in pesticide formulations)*
metallkeramisches H. ceramic-metal adhesive
Haftspannung *f* adhesive stress (tension)
Haftstelle *f (phys chem)* trap *(as for recombination of electrons and defect electrons)*
flache H. shallow trap
tiefe (tiefliegende) H. deep (recombination) trap
Haftstoff *m s.* Haftmittel
Haftung *f s.* Haften
Haftvermittler *m s.* Haftmittel
Haftvermögen *n s.* Haftfähigkeit
Hägglund-Verfahren *n* Hägglund process *(saccharification of wood)*
Hahn *m* cock, tap, plug valve (cock, bib)
H. mit hebelgelüftetem Küken lever-sealed plug cock
H. mit schräger Bohrung oblique cock
H. mit senkrechter Bohrung straight cock
Karlsruher H. T-shape 120° bore cock
Hahn-Aufsatz *m (distil)* Hahn head
Hahnenfuß *m (ceram)* [cock-]spur *(an item of kiln furniture)*
Hahnfett *n* tap (stopcock) grease
Hahnhülse *f* socket, cock shell (barrel)
Hahnkegel *m s.* Hahnküken
Hahnküken *n* [cock] plug, *(Am also)* stopper
Hahnrohr *n* **des Orsat-Apparates** Orsat gas manifold
Hahnschmiermittel *n* tap (stopcock) lubricant
Hahnsicherung *f* locking device
Hahnsystem *n* **des Orsat-Apparates** Orsat gas manifold
Hahnventil *n* plug valve (cock, bib) (*for compounds s.* Hahn)
Haidingerit *m (min)* haidingerite *(calcium hydrogenorthoarsenate)*
Haifisch[leber]tran *m* shark-liver oil
Halbanthrazit *m* semianthracite [coal], lean (dry steam) coal
Halbantigen *n (med)* hapten[e]
halbautomatisch semiautomatic
Halbazetal *n* hemiacetal
halbazetalartig hemiacetal-like
halbchemisch semichemical
Halbchinon *n* semiquinone
halbdirekt semidirect
halbdurchlässig semipermeable
halbdurchscheinend semitransparent
Halbedelstein *m* semiprecious stone

Halbelement *n (phys chem)* half-element, half-cell
Halbentbasten *n (text)* soupling
Halbester *m* semi-ester, half-ester
Halbfärbezeit *f* time of half-dyeing
halbfest semisolid
halbfeuerfest semirefractory
halbflächig *(cryst)* hemihedral
halbflüssig semiliquid, semifluid
Halbformal *n* $HOCH_2OR$ hemiformal *(any of the hemiacetals of formaldehyde)*
Halbfusinit *m* semifusinite *(a maceral of coal)*
halbgebleicht *(pap)* half-bleached, semibleached
halbgehärtet part-hydrogenated *(oils)*
halbgeleimt *(pap)* half-sized, $^1/_2$ sized
halbglasartig *(ceram)* semivitreous, semivitrified
halbhart semisolid
Halbhartgummi *m* semiebonite, half-hard rubber
Halbhydrat *n* hemihydrate
Halbkarton *m* cardboard
Halbkernseife *f* half-neat soap
Halbkoks *m* semicoke
Halbkolloid *n* semicolloid
Halbkonserve *f* partly preserved food
halbkontinuierlich semicontinuous, semibatch
Halbkristallglas *n* half-crystal
halbkristallin[isch] hemicrystalline, hypocrystalline
halbleitend semiconducting, semiconductive
Halbleiter *m* semiconductor
  H. mit Eigenleitfähigkeit intrinsic semiconductor
  elektronischer H. electronic semiconductor
  gemischter H. compensated semiconductor
  zusammengesetzter H. compound semiconductor
Halbleiterdehnungsmesser *m* semiconductor strain gauge
Halbleiterfotozelle *f* semiconductor photocell
Halbleitergleichrichter *m* barrier-layer rectifier
Halbleitermaterial *n* semiconducting material
Halbleiterschicht *f* semiconducting layer
Halbleitersperrschicht *f* semiconductor junction
Halbleiterteilchenzähler *m* semiconductor particle counter
Halbleiterübergang *m* semiconductor junction
Halbleiterverbindung *f* semiconducting compound
Halbleiterwiderstand *m* semiconductor resistor
Halblösung *f (pap)* weak acid
halbmatt *(phot)* semimat, semimatt[e], half-mat[t], semigloss[y]; *(text)* semidull
Halbmattglasur *f (ceram)* semimat glaze
Halbmetall *n* semimetal, crossroads element
Halbmetallglanz *m (min)* submetallic lustre
halbmetallisch semimetallic
Halbmikroanalyse *f* semimicroanalysis
Halbmikroansatz *m* semimicro batch
Halbmikrobestimmung *f* semimicrodetermination
Halbmikroextraktion *f* semimicro extraction
Halbmikromethode *f* semimicro method
  H. nach Kjeldahl semimicro Kjeldahl method
  H. nach Rast semimicro Rast method *(for determining molecular weights)*
Halbmikropräparation *f* semimicro preparation
Halbmikro-Torsionswaage *f* semimicro torsion balance
Halbmikroverfahren *n* semimicro method
Halbmikrowaage *f* semimicro balance
Halbmuffelofen *m* semimuffle kiln
halbnaß semidry
Halbneutralisationspunkt *m* half neutralization point
halbpolar half-polar, semipolar
Halbporzellan *n* semiporcelain, vitreous china
halbquantitativ semiquantitative
Halbsandwichverbindung *f* half-sandwich compound
Halbsäure *f (pap)* weak acid
Halbschatten *m* 1. half-shade; 2. half-shade angle *(polarimetry)*
Halbschattennicol *m* half-shade Nicol *(polarimetry)*
Halbschattenpolarimeter *n*, **Halbschattenpolarisator** *m* half-shade polarimeter
Halbschattenwinkel *m* half-shade angle *(polarimetry)*
Halbsesselform *f* half-chair form *(stereochemistry)*
Halbstoff *m (pap)* half-stuff, half-stock ✦ **zu H. aufschließen** *(pap)* to make into [a] pulp, to reduce to pulp, to pulp
  H. in Bogenform (Pappenform) half-stuff board, pulp (solid pulp) board, sheets (laps) of pulp, lap[ped] pulp
  textiler H. non-woody pulp
  trockener H. *s.* H. in Bogenform
Halbstoffholländer *m (pap)* half-stuff beater, breaking[-in] engine, breaker [engine], rag engine (breaker), Hollander washer
Halbstoffholländerwalze *f (pap)* breaker roll (drum)
Halbstoffsortierer *m (pap)* pulp screen
Halbstufenpotential *n* half-wave potential
Halbstundenlack *m* half-hour synthetic *(a nitrocellulose lacquer)*
halbtechnisch semitechnical, pilot-scale
Halbtrivialname *m* semitrivial (semisystematic) name
halbtrocken semidry
Halbtrockenpressen *n (ceram)* semidry pressing
halbtrocknend semidrying
Halbultrabeschleuniger *m (rubber)* semiultra accelerator
halbverglast *(ceram)* semivitreous, semivitrified
Halbverkokung *f* semicoking, semicarbonization
Halbwachs *n* propolis, bee glue, balm
Halbwassergas *n* semiwater gas
Halbweißöl half-white oil
Halbwelle *f* half wave
Halbwellenpotential *n* half-wave potential
Halbwert[s]breite *f (phys chem)* half-intensity width, half band width
Halbwerts[schicht]dicke *f (phys chem)* half-thickness

Halbwert[s]zeit *f* 1. [reaction] half-time, time (period) of half change, half change value *(reaction kinetics)*; *(nucl)* half-life; 2. *s.* Rückstands-Halbwertszeit
Halbwollfärben *n* union dyeing
Halbwollfarbstoff *m* union dye
Halbzellstoff *m* semichemical pulp
Halbzellstoffanlage *f s.* Halbzellstoffwerk
Halbzellstoffaufschluß *m* semichemical pulping
Halbzellstoffwerk *n* semichemical plant, semichemical-pulp mill
halbzersetzt semidecomposed
Halbzeug *n s.* Halbstoff
Halbzylinder *m* semicylinder
Halfagras *n* Alfa (Spanish) grass, esparto [grass], Stipa tenacissima L.
Halit *m (min)* halite, rock salt *(sodium chloride)*
Hall-Effekt *m* Hall effect *(a galvanomagnetic effect)*
Halloysit *m* halloysite *(a clay mineral)*
halluzinogen *(tox)* hallucinogenic
Halluzinogen *n (tox)* hallucinogen
Hall-Verfahren *n* Hall process *(1. for obtaining aluminium; 2. for gasification of oil)*
Halmyrolyse *f* halmyrolysis *(chemical destruction or rearrangement of a sediment on the sea floor)*
Halo *m* halo
Halochromie *f* halochromism *(phenomenon)*; halochromy *(property)*
Halochromieerscheinung *f* halochromic effect, halochromism
Haloform-Reaktion *f (org chem)* haloform (Einhorn) reaction
Halogen *n* halogen
Halogenabkömmling *m* halogen derivative
Halogenaddition *f* halogen addition
Halogenalkan *n* haloalkane, alkyl halide
Halogenanthrachinon *n* haloanthraquinone
Halogenäthan *n* haloethane, ethyl halide
Halogenbenzol *n* halobenzene
Halogenbrücke *f* halogen bridge
Halogenderivat *n* halogen derivative
Halogenelektrode *f* halogen electrode
Halogenentzug *m* dehalogenation
halogenhaltig halogen-containing
Halogenhydrin *n* halohydrin *(any of a class of glycerol derivatives)*
Halogenid *n* halide, halogenide
  siliziumorganisches H. *s.* Halogensilan / organisches
Halogenidphosphor *m* halide phosphor *(a halide exhibiting phosphorescence)*
halogenieren to halogenate
α-halogeniert α-halogenated, alpha-halogenated
halogeniert / mehrfach polyhalogenated
Halogenierung *f* halogenation
  H. in der Seitenkette side-chain halogenation
α-Halogenierung *f* alpha-halogenation
halogenisieren *s.* halogenieren
Halogenisierung *f s.* Halogenierung
α-Halogenkarbonsäure *f* α-halogenated acid, α-haloacid
Halogenketon *n* haloketone
Halogenkohlenwasserstoff *m* halogenated hydrocarbon, halocarbon
Halogenlampe *f* halide lamp
Halogen-Metall-Austausch *m* halogen-metal exchange
Halogenmethan *n* halomethane, methyl halide
Halogenmethyl *n s.* Halogenmethan
Halogenmethylaromat *m* halogenomethylaromatic
Halogenoform *n* CHX_3 haloform *(any of the trihalomethanes)*
Halogensilan *n* halogenosilane, halosilane
  organisches H. organohalogenosilane, organohalosilane, organosilicon halide
Halogensilber *n (phot)* silver halide
Halogensilberemulsion *f (phot)* silver-halide emulsion
Halogen-Stickstoff-Verbindung *f* halogen-nitrogen compound
halogensubstituiert halogen-substituted
Halogenüberträger *m* halogen carrier
Halogenverbindung *f* halo[gen] compound
Halogenwasserstoff *m* hydrogen halide (halogenide)
Halogenwasserstoffabspaltung *f* dehydrohalogenation
Halogenwasserstoffentzug *m* dehydrohalogenation
Halogenwasserstoffsäure *f* hydrohalic acid
Halometer *n* sali[ni]meter *(a hydrometer for salt solutions)*
Halophyt *m (bot)* halophyte
Halotrichit *m (min)* halotrichite, iron alum *(aluminium iron(II) sulphate-22-water)*
Hals *m* neck *(as of a bottle or shaft)*
haltbar durable, stable ✦ **h. machen** to preserve, to conserve, to prepare, to cure, *(food also)* to can
  h. verpackt packed for prolonged storage
  unbegrenzt h. sein to keep indefinitely
Haltbarkeit *f* durability, stability; *(coat, dye)* fastness; *(quantitatively)* life[time]; *(food)* storage (keeping) quality
Haltbarkeitsgrenze *f (plast)* endurance limit
Haltbarkeitsprüfung *f* durability test
Haltbarkeitszeit *f* shelf (storage) life
Haltbarmachung *f* preservation, conservation, preparation, curing, cure, *(food also)* canning
  H. für beschränkte Zeit temporary preservation
  H. von Lebensmitteln food preservation
Haltedruck *m* net positive suction head, NPSH *(of a pump)*
Halteklemme *f s.* Halteschelle
halten to support *(mechanically)*; to keep, to maintain *(e. g. a definite temperature)*
  am Kochen (Sieden) h. to keep at the boil
  im Gleichgewicht h. to equilibrate, to keep in equilibrium
  in Lösung h. to keep in solution
  in Suspension h. to keep in suspension
  konstant h. to maintain constant
  nahe am Sieden h. to keep near the boil

Haltepunkt *m* **[/ eutektischer]** [eutectic] halt
Halter *m* holder, clip, clamp
Halterollen *fpl* / **seitliche** edge rolls *(sheet glass manufacture)*
Halterung *f* bearing
Halteschelle *f* joint clamp, adapter *(for ground-glass joints)*
H. für Kegelschliffverbindungen cone-and-socket joint clamp, socket-to-cone adapter
H. für Kugelschliffverbindungen ball-and-socket joint clamp, socket-to-ball adapter
Haltevorrichtung *f* holder, clip, clamp
Haltewalze *f* nip roller
Haltezeit *f* residence (hold-up, holding, detention, retention) time
Häm *n (bioch)* haem, hem[e]
Hamamelitannin *n (tann)* hamameli-tannin
Hamamelose *f* hamamelose, 2-C-hydroxymethyl-ribose
Hämatin *n* haematin
Hämatit *m (min)* [red] haematite *(iron(III) oxide)*
Hämatit[roh]eisen *n* haematite [pig] iron
Hämiglobin *n* haemiglobin, methaemoglobin
Hämin *n* haemin
Häminchlorid *n* haemin chloride, protohaemin
Hammelfett *n*, **Hammeltalg** *m* mutton fat (tallow)
Hammer *m* beater *(of the hammer mill)*
Hammerbrecher *m* hammer grinder (crusher)
Hammermühle *f* hammer mill (grinder, disintegrator)
Hammerwalke *f (text)* fulling stocks
Hämochrom[ogen] *n* haemochromogen, haemochrome
Hämogen *n (bioch)* extrinsic factor, vitamin B_{12}
Hämogenase *f (bioch)* intrinsic factor
Hämoglobin *n* haemoglobin, Hb
reduziertes H. deoxygenated haemoglobin
Hämolymphe *f* haemolymph
Hämolyse *f* haemolysis
hämolysieren to haemolyze
Hämolysin *n* haemolysin *(toxic substance produced by certain bacteria)*
Hämostatikum *n s.* Hämostyptikum
hämostatisch *s.* hämostyptisch
Hämostyptikum *n (med)* haemostatic, styptic
hämostyptisch *(med)* haemostatic, styptic
Hämozyanin *n* haemocyanin *(a respiratory pigment of numerous invertebrate animals)*
Hämprotein *n* haem (hem, heme) protein
Handauflegeverfahren *n (plast)* hand (wet) lay-up technique, contact (impression) moulding
Handaustrag *m* hand scooping
Handbeschickung *f* hand charging
handbetätigt manually operated, hand-operated
Handbetrieb *m* manual (hand) operation
handbetrieben *s.* handbetätigt
Handbütten *n* vat (hand-made, genuine hand-made) paper
Handbüttenrand *m (pap)* deckle [edge]
Handcreme *f* hand cream
Handdruck *m* block printing
Handelsbenzol *n* commercial benzole
90er H. 90's benzole
Handelsbezeichnung *f s.* Handelsname
Handelschemiker *m* commercial chemist
Handelsdünger *m* commercial (artificial) fertilizer
Handelskarbid *n* commercial carbide
Handelskohle *f* commercial coal
Handelsmuster *n* trade sample
Handelsname *m* trade name (term), commercial name
Handelsprodukt *n* commercial product
formuliertes H. formulation
Handelsqualität *f* commercial (market) grade
Handelssorte *f* market type (grade)
Handelstannin *n* tannic acid of commerce *(a product consisting of gallic acid glucose esters with penta-m-digalloyl-β-glucose as chief constituent)*
handelsüblich commercial
Handentwicklung *f (phot)* see-saw development
Handfeuerlöscher *m* portable (household) fire extinguisher
Handform *f (plast)* hand mould
Handformen *n*, **Handformgebung** *f (ceram)* hand modelling (moulding)
Handformstein *m*, **Handformziegel** *m* hand-made brick
Handgebläse *n* hand-power air blower *(made of rubber bulbs)*
handgeformt *(ceram)* hand-modelled, hand-moulded
Handgraupappe *f* hand-made grey board
Handgriff *m* handle
handhaben to handle
sich h. lassen to handle
Handhabung *f* handling
Handholzpappe *f* hand-made wood board
Handklaubung *f (coal)* hand picking (cleaning, sorting)
Handlederpappe *f* hand-made leather board
Handleimung *f (pap)* hand sizing
Handloch *n* hand hole
Handlotion *f* hand lotion
Handmalerei *f (ceram)* hand painting
Handmuster *n (pap)* hand (pulp, test) sheet
Handpapier *n s.* Handbütten
Handpappe *f* cylinder board
Handpflegemittel *n (cosmet)* hand preparation
Handpresse *f (plast)* hand press
Handrad *n* handwheel
Handregelung *f* manual control
Handscheidung *f s.* Handklaubung
Handschuhbox *f (nucl)* glove box
Handschweißen *n* hand welding
Handschweißgerät *n* hand welding unit
Handsieben *n* hand sieving
handsortiert hand-sorted
Handsortierung *f* sorting by hand, hand sorting
Handspritze *f*, **Handspritzgerät** *n (agric)* hand sprayer
Handspritzpistole *f* hand [spray] gun
Handsprühgerät *n (agric)* hand sprayer
Handstäubegerät *n*, **Handstäuber** *m (agric)*

hand[-operated] duster; *(if pneumatically operated)* hand [dust] gun
Handsteuerung *f* manual control
Handtuchpapier *n* towelling paper
Handverstäuber *m s.* Handstäubegerät
Handwerkzeug *n (plast)* hand mould
Handzentrifuge *f* hand[-driven] centrifuge
Handzerstäuber *m (incorrectly for)* Handsprühgerät
Hanf *m* hemp, Cannabis L., *(specif)* [true] hemp, Cannabis sativa L.
Indischer H. Indian hemp, Cannabis indica Lam.; *(pharm)* Indian hemp *(dried summits from Cannabis indica Lam.)*
Hanfeibischfaser *f* kenaf (hibiscus, hemp-mallow) fibre *(from Hibiscus cannabinus L.)*
Hanffaser *f* hemp fibre *(from Cannabis sativa L.)*
Hanföl *n* hemp (hempseed, cannabic) oil *(from Cannabis sativa L.)*
Hanfpapier *n* hemp paper
Hanfsamenöl *n s.* Hanföl
Hängeäscher *m (tann)* rocker
Hängebandtrockner *m* festoon (loop) dryer
Hängedämpfer *m (text)* festoon ager
hängen to hang; to be attached *(as of atoms)*
hängenbleiben to stick, to hang up
Hängetrockner *m* festoon (loop) dryer
Hängezentrifuge *f* suspended (top-suspended, overdriven) centrifuge
Hansagelb *n* Hansa yellow *(any of various azo dyes)*
Hantelmodell *n* dumb--bell modell *(of a molecule)*
Hantelprüfkörper *m (rubber)* dumb-bell test piece, dumb-bell strip, dumb bell
Hapten *n (bioch)* hapten
Haptoglobin *n* haptoglobin *(any of a group of glycoproteids in blood)*
Harden-Young-Ester *m* Harden and Young ester, 1,6-fructofuranose diphosphate
Hardgrove-Maschine *f*, **Hardgrove-Mühle** *f* Hardgrove machine (mill) *(for determining grindability)*
Hardinge-Kaskadenmühle *f* Hardinge cascade mill
Hardinge-Mühle *f* Hardinge conical [ball] mill, Hardinge mill
Hard-Processing-Channel-Ruß *m* hard processing channel black, HPC black
Harfe *f (lab)* assembly
Hargreaves-Bird-Zelle *f* Hargreaves-Bird cell *(electrolysis)*
Hargreaves-Verfahren *n* Hargreaves process *(for obtaining HCl and Na_2SO_4 or K_2SO_4)*
Harmalaalkaloid *n* harmal[a] alkaloid
Harmalin *n* harmaline, 3,4-dihydro-7-methoxy-1-methyl-β-carboline *(a harmala alkaloid)*
Harmalraute *f*, **Harmelstaude** *f* harmal[a], harmel, Peganum harmala L.
Harmin *n* harmine, 7-methoxy-1-methyl-β-carboline *(a harmala alkaloid)*
harmlos innocuous
Harmotom *m (min)* harmotome *(a tectosilicate)*

Harn *m* urine
Harnanalyse *f* urinalysis, uranalysis, urine analysis
Harnantiseptikum *n* urinary antiseptic
Harngrieß *m* gravel
Harnsäure *f* uric acid, 2,6,8-trihydroxypurine
Harnsäurederivat *n* uric-acid derivative
Harnsäurestein *m (med)* uric-acid calculus
Harnstein *m (med)* urinary calculus
Harnstoff *m* $NH_2-CO-NH_2$ urea, carbamide
Harnstoffadditionsverbindung *f* urea addition compound
Harnstoffaddukt *n* urea adduct
Harnstoff-Aldehyd-Harz *n s.* Harnstoff-Formaldehyd-Harz
Harnstoffanlage *f (tech)* urea plant
Harnstoff-Bisulfit-Löslichkeit *f (text)* urea-bisulphite solubility
Harnstoffderivat *n* / **herbizides** urea herbicide
Harnstoffeinschlußverbindung *f* urea inclusion compound
Harnstoffentparaffinierung *f* urea dewaxing *(of lubricating-oil stocks)*
Harnstoff-Formaldehyd-Harz *n* urea[-formaldehyde] resin, polyurea
Harnstoff-Formaldehyd-Kondensat[ionsprodukt] *n* urea-formaldehyde condensation product, urea formaldehyde
Harnstoff-Formaldehyd-Leim *m* urea-formaldehyde glue
Harnstoffgitter *n* urea lattice
Harnstoffharz *n s.* Harnstoff-Formaldehyd-Harz
Harnstoffherbizid *n* urea herbicide
Harnstoffkalksalpeter *m s.* Harnstoff-Kalziumnitrat
Harnstoff-Kalziumnitrat *n* $Ca(NO_3)_2 \cdot 4CO(NH_2)_2$ calcium-nitrate-urea
Harnstoffkomplex *m*, **Harnstoffkomplexverbindung** *f* urea complex
Harnstoffmolekülverbindung *f* urea molecular compound
Harnstofformaldehyd *m s.* Harnstoff-Formaldehyd-Kondensat
harnstoffspaltend ureolytic
Harnstoffstickstoff *m (med)* urea nitrogen
Harnstofftrennung *f s.* Harnstoffentparaffinierung
Harnstoffzyklus *m (bioch)* urea cycle
harntreibend diuretic
Harnwaage *f* urinometer *(a hydrometer for determining the specific gravity of urine)*
Harristrip-Verfahren *n (text)* Harristrip process *(for decolourizing wool)*
Harris-Verfahren *n* Harris process *(for softening lead)*
Harrop-Ofen *m* Harrop kiln *(a tunnel kiln)*
hart hard *(as of metals, water, radiation)*; *(text)* crisp
h. werden to harden, to solidify, to chill, to set
Hartanodisierung *f* hard anodizing
Hartasphalt *m* hard asphalt
härtbar hardenable

Härtbarkeit *f* hardenability
Hartblei *n* hard[ened] lead, antimonial lead, regulus metal *(an alloy containing 90% Pb, 8% Sb, 2% Sn)*
Hartbrandstein *m (ceram)* hard-burned (hard-fired) brick
Härte *f* hardness *(as of metals, water, X-rays)*
  H. der [gehärteten] Randschicht *(met)* case hardness
  bleibende H. permanent (non-carbonate) hardness *(of water)*
  permanente H. *s.* bleibende H.
  schwindende H. *s.* temporäre H.
  temporäre H. temporary (carbonate) hardness, carbonate alkalinity *(of water)*
  vorübergehende H. *s.* temporäre H.
Härtebad *n* hardening (hardener) bath
Härtebestimmung *f* determination of hardness, hardness testing
Härtebildner *m* hardness element
Härtefaktor *m* hardness factor
Härtefixierbad *n (phot)* hardening fixer (fixing bath)
härtefrei hardness-free, zero-hardness *(water)*
Härtegrad *m* degree of hardness
Härtekasten *m (met)* case-hardening box
Härtekatalysator *m (plast)* curing catalyst
Härtelösung *f (phot)* hardening solution
Härtemesser *m s.* Härteprüfer
Härtemittel *n* hardening agent, hardener; *(coat, plast)* curing agent
härten *(met)* to harden; *(plast, coat)* to cure; *(plast)* to stove, *(Am)* to bake *(cast resins)*; *(glass)* to temper, to strengthen; *(food)* to hydrogenize, to hydrogenate, to harden *(oils)*
  durch Nitrierung h. to nitride *(steel)*
  im Einsatz[verfahren] h. to case-harden *(steel)*
  im Zyan[salz]bad h. to cyanide *(steel)*
  in Luft h. to air-harden *(steel)*
  in Öl h. to oil-harden *(steel)*
  in Wasser h. to water-harden *(steel)*
  oberflächlich h. to surface-harden
Härten *n (met)* hardening; *(plast, coat)* curing, cure; *(plast)* stoving, *(Am)* baking *(of cast resins)*; *(glass)* tempering, strengthening; *(food)* hydrogenation, hardening *(of oils)*
  selektives H. *(food)* selective hydrogenation
  vorzeitiges H. *(plast)* premature curing *(a moulding defect)*
Härteofen *m (ceram)* hardening furnace
Härteöl *n* hardening oil
Härteprüfer *m* hardness tester (meter); *(using a drill)* durometer; *(using the rebound of a ball)* scleroscope; *(using a stylus)* sclerometer
Härteprüfung *f* hardness test[ing]
  H. nach Knoop Knoop hardness test
Härtepulver *n (met)* case-hardening powder
Härter *m s.* Härtemittel
Härteschicht *f (met)* [hardened] case
Härteskala *f* scale of hardness
  H. nach Mohs Mohs' scale [of hardness]
Härtestufe *f* degree of hardness
Härtetiefe *f* depth of hardening
Härtezahl *f (rubber)* coefficient of hardness
Härtezeit *f (met)* hardening time; *(plast, coat)* curing (cure) time; *(plast)* stoving time *(of cast resins)*
Härtezyklus *m (plast)* curing cycle
Hartfaser *f* hard fibre
Hartfaserplatte *f* hard-board
Hartferrit *m (ceram)* hard ferrite
Hartfett *n* solid (hard) fat; hydrogenated (hardened) fat
Hartgewebe *n* laminated fabric, synthetic-resin-bonded fabric sheet
Hartgips *m* hard plaster
Hartglas *n* resistance (hard) glass
Hartglasgefäß *n* resistance-glass bottle
Hartglaskolben *m* resistance-glass flask
Hartgummi *m* hard rubber, ebonite, vulcanite
  zelliger H. cellular ebonite
Hartgummiartikel *m* hard-rubber article
Hartgummimischung *f* hard-rubber mix
Hartgummiplatte *f* hard-rubber sheet
Hartgummiwalze *f* hard-rubber-covered roll
Hartguß *m* chilled (white) cast iron, chill-cast iron
Hartgußwalze *f* chilled[-iron] roll
Hartharz *n* hard resin
Hartkarbid *n* hard [metal] carbide
Hartkautschuk *m s.* Hartgummi
hartkochen to undercook, to cook raw *(cellulose)*
Hartkochung *f* undercooking *(of cellulose)*
Hartkoks *m* hard coke
Hartlot *n* brazing (hard) solder
hartlöten to braze, to solder hard
  im Lötbad h. to dip-braze
Hartmasse *f (ceram)* hard paste, pâte dure
Hartmetall *n* **[/ gesintertes]** [cemented] hard metal, cemented [hard] carbide
Hartmetallegierung *f* hard-metal alloy
Hartmetallwerkstoff *m* cemented carbide material
Hartoxydation *f* hard anodizing *(of metals)*
Hartpapier *n* hard (bakelite, laminated) paper, synthetic-resin-bonded paper sheet
Hartpappe *f* hard-board, panel board
Hartparaffin *n* hard (solid) paraffin, ceresin [wax]
Hartpech *n* hard pitch
Hartpetrolat[um] *n* petrolatum
Hartporzellan *n* hard[-paste] porcelain
Hartpostpapier *n* bank paper (post), bond [paper]
Hart-PVC *n* unplasticized (rigid) PVC
Hartree-Einheiten *fpl* atomic units [Hartree] *(used in investigating the electronic structure of atoms and molecules)*
Hartsalz *n* hard salt *(crude potash salt containing $MgSO_4$)*
Hartseide *f* hard (ecru) silk
Hartseife *f* hard soap
Hartspiritus *m* solid (hard) spirit
Hartstoff *m* hard material
Härtung *f s.* Härten
Härtungsautoklav *m* hardening vessel *(for fats)*
Härtungsgeruch *m* hydrogenation (hardening) flavour

Härtungsgeschmack *m* hydrogenation (hardening) flavour
Härtungsmittel *n s.* Härtemittel
Härtungsperiode *f (plast)* curing cycle
Härtungstiefe *f* depth or hardening
Härtungszeit *f s.* Härtezeit
Hartverchromen *n* hard chrome-plating
Hartwachs *n* hard wax
Hartwasserbeständigkeit *f* resistance to hard water
Hartwerden *n* hardening, set[ting]
Hartzerkleinerung *f* crushing (size reduction) of hard material
Harz *n* resin; *(petrol)* gum; *(pap)* rosin *(colophony)*; *(as a deleterious component in paper pulp)* pitch
  H. aus Rohterpentin pine resin (rosin), [common] rosin, colophony
  H. für Handauflegeverfahren (Kontaktpreßverfahren, Niederdruckpreßverfahren) *(plast)* contact pressure resin
  H. im A-Zustand A-stage resin
  H. im B-Zustand B-stage resin
  H. im C-Zustand C-stage resin
  aktuelles H. *s.* vorgebildetes H.
  dispergiertes H. dispersion (grinding) resin
  fossiles H. fossil resin
  freies H. *(pap)* free rosin
  gehärtetes H. cured resin
  heißhärtendes H. hot-set resin
  lösliches H. soluble resin
  mögliches H. *(petrol)* ultimate (potential) gum
  natürliches H. natural resin
  ölmodifiziertes H. oil-modified resin
  ölreaktives H. oil-reactive resin
  potentielles H. *s.* mögliches H.
  schädliches H. *(pap)* pitch
  vorgebildetes (vorhandenes) H. *(petrol)* existent (preformed) gum
harzähnlich resin-like
Harzappretur *f (tann)* resin finish
harzartig resinous, resinoid, resiny
harzbildend *(bot)* resin-forming; *(petrol)* gum-forming
Harzbildnertest *m (petrol)* gum test
Harzbildung *f (bot)* resin formation; *(petrol)* gum formation, gumming
Harzeinschluß *m (plast)* resin pocket *(a moulding defect)*
Harzemulsion *f (pap)* rosin (size) milk, rosin size, size emulsion
  stabilisierte H. protected rosin size, high free protected size
Harzessenz *f s.* Harzgeist
Harzgang *m (bot)* resin canal
Harzgehalt *m* resin content; *(petrol)* gum content
Harzgeist *m* resin (rosin) spirit, pinolin[e], pinolene
harzgeleimt *(pap)* sized with rosin size
Harzgerbung *f* resin tannage
harzhaltig resiniferous, resinous; *(pap)* pitchy *(pulp)*
harzig resinous, resiny
Harzkanal *m (bot)* resin canal
Harzkiefer *f* red pine, Pinus resinosa Ait.
Harzkomponente *f* resin constituent
Harzkorn *n* resin bead *(ion-exchange material)*
Harzkörper *m* resinous body (matter, substance)
Harzlack *m* resinous varnish
Harzleim *m (pap)* rosin size
Harzleimpulver *n (pap)* dry rosin size
Harzleimung *f (pap)* rosin sizing
Harzlösung *f (coat)* resin solution; *(pap)* rosin (size) milk, size emulsion
Harzmasse *f s.* Harzkörper
Harzmilch *f s.* Harzemulsion
Harznest *n (plast)* resin pocket *(a moulding defect)*
Harzneubildung *f (petrol)* ultimate (potential) gum
Harzöl *n* resin oil, liquid resin
Harz-Öl-Farbe *f* oleoresinous paint
Harz-Öl-Lack *m* oleoresinous varnish
Harz-Öl-Verhältnis *n* resin-to-oil ratio, resin/oil ratio
Harz-Paraffin-Emulsion *f (pap)* rosin-wax emulsion
Harz-Paraffin-Leim *m (pap)* rosin-wax size
Harzpaste *f* paste resin
Harzpech *n* resin pitch
harzreich rich in resin, resinous
Harzsäure *f* oleoresin (resin, rosin) acid
Harzschwierigkeiten *fpl (pap)* pitch trouble[s]
Harzseife *f* resin soap *(salt of a resin acid)*
Harzsprit *m s.* Harzgeist
Harzstoff *m* resinous matter (substance)
Harztasche *f (plast)* resin pocket *(a moulding defect)*
Harzträger *m* resin[ous] binder
Harzvernetzung *f*, **Harzvulkanisation** *f (rubber)*, resin cure
Haschisch *m(n)* hashish, hasheesh, haschisch, marihuana, marijuana *(from Cannabis indica Lam.)*
Haselnußbutter *f*, **Haselnußöl** *n* hazelnut butter (oil)
Haspel *m(f) (tann)* paddle; *(text)* winch
Haspeläscher *m (tann)* paddle liming
Haspelfärbeapparat *m (text)* winch dyeing machine
Haspelgeschirr *n (tann)* paddle
Haspelkufe *f (text)* dye (winch) back
Haspelmaschine *f (text)* reeling machine
haspeln *(tann)* to paddle; *(text)* to reel [up]
Haspelseide *f* grege, greige, gum silk
Hatchettin *m (min)* hatchettine, hatchettite, mineral tallow *(a naturally occurring paraffin mixture)*
Haube *f (tech)* hood, head, cap, *(esp of a furnace)* dome, crown
Haubenofen *m (ceram)* top-hat kiln
Hauch *m (food)* bloom *(as on fruits or cocoa products)*
Hauchbildung *f* blooming *(esp of oil varnishes)* ✦
  H. zeigen to bloom

Hauerit *m (min)* hauerite *(manganese(IV) sulphide)*
Haufen *m* heap, pile
Haufenlaugung *f (met)* heap leaching
Haufenspeicher *m* pile
Häufigkeitsfaktor *m* frequency factor
Häufigkeitsverteilung *f* frequency distribution
Häufigkeitsverteilungskurve *f* frequency-distribution curve
Haufwerk *n* bed *(as of a filter)*
Haufwerkfilter *n* bed filter
Hauptalkaloid *n* main (major, principal, chief) alkaloid
Hauptbestandteil *m* main (principal, chief) component, main constituent (ingredient), base, basis
Hauptbrücke *f (nomencl)* main bridge
Hauptdampfleitung *f* main steam pipe
Hauptfarbstoff *m* principal colouring material
Hauptfluß *m (plast)* drag flow *(in an extruder)*
Hauptfraktion *f* main fraction
Hauptgärung *f* main (primary) fermentation
Hauptglukosid *n* main (chief) glucoside
Hauptgruppe *f* main group *(of the periodic system)*
Hauptkalkung *f (sugar)* main defecation
Hauptkette *f* main (fundamental, backbone) chain *(of a branched molecule)*
Hauptkolonne *f (distil)* main column
Hauptkomponente *f s.* Hauptbestandteil
Hauptlauf *m (distil)* main fraction
Hauptleitung *f* main
Hauptmasse *f*, **Hauptmenge** *f* bulk
Hauptnährstoff *m (agric)* macroelement, macronutrient, major element
Hauptname *m (pharm)* heading
Hauptperiode *f* reaction period *(of calorimetric measurements)*
Hauptprodukt *n* main (chief) product
Hauptquantenzahl *f* principal (total, first) quantum number
Hauptreaktion *f* main (principal, basic) reaction
Hauptring *m (nomencl)* main ring
Hauptrohr *n* main
Hauptsatz *m* **der Thermodynamik** law of thermodynamics
Hauptschale *f* main shell *(of an atom)*
Hauptscheidung *f (sugar)* main defecation
  kalte H. cold main defecation
Hauptschneide *f* principal (central, centre) knife edge *(of a balance)*
Hauptserie *f* principal series *(spectroscopy)*
Hauptsteinkohlenformation *f* coal measures
Hauptstrang *m* main *(as of a pipe system)*
Hauptstreifenart *f* lithotype *(of hard coal)*
Hauptsymmetrieebene *f (cryst)* unit (standard) plane
Hauptvalenz *f* primary (principal) valency
Hauptvalenzbindung *f* primary (major) valency bond
Hauptwürze *f (ferm)* first wort
Hausbrand *m s.* Hausbrandmaterial
Hausbrandkohle *f* domestic (household) coal
Hausbrandkoks *m* domestic coke
Hausbrandmaterial *n* domestic fuel
Hausenblasenleim *m* isinglass, ichthyocoll, fish glue (gelatine)
Hausenthärter *m* domestic softener *(for treating water)*
Haushaltbrennstoff *m* domestic fuel
Haushaltchemie *f* domestic chemistry
Haushalt-Enthärtungsapparat *m* domestic softener *(for treating water)*
Haushaltmargarine *f* household margarine
Haushaltporzellan *n* domestic porcelain, household china
Hausmannit *m (min)* hausmannite *(manganese(II,III) oxide)*
H-Austausch *m s.* Wasserstoffaustausch
Haut *f* 1. *(coat, met)* skin; 2. *(tann)* hide
  ungegerbte H. rawhide
Hautbildung *f (coat, met)* skin formation
Hautbräunungsmittel *n* suntan preparation (make-up)
Häutchen *n* film, membrane
Hautdesinfektionsmittel *n* skin disinfectant
Hauterweichungsmittel *n* skin softener
Hautleim *m* skin (hide, leather) glue
Hautlotion *f* skin lotion
Hautnährcreme *f* nourishing (lubricating) cream, skin food
Hautpergament *n* animal (skin, natural, writing) parchment
Hautpulver *n (tann)* hide powder
Hautreinigungscreme *f* cleansing cream
hautreizend skin-irritant
Hautreizstoff *m* skin irritant
Hauttonikum *n* skin tonic
Häutungshormon *n (biol)* juvenile (skin-shedding) hormone
Hautverhinderer *m*, **Hautverhinderungsmittel** *n s.* Hautverhütungsmittel
Hautverhütungsmittel *n (coat)* anticreaming (antiskinning) agent
Hautwolle *f* fellmongered wool
Haüyn *m (min)* hauyne, hauynite *(a tectosilicate)*
Havarie *f* upset
HB *s.* Brinellhärte
Hb *s.* Hämoglobin
Hb-CO *s.* Kohlenoxidhämoglobin
H-Bindung *f* hydrogen bonding (linkage) *(process)*; hydrogen bridge bond (linkage) *(state)*
HBL *s.* Harnstoff-Bisulfit-Löslichkeit
HBT *s.* Harzbildnertest
HCH *s.* Hexachlorzyklohexan
HCL-Gas *n* hydrochloric-acid gas
HCR-Mahlung *f (pap)* high-consistency refining, HCR
HD-Öl *n* HD (heavy duty) oil
HDS-Verfahren *n* HDS (hydrodesulphurization) process
Heater-Verfahren *n* heater process *(for regenerating rubber)*
Heavy-Duty-Öl *n s.* HD-Öl
Hebel *m* lever

Hebeleiste *f* lifting flight (plate), flight *(as in a drying cylinder)*
heben to lift, to elevate; to energize *(electrons into an excited state)*
Heber *m (lab)* siphon, syphon; *(tech)* lift
elektrolytischer H. salt bridge
Heberleitung *f* lift line (pipe) *(for lifting the catalyst in catcracking)*
hebern to siphon, to syphon
Hebestange *f* lever
Hebevorrichtung *f* lift[ing device], elevator
Hebewerk *n* drawworks
Hebezeug *n (petrol)* hoisting gear
Heckrolle *f*, **Hecktrommel** *f* tail pulley
Hectorit *m (min)* hectorite *(a phyllosilicate)*
Hedenbergit *m (min)* hedenbergite *(an inosilicate)*
Hederin *n* hederin *(a saponin)*
Hefe *f* yeast, barm
obergärige H. top[-fermentation] yeast
osmophile H. osmophilic yeast
untergärige H. bottom[-fermentation] yeast, low (lager) yeast
wilde H. wild yeast
Hefeadenylsäure *f* yeast adenylic acid *(a mixture of adenosine 2′-phosphate and adenosine 3′-phosphate)*
hefeartig yeast-like
Hefeaufziehapparat *m (ferm)* yeast-growing vat
Hefeautolysat *n* yeast autolysate (extract)
Hefebirne *f s.* Hefeaufziehapparat
Hefebottich *m*, **Hefebütte** *f (ferm)* yeast tub
Hefeextrakt *m s.* Hefeautolysat
Hefefabrik *f* yeast factory (works)
Hefeformmaschine *f* yeast-extruding machine
Hefegärung *f* yeast fermentation
hefegetrieben yeast-leavened, yeast-raised
Hefekultur *f* yeast culture
Hefekulturapparat *m* yeast propagator
Hefengut *n*, **Hefenmaische** *f (ferm)* yeast mash
Hefenukleinsäure *f* yeast nucleic acid
Hefepilz *m* yeast [plant]
Hefeschleuder *f*, **Hefeseparator** *m (ferm)* yeast separator
Hefetrub *m*, **Hefetrübung** *f (ferm)* yeast turbidity
Hefewanne *f (ferm)* yeast tub
Hefewasser *n (ferm)* yeast water
Hefewein *m* yeast wine
hefig yeast-like
Hefteisen *n (glass)* punty [iron]
heftig vigorous *(e. g. reaction)*
Heftpflaster *n* adhesive plaster (tape)
Heftschweißen *n (plast)* stitch welding
Heidehonig *m* heather honey
Heidemoorkrankheit *f (agric)* reclamation disease *(caused by cooper shortage)*
Heidetorf *m* heather peat
Heilbuttleberöl *n*, **Heilbuttlebertran** *m* halibut liver oil, haliver oil
Heildosis *f* therapeutic dose
Heilmittel *n* therapeutic agent, curative drug
antibiotisches H. antibiotic [agent, substance]
Heilquelle *f* medicinal (mineral) spring (well)
Heilwirkung *f* curative action
Heisenberg-Darstellung *f* Heisenberg representation *(quantum mechanics)*
heiß hot; *(nucl)* highly [radio]active, hot
h. [ver]pressen *s.* heißpressen
h. werden to become hot, to heat
Heißalkalisierung *f (pap)* hot [alkali] refining
Heißaufbereitung *f (ceram)* steam tempering, hot preparation
Heißblasen *n* blow[ing] *(in producing water gas)*
Heißblaseperiode *f* blow period *(in producing water gas)*
Heißchlorierung *f* hot chlorination
Heißchromatografie *f* hot chromatography
Heißdampf *m* superheated steam
Heißdampfregenerat *n (rubber)* steam reclaim
Heißdampfverfahren *n (rubber)* steam (thermal) process *(a reclaiming method)*
Heißfärben *n (text)* high-temperature dyeing
Heißfiltration *f* hot filtration
heißfixieren *(plast, text)* to heat-set
Heißgas *n* hot gas
Heißgaseintritt *m* hot-gas inlet
Heißgaserzeuger *m* hot-gas producer
Heißgasschweißen *n* hot-gas welding
Heißgassiegeln *n* hot-gas sealing *(of sheets)*
heißgereckt *(text)* hot-stretched, hot-drawn
Heißhaltezeit *f* retention (hold-up, holding) time, holding period *(of milk during pasteurization)*
Heiß-Kalt-Behandlung *f*, **Heiß-Kalt-Tränkung** *f* hot-and-cold open tank treatment *(wood preservation)*
Heiß-Kalt-Verfahren *n (nucl)* dual temperature process
Heißkanal-Spritzgießen *n (plast)* hot-runner moulding
Heißkanal-Spritzgießwerkzeug *n (plast)* hot-runner mould
Heißkanal-Spritzguß *m (plast)* hot-runner moulding
heißkleben ***(plast)*** **to heat-seal**
Heißkleber *m* hot-setting (hot-melt) adhesive, hot glue
Heißlauge *f* hot brine *(potash industry)*
Heißluft *f* hot (heated) air
Heißluftalterung *f* **[im Geer-Ofen]** *(rubber)* hot-air ageing, Geer (air) oven ageing
Heißluftheizung *f s.* Heißluftvulkanisation
Heißluftkammer *f*, **Heißluftraum** *m* hot-air chamber
Heißluftschrank *m* air-circulating oven
Heißluftsterilisator *m*, **Heißluftsterilisierschrank** *m* hot-air sterilizer
Heißluftstrom *m* hot-air stream (current)
Heißlufttrockenkammer *f* hot-air chamber
Heißlufttrockenmaschine *f (text)* hot flue
Heißluftvulkanisation *f* air (hot-air, dry-air, dry-heat) cure, hot-air vulcanization, HAV
kontinuierliche H. continuous [hot-]air cure
Heißmastikation *f*, **Heißmastizierung** *f* hot mastication

Heißmischen *n* hot mixing
Heißnebel *m (agric)* thermal aerosol
Heißnetzer *m*, **Heißnetzmittel** *n* hot wetting agent
Heißölfärben *n (text)* hot-oil dyeing
Heißpresse *f (pap)* hot press
heißpressen to hot-press, *(relating to powders also)* to sinter under pressure
Heißräuchern *n* hot smoking *(at 80 to 100 °C)*
Heißsäureverfahren *n* hot-acid process *(catalytic polymerization)*
Heißschleifen *n (pap)* hot grinding
Heißschliff *m (pap)* hot-ground pulp
heißsiegelfähig *(plast)* heat sealable
Heißsiegelfähigkeit *f (plast)* heat sealability
Heißsiegelkleber *m* hot-sealing adhesive
heißsiegeln *(plast)* to heat-seal
heißsintern to hot-press, to sinter under pressure
heißspritzen to hot-spray *(painting technology)*
Heißspritzlack *m* hot-spray lacquer
heißtauchen *(plast)* to dip-coat; *(with external moulds)* to dip-mould
Heißtrockenfarbe *f* heat-set ink
Heißtrub *m (ferm)* coarse sludge
heißveredeln *(pap)* to refine by the hot [alkali] process
Heißveredelung *f (pap)* hot [alkali] refining
Heißverlösen *n (fert)* hot dissolution
Heißverschweißen *n (plast)* heat welding, *(esp relating to films)* thermal (heat) sealing
heißverstreckt *(text)* hot-stretched, hot-drawn
Heißverstreckung *f (text)* hot stretching (drawing)
Heißvulkanisation *f* hot cure (vulcanization)
heißvulkanisierbar heat-curable
heißvulkanisierend heat-curing, hot-vulcanizing
heißvulkanisiert heat-cured, hot-cured, hot-vulcanized
Heißwasser *n* hot water
Heißwasserbehälter *m* hot-water tank (accumulator)
Heißwasserdekatur *f (text)* roll boiling
Heißwasserfixierung *f (text)* hydrosetting
Heißwasserpumpe *f* hot-water pump
Heißwasserrohr *n* hot-water pipe
Heißwassertrichter *m* hot-water funnel, heating funnel, funnel heater
Heißwind *m (met)* hot[-air] blast, heated air
Heißwindleitung *f* hot-[air-]blast main, hot-blast line
Heißwindring *m*, **Heißwindringleitung** *f (met)* bustle pipe
Heitler-London-Methode *f* Heitler-London method, HL method *(quantum chemistry)*
Heitler-London-Slater-Pauling-Methode *f* Heitler-London-Slater-Pauling method, HLSP method, spin (electron-pair, valence-bond) method, VB method *(quantum chemistry)*
Heitler-London-Slater-Pauling-Theorie *f* Heitler-London-Slater-Pauling theory, HLSP theory, electron-pair (valence-bond) theory, VB theory *(quantum chemistry)*
Heitler-London-Theorie *f* Heitler-London theory, HL theory *(quantum chemistry)*
Heizaggregat *n* heater assembly
Heizapparat *m* heater; *(rubber)* vulcanizer, vulcanizing apparatus
Heizbad *n* heating bath
Heizbalg *m (rubber)* diaphragm, bladder
Heizband *n* heating tape (band), strip (band) heater
heizbar heatable
Heizblock *m* heating block
Heizbrennstoff *m* fuel
Heizdampf *m* heating steam
Heizdampfeintritt *m* heating steam inlet
Heizeffekt *m* heating (calorific) effect
Heizeinsatzstück *n (plast)* adapter heater
Heizelement *n* heating element (unit), heater
Heizelementschweißen *n (plast)* heated-tool welding
heizen 1. to fire, to heat, to warm [up], to fuel; 2. *(rubber)* to cure, to vulcanize
Heizen *n* / **direktes** direct heating
indirektes H. indirect heating
heizend / langsam *(rubber)* slow-curing
rasch (schnell) h. *(rubber)* fast-curing, quick-curing
Heizer *m* 1. heater *(apparatus)*; 2. *(rubber)* vulcanizer, vulcanizing apparatus
Heizfläche *f* heating surface
Heizflächenofen *m* externally heated oven
Heizflansch *m* flange-type heater
Heizflüssigkeit *f* thermal liquid, heat transfer fluid, heat carrier
Heizgas *n* fuel (heating) gas
Heizgeflecht *n* heating blanket
Heizgerät *n* heater
Heizgeschwindigkeit *f (rubber)* cure (vulcanization) rate
Heizgruppe *f (pap)* dryer group (section)
Heizgut *n* heating load
Heizkammer *f* heating chamber; heating element, calandria *(of an evaporator)*
Heizkatode *f* hot cathode
Heizkeil *m (plast)* heated wedge
Heizkeilschweißen *n* heated-wedge (heated-tool) welding
Heizkörper *m* heating unit (element), heater
Heizkraft *f* calorivic power
heizkräftig of high calorific value
Heizleiter *m* heating resistor
Heizmantel *m* heating jacket (blanket, mantle)
ölgespeister H. oil jacket
Heizmittel *n* heating medium
Heizoberflächentemperatur *f (pap)* dryer surface temperature
Heizöl *n* fuel (heating) oil
H. auf Erdölbasis petroleum fuel oil
destilliertes H. distillate fuel oil
Heizpatrone *f* cartridge heater
Heizplatte *f* heating (hot) plate (platen)
Heizplattentrockner *m* jacketed shelf dryer
Heizpresse *f* hot press
Heizraum *m* heating chamber

Heizrohr *n* fire (heating) tube
Heizrohrkessel *m* fire-tube boiler
Heizschlange *f* heating coil
Heizschlauch *m* *(rubber)* curing bag (tube), air bag
Heizschlauchform *f* *(rubber)* air-bag mould
Heizschlauchmischung *f* *(rubber)* air-bag stock
Heizschnur *f* heating cord
Heizspirale *f* heating coil
Heizstrahler *m* radiant heater
Heiztellertrockner *m* rotary jacketed-shelf dryer
Heiztemperatur *f* *(rubber)* curing (cure, vulcanizing, vulcanization) temperature
Heiztisch *m* *(lab)* heating (hot) stage
Heiztischmikroskop *n* hot-stage microscope
Heizung *f* 1. heating; 2. *(rubber)* cure, vulcanization
  H. in Formen *(rubber)* mould cure (vulcanization)
  H. mit elektrischer Heizdecke *(plast)* electric blanket heating
  dielektrische H. dielectric (electronic, radio-frequency, high-frequency) heating
  elektrische H. electric heating
Heizvorrichtung *f* heating device, heater
Heizwand *f* heating wall
Heizwert *m* calorific value, c.v., heating (heat) value (power), thermal value, *(specif)* net calorific value, lower heating value
  oberer H. gross calorific value, higher heating value
  unterer H. net calorific value, lower heating value
heizwertarm of low calorific value
heizwertreich of high calorific value
Heizwertverlust *m* loss of calorific value
Heizwiderstand *m* heating resistor
Heizzeit *f* *(rubber)* curing (vulcanizing) time
Heizzone *f* heat[ing] zone
  hintere H. *(plast)* rear heat zone
  mittlere H. *(plast)* centre heat zone
  vordere H. *(plast)* front heat zone
Heizzug *m* [heating] flue
Heizzyklus *m* heating cycle
Heizzylinder *m* heating (heated) cylinder
Hektografentinte *f* hectographic ink
Helenin *n* helenin, alantolactone
Helianthin *n* $(CH_3)_2NC_6H_4N{=}NC_6H_4SO_3H$ 1. helianthin[e], *p*-(*p*-dimethylaminophenylazo)benzenesulphonic acid; 2. *(sometimes)* methyl orange, helianthin[e] *(sodium salt of 1.)*
Helioechtrot *n* Helio fast red, Harrison red *(a derivative of m-nitro-p-toluidine)*
Heliografie *f*, **Heliogravüre** *f* photogravure, asphalt (bitumen) process
Heliotrop *m* *(min)* heliotrope, bloodstone *(a subvariety of chalcedony)*
Heliotropin *n* heliotropin, 3,4-methylenedioxybenzaldehyde
Helium *n* He helium
Helium I *n* helium I *(a modification of liquefied helium)*
Helium II *n* helium II *(a modification of liquefied helium)*
Heliumkern *m* helium nucleus, α-particle
Heliumverflüssigung *f* helium liquefaction
γ-Helix *f* gamma helix *(a structural variety of polypeptide chains)*
Helixstruktur *f* helix (helical, helicoidal) structure *(of protein molecules)*
Helles *n s.* Bier / helles
Hellicht-Entwicklung *f* *(phot)* desensitization
Helligkeit *f* brightness, luminosity, *(esp quantitatively)* luminous intensity, intensity of light
Helligkeitsumfang *m* *(phot)* brightness range
Hell-Volhard-Zelinsky-Reaktion *f* Hell-Volhard-Zelinsky reaction *(α-halogenation of aliphatic carboxylic acids)*
Helm *m* *(tech)* head, *(distil also)* stillhead, still dome, distillation head
Helminthagogum *n* *(pharm)* helminthagogue, vermifuge
Helminthizid *n* helminthicide
Helvin *m* *(min)* helvin[e], helvite *(a tectosilicate)*
Helvolinsäure *f* helvolic acid, fumigacin *(antibiotic)*
Hemellithol *n* $C_6H_3(CH_3)_3$ hemimellitene, 1,2,3-trimethylbenzene
Hemellithsäure *f* $(CH_3)_2C_6H_3COOH$ hemellitic acid, 2,3-dimethylbenzoic acid
Hemialdol *n* *(org chem)* hemialdol
Hemiazetal *n* *(org chem)* hemiacetal
hemiedrisch *(cryst)* hemihedral
Hemikolloid *n* hemicolloid
Hemimellithsäure *f* $C_6H_3(COOH)_3$ hemimellitic acid, benzene-1,2,3-tricarboxylic acid
Hemimellitol *n s.* Hemellithol
Hemimellitsäure *f s.* Hemimellithsäure
Hemimorphit *m* *(min)* hemimorphite *(zinc dihydroxide disilicate)*
Hemipinsäure *f* $(CH_3O)_2C_6H_2(COOH)_2$ hemipic (hemipinic) acid, 3,4-dimethoxyphthalic acid
Hemiterpen *n* *(org chem)* hemiterpene
Hemizellulose *f* hemicellulose, pseudocellulose
hemmen to inhibit, to retard
hemmend inhibitory, retardant
Hemmstoff *m* inhibiting (retarding) substance (agent), inhibitor, retarder, anticatalyst, negative catalyst; *(biol)* growth inhibitor
Hemmung *f* inhibition, retardation, anticatalysis, negative catalysis
  kompetitive H. competitive inhibition *(of an enzymatic reaction)*
  nichtkompetitive H. non-competitive inhibition *(of an enzymatic reaction)*
Hemmungshof *m s.* Hemmungszone
Hemmungskurve *f* inhibition curve
Hemmungszone *f* zone of inhibition *(sterile zone in a penicillium culture)*
Hempel-Bürette *f* Hempel gas burette
Hempel-Pipette *f* Hempel gas pipette
Hendekan *n* $CH_3[CH_2]_9CH_3$ undecane, hendecane
Hendekanal *n* $CH_3[CH_2]_9CHO$ undecanal, hendecanal, undecyl aldehyde

Hendekandisäure *f* $HOOC[CH_2]_9COOH$ undecanedioic acid, hendecanedioic acid
Hendekanol *n* undecanol, hendecanol *(any of several isomeric alcohols $C_{11}H_{23}OH$)*
Hendekanon *n undecanone, hendecanone (any of several isomeric ketones $C_{11}H_{22}O$)*
Hendekansäure *f* $CH_3[CH_2]_9COOH$ undecanoic acid, hendecanoic acid
Hendezen *n* undecene, hendecene *(any of several isomeric alkenes $C_{11}H_{22}$)*
Hendezenol *n* undecenol, hendecenol *(any of several isomeric alcohols $C_{11}H_{21}OH$)*
Hendezensäure *f* undecenoic acid, hendecenoic acid *(any of several isomeric alkenoic acids $C_{11}H_{20}O_2$)*
Hendezen-(9)-säure *f* $CH_3CH=CH[CH_2]_7COOH$ 9-undecenoic acid, 9-hendecenoic acid
Hendezin *n* undecyne, hendecyne *(any of several isomeric alkynes $C_{11}H_{20}$)*
Hendezinsäure *f* undecynoic acid, hendecynoic acid *(any of several isomeric alkynoic acids $C_{11}H_{18}O_2$)*
Heneikosan *n* $C_{21}H_{44}$ heneicosane, *(specif)* $CH_3[CH_2]_{19}CH_3$ heneicosane, *n*-heneicosane
Heneikosandisäure *f* $HOOC[CH_2]_{19}COOH$ heneicosanedioic acid, Japanic acid
Henna *f* 1. henna *(a hair dye from Lawsonia inermis L.)*; 2. *s.* Hennastrauch
Hennastrauch *m* [Egyptian] henna, Egyptian privet, Lawsonia inermis L.
Hentriakontan *n* $C_{31}H_{64}$ hentriacontane, *(specif)* $CH_3[CH_2]_{29}CH_3$ hentriacontane, *n*-hentriacontane
Hepar sulfuris *n* hepar sulphuris *(technical potassium sulphide)*
Heparprobe *f* hepar test *(for detecting sulphur)*
Hepatokuprein *n* hepatocupreine *(a copper-containing protein)*
Heptachlor *n* heptachlor *(insecticide)*
Heptadekanon-(9) *n* $CH_3[CH_2]_7-CO-[CH_2]_7CH_3$ 9-heptadecanone, dioctyl ketone
Heptadekansäure *f* $CH_3[CH_2]_{15}COOH$ heptadecanoic acid
Heptafluorid *n* heptafluoride
Heptaldehyd *m s.* Heptanal
Heptamethylen *n s.* Zyklohexan
Heptamolybdat *n* $M_6^I[Mo_7O_{24}]$ heptamolybdate
Heptan *n* C_7H_{16} heptane, *(specif)* $CH_3[CH_2]_5CH_3$ heptane, *n*-heptane
***n*-Heptan** *n* $CH_3[CH_2]_5CH_3$ heptane *(proper)*, *n*-heptane
Heptanal *n* $CH_3[CH_2]_5CHO$ heptanal
Heptandikarbonsäure-(1,7) *f* $HOOC[CH_2]_7COOH$ *n*-heptane-1,7-dicarboxylic acid, nonanedioic acid, azelaic acid
Heptandisäure *f* $HOOC[CH_2]_5COOH$ heptanedioic acid, pimelic acid
Heptankarbonsäure-(1) *f* $CH_3[CH_2]_6COOH$ heptane-1-carboxylic acid, octanoic acid
Heptanol *n* $CH_3[CH_2]_6OH$ 1-heptanol, *n*-heptyl alcohol
Heptanon-(4) *n* $CH_3CH_2CH_2COCH_2CH_2CH_3$ 4-heptanone, butyrone
Heptansäure *f* $CH_3[CH_2]_5COOH$ heptanoic acid
Heptasulfid *n* heptasulphide
heptavalent heptavalent, septivalent
Heptavalenz *f* heptavalence, septivalence
Heptin-(1) *n* $CH \equiv C[CH_2]_4CH_3$ 1-heptyne
Heptose *f* heptose *(monosaccharide containing 7 carbon atoms per molecule)*
Heptoxid *n* heptaoxide
Heptoxotetraborat *n* $M_2^IB_4O_7$ heptaoxotetraborate, tetraborate
Heptoxotetraborsäure *f* $H_2B_4O_7$ heptaoxotetraboric acid, tetraboric acid
***n*-Heptylaldehyd** *m s.* Heptanal
***n*-Heptylalkohol** *m s.* Heptanol
n-**Heptylazetylen** *n* $CH \equiv C[CH_2]_6CH_3$ 1-nonyne, heptylacetylene
Heptylkarbinol *n s.* Oktanol -(1)
Heptylpenaldinsäure *f* heptylpenaldic acid, penaldic-K acid
Heptylpenillosäure *f* heptylpenilloic acid, penilloic-K acid
Heptylpenillsäure *f* heptylpenillic acid, penillic-K acid
Heptylpenizillin *n* heptylpenicillin, heptylpenicillinic acid, penicillin K
Heptylpenizilloinsäure *f* heptylpenicilloic acid, penicilloic-K acid
***n*-Heptylsäure** *f s.* Heptansäure
herabmindern *s.* herabsetzen
herabrieseln to trickle down
herabrinnen to trickle down
herabsetzen to lower, to reduce, to decrease; to slow down *(the speed)*; to relieve [down] *(the pressure)*
Herabsetzung *f* lowering, reduction, decrease; slowing-down *(of speed)*; relief *(of pressure)*
herabspülen to rinse (wash) down
herabtröpfeln to trickle down
herausdestillieren to distil out, to top
herausdrücken to push (blow) out *(as from a reactor)*; to squirt [out] *(as from a nozzle)*
herauslösen to pick out *(solid particles)*; to dissolve out, to lixiviate, to leach [out], *(esp relating to adsorbed substances)* to elute
herausnehmen to release *(from a mould)*
Herausnehmen *n* release *(from a mould)*
herauspressen to press (squeeze) out, to expel *(e. g. oil)*
herausschleppen *(distil)* to entrain out
herausspülen to rinse out; to eluate, to elute *(adsorbed substances from a solid adsorbent)*
herausstoßen to push out, to discharge *(e. g. coke from a coke oven)*
heraustreiben to expel, to drive out (off) *(e. g. gases)*
herb harsh, hard, sour; rough, dry *(wine)*
Herbar[ium]papier *n* herbarium paper
Herbe *f*, **Herbheit** *f* harshness, hardness, sourness; roughness, dryness *(of wine)*
herbizid herbicidal[ly active]
Herbizid *n* herbicide, weed-killer, weed control agent

H. gegen Gräser grass killer
nichtselektives H. non-selective herbicide
selektives (selektiv wirkendes) H. selective herbicide
staubförmiges H. herbicidal dust
systemisches H. systemic (translocated) herbicide, translocation weed-killer
total wirkendes H. *s.* nichtselektives H.
translokales (translokal wirkendes) H. *s.* systemisches H.
Herbizidwirkung *f* herbicidal action (effect)
Herbstdüngung *f* autumn fertilization
Hercynit *m (min)* hercynite *(iron(II) aluminate)*
Herd *m (min tech)* [concentrating, concentrate, concentrator] table; *(met)* ore hearth; hearth *(of an air furnace)*
Herdarbeit *f (min tech)* tabling
Herdflotation *f* table floatation
Herdfrischstahl *m* open-hearth (Siemens-Martin) steel
Herdfrischverfahren *n* open-hearth (Siemens-Martin) process, Siemens process
basisches H. basic open-hearth process
saures H. acid open-hearth process
Herdglas *n (glass)* slag
Herdofen *m* hearth furnace
Herdplatte *f* deck *(of a concentrating table)*
Herdsortieren *n (min tech)* tabling
Herdtafel *f s.* Herdplatte
Herdwagenofen *m (ceram)* bogie (truck chamber, truck, trolley hearth, car-bottom) kiln
Heringsöl *n*, **Heringstran** *m* herring oil
Herkunft *f* origin, source ✦ **pflanzlicher H.** of vegetable (plant) origin, plant-derived ✦ **tierischer H.** of animal origin
Herleitung *f* derivation
hermetisch [abgeschlossen, dicht, verschlossen] hermetic[al]
Heroin *n* heroin, diacetylmorphine *(a narcotic)*
Héroult-Lichtbogenofen *m* Héroult furnace
Herreshoff-Ofen *m* Herreshoff furnace
Herschel-Effekt *m (phot)* Herschel effect
Hershberg-Rührer *m* Hershberg stirrer
herstellen 1. *(esp tech)* to manufacture, to make, to produce, to process, *(esp lab)* to prepare; 2. to establish *(e. g. equilibrium or contact)*
eine Blaupause (Zyanotypie) h. to blueprint
gezielt h. to make to measure, to tailor[-make] *(e. g. polymers)*
großtechnisch h. to produce on the large scale
im Druckguß h. to [pressure-]diecast
im Sandguß h. to sand-cast *(foundry)*
Masterbatches h. *(rubber)* to masterbatch, to mix into a masterbatch
nach Maß h. *s.* gezielt h.
Vormischungen h. *s.* Masterbatches h.
Hersteller[betrieb] *m*, **Herstellerfirma** *f* manufacturer, maker, producer
Herstellung *f* 1. *(esp tech)* manufacture, make, production, processing, *(esp lab)* preparation; 2. establishment *(as of equilibrium or contact)*
H. des Gleichgewichts equilibration
H. von Formartikeln (Formteilen) moulding
H. von Gießkernen core making *(foundry)*
H. von Latexmischungen latex compounding
H. von Vormischungen *(rubber)* masterbatching
H. von Vormischungen auf nassem Wege *(rubber)* wet masterbatching
H. von Wasserzeichen *(pap)* watermarking
großtechnische H. large-scale production; *(as opposed to pilot-plant production)* full-scale factory production
Herstellungsdatum *n* date of manufacture
Herstellungsmethode *f* manufacturing method, method of production, *(esp lab)* method of preparation, preparative method
Herstellungsverfahren *n* manufacturing process
herunterkochen / weit to cook soft *(cellulose)*
herunterkühlen to cool down
herunterrieseln to trickle down
hervorrufen to bring about, to produce, to evolve
herzaktiv cardioactive
Herzenbergit *m (min)* herzenbergite *(tin sulphide)*
Herzgift *n* heart poison
Herzglykosid *n* cardiac glycoside
Herz-Verbindung *f (dye)* [intermediate] Herz compound
herzwirksam cardioactive
HE-Schweißen *n s.* Heizelementschweißen
Hesperetin *n* hesperetin, 3′,5,7-trihydroxy-4′-methoxyflavanone
Hesperetinsäure *f* $HO(CH_3O)C_6H_3CH{=}CHCOOH$ hesperetic acid, 3-hydroxy-4-methoxycinnamic acid
Hessit *m (min)* hessite *(silver telluride)*
Hessonit *m (min)* [h]essonite, cinnamon stone *(a variety of garnet)*
Hetaerolith *m (min)* hetaerolite *(a manganese zinc oxide)*
heteroanalog hetero-analogous
Heteroatom *n* heteroatom
Heteroauxin *n (bioch)* heteroauxin, 3-indolylacetic acid
heteroblastisch *(geol)* heteroblastic
heterofermentativ heterofermentative
heterogen heterogeneous, inhomogeneous
Heterogenität *f* heterogeneity, inhomogeneity
Heterolyse *f* heterolysis
heterolytisch heterolytic
Heterometrie *f* heterometry *(a method of titration)*
heteropolar heteropolar
heteropolymer heteropolymeric
Heteropolymer[es] *n* heteropolymer, heterogeneous polymer
Heteropolymerisation *f* heteropolymerization
Heteropolysäure *f* heteropoly acid
Heterosid *n (org chem)* heteroside
heterotroph *(biol)* heterotrophic
heterozyklisch heterocyclic
Heterozyklus *m (org chem)* heterocycle, heterocyclic [compound]

HETP-Wert *m (distil)* height equivalent to a theoretical plate, HETP
HET-Säure *f*, **Hetsäure** *f* chlorendic acid, hexachloroendomethylenetetrahydrophthalic acid
Heulandit *m (min)* heulandite *(a tectosilicate)*
Heuschreckenabwehrmittel *n* grasshopper repellent
Hevea-Kautschuk *m* hevea rubber *(from Hevea brasiliensis (H. B. K.) Muell. Arg.)*
Hevea-Latex *m* hevea latex
Hewettit *m (min)* hewettite *(a calcium vanadate)*
Hexaäthyltetraphosphat *n* $(C_2H_5)_6P_4O_{13}$ hexaethyltetraphosphate, HETP
Hexaboran *n* hexaborane
Hexaborid *n* hexaboride
Hexabromdisilan *n* Si_2Br_6 hexabromodisilane
Hexabromid *n* hexabromide
Hexabromoplatinat(IV) *n* $M_2^I[PtBr_6]$ hexabromoplatinate(IV), bromoplatinate(IV)
Hexabromoplatin(IV)-säure *f* $H_2[PtBr_6]$ hexabromoplatinic(IV) acid, bromoplatinic(IV) acid
Hexachloräthan *n* Cl_3CCCl_3 hexachloroethane, perchloroethane
Hexachlorbenzol *n* C_6Cl_6 hexachlorobenzene, perchlorobenzene
Hexachlordisilan *n* Si_2Cl_6 hexachlorodisilane
Hexachlorid *n* hexachloride
Hexachloroiridat(III) *n* $M_3^I[IrCl_6]$ hexachloroiridate(III), chloroiridate(III)
Hexachloroosmat(III) *n* $M_3^I[OsCl_6]$ hexachloroosmate(III)
Hexachloroosmat(IV) *n* $M_2^I[OsCl_6]$ hexachloroosmate(IV)
Hexachloropalladat(IV) *n* $M_2^I[PdCl_6]$ hexachloropalladate(IV)
Hexachlorophen *n* $Cl_3C_6H(OH)CH_2C_6H(OH)Cl_3$ hexachlorophene, 3,3′,5,5′,6,6′-hexachloro-2,2′-dihydroxydiphenylmethane
Hexachloroplatinat(IV) *n* $M_2^I[PtCl_6]$ hexachloroplatinate(IV), chloroplatinate(IV)
Hexachloroplatin(IV)-säure *f* $H_2[PtCl_6]$ hexachloroplatinic(IV) acid, chloroplatinic(IV) acid
Hexachlororhodat(III) *n* $M_3^I[RhCl_6]$ hexachlororhodate(III), chlororhodate(III)
Hexachlororuthenat(IV) *n* $M_2^I[RuCl_6]$ hexachlororuthenate(IV)
Hexachlorotellurat(IV) *n* $M_2^I[TeCl_6]$ hexachlorotellurate(IV), chlorotellurate(IV)
Hexachlorotitanat(IV) *n* $M_2^I[TiCl_6]$ hexachlorotitanate(IV), chlorotitanate(IV)
Hexachlorozinn(IV)-säure *f* $H_2[SnCl_6]$ hexachlorostannic acid
Hexachlorozirkonat(IV) *n* $M_2^I[ZrCl_6]$ hexachlorozirconate(IV), chlorozirconate(IV)
Hexachlorzyklohexan *n* hexachlorocyclohexane
Hexadekan *n* $C_{16}H_{34}$ hexadecane
Hexadekanol-(1) *n* $CH_3[CH_2]_{15}OH$ 1-hexadecanol
Hexadekansäure *f* $CH_3[CH_2]_{14}COOH$ hexadecanoic acid
1-Hexadekanthiol *n* $CH_3[CH_2]_{15}SH$ 1-hexadecanethiol
n-**Hexadezylalkohol** *m s.* Hexadekanol-(1)
n-**Hexadezylmerkaptan** *n s.* 1-Hexadekanthiol
n-**Hexadezylsäure** *f s.* Hexadekansäure
Hexadien-(1,5) *n* $CH_2{=}CHCH_2CH_2CH{=}CH_2$ 1,5-hexadiene
Hexadien-(2,4)-disäure *f* $HOOCCH{=}CHCH{=}CHCOOH$ 2,4-hexadien-1,6-dioic acid, muconic acid
Hexadien-(2,4)-säure *f* $CH_3CH{=}CHCH{=}CHCOOH$ 2,4-hexadienoic acid, sorbic acid
Hexaedrit *m* hexahedrite *(nickel-containing meteoric iron)*
Hexafluorid *n* hexafluoride
Hexafluoroferrat(III) *n* $M_3^I[FeF_6]$ hexafluoroferrate(III)
Hexafluorokieselsäure *f* $H_2[SiF_6]$ hexafluorosilicic acid, fluorosilicic acid
Hexafluoromanganat(IV) *n* $M_2^I[MnF_6]$ hexafluoromanganate(IV), fluoromanganate(IV)
Hexafluorophosphat *n* $M^I[PF_6]$ hexafluorophosphate
Hexafluorophosphorsäure *f* $H[PF_6]$ hexafluorophosphoric acid
Hexafluorosilikat *n* $M_2^I[SiF_6]$ hexafluorosilicate, fluorosilicate
Hexafluorostannat(IV) *n* $M_2^I[SnF_6]$ hexafluorostannate(IV), fluorostannate(IV)
hexagonal *(cryst)* hexagonal
Hexahydrat *n* hexahydrate
Hexahydrid *n* hexahydride
Hexahydrobenzol *n* hexahydrobenzene, cyclohexane
Hexahydrophenol *n* $C_6H_{11}OH$ hexahydrophenol, cyclohexanol
Hexahydrophthalsäure *f* $C_6H_{10}(COOH)_2$ hexahydrophthalic acid, cyclohexane-1,2-dicarboxylic acid
Hexahydrotoluol *n* hexahydrotoluene, methylcyclohexane
Hexahydroxoantimonat *n* $M^I[Sb(OH)_6]$ hexahydroxoantimonate, hydroxoantimonate
Hexahydroxoantimonsäure *f* $H[Sb(OH)_6]$ hexahydroxoantimonic acid, hydroxoantimonic acid
Hexahydroxostannat(IV) *n* $M_2^I[Sn(OH)_6]$ hexahydroxostannate(IV)
Hexahydroxybenzol *n* $C_6(OH)_6$ hexahydroxybenzene
Hexahydroxyzyklohexan *n* $C_6H_6(OH)_6$ hexahydroxycyclohexane, inositol
Hexahydrozymol *n* hexahydrocymene, menthane, 1-isopropyl-methylcyclohexane
Hexajodid *n* hexaiodide
Hexajodoplatin(IV)-säure *f* $H_2[PtI_6]$ hexaiodoplatinic(IV) acid, iodoplatinic(IV) acid
Hexakarbonyl *n* hexacarbonyl
1-Hexakosanol *n* $CH_3[CH_2]_{24}CH_2OH$ 1-hexacosanol, ceryl alcohol
n-**Hexakosansäure** *f* $CH_3[CH_2]_{24}COOH$ hexacosanoic acid, cerotic acid, cerinic acid
Hexamethylen *n s.* Zyklohexan
Hexamethylendiamin *n* hexamethylene diamine, 1,6-diaminohexane

Hexamethylendiaminadipat *n* hexamethylene diamine adipate, 6,6 salt, nylon salt
Hexamethylendiammoniumadipat *n* *s.* Hexamethylendiaminadipat
Hexamethylentetramin *n* hexamethylenetetramine, hexamine, metheneamine
Hexamin *n* 1. hexamine, metheneamine, hexamethylenetetramine; 2. di-2,4,6-trinitrophenylamine, hexanitrodiphenylamine, hexite
Hexammingallium(III)-chlorid *n* $[Ga(NH_3)_6]Cl_3$ hexaamminegallium(III) chloride
Hexamminkobalt(III)-chlorid *n* $[Co(NH_3)_6]Cl_3$ hexaamminecobalt(III) chloride
Hexamminnickel(II)-bromid *n* $[Ni(NH_3)_6]Br_2$ hexaamminenickel(II) bromide
Hexamminnickel(II)-chlorid *n* $[Ni(NH_3)_6]Cl_2$ hexaamminenickel(II) chloride
Hexamminnickel(II)-jodid *n* $[Ni(NH_3)_6]I_2$ hexaamminenickel(II) iodide
Hexamminplatin(IV)-chlorid *n* $[Pt(NH_3)_6]Cl_4$ hexaammineplatinum(IV) chloride
Hexan *n* C_6H_{14}hexane, *(specif)* $CH_3[CH_2]_4CH_3$ hexane, *n*-hexane
Hexanal *n* $CH_3[CH_2]_4CHO$ hexanal
Hexandikarbonsäure-(1,6) *f* $HOOC[CH_2]_6COOH$ hexane-1,6-dicarboxylic acid, octanedioic acid
Hexandisäure *f* $HOOC[CH_2]_4COOH$ hexanedioic acid, adipic acid
Hexanitrid *n* hexanitride
Hexanitroiridat(III) *n* $M_3^I[Ir(NO_2)_6]$ hexanitroiridate(III), nitroiridate(III)
Hexanitrokobaltat(II) *n* $M_4^I[Co(NO_2)_6]$ hexanitrocobaltate(II), nitrocobaltate(II)
Hexanitrokobaltat(III) *n* $M_3^I[Co(NO_2)_6]$ hexanitrocobaltate(III), nitrocobaltate(III)
Hexanitroniccolat(II) *n* $M_4^I[Ni(NO_2)_6]$ hexanitroniccolate(II), hexanitronickelate(II)
Hexanitrorhodat(III) *n* $M_3^I[Rh(NO_2)_6]$ hexanitrorhodate(III)
Hexanol-(1) *n* $CH_3[CH_2]_5OH$ 1-hexanol, *n*-hexyl alcohol
***n*-Hexanol** *n s.* Hexanol-(1)
Hexansäure *f* $CH_3[CH_2]_4COOH$ hexanoic acid
Hexansäureäthylester *m* $CH_3[CH_2]_4CO-OC_2H_5$ ethyl hexanoate
Hexasilan *n* Si_6H_{14} hexasilane
Hexasilikat *n* $M_{12}^ISi_6O_{18}$ hexasilicate
Hexasulfid *n* hexasulphide
Hexatantalat *n* $M_8^ITa_6O_{19}$ hexatantalate
Hexathionat *n* $M_2^I[S_6O_6]$ hexathionate
hexavalent hexavalent, sexavalent, sexivalent
Hexavalenz *f* hexavalence, sexavalende, sexivalence
Hexawolframat *n* hexawolframate, hexatungstate
Hexazyanochromat(II) *n* $M_4^I[Cr(CN)_6]$ hexacyanochromate(II), cyanochromate(II)
Hexazyanochromat(III) *n* $M_3^I[Cr(CN)_6]$ hexacyanochromate(III), cyanochromate(III)
Hexazyanoeisen(II)-säure *f* $H_4[Fe(CN)_6]$ hexacyanoferric(II) acid
Hexazyanoeisen(III)-säure *f* $H_3[Fe(CN)_6]$ hexacyanoferric(III) acid
Hexazyanoferrat(II) *n* $M_4^I[Fe(CN)_6]$ hexacyanoferrate(II), cyanoferrate(II)
Hexazyanoferrat(III) *n* $M_3^I[Fe(CN)_6]$ hexacyanoferrate(III), cyanoferrate(III)
Hexazyanoferrat(III)-komplex *m* hexacyanoferrate(III) complex, cyanoferrate(III) complex
Hexazyanokobaltat(III) *n* $M_3^I[Co(CN)_6]$ hexacyanocobaltate(III), cyanocobaltate(III)
Hexazyanomanganat(II) *n* $M_4^I[Mn(CN)_6]$ hexacyanomanganate(II), cyanomanganate(II)
Hexazyanomanganat(III) *n* $M_3^I[Mn(CN)_6]$ hexacyanomanganate(III), cyanomanganate(III)
Hexazyanomangan(II)-säure *f* $H_4[Mn(CN)_6]$ hexacyanomanganic(II) acid
Hexazyanoosmat(II) *n* $M_4^I[Os(CN)_6]$ hexacyanoosmate(II), cyanoosmate(II)
Hexazyanoplatinat(IV) *n* $M_4^I[Pt(CN)_6]$ hexacyanoplatinate(IV), cyanoplatinate(IV)
Hexenmehl *n* lycopodium powder
Hexin-(1) *n* $HC{\equiv}C[CH_2]_3$ 1-hexyne
Hexin-(2) *n* $CH_3C{\equiv}CCH_2CH_2CH_3$ 2-hexyne
Hexin-(3) *n* $CH_3CH_2C{\equiv}CCH_2CH_3$ 3-hexyne
Hexit *m* hexitol *(any of the hexahydroxy alcohols $HOCH_2[CHOH]_4CH_2OH$)*
Hexokinase *f* hexokinase
Hexonbase *f* hexone base
Hexose *f* hexose *(monosaccharide containing 6 oxygen atoms per molecule)*
Hexoxojodat(VII) *n* $M_5^IIO_6$ hexaoxoiodate(VII), orthoperiodate
Hexoxojod(VII)-säure *f* H_5IO_6 hexaoxoiodic(VII) acid, orthoperiodic acid
Hexoxotellursäure *f* H_6TeO_6 hexaoxotelluric acid, orthotelluric acid, telluric acid
***n*-Hexylaldehyd** *m s.* Hexanal
***n*-Hexylalkohol** *m s.* Hexanol-(1)
***n*-Hexylazetylen** *n s.* Oktin-(1)
***n*-Hexylchlorid** *n* $CH_3[CH_2]_4CH_2Cl$ 1-chlorohexane, 1-hexyl chloride
Hexylessigsäure *f s.* Oktansäure
Hexylsäure *f s.* Hexansäure
HF-... *s.* Hochfrequenz ...
HF-Alkylierung *f* HF alkylation, hydrofluoric-acid alkylation
HFS *s.* Hyperfeinstruktur
Hg-Destillationsapparat *m* mercury still
Hg-Fungizid *n* mercurial fungicide
Hgw *s.* Hartgewebe
Hiddenit *m (min)* hiddenite *(aluminium lithium disilicate)*
High-Abrasion-Furnace-Ruß *m (rubber)* high-abrasion furnace black, HAF black
High-Modulus-Furnace-Ruß *m (rubber)* high-modulus furnace black, HMF black
High-Structure-Ruß *m (rubber)* high-structure [carbon] black
High-Yield-Stoff *m (pap)* high-yield pulp
Hildebrandt-Extrakteur *m* Hildebrandt (U-tube) extractor
Hilfsausrüstung *f* ancillary equipment
Hilfselektrode *f* auxiliary electrode
H. zum Öffnen des Abstichlochs tapping electrode

Hilfsgerbstoff *m* **/ synthetischer** auxiliary (neutral) syntan
Hilfsknotenfänger *m (pap)* auxiliary strainer, back knotter
Hilfskolben *m (plast)* auxiliary ram
Hilfslöser *m s.* Hilfslösungsmittel
Hilfslösung *f* auxiliary solution
Hilfslösungsmittel *n* cosolvent, indirect (latent) solvent
Hilfsmittel *n* 1. auxiliary contrivance (device); 2. *s.* Hilfsstoff
Hilfspumpe *f* backing pump
Hilfsstandard *m* subsidiary standard
Hilfssteuerleitung *f* pilot supply line
Hilfssteuerung *f* pilot control ✦ **mit H.** pilot-controlled, pilot-operated
Hilfssteuerventil *n* pilot valve
Hilfsstoff *m* auxiliary (supplementary) agent, aid, *(esp pharm)* adjuvant; corrective *(in building-up active-substance mixtures)*
Hilfsthermometer *n* auxiliary thermometer
Hilfsvorrichtung *f* auxiliary contrivance (device)
  H. zum Ausdrücken *(plast)* extractor
Hill-Reaktion *f (bioch)* Hill reaction
Himmelblau *n* celestial (ethereal) blue *(any of several iron blue pigments)*
hinaufheben to lift
hinausdrücken to push out, to blow out *(as of a reactor)*; to squirt [out] *(as through a nozzle)*
hinauswandern to migrate out *(as of ions)*
Hinderung *f* hindrance
  H. der freien Drehbarkeit hindered rotation
  sterische H. steric hindrance (inhibition)
hindurchdrücken to press (force) through
hindurchlaufen to pass through *(as through a sieve)*
hindurchleiten to pass
hindurchperlen to bubble *(of a gas passing a liquid)*
  h. lassen to bubble *(a gas through a liquid)*
hindurchpressen to force through
hindurchsickern to soak
Hindurchwandern *n* **der Schmelzzone (Zone)** zone travel[ling] *(in zone melting)*
hineinfressen / sich to eat *(as of an acid)*
hineinwandern to migrate in *(as of ions)*
Hinokiflavon *n* hinokiflavone *(a biflavonyl)*
Hinokinin *n* hinokinin *(a lignan)*
Hinokiol *n* hinokiol *(a diterpene)*
Hinokisäure *f* hinokiic acid *(a sesquiterpene derivative)*
Hinreaktion *f* direct (forward) reaction
Hinsberg-Probe *f* Hinsberg [amine] test
Hintergrund *m (nucl)* background
Hintermauerung *f* backing-up *(as of a furnace)*
Hinterschneidung *f (plast)* counterdraft, undercut
hin- und herbewegen to reciprocate; *(tann)* to rock *(pelts in a rocker frame)*
  sich h. to reciprocate
hin- und hergehen to reciprocate
hinzufügen to add
hinzugeben to add
hinzuwandern to migrate in *(as of ions)*
H-Ion *n* hydrogen ion, H ion
H-Ionen ... *s.* Wasserstoffionen ...
Hippursäure *f* $C_6H_5CO-NHCH_2COOH$ hippuric acid, benzoylaminoacetic acid
HI-Presse *f (pap)* high-intensity press, HI press
Hiragonsäure *f* hiragonic acid, 6,10,14-hexadecatrienoic acid
Hirschhornsalz *n* [salt of] hartshorn, commercial ammonium carbonate, sal volatile *(a mixture consisting of ammonium hydrogencarbonate and ammonium carbamate)*
Hirschkolbensumach *m (tann)* staghorn sumac, Rhus typhina L.
His *s.* Histidin
Histamin *n* histamine, 4-(ω-aminoethyl)-glyoxaline
Histaminphosphat *n* histamine phosphate
Histidin *n* histidine, 2-amino-3-imidazolylpropionic acid
Histon *n* histone *(a protein)*
Histozym *n* histozyme *(an enzyme)*
Hitzdrahtanemometer *n* hot-wire anemometer
Hitzealterung *f* heat ageing, thermosenescence
Hitzebad *n* heating bath
Hitzebarriere *f* heat barrier
hitzebehandeln to heat-treat
Hitzebehandlung *f* heat treatment; *(text)* baking
hitzebeständig heat-resistant, heat-resisting, heat-proof, thermally stable, thermoresistant, *(relating to microorganisms also)* thermoduric
Hitzebeständigkeit *f* heat resistance, thermal stability (resistance, endurance), thermostability
Hitzedenaturierung *f* heat denaturation
hitzeempfindlich heat-sensitive
Hitzeempfindlichkeit *f* heat sensitivity
hitzefest *s.* hitzebeständig
Hitzefestigkeit *f s.* Hitzebeständigkeit
Hitzeflockung *f* heat flocculation
Hitzegrenze *f* heat barrier
hitzehärtbar *(plast, ceram)* thermosetting
Hitzeinaktivierung *f* heat inactivation
Hitzekoagulation *f* heat coagulation
Hitzemauer *f* heat barrier
hitzeresistent *s.* hitzebeständig
Hitzeresistenz *f s.* Hitzebeständigkeit
Hitzeschädigung *f* heat damage
Hitzespaltung *f* thermal decomposition, decomposition by heat
hitzestabil *s.* hitzebeständig
Hitzesterilisation *f* heat sterilization
hitzesterilisiert heat-sterilized
HLB-Wert *m (text)* hydrophilic-lipophilic balance, HLB
HLSP-Methode *f s.* Heitler-London-Slater-Pauling-Methode
HLSP-Theorie *f s.* Heitler-London-Slater-Pauling-Theorie
HMF-Ruß *m (rubber)* HMF black, high-modulus furnace black
hochaktiv highly active, *(nucl also)* high-level active, highly radioactive; *(rubber)* fully reinforcing

hocharomatisch highly aromatic
hochaschehaltig high-ash
Hochausbeutestoff *m (pap)* high-yield pulp
Hochausbeute-Sulfitzellstoff *m (pap)* high-yield sulphite pulp
Hochausbeutezellstoff *m (pap)* high-yield pulp
hochausraffiniert highly refined
hochbasisch highly basic
Hochbauschgarn *n* high-bulk yarn, bulky yarn
hochbeansprucht highly stressed
Hochbehälter *m* overhead (elevated) tank
hochbelastet highly (heavily) loaded
hochbleihaltig high-leaded *(e. g. alloy)*
Hochbunker *m* overhead hopper
hochchloren, hochchlorieren to superchlorinate *(water)*
Hochchlorierung *f*, **Hochchlorung** *f* superchlorination *(of water)*
hochdispers highly disperse
Hochdruck *m* 1. high pressure; 2. typographic (relief) printing
Hochdruckdampf *m* high-pressure steam, HP steam
Hochdruckdampferhärtung *f (build)* high-pressure steam curing
Hochdruckdampfverfahren *n (rubber)* high-pressure process, Palmer process *(a reclaiming process)*
Hochdruckdichtung *f* high-pressure packing
Hochdruckdüngelanze *f (agric)* high-pressure soil injector
Hochdruckfarbe *f* typographic [printing] ink
Hochdruckhydrierung *f* high-pressure hydrogenation
H. in flüssiger Phase high-pressure liquid-phase hydrogenation
Hochdruckhydrier[ungs]verfahren *n* **nach Bergius** Bergius process, berginization
Hochdruckjigger *m (text)* high-pressure jig
Hochdruckkochkessel *m (text)* high-pressure boiling kier
Hochdruckkompressor *m* high-pressure compressor
Hochdrucklaminieren *n* high-pressure laminating
Hochdruckleitung *f* high-pressure line
Hochdruckpackung *f* high-pressure packing
Hochdruckpolyäthylen *n* high-pressure-process polyethylene, branched polyethylene
Hochdruckpressen *n (plast)* high-pressure moulding
Hochdruckpumpe *f* high-pressure (high-head) pump
Hochdruckschicht[preß]stoff *m* high-pressure laminate
Hochdrucksprühgerät *n* high-pressure sprayer
Hochdruckstoffauflauf *m (pap)* high-pressure headbox, pressurized headbox
Hochdrucktechnik *f* high-pressure technology
Hochdruckverdichter *m* high-pressure compressor
Hochdruckverfahren *n* 1. high-pressure process; 2. typographic (relief) process
Hochdruck-Wasserstoff-Sauerstoff-Brennstoffelement *n* high-pressure hydrogen cell
Hochdruckwasserstrahl *m* high-pressure water jet
Hochdruckzelle *f* **von Bacon** Bacon high-pressure hydrogen cell
hocheisenhaltig high-iron
Hochelastizität *f* high elasticity
hochempfindlich highly sensitive
Hocherhitzer *m (food)* flash pasteurizer
Hocherhitzung *f (food)* flash pasteurization, flashing
hochevakuiert highly evacuated
hochexplosiv highly (violently) explosive
Hochfeuer *n (ceram)* full fire
hochfeuerfest highly refractory, superrefractory
hochflüchtig highly volatile, high-volatile
hochfördern to pass up[wards], to elevate
Hochfrequenzbeheizung *f s.* Hochfrequenzerhitzung
Hochfrequenzerhitzung *f*, **Hochfrequenzerwärmung** *f* high-frequency (radio-frequency) heating, dielectric (electronic) heating
Hochfrequenzheizung *f* 1. *(rubber)* high-frequency curing; 2. *s.* Hochfrequenzerhitzung
Hochfrequenzinduktionsofen *m* high-frequency induction furnace, coreless induction furnace
Hochfrequenzkoagulation *f* high-frequency coagulation
Hochfrequenzschweißen *n (plast)* high-frequency welding (bar sealing)
Hochfrequenzschweißgerät *n*, **Hochfrequenzsiegelgerät** *n (plast)* high-frequency sealing machine, bar sealer
Hochfrequenzsiegeln *n s.* Hochfrequenzschweißen
Hochfrequenzsirene *f* ultrasonic agglomerator *(gas purification)*
Hochfrequenztitration *f* high-frequency titration
Hochfrequenztitrator *m* high-frequency titrator
Hochfrequenztrockner *m* high-frequency dryer
Hochfrequenztrocknung *f* high-frequency drying
Hochfrequenzverleimung *f* high-frequency gluing
Hochfrequenzvorwärmung *f* high-frequency preheating
hochgebrannt *(ceram)* hard-fired, high-fired, hard-burned
hochgefüllt highly (heavily) loaded (filled)
Hochgehen *n* lifting *(of a coating by the action of a solvent)*
hochgekohlt high-carbon *(e. g. steel)*
hochgereinigt highly purified
hochgeschwefelt high-sulphur
hochgespannt highly strained *(e. g. ring system)*
Hochglanz *m* high gloss
hochglänzend high-gloss, high-lustrous, highly lustrous
Hochglanzpapier *n* bright enamel paper
hochheizen to heat up
hochhitzebeständig resistant to high temperature[s]

Hochhitzebeständigkeit *f* resistance to high temperature[s]
hochinkohlt high-rank
hochkapazitiv high-capacity, large-capacity
hochklopffest highly knockproof, high-octane *(carburetting fuel)*
hochkohlenstoffhaltig high-carbon *(e. g. steel)*
hochkolloidal highly colloidal
hochkomprimiert highly compressed
hochkonzentriert highly concentrated, high-concentration
Hochkräusen *pl (ferm)* rocky krausen
Hochkupferglanz *m (min)* high-chalcocite *(copper(I) sulphide)*
Hochkurzerhitzung *f*, **Hochkurzpasteurisation** *f*, **Hochkurzsterilisierung** *f (food)* short-time heat processing, high-temperature short-time pasteurization (heat treatment), HTST pasteurization
Hochleistungselement *n (agric)* microelement, micronutrient, minor [nutrient] element
Hochleistungsextruder *m* heavy-duty extruder
hochleistungsfähig high-capicity, large-capacity
Hochleistungskalander *m (pap)* supercalender
Hochleistungsöl *n* heavy-duty oil, HD oil
Hochleistungstrockner *m* high-duty dryer
Hochleistungszentrifuge *f* high-speed centrifuge
Hochmahlverfahren *n s.* Hochmüllerei
hochmolekular high-molecular
Hochmoortorf *m* moor peat
Hochmüllerei *f (food)* high (reduction) milling, high (open) grinding
Hochnaßmodulfaser *f* high-wet-modulus fibre, HWM
Hochnaßmodulfaserstoff *m* high-wet-modulus fibre, HWM
Hochofen *m* blast furnace
Hochofenanlage *f* blast-furnace plant
Hochofenfutter *n* blast-furnace lining
Hochofengas *n* blast-furnace gas
Hochofengestell *n* blast-furnace hearth
Hochofengicht *f* 1. throat, top *(of a furnace)*; 2. *s.* Hochofenmöller
Hochofengichtgas *n* blast-furnace gas
Hochofenkoks *m* [blast-]furnace coke
Hochofenmöller *m* charge [stock], burden, stock, feed
Hochofenschlacke *m* blast-furnace slag
Hochofenverfahren *n* blast-furnace process
Hochofenwerk *n* blast-furnace plant
Hochofenwind *m* furnace blast
Hochofenwinderhitzer *m* blast-furnace (air-blast, hot-blast) stove
Hochofenwürfel *m* blast-furnace cube
Hochofenzement *m* portland blast-furnace cement, slag cement
Hochoffsetdruck *m* dry offset printing
Hochoktanbenzin *n* high-octane gasoline (petrol)
hochoktanig high-octane
Hochoktankraftstoff *m* high-octane fuel
hochoktanzahlig high-octane
hochorientiert highly oriented *(bond)*
hochphosphorhaltig high-phosphorus
hochplastisch highly plastic
hochpolarisierbar highly polarizable
hochpolymer highly polymerized
Hochpolymer[es] *n* high polymer
hochporös highly porous
hochprozentig high-per-cent, high-percentage, high-analysis; *(relating to spirits)* strong, *(Am also)* hard
hochraffiniert highly refined
hochreaktionsfähig, hochreaktiv highly reactive
hochrein highly purified, high-purity
Hochreinigung *f* ultrapurification, superrefining
hochsauerstoffhaltig high-oxygen
hochschmelzend high-melting[-point], high-fusion
hochschrumpfend *(text)* high-shrinking
hochselektiv highly selective
hochsiedend high-boiling, heavy
Hochsieder *m* high boiler *(as of solvents)*
Hochsintern *n*, **Hochsinterung** *f* full (final) sintering *(of metals)*
Hochspannungselektrophorese *f* high-voltage electrophoresis
Hochspannungselektroporzellan *n* high-voltage (high-tension) electrical porcelain
Hochspannungspapierelektrophorese *f* high-voltage paper electrophoresis
Höchstdruckkessel *m* very-high-pressure boiler
Höchstdruckschmiermittel *n*, **Höchstdruckschmierstoff** *m* extreme-pressure lubricant, EP lubricant
Höchstdruckspritzen *n* **[/ druckluftloses]** *(coat)* airless spraying
Höchstlast *f* permitted load
Höchstmenge *f* **/ duldbare** *(tox)* [maximum] tolerance
zugelassene (zulässige) H. *s.* duldbare H.
Hochstruktur-Ruß *m (rubber)* high-structure [carbon] black
Hochtemperaturbehandlung *f* high-temperature treatment
hochtemperaturbeständig resistant to high temperature[s], stable at high temperature[s]
Hochtemperaturbeständigkeit *f* high-temperature resistance (stability, durability)
Hochtemperaturchemie *f* high-temperature chemistry
Hochtemperaturdämpfen *n (text)* high-temperature steaming
Hochtemperaturdämpfer *m (text)* high-temperature steamer
Hochtemperaturdestillation *f* high-temperature distillation
Hochtemperaturentgasung *f s.* Hochtemperaturverkokung
Hochtemperaturfärbemaschine *f (text)* high-temperature dyeing machine
Hochtemperaturfärben *n (text)* high-temperature dyeing
Hochtemperatur-Hitzebehandlung *f* high-temperature heat treatment

Hochtemperaturkochung *f (pap)* high-temperature digestion
Hochtemperaturkoks *m* high-temperature coke
Hochtemperaturkorrosion *f* high-temperature corrosion
Hochtemperaturlegierung *f* high-temperature alloy
Hochtemperaturofen *m* high-temperature furnace (kiln)
Hochtemperaturpolymer[es] *n*, **Hochtemperaturpolymerisat** *n* high-temperature polymer, hot (heat) polymer
Hochtemperaturpolymerisation *f* high-temperature polymerization, hot (heat) polymerization
Hochtemperaturpyrolyseverfahren *n* high-temperature pyrolysis process
Hochtemperaturreaktor *m (nucl)* high-temperature [gas-cooled] reactor
Hochtemperaturschmierfett *n* high-temperature grease
Hochtemperaturteer *m* high-temperature tar
Hochtemperaturtunnelofen *m (ceram)* high-temperature tunnel kiln
Hochtemperatur-Überdruckfärben *n (text)* high-temperature pressure dyeing
Hochtemperaturverkokung *f* high-temperature carbonization (coking)
Hochtemperaturvulkanisation *f* high-temperature cure (vulcanization)
hochtourig high-speed
hochtoxisch highly toxic
Hochvakuum *n* high vacuum
Hochvakuumbedampfung *f* vacuum coating [by evaporation]
Hochvakuumdestillation *f* high-vacuum distillation
Hochvakuumtechnik *f* high-vacuum technique
hochverstärkend *(rubber)* fully reinforcing
hochviskos highly viscous, high-viscosity
hochweiß extra white
hochwertig of high value (quality), high-grade
hochwirksam highly active
hochzähflüssig *s.* hochviskos
Hochzahl *f (nomencl)* [numeric] superscript, superscript numeral
Hochziehen *n (coat)* lifting *(by the action of solvents)*
Hof *m* halo *(as in spectrography and chromatography)*
Hoffmann-Ofen *m (ceram)* Hoffmann kiln
Hofmann-Abbau *m* Hofmann degradation *(of amides)*
Hofmann-Eliminierung *f* Hofmann elimination *(of quaternary ammonium hydroxides)*
Hofmann-Orientierung *f (org chem)* Hofmann orientation
Höhe *f* **einer Übertragungseinheit** *(distil)* height of one transfer unit, HTU
Höhenstrahlen *mpl* cosmic rays
Höhenstrahlung *f* cosmic radiation
höhergliedrig higher-membered
höherhalogeniert polyhalogenated
höhermolekular polymolecular
höherprozentig higher-percentage, higher-analysis
höherschmelzend higher-melting
höhersiedend high-boiling, higher-boiling, heavy
höherwertig of higher valence, higher-valent; polyhydric *(alcohol)*; higher-analysis *(as of a commercial product)*
Hohlblock *m (ceram)* hollow block
Hohlfaser *f* hollow fibre
Hohlform *f (plast)* die
Hohlglas *n* hollow [glass]ware, container glass
Hohlguß *m (ceram)* drain (hollow) casting
Hohlkegeldüse *f* hollow-cone nozzle
Hohlkörper *m (plast)* hollow article (body, part)
Hohlkörperblasen *n (plast)* blow moulding
Hohlprofil *n* hollow profile
Hohlraum *m* cavity, hollow (void) space; *(plast)* void *(a defect)*; *(cryst)* interstice, interstitial [lattice] site
  miarolitischer H. *(min)* miarolitic cavity, druse
Hohlraumbildung *f* cavitation *(in moving liquids)*
Hohlraumfilter *n* granular-bed separator *(gas cleaning)*
hohlraumfrei free from voids
Hohlraumstrahler *m* black-body radiator
Hohlraumstrahlung *f* black-body radiation
Hohlraumvolumen *n (soil)* volume of pore space
Hohlsog *m* cavitation *(in moving liquids)*
Hohlsprühkegel *m* hollow spray cone
Hohlsprühkegeldüse *f* hollow-cone nozzle
Hohlstein *m* hollow brick; hollow tile
Hohlware *f (ceram)* hollow ware
Hohlwelle *f* hollow shaft
Hohlzapfen *m* hollow journal
Hohlzapfenaustrag *m* trunnion discharge *(as of a mill)*
Hohlziegel *m* hollow brick; hollow tile
Holdcroft-Stäbe *mpl (ceram)* Holdcroft bars
Holländer *m (pap)* Hollander, Hollander beater (beating engine), pulp (stuff) engine, pulp grinder
  H. mit mehreren Grundwerken multiplate beater
  H. nach Jones-Bertrams Jones-Bertrams beater
Holländereintrag *m (pap)* furnish[ing]
Holländerfärbung *f (pap)* beater dyeing (colouring)
Holländerfüllung *f (pap)* furnish[ing]
Holländermesser *npl (pap)* knives, teeth, bars
Holländermüller *m (pap)* beaterman
Holländersaal *m (pap)* beater room (house)
Holländertrog *m (pap)* beater tub (vat, tank, pan)
Holländerwalze *f (pap)* Hollander (beater, beating, knive) roll, roll, cylinder
Höllenstein *m* caustic silver, lunar caustic *(silver nitrate)*
Holmes-Manley-Verfahren *n (petrol)* Holmes-Manley process
Holmium *n* Ho holmium
Holmiumchlorid *n* $HoCl_3$ holmium chloride
Holmiumhydroxid *n* $Ho(OH)_3$ holmium hydroxide

Holmiumoxid *n* Ho_2O_3 holmium oxide
Holmiumsulfat *n* $Ho_2(SO_4)_3$ holmium sulphate
holoedrisch *(cryst)* holohedral
Holoenzym *n* holoenzyme
holohyalin *(geol)* holohyaline
holokristallin holocrystalline
Holosid *n* holoside
Holosiderit *m* holosiderite *(meteoric iron)*
Holozellulose *f* holocellulose
Holst-Verfahren *n (pap)* Holst process *(for making chlorine-dioxide bleaching liquor)*
Holz *n* wood
  gerbstoffhaltiges H. tanwood
  verkieseltes H. silicified wood
Holzalkohol *m s.* Holzgeist
holzartig woody, xyloid, ligneous
Holzasche *f* wood ash[es]
Holzäther *m s.* Dimethyläther
Holzaufbereitung *f (pap)* wood preparation
Holzaufbereitungsanlage *f (pap)* wood room
Holzaufschluß *m (pap)* pulping of wood
  H. mit Salpetersäure nitric-acid pulping
Holzbearbeitung *f (pap)* wood preparation
Holzbeize *f* stain
Holzbottich *m*, **Holzbütte** *f* wooden tub (vat)
Holzchemie *f* wood chemistry
Holzdestillation *f* wood distillation
Holzdurchtränkung *f (pap)* penetration of wood (chips) *(with cooking liquor)*
Holzessig *m* wood vinegar, pyroligneous acid *(crude acetic acid obtained by wood distillation)*
Holzfaser *f* wood fibre
Holzfaserplatte *f* wood-fibre board, fibreboard
Holzfilter *n* wooden filter
holzfrei *(pap)* wood-free
Holzfülldichte *f (pap)* chip capacity
Holzfüllung *f (pap)* chip filling (packing)
  H. ohne Füllapparat gravity filling
Holzgas *n* wood[-distillation] gas; wood producer gas
Holzgefüge *n (coal)* woody structure
Holzgeist *m* wood alcohol (naphtha, spirit) *(crude methyl alcohol)*
Holzgewicht *n (pap)* wood weight
Holzgummi *n* xylan *(a pentosan)*
holzhaltig *(pap)* wood-containing, woody
holzig woody
Holzkalk *m (dye)* pyrolignite of lime *(crude calcium acetate)*
Holzkarton *m* wood-pulp cardboard *(from mechanical pulp)*
Holzkochung *f (pap)* cooking of wood
Holzkohle *f* [wood] charcoal
  aktive (aktivierte) H. active (activated) charcoal
  fossile (mineralische) H. fossil (mineral) charcoal, mother of coal
Holzkohleneisen *n* charcoal pig iron
  schwedisches H. Swedish iron
Holzkohlen[hoch]ofen *m* charcoal-fired [blast] furnace
Holzkohlenroheisen *n s.* Holzkohleneisen
Holzkonservierung *f* 1. wood preservation; 2. *(build)* timber proofing
Holzlagerplatz *m (pap)* wood yard
Holzlagerung *f (pap)* pulpwood storage
Holzleim *m* wood glue (adhesive)
Holzmasse *f* 1. *s.* Holzschliff; 2. *(pap)* wood weight
Holzmehl *n* / **feines** wood flour
  grobes H. wood meal
Holzöl *n* [/ **Chinesisches**] tung oil, China (Chinese) wood oil *(chiefly from the seeds of Aleurites fordii Hemsl.)*
Holzöl-Eisblumenbildung *f s.* Holzölerscheinung
Holzölerscheinung *f*, **Holzölkrankheit** *f (coat)* gas checking *(a defect)*
Holzopal *m (min)* wood opal
Holzpech *n* Stockholm pitch
Holzplatz *m (pap)* wood yard
Holzputzerei *f (pap)* wood room
Holzriffelwalze *f* wooden fluted roll
Holzrührer *m* wooden agitator
Holzsäure *f* wood acid *(formed when wood is heated to 120 °C)*
Holzschacht *m (pap)* magazine *(of a grinder)*
Holzschleifer *m (pap)* [pulpwood] grinder
Holzschleiferei *f (pap)* pulp (mechanical-pulp, groundwood) mill, grinder house (room)
Holzschleifmaschine *f (pap)* [pulpwood] grinder
Holzschliff *m (pap)* wood pulp, *(specif) s.* mechanischer H. ✦ **zu H. verschleifen** to reduce to pulp, to make into [a] pulp, to pulp
  H. in Bogenform (Pappenform) *s.* Holzschliffpappe
  brauner H. brown mechanical pulp
  chemischer H. chemigroundwood
  mechanischer H. mechanical pulp (wood pulp), MWP, groundwood [pulp]
Holzschliffaser *f (pap)* groundwood fibre
Holzschliffblätter *npl s.* Holzschliffpappe
Holzschliffbleiche *f (pap)* groundwood bleaching
Holzschliffentwässerungsmaschine *f (pap)* pulp (pulp-drying, half-stuff) machine, wet [press] machine, wet press, presse-pâte
Holzschlifferzeugung *f (pap)* manufacture of mechanical wood pulp, mechanical (groundwood) pulping
holzschliffhaltig *(pap)* wood-containing, woody
Holzschliffpapier *n* groundwood (wood-containing, woody) paper
Holzschliffpappe *f (pap)* wood-pulp board, board (sheets, laps) of mechanical wood pulp
Holzschliffverfahren *n (pap)* groundwood process
Holzschnitzel *npl (pap)* [wood] chips, chippings
Holzschutz *m* 1. wood preservation; 2. *(build)* timber proofing
Holzschutzmittel *n* wood preservative
  kombiniertes H. fire-retardant preservative
  wasserlösliches H. water-borne type preservative, WB-type preservative
Holzspäne *mpl s.* Holzschnitzel
Holzspiritus *m s.* Holzgeist

Holzsplitter *m* wood speck *(a defect in paper)*
Holzstoff *m* 1. lignin; 2. *s.* Holzschliff
Holzstruktur *f (coal)* woody structure
Holzsubstanz *f* ligneous substance
Holzteer *m* wood tar
Holzteerkreosot *n* wood[-tar] creosote
Holzverkohlung *f* wood carbonization, charcoal burning
Holzverkohlungsofen *m* char kiln
Holzverzuckerung *f* saccharification of wood
Holzvorbereitung *f (pap)* wood preparation
Holzvorschub *m (pap)* advance of wood *(in a grinder)*
Holzwalze *f* wooden roll
Holzwolle *f* wood wool
Holzzellstoff *m* 1. *(pap)* [chemical] wood pulp, CWP; 2. *s.* Holzzellulose
Holzzellstruktur *f* cell structure of wood
Holzzellulose *f* lignocellulose
Holzzerfaserungsmaschine *f s.* Holzschleifer
Holzzucker *m* wood sugar, xylose
Homatropin *n* homatropine, phenylglycollyltropine
Homilit *m (min)* homolite *(a neso-subsilicate)*
homodispers monodisperse
homofermentativ homofermentative
homogen homogeneous
Homogenisation *f s.* Homogenisierung
Homogenisator *m*, **Homogenisierapparat** *m* homogenizer
homogenisieren 1. to homogenize; 2. *(ceram)* to wedge
Homogenisierkopf *m* homogenizing head (valve) *(of a homogenizer)*
Homogenisiermaschine *f* homogenizer
Homogenisierung *f* 1. homogenization, homogenizing *(as of emulsions)*; 2. *(ceram)* wedging
Homogenisierungsdruck *m* homogenization pressure
Homogenisierungsglühen *n* homogenization, homogenizing *(of alloys)*
Homogenisierzone *f (plast)* metering section (zone) *(of an extrusion press)*
Homogenität *f* homogeneity
Homogenreaktion *f* homogeneous reaction
Homogenreaktor *m (nucl)* homogeneous reactor
homolog homologous
Homolog[es] *n* homologue, homolog
Homolyse *f* homolysis
homolytisch homolytic
homöoblastisch *(geol)* homoeoblastic
homöopolar covalent, homopolar, non-polar *(chemical-bond theory)*
homopolymer homopolymeric
Homopolymer[es] *n*, **Homopolymerisat** *n* homopolymer
Homopolymerisation *f* homopolymerization
homopolymerisieren to homopolymerize
4-Homosulfanilamid *n* 4-homosulphanilamide, α-aminotoluene-4-sulphonamide
homozyklisch homocyclic, isocyclic, carbocyclic
Honig *m* honey
honigend melliferous
honigerzeugend melliferous
Honigessig *m* honey vinegar
Honigpflanze *f* nectariferous (bee) plant
Honigstein *m (min)* mellite *(a hydrous aluminium mellitate)*
Honigsteinsäure *f* $C_6(COOH)_6$ mellitic acid, benzene-hexacarboxylic acid
Honigwein *m* mead
Hooker-Zelle *f* Hooker cell *(a diaphragm cell)*
Hoopes-Verfahren *n* Hoopes [electrolytic-refining] process
Hopeit *m (min)* hopeite *(zinc orthophosphate-4-water)*
hopfen to hop *(the wort)*
Hopfen *m* hop, Humulus L., *(specif) s.* Gemeiner H.
✦ **H. zusetzen** to hop *(the wort)*
Gemeiner H. common hop, Humulus lupulus L.
mittelfrüher H. mid-season hop
Hopfen *n* hopping *(of the wort)*
Hopfenanbau *m* hop growing
Hopfenaroma *n* hop aroma (flavour)
Hopfenbittere *f* bitterness of hops
Hopfenbittersäure *f* hop bitter acid
α-Hopfenbittersäure *f* α-lupulinic acid, α-bitter acid, humulone
β-Hopfenbittersäure *f* β-lupulinic acid, β-bitter acid, lupulone
Hopfenbitterstoff *m* hop bitter substance
Hopfendarre *f* hop dryer (kiln)
Hopfendolde *f* hop cone
Hopfenernte *f* hopping
Hopfenextrakt *m* hop extract[ive]
Hopfengabe *f* hop rate
Hopfenharz *n* hop resin
Hopfenkufe *f* hopper
Hopfenmehl *n* lupulin, hop flour
Hopfenöl *n* hop oil
Hopfenpektin *n* hop pectin
Hopfenpflanze *f* hop plant
hopfenreich hoppy
Hopfenseiher *m* hop back (jack, strainer)
Hopfentreber *m* spent hops
Hopfenzäpfchen *n* hop cone
Hopper *m* hopper *(in margarine making)*
Höppler-Kugelfallviskosimeter *n*, **Höppler-Viskosimeter** *n* Höppler [falling-ball] viscometer
Horde *f* [kiln] floor, tray *(as of a dryer)*
Hordein *n* hordein *(a prolamin)*
Hordenschwingtrockner *m* vibrating tray dryer
Hordenwagen *m* tray truck
Horizont *m (soil)* layer, horizon, level
Horizontalbeziehung *f* horizontal relationship *(in the periodic system)*
Horizontaldarre *f (ferm)* horizontal kiln
Horizontalkammer *f* horizontal chamber
Horizontalkammerofen *f* horizontal oven
H. mit senkrechten Heizzügen vertical-flue oven
Horizontalkanal *m* horizontal flue *(as for off-gas)*
Horizontalklärer *m* horizontal clarifyer *(waste-water treatment)*

Horizontalkolonne *f (distil)* horizontal still
Horizontalofen *m s.* Horizontalkammerofen
Horizontalretorte *f* horizontal retort
Horizontalrohrverdampfer *m* horizontal[-tube] evaporator
Horizontalzelle *f* horizontal [diaphragm] cell *(electrolysis)*
Horizontalziehverfahren *n (glass)* horizontal sheet drawing process
Horizontalzug *m* horizontal flue *(as for off-gas)*
Hormon *n* hormone
  adrenokortikotropes H. adrenocorticotropic hormone, corticotropin
  follikelstimulierendes H. follicle-stimulating hormone, FSH
  interstitielle Zellen stimulierendes H. *s.* luteinisierendes H.
  laktotropes H. lactogenic (luteotrophic) hormone, prolactin, luteotrop[h]in
  luteinisierendes H. luteinizing (interstitial-cell-stimulating) hormone, LH, ICSH, prolan B
  luteotropes H. *s.* luteinisierendes H.
  luteotrophes H. *s.* laktotropes H.
  östrogenes H. oestrus-producing hormone
  somatotropes H. somatotropic hormone, somatropin, STH
  thyreotropes H. thyrotrop[h]ic (thyroid-stimulating) hormone, thyrotrop[h]in
  zwischenzellenstimulierendes H. *s.* luteinisierendes H.
hornähnlich horn-like
Hornblende *f (min)* hornblende, amphibole *(an inosilicate)*
  Gemeine H. common hornblende
Hornblendeasbest *m (min)* amphibole asbestos
Hornfels *m (geol)* hornfels
Hornmehl *n (agric)* horn meal
Hornquecksilber *n (min)* horn mercury, calomel *(mercury(I) chloride)*
Hornsilber *n (min)* horn silver, chlorargyrite, cerargyrite *(silver chloride)*
Hornspatel *m* horn spatula
Hornstein *m (min)* hornstone, chert *(a mineral related to chalcedony)*
Hornsubstanz *f* keratin
Hosenrohr *n* wye
Hot-pit-Gerbung *f* hot pitting
Hottenroth-Zahl *f (text)* Hottenroth number
Houdresid-Verfahren *n* Houdresid process *(a catalytic cracking or reforming process)*
Houdriflow-Verfahren *n* Houdriflow [catalytic cracking] process
Houdriformen *n* houdriforming *(a variety of catalytic reforming)*
Houdriformer *m* houdriformer
Houdriformierung *f s.* Houdriformen
Houdriforming-Anlage *f* houdriformer
Houdry-Festbettverfahren *n* Houdry fixed-bed process
Houdry-Krackverfahren *n* / **katalytisches** Houdry catalytic cracking process
Howard-Kristallisator *m* Howard crystallizer
HOZ *s.* Hochofenzement
Hp *s.* Hartpapier
HPC-Ruß *m* hard processing channel black, HPC black
HR *s.* Rockwellhärte
H-Säure *f* $NH_2C_{10}H_4(OH)(SO_3H)_2$ H acid, 1-amino-8-naphthol-3,6-disulphonic acid
5-HT *s.* 5-Hydroxytryptamin
HT- ... *s.* Hochtemperatur ...
HTST-Erhitzung *f s.* Hochkurzerhitzung
HTU *s.* Höhe einer Übertragungseinheit
Huanaco-Koka *f* Huanuca coca *(from Erythroxylum coca Lam.)*
Hub *m* stroke [length] *(of a pump)*; vibration amplitude, stroke *(screening)*
Hubel *m (ceram)* clot, blank
Hubflügel *m* lifter *(in a tumbling mill)*
Hubkolbenpumpe *f* reciprocating pump
Hublänge *f s.* Hub
Hubleiste *f* lifting flight (plate) *(of a drying drum)*
Hübnerit *m (min)* huebnerite, hübnerite *(manganese(II) tungstate)*
Hubrückschlagventil *n* lift check valve
Hubseil *n* hoisting rope
Hubtür *f* tweel, tuille
Hubventil *n* globe valve
Hubvorrichtung *f* lifting device
Hubzahl *f* number of strokes
Hühneraugenmittel *n* corn remedy
Hülle *f* 1. shell, envelope, sheath *(of an atom)*; 2. *(tech)* jacket, sheath, cover, shell
Hüllenelektron *n* sheath (extranuclear) electron
Hüllpapier *n* wrapping (packing) paper, wrapper *(of high quality)*
Hülse *f* 1. *(tech)* socket, jacket; socket *(of a ground joint)*; 2. *(pap)* core, centre; 3. *(lab)* thimble *(for extracting)*
  keglige H. *(text)* cone
Hülsenlosfärben *n (text)* muff dyeing
Humat *n (soil)* humate
Humboldtin *m (min)* humboldtine *(iron oxalate)*
Hume-Rothery-Phase *f* Hume-Rothery phase
humid humid
Humifizierung *f* humification
Huminkohle *f* humic coal
Huminsäure *f* humic acid
Huminstoff *m*, **Huminsubstanz** *f* humic substance (material, matter)
Humit *m* 1. *(min)* humite *(a fluorine-containing magnesium silicate)*; 2. humic coal
Humulon *n* humulone, α-lupulinic acid
Humus *m* humus
Humusanreicherung *f* accumulation of humus
Humusboden *m* humus soil
Humuskarbonatboden *m* rendzina
Humuskohle *f* humic coal
Humusortstein *m (soil)* humic ortstein
Humussäure *f* humic acid
Humusstoff *m s.* Huminstoff
Humusstoffhorizont *m (soil)* H-layer
Humussubstanz *f s.* Huminstoff
Hund-Mulliken-Lennard-Jones-Hückel-Theorie *f*

Hund-Mulliken-Lennard-Jones-Hückel theory, molecular-orbital theory
Hunsdiecker-Reaktion *f* 1. Hunsdiecker reaction *(decarboxylation of the silver salt of an organic acid)*; 2. *s.* Hunsdiecker-Spaltung
Hunsdiecker-Spaltung *f* Hunsdiecker cleavage *(of esters)*
Huntilith *m (min)* huntilite *(silver arsenide)*
Huntington-Pendel[rollen]mühle *f* Huntington [ring-roll] mill
Huréaulith *m (min)* hureaulite *(an iron manganese phosphate)*
Hustenmittel *n* antitussive
Hut *m (geol)* cap
  Eiserner H. iron hat, gossan, gozzan
Hutmanschette *f* flange seal
Hütte *f (met)* refinery, smelting plant
Hüttenaluminium *n* primary aluminium pig
Hüttenbims *m* foamed slag
Hüttenchemie *f* metallurgical chemistry
Hüttenchemiker *m* metallurgical chemist
Hütteningenieur *m* metallurgical engineer
Hüttenkalk *m (agric)* smelting lime
Hüttenkoks *m* metallurgical (blast-furnace) coke
Hüttenkunde *f* metallurgy
hüttenmännisch metallurgical
Hüttentechnik *f* metallurgical technology
Hüttenwerk *n s.* Hütte
Hüttenwesen *n* metallurgy
Hüttenzement *m* portland blast-furnace cement, slag cement
Hutzucker *m* loaf sugar
HV *s.* Vickers-Härte
HWM-Faser *f* high-wet-modulus fibre, HWM
HWM-Faserstoff *m* high-wet-modulus fibre
hyalin hyaline
Hyalit *m (min)* hyalite *(silicon(IV) oxide)*
Hyalophan *m (min)* hyalophane *(a tectosilicate)*
hyalopilitisch *(geol)* hyalopilitic
Hyaluronidase *f* hyaluronidase, spreading factor *(any of a series of glycosidases)*
Hyaluronsäure *f* hyaluronic acid *(a mucopolysaccharide)*
Hyazinth *m (min)* hyacinth *(zirconium orthosilicate)*
hybrid hybrid
Hybrid *n* hybrid
  digonales H. digonal hybrid
sp-**Hybrid** *n sp*-hybrid
hybridisieren to hybridize
Hybridisierung *f* hybridization
  digonale H. digonal hybridization
  tetraedrische H. tetrahedral hybridization
  trigonale H. *s.* sp^2-Hybridisierung
sp^2-**Hybridisierung** *f* sp^2 hybridization, trigonal hybridization
Hybridorbital *n* hybrid [bond] orbital
sp-**Hybridorbital** *n* hybrid *sp* orbital
Hybridstruktur *f* hybrid structure
Hybridzustand *m* hybrid state
Hydantoinsäure *f* $NH_2CONHCH_2COOH$ hydantoic acid, ureidoacetic acid
hydatogen *(geol)* hydatogenous
Hydnokarpsäure *f* hydnocarpic acid, 11-(2-cyclopentenyl)undecanoic acid
Hydrakrylsäure *f* $HOCH_2CH_2COOH$ hydracrylic acid, 3-hydroxypropionic acid
Hydrangeasäure *f* hydrangeic acid, 3,4′-dihydroxystilbene-2-carboxylic acid
Hydrargillit *m (min)* hydrargillite, gibbsite *(aluminium trihydroxide)*
Hydrat *n* hydrate
Hydratation *f* hydration, aqua[tiza]tion
  H. der Ionen ionic hydration
Hydratationsenergie *f* hydration energy
Hydratationsgrad *m* degree of hydration
Hydratationswärme *f* heat of hydration
Hydratationszahl *f* hydration number
Hydratbildung *f* hydrate formation
Hydration *f (incorrectly for)* Hydratation
hydratisieren to hydrate, to aquate
Hydratisierung *f* hydration, aqua[tiza]tion
Hydratisierungs ... *s.* Hydratations ...
Hydrator *m* hydrator
Hydratwasser *n* hydrate (hydration) water
Hydratzellulose *f* hydrate (hydrated, regenerated) cellulose, cellulose hydrate
Hydraulikakkumulator *m* hydraulic accumulator
Hydraulikflüssigkeit *f* hydraulic fluid (medium)
Hydraulikkolben *m* hydraulic ram
Hydraulikspeicher *m* hydraulic accumulator
Hydraulikzylinder *m* hydraulic cylinder
hydraulisch angetrieben (betätigt, betrieben, bewegt) hydraulic-operated
Hydrazid *n* hydrazide
Hydrazin *n* N_2H_4 hydrazine
  wasserfreies H. anhydrous hydrazine
Hydrazingelb O *n* hydrazine yellow, tartrazine *(a pyrazole derivative)*
Hydrazinhydrat *n* $N_2H_4 \cdot H_2O$ hydrazine hydrate
Hydrazinolyse *f* hydrazinolysis
Hydrazinthiokarbonsäureamid *n* $NH_2-CS-NH-NH_2$ aminothiourea, thiosemicarbazide
Hydrazobenzol *n* $C_6H_5NHNHC_6H_5$ hydrazobenzene
Hydrazoverbindung *f* hydrazo compound
Hydrid *n* hydride
  interstitielles H. interstitial hydride
  komplexes H. complex hydride
  metallartiges (metallisches) H. transition-metal binary hydride
  salzartiges H. saline hydride
Hydrid-Ion *n* hydride ion
Hydridkomplex *m* complex hydride
Hydridoborat *n* $M^I[BH_4]$ hydridoborate, tetrahydridoborate
Hydridphase *f* hydride phase
Hydridverfahren *n* hydride process
Hydridverschiebung *f* hydride shift
Hydridverschiebungssatz *m* hydride displacement law
Hydrierautoklav *m* hardening vessel *(fat hardening)*

hydrierbar hydrogenable
Hydrierbarkeit *f* hydrogenability
Hydrierbenzin *n* hydrogenation gasoline (spirit)
hydrieren to hydrogenate, to hydrogenize
Hydrieren *n* hydrogenation
Hydriergas *n* hydrogenating gas
Hydrierkatalysator *m* hydrogenating (hydrogenation) catalyst
Hydrierofen *m* converter, *(food also)* hardening vessel
Hydrierreaktion *f* hydrogenating reaction
Hydrierung *f* hydrogenation
H. in flüssiger Phase *s.* H. in Sumpfphase
H. in Gasphase vapour-phase hydrogenation
H. in Sumpfphase sump-phase (liquid-phase) hydrogenation
abbauende (destruktive) H. destructive hydrogenation
katalytische H. catalytic hydrogenation
selektive H. selective hydrogenation
spaltende H. *s.* abbauende H.
Hydrierungs ... *s.* Hydrier ...
Hydrierverfahren *n* hydrogenation process, *(food also)* hardening process
Hydrierwärme *f* heat of hydrogenation
Hydrinden *n* hydrindene, indane
Hydrindon *n* hydrindone, indanone
Hydroaromaten *pl* hydroaromatic compounds
hydroaromatisch hydroaromatic
Hydroborazit *m* *(min)* hydroboracite *(an inoborate)*
Hydroborierung *f* hydroboration *(addition of diborane to olefines)*
Hydrobromid *n* hydrobromide
Hydrochemie *f* hydrochemistry, water chemistry
Hydrochinin *n s.* Dihydrochinin
Hydrochinon *n* $C_6H_4(OH)_2$ quinol, hydroquinone, *p*-dihydroxybenzene
Hydrochinonentwickler *m* *(phot)* hydroquinone developer
Hydrochinonklathrat *n* quinol clathrate
Hydrochlorid *n* hydrochloride
Hydrochlorierung *f* hydrochlorination
Hydrochlorkautschuk *m* rubber hydrochloride
Hydrocol-Verfahren *n* Hydrocol process *(for producing high-octane gasoline from natural gas)*
Hydrodesulfurierung *f* hydrodesulphurization, HDS
Hydrofiner *m* hydrofiner
Hydrofining *n* hydrofining *(a process for improving the quality of petroleum products by treating with hydrogen)*
Hydrofixierung *f (text)* hydrosetting
Hydrofluorid *n* hydrofluoride
Hydrofluorkautschuk *m* rubber hydrofluoride
Hydroformat *n (petrol)* hydroformate
Hydroformen *n s.* Hydroformieren
Hydroformer *m (petrol)* hydroformer
hydroformieren *(petrol)* to hydroform *(to reform by catalytic dehydrogenation and cyclization)*
Hydroformieren *n (petrol)* hydroforming
Hydroforming-Produkt *n (petrol)* hydroformate
Hydroformylierung *f* hydroformylation, oxo synthesis
Hydrogel *n* hydrogel
Hydrogenarsenat *n* $M_2^IHAsO_4$ hydrogenarsenate
Hydrogenase *f* hydrogenase
Hydrogenfluorid *n* M^IHF_2 hydrogendifluoride
Hydrogenkarbonat *n* M^IHCO_3 hydrogencarbonate
Hydrogenmonophosphat *n s.* Hydrogenorthophosphat
Hydrogenolyse *f* hydrogenolysis
Hydrogenorthophosphat *n* $M_2^IHPO_4$ hydrogenorthophosphate, hydrogenphosphate
Hydrogenoxalat *n* $M^IOOC-COOH$ hydrogenoxalate
Hydrogenperoxid *n* H_2O_2 hydrogen peroxide
hydrogenperoxidecht fast to hydrogen peroxide
Hydrogenphosphat *n s.* Hydrogenorthophosphat
Hydrogensalz *n* hydrogen salt
Hydrogensulfat *n* M^IHSO_4 hydrogensulphate
Hydrogensulfid *n* M^IHS hydrogensulphide
Hydrogensulfit *n* M^IHSO_3 hydrogensulphite
Hydroglimmer *m (min)* hydromica, hydrous mica
Hydroguttapercha *f(n)* hydro-gutta-percha
Hydrohalit *m (min)* hydrohalite *(sodium chloride)*
Hydrokautschuk *m* hydrogenated rubber, hydrorubber
Hydroklassieren *n* wet classification
Hydrokracken *n (petrol)* hydrocracking, hydrogenation cracking
Hydrokrackkatalysator *m (petrol)* hydrocracker
Hydrokumarsäure *f* $HOC_6H_4CH_2CH_2COOH$ hydrocoumaric acid, 3-hydroxyphenylpropionic acid
Hydrol *n* **/ Michlers** $[(CH_3)_2NC_6H_4]_2CHOH$ Michler's hydrol, di-(*p*-dimethylaminophenyl)methanol
Hydrolase *f* hydrolase
Hydrolysat *n* hydrolyzate
Hydrolyse *f* hydrolysis
H. der Fette fat hydrolysis
enzymatische H. enzymatic hydrolysis
partielle H. partial hydrolysis
saure H. acid hydrolysis
vorhergehende H. *(pap)* preimpregnation, preliminary impregnation (penetration) *(of the chips)*
Hydrolysebeständigkeit *f* hydrolytic stability, resistance to hydrolysis
Hydrolysegrad *m* degree of hydrolysis
Hydrolysenfällung *f* hydrolysis precipitation
Hydrolysenkonstante *f* hydrolysis constant
Hydrolyseresistenz *f*, **Hydrolysestabilität** *f s.* Hydrolysebeständigkeit
hydrolysierbar hydrolyzable
Hydrolysierbarkeit *f* hydrolyzability
hydrolysieren to hydrolyze
hydrolytisch hydrolytic
Hydromagnesit *m (min)* hydromagnesite *(magnesium carbonate hydroxide)*
Hydrometallurgie *f* hydrometallurgy, wet metallurgy

hydrometallurgisch hydrometallurgical
Hydronaphthalin *n* hydronaphthalene
Hydronephelit *m (min)* hydronepheline, hydronephelite
Hydroniumion *n* H_3O^+ hydronium ion
Hydroniumionenaktivität *f* hydronium-ion activity
Hydroniumionenkonzentration *f* hydronium-ion concentration ✦ **mit gleicher H.** isohydric
Hydroperoxid *n* hydroperoxide
Hydroperoxidumlagerung *f* hydroperoxide rearrangement
Hydrophan *m (min)* hydrophane *(silicon(IV) oxide)*
hydrophil hydrophilic, hydrophile
Hydrophilierung *f* hydrophiling
Hydrophilit *m (min)* hydrophilite *(calcium chloride)*
hydrophob hydrophobic, hydrophobe, water-repellent
Hydrophobiermittel *n* hydrophobing agent, water repellent
Hydrophobierung *f* hydrophobing
Hydrophobierungsmittel *n s.* Hydrophobiermittel
Hydroponik *f (agric)* hydroponic culture, hydroponics
Hydroseparator *m* hydroseparator *(a thickener)*
Hydrosol *n* hydrosol, aquasol
hydrostatisch hydrostatic
Hydrosulfitbleiche *f (pap)* sodium-hydrosulphite bleaching
hydrothermal hydrothermal
Hydrothermalsynthese *f* hydrothermal synthesis
Hydrotorf *m* hydro peat
Hydrotropie *f* hydrotropy
Hydroxamsäure *f* $C_nH_{2n+1}C(=O)NHOH$ hydroxamic acid
Hydroxid *n* hydroxide
Hydroxidion *n* hydroxide ion
Hydroxidionenaktivität *f* hydroxide ion activity
Hydroxidsalz *n* hydroxide salt
Hydroxoaluminat *n* hydroxoaluminate
Hydroxoantimonat *n s.* Hexahydroxoantimonat
Hydroxokomplex *m* hydroxo complex
Hydroxoniumion *n s.* Hydroniumion
Hydroxosalz *n* hydroxo salt
Hydroxostannat *n* hydroxostannate
Hydroxotrifluoroborat *n* $M^I[B(OH)F_3]$ trifluorohydroxoborate
Hydroxozinkat *n* hydroxozincate
Hydroxyaldehyd *m* hydroxyaldehyde
Hydroxyäthansäure *f s.* Hydroxyessigsäure
Hydroxyäthylzellulose *f* hydroxyethylcellulose
Hydroxyazetaldehyd *m* $HOCH_2CHO$ hydroxyacetaldehyde, glycollaldehyde
o-**Hydroxybenzoesäure** *f* $C_6H_4(OH)COOH$ *o*-hydroxybenzoic acid, salicylic acid
Hydroxybenzol *n* C_6H_5OH hydroxybenzene, phenol
o-**Hydroxybenzylalkohol** *m* $C_6H_4(OH)CH_2OH$ *o*-hydroxybenzyl alcohol, α,2-dihydroxytoluene, salicyl alcohol, saligenin
Hydroxybernsteinsäure *f* $HOOCCH(OH)CH_2COOH$ hydroxysuccinic acid, malic acid, hydroxybutanedioic acid
3-Hydroxybutanal *n* $CH_3CH(OH)CH_2CHO$ 3-hydroxybutanal, 3-hydroxybutyraldehyde, acetaldol, aldol
Hydroxybutandisäure *f s.* Hydroxybernsteinsäure
3-Hydroxybutanon-(2) *n* $CH_3CH(OH)COCH_3$ 3-hydroxy-2-butanone, dimethylketol, acetoin
3-Hydroxybutyraldehyd *m s.* 3-Hydroxybutanal
Hydroxychinolin *n* hydroxyquinoline
22-Hydroxydokosansäure *f* $HOCH_2[CH_2]_{20}COOH$ 22-hydroxydocosanoic acid, phellonic acid
Hydroxyessigsäure *f* $HOCH_2COOH$ hydroxyacetic acid, glycollic acid
Hydroxyfettsäure *f* hydroxy-fatty acid
Hydroxygruppe *f (incorrectly for)* Hydroxylgruppe
Hydroxykarbonsäure *f* hydroxycarboxylic acid, hydroxy acid
Hydroxyketon *n* hydroxy ketone, keto alcohol
Hydroxylamin *n* NH_2OH hydroxylamine
Hydroxylammoniumchlorid *n* $[H_3NOH]Cl$ hydroxylammonium chloride
Hydroxylammoniumnitrat *n* $[H_3NOH]NO_3$ hydroxylammonium nitrate
Hydroxylammoniumsulfat *n* $[H_3NOH]_2SO_4$ hydroxylammonium sulphate
Hydroxylaustausch *m* hydroxyl-cycle anion exchange *(water treatment)*
Hydroxylderivat *n* hydroxy derivative
Hydroxylgruppe *f* hydroxyl group, OH group ✦ **mit einer H.** monohydric *(alcohol, phenol)* ✦ **mit endständiger H.** hydroxy-terminated ✦ **mit mehreren Hydroxylgruppen** polyhydric *(alcohol, phenol)*
phenolische H. phenolic hydroxyl group
hydroxylhaltig containing hydroxyl, hydroxy
hydroxylieren to hydroxylate
Hydroxylierung *f* hydroxylation
Hydroxylsauerstoff *m* hydroxylic oxygen
Hydroxylzahl *f* hydroxyl value (number) *(of fats and fatty oils)*
Hydroxylzahlbestimmung *f* determination of the hydroxyl value
Hydroxymalonsäure *f* $HOCH(COOH)_2$ hydroxymalonic acid, tartronic acid
Hydroxymethylharnstoff *m* $HOCH_2-NH-CO-NH_2$ hydroxymethylurea, methylolurea
Hydroxynaphthalin *n* naphthol, hydroxynaphthalene
Hydroxynaphthalinkarbonsäure *f s.* Hydroxynaphthoesäure
Hydroxynaphthoesäure *f* $C_{10}H_6(OH)COOH$ hydroxynaphthoic acid
α-**Hydroxynitril** *n* alpha-hydroxy nitrile, cyanohydrin, cyanhydrin
Hydroxyölsäure *f s.* Rizinolsäure
2-Hydroxypropannitril *n* $CH_3CH(OH)CH$ 2-hydroxypropane nitrile, lactonitrile, acetaldehyde cyanohydrin

2-Hydroxypropansäure *f s.* 2-Hydroxypropionsäure

2-Hydroxypropionsäure *f* $CH_3CH(OH)COOH$ 2-hydroxypropionic acid, lactic acid

3-Hydroxypropionsäure *f* $HOCH_2CH_2COOH$ 3-hydroxypropionic acid, ethylenelactic acid

Hydroxysäure *f* hydroxycarboxylic acid, hydroxy acid

Hydroxysäureester *m* hydroxy ester

hydroxysubstituiert hydroxy-substituted

Hydroxytoluol *n* $CH_3C_6H_4OH$ hydroxytoluene, methylphenol, cresol

5-Hydroxytryptamin *n* serotonin, 5-hydroxytryptamine, 3-(2-aminoethyl)-5-hydroxyindole

Hydroxyxylol *n* $(CH_3)_2C_6H_3OH$ hydroxyxylene, xylenol, dimethylphenol

Hydroxyzimtsäure *f* $HOC_6H_4CH{=}CHCOOH$ hydroxycinnamic acid, coumaric acid, hydroxyphenylpropenoic acid

Hydrozellulose *f* hydrocellulose

Hydrozelluloseazetat *n (text)* secondary [cellulose] acetate, cellulose diazetate

Hydrozerussit *m (min)* hydrocerussite *(lead carbonate hydroxide)*

Hydrozimtaldehyd *m* $C_6H_5CH_2CH_2CHO$ hydrocinnamic aldehyde, hydrocinnamaldehyde, 3-phenylpropanal

Hydrozimtalkohol *m* $C_6H_5CH_2CH_2OH$ hydrocinnamyl alcohol, 3-phenyl-1-propanol

Hydrozimtsäure *f* $C_6H_5CH_2CH_2COOH$ hydrocinnamic acid, 3-phenylpropionic acid

Hydrozinkit *m (min)* hydrozincite *(a basic zinc carbonate)*

Hydrozyanit *m (min)* hydrocyanite *(copper(II) sulphate)*

Hydrozyklon *m* hydroclone, liquid cyclone, hydraulic cyclone separator, wet cyclone classifier

hygr. *s.* hygroskopisch

Hygrinsäure *f* hygrinic acid, 1-methylpyrrolidine-2-carboxylic acid

Hygromyzin *n* hygromycin *(antibiotic)*

hygroskopisch hygroscopic[al], water-absorbing, water-attracting

Hygroskopizität *f* hygroscopicity

Hylit *m(n)* xylite, [woody] lignite, woody brown coal

Hymatomelansäure *f (soil)* hymatomelanic acid

Hyoszin *n* hyoscine *(alkaloid)*

***DL*-Hyoszyamin** *n DL*-hyoscyamine, atropine *(alkaloid)*

Hypazidität *f (med)* subacidity

Hyperazidität *f (med)* superacidity

Hyperfeinspektrum *n* hyperfine spectrum

Hyperfeinstruktur *f* hyperfine structure

Hyperfiltration *f* hyperfiltration, reverse osmosis

Hyperformierung *f (petrol)* hyperforming

Hypergol *n* hypergol, hypergolic fuel (rocket propellant)

Hyperkonjugation *f* hyperconjugation, no-bond resonance

Hyperladung *f (nucl)* hypercharge

Hypernukleus *m* hypernucleus

Hyperon *n* hyperon *(a superheavy elementary particle)*

Hypersensibilisierung *f* hypersensitization

Hypersensibilität *f* hypersensitivity

Hypersorption *f* hypersorption

Hypersthen *m (min)* hypersthene *(an inosilicate)*

Hypervitaminose *f* hypervitaminosis

hypidiomorph *(min)* hypidiomorphic, subhedral

Hypnotikum *n* hypnotic, soporific, somnifacient, somnificant

Hypo *n* $Na_2S_2O_3$ hypo *(sodium thiosulphate)*

Hypoazidität *f (med)* subacidity

Hypobromit *n* M^IOBr hypobromite

Hypochlorit *n* M^IOCl hypochlorite

Hypochloritbehandlung *f (pap)* hypochlorite treatment (bleaching); *(petrol)* hypochlorite treatment (sweetening)

H. bei hoher Stoffdichte *(pap)* high-density hypochlorite treatment (bleaching)

Hypochloritbleiche *f (pap, text)* hypochlorite bleaching

H. bei hoher Stoffdichte *(pap)* high-density hypochlorite bleaching

Hypochloritbleichechtheit *f* fastness to hypochlorite bleaching

Hypochloritbleichlauge *f* hypochlorite bleach [liquor]

Hypochloritbleichstufe *f (pap)* hypochlorite bleaching stage

Hypochloritendbleiche *f (pap)* last (final) hypochlorite bleaching

Hypochloritsüßen *n (petrol)* hypochlorite sweetening (treatment)

hypogen *(geol)* hypogene

Hypoidöl *n* hypoid lubricant

Hypojodit *n* M^IOI hypoiodite

hypokristallin hypocrystalline

Hypomagma *n* hypomagma

Hyponitrit *n* $M_2^IN_2O_2$ hyponitrite

Hypophosphat *n* $M_4^IP_2O_6$ hypophosphate

Hypophosphit *n* $M^IPH_2O_2$ hypophosphite, *(better)* phosphinate

Hypophosphorsäure *f* $H_4P_2O_6$ hypophosphoric acid

Hypophysenhinterlappenextrakt *m* posterior-pituitary extract

Hypophysenhinterlappenpulver *n* posterior-pituitary powder

Hypotensivum *n* antihypertensive drug, blood pressure depressant

Hypothese *f* / **Avogadrosche** Avogadro hypothesis (law)

Goudsmit-Uhlenbecksche H. Goudsmit and Uhlenbeck assumption *(of rotating electrons)*

Proutsche H. Prout hypothesis

hypothetisch hypothetical

Hypovitaminose *f* hypovitaminosis

Hypoxanthin *n* hypoxanthine, 6-hydroxypurine

hypsochrom hypsochromic

Hypsochromie *f* hypsochromic shift

Hysterese *f* hysteresis

ferroelektrische H. ferroelectric hysteresis

ferromagnetische H. ferromagnetic hysteresis
Hysteresekurve *f*, **Hystereseschleife** *f* hysteresis loop
Hystereseverlust *m* hysteresis loss
Hysteresis *f s.* Hysterese

I

I. A. *s.* Ionenaustauscher
Ibogaalkaloid *n* iboga alkaloid
Ibogain *n* ibogaine *(alkaloid)*
Ibogamin *n* ibogamine *(alkaloid)*
iC$_4$-Kreislauf *m s.* Isobutankreislauf
ICSH *s.* Hormon / luteinisierendes
iC$_4$-Umlauf *m s.* Isobutankreislauf
Idaein *n* idaein *(a galactoside)*
Idealkristall *m* ideal crystal
Idealverhalten *n* ideal behaviour
Idealzustand *m* ideality
Identifikationsreaktion *f* identifying reaction
identifizierbar identifiable
identifizieren to identify
Identifizierung *f* identification
Identifizierungsreaktion *f* identifying reaction
Identitätsreaktion *f* identifying reaction
idiochromatisch idiochromatic
idiomorph *(min)* idiomorphic, idiomorphous, euhedral, automorphic *(having the proper crystal form)*
Idose *f* idose *(a monosaccharide)*
Idrialin *m (min)* idrialine, idrialite *(a naturally occurring hydrocarbon)*
I. E. *s.* Einheit / internationale
I-Effekt *m (phys chem)* I effect, induction (inductive) effect
IES *s.* Indolyl-3-essigsäure
I-Kalander *m* four-bowl stack type of calender
Ile *s.* Isoleuzin
i-Leitung *f* intrinsic semiconduction
Ilesit *m (min)* ilesite *(manganese sulphate-7-water)*
Ilhurinbalsam *m* Illorin gum *(from Daniella thurifera Bennett)*
Illingworth-Verfahren *n (coal)* Illingworth process *(a low-temperature carbonization process)*
Illinium *n s.* Promethium
Illit-Gruppe *f* illite series *(general term for micaceous clay minerals)*
Illustrationsdruckpapier *n* half-tone paper
Illuvialhorizont *m (soil)* illuvial (enrichment) horizon, B-horizon
Ilmenit *m (min)* ilmenite *(iron(II) metatitanate)*
Ilvait *m (min)* ilvaite, lievrite *(a sorosilicate)*
imbibieren to imbibe *(biological tissue with liquid)*
Imbibition *f* imbibition *(of biological tissue with liquid)*
Imen *n* imene *(a molecule fragment having only an electron sextet as outer shell of nitrogen)*
Imhoff-Brunnen *m* Imhoff tank *(water treatment)*
Imhoff-Trichter *m* Imhoff [sediment] cone
Imid *n* imide
Imidazol *n* imidazole, 1,3-diazole
Imidchlorid *n* imide chloride
Imidogruppe *f* imido group (residue)
Imidoharnstoff *m s.* Iminoharnstoff
Imidol *n s.* Pyrrol
Imidsäure *f* imidic acid
Imin *n* imine
Iminogruppe *f* =NH imino group (residue)
Iminoharnstoff *m* $HN{=}C(NH_2)_2$ iminourea, carbamidine, guanidine
Iminosäure *f* imino acid
Iminoverbindung *f* imino compound, imine
Immergan-Gerbung *f* Immergan process *(tanning with paraffin sulphochlorides)*
Immersion *f* immersion
Immersionsflüssigkeit *f* immersion liquid (fluid)
Immersionsöl *n* immersion oil
immobilisieren *(agric)* to immobilize *(nutrients)*
Immobilisierung *f (agric)* immobilization *(of nutrients by incorporation)*
immun immune
Immunbiologie *f* immunobiology
Immunglobulin *n* immune [serum] globulin
immunisieren to immunize
Immunisierung *f* immunization
Immunität *f* immunity
Immunkörper *m* antibody, immune body
Immunmilch *f* immune milk
Immunochemie *f* immunochemistry
immunochemisch immunochemical
Immunologie *f* immunology
Immunprotein *n* immune (antibody) protein
Immunserum *n* immune serum
impermeabel impermeable, impenetrable, impervious, tight
Impermeabilität *f* impermeability, impenetrability, imperviousness, tightness
impfen 1. *(cryst)* to seed *(solutions with nuclei of crystallization)*; 2. *(med)* to inoculate, *(relating to pocks)* to vaccinate, *(with antibodies)* to immunize
Impfkristall *m* seed crystal
Impfschlitzverfahren *n* gun injection *(wood preservation)*
Impfstoff *m* inoculating agent, inoculum, *(relating to pocks)* vaccine, *(containing antibodies)* immunizing agent
Impfung *f* 1. *(cryst)* seeding *(of solutions with nuclei of crystallization)*; 2. *(med)* inoculation, *(relating to pocks)* vaccination, *(with antibodies)* immunization
imprägnieren to impregnate, to imbibe; *(text)* to [water]proof, to impregnate; *(pap)* to penetrate, to impregnate *(chips)*
  flammfest (flammsicher) i. to flameproof
  mit CO$_2$ i. to carbonate, to aerate, to impregnate *(beverages)*
  mit Schwefel i. to sulphurize
  wasserabstoßend (wasserabweisend) i. to make water-repellent
  wasserdicht i. to waterproof

Imprägniergeschwindigkeit *f (pap)* rate of penetration *(of the chips with liquor)*
Imprägnierharz *n* impregnating resin
Imprägnierlösung *f* impregnating solution; *(rubber)* dope
Imprägniermasse *f* impregnating material
Imprägniermittel *n* impregnating agent, impregnant, proofing
Imprägnierung *f* 1. *(act)* impregnation, imbibition; *(text)* [water]proofing, impregnation; *(pap)* penetration, impregnation *(of chips)*; 2. *(state)* impregnation, finish
flammfeste (flammsichere) I. flameproof finish
wasserabstoßende (wasserabweisende) I. water-repellent finish
wasserdichte I. waterproof finish
Imprägnierungsmittel *n s.* Imprägniermittel
Imprägnierungsperiode *f (pap)* penetration period (time)
Impuls *m* momentum
Impulsmoment *n* moment of momentum, angular momentum
Impulsraum *m* momentum space
Impulsschmelzen *n* impulse rendering *(for extracting fats)*
Impulssiegeln *n* impulse sealing *(as of films)*
Impulsverteilungsgesetz *n* / **Maxwell-Boltzmannsches** Maxwell-Boltzmann distribution law
inaktiv inactive, inert, passive, *(rubber also)* non-reinforcing
optisch i. optically inactive
inaktivieren to deactivate, to inactivate, to block
Inaktivierung *f* deactivation, inactivation, blocking
Inaktivität *f* inactivity
Inaktivruß *m* inactive (inert, non-reinforcing) black
Inbetriebnahme *f* start-up, starting[-up]
inchromieren to chromize
Inchromierstahl *m* chromized steel
Inchromierung *f* chromizing
I. aus der Gasphase gas chromizing
Indan *n* indane, hydrindene
Indanon *n* indanone, hydrindone
Indanthron *n* indanthrone
Indanthron-Küpenfarbstoff *m* indanthrone vat dye
Index *m* index, value
chemotherapeutischer I. therapeutic index
hochgestellter (oberer) I. upper index, superscript
tiefgestellter (unterer) I. lower index, subscript
indifferent inert, indifferent, inactive, passive
Indifferenz *f* inertness, indifference
chemische I. chemical inertness
Indigblau *n s.* Indigoblau
Indigo *m(n)* indigo [blue]
natürlicher I. natural indigo
Indigoblau *n* indigo [blue]
Indigofarbstoff *m* indigoid [dye]
Indigolith *m (min)* indigolite, indicolite *(a variety of tourmaline)*
Indigopapier *n* indigo paper
Indigopflanze *f* indigo plant
Indigosol *n* indigosol *(any of several sulphuric-acid esters of leuco vat dyes)*
Indigotin *n* indigotin, indigo [blue]
Indikan *n* indican
Indikanreaktion *f* indican reaction *(for detecting glucosidase)*
Indikator *m (chem)* indicator, *(relating to isotopes preferably)* tracer; *(food)* indicator ingredient (substance)
basischer I. basic indicator
einfarbiger I. one-colour indicator
externer I. external (outside) indicator
interner I. internal indicator
isotoper I. isotopic tracer (indicator)
radioaktiver I. radioactive tracer (indicator), radiotracer
saurer I. acid indicator
zweifarbiger I. two-colour indicator
Indikatoratom *n* tracer atom
Indikatorbase *f* basic indicator
Indikatorbereich *m* indicator range
Indikatorchemie *f* tracer chemistry
Indikatordiagramm *n* indicator diagram
Indikatorelektrode *f* indicator electrode
Indikatorelement *n* tracer element
Indikatorfarbstoff *m* indicator dye
Indikatorgemisch *n* mixed indicator
Indikatorisotop *n* [isotopic] tracer
Indikatorkonstante *f* indicator constant
Indikatorlösung *f* indicator solution
Indikatormethode *f* tracer method
Indikatorpapier *n* indicator (test) paper, *(Am)* reaction paper
Indikatorsäure *f* acid indicator
Indikatorsubstanz *f* indicator, *(relating to isotopes preferably)* tracer
Indikatorumschlag *m* indicator change
Indischgelb *n* 1. Indian (cobalt) yellow, aureolin *(potassium hexanitrocobaltate)*; 2. *s.* echtes I.
echtes I. Indian yellow, piuri *(from Mangifera indica L.)*
Indium *n* In indium
Indium(I)-bromid *n* InBr indium(I) bromide, indium monobromide
Indium(II)-bromid *n* $InBr_2$ indium(II) bromide, indium dibromide
Indium(III)-bromid *n* $InBr_3$ indium(III) bromide, indium tribromide
Indium(I)-chlorid *n* InCl indium(I) chloride, indium monochloride
Indium(II)-chlorid *n* $InCl_2$ indium(II) chloride, indium dichloride
Indium(III)-chlorid *n* $InCl_3$ indium(III) chloride, indium trichloride
Indiumdi ... *s.* Indium(II)- ...
Indium(III)-fluorid *n* InF_3 indium(III) fluoride, indium trifluoride
Indium(III)-hydroxid *n* $In(OH)_3$ indium(III) hydroxide
Indium(III)-jodat *n* $In(IO_3)_3$ indium(III) iodate

Indium(I)-jodid *n* InI indium(I) iodide, indium monoiodide
Indium(II)-jodid *n* InI_2 indium(II) iodide, indium diiodide
Indium(III)-jodid *n* InI_3 indium(III) iodide, indium triiodide
Indiummono ... *s.* Indium(I)- ...
Indiummonoxid *n s.* Indium(II)-oxid
Indium(III)-nitrat *n* $In(NO_3)_3$ indium(III) nitrate, indium trinitrate
Indium(II)-oxid *n* InO indium(II) oxide, indium monooxide
Indium(III)-oxid *n* In_2O_3 indium(III) oxide, diindium trioxide
Indium(III)-perchlorat *n* $In(ClO_4)_3$ indium(III) perchlorate
Indium(III)-selenat *n* $In_2(SeO_4)_3$ indium(III) selenate, diindium triselenate
Indium(III)-sulfat *n* $In_2(SO_4)_3$ indium(III) sulphate, diindium trisulphate
Indium(II)-sulfid *n* InS indium(II) sulphide, indium monosulphide
Indium(III)-sulfid *n* In_2S_3 indium(III) sulphide, diindium trisulphide
Indiumtri ... *s.* Indium(III)- ...
Indium(III)-zyanid *n* $In(CN)_3$ indium(III) cyanide, indium tricyanide
Indizes *mpl* / **Bravaissche** *(cryst)* Bravais-Miller indices
Millersche I. Miller [crystal] indices
indizieren 1. *(nomencl)* to index, to indicate; 2. to label *(X-ray diffraction)*
Indizierung *f* 1. *(nomencl)* indexing, indication; 2. labelling *(X-ray diffraction)*
Indol *n* indole, 2,3-benzpyrrole
Indolalkaloid *n* indole alkaloid
Indolbrenztraubensäure *f s.* Indolylbrenztraubensäure
Indolbuttersäure *f s.* Indolylbuttersäure
Indolprobe *f* indole test
Indolsynthese *f* / **Fischersche** Fischer indole synthesis
Indolylbrenztraubensäure *f* indolylpyruvic acid
Indolylbuttersäure *f* indolylbutyric acid
Indolyl-3-essigsäure *f* 3-indolylacetic acid
Indopheninreaktion *f* / **Baeyers** Baeyer's indophenine reaction
Indophenol *n* indophenol
Indophenoloxidase *f* indophenoloxidase
Indoxyl *n* indoxyl, 3-hydroxyindole
Indoxylkarbonsäure *f* indoxylcarboxylic acid
Induktion *f* induction
Induktionseffekt *m* induction (inductive) effect, I effect
Induktionserwärmung *f* induction heating
Induktionsfaktor *m* induction factor
induktionshärten to induction-harden
Induktionshärtung *f* induction hardening
Induktionsheizgerät *n* induction heater
Induktionsheizung *f* induction heating
Induktionsoberflächenhärtung *f* induction surface hardening
Induktionsofen *m* induction (inductance, induction-heated) furnace
kernloser I. coreless (high-frequency) induction furnace
Induktionsperiode *f* induction period
Induktionsstoff *m s.* Induktor
Induktionszeit *f* induction time
Induktiv[be]heizung *f* induction heating
Induktor *m* inductor *(in chemical reactions)*; inductor, inductive agent *(enzymology)*
Industrie *f* / **chemische** chemical industry
feinkeramische I. fine ceramic industry
grobkeramische I. heavy-clay industry
keramische I. ceramic industry
petrolchemische I. petrochemicals industry
pharmazeutische I. pharmaceutic[al] industry
verarbeitende (weiterverarbeitende) I. processing industry
Industrieabfälle *mpl* industrial (trade) waste
Industrieabwasser *n* industrial waste water, factory (works, trade) effluent
Industriealkohol *m* commercial (industrial) alcohol (spirit)
Industrieanlage *f* industrial plant
Industriechemie *f* industrial (technical) chemistry
Industriechemiker *m* industrial chemist
Industriefilter *n* plant filter
Industriegas *n* industrial gas
Industriegasbrenner *m* industrial gas burner
Industriekohle *f* industrial coal
Industriemüll *m* industrial (trade) waste
Industrieofen *m* industrial kiln (furnace)
Industriestaub *m* industrial dust
inert inert, inactive, passive, indifferent
Inertgas *n* inert (inactive) gas
Inertia *f (phot)* inertia
inertieren *(coal)* to render inert
Inertinit *m* inertinite *(general term for some macerals of hard coal)*
inertisieren *s.* inertieren
Infiltration *f* infiltration
inflammabel inflammable, infl.
nicht i. non-[in]flammable
Inflammation *f* inflammation
infrarot infrared, ultrared
Infrarot *n* infrared [radiation]
Infrarotabsorption *f* infrared absorption
Infrarotabsorptionsspektrum *n* infrared absorption spectrum
Infrarotanalyse *f* infrared analysis
Infrarotbeheizung *f* infrared heating
Infrarotdunkelstrahler *m* far-infrared radiation element
infrarotdurchlässig infrared-transmitting, transparent to infrared
Infrarotdurchlässigkeit *f* infrared transmittancy
Infrarotfotografie *f* infrared photography
Infrarotfotometer *n* infrared photometer
Infrarotfrequenz *f* infrared frequency
Infrarotheizgerät *n* infrared heater
Infrarotheizung *f* infrared heating
Infrarothellstrahler *m* near-infrared radiation element

Infrarotkleber *m* infrared-drying adhesive
Infrarotlampe *f* infrared lamp (radiator)
Infrarotlicht *n* infrared light
Infrarotmikroskop *n* infrared microscope
Infrarotspektralfotometer *n* infrared spectrophotometer
Infrarotspektrometer *n* infrared spectrophotometer
Infrarotspektroskopie *n* infrared spectroscopy
Infrarotspektrum *n* infrared spectrum
Infrarotstrahler *m* infrared lamp (radiator)
Infrarotstrahlung *f* infrared radiation
Infrarottrockenofen *m* infrared drying oven
Infrarottrockner *m* infrared dryer
Infrarottrocknung *f* infrared drying
Infrarotuntersuchung *f* infrared study
Infrarotvorwärmung *f* infrared preheating
Infus *n (pharm)* infusion
Infusion *f* infusion
Infusionslösung *f (pharm)* infusion
Infusionsverfahren *n (ferm)* infusion mashing (process)
Infusorienerde *f* diatom (diatomaceous, infusorial) earth, kieselguhr
Infusum *n (pharm)* infusion
Ingangsetzen *n* start-up, starting[-up]
Ingenieurtechnik *f* / **chemische** chemical engineering technology
Ingenieurwesen *n* / **chemisches** chemical engineering
Inglasurdekor *n (ceram)* in-glaze (inter-glaze) decoration
Inglasurmalerei *f (ceram)* in-glaze (inter-glaze) painting
Ingot *m (met)* ingot
Ingrain-Farbe *f* ingrain dye
Ingrediens *n*, **Ingredienz** *f* ingredient
Ingwer *m* ginger *(from Zingiber officinale Rosc.)*
  Gelber I. curcuma, Indian saffron *(from Curcuma longa L.)*
  Japanischer I. Japanese (Mioga) ginger *(from Zingiber mioga (Thunb.) Rosc.)*
Ingwerbier *n* ginger ale (beer)
Ingwergrasöl *n* ginger-grass oil *(chiefly from Cymbopogon martini (Roxb.) Stapf var. sofia)*
Ingweröl *n* ginger oil
Ingwerwurzel *f* ginger *(from Zingiber officinale Rosc.)*
INH *s.* Isonikotinsäurehydrazid
Inhalationsnarkotikum *n* inhalation narcotic
Inhalt *m* content, *(quantitatively)* volume, capacity
inhibieren to inhibit, *(rubber also)* to shortstop *(e. g. polymerization)*
inhibierend inhibitory
Inhibition *f* inhibition
Inhibitor *m* inhibitor, inhibiting substance, retarder, retarding agent, *(rubber also)* stopper, shortstop, shortstopping agent
  anodischer I. anodic inhibitor
  katodischer I. cathodic inhibitor
inhomogen inhomogeneous, non-homogeneous, heterogeneous
Inhomogenität *f* inhomogeneity, non-homogeneity, heterogeneity
Initialsprengstoff *m* initiator, initiating (primary) explosive, primer, detonator, initial detonating agent
Initialtemperatur *f* initial temperature
Initialzündung *f* initial ignition
Initiator *m* initiator, initiating agent
Initiierbarkeit *f* sensitiveness to initiation *(of an explosive)*
initiieren to initiate, to start
Initiierung *f* initiation
Initiierungsstadium *n* initiation stage
Injektion *f* injection
  I. von flüssigem Ammoniak *(agric)* nitrojection
Injektionsgerät *n (agric)* injection apparatus *(for liquid fertilizers)*
Injektionslösung *f* solution for injection
Injektionsmetamorphose *f (geol)* injection metamorphism
Injektionspflugschar *n (agric)* injection plowshare *(for liquid fertilizers)*
Injektionsspritze *f* hypodermic (sample-charging) syringe *(gas chromatography)*
Injektionszerstäuber *m* gas-atomizing (two-fluid) nozzle
Injektor *m (agric)* [soil] injector, injector gun *(for soil fumigation)*
Injektormischer *m* injector mixer
Inklusion *f* inclusion
Inklusionsverbindung *f* inclusion compound
inkohlen *(geoch)* to coalify
Inkohlung *f (geoch)* coalification, carbonification
Inkohlungsband *n* coalification band
Inkohlungsgrad *m* degree of coalification, rank ✦ **von hohem I.** high-rank ✦ **von mittlerem I.** medium-rank ✦ **von niedrigem I.** low-rank
Inkohlungsmaßstab *m* rank parameter
Inkohlungsreihe *f* coalification series
Inkohlungsstadium *n* stage of coalification
Inkohlungsstreifen *m* coalification band
Inkohlungsstufe *f* stage of coalification
Inkohlungsvorgang *m* coalification process
inkompatibel incompatible *(as of pesticides or pharmaceuticals)*
Inkompatibilität *f* incompatibility *(as of pesticides or pharmaceuticals)*
Inkorporation *f (bioch)* incorporation
inkorporieren *(bioch)* to incorporate
inkromieren *s.* inchromieren
Inkromierung *f s.* Inchromierung
Inkrustation *f* incrustation, encrustation *(process or substance)*
Inkrusten *pl* incrustants, encrustants, incrusting material (matter, substance)
inkrustieren to incrust, to encrust
Inkrustierung *f s.* Inkrustation
Inkrustsubstanzen *fpl s.* Inkrusten
Inkubationsprobe *f (dye)* incubation test
Innenanstrichfarbe *f* interior paint
innenbürtig *(geol)* endogenous
Innenfilter *n* inside drum filter

Innenfläche *f* inner surface
Innengummi *m* innerliner *(of a tyre)*
Innenlack *m* interior paint, *(if transparent)* interior varnish
Innenmischer *m* closed mixer
Innenorbitalkomplex *m* inner orbital complex
Innenrohrschlange *f* tank coil
Innenrüttler *m (build)* immersion (poker, needle) vibrator
Innenschale *f* inner shell *(of an atom)*
Innenthermometer *n* internal thermometer
Innentrommelfilter *n s.* Innenfilter
Innenvibrator *m s.* Innenrüttler
Innenwasser *n* inherent moisture
Innenzellenfilter *n* inside drum filter
innenzentriert *(cryst)* space-centred, body-centred
Innerkomplex *m* inner complex
Innerkomplex-Anion *n* inner complex anion
Innerkomplex-Kation *n* inner complex cation
Innerkomplexsalz *n* inner-complex salt
innerlich *(pharm)* internal
innermolekular intramolecular
innersekretorisch endocrine
Innertherapeutikum *n* systemic [chemical] *(pest control)*
innertherapeutisch systemic *(pest control)*
innig intimate *(e. g. contact or mixture)*
Ino *s.* Inosin
Inosilikat *n (min)* inosilicate *(any of a class of polymeric silicates)*
Inosin *n* inosine *(a riboside)*
Inosinsäure *f* inosinic acid *(a nucleotide)*
Inosit *m* $C_6H_6(OH)_6$ inositol, hexahydroxycyclohexane
Insektenabwehrmittel *n* insect repellent, insectifuge
Insektenanlockmittel *n* insect attractant
Insektenbekämpfungsmittel *n* insecticide
Insektenfarbstoff *m* insect pigment
Insektenlockstoff *m* insect attractant
Insektenpuder *m*, **Insektenpulver** *n* insect powder
insektenschonend non-insecticidal *(e. g. fungicide)*
Insektenschutzmittel *n s.* Insektenabwehrmittel
Insektenstichmittel *n* insect bite remedy
insektentötend insecticidal
Insektentötungsmittel *n s.* Insektizid
Insektenvertilgungspapier *n* insecticide (poisoned) paper
insektenverträglich non-insecticidal *(e. g. fungicide)*
Insektenwachs *n* insect (Chinese tree) wax, vegetable spermaceti *(secreted by scales)*
insektizid insecticidal
Insektizid *n* insecticide
  endolytisches [systemisches] I. endolytic insecticide
  endometatoxisches [systemisches] I. endometatoxic insecticide
  pflanzliches I. insecticide of plant origin, botanical
  protektives (protektiv wirkendes) I. protective insecticide
  systemisches I. systemic insecticide
Insektizidaktivität *f* insecticidal power
Insektizidität *f* insecticidal efficiency
Insektizidnebel *m* insecticidal fog
Insektizidrauch *m* insecticidal smoke
Insektizidresistenz *f* insecticide resistance *(of animals)*
Insektizidsprühdose *f* insecticide bomb
insilizieren to siliconize *(metals for protection)*
Insilizierung *f* siliconization *(of metals for protection)*
in situ in place, in situ
In-situ-Laugung *f (met)* leaching in place, underground leaching
In-situ-Polymerisation *f* in-situ polymerization
instabil instable, unstable, transient
  thermisch i. thermolabile, heat-labile
Instabilität *f* instability
Instabilitätskonstante *f* instability constant *(as of complex ions)*
Instandhaltung *f* maintenance, upkeep
Instrument *n* / **registrierendes (selbstschreibendes)** graphic (recording) instrument, grapher, recorder
Instrumentalanalyse *f*, **Instrumentenanalyse** *f* instrumental analysis
Instrumentenfehler *m* instrumental error
Insulin *n* insulin ✦ **mit I. behandeln** to insulinize
Integraldetektor *m* integral detector *(gas chromatography)*
Intensität *f* intensity
Intensitätsgröße *f* intensive quantity *(being independent of the mass of the system concerned)*
Intensitätsregeln *fpl (phys chem)* intensity rules
Intensitätsverhältnis *n* transmittancy *(colorimetry)*
Intensitätsverteilung *f* intensity distribution
intensivieren to intensify
Intensivierung *f* intensification
Intensivkühler *m* jacketed coil condenser, high-efficiency condenser
Intensivkühlung *f* rapid cooling
interatomar interatomic
Interferenz *f (phys)* interference
Interferenzbild *n* interference figure (pattern)
Interferenzerscheinung *f* interference phenomenon
Interferenzfarbe *f* interference colour
Interferenzfigur *f s.* Interferenzbild
Interferenzfilter *n* interference filter
Interferenzmikroskop *n* interference microscope
Interferenzring *m* interference ring
Interferenzstreifen *m* interference fringe
interferieren to interfere
Interferometrie *f* interferometry
Interhalogen *n*, **Interhalogenverbindung** *f* interhalogen compound
interionisch interionic
interkristallin intercrystalline
interkrustal *(geol)* intercrustal

intermediär intermediate
Intermediärprodukt *n* intermediate product (substance), [reaction] intermediate
Intermediärstoffwechsel *m* *(bioch)* intermediary metabolism
Intermediärverbindung *f* intermediate [compound]
intermolekular intermolecular
intern *(pharm)* internal
Interstitiallösung *f s.* Interstitialmischkristall
Interstitialmischkristall *m* interstitial [solid] solution
interstitiell interstitial
Intervall *n* interval, space
Interzellularpigment *n* intercellular pigment *(in animals)*
Intoxikation *f* poisoning, *(med also)* intoxication
intramolekular intramolecular
intrazellulär intracellular, endocellular
Intrinsic-Leitfähigkeit *f* intrinsic semiconductivity (electrical conductivity)
Intrinsic-Leitung *f* intrinsic semiconduction (electrical conduction)
intrudieren *(geol, plast)* to intrude
Intrusion *f (geol, plast)* intrusion
Intrusionsgestein *n*, **Intrusivgestein** *n* intrusive (irruptive, contaminated) rock
Intussuszeption *f* *(biol)* intussusception *(interposition of new substance into growing cell membranes)*
Inulakampfer *m* alantolactone, helenin[e]
Inulin *n* inulin *(a polysaccharide)*
Inulinase *f* inulinase, inulase
invariabel invariable
invariant invariant, non-variant
Invariante *f* / **adiabatische** adiabatic invariant
Inversion *f* inversion
  Waldensche I. Walden inversion
Inversionsdrehachse *f (cryst)* inversion axis
Inversionspunkt *m (distil)* phase-inversion point
Inversionsschicht *f* inversion layer
Inversionstemperatur *f* inversion temperature
Inverspolarografie *f* anodic stripping
Invertase *f* invertase, saccharase, sucrase
Invertemulsion *f* invert emulsion *(as of pesticide formulations)*
invertieren to invert *(the configuration of a molecule)*
Invertierung *f* inversion *(of the configuration of a molecule)*
Invertin *n s.* Invertase
Invertseife *f* invert soap
Invertzucker *m* invert[ed] sugar
  durch Hydrolyse mit Säuren gewonnener I. acid-inverted sugar
Invertzuckersirup *m* invert syrup
Investmentguß *m* [precision] investment casting *(foundry)*
Ion *n* ion
  I. auf Zwischengitterplatz interstitial ion
  einfach geladenes I. mono-ion
  eintauschendes I. *(soil)* competitor ion
  einwertiges I. mono-ion
  hydratisiertes I. hydrated ion, aquo-ion
  komplexes I. complex ion
  negatives I. negative ion, anion
  positives I. positive ion, cation
  schnelles I. high-speed ion
ional ionic
Ionenadsorption *f* adsorption of ions
Ionenadsorptionsvermögen *n* ion-adsorbing capacity
Ionenaggregat *n* aggregate of ions
Ionenaktivität *f* ion[ic] activity
Ionenaktivitätskoeffizient *m* ion[ic] activity coefficient
Ionenantagonismus *m (bioch)* ion antagonism
Ionenäquivalentleitfähigkeit *f* equivalent ion[ic] conductance
Ionenäquivalentmasse *f* ionic equivalent
Ionenart *f* ionic species
Ionenassoziation *f* ion association
Ionenatmosphäre *f* *(phys chem)* ion[ic] atmosphere, ion cloud
Ionenaufnahme *f* ion absorption
Ionenausschlußverfahren *n* ion-exclusion process
Ionenaustausch *m* ion[ic] exchange
Ionenaustauschchromatografie *f* ion-exchange chromatography
Ionenaustauscher *m* ion-exchange material, ion exchanger
  I. auf Kunstharzbasis *s.* Ionenaustauschharz
Ionenaustauschermembran *f* ion-exchange membrane
Ionenaustauschersäule *f* ion-exchange column
Ionenaustauschgleichgewicht *n* ion-exchange equilibrium
Ionenaustauschharz *n* ion-exchange resin
Ionenaustauschmethode *f* ion-exchange method
Ionenaustauschreaktion *f* ion-exchange reaction
Ionenaustauschtechnik *f* ion-exchange technology
Ionenaustauschtrennung *f* ion-exchange separation
Ionenbeweglichkeit *f* ion[ic] mobility
Ionenbewegung *f* ion[ic] movement, ion[ic] motion
Ionenbeziehung *f* electrovalency, electrovalent (ionic, polar, heteropolar, electrostatic) bond
Ionenbildung *f* ionization, ionizing
Ionenbindung *f* 1. electrovalent (ionic, polar, heteropolar, electrostatic) linkage *(process)*; 2. *s.* Ionenbeziehung
Ionendichte *f* ion density
Ionendipolkomplex *m* ion-dipole complex
Ionendipolkräfte *fpl* ion-dipole forces
Ionendosis *f* ion dosage (dose)
ionenerzeugend ionogenic
Ionenfarbe *f* ion colour
Ionenformel *f* ionic formula
Ionengehalt *m* ionic content
Ionengeschwindigkeit *f* ionic speed (velocity)
Ionengetterpumpe *f* ionic getter pump
Ionengitter *n (cryst)* ionic [crystal] lattice
Ionengleichgewicht *n* ionic equilibrium (balance)

Ionengleichung *f* ionic equation
Ionengröße *f* ion[ic] size
Ionenhydrat *n* ion hydrate
Ionenhydratation *f* ionic hydration
Ionenhydration *f (incorrectly for)* Ionenhydratation
ioneninaktiv non-ionic, non-ionizing, non-ionogenic
Ionenkonzentration *f* ionic concentration
Ionenkräfte *fpl* ionic forces
Ionenkristall *m* ionic crystal
Ionenladung *f* ionic charge
Ionenladungszahl *f* ionic charge number
Ionenlawine *f* avalanche of ions
Ionenleitfähigkeit *f* ion[ic] conductivity, ion[ic] conductance
Ionenleitung *f* ionic conduction
Ionenmasse *f* ionic mass
Ionenmolekel *f*, **Ionenmolekül** *n* ionic molecule
Ionenpaar *n* ion pair
Ionenpolymerisation *f* ionic polymerization
Ionenprodukt *n* ion[ic] product
Ionenquelle *f* ion source
Ionenradius *m* ionic radius
Ionenreaktion *f* ion[ic] reaction
Ionenreihe *f* / **Hofmeistersche** lyotropic order (series)
Ionenrekombination *f* ion recombination
Ionenresonanz-Hochfrequenzspektrometer *n* ion-resonant spectrometer
Ionenrichtgitter *n* ion focus grid *(in a spectrometer)*
Ionensorte *f* ionic species
Ionenstärke *f* ionic strength
Ionenstrahl *m* ion beam
Ionenstruktur *f* ionic structure
Ionensuszeptibilität *f* ionic susceptibility *(susceptibility per gram ion)*
Ionentreibstoff *m* ionic propellant
Ionentrennung *f* ion separation
Ionenumtausch *m* ion exchange
Ionenverbindung *f* ionic (polar, heteropolar) compound
Ionenverzögerungsverfahren *n* ion-retardation process
Ionenwanderung *f* ionic migration, migration of ions
  unabhängige I. independent migration (mobility) of ions
Ionenwanderungsgeschwindigkeit *f* ionic speed (velocity)
Ionenwechselwirkung *f* ionic interaction, interionic action
Ionenwertigkeit *f* ionic valence
Ionenwind *m* ionic (electric) wind
Ionenwolke *f (phys chem)* ion cloud, ion[ic] atmosphere
Ionenzustand *m* ionic state
Ion-Ion-Komplex *m* ion-ion complex
Ion-Ion-Rekombination *f* ion-ion recombination
Ionisation *f* ionization
  differentielle I. *s.* spezifische I.
  lawinenartige I. cumulative ionization
  spezifische I. specific ionization
  thermische I. thermal ionization
Ionisationsarbeit *f s.* Ionisationsenergie
Ionisationsenergie *f* ionization energy
  erste I. first ionization potential *(expressed in eV)*
Ionisationsfähigkeit *f* ionizing power
Ionisationsgrad *m* degree of ionization
Ionisationskammer *f* ionization chamber
Ionisationspotential *n* ionization (ionizing) potential
  erstes I. first ionization potential
Ionisationsspannung *f s.* Ionisationspotential
Ionisationsspektrometer *n* ionization spectrometer
Ionisationsstärke *f* specific ionization
Ionisationsvakuummeter *n* ionization gauge
  I. mit heißer Katode thermionic (hot-filament) ionization gauge
Ionisationsvermögen *n* ionizing power
Ionisationswahrscheinlichkeit *f* probability of ionization
Ionisationswärme *f* heat of ionization
ionisch ionic
ionisierbar ionizable, ionogenic
ionisieren to ionize
ionisiert ionized, ionic
  einfach i. singly ionized
  zweifach i. doubly ionized
Ionisierung *f s.* Ionisation
Ionisierungsbereich *m* ionizing region *(spectrometry)*
ionogen ionogenic
ionoid ionic
Ionomer[es] *n* ionomer
ionometrisch ionometric
ionophil ionophilic
Ionophorese *f* ionophoresis
IPC *s.* Isopropyl-*N*-phenylkarbamat
Ipekakuanhaalkaloid *n* ipecacuanha alkaloid
Ipekakuanhawurzel *f (pharm)* ipecacuanha root *(from Cephaëlis ipecacuanha (Brot.) A. Rich. and C. acuminata Karsten)*
IP-Standardmethode *f* standard IP method
Irdengut *n*, **Irdenware** *f* earthenware
Iridium *n* Ir iridium
Iridosmium *n (min)* iridosmine, iridium-osmine
irisieren to be iridescent, to iridesce
Irisieren *n* iridescence
irisierend iridescent
Irispapier *n* iridescent paper, mother-of-pearl paper
irreversibel irreversible, non-reversible
I-Säure *f s.* J-Säure
I-Säure-Harnstoff *m s.* J-Säure-Harnstoff
i-s-Diagramm *n* enthalpy-entropy chart (diagram), Mollier chart
Isentrope *f (phys chem)* isentrope *(the representation of an isentropic process in a thermodynamic diagram)*

isentropisch *(phys chem)* isentropic
Islandspat *m (min)* Iceland spar *(calcium carbonate)*
Isoalkan *n* isoalkane
Isoalloxazin *n* isoalloxazine, flavin[e]
Isoamylaldehyd *m s.* Isovaleraldehyd
Isoamylalkohol *m* $(CH_3)_2CHCH_2CH_2OH$ isoamyl alcohol, 3-methyl-1-butanol
Isoamylnitrat *n* $(CH_3)_2CHCH_2CH_2ONO_2$ isoamyl nitrate
Isoamylvalerianat *n* $(CH_3)_2CHCH_2COOC_5H_{11}$ isoamyl valerianate (valerate)
Isoaskorbinsäure *f D-arabo*ascorbic acid, isoascorbic acid
Isobaldriansäure *f s.* Isovaleriansäure
isobar isobaric
Isobar *n* [nuclear] isobar
Isobare *f (phys chem)* isobar
Isobernsteinsäure *f* $CH_3CH(COOH)_2$ isosuccinic acid, methylmalonic acid, ethane-1,1-dicarboxylic acid
Isobutan *n* $(CH_3)_3CH$ isobutane, 2-methylpropane
Isobutankreislauf *m* isobutane recycle
Isobutanol *n s.* Isobutylalkohol
Isobuten *n s.* 2-Methylpropen
Isobuttersäure *f* $(CH_3)_2CHCOOH$ isobutyric acid, 2-methylpropionic acid
Isobutylalkohol *m* $(CH_3)_2CHCH_2OH$ isobutyl alcohol, 2-methyl-1-propanol
Isobutylen *n s.* 2-Methylpropen
Isobutylen-Isopren-Kautschuk *m* isobutylene-isoprene rubber, IIR
Isobutylessigsäure *f* $(CH_3)_2CHCH_2CH_2COOH$ 4-methylpentanoic acid, 4-methylvaleric acid, isobutyl acetic acid
Isobutylierung *f* isobutylation
Isobutylkarbinol *n s.* Isoamylalkohol
Isobutylmerkaptan *n s.* 2-Methyl-1-propanthiol
Isochinolin *n* isoquinoline, 2-benzazine
Isochinolinalkaloid *n* isoquinoline alkaloid
Isochore *f (phys chem)* isochore
isodiametrisch isodiametric
isodimensional isodimensional
Isodipren *n* (–)-3-carene, isodiprene, 3,7,7-trimethylbicyclo[2,2,1]hept-3-ene
isodispers isodisperse, monodisperse
Isodurol *n* $C_6H_2(CH_3)_4$ isodurene, 1,2,3,5-tetramethylbenzene
Isodurylsäure *f* $(CH_3)_3C_6H_2COOH$ isodurylic acid, trimethylbenzoic acid
isoelektrisch isoelectric
isoelektronisch isoelectronic, isosteric
Isoeugenol *n* $C_6H_3(OH)(OCH_3)CH{=}CHCH_3$ isoeugenol, 2-methoxy-4-propenylphenol
Isoferulasäure *f* $CH_3OC_6H_3(OH)CH{=}CHCOOH$ isoferulic acid, 3-hydroxy-4-methoxycinnamic acid
Isoflavon *n* isoflavone
Isogel *n (coll)* isogel
Isohemipinsäure *f* $(H_3CO)_2C_6H_2(COOH)_2$ isohemipinic acid, 4,5-dimethoxyisophthalic acid
Isokale *f* isocal, isocalorific line
Isokapronsäure *f s.* 4-Methylpentansäure
Iso-Kautschuk *m* isorubber
Isokrotonsäure *f* $CH_3CH{=}CHCOOH$ isocrotonic acid, *cis*-2-butenoic acid
Isolation *f* insulation
Isolations ... *s.a.* Isolier ...
Isolationswiderstand *m* insulation resistance
Isoleuzin *n* isoleucine *(an amino acid)*
isolierbar isolable
Isolierbeton *m* insulation concrete
isolieren 1. *(chem)* to isolate, to separate, to segregate; 2. to insulate *(against something)*
Isolierfestigkeit *f* insulation resistance
Isolierlack *m* insulating varnish
Isoliermasse *f* insulation compound
Isoliermaterial *n* insulation [material]
Isoliermischung *f (rubber)* insulating compound (stock)
Isoliermittel *n* insulating medium
Isolieröl *n* [electrical] insulating oil
Isolierpapier *n* [electrical] insulating paper
Isolierpappe *f* insulating (fuller) board
Isolierschlauch *m* insulation tubing
Isolierstein *m* insulating brick
Isolierstoff *m* insulation [material]
Isolierstreifen *m* insulation strip
Isolierung *f* 1. *(chem)* isolation, separation, segregation; 2. insulation *(against something)*
isomer isomeric
  optisch i. enantiomorphous, enantiomorphic, enantiomeric
Isomer *n* isomer, isomeride, isomeric compound
  geometrisches I. geometric[al] isomer
  optisches I. optical isomer, antimer
m-**Isomer** *n m*-isomer, meta isomer
o-**Isomer** *n o*-isomer, ortho isomer
p-**Isomer** *n p*-isomer, para isomer
Isomerase *f* isomerase
isomerenfrei free from isomers
Isomerengemisch *n* mixture of isomers
Isomerenpaar *n* isomeric pair
  optisch aktives I. pair of optical isomers
Isomeres *n s.* Isomer
Isomerie *f* isomerism
  geometrische I. geometrical (cis-trans) isomerism
  optische I. optical (mirror-image) isomerism, enantiomorphism
  räumliche (stereochemische) I. stereoisomerism, space isomerism
Isomerisation *f s.* Isomerisierung
isomerisieren to isomerize
  sich i. to isomerize
Isomerisierung *f* isomerization
  I. in der Dampfphase vapour-phase isomerization
  I. in der Flüssigphase liquid-phase isomerization
  I. in der Gasphase gas-phase isomerization
Isomerisierungsgleichgewicht *n* isomerization equilibrium
Isomerisierungsreaktion *f* isomerization reaction
Isomerisierungsverfahren *n* isomerization process

isomorph *(cryst)* isomorphic, isomorphous
Isomorphie *f (cryst)* isomorphism
Isomorphieregel *f* / **Mitscherlichsche** *(cryst)* Mitscherlich's law of isomorphism
Isomorphismus *m (cryst)* isomorphism
Isoniazid *n s.* Isonikotinsäurehydrazid
Isonikotinsäure *f* isonicotinic acid, pyridine-4-carboxylic acid
Isonikotinsäurehydrazid *n* isonicotinic acid hydrazide, isoniazid, INAH
Isonitril *n* R–N≡C isonitrile, isocyanide, carbylamine
Isooktan *n* isooctane, *(specif)* $(CH_3)_2CHCH_2C(CH_3)_3$ 2,2,4-trimethylpentane
Isooktanol *n s.* Isooktylalkohol
Isooktylalkohol *m* $(CH_3)_2CH[CH_2]_4CH_2OH$ 6-methylheptanol, isooctyl alcohol
Isopentan *n* $CH_3CH(CH_3)CH_2CH_3$ 2-methylbutane, isopentane
Isopersulfozyansäure *f* isoperthiocyanic acid
Isophthalsäure *f* $C_6H_4(COOH)_2$ isophthalic acid, *m*-phthalic acid, benzene-*m*-dicarboxylic acid
Isopolymorphie *f* isopolymorphism
Isopolysäure *f* isopoly acid
Isopren *n* $CH_2{=}CHC(CH_3){=}CH_2$ isoprene, 2-methyl-1,3-butadiene
Isoprenkautschuk *m* isoprene rubber, IR
Isoprenoid *n* isoprenoid *(general term for terpenes and steroids)*
Isopropylalkohol *m* $CH_3CH(OH)CH_3$ 2-propanol
Isopropylamin *n* $(CH_3)_2CHNH_2$ 2-propylamine, isopropylamine
Isopropyläther *m s.* Diisopropyläther
Isopropylbenzol *n* $C_6H_5CH(CH_3)_2$ isopropylbenzene, cumene, 2-phenylpropane
Isopropylchlorid *n* $CH_3CHClCH_3$ 2-chloropropane, isopropyl chloride
Isopropylessigsäure *f s.* Isovaleriansäure
Isopropylgruppe *f* $(CH_3)_2CH-$ isopropyl group (residue)
Isopropylidenazeton *n s.* Mesityloxid
Isopropylkarbinol *n s.* 2-Methylpropanol-(1)
Isopropyl-*N*-phenylkarbamat *n* $C_6H_5NHCOOCH(CH_3)_2$ isopropyl-*N*-phenylcarbamate, IPC *(a herbicide)*
isosmotisch is[o]osmotic, isotonic
isoster isosteric, isoelectronic
Isostere *f (phys chem)* isostere
Isosterie *f* isosterism
Isosynthese *f* isosynthesis
isotaktisch isotactic
Isotaktizität *f* isotacticity
Isoteniskop *n* isoteniscope *(a device for determining the saturation vapour pressure of liquids)*
isotherm isothermal
Isotherme *f* isotherm
  van-der-Waalssche I. van der Waals isotherm
isothermisch isothermal
Isoton *n (nucl)* isotone
isotonisch isotonic, is[o]osmotic
isotop isotopic
Isotop *n* isotope
  instabiles I. *s.* radioaktives I.
  künstlich erzeugtes I. artificial isotope
  nichtradioaktives I. *s.* stabiles I.
  radioaktives I. radioactive (unstable) Isotope, radioisotope
  schweres I. heavy isotope
  stabiles I. stable (non-radioactive) isotope
Isotopenanalyse *f* isotopic analysis
Isotopenaustausch *m* isotopic exchange
Isotopeneffekt *m* isotope effect
Isotopengemisch *n* mixture of isotopes, isotopic mixture
Isotopengewicht *n* physical (isotopic) atomic weight, mass value
Isotopenindikator *m* isotopic tracer (indicator)
Isotopenkasten *m (nucl)* glove box
Isotopentracer *m s.* Isotopenindikator
Isotopentrennung *f* isotope (isotopic) separation
Isotopenverbindung *f* isotopic compound
Isotopenverdünnung *f* isotopic dilution
Isotopenverdünnungsanalyse *f* isotopic dilution analysis
Isotopenverdünnungsverfahren *n* isotope dilution procedure
Isotopie *f* isotopism, isotopy
Isotopieeffekt *m* isotope effect
isotrop *(cryst)* isotropic
  optisch i. optically isotropic
Isotropie *f (cryst)* isotropism, isotropy
Isotypie *f (cryst)* isotypy
Isovaleraldehyd *m* $(CH_3)_2CHCH_2CHO$ isovaleraldehyde, 3-methylbutanal
Isovaleriansäure *f* $(CH_3)_2CHCH_2COOH$ isovaleric acid, 3-methylbutyric acid
Isovalersäure *f s.* Isovaleriansäure
Isovole *f* isovol *(line of equal volatile matter)*
Isoxylylsäure *f* $(CH_3)_2C_6H_3COOH$ isoxylylic acid, *p*-xylylic acid, 2,5-dimethylbenzoic acid
Isozitratdehydrogenase *f* isocitric acid dehydrogenase
Isozitronensäure *f* isocitric acid, 1-hydroxypropane-1,2,3-tricarboxylic acid
Isozyanat *n* R–N=C=O isocyanate
Isozyanatkleber *m* isocyanate adhesive
Isozyanatplast *m* isocyanate resin
Isozyanid *n* R–N≡C isocyanide, isonitrile, carbylamine
Isozyanin *n* isocyanine
Isozyaninkondensation *f* isocyanine condensation
Isozyanursäure *f* isocyanuric acid, fulminuric acid
Isozyklen *pl* isocyclic (homocyclic, carbocyclic) compounds
isozyklisch isocyclic, homocyclic, carbocyclic
Itakonsäure *f* $HOOCC(=CH_2)CH_2COOH$ itaconic acid, 2-propene-1,2-dicarboxylic acid
IT-Diagramm *n* IT diagram, heat-content/temperature diagram
It-Stoff *m* asbestos-rubber material *(for sealing)*
IUC-Regel *f (nomencl)* IUC rule
IUC-System *n (nomencl)* IUC system
IUPAC-Dyson-Notation *f (nomencl)* IUPAC-Dyson notation

IUPAC-Dyson-System *n (nomencl)* IUPAC-Dyson [notation] system
IUPAC-Regel *f (nomencl)* IUPAC rule
IUPAC-System *n (nomencl)* IUPAC system
Ivanoff-Reaktion *f* Ivanov reaction *(synthesis of hydroxy acids)*
Izod-Prüfung *f (plast)* Izod impact test
IZSH *s.* Hormon / luteinisierendes

J

Jaborandiöl *n* jaborandi oil *(from the leaves of Pilocarpus pennatifolius Lem.)*
Jade *m(f) (min)* jade *(a gemstone derived from jadeite or nephrite)*
Jadeit *m (min)* jadeite *(aluminium sodium metasilicate)*
Jalousiedrosselklappe *f* louvre
Jalousietrockner *m* louvre dryer
Jamaikapfeffer *m* allspice *(from Pimenta dioica (L.) Merr.)*
Jamesonit *m* jamesonite *(a sulphidic mineral)*
Japankampfer *m* Japan camphor *(from Cinnamomum camphora (L.) Sieb.)*
Japanlack *m* japan, Japan lacquer, *(specif)* Japanese (Chinese) lacquer *(from Rhus verniciflua Stokes)*
Japanpapier *n* Japan[ese] paper
Japansäure *f* $HOOC[CH_2]_{19}COOH$ Japanic acid, 1,21-heneicosanedioic acid
Japanseidenpapier *n* Japanese tissue paper
Japantalg *m* Japan tallow (wax) *(from Rhus succedanea L. and Rhus verniciflua Stokes)*
Japanwachs *n (incorrectly for)* Japantalg
Jargon *m (min)* jargo[o]n *(zirconium orthosilicate)*
Jaspis *m (min)* jasper *(a variety of quartz)*
Jaspisware *f (ceram)* jasper ware
Jaspopal *m (min)* jaspopal, jasper opal
Java-Kardamom *m(n)* Java cardamom *(from Amomum maximum Roxb.)*
Javakunstpapier *n* batik paper
Jensen-Turm *m (pap)* Jensen tower
Jequié-Kautschuk *m* Jequie rubber, mule gum *(from Manihot dichotoma Ule)*
Jervasäure *f* jervasic acid, chelidonic acid, 8-pyrone-2,6-dicarboxylic acid
Jet[t] *m(n)* jet, gagate *(a variety of lignite)*
J-Gefäß *n (text)* J box
Jigger *m (text)* jig[ger]
jj-Kopplung *f (nucl)* jj-coupling
Jod *n* I iodine
  radioaktives J. radioactive iodine, radioiodine, *(specif)* ^{131}I iodine-131
Jodargyrit *m (min)* iodargyrite *(silver iodide)*
Jodat *n* M^IIO_3 iodate
Jodäthan *n* CH_3CH_2I iodoethane
Jodäthanverbindung *f* ethiodide
Jodäthyl *n s.* Jodäthan
Jodatmethode *f* iodate method
Jodazid *n* IN_3 iodine azide
Jodbenzol *n* C_6H_5I iodobenzene
Jod(I)-bromid *n* IBr iodine(I) bromide, iodine monobromide
Jod(III)-bromid *n* IBr_3 iodine(III) bromide, iodine tribromide
Jodchinolin *n* iodoquinoline
Jod(I)-chlorid *n* ICl iodine(I) chloride, iodine monochloride
Jod(III)-chlorid *n* ICl_3 iodine(III) chloride, iodine trichloride
Joddampf *m* iodine vapour
Joddioxid *n* IO_2 iodine dioxide
Jodfärbevermögen *n* iodine staining power
Jod(V)-fluorid *n* IF_5 iodine(V) fluoride, iodine pentafluoride
Jod(VII)-fluorid *n* IF_7 iodine(VII) fluoride, iodine heptafluoride
Jodgorgosäure *f* $HOC_6H_2(I_2)CH_2CH(NH_2)COOH$ iodogorgoic acid
jodhaltig iodine-containing
Jodheptafluorid *n s.* Jod(VII)-fluorid
Jod(I)-hydroxid *n* IOH iodine hydroxide
Jodid *n* M^II iodide
jodieren to iodize, to iodinate
Jodierung *f* iodization, iodination
Jodkaliumstärkepapier *n* potassium-iodide-starch paper
Jodkohle *f* iodized active carbon
Jodlösung *f* iodine solution
  Lugolsche J. Lugol's solution
Jodmethan *n* CH_3I iodomethane, methyl iodide
Jodmethode *f* iodine method
Jodmethyl *n s.* Jodmethan
Jodmono ... *s.* Jod(I) ...
Jodoaurat(III) *n* $M^I[AuI_4]$ iodoaurate(III), tetraiodoaurate(III)
Jodobromit *m (min)* iodobromite *(silver bromide chloride iodide)*
Jodoform *n* CHI_3 iodoform, tri-iodomethane
Jodoformprobe *f* iodoform test
Jodomerkurat(II) *n* $M_2^I[HgI_4]$ iodomercurate(II), tetraiodomercurate(II)
Jodometrie *f* iodometry, iodimetry
jodometrisch iodometric, iodimetric
Jodoniumverbindung *f* iodonium compound
Jodoplatinsäure *f* $H_2[PtI_6]$ iodoplatinic acid, hexaiodoplatinic(IV) acid
Jod(V)-oxid *n* I_2O_5 iodine(V) oxide, diiodine pentaoxide
Jodpentafluorid *n s.* Jod(V)-fluorid
Jodpentoxid *n s.* Jod(V)-oxid
Jodsäure *f* HIO_3 iodic acid
Jodsilber *n (phot)* silver iodide
Jodspeisesalz *n* iodized salt
Jodstärkepapier *n* starch iodide paper
Jodtinktur *f* tincture of iodine
Jodtri ... *s.* Jod(III)- ...
Jodwasserstoffgleichgewicht *n* hydrogen-iodide equilibrium
Jodwasserstoffsäure *f* HI hydroiodic acid
Jodzahl *f* iodine [absorption] value, I. V., iodine (Huebl) number

rhodanometrische J. thiocyanogen number (value)
Jodzyanid *n* ICN iodine cyanide, cyanogen iodide
Johannisbrotgummi *n* carob (carob-seed, locust bean) gum, caroban *(from Ceratonia siliqua L.)*
Johannit *m (min)* johannite, uranvitriol
Jordan-Kegel[stoff]mühle *f*, **Jordan-Mühle** *f (pap)* Jordan engine (mill, refiner), jordan, refining (perfecting) engine
José-Papier *n* lens paper (tissue) *(for wiping optical lenses)*
Joule-Thomson-Effekt *m* Joule-Thomson effect
differentieller J. differential Joule-Thomson effect
Joule-Thomson-Koeffizient *m* Joule-Thomson coefficient
J-Säure *f* $NH_2C_{10}H_5(OH)SO_3H$ J acid, 2-amino-5-naphthol-7-sulphonic acid
J-Säure-Harnstoff *m* J acid urea
Juchtenleder *n* Russian leather
Juchtenöl *n (tann)* birch bark oil
Judäakaroben *fpl (dye)* carob (turpentine) galls *(from Pistacia terebinthus L.)*
Juglon *n* juglone, 5-hydroxy-1,4-naphthaquinone
Jungbier *n* new beer
Jungfernkautschuk *m* [caucho] virgin rubber *(from Sapium thomsoni God.)*
Jungfernöl *n* virgin [olive] oil *(obtained from the first light pressing in the cold)*
Jungfernquecksilber *n* native mercury
Jungfustik *m (dye)* young fustic, fustet *(from Cotinus coggygria Scop.)*
Junghopfen *m* green hop
Jungkräusen *pl (ferm)* low krausen
Jungwein *m* young (new) wine
Juniperinsäure *f* $HOCH_2[CH_2]_{14}COOH$ juniperic acid, 16-hydroxyhexadecanoic acid
Justage *f* rectification *(of an instrument)*
justieren to rectify *(instruments)*
justiert / auf Auslauf calibrated for delivery
auf Einguß j. calibrated for content
Justierung *f* rectification *(of an instrument)*
Jute *f* jute *(from Corchorus specc.)*
Jutedichtung *f* jute packing
Jutefaser *f* jute fibre
Jutepackung *f* jute packing
Juwelierborax *m* $Na_2B_4O_7 \cdot 5H_2O$ jeweller's borax, octahedral borax *(sodium tetraborate-5-water)*
JZ *s.* Jodzahl

K

KA *s.* Kaseinfaserstoff
Kabelbohranlage *f (petrol)* cable-tool installation (rig)
Kabelbohren *n (petrol)* cable-tool drilling
Kabelbohrgerät *n s.* Kabelbohranlage
Kabelbohrverfahren *n (petrol)* cable-tool method
Kabelisolieröl *n* cable oil
Kabelisolierpapier *n* cable paper
Kabelmantel *m* cable coating
Kabelöl *n* cable oil
Kabelpapier *n* cable paper
Kabelüberzug *m* cable coating
Kachel *f (ceram)* tile
trockengepreßte K. dust-pressed tile
Kachelpresse *f* pot press *(as for expression of oilseeds)*
Kadaverin *n* $NH_2CH_2[CH_2]_3CH_2NH_2$ cadaverine, pentamethylenediamine, 1,5-diaminopentane
Kadavermehl *n* animal (garbage) tankage
Kadinen *n* cadinene *(a bicyclic sesquiterpene)*
Kadinol *n* cadinol *(a bicyclic sesquiterpene alcohol)*
Kadmium *n* Cd cadmium
Kadmiumarsenid *n* Cd_3As_2 cadmium arsenide
Kadmiumazetat *n* $Cd(CH_3COO)_2$ cadmium acetate
Kadmiumbromat *n* $Cd(BrO_3)_2$ cadmium bromate
Kadmiumbromid *n* $CdBr_2$ cadmium bromide
Kadmiumchlorat *n* $Cd(ClO_3)_2$ cadmium chlorate
Kadmiumchlorid *n* $CdCl_2$ cadmium chloride
Kadmiumdiphosphat *n* $Cd_2P_2O_7$ cadmium diphosphate, cadmium pyrophosphate
Kadmiumdithionat *n* CdS_2O_6 cadmium dithionate
Kadmiumelektrode *f* cadmium electrode
Kadmiumfluorid *n* CdF_2 cadmium fluoride
Kadmiumgelb *n* cadmium (aurora, orient) yellow *(cadmium sulphide)*
kadmiumhaltig cadmium-bearing
Kadmiumhexazyanoferrat(II) *n* $Cd_2[Fe(CN)_6]$ cadmium hexacyanoferrate(II)
Kadmiumhydroxid *n* $Cd(OH)_2$ cadmium hydroxide
Kadmiumjodat *n* $Cd(IO_3)_2$ cadmium iodate
Kadmiumjodid *n* CdI_2 cadmium iodide
Kadmiumkarbonat *n* $CdCO_3$ cadmium carbonate
Kadmiummanganat(VII) *n* $Cd(MnO_4)_2$ cadmium manganate(VII)
Kadmiummetasilikat *n s.* Kadmiumsilikat
Kadmium-Nickel-Sammler *m* nickel-cadmium accumulator (cell)
Kadmiumnitrat *n* $Cd(NO_3)_2$ cadmium nitrate
Kadmiumorthophosphat *n* $Cd_3(PO_4)_2$ cadmium orthophosphate
Kadmium(II)-oxid *n* CdO cadmium oxide
Kadmiumpermanganat *n s.* Kadmiummanganat(VII)
Kadmiumpyrophosphat *n s.* Kadmiumdiphosphat
Kadmiumrot *n* cadmium red
Kadmiumselenat *n* $CdSeO_4$ cadmium selenate
Kadmiumselenid *n* CdSe cadmium selenide
Kadmiumsilikat *n* $CdSiO_3$ cadmium silicate, cadmium metasilicate, cadmium trioxosilicate
Kadmiumsulfat *n* $CdSO_4$ cadmium sulphate
Kadmiumsulfid *n* CdS cadmium sulphide
Kadmiumsulfit *n* $CdSO_3$ cadmium sulphite
Kadmiumtellurid *n* CdTe cadmium telluride
Kadmiumwolframat *n* $CdWO_4$ cadmium wolframate, cadmium tungstate
Kadmiumwolframatoborat *n* cadmium wolframoborate, cadmium tungstoborate

Kadmiumzyanid *n* $Cd(CN)_2$ cadmium cyanide
Kaffeebohne *f* coffee bean (nib)
Kaffee-Extrakt *m*, **Kaffee-Extraktpulver** *n* soluble coffee
Kaffeesahne *f* coffee cream
Kaffeesäure *f* $(HO)_2C_6H_3CH=CHCOOH$ caffeic acid, 3,4-dihydroxycinnamic acid
Kaffeesäureester *m* **der Chinasäure** caffeoylquinic acid, chlorogenic acid
Kaffeesäure-3-methyläther *m* caffeic acid 3-methyl ether, 3-hydroxy-3-methoxycinnamic acid, ferulic acid
Kaffeesäure-4-methyläther *m* caffeic acid 4-methyl ether, hesperetic acid
Kaffein *n* caffeine, 1,3,7-trimethylxanthine
Kaffeinsäure *f s.* Kaffeesäure
Käfig *m (cryst)* cage
Käfig[einschluß]verbindung *f* cage (clathrate) compound, clathrate [inclusion compound]
Käfigwand *f (cryst)* cage wall
Kahlappretur *f (text)* pileless finish
Kahm *m*, **Kahmhaut** *f* beeswing *(a film on liquids containing nutrients)*
Kahmhefe *f* scum yeast
kahmig ropy *(wine)*
Kahmpilz *m* scum yeast
Kainit *m (min)* kainit[e] *(magnesium potassium chloride sulphate)*
Kaisergrün *n s.* Grün / Schweinfurter
Kajeputöl *n (pharm)* cajeput oil *(from Melaleuca leucadendron L.)*
Kakao *m* cocoa, chocolate
Kakaobaum *m* cacao [tree], Theobroma cacao L.
Kakaobohne *f* cacao (cocoa) bean *(from Theobroma cacao L.)*
Kakaobruch *m* crushed cocoa
Kakaobutter *f*, **Kakaofett** *n*, **Kakaoöl** *n* cocoa (cacao) butter (oil)
Kakaopulver *n* cocoa [powder]
Kakaoschalen *fpl* cacao (cocoa) shells
Kakodyl *n* $(CH_3)_2AsAs(CH_3)_2$ cacodyl, tetramethyldiarsine
Kakodylchlorid *n* $(CH_3)_2AsCl$ cacodyl chloride, chlorodimethylarsine
Kakodyloxid *n* $[As(CH_3)_2]_2O$ cacodyl oxide
Kakodylsäure *f* $(CH_3)_2AsOOH$ cacodylic acid, dimethylarsinic acid
Kakoxen *m (min)* cacoxene, cacoxenite *(iron(III) trihydroxide orthophosphate)*
Kalabarbohne *f* calabar bean, Physostigma venenosum Balf.
Kalait *m s.* Kallait
Kalander *m* calender [machine]
K. in Tandemanordnung *(rubber)* tandem calender
K. mit Walzenschränkung *(rubber)* swivel-roll (crossed-axes) machine
K. zum Belegen von Geweben *(rubber)* coating (skim-coating, skimming) calender
K. zum Ziehen von Platten *(rubber)* sheeting calender
Kalandereffekt *m (rubber)* calender grain
Kalanderfärbung *f (pap)* calender staining (colouring), padding, stuffing
Kalanderführer *m* calender operator
Kalanderleimung *f (pap)* calender sizing
kalandern to calender
Kalanderplatte *f (rubber)* calendered sheet
Kalandersaal *m* calender department
Kalandersatz *m* calender stack
Kalanderschrumpfung *f (rubber)* calender shrinkage
Kalanderständer *m* calender frame
Kalanderwalze *f* calender roll (bowl)
Kalanderwalzenpapier *n* calender roll (bowl) paper, woollen paper
Kalanderwalzensatz *m* calender stack
kalandrieren *s.* kalandern
Kalbslederpapier *n* calf paper
Kaldo-Verfahren *n (met)* Kaldo process
Kalebassenalkaloid *n s.* Kalebassenkurare-Alkaloid
Kalebassenkurare *n* calabash (gourd) curare
Kalebassenkurare-Alkaloid *n* calabash-curare alkaloid
Kaledonischbraun *n* umber *(a naturally occurring brown pigment)*
Kaledonit *m (min)* caledonite *(a basic copper lead carbonate sulphate)*
Kali *n s.* 1. Kalisalz; 2. Kalidüngemittel; 3. Reinkali; 4. Kalium
schwefelsaures K. sulphate of potash *(a fertilizer chiefly consisting of K_2SO_4)*
Kaliabwasser *n* potash mine wastes
Kalialaun *m s.* Kalinit
Kaliapparat *m (lab)* potash bulb, alkalimeter *(for determining carbon dioxide)*
K. nach Geißler (Mohr) Geissler-Mohr absorption (potash) bulb
kalibrieren to calibrate *(measuring apparatus)*; to size *(e.g. tubing)*
Kalibriermaschine *f* calibrating machine
Kalibrierung *f* calibration *(of measuring apparatus)*; sizing *(as of tubing)*
Kalidüngemittel *n*, **Kalidünger** *m* potash (potassic) fertilizer
Kalifeldspat *m (min)* potash feldspar
Kalifornium *n* Cf californium
Kaliglas *n* potash glass
Kaliglimmer *m (min)* potash (potassium) mica, muscovite *(a phyllosilicate)*
kalihaltig potassiferous, potassic
Kaliindustrie *f* potash industry
Kalilager *n* potash deposit
Kalilauge *f* potash lye, caustic potash solution, potassium hydroxide solution
alkoholische K. alcoholic potash
Kalimagnesia *f* [single sulphate of] potash magnesia *(a fertilizer consisting of schoenite or leonite)*
Kalimangel *m (agric)* potassium shortage (deficiency)
Kalinit *m (min)* kalinite *(aluminium potassium sulphate-12-water)*

kalireich high-potassium
Kalirohsalz *n (fert)* potash ore, mine-run salt
Kalisalpeter *m* saltpetre *(proper)*, nitrate of potash *(potassium nitrate)*
Kalisalz *n* potash [salt], potassiferous salt
Kalisalzlager *n*, **Kalisalzlagerstätte** *f* potash deposit
Kalischmelze *f* potash fusion (melt)
Kaliseife *f* potassium soap, potash [soft] soap
Kalium *n* K potassium
Kaliumalaun *m* $KAl(SO_4)_2 \cdot 12H_2O$ potassium (potash) alum
Kaliumaluminat *n* $KAlO_2 \cdot 3H_2O$ potassium aluminate
Kaliumaluminiumalaun *m s.* Kaliumalaun
Kaliumaluminiumsulfat *n* $KAl(SO_4)_2$ aluminium potassium sulphate
Kaliumalumosilikat *n* $K[AlSi_3O_8]$ potassium aluminosilicate
Kaliumamalgam *n* potassium amalgam
Kaliumamid *n* KNH_2 potassium amide
Kaliumammoniumtartrat *n* $NH_4OOC[CHOH]_2COOK$ ammonium potassium tartrate
Kaliumantimonotartrat *n* $K[C_4H_2O_6Sb(OH_2)]$ potassium antimonotartrate
Kaliumantimonyltartrat *n s.* Kaliumantimonotartrat
Kaliumarsenat(III) *n s.* Kaliumorthoarsenat(III)
Kaliumäthylxanthogenat *n* C_2H_5OCSSK potassium ethyl xanthate
Kaliumaurat(III) *n* $KAuO_2$ potassium aurate(III)
Kaliumazetat *n* CH_3COOK potassium acetate
Kaliumazid *n* KN_3 potassium azide
Kaliumborotartrat *n* potassium borotartrate
Kaliumbromat *n* $KBrO_3$ potassium bromate
Kaliumbromid *n* KBr potassium bromide
Kaliumchlorat *n* $KClO_3$ potassium chlorate
Kaliumchlorid *n* KCl potassium chloride; *(in fertilizer analyses)* muriate
Kaliumchlorit *n* $KClO_2$ potassium chlorite
Kaliumchromalaun *m* $KCr(SO_4)_2 \cdot 12H_2O$ chrome [potash] alum, potassium chrome alum, common chrome alum
Kaliumchromat *n* K_2CrO_4 potassium chromate
Kaliumdichromat *n* $K_2Cr_2O_7$ potassium dichromate
Kaliumdihydrogenorthoarsenat *n* KH_2AsO_4 potassium dihydrogenorthoarsenate, potassium dihydrogenarsenate
Kaliumdihydrogenorthophosphat *n* KH_2PO_4 potassium dihydrogenorthophosphate, potassium dihydrogenphosphate
Kaliumdihydrogenphosphat *n s.* Kaliumdihydrogenorthophosphat
Kaliumdiphosphat *n* $K_4P_2O_7$ potassium diphosphate, potassium pyrophosphate
Kaliumdisulfat *n* $K_2S_2O_7$ potassium disulphate
Kaliumdisulfit *n* $K_2S_2O_5$ potassium disulphite
Kaliumdizyanoargentat *n* $K[Ag(CN)_2]$ potassium dicyanoargentate
Kaliumdizyanoaurat(I) *n* $K[Au(CN)_2]$ potassium dicyanoaurate(I)

Kaliumdodekawolframtosilikat *n* $K_4[SiW_{12}O_{40}]$ potassium dodecawolframosilicate, potassium dodecatungstosilicate
Kaliumeisen(III)-chlorid *n* $2KCl \cdot FeCl_3$ iron(III) potassium chloride, ferric potassium chloride
Kaliumeisen(III)-oxalat *n* $KFe(C_2O_4)_2$ iron(III) potassium oxalate, ferric potassium oxalate
Kaliumeisen(III)-sulfat-12-Wasser *n* $KFe(SO_4)_2 \cdot 12H_2O$ iron(III) potassium sulphate-12-water, ferric potassium sulphate-12-water, iron potassium alum
Kaliumfluorid *n* KF potassium fluoride
Kaliumfluoroberyllat *n* $K_2[BeF_4]$ potassium fluoroberyllate
Kaliumformiat *n* HCOOK potassium formate
Kaliumgallium(III)-sulfat *n* $KGa(SO_4)_2$ gallium potassium sulphate
kaliumhaltig potassium-containing, potassic
Kaliumhexabromoplatinat(IV) *n* $K_2[PtBr_6]$ potassium hexabromoplatinate(IV)
Kaliumhexachloroiridat(IV) *n* $K_2[IrCl_6]$ potassium hexachloroiridate(IV)
Kaliumhexachloroosmat(III) *n* $K_3[OsCl_6]$ potassium hexachloroosmate(III)
Kaliumhexachloroosmat(IV) *n* $K_2[OsCl_6]$ potassium hexachloroosmate(IV)
Kaliumhexachloropalladat(IV) *n* $K_2[PdCl_6]$ potassium hexachloropalladate(IV)
Kaliumhexachloroplatinat(IV) *n* $K_2[PtCl_6]$ potassium hexachloroplatinate(IV)
Kaliumhexafluorosilikat *n* $K_2[SiF_6]$ potassium hexafluorosilicate
Kaliumhexafluorotitanat(IV) *n* $K_2[TiF_6]$ potassium hexafluorotitanate(IV)
Kaliumhexafluorozirkonat(IV) *n* $K_2[ZrF_6]$ potassium hexafluorozirconate(IV)
Kaliumhexajodoplatinat(IV) *n* $K_2[PtI_6]$ potassium hexaiodoplatinate(IV)
Kaliumhexanitrokobaltat(III) *n* $K_3[Co(NO_2)_6]$ potassium hexanitrocobaltate(III)
Kaliumhexazyanoferrat(II) *n* $K_4[Fe(CN)_6]$ potassium hexacyanoferrate(II), yellow prussiate of potash, yellow potassium prussiate
Kaliumhexazyanoferrat(III) *n* $K_3[Fe(CN)_6]$ potassium hexacyanoferrate(III), red prussiate of potash, red potassium prussiate
Kaliumhexazyanokobaltat(II) *n* $K_4[Co(CN)_6]$ potassium hexacyanocobaltate(II)
Kaliumhexylxanthogenat *n* $C_6H_{13}OCSSK$ potassium hexylxanthate
Kaliumhydrid *n* KH potassium hydride
Kaliumhydrogenarsenat(V) *n s.* Kaliumhydrogenorthoarsenat(V)
Kaliumhydrogenfluorid *n* KHF_2 potassium hydrogenfluoride
Kaliumhydrogenkarbonat *n* $KHCO_3$ potassium hydrogencarbonate
Kaliumhydrogenorthoarsenat(V) *n* K_2HAsO_4 potassium hydrogenorthoarsenate, potassium hydrogenarsenate
Kaliumhydrogenorthophosphat *n* K_2HPO_4 potassium hydrogenorthophosphate, potassium hydrogenphosphate

Kaliumhydrogenoxalat *n* HOOC–COOK potassium hydrogenoxalate

Kaliumhydrogenphosphat *n s.* Kaliumhydrogenorthophosphat

Kaliumhydrogenphthalat *n* $C_6H_4(COOH)(COOK)$ potassium hydrogenphthalate

Kaliumhydrogensulfat *n* $KHSO_4$ potassium hydrogensulphate

Kaliumhydrogensulfid *n* KSH potassium hydrogensulphide

Kaliumhydrogensulfit *n* $KHSO_3$ potassium hydrogensulphite

Kaliumhydrogentartrat *n* $HOOC[CHOH]_2COOK$ potassium hydrogentartrate

Kaliumhydroxid *n* KOH potassium hydroxide

Kaliumhydroxidlösung *f* potassium hydroxide solution, potash lye, caustic potash solution

Kaliumhydroxidschmelze *f* 1. potassium hydroxide fusion, molten potassium hydroxide; 2. *(act)* potassium hydroxide fusion

alkoholische K. fusion with alcoholic potassium hydroxide

Kaliumhyperoxid *n* KO_2 potassium hyperoxide, potassium superoxide

Kaliumhypochlorit *n* KOCl potassium hypochlorite

Kaliumjodat *n* KIO_3 potassium iodate

Kaliumjodatstärkepapier *n* potassium-iodate-starch paper

Kaliumjodid *n* KI potassium iodide

Kaliumjodidstärkeindikator *m* potassium-iodide-starch indicator

Kaliumjodidstärkepapier *n* potassium-iodide-starch paper

Kaliumkalziumsulfat *n* $K_2Ca(SO_4)_2$ calcium potassium sulphate

Kaliumkarbonat *n* K_2CO_3 potassium carbonate

Kaliumkarbonyl *n* $K_6(CO)_6$ potassium carbonyl

Kaliumkobalt(II)-sulfat *n* $K_2SO_4 \cdot CoSO_4$ cobalt(II) potassium sulphate

Kaliumlaktat *n* $CH_3CH(OH)COOK$ potassium lactate

Kaliumlinie *f* potassium line

Kaliummagnesiumsulfat *n* $K_2SO_4 \cdot MgSO_4$ magnesium potassium sulphate

Kaliummanganat *n* K_2MnO_4 potassium manganate

Kaliummanganat(VII) *n s.* Kaliumpermanganat

Kaliummetaborat *n* KBO_2 potassium metaborate

Kaliummetaperjodat *n* KIO_4 potassium metaperiodate, potassium periodate, potassium tetraoxoiodate(VII)

Kaliummetaphosphat *n* $(KPO_3)_n$ potassium metaphosphate

Kaliummetasilikat *n* K_2SiO_3 potassium metasilicate, potassium silicate, potassium trioxosilicate

Kaliummethoxid *n s.* Kaliummethylat

Kaliummethylat *n* CH_3OK potassium methylate

Kaliummonosulfid *n* K_2S potassium monosulphide, potassium sulphide

Kaliumnatriumhexanitrokobaltat(III) *n* $K_2Na[Co(NO_2)_6]$ potassium sodium hexanitrocobaltate(III)

Kaliumnatriumkarbonat *n* $KNaCO_3$ potassium sodium carbonate

Kaliumnatriumtartrat *n* $KOOC[CHOH]_2COONa$ potassium sodium tartrate

Kaliumnitrat *n* KNO_3 potassium nitrate

Kaliumnitrid *n* K_3N potassium nitride

Kaliumnitrit *n* KNO_2 potassium nitrite

Kaliumnitroprussiat *n s.* Kaliumnitroprussid

Kaliumnitroprussid *n* $K_2[Fe(CN)_5(NO)]$ potassium nitroprusside, dipotassium pentacyanonitrosylferrate(II)

Kaliumoleat *n* potassium oleate, *(specif)* potassium 9-octadecenoate

Kaliumorthoarsenat(III) *n* K_3AsO_3 potassium orthoarsenite

Kaliumorthophosphat *n* K_3PO_4 potassium orthophosphate

Kaliumosmat(VI) *n* K_2OsO_4 potassium osmate(VI)

Kaliumoxalat *n* $K_2C_2O_4$ potassium oxalate

Kaliumoxid *n* K_2O potassium oxide

Kaliumpalmitat *n* $C_{15}H_{31}COOK$ potassium palmitate

Kaliumparawolframat *n* potassium parawolframate, potassium paratungstate

Kaliumpentachloroamminplatinat(IV) *n* $K[Pt(NH_3)Cl_5]$ potassium amminopentachloroplatinate(IV)

Kaliumpentasulfid *n* K_2S_5 potassium pentasulphide

Kaliumpentathionat *n* $K_2S_5O_6$ potassium pentathionate

Kaliumperchlorat *n* $KClO_4$ potassium perchlorate

Kaliumpermanganat *n* $KMnO_4$ potassium permanganate

Kaliumperoxid *n* K_2O_2 potassium peroxide

Kaliumperoxoborat *n* KBO_3 potassium peroxoborate

Kaliumperoxochromat *n* K_3CrO_8 potassium peroxochromate

Kaliumperoxodikarbonat *n* $K_2C_2O_6$ potassium peroxodicarbonate

Kaliumperoxodisulfat *n* $K_2S_2O_8$ potassium peroxodisulphate, potassium peroxosulphate

Kaliumperoxokarbonat *n s.* Kaliumperoxodikarbonat

Kaliumperrhenat *n* $KReO_4$ potassium perrhenate

Kaliumphenolat *n* C_6H_5OK potassium phenate, potassium phenoxide

Kaliumphosphat *n* potassium phosphate; *(specif) s.* Kaliumorthophosphat

Kaliumpyrophosphat *n s.* Kaliumdiphosphat

Kaliumpyrosulfit *n s.* Kaliumdisulfit

Kaliumquecksilberjodid *n* K_2HgI_4 mercury potassium iodide

Kaliumrhodanid *n s.* Kaliumthiozyanat

Kaliumseife *f s.* Kaliseife

Kaliumselenat *n* K_2SeO_4 potassium selenate

Kaliumselenid *n* K_2Se potassium selenide

Kaliumselenit *n* K_2SeO_3 potassium selenite

Kaliumselenozyanat *n* KSeCN potassium selenocyanate

Kaliumsilbernitrat *n* $KNO_3 \cdot AgNO_3$ potassium silver nitrate

Kaliumstearat *n* $C_{17}H_{35}COOK$ potassium stearate
Kaliumsulfat *n* K_2SO_4 potassium sulphate
Kaliumsulfid *n* potassium sulphide; *(specif)* s. Kaliummonosulfid
Kaliumsulfit *n* K_2SO_3 potassium sulphite
Kaliumtartrat *n* $KOOC[CHOH]_2COOK$ potassium tartrate
Kaliumtetraborat *n* $K_2B_4O_7$ potassium tetraborate
Kaliumtetrabromoaurat(III) *n* $K[AuBr_4]$ potassium tetrabromoaurate(III)
Kaliumtetrabromoplatinat(II) *n* $K_2[PtBr_4]$ potassium tetrabromoplatinate(II)
Kaliumtetrachloroaurat(III) *n* $K[AuCl_4]$ potassium tetrachloroaurate(III)
Kaliumtetrachloropalladat(II) *n* $K_2[PdCl_4]$ potassium tetrachloropalladate(II)
Kaliumtetrachloroplatinat(II) *n* $K_2[PtCl_4]$ potassium tetrachloroplatinate(II)
Kaliumtetrafluoroborat *n* $K[BF_4]$ potassium tetrafluoroborate
Kaliumtetraoxalat *n* $KHC_2O_4 \cdot H_2C_2O_4$ potassium tetraoxalate
Kaliumtetrasilikat *n* $K_2Si_4O_9$ potassium tetrasilicate
Kaliumtetrasulfid *n* K_2S_4 potassium tetrasulphide
Kaliumtetrathionat *n* $K_2S_4O_6$ potassium tetrathionate
Kaliumtetrazyanoaurat(III) *n* $K[Au(CN)_4]$ potassium tetracyanoaurate(III)
Kaliumtetrazyanomerkurat(II) *n* $K_2[Hg(CN)_4]$ potassium tetracyanomercurate(II)
Kaliumtetrazyanoniccolat(II) *n* $K_2[Ni(CN)_4]$ potassium tetracyanoniccolate(II)
Kaliumtetrazyanoplatinat(II) *n* $K_2[Pt(CN)_4]$ potassium tetracyanoplatinate(II)
Kaliumtetrazyanozinkat *n* $K_2[Zn(CN)_4]$ potassium tetracyanozincate
Kaliumtetroxojodat *n* s. Kaliummetaperjodat
Kaliumtetroxorhenat(VII) *n* s. Kaliumperrhenat
Kaliumthioarsenat(III) *n* K_3AsS_3 potassium thioarsenite
Kaliumthioarsenat(V) *n* K_3AsS_4 potassium thioarsenate
Kaliumthiokarbonat *n* K_2CS_3 potassium thiocarbonate, potassium trithiocarbonate
Kaliumthiosulfat *n* $K_2S_2O_3$ potassium thiosulphate
Kaliumthiozyanat *n* KSCN potassium thiocyanate, potassium rhodanide
Kaliumthiozyanatpapier *n* potassium thiocyanate paper
Kaliumtrioxosilikat *n* s. Kaliummetasilikat
Kaliumtrioxostannat(IV) *n* K_2SnO_3 potassium trioxostannate(IV)
Kaliumwolframat *n* K_2WO_4 potassium wolframate, potassium tungstate
Kaliumxanthogenat *n* potassium xanthate, *(specif)* C_2H_5OCSSK potassium ethylxanthate
Kaliumzitrat *n* $KOOCC(OH)(CH_2COOK)_2$ potassium citrate
Kaliumzyanat *n* KOCN potassium cyanate
Kaliumzyanid *n* KCN potassium cyanide
Kaliwasserglas *n* potassium (potash) water glass, soluble water (potash) glass
Kalk *m* lime ✦ **aus dem K. falzen** *(tann)* to green-shave ✦ **mit K. behandeln** to lime out *(manufacture of organic intermediates)* ✦ **mit K. düngen** *(agric)* to lime
an der Luft erhärtender K. non-hydraulic lime
durch feuchte Luft gelöschter K. air-slaked lime
essigsaurer K. vinegar salt
gebrannter K. burnt (burned, caustic) lime, quicklime *(calcium oxide)*
gelöschter K. hydrated (slaked, water-slaked) lime, slacklime, lime hydrate *(calcium hydroxide)*; *(agric)* agricultural hydrate
hochhydraulischer K. Roman (Parker's) cement
hydraulischer K. hydraulic (water) lime *(a lime which will harden under water)*
ungelöschter K. unslaked lime, quicklime *(calcium oxide)*
Wiener K. Vienna lime *(pulverized dolomite)*
Kalkalkaligestein *n* calc-alkali[c] rock
Kalkammoniak *n* kalkammon *(a fertilizer)*
Kalkammonsalpeter *m* nitrochalk
Kalkanreicherungshorizont *m* *(soil)* lime accumulation horizon, lime pan
kalkarm *(soil)* deficient in lime, sour
kalkartig limy, calcareous
Kalkäscher *m* 1. *(tann)* liming; lime pit; lime liquor; 2. *(text)* liming, lime boil
reiner K. *(tann)* straight lime liquor
Kalkäscherwolle *f* *(text)* slipe wool
Kalkbad *n* lime bath
Kalkbedarf *m* lime requirements
Kalkbeton *m* lime concrete
Kalkbeuche *f* *(text)* lime boil, liming
Kalkbilanz *f* *(agric)* lime balance
Kalkblau *n* copper blue, blue verditer, Bremen blue *(copper(II) hydroxide)*
Kalkboden *m* limy (calcareous) soil
Kalkbrei *m* *(tann)* lime cream (paint)
Kalkbrennen *n* lime burning
Kalkbrennofen *m* lime kiln
Kalkbrühe *f* *(agric)* lime wash, whitewash
Kalkchlorose *f* *(agric)* lime-induced chlorosis
Kalkchromgelb *n* gelbin *(calcium chromate)*
Kalkchromgranat *m* *(min)* lime-chrome garnet, calcium-chromium garnet
Kalkdüngemittel *n*, **Kalkdünger** *m* lime [fertilizer], *(specif)* agricultural limestone, agstone
kalkecht *(dye, coat)* fast to lime
Kalkechtheit *f* *(dye, coat)* fastness to lime
kalken 1. *(sugar, agric)* to lime; 2. to lime out *(manufacture of organic intermediates)*; 3. to whitewash *(e.g. walls)*
kalkfalzen *(tann)* to green-shave
Kalkfeldspat *m* *(min)* lime feldspar
Kalkflecken *mpl* 1. *(tann)* lime blasts (specks), blasting *(a result of incorrect liming)*; 2. *(phot)* drying marks
kalkfliehend calciphobous *(plant)*
Kalkgestein *n* calcareous rock
Kalkglas *n* lime glass

kalkhaltig limy, calcareous
Kalkhärte *f* calcium hardness *(of water)*
Kalkhydrat *n s.* Kalk / gelöschter
kalkig limy, calcareous
Kalk-Kohlensäure-Verfahren *n* *(sugar)* lime-carbon-dioxide process, carbonation process, defecosaturation, defecocarbonation
Kalkkruste *f* *(geol)* calcareous crust, caliche
Kalklicht *n* / **Drummondsches** Drummond's limelight
kalkliebend calciphilous *(plant)*
Kalklöschapparat *m* lime slaker
Kalklöschen *n* lime slaking (hydrating)
Kalklöscher *m* lime slaker
Kalklöschturm *m* slaking tower
kalkmeidend lime-intolerant *(plant)*
Kalkmergel *m* lime marl
Kalkmilch *f* milk of lime, lime milk; *(tann)* cream of lime; *(agric)* lime wash, whitewash *(a suspension of calcium hydroxide or hydrated lime in water)*
Kalkmilchscheidung *f* *(sugar)* defecation with milk of lime, wet liming
Kalkmilchsystem *n* *(pap)* milk-of-lime system
Kalkmörtel *m* lime mortar
Kalkmudde *f* *(geol)* calcareous mud
Kalknatronglas *n* soda-lime glass
Kalkofen *m* lime kiln
Kalkputz *m* lime plaster
kalkreich rich in lime, high-lime
Kalkringofen *m* lime ring furnace *(for lime burning)*
Kalksaccharat *n* lime saccharate
Kalksaccharatverfahren *n* **nach Steffen** Steffen [separation] process *(for recovering sugar from beet molasses)*
Kalksalpeter *m* $Ca(NO_3)_2$ nitrate of lime, lime saltpetre, calcium nitrate; *(min)* lime saltpetre, kalksaltpetre, nitrocalcite
Kalksandstein *m* sandy (psammitic) limestone
Kalkschachtofen *m* lime tunnel furnace
Kalkschatten *mpl* *(tann)* lime blasts (specks), blasting *(a result of incorrect liming)*
Kalkscheidung *f* *(sugar)* lime defecation
Kalkschlamm *m* 1. lime mud (sludge); 2. *(pap)* carbonate sludge, paper-mill sludge
Kalkschlammwäscher *m* lime mud washer
Kalkschwefelleber *f* liver of lime, sulphurated lime *(a mixture of calcium sulphides and calcium sulphate)*
Kalkschwefelnatriumäscher *m* *(tann)* sharpened lime *(cream of lime treated with sodium sulphide)*
Kalkseife *f* lime soap
Kalksilt *m* *(geol)* calcareous mud
Kalksinter *m* *(geol)* calcareous sinter
Kalk-Soda-Verfahren *n* lime-soda [softening] method *(water treatment)*
Kalkspat *m s.* Kalzit
Kalkstein *m* limestone
  dolomitischer K. dolomitic (high-magnesium) limestone
Kalksteineinlauf *m* *(pap)* limestone charging
Kalksteinfüllung *f* *(pap)* limestone packing
Kalkstickstoff *m* nitrolime, lime nitrogen *(calcium cyanamide)*
Kalkstickstoffverfahren *n* cyanamide process *(for producing ammonia)*
Kalktrichterofen *m* lime funnel furnace
Kalktuff *m* calcareous tufa, tufaceous limestone, tufa
Kalktünche *f* limewash
Kalküberschußverfahren *n* excess-lime process *(water conditioning)*
Kalkung *f* 1. *(sugar)* liming, defecation; 2. *(agric)* liming
Kalkversorgungsgrad *m* *(agric)* lime status
Kalkwasser *n* lime water *(an alkaline aqueous solution of calcium hydroxide)*
Kalkwasserpumpe *f* lime feeder *(water treatment)*
Kallait *m* *(min)* kalaite, turquois[e] *(a hydrous basic aluminium copper phosphate)*
Kallidin *n* kallidin *(a decapeptide)*
Kallikrein *n* kallikrein *(a pancreatic enzyme)*
Kalling-Domnarvet-Verfahren *n* *(met)* Kaldo process
Kallitypieverfahren *n* kallitype *(reprography)*
Kalmus *m* flag root, sweet flag *(Acorus calamus L.)*
Kalomel *m* *(min)* calomel, horn mercury *(mercury(I) chloride)*
Kalomel *n s.* Quecksilber(I)-chlorid
Kalomelelektrode *f* calomel electrode
Kalomelnormalelektrode *f* normal calomel electrode
Kalorie *f* calorie
  thermochemische K. thermochemical calorie
Kaloriengehalt *m* calorie content
Kalorienwert *m* *(food)* calorific value, c.v.
Kalorimeter *n* calorimeter
  K. für Gase gas calorimeter
  K. von Nernst Nernst calorimeter
  adiabatisches K. adiabatic calorimeter
  isothermes K. isothermal calorimeter
Kalorimeterbombe *f* calorimeter (calorimetric, explosion) bomb, bomb calorimeter
Kalorimeterflüssigkeit *f* calorimetric liquid (fluid)
Kalorimetergefäß *n* calorimeter (calorimetric) vessel
Kalorimeterschälchen *n* calorimeter fusion cup
Kalorimetrie *f* calorimetry
kalorimetrisch calorimetric[al]
kalorisch caloric
kalorisieren to calorize *(steel or cast iron)*
Kalotypie *f* *(phot)* calotype process
kaltabbindend cold-curing, cold-setting
Kaltalkalisierung *f* *(pap)* cold [alkali] refining
Kaltansatzlack *m* cold-cut varnish
Kaltasphalt *m* cold asphalt
kaltblasen to steam *(in producing water gas)*
Kaltblasen *n* steaming, run *(in producing water gas)*
Kaltbleiche *f* *(pap)* cold bleach[ing]
Kaltdampfverdichteranlage *f* vapour-compression system

Kälte *f* cold[ness]
Kälteanlage *f* refrigerating (cooling) plant
Kältebad *n* cold bath *(of a freeze dryer)*
kältebeständig cold-resistant, resistant to cold
Kältebeständigkeit *f* resistance to cold, cold (low-temperature) resistance
Kältebeständigkeitsprüfung *f* cold (low-temperature) test
Kältebiegeprüfung *f* cold-bend test
Kältebruchtemperatur *f* brittle-point temperature, brittleness (brittle) temperature
Kälteerzeugung *f* refrigeration, cooling
Kälteerzeugungsanlage *f* refrigerating (cooling) plant
kältefest *s.* kältebeständig
Kältefestigkeit *f s.* 1. Kältebeständigkeit; 2. Kältesprödigkeitspunkt
Kälteflexibilität *f* low-temperature flexibility
Kältegrad *m* degree of frost, degree below zero *(centigrade)*
Kälteisolierung *f* low-temperature insulation
Kälteleistung *f* refrigeration performance
Kältemaschine *f* refrigerating (cooling) machine, refrigerator
Kältemaschinenanlage *f* refrigerating (cooling) plant
Kältemischung *f* freezing (frigorific, cooling) mixture
Kältemittel *n* refrigerant, refrigerating medium, coolant, cooling medium (agent)
Kältemitteldampf *m* refrigerant vapour
Kälteprüfung *f* low-temperature test, cold test
Kältesprödigkeit *f* low-temperature brittleness
Kältesprödigkeitspunkt *m (rubber)* brittle point
Kältetechnik *f* cryogenic (refrigerating) engineering
Kälteträger *m s.* Kältemittel
Kältetrub *m (ferm)* cold sludge
Kältetrübung *f (ferm)* chill haze
Kälteverhalten *n* low-temperature behaviour characteristics, properties)
kaltfärbend cold-dyeing
Kaltfetten *n (tann)* dubbing, hand stuffing
kaltformen to cold-form, to cold-work
Kaltformgebung *f s.* Kaltformung
Kaltformteil *n (plast)* cold moulding
Kaltformung *f* cold forming (working)
Kaltgärhefe *f* cold-tolerant yeast
Kaltgas *n* cold gas
Kaltgasanlage *f* gas-cycle refrigeration system
kalthärten *(plast)* to cure cold; to work-harden *(metals)*
kalthärtend *(plast)* cold-curing, cold-setting; *s.* kaltvulkanisierend
Kalthärtung *f (plast)* cold curing; work hardening *(of metals)*
Kaltkammerdruckgießen *n*, **Kaltkammerdruckguß** *m* cold-chamber die (pressure) casting *(foundry)*
Kaltkautschuk *m* cold [polymerized] rubber, low-temperature polymer (rubber), LTP
Kaltkleber *m* cold adhesive
Kaltlack *m* cold-cut varnish
Kaltlagerung *f* cold storage
Kaltlatex *m* cold rubber latex
Kaltlauge *f* cool brine *(potash industry)*
Kaltleim *m* cold glue
Kaltlötmittel *n* cold-soldering flux
Kaltluft *f* cold air
Kaltmahlen *n* cold milling
Kaltmalerei *f (glass)* cold painting
Kaltmastikation *f*, **Kaltmastizierung** *f (rubber)* cold mastication
Kaltnatron-Halbzellstoff *m* cold soda pulp
Kaltnatronverfahren *n s.* Kaltsodaverfahren
Kaltnetzer *m* cold wetting agent
Kaltpolymerisation *f* cold polymerization
kaltpressen 1. *(met)* to cold-press; *(plast)* to cold-mould; 2. to cold-press, to cold-draw *(oils)*
Kaltpreßmasse *f (plast)* cold-moulding compound (material)
kaltpreßschweißen to cold-weld
Kaltpreßschweißen *n* cold welding
Kalträucherei *f (food)* cold smoking
kalträuchern *(food)* to cold-smoke
kaltrecken *(plast)* to cold-draw
Kaltrecken *n (plast)* cold drawing (stretching, orientation)
Kaltsäureverfahren *n* cold acid process *(for catalytic polymerization)*
kaltschlagen *s.* kaltpressen 2.
Kaltschleifen *n (pap)* cold grinding
Kaltschliff *m (pap)* cold-ground pulp
Kaltschmelzprozeß *m* cold impulse rendering [method] *(for recovering fats)*
Kaltschmieren *n s.* Kaltfetten
Kaltsodastoff *m (pap)* cold soda pulp
Kaltsodaverfahren *n (pap)* cold [caustic] soda process, cold caustic semichemical process
Kaltsterilisation *f* cold sterilization
Kaltstich *m* cold pass *(in powder rolling)*
Kaltstrecken *n s.* Kaltrecken
Kaltumformung *f* cold forming (working)
kaltverarbeiten to cold-work
Kaltverarbeitung *f* cold working
kaltveredeln *(pap)* to refine by the cold [alkali] process
Kaltveredelung *f (pap)* cold [alkali] refining
kaltverfestigen to work-harden *(metals)*
Kaltverformung *f (incorrectly for)* Kaltumformung
kaltverpressen *(plast)* to cold-mould
Kaltverstrecken *n s.* Kaltrecken
Kaltverweil-Färbeverfahren *n (text)* pad-batch process
Kaltvulkanisat *n* cold[-cured] vulcanizate
Kaltvulkanisation *f* cold curing (cure, vulcanization)
  K. nach dem Dunstverfahren vapour curing
kaltvulkanisieren to cure (vulcanize) at room temperature
kaltvulkanisierend room-temperature-curing, room-temperature-vulcanizing, RTV
Kaltwalzen *n* cold rolling

Kaltwalzstich *m s.* Kaltstich
Kaltwaschechtheit *f* fastness to cold washing
Kaltwasserextrakt *m* cold water extract
Kaltwellbehandlung *f (cosmet)* cold waving
Kaltwelle *f (cosmet)* cold wave
Kaltwerden *n* cooling[-down]
Kaltwind *m (met)* cold[-air] blast
kaltziehen to cold-draw *(metals)*
Kalzimeter *n* calcimeter
Kalzination *f s.* Kalzinierung
kalzinieren to calcine, to burn
Kalzinierofen *m* calcining kiln, calciner
Kalzinierung *f* calcination, calcining, burning
Kalzinierungsprodukt *n* calcine, calx
Kalzinierzone *f* calcining zone (compartment)
kalziphil calciphilous *(plant)*
Kalzit *m (min)* calcite, calc-spar, lime spar *(calcium carbonate)*
Kalzium *n* Ca calcium
Kalziumalginatfaser *f* calcium alginate fibre
Kalziumalginatfaserstoff *m* calcium alginate fibre
Kalziumaluminat *n* $Ca(AlO_2)_2$ calcium aluminate
Kalziumarsenat *n* calcium arsenate, *(specif)* $Ca_3(AsO_4)_2$ calcium orthoarsenate
Kalziumarsenid *n* Ca_3As_2 calcium arsenide
Kalziumazetat *n* $Ca(CH_3COO)_2$ calcium acetate
rohes (technisches) K. crude calcium acetate, grey acetate, grey lime
Kalziumazetylid *n s.* Kalziumkarbid
Kalziumbisulfit *n (incorrectly for)* Kalziumhydrogensulfit
Kalziumbisulfitkochsäure *f (pap)* calcium bisulphite cooking liquor, calcium-base acid (liquor)
Kalziumborid *n* CaB_6 calcium boride
Kalziumbromat *n* $Ca(BrO_3)_2$ calcium bromate
Kalziumbromid *n* $CaBr_2$ calcium bromide
Kalziumchelat *n* calcium chelate
Kalziumchlorat *n* $Ca(ClO_3)_2$ calcium chlorate
Kalziumchlorid *n* $CaCl_2$ calcium chloride
Kalziumchloridrohr *n (lab)* calcium chloride tube
Kalziumchromat *n* $CaCrO_4$ calcium chromate
Kalziumdichromat *n* $CaCr_2O_7$ calcium dichromate
Kalziumdihydrogenorthophosphat *n* $Ca(H_2PO_4)_2$ calcium dihydrogenorthophosphate
Kalziumdiphosphat *n* $Ca_2P_2O_7$ calcium diphosphate, calcium pyrophosphate
Kalziumdithionat *n* CaS_2O_6 calcium dithionate
Kalziumentzug *m (med)* decalcification
Kalziumfluorid *n* CaF_2 calcium fluoride
Kalziumglukonat *n* $Ca(HOCH_2[CHOH]_4COO)_2$ calcium gluconate
kalziumhart calcium-hard *(water)*
Kalziumhexafluorosilikat *n* $Ca[SiF_6]$ calcium hexafluorosilicate
Kalziumhexazyanoferrat(II) *n* $Ca_2[Fe(CN)_6]$ calcium hexacyanoferrate(II)
Kalziumhexazyanoferrat(III) *n* $Ca_3[Fe(CN)_6]_2$ calcium hexacyanoferrate(III)
Kalziumhydrid *n* CaH_2 calcium hydride
Kalziumhydrogenkarbonat *n* $Ca(HCO_3)_2$ calcium hydrogencarbonate
Kalziumhydrogenorthophosphat *n* $CaHPO_4$ calcium hydrogenorthophosphate
Kalziumhydrogensulfid *n* $Ca(SH)_2$ calcium hydrogensulphide
Kalziumhydrogensulfit *n* $Ca(HSO_3)_2$ calcium hydrogensulphite
Kalziumhydroxid *n* $Ca(OH)_2$ calcium hydroxide
Kalziumhypochlorit *n* $Ca(OCl)_2$ calcium hypochlorite
Kalziumhypochlorit-Bleichlauge *f (pap)* calcium hypochlorite bleach liquor
Kalziumhypophosphat *n* $Ca_2P_2O_6$ calcium hypophosphate
Kalziumhypophosphit *n* $Ca(PH_2O_2)_2$ calcium hypophosphite, *(better)* calcium phosphinate
Kalziumjodat *n* $Ca(IO_3)_2$ calcium iodate
Kalziumjodid *n* CaI_2 calcium iodide
Kalziumkarbid *n* CaC_2 calcium carbide, calcium acetylide, carbide *(proper)*
Kalziumkarbidherstellung *f* manufacture of calcium carbide
Kalziumkarbonat *n* $CaCO_3$ calcium carbonate
gefälltes (präzipitiertes) K. precipitated calcium carbonate
Kalziumlack *m* calcium lake
Kalziumligninsulfonat *n*, **Kalziumlignosulfonat** *n* calcium lignosulphonate
Kalziummanganat(VII) *n s.* Kalziumpermanganat
Kalziummangel *m (agric)* calcium shortage (deficiency)
Kalziummetaborat *n* $Ca(BO_2)_2$ calcium metaborate
Kalziummetaplumbat(IV) *n* $CaPbO_3$ calcium metaplumbate(IV), calcium trioxoplumbate(IV)
Kalziummetasilikat *n* $CaSiO_3$ calcium metasilicate, calcium trioxosilicate
Kalziummolybdat *n* $CaMoO_4$ calcium molybdate
Kalziumnitrat *n* $Ca(NO_3)_2$ calcium nitrate
Kalziumnitrid *n* Ca_3N_2 calcium nitride
Kalziumnitrit *n* $Ca(NO_2)_2$ calcium nitrite
Kalziumorthoarsenat *n* $Ca_3(AsO_4)_2$ calcium orthoarsenate
Kalziumorthophosphat *n* $Ca_3(PO_4)_2$ calcium orthophosphate
Kalziumorthoplumbat (IV) *n* Ca_2PbO_4 calcium orthoplumbate(IV), dicalcium tetraoxoplumbate(IV)
Kalziumorthosilikat *n* Ca_2SiO_4 calcium orthosilicate, dicalcium tetraoxosilicate
Kalziumoxid *n* CaO calcium oxide
Kalziumoxiderzeugnis *n (ceram)* lime refractory
Kalziumpektat *n* calcium pectate
Kalziumperchlorat *n* $Ca(ClO_4)_2$ calcium perchlorate
Kalziumpermanganat *n* $Ca(MnO_4)_2$ calcium permanganate
Kalziumperoxid *n* CaO_2 calcium peroxide
Kalziumphosphat *n* calcium phosphate, *(specif)* $Ca_3(PO_4)_2$ calcium orthophosphate
Kalziumphosphid *n* Ca_3P_2 calcium phosphide
Kalziumplumbat(II) *n* $CaPbO_2$ calcium plumbate(II)
Kalziumpyrophosphat *n s.* Kalziumdiphosphat

Kalziumrhodanid *n s.* Kalziumthiozyanat
Kalziumsaccharat *n* calcium saccharate
Kalziumseife *f* calcium soap
Kalziumselenat *n* $CaSeO_4$ calcium selenate
Kalziumsilikatschlacke *f* phosphate slag *(in manufacturing phosphorus in electric furnaces)*
Kalziumspiegel *m* calcium level *(of blood)*
Kalziumstearat *n* $Ca(CH_3[CH_2]_{16}COO)_2$ calcium stearate
Kalziumsulfat *n* $CaSO_4$ calcium sulphate
  gefälltes (präzipitiertes) K. precipitated calcium sulphate, precipitated gypsum
Kalziumsulfhydrat *n (tann)* calcium sulphydrate, *(better)* calcium hydrogensulphide
Kalziumsulfid *n* CaS calcium sulphide
Kalziumsulfit *n* $CaSO_3$ calcium sulphite
Kalziumtetraborat *n* CaB_4O_7 calcium tetraborate
Kalziumtetroxoplumbat(IV) *n s.* Kalziumorthoplumbat(IV)
Kalziumtetroxosilikat *n s.* Kalziumorthosilikat
Kalziumthiosulfat *n* CaS_2O_3 calcium thiosulphate
Kalziumthiozyanat *n* $Ca(SCN)_2$ calcium thiocyanate, calcium rhodanide
Kalziumtrioxoplumbat(IV) *n s.* Kalziummetaplumbat(IV)
Kalziumtrioxosilikat *n s.* Kalziummetasilikat
Kalziumwolframat *n* $CaWO_4$ calcium wolframate, calcium tungstate
Kalziumzyanamid *n* $CaCN_2$ calcium cyanamide
Kalziumzyanid *n* $Ca(CN)_2$ calcium cyanide
Kamazit *m (min)* kamacite *(a nickel-iron alloy occurring in meteoric iron)*
Kamerun-Kardamom *m(n)* Cameroon cardamom *(from Aframomum hanburgi Schum.)*
Kamin *m* [smoke]stack, chimney; *(distil)* riser [tube], chimney *(of a cap)*
Kaminklappe *f* stack valve
Kaminkühlturm *m* atmospheric (natural-draught) cooling tower
Kaminstummel *m (distil)* riser [tube], chimney *(of a cap)*
Kaminzug *m* draught, *(Am)* draft
Kammer *f* chamber, cabinet, compartment, cell, box; *(ceram)* corridor *(of a dryer)*; stall *(coal hydrogenation)*
  feuchte K. humidity chamber; *(lab)* moist chamber culture dish
  ringförmige K. annular chamber
Kammerabscheider *m* settling (gravity settling, fall-out) chamber, drop-out box *(gas cleaning)*
Kammerbegasung *f* chamber fumigation *(pest control)*
Kammerfilterpresse *f* chamber [filter] press, recessed-plate [filter] press
Kammerjäger *m* exterminator
Kammerofen *m (met)* chamber furnace; *(ceram)* chamber (box) kiln
Kammerofenkoks *m* oven coke
Kammerpresse *f s.* Kammerfilterpresse
Kammerraum *m* chamber space *(in manufacturing sulphuric acid)*
Kammerreaktion *f* chamber reaction *(in manufacturing sulphuric acid)*
Kammerringofen *m* annular chamber kiln
Kammer[schwefel]säure *f* chamber [sulphuric] acid
Kammertrockner *m* chamber dryer; cabinet[-type air] dryer *(with only one chamber)*; compartment dryer *(with multiple chambers)*
  K. mit Hordenwagen [tray-]truck dryer
Kammerverfahren *n* chamber process *(of manufacturing sulphuric acid)*
Kammervergasung *f* chamber fumigation *(pest control)*
Kammerwärmeaustauscher *m*, **Kammerwärmeübertrager** *m* plate heat exchanger
Kammerzentrifuge *f* multichamber centrifuge
Kammzug *m (text)* slubbing, top
Kammzugdruck *m (text)* vigoureux printing
Kammzugfärbeapparat *m (text)* top dyeing machine
Kammzugfärben *n (text)* top dyeing
Kampescheholz *n* Campe[a]chy wood, logwood *(from Haematoxylum campechianum L.)*
Kampfer *m* camphor, 1,7,7-trimethylbicyclo[2,2,1]-heptan-2-one
  künstlicher K. artificial (synthetic) camphor, 2-chlorobornane
  natürlicher K. natural camphor *(from Cinnamomum camphora (L.) Sieb.)*
***D*-Kampfer** *m* (+)-camphor, Japan camphor
***L*-Kampfer** *m* (−)-camphor, Matricaria camphor
kampferartig camphoraceous
Kampferblume *f*, **Kampferblüte** *f* flowers of camphor
Kampferliniment *n* camphorated oil, camphor liniment
Kampfermethode *f* camphor method *(for determining molecular weights)*
  K. nach Rast Rast camphor method, Rast [molecular weight] method, Rast micromethod
Kampferöl *n* camphor oil *(from Cinnamomum camphora (L.) Sieb.)*
  leichtes K. light camphor oil
  weißes K. white camphor oil
Kämpferol *n* kaempferol, 3,4′,5,7-tetrahydroxyflavone
Kampfersäure *f* camphoric acid, 1,2,2-trimethylcyclopentane-1,3-dicarboxylic acid
Kampfersäureanhydrid *n* camphoric anhydride, 1,2,2-trimethylcyclopentane-1,3-dicarboxylic anhydride
Kampfersäure-3-monoamid *n* camphoric acid α-monoamide, α-camphoramic acid, 3-carbamoyl-1,2,2-trimethylcyclopentanecarboxylic acid
Kampferweißöl *n* white camphor oil
Kampfmittel *n* / **biologisches** biological warfare agent
Kampfstoff *m* warfare agent
  biologischer K. *s.* Kampfmittel / biologisches
  blasenziehender K. blister gas (agent), vesicant
  chemischer K. chemical weapon (warfare agent), war gas

erstickender K. choking gas
flüchtiger K. non-persistent chemical agent
hautschädigender K. blister gas (agent), vesicant
kurzwirkender K. *s.* flüchtiger K.
lakrimogener K. *s.* tränenreizender K.
langwirkender K. *s.* seßhafter K.
lungenschädigender K. lung irritant (injurant)
psychotoxischer K. psychochemical
seßhafter K. persistent chemical agent
tränenreizender K. lachrymator, lacrimator
Kamphan *n s.* Bornan
Kamphanon-(2) *n* 2-camphanone, camphor, 1,7,7-trimethylbicyclo[2,2,1]heptan-2-one
Kamphen *n* camphene *(a bicyclic terpene hydrocarbon)*
Kampholsäure *f* campholic acid, 1,2,2,3-tetramethylcyclopentane-1-carboxylic acid
Kamphonansäure *f* camphonanic acid, 1,2,2-trimethylcyclopentane-1-carboxylic acid
α-Kamphoramsäure *f* α-camphoramic acid, 3-carbamoyl-1,2,2-trimethylcyclopentanecarboxylic acid
Kamphoronsäure *f* camphoronic acid, 2,3-dimethylbutane-1,2,3-tricarboxylic acid
Kamyr-Bleichturm *m (pap)* Kamyr bleacher
Kamyr-Schleifer *m (pap)* Kamyr grinder
Kanadabalsam *m (incorrectly for)* Kanadaterpentin
Kanadaterpentin *n(m)* Canada turpentine, balsam of fir *(from Abies balsamea (L.) Mill.)*
Kanadol *n* canadol *(a light ligroin)*
Kanal *m (tech)* canal, *(esp if tubular)* channel, duct; channel *(of an inclusion compound)*
Kanalbildung *f* channelling *(as in fluidized systems)*
Kanaleinschlußverbindung *f* channel inclusion compound
Kanalisation *f* sewerage
Kanalruß *m (rubber)* channel (impingement) black
mittelverarbeitbarer K. medium processing channel black, MPC black
Kanalrußverfahren *n (rubber)* channel method (process)
Kanalschwarz *n s.* Kanalruß
Kanalstrahl *m* canal (positive) ray
Kanalstrahlanalyse *f* canal-ray (positive-ray) analysis
Kanaltrockner *m* tunnel dryer (drying machine), canal dryer, drying tunnel; *(ceram)* corridor dryer
Kanangaöl *n* cananga oil *(from Cananga odorata (Lam.) Hook. f. et Thoms.)*
Kandelillawachs *n* candelilla wax *(from Pedilanthus pavonis (Klotzsch et Gcke.) Boiss.)*
Kandelit *m*, **Kandelkohle** *f s.* Kännelkohle
Kandelzucker *m s.* Kandiszucker
Kandis[zucker] *m* candy [sugar], sugar (rock) candy
Kaneel *m* / **Echter** Ceylon cinnamon *(from Cinnamomum zeylanicum Bl.)*
Kaneit *m (min)* kaneite *(manganese(III) arsenide)*
Kaneszin *n* deserpidine, 11-demethoxyreserpine, canescine
Kanister *m* can
Kanne *f (text)* can
Kännelkohle *f* cannel (candle, jet) coal
Kante *f* edge
Kanteneffekt *m (phot)* edge effect
Kantenschärfe *f (phot)* acutance
Kantenschliff *m (glass)* edging
Kantenversetzung *f (cryst)* edge dislocation
Kantenwinkel *m (cryst)* interfacial angle
kantig angular
Kanutillawachs *n s.* Kandelillawachs
Kanyabutter *f* kanya (Sierra Leone, lamy) butter *(from Pentadesma butyraceum Sabine)*
kanzerogen carcinogenic, cancerigenic, cancer-producing, cancer-causing
nicht k. non-carcinogenic
Kaolin *m* kaolin[e], china (porcelain) clay, white bole, bolus alba
K. für kautschuktechnische Zwecke rubber clay
geschlämmter K. washed kaolin, water-washed clay
harter K. hard [rubber] clay
kolloidaler K. colloidal kaolin
trockenaufbereiteter K. *(pap)* air-floated clay
weicher K. soft [rubber] clay
Zettlitzer K. Zettlitz kaolin
Kaolinbrei *m (pap)* clay[-water] slurry
Kaolindispersion *f (pap)* clay milk
Kaolinfüllstoff *m (pap)* clay filler
kaolingefüllt *(pap)* clay-filled
kaolingestrichen *(pap)* clay-coated
Kaolinisation *f* kaolinization
kaolinisieren to kaolinize
Kaolinisierung *f* kaolinization
Kaolinit *m (min)* kaolinite *(aluminium hydroxide silicate)*
kaolinitisch kaolinitic
Kaolinlager *n*, **Kaolinlagerstätte** *f* kaolin deposit
Kaolinmilch *f*, **Kaolintrübe** *f (pap)* clay milk
Kaolinvorkommen *n* kaolin deposit
Kaon *n (nucl)* kaon [particle], K meson
Kap-Aloe *f (pharm)* cape aloe *(from Aloe ferox Mill.)*
Kapbeerenwachs *n (incorrectly for)* Myrikatalg
Kapelle *f* 1. *(met)* cupel; 2. *s.* Abzugsschrank
Kapellenprobe *f (met)* cupel test, cupellation
Kapgummi *n* cape gum *(from Acacia specc.)*
kapillar capillary
kapillaraktiv capillary active, surface-active
Kapillaraktivität *f* capillary (surface) activity
Kapillaranalyse *f* capillary analysis
Kapillaraszension *f* capillary rise
Kapillarattraktion *f s.* Kapillarität
Kapillarbruch *m (text)* capillary breaking
Kapillardepression *f* capillary depression
Kapillardruck *m* capillary pressure
Kapillare *f* capillary [tube]
Kapillarelektrometer *n* capillary electrometer
Kapillarhahn *m* capillary stopcock (tap)
kapillarinaktiv capillary inactive

Kapillarität *f* capillarity
Kapillaritätskonstante *f* capillary constant
Kapillaritätstheorie *f* **der Gastrennung** capillary theory of separation
Kapillarkondensation *f* capillary condensation (sorption)
Kapillarkonstante *f* capillary constant
Kapillarmethode *f* capillary rise (tube) method *(for determining surface tension)*
Kapillarpipette *f* capillary pipette
Kapillarrohr *n*, **Kapillarröhrchen** *n* capillary [tube]
Kapillarrohre *npl* capillary tubing
Kapillarsäule *f* capillary column
Kapillärsirup *m* starch syrup
Kapillarsperrhahn *m* capillary stopcock (tap)
Kapillarviskosimeter *n* capillary viscometer, viscosity pipette
  K. nach Ostwald Ostwald [capillary] viscometer, Ostwald viscosity pipette
Kapillarwasser *n (soil)* capillary water
Kapillarwirkung *f* capillary action (attraction)
Kapok *m (text)* kapok, capoc *(fruit fibres esp from Ceiba pentandra Gaertn.)*
Kapoköl *n* kapok oil *(seed oil from Ceiba pentandra Gaertn.)*
Kappe *f* cap; *(glass)* moil
Kaprinaldehyd *m* $CH_3[CH_2]_8CHO$ decanal, capric aldehyde
Kaprinamid *n* $CH_3[CH_2]_8CONH_2$ decanamide, capric amide
Kaprinat *n* decanoate, caprate *(a salt or ester of decanoic acid)*
n-**Kaprinsäure** *f* $CH_3[CH_2]_8COOH$ decanoic acid, capric acid
Kaprinsäureanhydrid *n* $(C_9H_{19}CO)_2O$ decanoic anhydride, capric anhydride
Kaprinsäureäthylester *m* $CH_3[CH_2]_8COOC_2H_5$ ethyl decanoate, ethyl caprate
Kaprolaktam *n* caprolactam
n-**Kapronaldehyd** *m* $CH_3[CH_2]_4CHO$ hexanal, caproic aldehyde
Kapronat *n* hexanoate, capronate, caproate *(a salt or ester of hexanoic acid)*
Kaprononitril *n* $CH_3[CH_2]_4CN$ hexane nitrile, capronitrile
n-**Kapronsäure** *f* $CH_3[CH_2]_4COOH$ hexanoic acid, caproic acid
Kapronsäureanhydrid *n* $(C_5H_{11}CO)_2O$ hexanoic anhydride, caproic anhydride
Kapronsäureäthylester *m* $CH_3[CH_2]_4COOC_2H_5$ ethyl hexanoate, ethyl caproate
n-**Kaprylaldehyd** *m* $CH_3[CH_2]_6CHO$ octanal, caprylic aldehyde
n-**Kaprylalkohol** *m* $CH_3[CH_2]_6CH_2OH$ 1-octanol, caprylic alcohol
Kaprylat *n* octanoate, caprylate *(a salt or ester of octanoic acid)*
Kapryliden *n s.* Oktin-(1)
Kaprylonitril *n* $CH_3[CH_2]_6CN$ octane nitrile, caprylonitrile
n-**Kaprylsäure** *f* $CH_3[CH_2]_6COOH$ octanoic acid, *n*-caprylic acid
Kapsel *f* capsule; *(ceram)* saggar, sagger
Kapselgebläse *n* positive displacement (rotary) blower
kapseln to encase, to enclose; to waterproof; *(nucl)* to packet, to can *(fuel elements)*
Kapselpumpe *f* lobe pump
Kapselton *m (ceram)* saggar clay
Kapsenberg-Schmiere *f (lab)* Kapsenberg lubricant
Karakurin *n* caracurine *(alkaloid)*
Karamel *m* caramel, caramelized sugar
Karamelbier *n* malt beer
Karamelgeschmack *m* caramel flavour
Karamelisationsbräunung *f* caramelization browning
karamelisieren to caramelize
Karamelisierung *f* caramelization
Karamelmalz *n* caramel (crystal) malt
Karan *n* carane, 3,7,7-trimethylbicyclo[4,1,0]heptane
Karayagummi *n* karaya (sterculia) gum, [gum] karaya, Indian tragacanth *(chiefly from Sterculia urens Roxb.)*
Karayaschleim *m* karaya mucilage *(from Sterculia specc.)*
Karbachol *n* $NH_2COOCH_2CH_2N(CH_3)_3Cl$ carbachol, carbamylcholine chloride
Karbamat *n* NH_2COOM^I carbamate, aminoformate
Karbamid *n* NH_2CONH_2 carbamide, urea
Karbamidharz *n* urea resin
Karbamidkunststoff *m* urea plastic
Karbamidsäure *f* NH_2COOH carbamic acid, aminoformic acid
Karbamidsäureäthylester *m* $NH_2COOC_2H_5$ ethyl carbamate, ethyl aminoformate
Karbaminat *n s.* Karbamat
Karbaminsäure *f s.* Karbamidsäure
N-**Karbamoylglyzin** *n* $NH_2CONHCH_2COOH$ carbamoylglycine, hydantoic acid
Karbanilsäure *f* $C_6H_5NHCOOH$ carbanilic acid, phenylcarbamic acid
Karbanion *n s.* Karbeniat--Ion
Karbarson *n* $NH_2CONH-C_6H_4-AsO(OH)_2$ carbarsone, *N*-carbamoylarsanilic acid
Karbazol *n* carbazole, dibenzopyrrole
Karbazolfarbstoff *m* carbazole dye
Karbazolindophenol *n* carbazole indophenol
Karbazolring *m* carbazole ring
Karbazolsynthese *f* / **Graebe-Ullmannsche** Graebe-Ullmann synthesis of carbazoles
Karbazon *n* $R-N=N-CO-NH-NHR'$ carbazone, *(specif)* $H_2N-NH-CO-N=NH$ carbazone
Karben *n* $R_1-\overline{C}-R_2$ carbene, *(specif)* $|CH_2$ carbene, methylene
Karbeniat-Anion *n s.* Karbeniat-Ion
Karbeniatformel *f* carbeniate formula
Karbeniat-Ion *n* carbeniate ion, carbanion
Karbeniatstruktur *f* carbeniate structure
Karbeniumion *n*, **Karbeniumkation** *n* carbonium (carbenium) ion
Karbeniumsalz *n* carbonium (carbenium) salt

Karbeniumstruktur *f* carbonium (carbenium) structure
Karbid *n* carbide, *(specif)* CaC_2 calcium carbide, calcium acetylide
 gesintertes K. sintered [hard, metal] carbide
 hartes K. hard [metal] carbide
 interstitielles K. interstitial carbide
Karbidausscheidung *f* carbide precipitation
Karbidgerüst *n* carbide skeleton
Karbidhartmetall *n* cemented [hard] carbide, cemented hard metal
Karbidkalk *m* acetylene lime
Karbidofen *m* carbide furnace
 offener K. open carbide furnace
 rotierender K. rotating-hearth carbide furnace
Karbidschmelzofen *m s.* Karbidofen
Karbidskelett *n* carbide skeleton
Karbinol *n* 1. carbinol *(any of the branched-chain derivatives of methanol)*; 2. *s.* Methanol
Karboanhydr[at]ase *f* carbonic anhydrase
Karbodikarbonyl *n s.* Trikohlenstoffdioxid
karbofunktionell carbon-functional, organofunctional
Karbolin *n* carboline *(alkaloid)*
Karbolöl *n* carbolic oil
Karbolsäure *f* C_6H_5OH carbolic acid, phenol, hydroxybenzene
 rohe K. cresylic acid *(a crude mixture of phenols, cresols, and xylenols)*
Karbolschwefelsäure *f s.* Phenolsulfonsäure
Karbomyzin *n* carbomycin *(antibiotic)*
Karbonado *m* carbonado, black (carbon) diamond
Karbonat *n* $M_2^ICO_3$ carbonate
Karbonatanhydratase *f s.* Karboanhydrase
Karbonatbleiweiß *n* white lead, ceruse *(lead carbonate hydroxide)*
Karbonatgestein *n* carbonate rock
Karbonathärte *f* carbonate hardness (alkalinity), temporary hardness *(of water)*
Karbonatisation *f*, **Karbonatisierung** *f (geoch)* carbonatization
Karbonatochelat *n* carbonato chelate
Karbonatokomplex *m* carbonato complex
d-**Karbonatoverbindung** *f* dextro carbonato compound
Karbonatsediment *n* carbonate sediment
Karbonifikation *f (geoch)* carbonification, coalification
Karbonisation *f* 1. *(text)* carbonizing, carbonization *(removal of burrs from raw wool)*; 2. *(food)* carbonation, aeration, impregnation with carbon dioxide; 3. carbonation *(in sodium carbonate manufacture)*
 nasse K. *(text)* wet carbonizing
 trockene K. *(text)* dry carbonizing
Karbonisator *m s.* Karbonisierungskolonne
karbonisieren 1. *(text)* to carbonize *(wool)*; 2. *(food)* to carbonate, to aerate, to impregnate with carbon dioxide; 3. to carbonate *(in sodium carbonate manufacture)*
Karbonisierhilfsmittel *n*, **Karbonisiernetzmittel** *n (text)* carbonizing assistant
Karbonisierungskolonne *f* carbonating tower *(in sodium carbonate manufacture)*
karbonitrieren to carbonitride, to gas-cyanide, to dry-cyanide *(steel)*
Karbonitrierung *f* carbonitriding, gas (dry) cyaniding (cyanization), ni-carbing *(of steel)*
Karboniumion *n s.* Karbeniumion
Karboniumsalz *n s.* Karbeniumsalz
Karbonkohle *f* carboniferous coal
Karbonohydrazid *n* RNH−NH−CO−NH−NHR′ carbonohydrazide, carbazide, *(specif)* $H_2N-NH-CO-NH-NH_2$ carbonohydrazide, carbazide
Karbonpapier *n* carbon (carbonic, carbonized) paper
Karbonrohpapier *n*, **Karbonrohseide** *f* carbon base paper
Karbonsäure *f* carboxylic acid
 einbasige K. monocarboxylic acid
Karbonsäure-Abbau *m* / **Barbier-Wielandscher** Barbier-Wieland degradation
Karbonsäureamid *n* carboxylic acid amide, carboxazylic acid
Karbonsäure-Aufbau *m* / **Arndt-Eistertscher** Arndt-Eistert synthesis
Karbonsäurereduktion *f* / **McFadyen-Stevenssche** McFadyen-Stevens reduction
Karbonyl *n* carbonyl *(coordination chemistry)*
Karbonylaktivität *f* carbonyl activity
Karbonylbromid *n* $COBr_2$ carbonyl bromide, carbon dibromide oxide, bromophosgene
Karbonylchlorid *n* $COCl_2$ carbonyl chloride, carbon dichloride oxide, phosgene
Karbonyleisen *n* carbonyl iron
Karbonylgruppe *f* =C=O carbonyl (keto) group
 ketonartig gebundene K. ketocarbonyl (ketonic carbonyl) group
Karbonylsulfid *n* COS carbonyl sulphide, carbon oxide sulphide
Karbonylverbindung *f* carbonyl compound
 vinyloge K. vinylogous carbonyl compound
Karbonylverfahren *n (met)* carbonyl process
Karboxylase *f* carboxylase
Karboxylat *n* carboxylate *(a salt or ester of a carboxylic acid)*
Karboxylat-Ion *n* carboxylate ion
Karboxylatkautschuk *m* carboxylic (carboxylated, acid) rubber
Karboxylatkomplex *m* carboxylate complex
Karboxylende *n* C-terminal group (residue) *(in proteins)*
Karboxylgruppe *f* −COOH carboxyl group
karboxylieren to carboxylate
Karboxylierung *f* carboxylation
Karboxylkautschuk *m s.* Karboxylatkautschuk
Karboxymethoxy-essigsäure *f* $O(CH_2COOH)_2$ diglycolic (diglycollic) acid
Karboxymethylzellulose *f* carboxymethylcellulose, C. M. C.
Karboxypeptidase *f* carboxypeptidase

Karbozyanin *n (dye)* carbocyanine
Karbozyklen *pl* carbocyclic (homocyclic, isocyclic) compounds
karbozyklisch carbocyclic, homocyclic, isocyclic
Karburator *m* carburet[t]or *(for carburetting gases)*
karburieren to carburet *(gases)*
Karbylamin *n* R–N≡C carbylamine, isonitrile, isocyanide
Karbylaminreaktion *f* carbylamine test *(for detecting primary amines)*
Kardamom *m(n)* cardamom, cardamon *(from Elettaria and Amomum specc.)*
  Bengalischer K. Bengal (Nepal) cardamom *(from Amomum aromaticum Roxb. and Amomum subulatum Roxb.)*
  runder K. Siam cardamom *(from Elettaria cardamomum (L.) White et Maton)*
Kardamomöl *n* cardamom (cardamon) oil *(from Elettaria cardamomum (L.) White et Maton)*
Kardiotonikum *n (pharm)* cardiotonic, cardiac tonic
Δ^3-Karen *n* (−)-3-carene, 3,7,7-trimethylbicyclo-[2,2,1]hept-3-ene
Δ^4-Karen *n* (−)-4-carene, 3,7,7-trimethylbicyclo-[2,2,1]hept-2-ene
Karenzfrist *f* preharvest interval *(after application of pesticides)*
Karitebutter *f* shea (Bambuk, Galam) butter *(from Butyrospermum parkii (Don) Kotschy)*
Karkasse *f (rubber)* carcass, carcase, casing, case *(of a tyre)*
Karkasseneinlage *f s.* Karkaßlage
Karkassengummi *m* carcass (casing) rubber
Karkaßlage *f (rubber)* carcass (casing) ply
Karkaßmischung *f (rubber)* carcass (casing, body) stock, carcass compound
Karl-Fischer-Reagens *n* Karl Fischer reagent *(for determining the amount of water in various substances)*
Karminativum *n (pharm)* carminative
Karminazarin *n (dye)* carminazarin
Karminsäure *f* carminic acid
Karnallit *m (min)* carnallite *(magnesium potassium chloride)*
Karnaubawachs *n* carnauba (Brazil) wax *(from Copernicia prunifera (Muell.) H. E. Moore)*
Karneol *m (min)* carnelian, cornelian *(a chalcedony)*
Karotin *n* carotene
Karotinoid *n* carotinoid
Karpain *n* carpaine *(alkaloid)*
Karpamsäure *f* carpamic acid
Karpatenbalsam *m* Carpathian (Hungarian) turpentine *(from Pinus cembra L.)*
Karpholith *m (min)* carpholite *(an inosilicate)*
Karrag[h]een *n (pharm)* carrag[h]een, chondrus *(from marine algae Chondrus crispus (L.) Stackh. and Gigartina mamillosa (Gooden. et Woodw.) J. Agardh)*
Karragheenmoos *n (incorrectly for)* Karragheen
Karrenspritze *f*, **Karrenspritzgerät** *n (agric)* hand-propelled sprayer
Karrensprühgerät *n*, **Karrenzerstäuber** *m s.* Karrenspritze
Karteikarton *m* index board
Kartenpapier *n* map (chart, plan) paper, *(Am)* geography paper
Kartoffelmaische *f* potato mash
Kartoffelmehl *n (incorrectly for)* Kartoffelstärke
Kartoffelschlempe *f* potato slump
Kartoffelspiritus *m*, **Kartoffelsprit** *m* potato alcohol (spirits)
Kartoffelstärke *f* potato starch (flour), farina
Kartoffelwalzmehl *n* potato meal
Karton *m* cardboard, [paper]board
  gegautschter K. duplex cardboard
  geklebter K. pasteboard, pasted board
  gestrichener K. coated [card]board
Kartonagenpappe *f* [folding] boxboard, carton
Kartonfabrik *f* paperboard mill
Kartonmaschine *f* paperboard (vat) machine, board [making] machine
Kartonpapier *n* cardboard
Kartothekkarton *m* index board
Karusselltrockner *m (ceram)* dobbin *(a type of dryer)*
  K. mit Luftstromtrocknung jet drying dobbin
Karvakrol *n* $CH_3(OH)C_6H_3CH(CH_3)_2$ carvacrol, 2-hydroxy-4-isopropyl-1-methylbenzene
Karvon *n* carvone *(a monocyclic terpenoid ketone)*
Karyolymphe *f (bioch)* karyolymph, nuclear sap
Karyophyllin *n* caryophyllin, oleanoic acid
karzinogen carcinogenic, cancerigenic, cancer-producing, cancer-causing
  nicht k. non-carcinogenic
Karzinogen *n* carcinogen
kaschieren *(pap)* to laminate, to line, to paste, to paper; *(plast)* to back, to coat
Kaschieren *n* **mit stranggepreßter Folie** *(plast)* extrusion coating
Kaschierpapier *n* lining (pasting) paper, liners
kaschiert / mit Schaumstoff foam-backed
Kaschunuß *f* cashew nut *(from Anacardium occidentale L.)*
Käse *m* cheese
  gereifter K. ripened cheese
  grüner (ungereifter) K. green cheese
käseartig caseous *(e. g. precipitate)*
Käsebruch *m* [cheese] curd
Käsefarbe *f* cheese colouring
Käseherstellung *f* cheese making
Kasein *n* casein
Kaseinat *n* caseinate
Kaseindeckfarbe *f (tann)* casein coating colour
Kaseinfarbe *f* casein paint
Kaseinfaser *f* casein fibre (staple)
Kaseinfaserstoff *m* casein fibre
Kaseinleim *m* casein glue
Kaseinnatrium *n (pharm)* casein sodium
Kaseinogen *n* caseinogen
Kaseinsäure *f* caseinic acid *(a diaminotrihydroxydodecanoic acid)*

Kaseinspinnlösung *f (text)* casein dope
Käsemolke *f* cheese whey
Käsepulver *n* cheese powder
Käserei *f* 1. cheese making; 2. cheese factory, cheese-making plant
Käsereifung *f* cheese ripening
Käsereimilch *f* cheese milk
Käsereimolke *f* cheese whey
Käsewachs *n* cheese wax
käsig curd[l]y
Kaskadenmethode *f* cascade method *(gas liquefaction)*
Kaskadenmühle *f* cascade mill
Kaskadenofen *m* cascade burner
Kaskadenreaktor *m* cascade *(waste-water treatment)*
Kaskadenschaltung *f* cascade system
Kaskadentrockner *m* cascade dryer
Kaskadenverdampfer *m* cascade (multiple-effect) evaporator
Kaskararinde *f (pharm)* cascara sagrada [bark], sacred (Persian) bark, chittam (chittem, chittim) bark *(from Rhamnus purshiana DC.)*
Kaskarillarinde *f* cascarilla (eluteria, eluthera) bark *(from Croton eluteria Benn.)*
Kassavestärke *f* farinha *(from Manihot esculenta Crantz)*
Kassenrollenpapier *n* paper for calculating machines, tabulating paper
Kasserollenzange *f* casserole tongs
Kassiaöl *n* cassia (Chinese) oil *(from Cinnamomum aromaticum Nees)*
Kassiazimt *m* Chinese cinnamom *(from Cinnamomum aromaticum Nees)*
Kassiterit *m (min)* cassiterite, tin stone *(tin(IV) oxide)*
kastenaufkohlen to box-carburize *(steel)*
Kastenbandfilter *n* travelling-pan filter, TP filter
Kastenbeschicker *m* box feeder
kasteneinsetzen to box-carburize *(steel)*
kastenglühen to box-anneal *(castings)*
Kastenglühofen *m* box-annealing furnace
Kastenkristallisator *m* tank crystallizer
Kastenmälzerei *f* box malting
kastenzementieren to box-carburize *(steel)*
Kastor *m s.* Kastorzucker
Kastoröl *n* castor oil *(from Ricinus communis L.)*
Kastorzucker *m* castor sugar
Käswasser *n* lactoserum, whey, milk serum
Katabolismus *m (bioch)* katabolism, catabolism
Kataklase *f (geol)* cataclasis
Katalase *f* catalase
Katalaseaktivität *f* catalase activity
Katalasekomplex *m* catalase complex
Katalaseprobe *f* catalase test
Katalasewirkung *f* catalase action
Katalysator *m* [reaction] catalyst, catalyzer, cat
  K. in Pillenform pelletized catalyst
  K. ohne Stützmaterial (Träger) unsupported catalyst
  beweglicher (bewegter) K. moving catalyst
  fest angeordneter K. *s.* ruhender K.
  fester K. 1. solid catalyst; 2. *s.* ruhender K.
  festliegender K. *s.* ruhender K.
  gepulverter K. powdered catalyst
  heterogener K. contact catalyst
  negativer K. negative catalyst, anticatalyst, retarder, inhibitor
  perlförmiger K. bead catalyst
  pillenförmiger K. pelletized catalyst
  platinhaltiger K. platinum catalyst
  positiver K. positive catalyst
  pulverisierter (pulvriger) K. powdered catalyst
  ruhender K. static (fixed-bed) catalyst
  sich bewegender K. moving catalyst
  stereospezifischer K. stereospecific catalyst
Katalysatorabfall *m (petrol)* complex out *(in liquid-phase isomerization)*
Katalysatoraktivität *f* catalyst activity
Katalysatorauswaschkolonne *f* catalyst scrubber column
Katalysatorbett *n* catalyst bed
  bewegtes K. moving catalyst bed
  festes (festliegendes) K. *s.* ruhendes K.
  ruhendes (stationäres) K. static bed of catalyst, fixed catalyst bed
Katalysatorgift *n* catalyst (catalytic) poison, paralyzer
Katalysatorkammer *f* catalyst chamber
Katalysatorkreislauf *m* catalyst circulation (cycle, recycle)
Katalysatorleistung *f* catalyst performance
Katalysatoroberfläche *f* catalyst surface
Katalysator-Öl-Verhältnis *n* catalyst/oil ratio, catalyst-to-oil ratio
Katalysatorpille *f* catalyst pellet
Katalysatorpulver *n s.* Katalysatorstaub
Katalysatorregenerator *m* regenerator, kiln *(catalytic cracking)*
Katalysatorschicht *f* catalyst bed
  festliegende K. static bed of catalyst
Katalysatorschlamm *m* catalyst slurry
Katalysatorselektivität *f* catalyst selectivity
Katalysatorstaub *m* catalyst dust, powdered catalyst
Katalysatorstripper *m* catalyst removal column
Katalysatorträger *m* catalyst support (carrier)
Katalysatorumlauf *m s.* Katalysatorkreislauf
Katalysatorwirksamkeit *f* catalyst activity
Katalysatorwirkung *f* catalyst action
Katalyse *f* catalysis
  heterogene K. heterogeneous (contact, surface) catalysis
  homogene K. homogeneous catalysis
  negative K. negative catalysis, anticatalysis, inhibition
  positive K. positive catalysis
  stereospezifische K. stereospecific catalysis
Katalyseofen *m* catalytic reactor
Katalysewirkung *f* catalytic action
katalysieren to catalyze
katalysiert / basisch basically catalyzed
  sauer k. acid-catalyzed
katalytisch catalytic

Kataphorese *f* cataphoresis
kataphoretisch cataphoretic
Katapleit *m (min)* catapleiite *(a hydrous silicate of calcium, sodium, and zirconium)*
Katechin *n* catechin, catechol *(specif)* 3,3′,4′,5,7-pentahydroxyflavan, catechin
Katechingerbstoff *m* catechol tan
Katechu *n* cutch; *(specif) s.* Braunes K.
Braunes K. [black, dark] catechu, Pegu catechu (cutch) *(from Acacia catechu Willd.)*
Gelbes K. pale catechu, white cutch, gambi[e]r, catechu [gum] *(from Uncaria gambir Roxb.)*
Kathämoglobin *n (bioch)* kathaemoglobin
Katharometer *n* katharometer, hot-wire reference and detector cell *(a thermal-conductivity cell)*
Kathartikum *n (pharm)* cathartic
Kathedralglas *n* cathedral glass
Kathode *f s.* Katode
Kation *n* cation ✦ **von Kationen befreien** to decationize
komplexes K. complex cation
kationaktiv *s.* kationenaktiv
Kationcharakter *m* cationic nature
kationenaktiv cation-active, cationic
Kationenaustausch *m* cation exchange
Kationenaustauschadsorption *f* cation-exchange adsorption
Kationenaustauschbett *n* cation-exchange bed
Kationenaustauscher *m* cation (cationic) exchanger
Kationenaustauscher... *s.* Kationenaustausch...
Kationenaustauschfähigkeit *f* cation exchangeability
Kationenaustauschharz *n* cation-exchange resin
Kationenaustauschkapazität *f* cation-exchange capacity, CEC, total exchangeable bases
Kationenaustauschsäule *f* cation-exchange column
Kationenaustauschverfahren *n* cation-exchange method
Kationenfehlstelle *f*, **Kationenleerstelle** *f* cation vacancy (hole)
Kationenschwarm *m (coll)* cation swarm
Kationenüberführungszahl *f* cation transference number
Kationenumtausch *m* cation exchange
kationisch cationic
kationkapillaraktiv cation-active
kationoid cationoid, electrophilic, electron-attracting, electron-withdrawing
Kationsäure *f (phys chem)* cation acid
Kationseife *f* invert soap
Kationtensid *n* cationic surfactant
Katkracken *n (petrol)* cat (catalyst, catalytic) cracking
K. im Orthoflow-Verfahren *n* Orthoflow catalytic cracking
Katode *f* cathode, negative electrode
flüssige K. pool cathode
Katodendunkelraum *m* cathode (Crookes) dark space
Katodenglimmschicht *f* cathode-glow layer
Katodenkupfer *n* cathode copper
Katodenlumineszenz *f* cathode luminescence, cathodoluminescence
Katodennickel *n* electrolytic nickel
Katodenraum *m* cathode compartment
Katodenstrahl *m* cathode ray
Katodenstrahlröhre *f* cathode-ray tube
Katodenzerstäubung *f* cathode (cathodic) sputtering
katodisch cathodic
Katodolumineszenz *f s.* Katodenlumineszenz
Katolyt *m* catholyte *(electrolyte surrounding a cathode)*
Katzenauge *n (min)* cat's eye *(a variety of either quartz or chrysoberyl)*
Katzengold *n (min)* cat gold *(a partly weathered biotite)*
Katzensilber *n (min)* cat silver, potassium mica, muscovite *(a phyllosilicate)*
Kauren *n* kaurene *(a terpene)*
Kauri-Butanol-Wert *m*, **Kauri-Butanol-Zahl** *f (coat)* kauri-butanol number (value)
Kaurigum *m*, **Kauriharz** *n s.* Kaurikopal
Kaurikopal *m* kauri copal (gum, resin), cowrie *(from Agathis australis Salisb.)*
Kauriöl *n* kauri oil *(from Agathis australis Salisb.)*
Kaurireduktionsprüfung *f* kauri-reduction test
Kaustifizieranlage *f (pap)* causticizing department (plant, room)
Kaustifizierbehälter *m*, **Kaustifizierbottich** *m (pap)* causticizing tank, causticizer
kaustifizieren *(pap)* to [re]causticize
Kaustifizierung *f (pap)* causticization, [re]causticizing
Kaustikum *n (med)* caustic agent
kaustisch caustic
kaustizieren *s.* kaustifizieren
Kaustizierung *f s.* Kaustifizierung
Kauterisation *f (med)* cauterization, cautery *(burning of tissue with a caustic or heat)*
Kautschuk *m(n)* rubber, caoutchouc
K. mit geringem Spannungswert low-modulus rubber
K. mit hohem Spannungswert high-modulus rubber
K. mit mittlerem Spannungswert medium-modulus rubber
K. mit niederem Spannungswert low-modulus rubber
anorganischer K. PCl_2N phosphorus dichloride nitride
eiweißarmer (enteiweißter) K. deproteinized rubber
gefrorener K. frozen rubber
kalt polymerisierter K. cold [polymerized] rubber, low-temperature polymer (rubber), LTP
künstlicher K. *s.* synthetischer K.
ölgestreckter (ölhaltiger, ölplastizierter) K. oil-extended rubber
regenerierter K. reclaimed rubber, reclaim, shoddy

synthetischer K. synthetic [rubber], artificial (man-made, chemical) rubber
technisch klassifizierter K. technically classified rubber, T.C. rubber
totgewalzter (totmastizierter) K. dead-rolled (dead-milled) rubber, dead rubber
übermastizierter K. *s.* totgewalzter K.
universeller K. general-purpose rubber
vulkanisierter K. vulcanized (cured) rubber
zyklisierter K. cyclized rubber, cyclorubber
kautschukähnlich, kautschukartig rubber-like
Kautschukballen *m* bale of rubber
Kautschukbaum *m* rubber tree
Kautschukchemie *f* rubber chemistry
Kautschukchemiker *m* rubber chemist
Kautschukderivat *n* rubber derivative
Kautschukdibromid *n* $(C_5H_8Br_2)_n$ rubber dibromide
Kautschukelastizität *f* rubber elasticity
Kautschukfell *n* rubber sheet
Kautschuk-Füllstoff-Gel-Komplex *m* rubber-filler gel
Kautschuk-Füllstoff-Mischung *f* rubber-filler (rubber-pigment) mixture (stock)
Kautschukgehalt *m* rubber-hydrocarbon content, RHC
Kautschukgift *n* rubber poison
Kautschukhydrochlorid *n* rubber hydrochloride
Kautschukhydrofluorid *n* rubber hydrofluoride
Kautschukhydrogenchlorid *n s.* Kautschukhydrochlorid
Kautschukhydrogenfluorid *n s.* Kautschukhydrofluorid
Kautschukindustrie *f* rubber[-manufacturing] industry
Kautschukkohlenwasserstoff *m* rubber hydrocarbon
Kautschukkuchen *m (rubber)* slab
Kautschuk-KW *m s.* Kautschukkohlenwasserstoff
Kautschuklatex *m* rubber latex
synthetischer K. synthetic[-rubber] latex
kautschuklöslich rubber-soluble
Kautschuklösung *f* rubber solution (cement)
Kautschuklösungsmittel *n* rubber solvent
Kautschuklöwenzahn *m* kok-saghyz, Russian dandelion, Taraxacum bicorne Dahlst.
Kautschukmilch *f*, **Kautschukmilchsaft** *m* rubber latex (milk)
Kautschukmischung *f* rubber stock (mixture, compound)
Kautschukmolekül *n* rubber molecule
Kautschukpflanze *f* rubber-yielding (rubber-producing, rubber-bearing) plant, rubber plant
Kautschukplantage *f* rubber plantation
Kautschukplatte *f (rubber)* slab
Kautschukpulver *n* rubber powder
Kautschukqualität *f* grade of rubber
Kautschuk-Schwefel-Mischung *f* rubber-sulphur blend (compound, mixture, stock)
Kautschukspalter *m (rubber)* bale cutter (splitting machine), splitter
Kautschuktechnologe *m* rubber technologist
Kautschuktechnologie *f* rubber technology
Kautschukträger *m s.* Kautschukpflanze
Kautschuktrockengehalt *m* dry rubber content, DRC
Kautschuktrockensubstanz *f (rubber)* solid material, total solids, T. S.
Kautschuktyp *m* type of rubber
Kautschukverarbeitung *f* / **trockene** dry rubber manufacture
Kautschukvulkanisat *n* rubber vulcanizate, vulcanized (cured) rubber
kavernös drusy
Kavitation *f* cavitation
Kawaharz *n* kava resin *(from Piper methysticum G. Forst.)*
Kawain *n* kawain, 5,6-dihydro-4-methoxy-6-styrylpyran-2-one
Kawapfeffer *m* kava[kava], cava, kawa, Piper methysticum G. Forst.
Kawasäure *f* kawaic (kavaic) acid, 3-methoxy-7-phenyl-2,4,6-heptatrienoic acid
Kaysam-Prozeß *m (rubber)* Kaysam process
KD *(abbr)* Knockdown
KD-Effekt *m* knockdown effect *(of certain pesticides)*
Kefir *m (food)* kefir, kephir
Kefirkörner *npl (food)* kefir grains (seeds)
Kegel *m* 1. *(met)* cone *(of a blast furnace)*; 2. *s.* Schmelzkegel; 3. *s.* Kegelrotor
pyrometrischer K. *(ceram)* pyrometric cone
standardisierter K. *(ceram)* standard cone
Kegelbrecher *m* cone (conical, gyratory) crusher
Kegeldichtring *m* V-ring, V-seal
kegelförmig cone-shaped, conical, taper[ing]
Kegelgranulator *m* short-head cone crusher
Kegelmesser *npl (pap)* core bars, bars on the rotor, bars in the plug *(of a perfecting engine)*
Kegelmühle *f* 1. conical grinder, cone mill, rotary crusher; 2. *s.* Kegelstoffmühle
Kegelrefiner *m (pap)* conical refiner
Kegelring *m*, **Kegelringdichtung** *f* V-ring, V-seal
Kegelrotor *m (pap)* rotor, core, cone, plug *(of a perfecting engine)*
K. der Kegelstoffmühle jordan plug (rotor)
Kegelschliffverbindung *f* conical [ground-glass] joint, tapered joint
Kegelstoffmühle *f (pap)* perfecting engine, jordan, refining machine (engine), refiner
Kegelstrahldüse *f* swirl[-plate] nozzle, hollow-cone nozzle
Keilhauit *m (min)* keilhauite, yttrotitanite *(a variety of titanite)*
Keilküvette *f* wedge cell
Keilschieber *m* wedge gate valve
Keilspaltsieb *n* wedge-wire screen
Keim *m (cryst)* nucleus [of crystallization]
Keimbildner *m (cryst)* nucleation (nucleating) agent
Keimbildung *f (cryst)* nucleation, formation of nuclei
Keimbildungsgeschwindigkeit *f (cryst)* rate of nuclei forming, rate of formation of nuclei

keimfrei *(med)* sterile, aseptic
k. machen to sterilize, to degerm
Keimfreiheit *f (med)* sterility, asepsis
Keimfreimachung *f (med)* sterilization, degermation
Keimkristall *m* seed crystal
Keimlösung *f (coll)* nuclear solution
Keimöl *n (food)* germ oil
Keimschädigung *f* germination injury *(as by seed protectants)*
keimtötend sterilizing, germicidal, disinfectant
Keimtötung *f* sterilization, degermation, disinfection
Keimzahl *f (cryst)* number of nuclei
K-Einfang *m* K[-electron] capture
Kekulé-Formel *f* Kekulé formula
Kekulé-Grenzstrukturen *fpl*, **Kekulé-Strukturen** *fpl* Kekulé[-like] structures
K-Elektron *n* K electron
K-Elektroneneinfang *m s.* K-Einfang
Kelle *f (glass)* ladle
Kellerbehandlung *f (ferm)* cellar treatment
Kellog-Synthese *f* Kellog fluidized synthesis
Kelly-Filter *n* Kelly filter
Kelly-Stange *f (petrol)* kelly
Kelp *n* kelp *(the ashes of seaweed)*
Kelter *f (ferm)* [grape, wine] press
keltern to press *(grapes)*
Kelvin-Skale *f* Kelvin [temperature] scale
Kelvin-Temperatur *f* Kelvin (absolute) temperature
Kenab *m(n)*, **Kenaf** *m(n) s.* Kenaffaser
Kenaffaser *f* kenaf (hibiscus, hemp-mallow) fibre *(from Hibiscus cannabinus L.)*
Kennelkohle *f s.* Kännelkohle
Kennlinie *f* characteristic curve
Kennwert *m* [characteristic] value
Kennzahl *f*, **Kennziffer** *f* coefficient, value, index, characteristic
Kephalin *n* kephalin, cephalin *(a phosphatide)*
Kerametall *n* cermet, ceramal
Keramik *f* ceramics *(art, process, or product)*
K. für chemische Zwecke chemical ceramics
piezoelektrische K. piezoelectric ceramics
technische K. technical ceramics
Keramikchemiker *m* ceramic chemist
Keramiker *m* ceramist
Keramikfaser *f* ceramic fibre (staple)
Keramikfilter *n* ceramic filter
Keramikschüttung *f* ceramic packing
keramisch ceramic
Kerargyrit *m (min)* cerargyrite, chlorargyrite, horn silver *(silver chloride)*
Kerasin *n* kerasin *(a cerebroside)*
Keratin *n* keratin *(a scleroprotein)*
Keratinisation *f (biol)* keratinization
Kerbeinflußzahl *f* notch factor *(materials science)*
kerbempfindlich notch-sensitive
Kerbempfindlichkeit *f* notch sensitivity
Kerbschlagzähigkeit *f* notch impact resistance (strength)
K. nach Izod *(plast)* Izod impact strength
Kerbwirkungsfaktor *m*, **Kerbwirkungszahl** *f* notch factor
Kermes *m* 1. *(dye)* kermes [grains], kermes berries, grains of kermes, scarlet corns *(the dried bodies of the females of various scales, genus Kermes)*; 2. kermes scarlet, kermes [dye]; 3. kermes [mineral] *(a double compound of antimony trisulphide and antimony trioxide)*
mineralischer K. *s.* Kermes 3.
Kermeseiche *f* kermes [oak], Quercus coccifera L.
Kermesfarbstoff *m s.* Kermes 2.
Kermesit *m (min)* kermesite, red antimony *(antimony(III) oxide sulphide)*
Kermeskörner *npl s.* Kermes 1.
Kermessäure *f* kermesic acid *(an anthraquinone derivative)*
Kermesscharlach *m s.* Kermes 2.
Kern *m* nucleus *(of an atom)*; *(pap)* plug, rotor, core, cone *(of a perfecting machine)*; *(cryst)* nucleus; *(lab)* cone *(of a ground-glass joint)*; *(tann)* butt; *(rubber)* carcass; *(bot)* nucleus *(of a cell)*
K. der Kegelstoffmühle *(pap)* jordan plug (rotor)
anellierter K. *s.* kondensierter K.
aromatischer K. *(org chem)* aromatic (benzene) nucleus
dunkler K. *(pap)* burnt centre *(in a chip)*
kondensierter K. *(org chem)* condensed (fused) nucleus
schwarzer K. *(ceram)* black core (heart)
Kernabstand *m* nuclear (atomic, interatomic) distance; [inter]atomic spacing
Kernanregung *f* nuclear excitation
Kernaufbau *m* nuclear structure
Kernbaustein *m* nuclear constituent (particle), nucleon
Kernbildung *f (cryst)* nucleation, formation of nuclei
Kernbindemittel *n*, **Kernbinder** *m* core binder (binding agent) *(foundry)*
Kernbohren *n (petrol)* coring
Kernbohrer *m (petrol)* core drill
Kernbrennstoff *m* nuclear fuel.
Kernbruchstück *n* nuclear fragment
Kernchemie *f* nuclear chemistry
Kerndiagramm *n* / **elektrisches** *(petrol)* electric log *(in electrical coring)*
Kerndichte *f* nuclear density
Kerndrehimpuls *m* nuclear spin (angular momentum)
Kerneigenschaft *f* nuclear property
Kerneinfang *m* nuclear capture
Kerneisen *n* core iron *(reinforcement for heavy cores in foundry)*
Kernemulsion *f (nucl)* nuclear [track] emulsion
Kernemulsionstechnik *f (nucl)* nuclear emulsion technique
Kernen *n (geol)* coring, *(for measuring purposes also)* logging
elektrisches K. electrical coring
Kernenergie *f* nuclear (atomic) energy
kernenergiegetrieben nuclear-powered

Kernenergieniveau *n* nuclear energy level
Kernenergieniveaudichte *f* nuclear energy level density
Kernfaden *m (text)* core thread
Kernfeld *n* nuclear field
Kernfeldkräfte *fpl* nuclear forces
Kernfestigkeit *f* core strength *(foundry)*
Kernforschung *f* nuclear research
Kernfotoeffekt *m* nuclear photoeffect (photoelectric effect)
Kernfusion *f* nuclear fusion
Kerngarn *n* core spun yarn
Kerngrundsubstanz *f (bioch)* nuclear sap, karyolymph, enchylema
Kernguß *m (ceram)* solid casting
kernhalogeniert ring-halogenated
Kernhalogenierung *f* ring halogenation
Kernherstellung *f* core making *(foundry)*
Kernholz *n (bot)* heartwood, core
Kerninduktion *f* nuclear induction
Kernisobar *n* nuclear isobar
Kernisomeres *n* nuclear isomer
Kernisomerie *f* nuclear isomerism
Kernit *m (min)* kernite *(a hydrous basic sodium borate)*
Kernkettenreaktion *f* nuclear chain reaction
Kernkräfte *fpl* nuclear forces
Kernladung *f* nuclear charge
  effektive K. effective charge
Kernladungszahl *f* nuclear charge number, ordinal (atomic) number
  ungerade K. odd atomic number
Kernlochstift *m (plast)* plain core pin
Kernmagnetismus *m* nuclear magnetism
Kernmagneton *n* nuclear magneton
Kernmantelfaden *m* core-spun yarn
Kernmantelstruktur *f (text)* skin-core structure
Kernmasse *f* nuclear mass
Kernmaterie *f* nuclear matter
Kernmembran *f (bioch)* nuclear membrane
Kernmodell *n* nuclear model
Kernmoment *n* / **magnetisches** magnetic nuclear moment, nuclear magnetic moment
Kernnährstoff *m (agric)* primary nutrient (element)
Kernniveau *n* nuclear energy level
Kernniveaudichte *f* nuclear energy level density
Kernöl *n (food)* kernel oil; core oil *(foundry)*
Kernparamagnetismus *m* nuclear paramagnetism
Kernphysik *f* nuclear physics
  angewandte K. nucleonics
Kernpolarisation *f* nuclear polarization
Kernpotential *n* nuclear potential
Kernquadrupolmoment *n* nuclear quadrupole moment
Kernquadrupolresonanz *f* nuclear quadrupole resonance, NQR
Kernradius *m* nuclear radius
Kernreaktion *f* nuclear reaction
Kernreaktionsformel *f*, **Kernreaktionsgleichung** *f* nuclear reaction equation (formula), nuclear equation
Kernreaktor *m* [nuclear] reactor, [atomic] pile
  thermischer K. thermonuclear reactor
Kernresonanz *f* nuclear resonance
  magnetische K. nuclear magnetic resonance, NMR
Kernresonanzabsorption *f* nuclear resonance absorption
Kernresonanzspektrograf *m* nuclear resonance spectrograph
Kernresonanzspektroskopie *f* nuclear resonance spectroscopy
  magnetische K. nuclear magnetic resonance spectroscopy, NMR spectroscopy
Kernresonanzspektrum *n* / **magnetisches** nuclear magnetic resonance spectrum, NMR spectrum
Kernresonanzuntersuchung *f* nuclear resonance study
Kernrohr *n* inner pipe
Kernsaft *m (bioch)* nuclear sap, karyolymph, enchylema
Kernsand *m* core sand *(foundry)*
Kernsandbindemittel *n*, **Kernsandbinder** *m* core binder (binding agent) *(foundry)*
Kernseife *f* curd soap
Kernspaltung *f* **[/ gesteuerte]** nuclear fission (disintegration)
Kernspaltungsenergie *f* fission energy
Kernspaltungsspektrum *n* nuclear fission spectrum
Kernspektroskopie *f* nuclear spectroscopy
Kernspin *m* nuclear spin (angular momentum)
Kernspinmoment *n* nuclear spin moment
Kernspinquantenzahl *f* nuclear spin quantum number
Kernspinresonanz *f* nuclear magnetic resonance, NMR
Kernsprengstoff *m* nuclear explosive
Kernspur *f* nuclear track
Kernspuremulsion *f* nuclear [track] emulsion
Kernspuremulsionstechnik *f* nuclear emulsion technique
Kernstabilität *f* nuclear stability
Kernstatistik *f* nuclear statistics
Kernstoß *m* nuclear collision
Kernstrahlenchemie *f* [nuclear] radiation chemistry, radiochemistry
Kernstrahlung *f* nuclear radiation
Kernstruktur *f* nuclear structure
Kernstück *n (tann)* butt
Kernstückhälfte *f (tann)* bend
kernsubstituiert substituted in the ring
Kernsubstitution *f* substitution in the ring
Kernsynthese *f* [nuclear] fusion
Kerntechnik *f* nuclear technology
Kernteilchen *n* nuclear particle (constituent), nucleon
Kerntemperatur *f* nuclear temperature
Kernterm *m* nuclear energy level
Kerntheorie *f* nuclear theory
Kernübergang *m* nuclear transition
Kernumwandlung *f* [nuclear] transmutation, [nuclear] transformation

künstliche K. artificial transmutation (transformation)
Kernverdampfung *f* nuclear evaporation
Kernverschmelzung *f* [nuclear] fusion
Kernwicklung *f (pap)* centre rewind method
Kernzerfall *m* nuclear decay (disintegration)
Kernzersplitterung *f*, **Kernzertrümmerung** *f (nucl)* spallation
Kernzone *f* core *(foundry)*
Kernzusammenstoß *m* nuclear collision
Kerosen *n s.* Kerosin
Kerosin *n* kerosine, kerosene, paraffin [oil]
Kerosinfraktion *f* kerosine fraction
Kerosinschiefer *m* kerosine shale
Kerr-Effekt *m* Kerr effect
elektrooptischer K. electrooptical Kerr effect
magnetooptischer K. magnetooptical Kerr effect
Kerr-Konstante *f* Kerr constant
Kerze *f* candle; *(filtr)* tube, candle
Kerzenfilter *n* tube (tubular, candle) filter
Kerzenpapier *n* candle paper
Kessel *m* boiler; bowl *(of a centrifuge)*; *(rubber)* tank, pan
K. in Mantelform shell-type boiler
Kesselanlage *f* boiler plant
Kesselblech *n* boiler plate
Kesseldampf *m* boiler steam
Kesseldruckimprägnierung *f*, **Kesseldrucktränkung** *f* pressure process *(wood preservation)*
Kesselkohle *f* steam[-raising] coal
kokende K. coking steam coal
Kesselschlacke *f* boiler ash[es]
Kesselspeisepumpe *f* boiler feed pump
Kesselspeisewasser *n* boiler feed[ing] water
Kesselstein *m* [boiler] scale, fur ✦ **K. ansetzen** to scale ✦ **K. entfernen** to [de]scale
kesselsteinbildend scale-forming
Kesselsteinbildung *f* scale formation, scaling
Kesselsteinentfernung *f* boiler scale removal, [de]scaling
Kesselsteingegenmittel *n*, **Kesselsteinverhütungsmittel** *n* scale inhibitor, descaling (antinucleating) agent
Kessylalkohol *m* kessyl alcohol
Kessylketon *n* kessyl ketone
Kesting-Verfahren *n (pap)* Kesting electrolytic process *(for producing chlorine-dioxide bleaching liquor)*
Kestner-Verdampfer *m* Kestner long-tube evaporator, long-tube vertical-film evaporator, LTV evaporator
Kestose *f* kestose *(a trisaccharide)*
Ketal *n* ketal, ketone acetal
Ketazin *n* ketazine *(an azine formed from a ketone)*
Keten *n* keten[e] *(any of various derivatives of* $CH_2{=}C{=}O$*)*, *(specif)* $CH_2{=}C{=}O$ keten[e]
Ketenbase *f* keten[e] base
Ketimid *n*, **Ketimin** *n* ketimine, ketonimine
Ketimin-Enamin-Tautomerie *f* ketimine-enamine tautomerism
Ketin *n* ketine, 2,5-dimethylpyrazine
Ketipinsäure *f* ketipic acid, 3,4-dioxoadipic acid, 3,4-dioxohexanedioic acid
Keto ... *s. a.* Keton ...
Ketoamin *n* ketoamine
Ketobemidon *n* ketobemidone *(a piperidine derivative)*
Ketobernsteinsäure *f* $HOOC{-}CO{-}CH_2COOH$ oxosuccinic acid, oxalacetic acid
3-Ketobuttersäure *f* $CH_3{-}CO{-}CH_2COOH$ 3-oxobutyric acid, acetoacetic acid
Keto-Enol-Tautomerie *f* keto-enol tautomerism
Ketoester *m* keto (ketocarboxylic, keto-acid, ketonic) ester
Ketofettsäure *f* ketofatty acid
Ketoform *f* keto form
Ketofunktion *f* ketonic function
ketogen ketogenic
Ketogenese *f* ketogenesis
Ketoglutarsäure *f* oxoglutaric acid, ketoglutaric acid, *(specif)* 2-oxoglutaric acid, 2-oxopentanedioic acid
3-Ketoglutarsäure *f* $OC(CH_2COOH)_2$ 3-oxoglutaric acid, acetonedicarboxylic acid, ADA
Ketogruppe *f* oxo (keto, carbonyl) group, *(specif)* ketocarbonyl (ketonic carbonyl) group
Ketoheptose *f* ketoheptose
Ketohexose *f* ketohexose
Ketoindan *n s.* Hydrindon
Ketokarbonester *m s.* Ketokarbonsäureester
Ketokarbonsäure *f* keto (ketonic) acid
Ketokarbonsäureester *m* keto (keto-acid, ketonic) ester
Ketoketen *n* $(R)_2C{=}C{=}O$ ketoketene
Ketol *n* ketol, keto (ketone) alcohol, hydroxy ketone, *(specif)* monohydroxy ketone
Ketolaktam *n* keto lactam
Ketolyse *f* ketolysis
ketolytisch ketolytic
Ketomalonsäure *f* $OC(COOH)_2$ oxomalonic acid, mesoxalic acid, oxo-propanedioic acid
Keton *n* ketone
gemischtes K. $R'{-}CO{-}R''$ mixed ketone
makrozyklisches K. macrocyclic (macroring, large-ring) ketone
Michlers K. Michler's ketone, di-*p*-dimethylaminophenyl ketone
zyklisches K. cyclic ketone
Keton ... *s. a.* Keto ...
Ketonaldehyd *m* keto aldehyde
Ketonalkohol *m* keto alcohol, hydroxy ketone
Ketonämie *f (med)* ketonaemia
Ketonazetal *n* ketone acetal, ketal
Keton-Benzol-Verfahren *n (petrol)* benzole-ketone process *(dewaxing)*
Ketonbildung *f* ketone formation
Ketonharz *n* ketone (ketonic) resin
Ketonimid *n s.* Ketonimin
Ketonimin *n* ketonimine, ketimine
Ketoniminfarbstoff *m* ketonimine dye[stuff]
ketonisieren to ketonize

Ketonkarbonyl *n* ketocarbonyl [group], ketonic carbonyl [group]
Ketonkörper *m* ketone body
Ketonkörperbildung *f* , **Ketonkörperentstehung** *f* ketogenesis
Ketonmoschus *m (cosmet)* ketone musk, musk ketone, musk C *(a xylene derivative)*
Ketonperoxid *n* ketone peroxide
Ketonranzigkeit *f* ketonic rancidity
Ketonreaktion *f* ketone reaction
Ketonspaltung *f* ketonic cleavage (hydrolysis, fission)
Ketonstruktur *f* keto (ketonic) structure
Ketonsynthese *f* / **Friedel-Craftssche** Friedel-Crafts ketone synthesis
Ketonurie *f* ketonuria *(abnormal excretion of ketones via the urine)*
Ketopentose *f* ketopentose *(a ketonic sugar)*
Ketosäure *f* keto (ketonic, ketocarboxylic) acid
Ketosäureester *m s.* Ketokarbonsäureester
Ketose *f* 1. ketose, ketonic sugar; 2. *s.* Ketosis
Ketosid *n* ketoside *(a glycoside which on hydrolysis yields a ketose)*
Ketosis *f (med)* ketosis, ketoacidosis *(an abnormal increase of ketones in the body)*
Ketosteroid *n (org chem)* ketosteroid
4-Ketovaleriansäure *f s.* Lävulinsäure
Ketoverbindung *f* keto (ketonic) compound
Ketoxim *n* ketoxime
Ketozucker *m* ketose, ketonic sugar
Kette *f* chain; *(el chem)* cell, element; *(text)* warp ✦ **mit offener K.** *(org chem)* open-chain ✦ **mit verzweigter K.** *(org chem)* branched-chain
einsträngige K. single-strand chain *(conveying)*
elektrochemische K. *s.* galvanische K.
galvanische K. galvanic (voltaic) cell (element)
gerade K. *(org chem)* straight chain
geschlossene K. *(org chem)* closed chain
normale K. *(org chem)* straight chain
ringförmige K. *(org chem)* closed chain
sekundäre K. *(el chem)* secondary cell
unverzweigte K. *(org chem)* straight chain
verzweigte K. *(org chem)* branched (forked) chain
zweisträngige K. double-strand chain *(conveying)*
zyklische K. *(org chem)* closed chain
Kettenabbrecher *m* chain stopper
Kettenabbruch *m* chain termination (breakage)
Kettenabbruchmittel *n* chain stopper
Kettenabbruchreaktion *f* chain-terminating (chain-breaking, chain-stopping, break-off) reaction
Kettenachse *f* chain axis
kettenartig chain-like
Kettenbecherwerk *n* chain [and bucket] elevator
Kettenbeweglichkeit *f (org chem)* chain mobility
Kettenende *n (org chem)* chain end
Kettenentrinder *m (pap)* chain barker
K. nach Astrom Astrom chain barker
Kettenfaltung *f (org chem)* chain folding
Kettenförderer *m* chain conveyor
Kettenform *f (org chem)* chain form
offene K. open-chain form
Kettenfortpflanzung *f* chain propagation
Kettenfortpflanzungsreaktion *f* chain-propagating reaction
Kettenglied *n* link *(of a polymer or conveyor)*
Kettenisomerie *f* chain (nuclear) isomerism
Kettenklemme *(lab)* chain clamp
Kettenlänge *f* chain length
kinetische K. *(plast)* kinetic chain length
kettenlos chainless
Kettenmerzerisiermaschine *f (text)* chain mercerizer
Kettenmolekül *n* chain molecule
lineares K. linear chain molecule
starres K. rigid chain molecule
Kettenpolymerisation *f* chain polymerization
Kettenräumer *m* chain scraper *(waste-water treatment)*
Kettenreaktion *f* chain reaction
nukleare K. nuclear chain reaction
radikalische K. free-radical chain reaction
Kettenreduktion *f* chain reduction
Kettenrost *m* chain (travelling) grate
Kettenrostfeuerung *f* chain-grate stoker
Kettenschleifer *m (pap)* chain grinder
Kettensegment *n* chain segment
Kettenspaltung *f*, **Kettensprengung** *f* chain scission (splitting)
Kettenstart *m* chain initiation
K. durch Elektrolyse electrolytic initiation of polymerization
Kettenstrang *m* strand of chain *(conveying)*
Kettenstruktur *f* chain structure
kettentragend chain-carrying, chain-propagating
Kettenträger *m* chain carrier (initiator, propagator)
Kettenüberträger *m* chain-transfer agent
Kettenübertragung *f* chain transfer
Kettenverbindung *f* chain compound
Kettenverlängerung *f* chain prolongation (lengthening)
Kettenverzweigung *f* chain branching
Kettenverzweigungsmechanismus *m* branched-chain mechanism
Kettenwachstum *n* chain growth (propagation)
Kettfaden *m (text)* warp thread
Kettschlichte *f (text)* warp size
Kettschlichten *n (text)* warp sizing
Ketyl *n* [metal] ketyl
Keupermergel *m* keuper marl
kg-Molarität *f s.* Kilogramm-Molarität
KH *s.* Karbonathärte
Kharisalz *n (tann)* Khari salt *(a natural salt mixture containing chiefly sodium sulphate)*
Khellin *n* khellin *(a furanochromone derivative)*
Kiefernekrose *f (med)* phossy-jaw *(caused by phosphorus intake)*
Kiefernharz *n* pine resin
Kiefernholz *n* Scotch fir, pinewood *(from Pinus sylvestris L.)*
Kiefernholzteer *m* pine tar

Kiefernnadelöl *n* pine-needle oil, Scotch fir oil *(from Pinus sylvestris L.)*
Kiefernöl *n* tall oil, tallo[e]l, liquid resin (rosin)
Kiefernteer *m* pine tar
Kienöl *n* pine oil *(from Pinus specc.)*
Kies *m* gravel; *(mine)* pyrites
Kiesabbrand *m* roasted pyrites
Kiesabröstung *f* roasting (burning) of pyrites
Kiesel *m* pebble
Kieselfluorwasserstoffsäure *f s.* Fluorokieselsäure
Kieselgalmei *m s.* Kieselzinkerz
Kieselgel *n* silica (silicic-acid) gel, gelatinous silica
Kieselgestein *n* siliceous (silica) rock
Kieselglas *n* vitreous silica, silica glass
durchscheinendes K. fused (translucent vitreous) silica
durchsichtiges K. transparent [vitreous] silica, fused quartz, quartz glass
klares K. *s.* durchsichtiges K.
undurchsichtiges K. non-transparent vitreous silica
Kieselgur *f* kieselguhr, infusorial earth, diatom[aceous] earth
Kieselknolle *f (geol)* chert nodule
Kieselsäure *f* silicic acid, *(specif)* H_4SiO_4 orthosilicic acid, tetraoxosilicic acid
Kieselsäureester *m* silicic-acid ester, silicon ester
Kieselsäuregel *n s.* Kieselgel
Kieselsäuresol *n s.* Kieselsol
Kieselsediment *n* siliceous sediment
Kieselsinter *m (min)* siliceous sinter, geyserite
Kieselsol *n* silica sol
Kieselstein *m* pebble
Kieselzinkerz *n (min)* hemimorphite *(zinc dihydroxide disilicate)*
Kieserit *m (min)* kieserite *(magnesium sulphate-1-water)*
Kiesfilter *n* gravel (sand) filter
kiesig gravelly
Kiln *m* kiln, regenerator *(in catalytic cracking)*
Kilogramm-Molarität *f* molality, molal concentration, kilogram molarity
Kimberlit *m* kimberlite *(an agglomerate biotite-periodotite)*
Kinase *f* kinase
Kinderpuder *m* baby powder
Kinetik *f* / **chemische** chemical (reaction) kinetics
Kinetin *n* kinetin, 6-furfurylaminopurine
kinetisch kinetic
Kinin *n (bioch)* kinin
Kino *n* kino [gum]
Bengalisches K. Bengal kino, butea gum *(from Butea superba Roxb.)*
Indisches K. East India kino, Malabar kino *(from Pterocarpus marsupium Roxb.)*
Westindisches K. Jamaica kino *(from the bark of Coccoloba uvifera L.)*
Kinogerbsäure *f* kinotannic acid
Kinogummi *n*, **Kinoharz** *n s.* Kino
Kinotinktur *f* tincture of kino
Kippaufzug *m* skip hoist
kippen to tilt, to dump
Kipphorde *f* dumping floor
Kipphordenumlauftrockner *m* tilting (reversing) pan dryer
Kippkübel *m* skip [car]
Kipppresse *f* tilting head press
Kipptrommelmischer *m* tilting-drum mixer
Kippvorrichtung *f* tilting device, tilter
Kirnapparat *m s.* Kirne
Kirndauer *f* churning time, churning period *(margarine making)*
Kirne *f* [emulsion] churn *(margarine making)*
kirnen to churn *(margarine)*
Kirnmaschine *f s.* Kirne
Kirnung *f* churning *(margarine making)*
Kirschgummi *n* cherry gum
Kirschner-Zahl *f* Kirschner value
Kirschrotglut *f* cherry-red heat
kistenglühen to box-anneal *(metals)*
Kistenglühofen *m* box-annealing furnace
Kistenglühung *f* box-annealing *(of metals)*
Kistenpappe *f* container board
Kitt *m* cement, *(esp for sealing)* lute; *(for glass)* putty
hüpfender K. bouncing putty
kitten to cement; to putty *(glass)*
Kitten *n* cementation; puttying *(of glass)*
Kittharz *n (biol)* propolis, bee glue, balm
Ki-Z *s.* Kirschner-Zahl
Kjeldahl-Apparat *m* Kjeldahl digestion apparatus
Kjeldahl-Bestimmung *f* Kjeldahl determination
Kjeldahl-Kolben *m* Kjeldahl flask
Kjeldahl-Methode *f* Kjeldahl [nitrogen] method
Kjellin-Ofen *m* Kjellin furnace *(an induction furnace)*
Klammer *f* clamp, clip
Klammersicherung *f (lab)* joint clamp
Klang *m (pap)* rattle, snappiness, *(Am)* crackle
Klappe *f* lid, door; disk *(of a valve)*
Klappenboden *m* valve tray
Klappenrückschlagventil *n* swing check valve
Klapprost *m* dumping grate
klar clear *(e. g. liquid)*
k. werden to clarify, to [become] clear
Kläranlage *f* clarification (water purification) plant, *(municipally also)* sewage treatment (disposal) plant, sewage plant (works)
Klärapparat *m* clarifying (settling) apparatus, clarifier, settler
Klärbad *n* clearing bath
Klärbassin *n s.* Klärbecken
Klärbecken *n* clarifying (settling, precipitation, sedimentation) basin, clarifying tank, clarification unit, clarifier, settler, precipitator
K. mit horizontaler Durchströmung horizontal-flow clarification unit
K. mit vertikaler Durchströmung vertical-flow (up-flow) clarification unit
Klärbehälter *m s.* Klärgefäß
Klärbrunnen *m* clearwell

Kläre *f (sugar)* liquor
Klareis *n* clear (crystal) ice
klären to clarify, to clear, to purify, *(esp sugar)* to defecate
sich k. to clarify, to [become] clear
Klärfilter *n* clarifying (polishing) filter
Klarfiltration *f*, **Klärfiltration** *f* clarification, polishing [filtration]
Klärfläche *f* / **äquivalente** Σ value *(of a centrifuge)*
Klarflüssigkeit *f* clarified liquid (liquor); overflow product *(wet classification)*
Klärgefäß *n* clarifying (settling, precipitation, sedimentation) tank, clarifier, settler, precipitator
Klarglasur *f (ceram)* clear glaze
Klärgrube *f* settling pit, settler
Klarheit *f* clarity, clearness
Klärhilfsmittel *n* clarifying agent, clarifier, clarificant
Klarifikation *f* clarification
Klarifikator *m* clarifier
Klarit *m (coal)* clarain
Klarlack *m* varnish
Klärleistung *f* clarifying capacity
Klärmittel *n* clarifying agent, clarifier, clarificant; *(ferm)* fining agent
Klarodurit *m* clarodurite *(a banded variety of hard coal)*
Klarpunkt *m* clear point *(precipitation analysis)*
Klärrückstand *m* sewage sludge *(waste-water treatment)*
Klärschlamm *m* sewage sludge; *(agric)* sludge fertilizer
Klarsichtmittel *n* antifog[ging] agent, antifoggant
Klärspitze *f* [/ **konische**] cone classifier
Klärtank *m s.* Klärgefäß
Klärung *f* clarification, purification, *(esp sugar)* defecation; *(ferm)* fining
mechanische K. sedimentation
Klärvorrichtung *f* clarifier
Klärwanne *f* settling pit, settler
Klarwaschbad *n* clearing bath
Klarwasser *n*, **Klärwasser** *n* clarified (clear, filtered) water
Klärwerk *n s.* Kläranlage
Klärzeit *f* clarifying time, *(esp phot)* clearing time
Klärzentrifuge *f* centrifugal clarifier
Klärzone *f* clarification (clear-solution, free-settling) zone *(as in a thickener)*
Klassenbenennung *f (nomencl)* class name
Klassenkennzeichen *n (coal)* class parameter
Klassenname *m (nomencl)* class name
Klassennummer *f (coal)* class number
Klassenparameter *m (coal)* class parameter
Klassenziffer *f (coal)* class number
Klassierapparat *m s.* Klassierer
Klassierbecken *n* classifying pool
klassieren to classify, to class, to size, to grade [into size], to fractionate
Klassierer *m* classifier, classificator
hydraulischer K. water classifier
mechanischer K. mechanical classifier
Klassiergerät *n s.* Klassierer
Klassiergut *n* material being (*or* to be) classified (sized)
Klassierkegel *m* cone classifier
Klassiersieb *n* sizing screen
Klassiertrog *m* classifier trough
Klassierung *f* sizing, [size] classification, grading [into size]
K. nach dem Prinzip des freien Absetzens free-settling classification
K. nach dem Prinzip des gestörten Absetzens hindered-settling classification
hydraulische K. water classification
Klassierzyklon *m* cyclone classifier
Klassifikation *f (nomencl)* classification
Klassifikationssystem *n (nomencl)* system of classification
Klassifikator *m s.* Klassierer
klassifizieren *(nomencl)* to classify
Klassifizierung *f (nomencl)* classification
klastisch *(geol)* clastic
Klathrat *n* clathrate [inclusion compound], cage compound
Klathratbildung *f* clathrate formation
Klathratverbindung *f s.* Klathrat
Klaubeband *n (coal)* inspection belt
klauben *(min tech)* to pick, to sort
von Hand k. to hand-pick
Klauenöl *n* neatsfoot oil
Klebe ... *s. a.* Kleb ...
Klebeband *n* [cellulose] adhesive tape
Klebekarton *m* pasteboard
Klebelack *m* decorators' size
Klebelösung *f (rubber)* cement
K. auf Nitrilkautschukbasis nitrile cement
Klebemittel *n s.* Klebstoff
kleben to stick, to glue, to cement, to paste, to bond; to adhere, to stick
Kleben *n* sticking, gluing, cementation, pasting, bonding; adherence, adhesion
K. durch Anlösen (Anquellen) *(plast)* solvent welding
klebend sticky, adhesive, adherent
Kleber *m* 1. *(food)* gluten; 2. *s.* Klebstoff
Klebereiweiß *n* gluten
Klebestelle *f (glass)* tear *(defect)*
Klebetrockenverfahren *n (tann)* pasting process
Klebetrocknung *f (tann)* paste drying, pasting
Klebezement *m (rubber)* cement
klebfähig adhesive, adherent
Klebfähigkeit *f s.* Klebvermögen
Klebfilm *m* adhesive film
Klebfläche *f* bonded area
Klebfolie *f* adhesive film
klebfrei tack-free
klebfreudig *(rubber)* tacky
Klebfuge *f* glue line
Klebharz *n* adhesive resin
Klebkitt *m* bonding cement, putty
Klebkraft *f* adhesive capacity (power), adhesiveness, adherence
Klebpapier *n* adhesive (gummed) paper

klebrig sticky, tacky, viscid, glairy, tenaceous, ropy
k. machen to tackify
Klebrigkeit *f* stickiness, tackiness, viscidity, glairiness, tenacity, ropiness
Klebrigmacher *m (rubber)* tackifying (tack-producing) agent, tackifier
Klebstoff *m* adhesive [agent, substance], cementing material, cement, paste
K. auf Eiweißgrundlage glair[e]
K. für Sperrholz plywood adhesive
flüssiger K. solvent-based adhesive, solvent cement
härtender K. curing adhesive
heißabbindender (heißhärtender) K. hot-setting (holt-melt) adhesive, hot glue
hitzehärtbarer K. thermosetting adhesive
kalthärtender K. cold-setting (cold-cure) adhesive
thermoplastischer K. thermoplastic adhesive
wärmehärtbarer K. thermosetting adhesive
Klebstoffpulver *n* powder adhesive
Klebstoffschicht *f* adhesive layer
Klebstreifen *m* [cellulose] adhesive tape
Klebtechnik *f (plast)* cementing technique
Klebverbindung *f* adhesive bond
Klebvermögen *n* adhesive capacity (power), adhesiveness, adherence
Kleehonig *m* clover honey
Kleesalz *n* salt[s] of sorrel (lemon), sorrel salt, sal acetosella *(potassium tetraoxalate pure or mixed with potassium hydrogenoxalate)*
Kleesäure *f* HOOC−COOH oxalic acid, ethanedioic acid
Kleiderimprägnierung *f* clothing impregnation
Kleie *f* bran
Kleieköder *m* poison bran bait *(insect control)*
Kleiekulturmethode *f* bran culture method *(as for moulds producing penicillin)*
Kleiemehl *n* beeswing
Kleienbeize *f (tann)* bran drench[ing]
Kleieverfahren *n s.* Kleiekulturmethode
Klein-Bessemer-Birne *f*, **Klein-Bessemer-Konverter** *m* baby Bessemer converter
kleindrusig *(min)* miarolitic
Kleingefüge *n* microscopic structure, microstructure
Kleinkohle *f* small (small-sized) coal, smalls
Kleinkonverter *m* baby converter
kleinkristallin finely (minutely) crystalline
kleinlückig fine-pored
kleinmaschig fine-meshed, narrow-meshed, close-meshed
Kleinreihe *f* short run
Kleinringverbindung *f (org chem)* small-ring compound
Kleinschreiber *m* miniature recorder
Kleinschüttelgerät *n*, **Kleinschüttler** *m (lab)* minishaker
Kleinserie *f* short run
kleinstückig small-sized
Kleinvernebler *m (agric)* hand fogger
Kleinverstäuber *m (agric)* hand[-operated] duster
Kleinversuch *m* small-scale experiment
Kleinwinkelstreuung *f (phys chem)* small-angle scattering
Kleister *m* paste
Kleistertrübung *f (ferm)* starch turbidity
Klemme *f* clamp, clip
große K. *(lab)* condenser clamp
klemmen to clamp; to stick
Klemmring *m* locking ring
Klemmverbindung *f*, **Klemmverschraubung** *f* compression-type fitting
Kletterfilmverdampfer *m* rising-film evaporator, long-tube [vertical-film] evaporator, LTV evaporator
Kliachit *m (min)* kliachite, alumogel *(gel of aluminium hydroxide)*
Klimaanlage *f* air conditioning plant (system, apparatus)
Klimagerät *n* [air] conditioner
Klimakammer *f* climatic chamber
Klimaregelung *f* [air] conditioning
klimatisieren to [air-]condition
Klimatisierung *f* [air] conditioning
Klimatisierungsanlage *f s.* Klimaanlage
klingeln to pink, to knock *(of a carburettor engine)*
Klingeln *n* pinking, ping, knock[ing] *(of a carburettor engine)*
Klinker[stein] *m* clinker [brick], engineering brick
Klinkerverfahren *n* clinker process
Klinkerziegel *m s.* Klinker
Klinochlor *m (min)* clinochlore, clinochlorite *(a phyllosilicate)*
Klinoenstatit *m (min)* clinoenstatite *(magnesium metasilicate)*
Klinoklas *m (min)* clinoclase, clinoclasite *(a basic copper arsenate)*
Klinozoisit *m (min)* clinozoisite *(a sorosilicate)*
Klockmannit *m (min)* klockmannite *(copper(II) selenide)*
Klopfbremse *f* antiknock [additive, agent, compound], knock inhibitor (reducer, suppressor), octane improver
Klopfbremswirkung *f* antiknock effect
Klopfeigenschaft *f* knocking property
klopfempfindlich prone to knocking
Klopfempfindlichkeit *f s.* Klopfneigung
klopfen to knock, to pink *(of a carburetting fuel)*
Klopfen *n* knock[ing], pinking, ping *(of a carburetting fuel)*
Klopfer *m* rapper *(as for a gas-cleaning electrode)*
klopffest knockproof *(carburetting fuel)*
Klopffestigkeit *f* 1. antiknock quality, knock resistance; 2. *s.* Klopfwert
klopffrei 1. knock-free, knockless, non-knocking *(ignition of carburetting fuel)*; 2. *s.* klopffest
klopffreudig prone to knocking
Klopffreudigkeit *f*, **Klopfneigung** *f* knock proneness, knocking propensity
Klopfprüfmotor *m* knock-rating engine
Klopfprüfung *f* [anti]knock rating
klopfstark prone to knocking
Klopfstärke *f* knock intensity

Klopfvorrichtung *f* knocker; rapper *(as for a gas-cleaning electrode)*
Klopfwerk *n* knocker
Klopfwert *m* antiknock (knock) value (rating)
Klopfwertbestimmung *f*, **Klopfwertprüfung** *f* [anti]knock rating
Klotz *m*, **Klotzbad** *n s.* Klotzflotte
Klotzchassis *n (text)* pad box
Klotz-Dämpf-Färbeverfahren *n (text)* pad-steam process
klotzen *(text)* to pad
Klotzen *n*, **Klotzfärben** *n (text)* pad dyeing, [slop-]padding
Klotzfärbeverfahren *n (text)* pad dyeing process
Klotzfärbung *f (text)* pad dyeing
Klotz-Fixier-Verfahren *n (text)* pad-fix process
Klotzflotte *f (text)* [slop] pad liquor, pad[ding] bath
Klotzhilfsmittel *n (text)* padding auxiliary
Klotzmaschine *f (text)* padding machine (mangle), pad[der]
Klotz-Roll-Verfahren *n (text)* pad-roll method
Klotz-Thermofixier-Färbeverfahren *n (text)* pad thermofix dyeing
Klotz-Trocken-Kondensierverfahren *n (text)* pad-dry process
Klotztrog *m (text)* pad box
Klotz-Verweil-Verfahren *n (text)* pad-store process
klumpen to agglomerate, to clog, to clot, to lump, to agglutinate
Klumpen *m* lump, clot, bat
Klumpenbildung *f* agglomeration, clogging, clotting, lumping, agglutination
klumpig lumpy
  k. werden *s.* klumpen
Klumpigkeit *f* lumpiness
Klupanodonsäure *f* clupanodonic acid *(any of several polyunsaturated carboxylic acids, specif 4,7,11-docosatrien-18-ynoic acid)*
Klupein *n* clupeine *(a protamine)*
KM *s.* Kernmagneton
K-Meson *n* K meson
Knab-Ofen *m* sole-flue oven
Knallgas *n* detonating gas, *(specif)* oxyhydrogen gas
Knallgascoulometer *n* oxygen-hydrogen coulometer
Knallgaselement *n* oxyhydrogen (hydrogen-oxygen fuel) cell
Knallgasflamme *f* oxyhydrogen flame
Knallgasgebläse *n* oxyhydrogen blowpipe (burner, torch)
Knallgasreaktion *f* hydrogen-oxygen reaction
Knallquecksilber *n* $Hg(ONC)_2$ fulminating mercury, mercury (mercuric) fulminate
Knallsäure *f* C≡N−OH fulminic acid, carbonyl oxime
Knallzündschnur *f* detonating fuse
Knäuel *n* coil *(as of a macromolecule)*
Knäuelmolekül *n* coiled molecule
Knäuelung *f* coiling *(as of a macromolecule)*
Knetarm *m* mixing arm (blade)
  Z-förmiger K. sigma blade
kneten to knead
  zu Teig k. to dough [in]
Kneter *m s.* Knetwerk
Knetgut *n* material being (*or* to be) kneaded (mixed)
Knetlegierung *f* wrought alloy
Knetmaschine *f s.* Knetwerk
Knetmischer *m* kneader mixer
Knetorgan *n* kneading (mixing) element
Knetschaufel *f* mixing blade; *(rubber)* rotor *(of a closed mixer)*
Knetschnecke *f* mixing screw
Knetteller *m* [rotary] kneading table
Knettrog *m (rubber)* mixing chamber *(of a closed mixer)*
Knetwalze *f* kneading roll
Knetwerk *n* kneading machine, kneader, [dough] mixer, pug mill
Knickbeständigkeit *f s.* Knickfestigkeit
knicken *(text)* to roll *(flax)*
Knickfestigkeit *f (tann)* burst[ing] strength
Knickpunkt *m* break *(as of a curve)*; critical moisture point *(in drying processes)*
Knickpunktfeuchte *f*, **Knickpunkt-Feuchtebeladung** *f* critical moisture content *(theory of drying)*
Knie *n* elbow [fitting], ell
Kniehebelbackenbrecher *m* toggle (double toggle, Blake jaw) crusher
Kniehebelpresse *f* toggle [lever] press
Kniestück *n* elbow [fitting], ell
Knirschgriffappretur *f (text)* rustling finish
Knistersalz *n* crackling salt
knitterarm, knitterbeständig, knitterecht *s.* knitterfest
Knittererholungsvermögen *n (text)* crease recovery resistance
Knittererholungswinkel *m (text)* angle of crease recovery, crease angle
knitterfest *(text)* crease-resistant, non-creasing, non-creasable, *(Am)* noncrushable, crush-proof, crush-resistant
Knitterfestappretur *f*, **Knitterfestausrüstung** *f (text)* crease-resistant finish, anticrease (non-crease) finish (treatment), crush proofing
Knitterfestigkeit *f (text)* crease (creasing, wrinkle) resistance
Knitterfreiheit *f s.* Knitterfestigkeit
knittern *(text)* to crease, to wrinkle, to crush
Knitterpapier *n* creased paper
Knitterung *f (text)* creasing, wrinkling, crushing
Knitterwiderstand *m s.* Knitterfestigkeit
Knitterwinkel *m s.* Knittererholungswinkel
Knoblauchöl *n* garlic oil *(from Allium sativum L.)*
Knochenasche *f* bone ash
Knochenfett *n* bone fat (grease)
Knochengelatine *f* bone gelatin
Knochenkohle *f* animal char[coal], bone char[coal], spodium
Knochenleim *m* bone glue
Knochenöl *n* bone oil
Knochenporzellan *n* bone china

Knochenschwarz *n* animal (bone) black
Knochenteer *m* bone tar
knochentrocken *(ceram)* bone-dry, B.D., white-hard
Knockdown-Mittel *n* knockdown agent (poison) *(any of a class of immediately acting insecticides)*
Knockdown-Wirkung *f* knockdown effect *(of certain insecticides)*
Knock-out-Punkt *m* knock-out point *(in testing insecticides)*
Knoevenagel-Kondensation *f (org chem)* Knoevenagel condensation
Knöllchenbakterien *pl* nodule-forming bacteria, legume bacteria *(Rhizobium specc.)*
knollig *(min)* nodular
Knopflack *m* button lac
Knopfprobe *f*, **Knopfprüfung** *f (ceram)* [flow] button test, fusion flow test *(for testing the fusibility of an enamel frit)*
Knopfschellack *m* button lac
Knopper *f (tann)* knopper, knoppern nut, acorn gall
Knötchen *n (text)* burl, mote, nep
Knoten *m (pap, glass)* knot *(a defect)*
Knotenebene *f* nodal plane *(wave mechanics)*
Knotenfänger *m (pap)* knot screen, knotter [screen], jag-knotter
K. mit Stoffdurchgang von außen nach innen inward flow strainer
K. mit Stoffdurchgang von innen nach außen outward flow strainer
K. von Wandel Wandel strainer
rotierender K. rotary (rotating, revolving, revolving drum) strainer
Knotenfängerschlitz *m (pap)* screen slot
Knotenfläche *f* nodal plane *(wave mechanics)*
Knüppel *m (met)* billet
Koagel *n* coagel
Koagulans *n s.* Koagulationsmittel
Koagulans-Verfahren *n (rubber)* coagulant dipping process
Koagulase *f* coagulase
Koagulat *n* coagulate, coagulum, flocculate
Koagulation *f* coagulation, flocculation, flocculence, clotting, curdling
thixogene K. *(coll)* rheopexy
Koagulationsbad *n (text)* coagulating (coagulation) bath
Koagulationsdauer *f* coagulation time (period)
Koagulationsmittel *n* coagulant, coagulator, coagulating agent, coagulation chemical, flocculant
Koagulationstauchverfahren *n (rubber)* coagulant dipping process
Koagulationsvermögen *n* coagulating power
Koagulationsvitamin *n* coagulation (antihaemorrhagic) vitamin, phylloquinone
Koagulationswert *m (coll)* coagulation value
Koagulator *m s.* Koagulationsmittel
Koagulatplatte *f (rubber)* slab of coagulum
koagulierbar coagulable
Koagulierbarkeit *f* coagulability
koagulieren to coagulate, to flocculate, to curd[le], to clot
Koagulum *n s.* Koagulat
Koaleszenz *f* coalescence
koaleszieren, koal[is]ieren to coalesce
Koazervat *n* coacervate
Koazervation *f*, **Koazervierung** *f* coacervation
Kobalt *n* Co cobalt
radioaktives K. radioactive cobalt, radiocobalt, *(specif)* ^{60}Co cobalt-60
Kobalt-60 *n* ^{60}Co cobalt-60
Kobaltammin *n* cobalt ammine
Kobalt(II)-arsenat *n* cobalt(II) arsenate, *(specif)* $Co_3(AsO_4)_2$ cobalt(II) orthoarsenate
Kobaltat *n* cobaltate
Kobalt(II)-azetat *n* $Co(CH_3COO)_2$ cobalt(II) acetate
Kobalt(III)-azetat *n* $Co(CH_3COO)_3$ cobalt(III) acetate
Kobaltblau *n* cobalt (ultramarine, king's) blue *(cobalt aluminate)*
Kobaltblüte *f (min)* cobalt bloom, erythrine, erythrite *(cobalt(II) orthoarsenate)*
Kobalt(II)-bromat *n* $Co(BrO_3)_2$ cobalt(II) bromate
Kobalt(II)-bromid *n* $CoBr_2$ cobalt(II) bromide, cobalt dibromide
Kobaltchelat *n* cobalt chelate
Kobalt(II)-chlorat *n* $Co(ClO_3)_2$ cobalt(II) chlorate
Kobalt(II)-chlorid *n* $CoCl_2$ cobalt(II) chloride, cobalt dichloride
Kobaltchloridpapier *n* cobalt chloride test paper
Kobalt(II)-chromat *n* $CoCrO_4$ cobalt(II) chromate
Kobaltdi ... *s. a.* Kobalt(II)- ...
Kobaltdisulfid *n* CoS_2 cobalt disulphide
Kobalt(II)-fluorid *n* CoF_2 cobalt(II) fluoride, cobalt difluoride
Kobalt(III)-fluorid *n* CoF_3 cobalt(III) fluoride, cobalt trifluoride
Kobaltgelb *n* cobalt yellow, Indian yellow, aureolin *(potassium hexanitrocobaltate)*
Kobaltglanz *m s.* Kobaltin
Kobaltglas *n* cobalt (blue) glass
Kobaltgrün *n* cobalt (Rinmann's) green
Kobalt(II)-hexafluorosilikat *n* $Co[SiF_6]$ cobalt(II) hexafluorosilicate, cobalt fluorosilicate
Kobalt(II)-hexazyanoferrat(II) *n* $Co_2[Fe(CN)_6]$ cobalt(II) hexacyanoferrate(II)
Kobalt(II)-hexazyanoferrat(III) *n* $Co_3[Fe(CN)_6]_2$ cobalt(II) hexacyanoferrate(III)
Kobalt(II)-hydroxid *n* $Co(OH)_2$ cobalt(II) hydroxide
Kobalt(III)-hydroxid *n* $Co(OH)_3$ cobalt(III) hydroxide
Kobaltiak *n* cobalt(III) ammine
Kobaltin *m (min)* cobaltite, cobaltine, cobalt glance *(cobalt sulpharsenide)*
Kobalt(II)-jodat *n* $Co(IO_3)_2$ cobalt(II) iodate
Kobalt(II)-jodid *n* CoI_2 cobalt(II) iodide, cobalt diiodide
Kobalt(II)-karbonat *n* $CoCO_3$ cobalt(II) carbonate
Kobaltkatalysator *m* cobalt catalyst
Kobaltkies *m (min)* linnaeite, cobalt pyrites *(cobalt(II,III) sulphide)*

Kobaltkontakt *m* cobalt catalyst
Kobaltmonoselenid *n s.* Kobalt(II)-selenid
Kobaltmonosulfid *n s.* Kobalt(II)-sulfid
Kobaltmonoxid *n s.* Kobalt(II)-oxid
Kobalt(II)-nitrat *n* $Co(NO_3)_2$ cobalt(II) nitrate
Kobalt(II)-orthoarsenat(V) *n* $Co_3(AsO_4)_2$ cobalt(II) orthoarsenate, cobalt(II) arsenate
Kobalt(II)-orthophosphat *n* $Co_3(PO_4)_2$ cobalt(II) orthophosphate, cobalt(II) phosphate
Kobalt(II)-orthosilikat *n* Co_2SiO_4 cobalt(II) orthosilicate
Kobalt(II)-orthotitanat *n* Co_2TiO_4 cobalt(II) orthotitanate
Kobaltoxid *n* cobalt oxide, *(specif)* CoO cobalt(II) oxide
Kobalt(II)-oxid *n* CoO cobalt(II) oxide, cobalt monooxide
Kobalt(II,III)-oxid *n* Co_3O_4 cobalt(II,III) oxide, tricobalt tetraoxide
Kobalt(III)-oxid *n* Co_2O_3 cobalt(III) oxide, dicobalt trioxide
Kobalt(II)-perchlorat *n* $Co(ClO_4)_2$ cobalt(II) perchlorate
Kobalt(II)-perrhenat *n* $Co(ReO_4)_2$ cobalt(II) perrhenate, cobalt(II) tetraoxorhenate(VII)
Kobalt(II)-phosphat *n* cobalt(II) phosphate, *(specif)* $Co_3(PO_4)_2$ cobalt(II) orthophosphate
Kobalt(II)-phosphit *n* $CoPHO_3$ cobalt(II) phosphite, *(better)* cobalt phosphonate
Kobaltphthalozyanin *n* cobalt phthalocyanine
Kobaltpulver *n* cobalt powder
Kobalt(II)-rhodanid *n s.* Kobalt(II)-thiozyanat
Kobalt(II)-salz *n* cobalt(II) salt
Kobaltseife *f* cobalt soap
Kobalt(II)-selenat *n* $CoSeO_4$ cobalt(II) selenate
Kobalt(II)-selenid *n* CoSe cobalt(II) selenide
Kobaltsikkativ *n (coat)* cobalt drier
Kobalt(II)-sulfat *n* $CoSO_4$ cobalt(II) sulphate
Kobalt(III)-sulfat *n* $Co_2(SO_4)_3$ cobalt(III) sulphate
Kobalt(II)-sulfid *n* CoS cobalt(II) sulphide
Kobalt(II)-sulfit *n* $CoSO_3$ cobalt(II) sulphite
Kobalttetrakarbonyl *n* $Co(CO)_4$ cobalt tetracarbonyl
Kobalttetroxid *n s.* Kobalt(II,III)-oxid
Kobalt(II)-tetroxorhenat(VII) *n s.* Kobalt(II)-perrhenat
Kobalt(II)-tetroxosilikat *n s.* Kobalt(II)-orthosilikat
Kobalt(II)-thiozyanat *n* $Co(SCN)_2$ cobalt(II) thiocyanate, cobalt(II) rhodanide
Kobalttrifluorid *n s.* Kobalt(III)-fluorid
Kobalttrikarbonyl *n* $Co(CO)_3$ cobalt tricarbonyl
Kobalttrioxid *n s.* Kobalt(III)-oxid
Kobalttrockner *m (coat)* cobalt drier
Kobaltultramarin *n s.* Kobaltblau
Kobaltvitriol *m (min)* cobalt vitriol, bieberite *(cobalt(II) sulphate-7-water)*
Kobaltwolframat *n* $CoWO_4$ cobalt wolframate, cobalt tungstate
Kobalt(II)-zyanid *n* $Co(CN)_2$ cobalt(II) cyanide
Kochbedingungen *fpl (pap)* cooking conditions
kochbeständig resistant (fast) to boiling, boilproof, boilfast
Kochbeständigkeit *f* resistance (fastness) to boiling
Kochchemikalie *f (pap)* cooking agent (reagent, chemical)
Kochdauer *f* cooking time, *(pap also)* digestion time
kochecht *s.* kochbeständig
Kochechtheit *f s.* Kochbeständigkeit
kochen to cook, to boil; to be wild *(of rimming steel)*
auf Korn k. *(sugar)* to boil to grain, *(Am also)* to sugar off *(esp maple sap)*
im Autoklaven k. to autoclave
unter Rückfluß k. *(distil)* to reflux
unvollständig k. *(pap)* to undercook
Kochen *n* cooking, boiling, *(pap also)* digestion ✦ **am K. halten** to keep at the boil ✦ **zum K. bringen** to raise (bring) to the boil
K. auf Korn *(sugar)* boiling to grain, crystal boiling, evaporative crystallization
K. unter Druck boiling under pressure
nochmaliges K. recooking
Kocher *m* cooker, boiler, boiling vessel, *(pap also)* digester, kier
liegender K. *(pap)* horizontal digester
rotierender K. *(pap)* revolving (rotary) digester
stehender K. *(pap)* vertical (upright) digester
Kocherabgas *n (pap)* digester relief gas, release
Kocherabgaswärme *f (pap)* heat in the relief
Kocherausblasen *n (pap)* digester blow
Kocherdeckel *m (pap)* digester cover
Kocherdruck *m (pap)* digester pressure
Kocherei *f (pap)* digester (boiling, boil) house, boiling room
Kochereintrag *m s.* Kocherinhalt
Kocherführer *m s.* Kochermeister
Kocherfüllapparat *m (pap)* chip distributor (packer)
Kocherfüllung *f s.* Kocherinhalt
Kochergrube *f (pap)* blow pit (tank), wash (receiving) tank
Kocherinhalt *m (pap)* digester charge (contents), furnish, cook
ausgeblasener K. blow
Kocherleerung *f (pap)* digester blow
Kochermeister *m (pap)* cook[er], boilerman
Kocherraum *m s.* Kochervolumen
Kocherturnus *m*, **Kocherumtrieb** *m (pap)* digester cycle
Kochervolumen *n (pap)* digester capacity (space)
kochfertig ready-to-cook
kochfest resistant (fast) to boiling, boilproof, boilfast
Kochfestigkeit *f* resistance (fastness) to boiling
Kochfett *n* cooking fat
Kochflasche *f s.* Kochkolben
Kochflüssigkeit *f (pap)* digestion (cooking, pulping) liquor
Kochgeschmack *m* cooked flavour (taste)
Kochgut *n (pap)* digester charge, furnish, cook
Kochkäse *m* cook[ed] cheese
Kochkessel *m* cooking kettle (vat)

Kochkläre *f (sugar)* clearing liquor, clairce
Kochkolben *m* boiling (Florence) flask
Kochlauge *f (pap)* cooking (digestion, pulping) liquor
Kochlaugenanlage *f (pap)* liquor-making plant
Kochlaugenbehälter *m (pap)* cooking-liquor manufacture (preparation)
Kochlaugenleitung *f (pap)* liquor-circulating line
Kochlaugenvorratstank *m (pap)* liquor tank
Kochlaugenzusammensetzung *f (pap)* liquor composition
Kochlösung *f s.* Kochflüssigkeit
Kochmaische *f* decoction mash
Kochprobe *f* boil[ing] test; *(tann)* shrinkage test *(for determining the effectiveness of chrome tannings)*
Kochpunkt *m* boiling point, b.p., bp, boiling temperature
Kochraum *m s.* Kochervolumen
Kochsalz *n* common salt
Kochsalzlösung *f* solution of common salt, solution of sodium chloride
physiologische K. *(pharm)* physiological saline (salt solution), saline
Koch-Säure *f* $NH_2C_{10}H_4(SO_3H)_3$ Koch acid, 1-naphthylamine-3,6,8-trisulphonic acid
Kochsäure *f (pap)* cooking acid (liquor), digestion liquor
hochkonzentrierte (hochprozentige) K. *(pap)* high-strength (full-strength) cooking acid, strong acid
Kochsäureanlage *f (pap)* acid[-making] plant
Kochsäuredruckspeicher *m (pap)* pressure container (accumulator)
Kochsäuredurchtränkung *f (pap)* penetration of cooking acid
Kochsäureherstellung *f (pap)* cooking-acid manufacture (preparation)
Kochsäureleitung *f (pap)* acid-circulating line
Kochsäureumlauf *m (pap)* cooking-acid circulation
Kochsäureumwälzung *f*, **Kochsäurezirkulation** *f s.* Kochsäureumlauf
Kochsäurezusammensetzung *f (pap)* cooking-acid composition
Kochschnitzel *npl (pap)* [wood] chips, chippings
Kochstation *f (pap)* digester (boiling, boil) house, boiling room
Kochstoff *m (pap)* digester charge, furnish, cook
Kochtemperatur *f* cooking temperature, *(pap also)* digester temperature
Kochtrub *m (ferm)* coarse sludge
Kochung *f (pap)* cook[ing], boil[ing], digestion
alkalische K. alkaline cook
direkte K. direct (quick) cook
direkte K. nach Ritter-Kellner Ritter-Kellner cook process
indirekte K. indirect (slow) cook
indirekte K. nach Mitscherlich Mitscherlich cook process
unvollständige K. undercooking
Kochverfahren *n (ferm)* decoction process
Kochvorgang *m (pap)* cooking process
Kochzeit *f* cooking time, *(pap also)* digestion time
Kochzyklus *m (pap)* cooking cycle
Kodehydr[ogen]ase I *f s.* Nikotinamid-adenin-dinukleotid
Kodehydr[ogen]ase II *f s.* Nikotinamid-adenin-dinukleotidphosphat
Kodein *n* codeine, methylmorphine *(alkaloid)*
Kodeinon *n (org chem)* codeinone
Kodenummer *f* **der Gruppe** *(coal)* group number
K. der Klasse *(coal)* class number
K. der Untergruppe *(coal)* subgroup number
Köder *m* bait, baiting agent
Ködergift *n* poison for baits
Ködermittel *n s.* Köder
Kodestillation *f* codistillation
Kodeziffer *f* / **Dritte** *(coal)* subgroup number
Erste K. *(coal)* class number
Zweite K. *(coal)* group number
Koeffizient *m* / **osmotischer** osmotic coefficient
respiratorischer K. respiratory quotient (ratio)
Koenzym *n s.* Koferment
Koferment *n* coenzyme, prosthetic (active) group, agon
Koferment A *n* coenzyme A
Koferment I *n s.* Nikotinamid-adenin-dinukleotid
Koferment II *n s.* Nikotinamid-adenin-dinukleotidphosphat
Koffein *n* caffeine, 1,3,7-trimethylxanthine
koffeinarm decaffeinated
koffeinfrei caffeine-free, decaffeinated
koffeinhaltig caffeinic
Koffein-Natriumbenzoat *n (pharm)* caffeine and sodium benzoate
Koffein-Natriumsalizylat *n (pharm)* caffeine and sodium salicylate
Kogag-Ofen *m* Kogag oven *(for coking coal)*
Kogasinverfahren *n* Kogasin (Fischer-Tropsch) process (synthesis)
kohärent coherent
Kohärenz *f* coherence, coherency
kohärieren to cohere
Kohäsion *f* cohesion
Kohäsionsarbeit *f* work of cohesion
Kohäsionsdruck *m* cohesion (instrinsic) pressure
Kohäsionsfestigkeit *f* cohesive strength
Kohäsionskraft *f* cohesive (cohesion, cohesional) force
kohäsiv cohesive
Kohle *f* coal; *(el chem)* carbon [electrode]
K. der Kalksteingruppe limestone coal
K. mit erhöhtem Wasserstoffgehalt perhydrous coal
K. mit geringem Backvermögen weakly caking coal
aktive (aktivierte) K. active (activated) carbon (charcoal)
anthrazitische K. anthracitic coal
aufgegebene K. *(tech)* feed[-stock] coal
backende K. caking coal
bituminöse K. bituminous (soft) coal
brikettierte K. briquetted coal

eingesetzte K. *(tech)* feed[-stock] coal
fossile K. fossil (mineral, natural) coal
gasreiche bituminöse K. high-volatile bituminous coal
geringbituminöse K. low-volatile bituminous [steam] coal
glänzende K. bright coal
gut kokende (verkokbare) K. strongly coking coal
halbbituminöse K. semibituminous coal
hochflüchtige [bituminöse] K. high-volatile [bituminous] coal
humitische K. humic coal
karbonische K. carboniferous coal
kohlenstoffreiche K. carbonaceous coal
künstliche K. artificial coal; *(if made from wood, blood, or bones)* char[coal]
kurzflammige K. short-flame coal
langflammige K. long-flame coal
lignitische K. lignite
magere K. lean coal
mäßig backende K. medium-caking coal
mäßig kokende (verkokbare) K. medium coking coal
matte K. dull coal
metabituminöse K. metabituminous coal
mineralische K. *s.* fossile K.
mittelbackende K. medium-caking coal
mittelflüchtige [bituminöse] K. medium-volatile [bituminous] coal
natürliche K. *s.* fossile K.
nichtbackende K. non-caking coal
nichtkokende (nicht verkokbare) K. non-coking coal
niedrigflüchtige [bituminöse] K. low-volatile bituminous [steam] coal
orthobituminöse K. orthobituminous coal
parabituminöse K. parabituminous coal
perbituminöse K. perbituminous coal
perhydrierte K. perhydrous coal
pulverisierte K. powdered (pulverized) coal
pyrogene K. heat-altered coal *(having lost its caking power by the action of hot rock)*
reine K. pure coal material (substance); *(min tech)* pure (clean) coal *(with minimum ash content)*
sapropelitische K. sapropelic coal
schlecht kokende (verkokbare) K. weakly coking coal
schwachbackende K. weakly (feebly) caking coal
schwachkokende K. weakly coking coal
selbstbackende K. *(el chem)* self-baking electrode
semibituminöse K. semibituminous coal
starkbackende K. high-caking coal, strongly caking coal
streifige K. banded coal
subbituminöse K. subbituminous (lignitous) coal, bituminous (black) lignite
subhydrierte K. subhydrous coal
synthetische K. artificial coal
unverkokbare K. *s.* nichtkokende K.
vorgebrannte K. *(el chem)* prebaked electrode
wasserstoffarme K. subhydrous coal
wasserstoffreiche K. perhydrous coal
kohleähnlich coal-like
Kohleart *f* type (species) of coal
kohleartig coal-like
Kohleaufbereitung *f* coal preparation (dressing)
Kohleaufbereitungsanlage *f* coal-preparation plant
Kohleaufgabe *f* coal feed
Kohleaufgabeschleuse *f* coal-charging vessel *(as of a pressure gasifier)*
kohlebeheizt coal-fired
Kohlebeheizung *f* coal-fired heating
Kohlebeschickung *f* coal feed
Kohlebeschickungsmaschine *f* coal-charging machine
Kohlebestandteil *m* coal component
kohlebildend coal-forming
Kohlebildung *f* coal formation (genesis)
Kohlebildungsvorgang *m* coal-forming process
Kohlebogen *m* carbon arc
Kohlebohrer *m* *(lab)* charcoal borer
Kohlebraun *n* Cassel brown (earth), ulmin brown *(a naturally occurring pigment)*
Kohlebrei *m* coal slurry (paste)
Kohlebunker *m* coal bunker; coal [storage] hopper *(for bottom discharge)*
Kohlechemie *f* coal chemistry
Kohledestillation *f* distillation of coal
Kohleeinteilung *f* coal classification
Kohleelektrode *f* carbon [electrode]
Kohleentgasung *f* coal carbonization
Kohle-Erz-Brikett *n* coal-ore briquette
Kohle-Erz-Gemisch *n* coal-ore mixture
Kohleextraktion *f* coal extraction
Kohlefadenlampe *f* carbon filament lamp
Kohlefilter *n* char filter
Kohlefleck *m* carbon spot *(a defect in paper)*
Kohlefolgeprodukt *n* coal product
kohleführend coal-bearing
Kohlefüllmasse *f*, **Kohlefüllung** *f* coal charge
Kohlefüllwagen *m* coal-charging car
Kohlegefügebestandteil *m* coal maceral
kohlegeheizt coal-fired
Kohlegelbglas *n* amber glass
Kohlegemisch *n* coal blend (mixture)
Kohlegenesis *f* coal formation (genesis)
Kohlehydrat *n* *s.* Kohlenhydrat
Kohlehydrierung *f* coal hydrogenation
Kohlekammer *f* coal stall *(in coal hydrogenation)*
Kohleklasse *f* coal class
Kohleklassifikation *f* coal classification
K. nach dem Inkohlungsgrad rank classification of coals
Kohleklassifizierung *f* coal classification
Kohleklassifizierungsschaubild *n* coal chart
Seylers K. Seyler's [coal] chart
Kohleklein *n* small[-sized] coal, smalls
Kohlekomponente *f* coal component
Kohlelager *n* 1. *(geol)* coal deposit; 2. coal store
Kohlelichtbogen *m* carbon arc

Kohlemahlung *f* coal milling (grinding, pulverization)
Kohlemischung *f* coal blend (mixture)
Kohlemühle *f* coal pulverizer (mill), coal-pulverizing (coal-grinding, coal-dust) mill
kohlen to char; to carburize *(steel)*
kohlen ... *s. a.* kohle ...
Kohlen *n* charring; carburization, carburizing *(of steel)*
Kohlen ... *s. a.* Kohle ...
Kohlenband *n* coal band
Kohlenbrikett *n* coal briquette
Kohlencharge *f* coal charge
Kohlendioxid *n* CO_2 carbon dioxide
  festes K. solid carbon dioxide, dry ice
Kohlendioxidassimilation *f* carbon dioxide assimilation
Kohlendioxidlöscher *m* carbon dioxide fire extinguisher
Kohlendioxidschnee *m* carbon dioxide ice (snow)
Kohlendioxidüberträger *m* carbon dioxide carrier
Kohlendisulfid *n* CS_2 carbon disulphide
Kohlenentstehung *f* coal formation (genesis)
Kohlenfeld *n* panel of coal *(in underground gasification)*
Kohlenfüllhahn *m* coal inlet valve *(of a gasifier)*
Kohlengas *n* coal gas
Kohlengattung *f* species of coal
Kohlengestein *n* carbonaceous rock
Kohlengesteinskunde *f* coal petrology
Kohlengrus *m* coal breeze
kohlenhaltig carbonaceous
Kohlenhydrat *n* carbohydrate, saccharide
Kohlenhydratauslösung *f (pap)* dissolution of carbohydrates
Kohlenhydratchemie *f* carbohydrate chemistry
Kohlenhydratstoffwechsel *m* carbohydrate metabolism
Kohlenlademaschine *f* coal-load instrument
Kohlenladung *f* coal charge
Kohlenmazeral *n* coal maceral
Kohlenmonosulfid *n* CS carbon monosulphide
Kohlenmonoxid *n* CO carbon monooxide
Kohlenmonoxid-Hämoglobin *n s.* Kohlenoxidhämoglobin
Kohlenoxid *n s.* Kohlenmonoxid
Kohlenoxidbromid *n* $COBr_2$ carbon dibromide oxide, carbonyl bromide, bromophosgene
Kohlenoxidchlorid *n* $COCl_2$ carbon dichloride oxide, carbonyl chloride, phosgene
Kohlenoxidhämoglobin *n* carbon monooxyhaemoglobin, carboxyhaemoglobin, carbonylhaemoglobin
Kohlenoxidsulfid *n* COS carbon oxide sulphide, carbonyl sulphide
Kohlenpetrografie *f* coal petrography
Kohlenpetrologie *f* coal petrology
Kohlenpreßling *m* coal briquette
Kohlenrang *m*, **Kohlenrangstufe** *f* coal rank
Kohlensäule *f* column of coal *(as for determining its plasticity)*
Kohlensäure *f* H_2CO_3 carbonic acid
Kohlensäureanhydr[at]ase *f* carbonic anhydrase
Kohlensäurediamid *n s.* Karbamid
Kohlensäuredichlorid *n s.* Kohlenoxidchlorid
Kohlensäureerstarrungsverfahren *n* CO_2 process *(foundry)*
Kohlensäurelöscher *m* carbon dioxide fire extinguisher
Kohlensäuremonamid *n s.* Karbamidsäure
Kohlensäurepatrone *f* sparklet bulb
Kohlensäureschnee *m* carbon dioxide ice (snow), dry ice
Kohlensäureschreiber *m* CO_2 recorder
Kohlenschwelung *f* coal carbonization
Kohlensetzmaschine *f* coal jig
Kohlenstaub *m* coal dust, breeze; pulverized (powdered) coal
Kohlenstaubbrenner *m* pulverized-coal burner
Kohlenstaubfeuerung *f* suspension firing of coal
Kohlenstaubmahlung *f* coal milling (grinding, pulverization)
Kohlenstaubmühle *f s.* Kohlemühle
Kohlenstaub-Wasser-Trübe *f* coal/water slurry *(in elevated-pressure gasification)*
Kohlenstoff *m* C carbon
  aktiver (aktivierter) K. active (activated) carbon
  fester (fixer) K. fixed carbon, FC
  freier K. free carbon
  gelöster organischer K. total organic carbon, TOC *(waste-water evaluation)*
  radioaktiver K. *s.* Kohlenstoff-14
Kohlenstoff-14 *m* ^{14}C carbon-14, radioactive carbon, radiocarbon
Kohlenstoff-14-Alter *n* radiocarbon age
Kohlenstoff-14-Methode *f* ^{14}C method, radiocarbon method
Kohlenstoffablagerung *f* carbon deposit
Kohlenstoffalter *n* radiocarbon age
kohlenstoffarm poor in carbon, low-carbon *(e. g. steel)*
Kohlenstoffassimilation *f* carbon assimilation
Kohlenstoffatom *n* / **asymmetrisches** asymmetric carbon atom
  primäres K. primary carbon atom
  sekundäres K. secondary carbon atom
  tertiäres K. tertiary carbon atom
Kohlenstoffbilanz *f (agric)* carbon balance
Kohlenstoffbindung *f* / **doppelte** *s.* Kohlenstoff-Doppelbindung
  dreifache K. *s.* Kohlenstoff-Dreifachbindung
  einfache K. *s.* Kohlenstoff-Einfachbindung
Kohlenstoffblock *m (met)* carbon block
Kohlenstoffchemie *f* chemistry of the carbon compounds
Kohlenstoffdioxid *n s.* Kohlendioxid
Kohlenstoffdisulfid *n* CS_2 carbon disulphide
Kohlenstoff-Doppelbindung *f* carbon[-carbon] double bond, carbon-to-carbon double bond, C=C bond
Kohlenstoff-Dreifachbindung *f* carbon[-carbon] triple bond, carbon-to-carbon triple bond, C≡C bond
Kohlenstoffdreiring *m* three-membered carbon ring

Kohlenstoff-Einfachbindung *f* carbon[-carbon] single bond, carbon-to-carbon single bond, C–C bond
Kohlenstoffeinsatzstahl *m* carbon-carburizing steel, carbon case-hardening steel
kohlenstofffrei free from carbon
Kohlenstoff-Fünfring *m* five-membered carbon ring
Kohlenstoffgehalt *m* carbon content ✦ **mit hohem K.** high-carbon ✦ **mit mittlerem K.** medium-carbon ✦ **mit niedrigem K.** low-carbon
Kohlenstoffgerüst *n* carbon skeleton (framework)
Kohlenstoff-Halogen-Bindung *f* carbon-halogen bond
kohlenstoffhaltig carbonaceous
Kohlenstoffkette *f* carbon chain
geschlossene K. closed carbon chain
Kohlenstoff-Kohlenstoff-Bindung *f* carbon-carbon bond, C–C bond
Kohlenstoffmonosulfid *n* CS carbon monosulphide
Kohlenstoffnachweis *m* detection of carbon
Kohlenstoffpotential *n* carburizing level *(of steel)*
kohlenstoffreich rich in carbon, high-carbon
Kohlenstoffring *m* carbon ring
sechsgliedriger K. six-membered carbon ring, six-carbon[-atom] ring
siebengliedriger K. seven-membered carbon ring, seven-carbon[-atom] ring
Kohlenstoffrückstand *m* carbon residue; *(petrol)* coke (carbon) residue
K. nach Conradson *(petrol)* Conradson coke (carbon) residue
Kohlenstoffselenidsulfid *n* CSeS carbon selenide sulphide
Kohlenstoff-Silizium-Bindung *f* carbon silicon bond
Kohlenstoffstahl *m* carbon steel
reiner K. plain (straight) carbon steel
Kohlenstoffstein *m* carbon brick
Kohlenstoff-Stickstoff-Gerüst *n* carbon-nitrogen skeleton
Kohlenstoff-Stickstoff-Verhältnis *n* *(soil)* C/N ratio
Kohlenstofftelluridsulfid *n* CTeS carbon telluride sulphide
Kohlenstofftetrabromid *n* CBr_4 carbon tetrabromide, tetrabromomethane, perbromomethane
Kohlenstofftetrachlorid *n* CCl_4 carbon tetrachloride, tetrachloromethane, perchloromethane
Kohlenstoffverbindung *f* carbon compound
ringförmige K. carbocyclic compound
Kohlenstoff-Wasserstoff-Bindung *f* carbon-hydrogen bond
Kohlenstoff-Wasserstoff-Verhältnis *n* C/H ratio
Kohlenstoffzustellung *f* carbon lining *(of a blast furnace)*
Kohlensuboxid *n s.* Trikohlenstoffdioxid
Kohlensubsulfid *n s.* Trikohlenstoffdisulfid
Kohlenteer *m* coal tar
Kohlenteerdestillat *n* coal-tar distillate
Kohlenteeröl *n* coal-tar oil
Kohlenteerpech *n* coal-tar pitch
Kohlenteer-Solventnaphtha *n(f)* coal-tar naphtha
Kohlentrichter *m* coal hopper
Kohlenvorratsbunker *m* coal bunker; coal storage hopper *(for bottom discharge)*
Kohlenwäsche *f* 1. coal cleaning (washing); 2. coal-cleaning plant
Kohlenwasserstoff *m* hydrocarbon
K. mit gerader Kette straight-chain hydrocarbon
K. mit verzweigter Kette branched-chain hydrocarbon
aliphatischer K. aliphatic hydrocarbon
alizyklischer K. alicyclic (cycloaliphatic) hydrocarbon
aromatischer K. aromatic (benzene) hydrocarbon
azyklischer K. acyclic (aliphatic) hydrocarbon
chlorierter K. chlorinated hydrocarbon
fluorierter K. fluorinated hydrocarbon, fluorocarbon
geradkettiger K. straight-chain hydrocarbon
gesättigter K. saturated (paraffin) hydrocarbon, alkane
höherer K. higher (larger) hydrocarbon
kettenförmiger K. open-chain hydrocarbon, aliphatic hydrocarbon
offenkettiger K. *s.* kettenförmiger K.
paraffinischer K. *s.* gesättigter K.
polymerer K. hydrocarbon polymer, polyhydrocarbon
polyzyklischer K. polycyclic hydrocarbon, polycyclohydrocarbon
ringförmiger K. cyclic hydrocarbon
verzweigtkettiger K. branched-chain hydrocarbon
zyklischer K. cyclic hydrocarbon
zykloaliphatischer K. *s.* alizyklischer K.
Kohlenwasserstoffgas *n* hydrocarbon gas
Kohlenwasserstoffgemisch *n* hydrocarbon mixture
Kohlenwasserstoffgruppe *f* hydrocarbon group (residue)
Kohlenwasserstoffharz *n* hydrocarbon resin
Kohlenwasserstoffkette *f* hydrocarbon chain
kohlenwasserstofflöslich hydrocarbon-soluble
Kohlenwasserstofföl *n* hydrocarbon oil
Kohlenwasserstoffradikal *n* hydrocarbon radical, *(specif)* free hydrocarbon radical
Kohlenwasserstoffreihe *f* family of hydrocarbons
Kohlenwasserstoffrest *m s.* Kohlenwasserstoffgruppe
Kohlenwasserstoffsynthese *f* / **Kolbesche** Kolbe electrolysis (electrochemical) reaction
Kohlenwasserstoffwachs *n* hydrocarbon wax
Kohlepapier *n* carbon (carbonic, carbonized) paper
Kohlepartikel *f* coal particle
Kohlepaste *f* coal paste
Kohlepechbrikett *n* pitch-bound briquette
Kohleprobe *f*, **Kohleprobekörper** *m* coal sample
Kohlepulver *n* pulverized (powdered) coal
Kohleschiffchen *n* carbon boat

Kohleschlamm *m* coal slurry
Kohleschleuse *f* coal-charging vessel *(as of a pressure gasifier)*
Kohleschönung *f (food)* carbon treatment
Kohleschüttung *f* coal charge
Kohleschwarz *n* carbon black
Kohlestreifen *m* coal band
Kohlestreifenart *f* type of band, microlithotype
Kohlesubstanz *f* coal material (substance)
Kohleteilchen *n* coal particle
Kohletiegel *m* carbon crucible
Kohletyp *m* type of coal
Kohlevarietät *f* variety of coal
Kohleverflüssigung *f* coal liquefaction
Kohlevergasung *f* coal gasification
Kohleverkokung *f* coal carbonization
Kohlevermahlung *f* coal milling (grinding, pulverization)
Kohleverschwelung *f* coal carbonization
Kohlevorkommen *n* coal deposit
Kohlewertstoff *m* coal chemical, coal-derived chemical product
Kohlewertstoffgewinnung *f* recovery of coal chemicals
Kohlezeichenpapier *n* charcoal drawing paper
Kohlezerstäuber *m* coal atomizer
Kohlezufuhr *f* coal feed
Kohlezwischenbunker *m* auxiliary coal hopper
kohlig carbonaceous
Kohlrausch-Brücke *f (phys chem)* Kohlrausch bridge
Kohlsaatöl *n* rape[-seed] oil, colza oil
Kohlung *f* carburization *(of steel)*
Kohlungsbad *n* carburizing bath
Kohlungsgas *n* carburizing gas
Kohlungsmittel *n* carburizing agent, [case-hardening] carburizer
Kohlungsofen *m* carburizing furnace (oven)
Kohlungspegel *m* carburizing level
Kohlungspulver *n* carburizing powder
Kohlungssalz *n* carburizing salt
Kohlungstiefe *f* carburizing (carburization) depth, case depth
Kohobation *f (distil)* cohobation
kohobieren *(distil)* to cohobate
KOH-Zahl *f (rubber)* KOH number
koinzident coincident
Koinzidenz *f* coincidence
Koinzidenzanordnung *f (nucl)* coincidence arrangement
Koinzidenzmethode *f (nucl)* coincidence method
Koinzidenzzähler *m (nucl)* coincidence counter
Koinzidenzzählung *f (nucl)* coincidence counting
koinzidieren to coincide
Kojisäure *f* kojic acid, 5-hydroxy-2-hydroxymethyl-γ-pyrone
Koka *f s.* Kokastrauch
Kokaalkaloid *n* coca alkaloid
Kokain *n* cocaine
Kokainhydrochlorid *n* cocaine hydrochloride
Kokastrauch *m* coca, Erythroxylum P. Br.
Kokatalysator *m* co-catalyst
Kokatalyse *f* co-catalysis
koken to coke
Kokerei[anlage] *f* carbonizing (coking, coke) plant
K. mit bewegter Beschickung (Ladung) continuous carbonizing plant
K. mit ruhender Beschickung (Ladung) static carbonizing plant
Kokereibenzol *n* coke-oven benzole
Kokereigas *n* coke-oven gas
Kokereiindustrie *f* coke (carbonizing) industry
Kokereiofen *m* coke (coking) oven
Kokereiteer *m* coke-oven tar
Kokille *f* ingot mould *(for steel)*; [permanent] metal mould, gravity die *(for castings)*
Kokillengießmaschine *f* diecasting machine
Kokillenguß *m* ingot casting *(of steel)*; permanent-mould casting, gravity diecasting
Kokillengußstück *n*, **Kokillengußteil** *n* gravity die casting
Kokkolith *m (min)* coccolite *(a granular variety of pyroxene)*
Kokon *m* cocoon
Kokondensation *f (plast)* co-condensation
Kokonseide *f* florette (floss) silk
Kokosbutter *f s.* Kokosfett
Kokosfaser *f* coconut fibre, coir
Kokosfett *n* coconut (copra) butter (oil)
Kokoshartfett *n* hydrogenated coconut oil
Kokoskuchen *m* coconut (copra) cake
Kokoskuchenmehl *n* coconut (copra) meal
Kokosmilch *f* coconut milk (water)
Kokosnuß *f* coco[a]nut
Kokosöl *n s.* Kokosfett
Kokospreßkuchen *m s.* Kokoskuchen
Koks *m* coke
K. zur Wassergaserzeugung water-gas coke
hüttenfähiger (metallurgischer) K. metallurgical coke
Koksabwerframpe *f* coke wharf
Kok-Saghys-Kautschuk *m* kok-saghyz rubber *(from Taraxacum bicorne Dahlst.)*
Koksaschenbeton *m* breeze concrete
Koksausdrücken *n* coke discharge
Koksausdrückmaschine *f* coke-discharging machine, coke discharger (pusher)
Koksausdrückstange *f* coke pusher ram
Koksausstoß *m* coke discharge
Koksausstoßmaschine *f s.* Koksausdrückmaschine
Koksausstoßseite *f* coke[-discharge] side *(of a coke oven)*
Koksaustrag *m* coke discharge
Koksaustragevorrichtung *f* coke discharger
Koksaustragewalze *f* coke extractor
Koksband *n* coke belt conveyor
Koksbatterie *f s.* Koksofenbatterie
koksbeladen coke-contaminated *(catalyst)*
Koksbildung *f* coke formation
Koksbildungsvermögen *n* coking power
Koksbrecher *m* coke breaker
Koksbrikett *n* coke briquette
Koksbunker *m* coke bunker
Koksdrückmaschine *f s.* Koksausdrückmaschine

Kokseigenschaften *fpl* coke properties
Koksextraktor *m* coke extractor *(a discharging device)*
Koksfestigkeit *f* strength of coke
Koksführungsschild *m* coke guide
Koksführungswagen *m* coke guide
K. mit Türabhebemaschine coke guide and door machine
Koksgas *n* coke-oven gas
koksgefeuert coke-fired
Koksgenerator *m* coke producer
Koksgrus *m* coke breeze
Kokskammer *f* coke (coking) chamber *(of a coking plant or in cracking petrol)*
Kokskohle *f* coking coal
Kokskuchen *m* coke button
Kokskuchenführungswagen *m s.* Koksführungswagen
Kokskühlrampe *f* coke wharf
Kokskühlung *f s.* Kokslöschung
Kokslöschbeton *m* breeze concrete
Kokslöschturm *m* coke quenching tower
Kokslöschung *f* coke quenching
nasse K. wet quenching
trockene K. dry quenching
Kokslöschwagen *m* [coke-]quenching car
Koksofen *m* coke (coking) oven
K. mit Nebenproduktengewinnung by-product (chemical-recovery) coke oven
K. mit Unterbrennern underjet coke oven
Koksofenanlage *f s.* Kokereianlage
Koksofenbatterie *f* coke-oven battery, retort battery, carbonizing bench
Koksofenfüllwagen *m* coal-charging car
Koksofengas *n* coke-oven gas
Koksofenkammer *f* coking (coke) chamber
Koksofenteer *m* coke-oven tar
Koksrampe *f* coke wharf
Koksrückstand *m* **nach Conradson** *(petrol)* Conradson coke (carbon) residue
K. nach Ramsbottom *(petrol)* Ramsbottom coke (carbon) residue
Koksschleuse *f* coke extractor *(a discharging device)*
Koksschwarz *n* coke black
Koksseite *f* coke[-discharge] side *(of a coke oven)*
Koksstaub *m* coke dust
Kokstransportband *n* coke belt conveyor
Kokstyp[us] *m* coke type
K. nach Gray-King Gray-King [assay] coke type
Kokswagen *m* coke car, [coke-]quenching car
Kokswassergas *n* blue [water] gas
Kokumbutter *f* kokum (goa) butter *(from Garcinia indica Choisy)*
Kokungsdestillationsanlage *f* coking still
Kokungsofen *m* coking (coke) oven
Kokungsvermögen *n* coking power
Kolamin *n* $CH_2(OH)CH_2NH_2$ colamine, 2-aminoethanol, monoethanolamine, MEA
Kolanuß *f* cola (bissy, goora) nut *(from Cola specc.)*
Kolatur *f* colature
Kölbel *n (glass)* parison
Kolben *m* 1. *(lab)* flask, bulb; 2. *(tech)* piston, plunger, ram
Kolbendruckgießmaschine *f* piston-type diecasting machine
Kolbenfläche *f* piston area
Kolbenkompressor *m* reciprocating compressor
Kolbenpumpe *f* piston pump
K. mit hin- und hergehendem Kolben reciprocating pump
K. mit rotierendem Kolben rotary pump
Kolbenraum *m* plunger compartment
Kolbensetzmaschine *f (min tech)* plunger-type fixed-sieve jig
Harzer K. Harz [fixed-sieve] jig
Kolbenspritzgußmaschine *f (plast)* plunger-type injection machine
Kolbenstange *f* piston rod
Kolbenstangendichtung *f*, **Kolbenstangenpackung** *f* piston-rod packing
Kolbenstrangpresse *f (plast)* ram extruder
Kolbenstrangpressen *n (plast)* ram extrusion
Kolbenträger *m (lab)* flask holder (support)
Kolbenverbrennung *f* [oxygen] flask combustion *(analytical chemistry)*
Kolbenverdichter *m* reciprocating compressor
Kolbenvorplastizierung *f (plast)* ram-type preplastication
Kolbe-Schmitt-Synthese *f* Kolbe-Schmitt synthesis
Kolbe-Synthese *f* Kolbe synthesis
Kolchizin *n* colchicine *(alkaloid)*
kolieren to strain, to percolate
Kolieren *n* colation, straining
Kollagen *n* collagen *(a scleroprotein)*
Kollektivflotation *f* collective (bulk) floatation
Kollektor *m (min tech)* collector, promoter, collecting (promoting) agent
Kollergang *m* edge (pan, Chilean) mill, edge runner, mulling machine, kollergang, pan crusher
K. mit perforierter Mahlbahn perforated edge mill
Kollergangstein *m* runner [stone]
Kollermühle *f s.* Kollergang
kollern to disintegrate (grind) in an edge mill
Kollerstein *m* runner [stone]
Kollerstoff *m (pap)* [machine, mill] broke, brokes, broken [material, paper]
kollidieren to collide
α-Kollidin *n* α-collidine, 4-ethyl-2-methylpyridine
β-Kollidin *n* β-collidine, 3-ethyl-4-methylpyridine
γ-Kollidin *n* γ-collidine, 2,4,6-trimethylpyridine
symm.-**Kollidin** *n s.* γ-Kollidin
Kollinit *m (coal)* collinite
Kollision *f* collision
Kollodium *n* collodion
Kollodiummembran *f* collodion membrane
Kollodium-Ultrafilter *n* collodion ultrafilter
Kollodiumverfahren *n* **/ nasses** *(phot)* wet collodion process, wet-plate process
Kollodiumwolle *f* collodion cotton (wool), soluble nitrocellulose (guncotton, cotton), pyroxylin[e],

pyrocellulose *(a lower-nitrated cellulose)*
kolloid colloid[al]
Kolloid *n* colloid
festes K. solid colloid
globuläres K. globular colloid
heteropolares K. heteropolar colloid
lyophiles K. lyophile colloid
lyophobes K. lyophobe colloid
resolubles (reversibles) K. reversible colloid
kolloidal colloid[al]
Kolloidchemie *f* colloid chemistry, collochemistry
Kolloidchemiker *m* colloid chemist
kolloidchemisch colloid-chemical, colloidochemical
kolloiddispers colloid-disperse, colloid[al]
Kolloidelektrolyt *m* colloidal electrolyte
Kolloidik *f s.* Kolloidchemie
Kolloidkaolin *n* colloidal kaolin
Kolloidlösung *f* colloidal solution
Kolloidmühle *f* colloid mill
Oderberger K. Oderberg [colloid] mill
Plausonsche K. Plauson [colloid] mill
Kolloidpartikel *f* colloidal particle
Kolloidschwefel *m* colloidal sulphur
Kolloidsystem *n* colloidal system
Kolloidteilchen *n* colloidal particle
Kolloidzustand *m* colloidal state
Kolloxylin *n s.* Kollodiumwolle
Kölnischwasser *n* cologne [water]
Kolonne *f* column, tower
K. für Zweistoffgemische binary column
K. mit rotierendem Zylinder rotating-core column
K. mit rotierenden Scheiben rotary-disk tower (contactor, extractor)
K. mit schwingenden Siebböden reciprocating-plate column
K. zur Azeotropdestillation azeotropic column
atmosphärische K. atmospheric column
leere K. wetted-wall column
pulsierte K. pulse column, pulsed tower
Kolonnendestillation *f* column distillation
Kolonnendurchmesser *m* column diameter
Kolonnenhöhe *f* column height
Kolonnenkopf *m* top of the column
Kolonnenwirkungsgrad *m* column efficiency
Kolophonium *n* colophony, pine resin (rosin), rosin
Koloquinte *f (pharm)* colocynth, bitter apple (cucumber, gourd) *(fruit from Citrullus colocynthis (L.) Schrad.)*
kolorieren *(phot)* to colour
Kolorimeter *n* colorimeter
lichtelektrisches (objektives) K. photoelectric colorimeter
visuelles K. visual colorimeter
Kolorimeterrohr *n* colorimeter tube
Kolorimetrie *f* colorimetry
lichtelektrische (objektive) K. photocolorimetry
visuelle K. visual colorimetry
kolorimetrisch colorimetric
Kolorist *m (text)* colourist

Kolostralmilch *f*, **Kolostrum** *n* colostrum, beestings, first milk
Kolumbium *n s.* Niob
Kolzaöl *n* colza oil, rape[-seed] oil
Kombination *f* **/ lineare** linear combination *(as of atomic orbitals)*
Kombinationsdünger *m* compound (multinutrient, mixed) fertilizer
Kombinationsgerbung *f* combination tannage
Kombinationslack *m* combination lacquer
Kombinationsprinzip *n* **/ Ritz-Rydbergsches** *(phys chem)* Ritz-Rydberg combination principle, Ritz [combination] principle
Kombinationsschwingung *f* combination vibration
Kombinationstrockner *m (coat)* combination dryer
kombinieren to combine; *(dye)* to couple
kompakt compact, solid
Kompakt *m s.* Kompaktpuder
kompaktieren to compact
Kompaktiermaschine *f* compactor [mill], compacting mill
Kompaktierung *f* compaction
Kompaktpuder *m (cosmet)* compact powder
Komparator *m* comparator *(colorimetry)*
visueller K. visual comparator
kompatibel compatible
Kompatibilität *f* compatibility
Kompensation *f* compensation
Kompensationsmethode *f* compensation method, [null-]balance method
Poggendorffsche K. Poggendorff compensation method
Kompensations-pH-Meter *n* compensation pH-meter
Kompensationsschreiber *m* self-balancing recorder
Kompensationswägung *f* direct weighing
Kompensator *m* 1. compensator; 2. expansion joint *(of piping)*; 3. bias control *(as of a polarograph)*
kompensieren to compensate, to counterbalance; *(dye)* to offset
komplementär complementary
Komplementärfarbe *f* complementary colour
Komplementarität *f* complementarity
komplex complex
Komplex *m* complex ✦ **im K. binden** to complex
aktivierter K. activated complex, transition state *(reaction theory)*
koordinierter K. coordination complex
organomineralischer K. *(soil)* organo-clay (clay-humus) complex
schwach gebundener K. hypoligated complex
stark gebundener K. hyperligated complex
verbrauchter K. complex out *(in liquid-phase isomerization)*
Wernerscher K. Werner complex *(chemical-bond theory)*
1:1-Komplex *m* 1:1 complex
π-Komplex *m* π-complex
komplexbildend complex-forming, complexing

nicht k. non-complexing
Komplexbildner *m* complexing (sequestering) agent, sequestrant
Komplexbildung *f* complex formation, complexation, sequestration
Komplexbildungskonstante *f* complex-formation constant, stability constant
Komplexbildungstitration *f* complexation titration
Komplexbindung *f* complex bond
Komplexerz *n* complex ore
komplexieren to complex
Komplex-Ion *n* complex ion
Komplexisomer[es] *n* complex isomer
Komplexisomerie *f* complex isomerism
Komplexität *f (nomencl)* complexity
Komplexkoazervation *f* complex coacervation
Komplexometrie *f* complexometry
komplexometrisch complexometric, compleximetric
Komplexsalz *n* complex salt
inneres K. inner complex salt
offenes (offenkettiges) K. open-chain complex salt
Komplexverbindung *f* complex compound
innere K. inner complex compound
Kompliziertheit *f (nomencl)* complexity
Komponente *f* component, constituent, moiety
aktive K. *(dye)* diazo (diazonium, primary) component
dienophile K. dienophile
endständige K. *(nomencl)* end component
hochsiedende (höhersiedende) K. *(distil)* less volatile component, high-boiling component
leichterflüchtige (leichtersiedende) K. *s.* niedrigsiedende K.
leichtest siedende K. lighter-than-light component
leichtsiedende K. *s.* niedrigsiedende K.
mittlere K. middle component
niedrigsiedende K. *(distil)* more volatile component, M. V. C., low-boiling component, light[er] component
passive K. *(dye)* coupling (secondary) component
saure K. acid component
schwererflüchtige (schwerersiedende) K. *s.* hochsiedende K.
schwerflüchtige (schwersiedende) K. *s.* hochsiedende K.
tiefsiedende K. *s.* niedrigsiedende K.
Kompost *m (agric)* compost
Compound *n*, **Kompoundmasse** *f (plast)* compound *(mechanical polymer blend)*
kompressibel compressible
Kompressibilität *f* compressibility
Kompressibilitätskoeffizient *m* compressibility coefficient (factor)
Kompression *f* compression
Kompressionsanlage *f* compressor plant
Kompressionsarbeit *f* work of compression
Kompressionsdampfkälteanlage *f s.* Kompressionskälteanlage
Kompressionshahn *m* pet cock
Kompressionskälteanlage *f* compression refrigerating system, vapour-compression system
Kompressionskältemaschine *f* compression refrigerating machine, vapour-compression machine
Kompressionsmanometer *n* **nach McLeod** McLeod gauge
Kompressionsmaschine *f s.* Kompressionskältemaschine
Kompressionsmodul *m* compressive modulus
Kompressionsschrumpf *m (text)* compressive (compression) shrinkage
Kompressions-Verdrängungsverfahren *n (rubber, plast)* compression moulding
Kompressionsverhältnis *n* compression ratio
höchstes nutzbares K. highest useful compression ratio, H.U.C.R.
Kompressionswärme *f* heat of compression
Kompressionszone *f* compression zone
Kompressor *m* compressor
einstufiger K. single-stage compressor
mehrstufiger K. multistage compressor
ölfreier (ungeschmierter) K. non-lubricated compressor
komprimierbar compressible
Komprimierbarkeit *f* compressibility
komprimieren to compress, *(relating to solids also)* to compact
Konalbumin *n* conalbumin *(a glycoprotein obtained from egg albumen)*
Konche *f (food)* conching machine, [longitudinal] conche
konchieren *(food)* to conche, to mill
Konchinin *n* quinidine, conquinine, conchinine *(a cinchona alkaloid)*
Kondensat *n* condensate, condensation product
Kondensatabführung *f*, **Kondensatableitung** *f (pap)* condensate removal
Kondensation *f* 1. condensation *(of gas or vapour)*; 2. *(org chem)* condensation *(as of esters)*; fusion, anellation, annulization, condensation *(of cyclic compounds)*
Claisensche K. Claisen condensation *(of esters)*
extramolekulare K. self-condensation
kapillare K. capillary condensation
partielle K. partial condensation, *(distil also)* dephlegmation
retrograde K. retrograde condensation
Kondensationsanlage *f* condensing system
Kondensationsdruck *m* condensation (condensing) pressure
Kondensationsfläche *f* condensing surface
Kondensationsharz *n* condensation resin
Kondensationskalorimeter *n* steam calorimeter
Kondensationskammer *f* condensing chamber
Kondensationskeim *m*, **Kondensationskern** *m* condensation nucleus (centre)
Kondensationskolben *m* condensation bulb
Kondensationskurve *f*, **Kondensationslinie** *f (distil)* condensation (dew-point) curve
Kondensationsmethode *f* condensation method

Kondensationsmittel *n* condensing agent
Kondensationspolymer[es] *n* condensation polymer
Kondensationspolymerisation *f* condensation polymerization
Kondensationsprodukt *n* condensation product, condensate
Kondensationsreaktion *f* condensation reaction
Kondensationsrohr *n* condensing tube
Kondensationsschritt *m* condensation step
Kondensationsstelle *f* point (position, side) of fusion, common face *(of rings)*
Kondensationsturm *m* condensing tower
Kondensationswärme *f* heat of condensation
Kondensationszentrum *n s.* Kondensationskeim
Kondensationszwischenprodukt *n* half-condensation product
Kondensatlagerstätte *f (petrol)* condensate reservoir
Kondensator *m* 1. condenser; 2. *(el chem)* capacitor
K. mit Kühlschlange worm-type condenser
barometrischer K. barometric condenser
elektrolytischer K. electrolytic capacitor
Kondensatorkühler *m* condenser
Kondensatorkühlwasser *n* condenser water
Kondensator[seiden]papier *n* condenser paper, condenser tissue [paper]
Kondensatrücklauf *m* condensate returns
Kondensatrückleiter *m* lift steam trap, boiler return trap
Kondensatwasserableiter *m* steam trap
kondensierbar condensable
nicht k. non-condensable, uncondensable
Kondensierbarkeit *f* condensability
kondensieren 1. to condense, to precipitate *(gas or vapour)*; 2. *(org chem)* to condense *(e.g. esters)*; to fuse, to anellate, to annulize, to condense *(cyclic compounds)*; 3. *(food)* to condense, to inspissate; 4. *s.* sich k.
miteinander k. *(org chem)* to fuse together, to condense
sich k. to condense, to precipitate *(of gas or vapour)*
sich wieder k. to recondense
wieder k. to recondense
Kondensieren *n* / **festes** solidensing *(direct conversion of a gas into the solid state)*
Kondensmagermilch *f* condensed skim milk
Kondensmilch *f* condensed (concentrated, evaporated) milk
gezuckerte K. sweetened condensed milk
Kondensmilchfabrik *f* milk condensery (condensing plant, evaporating plant)
Kondensstelle *f s.* Kondensationsstelle
Kondenstopf *m* steam trap
Kondensvollmilch *f* condensed whole milk
Kondenswasser *n* condensation (condensed, condensate) water
Kondenswasserableiter *m* steam trap
Kondenswasserhahn *m* pet cock
Konditionieranlage *f* conditioning plant (unit)

Konditionierapparat *m* conditioning apparatus, conditioner
konditionieren to condition; *(relating to humidity)* to [air-]condition, to order
Kondorrinde *f s.* Kondurangorinde
Konduktanz *f* conductance
Konduktometrie *f* conductometry, conductimetry
konduktometrisch conductometric, conductimetric
Kondurangin *n* condurangin *(a bitter glycoside)*
Kondurangorinde *f (pharm)* cundurango bark *(from Marsdenia cundurango Rchb.fil.)*
Konfektion *f s.* Konfektionierung 2.
konfektionieren to formulate; *(rubber)* to assemble, to build
Konfektionierlaboratorium *n* regulatory laboratory *(for preparing pesticides)*
Konfektionierlösung *f (rubber)* assembling solution, cement
Konfektioniermaschine *f (rubber)* tyre-building machine (drum), lay-up machine
Konfektionierraum *m s.* Konfektionsabteilung
Konfektioniertisch *m (rubber)* assembling table
Konfektionierung *f* 1. formulation; 2. *(rubber)* assembly, building
Konfektionsabteilung *f (rubber)* assembling (assembly) department
Konfektionsklebrigkeit *f (rubber)* building tack
Konfektionslösung *f (rubber)* assembling solution
Konfiguration *f* [atomic] configuration, spatial (atom, atomic) arrangement
absolute K. absolute configuration
erzwungene K. forced configuration
geknickte K. puckered configuration
relative K. relative configuration
Konfigurationsbeweis *m* proof of configuration
Konfigurationserhaltung *f* retention of configuration
Konfigurationsformel *f* configurational (space) formula
Konfigurationssymbol *n* configurational symbol
Konfigurationsumkehr *f*, **Konfigurationswechsel** *m* inversion of configuration
konfigurativ configurational
Konformation *f* conformation
anti-clinale K. anti-clinal conformation
± **anti-periplanare K.** anti-periplanar conformation, anti (non-eclipsed, staggered) conformation
äquatoriale K. equatorial conformation
axiale K. axial conformation
ekliptische K. *s* ± syn-periplanare K.
gestaffelte K. *s.* ± anti-periplanare K.
polare K. *s.* axiale K.
syn-clinale K. syn-clinal conformation, gauche (skew) conformation
± **syn-periplanare K.** syn-periplanar conformation, syn (eclipsed) conformation
teilweise verdeckte K. *s.* anti-clinale K.
verdeckte K. *s.* ± syn-periplanare K.
windschiefe K. *s.* syn-clinale K.
Konformationsanalyse *f* conformational analysis
Konformationsformel *f* conformational formula
Konformationsisomer[es] *n* conformation isomer, conformer

Konformationsisomerie *f* conformational (rotational) isomerism
konformer conformational
Konformeres *n s.* Konformationsisomer
Konglomerat *n* conglomerate
Kongofarbstoff *m* Congo dye *(any of a group of azo dyes)*
Kongokopal *m* Congo copal (gum) *(a semifossil resin from Copaifera specc.)*
Kongokorinth *n (dye)* Congo corinth
Kongopapier *n* Congo paper
Kongorot *n* Congo red, direct red 28 *(an azo dye)*
Kongorubin *n* Congo rubin[e], direct red 17 *(an azo dye)*
Kongreßverfahren *n (ferm)* congress method
Kongreßwürze *f (ferm)* congress wort
Konhydrin *n* conhydrine, 2-1′-hydroxypropylpiperidine
Konidendrin *n* conidendrin *(a lignan)*
Koniferin *n*, **Koniferosid** *n* coniferin *(β-glucoside of coniferyl alcohol)*
Koniferylalkohol *m* $HOC_6H_3(OCH_3)CH{=}CHCH_2OH$ coniferyl alcohol, 4-γ-hydroxypropenyl-2-methoxyphenol
Königinnensubstanz *f* queen-bee's substance *(biologically active substance of the queen-bee's salivary glands)*
Königsgelb *n* king's yellow (gold), royal yellow, yellow arsenic [sulphide], arsenic yellow, orpiment [yellow] *(arsenic(III) sulphide)*
Königswasser *n* aqua regia, aq.reg., chloronitrous (chloroazotic, nitrohydrochloric) acid
Koniin *n* coniine, 2-propylpiperidine *(alkaloid)*
konisch conical, cone-shaped ✦ **k. zulaufen** to taper ✦ **k. zulaufend** taper[ing], tapered
Konizität *f* conicity; taper *(of piping)*
Konjugation *f* conjugation
Konjugationsenergie *f* conjugation (resonance, delocalization, mesomeric) energy
konjugieren to conjugate
konjugiert conjugate[d]
konkav concave, dished
Konkavgitter *n* concave grating *(spectrography)*
Konkavgitterspektrograf *m* concave-grating spectrograph
Konkrement *n (med)* calculus
konkret *(cosmet)* concrete *(e.g. oil)*
Konkret *n (cosmet)* concrete [oil]
Konkretion *f (geol)* concretion
Konkurrenzreaktion *f* competitive (competing, concurrent) reaction
Konnellit *m* connellite *(a copper mineral)*
Konserve *f (food)* preserve
Konservendose *f* preserve can (tin), tin [can], can
  verzinnte K. tin [can]
Konservenfabrik *f* canning (preserving) plant, cannery, food-canning factory, *(Am)* packing house
Konservenfabrikation *f* canning
Konservenglas *n* preserve (preserving) jar
Konservenherstellung *f* canning
Konservenindustrie *f* canning (canned foods) industry
konservieren *(tech)* to conserve, to preserve; *(food)* to preserve, *(esp by drying, salting, or smoking)* to cure; *(tann)* to cure
  durch Kälte k. to deep-freeze
  in Dosen k. to can, to tin
konservierend preservative
konserviert / mit Ammoniak *(rubber)* ammonia-preserved
Konservierung *f (tech)* conservation, preservation; *(food)* preservation, *(esp by drying, salting, or smoking)* cure, curing; *(tann)* cure, curing
  K. durch Salzlakenbehandlung *(tann)* brine cure (curing)
  K. in Dosen canning, tinning
Konservierungsmittel *n (tech, food)* preservative [agent], preserving agent; *(food, tann)* curing agent
Konservierungssalz *n* curing salt
konsistent consistent
Konsistenz *f* consistency, consistence; *(coat)* body
  pastöse (teigige) K. doughiness
Konsistenzregler *m (pap)* consistency regulator
Konsistometer *n* consistometer
konstant constant ✦ **k. halten** to maintain constant
Konstante *f* constant [quantity]
  K. der inneren Reibung viscosity coefficient
  Boltzmannsche K. Boltzmann constant
  ebullioskopische K. ebullioscopic (boiling-point) constant, molal boiling-point[-elevation] constant, molal (molar, molecular) elevation constant
  kryoskopische K. cryoscopic constant, molal freezing-point[-depression] constant, molal (molar, molecular) depression constant
  Madelungsche K. Madelung constant
  Plancksche K. Planck (action) constant, [Planck] quantum of action
  Poissonsche K. Poisson's ratio
  Rydbergsche K. Rydberg constant (number)
  van-der-Waalssche K. van der Waals constant
Konstanten *fpl* / **kritische** critical constants (data)
Konstantpumpe *f* constant-displacement pump
konstantsiedend constant-boiling
Konstellation *f s.* Konformation
Konstitution *f* constitution, structure
Konstitutionsaufklärung *f*, **Konstitutionsbestimmung** *f* determination of constitution, structure elucidation
Konstitutionsbeweis *m* proof of constitution
Konstitutionserforschung *f*, **Konstitutionsermittlung** *f s.* Konstitutionsaufklärung
Konstitutionsformel *f* constitutional (structural, graphic) formula
Konstitutionswasser *n* constitutional water, water of constitution
konstitutiv constitutive
Konstruktionsmerkmal *n* design feature
Konstruktionswerkstoff *m* material of construction
Kontakt *m* 1. contact; 2. [contact] catalyst, cat ✦ **in K. bringen** to contact ✦ **miteinander in K. kommen** to contact

beweglicher (bewegter) K. moving catalyst
fester (festliegender) K. *s.* ruhender K.
perlförmiger K. bead catalyst
ruhender K. static (fixed-bed) catalyst
Kontaktabzug *m (phot)* contact print
Kontaktaureole *f (geol)* contact aureole (zone)
Kontaktautoklav *m* contactor
Kontaktbacke *f* contact shoe *(for electrodes)*
Kontaktbaustoff *m* electrical contact material
Kontaktbett *n* catalyst bed
festes (festliegendes) K. *s.* ruhendes K.
ruhendes (stationäres) K. static bed of catalyst
Kontaktdauer *f* time of contact
Kontaktdüngung *f* contact fertilization
Kontaktfiltration *f* contact filtration
Kontaktfläche *f* surface (area) of contact
Kontaktgefrieren *n* contact freezing
Kontaktgetterung *f* contact gettering *(vacuum technology)*
Kontaktgift *n* 1. catalyst (catalytic) poison, paralyzer; 2. *(agric)* [direct] contact poison, contact toxicant
Kontaktherbizid *n* contact herbicide (weed-killer)
Kontakthof *m s.* Kontaktaureole
Kontaktinsektizid *n* contact insecticide
K. mit Dauerwirkung residual contact insecticide
protektives K. protective contact insecticide
Kontaktkammer *f* catalyst chamber
Kontaktkatalysator *m* [contact] catalyst, cat
Kontaktkatalyse *f* contact (surface, heterogeneous) catalysis
kontaktkatalytisch contact-catalytic
Kontaktkleber *m*, **Kontaktklebstoff** *m* contact adhesive (cement)
Kontaktkopie *f (phot)* contact print
Kontaktkopieren *n (phot)* contact printing
Kontaktkopiergerät *n (phot)* contact printer
Kontaktkorrosion *f* contact corrosion
Kontaktmetamorphose *f (geoch)* contact metamorphism (metamorphosis)
Kontaktmineral *n* contact mineral
Kontaktmittelemulsion *f* contact emulsion *(for weed control)*
Kontaktofen *m* converter
Kontakt-Öl-Verhältnis *n* catalyst-to-oil ratio, catalyst/oil ratio
Kontaktor *m* contactor *(1. an autoclave; 2. a solvent extractor)*
Kontaktpapier *n (phot)* contact [printing] paper, silver-chloride paper
Kontaktpille *f* catalyst pellet
Kontaktpotentialdifferenz *f* difference of potential on direct contact
Kontaktpressen *n (plast)* contact (impression) moulding, hand lay-up [technique]
Kontaktraffination *f (petrol)* contact treatment
Kontaktraum *m* contact space
Kontaktsäure *f* contact [sulphuric] acid
Kontaktschicht *f* catalyst bed
Kontaktschwefelsäure *f* contact [sulphuric] acid
Kontaktschwefelsäureverfahren *n* contact process (method)
Kontaktstaub *m* catalyst dust, powdered catalyst
Kontaktstoff *m s.* Kontaktkatalysator
Kontaktthermometer *n* contact thermometer
Kontaktträger *m* catalyst carrier (support)
Kontakttrocknung *f* contact (conduction, indirect) drying, drying by contact
Kontaktumlauf *m* catalyst circulation (cycle, recycle)
Kontaktverfahren *n* 1. contact process; 2. *(pap)* cast coating; 3. *(phot)* contact printing
Kontaktwerkstoff *m* electrical contact material
Kontaktwinkel *m* contact angle
Kontaktwirkung *f* contact action
endomorphe K. *(geol)* endomorphism
exomorphe K. *(geol)* exomorphism
Kontaktzeit *f* time of contact
Kontamination *f* contamination
kontaminieren to contaminate
Kontinuebetrieb *m (dye)* continuous operation (working, processing) ✦ **im K.** by a continuous process
Kontinuebleiche *f (text)* continuous bleaching
Kontinue-Breitbleichanlage *f (text)* continuous open-width bleaching machine
Kontinuedämpfer *m (text)* continuous steamer
Kontinuefärbemaschine *f (text)* continuous-dyeing machine
Kontinuefärben *n (text)* continuous dyeing
Kontinuespinnverfahren *n (text)* continuous spinning
Kontinueverfahren *n (dye)* continuous process
kontinuierlich continuous
k. arbeitend continuous
Kontinuum *n* continuum
kontrahieren to contract
Kontraktion *f* contraction
Kontraktionsberührungswinkel *m* receding contact angle *(testing of tension depressors)*
Kontraktionskoeffizient *m* coefficient of contraction
Kontraktionsstelle *f* vena contracta *(flow measurement)*
Kontrast *m (phot)* contrast
weicher K. soft contrast
kontrastarm *(phot)* low-contrast, thin
Kontrastentwickler *m (phot)* high-contrast developer
Kontrastfaktor *m (phot)* development factor, gamma value
Kontrastfärbung *f (text)* differential dyeing
Kontrastlosigkeit *f (phot)* flatness
Kontrastmittel *n (med)* contrast medium
kontrastreich *(phot)* high-contrast, contrasty
Kontrastumfang *m (phot)* contrast (brightness) range
Kontrastverminderung *f (phot)* decrease in contrast, reduction in image contrast
Kontrolle *f* check[ing], inspection, control ✦ **außer K. geraten** to get out of control (hand), to run wild
kinetische K. kinetic control
Kontrollfläche *f (agric)* check plot

Kontrollhahn *m*, **Kontrollhahnventil** *n* test cock
kontrollieren to check, to inspect, to control
Kontrollöffnung *f* inspection door
Kontrollstab *m (nucl)* control rod
Konturenschärfe *f (phot)* acutance; *(text)* sharpness in print outline
Konus *m* cone; *(pap)* plug, rotor, cone, core *(of a perfecting engine)*
  K. der Kegelstoffmühle *(pap)* jordan plug
konusartig cone-shaped, conical
Konusfärbeapparat *m* cone dyeing apparatus
Konusmühle *f* cone mill, conical [ball] mill
Konvallamarin *n* convallamarin *(a glycoside)*
Konvallarin *n* convallarin *(a glycoside)*
Konvallatoxin *n* convallatoxin *(a glycoside)*
Konvektion *f* convection
Konvektionsheizung *f* convection heating
Konvektionsmischen *n* convective mixing
Konvektionsstrom *m* convection current
Konvektionstrockner *m* convection dryer
Konvektionstrocknung *f* convection (direct) drying
Konvektionsvermischen *n* convective mixing
Konvektionswärme *f* convection (convected) heat
konvektiv convective
Konvergenz *f* convergence, convergency
Konversion *f* conversion
Konversionsreaktor *m (nucl)* conversion reactor, converter
Konversionssalpeter *m* conversion (converted) saltpetre
Konverter *m* 1. *(met, text)* converter; 2. *s.* Konversionsreaktor ✦ **im K. verblasen** *(met)* to convert, to bessemerize
  bodenblasender K. bottom-blown converter
  drehbarer K. rotating converter
  normal blasender K. bottom-blown converter
  rotierender K. rotating converter
  seitlich blasender K. side-blown converter
Konverterauskleidung *f (met)* converter lining
Konverterfrischverfahren *n (met)* converter (converting, Bessemer) process
Konverterfutter *n (met)* converter lining
Konvertermittelstück *n* body *(the cylindrical part of a steel converter)*
Konverterprozeß *m s.* Konverterfrischverfahren
Konverterreaktor *m s.* Konversionsreaktor
Konverterstahl *m* converter (Bessemer) steel
  sauerstoffgefrischter K. basic oxygen [furnace] steel
Konverterverfahren *n* 1. *s.* Konverterfrischverfahren; 2. *(text)* tow-to-top process
Konverterzustellung *f (met)* converter lining
konvertieren to convert
Konvertierung *f* conversion
Konvertierungsverfahren *n* conversion process (method)
konvex convex, domed
Konvolvulinolsäure *f* convolvulinolic acid, 11-hydroxytetradecanoic acid
Konvolvulinsäure *f* convolvulinic acid *(a glucoside)*
konz. *s.* konzentriert
Konzentrat *n* concentrate
  emulgierbares K. emulsifiable concentrate *(pest control)*
Konzentrataustrag *m* concentrate discharge
Konzentration *f* concentration, *(relating to solutions preferably)* strength ✦ **zur ursprünglichen K. lösen** to reconstitute
  K. in Prozent per cent concentration
  K. Null zero concentration
  gesamte ionale K. total ionic concentration
  höchstzulässige K. *(tox)* maximum permissible concentration
  ionale K. ionic concentration
  molale K. molal concentration, molality
  molare K. molar concentration, molarity
  prozentuale K. per cent concentration
Konzentrationsänderung *f* concentration change
Konzentrationsanlage *f* concentration plant
Konzentrationselement *n (el chem)* concentration cell
  K. mit Überführung concentration cell with transference
  K. ohne Überführung concentration cell without transference
Konzentrationsgefälle *n*, **Konzentrationsgradient** *m* concentration gradient
Konzentrationskette *f (el chem)* concentration cell
Konzentrationspolarisation *f* concentration polarization
Konzentrationsprofil *n* concentration profile *(as in a column)*
Konzentrationsveränderung *f* concentration change
Konzentrationsverhältnis *n* ratio of concentrations
Konzentrationszelle *f (el chem)* concentration cell
Konzentratschaum *m (min tech)* concentrate-laden froth
konzentrieren to concentrate
  durch Abdunsten k. to graduate
Konzentrieren *n* concentration *(of a solution)*
  K. durch Abdunsten graduation
konzentriert concentrated, conc.
  doppelt k. double-strength
Konzentrierungsanlage *f* concentration plant
Koordinaten *fpl* **der Atomschwerpunkte im Gitter** *(cryst)* atomic parameters
Koordination *f* coordination *(chemical-bond theory)*
  K. im Sinne von Werner Werner-type coordination
Koordinationsbestreben *n* coordination tendency
Koordinationschemie *f* coordination chemistry, chemistry of coordination compounds
Koordinationseinheit *f* coordination unit
Koordinationsformel *f* coordination formula
Koordinationsgebilde *n* coordination entity
Koordinationsgitter *n* coordination lattice

Koordinationsgruppe *f* coordination (coordinated) group
Koordinationskomplex *m* coordination complex
Koordinationskörper *m* coordination entity
Koordinationslehre *f* coordination theory
Wernersche K. Werner [coordination] theory
Koordinationspolymerisation *f* coordination polymerization
Koordinationsstelle *f* coordination position (site)
Koordinationstheorie *f* coordination theory
Koordinationsverbindung *f* coordination compound
Koordinationszahl *f* coordination (covaleny) number, ligancy
koordinativ coordinative, coordinate
koordiniert / dreifach three-coordinate
Kopaivabalsam *m (incorrectly for)* Kopaivaterpentin
Kopaivaöl *n* copaiba oil *(from Copaifera specc.)*
Kopaivaterpentin *n(m)* copaiba resin, Jesuit's balsam *(from Copaifera specc.)*
Kopal *m* copal [resin], [gum] copal *(collectively for high-melting resins esp of fossil origin)*
Amerikanischer K. Colombia (Brazil) copal *(from Hymenaea courbaril L.)*
Kopalin *m (min)* copalite, copaline, highgate resin *(an amber-like fossil resin)*
Kopf *m (tech)* head; *(met)* top
schwimmender K. floating head *(of a heat exchanger)*
Kopfdünger *m* top-dressing, direct-application fertilizer ✦ **mit K. behandeln** to top-dress
Kopfdüngung *f* top-dressing
Kopfform *f (glass)* ring (finish) mould, neckring
Kopffraktion *f (distil)* top fraction
Kopfkalkung *f (agric)* top liming
Kopf-Kopf-Polymerisation *f* head-to-head polymerization
Kopf-Kopf-Struktur *f*, **Kopf-Kopf-Verkettung** *f* head-to-head structure *(of polymers)*
Kopfprodukt *n (distil)* overhead [product], top product
Kopfrolle *f* head pulley *(of a conveyor)*
angetriebene K. motorized-head pulley
Kopf-Schwanz-Polymerisation *f* head-to-tail polymerization
Kopf-Schwanz-Struktur *f*, **Kopf-Schwanz-Verkettung** *f* head-to-tail structure *(of polymers)*
Kopftemperatur *f (distil)* overhead temperature
Kopf- und Bodenschmelzen *n (met)* top-and-bottom smelting
Kopfwalze *f (pap)* bottom couch-press roll
Kopiapit *m (min)* copiapite *(a basic iron sulphate)*
Kopie *f (phot)* print
positive K. positive print
Kopierdruckfarbe *f* copying ink
kopieren *(phot)* to print
Kopiergerät *n (phot)* contact printer
Kopierpapier *n (phot)* print[ing] paper
Kopierseidenpapier *n* copying tissue paper
Kopierstift *m* copying (indelible) pencil
Kopiertinte *f* copying ink
Kopigment *n (bioch)* copigment
Kopolyaddition *f* copolyaddition
Kopolykondensation *f* copolycondensation
Kopolymer[es] *n* copolymer, mixed polymer
K. mit hohem Styrolgehalt high-styrene copolymer (polymer, resin)
alternierendes K. alternating copolymer
statistisches K. random copolymer
Kopolymerisat *n s.* Kopolymer
Kopolymerisatfaser *f* copolymer fibre
Kopolymerisatfaserstoff *m* copolymer fibre
Kopolymerisation *f* copolymerization
K. mit Vernetzung copolymerization with cross-linking
alternierende K. alternating copolymerization
statistische K. random copolymerization
kopolymerisieren to copolymerize
Koppers-Becker-Ofen *m*, **Koppers-Becker-Verbundkoksofen** *m* Koppers-Becker [combination coke] oven
K. mit Unterbrennern Koppers-Becker [combination] underjet coke oven
Koppers-Ofen *m (coal)* Koppers oven
Koppers-Totzek-Verfahren *n* Koppers process *(for gasifying pulverized coal)*
Koppers-Verfahren *n* Koppers process *(for gas cleaning)*
Kopplung *f (dye, phys chem)* coupling
Russell-Saunderssche K. Russell-Saunders coupling, LS coupling *(of spins and moments of momentum)*
Kopplungskoeffizient *m* / **piezoelektrischer** piezoelectric coupling coefficient
Kopra *f* copra *(dried coconut meat)*
Koprogen *n* coprogen *(a sideramine)*
Koprolith *m (geol)* coprolite *(a fossile excrement)*
Kops *m (text)* cop
Kopsfärbeapparat *m (text)* cop dyeing machine
Kopsin *n* kopsin *(alkaloid)*
Korallenkalk *m* coral lime
Korallenschlick *m* coral mud
Korarima-Malagetta *m (food)* Madagascar cardamom *(from Aframomum angustifolium Schum.)*
Korbflasche *f* basket bottle, *(holding from 1 to 10 gallons)* demijohn, *(esp for acids)* carboy
Korbpresse *f* basket (curb) press
Kord[faden] *m (rubber)* cord
Kordgewebe *n (rubber)* cordage, cord fabric
Kordherd *m (min tech)* corduroy blanket table
Kordierit *m (min)* cordierite *(a silicate of aluminium, iron, and magnesium)*
Kordieritkeramik *f* cordierite ceramics
Kordieritporzellan *n* cordierite porcelain
Kordieritweißware *f* cordierite whiteware
Kordlage *f (rubber)* carcass (casing) ply
Koriander *m* 1. coriander, Coriandrum sativum L.; 2. coriander [seed] *(from 1.)*
Korilagin *n* corilagin *(an ellagitannin)*
Korium *n (tann)* corium
Kork *m* 1. [natural] cork; 2. *(bot)* [bark] cork, phellem; 3. *s.* Korken ✦ **aus K.** subereous

korkähnlich, korkartig cork-like, corky, suberose, suberous
Korkbohrer *m (lab)* cork borer
Korkbohrerschärfer *m (lab)* cork borer sharpener
Korkdichtung *f* cork gasket
Korkeiche *f* cork oak, Quercus suber L.
Korken *m* cork *(stopper)*
Korkgewebe *n (bot)* cork, phellem
Korkmehl *n* cork powder
Kork[mehl]papier *n* cork paper
Korkpresse *f (lab)* cork press (softener, squeezer)
Korkrinde *f s.* Kork 1.
Korkring *m* cork ring
Korksäure *f* $HOOC[CH_2]_6COOH$ suberic acid, octanedioic acid
Korkstopfen *m* cork
Korkwachs *n* cork wax
Korn *n* grain, particle; *(phot, cryst)* grain; *(rubber)* pellet ✦ **auf K. [ver]kochen** *(sugar)* to boil to grain, *(Am also)* to sugar off *(esp maple sap)*
grobes K. *(phot)* coarse grain
Kornbildung *f (sugar)* graining
Kornbranntwein *m* grain alcohol
Kornbrennerei *f* grain-alcohol plant
Körnchen *n* granule, [small] grain; *(rubber)* pellet
Körnchenbildung *f* granulation
Korndichte *f* grain (granule, lump, apparent) density *(of bulk material)*
körnen to grain, to granulate; *(rubber)* to pellet[ize] *(soot)*
Körnergelatine *f* kibbled gelatin
Körnerhaufwerk *n* granular bed, bed of granular solids
Körnerlack *m* grained (seed) lac
Körnerschüttung *f s.* Körnerhaufwerk
Körnerzinn *n* grain tin
Kornform *f* grain (particle) shape
Kornfraktion *f* size fraction *(classifying)*
feine K. fine sizes, fines
Korngestalt *f s.* Kornform
Korngrenze *f (cryst)* grain boundary
Korngrenzenbruch *m (cryst)* intercrystalline cracking (failure, fracture)
Korngrenzenkorrosion *f (cryst)* intergranular attack (corrosion)
Korngröße *f* particle (grain) size, *(in screening also)* grade, screen size
mittlere K. average particle size
Korngrößenanalyse *f* particle-size analysis, granulometric analysis
Korngrößenbereich *m* size range, range of particle size[s]
Korngrößenbestimmung *f s.* Korngrößenanalyse
Korngrößenklasse *f* size fraction
Korngrößenverteilung *f* particle-size distribution
Korngruppe *f* size fraction
körnig granular, granulate[d], grainy, grained ✦ **k. machen** to granulate, to grain
gleichmäßig k. equigranular
Körnigkeit *f* granularity, graininess; *(phot)* grain
Körnigmachen *n* granulation, graining
Kornkennlinie *f* particle-size distribution curve
Kornklasse *f* [size] fraction
Kornkochen *n (sugar)* boiling to grain, crystal boiling
kornlos grain-free
Kornmittel *n* average particle size
Kornoberfläche *f* grain surface
Kornpolymerisation *f* bead (suspension) polymerization
Kornscheide *f* size of separation, cut size (point), critical diameter *(classifying)*
Kornspanne *f s.* Korngrößenbereich
Körnung *f* granularity, grain, *(in screening also)* grade, screen size
Körnungsanalyse *f* size[-frequency] analysis
Körnungsbereich *m s.* Korngrößenbereich
Körnungsgesetz *n* law of size distribution
Körnungslinie *f* particle-size distribution curve
Körnungsmodul *m* fineness modulus, *(Am)* fineness module
Körnungsschwankung *f (geol)* granular variation
Kornverteilung *f* particle-size distribution
Kornverteilungsgesetz *n* law of size distribution
Rosin-Rammlersches K. Rosin-Rammler exponential law
Kornverteilungskurve *f* particle-size distribution curve
Koronabeständigkeit *f* corona resistance, resistance to corona [discharge]
Koronaentladung *f* corona [discharge]
koronisieren to coronize *(glass cloth for rendering it crease-resistant)*
Körper *m (tech)* body; effect *(of an evaporator)* ✦ **K. geben** *(coat)* to body
K. des Weines wine body
diamagnetischer K. diamagnetic material (substance)
fester K. solid [matter]
grauer K. *(phys chem)* grey body
keramischer K. ceramic body
paramagnetischer K. paramagnetic material (substance)
schwarzer K. *(phys chem)* black body
Körperfarbe *f* pigment
Körperfett *n* body fat
Körperflüssigkeit *f* body fluid
körpergebend *(coat)* bodying
Körpergehalt *m* / **scheinbarer** *(plast)* false body
Körperpflegemittel *n* cosmetic
Körperpuder *m* body powder
Korpuskel *n (nucl)* corpusc[u]le, particle
Korpuskularstrahlung *f* corpuscular (particle) radiation
Korrektionsgröße *f* correction term
Korrekturfaktor *m* 1. correction factor; 2. normality (titrimetric, volumetric) factor *(the numerical value of the normality of a solution)*
Korrekturglied *n* correction term
Korrekturkoeffizient *m* correction factor
Korrelation *f* correlation
korrespondierend conjugate *(acid, base)*
Korrigens *n (pharm)* corrigent, corrective

korrodierbar corrodible
Korrodierbarkeit *f* corrodibility
korrodieren to corrode
Korrosion *f* corrosion
K. durch das Erdreich soil corrosion
K. durch Fremdstrom electrocorrosion
K. durch Luftsauerstoff atmospheric corrosion
allgemeine K. general corrosion
ebenmäßige K. uniform corrosion
elektrochemische K. durch Streuströme electrocorrosion
galvanische K. galvanic corrosion
interkristalline K. intercrystalline (intergranular) corrosion
örtliche K. localized corrosion
selektive K. selective corrosion
korrosionsanfällig corrodible
Korrosionsanfälligkeit *f* corrodibility
korrosionsbeständig corrosion-resistant, resistant to corrosion, non-corroding
Korrosionsbeständigkeit *f* corrosion resistance, resistance to corrosion
Korrosionselement *n (el chem)* corrosion cell
korrosionsempfindlich corrodible
Korrosionsempfindlichkeit *f* corrodibility
Korrosionsermüdung *f* corrosion fatigue
korrosionsfest *s.* korrosionsbeständig
Korrosionsgeschwindigkeit *f* rate of corrosion
Korrosionshemmer *m*, **Korrosionshemmstoff** *m*, **Korrosionsinhibitor** *m s.* Korrosionsschutzmittel
Korrosionsmedium *n s.* Korrosionsmittel
Korrosionsmittel *n* corrosive agent, corroding medium
Korrosionsprodukt *n* corrosion product
Korrosionsprüfgerät *n* corrosion meter
Korrosionsprüfung *f* corrosion test
Korrosionsschutz *m* corrosion control (protection), protection from corrosion
katodischer K. durch (mit) Opferanoden sacrificial protection
korrosionsschützend anticorrosive
Korrosionsschutzmittel *n* corrosion inhibitor, anticorrosive agent
Korrosionsschutzpapier *n* anticorrosive (antitarnish) paper
Korrosionstest *m* corrosion test
korrosionsverhütend anticorrosive
Korrosionsversuch *m* corrosion test
Korrosionsverzögerer *m s.* Korrosionsschutzmittel
Korrosionswiderstand *m* corrosion resistance
korrosiv corrosive
Korrosivität *f* corrosiveness, corrosivity
Kortikoid *n (bioch)* corticoid
Kortikosteroid *n (bioch)* corticosteroid
Kortikosteron *n* corticosterone *(an adrenocortical hormone)*
Kortikotropin *n* corticotropin, adrenocorticotropic hormone, ACTH
Kortin *n* cortin *(collectively for a group of adrenocortical hormones)*
Kortisol *n* cortisol *(an adrenocortical hormone)*
Kortison *n* cortison *(an adrenocortical hormone)*
Korund *m (min)* corundum *(α-aluminium oxide)*
Korydin *n* corydine *(alkaloid)*
Koschenille *f (dye)* cochineal
Koschenillesäure *f* cochenillic acid, 5-hydroxytoluene-2,3,4-tricarboxylic acid
Koseiseide *f* Kosey silk *(from regenerated fibroin)*
Kosmetikchemiker *m* cosmetic chemist
Kosmetikpräparat *n* cosmetic preparation
Kosmetikum *n* cosmetic
Kosmetologie *f* cosmetology
Kosmochemie *f* cosmochemistry, cosmic (space) chemistry
kosten to taste *(a substance)*
Koster *m* taster *(profession)*
Kostinky-Effekt *m (phot)* Kostinky effect
Kot *m* faecal matter, faeces
Kotoin *n* cotoin, *(specif)* $C_6H_2(OH)_2(OCH_3)COC_6H_5$ cotoin, 2,6-dihydroxy-4-methoxybenzophenone
kotonisieren to cottonize *(flax or hemp)*
Kotunnit *m (min)* cotunnite *(lead(II) chloride)*
Kötzer *m (text)* cop
kovalent covalent, homopolar
Kovalenz *f* 1. covalence, covalency; 2. *s.* Bindung / homöopolare
Kovalenzbindungswinkel *m* covalent bond angle
Kovellin *m (min)* covellite, covelline, indigo copper, blue copper *(copper(II) sulphide)*
Kovolumen *n (phys chem)* covolume
Kozymase *f s.* Nikotinamid-adenin-dinukleotid
KP. *s.* Kochpunkt
krabben *(text)* to crab
Krabbmaschine *f (text)* crab
krachen (text) to rustle
Krachen *n (text)* rustle, scroop
krachend *(text)* scroopy *(feel)*
Krackanlage *f* cracking plant, cracker
katalytische K. catalytic cracking plant, catalytic (catalyst, cat) cracker
Krackbedingungen *fpl* cracking conditions
milde K. mild cracking conditions
scharfe K. severe cracking conditions
Krackbehandlung *f* cracking treatment
Krackbenzin *n* cracked gasoline
katalytisches K. cat-cracked gasoline
Krackdestillat *n* cracked distillate
Krackeinsatz *m s.* Krackgut
kracken to crack
gelinde k. *s.* mild k.
katalytisch k. to cat-crack
mild k. to give a mild cracking treatment
scharf k. to give a severe cracking treatment
Kracken *n* cracking
K. am Katalysator *s.* katalytisches K.
K. auf flüssigen Rückstand residue cracking, flashing
K. auf Koks[rückstand] non-residue cracking, coking
K. in der Dampfphase (Gasphase) vapour-phase cracking

K. in Flüssigphase (flüssiger Phase) liquid-phase cracking
K. in gemischt flüssiger und dampfförmiger Phase mixed-phase cracking
K. in Wirbelschicht mit Fließbettkatalysator fluid catalytic cracking
K. mit Katalysatorbett bed cracking
K. mit Kokungsarbeitsweise *s.* K. auf Koks
K. mit Rückstandsarbeitsweise *s.* K. auf flüssigen Rückstand
K. mit suspendiertem Katalysator suspensoid [catalytic] cracking
K. nach dem Airliftverfahren riser cracking
K. nach der Entspannungsfahrweise *s.* K. auf flüssigen Rückstand
K. nach der Verkokungsfahrweise *s.* K. auf Koks
K. über Ionen *s.* katalytisches K.
fluidkatalytisches K. fluid catalytic cracking
hydrierendes K. hydrogenation cracking, hydrocracking
ionisches K. *s.* katalytisches K.
katalytisches K. catalytic (catalyst, cat) cracking
katalytisches K. im Orthoflow-Verfahren Orthoflow catalytic cracking
katalytisches K. in [der] Wirbelschicht fluid catalytic cracking
rückstandsloses K. *s.* K. auf Koks
selektives K. selective cracking
thermisches K. thermal cracking
thermisch-katalytisches (thermokatalytisches) K. thermal-catalytic cracking
Krackfraktion *f* cracked fraction
Krackgas *n* cracker (cracking) gas
Krackglasur *f (ceram)* crackle glaze
Krackgut *n* cracking feedstock (stock, feed)
Krackmittelöl *n* cycle stock
Krackofen *m* cracking furnace
Krackraum *m* cracking chamber
Krackreaktion *f* cracking reaction
Krackreaktor *m* cracking chamber
Krackröhrenerhitzer *m*, **Krackröhrenofen** *m* tubular cracking furnace
Krackrohstoff *m s.* Krackgut
Krackung *f s.* Kracken
Krackverfahren *n* cracking process
K. nach Cross Cross process
K. nach Holmes und Manley Holmes-Manley process
K. nach Winkler und Koch Winkler-Koch process
Krackvorgang *m* cracking process
Krackzone *f* cracking zone
Kraft *f* power; *(phys, phys chem)* force
abstoßende K. repulsive force
abweisende K. repellency *(as of a surface for water)*
bewegende K. *(tech)* momentum
Coulombsche K. Coulomb (coulombic) force *(electrostatics)*
elektromotorische K. electromotive force, e.m.f., EMF
flockende K. flocculating (flocculation) power
gegenelektromotorische K. counter-electromotive force, back electromotive force, back e.m.f.
magnetomotorische K. magnetomotive force, m.m.f.
molekulare K. [inter]molecular force
nachschaffende K. *(agric)* supplying power *(of a soil)*
nukleophile K. nucleophilicity
ölabweisende K. oil repellency
osmotische K. osmotic force
rücktreibende K. restoring force *(in oscillating molecules)*
treibende K. moving force, agency
wasserabweisende K. water repellency
zwischenmolekulare K. [inter]molecular force
Kraftdeckenpapier *n* kraft liner paper
Kraft-Dehnungs-Kurve *f* stress-strain curve *(expressed as tons or pounds per squ.in.)*; load-elongation (load-extension) curve *(expressed as inches per inch)*
Kräfte *fpl* **/ van-der-Waalssche** van der Waals forces [of attraction], van der Waals attractive forces
Kräftefeld *n* **des Atoms** atomic field
Kraftfeld *n* force field
Kraftgas *n s.* Generatorgas
kräftigend *(pharm)* tonic, roborant
Kräftigungsmittel *n (pharm)* tonic, roborant
Kraftkarton *m* kraft cardboard
Kraftkonstante *f* force constant
Kraft-Längenänderungs-Diagramm *n* stress-strain diagram *(expressed as tons or pounds per square inch)*; load-elongation (load-extension) diagram *(expressed as inches per inch)*
Kraftlinie *f* line of force
Kraft[pack]papier *n* kraft (strong) paper, kraft wrapping paper, kraft
imitiertes K. imitation kraft paper
Kraftpapiermaschine *f* kraft paper machine
Kraftquelle *f* source of energy
Kraftschaufel *f* power shovel *(conveying)*
Kraftspiritus *m s.* Kraftsprit
Kraftsprit *m* power (fuel) alcohol
Kraftstoff *m* [power] fuel
K. für Fahrdieselmotoren automotive diesel fuel, diesel fuel (oil) for road vehicles
fester K. solid fuel
gebleiter K. leaded (ethylized) fuel
hochklopffester K. high-octane fuel
hochoktaniger (hochoktanzahliger) K. high-octane fuel
klopffester K. antiknock fuel
verbleiter K. *s.* gebleiter K.
Kraftstoffadditiv[e] *n* fuel additive
Kraftstoffbehälter *m* fuel tank
Kraftstoffempfindlichkeit *f* sensitivity *(difference between octane numbers obtained by F1 method and F2 method)*
Kraftstoffklopfen *n* fuel knock
Kraft-Verlängerungs-Kurve *f s.* Kraft-Dehnungs-Kurve
Kraft-Verlängerungs-Schaubild *n s.* Kraft-Längenänderungs-Diagramm

Kraftwirkung *f* force effect
Kraftzellstoff *m (pap)* kraft pulp
Kraftzellstoffkocher *m (pap)* kraft digester
Kraftzellstoffverfahren *n (pap)* kraft process
Krähenaugen *npl (pharm)* nux vomica, strychnos seed *(from Strychnos nux-vomica L.)*
Krählarm *m* raking (rake, agitator, rabble) arm, rabble[r]
krählen to rake, to rabble
Krähler *m s.* Krählarm
Krählwerk *n* raking mechanism
Krakeleeglas *n* crackled glass
Krakeleeglasur *f (ceram)* crackle glaze
Krämer-Mühle *f* Krämer mill *(a beater mill)*
Krampfgift *n* tetanic poison
krampflindernd anticonvulsant
krampflösend spasmolytic
krankheitserregend, krankheitserzeugend pathogenic
krappen *(text)* to crab
Krappfarbstoff *m* madder *(from Rubia tinctorum L.)*
Krapplack *m* madder lake
Krapprot *n* Turkey (alizarine) red, madder
Krappwurzel *f* madder [root] *(from Rubia tinctorum L.)*
Krater *m* 1. *(plast)* crater; 2. [charge] crucible *(as in manufacturing calcium carbide)*
Kratzband *n* drag classifier; *s.* Kratzerförderer
Krätzblei *n* slag lead
Kratze *f s.* Krählarm
Krätze *f (met)* blue dust (powder) *(by-product of zinc reduction)*
kratzen to scrape, to scratch, to rabble
Kratzenband *n s.* Kratzerförderer
Kratzer *m* 1. scratch; *(glass)* cat scratch *(a defect)*; 2. *s.* Kratzerförderer; 3. *s.* Krählarm
Kratzerförderer *m* flight[ed] conveyor
  einsträngiger K. single-strand flight conveyor
  zweisträngiger K. double-strand flight conveyor
Kratzerkette *f s.* Kratzerförderer
kratzfest scratch-proof, scratch-resistant, marproof, mar-resistant
Kratzfestigkeit *f* scratch (mar) proofness (resistance); *(ceram)* scratch hardness
Kratzprobe *f* scratch test
Kratz-Rohrkristallisator *m* scraped-pipe crystallizer
Kräuselfaden *m s.* Kräuselgarn
Kräuselfestigkeit *f s.* Kräuselungsbeständigkeit
Kräuselgarn *n* crimped (crinkled) yarn
Kräusellack *m* wrinkle varnish (finish)
kräuseln *(text)* to crimp, to crinkle, to crêpe; *(pap)* to curlate; *(coat)* to wrinkle
  sich k. *(text)* to crimp, to crinkle; *(phot)* to frill
Kräuseln *n (ceram)* curling *(a defect)*; *(pap)* curlation; *s.* Kräuselung 1.
Kräuselung *f* 1. *(text)* crimping, crinkling, crêp[e]ing; *(coat)* wrinkling; 2. *(text)* crimp
  latente K. *(text)* latent crimp
Kräuselungsbeständigkeit *f (text)* crimp rigidity (stability)
Kräusen *pl (ferm)* krausen, bloom
  hohe K. rocky krausen
Kreatin *n (bioch)* creatine
Kreatinin *n (bioch)* creatinine
krebsauslösend *s.* krebserregend
krebserregend carcinogenic, cancerigenic, cancer-causing, cancer-producing
  nicht k. non-carcinogenic
krebserzeugend *s.* krebserregend
Krebszement *m* grappier cement
Krebs-Zyklus *m* Krebs cycle, tricarboxylic-acid (citric-acid) cycle, TCA cycle
Kreide *f* chalk
  gefällte K. precipitated chalk (whiting)
  gemahlene K. whiting
  geschlämmte und gemahlene K. *s.* präparierte K.
  lithografische K. lithographic crayon
  präparierte K. prepared chalk
  präzipitierte K. *s.* gefällte K.
kreideartig chalky
kreiden *(coat)* to become chalky, to chalk
Kreidepapier *n* chalk paper; *s.* Kreidereliefpapier
Kreidepulver *n* whitening dust (powder)
Kreidereliefpapier *n*, **Kreidezurichtepapier** *n* chalk overlay paper
kreidig chalky
Kreidigkeit *f* chalkiness
Kreisbahn *f* [circular] orbit
Kreisblattschreiber *m* circular-chart recorder
Kreisel *m* / **[schnell]rotierender** spinning top *(as for oil burners)*
Kreiselbrecher *m* gyratory crusher
Kreiselerhitzer *m (food)* cylindrical batch pasteurizer
Kreiselgebläse *n* centrifugal (turbine) blower, turboblower
Kreiselkompressor *m* centrifugal compressor
Kreiselkraftdüse *f* swirl[-plate] nozzle
Kreiselpumpe *f* centrifugal pump
Kreiselpumpenmischapparat *m* centrifugal pump mixer
Kreiselradkompressor *m* centrifugal compressor
Kreiselradlüfter *m* centrifugal fan
Kreiselradverdichter *m* centrifugal compressor
Kreiselsichter *m* whizzer classifier
Kreiselversprüher *m*, **Kreiselzerstäuber** *m* spinning-top atomizer (sprayer)
kreisen to rotate, to revolve, to circulate, to spin
Kreislauf *m* 1. cycle, circuit *(scheme)*; 2. circulation *(of liquid or gas)* ✦ **im K. führen** to [re]circulate, to recycle ✦ **in den K. zurückführen** to return to the circuit, to recycle, to recirculate
  geschlossener K. closed cycle (circuit)
Kreislaufchlorwasserstoff *m* recycle hydrogen chloride
Kreislaufführung *f* [re]cycling, recirculating
Kreislaufgas *n* recycle gas
Kreislauföl *n* cycle oil
Kreislaufpumpe *f* circulating pump, circulator
Kreislaufrückwasser *n (pap)* white water

Kreislaufsystem *n* circulation system
Kreislaufverfahren *n* recycling procedure *(in extracting)*
Kreislaufwasserstoff *m* recycle hydrogen
Kreismesser *n* *(pap)* disk knife, circular [slitting] knife, slitter
Kreisprozeß *m* cycle, cyclic process
 Born-Haberscher K. Born-Haber [thermochemical] cycle
 Carnotscher K. Carnot cycle
 Szent-Györgyi-Krebsscher K. Krebs cycle, tricarboxylic-acid (citric-acid) cycle, TCA cycle
 thermodynamischer K. thermodynamic cycle
Kreisrost *m* circular grate
Kreisschwingsieb *n* circle-throw screen
Krem *f* cream *(for compounds s.* Creme)
Krempelband *n* *(text)* sliver
Krennerit *m* *(min)* krennerite *(gold(III) telluride)*
Krensäure *f* *(soil)* crenic acid *(a fulvic acid)*
Kreosot *n* creosote, *(specif)* wood[-tar] creosote
Kreosotal *n* creosote carbonate
kreosotieren to creosote
Kreosotkarbonat *n* creosote carbonate
Kreosotöl *n* creosote oil, *(specif)* coal-tar creosote oil
Krepeninsäure *f* crepenynic acid, *cis*-9-octadecen-12-ynoic acid
Kreponierbad *n* *(text)* crêp[e]ing bath (liquor)
kreponieren *(text)* to crêpe, to crimp, to crinkle
Krepp *m* 1. *(text)* crêpe; 2. *s.* Kreppkautschuk
Kreppapier *n* crêpe paper
Kreppbad *n s.* Kreponierbad
kreppen *(pap, text)* to crêpe, *(text also)* to crimp, to crinkle
Kreppkautschuk *m* crêpe [rubber]
Krepp-Pack[papier] *n* crêpe wrapping paper
Kreppseidenpapier *n* crêpe tissue paper
Kreppstoff *m* *(text)* crêpe
Kresol *n* $CH_3C_6H_4OH$ cresol, hydroxytoluene, methylphenol
Kresolharz *n* cresol (cresylic) resin
Kresolphthalein *n* cresolphthalein
m-**Kresolpurpur** *m* cresol purple, *m*-cresolsulphonephthalein
Kresolrot *n* cresol red, *o*-cresolsulphonephthalein
Kresolsulfophthalein *n* cresolsulphonephthalein
Kresotinsäure *f* $C_6H_3(CH_3)(OH)COOH$ cresotic (cresotinic) acid, hydroxytoluic acid
Kresylsäure *f* coal-tar-derived cresylic acid *(a mixture of o-, m-, and p-cresol)*
Kreuzbalkenrührer *m* cross-arm paddle mixer
Kreuzband[magnet]scheider *m* cross-belt [magnetic] separator
Kreuzkopfbohrer *m* *(lab)* charcoal borer
Kreuzresistenz *f* *(tox)* cross resistance
Kreuzschichtstoff *m* cross-laminate
Kreuzspule *f* *(text)* cheese
 konische K. cone
Kreuzspulfärbeapparat *m* *(text)* cheese dyeing machine
Kreuzstrom *m* cross flow
Kreuzstromboden *m* *(distil)* cross-flow tray
Kreuzstromkühlturm *m* cross-flow cooling tower
Kreuzstück *n* cross
kriechen to creep; *(ceram)* to crawl *(of glaze)*
Kriechen *n* creep; *(ceram)* crawling *(a defect during glazing)*
Kriechfestigkeit *f* creep resistance
Kriechpunkt *m* creep point
Kriechstromfestigkeit *f* track[ing] resistance
Kriechwegbildung *f* tracking
 K. durch leitende Ablagerungen deposit tracking
 K. durch Lichtbogen arc tracking
Kriechwiderstand *m* creep resistance
Krinkelgarn *n* crimped (crinkled) yarn
Krispelmaschine *f* *(tann)* boarding machine
krispeln *(tann)* to board, to grain, to pommel
Kristall *m* crystal
 flüssiger K. liquid crystal, crystalline liquid
 gestörter K. imperfect crystal
 homöopolarer K. *s.* kovalenter K.
 kovalenter K. covalent crystal
 optisch positiver K. positive crystal
 piezoelektrischer K. piezoelectric crystal
 pseudomorpher K. pseudomorph
 realer K. imperfect crystal
 rotierender K. rotating (rotation) crystal
 xenomorpher K. xenomorphic (allotriomorphic) crystal, anhedron
Kristallabscheider *m*, **Kristallabscheideraum** *m* crystallizing chamber
Kristallachse *f* crystal (crystallographic) axis
Kristallaggregat *n* crystal[line] aggregate
 van-der-Waalssches K. van der Waals crystal aggregate
kristallartig crystal-like, crystalline
Kristallaustrag *m* crystal discharge
Kristallbau *m* crystal structure
Kristallbaufehler *m* crystal defect, lattice defect (imperfection)
Kristallbett *n* crystal bed
Kristallchemie *f* crystal chemistry
kristallchemisch crystallochemical
Kristalldruse *f* *(geoch)* [crystal] druse, geode
Kristallebene *f* crystal[lographic] plane
Kristalleigenschaft *f* crystal property
Kristalleis *n* crystal (clear) ice
kristallen crystalline
Kristaller *m* crystallizer
Kristallfehler *m s.* Kristallbaufehler
Kristallfeldtheorie *f* crystal field theory
Kristallfläche *f* crystal face
Kristallform *f* crystal form (shape) ✦ **in K.** in crystalline form
Kristallgitter *n* crystal lattice (grating), space lattice
Kristallgitterspektrograf *m* crystal grating spectrograph
Kristallgittertyp *m* *(cryst)* lattice type
Kristallglas *n* crystal glass, *(Am)* rock crystal
Kristallglasur *f* *(ceram)* crystal[line] glaze
Kristallgummi *n* $(C_6H_{10}O_5)_x$ starch (artificial) gum

Kristallhabitus *m* crystal habit
Kristallhaufwerk *n* crystal[line] aggregate
Kristallierer *m* crystallizer
kristallin crystalline
nicht k. non-crystalline, amorphous
kristallinisch *s.* kristallin
Kristallinität *f* crystallinity
Kristallinitätsgrad *m* degree of crystallinity
Kristallisat *n* crystallizate, crop of crystals, crystalline crop; solid *(in zone melting)*
Kristallisation *f* crystallization ✦ **die K. anregen** to induce crystallization ✦ **zur K. bringen** to crystallize out
K. in Bewegung *(sugar)* crystallization in motion
extraktive K. extractive crystallization
fraktionierte K. fractional crystallization
ungeleitete K. uncontrolled crystallization
Kristallisationsbedingung *f* condition of crystallization
Kristallisationsdifferentiation *f* *(geol)* crystallization differentiation, fractional crystallization
kristallisationsfähig crystallizable
Kristallisationsfähigkeit *f* crystallizability
Kristallisationsgefäß *n* crystallizing vessel
Kristallisationsgeschwindigkeit *f* rate of crystallization
Kristallisationskeim *m* crystal nucleus, centre of crystallization ✦ **als K. wirken** to nucleate
Kristallisationskeimbildung *f* nucleation, creation of nucleation centres, formation of nuclei
Kristallisationskeimzahl *f* number of nuclei
Kristallisationskern *m* *s.* Kristallisationskeim
Kristallisationskraft *f* force of crystallization
Kristallisationslösung *f* crystallizing solution
Kristallisationsneigung *f* tendency to crystallize
geringe K. reluctance to crystallize
Kristallisationspapier *n* ice paper
Kristallisationsprodukt *n* crystalline crop, crop of crystals, crystallizate
Kristallisationsraum *m* crystallizing chamber
Kristallisationstendenz *f* tendency to crystallize
Kristallisationsvermögen *n* crystallizability
Kristallisationswagen *m* crystallization truck [for wet cooling], crystallization wag[g]on *(margarine making)*
Kristallisationswärme *f* heat of crystallization
Kristallisationswiderstand *m* resistance to crystallization
Kristallisationszentrum *n* *s.* Kristallisationskeim
Kristallisator *m* crystallizer; *(ceram)* accelerator
klassifizierender K. classifying crystallizer
offener feststehender K. tank crystallizer, crystallizing tank
kristallisch crystalline
Kristallisier ... *s.a.* Kristallisations ...
Kristallisierapparat *m* crystallizer
kristallisierbar crystallizable
Kristallisierbarkeit *f* crystallizability
Kristallisierbecken *n* crystallizing pond *(as in a saltern)*
Kristallisierbehälter *m* crystallizing tank, tank crystallizer
kristallisieren to crystallize [out]
Kristallisiermulde *f*, **Kristallisierpfanne** *f* crystallizing tank, tank crystallizer
Kristallisierschale *f* crystallization (crystallizing) dish
Kristallisierung *f* crystallization
Kristallisierwiege *f* **[/Wulff-Bocksche]** Wulff-Bock crystallizer
Kristallit *m* crystallite
Kristallittheorie *f* *(glass)* crystallite (microheterogeneity) theory
Kristallkeim *m*, **Kristallkern** *m* *s.* Kristallisationskeim
Kristallklasse *f* crystallographic class, class of crystal symmetry
Kristallkörnchen *n* *(sugar)* grain
Kristalloberfläche *f* crystal surface
Kristallografie *f* crystallography
kristallografisch crystallographic
Kristalloid *n* crystalloid
Kristallolumineszenz *f* crystalloluminescence
Kristallose *f* soluble saccharin *(sodium salt of saccharin)*
Kristallphase *f* crystalline phase
Kristallphosphor *m* crystal phosphor
Kristallphosphoreszenz *f* crystal phosphorescence
Kristallpolster *n* crystal bed
Kristallpulver *n* crystal (crystalline) powder
Kristall[schütt]schicht *f*, **Kristallschüttung** *f* crystal bed
Kristallsoda *f* salt of soda, sal soda, soda [crystals], washing soda, natron *(sodium carbonate-10-water)*
Kristallspektrograf *m* crystal grating spectrograph
Kristallstörung *f* crystal imperfection
Kristallstruktur *f* crystal structure
Kristallstrukturanalyse *f* crystal[-structure] analysis
röntgenografische K. X-ray crystal[-structure] analysis, X-ray crystallographic analysis
Kristallstrukturbestimmung *f* crystal-structure determination
Kristallsystem *n* crystal (crystallographic) system
hexagonales K. hexagonal [crystal] system
kubisches K. cubic [crystal] system, regular system
monoklines K. monoclinic [crystal] system
orthorhombisches K. [ortho]rhombic [crystal] system
reguläres K *s.* kubisches K.
rhombisches K. *s.* orthorhombisches K.
rhomboedrisches K. rhombohedral [crystal] system, trigonal system
tetragonales K. tetragonal [crystal] system
trigonales K. *s.* rhomboedrisches K.
triklines K. triclinic [crystal] system
Kristalltuff *m* *(geol)* crystal tuff
Kristallversetzung *f* crystal dislocation
Kristallviolett *n* crystal violet, hexamethyl-*p*-rosaniline hydrochloride

Kristallwachstum *n* crystal growth
Kristallwasser *n* water of crystallization
kristallwasserfrei free from crystal water, anhydrous
Kristallwinkel *m* crystal angle
Kristallzüchtung *f* crystal growth
Kristallzucker *m* crystallized (granulated) sugar, white crystals
Kröhnkit *m (min)* kröhnkite *(copper sodium sulphate)*
Krokoit *m (min)* crocoite, crocoisite, red lead ore *(lead(II) chromate)*
Krokydolith *m* crocidolite, cape (blue) asbestos, cape blue *(a mineral of the amphibole group)*
Kroll-Verfahren *n (met)* Kroll process *(a reduction process)*
Kron-Effekt *m (phot)* Kron effect
Kronenblock *m* crown block *(of a rotary-drilling installation)*
Kronenverschließmaschine *f (food)* crown corking machine, crowner
Kronenverschluß *m (food)* crown cap (closure, cork)
Kronflintglas *n* crown flint glass, lead crown glass
Kronglas *n* crown [optical] glass, crown
K-Röntgenstrahlung *f* K radiation
Kropf *m (pap)* backfall, descent plate, weir *(of a beater)*
Kropfkrone *f (pap)* backfall crest (crown) *(of a beater)*
Krötengift *n* toad venom (poison)
Krotonaldehyd *m* $CH_3CH{=}CHCHO$ crotonaldehyde, 2-butenal
Krotonöl *n* croton (tiglium) oil *(from Croton tiglium L.)*
Krotonsäure *f* $CH_3CH{=}CHCOOH$ crotonic acid, 3-methylacrylic acid, 2-butenoic acid, *(specif) trans*-2-butenoic acid
α-**Krotonsäure** *f* crotonic acid *(proper)*, α-crotonic acid, *trans*-2-butenoic acid
β-**Krotonsäure** *f* isocrotonic acid, β-crotonic acid, *cis*-2-butenoic acid
Krotonsäureäthylester *m* $CH_3CH{=}CHCO{-}OC_2H_5$ ethyl crotonate
Krotonylen *n* $CH_3{-}C{\equiv}C{-}CH_3$ crotonylene, 2-butyne
Krozeinsäure *f* crocein acid, 2-naphthol-8-sulphonic acid
Krücke *f* agitator (raking, rabble) arm, rabble[r]
Krume *f (soil)* topsoil
Krümel *m(n) (soil)* ped, aggregate, crumb; *(rubber)* pellet, *(sometimes also)* crumb
Krümelbildung *f (soil)* flocculation
Krümelbuna *m(n)* crumbs of buna synthetic rubber
Krümelgefüge *n (agric)* crumb structure
krümelig crumbly, friable
krümeln to crumble; *(by means of a liquid)* to agglomerate
Krümelstruktur *f (agric)* crumb structure
Krümelung *f* crumbling; *(by means of a liquid)* agglomeration; *(soil)* flocculation
krümmen to bend, to curve, to camber
Krümmer *m (tech)* elbow [fitting], ell
K. mit rechtwinkliger Ablenkung right-angle elbow
Krümmer-Durchfluß[mengen]messer *m* elbow meter *(flow measurement)*
Krümmung *f* bend, curvature, camber
krumpfbeständig *s.* krumpffest
krumpfen *(text)* to shrink
krumpffest *(text)* shrink-resistant, shrinkproof, unshrinkable
Krumpffestausrüstung *f (text)* shrink-resist finish, unshrinkable finish
Krumpffestigkeit *f (text)* shrink resistance, unshrinkability
Krumpffestmachen *n (text)* shrinkproofing
krumpffrei *(text)* non-shrinking
Krumpfmaschine *f (text)* shrinking machine
Krumpfung *f (text)* shrinkage, shrinking
erzwungene (kompressive) K. compressive (compression) shrinkage
Krupon *m (tann)* butt
kruponieren *(tann)* to butt
Krupp-Lurgi-Schwelverfahren *n* Krupp-Lurgi process
Krupp-Rennverfahren *n (met)* Krupp Renn process
Kruste *f* crust, incrustation, encrustation; *(food)* rind ✦ **eine K. bilden** to form a crust, to encrust, to incrust
Kryogenik *f* cryogenics
Kryohydrat *n* cryohydrate
kryohydratisch cryohydric
Kryolith *m (min)* cryolite, ice stone, Greenland spar *(sodium fluoroaluminate)*
Kryoskop *n* cryoscope
Kryoskopie *f* cryoscopy
Kryptobase *f* cryptobase
Kryptoionenreaktion *f* crypto-ionic reaction
kryptoionisch crypto-ionic
kryptokristallin cryptocrystalline
Kryptolepin *n* cryptolepine, methylquindolanol
Kryptomeren *n* cryptomerene *(a diterpene derivative)*
Krypton *n* Kr krypton
Kryptozyanin *n* kryptocyanine, 1,1-diethyl-4,4′-carbocyanine iodide
K-Säure *f* $NH_2C_{10}H_4(OH)(SO_3H)_2$ K acid, 1-amino-8-naphthol-4,6-disulphonic acid
K-Schale *f* K-shell *(of an atom)*
KU *s.* Kuoxamfaserstoff
Kuba[gelb]holz *n* Cuba wood *(a sort of the dyewood from Chlorophora tinctoria Gaud.)*
Kubeben *fpl (pharm)* cubebs *(from Piper cubeba L.f.)*
Kubebenpfeffer *m (pharm)* cubeba, Piper cubeba L.f.
Kübel *m* tub, vat; bucket *(esp in conveying)*
Kübelaufzug *m* skip hoist
Kubierschky-Nitrierapparat *m* Kubierschky nitrator
Kubierschky-Turm *m* Kubierschky tower *(for re-*

covering bromine from brines containing magnesium bromide)

kubisch cubic

kubisch-flächenzentriert *(cryst)* face-centred cubic

kubisch-innenzentriert *s.* kubisch-raumzentriert

kubisch-raumzentriert *(cryst)* body-centred cubic

Kuchen *m (chem, tech)* cake; *(glass)* shear cake

Kuchenabnahme *f*, **Kuchenaustrag** *m (filtr)* cake discharge

Kuchendicke *f (filtr)* cake thickness

Kuchendickefühler *m*, **Kuchendickenmeßeinrichtung** *f (filtr)* cake thickness detector (sensing device)

Kuchenfeuchte *f*, **Kuchenfeuchtigkeit** *f* cake moisture

Kuchenführungswagen *m (coal)* coke guide

Kuchenhöhe *f*, **Kuchenstärke** *f s.* Kuchendicke

Kuchenwiderstand *m (filtr)* cake resistance

Kufe *f* tub, bark, beck, vat

Kugel *f* sphere, globe; *(tech)* ball *(as in bearings or mills)*; *(met)* pellet; *(lab)* bulb

Kügelchen *n* spherule, globule; *(met)* pellet; *(plast)* bead

Kugeldruckhärte *f* ball-puncture resistance, indentation hardness

Kugeldruckprüfung *f*, **Kugeldrucktest** *m* ball test

Kugelfallmethode *f* falling-ball (falling-sphere) method

Kugelfallprüfung *f* falling-ball [impact] test

Kugelfallviskosimeter *n* falling-ball (falling-sphere, dropping-ball) viscometer, ball viscometer

Kugelfallwerk *n* drop-weight device

Kugelform *f* spherical (globular) form

kugelförmig spheric[al], globular

Kugelförmigkeit *f* sphericity

Kugelfüllung *f* ball charge

Kugelgraphit *m* nodular (spheroidal) graphite

Kugelgraphit[grau]guß *m* nodular cast iron

Kugelhaufenreaktor *m (nucl)* pebble [bed] reactor, PBR

kugelig spheric[al], globular; *(min)* nodular

Kugelkocher *m (pap)* spherical boiler (cooker, digester, rotary cooker)

Kugelkühler *m* ball (bulb) condenser

K. nach Allihn Allihn condenser

kugelmahlen to ball-mill

Kugelmühle *f* ball (globe) mill, *(specif)* pebble mill *(filled with flint pebbles or porcelain balls)* ✦ **in der K. mahlen** to ball-mill

konische K. conical ball mill

schwindende K. vibrating (vibratory, vibration, oscillating, oscillatory) ball mill

zylindrisch-konische K. *s.* konische K.

Kugelmühlenaufbereitung *f* ball milling

Kugelmühlenmethode *f* ball-mill method

Kugelpackung *f (cryst)* sphere packing

dichteste K. closest (close) packing

hexagonal dichteste K. hexagonal closest (close) packing

Kugelprotein *n* globular protein

Kugelpulver *n* ball powder *(a propellant powder in the form of spherules)*

Kugelringmühle *f* ball-and-ring (ball-and-race) mill, ball roller mill

K. für Kohlenstaubmahlung ball-and-ring coal pulverizer

Kugelrohr *n* bulb tube *(tube with bulbar enlargements)*

Kugelröhre *f s.* Kugelrohr

Kugelrohrmühle *f* bulb-tube mill

Kugelrückschlagventil *n* ball-check valve, ball check

Kugelschale *f* spherical shell

Kugelschliffverbindung *f* spherical [ground-glass] joint, ball-and-cup (ball-and-socket) joint

kugelsintern to pelletize, to nodulize *(ore)*

Kugelsinterung *f* pelletizing, nodulizing *(of ore)*

Kugelsymmetrie *f* spherical symmetry

kugelsymmetrisch spherically symmetrical

Kugelventil *n* ball valve

Kugelverschlußdüse *f* ball-check nozzle

Kuhbutterfett *n* cow milk fat

kühl cool

Kühlanlage *f* refrigerating (cooling) plant

Kühlapparat *m* cooling apparatus, cooler; *(for temperatures below 0 °C)* chiller

Kühlauto *n* refrigerator truck

Kühlbad *n* cooling bath

Kühlbereich *m (glass)* annealing range

Kühlcreme *f (cosmet)* cold cream

kühlen to cool, *(esp food)* to refrigerate; *(quickly by immersion)* to chill, to quench

mit Wasser k. to water-cool

Kühler *m* 1. *(lab)* [vapour, steam] condenser; 2. sheet cooler *(for sheet glass)*; 3. *s.* Kühlvorrichtung

K. nach Friedrichs Friedrichs [reflux] condenser

K. nach West West condenser

Allihnscher K. Allihn condenser

Liebigscher K. Liebig condenser

Kühlerklemme *f (lab)* condenser clamp

Kühlermantel *m (lab)* condenser jacket

Kühlerschweinchen *n (lab)* condenser jacket

Kühlerwascher *m* washer-cooler

Kühlfalle *f* cold trap

Kühlfalte *f (glass)* chill (settle) mark *(a surface defect)*

Kühlfeld *n (text)* cooling zone

Kühlfinger *m* cold finger [condenser]

Kühlfläche *f* cooling surface

Kühlflüssigkeit *f* cooling liquid

Kühlgeschwindigkeit *f* cooling rate

Kühlgrenztemperatur *f* wet-bulb (wet-surface) temperature

Kühlhalle *f*, **Kühlhaus** *n* cold store (storage), cold-storage house, coolhouse

Kühlhausaufbewahrung *f* cold storage

Kühlkammer *f* cooling chamber

Kühlkanal *m* cooling channel; *(glass)* lehr

Kühllagerung *f* cold storage

Kühlluft *f* cooling air (wind)

Kühlmantel *m* cooling jacket

Kühlmedium *n*, **Kühlmittel** *n* cooling agent (medium), coolant, refrigerating medium, refrigerant
Kühloberfläche *f* cooling surface
Kühlofen *m (glass)* [annealing] lehr, annealing oven, leer
kontinuierlich arbeitender K. continuous-annealing lehr
Kühlofenbeschicker *m (glass)* lehr loader, stacker
Kühlplatte *f* cool plate
Kühlpunkt *m* / **oberer** *s.* Kühltemperatur / obere
unterer K. *s.* Kühltemperatur / untere
Kühlraum *m* cold-storage room
Kühlraumaufbewahrung *f*, **Kühlraumlagerung** *f* cold storage
Kühlrippe *f* cooling fin
Kühlriß *m (ceram)* dunt, cooling crack
Kühlrißbildung *f (ceram)* dunting
Kühlrohr *n* cooling (chilling) tube, chilling cylinder; *(distil)* condensing tube
Kühlschacht *m (glass)* vertical lehr
Kühlschiff *n (ferm)* coolship, cooler
Kühlschlange *f* cooling (refrigerating) coil; attemperator *(in a fermenter)*
Kühlschrank *m* refrigerator
Kühlsole *f* cooling (refrigerating, cold) brine
Kühlsystem *n* cooling (refrigeration) system
K. mit einer gekühlten Walze single-drum system *(margarine making)*
K. mit Walzenpaar double-drum system *(margarine making)*
Kühltank *m* cooling tank
Kühltankwagen *m* refrigerated-tank truck
Kühlteich *m* cooling pond
Kühltemperatur *f* / **obere** *(glass)* annealing temperature (point), A.P., 13.0 temperature *(at which the viscosity is* 10^{13} *poises)*
untere K. *(glass)* strain temperature (point), St.P.
Kühltrommel *f* cooling drum, cooling (chill) roll
Kühltrommelverfahren *n* dry [drum-]cooling process, chill-roll method *(margarine making)*
Kühltrub *m (ferm)* cold sludge
Kühlturm *m* cooling tower; chilling tower *(petroleum dewaxing)*
K. mit natürlichem Zug natural-draught cooling tower, atmospheric cooling tower
selbstbelüfteter K. *s.* K. mit natürlichem Zug
Kühlturmkamin *m* cooling-tower casing
Kühlung *f* cooling, *(esp food)* refrigeration; *(quickly by immersion)* chilling, quench[ing]
K. mit Wasser water cooling
Kühlungskristallisator cooling (cooler) crystallizer
Kühlvorrichtung *f* cooling facility, cooler
Kühlwagen *m*, **Kühlwaggon** *m* refrigerator car (wagon)
Kühlwalze *f* cooling roll (drum)
Kühlwanne *f* cooling vat
Kühlwascher *m* washer-cooler
Kühlwasser *n* cooling water
Kühlwasseraustritt *m* cooling-water outlet
Kühlwirkung *f* cooling effect
Kühlzentrifuge *f* refrigerated centrifuge
Kühlzone *f* cooling zone (compartment, section)
Kühlzylinder *m* cooling cylinder (roll); *(pap)* sweat cylinder (roll)
Kuhmilch *f* cow milk
Kuhn-Roth-Bestimmung *f* Kuhn-Roth determination *(of terminal methyl groups)*
Kuhpockenlymphe *f* vaccine
KUK *(abbr)* Kationenumtauschkapazität, *s.* Kationenaustauschkapazität
Küken *n (lab)* stopper, plug
massives K. solid stopper
Külbel *n (glass, plast)* parison
spritzgegossenes K. *(plast)* injection-moulded parison
Kulör *f* sugar colouring (dye)
kultivieren to cultivate
kultiviert / in Nährlösung solution-grown
Kultur *f* culture, *(act also)* cultivation *(as of microorganisms)*
K. im hängenden Tropfen [hanging-]drop culture
submerse K. submerged culture
Kulturfiltrat *n* culture filtrate
Kulturflüssigkeit *f* nutrient broth *(as for microorganisms)*
Kulturgefäß *n (agric)* pot
Kulturhefe *f* cultivated (culture) yeast
Kulturkolben *m* culture bottle (flask)
K. nach Roux Roux culture bottle
Kulturlösung *f* nutrient broth *(as for microorganisms)*
Kulturmedium *n* culture medium
Kulturplatte *f* culture plate
Kulturröhrchen *n* culture tube
Kulturstamm *m* strain *(as of microorganisms)*
Kumalinsäure *f* coumalic acid, coumalinic acid, 2-oxo-2H-pyran-5-carboxylic acid
Kumarilsäure *f* coumarilic acid, benzofuran-2-carboxylic acid
Kumarin *n* coumarin, cumarin, 1,2-benzopyrone
Kumarinsäure *f* coumarinic acid, *cis-o*-coumaric acid
Kumarinsäurelakton *n* coumarinic lactone
Kumarinsynthese *f* / **Pechmannsche** Pechmann condensation (coumarin synthesis)
Kumaron *n* coumarone, cumarone, benzofuran
Kumaronharz *n*, **Kumaron-Inden-Harz** *n* coumarone[-indene] resin
Kumaron-2-karbonsäure *f s.* Kumarilsäure
Kumarsäure *f* $HOC_6H_4CH{=}CHCOOH$ coumaric (cumaric) acid, hydroxycinnamic acid, *(specif) o*-coumaric acid, *trans-o*-hydroxycinnamic acid
Kumarylchinasäure *f* coumaroylquinic acid *(a depside)*
Kümmelöl *n* caraway oil *(from Carum carvi L.)*
Kumol *n* $C_6H_5CH(CH_3)_2$ cumene, 2-phenylpropane
kumulativ cumulative
Kumulen *n* cumulene *(any of a class of compounds having three or more cumulated double bonds)*
kumulierend cumulative

Kumys *m*, **Kumyß** *m* koumis[s], koumyss
Kunstbutter *f* artificial butter
Kunstdruckkarton *m* art cardboard
Kunstdruckpapier *n* art paper
Kunstdruckrohpapier *n* art base paper
Kunstdünger *m s.* Handelsdünger
Kunsteis *n* artificial (manufactured) ice
Kunstfaser *f s.* Chemiefaser
Kunstfaserstoff *m s.* Chemiefaserstoff
Kunstfaserzellstoff *m* rayon (dissolving) pulp
Kunstfett *n* artificial (synthetic) fat
Kunsthaar *n* artificial hair
Kunstharz *n* artificial (synthetic) resin
K. zur Naßfestleimung *(pap)* wet-strength resin
Kunstharzappretur *f s.* Kunstharzausrüstung
Kunstharzausrüstung *f (text)* [synthetic-]resin finish
K. mit verzögerter Formfixierung deferred-curing finish
Kunstharzaustauscher *m* resinous exchanger
Kunstharzbehandlung *f (text)* resin treatment
Kunstharzdispersion *f* latex
Kunstharzionenaustauscher *m s.* Kunstharzaustauscher
Kunstharzkitt *m* synthetic-resin cement
Kunstharzkleber *m*, **Kunstharzklebstoff** *m* synthetic-resin adhesive
Kunstharzlack *m* synthetic-resin varnish
Kunstharzpulver *n* synthetic-resin powder
Kunstharzsperrholz *n* resin-bonded plywood
Kunsthonig *m* artificial honey
Kunsthorn *n* artificial horn, casein plastic
Kunstkautschuk *m* synthetic (man-made, artificial, chemical) rubber
Kunstkautschukklebstoff *m* synthetic-rubber adhesive
Kunstkautschuklatex *m* synthetic[-rubber] latex
Kunstkautschukmischung *f* synthetic-rubber mix (stock, compound)
Kunstkohle *f* artificial coal
Kunstleder *n* artificial (imitation) leather
Künstlerfarbe *f* artists' colour
künstlich artificial, factitious, non-natural, manufactured, synthetic, man-made
Kunstlicht *n* artificial light
Kunstmist *m* artificial manure
Kunstrahm *m* artificial cream
Kunstseide *f s.* Chemieseide
Kunststoff *m* plastic [material] *(for compounds s.a.* Plast)
K. aus Harnstoff-Formaldehyd-Harz urea-formaldehyde plastic
K. aus Harnstoffharz urea plastic
duroplastischer K. thermosetting plastic (resin), thermoset [resin]
glasfaserverstärkter K. glass-fibre reinforced plastic, G.R.P.
harter K. rigid plastic
thermoplastischer K. thermoplastic [material]
verstärkter K. reinforced plastic
Kunststofferzeugnis *n* plastic product
kunststoffimprägniert plastic-proofed
Kunststoffolie *f (plast)* [self-supporting] film *(esp if thickness less than 0.01 inch)*; *(if thickness greater than 0.01 inch)* sheeting, *(pieces)* sheet
Kunststoffrohr *n* plastic pipe
Kunststoffrohrmaterial *n* plastic piping
Kunststoffsack *m* plastic bag
Kunststoffschweißen *n* welding of plastics
Kunststoffüberzug *m* plastic covering
Kunstumblattpapier *n* cigar wrapping paper
Kuoxam *n* $[Cu(NH_3)_4](OH)_2$ cuprammonium hydroxide solution, cuprammonia
Kuoxamfaser *f* cuprammonium rayon staple fibre, cuprammonium (cupro) staple
Kuoxamfaserstoff *m* cuprammonium rayon
Kuoxamseide *f* continuous-filament cuprammonium rayon
Kuoxamzellulose *f* cuprammonium cellulose
Kuparen *n* cuparene, *p*-(1′,2′,2′-trimethylcyclopentyl-)toluene
Kuparensäure *f* cuparenic acid, *p*-(1′,2′,2′-trimethylcyclopentyl-)benzoic acid
Küpe *f (dye)* vat ✦ **in der K. behandeln** to vat
ammoniakalische K. ammonia vat
blinde K. blank vat
Kupellation *f (met)* cupellation
kupellieren *(met)* to cupel
küpen *(dye)* to vat
Küpenfärberei *f* vat dyeing
Küpenfarbstoff *m* vat dye[stuff], vat colour
Küpenflüssigkeit *f (dye)* vat liquor
Küpensäure *f (dye)* vat acid
Küpensäureverfahren *n (dye)* vat-acid process
Kupfer *n* Cu copper
K. hoher Leitfähigkeit high-conductivity copper, H.C. copper
elementares K. elemental copper
gediegen[es] K. native copper
hammergares K. tough-pitch copper
metallisches K. metallic copper
sauerstofffreies K. oxygen-free copper, O.F. copper
stranggepreßtes K. coalesced copper
zähgepoltes K. tough-pitch copper
Kupferamalgamelektrode *f* copper amalgam electrode
Kupfer(I)-ammin *n* copper(I) ammine, cuprous ammine
Kupfer(II)-ammin *n* copper(II) ammine, cupric ammine
Kupfer(I)-antimonid *n* Cu_3Sb copper(I) antimonide, cuprous antimonide
Kupfer(II)-arsenat(III) *n* $Cu_3(AsO_3)_2$ copper(II) arsenite, cupric arsenite
Kupfer(II)-arsenat(V) *n* $Cu_3(AsO_4)_2$ copper(II) arsenate, copper(II) orthoarsenate, cupric arsenate
Kupfer(I)-arsenid *n* Cu_3As copper(I) arsenide, cuprous arsenide
Kupferarsenitazetat *n* copper aceto-arsenite *(approximately* $Cu(CH_3COO)_2 \cdot 3\,Cu(AsO_2)_2$*)*
Kupferäthylendiamin *n (text)* cupriethylenediamine
Kupferätze *f (glass)* copper stain

Kupferaventurin *m* gold aventurine
Kupfer(II)-azetat *n* $Cu(CH_3COO)_2$ copper(II) acetate, cupric acetate
Kupferazetatarsenit *n s.* Kupferarsenitazetat
Kupfer(I)-azetylid *n* Cu_2C_2 copper(I) acetylide, copper(I) carbide, cuprous acetylide
Kupferbad *n* copper[plating] bath
Kupferbeize *f (glass)* copper stain
Kupfer(II)-benzoat *n* $Cu(C_6H_5COO)_2$ copper(II) benzoate, cupric benzoate
Kupferblau *n* verditer blue *(a basic copper carbonate)*
Kupfer(II)-borid *n* Cu_3B_2 copper(II) boride
Kupfer(II)-bromat *n* $Cu(BrO_3)_2$ copper(II) bromate, cupric bromate
Kupfer(II)-bromid *n* $CuBr_2$ copper(II) bromide, copper dibromide, cupric bromide
Kupferbruch *m* copper casse, cuprous-sulphide cloud *(a disorder in wine)*
Kupferbrühe *f (agric)* copper spray
Kupfer(II)-butyrat *n* $Cu(C_3H_7COO)_2$ copper(II) butyrate, cupric butyrate
Kupfer(II)-chelat *n* copper(II) chelate, cupric chelate
Kupfer(I)-chlorid *n* CuCl copper(I) chloride, copper monochloride, cuprous chloride
Kupfer(II)-chlorid *n* $CuCl_2$ copper(II) chloride, copper dichloride, cupric chloride
Kupferchloridverfahren *n (petrol)* copper chloride [sweetening] process
Kupfercoulometer *n* copper coulometer (voltameter)
Kupferdi . . . *s.a.* Kupfer(II)- . . .
Kupferdichromat *n* $CuCr_2O_7$ copper dichromate
Kupferdichtung *f* copper gasket
Kupferdrehspäne *mpl* copper turnings
Kupferdruck *m* copperplate printing
Kupferdruckfarbe *f* copperplate ink
Kupferdruckpapier *n* [soft] plate paper, etching paper
Kupferelektrode *f* copper electrode
Kupferenzym *n* copper enzyme
Kupfererz *n* copper ore
Kupfer-Farbstoffchelat *n* copper-dye chelate
Kupferfaser *f* 1. copper fibre; 2. *s.* Kuoxamfaser
Kupferfaserstoff *m* 1. copper fibre; 2. *s.* Kuoxamfaserstoff
Kupferfeilspäne *mpl* copper filings
Kupfer(II)-fluorid *n* CuF_2 copper(II) fluoride, copper difluoride, cupric fluoride
Kupferfolie *f* copper foil
Kupfer(II)-formiat *n* $Cu(HCOO)_2$ copper(II) formate, cupric formate
Kupferfungizid *n* copper fungicide
Kupferglanz *m* copper glance *(a mineral group)*
Kupferglyzinchelat *n* copper-glycine chelate
Kupfergraphit *m* copper graphite
kupferhaltig copper-bearing, cupriferous
Kupfer(I)-hexafluorosilikat *n* $Cu_2[SiF_6]$ copper(I) hexafluorosilicate, cuprous fluorosilicate
Kupfer(II)-hexafluorosilikat *n* $Cu[SiF_6]$ copper(II) hexafluorosilicate, cupric fluorosilicate
Kupfer(I)-hexazyanoferrat(II) *n* $Cu_4[Fe(CN)_6]$ copper(I) hexacyanoferrate(II)
Kupfer(I)-hexazyanoferrat(III) *n* $Cu_3[Fe(CN)_6]$ copper(I) hexacyanoferrate(III)
Kupfer(II)-hexazyanoferrat(II) *n* $Cu_2[Fe(CN)_6]$ copper(II) hexacyanoferrate(II)
Kupfer(II)-hexazyanoferrat(III) *n* $Cu_3[Fe(CN)_6]_2$ copper(II) hexacyanoferrate(III)
Kupfer(I)-hydrid *n* CuH copper(I) hydride
Kupfer(II)-hydrogenarsenat(III) *n* $CuHAsO_3$ copper(II) hydrogenarsenite, cupric hydrogenarsenite
Kupfer(II)-hydroxid *n* $Cu(OH)_2$ copper(II) hydroxide, cupric hydroxide
Kupferhydroxidlösung *f* **/ ammoniakalische** *s.* Kupferoxidammoniak
Kupferindig[o] *m (min)* indigo copper, blue copper, covellite, covelline *(copper(II) sulphide)*
Kupfer(I)-ion *n* copper(I) ion, cuprous ion
Kupfer(II)-ion *n* copper(II) ion, cupric ion
Kupfer(I)-ionen-Verfahren *n* cuprous ion method
Kupfer(II)-jodat *n* $Cu(IO_3)_2$ copper(II) iodate, cupric iodate
Kupferkalkbrühe *f* Bordeaux mixture *(fungicide)*
Kupferkarbid *n* Cu_2C_2 copper(I) carbide, copper(I) acetylide, cuprous acetylide
Kupfer(I)-katalysator *m* cuprous catalyst
Kupferkies *m (min)* chalcopyrite, chalkopyrite, copper pyrites *(copper(II) iron(II) sulphide)*
Kupferkomplex *m* copper complex
Kupferkonverter *m* copper converter
Kupferkonzentrationszelle *f* copper concentration cell
Kupferkopf *m* copper head *(a defect in enamel)*
Kupferkunstseide *f s.* Kuoxamseide
Kupfer(II)-laktat *n* $Cu(CH_3CH(OH)COO)_2$ copper(II) lactate, cupric lactate
Kupferlasur *m (min)* blue copper ore, azurite
Kupferlegierung *f* copper alloy
Kupfermangel *m* copper deficiency
Kupfer(II)-metaborat *n* $Cu(BO_2)_2$ copper(II) metaborate
Kupfermineral *n* copper mineral
Kupfermonoxid *n s.* Kupfer(II)-oxid
kupfern *(dye, text)* to copperize
Kupfernachbehandlung *f* copper aftertreatment
Kupfernaphthenat *n* copper naphthenate
Kupfer-Nickel-Legierung *f* cupro-nickel alloy
Kupfer(II)-nitrat *n* $Cu(NO_3)_2$ copper(II) nitrate, cupric nitrate
Kupfernitrid *n* Cu_3N copper nitride
Kupfer(II)-nitroprussiat *n*, **Kupfer(II)-nitroprussid** *n s.* Kupfer(II)-pentazyanonitrosylferrat
Kupfer(II)-oleat *n* $Cu(C_{17}H_{33}COO)_2$ copper(II) oleate, cupric oleate
Kupfer(II)-orthophosphat *n* $Cu_3(PO_4)_2$ copper(II) orthophosphate, copper(II) phosphate, cupric phosphate
Kupfer(II)-oxalat *n* $Cu(COO)_2$ copper(II) oxalate, cupric oxalate
Kupferoxid *n* copper oxide; *(specif) s.* Kupfer(II)-oxid

Kupfer(I)-oxid *n* Cu_2O copper(I) oxide, cuprous oxide, red copper oxide
Kupfer(II)-oxid *n* CuO copper(II) oxide, copper monooxide, cupric oxide
Kupferoxidammoniak *n* $[Cu(NH_3)_4](OH)_2$ cuprammonium hydroxide solution, cuprammonia, ammoniacal copper hydroxide, tetraamminecopper hydroxide
Kupferoxidammoniakzellulose *f* cuprammonium cellulose
Kupfer(II)-oxidchlorid *n* copper(II) chloride oxide, cupric chloride oxide
Kupfer(I)-oxid-Element *n* cuprox cell
Kupfer(I)-oxid-Gleichrichter *m* copper-oxide rectifier, cuprous oxide rectifier, cuprox
Kupfer(I)-oxid-Zelle *f* cuprox cell
Kupferoxydase *f* copper oxidase
Kupfer(II)-pentazyanonitrosylferrat *n* $Cu[Fe(CN)_5NO]$ copper(II) pentacyanonitrosylferrate, cupric nitroprusside, cupric nitroprussiate
Kupferperoxid *n* CuO_2 copper peroxide
Kupfer(II)-phosphat *n s.* Kupfer(II)-orthophosphat
Kupfer(I)-phosphid *n* Cu_3P copper(I) phosphide, tricopper monophosphide, cuprous phosphide
Kupfer(II)-phosphid *n* Cu_3P_2 copper(II) phosphide, tricopper diphosphide, cupric phosphide
Kupfer(II)-phosphit *n* $CuPHO_3$ copper(II) phosphite, cupric phosphite, *(better)* copper(II) phosphonate
Kupferphthalozyanin *n* copper phthalocyanine
Kupferpulver *n* copper powder
Kupfer(I)-rhodanid *n s.* Kupfer(I)-thiozyanat
Kupfer(II)-rhodanid *n s.* Kupfer(II)-thiozyanat
Kupferron *n* $[C_6H_5N(NO)O]NH_4$ cupferron, ammonium *N*-nitrosophenylhydroxylamine
kupferrot copper-red
Kupferrubinglas *n* copper ruby glass
Kupfer(II)-salizylat *n* $Cu(HOC_6H_4COO)_2$ copper(II) salicylate, cupric salicylate
Kupferschachtofen *m* copper blast furnace
Kupferschaum *m (min)* copper froth, froth copper, tyrolite
Kupferschmelzofen *m* copper blast furnace
Kupferseide *f s.* Kuoxamseide
Kupferseife *f* copper soap
Kupfer(II)-selenat *n* $CuSeO_4$ copper(II) selenate, cupric selenate
Kupfer(I)-silizid *n* Cu_4Si copper(I) silicide, cuprous silicide
Kupfersodabrühe *f* soda bordeaux, Burgundy mixture *(fungicide)*
Kupferspirale *f* copper spiral
Kupferstaub *m* copper dust *(a fungicide)*
Kupfer(II)-stearat *n* $Cu(CH_3[CH_2]_{16}COO)_2$ copper(II) stearate, cupric stearate
Kupferstein *m* **/ armer** *(met)* copper matte
reicher K. *(met)* copper bottom
Kupfersteinkonverter *m* copper converter
Kupfersteinverblasen *n* copper converting
Kupferstreifenkochprobe *f* copper strip test *(for determining active sulphur)*
Kupfer(I)-sulfat *n* Cu_2SO_4 copper(I) sulphate, cuprous sulphate
Kupfer(II)-sulfat *n* $CuSO_4$ copper(II) sulphate, cupric sulphate
Kupfer(II)-sulfid *n* CuS copper(II) sulphide, cupric sulphide
Kupfer(I)-sulfit *n* Cu_2SO_3 copper(I) sulphite, cuprous sulphite
Kupfersüßen *n (petrol)* copper sweetening
Kupfer(II)-tartrat *n* Cu(OOC–CH(OH)CH(OH)–COO) copper(II) tartrate, cupric tartrate
Kupfer(II)-tetramminhydroxid *n* $[Cu(NH_3)_4](OH)_2$ tetramminecopper hydroxide, ammoniacal copper hydroxide
Kupfer(I)-thiozyanat *n* Cu(SCN) copper(I) thiocyanate, copper(I) rhodanide, cuprous thiocyanate
Kupfer(II)-thiozyanat *n* $Cu(SCN)_2$ copper(II) thiocyanate, copper(II) rhodanide, cupric thiocyanate
Kupfertrübung *f s.* Kupferbruch
Kupferung *f (dye)* copperization, treatment with copper
oxydative K. oxidative copperization
Kupferverfahren *(petrol)* copper sweetening process
Kupfervitriol *m (min)* copper (blue) vitriol, chalcanthite *(copper(II) sulphate-5-water)*
Kupferwasserstoff *m* CuH copper hydride
Kupfer(II)-wolframat *n* $CuWO_4$ copper(II) wolframate, cupric wolframate, cupric tungstate
Kupferzahl *f (sugar)* copper reducing power, K value; *(pap, text)* copper number (index, value)
Kupfer-Zinn-Legierung *f* copper-tin alloy
Kupfer(II)-zitrat *n* copper(II) citrate, cupric citrate
Kupfer(I)-zyanid *n* CuCN copper(I) cyanide, copper monocyanide, cuprous cyanide
Kupfer(II)-zyanid *n* $Cu(CN)_2$ copper(II) cyanide, copper dicyanide, cupric cyanide
kupieren *(food)* to blend
Kupolofen *m* cupola [furnace]
Kupolofenausmauerung *f*, **Kupolofenfutter** *n* cupola linings
Kupolofenstein *m* cupola brick
Kuppel *f* crown *(of a glass furnace)*
kuppeln *(dye)* to couple
alkalisch k. to couple in alkaline solution
sauer k. to couple in acid solution
Kuppelofen *m s.* Kupolofen
Kupplung *f (dye)* coupling
oxydative K. oxidative coupling
Kupplungsbottich *m*, **Kupplungsbütte** *f (dye)* coupling vat
Kupplungsgeschwindigkeit *f (dye)* coupling rate
Kupplungskomponente *f (dye)* coupling (secondary) component
Kupplungskufe *f (dye)* coupling vat
Kuprat *n* cuprate
Kuprit *m (min)* cuprite, red (ruby) copper ore *(copper(I) oxide)*
Kuproionenmethode *f s.* Kupfer(I)-ionen-Verfahren
Kuproxgleichrichter *m s.* Kupfer(I)-oxid-Gleichrichter

Kurare *n* curare, curara *(an arrow poison from several menispermaceae and loganiaceae)*
Kurarealkaloid *n* curare alkaloid
Kurbelpresse *f* crank press
Kurbelwinkel *m* crank angle *(a measure for ignition delay)*
Kürbiskernöl *n* pumpkin [seed] oil *(from Cucurbita pepo L.)*
Kurchirinde *f (pharm)* kurchee (kurchi) bark *(from Holarrhena antidysenterica Wall.)*
Kurin *n* curine, bebeerine *(a curare alkaloid)*
Kurkuma *f* turmeric, curcuma, Indian saffron, Curcuma longa L.
Kurkumapapier *n* turmeric paper *(an indicator paper)*
Kurkumaprobe *f* turmeric test
Kurkumin *n* curcumin, turmeric yellow *(colouring principle of turmeric)*
Kurrunjeöl *n* Pongam (Hongay) oil *(from Pongamia pinnata (L.) Merr.)*
Kurtschatovium *n* Ku kurchatovium
Kurve *f* curve; graph, diagram; trace *(the marking made by a recording instrument)*
K. gleichen Gehalts an flüchtiger Substanz line of equal volatile matter, isovol
K. gleichen Heizwertes isocalorific line, isocal
K. gleichen Kohlenstoffgehaltes isocarbon line
binodale K. binodal curve
charakteristische K. *(phot)* characteristic curve, Hurter and Driffield curve, H and D curve
polarografische K. polarographic curve, polarogram
sensitometrische K. *s.* charakteristische K.
Kurvendurchhang *m (phot)* foot, toe, region of underexposure *(of the characteristic curve)*
kurz *(ceram)* short *(clay body)*
Kurzalterung *f* accelerated (artificial) ageing *(materials testing)*
Kurzbezeichnung *f* / **chemische** abbreviated chemical name
Kurzhalskolben *m* short-neck[ed] flask
Kurzhalsrundkolben *m* short-neck round-bottom flask, balloon flask
Kurzhalsstehkolben *m* short-neck flat-bottom flask
kurzkettig short-chain
kurzlebig *(nucl)* short-lived
Kurzmalz *n (ferm)* chit (slightly grown) malt
Kurzname *m* / **chemischer** abbreviated chemical name
Kurznaßbeize *f (agric)* instant dip
Kurzperiode *f* short (small) period *(in the periodic system)*
Kurzprüfung *f* accelerated test
Kurzrohrverdampfer *m* short-tube evaporator
Kurzschleifentrockner *m (text)* short-loop dryer, roller dryer
Kurzventuridüse *f* short-tube venturi
Kurzversuch *m* accelerated test
Kurzwegdestillation *f* short-path [high-vacuum] distillation
Kurzwegdestillierapparat *m* short-path still
Kurzzeiterhitzer *m (food)* high-temperature short-time pasteurizer
Kurzzeiterhitzung *f (food)* short-time heat treatment, high-temperature short-time pasteurization, HTST pasteurization
Kurzzeitmesser *m* timing device
Kurzzeitmethode *f* quick-assay method
Kurzzeitpasteurisation *f s.* Kurzzeiterhitzung
Kurzzeitpasteurisierapparat *m s.* Kurzzeiterhitzer
Kurzzeitprüfung *f* short-time test (assay), S.T.T.
Kurzzeittrockner *m* short-retention-time dryer
Kurzzeitwecker *m (lab)* interval timer
Kuskhygrin *n* cuskhygrine *(alkaloid)*
Kuteragummi *n* kuteera (kateera, kateira) gum, gum kuteera *(from Sterculia, Cochlospermum, or Astragalus specc.)*
Kutinit *m (coal)* cutinite
Kuvertpapier *n* envelope paper
Küvette *f* cuvet[te], cell
Küvettenwechselvorrichtung *f* cuvette well adapter *(of a photometer)*
Kw. *s.* Königswasser
K-Wert *m (plast)* K value, K factor
K. nach Fikentscher Fikentscher K-value *(characterizing the molecular weight of high polymers)*
KW-Stoff *m* hydrocarbon
kyanisieren to kyanize *(to protect wood by saturating it with aqueous mercuric chloride)*
Kyanisierung *f* kyanizing *(protection of wood by saturating it with aqueous mercuric chloride)*
Kyanit *m (min)* kyanite, cyanite, disthene *(aluminium oxide orthosilicate)*
Kynurenin *n* kynurenine, 3-anthraniloyl-*L*-alanine
Kynurensäure *f* kynurenic acid, 4-hydroxyquinoline-2-carboxylic acid
Kynurin *n* kynurine, 4-hydroxyquinoline
Kynursäure *f* $HOOC-CO-NHC_6H_4COOH$ kynuric acid, *o*-carboxyoxanilic acid

L

l. *s.* löslich
l *s.* linksdrehend
L s. 1. Leuchtdichte; 2. Löslichkeitsprodukt
Lab *n* rennet ✦ **mit L. versetzen** to rennet *(milk)*
labbehandelt rennet-treated *(e.g. casein)*
Labbruch *m* rennet curd
Labdan[um] ... *s.* Ladanum ...
laben to rennet *(milk)*
Labessenz *f*, **Labextrakt** *m* rennet extract
Labfähigkeit *f* rennetability, renneting ability, rennet coagulability
Labferment *n* rennin, rennase, rennet [ferment]
Labgärprobe *f* rennet[-fermentation] test
Labgärung *f* rennet fermentation
Labgerinnung *f* rennet clotting (coagulation)
Labgerinnungsfähigkeit *f s.* Labfähigkeit
Labgerinnungszeit *f* rennet-clotting time, renneting time
labil labile, unstable, instable
thermisch l. thermolabile, heat-labile

Labilität *f* lability, unstableness, instability
Labkäse *m* rennet cheese
Labkäsebruch *m* rennet curd
Labkasein *n* rennet[-precipitated] casein
Labkoagulation *f* rennet clotting (coagulation)
Labmagen *m* rennet stomach (bag), abomasum, abomasus
Labor *n s.* Laboratorium
Laborant *m* laboratory assistant
Laborantin *f* [female] laboratory assistant
Laboratorium *n* laboratory, lab
heißes L. *(nucl)* hot laboratory
Laboratoriumsabriebprüfung *f* laboratory abrasion test
Laboratoriumsapparat *m* laboratory apparatus (instrument)
Laboratoriumsausstattung *f* laboratory equipment
Laboratoriumsbecken *n* laboratory sink
Laboratoriumsbestimmung *f* laboratory determination
Laboratoriumseinrichtung *f* laboratory equipment
Laboratoriumsgerät *n* laboratory apparatus
Laboratoriumsglas *n* chemically resistant glass
Laboratoriumskolonne *f* laboratory column
Laboratoriumsmethode *f* laboratory method
Laboratoriumsofen *m* laboratory furnace
Laboratoriums-pH-Meter *n* laboratory pH meter
Laboratoriumspotentiometer *n* laboratory potentiometer
Laboratoriumspraktikum *n* laboratory period *(during the study of chemistry)*
Laboratoriumspresse *f* laboratory press
Laboratoriumsprüfung *f* laboratory test
Laboratoriumstagebuch *n* laboratory manual
Laboratoriumstest *m* laboratory test
Laboratoriumstisch *m* laboratory table (bench, desk)
Laboratoriumstitrimeter *n* laboratory titrimeter
Laboratoriumsverfahren *n* laboratory process
Laboratoriumsversuch *m* laboratory test
Labpulver *n* rennet powder
Labquark *m* rennet curd
Labradorit *m (min)* labradorite *(a felspar)*
Labstärke *f* rennet strength
labträge slow-renneting *(milk)*
Labung *f* renneting *(of milk)*
Labungsfähigkeit *f s.* Labfähigkeit
Labyrinthdichtung *f* labyrinth seal
Labyrinthkondenswasserableiter *m* labyrinth trap
Labyrinthspaltdichtung *f* labyrinth seal
Labzeit *f* rennet-clotting time, renneting time
Lachgas *n* laughing gas *(nitrogen(I) oxide)*
Lachsfett *n*, **Lachsöl** *n* salmon oil
Lack *m (if drying solely by evaporation of the solvent)* lacquer, *(more generally)* paint; *(if transparent)* varnish; *(of animal or vegetable origin)* lac[k]; *(organic compounds precipitated on a carrier, esp alumina)* lake; *(yielding a very hard and glossy coating)* enamel
L. für Außenanstriche exterior paint (varnish)
L. für Innenanstriche interior paint (varnish)
fetter L. long-oil varnish
halbfetter L. medium-oil varnish
kalthärtender L. cold-hardening varnish
magerer L. short-oil varnish
mittelfetter L. medium-oil varnish
ofentrocknender L. stoving lacquer (varnish), *(Am)* baking finish
pigmentierter L. lacquer, paint
überfetter L. extra long-oil varnish
Lackbehälter *m* sump *(in flow-coating process)*
Lackbenzin *n* varnish-makers' [and painters'] naphtha, V.M.P. naphtha, painter's naphtha
Lackbildung *f* varnishing
Lackentferner *m* paint (lacquer, varnish) remover
Lackfarbe *f* 1. lacquer, paint; 2. *s.* Lackfarbstoff
Lackfarbstoff *m* lacquer dye; *(if adsorbed on a carrier as on alumina)* lake colour
roter L. lac dye *(obtained from stick lac)*
Lackgewebe *n* varnished fabric
Lackgießen *n* curtain coating
Lackharz *n* varnish (coating) resin
lackieren 1. *(using products drying by evaporation)* to lacquer, *(more generally)* to paint; *(using transparent products)* to varnish; *(using products yielding very hard coatings)* to enamel; 2. *(cosmet)* to enamel *(e.g. nails)*
Lackiererei *f* paint shop
Lackiermaschine *f* coating machine, coater
Lackiertrommel *f* coating pan, tumbling barrel
Lack-Lack *m* lac (Indian) lake, lac (lake) lac *(a product prepared from lac dye)*
Lacklaus *f* lac insect, Tachardia lacca
Lackleder *n* varnished (patent, japanned, Japan, enamelled) leather
Lacklösungsmittel *n* lacquer solvent
Lackmoid *n* $(HO)_2C_6H_3N[C_6H_2(OH)_3]_2$ lacmoid, lackmoid
Lackmus *n(m)* litmus, lacmus, lakmus, lichen blue *(a lichen dye)*
Lackmuspapier *n* litmus [test] paper
Lackmustinktur *f* litmus tincture
Lacköl *n* varnish oil
Lackpapier *n* varnished (varnishing) paper
Lackschildlaus *f* lac insect, Tachardia lacca
Lacküberzug *m* organic coating
Lackvorhang *m* curtain
Lackvorratsbehälter *m* sump *(in flow-coating process)*
Ladangummi *n (incorrectly for)* Ladanumharz
Ladanharz *n s.* Ladanumharz
Ladanum[harz] *n* ladanum, labdanum [resin] *(an oleoresin from Cistus specc.)*
Ladanumöl *n* la[b]danum oil *(from Cistus specc.)*
laden 1. to load, to charge, to prime *(filling material)*; 2. *(phys chem)* to charge
Ladung *f* 1. load, charge, batch; 2. *(phys chem)* charge
elektrostatische L. electrostatic charge
formale L. formal charge
gleichnamige L. like charge

ruhende L. static charge *(of an intermittent gas-making retort)*
ungleichnamige L. unlike charge
Ladungsabtrennung *f (phys chem)* charge separation
Ladungsdichte *f* 1. loading density *(of explosives)*; 2. *(phys chem)* charge density
Ladungsdichteverteilung *f (phys chem)* charge distribution
Ladungsdifferenz *f s.* Ladungsunterschied
Ladungsdosis *f (phys chem)* charge dosage
Ladungskonzentration *f* loading density *(of explosives)*
ladungslos uncharged
Ladungsmenge *f* amount of charge
Ladungsträger *m (phys chem)* charge carrier
Ladungsträgerrekombination *f* recombination of carriers
Ladungstrennung *f* separation of charge
Ladungsunterschied *m (phys chem)* difference in charge, charge difference
Ladungsverteilung *f (phys chem)* charge distribution
Ladungswolke *f (phys chem)* charge cloud
L. einer π-Bindung π cloud
Ladungszahl *f (phys chem)* charge number
Lage *f* 1. position; 2. layer; ply *(as of laminated material)*; 3. *(geol)* stratum ✦ **in natürlicher L.** *(geol)* autochthonous, in place, in situ
ungeordnete L. *(phys chem)* random (disordered) fashion
Lageenergie *f* potential energy
Lagenlösung *f (rubber)* ply separation
Lagentextur *f (geol)* banded structure
Lager *n* 1. storage, store[house], stock house (room); 2. *(geol)* layer, bed, deposit, lode, seam; 3. *(tech)* bearing *(as of shafts)*
ölfreies (ölloses) L. oilless bearing
Lagerbehälter *m* storage tank (vessel), storage
lagerbeständig stable in storage, resistant to storage
Lagerbeständigkeit *f* stability in storage, resistance to storage, storage stability (resistance, quality, life), *(plast, rubber also)* shelf life
Lagerbier *n* stock beer, *(esp)* lager [beer]
lagerfähig *s.* lagerbeständig
Lagerfähigkeit *f s.* Lagerbeständigkeit
Lagergefäß *n s.* Lagerbehälter
Lagerhalle *f* store[house], storage, stock house (room)
Lagerhaltbarkeit *f s.* Lagerbeständigkeit
Lagerhaus *n s.* Lagerhalle
Lagerkeller *m* storage cellar
Lagermetall *n* bearing metal
lagern 1. to store, *(bulk material also)* to stockpile; 2. *(ceram)* to age, to sour *(moistened clay)*
im Tank l. to tank
in Borke l. *(tann)* to age
kalt (kühl) l. to cold-store
Lagerraum *m* store (stock) room, storage
Lagerstätte *f (geol)* layer, bed, deposit, lode, seam; *(petrol)* field, reservoir
L. mit Gasentlösungsdruck *(petrol)* depletion-type field (reservoir, solution gas-drive field (reservoir)
L. mit Gaskappe *(petrol)* gas cap-drive field (reservoir)
L. mit Wassertrieb *(petrol)* water-drive field (reservoir)
unter Gasdruck (Gastrieb) stehende L. *(petrol)* gas-drive field (reservoir)
unter Schwerkraft entölende L. *(petrol)* gravity drainage reservoir
Lagerstättenvergasung *f (coal)* underground gasification
Lagertank *m* storage tank (vessel)
L. für Abfallsäure waste-acid-storage tank
Lagertemperatur *f* storage (stock, holding) temperature
Lagerung *f* 1. storage, *(of bulk material also)* stockpiling; 2. *(ferm)* afterfermentation, secondary fermentation; 3. *(tech)* bearing
L. unter Wasser *(pap)* water storage
unterirdische L. underground storage
Lagerungstemperatur *f s.* Lagertemperatur
Lagerungsvorschrift *f* storage regulation
Lagerungszeit *f* storage period
Lagervorrat *m* stock
Lagerweißmetall *n* white metal
Lagerzeit *f* storage period
Lainer-Effekt *m (phot)* Lainer effect
Lake *f* [salt] brine ✦ **mit L. behandeln** to brine
Lakenbehandlung *f* brining
lakenkonserviert *(tann)* brine-cured
Lakenkonservierung *f (tann)* brine curing (cure)
Lakkainsäure *f* laccainic acid
Lakkase *f* laccase
Lakmoid *n s.* Lackmoid
lakrimogen lachrymatory
Lakritze *f* liquorice, licorice *(from Glycyrrhiza glabra L.)*
Laktagogum *n* galactagogue, lactagogue *(an agent promoting secretion of milk)*
Laktalbumin *n* lactalbumin, milk albumin
Laktam *n* lactam
Laktam-Laktim-Tautomerie *f* lactam-lactim tautomerism
β-Laktamring *m* β-lactam ring
Laktarinsäure *f* lactarinic acid, 6-oxo-octadecanoic acid
Laktarsäure *f s.* Stearinsäure
Laktase *f* β-galactosidase, lactase
Laktat *n* lactate
Laktatdehydrase *f s.* Laktatdehydrogenase
Laktatdehydrogenase *f* lactic dehydrogenase
Laktationshormon *n* lactogenic (luteotrophic) hormone, lactogen, prolactin, luteotrophin
Laktatmethode *f (soil)* double-lactate method
Laktazidogen *n* lactacidogen
Laktid *n* lactide *(any of a group of dilactones)*
Laktikodehydrase *f s.* Laktatdehydrogenase
Laktim *n* lactim
Laktinsäure *f s.* Milchsäure
Laktobionsäure *f* lactobionic acid

Laktobiose *f* lactobiose, lactose, milk sugar
Laktobutyrometer *n* lactobutyrometer
Laktodensimeter *n* lactodensimeter
Laktogen *n s.* Laktationshormon
Laktoglobulin *n* lactoglobulin, milk globulin
Laktometer *n* galactometer, lactometer
Lakton *n* lactone
Laktonbildung *f* lactonization, lactone formation
Laktonbindung *f* lactonic linkage
laktonisieren to lactonize
Laktonisierung *f* lactonization
Laktonitril *n* $CH_3CH(OH)CN$ lactonitrile, acetaldehyde cyanhydrin, 2-hydroxypropane nitrile
Laktonregel *f* **[/ Hudsonsche]** [Hudson] lactone rule *(of optical rotation)*
Laktonring *m* lactone ring
Laktonsäure *f* 1. lactone acid, lactonic acid *(any of several acids with a lactone ring bearing the carboxyl group)*; 2. *s.* Galaktonsäure
Laktoperoxydase *f* lactoperoxidase
Laktose *f s.* Laktobiose
Laktosesirup *m* lactose syrup
Laktoskop *n* lactoscope
Laktosurie *f (med)* lactosuria
laktotrop lactogenic
Laktotropin *n s.* Laktationshormon
Laktoylgruppe *f* $CH_3CH(OH)CO-$ lactoyl group (residue)
Laktylharnstoff *m* lactylurea
Laktylmilchsäure *f* lactyllactic acid
lakustrisch *(geoch)* lacustrine
lamellar lamellar
Lamelle *f* lamella
Lamellenmethode *f (phys chem)* detachment method *(for determining surface tension)*
laminar laminar
Laminarbewegung *f*, **Laminarströmung** *f* laminar (streamlined) flow
Laminat *n* laminate, laminated material (plastic)
laminieren to laminate
Laminierharz *n* laminating resin
laminiert / mit Schaumstoff foam-backed
Laminierung *f* lamination
Lampe *f (glass)* [glass blower's] lamp, blowtorch, bench blowpipe ✦ **vor der L. geblasen** lamp-blown, lampworked
Lampenarbeit *f (glass)* lampworking
Lampenbläser *m (glass)* lampworker
Lampenbläserei *f (glass)* lampworking
lampengeblasen *(glass)* lamp-blown, lampworked
Lampenmethode *f (petrol)* lamp method *(for determining sulphur content)*
Lampenpetroleum *n* lamp (illuminating) oil
Lampenruß *m*, **Lampenschwarz** *n* lampblack
Lana philosophica *f* philosopher's wool, flowers of zinc *(woolly zinc oxide)*
Lanarkit *m (min)* lanarkite *(lead(II) oxide sulphate)*
Lanatosid *n* lanatoside *(a glycoside)*
Lancashire-Kessel *m* Lancashire boiler *(an internally fired boiler having two flues)*
Landé-Faktor g *m* Landé g-factor, [Landé] splitting factor g *(chemical-bond theory)*
Landkartenpapier *n* map (chart, plan) paper, *(Am)* geography paper
Landwirtschaftschemie *f* agricultural chemistry, agrochemistry
Langarmzentrifuge *f* long-arm centrifuge
längen to stretch
Längenänderung *f* strain *(materials testing)*
  L. bei der Bruchkraft (Reißkraft) extension at break
  bleibende L. secondary creep
Längenzunahme *f* extension
langfas[e]rig long-fibre
Langfilz *m (pap)* long felt *(in cylinder board machines)*
langhalsig long-neck[ed]
Langhalskolben *m* long-neck[ed] flask
Langhalsrundkolben *m* round-bottom long-neck flask
Langhalsstehkolben *m* flat-bottom long-neck flask, Florence (boiling) flask
Langholz *n (pap)* log
Langit *m (min)* langite *(copper(II) hexahydroxide sulphate)*
langkettig long-chain
Langlochziegel[stein] *m* horizontally perforated brick
Langmaschengewebe *n* oblong-mesh (rectangular-opening) cloth
Langperiode *f* big series, long period *(in the periodic system)*
Langrohr[vertikal]verdampfer *m* long-tube vertical-film evaporator, LTV evaporator
Langsamfilter *n* slow [sand] filter, low-rate filter, English filter *(water treatment)*
Langsamfiltration *f* slow [sand] filtration
langsamflüchtig slow-evaporating
Langsamkochung *f (pap)* slow cook
langsamwirkend slow-acting
langsamziehend *(dye)* slow-striking
Längsbecken *n* rectangular basin (tank)
Längsdehnung *f* linear expansion
Langsieb *n* **der Papiermaschine** Fourdrinier wire
Langsieb[papier]maschine *f* Fourdrinier [paper machine]
Langsiebpartie *f (pap)* Fourdrinier part (section)
Längsreibe[maschine] *f (food)* conching machine, [longitudinal] conche
Längsrippenrohr *n* long-fin tube
Längsschneidemaschine *f (pap)* reel-slitting (roll-slitting) machine, rereeling (rewinding, slitting) machine, slitter, rewinder
längsschneiden *(pap)* to slit, to cut lengthways
Längsschneiden *n* **nach dem Druckprinzip** *(pap)* score cutting
  L. nach dem Scherenprinzip shear cutting
  L. und Umrollen *n* rewinding
Längsschneider *m s.* Längsschneidemaschine
Längsspritzkopf *m (plast)* horizontal (axial extruder) head
längstrennen *s.* längsschneiden

Langzeitbrennöl *n* long-time burning oil, signal oil
Langzeitkonservierung *f (food)* long-range (long-term) preservation
Langzeittrockner *m* long-retention-time dryer
Lanolin *n* lanolin[e] *(refined wool grease)*
Lanopalmitinsäure *f* $HOC_{15}H_{30}COOH$ lanopalminic acid
Lanosterin *n* lanosterol
Lanozerinsäure *f* $(HO)_2C_{29}H_{57}COOH$ lanoceric acid
Lanthan *n* La lanthanum
Lanthanazetylid *n s.* Lanthankarbid
Lanthanbromid *n* $LaBr_3$ lanthanum bromide
Lanthanchlorid *n* $LaCl_3$ lanthanum chloride
Lanthanerde *f s.* Lanthan(III)-oxid
Lanthanfluorid *n* LaF_3 lanthanum fluoride
Lanthanhydroxid *n* $La(OH)_3$ lanthanum hydroxide
Lanthanid *n s.* Lanthanoid
Lanthaniden ... *s.* Lanthanoiden ...
Lanthanit *m (min)* lanthanite *(cerium dysprosium lanthanum carbonate)*
Lanthanjodat *n* $La(IO_3)_3$ lanthanum iodate
Lanthankarbid *n* LaC_2 lanthanum carbide
Lanthankarbonat *n* $La_2(CO_3)_3$ lanthanum carbonate
Lanthannitrat *n* $La(NO_3)_3$ lanthanum nitrate
Lanthanoid *n*, **Lanthanoidenelement** *n* lanthanoid [element]
Lanthanoidengruppe *f s.* Lanthanoidenreihe
Lanthanoidenkontraktion *f* lanthanoid contraction
Lanthanoidenmetall *n* lanthanoid metal
Lanthanoidenreihe *f* lanthanoid series (group)
Lanthan(III)-oxid *n* La_2O_3 lanthanum(III) oxide, lanthanum trioxide
Lanthanreihe *f s.* Lanthanoidenreihe
Lanthansalz *n* lanthanum salt
Lanthansulfat *n* $La_2(SO_4)_3$ lanthanum sulphate
Lanthantrioxid *n s.* Lanthan(III)-oxid
Lanthionin *n* $S[CH_2CH(NH_2)COOH]_2$ lanthionine, ββ'-diamino-ββ'-dicarboxydiethyl sulphide
Lanzendüngung *f* fertilization by soil injection
Lanzettnadel *f (lab)* lancet-point dissecting needle
Lapachoholz *n (dye)* lapacho *(from Tabebuia and Tecoma specc.)*
Lapachol *n* lapachol *(a naphthoquinone derivative)*
Lapachosäure *f s.* Lapachol
Lapillituff *m (geol)* lapilli tuff
Lapislazuli *m (min)* lapis [lazuli], lazurite (a tectosilicate)
Lardöl *n* lard (grease) oil
Larixinsäure *f* larixinic acid, maltol
Larizin *n s.* Koniferin
Larmor-Präzession *f* Larmor precession *(of atoms and electrons in a magnetic field)*
Larnit *m (min)* larnite *(calcium orthosilicate)*
Larvengift *n*, **Larvizid** *n* larvicide, larvacide
Lassaigne-Probe *f* Lassaigne test *(for detecting nitrogen)*
Last *f* load
lastbetätigt gravity-operated, weight-operated
Lastdurchbiegungskurve *f (plast)* load deflection curve
Lastschale *f* left-hand pan *(of a balance)*
Lasttrum *m(n)* carrying side, top strand, *(relating to belt conveyors also)* drive belt
Lastwiderstand *m* load resistance
Lasurit *m s.* Lapislazuli
Latensifikation *f (phot)* latensification
Latenzzeit *f* period of latency (induction)
Lateritboden *m* lateritic soil
Lateritisierung *f* lateritization *(of rocks)*
Latex *m* latex
  L. mit niedrigem Ammoniakgehalt low-ammonia latex
  aufgerahmter L. creamed latex
  eingedampfter (eingedickter) L. evaporated latex
  frisch gezapfter L. field latex
  konzentrierter L. concentrated latex
  künstlicher L. synthetic[-rubber] latex
  normaler L. normal latex
  zentrifugierter L. centrifuged latex
Latexanstrichfarbe *f s.* Latexfarbe
Latexbecher *m (rubber)* collection (tapping) cup
Latex-Chlorkautschuk *m* latex-chlorinated rubber
Latexfaden *m* latex thread
Latexfarbe *f* latex (water) paint
latexführend latex-bearing, laticiferous
Latexgummifaden *m* latex thread
Latexkonzentrat *n* latex concentrate
Latexmischung *f* latex compound
Latexschaum[gummi] *m* latex foam [rubber], foamed latex rubber
Latexschwamm *m* latex foam sponge
Latextechnologie *f* latex technology
Latschenkiefernöl *n* dwarf pine-needle oil *(from Pinus mugo Turra)*
Lattentrommel *f (tann)* slatted drum ✦ **in der L. behandeln (durcharbeiten)** to drum
Latwerge *f (pharm)* electuary
Laubgrün *n* chrome [oxide] green, green cinnabar *(chromium(III) oxide)*
Laubholz *n* hardwood, angiospermous wood
Laubholzschliff *m* hardwood groundwood
Laubholzzellstoff *m* hardwood pulp
Laudanosin *n* laudanosine, laudanine methyl ether *(an opium alkaloid)*
Laudanum *n (pharm)* laudanum *(a tincture of opium)*
Laue-Aufnahme *f*, **Laue-Diagramm** *n (cryst)* Laue diagram (pattern, photograph)
Laue-Verfahren *n (cryst)* Laue [photograph] method, Laue X-ray method
Laue-Versuch *m* Laue experiment
Lauf *m* 1. travel *(as of a reactant)*; course *(of a reaction)*; 2. run *(of a machine)*
Laufbandtrockner *m* festoon (loop) dryer
Laufdauer *f* wear life
Laufeigenschaften *fpl* runnability *(of paper)*
laufen 1. to run, to flow *(of a liquid)*; to sag, to

curtain *(of surface coatings)*; 2. to run *(of an experiment)*; 3. to leak, to run *(of a vessel)*; 4. to run *(of a machine)*; to travel *(of the paper machine wire)*

Läufer *m* 1. *(coat)* curtain *(faulty film)*; 2. *s.* Laufstein

Läuferbildung *f* sagging, curtaining *(of surface coatings)*

Läuferstein *m s.* Laufstein

Lauffläche *f (rubber)* tread, wearing surface

Laufflächenabnutzung *f*, **Laufflächenabrieb** *m (rubber)* tread wear

Laufflächengummi *m* tread rubber

Laufflächenmischung *f (rubber)* tread stock (compound, mix)

Laufflächenspritzkopf *m (rubber)* tread head

Laufflächenspritzmaschine *f (rubber)* tread extruder

Laufgeschwindigkeit *f* travel rate

Laufmittel *n* mobile solvent *(chromatography)*

Laufmittelfront *f* solvent front *(chromatography)*

Laufmittelgemisch *n* solvent system *(chromatography)*

Laufrad *n* impeller

geschlossenes L. enclosed (closed, shrouded) impeller *(of a centrifugal pump)*

halboffenes L. semienclosed (semiopen) impeller *(of a centrifugal pump)*

Laufrichtung *f* 1. *(pap)* running (making, machine, grain, long) direction, direction of travel; 2. flow direction *(chromatography)*

Laufrolle *f* idler

Laufschaufel *f* impeller blade (vane)

Laufsteg *m (tech)* walkway

Laufstein *m* runner [stone], mill runner, muller

Laufstreifen *m* / **roher** *(rubber)* camelback *(for retreading tyres)*

Laufstreifenspritzmaschine *f (rubber)* tread extruder

Laufterm *m*, **Laufzahl** *f* variable (current) term

Laufzeit *f* 1. [wear] life; 2. running time *(of a chromatogram)*

L. des Filzes *(pap)* felt life

Laufzeitspektrometer *n* time-of-flight spectrometer

Lauge *f* 1. lye *(alkaline solution)*; 2. *(tech)* liquor; leach[ate] *(solution obtained by leaching)*; *(text)* buck *(for washing or bleaching)*

eingestellte L. standard base

Javellesche L. eau de Javel[le], Javelle water *(a bleaching agent)*

standardisierte L. standard base

Laugemittel *n* leaching agent

laugen *(met)* to leach [out], to lixiviate; *(text)* to buck, to mercerize

Laugen *n* 1. *(text)* mercerizing, mercerization; 2. *s.* Laugung

L. unter Spannung mercerization with tension

spannungsloses L. mercerization without tension

Laugenaustritt *m* liquor outlet *(as on an evaporator)*

Laugenbehandlung *f* 1. lye treating; 2. *(petrol)* caustic[-soda] wash, alkali wash

laugenbeständig resistant (stable) to alkali[es], alkali-resistant, lye-proof, caustic-proof

Laugenbeständigkeit *f* resistance (stability) to alkali[es], alkali resistance

Laugenbrüchigkeit *f (met)* caustic cracking (embrittlement)

Laugeneintritt *m* liquor inlet *(as on an evaporator)*

Laugenentsäuerung *f* neutralization [with alkali solutions]

laugenfest *s.* laugenbeständig

Laugenfestigkeit *f s.* Laugenbeständigkeit

Laugengehalt *m* alkali content

Laugenphase *f* lye phase

Laugenregeneration *f (pap)* liquor (waste-liquor, spent-liquor) recovery

Laugensalz *n s.* Hirschhornsalz

Laugensprödigkeit *f (met)* caustic cracking (embrittlement)

Laugenstation *f (pap)* liquor-making plant

Laugenturm *m (pap)* reaction tower *(in pulping with chlorine)*

Laugenumlauf *m*, **Laugenumwälzung** *f (pap)* circulation of liquor

Laugenverhältnis *n* (pap) liquor[-to-wood] ratio

Laugenwäsche *f (petrol)* caustic[-soda] wash, alkali wash

Laugenzirkulation *f s.* Laugenumlauf

laugieren *(text)* to buck, to mercerize

Laugung *f* 1. *(met)* leach[ing], lixiviation; 2. *(petrol)* caustic[-soda] wash, alkali wash

L. in der Grube *(met)* underground leaching, leaching in place

L. in situ *s.* L. in der Grube

Laugungsmittel *n* leaching agent

Laumontit *m (min)* laumon[t]ite, lomonite *(a hydrous calcium and aluminium silicate)*

Lauraldehyd *m* $CH_3[CH_2]_{10}CHO$ lauraldehyde, lauric aldehyde, aldehyde C-12, dodecanal

Laurat *n* laurate *(salt or ester of lauric acid)*

Laurent-Säure *f* $NH_2C_{10}H_6SO_3H$ Laurent's acid, 1-naphthylamine-5-sulphonic acid

Laurinaldehyd *m s.* Lauraldehyd

Laurinsäure *f* $CH_3[CH_2]_{10}COOH$ lauric acid, dodecanoic acid

Laurinsäureäthylester *m* $CH_3[CH_2]_{10}CO\text{-}OC_2H_5$ ethyl laurate

Lauroleinsäure *f* $CH_3CH_2CH{=}CH[CH_2]_7COOH$ lauroleic acid, 9-dodecenoic acid

Laurylalkohol *m* $CH_3[CH_2]_{10}CH_2OH$ lauryl alcohol, alcohol C-12, 1-dodecanol

Laurylamin *n* $CH_3[CH_2]_{10}CH_2NH_2$ laurylamine, dodecylamine

Laurylmerkaptan *n s.* 1-Dodekanthiol

Läuse[bekämpfungs]mittel *n (med)* pediculicide; *(agric)* lousicide

Läusepulver *n* louse powder

Läuterboden *m* false bottom; *(ferm)* strainer [bottom]

Läuterbottich *(ferm)* lauter tub (tun)

Lautermaische *f*, **Läutermaische** *f (ferm)* lauter mash
Läutermittel *n (glass)* [re]fining agent
läutern to purify, to clarify, to refine; *(min tech)* to wash, to scavenge; *(ferm)* to lauter; *(glass)* to plain, to refine, to found *(to free from bubbles)*
  sich l. to clarify
Läuterung *f* purification, clarification, refining [treatment], refinement; *(min tech)* washing, scavenging; *(ferm)* lautering; *(glass)* plaining, [re]fining, founding
Läuterungsmittel *n s.* Läutermittel
Läuterwanne *f (glass)* plaining (refining) chamber (end), refiner, nose
Läuterzone *f (glass)* refining zone
Lava *f* lava
  erstarrte L. frozen lava
Lavandinöl *n* lavandin oil *(from a hybrid Lavandula angustifolia Mill.* × *L. latifolia (L.fil.) Medik.)*
Lavandulol *n* lavandulol *(a terpenoid alcohol)*
Lavendelöl *n* lavender [flower] oil *(from Lavandula specc.)*
Lavendelwasser *n (cosmet)* lavender water
Låvenit *m (min)* la[a]venite *(a sorosilicate containing zirconium)*
Laves-Phase *f* Laves phase *(an intermetallic structure)*
Lävopimarsäure *f* (-)-pimaric acid, laevopimaric acid, (-)-sapietic acid
Lävulinsäure *f* $CH_3-CO-CH_2CH_2COOH$ laevulinic acid, 4-oxovaleric acid, 4-oxopentanoic acid
Lävulose *f* laevulose, fructose *(a monosaccharide)*
Lawine *f (phys chem)* avalanche
Lawrentium *n* Lr lawrencium
Lawson *n* lawsone, 2-hydroxy-1,4-naphthoquinone
Lawsonit *m (min)* lawsonite *(a sorosilicate)*
L-Band *n s.* Leitfähigkeitsband
LCAO-Methode *f* linear-combination-of-atomic-orbitals method, LCAO method
  L. der Molekülorbitale LCAO molecular-orbital method, LCAO MO method
  selbstkonsistente L. LCAO self-consistent method
LCAO-Molekülorbital *n* LCAO molecular orbital, LCAO MO
LCAO-MO-Methode *f* LCAO molecular-orbital method, LCAO MO method
LCAO-Näherung *f* linear-combination-of-atomic-orbitals approximation, LCAO approximation
LCM-Vulkanisation *f* liquid curing
LD *s.* Dosis / letale
LD 50, LD_{50} *s.* Dosis / mittlere letale
LD-Aufblaseverfahren *n*, **LD-Blasstahlverfahren** *n* Linz-Donawitz process, L-D process
LDH *s.* Laktatdehydrogenase
LD-Verfahren *n s.* LD-Aufblaseverfahren
Lea-Zahl *f* Lea [peroxide] value *(for characterizing oils and fats)*
Lebedew-Verfahren *n* Lebedev process *(for obtaining butadiene)*
Lebensdauer *f* lifetime, life [period], durability
  L. eines Austauscherharzes resin life
  mittlere L. *(phys chem)* average (mean) life (lifetime)
Lebensmittel *npl* edibles, eatables, comestibles
  diätetische L. dietary food
Lebensmittelchemie *f* food chemistry
Lebensmittelfarbe *f s.* Lebensmittelfarbstoff
Lebensmittelfarbstoff *m* food dye (colourant)
Lebensmittelindustrie *f* food (food-processing, foodstuff, provisions) industry
Lebensmittelkonservierung *f* food preservation
Lebensmitteltechnologie *f* food technology
Lebensmittelüberwachung *f* food control
Lebensmittelverarbeitung *f* food processing
Lebensmittelzusatz[stoff] *m* food additive
lebensnotwendig *(biol)* essential
Leberöl *n* liver oil
Leberstärke *f* animal (liver) starch, glycogen
Lebertran *m* liver oil, *(specif)* cod-liver oil
Lebertranemulsion *f* cod-liver oil emulsion
lebhaft vigorous, brisk *(reaction)*; bright, vivid *(colour)*
Lebhaftigkeit *f* vigorousness, briskness *(of a reaction)*; brightness, vividness *(of colour)*
Leblanc-Soda *f* Leblanc soda
Leblanc-Verfahren *n* Leblanc process *(for obtaining soda)*
leck leaking
  l. sein to leak, to run
Leck *n* leak
Leckage *f* leakage
lecken to leak, to run
Lecken *n* leak[age]
Leckflüssigkeit *f* leakage
leckfrei leakproof, leaktight
Leckluft *f* leakage; inleakage *(vacuum technology)*
Lecksaft *m (pharm)* linctus
lecksicher leakproof, leaktight
Leckströmung *f* leakage flow
Lecksuche *f* leak testing
Lecksucher *m*, **Lecksuchgerät** *n* leak detector
Leckverlust *m* leakage, slippage loss, slip
Leckweg *m* leakage path
Leclanché-Element *n* Leclanché cell
Leder *n* leather
  leeres L. empty leather *(result of incorrect tanning)*
  pflanzlich gegerbtes L. bark leather
  synthetisches L. man-made leather
  weißgares L. white leather
lederähnlich, lederartig leathery
Lederausschlag *m (tann)* bloom, exudation
Lederaustauschstoff *m* leathercloth
Lederfett *n* leather grease
lederhart *(ceram)* leather-hard
Lederhaut *f (tann)* corium
Lederindustriechemiker *m* leather chemist
Lederkitt *m* leather cement
Lederkohle *f* leather charcoal
Lederleim *m* leather (skin, hide) glue

Lederpappe *f* [/ **braune]** leather board
  imitierte L. imitation leather board
Lederpflegemittel *n* leather-dressing agent
Lederschmiere *f (tann)* stuffing [mixture]
Lederwalze *f (tann)* bend (butt) roller
Lederzurichter *m (tann)* currier
leer 1. empty; evacuated; 2. *(cryst)* empty, vacant; 3. vacant *(orbital)*
leerblasen *(pap)* to blow [off] *(a digester)*
Leerblasen *n* **des Zellstoffkochers** *(pap)* digester blow
leeren to empty
leerlaufen to drain
  l. lassen to run dry *(e.g. a vessel)*
Leerlaufen *n* drainage
Leerlaufregulierung *f* inlet-valve control unloading *(of a compressor)*
Leerlaufspannung *f* self-stress *(in the wall of a centrifuge)*
Leerplatz *m s.* Leerstelle 1.
Leerstelle *f* 1. *(cryst)* empty place, hole [position], lattice vacancy (hole), vacant lattice position (site); 2. void *(as in bulk material)*
Leerstellenpaar *n* vacancy pair
Leertrum *m(n)* slack side, return strand (side), *(relating to belt conveyors also)* return belt
Leerventil *n* discharge valve
Leerversuch *m* blank experiment (trial, test), blank
Leerwert *m* blank reading; *(nucl)* background
legieren to alloy
Legierung *f* alloy
  L. nach Rose Rose's metal (alloy)
  Arndtsche L. Arndt's alloy *(a reductant consisting of Cu and Mn)*
  binäre L. binary alloy
  Devardasche L. Devarda's alloy *(a reductant consisting of Cu, Al, and Zn)*
  eutektische L. eutectic alloy
  leichtschmelzende L. low-melting alloy, fusible alloy
  Lipowitzsche L. Lipowitz's metal (alloy)
  niedrigschmelzende L. *s.* leichtschmelzende L.
  pyrophore L. pyrophoric alloy
  quaternäre L. quaternary alloy
  ternäre L. ternary alloy
  Woods L. Wood's metal (alloy)
Legierungsbestandteil *m*, **Legierungselement** *n*, **Legierungskomponente** *f* alloying element (agent, constituent, ingredient)
Lehm *m* loam, clay
lehmartig loamy, clayey, clayish
Lehmform *f* loam mould *(foundry)*
Lehmformen *n* loam moulding *(foundry)*
Lehmglasur *f (ceram)* slip glaze
lehmhaltig loamy, clayey, clayish, *(of rocks also)* argilliferous
lehmig loamy, clayey, clayish, argillaceous
Leichengift *n* ptomaine
Leichtbauplatte *f* building board
Leichtbenzin *n* light gasoline (benzine, naphtha, spirit)
Leichtbenzol *n* light benzole
Leichtbeton *m* lightweight concrete
  L. mit Koksaschenzusatz breeze concrete
Leichterflüchtiges *n s.* Leichtersiedendes
leichtersiedend lower-boiling, light
Leichtersiedendes *n* more volatile component, M.V.C., lighter (low-boiling) component, light phase
Leichtflintglas *n* light flint glass
leichtflüchtig highly (readily) volatile, high-volatile
Leichtflüchtigkeit *f* high volatility
Leichtgut *n* light material; light fraction
leichtlöslich readily (freely) soluble
Leichtmetall *n* light metal
Leichtmineral *n* light mineral
Leichtöl *n* light oil
Leichtparaffin *n* light liquid paraffin
leichtschmelzbar, leichtschmelzend low-melting, low-melting-point, low-fusion
leichtsiedend low-boiling, light
Leichtsiedendes *n s.* Leichtersiedendes
Leichtstein *m* lightweight brick
Leichtstoffabscheider *m* light-solids remover *(waste-water treatment)*
leichtviskos low-viscosity
Leichtwaschmittel *n* light-duty detergent
Leichtwasserreaktor *m* light-water reactor
leichtzersetzlich easily decomposable
Leichtzuschlagstoff *m* lightweight aggregate
Leim *m* glue; *(pap)* sizing material (agent), size ✦
  mit L. bestreichen to glue
  flüssiger L. liquid glue
  freiharzreicher (hochfreiharzhaltiger) L. *(pap)* high free rosin size
  pflanzlicher L. vegetable glue
  tierischer L. animal gelatine (glue); *(pap)* animal size
Leimaufnahme *f (pap)* pickup of size
Leimauftragmaschine *f* spreader
Leimbad *n (pap)* size bath
leimbar *(pap)* sizable
Leimbarkeit *f (pap)* sizability
Leimbrühe *f s.* Leimlösung
Leimbütte *f s.* Leimtrog
leimen to glue; *(pap)* to size
Leimfarbe *f* calcimine, *(if suspended)* distemper
Leimfleck *m (pap)* size speck (spot) *(a defect)*
Leimgürtel *m (agric)* greaseband
Leimkocher *m (pap)* size cooker
Leimleder *n (tann)* hide scrapings (shavings)
Leimlösung *f (pap)* size (sizing) solution
Leimmilch *f (pap)* size emulsion (milk)
Leimmittel *n s.* Leim
Leimpresse *f (pap)* [surface] sizing press
Leimpressenwalze *f (pap)* sizing press roll
Leimring *m (agric)* greaseband
Leimstoff *m s.* Leim
Leimsüß *n s.* Glykokoll
Leimtrog *m (pap)* size (sizing) vat (tub)
Leimüberschuß *m (pap)* excess size
Leimung *f* gluing; *(pap)* sizing ✦ **mit mittlerer L.**

half-sized, $^1/_2$ sized ✦ **mit schwacher L.** soft-sized, slack-sized, S.S. ✦ **mit starker L.** hard-sized, H.S., strongly sized
L. im Stoff *(pap)* engine (beater, pulp) sizing, E.S., sizing in the engine (stuff)
L. in der Masse *s.* L. im Stoff
L. mit Gelatine *(pap)* animal tub sizing, A.T.S., gelatin sizing
L. mit Natronwasserglas *(pap)* silicate sizing
L. mit Tierleim *s.* L. mit Gelatine
Leimungsgrad *m (pap)* degree of sizing
Leimwalze *f (pap)* sizing press roll
Leimwanne *f s.* Leimtrog
Leimzucker *m s.* Glykokoll
Lein *m* flax, Linum L., *(specif)* Linum usitatissimum L.
Leindotteröl *n* cameline (dodder) oil *(from the seeds of Camelina sativa Crantz)*
Leinen[gewebe] *n* linen [fabric]
Leinenhadern *mpl*, **Leinenlumpen** *mpl (pap)* linen rags
Leinenpapier *n* 1. reinforced (linen, cloth-mounted) paper, papyrolin; 2. linen (linen-embossed, linen-finished, linen-faced) paper
Leinenpostpapier *n* linen-embossed writing paper
Leinenprägekalander *m (pap)* linenizing calender
Leinenprägen *n (pap)* linenizing
Leinenprägepresse *f (pap)* linenizing calender
Leinenprägung *f (pap)* linen finish, *(Am)* cloth (Damask) finish
Leinenstoff *m* linen [fabric]
Leinkuchen *m* linseed cake
Leinöl *n* linseed (flax-seed, flax) oil
Leinölfirnis *m* boiled linseed oil
Leinöllack *m* linseed-oil varnish
Leinölsäure *f s.* Linolsäure
Leinöl-Standöl *n* calorized linseed oil
Leinsaat *f*, **Leinsamen** *m* linseed
Leinsamenschleim *m* linseed mucilage
Leinwandgewebe *n* / **geteertes** tarpaulin
Leiste *f* / **dreieckige** *(ceram)* saddle *(a piece of kiln furniture)*
Leistung *f (tech)* performance, output; *(phys)* power
katalytische L. catalytic performance
spezifische L. power number
Leistungsbrutreaktor *m (nucl)* power breeder reactor
Leistungsdüngung *f* output fertilization
leistungsfähig *(tech)* efficient
Leistungsfähigkeit *f (tech)* efficiency, capacity
katalytische L. catalytic efficiency
Leistungsfaktor *m* power factor
Leistungskennlinie *f*, **Leistungskurve** *f* performance curve
Leistungsreaktor *m (nucl)* power reactor
Leistungssteigerung *f* increase of efficiency
Leistungsvermögen *n* capacity, power
Leistungszahl *f* performance coefficient (number)
Leitblech *n* deflector [plate], baffle [plate] ✦ **mit Leitblechen [versehen]** baffled ✦ **ohne Leitbleche** unbaffled
Leitelektrolyt *m* supporting electrolyte *(polarography)*
leiten 1. to lead, to pipe; to direct *(a liquid or gas stream towards something)*; 2. to control *(a reaction)*; to conduct *(electricity, heat)*
in Rohrleitungen l. to pipe
leitend conductive
schlecht l. poorly conducting
Leiter *m* conductor
L. erster Ordnung electronic conductor
L. zweiter Ordnung electrolytic conductor
elektrischer L. conductor of electricity, electrical conductor
elektrolytischer L. electrolytic conductor
schlechter L. poor conductor
Leiterpolymer[es] *n* ladder polymer
leitfähig conductive
Leitfähigkeit *f* conductivity, conductance, conducting power
elektrische L. electrical conductivity
elektrolytische L. electrolytic conductivity
elektronische L. electronic conductivity
fotoelektrische (lichtelektrische) L. photoconductivity
molare L. molar conductance
spezifische L. specific conductivity
thermische L. thermal conductivity
Leitfähigkeitsband *n* conductivity (conduction, conductance) band
leeres (unbesetztes) L. empty band
Leitfähigkeitselektron *n* conduction electron
Leitfähigkeitserhöhung *f* increase in conductivity (conductance)
Leitfähigkeitskoeffizient *m* conductivity (conductance) ratio
Leitfähigkeits-Konzentrations-Kurve *f* conductivity-concentration curve
Leitfähigkeitskurve *f* conductivity curve
Leitfähigkeitsmeßbrücke *f* conductivity bridge
Leitfähigkeitsmeßgerät *n* conductivity apparatus, conductance meter, conductometer, conductimeter
Leitfähigkeitsmessung *f* conductivity (conductance] measurement
Leitfähigkeitsmethode *f* conductometric (conductance) method
Leitfähigkeitsstrom *m* conduction current
Leitfähigkeitstitration *f* conductometric (conductance) titration
Leitfähigkeitswasser *n* conductivity (conductance) water
Leitfähigkeitszelle *f* conductivity (conductance) cell
Leitisotop *n* [isotopic] tracer
Leitrad *n*, **Leitring** *m* diffusion ring *(as of a pump)*
Leitrohr *n* shroud ring, draught tube *(of an agitator)*
Leitrolle *f* guide roll, snub pulley
Leitsalz *n* supporting electrolyte *(polarography)*
Leitschaufel *f* guide vane *(as of a turbine)*

Leitung *f* 1. line, duct, pipe[line]; 2. cable, wire *(in an electric circuit)*; 3. conduction *(of electricity or heat)*
L. für den Schlammaustrag sludge-discharge line
elektrolytische L. electrolytic conduction
erdverlegte (unterirdische) L. underground line, buried duct
Leitungsband *n s.* Leitfähigkeitsband
Leitungselektron *n* conduction electron
Leitungsrohr *n* duct, line, pipe[line]
Leitungsvermögen *n s.* Leitfähigkeit
Leitungswasser *n* tap water
Leitvermögen *n s.* Leitfähigkeit
Leitwalze *f* 1. lead[ing] roll, guide roll; 2. *(pap)* dipping (size) roll *(in vat sizing)*; wire (wire-guide, wire-leading) roll
Leitwert *m* / **elektrischer** electrical conductance
Lekanorsäure *f* lecanoric acid *(a lichen acid)*
Lemongrasöl *n (cosmet)* lemon-grass oil, East Indian verbena oil *(from Cymbopogon flexuosus Stapf and C. citratus (DC.) Stapf)*
Lenkblech *n* baffle [plate], deflector [plate]
lenken to control *(a process)*; to direct *(a substituent)*
Lenkung *f* control *(of a process)*; direction *(of a substituent)*
Lennard-Jones-Potential *n* Lennard-Jones potential
Lenzpumpe *f* sump pump
Lepidin *n* lepidine, 4-methylquinoline
Lepidinäthyljodid *n* lepidine ethiodide
lepidoblastisch lepidoblastic *(rock fabric)*
Lepidokrokit *m (min)* lepidocrocite *(iron hydroxide oxide)*
Lepidolith *m (min)* lepidolite *(a mica)*
Lepton *n* lepton *(any of a family of light elementary particles)*
Leseband *n (min tech)* picking (inspection) belt
lesen (min tech) to pick, to sort
Lessing-Ring *m* Lessing ring *(a variety of a Raschig ring)*
Letternmetall *n* type metal
Leuchtbakterien *pl* luminous bacteria
Leuchtdichte *f* luminance, brightness
Leuchtdichtepyrometer *n* [partial-]radiation pyrometer
Leuchtdraht *m* incandescent filament
Leuchtelektron *n* optical (valency, outermost) electron
leuchten to give [off] light, to radiate [light], to glow, to luminesce
Leuchten *n* radiation [of light], glow, luminosity
kaltes L. luminescence
leuchtend luminous, luminiferous; brilliant, bright *(colour)*
schwach l. faintly luminous
stark l. highly luminous
Leuchterscheinung *f* luminous effect
Leuchtfaden *m* incandescent filament
Leuchtfarbe *f* luminous (luminescent) paint
fluoreszierende L. fluorescent paint
nachleuchtende (phosphoreszierende) L. phosphorescent paint
Leuchtflammenbrenner *m* nozzle-mix burner
Leuchtfleck *m (phot)* light spot
Leuchtgas *n s.* Stadtgas
Leuchtkraft *f* luminosity; brilliance, brightness *(of colours)*
Leuchtmasse *f* / **radioaktive** radioactive phosphorescent material
Leuchtmittel *n* illuminant
Leuchtöl *n*, **Leuchtpetroleum** *n* illuminating (lamp) oil
Leuchtstärke *f* luminosity
Leuchtstoff *m* illuminant; *(phys)* luminophore, luminescent substance
aufhellender L. *(text)* fluorescent brightener (brightening agent, whitening agent, white dye)
Leuchtwirkung *f* luminous effect
Leuckart-Reaktion *f* Leuckart reaction *(alkylation of amines)*
Leukindigo *m* leucoindigo
Leukoalizarin *n* leucoalizarin
Leukoanthozyan *n* leucoanthocyanin, anthocyanogen
Leukobase *f* leuco base
Leukochinizarin *n* leucoquinizarine
Leukoester *m* leuco ester
Leukoform *f* leuco form
Leukogen *n* leucogen *(a solution of sodium hydrogensulphite in water)*
Leukoindigo *m* leucoindigo
leukokrat *(min)* leucocratic
Leukomalachitgrün *n* leucomalachite green
Leukopararosanilin *n* leucopararosaniline
Leukophan *m (min)* leucophanite, leucophane *(a sorosilicate)*
Leukorosanilin *n* leucorosaniline
Leukosalz *n* leuco salt
Leukoschwefelsäureester *m* leucosulphuric acid ester
Leukotetraschwefelsäureester *m* leucotetrasulphuric acid ester
Leukoverbindung *f* leuco compound
Leukovorin *n* leucovorin, citrovorum factor, N^5-formyltetrahydrofolic acid, folinic acid SF
Leukozyt *m* / **basophiler** *(med)* basophil[e]
Leuna-Verfahren *n* Bergius process, berginization
Leuzenin *n* leucenine *(an amino acid deriving from pyridine)*
Leuzenol *n s.* Leuzenin
Leuzin *n* $(CH_3)_2CHCH_2CH(NH_2)COOH$ leucine, 2-amino-4-methylvaleric acid
Leuzit *m (min)* leucite *(potassium aluminodisilicate)*
Leviathan-Wollwaschmaschine *f* Leviathan washer
Levyn *m (min)* levyne, levynite *(calcium dialuminotrisilicate)*
Lewis-Base *f* Lewis base
Lewis-Säure *f* Lewis acid
Lezithin *n* lecithin

Lezithinase *f* lecithinase
Lezithinnaßschlamm *m* wet lecithin sludge
LH *s.* Luteinisierungshormon
Libbey-Owens-Verfahren *n* Libbey-Owens[-Ford Colburn] process, LOF-Colburn process, Colburn process *(for manufacturing sheet glass)*
Libethenit *m (min)* libethenite *(copper(II) hydroxide orthophosphate)*
Libidibi *pl (tann)* divi-divi *(from Caesalpinia coriaria (Jacq.) Willd.)*
Lichenin *n* lichenin *(a polysaccharide)*
licht light *(colour)*
Licht *n* light ✦ **am (im) L.** under illumination, on exposure to light ✦ **vor L. geschützt** protected from light
auffallendes L. incident light
diffuses L. diffused (scattered) light
durchfallendes L. transmitted light
einfallendes L. incident light
gestreutes L. scattered light
künstliches L. artificial light
lichtabsorbierend light-absorbing
Lichtabsorption *f* light absorption
Lichtalterung *f* light ageing
lichtbeständig insensitive (resistant) to light, nonfading, *(Am also)* lightfast
Lichtbeständigkeit *f* resistance to light, light resistance, *(Am also)* lightfastness
Lichtbeugung *f* diffraction of light
Lichtbogen *m* **[/ elektrischer]** [electric] arc
lichtbogenbeständig arc-resistant, arc-proof
Lichtbogenbeständigkeit *f* arc resistance
Lichtbogenelektroofen *m s.* Lichtbogenofen
Lichtbogenentladung *f* arc discharge
lichtbogenfest *s.* lichtbogenbeständig
Lichtbogenheizung *f* arc heating
Lichtbogenkohle *f* arc carbon
Lichtbogenlampe *f* arc lamp
Lichtbogenofen *m* arc (arc-heated) furnace, electric-arc furnace
L. mit direkter Beheizung direct arc (arc-heated) furnace
L. mit indirekter Beheizung indirect arc (arc-heated) furnace
L. mit reiner Strahlungsbeheizung *s.* L. mit indirekter Beheizung
direkter L. *s.* L. mit direkter Beheizung
indirekter L. *s.* L. mit indirekter Beheizung
Lichtbogenschweißen *n* arc welding
atomares L. atomic hydrogen [arc] welding
Lichtbogenstrahlungsofen *m* indirect arc (arc-heated) furnace
Lichtbogenverfahren *n* luminous arc process *(Birkeland-Eyde process)*
Lichtbogenwiderstandsofen *m* arc resistance furnace
lichtbrechend refractive
Lichtbrechung *f* refraction of light
Lichtbündel *n* beam of light
Lichtchlorierung *f* photochemical chlorination
lichtdicht lightproof, lighttight
Lichtdichtheit *f* lightproofness, lighttightness
Lichtdruck *m* photomechanical printing
Lichtdruckfarbe *f* photogelatin ink
Lichtdruckgelatine *f* photogelatin
Lichtdruckkarton *m* phototyping cardboard
Lichtdruckpapier *n* phototyping (phototype, collotype) paper
Lichtdruckverfahren *n* photogelatin process
lichtdurchlässig transparent, translucent
Lichtdurchlässigkeit *f (qualitatively)* transparency, light transmission, clarity; *(quantitatively)* light transmittance
lichtecht *s.* lichtbeständig
Lichtechtheit *f s.* Lichtbeständigkeit
Lichtechtheitsmesser *m (text)* fad[e]ometer
Lichtechtheitsprüfung *f (text)* light test
Lichteinfluß / unter on exposure to light, under illumination
lichtelektrisch photoelectric
lichtempfindlich sensitive to light, light-sensitive, photosensitive ✦ **l. machen** *(phot)* to sensitize
Lichtempfindlichkeit *f* sensitivity to light, light sensitivity, photosensitivity
Lichtempfindlichmachen *n (phot)* sensitization
Lichtenergie *f* light (luminous) energy
Lichterscheinung *f* luminous phenomenon (effect)
lichterzeugend luminiferous, photogenic
Lichtfilter *n* light filter
Lichtfleck *m (phot)* light spot
lichtgeschützt protected from light
Lichtgeschwindigkeit *f* velocity of light
Lichthof *m* halo
Lichthofschutzschicht *f (phot)* antihalation (antihalo) backing (coating), anti-halo layer
Lichtintensität *f* intensity of light
lichtkatalysiert light-catalyzed, photocatalyzed
Lichtmikroskop *n* light microscope
Lichtpausautomat *m* blueprinter
Lichtpause *f* blueprint, blue print, *(if based on diazo compounds also)* diazo copy, *(if based on ferricyanide also)* blue negative print
lichtpausen to make (take) a blue print, to blueprint
Lichtpausen *n* blueprint, blue-printing, *(if based on diazo compounds also)* diazo print[ing]
Lichtpausgewebe *n* translucent tracing cloth
Lichtpauspapier *n* blue print paper, *(if based on diazo compounds also)* diazo paper
Lichtpolarisation *f* polarization of light
Lichtquant *n* photon, light quant[um]
Lichtquelle *f* light (luminous) source
Lichtreaktion *f* photochemical reaction, photoreaction
lichtschluckend light-absorbing
Lichtschutz *m* protection from light
lichtspendend luminiferous
Lichtstabilisator *m* light stabilizer
Lichtstärke *f* luminosity, luminous intensity
Lichtstärkemessung *f* photometric analysis, photometry
Lichtstrahl *m* light ray; *(if bundled)* beam of light
Lichtstreuung *f* light scattering

Lichtstreuungsmethode *f* light scattering method *(for determining molecular weights)*
Lichtstrom *m* luminous flux
Lichtstromdichte *f* luminous-flux density
lichtundurchlässig opaque, lightproof, lighttight
Lichtundurchlässigkeit *f* opacity, lightproofness, lighttightness
Lichtwelle *f* light wave
Lichtwiderstand *m* photoresistor, photoconductive cell
Lichtwirkung *f* effect of light, luminous effect
Licker *m (tann)* fat liquor
Lick-up-Bahnabnahme *f (pap)* lick-up
Lick-up-Filz *m (pap)* lick-up overfelt (wet felt, felt)
Liderung *f* packing *(of a pump)*
Lidschatten *m (cosmet)* eye-shadow
Liebermann-Storch-Test *m* Liebermann-Storch test *(for detecting rosin)*
Liebig-Kühler *m* Liebig condenser
Liebstöckelöl *n* lovage (levisticum) oil *(from Levisticum officinale W.D.I. Koch)*
Lieferbeton *m* ready-mixed concrete
Lieferdruck *m* discharge (delivery) pressure
Liefergeschwindigkeit *f* delivery speed (rate)
liefern to deliver, to yield *(as in a reaction)*; to donate, to contribute *(electrons)*
Lieferseite *f* delivery side
Lieferstrom *m* delivery
liegen / zutage *(geol)* to crop out, to outcrop
Liegepresse *f (pap)* straight-through press
Liegerohr *n* horizontal pipe
Liesegang-Ringe *mpl (coll)* Liesegang rings
Lievrit *m (min)* lievrite, ilvaite *(a sorosilicate)*
Liftbehälter *m* lift tank
liften to lift *(catalysts)*
Liften *n* lifting *(as of catalysts)*
L. mit Druckluft air lifting
Liftleitung *f* lift line (pipe) *(for lifting catalysts)*
Lifttopf *m* lift pot
Ligand *m* ligand, addend
Ligandenfeldtheorie *f* ligand field theory
Ligandenkonzentration *f* ligand concentration
Ligatur *f* joint clamp *(for ground-glass joints)*
L. für Kugelschliffe ball-and-socket joint clamp
Lignan *n* lignan
lignifizieren to lignify
Lignifizierung *f* lignification
Lignin *n* lignin
chloriertes L. *(pap)* chlorinated lignin, chlorolignin
restliches L. *(pap)* residual lignin, lignin residues
Ligninase *f* ligninase
Ligninauslösung *f*, **Ligninentfernung** *f (pap)* dissolution (removal) of lignin
Ligningehalt *m* lignin content
Ligninherauslösung *f s.* Ligninauslösung
Ligninkohle *f* lignin coal
Ligninpech *n (pap)* lignin pitch *(concentrated sulphite waste liquor)*
ligninreich rich in lignin
Ligninreste *mpl (pap)* lignin residues, residual lignin
Ligninrestentfernung *f (pap)* removal of lignin residues
Ligninsulfonsäure *f* lignosulphonic acid
nach Hägglund bestimmte feste L. Hägglund's solid lignosulphonic acid
nach Kullgren bestimmte wasserlösliche L. Kullgren lignosulphonic acid
Lignit *m s.* Xylit
Lignosulfonsäure *f s.* Ligninsulfonsäure
Lignozerinsäure *f* lignoceric acid, tetracosanoic acid
Ligroin *n* ligroin[e] *(with boiling range from 90°C to 120°C)*
Likansäure *f* licanic acid, couepic acid
limnisch *(geoch)* limnic, lacustrine
Limonen *n* limonene, (+)-4-isopropenyl-1-methylcyclohexene
Limonit *m (min)* limonite, brown iron ore (stone)
Linarit *m (min)* linarite *(copper(II) lead(II) dihydroxide sulphate)*
Lindemann-Glas *n* Lindemann glass *(transparent to X-rays)*
Lindenholz *n* basswood
Linderungsmittel *n (pharm)* palliative
Linde-Verfahren *n* Linde process *(for air liquefaction)*
Lindgrenit *m (min)* lindgrenite *(copper(II) hydroxide molybdate)*
linear linear, *(relating to chain molecules also)* unbranched
l. polarisiert linearly polarized, plane-polarized
Linearbeschleuniger *m (nucl)* linear accelerator
Linearkolloid *n* linear (fibrous) colloid
Linearkombination *f* linear combination
L. von Atomorbitalen linear combination of atomic orbitals, LCAO
Linearpolyäthylen *n* linear (low-pressure) polyethylene
Linearpolymerisation *f* linear polymerization
Linearprotein *n* fibrous protein
Linie *f* **gleichen Gehalts an flüchtiger Substanz** isovol, line of equal volatile matter
L. gleichen Heizwertes isocal, isocalorific line
L. gleichen Kohlenstoffgehalts isocarbon line
Anti-Stokessche L. anti-Stokes line *(in Raman spectra)*
beständige (letzte) L. persistent (ultimate) line *(spectral analysis)*
Stokessche L. Stokes line *(in Raman spectra)*
verbotene L. forbidden line *(in a spectrum)*
Linien *fpl* / **Fraunhofersche** Fraunhofer lines
Neumannsche L. *(cryst)* Neumann lines
Linienbreite *f* line width *(spectroscopy)*
natürliche L. natural line width
Liniendruck *m (pap)* nip pressure, pressure at the roll nips
Liniengitter *n* line grating *(in X-ray spectrography)*
Linienintensität *f* line intensity *(spectroscopy)*
Linienschreiber *m* continuous-line recorder
Linienserie *f* series of spectral lines
Linienspektrum *n* line (discrete) spectrum

Linienverbreiterung *f* broadening of spectral lines
Liniment *n (pharm)* liniment
Linksdraht *m (rubber)* S twist
linksdrehend laevorota[to]ry, laevorotating, laevo, laevogyrate, laevogyratory, laevogyre, laevogyrous, *(esp cryst)* left-handed
Linksdrehung *f* laevorotation, *(esp cryst)* left-handed rotation
Linksform *f* laevorotatory (laevorotating, laevo) form, (-)form
Linksmilchsäure *f* $CH_3CH(OH)COOH$ D(-)-lactic acid, laevolactic acid
Linkssäure *f* laevorotatory (laevo) acid
Linksweinsäure *f s.* *D*(-)-Weinsäure
linkszirkular polarisiert left-circularly (left-hand circularly) polarized
Linneit *m (min)* linn[a]eite, cobalt pyrites *(cobalt(II,III) sulphide)*
Linolensäure *f* linolenic acid, *(specif)* α-linolenic acid, 9-*cis*, 12-*cis*, 15-*cis*-octadecatrienoic acid
Linolsäure *f* linoleic acid, 9,12-octadecadienoic acid
Linoxyn *n* linoxyn, linoxylin *(a substance obtained by oxidation and polymerisation of linseed oil)*
Linsenapertur *f* aperture of a lens
Linsenerz *n s.* Lirokonit
Linsenfläche *f* lens surface
linsenförmig lenticular
Linsenglas *n* optical glass
Linsenöffnung *f* aperture of a lens
Lint *m*, **Lintbaumwolle** *f* lint cotton
Linters *pl* linters
Lintershalbstoff *m (pap)* linters pulp
Lintwolle *f* lint cotton
Linz-Donawitz-Verfahren *n* Linz-Donawitz process, L-D process *(for steelmaking)*
Liparit *m* liparite, rhyolite *(an acid volcanic rock)*
Lipase *f* lipase
Lipid *n* lipid[e] *(any of a group of fat-like substances)*
  einfaches L. simple lipid
  komplexes L. compound lipid
Lipoid *n* lipoid
Lipolyse *f* lipolysis, fat splitting
lipolytisch lipolytic, fat-splitting
α-**Liponsäure** *f* α-lipoic acid, 3-(4-carboxybutyl)-1,2-dithiolane
lipophil lipophilic, lipophile
Lipoproteid *n* lipoprotein, lipoproteid[e]
Lippenglanz *m (cosmet)* lip-gloss
Lippenpomade *f (cosmet)* lipsalve
Lippenstift *m (cosmet)* lipstick
Lippentupfpapier *n (cosmet)* facial (cleansing) tissue
Liquation *f (geol)* liquation [differentiation] *(process of separating of magmatic fusions)*
Liquid-Polymer[es] *n (rubber)* [polysulphide] liquid polymer
Liquiduskurve *f*, **Liquiduslinie** *f* liquidus [curve, line] *(of a melting diagram)*
Liquidus-Liquidus-Chromatografie *f* liquid-liquid chromatography, L.L.C.
Liquidus-Solidus-Chromatografie *f* liquid-solid chromatography, L.S.C.
Liquidustemperatur *f* liquidus (limiting crystallization) temperature, *(glass also)* limiting devitrification temperature
Lirokonit *m (min)* liroconite
Lisseuse *f (text)* backwashing machine
Lissieren *n (text)* backwashing
Liter *n* / **Mohrsches** Mohr litre *(gas volumetry)*
Liter-Molarität *f* molarity, molar concentration
Lithidionit *m s.* Litidionit
Lithifikation *f (geol)* lithification, induration
Lithiophilit *m (min)* lithiophilite *(a phosphate of lithium containing manganese and iron)*
Lithiophorit *m (min)* lithiophorite *(a hydrous oxide of manganese, aluminium, and lithium)*
Lithium *n* Li lithium
Lithiumalanat *n s.* Lithiumaluminiumhydrid
Lithiumalkyl *n* alkyl lithium, lithium alkyl
Lithiumaluminat *n* $LiAlO_2$ lithium aluminate
Lithiumaluminiumhydrid *n* $Li[AlH_4]$ lithium aluminium hydride, lithium tetrahydridoaluminate
Lithiumamid *n* $LiNH_2$ lithium amide
Lithiumaryl *n* aryl lithium, lithium aryl
Lithiumazetat *n* CH_3COOLi lithium acetate
Lithiumazetylid *n s.* Lithiumkarbid
Lithiumbromid *n* LiBr lithium bromide
Lithiumchlorat *n* $LiClO_3$ lithium chlorate
Lithiumchlorid *n* LiCl lithium chloride
Lithiumchromat *n* Li_2CrO_4 lithium chromate
Lithiumdichromat *n* $Li_2Cr_2O_7$ lithium dichromate
Lithiumdihydrogenorthophosphat *n* LiH_2PO_4 lithium dihydrogenorthophosphate, lithium dihydrogenphosphate
Lithiumfluorid *n* Li_2F_2 lithium fluoride
Lithiumhexachloroplatinat(IV) *n* $Li_2[PtCl_6]$ lithium hexachloroplatinate(IV), lithium chloroplatinate(IV)
Lithiumhexafluorosilikat *n* $Li_2[SiF_6]$ lithium hexafluorosilicate, lithium fluorosilicate
Lithiumhydrid *n* LiH lithium hydride
Lithiumhydrogenkarbonat *n* $LiHCO_3$ lithium hydrogencarbonate
Lithiumhydrogensulfat *n* $LiHSO_4$ lithium hydrogensulphate
Lithiumhydrogensulfid *n* LiHS lithium hydrogensulphide
Lithiumhydroxid *n* LiOH lithium hydroxide
Lithiumjodat *n* $LiIO_3$ lithium iodate
Lithiumkarbid *n* Li_2C_2 lithium carbide, lithium acetylide
Lithiumkarbonat *n* Li_2CO_3 lithium carbonate
Lithiummanganat(VII) *n s.* Lithiumpermanganat
Lithiummetasilikat *n* Li_2SiO_3 lithium metasilicate, lithium trioxosilicate
Lithiummetavanadat *n* $LiVO_3$ lithium metavanadate, lithium trioxovanadate
Lithiummethyl *n* CH_3Li methyllithium
Lithiummolybdat *n* Li_2MoO_4 lithium molybdate
Lithiumnitrat *n* $LiNO_3$ lithium nitrate
Lithiumnitrid *n* Li_3N lithium nitride

Lithiumnitrit *n* $LiNO_2$ lithium nitrite
Lithiumorthoarsenat(V) *n* Li_3AsO_4 lithium orthoarsenate
Lithiumorthosilikat *n* Li_4SiO_4 lithium orthosilicate, lithium tetraoxosilicate
Lithiumoxid *n* Li_2O lithium oxide
Lithiumperchlorat *n* $LiClO_4$ lithium perchlorate
Lithiumpermanganat *n* $LiMnO_4$ lithium permanganate
Lithiumperoxid *n* Li_2O_2 lithium peroxide
Lithiumphosphat *n* lithium phosphate, *(specif)* Li_3PO_4 lithium orthophosphate
Lithiumrhodanid *n s.* Lithiumthiozyanat
Lithiumsalizylat *n* HOC_6H_5COOLi lithium salicylate
Lithiumselenid *n* Li_2Se lithium selenide
Lithiumsilizid *n* Li_6Si_2 lithium silicide
Lithiumsulfat *n* Li_2SO_4 lithium sulphate
Lithiumsulfid *n* Li_2S lithium sulphide
Lithiumsulfit *n* Li_2SO_3 lithium sulphite
Lithiumtetroxosilikat *n s.* Lithiumorthosilikat
Lithiumthiozyanat *n* LiSCN lithium thiocyanate
Lithiumtrioxosilikat *n s.* Lithiummetasilikat
Lithiumtrioxovanadat(V) *n s.* Lithiummetavanadat
Lithiumwolframat *n* Li_2WO_4 lithium wolframate (tungstate)
Lithocholsäure *f* lithocholic acid
Lithografenfirnis *m* lithographic (litho) varnish
Lithografenkalk *m*, **Lithografenschiefer** *m* lithographic limestone
Lithografie *f* lithography, lithographic printing
Lithografiefarbe *f* lithographic [printing] ink, litho ink
Lithografiekreide *f* lithographic crayon
Lithografiepapier *n* lithographic [printing] paper, litho
Lithografiestein *m* lithographic limestone
Litholrot *n* lithol red
Lithopon *n*, **Lithopone** *f* lithopone, zinc baryta white, Orr's white *(consisting of ZnS and $BaSO_4$)*
Lithosphäre *f (geol)* lithosphere
Lithotype *f (coal)* lithotype, rock type
Litidionit *m (min)* lithidionite, neocyanite *(a copper-containing phyllosilicate)*
Littleton-Punkt *m (glass)* Littleton [softening] point, seven-point-six temperature, 7.6 temperature *(at which the viscosity is $10^{7.6}$ poises)*
Littrow-Spektrograf *m* Littrow spectrograph
Livingstonit *m (min)* livingstonite *(antimony(III) mercury(II) sulphide)*
Ljungström-Regenerator *m*, **Ljungström-Vorwärmer** *m* Ljungstrom heater (regenerator)
L-Kalander *m* L type of calender
L-Konfiguration *f* L configuration
ll *s.* leichtlöslich
LLC *s.* Liquidus-Liquidus-Chromatografie
Lobarsäure *f* lobaric acid *(a lichen acid)*
Lobeliaalkaloid *n* lobelia alkaloid
Lobelienkraut *n (pharm)* asthma weed *(from Lobelia inflata L.)*
Lobelin *n* lobeline *(a piperidine derivative)*
Loch *n* 1. hole; 2. *(phys chem)* [electron] hole, hole position, defect electron; *(cryst)* [lattice] vacancy, vacant lattice position (site), hole; 3. *(plast)* pinhole, pit *(a moulding defect)*
Lochblech *n* perforated (punched) plate
Lochblechsieb *n* perforated-metal (punched-plate) screen
Lochdüngung *f (agric)* hole dressing
Loch-Elektronen-Paar *n* hole-electron pair
Löcherleitung *f* hole conduction
Löchertheorie *f* **der Flüssigkeiten** hole theory of liquids
Lochfraß *n* pitting
Lochfraßkorrosion *f* pitting corrosion
Lochleitung *f* hole conduction
Lochmaß *n* size of aperture *(classifying)*
Lochplatte *f* perforated plate; strainer *(filtration)*
Lochpresse *f* punch press
Lochscheibe *f (plast)* breaker plate
Lochstanze *f* punch press
Lochstein *m* perforated brick
Lochstift *m* plain core pin
Lochtrommel *f* perforated basket *(of a centrifuge)*
Lochwalze *f* perforated (holey) roll
Lochweite *f* orifice
Lochwerte *mpl* perforation *(of a screen)*
Lochziegel[stein] *m* perforated (hollow) brick; hollow tile
Lockerstelle *f (cryst)* loose place (position), flaw
Lockerung *f* antibond[ing], loosening *(chemical-bond theory)*
Lockmittel *n* attractant
Lockspeise *f* food lure *(insect control)*
Lockstoff *m* attractant
  L. zur Eiablage oviposition lure *(insect control)*
Lockstoffeigenschaften *fpl* attractive properties
Lockwirkung *f* attractant (attractive) action
Loesche-Kohlen[staub]mühle *f* Loesche coal mill
Loeweit *m s.* Löweit
Löffel *m* 1. *(glass)* ladle; 2. bucket *(of an elevator)*; 3. *(lab)* spoon
Logarithmenpapier *n* logarithmic paper
Lohbrühe *f (tann)* bark liquor
  ausgelaugte L. spent bark liquor
Lohe *f (tann)* bark [of tan]
Lohgerbung *f* bark tannage (tanning)
Lohgrube *f (tann)* tan[ning] pit, handler
Lohmühle *f (tann)* bark mill
Lokalanalgetikum *n* local analgesic
Lokalanästhetikum *n* local anaesthetic
Lokalelement *n* local cell, local galvanic element *(a very small corrosion cell)*
Lokalisation *f* localization [of position], determination of position
lokalisieren to localize
Lokao *n* locao, Chinese green *(natural dye from Rhamnus specc.)*
Löllingit *m (min)* loellingite, löllingite *(iron arsenide)*
Longifolen *n* longifolene *(a tricyclic sesquiterpene)*

Lorbeerbutter *f*, **Lorbeeröl** *n* laurel oil, [sweet] bay oil, bay fat *(from Laurus nobilis L.)*
Lorbeerwachs *n (incorrectly for)* Myrikatalg
Loröl *n s.* Lorbeerbutter
lösbar 1. breakable *(chemical bond)*; detachable *(esp from a surface)*; 2. resolvable *(problem)*; 3. *s.* löslich
leicht l. easy to disconnect *(e.g. ground-glass joints)*
Lösbarkeit *f* 1. detachability *(esp from a surface)*; 2. resolvability *(of a problem)*; 3. *s.* Löslichkeit
Löschanlage *f* quenching station *(for coke)*
Löschbrause *f* emergency (safety, drench) shower
Lösche *f* breeze
löschen to extinguish *(fire)*; to quench *(coke)*; to slake, to hydrate *(lime)*; to quench *(an electric arc)*
Löscher *m* hydrator, slaker *(for lime)*
Löschgas *n* quench[ing] gas
Löschkalk *m* slaked (water-slaked, hydrated) lime, slacklime, lime hydrate, *(agric also)* agricultural hydrate *(calcium hydroxide)*
Löschkarton *m* absorbent (blotting) board
Löschmaschine *f* hydrator, slaker *(for lime)*
Loschmidt-Zahl *f* Avogadro number (constant) *(the number of molecules in a gram molecule of any substance)*
Löschpapier *n* blotting paper
Löschpapierprüfgerät *n* blotting paper tester
Löschstation *f* quenching station *(for coke)*
Löschturm *m* quenching tower *(for coke)*
Löschwagen *m* quenching car *(for coke)*
Lösebehälter *m s.* Lösetank
Lösefähigkeit *f s.* Lösekraft
Lösegeschwindigkeit *f* rate of dissolution
Lösegut *n* material being (*or* to be) dissolved; material being (*or* to be) extracted
Lösekraft *f* solvent (solubilizing) power, solvency
Lösemittel *n s.* Lösungsmittel
lösen 1. to dissolve *(a chemical)*; 2. to break, to crack *(a chemical bond)*; to detach *(an electron from its shell)*; to disconnect, to undo *(a joint)*; to detach *(from a surface)*; 3. to resolve *(a problem)*
sich l. to dissolve, to go into solution
wieder l. to redissolve
zur ursprünglichen Konzentration l. to reconstitute
lösend dissolving, solvent
Löser *m s.* Lösungsmittel
Lösetank *m (pap)* [smelt] dissolving tank, dissolving chest, dissolver
Lösevermögen *n* solvent (solubilizing) power, solvency
latentes L. latent solvency
Lösewärme *f s.* Lösungswärme
lösl. *s.* löslich
löslich soluble, dissoluble ✦ **l. machen** to solubilize
größtenteils l. substantially soluble
gut l. *s.* leicht l.
leicht l. readily (freely) soluble
nicht l. insoluble, non-soluble, indissoluble
schwach l. *s.* schwer l.
schwer l. slightly (sparingly) soluble
sehr leicht l. very (highly) soluble, v.s.
sehr schwer (wenig) l. extremely (very) slightly soluble, v.s.s., extremely insoluble
teilweise l. partially soluble
wenig l. *s.* schwer l.
Löslichkeit *f* solubility
L. in festem Zustand solid solubility *(of metals)*
L. in Wasser aqueous (water) solubility
gegenseitige L. mutual solubility
scheinbare L. apparent solubility
Löslichkeitsdiagramm *n* solubility curve
Löslichkeitseigenschaften *fpl* solubility properties
Löslichkeitserniedrigung *f* decrease in solubility
Löslichkeitskoeffizient *m*, **Löslichkeitskonstante** *f* solubility coefficient
Löslichkeitskurve *f* solubility curve
Löslichkeitsprodukt *n* solubility product
Löslichkeitsunterschied *m* difference in solubility
Löslichkeitsverbesserer *m s.* Löslichkeitsvermittler
Löslichkeitsverhalten *n* solubility behaviour
Löslichkeitsverminderung *f* decrease in solubility; common ion effect *(in the presence of a second electrolyte with a common ion)*
Löslichkeitsvermittler *m* solutizer, solubilizer, solubility promoter, solutizing agent
Löslichmachen *n* solubilizing
loslösen to release *(electrons)*
losnarbig *(tann)* loose-grained, pipey
Losnarbigkeit *f (tann)* grain pipeyness
Löß *m* loess
Lößboden *m* loess soil
Lößlehm *m* loess loam
Lost *n s.* Dichlordiäthylsulfid
Losttherapie *f* mustard therapy
Lösung *f* 1. solution; 2. *(act or process)* dissolution ✦ **in L. bringen** to bring (put) into solution, to render soluble ✦ **in L. gehen** to go into solution, to dissolve ✦ **in L. halten** to keep in solution
alkalische L. alkaline (alkali) solution
alkoholische L. alcoholic solution
äquimolare L. equimolar solution
Benedictsche L. Benedict solution *(for deteoting reducing sugars)*
Bialsche L. Bial reagent *(for detecting pentoses)*
Cramersche L. Cramer solution *(for detecting reducing sugars)*
echte L. real (true) solution
eingestellte L. standard solution
Fehlingsche L. Fehling's solution (reagent), Fehlings's
feste L. solid solution
Flemmingsche L. *(biol)* Flemming solution *(a fixative)*
Flicksche L. Flick solution *(of HCl and H_2F_2 for etching aluminium)*
Fowlersche L. *(med)* Fowler solution *(potassium arsenite)*

geimpfte L. seeded solution
gepufferte L. buffered solution
gesättigte L. saturated solution
gewichtsmolare L. molal solution
Hainesche L. Haine reagent *(for detecting glucose)*
heiß gesättigte L. hot-saturated solution
hydrotrope L. hydrotropic solution
hypertonische L. hypertonic solution
hypotonische L. hypotonic solution
ideale L. ideal solution
interstitielle feste L. interstitial [solid] solution
isosmotische (isotonische) L. isosmotic (isotonic) solution
Knappsche L. Knapp solution *(of $Hg(CN)_2$ and NaOH for determining glucose)*
Knopsche L. *(agric)* Knop's solution
kolloidale L. *s.* kolloide L.
kolloide L. colloidal solution
Lugolsche L. Lugol's solution *(aqueous solution of potassium iodide and iodine)*
molale L. molal solution
molare L. molar solution
nichtwäßrige L. non-aqueous solution
normale L. standard solution, *(specif)* normal solution, N solution
Ostsche L. Ost's solution *(of $CuSO_4$, Na_2CO_3, and $NaHCO_3$ for detecting glucose)*
Pavysche L. Pavy solution *(for detecting glucose)*
Ringersche L. *(med)* Ringer solution (fluid), Ringer artificial serum
Sachssesche L. Sachsse solution *(for determining glucose)*
saure L. acid solution
selbstvulkanisierende L. *(rubber)* self-curing (self-vulcanizing) cement
standardisierte L. standard solution
übersättigte L. supersaturated solution
ungeimpfte L. unseeded solution
verdünnte L. dilution
volumenmolare L. molar solution
wäßrige L. aqueous (water) solution
Wijssche L. Wijs [iodine monochloride] solution *(for determining the iodine number)*
m-**Lösung** *f s.* Lösung / molare
1m-**Lösung** *f* 1.0 molar solution
0,1m-**Lösung** *f, m/10*-**Lösung** *f* decimolar (tenth molar) solution
0,01m-**Lösung** *f, m/100*-**Lösung** *f* centimolar solution
0,001m-**Lösung** *f, m/1000*-**Lösung** *f* millimolar solution
n-**Lösung** *f s.* Lösung / normale
1n-**Lösung** *f* N solution, normal solution
0,1n-**Lösung** *f, n/10*-**Lösung** *f* decinormal (tenth normal) solution
0,01n-**Lösung** *f, n/100*-**Lösung** *f* centinormal solution
0,001n-**Lösung** *f, n/1000*-**Lösung** *f* millinormal solution

Lösungen *fpl* **gleichen Dampfdrucks** isopiestic solutions
Lösungsaustritt *m* liquor outlet *(as on an evaporator)*
Lösungsbehälter *m (rubber)* dip[ping] tank
Lösungsbenzin *n* mineral (petroleum) spirit[s]
Lösungsbeschleuniger *m s.* Lösungsvermittler
Lösungschromatografie *f* solubilization chromatography
Lösungsdruck *m* solution pressure (tension)
elektrolytischer L. electrolytic solution pressure
Lösungseintritt *m* liquor inlet *(as on an evaporator)*
Lösungsenthalpie *f* enthalpy of solution
Lösungsfähigkeit *f s.* Lösungsvermögen
Lösungsfigur *f (cryst)* corrosion (etch) figure
Lösungsgeschwindigkeit *f* rate of dissolution
Lösungsgleichgewicht *n* [dis]solution equilibrium
Lösungsglühen *n* solution [heat] treatment
Lösungshilfsmittel *n* solvent assistant
Lösungskasten *m (rubber)* dip[ping] tank
Lösungskraft *f s.* Lösungsvermögen
Lösungsmittel *n* solvent, dissolver, dissolvent
L. für Chemischreinigung dry-cleaning solvent
aktives (echtes) L. active (true) solvent
gemischtes L. mixed solvent
hochsiedendes L. high-boiling solvent, high boiler
latentes L. latent solvent, cosolvent
leichtflüchtiges L. fast solvent
mittelbares L. *s.* latentes L.
mittelsiedendes L. medium-boiling solvent, medium boiler
nichtwäßriges L. non-aqueous solvent
niedrigsiedendes L. low-boiling solvent, low-boiler
organisches L. organic solvent
polares L. polar solvent
schlechtes L. poor solvent
schnellflüchtiges L. fast solvent
selektives L. selective solvent
lösungsmittelabstoßend lyophobe, lyophobic
lösungsmittelanziehend lyophile, lyophilic
Lösungsmittelbehandlung *f* solvent treatment
lösungsmittelbeständig fast to solvents, solvent-resisting
Lösungsmittelbeständigkeit *f* fastness to solvents, solvent resistance
Lösungsmitteldampf *m* solvent vapour
Lösungsmitteleffekt *m* solvent effect
Lösungsmittelentparaffinierung *f (petrol)* solvent dewaxing
Lösungsmittelextraktion *f* solvent extraction
lösungsmittelfest *s.* lösungsmittelbeständig
lösungsmittelfrei solventless
Lösungsmittelfront *f* solvent front *(chromatography)*
Lösungsmittelgemisch *n* mixed solvent
Lösungsmittelgerbung *f* solvent tannage
Lösungsmittelgleichgewicht *n* solvent balance

Lösungsmittelkleber *m* solvent-based adhesive, solvent cement
Lösungsmittelphase *f* solvent phase
Lösungsmittelpolymerisation *f s.* Lösungspolymerisation
Lösungsmittelraffination *f* solvent refining
lösungsmittelraffiniert solvent-refined
Lösungsmittelretention *f* solvent retention
Lösungsmittelrückgewinnung *f* solvent recovery
Lösungsmittelrückgewinnungsanlage *f* solvent-recovery plant
Lösungsmittelschale *f* solvent dish
Lösungsmittelschweißen *n* solvent welding
Lösungsmittelspektrum *n* solvent spectrum
Lösungsmittelsystem *n* solvent system
Lösungsmitteltrog *m* solvent trough *(chromatography)*
Lösungsmittelverfahren *n (text)* solvent process
Lösungsmittelwäsche *f (text)* solvent scouring
Lösungsmittelwiedergewinnung *f* solvent recovery
Lösungspolymerisation *f* solvent (solution) polymerization
Lösungspunkt *m* / **kritischer** *s.* Lösungstemperatur / kritische
Lösungsraffination *f* solvent refining
Lösungsreaktion *f* reaction in solution
Lösungsregenerierverfahren *n* solution reclaiming process
Lösungsschweißen *n* solvent welding
Lösungsspinnen *n (text)* solution (solvent) spinning
Lösungstemperatur *f* solution temperature
kritische L. consolute (critical solution) temperature
obere kritische L. upper consolute (critical solution) temperature
untere kritische L. lower consolute (critical solution) temperature
Lösungstension *f* solution pressure (tension)
Lösungstrog *m (rubber)* dip[ping] tank
Lösungsverbesserer *m s.* Lösungsvermittler
Lösungsverhalten *n* solution behaviour
Lösungsvermittler *m* solutizer, solubilizer, solubility promoter, solutizing agent, solution assistant
Lösungsvermögen *n* solvency, solvent (solubilizing) power
latentes L. latent solvency
Lösungsverwitterung *f (soil)* disintegration by solution
Lösungsvorgang *m* dissolving process
Lösungswärme *f* heat of solution
differentiale (differentielle) L. *s.* partielle L.
integrale L. integral (total) heat of solution
partielle L. partial (differential) heat of solution
Lot *n* solder
Lötbad *n* solder bath
lötbar solderable
Lötdraht *m* solder wire
löten to solder
lötfähig *s.* lötbar
Lötfett *n* soldering paste
Lötglas *n* solder (sealing, solder sealing) glass
Löthilfsmittel *n* soldering agent, flux
Lotion *f (cosmet)* lotion, wash; *(pharm)* lotion
desodorierende L. deodorant lotion
transpirationsverringernde L. antiperspirant lotion
Lötlampe *f* [blow]torch, blowlamp
Lotlegierung *f* solder alloy
Lötmetall *n* solder
Lötmittel *n s.* 1. Lötmetall; 2. Löthilfsmittel
Lötpaste *f* soldering paste; paste solder *(containing all components for soldering)*
Lötrohr *n* [mouth] blowpipe ✦ **mit dem L. arbeiten** to blowpipe
Lötrohranalyse *f* blowpipe analysis
Lötrohrflamme *f* blowpipe flame
Lötrohrkohle *f* blowpipe charcoal
Lötrohrmundstück *n* blowpipe mouthpiece
Lötrohrprobe *f* blowpipe test (assay, proof)
Lötrohrprobierkunde *f s.* Lötrohranalyse
Lötsäure *f* soldering acid
Lötstelle *f* soldered joint, soldering
Lötverbindung *f* soldered joint, soldering
Lötwasser *n* soldering fluid (liquid)
Lötzinn *n* soldering tin, plumber's solder
Löweit *m (min)* loeweite *(a hydrous magnesium sodium sulphate)*
Low-structure-Ruß *m (rubber)* low-structure [carbon] black
LP-Beton *m* air-entrained (air-entraining) concrete
LP-Bildner *m (build)* air-entraining additive (admixture, agent, compound)
LSC *s.* Liquidus-Solidus-Chromatografie
L-Schale *f* L-shell *(of an atom)*
LSD *s.* Lysergsäurediäthylamid
Lsgm. *s.* Lösungsmittel
LS-Kopplung *f (nucl)* LS coupling, Russell-Saunders coupling
LTH *s.* Hormon / laktotropes
Lücke *f (cryst)* [lattice] vacancy, vacant lattice position (site), hole; gap, interstice, hole position *(between the regular lattice sites)*
Ludlamit *m (min)* ludlamite *(a hydrous iron(II) orthophosphate)*
Luft *f* air ✦ **an der L.** on exposure to air
atmosphärische L. atmospheric air
eingeblasene L. draught air
flüssige L. liquid air
mit Feuchtigkeit beladene L. moisture-laden air
überschüssige L. excess air
Luftablaß *m* air relief
Luftabschluß *m* exclusion of air ✦ **unter L.** out of contact with air, in the absence of oxygen, *(biol also)* under anaerobic conditions
Luftabschreckung *f (glass)* air quenching
luftangetrieben air-driven
Luftaufbereitung *f (min tech)* dry (pneumatic) cleaning
Luftauftrieb *m* buoyancy of the air

Luftbad *n* air bath (jacket)
Luftbedarf *m* air requirements
Luftbefeuchtung *f* air moistening
luftbeständig stable in air
luftbewegt air-operated
Luftblase *f* air bubble, *(on the surface)* blister, air bell, *(pap also)* foam mark
Luftbombenalterung *f (rubber)* air bomb ageing (test), air pressure [heat] test
Luftbürste *f* air brush (knife)
Luftbürstenauftragmaschine *f (plast)* air knife coater
Luftbürstenstreichmaschine *f (pap)* air brush coater
Luftdämpfungseinrichtung *f* air damping device *(on precision balances)*
luftdicht airtight, air-impermeable
Luftdichtigkeit *f s.* Luftundurchlässigkeit
Luftdruck *m* air pressure
Luft-Druckalterung *f s.* Luftbombenalterung
Luftdruckmesser *m* barometer
Luftdruckmessung *f* barometry
Luftdruckregler *m* air pressure regulator
Luftdruckschreiber *m* barograph
luftdurchlässig air-permeable
Luftdurchlässigkeit *f* air permeability, permeability to air
Luftdurchlässigkeitsprüfer *m* densimeter, densometer
L. nach Schopper *(pap)* Schopper densimeter
Luftdüsenblasverfahren *n (glass)* air-blowing process
Lufteingang *m* air intake, blast inlet *(in underground gasification)*
Lufteinschluß *m* inclusion of air, [en]trapped air, *(if material fault also)* air pocket, [air] blister, air void
Lufteinwirkung / unter on exposure to air
Luftelektrizität *f* atmospheric electricity
luftempfindlich air-sensitive
Lüften *n (plast)* venting, breathing, degassing *(of the mould)*
Lüfter *m* fan, blower
L. mit geraden Schaufeln straight-blade fan
L. mit rückwärtsgekrümmten Schaufeln backward-curved-blade fan
L. mit vorwärtsgekrümmten Schaufeln forward-curved-blade fan
Lufterhitzer *m* air heater
Luftfeuchte *f*, **Luftfeuchtigkeit** *f* atmospheric humidity
Luftfeuchtigkeitsmesser *m* hygrometer
Luftfilter *n* air filter
Luftförderer *m* air conveyor
Luftführung *f* **im Kreislauf (Umluftbetrieb)** air recirculation
Luftgas *n s.* Generatorgas
Luftgebläse *n* forced-draught fan (blower), air blower
Luftgefrierapparat *m* [air-]blast freezer
luftgefüllt air-filled
luftgekühlt air-cooled
Luftgeschwindigkeit *f* air velocity
luftgesteuert air-operated
luftgetrieben air-driven
luftgetrocknet air-dried; *(tann)* air-conditioned
Luft-Glas-Fläche *f (phot)* glass-air interface
Lufthahn *m* air cock
lufthärten to air-harden
Lufthärter *m*, **Lufthärtestahl** *m* air-hardening steel
Lufthärtung *f* air hardening
Luftheber *m* air-lift [pump], mammoth pump
Luftherd *m (min tech)* air (dry) table
Lufthülle *f* atmosphere
Luftkalk *m* non-hydraulic lime
Luftkammer *f (min tech)* air chamber; *(glass)* air regenerator chamber; plenum chamber *(for gas cleaning)*
Luftkanal *m* air duct
Luftklappe *f* air register
Luftkolben *m* air slug *(of an air-lift pump)*
Luftkonditionieranlage *f* air conditioning plant
Luftkonditionierung *f* air conditioning
Luftkorrosion *f* atmospheric corrosion
Luftkühler *m (lab)* air condenser
Luftkühlung *f* air cooling
luftleer evacuated, exhausted ✦ **l. machen** to evacuate, to exhaust
Luftleitung *f* air line; air-blast main *(of a producer)*
Luftmantel *m* air jacket
Luftmenge *f* **/ kritische** *(coal)* critical air blast, C.A.B.
Luftmesser *n (pap)* air knife
Luftmesserstreichmaschine *f (pap)* air knife coater
Luftmesserstreichverfahren *n (pap)* air knife coating
Luftmörtel *m* non-hydraulic mortar
Luftmotor *m (lab)* air motor
Luftoxydation *f* oxidation by air
Luftpore *f* air void
Luftporenbeton *m* air-entrained (air-entraining) concrete
Luftporenbildner *m (build)* air-entraining additive (admixture, agent, compound)
Luftporenbildung *f* air entrainment
Luftporenzement *m* air-entraining cement
Luftpostpapier *n* air-mail paper
Luftrakelauftragmaschine *f* air blade coater
Luftregenerativkammer *f (glass)* air regenerator chamber
Luftreifen *m (rubber)* pneumatic [tyre]
Luftreiniger *m* air purifier (cleaner)
Luftreinigung *f* cleaning of air
Luftrückführung *f* air recirculation
Luftsack *m (glass)* bubble *(in a parison)*
Luftsauerstoff *m* atmospheric oxygen
Luftschieber *m* air register
Luftschlauch *m (rubber)* inner (air) tube
Luftschlauchmischung *f (rubber)* inner tube compound
Luftschlauchregenerat *n (rubber)* inner tube reclaim

Luftschleier *m* aerial fog
Luftsetzapparat *m (min tech)* dry cleaner
Luftspalt *m* air gap
Luftstickstoff *m* atmospheric nitrogen
Luftstrahlgebläse *n,* **Luftstrahlpumpe** *f,* **Luftstrahlverdichter** *m* compressed-air ejector
Luftstrom *m* draught (blast, current) of air, air flow (stream)
Luftstrommühle *f* air-swept mill
Luftstromsichter *m* air (pneumatic) classifier, air separator
Luftstromsichtung *f* air-flow classification
Luftstromtexturieren *n (text)* air-jet crimping (texturing)
Luftstromtrockner *m* [air-]jet dryer
Luftstromtrocknung *f* [air-]jet drying
Lufttaupunkt *m* air dew point
Lufttrennung *f* air separation
lufttrocken air-dry
l. und mineral[stoff]frei dry and mineral-matter-free, d.m.m.f., D.M.F.
Lufttrockner *m* air (atmospheric) dryer; *(lab)* balance desiccator
Lufttrocknung *f* air (open-air, atmospheric) drying
Luftüberschuß *m* excess of air
Luftüberwachungsanlage *f* dust monitor; *(nucl)* radiation monitor
luftundurchlässig air-impermeable, airtight
Luftundurchlässigkeit *f* air impermeability (tightness)
Lüftung *f* aeration
Lüftungsöffnung *f* air vent (relief), vent
Lüftungssystem *n* ventilation system
Luftventilator *n* air blower
Luftverbesserungsmittel *n* air improver
Luftverflüssigung *f* air liquefaction
Luftverschmutzung *f,* **Luftverunreinigung** *f* air pollution
Luftvorwärmer *m* air preheater
Luftwäsche *f (min tech)* dry (pneumatic) cleaning
Luftzerlegung *f* air separation
Luftzirkulation *f* air circulation
Luftzufuhr *f* air supply
Luftzuführung *f* 1. *s.* Luftzufuhr; 2. air inlet (supply) *(point)*
Luftzutritt *m* access of air ✦ **unter L.** in the presence of oxygen, *(biol also)* under aerobic conditions
Lukas-Test *m* Lukas test *(for detecting alcohol)*
lumineszent luminescent
Lumineszenz *f* luminescence
Lumineszenzanalyse *f* luminescent analysis
Lumineszenzfarbe *f* luminescent (luminous) paint
Lumineszenzindikator *m* luminescent indicator
Lumineszenzstrahler *m s.* Luminophor
lumineszieren to luminesce
lumineszierend luminescent
Luminophor *m* luminophor[e], luminescent substance
Luminosität *f* luminosity
Lumpen *pl (pap)* rags
Lumpenhalbstoff *m (pap)* rag pulp (stuff, stock), all-rag furnish
Lumpenpapier *n* [all-]rag paper
Lumpenschneider *m (pap)* rag cutter
Lungengift *n* lung injurant
Lunker *m* bubble, pipe *(foundry)*; *(plast)* bubble
primärer L. primary pipe
sekundärer L. secondary pipe
Lunker[aus]bildung *f,* **Lunkern** *n,* **Lunkerung** *f* piping *(foundry)*
Lunkerverhütungsmittel *n* anticavitation (cavity-preventing) agent *(foundry)*
Lunte *f* 1. *(glass)* sliver; 2. fuse *(for setting off explosives)*
Lupinenalkaloid *n* lupin[e] alkaloid
Lüpke-Pendel *n (rubber)* Lüpke pendulum (resiliometer) *(a testing instrument)*
Luppe *f (met)* loup, ball ✦ **Luppen bilden** to ball [up]
Luppenbildung *f (met)* balling
Lupulin *n* lupulin, hop flour
Lupulinbecher *m* lupulin gland
α-Lupulinsäure *f* α-lupulinic acid, humulone
β-Lupulinsäure *f* β-lupulinic acid, lupulone
Lupulon *n s.* β-Lupulinsäure
Lurgi-Druckgasverfahren *n* Lurgi pressure-gasification (high-pressure) process
Lurgi-Druckvergaser *m* Lurgi pressure gasifier
Lurgi-Druckvergasungsanlage *f* Lurgi gasification plant
Lurgi-Druckvergasungsverfahren *n s.* Lurgi-Druckgasverfahren
Lurgi-Spülgas[schwel]verfahren *n* Lurgi Spülgas low-temperature carbonization process
Lüsterfarbe *f (ceram)* lustre colour
Lüsterglasur *f (ceram)* lustre glaze
Lustgas *n s.* Lachgas
Lüstriermittel *n (text)* lustring agent
Lutein *n* lutein *(a yellow natural pigment)*
luteinisieren *(bioch)* to luteinize
Luteinisierungshormon *n* luteinizing hormone, LH, interstitial-cell-stimulating hormone, ICSH
Luteolin *n* luteolin *(a flavone pigment)*
Luteotrophin *n* luteotrophin, prolactin, lactogen, lactogenic (luteotrophic) hormone
Lutetium *n* Lu lutetium, lutecium
Lutetiumchlorid *n* $LuCl_3$ lutetium chloride
Lutetiumerde *f s.* Lutetiumoxid
Lutetiumoxid *n* Lu_2O_3 lutetium oxide
Lutetiumsulfat *n* $Lu_2(SO_4)_3$ lutetium sulphate
Lutidin *n* lutidine, dimethylpyridine
Lutoide *npl (rubber)* lutoids
lutro *s.* lufttrocken
Luvo *m s.* Luftvorwärmer
Lux-Masse *f* luxmasse
Luxusaufnahme *f,* **Luxuskonsum** *m* luxury consumption *(uptake of unnecessary amounts of nutrients by plants)*
L-Walzenkalander *m* L type of calender
Lydit *m (min)* lydite, lydian stone, touchstone
Lykopen *n* lycopene *(a natural dye)*
Lykorin *n* lycorine *(alkaloid)*

Lyman-Serie *f* Lyman series *(spectroscopy)*
Lyoenzym *n*, **Lyoferment** *n* lyo-enzyme
Lyogel *n* lyogel
Lyolysis *f* lyolysis, solvolysis
lyophil lyophile, lyophilic
Lyophilisation *f s.* Lyophilisierung
lyophilisieren to lyophilize, to freeze-dry
Lyophilisierung *f* lyophilization, freeze drying
lyophob lyophobe, lyophobic
Lysergsäure *f* lysergic acid
Lysergsäurediäthylamid *n* lysergic acid diethylamide, LSD
Lysin *n* $NH_2[CH_2]_4CH(NH_2)COOH$ lysine, 2,6-diaminohexanoic acid
Lysozym *n* lysozyme *(a bacteriolytic enzyme)*
Lyzin *n* lycine, betaine *(proper)*, trimethylglycine, oxyneurine
LZ *s.* Leistungszahl

M

M *s.* Massenzahl
Macassar-Öl *n* Macassar (kussum) oil *(from Schleichera specc.)*
Mac-Dougall-Ofen *m* McDougall furnace *(a multi-hearth furnace)*
machen:
  alkalisch m. to make alkaline, to alkalify, to alkal[in]ize
  brandsicher m. *s.* feuerbeständig m.
  durch Gefriertrocknung haltbar m. *(food)* to dehydrofreeze
  einen Blindversuch m. to run a blank
  feuerbeständig m. to fireproof
  geruchlos m. to deodorize
  haltbar m. to preserve, to conserve, to prepare, to cure, *(food also)* to can
  keimfrei m. *(med)* to sterilize, to degerm
  körnig m. to granulate, to grain
  lichtempfindlich m. *(phot)* to sensitize
  löslich m. to solubilize
  luftleer m. to evacuate, to exhaust
  mottenecht m. *(text)* to mothproof
  spannungsfrei m. *(glass, plast)* to temper, to anneal
  stückig m. to agglomerate
  unempfindlich m. *(phot)* to desensitize
  unlöslich m. to unsolubilize
  unwirksam m. to inactivate, to block
  verfallen m. *(tann)* to bring down, to deplete, to fall *(pelts)*
  wasserdicht (wasserundurchlässig) m. *(text)* to waterproof
  weich m. to soften *(e.g. water, plastics)*
  zähflüssig m. to thicken
mächtig *(mine)* thick
Mächtigkeit *f (mine)* thickness
Mackie-Linie *f (phot)* Mackie line
Madelung-Konstante *f (cryst)* Madelung constant *(of lattice energy)*
Madiöl *n* madia oil *(from Madia sativa Mol.)*

Magazin *n* 1. storehouse, storage, store; 2. *(pap)* magazine
Magazinschleifer *m (pap)* magazine grinder
  hydraulischer M. hydraulic magazine grinder
Magengift *n* stomach poison
Magensaft *m* gastric juice
Magensäure *f* gastric acid
Magenta *n* magenta, fuchsin, rosaniline
mager lean *(e.g. ore, concrete, coal)*; *(food)* fatless; *(soil)* poor, infertile, thin
Magerbeton *m* lean[-mixed] concrete, poor concrete
Magererz *n* lean (low-grade) ore
Magerkalk *m* poor lime
Magerkäse *m* skim-milk cheese
Magerkohle *f* lean (dry steam) coal, semianthracite [coal]
Magermilch *f* skim[med] milk, separated (fat-free) milk
  eingedickte (kondensierte) M. condensed skim milk
Magermilchpulver *n* skim-milk powder, dry skim milk, dried fat-free milk
magern *(ceram)* to shorten, *(esp using crushed firebricks)* to grog
Mageröl *n* lean oil *(as for an absorption column)*
Magerton *m* lean clay
Magerungsmittel *n* leaning material; *(ceram)* shortening (non-plastic) material, *(esp crushed firebricks)* grog
Magma *n* magma
  feuerflüssiges M. igneous magma
  palingenetisches M. palingen[et]ic magma, neomagma
Magmagestein *n* magmatic (igneous) rock
Magmaherd *m* magmatic hearth, pocket of magma
Magmaintrusion *f* magmatic intrusion
magmatisch magmatic, igneous
Magmatit *m s.* Magmagestein
Magnesia *f* MgO magnesium oxide
  gebrannte (kalzinierte) M. calcined magnesium oxide
Magnesiabinder *m (build)* magnesia (magnesium oxychloride, Sorel) cement
Magnesiaeisenglimmer *m (min)* black (dark) mica, biotite
magnesiahaltig containing magnesia, magnesian, magnesial
Magnesiamilch *f* milk of magnesia, magnesia magma
Magnesiamixtur *f* magnesia mixture *(aqueous solution of NH_4Cl, NH_4OH, and $MgCl$)*
Magnesiasalpeter *m (min)* nitromagnesite *(magnesium nitrate)*
Magnesiazement *m s.* Magnesiabinder
Magnesiochromit *m (min)* magnesiochromite *(magnesium chromate(III))*
Magnesiospinell *m (min)* spinel *(proper) (magnesium aluminate)*
Magnesit *m (min)* magnesite, bitter spar *(magnesium carbonate)*

Magnesitbinder *m s.* Magnesiabinder
Magnesitstein *m* **[/ feuerfester]** magnesite brick
Magnesium *n* Mg magnesium
Magnesiumazetat *n* $Mg(CH_3COO)_2$ magnesium acetate
Magnesiumband *n* magnesium ribbon
Magnesiumbasis / auf magnesium-base
Magnesiumbisulfit *n (incorrectly for)* Magnesiumhydrogensulfit
Magnesiumbisulfitkochsäure *f (pap)* magnesium-base acid (liquor, sulphite liquor)
Magnesiumbranntkalk *m* dolomitic (dolomite) lime *(calcium magnesium oxide)*
Magnesiumbromat *n* $Mg(BrO_3)_2$ magnesium bromate
Magnesiumbromid *n* $MgBr_2$ magnesium bromide
Magnesiumchlorat *n* $Mg(ClO_3)_2$ magnesium chlorate
Magnesiumchlorid *n* $MgCl_2$ magnesium chloride
Magnesiumchromat *n* $MgCrO_4$ magnesium chromate
Magnesiumdiäthyl *n s.* Diäthylmagnesium
Magnesiumdihydrogenorthophosphat *n* $Mg(H_2PO_4)_2$ magnesium dihydrogenorthophosphate, magnesium dihydrogenphosphate
Magnesiumdiphosphat *n* $Mg_2P_2O_7$ magnesium diphosphate, magnesium pyrophosphate
Magnesiumdrehspäne *mpl* magnesium turnings
Magnesiumfluorid *n* MgF_2 magnesium fluoride
Magnesiumgrundlage / auf magnesium-base
Magnesiumhexachloropalladat(IV) *n* $Mg[PdCl_6]$ magnesium hexachloropalladate(IV)
Magnesiumhexachloroplatinat(IV) *n* $Mg[PtCl_6]$ magnesium hexachloroplatinate(IV)
Magnesiumhexachlorostannat(IV) *n* $Mg[SnCl_6]$ magnesium hexachlorostannate(IV)
Magnesiumhexafluorosilikat *n* $Mg[SiF_6]$ magnesium hexafluorosilicate
Magnesiumhexazyanoferrat(II) *n* $Mg_2[Fe(CN)_6]$ magnesium hexacyanoferrate(II)
Magnesiumhydrid *n* MgH_2 magnesium hydride
Magnesiumhydrogenkarbonat *n* $Mg(HCO_3)_2$ magnesium hydrogencarbonate
Magnesiumhydrogenorthoarsenat(V) *n* $MgHAsO_4$ magnesium hydrogenorthoarsenate, magnesium hydrogenarsenate
Magnesiumhydrogenorthophosphat *n* $MgHPO_4$ magnesium hydrogenorthophosphate, magnesium hydrogenphosphate
Magnesiumhydrogensulfit *n* $Mg(HSO_3)_2$ magnesium hydrogensulphite
Magnesiumhydroxid *n* $Mg(OH)_2$ magnesium hydroxide
Magnesiumhypophosphit *n* $Mg(PH_2O_2)_2$ magnesium hypophosphite, *(better)* magnesium phosphinate
Magnesiumjodat *n* $Mg(IO_3)_2$ magnesium iodate
Magnesiumjodid *n* MgI_2 magnesium iodide
Magnesiumkarbonat *n* $MgCO_3$ magnesium carbonate
Magnesiumlöschkalk *m (agric)* hydrated magnesium lime
Magnesiummanganat(VII) *n s.* Magnesiumpermanganat
Magnesiummangel *m (agric)* magnesium shortage
Magnesiumnitrat *n* $Mg(NO_3)_2$ magnesium nitrate
Magnesiumnitrid *n* Mg_3N_2 magnesium nitride
Magnesiumorthoarsenat(III) *n* $Mg_3(AsO_3)_2$ magnesium orthoarsenite
Magnesiumoxid *n* MgO magnesium oxide
Magnesiumperchlorat *n* $Mg(ClO_4)_2$ magnesium perchlorate
Magnesiumpermanganat *n* $Mg(MnO_4)_2$ magnesium permanganate
Magnesiumperoxid *n* MgO_2 magnesium peroxide
Magnesiumphosphit *n* $MgPHO_3$ magnesium phosphite, *(better)* magnesium phosphonate
Magnesiumphthalozyanin *n* magnesium phthalocyanine
Magnesiumpyrophosphat *n s.* Magnesiumdiphosphat
Magnesiumsulfat *n* $MgSO_4$ magnesium sulphate
Magnesiumsulfid *n* MgS magnesium sulphide
Magnesiumsulfit *n* $MgSO_3$ magnesium sulphite
Magnesiumthiosulfat *n* MgS_2O_3 magnesium thiosulphate
Magnetband *n* magnetic tape
Magnetbandrolle *f* magnetic pulley *(in belt conveyors)*
Magneteisenerz *n*, **Magneteisenstein** *m s.* Magnetit
Magnetfeld *n* magnetic field
Magnetfilter *n* magnetic filter
magnetisch magnetic
magnetisierbar magnetizable
magnetisieren to magnetize
Magnetisierung *f* magnetization
Magnetismus *m* magnetism
Magnetit *m (min)* magnetite, magnetic iron [ore] *(iron(II,III) oxide)*
Magnetkies *m (min)* magnetic pyrites, pyrrhotite, pyrrhotine *(iron(II) sulphide)*
Magnetochemie *f* magnetochemistry
Magneton *n* magneton *(a unit of the magnetic moment)*
Bohrsches M. Bohr magneton
Magnetorotation *f* magnetic rotation, Faraday effect
Magnetrolle *f s.* Magnetbandrolle
Magnetrührer *m* magnetic stirrer
Magnetscheiden *n* magnetic separation
Magnetscheider *m* magnetic separator
Magnetsortieren *n* magnetic separation
Magnettrommel *f* magnetic drum
Magnettrommelscheider *m* magnetic drum [separator], induced-roll [magnetic] separator, rotor separator
Magnetvibrator *m* electromagnetic vibrator
Magnolit *m (min)* magnolite *(mercury tellurate)*
Mahlbahn *f* grinding surface

mahlbar grindable
Mahlbarkeit *f* grindability
Mahlbarkeitsindex *m* **nach Hardgrove** Hardgrove grindability index
Mahlbarkeitsprüfung *f*, **Mahlbarkeitstest** *m* grindability test
Mahlbarkeitszahl *f* grindability index (value)
Mahldruck *m (pap)* beating pressure *(in a Hollander beater)*; *(pap)* plug pressure *(in perfecting engines)*
Mahleffekt *m (pap)* effect of beating
mahlen to grind, to mill; *(pap)* to beat *(in a Hollander beater)*; *(pap)* to refine, to clear, to brush out *(in a refiner)*
auf Staubfeinheit m. to pulverize
in der Kugelmühle m. to ball-mill
wieder m. *(plast)* to regrind
Mahlfeinheit *f*, **Mahlfeinheitsgrad** *m* fineness of grind[ing]
Mahlfläche *f s.* Mahlbahn
Mahlgang *m s.* Mahlscheibenmühle
Mahlgeschirr *n (pap)* beating engine, beater
Mahlgrad *m* 1. *(pap)* freeness [value], degree of beating; 2. *s.* Mahlfeinheit
Mahlgradbestimmung *f (pap)* freeness test
Mahlgradprüfer *m (pap)* freeness (beaten stuff) tester
M. nach Schopper-Riegler Schopper-Riegler apparatus
Mahlgradprüfung *f (pap)* freeness test
Mahlgut *n* 1. material being (*or* to be) ground; 2. *(pap)* beating material, beater charge, [fibrous, pulp] furnish
Mahlhilfe *f* grinding aid
Mahlhilfsmittel *n (pap)* beater additive
Mahlhilfsstoff *m* grinding aid
Mahlholländer *m (pap)* Hollander [beater, beating engine], pulp engine (grinder), stuff engine
Mahlkammer *f* grinding (pulverizing) chamber
Mahlkörper *m* grinding medium
Mahlkugel *f* grinding ball
Mahlmaschine *f* grinding machine; *(pap)* beating engine, beater
Mahlorgan *n* grinding element
Mahlraum *m s.* Mahlkammer
Mahlring *m* grinding (pulverizing, bull) ring *(in a roller mill)*
Mahlscheibe *f* grinding disk *(of a disk mill)*
Mahlscheibenmühle *f* disk (attrition) mill, disk grinder
Mahlschüssel *f* grinding pan, bowl
Mahlstein *m* grindstone, millstone
Mahlstoff *m s.* Mahlgut 2.
Mahlteller *m (ceram)* grinding pan
Mahltrocknung *f* mill drying
Mahlung *f* grinding, milling; *(pap)* beating *(in a Hollander beater)*; *(pap)* refining, clearing, brushing-out *(in a refiner)*
M. auf Staubfeinheit pulverizing
M. im geschlossenen Kreislauf closed-circuit grinding
M. in der Kugelmühle mill mixing
autogene M. autogenous grinding
feine M. fine grinding
grobe M. coarse grinding
rösche M. *(pap)* free beating
schmierige M. *(pap)* wet (slow) beating
Mahlwalze *f* grinding roll, *(in edge-runner mills)* muller; *(pap)* beater (beating, Hollander, knife) roll
Mahlwiderstand *m* grinding resistance
Mahlwirkung *f (pap)* beating action
Mahlzeug *n (pap)* beating material *(active portion of the beating apparatus)*
Maillard-Reaktion *f* Maillard reaction *(for detecting reducing sugars)*
Maischapparat *m* masher
Maischbottich *m* mash[ing] tub
Maische *f (ferm)* mash *(in producing beer)*; must *(in producing wine)* ✦ **M. abziehen** to remove the mash
süße M. sweet mash
Maischebottich *m s.* Maischbottich
Maischefilter *n* mash filter
Maischefilterplatte *f* mash filter plate
Maischekessel *m s.* Maischbottich
Maischekochkessel *m s.* Maischepfanne
maischen to mash, to dough [in]
Maischepfanne *f* mash tun (copper, kettle)
Maischverfahren *n* mashing process
Maiskeimöl *n* maize [germ] oil, *(Am)* corn oil
Maismehl *n* maize meal, *(Am)* corn meal
Maisöl *n s.* Maiskeimöl
Maisprotein *n* maize protein
Maisstärke *f* maize starch, *(Am)* corn starch
Maiszucker *m* maize sugar, *(Am)* corn sugar
Majolika *f (ceram)* majolica, maiolica
Majoranöl *n* marjoram oil *(from Majorana hortensis Moench)*
Majoritäts[ladungs]träger *m* majority carrier
Makajabutter *f* macaja (micauba) oil *(a palm kernel oil from Acrocomia sclerocarpa Mart.)*
Make-up-Creme *f* cream make-up
Makroanalyse *f* macroanalysis
Makroansatz *m (lab)* macro batch
makroätzen to macroetch
Makroätzung *f* macroetching
Makroaufnahme *f* photomacrograph
Makrobestandteil *m s.* Makrokomponente
Makrochemie *f* macrochemistry
Makrofibrille *f* macrofibrille
Makrofoto *n* photomacrograph
Makrofotografie *f* 1. photomacrography; 2. photomacrograph
Makrogefüge *n* macrostructure
Makrokomponente *f* macrocomponent, macroconstituent
makrokristallin macrocrystalline
Makrolid *n (org chem)* macrolide
Makromethode *f* macromethod
Makromolekül *n* macromolecule, giant (large) molecule
makromolekular macromolecular
Makronährstoff *m (agric)* macroelement, macronutrient, major element

Makroradikal *n* macroradical
Makroring *m* macroring, large ring *(consisting of 13 or more members)*
makroskopisch macroscopic
Makrostruktur *f* macrostructure
Makrotetrolid *n (org chem)* macrotetrolide
makrozyklisch macrocyclic
Makulatur *f* waste (old) paper
Malabarkino *n* Malabar (East India) kino *(kino gum from Pterocarpus marsupium Roxb.)*
Malabartalg *m* piney tallow, Dhupa fat *(seed fat from Vateria indica L.)*
Malachit *m (min)* malachite *(copper(II) carbonate dihydroxide)*
Malachitgrün *n* malachite (Victoria, benzal, benzaldehyde) green, basic green 4 *(a triphenylmethane dye)*
Malagetta *m* / **Abessinischer (Madagassischer)** Madagascar cardamom *(from Aframomum angustifolium Schum.)*
Malagettapfeffer *m* grains of paradise *(from Aframomum melegueta (Rosc.) Schum.)*
Malakon *m (min)* malacon *(a nesosilicate containing zirconium)*
Malaria[bekämpfungs]mittel *n* antimalarial [drug]
Malariawirksamkeit *f (pharm)* antimalarial activity
Malat *n* malate *(salt or ester of malic acid)*
Malatdehydrogenase *f* malic acid dehydrogenase
Malatenzym *n* malic enzyme
Maldonit *m (min)* maldonite, bismuth gold *(a natural alloy consisting of gold and bismuth)*
Maleat *n*, **Maleinat** *n* maleate, maleinate *(salt or ester of maleic acid)*
Maleinsäure *f* HOOC−CH=CH−COOH maleic acid, *cis*-butenedioic acid, *cis*-ethylene-1,2-dicarboxylic acid
Maleinsäureanhydrid *n* maleic anhydride, *cis*-butenedioic anhydride, 2,5-furandione
Maleinsäurehydrazid *n (incorrectly for)* Maleinylhydrazin
Maleinylhydrazin *n* maleic hydrazide
Malergold *n* mosaic gold, ormolu *(tin(IV) sulphide)*
Malerpappe *f s.* Malkarton
Malett[o]rinde *f (tann)* mallet bark *(from Eucalyptus specc., esp Eucalyptus occidentalis Endl.)*
Malinsäure *f s.* Äpfelsäure
Malkarton *m* painters' (artists') cardboard, painting board
Mallardit *m (min)* mallardite *(manganese(II) sulphate-7-water)*
Malonat *n* malonate *(salt or ester of malonic acid)*
Malonestersynthese *f* malonic ester synthesis
Malonsäure *f* $CH_2(COOH)_2$ malonic acid, methanedicarboxylic acid
Malonsäurediäthylester *m* $CH_2(COOC_2H_5)_2$ diethyl malonate, malonic ester
Malonsäuredinitril *n* $CH_2(CN)_2$ malonitrile
Malonsäureester *m s.* Malonsäurediäthylester
N,N′-**Malonylharnstoff** *m N,N′*-malonylurea, barbituric acid
Malpappe *f s.* Malkarton
Maltase *f* maltase
Maltodextrin *n* maltodextrin
Maltol *n* maltol, larixinic acid
Maltonsäure *f s. D*-Glukonsäure
Maltose *f* maltose, malt sugar
Maltosedextrin *n* maltodextrin
Maltosesirup *m* malt extract
Malz *n* malt
  dunkles M. dark malt
  geröstetes M. roasted malt
  geschrotetes M. ground (crushed) malt, malt meal, grist
  helles M. light (white, ordinary) malt
  Münchner M. Munich malt
  Pilsner M. Pilsen malt
Malzamylase *f* malt amylase
Malzanalyse *f* malt analysis
malzartig malty
Malzbereitung *f* malting
Malzbier *n* malt beer
Malzdarre *f* malt [drying] kiln
malzen, mälzen to malt
Mälzer *m* maltster
Mälzerei *f* 1. malting; 2. *s.* Malzfabrik
  mechanische (pneumatische) M. pneumatic malting
Mälzereiabwasser *n* malt-house waste
Malzessig *m* malt vinegar
Malzextrakt *m* malt extract
Malzfabrik *f* malt factory, malt-house, malting plant
Malzgerste *f* malting barley
Malzkaffee *m* malt coffee
Malzkeim *m* malt rootlet
Malzmeister *m* maltster
Malzmilch *f* malt slurry
Malzquetsche *f*, **Malzquetscher** *m* malt mill (crusher)
Malzrumpf *m* malt hopper
Malzschrot *n(m)* ground (crushed) malt, malt meal, grist
Malzstärke *f* malt starch
Malztenne *f* malting (malt, malt-house) floor
Malztreber *pl* malt spent grains
Malztrichter *m* malt hopper
Mälzung *f* malting
Mälzungsschwund *m* malting loss
Malzwender *m*, **Malzwendevorrichtung** *f* malt turning device, malt plough, maltomobile
Malzwürze *f* malt wort
Malzzerkleinerungsapparat *m s.* Malzquetsche
Malzzucker *m* malt sugar, maltose
Mammutbaum *m* [California] redwood, Sequoia sempervirens (D. Don) Endl.
Mammutpumpe *f* air lift, air-lift pump, mammoth pump
Mammutrührwerk *n* air-lift mixer
Manchesterbraun *n* Manchester brown
Manchestergelb *n* Manchester (Martius) yellow

Manchester-Ofen *m (ceram)* Manchester kiln
Mandel *f* 1. *(food)* almond *(from Prunus amygdalus Batsch)*; 2. *(geol)* amygdale, amygdule, geode
Mandelöl *n* almond oil
Mandelsäure *f* $C_6H_5CH(OH)COOH$ mandelic acid, 2-hydroxy-2-phenylacetic acid
para-**Mandelsäure** *f DL*-mandelic acid
Mandelsäurebenzylester *m* $C_6H_5CH(OH)CO{-}OCH_2C_6H_5$ benzyl mandelate
Mandelsäuretropylester *m* mandelyltropine, homatropine, phenylglycollyltropine
Mandelstein *m (geol)* amygdaloid
Mangabeirakautschuk *m* Mangabeira rubber *(from Hancornia speciosa Gomez)*
Mangan *n* Mn manganese
Manganat *n* manganate
Manganat(IV) *n* manganate(IV), manganite
Manganat(VII) *n* permanganate
Mangan(II)-azetat *n* $Mn(CH_3COO)_2$ manganese(II) acetate
Manganbister *m(n)* manganese brown
Manganblende *f (min)* manganblende, alabandite *(manganese(II) sulphide)*
Manganborat *n* manganese borate
Manganbraun *n* manganese brown *(manganese(III) hydroxide)*
Mangan(II)-bromid *n* $MnBr_2$ manganese(II) bromide, manganese dibromide
Manganbronze *f* manganese bronze
Mangan(II)-chlorid *n* $MnCl_2$ manganese(II) chloride, manganese dichloride
Mangan(IV)-chlorid *n* $MnCl_4$ manganese(IV) chloride, manganese tetrachloride
Mangandichlorid *n s.* Mangan(II)-chlorid
Mangandifluorid *n s.* Mangan(II)-fluorid
Mangan(II)-dihydrogenorthophosphat *n* $Mn(H_2PO_4)_2$ manganese(II) dihydrogenorthophosphate, manganese(II) dihydrogenphosphate
Mangandijodid *n s.* Mangan(II)-jodid
Mangandioxid *n s.* Mangan(IV)-oxid
Mangan(II)-diphosphat *n* $Mn_2P_2O_7$ manganese(II) diphosphate, manganese(II) pyrophosphate
Mangandisilizid *n* $MnSi_2$ manganese disilicide
Mangandisulfid *n s.* Mangan(IV)-sulfid
Mangan(II)-dithionat *n* MnS_2O_6 manganese(II) dithionate
Mangan-Epidot *m (min)* manganepidote
Manganerz *n* manganese ore
Mangan(II)-fluorid *n* MnF_2 manganese(II) fluoride, manganese difluoride
Mangan(III)-fluorid *n* MnF_3 manganese(III) fluoride, manganese trifluoride
Mangangrün *n* manganese (Rosenstiehl's, Cassel) green *(barium manganate)*
Manganhartstahl *m* manganese steel
Manganheptoxid *n s.* Mangan(VII)-oxid
Mangan(II)-hexafluorosilikat *n* $Mn[SiF_6]$ manganese(II) hexafluorosilicate, manganese(II) fluorosilicate
Mangan(II)-hexazyanoferrat(II) *n* $Mn_2[Fe(CN)_6]$ manganese(II) hexacyanoferrate(II)
Mangan(II)-hydrogenorthophosphat *n* $MnHPO_4$ manganese(II) hydrogenorthophosphate, manganese(II) hydrogenphosphate
Mangan(II)-hydroxid *n* $Mn(OH)_2$ manganese(II) hydroxide
Mangan(II)-hypophosphit *n* $Mn(PH_2O_2)_2$ manganese(II) hypophosphite, *(better)* manganese(II) phosphinate
Manganit *m (min)* manganite *(manganese(III) hydroxide oxide)*
Manganit *n s.* Manganat(IV)
Mangan(II)-jodid *n* MnI_2 manganese(II) iodide
Mangankarbid *n* manganese carbide
Mangan(II)-karbonat *n* $MnCO_3$ manganese(II) carbonate
Manganknolle *f (geol)* manganese nodule
Mangan(II)-metasilikat *n* $MnSiO_3$ manganese(II) metasilicate, manganese trioxosilicate
Manganmonosulfid *n s.* Mangan(II)-sulfid
Manganmonoxid *n s.* Mangan(II)-oxid
Mangan(II)-nitrat *n* $Mn(NO_3)_2$ manganese(II) nitrate
Manganometrie *f* permanganometry
Mangan(II)-orthophosphat *n* $Mn_3(PO_4)_2$ manganese(II) orthophosphate, manganese(II) phosphate
Mangan(II)-orthosilikat *n* Mn_2SiO_4 manganese(II) orthosilicate, dimanganese tetraoxosilicate
Manganosit *m (min)* manganosite *(manganese(II) oxide)*
Mangan(II)-oxid *n* MnO manganese(II) oxide, manganese monooxide
Mangan(III)-oxid *n* Mn_2O_3 manganese(III) oxide, dimanganese trioxide
Mangan(IV)-oxid *n* MnO_2 manganese(IV) oxide, manganese dioxide
Mangan(VI)-oxid *n* MnO_3 manganese(VI) oxide, manganese trioxide
Mangan(VII)-oxid *n* Mn_2O_7 manganese(VII) oxide, dimanganese heptaoxide
Mangan(II,IV)-oxid *n* Mn_3O_4 manganese(II,IV) oxide, trimanganese tetraoxide, red manganese oxide
Manganoxid *n* / **rotes** *s.* Mangan(II,IV)-oxid
Mangan(III)-oxidhydrat *n* $Mn_2O_3 \cdot nH_2O$ hydrated manganese(III) oxide
Manganphosphid *n* manganese phosphide
Mangan(II)-phosphit *n* $MnPHO_3$ manganese(II) phosphite, *(better)* manganese(II) phosphonate
Mangan(II)-pyrophosphat *n s.* Mangan(II)-diphosphat
Manganresinat *n* manganese resinate
Mangan(II)-rhodanid *n s.* Mangan(II)-thiozyanat
Mangansäure *f* H_2MnO_4 manganic acid
Mangan(VII)-säure *f* $HMnO_4$ permanganic acid
Manganschwarz *n* manganese black *(manganese(IV) oxide)*
Manganseife *f* manganese soap
Mangan(II)-selenat *n* $MnSeO_4$ manganese(II) selenate
Mangan(II)-selenid *n* MnSe manganese(II) selenide

Mangan(II)-silizid *n* Mn_2Si manganese(II) silicide, dimanganese silicide
Mangansiliziumstahl *m* silicon-manganese steel, silicomanganese steel
Manganspat *m (min)* dialogite, rhodochrosite *(manganese(II) carbonate)*
Manganstahl *m* manganese steel
Mangan(II)-sulfat *n* $MnSO_4$ manganese(II) sulphate
Mangan(III)-sulfat *n* $Mn_2(SO_4)_3$ manganese(III) sulphate
Mangan(II)-sulfid *n* MnS manganese(II) sulphide, manganese monosulphide
Mangan(IV)-sulfid *n* MnS_2 manganese(IV) sulphide, manganese disulphide
Mangantetrachlorid *n s.* Mangan(IV)-chlorid
Mangan(II)-tetroxosilikat *n s.* Mangan(II)-orthosilikat
Mangan(II)-thiozyanat *n* $Mn(SCN)_2$ manganese(II) thiocyanate, manganese(II) rhodanide
Mangantrifluorid *n s.* Mangan(III)-fluorid
Mangantrioxid *n s.* Mangan(VI)-oxid
Mangan(II)-trioxosilikat *n s.* Mangan(II)-metasilikat
Manganvitriol *m (min)* mallardite *(manganese(II) sulphate-7-water)*
Mangan(II)-zyanwasserstoffsäure *f s.* Hexazyanomangan(II)-säure
Mangel *m* 1. deficiency, *(agric also)* starvation; 2. *(tech)* fault, defect
Mangelelektron *n (phys chem)* electron hole
Mangelerscheinung *f (agric)* deficiency symptom
Mangelleiter *m (phys chem)* p-type conductor, hole conductor
Mangelleitung *f (phys chem)* p-type conduction, hole conduction
Mangelsymptom *n (agric)* deficiency symptom
Mangeltrockner *m (ceram)* mangle [dryer]
Mangroverinde *f (tann)* mangrove bark
Mangroverindenextrakt *m(n) (tann)* mangrove cutch, kutch
Manicoba-Kautschuk *m* manicoba (Ceará) rubber *(from Manihot specc.)*
Manilahanf *m* Manila (Manilla) hemp (fibre), abaca [fibre] *(leaf fibres from Musa textilis Née)*
Manilakopal *m* Manila copal *(from Agathis specc.)*
halbfossiler M. pontianac, pontianak [gum], gum pontianak *(a copal from Agathis alba (Lam.) Foxw.)*
Manilakraftpapier *n*, **Manila[pack]papier** *n* Manila (Manilla) paper
Maniok *m* cassava, Manihot esculenta Crantz
Süßer M. aipi, Manihot dulcis (J.F. Gmel.) Pax
Mankettinußöl *n* Manketti nut oil *(from Ricinodendron rautaneni Schinz)*
Manna *n(f)* manna *(from Fraxinus ornus L.)*
Mannan *n* mannan *(a polysaccharide)*
Mannazucker *m s.* Mannit
Mannich-Base *f* Mannich base
Mannich-Reaktion *f* Mannich reaction *(aminomethylation and variations of it)*
Mannit *m* $HOCH_2[CHOH]_4CH_2OH$ mannitol *(a sugar alcohol)*
Mannloch *n* manhole, manway
Mannlochdeckel *m* manhole cover
Mannogalaktan *n* mannogalactan, galactomannan
Mannose *f* mannose *(a monosaccharide)*
Manometer *n* manometer, pressure gauge
manometrisch manometric
Manool *n* manool *(a bicyclic diterpene)*
Manschette *f (tech)* collar, sleeve
drehbare M. movable collar *(of a Bunsen burner)*
Manschettendichtung *f* oil seal ring
Mantel *m (tech)* jacket, shell, mantle, casing; cover, sheath *(a protective layer)*; *(rubber)* cover *(of a tyre)*; *(text)* skin, sheath *(of core-spun yarn)*
Mantelbehälter *m*, **Mantelgefäß** *n* jacketed vessel (kettle)
Mantelkessel *m* shell-type boiler
Mantelkühler *m* jacket cooler
Mantelmesser *npl (pap)* shell bars, bars in the shell, bars on the casing *(of a perfecting engine)*
Mantelmischung *f (rubber)* sheath[ing] compound *(as for cables)*
Mantelraummedium *n* shell-side medium
Mantelrohr *n* jacket[ed] pipe
Marakaibobalsam *m* Maracaibo resin *(from Copaifera specc.)*
Marantastärke *f* arrowroot *(from Maranta arundinacea L. and related specc.)*
Marbel *f (glass)* marver [plate]
marbeln *(glass)* to marver *(a gather on a flat plate)*
Marbelplatte *f (glass)* marver [plate]
Marcy-Mühle *f* Marcy [ball] mill
Margarine *f* margarine
aus Pflanzenfetten hergestellte M. vegetable margarine
aus Tierfetten hergestellte M. animal fat margarine
mit Molke hergestellte M. whey margarine
zum Tränen neigende M. weeping (leaking) margarine
Margarineanlage *f* margarine[-making] plant
Margarinearoma *n* margarine flavour
Margarineemulsion *f* margarine emulsion
Margarinefabrik *f* margarine factory (works)
Margarinefarbe *f* margarine colouring *(substance)*
Margarinefärbung *f* margarine colouring
Margarinefett *n* margarine fat
Margarineindustrie *f* margarine industry
Margarinekonservierungsmittel *n* margarine preservative
Margarinekonsistenz *f* consistency of margarine
Margarinerezeptur *f* margarine blend, blend for margarine
Margarineschmalz *n* margarine fat
Margarinestrang *m* strand (bar) of margarine
Margarinsäure *f* $CH_3[CH_2]_{15}COOH$ margaric acid, heptadecanoic acid
Margarit *m (min)* margarite, pearl-mica *(a phyllosilicate)*

Marialith *m (min)* marialite *(a tectosilicate)*
Marihuana *n* marihuana, marijuana, hasheesh, hashish *(from Cannabis indica Lam.)*
Marinade *f (food)* pickle, souse
Marineblau *n* navy blue
Marinekohle *f* navigation coal
Marineleim *m* marine glue
Marineöl *n* marine [animal] oil
marinieren *(food)* to pickle, to souse
Mark *n (food)* pulp; *(bot)* pith
Markasit *m (min)* marcasite *(iron disulphide)*
Marke *f* mark *(as of calibration)*; brand
Markenbezeichnung *f*, **Markenname** *m* brand name
Markenspitze *f* levelling wire *(of a Redwood viscometer)*
markieren to mark; *(nucl)* to label, to tag
mit Brandzeichen m. to brand *(leather)*
Markierfilz *m (pap)* marking (ribbed, ribbing) felt
markiert / radioaktiv radioactively labelled, radiolabelled
Markierung *f* 1. marking; *(nucl)* labelling, tagging; 2. mark
M. der Siebnaht seam mark *(a defect in paper)*
M. durch die Heizplatte *(plast)* platen mark *(a defect)*
M. durch überfließendes Material skid *(a defect in injection-moulded plastics)*
spritzerförmige M. splash *(a defect in injection-moulded plastics)*
Markierungselement *n (nucl)* tracer element
Markownikow-Addition *f* Markovnikov addition
Markownikow-Regel *f* Markovnikov's rule
Markpapier *n* / **Chinesisches** rice paper *(from the pith of Tetrapanax papyriferum (Hook.) K. Koch)*
Marmor *m* marble
Marmorpapier *n* marble[d] paper
Marmorpappe *f* marble[d] board
Marsgestein *n* Martian rock
Marshit *m (min)* marshite *(copper(I) iodide)*
Martensit *m (met)* martensite *(the hard constituent of which quenched steel is chiefly composed)*
martensitisch *(met)* martensitic
Martit *m (min)* martite *(a pseudomorph of haematite after magnetite)*
Martiusgelb *n* Martius (Manchester) yellow, acid yellow 24
Marzetti-Plastometer *n (rubber)* Marzetti plastometer
Mascagnit *m (min)* mascagnite *(ammonium sulphate)*
Mascara *m* mascara *(a cosmetic for colouring the eyebrows)*
Masche *f* mesh *(screening)*
Maschendrahtfüllkörper *mpl* mesh packings
Maschenweite *f* mesh size, screen aperture, clear opening
Maschenzahl *f* mesh *(number of openings per linear inch)*
Maschinenausfallzeit *f* downtime, down period
Maschinenbreite *f (pap)* width of the machine
Maschinenbütte *f (pap)* service (machine, pulp, supply, stuff) chest
Maschinenbüttenpapier *n* machine-made (cylinder-made) deckle-edge paper, mouldmade paper
Maschinenformen *n* machine moulding
Maschinengeschwindigkeit *f (pap)* machine speed
maschinengestrichen *(pap)* machine-coated
maschinengetrocknet *(pap)* machine-dried, cylinder-dried, steam-dried
maschinenglatt *(pap)* machine-finished, MF
Maschinenglätte *f (pap)* machine finish, MF
Maschinenglättwerk *n* calender [machine]
Maschinengraupappe *f* chip board
Maschinenkalander *m s.* Maschinenglättwerk
Maschinenöl *n* machine (machinery) oil
Maschinenpapier *n* machine[-made] paper
Maschinenpappe *f* mill board
Maschinenrichtung *f (pap)* machine direction (way), making (grain, long) direction
Maschinenrolle *f (pap)* machine (mill, jumbo) roll
Maschinenschmieröl *n s.* Maschinenöl
Maschinensieb *n (pap)* Fourdrinier wire
Maschinenstrich *m* [paper] machine coating, on-machine coating
Maschinentorf *m* machine[-cut] peat
maschinentrocken *s.* maschinengetrocknet
Maserpappe *f* grained board
Maskenform *f* shell mould *(foundry)*
Maskenformen *n* shell moulding *(foundry)*
Maskenformverfahren *n* shell-moulding process, Croning process, C process *(foundry)*
maskieren *(chem, tann)* to mask, to sequester *(ions)*; *(ceram)* to mask
Maskierung *f (chem, tann)* masking, sequestration *(of ions)*; *(ceram)* masking
M. mit Formiaten *(tann)* formate masking *(of chrome liquors)*
Maskierungsmittel *n*, **Maskierungsreagens** *n* masking (sequestering) agent, sequestrant
Maskierungsvermögen *n (chem, tann)* masking (sequestering) power; *(ceram)* masking power *(of a glaze)*
Masonite-Verfahren *n (pap)* Masonite (explosion) process *(chemigroundwood process)*
Maß *n* measure ✦ **nach M. aufbauen** to tailor[-make], to make to measure *(e. g. polymers)*
Massagecreme *f* massage cream
Maßanalyse *f* mensuration (volumetric, titrimetric) analysis
elektrometrische M. electrometric titration
konduktometrische M. conductometric titration
potentiometrische M. potentiometric titration
maßanalytisch volumetric, titrimetric
Maßbeständigkeit *f* dimensional stability
Masse *f* 1. mass, *(if loose also)* bulk; *(ceram)* body, paste; 2. *(quantitatively)* mass, *(chem, tech esp in word compounds often loosely)* weight
M. des feuchten Stoffs wet weight
M. je Bogen *(pap)* weight of a sheet of paper

M. je Flächeneinheit *(pap)* substance, substance weight (number), basis (basic) weight
M. je Ries *(pap)* weight per ream
M. mit hohem Aluminiumoxidgehalt *(ceram)* high-alumina body
M. mit hohem Berylliumoxidgehalt *(ceram)* high-beryllia body
M. mit hohem Magnesiumoxidgehalt *(ceram)* high-magnesia body
M. mit hohem Titanoxidgehalt *(ceram)* high-titania body
M. mit hohem Tonerdegehalt *(ceram)* high-alumina body
M. mit hohem Zirkoniumoxidgehalt *(ceram)* high-zirconia body
M. mit niedrigem Verlustfaktor *(ceram)* low-loss body
aktive M. *(phys chem)* active mass
atomare M. atomic mass
bildsame M. *(ceram)* plastic body
breiige M. pulp
gebrannte M. *(ceram)* fired body
halbplastische M. *(ceram)* stiff-plastic body
keramische M. ceramic body (paste, mix)
molare M. molar mass
nicht schwindende M. *(ceram)* non-shrinking body
plastische M. plastic material; *(ceram)* plastic body
reduzierte M. *(phys chem)* reduced mass
tonfreie M. *(ceram)* non-clay body
ungebrannte M. *(ceram)* raw body
weichplastische M. *(ceram)* soft plastic body
Masse ... *s. a.* Massen ...
Masseabfall *m (ceram)* body scrap
Masseäquivalent *n* mass equivalent
Masseaufbereitung *f (ceram)* body preparation
Massedefekt *m (nucl)* mass defect
Massedosiervorrichtung *f (plast)* weight feeder (feeding device)
Masseeinheit *f* mass unit
atomare M. atomic mass unit, amu
Masse-Energie-Äquivalenzprinzip *n* mass-energy equivalence principle
Masse-Energie-Beziehung *f*, **Masse-Energie-Gleichung** *f* mass-energy relation
Massefärbung *f (pap)* beater dyeing (colouring), dyeing (colouring) in the pulp
massegeleimt *(pap)* beater-sized, pulp-sized, engine-sized, E.S., sized in the engine (stuff)
Maßeinteilung *f* scale *(of a measuring instrument)*
Massekammer *f (plast)* plenum chamber
Massekeller *m (ceram)* maturing cellar
Massekonstanz *f* constant weight ✦ **bis zur M. glühen** to ignite to constant weight
Massekuchen *m (filtr)* pulp disk
Massel *f (met)* pig
Masselbeet *n*, **Masselbett** *n (met)* pig bed
Masseleimung *f (pap)* beater (pulp, engine) sizing, E.S., sizing in the engine (stuff)
Masseleisen *n* pig iron, ferrocarbon
Masselgießmaschine *f (met)* pig-casting machine
Massen ... *s. a.* Masse ...
Massenabsorptionskoeffizient *m* mass-absorption coefficient
Massenanalyse *f* routine analysis
Massenanziehung *f* gravitation, gravity
Massenäquivalent *n* mass equivalent
Massenbeton *m* mass concrete
Massendefekt *m* mass defect
Massendichte *f* mass density
Masseneffekt *m* mass effect
Massenerhaltung *f* conservation of mass (matter)
Massenerhaltungssatz *m* law of conservation of mass (matter)
Massenkonzentration *f* mass per unit volume
Massenpeak (Massenpik) *m* **im Massenspektrum** parent [mass] peak
Massenschwund *m s.* Massendefekt
Massenskala *f* mass scale
Massenspektrograf *m* mass spectrograph
Astonscher M. Aston mass spectrograph
doppel[t]fokussierender M. double-focus[s]ing mass spectrograph
Massenspektrogramm *n* mass spectrogram
Massenspektrometer *n* mass spectrometer
Massenspektrometrie *f* mass spectrometry
Massenspektroskopie *f* mass spectroscopy
Massenspektrum *n* mass spectrum
Massensuszeptibilität *f* mass (specific) susceptibility, susceptibility per gram *(in magnetization)*
Massenverhältnis *n* mass ratio
M. flüssig zu fest mass ratio of liquid to solid
Massenwert *m* mass value, physical (isotopic) atomic weight
Massenwirkungsgesetz *n* law (principle) of mass action, mass-action expression (law), *(sometimes also)* reaction isotherm
Massenwirkungskonstante *f* equilibrium constant
Massenzahl *f* [atomic] mass number, nuclear number
Masseplatte *f (filtr)* pulp disk
Massepolymerisation *f* mass (bulk) polymerization
Masseprozent *n* percentage (per cent) by weight
Massequirl *m (ceram)* blunger
Massescherben *mpl (ceram)* pitchers
Masseschlagmaschine *f (ceram)* kneading machine (table), kneader
Masseschlicker *m (ceram)* body slip
Massestrang *m (ceram)* clay column
Massestück *n* balance weight
Masseteil *m* part by weight
Masseversatz *m (ceram)* batch
Massey-Papier *n* Massey [process-coated] paper
Massezusammensetzung *f (ceram)* body composition
Massezylinder *m (plast)* injection (shooting, plasticating) cylinder
Massicot *m* massicot, lead ochre *(a yellow powder consisting of lead(II) oxide)*

massiv massive, solid
Massivguß *m (ceram)* solid casting
Massivreifen *m* solid[-rubber] tyre, *(Am)* band tire
Maßkolben *m (lab)* measuring (volumetric, graduated) flask
Maßlöffel *m (lab)* measuring spoon
Maßlösung *f* standard solution
M. einer Lauge standard base
M. einer Säure standard acid
maßschneidern *s.* aufbauen / gezielt
Maßstab *m* scale ✦ **in großem (großtechnischem) M.** large-scale, *(as opposed to pilot-plant-scale also)* full-scale ✦ **in großtechnischem M. hergestellt** produced on the large scale ✦ **in halbtechnischem M.** pilot-plant-scale, semicommercial-scale
logarithmischer M. log[arithmic] scale
Maßzylinder *m* measuring (graduated) cylinder
Masterbatch *m (rubber)* masterbatch, mother stock ✦ **Masterbatches herstellen** to mix into a masterbatch, to masterbatch ✦ **mit Masterbatches mischen** to masterbatch
Mastikation *f (rubber)* mastication
M. auf Walzwerken open-mill mastication
heiße M. hot mastication
kalte M. cold mastication
Mastikator *m (rubber)* masticator
Mastix *m* mastic [gum], mastix, gum (Chios) mastic, pistachia galls *(from Pistacia lentiscus L.)*
Amerikanischer M. American mastic *(from Schinus molle L.)*
Mastixharz *n s.* Mastix
mastizieren *(rubber)* to masticate
Mastiziermaschine *f (rubber)* masticator
Masurium *n s.* Technetium
Masut *m* maz[o]ut, masut
Matairesinol *n* matairesinol *(a lignan)*
Material *n* material, substance, matter
abgeröstetes M. *(met)* calcine
aufgearbeitetes M. *(plast)* reground material
basisches feuerfestes M. basic refractory [material]
ferroelektrisches M. ferroelectric [material, substance]
ferromagnetisches M. ferromagnetic [material, substance]
feuerfestes M. refractory [material]
filmbildendes M. film-forming material (substance, component), film former
frisch hergestelltes M. *(plast)* virgin material
halbleitendes M. semiconducting material
lichtempfindliches M. *(phot)* sensitive (sensitized) material
mineralisches M. mineral matter (substance)
neutrales feuerfestes M. neutral refractory [material]
saures feuerfestes M. acid refractory [material]
spaltbares M. *(nucl)* fissionable material
thermoadhäsives M. thermoadhesive [material]
zu reformierendes M. *(petrol)* reformer feedstock
zu verarbeitendes M. [feed]stock
Materialbilanz *f* material balance
Materialfehler *m* fault, defect, flaw
Materie *f* matter, substance
feste M. solid [matter]
materiell material
Materieteilchen *n* particle of matter, material particle
Materiewellen *fpl* matter waves
Mathieson-Zelle *f* Mathieson cell *(a chlor-alkali cell)*
Matlockit *m (min)* matlockite *(lead chloride fluoride)*
Matrix *f (ceram)* matrix, ground-mass
Matrize *f* die; *(plast)* female die (mould), negative die, cavity retainer (block)
zweiteilige M. two-piece die
Matrizeneinsatz *m (plast)* bottom plug
Matrizenmechanik *f* matrix (quantum) mechanics
Matrizenname *m (nomencl)* replacement name
Matrizenplatte *f (plast)* retainer plate
matt mat[t], dull, lustreless *(esp surfaces)*; flat, dead *(esp colours)*
m. werden to dull, *(Am also)* to blind
Mattätze *f (glass)* frosting
mattätzen *(glass)* to frost
Mattdruck *m* dull print
Mattenpreßverfahren *n (plast)* mat moulding
Mattglanz *m* dead lustre (gloss), *(tann also)* dead finish
Mattglasur *f (ceram)* mat glaze
Mattheit *f* dullness *(esp of surfaces)*; flatness, deadness *(esp of colours)*
mattieren to mat, to dull *(esp surfaces)*; to flat, to deaden *(esp colours)*; *(glass)* to frost; *(text)* to delustre
mit Sandstrahl m. *(glass)* to sandblast
Mattierungsmittel *n (coat)* flatting agent; *(text)* delustrant, delustring (dulling) agent
Mattkohle *f* dull coal
Mattlack *m* flat varnish
Mattsalz *n (glass)* frosting agent
Mauerwerk *n* brickwork
Mauerziegel *m* [building] brick
mauken *(ceram)* to mature, to age, to sour
Maukkeller *m (ceram)* maturing cellar
Maul *n* inlet *(of a jaw breaker)*
Maulpresse *f* jaw (gap) type press, open-side (open-gap, C-frame) press
Maulwurfpumpe *f* close-coupled pump
Mauvein *n* **[/ Perkins]** mauvein[e], Perkin's mauve (purple, violet) *(a quinone dye)*
Mavakurin *n* mavacurine *(a calabash curare alkaloid)*
Maxecon-Kent-Mühle *f* Kent Maxecon mill *(a ring-roll mill)*
Maximadämpfer *m* maximum suppressor *(polarography)*
Maximaldosis *f* maximal dose
Maximalschwärzung *f (phot)* maximum density
Maximalvalenz *f* maximum valency
negative M. maximum negative valency

positive M. maximum positive valency
Maximum *n* maximum, peak
verdecktes M. *(cryst)* hidden maximum
Maximum-Minimum-Thermometer *n* maximum and minimum thermometer
Maximum-Siedepunkt *m* maximum boiling point
Maxwell-Boltzmann-Statistik *f* Maxwell-Boltzmann statistics
Mazeral *n (coal)* maceral, constituent
Mazeralgruppe *f (coal)* maceral group
Mazeration *f* maceration
Mazerator *m* macerator, blendor
mazerieren to macerate
Mazis *m* mace *(from Myristica fragrans Houtt.)*
McDougall-Ofen *m* McDougall furnace (roaster) *(a multihearth roaster)*
McLeod-Manometer *n*, **McLeod-Vakuummeter** *n* McLeod gauge
MCPA *s.* 2-Methyl-4-chlorphenoxyessigsäure
McQuaid-Ehn-Probe *f* McQuaid-Ehn test *(for determining particle sizes)*
Mechanochemie *f* mechanochemistry
mechanochemisch mechanochemical
Mechloräthamin *n* $(ClCH_2CH_2)_2NCH_3$ chlormethine, mechlorethamine, mustine, *N*-di-(2-chloroethyl)methylamine hydrochloride
Medikament *n* medicament, remedy, *(if for internal use also)* medicine
Medium *n* medium
Medium-Processing-Channel-Ruß *m (rubber)* medium processing channel black, MPC black
Medium-Thermal-Ruß *m* medium thermal black, MT black
Medizinalöl *n* medicinal oil
medizinisch medicinal
Meerrettichperoxydase *f* horse radish peroxidase
Meersalz *n* sea (marine) salt
Meerschaum *m (min)* sea foam, meerschaum, sepiolite *(a phyllosilicate)*
Meerwasser *n* sea water
Meerwasserentsalzung *f* desalin[iz]ation of sea water
Meerwein-Ponndorf-Verley-Karbonylreduktion *f* Meerwein-Ponndorf-Verley reduction
Meerzwiebel *f* red squill, Urginea maritima (L.) Bak.
M-Effekt *m s.* Mesomerieeffekt
megaskopisch megascopic
Mehl *n* 1. *(food) (if finely ground)* flour, *(if coarsely ground)* meal; 2. *(min tech, chem)* flour, powder, meal
mehlartig *s.* mehlig
Mehlbleichmittel *n* flour-bleaching agent
Mehlbleichung *f* flour bleaching
mehlig floury, mealy, farinaceous
Mehligkeit *f* mealiness
Mehlmischer *m* flour mixer
Mehlstoff *m (pap)* flour
Mehlverbesserungsmittel *n* flour improver *(for increasing the baking qualities)*
mehratomig polyatomic
Mehrbahnofen *m (ceram)* multipassage kiln
Mehrbandtrockner *m* multistage belt dryer, multiple-belt [tunnel] dryer, multiconveyor [tunnel] dryer
mehrbasig polybasic *(acid)*
Mehrbereichsöl *n* multigrade oil
Mehrdeckersiebmaschine *f* multideck (multiple-deck) screen
Mehretagenofen *m* multihearth (multiple-hearth) furnace
Mehretagenpresse *f* multidaylight (multiple-daylight, multiplaten) press
Mehretagenröstofen *m* multihearth (multiple-hearth) roaster (roasting furnace)
M. nach Herreshoff Herreshoff roaster (furnace)
M. nach McDougall McDougall roaster (furnace)
M. nach Nichols Nichols roaster (furnace)
Mehrfachbindung *f* multiple bond, multiple link[age]
Mehrfachbindungssystem *n* multiple-bond system
Mehrfachelektrode *f* multiple electrode, polyelectrode
Mehrfachform *f s.* Mehrfachwerkzeug
Mehrfachpunktschreiber *m* multipoint recorder
Mehrfachreflexion *f* multiple reflection
Mehrfachschicht *f* multilayer, multimolecular [adsorbed] layer
Mehrfachstreuung *f (phys chem)* plural scattering
Mehrfachsubstitution *f* polysubstitution
Mehrfachverdampfer *m* multiple-effect evaporator
Mehrfachwerkzeug *n (plast)* multi-impression mould, *(Am)* multi-cavity mold; *(plast)* composite mould *(containing dissimilar impressions within a common bolster)*
M. mit getrennten Füllräumen separate pot mould
Mehrfachzerfall *m (nucl)* multiple (branched) disintegration (decay), branching
Mehrfachzyklon *m* multiple[-unit] cyclone
Mehrfarbendruck *m* [multi]colour printing
Mehrfarbeneffekt *m* multicolour[ed] effect
mehrfarbig multicolour[ed]
Mehrfarbigkeit *f (cryst)* pleochro[mat]ism, polychroism
mehrfunktionell polyfunctional
Mehrgutapparat *m* multiproduct unit
Mehrhalskolben *m (lab)* multinecked flask
Mehrkammereindicker *m* tray thickener
M. mit parallelgeschalteten Kammern balanced tray thickener
Mehrkammermühle *f* [multi]compartment mill, compound mill
Mehrkammerofen *m* multichamber kiln
Mehrkammerrohrmühle *f s.* Mehrkammermühle
Mehrkammerzentrifuge *f* multichamber centrifuge

Mehrkanaldurchschubofen *m (ceram)* multipassage kiln
mehrkernig multinuclear, polynuclear
Mehrkolbenpumpe *f* multipiston pump
Mehrkomponentengemisch *n* multicomponent mixture
Mehrkomponentensystem *n* multicomponent system
Mehrkörperextraktionsanlage *f* pot plant *(solvent extraction)*
Mehrkörperkräfte *fpl (phys chem)* many-body forces
Mehrkörperproblem *n (phys chem)* many-body problem
Mehrkörperverdampfer *m s.* Mehrstufenverdampfer
Mehrkörperverdampfung *f s.* Mehrstufenverdampfung
mehrlagig multi-ply
Mehrmulden-Unterschubrost *m* multiple-retort [underfeed] stoker
Mehrnährstoffdünger *m* mixed (multinutrient, compound) fertilizer
Mehrpendelmühle *f* multiroll mill
Mehrphasensystem *n* multiphase system
mehrphasig multiphase
Mehrplatten[schnell]gefrierapparat *m* multiplate freezer
Mehrpressenschleifer *m (pap)* pocket grinder
Mehrproduktenapparat *m* multiproduct unit
Mehrrundsiebmaschine *f (pap)* multicylinder (multivat) machine
mehrsäurig polyacid *(base)*
Mehrscheiben[sicherheits]glas *n*, **Mehrschichten-[sicherheits]glas** *n* laminated safety [sheet] glass, laminated glass
mehrschichtig multi-ply
Mehrschneckenextruder *m (plast)* multiscrew extruder
Mehrstoffgemisch *n* multicomponent mixture
Mehrstoffkatalysator *m* mixed catalyst
Mehrstoffsystem *n* multicomponent system
Mehrstufenbleiche *f (pap)* multistage bleaching
Mehrstufeneindampfung *f s.* Mehrstufenverdampfung
Mehrstufenkompressor *m* multistage compressor
Mehrstufenpolarogramm *n* multiple polarogram
Mehrstufenschubzentrifuge *f* multistage reciprocating-pusher centrifuge
Mehrstufenseparator *m (petrol)* multistage separator
Mehrstufensynthese *f* many-step synthesis
Mehrstufenverdampfer *m* multieffect (multiple-effect, multistage, cascade) evaporator
Mehrstufenverdampfung *f* multieffect (multiple-effect, multistage) evaporation
Mehrstufenverdichter *m* multistage compressor
Mehrstufenwäscher *m* multistage washer
mehrstufig multistage, multiple-stage, multistep
Mehrwalzenbrecher *m* multiroll crusher
Mehrwalzenmühle *f* multiroll mill
Mehrwalzenstuhl *m* multiroll mill
Mehrweg[e]hahn *m* multiport plug valve
mehrwertig multivalent, polyvalent, polyad
Mehrwertigkeit *f* multivalence, polyvalence
mehrzählig *s.* mehrzahnig
mehrzahnig, mehrzähnig multidentate, polydentate *(coordination chemistry)*
Mehrzellenelektrodialysator *m* multimembrane electrodialyzer
Mehrzellenflotationsgerät *n*, **Mehrzellenflotationsmaschine** *f* multicell floatation machine
Mehrzentrenbindung *f* multicentre bond
Mehrzweckreinigungsmittel *n* all-purpose cleaner
Meiler[haufen] *m* pile, heap *(for producing charcoal)*
M-Einheit *f (nomencl)* monofunctional unit, M unit
Meißel *m* **für hartes Gebirge (Gestein)** *(petrol)* hard-formation bit, rock bit
M. für lockeres Gebirge (Gestein) *(petrol)* soft-formation bit
Mejonit *m (min)* meionite *(a tectosilicate)*
MEK *s.* Methyläthylketon
MEK-Benzol-Entparaffinierungsanlage *f (petrol)* MEK-benzene dewaxing plant
MEK-Benzol-Verfahren *n (petrol)* MEK-benzene [dewaxing] process
MEK-Entparaffinierung *f (petrol)* MEK dewaxing
Méker-Brenner *m (lab)* Meker burner
Mekkabalsam *m* Mecca balsam, balm of Gilead *(from Commiphora opobalsamum (L.) Engl.)*
Mekonsäure *f* meconic acid, 3-hydroxy-4-pyrone-2,6-dicarboxylic acid
MEK-Verfahren *n (petrol)* MEK [dewaxing] process
Melamalz *n s.* Melanoidinmalz
Melamin *n* melamine, triaminotriazine
Melamin-Formaldehydharz *n*, **Melaminharz** *n* melamine[-formaldehyde] resin
Melamin-Phenolharz *n* melamine-phenolic resin
Melampyrin *m*, **Melampyrit** *m s.* Dulzit
Melangedruck *m (text)* vigoureux printing
Melangegarn *n* mixture yarn *(made from fibres of different colour)*
Melanit *m (min)* melanite *(a nesosilicate)*
Melanoidin *n* melanoidin[e] *(colouring matter and aromatic ingredient of malt)*
Melanoidinmalz *n* melanoidin[e] malt
Melanozerit *m (min)* melanocerite *(a nesosilicate)*
Melanterit *m (min)* melanterite, iron vitriol, copperas *(iron(II) sulphate-7-water)*
Melasse *f (sugar)* [sugar house] molasses, treacle
melassebildend *(sugar)* molasses-forming, melassigenic
Melassebildner *m (sugar)* molasses-forming substance
Melassebildung *f (sugar)* molasses formation
Melasseentzuckerung *f* desugarizing of molasses
Melassesirup *m s.* Melasse
Melassezucker *m* molasses sugar

Meldeeinrichtung *f*, **Melder** *m* alarm
Meldolablau *n* Meldola's blue, new blue R, basic blue 6
Melibiose *f* melibiose *(a disaccharide)*
Melierung *f (pap)* mottling
Melilith *m (min)* melilite *(a sorosilicate)*
Melilotsäure *f* $C_6H_4(OH)CH_2CH_2COOH$ melilotic acid, *o*-hydrocoumaric acid, 3-*o*-hydroxyphenylpropionic acid
Melinophan *m (min)* melinophane *(a sorosilicate)*
Melioration *f (agric)* amelioration, amendment
Meliorationslanze *f (agric)* amelioration injector *(for liquid fertilizers)*
meliorieren *(agric)* to ameliorate
Melissenöl *n* melissa oil, [lemon] balm oil *(from Melissa officinalis L.)*
Melissinsäure *f* $CH_3[CH_2]_{28}COOH$ melissic acid, triacontanoic acid
Melissylalkohol *m* melissyl alcohol *(1.* $C_{30}H_{61}OH$ *triacontanol; 2.* $C_{31}H_{63}OH$ *hentriacontanol)*
Melitose *f*, **Melitriose** *f s.* Raffinose
Melkfett *n* milking grease
Mellit *m (incorrectly for)* Mellith
Mellith *m (min)* mellite *(a hydrous aluminium mellitate)*
Mellithsäure *f* $C_6(COOH)_6$ mellitic acid, benzenehexacarboxylic acid
Melonit *m (min)* melonite *(nickel telluride)*
Membran *f* membrane, diaphragm, partition [wall]
halbdurchlässige M. *s.* semipermeable M.
ionenaustauschende M. ion-exchange membrane
semipermeable M. semipermeable membrane
Membranfilter *n* membrane filter
Membrangleichgewicht *n* membrane equilibrium
Donnansches M. Donnan [membrane] equilibrium
Membranhydrolyse *f* membrane hydrolysis
Membrankolbensetzmaschine *f (min tech)* diaphragm-actuated jig
Membrankompressor diaphragm compressor
Membranpotential *n* membrane potential
Membranpumpe *f* diphragm pump
Membransortierer *m* diaphragm screen
Membranventil *n* diaphragm valve
Membranverdichter *m* diaphragm compressor
Memory-Effekt *m (nucl)* memory effect
Mendelejew-System *n* Mendeléeff [periodic] system (table)
Mendelevium *n* Md mendelevium
Mendheim-Ofen *m (ceram)* Mendheim kiln *(a gas-fired chamber kiln)*
Mendipit *m (min)* mendipite *(lead(II) chloride oxide)*
Mendozit *m (min)* mendozite *(aluminium sodium sulphate-11-water)*
Meneghinit *m (min)* meneghinite *(antimony(III) lead(II) sulphide)*
Menge *f* quantity, quantum, amount
äquivalente M. equivalent
heilende M. *(pharm)* therapeutic dose
schädigende M. *(pharm)* toxic dose
theoretische (theoretisch nötige) M. theoretical quantity
mengen to mix, to mingle, to blend
Mengenmesser *m* quantity meter
Mengenstrommesser *m* flow (rate, stream) meter, flowmeter
M. für Flüssigkeiten liquid meter
Mengenstrommessung *f* flow measurement
Mengenverhältnis *n* quantity ratio
Menhadenöl *n* menhaden oil *(a fish oil)*
Menilit *m (min)* menilite *(a variety of opal)*
Meni-Öl *n* Meni oil *(from Lophira alata Banks)*
Meniskus *m* meniscus
Meniskusvisierblende *f* meniscus reader *(titration)*
Mennige *f* minium, red lead [oxide] *(lead(II) orthoplumbate)*
Mennigepaste *f* red-lead paste
Mensur *f* measuring (graduated) cylinder, graduate
Menthadien *n (org chem)* menthadiene
p-**Menthan** *n* *p*-menthane, 1-isopropyl-4-methylcyclohexane
Menthen *n* menthene, *(specif)* 1-isopropyl-4-methylcyclohexene
Menthol *n* menthol, 2-isopropyl-5-methylcyclohexanol
Menthon *n* menthone, 2-isopropyl-5-methylcyclohexanone
Mercapsol-Verfahren *n* Mercapsol process *(for desulphurizing petrol distillates)*
Mergel *m* marl
mergelig marly
Mergelton *m* marl clay
Merichinon *n (org chem)* semiquinone
Merkaptal *n* mercaptal *(any of a class of condensation products of thiols with aldehydes)*
Merkaptan *n* mercaptan, *(better)* thiol
merkaptanarm poor in mercaptan
Merkaptanentfernung *f* mercaptan removal
Merkaptanextraktion *f* mercaptan extraction
merkaptanreich rich in mercaptan, mercaptan-rich
Merkaptanumwandlung *f* mercaptan conversion
Merkaptid *n* mercaptide *(a metallic derivative of a thiol)*
Merkaptoäthansäure *f s.* Merkaptoessigsäure
Merkaptobenzoesäure *f* $C_6H_4(SH)COOH$ mercaptobenzoic acid
Merkaptoessigsäure *f* $HSCH_2COOH$ mercaptoacetic acid, thioglycollic acid
Merkaptogruppe *f* $-SH$ mercapto group, thiol group, sulphydryl group
Merkaptol *n* mercaptol[e] *(any of a class of condensation products of thiols with ketones)*
6-Merkaptopurin *n* 6-mercaptopurin, 6-purinethiol, leukerin
Merkaptorest *m s.* Merkaptogruppe
Merkapto-Schwefel *m* mercaptan sulphur
Merkurierung *f* mercur[iz]ation *(treatment of an organic compound with a mercury salt)*

Merkurimetrie *f* mercurimetry *(titration with a mercury(II) nitrate solution)*
merkurimetrisch mercurimetric
Merkurometrie *f* mercurometry *(titration with a mercury(I) nitrate solution)*
merkurometrisch mercurometric
Meroxen *m (min)* meroxene *(the most common variety of biotite)*
Merzerisation *f (text)* mercerization
Merzerisierechtheit *f (text)* fastness to mercerization (mercerizing)
merzerisieren *(text)* to mercerize
Merzerisieren *n (text)* mercerization
  M. unter Spannung mercerization with tension
  spannungsloses M. mercerization without tension, slack mercerization
Merzerisierhilfsmittel *n (text)* mercerizing assistant
Merzerisiermaschine *f (text)* mercerizing machine
  kettenlose M. chainless mercerizing machine
Mesitinspat *m*, **Mesitit** *m (min)* mesitine [spar], mesitite *(ferroan magnesite)*
Mesitylen *n* $C_6H_3(CH_3)_3$ mesitylene, 1,3,5-trimethylbenzene
Mesitylen-2-karbonsäure *f* $(CH_3)_3C_6H_2COOH$ mesitoic acid, 2,4,6-trimethylbenzoic acid
Mesityloxid *n* $(CH_3)_2C{=}CH{-}CO{-}CH_3$ mesityl oxide, 4-methyl-3-penten-2-one
Meskalin *n* $(CH_3O)_3C_6H_2CH_2CH_2NH_2$ mescaline, β-[3,4,5-trimethoxyphenyl]ethylamine
Mesoatom *n s.* Mesonenatom
Meso-Form *f (org chem)* meso form
mesoionisch mesoionic
Mesokolloid *n* mesocolloid
Mesolith *m (min)* mesolite *(a tectosilicate)*
mesomer mesomeric
Mesomerie *f* mesomerism, resonance
Mesomeriebegriff *m* concept of mesomerism (resonance)
Mesomerieeffekt *m* mesomeric (resonance) effect
Mesomerieenergie *f* mesomeric (resonance) energy
mesomeriefrei free from mesomerism, resonance-free
Mesomeriepfeil *m* double-headed arrow
Mesomerievorstellung *f s.* Mesomeriebegriff
mesomorph mesomorphic, mesomorphous
Meson *n (nucl)* meson *(an elementary particle)*
  neutrales M. neutral meson, neutretto
μ-Meson *n* μ-meson, mu meson, muon
π-Meson *n* π-meson, pi meson, pion
Meson[en]atom *n* mesonic atom
Mesonenfeld *n* meson field
Mesonentheorie *f* meson theory
mesonisch mesonic
Mesoperjodat *n* mesoperiodate, pentaoxoiodate(VII)
Mesotartarsäure *f s.* Mesoweinsäure
Mesothorium *n* mesothorium
Mesotron *n s.* Meson

Mesoverbindung *f (org chem)* meso compound
Mesoweinsäure *f* $HOOC[CHOH]_2COOH$ mesotartaric acid
Mesoxalsäure *f* $OC(COOH)_2$ mesoxalic acid, oxomalonic acid, oxo-propanedioic acid
Meßbehälter *m* measuring vessel
Meßbereich *m* measuring range
Meßblende *f* orifice meter (flowmeter, plate)
Meßbrücke *f* bridge connection
Meßdüse *f* flow nozzle *(flow measurement)*
Meßeinrichtung *f* measuring device, measurement element
Meßelektrode *f* measuring (indicating, indicator) electrode
Messenger-Ribonukleinsäure *f*, **Messenger-RNS** *f* messenger ribonucleic acid, messenger RNA
Messer *m* gauge, meter
Messer *n* knife, *(pap also)* bar
  M. der Kegelstoffmühle *(pap)* jordan bar
  feststehendes M. *(pap)* dead knife
Messerabstand *m (pap)* spacing between bars
Messerblock *m (pap)* beater (dead) plate, bedplate *(of a Hollander beater)*
Messerentrinder *m (pap)* knife (disk) barker (barking machine)
Messergarnierung *f (pap)* filling, tackle *(of a Hollander beater)*
Messerholländer *m (pap)* Hollander, Hollander beater (beating engine), stuff (pulp) engine, pulp grinder
Messernarbe *f (glass)* shear mark *(a defect)*
Messerscheibenentrinder *m s.* Messerentrinder
Messerstreichmaschine *f (plast)* knife coater
Messerwalze *f (pap)* Hollander (beater, beating, knife) roll
Messerwellenquerschneider *m (pap)* revolving-knife cutting machine, rotary [knife] cutter
Messerwerk *n s.* Messerblock
Meßfehler *m* error of measurement
Meßflasche *f s.* Meßkolben
Meßfühler *m* sensing device, sensor, detector
Meßgefäß *n* measuring vessel (pot), graduate
  M. für Oleum oleum-measuring vessel
Meßgerät *n* measuring instrument, gauge, meter
  anzeigendes M. indicating instrument
  registrierendes M. recording instrument, recorder
Meßgerinne *n* flume *(flow measurement)*
Meßglas *n* measuring glass, glass measure
Meßglied *n* measurement element
Meßgröße *f* quantity being (*or* to be) measured
Messing *n* [yellow] brass
  M. mit hohem Zinkgehalt high-zinc brass
  M. mit niedrigem Zinkgehalt low-zinc brass
α-Messing *n* alpha brass
β-Messing *n* beta brass
Messingbindung *f (rubber)* rubber-brass bond
Messingblüte *f (min)* aurichalcite *(a basic copper zinc carbonate)*
Messingdichtung *f* brass gasket
Messinggewicht *n (incorrectly for)* Messingwägestück

Messingwägestück *n* brass weight
Meßinstrument *n* measuring instrument, meter
Meßkammer *f* measuring chamber
Meßkelch *m (lab)* measuring cup
Meßkolben *m (lab)* measuring (graduated, volumetric, delivery) flask
M. mit [einer] Marke one-mark volumetric flask
Meßlatte *f* gauge stick
Meßlöffel *m (lab)* measuring spoon
Meßmethode *f* method of measurement
Meßmikroskop *n* scanning microscope
Meßpipette *f* measuring (graduated) pipette
M. nach Mohr Mohr measuring pipette
Meßreihe *f* series of measurements
Meßstelle *f* measuring point; measuring junction *(of a thermocouple)*
Messung *f* measurement
M. der Dielektrizitätskonstante measurement of dielectric constant, dielectrometry
M. des Dipolmoments dipole measurement
M. des Formänderungsrestes *(rubber)* permanent-set test *(in tension or compression)*
M. nach einer Skale scaling
elektrometrische M. electrometric measurement
potentiometrische M. potentiometric measurement
röntgenografische M. X-ray measurement
Meßungenauigkeit *f* measuring accuracy, accuracy in (of) measurement
Meßvorrichtung *f s.* Meßeinrichtung
Meßwagen *m* measuring van
Meßwalze *f (pap)* metering roll
Meßwert *m* measured value
Meßzelle *f* measuring cell
Meßzylinder *m* measuring (graduated) cylinder
M. nach Crow Crow receiver *(a receiver with a conical base)*
Mesylat *n*, **Mesylester** *m* mesylate, methane sulphonate
Metaaluminat *n* M^IAlO_2 metaaluminate
Meta-Anthrazit *m* meta-anthracite
Metaarsenat(III) *n* M^IAsO_2 metaarsenite
Metaarsenat(V) *n* M^IAsO_3 metaarsenate
Metaarsen(V)-säure *f* $HAsO_3$ metaarsenic acid
Metaaurat(III) *n* M^IAuO_2 metaaurate(III)
Metabolismus *m* metabolism
Metabolit *m* metabolite *(a substance essential to metabolism)*
Metaborat *n* M^IBO_2 metaborate, dioxoborate
Metaborsäure *f* HBO_2 metaboric acid, dioxoboric acid
Metachromverfahren *n (text)* metachrome [dyeing] method, chromate [dyeing] method, chromate process
meta-dirigierend meta-directing
Metahalloysit *m (min)* metahalloysite *(a phyllosilicate)*
Meta-Isomer[es] *n* metaisomer, *m*-isomer
Metakaolin *m* metakaolin
Metakieselsäure *f* H_2SiO_3 metasilicic acid, trioxosilicic acid

Metaldehyd *m* $(CH_3CHO)_n$ metaldehyde
Metall *n* metal
aufgetragenes M. deposited metal
edles M. noble metal
gediegenes M. native [metal]
gelochtes M. perforated metal
gepulvertes M. powder[ed] metal, metal powder
passives M. passive metal
reines M. pure metal
Rosesches M. Rose's metall (alloy)
schmelzflüssiges M. molten metal
unedles M. base metal
Woodsches M. Wood's metal
Metallabscheidung *f* metal deposition
katodische M. cathodic deposition of metals
metallaktiviert metal-activated
Metallalkyl *n* metal alkyl
Metallamid *n* M^INH_2 [metal] amide
Metallatom *n* metal[lic] atom
Metallauftrag *m* 1. deposition (production) of metallic coatings; 2. metallic coating *(result)*
Metallausdehnungsthermometer *n* solid expansion thermometer
Metallbad *n* metal bath
Metallbadfärbeverfahren *n (text)* molten-metal [dyeing] process, Standfast molten-metal process
Metallbadfärbung *f (text)* molten-metal dyeing
Metallbadverfahren *n s.* Metallbadfärbeverfahren
Metallbeize *f* metallic mordant
Metallbeschlag *m* metal crust
Metallbindung *f* metal[lic] bond, metal-metal bond
Metallborhydrid *n*, **Metallborwasserstoff** *m* $M^I[BH_4]$ tetrahydridoborate, hydridoborate
Metallchelat *n s.* Metallchelatkomplex
Metallchelatbindung *f* metal-chelate bond
Metallchelatkomplex *m*, **Metallchelatverbindung** *f* metal-chelate complex (compound)
Metallchemie *f* metal chemistry
Metallde[s]aktivator *m* metal deactivator
Metall-Dewar-Gefäß *n* Dewar [vessel] *(container for liquid gases)*
Metalldichtung *f (for moving parts)* metal[lic] packing, metal seal; *(for parts without relative motion)* metal gasket
Metall-Donatorbindung *f* metal-donor bond (linkage)
Metalleiweißtrübung *f* protein-metal turbidity
Metallelektrode *f* metal[lic] electrode
metallen metallic
Metallenzym *n* metallo enzyme
Metallerz *n* metal[lic] ore
Metallfadenlampe *f* metal-filament lamp
Metallfarbe *f* metallic ink
Metallfärbung *f* metal colouring
Metallfaser *f* metallic fibre
Metallfaserstoff *m* metallic fibre
Metallfilter *n* porous-metal filter
Metallflansch *m* metal flange

Metallfolie *f* metal foil
Metallform *f* [permanent] metal mould, gravity die *(foundry)*
metallfrei metal-free
metallführend metalliferous
Metallgarn *n* metallic yarn
Metallgehalt *m* metal content
  M. des Roherzes tenor of ore
Metallgewebe *n* metal fabric (gauze), wire cloth (gauze) ✦ **mit M. abgedeckt** *(filtr)* screen-covered
Metallgewinnung *f* metal extraction
Metallgitter *n* metallic lattice
Metallglanz *m* metallic lustre
Metall-Halogen-Austausch *m* metal-halogen exchange
metallhaltig metal-containing, metalliferous
Metallhüttenwesen *n* non-ferrous metallurgy
Metallhydrid *n* metal hydride
Metallic-Papier *n* metallic paper
Metallierung *f* metalation *(attachment of a metal atom to a carbon atom of an organic molecule)*
Metall-Ion *n* metal[lic] ion
metallisch metallic
metallisieren to metallize
Metallisierung *f* metallization
Metallkalorimeter *n* / **Nernstsches** Nernst calorimeter
Metallkarbid *n* metal carbide
Metallkarbonyl *n* metal carbonyl
Metallkatalysator *m* metal[lo] catalyst
metallkatalysiert metal-catalyzed
Metallkation *n* metal cation
Metallkeramik *f* metal ceramics
metallkeramisch metal-ceramic, cerametallic
Metallketyl *n* [metal] ketyl
Metallkleben *n* adhesive bonding of metals
Metallklebstoff *m* metal-bonding adhesive
Metallkomplex *m* metal complex
Metallkönig *m s.* Regulus
Metallkontakt *m s.* Metallkatalysator
Metall-Metall-Austausch *m* metal-metal exchange
metallogen *(geoch)* metallogen[et]ic
Metallografie *f* metallography
metallografisch metallographic
Metalloid *n s.* Nichtmetall
metallorganisch metallo-organic, metalorganic
Metallorganyl *n s.* Verbindung / metallorganische
Metalloxid *n* metal[lic] oxide
Metalloxidvernetzung *f* *(rubber)* metallic-oxide cure
Metallpackung *f* metal[lic] packing
Metallpapier *n* metal[lized] paper
Metallphthalozyanin *n* metal phthalocyanine
Metallpuffer *m* metal buffer
Metallpulver *n* metal powder, powder[ed] metal
  M. mit einer Komponente single-metal powder
Metallpulverfilter *n* metal-powder filter
Metallpulverpreßling *m* powder-metal compact *(in powder metallurgy)*
Metallsalzbeize *f* metallic mordant
Metallschaum *m* dross
Metallschlauch *m* [flexible] metal hose
Metallschmelze *f* molten (fused) metal
Metallseife *f* metallic soap
Metallsieb *n* metal screen, wire screen (sieve)
Metallsol *n* metal sol
Metallspatel *m* *(lab)* metal spatula
Metallspritzen *n* metal spraying, spray metallization, Schoop metallizing, schooping
Metallspritzverfahren *n* metal spraying process, Schoop process
Metallstearat *n* metal stearate
Metallsulfid *n* metal sulphide
Metalltitration *f* metal titration
Metallüberzug *m* metallic coating
Metallurgie *f* metallurgy
metallurgisch metallurgical
Metallvalenzorbital *n* metal-valence-bond orbital
Metallverbindung *f* 1. metal[lic] compound; 2. bonding (joining) of metals; *(rubber)* bonding to metals; 3. metal joint (seal) *(result of 2.)*
Metallverklebung *f*, **Metallverleimung** *f* adhesive bonding of metals
Metall-Wasserstoff-Austausch *m* metal-hydrogen exchange
Metall-Weichstoffdichtung *f* semimetallic packing
Metallzustand *m* metallic state
metamer metameric
Metamer[es] *n* metamer
Metamerie *f* metamerism *(one form of structural isomerism)*
Metametall *n* metametal
metamorph *(geol)* metamorphic
Metamorphismus *m s.* Metamorphose
Metamorphit *m* metamorphite, metamorphic (metamorphosed) rock
Metamorphose *f* *(geol)* metamorphism, metamorphosis, transition
  kinetische (mechanische) M. dynamometamorphism, dynamic metamorphism
  regionale M. regional metamorphism
Metamorphosegrad *m* *(geol)* degree of metamorphosis
Metanilsäure *f* $C_6H_4(NH_2)SO_3H$ metanilic acid, aniline-*m*-sulphonic acid
Metaperjodat *n* M^IIO_4 metaperiodate, periodate, tetraoxoiodate(VII)
Metaperjodsäure *f* HIO_4 metaperiodic acid, periodic acid, tetraoxoiodic(VII) acid
Metaphosphat *n* $(M^IPO_3)_n$ metaphosphate
Metaphosphit *n* M^IPO_2 metaphosphite
Metaphosphorsäure *f* $(HPO_3)_n$ metaphosphoric acid
Metaplumbat(IV) *n* $M_2^IPbO_3$ metaplumbate(IV), trioxoplumbate(IV)
Metasilikat *n* $M_2^ISiO_3$ metasilicate, trioxosilicate
metastabil metastable
metaständig meta, in meta position
  m. sein to be [located, situated] meta
meta-Stellung *f* meta position ✦ **in meta-Stellung**

in meta position, meta ✦ **nach der meta-Stellung dirigierend** meta-directing
metasubstituiert meta-substituted, *m*-substituted
Metatellurat(VI) *n* M_2TeO_4 metatellurate(VI), tetraoxotellurate(VI)
Metathese[reaktion] *f* metathesis, metathetical (double-decomposition) reaction, double decomposition
metathetisch metathetical
Metathioarsenat(III) *n* M^IAsS_2 metathioarsenite, dithioarsenate(III)
Metathioarsenat(V) *n* M^IAsS_3 metathioarsenate, trithioarsenate(V)
Metathiostannat(IV) *n* $M_2{}^ISnS_3$ metathiostannate(IV), trithiostannate(IV)
Metatitanat(IV) *n* $M_2{}^ITiO_3$ metatitanate(IV), trioxotitanate(IV)
Metavanadat *n* M^IVO_3 metavanadate, trioxovanadate(V)
meta-Verbindung *f* meta compound
Metawolframat *n* *s.* Dihydrogendodekawolframat(VI)
Metawolframsäure *f* *s.* Dihydrogendodekawolframsäure
Metazinnabarit *m (min)* metacinnabar[ite] *(mercury(II) sulphide)*
Metazinnsäure *f* metastannic acid
Metazirkonat(IV) *n* $M_2{}^IZrO_3$ metazirconate(IV), trioxozirconate(IV)
Meteoreisen *n* meteoric iron, iron meteorite, [holo]siderite
Meteorit *m* meteorite
Meteorstein *m (geol)* meteoric stone, stone (stony) meteorite, aerolite, aerolith
Methakrolein *n*, **Methakrylaldehyd** *m* $CH_2{=}C(CH_3)CHO$ methacrolein, methacrylaldehyde
Methakrylat *n* methacrylate
2-Methakrylsäure *f* $CH_2{=}C(CH_3)COOH$ 2-methylacrylic acid, 2-methylpropenoic acid
Methakrylsäuremethylester *m* $CH_2{=}C(CH_3)COOCH_3$ methyl methacrylate, methyl-2-methylacrylate
Methallylchlorid *n* $CH_2{=}C(CH_3){-}CH_2Cl$ methallyl chloride
Methämoglobin *n* methaemoglobin, haemiglobin
Methan *n* CH_4 methane
Methanal *n* HCHO methanal, formaldehyde
Methanamid *n* $H{-}CO{-}NH_2$ formamide
Methandikarbonsäure *f* $CH_2(COOH)_2$ methanedicarboxylic acid, propanedioic acid, malonic acid
Methangewinnung *f* methane recovery
Methanisierung *f* methanation *(as of synthesis gas)*
Methanol *n* CH_3OH methanol, methyl alcohol
methanolisch methanolic
Methanolyse *f* methanolysis
methanreich rich in methane
Methansäure *f* HCOOH methanoic acid, formic acid
Methansulfonat *n*, **Methansulfonsäureester** *m* methane sulphonate, mesylate
Met-Hb *s.* Methämoglobin
Methenamin *n* metheneamine, hexamethylenetetramine
Methid *n (org chem)* methide
Methin *n (dye)* methine
Methinbrücke *f (org chem)* methine bridge
Methinfarbstoff *m* methine dye
Methingruppe *f* $={CH}{-}$ methine group
Methionin *n* $CH_3SCH_2CH_2CH(NH_2)COOH$ methionine, 4-methylmercapto-2-aminobutyric acid
Methode *f* method, procedure, technique
M. champenoise *(ferm)* bootle champagnization
M. der Chemical Abstracts *(nomencl)* Chemical Abstracts method
M. der Dampfdichtebestimmung vapour-density method *(for estimating molecular weights)*
M. der freischwebenden Zone floating-zone method *(zone melting)*
M. der Gefrierpunktserniedrigung freezing-point method *(for determining molecular weights)*
M. der geneigten Platte *(phys chem)* tilting plate method *(for measuring the contact angle)*
M. der kritischen Luftmenge *(coal)* critical air blast method
M. der linearen Kombination von Atomorbitalen linear-combination-of-atomic-orbitals method, LCAO method
M. der Molekularstrahlen *(phys chem)* molecular-beam method (technique)
M. der Molekülorbitale molecular-orbital method, Hund-Mulliken-Lennard-Jones-Hückel method
M. der schwebenden Zone floating-zone method *(zone melting)*
M. der schwingenden Scheibe oscillating disk method *(for determining the viscosity of gases)*
M. der Tiegelverkokung *(coal)* crucible method
M. der Valenzstrukturen valence-bond method, VB method, method of valence-bond structures
M. der wandernden Grenzflächen *(phys chem)* moving-boundary method
M. des selbstkonsistenten Feldes self-consistent-field method
M. des verlorenen Wachsmodells *(met)* lost-wax process
M. von Manning-Shepperd Manning-Shepperd method *(for determining alkanes)*
M. von Rabi *(nucl)* Rabi method
M. von Tschugajew-Zerewitinow *(org chem)* Chugaev-Zerevitinov (Tschugaeff-Zerewitinoff) method *(for determining the number of hydroxyl groups)*
Beckmannsche M. Beckmann method *(for determining molecular weights)*
Curtiussche M. Curtius method (reaction, rearrangement) *(of decomposing acid azides for preparing primary amines)*
dynamische M. dynamic method *(for determining vapour pressures)*
gravimetrische M. gravimetric method

Heumannsche M. Heumann method *(for synthesizing indigo)*
isopiestische M. isopiestic method *(for determining molecular weights)*
kolorimetrische M. colorimetric method
konduktometrische M. conductometric (conductance) method
lichtelektrische kolorimetrische M. photocolorimetric method
magnetische M. *(petrol)* magnetic method
potentiometrische M. potentiometric method
pulvermetallurgische M. powder-metallurgical method, powder-metallurgy (powdered-metal) technique
röntgenografische M. X-ray method
seismische M. *(petrol)* seismic method
selbstkonsistente M. *(phys chem)* self-consistent method
standardisierte M. standard method (procedure, technique)
statische M. static method *(for measuring vapour pressures)*
Stelznersche M. *(nomencl)* Stelzner method
turbidimetrische M. turbidimetric method
Methodik *f* method, experimental *(in treatises)*
Methoxid *n* CH_3OM^I methoxide, methylate
***p*-Methoxybenzoesäure** *f* $CH_3OC_6H_4COOH$ *p*-methoxybenzoic acid
Methoxybenzol *n* $CH_3OC_6H_5$ methoxybenzene, methylphenyl ether, anisole
Methoxylbestimmung *f* / **Zeiselsche** Zeisel [methoxyl] determination
Methoxylgruppe *f*, **Methoxylrest** *m* CH_3O- methoxyl group (residue)
Methoxymethan *n s.* Dimethyläther
Methyl *n* methyl
3-Methylakrolein *n* $CH_3CH=CHCHO$ 3-methylacrolein, crotonaldehyde, 2-butenal
Methylakrylat *n* $CH_2=CHCOOCH_3$ methyl acrylate
Methylal *n* $H_2C(OCH_3)_2$ methylal, dimethoxymethane, formaldehyde dimethyl acetal
Methylalkohol *m* CH_3OH methyl alcohol, methanol
Methylamin *n* CH_3NH_2 methylamine
Methylaminoessigsäure *f* CH_3NHCH_2COOH methylaminoacetic acid, sarcosine
***N*-Methylanilin** *n* $C_6H_5NHCH_3$ *N*-methylaniline
Methylat *n* CH_3OM^I methylate, methoxide
Methyläthen *n s.* Propen
Methyläther *m s.* Dimethyläther
Methyläthin *n* $CH_3C\equiv CH$ propyne, methylacetylene
Methyläthylkarbinol *n s.* Butanol-(2)
Methyläthylketon *n* $CH_3COC_2H_5$ methyl ethyl ketone, MEK, ethyl methyl ketone, 2-butanone
Methyläthylsulfid *n* $CH_3SC_2H_5$ ethyl methyl sulphide, methyl ethyl sulphide
Methylazetaldehyd *m s.* Propanal
Methylazetat *n* CH_3COOCH_3 methyl acetate
Methylazetylen *n* $CH_3C\equiv CH$ propyne, methylacetylene
Methylbenzoat *n* $C_6H_5COOCH_3$ methyl benzoate
Methylbenzoesäure *f* $CH_3C_6H_4COOH$ methylbenzoic acid, toluic acid
Methylbenzol *n* $C_6H_5CH_3$ methylbenzene, toluene
Methylbenzolkarbonsäure *f s.* Methylbenzoesäure
Methylbernsteinsäure *f* $HOOC-CH(CH_3)-COOH$ methylsuccinic acid, 2-methyl-1,4-butanedioic acid
Methylbromid *n* CH_3Br bromomethane, methyl bromide
2-Methylbutan *n* $CH_3CH(CH_3)CH_2CH_3$ 2--methylbutane, isopentane
3-Methylbutanal *n* $(CH_3)_2CHCH_2CHO$ 3-methylbutanal, isovaleraldehyde
Methylbutandisäure $HOOC-CH(CH_3)CH_2-COOH$ 2-methyl-1,4-butanedioic acid, methylsuccinic acid
2-Methylbutanol-(1) *n* $CH_3CH_2CH(CH_3)CH_2OH$ 2-methyl-1-butanol
2-Methylbutanol-(2) *n* $CH_3CH_2C(CH_3)(OH)CH_3$ 2-methyl-2-butanol
3-Methylbutanol-(1) *n* $(CH_3)_2CHCH_2CH_2OH$ 3-methyl-1-butanol
3-Methylbuttersäure *f* 3-methylbutyric acid, isovaleric acid
3-Methylbutyraldehyd *m s.* 3-Methylbutanal
2-Methylchinolin *n* 2-methylquinoline, quinaldine
4-Methylchinolin *n* 4-methylquinoline, lepidine
Methylchlorid *n* CH_3Cl chloromethane, methyl chloride
2-Methyl-4-chlorphenoxyessigsäure *f* 2-methyl-4-chlorophenoxyacetic acid, MCPA *(a herbicide)*
Methylchlorsilan *n* methylchlorosilane
Methylen *n* 1. $=CH_2$ methylene *(radical)*; 2. $R_1-\overline{C}-R_2$ carbene, *(specif)* $|CH_2$ carbene, methylene
Methylenbernsteinsäure *f* methylene-succinic acid, itaconic acid, 2-propene-1,2-dicarboxylic acid
Methylenblau *n* methylene blue
Methylenblau[reduktions]probe *f* methylene-blue test
Methylenbromid *n* CH_2Br_2 dibromomethane, methylene dibromide
Methylenbrücke *f (org chem)* methylene bridge
Methylenchlorid *n* CH_2Cl_2 dichloromethane, methylene dichloride
Methylenhalogenid *n* methylene halogenide (halide)
Methylenjodid *n* CH_2I_2 diiodomethane, methylene diiodide
Methylenzyanid *n* $CH_2(CN)_2$ methylene cyanide, malonitrile
Methylester *m* methyl ester
Methylformiat *n* $HCOOCH_3$ methyl formate
***N*-Methylglykokoll** *n* CH_3NHCH_2COOH *N*-methylglycine, sarcosine
Methylglyoxal *n* $CH_3-CO-CHO$ methylglyoxal, pyruvic aldehyde, 2-oxopropionaldehyde
Methylgruppe *f* CH_3- methyl group (residue)
ringständige M. ring-methyl group

Methylhalogenid *n* methyl halogenide (halide)
methylieren to methylate
Methylierung *f* methylation
erschöpfende M. exhaustive methylation
Methyljodid *n* CH_3I iodomethane, methyl iodide
Methylkautschuk *m* methyl rubber
Methyllithium *n* CH_3Li methyllithium
Methylmagnesiumjodid *n* CH_3MgI methylmagnesium iodide *(a Grignard reagent)*
Methylmaleinsäure *f* $CH_3C(COOH)=CHCOOH$ methylmaleic acid, citraconic acid, *cis*-methylbutenedioic acid
Methylmethakrylat *n* $CH_2=C(CH_3)COOCH_3$ methyl methacrylate, methyl 2-methylacrylate
Methylmethan *n* CH_3CH_3 methylmethane, ethane
Methylnaphthalin *n* $C_{10}H_7CH_3$ methylnaphthalene
Methylnatrium *n* CH_3Na methylsodium
Methylnitroanilin *n* $CH_3C_6H_3(NO_2)NH_2$ methylnitroaniline
Methylolharnstoff *m* $HOCH_2NH{-}CO{-}NH_2$ methylolurea, hydroxymethylurea
Methylorange *n* methyl orange, sodium *p*-(*p*-dimethylaminophenylazo)benzenesulphonate
2-Methylpentanol-(3) *n* $(CH_3)_2CH-CHOH-CH_2CH_3$ 2-methyl-3-pentanol
4-Methylpentansäure *f* $(CH_3)_2CHCH_2CH_2COOH$ 4-methylpentanoic acid, 4-methylvaleric acid
Methylphenol *n* $CH_3C_6H_4OH$ methylphenol, hydroxytoluene, cresol
Methylphenyläther *m* $CH_3OC_6H_5$ methyl phenyl ether, methoxybenzene, anisole
Methylphenylketon *n* $C_6H_5-CO-CH_3$ methyl phenyl ketone, acetophenone, acetylbenzene
Methylphenylsilikon *n* methyl phenyl silicone
Methylphenylsilikonharz *n* methyl phenyl silicone resin
Methylphenylsilikonöl *n* methyl phenyl silicone fluid
2-Methylpropan *n* $(CH_3)_2CHCH_3$ 2-methylpropane, isobutane
2-Methylpropandisäure *f* $CH_3CH(COOH)_2$ 2-methylpropanedioic acid, methylmalonic acid, ethane-1,1-dicarboxylic acid
2-Methylpropanol-(1) *n* $(CH_3)_2CHCH_2OH$ 2-methyl-1-propanol
2-Methylpropanol-(2) *n* $(CH_3)_3COH$ 2-methyl-2-propanol
2-Methyl-1-propanthiol *n* $CH_3CH(CH_3)CH_2SH$ 2-methyl-1-propanethiol
2-Methylpropen *n* $(CH_3)_2C=CH_2$ 2-methylpropene
2-Methylpropionsäure *f* $(CH_3)_2CHCOOH$ 2-methylpropionic acid, isobutyric acid
Methylpropylazetylen *n* $CH_3C{\equiv}CCH_2CH_2CH_3$ 2-hexyne, methylpropylacetylene
Methylpropylketon *n* $CH_3-CO-CH_2CH_2CH_3$ methyl propyl ketone, 2-pentanone
Methylradikal *n* methyl radical, *(specif)* methyl free radical
Methylrest *m s.* Methylgruppe
Methylsalizylat *n* $HOC_6H_4COOCH_3$ methyl salicylate
Methylsilikon *n* methyl silicone
Methylsilikongummi *m* methyl silicone rubber
Methylsilikonharz *n* methyl silicone resin
Methylsilikonöl *n* methyl silicone fluid (oil)
Methylsubstituent *m* methyl substituent
Methylsulfat *n s.* Dimethylsulfat
Methylsulfoxid *n s.* Dimethylsulfoxid
Methylthioäthan *n* $CH_3SC_2H_5$ methyl thioethane, methyl ethyl sulphide
Methylthiophen *n* methylthiophene, thiotolene
Methyltrichlorsilan *n* CH_3SiCl_3 methyltrichlorosilane
Methylviolett *n* methyl violet
Methylzellosolve *n* methylcellosolve, ethylene glycol monomethyl ether
Methylzellulose *f* methyl cellulose
Methylzyklohexan *n* $CH_3C_6H_{11}$ methylcyclohexane, hexahydrotoluene
Methymyzin *n* methymycin *(antibiotic)*
Metol-Hydrochinon-Entwickler *m (phot)* metolhydroquinone developer, M.Q. developer
Mevalonsäure *f* $HOH_2CCH_2C(CH_3)(OH)CH_2COOH$ mevalonic acid, MVA, 3,5-dihydroxy-3-methylvaleric acid
Mezkalin *n s.* Meskalin
MF-Induktionsofen *m* coreless induction furnace
mgl *s.* maschinenglatt
Mgl *s.* Maschinenglätte
MH *s.* Maleinylhydrazin
Miargyrit *m (min)* miargyrite *(antimony(III) silver sulphide)*
miarolitisch *(min)* miarolitic, drusy
Miazin *n* pyrimidine, 1,3-diazine, miazine
Michael-Addition *f (org chem)* Michael condensation (reaction)
Miemit *m (min)* miemite *(a variety of dolomite)*
Migma *n (geol)* migma
Migmabildung *f*, **Migmatisierung** *f (geol)* migmatization
Migränemittel *n* anticephalalgic
Migration *f* migration *(as of ions or petroleum)*
Migrationsfläche *f (nucl)* migration area
Migrationslänge *f (nucl)* migration length
Migrationsstrom *m* migration current
migrieren to migrate *(as of ions or petroleum)*
Mikrinit *m (coal)* micrinite *(a maceral)*
Mikroanalyse *f* microanalysis
mikroanalytisch microanalytic[al]
Mikroaufnahme *f* photomicrograph
Mikrobe *f* microbe, microorganism ✦ **von Mikroben erzeugt** microbial-derived
Mikrobestandteil *m* microconstituent, microcomponent
Mikrobestimmung *f* microdetermination, microestimation
mikrobiell microbial, microbian, microbic ✦ **mikrobiellen Ursprungs** microbial-derived
Mikrobiologie *f* microbiology
industrielle (technische) M. industrial microbiology
mikrobiologisch microbiologic[al]
mikrobizid microbicidal *(killing microorganisms)*
Mikrobombe *f (lab)* microbomb

Mikrobrenner *m* microburner
Mikrobürette *f* microburette
Mikrochemie *f* microchemistry
mikrochemisch microchemical
Mikrochromatografie *f* microchromatography
Mikrodehnung *f (text)* micro length stretch, ML *(before resin treatment)*
Mikrodestillation *f* microdistillation
Mikrodichtemesser *m* microdensitometer
Mikroelektrode *f* microelectrode
Mikroelektrophorese *f* microelectrophoresis
Mikroelement *n s.* Mikronährstoff
mikrofibrillär microfibrillar
Mikrofibrille *f* microfibril
Mikrofoto *n* photomicrograph
Mikrofotografie *f* 1. photomicrography; 2. photomicrograph
Mikrogasanalyse *f* micro gas analysis
Mikrogefüge *n* microstructure, microscopic structure
Mikrogel *n (coll)* microgel
Mikrogramm-Methode *f* microgram method *(analytical chemistry)*
Mikrohärte *f* microhardness
Mikroheterogenität *f* microheterogeneity
Mikrohydrierung *f* microhydrogenation
Mikrokalorimeter *n* microcalorimeter
mikrokalorimetrisch microcalorimetric
Mikroklin *m (min)* microcline, amazonite, amazonstone *(aluminium potassium silicate)*
Mikrokolorimeter *n* microcolorimeter
Mikrokomponente *f* microcomponent
mikrokristallin[isch] microcrystalline
Mikroküvette *f* microcell
Mikrolith *m (min)* microlite *(an oxidic mineral containing tantalum and niobium)*
Mikrolithotype *f (coal)* microlithotype, banded component (constituent)
Mikromanipulator *m* micromanipulator
Mikromanometer *n* micromanometer
Mikrometerschraube *f* [micrometer] caliper
Mikromethode *f* micromethod
  M. von Rast Rast micromethod, Rast [molecular weight, camphor] method
mikromolekular micromolecular
Mikron *n* micron *(dispersed particle visible in an ordinary microscope)*
Mikronährstoff *m (agric)* micronutrient, microelement, minor [nutrient] element, [nutritional] trace element
Mikroorganismus *m* microorganism, microbe
Mikroparaffin *n* microparaffin, micro wax
Mikrophysik *f* microphysics
Mikropipette *f* micropipette, teat pipette
Mikropore *f* micropore
mikroporös microporous
Mikropyrometer *n* micropyrometer
Mikroradiometer *n* microradiometer
Mikroreagenzglas *n* micro test tube
Mikrosensor *m* microsensor *(e. g. an automatic gas detector)*
Mikroskop *n* microscope
Mikroskopie *f* microscopy
  M. mit Röntgenstrahlen X-ray microscopy
mikroskopisch microscopic[al]
Mikroskopobjektträger *m* microscope slide
Mikrospatel *m* micro-spatula
Mikrostruktur *f* microstructure, microscopic structure
Mikrosublimation *f* microsublimation
Mikrotechnik *f* microtechnique, microtechnic, microscopic technique, micrology
Mikrotitration *f* microtitration, microanalytic[al] titration
Mikrountersuchung *f* microexamination
Mikrovitrain *m*, **Mikrovitrit** *m (coal)* microvitrain
Mikrowaage *f* microbalance, microchemical balance
Mikrowachs *n (incorrectly for)* Mikroparaffin
Mikrowelle *f* microwave
Mikrowellenspektroskop *n* microwave spectroscope
Mikrowellenspektroskopie *f* microwave spectroscopy
Mikrowellenspektrum *n* microwave spectrum
Mikrozustand *m (phys chem)* microstate
Milbenbekämpfungsmittel *n*, **Milbengift** *n* acaricide, miticide
milbentötend acaricidal, miticidal
Milch *f* milk
  dickgelegte M. soured (fermented, cultured) milk
  eingedampfte (eingedickte) M. *s.* kondensierte M.
  entrahmte M. skim (skimmed, separated, fatfree) milk
  evaporierte M. *s.* kondensierte M.
  gereifte M. ripened (acidified) milk
  geronnene M. curd[s]
  gesäuerte M. *s.* gereifte M.
  kondensierte M. condensed (evaporated, concentrated) milk
  mit Säureweckern gesäuerte (versetzte) M. *s.* dickgelegte M.
  rekonstituierte M. reconstituted milk *(redissolved milk powder)*
  saure M. sour milk
  spontan gesäuerte M. spontaneously (naturally) soured milk
  sterilisierte M. sterilized milk
  teilentrahmte M. partly skimmed milk
  walzengetrocknete M. roller[-dried] milk powder
  weichgerinnende M. soft-curd milk
Milchabsonderung *f* milk secretion
Milchalbumin *n* lactalbumin, milk albumin
Milchannahmebehälter *m* milk storage vessel
milchartig milk-like, milky
Milchbehandlung *f* milk treatment
Milchbestandteile *mpl* milk constituents
Milchbutyrometer *n* lactobutyrometer
Milchchemie *f* dairy chemistry
Milchdauerwaren *fpl* milk preserves
Milcheiweiß *n* milk protein
Milchenzym *n* milk enzyme

Milcherhitzer *m*, **Milcherhitzungsapparat** *m* [milk] pasteurizer
Milcherzeugnis *n* milk product
Milcherzeugung *f* milk production
Milchfehler *m* milk defect
Milchferment *n* milk enzyme
Milchfett *n* milk fat
  wasserfreies M. anhydrous milk fat
Milchfettsynthese *f* milk fat synthesis
Milchgefäß *n (bot)* laticiferous (latex) vessel
Milchglas *n* milk glass
Milchglobulin *n* milk globulin
milchhaltig lactiferous, *(bot also)* laticiferous
milchig milky, *(as opposed to translucent also)* opaque
Milchindustrie *f* dairy industry
Milchkatalase *f* milk catalase
Milchkühler *m* milk cooler
Milchleistung *f* milk yield
Milchlipase *f* milk lipase
Milchmargarine *f* milk margarine
Milchopal *m (min)* milk opal *(silicon(IV) oxide)*
Milchphase *f* milk phase *(in margarine making)*
Milchprodukt *n* dairy (milk) product
  gesäuertes M. cultured dairy (milk) product
Milchprotein *n* milk protein
Milchpulver *n* milk powder, dry (dried, desiccated, powdered) milk
Milchpulvermilch *f* reconstituted milk
Milchquarz *m* milky quartz
Milchreifung *f* milk ripening
Milchröhre *f (bot)* laticiferous (latex) tube
  gegliederte M. laticiferous (latex) vessel
  ungegliederte M. laticiferous (latex) cell
Milchsaft *m (bot)* latex, milky sap (juice), milk
milchsaftführend *(bot)* laticiferous, latex-bearing
Milchsaftgefäß *n (bot)* laticiferous (latex) vessel
Milchsaftschlauch *m (bot)* laticiferous (latex) tube
Milchsaftzelle *f (bot)* laticiferous (latex) cell
Milchsalz *n* milk salt
Milchsäuerung *f* souring of milk
Milchsäure *f* lactic acid, hydroxypropionic acid, *(specif)* $CH_3CH(OH)COOH$ lactic acid, 2-hydroxypropionic acid
  gewöhnliche M. *s. DL*-Milchsäure
  linksdrehende M. *s. D*-Milchsäure
d-**Milchsäure** *f s. L*-Milchsäure
D-**Milchsäure** *f D*(-)-lactic acid, laevolactic acid
dl-**Milchsäure** *f s. DL*-Milchsäure
DL-**Milchsäure** *f DL*-lactic acid, ordinary lactic acid
l-**Milchsäure** *f s. D*-Milchsäure
L-**Milchsäure** *f L*(+)-lactic acid, dextrolactic acid
Milchsäureanhydrid *n* $[CH_3CH(OH)CO]_2O$ lactic anhydride
Milchsäureäthylester *m* $CH_3CH(COOC_2H_5$ ethyl lactate
Milchsäurebakterien *pl* lactic-acid[-producing] bacteria
Milchsäurebazillus *m* lactobacillus
Milchsäurebildner *m* lactic-acid organism
Milchsäuredehydrase *f s.* Milchsäuredehydrogenase
Milchsäuredehydrogenase *f* lactic dehydrogenase
Milchsäureerzeuger *m s.* Milchsäurebildner
Milchsäuregärung *f* lactic[-acid] fermentation
Milchsäuremikrobe *f* lactic-acid organism
Milchsäurenitril *n s.* Laktonitril
Milchschleuder *f s.* Milchzentrifuge
Milchschokolade *f* milk chocolate
Milchsekretion *f* milk secretion
Milchseparator *m s.* Milchzentrifuge
Milchserum *n* milk serum, lactoserum, whey
Milchserumprotein *n* whey (milk serum) protein
Milchspindel *f* lactodensimeter
Milchstein *m* milk stone
Milchsteinentferner *m* milk stone remover
Milchtrockenmasse *f* milk solids
Milchverarbeitung *f* milk processing
Milchvitamin[is]ierung *f* milk vitaminizing
Milchvorratstank *m* milk storage vessel
Milchwirtschaft *f* dairying
milchwirtschaftlich dairy
Milchzelle *f (bot)* laticiferous (latex) cell
Milchzentrifuge *f* milk centrifuge, milk (cream) separator, skimming machine
Milchzucker *m* lactose, lactobiose, milk sugar
Milchzuckergärung *f* lactose fermentation
mild mild, bland *(e.g. taste or remedy)*; mild *(reaction conditions)*
Milieu *n* environment
Miller-Indizes *mpl* Miller [crystal] indices
Millerit *m (min)* millerite *(nickel(II) sulphide)*
Milliäquivalent *n s.* Milligrammäquivalent
Milligrammäquivalent *n* milliequivalent, mequiv, meq
Milligramm-Methode *f* milligram method *(analytical chemistry)*
Milliliter *n* milliliter, ml., mil
Millival *n s.* Milligrammäquivalent
Mills-Packard-Kammer *f* Mills-Packard chamber *(for producing sulphuric acid)*
Mimeografenfarbe *f* mimeograph ink
Mimetesit *m (min)* mimet[es]ite, mimetene *(a lead arsenate and chloride)*
Mimosengummi *n (incorrectly for)* Akaziengummi
Mimosin *n* mimosine, leucenine *(an amino acid deriving from pyridine)*
Minderheitsträger *m* minority carrier *(in a semiconductor)*
mindern to lower, to reduce, to decrease, to diminish
  im Wert m. to deteriorate
Minderung *f* lowering, reduction, decrease, diminution
minderwertig low-grade
Mindestbodenzahl *f (distil)* minimum number of plates (trays)
Mindestdosis *f* minimum dosage (dose)
Mindestfettgehalt *m* minimum fat content
Mindestrücklaufverhältnis *n* minimum reflux ratio
Mindesttrennstufenzahl *f s.* Mindestbodenzahl

Mine *f* mine
Mineral *n* mineral
akzessorisches M. accessory mineral (component, constituent)
allothigenes M. allothigenic mineral
authigenes M. authigene mineral
beigemengtes M. *s.* akzessorisches M.
gesteinsbildendes M. rock-forming mineral
kritisches M. critical mineral
nichtmetallisches M. non-metallic mineral
primäres M. primary mineral
schweres M. heavy mineral
sekundäres M. secondary mineral
typomorphes M. typomorphic mineral
Mineralaggregat *n* mineral aggregate
Mineralassoziation *f* mineral association
Mineralaufbereitung *f* mineral dressing
Mineralaustauscher *m* / **künstlicher** *s.* Zeolith / künstlicher
Mineralchemie *f* mineral (mineralogical) chemistry
Mineraldüngemittel *n*, **Mineraldünger** *m* mineral (inorganic) fertilizer
Mineralfaser *f* mineral fibre
Mineralfaserstoff *m* mineral fibre
Mineralfazies *f* mineral facies
mineralfrei mineral-matter-free, mmf
Mineralgang *m (geol)* mineralized lode
Mineralgerbung *f* mineral tanning
Mineralhefe *f* mineral yeast
Mineralisation *f (soil)* mineralization
Mineralisator *m (ceram)* mineralizer, accelerator
mineralisch mineral
mineralisieren *(soil)* to mineralize
Mineralisierung *f (soil)* mineralization
Mineralkermes *m* kermes mineral *(a double compound of antimony trisulphide and antimony trioxide)*
Mineralkohle *f* mineral (natural, fossil) coal
Mineralkombination *f (geol)* mineral association
Mineralkortikoid *n* mineralocorticoid *(a hormone)*
Mineralkunde *f* mineralogy
Mineralmörser *m* diamond (percussion) mortar
Mineralneubildung *f* neomineralization
Mineralog[e] *m* mineralogist
Mineralogie *f* mineralogy
Mineralöl *n* mineral oil
Mineralöltechnologie *f* mineral-oil technology
Mineralphosphat *n* rock (mineral) phosphate
Mineralpigment *n* synthetic inorganic pigment
Mineralprovinz *f (geol)* mineral province
Mineralquelle *f* mineral spring
Mineralsalz *n* mineral salt
Mineralsalzernährung *f (agric)* mineral nutrition
Mineralsäure *f* mineral acid
Mineralschwarz *n* slate black, black chalk *(a natural pigment)*
Mineralstoff *m* mineral matter (substance)
Mineralstoffdüngung *f* mineral (inorganic) fertilization
mineralstofffrei mineral-matter-free, mmf
Mineralstoffgehalt *m* mineral content
Mineralstoffwechsel *m* mineral metabolism
Mineralsubstanz *f s.* Mineralstoff
Mineralvergesellschaftung *f (geol)* mineral association
Mineralverwitterung *f* mineral weathering
Mineralwasser *n* mineral water
natürliches M. natural mineral water
Mineralwolle *f* mineral wool (cotton), rock wool
Minimaltemperatur *f* minimum temperature
Minimum *n* minimum
Minimum-Siedepunkt *m* minimum boiling point
Minoritäts[ladungs]träger *m* minority carrier *(in a semiconductor)*
Minton-Ofen *m (ceram)* Minton oven
Miotikum *n (pharm)* miotic, myotic
miotisch *(pharm)* miotic, myotic
Mirabilit *m (min)* mirabilite *(sodium sulphate-10-water)*
Mirbanessenz *f*, **Mirbanöl** *n (cosmet)* mirbane (myrbane) oil, essence of mirbane (myrbane) *(nitrobenzene)*
Mischanilinpunkt *m* mixed aniline point
Mischanlage *f* blending (mixing) plant
Mischapparat *m* mixer
Mischarm *m* agitator (raking, rabble) arm, rabbler
mischbar miscible
begrenzt m. incompletely (partially) miscible
beliebig (in jedem Verhältnis) m. miscible in all proportions
leicht m. freely miscible
nicht m. immiscible, non-miscible
teilweise m. partially (incompletely) miscible
Mischbarkeit *f* miscibility
begrenzte (teilweise) M. incomplete (partial) miscibility
Mischbehälter *m* mixing (blending) tank, mixing vessel (vat, chest, box), mixer bowl
Mischbett *n* mixed bed
Mischbettaustauscher *m*, **Mischbettfilter** *n* mixed-bed [ion] exchanger, mixed-bed deionizer (demineralizer)
Mischblende *f* orifice mixer
Mischbottich *m* mixing chest (vat)
Mischbunker *m* mixing bin
Mischbütte *f s.* Mischbottich
Mischdünger *m* 1. mixed (dry-blended) fertilizer; 2. *(incorrectly for)* Mehrnährstoffdünger
Mischdüngerfabrik *f*, **Mischdüngerwerk** *n* mixed-fertilizer (bulk mixing) plant
Mischdüse *f* mixing nozzle
Mischelement *n* mixed element
mischen to mix, to blend; *(in accordance with a recipe)* to compound; *(esp of solids with a liquid to obtain a desired consistency)* to temper
auf dem Walzwerk m. *(rubber)* to mill[-mix]
im Bleichholländer m. *(pap)* to potch, to poach
mit Masterbatches (Vormischungen) m. *(rubber)* to masterbatch
Mischen *n* **im Fertigtank** *(petrol)* batch blending
M. in der Pumpleitung *(petrol)* in-line blending
M. mit Masterbatches (Vormischungen) *(rubber)*

masterbatch method of mixing, masterbatching

M. von Feststoffkomponenten dry (solid-solid) mixing (blending)

Mischer *m* 1. mixer, blender; 2. *s.* Mischbehälter

M. mit Wechselbehälter change-can mixer

geschlossener M. closed mixer

Mischer-Abscheider *m* mixer-settler *(for solvent extraction)*

Mischerbehälter *m s.* Mischbehälter

Mischerschaufel *f* mixing blade

Mischerz *n* complex ore

Mischfarbe *f* 1. mixed colour *(phenomenon)*; 2. *(incorrectly for)* Mischfarbstoff

Mischfarbstoff *m* mixed colouring matter; *(if soluble)* mixed dye

Mischfaserfarbstoff *m* union dye

Mischfett *n* compound fat (lard) *(a mixture of solid animal fats with liquid animal fats or vegetable oils)*

Mischflügel *m* mixing blade, agitator [blade]

Mischfolge *f (rubber)* order of adding materials (compounding ingredients)

Mischgarn *n* mixture yarn

Mischgas *n* mixed gas

Mischgefäß *n s.* Mischbehälter

Mischgeschwindigkeit *f* rate of mixing

Mischgestein *n* hybrid rock, migmatite

Mischgewebe *n* union fabric

Mischgewebefärben *n* union dyeing

Mischgut *n* material being (*or* to be) mixed (blended)

Mischholländer *m (pap)* [mixing] potcher, poacher, potching (poaching) engine

Mischindikator *m* mixed indicator

Mischkalk *m* compound lime fertilizer

Mischkammer *f* 1. mixing chamber *(as of a condenser)*; 2. *(rubber)* compounding room

Mischkatalysator *m* mixed catalyst

Mischkeramik *f* cermet, ceramal, ceramel, ceramet

Mischkessel *m* mixing vessel

Mischklebstoff *m* mixed adhesive

Mischkneter *m* kneader-mixer

Mischkollergang *m* muller mixer

Mischkomplex *m* mixed complex *(coordination chemistry)*

Mischkomponente *f* blend component

Mischkondensation *f (plast)* co-condensation

Mischkondensator *m* [direct-]contact condenser

barometrischer M. barometric condenser

nasser M. wet (jet) condenser

Mischkristall *m* mixed crystal, mix-crystal, solid solution

Mischkristallbildung *f* mixed-crystal formation

Mischkultur *f (biol)* mixed culture

Mischleiter *m* mixed conductor

Mischmaschine *f* mixing machine, mixer, blender

Mischmilch *f* mixed milk

Mischmolekül *n (nucl)* mixed molecule

Mischoktanzahl *f* blending octane number, blending value

Mischöl *n* mixed oil; mixed-base petroleum (crude oil, crude)

Mischoxid *n* mixed oxide

Misch-OZ *f s.* Mischoktanzahl

Mischpappe *f* mixed board

Mischphase *f* mixed phase

Mischpolymer[es] *n s.* Kopolymeres

Mischpolymerisat *n s.* Kopolymerisat

Mischpolymerisation *f s.* Kopolymerisation

mischpolymerisieren *s.* kopolymerisieren

Mischquirl *m* mixing blunger

Mischsalz *n* mixed salt

Mischsalzkatalysator *m*, **Mischsalzkontakt** *m* mixed-salt catalyst

Mischsäure *f (tech)* mixed acid, nitrating acid *(consisting of concentrated nitric acid and sulphuric acid)*

Mischschmelzpunkt *m* mixed melting point

Mischstromführung *f* mixed-feed operation *(as in a multiple-effect evaporator)*

Mischsystem *n* combined system of sewerage, combined sewer system *(municipal water treatment)*

Mischtank *m* mixing (blending) tank

Misch-Trenn-Behälter *m* mixer-settler *(for solvent extraction)*

Mischtrommel *f* blending drum

Mischung *f* 1. mixing, blending; 2. mixture, mix, blend; *(rubber)* compound, stock, composition

M. für die Schlauchseele *(rubber)* inner tube compound

M. für Kordgummierung *(rubber)* cord-rubberizing compound

M. für Walzen *(rubber)* roll compound

azeotrope M. azeotropic mixture, azeotrope

eutektische M. eutectic mixture, eutectic

konjugierte M. *(phys chem)* conjugate solution

rußgefüllte (rußhaltige) M. *(rubber)* [carbon] black compound

schwach gefüllte M. *(rubber)* low-load compound

ungefüllte M. *(rubber)* pure gum compound

unvulkanisierte M. *(rubber)* green compound

Mischungsanteil *m* blend component

Mischungsbestandteil *m* ingredient of a mixture; *(rubber)* compounding ingredient

Mischungsentropie *f* entropy of mixing

Mischungsentwickler *m*, **Mischungsfachmann** *m (rubber)* compound designer, compounder

Mischungsgrad *m* degree of mixing

Mischungsherstellung *f (rubber)* compounding

Mischungskalorimeter *n* water calorimeter

Mischungslücke *f* miscibility gap

Mischungsmethode *f (phys chem)* method of mixture

Mischungspunkt *m* / **kritischer** *s.* Mischungstemperatur / kritische

Mischungsrezept *n*, **Mischungsrezeptur** *f (rubber)* compound[ing] formula, mix[ing] formula, compounding recipe, recipe of mix

Mischungstemperatur *f* / **kritische** consolute (critical solution) temperature

obere kritische M. upper consolute temperature
untere kritische M. lower consolute temperature
Mischungsverhältnis *n* mixing (blending) ratio, formula
Mischungswärme *f* heat of mixing
Mischverfahren *n* 1. mixing process; 2. *s.* Mischsystem
Mischvorgang *m* mixing process
Mischvorrichtung *f* mixing device
Mischvorschrift *f* mixing instruction, *(rubber also)* order of milling
Mischwalze *f* mixing (homogenizing) roll
Mischwalzwerk *n* mixing mill (rolls)
Mischwerk *n* mixer
Mispickel *m (min)* mispickel, arsenic[al] iron, arsenopyrite, arsenical pyrite *(iron sulpharsenide)*
Mißbildung *f* malformation
Mißfärbung *f (glass)* discoloration
Mistgas *n* fermentation gas
Miszella *f* miscella *(an extractant containing an extracted oil or grease)*
Miszellaschleuder *f* miscella filter
mitfallen to coprecipitate
Mitfallen *n* coprecipitation
mitfällen to coprecipitate
Mitfällung *f* **[/ induzierte]** coprecipitation
mitführen to entrain, to carry over (off) *(as in a gas stream)*
Mitführen *n* entrainment, carry-over *(as in a gas stream)*
Mitisgrün *n s.* Grün / Schweinfurter
Mitläufer *m (rubber)* wrapper, leader; *(text)* back [grey] cloth
Mitläufergewebe *n*, **Mitläuferstoff** *m (rubber)* lining, liner
Mitnehmer *m* 1. *(distil)* entrainer, entraining (azeotroping) agent, azeotrope-former; 2. flight *(conveying)*
Mitnehmerblech *n* lifting flight (plate) *(as of a drying drum)*
Mitnehmerstange *f (petrol)* kelly
Mitomyzin *n* mitomycin *(antibiotic)*
Mitosegift *n (biol)* mitotic poison
Mitoseindex *m (biol)* mitotic index
mitreißen to entrain, to carry over (off) *(as in a gas stream)*
Mitreißen *n* entrainment, carry-over *(as in a gas stream)*
Mittel *n* 1. agent; *(pharm)* preparation, remedy; 2. *s.* Mittelwert
M. gegen Beschlagen antidimming agent, antidimmer
M. gegen Bluthochdruck *s.* blutdrucksenkendes M.
M. gegen depressive Verstimmung antidepressant
M. gegen Durchfall antidiarrhoeal, styptic
M. gegen Epilepsie antiepileptic
M. gegen Erbrechen antiemetic
M. gegen Festfressen *(plast)* antiseize agent
M. gegen Gicht antiarthritic
M. gegen Kesselstein descaling agent
M. gegen Krätze scabicide, scabieticide
M. gegen Malaria antimalarial
M. gegen Nervenschmerzen antineuralgic
M. gegen Rheuma[tismus] antirheumatic
M. gegen Rückvergrauung *(text)* antiredeposition agent
M. gegen Schaumbildung antifoam[ing] agent, antifoamer, antifrothing (froth-preventing) agent, foam inhibitor (killer, breaker, destroyer)
M. gegen Schnecken molluscacide, molluscide
M. gegen Wasserschnecken aquatic molluscacide
M. gegen Zuckerkrankheit antidiabetic
M. zur Herabsetzung der Viskosität viscosity depressant
M. zur pH-Regelung pH regulator *(floatation)*
absetzverhinderndes M. antisettling agent
absorptionsbeschleunigendes (absorptionsförderndes) M. absorbefacient
adstringierendes M. astringent, styptic
aktivierendes M. 1. activator *(floatation)*; 2. *(met)* energizer *(for promoting carburization)*
alkylierendes M. alkylating agent
analgetisches M. analgesic, pain-reliever, pain-killer
anregendes M. *(pharm)* analeptic, central nervous system stimulant
antikonzeptionelles M. contraceptive
antiperspirierendes M. *s.* schweißhemmendes M.
antiseptisches M. antiseptic [agent]
antistatisches M. antistatic [agent]
appetitanregendes M. stomachic
arithmetisches M. arithmetic mean *(statistics)*
aufkohlendes M. *(met)* carburizing agent, [case-hardening] carburizer
auswurfförderndes M. *(pharm)* expectorant
bakteriostatisches M. bacteriostat[ic]
belebendes M. activator *(floatation)*
blähungstreibendes M. *(pharm)* carminative
blasenziehendes M. *(pharm)* vesicant
blutdrucksenkendes M. antihypertensive (hypotensive) drug, blood pressure depressant
blutstillendes M. haemostatic, styptic
chemotherapeutisches M. chemotherapeutic agent
depilierendes M. depilator, depilatory [agent], depilitant, hair remover
desinfizierendes M. disinfectant
desodor[is]ierendes M. deodorizer, deodorant
die Milchsekretion förderndes M. milk-ejecting agent, galactagogue
dispergierendes M. dispersing agent, dispersant, dispergator
diuretisches M. diuretic
drückendes M. depressant *(floatation)*
empfängnisverhütendes M. contraceptive
endometatoxisches M. *(agric)* systemic poison
entzündungshemmendes M. antiphlogistic

erweichendes M. *(tech)* softener, softening agent; *(cosmet)* emollient, softener
feuerhemmendes M. fire-retardant agent, fire retardant
fiebersenkendes M. antipyretic
galenisches M. *(pharm)* galenical
gefäßerweiterndes M. vasodilator, vasodepressor
gefäßkontrahierendes (gefäßverengendes) M. vasoconstrictor, vasoexcitor
geometrisches M. geometric mean *(statistics)*
gerinnungshemmendes (gerinnungsverzögerndes) M. anticoagulant
geruchsbeseitigendes (geruchszerstörendes) M. deodorizer, deodorant
gewogenes M. weighted mean *(statistics)*
harntreibendes M. diuretic
hauterweichendes M. skin softener
hydrophobierendes M. hydrophobing agent, water repellent
innertherapeutisches M. *s.* systemisches M.
insektenabschreckendes (insektenvertreibendes) M. insect repellent, insectifuge
keimbildendes M. *(cryst)* nucleation (nucleating) agent
keimtötendes M. disinfectant, germicide
koagulierendes M. coagulating agent, coagulant, coagulator
kohlendes M. *s.* aufkohlendes M.
Konvulsionen auslösendes (erregendes) M. convulsant
korrodierendes M. corrosive agent
kräftigendes M. tonic
krampflösendes M. antispasmodic, spasmolytic
maskierendes M. masking agent
mikrobizides M. microbicide
oberflächenaktives M. surface-active agent, surfactant
örtlich schmerzstillendes M. local analgesic
oxydierendes M. oxidizing agent, oxidant
passivierendes M. 1. depressant *(floatation)*; 2. passivator *(for metals)*
pharmazeutisches M. pharmaceutic[al]
pilztötendes M. fungicide
protektives (protektiv wirkendes) M. protective
pupillenerweiterndes M. mydriatic
pupillenverengendes M. miotic, myotic
reduzierendes M. reducing agent, reductant, reducer, reductive
regelndes M. modifying agent, modifier *(floatation)*
schaumerzeugendes M. foaming (frothing, froth-forming) agent, foamer, frother
schleierdämpfendes (schleierverhütendes, schleierwidriges) M. *(phot)* antifog[ging] agent, antifoggant
schleimlösendes M. *(pharm)* expectorant
schmerzlinderndes (schmerzstillendes) M. analgesic, pain-reliever, pain-killer
Schüttelkrämpfe auslösendes (erregendes) M. convulsant
schwangerschaftsverhütendes M. contraceptive
schweißhemmendes (schweißlinderndes) M. antiperspirant, antihidrotic, perspiration check
schweißtreibendes M. sudorific, diaphoretic
spermienabtötendes (spermizides) M. *(pharm)* spermatocide, spermicide
stärkendes M. tonic
stopfendes M. *(pharm)* styptic
sulfonierendes M. sulphonating agent
systemisches M. systemic [chemical, poison, insecticide]
tonisierendes M. tonic
verdauungsförderndes M. digestive stimulant, digester
virentötendes (viruzides) M. virucide, viricide, viricidal agent
vorbeugendes (vorbeugend wirkendes) M. prophylactic, protective
wasserabspaltendes M. dehydrating agent, dehydrator
wasserabstoßendes (wasserabweisendes) M. water repellent
wasserdichtmachendes M. waterproofing agent
wasserentziehendes M. dehydrating agent, dehydrator
wirksames M. agent
wurmabtreibendes M. helminthagogue, vermifuge
wurmtötendes M. helminthicide, vermicide
wurmwidriges M. anthelmint[h]ic
zusammenziehendes M. *(pharm)* astringent, styptic

mittelaktiv *(nucl)* medium-level active
Mittelbenzin *n* medium[-heavy] gasoline
Mittelbrechen *n* intermediate crushing
Mitteldestillat *n* middle distillate
Mitteldrucksynthese *f* medium-pressure synthesis
mittelfein moderately fine
mittelfeinkörnig medium-grained
Mittelfraktion *f* middle fraction
Mittelfrequenzinduktionsofen *m* high-frequency induction furnace, coreless induction furnace
mittelgekohlt medium-carbon *(e. g. steel)*
mittelgrob moderately coarse
Mittelgut *n* middlings product, intermediate material *(classifying)*
Mittelhartzerkleinerung *f* size reduction of medium-hard materials
Mittelkammer *f s.* Mittelraum
Mittelkorn *n* middle grain
mittelkörnig medium-grained
Mittelöl *n* middle oil
Mittelpech *n* medium-soft pitch
Mittelprodukt *n* *(sugar)* intermediate product; *s.* Mittelgut
Mittelproduktfüllmasse *f* *(sugar)* intermediate massecuite, second fillmass
Mittelproduktzucker *m* *(sugar)* intermediate product

Mittelraum *m* middle chamber (compartment) *(as of an electrolytic cell)*
Mittelsand *m* medium-grained sand
Mittelschneide *f* principal (central, centre) knife-edge *(of a balance)*
Mittelschwerbenzin *n s.* Mittelbenzin
Mittelsieder *m* medium boiler *(e. g. a solvent)*
mittelständig centrally located
Mittelstellung *f* intermediate position
Mitteltemperaturentgasung *f s.* Mitteltemperaturverkokung
Mitteltemperaturkoks *m* medium-temperature coke
Mitteltemperaturverkokung *f* medium-temperature carbonization
Mitteltöne *mpl (phot)* middle tones
mittelviskos moderately viscous, medium-viscosity
Mittelwand *f (pap)* midfeather, mid-wall, midriff, centre division *(of a Hollander beater)*
Mittelwert *m* average, av., mean [value]
mitten to centre
Mitten *n* centr[e]ing
mittig on-centre
Mixer *m* mixer
Mixer-Settler-Anlage *f,* **Mixer-Settler-Extraktor** *m* mixer-settler unit
Mixit *m (min)* mixite *(a hydrous basic bismuth copper arsenate)*
Mixtur *f (pharm)* mixture
Mizell *n s.* Mizelle
mizellar micellar
Mizellarstrang *m* micellar string
Mizellarstruktur *f* micellar structure
Mizellartheorie *f* micellar theory (hypothesis)
Mizellbildungskonzentration *f* **/ kritische** *(coll)* critical micelle concentration, c.m.c.
Mizelle *f* micelle, micell[a], *(in fibrous material also)* crystallite
Mizellkolloid *n* micellar colloid
Mizzonit *m (min)* mizzonite *(a tectosilicate)*
m-Lösung *f* molar solution (*for further compounds s.* Lösung)
MMK *s.* Kraft / magnetomotorische
MO *s.* Molekülorbital
mobil mobile
Modakrylfaser *f* modacrylic (modified acrylic) fibre
Modakrylfaserstoff *m* modacrylic (modified acrylic) fibre
Modakrylnitrilfaser *f* modacrylonitrile fibre
Modakrylnitrilfaserstoff *m* modacrylonitrile fibre
Modeldruck *m (text)* block printing
Modell *n* model
  ausschmelzbares (verlorenes) M. investment (fusible alloy) pattern *(foundry)*
Modellplatte *f* pattern plate
Modellverbindung *f* model compound
Moder *m (soil)* duff *(one form of humus)*
Moderator *m,* **Moderatorsubstanz** *f (nucl)* moderator
moderieren *(nucl)* to moderate, to slow down
Modifikation *f* modification
  allotrope M. allotrope, allotropic form
  geometrisch isomere M. geometric[al] isomer
Modifikationsmittel *n* modifying agent, modifier
Modifikator *m (text)* modifier, modifying agent
modifizieren to modify
Modifizierer *m (glass)* network modifier
Modifizierung *f* modification
Modul *m* modulus
  Youngscher M. Young's modulus [of elasticity], elastic modulus
Moellon *n* moellon, degras *(a fatty substance used in dressing leather)*
Mohnalkaloid *n* opium alkaloid
Mohnöl *n* poppy[-seed] oil
Mohnsäure *f* meconic acid, 3-hydroxy-4-pyrone-2,6-dicarboxylic acid
Mohs-Skala *f* Mohs' [hardness] scale
Moiré-Effekt *m (text)* moiré effect
moirieren *(text)* to cloud
moiriert *(text)* cloudy
Mojonnier-Test *m (food)* Mojonnier [solids] test; Mojonnier [fat] test
Mokkastein *m (min)* Mocha stone *(chalcedony containing dendritic inclusions)*
Mol *n* [gram] mole, mol, gram molecule, gram-molecular weight
molal molal
Molalität *f* molality, molal concentration, kilogram molarity
molar molar
Molardispersion *f s.* Molekulardispersion
Molarität *f* molarity, molar concentration
Molarkonzentration *f s.* Molarität
Molch *m (petrol)* go-devil
Molekel *f* molecule *(for compounds s.* Molekül*)*
Molekül *n* molecule
  aktives (aktiviertes) M. active (activated) molecule
  angeregtes M. excited molecule
  energiereicheres M. *s.* aktives M.
  geknäueltes M. coiled molecule
  homöopolares (homöopolar gebundenes) M. homopolar molecule
  langgestrecktes M. long molecule
  mehratomiges M. polyatomic molecule
  nichtpolares M. non-polar molecule
  polares M. polar molecule
  unpolares M. non-polar molecule
  van-der-Waalssches M. van der Waals molecule
  vernetztes M. crosslinked molecule
  verzweigtes M. branched molecule
  vielatomiges M. polyatomic molecule
Molekülaggregat *n,* **Molekülaggregation** *f* aggregate of molecules, molecular aggregate
Molekülaktivierung *f* molecular activation
Molekülanziehung *f* molecular attraction
molekular molecular
Molekularaggregat *n,* **Molekularaggregation** *f s.* Molekülaggregat
Molekularasymmetrie *f* molecular asymmetry (dissymmetry)

Molekularattraktion *f* molecular attraction
Molekularbewegung *f* molecular motion (movement)
Brownsche M. Brownian motion (movement)
Molekularbiologie *f* molecular biology
Molekulardestillation *f* molecular distillation
Molekulardestillierapparat *m* molecular still
molekulardispers molecularly disperse
Molekulardispersion *f* molecular (molar) dispersion (dispersivity) *(difference in molar refraction)*
Molekulardrehung *f* molecular rotation
Molekularformel *f* molecular formula
empirische M. empirical molecular formula
Molekulargewicht *n* molecular weight
Molekulargewichtsbestimmung *f* molecular-weight determination
Molekulargewichtsverteilung *f* molecular weight distribution
Molekularität *f* molecularity *(of a reaction)*
Molekularkraft *f* [inter]molecular force
Molekularmasse *f s.* Molekülmasse
Molekularorbital *n s.* Molekülorbital
Molekularpolarisation *f* molecular (molar) polarization
Molekularpolarisierbarkeit *f* molecular (molar) polarizability
Molekularrefraktion *f* molecular (molar) refraction (refractivity)
Molekularrotation *f* molecular rotation
Molekularsieb *n* molecular sieve
Molekularstrahl *m* molecular beam
Molekularstrahlapparatur *f* molecular-beam apparatus
Molekularstrahlmessung *f* molecular-beam measurement
Molekularstrahlmethode *f* molecular-beam method (technique)
Molekularstrahlspektroskopie *f* molecular-beam spectroscopy
Molekularstrahlversuch *m* molecular-beam experiment
Molekularstruktur *f* molecular structure
Molekularverbindung *f* molecular compound
Molekülassoziation *f* molecular association
Molekülasymmetrie *f* molecular asymmetry (dissymmetry)
Moleküldissoziation *f* molecular dissociation
Moleküldurchmesser *m* molecular diameter
Moleküleigenfunktion *f* molecular eigenfunction
Molekülelektronenwolke *f* molecular electron cloud
Molekülformel *f* molecular formula
Molekülgeschwindigkeit *f* molecular velocity
Molekülgitter *n* molecular lattice
Molekülgruppe *f* molecular group
Molekülkette *f* molecular chain
Molekülkolloid *n* molecular colloid
Molekülkomplex *m* molecular complex
Molekülkristall *m* molecular crystal
Molekülmasse *f* 1. mass of a molecule; 2. *s.* relative M. ✦ **von geringer M.** low-molecular-weight
mittlere relative M. average molecular weight
relative M. molecular weight, M.W.
relative M. nach Staudinger Staudinger molecular weight
Molekülmodell *n* molecular model
Molekülorbital *n* molecular orbital, MO
Hückelsches M. Hückel molecular orbital, HMO
polyzentrisches M. polycentric molecular orbital
Molekülorbitalmethode *f* molecular-orbital method, Hund-Mulliken-[Lennard-Jones-]Hückel method
Molekülorbitalrechnung *f* molecular-orbital calculation
Molekülorbitaltheorie *f* molecular-orbital theory, Hund-Mulliken-[Lennard-Jones-]Hückel theory
Molekülorientierung *f* molecular orientation
Molekülphosphoreszenz *f* molecular phosphorescence
Molekülschicht *f* molecular layer
Molekülschwarm *m* swarm (bundle) of molecules
Molekülsieb *n* molecular sieve
Molekülspaltung *f* molecular cleavage
Molekülspektrum *n* molecular (band, banded) spectrum
Molekülspinorbital *n* molecular spin orbital
Molekülstrahl *m s.* Molekularstrahl
Molekülstruktur *f* molecular structure
Molekülverbindung *f* molecular compound
Molekülwellenfunktion *f* molecular wave function
Molenbruch *m* mole (molar) fraction
Moler *m*, **Molererde** *f* moler *(a kind of diatomaceous earth)*
Molettewasserzeichen *n (pap)* impressed (rubber-stamp) mark, *(Am)* press mark
Molgewichtsverteilung *f s.* Molekulargewichtsverteilung
Molisch-Reaktion *f* Molisch reaction *(for detecting carbohydrates)*
Molke *f* whey, [milk] serum, lactoserum
süße M. sweet whey
Molken *m s.* Molke
Molkenbutter *f* whey (serum) butter
Molkenkäse *m* whey cheese
Molkenmargarine *f* whey margarine
Molkenprotein *n* whey (milk serum) protein
Molkenpulver *n* whey powder, powdered (dried, dry) whey
Molkerei *f*, **Molkereibetrieb** *m* dairy, *(Am also)* creamery
Molkereibutter *f* dairy butter
Molkereierzeugnisse *npl*, **Molkereiprodukte** *npl* dairy products (foods)
Molkonzentration *f* molar concentration, molarity
Molleharz *n* American mastic *(a gum from Schinus molle L.)*
Möller *m (met)* burden
möllern *(met)* to burden
Möllersonde *f* stock level (line) indicator *(of a blast furnace)*

Möllerwagen *m* scale car *(for charging blast furnaces)*
Mollier-Diagramm *n* Mollier (enthalpy-entropy) chart (diagram)
Molluskizid *n (agric)* molluscacide, molluscide
Molmenge *f* mole quantity
Molnormvolumen *n* gram-molecular volume
Molpolarisation *f* molar (molecular) polarization
Molpolarisierbarkeit *f* molar (molecular) polarizability
Molprozent *n* **Ungesättigtheit** *(rubber)* mole per cent unsaturation
Molrefraktion *f* molar (molecular) refraction (refractivity)
Molsieb *n s.* Molekularsieb
Molsuszeptibilität *f* molar susceptibility
Molverhältnis *n* molar (mole) ratio
Molvolum[en] *n* molar (molecular, gram-molecular) volume
  kritisches M. critical volume
  partielles M. partial molar (molal) volume
Molwärme *f* molar heat [capacity], molecular heat
Molybdän *n* Mo molybdenum
Molybdänblau *n* molybdenum blue *(molybdenum(V,VI) oxide)*
Molybdän(II)-dihydroxidtetrabromid *n* $Mo_3Br_4(OH)_2$ molybdenum tetrabromide dihydroxide
Molybdän(II)-dihydroxidtetrachlorid *n* $Mo_3Cl_4(OH)_2$ molybdenum tetrachloride dihydroxide
Molybdändioxiddibromid *n* MoO_2Br_2 molybdenum dibromide dioxide
Molybdändisulfid *n s.* Molybdän(IV)-sulfid
Molybdänerz *n* molybdenum ore
Molybdänglanz *m s.* Molybdänit
molybdänhaltig molybdeniferous
Molybdänit *m (min)* molybdenite *(molybdenum(IV) sulphide)*
Molybdänkarbid *n* molybdenum carbide
Molybdänocker *m (min)* 1. molybdic ochre, molybdite *(molybdenum(VI) oxide)*; 2. molybdic ochre, ferrimolybdite *(a hydrous iron(III) molybdate)*
Molybdän(III)-oxid *n* Mo_2O_3 molybdenum(III) oxide, dimolybdenum trioxide
Molybdän(VI)-oxid *n* MoO_3 molybdenum(VI) oxide, molybdenum trioxide
Molybdänoxidtetrachlorid *n* $MoOCl_4$ molybdenum tetrachloride oxide
Molybdänoxidtetrafluorid *n* $MoOF_4$ molybdenum tetrafluoride oxide
Molybdänpulver *n* molybdenum powder
Molybdänsäure *f* H_2MoO_4 molybdic acid
Molybdänstahl *m* molybdenum steel
Molybdän(III)-sulfid *n* Mo_2S_3 molybdenum(III) sulphide, dimolybdenum trisulphide
Molybdän(IV)-sulfid *n* MoS_2 molybdenum(IV) sulphide, molybdenum disulphide
Molybdäntrioxid *n s.* Molybdän(VI)-oxid
Molybdäntrioxidhexachlorid *n* $Mo_2O_3Cl_6$ molybdenum hexachloride trioxide
Molybdäntrioxidpentachlorid *n* $Mo_2O_3Cl_5$ molybdenum pentachloride trioxide
Molybdat(VI) *n* $M_2^IMoO_4$ molybdate(VI)
Molybdatophosphat *n* molybdophosphate, *(specif)* $M_3^I[PMo_{12}O_{40}]$ dodecamolybdatophosphate
Molybdatophosphorsäure *f* molybdophosphoric acid, *(specif)* $H_3[PMo_{12}O_{40}]$ dodecamolybdatophosphoric acid
Molybdit *m (min)* molybdite, molybdic ochre *(molybdenum(VI) oxide)*
Molzahl *f* number of moles
Moment *n* moment
  kernmagnetisches M. nuclear magnetic moment
  magnetisches M. magnetic moment
  orbitalmagnetisches M. orbital [magnetic] moment
Momentanpasteurisation *f*, **Momenterhitzung** *f* *(food)* flash pasteurization, flashing
MO-Methode *f s.* Molekülorbitalmethode
MO-Näherung *f* molecular-orbital approximation (approach)
Monamid *n s.* Monoamid
Monamin *n s.* Monoamin
Monammingallium(III)-chlorid *n* $[Ga(NH_3)]Cl_3$ monoamminegallium(III) chloride
Monardaöl *n* monarda (horsemint) oil *(essential oil from Monarda specc.)*
Monazit *m (min)* monazite *(cerium phosphate)*
Monazitsand *m* monazite sand
Mönchspergament *n* vellum [paper]
Mond-Gas *n* Mond gas *(a producer gas)*
Mond-Gasgenerator *m* Mond [gas] producer
Mond-Gasverfahren *n* Mond process
Mond-Generator *m s.* Mond-Gasgenerator
Mondgestein *n* lunar rock
Mondglas *n* crown glass
Mondmineral *n* lunar mineral
Mond-Niederdruckkarbonylverfahren *n* *(met)* Mond [carbonyl] process
Mondstaub *m* lunar dust
Mondstein *m (min)* moonstone *(a feldspar)*
Mond-Verfahren *n s.* 1. Mond-Gasverfahren; 2. Mond-Niederdruckkarbonylverfahren
Mong Yu *n* stillingia (tallow-seed) oil *(from Sapium sebiferum (L.) Roxb.)*
Monnier-Ofen *m (ceram)* Monnier kiln
Monoalkylbenzol *n* monoalkylbenzene
monoalkylieren to monoalkylate
Monoalkylierung *f* monoalkylation
Monoamid *n* monoamide
Monoamin *n* monoamine
Monoaminoxydase *f* monoamine oxidase
Monoarsin *n* AsH_3 monoarsine, arsine, arsenic trihydride
Monoäthanolamin *n* $CH_2(OH)CH_2NH_2$ 2-aminoethanol, monoethanolamine, MEA, colamine
Monoäthylamin *n* $C_2H_5NH_2$ ethylamine
monoatomar mon[o]atomic
Monoazetin *n* monoacetin, glycerol monoacetate
Monoazofarbstoff *m* monoazo dye
Monoborid *n* monoboride
Monoborin *n* BH_3 borane(3), monoborane(3)
Monoborsäure *f* H_3BO_3 orthoboric acid, boric acid, trioxoboric acid

Monobrombenzol *n* C_6H_5Br bromobenzene
Monobromid *n* monobromide
monobromieren to monobrominate
Monobromierung *f* monobromination
Monobromkampfer *m* bromocamphor, brominated (monobrominated) camphor
Monobrommethan *n* CH_3Br bromomethane
Monochloralkan *n* monochloroalkane
Monochloräthan *n* CH_3CH_2Cl chloroethane
Monochlorderivat *n* monochloro derivative
Monochlorid *n* monochloride
monochlorieren to monochlorinate
Monochlorierung *f* monochlorination
Monochlorsilan *n* SiH_3Cl chlorosilane, monochlorosilane
Monochromarsenid *n* CrAs monochromium arsenide
monochromatisch monochromatic, monochrome
Monochromator *m* monochromator, monochromatic illuminator
Monochromborid *n* CrB monochromium boride
Monochromsäure *f* H_2CrO_4 chromic acid
Monoderivat *n* mono derivative
monodispers monodisperse
Monoester *m* monoester
Monofil[garn] *n* monofil, monofilament [yarn]
Monofluorid *n* monofluoride
monofunktionell monofunctional
Monogerman *n* GeH_4 monogermane, germane, germanium tetrahydride
Monoglyzerid *n* monoglyceride
Monohalogenalkan *n* monohalogen alkane, alkyl monohalide
Monohalogenderivat *n* monohalogen derivative
Monohalogenid *n* monohalogenide, monohalide
monohalogenieren to monohalogenate
Monohalogenierung *f* monohalogenation
Monohydrat *n* monohydrate
Monohydrid *n* monohydride
Monohydrogenphosphat *n* $M_2^IHPO_4$ monohydrogenphosphate
Monohydrogensalz *n* monohydrogen salt
Monohydroxyverbindung *f* monohydroxy compound
monoisotop monoisotopic
Monojodid *n* monoiodide
Monokaliumoxalat *n* HOOC−COOK potassium hydrogen oxalate
Monokarbonsäure *f* monocarboxylic acid
Monoketon *n* monoketone
Monoketonimid *n* monoketone imide
monokl. *s.* monoklin
monoklin *(cryst)* monoclinic, mon., mn.
monolithisch monolithic
monomer monomeric
Monomer[es] *n* monomer
monomineralisch monomineral
monomolekular monomolecular, unimolecular
Monomolekularfilm *m* monomolecular (unimolecular) film (layer), monolayer
Mononatriumglutamat *n* NaOOC−$CH_2CH_2CH(NH_2)$−COOH monosodium glutamate
mononitrieren to mononitrate
Mononitrierung *f* mononitration
Mononitrokörper *m* mononitro body
Mononitroverbindung *f* mononitro compound
Mononukleotid *n* mononucleotide
Monoolefin *n* monoolefin, alkene, ethylenic hydrocarbon
Monoperphthalsäure *f* monoperphthalic acid
Monophosphat *n* $M_3^IPO_4$ monophosphate, orthophosphate
Monophosphid *n* monophosphide
Monophosphin *n* PH_3 monophosphine, phosphine, phosphorus trihydride
Monophosphorsäure *f* H_3PO_4 monophosphoric acid, orthophosphoric acid, phosphoric acid
Monosa[c]charid *n* monosaccharide
Monoschicht *f* monolayer, monomolecular (unimolecular) film (layer)
Monoschwefelwasserstoff *m* H_2S hydrogen sulphide, monosulphane
Monose *f s.* Monosaccharid
Monosilan *n* SiH_4 monosilane
Monosolverfahren *n (petrol)* single-solvent process
Monospiro-Verbindung *f* monospiro compound (hydrocarbon)
Monostearin *n* monostearin, glycerol monostearate, GMS, glycerol octadecanoate
monosubstituieren to monosubstitute
Monosubstitution *f* monosubstitution
Monosubstitutionsprodukt *n* monosubstitution product
Monosulfan *n* H_2S monosulphane, hydrogen sulphide
Monosulfid *n* monosulphide
Monosulfidbrücke *f* monosulphide bridge (crosslink)
Monosulfitaufschluß *m (pap)* pulping with sodium sulphite
Monosulfonsäure *f* monosulphonic acid
Monoterpen *n* monoterpene
Monotreibstoff *m* monofuel, monopropellant
monotrop monotropic
Monotropie *f* monotropy
monovalent monovalent, univalent
monovariant monovariant, univariant
Monowasserstoff *m* monohydrogen, atomic hydrogen
Monoxid *n* monooxide
monozyklisch monocyclic
Montansäure *f* $C_{27}H_{55}COOH$ montanic acid, octacosanoic acid
Montanwachs *n* montan[in] wax
Montanwachsleim *m (pap)* montan-wax size
Montanwachsleimung *f (pap)* sizing with montan wax
Montanwachspech *n* montan-wax pitch
Mont-Cenis-Verfahren *n* Mont Cenis process *(for producing ammonia)*

Montejus *n* montejus, acid egg, blowcase *(an apparatus for lifting liquids)*
Monticellit *m (min)* monticellite *(calcium magnesium orthosilicate)*
Monuron *n* $ClC_6H_4NH-CO-N(CH_3)_2$ monuron, 3-(*p*-chlorophenyl)-1,1-dimethylurea *(a herbicide)*
Mooney-Anvulkanisationszeit *f* Mooney scorch [time]
Mooney-Grad *m (rubber)* Mooney unit
Mooney-Plastizität *f (rubber)* Mooney plasticity
Mooney-Plastometer *n s.* Mooney-Viskosimeter
Mooney-Viskosimeter *n (rubber)* Mooney viscometer (plastometer, instrument)
Mooney-Viskosität *f (rubber)* Mooney viscosity
Mooney-Wert *m*, **Mooney-Zahl** *f (rubber)* Mooney [value]
Moore-Campbell-Ofen *m* Moore-Campbell kiln *(an electric tunnel kiln)*
Moore-Filter *n* Moore filter
Moortorf *m* bog peat
Moos *n* / **Irländisches** *(incorrectly for)* Karrag[h]een
  Isländisches M. *(pharm)* Iceland moss, Cetraria islandica (L.) Acharius *(a lichen)*
Moosachat *m (min)* moss agate
Moosgummi *m* microcellular rubber
Moostorf *m* moss peat
MO-Rechnung *f* molecular-orbital calculation
Morenosit *m (min)* morenosite, nickel vitriol *(nickel sulphate-7-water)*
Morgan-Gaserzeuger *m*, **Morgan-Generator** *m* Morgan [gas] producer
Morin *n* morin, 2′,3,4′,5,7-pentahydroxyflavone
Morphin *n* morphine *(alkaloid)*
morphinähnlich morphine-like
Morphinalkaloid *n* morphine alkaloid
Morphinhydrochlorid *n* morphine hydrochloride
Morphinsulfat *n* morphine sulphate
Morphol *n* morphol, 3,4-dihydroxyphenanthrene
Morpholin *n* morpholine, tetrahydro-1,4-oxazine
morsch werden *(text)* to tender
Mörser *m* mortar
Mörserkeule *f s.* Pistill
Mörtel *m* mortar
  an der Luft erhärtender M. non-hydraulic mortar
  fetter M. rich mortar
  feuerfester M. refractory mortar
  hydraulischer M. hydraulic mortar
Mörtelstruktur *f (geol)* mortar structure
Mosaikblock *m*, **Mosaikblöckchen** *n (cryst)* mosaic bloc, crystallite, domain, secondary structure
Mosaikgold *n* mosaic gold, ormolu *(1. tin(IV) sulphide; 2. a sort of brass)*
Mosaikkristall *m* mosaic crystal
Mosaikstruktur *f*, **Mosaiktextur** *f (cryst)* mosaic structure (texture)
Mosandrit *m (min)* mosandrite *(a sorosilicate)*
Moschus *m* musk
  M. Ambrette musk ambrette, 2,6-dinitro-3-methoxy-4-*tert*-butyltoluene
  M. Baur Baur musk *(a synthetic musk)*
  M. C *s.* M. Keton
  M. Keton musk ketone, ketone musk, musk C *(an acetophenone derivative)*
  M. Xylol musk xylol, musk xylene, 1,3-dimethyl-5-*tert*-butyl-2,4,6-dinitrobenzene
  echter M. natural musk
  künstlicher M. *s.* synthetischer M.
  natürlicher M. natural musk
  synthetischer M. synthetic (artificial) musk *(common name of several organic compounds)*
moschusartig musk-like
Moschusketon *n s.* Moschus C
Moschuskörneröl *n* ambrette (amber seed) oil *(from Hibiscus abelmoschus L.)*
Moschusöl *n* sumbul oil *(from Ferula sumbul Hook.)*
Moseley-Diagramm *n* Moseley diagram
Mößbauer-Effekt *m* Mössbauer effect
Mößbauer-Spektroskopie *f* Mössbauer spectroscopy
Mößbauer-Spektrum *n* Mössbauer spectrum
Most *m* must, stum
  abgepreßter M. press must
Mostaräometer *n* must gauge
Mostrich *m* mustard
Mostwaage *f* must gauge
MO-Theorie *f s.* Molekülorbitaltheorie
Motor[en]benzin *n* motor gasoline (spirit)
Motorenbenzol *n* motor benzole, benzole mixture (motor spirit)
Motorenkraftstoff *m* motor (power) fuel
Motorenöl *n* motor oil
Motorenpetroleum *n* power kerosine (vaporizing oil)
Motorkraftstoff *m* motor (power) fuel
Motormethode *f* motor method, F2 method *(for octane rating)*
Motor-Oktanzahl *f* motor[-method] octane number, MON, F2 octane
Motorverfahren *n s.* Motormethode
Motorverstäuber *m (agric)* power duster
mottenbeständig *s.* mottenecht
mottenecht mothproof ✦ **m. machen** to mothproof
Mottenechtappretur *f*, **Mottenechtausrüstung** *f* mothproof finish
Mottenechtheit *f* resistance to moth
mottenfest *s.* mottenecht
Mottenkugel *f* moth ball
Mottenpapier *n* mothproof paper
Mottenschutz *m* mothproofing
Mottenschutzmittel *n* mothproofing agent, mothproofer, moth repellent
mottensicher *s.* mottenecht
Mottramit *m (min)* mottramite *(a basic vanadate of lead, copper, and zinc)*
motzen *(glass)* to marver *(a gather in an ovoid mould)*
moussieren to effervesce, to sparkle
Moussieren *n* effervescence
moussierend effervescent

Mowra[h]butter *f*, **Mowra[h]öl** *n* mowra (mowrah, moura) butter (fat, oil) *(from Madhuca specc.)*
Moyno-Pumpe *f* Moyno pump *(a single-rotor screw pump)*
MPC-Ruß *m* medium processing channel black, MPC black
Ms *s.* Messing
M-Schale *f* M-shell *(of an atom)*
MT *s.* Metallfaserstoff
MT-Ruß *m* medium thermal black, MT black
Mücken[schutz]mittel *n* mosquito repellent
Mud *m* 1. *(petrol)* mud; 2. *(tann)* bloom, exudation
Muffe *f* socket, boss
Muffel *f* muffle
Muffelofen *m* muffle furnace (kiln)
Muffeltunnelofen *m* muffle tunnel kiln
Muffenrohr *n* socket[ed] pipe
Muffenschweißverbindung *f* socket-weld joint *(a pipe connection)*
Muffenverbinder *m*, **Muffenverbindungsstück** *n* socket fitting
Mühle *f* 1. mill *(works)*; 2. [grinding] mill, grinder *(apparatus)*
M. zur Naß[ver]mahlung wet-grinding mill
autogen arbeitende M. autogenous mill
chilenische M. Chilean (Chile) mill, pan (pan-type roller, edge-runner, edge) mill, edgerunner
Mühlenzusatz *m (ceram)* mill addition
Mühlstein *m* grindstone, millstone, bur[r]stone, buhr[stone]
Mukobromsäure *f* OHC−C(Br)=C(Br)COOH mucobromic acid, dibromoaldehydoacrylic acid
Mukochlorsäure *f* OHC−C(Cl)=C(Cl)COOH mucochloric acid, dichloroaldehydoacrylic acid
Mukoid *n* mucoid *(a glycoprotein)*
Mukoitinschwefelsäure *f* mucoitinsulphuric acid, mucoitin sulphate *(an acidic polysaccharide)*
Mukonsäure *f* HOOC−CH=CHCH=CH−COOH muconic acid, 2,4-hexadien-1,6-dioic acid
Mukopeptid *n* mucopeptide
Mukopolysaccharid *n* mucopolysaccharide
Mukoproteid *n*, **Mukoprotein** *n* mucoprotein
mukos, mukös slimy
Mulde *f* trough, pan
muldenförmig trough-shaped
Muldenmischer *m* trough mixer
Muldenrolle *f* troughing idler
Muldentrockner *m* trough conveyor dryer
Müll *m* waste [material, product], refuse, rubbish, *(Am also)* garbage
Müllaufbereitungsanlage *f* waste-treatment plant
Müllbeseitigung *f* waste disposal
Müllerei *f (food)* milling
Müller-Rochow-Synthese *f* Müller-Rochow synthesis *(for producing chlorosilanes)*
Müller-Rochow-Verfahren *n* Müller-Rochow process *(for producing chlorosilanes)*
Mullit *m (min)* mullite *(aluminium silicate)*
Mullitporzellan *n* mullite porcelain
Müllkompost *m* refuse compost
Müllverbrennungsanlage *f* incinerating plant
Müllverbrennungsofen *m* incinerator, [refuse] destructor
Müllverträglichkeit *f (agric)* tolerance to waste
Multiaerozyklon *m* multitube cyclone separator
Multieinheit *f* multiple [unit]
Multifil[garn] *n* multifil, multifilament [yarn]
Multiklon *m s.* Multiaerozyklon
multipel multiple
Multiples *n* multiple
Multiplett *n* multiplet *(spectroscopy)*
normales (regelrechtes) M. normal (regular) multiplet
verkehrtes M. inverted multiplet
Multiplettaufspaltung *f* multiplet splitting
Multipletterm *m* multiplet level
Multiplettniveau *n* multiplet level
Multiplettstruktur *f* multiplet structure
Multiplexanlage *f* multiplex-roll plant *(a system with more than two pairs of rolls)*
Multiplexpappe *f* multi-ply board
Multiplexwalze *f* multiplex [roll]
Multiplikationsfaktor *m (nucl)* multiplication constant
Multiplikativzahl *f* multiplying prefix, multiplicative numer[ic]al prefix
Multiplizität *f* multiplicity
maximale M. maximum multiplicity
Multiplizitätsprinzip *n* / **Hundsches** Hund maximum-multiplicity principle (rule), Hund's first rule
Multirotation *f* multirotation, mutarotation
Mu-Meson *n s.* Myon
Mundblaseglas *n* hand-blown (hand-made) glass
Mundblasverfahren *n (glass)* hand-blown process
mundgeblasen *(glass)* hand-blown
Mundstück *n* mouthpiece; *(ceram)* die *(of an extruder)*
Mündung *f* orifice
versetzte M. *(glass)* offset finish *(a defect)*
Mündungsbär *m (met)* skull *(in a converter)*
Mundwasser *n* mouthwash
Munkettinußöl *n* Manketti nut oil *(from Ricinodendron rautaneni Schinz)*
Münzbronze *f* coinage bronze
Münzgold *n* coin gold
Münzlegierung *f* coinage alloy
Münzmetall *n* coinage metal
Muon *n*, **Müon** *n s.* Myon
Murakami-Reagens *n* Murakami's reagent *(for etching metals)*
Muraminsäure *f* muramic acid, 3-*O*-α-carboxyethyl-*D*-glucosamine
Murexid *n* murexide *(ammonium salt of purpuric acid)*
Murexidprobe *f* murexide test *(for detecting uric acid)*
Muschelgold *n* [/ **unechtes**] artificial gold
muschelig *(min)* conchoidal *(surface produced by fracture)*
Muschelkalk *m* shell lime, coquina
Muschelkalkstein *m* shell (shelly, coquinoid) limestone

Musivgold *n* mosaic gold, ormolu *(1. tin(IV) sulphide; 2. a sort of brass)*
Muskarin *n* muscarine *(a tetrahydrofuran derivative)*
Muskatbalsam *m s.* Muskatnußbutter
Muskatblüte *f* mace *(from Myristica fragrans Houtt.)*
Muskatnußbutter *f* nutmeg butter (oil) *(from Myristica fragrans Houtt.)*
Muskatnußöl *n s.* 1. Muskatnußbutter; 2. Muskatöl / ätherisches
Muskatöl *n* / **ätherisches** nutmeg (myristica) oil *(from Myristica fragrans Houtt.)*
Muskeladenylsäure *f* [muscle] adenylic acid, adenosine 5′-phosphate
Muskelinosinsäure *f s.* Inosinsäure
Muskelöl *n (cosmet)* muscle oil
Muskon *n* muscone, 3-methylcyclopentadecanone
Muskovit *m (min)* muscovite, potassium (potash) mica *(a phyllosilicate)*
Muster *n* 1. sample; 2. pattern ✦ **nach M. färben** *(text)* to match
Musterfärbejigger *m* sample-dye[ing] jig
Musterfärbung *f* sample dyeing
Mutagen *n (biol)* mutagen, mutagenic agent
Mutarotation *f* mutarotation, multirotation
Mutase *f* mutase
Mutterboden *m* topsoil
Mutterelement *n* parent element
Mutterform *f (ceram)* master mould
Muttergestein *n* 1. *(geol)* parent (mother, source) rock *(as of sediments)*; 2. *(soil)* bedrock; 3. matrix, groundmass *(in which larger crystals are embedded)*
Mutterhefe *f* inoculating (seed, pitching) yeast
Mutterkorn *n (pharm)* ergot
Mutterkornalkaloid *n* ergot alkaloid
Mutterkornpilz *m* ergot, Claviceps purpurea (Fr.) Tul.
Mutterkornvergiftung *f* ergotism
Mutterkultur *f* mother culture (starter), original starter culture *(margarine making)*
Mutterlauge *f* mother liquor
Muttermagma *n* parental magma
Muttermilch *f* breast (human) milk
Mutterpause *f* transparent (translucent) master *(reprography)*
Muttersaft *m (food)* mother (natural) juice
Muttersäurekultur *f s.* Mutterkultur
Muttersubstanz *f* mother (parent) substance
Mutungsbohrung *f (petrol)* exploration drilling, wildcat
Muzin *n* mucin *(a glycoprotein)*
Muzinsäure *f* $HOOC[CHOH]_4COOH$ mucic acid, galactosaccharic acid
MWG *s.* Massenwirkungsgesetz
Mydriatikum *n (med)* mydriatic
mydriatisch *(med)* mydriatic
Myelin *n (bioch)* myelin[e]
Myeloperoxydase *f* myeloperoxidase
Mykarose *f* mycarose *(a monosaccharide)*
Mykolipensäure *f* mycolipenic acid, *trans*-2,4,6-trimethyl-2-tetracosenoic acid
Mykolsäure *f* mycolic acid *(any of a group of acids occurring in tubercle bacilli)*
Mykomyzin *n* mycomycin, 2,4,6,7-tridecatetraene-9,11-di-ynoic acid
Mykophenolsäure *f* mycophenolic acid *(antibiotic)*
Mykosamin *n* mycosamine *(an amino-sugar)*
Mykosterin *n* mycosterol
Myogen *n* myogen *(a mixture of albumins found in muscle)*
Myoglobin *n*, **Myohämoglobin** *n (bioch)* myoglobin
Myon *n (nucl)* muon, mu meson, μ-meson *(an elementary particle)*
Myosin *n* myosin *(a protein of muscle)*
Myrikatalg *m* myrica (bayberry) tallow *(from Myrica specc.)*
Myrikawachs *n (incorrectly for)* Myrikatalg
Myristaldehyd *m* $CH_3[CH_2]_{12}CHO$ myristaldehyde, tetradecanal
Myristat *n* myristate *(a salt or ester of myristic acid)*
Myristinaldehyd *m s.* Myristaldehyd
Myristinalkohol *m s.* Myristylalkohol
Myristinsäure *f* $CH_3[CH_2]_{12}COOH$ myristic acid, tetradecanoic acid
Myristinsäureglyzerylester *m* glycerol trimyristate, trimyristin
Myristizin *n* myristicin, 1-allyl-3-methoxy-4,5-methylenedioxybenzene
Myristizinaldehyd *m* myristicinaldehyde, 3-methoxy-4,5-methylenedioxybenzaldehyde
Myristizinsäure *f* myristicic acid, 3-methoxy-4,5-methylenedioxybenzoic acid
Myristoleinsäure *f* $CH_3[CH_2]_3CH{=}CH[CH_2]_7COOH$ myristoleic acid, 9-tetradecenoic acid
Myriston *n* $CH_3[CH_2]_{12}-CO-[CH_2]_{12}CH_3$ myristone, 14-heptacosanone
Myristylaldehyd *m s.* Myristaldehyd
Myristylalkohol *m* $CH_3[CH_2]_{12}CH_2OH$ myristyl alcohol, 1-tetradecanol
Myrizetin *n* myricetin *(a hexahydroxyflavone)*
Myrizitrin *n* myricitrin *(a glycosidic colouring matter)*
Myrizylalkohol *m* myricyl alcohol *(1. $C_{30}H_{61}OH$ triacontanol; 2. $C_{31}H_{63}OH$ hentriacontanol)*
Myrizylpalmitat *n* myricyl palmitate *(palmitic acid ester of either triacontanol or hentriacontanol)*
Myrobalane *f (tann)* myrobalan, myrab[olan] *(from Terminalia specc.)*
Myronsäure *f* myronic acid
Myrosin *n* myrosin *(an enzyme occurring in various brassicaceous plants)*
Myrrhe *f*, **Myrrhenharz** *n* myrrh gum *(a gum resin from Commiphora specc.)*
Myrrhenöl *n* myrrh oil *(from Commiphora specc.)*
Myrtenal *n* myrtenal, 2-formyl-6,6-dimethyl-2-norpinene
Myrtenol *n* myrtenol, 2-pinen-10-ol
Myrtenöl *n* myrtle oil *(from Myrtus communis L.)*

Myrtensäure *f* myrtenic acid, 6,6-dimethyl-2-norpinene-2-carboxylic acid
Myrtenwachs *n (incorrectly for)* Myrikatalg
Myrtol *n* myrtol *(a fraction of myrtle oil distilling between 160 und 180 °C)*
Myrzen *n* myrcene, 7-methyl-3-methylene-1,6-octadiene
Myzel[ium] *n* mycelium

N

nachappretieren *(text)* to resize
Nachappretur *f (text)* additional finish, resizing
Nachäscher *m (tann)* fresh lime
Nachauflaufbehandlung *f (agric)* post-emergence treatment
Nachauflaufherbizid *n (agric)* post-emergence herbicide
Nachbaratom *n* neighbouring (adjacent) atom
Nachbareffekt *m (phot)* adjacency effect
Nachbargestein *n* adjacent (adjoining, wall) rock
Nachbargruppeneffekt *m* neighbouring effect
Nachbarmolekül *n* adjacent molecule
nachbarständig vicinal, adjacent, neighbouring
Nachbarstellung *f* vicinal (adjacent, neighbouring) position
nachbearbeiten *(tech)* to finish
Nachbearbeitung *f (tech)* finish[ing]
nachbehandeln to aftertreat, to re-treat, to cure
Nachbehandlung *f* aftertreatment, re-treatment, secondary treatment, cure
N. mit Dampf steam-curing *(of concrete)*
N. mit Doktorlauge (Doktorlösung) *(petrol)* doctor treatment (sweetening)
reduktive N. *(text)* reduction clearing
Nachbehandlungsfarbstoff *m* aftertreated dye
Nachbelichtung *f (phot)* post-exposure, re-exposure
nachbessern *(tann)* to mend *(the lime liquor)*; to feed [in] *(tanning agents)*
nachbilden / sich to recover *(as of isotopes)*
Nachbildung *f* recovery *(as of isotopes)*
nachblasen *(met)* to afterblow
Nachblasen *n (met)* afterblow
Nachbleiche *f* final bleaching
Nachblütenspritzung *f (agric)* post-blossom spray
Nachbrand *m (ceram)* refiring
Nachbrecher *m* secondary crusher
nachbrennen *(ceram)* to refire
nachchloren to post-chlorinate *(water)*
nachchlorieren to post-chlorinate, to after-chlorinate
Nachchlorierung *f* post-chlorinating, post-chlorination, after-chlorination
Nachchlorung *f* post-chlorinating, post-chlorination *(of water)*
nachchromieren *(text)* to afterchrome
Nachchromierfarbstoff *m (text)* afterchrome (top chrome, chrome-developed) dye
nachchromiert *(text)* afterchromed, chrome topped
Nachchromierung *f (text)* afterchroming, topchroming
Nachchromierungsfarbstoff *m s.* Nachchromierfarbstoff
Nachchromier[ungs]verfahren *n* afterchrome (top chrome) dyeing process
nachdecken *(text)* to fill up, to top
nachdiazotieren to rediazotize
nachdunkeln to darken
nachentfleischen *(tann)* to re-flesh
nachentwickeln *(phot)* to redevelop
Nachentwicklung *f (phot)* redevelopment
Nachfällung *f* postprecipitation
nachfärben to redye, to top *(one component in textile-fibre mixtures)*
Nachfärbung *f* redyeing, topping, cross dyeing *(of one component in textile-fibre mixtures)*
Nachfaulbecken *n*, **Nachfaulbehälter** *m*, **Nachfaulraum** *m* secondary digestion tank *(waste-water treatment)*
Nachfiltration *f* afterfiltration
Nachflotation *f* second-stage floatation
Nachformung *f* aftershaping
nachfüllen to replenish, to fill up, to refill
Nachfüllösung *f (phot)* replenisher [solution]
Nachfüllung *f* replenishment
Nachgärung *f* afterfermentation, secondary fermentation
nachgerben to retan, to fill
Nachgerbung *f* retannage, filling
Nachgeschmack *m* aftertaste
Nachgiebigkeit *f (text)* compliance; *(tann)* run *(of leather)*
elastische N. *(text)* elastic compliance; *(rubber)* [elastic] resilience
Nachgiebigkeitsverhältnis *n (text)* compliance ratio
Nachglimmen *n*, **Nachglühen** *n* afterglow
Nachguß *m (ferm)* sparge liquor
nachhärten *(plast)* to afterbake
Nachhärtung *f (plast)* afterbake, postcure, second cure
Nachheizung *f s.* Nachvulkanisation
Nachhydrolyse *f* afterhydrolysis
Nachklärbecken *n* secondary sedimentation basin (tank)
nachklassieren to rescreen
Nachklassierung *f* secondary (fine) screening, rescreening
Nachkristallisation *f* after-crystallization, post-crystallization
Nachkühlung *f* aftercooling, secondary cooling
Nachkultur *f* subculture *(bacteriology)*
nachkupfern to aftercopper
Nachkupferung *f* aftercoppering, copper aftertreatment
nachlassen to die down (away), to quieten down *(of a reaction)*
Nachlauf *m (distil)* tailing[s], tail[s], back end, foots
Nachleuchten *n* afterglow
Nachmehl *n (food)* middlings

Nachperiode *f* final period, postperiod *(in calorimetric measurements)*
Nachpolymerisation *f* after-polymerization, postpolymerization
Nachprodukt *n (sugar)* afterproduct
Nachproduktfüllmasse *f (sugar)* low-grade massecuite, aftermassecuite
Nachproduktzucker *m* low (low-grade, low-raw, second raw) sugar
nachprüfen to recheck
Nachprüfung *f* recheck
nachreifen *(fert)* to cure *(as of superphosphate)*
Nachreifen *n (fert)* curing, cure *(as of superphosphate)*
Nachreinigung *f* afterpurification
nachsalzen *(tann)* to resalt
Nachsaturation *f (sugar)* final saturation
nachschäumen *(glass)* to reboil
Nachscheidung *f (sugar)* post-defecation, postliming
Nachschwaden *m (mine)* afterdamp, chokedamp *(after explosions of firedamp)*
Nachschwinden *n (ceram)* aftershrinkage, aftercontraction; *(plast)* aftershrinkage, postmoulding deformation
Nachseifen *n (text)* soaping aftertreatment
nachsetzen to slip *(a Söderberg electrode)*
Nachsetzvorrichtung *f* slipping device *(for Söderberg electrodes)*
nachsichten to rescreen
nachsintern to resinter
nachsortieren to rescreen
Nachsortierer *m* secondary (second, fine) screen
Nachsortierung *f* secondary (fine) screening, rescreening
Nachspülbad *n* clearing bath
nachspülen to [re]rinse
Nachspülen *n* rinsing, rinse
Nachspülmittel *n* rinsing agent, rinse
Nachtcreme *f* night cream
nachtönen to tint, to tone
Nachtrockenzylinder *m*, **Nachtrockner** *m* afterdryer
Nachturm *m (pap)* weak[-acid] tower *(of a two-tower system)*
nachverarbeiten to reprocess
Nachverarbeitung *f* reprocessing
Nachverbrennungskammer *f (pap)* combustion chamber
nachverdichten to seal *(in anodic oxidation)*
Nachverformung *f s.* Nachschwinden
Nachverstrecken *n (text)* afterstretching, poststretching
Nachvulkanisation *f (rubber)* after-vulcanization, after-cure, post-vulcanization, post-cure
  N. im Ofen post-oven cure
nachvulkanisch postvolcanic
Nachwachsen *n (ceram)* afterexpansion
nachwaschen to rewash
nachwässern *(phot)* to rewash
Nachweis *m* detection, identification, [confirmatory] test
nachweisbar detectable
Nachweisbarkeit *f* detectability
Nachweisempfindlichkeit *f* detection sensitivity
nachweisen to detect, to identify
Nachweisgerät *n* detection unit
Nachweislinien *fpl* persistent (ultimate) lines, raies ultimes *(spectral analysis)*
Nachweismethode *f* detection method
Nachweismittel *n* **[/ chemisches]** *s.* Nachweisreagens
Nachweisreagens *n* analytical reagent, A. R.
Nachweisreaktion *f* test reaction
Nachwirkung *f* aftereffect, residual action (effect)
  elastische N. elastic aftereffect, delayed elasticity, memory effect
Nachwürze *f (food)* afterwort, last wort
Nacktgerste *f (ferm)* hulless (naked) barley
NAD *s.* Nikotinamid-adenin-dinukleotid
Nadel *f (cryst)* needle
Nadelausreißfestigkeit *f (rubber)* stitch-tear strength, needle-tear resistance
Nadelausreißprüfung *f (rubber)* stitch-tear test
Nadelausreißwiderstand *m s.* Nadelausreißfestigkeit
Nadelblei *n* needle lead
Nadeleisenerz *n (min)* goethite, göthite *(iron hydroxide oxide)*
nadelförmig needle-like, needle-shaped, acicular
Nadelholz *n* softwood, coniferous wood
Nadelholzzellstoff *m* softwood pulp
nadelig *s.* nadelförmig
Nadelpunktanguß *m (plast)* pinpoint (pinhole) gate
Nadelstich *m (ceram)* pinhole *(a defect)*
Nadelventil *n* needle valve
Nadelverfahren *n (cosmet)* ecuelle method *(for expressing lemon oil)*
Nadelwärmeaustauscher *m*, **Nadelwärmeübertrager** *m* bayonet-tube heat exchanger
Na-D-Linie *f* sodium D line
Nadorit *m (min)* nadorite *(antimony(III) lead(II) chloride dioxide)*
NADP$^+$ *s.* Nikotinamid-adenin-dinukleotidphosphat
Nagelfang *m (pap)* button trap (catcher)
Nagelhautentferner *m (cosmet)* cuticle remover
Nagellack *m (cosmet)* nail lacquer
Nagellackentferner *m (cosmet)* nail lacquer remover
Nageln *n* diesel knock
Nagelpflegemittel *n (cosmet)* manicure preparation
Nagelpoliermittel *n*, **Nagelpolitur** *f (cosmet)* nail polish
Nagetiergift *n* rodenticide
Nagyagit *m (min)* nagyagite *(a sulphide of lead, gold, and tellurium)*
Näherung *f* approximation
  Bornsche N. Born approximation *(for computing wave functions)*
Näherungsformel *f* approximate formula
Näherungsverfahren *n* approximation method, approximate procedure

empirisches N. trial-and-error procedure
Näherungswert *m* approximate value
Nahordnung *f*, **Nahordnungsgrad** *m (cryst)* short-range order
Nähragar *m(n)* nutrient agar
Nährboden *m* nutrient (nutritive, culture, growth) medium, substrate *(for microorganisms)*
fester N. solid medium
synthetischer N. synthetic medium
Nährbouillon *f* nutrient broth
Nährcreme *f (cosmet)* nourishing cream, skin food
Nährelement *n s.* Nährstoffelement
nähren to nourish
nährend nutrient, nutritive, nutritious, nourishing
Nährflüssigkeit *f* nutrient broth
Nährhefe *f* nutrient (food) yeast
Nährhumus *m* friable (nutritive) humus
Nährkartonscheibe *f* cardboard culture disk
Nährlösung *f* nutrient solution ✦ **in N. kultiviert** solution-grown
N. nach Hoagland Hoagland solution
N. nach Johnson Johnson solution
N. nach Knop Knop solution
Nährmedium *n s.* Nährboden
Nährsalz *n* nutrient salt
Nährstoff *m* nutrient, nutritive, foodstuff
unentbehrlicher N. *(agric)* essential element
Nährstoffanreicherung *f* nutrient enrichment
Nährstoffansprüche *mpl* nutrient requirements
nährstoffarm poor in nutrients, *(soil also)* infertile
Nährstoffarmut *f* nutrient deficiency (lack), *(soil also)* infertility
Nährstoffaufnahme *f* nutrient uptake (adsorption)
Nährstoffausnutzung *f* nutrient utilization
Nährstoffauswaschung *f* nutrient elution (leaching)
Nährstoffbedarf *m* nutrient demand (requirements)
Nährstoffbilanz *f* nutrient balance
Nährstoffelement *n (agric)* nutrient (food) element
Nährstoffentzug *m* nutrient withdrawal
Nährstoffgehalt *m* nutrient content, *(if expressed in percentage* $N-P_2O_5-K_2O$*)* grade
Nährstofflinie *f* nutrient line
Nährstoffmangel *m* nutrient deficiency (lack)
Nährstoffmangelerscheinung *f (agric)* nutrient deficiency symptom, hunger sign
Nährstoffnachlieferung *f* nutrient supply
nährstoffreich nutritious, rich in nutrients, *(soil also)* fertile
Nährstoffreichtum *m* nutritiousness, *(soil also)* fertility
Nährstoffreserve *f* nutrient reserve
Nährstoffrückwanderung *f* nutrient remigration
Nährstoffträger *m* nutrient carrier *(in fertilizers)*
Nährstoffverhältnis *n* nutrient ratio
Nährstoffverlagerung *f (bot)* nutrient displacement (translocation)
Nährstoffversorgung *f* nutrient supply
Nährstoffvorrat *m* nutrient reserve
Nährstoffzufuhr *f* nutrient supply
Nährsubstrat *n s.* Nährboden
Nahrung *f* diet, nourishment
Nahrungsfett *n* dietary fat
Nahrungslockstoff *m* food lure *(as for insect control)*
Nahrungsmittel *n* food[stuff], nourishment
Nahrungsmittelchemie *f* food chemistry
Nahrungsmitteleiweiß *n* food protein
Nahrungsmittelindustrie *f* food (food-processing, foodstuff, provisions) industry
Nahrungsmittelvergiftung *f* food poisoning
Nährwert *m* nutritive (nutritional, food) value
Naht *f* seam
geschweißte N. weld
Nahtstelle *f* **der Form** *(plast)* mould-parting line
Nahwirkung *f (phys chem)* contact (close-up) effect
Nakrit *m (min)* nacrite *(a phyllosilicate)*
Name *m (nomencl)* name, term
N. nach Patterson Patterson name
additiver N. additive name
allgemeiner N. generic (non-proprietary) name
falscher N. misnomer
generischer N. *s.* allgemeiner N.
Genfer N. Geneva name
geschützter N. proprietary name
halbsystematischer (halbtrivialer) N. semi-systematic (semitrivial) name
kommerzieller N. trade (commercial) name
konjunktiver N. conjunctive name
nach dem Ring-Index gebildeter N. Ring Index name
nach den IUPAC-Regeln gebildeter N. IUPAC name
nichtgeschützter N. *s.* allgemeiner N.
nichtsystematischer N. *s.* trivialer N.
offizieller N. official (approved) name
radikofunktioneller N. radicofunctional name
rationeller (systematischer) N. systematic name
trivialer (unsystematischer) N. trivial (unsystematic) name
zusammengesetzter N. conjunctive name
Namengebung *f (nomencl)* naming
Namenreaktion *f* name reaction
Nantokit *m (min)* nantokite *(copper(I) chloride)*
Napalm *n* napalm
Napfmanschette *f* cup [ring]
Naphtha *n(f)* 1. [heavy] naphtha *(boiling range 150 to 210°C)*; 2. *s.* Erdöl
leichtes N. light gasoline (naphtha)
Naphthalen *n s.* Naphthalin
Naphthalin *n* $C_{10}H_8$ naphthalene
durch Abpressen gereinigtes N. hot-pressed naphthalene
durch Zentrifugieren gereinigtes N. whizzed naphthalene
hochgereinigtes N. pure flake naphthalene
Naphthalindampf *m* naphthalene vapour
Naphthalindikarbonsäure *f* $C_{10}H_6(COOH)_2$ naphthalene dicarboxylic acid

Naphthalindisulfonsäure *f* $C_{10}H_6(SO_3H)_2$ naphthalenedisulphonic acid
Naphthalinindigo *m(n)* naphthindigo
Naphthalinkarbonsäure *f* naphthalene-carboxylic acid, naphthoic acid
Naphthalinpfanne *f* naphthalene tray
Naphthalinreihe *f* naphthalene series
Naphthalinsulfochlorid *n* $C_{10}H_7SO_2Cl$ naphthalene sulphonyl chloride
Naphthalinsulfonsäure *f* $C_{10}H_7SO_3H$ naphthalenesulphonic acid
Naphthalinsulfonylchlorid *n s.* Naphthalinsulfochlorid
Naphthalol *n* naphthalol, betol, 2-naphthyl salicylate
Naphthalsäure *f* $C_{10}H_6(COOH)_2$ naphthalic acid, naphthalene-1,8-dicarboxylic acid
Naphthan *n s.* Dekahydronaphthalin
Naphthazen *n* naphthacene, 2,3-benzanthracene
Naphthen *n* naphthene, cycloalkane, cyclane, cycloparaffin
naphthenartig naphthenic
Naphthenat *n* naphthenate
Naphthenbasis *f* naphthene base *(of a crude oil)*
Naphthenbasisöl *n*, **Naphthenerdöl** *n* naphthene-base petroleum (crude oil), naphthenic petroleum
naphthenisch naphthenic
Naphthenöl *s.* Naphthenbasisöl
Naphthensäure *f* naphthenic acid
Naphth[indol]indigo *m(n)* naphthindigo
Naphthionsäure *f* $NH_2C_{10}H_6SO_3H$ naphthionic acid, 1-naphthylamine-4-sulphonic acid
α-**Naphthochinolin** *n s.* 7,8-Benzochinolin
β-**Naphthochinolin** *n s.* 5,6-Benzochinolin
Naphthochinon *n* naphthoquinone
Naphthoesäure *f* $C_{10}H_7COOH$ naphthoic acid, naphthalene-carboxylic acid
1-Naphthol *n* $C_{10}H_7OH$ 1-naphthol, 1-hydroxynaphthalene
α-**Naphthol** *n s.* 1-Naphthol
Naphtholblauschwarz B *n* naphthol blue black B
Naphtholdisulfonsäure *f* $HOC_{10}H_5(SO_3H)_2$ naphtholdisulphonic acid
Naphtholgelb S *n* naphthol yellow S, acid yellow 1
Naphtholkomponente *f* naphthol component
α-**Naphtholorange** *n* $NaSO_3C_6H_4N{=}NC_{10}H_6OH$ α-naphthol orange, orange I, sodium-azo-α-naphthol sulphanilate
Naphtholpech *n* naphthol pitch
Naphtholsulfonsäure *f* $HOC_{10}H_6SO_3H$ naphtholsulphonic acid
β-**Naphtholsulfonsäure** *f s.* 2-Naphthol-7-sulfonsäure
1-Naphthol-4-sulfonsäure *f* 1-naphthol-4-sulphonic acid, Nevile-Winther acid
2-Naphthol-6-sulfonsäure *f* 2-naphthol-6-sulphonic acid, Schäffer acid
2-Naphthol-7-sulfonsäure *f* 2-naphthol-7-sulphonic acid
Naphthylamin *n* $C_{10}H_7NH_2$ naphthylamine
Naphthylamindisulfonsäure *f* $NH_2C_{10}H_5(SO_3H)_2$ naphthylaminedisulphonic acid
Naphthylamindisulfonsäure S *f* 1-naphthylamine-4,8-disulphonic acid
Naphthylaminsulfat *n* $(C_{10}H_7NH_2)_2 + H_2SO_4$ naphthylamine sulphate
Naphthylaminsulfonsäure *f* $NH_2C_{10}H_6SO_3H$ naphthylaminesulphonic acid
Naphthylessigsäure *f* $C_{10}H_7CH_2COOH$ naphthylacetic acid
Naphthylgruppe *f*, **Naphthylrest** *m* $C_{10}H_7-$ naphthyl group (residue)
α-**Naphthylthioharnstoff** *m* 1-naphthylthiourea *(a rodenticide)*
Narben *m (tann)* grain ✦ **den N. abschleifen (abziehen)** to buff
geschliffener N. corrected grain
gezogener N. drawn (distorted) grain *(a defect)*
loser N. loose grain *(a defect)*
rinnender N. pipey (empty) grain *(a defect)*
Narbenfestigkeit *f (tann)* grain crack resistance
Narbenpressen *n (tann)* embossing
narbenrein *(tann)* clean-grained
Narbenseite *f (tann)* grain layer
Narbenspalt *m (tann)* grain [split]
narbig pitted
Narkoseäther *m* anaestesia (anaesthetic) ether
Narkosechloroform *n* anaesthesia (anaesthetic) chloroform
Narkosemittel *n* narcotic
Narkotikum *n* narcotic
Narkotin *n* narcotine *(alkaloid)*
narkotisch narcotic
Nasenstein *m* tuckstone *(in a glass furnace)*
naß wet, moist, damp
Naßabscheiden *n* wet collecting (collection)
Naßabscheider *m* wet collector, scrubber
Naßanalyse *f* wet analysis
naßaufbereiten *(min tech)* to wet-clean, to wash
Naßaufbereitung *f (min tech)* wet cleaning, washing; *(ceram)* wet preparation (mixing)
Naßaufbereitungsanlage *f* wet-process plant
Naß-auf-Naß-Druckverfahren *n (text)* wet-on-wet printing method
Naßausschuß *m (pap)* wet broke
Naßaustrag *m* wet discharge
Naßbehandlung *f* wet treatment
Naßbeize *f*, **Naßbeizung** *f (agric)* wet treatment *(of seed)*
Naßbetrieb *m* steaming *(in gas production)*
Naßdampf *m* wet steam (vapour)
Naßdekatur *f (text)* wet decatizing, roll boiling
Nässe *f* wetness, moisture, dampness
Naßechtheit *f (text)* wet strength (fastness), fastness to wetting
Naßelektroabscheider *m*, **Naßelektrofilter** *n* wet (film) precipitator *(for electrical gas cleaning)*
Naßentrindungsanlage *f (pap)* waterous barker
Naßfäule *f* wet rot
Naßfestigkeit *f* wet strength (fastness)
Naßfestleim *m (pap)* wet-strength resin
Naßfilz *m (pap)* wet felt
wollener N. wool (woollen) felt
Naßfilzleitwalze *f (pap)* wet-felt roll

Naßgas *n* wet gas
Naßgasreinigung *f* wet gas cleaning
Naßguß *m* green-sand casting
Naßgußform *f* green-sand mould
Naßgußformen *n* green-sand moulding
Naßguß[form]sand *m* green [moulding] sand
Naßgut *n* wet product; *(relating to dryers)* wet feed
Naßgutaufgabe *f* wet feeding
Naßgutaufgabevorrichtung *f* wet feeder
Naßherd *m (min tech)* wet table
Naß-in-Naß-Druckverfahren *n (text)* wet-on-wet printing method
Naßkarbonisation *f (text)* wet carbonizing *(of wool)*
Naßklassieren *n* wet classification
Naßklassierer *m* wet classifier
Naßkleber *m* wet adhesive
Naßknitterarm-Ausrüstung *f (text)* no-iron (smooth-drying) finish
Naßkoller[gang] *m* wet pan
Naßkollodiumplatte *f (phot)* wet collodion plate
Naßkollodiumverfahren *n (phot)* wet collodion process, wet-plate process
Naßlöschen *n* wet quenching *(of coke)*
Naßmagnetscheider *m* wet magnetic separator
Naßmahlung *f* wet milling (grinding)
Naßmetallurgie *f* hydrometallurgy, wet metallurgy
naßmetallurgisch hydrometallurgical
Naßmühle *f* wet-grinding mill
Naßpartie *f (pap)* wet part (end)
Naßphosphorsäure *f* wet-process phosphoric acid, green acid
Naßplatte *f (phot)* wet collodion plate
naßpökeln *(food)* to brine, to pickle
Naßpökelung *f (food)* brining, brine curing (cure, salting), pickling, pickle curing (cure), wet salting
Naßpressen *n (ceram)* wet (plastic) pressing
Naßpressenpartie *f (pap)* press part (section)
Naßprobe *f (met)* wet assay
Naßreibechtheit *f (text)* fastness to wet rubbing
naßreinigen to wet-clean
Naßreiniger *m* wet cleaner
Naßreinigung *f* wet cleaning
  N. eines Gases wet gas cleaning
Naßrühren *n (ceram)* blunging
naßsalzen *(food)* to brine
Naßscheidung *f (sugar)* wet liming, defecation with milk of lime
Naßschlamm *m* wet sludge
Naßschmelze *f (food)* wet rendering
  N. auf Dampf steam rendering
Naßschmelzen *n s.* Naßschmelze
Naßschmelzgrieben *fpl (food)* wet-rendered fat
Naßschnitzel *npl (sugar)* wet [beet] pulp
Naßsetzmaschine *f (min tech)* wet jig, jig washer
Naßsieben *n* wet screening
Naßspinnen *n* wet spinning
Naßtreber *pl (ferm)* wet spent grains
Naß-Trocken-Verfahren *n (text)* wet-on-dry technique
Naßverbrennung *f* wet combustion
Naßverfahren *n* wet process, wet-processing method; *(ceram)* slip process; ice-water (wet-cooling) method *(margarine making)*
Naßvermahlung *f* wet milling (grinding)
Naßwäsche *f* wet cleaning (washing)
Naßzyklon *m* liquid cyclone, wet-cyclone classifier, hydraulic cyclone separator
naszierend nascent
nativ native
Nativserum *n* native serum
Natrit *m (min)* natrite, natron, soda *(sodium carbonate-10-water)*
Natrium *n* Na sodium
  N. in Bandform sodium ribbon
  radioaktives N. radioactive sodium, radiosodium, *(specif)* ^{24}Na sodium-24
Natrium-24 *n* ^{24}Na sodium-24
Natriumabietat *n* sodium abietate
Natriumalaun *m* $NaAl(SO_4)_2 \cdot 12H_2O$ sodium alum, soda alum
Natriumalginat *n* sodium alginate
Natriumalkoholat *n* NaOR sodium alkoxide
Natriumaluminat *n* sodium aluminate
Natriumaluminiumchlorid *n* $AlCl_3 \cdot NaCl$ aluminium sodium chloride
Natriumaluminiumfluorid *n s.* Natriumhexafluoroaluminat
Natriumaluminiumsulfat *n* $NaAl(SO_4)_2$ aluminium sodium sulphate
Natriumalumosilikat *n* $Na[AlSi_3O_8]$ sodium aluminosilicate
Natriumamalgam *n* sodium amalgam
Natriumamid *n* $NaNH_2$ sodium amide
Natriumammoniumhydrogenphosphat-4-Wasser *n* $Na(NH_4)HPO_4 \cdot 4H_2O$ ammonium sodium hydrogenphosphate-4-water, phosphorus salt
Natriumammoniumsulfat *n* $Na_2SO_4 \cdot (NH_4)_2SO_4$ ammonium sodium sulphate
Natriumanthrachinon-2-sulfonat *n* sodium anthraquinone-2-sulphonate, silver salt
Natriumarsenit *n* Na_3AsO_3 sodium arsenite
Natriumäthoxid *n s.* Natriumäthylat
Natriumäthylat *n* C_2H_5ONa sodium ethoxide
Natriumäthylxanthogenat *n* sodium ethyl xanthate
Natriumaustauscher *m* sodium-cycle cation exchanger
Natriumazetat *n* CH_3COONa sodium acetate
Natriumazetessigester *m* $CH_3COCHCOOC_2H_5Na$ sodioacetoacetic ester
Natriumazetylid *n* Na_2C_2 sodium acetylide (carbide)
Natriumazid *n* NaN_3 sodium azide
Natriumbenzoat *n* C_6H_5COONa sodium benzoate
Natriumbenzolsulfonat *n* $C_6H_5SO_3Na$ sodium benzene sulphonate
Natriumberylliumfluorid *n s.* Natriumtetrafluoroberyllat
Natriumbisulfitbleiche *f (pap)* sodium bisulphite bleaching

Natriumbisulfitkochsäure *f (pap)* sodium bisulphite cooking liquor, sodium-base [sulphite] liquor, sodium-base acid
Natriumbisulfitlösung *f (pap)* sodium bisulphite liquor
Natriumboranat *n* $Na[BH_4]$ sodium tetrahydridoborate, sodium hydridoborate
Natriumborhydrid *n*, **Natriumborwasserstoff** *m s.* Natriumboranat
Natriumbromat *n* $NaBrO_3$ sodium bromate
Natriumbromid *n* NaBr sodium bromide
Natrium-Butadienkautschuk *m* sodium-butadiene rubber
Natriumbutyrat *n* $CH_3CH_2CH_2COONa$ sodium butyrate
Natriumchlorat *n* $NaClO_3$ sodium chlorate
Natriumchlorid *n* NaCl sodium chloride
Natriumchloridgitter *n* sodium chloride lattice
Natriumchlorit *n* $NaClO_2$ sodium chlorite
Natriumchloritbleiche *f (pap)* sodium chlorite bleaching
Natriumchloritbleichlauge *f (pap)* sodium chlorite bleaching liquor
Natriumchromat *n* Na_2CrO_4 sodium chromate
Natriumdampflampe *f* sodium-vapour [discharge] lamp, sodium lamp
Natriumderivat *n (org chem)* sodio derivative
Natriumdichromat *n* $Na_2Cr_2O_7$ sodium dichromate
Natriumdihydrogenarsenat *n s.* Natriumdihydrogenorthoarsenat
Natriumdihydrogendiphosphat *n* $Na_2H_2P_2O_7$ sodium dihydrogendiphosphate, sodium dihydrogenpyrophosphate
Natriumdihydrogenorthoarsenat *n* NaH_2AsO_4 sodium dihydrogenorthoarsenate, sodium dihydrogenarsenate
Natriumdihydrogenorthophosphat *n* NaH_2PO_4 sodium dihydrogenorthophosphate, sodium dihydrogenphosphate
Natriumdihydrogenphosphat *n s.* Natriumdihydrogenorthophosphat
Natriumdihydrogenpyrophosphat *n s.* Natriumdihydrogendiphosphat
Natriumdinitrophenolat *n* sodium dinitrophenate
Natriumdiphosphat *n* $Na_4P_2O_7$ sodium diphosphate, sodium pyrophosphate
Natriumdisilikat *n* $Na_2Si_2O_5$ sodium disilicate
Natriumdisulfat *n* $Na_2S_2O_7$ sodium disulphate
Natriumdisulfit *n* $Na_2S_2O_5$ sodium disulphite
Natriumdithionat *n* $Na_2S_2O_6$ sodium dithionate
Natriumdithionit *n* $Na_2S_2O_4$ sodium dithionite
Natriumdithiosulfatoaurat(I) *n* $Na_3[Au(S_2O_3)_2]$ sodium dithiosulphatoaurate(I)
Natriumdiuranat *n* $Na_2U_2O_7$ sodium diuranate
Natriumdivanadat(V) *n* $Na_4V_2O_7$ sodium divanadate(V)
Natrium-D-Linie *f* sodium D line
Natriumdodezylsulfat *n s.* Natriumlaurylsulfat
Natriumdraht *m* sodium wire
Natriumeisen(III)-oxalat *n* $Na_3[Fe(C_2O_4)_3]$ iron(III) sodium oxalate, ferric sodium oxalate, sodium trioxalatoferrate(III)
Natriumferrat(III) *n* $NaFeO_2$ sodium ferrate(III)
Natriumflamme *f* sodium flame
Natriumfluorid *n* sodium fluoride
Natriumfluoroberyllat *n s.* Natriumtetrafluoroberyllat
Natriumformaldehydsulfoxylat *n* $HO-CH_2-SO_2Na$ sodium formaldehyde-sulphoxylate
Natriumformiat *n* HCOONa sodium formate
Natriumglutamat *n* $NaOOC-CH_2CH_2CH(NH_2)COOH$ sodium glutamate
Natriumgold(III)-chlorid *n s.* Natriumtetrachloroaurat(III)
Natriumgold(I)-sulfid *n* NaAuS gold sodium sulphide
Natrium-Graphit-Reaktor *m (nucl)* sodium graphite reactor
Natriumheptoxodivanadat(V) *n s.* Natriumdivanadat(V)
Natriumhexachloroiridat(III) *n* $Na_3[IrCl_6]$ sodium hexachloroiridate(III)
Natriumhexachloroiridat(IV) *n* $Na_2[IrCl_6]$ sodium hexachloroiridate(IV)
Natriumhexachloroosmat(IV) *n* $Na_2[OsCl_6]$ sodium hexachloroosmate(IV)
Natriumhexachloroplatinat(IV) *n* $Na_2[PtCl_6]$ sodium hexachloroplatinate(IV)
Natriumhexachlororhodat(III) *n* $Na_3[RhCl_6]$ sodium hexachlororhodate(III)
Natriumhexafluoroaluminat *n* $Na_3[AlF_6]$ sodium hexafluoroaluminate
Natriumhexafluoroantimonat(V) *n* $Na[SbF_6]$ sodium hexafluoroantimonate
Natriumhexafluorosilikat *n* $Na_2[SiF_6]$ sodium hexafluorosilicate
Natriumhexahydroxostannat(IV) *n* $Na_2[Sn(OH)_6]$ sodium hexahydroxostannate(IV), preparing salt
Natriumhexajodoplatinat(IV) *n* $Na_2[PtI_6]$ sodium hexaiodoplatinate(IV)
Natriumhexametaphosphat *n* $(NaPO_3)_6$ sodium hexametaphosphate
Natriumhexanitrokobaltat(III) *n* $Na_3[Co(NO_2)_6]$ sodium hexanitrocobaltate(III)
Natriumhexazyanoferrat(II) *n* $Na_4[Fe(CN)_6]$ sodium hexacyanoferrate(II)
Natriumhexazyanoferrat(III) *n* $Na_3[Fe(CN)_6]$ sodium hexacyanoferrate(III)
Natriumhydrid *n* NaH sodium hydride
Natriumhydrogenfluorid *n* $NaHF_2$ sodium hydrogenfluoride
Natriumhydrogenkarbonat *n* $NaHCO_3$ sodium hydrogencarbonate
Natriumhydrogenorthophosphat *n* Na_2HPO_4 sodium hydrogenorthophosphate, sodium hydrogenphosphate
Natriumhydrogenperoxid *n* NaOOH sodium hydrogenperoxide
Natriumhydrogenphosphat *n s.* Natriumhydrogenorthophosphat

Natriumhydrogensulfat *n* $NaHSO_4$ sodium hydrogensulphate
Natriumhydrogensulfid *n* NaHS sodium hydrogensulphide
Natriumhydrogensulfit *n* $NaHSO_3$ sodium hydrogensulphite
Natriumhydrogentartrat *n* $NaOOC[CHOH]_2COOH$ sodium hydrogentartrate
Natriumhydroxid *n* NaOH sodium hydroxide
Natriumhypochlorit *n* NaOCl sodium hypochlorite
Natriumhypochloritbleiche *f (pap)* sodium hypochlorite bleaching
Natriumhypochloritbleichlauge *f (pap)* sodium hypochlorite bleaching liquor
Natriumhyponitrit *n* $Na_2N_2O_2$ sodium hyponitrite
Natriumhypophosphat *n* $Na_4P_2O_6$ sodium hypophosphate
Natriumhypophosphit *n* $NaPH_2O_2$ sodium hypophosphite, *(better)* sodium phosphinate
Natriumjodat *n* $NaIO_3$ sodium iodate
Natriumjodid *n* NaI sodium iodide
Natriumkarbid *n* Na_2C_2 sodium carbide (acetylide)
Natriumkarbonat *n* Na_2CO_3 sodium carbonate, soda
wasserfreies N. anhydrous sodium carbonate, *(tech also)* calcined soda, [soda] ash
Natriumkarboxymethylzellulose *f* sodium carboxymethyl cellulose
natriumkatalysiert sodium-catalyzed
Natriumkondensation *f (org chem)* sodium condensation
Natriumkuchen *m* nitre cake *(consisting of sodium sulphate and sodium hydrogensulphate)*
Natriumlaktat *n* $CH_3CH(OH)COONa$ sodium lactate
Natriumlampe *f s.* Natriumdampflampe
Natriumlaurylsulfat *n* $CH_3[CH_2]_{11}SO_4Na$ sodium lauryl sulphate
Natriumlicht *n* sodium light
Natriumlöffel *m* sodium spoon
Natriummanganat(VI) *n* Na_2MnO_4 sodium manganate(VI)
Natriummanganat(VII) *n s.* Natriumpermanganat
Natriummetaarsenat(III) *n* $NaAsO_2$ sodium metaarsenite
Natriummetaarsenat(V) *n* $NaAsO_3$ sodium metaarsenate
Natriummetaborat *n* $NaBO_2$ sodium metaborate
Natriummetaperjodat *n s.* Natriumtetroxojodat(VII)
Natriummetaphosphat *n* $(NaPO_3)_n$ sodium metaphosphate
Natriummetaphosphatperle *f* sodium phosphate bead
Natriummetasilikat *n* Na_2SiO_3 sodium metasilicate
Natriummetavanadat *n* $NaVO_3$ sodium metavanadate, sodium trioxovanadate(V)
Natriummethoxid *n s.* Natriummethylat
Natriummethyl *n s.* Methylnatrium
Natriummethylat *n* $NaOCH_3$ sodium methylate, sodium methoxide
Natriummethylsilikonat *n* $CH_3SiOONa$ *or* $CH_3Si(OH)_2ONa$ sodium methylsiliconate
Natriummineral *n* sodium mineral
Natriummolybdat *n* Na_2MoO_4 sodium molybdate
Natriummolybdat-2-Wasser *n* $Na_2MoO_4 \cdot 2H_2O$ sodium molybdate-2-water, sodium molybdate crystals
Natriummonouranat *n* Na_2UO_4 sodium uranate
Natriummyristat *n* $CH_3[CH_2]_{12}COONa$ sodium myristate
Natriumnitrat *n* $NaNO_3$ sodium nitrate
Natriumnitrid *n* Na_3N sodium nitride
Natriumnitrit *n* $NaNO_2$ sodium nitrite
Natriumnitroprussiat *n*, **Natriumnitroprussid** *n s.* Nitroprussidnatrium
Natriumoleat *n* $CH_3[CH_2]_{16}COONa$ sodium oleate
Natriumorthophosphat *n* Na_3PO_4 sodium orthophosphate
Natriumorthosilikat *n* Na_4SiO_4 sodium orthosilicate, sodium tetraoxosilicate
Natriumorthovanadat(V) *n* Na_3VO_4 sodium orthovanadate, sodium tetraoxovanadate(V)
Natriumoxalat *n* NaOOC−COONa sodium oxalate
Natriumoxid *n* Na_2O sodium oxide
Natriumpalmitat *n* $CH_3[CH_2]_{14}COONa$ sodium palmitate
Natriumperchlorat *n* $NaClO_4$ sodium perchlorate
Natriumpermanganat *n* $NaMnO_4$ sodium permanganate
Natriumperoxid *n* Na_2O_2 sodium peroxide
Natriumperoxoborat *n* $NaBO_3$ sodium peroxoborate
Natriumperoxochromat *n* Na_3CrO_8 sodium peroxochromate
Natriumperoxodisulfat *n* $Na_2S_2O_8$ sodium peroxodisulphate
Natriumperrhenat *n* $NaReO_4$ sodium perrhenate, sodium tetraoxorhenate(VII)
Natriumphenolat *n* $NaOC_6H_5$ sodium phenate
Natriumphosphat *n* sodium phosphate; *(specif) s.* Natriumorthophosphat
Natriumphosphid *n* Na_3P sodium phosphide
Natriumphosphit *n* Na_2PHO_3 sodium phosphite, *(better)* sodium phosphonate
Natriumpolybutadien *n* sodium polybutadien
Natriumpolymerisation *f* sodium[-catalyzed] polymerization
Natriumpolysulfid *n* sodium polysulphide
Natriumpresse *f* sodium [wire] press
Natriumpyrophosphat *n s.* Natriumdiphosphat
Natriumpyrosulfit *n s.* Natriumdisulfit
Natriumpyrovanadat(V) *n s.* Natriumdivanadat(V)
Natriumrhodanid *n s.* Natriumthiozyanat
Natriumsalizylat *n* HOC_6H_4COONa sodium salicylate
Natriumsalz *n* sodium salt
Natriumseife *f* sodium (soda) soap
Natriumselenat *n* Na_2SeO_4 sodium selenate

Natriumselenid *n* Na_2Se sodium selenide
Natriumselenit *n* Na_2SeO_3 sodium selenite
Natriumsesquisilikat *n* sodium sesquisilicate
Natriumsilikat *n* sodium silicate
Natriumsilikatglas *n* soda-silica glass
Natriumspektralleuchte *f* sodium-spectrum lamp
Natriumstearat *n* $CH_3[CH_2]_{16}COONa$ sodium stearate
Natriumsukzinat *n* $NaOOC-CH_2CH_2-COONa$ sodium succinate
Natriumsulfat *n* Na_2SO_4 sodium sulphate
Natriumsulfat-10-Wasser *n* $Na_2SO_4 \cdot 10H_2O$ sodium sulphate-10-water, Glauber salt
Natriumsulfhydrat *n (tann)* sodium hydrogensulphide, sodium sulphydrate
Natriumsulfid *n* Na_2S sodium sulphide
Natriumsulfit *n* Na_2SO_3 sodium sulphite
Natriumsulfonat *n* sodium sulphonate
Natriumsulfoxylat *n* Na_2SO_2 sodium sulphoxylate
Natriumtartrat *n* $NaOOC[CHOH]_2COONa$ sodium tartrate
Natriumtetraborat *n* / **wasserfreies** $Na_2B_4O_7$ sodium tetraborate, *(tech also)* calcined (burnt, anhydrous, dehydrated) borax
Natriumtetraborat-5-Wasser *n* $Na_2B_4O_7 \cdot 5H_2O$ sodium tetraborate-5-water, octahedral borax
Natriumtetraborat-10-Wasser *n* $Na_2B_4O_7 \cdot 10H_2O$ sodium tetraborate-10-water, borax
Natriumtetrachloroaurat(III) *n* $Na[AuCl_4]$ sodium tetrachloroaurate(III)
Natriumtetrachloropalladat(II) *n* $Na_2[PdCl_4]$ sodium tetrachloropalladate(II)
Natriumtetrachloroplatinat(II) *n* $Na_2[PtCl_4]$ sodium tetrachloroplatinate(II)
Natriumtetrafluoroberyllat *n* $Na_2[BeF_4]$ sodium tetrafluoroberyllate
Natriumtetrafluoroborat *n* $Na[BF_4]$ sodium tetrafluoroborate
Natriumtetrahydridoborat *n s.* Natriumboranat
Natriumtetrasulfid *n* Na_2S_4 sodium tetrasulphide
Natriumtetrathionat *n* $Na_2S_4O_6$ sodium tetrathionate
Natriumtetroxojodat(VII) *n* $NaIO_4$ sodium tetraoxoiodate(VII), sodium periodate, sodium metaperiodate
Natriumtetroxorhenat(VII) *n s.* Natriumperrhenat
Natriumtetroxosilikat *n s.* Natriumorthosilikat
Natriumthioantimonat(V) *n* Na_3SbS_4 sodium thioantimonate
Natriumthiokarbonat *n* Na_2CS_3 sodium thiocarbonate, sodium trithiocarbonate
Natriumthiosulfat *n* $Na_2S_2O_3$ sodium thiosulphate, *(phot also)* hyposulphite
Natriumthiozyanat *n* NaSCN sodium thiocyanate, sodium rhodanide
Natriumtrioxobismutat(V) *n* $NaBiO_3$ sodium trioxobismuthate(V)
Natriumtrioxovanadat(V) *n s.* Natriummetavanadat
Natriumtriphosphat *n* $Na_5P_3O_{10}$ sodium triphosphate
Natriumuranat *n s.* Natriumdiuranat
Natriumuranylazetat *n* $NaUO_2(CH_3COO)_3$ sodium uranyl acetate
Natriumwolframat *n* Na_2WO_4 sodium wolframate, sodium tungstate
Natriumzange *f* sodium tongs
Natriumzelluloseglykolat *n s.* Natriumkarboxymethylzellulose
Natriumzellulosexanthogenat *n* sodium cellulose xanthate
Natriumzitrat *n* $NaOOC-C(OH)(CH_2COONa)_2$ sodium citrate
Natriumzyanat *n* NaOCN sodium cyanate
Natriumzyanid *n* NaCN sodium cyanide
Natriumzyklus *m* sodium cycle *(ion exchange)*
Natrolith *m (min)* natrolite, needle zeolite *(a hydrous aluminium sodium silicate)*
Natron *n* 1. *(min)* soda, natron, natrite *(sodium carbonate-10-water)*; 2. *s.* Natriumhydrogenkarbonat
Natronalaun *m s.* Natriumalaun
Natronaufschluß *m (pap)* soda pulping
Natronbleichlauge *f* Labarraque's solution, eau de Labarraque *(aqueous solution containing sodium hypochlorite)*
Natronglas *n* soda (soft) glass
Natronglimmer *m (min)* soda mica, paragonite *(a phyllosilicate)*
Natronkalk *m* soda lime *(a mixture of caustic soda with caustic lime)*
Natronkalkglas *n* soda-lime glass
Natron-Kalk-Kieselsäureglas *n* soda-lime-silica glass
Natronkalkrohr *n* soda lime tube
Natronkochlauge *f (pap)* soda cooking (digestion) liquor, soda liquor (lye)
Natronkochung *f (pap)* soda cook
Natronlauge *f* sodium hydroxide solution, caustic-soda solution, caustic lye of soda
Natronlauge-Aluminatlösung *f* sodium aluminate solution, aluminate liquor *(used in the Bayer process)*
Natronsalpeter *m* soda nitre, Chile saltpetre (nitre, nitrate), Chilean (Chilian) nitrate *(sodium nitrate)*
Natronseife *f* soda (sodium) soap
Natronstoff *m s.* Natronzellstoff
Natronverfahren *n (pap)* soda process, soda [wood]pulp process
Natronwasserglas *n* [soda] water glass, sodium silicate ✦ **mit N. geleimt** *(pap)* silicate-sized
Natronweinstein *m* Rochelle salt, potassium sodium tartrate-4-water
Natronzellstoff *m* soda pulp
Natronzellstoffabrik *f* soda mill
Natronzellstoffkocher *m* soda digester
Naturambra *f* ambergris, ambergrease
Naturasphalt *m* rock (native, natural) asphalt
Naturbenzin *n s.* Naturgasbenzin
Naturbleiche *f (text)* natural (grass) bleaching, grassing

Naturbleicherde *f* natural (naturally occurring) clay
Natureis *n* natural ice
Naturfarbstoff *m* natural colouring matter, *(esp text)* natural dyestuff; *(bioch)* biochrome
Naturfaser *f* natural fibre
Naturfaserstoff *m* natural fibre
Naturfett *n* natural fat
Naturformsand *m* natural [moulding] sand, naturally bonded sand
Naturgas *n* natural gas
feuchtes (nasses) N. wet natural gas
trockenes N. dry natural gas
Naturgasbenzin *n* natural gasoline, casing-head gasoline (spirit)
durch Absorption gewonnenes N. absorption gasoline
Naturglas *n* natural glass
Naturgummi *n* natural gum
Naturharz *n* natural resin
Naturindigo *m(n)* natural indigo
Naturkautschuk *m* natural rubber, NR
Naturkautschuklatex *m* natural-rubber latex
Naturkautschukmischung *f* natural-rubber compound (mix, stock)
Naturkautschukvulkanisat *n* natural-rubber vulcanizate
Naturkohle *f* natural (mineral, fossil) coal
Naturkohlenwasserstoff *m* natural hydrocarbon
Naturkork *m* bark cork
Naturkunstdruckpapier *n* imitation art paper
Naturlatex *m s.* Naturkautschuklatex
Naturlegierung *f* natural alloy
natürlich [vorkommend] naturally occurring, found in nature, from natural sources, native
Naturmoschus *m* natural musk
Naturprodukt *n* natural product
Naturriechstoff *m* natural perfume
Natursand *m s.* Naturformsand
Naturseide *f* natural silk
Naturstoff *m* natural product (material)
insektizider N. botanical
organischer N. natural organic product
plastischer N. natural plastic
Naturstoffchemie *f* chemistry of natural products
Naturton *m* natural (naturally occurring) clay
Naturumlauf *m* natural circulation, gravity return
Naturumlaufsystem *n* natural-circulation (gravity-return) system
Naturversuch *m* field experiment (test, trial)
Naturzement *m* natural cement
Naumannit *m (min)* naumannite *(silver selenide)*
NBS *s. N*-Brombernsteinsäureimid
Nc-Lack *m*, **N.C.-Lack** *m s.* Zellulosenitratlack
Neamin *n* neamine *(antibiotic)*
Neapelgelb *n* Naples (antimony) yellow
Nebel *m* fog, mist
Nebelabscheider *m* mist eliminator
Nebelblaser *m*, **Nebelgerät** *n (agric)* fog generator (appliance), fogging machine, fogger, nebulizer, aero-mist sprayer
Nebelkammer *f* **[/ Wilsonsche]** [Wilson] cloud chamber, [Wilson] cloud-track apparatus
Nebelkammeraufnahme *f*, **Nebelkammerbild** *n* cloud-chamber photograph
Nebeln *n (agric)* fogging
Nebelspur *f (phys chem)* cloud (fog) track
Nebeltröpfchen *n* fog droplet
Nebenalkaloid *n* companion alkaloid
Nebenbild *n (phot)* ghost image
Nebenerzeugnis *n s.* Nebenprodukt
Nebengemengteile *mpl* accessory minerals (components, constituents)
Nebengestein *n* adjacent (adjoining, enclosing, inclosing, wall) rock
Nebengewinnungsofen *m s.* Nebenproduktenofen
Nebengruppe *f* subgroup, auxiliary group *(of the periodic system)*
Nebenmineral *n* minor mineral
Nebennierenrinde *f* adrenal cortex
Nebennierenrindenhormon *n* adrenocortical (adrenal-cortical, corticoid, cortical) hormone
Nebenprodukt *n* by-product
Nebenproduktenanlage *f* by-product plant
Nebenproduktengewinnung *f* by-product recovery
Nebenproduktenofen *m* by-product (chemical-recovery) oven
Nebenquantenzahl *f* subsidiary quantum number
Nebenreaktion *f* side (secondary) reaction
Nebenreihe *f* auxiliary series *(of the periodic system)*
Nebenserie *f* **/ diffuse (erste)** diffuse series *(spectroscopy)*
scharfe (zweite) N. sharp series *(spectroscopy)*
Nebenvalenz *f* secondary valency
Nebenvalenzbindung *f* secondary [valency] bond
Nebenwirkung *f (pharm)* side effect
Nebligwerden *n* blooming *(of oil varnishes)*
Nebulit *m* nebulite *(a migmatite)*
negativ negative
n. geladen negatively charged
einfach n. uninegative
zweifach n. binegative, dinegative
Negativ *n (phot)* negative
Negativbild *n (phot)* negative image
Negativemulsion *f (phot)* negative emulsion
Negativentwickler *m (phot)* negative developer
Negativentwicklung *f (phot)* negative development
Negativkopie *f (phot)* negative copy
Negativladung *f* negative charge
Negativmaterial *n (phot)* negative material
Negativpapier *n (phot)* negative paper
abziehbares N. stripping paper
Negatron *n* negatron, [negative] electron
Neigung *f* 1. inclination, *(quantitatively)* gradient, slope; 2. tendency, propensity *(as to chemical reaction)*
Neigungswinkel *m* angle of inclination (incline)
nektarführend melliferous
Nektarlieferant *m*, **Nektarspender** *m* nectariferous (bee) plant

NE-Legierung *f* non-ferrous alloy
Nelkenöl *n* caryophyllus (clove) oil, oil of cloves *(from Syzygium aromaticum (L.) Merr. et L.M. Perry)*
Nelkenpfeffer *m* allspice *(from Pimenta dioica (L.) Merr.)*
Nelkenrinde *f*, **Nelkenzimt** *m* clove bark *(from Dicypellum caryophyllatum Nees)*
Nelson-Zelle *f* Nelson cell *(used in the electrolysis of alkali-metal chlorides)*
Nemalith *m (min)* nemalite *(magnesium hydroxide)*
nematizid nematocidal, nema[ti]cidal
Nematizid *n* nematocide, nema[ti]cide
nematoblastisch *(geol)* nematoblastic
Namatozid *n s.* Nematizid
NE-Metall *n* non-ferrous metal
N-endständig N-terminal
Neodym *n* Nd neodymium
Neodymazetat *n* $Nd(CH_3COO)_3$ neodymium acetate
Neodymbromat *n* $Nd(BrO_3)_3$ neodymium bromate
Neodymbromid *n* $NdBr_3$ neodymium bromide
Neodymchlorid *n* $NdCl_3$ neodymium chloride
Neodymglas *n* neodymium glass
Neodymjodid *n* NdI_3 neodymium iodide
Neodymmolybdat(VI) *n* $Nd_2(MoO_4)_3$ neodymium molybdate(VI)
Neodymnitrat *n* $Nd(NO_3)_3$ neodymium nitrate
Neodymnitrid *n* NdN neodymium nitride
Neodymoxid *n* Nd_2O_3 neodymium oxide
Neodymsulfat *n* $Nd_2(SO_4)_3$ neodymium sulphate
Neodymsulfid *n* Nd_2S_3 neodymium sulphide
Neomethymyzin *n* neomethymycin *(antibotic)*
neomorph *(geol)* neomorphic
Neomyzin *n* neomycin *(antibiotic)*
Neon *n* Ne neon
Neonröhre *f* neon discharge tube
Neopentan *n* $C(CH_3)_4$ neopentane, 2,2-dimethylpropane
Neoprenkautschuk *m* neoprene rubber
Neoprenklebezement *m* neoprene cement
Neoprenlatex *m* neoprene latex
Neoprenmischung *f* neoprene compound (stock)
Neoprenverseilmaschine *f* neoprene roper
Neoprenvulkanisat *n* neoprene vulcanizate
Neoprenzement *m* neoprene cement
neovulkanisch neovolcanic
Neozyanit *m (min)* neocyanite, lithidionite *(a copper-containing phyllosilicate)*
Nepalkardamom *m(n)* Nepal (Bengal) cardamom *(from Amomum aromaticum Roxb. and A. subulatum Roxb.)*
Nephelin *m (min)* nepheline, nephelite *(a tectosilicate)*
Nephelometer *n* nephelometer
Nephelometrie *f* nephelometry
nephelometrisch nephelometric
Nephrit *m (min)* nephrite, greenstone *(an inosilicate)*
nephrotoxisch nephrotoxic *(poisonous to the kidney)*
Neptunium *n* Np neptunium
Neral *n* neral, citral b, *cis*-3,7-dimethyl-2,6-octadienal
Nernst-Brenner *m*, **Nernst-Lampe** *f* Nernst lamp (glower)
Nerol *n* nerol, 3,7-dimethyl-2,6-octadiene-1-ol
Neroliöl *n* neroli oil *(from Citrus aurantium L. ssp. aurantium)*
Nerv *m (rubber)* nerve, snap
Nervengas *n* nerve gas
nervenschädigend neurotoxic
nervig *(rubber)* nervy, snappy
Nervonsäure *f* $CH_3[CH_2]_7CH{=}CH[CH_2]_{13}COOH$ nervonic acid, *cis*-15-tetracosenoic acid
Nesosilikat *n* nesosilicate *(a silicate containing independent SiO_4 tetrahedra)*
Nessel *f* / **Indische** China (Chinese) grass, Boehmeria nivea (L.) Gaudich.
Nesselfaser *f* nettle fibre
Neßler-Gefäß *n*, **Neßler-Rohr** *n*, **Neßler-Röhre** *f* Nessler glass (tube) *(colorimetry)*
Nestdüngung *f (agric)* nest fertilization
Nesterbehandlung *f (agric)* spot treatment *(as with herbicides)*
netzbar wettable
Netzbarkeit *f* wettability
Netzbildung *f* 1. *(phot)* reticulation; 2. *(coat)* stringing
Netzbottich *m (text)* steeping pan
Netzebene *f (cryst)* lattice (net, atomic) plane
Netzebenenabstand *m (cryst)* lattice distance (spacing), spacing of the planes
Netzeigenschaften *fpl* wetting properties (characteristics)
Netzelektrode *f* gauze electrode
netzen to wet
Netzen *n* wetting, *(text also)* dewing, damp[en]ing
Netzer *m s.* Netzmittel
Netzfähigkeit *f* wetting ability
Netzkatode *f* gauze cathode
Netzkraft *f* wetting power
Netzmittel *n* wetting agent (aid), wetter, spreading agent, spreader, humectant
Netzpulver *n* wettable powder *(pest control)*
Netzschwefel *m* wettable sulphur *(pest control)*
Netzspirale *f* gauze plug *(in a combustion tube)*
Netzvermögen *n* wetting ability
Netzwärme *f* heat of wetting
Netzwerk *n* network
  räumliches N. space network
Netzwerkanalyse *f* network analysis
Netzwerkbildner *m (glass)* network former, network-forming ion
  N. und -wandler *m* network co-former
Netzwerkformer *m s.* Netzwerkbildner
Netzwerkhypothese *f* **von W.H. Zachariasen** *(glass)* Zachariasen's theory
Netzwerkwandler *m (glass)* network modifier, network-modifying ion

Netzwerkzwitter *m (glass)* net intermediate
Netzwirkung *f* wetting action (effect)
neuappretieren *(text)* to resize
Neuappretur *f (text)* resizing, additional finish
Neubestimmung *f* redetermination
Neubildungsgeschwindigkeit *f (bioch)* turnover rate
Neubildungszeit *f (bioch)* turnover time
Neublau *n* new blue *(any of several blue dyes and pigments)*
Neublau R *n* new blue R, Meldola's blue, basic blue 6
Neubohrung *f (petrol)* wildcat
Neufuchsin *n* new fuchsine, basic violet 2 *(a triphenylmethane dye)*
Neugewürz *n* allspice *(from Pimenta dioica (L.) Merr.)*
Neugrün *n* malachite (fast) green, basic green 4 *(a triphenylmethane dye)*; new green *(copper(II) acetate arsenite)*
Neuraminsäure *f* neuraminic acid *(an amino sugar)*
Neurin *n* $CH_2{=}CHN(CH_3)_3OH$ neurine, trimethylvinylammonium hydroxide
Neuroleptikum *n (pharm)* neuroleptic [drug], CNS-depressant
neuroleptisch *(pharm)* neuroleptic
Neurot *n* scarlet red, Biebrich [scarlet] red
neurotoxisch neurotoxic
Neurotoxizität *f* neurotoxicity
Neuseelandflachs *m* phormium, Phormium tenax J.R. et G. Forst.
Neusilber *n* nickel silver, *(Am)* nickel brass
neutral neutral ✦ **n. reagieren** to react neutral
Neutralfett *n* neutral fat
Neutralglyzerid *n* triglyceride
Neutralisation *f* neutralization
Neutralisationsaktivität *f* neutralizing activity
Neutralisationsanalyse *f* [volumetric] neutralization titration
Neutralisationsanlage *f* neutralizing plant
Neutralisationsbehälter *m* neutralizer
Neutralisationsindikator *m* neutralization indicator
Neutralisationskurve *f* neutralization curve
Neutralisationsmittel *n* neutralizing agent, neutralizer
Neutralisationsreaktion *f* neutralization reaction
Neutralisationsschritt *m* neutralization step
Neutralisationstitration *f* neutralization titration, acid-base titration
Neutralisationswärme *f* heat of neutralization
Neutralisationszahl *f* neutralization value, acid value (number), A.V.
Neutralisator *m s.* Neutralisationsmittel
neutralisieren to neutralize
Neutralisierungs ... *s.* Neutralisations ...
Neutralität *f* neutrality
Neutralkolloid *n* neutral colloid
Neutrallard *n* neutral lard
Neutralöl *n* neutral oil
Neutralpunkt *m* neutral point, point of neutrality
Neutralrot *n* neutral red *(an oxidation-reduction indicator)*
Neutralsalz *n* neutral (normal) salt
Neutralsalzeffekt *m s.* Neutralsalzwirkung
Neutralsalzfehler *m* [neutral-]salt error *(in pH determinations)*
Neutralsalzquellung *f (tann)* osmotic swelling
Neutralsalzverfahren *n (rubber)* neutral [reclaiming] process
Neutralsalzwirkung *f* neutral-salt effect *(acidimetry)*
Neutralschmalz *n* neutral lard
Neutralsulfitablauge *f (pap)* neutral sodium sulphite waste liquor, neutral sulphite semichemical spent liquor
Neutralsulfit-Halbzellstoff *m* neutral sulphite semichemical pulp, NSSC pulp
Neutralsulfit-Halbzellstoffaufschluß *m* neutral sulphite semichemical pulping, NSSC pulping
Neutralsulfitkochlauge *f (pap)* neutral sulphite semichemical liquor, NSSC liquor, neutral sodium sulphite cooking liquor, semichemical pulping liquor
Neutralsulfitstoff *m s.* Neutralsulfit-Halbzellstoff
Neutralsulfitverfahren *n (pap)* neutral sulphite [semichemical] process, NSSC process, neutral sodium sulphite process
Neutralverfahren *n (rubber)* neutral [reclaiming] process
Neutretto *n* neutretto, neutral meson
Neutrino *n* neutrino
Neutron *n* neutron
  energiereiches N. high-energy neutron
  gestreutes N. scattered neutron
  kaltes N. cold neutron
  langsames N. slow neutron
  schnelles N. fast (high-speed) neutron
  thermisches N. thermal neutron
  unterthermisches N. *s.* kaltes N.
  unverzögertes N. *s.* schnelles N.
  vagabundierendes N. stray neutron
  verzögertes N. delayed neutron
Neutronenabsorber *m* neutron absorber
Neutronenaktivierungsanalyse *f* neutron-activation analysis
Neutronenbeschuß *m* neutron bombardment
Neutronenbeugung *f* neutron diffraction
Neutronenbeugungsuntersuchung *f* neutron diffraction study
neutronenbestrahlt neutron-irradiated
Neutronenbestrahlung *f* neutron irradiation
Neutronenbindungsenergie *f* neutron-binding energy
Neutronenbremsung *f* neutron moderation
Neutronendichte *f* neutron density
Neutroneneinfang *m* neutron capture
  parasitärer N. parasitic neutron capture
Neutroneneinfangquerschnitt *m* neutron-capture cross section
Neutronenemission *f* neutron emission
  verzögerte N. delayed neutron emission
Neutronenfänger *m* neutron absorber
Neutronenfluß *m* neutron flux
Neutronennachweis *m* neutron detection

Neutronenquelle *f* neutron source
Neutronenruh[e]masse *f* neutron rest mass
Neutronenspektrum *n* neutron spectrum
Neutronenstrahl *m* neutron ray; *(if bundled)* neutron beam
Neutronenstrahlenquelle *f* neutron source
Neutronenstrahlung *f* neutron radiation
Neutronenüberschuß *m* neutron excess
Neutronenzerfall *m* neutron decay
Neuverteilung *f* redistribution
Neuverteilungsreaktion *f* redistribution reaction
Nevile-Winther-Säure *f* $C_{10}H_6(OH)SO_3H$ Nevile-Winther acid, NW acid, 1-naphthol-4-sulphonic acid
Newcastle-Ofen *m* Newcastle kiln *(a horizontal-draught kiln)*
New-Jersey-Zinkverfahren *n* New Jersey zinc[-recovery] process
Nezinsäure *f* necic acid *(acid component of senecio alkaloids)*
NF-Ofen *m s.* Nichols-Freeman-Ofen
Ngaikampfer *m* ngai camphor *(chemically nearly pure L-borneol)*
NH_3-Wasser *n s.* Ammoniakwasser 1.
Niazin *n s.* Nikotinsäure
Niazinamid *n s.* Nikotinamid
Niccolit *m (min)* niccolite, arsenical nickel *(nickel arsenide)*
Nichols-Freeman-Ofen *m* Nichols-Freeman flash roaster
Nicholson-Blau *n* Nicholson blue
nichtaktiviert non-activated
nichtalkoholisch non-alcoholic
nichtaromatisch non-aromatic
nichtaustauschbar non-exchangeable
nichtbenzoid non-benzoid
nichtchinoid non-quinonoid
Nichtedelmetall *n* base metal
Nichtedelmetallkatalysator *m* base-metal catalyst
nichteinlaufend *s.* nichtkrumpfend
Nichteisenhüttenwesen *n* non-ferrous metallurgy
Nichteisenlegierung *f* non-ferrous alloy
Nichteisenmetall *n* non-ferrous metal
Nichteisenmetallurgie *f* non-ferrous metallurgy
Nichteiweißstickstoff *m* non-protein nitrogen
Nichtelektrolyt *m* non-electrolyte
Nichtelektrolytchelat *n* non-electrolyte chelate
nichtentflammbar non-[in]flammable, uninflammable, flameproof
Nichtentflammbarkeit *f* non-[in]flammability, uninflammability, flameproofness
nichtenzymatisch non-enzymatic
Nichterz *n* gangue [mineral], gang, waste rock, matrix
Nichtexistenz *f* non-existence
nichtfarbig colourless
nichtfettig non-greasy
nichtflüchtig non-volatile
Nichtflüchtiges *n* non-volatile matter
Nichtflüchtigkeit *f* non-volatility
Nichtgerbstoff *m* non-tan[nin]
nichtgilbend *(pap)* non-fading, non yellowing
nichthybridisiert unhybridized
nichthydratisiert non-hydrated
nichtideal non-ideal
nichtionisch non-ionic
nichtionisierend *s.* nichtionogen
nichtionisiert unionized
nichtionogen non-ionogenic, non-ionizing
Nichtkarbonathärte *f* non-carbonate hardness, permanent hardness *(water chemistry)*
Nichtkautschukbestandteil *m* non-rubber constituent
Nichtkautschuksubstanz *f* non-rubber substance (material)
nichtklebend non-stick
nichtklopfend knock-free, knockless, non-knocking *(carburetting fuel)*
nichtkondensierbar non-condensable
Nichtkondensierbarkeit *f* non-condensability
nichtkonjugiert non-conjugated, unconjugated
nichtkristallin[isch] non-crystalline
nichtkrumpfend *(text)* shrink-resistant, non-shrinking, shrinkproof, unshrinkable
nichtkumuliert non-cumulative *(double bond)*
nichtleitend non-conducting
Nichtleiter *m* non-conductor, dielectric [material], electrical insulator
nichtleuchtend non-luminous
nichtlinear non-linear
Nichtlinearität *f* non-linearity
nichtlokalisiert delocalized, non-localized
Nichtlokalisierung *f* delocalization
Nichtlöser *m* non-solvent
nichtlöslich insoluble, non-soluble, indissoluble
Nichtlöslichkeit *f* insolubility, insolubleness
nichtmagnetisch non-magnetic
nichtmarkiert unlabelled
Nichtmetall *n* non-metal, non-metallic element
nichtmetallisch non-metallic
Nichtmischbarkeit *f* immiscibility
nichtmodifiziert unmodified
Nichtnetzer *m* non-wetter
nichtoxydierend non-oxidizing
nichtphenolisch non-phenolic
nichtpigmentiert unpigmented
nichtplastisch non-plastic
nichtpolar non-polar, apolar
nichtpolarisierbar non-polarizable
Nichtproteinstickstoff *m* non-protein nitrogen
nichtradioaktiv non-radioactive
nichtreaktionsfähig non-reactive, unreactive, inactive
nichtreduzierend non-reducing
nichtregenerativ non-regenerative
nichtrelativistisch non-relativistic
nichtrostend stainless, rustproof, rustless *(steel)*
Nichts *n* / **weißes** nihilum album, nix alba *(white woolly zinc oxide)*
Nichtsättigung *f* unsaturation
nichtschäumend foamless; *(cosmet)* non-lathering
nichtschrumpfend *s.* nichtkrumpfend

nichtspaltbar non-fissile, non-fissionable
nichtstarr non-rigid
nichtstaubend dustless, dust-free
nichtstöchiometrisch non-stoichiometric
nichtsubstituiert unsubstituted
nichtsulfoniert unsulphonated
nichttitrimetrisch non-titrative
nichttoxisch non-toxic, non-poisonous
nichttrocknend non-drying
nichtumgesetzt unreacted
nichtverfestigt unconsolidated
nichtvergilbend *(pap)* non-fading, non-yellowing
nichtvernetzt *(rubber)* unvulcanized
nichtverseifend non-saponifying
nichtwandernd non-migrating
nichtwässerig *s.* nichtwäßrig
nichtwasserlöslich water-insoluble, insoluble in water
nichtwäßrig non-aqueous
Nichtzellulosebestandteile *mpl (pap)* non-cellulosic constituents
nichtzentral non-central
Nichtzuckeranteil *m* non-sugar portion
nichtzyklisch non-cyclic[al], acyclic
Nickel *n* Ni nickel
Nickelazetat *n* $Ni(CH_3COO)_2$ nickel acetate
Nickelblüte *f (min)* nickel bloom, annabergite *(nickel orthoarsenate)*
Nickel(II)-bromid *n* $NiBr_2$ nickel(II) bromide, nickel dibromide
Nickelchelat *n* nickel chelate
Nickel(II)-chlorid *n* $NiCl_2$ nickel(II) chloride, nickel dichloride
Nickeldi ... *s. a.* Nickel(II)- ...
Nickeldiazetyldioxim *n s.* Nickeldimethylglyoxim
Nickeldimethylglyoxim *n* $Ni[(CH_3)_2(CNO)_2H]_2$ nickel dimethylglyoxime
Nickeldithionat *n* NiS_2O_6 nickel dithionate
Nickel-Eisen-Akkumulator *m* nickel-iron accumulator (battery, cell), Edison accumulator
Nickeleisenkern *m (geoch)* iron-nickel core
Nickelelektrode *f* nickel electrode
Nickel(II)-fluorid *n* NiF_2 nickel(II) fluoride, nickel difluoride
Nickelformiat *n* $Ni(HCOO)_2$ nickel formate
nickelhaltig nickeliferous
Nickelhexafluorosilikat *n* $Ni[SiF_6]$ nickel hexafluorosilicate
Nickel(II)-hexazyanoferrat(II) *n* $Ni_2[Fe(CN)_6]$ nickel(II) hexacyanoferrate(II)
Nickel(II)-hydroxid *n* $Ni(OH)_2$ nickel(II) hydroxide
Nickelin *m s.* Niccolit
Nickel(II)-jodid *n* NiI_2 nickel(II) iodide, nickel diiodide
Nickel-Kadmium-Akkumulator *m* nickel-cadmium accumulator (cell)
Nickelkarbid *n* Ni_3C nickel carbide
Nickelkarbonat *n* $NiCO_3$ nickel carbonate
Nickelkatalysator *m* nickel catalyst
Nickelkugeln *fpl (met)* nickel pellets *(as produced by the Mond carbonyl process)*
Nickellegierung *f* nickel alloy
Nickelmonosulfid *n s.* Nickel(II)-sulfid
Nickelmonoxid *n s.* Nickel(II)-oxid
Nickelnitrat *n* $Ni(NO_3)_2$ nickel nitrate
Nickelorthoarsenat(V) *n* $Ni_3(AsO_4)_2$ nickel orthoarsenate
Nickelorthophosphat *n* $Ni_3(PO_4)_2$ nickel orthophosphate
Nickel(II)-oxid *n* NiO nickel(II) oxide, nickel monooxide
Nickel(II,III)-oxid *n* Ni_3O_4 nickel(II,III) oxide, trinickel tetraoxide
Nickel(III)-oxid *n* Ni_2O_3 nickel(III) oxide, dinickel trioxide
Nickelperchlorat *n* $Ni(ClO_4)_2$ nickel perchlorate
Nickelphthalozyanin *n* nickel phthalocyanine
Nickelschwamm *m* spongy nickel
Nickelschwammkatalysator *m* spongy-nickel catalyst
Nickelskutterudit *m (min)* nickel-skutterudite, white nickel, chloanthite *(nickel arsenide)*
Nickelstahl *m* nickel [alloy] steel
Nickelstein *m (met)* nickel matte
Nickelsulfat *n* $NiSO_4$ nickel sulphate
Nickel(II)-sulfid *n* NiS nickel(II) sulphide, nickel monosulphide
Nickel(II,III)-sulfid *n* Ni_3S_4 nickel(II,III) sulphide, trinickel tetrasulphide
Nickeltetrakarbonyl *n* $Ni(CO)_4$ nickel tetracarbonyl
Nickeltiegel *m* nickel crucible
Nickelvitriol *m (min)* nickel vitriol, morenosite *(nickel sulphate-7-water)*
Nickelzyanid *n* $Ni(CN)_2$ nickel cyanide
Nickschwingung *f* wagging vibration *(spectroscopy)*
Nicol-Prisma *n* Nicol prism
niederblasen *(glass)* to blow down
Niederblasen *n (glass)* settle blow
niederbringen *(mine)* to sink *(a shaft)*
eine Bohrung n. *(petrol)* to sink a bore
Niederdruck *m* low pressure
Niederdruckdampf *m* low-pressure steam, LP steam, ordinary-pressure steam
Niederdruckharz *n* low-pressure resin
Niederdrucklaminieren *n* low-pressure laminating
Niederdruckleitung *f* low-pressure line
Niederdruckpolyäthylen *n* low-pressure (high-density, linear) polyethylene, H.D. polythene
Niederdruckpreßverfahren *n (plast)* low-pressure moulding method
Niederdruckschichtpressen *n* low-pressure laminating
Niederdruckschichtstoff *m* low-pressure laminate
Niederdrucksprühgerät *n (agric)* low-pressure sprayer
Niederdruckverfahren *n* **nach Ziegler** Ziegler process *(for polymerization)*
niederinkohlt low-rank
Niederkräusen *pl (ferm)* low krausen
niedermolekular low-molecular

Niedermoortorf *m* fen peat
Niederschlag *m* precipitate, [bottom] sediment, [bottom] settlings, B.S., bottoms, deposit, foots, *(esp ferm)* lees; *(from vapour)* condensate ✦
einen N. bilden *s.* niederschlagen / sich
flockiger N. flocculate, flocculation
galvanischer N. electrodeposit
käsiger N. curdy precipitate
radioaktiver N. [radioactive] fallout
niederschlagbar precipitable; condensable *(vapour)*
Niederschlagbarkeit *f* precipitability; condensability *(of vapour)*
niederschlagen to precipitate, to sediment[ate], to set, to deposit; to condense *(vapour)*
elektrochemisch (elektrolytisch, galvanisch) n. to electrodeposit
sich n. to precipitate, to sediment, to set, to settle [down, out], to deposit, to subside; *(of vapour)* to condense
sich elektrochemisch (elektrolytisch, galvanisch) n. to plate out
Niederschlagselektrode *f* precipitating (collecting, receiving) electrode
Niederschlagsmenge *f* quantity (amount) of precipitate
Niederschlagsmittel *n* precipitant, precipitating agent, precipitator
Niederschlagsplatte *f* collecting plate
Niederschlagsreaktion *f* precipitation reaction
Niederschlagswasser *n* condensation (condensed) water, condensate [water]
niederschmelzen to melt down
Niederschmelzen *n* melting-down
Niederspannung *f* low voltage, low tension
Niederspannungselektrophorese *f* low-voltage electrophoresis
Niederspannungs[-Elektro]porzellan *n* low-tension [electrical] porcelain
Niederstruktur-Ruß *m (rubber)* low-structure [carbon] black
Niedertemperaturofen *m* low-temperature kiln
Niederungsmoortorf *m* fen peat
Niedervakuum *n* low vacuum
niederwertig low-grade; lower-valent
niedrigaktiv *(nucl)* low-level active
Niedrigfeuer *n* slow fire
niedriggekohlt low-carbon *(e.g. steel)*
niedriginkohlt low-rank
niedrigmolekular low-molecular
niedrignitriert low-nitrated
niedrigoktanig low-octane
niedrigphosphorhaltig low-phosphorus
niedrigschmelzend low-melting[-point], low-fusion
niedrigsiedend low-boiling, light
Niedrigsieder *m* low boiler *(as of solvents)*
niedrigviskos of low viscosity, low-viscosity
Nierenfett *n* leaf fat, suet
nierenschädigend nephrotoxic
Niesmittel *n (pharm)* sternutator
NiFe-Akkumulator *m* nickel-iron accumulator (cell, battery), Edison accumulator
NiFe-Kern *m (geoch)* iron-nickel core
Nigeröl *n* niger-seed oil, Ramtilla oil *(from Guizotia abyssinica (L.f.) Cass.)*
Nigrotinsäure *f* nigrotic acid, 3,5-dihydroxy-7-sulpho-2-naphthoic acid
Nihilum album *n s.* Nichts / weißes
Nikaragua-Brechwurzel *f* Nicaragua (Panama, Cartagena) ipecacuanha *(from Cephaelis acuminata Karsten)*
Nikotin *n* nicotine
Nikotinamid *n* nicotinamide, nicotinic acid amide, niacin amide, pyridine-3-carboxamide
Nikotinamid-adenin-dinukleotid *n* nicotinamide-adenine dinucleotide, NAD
Nikotinamid-adenin-dinukleotidphosphat *n* nicotinamide-adenine dinucleotide phosphate, NADP, NADPH
Nikotinsäure *f* nicotinic acid, niacin, pyridine-3-carboxylic acid
Nikotinsäureamid *n s.* Nikotinamid
Nikotinsäurebenzylester *m* benzyl nicotinate
Nikotinvergiftung *f* nicotine poisoning
Nilgummi *n* Somali gum *(from several Acacia specc.)*
Ninhydrinreaktion *f* ninhydrin reaction (test) *(for detecting proteins and amino acids)*
Niob *n* Nb niobium
Niobat *n* niobate
Niob(V)-bromid *n* $NbBr_5$ niobium(V) bromide, niobium pentabromide
Niob(V)-chlorid *n* $NbCl_5$ niobium(V) chloride, niobium pentachloride
Niobdioxid *n s.* Niob(IV)-oxid
Niobeöl *n* $C_6H_5COOCH_3$ niobe oil, methyl benzoate
Niob(V)-fluorid *n* NbF_5 niobium(V) fluoride, niobium pentafluoride
Niobhydrid *n* NbH niobium hydride
Niobit *m* niobite *(a mineral containing niobium and tantalum)*
Niobkarbid *n* NbC niobium carbide
Niobmonoxid *n s.* Niob(II)-oxid
Niob(II)-oxid *n* NbO niobium(II) oxide, niobium monooxide
Niob(IV)-oxid *n* NbO_2 niobium(IV) oxide, niobium dioxide
Niob(V)-oxid *n* Nb_2O_5 niobium(V) oxide, niobium pentaoxide
Niobpenta ... *s.* Niob(V)- ...
Niobpentoxid *n s.* Niob(V)-oxid
Niobsäure *f* niobic acid
Nisinsäure *f* nisinic acid *(a tetracosahexaenoic acid)*
Niter *m s.* Nitronatrit
Niton *n s.* Radon
Nitramid *n* NH_2NO_2 nitramide
Nitramin *n* nitramine *(general formula $RHN \cdot NO_2$ or $R_1R_2N \cdot NO_2$)*
Nitranilin *n* $NH_2C_6H_4NO_2$ nitroaniline
Nitranilinrot *n* $C_{10}H_6(OH)N{=}NC_6H_4NO_2$ paranitraniline red, para red
Nitranilsäure *f* $(NO_2)_2C_6O_2(OH)_2$ nitranilic acid

Nitrat *n* nitrate
Nitratbakterien *pl*, **Nitratbildner** *mpl* nitrate (nitric) bacteria, nitrobacteria *(genus Nitrobacter)*
Nitrator *m s.* Nitrierapparat
Nitratstickstoff *m* nitrate nitrogen
Nitratzellulose *f (incorrectly for)* Zellulosenitrat
Nitren *n* nitrene, imene *(a molecule fragment having only an electron sextet as outer shell of nitrogen)*
Nitrid *n* nitride
Nitridhärtung *f s.* Nitrierhärtung
Nitrierabteilung *f* nitrating department
Nitrieragens *n* nitrating agent
Nitrieranlage *f* nitration plant, nitrating unit
Nitrierapparat *m* nitrator
N. nach Kubierschky Kubierschky's nitrator
N. nach Weiler ter Meer ter Meer's nitrator
Nitrierbad *n (met)* nitriding bath
nitrieren to nitrate; *(met)* to nitride
Nitriergefäß *n* nitrating pan
Nitriergemisch *n* nitration mixture
Nitriergeschwindigkeit *f* speed (rate) of nitration
nitrierhärten to nitride *(steel)*
Nitrierhärtung *f* nitride hardening, nitriding, nitridation, nitrogen [case-]hardening *(of steel)*
Nitrierkasten *m* nitriding box *(for treating steel)*
Nitrierkessel *m* nitrating pan
Nitrierkrepp *m* nitrated (nitrate, nitrating) paper
Nitriermittel *n* nitrating agent
Nitrierofen *m* nitriding furnace *(for treating steel)*
Nitrierprodukt *n* nitration product
Nitriersalzbad *n* nitriding bath *(for treating steel)*
Nitriersäure *f* nitrating acid, mixed acid *(a mixture of concentrated nitric and sulphuric acid)*
Nitrierschicht *f* nitride (nitrided) case (layer) *(on steel)*
Nitrierstahl *m* nitriding steel
Nitriertiefe *f* nitriding depth *(in steel)*
Nitrierung *f* 1. nitration; 2. *s.* Nitrierhärtung
direkte N. direct nitration
diskontinuierliche N. batch nitration
kontinuierliche N. continuous nitration
Nitrierungs... *s.* Nitrier...
Nitrierverfahren *n* 1. nitration process; 2. nitriding process, nitrogen case-hardening process *(for treating steel)*
Nitrifikanten *mpl s.* Nitratbakterien
Nitrifikation *f (agric)* nitrification
Nitrifikationsbakterien *pl* nitrifying bacteria, nitrobacteria *(collectively for nitrite and nitrate bacteria)*
Nitrifizierung *f (agric)* nitrification
Nitril *n* R–C≡N nitrile
Nitril-Chloroprenkautschuk *m* nitrile-chloroprene rubber, NCR
Nitrilgruppe *f* –C≡N nitrile group (residue)
Nitrilkautschuk *m* [acrylo]nitrile-butadiene rubber, nitrile rubber, NBR
karboxylgruppenhaltiger (karboxylierter) N. carboxynitrile rubber, carboxy-modified nitrile rubber
Nitrilotriessigsäure *f* $N(CH_2COOH)_3$ nitrilo-triacetic acid, tri-(carboxymethyl)amine
Nitrilsilikongummi *m s.* Nitrilsilikonkautschuk
Nitrilsilikonkautschuk *m* nitrile-silicone rubber, NSR, cyano silicone rubber
Nitrilsynthese *f* nitrile synthesis
Nitrit *n* nitrite
Nitritbakterien *pl*, **Nitritbildner** *mpl* nitrite[-forming] bacteria, nitrosobacteria, nitrous bacteria
nitrithaltig nitrite-containing
3-Nitroalizarin *n* 3-nitroalizarin, alizarin orange, 1,2-dihydroxynitroanthraquinone
Nitroalkan *n* nitroalkane, nitroparaffin
Nitroanilin *n* $NH_2C_6H_4NO_2$ nitroaniline
Nitroanthrachinon *n* nitroanthraquinone
Nitroäthan *n* $CH_3CH_2NO_2$ nitroethane
Nitrobakterien *pl s.* Nitrifikationsbakterien
Nitrobarit *m (min)* nitrobarite *(barium nitrate)*
Nitrobenzaldehyd *m* $NO_2C_6H_4CHO$ nitrobenzaldehyde
Nitrobenzoesäure *f* $NO_2C_6H_4COOH$ nitrobenzoic acid
Nitrobenzol *n* $C_6H_5NO_2$ nitrobenzene
Nitrobenzolkarbonsäure *f s.* Nitrobenzoesäure
***p*-Nitrobenzolsulfochlorid** *n*, ***p*-Nitrobenzolsulfonylchlorid** *n* $NO_2–C_6H_4SO_2Cl$ *p*-nitrobenzenesulphonyl chloride
Nitrochlorbenzol *n s.* Chlornitrobenzol
Nitrochloroform *n* CCl_3NO_2 nitrochloroform, chloropicrin, trichloronitromethane
Nitroderivat *n* nitro derivative
Nitroechtfarbstoff *m* fast nitro dye
Nitroessigsäure *f* NO_2CH_2COOH nitroacetic acid
Nitrofarbstoff *m* nitro dye
Nitroglyzerin *n (incorrectly for)* Glyzerintrinitrat
Nitrogruppe *f* $–NO_2$ nitro group
Nitro-Isonitro-Tautomerie *f* nitro-isonitro tautomerism
Nitrojektion *f (agric)* nitrojection
Nitrokalzit *m (min)* nitrocalcite, kalksaltpetre, lime saltpetre *(calcium nitrate)*
Nitrokobaltat *n* nitrocobaltate
Nitrokombinationslack *m* combination cellulose nitrate lacquer
Nitrokörper *m* nitro body
Nitrolack *m* cellulose nitrate lacquer
Nitrolsäure *f* nitrolic acid *(any of a class of compounds $RC(=NOH)NO_2$)*
Nitromagnesit *m (min)* nitromagnesite *(magnesium nitrate)*
Nitrometer *n* nitrometer, azetometer
Nitromethan *n* CH_3NO_2 nitromethane
Nitromoschus *m* nitro musk
Nitronaphthalin *n* $C_{10}H_7NO_2$ nitronaphthalene
Nitronatrit *m (min)* nitronatrite, [soda] nitre *(sodium nitrate)*
Nitroniumion *n s.* Nitrylion
Nitroniumperchlorat *n s.* Nitrylperchlorat
Nitronsäure *f* nitronic acid *(any of a class of compounds $R=NO(OH)$)*
Nitroparaffin *n* nitroparaffin, nitroalkane
Nitroperbenzoesäure *f* $NO_2C_6H_4CO_3H$ nitroperbenzoic acid

Nitrophosphat *n* nitrophosphate *(any of a group of nitrogen-phosphorus fertilizers)*
Nitroprussiat *n s.* Nitroprussid
Nitroprussid *n* $M_2^I[Fe(CN)_5(NO)]$ nitroprusside, nitroprussiate, pentacyanonitrosylferrate
Nitroprussidnatrium *n* $Na_2[Fe(CN)_5(NO)]$ sodium nitroprusside, disodium pentacyanonitrosylferrate
Nitroprussidnatriumpapier *n* sodium nitroprusside paper
nitros nitrous *(containing nitrogen oxides)*
Nitrosamin *n* nitrosamine *(any of a class of compounds R_1R_2N-NO)*
Nitrosaminrot *n* nitrosamine red
Nitrosat *n (org chem)* nitrosate
Nitrose *f* nitrous vitriol *(an intermediate in manufacturing sulphuric acid)*
Nitrosebakterien *pl s.* Nitritbakterien
nitrosieren to nitrosate
Nitrosiermittel *n* nitrosating agent
Nitrosierung *f* nitrosation
Nitrosit *n (org chem)* nitrosite
***p*-Nitrosoanilin** *n* $NH_2C_6H_4NO$ *p*-nitrosoaniline
Nitrosobenzol *n* C_6H_5NO nitrosobenzene
Nitrosofarbstoff *m* nitroso dye
Nitrosogruppe *f* $-N=O$ nitroso group
Nitroso-Isonitroso-Tautomerie *f* nitroso-isonitroso tautomerism
Nitrosokautschuk *m* nitroso rubber
Nitrosokresol *n* $ONC_6H_3(CH_3)OH$ nitrosocresol
Nitrosonaphthol *n* nitrosonaphthol
Nitrosoverbindung *f* nitroso compound
Nitrostärke *f (incorrectly for)* Stärkenitrat
Nitrosylchlorid *n* NOCl nitrosyl chloride
Nitrosylfluorid *n* NOF nitrosyl fluoride
Nitrosylhydrogensulfat *n* $NOHSO_4$ nitrosyl hydrogensulphate
Nitrosylion *n* NO^+ nitrosyl ion
Nitrotoluidin *n* $CH_3C_6H_3(NO_2)NH_2$ nitrotoluidine
Nitroverbindung *f* nitro compound
Nitroxylol *n* nitroxylene
Nitrozellulose *f (incorrectly for)* Zellulosenitrat
Nitrozelluloselack *m s.* Zellulosenitratlack
Nitrylchlorid *n* NO_2Cl nitryl chloride
Nitrylion *n* nitryl ion
Nitrylperchlorat *n* $NO_2[ClO_4]$ nitryl perchlorate
Niveau *n (phys chem, tech)* level
  angeregtes N. excited (excitation) level
  hängendes N. suspended level *(as in viscometers)*
Niveaubirne *f* [gas] levelling bulb
Niveaufläche *f* equipotential (potential energy) surface
Niveauflasche *f*, **Niveaugefäß** *n* levelling bottle
Niveaukonstanthalter *m* constant-level device; constant-level tank
Niveaukugel *f* [gas] levelling bulb
Niveaumessung *f* level measurement
Niveauregler *m* level controller
Niveaurohr *n* levelling tube
NK *s.* Naturkautschuk
NKH *s.* Nichtkarbonathärte
nl *s.* nichtlöslich
N-Lost *n s.* Stickstoffyperit
n-Lösung *f* N solution, normal solution (*for further compounds s.* Lösung)
NMR *s.* Resonanz / kernmagnetische
NMR-Spektrum *n s.* Resonanzspektrum / kernmagnetisches
Nobelium *n* No nobelium
No-iron-Ausrüstung *f (text)* no-iron finish
Nomenklatur *f* nomenclature
  chemische N. chemical nomenclature
  Genfer N. Geneva nomenclature
  Stocksche N. Stock nomenclature
Nomenklaturkommission *f* commission on nomenclature, nomenclature commission
Nomenklaturregel *f* nomenclature rule
Nomenklatursystem *n* nomenclature system
  Genfer N. Geneva system of nomenclature (naming)
Nonakosan *n* $C_{29}H_{60}$ nonacosane, *(specif)* $CH_3[CH_2]_{27}CH_3$ nonacosane, *n*-nonacosane
Nonaktin *n* nonactin *(antibiotic)*
Nonaktinsäure *f* nonactinic acid *(a furan derivative)*
Nonandisäure *f* $HOOC[CH_2]_7COOH$ nonanedioic acid, azelaic acid
1-Nonanol *n* $CH_3[CH_2]_7CH_2OH$ 1-nonanol, *n*-nonyl alcohol
Nonansäure *f* $CH_3[CH_2]_7COOH$ nonanoic acid, pelargonic acid
Nonin-(1) *n* $CH_3[CH_2]_6C{\equiv}CH$ 1-nonyne
nonvariant non-variant, invariant
***n*-Nonylaldehyd** *m* $CH_3[CH_2]_7CHO$ nonanal, *n*-nonylic aldehyde
***n*-Nonylalkohol** *m* $CH_3[CH_2]_7CH_2OH$ 1-nonanol, *n*-nonyl alcohol
Nonylon *n* $CH_3[CH_2]_7CO[CH_2]_7CH_3$ 9-heptadecanone, nonylone
***n*-Nonylsäure** *f s.* Nonansäure
Nootkaten *n* nootkatene *(a sesquiterpene)*
Nootkatin *n* nootkatin *(a tropolone)*
N-O-Peptidylverschiebung *f (text)* N-O peptidyl shift
Noppenfärben *n (text)* burl dyeing
2-Norbornen *n* norbornene, bicyclo-[2,2,1]hept-2-ene
Norekgonin *n* norecgonine
Norgesalpeter *m* Norway (Norwegian) saltpetre, Norge nitre *(calcium nitrate)*
Norm ... *s.a.* Normal ...
normal normal
  n. anfärbend regular-dyeing
Normal *n* standard
Normal ... *s.a.* Norm ...
Normalalkali *n* standard alkali
Normalalkohol *m* proof spirit
Normalatmosphäre *f* standard atmosphere
Normalausrüstung *f* standard equipment
Normalbedingungen *fpl* normal conditions
Normalbutan *n* $CH_3CH_2CH_2CH_3$ butane *(proper)*, *n*-butane
Normaldosis *f* normal dose
Normaldruck *m* normal (standard) pressure

Normaldrucktrockner *m* atmospheric dryer
Normaleinheit *f* standard (normal) unit
Normalelektrode *f* normal electrode
Normalelement *n* standard cell
Westonsches N. Weston [normal] cell, standard Weston cell
Normalentwickler *m (phot)* normal developer
normalglühen to normalize *(steel)*
Normalglühen *n* normalizing, normalization *(of steel)*
Normalglühofen *m* normalizing furnace *(for treating steel)*
Normalheptan *n* $CH_3[CH_2]_5CH_3$ heptane *(proper)*, *n*-heptane
Normalhexan *n* $CH_3[CH_2]_4CH_3$ hexane *(proper)*, *n*-hexane
normalisieren *s.* normalglühen
Normalisierofen *m s.* Normalglühofen
Normalisierungsglühen *n s.* Normalglühen
Normalität *f* normality *(of solutions)*
Normalitätsfaktor *m* normality (volumetric, titrimetric) factor
Normal-Kalilauge *f* normal KOH solution
Normalkalomelelektrode *f* normal calomel electrode
Normalklima *n (text)* standard[ized] conditions
Normalkomplex *m* normal (addition) complex
Normallauge *f* standard base
Normallösung *f* standard (normal) solution, *(specif)* N solution, normal solution *(containing one gram equivalent per litre)*
$^1/_{10}$-Normallösung *f* decinormal solution (*for similar constructions s. under* Lösung)
Normalluftdruck *m* standard (normal) pressure
Normal-Nitritlösung *f (dye)* standard solution of sodium nitrite
Normaloktan *n* $CH_3[CH_2]_6CH_3$ octane *(proper)*, *n*-octane
Normalpentan *n* $CH_3[CH_2]_3CH_3$ pentane *(proper)*, *n*-pentane
Normalpotential *n* normal potential
Normalprobe *f* standard sample
Normalsalz *n* normal (neutral) salt
Normalsäure *f* standard acid
Normalschwingung *f (phys chem)* normal vibration
Normalseifenlösung *f* standard soap solution
Normalsieb *n* standard sieve (screen)
Normalsiebreihe *f*, **Normalsiebskala** *f* standard sieve scale (series)
Normal-Silber-Silberchloridelektrode *f* normal silver-silver chloride electrode
Normalsintern *n* pressureless sintering
Normalspannung *f* normal voltage
Normalspektrum *n* normal spectrum
normalstark proof *(of liquids containing alcohol)*
Normalstärke *f* proof *(of liquids containing alcohol)*
Normaltemperatur *f* normal temperature
Normalthermometer *n* standard thermometer
Normalton *m (dye)* standard shade
Normaltontiefe *f (dye)* standard depth [of shade]
Normaltropfenzähler *m* standard dropper
Normalverbindung *f* normal compound
Normalverdampfer *m* standard evaporator
Normalverkokung *f* normal (high-temperature) carbonization (coking)
Normalvolumen *n* standard (normal) volume
Normalwasserstoffelektrode *f* normal hydrogen electrode
Normalweingeist *m* proof spirit
Normalwert *m* standard value
Normalwiderstand *m* standard resistance
Normalwiderstandsthermometer *n* standard resistance thermometer
Normalzustand *m* 1. normal (ground) state, ground term (level), basic term *(of an atom)*; 2. *s.* Normzustand
Normblende *f* standard orifice *(flow measurement)*
Normdichte *f* normal density
Normdüse *f* standard nozzle *(flow measurement)*
Normprüfsieb *n* standard (normal) test sieve (screen)
Normschliff *m (lab)* standard ground joint
Normschliffhahn *m* standard ground stopcock
Normzustand *m* 1. standard state; 2. normal (standard) temperature and pressure, NTP, STP *(0°C and 760 mm Hg)*
Nornikotin *n* nornicotine *(a tobacco alkaloid)*
3-Nortropanol *n s.* Nortropin
Nortropin *n* nortropine, 3-nortropanol, 8-azabicyclo-[3,2,1]octan-3-ol
Nosean *m (min)* nosean, noselite *(a feldspar)*
Notation *f (nomencl)* notation
Notationssystem *n (nomencl)* notation system
N. von Wiswesser Wiswesser [notation] system
Notbrause *f* safety (emergency, drench) shower
Notenpapier *n* music paper
Novolack *m s.* Novolak
Novolak *m*, **Novolakharz** *n* novolak [resin], two-stage resin
N-Oxid *n s.* Aminoxid
N-Schale *f* N-shell *(of an atom)*
NSSC-Stoff *m* neutral sodium sulphite semichemical pulp, NSSC pulp
NSSC-Verfahren *n (pap)* neutral sulphite [semichemical] process, NSSC process, neutral sodium sulphite process
NST-Wert *m (plast)* no-strength temperature
NTE *s.* Nitrilotriessigsäure
N-terminal N-terminal
Nuance *f* shade, tint, tone, hue, *(text also)* cast
nuancieren to shade, to tint, to tone, *(text also)* to cast
Nuancierung *f* 1. shading, tinting, toning, *(text also)* casting; 2. *s.* Nuance
Nugget *n* gold nugget
Nuklealreaktion *f* / **Feulgensche** Feulgen reaction (staining method, test) *(for detecting deoxyribonucleic acid)*
nuklear nuclear
Nuklease *f* nuclease
Nuklein *n* 1. nuclein; 2. *s.* Nukleinsäure
Nukleinsäure *f (bioch)* nucleic acid
Nukleinsäuresynthese *f* nucleic acid synthesis
Nukleinstoff *m* nuclein

Nukleon *n* nucleon, nuclear particle (constituent)
Nukleonenkomponente *f* nucleonic component
nukleophil nucleophilic
Nukleophil *n* nucleophile
Nukleophilie *f* nucleophilicity
Nukleoplasma *n (bioch)* nuclear sap
Nukleoproteid *n* nucleoprotein
Nukleosid *n* nucleoside
Nukleosidase *f* nucleosidase
Nukleotid *n* nucleotide
Nukleotidase *f* nucleotidase
Nukleus *m* nucleus
Nuklid *n* nuclide
  radioaktives N. radioactive nuclide, radionuclide
Nulladung *f* zero charge
Nullage *f* rest point *(of a balance)*
Nulleffekt *m (nucl)* background
Nullelektrode *f* null electrode
Nullfläche *f (agric)* check plot, nil *(as in testing fertilizer effects)*
Nullinie *f* zero (null) line *(in the band spectrum)*
Nullinstrument *n* null instrument
Nullmethode *f* null method [of measurement]
Nullparzelle *f s.* Nullfläche
Nullporosität *f* zero porosity
Nullpunkt *m* zero [point] *(of a scale)*
  absoluter N. absolute zero
Nullpunktsenergie *f* zero-point energy
Nullpunktsentropie *f* zero-point entropy, entropy of absolute zero
Nullpunktskonfiguration *f* zero-point configuration
Nullpunktsschwingung *f* zero-point vibration
Nullrate *f (nucl)* background
Nullstelle *f* origin (centre) of the band *(spectroscopy)*
Nullstellung *f* zero position
Nulltoleranz *f (tox)* zero tolerance
Nullvariante *f (agric)* nil *(as in testing fertilizer effects)*
Nullversuch *m* 1. blank, blank experiment (test, trial); 2. *s.* Nullvariante
nullwertig zero-valent, non-valent, avalent
Nullwertigkeit *f* zero valence
Nullzweig *m* Q-branch *(of a band spectrum)*
numerieren *(nomencl)* to number
Numerierung *f (nomencl)* numbering
  N. im Uhrzeigersinn clockwise numbering
Numerierungssystem *n (nomencl)* numbering system
Nummer *f (text)* number, count
Nur-Glas-Papier *n* glass paper *(an insulating material)*
Nußöl *n* nut oil
Nutringdichtung *f* U-seal
Nutsche *f* nutsch[e], nutsch filter
nutschen to filter [off] by suction, to filter under suction
Nutzarbeit *f (phys chem)* useful work
  maximale N. useful maximum work
nutzbar effective
Nutzleistung *f* efficiency
Nutzraum *m (ceram)* setting space *(of a kiln)*
Nutzwasser *n* industrial (service, process) water, water for industrial use
n-Verbindung *f* normal compound *(as opposed to iso compounds)*
Nylonfarbstoff *m* nylon dye
Nylonfilz *m (pap)* nylon felt
N-Yperit *n s.* Stickstoffyperit

O

Oakes-Maschine *f (rubber)* Oakes frother
O_2-Aufblaskonverter *m s.* Oberwindkonverter
O_2-Aufblasverfahren *n s.* Oberwindfrischverfahren
Obenaufgabe *f (tech)* top feed
Oberbainit *m (met)* upper bainite
Oberbauseitenwand *f* casement (casing, jamb, breast) wall *(of a glass-melting furnace)*
Oberboden *m (soil)* eluvial horizon, A-horizon
Oberdruckpresse *f (plast)* top ram press, down stroke press
Oberfilz *m (pap)* top felt, overfelt
Oberfläche *f* surface ✦ **die O. behandeln** to surface, to finish ✦ **mit glatter O.** smooth-surfaced
  gehämmerte O. batter *(a defect in glass)*
  innere O. inner surface
  polierte O. polished surface
  rauhe O. rough surface; pulled surface *(a defect in plastics)*
  spezifische O. specific surface
  wirksame O. effective surface
oberflächenaktiv surface-active
Oberflächenaktivität *f* surface activity
  spezifische O. specific surface activity
Oberflächenanästhetikum *n* surface anaesthetic
Oberflächenarbeit *f* free surface energy
Oberflächenatom *n* surface atom
Oberflächenausdehnung *f* surface extent
Oberflächenbehandlung *f* surface treatment, surfacing, finish[ing]
Oberflächenbelüftung *f* surface aeration
Oberflächenbeschaffenheit *f* surface properties
Oberflächenbild *n (phot)* surface image
  latentes O. surface latent image
Oberflächenblase *f* skin blister *(a defect in glass)*
Oberflächenchemie *f* surface chemistry
Oberflächendiffusion *f* surface diffusion
Oberflächendruck *m* surface pressure
Oberflächeneffekt *m* surface effect
Oberflächeneigenschaften *fpl* surface properties
Oberflächenenergie *f* surface energy
  freie O. free surface energy
Oberflächenenthalpie *f* surface enthalpy
Oberflächenentropie *f* surface entropy
Oberflächenentwickler *m (phot)* surface developer
Oberflächenentwicklung *f (phot)* surface development
Oberflächenerscheinung *f* surface phenomenon
Oberflächenfärbung *f (pap)* surface colouring (staining), tub colouring, dipping
  O. im Kalander calender colouring (staining), padding, stuffing

Oberflächenfilm *m* [surface] film
  gasanaloger O. gaseous film
  kondensierter O. condensed film
  monomolekularer O. monomolecular (unimolecular) surface film
Oberflächenfilter *n* surface (edge) filter
Oberflächenfiltration *f* surface (edge) filtration
oberflächengefärbt *(pap)* surface-coloured
oberflächengeleimt *(pap)* surface-sized, top-sized
  im Leimbadtauchverfahren o. surface-sized with size tub, tub-sized, T.S., vat-sized
  in der Leimpresse o. surface-sized with size press
  mit Gelatine o. *s.* mit Tierleim o.
  mit Stärke o. surface-sized with starch
  mit Tierleim o. surface-sized with animal glue, animal-sized, gelatin-sized
Oberflächengestein *n* effusive rock
Oberflächenglanz *m* surface lustre, gloss, glaze
Oberflächengüte *f* surface finish
Oberflächenhaftung *f* surface attachment
Oberflächenhärte *f* surface hardness
oberflächenhärten to surface-harden *(metals)*
  durch Diffusion o. to cement
Oberflächenhärten *n* surface hardening *(of metals)*
  O. durch Diffusion cementation, cementing
Oberflächenhaut *f* surface film
Oberflächenkatalyse *f* surface (contact, heterogeneous) catalysis
Oberflächenkondensator *m* surface condenser
Oberflächenkonzentration *f* surface concentration
Oberflächenkraft *f* surface force
Oberflächenkultur *f (bioch)* surface culture
Oberflächenlebensdauer *f* surface lifetime *(of a semiconducting crystal)*
Oberflächenleim *m (pap)* surface-sizing agent, tub size
Oberflächenleimmaschine *f (pap)* surface-sizing (tub-sizing) machine
Oberflächenleimung *f (pap)* surface (top) sizing
  O. im Leimbadtauchverfahren surface sizing with size tub, tub (vat) sizing, T.S., size-tub treatment
  O. in der Leimpresse surface sizing with size press
  O. mit Gelatine *s.* O. mit Tierleim
  O. mit Stärke surface sizing with starch
  O. mit Tierleim surface sizing with animal glue, animal tub sizing, A.T.S., gelatin sizing
Oberflächenleitfähigkeit *f*, **Oberflächenleitvermögen** *n* surface conductivity
Oberflächenlösung *f* surface solution
Oberflächenlüftung *f* surface aeration
Oberflächenmatte *f (plast)* surfacing mat, overlay sheet
Oberflächenniveau *n* surface level
Oberflächenpotential *n* surface potential
Oberflächenrauhigkeit *f* surface roughness
Oberflächenreaktion *f* surface reaction
Oberflächenriß *m* surface crack; *(glass)* check, vent, *(in the neck of a bottle)* smear
Oberflächenrückstand *m* extrasurface residue *(of pesticides)*
Oberflächenrüttler *m* surface vibrator
Oberflächenrüttlung *f* surface vibration
Oberflächenschicht *f* surface layer
  aufgekohlte (eingesetzte, zementierte) O. carburized case *(on steel)*
Oberflächenschliere *f* surface cord *(a defect in glass)*
Oberflächenschutz *m* surface protection
Oberflächenspannung *f* surface tension
oberflächenspannungsvermindernd surface-tension-depressing
Oberflächentemperatur *f* surface temperature
Oberflächentextur *f* surface texture
Oberflächentrockner *m (coat)* surface drier
Oberflächenverbindung *f* surface compound
Oberflächenverbrennung *f* surface combustion
Oberflächenverdichter *m* surface condenser
Oberflächenverdichtung *f (build)* surface compaction
Oberflächenverdunstung *f* surface evaporation
Oberflächenvergrößerung *f* increase of surface
Oberflächenverwitterung *f* surface weathering
Oberflächenwachstum *n* surface growth
Oberflächenwärmeaustauscher *m* surface exchanger
Oberflächenwasser *n* surface water
Oberflächenwiderstand *m* surface resistance
  spezifischer O. surface resistivity
oberflächenwirksam surface-active
Oberflächenwirkung *f* surface effect
Oberflächenzone *f s.* Oberflächenschicht
Oberflächenzustand *m* surface conditions
obergärig top-fermenting, top-fermented
Obergärung *f* top (surface) fermentation
Oberhefe *f* top[-fermentation] yeast
Oberkolben *m (plast)* top ram (force) *(of a press)*
Oberkolbenpresse *f (plast)* top ram press, down stroke press
Oberleder *n* upper (dressing) leather
Obermesser *n (pap)* revolving (fly) knife *(of a cross-cutter)*
Oberphos-Verfahren *n (fert)* Oberphos process
Oberschale *f* **der Petrischale** *(lab)* Petri dish top
Oberschicht *f* top layer
Oberschwingungsbande *f* overtone band *(in molecular spectra)*
Oberseite *f (pap)* top (felt) side
Oberstempel *m (plast)* top ram (force) *(of a press)*; *(ceram)* top punch; *(met)* upper plunger *(of a die)*
Oberteil *n* upper part, top
Obertuch *n (pap)* top felt, overfelt
Oberwalze *f* top roll
  O. der Leimpresse *(pap)* top size press roll
Oberwindfrischverfahren *n* basic (top-blown) oxygen converter (furnace) process, basic oxygen [steel] process
Oberwindkonverter *m* basic (top-blown) oxygen converter (furnace)
Objektträger *m* [microscope] slide
Objektträgerzellenmethode *f* slide-cell method *(for testing antibiotics)*
Objektträgerzellentechnik *f* / **Wrightsche** Wright slide-cell technique *(for testing antibiotics)*

Obstbaumspritzmittel *n* fruit tree spray
fungizides O. fruit-fungicide spray
Obstessig *m* fruit vinegar
Obstpülpe *f* fruit pulp
Obstsaft *m* [fruit] juice
Obstwein *m* fruit wine
Ocker *m (min)* ochre
Gelber O. yellow ochre *(a mixture of limonite with clay and silica used as a pigment)*
Italienischer O. Italian red *(a pigment consisting of iron(III) oxide)*
Roter O. red ochre (rudd), stone red *(a red haematite used as a pigment)*
ockerhaltig ochreous
Oderberg-Mühle *f* Oderberg [colloid] mill
Odontolith *m (min)* odontolite
Odorans *n* odorant
odorieren to odorize *(toxic gases)*
Odoriermittel *n* odorant
Odorierung *f* odorization *(of toxic gases)*
Odorierungsmittel *n* odorant
Odorimetrie *f* 1. odorimetry *(measurement of the intensity of odours)*; 2. *s.* Olfaktometrie
odorisieren *s.* odorieren
OE *s.* Oxford-Einheit
Oechsle-Waage *f* must gauge, saccharometer for grapes
Ofen *m (tech)* furnace, *(esp ceram)* kiln, *(esp for lower temperatures)* oven ✦ **im O. trocknen** to kiln-dry, to oven-dry
O. für durchlaufenden (kontinuierlichen) Betrieb continuous furnace
O. mit aufsteigender Flamme *(ceram)* updraught kiln
O. mit Außenbeheizung *s.* außenbeheizter O.
O. mit elektrischer Beheizung *s.* elektrischer O.
O. mit horizontaler Flammenführung *(ceram)* horizontal-draught kiln
O. mit indirekter Beheizung *s.* außenbeheizter O.
O. mit Längsgewölbe *(ceram)* longitudinal arch kiln
O. mit rascher Brandfolge *(ceram)* short-cycle kiln
O. mit Sohlebeheizung sole-flue oven
O. mit überschlagender Flamme *(ceram)* downdraught kiln
O. mit U-Flammenführung *(glass)* end-fired (end-port) furnace
O. mit waagerechter Flammenführung *(ceram)* horizontal-draught kiln
O. mit wanderndem Feuer *(ceram)* moving-fire kiln
O. mit Widerstandserhitzung (Widerstandsheizung) resistance (resistance-heated, resistor) furnace
außenbeheizter O. indirect-fired (indirect-heat, externally heated) furnace
belgischer O. *(ceram)* Belgian kiln
brennstoffbeheizter O. fuel-heated furnace
deckenbeheizter O. *s.* von oben beheizter O.
direkt beheizter O. direct-fired furnace
elektrischer (elektrisch beheizter, elektrothermischer) O. electric furnace
gemuffelter O. muffle furnace; *(ceram)* muffle kiln
halbgemuffelter O. *(ceram)* semimuffle kiln
holzgefeuerter O. wood-fired furnace; *(ceram)* wood-fired kiln
holzkohlengefeuerter O. charcoal-fired furnace
indirekt beheizter O. indirect-fired furnace
induktionsbeheizter O. induction (inductance, induction-heated) furnace
intermittierender O. *s.* periodischer O.
Kasseler O. *(ceram)* Kassel[er] kiln
kontinuierlicher (kontinuierlich arbeitender) O. continuous furnace
Mannheimer O. Mannheim furnace *(for producing hydrochloric acid)*
mehretagiger (mehrherdiger) O. multihearth (multiple-hearth) furnace
periodischer (periodisch arbeitender) O. batch-type furnace; *(ceram)* periodic kiln
rotierender O. rotary furnace; *(for sintering or calcining)* rotary kiln
von oben beheizter O. top-fired furnace, overfired furnace; *(ceram)* top-fired kiln
widerstandsbeheizter O. *s.* O. mit Widerstandserhitzung
zweietagiger O. *(ceram)* two-tier kiln
Öfen *mpl* / **gekoppelte** *(ceram)* linked kilns
Ofenabwärme *f* furnace waste heat
Ofenalterung *f (rubber)* [air] oven ageing
Ofenatmosphäre *f* furnace atmosphere
Ofenauskleidung *f*, **Ofenausmauerung** *f* furnace lining
Ofenausschuß *m (ceram)* kiln loss
Ofenbeschickung *f* furnace charge
Ofenbetrieb *m* furnace operation
Ofenboden *m s.* Ofensohle
Ofencharge *f* furnace charge
Ofendrehwerk *n* furnace-rotating mechanism
Ofeneinsatz *m* furnace charge
Ofenfutter *n* furnace lining
basisches O. *(met)* basic lining
saures O. *(met)* acid lining
Ofengas *n* furnace (flue) gas
Ofengestell *n* furnace hearth
ofengetrocknet kiln-dried, oven-dried
Ofenlack *m* stoving lacquer; *(if unpigmented)* stoving varnish
Ofenladung *f* furnace charge
Ofenmantel *m* furnace shell
Ofenraum *m* furnace chamber
Ofenruß *m (rubber)* furnace [combustion] black
Ofenschacht *m* furnace shaft
Ofenschlacke *f* furnace slag (clinker)
Ofenschwarz *n s.* Ofenruß
Ofensohle *f* furnace bottom; *(ceram)* kiln floor; *(glass)* bench, siege *(of a pot furnace)*
Ofenstützmaterial *n (ceram)* kiln furniture
Ofentransformator *m* furnace transformer
ofentrocken kiln-dried, oven-dried
Ofentrocknung *f* kiln (oven) drying
Ofentür *f* / **gemauerte** *(ceram)* [kiln] wicket
Ofenverfahren *n (rubber)* furnace [combustion]

process, continuous-furnace method *(for producing carbon black)*
Ofenvorlage *f* gas collecting main *(of a coke-oven battery)*
Ofenwagen *m (ceram)* kiln car
Ofenwanne *f* crucible *(of an arc furnace)*
Ofenzug *m* flue
Ofenzustellung *f (met)* furnace lining
offenkettig open-chain
offenzellig open-cell *(foamed plastics)*
offizinell *(pharm)* official, pharmacopoeial
öffnen 1. to open; 2. to dismantle *(a filter press)*
Öffnung *f* aperture, opening, orifice, hole
Öffnungsdruck *m* opening pressure *(of a valve)*
Offsetdruck *m* offset printing
Offsetdruckfarbe *f* offset [printing] ink
Offsetdruckpapier *n* offset [printing] paper
Offsetdruckverfahren *n* offset [printing] process
Offsetkarton *m* offset [printing] cardboard
Offsetpapier *n* offset [printing] paper
Offsetpresse *f (pap)* offset (smoothing) press
OFHC-Kupfer *n* oxygen-free high-conductivity copper, O.F.H.C. copper
Ogia-Kopal *m* Accra copal *(from Daniella ogea Rolfe)*
OH-Gruppe *f* OH group, hydroxyl group (residue)
OH-Radikal *n* OH radical, *(specif)* free OH radical
OH-Rest *m s.* OH-Gruppe
OHZ *s.* Hydroxylzahl
Oiazin *n s.* Pyridazin
Oiticicaöl *n* oiticica oil *(from Licania rigida Benth.)*
okkludieren to occlude
Okklusion *f* occlusion
Oktadekadien-(9,12)-säure *f* 9,12-octadecadienoic acid
Oktadekan *n* $C_{16}H_{38}$ octadecane
Oktadekanal *n* $CH_3[CH_2]_{16}CHO$ octadecanal, stearaldehyde
Oktadekanamid *n* $CH_3[CH_2]_{16}CONH_2$ octadecanoamide, stearamide, stearic acid amide
Oktadekananilid *n* $CH_3[CH_2]_{16}CONHC_6H_5$ octadecanoanilide, stearanilide
Oktadekannitril *n* $CH_3[CH_2]_{16}CN$ octadecanonitrile, stearonitrile
Oktadekanol-(1) *n* $CH_3[CH_2]_{16}CH_2OH$ 1-octadecanol, octadecyl alcohol
Oktadekansäure *f* $CH_3[CH_2]_{16}COOH$ octadecanoic acid, stearic acid
Oktadekatriensäure *f* octadecatrienoic acid
Oktadezen-(11)-in-(9)-säure *f* 11-octadecen-9-ynoic acid, santalbic acid
Oktadezensäure *f* octadecenoic acid
Oktadezin-(9)-säure *f* 9-octadecynoic acid, stearolic acid
***n*-Oktadezylalkohol** *m s.* Oktadekanol-(1)
Oktaeder *n (cryst)* octahedron
oktaedrisch *(cryst)* octahedral, oct.
Oktafluorid *n* octafluoride
Oktahydrat *n* octahydrate
Oktahydrid *n* octahydride
Oktahydronaphthalin *n* octahydronaphthalene, octalin
Oktakosansäure *f* $C_{27}H_{55}COOH$ octacosanoic acid, montanic acid
Oktalin *n s.* Oktahydronaphthalin
Oktamolybdat *n* $M_4^IMo_8O_{26}$ octamolybdate
Oktan *n* C_8H_{18} octane, *(specif)* $CH_3[CH_2]_6CH_3$ octane, *n*-octane
Oktanal *n* $CH_3[CH_2]_6CHO$ octanal, *n*-octyl aldehyde
Oktandikarbonsäure-(1,8) *f* $HOOC[CH_2]_8COOH$ octane-1,8-dicarboxylic acid, decanedioic acid
Oktandisäure *f* $HOOC[CH_2]_6COOH$ octanedioic acid
Oktanol-(1) *n* $CH_3[CH_2]_6CH_2OH$ 1-octanol, *n*-octyl alcohol
Oktansäure *f* $CH_3[CH_2]_6COOH$ octanoic acid
Oktanzahl *f* octane number (rating, value) ✦ **mit hoher O.** high-octane ✦ **mit niedriger O.** low-octane
Oktanzahlbestimmung *f* octane rating
oktavalent octavalent
Oktavalenz *f* octavalency
Oktawolframat *n* octawolframate, octatungstate
Oktazyanowolframat(IV) *n* $M_4^I[W(CN)_8]$ octacyanowolframate(IV), octacyanotungstate(IV)
Oktazyanowolframat(V) *n* $M_3^I[W(CN)_8]$ octacyanowolframate(V), octacyanotungstate(V)
Oktett *n* octet
Oktettlücke *f* octet gap
Oktettregel *f* octet rule
Oktin-(1) *n* $HC{\equiv}C[CH_2]_5CH_3$ 1-octyne
Oktose *f* octose *(any of a class of monosaccharides)*
Oktoxid *n* octaoxide
***n*-Oktylaldehyd** *m s.* Oktanal
***n*-Oktylalkohol** *m s.* Oktanol-(1)
***n*-Oktylazetylen** *n* $HC{\equiv}C[CH_2]_7CH_3$ 1-decyne, octylacetylene
***n*-Oktylsäure** *f s.* Oktansäure
Okular *n* eyepiece, ocular *(of a microscope)*
Öl *n* oil ✦ **in Ö. ablöschen (abschrecken)** to oil-quench ✦ **in Ö. härten** to oil harden *(steel)* ✦ **mit Ö. beheizt** oil-fired, oil-heated ✦ **mit Ö. getränkt** oil-impregnated
Ö. auf Erdölbasis petroleum oil
Ö. aus Rückstandsaufarbeitung recovered oil *(in coal hydrogenation)*
Ö. mit negativem Doktortest *(petrol)* sweet oil
Ö. mit positivem Doktortest *(petrol)* sour oil
abgepreßtes Ö. *(petrol)* pressed distillate, blue oil
abgetopptes Ö. *s.* getopptes Ö.
absolutes Ö. *(cosmet)* absolute essence, absolute [from concrete]
ätherisches Ö. essential (volatile) oil
compoundiertes Ö. compounded oil
destilliertes Ö. distilled oil
Dippelsches Ö. Dippel's (bone) oil
doktor-negatives Ö. *(petrol)* sweet oil
doktor-positives Ö. *(petrol)* sour oil
eingedicktes Ö. thickened oil
emulgierbares Ö. emulsifiable (soluble, miscible) oil
fettes Ö. fat[ty] oil, fixed oil *(as opposed to essential oil)*; *(petrol)* rich oil *(an absorption oil for light hydrocarbons)*
geblasenes Ö. blown oil

gebranntes Ö. distilled oil
gefettetes Ö. compounded oil
gehärtetes Ö. hardened (hydrogenated) oil
gesäuertes Ö. *s.* saures Ö.
getopptes Ö. topped crude [petroleum], reduced crude [oil], reduced oil
halbtrocknendes Ö. semidrying oil
helles Ö. pale oil *(a lubricating-oil distillate)*
hydriertes Ö. *s.* gehärtetes Ö.
kaltgepreßtes (kaltgeschlagenes) Ö. cold-drawn (cold-pressed) oil
konkretes Ö. *(cosmet)* concrete [oil]
leichtes Ö. light oil
lösliches Ö. soluble oil
medizinisches Ö. medicinal oil
mineralisches Ö. mineral oil
mischbares Ö. 1. *(agric)* emulsifiable concentrate *(used as a pesticide)*; 2. *s.* emulgierbares Ö.
naphthenhaltiges (naphthenisches) Ö. naphthenic oil
neutrales Ö. neutral oil
nichttrocknendes Ö. non-drying oil, permanent oil
oxydiertes Ö. *(coat)* blown oil
Paalsgardsches Ö. Paalsgard emulsion oil, P.E.O.
pflanzliches Ö. vegetable oil
reduziertes Ö. *s.* getopptes Ö.
rohes Ö. *(petrol)* crude [oil]
rostschützendes Ö. rust-inhibiting oil
saures Ö. *(petrol)* sour oil *(acid-treated oil before neutralization)*
schwach trocknendes Ö. *s.* halbtrocknendes Ö.
schweres Ö. heavy oil
staubbindendes Ö. dust-laying oil
sulfatiertes (sulfiertes) Ö. sulphated oil
sulfoniertes (sulfuriertes) Ö. sulphonated oil
trocknendes Ö. drying oil
wasserlösliches Ö. *s.* emulgierbares Ö.
zurückgewonnenes Ö. recovered oil
Ölabfluß *m* drainage of oil
Öablöschung *f* oil quenching
Ölabscheider *m* oil separator
Ölabscheidung *f* oil separation
Ölabschreckung *f* oil quenching
Ölabsorption *f* oil absorption
ölabweisend oil-repellent
Ölabweisungsvermögen *n* oil repellency
Ölanreicherung *f*, **Ölansammlung** *f* oil accumulation
ölartig oily, oleaginous
Ölausbruch *m (petrol)* blow-out
Ölbad *n* oil bath
Ölbatch *m (rubber)* oil masterbatch
Ölbehälter *m* oil tank (container); oil cup *(of a viscometer)*
ölbeheizt oil-fired, oil-heated
Ölbeheizung *f s.* Ölheizung
Ölbeize *f (coat)* oil stain
ölbeständig resistant (fast) to oil, oil-resistant, oil-resisting
Ölbeständigkeit *f* resistance to oil, oil resistance
ölbildend oil-forming
Ölbitumen *n* oily bitumen
Ölbohne *f* soybean, soya[-bean], soja, soy, Glycine max (L.) Merr.
Ölbohrloch *n*, **Ölbohrung** *f* oil (petroleum) well
Ölbrenner *m* oil burner (gun)
Olbrücke *f* ol (olation) bridge
Ölbrunnen *m* oil (petroleum) well
pumpender Ö. pumping well
öldicht oiltight, oilproof
Öldichtigkeit *f* oiltightness, oilproofness
Oleanolsäure *f* oleanoic acid *(a triterpene derivative)*
Oleat *n* oleate
Olefin *n* olefin, alkene, ethylenic hydrocarbon
Olefinierung *f* olefination
Olefinierungsreagens *n* olefin-forming reagent
olefinisch olefinic
Olefinoxidpolymerisation *f* olefin-oxide polymerization
Olefinpolymerisation *f* olefin polymerization
Olefinreihe *f* olefin series, alkene family
Olein *n* olein *(commercial oleic acid)*
Öleinsatz *m* oil feed
dampfförmiger Ö. vapour feed *(in Thermofor catalytic cracking)*
flüssiger Ö. liquid feed *(in Thermofor catalytic cracking)*
Oleinsäure *f s.* Ölsäure
ölen to oil
Oleo[margarin] *n* oleomargarine
Oleoresin *n* oleoresin *(a mixture of an essential oil and a resin)*
Oleostearin *n*, **Oleostock** *n (food)* oleostearin
Oleum *n* oleum, fuming sulphuric acid
Oleylalkohol *m* $C_8H_{17}CH{=}CH[CH_2]_8OH$ oleyl alcohol, *cis*-9-octadecenol
1-Oleylglyzeryläther *m* 1-oleyl glycerol ether, selachyl alcohol
Olfaktometer *n (med)* olfactometer *(an instrument for measuring the keenness of the sense of smell)*
Olfaktometrie *f (med)* olfactometry
Ölfalle *f* oil trap
Ölfänger *m* oil separator
Ölfarbe *f* oil paint, oil-base[d] paint, *(esp in art)* oil colour
Ölfeld *n* oil field (pool)
Ölfeldentwicklung *f* oil-field development
Ölfeldwasser *n (petrol)* edge water
ölfest *s.* ölbeständig
Ölfeuerung *f s.* Ölheizung
Ölfilter *n* oil filter
Ölfirnis *m* boiled oil; oil varnish *(for printing)*
Ölfleck *m* oil spot
ölfrei free from oil, oil-free
Ölfrucht *f* oil plant
ölführend oil-bearing, petroleum-bearing, petroliferous
Öl-Furnace-Anlage *f (rubber)* oil-furnace plant
Öl-Furnace-Ruß *m (rubber)* oil-furnace black
Öl-Furnace-Verfahren *n (rubber)* oil-furnace process
Ölgas *n* oil (fatty) gas

ölgefeuert oil-fired, oil-heated
Ölgehalt *m* oil content; *(coat)* oil length *(related to resin)*
Ölgemisch *n* oil mixture
ölgestreckt *(rubber)* oil-extended, oil-filled
ölgetränkt oil-impregnated
Ölgrün *n* oil green *(chromium(III) oxide)*; chrome green *(consisting of iron blue and chrome yellow)*
ölhaltig 1. oil-containing, oily, oleaginous; 2. *(rubber)* oil-extended, oil-filled; 3. *s.* ölführend
ölhärten to oil-harden *(steel)*
Ölhärter *m*, **Ölhärtestahl** *m* oil-hardening steel
Ölhärtung *f* 1. oil hardening *(of steel)*; 2. *(food)* hydrogenation of oil
Ölharz *n* oleoresin
Ölharzfarbe *f* oleoresinous paint
Ölharzlack *m* oleoresinous varnish
Öl-Harz-Verhältnis *n* oil/resin ratio, oil-to-resin ratio
Ölheizung *f* oil-fired heating
Ölhorizont *m (geol)* oil layer (horizon)
Olibanum *n* [frank]incense, olibanum *(from Boswellia specc.)*
Olibanumöl *n (pharm)* frankincense (olibanum) oil *(from Boswellia specc.)*
ölig oily, oleaginous, *(esp of crystals)* unctuous
Oligase *f* oligase
Öligkeit *f* oiliness, *(esp of crystals)* unctuousness
Oligoklas *m (min)* oligoclase *(a tectosilicate)*
Oligomer[es] *n* oligomer
Oligopeptid *n* oligopeptide *(containing up to 10 amino acids)*
Oligosaccharid *n* oligosaccharide
oligotroph oligotrophic *(deficient in dissolved plant nutrients)*
Ölindustrie *f* oil industry
Öl-in-Wasser-Emulsion *f* oil[-in]-water emulsion, O/W emulsion
Öl-in-Wasser-Typ *m* oil-in-water type, O/W type *(of emulsions)*
Olive *f* 1. olive *(from Olea europaea L.)*; 2. *(lab)* hose coupling (connection, connector)
Olivenit *m (min)* olivenite *(copper(II) hydroxide orthoarsenate)*
Olivenöl *n* olive oil *(from Olea europaea L.)*
Oliver-Filter *n* Oliver filter
Olivin *m (min)* olivine *(a nesosilicate)*
Olivinerzeugnis *n (ceram)* olivine refractory
Ölkarton *m* painters' (artists') cardboard
Ölkautschuk *m s.* Faktis
Ölkern *m s.* Ölsandkern
Ölkracken *n* oil cracking
Ölkuchen *m* oil (mill) cake
Ölkuchenbrecher *m* oil-cake breaker (crusher), cake mill
Ölkuchenmehl *n* oil-cake meal
Öllack *m* oil varnish
 fetter Ö. long-oil varnish
 halbfetter Ö. medium-oil varnish
 magerer Ö. short-oil varnish
 mittelfetter Ö. medium-oil varnish
 überfetter Ö. extra long-oil varnish
Öllein *m* oil flax *(a variety of Linum usitatissimum L.)*
Öllos-Lager *n (tech)* oilless bearing
öllöslich oil-soluble
Öllöslichkeit *f* oil solubility, solubility in oil
Ölmischung *f* oil mixture
ölmodifiziert oil-modified
Ölmühle *f* oil mill
Öl-Naturharz-Farbe *f* oleoresinous paint
Öl-Naturharz-Lack *m* oleoresinous varnish
Ölniveau *n* oil level
Öl[pack]papier *n* oiled (oil wrapping) paper
Ölpergament *n* oil parchment
Ölpflanze *f* oil plant
ölplastiziert *s.* ölgestreckt
Ölplastizierung *f s.* Ölstreckung
Ölpräserve *f* oil preserve
Ölpumpe *f* oil pump
OLP-Verfahren *n (met)* O.L.P. process *(an oxygen-lance process)*
Ölquelle *f* oil (petroleum) well
Ölraffination *f* 1. *(petrol)* oil refining; 2. *(food)* oil degumming
Ölraps *m* rape, Brassica napus L. em. Metzger var. napus
ölreaktiv *(coat)* oil-reactive
Öl-Ruß-Batch *m (rubber)* oil/carbon black masterbatch
Ölsand *m (geol, met)* oil sand
Ölsandkern *m* oil-sand core *(foundry)*
Ölsäure *f* $CH_3[CH_2]_7CH{=}CH[CH_2]_7COOH$ oleic acid, *cis*-9-octadecenoic acid
Ölscheidung *f* oil liberation *(in emulsions)*
Ölschicht *f* 1. oil layer, *(if thin)* oil film; 2. *(geol)* oil layer (horizon)
Ölschiefer *m (geol)* oil shale
Ölschlamm *m (petrol)* oil sludge; engine sludge *(in an internal combustion engine)*
Ölschwarz *n* black chalk, slate black *(a natural pigment)*
Ölspachtel *m(f) (coat)* oil filler
Ölspaltung *f (petrol)* oil cracking
Ölspiegel *m* oil level
Ölspülung *f (petrol)* oil-base mud
Ölspur *f (geol)* oil trace
Ölstand *m* oil level
Ölstreckung *f (rubber)* oil extension
Ölsüß *n s.* Glyzerin
Ölteer *m* oil tar
Ölträger *m (petrol)* producing formation
Öltränkung *f* oil impregnation
Öltröpfchen *n* oil droplet
Öltropfenmethode *f* oil-drop method *(as for determining the charge of electrons)*
Ölturbinen[ultra]zentrifuge *f* oil turbine ultracentrifuge
Ölumlauf *m* oil circulation
ölundurchlässig oilproof, oiltight
Ölundurchlässigkeit *f* oilproofness, oiltightness
Ölvergasung *f* oil gasification
Ölvormischung *f (rubber)* oil masterbatch
Ölwäsche *f* oil washing

Ölwäscher *m* oil scrubber
Ommatin *n* ommatine *(an eye pigment)*
Ommin *n* ommine *(an eye pigment)*
Ommochrom *n* ommochrome *(any of a class of eye pigments)*
Omskriptpapier *n* metallic paper
Omunketenußöl *n* Manketti nut oil *(from Ricinodendron rautaneni Schinz)*
Önanthalkohol *m* $CH_3[CH_2]_5CH_2OH$ 1-heptanol, oenanthic alcohol
Önanthat *n* heptanoate, oenanthate *(salt or ester of heptanoic acid)*
Önanthsäure *f* $CH_3[CH_2]_5COOH$ heptanoic acid, oenanthic acid
Önanthyliden *n s.* Heptin-(1)
Onia-Gegi-Verfahren *n (petrol)* Onia-Gegi process
Oniumstruktur *f (org chem)* onium structure
Oniumverbindung *f (org chem)* onium compound
Önologie *f (food)* oenology
önologisch *(food)* oenological
Önometer *n* oenometer, wine gauge
Onsäure *f* $HOCH_2[CHOH]_nCOOH$ aldonic acid *(any of a class of acids derived from aldoses)*
Onyx *m (min)* onyx *(silicon(IV) oxide)*
Oolith *m (min)* oolite
oolithisch oolitic
opak opaque
Opakglasur *f (ceram)* opaque glaze
Opaksubstanz *f* opaque matter
Opal *m (min)* opal *(silicon(IV) oxide)*
Opaleszenz *f* opalescence
Opaleszenzfarbe *f* opalescence colour
opaleszieren to opalesce
opaleszierend opalescent
Opalglas *n* opal glass
opalisieren to opalesce
opalisierend opalescent
Opazität *f* opacity
Operment *n* orpiment [yellow], yellow arsenic [sulphide], arsenic (royal) yellow, king's yellow (gold) *(technically pure arsenic(III) sulphide)*
Opferanode *f* sacrificial anode
Opium *n* opium
Opiumalkaloid *n* opium alkaloid
Opiumpulver *n* powdered opium
  eingestelltes O. standardized powdered opium
Opiumsäure *f s.* Mekonsäure
Opiumtinktur *f* tincture of opium
  benzoesäurehaltige O. benzoated tincture of opium
OP-Kautschuk *m* oil-extended styrene-butadiene rubber, OE-SBR, oil-extended (oil-masterbatched) polymer, OEP
Oppenauer-Reaktion *f* Oppenauer oxidation (reaction) *(for dehydrogenating secondary alcohols)*
Oppenheimer-Phillips-Prozeß *m (nucl)* Oppenheimer-Phillips process
optisch aktiv optically active
  o. einachsig [optically] uniaxial
  o. inaktiv optically inactive
  o. isomer enantiomorphous, enantiomorphic, enantiomeric
  o. isotrop optically isotropic
  o. zweiachsig biaxial
Orange I *n* orange I, α-naphthol orange, sodium-azo-α-naphthol sulphanilate, acid orange 20
Orange II *n* orange II, sodium-azo-β-naphthol sulphanilate, acid orange 7
Orange III *n* orange III, methyl orange, sodium *p*-(*p*-dimethylaminophenylazo)benzenesulphonate, acid orange 52
Orange IV *n* orange IV, tropaeolin 00, sodium *p*-diphenylaminoazobenzenesulphonate, acid orange 5
Orange N *n s.* Orange IV
Orangemennige *f* orange lead (mineral)
Orangenschaleneffekt *m (coat)* orange peel, bad flow
Orangenschalenöl *n* orange-peel oil
Orangenschellack *m* orange shellac (lac)
Orangit *m (min)* orangite *(thorium orthosilicate)*
Orbital *n* orbital
  antibindendes O. antibonding orbital
  atomares O. *s.* Atomorbital
  bindendes O. bonding orbital
  elektronisches O. electron orbital
  für Bindungen benötigtes O. bond orbital
  ***sp*-hybridisiertes O.** hybrid *sp* orbital
  lockerndes O. antibonding orbital
  molekulares O. *s.* Molekülorbital
  nichtbindendes O. non-bonding orbital
***p*-Orbital** *n p* orbital
***s*-Orbital** *n s* orbital
***sp*-Orbital** *n sp* orbital
π-Orbital *n* π orbital, pi orbital
Orbitalbewegung *f* orbital motion
Orbitaldrehimpuls *m* orbital angular momentum
Orbitaldrehimpulsquantenzahl *f* orbital (azimuthal, secondary) quantum number
Orbitalelektron *n* orbital electron
Orbitalmoment *n* orbital [magnetic] moment
Orbitaltheorie *f* orbital theory
Ordnung *f* order
  gebrochene O. fractional order *(reaction kinetics)*
  laterale O. lateral order
  weitreichende O. *(cryst)* long-range order
Ordnungsgrad *m* degree of order (orientation) *(as of polymers)*
Ordnungszahl *f* atomic (ordinal) number, nuclear charge number
  effektive O. effective atomic number, E.A.N.
  ungerade O. odd atomic number
Ordnung-Unordnung-Umwandlung *f (cryst)* order-disorder transformation (transition)
Orford-Verfahren *n (met)* Orford process, top-and-bottom smelting process
Organiker *m* organic chemist
organisch organic
organisch-chemisch organic-chemical
Organoberylliumverbindung *f* organoberyllium compound
Organoborverbindung *f* organoboron compound
Organochemiker *m* organic chemist
Organochlorsilan *n* organochlorosilane

Organoderivat *n* organoderivative
organofunktionell organofunctional, carbon-functional
Organogel *n* organogel
organogen organogenic
Organogen *n* organogen *(any of the elements characteristic of organic compounds)*
Organogruppe *f* organic group
Organohalogensilan *n* organosilicon halogen (halide), organohalogenosilane, organohalosilane
Organokadmiumverbindung *f* organocadmium compound
organoleptisch organoleptic
Organolithiumverbindung *f* organolithium compound
Organometallverbindung *f* organometallic (metallo-organic) compound
Organophosphorverbindung *f* organophosphorus compound
Organopolysiloxan *n* organopolysiloxane, polyorganosiloxane, polymeric organosiloxane
Organoquecksilberbeize *f (agric)* organomercury dressing
Organoquecksilberton *m* organomercury clay
Organoquecksilberverbindung *f* organomercury compound
Organosilan *n* organosilane, organic silane
Organosilazan *n* organosilazane
Organosilikon *n* organosilicone
Organosiliziumchemie *f* organosilicon chemistry
Organosiliziumhalogenid *n s.* Organohalogensilan
Organosiliziumoxid *n s.* Organosiloxan
Organosiliziumpolymer[es] *n* organosilicon polymer
Organosiliziumverbindung *f* organosilicon compound
Organosiloxan *n* organosiloxane, organic siloxane, organosilicon oxide
  polymeres O. *s.* Organosiloxanpolymer
Organosiloxanpolymer[es] *n* organosiloxane polymer, organopolysiloxane, polyorganosiloxane
Organosol *n* organosol
Organozinnmerkaptid *n* organotin mercaptide
Organozinnstabilisator *m* organotin stabilizer
Organozinnverbindung *f* organotin compound
orientieren to orient[ate], *(in linear direction also)* to align
Orientierung *f* orientation, *(in linear direction also)* alignment
  axiale O. monoaxial orientation
  biaxiale O. biaxial orientation
  einachsige O. monoaxial orientation
  molekulare O. molecular orientation
  nichtbevorzugte (regellose) O. random orientation
Orientierungseffekt *m* orientation effect *(as with molecules)*
Orientierungserscheinung *f (rubber)* grain effect *(in sheets)*
Orientierungsgrad *m* degree of order (orientation)
Orientierungspolarisation *f* orientation polarization
Originalbehälter *m* original container
Originalbestimmung *f* original assay
O-Ring *m* O-ring *(seal)*
Orlean *m* annatto, annotta, arnatto, arnotta *(a colouring matter from Bixa orellana L.)*
Ornamentglas *n* patterned (figured) glass
Ornithin *n* $NH_2[CH_2]_3CH(NH_2)COOH$ ornithine, 2,5-diamino-*n*-valeric acid
Ornithinzyklus *m (bioch)* ornithine (urea) cycle
Orotsäure *f* orotic acid, uracil-4-carboxylic acid
Orsat-Apparat *m*, **Orsat-Gerät** *n* Orsat analyzer, Orsat gas [analysis] apparatus
Orseille *f* orchil, archil *(a lichen dye)*
Orsellinsäure *f* $(HO)_2C_6H_2(CH_3)COOH$ orsellic (orsellinic) acid, 4,6-dihydroxy-*o*-toluic acid
Ortbeton *m* cast-in-place (poured-in-place) concrete, in-situ (cast-in-situ) concrete
Orthanilsäure *f* $NH_2C_6H_4SO_3H$ orthanilic acid, aniline-*o*-sulphonic acid
Orthit *m (min)* orthite, allanite *(a sorosilicate)*
Orthoameisensäureäthylester *m* $HC(OC_2H_5)_3$ ethyl orthoformate
orthoanelliert *(org chem)* ortho-fused
Orthoanellierung *f (org chem)* ortho fusion
Orthoarsenat(III) *n* $M_3^IAsO_3$ orthoarsenite, arsenite
Orthoarsenat(V) *n* $M_3^IAsO_4$ orthoarsenate, arsenate
Orthoborat *n* $M_3^IBO_3$ orthoborate, borate, trioxoborate
Orthoborsäure *f* H_3BO_3 orthoboric acid, boric acid, trioxoboric acid
orthochromatisch *(phot)* orthochromatic
ortho-dirigierend ortho-directing
Orthoflow-Verfahren *n (petrol)* Orthoflow [catalytic cracking] process
Orthogestein *n* ortho rock
ortho-Isomer[es] *n* ortho isomer, *o*-isomer
Orthokarbonat *n* $M_4^ICO_4$ orthocarbonate
Orthokieselsäure *f* H_4SiO_4 orthosilicic acid, silicic acid, tetraoxosilicic acid
Orthoklas *m (min)* orthoclase *(a feldspar)*
Orthokondensation *f* ortho fusion
orthokondensiert ortho-fused
Orthokortex *m (text)* orthocortex
ortho-Molekül *n* ortho molecule
Ortho-Öl *n* ortho oil
ortho-Orientierung *f* ortho orientation
ortho-para-dirigierend ortho-para-directing
ortho-peri-anelliert, ortho-peri-kondensiert ortho[-and]-peri-fused
Orthoperjodat *n* $M_5^I[IO_6]$ orthoperiodate, hexaoxoiodate(VII)
Orthoperjodsäure *f* H_5IO_6 orthoperiodic acid, hexaoxoiodic(VII) acid
Orthophosphat *n* $M_3^IPO_4$ orthophosphate, phosphate
  neutrales O. $M_3^IPO_4$ neutral (normal) orthophosphate (phosphate)
  primäres O. *s.* Dihydrogenorthophosphat
  sekundäres O. *s.* Hydrogenorthophosphat
  tertiäres O. *s.* neutrales O.
Orthophosphit *n* $M_2^IPHO_3$ orthophosphite, phosphite, *(better)* phosphonate

Orthophosphorsäure *f* H_3PO_4 orthophosphoric acid, phosphoric acid
Orthoplumbat *n* $M_4^IPbO_4$ orthoplumbate, tetraoxoplumbate(IV)
Ortho-Positronium *n (nucl)* orthopositronium
orthorhombisch *(cryst)* orthorhombic, o-rh.
Orthosilikat *n* $M_4^ISiO_4$ orthosilicate, silicate, tetraoxosilicate
orthoständig ortho, in ortho position
o. sein to be [located, situated] ortho
Orthostannat *n* $M_4^ISnO_4$ orthostannate, tetraoxostannate(IV)
ortho-Stellung *f* ortho position ✦ **in ortho-Stellung** in ortho position, ortho ✦ **nach der ortho-Stellung dirigierend** ortho-directing
orthosubstituiert ortho-substituted, *o*-substituted
Orthotellurat *n* $M_6^ITeO_6$ orthotellurate, tellurate, hexaoxotellurate
Orthotellursäure *f* H_6TeO_6 orthotelluric acid, telluric acid, hexaoxotelluric acid
Orthotitanat *n* $M_4^ITiO_4$ orthotitanate, tetraoxotitanate(IV)
Orthotitansäure *f* H_4TiO_4 orthotitanic acid, tetraoxotitanic acid
Orthovanadat *n* $M_3^IVO_4$ orthovanadate, tetraoxovanadate(V)
ortho-Verbindung *f* ortho compound
Orthowasserstoff *m* ortho hydrogen
Orthozinnsäure *f* H_4SnO_4 orthostannic acid, tetraoxostannic acid
ortho-Zustand *m* ortho state
Ortonkegel *m (ceram)* Orton cone
Ortsbestimmung *f* localization [of position], determination of position
ortsbeweglich movable, mobile
ortseigen *(geol)* autochthonous
ortsfest non-mobile, stationary
ortsfremd *(geol)* allochthonous
Ortsisomerie *f* position (place) isomerism
Ortstein *m (soil)* ironpan, hardpan, ortstein, boundary stone
Orzein *n (dye)* orcein
Orzin *n* $CH_3C_6H_3(OH)_2$ orcinol, 3,5-dihydroxytoluene
Osazon *n (org chem)* osazone
Osazonbildung *f (org chem)* osazone formation
Osmat(VI) *n* $M_2^IOsO_4$ osmate(VI), tetraoxoosmate(VI)
Osmium *n* Os osmium
Osmium(II)-chlorid *n* $OsCl_2$ osmium(II) chloride, osmium dichloride
Osmium(III)-chlorid *n* $OsCl_3$ osmium(III) chloride, osmium trichloride
Osmium(IV)-chlorid *n* $OsCl_4$ osmium(IV) chloride, osmium tetrachloride
Osmiumdichlorid *n s.* Osmium(II)-chlorid
Osmiumdioxid *n s.* Osmium(IV)-oxid
Osmiumdisulfid *n s.* Osmium(IV)-sulfid
Osmium(IV)-fluorid *n* OsF_4 osmium(IV) fluoride, osmium tetrafluoride
Osmium(VI)-fluorid *n* OsF_6 osmium(VI) fluoride, osmium hexafluoride
Osmium(VIII)-fluorid *n* OsF_8 osmium(VIII) fluoride, osmium octafluoride
Osmiumhexafluorid *n s.* Osmium(VI)-fluorid
Osmiummonoxid *n s.* Osmium(II)-oxid
Osmiumoktafluorid *n s.* Osmium(VIII)-fluorid
Osmium(II)-oxid *n* OsO osmium(II) oxide, osmium monooxide
Osmium(III)-oxid *n* Os_2O_3 osmium(III) oxide, diosmium trioxide
Osmium(IV)-oxid *n* OsO_2 osmium(IV) oxide, osmium dioxide
Osmium(VIII)-oxid *n* OsO_4 osmium(VIII) oxide, osmium tetraoxide
Osmium(IV)-sulfid *n* OsS_2 osmium(IV) sulphide, osmium disulphide
Osmium(VIII)-sulfid *n* OsS_4 osmium(VIII) sulphide, osmium tetrasulphide
Osmiumtetrachlorid *n s.* Osmium(IV)-chlorid
Osmiumtetrafluorid *n s.* Osmium(IV)-fluorid
Osmiumtetrasulfid *n s.* Osmium(VIII)-sulfid
Osmiumtetroxid *n s.* Osmium(VIII)-oxid
Osmiumtrichlorid *n s.* Osmium(III)-chlorid
Osmometer *n* osmometer
Osmometrie *f* osmometry
Osmose *f* osmosis
umgekehrte O. reverse osmosis, hyperfiltration *(as in waste-water treatment)*
Osmotierung *f* diffusion treatment *(wood preservation)*
osmotisch osmotic
Oson *n (org chem)* osone
Osteolith *m (min)* osteolite *(a massive earthy apatite)*
Östradiol *n* oestradiol *(a sex hormone)*
östrogen *(pharm)* oestrogenic, oestrus-producing
Östrogen *n* oestrogen *(any of a class of sex hormones)*
Ostwald-Reifung *f (coll)* Ostwald ripening
Ostwald-Verfahren *n* Ostwald process *(for obtaining nitric acid)*
Ostwald-Viskosimeter *n* Ostwald viscometer
Oszillation *f* oscillation, vibration
Oszillationsenergie *f* oscillational (vibrational) energy
Oszillator *m* oscillator
oszillatorisch oscillatory, vibratory
oszillieren to oscillate, to vibrate
Oszillometrie *f* oscillometry
oszillopolarografisch oscillo-polarographic
Oszin *n* scopoline, oscine *(alkaloid)*
o. T. *s.* Tiegel / offener
Otto-Hoffmann-Koksofen *m* Otto-Hoffmann coke oven
Ouabain *n* ouabain, *g*-strophantin *(glycoside)*
Ouricury-Wachs *n* ouricury wax *(from Cocos coronata Mart.)*
Ovalbumin *n* egg albumin
ovizid ovicidal
Ovizid *n* ovicide *(pest control)*
Ovomukoid *n* ovomucoid *(a protein)*
Ovovitellin *n* ovovitellin, egg vitellin *(a phosphoprotein)*
Ovulationshemmer *m* ovulation inhibitor, antifertility agent

Ö/W *s.* Öl-in-Wasser-Emulsion
Owens-Maschine *f (glass)* Owens [bottle] machine
Owens-Verfahren *n* Owens process *(for making bottles)*
Ö/W-Seifenemulsion *f* soap-stabilized O/W emulsion
Oxalaldehyd *m* OHC–CHO oxaldehyde, glyoxal, ethanedial
Oxalaldehydsäure *f s.* Oxoessigsäure
Oxalat *n* oxalate ✦ **mit Oxalaten behandeln** to oxalate *(e. g. blood)*
saures O. *s.* Hydrogenoxalat
Oxalatchelat *n* oxalate chelate
Oxalatoaluminat *n* oxalatoaluminate
Oxalatoboratkomplex *m* oxalate complex of boron
Oxalatochromat(III) *n* $M_3^I[Cr(C_2O_4)_3]$ oxalatochromate(III), trioxalatochromate(III)
Oxalatokobaltat(III) *n* $M_3^I[Co(C_2O_4)_3]$ oxalatocobaltate(III), trioxalatocobaltate(III)
Oxalatokomplex *m* oxalate complex
Oxalbernsteinsäure *f* $HOOCCH_2CH(COOH)COCOOH$ oxalosuccinic acid
Oxalbernsteinsäurekarboxylase *f* oxalsuccinic carboxylase
Oxalessigester *m s.* Oxalessigsäurediäthylester
Oxalessigsäure *f* $HOOC–COCH_2–COOH$ oxalacetic acid, oxosuccinic acid
Oxalessigsäurediäthylester *m* diethyl oxalacetate
Oxalessigsäurekarboxylase *f* oxalacetic carboxylase
Oxalsäure *f* HOOC–COOH oxalic acid, ethanedioic acid
Oxalsäuredialdehyd *m s.* Oxalaldehyd
Oxalsäurediäthylester *m* $C_2H_5O–CO–CO–OC_2H_5$ diethyl oxalate, oxalic acid diethyl ester, oxalic ester
Oxalsäuremonoureid *n s.* Oxalursäure
Oxalsäurenitril *n* N≡C–C≡N oxalonitrile, cyanogen
Oxalurie *f (med)* oxaluria
Oxalursäure *f* $NH_2CONHCOCOOH$ oxaluric acid, mono-oxalylurea
Oxalylharnstoff *m* oxalylurea, parabanic acid, imidazolidine-2,4,5-trione
Oxamid *n* $NH_2COCONH_2$ oxamide, oxalic acid diamide
Oxamidsäure *f* $NH_2COCOOH$ oxamic acid, oxalic acid monoamide
Oxazinfarbstoff *m* oxazine dye
Oxford-Einheit *f* Oxford (Florey) unit *(formerly a biological measure for penicillin)*
Oxford-Methode *f* Oxford (cup) method *(for evaluating penicillin)*
Oxid *n* oxide
basisches O. basic oxide
saures O. acidic oxide
siliziumorganisches O. organosilicon oxide, organosiloxane, organic siloxane
Oxidbelag *m* oxide layer, *(if thin)* oxide film
Oxidchlorid *n* chloride oxide
Oxidelektrode *f* oxide electrode
Oxidfilm *m*, **Oxidhaut** *f* oxide film
Oxidhydrat *n* hydrated oxide
Oxidhydratsol *n* sol of hydrated oxide
oxidisch oxidic
Oxidkatode *f* oxide[-coated] cathode
Oxidkeramik *f* oxide ceramics, oxide-ceramic products
Oxidphosphor *m* oxide phosphor *(any of a class of compounds which exhibit phosphorescence)*
Oxidrot *n (coat)* red oxide
Oxidsalz *n* oxide salt
Oxidschicht *f* oxide layer, *(if thin)* oxide film
Oxidsinterung *f (ceram)* oxide sintering
Oxim *n* oxime *(any of a class of compounds containing the group >C=NOH)*
Oxin *n* oxine, 8-hydroxyquinoline
Oxinat *n* oxinate *(any of the complex compounds of 8-hydroxyquinoline)*
Oxinchelat *n* oxine chelate
Oxindol *n* oxindole, 2-hydroxyindole
Oxindolalkaloid *n* oxindole alkaloid
Oxiran *n* oxiran, ethylene oxide, epoxyethane
Oxoäthansäure *f s.* Oxoessigsäure
Oxobrücke *f* oxo bridge
Oxoessigsäure *f* OHC–COOH oxoethanoic acid, glyoxylic acid
Oxoform *f* oxo form
Oxofunktion *f* oxo function
Oxoglutarsäure *f* oxoglutaric acid, *(specif)* 2-oxoglutaric acid, 2-oxopentanedioic acid
Oxogruppe *f* oxo group
Oxokomplex *m* oxo complex
Oxomalonsäure *f* $OC(COOH)_2$ oxomalonic acid, mesoxalic acid, oxopropanedioic acid
Oxomethan *n* oxomethane, methanal, formaldehyde
Oxonium-Ion *n* H_3O^+ oxonium ion
Oxoniumsalz *n* oxonium salt
Oxoniumverbindung *f* oxonium compound
Oxopentandisäure *f* oxopentanedioic acid, oxoglutaric acid
4-Oxopentansäure *f* $CH_3COCH_2CH_2COOH$ 4-oxopentanoic acid, 4-oxovaleric acid, laevulinic acid
2-Oxopropanal *n* CH_3COCHO 2-oxopropionaldehyde, pyruvic aldehyde
Oxopropandisäure *f* $OC(COOH)_2$ oxopropanedioic acid, oxomalonic acid
2-Oxopropionsäure *f* $CH_3COCOOH$ 2-oxopropionic acid, pyruvic acid
Oxoreaktion *f* oxo reaction
Oxosalz *n* oxo salt *(salt of an oxo acid)*
Oxosäure *f* oxo acid
Oxosynthese *f* oxo synthesis, hydroformylation
Oxosynthesereaktion *f* oxo reaction
Oxoverbindung *f* oxo compound
Oxo-Zyklo-Tautomerie *f* oxo-cyclo-tautomerism
Oxy... *s. a.* Hydroxy...
oxybiontisch *(biol)* oxybiotic, aerobiotic *(living in the presence of air oxygen)*
Oxybiose *f (biol)* oxybiosis, aerobiosis *(life in the presence of air oxygen)*
oxydabel oxidizable
Oxydans *n* oxidant, oxidizing agent
Oxydase *f* oxidase

kupferhaltige O. copper oxidase
Oxydation *f* oxidation
O. durch Luftsauerstoff air (aerial) oxidation
O. von Kohle im Wirbelbett fluidized oxidation of coal
anodische O. anodic (electrolytic) oxidation, *(met also)* anodizing, anodization, anodic coating (treatment)
elektrochemische (elektrolytische) O. *s.* anodische O.
fotochemische O. photochemical oxidation, photooxidation
katalytische O. catalytic oxidation
milde O. mild oxidation
oberflächliche O. surface oxidation
Oppenauersche O. Oppenauer oxidation *(for dehydrogenating secondary alcohols)*
selektive O. selective (preferential) oxidation *(as in fire refining)*
spontane O. spontaneous oxidation
Oxydationsätze *f (text)* oxidation discharge
Oxydationsbad *n* oxidizing bath
oxydationsbeständig resistant (stable) to oxidation, oxidatively stable
Oxydationsbeständigkeit *f* resistance (stability) to oxidation, oxidation resistance, oxidative stability
Oxydationsbleiche *f* oxidation bleaching, oxidizing bleach
Oxydationsferment *n* oxidative enzyme; *(specif)* *s.* gelbes O.
gelbes O. yellow enzyme, flavin[e] enzyme *(any of a class of redoxases)*
Warburgsches gelbes O. Warburg's enzyme
Oxydationsflamme *f* oxidizing flame
Oxydationsgeschmack *m* oxidized flavour
Oxydationsgeschwindigkeit *f* rate of oxidation
Oxydationsgrad *m* degree of oxidation
Oxydationsinhibitor *m* oxidation inhibitor, antioxidant, antioxidizer, antioxidizing agent, antioxygen, *(petrol also)* gum inhibitor
Oxydationskatalysator *m* oxidation (oxidizing) catalyst
Oxydationsmittel *n* oxidizing agent, oxidant
Oxydationspotential *n* oxidation potential
Oxydationsprodukt *n* oxidation product
Oxydationsraum *m* oxidizing zone *(of a burner flame)*
Oxydations-Reduktions-Elektrode *f* oxidation-reduction electrode
Oxydations-Reduktions-Katalysator *m* oxidation-reduction catalyst
Oxydations-Reduktions-Potential *n* oxidation-reduction potential, redox potential
Oxydations-Reduktions-Reaktion *f* oxidation-reduction reaction, redox reaction
Oxydations-Reduktions-System *n* oxidation-reduction system, redox system
Oxydations-Reduktions-Titration *f* oxidation-reduction titration, redox titration
Oxydationsschicht *f* oxide layer
Oxydationsstabilität *f s.* Oxydationsbeständigkeit
Oxydationsstufe *f s.* Oxydationszahl
Oxydationsverfahren *n* oxidation process
anodisches O. *(met)* anodizing process
Oxydationsverhinderer *m s.* Oxydationsinhibitor
Oxydationsvorgang *m* oxidation process
Oxydationswirkung *f* oxidizing action (effect)
Oxydationszahl *f* oxidation number
Oxydationszone *f* oxidation (oxidizing) zone; combustion zone *(of a blast furnace)*
Oxydationszustand *m* oxidation state
oxydativ oxidative
Oxydiäthansäure *f s.* Diglykolsäure
oxydierbar oxidizable
leicht o. readily oxidizable
Oxydierbarkeit *f* oxidizability
oxydieren to oxidize
anodisch o. *(met)* to anodize
elektrochemisch (elektrolytisch) o. *s.* anodisch o.
Oxydiessigsäure *f s.* Diglykolsäure
Oxydimetrie *f* oxidimetry
oxydimetrisch oxidimetric
Oxydoreduk[t]ase *f* oxido-reductase, oxidation-reduction enzyme, redoxase
Oxydoreduktion *f* oxidation-reduction, redox reaction
Oxygenase *f* oxygenase
Oxyhämoglobin *n* oxyhaemoglobin
Oxyhämozyanin *n* oxyhaemocyanin *(a copper-containing blood pigment)*
Oxyliquit *n* oxyliquit *(an explosive)*
Oxyn *n* oxyn *(any of the solid oxidation products of drying oils)*
Oxyneurin *n* oxyneurine, trimethylglycine, lycine, betaine *(proper)*
α-Oxypropionsäure *f s.* 2-Oxopropionsäure
Oxytozin *n* oxytocin *(a polypeptide hormone)*
Oxytozinwirksamkeit *f (pharm)* oxytocic activity
oxytozisch *(pharm)* oxytocic, accelerating parturition
Oxyzellulose *f* oxycellulose
OZ *s.* 1. Ordnungszahl; 2. Oktanzahl
OZ-Bestimmung *f* octane rating
Ozimen *n* ocimene, 3,7-dimethyl-1,3,6-octatriene
Ozokerit *m (min)* ozokerite, earth (ader) wax, native paraffin
Ozokeritgestein *n* ozokerite rock
Ozon *n* O_3 ozone, trioxygen ✦ **sich in O. verwandeln** to ozonize, to ozonify
Ozonabbau *m s.* Ozonspaltung
Ozonalterung *f (rubber, plast)* ozone [exposure] test
ozonartig ozone-like, ozonic, ozonous
ozonbeständig ozone-resistant, ozone-resisting, resistant to ozone
Ozonbeständigkeit *f* ozone resistance, resistance to ozone
ozonfest *s.* ozonbeständig
Ozongenerator *m s.* Ozonisator
ozonhaltig containing ozone, ozoniferous, ozonic, ozonous
Ozonid *n* ozonide
Ozonidspaltung *f s.* Ozonspaltung
Ozonisation *f s.* Ozonisierung

Ozonisator *m* ozonizer, ozone generator
ozonisieren to ozonize, to ozonify
Ozonisierung *f* ozonization, ozonification
Ozonisierungsapparatur *f* ozonization apparatus
Ozonkonzentration *f* ozone concentration
Ozonolyse *f s.* Ozonspaltung
Ozonosphäre *f* ozonosphere, ozone layer
Ozonpapier *n* ozone [test] paper, potassium-iodide-starch paper
Ozonprüfung *f (rubber, plast)* ozone [exposure] test
ozonresistent *s.* ozonbeständig
Ozonresistenz *f s.* Ozonbeständigkeit
Ozonriß *m (rubber)* ozone crack (cut)
Ozonrißbildung *f (rubber)* ozone cracking
Ozonschicht *f s.* Ozonosphäre
Ozonschutzmittel *n* antiozonant, antiozidant, sunproofing agent
Ozonspaltung *f* ozonolysis, cleavage by ozone

P

p *s.* Proton
p̄ *s.* Antiproton
P *s.* Parachor
p. a. *(abbr)* pro analysi, *s.* analysenrein
PA *s.* 1. Polyamid; 2. Polyamidfaserstoff
Paalsgard-Emulsionsöl *n* Paalsgard emulsion oil, P.E.O.
Paarbildung *f*, **Paarerzeugung** *f* pair formation (creation, production) *(esp of an electron and a positron)*; pairing *(of electrons)*
Paarspektrometer *n* pair spectrometer
Paarspektroskop *n* pair spectroscope
Paarung *f s.* Paarbildung
Paarvernichtung *f* pair annihilation *(esp of an electron and a positron)*
paarweise zusammentreten to pair *(of electrons)*
Pachnolith *m (min)* pachnolite *(calcium sodium fluoroaluminate)*
Pachuca-Tank *m (met)* Pachuca tank, cyanidation vat
Pack *m (text)* package
packen to pack[age]
Packen *n* packing, package, packaging
Packer *m (petrol)* packer *(as for sealing part of a borehole)*
Packfärbeapparat *m (text)* package dyeing machine
Packfärben *n*, **Packfärberei** *f (text)* pack[age] dyeing
packfertig ready (suitable) for packing
packgefärbt *(text)* package-dyed
Packgewebe *n* reinforced paper, papyrolin *(cloth-faced or cloth-centred paper)*
Packkrepp *m* crêpe wrapping paper
Packleinen *n* burlap, gunny, hessian
Packpapier *n* wrapping (packing, package, packaging) paper
  P. für Papierrollen mill wrapper (wrapping)
Packpresse *f* platen press
Packseidenpapier *n* packing (wrapping, commercial) tissue, tissue wrapper (wrapping)
Packstoff *m* adherend *(a body to be attached to another one by an adhesive)*
Packung *f* packing *(as of a column)*; *(cryst)* packing; seal *(as in a valve)*; pack[age] *(a packed quantity)*; *(med)* pack
  P. mit radioaktivem Heilschlamm radium pack
  dichteste P. *(cryst)* closest (close) packing (package)
  halbmetallische P. semimetallic packing *(as in a valve)*
  hexagonal dichteste P. *(cryst)* hexagonal closest (close) packing
Packungsanteil *m (nucl)* packing fraction
Packungsdichte *f* packing density
  P. der Hackschnitzel *(pap)* chip capacity
Packungseffekt *m (nucl)* packing effect
Packungsmaterial *n* packing material *(as for valves)*
Packungsring *m* packing (seal, sealing) ring
Paddel *n* paddle *(as of a mixer)*
Paddelfärbemaschine *f (text)* paddle [wheel] dyeing machine
Paddelrad *n* paddle wheel *(waste-water treatment)*
Paddelrührer *m* paddle agitator (mixer), blade (arm) mixer
Pad-Jig-Verfahren *n (text)* pad-jig process
Pad-Roll-Verfahren *n (text)* pad-roll process
Pad-Steam-Verfahren *n (text)* pad-steam process
paketieren to pack[age]
Paketstahl *m* refined steel (iron, bar), shear steel, merchant bar
Paläobiochemie *f* palaeobiochemistry
Palette *f (tech)* pallet
Palingenese *f*, **Palingenesis** *f (geol, biol)* palingenesis
Palladium *n* Pd palladium
Palladiumasbest *m* palladinized asbestos
Palladium(II)-bromid *n* $PdBr_2$ palladium(II) bromide, palladium dibromide
Palladium(II)-chlorid *n* $PdCl_2$ palladium(II) chloride, palladium dichloride
Palladiumchloridpapier *n* palladium chloride paper
Palladiumdi... *s. a.* Palladium(II)-...
Palladiumdioxid *n s.* Palladium(IV)-oxid
Palladiumdisulfid *n s.* Palladium(IV)-sulfid
Palladium(II)-fluorid *n* PdF_2 palladium(II) fluoride, palladium difluoride
Palladium(III)-fluorid *n* PdF_3 palladium(III) fluoride, palladium trifluoride
Palladiumgold *n (min)* palladium gold, porpezite *(a natural alloy)*
Palladium(II)-jodid *n* PdI_2 palladium(II) iodide, palladium diiodide
Palladiummohr *n* palladium black
Palladiummonosulfid *n s.* Palladium(II)-sulfid
Palladiummonoxid *n s.* Palladium(II)-oxid
Palladium(II)-nitrat *n* $Pd(NO_3)_2$ palladium(II) nitrate
Palladium(II)-oxid *n* PdO palladium(II) oxide, palladium monooxide
Palladium(IV)-oxid *n* PdO_2 palladium(IV) oxide, palladium dioxide
Palladiumröhre *f* palladium tube *(for separating hydrogen)*
Palladiumschwarz *n* palladium black
Palladium(II)-sulfat *n* $PdSO_4$ palladium(II) sulphate

Palladium(II)-sulfid *n* PdS palladium(II) sulphide, palladium monosulphide
Palladium(IV)-sulfid *n* PdS_2 palladium(IV) sulphide, palladium disulphide
Palladiumtrifluorid *n s.* Palladium(III)-fluorid
Palladium(II)-zyanid *n* $Pd(CN)_2$ palladium(II) cyanide
Pallasit *m* pallasite, pallas iron *(a meteorite)*
Palliativum *n (pharm)* palliative
Pall-Ring *m (distil)* Pall ring *(a kind of packing)*
Palmarosaöl *n* palmarosa oil, Indian geranium (grass) oil *(from Cymbopogon martini (Roxb.) Stapf)*
Palmensago *m*, **Palmenstärke** *f* palm starch, sago *(from Metroxylon specc.)*
Palmer-Verfahren *n (rubber)* Palmer (high-pressure) process *(a reclaiming process)*
Palmfett *n s.* Palmöl
Palmitat *n* palmitate *(a salt or ester of palmitic acid)*
Palmitinsäure *f* $CH_3[CH_2]_{14}COOH$ palmitic acid, hexadecanoic acid
Palmitinsäurechlorid *n s.* Palmitoylchlorid
Palmitinsäurezetylester *m* $C_{15}H_{31}COOC_{16}H_{33}$ cetyl palmitate, cetin
Palmitoleinsäure *f* $CH_3[CH_2]_5CH{=}CH[CH_2]_7COOH$ palmitoleic acid, zoomaric acid, 9-hexadecenoic acid
Palmitoylchlorid *n* $CH_3[CH_2]_{14}COCl$ palmitoyl chloride, hexadecanoyl chloride
Palmkernfett *n s.* Palmkernöl
Palmkernhartfett *n* hydrogenated palm kernel oil
Palmkernöl palm kernel (nut) oil *(from Elaeis guineensis Jacq.)*
Palmöl *n* palm oil (butter) *(from Elaeis guineensis Jacq.)*
Palmwein *m* palm wine
Palmzucker *m* palm sugar
Palygorskit *m* palygorskite *(a phyllosilicate)*
Pamaquin *n* pamaquin[e] *(a quinoline derivative)*
PAN *s.* Polyakrylnitril
Panama-Brechwurzel *f (pharm)* Panama (Cartagena, Nicaragua) ipecacuanha *(from Cephaelis acuminata Karsten)*
Panamarinde *f* Panama (soap) bark *(from Quillaja saponaria Mol.)*
Panazee *f (pharm)* panacea, cure-all
panchromatisch *(phot)* panchromatic ✦ **p. machen (sensibilisieren)** to panchromatize
Pandermit *m (min)* pandermite *(a soroborate)*
panidiomorph[körnig] *(geol)* panidiomorphic, automorphic-granular
Pankreasbeize *f (tann)* pancreatic bate
Pankreassaft *m (bioch)* pancreatic juice
pantoffeln *(tann)* to grain
Pantothensäure *f* pantothenic acid
Pan-Verfahren *n (rubber)* pan [reclaiming] process, pan devulcanization
Panzerglas *n* bullet-proof (bullet-resistant) glass
Päonin *n* paeonin *(an anthocyanin)*
Papageiengrün *n* parrot (Paris, emerald) green *(copper aceto-arsenite)*
Papain *n* papain *(an enzyme)*
Papainase *f* papainase *(any of several proteinases)*
Papaverin *n* papaverine *(alkaloid)*
Papier *n* paper ✦ **auf (zu) P. verarbeiten** to make into paper
P. mit schwacher Leimung weakly sized paper
auf Format geschnittenes P. sheeted (sheet, ream) paper, paper in sheets
azetyliertes P. acetylated paper *(chromatography)*
bituminiertes P. tar (tarred, tarred brown, tar-impregnated, pitch, asphalt) paper
Chinesisches P. China (Chinese, India, Indian) paper
chromatografisches P. chromatographic paper
elektrisch leitendes P. electrical conducting paper
farbig gestrichenes P. coloured coated paper
farbiges P. coloured paper
feuerfestes (feuersicheres) P. fireproof (fire-resistant, fire-resisting) paper
flammsicheres P. flameproof paper
fotografisches P. photographic (photo) paper
gaufriertes P. embossed paper
gegautschtes P. duplex paper
gehämmertes P. hammer-finished paper
gekrepptes P. crêpe paper
geprägtes P. embossed paper
geräuschloses P. noiseless (programme) paper
gestrichenes P. coated (surfaced) paper
gummiertes P. gummed (adhesive) paper
hadernhaltiges P. rag content paper
handgeschöpftes P. [genuine] hand-made paper, vat paper
holzfreies P. wood-free paper
holzhaltiges P. wood-containing paper, groundwood (woody) paper
konservierendes P. preservative (preserving) paper
korrosionsschützendes P. anticorrosive (anti-tarnish) paper
leinengeprägtes P. linen-embossed (linen-finished, linen-faced, linen) paper
mit Wasserzeichen versehenes P. watermarked paper
naßfestes P. wet-strength paper
paraffiniertes P. paraffin (paraffined, paraffin-waxed, waxed, wax) paper
satiniertes P. [super]calendered paper, glazed paper
säurefreies P. acid-free paper
technisches P. technical paper, paper for technical purposes
textilverstärktes P. reinforced paper
ungeleimtes P. waterleaf paper
veredeltes P. processed paper
wasserfestes P. wet-strength paper
Papierabfälle *mpl* waste (old) paper
Papierabzug *m (phot)* paper print
Papieraufrolltrommel *f (pap)* reel (reeling) drum (cylinder)
Papieraufrollung *f*, **Papieraufwicklung** *f (pap)* winding, reeling

Papierausrüstung *f* paper finishing
Papierausschuß *m* broken [material, paper], [mill, machine] broke, brokes, waste stuff
Papierbahn *f* [paper] web, mat
Papierbahn[ab]riß *m (pap)* break in the web, sheet break
Papierbahnspannung *f s.* Papierzugspannung
Papierbild *n (phot)* paper print
Papierbirke *f* paper birch, Betula papyrifera Marsh.
Papierbrei *m* paper[making] stock
Papierchromatografie *f* paper chromatography
absteigende P. descending [paper] chromatography
aufsteigende P. ascending [paper] chromatography
eindimensionale P. one-dimensional (one-way) paper chromatography
zweidimensionale P. two-dimensional (two-way) paper chromatography
Papierchromatogramm *n* paper chromatogram
Papiereinlage *f* paper mat
Papierelektrophorese *f* paper electrophoresis, electrophoresis on paper
Papierentwickler *m (phot)* paper developer
Papierentwicklung *f (phot)* paper development
Papierfabrik *f* paper mill (factory)
Papierfabrikabwasser *n* paper mill wastes
Papierfärben *n*, **Papierfärberei** *f* paper dyeing, *(Am also)* paper coloring
Papierfaser *f* papermaking fibre
Papierfaserstoff *m* [paper, raw] stock, raw papermaking material
Papierfehler *m* defect in [the] paper, defect of paper
Papierfilter *n* paper filter
Papierformat *n* paper size
Papierfüllstoff *m* [paper] filler, loader
Papiergarn *n* paper yarn
Papiergewicht *n* paper weight, weight of paper
Papiergradation *f*, **Papierhärtegrad** *m (phot)* paper grade
Papierherstellung *f* papermaking
Papierholz *n* pulpwood
Papierindustrie *f* paper industry
Papierkalander *m (pap)* calender [machine], machine calender stack
Papierkalanderwalze *f* paper bowl
Papierleimung *f* paper sizing
Papierleitwalze *f* fly roll
Papiermacher *m* papermaker
Papiermacheralaun *m* papermaker's alum *(technical aluminium sulphate)*
Papiermacherei *f* papermaking
Papiermaschine *f* paper[making] machine
P. mit zwei Langsieben twin-wire paper machine
P. mittlerer Geschwindigkeit moderate-speed paper machine
langsamlaufende P. slow-speed paper machine
schnellaufende P. high-speed paper machine
Papiermaschinenfilz *m* paper machine felt, papermaker's felt
Papiermaschinengeschwindigkeit *f* paper machine speed
Papiermaschinensieb *n* [paper] machine wire, travelling wire
Papiermasse *f* 1. paper[making] stock; 2. paper weight, weight of paper
Papiermaulbeerbaum *m* paper-mulberry, cloth tree *(Broussonetia papyrifera (L.) L'Hérit.)*
Papiermühle *f* paper mill
Papierrohstoff *m* [paper, raw] stock, raw papermaking material
Papierrolle *f* roll [of paper], reel [of paper]
von der Maschine kommende P. machine (mill, jumbo) roll
Papierrollstange *f (pap)* winder (rewind) shaft
Papiersack *m* paper bag (sack)
mehrlagiger P. multiwall paper bag
Papierschnitzel *npl(mpl) (pap)* shavings
Papiersorte *f* grade of paper
Papierstoff *m* [**/ fertiger**] paper[making] stock, [finished] stuff
Papiertambour *m s.* Papierrolle
Papiertrockenzylinder *m* paper dryer
Papiertrocknung *f* paper drying
Papierunterlage *f (phot)* paper base
Papiervlies *n s.* Papierbahn
Papierwalze *f s.* 1. elastische P.; 2. Papierkalanderwalze
elastische P. paper (resilient, filled) roll *(of a supercalender)*
Papierwolle *f* paper wool
Papierzellstoff *m* paper[making] pulp
Papierzugspannung *f* sheet (web, paper) tension, pull (tension) of the web
Pappe *f* [paper]board, cardboard
beklebte P. lined board
gedeckte (gegautschte) P. couched (lined, vatlined) board, *(Am)* nonpasted board
gehärtete P. hardboard, panel board
gemaserte P. grained board
im Format geklebte (kaschierte) P. sheet-lined board
in der Rolle geklebte (kaschierte) P. mill-lined board
kaschierte P. lined board
mehrschichtige P. multi-ply board
zweischichtige P. two-ply board
Pappebogen *m* sheet of board
Pappenfabrik *f* paperboard mill
Pappenmaschine *f* board [making] machine
Pappenrundsiebmaschine *f* cylinder board machine
Pappkarton *m*, **Pappschachtel** *f* [paperboard] carton, cardboard box
Papyrolin *n* papyrolin, reinforced paper *(cloth-faced or cloth-centred paper)*
Para *m s.* Parakautschuk
Paraaminophenolentwickler *m (phot)* paraminophenol (para-aminophenol) developer
Paraazetaldehyd *m s.* Paraldehyd
Parabansäure *f* parabanic acid, imidazolidine-2,4,5-trione
Parachor *m (phys chem)* parachor
Parachormessung *f* parachor measurement

Paradamin *m (min)* paradamite *(a basic zinc arsenate)*
Paradichlorbenzol *n* $C_6H_4Cl_2$ paradichlorobenzene, *p*-dichlorobenzene
Paradieskörner *npl* grains of paradise *(from Aframomum melegueta Schum.)*
para-dirigierend para-directing
Paraffin *n* 1. paraffin [wax] *(mixture of high-molecular-weight hydrocarbons)*; 2. *s.* Alkan
  P. für Sprühzwecke spray (light liquid) paraffin
  amorphes P. *s.* mikrokristallines P.
  festes P. solid paraffin
  flüssiges P. liquid (medicinal) paraffin, liquid petrolatum, paraffin oil
  hartes P. hard paraffin
  mikrokristallines P. microparaffin, micro wax
  weiches P. soft paraffin
Paraffinalkohol *m* paraffinic alcohol
Paraffinanteil *m* paraffinicity *(percentage of alkanes as in insecticidal oils)*
Paraffinbasis *f* paraffin base *(of crude petroleum)*
Paraffinbasisöl *n (petrol)* paraffin-base petroleum, paraffinic petroleum, paraffin-base crude [oil]
Paraffindestillat *n* paraffin distillate
Paraffinemulsion *f (pap)* wax emulsion
Paraffineur *m (text)* waxing device
Paraffingatsch *m* [paraffin] slack wax
paraffinhaltig containing paraffin, paraffinic
paraffiniert paraffin-coated
paraffinisch paraffinic
Paraffinität *f* paraffinicity *(percentage of alkanes as in insecticidal oils)*
Paraffinkohlenwasserstoff *m* paraffin [hydrocarbon], saturated hydrocarbon, alkane
  monochlorierter P. monochloroparaffin
Paraffinkuchen *m* [paraffin] wax cake
Paraffinleim *m (pap)* [paraffin] wax size
Paraffinleimung *f (pap)* paraffin wax sizing, sizing with wax emulsions
Paraffinöl *n* 1. paraffin[ic] oil, liquid paraffin (petrolatum); 2. *s.* Paraffinbasisöl
Paraffinpapier *n* paraffin[ed] paper, paraffin-waxed paper, wax paper
Paraffinreihe *f s.* Alkanreihe
Paraffinschuppen *fpl* paraffin scale, scale wax
Paraffinwachs *n* paraffin wax
  P. aus Erdöl petroleum wax
  mikrokristallines P. *s.* Paraffin / mikrokristallines
Paraform *n s.* Paraformaldehyd
Paraformaldehyd *m* $(CH_2O)_n$ paraformaldehyde, polyoxymethylene
Parafuchsin *n* parafuchsine
Paragonit *m (min)* paragonite, soda mica *(a phyllosilicate)*
Paragummi *m s.* Parakautschuk
Parahelium *n* parahelium, parhelium
Para-Isomer[es] *n* para isomer, *p*-isomer
Parakasein *n* paracasein *(insoluble casein)*
Parakautschuk *m* Para rubber *(from Hevea brasiliensis (H.B.K.) Muell. Arg.)*
parakristallin paracrystalline
Paraldehyd *m* $(CH_3CHO)_3$ paraldehyde, paraacetaldehyde, 2,4,6-trimethyl-1,3,5-trioxane
Paraleukorosanilin *n* leucopararosaniline
Parallaxenfehler *m* parallax error *(as in reading burettes)*
Parallelplattendruckgerät *n*, **Parallelplattenplastometer** *n (rubber)* parallel plate plastometer
Parallelreaktion *f* competitive (competing, concurrent) reaction
Parallelschieber *m* parallel-seat gate valve
Parallelstrom *m* parallel (co-current, concurrent) flow ✦ **im P. [geführt]** co-current
Parallelverschiebung *f (cryst)* translation
Paramagnetikum *n* paramagnetic [material, substance]
paramagnetisch paramagnetic
Paramagnetismus *m* paramagnetism
Parameter *m* parameter, *(cryst also)* intercept
Paramilchsäure *f* $CH_2(OH)CH(OH)COOH$ paralactic acid, *L*(+)-lactic acid, dextrorotatory lactic acid
Paraminophenolentwickler *m s.* Paraaminophenolentwickler
Paramolybdat *n* paramolybdate
Paramorphin *n* thebaine, paramorphine *(alkaloid)*
Paranitranilin *n* $NH_2C_6H_4NO_2$ *p*-nitroaniline
Paranitranilinrot *n* $C_{10}H_6(OH)NNC_6H_4NO_2$ paranitraniline red, para red *(a pigment dye)*
Paranuß *f* para (Brazil, cream) nut *(seed from Bertholletia excelsa Humb. et Bonpl.)*
Paraplasma *n (biol)* paraplasm, ergastoplasm *(collectively for the reserve and waste inclusions of protoplasm)*
Para-Positronium *n* parapositronium
Pararosanilin *n* parafuchsine, pararosaniline
Pararosolsäure *f* $(HOC_6H_4)_2C{=}C_6H_4{=}O$ para-rosolic acid, rosolic acid
Pararot *n s.* Paranitranilinrot
Parasorbinsäure *f* parasorbic acid
paraständig para, in para position
  p. sein to be [located, situated] para
para-Stellung *f* para position ✦ **in para-Stellung** in para position, para ✦ **nach der para-Stellung dirigierend** para-directing
parasubstituiert para-substituted, *p*-substituted
Parasympathikomimetikum *n (pharm)* parasympathomimetic agent
parasympathikomimetisch *(pharm)* parasympathomimetic
Parathion *n* parathion, diethyl-*p*-nitrophenylphosphorothionate *(an insecticide)*
Parathormon *n (bioch)* parathormone
para-Verbindung *f* para compound
Parawasserstoff *m* para hydrogen
Paraweinsäure *f* $HOOC[CHOH]_2COOH$ (±)-tartaric acid, racemic (inactive) tartaric acid
Parawolframat *n* parawolframate, paratungstate
Parazetaldehyd *m s.* Paraldehyd
para-Zustand *m* para state
Parelleflechte *f (dye)* crabeye lichen, Lecanora parella Mass.
Parfüm *n* perfume, essence
parfümieren to perfume, *(relating to tobacco also)* to flavour
Parfümpapier *n* perfumed paper

Parfümranzigkeit *f (food)* ketonic rancidity
Parhelium *n s.* Parahelium
Pariangips *m* Parian cement *(a hard plaster made up of gypsum and borax)*
Parica parica, cohoba [snuff] *(narcotic seed meal from Piptadenia peregrina (L.) Benth.)*
Parität *f (phys chem)* parity
parkerisieren to parkerize *(corrosion inhibition)*
Parker-Verfahren *n (coal)* Parker process *(low-temperature carbonization)*
Parkes-Reagens *n* Parkes reagent *(for detecting artificial colourants in fats)*
Parkes-Verfahren *n* Parkes process *(for removing noble metals from lead)*
Partialdampfdruck *m* partial vapour pressure
Partialdruck *m* partial pressure
Partialdruckgesetz *n* / **Daltonsches** Dalton's law of partial pressures
Partialdruckkurve *f* partial pressure curve
Partialhydrolyse *f* partial hydrolysis
Partialvalenz *f* partial valency (valence)
Partialvalenztheorie *f* partial-valence theory
Partie *f (dye)* batch
partiell partial
Partikel *f* particle
Partikelgröße *f* particle size
Partner *m* participant *(of a reaction)*, reactant
Paschen-Back-Effekt *m* Paschen-Back effect *(splitting up of spectral lines in a strong magnetic field)*
Paschen-Serie *f* Paschen series *(of the hydrogen spectrum)*
Passage *f* pass *(in zone melting)*
Passageofen *m (ceram)* multipassage kiln
Paßburg-Trockenschrank *m* Passburg dryer
passieren to pass, to penetrate
Passiermaschine *f (food)* pulping machine
passiv passive
Passivator *m* passivator *(for metals)*; inhibitor, negative catalyst, anticatalyst, retarder *(of a chemical reaction)*; *(min tech)* depressant
passivieren *(phys chem)* to passivate *(metals)*; to inhibit, to retard *(a chemical reaction)*; *(min tech)* to depress
Passivierung *f (phys chem)* passivation *(of metals)*; inhibition, retardation *(of a chemical reaction)*; *(min tech)* depressing
Passivierungsmittel *n* passivator *(for metals)*
Passivität *f (phys chem)* passivity
  anodische P. anodic passivity
  chemische P. chemical passivity
Paßstück *n* adapter
Paste *f* paste
Pastellkreide *f* crayon
Pastellstift *m* crayon
pasten *(tann)* to paste-dry
pastenartig paste-like, pasty
Pastengießen *n (plast)* slush moulding
Pastenharz *n* paste resin
Pastenmischer *m* paste mixer (mill)
Pasteur-Effekt *m* Pasteur effect *(the inhibition of anaerobic fermentation by oxygen)*
Pasteurisation *f s.* Pasteurisierung
Pasteuriseur *m*, **Pasteurisierapparat** *m* pasteurizer
pasteurisieren to pasteurize
  wiederholt p. to repasteurize
Pasteurisierung *f* pasteurization, pasteurizing
  kontinuierliche P. continuous pasteurization
  nochmalige (wiederholte) P. repasteurization
Pasteurisierungsabteilung *f* pasteurization section
Pasteurisierungsgeschmack *m* pasteurization flavour
Pasteur-Kolben *m* Pasteur flask
pastieren to paste *(lead-acid accumulators)*
pastig *s.* pastös
Pastille *f (pharm)* pastille
Pastillenpresse *f* briquetting (briquette) press *(for calorimetric tests)*
Pasting-Verfahren *n (tann)* pasting process
pastös pasty, paste-like
Patentgrün *n s.* Grün / Schweinfurter
patentieren to patent *(1. an invention; 2. wire by heat)*
Patentierofen *m* patenting furnace *(for heat treatment of wire)*
Patentpapier *n* machine[-made] paper
pathogen pathogenic
Pathogen *n* pathogen
Patina *f* patina
Patrize *f (plast)* patrix, moulding (force) plug, male form (mould)
Patrone *f* cartridge
Patronenfilter *n* cartridge filter
Patronenheizkörper *m* cartridge heater
patschen *(glass)* to paddle
Patschoulialkohol *m* patchouli alcohol
Patschouliöl *n* patchouli oil *(from Pogostemon cablin (Blanco) Benth.)*
Patschuli... *s.* Patschouli...
Pattinson-Bleiweiß *n* Pattinson's white lead
Pattinson-Verfahren *n* Pattinson process *(for separating silver from lead)*
Pauli-Prinzip *n* [Pauli] exclusion principle
pauschen *(met)* to liquate
Pauschen *n (met)* liquating, liquation
Pausleinen *n* tracing cloth *(reprography)*
Pauspapier *n* tracing paper
PC *s.* Polykarbonat
PCE *(abbr)* pyrometric cone equivalent, *s.* Schmelzkegeläquivalent
PCP *s.* Pentachlorphenol
PCTFE *s.* Polychlortrifluoräthylen
PCV *s.* Polyvinylkarbazol
PE *s.* 1. Polyäthylen; 2. Polyesterfaserstoff
Peachey-Prozeß *m* Peachey process *(of curing rubber with SO_2 and H_2S)*
Peak *m* peak *(as of a chromatogram)*
Peakbreite *f* peak width *(as in a chromatogram)*
Pebble-Heater-Pyrolyse *f (petrol)* pebble-heater pyrolysis
Pebble-Reaktor *m (nucl)* pebble [bed] reactor, PBR
Pech *n* pitch
  Kanadisches P. Canada (hemlock) pitch *(usually from Tsuga canadensis (L.) Carr.)*
pechartig pitchy, pitch-like

Pechblende *f (min)* pitchblende *(uranium(IV) oxide)*
Pechkohle *f* pitch coal
Pechpolitur *f (glass)* pitch polishing
Pechsee *m (geol)* pitch (asphalt) lake
Pechspitze *f (text)* pitch tip
Pedersen-Verfahren *n* Pedersen process *(for obtaining aluminium)*
Pedologie *f* pedology, soil science
pedologisch pedologic[al]
Pegel *m* 1. *(tech)* level *(as of a liquid)*; 2. *(glass)* tip
Pegukatechu *n* Pegu (black, dark) catechu, Pegu cutch *(from Acacia catechu Willd.)*
peilen to dip *(to measure the oil level in a tank)*
Peilen *n* dip *(measuring of the oil level in a tank)*
Peilstab *m* gauge stick
Peirce-Smith-Konverter *m (met)* Peirce-Smith converter
Pektase *f* pectase, pectinesterase
Pektat *n* pectate *(a salt or ester of a pectic acid)*
Pektin *n* 1. pectin; 2. *s.* Pektinstoff
  eigentliches P. pectin proper
  hochverestertes P. high-ester pectin
  leichtverestertes P. *s.* niederverestertes P.
  methoxylarmes P. *s.* niederverestertes P.
  niederverestertes (niedrig verestertes) P. low-ester (low-methoxyl) pectin
Pektin H *n* high-ester pectin
pektinabbauend pect[in]olytic
Pektinase *f* pectinase, pectic enzyme
Pektinat *n* pectinate *(a salt or ester of a pectinic acid)*
Pektinchemie *f* pectin chemistry
Pektinesterase *f s.* Pektase
Pektingel *n*, **Pektingelee** *n* pectin jelly
Pektinglykosidase *f* pectinase, pectic enzyme
Pektinmethylesterase *f s.* Pektase
pektinolytisch pect[in]olytic
Pektinpräparat *n* pectin preparation
Pektinsäure *f* pectic acid *(a high-molecular-weight polymer of D-galacturonic acid)*
pektinspaltend pect[in]olytic
Pektinstoff *m* pectic (pectinous) substance, pectin
Pektinzucker *m* pectin sugar, pectinose
Pektisation *f (coll)* pectization, flocculation
Pektolith *m (min)* pectolite *(an inosilicate)*
pektolytisch pect[in]olytic
Pektose *f s.* Protopektin
Pelagit *m (min)* pelagite *(pelagic manganese oxide)*
Pelargon *n* $CH_3[CH_2]_7CO[CH_2]_7CH_3$ pelargone, dioctyl ketone, 9-heptadecanone
Pelargonaldehyd *m* $CH_3[CH_2]_7CHO$ pelargonic aldehyde, *n*-nonyl aldehyde, nonanal
Pelargonsäure *f* $CH_3[CH_2]_7COOH$ pelargonic acid, nonanoic acid
Pellagrapräventivvitamin *n* pellagra-preventive factor, PP factor *(either nicotinic acid or nicotinamide)*
Pellet *n* pellet
Pelletformeinrichtung *f* pelletizing (balling) device
Pelletier... *s.* Pelletisier...
pelletieren *s.* pelletisieren
Pelletisieranlage *f* pelletizing plant
pelletisieren to pellet[ize], to ball
Pelletisierkonus *m* pelletizing (balling) cone
Pelletisiermaschine *f* pellet[izing] machine, pelletizer
Pelletisierteller *m* pelletizing (balling) disk
Pelletisiertrommel *f* pelletizing (balling) drum
Pelletisierungsanlage *f* pelletizing plant
Pelletisier[ungs]verfahren *n* pelletizing process
Peltier-Effekt *m* Peltier effect *(a thermoelectric phenomenon)*
Pelzfarbstoff *m* fur dye
Penaldin-F-Säure *f* penaldic-F acid, 2-pentenylpenaldic acid
Penaldin-G-Säure *f* penaldic-G acid, benzylpenaldic acid
Penaldin-K-Säure *f* penaldic-K acid, heptylpenaldic acid
Penaldinsäure *f* penaldic acid *(any of a group of acids RCONHCH(CHO)COOH)*
Penaldin-X-Säure *f* penaldic-X acid, *p*-hydroxybenzylpenaldic acid
Penaldsäure *f s.* Penaldinsäure
Pendelbecherwerk *n* swing-bucket (swing-tray) elevator, pivoted-bucket conveyor
Pendelfallmethode *f* pendulum method *(for testing rubber)*
Pendelmühle *f s.* Pendelrollenmühle
pendeln to oscillate, to pendulate
Pendelrollenmühle *f* pendulum roller mill, ring-roll mill *(with horizontal grinding ring)*
Pendelschlagwerk *n* impact pendulum machine
Pendelzentrifuge *f* link-suspended centrifuge
Penetration *f* 1. penetration; 2. *(cryst)* intergrowth
penetrieren 1. to penetrate; 2. *(cryst)* to intergrow
Penetrometer *n* penetrometer
Penetrometerverfahren *n (coal)* penetrometer method
Penill-F-Säure *f* penillic-F acid, 2-pentenylpenillic acid
Penill-G-Säure *f* penillic-G acid, benzylpenillic acid
Penill-K-Säure *f* penillic-K acid, heptylpenillic acid
Penilloaldehyd *m* $RCONHCH_2CHO$ penilloaldehyde
Penillo-G-Säure *f* penilloic-G acid, benzylpenilloic acid
Penillo-K-Säure *f* penilloic-K acid, heptylpenilloic acid
Penillosäure *f* penilloic acid *(any of a group of acids produced by Penicillium specc.)*
Penillo-X-Säure *f* penilloic-X acid, *p*-hydroxybenzylpenilloic acid
Penillsäure *f* penillic acid *(any of a group of inactivation products of penicillins)*
Penill-X-Säure *f* penillic-X acid, *p*-hydroxybenzylpenillic acid
Penizillamin *n* $(CH_3)_2C(SH)CH(NH_2)COOH$ penicillamine
Penizillaminsäure *f* $(CH_3)_2C(SO_3H)CH(NH_2)COOH$ penicillaminic acid
Penizillansäure *f* penicillanic acid *(a building brick of penicillins)*
Penizillase *f* penicill[in]ase
Penizillin *n* penicillin

Penizillinamidase *f* penicillin amidase
Penizillinase *f* penicill[in]ase
penizillinempfindlich penicillin-sensitive, sensitive to penicillin
penizillinresistent penicillin-resistant
Penizillinsäure *f s.* Penizillsäure
Penizillinwirksamkeit *f* penicillin activity
Penizilloat *n* penicilloate
Penizilloinsäure *f s.* Penizillosäure
Penizillosäure *f* penicilloic acid *(any of a group of amidodicarboxylic acids $RCONH(C_6H_{10}NS)(COOH)_2$)*
Penizillsäure *f* penicillic acid, 3-methoxy-5-methyl-4-oxo-2,5-hexadienoic acid
Pennin *m (min)* penninite, pennine *(a phyllosilicate)*
Pensky-Martens-Gerät *n* Pensky-Martens [flash-point] apparatus
Pentaboran *n* pentaborane
Pentaborat *n* pentaborate
Pentabromid *n* pentabromide
Pentachlorid *n* pentachloride
Pentachlorphenol *n* C_6Cl_5OH pentachlorophenol
Pentadekan *n* $C_{15}H_{32}$ pentadecane
Pentadekanol *n* $C_{15}H_{31}OH$ pentadecanol, *(specif)* $CH_3[CH_2]_{13}CH_2OH$ 1-pentadecanol
Pentadekansäure *f* $CH_3[CH_2]_{13}COOH$ pentadecanoic acid
Pentadezylalkohol *m* $CH_3[CH_2]_{13}CH_2OH$ pentadecyl alcohol, 1-pentadecanol
Pentadezylsäure *f s.* Pentadekansäure
Pentadien *n* pentadiene, *(specif)* $CH_2{=}CHCH{=}CHCH_3$ 1,3-pentadiene, piperylene
Pentadien-(2,4)-säure *f* $CH_2{=}CHCH{=}CHCOOH$ 2,4-pentadienoic acid, 3-vinylacrylic acid
Pentaerythrit *m* $C(CH_2OH)_4$ pentaerythritol, PE, tetrahydroxytetramethylmethane
Pentaerythrit-tetranitrat *n* $C(CH_2ONO_2)_4$ pentaerythritol tetranitrate, penthrite, PETN *(an explosive)*
Pentaglyzerin *n* $CH_3C(CH_2OH)_3$ pentaglycerol, 1,1,1-trimethylolethane
Pentahydrat *n* pentahydrate
Pentajodid *n* pentaiodide
Pentakontan *n* $C_{50}H_{102}$ pentacontane
Pentakosan *n* $C_{25}H_{52}$ pentacosane
pentakovalent pentacovalent
Pentamethonium *n* $(CH_3)_3N(CH_2)_5N(CH_3)_3$ pentamethonium
Pentamethylbenzoesäure *f* $C_6(CH_3)_5COOH$ pentamethylbenzoic acid
Pentamethylbenzol *n* $C_6H(CH_3)_5$ pentamethylbenzene
Pentamethylen *n s.* Zyklopentan
Pentamethylendiamin *n* $H_2N[CH_2]_5NH_2$ pentamethylenediamine, cadaverine
Pentammin *n* pentammine
Pentan *n* C_5H_{12} pentane, *(specif)* $CH_3[CH_2]_3CH_3$ pentane, *n*-pentane
Pentanal *n* $CH_3[CH_2]_3CHO$ pentanal, valeraldehyde
Pentandial *n* $OCH[CH_2]_3CHO$ pentanedial, glutaraldehyde
Pentandikarbonsäure *f* pentanedicarboxylic acid
Pentandiol *n* pentanediol
Pentandisäure *f* $HOOC[CH_2]_3COOH$ pentanedioic acid
Pentanitroosmat(III) *n* $M_2^I[Os(NO_2)_5]$ pentanitroosmate(III), nitroosmate(III)
Pentankarbonsäure *f* pentanecarboxylic acid
Pentanol *n* $C_5H_{11}OH$ pentanol
Pentanol-(1) *n* $CH_3[CH_2]_3CH_2OH$ 1-pentanol
Pentaquin *n* pentaquin *(a quinoline derivative)*
Pentaschwefelwasserstoff *m s.* Pentasulfan
Pentaselenid *n* pentaselenide
Pentasilan *n* Si_5H_{12} pentasilane
Pentasulfan *n* H_2S_5 pentasulphane, hydrogen pentasulphide
Pentasulfid *n* pentasulphide
Pentathionat *n* $M_2^IS_5O_6$ pentathionate
Pentathionsäure *f* $H_2S_5O_6$ pentathionic acid
Pentatriakontan *n* $C_{35}H_{72}$ pentatriacontane
pentavalent pentavalent, quinquevalent
Pentavalenz *f* pentavalency, quinquevalency
Pentawolframat *n* pentawolframate, pentatungstate
Pentazen *n* pentacene, dibenz[*b,i*]anthracene
Pentazolyl *n* pentazolyl *(a heterocyclic group of atoms)*
pentazyklisch pentacyclic
Penten-(1) *n* $CH_3CH_2CH_2CH{=}CH_2$ 1-pentene
2-Pentenylpenillsäure *f* 2-pentenylpenillic acid, penillic-F acid
2-Pentenylpenizilloinsäure *f* 2-pentenylpenicilloic acid, penicilloic-F acid
Pentin-(1) *n* $CH{\equiv}CCH_2CH_2CH_3$ 1-pentyne
Pentlandit *m (min)* pentlandite *(an iron nickel sulphide)*
Pentosan *n* pentosan *(any of a class of polysaccharides)*
Pentose *f* pentose *(monosaccharide containing 5 carbon atoms per molecule)*
Pentoxid *n* pentaoxide
Pentoxojodat(VII) *n* $M_3^I[IO_5]$ pentaoxoiodate(VII), mesoperiodate
Pentoxorhenat(VII) *n* $M_3^I[ReO_5]$ pentaoxorhenate(VII)
Pentylalkohol *m s.* Pentanol-(1)
Pepsin *n* pepsin *(a proteinase)*
Pepsinogen *n* pepsinogen *(inactive precursor of pepsin)*
Pepsinverdauung *f* peptic digestion
Peptid *n* peptide
  zyklisches P. cyclic peptide
Peptidanteil *m* peptide moiety (portion)
Peptidase *f* peptidase
Peptidbindung *f* peptide bond (linkage)
Peptidgruppe *f* peptide group
Peptidkette *f* peptide chain
Peptidkomponente *f s.* Peptidanteil
Peptidsynthese *f* peptide synthesis
Peptidteil *m s.* Peptidanteil
Peptisation *f* peptization, deflocculation
Peptisationsmittel *n s.* Peptisator
Peptisator *m* peptizer, peptizing (deflocculating) agent, deflocculant, deflocculator; *(rubber)* [chemical] plasticizer, [chemical] plasticizing agent

peptisieren to peptize, to deflocculate; *(rubber)* to plasticize
Peptisiermittel *n s.* Peptisator
Peptisierung *f* peptization
Peptisierungsmittel *n s.* Peptisator
Peptolyse *f* peptolysis
peptolytisch peptolytic
Pepton *n* peptone *(any of various high-molecular protein derivatives)*
peptonisieren to peptonize
Peptonisierung *f* peptonization
Perameisensäure *f* HCOOOH performic acid
Peräthansäure *f s.* Peressigsäure
Perbenzoesäure *f* C_6H_5COOOH perbenzoic acid
Perbenzolkarbonsäure *f s.* Perbenzoesäure
Perchlorat *n* M^IClO_4 perchlorate
Perchloräthan *n* Cl_3CCCl_3 perchloroethane, hexachloroethane
Perchlormethan *n* CCl_4 perchloromethane, tetrachloromethane, carbon tetrachloride
Perchlormethylmerkaptan *n s.* Trichlormethansulfenylchlorid
Perchlorsäure *f* $HClO_4$ perchloric acid
Peressigsäure *f* CH_3COOOH peracetic acid
Perester *m* perester
Perezon *n* perezone, pipitzahoic acid *(a benzoquinone derivative)*
Perfektkautschuk *m* superior processing rubber, S.P. rubber
Perfluoräthen *n*, **Perfluoräthylen** *n* $F_2C{=}CF_2$ perfluoroethylene, tetrafluoroethylene, T.F.E.
perfluorieren to perfluorinate
Perfluorierung *f* perfluorination
Perfluorvinylchlorid *n s.* Chlortrifluoräthylen
Perforation *f* perforation
Perforationsgerät *n*, **Perforator** *m* perforator *(continuous liquid extraction apparatus)*
perforieren to perforate
Perforieren *n* perforation
Pergament *n* parchment
  animalisches (tierisches) P. animal (skin, natural, writing) parchment
  vegetabilisches P. vegetable parchment, parchment paper
pergamentartig parchmenty
Pergamentersatz *m*, **Pergamentersatzpapier** *n* artificial (imitation) parchment
pergamentieren *(pap)* to parchmentize
Pergamentierung *f (pap)* parchmentization
Pergamentierungsmittel *n (pap)* parchmentizing agent
Pergamentkarton *m* parchment cardboard
Pergamentpapier *n* **[/ echtes]** parchment paper, vegetable parchment
Pergamentrohstoff *m* paper for parchmentizing
Pergamin[papier] *n*, **Pergamyn** *n* pergamyn, glassine [paper]
perhydrieren to perhydrogenate, to perhydrogenize
Perhydrierung *f* perhydrogenation
perianelliert peri-fused
Perianellierung *f* peri fusion
Peridot *m (min)* peridot[e] *(a variety of olivine)*
Periklas *m (min)* periclas[it]e *(magnesium oxide)*
Periklin *m (min)* pericline *(a tectosilicate)*
Perikondensation *f* peri fusion
perikondensiert peri-fused
Perillaalkohol *m* perilla (perillic, perillyl) alcohol, 1-hydroxymethyl-4-isopropenylcyclohexane
Perillaldehyd *m* perillaldehyde, perillic (perillyl) aldehyde, 4-isopropenyl-1-cyclohexene-1-aldehyde
Perillaöl *n* perilla oil *(from Perilla specc.)*
Perillasäure *f* perillic acid, 4-isopropenyl-1-cyclohexene-1-carboxylic acid
Periode *f* period *(as of the periodic table)*
  große P. *s.* lange P.
  kleine P. *s.* kurze P.
  kurze P. short period
  lange P. long period
Periodengesetz *n* periodic law
Periodenmischer *m* batch mixer
Periodensystem *n* **[der Elemente]** periodic system (table)
  Mendelejewsches P. Mendeléeff periodic system (table)
periodisch periodic, *(relating to operations preferably)* batch[wise]
  p. arbeitend batch *(apparatus)*
Periodizität *f* periodicity
Peripherie *f* periphery
Perisäure *f* $NH_2C_{10}H_6SO_3H$ peri acid, 1-naphthylamine-8-sulphonic acid
peri-Stellung *f* peri position
Peristerit *m (min)* peristerite *(any of a group of tectosilicates)*
Peritektikum *n (phys chem)* peritectic (transition-type) system
peritektisch *(phys chem)* peritectic
Perjodat *n* periodate; *(specif) s.* Metaperjodat
Perjodsäure *f* periodic acid; *(specif) s.* Metaperjodsäure
Perkin-Kondensation *f*, **Perkin-Reaktion** *f* Perkin condensation (reaction)
Perkinviolett *n* Perkin's mauve (purple, violet), aniline purple
Perkolat *n* percolate, leachate
Perkolation *f* percolation
  P. unter Druck diacolation
  kontinuierliche P. continuous percolation
Perkolationsverfahren *n* percolation process
  kontinuierliches P. continuous-percolation process
Perkolator *m* percolator
perkolieren to percolate
Perlanlage *f (rubber)* pelletizing equipment
Perlasche *f* pearl ash *(potassium carbonate)*
Perle *f* bead
perlen 1. to bubble, to effervesce, *(of wine also)* to sparkle; 2. to pelletize *(soot)*
Perlen *fpl* shot, slug *(a defect in fibre-glass products)*
perlend bubbling, effervescent, *(of beverages also)* sparkling, brisk

Perlenessenz *f (coat, cosmet)* [natural] pearl essence
perlenförmig beaded
Perlenprobe *f*, **Perlenreaktion** *f* bead test *(analytical chemistry)*
Perlessenz *f s.* Perlenessenz
Perlglimmer *m (min)* pearl-mica, margarite *(a phyllosilicate)*
Perlit *m* 1. *(met)* pearlite *(a microconstituent of steel)*; 2. *(geol)* perlite *(a natural glass)*
perlitisch 1. *(met)* pearlitic; 2. *(geol)* perlitic
Perlkontakt *m* bead catalyst
Perlkontaktmasse *f* beaded material *(in catalytic cracking)*
Perlmutt *n s.* Perlmutter
perlmutt... *s.* perlmutter...
Perlmutteffekt *m* nacreous (mother-of-pearl) effect
Perlmutter *f* nacre, mother of pearl ✦ **aus P. [bestehend]** nacr[e]ous
perlmutterähnlich, perlmutterartig nacr[e]ous, pearly
Perlmutterglanz *m* pearly lustre
perlmutterglänzend nacr[e]ous, pearly
Perlmutterpapier *n* mother-of-pearl paper, iridescent [paper]
Perlpolymerisat *n* bead polymerizate
Perlpolymerisation *f* bead (suspension) polymerization
Perlruß *m* beaded (pelletized) black
Perlspat *m (min)* pearl spar *(loosely for dolomite and aragonite)*
Perlstein *m (geol)* perlite *(a natural glass)*
Perlwein *m* carbonated wine
Perlweiß *n* pearl (bismuth, Spanish) white, cosmetic bismuth *(consisting of bismuth chloride oxide or bismuth nitrate oxide)*
Permanentappretur *f*, **Permanentausrüstung** *f (text)* permanent (durable) press, PP
Permanentblau *n* permanent (French) blue *(an artificially prepared pigment)*
Permanentgrün *n* permanent green *(a pigment consisting of barium sulphate and guignet green)*
Permanenthärte *f* permanent hardness *(water chemistry)*
Permanentrot *n* permanent red *(an azo dye)*
Permanentweiß *n* permanent (fixed) white, blanc fixe *(precipitated barium sulphate)*
Permanganat *n* M^IMnO_4 permanganate
Permanganatoxydation *f* permanganate oxidation
Permanganatprobe *f* permanganate test *(for detecting alkenes)*
Permanganatzahl *f* permanganate number *(for characterizing the organic pollution of water)*
Permanganometrie *f* permanganometry
Permangansäure *f* $HMnO_4$ permanganic acid
permeabel permeable
Permeabilität *f* permeability
Permeation *f* permeation
permeieren to permeate
Permeiervermögen *n* permeativity
Permethansäure *f s.* Perameisensäure
Pernambukholz *n (dye)* brazilwood *(from Caesalpinia specc., specif from C. echinata Lam.)*
Pernambukokautschuk *m s.* Mangabeirakautschuk
Perowskit *m (min)* perovskite, perofskite *(a calcium titanate)*
Peroxid *n* peroxide
Peroxidbindung *f* peroxide bond (link)
Peroxidbleiche *f* peroxide bleaching
P. bei hoher Stoffdichte *(pap)* high-density peroxide bleaching
Peroxidbleichlösung *f* peroxide bleaching solution
Peroxideffekt *m* peroxide effect *(directive influence of peroxides in the addition of HBr to alkanes)*
Peroxidhydrat *n s.* Peroxyhydrat
Peroxidigkeit *f s.* Peroxidranzigkeit
peroxidisch peroxidic
Peroxidranzigkeit *f* peroxide (oxidative) rancidity, oiliness, tallowiness
Peroxidvernetzung *f* peroxide cross-linking
Peroxidvulkanisation *f (rubber)* peroxide cure (vulcanization)
Peroxidwert *m*, **Peroxidzahl** *f* peroxide number (value) *(measure of rancidity)*
Peroxoborat *n* peroxoborate
echtes (wahres) P. true peroxoborate
Peroxoborsäure *f* peroxoboric acid
Peroxochromat *n* peroxochromate
Peroxochromsäure *f* peroxochromic acid
Peroxoderivat *n* peroxo derivative
Peroxodischwefelsäure *f* $H_2S_2O_8$ peroxodisulphuric acid
Peroxodisulfat *n* $M_2^IS_2O_8$ peroxodisulphate
Peroxogruppe *f* $-O-O-$ peroxo group
Peroxohydrat *n* peroxohydrate, hydroperoxidate
Peroxokarbonat *n* $M_2^IC_2O_6$ peroxocarbonate
Peroxokarbonat-Peroxyhydrat *n* peroxyhydrated peroxocarbonate
Peroxokohlensäure *f* $H_2C_2O_6$ peroxocarbonic acid
Peroxomonophosphat *n* $M_3^IPO_5$ peroxomonophosphate
Peroxomonophosphorsäure *f* H_3PO_5 peroxomonophosphoric acid
Peroxomonoschwefelsäure *f* H_2SO_5 peroxomonosulphuric acid, Caro's acid
Peroxomonosulfat *n* $M_2^ISO_5$ peroxomonosulphate
Peroxonitrat *n* M^INO_4 peroxonitrate
Peroxonitrit *n* $M^I[OON=O]$ peroxonitrite
Peroxophosphat *n* peroxophosphate; *(specif) s.* Peroxomonophosphat
Peroxophosphorsäure *f* peroxophosphoric acid; *(specif) s.* Peroxomonophosphorsäure
Peroxosalpetersäure *f* HNO_4 peroxonitric acid
Peroxosalz *n* peroxo salt
Peroxosäure *f* peroxo acid
Peroxoschwefelsäure *f* peroxosulphuric acid; *(specif) s.* Peroxodischwefelsäure
Peroxosulfat *n* peroxosulphate; *(specif) s.* Peroxodisulfat
Peroxouranat *n* $M_2^IUO_6$ peroxouranate
Peroxovanadat *n* peroxovanadate
Peroxovanadinsäure *f* peroxovanadic acid
Peroxoverbindung *f* peroxo compound
Peroxy... *(with inorganic compounds incorrectly for)* Peroxo...

Peroxyameisensäure *f s.* Perameisensäure
Peroxybenzoesäure *f s.* Perbenzoesäure
Peroxydase *f* peroxidase
Peroxydaseprobe *f*, **Peroxydasetest** *m* peroxidase test *(liberation of iodine from potassium iodide by peroxidases in presence of H_2O_2)*
Peroxydation *f* peroxidation
peroxydatisch peroxidatic *(of or relating to peroxidase)*
peroxydieren to peroxidize, to peroxidate
Peroxydierung *f* peroxidation
Peroxyessigsäure *f s.* Peressigsäure
Peroxyester *m* peroxy ester
Peroxygruppe *f (org chem)* –0–0– peroxy group; *(with inorganic compounds incorrectly for)* Peroxogruppe
Peroxyhydrat *n* peroxyhydrate
Peroxykarbonsäure *f* peroxycarboxylic acid
Peroxysäure *f (org chem)* RC(=O)OOH peroxy acid; *s.* Peroxosäure
Peroxyverbindung *f (org chem)* peroxy compound; *s.* Peroxoverbindung
Perpetuum mobile *n* **erster Art** *(phys chem)* perpetual motion of the first kind
P. mobile zweiter Art perpetual motion of the second kind
Perrhenat *n* M^IReO_4 perrhenate, tetraoxorhenate(VII)
Perrheniumsäure *f* $HReO_4$ perrhenic acid, tetraoxorhenic(VII) acid
Perrin-Verfahren *n* Perrin process *(for producing ingot steel)*
Persalz *n* persalt
Persäure *f* peracid; *(org chem)* RC(=O)OOH peroxy acid
Persio *m* persis, persio, cudbear *(dried paste of archil, a lichen dye)*
Persischrot *n* Persian red, chrom[at]e red, Austrian cinnabar *(a basic lead chromate)*
persistent persistent
p. sein to persist
Persistenz *f* persistence
Persoz-Reagens *n (text)* Persoz's reagent *(for detecting silk in presence of wool)*
Persubstitution *f* per substitution
Persulfatoxydation *f* / **Elbssche** Elbs persulphate oxidation *(of phenols)*
Pertechnetat *n* M^ITcO_4 pertechnate
Pertechnetiumsäure *f* $HTcO_4$ pertechnetic acid
Perthiokarbonat *n* $M_2^ICS_4$ perthiocarbonate
Perthiokohlensäure *f* H_2CS_4 perthiocarbonic acid
Perthit *m (min)* perthite *(any of a group of tectosilicates)*
Perubalsam *m* Peru (Peruvian, Indian, black) balsam, China oil *(from Myroxylon balsamum (L.) Harms var. pereirae)*
Perverbindung *f* per compound
Perylen *n* perylene *(a polycyclic hydrocarbon)*
Perylenchinon *n* perylenequinone
Perylenkarbonsäure *f* perylenecarboxylic acid
Perylenringsystem *n* perylene ring system
Pestizid *n* pesticide, pest control agent
Pestizidrückstand *m* pesticide residue
Petalit *m (min)* petalite *(a lithium aluminium silicate)*
Petersen-Verfahren *n* Petersen process *(for manufacturing sulphuric acid)*
Petersilienkampfer *m* parsley camphor, apiol, 2,5-dimethoxy-3,4-methylenedioxy-1-allylbenzene
Petersilienöl *n* parsley oil *(from Petroselinum crispum (Mill.) Nym.)*
PETN *s.* Pentaerythrit-tetranitrat
PETP *s.* Polyäthylenterephthalat
Petrifikation *f (geol)* petrifaction
petrifizieren *(geol)* to petrify
Petrischale *f* Petri [culture] dish
Petrischalenbüchse *f* Petri dish box
Petri-Unterschale *f* Petri dish bottom
Petrochemie *f* 1. petrochemistry, chemistry of rocks; 2. *(incorrectly for)* Petrolchemie
Petrochemikalie *f (incorrectly for)* Petrolchemikalie
petrochemisch 1. petrochemical *(relating to the chemistry of rocks)*; 2. *(incorrectly for)* petrolchemisch
petrogen petrogenic, rock-forming
Petrogenese *f*, **Petrogenesis** *f (geol)* petrogenesis
petrogenetisch petrogen[et]ic *(relating to the formation of rocks)*
Petrografie *f* petrography
petrografisch petrographic[al]
Petrol *n s.* Petroleum
Petrolasphalt *m s.* Petroleumasphalt
Petrolat *n s.* Petrolatum
Petroläther *m* petroleum ether, ligroin[e], light petroleum (ligroin), benzin[e] *(with a boiling point range of 40 to 70°C)*
Petrolatum *n* petrolatum, petroleum (mineral) jelly, mineral fat
Petrolchemie *f* petrochemistry
Petrolchemikalie *f* petrochemical
petrolchemisch petrochemical, petroleum-chemical
Petroleum *n* kerosene, kerosine, paraffin oil
Petroleumasphalt *m* petroleum asphalt
Petroleumäther *m s.* Petroläther
Petroleumbenzin *n* petroleum benzin
Petroleumemulsion *f* petroleum wash, white-oil spray *(for pest control)*
Petroleumfraktion *f* kerosene fraction
Petroleumgefäß *n* oil cup *(of a flash-point tester)*
Petrolfraktion *f* kerosene fraction
Petrolharz *n* petroleum resin
Petrolkoks *m* petroleum (still) coke
Petrolnaphtha *n(f)* petroleum naphtha
Petrologie *f* petrology
petrologisch petrologic[al]
Petrolpech *n* petroleum pitch
Petroselinsäure *f* petroselic (petroselinic) acid, *cis*-6-octadecenoic acid
Pe-tun-tse *m (ceram)* petun[t]se, petun[t]ze, china stone
Petzit *m (min)* petzite *(gold silver telluride)*
Pfaff *m* punch, *(plast also)* hob

Pfahlrohr *n (pap)* scriptural reed, Arundo donax cane
Pfanne *f* pan; ladle *(as for conveying molten metal)*; plate *(of a balance)*
Pfannenamalgamation *f*, **Pfannenamalgamierung** *f* pan amalgamation *(of noble metals)*
Pfannenbär *m (met)* skull
Pfannenfutter *n* ladle lining
Pfannenkristallisator *m* tank crystallizer
PFC *s.* Polychlortrifluoräthylen
Pfeffer *m* 1. pepper, Piper L., *(specif)* black pepper, Piper nigrum L.; 2. pepper *(seeds from 1.)*
  Echter P. black pepper, Piper nigrum L.
  Langer P. long pepper *(fruit from Piper longum L.)*
  Schwarzer P. black pepper *(the shrub Piper nigrum L. or its seeds)*
  Spanischer P. hot pepper, chili, chilli *(fruit from Capsicum specc.)*
Pfefferöl *n* [black] pepper oil *(from Piper nigrum L.)*
Pfefferstrauch *m* / **Schwarzer** black pepper, Piper nigrum L.
Pfefferwurzel *f (pharm)* kava, cava, kawa, kavakava *(from Piper methysticum G. Forst.)*
Pfeife *f (glass)* blowpipe, blow[ing] iron, *(in the tube-drawing process also)* mandrel
Pfeifenende *n*, **Pfeifenkopf** *m (glass)* nose-piece, nose
Pfeifenton *m* pipeclay
Pfeilgift *n* arrow poison
Pfeilwurzelmehl *n* arrowroot *(from Maranta specc.)*
PF-Harz *n s.* Phenolformaldehydharz
Pfirsichkernöl *n* peach-kernel oil
Pflanze *f* plant
  gerbstoffhaltige P. tanniferous (tanning) plant
  honigende P. nectariferous plant
  kautschukführende (kautschukhaltige) P. rubber-bearing (rubber-producing, rubber) plant
  kautschukliefernde P. rubber-yielding plant
  stärkeliefernde P. starch-yielding plant
Pflanzenauszug *m* plant extract
Pflanzenbiochemie *f* plant biochemistry
Pflanzenblindwert *m (tox)* plant blank value
Pflanzenbutter *f* vegetable butter
Pflanzenernährung *f* plant (crop) nutrition
Pflanzenextrakt *m* plant extract
Pflanzenfarbstoff *m* plant pigment
Pflanzenfaser *f* vegetable fibre
Pflanzenfett *n* vegetable (plant) fat
Pflanzengift *n* plant poison
Pflanzengummi *n* [plant, tree, vegetable] gum
Pflanzenindigo *m(n)* natural indigo
Pflanzenindikan *n* indican [of plants], indoxyl-β-glucoside
Pflanzenleim *m* vegetable glue (adhesive)
Pflanzenlezithin *n* vegetable lecithin
Pflanzenmargarine *f* vegetable margarine
Pflanzennährstoff *m* plant nutrient
Pflanzenöl *n* vegetable oil
Pflanzenpreßpapier *n* herbarium paper
Pflanzenresistenz *f* plant resistance
Pflanzensaft *m* [plant] juice; *(bot)* sap; *(pharm)* succus
pflanzenschädigend phytotoxic
Pflanzenschleim *m* mucilage
Pflanzenschutz *m* crop protection
Pflanzenschutzgerät *n (agric)* crop protection apparatus, applicator, application apparatus
Pflanzenschutzmittel *n* agricultural pesticide (control chemical), economic poison, crop protection chemical, plant protection product
  flüssiges P. plant spray
  nicht persistentes P. non-persistent pesticide
  protektives (vorbeugend wirkendes) P. plant protective, protective toxicant
Pflanzenschutzpräparat *n s.* Pflanzenschutzmittel
Pflanzensterin *n* phytosterol, plant sterol
Pflanzentalg *m* vegetable tallow
pflanzentötend phytocidal
pflanzenverfügbar *(agric)* available *(nutrients)*
Pflanzenverfügbarkeit *f (agric)* availability *(of nutrients)*
pflanzenverträglich non-phytotoxic
pflanzlich vegetable
Pflanzungskautschuk *m* plantation (estate) rubber
Pflatschen *n (text)* slop-padding
pflegeleicht *(text)* easy-care, minicare
Pflegeleicht-Ausrüstung *f (text)* easy-care finish
Pflegeleichtigkeit *f (text)* ease of care, easy care
Pflegemittel *n* / **kosmetisches** treatment cosmetic (preparation)
Pflugschararm *m* plough bar *(of a dryer)*
Pfropf *m s.* Pfropfen
Pfropfelastomer[es] *n* graft elastomer
pfropfen *(plast)* to graft
Pfropfen *m* stopper, plug, cork
  kalter P. *(plast)* cold slug
Pfropfenbildung *f (plast)* plug-up *(a defect in injection moulding)*
Pfropfgrad *m (plast)* grafting degree
Pfropfkopolymer[es] *n*, **Pfropfkopolymerisat** *n* graft copolymer
Pfropfpolymer[es] *n* graft polymer
  P. von Naturkautschuk natural-rubber graft [polymer]
Pfropfpolymerisat *n s.* Pfropfpolymeres
Pfropfpolymerisation *f* graft polymerization
Pfropfreaktion *f (plast)* grafting reaction
Pfropfung *f (plast)* grafting
PFT *s.* Polytetrafluoräthylenfaserstoff
Pfund-Serie *f* Pfund series *(of the hydrogen spectrum)*
PH *s.* Polyharnstoffaserstoff
pH-Abfall *m*, **pH-Abnahme** *f* pH decrease (drop)
Phage *m* [bacterio]phage
Phakolith *m (min)* phacolite *(a tectosilicate)*
pH-Änderung *f* pH change
phanerokristallin phanerocrystalline
Phänomen *n* phenomenon
  P. der Liesegangschen Ringe Liesegang phenomenon *(formation of ring-shaped precipitates in gels)*
  Purkinjesches P. *(phot)* Purkinje effect *(shifting in the spectral sensitivity of emulsions)*
pH-Anstieg *m* pH increase

Phantombild *n (phot)* ghost image
pH-Anzeigegerät *n* pH indicator
Pharmakochemie *f* pharmaceutic[al] chemistry
Pharmakochemiker *m* pharmaceutic[al] chemist
Pharmakodynamie *f*, **Pharmakodynamik** *f* pharmacodynamics
pharmakodynamisch pharmacodynamic
Pharmakognosie *f* pharmacognosy, pharmacognosia
pharmakognostisch pharmacognostic
Pharmakolith *m (min)* pharmacolite *(calcium hydrogenorthoarsenate)*
Pharmakologie *f* pharmacology, pharmacologia
pharmakologisch pharmacologic[al]
Pharmakon *n* pharmacon
  psychotropes P. psychotropic agent
Pharmakopöe *f* pharmacopoeia
Pharmakosiderit *m (min)* pharmacosiderite *(a hydrous basic iron(III) arsenate)*
Pharmakotherapeutik *f* pharmacotherapeutics
pharmakotherapeutisch pharmacotherapeutic[al]
Pharmakotherapie *f* pharmacotherapy
Pharmazeut *m* pharmac[eut]ist
Pharmazeutik *f* pharmaceutics
Pharmazeutikum *n* pharmaceutic[al]
pharmazeutisch pharmaceutic[al]
Pharmazie *f* pharmacy
Phase *f* phase, stage ✦ **in homogener P.** homogeneously
  amorphe P. amorphous phase
  bewegliche P. *s.* mobile P.
  cholesterische P. cholesteric phase
  dampfförmige P. vapour phase
  dispergierte (disperse) P. disperse (dispersed, discontinuous, internal) phase
  feste P. solid phase; solid *(in zone melting)*
  flüssige P. liquid phase; liquid *(in zone melting)*
  gasförmige P. gas phase
  gemischte P. mixed phase
  geochemische (geologische) P. geochemical stage *(of coalification)*
  geschlossene P. continuous phase
  glasige P. glassy (vitreous) phase
  Hume-Rotherysche P. *(cryst)* Hume-Rothery phase
  innere P. *s.* disperse P.
  intermetallische P. intermetallic (metal) compound, intermediate constituent (phase)
  kristalline P. crystalline phase
  metallische intermediäre P. *s.* intermetallische P.
  mobile P. mobile (moving) phase *(chromatography)*
  nematische P. *(coll)* nematic phase
  offene P. *s.* disperse P.
  smektische P. *(coll)* smectic phase
  stationäre (unbewegliche) P. stationary (immobile, non-mobile) phase *(chromatography)*
  wäßrige P. aqueous (water) phase
  zusammenhängende P. continuous phase
Phasenänderung *f* phase change (transition)
Phasenbeziehung *f* phase relation
Phasendiagramm *n* phase (equilibrium) diagram
Phasendifferenz *f* phase difference
Phasendrehung *f* phase shift
Phasengeschwindigkeit *f* phase velocity
Phasengesetz *n s.* Phasenregel
Phasengleichgewicht *n* phase equilibrium
Phasengrenze *f* phase boundary
  P. fest-flüssig solid-liquid boundary
Phasengrenzfläche *f* interface, boundary (bounding) surface, junction
Phasengrenzpotential *n* phase-boundary potential
Phasenkontrastmikroskop *n* phase-contrast microscope
Phasenregel *f* **[von Gibbs]** phase rule [of Gibbs]
Phasenschiebung *f* phase shift
Phasenstabilität *f* phase stability
Phasentrennung *f* phase separation
Phasenübergang *m s.* Phasenumwandlung
Phasenumkehr *f* phase inversion, reversal of phases
Phasenumwandlung *f* phase transition (change)
  P. erster Ordnung first-order transition
  P. zweiter Ordnung second-order transition
Phasenunterschied *m* phase difference
Phasenverschiebung *f* phase shift
Phasenwinkel *m* phase angle
pH-Bereich *m* pH range
pH-Bestimmung *f* pH determination
pH-Bestimmungsapparat *m s.* pH-Meßgerät
pH-Einheit *f* pH unit
Phellandren *n* phellandrene *(a monocyclic terpene)*
Phellem *n (bot)* phellem, cork
Phellonsäure *f* $HOCH_2[CH_2]_{20}COOH$ phellonic acid, 22-hydroxydocosanoic acid
Phenakit *m (min)* phenakite, phenacite *(beryllium orthosilicate)*
Phenanthren *n* phenanthrene
Phenanthrenchinon *n* phenanthraquinone, *(specif)* 9,10-phenanthraquinone
Phenanthrenringsystem *n* phenanthrene ring system
Phenanthrochinolizidin-Alkaloid *n* phenanthroquinolizidine alkaloid
Phenanthroindolizidin-Alkaloid *n* phenanthroindolizidine alkaloid
Phenanthrol *n* phenanthrol, hydroxyphenanthrene
Phenarsazinchlorid *n (tox)* 10-chloro-5,10-dihydrophenarsazine, phenarsazine chloride
Phenarsazinsäure *f* phenarsazinic acid
Phenat *n* phenate, phenolate, phenoxide
Phenäthylalkohol *m* $C_6H_5CH_2CH_2OH$ phenethyl alcohol, 2-phenylethanol
Phenäthylamin *n* phenethylamine, phenylethylamine, aminoethylbenzene
Phenäthylgruppe *f* $C_6H_5CH_2CH_2-$ phenethyl group
Phenazarsinsäure *f s.* Phenarsazinsäure
Phenazetaminoessigsäure *f s.* Phenazetursäure
Phenazetin *n* phenacetin, acet-*p*-phenetidide
Phenazetursäure *f* $C_6H_5CH_2-CO-NHCH_2COOH$ phenaceturic acid, phenylacetylglycine
Phenazetylglykokoll *n*, **Phenazetylglyzin** *n s.* Phenazetursäure
Phenazetylharnstoff *m* $C_6H_5CH_2-CO-NHCONH_2$ phenacetylurea

Phenazin *n* phenazine, azophenylene
Phenazon *n* phenazone, antipyrine
Phenazozin *n* phenazocine *(a benzomorphan derivative)*
Phenazylalkohol *m* $C_6H_5COCH_2OH$ phenacyl alcohol, α-hydroxyacetophenone
Phenazylchlorid *n* $C_6H_5COCH_2Cl$ phenacyl chloride, α-chloroacetophenone
Phenazylgruppe *f* $C_6H_5COCH_2-$ phenacyl group
Phenazylidengruppe *f* $C_6H_5COCH=$ phenacylidene group
Phenetidin *n* $C_6H_4(NH_2)OC_2H_5$ phenetidine, aminophenol ethyl ether
Phenetol *n* $C_2H_5OC_6H_5$ phenetol, ethoxybenzene, ethyl phenyl ether
Phengit *m (min)* phengite *(a phyllosilicate)*
Phenindion *n* phenindione, 2-phenyl-1,3-indanedione
Phenmetrazin *n* phenmetrazine, tetrahydro-3-methyl-2-phenyl-1,4-oxazine
Phenol *n* phenol, *(specif)* C_6H_5OH phenol, benzophenol, hydroxybenzene
einwertiges P. monohydric phenol
dreiwertiges P. trihydric phenol
mehrwertiges P. polyhydric phenol
verflüssigtes P. *(pharm)* liquefied phenol
zweiwertiges P. dihydric phenol
Phenolabscheider *m* dephenolizer
Phenolabwasser *n* phenolic wastes
Phenolaldehyd *m* phenolic aldehyde
Phenolalkohol *m* phenolic alcohol
Phenolase *f s.* Phenoloxydase
Phenolat *n* phenolate, phenate, phenoxide
Phenoläther *m* phenolic ether
Phenolatverfahren *n* phenolate process *(for removing H_2S from gas)*
Phenolester *m* phenolic ester
Phenolformaldehyd *m (plast)* phenolformaldehyde, PF
Phenolformaldehydharz *n* phenolformaldehyde resin
Phenolformaldehydkondensation *f* phenolformaldehyde condensation
phenolfrei free from phenol
Phenolfurfuralharz *n* phenol-furfural resin
Phenolgeschmack *m* phenolic taste
Phenolharz *n* phenolic resin
P. im A-Zustand A-stage resin, resol
P. im B-Zustand B-stage resin, resitol
P. im C-Zustand C-stage resin, resite
Phenolharzkitt *m* phenolic cement
Phenolharzklebstoff *m* phenolic adhesive
Phenolharzkunststoff *m* phenoplast, phenolic plastic
Phenolharzlack *m* phenolic varnish
Phenolharzlaminat *n* phenolic laminate
Phenolharzpreßmasse *f* phenolic moulding compound
Phenolharzschaum[stoff] *m* phenolic[-resin] foam
Phenolhydroxyl *n* phenolic hydroxyl
phenolisch phenolic
phenolisieren to phenolate, to phenolize
Phenolkalium *n s.* Kaliumphenolat
Phenolkarbonsäure *f* phenolic acid *(any of a group of aromatic hydroxycarboxylic acids)*
Phenolkern *m* phenol nucleus
Phenolkitt *m s.* Phenolharzkitt
Phenolkleber *m s.* Phenolharzklebstoff
Phenolkoeffizient *m* phenol coefficient *(of disinfectants)*
2-Phenolmethylal *n s.* Salizylaldehyd
Phenolnatrium *n s.* Natriumphenolat
Phenoloxydase *f* phenol oxidase, phenolase, tyrosinase
Phenolphthalein *n* phenolphthalein
Phenolreaktion *f* / **Liebermannsche** Liebermann reaction for phenols
Phenolring *m* phenol ring
Phenolrot *n* phenol red *(phenolsulphonphthalein)*
Phenolschaum *m s.* Phenolharzschaum
Phenolsulfonphthalein *n* phenolsulphonphthalein
Phenolsulfonsäure *f* $C_6H_4(OH)SO_3H$ phenolsulphonic acid
Phenonium-Ion *n* phenonium ion
Phenoplast *m* phenoplast, phenolic plastic
Phenoxid *n* phenoxide, phenolate, phenate
Phenoxyessigsäure *f* $C_6H_5O-CH_2COOH$ phenoxyacetic acid
Phenoxygruppe *f* phenoxy group (residue)
Phenoxyharz *n* phenoxy resin
Phenoxyrest *m s.* Phenoxygruppe
Phenylakrylsäure *f* phenylacrylic acid
α-Phenylalanin *n* $CH_3C(NH_2)(C_6H_5)COOH$ α-phenylalanine, 2-amino-2-phenylpropionic acid
β-Phenylalanin *n* $C_6H_5CH_2CH(NH_2)COOH$ β-phenylalanine, 1-amino-2-phenylpropionic acid
Phenylalkan *n* phenylalkane
1-Phenylallylalkohol *m* $C_6H_5CH(OH)CH=CH_2$ 1-phenylallyl alcohol
Phenylamin *n* $C_6H_5NH_2$ phenylamine, aminobenzene, aniline
Phenyläthan *n* $C_6H_5C_2H_5$ phenylethane, ethylbenzene
1-Phenyläthanol *n* $C_6H_5CH(OH)CH_3$ 1-phenylethanol
2-Phenyläthanol *n* $C_6H_5CH_2CH_2OH$ 2-phenylethanol, phenethyl alcohol
Phenyläther *m s.* Diphenyläther
α-Phenyläthylalkohol *m s.* 1-Phenyläthanol
β-Phenyläthylalkohol *m s.* 2-Phenyläthanol
Phenyläthylamin *n* aminoethylbenzene, phenylethylamine
Phenyläthylen *n* $C_6H_5CH=CH_2$ phenylethylene, styrene
***N*-Phenylazetamid** *n* $C_6H_5NHCOCH_3$ *N*-phenylacetamide, acetanilide
Phenylazetat *n* $CH_3COOC_6H_5$ phenyl acetate
Phenylazetonitril *n* $C_6H_5CH_2CN$ phenylacetonitrile, benzyl cyanide
***N*-Phenylazetursäure** *f* $C_6H_5N(COCH_3)CH_2COOH$ *N*-phenylaceturic acid, *N*-acetylphenylglycine
Phenylazetylen *n* $C_6H_5C{\equiv}CH$ phenylacetylene, acetylenylbenzene, ethynylbenzene
Phenylazetylharnstoff *m* $C_6H_5CH_2CONHCONH_2$ phenacetylurea

Phenylbenzoat *n* $C_6H_5COOC_6H_5$ phenyl benzoate
Phenylbenzol *n* $C_6H_5C_6H_5$ phenylbenzene, biphenyl
Phenylbenzylkarbinol *n s.* 1,2-Diphenyläthanol
Phenylbernsteinsäure *f* $HOOCCH(C_6H_5)CH_2-COOH$ phenylsuccinic acid
Phenylboronsäure *f s.* Phenylborsäure
Phenylborsäure *f* $C_6H_5B(OH)_2$ benzeneboronic acid, phenylboric acid
Phenylbrenztraubensäure *f* $C_6H_5CH_2COCOOH$ phenylpyruvic acid
Phenylbromid *n* C_6H_5Br bromobenzene, phenyl bromide
Phenylbutansäure *f s.* Phenylbuttersäure
Phenylbuttersäure *f* phenylbutyric acid
Phenylchlorid *n* C_6H_5Cl chlorobenzene, phenyl chloride
Phenylchloroform *n* $C_6H_5CCl_3$ α,α,α-trichlorotoluene, phenylchloroform
Phenylchlorsilan *n* phenylchlorosilane
Phenyldisulfid *n s.* Diphenyldisulfid
Phenyldithiobenzol *n s.* Diphenyldisulfid
Phenylenblau *n* phenylene blue *(an indamine dye)*
Phenylendiamin *n* $C_6H_4(NH_2)_2$ phenylenediamine, diaminobenzene
Phenylengruppe *f*, **Phenylenrest** *m* $-C_6H_4-$ phenylene group (residue)
Phenylessigsäure *f* $C_6H_5CH_2COOH$ phenylacetic acid, α-toluic acid
Phenylessigsäureäthylester *m* $C_6H_5CH_2CO-OC_2H_5$ ethyl phenylacetate
Phenylessigsäurenitril *n s.* Phenylazetonitril
Phenylester *m* phenyl ester
Phenylglykokoll *n s.* Phenylglyzin
Phenylglykolsäure *f* $C_6H_5CH(OH)COOH$ phenylglycollic acid, mandelic acid, 2-hydroxy-2-phenylacetic acid
***O*-Phenylglykolsäure** *f* $C_6H_5OCH_2COOH$ glycollic acid phenyl ether, phenoxyacetic acid
Phenylglyoxylsäure *f* $C_6H_5COCOOH$ phenylglyoxylic acid, benzoylformic acid
Phenylglyzin *n* phenylglycine, *(specif)* $C_6H_5NHCH_2COOH$ *N*-phenylglycine, anilinoacetic acid
Phenylgruppe *f* C_6H_5- phenyl group (residue)
Phenylharnstoff *m* $C_6H_5NHCONH_2$ phenylurea, phenylcarbamide
2-Phenylhydrakrylsäure *f* $HOCH_2CH(C_6H_5)COOH$ 2-phenylacrylic acid, tropic acid, 3-hydroxy-2-phenylpropionic acid
Phenylhydrazin *n* $C_6H_5NHNH_2$ phenylhydrazine
Phenylhydrazon *n* phenylhydrazone *(any of several phenylhydrazine derivatives $C_6H_5NHN=CR_1R_2$)*
phenylieren to phenylate
Phenylierung *f* phenylation
Phenyl-I-Säure *f s.* Phenyl-J-Säure
Phenylisozyanid *n* C_6H_5NC phenylisocyanide, phenylcarbylamine
Phenyljodid *n* C_6H_5I iodobenzene, phenyl iodide
Phenyl-J-Säure *f* $HOC_{10}H_5(NHC_6H_5)SO_3H$ phenyl-J acid, 7-anilino-4-hydroxynaphthalene-2-sulphonic acid
Phenylkarbamid *n* $C_6H_5NHCONH_2$ phenylcarbamide, phenylurea
Phenylkarbamidsäure *f* $C_6H_5NHCOOH$ phenylcarbamic acid, carbanilic acid
Phenylkarbinol *n s.* Benzylalkohol
Phenylkarbylamin *n* C_6H_5NC phenylcarbylamine, phenyl isocyanide
Phenylketon *n s.* Diphenylketon
Phenylketonurie *f (med)* phenylketonuria
Phenylmagnesiumbromid *n* C_6H_5MgBr phenylmagnesium bromide
Phenylmethan *n* $C_6H_5CH_3$ phenylmethane, methylbenzene, toluene
2-Phenylmilchsäure *f* $CH_3C(C_6H_5)(OH)COOH$ 2-phenyl-lactic acid, atrolactinic acid, 2-hydroxy-2-phenylpropionic acid
Phenylperisäure *f* $C_6H_5NHC_{10}H_6SO_3H$ phenyl-peri acid, *N*-phenyl-1-naphthylamine-8-sulphonic acid
Phenylphenazoniumsalz *n* phenylphenazonium salt
Phenylphosphinsäure *f* $C_6H_5PO_2H_2$ phenylphosphinic acid, benzenephosphinic acid
Phenylphosphonsäure *f* $C_6H_5PO(OH)_2$ phenylphosphonic acid, benzenephosphonic acid
2-Phenylpropansäure *f s.* 2-Phenylpropionsäure
Phenylpropinsäure *f s.* Phenylpropiolsäure
Phenylpropiolsäure *f* $C_6H_5C\equiv C-COOH$ phenylpropiolic acid, 3-phenyl-2-propynoic acid
2-Phenylpropionsäure *f* $CH_3CH(C_6H_5)COOH$ 2-phenylpropionic acid, hydratropic acid
Phenylquecksilberazetat *n* $CH_3COOHgC_6H_5$ phenylmercuric acetate, PMA *(a pesticide)*
Phenylquecksilberharnstoff *m* $C_6H_5HgNHCONH_2$ phenylmercury urea
Phenylradikal *n* phenyl radical, *(specif)* phenyl free radical
Phenylrest *m s.* Phenylgruppe
Phenylsalizylat *n* $C_6H_4(OH)CO-OC_6H_5$ phenyl salicylate, salol
Phenyl-γ-Säure *f* $HOC_{10}H_5(NHC_6H_5)(SO_3H)$ phenyl-gamma acid, 6-anilino-4-hydroxynaphthalene-2-sulphonic acid
Phenylschwefelsäure *f* $C_6H_5OSO_3H$ phenylsulphuric acid
Phenylsilikon *n* phenyl silicone
Phenylsilikonflüssigkeit *f*, **Phenylsilikonöl** *n* phenyl silicone fluid
Phenylsulfamidsäure *f*, **Phenylsulfaminsäure** *f* $C_6H_5NHSO_3H$ phenylsulphamic acid
Phenylsulfonsäure *f s.* Benzolsulfonsäure
Pheromon *n (bioch)* pheromone
Pheron *n* pheron *(colloid carrier of an enzyme)*
pH-Gebiet *n* pH range
Phillipsit *m (min)* phillipsite *(a tectosilicate)*
Phillips-Verfahren *n* Phillips process *(for producing high-density polyethylene)*
philodien dienophilic
Philodien *n* dienophile
Philosophenwolle *f s.* Zinkblumen
pH-Indikator *m* pH indicator
pH-Intervall *n* pH interval
pH-Kontrolle *f* pH control
Phlegma *n (distil)* less volatile component

phlegmatisieren to desensitize *(an explosive)*
Phlegmatisierung *f* desensitization *(of an explosive)*
Phlobaphen *n* phlobaphene, tanner's red
Phlogopit *m (min)* phlogopite *(a phyllosilicate)*
Phloridzin *n* phlori[d]zin *(β-glucoside of phloretin)*
Phlorogluzin *n* $C_6H_3(OH)_3$ phloroglucinol, 1,3,5-trihydroxybenzene
Phlorogluzinkarbonsäure *f* $(HO)_3C_6H_2COOH$ phloroglucinol-carboxylic acid, 2,4,6-trihydroxybenzoic acid
Phlorrhizin *n s.* Phloridzin
pH-Meßgerät *n* pH meter (instrument, determination apparatus)
pH-Meßsystem *n* pH-measuring system
pH-Messung *f* pH measurement
pH-Meter *n s.* pH-Meßgerät
Phönikochroit *m (min)* phoenicochroite *(a basic lead chromate)*
Phönizin *n* phoenicin[e], 2,2′-dihydroxy-4,4′-ditoluquinone
Phoron *n* $(CH_3)_2C{=}CHCOCH{=}C(CH_3)_2$ phorone, 2,6-dimethyl-2,5-heptadien-4-one
Phosgen *n* $COCl_2$ phosgene, carbonyl chloride, carbon dichloride oxide
phosgenieren to phosgenate
Phosgenierung *f* phosgenation
Phosgenit *m (min)* phosgenite *(lead carbonate chloride)*
Phosphat *n* phosphate; *(specif) s.* Orthophosphat
  neutrales P. neutral (normal) phosphate
  primäres P. *s.* Dihydrogenorthophosphat
  sekundäres P. *s.* Hydrogenorthophosphat
  tertiäres P. *s.* Orthophosphat / neutrales
Phosphatase *f* phosphatase
  alkalische P. alkaline phosphatase
  saure P. acid phosphatase
Phosphataseprobe *f (food)* phosphatase test
Phosphatdüngemittel *n* phosphate fertilizer
  stickstoffhaltiges P. nitrogen-phosphorus fertilizer
Phosphatentschwefelungsverfahren *n (petrol)* phosphate process *(esp for desulphurizing natural gas)*
Phosphaterz *n* phosphate ore (rock)
phosphatgepuffert phosphate-buffered
Phosphatglas *n* phosphate glass
Phosphatid *n* phosphatide, phospholipid[e]
Phosphatidsäure *f* phosphatidic acid
phosphatieren to phosphate, to phosphatize, to coslettize *(for preventing corrosion)*
Phosphatierung *f* phosphation, phosphating, phosphatizing, coslettizing *(anticorrosive treatment)*
Phosphatimpfverfahren *n* threshold treatment *(for softening water)*
Phosphatlöslichkeit *f (agric)* phosphate solubility
Phosphatpuffer *m* phosphate buffer
Phosphatquelle *f (agric)* phosphate source
Phosphatträger *m (agric)* phosphate carrier
Phosphatverfahren *n s.* Phosphatentschwefelungsverfahren
Phosphatweichmacher *m* phosphate plasticizer
Phosphensäure *f* $HOPO_2$ phosphenic acid; *(org chem)* $R_1R_2P(OH)$ phosphinous acid
Phosphid *n* phosphide
Phosphin *n* phosphine, phosphorus hydride; *(specif) s.* Monophosphin
Phosphinigsäure *f* 1. RPHOH *or* $RPH_2{=}O$ phosphinous acid *(IUPAC nomenclature)*; 2. *(formerly in Beilstein nomenclature for)* Phosphinsäure
Phosphinsäure *f* RPH=O(OH) phosphinic acid, *(specif)* HPH_2O_2 phosphinic acid, hypophosphorous acid
Phosphit *n* 1. phosphite, *(specif)* $M_2{}^IPHO_3$ orthophosphite, *(better)* phosphonate; 2. phosphite *(an ester of the hypothetical acid $P(OH)_3$)*
Phosphoglyzerinsäure *f* phosphoglyceric acid
Phosphokinase *f* phosphokinase
Phospholipase *f* phospholipase, lecithinase
Phospholipid *n*, **Phospholipoid** *n* phospholipid[e], phosphatide
Phosphonat *n* phosphonate
Phosphonigsäure *f* $RP(OH)_2$ phosphonous acid
Phosphoniumbase *f* phosphonium base
Phosphoniumbromid *n* $[PH_4]Br$ phosphonium bromide
Phosphoniumchlorid *n* $[PH_4]Cl$ phosphonium chloride
Phosphoniumion *n* $[PH_4]^+$ phosphonium ion
Phosphoniumjodid *n* $[PH_4]I$ phosphonium iodide
Phosphoniumsulfat *n* $[PH_4]_2SO_4$ phosphonium sulphate
Phosphonsäure *f* $RP{=}O(OH)_2$ phosphonic acid, *(specif)* H_2PHO_3 phosphonic acid
Phosphonsäureester *m* phosphonate
Phosphoproteid *n* phosphoprotein
Phosphor *m* 1. P phosphorus; 2. phosphor *(a substance which exhibits phosphorescence)*
  amorpher P. amorphous phosphorus
  farbloser (gelber) P. *s.* weißer P.
  hexagonal kristallisierter weißer P. β-white phosphorus
  Hittorfscher P. *s.* violetter P.
  kubisch kristallisierter weißer P. α-white phosphorus
  radioaktiver P. radioactive phosphorus, radiophosphorus, *(specif)* ^{32}P phosphorus-32
  roter P. red (amorphous) phosphorus
  schwarzer P. black (β-metallic) phosphorus, phosphorus IV
  violetter P. violet (α-metallic) phosphorus, phosphorus III
  weißer P. white (yellow) phosphorus, tetraphosphorus
Phosphor-32 *m* ^{32}P phosphorus-32
Phosphorandisäure *f* $H_3P(OH)_2$ phosphoranedioic acid
Phosphoranpentasäure *f* $P(HO)_5$ phosphoranepentoic acid
Phosphoransäure *f* H_4POH phosphoranoic acid
Phosphorantetrasäure *f* $HP(OH)_4$ phosphoranetetroic acid
Phosphorantrisäure *f* $H_2P(OH)_3$ phosphoranetrioic acid

phosphorarm poor in phosphorus, low-phosphorus
Phosphorbedarf *m (agric)* phosphorus needs
Phosphor(III)-bromid *n* PBr_3 phosphorus(III) bromide, phosphorus tribromide
Phosphor(V)-bromid *n* PBr_5 phosphorus(V) bromide, phosphorus pentabromide
Phosphorbronze *f* phosphor bronze
Phosphor(III)-chlorid *n* PCl_3 phosphorus(III) chloride, phosphorus trichloride
Phosphor(V)-chlorid *n* PCl_5 phosphorus(V) chloride, phosphorus pentachloride
Phosphor(V)-dibromidtrichlorid *n* PBr_2Cl_3 phosphorus dibromide trichloride
Phosphor(V)-dibromidtrifluorid *n* PBr_2F_3 phosphorus dibromide trifluoride
Phosphordichlorid *n* PCl_2 phosphorus dichloride
Phosphor(V)-dichloridtrifluorid *n* PCl_2F_3 phosphorus dichloride trifluoride
Phosphoreszenz *f* phosphorescence
Phosphoreszenzfarbe *f* phosphorescent paint
Phosphoreszenzspektrum *n* phosphorescence spectrum
phosphoreszieren to phosphoresce
phosphoreszierend phosphorescent, phosphoric, phosphorous
Phosphorfestlegung *f (agric)* phosphorus fixation
Phosphor(III)-fluorid *n* PF_3 phosphorus(III) fluoride, phosphorus trifluoride
Phosphor(V)-fluorid *n* PF_5 phosphorus(V) fluoride, phosphorus pentafluoride
Phosphorgehalt *m* phosphorus content ✦ **mit hohem P.** rich in phosphorus, high-phosphorus ✦ **mit niedrigem P.** poor in phosphorus, low-phosphorus
Phosphorhalogenid *n* phosphorus halide
phosphorhaltig containing phosphorus, phosphorus-bearing
Phosphorheptasulfid *n* P_4S_7 phosphorus heptasulphide, tetraphosphorus heptasulphide
Phosphor(III)-hydrid *n* PH_3 phosphorus trihydride, monophosphine, phosphine
Phosphorigsäure-alkylester-alkylimid *n* (RO)P=NR alkoxyalkyliminophosphine, alkyl *N*-alkylphosphenimidite, *(Kosolapoff's nomenclature)* alkylalkylimidophosphite
Phosphorigsäureester *m* phosphite
Phosphorinsektizid *n* / **organisches** O–P insecticide
Phosphorit *m* phosphorite, phosphate rock
Phosphor(III)-jodid *n* PI_3 phosphorus(III) iodide, phosphorus triiodide
Phosphormonoselenid *n* P_2Se phosphorus monoselenide
Phosphornekrose *f (med)* phossy jaw
Phosphornitrid *n* P_3N_5 phosphorus nitride, triphosphorus pentanitride
Phosphornitriddibromid *n* PBr_2N phosphorus dibromide nitride
Phosphornitriddichlorid *n* $(PCl_2N)_n$ phosphorus dichloride nitride
Phosphorochalzit *m (min)* phosphorochalcite, pseudomalachite, dihydrite *(a hydrous basic copper phosphate)*
Phosphorofen *m* phosphorus (phosphate) furnace *(for manufacturing phosphoric acid)*
Phosphorogen *n* phosphorogen *(a substance which produces or induces phosphorescence)*
Phosphororganikum *n* / **toxisches** *(agric)* O–P toxicant
Phosphor(III)-oxid *n* P_2O_3 phosphorus(III) oxide, phosphorus trioxide, diphosphorus trioxide
Phosphor(V)-oxid *n* P_2O_5 phosphorus(V) oxide, phosphorus pentaoxide, diphosphorus pentaoxide
Phosphor(V)-oxidbromid *n s.* Phosphorylbromid
Phosphor(V)-oxidchlorid *n s.* Phosphorylchlorid
Phosphor(V)-oxidfluorid *n s.* Phosphorylfluorid
Phosphoroxidtriamid *n s.* Phosphoryltriamid
Phosphorpenta..., **Phosphorpent...** *s.a.* Phosphor(V)-...
Phosphorproteid *n* phosphoprotein
phosphorreich rich in phosphorus, high-phosphorus
Phosphor(III)-rhodanid *n s.* Phosphor(III)-thiozyanat
Phosphorsalz *n* $Na(NH_4)HPO_4 \cdot 4H_2O$ phosphorus salt, microcosmic salt, ammonium sodium hydrogenphosphate-4-water
Phosphorsalzperle *f* [sodium] phosphate bead, microcosmic [salt] bead
Phosphorsäure *f* phosphoric acid; *(specif)* H_3PO_4 orthophosphoric acid, phosphoric acid
auf nassem Wege hergestellte P. wet-process phosphoric acid
auf trockenem Wege hergestellte P. furnace acid
Phosphorsäurepolymerisationsverfahren *n* phosphoric-acid polymerization process
Phosphorsäuretriamid *n s.* Phosphoryltriamid
Phosphorsäuretriäthylester *m* $(C_2H_5)_3PO_4$ triethyl phosphate
Phosphorsäuretrikresylester *m* $[CH_3C_6H_4O]_3PO$ tritolyl phosphate, tricresyl phosphate, TCP
Phosphorsäuretriphenylester *m* $(C_6H_5O)_3PO$ triphenyl phosphate
Phosphorsäureverfahren *n* phosphoric-acid process *(catalytic polymerization)*
Phosphorstickstoff *m s.* Phosphornitrid
Phosphorstoffwechsel *m (bioch)* phosphorus metabolism
Phosphor(III)-sulfid *n* P_2S_3 phosphorus(III) sulphide, diphosphorus trisulphide
Phosphor(V)-sulfid *n* P_2S_5 phosphorus(V) sulphide, diphosphorus pentasulphide
Phosphor(V)-tetrabromidtrichlorid *n* PBr_4Cl_3 phosphorus tetrabromide trichloride
Phosphortetroxid *n* P_2O_4 phosphorus tetraoxide, diphosphorus tetraoxide
Phosphor(III)-thiozyanat *n* $P(SCN)_3$ phosphorus(III) thiocyanate, phosphorus(III) rhodanide
Phosphortri... *s.* Phosphor(III)-...
Phosphorwasserstoff *m* phosphorus hydride, phosphine
Phosphorylase *f* phosphorylase
Phosphorylbromid *n* $POBr_3$ phosphoryl bromide, phosphorus tribromide oxide
Phosphorylchlorid *n* $POCl_3$ phosphoryl chloride, phosphorus trichloride oxide

Phosphorylfluorid *n* POF_3 phosphoryl fluoride, phosphorus trifluoride oxide
Phosphorylgruppe *f* $\equiv PO$ phosphoryl group
Phosphoryltriamid *n* $OP(NH_2)_3$ phosphoryl triamide, phosphoric triamide, triamidophosphoric acid
Phosphuranylit *m (min)* phosphuranylite *(uranyl phosphate-6-water)*
Phosvitin *n* phosvitin *(a phosphoprotein)*
Photo... *s.* Foto...
Photon *n* photon
virtuelles P. virtual photon
phototrop *(dye)* phototropic
nicht p. non-phototropic
pH-Regelgerät *n* pH controller
pH-Regelsystem *n* pH-control system
pH-Regelung *f* pH control
pH-Regler *m* 1. pH controller *(apparatus)*; 2. pH regulator *(floatation agent)*
pH-Skale *f* pH scale
pH-Standard *m* pH standard
Phthalamid *n* $C_6H_4(CONH_2)_2$ phthalamide, phthalic acid diamide
Phthalamidsäure *f* $C_6H_4(CONH_2)COOH$ phthalamic acid, phthalic acid monoamide
Phthalaminsäure *f s.* Phthalamidsäure
Phthalat *n* phthalate
Phthalatweichmacher *m* phthalate plasticizer
Phthalein *n* phthalein
Phthalimid *n* phthalimide
Phthalodinitril *n* $C_6H_4(CN)_2$ phthalonitrile, phthalic acid dinitrile
Phthalogenbrillantblau *n (dye)* phthalogen brilliant blue
Phthalomonopersäure *f* monoperphthalic acid
Phthalophenon *n* phthalophenon
Phthaloylchlorid *n* $C_6H_4(COCl)_2$ phthaloyl chloride
Phthalozyanin *n* phthalocyanine
Phthalozyaninfarbstoff *m* phthalocyanine dye
Phthalozyaninpigment *n* phthalocyanine pigment
Phthalozyaninreihe *f* phthalocyanine series
Phthalsäure *f* $C_6H_4(COOH)_2$ phthalic acid; *(specif) s.* *o*-Phthalsäure
***m*-Phthalsäure** *f m*-phthalic acid, isophthalic acid, benzene-*m*-dicarboxylic acid
***o*-Phthalsäure** *f o*-phthalic acid, phthalic acid *(proper)*, benzene-*o*-dicarboxylic acid
***p*-Phthalsäure** *f p*-phthalic acid, terephthalic acid, benzene-*p*-dicarboxylic acid
Phthalsäureanhydrid *n* phthalic anhydride
Phthalsäurediamid *n s.* Phthalamid
Phthalsäurediäthylester *m* $C_6H_4(COOCH_2CH_3)_2$ diethyl phthalate, D.E.P.
Phthalsäuredimethylester *m* $C_6H_4(COOCH_3)_2$ dimethyl phthalate, D.M.P.
Phthalsäuredinitril *n s.* Phthalodinitril
Phthalsäure-di-*n*-oktylester *m* $(COOC_8H_{17})_2$ dioctyl phthalate
Phthalsäureimid *n s.* Phthalimid
Phthalsäuremonoamid *n s.* Phthalamidsäure
Phthiokol *n* phthiocol, 2-hydroxy-3-methyl-1,4-naphthoquinone
pH-Wert *m* pH value (number), hydrogen ion exponent ✦ **von gleichem pH-Wert** isohydric
pH-Wert-Bestimmung *f* pH determination
pH-Wert-Messer *m*, **pH-Wert-Meßgerät** *n s.* pH-Meßgerät
pH-Wert-Meßsystem *n* pH-measuring system
pH-Wert-Messung *f* pH measurement
pH-Wert-Regelsystem *n* pH-control system
pH-Wert-Regelung *f* pH control
pH-Wert-Regler *m s.* pH-Regler
Phykoerythrin *n* phycoerythrin *(a protein pigment of the red algae)*
Phykokolloid *n* phycocolloid *(a polysaccharide of brown and red algae)*
Phykozyan *n* phycocyan[in] *(a protein pigment of the blue-green algae)*
Phyllochinon *n* phylloquinone, coagulation (antihaemorrhagic) vitamin
Phyllosilikat *n (min)* phyllosilicate, sheet-silicate *(any of a class of polymeric silicates)*
Physetölsäure *f s.* Zoomarinsäure
physikalisch physical
physikalisch-chemisch, physikochemisch physicochemical
physiologisch physiologic[al]
physisch physical
Physostigmin *n* physostigmine, eserine *(an indole alkaloid)*
Physostigmol *n* physostigmol, 5-hydroxy-1,3-dimethylindole *(alkaloid)*
Phytase *f* phytase *(an enzyme)*
Phyteral *n* phyteral *(a vegetable structural element of coal)*
Phytin *n* phytin *(calcium magnesium potassium salt of phytic acid)*
Phytinsäure *f* phytic acid
Phytoalexin *n (bioch)* phytoalexin
Phytochemie *f* phytochemistry
Phytol *n* phytol, 3,7,11,15-tetramethyl-2-hexadecen-1-ol
Phytosterin *n* phytosterol, plant sterol
phytotoxisch phytotoxic
Phytotoxizität *f* phytotoxicity
pH-Zahl *f s.* pH-Wert
Piazin *n s.* Pyrazin
PIB *s.* Polyisobutylen
Pi-Bindung *f* pi bond, π bond
pichen to pitch
Pickel *m* 1. *(tann)* pickle [liquor, solution], pickling solution; 2. *(plast)* pimple *(a moulding defect)*
pickeln *(tann)* to pickle, to sour
Pickeringit *m (min)* pickeringite *(aluminium magnesium sulphate-22-water)*
Pick-up-Walze *f (pap)* pickup roll
Pidgeon-Verfahren *n* Pidgeon [vacuum] process *(for producing metallic magnesium)*
Piemontit *m (min)* piedmontite *(a sorosilicate)*
Piezochemie *f* piezochemistry
Piezochromie *f* piezochromism
Piezodruckmesser *m* piezometer
Piezoeffekt *m* piezo[electric] effect
piezoelektrisch piezoelectric

Piezoelektrizität *f* piezoelectricity
Piezokristall *m* piezoelectric crystal
Piezokristallisation *f* piezocrystallization
Piezometer *n* piezometer
Pigment *n* pigment
anorganisches P. inorganic pigment
geflushtes P. flushed pigment, *(Am also)* flushed color
künstliches anorganisches P. synthetic inorganic pigment
metallisches P. metallic pigment
mikronisiertes P. micronized pigment
natürliches P. natural pigment
natürliches anorganisches P. earth (mineral) pigment
organisches P. organic pigment
respiratorisches P. *(bioch)* respiratory pigment
Pigmentanreibung *f* pigment grinding
Pigmentation *f* pigmentation
Pigment-Bindemittel-Verhältnis *n* pigment-binder ratio
Pigmentdispergierung *f*, **Pigmentdispersion** *f* pigment dispersion
Pigmentdruck *m* pigment printing
Pigmentfarbstoff *m* pigment [dyestuff, dye], *(Am also)* pigment color; *(biol)* pigment; pigment colour *(for artists)*
pigmentieren to pigment
Pigmentklotzung *f (text)* pigment padding
Pigmentrot *n* pigment (para) red, paranitraniline red
Pigmentvolumenkonzentration *f* pigment volume concentration, p.v.c., PVC
Pigmentwanderung *f* pigment migration
Pik *m* peak *(as of a chromatogram)*
Pikolin *n* picoline, methylpyridine
Pikolinsäure *f* picolinic acid, pyridine-2-carboxylic acid
Pikotit *m (min)* picotite *(a variety of spinel)*
Pikraminsäure *f* $C_6H_2(NO_2)_2(NH_2)OH$ picramic acid, 2-amino-4,6-dinitrophenol
Pikrat *n* picrate
Pikrinsäure *f* $C_6H_2(NO_2)_3OH$ picric acid, 2,4,6-trinitrophenol
Pikromyzin *n* picromycin *(antibiotic)*
Pile *m (nucl)* pile
Pile *f* [amalgam] decomposer *(electrolysis)*
Pilinußöl *n* pili nut oil *(from Canarium specc.)*
Pilkington-Verfahren *n* Pilkington process *(plate-glass manufacture)*
Pillbildung *f (text)* pilling
Pille *f* pellet *(as of a catalyst)*; *(med)* pill
pillieren to pill
Pillingeffekt *m (text)* pilling
Pilokarpin *n* pilocarpine *(alkaloid)*
Pilokarpinnitrat *n* pilocarpine nitrate
Pilotanlage *f* pilot plant, semiworks
Pilzdiastase *f* fungal diastase
Pilzfarbstoff *m* fungus (fungal) pigment
pilzfest fungus-proof
Pilzmaischverfahren *n* amylo fermentation process *(use of fungal amylases for fermenting starchy materials)*
Pilzmischer *m* mushroom mixer
pilztötend fungicidal
Pilzventil *n* mushroom-seated valve
pilzwidrig, pilzwirksam antifungal
Pimarizin *n* pimaricin *(antibiotic)*
(+)-Pimarsäure *f* (+)-pimaric acid, dextropimaric acid
(−)-Pimarsäure *f* (−)-pimaric acid, laevopimaric acid, *(proposed name)* (−)-sapietic acid
Pimelinsäure *f* $HOOC[CH_2]_5COOH$ pimelic acid, heptanedioic acid
Pimelit *m (min)* pimelite *(a phyllosilicate)*
Piment *n* allspice *(from Pimenta dioica (L.) Merr.)*
Pinakoid *n (cryst)* pinacoid
pinakoidal *(cryst)* pinacoidal
Pinakol *n* $(CH_3)_2C(OH)C(OH)(CH_3)_2$ pinacol, 2,3-dimethyl-2,3-butanediol
Pinakolin *n s.* Pinakolon
Pinakolinumlagerung *f s.* Pinakol-Pinakolon-Umlagerung
Pinakolon *n* $CH_3COC(CH_3)_3$ pinacolone, 3,3-dimethyl-2-butanone, 1,1,1-trimethylacetone
Pinakol-Pinakolon-Umlagerung *f* pinacol rearrangement
Pinakon *n s.* Pinakol
Pinakryptolgelb *n* pinacryptol yellow *(an isocyanine)*
Pinakryptolgrün *n* pinacryptol green *(an isocyanine)*
Pinen *n* pinene *(a bicyclic monoterpene)*
Pineyharz *n* piney resin, piney (white) dammar, Indian copal *(from Vateria indica L.)*
Pineytalg *m* piney tallow, Dhupa fat *(from Vateria indica L.)*
Pinit *m* 1. $C_6H_6(OH)_5OCH_3$ pinitol, inositol 3-monomethyl ether; 2. *(min)* pinite *(any of several pseudomorphs of mica-like minerals)*
Pinksalz *n* $(NH_4)_2[SnCl_6]$ pink salt, ammonium hexachlorostannate(IV)
Pinne *f (ceram)* pin
Pinobanksin *n* pinobanksin, 3,5,7-trihydroxyflavanone
Pinolin *n* pinolin[e], pinolene, rosin (resin) spirit
Pinonen *n* pinonene, (−)-4-carene, 3,7,7-trimethylbicyclo[2,2,1]hept-2-ene
Pinoresinol *n* pinoresinol *(a derivative of coniferyl alcohol)*
Pinosylvin *n* $(HO)_2C_6H_3CH{=}CHC_6H_5$ pinosylvin, 3,5-dihydroxystilbene
Pintsch-Gas *n* Pintsch gas *(an oil gas)*
Pinzette *f* pincers, tweezers
Pion *n (nucl)* pion, pi meson, π-meson *(an elementary particle)*
Pioskop *n* pioscope *(for determining the fat content of milk)*
Pipekolinsäure *f* pipecolic acid, piperidine-2-carboxylic acid
Pipeline *f* **für Fertigerzeugnisse** *(petrol)* refined-product pipeline
Piperazin *n* piperazine
Piperidin *n* piperidine, hexahydropyridine
Piperidon *n* piperidone

Piperinsäure *f* piperic acid, 5-(3,4-methylenedioxyphenyl)-2,4-pentadienoic acid
Piperonal *n* $CH_2(O_2)C_6H_3CHO$ piperonal, 3,4-methylenedioxybenzaldehyde
Piperonylsäure *f* $CH_2(O_2)C_6H_3COOH$ piperonylic acid, 3,4-methylenedioxybenzoic acid
Piperylen *n* $CH_2{=}CHCH{=}CHCH_3$ piperylene, 1,3-pentadiene
Pipestill-Anlage *f (petrol)* pipe-still (tube-still) plant (unit), pipe (tube) still
Pipette *f* pipette, *(if small also)* dropper
P. mit Farbmarkierung colour-code pipette
Pipettenbürste *f* pipette brush
Pipetten-Etagere *f*, **Pipettengestell** *n s.* Pipettenständer
Pipettenhütchen *n* dropper teat
Pipettenspitze *f* pipette tip
Pipettenständer *m* pipette rack (stand, support)
pipettieren to pipette
Pipettmethode *f* pipetting method
Pipitzahoinsäure *f* pipitzahoic acid, perezone *(a benzoquinone derivative)*
Pirani-Manometer *n*, **Pirani-Vakuummeter** *n* Pirani gauge
Pisolith *m (min)* pisolite *(calcium carbonate)*
Pisolithtuff *m* pisolitic tuff
Pistaziengallen *fpl (coat)* carob (turpentine) galls *(from leaves of Pistacia terebinthus L.)*
Pistill *n (lab)* pestle
Pitot-Rohr *n* Pitot tube *(for measuring the velocity of a flowing medium)*
Pittsburgh-Verfahren *n* Pittsburgh [sheet] process, Pennvernon process *(for the vertical drawing of sheet glass)*
Piuri *n (dye)* piuri, Indian yellow *(from Mangifera indica L.)*
Pivalaldehyd *m* $(CH_3)_3CCHO$ pivalic aldehyde, pivaldehyde, 2,2-dimethylpropanal
Pivalinsäure *f* $(CH_3)_3CCOOH$ pivalic acid, 2,2-dimethylpropionic acid
Pi-Yu *n s.* Chinatalg
Pizeatannol *n* piceatannol *(a stilbene derivative)*
Pizein *n* picein, piceoside, *p*-hydroxyacetophenone-β-glucoside
Pizen *n* picene, 1,2,7,8-dibenzphenanthrene
PK *s.* Polykarbonatfaserstoff
Plachenherd *m*, **Plachentisch** *m (min tech)* vanner
Placierung *f* placement *(of fertilizers or pesticides)*
Plagioklas *m (min)* plagioclase *(any of a series of tectosilicates)*
Plagionit *m (min)* plagionite *(antimony(III) lead(II) sulphide)*
Plakatfarbe *f* poster paint (colour)
Plakatpapier *n* poster paper, posters
planar planar
Plancheit *m (min)* planchéite *(a cyclosilicate)*
Planetenrührer *m*, **Planetenrührwerk** *n* planet[ary] stirrer
Planfilm *m (phot)* sheet film
Planfilter *n* table filter
planieren to level
Planierstange *f* levelling (coal leveller) bar *(in a coke chamber)*
Planiervorrichtung *f* levelling device
Planknotenfänger *m (pap)* flat strainer (screen)
Planlager *n* plane bearing *(of a balance)*
Planschliffverbindung *f* plane (flat-flange) joint
Planschneiden *n (pap)* guillotine cutting (trimming)
Planschneider *m (pap)* guillotine cutting (trimming) machine, guillotine cutter (trimmer, press), trimmer
Planschwingsiebmaschine *f* oscillating screen (strainer)
Plansichter *m* gyratory screen (sifter)
Plansortierer *m* flat screen
Plantagenkautschuk *m* plantation (estate) rubber
Plantagenlatex *m* plantation latex
Plasma *n* 1. *(phys chem)* plasma; 2. *(min)* plasma *(variety of chalcedony)*
Plasmabrenner *m* plasma burner
Plasmachemie *f* plasma chemistry
Plasmamembran *f (biol)* plasma membrane
Plasmaphysik *f* plasma physics
Plasmaschwingungen *fpl* plasma [electron] oscillations
Plasmazustand *m (phys chem)* plasma state
Plast *m* plastic [material] *(for compounds s. a.* Kunststoff*)*
flexibler P. flexible plastic
glasfaserverstärkter P. glass-fibre reinforced plastic, G.R.P.
harter P. rigid plastic
hitzehärtbarer P. *s.* wärmehärtbarer P.
verstärkter P. reinforced plastic
wärmehärtbarer P. thermosetting plastic (resin), thermoset [resin]
weicher (weichgestellter) P. non-rigid plastic
Plastansatz *m* plastic composition
Plasterzeugnis *n* plastic product
Plastfolie *f* plastic foil (film)
Plastfolienschweißgerät *n* plastic foil welder
Plastifikation *f s.* Plastizierung
Plastifikationsmittel *n*, **Plastifikator** *m s.* Plastizierungsmittel
Plastifizier... *s.* Plastizier...
plastifizieren *s.* plastizieren
Plastigel *n* plastigel
Plastikator *m* 1. plasticator *(a machine for plasticizing rubber or plastics)*; 2. *s.* Plastizierungsmittel
plastisch plastic
plastizieren to plasticize, to plastify, to plasticate, to soften, *(rubber also)* to break down
mit Peptisiermitteln p. to peptize
thermisch p. to heat-soften
Plastizierleistung *f* plasticizing capacity
Plastiziermaschine *f* plasticator
Plastizierung *f* plasticization, plasti[fi]cation, softening, *(rubber also)* breakdown
P. mit Peptisiermitteln peptization
chemische P. chemical plasticization
mechanische P. *(rubber)* mechanical (mill) breakdown
thermische P. thermal plasticization, thermal (heat) softening
Plastizierungsmittel *n* plasticizer, plasticizing agent

chemisches P. chemical plasticizer, *(rubber also)* peptizer, peptizing agent
Plastizität *f* plasticity
Plastizitätsbereich *m* plastic range
Plastizitätsmessung *f* plasticity measurement
Plastizitätsprüfgerät *n s.* Plastometer
Plastizitätsprüfung *f* plasticity test
Plastizitätswasser *n (ceram)* water of plasticity
Plastmischung *f* plastic composition
Plastograf *m* plastograph
Plastomer[es] *n* plastomer
Plastometer *n* plastometer, plastimeter
P. von Williams Williams plastometer
Gieselersches P. Gieseler plastometer
Plastrohr *n* plastic pipe
Plastrohrmaterial *n* plastic piping
Plastsack *m* plastic bag
Plastschmelze *f* plastic melt
Plastschweißen *n* welding of plastics
Plastüberzug *m* plastic covering
Plastverarbeiter *m* plastics processor
Plastverarbeitung *f* plastics processing
Plateau *n (rubber)* cure plateau ✦ **mit breitem P.** flat-curing *(having a wide optimum range of vulcanization)* ✦ **mit kurzem P.** peaky[-curing] *(having a narrow optimum range of vulcanization)*
Plateaueffekt *m (rubber)* plateau (flat-curing) effect
Platformat *n (petrol)* platformate
Platformer *m s.* Platforming-Anlage
Platformerprodukt *n s.* Platformat
Platformieren *n (petrol)* platforming, platinum reforming
Platforming-Anlage *f (petrol)* platforming unit, platformer
Platforming-Produkt *n s.* Platformat
Platforming-Verfahren *n (petrol)* platforming process
Platiak *n* platinum ammine, platinammine
Platin *n* Pt platinum
Platinasbest *m* platinized asbestos
Platin(II)-chlorid *n* $PtCl_2$ platinum(II) chloride, platinum dichloride
Platin(III)-chlorid *n* $PtCl_3$ platinum(III) chloride, platinum trichloride
Platin(IV)-chlorid *n* $PtCl_4$ platinum(IV) chloride, platinum tetrachloride
Platinchlorwasserstoffsäure *f s.* Hexachloroplatin(IV)-säure
Platindichlorid *n s.* Platin(II)-chlorid
Platindioxid *n s.* Platin(IV)-oxid
Platindiphosphat *n* PtP_2O_7 platinum diphosphate, platinum pyrophosphate
Platindisulfid *n s.* Platin(IV)-sulfid
Platindraht *m* platinum wire
Platindruck *m* platinotype *(reprography)*
Platinelektrode *f* platinum electrode
platinhaltig platinum-bearing, platiniferous
platinieren to platinize, to platinate
Platinieren *n* platinization, platinum plating
Platinkatalysator *m* platinum catalyst
Platinkontakt *m* platinum catalyst
Platinmikroelektrode *f* platinum microelectrode
Platinmohr *n* platinum black, platina mohr
Platinmonosulfid *n s.* Platin(II)-sulfid
Platinmonoxid *n s.* Platin(II)-oxid
Platin(II)-oxid *n* PtO platinum(II) oxide, platinum monooxide
Platin(IV)-oxid *n* PtO_2 platinum(IV) oxide, platinum dioxide
Platin(VI)-oxid *n* PtO_3 platinum(VI) oxide, platinum trioxide
Platinoxidkatalysator *m* platinum oxide catalyst
Platinpapier *n (phot)* platinotype paper
Platinpyrophosphat *n s.* Platindiphosphat
Platinschale *f* platinum dish
Platinschiffchen *n* platinum boat
Platinschwamm *m* platinum sponge, spongy platinum
Platinschwarz *n s.* Platinmohr
Platinsol *n* platinum sol
Platin(IV)-sulfat *n* $Pt(SO_4)_2$ platinum(IV) sulphate
Platin(II)-sulfid *n* PtS platinum(II) sulphide, platinum monosulphide
Platin(III)-sulfid *n* Pt_2S_3 platinum(III) sulphide, platinum trisulphide, diplatinum trisulphide
Platin(IV)-sulfid *n* PtS_2 platinum(IV) sulphide, platinum disulphide
Platintetrachlorid *n s.* Platin(IV)-chlorid
Platintiegel *m* platinum crucible
Platintonung *f (phot)* platinum toning
Platintrichlorid *n s.* Platin(III)-chlorid
Platintrioxid *n s.* Platin(VI)-oxid
Platinwiderstandsthermometer *n* platinum resistance thermometer
Platin(II)-zyanid *n* $Pt(CN)_2$ platinum(II) cyanide
Platte *f* plate, *(if thick)* slab, *(if thin)* sheet; tile *(as of fired clay or concrete)*; board *(as of wood pulp)*; *(filtr)* disk, pad; *(rubber)* sheet, slab
amalgamierte P. amalgamated plate
bewegliche P. *(plast)* floating plate *(of a press)*
fotografische P. photographic plate
stranggepreßte P. *(plast)* extruded sheet
xerografische P. xerographic plate
Plattenabscheider *m* plate precipitator *(gas cleaning)*
Plattenbandförderer *m* apron conveyor
Platteneis *n* plate (slice) ice
Plattenelektroabscheider *m*, **Plattenelektrofilter** *n* plate precipitator *(gas cleaning)*
Plattenerhitzer *m (food)* plate pasteurizer
Plattenförderer *m* apron conveyor
plattenförmig plate-like; *(cryst)* lamellar, flat
Plattenformung *f (plast)* sheet forming
Plattenkühler *m* plate cooler
Plattenpresse *f* platen press
Plattenschieber *m* parallel-seat gate valve
Plattentrockner *m* tray (shelf) dryer
Plattenwärmeaustauscher *m* plate heat exchanger
berippter P. plate-fin heat exchanger
Plattenziehen *n (rubber)* sheet calendering, sheeting[-out]
plattieren to plate *(metal)*, *(by bonding or welding)* to clad; *(tann)* to strike (set) out; *(text)* to plate

Plattierung *f* plating *(of metal), (result also)* plate; *(by bonding or welding)* cladding; *(text)* plating
Plattierungswerkstoff *m* plating (cladding) material
plattstengelig *(cryst)* bladed
Platzbedarf *m (cryst)* size requirements
Plätzchen *n* pellet *(as of potassium hydroxide)*
platzen to burst, to break, to explode *(by the force of internal pressure)*; to crack *(as of a glass plate)*
Platzscheibe *f* bursting (rupture) disk
Plauson-Mühle *f* Plauson [colloid] mill
Pleochroismus *m (cryst)* pleochro[mat]ism, polychroism
pleochroitisch *(cryst)* pleochro[mat]ic, polychroic
Pleonast *m (min)* pleonaste, ceylonite, ceylanite *(a spinel containing iron(II))*
Plessit *m (min)* plessite *(a natural alloy occurring in meteorites)*
pl-Phase *f (coll)* nematic phase
Plumban *n* PbH_4 plumbane, lead hydride
Plumbat *n* $M_2^IPb(OH)_6$ *or* $M_4^IPbO_4$ plumbate
Plumbitverfahren *n (petrol)* plumbite [sweetening] process
Plumbogummit *m (min)* plumbogummite *(a hydrous basic phosphate of lead and aluminium)*
Plundermilch *f* [naturally, spontaneously] soured milk, sour (set) milk
Plunger *m* plunger; *(glass)* needle *(of a feeder)*
Plungerkolben *m* plunger
Plungerpumpe *f* plunger (ram) pump
Plüschpapier *n* velour (flock) paper
Plusplatte *f (el chem)* positive plate
Plutonit *m (geol)* plutonite, plutonic (hypogene, irruptive) rock
Plutonium *n* Pu plutonium
PMA, PMAS *s.* Phenylquecksilberazetat
PMMA *(abbr)* Polymethylmethakrylat, *s.* Polymethakrylat
Pneumatikreifen *m* pneumatic [tyre]
Pneumatikventil *n* pneumatic (air) valve
pneumatisch pneumatic
pn-Übergang *m* p-n junction *(in semiconductors)*
PO *s.* 1. Polyolefin; 2. Polyolefinfaserstoff
Pochwerk *n (min tech)* stamp battery, stamp[ing] mill
Podbielniak-Extraktor *m*, **Podbielniak-Kontaktor** *m*, **Podbielniak-Zentrifuge** *f* Podbielniak contactor (machine), Podbielniak centrifugal [countercurrent] contactor
Podokarpren *n* podocarprene *(a diterpene)*
Podsol[boden] *m* podzol [soil], podsol [soil], podzolic soil
Podsolierung *f (soil)* podzolization
Poise *n* poise *(a unit of dynamic viscosity)*
Pökelfaß *n (food)* curing vat
Pökelfleisch *n* cured meat
Pökelflüssigkeit *f*, **Pökellake** *f (food)* curing (pickling) solution, pickle [liquor, solution], souse
pökeln *(food)* to cure
Pökeln *n (food)* curing, cure
Pökelsalzlösung *f s.* Pökelflüssigkeit
polar polar
Polarimeter *n* polarimeter, polariscope
Polarimetrie *f* polarimetry
Polarisation *f* polarization
  P. des Vakuums vacuum polarization
  dielektrische P. dielectric polarization
  elektrolytische P. electrolytic polarization
  elliptische P. elliptical polarization
  galvanische P. electrolytic polarization
  lineare P. linear (plane) polarization
  zirkulare P. circular polarization
Polarisationsebene *f* polarization plane
Polarisationseffekt *m* polarization (polarizing) effect
Polarisationselektrode *f* polarization electrode
Polarisationsellipsoid *n* polarization ellipsoid
Polarisationsgrad *m* degree of polarization
Polarisationsmikroskop *n* polarizing microscope
Polarisationssättigung *f* polarization saturation
Polarisationsspannung *f* polarization voltage (potential)
Polarisationsstrom *m* polarization (polarizing) current
Polarisationswinkel *m* angle of polarization, polarizing (Brewster) angle
Polarisationswirkung *f s.* Polarisationseffekt
Polarisator *m* polarizer
polarisierbar polarizable
Polarisierbarkeit *f* polarizability
Polarisierbarkeitsellipsoid *n* polarization ellipsoid
polarisieren to polarize
polarisiert / elliptisch elliptically polarized
  linear p. linearly polarized, plane-polarized
  linkszirkular p. left-circularly (left-hand circularly) polarized
  rechtszirkular p. right-circularly (right-hand circularly) polarized
  zirkular p. circularly polarized
Polarisierung *f* polarization
Polarisierungskraft *f* polarizing power
Polaristrobometer *n* polaristrobometer
Polarität *f* polarity
Polarograf *m* polarograph
  P. mit fotografischer Registrierung photographic recording polarograph
Polarografie *f* polarography
  oszillografische P. oscillographic polarography
  oszillografische P. mit Wechselstrom multisweep polarography
polarografisch polarographic
Polarogramm *n* polarogram
Poleiöl *n* European pennyroyal oil *(essential oil from Mentha pulegium L.)*
polen *(met)* to pole
  zu weit p. to overpole
Polenske-Zahl *f* Polenske number (value) *(indicating the content of volatile water-insoluble acids in fat)*
Polgewebe *n (text)* pole fabric
Polianit *m (min)* polianite *(manganese(IV) oxide)*
Polierblech *n* polishing plate
polieren to polish
  elektrolytisch p. to electropolish
Polierfilter *n* polishing filter

Polierfiltration *f* polishing [filtration]
polierfiltrieren to polish
Poliergold *n (ceram)* burnish[ed] gold, best gold
Poliermasse *f* polishing compound
Poliermittel *n* polishing agent (material)
Polieröl *n* polishing oil
Polierpech *n* polishing pitch
Polierrot *n* polishing rouge *(iron(III) oxide)*
Polierstein *m*, **Poliertisch** *m (glass)* polisher block
Politur *f* 1. polish[ing agent]; 2. lustre, gloss, polish
Polluzit *m (min)* pollucite *(a tectosilicate)*
Polonium *n* Po polonium
Polpapier *n* pole[-finding] paper, pole reagent paper, *(Am)* polarity paper
Polprüfer *m* polarity tester
Polreagenzpapier *n s.* Polpapier
Polstermischung *f (rubber)* cushion stock
Polyaddition *f* polyaddition, addition polymerization
Polyaddukt *n* addition polymer
Polyaffinität *f* polyaffinity
Polyaffinitätstheorie *f* polyaffinity theory
Polyakrylat *n* polyacrylate; *s.* Polyakrylharz
Polyakrylat-Elastomer[es] *n* polyacrylate elastomer, acrylate (acrylic) elastomer
Polyakrylharz *n* polyacrylate, acrylate (acrylic-acid, acrylic) resin
Polyakrylnitril *n* polyacrylonitrile
Polyakrylnitrilfaser *f* polyacrylonitrile (acrylic) fibre
Polyakrylnitrilfaserstoff *m* polyacrylonitrile (acrylic) fibre
Polyakrylsäure *f* polyacrylic acid
Polyakrylsäureester *m* polyacrylate
Polyalkohol *m* polyalcohol, polyol, polyfunctional (polyhydric, polyhydroxy) alcohol
Polyalkylsiloxan *n* alkyl polysiloxane
Polyallomer[es] *n* polyallomer
Polyamid *n* polyamide
Polyamidfaser *f* polyamide fibre
Polyamidfaserstoff *m* polyamide fibre
Polyampholyt *m* polyampholyte *(a polymer reacting with acids as well as with bases)*
Polyanion *n* polyanion
Polyaryläther *m* polyaryl ether
Polyäthen *n s.* Polyäthylen
Polyäther *m* polyether
Polyäthylen *n* polyethylene, polythene, PE
  P. hoher Dichte high-density polyethylene, H.D. polythene
  P. mittlerer Dichte medium-density polyethylene
  P. niedriger Dichte low-density polyethylene, L.D. polythene
  chlorsulfoniertes P. chlorosulphonated polyethylene
  sulfochloriertes P. sulphochlorinated polyethylene
  unverzweigtes P. unbranched (linear, low-pressure) polyethylene
  verzweigtes P. branched (high-pressure) polyethylene
Polyäthylenadipat *n* polyethylene adipate
Polyäthylenfaser *f* polyethylene fibre
Polyäthylenfaserstoff *m* polyethylene fibre
Polyäthylenglykol *n s.* Polyäthylenoxid
Polyäthylenimin *n* polyethylene imine
Polyäthylenoxid *n* polyethylene oxide, polyethylene glycol
Polyäthylenschaum[stoff] *m* polyethylene foam
Polyäthylenterephthalat *n* polyethylene terephthalate
polyatomar polyatomic
Polyazetal *n* polyacetal
Polybasit *m (min)* polybasite *(a sulphidic silver ore)*
Polybenzimidazol *n* polybenzimidazole, PBI
Polybenzothiazol *n* polybenzothiazole, PBT
Polyblend *n* polyblend, polymer blend *(a mixture of several thermoplastics)*
Polybutadien *n* polybutadiene, butadiene polymer
Polybuten *n*, **Polybutylen** *n* polybutene, polybutylene
Polychloräthan *n* polychloroethane
Polychloropren *n* polychloroprene, PCP
Polychlorstyrol *n* polychlorostyrene
Polychlortrifluoräthylen *n* polychlorotrifluoroethylene, PCTFE
polydispers polydisperse
Polydispersität *f* polydispersity
Polydymit *m (min)* polydymite *(a nickel sulphide)*
Polyeder *n (cryst)* polyhedron
polyedrisch *(cryst)* polyhedral
Polyelektrolyt *m* polyelectrolyte
Polyen *n (org chem)* polyene
Polyen-Antibiotikum *n* polyene antibiotic
Polyepoxid *n* polyepoxide
Polyester *m* polyester
Polyesterfaser *f* polyester fibre
Polyesterfaserstoff *m* polyester fibre
Polyesterharz *n* polyester resin
Polyesterkautschuk *m* polyester rubber
Polyfluoräthylenpropylen *n* polyfluoroethylene propylene, PFEP
Polyformaldehyd *m* polyformaldehyde
Polyformen *n (petrol)* polyforming
Polyform[ing]-Verfahren *n (petrol)* polyform[ing] process
polyfunktionell polyfunctional, multifunctional, multiple-function
Polygalakturonase *f* polygalacturonase
Polygalakturonsäure *f* polygalacturonic acid
Polyglykol *n s.* Polyäthylenoxid
Polyglyzerin *n* polyglycerol
polygonal polygonal
Polyhalogenid *n* polyhalogenide, polyhalide
Polyharnstoff *m* polyurea
Polyharnstoffaser *f* polyurea fibre
Polyharnstoffaserstoff *m* polyurea fibre
Polyhydrat *n* polyhydrate
Polyhydroxyaldehyd *m* polyhydroxy aldehyde
Polyhydroxyanthrachinon *n* polyhydroxyanthraquinone
Polyhydroxyketon *n* polyhydroxy ketone
Polyhydroxyverbindung *f* polyhydroxy compound
Polyimid *n* polyimide, PI
Polyion *n* polyion

Polyisobuten *n*, **Polyisobutylen** *n* polyisobutene, polyisobutylene
Polyisopren *n* polyisoprene
Polyjodid *n* polyiodide
Polykaprolaktam *n* polycaprolactam
Polykarbamid *n* polyurea
Polykarbonat *n* polycarbonate
Polykarbonatfaser *f* polycarbonate fibre
Polykarbonatfaserstoff *m* polycarbonate fibre
Polykarbonatharz *n* polycarbonate resin
Polykarbonsäure *f* polycarboxylic acid
Polyketon *n* polyketone
Polykieselsäure *f* polysilicic acid
Polykondensat *n* polycondensate, condensation polymer
Polykondensation *f* polycondensation, condensation polymerization
Polykras *m* polycrase *(an oxidic rare-earth mineral)*
polymer polymeric
Polymer *n* polymer, *(tech also)* polymerizate
amorphes P. amorphous polymer
anorganisches P. inorganic polymer
ataktisches P. atactic polymer
eindimensionales P. linear polymer
[elementar]fadenbildendes P. fibrous polymer
flüssiges P. *(rubber)* [polysulphide] liquid polymer
hochstyrolhaltiges P. *(rubber)* high-styrene polymer (copolymer, resin), self-reinforced elastomer
isomeres P. isomeric polymer
isotaktisches P. isotactic polymer
langkettiges P. long-chain polymer
lebendes P. living polymer
lineares P. linear polymer
organisches P. organic polymer
räumlich vernetztes P. space-network polymer
siliziumorganisches P. organosilicon polymer
stereoreguläres P. stereoregular polymer
stereospezifisches P. stereospecific polymer
syndiotaktisches P. syndiotactic polymer
tritaktisches P. tritactic polymer
vernetztes P. cross-linked polymer
verzweigtes P. branched polymer
Polymerbenzin *n* polymer gasoline
polymereinheitlich polymer-homologous
Polymeres *n s.* Polymer
polymerhomolog polymer-homologous
Polymerhomolog[es] *n* polymer homologue
Polymerisat *n* polymerizate, polymer (*for compounds s.* Polymer)
Polymerisatbinder *m* polymer binder
Polymerisation *f* polymerization
P. an Ort und Stelle in situ polymerization
P. in der Gasphase gas (gaseous) polymerization
P. in Emulsion emulsion polymerization
P. in Masse (Substanz) bulk (mass) polymerization
anionische P. anionic polymerization
durch freie Radikale ausgelöste P. [free-]radical polymerization
ionische P. ionic polymerization
katalytische P. catalytic polymerization
kationische P. cationic polymerization
kontinuierliche P. in einer Druckschnecke continuous screw-feed process of polymerization
koordinative P. coordination polymerization
radikalische P. [free-]radical polymerization
stereospezifische P. stereospecific polymerization
stöchiometrische P. living polymerization
Polymerisationsabstoppmittel *n (rubber, plast)* shortstopping agent, shortstop, stopper
Polymerisationsaktivator *m* activator of polymerization
Polymerisationsansatz *m* polymerization recipe
Polymerisationsbenzin *n* polymer gasoline
Polymerisationserreger *m* polymerization initiator
polymerisationsfähig polymerizable
Polymerisationsfähigkeit *f* polymerizability
Polymerisationsgerbung *f* polymerization (polymeric) tannage
Polymerisationsgeschwindigkeit *f* rate of polymerization
Polymerisationsgrad *m* degree of polymerization
Polymerisationsinitiator *m* polymerization initiator
Polymerisationskessel *m* polymerization kettle
Polymerisationsprodukt *n* polymerization product
Polymerisationsstopper *m s.* Polymerisationsabstoppmittel
Polymerisationstemperatur *f* polymerization temperature
polymerisierbar polymerizable
Polymerisierbarkeit *f* polymerizability
polymerisieren to polymerize
Polymerisierung *f s.* Polymerisation
Polymerkette *f* polymer chain
Polymermischung *f* polyblend, polymer blend *(a mixture of several thermoplastics)*
Polymer-Ruß-Batch *m (rubber)* carbon black [master]batch, black batch
Polymetamorphose *f (geol)* polymetamorphism
Polymethakrylat *n* polymethacrylate, polymethyl acrylate (methacrylate)
Polymethakrylsäure *f* polymethacrylic acid
Polymethakrylsäureester *m s.* Polymethakrylat
Polymethinfarbstoff *m* polymethine dye
Polymethylen *n* polymethylene
Polymethylmethakrylat *n s.* Polymethakrylat
Polymignit *m* polymignyte, polymignite *(an oxidic rare-earth mineral)*
polymineralisch polymineral
polymolekular polymolecular
Polymolekularität *f* polymolecularity
polymorph polymorphic, polymorphous
Polymorphie *f*, **Polymorphismus** *m* polymorphism
Polymyxin *n* polymyxin *(any of a group of polypeptide antibiotics)*
Polynitroderivat *n* polynitro derivative
Polynosefaser *f* polynosic fibre
Polynosefaserstoff *m* polynosic fibre
Polyol *n s.* Polyalkohol
Polyolefin *n* polyolefin
Polyolefinfaser *f* polyolefin fibre

Polyolefinfaserstoff *m* polyolefin fibre
Polyolefinkautschuk *m* polyolefin rubber
Polyorganosiloxan *n* polyorganosiloxane, organopolysiloxane, polymeric organosiloxane, silicone
Polyorganosiloxanchemie *f s.* Silikonchemie
Polyose *f* polyose, polysaccharose
Polyoxazyklobutan *n* polyoxacyclobutane
Polyoxymethylen *n* polyoxymethylene
Polypeptid *n* polypeptide
Polypeptidantibiotikum *n* polypeptide antibiotic
Polypeptidkette *f* polypeptide chain
Polyphenol[oxyd]ase *f* polyphenol oxydase, polyphenolase
Polyphenylenoxid *n* polyphenylene oxide, PPO
Polyphenylsulfid *n* polyphenyl sulphide
Polyphosphat *n* polyphosphate
Polyporsäure *f* polyporic acid, 3,6-dihydroxy-2,5-diphenyl-*p*-benzoquinone
Polypropen *n*, **Polypropylen** *n* polypropene, polypropylene
P. mit vorgebildeter Faserstruktur fibrillated polypropene
Polypropylenfaser *f* polypropylene fibre
Polypropylenfaserstoff *m* polypropylene fibre
Polypropylenschaum[stoff] *m* polypropylene foam
Polyreaktion *f* polyreaction *(polymerization, polycondensation, or polyaddition)*
Polysaccharid *n* polysaccharide
Polysiloxan *n* polysiloxane, polymeric siloxane, *(specif) s.* Polyorganosiloxan
organisches P. *s.* Polyorganosiloxan
Polystyrol *n* polystyrene
aufschäumbares P. expandable polystyrene
geschäumtes P. foamed (expanded) polystyrene
hochschlagfestes P. high-impact polystyrene
schlagfestes P. impact polystyrene
Polystyrolfaser *f* polystyrene fibre
Polystyrolfaserstoff *m* polystyrene fibre
Polystyrolperle *f* / **aufschäumbare** expandable polystyrene bead
Polystyrolschaum[stoff] *m* polystyrene foam
Polystyrolspritzgußmasse *f* injection moulding polystyrene
polysubstituiert polysubstituted
Polysubstitution *f* polysubstitution
Polysulfid *n* polysulphide
Polysulfidbrücke *f (rubber)* polysulphide bridge (cross-link, link)
polysulfidisch polysulphidic
Polysulfidkautschuk *m* polysulphide rubber
Polysulfon *n* polysulphone
Polysulfonharz *n* polysulphone resin
Polyterpen *n* polyterpene
Polytetrafluoräthen *n s.* Polytetrafluoräthylen
Polytetrafluoräthylen *n* polytetrafluoroethylene, PTFE
Polytetrafluoräthylenfaser *f* polytetrafluoroethylene fibre
Polytetrafluoräthylenfaserstoff *m* polytetrafluoroethylene fibre
Polythiokarbamid *n* polythiourea
Polythionat *n* $M_2^I[S_xO_6]$ polythionate
Polythionsäure *f* polythionic acid
Polytriazol *n* polytriazole
Polytrifluorchloräthylen *n s.* Polychlortrifluoräthylen
polytropisch *(phys chem)* polytropic
Polyurethan *n* polyurethan[e]
Polyurethanelastomer[es] *n* polyurethane elastomer
walzbares P. millable polyurethane elastomer
Polyurethanfaser *f* polyurethane fibre
Polyurethanfaserstoff *m* polyurethane fibre
Polyurethanharz *n* polyurethane (isocyanate) resin
Polyurethankautschuk *m* polyurethane (isocyanate) rubber
Polyurethanschaum[stoff] *m* polyurethane foam
Polyuronid *n* polyuronid[e]
Polyuronsäure *f* polyuronic acid
polyvalent 1. polyvalent, multivalent *(chemical-bond theory)*; 2. *(agric)* broad-scale, broad-spectrum *(insecticide)*
Polyvalenz *f* polyvalency, multivalency
Polyvinylalkohol *m* polyvinyl alcohol
Polyvinylalkoholfaser *f* polyvinyl alcohol fibre
Polyvinylalkoholfaserstoff *m* polyvinyl alcohol fibre
Polyvinyläther *m* polyvinyl ether
Polyvinyläthyläther *m* polyvinyl ethyl ether
Polyvinylazetal *n* polyvinyl acetal
Polyvinylazetat *n* polyvinyl acetate, PVA
Polyvinylbenzolfaser *f s.* Polystyrolfaser
Polyvinylbutyral *n* polyvinyl butyral
Polyvinylchlorid *n* polyvinyl chloride, PVC
hochschlagfestes P. high-impact polyvinyl chloride
unplastifiziertes (weichmacherfreies) P. unplasticized (rigid) polyvinyl chloride
Polyvinylchloridazetat *n* polyvinyl chloride acetate
Polyvinylchloridfaser *f* polyvinyl chloride fibre
Polyvinylchloridfaserstoff *m* polyvinyl chloride fibre
Polyvinylfaser *f* polyvinyl fibre
Polyvinylfaserstoff *m* polyvinyl fibre
Polyvinylfluorid *n* polyvinyl fluoride, PVF
Polyvinylformal *n* polyvinyl formal
Polyvinyl-Formaldehydazetal *n s.* Polyvinylformal
Polyvinylidenchlorid *n* polyvinylidene chloride
Polyvinylidenchloridfaser *f* polyvinylidene chloride fibre
Polyvinylidenchloridfaserstoff *m* polyvinylidene chloride fibre
Polyvinylidenchloridharz *n* polyvinylidene chloride resin
Polyvinylidendinitrilfaser *f* polyvinylidene dinitrile fibre
Polyvinylidendinitrilfaserstoff *m* polyvinylidene dinitrile fibre
Polyvinylidenfluorid *n* polyvinylidene fluoride
Polyvinylidenzyanidfaser *f* polyvinylidene cyanide fibre
Polyvinylidenzyanidfaserstoff *m* polyvinylidene cyanide fibre
Polyvinylkarbazol *n* polyvinyl carbazole

Polyvinylmethyläther *m* polyvinyl methyl ether
Polyvinylpropionat *n* polyvinyl propionate
Polyvinylpropionatharz *n* polyvinyl propionate resin
Polyvinylpyrrolidin *n* polyvinylpyrrolidine, PVP
Polyvinylpyrrolidon *n* polyvinylpyrrolidone, PVP
Polyvinyltoluol *n* polyvinyl toluene
Polyvinylzyklohexan *n* polyvinyl cyclohexane
polyzentrisch polycentric, multicentre
polyzyklisch polycyclic
Polyzyklochinon *n* polycyclic quinone
POM *s.* Polyoxymethylen
Pomade *f (cosmet)* pomade, pomatum
Pomeranzenblütenöl *n* orange-flower oil, neroli oil *(from Citrus aurantium L. ssp. aurantium)*
Pomeranzenschalenöl *n* orange-peel oil
  bitteres P. bitter orange-peel oil *(from Citrus aurantium L. ssp. aurantium)*
  süßes P. sweet orange-peel oil *(from Citrus sinensis (L.) Osbeck)*
Pomilio-Verfahren *n (pap)* Pomilio process *(pulping with chlorine)*
Pompejanischrot *n* Pompeian red *(a pigment consisting of Fe_2O_3)*
Pontianak *m s.* Djelutungharz
Poort *m* bort, boort, boart *(collectively for diamonds of inferior quality)*
Pore *f* pore, void; *(rubber)* cell
Porenbeton *m* cellular[-expanded] concrete
porenbildend pore-forming
Porenbildung *f* pore formation; *(build)* air entrainment *(for improving the properties of concrete)*; *(coat)* pinholing *(a defect)*
Porendurchmesser *m* pore diameter
porenfrei non-porous
Porenfüller *m*, **Porenfüllmittel** *n* pore filler; *(coat)* sealer, sealing paint
Porengröße *f* pore size
Porengrößenverteilung *f* pore-size distribution
Porengummi *m* microcellular rubber
Porenraum *m* pore space
  relativer P. *(filtr)* porosity, voidage
Porensaugwasser *n (build)* water of capillarity
Porenschließer *m (coat)* sealer, sealing paint
Porensinter *m* [lightweight] expanded clay aggregate; expanded shale
Porenvolumen *n* pore volume; *(soil)* volume of pore space
  relatives P. *(filtr)* porosity, voidage
Porenwasser *n (ceram)* pore water; *(build)* water of capillarity
Porenweite *f* pore size
porig pored
porös 1. porous, porose, poriferous; 2. *s.* porig
Porosimeter *n* porosimeter
Porosität *f* porosity, *(quantitatively)* voidage
  scheinbare P. *(ceram)* apparent porosity
  wahre P. *(ceram)* true porosity
Porpezit *m (min)* porpezite, palladium gold
Porphin *n* porphin[e] *(a pyrrole pigment)*
Porphinskelett *n* porphin[e] ring
Porphyr *m* porphyry
Porphyrglattwalze *f* smooth porphyry roll
Porphyrin *n* porphyrin
Porphyrinchelat *n* porphyrin chelate
porphyrisch porphyritic
Porphyroblast *m (geol)* porphyroblast
porphyroblastisch *(geol)* porphyroblastic
Porphyrwalze *f* porphyry roll
Portion *f* portion
portionsweise in portions
Portlandzement *m* portland cement
Portlandzementklinker *m* portland cement clinker
Portugalöl *n s.* Pomeranzenschalenöl / süßes
Porzellan *n* 1. porcelain, *(for non-technical use also)* china; 2. *s.* Porzellanware ✦ **aus P.** *s.* porzellanen
  P. für hohe Temperaturen high-temperature porcelain
  P. mit hohem Aluminiumoxidgehalt (Tonerdegehalt) high-alumina porcelain
  chemisches P. chemical porcelain
  elektrotechnisches P. electrical porcelain
  lithiumoxidhaltiges P. lithia porcelain
  nichttechnisches P. china
Porzellanabdampfschale *f* porcelain evaporating basin (dish)
Porzellanbehälter *m* porcelain tank
Porzellandreieck *n (lab)* porcelain triangle
Porzellandruck-Seidenpapier *n* porcelain (pottery) tissue paper
porzellanen porcelan[e]ous, porcelainous
Porzellanerde *f* porcelain (china) clay, kaolin[e], bolus alba, white bole
  geschlämmte P. [china] clay, kaolin[e]
Porzellanfilter *n* porcelain filter
Porzellanfiltertiegel *m* porous porcelain crucible
Porzellanfliese *f* porcelain tile
Porzellangut *n* vitreous china
Porzellanherstellung *f* porcelain manufacture
Porzellanisolator *m* porcelain insulator
Porzellanjaspis *m* porcelain jasper *(a variety of porcellanite)*
Porzellankasserolle *f* porcelain casserole
Porzellankitt *m* porcelain cement
Porzellanmehl *n* powdered porcelain
Porzellanplatte *f* porcelain plate
Porzellanrohr *n* porcelain pipe
Porzellanschale *f* porcelain basin (dish)
Porzellanscharffeuerglasur *f* high-firing porcelain glaze
Porzellanscherben *m (ceram)* porcelain body
Porzellanschiffchen *n (lab)* porcelain boat
Porzellanseidenpapier *n* porcelain (pottery) tissue paper *(a wrapping paper)*
Porzellantiegel *m (lab)* porcelain crucible
Porzellanware *f* porcelain [ware], *(for non-technical use also)* china[ware]
Position *f* position
positiv positive
  p. geladen positively charged
  einfach p. unipositive
  zweifach p. dipositive
Positiv *n (phot)* positive
Positivbild *n (phot)* positive image

Positivemulsion *f (phot)* positive emulsion
Positiventwickler *m (phot)* positive developer
Positivfilm *m (phot)* positive film
Positivform *f (plast)* positive mould, *(in drape and vacuum forming)* male mould
Positivladung *f* positive charge
Positiv-Lichtpausverfahren *n* autopositive photocopying process
Positivmaterial *n (phot)* positive material
Positron *n* positron, positive electron, antielectron
Positron-Elektron-Paar *n* positron-electron pair, positive-negative electron pair
Positronenbildung *f* positron formation
Positronenemission *f* positron emission
Positronenstrahler *m* positron radiator
Positronenstrahlung *f* positron radiation
Positronenzerfall *m* positron decay (disintegration)
Positronenzerstrahlung *f* destruction of positrons
Positronium *n* positronium *(a system consisting of a positron and an electron)*
Posten *m (tech)* batch; *(glass)* [glass] gob, gather [of glass]
Postenform *f (glass)* gob shape
Postengewicht *n (glass)* gob weight
Postenspeiser *m (glass)* gob feeder
Postenspeisung *f (glass)* gob feeding
Postpapier *n* letter (note) paper, *(Am)* correspondence paper
postvulkanisch postvolcanic
Potential *n* potential
  chemisches P. chemical potential
  elektrisches P. electric potential
  elektrochemisches P. electrochemical potential
  elektrokinetisches P. electrokinetic (double-layer, zeta) potential
  retardiertes P. retarded potential
  thermodynamisches P. thermodynamic potential
ζ-Potential *n s.* Potential / elektrokinetisches
Potentialabfall *m* potential drop, fall of potential
Potentialbarriere *f*, **Potentialberg** *m s.* Potentialwall
potentialbildend potential-forming
Potentialdifferenz *f* potential difference
Potentialenergie *f* potential energy
Potentialenergiekurve *f* potential-energy curve
Potentialfeld *n* potential field
Potentialfläche *f* equipotential (potential-energy) surface
Potentialfunktion *f* potential function
Potentialgefälle *n* potential gradient
Potentialgleichung *f* potential[-energy] equation
Potentialgradient *m* potential gradient
Potentialkasten *m s.* Potentialtopf
Potentialkurve *f* potential-energy curve
Potentialmessung *f* potential measurement
Potentialmulde *f s.* Potentialtopf
Potentialrückgang *m* potential drop, drop of potential
Potentialschwelle *f s.* Potentialwall
Potentialsprung *m* potential jump
Potentialstreuung *f* potential scattering
Potentialtopf *m* potential well (hole)
Potentialunterschied *m* potential difference
Potentialvermittler *m* potential mediator
Potentialwall *m* potential[-energy] barrier, potential threshold (hill, wall)
potentiell potential
Potentiometer *n* potentiometer
Potentiometerverfahren *n* potentiometric method
Potentiometrie *f* potentiometry
potentiometrisch potentiometric
Potenz *f* / **nukleophile** nucleophilicity
Pottasche *f* K_2CO_3 potash, potassa, carbonate of potash, potassium carbonate
Pott-Broche-Verfahren *n* Pott-Broche process *(coal extraction)*
Potten *n (text)* potting
Pourpoint *m* pour point *(as of a lubricating oil)*
Pourpoint-Depressor *m* pour-point depressant
Pourpre Française *f* archil, orchil *(a lichen dye)*
Po-Z *s.* Polenske-Zahl
Pozz[u]olanerde *f s.* Puzzolanerde
PP *s.* 1. Polypropylen; 2. Polypropylenfaserstoff
PP-Faktor *m s.* Pellagrapräventivvitamin
PPO *s.* Polyphenylenoxid
Prädissoziation *f* predissociation
Prädissoziationsspektrum *n* predissociation spectrum
Prädissoziationszustand *m* predissociation level
Präfix *n (nomencl)* prefix
  multiplizierendes P. *s.* vervielfachendes P.
  numerisches P. numer[ic]al prefix
  vervielfachendes P. multiplying (multiplicative numeral) prefix
Prägekalander *m* embossing (goffering) calender
prägen to emboss, to goffer
Prägen *n* embossing, goffering
Prägepapier *n* embossing paper
Prägepresse *f (plast)* hobbing press
Prägestempel *m (plast)* punch, hob
Prägung *f* 1. embossing, goffering, embossed design; 2. *s.* Prägen
Präionisation *f* preionization, autoionization
Präkursor *m* precursor, progenitor
präkursorfrei precursor-free
Prallabscheiden *n* impingement separation
Prallabscheider *m* impingement (inertial, momentum) separator, impingement collector
Prallblech *n s.* Prallfläche
Prallbrecher *m* impact crusher
Pralldüse *f* impact (deflector) nozzle
Prallfläche *f* baffle, baffle (impingement, deflector) plate ✦ **mit Prallflächen ausstatten** to baffle ✦ **ohne Prallflächen** unbaffled
  P. aus feuerfestem Ton fireclay baffle
Prallmühle *f* impact (reflection) mill
Prallplatte *f s.* Prallfläche
Prallscheider *m s.* Prallabscheider
Prallwand *f s.* Prallfläche
Prallzerkleinerung *f* impact crushing
Präparat *n* preparation
  galenisches P. *(pharm)* galenical
  kosmetisches P. cosmetic [preparation]
  pharmazeutisches P. pharmaceutic[al] preparation

virentötendes (viruzides) P. virucide, viricide, viricidal agent
Präparatenchemie *f* preparative chemistry
Präparatenglas *n* specimen (preservation, museum) jar, show (inverted) bottle
Präparationsgalette *f (text)* sizing pad
präparativ preparative
präparieren to prepare
Präpariermikroskop *n* dissecting microscope
Präpariernadel *f* dissecting needle
lanzettenförmige P. lancet-point dissecting needle
Präparierpinzette *f* pinning forceps
Präpariersalz *n* $Na_2[Sn(OH)_6]$ preparing salt, sodium hexahydroxostannate(IV)
Prasem *m (min)* prase *(a variety of chalcedony)*
Praseodym *n* Pr praseodymium
Praseodymchlorid *n* $PrCl_3$ praseodymium chloride
Praseodymdioxid *n s.* Praseodym(IV)-oxid
Praseodymkarbonat *n* $Pr_2(CO_3)_2$ praseodymium carbonate
Praseodym(III)-oxid *n* Pr_2O_3 praseodymium(III) oxide, praseodymium trioxide
Praseodym(IV)-oxid *n* PrO_2 praseodymium(IV) oxide, praseodymium dioxide
Praseodymsulfat *n* $Pr_2(SO_4)_3$ praseodymium sulphate
Praseodymsulfid *n* Pr_2S_3 praseodymium sulphide
Praseodymtrioxid *n s.* Praseodym(III)-oxid
Präserve *f (food)* preserve
Prayon-Filter *n* Prayon [continuous] filter, tilting-pan filter
Präzipitat *n* 1. precipitate *(any of several mercury compounds)*; 2. *s.* Niederschlag
gelbes P. HgO yellow precipitate, yellow mercuric oxide
rotes P. HgO red precipitate, red mercuric oxide
schmelzbares weißes P. $[Hg(NH_3)_2]Cl_2$ fusible white precipitate, diamminemercury(II) chloride, diammine mercuric chloride
unschmelzbares weißes P. $[Hg(NH_2)]Cl$ infusible white precipitate, amidomercury(II) chloride
Präzipitation *f* precipitation
präzipitieren to precipitate
präzipitierend precipitative
Präzipitiermittel *n* precipitant, precipitating agent
Präzipitin *n (med)* precipitin *(an antibody)*
Präzisionsapothekerwaage *f* prescription balance
Präzisionsfotometrie *f* precision photometry
Präzisionsguß *m (met)* precision casting
P. nach dem Ausschmelzverfahren precision investment casting
Präzisionspolarimeter *n* precision polarimeter
Präzisionswaage *f* precision balance
Precoatfilter *n* precoat[ed] filter
Precoatschicht *f (filtr)* precoat [layer, bed]
Prehnit *m (min)* prehnite *(a sorosilicate)*
Prehnitol *n* $C_6H_2(CH_3)_4$ prehnitene, prehnitol, 1,2,3,4-tetramethylbenzene
Prehnitsäure *f* $C_6H_2(COOH)_4$ prehnitic acid
Prehnitylsäure *f* $(CH_3)_3C_6H_2COOH$ prehnitylic acid, 2,3,4-trimethylbenzoic acid
Premier jus *m* premier jus *(fine edible tallow)*
Premier-Kolloidmühle *f* Premier [colloid] mill
Premium-Benzin *n*, **Premium-Kraftstoff** *m* premium gasoline (motor fuel, spirit)
Premium-Öl *n* premium oil *(a lubricant)*
Prepaktbeton *m* prepacked (grouted) concrete
Prephensäure *f* $HOC_6H_5(COOH)CH_2COCOOH$ prephenic acid *(a cyclohexadiene derivative)*
Prepolymer-Verfahren *n (plast)* prepolymer process *(foaming)*
Prepreg *n* prepreg *(preimpregnated glass-fibre material)*
Preßautomat *m (plast)* automatic moulding machine
Preßband *n* compacted strip *(in powder rolling)*
Preßblasmaschine *f (glass)* press-and-blow machine
Preßblasverfahren *n (glass)* press-and-blow process
Preßdauer *f (plast)* moulding cycle
preßdicht compact
Preßdruck *m* pressing pressure, *(relating to solid particles also)* compacting pressure; *(plast)* moulding pressure; *(coal)* briquetting pressure
P. beim Formpressen *(plast)* compression-moulding pressure
P. beim Preßspritzen (Spritzpressen) *(plast)* transfer moulding pressure
Presse *f* press; *(food)* press, squeezer; *(plast)* moulding press
P. mit Einzelantrieb self-contained press
beheizbare P. hot press
einhüftige P. open-side (open-gap, gap-type, jaw-type, C-frame) press
filzlose (glättende) P. *(pap)* smoothing (offset) press
kippbare P. tilting-head press
pressen to press, to compress, to compact; to mould; *(food)* to press, to express, to squeeze [out]
heiß p. to hot-press, *(relating to powders also)* to sinter under pressure
nochmals p. to re-press
zu Briketts p. to briquette
Pressen *n* **in halbtrockenem Zustand** *(ceram)* semi-dry pressing
P. mit Gummisack *(plast)* [rubber-]bag moulding
P. von Faserbreiformteilen *(plast)* pulp moulding
plastisches P. *(ceram)* plastic (wet) pressing
Pressenanordnung *f* press arrangement
Pressenheizung *f (rubber)* press cure (curing, vulcanization)
Pressenpartie *f (pap)* press part (section)
Pressenschleifer *m (pap)* pocket grinder
3-Pressen-Schleifer *m (pap)* three-pocket grinder
Pressentisch *m* press table, table press, ram, *(Am)* [press] platen
Preßfett *n* expressed fat
Preßfilz *m (pap)* press[ing] felt
Preßfläche *f (plast)* projected area
Preßform *f* mould; die *(of an extruder)*
Preßglanzdekatur *f (text)* pressure decatizing

Preßglas *n* pressed glass
Preßgrat *m (plast)* fin, fan
Preßguß *m s.* Kaltkammerdruckgießen
Preßgut *n* 1. material being (*or* to be) compacted; 2. material being (*or* to be) pressed (expressed)
Preßharz *n* [compression-]moulding resin
Preßhefe *f* compressed yeast
Preßholz *n* compressed (compression) wood
Preßkasten *m (pap)* pocket *(of a pulpwood grinder)*
Preßkohle *f* briquetted coal
Preßkopf *m* extrusion tube
Preßkorb *m* curb *(of a wine press)*
Preßkörper *m* 1. *s.* Preßling; 2. compressed cartridge *(technology of explosives)*
Preßkuchen *m (tech)* press[ed] cake; *(food)* oil (mill) cake; *(plast)* biscuit *(for pressing disk records)*
Preßlauge *f* press liquor
Preßling *m (plast, ceram, glass)* moulding; compact *(powder metallurgy)*; *(coal)* briquet[te]
gesinterter P. sintered[-powder metal] compact
Preßlingsfläche *f (plast)* projected area
Preßluft *f* compressed (compression) air
Preßluftleitung *f* compressed-air line
Preßlufttrüttler *m* pneumatic (air-driven) vibrator
Preßluftventil *n* pneumatic (air) valve
Preßmasse *f* 1. *(ceram)* press body (mix); press[ing] dust *(in dry pressing)*; 2. *(plast)* compression-moulding material
pulvrige P. *(plast)* moulding powder
Preßöl *n (food)* expressed oil; *(petrol)* pressed distillate, blue oil *(obtained by dewaxing)*
pressorisch *(pharm)* pressor
Preßplatte *f (pap)* pressure foot *(of a pocket grinder)*
Preßpumpe *f* high-pressure (high-head) pump
Preßring *m* clamp ring *(as on a Söderberg electrode)*
Preßrißbildung *f (ceram)* pressure cracking
Preßrückstand *m* expressed residue
Preßrunzel *f (glass)* flow line *(a surface defect)*
Preßschichtholz *n* compreg, compressed resin-impregnated wood
Preßsintern *n* sintering under pressure, hot pressing *(of metal powder)*
Preßspan *m* pressboard, pressing board, press[s]pahn
P. für Elektrotechnik electrical pressboard
Preßspanersatz *m* imitation pressboard
Preßspritzen *n (plast)* transfer (flow) moulding, *(Am also)* plunger molding
Preßspritzwerkzeug *n (plast)* transfer mould
Preßstaub *m (ceram)* press[ing] dust
Preßstempel *m (plast)* male form (mould), moulding (force) plug, patrix
Preßstück *n s.* Preßling
Preßtalg *m (food)* pressed tallow, oleostearin[e]
Preßtasche *f (pap)* pocket *(of a pulpwood grinder)*
Preßteil *n (plast)* moulding
P. aus Schichtstoff moulded laminate
P. mit Schnitzelfüllstoff macerate moulding
ausgehärtetes P. cured moulding
Preßtuch *n* filter cloth
Preßtuchmatte *f* filter mat
preßverdichten to compact
Preßverdichtung *f* compaction
Preßvollholz *n s.* Preßholz
Preßvulkanisation *f* press cure (curing, vulcanization)
Preßwalze *f* press (pressure, compression, squeeze) roll; *(coal)* briquetting roll
Preßwasserreaktor *m* pressurized-water reactor, PWR
Preßwerkzeug *n* pressing tool; *(plast)* mould
P. mit vertieft liegendem Abquetschrand *(plast)* semipositive mould
zusammengesetztes P. *(plast)* composite mould
Preßzyklus *m (plast)* moulding cycle
Preußischblau *n* Prussian blue
Preventer *m (petrol)* blow-out preventer
Priceit *m (min)* priceite *(a soroborate)*
prillen *(fert)* to prill
Prillturm *m (fert)* prilling tower
Primaquin *n* primaquine *(a quinoline derivative)*
primär primary
Primärakt *m* primary reaction
Primärazetat *n (text)* primary [cellulose] acetate
Primärbatterie *f* primary battery
Primärbeschleuniger *m (rubber)* primary accelerator
Primärbezugskraftstoff *m* primary reference fuel
Primärcharge *f* / **zusätzlich verdünnte** *(petrol)* primary dilute charge *(in dewaxing)*
Primärdestillation *f* primary distillation
Primärelektron *n* primary (initiating) electron
Primärelement *n (el chem)* primary cell
Primärfaden *m (glass)* basic fibre
Primärgraphit *m (met)* kish, keesh
Primärionisation *f* initial ionization
Primärlagerstätte *f* primary deposit
Primärluft *f* primary air
Primärlunker *m* [/ **offener**] *(met)* primary pipe *(in an ingot)*
Primärmetall *n* primary (virgin) metal
Primärprodukt *n* primary product
Primärreaktion *f* primary (initiating) reaction
fotochemische P. primary photoreaction (photochemical reaction)
Primärschicht *f* primary layer
Primärsprengstoff *m* initial detonating agent
Primärspule *f (text)* winding head
Primärstandard *m* primary standard *(pH measurement)*
Primärstufe *f* initiating step
Primärteer *m* low-temperature tar
Primärton *m* primary (residual) clay
Primärwand *f* primary (outside, outer) wall *(of a vegetable fibre)*
Primärzentrifuge *f* primary centrifuge
Primulin *n* primuline *(an anthraquinone derivative)*
Primulinbase *f* primuline-type base
Primulingelb *n* primuline yellow
Primulinrot *n* primuline red
Prins-Reaktion *f* Prins reaction *(addition of formaldehyde)*
Prinzip *n* 1. principle, rule, law; 2. principle, fundamental constituent, base

P. der Erhaltung der Energie law of conservation of energy, energy principle
P. der größten Multiplizität [Hund] maximum-multiplicity principle, Hund's first rule
P. des beweglichen Gleichgewichtes *s.* P. des kleinsten Zwanges
P. des kleinsten Zwanges principle of least resistance (restraint), Le Chatelier[-Braun] principle, principle of mobile equilibrium
aktives P. *s.* wirksames P.
Babinetsches P. *(cryst)* Babinet absorption rule
Berthelot-Thomsensches P. Thomsen-Berthelot principle
Carnotsches P. Carnot theorem
färbendes (färberisches) P. colouring principle
Fermatsches P. Fermat's principle
fluoreszierendes P. fluorophore, fluorogen *(a group of atoms which give a molecule fluorescent properties)*
Franck-Condonsches P. Franck-Condon principle
Hundsches P. *s.* P. der größten Multiplizität
Le Chatelier-Braunsches P. *s.* P. des kleinsten Zwanges
Paulisches P. [Pauli] exclusion principle
toxisches P. toxic principle
wirksames P. active principle (ingredient, agent, substance)

Prisma *n* / **Nicolsches** Nicol prism
Prismenspektrograf *m* prism spectrograph
Prismenspektroskop *n* prism spectroscope
Pro *s.* Prolin
Probe *f* 1. sample, specimen; *s.* Probekörper; 2. *(met)* assay; *s.* Prüfung ✦ **eine P. [ent]nehmen** to sample
Baudouinsche P. Baudouin test (reaction) *(for detecting sesame oil)*
Bettendorfsche P. Bettendorf's test [for arsenic]
Delfter P. *(rubber)* Delft test piece
Gutzeitsche P. Gutzeit test *(for detecting arsenic)*
Lassaignesche P. Lassaigne['s] test *(for detecting nitrogen in organic substances)*
Marshsche P. Marsh's arsenic test, Marsh's test [for arsenic]
nasse P. *(met)* wet assay
Reinschsche P. Reinsch test [for arsenic]
trockene P. *(met)* dry assay

Probeabzug[s]papier *n* [galley] proof paper, proofing paper
Probeentnahme *f* sampling
Probefärbung *f* trial dyeing
probehaltig proof *(of standard alcoholic content)*
Probekörper *m* test piece (specimen), sample
bogenförmiger P. *(rubber)* crescent test piece
ringförmiger P. *(rubber)* ring test piece, ring sample
stabförmiger P. test rod; *(rubber)* dumb-bell test piece, dumb-bell strip, dumb bell

Probelauf *m* test (dry) run
Probelösung *f* test (experimental) solution, solution to be analyzed
Probemischung *f* trial mix[ture]
Probenahme *f* sampling
Probenahmegerät *n*, **Probenehmer** *m* sampling device (tool), sampler
Probenehmerhahn *m* sampling cock
Probenrohr *n* sample tube
Probestab *m* test rod
Probestück *n* *s.* Probekörper
Probesubstanz *f* test (experimental) substance (material), substance under investigation, substance being (*or* to be) investigated
probieren 1. *(met)* to assay; *s.* prüfen; 2. *(food)* to taste
Probierglas *n* test tube
Probiergläschen *n* taster *(as for wine)*
Probierglasgestell *n* test-tube rack (stand, support)
Probierglashalter *m* test-tube holder
Probierhahn *m* sampling cock
Probierstein *m* touchstone
Probitmortalität *f (tox)* probit mortality
Prochlorit *m (min)* prochlorite, ripidolite *(a phyllosilicate)*
Proctor-Trockner *m (ceram)* Proctor dryer *(a tunnel dryer)*
Prodigiosin *n* prodigiosin[e] *(a red pigment produced by bacteria)*
Produkt *n (chem, tech)* product
disubstituiertes P. disubstitution product
feuerfestes P. refractory [product]
Habersches P. *(tox)* ct product
helles P. *(petrol)* white product
kernsubstituiertes P. nuclear substitution product
monosubstituiertes P. monosubstitution product
primäres P. primary product
reformiertes P. *(petrol)* reformate
verperltes P. shot
vulkanisiertes P. vulcanized product
weißes P. *(petrol)* white product

Produktenleitung *f (petrol)* refined-product pipeline
Produktion *f* 1. manufacture, make, production; 2. *s.* Produktionsleistung
P. in großtechnischem Maßstab large-scale production
P. in halbtechnischem Maßstab pilot-plant-scale production, pilot production

Produktionsabwasser *n* factory (works, trade) effluent, industrial waste-water
Produktionsbohrung *f (petrol)* exploitation (development) well
Produktionskraft *f (agric)* productive capacity *(of soils)*
Produktionsleistung *f*, **Produktionsmenge** *f* production, output, make
Produktionsphase *f* make (run) part *(of a production cycle)*
Produktionsprogramm *n* production pattern
Produktionsverfahren *n* manufacturing process
Produktionsvermögen *n* capacity
Produktkühler *m (distil)* product condenser
Proenzym *n*, **Proferment** *n* proenzyme, proferment, zymogen

Profil *n* profile, shape; *(plast)* section
gespritztes (stranggepreßtes) P. *(plast)* extruded section
profilieren to profile, to shape
Profilkalander *m* *(rubber)* profiling calender
Profilpapier *n* profile (scale, sectional) paper
Profilseide *f* bulky yarn *(chemical fibre)*
Progesteron *n* progesterone, progestine *(a hormone)*
Programmpapier *n* programme (noiseless) paper
Proguanil *n* proguanil, *N*[1]-*p*-chlorophenyl-*N*[5]-isopropyldiguanide
Projektionsformel *f* projection formula
Fischersche P. Fischer projection formula
Projektionsgalvanometer *n* projection galvanometer
Prokain *n* $NH_2C_6H_4CO–OCH_2CH_2N(C_2H_5)_2$ procaine, β-diethylaminoethyl *p*-aminobenzoate
Prolaktin *n* prolactin (lactogenic, luteotrophic) hormone
Prolamin *n* prolamine *(a simple vegetable protein)*
Prolin *n* proline, pyrrolidine-2-carboxylic acid
Prolinase *f* prolinase
Promethium *n* Pm promethium
Promotion *f* *(phys chem)* promotion
Promotionsenergie *f* *(phys chem)* promotion energy
Promotor *m* promoter, promoting agent, activator, activating substance
Promotorwirkung *f* promoter action
Promovierung *f* *(phys chem)* promotion
Prooxygen *n* prooxidant *(a substance which accelerates oxidation)*
Propadien *n* $CH_2{=}C{=}CH_2$ propadiene, allene
Propan *n* $CH_3CH_2CH_3$ propane
Propanabtrennung *f* depropanization
Propanal *n* CH_3CH_2CHO propanal, propionaldehyde
1,2-Propandiamin *n* $CH_3CH(NH_2)CH_2NH_2$ 1,2-propanediamine, propylenediamine
Propandiol-(1,2) *n* $CH_3CH(OH)CH_2OH$ 1,2-propanediol, 1,2-dihydroxypropane
Propandisäure *f* $CH_2(COOH)_2$ propanedioic acid, malonic acid
Propanentasphaltierung *f* propane deasphalting
Propanentasphaltierungsanlage *f* propane-deasphalting plant
Propanentparaffinierung *f* propane dewaxing
Propanentparaffinierungsanlage *f* propane-dewaxing plant
Propannitril *n s.* Propionitril
Propanol-(1) *n* $CH_3CH_2CH_2OH$ 1-propanol
Propanol-(2) *n* $CH_3CH(OH)CH_3$ 2-propanol
Propanolyse *f* propanolysis
Propanon *n* CH_3COCH_3 propanone, acetone, dimethylketone
Propansäure *f s.* Propionsäure
1-Propanthiol *n* $CH_3CH_2CH_2SH$ 1-propanethiol
Propantrikarbonsäure-(1,2,3) *f* 1,2,3-propanetricarboxylic acid
Propantriol-(1,2,3) *n* $HOCH_2CH(OH)CH_2OH$ 1,2,3-propanetriol, 1,2,3-trihydroxypropane, glycerol
Propargylalkohol *m* $CH{\equiv}CCH_2OH$ propargyl alcohol, 2-propyn-1-ol
Propargylsäure *f s.* Propiolsäure
Propektin *n s.* Protopektin
Propeller *m* propeller
Propellermischer *m* propeller mixer (agitator)
Propellerpumpe *f* propeller (axial-flow) pump
Propellerrührer *m*, **Propellerrührwerk** *n s.* Propellermischer
Propellerventilator *m* propeller (axial-flow) fan
Propen *n* $CH_3CH{=}CH_2$ propene
Propenal *n* $CH_2{=}CHCHO$ propenal, acraldehyde, acrolein
Propenamid *n* $CH_2{=}CHCONH_2$ propenamide, acrylamide
Propennitril *n* $CH_2{=}CHCN$ propenenitrile, acrylonitrile, vinyl cyanide
Propen-(2)-ol-(1) *n* $CH_2{=}CHCH_2OH$ 2-propen-1-ol, allyl alcohol
Propenol-(3) *n s.* Propen-(2)-ol-(1)
Propensäure *f* $CH_2{=}CHCOOH$ propenoic acid, acrylic acid
Propenylbromid *n* $CH_3CH{=}CHBr$ 1-bromopropene, propenyl bromide
Propenylgruppe *f*, **Propenylrest** *m* $CH_3CH{=}CH–$ propenyl group (residue)
Propepsin *n s.* Pepsinogen
Prophylaktikum *n* prophylactic
prophylaktisch prophylactic
Propin *n* $CH_3C{\equiv}CH$ propyne
Propin-(2)-ol-(1) *n* $CH{\equiv}CCH_2OH$ 2-propyn-1-ol, propargyl alcohol
β-Propiolakton *n* β-propiolactone, BPL
Propiolalkohol *m s.* Propin-(2)-ol-(1)
Propiolsäure *f* $HC{\equiv}CCOOH$ propiolic acid, propargylic acid, propynoic acid
Propionaldehyd *m* CH_3CH_2CHO propionaldehyde, propanal
Propionat *n* propionate *(salt or ester of propionic acid)*
Propionitril *n* CH_3CH_2CN propionitrile, ethyl cyanide
Propionsäure *f* CH_3CH_2COOH propionic acid, propanoic acid
Propionsäureäthylester *m* $CH_3CH_2CO–OC_2H_5$ ethyl propionate
Propionsäurebenzylester *m* $C_2H_5CO–OCH_2C_6H_5$ benzyl propionate
Propionsäuregärung *f* propionic-acid fermentation
Propionsäurenitril *n s.* Propionitril
Propionylbenzol *n* $C_6H_5COCH_2CH_3$ propionylbenzene, propiophenone
Propionylgruppe *f*, **Propionylrest** *m* $C_2H_5CO–$ propionyl group (residue)
Propiophenon *n* $C_6H_5COCH_2CH_3$ propiophenone, ethyl phenyl ketone
Propolis *f* propolis, bee glue, balm
Proportion *f* proportion
Proportionalitätsbereich *m* proportional region
Proportionalitätsfaktor *m* proportionality constant
Proportionalitätsgrenze *f* proportional limit; *(plast)* offset yield strength (stress)
Proportional[itäts]wägung *f* direct weighing

Proportionalzähler *m*, **Proportionalzählrohr** *n* proportional counter
Propylalkohol *m* propanol, propyl alcohol, *(specif)* $CH_3CH_2CH_2OH$ 1-propanol
sek.-**Propylalkohol** *m s.* Propanol-(2)
Propylamin *n* $CH_3CH_2CH_2NH_2$ propylamine
n-**Propyläthen** *n*, *n*-**Propyläthylen** *n s.* Penten-(1)
Propylazetat *n* $CH_3COOC_3H_7$ propyl acetate
n-**Propylazetylen** *n s.* Pentin-(1)
Propylchlorid *n* $CH_3CH_2CH_2Cl$ 1-chloropropane, propyl chloride
Propylen *n s.* Propen
Propylenaldehyd *m s.* Krotonaldehyd
Propylenbromid *n s.* Propylendibromid
Propylendiamin *n* $CH_3CH(NH_2)CH_2NH_2$ 1,2-propanediamine, propylenediamine
Propylendibromid *n* $CH_3CHBrCH_2Br$ 1,2-dibromopropane, propylene dibromide
Propylendikarbonsäure *f* $HOOCC(=CH_2)CH_2COOH$ 2-propene-1,2-dicarboxylic acid, itaconic acid
Propylenglykol *n* $HOCH_2CH(OH)CH_3$ 1,2-propanediol, 1,2-propylene glycol
Propylenoxid *n* propylene oxide, 2-methyloxiran
propylenoxidisch propylene-oxidic
Propylessigsäure *f* $CH_3[CH_2]_3COOH$ *n*-valeric acid, pentanoic acid, propyl acetic acid
Propylit *m* propylite *(an extrusive rock)*
Propylkarbinol *n s.* Butanol-(1)
n-**Propylmagnesiumbromid** *n* C_3H_7MgBr *n*-propylmagnesium bromide
Propylmerkaptan *n* / **primäres** *s.* 1-Propanthiol
Prosize-Verfahren *n (pap)* Prosize process
prospektieren *(petrol, mine)* to prospect
Prospektierung *f (petrol, mine)* prospecting
prosthetisch prosthetic *(enzymology)*
Protaktinium *n* Pa protactinium
Protamin *n* protamine *(any of a class of simple proteins)*
Protease *f* protease, proteolytic (protein-digesting) enzyme
Proteid *n* conjugated protein, proteid[e]
Protein *n* protein, *(specif)* simple protein
  faserartiges (fibrilläres) P. fibrous protein
  globuläres (kugelförmiges) P. globular protein
Proteinabbau *m* protein degradation (breakdown)
Proteinanteil *m* protein moiety
proteinartig proteinaceous
Proteinase *f* proteinase
Proteindenaturierung *f* protein denaturation
Proteinfaser *f (text)* protein [man-made] fibre
Proteinfaserstoff *m (text)* protein [man-made] fibre
proteinhaltig containing protein
Proteinkomponente *f* protein moiety
Proteinkunststoff *m* protein plastic
Proteinlösung *f* protein solution
Proteinschicht *f* layer of protein
Proteinsilber *n (pharm)* silver protein[ate]
Proteinsol *n* protein sol
proteinspaltend *s.* proteolytisch
Proteinteil *m s.* Proteinanteil
Protektor *m (rubber)* tread, wearing surface
Protektorgummi *m* tread rubber
Protektorspritzkopf *m (rubber)* tread head
Protektorspritzmaschine *f (rubber)* tread extruder
Proteohormon *n* proteohormone
Proteolyse *f* proteolysis
Proteolyten *mpl* proteolytic bacteria
proteolytisch proteolytic, proteoclastic, protein-digesting
Proteose *f* proteose *(any of various cleavage products formed by partial hydrolysis of proteins)*
Prothrombin *n* prothrombin *(a glycoprotein)*
Protium *n* ^1H protium, light hydrogen
Protohämin *n* protohaemin, haemin chloride
Protokatechualdehyd *m* $(HO)_2C_6H_3CHO$ protocatechuic aldehyde, 3,4-dihydroxybenzaldehyde
Protokatechusäure *f* $(HO)_2C_6H_3COOH$ protocatechuic acid, 3,4-dihydroxybenzoic acid
Protolyse *f* protolysis
Protolyt *m* protolyte
protolytisch protolytic
Proton *n* proton
protonenabspaltend protogenic
Protonenaffinität *f* proton affinity
Protonenakzeptor *m* proton acceptor
Protonenbindungsenergie *f* proton binding energy
Protonendon[at]or *m* proton don[at]or
Protonenfänger *m* proton catcher
protonenfrei proton-free
Protonenresonanz *f* proton resonance
Protonenruh[e]masse *f* proton rest mass
Protonensäure *f* protonic acid
Protonenübertragung *f* proton transfer
Protonenverschiebung *f* proton shift
protonieren to protonate
Protonierung *f* protonation
protonogen protogenic
protonophil protophilic
Protonsäure *f* protonic acid
Protopektin *n* protopectin, pectinogen *(any of a group of water-insoluble pectic substances)*
Protopin *n* protopine, fumarine *(an opium alkaloid)*
Protoplasma *n* protoplasm
prototrop prototropic
Prototropie *f* prototropism, prototropy *(one form of tautomerism)*
Proustit *m (min)* proustite, light red silver ore *(arsenic(III) silver sulphide)*
Provitamin *n* provitamin
Prozentgehalt *m* percentage
Prozeß *m* process
Prozeßdampf *m* process steam
Prozeßgas *n* process gas
Prozeßpumpe *f* process pump
Prozeßwasser *n* [industrial] process water
Prüfbecher *m (plast)* flow cup
Prüfbogen *m (pap)* test (hand, pulp) sheet
Prüfdauer *f* duration of test
prüfen to test; to control, to check, *(using an electronic device)* to scan
  auf Gehalt p. *(pharm, met)* to assay
  auf Geschmack p. to taste
Prüfer *m* 1. tester *(person)*; *(food)* taster; 2. *s.* Prüfgerät

Prüfgerät *n* testing instrument, tester, meter
Prüfglas *n* test glass
Prüfkörper *m*, **Prüfling** *m* test piece (specimen)
Prüflösung *f* test solution
Prüfmethode *f* test[ing] method
  standardisierte P. standard test[ing] method
Prüfmischung *f (rubber)* test compound (mix, mixture)
Prüfmittel *n* testing agent
Prüfmuster *n s.* Prüfkörper
Prüfpapier *n* test (indicator) paper, *(Am)* reaction paper
Prüfprotokoll *n* testing protocol
Prüfsieb *n* test (testing) sieve (screen)
  standardisiertes P. standard (normal) test sieve
Prüfsiebreihe *f* test sieve series
Prüfstab *m* test rod
Prüftiegel *m* test cup *(of a flash-point tester)*
Prüfung *f* test; *(met)* assay
  P. auf Ameisensäure formic-acid test
  P. auf Biegefestigkeit cross-bend[ing] test
  P. auf Formaldehyd formaldehyde test
  P. auf Formaldehyd nach Arnold und Mentzel Arnold-Mentzel formaldehyde test
  P. auf Fuselöl fusel-oil test
  P. auf Kohlendioxid test for carbon dioxide
  P. auf Lichtbeständigkeit (Lichtechtheit) light test
  P. auf Phenol phenol test
  P. auf Salizylsäure salicylic-acid test
  P. auf Sterilität sterility test
  P. der Anvulkanisation *(rubber)* scorch test
  P. der Biegerißfestigkeit flex-cracking test
  P. des Anbrennens (Anspringens) *(rubber)* scorch test
  P. durch Außenbewetterung outdoor weathering test
  P. im Laboratoriumsmaßstab laboratory-scale test
  P. mit dem ballistischen Pendel ballistic pendulum test *(of explosives)*
  P. mit Röntgenstrahlen X-ray test
  dynamische P. dynamic test
  organoleptische P. organoleptic (sensory) estimation (evaluation)
  orientierende P. preliminary (basic) test *(as of pesticides)*
  statische P. static test
  zerstörungsfreie P. non-destructive test
Prüfungsdosis *f* test dose
Prüfungstoxin *n* test toxin
Prüfungsvorschrift *f* specification
Prüfverfahren *n* test[ing] method
Prüfversuch *m* test
Prüfvorrichtung *f s.* Prüfgerät
Prüfzeit *f* duration of test
Prussiat *n*, **Prussid** *n* prussiate, prussate
PS *s.* Polystyrol
Psammit *m* psammite *(a rock composed of sandy particles)*
Pschorr-Synthese *f* Pschorr reaction (synthesis) *(for obtaining phenanthrene derivatives)*
PSE *s.* Periodensystem [der Elemente]
Pseudoanthrazit *m* pseudo-anthracite
Pseudoaromaten *pl* pseudo-aromatics
Pseudoasymmetrie *f* pseudo-asymmetry
pseudoasymmetrisch pseudo-asymmetric[al]
Pseudobase *f* pseudo base
Pseudobenzol *n* $B_3N_3H_6$ inorganic benzene, borazole, triborine triamine
Pseudobrookit *m (min)* pseudobrookite *(an iron(II) titanium oxide)*
pseudodimolekular pseudo-dimolecular
Pseudoekgonin *n* Ψ-ecgonine, pseudo-ecgonine
Pseudohalogen *n* pseudo-halogen
Pseudokannelkohle *f* pseudocannel [coal]
Pseudokatalyse *f* pseudo-catalysis
pseudokristallin pseudo-crystalline
Pseudokumol *n* $C_6H_3(CH_3)_3$ Ψ-cumene, pseudocumene, 1,2,4-trimethylbenzene
Pseudolösung *f* pseudo-solution
Pseudomalachit *m (min)* pseudomalachite, dihydrite *(a hydrous basic copper phosphate)*
pseudomonomolekular pseudo-monomolecular
pseudomorph *(cryst)* pseudomorphic, pseudomorphous
Pseudomorphie *f (cryst)* pseudomorphism
Pseudomorphose *f (cryst)* 1. pseudomorphosis *(process)*; 2. pseudomorph *(substance)*
Pseudomorphosierung *f (cryst)* pseudomorphosis
pseudorazemisch pseudo-racemic
Pseudosalz *n* pseudo salt
Pseudosäure *f* pseudo acid
pseudostabil pseudo-stable
Pseudosymmetrie *f* pseudo-symmetry
pseudosymmetrisch pseudo-symmetric[al]
pseudotrimolekular pseudo-trimolecular
Pseudowollastonit *m (min)* pseudowollastonite *(calcium metasilicate)*
Psilomelan *m (min)* psilomelane *(manganese(IV) oxide)*
Psilozybin *n* psilocybin *(a tryptamine derivative)*.
Psychochemikalie *f* psychochemical
Psychomimetikum *n* psycho[to]mimetic
psychomimetisch psycho[to]mimetic
Psychopharmakon *n* psychotropic agent
Psychrometer *n* psychrometer, wet-and-dry-bulb hygrometer (thermometer)
PT *s.* Polyäthylenfaserstoff
Pteridin *n* pteridine, pyrimido-[4,5-*b*]-pyrazine
Pterin *n* pterin *(any of the pteridine derivatives)*
Pteroylglutaminsäure *f* pteroylglutamic acid
PTFE *s.* Polytetrafluoräthylen
Ptomain *n* ptomaine
Ptyalin *n* ptyalin *(salivary amylase)*
PU *s.* Polyurethanfaserstoff
Pucherit *m (min)* pucherite *(bismuth(III) orthovanadate)*
Puddeleisen *n* puddle iron
puddeln *(met)* to puddle
Puddeln *n (met)* puddling
Puddelofen *m (met)* puddling furnace
Puddelroheisen *n* forge pig iron
Puddelstahl *m* puddle steel

Puddelverfahren *n (met)* puddling process
Puder *m* powder
desodor[is]ierender P. deodorant powder
festgepreßter P. compressed powder
flüssiger P. liquid powder
gepreßter P. compressed powder
kompakter P. compact powder
Pudermittel *n* dusting powder, dusting (powdering) agent, *(rubber also)* chalk; coating substance *(as for conditioning fertilizers)*
pudern to dust, to powder, *(rubber also)* to chalk; to coat *(e.g. fertilizers)*
mit Talkum p. *(rubber)* to soapstone
Puderpapier *n (cosmet)* powder paper
Puderstoff *m* coating substance *(as for conditioning fertilizers)*
Puderzucker *m* powdered (castor, icing) sugar
Puffer *m* 1. *(chem)* buffer; 2. *(petrol)* surge tank *(clay contacting process)*
Pufferbereich *m* buffer region
Puffergefäß *n* buffer vessel
Puffergemisch *n* buffer mixture
Pufferion *n* buffer ion
Pufferkapazität *f* buffering capacity (power)
Pufferlösung *f* buffer solution; buffered solution
standardisierte P. standard buffer [solution]
Puffermischung *f* buffer mixture
puffern to buffer
Puffersubstanz *f* buffering agent (substance), buffer reagent
Puffersystem *n* buffer system
Pufferung *f* buffering
Pufferungsvermögen *n* buffering capacity (power)
Pufferwert *m* buffer value (index), BI
Pufferwirkung *f* buffering effect (action), buffer action
Pufferzone *f* buffer region
Pulegon *n* pulegone *(a monocyclic ketone)*
Pulfrich-Refraktometer *n* Pulfrich refractometer
Pulp *m*, **Pulpe** *f s.* Pülpe
Pülpe *f (food)* [fruit] pulp, pomace, pummace
Pülpefänger *m (sugar)* beet pulp catcher
Pulper *m (pap)* pulper, pulping engine, hydrapulper
P. mit vier Auflösescheiben quatropulper
P. mit zwei Auflösescheiben duopulper
Pulsation *f* pulsation
Pulsationsdämpfer *m* pulsation damper, [pulsation] snubber
pulsationsfrei non-pulsating
Pulsationskolonne *f* pulsed column (tower)
Pulsator *m* pulser, pulsing device
pulsieren to pulse, to pulsate
Pulsometer *n* pulsometer [pump], acid egg, blow case
Pulspolarografie *f* pulse polarography
Pulver *n* powder; meal *(of rock)*
einbasiges P. single-base powder *(technology of explosives)*
elektrolytisches (katodisches) P. electrolytic powder
metallisches P. metal powder, powder[ed] metal
oberflächenaktives P. wettable powder *(crop protection)*
rauchloses (rauchschwaches) P. smokeless powder
zweibasiges P. double-base powder *(technology of explosives)*
pulverartig powdery, pulverulent
pulveraufkohlen *(met)* to pack-carburize
Pulveraufkohlen *n (met)* pack (solid-pack, solid) carburizing
Pulveraufnahme *f s.* Pulverdiagramm
Pulverband *n* / **dichtgepreßtes** compacted strip *(in powder rolling)*
Pulverdiagramm *n* powder diagram, X-ray (back reflection) powder pattern
pulvereinsetzen *s.* pulveraufkohlen
pulverförmig, pulverig powdery, pulverulent
Pulverisieranlage *f* pulverizing equipment
pulverisieren to pulverize, to reduce to powder, to powder, to triturate
Pulverisiermühle *f* pulverizing mill, pulverizer
Pulverkamera *f* [X-ray] powder camera
Pulvermetallurgie *f* powder metallurgy, metal ceramics
pulvermetallurgisch powder-metallurgical, metal-ceramic
Pulvermethode *f* [/ **Debye-Scherrersche**] *(cryst)* [X-ray] powder method, Debye-Scherrer-Hull method
pulvern *s.* pulverisieren
Pulverpreßkörper *m*, **Pulverpreßling** *m* [powder metal] compact
Pulverseele *f* [powder] core *(of a blasting fuse)*
Pulversinterverfahren *n (plast)* powder moulding (sintering) process
Pulversprengstoff *m* powder (low) explosive, deflagrating powder, propellant [explosive]
Pulvertrichter *m* powder (filling, wide-stemmed) funnel
Pulververfahren *n s.* Pulvermethode
Pulververstäuber *m (agric)* dusting machine, duster
Pulverwalzen *n* powder (direct) rolling *(of powder metal)*
pulverzementieren *s.* pulveraufkohlen
pulvrig *s.* pulverförmig
Pumpbeton *m* pumping (pumped) concrete
Pumpe *f* pump
P. für Gase gas pump
P. für heiße Medien hot-charge pump
P. in diagonaler Bauart *s.* halbaxiale P.
P. mit einseitigem Flüssigkeitseintritt single-suction pump
P. mit Leitrad diffuser (turbine) pump
P. mit zweiseitigem Flüssigkeitseintritt double-suction pump
dreistufige P. three-stage pump
einseitig saugende P. single-suction pump
einstufige P. single-stage pump
elektromagnetische P. electromagnetic pump
halbaxiale P. mixed-flow pump, turbine pump
mehrstufige P. multistage pump

rotierende P. rotary pump
vierstufige P. four-stage pump
zweiseitig saugende P. double-suction pump
zweistufige P. two-stage pump
pumpen to pump
Pumpengehäuse *n*, **Pumpenkörper** *m* pump casing
Pumpenwirkungsgrad *m* pump efficiency
pumpfähig pumpable
Pumpfähigkeit *f* pumpability
Pumpgrenze *f* pumping limit (point) *(of a compressor)*
Punkt *m* / **azeotrop[isch]er** azeotropic point
dunkler P. *(plast)* dark speck, hull *(as in laminated fibres)*
elektrisch neutraler P. isoelectric point
eutektischer P. eutectic [point]
isoelektrischer P. isoelectric point
isoionischer P. isoionic point
kritischer P. critical (plait) point, point of criticality *(as in three-component systems)*; critical moisture point *(in drying processes)*
kryohydratischer P. cryohydric point
peritektischer P. peritectic (transition) point
Punktbeschichten *n (plast)* spot coating
Punktgruppe *f s.* Punktsymmetriegruppe
Punktschweißen *n* spot welding
Punktsymmetriegruppe *f (cryst)* point group
pupillenerweiternd mydriatic
pupillenverengend myotic, miotic
Puppe *f (rubber, plast)* puppet, billet
PUR *s.* Polyurethan
Purgativum *n (pharm)* purgative
Purgiernuß *f* physic nut, Jatropha curcas L.
Purifikation *f* purification *(of emulsions)*
Purin *n* purine
Purinbase *f* purine base
Puringruppe *f* purine group
6-Purinon *n* 6(1H)-purinone, hypoxanthine, 6-hydroxypurine
6-Purinthiol *n* 6-purinethiol, 6-mercaptopurine, leukerin
Purkinje-Phänomen *n (phot)* Purkinje effect
Purpur *m* purple
P. der Alten *s.* Antiker P.
Antiker (Byzantinischer, Tyrischer) P. Phoenician (Tyrian) purple, purple of the ancients *(6,6'-dibromoindigo)*
Purpurerz *n* purple ore, blue billy *(leached residue of roasted pyrites containing copper)*
Purpurin *n* purpurin, 1,2,4-trihydroxyanthraquinone
Purpursäure *f* purpuric acid
Purpursäurechelat *n* purpureate chelate
Purpurschnecke *f* purple snail
Putreszin *n* $H_2N[CH_2]_4NH_2$ putrescine, 1,4-diaminobutane
Putride *npl* putrefactive bacteria
putzen to fettle, to trim *(metal castings, plastics mouldings, or excess body in pottery-ware)*; *(build)* to plaster
Putzen *n* **von Hand** hand fettling
Putzmaschine *f* fettling machine
Putzmittel *n* scouring agent
Putzmörtel *m* plaster [mortar]
Putztrommel *f (plast)* tumbling barrel
Puzzolanerde *f (build)* pozzolana, puzzolan[a]
Puzzolanzement *m* pozzolanic (puzzolanic) cement
PV *s.* Polyvinylfaserstoff
PVA *s.* 1. Polyvinylalkohol; 2. Polyvinylalkoholfaserstoff
PVAC *s.* Polyvinylazetat
PVB *s.* 1. Polystyrolfaserstoff; 2. Polyvinylbutyral
PVC *s.* 1. Polyvinylchlorid; 2. Polyvinylchloridfaserstoff
PVCA *s.* Polyvinylchloridazetat
PVC-H, PVC-hart *n* unplasticized (rigid) PVC
PVC-W, PVC-weich *n* plasticized (flexible) PVC
PVD *s.* Polyvinylidenchloridfaserstoff
PVDC *s.* Polyvinylidenchlorid
PVK *s.* Pigmentvolumenkonzentration
PVP *s.* Polyvinylpyrrolidon
PVY *s.* Polyakrylnitrilfaserstoff
Pyknometer *n* pycnometer, density (specific-gravity) bottle (flask)
pyknometrisch pycnometric
Pyozyanin *n* pyocyanin[e] *(antibiotic)*
Pyran *n* pyran
Pyranose *f* pyranose *(a cyclic monosaccharide)*
Pyranosering *m* pyranose ring
Pyranosestruktur *f* pyranose structure
Pyranosid *n* pyranoside
Pyrargyrit *m* pyrargyrite, dark red silver [ore], red silver ore, silver ruby *(antimony(III) silver sulphide)*
Pyrazin *n* pyrazine, 1,4-diazine
Pyrazol *n* pyrazole, 1,2-diazole
Pyrazolin *n* pyrazoline
Pyrazolon *n* pyrazolone
Pyrazolon-Azofarbstoff *m*, **Pyrazolonfarbstoff** *m* [azo-]pyrazolone dye
Pyren *n* pyrene *(a tetracyclic compound)*
Pyrethrin *n* pyrethrin
Pyrethrineinwirkung *f (tox)* pyrethrinization
Pyrethroid *n* pyrethroid *(any of a series of insecticidal compounds)*
Pyridazin *n* pyridazine, 1,2-diazine
Pyridin *n* pyridine
Pyridinalkaloid *n* pyridine-type alkaloid
Pyridin-Butadienkautschuk *m* pyridine-butadiene rubber, PBR
Pyridindikarbonsäure *f* pyridinedicarboxylic acid
Pyridinextraktion *f* pyridine extraction
Pyridin-3-karbonsäure *f* pyridine-3-carboxylic acid, nicotinic acid
Pyridinring *m* pyridine ring
Pyridinsynthese *f* / **Hantzschsche** Hantzsch pyridine synthesis
Pyridin-2,3,5-trikarbonsäure *f* pyridine-2,3,5-tricarboxylic acid, carbodinicotinic acid
Pyridin-2,4,5-trikarbonsäure *f* pyridine-2,4,5-tricarboxylic acid, berberonic acid
Pyridon *n* pyridone, hydroxypyridine
Pyridoxalphosphat *n* pyridoxal phosphate *(coenzyme of decarboxylase)*

Pyridoxin *n* pyridoxin, adermin, 3-hydroxy-4,5-di-(hydroxymethyl)-2-methylpyridine *(vitamin B_6)*
Pyrimidin *n* pyrimidine, 1,3-diazine
Pyrit *m (min)* [iron] pyrite[s], mundic *(iron disulphide)*
Pyritkonzentrat *n* pyrite[s] concentrate
Pyritröstung *f* roasting of pyrite[s]
Pyritschmelzen *n (met)* pyritic smelting
Pyroantimonat(V) *n s.* Diantimonat(V)
Pyroantimonsäure *f s.* Diantimon(V)-säure
Pyroarsenat(III) *n s.* Diarsenat(III)
Pyroarsenat(V) *n s.* Diarsenat(V)
Pyroarsensäure *f s.* Diarsen(V)-säure
Pyroborat *n s.* Tetraborat
Pyroborax *m* dehydrated (calcined, burnt, anhydrous) borax *(sodium tetraborate)*
Pyroborsäure *f s.* Tetraborsäure
Pyrochlor *m (min)* pyrochlore
Pyrochroit *m (min)* pyrochroite *(manganese(II) hydroxide)*
Pyrogallol *n* $C_6H_3(OH)_3$ pyrogallol, pyrogallic acid, 1,2,3-trihydroxybenzene
Pyrogallolgerbstoff *m* pyrogallol tan
Pyrogallussäure *f s.* Pyrogallol
pyrogen pyrogenic
Pyrokatechol *n s.* Brenzkatechin
Pyroligninsäure *f* pyroligneous acid *(crude acetic acid obtained by wood distillation)*
Pyrolusit *m (min)* pyrolusite *(manganese(IV) oxide)*
Pyrolyse *f* pyrolysis
  P. mit Pebbles pebble-heater pyrolysis
  gelenkte P. controlled pyrolysis
Pyrolyseprodukt *n* pyrolyzate
pyrolysieren to pyrolyze
pyrolytisch pyrolytic
Pyromellithsäure *f* $C_6H_2(COOH)_4$ pyromellitic acid, benzene-1,2,4,5-tetracarboxylic acid
Pyromellithsäuredianhydrid *n* pyromellitic dianhydride
Pyrometallurgie *f* pyrometallurgy
pyrometallurgisch pyrometallurgical
Pyrometer *n* pyrometer
  fotoelektrisches P. photoelectric pyrometer
  optisches P. optical pyrometer
  thermoelektrisches P. thermoelectric pyrometer
Pyrometrie *f* pyrometry
pyrometrisch pyrometric
Pyromorphit *m (min)* pyromorphite, green (brown) lead ore
Pyron *n* pyrone *(any of a class of heterocyclic compounds containing the oxo group)*
Pyrop *m (min)* pyrope *(a nesosilicate)*
Pyrophanit *m* pyrophanite *(manganese metatitanate)*
pyrophor pyrophoric, pyrophorous
Pyrophosphat *n s.* Diphosphat
Pyrophosphatase *f* pyrophosphatase
Pyrophosphit *n s.* Diphosphit
Pyrophosphorsäure *f s.* Diphosphorsäure
Pyrophyllit *m (min)* pyrophyllite *(a phyllosilicate)*
Pyrophysalit *m (min)* pyrophysalite *(a variety of topaz)*
Pyrosäure *f* pyroacid
Pyroschleimsäure *f s.* Brenzschleimsäure
Pyroskop *n (ceram)* pyroscope
Pyrosol *n (coll)* pyrosol *(a type of solid sols)*
Pyrostilpnit *m (min)* pyrostilpnite *(antimony(III) silver sulphide)*
Pyrosulfit *n s.* Disulfit
Pyrosulfurylchlorid *n s.* Disulfurylchlorid
Pyrotechnik *f* pyrotechnics
Pyrotechniker *m* pyrotechnist, pyrotechnician
pyrotechnisch pyrotechnic[al]
Pyrotritarsäure *f* pyrotritaric acid, 2,5-dimethylfuran-3-carboxylic acid
Pyrovanadat *n s.* Divanadat
Pyroweinsäure *f* $HOOCCH(CH_3)CH_2COOH$ pyrotartaric acid, methylsuccinic acid, 2-methyl-1,4-butanedioic acid
Pyroxen *m (min)* pyroxene *(a member of a group of inosilicates)*
Pyroxenfamilie *f (min)* pyroxene group
Pyrrhotin *m (min)* pyrrhotite, pyrrhotine, magnetic pyrites *(iron(II) sulphide)*
Pyrrol *n* pyrrole
Pyrrolfarbstoff *m* pyrrole pigment
Pyrrolidin *n* pyrrolidine, tetrahydropyrrole
Pyrrolidinalkaloid *n* pyrrolidine alkaloid
Pyrrolidin-2-karbonsäure *f* pyrrolidine-2-carboxylic acid, proline
Pyrrolring *m* pyrrole ring
Pyrrolsynthese *f* / **Knorrsche** Knorr pyrrole synthesis
Pyrrolylen *n s.* Butadien-(1,3)
Pyruvat *n* pyruvate *(salt or ester of pyruvic acid)*
Pyruvinaldehyd *m s.* Methylglyoxal
Pyruvinsäure *f s.* Brenztraubensäure
PZ *s.* 1. Portlandzement; 2. Polenske-Zahl
P-Zweig *m* P-branch *(of a band spectrum)*

Q

Q-Einheit *f* Q unit, tetrafunctional unit *(a structural unit)*
Quaderkalk *m* ashlar lime
Quadratbecken *n* square basin (tank)
Quadratmetergewicht *n s.* Masse je Flächeneinheit
Quadrupeleffektverdampfer *m* quadruple-effect evaporator
Quadrupelpunkt *m (phys chem)* quadruple point
Quadrupol *m (phys chem)* quadrupole
Quadrupolkräfte *fpl* quadrupole forces
Quadrupolmoment *n* quadrupole moment
  Q. des Kerns nuclear quadrupole moment
Quadrupolstrahlung *f* quadrupole radiation
Qualität *f* grade, quality ✦ **von minderer Q.** low-grade
  handelsübliche Q. commercial grade
Qualitätsanforderungen *fpl* quality demands
Qualitätsdüngung *f* quality fertilization
Qualitätskontrolle *f* quality control
Qualitätsminderung *f* deterioration in quality
Qualitätsprüfung *f* quality test

Qualitätsverbesserung *f* improvement in quality, upgrading
Quant *n (phys chem)* quant[um]
quanteln *(phys chem)* to quantize
Quantelung *f (phys chem)* quantization
Quantenäquivalenz *f* quantum equivalence
Quantenäquivalenzgesetz *n* law of the photochemical equivalent, Stark-Einstein law
Quantenausbeute *f* quantum efficiency (yield)
Quantenbedingung *f* quantum condition
Quantenchemie *f* quantum chemistry
Quantenelektrodynamik *f* quantum electrodynamics
Quantenenergie *f* quantum energy
Quantenflüssigkeit *f* quantum liquid
Quantenmechanik *f* quantum (matrix) mechanics
quantenmechanisch quantum-mechanical
Quantensprung *m* quantum jump (transition)
Quantenstatistik *f* quantum statistics
Quantentheorie *f* quantum theory
Quantenzahl *f* quantum number
  azimutale Q. azimuthal (secondary, orbital) quantum number
  innere Q. inner quantum number
  magnetische Q. magnetic quantum number
  radiale Q. radial quantum number
  räumliche Q. *s.* magnetische Q.
  sekundäre Q. *s.* azimutale Q.
Quantenzähler *m* quantum counter
Quantenzustand *m* quantum state
quantisieren to quantize
Quantisierung *f* quantization
Quantitätseigenschaft *f (phys chem)* extensive property (quantity)
Quantum *n* quantum, quantity, amount; portion
Quark *n* quark *(a hypothetical particle)*
quartär quaternary
Quartärsalz *n* quaternary salt
Quartärstruktur *f* quaternary structure *(of proteins)*
Quartation *f* quartation *(separating gold and silver by hot nitric acid)*
Quarz *m* quartz ✦ **aus Q.** *s.* quarzig
  linker Q. left-handed quartz
  rechter Q. right-handed quartz
Quarzboot *n* quartz boat
Quarzfilternutsche *f* quartz filter funnel
quarzführend *s.* quarzhaltig
Quarzgefäß *n* quartz vessel
Quarzglas *n* quartz glass, transparent [vitreous] silica, fused quartz
Quarzgut *n* translucent vitreous silica, fused silica
quarzhaltig quartzic, quartziferous, quartzose, quartzous, quartzy
quarzig quartzose, quartzous, quartzy
Quarzit *m (min)* quartzite
quarzitisch quartzitic
Quarzkeil *m* quartz wedge
Quarzkorn *n* quartz grain
Quarzkristall *m* quartz crystal
Quarzlampe *f* quartz lamp
Quarzporphyr *m* quartz porphyry
Quarzsand *m* quartz sand
Quarzschiffchen *n* quartz boat
Quarzspektrograf *m* quartz spectrograph
Quarztiegel *m* quartz crucible
Quarzuhr *f* quartz[-crystal] clock
Quasi-Emulgator *m* quasi-emulsifier
Quasi-Fermi-Niveau *n* quasi-Fermi level
Quasifließen *n* pseudo-plastic flow
quasikristallin quasi-crystalline
quasiplastisch pseudo-plastic
Quasirazemat *n* quasi-racemate
quasistabil quasi-stable
quasistationär quasi-stationary
quasistatisch quasi-static
quasiviskos quasi-viscous
Quassiaholz *n (pharm)* quassia, bitterwood, bitter ash *(from Quassia amara L. or Picrasma excelsa (Swartz) Planchon)*
quaternär quaternary
Quaternisierung *f* quaternization *(as of amines)*
Quaternisierungsmittel *n* quaternizing agent
Quebrachin *n* quebrachine, yohimbine, aphrodine *(a rauvolfia alkaloid)*
Quebrachoextrakt *m (tann)* quebracho extract
Quebrachoholz *n (tann)* quebracho *(from Quebrachia lorentzii Griseb. and Schinopsis balansae Engl.)*
Quebrachorinde *f (pharm)* quebracho bark *(from Aspidosperma quebracho-blanco Schlechtend.)*
Quecksilber *n* Hg mercury
  gediegenes Q. native mercury
Quecksilberalkyl *n s.* Alkylquecksilberverbindung
Quecksilber(II)-amidochlorid *n* $[Hg(NH_2)]Cl$ amidomercury(II) chloride, infusible white precipitate
Quecksilber(I)-azetat *n* CH_3COOHg mercury(I) acetate, mercurous acetate
Quecksilber(II)-azetat *n* $(CH_3COO)_2Hg$ mercury(II) acetate, mercuric acetate
Quecksilber(II)-azetylid *n* HgC_2 mercury(II) acetylide, mercuric acetylide
Quecksilber(I)-azid *n* HgN_3 mercury(I) azide, mercurous azide
Quecksilberbarometer *n* mercury (mercurial) barometer
Quecksilber(I)-bromat *n* $HgBrO_3$ mercury(I) bromate, mercurous bromate
Quecksilber(II)-bromat *n* $Hg(BrO_3)_2$ mercury(II) bromate, mercuric bromate
Quecksilber(I)-bromid *n* Hg_2Br_2 mercury(I) bromide, mercurous bromide
Quecksilber(II)-bromid *n* $HgBr_2$ mercury(II) bromide, mercuric bromide
Quecksilber(II)-bromidjodid *n* HgBrI mercury(II) bromide iodide, mercuric bromide iodide
Quecksilber(I)-chlorat *n* $HgClO_3$ mercury(I) chlorate, mercurous chlorate
Quecksilber(II)-chlorat *n* $Hg(ClO_3)_2$ mercury(II) chlorate, mercuric chlorate
Quecksilber(I)-chlorid *n* Hg_2Cl_2 mercury(I) chloride, mercurous chloride, calomel
Quecksilber(II)-chlorid *n* $HgCl_2$ mercury(II) chloride, mercuric chloride, sublimate

Quecksilber(II)-chloridjodid *n* HgClI mercury(II) chloride iodide, mercuric chloride iodide
Quecksilber(I)-chromat *n* Hg_2CrO_4 mercury(I) chromate, mercurous chromate
Quecksilber(II)-chromat *n* $HgCrO_4$ mercury(II) chromate, mercuric chromate
Quecksilberdampf *m* mercury vapour
Quecksilberdampflampe *f* mercury[-vapour] lamp, mercury arc (discharge) lamp
Quecksilberdestillationsapparat *m* mercury still
Quecksilberdi ... *s.* Quecksilber(II)- ...
Quecksilberdichtung *f* mercury seal *(as on a stirrer)*
Quecksilberdiffusionspumpe *f* mercury diffusion pump, mercury[-vapour] pump
Quecksilberelektrode *f* mercury electrode
strömende Q. venous mercury electrode
Quecksilbererz *n* mercury ore
Quecksilberfahlerz *n (min)* schwazite *(a tetrahedrite containing mercury)*
Quecksilberfalle *f* mercury trap
Quecksilber(I)-fluorid *n* Hg_2F_2 mercury(I) fluoride, mercurous fluoride
Quecksilber(II)-fluorid *n* HgF_2 mercury(II) fluoride, mercuric fluoride
Quecksilber(II)-fulminat *n* $Hg(ONC)_2$ mercury fulminate, fulminating mercury, mercuric fulminate
quecksilberhaltig containing mercury, mercurial
Quecksilber(I)-hexafluorosilikat *n* $Hg_2[SiF_6]$ mercury(I) hexafluorosilicate, mercurous fluorosilicate
Quecksilber(II)-hexafluorosilikat *n* $Hg[SiF_6]$ mercury(II) hexafluorosilicate, mercuric fluorosilicate
Quecksilber(II)-hexoxotellurat(VI) *n* Hg_3TeO_6 mercury(II) hexaoxotellurate(VI), mercury(II) orthotellurate, mercuric orthotellurate
Quecksilber(II)-hydrogenarsenat(V) *n* $HgHAsO_4$ mercury(II) hydrogenarsenate, mercuric hydrogenarsenate
Quecksilber(I)-jodat *n* $HgIO_3$ mercury(I) iodate, mercurous iodate
Quecksilber(II)-jodat *n* $Hg(IO_3)_2$ mercury(II) iodate, mercuric iodate
Quecksilber(I)-jodid *n* Hg_2I_2 mercury(I) iodide, mercurous iodide
Quecksilber(II)-jodid *n* HgI_2 mercury(II) iodide, mercuric iodide
gelbes Q. yellow mercuric iodide
rotes Q. red mercuric iodide
Quecksilber(I)-karbonat *n* Hg_2CO_3 mercury(I) carbonate, mercurous carbonate
Quecksilberkatode *f* mercury cathode
Quecksilberlampe *f s.* Quecksilberdampflampe
Quecksilberlichtbogen *m* mercury arc
Quecksilbermanometer *n* mercury (mercurial) manometer, mercury gauge
Quecksilbermikroelektrode *f* mercury microelectrode
Quecksilber(I)-nitrat *n* $Hg_2(NO_3)_2$ mercury(I) nitrate, mercurous nitrate
Quecksilber(II)-nitrat *n* $Hg(NO_3)_2$ mercury(II) nitrate, mercuric nitrate
Quecksilber(II)-nitrid *n* Hg_3N_2 mercury(II) nitride, mercuric nitride
Quecksilber(I)-nitrit *n* $Hg_2(NO_2)_2$ mercury(I) nitrite, mercurous nitrite
Quecksilber(I)-orthoarsenat(V) *n* Hg_3AsO_4 mercury(I) orthoarsenate, mercurous orthoarsenate
Quecksilber(II)-orthoarsenat(V) *n* $Hg_3(AsO_4)_2$ mercury(II) orthoarsenate, mercuric orthoarsenate
Quecksilber(I)-orthophosphat *n* Hg_3PO_4 mercury(I) orthophosphate, mercurous orthophosphate
Quecksilber(II)-orthophosphat *n* $Hg_3(PO_4)_2$ mercury(II) orthophosphate, mercuric orthophosphate
Quecksilber(I)-oxid *n* Hg_2O mercury(I) oxide, mercurous oxide
Quecksilber(II)-oxid *n* HgO mercury(II) oxide, mercuric oxide
gefälltes (gelbes) Q. yellow mercuric oxide, yellow precipitate
rotes Q. red mercuric oxide, red precipitate
Quecksilberpumpe *f s.* Quecksilberdiffusionspumpe
Quecksilber(I)-rhodanid *n s.* Quecksilber(I)-thiozyanat
Quecksilber(II)-rhodanid *n s.* Quecksilber(II)-thiozyanat
Quecksilbersalbe *f* mercurial ointment
Quecksilber(I)-salz *n* mercury(I) salt, mercurous salt
Quecksilber(II)-salz *n* mercury(II) salt, mercuric salt
Quecksilbersäule *f* mercury column
Quecksilber(II)-selenid *n* HgSe mercury(II) selenide, mercuric selenide
Quecksilberstand *m* mercury level
Quecksilber(I)-sulfat *n* Hg_2SO_4 mercury(I) sulphate, mercurous sulphate
Quecksilber(II)-sulfat *n* $HgSO_4$ mercury(II) sulphate, mercuric sulphate
Quecksilber(I)-sulfid *n* Hg_2S mercury(I) sulphide, mercurous sulphide
Quecksilber(II)-sulfid *n* HgS mercury(II) sulphide, mercuric sulphide
rotes Q. red mercuric sulphide
Quecksilbertauchlampe *f* mercury immersion lamp
Quecksilber(II)-tetroxotellurat(VI) *n* $HgTeO_4$ mercury(II) tetraoxotellurate(VI), mercury(II) metatellurate, mercuric metatellurate
Quecksilberthermometer *n* mercury-in-glass thermometer
Quecksilber(I)-thiozyanat *n* HgSCN mercury(I) thiocyanate, mercury(I) rhodanide, mercurous thiocyanate (rhodanide)
Quecksilber(II)-thiozyanat *n* $Hg(SCN)_2$ mercury(II) thiocyanate, mercury(II) rhodanide, mercuric thiocyanate (rhodanide)
Quecksilberton *m* mercury clay
Quecksilbertropfelektrode *f* dropping mercury electrode, DME
Quecksilbertropfkatode *f* dropping mercury cathode
Quecksilberverfahren *n* mercury-cell process *(electrolysis)*

Quecksilbervergiftung *f* mercurialism, hydrargyrism
Quecksilberverstärker *m (phot)* mercury intensifier
Quecksilberverstärkung *f (phot)* mercury intensification
Quecksilber(I)-wolframat *n* Hg_2WO_4 mercury(I) wolframate, mercury(I) tungstate, mercurous wolframate (tungstate)
Quecksilber(II)-wolframat *n* $HgWO_4$ mercury(II) wolframate, mercury(II) tungstate, mercuric wolframate (tungstate)
Quecksilberzange *f* mercury tongs
Quecksilberzelle *f* mercury cell
Quecksilber(II)-zyanid *n* $Hg(CN)_2$ mercury(II) cyanide, mercuric cyanide
quellbar capable of swelling
Quellbarkeit *f* capability of swelling, swelling capacity
quellbeständig resistant to swelling, swell-resistant, non-swelling
Quellbeständigkeit *f* resistance to swelling
Quellbottich *m (ferm)* steep tank, steeping vat, steeper
Quelle *f* 1. source; 2. spring
  muriatische Q. brine spring
quellen 1. to swell; 2. to steep, to [cause to] swell
quellfähig capable of swelling
Quellfähigkeit *f* 1. capability of swelling, swelling capacity; 2. swelling power
quellfest resistant to swelling, swell-resistant, non-swelling
Quellfestmittel *n* non-swelling agent
Quellgrad *m s.* Quellungsgrad
Quellprüfung *f* swelling test
Quellpunkt *m (glass)* hot spot
Quellschweißen *n* solvent welding
Quellstock *m s.* Quellbottich
Quellung *f* 1. *(of a substance)* swelling; 2. *(of human agent)* steeping, steepage, swelling
Quellungsbetrag *m s.* Quellungsgrad
Quellungsdruck *m* swelling pressure
Quellungsgeschwindigkeit *f* rate of swelling
Quellungsgrad *m* degree (extent, amount) of swelling
Quellungsmittel *n* swelling agent
Quellungsprüfung *f*, **Quellungsversuch** *m* swelling test
Quellungswärme *f* heat of swelling
Quellverhalten *n* swelling behaviour (properties)
Quellvermögen *n s.* Quellfähigkeit
Quellwasser *n* spring water
Quellwirkung *f* swelling effect
Quellzement *m* expanding (expansive) cement
Quenchöl *n (met)* quench[ing] oil
Querbruchfestigkeit *f* transverse strength
Querflammenwanne *f (glass)* side-fired (side-port) furnace
Querlauf *m*, **Querrichtung** *f (pap)* cross direction (way)
Querschlag *m* cross-gallery *(underground gasification)*
querschleifen *(pap)* to grind across the grain
Querschleifen *n (pap)* cross-grinding
Querschleifer *m (pap)* cross-grinder
querschneiden *(pap)* to cross-cut, to cut across, to sheet
Querschneiden *n (pap)* cross-cutting, sheeting
  Q. mit rotierenden Messern rotary cutting
Querschneider *m (pap)* cross-cutter, sheet cutter, sheeter
  Q. mit rotierenden Messern rotary [knife] cutter, revolving-knife cutting machine
Querschnitt *m* cross section
  Q. im Augenblick des Bruchs *s.* wirklicher Q.
  effektiver Q. effective cross section
  ursprünglicher Q. original cross section
  wirklicher Q. *(rubber)* cross section at break
Querschnittsfläche *f* cross-sectional area
Querschnittsverminderung *f (text)* necking *(on stretching filaments)*
Querspritzkopf *m (plast)* cross (transversal) extruder head, crosshead
Querstrom-Kühlturm *m* crossflow cooling tower
quervernetzen to cross-link
Quervernetzung *f* cross-linkage *(process and result)*
Querzahl *f* Poisson's ratio *(of transverse to longitudinal strain in a material under tension)*
Querzetin *n* quercetin, 3,3′,4′,5,7-pentahydroxyflavone
Querzitol *n* quercitol, pentahydrocyclohexane
Querzitrin *n* quercitrin *(rhamnoside of quercetin)*
Querzitron *n*, **Querzitronrinde** *f (dye)* quercitron [bark] *(from several Quercus specc.)*
Quetsche *f* press, squeezer
quetschen to press, to squeeze
Quetschfalten *fpl (pap)* calender cuts *(a defect)*
Quetschhahn *m* pinchcock, hose cock, pinch clamp
  Q. nach Day Day pinchcock
  Q. nach Hoffmann Hoffmann clamp, Bunsen screw clip
  Q. nach Mohr Mohr pinchcock (clip)
Quetschklemme *f s.* Quetschhahn
Quetschkolben *m* compressor *(of a diaphragm valve)*
Quetschwalze *f* press (pressure, compression) roll, squeeze (squeezing) roll, squeegee
Quillajarinde *f* quillaja, soap (China, Panama) bark *(from Quillaja saponaria Mol.)*
Quirl *m (ceram)* blunger
quirlen *(ceram)* to blunge
Quitten[samen]schleim *m (pharm)* quince seed mucilage
Quotient *m* / **respiratorischer** *(bioch)* respiratory quotient (ratio)
Q-Wert *m* Q-value *(of a nuclear reaction)*
Q-Zweig *m* Q-branch *(of a band spectrum)*

R

R *s.* Röntgen-Einheit
Rabi-Methode *f (nucl)* Rabi method

Rad *n*, **Rad-Einheit** *f* rad *(a unit of absorbed dose of ionizing radiation)*
Radialbeschleunigung *f* angular acceleration
Radialkolbenpumpe *f* radial-piston pump
Radialschaufel *f* radial blade
radialstengelig, radialstrahlig *(cryst)* radiating, divergent
Radialventilator *m* centrifugal fan
Radierbarkeit *f*, **Radierfestigkeit** *f* *(pap)* erasability
Radikal *n* radical, *(specif)* free radical
freies R. free radical
kettentragendes R. chain carrier (initiator, propagator)
radikalartig radical-like
Radikalbildung *f* radical formation
Radikaldissoziation *f* dissociation to radicals
Radikalfänger *m* radical trap
Radikal-Ion *n* radical ion
radikalisch radical
Radikalkettenpolymerisation *f* radical[-chain] polymerization, free-radical [chain] polymerization
Radikalkettenreaktion *f* radical-chain (free-radical chain) reaction
Radikalname *m* radical name
Radikalpolymerisation *f* *s.* Radikalkettenpolymerisation
Radikalreaktion *f* radical reaction
Radikalübertragung *f* radical transfer
Radikalwanderung *f* radical migration
Radioadapt[at]ion *f* *(biol)* radioadap[ta]tion
Radioaktinium *n* ^{227}Th, RdAc radioactinium
radioaktiv radioactive ✦ **r. machen** to radioactivate;
r. markiert radio-labelled
hochgradig r. *s.* stark r.
nicht r. non-radioactive, cold
stark r. highly radioactive, hot
Radioaktivität *f* radioactivity
induzierte (künstliche) R. induced radioactivity
Radioautografie *f* autoradiography, radioautography
Radiobiologie *f* radiobiology, radiation biology
Radiochemie *f* radiochemistry
radiochemisch radiochemical
Radiochromatografie *f* radiochromatography
Radioelement *n* radioelement, radioactive element
radiogen radiogenic
Radiografie *f* radiography *(materials testing)*
radiografisch radiographic
Radioindikator *m* radiotracer, radioactive indicator (tracer)
radioindiziert radio-labelled
Radioisotop *n* *(incorrectly for)* Radionuklid
Radiojod *n* radioiodine, radioactive iodine, *(specif)* ^{131}I iodine-131
Radiokarbonmethode *f* radiocarbon method, ^{14}C method
Radiokobalt *n* radiocobalt, radioactive cobalt, *(specif)* ^{60}Co cobalt-60
Radiokohlenstoff *m* radiocarbon, radioactive carbon, *(specif)* ^{14}C carbon-14
Radiokohlenstoffmethode *f* *s.* Radiokarbonmethode
Radiokolloid *n* radiocolloid
Radiologie *f* radiology
Radiolumineszenz *f* radioluminescence
Radiolyse *f* radiolysis
Radiometer *n* radiometer
Radiometrie *f* radiometry
radiometrisch radiometric
Radiomimetikum *n* *(biol)* radiomimetic
radiomimetisch *(biol)* radiomimetic
Radionatrium *n* radiosodium, radioactive sodium, *(specif)* ^{24}Na sodium-24
Radionuklid *n* radionuclide, radioactive nuclide
Radiophosphor *m* radiophosphorus, radioactive phosphorus, *(specif)* ^{32}P phosphorus-32
Radioschwefel *m* radiosulphur, radioactive sulphur, *(specif)* ^{35}S sulphur-35
Radiostrahlung *f* radio radiation
Radiostrontium *n* radiostrontium, radioactive strontium, *(specif)* ^{90}Sr strontium-90
Radiothorium *n* ^{228}Th, RdTh radiothorium
Radiotoxizität *f* radiotoxicity
Radium *n* Ra radium
Radiumbehandlung *f* *(med)* radium therapy
Radium-Beryllium-Neutronenquelle *f* radium-beryllium neutron source
Radiumbromid *n* $RaBr_2$ radium bromide
Radiumchlorid *n* $RaCl_2$ radium chloride
Radiumeinlage *f* *(med)* radium pack
Radium-Emanation *f* *s.* Radon-222
Radiumgehalt *m* radium content
Radiumisotop *n* radium isotope
Radiumjodat *n* $Ra(IO_3)_2$ radium iodate
Radiumkarbonat *n* $RaCO_3$ radium carbonate
Radiummoulage *f* *(med)* radium mould
Radiumnadel *f* *(med)* radium needle
Radiumpräparat *n* radium preparation
Radiumreihe *f* radium series
Radiumsulfat *n* $RaSO_4$ radium sulphate
Radiumtherapie *f* radium therapy
Radius *m*/ **van-der-Waalsscher** van der Waals radius
Radon *n* Rn radon
Radon-218 *n* ^{218}Rn radon-218
Radon-219 *n* ^{219}Rn radon-219
Radon-220 *n* ^{220}Rn radon-220
Radon-222 *n* ^{222}Rn radon-222
Radonfluorid *n* radon fluoride
Raffinade *f* refined sugar
Raffinadekläre *f* *(sugar)* refined liquor
Raffinadekochkläre *f* *(sugar)* clearing liquor, clairce
Raffinadezucker *m* refined sugar
Raffinase *f* raffinase
Raffinat *n* raffinate *(solvent extraction)*
Raffinatblei *n* refined lead
Raffinatende *n* raffinate end *(solvent extraction)*
Raffination *f* refining, refinement, purification; *(petrol)* [refining] treatment; *(food)* degumming *(of oil)*
R. im Schmelzfluß fire refining

elektrolytische R. electrolytic refining, electrorefining
hydrierende R. hydrorefining
Raffinationsanlage *f* refinery, refining plant
Raffinationsrückstand *m* refinery residue
Raffinatkupfer *n* refined (casting) copper
Raffinatphase *f* raffinate phase *(solvent extraction)*
Raffinatseite *f* raffinate end *(solvent extraction)*
Raffinatstripper *m* raffinate stripper *(solvent extraction)*
Raffinatverdampfer *m* raffinate evaporator *(solvent extraction)*
Raffinerie *f* refinery, refining plant
Raffineriegas *n* refinery gas
Raffineriemelasse *f (sugar)* refinery molasses
raffinieren to refine, to purify; *(petrol)* to refine, to treat; *(food)* to degum *(oil)*
im Schmelzfluß r. to fire-refine
Raffiniergas *n* refinery gas
Raffinierstahl *m* refined steel (iron, bar), shear steel, merchant bar
raffiniert / mit Lösungsmitteln solvent-refined
Raffinose *f* raffinose *(a trisaccharide)*
Ragmischung *f (rubber)* rag mix (stock)
Rahm *m* cream
geschlagener R. whipped cream
saurer R. cultured cream
Rahmeis *n* ice cream, cream ice
Rahmen *n* creaming *(of emulsions)*
Rahmenfilter *n* 1. *(for gases)* screen (envelope) filter; 2. *(for liquids)* plate-and-frame filter
Rahmen[filter]presse *f* plate-and-frame press (filter press, filter), plate-and-ring filter press, flush-plate [filter] press
Rahmenspannmaschine *f (text)* tenter [frame], stenter
Rahmkäse *m* cream cheese
Rahmkelle *f* skimmer
Rahmkühlwanne *f* cream chilling (cooling) vat
Rahmlöffel *m* skimmer
Rahmplasma *n* cream plasma
Rahmpulver *n* dried (dry) cream
Rahmreifer *m* cream (milk) ripener, cream ripening (ageing) tank
Rahmreifung *f* cream ripening
Rahmseparator *m* cream separator, skimming machine
Rahnwerden *n* darkening of wine *(a disorder)*
Rainfarnöl *n* tansy oil *(from Chrysanthemum vulgare (L.) Bernh.)*
Rakel *f (text, pap)* doctor blade (knife) *(in printing);* squeegee *(for screen printing)*
Rakelappretur *f (text)* doctor finish
Rakelauftragmaschine *f s.* Rakelstreichmaschine
Rakelmesser *n (plast, rubber)* doctor blade (knife)
Rakelstreichmaschine *f* doctor [kiss] coater, knife coater
Rakelstreifen *m* doctor streak
Raketentreibstoff *m* rocket propellant (fuel)
hypergoler R. hypergolic fuel (rocket propellant), hypergol *(self-igniting upon contact of components)*
ramanaktiv Raman-active
Raman-Effekt *m* Raman effect
Raman-Frequenz *f* Raman frequency
ramaninaktiv Raman-inactive
Raman-Linie *f* Raman line
Raman-Spektroskopie *f* Raman spectroscopy
Raman-Spektrum *n* Raman spectrum
Raman-Streuung *f* Raman scattering
Raman-Untersuchung *f* Raman study
Raman-Verschiebung *f* Raman shift
Ramiedichtung *f* ramie packing
Ramiefaser *f* ramie [fibre] *(from Boehmeria nivea (L.) Gaudich.)*
Ramiepackung *f* ramie packing
Rammelkamp-Methode *f* Rammelkamp method *(for testing penicillin)*
Rampe *f* bench, platform
schräge R. slope
Rampenbildung *f* casing (outer layer) effect *(a defect in glass)*
Ramsbottom-Carbon-Wert *m s.* Ramsbottom-Verkokungswert
Ramsbottom-Methode *f* Ramsbottom [coking] method, Ramsbottom carbon residue method
Ramsbottom-Test *m* Ramsbottom test *(for determining the carbon residue of oils)*
Ramsbottom-Verkokungswert *m,* **Ramsbottom-Verkokungszahl** *f* Ramsbottom value (coke number)
Rand *m* edge, border, brim; *(met)* case, surface layer
aufgekohlter R. *(met)* carburized case
nitrierter R. *(met)* nitride[d] case
Randabfall *m (pap)* trim
Randabspritzeinrichtung *f (pap)* squirt trim
randaufkohlen *(met)* to case-carburize
Randaufkohlung *f (met)* case carburization
Randbeschnitt *m (pap)* trim
Randblase *f* subcutaneous blowhole *(foundry)*
Randeffekt *m* border effect
Rändern *n (ceram)* banding
randkohlen *s.* randaufkohlen
Randschicht *f (met)* case, surface layer
aufgekohlte R. carburized case
eingesetzte R. *s.* aufgekohlte R.
gehärtete R. hardened case
nitrierte R. nitride[d] case
zementierte R. *s.* aufgekohlte R.
Randspritzer *m (pap)* squirt trim
Randspritzstoff *m (pap)* squirt-trimmed stock, squirt trim
Randstreifen *m (pap)* [cutter] trim
Randwalzen *fpl* edge rolls *(sheet-glass manufacture)*
Randwasser *n (petrol)* edge water
Randwassergrenze *f (petrol)* water table
Randwasserzeichen *n (pap)* edge (marginal) watermark
Randwinkel *m* contact angle *(in testing surface-active substances)*
vorwärtsschreitender R. advancing contact angle
Randwinkelmessung *f* measurement of the contact angle

Randzone *f s.* Randschicht
Raney-Katalysator *m* Raney (skeletal) catalyst
Raney-Nickel *n* Raney nickel
Rang *m (coal)* rank
Rangklassifikation *f* **der Kohlen** rank classification of coals
Rangordnung *f (nomencl)* seniority
Rangstufe *f (coal)* rank
Rankine-Skale *f* Rankine temperature scale
Ranzidität *f (food)* rancidity, rancidness
Ranziditätsprüfung *f (food)* rancidity test
ranzig rancid
r. werden to rancidify
Ranzigkeit *f* rancidity
hydrolytische R. *(food)* hydrolytic (lipolytic) rancidity
Ranzigwerden *n* rancidification
Rapidnetzer *m* rapid wetting agent
Rapidpolarografie *f* rapid polarography
Rapshartfett *n* hydrogenated rape-seed oil
Rapsöl *n* rape[-seed] oil *(from Brassica napus L. em. Metzg.)*
Rasamalaharz *n* Rasamala resin (wood oil) *(from Altingia excelsa Noron.)*
Raschig-Ring *m* Raschig ring *(a packing material)*
Raschig-Synthese *f* Raschig [hydrazine] synthesis
Raschig-Verfahren *n* Raschig process (method) *(catalytic conversion of benzene to phenol)*
raschwirkend quick-acting, short-action
Raseneisenerz *n (min)* bog iron [ore], bog ore
Rasenröste *f (text)* dew-ret[ting]
Rasiercreme *f* shaving cream
schäumende R. lather shaving cream
Rasier[hilfs]mittel *n* shaving preparation
Rasierstein *m* styptic pencil
Rasierwasser *n* / **nachbehandelndes** after-shave lotion
vorbehandelndes R. pre-shave lotion
Rast *f* bosh *(of a blast furnace)*
Rast-Methode *f* Rast method, Rast's molecular weight determination
Rationalitätsgesetz *n (cryst)* rational index law, law of rational indices, Haüy law
Rattenbekämpfung *f* rat destruction
Rattenbekämpfungsmittel *n*, **Rattengift** *n* rat poison, raticide
Rattenvertilgung *f* rat destruction
Rätter *m* gyratory screen (sifter), riddle
Rauch *m* smoke *(as of a stack)*; fume *(as of concentrated acids)*
rauchen to smoke *(as of a stack)*; to fume *(as of concentrated acids)*
rauchend fuming *(e.g. acids)*
Räucherapparat *m* fumigator *(pest control)*
Räucherei *f (food)* 1. smoke curing (drying), smoking; 2. smokery, smokehouse
Räucherhaus *n (food)* smokehouse, smokery
Räucherkammer *f (food)* smoke chamber
Räucherkautschuk *m* smoked sheet [rubber]
Räucherkerze *f* fumigating candle
Räuchermittel *n* [dry] fumigant *(pest control)*
räuchern *(food)* to smoke[-dry]; to fume *(timber)*; to fumigate *(e.g. rooms or plants for pest control)*
mit Schwefel r. to sulphur
Räuchern *n (food)* smoking, smoke drying (curing); fuming *(of timber)*; fumigation *(for pest control)*
Räucherofen *m* smoking kiln
Räucherpapier *n* fumigating paper *(for destroying insects)*; incense paper
Räucherpulver *n (agric)* fumigating (combustible) powder
Räucherung *f s.* Räuchern
Rauchfangdach *n* fume (canopy) hood
Rauchfeuer *n* soft fire
Rauchgas *n* flue (waste, combustion) gas, fume
Rauchgasanalyse *f* flue-gas analysis
Rauchgasbeständigkeit *f* resistance to fumes
Rauchgasechtheit *f (text)* gas-fume fastness, fastness to gas fading
Rauchgasentstauber *m* fly-ash precipitator
Rauchgasprüfer *m* flue-gas analyzer
Rauchgasresistenz *f* resistance to fumes
Rauchgassammelkanal *m* waste-gas flue
Rauchgasverlust *m* stack loss
Rauchgasvorwärmer *m* flue-gas preheater, economizer
Rauchgenerator *m* smoke generator
pyrotechnischer R. pyrotechnic smoke generator
Rauchkammer *f* smoke chamber
Rauchmeldeanlage *f* smoke (flue dust) monitor
Rauchnebel *m* smaze *(a type of air pollution)*
Rauchopium *n* chandoo
Rauchquarz *m (min)* smoky quartz, cairngorm [stone]
Rauchrohr *n* fire tube
Rauchrohrkessel *m* fire-tube boiler
Rauchsäule *f* column of smoke
Rauchschirm *m* smoke screen
Rauchsignal *n* smoke signal
Rauchwacke *f* cellular dolomite
rauh rough
Rauhbrand *m (ceram)* biscuit firing, biscuitting
Rauheit *f* roughness
Rauhigkeit *f* roughness; *(plast)* pulled surface *(a defect)*
Rauhigkeitsgrad *m*, **Rauhigkeitszahl** *f* coefficient of roughness
Rauhschliff *m (glass)* grey cutting
Rauhwacke *f* cellular dolomite
Raum *m* 1. space; 2. chamber, cabinet
feldfreier R. field-free region *(as in spectrographs)*
luftleerer R. absolute vacuum
luftverdünnter R. partial vacuum
schädlicher R. 1. clearance volume *(of a pump)*; 2. *s.* toter R.
toter R. dead volume (space) *(chromatography)*
Raumbegasungsmittel *n* space fumigant
raumbeständig volume-stable
Raumbeständigkeit *f* volume stability

Raumchemie *f* stereochemistry
Raumelement *n* volume element
Raumformel *f* space formula
Raumgeschwindigkeit *f* space velocity
Raumgewicht *n s.* Raummasse
Raumgitter *n (cryst)* space lattice
Raumgruppe *f (cryst)* space group
Raumhundertstel *n* volume percentage
Rauminhalt *m* volume
raumisomer stereoisomeric
Raumisomer[es] *n* stereo[iso]mer
Raumisomerie *f* stereoisomerism, space isomerism
Raumladung *f* space charge
räumlich spatial, steric
Raummasse *f* bulk density, B. D.
Raummodell *n* space model
Raumquantelung *f* quantization of direction, directional quantization
Raumrichtung *f* direction in space
Raumspray *m* space (household) spray
Raumsymmetriegruppe *f (cryst)* space group
Raumteil *m* part by volume
Raumtemperatur *f* room (ordinary) temperature
Raumtemperaturvernetzung *f (rubber)* room cure, room-temperature vulcanization
Raumverhältnis *n* proportion by volume
Raum-Zeit-Ausbeute *f* space-time yield
raumzentriert *(cryst)* space-centred, body-centred
Raupenleim *m* banding grease
Rauschgelb *n* yellow arsenic [sulphide], arsenic [sulphide], arsenic yellow, king's yellow (gold), royal yellow, orpiment [yellow] *(technically pure arsenic(III) sulphide)*
Rauschgift *n* narcotic, drug
Rauschpfeffer *m* cava, kawa, kava[kava], Piper methysticum G. Forst.
Rauschpilz *m* / **mexikanischer** Mexican hallucinogenic mushroom, Psilocybe mexicana Heim
Rauschrot *n* ruby arsenic, red orpiment *(tetraarsenic tetrasulphide)*
Rayleigh-Interferometer *n* Rayleigh refractometer
Raymond-Pendel[rollen]mühle *f* Raymond [roller] mill, Raymond ring-roll[er] mill
Raymond-Schüsselmühle *f* Raymond bowl mill
RaZ *s.* Rhodanzahl
Razemase *f* racemase
Razemat *n* racemate *(1. optically inactive compound; 2. salt or ester of DL-tartaric acid)*
  kristallisiertes R. solid racemate
  partielles R. partial racemate
Razematspaltung *f*, **Razemattrennung** *f* resolution of racemates
Razematverbindung *f* racemic compound
Razemform *f* racemic form
razemisch racemic
razemisierbar racemizable
Razemisierbarkeit *f* racemizability
razemisieren to racemize
Razemisierung *f* racemization
  partielle R. partial racemization
Razemisierungsgeschwindigkeit *f* velocity (rate) of racemization

RBW *s.* Wirksamkeit / relative biologische
R. D. *s.* Rotationsdispersion
Reagens *n* reagent
  R. nach Bey Bey reagent *(for detecting cadmium and tin)*
  R. nach Bezssonow Bezssonov reagent *(for detecting polyphenols)*
  R. nach Florence Florence reagent *(for detecting blood and bile pigments in urine)*
  R. nach Molisch Molisch reagent *(for detecting albumin and peptone)*
  R. nach Nadi Nadi reagent *(for detecting indophenol oxidase)*
  Abels R. Abel reagent *(for etching metals)*
  anionisches R. anionoid (nucleophilic) reagent, nucleophile
  Barfoeds R. Barfoed reagent *(for determining monosaccharides)*
  Barnardsches R. Barnard reagent *(for detecting aldehydes)*
  Bealesches R. Beale reagent *(a biological stain)*
  Benedictsches R. Benedict solution *(for detecting reducing sugars)*
  Bials R. Bial reagent *(for detecting pentoses)*
  Brückesches R. Brücke reagent [for proteins]
  chelatbildendes R. chelating agent
  chemisches R. reagent chemical
  Coopersches R. Cooper reagent *(for detecting trivalent iron ions)*
  Dragendorffs R. Dragendorff [alkaloid] reagent
  drückendes R. *(min tech)* depressant
  elektrophiles R. electrophilic reagent, electrophile
  Fleigsches R. Fleig reagent [for blood]
  Fröhdes R. Fröhde reagent *(for detecting alkaloids)*
  Frysches R. Fry reagent *(for etching steel)*
  Gibbs' R. Gibbs reagent *(for detecting hydroxyflavones)*
  Giemsasches R. Giemsa reagent *(for detecting quinine)*
  Grignardsches R. Grignard reagent *(for synthesizing various organic compounds)*
  Mandelins R. Mandelin reagent *(for detecting alkaloids)*
  Marmes R. Marme reagent (solution) *(for detecting alkaloids)*
  Marquis' R. Marquis solution *(for detecting alkaloids)*
  Mayers R. Mayer reagent *(for detecting alkaloids)*
  Meckes R. Mecke solution *(for detecting alkaloids)*
  Millons R. Millon reagent *(for detecting proteins)*
  Mohlers R. Mohler solution *(for detecting tartaric acid)*
  Neßlers R. Nessler reagent (solution) *(for detecting ammonia and various amines)* ✦ **mit Neßlers R. versetzen** to nesslerize
  nukleophiles R. nucleophilic reagent, nucleophile
  Nylanders R. Nylander solution *(for detecting glucose in urine)*

passivierendes R. *(min tech)* depressant
regelndes R. *(min tech)* modifying agent, modifier
Rieglers R. *(med, food)* Riegler reagent
Sangersches R. Sanger reagent *(for detecting amino-acids and proteins)*
Scheiblers R. Scheibler reagent *(for detecting alkaloids)*
Schiffsches R. Schiff reagent *(for detecting aldehydes)*
Schweizers R. Schweizer reagent, cuprammonium [hydroxide] solution, cuprammonia *(for dissolving or detecting cellulose)*
spezifisches R. special reagent
Steadsches R. Stead reagent *(for detecting phosphorus segregation in steel)*
Vervens R. Verven solution *(for detecting alkaloids)*
Zerewitinows R. Zerewitinoff reagent *(methyl magnesium chloride in butyl ether)*
Reagenserzeugung *f* **außerhalb der Titrierzelle** external generation *(coulometry)*
R. innerhalb der Titrierzelle internal generation *(coulometry)*
Reagenslösung *f* reagent solution
Reagenz *f* reagency
Reagenzglas *n* test tube, proof
Reagenzglasbürste *f* test-tube brush
Reagenzglasgestell *n* test-tube rack (stand, support)
R. für Wasserbäder water-bath rack
Reagenzglashalter *m* test-tube holder
Reagenzglasklammer *f*, **Reagenzglasklemme** *f* test-tube clamp
Reagenzglasversuch *m* test-tube experiment, t.t. experiment
Reagenzienflasche *f* reagent bottle
Reagenzienraum *m* reagent room
Reagenzpapier *n* test (indicator) paper, *(Am)* reaction paper
reagieren to react
alkalisch r. to react alkaline
auf Düngung r. *(agric)* to respond to fertilizing
neutral r. to react neutral
sauer r. to react (be) acid
Reaktant *m s.* Reaktionspartner
Reaktantharz *n* reactant-type resin
Reaktion *f* reaction ✦ **zur R. bringen** to react
R. dritter Ordnung third-order reaction
R. erster Ordnung first-order reaction
R. höherer Ordnung higher-order reaction
R. mit wechselseitiger Substitution displacement reaction
R. nach Arnold und Mentzel Arnold-Mentzel formaldehyde test
R. nach Claisen-Tischtschenko Claisen-Tishchenko reaction *(for converting aldehydes into esters)*
R. nach Zimmermann Zimmermann reaction *(for determining ketonic steroids)*
R. nullter Ordnung zero-order reaction
R. zweiter Ordnung second-order reaction
alkalische R. alkaline reaction
anionoide R. anionoid (nucleophilic) reaction
autokatalytische R. autocatalytic reaction
Bartsche R. Bart reaction *(for preparing aromatic arsonic acids)*
basische R. alkaline reaction
bimolekulare R. bimolecular reaction
Blancsche R. Blanc [chloromethylation] reaction
Bouveault-Blancsche R. Bouveault-Blanc reaction *(for reducing esters to alcohols)*
Buchererscha R. Bucherer reaction *(the conversion of a naphthylamine to a naphthol or vice versa)*
Cannizzarosche R. Cannizzaro reaction *(aldehyde dismutation)*
chemische R. chemical reaction
dimolekulare R. bimolecular reaction
Doebnersche R. Doebner synthesis *(of substituted cinchoninic acids)*
einfache R. simple reaction
elektrophile R. electrophilic reaction
endotherme R. endothermic reaction
enzymatische (enzymkatalysierte) R. enzymatic reaction
epithermische R. epithermal reaction
Étardsche R. Étard reaction *(for preparing aromatic aldehydes)*
exotherme R. exothermic reaction
Feiglsche R. Feigl microreaction
Feulgensche R. Feulgen reaction (staining method, test) *(for detecting deoxyribonucleic acid)*
fotochemische R. photochemical reaction, photoreaction
Friedel-Craftssche R. Friedel-Crafts reaction *(for synthesizing aromatic hydrocarbon derivatives)*
gekoppelte R. coupled (linked) reaction
Hehnersche R. Hehner [formaldehyde] test
heterogene R. heterogeneous reaction
heterolytische R. heterolytic reaction
Hillsche R. *(bioch)* Hill reaction
Hinsbergsche R. Hinsberg [amine] test
homogene R. homogeneous reaction
homolytische R. homolytic reaction
im Neutralbereich liegende R. circumneutral reaction
induzierte R. induced (sympathetic) reaction
irreversible R. irreversible (one-way) reaction
isolierte R. isolated reaction
kationoide R. cationoid (electrophilic) reaction
Kolbesche R. Kolbe reaction *(for synthesizing aromatic hydroxy acids)*
Kolbe-Schmittsche R. Kolbe-Schmitt reaction *(for synthesizing aromatic hydroxy acids)*
komplexe R. complex reaction
Landoltsche R. Landolt reaction *(liberation of iodine by oxidizing sulphurous acid with iodates)*
mechanochemische R. mechanochemical reaction
Millonsche R. Millon reaction *(for detecting proteins)*

Molischsche R. Molisch reaction *(for detecting carbohydrates)*
monomolekulare R. monomolecular (unimolecular) reaction
nichtumkehrbare R. *s.* irreversible R.
nukleophile R. nucleophilic reaction
Paulysche R. Pauly [protein] reaction
periodische R. periodic reaction
Perkinsche R. Perkin [condensation] reaction *(for synthesizing unsaturated carboxylic acids)*
Pfitzingersche R. Pfitzinger reaction *(for preparing quinoline)*
Pictet-Spenglersche R. Pictet-Spengler reaction *(isoquinoline ring closure)*
protolytische R. protolytic reaction
pseudomonomolekulare R. pseudomonomolecular (pseudounimolecular) reaction
radikalische R. free-radical reaction
Reedsche R. Reed reaction *(photocatalytic sulphochlorination of hydrocarbons)*
Reformatskysche R. Reformatsky reaction *(of 3-hydroxycarboxylic-acid esters)*
reversible R. reversible (balanced) reaction
rhythmische R. periodic reaction
Sabatier-Senderenssche R. Sabatier-Senderens reaction *(for reducing organic compounds)*
Sandmeyersche R. Sandmeyer [diazo] reaction
Schiffsche R. Schiff's test *(for detecting aldehydes)*
Schmidtsche R. Schmidt reaction *(between hydrazoic acid and carbonyl compounds)*
Schotten-Baumannsche R. Schotten-Baumann reaction *(acylation)*
Schwarzsche R. Schwarz reaction *(for detecting naphthalene or chloroform)*
Seliwanowsche R. Seliwanoff reaction *(for detecting hexoses)*
stereospezifische R. stereospecific reaction
thermonukleare R. thermonuclear reaction
trimolekulare R. trimolecular reaction
Tschugajewsche R. Chugaev reaction *(for preparing olefins from alcohols)*
Ullmannsche R. Ullmann reaction *(for synthesizing diaryls)*
umkehrbare R. *s.* reversible R.
umkehrbare R. erster Ordnung reversible first-order reaction
unimolekulare R. *s.* monomolekulare R.
Wurtzsche R. Wurtz reaction *(for synthesizing alkanes)*
Zeiselsche R. Zeisel reaction *(demethylation)*
zusammengesetzte R. complex reaction

Reaktionsabbruch *m* stoppage of the reaction
Reaktionsablauf *m* reaction course (sequence)
Reaktionsapparat *m* reactor
Reaktionsarbeit *f (phys chem)* work of reaction
Reaktionsbehälter *m* reaction vessel
Reaktionsbestreben *n* tendency to react
Reaktionsdrehofen *m* rotary kiln, *(esp met)* rotary furnace
Reaktionsenergie *f* reaction energy
Reaktionsenthalpie *f* reaction enthalpy
Reaktionsentropie *f* reaction entropy
reaktionsfähig reactive
Reaktionsfähigkeit *f* reactivity
R. gegen[über] Sauerstoff reactivity to (with) oxygen
Reaktionsfähigkeitsindex *m* reactivity index
Reaktionsfolge *f* reaction sequence
reaktionsfreudig reactive
Reaktionsfreudigkeit *f* reactivity
Reaktionsgas *n* reaction gas
Reaktionsgaschromatografie *f* reaction gas chromatography
Reaktionsgefäß *n* reaction vessel
Reaktionsgemisch *n* reaction mixture
Reaktionsgeschwindigkeit *f* reaction rate (velocity)
spezifische R. *s.* Reaktionsgeschwindigkeitskonstante
Reaktionsgeschwindigkeitskonstante *f* reaction-rate constant, [reaction-]velocity constant, specific reaction rate (velocity)
Reaktionsgleichung *f* **[/ chemische]** reaction (chemical) equation
Reaktionsgrundierung *f (coat)* reaction (wash, self-etch pretreatment, self-etching) primer
Reaktionshemmung *f* reaction inhibition
Reaktionsisobare *f* **[/ van't-Hoffsche]** van't Hoff [reaction] isobar
Reaktionsisochore *f* **[/ van't-Hoffsche]** van't Hoff [reaction] isochore
Reaktionsisotherme *f* **[/ van't-Hoffsche]** van't Hoff [reaction] isotherm
Reaktionskammer *f* reaction chamber; *(petrol)* soaking chamber (drum), soaker *(in the tube-and-tank process)*
Reaktionskessel *m* reaction vessel
Reaktionskette *f* reaction chain
Reaktionskinetik *f* **[/ chemische]** reaction (chemical) kinetics
Reaktionskleber *m* two-component adhesive, mixed adhesive
Reaktionskoordinate *f* reaction coordinate
Reaktionskurve *f* reaction curve
reaktionslos reactionless
Reaktionsmechanismus *m* reaction mechanism
Reaktionsmolekularität *f* molecularity of reaction
Reaktionsofen *m* reactor, converter
R. für Flammenreaktionen flame reactor
Reaktionsordnung *f* reaction order
Reaktionspartner *m* participant in a reaction, reactant, coreactant
Reaktions-Primer *m s.* Reaktionsgrundierung
Reaktionsprodukt *n* reaction product
Reaktionsraum *m* 1. reaction space; 2. reaction chamber
Reaktionsschema *n* reaction scheme
Reaktionstechnik *f* **/ chemische** chemical reaction engineering
Reaktionsteilnehmer *m s.* Reaktionspartner
Reaktionstemperatur *f* reaction temperature
reaktionsträge [chemically] inert, chemically indifferent, unreactive, non-reactive, inactive
Reaktionsträgheit *f* **[/ chemische]** [chemical] inertness

Reaktionsturm *m* reaction (reacting) tower; *(pap)* retention tower
Reaktionsverlauf *m* reaction course (sequence)
Reaktionsvermögen *n* reactivity
Reaktionswärme *f* reaction heat, heat of reaction
Reaktionsweg *m* reaction path
Reaktionszone *f* reaction zone
Reaktionszwischenstufe *f* reaction intermediate
reaktiv reactive
Reaktivfarbstoff *m* reactive dye
Reaktivgruppe *f* reactive group
reaktivieren to reactivate, to revivify *(esp charcoal)*, to revive *(esp metals)*
Reaktivierung *f* reactivation, *(esp relating to metals or charcoal)* revivification
Reaktivität *f* reactivity
Reaktor *m* 1. reactor, converter; 2. [nuclear] reactor, pile
  R. mit Flüssigmetallbrennstoff liquid metal fuelled reactor, LMFR
  graphitmoderierter R. graphite-moderated reactor
  heterogener R. heterogeneous reactor
  homogener R. homogeneous reactor
  natriumgekühlter R. sodium[-cooled] reactor
  schneller R. fast [neutron] reactor
  thermonuklearer R. thermonuclear reactor
  überkritischer R. supercritical reactor
  unterkritischer R. subcritical reactor
  wassermoderierter R. water-moderated reactor
Reaktorabschirmung *f (nucl)* reactor shielding
Reaktorgift *n (nucl)* fission (neutron) poison
Reaktorprodukt *n* reactor product
Reaktortechnik *f (nucl)* reactor engineering
real real *(e. g. gas)*; actual *(as opposed to theoretical)*
Realgar *m (min)* realgar *(tetraarsenic tetrasulphide)*
Realkristall *m* real (imperfect) crystal
Rebromierung *f (phot)* rebromination
Rechen *m (tech)* rake
Rechenanlage *f* trash removal facility *(waste-water treatment)*
Rechengut *n* screenings *(in waste water)*
Rechengutzerkleinerer *m* screenings disintegrator *(waste-water treatment)*
Rechenklassierer *m* rake classifier
Rechteckbecken *n* rectangular [sedimentation] tank *(waste-water treatment)*
Rechteckofen *m* rectangular kiln
Rechteckwellenpolarografie *f* square-wave polarography
Rechtsdraht *m (rubber)* Z twist
rechtsdrehend dextrorota[to]ry, dextrorotating, dextro, dextrogyrate, dextrogyratory, dextrogyre, dextrogyrous, *(esp cryst)* right-handed
Rechtsdrehung *f* dextrorotation, *(esp cryst)* right-handed rotation
Rechtsform *f* dextrorotatory (dextrorotating, dextro) form, (+) form
Rechtsmilchsäure *f* $CH_3CH(OH)COOH$ L(+)-lactic acid, dextrolactic acid
Rechtsquarz *m* right-handed quartz
Rechtssäure *f* dextro[rotatory] acid
Rechtsweinsäure *f* $HOOC[CHOH]_2COOH$ L(+)-tartaric acid, dextrorotatory tartaric acid
rechtszirkular polarisiert right-circularly (right-hand circularly) polarized
recken *(text, plast)* to stretch
Recken *n (text, plast)* stretch[ing]
  zweiachsiges R. biaxial stretching
Recker *m (tann)* slicker
Reckerwalze *f (tann)* setting-out cylinder
Reckfestigkeit *f (text)* stretch resistance
Reckgrad *m* degree of stretching
Reckung *f s.* Recken
Reckverhältnis *n* **beim Extrudieren** extrusion ratio
Reclamator-Verfahren *n (rubber)* reclamator process *(a reclaiming process)*
Reddingit *m (min)* reddingite *(a hydrous manganese phosphate)*
Redestillat *n (petrol)* rerun oil
Redestillation *f* redistillation, rerunning
Redestillationskolonne *f (petrol)* rerun tower
redestillieren to redistil, to rerun
Redler *m*, **Redler-Band** *n s.* Redler-Förderer
Redler-Förderer *m*, **Redler-Kettenförderer** *m* Redler conveyor, skeleton-flight conveyor
Redoxanalyse *f* redox analysis
Redoxase *f* redoxase, oxido-reductase, oxidation-reduction enzyme
Redoxaustauscher *m s.* Redox-Ionenaustauscher
Redoxelektrode *f* redox electrode
Redoxgleichgewicht *n* redox equilibrium
Redoxharz *n* redox resin
Redoxindikator *m* redox indicator
Redox-Ionenaustauscher *m* redox ion exchanger
Redoxpolymerisat *n* redox polymer
Redoxpotential *n* redox (oxidation-reduction) potential
Redoxreaktion *f* redox (oxidation-reduction) reaction
Redoxsystem *n* redox (oxidation-reduction) system
  eisenfreies R. iron-free redox system
Redoxtitration *f* redox titration
Reduktase *f* reductase
Reduktaseprobe *f* reductase test, methylene-blue reductase fermentation test
Reduktion *f* reduction
  R. mit Wasserstoff hydrogen reduction
  R. nach Wolff-Kishner Wolff-Kishner reduction
  Bouveault-Blancsche R. Bouveault-Blanc reduction *(of esters to alcohols)*
  direkte R. direct reduction
  elektrolytische R. electrolytic reduction
  Sabatier-Senderenssche R. Sabatier-Senderens reduction
Reduktionsanlage *f* reduction plant
Reduktionsapparatur *f* reduction unit
Reduktionsbad *n* reducing bath
Reduktionsbleiche *f* reduction bleaching

Reduktionsbrühe *f (dye)* reduction liquor *(Béchamp reduction)*
Reduktionsflamme *f* reducing flame
Reduktionsflüssigkeit *f* reduction liquor
Reduktionsgemisch *n* reduction mixture
Reduktionskolben *m* reduction flask
Reduktionskraft *f* reductive capacity
Reduktionslauge *f* reduction liquor
Reduktionsmethode *f* reduction method
Reduktionsmischung *f* reduction mixture
Reduktionsmittel *n* reducing agent, reductant, reducer, reductive
Reduktions-Oxydations- ... *s.* Redox ...
Reduktionspotential *n* reduction potential
Reduktionsverfahren *n* reduction process, *(lab also)* reduction technique
Reduktionsvermögen *n* reducing power
Reduktionsvorgang *m* reduction process
Reduktionswirkung *f* reducing action
Reduktionszone *f* reduction (reducing) zone
reduktiv reductive
Reduktor *m* 1. reduction pan; 2. *s.* Reduktionsmittel
reduzierbar reducible
Reduzierbarkeit *f* reducibility
reduzieren *(chem)* to reduce; to reduce, to lower, to decrease *(e.g. pressure or temperature)*
auf ein Mindestmaß r. to minimize
sich r. lassen to reduce
reduzierend reductive
Reduzierstück *n* reducer, reducing fitting, adapter
Reduzierung *f* reduction, lowering, decrease *(as of pressure or temperature)*
Reduzierventil *n* reducing valve
Redwood-Viskosimeter *n* Redwood viscometer
Referateblatt *n*, **Referateorgan** *n*, **Referatezeitschrift** *f* abstract[ing] journal
Refiner *m* 1. *(rubber)* refiner, refining mill; 2. *(pap)* refiner, refining machine (engine), perfecting engine
refinern *(rubber)* to refine
Refinern *n (rubber)* refining [treatment], refinement
Refinerscheibe *f (pap)* refiner disk
Refiner-Walzwerk *n (rubber)* refining mill, refiner
reflektieren to reflect [back]
Reflexbild *n (phot)* ghost image
Reflexion *f* reflection
R. von Röntgenstrahlen X-ray reflection
mehrfache R. multiple reflection
Reflexionsfähigkeit *f* reflectivity, reflecting power
Reflexionsfleck *m (phot)* flare *(on a negative)*
Reflexionsgesetz *n* / **Braggsches** *(cryst)* Bragg law
Reflexionsgleichung *f* / **Braggsche** *(cryst)* Bragg equation
Reflexionsgoniometer *n (cryst)* reflection goniometer
Reflexionsgrad *m*, **Reflexionskoeffizient** *m* reflectance, reflection coefficient (factor)
Reflexionskraft *f s.* Reflexionsfähigkeit
Reflexionslage *f* reflecting position
Reflexionsmessung *f* reflection (reflectance) measurement
Reflexionsseismik *f (petrol)* reflection shooting
Reflexionsvermögen *n* reflectivity, reflecting power; *(quantitatively)* reflectance, reflection coefficient (factor)
Reflexionswinkel *m* reflection angle
Reflexionszahl *f s.* Reflexionsgrad
Reflexkopierverfahren *n (phot)* reflex copying
Reflux *m (distil)* reflux [stream]
Reformanlage *f (petrol)* reforming plant, reformer
Reformat *n (petrol)* reformate
Reformatsky-Reaktion *f* Reformatsky reaction *(of 3-hydroxycarboxylic-acid esters)*
Reformbenzin *n* reformed gasoline
Reformen *n s.* Reformieren
Reformer *m s.* Reformieranlage
Reformieranlage *f* reforming plant, reformer
Reformierbenzin *n* reformed gasoline
reformieren to reform *(e.g. knocking gasoline)*
Reformieren *n* **an Platinkatalysatoren (Platinkontakten)** platinum reforming, platforming
katalytisches R. catalytic reforming, cat-forming
thermisches R. thermal reforming
reformiert / katalytisch cat-reformed
Reformierung *f* reforming *(as of knocking gasoline)*
Reformierungsanlage *f s.* Reformieranlage
Reformierungsprozeß *m* **an Platinkatalysatoren (Platinkontakten)** platforming process
Reformierungsreaktion *f* reforming reaction
Reformier[ungs]verfahren *n* reforming process
Reforming-Anlage *f s.* Reformieranlage
Reforming-Benzin *n* reformed gasoline
Reformingstock *m s.* Reformstock
Reformreaktion *f* reforming reaction *(of hydrocarbons)*
Reformstock *m* reformer feedstock *(of hydrocarbons)*
Refraktion *f* refraction
spezifische R. specific refraction (refractive index, refractivity)
Refraktionsmethode *f* refraction method
Refraktionsseismik *f (petrol)* refraction shooting
Refraktionsvermögen *n* refractivity
Refraktometer *n* refractometer
Refraktometrie *f* refractometry
refraktometrisch refractometric
Regel *f* rule, principle
R. der größten Multiplizität *s.* erste Hundsche R.
R. der IUC IUC rule
R. der IUPAC IUPAC rule
Antonowsche R. Antonoff rule *(of interfacial tension)*
Astonsche R. Aston whole number rule *(of the atomic weights of isotopes)*
Auwers-Skitasche R. Auwers-Skita rule *(of catalytic hydrogenation)*
Avogadrosche R. Avogadro hypothesis (law) *(of the number of molecules in gases)*
Babinetsche R. *(cryst)* Babinet absorption rule

Blancsche R. Blanc rule *(of the dehydration of dicarboxylic acids)*
Braggsche R. *(nucl)* Bragg rule
Bredtsche R. Bredt rule *(of bridged polycyclic systems)*
Drapersche R. Draper law *(of chemically effective radiation)*
Dühringsche R. *(phys chem)* Dühring's rule
Dulong-Petitsche R. Dulong and Petit's law *(of atomic heats)*
Eötvössche R. Eötvös rule *(of molar surface energy)*
erste Hundsche R. Hund's first rule, maximum-multiplicity principle (rule) *(of unpaired electrons)*
Friessche R. Fries rule *(of the bond structure of polynuclear compounds)*
Geiger-Nuttallsche R. *(nucl)* Geiger-Nuttall rule
Hildebrandsche R. Hildebrand rule *(of constant entropy of vaporization)*
Hiltsche R. *(coal)* Hilt's law (rule)
Hume-Rotherysche R. Hume-Rothery rule *(of alloy sytems)*
Hundsche R. *(phys chem)* Hund's rule
Hundsche R. der größten Multiplizität *s.* erste Hundsche R.
Konowalowsche R. *(phys chem)* Konowaloff rule
Koppsche R. Kopp law *(of molar heats)*
Markownikowsche R. Markovnikov's rule *(of the addition of compounds to olefins)*
Mattauchsche R. *(nucl)* Mattauch rule
Matthiessensche R. *(phys chem)* Matthiessen rule
Ramsay-Youngsche R. *(phys chem)* Ramsay-Young rule
Schürmannsche R. *(coal)* Schürmann's rule
Stokessche R. Stokes rule *(of luminescence)*
Traubesche R. Traube rule *(of the surface tension of water)*
Troutonsche R. Trouton rule (law) *(of the heat of evaporation)*
Vegardsche R. *(cryst)* Vegard's rule (law)
Waldensche R. *(phys chem)* Walden rule (law)

Regelanlage *f*, **Regeleinrichtung** *f* controlling device, controller
regellos random
regelmäßig regular
sterisch r. stereoregular
Regelmäßigkeit *f* regularity
sterische R. stereoregularity
regeln to control, to adjust
Regelstab *m (nucl)* control rod
Regelung *f* control, adjustment; *(specif) s.* automatische R.
automatische R. automatic (feedback) control
nichtselbsttätige R. manual control
selbsttätige R. *s.* automatische R.
Regelungsrechner *m* control computer
Regelungssystem *n* control system
Regelventil *n* control valve
regelwidrig anomalous, abnormal
Regelwidrigkeit *f* anomaly, abnormality
regendicht rainproof, rain-tight, shower-proof
Regenerat *n (rubber)* reclaim, reclaimed rubber, shoddy; *(plast)* reclaim, reground material
Regeneratdispersion *f* reclaim [rubber] dispersion, dispersed reclaimed rubber
Regeneratfaser *f* semisynthetic (manufactured) fibre
Regeneratfaserstoff *m* semisynthetic (manufactured) fibre
Regeneratherstellen *m (rubber, plast)* reclaimer
Regeneration *f* regeneration; *(rubber, plast)* reclaiming, reclamation, recovery; *(pap)* remanufacture; revivification *(as of catalysts)*
regenerativ regenerative
Regenerativfeuerung *f* regenerative firing
Regenerativgummi *m* reclaimed rubber, reclaim, shoddy
Regenerativkammer *f* regenerator chamber
Regenerativofen *m* regenerative furnace
Regenerativprinzip *n* regenerative principle
Regenerativschmelzofen *m* regenerative melting furnace
Regenerativsystem *n* regenerative system
Regenerativ-Verbund[koks]ofen *m* **nach Still** Still oven
Regeneratmischung *f (rubber)* reclaim compound (mix)
Regeneratmischwalzwerk *n (rubber)* reclaim mixing mill
Regenerator *m* 1. regenerator, *(in catalytic cracking also)* kiln; 2. revivifier *(of catalysts)*; 3. *(phot)* replenisher [solution]
Regeneratorkammer *f* regenerator chamber
Regeneratorlösung *f (phot)* replenisher [solution]
Regeneratorraum *m* regenerator chamber
Regeneratpulver *n (rubber)* powdered reclaim
Regeneratwolle *f* reclaimed (recovered) wool
Regeneratzellulosefaser *f* regenerated cellulose fibre
Regeneratzellulosefaserstoff *m* regenerated cellulose fibre
regenerieren to regenerate; *(rubber, plast)* to reclaim, to recover; *(pap)* to remanufacture; to revivify *(e. g. catalysts)*
Regenerierlösung *f* brine regenerant *(ion exchange)*
Regeneriermittel *n* regenerant [chemical]; *(rubber)* reclaiming agent
Regenerierofen *m* regenerator, kiln *(in catalytic cracking)*
Regenerierung *f s.* Regeneration
Regenerierungsmittel *n* regenerant [chemical]
Regenerierverfahren *n (rubber, plast)* reclaiming process
Regenschirmanguß *m (plast)* disk gate
Regenwasser *n* rain water
Regionalmetamorphose *f (geoch)* regional metamorphism
Registerpartie *f (pap)* table-roll section
Registerrollenpapier *n* tabulating paper
Registerschienen *fpl (pap)* shake rails
Registerteil *m (pap)* table-roll section

Registerwalze *f (pap)* table (tube, wire-cloth) roll
registrieren to index; to record *(measuring values)*
Registriergerät *n* recording instrument
Registrierkassenpapier *n s.* Registerrollenpapier
Registriername *m (nomencl)* index name
Registrierpapier *n* recorder chart; *s.* Registerrollenpapier
Registrierung *f* indexing; recording *(of measuring values)*
  fotografische R. photographic record
Registrierungssystem *n (nomencl)* indexing system
Regler *m* 1. [automatic] controller, control[ling] unit; 2. regulator *(in polymerization processes)*; 3. modifier, modifying agent *(in floatation processes)*
Reglermembran *f* governor diaphragm
Reglersubstanz *f* regulator *(in polymerization)*
regulär regular
Reguliergefäß *n* regulating vessel
Regulus *m (lab)* regulus, prill
Regulusmetall *n* regulus metal *(an alloy containing 90% lead, 8% antimony, and 2% tin)*
Rehalogen[is]ierung *f (phot)* rehalogenation
Reib ... *s. a.* Reibungs ...
Reibechtheit *f* (text) fastness to rubbing, rub[bing] fastness, *(relating to dyes also)* crock fastness
reiben to rub; to grind, to triturate *(to a fine powder)*
Reiberwalze *f* triturating roll *(as for preparing oil paints)*
Reibkorrosion *f* fretting (chafing) corrosion
Reibmittel *n (text)* abradant
Reibmühle *f* attrition mill
Reibschale *f* mortar
Reibung *f* friction
  innere R. internal friction
Reibungseffekt *m* / **richtungsabhängiger** directional frictional effect
Reibungsempfindlichkeit *f* sensitiveness to friction
Reibungsentrinder *m* nach Thorne *(pap)* Thorne (waterous) barker
Reibungsfaktor *m* coefficient of friction
reibungsfrei frictionless
Reibungshöhe *f* friction head *(in pumps)*
Reibungskalander *m (pap, rubber)* friction[ing] calender
Reibungskoeffizient *m* coefficient of friction
Reibungskraft *f* friction force; viscous force *(of a liquid)*
reibungslos frictionless
Reibungslöten *n* friction soldering
Reibungslumineszenz *f* triboluminescence
Reibungsschweißen *n* friction (spin) welding
Reibungsverhältnis *n* friction ratio
Reibungsverlust *m* friction loss
Reibungsverschleiß *m* abrasive wear, abrasion
Reibungswärme *f* friction heat
Reibungswiderstand *m* resistance to friction, friction resistance; friction drag *(acting on a body immersed in a moving fluid)*
Reibwert *m* coefficient of friction
Reichert-Meissl-Zahl *f* Reichert-Meissl number (value), R-M number *(for evaluating oils and fats)*
Reichgas *n* rich gas
Reichölerhitzer *m* rich-oil heater
Reichweite *f* range
Reichweite-Energie-Beziehung *f* range-energy relation
Reid-Dampfdruck *m* Reid vapour pressure, R.V.P.
reif ripe, mature
Reif *m (food)* bloom *(as on fruit or chocolate)*
Reife *f* 1. ripeness, maturity; 2. *s.* Reifung
Reifebeschleuniger *m (food)* ripening agent
Reifegrad *m (text)* maturity level, degree of ripeness
Reifegradbestimmung *f (text)* maturity test
reifen to ripen, to mature
Reifen *m (rubber)* tyre
  schlauchloser R. tubeless tyre
Reifenaufbaumaschine *f (rubber)* tyre-building machine (drum), lay-up machine
Reifeneinzelheizer *m (rubber)* single tyre press, unit vulcanizer for tyres
Reifenform *f (rubber)* tyre mould
Reifenheizung *f (rubber)* tyre curing
Reifenindustrie *f (rubber)* tyre industry
Reifenkord *m (rubber)* tyre cord
Reifenmischung *f (rubber)* tyre compound
Reifenrohling *m (rubber)* green tyre
Reifenwerk *n (rubber)* tyre factory (plant)
Reifenwickelmaschine *f s.* Reifenaufbaumaschine
Reifenwickeltrommel *f (rubber)* tyre-building drum, building (case-making) drum
Reifenwulst *m (rubber)* tyre bead
Reifezahl *f s.* Reifegrad
Reifung *f* 1. ripening, maturing, maturation; 2. *(phot)* digestion *(of an emulsion)*; 3. *(filtr)* ripening
Reifungsbeschleuniger *m (food)* ripening agent
Reifungsgrad *m (food)* degree of ripening
Reifungsvorgang *m* ripening process
Reifwerden *n* ripening, maturing, maturation
Reihe *f* series; family *(as of hydrocarbons)*
  Hofmeistersche R. Hofmeister series, lyotropic order (series)
  homologe R. homologous series
  idioblastische (kristalloblastische) R. *(min)* crystalloblastic series
  lyotrope R. *s.* Hofmeistersche R.
  radioaktive R. radioactive series (family, chain), radioactive decay (disintegration) series
D-**Reihe** *f D*-series *(stereochemistry)*
L-**Reihe** *f L*-series *(stereochemistry)*
Reihenanalyse *f* routine analysis
Reihendüngung *f* row dressing
Reihenfolge *f* sequence
  R. der [Verknüpfung der] Aminosäuren sequence of amino-acid residues, amino-acid sequence
Reihenverdünnung *f* serial dilution
Reihenverdünnungsmethode *f* serial-dilution method *(as for evaluating antibiotics)*

Reimer-Tiemann-Synthese *f* Reimer-Tiemann synthesis

rein pure, *(of noble metals also)* fine, *(of chemical elements also)* elemental; plain, unalloyed *(steel)*; absolute *(alcohol)*; neat *(wine)* ✦ **r. darstellen** to isolate

chemisch r. chemically pure, CP

nicht r. impure

technisch r. technical

Reinaluminium *n* pure aluminium

Reinbenzol *n* pure benzene

Reinblau *n* celestial (ethereal) blue *(any of several iron-blue pigments)*

Reinblei *n* chemical lead

Reinchemikalie *f* pure chemical

Reindarstellung *f* isolation

Reineck[e]at *n* reineckate *(a salt of Reinecke acid)*

Reinecke-Salz *n* $NH_4[Cr(SCN)_4(NH_3)_2] \cdot H_2O$ Reinecke salt, ammonium tetrathiocyanodiammonochromate

Reinelement *n* monoisotopic element

Reinerzeugnis *n* pure product

Reingas *n* clean[ed] gas

Reinheit *f* purity *(as of chemicals)*; clean[li]ness *(as of surfaces)* ✦ **von höchster R.** superpure

Reinheitsgrad *m* degree of purity

Reinheitsquotient *m* *(sugar)* purity quotient (coefficient)

reinigen to clean[se]; to purify *(e.g. chemicals)*; to refine *(e.g. metals)*; to clarify, to defecate, to clear *(liquids)*

chemisch r. to dry-clean *(clothing)*

durch Zonenschmelzen r. to zone-purify, to zone-refine

mit einem Schaber r. to doctor *(e.g. a roll)*

mit einer Bürste r. to scrub

trocken r. to dry-clean *(e.g. gas or clothing)*

reinigend detergent

Reiniger *m* *s.* 1. Reinigungsanlage; 2. Reinigungsmittel

Reinigung *f* clean[s]ing; purification *(as of chemicals)*; refining, refinement *(as of metals)*; clarification, defecation, clearing *(of liquids)*

elektrische (elektrostatische) R. electrical (electrostatic) precipitation

extreme R. superrefining, ultrapurification *(as of metals)*

nasse R. wet cleaning *(as of gas)*

trockene R. dry cleaning *(as of gas)*

Reinigungsanlage *f* purification plant; *(coal, pap)* cleaning plant

Reinigungsapparat *m* purifier

Reinigungsbad *n* *(text)* clearing (scouring) bath

Reinigungsbenzin *n* cleaner's naphtha (solvent)

Reinigungscreme *f* *(cosmet)* cleansing cream

Reinigungseffekt *m* *(text)* cleaning (detergent) effect (action); *(filtr)* clarification efficiency *(per cent)*

Reinigungsflüssigkeit *f* *(filtr)* washing liquid (liquor), wash [solvent]

Reinigungsgrad *m* degree of purification

Reinigungskraft *f* *s.* Reinigungsvermögen

Reinigungsleistung *f* *(text)* cleaning efficiency, detergent performance

Reinigungslotion *f* *(cosmet)* cleansing lotion

Reinigungsmittel *n* cleansing agent, clean[s]er, purifier, detergent

synthetisches R. [synthetic] detergent, syndet, soapless soap

Reinigungsturm *m* tower purifier

Reinigungsvermögen *n* *(text)* cleaning efficiency, detergency, detergent power (properties)

Reinigungsverstärker *m* *(text)* cleaning promoter (intensifier)

Reinigungswirkung *f* *s.* Reinigungseffekt

Reinigungszusatz *m* detergent additive

Reinkali *n* potassium oxide, [soluble] potash *(in fertilizer analyses)*

Reinkohle *f* pure (clean) coal *(a coal of minimum ash content)*

Reinkohlensubstanz *f* pure coal material (substance)

Reinkultur *f* pure culture

Reinkupfer *n* pure copper

Reinlezithin *n* pure lecithin

reinmachen *(tann)* to scud *(to remove remaining hairs or lime from hides or skins)*

Reinmetall *n* pure metal

Reinprodukt *n* pure product

Reinprodukte *npl* pures

Reinprotein *n* pure (true) protein

Reinsole *f* pure brine

reinst superpure

Reinstaluminium *n* superpure aluminium

Reinsubstanz *f* pure substance

Reinvulkanisat *n* *(rubber)* [pure] gum vulcanizate

Reinwasser *n* clean water

Reinzucht *f* *s.* Reinkultur

Reisbier *n* rice beer

Reisglas *n* alabaster glass

Reismehl *n* rice flour

Reisöl *n* rice oil

Reispapier *n* **[/ Chinesisches]** rice paper *(from the pith of Tetrapanax papyriferum (Hook.) K. Koch)*

Reispuder *m* rice powder

Reisschleifmehl *n* rice polish (dust) *(removed from rice in polishing)*

Reißdehnung *f* elongation (strain, extension) at break, breaking elongation (extension)

reißen 1. to break, to rupture, to tear; 2. to break, to rupture *(as of paper webs)*; to crack, to slit; *(ceram)* to craze *(of glazes)*

Elementarfäden auf Stapel (bestimmte Länge) r. *(text)* to staple

Reißfestigkeit *f* [ultimate] tensile strength, breaking strength (tenacity), resistance to tearing, tear resistance (strength)

R. in trockenem Zustand *(text)* dry tensile strength

Reißkonverterverfahren *n* *(text)* tow-to-top breaking system

Reißlänge *f* *(text)* breaking length, strength-to-weight ratio

Reißscheibe *f* rupture (bursting) disk
Reisstärke *f* rice starch
Reißverschlußreaktion *f (plast)* chain unzipping reaction
Reißwerk *n* macerator *(as for peat)*
Reißwolle *f* reclaimed (recovered) wool, *(if recovered from heavily felted wool goods or wastes)* mungo
Reiter *m*, **Reiterchen** *n* [balance] rider, slide *(of an analytical balance)*
Reiterlineal *n* rider bar (carrier) *(of an analytical balance)*
Reiterwägestück *n s.* Reiter
reizen *(med)* to irritate
  zu Tränen r. to produce (prompt) tears
  zum Husten r. to provoke cough
  zur Blasenbildung r. to vesicate, to blister
reizend *(med)* irritant, irritating
  zu Tränen r. lachrymatory, lacrimatory
  zum Husten r. cough-provoking
  zur Blasenbildung r. vesicant, vesicatory
reizlos *(cosmet)* non-irritant
  physiologisch r. physiologically inert
Reizlosigkeit *f (cosmet)* non-irritance, freedom from irritation
  physiologische R. physiological inertness
Reizmittel *n (pharm)* stimulant, stimulatory drug, *(esp if used externally)* irritant
Reizschwellenwert *m (tox)* activation threshold
Reizstoff *m s.* Reizmittel
Reizwirkung *f* irritant action
Rekaleszenz *f (cryst)* recalescence
Rekombination *f* recombination
Rekombinationsgeschwindigkeit *f* rate of recombination
Rekombinationskoeffizient *m* coefficient of recombination
Rekombinationswärme *f* heat of recombination
Rekombinationszentrum *n (nucl)* recombination (deep) trap
rekombinieren to recombine
  sich r. to recombine
rekonstruieren to reconstruct, to remodel
Rekonstruktion *f* reconstruction
Rekristallisation *f* recrystallization
Rekristallisationsglühen *n* recrystallization annealing
rekristallisieren to recrystallize
Rektifikation *f* rectification
  diskontinuierliche R. batch rectification
  kontinuierliche (stetige) R. continuous rectification
  unstetige R. *s.* diskontinuierliche R.
Rektifikationsanlage *f* rectifying plant
Rektifikationsapparat *m* rectifying apparatus, rectification still, rectifier
Rektifikationskolonne *f* rectifying (rectification) column
Rektifikationssäule *f s.* Rektifikationskolonne
Rektifikationsteil *m*, **Rektifikationszone** *f* rectifying (enriching, enrichment) section
Rektifizier... *s.a.* Rektifikations...
Rektifizierboden *m* exchange plate
rektifizieren to rectify
rektifiziert / doppelt (zweimal) twice-rectified
Rektifizierung *f* rectification
rekuperativ recuperative
Rekuperativfeuerung *f* recuperative firing
Rekuperativofen *m* recuperative furnace
Rekuperativsystem *n* continuous-recuperative system
Rekuperator *m* recuperator
Relativfotometrie *f* relative photometry
relativistisch relativistic
Relativität *f* relativity
Relativitätstheorie *f* relativity theory
  allgemeine R. general relativity theory
  spezielle R. special relativity theory
Relaxation *f* relaxation
Relaxationseffekt *m* relaxation (asymmetry) effect
Relaxationsgeschwindigkeit *f* relaxation rate
Relaxationsperiode *f* relaxation time
Relaxationsverfahren *n* relaxation process *(reaction kinetics)*
Relaxationszeit *f* relaxation time
Reliktmineral *n* relict
Rendzina *f (soil)* rendzina
Reniérit *m* renierite *(a mineral containing germanium)*
Rennin *n* rennin, rennase, rennet [ferment]
Rennkraftstoff *m* racing fuel
Reoxydation *f* reoxidation
Reparaturlack *m* touch-up paint
Reparaturplatte *f* repair patch, patching rubber
Repellent *n (agric)* repellent
  R. gegen Heuschrecken grasshopper repellent
  R. gegen Nagetiere rodent repellent
Repellentstoff *m s.* Repellent
Replastizieren *n* premilling *(of silicone rubber mixtures)*
Reppe-Chemie *f* Reppe chemistry
Reppe-Verfahren *n* Reppe process *(for synthesizing high polymers derived from acetylene)*
Reproduktion *f* reproduction
Reproduktionsfaktor *m (nucl)* multiplication constant
reproduzierbar reproducible
Reproduzierbarkeit *f* reproducibility
Reprografie *f* reprography
Repulsion *f (phys chem)* repulsion
Resazurinprobe *f* resazurin [reduction] test *(for testing the keeping quality of milk)*
Research-Methode *f* research method, F1 method *(for determining the octane number)*
Research-Oktanzahl *f* research octane number, research-method rating, RON
Reserpin *n* reserpine *(a rauvolfia alkaloid)*
Reserpinsäure *f*, **Reserpsäure** *f* reserpic acid
Reservagedruck *m s.* Reservedruck
Reserve *f* 1. store; 2. *s.* Reservierungsmittel
Reservedruck *m (text)* reserve (resist) printing
Reservekohlenhydrat *n* reserve carbohydrate
Reservemittel *n s.* Reservierungsmittel
reservieren *(text)* to reserve, to resist

Reservierung *f (text)* reservation, resisting
Reservierungsmittel *n (text)* reserve, resist[ing agent]
Reservierungspaste *f (text)* resist paste
Reservoir *n* reservoir, tank
Residualaffinität *f* residual affinity
Residualfungizid *n* protective fungicide
Residualöl *n (petrol)* residual oil (stock)
Residualton *m (geoch)* residual (primary) clay
Residualwirkung *f* residual action (effect)
Residuum *n* residue, *(petrol also)* resid[uum]
Resiliometer *n* resiliometer, resilience meter
Resinat *n* resinate *(resin soap or resin ester)*
Resinoid *n* resinoid
Resinosäure *f* resin acid
resistent resistant, resisting, stable
Resistenz *f* resistance, stability
R. gegen Chemikalien resistance to chemicals
chemische R. chemical resistance (stability), resistance (stability) to chemical attack
mikrobiologische R. microbiological resistance
pflanzliche R. plant resistance *(as to pesticides)*
Resit *n* resite, C-stage resin
Resitol *n* resitol, B-stage resin
Resol *n* resol, A-stage (one-stage) resin
Resonanz *f* 1. *(org chem)* resonance, mesomerism; 2. *(phys)* resonance
kernmagnetische R. nuclear magnetic resonance, NMR
paramagnetische R. paramagnetic [electronic] resonance, PMR, electron spin resonance, ESR
quantenmechanische R. quantum-mechanical resonance
Resonanzabsorption *f* resonance absorption
Resonanzbastard *m* resonance hybrid
Resonanzbegriff *m* concept of resonance
Resonanzbereich *m* resonance region
Resonanzeffekt *m* resonance effect
Resonanzeinfang *m* resonance capture
Resonanzenergie *f* resonance (mesomeric, delocalization) energy
Resonanzentkommwahrscheinlichkeit *f (nucl)* resonance escape probability
Resonanzerscheinung *f* resonance phenomenon
Resonanzfluoreszenz *f s.* Resonanzstrahlung
Resonanzformel *f* resonance formula
resonanzfrei resonance-free
Resonanzfrequenz *f* resonance (resonant) frequency
Resonanzhybrid *n* resonance hybrid
Resonanzintegral *n* resonance integral *(quantum chemistry)*
Resonanzlinie *f* resonance [spectral] line
Resonanzmethode *f* resonance method
Resonanzneutron *n* resonance neutron
Resonanzniveau *n* resonance level (state)
Resonanzpfeil *m* double-headed arrow
Resonanzphänomen *n* resonance phenomenon
Resonanzpotential *n* resonance potential
Resonanzsignal *n* resonance signal *(spectroscopy)*
Resonanzspektroskopie *f* resonance spectroscopy
kernmagnetische R. nuclear magnetic resonance spectroscopy, NMR spectroscopy
magnetische R. magnetic resonance spectroscopy
paramagnetische R. electron paramagnetic (spin) resonance spectroscopy, EPR spectroscopy, ESR spectroscopy
Resonanzspektrum *n* resonance spectrum
kernmagnetisches R. nuclear magnetic resonance spectrum, NMR spectrum
paramagnetisches R. electron paramagnetic (spin) resonance spectrum, EPR spectrum, ESR spectrum
resonanzstabilisiert resonance-stabilized
Resonanzstabilisierung *f* resonance stabilization, stabilization through resonance
Resonanzstrahlung *f* resonance radiation (fluorescence)
Resonanzstreuung *f* resonance scattering
Resonanzstruktur *f* resonance (resonating) structure
Resonanztheorie *f* resonance theory
Resonanzübergang *m* resonance transition
Resonanzvalenzbindungssystem *n* resonating valence bond system
Resonanzvorstellung *f* concept of resonance
Resonanzwechselwirkung *f* resonance interaction
resorbieren to resorb
resorbiert werden to resorb, to undergo resorption
Resorption *f* resorption
Resorzin *n* $C_6H_4(OH)_2$ resorcinol, *m*-dihydroxybenzene
Resorzinblau *n* resorcinol blue, lac[k]moid
Resorzingelb *n* resorcinol yellow, tropaeolin 0 *(sodium azoresorcinol-sulphanilate)*
Resorzinharz *n* resorcinol-formaldehyde resin
Resorzinmonoäthyläther *m* $C_2H_5OC_6H_4OH$ resorcinol monoethyl ether, *m*-ethoxyphenol
Resorzinphthalein *n* resorcinolphthalein, fluorescein
Respiration *f* respiration
respiratorisch respiratory
Rest *m* residue, group *(of a molecule)*
C-terminaler R. C-terminal residue (group) *(in proteins)*
Restaffinität *f* residual affinity
Restalkaligehalt *m (pap)* residual alkalinity
Restbrühe *f (tann)* tailing[s], tails
Restdextrin *n* residual (limit) dextrin
Restfeuchte *f* residual moisture
Restfeuchtebeladung *f* residual moisture content
Restfeuchtegehalt *m* residual moisture content
Restfeuchtigkeit *f* residual moisture
Restflüssigkeit *f* residual liquor
Restgas *n* residual (residue) gas
Resthärte *f* residual hardness *(of water)*
Restkrumpfung *f (text)* residual shrinkage
Restlignin *n (pap)* residual lignin, lignin residues
Restlinien *fpl* persistent (ultimate) lines, raies ultimes *(spectral analysis)*
Restmenge *f* / **duldbare (zulässige)** *(tox)* [residue] tolerance

Restöl *n (petrol)* residual oil (stock)
Restparamagnetismus *m* residual paramagnetism
Restschmutz *m (text)* soil residue
Restschrumpf *m*, **Restschrumpfung** *f (text)* residual shrinkage
Restspannung *f* 1. *(tech)* residual stress; 2. *(el chem)* residual voltage
Reststickstoff *m* residual (non-protein) nitrogen
Reststrahlen *mpl* residual rays, reststrahlen
Reststrahlung *f* residual radiation
Reststrom *m (el chem)* residual current
Restsüße *f (ferm)* residual sugar
Restvalenz *f* residual valency
Restwiderstand *m* residual resistance
Restzucker *m (ferm)* residual sugar
Reszidin *n* rescidine *(a rauvolfia alkaloid)*
Reszinnamin *n* rescinnamine *(3,4,5-trimethoxycinnamoyl ester of reserpic acid)*
Retardiermittel *n* retarder, retarding agent (material), *(text also)* dye retardant
Reten *n* retene, 7-isopropyl-1-methylphenanthrene
Retention *f* retention
  R. der Konfiguration retention of configuration
  relative R. relative retention *(gas chromatography)*
Retentionsvermögen *n* retention power *(gas chromatography)*
Retentionsvolumen *n* retention volume
Retentionszeit *f* retention (hold-up) time *(gas chromatography)*; *(tech)* residence (retention, hold-up, holding) time
Retinit *m (min)* retinite *(collectively for a series of fossil resins)*
Retorte *f* retort ✦ **in der R. erhitzen** to retort
  R. mit ruhender Beschickung (Ladung) static retort
  gemauerte R. brick retort
  gußeiserne R. cast-iron retort
  horizontale R. horizontal retort
  keramische R. brick retort
  liegende R. horizontal retort
  steinerne R. brick retort
  vertikale R. vertical retort
Retortenbatterie *f* retort battery
Retorteneinheit *f s.* Retortengruppe
Retortengas *n* gas-retort gas
Retortengraphit *m* gas carbon
Retortengruppe *f* retort setting
Retortenhals *m* retort neck
Retortenkohle *f* gas carbon
Retortenkoks *m* retort coke
Retortenofen *m* retort furnace (oven)
Retortenschwelen *n* retorting *(of oil shale)*
Retortenverfahren *n (met)* distillation process
retrograd retrograde
Retrogradation *f (coll)* retrogradation *(esp of starch solutions)*
Retropinakolin-Umlagerung *f (org chem)* retropinacolin rearrangement
reversibel reversible
  thermisch r. thermally reversible
Reversibilität *f* reversibility
  mikroskopische R. microscopic reversibility
  thermische R. thermal reversibility
Reversierventil *n* reversing valve
Reversion *f (rubber)* reversion ✦ **zur R. neigen** to tend to revert
Reversions-Gaschromatografie *f* reversion gas chromatography
Reversionsneigung *f*, **Reversionstendenz** *f (rubber)* reversion tendency
Reversosmose *f (filtr)* reverse osmosis
Revolverpresse *f (ceram)* revolver press
Reynolds-Zahl *f* Reynolds number
  R. des Rührvorgangs impeller Reynolds number
rezent recent
Rezept *n* recipe, formula
Rezeptaufstellung *f (rubber)* design of compound, compounding
Rezeptur *f* recipe, formulation
Rezipient *m (pap)* pressure container (accumulator)
Rezipientenglocke *f (lab)* bell jar
Reziprozitätsgesetz *n*, **Reziprozitätsregel** *f (phot)* reciprocity law, Bunsen-Roscoe [reciprocity] law
rezirkulieren to return to the circuit, to recirculate, to recycle
Rf-Wert *m* retention (retardation) factor, Rf value *(chromatography)*
Rhabdophan *m (min)* rhabdophane, rhabdophanite *(a cerium phosphate)*
Rhamnazin *n* rhamnazin *(a natural dye)*
Rhamnetin *n* rhamnetin *(a natural dye)*
Rhamnose *f* rhamnose *(a monosaccharide)*
Rhein *n* rhein, 4,5-dihydroxyanthraquinone-2-carboxylic acid
Rhenat(IV) *n* $M_2^IReO_3$ rhenate(IV), trioxorhenate(IV)
Rhenat(VI) *n* $M_2^IReO_4$ rhenate (VI), rhenate, tetraoxorhenate(VI)
Rhenat(VII) *n* M^IReO_4 perrhenate, tetraoxorhenate(VII)
Rhenium *n* Re rhenium
Rhenium(V)-chlorid *n* $ReCl_5$ rhenium(V) chloride, rhenium pentachloride
Rhenium(VI)-chlorid *n* $ReCl_6$ rhenium(VI) chloride, rhenium hexachloride
Rheniumdioxid *n s.* Rhenium(IV)-oxid
Rheniumheptoxid *n s.* Rhenium(VII)-oxid
Rheniumhexachlorid *n s.* Rhenium(VI)-chlorid
Rhenium(III)-oxid *n* Re_2O_3 rhenium(III) oxide, dirhenium trioxide
Rhenium(IV)-oxid *n* ReO_2 rhenium(IV) oxide, rhenium dioxide
Rhenium(VI)-oxid *n* ReO_3 rhenium(VI) oxide, rhenium trioxide
Rhenium(VII)-oxid *n* Re_2O_7 rhenium(VII) oxide, dirhenium heptaoxide
Rheniumpentachlorid *n s.* Rhenium(V)-chlorid
Rheniumperoxid *n* Re_2O_8 rhenium peroxide
Rheniumsäure *f* H_2ReO_4 rhenic acid, tetraoxorhenic(VI) acid
Rhenium(VII)-säure *f* $HReO_4$ perrhenic acid, tetraoxorhenic(VII) acid

Rheniumtrioxid *n s.* Rhenium(VI)-oxid
Rheologie *f* rheology
rheologisch rheological
Rheomorphose *f (geoch)* rheomorphism
Rheopexie *f (coll)* rheopexy
Rheorinne *f* trough washer
Rheotron *n s.* Betatron
Rhesus-Faktor *m,* **Rh-Faktor** *m (med)* rhesus factor (antigen), rh [factor]
Rhipidolith *m (min)* ripidolite, prochlorite *(a phyllosilicate)*
Rh-negativ *(med)* rh-negative
Rhodamin *n (dye)* rhodamine *(any of a class of fluorescein derivatives)*
Rhodanese *f* rhodanese *(an enzyme belonging to the transferases)*
Rhodanid *n s.* Thiozyanat
Rhodano ... *s. a.* Thiozyanato ...
Rhodanometrie *f* rhodanometry, thiocyanate method
Rhodanwasserstoffsäure *f s.* Thiozyansäure
Rhodanzahl *f* rhodanic (thiocyanogen) number (value) *(a measure of unsaturation of fats)*
Rhodat *n* $M_2^IRhO_4$ rhodate
rhodinieren [/ galvanisch] to rhodanize *(to plate with rhodium)*
Rhodinieren *n* **[/ galvanisches]** rhodanizing, rhodium plating
Rhodinsäure *f* rhodinic acid, 3,7-dimethyl-6-octenoic acid
Rhodium *n* Rh rhodium
Rhodium(III)-chlorid *n* $RhCl_3$ rhodium(III) chloride, rhodium trichloride
Rhodiumdioxid *n s.* Rhodium(IV)-oxid
Rhodium(III)-fluorid *n* RhF_3 rhodium(III) fluoride, rhodium trifluoride
Rhodiumholz *n* red gum *(heartwood of Liquidambar styraciflua L.)*
Rhodium(III)-hydrogensulfid *n* $Rh(SH)_3$ rhodium(III) hydrogensulphide
Rhodium(III)-hydroxid *n* $Rh(OH)_3$ rhodium(III) hydroxide, rhodium trihydroxide
Rhodium(IV)-hydroxid *n* $Rh(OH)_4$ rhodium(IV) hydroxide, rhodium tetrahydroxide
Rhodiummonosulfid *n s.* Rhodium(II)-sulfid
Rhodiummonoxid *n s.* Rhodium(II)-oxid
Rhodium(III)-nitrat *n* $Rh(NO_3)_3$ rhodium(III) nitrate
Rhodium(II)-oxid *n* RhO rhodium(II) oxide, rhodium monooxide
Rhodium(III)-oxid *n* Rh_2O_3 rhodium(III) oxide, dirhodium trioxide
Rhodium(IV)-oxid *n* RhO_2 rhodium(IV) oxide, rhodium dioxide
Rhodium(VI)-oxid *n* RhO_3 rhodium(VI) oxide, rhodium trioxide
Rhodium(III)-sulfat *n* $Rh_2(SO_4)_3$ rhodium(III) sulphate
Rhodium(II)-sulfid *n* RhS rhodium(II) sulphide, rhodium monosulphide
Rhodium(III)-sulfid *n* Rh_2S_3 rhodium(III) sulphide, dirhodium trisulphide
Rhodiumtri ... *s.a.* Rhodium(III)- ...
Rhodiumtrioxid *n s.* Rhodium(VI)-oxid
Rhodizit *m (min)* rhodizite *(a tectoborate)*
Rhodochrosit *m (min)* rhodochrosite, dialogite *(manganese(II) carbonate)*
Rhodommatin *n* rhodommatine *(an animal eye pigment)*
Rhodonit *m (min)* rhodonite *(an inosilicate)*
Rhodopsin *n* rhodopsin, visual purple
rhombisch *(cryst)* [ortho]rhombic, o-rh., rhomb.
rhombisch-pyramidal *(cryst)* rhombic-pyramidal
Rhomboeder *n (cryst)* rhombohedron
rhomboedrisch *(cryst)* rhombohedral
Rh-positiv *(med)* rh-positive
rH-Wert *m* rH [value]
Rhyolit *m* rhyolite, liparite *(an acid volcanic rock)*
RhZ *s.* Rhodanzahl
Ribbonisation *f* ribbonization *(of glass-fibre reinforced plastics)*
2-Ribodesose *f* 2-ribodesose, 2-deoxyribose
Riboflavin *n* riboflavin[e], lactoflavin[e]
Riboflavinphosphat *n* riboflavin[e] phosphate
Ribonuklease *f* ribonuclease
Ribonukleat *n* ribonucleate *(a salt of a ribonucleic acid)*
Ribonukleinsäure *f* ribonucleic acid, RNA
  lösliche R. soluble (transfer) ribonucleic acid, s-RNA
Ribonukleoproteid *n* ribonucleoprotein
Ribose *f* ribose, Rib *(a pentose)*
Richardson-Effekt *m* Richardson effect *(emission of electrons from hot metallic surfaces)*
Richtbohren *n (petrol)* directional drilling
Richteffekt *m* orientation effect
Richtkeil *m (petrol)* whipstock
Richttyptiefe *f (text)* standard depth [of shade]
Richtung *f* direction
  R. der Spaltebene *(min)* cleavage
  R. des Zonendurchgangs direction of zoning (zone travel) *(in zone melting)*
Richtungsfokussierung *f* direction focus[s]ing
Richtungsquantelung *f* directional quantization
Richtwert *m* guide value
  konventioneller R. *(tox)* working level *(for maximum tolerances)*
Rickardit *m (min)* rickardite *(copper telluride)*
Riebeckit *m (min)* riebeckite *(an inosilicate)*
riechen to smell
riechend / angenehm pleasant-smelling
  aromatisch r. fragrant
  aufdringlich r. obnoxious
  stechend r. pungent-smelling
  süßlich r. sweet-smelling
  unangenehm (widerlich) r. ill-smelling, foul-smelling, obnoxious
Riechsalz *n* smelling salt
Riechstoff *m* odoriferous substance, perfume
Riemen *m* belt
Riemenwerkstoff *m* belting
Ries *n (pap)* ream
Riesbeschneidemaschine *f (pap)* guillotine cutter (cutting machine, press, trimmer)
Rieseinschlagpapier *n* ream (mill) wrapper (wrapping)

Rieselblech *n*, **Rieselboden** *m* shower tray
Rieseleinbauten *mpl* film fill (pack) *(as of cooling towers)*
Rieselfähigkeit *f s.* Rieselvermögen
Rieselfilmkolonne *f* wetted-wall column
Rieselkühler *m* trickle (spray, film) cooler
rieseln to trickle
Rieselvermögen *n* flowability, free flowing *(of bulk material)*
Riesenfeld-Probe *f* Riesenfeld test *(for detecting peroxo acids)*
Riesenmolekül *n* giant molecule, macromolecule
Riesenschilf *n* scriptural reed, Arundo donax L.
Riesgewicht *n (pap)* weight per ream
Riffel *f* riffle
Riffelkneter *m* kneading table with fluted roll *(in margarine making)*
riffeln to flute, to corrugate
Riffelwalze *f* fluted (corrugated) roll
Riffkalk *m* reef lime
Rille *f* groove, riffle
Rillenwalzentrockner *m* fin drum dryer
Rinde *f (bot)* bark, *(esp)* inner bark; *(pharm)* cortex; *(chem)* crust *(as on sodium metal)*
  ausgelaugte R. *(tann)* spent bark
Rindenfleck *m* bark speck (spot) *(a defect in paper)*
Rinderfußöl *n* neatsfoot oil
  geklärtes (gereinigtes) R. cold-tested neatsfoot oil
Rinderklauenöl *n s.* Rinderfußöl
Rindertalg *m*, **Rindstalg** *m* beef fat (tallow)
Ring *m (org chem)* ring, nucleus; *(tech)* ring, *(esp for tightening)* washer
  anellierter R. fused ring
  benzoider R. benzenoid ring
  einzelner R. single ring
  gewöhnlicher R. common ring *(5 to 7 members)*
  großer R. large ring, macroring *(13 or more members)*
  kleiner R. small ring *(3 and 4 members)*
  kondensierter R. fused ring
  mittelgroßer (mittlerer) R. medium[-size] ring *(8 to 12 members)*
  nicht ebener R. puckered ring
  normaler R. *s.* gewöhnlicher R.
  sechsgliedriger R. six-membered ring
Ringatom *n* ring atom
Ringaufspaltung *f* ring fission (opening, scission)
Ringbildung *f* ring formation
Ringbrenner *m* ring burner
Ringchelat *n* ring chelate
Ringchromatografie *f* circular [paper] chromatography
Ringdichtung *f* ring packing (seal)
Ringdüse *f* ring nozzle, *(plast also)* circular (ring-shaped) die
Ringe *mpl* / **Liesegangsche** Liesegang rings
  Newtonsche R. Newton rings
Ringebene *f* ring plane
Ringer-Lösung *f (med)* Ringer's solution, Ringer artificial serum
Ringerweiterung *f* ring enlargement
ringförmig ring-shaped, annular, cyclic
Ringgerüst *n* ring skeleton
Ringglied *n* ring link (member)
Ringgröße *f* ring size
Ring-Index *m (nomencl)* Ring Index
Ring-Index-System *n (nomencl)* Ring Index system
Ringkammer *f* annular chamber *(as in gas manufacture)*
Ringketon *n* cyclic ketone
  großes R. macrocyclic (macroring, large-ring) ketone
Ring-Ketten-Tautomerie *f* ring-chain tautomerism
Ringkohlenstoffatom *n* ring-carbon atom
Ringkohlenwasserstoff *m* cyclic hydrocarbon
  anellierter (kondensierter) R. fused-ring hydrocarbon
Ringkomplex *m* ring aggregate
Ringkondensation *f* fusion, anellation, annulization
Ring-Kugel-... *s.* Ring-und-Kugel-...
Ringlüfter *m* tubeaxial (duct) fan
Ringmühle *f s.* Ringrollenmühle
Ringofen *m* ring (annular) kiln
  Hoffmannscher R. *(ceram)* Hoffmann kiln
Ringöffnung *f s.* Ringspaltung
Ringpolymer[es] *n* ring polymer
Ringpresse *f* pot press
Ringprobe *f* 1. *(rubber)* ring sample (test piece); 2. brown-ring test *(for detecting nitrate ions)*
Ringprüfung *f* ring test *(for glazes)*
Ringraum *m s.* 1. Ringkammer; 2. freier R.
  freier R. annular space *(as in piston pumps)*
Ringrohr *n* annulus, collar
Ringrollenmühle *f* ring-roll mill (pulverizer), centrifugal grinder (attrition mill), channel-roller pulverizer
Ringschleifer *m (pap)* rotary grinder
Ringschluß *m* ring closure, cyclization
  doppelter R. double ring closure
Ringschlußreaktion *f* cyclization (ring closure) reaction
Ringsequenz *f* ring assembly
Ringskelett *n* ring skeleton
Ringspalt-Tellerzentrifuge *f* annular solids-discharge disk centrifuge
Ringspaltung *f* ring fission (opening, scission)
Ringspannung *f* ring (angle) strain *(chemical-bond theory)*
Ringsprengung *f s.* Ringspaltung
Ringstruktur *f* ring (annular) structure
Ringsystem *n* ring system
  anelliertes R. *s.* kondensiertes R.
  kondensiertes R. fused (fused-ring, condensed-ring, anellated) system
  kondensiertes aromatisches R. fused-ring aromatic system
Ring-und-Kugel-Gerät *n (plast)* ring-and-ball apparatus
Ring-und-Kugel-Methode *f (plast)* ring-and-ball method
Ringverbindung *f* ring (cyclic) compound

R. **mit großer Gliederzahl** large-ring compound
R. **mit mehreren Heteroatomen** polyheteroatomic-ring compound
Ringverengung *f* ring contraction
Ringverformung *f (incorrectly for)* Streckformen mit Ring
Ringverzweigungsstelle *f* ring-branching position
Ringwalzenmühle *f s.* Ringrollenmühle
Ringwalzenpresse *f* ring-roll press
Rinkit *m (min)* rinkite *(a sorosilicate)*
Rinne *f* channel, gutter, trough; *(met)* runner; *(min tech)* launder; chute, trough *(for transporting bulk material)*
pneumatische R. gravity fluidizing conveyor
rinnen to trickle, to run
rinnend *(tann)* pipey *(grain)*
Rio-Brechwurz[el] *f* Rio (Brazilian) ipecacuanha *(from Cephaelis ipecacuanha (Brot.) A. Rich.)*
Rippe *f (tech)* fin
Rippenglas *n* fluted (ribbed) glass
Rippenheizelement *n*, **Rippenheizkörper** *m* finned heater
Rippenrohr *n* fin[ned] tube
Rippentrichter *m* ribbed (fluted) funnel
Rippenziegel *m* rib tile
Rippfilz *m (pap)* ribbed (ribbing) felt
Riß *m* crack, flaw, crevice
interkristalliner R. intercrystalline (intergranular) crack
transkristalliner R. transcrystalline (transgranular) crack
Rißbeständigkeit *f* resistance to cracking (tearing), tear resistance (strength)
Rißbildung *f* crack initiation, cracking; *(ceram)* crazing *(a defect in glazes)*
R. infolge Korrosion corrosion cracking
Rissebildung *f s.* Rißbildung
rissig cracked, flawy
r. werden to crack
Rissigwerden *n* cracking
Rißwachstum *n* crack growth, *(rubber also)* cut growth
Rittinger-Gesetz *n* Rittinger's law *(of size reduction)*
ritzen to scratch
Ritzfestigkeit *f s.* Ritzhärte
Ritzhärte *f* scratch hardness (resistance), resistance to scratching
Ritzhärteprüfer *m* sclerometer
Ritzprobe *f*, **Ritzprüfung** *f* scratch test
Rizin *n* ricin *(a vegetable protein)*
Rizinelaidinsäure *f* ricinelaidic acid, *trans*(+)-12-hydroxy-9-octadecenoic acid
Rizinin *n* ricinine *(alkaloid)*
Rizinoleat *n* ricinoleate *(a salt of ricinoleic acid)*
Rizinoleinsäure *f s.* Rizinolsäure
Rizinolsäure *f* ricinoleic acid, *cis*(+)-12-hydroxy-9-octadecenoic acid
Rizinusöl *n* castor oil
sulfatiertes (sulfiertes) R. sulphated castor oil
sulfoniertes R. *(incorrectly for)* sulfatiertes R.
sulfuriertes R. *s.* sulfatiertes R.
Rizinusölsäure *f s.* Rizinolsäure
Rizinuspreßkuchen *m*, **Rizinussaatkuchen** *m* castor pomace (cake, meal)
Rizinussäure *f s.* Rizinolsäure
RKS *s.* Röntgenkleinwinkelstreuung
RL_{50} *(abbr residue-life 50 per cent) s.* Rückstands-Halbwertszeit
RMZ, R-M-Z *s.* Reichert-Meissl-Zahl
RNS *s.* Ribonukleinsäure
Robbenöl *n* seal oil
mineralisches R. mineral seal [oil]
Robbentran *m s.* Robbenöl
Roberts-Ringschleifer *m*, **Roberts-Schleifer** *m (pap)* Roberts grinder
Robert-Verdampfer *m* Robert (calandria) evaporator
Robison[-Embden]-Ester *m* Robison-Embden ester *(glucose-6-phosphate)*
Roborans *n (pharm)* roborant
roborierend *(pharm)* roborant
Rochellesalz *n* $KOOC[CHOH]_2COONa \cdot 4H_2O$ Rochelle salt, potassium sodium tartrate-4-water
Rockwellhärte *f* Rockwell hardness, R.H.
Rockwell-Härteprüfung *f* Rockwell hardness test
rodentizid rodenticidal
Rodentizid *n* rodenticide
Roelig-Maschine *f* Roelig hysteresis apparatus
Roè-Zahl *f (pap)* Roè chlorine number
Roga-Backzahl *f*, **Roga-Index** *m (coal)* Roga index
Roga-Test *m* Roga test *(for determining the caking properties of coal)*
Rogenstein *m (min)* oolite *(consisting of minute spherical concretions)*
Roggenmehl *n* rye flour, *(if coarse)* rye meal
Roggenstärke *f* rye starch
roh crude *(esp chemicals)*, raw, untreated, unprocessed *(material)*, *(pap also)* uncooked; *(tann)* raw, green; *(ceram)* unfired, green; *(met)* unwrought
Rohabwasser *n* crude (raw) sewage
Rohbaumwolle *f* raw (grey) cotton
Rohbenzin *n* raw (virgin) gasoline
Rohbenzol *n* crude benzole
Rohbenzolabtreiber *m* crude-benzole still
Rohbenzolanlage *f* crude-benzole plant
Rohbenzoldestillieranlage *f* crude-benzole still
Rohblei *n* crude (pig) lead
Rohblock *m (met)* ingot
Rohbramme *f (met)* slab ingot
Rohbrand *m (ceram)* biscuit firing, biscuitting
Rohbranntwein *m* crude alcohol
Rohbraunkohle *f* raw lignite
Rohbruchfestigkeit *f (ceram)* green strength
Rohcharge *f* raw stock
Rohdestillat *n* crude distillate
Rohdichte *f (pap)* bulk; bulk density, B.D. *(of timber)*
Roheisen *n* pig iron
graues R. grey pig iron
heißerblasenes R. hot-blast pig iron
kalterblasenes R. cold-blast pig iron
meliertes R. mottled pig iron

weißes R. white pig iron
Roheisenpfanne *f* hot-metal ladle
Roherde *f* natural earth
Roherdöl *n* crude [oil, petroleum], mineral-oil crude
Roherz *n* crude (raw, run-of-mine, as-mined) ore
Rohextrakt *m* crude extract
Rohfaser *f* crude fibre
Rohfett *n* crude (raw) fat
Rohfettzerkleinerungsmaschine *f* fat hasher
Rohförderkohle *f s.* Rohkohle
Rohgas *n* crude (raw) gas, *(in gas cleaning also)* dust-laden gas
Rohgemenge *n (glass)* raw batch
Rohglas *n* rough[-cast] glass
Rohglasplatte *f* rough-cast plate
Rohglasur *f (ceram)* raw glaze
Rohgut *n* crude
Rohharz *n* crude resin
Rohhaut *f (tann)* green hide, rawhide
Rohholz *n* rough wood
Rohhumus *m* raw humus, mor
Rohkaolin *m* crude (raw) kaolin
Rohkautschuk *m* crude (raw) rubber
Rohkern *m (food)* suet *(from beef)*
Rohkohle *f* raw (run-of-mine) coal
Rohkonzentrat *n* crude concentrate
Rohkreosot *n* crude creosote
Rohlaufstreifen *m (rubber)* camelback *(for retreading tyres)*
Rohlaufstreifenmischung *f (rubber)* camelback compound
Rohling *m* blank; *(plast)* parison, blank, *(for manufacturing records also)* biscuit, bisque
vorgeformter (vorkonfektionierter) R. *(rubber)* preform
Rohmaterial *n* raw material
Rohmilch *f* raw milk
Rohmischung *f* raw mixture; *(rubber)* green compound
Rohmodell *n* basic model
Rohöl *n* crude oil; *(petrol)* crude [oil, petroleum], mineral-oil crude
abgetopptes R. *s.* getopptes R.
asphaltbasisches (asphaltisches) R. asphalt-base crude, asphaltic petroleum
gemischtbasisches R. mixed-base crude (petroleum)
getopptes R. topped (reduced) crude
naphthenbasisches R. naphthene-base crude, naphthenic petroleum
naphthenisch-aromatisches R. naphthenic-aromatic crude (petroleum)
naphthenisches R. *s.* naphthenbasisches R.
paraffinbasisches (paraffinisches) R. paraffin-base crude, paraffinic petroleum
paraffinisch-naphthenisches R. paraffinic-naphthenic crude (petroleum)
reduziertes R. *s.* getopptes R.
Rohölkühler *m* crude-oil cooler
Rohöllagergefäß *n* crude-oil tank
Rohölleitung *f*, **Rohölpipeline** *f* crude-oil pipeline
Rohöltank *m* crude-oil tank
Rohopium *n* crude (raw) opium
Rohpapier *n* base (raw, body) paper
fotografisches R. photographic base paper
Rohpappe *f* raw (body) board
Rohphosphat *n* rock phosphate
Rohprodukt *n* crude (raw) product
Rohprotein *n* crude protein
Rohr *n* tube, pipe
R. für direkten Dampf (Wasserdampf) live-steam tube
R. mit Klebnaht *(plast)* cemented tube
blindes R. dummy tube
extrudiertes R. extruded tube
gegossenes R. cast tube
geschweißtes R. welded tube
gewickeltes R. rolled tube
glattes R. bare tube
nahtloses R. seamless tube
stranggepreßtes R. extruded tube
Rohraufhänger *m* tube hanger
Rohrboden *m* tube sheet
beweglicher R. floating head *(of a heat exchanger)*
Rohrbündel *n* tube bundle
Röhre *f* tube, pipe
Bourdonsche R. Bourdon [pressure] gauge
Geißlersche R. Geissler tube
Pitotsche R. Pitot tube
Röhrenabscheider *m* tube (pipe) precipitator
röhrenartig tubular
Röhrenbündelverdampfer *m* forced-circulation reboiler
Röhrendestillation *f* pipe-still distillation
Röhrendestillationsanlage *f* pipe-still distillation unit
Röhreneis *n* tube ice
Röhrenerhitzer *m* 1. tubular pasteurizer *(for milk)*; 2. *s.* Röhrenofen
Röhrenfedermanometer *n* Bourdon [pressure] gauge
röhrenförmig tubular
Röhrenglas *n* glass piping
Röhrengutti *n* pipe gamboge *(from Garcinia specc.)*
Röhrenkessel *m* tubular boiler
Röhrenkonverter *m* tubular converter
Röhrenkühler *m* tubular (pipe) cooler, *(esp lab)* tubular condenser
Röhrenofen *m (distil)* tube (pipe) still, tubular (pipe) furnace (heater)
Röhrenofenanlage *f* tube-still (pipe-still) plant (unit), tube (pipe) still
Röhrenofendestillation *f* tube-still (pipe-still) distillation
Röhrenofendestillationsanlage *f s.* Röhrenofenanlage
Röhrenpresse *f (ceram)* pipe press (machine)
Röhrentrockner *m* tube rotary dryer; *(lab)* drying pistol
dampfbeheizter R. steam-tube rotary dryer
Röhrentrommeltrockner *m* indirect rotary dryer

Röhrenvoltmeter *n* valve voltmeter
Röhrenwachs *n* [sucker-]rod wax
Röhrenwärme[aus]tauscher *m* tubular heat exchanger
R. mit Mantel shell-and-tube heat exchanger
Röhrenwärmeübertrager *m s.* Röhrenwärme[aus]tauscher
Röhrenzentrifuge *f* tubular [bowl] centrifuge
Röhrenziehverfahren *n (glass)* tube-drawing process
Rohrflansch *m* tube (pipe) flange
Rohrgewinde *n* tube (pipe) thread
gerades R. straight tube thread
kegliges R. taper tube thread
Rohrheizelement *n*, **Rohrheizkörper** *m* tubular heater
Rohrkopf *m (petrol)* casing head
Rohrkopfbenzin *n* casing-head gasoline (spirit)
Rohrkorbverdampfer *m* basket evaporator
Rohrkrümmer *m* elbow [fitting], ell
R. mit rechtwinkliger Ablenkung right-angle elbow
Rohrleitung *f* tubing, piping; *(petrol)* pipeline
R. für Fertigerzeugnisse (Fertigprodukte) refined-product pipeline
Rohrleitungsflansch *m* tube (pipe) flange
Rohrleitungsrührwerk *n* pipeline (in-line) agitator, agitated-line mixer
Rohrmaterial *n* tubing, piping
Rohrmedium *n* tube-side medium
Rohrmelasse *f* [sugar] cane molasses
Rohrmühle *f* 1. tube (tubular) mill; 2. sugar crusher (mill)
Rohrofen *m (ceram, met)* tube (tubular) furnace
Rohrpresse *f (ceram)* pipe press
Rohrreibung *f* pipe friction
Rohrreiniger *m* tube cleaner
Rohrrohzucker *m* raw cane sugar
Rohrsaft *m (sugar)* cane juice
Rohrsaturation *f (sugar)* tube saturation
Rohrsaug[er]filzwäsche *f (pap)* suction pipe felt cleaner
Rohrschelle *f* tube (pipe) clamp
R. für Gasflaschen gas-cylinder support
Rohrschlange *f* pipe coil
grätenförmige R. herringbone coil
Rohrschlangenkondensator *m* multicoil condenser
Rohrschlangenmantel *m* external coil
Rohrschleuder *f* centrifugal cleaner, *(Am)* centrifiner
Rohrschweißgerät *n (plast)* tube-welding machine, tube welder
Rohrspirale *f* pipe coil
Rohrstutzen *m* socket
Rohrverbinder *m*, **Rohrverbindung** *f*, **Rohrverbindungsstück** *n* tube (pipe) joint (connection), union
Rohrverschraubung *f* screwed fitting
Rohrverteiler *m* manifold
Rohrwand[ung] *f* tube (pipe) wall
Rohrwärmeaustauscher *m* tubular heat exchanger
Rohrzange *f* gas pliers, pipe wrench
Rohrzucker *m* cane sugar *(sucrose, esp from Saccharum officinarum L.)*
Rohrzuckerfabrik *f* cane sugar factory
Rohrzuckerinversion *f* inversion of sucrose
Rohrzunder *m* pipe scale
Rohsaft *m* crude (raw) juice, *(sugar also)* diffusion juice
Rohsaftpumpe *(sugar)* raw juice pump
Rohsalz *n* crude (mine-run) salt
Rohsäure *f* crude (raw) acid, *(pap also)* raw sulphite cooking acid, storage (tower) acid
Rohschellack *m* raw lac
Rohschieferöl *n* crude shale oil
Rohschwefel *m* crude sulphur
Rohseide *f* raw (gum) silk, grege, greige
unentbastete R. hard silk
Rohsoda *f* black ash
Rohspiritus *m*, **Rohsprit** *m* crude alcohol, raw spirit
Rohstahlblock *m* steel ingot
Rohstärke *f* raw (wet-end) starch
Rohstoff *m* raw material
R. für die Papiererzeugung papermaker's furnish, raw papermaking material, raw stock
R. für Textilindustrie textile material
pflanzlicher R. plant material
Rohstoffbedarf *m* requirements of raw materials
Rohstoffkosten *pl* raw-material cost[s]
Rohstofflager *n* stock house
Rohstoffquelle *f* source of raw material
Rohstück *n* blank
Rohsulfat *n* salt cake *(sodium sulphate)*
Rohsulfatzusatz *m (pap)* salt-cake makeup *(for replacing lost alkali)*
Rohtalg *m* raw tallow
Rohteil *n* blank
Rohton *m* crude (raw) clay
Rohtorf *m* raw peat
Rohvaseline *f* petrolatum, petroleum jelly
Rohvolumen *n (ceram)* bulk volume
Rohware *f (text)* grey goods
Rohwasser *n* raw (untreated) water
Rohweinstein *m* wine stone, argol, argal *(potassium hydrogen tartrate)*
Rohwolle *f* raw (grease) wool
Rohwollfett *n* wool wax
Rohzink *n* virgin (primary) zinc
Rohzucker *m* crude (raw) sugar
brauner R. brown sugar
Rohzucker I *m* first raw sugar, high-grade sugar
Rohzucker II *m* second raw sugar
Rohzuckererstprodukt *n s.* Rohzucker I
Rohzuckernachprodukt *n s.* Rohzucker II
Rohzustand *m* crude (raw) state; *(text)* grey state
Rollapparat *m s.* Rollmaschine
Rollbutterfertiger *m* roller-type churn
Rolle *f* roll[er]; *(rubber)* puppet
rollen to roll
Rollenaufwicklung *f (pap)* reeling
Rollenbahn *f* gravity-roller conveyor
Rollenbreite *f (pap)* width of a roll
Rollendruckmaschine *f (pap)* web-fed press

rollengeglättet *(pap)* web-calendered
Rollenkette *f* roller chain
Rollenkufe *f (text)* back with rollers, roller vat
Rollenkühlofen *m (glass)* roller lehr
Rollenlager *n* roller bearing
Rollenmeißel *m (petrol)* roller bit
Rollenpackmaschine *f (pap)* reel-packing machine, wrapper
Rollenpapier *n* reeled (continuous, roll-finished) paper
Rollenreckmaschine *f (plast)* roller stretching machine
Rollensatinage *f (pap)* web calendering
rollensatiniert *(pap)* web-calendered
Rollensatz *m (pap)* set of rolls
Rollenschneider *m s.* Rollenschneidmaschine
Rollenschneidmaschine *f (pap)* reel-slitting (roll-slitting) machine, rereeling (rewinding, slitting) machine, rewinder, slitter
R. mit Kernwicklung (axialer Papieraufwicklung) centre rewinder
Rollenwälzmühle *f s.* Ringrollenmühle
Roller *m s.* Rollmaschine
Rollfilm *m* roll film
Rollgang *m* roll train *(in a rolling mill)*
Rollmaschine *f (pap)* reeling machine, reel, reeler, winder
Rollmühle *f* roller mill
Roll-Schicht-Frosten *n* shell freezing *(a freeze-drying process)*
Rollstange *f (pap)* winder (rewind) shaft, reel
Romankalk *m,* **Romanzement** *m* Roman cement, Parker's cement
Römischbraun *n* umber *(an earth pigment)*
rommeln *(tech)* to tumble
röntgen to roentgenize, to X-ray
Röntgen *n* 1. roentgenization; 2. *s.* Röntgen-Einheit
Röntgenabsorption *f* X-ray absorption
Röntgenabsorptionsspektrum *n* X-ray absorption spectrum
röntgenamorph X-amorphous
Röntgenanalyse *f* X-ray analysis
röntgenanalytisch X-ray-analytical
Röntgenapparat *m* X-ray apparatus, fluoroscope
Röntgenäquivalent *n* roentgen equivalent, equivalent roentgen
Röntgenaufnahme *f* X-ray photograph (image, picture), radiograph, radiogram
Röntgenbefund *m* X-ray result
Röntgenbestrahlung *f* X[-ray] irradiation
Röntgenbeugung *f* X-ray diffraction
Röntgenbeugungsaufnahme *f* X-ray diffraction pattern (diagram)
Röntgenbeugungsbild *n,* **Röntgenbeugungsdiagramm** *n s.* Röntgenbeugungsaufnahme
Röntgenbeugungsmethode *f* X-ray diffraction method
Röntgenbild *n s.* Röntgenaufnahme
Röntgenbildschirm *m* fluorescent screen
Röntgenbündel *n* X-ray beam
Röntgenchemikalie *f* radiophotographic chemical
Röntgendaten *pl* X-ray data
Röntgendiagramm *n* X-ray diagram (pattern)
Röntgendiffraktion *f* X-ray diffraction *(for compounds s.* Röntgenbeugung)
Röntgendiffraktometer *n* X-ray diffractometer
Röntgendrehaufnahme *f* X-ray rotation photograph
Röntgendurchleuchtung *f* fluoroscopy
Röntgen-Einheit *f* roentgen [unit]
Röntgeneinrichtung *f* X-ray equipment
Röntgenemission *f* X-ray emission
Röntgenemulsion *f* X-ray emulsion
Röntgenfilm *m* X-ray film
Röntgenfluoreszenz *f* X-ray fluorescence
Röntgenfluoreszenzanalyse *f* X-ray fluorescence analysis
Röntgenfotografie *f s.* Röntgenografie
Röntgenfotogramm *n s.* Röntgenaufnahme
Röntgengebiet *n* X-ray region (range)
Röntgengerät *n s.* Röntgenapparat
Röntgengoniometer *n* X-ray goniometer
Röntgenintensität *f* X-ray intensity
Röntgeninterferenz *f* X-ray interference
Röntgenkamera *f* X-ray camera
Röntgenkleinwinkelstreuung *f* X-ray small-angle (low-angle) scattering
Röntgenkontrastmittel *n (med)* contrast medium
positives R. radiopaque contrast medium
Röntgenkristallografie *f* X-ray crystallography
Röntgenkristallstrukturanalyse *f* X-ray crystal-structure analysis, crystal (crystallographic) analysis
Röntgenkunde *f* roentgenology
Röntgenmessung *f* X-ray measurement
Röntgenmetallografie *f* X-ray metallography
Röntgenmethode *f* X-ray method
Röntgenmikroanalyse *f* X-ray microanalysis
Röntgenmikroskop *n* X-ray microscope
Röntgenmikroskopie *f* X-ray microscopy
Röntgenniveau *n* X-ray level
Röntgenografie *f* X-ray photography, radiography
röntgenografisch X-ray-photographic, radiographic
Röntgenogramm *n s.* Röntgenaufnahme
Röntgenologie *f* roentgenology
röntgenologisch roentgenologic[al]
Röntgenometrie *f (cryst)* roentgenometry
Röntgenoptik *f* X-ray optics
Röntgenprüfung *f* X-ray test[ing]
Röntgenpulverkamera *f* X-ray powder camera
Röntgenquant *n* X-ray quantum
Röntgenreflexion *f* X-ray reflection
Röntgenröhre *f* X-ray tube
Röntgenschirm *m* fluorescent screen
Röntgenschutz *m* X-ray protection
Röntgenschutzglas *n* X-ray-protective glass
Röntgenspektralanalyse *f* X-ray spectrum analysis
Röntgenspektrograf *m* X-ray spectrograph
Röntgenspektrografie *f* X-ray spectrography
Röntgenspektrogramm *n* X-ray spectrogram
Röntgenspektrometer *n* X-ray spectrometer
Röntgenspektrometrie *f* X-ray spectrometry

Röntgenspektroskopie *f* X-ray spectroscopy
R. nach dem Debye-Scherrer-Verfahren, R. nach dem Pulververfahren Debye-Scherrer-X-ray method, X-ray powder spectroscopy
Röntgenspektrum *n* X-ray spectrum, roentgen spectrum
Röntgenstrahl *m* X-ray
Röntgenstrahlanalyse *f* X-ray analysis
Röntgenstrahlbeugung *f* X-ray diffraction
Röntgenstrahlbündel *n* beam of X-rays
Röntgenstrahlenabsorption *f* X-ray absorption
Röntgenstrahlenbeugung *f* X-ray diffraction
Röntgenstrahlenemissionsspektrum *n* X-ray [emission] spectrum
Röntgenstrahlenkunde *f* roentgenology
Röntgenstrahlenmethode *f* X-ray method
Röntgenstrahlenquelle *f* X-ray source
Röntgenstrahlenschutz *m* X-ray protection
Röntgenstrahlenspektrum *n* X-ray [emission] spectrum
Röntgenstrahlintensität *f* X-ray intensity
Röntgenstrahlinterferenz *f* X-ray interference
Röntgenstrahlmikroanalyse *f* X-ray microanalysis
Röntgenstrahlmikroskop *n* X-ray microscope
Röntgenstrahlstreuung *f* X-ray scattering
Röntgenstrahltechnik *f* X-ray technique
Röntgenstrahlung *f* X-ray radiation, X-radiation
Röntgenstreuung *f* X-ray scattering
Röntgenstruktur *f* X-ray structure
Röntgenstrukturanalyse *f* X-ray [structure, structural] analysis
R. von Kristallen *s.* Röntgenkristallstrukturanalyse
Röntgentechnik *f* X-ray technique
Röntgenterm *m* X-ray level
Röntgenuntersuchung *f* X-ray investigation (examination, study)
Röntgenverfahren *n* X-ray method
Röntgenwellenlänge *f* X-ray wavelength
Roots-Gebläse *n* Roots (cycloidal) blower, straight-lobe compressor
Rosanilin *n* rosaniline, fuchsin[e], magenta
Rosanilinchlorhydrat *n s.* Rosanilinhydrochlorid
Rosanilinfarbstoff *m* rosaniline dye
Rosanilinhydrochlorid *n* rosaniline hydrochloride
Roscoelith *m (min)* roscoelite, vanadium mica *(a phyllosilicate)*
Roselith *m (min)* roselite *(calcium cobalt(II) orthoarsenate)*
Rosendammar *n* rose dammar *(from Vatica rassak Blume)*
Rosenmund-[Saizew-]Reaktion *f* Rosenmund reaction (reduction) *(catalytic hydrogenation of acid chlorides)*
Rosenöl *n* rose oil, oil (otto, attar, essence) of roses
Rosenquarz *m (min)* rose quartz
Rosenwasser *n* rose water
Roseokobaltchlorid *n s.* Aquopentamminkobalt(III)-chlorid
Rose-Tiegel *m* Rose crucible
Rosolsäure *f* rosolic acid *(any of several related compounds, esp aurine)*
p-**Rosolsäure** *f* $(HOC_6H_4)_2C{=}C_6H_4{=}O$ *p*-rosolic acid, pararosolic acid
Rosolsäurefarbstoff *m* rosolic acid dye[stuff], aurine dyestuff
Ross-Effekt *m (phot)* Ross effect
Rossit *m (min)* rossite *(a calcium vanadate)*
Rost *m* 1. *(chem)* rust; 2. *(tech)* grid, grate, grating
weißer R. white rust *(consisting of zinc carbonate)*
Röstanlage *f* roasting plant
Rostantrieb *m* grate drive
Rostaustrag *m* grating discharge
Röstbakterien *pl* retting bacteria
Röstbassin *n (text)* retting vat
Rostbelag *m* 1. rust layer; 2. sinter cake *(of a sintering machine)*
rostbeständig stainless, rustless, rust-resistant, rust-resisting
Rostbeständigkeit *f* stainlessness, rust resistance, resistance to rusting
Röstbetriebsdauer *f* roasting time
Rostbildung *f* rust formation
Röstdextrin *n* torrefaction dextrin
Röste *f (text)* ret[ting]
R. im Wasserbehälter tank retting
biologische R. biological retting
chemische R. chemical retting
unvollständige R. underretting
rosten to rust, to corrode
rösten 1. *(min tech, food)* to roast, to burn, to torrefy, *(relating to ores also)* to calcine; 2. *(text)* to ret, to rot
im Ofen r. to kiln[-dry]
Rösten *n* 1. *(min tech, food)* roasting, burning, torrefaction, *(of ores also)* calcination; 2. *(text)* ret[ting], rotting
chlorierendes R. *(min tech)* chloridizing roasting
oxydierendes R. *(min tech)* oxidizing roasting
reduzierendes R. *(min tech)* reducing roasting
sulfatisierendes R. *(min tech)* sulphat[iz]ing roasting
rostentfernend rust-removing
Rostentferner *m*, **Rostentfernungsmittel** *n* rust-removing agent, rust remover
Rostfeuerung *f* grate (fuel-bed) firing
Rostfilm *m* rust film
Rostfleck *m* rust spot
rostfrei stainless, rustless
Röstgas *n* roaster gas
Rostgelb *n (dye)* iron buff *(hydrated ferric oxide)*
Röstgut *n* roasted material, calcine
Röstgutaustrag *m* calcine discharge
Rösthorde *f* kiln floor
rostinhibierend *s.* rostschützend
Rostinhibitor *m s.* Rostschutzmittel
Röstkaffee *m* roasted coffee
Rostkitt *m* rust (iron, iron-rust) cement
Röstofen *m* roasting furnace (kiln, oven), roaster, calcining kiln, calciner
mehretagiger (mehrherdiger) R. multihearth (multiple-hearth) roaster (roasting furnace)

Röstofenanlage *f* roasting plant
Röstprodukt *n* roasted product, product of roasting
Röstprozeß *m* roasting process
Röstreaktion *f* roast reaction
Röstreaktionsarbeit *f* roast-reaction process
Röstreaktionsverfahren *n* roast-reaction process
Röstreduktionsarbeit *f* roast-reduction process
Röstreduktionsverfahren *n* roast-reduction process
Röstreife *f (text)* retting maturity
Röstschale *f* roasting dish
Rostschicht *f* rust layer, *(if thin)* rust film
Rostschutz *m* rust inhibition (prevention, protection)
Rostschutzadditiv *n* rust-inhibiting (rust-preventing) additive
rostschützend rust-inhibiting, rust-preventing, rust-preventive, rust-protective, antirust
Rostschutzfarbe *f* rust-protective (rust-resisting) paint, antirust paint
Rostschutzfett *n* rust-inhibiting grease
Rostschutzmittel *n* rust inhibitor (preventive, preventer), rust-preventing agent
Rostschutzöl *n* rust-inhibiting oil, slushing oil
Rostschutzpapier *n* anticorrosive (antitarnish) paper
rostsicher *s.* rostfrei
Rostsieb *n*, **Rostsiebmaschine** *f* grizzly [screen], bar screen (grating)
Roststange *f* grate bar
Röststärke *f* roasted starch
Roststelle *f* rust spot
Rösttemperatur *f* roasting temperature
Rösttrommel *f* roasting cylinder
rostverhindernd *s.* rostschützend
rostverhütend *s.* rostschützend
Rostwärmebelastung *f* grate heat release
Röstwasser *n (text)* retting water
Rot *n* / **Pariser** Paris red *(1. minium; 2. iron(III) oxide)*
Rotameter *n* rotameter, tube-and-float meter *(a flowmeter)*
Rotary[bohr]anlage *f* rotary-drilling installation, rotary rig
Rotarybohren *n* rotary drilling
Rotarybohrtisch *m* rotary table
Rotarybohrverfahren *n* rotary[-drilling] method
Rotarysystem *n* rotary system
Rotation *f* rotation, spinning
  freie R. free rotation
  optische R. optical rotation
Rotationsabsorber *m* rotary absorber
Rotationsachse *f* axis of rotation
Rotationsanteil *m* rotatory contribution
Rotationsbewegung *f* motion of rotation
Rotationsdispersion *f* [optical] rotatory dispersion, rotational dispersion
  anomale R. anomalous rotatory dispersion
  normale R. normal rotatory dispersion
Rotationsdispersionskurve *f* rotational-dispersion curve
Rotationsdruckfarbe *f* rotary-press ink
Rotationsdruckpapier *n* newsprint [paper]
Rotationsenergie *f* rotational energy
Rotationsfeinstruktur *f* rotational fine structure
Rotationsfilmdruck *m (text)* rotary screen printing
Rotationsformen *n (plast)* rotation[al] moulding, rotomoulding
Rotationsfreiheitsgrad *m* degree of rotational freedom
Rotationsfrequenz *f* rotational frequency
Rotationsgebläse *n* rotary blower
Rotationsguß *m (plast)* rotational casting; *(glass)* centrifugal casting
Rotationsisomer[es] *n* rotational isomer
Rotationsisomerie *f* rotational isomerism
Rotationskolonne *f* rotary (centrifugal) still (column)
Rotationskompressor *m* rotary compressor
Rotationsniveau *f* rotational level
Rotationspressen *n s.* Rotationsformen
Rotationspumpe *f* rotary pump
Rotationsquantenzahl *f* rotation[al] quantum number
Rotationsquerschneider *m (pap)* revolving-knife cutting-machine, rotary [knife] cutter
Rotationsschwingungsbande *f* rotation-vibration band
Rotationsschwingungsspektrum *n* rotation-vibration[al] spectrum
Rotationsspektrum *n* rotation[al] spectrum
Rotationstiefdruckfarbe *f* rotogravure ink
Rotationstrommel *f* rotary drum
Rotationsübergang *m* rotational transition
Rotationsverdampfer *m* rotary [film] evaporator
Rotationsverdichter *m* rotary compressor
Rotationsversprüher *m* rotary[-cup] atomizer, spinning[-cup] atomizer, rotating atomizer (nozzle)
Rotationsverteilungsfunktion *f* rotational partition function
Rotationsviskosimeter *n* rotating-cylinder viscometer, rotational viscometer
Rotationsvulkanisation *f* continuous rotary cure
Rotationswalkmaschine *f (text)* rotary milling machine
Rotationswärme *f* rotational heat
Rotationszerstäuber *m s.* Rotationsversprüher
Rotationszustand *m* rotational state
Rotätze *f s.* Rotbeize 2.
Rotbeize *f* 1. *(text)* red mordant (acetate), red liquor, mordant rouge *(a solution of aluminium acetate in acetic acid)*; 2. *(glass)* red (copper) stain
Rotbleierz *n (min)* red lead ore, crocoite, crocosite *(lead(II) chromate)*
Roteisenerz *n*, **Roteisenstein** *m (min)* red iron ore, blood-stone *(iron(III) oxide)*
Rötel *m (min)* ruddle, reddle, red bole *(iron(III) oxide)*
Rotenoid *n (org chem)* rotenoid
Rotenon *n* rotenone *(an insecticide)*
Rotenonsäure *f* rotenonic acid
Roterde *f (soil)* krasnozem; *(ceram)* terra rossa
Rotglas *n* red arsenic glass *(glass-like arsenic sulphide)*

rotglühen *s.* auf Rotglut erhitzen
rotglühend red-hot
Rotglut *f* red heat (glow), R.H., redness ✦ **auf R. erhitzen** to heat to redness, to make red-hot
dunkle R. dull red heat
Rotgültigerz *n (min)* red silver ore *(either of two silver sulphides)*
dunkles R. dark red silver ore, silver ruby, pyrargyrite *(antimony(III) silver sulphide)*
lichtes R. light red silver ore, proustite *(arsenic(III) silver sulphide)*
Rotgummi *n* red (eucalyptus) gum, Australian kino *(kino gum from Eucalyptus camaldulensis Dehnh.)*
Rotguß *m* red brass
Rotholz *n* redwood *(collectively for various dyewoods)*
Afrikanisches R. 1. camwood *(from Baphia nitida Afz.)*; 2. barwood *(any of several African dyewoods)*
Indisches R. sappan[wood] *(from Caesalpinia sappan L.)*
rotieren to rotate, to spin
Rotkupfererz *n (min)* red (ruby) copper ore, cuprite *(copper(I) oxide)*
Rotmetall *n* red brass
Rotnickelkies *m (min)* arsenical nickel, niccolite *(nickel arsenide)*
Rotocker *m s.* Rötel
Rotor *m* rotor; *(pap)* rotor, core, plug, cone, jordan rotor (plug) *(of a perfecting engine)*
R. einer Horizontalzentrifuge *s.* ausschwingender R.
ausschwingender R. swing[ing]-bucket rotor, swingout centrifuge head
Rotor-Blasstahlverfahren *n* Rotor process
Rotormesser *n* rotor (fly) knife *(of a rotary cutter)*
Rotormesser *npl (pap)* core bars, bars on the rotor, bars in the plug *(of a perfecting engine)*
Rotorscheibe *f* rotor disk *(as of an extraction column)*
Rotorschneidmaschine *f* rotary cutter
Rotorverfahren *n* **[/ Oberhausener]** *(met)* Rotor process
Rotschlamm *m (met)* red mud
Rotsekt *m* pink champagne
Rotspießglanz *m (min)* red antimony, kermesite *(antimony(III) oxide sulphide)*
Rotstein *m s.* Rötel
Rotte *f (text)* ret[ting] *(for compounds s.* Röste)
rotten *(text)* to ret, to rot *(esp flax)*; to ferment *(cocoa beans)*
im Haufen r. to ferment in heaps *(cocoa beans)*
Röttisit *m (min)* röttisite *(a phyllosilicate)*
Rottung *f* fermentation *(of cocoa beans)*
R. im Haufen heap fermentation
Rotverschiebung *f* shift to the red
Rotwein *m* red wine
Rotzinkerz *n (min)* red zinc ore, zincite *(zinc oxide)*
Rouge *n (cosmet)* rouge
Rouleauxdruck *m (text)* roller printing
Rouleauxdruckmaschine *f (text)* roller printing machine
Routineanalyse *f* routine analysis
Roux-Flasche *f*, **Roux-Kolben** *m* Roux bottle
Roving *m* [glass-fibre] roving; *(text)* roving
Rovinggewebe *n (text, glass)* roving fabric, woven roving
Rowlandit *m (min)* rowlandite *(an yttrium silicate containing iron, fluorine, and cerium)*
RQ *s.* Quotient / respiratorischer
RR-Säure *f* RR acid, 2R acid, 2-amino-8-naphthol-3,6-disulphonic acid
R-Salz *n* R salt *(disodium salt of 2-naphthol-3,6-disulphonic acid)*
R-Säure *f* $C_{10}H_5(OH)(SO_3H)_2$ R acid, 2-naphthol-3,6-disulphonic acid
Rubeanwasserstoff *m*, **Rubeanwasserstoffsäure** *f* $H_2NSCCSNH_2$ rubeanic acid, dithiooxamide
Rubellit *m (min)* rubellite *(a cyclosilicate)*
Rübenmelasse *f* beet molasses
Rübenpektin *n* beet pectin
Rübenrohsaft *m* crude beet juice
Rübenrohzucker *m* raw beet sugar
Rübensaft *m* beet juice
Rübenschneidmaschine *f* beet slicing machine, beet slicer (cutter)
Rübenschnitzel *npl* beet cossettes (slices), sugar beet chips
ausgelaugte R. [exhausted, sugar] beet pulp, exhausted (leached) cossettes
Rübenschnitzelmaschine *f s.* Rübenschneidmaschine
Rübenschwanzfänger *m* beet tail catcher
Rübenwäsche *f* 1. beet cleaning; 2. *s.* Rübenwaschmaschine
Rübenwaschmaschine *f* beet washer (washing machine)
Rübenzucker *m* beet sugar *(sucrose, esp from Beta vulgaris var. altissima Doell)*
Rübenzuckerfabrik *f* beet sugar factory (house), sugar beet mill
Rübenzuckerfabrikation *f* beet sugar manufacture
Rübenzuckerindustrie *f* beet sugar industry
Ruberythrinsäure *f* ruberythric acid, alizarin primveroside
Rubidium *n* Rb rubidium
Rubidiumalaun *m* rubidium alum; *(specif) s.* Rubidiumaluminiumalaun
Rubidiumaluminiumalaun *m* $RbAl(SO_4)_2 \cdot 12H_2O$ rubidium alum
Rubidiumchlorat *n* $RbClO_3$ rubidium chlorate
Rubidiumchlorid *n* RbCl rubidium chloride
Rubidiumchromat *n* Rb_2CrO_4 rubidium chromate
Rubidiumchrom(III)-sulfat *n* $RbCr(SO_4)_2$ chromium rubidium sulphate
Rubidiumdichromat *n* $Rb_2Cr_2O_7$ rubidium dichromate
Rubidiumfluorid *n* RbF rubidium fluoride
Rubidiumhexachloroplatinat(IV) *n* $Rb_2[PtCl_6]$ rubidium hexachloroplatinate(IV)
Rubidiumhydrogensulfat *n* $RbHSO_4$ rubidium hydrogensulphate
Rubidiumhydroxid *n* RbOH rubidium hydroxide
Rubidiumjodat *n* $RbIO_3$ rubidium iodate

Rubidiumjodid *n* RbI rubidium iodide
Rubidiumkarbonat *n* Rb_2CO_3 rubidium carbonate
Rubidiummetaperjodat *n s.* Rubidiumperjodat
Rubidiumnitrat *n* $RbNO_3$ rubidium nitrate
Rubidiumperjodat *n* $RbIO_4$ rubidium periodate, rubidium metaperiodate, rubidium tetraoxoiodate(VII)
Rubidiumpermanganat *n* $RbMnO_4$ rubidium permanganate
Rubidiumperoxid *n* Rb_2O_2 rubidium peroxide
Rubidiumsulfat *n* Rb_2SO_4 rubidium sulphate
Rubidiumtetroxojodat(VII) *n s.* Rubidiumperjodat
Rubin *m (min)* [oriental] ruby
Rubinglas *n* ruby glass
Rubinglimmer *m (min)* lepidocrocite *(iron hydroxide oxide)*
Rubin[schel]lack *m* garnet lac
Rubinschwefel *m* ruby arsenic, red orpiment *(tetraarsenic tetrasulphide)*
Rubinspinell *m (min)* ruby spinel, spinel ruby
Rubinzahl *f* ruby (rubin, rubine) number, Congo rubin number *(a measure for the protective action of a colloid)*
Rubizen *n* rubicene *(an anthrylene derivative)*
Rüböl *n* colza (rape-seed, rape) oil *(from Brassica rapa L. em. Metzg.)*
mineralisches R. mineral colza [oil]
Rübsenöl *n s.* Rüböl
Rückbildung *f* re-formation
Rückbrennen *n (pap)* reburning *(of lime mud)*
Rückdiffusion *f* back-diffusion
Rückdruck *m* back pressure
Rückdrückstift *m (plast)* return pin
Rückenappretur *f (text)* back-sizing, backing
Rückenbegießung *f (agric)* pour-on method *(for bot control)*
Rückenbeschichtung *f (text)* back coating
Rückenspritze *f s.* Rückensprüher
Rückensprüher *m*, **Rückensprühgerät** *n* knapsack sprayer *(for pesticides)*
Rückenstäubegerät *n*, **Rückenstäuber** *m* knapsack duster *(for pesticides)*
Rückenverstäuber *m s.* Rückenstäubegerät
rückerwärmen *(glass)* to reheat *(the parison)*
Rückerwärmung *f (glass)* reheat *(of the parison)*
rückfedern *(rubber)* to recover, to rebound
Rückfederung *f (rubber)* [elastic] recovery, rebound
Rückfluß *m* 1. back (return) flow, *(esp distil)* reflux; 2. *s.* Rücklauf 2. ✦ **unter R. erhitzen (kochen)** *(distil)* to reflux
totaler R. *(distil)* total reflux
Rückflußabscheider *m (distil)* reflux separator
Rückflußbehälter *m (distil)* reflux tank
Rückflußkühler *m* reflux condenser
Rückflußrohr *n (distil)* downpipe, downspout, downtake, downcomer
Rückflußsammelbehälter *m*, **Rückflußtank** *m (distil)* reflux tank
Rückflußteiler *m (distil)* reflux divider
Rückflußventil *n* return valve
Rückflußverhältnis *n (distil)* reflux ratio
rückführen to return, to recycle, to feed back
Rückführöl *n* recycle oil (stock), return oil
Rückführung *f* return, recycling, feedback
Rückgang *m* reduction, diminution, decrease
R. des Backvermögens *(coal)* reduction of caking power
R. des latenten Bildes *(phot)* latent-image regression
rückgewinnen to recover, to reclaim
Rückgewinnung *f* recovery, reclaim
Rückgewinnungsanlage *f* recovery plant
Rückhaltefaktor *m* retention (retardation) factor, Rf value *(chromatography)*
Rückhalteträger *m (nucl)* hold-back carrier
Rückhaltevolumen *n* retention volume *(chromatography)*
Rückhaltezeit *f* retention (residence, hold-up, holding) time
Rückhub *m* return stroke
Rückkondensation *f* retrograde condensation
Rückkoppelungshemmung *f (bioch)* feedback inhibition
Rückkreisöl *n s.* Rücklauföl
Rückkühlzeit *f* cooling-down time
Rücklauf *m* 1. *(process)* back (return) flow, *(esp distil)* reflux; *(pap)* return journey *(as of the sieve)*; 2. *(material)* return flowage; *(distil)* reflux [stream, liquid, liquor]
totaler R. *(distil)* total reflux
Rücklaufbehälter *m (pap)* reclaiming tank
Rücklaufflüssigkeit *f (distil)* reflux liquid (liquor)
rückläufig retrograde
Rücklaufkondensator *m (distil)* reflux condenser
Rücklauföl *n* recycle oil (stock), return oil
Rücklaufrohr *n s.* Rückflußrohr
Rücklaufstrom *m (distil)* downflow
Rücklaufteiler *m (distil)* reflux divider
Rücklaufverhältnis *n (distil)* reflux ratio
Rücklaufwasser *n (pap)* white water
Rücklauge *f (pap)* relief liquor, release, blow-off
Rückleitung *f* 1. return [pipe, line]; 2. *s.* Rückführung
Rückoxydation *f* reoxidation
Rückprall *m* rebound, recoil
Rückprallelastizität *f (rubber)* rebound [elasticity, resilience], impact resilience
Rückprallhöhe *f (rubber)* rebound height
Rückprallkitt *m* bouncing putty
Rückprallpendel *n* **nach Lüpke** Lüpke pendulum (resiliometer)
Rückpralltest *m*, **Rückprallversuch** *m (rubber)* rebound (resilience) test
Rückreaktion *f* reverse (backward, back, opposing) reaction
Rückschicht *f (phot)* backing [layer]
Rückschlag *m* blowback, recoil
Rückschlagklappe *f* swing check valve
Rückschlagventil *n* check (non-return) valve
rückschwefeln *(met)* to resulphurize
Rückschwefelung *f (met)* resulphurization
Rücksprunghärteprüfer *m* scleroscope
rückspülen to backwash, to flush back

Rückspülung *f* backwash[ing], back-flushing
Rückstand *m* residue, remainder, back; *(petrol)* residuum, resid; *(min tech)* underflow ✦ **Rückstände aufarbeiten** *(petrol)* to run resid
R. der Vakuumdestillation vacuum residue
R. im Boden soil residue *(of pesticides)*
atmosphärischer R. *(distil)* long residue (residuum)
fester R. residual solid matter, dry residue
kutikulärer R. cuticular (cuticle) residue *(of pesticides)*
subkutikulärer R. subcuticular residue *(of pesticides)*
unlöslicher R. insoluble residue
Rückstandsabscheider *m* residue separator
Rückstandsanalytiker *m* residue chemist
Rückstandsasphalt *m* residual asphalt
Rückstandsbrennstoff *m* residue fuel
Rückstandsfilter *n* cake filter
rückstandsfrei residue-free, non-residue
Rückstandsgas *n* residue (residual) gas
Rückstands-Halbwertszeit *f* *(tox)* residue-life 50 percent, RL_{50}, biological half-life
Rückstandsheizöl *n* residual fuel oil
Rückstandskracken *n* resid operation *(working-up of heavy long residues)*
rückstandslos *s.* rückstandsfrei
Rückstandsmenge *f* / **duldbare** residue tolerance *(of a pesticide)*
Rückstandsöl *n* residual oil (stock)
Rückstandsschmieröl *n* residual lubricating oil
Rückstandstoxizität *f* residual toxicity *(of pesticides)*
Rückstandswirkung *f* residual action (effect) *(of pesticides)*
Rückstandszone *f* sludge zone *(of a thickener)*
Rückstoß *m* recoil, rebound, blowback
Rückstoßatom *n* recoil atom
Rückstoßelektron *n* recoil (Compton) electron
Rückstoßion *n* recoil ion
Rückstoßkern *m* recoil nucleus
Rückstoßstift *m* *(plast)* return pin
Rückstoßstrahlung *f* recoil radiation
Rückstoßteilchen *n* recoil particle
Rückstrahl[pulver]diagramm *n* back reflection powder diagram (pattern)
Rückstrahlung *f* reflection
Rückstrahlungsvermögen *n* reflectivity, reflecting power
Rückstreuung *f* *(nucl)* backscattering
Rückstrom *m* back flow; *(distil)* down[ward] flow
Rückstromsperre *f* back-flow valve
Rückstromventil *n* back-flow valve
Rücktitration *f* back titration
Rückverdampfer *m* *(petrol)* reboiler [furnace]
Rückverdampfung *f* re-evaporation
Rückverformung *f* *(rubber)* [elastic] recovery, rebound
Rückvergrauung *f* *(text)* soil redeposition
Rückwärtsgasung *f* back run *(in producer-gas manufacture)*
Rückwasch *m* *(petrol)* backwash
Rückwäsche *f* backwash[ing]
Rückwasser *n* circulating water; *(pap)* white water
faser- und füllstoffreiches R. *(pap)* rich white water
Rückwasserbehälter *m* *(pap)* white-water chest
Rückwasserpumpe *f* *(pap)* white-water pump
Rückwassersammelbehälter *m* *(pap)* white-water chest
Rückzugfeder *f* *(plast)* return spring
Rückzugkolben *m* *(plast)* pull-back ram
Rückzugsrandwinkel *m* receding contact angle *(in testing surfactants)*
Ruheenergie *f* rest energy
Ruheinhalt *m* [liquid] hold-up *(as of a column)*
Ruhelage *f* position at rest, rest position
Ruhemasse *f* rest mass
ruhen to rest
Ruheperiode *f* rest[ing] period
Ruherohr *n* resting (recrystallization) tube (cylinder) *(in margarine making)*
Ruhestellung *f s.* Ruhelage
Ruhewert *m* **[/ idealer]** equilibrium value *(theory of zone melting)*
Ruhezustand *m* state of rest
Ruhmasse *f* rest mass
Rühranker *m* stirring bar
Rührapparat *m* agitator, stirring apparatus, stirrer, mixer (*for compounds s.* Rührwerk)
Rührarm *m* agitator (stirring, raking, rabble) arm
Rührautoklav *m* agitated (stirred) autoclave
rührbar agitable, stirrable
Rührbehälter *m s.* Rührgefäß
Rührdauer *f* agitating (stirring, mixing) time
Rühreinrichtung *f s.* Rührwerk
rühren to agitate, to stir, to mix
Rühren *n* agitation, stirring, mixing
pneumatisches R. air (gas) agitation
Rührer *m* agitator, stirrer, mixer (*s. a.* Rührwerk)
Rührerflügel *m s.* Rührflügel
Rührerführung *f* stirrer guide
Rührerschaufel *f s.* Rührflügel
Rührerwelle *f* agitator shaft
Rührextrakteur *m*, **Rührextraktor** *m* agitated extractor
Rührflügel *m* agitator (mixing) blade
Rührgefäß *n* agitated (stirred) tank (vessel), mixing vat (vessel), mixer bowl
Rührgeschwindigkeit *f* stirring rate
Rührgut *n* material being (*or* to be) stirred (mixed)
Rührkessel *m s.* Rührgefäß
Rührkolonne *f* stirred [pot] still *(as in molecular distillation)*
R. nach Scheibel Scheibel column
Rührkristaller *m*, **Rührkristallisator** *m* agitated (stirred) crystallizer
Rührlaugung *f* leaching by agitation
Rührleistung *f* agitator power
Rührmaschine *f s.* Rührwerk
Rührmotor *m* stirrer motor
Rührorgan *n* agitating (stirring, mixing) element
Rührschaufel *f* agitator (mixing) blade

Rührstab *m* stirring pole; *(lab)* stirring rod
Rührtrockner *m* agitated pan dryer
Rührverschluß *m* stirrer seal
Rührvorrichtung *f s.* Rührwerk
Rührwelle *f* agitator shaft
Rührwerk *n* agitator, stirring apparatus, stirrer, mixer; raking mechanism *(as of a sedimentation tank)*
R. mit schräger Rührwelle angular agitator
gegenläufiges R. double-motion agitator
magnetisches R. magnetic stirrer
pneumatisches R. gas-agitated stirrer
zweiachsiges R. *s.* gegenläufiges R.
Rührwerkbehälter *m* mixing vat (vessel), mixer bowl
Rührwerkextrakteur *m* mechanically agitated extractor
Rührwerkkessel *m s.* Rührwerkbehälter
Rührwerkmischextrakteur *m* mechanically agitated extractor
Rührwerksautoklav *m* agitated (stirred) autoclave
RuK *s.* Ring-und-Kugelmethode
Rummel-Schlackenbadverfahren *n* Otto-Rummel process *(for gasifying coal)*
Rumpeln *n* rumble *(in motors with a compression ratio of more than 10:1)*
Rumpf *m* core, rumpf *(of an atom)*; backbone *(of a molecule)*
Rundabsetzbecken *n* circular (radial-flow) sedimentation tank *(waste-water treatment)*
Rundbecken *n* circular tank
Rundbrecher *m* gyratory crusher
Runddrahtrost *m* rod deck *(of a grizzly screen)*
runderneuern to retread, to recap *(tyres)*
Runderneuerung *f* retreading, recapping *(of tyres)*
Rundfilter *n (lab)* round (plain) filter, filter paper disk
Rundfilterchromatografie *f* circular [paper] chromatography
rundführen to recycle
Rundglocke *f (distil)* circular (bell) cap
Rundklärbecken *n*, **Rundklärer** *m* round (circular) clarifier
Rundkolben *m* round-bottom[ed] flask
Rundmesser *n* circular knife
Rundofen *m (ceram)* round (beehive) kiln
Rundpumpverfahren *n (food)* generator method *(of acetification)*
Rund[schnur]ring *m* O-ring *(seal)*
Rundsieb *n* rotary (revolving) screen, trommel [screen]
Rundsiebbütte *f* [cylinder] vat *(of the cylinder paper machine)*
Rundsieb-Büttenpapier *n* cylinder-made (machine-made) deckle-edge paper, cylinder machine-made paper, mould-made paper
Rundsiebkartonmaschine *f* cylinder board machine
Rundsiebmaschine *f (pap)* cylinder [vat] machine, vat (mould) machine
Rundsiebpapiermaschine *f* cylinder paper machine
Rundsiebzylinder *m (pap)* cylinder mould
Rundsortierer *m* rotary screen
Rundtischverfahren *n (glass)* round-table system
Runzel *f* chill mark, flow line *(a surface defect on glass)*; wrinkle *(of a paint)*
Runzelkorn *n (phot)* reticulation
Runzellack *m* wrinkle varnish (finish)
runzeln to wrinkle *(painting technology)*
Runzelung *f* wrinkling *(painting technology)*
rupfen *(pap)* to pick, to pluck, to lift, to pull up, *(Am)* to flake
Rupffestigkeit *f (pap)* picking (plucking) resistance
Rupffestigkeitsprüfung *f (pap)* picking-resistance (plucking-resistance) test
Rupfwiderstand *m s.* Rupffestigkeit
Rüping-Spartränkverfahren *n* Rueping process *(wood preservation)*
Rusagrasöl *n (cosmet)* rusa oil *(from Cymbopogon martini Stapf)*
Ruß *m* soot, *(for technical purposes preferably)* [carbon] black
R. für kautschuktechnische Zwecke rubber [carbon] black
aktiver R. active (reinforcing) black
geperlter R. beaded (pelletized) black
halbaktiver (halbverstärkender) R. *(rubber)* semi-reinforcing black
inaktiver R. *(rubber)* inactive (inert, non-reinforcing) black
loser R. *(rubber)* fluffy black
thermatomischer R. *(rubber)* thermatomic black
Rußbatch *m (rubber)* [carbon] black batch
Rußdispergierung *f* [carbon] black dispersion
Rußdosierung *f (rubber)* [carbon] black loading
Russel-Effekt *m (phot)* Russel effect
Russell-Saunders-Kopplung *f (nucl)* Russell-Saunders coupling, LS coupling
Rußfabrik *f* [carbon] black plant
rußfrei *(rubber)* non-black
rußgefüllt *(rubber)* [carbon-]black-filled, black-loaded, black-reinforced, black-pigmented
Rußgel *n* [rubber] carbon gel
rußhaltig *s.* rußgefüllt
rußig sooty
Rußpunkt *m (petrol)* smoke point
Rußschwarz *n* carbon black
Rußvormischung *f (rubber)* [carbon] black batch
Rußzusatz *m*, **Rußzuschlag** *m (rubber)* [carbon] black loading
Ruthenat *n* ruthenate
Ruthenium *n* Ru ruthenium
Ruthenium(II)-chlorid *n* $RuCl_2$ ruthenium(II) chloride, ruthenium dichloride
Ruthenium(IV)-chlorid *n* $RuCl_4$ ruthenium(IV) chloride, ruthenium tetrachloride
Rutheniumdichlorid *n s.* Ruthenium(II)-chlorid
Rutheniumdioxid *n s.* Ruthenium(IV)-oxid
Ruthenium(III)-hydroxid *n* $Ru(OH)_3$ ruthenium(III) hydroxide
Ruthenium(III)-oxid *n* Ru_2O_3 ruthenium(III) oxide, diruthenium trioxide

Ruthenium(IV)-oxid *n* RuO_2 ruthenium(IV) oxide, ruthenium dioxide
Rutheniumrot *n* ruthenium red
Rutheniumtetrachlorid *n* s. Ruthenium(IV)-chlorid
Ruthenrot *n* s. Rutheniumrot
Rutherfordin *m (min)* rutherfordine *(uranyl carbonate)*
Rutil *m (min)* rutile *(titanium(IV) oxide)*
Rutilmasse *f (ceram)* rutile body
Rutilporzellan *n* rutile porcelain
Rutin *n* rutin *(a glycoside)*
Rutinose *f* rutinose, 6-(α-*L*-rhamnosyl)-*D*-glucose
Rutsche *f* chute
Rutschvermögen *n (pap)* slipperiness
Rutschwinkel *m* angle of repose *(of bulk material)*
Rüttelbeton *m* vibrated concrete
Rüttelformmaschine *f (met)* jolt-ram machine
Rüttelgrobbeton *m* vibrated coarse concrete
rütteln to rap, to vibrate, to shake
Rüttelsieb *n* riddle
Rüttelvorrichtung *f*, **Rüttler** *m* rapper, vibrator
Rydberg-Konstante *f*, **Rydberg-Zahl** *f* Rydberg constant (number)
RZ s. Regeneratzellulosefaserstoff
R-Zweig *m* R-branch *(of a band spectrum)*

S

Saatbeize *f* 1. seed protection, brining of seed; 2. s. Saatbeizmittel
Saatbeizmittel *n* seed protectant (disinfectant), seed-treatment fungicide
Saatbeizung *f* s. Saatbeize 1.
Saatgutbehandlung *f* seed treatment
Saatgutbeize *f* s. Saatbeize
Saatgutbeizmittel *n* s. Saatbeizmittel
Saatkristall *m* seed crystal
Saatschutzmittel *n* seed protectant
  S. gegen Vögel bird repellent
Sabadillsamen *mpl* sabadilla seeds, cevadilla [seeds] *(from Schoenocaulon officinale (Schl. et Ch.) A. Gray; an insecticide)*
Sabattier-Effekt *m (phot)* Sabattier effect
Säbelkolben *m* **[nach Anschütz]** sausage flask
Sabinen *n* sabinene, 4(10)-thujene *(a bicyclic terpene)*
Sabininsäure *f* $HOCH_2[CH_2]_{10}COOH$ sabinic acid, 12-hydroxydodecanoic acid
Saccharase *f* saccharase, sucrase, invertase
Saccharat *n* saccharate, sucrate
Saccharatverfahren *n (sugar)* saccharate process
Saccharid *n* saccharide, carbohydrate
Saccharifikation *f* saccharification, saccharization
saccharifizieren to saccharify, to saccharize
Saccharifizierung *f* saccharification, saccharization
Saccharimeter *n* saccharimeter *(a polarimeter)*
Saccharimetrie *f* saccharimetry
Saccharin *n* saccharin, benzoic sulphimide
Saccharinsäure *f* saccharinic acid *(any of several tetrahydroxycarboxylic acids), (specif)* $HOCH_2[CHOH]_2C(OH)(CH_3)COOH$ saccharinic acid
Saccharogenamylase *f* saccharogenic amylase
Saccharometer *n* saccharometer *(a hydrometer)*
Saccharometrie *f* saccharometry
Saccharomyzet *m* saccharomycete *(a yeast fungus)*
Saccharose *f* sucrose, saccharose *(a disaccharide)*
Saccharoseoktaazetat *n* sucrose octaacetate
Saccharosephosphat *n* sucrose phosphate
Saccharosephosphorylase *f* sucrose phosphorylase
Saccharosespaltung *f* sucrose splitting
Sachar ... s. Sacchar ...
Sachse-Mohr-Theorie *f (org chem)* Sachse-Mohr concept of strainless rings
Sächsischblau *n* s. Smalte
Sackfilter *n* bag filter
Sackleinwand *f* burlap
Sackpapier *n* bag paper
Sadebaumöl *n* savin oil *(from Juniperus sabina L.)*
Safflor *m* s. Saflor
Safflorit *m (min)* safflorite *(cobalt arsenide)*
Saflor *m* 1. safflower, safflor, Carthamus tinctorius L.; 2. safflower, safflor *(blossoms from 1.)*; 3. zaffre[e], zaffar, zaffer, zaffir *(crude cobalt oxide)*
Safloröl *n* safflower (carthamus) oil *(from the seeds of Carthamus tinctorius L.)*
Saflorrot *n* safflor [red], safflower
Safran *m* saffron *(from Crocus sativus L.)*
Safranin *n (dye)* safranin[e], saffranine
Safrol *n* safrol, 3,4-methylenedioxyallylbenzene
SAF-Ruß *m (rubber)* super abrasion furnace black, SAF black
Saft *m* juice, sap, *(esp of plants)* succus
  eingedickter S. condensed (inspissated) juice
  geschiedener S. *(sugar)* defecated (limed) juice
  naturreiner S. natural (mother) juice
  vorgeschiedener S. *(sugar)* predefecated (predefecation, prelimed) juice
Safterhitzer *m* juice heater
Saftfänger *m (sugar)* juice catcher
Saftpresse *f* squeezer
Saftpumpe *f* juice pump
Saftreinigung *f* juice clarification
Saftverdrängungsverfahren *n* Boucherie process *(wood preservation)*
Sägeegreniermaschine *f (text)* saw gin
Sägemehl *n* sawdust
Sägengin *m (text)* saw gin
Sagenit *m (min)* sagenite *(a variety of rutile)*
Sägespäne *pl* sawdust
Sagostärke *f* sago *(from Metroxylon sagu Rottb.)*
Sagradarinde *f (pharm)* cascara sagrada [bark], sacred (Persian, chittam, chittem, chittim) bark *(from Rhamnus purshianus DC.)*
Sahne *f* cream
  saure S. cultured cream

sterilisierte S. sterilized cream
Sahneeis *n* ice cream, cream ice
Sahnemargarine *f* cream margarine
Sahnepulver *n* dried (dry) cream
Saitengalvanometer *n* string galvanometer
S. nach Einthoven Einthoven galvanometer
Saizew-Regel *f (org chem)* Saytzeff rule
Saladin-Mälzerei *f (ferm)* Saladin malting
Saladin-Weiche *f (ferm)* Saladin steep
Salammoniak *m s.* Salmiak 2.
Salatöl *n* salad oil
Salbe *f (pharm)* ointment
salbenartig paste-like, pasty
Salbengrundlage *f (pharm)* ointment base
salbig *s.* salbenartig
Saligenin *n* $C_6H_4(OH)CH_2OH$ saligenin, salicyl alcohol, α,2-dihydroxytoluene
salin[ar] *s.* 1. salzführend; 2. salzig
Saline *f* saline, salina, saltern, salt works; salt garden
Salit *m (min)* salite *(an inosilicate)*
Salizin *n* salicin, α,2-dihydroxytoluene glucoside
Salizylaldehyd *m* $C_6H_4(OH)CHO$ salicylaldehyde, *o*-hydroxybenzaldehyde
Salizylalkohol *m* $C_6H_4(OH)CH_2OH$ salicyl alcohol, saligenin, α,2-dihydroxytoluene
Salizylamid *n s.* Salizylsäureamid
Salizylanilid *n s.* Salizylsäureanilid
Salizylat *n* salicylate *(salt or ester of salicylic acid)*
Salizyl-β-naphthylester *m* 2-naphthyl salicylate, naphthalol, betol
O-Salizyloylsalizylsäure *f* $HOC_6H_4COOC_6H_4COOH$ *O*-salicyloylsalicylic acid
Salizylsäure *f* HOC_6H_4COOH salicylic acid, *o*-hydroxybenzoic acid
Salizylsäureamid *n* $HOC_6H_4CONH_2$ salicylamide
Salizylsäureanilid *n* $HOC_6H_4CONHC_6H_5$ salicylanilide
Salizylsäurebenzylester *m* $HOC_6H_4COOCH_2C_6H_5$ benzyl salicylate
Salizylsäuremethylester *m* $HOC_6H_4COOCH_3$ methyl salicylate
Salizylsäure-2-naphthylester *m* 2-naphthyl salicylate, naphthalol
Salizylsäurephenylester *m* $HOC_6H_4COOC_6H_5$ phenyl salicylate, salol
Salizylsäure-5-sulfonsäure *f* 5-sulphosalicylic acid, 2-hydroxy-5-sulphobenzoic acid
Salmiak *m* 1. NH_4Cl salmiac, sal ammoniac, ammonium chloride; 2. *(min)* salmiac *(ammonium chloride)*
natürlicher S. *s.* Salmiak 2.
Salmiakgeist *m* household [aqua] ammonia, ammonia water (solution, spirit)
Salmiaksalz *n s.* Salmiak 1.
Salmin *n* salmine *(a protamine)*
Salol *n* $C_6H_4(OH)COOC_6H_5$ salol, phenyl salicylate
Salpeter *m* saltpetre, nitre
kubischer S. *s.* Natronsalpeter
salpeterhaltig containing saltpetre, nitrous
Salpetersäure *f* HNO_3 nitric acid
konzentrierte S. concentrated nitric acid
rauchende S. fuming nitric acid, *(esp)* white fuming nitric acid, WFNA *(containing about 98% HNO_3)*
rote rauchende S. red fuming nitric acid, RFNA *(containing more than 86% HNO_3 and 6 to 15% nitric oxides)*
Salpetersäureäthylester *m* $C_2H_5NO_3$ ethyl nitrate
Salpetersäureaufschluß *m (pap)* nitrate (nitric acid) pulping
Salpetersäureherstellung *f* manufacture of nitric acid
Salpetersäureisoamylester *m* $(CH_3)_2CHCH_2CH_2ONO_2$ isoamyl nitrate
Salpetersäureoxydation *f* nitric acid oxidation
Salpetersäure[zellstoff]verfahren *n (pap)* nitrate (nitric acid) pulping
Salpetersiederei *f* saltpetre refinery
salpetrig nitrous
Salpetrigsäureamylester *m* $C_5H_{11}ONO$ amyl nitrite, *(specif)* $(CH_3)_2CHCH_2CH_2ONO$ [ordinary] amyl nitrite, 3-methyl-1-butyl nitrite
Salpetrigsäureäthylester *m* C_2H_5ONO ethyl nitrite
Salsolin *n* salsoline *(an isoquinoline derivative)*
Salz *n* salt, *(food specif)* common salt *(sodium chloride)* ✦ **in ein S. überführen** to salify *(an acid or a base)*
aus Meer[salz]salinen gewonnenes S. bay salt
basisches S. basic salt, *(better)* oxide salt *(or)* hydroxide salt
Buntesches S. Bunte salt *(sodium salt of ethyl thiosulphate)*
einfaches S. simple (single) salt
Englisches S. *s.* Hirschhornsalz
fettsaures S. fatty acid salt
flüchtiges S. *s.* Hirschhornsalz
Frémysches S. Frémy's salt *(1. potassium hydrogenfluoride; 2. potassium nitrosodisulphonate)*
Grahamsches S. Graham's salt *(a sodium polyphosphate)*
inneres S. inner salt
intramolekulares S. intramolecular salt
komplexes S. complex salt
Kurrolsches S. Kurrol salt *(a potassium metaphosphate)*
Mohrsches S. Mohr's salt *(diammonium iron(II) sulphate-6-water)*
neutrales (normales) S. neutral (normal) salt
Peligotsches S. Peligot's salt *(potassium chlorochromate)*
quartäres (quaternäres) S. quaternary salt
saures S. acid salt, *(better)* hydrogen salt
Schlippesches S. Schlippe's salt *(sodium thioantimonate-9-water)*
Salzablagerung *f (geol)* salt deposit; salting *(as on evaporator walls)*
salzähnlich salt-like
Salzanreicherung *f (agric)* salt accumulation
salzartig saline, salt-like
Salzartigkeit *f* salinity
Salzbad *n* salt bath
aufkohlendes S. *(met)* carburizing bath

zyan[id]haltiges S. *(met)* cyanide [salt] bath
Salzbadaufkohlen *n (met)* bath (molten-salt) carburizing, liquid[-salt] carburizing
Salzbadchromieren *n (met)* salt-bath chromizing
Salzbadeinsatzhärtung *f (met)* salt-bath [case-] hardening
Salzbadinchromieren *n (met)* salt-bath chromizing
Salzbadzementieren *n s.* Salzbadaufkohlen
Salzbeet *n* salt meadow
salzbildend salt-forming
Salzbildung *f* salt formation, salification
salzbildungsfähig salifiable
Salzbindung *f* salt link *(chemical-bond theory)*
Salzboden *m* saline soil, *(specif)* solonchak
Salzbrücke *f (el chem)* salt bridge
Salzdenaturierung *f* denaturation (denaturing) of salt
Salzdom *m (geol)* salt dome (plug)
Salzdomfalle *f (geol)* salt-dome trap
Salzeffekt *m (phys chem)* salt effect
primärer S. primary salt effect
sekundärer S. secondary salt effect
salzen to salt
Salzfehler *m* [neutral-]salt error *(in pH determinations)*
Salzfisch *m* salt fish
Salzfleck *m (tann)* salt stain (spue)
Salzfleisch *n* salt meat
salzführend saliferous
Salzgarten *m* salt garden
Salzgehalt *m* salt content
Salzgehaltmesser *m* salinometer, salinimeter *(a hydrometer for salt solutions), (specif)* salimeter *(for indicating directly the percentage of salt)*
Salzgemisch *n* salt mixture
flüssiges (geschmolzenes) S. molten-salt mixture
Salzgeschmack *m* salt (salty) flavour (taste)
Salzgewicht *n s.* Salzmasse
salzglasieren *(ceram)* to salt-glaze
Salzglasieren *n (ceram)* salt glazing
Salzglasur *f (ceram)* salt glaze, smear
salzhaltig saline, salty, briny *(liquid)*; saliferous, salt-bearing *(rock)*
Salzhaltigkeit *f* salinity
Salzhut *m (geol)* salt dome (plug)
salzig salty, saline, briny
Salzigkeit *f* saltiness, salinity
Salzkristall *m* salt crystal
Salzlager *n* salt deposit, saline
Salzlake *f* brine, pickle ✦ **mit S. behandeln** *(food, tann)* to brine
Salzlakenbehandlung *f (food, tann)* brining
salzlakenkonserviert *(tann)* brine-cured
Salzlakenkonservierung *f. (tann)* brine curing (cure)
Salzlauge *f s.* Salzlake
Salzlöser *m* salt dissolver
Salzlösung *f* salt (saline) solution
Salzmasse *f (tann)* cured weight *(of hides)*
Salzmesser *m s.* Salzgehaltmesser
Salzmischung *f s.* Salzgemisch
Salzpaar *n* salt pair
Salzpflanze *f* halophyte
Salzprobe *f* specimen of salt
Salzreihe *f* series of salts
Salzsäure *f* HCl hydrochloric acid
Salzsäureauszug *m* hydrochloric-acid extract
Salzsäuregas *n* hydrochloric-acid gas
Salzsäureofen *m* hydrochloric-acid furnace
Salzsäureverfahren *n* / **Mannheimer** Mannheim furnace process
Salzschmelze *f* salt melt, molten (fused) salt; *(met)* salt bath
Salzschmelzenreaktor *m* molten-salt reactor
Salzschmelzflußextraktion *f* molten-salt extraction
Salzsiederei *f s.* Salzwerk
Salzsole *f* [salt] brine, brine solution
Salzsprühversuch *m* salt spray test *(a corrosion test)*
Salztoleranz *f (biol)* salt tolerance
Salzturm *m* salt tower
Salzwaage *f s.* Salzgehaltmesser
Salzwasser *n* salt (saline) water
Salzwerk *n* salt works, saline, salina, saltern
Samarium *n* Sm samarium
Samarium(II)-chlorid *n* $SmCl_2$ samarium(II) chloride, samarium dichloride
Samarium(III)-chlorid *n* $SmCl_3$ samarium(III) chloride, samarium trichloride
Samariumdichlorid *n s.* Samarium(II)-chlorid
Samariumdijodid *n s.* Samarium(II)-jodid
Samariumerde *f s.* Samarium(III)-oxid
Samarium(III)-hydroxid *n* $Sm(OH)_3$ samarium(III) hydroxide
Samarium(II)-jodid *n* SmI_2 samarium(II) iodide, samarium diiodide
Samarium(III)-jodid *n* SmI_3 samarium(III) iodide, samarium triiodide
Samarium(III)-oxid *n* Sm_2O_3 samarium(III) oxide
Samariumtrichlorid *n s.* Samarium(III)-chlorid
Samariumtrijodid *n s.* Samarium(III)-jodid
Samarskit *m* samarskite *(a rare-earth mineral)*
Samenfaser *f* seed fibre
Samenfett *n* seed fat
Samenhaar *n* seed hair
Samenlack *m* seed (grained) lac
Samenöl *n* seed oil
sämischgar, sämischgegerbt chamois, oil-tanned
Sämischgerber-Degras *m* sod oil *(a by-product from chamois tannage)*
Sämischgerbung *f* chamois (oil) tannage, chamois[ing] process
Sämischleder *n* chamois [leather], shammy, shamoy
Sammelablaßleitung *f* discharge manifold
Sammelbecher *m (rubber)* collection (tapping) cup
Sammelbecken *n* **des Stoffwechsels** *(bioch)* metabolic pool
Sammelbehälter *m* collecting (receiving) tank, receiver, receptacle
Sammelbezeichnung *f* generic name (term)

Sammelelektrode *f* collecting (receiving) electrode, collector [electrode]
Sammelgefäß *n s.* Sammelbehälter
Sammelkanal *m s.* Sammelleitung
Sammelkristallisation *f* coarsening crystallization
Sammelleitung *f* collecting main, manifold
S. für Schwachgas lean fuel gas main *(of a coke oven)*
S. für Starkgas rich fuel gas main *(of a coke oven)*
Sammelmilch *f* bulk milk
sammeln to collect, to receive, to accumulate
sich s. to collect, to accumulate
Sammelname *m* generic name (term)
Sammelplatte *f* collecting plate
Sammelprobe *f* cumulative sample
Sammelrinne *f* gutter; *(pap)* stock sewer (line)
Sammelrohr *n s.* Sammelleitung
Sammeltrog *m* collecting trough
Sammelvorlage *f* gas collecting main *(of a coke-oven battery)*
Sammler *m* 1. *(min tech)* collecting (promoting) agent, collector, promoter; 2. *(el chem)* [storage] battery, accumulator
drückender S. *(min tech)* depressant
kationaktiver (kationischer) S. *(min tech)* cationic collector
Samtpapier *n* velour (flock) paper
Sand *m* sand
S. geringer Gasdurchlässigkeit close (poor-venting) sand *(foundry)*
S. guter (hoher) Gasdurchlässigkeit open (free-venting) sand *(foundry)*
bitumenhaltiger (bituminöser) S. bituminous sand
feinster S. flour (very fine) sand
grüner S. green [moulding] sand *(foundry)*
nasser S. *s.* grüner S.
natürlicher (natürlich vorkommender) S. natural [moulding] sand *(foundry)*
ölhaltiger S. oil sand
synthetischer S. synthetic [moulding] sand *(foundry)*
Sandarach *n* ruby arsenic, red orpiment *(tetraarsenic tetrasulphide)*
Sandarak *m*, **Sandarakharz** *n* gum sandarac (juniper) *(from Tetraclinis articulata (Vahl.) Mast.)*
Sandaustrag *m* sand discharge *(of a classifier)*
Sandbad *n* sand bath
Sandel[holz]öl *n* sandalwood oil *(from Santalum album L.)*
Sanden *n (ceram)* sanding
Sandfang *m*, **Sandfänger** *m* sand trap (well, box, catcher, sifter, grate); *(pap)* sand (settling) table, bed-washer
Sandfilter *n* sand filter
Sandflotation *f* sand floatation
Sandform *f* sand mould *(foundry)*
getrocknete (trockene) S. dry-sand mould
Sandguß *m* sand casting *(foundry)* ✦ **im S. herstellen** to sand-cast
Sandgußstück *n* sand casting *(foundry)*
sandhaltig, sandig arenaceous, arenarious, sandy
Sandigkeit *f* sandiness
Sandkohle *f* sand coal
Sandmeyer-Reaktion *f* Sandmeyer [diazo] reaction
Sandpapier *n* sand paper
Sandrinne *f (met)* runner
Sandschleudermaschine *f* sandslinger *(foundry)*
Sandschwimmverfahren *n* sand-floatation process
Sandstein *m* sandstone
sandstrahlen to sandblast
Sandstrahlmattieren *n (glass)* sand carving
Sandstrahlreinigung *f* sandblasting
Sandwichbauweise *f (plast)* sandwich construction
Sandwichbindung *f* sandwich bond *(chemical-bond theory)*
Sandwichkomplex *m* sandwich complex *(chemical-bond theory)*
Sandwichmolekül *n* sandwich molecule, molecular sandwich
Sandwichofen *m (ceram)* sandwich kiln
Sandwichplatte *f (build)* sandwich panel
Sandwichstruktur *f* sandwich structure *(chemical-bond theory)*
Sandwichverbindung *f* sandwich[-bonded] compound *(chemical-bond theory)*
Sandzucker *m* sand sugar
sanforisieren *(text)* to sanforize
Sanidin *m (min)* sanidine *(a feldspar)*
Sanitärkeramik *f* sanitaryware
Sanitärporzellan *n* vitreous china sanitaryware
Sansibar-Kopal *m* Zanzibar (Madagascar) copal, gum zanzibar *(from Trachylobium specc.)*
Santal *n* santal *(an isoflavone)*
Santalbsäure *f* $CH_3[CH_2]_5CH{=}CHC{\equiv}C[CH_2]_7COOH$ santalbic acid, 11-octadecen-9-ynoic acid
Santalen *n* santalene *(a sesquiterpene)*
Santen *n* santene, 2,3-dimethyl-2-norbornene
Santenonalkohol *m* santenone alcohol, 1,7-dimethyl-2-norbornanol
Santensäure *f* santenic acid *(one form of 1,2-dimethyl-cyclopentane-1,3-dicarboxylic acid)*
Santonin[lakton] *n* santonin, santonic lactone
Santoninsäure *f* santoninic acid
Santonsäure *f* santonic acid
São-Francisco-Kautschuk *m* São Francisco rubber *(from Manihot heptaphylla Ule)*
SAP *s.* Sinteraluminiumpulver
Saphir *m (min)* [true] sapphire *(aluminium oxide)*
saphirartig sapphirine
Saphirin *m (min)* sapphirine *(a neso-subsilicate)*
Saponifikation *f* saponification
Saponin *n* saponin *(any of various plant glucosides)*
Sappanholz *n* sappan[wood], sapanwood *(from Caesalpinia sappan L.)*
Sapphir *m s.* Saphir
saprogen saprogenic, saprogenous
Sapropel *n(m)* sapropel
Sapropelgestein *n*, **Sapropelit** *m* sapropelite

Sapropel[it]kohle *f* sapropelic coal
Sapropelwachs *n* sapropel wax
Saprophyt *m* saprophyte, saprophytic organism
saprophytisch saprophytic
Sardinenöl *n*, **Sardinentran** *m* sardine oil
Sardonyx *m* *(min)* sardonyx *(a variety of chalcedony)*
Sarin *n* sarin, isopropyl methylphosphonofluoridate
Sarkin *n* sarcine, hypoxanthine, 6-hydroxypurine
Sarkosin *n* CH_3NHCH_2COOH sarcosine, methylaminoacetic acid
Sassafrasöl *n* sassafras (saxifrax) oil *(from Sassafras albidum (Nutt.)Nees)*
Sassolin *m* *(min)* sassolite, sassolin *(orthoboric acid)*
Satinage *f* *(pap)* [super]calendering, glazing
  S. von Bogenpapieren sheet calendering
  S. von Rollenpapieren web calendering
  scharfe S. strong glazing
satinieren *(pap)* to [super]calender, to glaze, to plate, to enamel; *(tann)* to satine
Satinierfalten *fpl* calender cuts *(a defect in paper)*
Satinierflecken *mpl* calender spots *(a defect in paper)*
Satinierkalander *m* *(pap)* calender [machine], calender section, machine calender stack
Satinierung *f s.* Satinage
Satinier[walz]werk *n s.* Satinierkalander
Satinweiß *n* satin white (spar) *(a pigment)*
Sattdampf *m* saturated steam
Sattel *m* *(pap)* backfall crest (crown) *(of a Hollander beater)*
Sattel[füll]körper *m* *(distil)* saddle
sättigen to saturate *(solutions)*
  mit Benzol s. to benzolize
  mit CO_2 s. to carbonate, to impregnate with carbon dioxide
Sättiger *m* saturator
Sättigung *f* saturation *(of solutions)*
Sättigungsapparat *m* saturator
Sättigungs[dampf]druck *m* saturation (saturated) vapour pressure
Sättigungsfeuchte *f*, **Sättigungsfeuchtigkeit** *f* saturation humidity
Sättigungsgefäß *n* saturator
Sättigungsgrad *m* degree of saturation; *(soil)* base-saturation percentage
Sättigungsgrenze *f* saturation limit
Sättigungskonzentration *f* saturation concentration
Sättigungspunkt *m* saturation point
Sättigungsstrom *m* *(phys chem)* limiting (maximum) diffusion current
Sättigungsvorrichtung *f* saturator
Sättigungswassergehalt *m* saturation humidity
Sättigungswert *m* saturation value
Saturateur *m* saturator
Saturation *f* *(sugar)* carbonation, saturation *(for removing excess lime)*
  S. mit schwefliger Säure sulphitation
  periodische S. batch carbonation
Saturationsgas *n* *(sugar)* saturation gas
Saturationsgefäß *n* **für CO_2** *(sugar)* carbonation (carbonating, saturation) tank
Saturationsmethode *f* transpiration (gas-saturation) method *(for measuring vapour pressure)*
Saturationspapier *n* saturation paper
Saturationspfanne *f* *(sugar)* carbonation (saturation) pan
Saturationssaft *m* *(sugar)* carbonation (saturation) juice
  erster S. first carbonation juice, scum juice
  zweiter S. second carbonation juice
Saturationsschlamm *m* *(sugar)* carbonation (saturation) mud (scum), sugar-factory lime
Saturator *m* saturator
saturieren *(sugar)* to carbonate *(for removing excess lime)*
  mit schwefliger Säure s. to sulphite
saturnin *(med)* saturnine *(relating to lead poisoning)*
Saturnismus *m* *(med)* saturnism *(lead poisoning)*
Saturnzinnober *m* orange lead (mineral) *(obtained by roasting white lead)*
Satz *m* 1. sediment, subsidence, settling[s], bottom settlings (sediment), B.S., bottoms, lees, dregs, foots; 2. set, nest *(as of laboratory appliances)*; 3. charge [stock], charging stock, batch, feed[stock]; 4. principle, theorem, law
  S. von der Erhaltung der Energie law of conservation of energy, energy principle
  Abeggscher S. Abegg rule
  Hessscher S. law of Hess, Hess's law of heat summation, law of constant heat summation
  pyrotechnischer S. pyrotechnic mixture
Satzbetrieb *m* batch processing
satzweise batch[wise]
säubern to clean[se], to scavenge
Säuberung *f* clean[s]ing
sauer 1. acid, acidic; sour *(soil)*; sour *(petroleum distillate)*; 2. acid, sour, hard *(taste)* ✦ **s. ausgekleidet (zugestellt)** *(met)* acid-lined
  s. werden to [become] sour *(relating to milk, beer, wine)*
  nicht s. non-acidic
  schwach s. weakly (feebly, faintly) acid, subacid
  stark s. strongly acid, superacid
Sauer *m s.* Sauerteig
Sauerbier *n* sour beer
Sauerbrunnen *m* naturally carbonated water
Sauerkleesalz *n s.* Kleesalz
Sauerkraut *n s.* Spuckstoff
säuerlich sour[ish], subacid
Säuerling *m* naturally carbonated water
Sauermilch *f* sour milk
Sauermilchkäse *m* sour-milk cheese
säuern *(food)* 1. to pickle, to souse *(vegetables)*; to sour, to ripen *(milk)*; to leaven *(dough)*; 2. to sour, to ripen *(of milk)*
Saueröl *n* *(petrol)* sour oil
Sauerrahm *m* sour cream
Sauerrahmbutter *f* ripened (starter-ripened, cultured) butter

Sauerstoff *m* O oxygen ✦ **mit S. anreichern** to oxygenate, to oxygenize, to enrich by oxygen
aktiver S. available oxygen
atmosphärischer S. atmospheric oxygen
flüssiger S. liquid oxygen
Sauerstoffabscheidung *f* oxygen evolution
Sauerstoffabsorption *f* oxygen absorption
Sauerstoffabsorptionsmittel *n* oxygen absorbent
Sauerstoffabsorptionsprüfung *f (rubber)* oxygen absorption test *(an ageing test)*
Sauerstoffabsorptionswärme *f* heat of oxygenation
Sauerstoffabspaltung *f* oxygen release, deoxygenation
Sauerstoffaffinität *f* affinity for oxygen
sauerstoffähnlich resembling oxygen, oxygenic, oxygenous
Sauerstoffalterung *f (rubber)* oxygen ageing
sauerstoffangereichert oxygen-enriched
sauerstoffarm low-oxygen
Sauerstoffaufblaskonverter *m* top-blown basic oxygen converter (furnace), basic oxygen converter (furnace)
Sauerstoffaufblas-Konverterprozeß *m*, **Sauerstoffaufblasverfahren** *n* top-blown oxygen converter process, basic oxygen [converter, furnace, steel] process, oxygen process of steelmaking, oxygen lance process
Sauerstoffaufnahme *f* oxygen uptake
Sauerstoffbad *n* **/ flüssiges** liquid-oxygen bath
Sauerstoffbedarf *m* oxygen demand
biochemischer S. biochemical oxygen demand, B.O.D. *(of waste-water)*
chemischer S. chemical oxygen demand, C.O.D. *(of waste-water)*
fünftägiger biochemischer S. five days biochemical oxygen demand
sauerstoffbeladen oxygenated
Sauerstoffbilanz *f* oxygen balance *(as of explosives)*
Sauerstoffblaskonverter *m s.* Sauerstoffaufblaskonverter
Sauerstoffblasstahl *m* basic oxygen [furnace] steel
Sauerstoffblasstahlverfahren *n s.* Sauerstoffaufblas-Konverterprozeß
Sauerstoffblaswerk *n* [basic] oxygen steel plant
Sauerstoffbombe *f* oxygen bomb
S. nach Bierer und Davis Bierer-Davis oxygen bomb
Sauerstoffbombenalterung *f (rubber)* oxygen bomb ageing, oxygen pressure method
Sauerstoffdifluorid *n* OF_2 oxygen difluoride
Sauerstoffdon[at]or *m* oxygen donor
Sauerstoffdruckalterung *f s.* Sauerstoffbombenalterung
Sauerstoffelektrode *f* oxygen electrode
Sauerstoffentwicklung *f* oxygen evolution
Sauerstoffentwicklungsapparat *m* oxygenerator
Sauerstoffentzug *m* deoxid[iz]ation, disoxidation, deoxygenation, disoxygenation
Sauerstofffluorid *n s.* Sauerstoffdifluorid
sauerstofffrei free from oxygen, oxygen-free
Sauerstofffreiheit *f* freedom from oxygen
Sauerstofffrischverfahren *n s.* Sauerstoffaufblas-Konverterprozeß
sauerstoffgebunden oxygen-linked
Sauerstoffgehalt *m* oxygen content ✦ **mit geringem S.** low-oxygen ✦ **mit hohem S.** high-oxygen
sauerstoffgesättigt oxygenated
sauerstoffhaltig oxygen-containing, oxygenic, oxygenous
Sauerstoffkonverter *m s.* Sauerstoffaufblaskonverter
Sauerstofflanze *f (met)* oxygen lance
Sauerstoffmangel *m* oxygen deficiency
Sauerstoffpunkt *m (phys chem)* oxygen point
Sauerstoffquelle *f* oxygen source
sauerstoffreich hig-oxygen
Sauerstoffsäure *f* oxo acid
Sauerstoffschuld *f (med)* oxygen debt
Sauerstoffstrom *m* current (stream) of oxygen
sauerstofftragend oxygen-carrying
Sauerstoffträger *m* oxygen carrier; oxidizer *(for explosives)*
Sauerstoffüberspannung *f (el chem)* oxygen overvoltage
Sauerstoffüberträger *m* oxygen carrier
Sauerstoffverbindung *f* oxygen (oxy) compound
Sauerstoffverbrauch *m* oxygen consumption
chemischer S. chemical oxygen demand, C.O.D. *(of waste-water)*
sauerstoffzehrend oxygen-consuming, oxygen-depleting
Sauerteig *m* sour[dough], leaven, leavening [agent]
Säuerung *f (food)* pickling, pickle cure (curing) *(of vegetables)*; souring *(of milk)*; leavening *(of dough)*
Säuerungsbakterien *pl* acid-forming (acid-producing) bacteria, acid-formers, acid-producers
Säuerungsgefäß *n* souring (ripening, ageing, cream holding) tank, souring vat, cream (milk) ripener, ripener
Säuerungsvorgang *m* souring (ripening) process *(in milk treatment)*
Säuerungswanne *f s.* Säuerungsgefäß
Sauerwerden *n* souring *(as of milk)*
Saug-Absperr-Regulierung *f* closed-suction control *(of a compressor)*
Saugansatz *m* exit tube *(of a suction flask)*
Saugapparat *m* suction extractor
Saugbassin *n*, **Saugbehälter** *m* suction tank
Saugblasmaschine *f (glass)* suck-and-blow machine, suction-type machine
Saugblasverfahren *n (glass)* suck-and-blow (vacuum-and-blow) process, suction process
Saugdruck *m* suction pressure
saugen to suck, to draw
Saugen *n* suction, drawing
Sauger[kasten] *m (pap)* suction (vacuum, pump) box
Saugerkastendeckel *m (pap)* suction box cover
Saugerwasser *n (pap)* suction water

saugfähig absorptive, absorbent; *(pap)* bibulous
Saugfähigkeit *f* absorptivity, absorptive (absorbing, absorption) capacity, absorptive power, absorbancy, absorbency; *(pap)* bibulousness
Saugfähigkeitsprüfgerät *n (pap)* bibliometer
Saugfilter *n* suction (vacuum) filter
Saugflasche *f* suction (filter, aspirator) bottle
Saugförderer *m* suction conveyor
Saugform *f* preform screen *(for glass-fibre reinforced plastics)*
Sauggasgenerator *m* suction gas producer, suction generator
Sauggautsche *f (pap)* suction couch [roll], suction roll
Sauggebläse *n* aspirator
Sauggrube *f* suction pit
Saugheber *m* siphon, syphon
Saughöhe *f* suction head
größtmögliche S. net positive suction head, NPSH
Saughöhenprüfgerät *n (pap)* bibliometer
Saughub *m* suction stroke
Saugkasten *m* 1. wind box *(of a downdraught sintering machine)*; 2. *s.* Saugerkasten
Saugkraft *f (biol)* suction pressure (force, tension)
Saugleitung *f* suction line (pipe)
Saugmund *m* suction port
Saugnutsche *f* vacuum nutsche
Saugpapier *n* absorbent paper
Saugpappe *f* absorbent (coaster) board
Saugpipette *f* suction pipette
Saugpresse *f (pap)* suction press
Saugpreßwalze *f (pap)* suction press roll
Saugpumpe *f* suction pump
Saugreibungshöhe *f* suction friction head
Saugrohr *n* suction line (pipe)
Saugseite *f* suction side
Saugspeiser *m (glass)* suction feeder
Saugspeisung *f (glass)* suction feeding
Saugstelle *f* suction point
Saugstrahlgebläse *n*, **Saugstrahlpumpe** *f* ejector, eductor, siphon, syphon
Saugstutzen *m* suction port
Saugtopf *m* suction pot
Saugtrockner *m* suction dryer
Saugtrommeltrockner *m* suction-drum dryer
Saugventil *n* suction valve
Saugventilator *m* aspirator
Saugvermögen *n s.* Saugfähigkeit
Saugwalze *f (pap)* suction couch [roll], suction roll
Saugwirkung *f* sucking action, suction effect
S. der Kapillaren capillary suction
Saugzellenfilter *n (pap)* rotary vacuum drum-type save-all
Saugzone *f* suction area
Saugzug *m* induced (forced) draught, downdraught, suction
Saugzuggebläse *n* induced-draught fan
Saugzuglüfter *m* exhauster
Saugzugsinterung *f* downdraught sintering
Säule *f* 1. *(tech)* column, tower; 2. *(cryst)* prism; 3. column *(chromatography)*
S. mit rotierenden Scheiben rotating-disk (rotation-disk, rotary-disk) tower, rotating-disk contactor (extractor)
chromatografische S. chromatographic column
positive S. *(phys chem)* positive column
Säulenadsorptionschromatografie *f* column adsorption chromatography
säulenartig *(cryst)* columnar
Säulenbetrieb *m* column operation *(as in the ion-retardation process)*
Säulenchromatografie *f* column chromatography
säulenförmig *(cryst)* columnar
Säulenfüllung *f* column packing
Säulenhöhe *f* column height
Säulentrennung *f* column separation *(chromatography)*
Säulenwand *f* column wall
Saumeffekt *m (phot)* fringe effect
Säure *f* 1. acid; 2. sourness, acidity
anorganische S. inorganic (mineral) acid
arsenige S. H_3AsO_3 arsenious acid
bromige S. $HBrO_2$ bromous acid
Carosche S. H_2SO_5 Caro's acid, peroxomonosulphuric acid
Cassellasche S. Cassella's acid *(2-naphthol-7-sulphonic acid or 2-naphthylamine-4,8-disulphonic acid)*
chlorige S. $HClO_2$ chlorous acid
Clevesche S. Cleve's acid *(1-naphthol-5-sulphonic acid and several 1-naphthylamine sulphonic acids)*
diphosphorige S. $H_2P_2H_2O_5$ diphosphorous acid, *(better)* diphosphonic acid
dischweflige S. $H_2S_2O_5$ disulphurous acid
dithionige S. $H_2S_2O_4$ dithionous acid
dreibasige (dreiwertige) S. tribasic acid, triacid
einbasige S. monobasic acid, monoacid
eingestellte S. standard acid
einwertige S. *s.* einbasige S.
Freundsche S. $NH_2C_{10}H_5(SO_3H)_2$ Freund's acid, 1-naphthylamine-3,6-disulphonic acid
frische S. fresh acid
fuchsinschweflige S. fuchsinesulphurous acid
hochkonzentrierte (hochprozentige) S. high-strength acid
hypobromige S. HOBr hypobromous acid
hypochlorige S. HOCl hypochlorous acid
hypojodige S. HOI hypoiodous acid
hypophosphorige S. HPH_2O_2 hypophosphorous acid, *(better)* phosphinic acid
hyposalpetrige S. $H_2N_2O_2$ hyponitrous acid
isomere S. iso acid
Kochsche S. $NH_2C_{10}H_4(SO_3H)_3$ Koch's acid, 1-naphthylamine-3,6,8-trisulphonic acid
konzentrierte S. concentrated (strong) acid
korrespondierende S. conjugate acid
Laurentsche S. $NH_2C_{10}H_6SO_3H$ Laurent's acid, 1-naphthylamine-5-sulphonic acid
linksdrehende S. laevo[rotatory] acid
mehrbasige S. polybasic (polyprotic) acid, polyacid

metaphosphorige S. HPO_2 metaphosphorous acid
mittelstarke S. moderately strong (weak) acid
monophosphorige S. *s.* orthophosphorige S.
nitrose S. nitrous vitriol *(in the chamber process)*
normale S. standard acid
organische S. organic acid
orthophosphorige S. H_2PHO_3 [ortho]phosphorous acid, *(better)* phosphonic acid
pektinige S. pectinic acid
peroxosalpetrige S. HOON=O peroxonitrous acid
phosphonige S. $RP(OH)_2$ phosphonous acid
phosphorige S. phosphorous acid; *(specif) s.* orthophosphorige S.
pyrophosphorige S. *s.* diphosphorige S.
pyroschweflige S. *s.* dischweflige S.
razemische S. racemic acid
rechtsdrehende S. dextro[rotatory] acid
salpetrige S. HNO_2 nitrous acid
Schäffersche S. $HOC_{10}H_6SO_3H$ Schäffer acid, 2-naphthol-6-sulphonic acid
schwache S. weak acid
schweflige S. H_2SO_3 sulphurous acid
selenige S. H_2SeO_3 selenious acid
standardisierte S. standard acid
starke S. strong acid
ungesättigte S. unsaturated acid
α,β-ungesättigte S. α,β-unsaturated acid
unterbromige S. *s.* hypobromige S.
unterchlorige S. *s.* hypochlorige S.
unterhalogenige S. hypohalous acid
unterjodige S. *s.* hypojodige S.
zweibasige (zweiwertige) S. dibasic acid, diacid

***d*-Säure** *f s.* Säure / rechtsdrehende
***l*-Säure** *f s.* Säure / linksdrehende
γ-Säure *f* γ-acid *(2-amino-8-naphthol-6-sulphonic acid)*
δ-Säure *f* δ-acid *(1-naphthol-4,8-disulphonic acid or 1-naphthylamine-4,8-disulphonic acid)*
ε-Säure *f* ε-acid *(1-naphthylamine-3,8-disulphonic acid or 1-naphthol-3,8-disulphonic acid)*
Säureabscheider *m* acid separator
Säureabsorptionsturm *m* acid-absorption tower
Säureakkumulator *m* lead (lead-acid) accumulator (battery)
Säureamid *n* $RCONH_2$ acid amide
Säureamidabbau *m* degradation of amides
Hofmannscher S. Hofmann degradation of amides, Hofmann reaction
Säureanhydrid *n* acid anhydride
Säureanthrazenbraun *n* acid anthracene brown
Säureäquivalent *n* acid equivalent
Säureätzung *f* acid etching
Säureaufschluß *m* decomposition (digestion) by acids; *(fert)* acidulation; *(pap)* acid pulping
Säureaustausch *m* acid exchange
Säureazid *n* RCO−N=N=N acid azide
Säurebad *n* acid bath
Säureballon *m* acid carboy
Säure-Basen-Dissoziationskurve *f* acid-base dissociation curve *(of proteins)*
Säure-Basen-Gleichgewicht *n* acid-base equilibrium (balance)
Säure-Basen-Indikator *m* acid-base indicator
Säure-Basen-Katalyse *f* acid-base catalysis
säure-basen-katalysiert acid-base catalyzed, catalyzed by acids and bases
Säure-Basen-Paar *n* acid-base pair
Säure-Basen-Reaktion *f* acid-base reaction
Säure-Basen-Theorie *f* acid-base theory
Säure-Basen-Titration *f* acid-base titration, neutralization titration
Säurebehälter *m* acid tank
Säurebehandlung *f* acid treatment, *(text also)* acid steeping, grey souring
Säurebereitung *f* acid preparation (making), *(pap also)* cooking-liquor manufacture
säurebeständig acid-resistant, acid-resisting, acid-fast, acid-stable, acid-proof, resistant (fast, stable) to acids
Säurebeständigkeit *f* acid resistance (fastness, stability, endurance), resistance (stability) to acids
säurebildend acid-forming
Säurebildner *mpl* acid formers, acid-producers *(esp acid-forming bacteria)*
säurebindend acid-binding
Säureblau *n* acid blue
Säurebraun *n* acid brown
Säurebromid *n* acid bromide
Säurecharakter *m* acidic character
Säurechlorid *n* acid chloride
Säurechlorid-Reduktion *f* / **Rosenmund-Saizewsche** Rosenmund (acid chloride) reduction, Rosenmund reaction
Säuredampf *m* acid vapour (fume)
Säuredämpfen *n (text)* acid ageing
Säuredämpfer *m (text)* acid ager
Säuredissoziation *f* acid[ic] dissociation
Säuredissoziationskonstante *f* acid[ic] dissociation constant
Säuredruckvorlage *f* acid egg
säureecht *(dye)* acid-fast, fast to acid
Säureechtheit *f (dye)* acid fastness
säureempfindlich acid-sensitive, acid-labile, sensitive to acids
Säureempfindlichkeit *f* sensitivity to acids
Säurefällung *f* acid precipitation
Säurefarbstoff *m* acid[ic] dye
säurefest *s.* säurebeständig
Säureform *f* aci form *(of nitro compounds)*
Säurefraktion *f* acidic fraction
säurefrei acid-free, free from acidity
Säurefuchsin *n* acid fuchsine
Säurefunktion *f* acidic function
säuregefällt acid-precipitated
Säuregehalt *m* 1. acid content; 2. *s.* Säuregrad
Säureglasballon *m* acid carboy
Säuregoudron *m* acid tar
Säuregrad *m* [degree of] acidity
S. nach Soxhlet-Henkel degree Soxhlet-Henkel, degree S/H

Säuregradbestimmung *f* acidity test, test for acidity
Säurehalogenid *n* acid halide
säurehaltig acid-containing, acidic, acidiferous
Säurehaltigkeit *f s.* Säuregrad
Säurehärtung *f (coat, plast)* acid-catalyzed cure, acid hardening
Säureheber *m* acid siphon
Säurehydrolyse *f* acid hydrolysis
Säureion *n* acid[ic] ion
Säurekatalysator *m* acid[ic] catalyst
Säurekatalyse *f* acid[ic] catalysis
allgemeine S. general acid catalysis
spezifische S. specific acid catalysis
säurekatalysiert acid-catalyzed, catalyzed by acid[s]
Säurekochechtheit *f (text)* fastness to cross-dyeing, fastness to topping
Säurekomponente *f* acid component
Säurekonstante *f* acidity constant
Säurekonzentration *f* acid concentration
säurelabil acid-labile, acid-sensitive, sensitive to acid
Säureleitung *f* acid-proof pipe; *(pap)* liquor-circulating line
säureliebend acidophilic, acidophilous, oxyphil[e], oxyphilic, oxyphilous
säurelöslich acid-soluble, soluble in acids
Säurelöslichkeit *f* solubility in acids
Säurelösung *f* acid solution
Säuremattierung *f* acid etching
Säurenebel *m* acid mist
Säureorange *n* acid orange
Säurepergament *n* vegetable parchment, parchment paper
Säurepolieren *n*, **Säurepolitur** *f* acid polishing
Säureprobe *f*, **Säureprüfung** *f* acid test
Säurepumpe *f* acid pump
Säureregenerat *n (rubber)* acid reclaim
säureresistent *s.* säurebeständig
Säureresistenz *f s.* Säurebeständigkeit
Säurerest *m* acid residue
Säurescharlach *m* acid scarlet
Säureschicht *f* acid layer
Säureschlamm *m* acid sludge
Säureschockverfahren *n (text)* pad-acid develop process
Säureschutzschürze *f* acid apron
Säureschwarz *n* acid black
Säurespaltung *f* acid cleavage, cleavage by acids
säurestabil *s.* säurebeständig
Säurestabilität *f s.* Säurebeständigkeit
Säurestärke *f* acid[ic] strength
Säurestation *f (pap)* acid [preparation] plant, liquor-making plant
Säuretank *m* acid tank
Säureteer *m* acid tar
Säureturm *m (pap)* 1. acid[-making] tower; 2. absorption (reaction, limestone) tower
S. aus Beton concrete acid tower
Säureüberschuß *m* excess of acid
säureunbeständig acid-labile
säureunlöslich acid-insoluble, insoluble in acids
Säureverfahren *n (rubber)* acid [reclaiming] process
Säureverhalten *n* acid behaviour
Säureverhältnis *n (pap)* liquor[-to-wood] ratio *(ratio of cooking liquor to wood weight)*
Säurevorratsbehälter *m* acid storage tank
Säurewäsche *f* acid washing
Säurewasserstoff *m* acidic hydrogen
Säurewecker *m (food)* starter, seeding material
Säureweckerapparat *m (food)* starter heater
Säureweckerdestillat *n (food)* starter distillate
Säureweckerkultur *f (food)* starter culture
säurewidrig antacid
Säurezahl *f* acid value (number), A.V.
säurezersetzlich acid-labile, acid-sensitive, sensitive to acids
Saussurit *m (min)* saussurite *(a tectosilicate)*
Saussuritbildung *f*, **Saussuritisierung** *f (geol)* saussuritization
Saybolt-Kolben *m* Saybolt distilling flask
Saybolt-Universalviskosimeter *n (petrol)* Saybolt visco[si]meter
SB *s.* Siedebeginn
S-Benzin *n* liquid-phase gasoline (petrol)
SBK *s.* Styrol-Butadien-Kautschuk
Sbp. *s.* Sublimationspunkt
SBR-Latex *m* SBR latex, styrene-butadiene latex
SCF-LCAO-Rechnung *f* LCAO SCF calculation
SCF-Methode *f* SCF method, self-consistent-field method
SCF-Molekülorbital *n* self-consistent-field molecular orbital
SCF-Rechnung *f* SCF calculation, self-consistent-field calculation
Schäbe *f (text)* shive
Schabeisen *n (ceram)* scraper
Schabemesser *n* scraper knife (blade); *(for rolls)* doctor knife (blade)
Schaber *m* scraper; *(for rolls)* doctor
S. am Trockenzylinder *(pap)* dryer doctor
S. an der Brustwalze *(pap)* breast roll doctor
S. an der [oberen] Preßwalze *(pap)* [top] press roll doctor
S. an der Siebleitwalze *(pap)* wire-roll doctor
traversierender S. *(pap)* vibrating (oscillating) doctor
Schaberblech *n* raking blade
Schaberklinge *f*, **Schabermesser** *n s.* Schabemesser
Schaberwalze *f* doctor roll
Schabkarton *m* scraper (scratch) board
Schablone *f* pattern; *(for painting or printing)* stencil
Schablonenätzung *f (glass)* plate etching
Schablonenseide *f* stencil silk
Schabmesser *n s.* Schabemesser
Schacht *m* shaft, duct; stack *(of a blast furnace)*; pocket *(of a pulpwood grinder)*
Schachtelkarton *m* [folding] boxboard, carton
Schachtelpresse *f* box (pot) press
Schacht-Kettenrost *m* shaft-chain grate

S. von Makarjew Makarev shaft-chain grate
Schachtmauerwerk *n* inwall
Schachtofen *m* shaft furnace (kiln), vertical (upright) kiln
Schachttrockner *m* shaft (tower) dryer
Schachtwasser *n* mine water
Schädigungseffekt *m (text)* tendering effect
Schädigungsfaktor *m* damage factor
schädlich noxious, harmful, injurious, deleterious, detrimental
Schädlichkeit *f* noxiousness, harmfulness, injuriousness, deleteriousness, detrimentalness
Schädling *m (agric)* pest
Schädlingsbekämpfung *f* pest control
Schädlingsbekämpfungsmittel *n* pesticide, [pest] control agent
chemisches S. pesticide (pesticidal) chemical
Schädlingsbekämpfungsmittelrückstand *m* pesticide residue
Schadraum *m* clearance volume *(in a reciprocating engine)*
relativer S. clearance *(ratio of clearance volume to swept volume)*
Schadstoff *m* pollutant, polluting agent
Schäffer-Salz *n* Schäffer salt *(sodium salt of Schäffer acid)*
Schäffer-Säure *f* $HOC_{10}H_6SO_3H$ Schäffer acid, 2-naphthol-6-sulphonic acid
Schafgarbenöl *n* milfoil oil
Schafmilch *f* sheep milk
schal stale, flat
Schale *f* 1. *(phys chem)* shell *(of an atom)*; 2. *(lab)* dish; *(tech)* pan, tray; 3. *(bot)* peel, rind, *(if dry)* husk, hull, *(if hard or fibrous)* shell
abgeschlossene S. closed shell *(of an atom)*
äußere S. external (outermost, outer) shell *(of an atom)*
innere S. inner shell *(of an atom)*
schälen 1. *(pap)* to peel, to remove the bark, to [de]bark, *(Am also)* to ross; 2. *(food)* to decorticate, to excorticate *(cereals)*; 3. *(tech)* to skim *(as in the knife-discharge centrifuge)*
Schalenaufbau *m* shell structure *(of an atom)*
Schalendestillierapparat *m* pot still
Schalenelektron *n* shell electron
Schalenentwicklung *f (phot)* dish development
Schalenguß *m* 1. chill casting; 2. chill-cast (chilled cast) iron, chill casting
Schalengußkern *m* chill[ed] core
Schalengußstück *n s.* Schalenguß 2.
Schalenhartguß *m s.* Schalenguß
Schalenmehl *n (plast)* shell flour *(a filler)*
Schalenmodell *n* shell model
Schalenstruktur *f* shell structure
Schäler *m s.* 1. Schälmaschine; 2. Schälmesser
Schälfestigkeit *f (plast)* peel strength
Schälfolie *f* sliced sheet, *(if thin)* sliced film
Schallaerozyklon *m* sonic cyclone *(gas cleaning)*
Schalldämmplatte *f* sound-insulation board
Schalldämmstoff *m* sound-absorbing material
Schallenergie *f* sonic energy
Schallfeld *n* sound field
Schällöffel *m (food)* skimmer
Schallsirene *f s.* Schallaerozyklon
Schallwelle *f* sound wave
Schälmaschine *f (pap)* barking (peeling) machine, barker; *(food)* peeling machine
Schälmesser *n* unloader knife, skimmer *(as of a centrifuge)*
Schälprüfung *f* peeling test *(for testing the strength of an adhesive-bonded joint)*
Schälrohr *n* skimming tube, skimmer [pipe] *(of a centrifuge)*
Schälschleuder *f s.* Schälzentrifuge
Schalter *m* switch
Schalteröl *n* switch oil
Schäl- und Haftprüfung *f (plast)* peel bond test
Schälung *f (pap)* peeling, [de]barking, *(Am also)* rossing
Schälzentrifuge *f* skimmer (knife-discharge) centrifuge
Schamotte *f* chamotte
zerkleinerte S. grog
Schamotteerzeugnis *n* fireclay refractory [material]
SiO_2-reiches S. siliceous refractory *(containing 78 to 92% SiO_2)*; silica refractory *(containing more than 92% SiO_2)*
Schamottekapsel *f (ceram)* saggar, fireclay box
Schamotteplatte *f* fireclay plate
Schamottestein *m* refractory (fireclay) brick, firebrick
Schamotteton *m* refractory clay, fireclay
Schamotteziegel *m s.* Schamottestein
Schampun *n (cosmet)* shampoo
Schappe[seide] *f* chappe (schappe) silk, [s]chappe
Schardinger-Dextrin *n* Schardinger dextrin
Schardinger-Enzym *n* Schardinger enzyme, xanthine oxidase
scharf 1. severe, rigorous, drastic *(reaction conditions)*; 2. sharp, acrid, pungent *(taste)*
Schärfe *f* 1. severity, severeness, rigorousness *(of reaction conditions)*; 2. sharpness, acridity, pungency *(of taste)*
S. der Absiebung cleanliness of cut *(classifying)*
S. der Küpe *(text)* sharpness of the vat
schärfen *(text)* to sharpen *(the vat)*
Scharffeuer *n (ceram)* hard (quick) fire
scharfgebrannt *(ceram)* hard-burned, hard-fired
Scharlach *m (dye)* scarlet
Biebricher S. Biebrich (scarlet) red
Venezianischer S. Venetian scarlet
Scharlachkörner *npl (dye)* scarlet corns, kermes berries (grains), kermes *(dried bodies of the females of various scales, genus Kermes)*
Schatten *mpl (tann)* blasting
Schattenwand *f* shadow (baffle) wall *(of a glass-melting furnace)*
Schattenzeichnung *f (phot)* shadow detail
schattieren to shade
Schattierung *f* shade; shading
Schauer *m (nucl)* shower
Schauerteilchen *n (nucl)* shower particle
Schaufel *f* 1. shovel, scoop; flight *(as in a drying cylinder)*; 2. *s.* Schaufelblatt

Schaufelblatt *n* shovel, vane, blade, paddle
Schaufelelement *n* mixing element *(of an agitator)*
Schaufelkneter *m* kneading machine, kneader, double-arm mixer; paddle kneading table *(margarine making)*
Schaufelmischer *m* blade (arm) mixer
Schaufelradfärbemaschine *f (text)* paddle [wheel] dyeing machine
Schaufelrührer *m*, **Schaufelrührwerk** *n* blade (arm) mixer
Schaufelwinkel *m* vane angle
Schauglas *n* 1. inspection (sight, gauge) glass; 2. show (inverted) bottle, specimen (museum, preservation) jar
Schaukelbecherwerk *n*, **Schaukelförderer** *m* swing-bucket (swing-tray) elevator, pivoted-bucket conveyor
schaukeln to swing, to rock
Schaukelofen *m* rocking furnace
Schaukelrahmen *m (tann)* rocker [frame] *(for moving pelts in suspender pits)*
Schaukelschwingung *f* rocking vibration *(spectroscopy)*
Schaukeltrockner *m* tilting (reversing) pan dryer
Schauloch *n* inspection hatch (port), sight (spy) hole, peephole, viewing aperture
Schaum *m* 1. foam, froth, spume, *(if undesired)* scum, skimmings; 2. *(glass)* scum[ming]; 3. *(met)* dross; 4. *(ferm)* head; 5. *(cosmet)* lather ✦ **sich mit S. bedecken** to foam, to froth, to spume
fester S. *(plast)* rigid foam
halbharter S. *(plast)* semirigid foam
zweiphasiger S. *(coll)* two-phase foam
Schaumabstreifer *m* scum skimmer (baffle, remover), skimmer blade; *(min tech)* froth skimmer
schaumartig foam-like, foamy, frothy, spumescent
Schaumbekämpfungsöl *n (pap)* [anti]froth oil
Schaumbeständigkeit *f* foam stability (persistence), foam-holding capacity, lifetime of the foam; *(ferm)* firmness of the head, head retention
Schaumbeton *m* foamed (aerated, cellular-expanded) concrete
schaumbildend foam-producing
Schaumbildner *m* foaming (frothing, froth-forming, gasifying) agent, foamer, frother
Schaumbildung *f* foam (froth) formation, foaming, frothing; *(ferm)* head formation
Schaumbildungsfähigkeit *f s.* Schaumbildungsvermögen
Schaumbildungsmittel *n s.* Schaumbildner
Schaumbildungsvermögen *n* foaming power (ability), foaminess, frothiness
Schaumdämpfer *m*, **Schaumdämpfungsmittel** *n s.* Schaumverhütungsmittel
Schaumdauer *f s.* Schaumbeständigkeit
Schäumeigenschaft *f* foaming property; *(cosmet)* lathering property
Schaumeindicker *m* scum thickener
schäumen 1. to foam, to expand *(e. g. plastics)*; 2. to foam, to froth, to spume, to effervesce; *(cosmet)* to lather
schäumend *(chem)* effervescent, spumescent; *(ferm)* effervescent, brisk
Schaumentwässerung *f* foam draining
Schaumentwicklung *f s.* Schaumbildung
Schäumer *m*, **Schaumerzeuger** *m s.* Schaumbildner
Schaumfaden *m (text)* foam filament
schaumfähig *(incorrectly for)* 1. schaumbildend; 2. verschäumbar
Schaumfleck *m* froth spot (mark) *(a defect in paper)*
Schaumflotation *f* froth floatation
schaumfördernd foam-promoting
Schaumgärung *f* foam (froth) fermentation
Schaumglas *n* foam (foamed, cellular, sponge) glass
Schaumglasblock *m* foam glass block
Schaumgummi *m* foam (foamed) rubber (latex), latex foam [rubber]
Schaumhaltigkeit *f s.* Schaumstabilität
Schaumherstellungsmaschine *f* frother
schaumig, schäumig foamy, frothy, spumescent
Schaumigkeit *f*, **Schäumigkeit** *f* foaminess, frothiness
Schaumkelle *f* [scum] skimmer
Schaumkonzentrat *n (min tech)* concentrate-laden froth, froth product
Schaumkraft *f* foaming power (ability)
Schaumkrone *f* head *(on beer)*
Schaumkunststoff *m s.* Schaumstoff
Schaumlamelle *f (text)* foam film
Schaumlinie *f (glass)* foam line
Schaumlöffel *m* [scum] skimmer
Schaumlöscher *m* foam extinguisher
Schaummaschine *f s.* Schaumschlagmaschine
Schaummeßgerät *n* foam meter
Schaummittel *n*, **Schäummittel** *n s.* Schaumbildner
Schaumöl *n (pap)* [anti]froth oil
Schaumpolystyrol *n* foamed (expanded) polystyrene
Schaumpolyvinylchlorid *n* foamed (expanded) polyvinyl chloride
Schaumprodukt *n (min tech)* froth product
Schaum-PVC *n s.* Schaumpolyvinylchlorid
Schaumschlacke *f* foamed slag
Schaumschlagmaschine *f (rubber)* foaming (frothing, beating, whisking) machine, frother
Schaumschwimmaufbereitung *f* froth floatation
Schaumstabilisator *m* foam stabilizer (stabilizing compound)
Schaumstabilität *f* foam persistence (stability), foam-holding capacity, lifetime of the foam; *(ferm)* firmness of the head, head retention
Schaumstoff *m* foamed (expanded, cellular) plastic ✦ **mit S. beschichtet** foam-backed
chemisch getriebener S. chemically foamed plastic
harter S. rigid foam
offenzelliger S. open-cell foam

physikalisch getriebener S. mechanically foamed plastic
„schaumloser" S. foamless foamback *(obtained by melting thin laminates)*
weicher (weich-elastischer) S. flexible foam
Schaumstoffaden *m (text)* foam filament
schaumstoffbeschichtet foam-backed
Schaumstoffdämmung *f*, **Schaumstoffisolierung** *f* rigid-foam insulation
schaumstoffkaschiert *(text)* foam-backed
Schaumstoffolie *f* expanded sheet
Schaumstofformen *n* foam moulding
Schaumstoffschnittfaden *m (text)* foam filament
Schaumton *m* foamed clay, foamclay
Schaumverbesserer *m* foam improver
Schaumverhinderungsmittel *n s.* Schaumverhütungsmittel
Schaumverhütung *f* foam (froth) suppression (prevention), inhibition of foaming
Schaumverhütungsmittel *n* defoaming (antifoaming, antifoam) agent, antifoam[er], defoamer, foam inhibitor (killer, breaker, destroyer), antifrothing (antifroth, froth-preventing) agent, froth suppressor
Schaumverhütungsöl *n (pap)* [anti]froth oil
Schäumverlust *m* foaming loss
Schaumvermögen *n (incorrectly for)* Schaumbildungsvermögen
Schaumvolumen *n* foam volume
Schaumwein *m* sparkling (effervescent) wine, champagne
im Flaschengärverfahren hergestellter S. bottle-fermented champagne
im Tankgärverfahren hergestellter S. bulk process champagne
Schaumweinherstellung *f* champagnization
Schaumzahl *f* foam number
Schaumzerstörung *f* breaking of foams
Schaumzerstörungsmittel *n s.* Schaumverhütungsmittel
Schauöffnung *f* inspection hatch (port), viewing aperture
Scheckpapier *n* safety (cheque) paper
Scheelit *m (min)* scheelite *(calcium wolframate)*
Scheibe *f* disk; *(filtr)* leaf, disk; washer *(as for tightening joints)*
S. mit Flügeln (Schaufeln) vaned disk
rotierende S. rotating disk
Scheibel-Kolonne *f* Scheibel column
Scheibenaufschläger *m (pap)* disk refiner
Scheibenbrecher *m* disk crusher
Scheibenfilter *n* leaf (disk) filter
scheibenförmig disk-shaped
Scheibengasbehälter *m* dry seal gasholder
Scheiben[kolloid]mühle *f* disk [attrition] mill, disk grinder
Scheibenrefiner *m (pap)* disk refiner
Scheibenrührer *m* disk agitator
Scheibenseparator *m*, **Scheibentrieur** *m* disk separator
Scheibenverdampfer *m* disk evaporator
Scheibenversprüher *m* disk (centrifugal-disk, spinning-disk, rotary-disk, spray-disk) atomizer
Scheibenversprühung *f* disk (centrifugal-disk, spinning-disk, rotary-disk, spray-disk) atomization
Scheibenzähler *m* disk meter, nutating-piston meter *(flow measurement)*
Scheibenzerstäuber *m s.* Scheibenversprüher
Scheibenzerstäubung *f s.* Scheibenversprühung
Scheibler-Exsikkator *m* Scheibler desiccator
Scheidebehälter *m s.* Scheidegefäß
Scheidefiltration *f* cake filtration, cake[-filter] operation
Scheidegefäß *n* separating (separatory) vessel, settler
Scheidegut *n* material being (*or* to be) separated
Scheidekalk *m* sugar-factory lime
Scheidemittel *n* separating agent
scheiden 1. to separate, to segregate, *(relating to size)* to classify, to make a cut; 2. *(sugar)* to defecate, to lime
Scheidepfanne *f (sugar)* defecation tank (pan), defecator
Scheidepresse *f* wringer, press
Scheider *m* separator, *(relating to size)* classifier
Scheidesaft *m (sugar)* defecated juice
Scheidesaturation *f (sugar)* carbonation process, lime-carbon-dioxide process
Scheideschlamm *m* 1. *(sugar)* defecation mud (scum); 2. *(agric)* sugar-factory lime
Scheidetrichter *m* separatory funnel
Scheidewand *f* 1. diaphragm, membrane, septum; partition [wall]; 2. *(pap)* centre division, mid-feather, mid-wall, midriff *(of a Hollander beater)*
halbdurchlässige S. *s.* semipermeable S.
poröse S. porous diaphragm (membrane)
semipermeable S. semipermeable diaphragm (membrane)
Scheidewasser *n* aqua fortis *(concentrated nitric acid)*
Scheidung *f* 1. separation, segregation, *(relating to size)* classification; 2. *(sugar)* defecation, liming
elektrostatische S. electrostatic separation
heiße S. *(sugar)* hot defecation (liming)
kalte S. *(sugar)* cold defecation (liming)
nasse S. *(sugar)* defecation with milk of lime, wet liming
trockene S. *(sugar)* defecation with dry lime, dry liming
warme S. *s.* heiße S.
Scheinlöslichkeit *f* apparent solubility
Scheinporosität *f (ceram)* apparent porosity
Scheinviskosität *f* apparent viscosity
Scheitelpunkt *m* peak *(as of a curve)*
Scheitelwert *m* peak
Scheitermost *m* last pressed juice
Schellack *m* shellac
gebleichter S. bleached (white) shellac
weißer (weißgebleichter) S. *s.* gebleichter S.
Schellacklösung *f* shellac varnish
Schellbach-Bürette *f* Schellbach burette
Schelle *f* clamp, clip
Schellolsäure *f* shellolic acid *(a dicarboxylic acid present in shellac)*

Schenkel *m* leg *(of a U-tube)*
Scherarbeit *f* work done in shear
Scherben *m (ceram)* [ceramic] body
  geschrühter (verschrühter) S. biscuitted body
Scherben *mpl* 1. *(glass)* cullet; 2. *(ceram)* pitchers
  fabrikeigene S. *(glass)* factory (domestic) cullet
  fremde S. *(glass)* foreign cullet
Scherbeneis *n* flake ice
Scherbengemenge *n (glass)* raw cullet
Scherbenoberfläche *f (ceram)* body surface
Scherdegen *m (tann)* fleshing knife
scheren to shear
Scheren *n* shear[ing]
Scherenarm *m* chelate arm *(coordination chemistry)*
Scherenbildung *f* chelation, chelate formation *(coordination chemistry)*
Scherenverbindung *f* chelate, chelate complex (compound), crab's claw complex, scissor compound
Scherfestigkeit *f* shear[ing] strength, shear stability
Schergeschwindigkeit *f* shearing rate (speed)
Schermaschine *f* 1. *(text)* shearing machine, shearer; 2. longitudinal covering (insulating) machine *(for cable manufacture)*
Scherscheibenviskosimeter *n* shearing-disk viscometer
Scherstift *m* shear pin
Scherung *f* shear[ing]
Scherverfahren *n* long covering (insulating) process *(for cable manufacture)*
Scherwirkung *f* shearing effect (action)
scherzerkleinern to shear
Scherzerkleinerung *f* shear[ing]
Scherzylinder *m* shearing cylinder
scheuerbeständig *(text)* abrasion-resistant, wear-resistant, resistant to abrasion (wear)
Scheuerbeständigkeit *f (text)* abrasion (wear) resistance
scheuerfest *s.* scheuerbeständig
Scheuerfestigkeit *f s.* Scheuerbeständigkeit
Scheuermittel *n* 1. scouring agent; 2. abrasive, abradant *(for testing textiles)*
scheuern 1. to scour, to scrub; 2. to abrade, to gall
Scheuerprüfung *f (text)* abrasion test
Scheuerpulver *n* scouring powder
Scheuerseife *f* scouring soap
Scheuerverschleiß *m* abrasive wear, abrasion
Schibutter *f* shea (Bambuk, Galam) butter *(from Butyrospermum parkii (Don) Kotschy)*
Schicht *f* 1. layer, *(esp relating to bulk material)* stratum; coat[ing], cover, *(if thin)* film; ply *(as of laminated fabric or paperboard)*; 2. *(geol)* layer, stratum, bed
  angereicherte S. *(phys chem)* enriched layer
  aufgekohlte (eingesetzte) S. *(met)* carburized case
  elektrochemisch (elektrolytisch) aufgebrachte S. electrodeposit
  fotografische S. emulsion, sensitive (photographic) layer
  galvanisch aufgebrachte S. electrodeposit
  gekohlte S. *s.* aufgekohlte S.
  lichtempfindliche S. *s.* fotografische S.
  molekulare S. molecular layer
  monomolekulare S. monomolecular (unimolecular) film (layer), monolayer
  multimolekulare S. multimolecular layer, multilayer
  nitrierte S. *(met)* nitrided case
  obere S. top layer
  ölführende (produzierende) S. *(petrol)* producing formation
  reflexmindernde S. *(phot)* antireflection (antiflare) coating
  trägerverarmte S. *(phys chem)* depletion layer
  untere S. bottom layer
  verarmte S. *s.* trägerverarmte S.
  zementierte S. *s.* aufgekohlte S.
Schichtdicke *f* layer (film) thickness; bed depth *(of bulk material)*; length of path *(colorimetry)*
Schichtempfindlichkeit *f (phot)* emulsion sensitivity (speed)
schichten to stratify *(bulk material)*; *(plast)* to laminate
Schichtenbildung *f (geol)* stratification, lamination
Schichtenfilter *n* sheet filter, plate press
Schichtengitter *n (cryst)* layer lattice
Schichtenspaltung *f* delamination
Schichtenströmung *f* laminar (streamlined) flow
Schichtentrennung *f* delamination
Schichtfolie *f* laminated sheet
Schichtgestein *n* sedimentary rock, sediment
Schichtgitter *n (cryst)* layer lattice
Schichthöhe *f* depth of bed *(as of bulk material)*
Schichtlinie *f (cryst)* layer line
Schichtoberfläche *f s.* Schichtseite
Schichtpressen *n (plast)* laminated moulding
Schichtpreßstoff *m* laminated material (plastic), laminate
Schichtpreßstofferzeugnis *n* laminated product
Schichtpreßstoffplatte *f* laminated sheet
Schichtprobe *f* / **Hellersche** Heller layer test *(for detecting proteins in urine)*
Schichtseite *f (phot)* emulsion surface (side)
Schichtspaltung *f* delamination
Schichtstoff *m s.* Schichtpreßstoff
Schichtstoffbauweise *f (plast)* sandwich construction
Schichtstoffpreßteil *n* laminated moulding, moulded laminate
Schichtstoffprofil *n* laminated section
  [form]gepreßtes S. moulded laminated section
  nachgeformtes S. postformed laminated section
Schichtstoffrohr *n* laminated tube
  [form]gepreßtes S. moulded laminated tube
  gewickeltes S. rolled laminated tube
Schichtstruktur *f* layer structure
Schichtträger *m* 1. substrate, adherend *(material on which an adhesive is spread)*; 2. *(phot)* [film] base, [emulsion] support

Schichtung *f* 1. *(plast)* lamination; 2. *(geol)* striation, lamination
schiebefest *(text)* slip-resistant
Schiebefestappretur *f*, **Schiebefestausrüstung** *f* *(text)* antislip (non-slip) finish
schieben to push *(a reaction)*
Schieber *m* 1. gate, disk *(of a valve)*; 2. *s.* Schieberventil; 3. *(ceram)* damper
Schieberaustrag *m* gate discharge
Schiebersteuerung *f* gate control
Schieberventil *n* gate valve
Schiebestempel *m* *(plast)* sliding punch
Schiebungswinkel *m* angle of slide
Schiedsanalyse *f*, **Schiedsprobe** *f* referee check *(forensic chemistry)*
Schiefer *m* *(geol)* schist, slate, *(if argillaceous)* shale
  bituminöser S. bituminous (oil) shale
  kristalliner S. crystalline schist
schieferartig schistous, slaty, shaly
Schiefergips *m* foliated gypsum
Schiefermehl *n* slate flour
Schieferöl *n* shale oil
Schieferpapier *n* slate paper
Schieferplatte *f* slate
Schieferrohöl *n* crude shale oil
Schieferschwarz *n* slate black, black chalk *(a natural pigment)*
Schieferteer *m* shale tar
Schieferton *m* shale clay
Schieferweiß *n* flake white *(a lead pigment)*
Schieler *m s.* Schillerwein
Schiemann-Reaktion *f* Schiemann reaction *(for preparing aryl fluorides)*
Schierling *m* hemlock, Conium L.
  Gefleckter S. poison hemlock (parsley), Conium maculatum L.
Schießbaumwolle *f* gun-cotton, nitrocotton
Schießmittel *n* low explosive, propellant [explosive], deflagrating powder
Schießofen *m* *(lab)* Carius (tube, tubular) furnace
Schießpulver *n* gunpowder
Schießpunkt *m* *(petrol)* shot point *(in refraction shooting)*
Schießrohr *n* tube for sealing, sealing (sealed) tube
  S. nach Carius Carius tube
Schießstoff *m s.* Schießmittel
Schießwolle *f* gun-cotton, nitrocotton
Schiffchen *n* *(lab)* [combustion] boat
Schiffsbodenfarbe *f* ship-bottom paint
Schiffskesselkohle *f* navigation coal
Schilddrüsenhormon *n* thyroid hormone
Schildlausbekämpfungsmittel *n* scalicide
schillern to iridesce
Schillern *n* iridescence
Schillerspat *m* *(min)* schillerspar, bastite *(an inosilicate)*
Schillerwein *m* rose wine
Schimmel *m* mould *(as on food)*; mildew *(as on cloth or leather)*
schimmelbeständig mildew-resistant, mildew-proof
Schimmelbeständigkeit *f* mildew resistance
schimmelfest *s.* schimmelbeständig
Schimmelfestappretur *f*, **Schimmelfestausrüstung** *f* mildew-proofing
schimmelgereift *(food)* mould-ripened
Schimmelgeruch *m* mouldy smell
Schimmelgeschmack *m* mouldy taste (flavour)
schimmeln to mould
Schimmelpilz *m* mould [fungus]
Schimmelpilzfarbstoff *m* mould pigment
Schipprigfärben *n* *(text)* tippy dyeing
Schirm *m* *(tech)* screen
Schirmwand *f* screen wall
Schlacke *f* *(met)* slag, cinder, scum, scoria, dross; *(coal)* clinker; *(geol)* scoria ✦ **S. bilden** to slag; *(coal)* to clinker
  basische S. basic slag
  saure S. acid slag
schlacken to slag; to scorify
Schlackenabstich *m* 1. tapping *(of slag)*; 2. slagging hole, slag hole (notch, tap), slag-tap hole, cinder notch (tap)
Schlackenabstichgenerator *m* slagging[-ash] producer
  S. nach Würth Würth producer
Schlackenabstichloch *n s.* Schlackenabstich 2.
Schlackenabstichrinne *f* slag runner (spout)
Schlackenangriff *m* *(met)* slag attack
schlackenartig slaggy, scoriaceous
Schlackenbeständigkeit *f* slag resistance
Schlackenbeton *m* slag (clinker) concrete
schlackenbildend slag-forming
Schlackenbildung *f* 1. *(met)* slagging, scorification; 2. *(coal)* clinkering, clinker formation
Schlackenbrecher *m* *(coal)* clinker grinder
Schlackendamm *m* *(met)* skimmer dam (block)
Schlackenebene *f* *(met)* slag line
Schlackenfaser *f* slag wool
Schlackenkegel *m* *(met)* scorifier
Schlackenmetall *n* prill[i]on *(tin)*
Schlackenmühle *f* *(coal)* clinker grinder
Schlackenpfanne *f* slag ladle
Schlackenreaktionsverfahren *n* *(met)* Perrin process
schlackenreich slaggy
Schlackenrinne *f* slag runner (spout)
Schlackenscherben *m* *(met)* scorifier
Schlackenstein *m* slag brick (stone)
Schlackenstich *m*, **Schlackenstichloch** *n s.* Schlackenabstich 2.
Schlackenwolle *f* slag wool
Schlackenzement *m* slag cement
Schlackenziegel *m* slag brick (stone)
Schlackenziehen *n* *(coal)* clinker discharge
Schlackenzinn *n* prill[i]on
schlackig *s.* schlackenartig
Schlafmittel *n* hypnotic, somnifacient, somnificant, soporific
Schlafmohn *m* [opium] poppy, Papaver somniferum L.
Schlag *m* impact
Schlagbiegefestigkeit *f* impact bending strength

S. nach Izod *(plast)* Izod impact strength
Schlagbiegezähigkeit *f* impact bending strength
Schlagbohrverfahren *n* *(petrol)* percussion method
schlagempfindlich sensitive to impact (shock)
Schlagempfindlichkeit *f* sensitiveness to impact (shock)
Schläger *m* beater *(of a hammer mill)*
Schlägermühle *f* 1. beater mill; 2. *(pap)* chip crusher (breaker), rechipper
schlagfest impact-resistant, shock-resistant
Schlagfestigkeit *f* resistance to impact (shock), impact (shock) resistance, impact strength
Schlagfestigkeitsprüfung *f* **nach Charpy** Charpy test
Schlagfiguren *fpl (cryst)* impact figures
Schlaghärte *f* impact hardness
Schlaghärteprüfung *f* impact test
Schlagkorbmühle *f* cage (bar) mill, [squirrel-]cage disintegrator
Schlaglot *n* hard solder
Schlagmühle *f s.* Schlägermühle
Schlagpressen *n* impact moulding
Schlagprobe *f*, **Schlagprüfung** *f* impact test
Schlagsahne *f* whipped cream
Schlagstiftmühle *f* pinned-disk disintegrator (mill), pin[-type] mill
Schlagwetter *pl (mine)* firedamp, filty
Schlagwirkung *f* impact effect
Schlagzähigkeit *f s.* Schlagfestigkeit
Schlamm *m* sludge, mud, slurry, slush, *(if viscous)* slime, *(esp geol)* ooze
aktivierter S. activated (biological) sludge *(waste-water treatment)*
ausgefaulter S. digested sludge *(waste-water treatment)*
belebter (biologischer) S. *s.* aktivierter S.
konzentrierter S. thickener pulp
Schlammablaßrohr *n* sludge pipe
Schlammablaßschieber *m* desludging valve
Schlammabscheider *m* sludge pocket, slurry settler
Schlammabzug *m* sludge removal
Schlämmanalyse *f* elutriation analysis
Schlämmapparat *m* elutriator
Schlammaustrag *m* sludge discharge
Schlammbeize *f (agric)* slurry [method of seed] treatment
Schlammbelebung *f* sludge activation, bioaeration *(waste-water treatment)*
Schlammbelebungsverfahren *n* activated-sludge method *(waste-water treatment)*
Schlammberäumung *f* sludge removal
Schlämmbeton *m* prepacked (grouted) concrete
Schlammblockierungsmittel *n (min tech)* blinding agent
Schlämme *f* slurry
Schlammeindicker *m* sludge thickener (concentrator)
Schlammeindickung *f* sludge thickening (concentration)
schlämmen to elutriate, to wash
Schlämmen *n* elutriation, washing
Schlammentwässerung *f* sludge drying
Schlammfang *m* sludge pocket, slurry settler
Schlammfaulbehälter *m* sludge-digestion tank, digester
Schlammfaulraum *m* sludge-digestion chamber (compartment)
Schlammfaulung *f* sludge digestion
Schlammfilter *n* cake filter
Schlammfiltration *f* cake filtration, cake[-filter] operation
Schlammgrube *f (petrol)* mud pit *(of a rotary-drilling installation)*
schlammig sludgy, muddy, slimy
Schlammigkeit *f* muddiness, sliminess
Schlamminhibitor *m* sludge dispersant, detergent
Schlämmkaolin *m* washed kaolin, water-washed clay
Schlammkonzentrat *n* thickener pulp
Schlammkonzentration *f* sludge concentration (thickening)
Schlämmkreide *f* prepared (drop) chalk, prepared calcium carbonate, *(for technical purposes)* whiting
Schlammöl *n (petrol)* slurry oil
Schlammpumpe *f* 1. sludge (mud) pump; 2. *(sugar)* scum pump
Schlammraum *m* dirt-holding space *(of a bowl centrifuge)*
Schlammräumer *m*, **Schlammräumvorrichtung** *f* sludge-removal device (facility), sludge scraper
Schlammsaft *m (sugar)* carbonation (saturation) juice
erster S. first carbonation juice, scum juice
zweiter S. second carbonation juice
Schlammsammelschacht *m s.* Schlammsumpf
Schlammschild *m* raking blade *(of a thickener)*
Schlammschleuse *f* sludge gate
Schlammstein *m (geol)* mudstone
Schlammsumpf *m* sludge well *(waste-water treatment)*
Schlammtrichter *m* sludge hopper *(waste-water treatment)*
Schlammüberlauf *m (min tech)* slime overflow
Schlammzone *f* sludge zone
Schlange *f (tech)* coil
Schlangengift *n* snake poison
Schlangenhautglasur *f (ceram)* snakeskin glaze
Schlangenkühler *m* coil (coiled-tube, spiral) condenser
Schlangenrohr *n* coiled (spiral) tube, coil
Schlangenwärmeaustauscher *m* coil heat exchanger
Schlappgurt *m* slack belt
Schlauch *m* hose, *(without textile casing)* tubing; *(filtr)* bag
S. mit Gewebeeinlage reinforced hose
geklöppelter S. braided hose
Schlauchabscheider *m s.* Schlauchfilter
Schlauchbandförderer *m* closed-belt conveyor
Schlaucheinzelheizer *m (rubber)* unit vulcanizer for tubes

Schlauchfilter *n* bag filter, *(gas cleaning also)* bag collector
Schlauchfolie *f (plast)* blown tubing, tubular film, *(if slit)* blown film
Schlauchkammerfilter *n* bag collector *(for gas cleaning)*
Schlauchklemme *f* pinchcock, hose cock, pinch clamp
S. nach Hoffmann Hoffmann clamp, Bunsen screw clip
S. nach Mohr Mohr pinchcock (clip)
Schlauch[quetsch]pumpe *f* flow inducer
Schlauchschelle *f* hose clamp (clip)
Schlauchseele *f (rubber)* inner tube
Schlauchspritzmaschine *f* tube-extruding (tube-extrusion) press
Schlauchspritzmundstück *n* die for tubing
Schlauchstück *n* piece of tubing
Schlauchtülle *f* hose connector
Schlauchventil *n* pinch valve
Schlauchverbindung *f* hose connection
Schlauchverbindungsstück *n* hose coupling (connection, connector)
schleichend wirkend *(tox)* slow-acting
Schleier *m* 1. *(phot)* fog, haze; 2. *(coat)* bloom
dichroitischer S. dichroic fog
Schleierbildung *f* 1. *(phot)* fogging; 2. *(coat)* blooming
Schleierdichte *f (phot)* fog density
schleierfrei, schleierlos *(phot)* fog-free
schleiern *(phot)* to fog
Schleierschwärzung *f s.* Schleier 1.
Schleierverhinderung *f*, **Schleierverhütung** *f (phot)* fog prevention (inhibition)
schleifbar grindable
Schleifbarkeit *f* grindability
Schleifdruck *m (pap)* grinding pressure
schleifen to grind, *(finely)* to polish; to buff *(leather)*
Schleifengalvanometer *n* torsion loop (string) galvanometer
Schleifer *m (pap)* grinder
stetiger S. continuous grinder
Schleiferei *f (pap)* grinder house (room); *s.* Schleifereibetrieb
Schleifereibetrieb *m (pap)* groundwood mill, [mechanical] pulp mill
Schleiferstein *m (pap)* grindstone, pulpstone, abrasive stone
Schleifertrog *m (pap)* grinder pit
Schleiffläche *f (pap)* grinding surface (area)
Schleifholz *n (pap)* pulp wood
Schleifmaschine *f* grinding machine, grinder; *(tann)* buffing machine
Schleifmasse *f (pap)* mechanical [wood-]pulp, M.W.P., mechanical wood, groundwood [pulp]
Schleifmittel *n* abrasive
gebundenes S. bonded abrasive
Schleifpapier *n* abrasive paper, *(Am)* sander
Schleifpulver *n* abrasive powder
Schleifrohpapier *n* abrasive body paper
Schleifsand *m* grinding sand
Schleifscheibe *f* grinding (grinder) wheel
Schleifstein *m* grindstone, pulpstone, abrasive stone
Schleiftisch *m* grinding table
Schleifzone *f s.* Schleiffläche
Schleim *m (tech)* slime; *(bot, pharm)* mucilage; *(of animal origin)* mucus
Schleimansammlung *f (pap)* accumulation of slime
Schleimappretur *f (tann)* mucilage dressing
schleimartig *s.* schleimig
Schleimbatzen *m s.* Schleimansammlung
Schleimbildung *f (tech)* slime formation; *(bot)* mucilage formation; mucus formation *(by animals)*
Schleimfleck *m* slime spot *(a defect in paper)*
Schleimgärung *f* slime (ropy) fermentation
schleimhaltig *(bot, pharm)* mucilaginous
schleimig *(tech)* slimy; *(bot)* mucilaginous; *(relating to animal secretions)* mucous
Schleimigkeit *f (tech)* sliminess; *(biol)* mucosity
Schleimkontrolle *f (pap)* control of slime
Schleimsäure *f* $HOOC[CHOH]_4COOH$ mucic acid, galactosaccharic acid
Schleimstoff *m*, **Schleimsubstanz** *f (tech)* slimy substance (material); *(bot, pharm)* mucilaginous substance
Schleimverhütungsmittel *n (pap)* slimicide
Schlempe *f (ceram)* slip; *(ferm)* vinasse, slop, distillers' wash, spent mash (wash), stillage, pot ale
Schlempekohle *f (ferm)* vinasse cinder
Schlepper *m s.* Schleppmittel
Schleppgas *n* carrier gas *(gas chromatography)*
Schleppklingenstreichmaschine *f (pap)* trailing blade coater
Schlepplöffelbagger *m* dragline [excavator]
Schleppmittel *n (distil)* entrainer, entraining (azeotroping, azeotropic-distillation) agent, azeotrope-former
Schlepprakelstreichmaschine *f (pap)* trailing blade coater
Schleppschaufelbagger *m* dragline [excavator]
Schleppströmung *f (plast)* drag flow *(in extruders)*
Schleuder *f* centrifuge, *(tech esp for filtering)* centrifugal (*s. a.* Zentrifuge)
Schleuderbeton *m* centrifugally cast concrete, centrifugal (spun) concrete
Schleuderdüngerstreuer *m* centrifugal fertilizer distributor
Schleudereffekt *m* relative centrifugal force
Schleuderformmaschine *f (met)* sandslinger
Schleudergießverfahren *n* centrifugal-casting process *(foundry)*; *(plast)* centrifugal-moulding process
Schleuderguß *m* centrifugal casting *(foundry)*; *(plast)* centrifugal moulding
Schleudergußrohr *n* centrifugally cast pipe
Schleudergußteil *n* centrifugal casting *(foundry)*; *(plast)* centrifugal moulding

Schleuderhonig *m* centrifugal (extracted, run) honey
Schleuderkraftabscheider *m* centrifugal collector (separator)
Schleudermaschine *f (met)* sandslinger
Schleudermühle *f* cage (squirrel-cage) disintegrator (mill), centrifugal (bar) mill; *(pap)* chip crusher (breaker), rechipper
schleudern to centrifuge, to centrifugate; *(for drying)* to hydroextract, to whiz
Schleudern *n* centrifuging, centrifugation; *(for drying)* hydroextraction, whizzing
Schleuderscheibe *f* centrifugal disk
Schleudersortierer *m (pap)* centrifugal strainer, erkensator *(for the clean-up of the stock)*
Schleuderstreuer *m* centrifugal fertilizer distributor
Schleudertrockner *m* hydroextractor, whizzer
Schleuderverfahren *n* centrifugal process *(fibreglass manufacturing); s.* Schleudergießverfahren
Schleuderzahl *f* relative centrifugal force
Schleuse *f* sluice; *(esp for waste water)* sewer; [air] lock
Schlichtanlage *f (text)* sizing machine
Schlichtbaum *m (tann)* perch
Schlichte *f (text)* size, sizing [material, substance], dressing
  haftmittelfreie S. textile size
  haftmittelhaltige S. reinforcement size
  textile S. textile size
Schlichteauftragvorrichtung *f (text)* sizing pad
Schlichtebad *n (text)* sizing bath
Schlichteflotte *f (text)* sizing bath
Schlichtekocher *m (text)* size cooker
Schlichtemittel *n s.* Schlichte
schlichten 1. *(text)* to size, to dress, to slash; 2. *(tann)* to perch
Schlichten *n* **der Kettfäden** *(text)* warp sizing
  S. im Strang, S. in Strangform *(text)* hank sizing
Schlichtmaschine *f (text)* sizing machine, slasher
Schlichtmond *m (tann)* moon knife
Schlick *m* slime, mud, ooze
Schlicker *m* 1. *(ceram)* slip, slop, slurry; 2. *(tann)* slicker
  flüssiger S. *(ceram)* liquid slip
  glasartiger S. *(ceram)* vitreous slip
Schlickergießen *n (ceram, met)* slip casting
Schlickerglasur *f (ceram)* slip glaze
Schlickerguß *m (ceram, met)* slip casting
Schlickermalerei *f (ceram)* trailing
Schlickermilch *f s.* Schlippermilch
Schlickerofen *m (ceram)* slip kiln
Schlickerüberschuß *m (ceram)* surplus slip
Schlickerüberzug *m (ceram)* slip coating
Schlickerverfahren *n (ceram)* slip (wet) process
Schliere *f (glass)* stria, vein, cord; *(plast)* stria; schliere *(in a fluid)*
Schlierenaufnahme *f* schlieren photo[graph]
Schlierenbild *n* schlieren picture
Schlierenbildung *f (glass, plast)* striation
Schlierenfotografie *f* schlieren photography
Schlierengerät *n* schlieren apparatus
Schlierenverfahren *n* schlieren method
  Toeplersches S. Toepler schlieren method
Schlierigkeit *f* cordiness *(a defect in glass)*
Schließdruck *m (plast)* clamping (locking, mould-locking) pressure
schließen / sich zum Ring to cyclize
Schließring *m (plast)* locking ring
Schließweg *m* closing travel *(of a press)*
Schließzeitbestimmung *f (plast)* cup flow test, flow cup test
Schliff *m* 1. grinding, *(glass also)* cutting; 2. *(result)* cut *(of glass or gems)*; 3. *(substance)* grindings; *(pap)* groundwood; section *(of rocks for microscopy)*; 4. *s.* Schliffverbindung
  chemischer S. *(pap)* chemigroundwood
Schliffball *m s.* Schliffkugel
Schliffläche *f* ground surface
Schliffgeräte *npl* ground-glass equipment
Schliffhülse *f* socket, female tapered joint
Schliffkern *m* cone, male tapered joint
Schliffkette *f* chain of joints
Schliffkugel *f* ball, male spherical joint
Schliffpfanne *f* socket, female spherical joint
Schliffschale *f s.* Schliffpfanne
Schliffstopfen *m* ground[-glass] stopper
Schliffverbindung *f* ground[-glass] joint
Schlippermilch *f* [naturally, spontaneously] soured milk, sour (set) milk
Schlitz *m* slit, slot
Schlitzaufsatz *m* [burner] flame spreader, burner wingtop
Schlitzbreite *f* slot width
Schlitzbrenner *m* bats-wing (flat-flame, fish-tail) burner
Schlitzdüse *f* slot nozzle; *(plast)* slit die (orifice)
Schlitzdüsenauftragmaschine *f (plast)* air knife coater
Schlitzform *f* slot die *(for manufacturing plastic films)*
Schlitzglocke *f (distil)* slotted cap
Schlot *m* [smoke]stack, chimney
Schluff *m* silt *(grain size 0.02 to 0.002 mm)*
Schlumberger-Methode *f* Schlumberger method *(for measuring oil wells)*
Schlupf *m* slip[page]
Schlüpffaktor *m (biol)* hatching factor *(root excretion causing hatching of nematodes)*
Schluß *m* closure *(as of rings)*
Schlußanstrich *m* 1. finish[ing]; 2. finish, finishing coat
Schlußbütte *f (dye)* final vat
Schlüsselatom *n* key atom *(for polarizing a molecule)*
Schlüsselkomponente *f (distil)* key [component]
  höhersiedende S. heavy key
  niedrigsiedende S. light key
Schlüsselstoff *m* key substance (chemical)
Schlußkupplung *f (dye)* final coupling
schmackhaft tasteful, tasty
Schmackhaftigkeit *f* tastiness

Schmalte *f* smalt, powder blue *(cobalt(II) potassium silicate)*
Schmalz *n* melted (rendered) fat, *(esp)* lard, pig fat
Schmälze *f (text)* spinning oil, lubricant, lube; *(glass)* binder
schmälzen *(text)* to oil, to lubricate
Schmälzmittel *n s.* Schmälze
Schmälznebel *m (glass)* binder spray
Schmalzöl *n* lard oil
Schmälzöl *n (text)* spinning oil
schmauchen *(ceram)* to water-smoke
Schmauchen *n (ceram)* water-smoking
Schmauchfeuer *n (ceram)* prefire
Schmauchperiode *f (ceram)* water-smoking period
schmecken to taste
Schmelz *m (biol)* enamel
Schmelzaufschluß *m* fusion *(of sparingly soluble substances)*
Schmelzausdehnung *f* melting dilatation
Schmelzausdehnungskurve *f* melting dilatation curve
Schmelzbad *n* molten bath
schmelzbar fusible, meltable
Schmelzbarkeit *f* fusibility, meltability
Schmelzbarren *m* ingot *(in zone melting)*
Schmelzbereich *m* melting range
Schmelzbeschichtung *f (text)* flame lamination
Schmelzbutter *f* rendered butter, butterfat
Schmelzdiagramm *n* melting[-point] diagram
Schmelzdilatation *f* melting dilatation
Schmelzdruckkurve *f* melting-point [pressure] curve
Schmelze *f* 1. *(act)* fusion; *(food)* rendering; 2. *(substance)* melt, fusion; *(met)* smelt; *(geol)* fused (molten) rock; liquid *(in zone melting)*
Schmelzelektrolyse *f s.* Schmelzflußelektrolyse
schmelzen to melt, to fuse, to flux, to liquefy; to smelt *(ore)*; to render *(fat)*; to thaw *(of ice)*
  auf Stein s. *(met)* to matte-smelt
Schmelzen *n* **auf Stein** *(met)* matte smelting
  S. mit Alkali alkali[ne] fusion
  S. mit alkoholischem Kaliumhydroxid *(dye)* fusion with alcoholic potassium hydroxide
  S. unter Vakuum vacuum melting *(of metals for refining or alloying)*
  autogenes S. *(met)* autogenous smelting
  direktes S. *(met)* direct smelting
  pyritisches S. *(met)* pyritic smelting
  unmittelbares S. *(met)* direct smelting
Schmelzentropie *f* entropy of fusion
Schmelzer *m (met)* smelter; melter *(foundry)*; *(glass)* teaser, founder, melter
Schmelzextraktor *m (plast)* melt extractor *(of a plunger-type injection machine)*
Schmelzfarbe *f (ceram)* enamel (vitrifiable, overglaze, fused-on) colour
Schmelzfeuerung *f* slag-tap (wet-bottom) furnace *(with separation of slag in a molten condition)*
Schmelzfluß *m* melt, fusion; *(met)* smelting flux, smelt; *(geol)* fused (molten) rock
Schmelzflußelektrolyse *f* fused-salt (molten-salt) electrolysis
schmelzflüssig molten
Schmelzformen *n (ceram)* fusion casting
Schmelzgefäß *n s.* Schmelzkessel
Schmelzgießen *n (ceram)* fusion casting
Schmelzgut *n* material being (*or* to be) melted; molten material; *(met)* material being (*or* to be) smelted; smelted material
Schmelzharz *n* cast[ing] resin
Schmelzhütte *f (met)* smelting plant
Schmelzindex *m (plast)* melt [flow] index, M.F.I.
Schmelzkäse *m* process[ed] cheese
Schmelzkegel *m (ceram)* pyrometric (fusion) cone
  S. nach Seger *(ceram)* Seger cone
Schmelzkegeläquivalent *n (ceram)* pyrometric-cone equivalent, PCE
Schmelzkessel *m* melting pot (vessel)
Schmelzkleber *m* hot-setting (hot-melt) adhesive
Schmelzkurve *f* melting curve
Schmelzling *m* ingot *(in zone melting)*
Schmelzlöser *m (pap)* [smelt] dissolving tank, dissolving chest, dissolver
Schmelzmargarine *f* rendered margarine, margarine fat
Schmelzmittel *n* fluxing agent, flux
Schmelzofen *m* melting furnace, melter; *(met)* smelting furnace, smelter
  S. zur Verbrennung der Schwarzlauge *(pap)* recovery furnace
  vollelektrisch beheizter S. all-electric furnace
Schmelzpfanne *f* fusion pan, melt[ing] pan
Schmelzprodukt *n* product of fusion
Schmelzpunkt *m* melting point, m.p. ✦ **mit hohem S.** high-melting[-point], high-fusion ✦ **mit niedrigem S.** low-melting[-point], low-fusion
Schmelzpunktapparat *m* melting-point apparatus
  S. nach Thiele Thiele melting-point tube
Schmelzpunktbad *n* melting-point bath
Schmelzpunktbestimmung *f* melting-point determination
Schmelzpunktbestimmungsapparat *m*, **Schmelzpunktbestimmungsgerät** *n s.* Schmelzpunktapparat
Schmelzpunktdepression *f* melting-point depression (lowering)
Schmelzpunktkapillare *f*, **Schmelzpunktröhrchen** *n* melting-point capillary (tube)
Schmelzpunktserniedrigung *f s.* Schmelzpunktdepression
Schmelzraum *m (glass)* melting end (chamber), melter
Schmelzsoda *f (pap)* [soda] smelt, black ash
Schmelzspinnanlage *f* melt spinning line
Schmelzspinnen *n* melt spinning (extrusion)
Schmelzspinnverbundstoff *m (text)* spun-bonded product
Schmelzteil *m s.* Schmelzraum
Schmelztemperatur *f* melting temperature
Schmelztiegel *m* melting (ignition, fusion) crucible; *(met)* smelting pot

S. nach Rose Rose crucible
Schmelzvergasung *f (met)* gassing
Schmelzviskosität *f* melt[ing] viscosity
Schmelzwanne *f* melting tank, melter, *(glass also)* melting chamber (end)
Schmelzwärme *f* heat of fusion (melting)
Schmelzzone *f* melting zone; *(met)* smelting zone; molten zone *(in zone melting)*
Schmelzzonenbreite *f* zone width *(in zone melting)*
Schmelzzonenlänge *f* zone length *(in zone melting)*
Schmelzzonenrichtung *f* direction of zoning (zone travel) *(in zone melting)*
Schmerzlinderungsmittel *n* pain-reliever, analgesic
schmerzstillend pain-relieving, analgesic
Schmidt-Reaktion *f* Schmidt reaction *(between hydrazoic acid and carbonyl compounds)*
Schmiedekohle *f* fat coal *(proper)*
Schmiedeofen *m* forging furnace
Schmiere *f s.* Schmierstoff
Schmiereigenschaften *fpl* lubricating properties
schmieren 1. to smear; to lubricate, to grease, to oil; *(tann)* to oil, to stuff, to dub; 2. to smear *(of a substance)*
auf der Tafel s. *(tann)* to dub
Schmierergiebigkeit *f s.* Schmiergüte
Schmierfähigkeit *f* lubricating power, lubricity
Schmierfett *n* lubricating grease
Schmierflüssigkeit *f* lubricating fluid
Schmiergüte *f* oiliness *(of a lubricating oil)*
schmierig smeary, greasy, slippery, unctuous, slimy
Schmierigkeit *f* smeariness, greasiness, slipperiness, unctuousness, sliminess; *(pap)* softness, wetness, slowness
Schmierigmahlung *f (pap)* wet (slow) beating
Schmiermittel *n s.* Schmierstoff
Schmieröl *n* lubricating (lubrication, lube) oil
S. auf Erdölbasis petroleum lubricating oil
dunkles S. black oil
Schmierölausgangsstoff *m* lubricating[-oil] stock
Schmieröldestillat *n* lubricating-oil distillate
Schmierölextrakt *m* lubricating-oil extract
Schmierölfraktion *f* lubricating-oil fraction
Schmierölraffination *f* lubricating-oil refining
Schmierseife *f* soft soap
Schmierstelle *f (plast)* smudge *(on injection-moulded parts)*
Schmierstoff *m* lubricant, lubricating agent, lube
Schmierstoffschicht *f* lubricating film
Schmierung *f* lubrication, greasing, oiling
Schmierwert *m s.* Schmiergüte
Schminkrot *n* rouge
Schminkweiß *n* pearl (bismuth, Spanish) white, cosmetic bismuth *(bismuth chloride oxide)*
Schmirgel *m* emery
Schmirgelleinen *n* emery cloth
Schmirgelpapier *n* emery paper
Schmutz *m* dirt, soil, grime
künstlicher S. *(text)* artificial soil
standardisierter S. *(text)* standard soil
schmutzabstoßend, schmutzabweisend dirt-repellent, soil-repellent, antisoiling
Schmutzauswaschbarkeit *f (text)* soil release, SR
schmutzen to soil
Schmutzentfernungsvermögen *n (text)* soil removing capacity
Schmutzfang *m*, **Schmutzfänger** *m (pap)* dirt (junk) remover
Schmutzfleck *m* stain, soil, spot, smear, smudge; *(pap)* dirt speck
Schmutzflotte *f (text)* used detergent solution
Schmutzgehalt *m* dirt content *(as of paper)*
Schmutzhaftung *f (text)* soil adherence
schmutzig dirty
Schmutzlösevermögen *n (text)* soil removing capacity
Schmutzraum *m* dirt-holding space *(of a sedimentation centrifuge)*
Schmutzrückhaltevermögen *n* power of holding dirt
Schmutzschleuse *f (pap)* dirt (junk) remover
Schmutzstoff *m* contaminant, *(esp in environment)* polluting matter, pollutant; *(pap)* junk
Schmutzteilchen *n* dirt particle
Schmutztragevermögen *n* dirt-suspending (soil-suspending) power *(as of detergents)*
Schmutzwasser *n* dirty water, sewage, slops
Schmutzwasserkanalisation *f* sewerage [system], *(sanitary engineering also)* water-carriage system
Schmutzwasserpumpe *f* sewage (sludge) pump
Schmutzwolle *f* raw (grease) wool
Schnauze *f* spout *(as of a beaker)*
Schnecke *f (tech)* 1. screw, scroll, worm; 2. *s.* Schneckenförderer
S. mit Homogenisier[ungs]zone *(plast)* metering screw
tiefgeschnittene S. deep-cut screw
Schneckenaufgabegerät *n* screw (worm) feeder
Schneckenaustragzentrifuge *f* helical-conveyor centrifuge
Schneckendrehzahl *f* screw speed
Schneckenförderer *m* screw (scroll, worm, spiral, helix) conveyor
Schneckengang *m* screw flight
Schneckengetriebe *n* worm gear
Schneckenkneter *m* screw mixer
Schneckenkompressor *m* screw compressor
Schneckenmischer *m* screw mixer
senkrechter S. vertical screw mixer
Schneckenpresse *f* screw (worm) press, screw extruder; *(ceram)* auger
Schneckenpumpe *f* screw pump
Schneckenrührer *m* screw mixer
Schneckenschleuder *f* helical-conveyor centrifuge
Schneckenspeiser *m* screw (worm) feeder
Schneckenspritzgießmaschine *f (plast)* screw injection [moulding] machine
Schneckenstrangpresse *f s.* Schneckenpresse
Schneckentrieb *m* worm gear

Schneckentrockner *m* screw-conveyor dryer
Schneckenverdichter *m* screw compressor
Schneckenvorplastizierung *f (plast)* screw pre-plastication
Schneckenwelle *f* worm shaft
Schneckenzentrifuge *f* helical-conveyor centrifuge
Schneidapparat *m* cutter
schneidbar sectile
Schneidbrenner *m* dissecting blowpipe
Schneide *f* knife edge *(as of an analytic balance)*
Schneide ... *s.a.* Schneid ...
Schneidegranulator *m* rotary cutter
schneiden to cut
  autogen s. to cut autogenously
  einlagig s. *(pap)* to cut single sheet
  Elementarfäden auf Stapel s. *(text)* to staple
  in Bogen (Format) s. *(pap)* to cut into sheets, to sheet
  in Würfel s. to dice
  mehrlagig s. *(pap)* to group-cut *(sheets)*
  von der Walze s. to cut (slab) off *(rubber sheet)*
Schneiden *n* / **autogenes** autogenous (oxygen) cutting
Schneider *m s.* 1. Schneidvorrichtung; 2. Schneidmaschine
Schneidflüssigkeit *f* cutting fluid
Schneidkeramik *f* cutting ceramics
Schneidkonverterverfahren *n (text)* tow-to-top cutting system
Schneidmahlung *f (pap)* free beating
Schneidmaschine *f* cutting machine, cutter
Schneidmühle *f* cutting mill
Schneidöl *n* cutting oil
  wasserlösliches S. soluble cutting oil
Schneidplatte *f* die
Schneidriefen *fpl (plast)* sheeter lines
Schneidringverbindung *f*, **Schneidringverschraubung** *f* bite-type fitting joint
Schneidvorrichtung *f* cutting device
schnellaufend high-speed
Schnelläufer *m* high-speed (fast-running) machine
Schnellaufschluß *m (pap)* fast pulping
Schnellaufzahl *f* specific speed *(as of a pump)*
Schnellbestimmung *f* rapid determination
schnellbindend rapid-hardening, fast-setting *(e.g. cement)*
Schnellbleiche *f* quick bleach
schnellbleichend quick-bleaching
Schnelldämpfen *n (text)* flash ageing
Schnelldämpfer *m (text)* flash ager
Schnellentwickler *m (phot)* rapid (high-speed) developer
Schnellessigverfahren *n* rapid acetification process, quick vinegar process, German method of acetification
Schnellfilter *n* rapid (fast, high-rate) filter
Schnellfiltration *f* rapid (fast, high-rate) filtration
Schnellfixierbad *n (phot)* rapid (high-speed) fixing bath, rapid fixer
Schnellfixierung *f (phot)* rapid fixing
Schnellgefrieranlage *f* quick-freezing plant
Schnellgefrierapparat *m* quick-freezer
schnellgefrieren *(food)* to quick-freeze
Schnellgefrieren *n (food)* quick (rapid, sharp) freezing
Schnellgerbung *f* rapid (accelerated) tannage
schnellhärtend rapid hardening *(e.g. cement)*; *(plast)* quick-curing, fast-setting
Schnellkochung *f (pap)* quick cook
Schnellmethode *f* quick-assay method
Schnellpökeln *n*, **Schnellpökelung** *f (food)* quick curing, injection cure
Schnellpreßmasse *f (plast)* quick-curing moulding compound
schnellrotierend *s.* schnellumlaufend
Schnellrührer *m* impeller
Schnellsandfilter *n* rapid sand filter
Schnellsandfilterung *f*, **Schnellsandfiltration** *f* rapid sand filtration
Schnellschlußventil *n* quick-operating valve
Schnellschreiber *m* high-speed recorder
Schnellspaltung *f (nucl)* fast (high-energy) fission
Schnellspaltungseffekt *m (nucl)* fast [fission] effect
Schnelltest *m* quick assay, short-time test (assay), S.T.T.
schnelltrocknend quick-drying
Schnelltrocknung *f* quick (rapid, fast, high-speed) drying
Schnellüberzug *m* flash plate *(electroplating)*
schnellumlaufend spinning, high-speed
Schnellverband *m* adhesive bandage, band aid
Schnellverdampfer *m* flash evaporator, flasher
Schnellverfahren *n* short-cut method
schnellvulkanisierend *(rubber)* quick-curing, fast-curing
schnellwirkend quick-acting
Schnitt *m* cut; *(distil)* cut; [micro]section *(microscopy)*
Schnittbrenner *m* fish-tail (bats-wing, flat-flame) burner
Schnittbrenneraufsatz *m* burner wingtop (flame spreader)
Schnittfläche *f* surface of cut
Schnittlänge *f (pap)* cutting (chop) length
Schnittnarbe *f* shear mark *(a defect in glass)*
Schnittplatte *f* die
Schnittwachstum *n (rubber)* cut growth
Schnittwerkzeug *n (plast)* die
Schnittzahl *f (pap)* number of cuts
Schnitzel *npl s.* 1. Kochschnitzel; 2. Zuckerrübenschnitzel
Schnitzelmaschine *f* dicing cutter (machine), dicer
Schnitzelmaterial *n (plast)* macerate *(filler)*
schnitzeln to chop
Schnitzelpreßmasse *f (plast)* macerate moulding compound
Schnitzelpreßwasser *n (sugar)* [beet] pulp press water
Schnitzelpumpe *f (sugar)* [beet] pulp pump
Schnitzeltrockner *m (sugar)* [beet] pulp dryer

Schnitzeltrocknung *f (sugar)* [beet] pulp drying
Schnupftabak *m* snuff[ing] tobacco
Schnurabnahme *f*, **Schnürenabnahme** *f (filtr)* string discharge
Schnurfang *m (pap)* string catcher
Schokolade *f* chocolate
Schokoladenbraun *n* chocolate
Schokoladenindustrie *f* chocolate industry
Schollenlack *m* shellac
Schöllkopf-Säure *f* Schöllkopf acid *(1. $NH_2C_{10}H_5SO_3H$ 1-naphthylamine-8-sulphonic acid; 2. $HOC_{10}H_5(SO_3H)_2$ 1-naphthol-4,8-disulphonic acid)*
Schöne *f (ferm)* fining [agent]
schönen *(ferm)* to clarify, to clear, to fine
schonend mild *(e.g. oxidation)*
Schönheitsmittel *n* cosmetic, [makeup] preparation
Schönherr-Verfahren *n* Schönherr process *(for fixing atmospheric nitrogen)*
Schönseite *f* top (felt) side *(of paper)*
Schönung *f (ferm)* clarification, clearing, fining
Schönungsmittel *n (ferm)* fining [agent]
schoop[is]ieren to metallize, to sputter
Schoop[is]ieren *n* schooping, [Schoop] metallizing, [spray] metallization, metal spraying, sputtering
Schöpfbecherwerk *n* scooping bucket elevator
Schöpfbütte *f (pap)* dipping (working) vat
schöpfen *(pap)* to dip out, to mould
Schöpfer *m* 1. *(pap)* dipper, vatman, moulder; 2. *(pap)* dipper *(apparatus)*
Schöpfform *f (pap)* [hand] mould, paper-mould
Schöpfgefäß *n* dipper, scoop
S. für Chemikalien chemical dipper
Schöpfkelle *f (glass)* ladle
Schöpfpapier *n* vat (hand-made) paper
Schöpfpapiermuster *n (pap)* pulp (test, hand) sheet
Schöpfprobe *f (glass)* spoon proof
Schöpfrad *n* flighted wheel
Schöpfrahmen *m* deckle, hand-mould, [paper-] mould
Schöpfrand *m (pap)* deckle [edge] ✦ **mit S.** deckled
Schöpfwerk *n* scooping bucket elevator
Schopper-Dalen-Maschine *f (rubber)* Schopper machine
Schörl *m (min)* schorl, shorl, schorlite *(a cyclosilicate)*
Schorlomit *m (min)* schorlomite *(a variety of melanite)*
Schornstein *m* chimney, [smoke]stack; *(lab)* burner chimney
S. zum Anlassen starting stack *(of a FluoSolids reactor)*
Schornsteinaufsatz *m (lab)* burner chimney
Schornsteinzug *m* draught, chimney pull
Schotten-Baumann-Reaktion *f (org chem)* Schotten-Baumann reaction
Schotter *m* gravel
Schrägagarkultur *f* slant culture *(microbiology)*
Schrägaufzug *m* skip (inclined) hoist
Schrägbecherwerk *n* inclined bucket elevator
Schrägbeziehung *f* diagonal relationship *(in the periodic system)*
Schrägbrücke *f* hoist bridge
S. mit Kippkübel skip bridge *(of a blast furnace)*
Schrägkultur *f* slant culture *(microbiology)*
Schrägrinne *f* trough washer
Schrägrohrverdampfer *m* inclined evaporator
Schrägrost *m* sloping grate
Schrägschnittpapier *n* angle (angular, angle-cut, cater-cornered) paper
Schrägsitzventil *n* inclined-seat valve, Y valve
Schrägspritzkopf *m (plast)* oblique (angular, angle) head *(of an extruder)*
Schrapper *m* scraper
Schraube *f* 1. screw; 2. propeller
Schraubenachse *f (cryst)* screw axis
schraubenförmig spiral, helical
Schraubenformstück *n* screwed fitting
Schraubenkappe *f* screw-on-type cap
Schraubenklassierer *m* spiral classifier
Schraubenkompressor *m* screw compressor
Schraubenkühler *m* Friedrichs [reflux] condenser
Schraubenpumpe *f* screw pump
Schraubenquetschhahn *m (lab)* screw [compressor] clamp, Hoffmann clamp, [Bunsen] screw clip
Schraubenquirl *m* propeller blunger
Schraubenrührer *m* propeller mixer (agitator)
Schraubenspindel *f* screw
Schraubentute *f (petrol)* overshot
Schraubenverbindungsstück *n* screwed fitting
Schraubenverdichter *m* screw compressor
Schraubenversetzung *f (cryst)* screw dislocation
Schraubfitting *m(n)* screwed fitting
Schraubklemme *f* **[nach Hoffmann]** *s.* Schraubenquetschhahn
Schreckplatte *f* chill *(foundry)*
Schreckschicht *f* chilled portion, chill *(foundry)*
Schreiber *m* recording instrument, recorder
Schreibgalvanometer *n* galvanometer recorder
Schreibinstrument *n s.* Schreiber
Schreibkreide *f* chalk *(calcium carbonate)*
Schreibmaschinenbänderfarbe *f* typewriter-ribbon ink
Schreibmaschinenpapier *n* typewriting (typewriter) paper, T.W.
Schreibpapier *n* writing paper
Schreibpergament *n* vellum [paper]
Schreibspur *f* trace *(of a recorder)*
Schreibtinte *f* writing ink
Schreibwerk *n* recorder
Schrenzkarton *m* screenings
Schrenzpapier *n* screenings
Schriftmetall *n* type metal
Schritt *m* stage, step
schrittweise stepwise, step-by-step
Schrödinger-Gleichung *f* Schrödinger [wave] equation
Schrotbeize *f (tann)* bran drench[ing] ✦ **in S. behandeln** to drench
Schrotmehl *n* coarse meal, grout

Schrotmühle *f* corn crusher; malt mill (crusher)
Schrott *m* scrap [metal]
schrubben to scrub
Schrühbrand *m (ceram)* biscuit firing, biscuitting
Schrühbrandofen *m s.* Schrühofen
Schrühbrandscherben *m (ceram)* biscuitted body
schrühen *(ceram)* to bake
Schrühofen *m (ceram)* biscuit kiln (oven)
Schrühware *f (ceram)* biscuit (biscuit-fired, biscuitted) ware, biscuit earthenware, bisque
schrumpfbeständig *s.* schrumpffest
schrumpfen to shrink, to contract
schrumpffest *(text)* shrink-resistant, shrinkproof, unshrinkable
Schrumpffestausrüstung *f (text)* shrink-resist finish, unshrinkable finish
Schrumpffestigkeit *f (text)* shrink resistance
Schrumpffestmachen *n (text)* shrinkproofing
Schrumpfriß *m* shrinkage crack
Schrumpfung *f* shrinkage, contraction
differentielle S. *(text)* differential shrinkage
erzwungene (kompressive) S. *(text)* compressive (compression) shrinkage
Schrumpfungsgrad *m* degree of shrinkage
Schrumpfverhältnis *n* shrink ratio
Schrumpfvorrichtung *f (plast)* shrinkage (cooling) jig, shrink (cooling) fixture *(for mouldings)*
Schub *m* 1. pushing force, push; 2. shear[ing]
Schubboden *m* pusher plate (disk)
Schubschleuder *f s.* Schubzentrifuge
Schubstange *f (min tech)* pitman *(as of a Wilfley table)*
Schubzentrifuge *f* push-type (reciprocating-conveyor, reciprocating-pusher) centrifuge
Schuhkrem *f* shoe polish
Schuller-Verfahren *n (glass)* Schuller (updraw) process
Schulter *f (phot)* shoulder, knee, region of overexposure *(of the characteristic curve)*
Schulterstab *m* / **hantelförmiger** *(rubber)* dump bell [test piece, strip]
Schumann-Platte *f* Schumann plate *(ultraviolet photography)*
Schüppchen *n (min)* spangle
Schuppe *f* scale, flake
Schuppenbildung *f* scaling, flaking
schuppenförmig scale-like, scaly
Schuppenglas *n* scaly glass
Schuppenparaffin *n* scale wax, paraffin scale
schuppig scaly, foliated; lepidoblastic *(texture of rocks)*
schüren to stoke
Schüren *n* **von Hand** hand stoking
schürfen *(mine)* to prospect, to explore
Schürfkübelbagger *m* dragline [excavator]
Schürfkübelraupe *f* caterpillar-powered scraper
Schürfkübelwagen *m* dragline [excavator]
Schürfung *f (mine)* prospecting, exploration
Schürloch *n* stoke (fire, fuel) hole
Schurre *f* chute
Schürvorrichtung *f* poker
Schurwolle *f* shorn (clip) wool
Schuß *m* 1. *(plast)* shot; 2. shot *(explosives)*
Schüssel *f* dish, bowl, basin
Schüsselflechte *f (dye)* [dark] crottle, crottal, Parmelia physodes (L.)Ach.
Schüssel-Kegel-Mühle *f* ring-roll mill *(with horizontal grinding ring)*
Schüsselklassierer *m* bowl classifier
S. mit seitlichem Austrag bowl desiltor
Schüsselmühle *f* [roller-and-]bowl mill, bowl ring-roller mill
S. für Kohlenstaubmahlung bowl-mill coal pulverizer
Schußfaden *m (text)* weft thread
Schußgewicht *n s.* Schußmasse
Schußmasse *f (plast)* shot size (weight)
maximale S. *(plast)* shot capacity
Schußmoment *n (petrol)* moment of discharge
Schußperforierung *f (petrol)* gun perforating
Schußpunkt *m (petrol)* shot point *(in refraction shooting)*
Schuß[roh]seide *f* tram silk
Schüttbeton *m* no-fines concrete
Schüttdichte *f (chem, mine)* bulk density, B.D., apparent density
S. von Pulver powder density
Schütte *f* chute
Schüttelapparat *m (pap)* shake [apparatus]; *s.* Schüttelmaschine
Schüttelautoklav *m* shaker autoclave
Schüttelbock *m (pap)* shake [apparatus]
Schüttelflasche *f* shaking bottle
Schüttelgefäß *n* shaking (shaker) flask
Schüttelherd *m (min tech)* shaking table
Schüttelmaschine *f* shaking machine, [mechanical] shaker
schütteln to shake
Schütteln *n* shake, shaking ✦ **unter ständigem S.** with constant shaking
Schüttelrinne *f*, **Schüttelrutsche** *f* shaking chute
Schüttelsieb *n* 1. shaking (shaker) screen (sieve); 2. *(petrol)* vibrating mudscreen *(of a rotary-drilling installation)*
Schüttelsortierer *m s.* Schüttelsieb 1.
Schütteltrichter *m* separatory (separating) funnel
Schüttelung *f* shake, shaking
Schüttelzylinder *m* shaking cylinder
schütten to pour
Schüttgewicht *n s.* Schüttdichte
Schüttgut *n* bulk material
Schüttgutbett *n*, **Schütt[gut]schicht** *f* bed
Schüttschicht-Staubabscheider *m* granular-bed separator
Schütttrichter *m* [feed, charge] hopper, feeding funnel; cup *(of a blast furnace)*
Schüttung *f* 1. *(act)* pouring; 2. *(substance)* bed; feed[stock], charge, charge (charging) stock *(as of a blast furnace)*; *(distill)* packing *(of a column)* ✦ **in loser S.** in bulk
keramische S. *(distil)* ceramic packing
körnige S. granular bed
ruhende S. fixed (static) bed
sich bewegende S. fluid bed

statische S. *s.* ruhende S.
Schüttungshöhe *f* bed depth; *(distil)* packing depth
Schüttvolum[en] *n* bulk (dry) volume
Schüttwinkel *m* angle of repose
Schutz *m* protection; safeguard
katodischer S. cathodic protection
katodischer S. durch Opferanoden sacrificial protection
Schütz *n* 1. sluice *(in a liquid stream)*; 2. relay *(in an electric circuit)*
Schutzanstrich *m* protective coating, protecting (protection) layer, pro-coating
Schutzatmosphäre *f* protective atmosphere
Schutzbekleidung *f s.* Schutzkleidung
Schutzbrett *n (pap)* spatter (baffle) board
Schutzbrille *f* protective (safety) goggles
schützend protective
Schutzengobe *f (ceram)* protective engobe
Schutzfähigkeit *f* protective capacity
Schutzfilm *m* protective (protecting) film
Schutzgas *n* protective (inert) gas, *(esp welding)* shielding gas
Schutzgasatmosphäre *f* protective (inert) atmosphere
kontrollierte S. controlled atmosphere
Schutzgasglühen *n* protective-gas annealing, bright annealing
Schutzgas-Lichtbogenschweißen *n* inert-gas-shielded arc-welding
Schutzglas *n* safety glass
Schutzgruppe *f* protective (protecting) group
Schutzhaube *f* **für Exsikkatoren** desiccator cage (guard)
Schutzhaut *f* protective (protecting) film
Schutzhülle *f* protective covering (sheath)
Schutzhülse *f* protective (protecting) tube
Schutzkleidung *f* protective (safety) clothing
Schutzkolloid *n* protective colloid
Schutzkorb *m* **für Exsikkatoren** desiccator cage (guard)
Schutzlack *m* protective lacquer; *(if unpigmented)* protective varnish
Schutzmagnet *m* tramp-iron magnet (magnetic separator)
Schutzmittel *n* protecting agent, protective, protectant; *(pharm)* prophylactic, protective
S. mit abstoßender Wirkung *(text)* repellent
Schutzrohr *n* protective (protecting) tube
Schutzscheibe *f (lab)* explosion screen *(made from wire glass)*
Schutzschicht *f* protective layer
S. gegen Abrieb antiabrasion layer
dünne S. protective (protecting) film
Schutzschirm *m* explosion screen
Schutzüberzug *m* protective coating, protecting (protection) layer
Schutzwirkung *f* protective action
schwach 1. weak *(e.g. acid, lye)*; 2. faint *(reaction)*
Schwachbrandstein *m (ceram)* soft[-fired] brick
schwächen 1. to weaken *(e.g. the basicity)*; 2. *(text)* to tender
schwächer werden to faint *(as of reactions)*
schwachfarbig feebly (weakly) coloured
Schwachfeldscheidung *f (min tech)* low-intensity magnetic separation
Schwachgas *n* poor (lean) gas
schwachgeleimt *(pap)* soft-sized, slack-sized, S.S.
Schwachlauge *f (pap)* weak liquor
schwachlöslich slightly (sparingly) soluble, low-solubility
Schwachsäure *f (pap)* weak acid
Schwächung *f* 1. weakening *(as of the basicity)*; 2. *(text)* tendering
Schwaden *m* swathe, damp
Schwadenhaube *f* hood
schwammartig sponge-like, spongy
Schwammeisen *n* sponge iron
Schwammeisenpulver *n* sponge-iron powder
Schwammgummi *m* sponge rubber
schwammig spongy, porous, porose
Schwammkunststoff *m* sponge plastic
Schwammverfahren *n* sponge process *(for squeezing off essential oil by hand)*
schwammverputzen *(ceram)* to sponge
schwangerschaftsverhütend contraceptive
Schwankung *f* variation
Schwanz *m* tail (1. *chromatography*; 2. *of a monomer*)
Schwanzbildung *f* tailing, trailing *(chromatography)*
Schwanzhahn *m* tailkey (tailed) stopcock
Schwanz-Schwanz-Polymerisation *f* tail-to-tail polymerization
Schwanz-Schwanz-Struktur *f* tail-to-tail structure
Schwarm *m* swarm, bundle *(of molecules)*
Schwarmwasser *n (soil)* hydration water of exchangeable ions
Schwarz *n* / **Frankfurter** Frankfort black
Kölner S. bone (animal) black
Schwarzalkaliboden *m (soil)* solonetz
Schwarzbeize *f (dye)* black mordant (liquor), iron [acetate] liquor
Schwarzblech *n* black plate
Schwärze *f* black
schwärzen to blacken; to darken, to affect *(the photosensitive layer)*
sich s. to blacken; *(phot)* to darken
Schwarzerde *f (soil)* chernozem, black earth
Schwarzfärbung *f* blackening
Schwarzfäule *f* black rot
Schwarzglas *n* black glass
schwarzkochen *(pap)* to burn *(the cook)*
Schwarzkochung *f (pap)* 1. burning *(process)*; 2. burned (burnt, black) cook
Schwarzkohle *f* bituminous (black) coal
Schwarzkopie *f (phot)* negative copy
Schwarzkümmel *m* black cummin, Nigella sativa L.
Schwarzkupfer *n (met)* black (coarse) copper
Schwarzlack *m* black varnish
Schwarzlauge *f (pap)* black liquor (lye)

eingedickte S. *(pap)* concentrated (evaporated, thick) black liquor
Schwarzlaugenbehälter *m*, **Schwarzlaugenvorratstank** *m (pap)* black liquor storage [tank]
Schwarzmaterial *n coal or coke in calcium carbide manufacture*
Schwarzmetallurgie *f* ferrous (iron) metallurgy
Schwarzpulver *n* gunpowder, black (blasting) powder
Schwarzschild-Effekt *m (phot)* Schwarzschild effect, reciprocity[-law] failure
Schwarzschmelze *f*, **Schwarzsoda** *f (pap)* black ash
Schwarztorf *m (agric)* black peat
Schwärzung *f* 1. blackening; *(phot)* darkening; fog *(in the unexposed part of an image)*; 2. *s.* Schwärzungsdichte
Schwärzungsabstufung *f (phot)* gradation
Schwärzungsbereich *m (phot)* density range
Schwärzungsdichte *f (phot)* extinction, [optical, photographic] density
Schwärzungshof *m (phot)* halo
Schwärzungskurve *f* **[/ fotografische]** *(phot)* characteristic curve, Hurter and Driffield curve, H and D curve
Schwärzungsmesser *m (phot)* densitometer
Schwärzungsmessung *f (phot)* densitometry
Schwärzungsring *m (phot)* halo
Schwärzungsschwelle *f (phot)* threshold [point]
Schwarzweißkopie *f* black-and-white print
Schwazit *m (min)* schwazite *(tetrahedrite containing mercury)*
Schwebekörper *m* plummet *(in flowmeters)*
Schwebekörpermesser *m* rotameter, tube-and-float meter *(a flowmeter)*
Schwebemittel *n* antisettling agent
Schweberöstung *f (met)* suspension roasting
Schwebeschmelzen *n* **/ autogenes** *(met)* autogenous smelting
Schwebestoff *m s.* Schwebstoff
Schwebeteilchen *n* suspended particle
Schwebetrockner *m* moving-product dryer
Schwebezone *f* floating zone *(in zone melting)*
Schwebezonenapparatur *f* floating-zone apparatus (unit)
Schwebezonenschmelzen *n* floating-zone melting
Schwebezonenverfahren *n* floating-zone method *(of zone melting)*
Schwebstoff *m* suspended material (matter)
schwebstofffrei free from suspended matter
Schwefel *m* S sulphur ✦ **aus S. bestehend** sulphur[e]ous, sulphuric ✦ **mit S. behandeln** to sulphur ✦ **mit S. imprägnieren** to sulphurize ✦ **mit S. räuchern** to sulphur ✦ **mit S. vernetzbar** *(rubber)* sulphur-curable, sulphur-curing, sulphur-vulcanizable ✦ **mit S. vernetzt** *(rubber)* sulphur-cured, sulphur-vulcanized
amorpher S. amorphous sulphur
anorganisch gebundener S. mineral sulphur
extrahierbarer S. *(rubber)* [total] extractable sulphur
freier S. free sulphur, *(rubber also)* true free sulphur
gebundener S. bound (combined) sulphur, *(rubber also)* rubber-combined sulphur
gediegener S. native (virgin) sulphur
gefällter S. precipitated sulphur, milk of sulphur
kolloid[al]er S. colloidal sulphur
monokliner S. monoclinic sulphur, β-sulphur
natürlicher S. *s.* gediegener S.
perlmutt[er]artiger S. mother-of-pearl sulphur, nacreous sulphur *(a modification)*
plastischer S. plastic sulphur, γ-sulphur
polysulfidisch gebundener S. polysulphidic sulphur
radioaktiver S. radioactive sulphur, radiosulphur, *(specif)* ^{35}S sulphur-35
rhombischer S. rhombic sulphur, α-sulphur
sublimierter S. sublimed sulphur, sulphur flowers
totaler S. *(rubber)* total sulphur
unlöslicher S. insoluble sulphur
Schwefel-35 *m* ^{35}S sulphur-35
α-**Schwefel** *m* α-sulphur, rhombic sulphur
β-**Schwefel** *m* β-sulphur, monoclinic sulphur
γ-**Schwefel** *m* γ-sulphur, plastic sulphur
λ-**Schwefel** *m* λ-sulphur *(an amorphous modification)*
μ-**Schwefel** *m* μ-sulphur *(an amorphous modification)*
Schwefelabdruck *m* sulphur print *(for detecting sulphur)*
Schwefelablagerung *f* sulphur deposit
Schwefelaffinität *f* affinity for sulphur
Schwefelantimon *n s.* Antimon(III)-sulfid
schwefelarm poor (low) in sulphur, low-sulphur
Schwefelarsen *n s.* Arsensulfid
schwefelartig sulphur[e]ous
Schwefelausblühung *f (rubber)* sulphur bloom
Schwefelbakterien *pl* sulphur bacteria
Schwefelbestimmung *f* determination (estimation) of sulphur
Schwefelblume *f*, **Schwefelblüte** *f* sulphur flowers, sublimed sulphur
Schwefelbrücke *f (rubber)* sulphur bridge (link, cross-linkage, cross-link)
Schwefel(II)-chlorid *n* SCl_2 sulphur(II) chloride, sulphur dichloride
Schwefel(IV)-chlorid *n* SCl_4 sulphur(IV) chloride, sulphur tetrachloride
Schwefeldampf *m* sulphur vapour
Schwefeldichlorid *n s.* Schwefel(II)-chlorid
Schwefeldioxid *n* SO_2 sulphur dioxide, sulphur(IV) oxide
gebundenes S. *(pap)* combined (non-available) sulphur dioxide
Schwefeldonator *m* sulphur donor
Schwefeldosierung *f* sulphur dosage (level)
Schwefelerz *n* sulphide ore
Schwefelfarbstoff *m* sulphur (sulphide) dye (dyestuff), *(Am also)* sulfur color
Schwefel(IV)-fluorid *n* SF_4 sulphur(IV) fluoride, sulphur tetrafluoride

Schwefel(VI)-fluorid *n* SF_6 sulphur(VI) fluoride, sulphur hexafluoride
schwefelfrei sulphur-free, sulphurless, non-sulphur
schwefelführend sulphur-bearing
Schwefelgehalt *m* sulphur content
schwefelhaltig sulphur-containing, sulphur[e]ous, *(min also)* sulphur-bearing
Schwefelharnstoff *m s.* Thioharnstoff
Schwefelheptoxid *n* S_2O_7 sulphur heptaoxide, disulphur heptaoxide
Schwefelhexafluorid *n s.* Schwefel(VI)-fluorid
Schwefelhexajodid *n s.* Schwefel(VI)-jodid
Schwefel(II)-hydroxid *n s.* Sulfoxylsäure
Schwefelindigoblau *n* sulphur indigo blue
Schwefel(VI)-jodid *n* SI_6 sulphur(VI) iodide, sulphur hexaiodide
Schwefelkalkbrühe *f* lime sulphur *(a pesticide)*
Schwefelkies *m (min)* iron pyrite[s], pyrite, mundic *(iron disulphide)*
Schwefelkohlenstoff *m* CS_2 carbon disulphide
Schwefellager *n (geol)* sulphur deposit
Schwefelleber *f* liver of sulphur, hepar sulfuris *(technical potassium sulphide)*
Schwefellost *n* mustard gas, sulphur mustard *(di-2-chlorodiethylsulphide)*
Schwefellösung *f* sulphur solution
Schwefelmehl *n* flour sulphur
Schwefelmilch *f* milk of sulphur, precipitated sulphur
schwefelmodifiziert sulphur-modified
Schwefelmonoxid *n s.* Schwefel(II)-oxid
schwefeln 1. to sulphur[ize], *(text also)* to stove; 2. *(sugar)* to sulphite; 3. to thionate *(organic compounds for producing dyestuffs)*
Schwefelnatriumäscher *m (tann)* sulphide lime
Schwefelnitrid *n* tetrasulphur tetranitride
Schwefelofen *m* sulphur burner
rotierender S. rotary sulphur burner
Schwefel(II)-oxid *n* SO sulphur(II) oxide, sulphur monooxide
Schwefel(IV)-oxid *n* SO_2 sulphur(IV) oxide, sulphur dioxide
Schwefel(VI)-oxid *n* SO_3 sulphur(VI) oxide, sulphur trioxide
Schwefelpunkt *m* boiling point of sulphur
Schwefelsaturation *f (sugar)* sulphitation
Schwefelsäure *f* H_2SO_4 sulphuric acid ✦ **mit S. beizen** *(met)* to vitriol
S. aus Rohschwefel sulphur (brimstone) acid
nitrose S. nitrous vitriol *(in the chamber process of manufacturing sulphuric acid)*
rauchende S. fuming sulphuric acid, oleum
Schwefelsäurealkylierung *f* sulphuric-acid alkylation
Schwefelsäureäthylester *m s.* Schwefelsäurediäthylester
Schwefelsäurebad *n* bath of sulphuric acid
Schwefelsäurebehandlung *f (petrol)* sulphuric-acid refining
Schwefelsäurediäthylester *m* $(C_2H_5)_2SO_4$ diethyl sulphate
Schwefelsäuredimethylester *m* $(CH_3)_2SO_4$ dimethyl sulphate
Schwefelsäureindustrie *f* sulphuric-acid industry
Schwefelsäurekontaktverfahren *n* contact process (method)
Schwefelsäuremonoäthylester *m* $C_2H_5HSO_4$ ethyl hydrogensulphate, ethylsulphuric acid
Schwefelsäurepolymerisationsverfahren *n* sulphuric-acid polymerization process
Schwefelsäureraffinage *f (petrol)* sulphuric-acid refining
Schwefelsäurewaschprobe *f (dye)* sulphuric-acid test
Schwefelschwarz *n* sulphur black *(any of several sulphur dyes)*
Schwefelschwarzpaste *f* sulphur black paste
Schwefelschwarzpulver *n* sulphur black powder
Schwefelsensibilisator *m (phot)* sulphur sensitizer
Schwefelsol *n* sulphur sol
Schwefelspender *m* sulphur donor
Schwefeltetrachlorid *n s.* Schwefel(IV)-chlorid
Schwefeltetrafluorid *n s.* Schwefel(IV)-fluorid
Schwefeltetroxid *n* SO_4 sulphur tetraoxide
Schwefeltonung *f (phot)* sulphur (sulphite) toning
Schwefeltrioxid *n s.* Schwefel(VI)-oxid
Schwefelung *f* 1. sulphurization, *(text also)* stoving; 2. *(sugar)* sulphitation; 3. thionation *(of organic compounds for producing dyestuffs)*
Schwefelverbindung *f* sulphur compound
Schwefelverbrennungen *fpl (agric)* sulphur scald
Schwefelverbrennungsofen *m* sulphur burner
rotierender S. rotary sulphur burner
Schwefelverlust *m (pap)* loss of sulphur
Schwefelvernetzungsbrücke *f (rubber)* sulphur bridge (link, cross-linkage, cross-link)
Schwefelvulkanisat *n* sulphur vulcanizate
Schwefelvulkanisation *f* [elemental] sulphur cure, sulphur vulcanization
Schwefelvulkanisationssystem *n* sulphur curing (vulcanization) system
Schwefelwasserstoff *m* hydrogen sulphide, sulphane, *(specif)* H_2S hydrogen sulphide, monosulphane
Schwefelyperit *n* mustard gas, sulphur mustard, yperite, di-2-chloroethyl sulphide
schweflig sulphur[e]ous
Schwefligsäureester *m* sulphite ester
Schweif *m* tail *(chromatography)*
Schweifbildung *f* tailing, trailing *(chromatography)*
Schweinchen *n (lab)* 1. weighing piggy *(weighing bottle with feet)*; 2. condenser jacket
Schweinefett *n*, **Schweineschmalz** *n* hog (pig) fat, lard
Schweinsgummi *n (pharm)* 1. hog gum *(from Clusia flava L. and other trees confused with it)*; 2. doctor gum *(from Symphonia globulifera L.)*
Schweiß *m* sweat
schweißbar weldable
Schweißbarkeit *f* weldability

Schweißbrenner *m* [welding] torch, blowpipe
Schweißdraht *m* welding wire
schweißecht *(tann, text)* fast to perspiration, perspiration-resistant
Schweißechtheit *f* *(tann, text)* fastness to perspiration, perspiration resistance
Schweißelektrode *f* welding electrode
schweißen to weld, *(plast also)* to seal
Schweißen *n* **durch Anlösen (Anquellen)** solvent welding
autogenes S. autogenous (gas) welding
Schweißerglas *n* welding glass
Schweißflußmittel *n* welding flux
Schweißgerät *n* welding apparatus, *(plast also)* sealing unit
schweißhemmend *(cosmet)* antiperspirant, antihidrotic
Schweißhemmungsmittel *n* *(cosmet)* antiperspirant, antihidrotic, perspiration check
schweißlindernd *s.* schweißhemmend
Schweißlineal *n* *(plast)* heated bar
Schweißmittel *n* 1. welding flux; 2. *(pharm)* sudorific; 3. *s.* Schweißhemmungsmittel
Schweißpistole *f* welding gun
Schweißpulver *n* welding powder
Schweißstab *m* welding (filler) rod
Schweißstahl *m* wrought iron
schweißtreibend *(pharm)* sudorific
Schweißung *f* weld[ing]
Schweißwärme *f* welding heat
Schweißwolle *f* raw (grease) wool
Schwelanlage *f* low-temperature carbonization (carbonizing) plant
Schwelaufbau *m*, **Schwelaufsatz** *m* superimposed carbonizing chamber
Schwelbrikett *n* carbonized briquette
schwelen to carbonize *(e.g. coal at low temperature)*; to smoulder
in der Retorte s. to retort *(e.g. oil shale)*
Schwelen *n* low-temperature carbonization, *(coal also)* low-temperature coking
Schweler *m* carbonizer
Schwelerei *f* *s.* Schwelanlage
Schwelgas *n* carbonization (low-temperature) gas; distillation gas *(obtained in a predistillation gas producer)*
Schwelgasleitung *f* distillation gas main *(of a predistillation gas producer)*
Schwelgenerator *m* predistillation [gas] producer
Schwelindustrie *f* carbonizing industry
Schwelkoks *m* low-temperature coke
Schwelläscher *m* *(tann)* fresh lime
Schwelle *f* *(phys)* threshold; *(phot)* threshold [point]
schwellen to swell [up], to expand; *(tann)* to plump *(pelts)*
Schwellenenergie *f* threshold energy
Schwellenverfahren *n* *s.* Schwellenwertbehandlung
Schwellenwert *m* threshold value
Schwellenwertbehandlung *f* threshold treatment *(of water)*
Schwellmittel *n* *(tann)* plumping agent, plumper
Schwellvorgang *m* swelling, expansion
Schwellwirkung *f* *(tann)* plumping power
Schwellzement *m* expanding (expansive) cement
Schwelofen *m* carbonizer
Schwelretorte *f* low-temperature retort
Schwelschacht *m* carbonizing (carbonization) chamber *(of a predistillation gas producer)*
aufgesetzter S. superimposed carbonizing chamber
Schwelteer *m* low-temperature tar
Schwelung *f* low-temperature carbonization, *(coal also)* low-temperature coking
S. im Wirbelbett fluidized carbonization
Schwelverfahren *n* low-temperature carbonization (carbonizing) process
Schwelvorgang *m* low-temperature carbonization (carbonizing) process
Schwelwerk *n* low-temperature carbonization (carbonizing) plant
Schwelzone *f* distillation (carbonization) zone *(of a gas producer)*
Schwemmtorf *m* hydro peat
Schwemmwasser *n* flume (fluming) water
schwenkbar turning, turnable, mobile
schwenken to swirl *(e.g. a beaker)*; *(tech)* to turn, to pivot, *(horizontally)* to swivel, *(to and fro)* to oscillate
Schwenkmethode *f* oscillating crystal method *(for investigating crystal structures)*
Schwerbenzin *n* heavy gasoline (benzine, naphtha) *(boiling range 150 to 210 °C)*
Schwerbenzol *n* heavy benzole
Schwerbeton *m* heavy concrete
Schwerchemikalien *fpl* heavy chemicals *(acids, salts, and alkalies produced on a large scale)*
schwererflüchtig *(distil)* less volatile
schwererschmelzbar higher-melting
schwerersiedend higher-boiling
Schwerersiedendes *n* higher-boiling component, less volatile component
Schweretrennung *f* *s.* Schwerkraftabscheidung
Schweretrübe *f* *s.* Schwerflüssigkeit
schwerflüchtig difficultly volatile, slow-evaporating, heavy
Schwerflüssigkeit *f* *(min tech)* heavy (dense) medium, high-gravity medium
Schwerflüssigkeitsanlage *f* *(min tech)* heavy-medium (dense-medium) separation plant, gravity concentrating (concentration) apparatus
Schwerflüssigkeitsaufbereitung *f* *(min tech)* heavy-medium (dense-medium) separation (cleaning)
Schwerflüssigkeitsaufbereitungsanlage *f* *s.* Schwerflüssigkeitsanlage
Schwerflüssigkeitsscheider *m* *(min tech)* heavy-medium (dense-medium) separator (vessel, washer)
Schwerflüssigkeitssortieren *n* *s.* Schwerflüssigkeitsaufbereitung

Schwerflüssigkeitsverfahren *n (min tech)* sink-and-float process, sink-float method
Schwergut *n* heavy material
Schwerkraftabscheider *m* gravity (gravitational) separator
Schwerkraftabscheidung *f* gravity (gravitational) separation
Schwerkraftabsetzbehälter *m*, **Schwerkraftabsetzer** *m* gravity settling tank
Schwerkraftabsetzung *f s.* Schwerkraftabscheidung
Schwerkraftaufbereitung *f (min tech)* gravity concentration
Schwerkraftfilter *n* gravity (hydrostatic head) filter
Schwerkrafttrennung *f s.* Schwerkraftabscheidung
Schwerkraftzuführung *f* gravity feed
schwerkraftzugeführt gravity-fed
Schwerkron[glas] *n* heavy crown glass
Schwerlegierung *f* heavy [metal] alloy
schwerlöslich slightly (sparingly) soluble
Schwermetall *n* heavy metal
Schwermetallegierung *f* heavy [metal] alloy
Schwermetallsalz *n* heavy metal salt
Schwermineral *n* heavy mineral
Schweröl *n* heavy oil
Schwerschlamm *m* sludge *(waste-water treatment)*
schwerschmelzbar difficultly fusible, high-melting, high-fusion
schwersiedend high-boiling
Schwersiedendes *n* high-boiling component
Schwerspat *m (min)* heavy spar, barite, baryte[s] *(barium sulphate)*
Schwerstange *f* drill collar *(of a rotary-drilling installation)*
Schwerstbeton *m* super-heavy concrete
Schwerstkron[glas] *n* heaviest crown glass
Schwerstoff *m* high-gravity solid
Schwertkolben *m (lab)* sausage flask
Schwertrübe *f s.* Schwerflüssigkeit
Schwertrübe[wasch]zyklon *m* heavy-medium (dense-medium) cyclone
Schwerwasserdruckreaktor *m (nucl)* pressurized heavy-water reactor
Schwerwasserherstellung *f* heavy-water production
Schwimmanteil *m (min tech)* floating fraction
Schwimmäscher *m (tann)* floating lime
Schwimmaschine *f (min tech)* floatation machine (apparatus)
Schwimmaufbereitung *f* flo[a]tation
Schwimmdeckenabstreifer *m* surface skimming device
Schwimmdüse *f (glass)* debiteuse *(in the Fourcault process)*
Schwimmer *m* float[er]
Schwimmerdruckmesser *m* float-type manometer
Schwimmerkondenstopf *m*, **Schwimmerkondenswasserableiter** *m* ball float trap
Schwimmerkörper *m* float[er]
Schwimmermanometer *n* float-type manometer
Schwimmermesser *m* 1. tube-and-float meter, rotameter *(a flowmeter)*; 2. float gauge *(for liquid-level measurement)*
Schwimmerventil *n* float valve
schwimmfähig floatable
Schwimmfähigkeit *f* floatability
Schwimmgerät *n (min tech)* floatation apparatus (machine)
Schwimmgut *n (min tech)* float[ing] material, floats, floating fraction
Schwimmittel *n (min tech)* floatation agent
drückendes S. depressant
regelndes S. modifying agent, modifier
Schwimmkiesel *m (min)* float stone *(a spongy variety of opal)*
Schwimmkopf *m* floating head *(of a heat exchanger)*
Schwimmkurve *f (min tech)* floats curve
Schwimmschicht *f s.* Schwimmschlamm
Schwimmschlamm *m* surface mat, [top] scum *(on waste water)*
Schwimmschlammräumschild *m* skimmer blade *(waste-water treatment)*
Schwimmseife *f* floating soap
Schwimm-Sink-Aufbereitung *f (min tech)* sink-float (heavy-medium, dense-medium) separation
Schwimm-Sink-Verfahren *n (min tech)* sink-float process
Schwimmverfahren *n (min tech)* floatation process
schwinden to shrink, to contract
schwindend / nicht non-shrinking
Schwindmaß *n* measure of shrinkage (contraction)
Schwindriß *m (ceram)* shrinkage crack
Schwindung *f* shrinkage, contraction
kubische S. *(ceram)* cubic (volume) shrinkage
lineare S. *(ceram)* linear shrinkage
räumliche S. *s.* kubische S.
Schwindungshohlraum *m* shrinkage (contraction) cavity
Schwindungskoeffizient *m (ceram)* sintering coefficient
Schwindungsriß *m (ceram)* shrinkage crack
Schwingamplitude *f* vibration amplitude
Schwing[backen]brecher *m* balanced-jaw crusher
schwingen to vibrate, to oscillate; *(text)* to scutch *(flax)*
Schwingerbrechbacke *f* moving jaw *(of a jaw crusher)*
Schwingfeuer[nebel]gerät *n* swingfog *(crop protection)*
Schwingfrequenz *f s.* Schwingungsfrequenz
Schwingmaschine *f (text)* scutching mill
Schwingmühle *f* vibratory (vibrating, oscillatory, oscillating) mill
Schwingrechen *m* swing rake
Schwingsieb *n*, **Schwingsiebmaschine** *f* vibrating (oscillating) screen

Schwingsiebschleuder *f*, **Schwingsiebzentrifuge** *f* oscillating-screen (oscillating-basket) centrifuge
Schwingsortierer *m s.* Schwingsiebmaschine
Schwingtrockner *m* vibrating conveyor dryer
Schwingung *f* vibration, oscillation
Schwingungsamplitude *f* vibration amplitude
Schwingungsbande *f* vibration[al] band
Schwingungseinrüttler *m* vibrator
Schwingungsenergie *f* vibrational energy
schwingungsfrei vibrationless
Schwingungsfreiheitsgrad *m* vibrational degree of freedom
Schwingungsfrequenz *f* vibration (oscillation) frequency
Schwingungsgrundzustand *m* vibrational ground state
Schwingungsniveau *n* vibrational level
Schwingungsquantenzahl *f* vibrational quantum number
Schwingungsrichtung *f* direction of vibration
Schwingungsrüttler *m* vibrator
Schwingungsspektrum *n* vibration[al] spectrum
Schwingungssystem *n* vibrating system
Schwingungsübergang *m* vibrational transition
Schwingungsverteilungsfunktion *f* vibrational partition function
Schwingungszahl *f* vibration number
Schwingungszustand *m* vibrational state
Schwingweg *m*, **Schwingweite** *f* vibration amplitude
Schwitzapparat *m* sweating stove *(for refining paraffin)*
  S. von Henderson Henderson stove
Schwitze *f (tann)* sweating *(for removing hairs)*
schwitzen 1. to sweat; *(relating to cement)* to weep, to bleed; 2. to sweat *(crude paraffin for removing oil)*
Schwitzkammer *f* sweating room (chamber) *(for refining paraffin)*
Schwitzöl *n (petrol)* sweats (foots) oil, sweats
Schwitzpfanne *f s.* Schwitzwanne
Schwitzraum *m s.* Schwitzkammer
Schwitztasse *f s.* Schwitzwanne
Schwitzung *f* sweating *(for refining paraffin)*
Schwitzverfahren *n* sweating process *(for refining paraffin)*
Schwitzwanne *f* sweating pan (tray), sweat pan *(for refining paraffin)*
Schwitzwasser *n* condensation (condensed) water, condensate [water]
Schwitzwolle *f* sweated wool
Schwöde *f (tann)* painting, paint unhairing
Schwödebrei *m (tann)* lime cream (paint)
schwöden *(tann)* to paint
Schwödewolle *f* sulphide-painted wool
Schwund *m* 1. loss, wastage, outage *(a quantity lost in transportation or storage)*; 2. shrinkage, contraction *(relating to dimensions)*
Schwundriß *m* shrinkage crack
scorchanfällig *(rubber)* scorchy
scorchbeständig *(rubber)* scorch-resistant
Scorchbeständigkeit *f (rubber)* scorch resistance
Scorchcharakteristik *f (rubber)* scorch characteristic
Scorchkurve *f (rubber)* scorch curve
Scorchneigung *f (rubber)* scorch tendency, scorchiness
Scorchperiode *f (rubber)* scorch period
Scorchprüfung *f (rubber)* scorch test
Scorchpunkt *m (rubber)* scorch point
scorchresistent *(rubber)* scorch-resistant
Scorchresistenz *f (rubber)* scorch resistance
Scorchtendenz *f s.* Scorchneigung
Scorchzeit *f (rubber)* scorch time
S-Drehung *f (rubber)* S twist
SE *s.* 1. Sekundärelektron; 2. Siedeendpunkt
Sebazinsäure *f* $HOOC[CH_2]_8COOH$ sebacic acid, decanedioic acid
Sebazinsäurediäthylester *m* $C_2H_5O-CO[CH_2]_8CO-OC_2H_5$ diethyl sebacate
Sebazylsäure *f s.* Sebazinsäure
sechsatomig hexatomic
sechseckig *(cryst)* hexagonal
Sechserring *m s.* Sechsring
sechsfach koordiniert hexacoordinated
sechsfachpositiv hexapositive
sechsgliedrig six-membered
Sechskohlenstoffkern *m* six-carbon nucleus
Sechskörperverdampfer *m* sextuple evaporator
Sechsring *m* six-membered ring
Sechsstufenbleiche *f (pap)* six-stage bleaching
Sechsstufenverdampfer *m* sextuple evaporator
sechswertig sexivalent, sexavalent, hexavalent
Sechswertigkeit *f* sexivalence, sexavalence
sechszählig 1. *(cryst)* hexad, sixfold; 2. *s.* sechszähnig
sechszähnig sexadentate *(ligand)*
Sedativum *n (pharm)* sedative
Sediment *n* sediment, subsidence, settling[s], bottom settlings (sediment), B.S., bottoms, lees, dregs, foots
  biogenes (organogenes) S. organic (biogenic) rock, biolith
sedimentär sedimentary
Sedimentation *f* sedimentation, settling
  S. im Schwerefeld gravity (gravitational) sedimentation (settling)
  freie S. *(min tech)* free settling
  gestörte S. *(min tech)* hindered settling
Sedimentationsanalyse *f* sedimentation analysis
Sedimentationsgeschwindigkeit *f* sedimentation (settling) velocity (rate)
Sedimentationsgleichgewicht *n* sedimentation equilibrium
Sedimentationskonstante *f* sedimentation constant (value, coefficient)
Sedimentationspotential *n* sedimentation potential
Sedimentationsstoffänger *m (pap)* sedimentation (gravity) save-all
Sedimentationstest *m* sedimentation test
Sedimentationswaage *f* sedimentation balance
Sedimente *npl* / **grobklastische (grobkörnige kla-**

stische) *(geol)* macroclastics, coarse-grained clastics
klastische S. *(geol)* clastic sediments, clastics
makroklastische S. *s.* grobklastische S.
Sedimentgestein *n* sedimentary rock, sediment
sedimentieren 1. *(of human agent)* to sediment[ate], to precipitate, to set; 2. *(of a precipitate)* to deposit, to sediment, to subside, to settle [down, out]
Sedimentierzentrifuge *f* sedimentation centrifuge
Sedimentpetrografie *f* sedimentary petrography, sedimentography
Sedoheptose *f s.* Sedoheptulose
Sedoheptulose *f* $HOCH_2[CHOH]_4COCH_2OH$ sedoheptulose *(a ketoheptose)*
Seehundstran *m* seal oil
Seekreide *f (min)* chalk *(calcium carbonate)*
Seele *f* core *(as of a cable)*
Seelenmischung *f (rubber)* inner tube compound
Seesalz *n* marine (sea, solar, bay) salt
Seesand *m* sea-shore sand
Seetang *m* [sea]tang
Seetieröl *n* marine [animal] oil
Seewasser *n* sea water
Seewasserfestigkeit *f* resistance to sea water
Segas-Verfahren *n (petrol)* Segas process
Seger-Formel *f (ceram)* Seger formula
Seger-Kegel *m (ceram)* Seger cone *(a pyrometric cone)*
Seger-Porzellan *n* Seger porcelain
Seggentorf *m* sedge peat
Segregation *f* segregation
Segregationskonstante *f* segregation (partition, distribution) coefficient *(in zone-melting theory)*
Sehpurpur *m* visual purple, rhodopsin
Sehstoff *m* visual pigment
Seide *f* 1. silk; 2. continuous-filament yarn *(man-made fibre)*
echte S. natural silk
gereckte S. drawn yarn
gesponnene S. spun silk
halbentbastete S. souple silk
monofile S. monofil, monofilament [yarn]
multifile (polyfile) S. multifil, multifilament [yarn]
reine S. natural silk
souplierte S. souple silk
texturierte S. bulked yarn
wilde S. wild silk *(e.g. tussah silk)*
seidenartig silk-like, silky
Seidenbast *m* silk gum, sericin *(a protein occurring in silk)*
Seidenfibroin *n* silk fibroin
Seidengewebe *n* silk fabric
Seidenglanz *m* silky lustre *(of crystals)*
Seidenkautschuk *m* silk rubber *(from Funtumia elastica Stapf)*
Seidenkokon *m* cocoon
Seidenkreppapier *n* crêpe tissue paper
Seidenleim *m s.* Seidenbast
Seidenpapier *n* tissue [paper]
gekrepptes S. crêpe tissue paper
Seidenraster *m* silk screen
Seidenrasterdruck *m* silk-screen printing
Seidensieb *n* silk screen
Seidenstoff *m* 1. silk fabric; 2. fibroin *(the fibrous component of natural silk)*
seidig silky
Seife *f* 1. soap; 2. *(geol)* placer [deposits]
alluviale S. *(geol)* alluvial placer
äolische S. *(geol)* eolian placer
aufgebaute S. built soap
feste S. hard soap
flüssige S. liquid soap
fluviatile S. *(geol)* fluviatile placer
harte S. hard soap
kastilianische S. [olive-oil] castile soap, castile
marine S. *(geol)* marine placer
Marseiller S. Marseilles soap
medizinische S. medicated (medicinal) soap
neutrale S. neutral soap
transparente S. transparent soap
Venezianer (venezianische) S. Venetian soap
weiche S. soft soap
seifecht fast to soaping
Seifechtheit *f* fastness to soaping
Seifenbad *n* soaping bath
Seifenblase *f* soap bubble
Seifenblätter *npl* shavings
Seifenfabrikation *f* soap manufacture
Seifenflocken *fpl* soap flakes
Seifenfluß *m* soapstock
seifenfrei non-soapy
Seifengold *n* placer (stream) gold
Seifenherstellung *f* soap manufacture
seifenlos soapless, non-soapy
Seifenmizell *n*, **Seifenmizelle** *f* soap micelle
Seifennachbehandlung *f (text)* soaping aftertreatment
Seifenpapier *n* soap paper (tissue)
Seifenpulver *n* soap powder
Seifenriegel *m* soap bar
Seifenrinde *f* soap (Panama, China) bark *(from Quillaja saponaria Mol.)*
Seifenschampun *n* soap shampoo
Seifenschaum *m* lather
Seifenschnitzel *npl* soap chippings
Seifenshampoon *n s.* Seifenschampun
Seifenspäne *mpl* shavings
Seifenstange *f* soap bar
Seifenstock *m* soapstock
Seifenstrang *m* soap bar
Seifenstück *n* cake *(in soap manufacture)*
Seifenzinn *n* stream tin
seigern 1. to liquate, to segregate *(metal)*; 2. to liquate [out], to segregate *(said of the metal)*
Seigerraffination *f* liquation refining *(for removing impurities from metallic fusions)*
Seigerung *f (met)* liquation, segregation
Seignettesalz *n* $KOOC[CHOH]_2COONa \cdot 4H_2O$ Rochelle (Seignette) salt, potassium sodium tartrate-4-water

seihen to strain
Seiher *m* 1. strainer; 2. *s.* Seiherkörper
Seiherkörper *m* cage, curb *(of a cage press)*
Seiherpresse *f* cage (curb) press
Seiherverfahren *n (petrol)* percolation process *(for refining bleaching earth)*
Seihfilter *n* bag filter
Seihtuch *n* straining cloth
Seilbohranlage *f* cable tool installation (rig)
Seilbohren *n* **[/ pennsylvanisches]** cable tool drilling
Seilbohrgerät *n* cable tool installation (rig)
Seilbohrloch *n* cable tool well
Seilbohrung *f* 1. cable tool well; 2. *s.* Seilbohren
Seilbohrwerkzeug *n* cable tool
Seilschlagbohrung *f* cable tool well
Seilschlag[bohr]verfahren *n* cable tool method (system)
Seite *f* side *(as of a molecule), (relating to rings also)* face
gemeinsame S. common face, side of fusion *(in a ring system)*
Seitenablauf *m (text)* side-to-centre shading
Seitenkette *f* side (lateral) chain
Seitenkettenabbau *m* side-chain degradation
Seitenkettenchlorierung *f* side-chain chlorination
Seitenkettenhalogenierung *f* side-chain halogenation
Seitenkettenmetallierung *f* side-chain metalation
Seitenkolonne *f (distil)* side stripper
Seitenprodukt *n (distil)* side product
Seitenschneide *f* terminal knife edge *(as of an analytical balance)*
Seitenstreifenmischung *f (rubber)* sidewall compound (stock)
Seitenstrom *m* side stream
Seitenwand *f* sidewall
Seitenwindkonverter *m* side-blow[n] converter
Seitz-Entkeimungsfilter *n*, **Seitz-Filter** *n* Seitz [germ-proofing] filter
Sekalonsäure *f* secalonic acid *(a fungal pigment from Claviceps purpurea (Fr.)Tul.)*
Sekretion *f* secretion
Sekretionsvorgang *m* secretory process
sekretorisch secretory
Sekt *m* champagne, sparkling wine
Sektorverfahren *n* sector process *(paper chromatography)*
sekundär secondary
Sekundärazetat *n s.* Sekundärzelluloseazetat
Sekundärbatterie *f* secondary battery
Sekundärbeschleuniger *m (rubber)* secondary accelerator
aktivierender S. activating accelerator, booster
Sekundärbrücke *f (nomencl)* secondary bridge
Sekundäreffekt *m* secondary effect
Sekundärelektron *n* secondary electron
Sekundärelektronenvervielfacher *m* photomultiplier [tube], multiplier phototube, secondary-emission electron multiplier
Sekundärelement *n* secondary cell
Sekundäremission *f* secondary emission
Sekundärförderung *f (petrol)* secondary recovery
Sekundärluft *f* secondary air
Sekundärlunker *m* secondary pipe *(foundry)*
Sekundärreaktion *f* secondary reaction
fotochemische S. secondary photochemical reaction
Sekundärstandard *m* secondary standard
Sekundärwand *f* secondary wall *(of a vegetable fibre)*
Sekundärzelluloseazetat *n (text)* secondary [cellulose] acetate, cellulose diacetate
Sekundärzentrifuge *f* secondary centrifuge
Selachensäure *f* selacholeic acid, nervonic acid, *cis*-15-tetracosenoic acid
Selachylalkohol *m* selachyl alcohol, glycerol 1-octadecenyl ether
Selbstabnahmefilz *m (pap)* lick-up overfelt (wet felt, felt)
Selbstabnahmemaschine *f s.* Selbstabnahmepapiermaschine
Selbstabnahmeoberfilz *m*, **Selbstabnahmeobertuch** *n s.* Selbstabnahmefilz
Selbstabnahmepapiermaschine *f* lick-up (single-cylinder) machine, Yankee machine
Selbstabnahmewalze *f (pap)* pickup roll
Selbstabsorption *f* self-absorption
Selbstalkylierung *f* self-alkylation
selbstansaugend self-priming *(pump)*
selbstauflösend *(biol)* autolytic
Selbstauflösung *f (biol)* autolysis, autolytic decomposition
selbstauslöschend self-extinguishing
Selbstdiffusion *f* self-diffusion
selbstemulgierend self-emulsifying
selbstentzündlich self-igniting, pyrophoric, pyrophorous
Selbstentzündung *f* self-ignition, spontaneous (autogenous) ignition
Selbstentzündungstemperatur *f* self-ignition (spontaneous-ignition) temperature, S.I.T., autogenous ignition temperature
Selbsterhitzung *f* self-heating, spontaneous heating
Selbsterwärmung *f* self-heating, spontaneous heating
Selbstfarbstoff *m* self-colour
selbstgängig, selbstgehend self-fluxing *(ore)*
selbsthärtend *(plast)* self-curing
Selbstionisation *f* autoionization, preionization
Selbstklebefolie *f* self-adhesive film
selbstklebend self-adherent
Selbstkleber *m* pressure-sensitive adhesive
Selbstkondensation *f* self-condensation
selbstlöschend self-extinguishing
Selbstreinigung *f* self-purification, autopurification, self-cleansing
Selbstreinigungskraft *f*, **Selbstreinigungsvermögen** *n* self-purification power, autopurification power
selbstschmierend self-lubricating

Selbstschmierung *f* self-lubrication
Selbstschreiber *m* recording instrument, recorder
Selbstspalten *n (tann)* blistering *(of hides)*
selbsttätig automatic
selbsttonend *(phot)* self-toning
selbstverlöschend self-extinguishing
selbstvulkanisierend *(rubber)* self-curing, self-vulcanizing
Selbstzündpunkt *m*, **Selbstzündtemperatur** *f* self-ignition (spontaneous-ignition) temperature, S.I.T., autogenous ignition temperature
Selbstzündung *f* self-ignition, spontaneous ignition
selektieren to select
Selektion *f* selection
selektiv selective
Selektivaustauscher *m* selective exchanger
Selektivaustauscherharz *n* selective resin
Selektivflotation *f* selective (differential) floatation
Selektivherbizid *n* selective herbicide (weed-killer)
Selektivinsektizid *n* selective insecticide
Selektivität *f* selectivity
Selektivkrackung *f* selective cracking
Selektivlösungsmittel *n* selective solvent
Selektivnährboden *m* selective medium *(microbiology)*
Selen *n* Se selenium
Selenat *n* $M_2^ISeO_4$ selenate
Selen(IV)-bromid *n* $SeBr_4$ selenium(IV) bromide, selenium tetrabromide
Selen(IV)-chlorid *n* $SeCl_4$ selenium(IV) chloride, selenium tetrachloride
Selendioxid *n s.* Selen(IV)-oxid
Selendisulfid *n s.* Selen(IV)-sulfid
Selenfilter *n* selenium glass
Selen(IV)-fluorid *n* SeF_4 selenium(IV) fluoride, selenium tetrafluoride
Selen(VI)-fluorid *n* SeF_6 selenium(VI) fluoride, selenium hexafluoride
Selengleichrichter *m* selenium rectifier
Selenhalogenid *n* selenium halogenide, selenium halide
Selenhexafluorid *n s.* Selen(VI)-fluorid
Selenid *n* M_2^ISe selenide
Selenit *m (min)* selenite *(a variety of gypsum)*
Selenit *n* $M_2^ISeO_3$ selenite
Selenmonosulfid *n s.* Selen(II)-sulfid
Selennitrid *n* Se_4N_4 selenium nitride, tetraselenium tetranitride
Selenoniumverbindung *f* selenonium compound
Selen(IV)-oxid *n* SeO_2 selenium(IV) oxide, selenium dioxide
Selen(VI)-oxid *n* SeO_3 selenium(VI) oxide, selenium trioxide
Selenoxidbromid *n* $SeOBr_2$ selenium dibromide oxide
Selenoxidchlorid *n* $SeOCl_2$ selenium dichloride oxide
Selenoxidfluorid *n* $SeOF_2$ selenium difluoride oxide
Selenrubinglas *n* selenium ruby glass
Selensäure *f* H_2SeO_4 selenic acid
Selenstickstoff *m s.* Selennitrid
Selen(II)-sulfid *n* SeS selenium(II) sulphide, selenium monosulphide
Selen(IV)-sulfid *n* SeS_2 selenium(IV) sulphide, selenium disulphide
Selentetra ... *s.* Selen(IV)- ...
Selentrioxid *n s.* Selen(VI)-oxid
Selenwasserstoff *m* H_2Se hydrogen selenide, selenium hydride
Selenwismutglanz *m (min)* guanajuatite *(bismuth selenide)*
Selenzelle *f* selenium cell
Seliwanow-Reaktion *f* Seliwanoff reaction *(for detecting hexoses)*
Sellait *m (min)* sellaite *(magnesium fluoride)*
Sellerieöl *n* celery [seed] oil *(from Apium graveolens L.)*
Selleriesalz *n* celery salt
Seltenerden *pl* rare earths
Seltenerdmetalle *npl* rare-earth elements (metals)
Selterswasser *n* seltzer [water], selter, selters water
Semet-Solvay-Ofen *m* Semet-Solvay oven *(a coke oven)*
Semianthrazit *m* semianthracite [coal]
Semichemical-Zellstoff *m* semichemical pulp
Semichinon *n (org chem)* semiquinone
Semichromgerbung *f* semichrome tannage
Semidin *n* $C_6H_5NHC_6H_4NH_2$ semidine
Semidinumlagerung *f* semidine rearrangement (transformation)
Semiebonit *n* semiebonite, half-hard rubber
Semifusinit *m* semifusinite *(a maceral of coal)*
Semikarbazid *n* $NH_2CONHNH_2$ semicarbazide
Semikarbazon *n (org chem)* semicarbazone
Semikolloid *n* semicolloid
Semikraft-Verfahren *n (pap)* kraft semichemical process
Semimikroanalyse *f* semimicroanalysis
semipermeabel semipermeable
Semipermeabilität *f* semipermeability
semipolar semipolar, half-polar
semiquantitativ semiquantitative
Semi-Reinforcing-Furnace-Ruß *m (rubber)* semireinforcing furnace black, SRF black
semizyklisch semicyclic
Senarmontit *m (min)* senarmontite *(antimony(III) oxide)*
Senezioalkaloid *n* senecio alkaloid
Seneziosäure *f* $(CH_3)_2C{=}CHCOOH$ senecioic acid, 3,3-dimethylacrylic acid
Senf *m* 1. mustard, *(esp genus)* Sinapis L.; 2. *(food)* [prepared] mustard
  Brauner S. *s.* Schwarzer S.
  Indischer S. Indian mustard, Brassica juncea (L.) Coss.
  Roter S. *s.* Schwarzer S.
  Schwarzer S. black (brown) mustard, Brassica nigra (L.) W.D.J. Koch

Weißer S. white mustard, Sinapis alba L.
Senfgas *n* mustard gas, di-2-chloroethyl sulphide
Senföl *n* mustard[seed] oil; mustard oil *(any of a class of compounds RNCS)*
Senfölglykosid *n* mustard glycoside
Senfpapier *n (med)* mustard paper
Senkblei *n* plummet, sounding lead
senken to lower, to reduce, to depress
Senkgrube *f* sink, sump
Senkrechtbecherwerk *n* vertical bucket elevator
Senkrechtförderer *m* vertical elevator
Senkrechtkammer *f* vertical chamber *(as of an oven)*
Senkrechtofen *m* vertical oven *(as for manufacturing gas)*
Senkrechtstellung *f* perpendicular position
Senkrechtziehverfahren *n* **für Tafelglas** vertical sheet drawing process
Senkspindel *f* densi[to]meter, araeometer, hydrometer
Senkung *f* lowering, reduction, depression
Senkwaage *f s.* Senkspindel
Sennaar-Gummi *n* Sennaar gum *(mainly from Acacia senegal (L.) Willd.)*
Sensibilisator *m* sensitizer
chemischer S. chemical sensitizer
optischer (spektraler) S. spectral (optical) sensitizer
sensibilisieren to sensitize
panchromatisch s. to panchromatize
Sensibilisierung *f* sensitization
chemische S. chemical sensitization
optische (spektrale) S. spectral (optical) sensitization
Sensibilisierungsfarbstoff *m* sensitizing dye
Sensibilität *f* sensitivity, sensitiveness
Sensitometer *n* sensitometer
Sensitometrie *f* sensitometry
Separation *f* separation
Separator *m* separator
mehrstufiger S. multistage separator
Separator-Nobel-Verfahren *n* Separator-Nobel process *(for deparaffinizing petroleum)*
Separatstreichen *n*, **Separatstrich** *m (pap)* separate (conversion, off-machine) coating
separieren to separate
Separierung *f* separation
Sepiabraun *n* umber *(an earth pigment)*
Sepialichtpause *f* brown print
Sepia-Lichtpausverfahren *n* sepia negative process, silver-iron process, Vandyke process, brownprint
Sepiatonung *f (phot)* sepia toning
Sepiolith *m (min)* sepiolite, sea foam, meerschaum *(a phyllosilicate)*
Sequenz *f* sequence
Sequestiermittel *n* sequestering agent, sequestrant
Ser *s.* Serin
Serie *f* 1. series; 2. battery, bank *(as of apparatus)*
Serienanalyse *f* routine analysis
Seriengrenze *f* series (convergence) limit *(of a spectral series)*
Seriengrenzfrequenz *f* convergence frequency *(of a spectral series)*
Serin *n* $HOCH_2CH(NH_2)COOH$ serine, 2-amino-3-hydroxypropionic acid
Serizin *n* sericin, silk gum *(a protein occurring in silk)*
Serizit *m (min)* sericite *(a phyllosilicate)*
Serotonin *n* serotonin, 3-(2-aminoethyl)-5-hydroxyindole
Serpentin *m (min)* serpentine *(a phyllosilicate)*
Serpentinasbest *m (min)* serpentine asbestos
Serum *n* serum; *(rubber)* skim
natives S. native serum
Serumalbumin *n* [blood] serum albumin, seralbumin
Serumglobulin *n* serum globulin, serglobulin
Serumprotein *n* serum protein
Sesamöl *n* sesame (benne, gingelly, teel) oil *(from Sesamum indicum L.)*
Sesci-Ofen *m* Sesci furnace
Sesquioxid *n* sesquioxide *(deprecated term for an oxide of the type $M_2^{III}O_3$)*
Sesquiterpen *n* sesquiterpene
Sesquiterpenalkohol *m* sesquiterpene alcohol
Sesselform *f* chair form, chair [conformation] *(stereochemistry)*
Setzapparat *m (min tech)* jig
hydraulischer S. hydraulic (wet) jig
Setzarbeit *f (min tech)* jigging
Setzbett *n (min tech)* jig bed
setzen *(min tech)* to jig; *(ceram)* to place, to set *(material to be burned)*
außer Betrieb s. 1. to put out of operation, to cut out of service, to stop *(a machine)*; 2. to shut [down], to close down *(a factory)*
außer Gang s. *s.* außer Betrieb s. 1.
in Betrieb s. to put (set) in operation, to set in action, to start [up], to prime *(a machine)*
in Freiheit s. to release, to liberate
in Gang s. 1. to initiate, to start up *(a reaction)*; 2. *s.* in Betrieb s.
ins Gleichgewicht s. to equilibrate, to bring into equilibrium
sich s. to settle [out], to deposit, to sediment, to subside, to set
unter Druck s. to pressurize
Setzfaß *n s.* Setzkasten
Setzgut *n* material being (*or* to be) separated
Setzherd *m (min tech)* table
Setzkasten *m (min tech)* wash box, hutch *(of a jig washer)*
Baumscher S. Baum wash box
Setzmaschine *f (min tech)* jig [washer]
S. mit bewegtem Sieb movable-sieve jig
S. mit festem Sieb fixed-sieve jig
Baumsche S. Baum jig [washer] *(an air-operated jig)*
Harzer S. Harz [fixed-sieve] jig
hydraulische S. hydraulic (wet) jig
luftbewegte (luftgesteuerte) S. air-operated jig
Setzmilch *f* [naturally, spontaneously] soured milk, sour (set) milk

Setzraum *m (min tech)* jigging compartment, ore box *(of a jig washer)*
Setzsieb *n (min tech)* jig[ging] screen
Setzweise *f* / **offene** *(ceram)* open setting *(the arrangement of ware in a kiln without saggars)*
SEV *s.* Sekundärelektronenvervielfacher
Sexagen *n*, **Sexualhormon** *n* sex hormone
Sexuallockstoff *m* sex[ual] attractant, sex lure
sezernieren *(biol)* to secrete
S-Finish *n (text)* S finishing, surface saponification
Shampoo[n] *n* shampoo
Sharples-Verfahren *n (petrol)* Sharples [two-stage dewaxing] process
Shatter-Test *m (coal)* shatter test
Shattuckit *m (min)* shattuckite *(a cyclosilicate)*
Sheet *m* / **luftgetrockneter** *(rubber)* air-dried sheet, A.D.S.
Sheet-Mangel *f (rubber)* sheeting mill
sherardisieren to sherardize *(to heat-treat with zinc dust)*
Sherardisierung *f* sherardizing *(heat-treatment with zinc dust)*
Shikimisäure *f* shikimic acid, 3,4,5-trihydroxy-1-cyclohexene-1-carboxylic acid
Shoddywolle *f (text)* shoddy
Shonansäure *f* shonanic acid *(a tropolone derivative)*
Shore-Härte *f* Shore hardness
Shore-Härtemesser *m*, **Shore-Härteprüfer** *m* Shore durometer
Showerdeck *n (petrol)* shower deck
SHZ *s.* Sulfathüttenzement
Siaktalg *m* Siak fat (tallow) *(from Palaquium oleiferum Blanco)*
Sial *n (geoch)* sial
Sialsäure *f* sialic acid *(any of a group of acylated neuraminic acids)*
Sialzone *f (geoch)* sial
Siambenzoe *f* Siam benzoin *(from Styrax specc.)*
Siamkardamom *m(n)* Siam cardamom *(from Elettaria cardamomum (L.) White et Maton)*
Siaresinolsäure *f* siaresinolic acid *(a polycyclic compound isolated from Siam benzoin)*
Sicherheitsabsperrventil *n* safety cut-off
Sicherheitsbeiwert *m* safety factor
Sicherheitsbeschleuniger *m (rubber)* delayed-action accelerator
Sicherheitsbestimmung *f* safety regulation
Sicherheitsfaktor *m* safety factor
Sicherheitsfarbe *f* safety (sensitive) ink *(printing ink for safety papers)*
Sicherheitsgefäß *n* safety vessel
Sicherheitsglas *n* safety (shatterproof, non-shattering) glass
  geschichtetes S. laminated safety [sheet] glass
Sicherheitsglas-Zwischenschicht *f* safety-glass interlayer (interleaver)
Sicherheitsingenieur *m* safety engineer
Sicherheitslicht *n* safelight
Sicherheitsmaßnahme *f* safety measure, precaution
Sicherheitspackung *f* safety package
Sicherheitspapier *n* safety (cheque, security) paper
Sicherheitspipette *f* safety pipette
Sicherheitssprengstoff *m* safety explosive, *(Am)* permissible [explosive]
Sicherheitsventil safety[-relief] valve, [pressure] relief valve
  entlastetes S. balanced relief valve
  federbelastetes S. spring-actuated relief valve
Sicherheitsvorkehrungen *fpl* safety precautions
Sicherheitsvorschrift *f* safety instruction
Sicherheitswaschflasche *f* safety wash-bottle
Sicherheitszündholz *n* safety match
sichern to secure; to establish *(e.g. the structure of a compound)*; to check *(results)*
Sicherung *f* 1. securing; establishing *(of the structure of a compound)*; checking *(of results)*; 2. safeguard, security device
sichten to classify pneumatically (by air)
Sichten *n* pneumatic (air) classification, air sizing
Sichtentwicklung *f (phot)* development by inspection
Sichter *m* pneumatic (air) classifier
Sichtermühle *f* classifier mill
Sichtfeines *n* undersize [material], minus material *(air classification)*
Sichtfenster *n*, **Sichtglas** *n* inspection (sight) glass
Sichtgrobes *n* oversize [material], tailing[s], tails *(air classification)*
Sichtkammer *f* classifying chamber
Sichtmühle *f* classifier mill
Sichtung *f s.* Sichten
sickern to trickle
  s. lassen to [allow to] trickle
Sickerverlust *m* leakage, slippage loss, slip
Sickerwasser *n* drainage water
Sideramin *n* sideramine *(any of a class of compounds promoting bacterial growth)*
Siderazot *m (min)* siderazot[e], silvestrite *(an iron nitride)*
Sideringelb *n* siderin yellow *(iron(III) chromate)*
Siderit *m* 1. [holo]siderite, iron meteorite, meteoric iron; 2. *(min)* siderite, spathic iron [ore] *(iron(II) carbonate)*
Siderografie *f* siderography, steel-plate printing
Siderolith *m* siderolite, sideraerolite *(a kind of iron meteorite)*
Sideromyzin *n* sideromycin *(antibiotic)*
Siderosphäre *f* siderosphere, earth's core (nucleus)
Sieb *n* 1. sieve, *(esp tech)* screen, *(for liquids containing solid particles)* strainer; *(pap)* wire; 2. *s.* Siebapparat
  endloses S. *(pap)* endless wire
  standardisiertes S. standard sieve (screen)
Siebabwasser *n (pap)* tray (pulp, free, loose, saveall) water, [machine] white water, pulpwater, backwater
Siebanalyse *f* sieve (screen) analysis
Siebanlage *f* screening plant

Siebapparat *m* screening machine, screen [classifier], sifter
Siebaustrag *m* screen discharge
Siebblech *n* perforated plate (tray)
Siebboden *m* sieve (perforated) bottom, screen deck; *(distil)* sieve (perforated) plate (tray); *(ferm)* strainer [bottom]
Siebbodenkolonne *f (distil)* sieve-plate (perforated-plate) column (tower)
pulsierte S. pulsed sieve-plate column
Siebbreite *f (pap)* width of the wire
Siebdruck *m* [silk-]screen printing, stencil printing
Siebdruckfarbe *f* silk-screen (screen-process) ink
Siebdruckverfahren *n* silk-screen process
Siebdurchgang *m*, **Siebdurchlauf** *m* underflow
Siebeinlage *f* strainer
Siebeinsatz *m* strainer; *(plast)* screen pack *(as of an extruder)*
sieben to sieve, *(esp tech)* to screen, to sift
Siebenerring *m s.* Siebenring
siebengliedrig seven-membered
Siebenring *m* seven-membered ring
Siebenstufenbleiche *f (pap)* seven-stage bleaching
siebenwertig septivalent, heptavalent
Siebenwertigkeit *f* septivalence
Sieberei *f* screening station
Siebereianlage *f* screening plant
Sieberfolg *m* screen[ing] efficiency
Siebfeines *n* undersize [material], minus material
Siebfeinheit *f* sieve fineness, grade, *(quantitatively)* mesh
Siebfilter *n* screen filter (strainer)
Siebfläche *f* screening (screen) area (surface), sieve area
nützliche S. active screen area
Siebgeschwindigkeit *f (pap)* speed of the wire
Siebgewebe *n* straining cloth, gauze; *(pap)* wire cloth (gauze) ✦ **mit S. abgedeckt** *(filtr)* screen-covered
Siebgrobes *n* [screen] oversize, overs, tailings
Siebgut *n* material being (*or* to be) screened, screen feed
Siebgütegrad *m* screen[ing] efficiency
Siebkante *f (pap)* wire edge
Siebkasten *m* screen frame
siebklassieren to screen
Siebklassierung *f* screen classification, screening
Siebkopf *m* screen (straining) head
Siebkopf-Spritzmaschine *f (plast, rubber)* screen head extruder, straining machine, strainer
Siebkurve *f* grading curve (limit)
Sieblänge *f (pap)* wire length
Sieblauf *m (pap)* run of the wire
Sieblaufregler *m (pap)* wire guide
Sieblaufregulierwalze *f (pap)* wire-guide roll, wire-leading roll
Siebleder *n (pap)* apron
Siebleistung *f* screening capacity
Siebleitwalze *f (pap)* wire-guide roll, wire-leading roll
Sieblinie *f s.* Siebkurve
Sieblochung *f* perforation *(of a screen)*, *(relating to diameter also)* orifice, screen aperture (size)
Siebmantel *m (pap)* cylinder cover *(of a cylinder mould)*
siebmarkiert *(pap)* wire-marked
Siebmarkierung *f (pap)* wire mark ✦ **mit S.** wire-marked *(paper)*
Siebmaschine *f s.* Siebapparat
Siebmühle *f* screen-type mill
Siebnaht *f (pap)* seam of the machine wire
Siebnummer *f* mesh *(number of openings per linear inch)*; *(pap)* number
Siebnutzfläche *f* active screen area
Sieboberfläche *f* screen[ing] surface
Sieböffnung *f* screen aperture (size), orifice
Siebpartie *f (pap)* wire part (end), Fourdrinier part (section)
S. der Rundsiebpapiermaschine cylinder part
ausfahrbare S. removable Fourdrinier [part]
Siebplättchen *n* perforated plate *(of a Gooch crucible)*
Siebplatte *f* perforated plate; *(plast)* screen pack *(of an extruder)*
Siebpresse *f s.* Siebkopf-Spritzmaschine
Siebrand *m (pap)* wire edge
Siebrandspritzwasser *n (pap)* squirt-trim water
Siebrätter *m* gyratory screen (sifter)
Siebreihe *f* screen (sieve) series (scale)
Siebrost *m* bar screen
Siebrückstand *m s.* Siebgrobes
Siebrüttelmaschine *f* sieve shaker
Siebsatz *m* set of screens (sieves)
Siebsaugwalze *f (pap)* suction [couch] roll
Siebschleuder *f s.* Siebzentrifuge
Siebschneckenaustragzentrifuge *f*, **Siebschnekkenschleuder** *f* screen-conveyor centrifuge
Siebschüttelung *f (pap)* shake of the wire
Siebseite *f* wire side *(of paper)*
Siebskala *f s.* Siebreihe
Siebspannung *f (pap)* wire tension
Siebspannwalze *f (pap)* wire-stretch roll
Siebstation *f* screening station
Siebtisch *m (pap)* wire table (frame), forming table
Siebtrog *m* [cylinder] vat *(of a cylinder paper machine)*
Siebtrommel *f* 1. revolving screen, trommel [screen]; 2. perforated basket (bowl) *(of a centrifugal)*
Siebtrum *m (pap)* wire run
rücklaufender (rückläufiger, unterer) S. return wire run
vorlaufender S. wire run
Siebtuch *n* straining cloth, gauze
Siebtuchpresse *f (pap)* fabric press
Siebübergang *m*, **Siebüberlauf** *m s.* Siebgrobes
Siebüberzug *m (pap)* cylinder cover *(of a cylinder mould)*
Siebunterlauf *m s.* Siebfeines
Siebvorrichtung *f* screen classifier, screening device

angebaute (äußere) S. external screen classifier

eingebaute S. internal screen classifier

Siebwasser *n (pap)* backwater, tray (pulp, free, loose, save-all) water, [machine] white water, pulpwater

Siebwasserbehälter *m (pap)* backwater tank (box), hog (wire, machine-wire) pit, save-all tray

Siebwasserpumpe *f (pap)* backwater pump

Siebwassersammelbecken *n s.* Siebwasserbehälter

Siebwechsel *m (pap)* wire changing

Siebweite *f* screen size (aperture), screen-size opening, orifice

Siebwirkung *f* screening action

Siebwirkungsgrad *m* screen[ing] efficiency

Siebzentrifuge *f* screen (perforate bowl) centrifuge, centrifugal

S. mit konischer Trommel conical-screen centrifuge

S. mit zylindrischer Trommel cylindrical-screen centrifuge

Siebzylinder *m (pap)* cylinder mould

Siedeanalyse *f* distillation analysis

Siedebeginn *m* initial boiling point, I.B.P.

Siedebereich *m* boiling[-point] range

Siedediagramm *n* boiling-point diagram, liquid-vapour equilibrium diagram

S. mit Maximum maximum boiling-point system

S. mit Minimum minimum boiling-point system

Siedeendpunkt *m* final boiling point, F.B.P., end point

Siedefläche *f* plane of the boiling-point diagram

Siedegrenze *f* boiling limit

Siedegrenzenbenzin *n* special boiling-point spirit, SBP spirit

Siedehitze *f* boiling heat

Siedeintervall *n* boiling[-point] range

Siedekapillare *f* capillary air bleed, whipping (air-leak) tube

Siedekonstanz *f* regular boiling

Siedekurve *f s.* 1. Siedelinie; 2. Siedepunktskurve

Siedelinie *f* boiling curve *(of a two-component system)*

sieden to boil

Sieden *n* boil[ing] ✦ **am S. halten** to keep at the boil ✦ **nahe am S. halten** to keep near the boil

konstantes S. regular boiling

siedend boiling

höher s. higher-boiling

konstant s. constant-boiling

leichter s. lower-boiling

niedriger s. lower-boiling

schwerer s. higher-boiling

Siedepfanne *f* pan *(for obtaining table salt)*

Siedepunkt *m* boiling point (temperature), b.p.

mittlerer S. mid-boiling point

Siedepunktsbestimmung *f* boiling-point determination

Siedepunktserhöhung *f* boiling-point elevation (rise)

molale (molare) S. molal boiling-point[-elevation] constant, molal (molar, molecular) elevation constant, ebullioscopic constant

Siedepunktshöchstwert *m* maximum boiling point

Siedepunktskurve *f* boiling-point curve *(of a multicomponent system)*

S. bei geschlossener Verdampfung [single-]flash curve

Siedepunktsmaximum *n* maximum boiling point

Siedepunktsminimum *n* minimum boiling point

Siedereaktor *m (nucl)* boiling [water] reactor

Siederohr *n* boiling tube

Siederohrdampferzeuger *m*, **Siederohrkessel** *m* water-tube boiler

Siedesalz *n* evaporated salt

Siedeschaubild *n* boiling-point diagram

Siedestab *m (lab)* bumping stick

Siedestein *m*, **Siedesteinchen** *n* boiling stone

Siedetemperatur *f* boiling temperature (point)

Siedeverfahren *n* boiling-off process *(for making butter)*

Siedeverzögerung *f s.* Siedeverzug

Siedeverzug *m* delay in boiling, delayed boiling

Siedewasserreaktor *m (nucl)* boiling [water] reactor

Siedezone *f* boiling zone

Siegelgerät *n (plast)* sealing unit

Siegellack *m* sealing wax

siegeln *(plast)* to seal

Siegeln *n (plast)* sealing

S. mit gespritztem Zusatzdraht extruded bead sealing

S. mit Heizstab heated-bar sealing

S. mit Schweißlineal heated-bar sealing

S. mit stranggepreßtem Zusatzdraht extruded bead sealing

Siegler *m (plast)* sealer

Siemens-Martin-Ofen *m* open-hearth furnace, OH-furnace, Siemens-Martin furnace

Siemens-Martin-Schlacke *f* open-hearth slag

Siemens-Martin-Stahl *m* open-hearth (Siemens-Martin) steel

Siemens-Martin-Stahlschrott *m* open-hearth steel scrap

Siemens-Martin-Stahlwerk *n* open-hearth (Siemens-Martin) steel plant

Siemens-Martin-Verfahren *n* open-hearth (Siemens-Martin) process

Sienaerde *f* sienna *(hydrous iron oxide)*

Sigmabindung *f* sigma bond, σ bond

Sigmaelektron *n* sigma electron, σ electron

Sigmaphase *f (met)* sigma phase

Signalöl *n* signal (long-time burning) oil

Sikkativ *n* siccative, desiccant, drying agent, drier

Silan *n* Si_nH_{2n+2} silane, *(specif)* SiH_4 monosilane

organisches S. organic silane, organosilane

Silanol *n* silanol *(a silicon compound which contains OH groups, specif H_3SiOH)*

Silazan *n* $H_3Si(NHSiH_2)_nNHSiH_3$ silazane

organisches S. organosilazane
Silber *n* Ag silver
gediegen[es] S. native silver
knallsaures S. *s.* Silberfulminat
metallisches S. metallic silver
Silberabscheidung *f* 1. deposition of silver; 2. silver deposit
Silberarsenat(III) *n* Ag_3AsO_3 silver arsenite
Silberarsenat(V) *n* Ag_3AsO_4 silver arsenate
Silberätze *f (glass)* silver stain
Silberazetylid *n* C_2Ag_2 silver acetylide (carbide)
Silberbad *n (phot)* silver bath; silver-plating bath *(electroplating)*
Silberbeize *f (glass)* silver stain
Silberbelag *m* silver deposit
Silberbild *n (phot)* silver image
Silberbromid *n* AgBr silver bromide
Silberbromidelektrode *f* silver-bromide electrode
Silberbromidkorn *n (phot)* silver-bromide grain
Silberchlorid *n* AgCl silver chloride
Silberchloridelektrode *f* silver-chloride electrode
Silberchromat *n* Ag_2CrO_4 silver chromate
Silbercoulometer *n* silver coulometer (voltameter)
Silberdifluorid *n s.* Silber(II)-fluorid
Silberdiphosphat *n* $Ag_4P_2O_7$ silver diphosphate, silver pyrophosphate
Silberelektrode *f* silver electrode
Silber(I)-fluorid *n* AgF silver(I) fluoride, silver monofluoride
Silber(II)-fluorid *n* AgF_2 silver(II) fluoride, silver difluoride
silberführend silver-bearing, argentiferous
Silberfulminat *n* AgONC silver fulminate, fulminating silver
Silbergehalt *m* silver content
Silberglanz *m (min)* silver glance, vitreous silver *(silver sulphide)*
silberglänzend silvery
Silberhalogenid *n* silver halide
Silberhalogenidemulsion *f (phot)* silver-halide emulsion
Silberhalogenidkorn *n (phot)* silver-halide grain
silberhaltig containing silver, argentiferous
Silberhexafluorosilikat *n* $Ag_2[SiF_6]$ silver hexafluorosilicate
Silberhexazyanoferrat(II) *n* $Ag_4[Fe(CN)_6]$ silver hexacyanoferrate(II)
Silberhexazyanoferrat(III) *n* $Ag_3[Fe(CN)_6]$ silver hexacyanoferrate(III)
Silberhydrogenorthophosphat *n* Ag_2HPO_4 silver hydrogenorthophosphate, silver hydrogenphosphate
silberig silvery
Silberjodid *n* AgI silver iodide
Silberkarbid *n s.* Silberazetylid
Silberkeim *m (phot)* silver nucleus
Silberkorn *n (phot)* silver grain
Silberlot *n* silver solder
Silbermolybdat *n* Ag_2MoO_4 silver molybdate
Silbermonofluorid *n s.* Silber(I)-fluorid
Silberniederschlag *m* silver deposit
Silbernitrat *n* $AgNO_3$ silver nitrate
Silbernitratpapier *n* silver-nitrate paper
Silbernitroprussiat *n s.* Silbernitroprussid
Silbernitroprussid *n* $Ag_2[Fe(CN)_5NO]$ silver nitroprusside
Silberorthophosphat *n* Ag_3PO_4 silver orthophosphate
Silberoxid *n* silver oxide; *(specif) s.* Silber(I)-oxid
Silber(I)-oxid *n* Ag_2O silver(I) oxide
Silber(II)-oxid *n* AgO silver(II) oxide
Silber[pack]papier *n* aluminium (silver) paper
silberplattieren to silver-clad
Silberplattierung *f* silver cladding
Silberpyrophosphat *n s.* Silberdiphosphat
Silberrhodanid *n s.* Silberthiozyanat
Silberrückgewinnung *f* silver recovery
Silbersalz *n* 1. silver salt; 2. *(org chem)* silver salt, sodium anthraquinone-2-sulphonate *(for detecting alkaloids)*
Silbersalzdiffusion *f (phot)* silver-salt diffusion
Silber-Silberbromid-Elektrode *f* silver-silver bromide electrode
Silber-Silberchlorid-Elektrode *f* silver-silver chloride electrode
Silberspiegel *m* silver mirror
Silberstrichbildung *f (ceram)* silver (cutlery) marking
Silbersulfat *n* Ag_2SO_4 silver sulphate
Silbertetrajodomerkurat(II) *n* $Ag_2[HgI_4]$ silver tetraiodomercurate(II)
Silberthiozyanat *n* AgSCN silver thiocyanate, silver rhodanide
Silberverstärker *m (phot)* silver intensifier
Silberwolle *f* silver wool
Silexknolle *f (geol)* chert nodule
Silifizierung *f* silification *(thickening of silicate paint)*
Silika *f s.* Silikamasse
Silikabaustein *m* silica brick
Silikamasse *f*, **Silikamaterial** *n* silica
Silikamörtel *m* silica cement
Silikan *n s.* Silan
Silikasand *m* silica sand
Silikastein *m* silica brick
Silikat *n* silicate
Silikatbildung *f* silicate formation
Silikatbindung *f* silicate bond
Silikatgestein *n* silicate rock
Silikatglas *n* silicate glass
Silikatphosphor *m* silicate phosphor
Silikatschlacke *f* silicate slag
Silikoameisensäure *f* HSiOOH silicoformic acid
Silikoameisensäureanhydrid *n* $[Si_2O_3H_2]_x$ silicoformic anhydride, dioxodisiloxane
Silikoäthan *n s.* Disilan
Silikobenzoesäure *f* C_6H_5SiOOH silicobenzoic acid
Silikobromoform *n* $SiHBr_3$ tribromosilane, silicobromoform
Silikochloroform *n* $SiHCl_3$ trichlorosilane, silicochloroform
Silikoessigsäure *f* CH_3SiOOH silicoacetic acid

Silikofluorid *n s.* Hexafluorosilikat
Silikofluoroform *n* $SiHF_3$ trifluorosilane, silicofluoroform
Silikojodoform *n* $SiHI_3$ triiodosilane, silicoiodoform
Silikomesoxalsäure *f* $H_4Si_3O_6$ silicomesoxalic acid
Silikomethan *n s.* Monosilan
Silikomethyläther *m s.* Disiloxan
Silikon *n* silicone, organopolysiloxane, polyorganosiloxane ✦ **mit Silikonen behandeln** to silicone-treat, to siliconize
elastomeres S. silicone elastomer
karbofunktionelles S. carbon-functional silicone
organofunktionelles S. organofunctional silicone
Silikonanlage *f* silicone plant
Silikonanstrichfarbe *f* silicone paint
Silikonantischaummittel *n* silicone antifoam
Silikonat *n* siliconate
Silikonausrüstung *f* silicone finish
Silikonbasis / auf silicone-base[d]
silikonbehandelt silicone-treated, siliconized
Silikonbehandlung *f* silicone treatment, siliconization
silikonbeschichtet silicone-coated
Silikonchemie *f* silicone chemistry
Silikonelastomer[es] *n* silicone elastomer
Silikonemulsion *f* silicone emulsion
Silikonfarbe *f* silicone paint
Silikonfett *n* silicone grease
Silikonfilm *m* silicone film
Silikonfluid *n*, **Silikonflüssigkeit** *f* silicone fluid (liquid, oil)
Silikonform[en]trennmittel *n* silicone mould-release agent
Silikonfüllstoff *m* silicone filler
Silikongrundlage / auf silicone-base[d]
Silikongummi *m* silicone rubber *(vulcanized product)*
Silikongummimischung *f s.* Silikonkautschukmischung
Silikonhahnfett *n* silicone stopcock grease
Silikonharz *n* silicone resin
Silikonhaut *f* silicone film
silikonisieren to siliconize, to silicone-treat
Silikonisierung *f* siliconization, silicone treatment
Silikonkautschuk *m* silicone [rubber] gum
Silikonkautschukmischung *f* silicone rubber compound (mixture, stock)
Silikonkitt *m* silicone putty
Silikonklebstoff *m* silicone adhesive
Silikonkunstharz *n* silicone resin
Silikonlack *m* silicone varnish, *(if pigmented)* silicone lacquer
silikonmodifiziert silicone-modified
Silikonöl *n* silicone oil (liquid, fluid)
Silikononan *n s.* Tetraäthylsilan
Silikonpaste *f* silicone compound
Silikonpolymer[es] *n* silicone polymer
Silikonrohkautschuk *m* silicone [rubber] gum
Silikonschmiermittel *n* silicone lubricant
Silikonspringkitt *m* bouncing putty
Silikontrennmittel *n* silicone abherent (release agent)
silikonüberzogen silicone-coated
Silikonüberzug *m* silicone coating
Silikonwerk *n* silicone plant
Silikooxalsäure *f* $[Si_2O_4H_2]_x$ silicooxalic acid
Silikopropan *n s.* Trisilan
Silikose *f (med)* silicosis
Silikospiegel *m* silicospiegel *(an iron-base alloy containing Mn and Si)*
silikothermisch silicothermic
Silikowolframat *n s.* Wolframatosilikat
Silizid *n* silicide
silizieren to siliconize *(metals for protecting them)*
Silizierung *f* siliconization *(of metals)*
Silizifikation *f*, **Silizifizierung** *f (geol)* silicification
Silizium *n* Si silicon
Silizium(IV)-bromid *n* $SiBr_4$ silicon(IV) bromide, silicon tetrabromide
Siliziumbromoform *n s.* Silikobromoform
Siliziumbronze *f* silicon bronze
Silizium(IV)-chlorid *n* $SiCl_4$ silicon(IV) chloride, silicon tetrachloride
Siliziumchloroform *n s.* Silikochloroform
Siliziumdioxid *n s.* Silizium(IV)-oxid
siliziumdioxidhaltig siliciferous, siliceous, silicic
Siliziumdisulfid *n s.* Silizium(IV)-sulfid
Siliziumeisen *n* silicon iron
Siliziumester *m* silicic-acid ester, silicon ester
Silizium(IV)-fluorid *n* SiF_4 silicon(IV) fluoride, silicon tetrafluoride
Siliziumfluoroform *n s.* Silikofluoroform
Siliziumfluorwasserstoffsäure *f s.* Hexafluorokieselsäure
siliziumfunktionell silicon-functional
Siliziumgleichrichter *m* silicon rectifier
Siliziumguß *m* silicon cast iron
siliziumhaltig containing silicon; *s.* siliziumdioxidhaltig
Siliziumhexabromid *n* Si_2Br_6 hexabromodisilane, disilicon hexabromide
Siliziumhexachlorid *n* Si_2Cl_6 hexachlorodisilane, disilicon hexachloride
Siliziumhydrid *n s.* Siliziumwasserstoff
Silizium(IV)-jodid *n* SiI_4 silicon(IV) iodide, silicon tetraiodide
Siliziumjodoform *n s.* Silikojodoform
Siliziumkarbid *n* SiC silicon carbide
Silizium-Kohlenstoff-Bindung *f* silicon-carbon bond
Silizium(IV)-oxid *n* SiO_2 silicon(IV) oxide, silicon dioxide
Siliziumpolymer[es] *n* silicon polymer
organisches S. organosilicon polymer
Silizium(IV)-sulfid *n* SiS_2 silicon(IV) sulphide, silicon disulphide
Siliziumtetra ... *s.a.* Silizium(IV)- ...
Siliziumtetraäthyl *n s.* Tetraäthylsilan
Siliziumtetramethyl *n s.* Tetramethylsilan
Siliziumwasserstoff *m* Si_nH_{2n+2} silicon hydride, silane

Silizylengruppe *f s.* Silylengruppe
Silizylgruppe *f s.* Silylgruppe
Silizyloxid *n s.* Disiloxan
Sillimanit *m (min)* sillimanite *(a neso-subsilicate)*
Sillimaniterzeugnis *n (ceram)* sillimanite refractory
Sillimanitstein *m* [/ **feuerfester**] sillimanite refractory brick
Silo *n(m)* silo, bin, [storage] hopper
Siloxan *n* siloxane
organisches S. organic siloxane, organosiloxane, organosilicon oxide
polymeres S. polymeric siloxane, polysiloxane
ringförmiges (zyklisches) S. cyclic siloxane, cyclosiloxane
Siloxanbindung *f* siloxane bond
Siloxaneinheit *f* siloxane unit
Siloxanpolymer[es] *n* siloxane polymer
Siloxen *n* $Si_6O_3H_6$ siloxen[e]
Silt *m (geol)* silt
Silthian *n* $H_3Si(SSiH_2)_nSSiH_3$ silthiane
Silvan *n* silvan, 2-methylfuran
Silvestrit *m (min)* silvestrite, siderazot[e] *(an iron nitride)*
Silylengruppe *f* $H_2Si=$ silylene group (residue)
Silylgruppe *f* H_3Si- silyl group (residue)
Silylidingruppe *f* $HSi\equiv$ silylidyne group (residue)
Sima *n (geoch)* sima
Simazin *n* simazine *(a herbicidal triazine derivative)*
Simazone *f (geoch)* sima
Simplexpumpe *f* simplex pump
Simultanreaktion *f* simultaneous reaction
Sinalbin *n* sinalbin *(a mustard glucoside)*
Sinapin *n* sinapin[e], 4-hydroxy-3,5-dimethoxycinnamic acid choline ester
Sinapinalkohol *m* $HO(CH_3O)_2C_6H_2CH=CHCH_2OH$ sinapic alcohol
Sinapinsäure *f* $HO(CH_3O)_2C_6H_2CH=CHCOOH$ sinapic acid, 3-(4-hydroxy-3,5-dimethoxyphenyl)acrylic acid
Singulett *n* singlet
Singulettbindung *f* one-electron (single-electron) bond, singlet link[age]
Singulettzustand *m* singlet state
Singulosilikat *n* monosilicate
Singulosilikatschlacke *f* monosilicate slag
Sinigrin *n* sinigrin *(a mustard glycoside)*
Sinkanteil *m (min tech)* sinking fraction
sinken 1. to sink, to fall, to subside; *(if bottoms are formed)* to sediment, to settle; 2. *(of physical data)* to drop, to decrease
Sinken *n* 1. sinking, fall, subsidence; *(if bottoms are formed)* sedimentation, settling; 2. drop, decrease *(of physical data)*
Sinkerviskosimeter *n* sinking-body viscometer
Sinkfraktion *f (min tech)* sinking fraction
Sinkgeschwindigkeit *f* rate (velocity) of fall; sedimentation (settling) rate (velocity)
S. im Gravitationsfeld rate of fall under gravity
Sinkgut *n* settling (sinking) material; settled material; *(min tech)* underflow, sinks
Sinkkurve *f* sink curve
Sinkscheider *m (min tech)* dense-medium separator (vessel, washer), heavy-medium separator
Sinkscheideranlage *f (min tech)* dense-medium (heavy-medium) separation plant, gravity concentration (concentrating) apparatus
Sinkscheideverfahren *n (min tech)* dense-medium (heavy-medium) separation, dense-medium process
Sinkschlamm *m* bottom sludge
Sinkstoff *m* suspended matter, settling (sinking) material; deposited matter, settled material, settlings
Sinnenprüfung *f* sensory (organoleptic) estimation (evaluation)
Sinter *m (ceram, geol)* sinter; *(met)* scale
Sinteraluminiumpulver *n* sintered aluminium powder, S.A.P.
Sinteranlage *f* sintering plant
Sinterapparat *m* sintering machine
Sinterband *n* sinter[ing] strand
Sinterberyllerde *f (ceram)* sintered beryllia
Sinterbrand *m (ceram)* sinter firing
Sinterdolomit *m (ceram)* sintered (dead-burned) dolomite
Sintereisen *n* sintered iron
Sintererzeugnis *n* sintered product, sinter
Sintergut *n* material being (*or* to be) sintered; sinter, sintered product
Sinterhartmetall *n* [cemented] hard metal, cemented [hard] carbide
Sinterkarbid[metall] *n s.* Sinterhartmetall
Sinterkörper *m* sintered[-powder metal] compact
Sinterkorund *m (ceram)* sintered alumina
Sintermaschine *f* sintering machine
Sintermetall *n* sintered[-powder] metal
sintern to sinter, *(esp glass)* to frit
Sintern *n* sintering
S. ohne Druckanwendung pressureless sintering
aktiviertes S. activated sintering
druckloses S. pressureless sintering
Sinterofen *m* sintering furnace
Sinterstahl *m* sintered steel
Sinterstoff *m s.* Sintergut
Sintertechnik *f* sintering technique
Sintertonerde *f (ceram)* sintered alumina
Sinterung *f s.* Sintern
Sinterverfahren *n* sintering process
Sintervorgang *m* sintering process
Sinterwanne *f (ceram)* sintering tray
Sinterwerkstoff *m* sintered material
Sinterzone *f* sintering zone
Siphon *m (lab)* siphon, syphon *(for handling liquids)*; [siphon] trap *(as of a sink)*; *(tech)* siphon, syphon *(as for condensate removal in gas mains)*
feststehender S. *(tech)* stationary-type siphon
rotierender S. *s.* umlaufender S.
umlaufender S. *(tech)* revolving-type siphon
Siphonrohr *n* siphon pipe
Sirup *m* syrup, sirup; *(sugar)* treacle, *(if dark also)* molasses, *(if bright also)* golden syrup

sirupartig, sirupös syruplike, syrupy
Sisal *m*, **Sisalfaser** *f* sisal [hemp] *(leaf fibre from various Agave specc., esp from Agave sisalana Perr.)*
Sitz *m* seat *(as of a valve)*
Sitzring *m* seat ring
Sitzventil *n* globe valve
SK *s.* Synthesekautschuk
Skala *f* scale
Skale *f* scale *(on a measuring device)*
Skaleneinteilung *f* scale graduation
Skalenlänge *f* scale length
Skalenteil *m* scale division (interval)
Skalenteilstrich *m* scale division
Skalenteilung *m* scale graduation
Skalenwert *m* reading
Skammoniumharz *n* scammony [resin], resin of ipomoea *(from Ipomoea orizabensis Ledanois)*
Skandium *n* Sc scandium
Skandiumbromid *n* $ScBr_3$ scandium bromide
Skandiumchlorid *n* $ScCl_3$ scandium chloride
Skandiumfluorid *n* ScF_3 scandium fluoride
Skandiumhydroxid *n* $Sc(OH)_3$ scandium hydroxide
Skandiumhydroxidnitrat *n* $Sc(OH)(NO_3)_2$ scandium hydroxide nitrate
Skandiumjodid *n* ScI_3 scandium iodide
Skandiumkarbonat *n* $Sc_2(CO_3)_3$ scandium carbonate
Skandiumnitrat *n* $Sc(NO_3)_3$ scandium nitrate
Skandiumoxid *n* Sc_2O_3 scandium oxide
Skandiumsulfid *n* Sc_2S_3 scandium sulphide
Skapolith *m (min)* scapolite, wernerite *(a tectosilicate)*
Skarn *m* skarn *(contact metamorphic high-iron hornfels)*
Skatol *n* skatole, 3-methylindole
Skelett *n (org chem)* skeleton, backbone
Skelettkatalysator *m*, **Skelettkontakt** *m* skeleton (skeletal, Raney) catalyst
Skidmore-Eisentiegel *m* Skidmore iron crucible
Skimkautschuk *m* skim rubber
Skimlatex *m (rubber)* skim latex
skimmen *(rubber)* to skim[coat] *(frictioned tissue)*; to skim *(petroleum)*
Skimmen *n (rubber)* skim coating, skimming *(of frictioned tissue)*; skimming *(of petroleum)*
Skim[m]rinne *f* skimming slot *(waste-water technology)*
Skipgefäß *n* skip [car]
Sklerometer *n* sclerometer
Skleroprotein *n* scleroprotein, skeletal protein
Skleroskop *n* scleroscope
SK-Mischung *f s.* Synthesekautschukmischung
Skolezit *m (min)* scolecite *(a tectosilicate)*
Skopolamin *n* scopolamine, hyoscine *(alkaloid)*
Skopoletin *n* scopoletin, 7-hydroxy-6-methoxycoumarin
Skopolin *n* scopoline *(alkaloid)*
Skorodit *m (min)* scorodite *(iron(III)* orthoarsenate)
Skraup-Synthese *f* Skraup [quinoline] synthesis, Skraup reaction
Skrubber *m* scrubber, scrubbing tower, gas washer, wet collector
Skutterudit *m (min)* skutterudite, smaltite, smaltine *(cobalt triarsenide)*
Slack-Merzerisation *f (text)* slack mercerization
slö *s.* schwerlöslich
Slop-Wax *n* slop wax *(non-pressable wax from heavy cracked distillates)*
S-Lost *n s.* Schwefellost
SM- ... *s.* Siemens-Martin- ...
Smalte *f* smalt, powder blue *(cobalt(II) potassium silicate)*
Smaltin *m s.* Skutterudit
Smaragd *m (min)* emerald *(a cyclosilicate)*
Smaragdgrün *n* emerald (chrome, Guignet's, Mittler's) green *(hydrated chromium oxide)*
smektisch smectic
Smithsonit *m (min)* smithsonite, zinc spar, dry-bone ore *(zinc carbonate)*
SM-Verfahren *n* open-hearth (Siemens-Martin) process
S.M.-Verfahren *n (coal)* Dutch State Mines process
S_N-Reaktion *f* S_N reaction, nucleophilic substitution
S_N1-Reaktion *f* S_N1 reaction *(monomolecular nucleophilic substitution)*
S_N2-Reaktion *f* S_N2 reaction *(bimolecular nucleophilic substitution)*
SN-Verfahren *n* Separator-Nobel process *(a deparaffinization process)*
Soak-Sektion *f (petrol)* soaking section *(of a pipe furnace)*
Sockel *m* base
Soda *f* 1. Na_2CO_3 soda, sodium carbonate; 2. *(min)* soda, natron, natrite *(sodium carbonate-10-water)*
  kalzinierte S. calcined soda, [soda] ash, anhydrous sodium carbonate
  kaustische S. caustic soda
  [kristall]wasserfreie S. *s.* kalzinierte S.
Sodaablauge *f* spent soda
Sodaalaun *m* soda (sodium) alum *(crystalline aluminium sodium sulphate)*
Sodaaufschluß *m (lab)* fusion with sodium carbonate; *(pap)* soda pulping
Sodaenthärtung *f* soda softening *(of water)*
Sodakalkglas *n* soda-lime glass
Sodakochung *f (pap)* soda cook
Sodalith *m (min)* sodalite *(a tectosilicate)*
Sodaschmelze *f (pap)* soda smelt, black ash
Sodastein *m* caustic soda
Sodaverfahren *n (pap)* soda process, soda [wood] pulp process
Sodawäsche *f (petrol)* soda washing
Sodawasser *n* soda [water], carbonated water
Sodazellstoff *m* soda pulp
Söderberg-Elektrode *f* Söderberg [continuous, self-baking] electrode
Söderberg-Elektrodenmasse *f s.* Söderberg-Masse
Söderberg-Masse *f* Söderberg paste

grüne (rohe) S. green [Söderberg] paste
Sofiaöl *n (cosmet)* ginger-grass oil *(chiefly from Cymbopogon martini (Roxb.) Stapf var. sofia)*
Sohlenimprägniermittel *n (tann)* sole impregnating agent
Sohlenkalander *m (rubber)* soling calender
Sohlenkrepp *m* sole crepe
Sohlenmischung *f (rubber)* soling compound
Sojabohne *f* soybean, soya[-bean], Glycine (L.)Merr.
Sojabohnenmehl *n* soybean flour
Sojabohnenöl *n* soy[bean] oil, Chinese bean oil
Sojaeiweiß *n* soybean protein
Sojaeiweißfaser *f (text)* soybean fibre
Sojaeiweißfaserstoff *m (text)* soybean fibre
Sojaeiweißleim *m* soybean glue
Sojahartfett *n* hydrogenated soybean oil
Sojalezithin *n* soybean lecithin
Sojamehl *n* soybean flour
Sojaöl *n s.* Sojabohnenöl
Sojaprotein *n* soybean protein
Sokotra-Aloe *f (pharm)* Socotra (Zanzibar) aloe *(from Aloe perryi Baker)*
Sol *n (coll)* sol
irresolubles (irreversibles) S. irreversible sol
lyophiles S. lyophilic sol
lyophobes S. lyophobic sol
resolubles (reversibles) S. reversible sol
Solanin *n* solanin[e] *(a steroid alkaloid)*
Solarisation *f (phot, glass)* solarization
Solaröl *n* solar oil
Solarstearin *n* solar stearin
Sole *f* [salt] brine, salt (saline) water
ablaufende S. spent brine
eutektische S. eutectic brine
verbrauchte S. spent brine
solehaltig briny
Soleil-Doppelplatte *f* biquartz *(polarimetry)*
Solekühler *m* brine cooler (refrigerator)
Sohlekühlung *f* brine cooling (refrigeration)
Sol-Gel-Übergang *m (coll)* sol-gel transformation
Solidensieren *n* desublimation *(direct conversion of a gas into the solid state)*
Solidgrün *n s.* Malachitgrün
Soliduskurve *f*, **Soliduslinie** *f* solidus [curve, line] *(of a melting diagram)*
Soliduspunkt *m* solidus point
Sollwert *m* set point
Solonetz *m (soil)* solonetz
Solontschak *m (soil)* solonchak
Solquelle *f* brine spring
Solubilisation *f*, **Solubilisierung** *f* solubilization
Solutizerverfahren *n* solutizer process *(for desulphurizing petroleum distillates)*
Solvat *n* 1. solvate *(solvent and solute or dispersed phase and dispersion medium)*; 2. raffinate *(less soluble residue remaining after solvent extraction)*
Solvatation *f* solvation
Solvatationseffekt *m* solvation effect
Solvatationsenergie *f* solvation energy
Solvatationsgrad *m* degree of solvation
Solvatationsreaktion *f* solvation reaction
Solvathülle *f* sheath of solvent molecules
solvatisieren to solvate
Solvatisierung *f* solvation
Solvatochromie *f* solvatochromism
Solvay-Verfahren *n* Solvay process, [Solvay's] ammonia soda process
Solvay-Zelle *f* Solvay cell *(a mercury cell)*
Solvens *n* [dis]solvent, dissolver
selektiv wirkendes S. selective solvent
Solvententölung *f (petrol)* solvent deoiling
Solvententparaffinierung *f (petrol)* solvent dewaxing
Solvententwachsung *f s.* Solvententparaffinierung
Solventextraktion *f* solvent (liquid-liquid) extraction
Solventnaphtha *n(f)* solvent naphtha *(a fraction of middle and high-boiling benzene hydrocarbons)*
Solventraffination *f* solvent refining
Solventtrocknung *f* solvent drying
Solvolyse *f* solvolysis, lyolysis
Solzustand *m (coll)* sol state (condition) ✦ **in den S. übergehen** to solate
Soman *n* soman, methyl-1,2,2-trimethylpropoxyfluorophosphine oxide
Somatropin *n* somatotropic hormone, somatropin
SO_2-Meßgerät *n* sulphur dioxide meter
Sommelet-Umlagerung *f* Sommelet rearrangement *(isomerization of quaternary benzylammonium compounds)*
sommern *(ceram)* to weather
Sommerspritzmittel *n (agric)* summer spray
Sommerspritzung *f (agric)* summer spraying
Sonde *f (petrol)* sonde *(in electrical coring)*
Sonderenergie *f (phys chem)* resonance energy
Sonderkoks *m* special coke
Sonderpapier *n* / **technisches** technical paper
Sonderpappe *f* special board
Sonderzement *m* special cement
Sonnenblumenhartfett *n* hydrogenated sunflower oil
Sonnenblumenöl *n* sunflower [seed] oil
Sonnenbräunungsmittel *n (cosmet)* suntan makeup, suntan preparation
Sonnendestillationsanlage *f* solar still (evaporator)
Sonnenenergie *f* solar energy
Sonnenlichtbeständigkeit *f* resistance to sunlight, sunlight resistance (stability)
Sonnenofen *m* solar furnace
Sonnenschutzmittel *n* sunscreen [agent]
Sonnenschutzöl *n* sunscreen oil
Sonnenspektrum *n* solar spectrum
Sonnenstein *m (min)* sunstone *(a tectosilicate)*
Sonnenstrahlung *f* solar radiation
Sonnentrocknung *f* solar (sun) drying
Sonolumineszenz *f* sonoluminescence, sonic luminescence
Sophia-Jacoba-Verfahren *n (coal)* Barvoys process

Sophorin *n* cytisine, sophorine *(a lupine alkaloid)*
Sorbend *m* sorbate
Sorbens *n* sorbent [material]
sorbieren to sorb
Sorbinsäure *f* $CH_3CH=CHCH=CHCOOH$ sorbic acid, 2,4-hexadienoic acid
Sorbit *m* 1. $CH_2OH[CHOH]_4CH_2OH$ sorbitol *(a sugar alcohol)*; 2. *(met)* sorbite
sorbitisch *(met)* sorbitic
Sorbose *f* sorbose *(a monosaccharide)*
Sorelzement *m* Sorel cement
Sörensen-Titration *f* Sörensen's formol titration
Sorosilikat *n (min)* sorosilicate *(any of a class of polymeric silicates)*
Sorption *f* sorption
Sorptionskapazität *f (soil)* total exchangeable bases
Sorptionskomplex *m (soil)* exchange complex
Sorptionskurve *f* sorption curve
Sorptionsmittel *n* sorbent [material]
Sorptionswaage *f* sorption balance
Sorptiv *n* sorbate
Sorte *f* grade *(as of chemicals)*
Sortierblech *n (pap)* screen plate
sortieren to sort, to grade; *(min tech)* to classify, to sort, to separate, *(by hand)* to pick, *(according to diameter)* to size
Sortiergut *n* material being (*or* to be) sorted; *(according to diameter)* material being (*or* to be) sized
Sortiermaschine *f* grading machine, grader
Sortierplatte *f (pap)* screen plate
Sortiersaal *m (pap)* [rag-]sorting room
Sortierung *f* sorting, grading; *(min tech)* classification, sorting, separation, *(by hand)* picking, *(according to diameter)* sizing
S. der Hackschnitzel *(pap)* chip screening
S. des Bogenpapiers *(pap)* sheet sorting
S. von Hand hand sorting, *(min tech also)* picking
statistische S. *(pap)* sampling
SO_2-Rückgewinnung *f (pap)* sulphur dioxide recovery
Souple *m*, **Soupleseide** *f* souple silk
souplieren *(text)* to souple
SO_2-Verlust *m (pap)* sulphur dioxide loss
Soxhlet[-Apparat] *m*, **Soxhlet-Extraktor** *m* Soxhlet [extractor]
Spachtel *m(f) s.* Spachtelmasse
Spachtellack *m* flatting varnish
Spachtelmasse *f* filler, surfacer
Spallation *f (nucl)* spallation
Spalt *m* slit, *(between rolls)* gap; crack, crevice, fissure
Spaltanlage *f (petrol)* cracking plant; *(agric)* split plot design *(as for fertilizer testing)*
Spaltausbeute *f (nucl)* fission yield
Spaltausbeutekurve *f (nucl)* fission yield curve
spaltbar *(min, cryst)* cleavable, fissile; *(nucl)* fissile, fissionable; divisible, resolvable *(optically active compounds)*
Spaltbarkeit *f (min, cryst)* cleavability, fissility; *(nucl)* fissility, fissionability; divisibility, resolvability *(of optically active compounds)*
unvollkommene S. *(min, cryst)* imperfect cleavage
vollkommene S. *(min, cryst)* perfect cleavage
Spaltbenzin *n* cracked gasoline
Spalte *f s.* Spalt
Spaltebene *f (min, cryst)* cleavage plane (surface, face)
Spalteinfang *m (nucl)* fission capture
spalten to cleave, to crack, to break down, to decompose *(chemical compounds)*; to cleave, to break, to crack, to split *(a chemical bond)*; *(nucl)* to fission, to split; to break, to crack, to demulsify *(emulsions)*; to resolve *(a racemate)*; to cut *(a bale of rubber)*
katalytisch s. to cat-crack
sich s. to crack, to break down, to decompose, to split up *(of chemical compounds)*; *(nucl)* to split, to break up, to [undergo] fission
Spaltenergie *f (nucl)* fission energy
Spaltereignis *n (nucl)* fission event
spaltfähig *s.* spaltbar
Spaltfaser *f* split fibre
Spaltfestigkeit *f (plast)* interlaminar strength
Spaltfilter *n* edge filter
Spaltfläche *f* cleavage plane (surface, face)
Spaltgift *n (nucl)* fission poison
Spaltgut *n (petrol)* cracking feed (stock, feedstock)
Spaltkette *f (nucl)* fission [decay] chain
Spaltkorrosion *f* crevice corrosion
Spaltmaschine *f (tann)* splitting machine
Spaltmaterial *n (nucl)* fissionable material
Spaltmethode *f* method of resolution *(stereochemistry)*
Spaltprodukt *n* cleavage (breakdown) product; *(nucl)* fission product
Spaltproduktausbeute *f (nucl)* fission yield
Spaltproduktausbeutekurve *f (nucl)* fission yield curve
Spaltproduktreihe *f (nucl)* fission [decay] chain
Spaltprozeß *m s.* Spaltungsreaktion
Spaltrohrpumpe *f* canned-motor pump
Spaltruß *m* **[/ thermischer]** thermal [carbon] black, thermal-decomposition black, furnace thermal black
Spaltsieb *n* wedge-wire screen
Spaltstoff *m (nucl)* fissionable material
Spaltstoffelement *n (nucl)* fuel element
Spaltstück *n* fragment, *(nucl also)* fission fragment
Spaltultramikroskop *n* slit ultramicroscope
Spaltung *f* cleavage, cracking, breakdown, fission, decomposition *(of chemical compounds)*; cleavage, breaking, splitting[-up], fission, scission *(of chemical bonds)*; breaking, cracking, demulsification *(of emulsions)*; *(nucl)* fission; resolution *(of a racemate)*
S. durch schnelle Neutronen high-energy fission
enzymatische (fermentative) S. enzymatic de-

composition
homolytische S. homolytic cleavage
hydrierende S. *(petrol)* hydrogenation cracking, hydrocracking
hydrogenolytische S. hydrogenolysis
hydrolytische S. hydrolytic cleavage (decomposition)
ionische S. *s.* katalytische S.
katalytische S. catalytic (cat, catalyst) cracking
radikalische S. *s.* thermische S.
schnelle S. *(nucl)* high-energy fission
selektive S. selective cracking
thermische S. thermal decomposition; *(petrol)* thermal cracking
Spaltungseinfang *m (nucl)* fission capture
Spaltungsenergie *f (nucl)* fission energy
Spaltungskammer *f (nucl)* fission chamber
Spaltungsprodukt *n s.* Spaltprodukt
Spaltungsquerschnitt *m (nucl)* fission cross section
Spaltungsreaktion *f* cleavage reaction; *(nucl)* fission reaction
Spaltverfahren *n (petrol)* cracking process
S. nach Burton Burton [cracking] process
S. nach de Florez deFlorez [cracking] process
S. nach Holmes und Manley Holmes-Manley [cracking] process
katalytisches S. catalytic-cracking process
selektives S. selective-cracking process
thermisches S. thermal-cracking process
Spaltvorgang *m (petrol)* cracking process; *(nucl)* fission process
Spaltzone *f (petrol)* cracking zone; *(nucl)* core
Span *m* splinter, chip, shaving, *(esp of wood)* sliver
Spänehaus *n (pap)* chip loft
Spanholzplatte *f* chipboard
Spanischweiß *n* Spanish (bismuth, pearl) white, cosmetic bismuth *(bismuth nitrate oxide or bismuth chloride oxide)*
Spannbacke *f (lab)* gripping jaw *(of an apparatus clamp)*
Spannbeton *m* prestressed concrete
Spanndruck *m (plast)* [mould-]locking pressure, clamping pressure
Spanneinrichtung *f* tensioning device *(as of a conveyor)*
spannen to strain, to stress, *(esp in length)* to stretch
Spannrahmen *m (plast, text)* tenter [frame], stenter, *(plast also)* clamping frame
Spannrahmentrockner *m* tenter [frame] dryer
Spannring *m* locking ring
Spannrolle *f* tension pulley
Spanntrommel *f* tension pulley
Spannung *f* 1. *(mechanically)* strain, stress, tension; 2. *(el chem)* voltage; 3. strain *(of ring molecules)*
eingefrorene S. *(plast)* frozen-in stress
kritische S. sparking potential *(as in electrical precipitation)*
Spannungsabfall *m* drop in voltage
spannungsarm low-strain
Spannungs-Dehnungs-Diagramm *n* stress-strain diagram; load-elongation (load-extension) diagram
Spannungs-Dehnungs-Linie *f* stress-strain curve *(expressed as tons or pounds per squ.in.)*; load-elongation (load-extension) curve *(expressed as inches per inch)*
spannungsfrei strainless, strain-free, *(org chem also)* free from angle strain ✦ **s. machen** *(glass, plast)* to temper, to anneal
Spannungskorrosion *f* stress corrosion
spannungslos *(el chem)* dead
spannungsoptisch stress-optical
Spannungsreihe *f* / **elektrochemische** electrochemical series, electromotive [force] series
Spannungsrelaxation *f (rubber)* stress relaxation
Spannungsrelaxationsverfahren *n (rubber)* stress relaxation method
Spannungsrißbildung *f* stress cracking
Spannungsrißkorrosion *f* environmental stress cracking
Spannungsrückgang *m* drop in voltage
Spannungsscheibe *f (glass)* strain disk
Spannungstheorie *f* theory of strain
Baeyersche S. Baeyer strain (angle strain, tension) theory
Sachse-Mohrsche S. Sachse-Mohr concept of strainless rings
Spannungsverlust *m* loss of tension; *(el chem)* loss of voltage
Spannungswert *m (rubber)* [tensile] modulus
Spannungswertcharakteristik *f* / **ansteigende** *(rubber)* marching modulus
Spannwalze *f (pap)* stretch (stenting, tension, hitch) roll, tightener
Spannweite *f* support span
Spanplatte *f* particle board, chipboard
Spanpressen *n (text)* paper pressing
Sparbeton *m* lean (lean-mixed, poor) concrete
Sparflamme *f* small flame
Sparkapsel *f (ceram)* crank *(a type of support in kilns)*
Spartein *n* sparteine *(a lupine alkaloid)*
Sparträn kverfahren *n* **nach Rüping** Rueping process *(wood preservation)*
Spasmolytikum *n (pharm)* spasmolytic, antispasmodic
spasmolytisch *(pharm)* spasmolytic
spatartig *(min)* spathic, spathose
Spatel *m (lab)* spatula
Spatelspitze voll / **eine** a spatula-tipfull
Spatenwischer *m (lab)* wing-shape policeman
spatig *s.* spatartig
Specköl *n* grease (lard) oil
Speckstein *m (min)* soapstone, steatite *(a variety of talc)*
Speerkies *m (min)* spear pyrite *(a variety of marcasite)*
Speichel *m* saliva
Speicher *m* store[house], storage, reservoir; accumulator *(hydraulics)*

unterirdischer S. underground reservoir (storage tank)
Speicherbehälter *m* storage tank (vessel), hold tank
Speichergestein *n* reservoir rock, carrier bed, pool *(a sedimentary rock containing petroleum or gas)*
Speichergewebe *n (biol)* storage tissue
speichern to store [up], to accumulate
Speichertank *m s.* Speicherbehälter
Speicherung *f* storage, accumulation
unterirdische S. underground storage
Speirohr *n* spout
Speise *f* 1. food; 2. *(met)* speiss, speise
Speisebehälter *m* feed tank
Speiseboden *m (distil)* feed tray (plate)
Speiseeis *n* ice cream
Speiseeisbereitung *f* ice-cream making
Speiseessig *m* table vinegar
Speisefett *n* edible fat
Speisegelatine *f* edible gelatin
Speisehartfett *n* shortening *(for baked goods)*
Speiseleitung *f* supply line
speisen 1. *(tech)* to feed, to charge; 2. to power *(an electrical circuit)*
Speiseöl *n* edible oil
Speisepumpe *f* feed pump
Speiser *m* feeder
Speiserbecken *n (glass)* feeder bowl (nose), [feeder] spout
Speiserkanal *m (glass)* feeder channel
Speiserkopf *m s.* Speiserbecken
Speisermaschine *f (glass)* gob-fed machine
Speiserring *m (glass)* orifice ring
Speiserrinne *f (glass)* feeder channel
Speiserrohr *n (glass)* feeder sleeve (tube)
Speisertropfen *m (glass)* [glass] gob
Speiserverfahren *n (glass)* gob (feeder) process
Speisesalz *n* table salt
Speisesenf *m* prepared mustard
Speisesirup *m* edible (table) syrup
Speisetalg *m* edible tallow
Speisewasser *n* feed water
Speisewasserverdampfer *m* boiler make-up evaporator
Speisewasservorwärmer *m* boiler feed preheater, feed-water heater, economizer
Speisezone *f (plast)* feed zone (section)
Speisezwecke / für for edible purposes
Speiskobalt *m s.* Skutterudit
Speisung *f* feed[stock], charge, charging stock
spektral spectral
Spektralanalyse *f* spectral (spectroscopic) analysis
Spektralbereich *m* spectral region (range)
kurzwelliger S. short-wavelength region
langwelliger S. long-wavelength region
Spektralfarbe *f* spectral colour
Spektralfotometer *n* spectrophotometer
Bracesches S. Brace-Lemon spectrophotometer
Spektralfotometrie *f* spectrophotometry
spektralfotometrisch spectrophotometric
Spektralgebiet *n s.* Spektralbereich
Spektrallinie *f* spectral (spectrum) line
Spektralliniendublett *n* doublet
spektralrein spectroscopically (spectrally) pure
Spektralserie *f* spectral series
Spektralterm *m* spectral (series) term
spektrochemisch spectrochemical
Spektrofotometer *n* spectrophotometer
Spektrofotometrie *f* spectrophotometry
spektrofotometrisch spectrophotometric
Spektrograf *m* spectrograph
spektrografisch spectrographic
Spektrogramm *n* spectrogram
Spektrometer *n* spectrometer
Braggsches S. Bragg spectrometer
Spektrometrie *f* spectrometry
Spektropolarimeter *n* spectropolarimeter
Spektroskop *n* spectroscope
geradsichtiges S. direct-vision spectroscope
Spektroskopie *f* spectroscopy
spektroskopisch spectroscopic[al]
Spektrum *n* spectrum
diskretes S. discrete spectrum
kontinuierliches S. continuous spectrum
optisches S. optical spectrum
sichtbares S. visible spectrum
α-**Spektrum** *n* alpha-particle spectrum, α spectrum
β-**Spektrum** *n* beta-ray spectrum, β spectrum
Spekularit *m (min)* specularite, specular iron [ore] *(a variety of haematite)*
spenden to donate *(e.g. electrons)*
Spermazet[i]öl *n s.* Spermöl
spermienabtötend, spermizid *(pharm)* spermatocidal, spermicidal
Spermöl *n* sperm [whale] oil
mineralisches S. mineral sperm [oil]
Sperre *f* arrest device
sperren to arrest, to lock, to block; *(coat)* to seal
Sperrflüssigkeit *f* sealing liquid (fluid)
Sperrflüssigkeitsdichtung *f* liquid-buffered seal
Sperrgrund *m (coat)* sealing paint, sealer
Sperrhahn *m* stopcock
Sperrholz *n* plywood
Sperrholzpappe *f* plywood board
sperrig bulky
Sperring *m* locking ring
Sperröldichtung *f* oil-buffered seal
Sperrschicht *f* barrier layer, interlining
Sperrschichteffekt *m* photovoltaic effect
Sperrschichtelement *n* photovoltaic (photochemical, photoelectrolytic, photobarrier, barrier-layer) cell
Sperrschichtfotoeffekt *m* photovoltaic effect
Sperrschichtfotoelement *n*, **Sperrschichtfotozelle** *f s.* Sperrschichtelement
Sperrschichtgleichrichter *m* barrier-layer rectifier
Sperrschichtzelle *f s.* Sperrschichtelement
Sperrvorrichtung *f* arrest device
Spessartin *m (min)* spessartite, spessartine *(aluminium manganese(II) orthosilicate)*
Spezialdünger *m* specialty fertilizer

Spezialkoks *m* special coke
Spezialkokskohle *f* / **bituminöse** prime coking coal
Spezialpapier *n* special paper
Spezialpappe *f* special board
Spezialwirksamkeit *f* selectivity *(as of pesticides)*
Spezialzement *m* special cement
Spezifikation *f* specification
Spezifität *f* / **stereochemische** stereospecificity, stereoselectivity
Sphagnumtorf *m* sphagnum peat
Sphalerit *m (min)* sphalerite, zinc blende, *(Am also)* black jack
sphärisch spheric[al]
Sphärizität *f* sphericity
sphäroidal spheroidal
Sphärokolloid *n* spherocolloid, globular colloid
Sphärolith *m (geol)* spherulite
sphärolithisch *(geol)* spherulitic
Sphäroprotein *n* globular protein
Sphen *m (min)* sphene *(a variety of titanite)*
Sphingolipoid *n (bioch)* sphingolipid[e]
Sphingomyelin *n* sphingomyelin *(any of a group of crystalline phosphatides)*
Sphingosin *n* sphingosine *(an amino alcohol)*
sp-Hybrid *n* sp hybrid *(chemical-bond theory)*
sp-Hybridisierung *f* sp hybridization *(chemical-bond theory)*
Spiegel *m* level *(of a liquid)*
Spiegelamalgam *n* mirror amalgam *(consisting of 77% Hg and 23% Sn)*
Spiegelbild *n* mirror image *(stereochemistry)*
spiegelbildisomer enantiomorphous, enantiomorphic, enantiomeric
Spiegelbildisomer[es] *n* enantiomorph, enantiomer, enantiomorphous form (isomer), optical antipode (isomer), antimer, mirror-image isomer
Spiegelbildisomerie *f* enantiomorphism, optical isomerism, mirror-image isomerism (relationship)
Spiegelebene *f (cryst)* plane of symmetry, mirror plane
Spiegeleisen *n* spiegel [iron], spiegeleisen, mirror iron *(a pig iron containing manganese)*
Spiegelfleck *m (phot)* flare *(on a negative)*
Spiegelgalvanometer *n* mirror galvanometer
Spiegelglas *n* [polished] plate glass
Spiegelglaswanne *f* plate-glass furnace
Spiegellackierung *f* mirror varnishing
Spiegelmethode *f* **nach Paneth** Paneth's mirror method, Paneth technique *(for detecting free radicals)*
spiegeln to reflect [back]
Spiegelreflexion *f* mirror reflection
Spiegelung *f* reflection
Spielkartenkarton *m* playing cardboard
Spiköl *n* [lavender-]spike oil, aspic (Spanish lavender) oil *(from Lavandula latifolia (L.fil.)Medik.)*
Spin *m* spin, *(of elementary particles also)* [intrinsic] angular momentum
nichtkompensierter S. uncoupled spin
nichtverschwindender (von Null verschiedener) S. non-zero spin
spinabhängig spin-dependent
Spinauslöschung *f* quenching of orbital angular momentum
Spindel *f* 1. araeometer, hydrometer; 2. screw, stem *(as of a valve)*
spindelbetätigt screw-operated
Spindelöl *n* spindle oil
Spindelpumpe *f* screw pump
Spinell *m* 1. *(min)* spinel *(magnesium aluminate)*; 2. spinel *(any of a class of aluminates of the type* $M^{II}Al_2O_4$*)*
Spinmethode *f* spin (electron-pair, valence-bond, VB, spin-state) method, Heitler-London-Slater-Pauling (HLSP) method
Spinmoment *n* spin moment
Spinmultiplizität *f* spin multiplicity
Spinnbad *n* spinning bath
Spinnband *n (text)* slubbing
Spinnbrause *f s.* Spinndüse
Spinndüse *f (text)* spinneret, spinning jet (shower)
spinnen to spin *(yarn)*
Spinnen *n* **aus der Schmelze** melt spinning (extrusion)
S. aus Lösungen solution (solvent) spinning
Spinnfaden *m (text)* strand
Spinnfärbung *f* spin (dope) dyeing
Spinnfaser *f (text)* [staple] fibre *(a natural or man-made object of relatively short length)*
spinngefärbt *(text)* spin-dyed, spun-dyed, dope-dyed
Spinnkabel *n (text)* tow
Spinnkanne *f (text)* spinning can
Spinnkopf *m* spinning head
Spinnkuchen *m* cake *(in a centrifuge)*
Spinnlösung *f (text)* spinning solution, dope
Spinnmaschine *f* spinning machine
spinnmattiert *(text)* dull-spun
Spinnpapier *n* spinning paper
Spinnpumpe *f* spinning (metering) pump
Spinnroving *m* spun roving
Spinnschacht *m* spinning cabinet (cell, tube)
Spinnstrecken *n* spinning stretch
Spinntisch *m* spinning table
Spinntopf *m* spinning pot (vessel), Topham box
Spinntopfverfahren *n* [centrifugal] pot spinning
Spinnviskosität *f* dope viscosity
Spinnwert *m* / **technischer** spinnability
Spinnzentrifuge *f s.* Spinntopf
Spinnzwiebel *f* [spinning] bulb
Spinoperator *m* spin operator
Spinorbital *n* spin orbital, SO
molekulares S. *s.* Molekülspinorbital
Spinquantenzahl *f* spin quantum number
Spin-Spin-Wechselwirkung *f* spin-spin interaction
Spinthariskop *n*, **Spintheriskop** *n* spinthariscope *(a device for investigating* α *rays)*
Spiralabscheider *m* dust collecting fan, fan impeller-type collector

Spiralausbreiter *m*, **Spiralbreithalter** *m (text)* worm expander
Spirale *f* 1. spiral, coil, helix; 2. *(distil)* beehive *(a device for retaining packing in position)*; spiral tile *(a kind of packing)*; 3. vortex *(of fluid as in a hydrocyclone)*
spiralförmig spiral
Spiralgehäuse *n* volute casing *(of pumps)*
Spiralgehäusepumpe *f* volute pump
spiralig spiral
Spiralklassierer *m* spiral classifier
Spiralkühler *m* **[nach Friedrichs]** Friedrichs condenser
Spiralrohrbündelwärme[aus]tauscher *m*, **Spiralrohrbündelwärmeübertrager** *m* spiral-tube heat exchanger
Spiralrohrschlange *f* spiral coil
Spiralscheider *m s.* Spiralabscheider
Spiralschlauch *m (rubber)* wire-spiral-reinforced hose
Spiraltest *m (plast)* spiral flow test
Spiralwärme[aus]tauscher *m*, **Spiralwärmeübertrager** *m* spiral[-plate heat] exchanger
Spiran *n (org chem)* spiran[e], spiro-compound, spiro hydrocarbon
Spiranringsystem *n (org chem)* spiro-ring system
Spiranverbindung *f s.* Spiran
spirituos, spirituös alcoholic, spirituous
Spirituosen *pl* spirits
Spiritus *m* spirit[s] *(industrially manufactured ethanol)*
  S. absolutus dehydrated alcohol
  S. denaturatus *s.* vergällter S.
  denaturierter S. *s.* vergällter S.
  vergällter S. methylated spirit, denatured alcohol
Spiritusbeize *f (coat)* spirit stain
Spiritusbrenner *m (lab)* alcohol burner
Spiritusbrennerei *f* alcohol plant, distillery
Spiritusindustrie *f* distilling industry
Spirituslack *m* spirit varnish
Spirituslampe *f (lab)* alcohol lamp
Spiroatom *n (org chem)* spiro atom
Spirobindung *f* spiro union (junction) *(of atoms)*
Spirostellung *f (org chem)* spiro position
Spiroverbindung *f (org chem)* spiro-compound, spiran[e]
Spiroverknüpfung *f s.* Spirobindung
Spitze *f* top; vertex *(as of a tetrahedral molecule)*; ✦ **bis zur S. geteilt** calibrated to jet *(pipette)*
Spitzentemperatur *f* peak temperature
Spitzenwert *m* extreme [value], extremum
Spitzigfärben *n* tippy dyeing
Spitzigkeit *f (dye)* tippiness
Spitztrichter *m* cone classifier
SP-Kautschuk *m* superior processing rubber, S.P. rubber
Splint *m*, **Splintholz** *n* sapwood
Splintkohle *f* splint coal
Splitt *m* grit
Splitter *m* 1. splinter, chip, sliver; 2. *(pap)* shim; wood speck *(a defect in paper)*
Splittereis *n* flake (shaved, chipped) ice
Splitterfänger *m (pap)* sliver (shim, bull) screen
splitterfest shatterproof *(glass)*
Splitterfestigkeit *f* shatter resistance (strength) *(of glass)*
Splitterkohle *f* splint coal
splittern to shatter, to spall
splittersicher shatterproof *(glass)*
Splittersicherheit *f* shatter resistance (strength) *(of glass)*
Spodium *n* spodium, bone char[coal]
Spodumen *m (min)* spodumene *(aluminium lithium silicate)*
spontan spontaneous
Spontangärung *f* spontaneous (wild) fermentation, autofermentation
Spontansäuerung *f (food)* spontaneous (natural) souring
Spontanzündung *f* spontaneous ignition
Sporinit *m* sporinite *(a maceral of coal)*
spratzen *(met)* to sputter, to spatter, to spit
spreitbar spreadable
Spreitbarkeit *f* spreadability
spreiten to spread [out]
Spreitungsdruck *m* spreading pressure
Spreitungskoeffizient *m* spreading coefficient, SC
Spreitungsvermögen *n* spreading ability
Sprengel-Pumpe *f* Sprengel pump
sprengen 1. to break, to crack, to rupture *(a chemical bond)*; 2. to blast *(by means of explosives)*; 3. to water
Sprengfähigkeit *f* explosibility, explosiveness, explosivity
Sprenggelatine *f* blasting (explosive) gelatin, gelatin dynamite
Sprengkapsel *f* blasting (detonating) cap, primer, detonator
Sprengkraft *f* explosive power (force), shattering power, brisance
Sprengladung *f* blasting (explosive) charge
Sprengluft *f* oxyliquit *(an explosive)*
Sprengmittel *n* blasting agent
Sprengöl *n* blasting (explosive) oil *(glycerol trinitrate or ethyleneglycol dinitrate or a mixture of both)*
Sprengpulver *n* blasting (black) powder, gunpowder
Sprengpunkt *m (petrol)* shot point
Sprengsalpeter *m* explosive saltpetre
Sprengschnur *f* detonating fuse
Sprengstelle *f (petrol)* shot point *(in refraction shooting)*
Sprengstoff *m* explosive
  brisanter S. high explosive, H.E.
  initiierender S. initiating explosive, initiator
Sprengstoffeigenschaften *fpl* explosive properties
Sprengstoffgemisch *n* explosive mixture
Sprengstoffladung *f s.* Sprengladung
Sprengung *f* 1. breaking, cracking, rupture, fission *(of a chemical bond)*; 2. blast[ing], detonation, shot

S. der C–C-Bindung C–C [bond] rupture
Sprengwirkung *f* explosive action (effect)
Sprenkpapier *n* marble[d] paper
springen to burst, to crack, to break
Springkitt *m* bouncing putty
Sprit *m s.* Spiritus
Spritbeize *f (coat)* spirit stain
Spritblau *n* spirit (aniline) blue *(triphenylrosaniline hydrochloride)*
Spritgelb *n* spirit yellow G, solvent yellow 1 *(p-aminoazobenzene)*
spritlöslich spirit-soluble, alcohol-soluble
Spritzartikel *m (plast)* extruded article; injection-moulded article
Spritzauftrag *m* spray coat; splash feed *(in roller dryers)*
Spritzbarkeit *f (plast)* extrudability, extrusion property; injecting property *(injection moulding)*
Spritzbelag *m (agric)* spray residue (load)
Spritzbeton *m* gunned concrete, gunite
Spritzbrett *n (pap)* spatter (baffle) board
Spritzbrühe *f (agric)* mixture for spraying, wash
Spritzdorn *m (plast)* extruder core
Spritzdruck *m* spraying pressure; *(plast)* injection pressure
S. beim Spritzgießen *(plast)* injection moulding pressure
Spritzdüse *f* spray nozzle; *(plast)* injection nozzle
Spritzeigenschaften *fpl (plast)* extrusion properties; injecting properties *(injection moulding)*
spritzen 1. to spray; 2. *(of a liquid)* to splash, to spatter, to sputter; 3. *(plast)* to extrude; to inject; 4. *(coat)* to spray
Spritzen *n* 1. spraying; 2. splashing, spattering, sputtering *(of a liquid)*; 3. *(plast)* extruding, extrusion [moulding]; injection [moulding]; 4. *(coat)* spraying, spray painting; 5. *(agric)* high-volume spraying *(application of pesticides with abundant water as a carrier and dispersant)*
druckluftfreies (druckluftloses) S. *(coat)* airless spraying
elektrostatisches S. *(coat)* electrostatic spraying
luftloses S. *s.* druckluftfreies S.
Spritzer *m* splash
Spritzfärbung *f (tann)* spray dyeing
Spritzfehler *m (plast)* injection defect
Spritzflasche *f (lab)* wash[ing] bottle
S. aus Weichplast squeeze bottle
Spritzflüssigkeit *f* spray, *(agric also)* plant spray
Spritzgehäuse *n (plast)* cylinder *(of an extruder)*; barrel *(of an injection moulding machine)*
Spritzgerät *n (agric)* spray machine, sprayer
Spritzgeräte *npl (agric)* spray equipment
Spritzgießen *n (plast)* injection moulding
Spritzgießer *m (plast)* injection moulder
Spritzgießmaschine *f (plast)* injection moulding machine
S. mit Schneckenkolben screw injection [moulding] machine
S. mit Schneckenvorplastizierung screw pre-plasticizing machine
Spritzgießteil *n s.* Spritzgußteil
Spritzgießwerkzeug *n (plast)* injection mould
S. mit Zentralanguß centre-gated mould
angußloses S. runnerless mould
Spritzgießzylinder *m (plast)* injection (plasticating, shooting) cylinder
Spritzguß *m (plast)* injection moulding
Spritzgußform *f (plast)* injection mould
Spritzgußkunststoff *m* injection-moulded plastic
Spritzgußmasse *f* injection-moulding material, injection compound
Spritzgußteil *n (plast)* injection moulding, injection-moulded part
unvollständiges S. short shot
Spritzgußverfahren *n* injection-moulding process
Spritzgußwerkzeug *n (plast)* injection mould
Spritzhilfsmittel *n (agric)* auxiliary (supplementary) spray material, spray supplement
Spritzkabine *f (ceram)* spray booth
Spritzkammer *f (plast)* plenum chamber
Spritzkolben *m (plast)* injection plunger (piston, ram) *(injection moulding)*; pot plunger *(transfer moulding)*
Spritzkonzentrat *n (agric)* spray concentrate
Spritzkonzentration *f (agric)* spray strength
Spritzkopf *m (plast)* extruder (extrusion, die) head, die box
spritzlackieren to spray, *(relating to wooden material also)* to varnish by spraying, to spray-varnish
Spritzlackieren *n* / **elektrostatisches** electrostatic spraying
Spritzling *m (plast)* injection moulding (moulded part)
Spritzmaschine *f* 1. *(plast)* extruding (forcing) machine, extruder; injection-moulding machine; 2. *(ceram)* extrusion auger
Spritzmetallisieren *n* [spray, Schoop] metallizing, metallization, schooping, metal spraying
Spritzmittel *n* spray, *(agric also)* plant spray
fungizides S. fungicide spray
Spritzmundstück *n (plast)* extruder (extrusion) die
Spritznarben *fpl (coat)* orange peel, bad flow
Spritznebel *m (coat)* spray dust
Spritzniederschlag *m (agric)* spray residue (load)
Spritzöl *n (agric)* spray oil
Spritzpfahl *m* injector gun, soil injector *(for applying liquid pesticides)*
Spritzpistole *f (coat)* spray[ing] gun, paint sprayer (spray gun); schooping gun *(for metal spraying)*
elektrostatische S. electrostatic spray gun
Spritzplastometer *n* **nach Dillon-Firestone** *(rubber)* Firestone[-Dillon] plastometer
Spritzpresse *f (plast)* transfer moulder
Spritzpressen *n (plast)* transfer (flow) moulding, *(Am also)* plunger molding
Spritzpreßkolben *m* pot plunger

Spritzpreßmaschine *f s.* Spritzpresse
Spritzpreßteil *n (plast)* transfer moulding
Spritzpreßwerkzeug *n (plast)* transfer mould
Spritzpulver *n (agric)* wettable powder
Spritzquellung *f (plast)* swelling from the die, swelling after extrusion, die swell
Spritzrohrfeuchter *m (pap)* spray damper
Spritzrückstand *m (agric)* spray residue (load)
Spritzschaden *m (agric)* spray injury
Spritzschleier *m (agric)* spray swathe
Spritzstreichen *n (pap)* spray coating
Spritzstreichmaschine *f (pap)* spray coater
Spritzteil *n s.* Spritzgußteil
Spritzteller *m* splash plate
Spritztopf *m (plast)* transfer pot
Spritztorf *m* hydro peat
Spritzung *f (agric)* high-volume spraying *(application of pesticides with abundant water as a carrier and dispersant)*
Spritzverlust *m* splashing loss
Spritzvolumen *n (plast)* injection capacity
Spritzwasser *n (pap)* shower (flush) water
Spritzwasserbehälter *m (pap)* shower supply tank
Spritzwasserfalle *f* steam trap *(of an evaporator)*
Spritzwasserpumpe *f (pap)* shower pump
Spritzwechsel *m (agric)* spray rotation *(systematic change of pesticides applied)*
Spritzwinkel *m* angle of spray
Spritzzylinder *m (plast)* injection (shooting, plasticating) cylinder
Sprödbruch *m* brittle failure
Sprödbruchtemperatur *f* brittle (brittle-point, brittleness) temperature
spröde brittle ✦ **s. machen** to embrittle, to make brittle ✦ **s. werden** to embrittle, to become brittle
Spröde *f* brittleness
Sprödglimmer *m* brittle mica
Sprödigkeit *f* brittleness
Steadsche S. Stead's brittleness
Sprödigkeitspunkt *m* brittle point
Sprödwerden *n* embrittlement
Sprout-Waldron-Mühle *f (pap)* Sprout-Waldron refiner
Sprudel *m* sparkling water
sprudeln to bubble, to effervesce
Sprudeln *n* effervescence
sprudelnd effervescent, effervescing
Sprühbeschichten *n (plast)* spray coating
elektrostatisches S. electrostatic spray coating
Sprühblaser *m (agric)* low-volume mist blower, pneumatic (air-blast) sprayer
Sprühdose *f* aerosol bomb, pocket sprayer
Sprühdraht *m* ionizer wire *(as of an electrical precipitator)*
Sprühdüse *f* spray (atomizing) nozzle
Sprühdüsenkühler *m* spray-type cooler
Sprüheinrichtung *f* sprayer, atomizer
rotierende S. spinning atomizer
Sprühelektrode *f* discharge (ionizing, ionic) electrode
sprühen to spray
Sprühen *n* / **elektrostatisches** electrostatic spraying
Sprüher *m* sprayer, atomizer; chromatosprayer *(chromatography)*
Sprühflüssigkeit *f* spray
Sprühgefrierverfahren *n (food)* spray freezing process
Sprühgerät *n (agric)* spray machine, [crop] sprayer
Sprühgeräte *npl* spray equipment
Sprühkautschuk *m* [latex-]sprayed rubber
Sprühkegel *m* spray cone
Sprühkegelwinkel *m* angle of spray
Sprühkorb *m* rotary-cup atomizer
Sprühkörper *m (petrol)* shower-deck tray
Sprühkristallisation *f* prilling
sprühkristallisieren to prill
Sprühmilchpulver *n* spray-dried milk [powder], spray powder milk
Sprühmittel *n* spray, *(agric also)* plant spray
S. für den Gartenbau horticultural spray
S. gegen Fliegen fly spray
Sprühnebel *m (agric)* spray mist
Sprühnebler *m s.* Sprühblaser
Sprühparaffin *n* spray (light liquid) paraffin
Sprühpistole *f* spray[ing] gun
Sprühreagens *n* spray reagent
Sprührohr *n* spray pipe
Sprührückstand *m (agric)* spray residue (load)
Sprühsättiger *m* spray-type ammonia absorber
Sprühscheibe *f* spray disk, rotating-disk atomizer
S. mit Schaufeln vaned-disk atomizer
Sprühschleier *m (agric)* spray swathe
Sprühstrahl *m (agric)* directed spray
Sprühtank *m* spray tank
Sprühteller *m* splash plate
Sprühtrockner *m* spray (flash) dryer
Sprühtrocknung *f* spray (flash) drying
Sprühturm *m* spray tower (column, washer)
Sprühwasser *n* spray[ing] water
Sprühwinkel *m* angle of spray
Sprung *m* 1. crack; *(ceram)* flaw; 2. crack of the grain *(a property of leather relating to its elasticity)*
Sprungpunkt *m*, **Sprungtemperatur** *f* superconductive (transition) temperature *(of a superconductor)*
spucken to prime *(of kettles or distillation columns)*, *(distil also)* to puke
Spuckgrenze *f (distil)* flooding point
Spuckstoff *m (pap)* groundwood (screen) rejects, screen[ing]s, tail[ing]s, coarse (waste) material, waste [product], rejected stock *(classification of mechanical pulp)*
Spülapparat *m* rinser
Spülbad *n* rinsing (scouring) bath
Spülbecken *n* [bench] sink
Spülboden *m (filtr)* scavenger plate
Spüldampf *m (petrol)* stripping (purge) steam
Spüldüse *f* scour (sluicing) nozzle
Spule *f (text)* bobbin

keglige S. cone
spulen *(text)* to wind
spülen to rinse, to scour, to scavenge, to wash, to swill, to flush
Spulenspinnverfahren *n* bobbin spinning
Spüler *m* rinser
Spülflüssigkeit *f* rinsing (purging, scouring) liquid, rinsing[s]
Spülflüssigkeitsschieber *m* rinse (scour) valve
Spülgas *n* purge gas *(chromatography)*
Spülgas[schwel]verfahren *n* Spülgas process *(gas making)*
Spülgrube *f (petrol)* mud pit
Spülherd *m (min tech)* wet table
Spülkopf *m (petrol)* swivel
Spülluft *f* scavenging air
Spulmaschine *f (text)* winding machine, winder
Spülmittel *n* rinsing (scouring) agent, rinse, scavenger
Spülmud *m (petrol)* drilling mud
Spülpumpe *f (petrol)* mud pump
Spülsäule *f (petrol)* mud column
Spülschieber *m* scour valve
Spülschlamm *m (petrol)* drilling mud
S. auf Erdölbasis oil-base mud
Spülschlauch *m (petrol)* mud (rotary) hose
Spülung *f* 1. rinse, rinsing, scouring, scavenging, wash[ing], swilling, flushing; 2. *s.* Spülschlamm
Spülungsbehälter *m (petrol)* mud pit
Spülungssäule *f (petrol)* mud column
Spülungstank *m (petrol)* mud pit
Spülungsumlaufsystem *n (petrol)* drilling fluid circulating system
Spülventil *n* rinse (scour) valve
Spülwasser *n* rinsing (rinse, wash, swill) water, wash [liquid, liquor]; *(filtr)* backwash water
Spund *m* plug, stopper, bung
spunden, spünden to bung
Spundloch *n* bunghole
Spur *f* 1. trace *(analytical chemistry)*; remnant *(as of an undesired substance)*; 2. trace *(of a recording instrument)*
Spurenanalyse *f* trace analysis
Spurenelement *n (biol, agric)* [nutritional] trace element, minor [nutrient] element, microelement
Spurenmaterial *n* tracer *(for investigating the pathway of reactions or nutrients)*
Spurenmenge *f* trace amount
Spurenmetall *n* trace metal
Spurennährstoff *m* micronutrient
Spurensucher *m s.* Spurenmaterial
Spürmethode *f* / **magnetische** *(petrol)* magnetic method
seismische S. seismic method
Spurstein *m* white metal *((product obtained by removing iron from copper matte)*
S-PVC *s.* Suspensionspolyvinylchlorid
Squalen *n* squalene *(an aliphatic triterpene)*
SR-Benzin *n* straight-run gasoline (benzine, naphtha, spirit), S.R.B.
SRF-Ruß *m (rubber)* semireinforcing furnace black, SRF black
S-Säure *f* $HOC_{10}H_5(NH_2)(SO_3H)$ S acid, 1-amino-8-naphthol-4-sulphonic acid
2S-Säure *f s.* SS-Säure
S-S-Bindung *f* disulphide bond (crosslink, link)
S-S-Brücke *f* disulphide bridge
ssl. *s.* löslich/sehr schwer
SS-Säure *f* $HOC_{10}H_4(NH_2)(SO_3H)_2$ SS acid, 2S acid, Chicago acid, 1-amino-8-naphthol-2,4-disulphonic acid
Stababgabe *f (pap)* stick downtake *(in a festoon dryer)*
Stabaufnahme *f (pap)* stick uptake *(in a festoon dryer)*
stabil stable, persistent, resistant *(chemically)*; stable, rigid *(physically)*
thermisch s. *(relating to decomposition)* thermally stable; *(relating to deformation)* heat-resistant
Stabilbenzin *n* stabilized (stable) gasoline
stabilglühen *(met)* to stabilize
Stabilglühen *n (met)* stabilizing [anneal]
Stabilisation *f* stabilization
Stabilisationsanlage *f (petrol)* stabilizer plant
Stabilisationskolonne *f (petrol)* stabilizer
Stabilisator *m* stabilizing agent, stabilizer, *(relating to suspensions also)* suspending agent; *(rubber)* preserving agent, preservative; *(petrol)* stabilizer
Stabilisieranlage *f (petrol)* stabilizer plant
stabilisieren to stabilize
Stabilisierkolonne *f (petrol)* stabilizer
Stabilisierung *f* stabilization
Stabilisierungsbad *n* stabilizing bath
Stabilisierungsglühen *n (met)* stabilizing [anneal]
Stabilisierungsmittel *n s.* Stabilisator
Stabilität *f* stability, persistence, resistance *(chemically)*; stability, rigidity *(physically)*
thermische S. *(relating to decomposition)* thermal stability, thermostability; *(relating to deformation)* resistance to heat
Stabilitätsfaktor *m* stability factor
Stabilitätskonstante *f* stability constant
Stabilitätszone *f (coll)* stability zone
Stabmühle *f* rod mill
Stab[sieb]rost *m* rod deck *(as of a screening machine)*; grizzly [screen], bar screen (grizzly) *(a kind of screening machine)*
Stabspritze *f (agric)* [hand] lance
Stabziehverfahren *n (glass)* drawn-rod method
Stachelwalzenbrecher *m* toothed-roll crusher
Stachydrin *n* stachydrine *(a betaine)*
Stachyose *f* stachyose *(a tetrasaccharide)*
Stadiendüngung *f* fertilization in stages
Stadium *n* stage
hydrothermales S. *(geol)* hydrothermal stage
liquidmagmatisches S. *(geol)* orthomagmatic stage
Stadtentwässerung *f* sewerage
Stadtgas *n* town (city) gas
Stadtmüll *m* town refuse, city garbage
Stadtnebel *m* smog *(a type of air pollution)*

Staffelform *f* staggered form *(stereochemistry)*
Staffordshire-Ofen *m (ceram)* Staffordshire kiln
Staffordshire-Seger-Kegel *m (ceram)* Staffordshire [Seger] cone
Stahl *m* steel
S. für Einsatzhärtung case-hardening steel, carburizing steel
S. für Nitrierhärtung nitriding steel
aufgekohlter S. *s.* einsatzgehärteter S.
basischer S. basic steel
beruhigter (beruhigt vergossener) S. killed steel
einsatzgehärteter S. case-hardened steel, carburized steel
feuerverzinkter S. hot-dip galvanized steel
halbberuhigter S. semikilled steel
herdgefrischter S. open-hearth (Siemens-Martin) steel
hochlegierter S. high-alloy steel
inchromierter (inkromierter) S. chromized steel
kaltgezogener S. cold-drawn steel
knetbarer S. plastic steel
legierter S. alloy steel
lufthärtender S. air-hardening steel
nickellegierter S. nickel alloy steel
niedriglegierter S. low-alloy steel
nitriergehärteter (nitrierter) S. nitrided steel
ölhärtender S. oil-hardening steel
ruhig vergossener S. killed steel
sauerstoffgefrischter S. basic oxygen [furnace] steel
saurer S. acid steel
unberuhigter (unberuhigt vergossener) S. rimmed (rimming, wild) steel
verzinnter S. tinned steel
wasserhärtender S. water-hardening steel
weicher S. mild (soft) steel
windgefrischter S. Bessemer (converter) steel
zementierter S. *s.* einsatzgehärteter S.
Stahlakkumulator *m* Edison accumulator (cell)
Stahlarmierung *f* steel armouring
Stahlautoklav *m* steel autoclave
Stahlband *n* steel ribbon
Stahlbeton *m* reinforced concrete
Stahlblech *n* steel plate; *(if less than 0.25 inch in thickness)* steel sheet
Stahlblechmantel *m* steel jacket
Stahlblock *m* steel ingot
Stahlbombe *f* steel cylinder
Stahldruck *m* steel-plate printing
Stahldruckfarbe *f* steel-plate[-engraving] ink
Stahleisen *n* basic (steelmaking) pig iron
Stahlerzeugung *f* steelmaking
Stahlflasche *f* steel cylinder
Stahlformguß *m* cast steel
Stahlgewinnung *f* steelmaking
Stahlguß *m* 1. *(act)* steel casting; 2. *(product)* cast steel, steel casting
Stahlgußstück *n* steel casting
Stahlherstellung *f* steelmaking
Stahlmantel *m* steel jacket
Stahlmörser *m* percussion (diamond) mortar
Stahlpanzer *m* steel shell
Stahlroheisen *n* basic (steelmaking) pig iron
Stahlrohr *n* steel pipe
nahtloses S. line pipe
Stahlröhre *f* steel pipe
Stahlschmelzofen *m* steelmaking furnace
Stahlschrott *m* steel scrap
Stahlstich[druck]farbe *f* steel-plate[-engraving] ink
stahlummantelt steel-cased, steel-jacketed
Stahlunterbau *m* steel substructure
Stahlwalze *f* steel roll
Stahlwerksofen *m* steelmaking furnace
Stalagmit *m (min)* stalagmite
stalagmitisch *(min)* stalagmitic
Stalagmometer *n* stalagmometer *(for measuring surface tensions)*
Stalagmometrie *f* stalagmometry *(a method for measuring surface tensions)*
Stalaktit *m (min)* stalactite
stalaktitisch *(min)* stalactitic
Stamm *m* 1. *(nomencl)* parent; 2. *(biol)* strain *(of microorganisms)*
Stammansatz *m (dye)* stock liquor
Stammbase *f* parent base
Stammbaum *m* 1. family tree *(as of hydrocarbons)*; 2. *s.* Stammreihe / radioaktive
Stammemulsion *f* stock emulsion
Stammflotte *f (dye)* stock liquor
Stammgesenk *n (plast)* fixed plate
Stammkohlenwasserstoff *m* parent hydrocarbon
Stammkörper *m* parent (mother) substance
Stammküpe *f (dye)* stock vat
Stammlauge *f* mother liquor
Stammlösung *f* stock solution
Stamm-Magma *n* parental magma
Stammname *m (nomencl)* parent name
Stammplatte *f (plast)* fixed plate
Stammreihe *f* **/ radioaktive** *(nucl)* radioactive decay (disintegration) series, radioactive series (family, chain)
Stammringsystem *n (nomencl)* parent ring system
Stammsäure *f* parent acid
Stammsubstanzname *m (nomencl)* semisystematic (semitrivial) name
Stammverbindung *f* parent compound
Stammwürze *f (ferm)* original wort
stampfen to stamp, to ram, to tamp
Stampfer *m* 1. *(ceram)* tamp; 2. *(tech)* beater *(of a hammer mill)*
Stampfgemisch *n s.* Stampfmasse 1.
Stampfkalander *m (text)* beetler, beetling machine
Stampfmaschine *f (coal)* stamper
Stampfmasse *f* 1. *(ceram)* ramming mass (material, mix, mixture); 2. electrode material, paste
feuerfeste S. refractory ramming material
Stampfmischung *f s.* Stampfmasse 1.
Stampfmühle *f* stamp mill
Stampfwerk *n (pap)* stamping (hammer) mill,

stamper, stamps, stocks

Stand *m* 1. level, height *(as of liquid in a vessel)*; 2. *(tann)* firmness *(a property of leather)*

Standard *m* standard

innerer S. internal standard *(resonance spectroscopy)*

radioaktiver S. *s.* Standardpräparat / radioaktives

sekundärer S. secondary standard

Standardabweichung *f* standard deviation *(statistics)*

Standardausrüstung *f* standard equipment

Standardbedingungen *fpl* standard[ized] conditions, normal conditions

Standardbezugs ... *s.* Standard ...

Standardbildungsenthalpie *f*, **Standardbildungswärme** *f* standard heat of formation

Standarddruck *m* standard (normal) pressure

Standardeinheit *f* standard (normal) unit

Standardelektrode *f* standard electrode (half-cell)

sekundäre S. secondary standard electrode

Standardelektrodenpotential *n* standard [electrode] potential, S.E.P.

Standardelement *n (el chem)* standard cell

Standard-EMK *f* standard electromotive force, standard [reference] emf

Standardenthalpie *f* standard enthalpy

Standardentropie *f* standard entropy

Standardfarbe *f* standard colour

Standardgold *n* standard gold *(for coinage)*

Standardgoldsol *n* standard gold sol

Standardhalbelement *n*, **Standardhalbzelle** *f s.* Standardelektrode

Standardkatalysator *m* standard catalyst

Standardlösung *f* standard solution

Standardmethode *f* standard method (procedure, technique)

Standardmineral *n* standard mineral

Standardnormalpotential *n* standard potential

Standardopaleszenz *f* standard opalescence (turbidity)

Standardoxydationspotential *n* standard oxidation potential

Standardpenizillin *n* standard penicillin

Standardpotential *n* standard [electrode] potential, S.E.P.

S. des Halbelements standard-half-cell potential

S. des Redoxsystems standard redox (oxydation-reduction) potential

chemisches S. standard chemical potential

Standardpräparat *n* standard preparation

radioaktives S. radioactive (radioactivity) standard

Standardprobe *f* standard sample

Standardprüfsieb *n* standard (normal) test (testing) screen

Standardprüfung *f* standard test

Standardpuffer *m*, **Standardpufferlösung** *f* standard buffer [solution]

Standardradius *m* standard radius

Standardreagens *n* standard reagent

Standardreaktionsarbeit *f* standard work of reaction

Standardreaktionsenthalpie *f* standard enthalpy of reaction

Standardredoxpotential *n* standard redox (oxidation-reduction) potential

Standardreduktionspotential *n* standard reduction potential

Standardschliff *m* standard ground joint

Standardschliffhahn *m* standard ground stopcock

Standardschliffverbindung *f* standard ground joint

Standardsieb *n* standard sieve (screen)

Standardsiebreihe *f* standard sieve scale (series), standard series of screens

Standardsilber *n* standard silver *(for coinage)*

Standardspannung *f* standard [reference] voltage

Standardsubstanz *f* standard [substance]

Standardtemperatur *f* standard temperature

Standardthermometer *n* standard thermometer

Standardthermopaar *n* standard thermocouple

Standardtrübung *f* standard opalescence (turbidity)

Standardverfahren *n* standard method (procedure, technique)

Standardversuch *m* standard test

Standardvorschrift *f* standard specification

Standardwasserstoffelektrode *f* standard hydrogen electrode, SHE

Standardwert *m* standard value

Standardwiderstandsthermometer *n* standard resistance thermometer

Standardzelle *f (el chem)* standard cell

Standardzustand *m* standard state

Standbad *n (text)* standing bath

Standentwicklung *f (phot)* stand development

Ständer *m* rack; pedestal

Standfast-Färbemaschine *f (text)* Standfast molten-metal machine

Standflasche *f* storage (laboratory) bottle

Standglasverfahren *n* long-tube method *(in determining sedimentation rates)*

***m*-ständig** in meta position, meta

***m*-ständig sein** to be [located, situated] meta

***o*-ständig** in ortho position, ortho

***p*-ständig** in para position, para

Standöl *n* stand oil

Standrohr *n* standpipe

Standzeit *f* residence (hold-up, holding, detention, retention) time; *(plast)* pressing time; *(coat)* pot life

Standzylinder *m* gas jar

Stange *f* 1. *(tech)* rod, bar, pole; 2. roll *(of sulphur)*

Stangenrost *m* rod deck

Stangenrostsieb *n* bar screen (grizzly), grizzly [screen]

Stangenschwefel *m* roll sulphur

Stangensiebrost *m s.* Stangenrostsieb

Stangentusche *f* India ink *(in the form of sticks)*

Stannan *n* SnH_4 stannane, tin hydride, stannic hydride
Stannat *n* stannate
Stannin *m (min)* stannite, tin pyrites, bell-metal ore
Stanniol *n*, **Stanniolfolie** *f* tin foil
Stanniolpapier *n* tin-foil paper
stanzen to punch, to blank
Stanzform *f* die
Stanzmasse *f (ceram)* dust
Stanzwerkzeug *n* die
Stapel *m* 1. stack; *(ceram)* bung *(as of bricks)*; 2. *(text)* staple [length], fibre staple ✦ **Elementarfäden auf S. reißen (schneiden)** *(text)* to staple
Stapelbecken *n* detention (holding) basin (pond) *(waste-water treatment)*
Stapelfaser *f (text)* staple [fibre]
Stapelfehler *m (cryst)* stacking fault
Stapelglasseide *f* chopped strands
Stapellänge *f (text)* staple [length], fibre staple
Stapelmaschine *f (ceram)* stacking machine
stapeln to stack
Stapelplatte *f* pallet
Stapelsalzung *f (tann)* green salting *(of hides)*
Stapelung *f (cryst)* stacking
stark 1. *(relating to beverages)* potent, strong, spirituous, intoxicating, *(Am also)* hard; 2. strong *(acid or base)*; powerful, vigorous *(reaction)*; 3. strong *(remedy)*
Stärke *f* 1. starch *(a polysaccharide)*; 2. strength *(of an acid or base)*; intensity, power *(of a reaction)*; intensity *(of irradiation)*
S. der Röntgenstrahlen X-ray intensity
S. des Schleiers *(phot)* fog level
enzymatisch (fermentativ) abgebaute S. enzyme-converted starch
lösliche S. soluble starch
native S. native starch
tierische S. animal (liver) starch
Stärkeabbau *m* starch breakdown
stärkeartig amyloid[al]
Stärkechemie *f* starch chemistry
stärkeführend *s.* stärkehaltig
Stärkegehalt *m* starch content
Stärkegranulose *f* amylopectin
Stärkegummi *n* starch (artificial, British) gum, dextrin[e] *(a polysaccharide)*
stärkehaltig starch-containing, starchy, amylaceous, amyliferous, farinaceous
Stärkeindustrie *f* starch industry
Stärke-Jodat-Papier *n* starch-iodate paper
Stärkekette *f* starch chain
Stärkekleister *m* starch paste
Stärkekorn *n* starch grain (granule, kernel)
Stärkeleim *m* starch glue (adhesive)
Stärkeleimung *f (pap)* starch sizing
Stärkelieferant *m* starch-yielding (starch-bearing) plant
Stärkelösung *f* starch solution
Stärkemais *m* soft maize, *(Am)* soft corn *(Zea mays L. var. amylacea)*
Stärkemehl *n* starch flour
Stärkemilch *f* starch milk (slurry)
stärken *(text)* to starch
stärkend *(pharm)* roborant, tonic
Stärkenitrat *n* starch nitrate, nitrostarch
Stärkepapier *n* starch paper
Stärkepflanze *f* starch plant *(a plant which converts excess assimilation products into starch)*
stärkereich rich in starch, starchy, farinaceous
Stärkesirup *m* starch (grain, glucose) syrup
stärkespaltend amylolytic, amyloclastic, starch-splitting
Stärkeverzuckerung *f* saccharification of starch, starch conversion
Stärkewert *m (agric)* starch equivalent
Stärkezucker *m* starch (corn) sugar, dextrose, *D*-glucose *(a monosaccharide)*
Starkfeldscheidung *f (min tech)* high-intensity magnetic separation
Starkgas *n* rich gas
starkgeleimt *(pap)* strongly sized, hard-sized, H.S.
Starkgift *n* / **rodentizides** acute (single-dose) rodenticide
Starkreiniger *m* heavy-duty detergent
Starksäure *f (pap)* strong acid
starr rigid *(e.g. molecule, gel)*
Starrheit *f* rigidity *(as of molecules or gels)*
Starrschmiermittel *n* sett grease *(consisting of lime, rosin oil, and mineral oil)*
Start *m* initiation *(as of a reaction)*
Startblech *n* starting sheet
Startdünger *m* starter [fertilizer]
starten to initiate *(e.g. a reaction)*
Starter *m (plast)* polymerization initiator
Startreaktion *f* start (initiating) reaction
Startreaktionsstadium *n* initiation stage
stationär stationary, immobile, non-mobile
Statistik *f* / **Boltzmannsche (klassische)** Boltzmann statistics
statistisch verteilt *(cryst)* randomly distributed
Stativ *n (lab)* stand, retort (support) stand
Stativfuß *m (lab)* retort stand base
Stativhaken *m* **mit Muffe** *(lab)* suspension clamp
Stativharfe *f (lab)* assembly
Stativklemme *f (lab)* apparatus clamp
S. mit halbrunden Spannbacken flask (condenser) clamp
Stativstab *m (lab)* retort stand rod
Statormesser *n* bed knife *(of a rotary cutter)*
Statorring *m (distil)* stator ring *(of a rotating-disk contactor)*
Staub *m* 1. dust; *(intentionally made as from stone)* powder; 2. *(pap)* fines *(in classifying wood flour)*
Staubabscheider *m* dust collector (separator, settler, catcher)
akustischer S. sonic agglomerator
Staubabscheidezyklon *m* cyclone dust collector
Staubabscheidung *f* dust elimination (extraction, removal), dedusting
Staubaufnahme *f* dust absorption
Staubaufnahmevermögen *n* dust-holding capacity

Staubaustrag *m* dust discharge; dust outlet
Staubaustragöffnung *f* dust outlet
Staubaustritt *m* dust outlet
Staubbekämpfung *f* dust control
staubbeladen dust-laden, dust-bearing
Staubbeladung *f* dust loading
Staubbelag *m* dust covering, film of dust
Staubbildung *f* dust formation
Staubbindeflüssigkeit *f* dust-collecting liquid
Staubbindeöl *n* dust-laying oil
Staubbrenner *m* pulverized-coal burner
Staubbunker *m* dust hopper
staubdicht dustproof
Stäubebelag *m* dust residue *(plant protection)*
Stäubebeutel *m* dust-bag *(plant protection)*
Stäubegerät *n (agric)* dusting machine, duster
Stäubemittel *n* dusting agent, dust
  herbizides S. herbicidal dust
  insektizides S. insecticidal dust
  kombiniertes S. *(agric)* dust formulation
stauben to dust
stäuben 1. *(agric)* to dust *(with pesticides)*; 2. to dust *(as of heavily filled papers)*
Stäuber *m (pap)* dusting machine, duster *(for cleaning rags)*
Stäubeschwefel *m* winnowed (wind-blown) sulphur
Staubexplosion *f* dust explosion
Staubfallraum *m* gravity chamber *(gas cleaning)*
Staubfänger *m s.* Staubabscheider
Staubfeuerung *f* pulverized-fuel firing, [solid-fuel] suspension firing
Staubfilter *n* dust filter
Staubfließkatalysator *m* fluidized (fluid-bed, fluid) catalyst
Staubfließsystem *n* fluid[ized] system
Staubfließtechnik *f* fluid-bed (fluidized-bed, boiling-bed) technique
Staubfließverfahren *n* fluid (fluidized) process (operation)
staubfrei dust-free, dustless
Staubgas *n* dust-laden gas
staubgefeuert pulverized-coal-fired, pulverized-fuel-fired
Staubgehalt *m* dust loading (content)
staubhaltig dust-laden
Staubhefe *f* non-flocculating yeast
staubig dusty, dust-laden, pulverulent
Staubkammer *f* dust (fall-out, settling, settlement, gravity) chamber, drop-out box *(gas cleaning)*
  S. mit Prallflächen baffle separator
Staubkohle *f* powdered (fluid) coal
Staubkonzentrat *n* dry concentrate *(a kind of pesticide formulation)*
Staubkorn *n* dust particle
Staubluft *f* dust-laden air
Staubmühle *f* pulverizing mill, pulverizer
Staubniederschlag *m* dust deposition *(act)*; dust residue (deposit) *(substance)*
Staubpartikel *f* dust particle
Staubprobensammler *m* dust impinger (apparatus)
Staubrett *n (pap)* dam *(groundwood pulping)*
Staubrückstand *m* dust residue
Staubsack *m* 1. gravity dust catcher *(of a blast furnace)*; 2. *s.* Staubabscheider
Staubsammelbunker *m* dust hopper
Staubsammler *m s.* Staubabscheider
Staubsand *m* flour (very fine) sand
Staubschutz *m* dust shield
Staubtasche *f* dust chamber
Staubteilchen *n* dust particle
Staubtuff *m (geol)* dust tuff
staubundurchlässig dustproof
Staubvergasung *f* entrainment gasification
Staubvergasungsverfahren *n* entrained gasification process
  S. nach Koppers[-Totzek] Koppers process [for gasification of pulverized coal]
Staubzucker *m* powdered (icing, castor) sugar
Staubzyklon *m* cyclone dust collector
Stauchkammer *f (text)* stuffing (stuffer) tube (box)
Stauchkammertexturieren *n (text)* stuffing box crimping
Stauchsetzapparat *m (min tech)* moving-sieve-type jig
Staudinger-Einheit *f (plast)* structural element (unit), base (repeating) unit, base molecule
Staudruck *m* stagnation (dynamic) pressure
Stauer *m* expansion trap *(a kind of steam trap)*
Stauhöhe *f (pap)* head of stock *(in the headbox)*
Staupunkt *m (distil)* loading (phase-inversion) point
Staurohr *n* Pitot tube *(flow measurement)*
Staurolith *m (min)* staurolite *(a neso-subsilicate)*
Stauscheibe *f (plast)* breaker plate
Steapsin *n* steapsin *(a lipase)*
Stearaldehyd *m* $CH_3[CH_2]_{16}CHO$ stearaldehyde, octadecanal
Stearamid *n* $CH_3[CH_2]_{16}CONH_2$ stearamide, stearic acid amide, octadecanoamide
Stearanilid *n* $CH_3[CH_2]_{16}CONHC_6H_5$ stearanilide, octadecanoanilide
Stearat *n* stearate
Stearin *n* stearin, *(specif)* tristearin, glyceryl tristearate
Stearinaldehyd *m s.* Stearaldehyd
Stearinpech *n* stearin pitch
Stearinsäure *f* $CH_3[CH_2]_{16}COOH$ stearic acid, octadecanoic acid
Stearinsäureamid *n s.* Stearamid
Stearinsäureäthylester *m* $CH_3[CH_2]_{16}CO{-}OC_2H_5$ ethyl stearate
Stearinsäurechlorid *n s.* Stearoylchlorid
Stearinseife *f* stearin soap
Stearolsäure *f* $CH_3[CH_2]_7C{\equiv}C[CH_2]_7COOH$ stearolic acid, 9-octadecynoic acid
Stearon *n* $CH_3[CH_2]_{16}CO[CH_2]_{16}CH_3$ stearone, diheptadecyl ketone, 18-pentatriacontanone
Stearonitril *n* $CH_3[CH_2]_{16}CN$ stearonitrile, octadecanonitrile
Stearophansäure *f s.* Stearinsäure
Stearoxylsäure *f* $CH_3[CH_2]_7COCO[CH_2]_7COOH$

stearoxylic acid, 9,10-dioxo-octadecanoic acid
Stearoylchlorid *n* $CH_3[CH_2]_{16}COCl$ stearoyl chloride
Stearoylgruppe *f* stearoyl group
Stearylalkohol *m* $CH_3[CH_2]_{16}CH_2OH$ stearyl alcohol, 1-octadecanol
Stearylamin *n* $CH_3[CH_2]_{16}CH_2NH_2$ stearylamine, 1-amino-octadecane
Stearylchlorid *n (incorrectly for)* Stearoylchlorid
Steatit *m (min)* steatite, soapstone *(a variety of talc)*
Steatitkeramik *f* steatite ceramics
Steatitporzellan *n* steatite porcelain
Steatitweißware *f* steatite whiteware
stechend pungent *(smell)*
Stechpipette *f* dropping (teat) pipette
Stedman-Körper *m (distil)* Stedman cone
Steffen-Abwasser *n (sugar)* Steffen waste water
Steffen-Melasse *f (sugar)* Steffen molasses
Steffen-Verfahren *n* Steffen [separation] process *(for recovering sugar from beet molasses)*
Stegkettenförderer *m* scraper[-chain] conveyor, *(specif)* drag (drag-chain, drag-link, slat) conveyor
stehen / äquatorial to be equatorial
cis (in cis-Stellung) s. to be cis
in Wechselwirkung s. to interact
s. lassen to allow to stand
Stehkocher *m (pap)* upright (vertical) digester
Stehkolben *m* flat-bottom[ed] flask
Stehzentrifuge *f* underdriven (bottom-driven, base-bearing) centrifuge
Steifheit *f* rigidity, stiffness
Steifwerden *n (coll)* gelation
steigen 1. to increase, to rise *(as of temperature)*; 2. to pass up[wards] *(as of fluid)*
Steigen *n* 1. increase, rise *(as of temperature)*; 2. passing up[wards] *(as of fluid)*
steigern 1. to increase *(e.g. number, quantity)*; to raise, to elevate *(e.g. temperature, boiling point)*; 2. to potentiate *(the effect esp of drugs or pesticides by adding another agent)*
im Heizwert s. to enrich *(fuels)*
Steigerohr *n s.* Steigrohr
Steigerung *f* 1. increase *(as of number or quantity)*; raise, elevation *(as of temperature, boiling point)*; 2. potentiation *(of the effect esp of drugs or pesticides by adding another agent)*
Steiggefäß *n* ascending chromatography tank, developing chamber
Steiggeschwindigkeit *f* rising velocity, upward-flow rate
Steigleitung *f* rising main
Steiglochkreisradius *m* interface radius *(of a separator)*
Steigpresse *f (pap)* reverse press
Steigpreßfilz *m (pap)* reverse-press felt
Steigrohr *n* rising (ascension, lift) pipe, riser [tube], upspout; *(distil)* riser [tube], chimney *(of a tray)*
Steigungswinkel *m* helix angle *(as of an extruder screw)*
Steilbrustflasche *f* wide-necked reagent bottle with conical shoulder
Steilförderer *m* elevator
Steilheit *f* gradient
Steilrohr-Berieselungsverflüssiger *m* vertical shell-and-tube condenser
Steilwurfsieb *n* reciprocating (circle-throw) screen
Stein *m* 1. stone; 2. *(ceram)* brick; 3. *(glass)* [glass] stone *(a defect)*; 4. *(met)* matte *(artificially smelted mixture of metal sulphides)* ✦ **auf S. [ver]schmelzen** *(met)* to matte-smelt
feuerfester S. refractory (fireclay) brick, firebrick
hochfeuerfester S. highly refractory brick
lithografischer S. lithographic limestone
scharfgebrannter S. *(ceram)* hard-fired (hard-burned) brick
schwachgebrannter S. *(ceram)* soft[-fired] brick
Steinabstich *m*, **Steinabstichloch** *n (met)* taphole for matte
Steinbildung *f* scaling, scale formation *(in a vessel)*
Steinchen *n* [glass] stone *(a defect)*
Steindruck *m* lithographic printing, lithography
Steindruckfarbe *f* lithographic [printing] ink, litho ink
Steindruckpapier *n* lithographic [printing] paper, litho
Steinerhitzer *m* pebble heater (stove)
Steinerhitzerverfahren *n* pebble-heater process
Steinfang *m*, **Steinfänger** *m* stone (rock) catcher
Steingitterwerk *n* chequer brickwork, chequerwork
Steinglätte *f (pap)* flint-glazing machine, flint glazer, stone burnisher
Steingut *n* earthenware
Steingutfilter *n* earthenware filter
Steinguthahn *m* earthenware cock
Steinkohle *f* [hard] coal
kurzflammige (kurzflammig brennende) S. short-flame coal
langflammige (langflammig brennende) S. long-flame coal
Steinkohlehydrierung *f* coal hydrogenation
Steinkohlenaufbereitung *f* coal preparation (cleaning)
Steinkohlenbildung *f* coal formation
Steinkohlenbrikett *n* [hard-]coal briquette
Steinkohlenentgasung *f* coal carbonization
Steinkohlenentstehung *f* coal formation
Steinkohlenformation *f* / **produktive** coal measures
Steinkohlengas *n* coal gas
Steinkohlengefügebestandteil *m* coal maceral (constituent)
Steinkohlenhydrierung *f* coal hydrogenation
Steinkohlenkampfer *m s.* Naphthalin
Steinkohlenklassifizierung *f* classification of [hard] coals
Steinkohlenklein *n* small[-sized] coal, smalls
Steinkohlenkoks *m* coal coke
Steinkohlenkreosot *n* coal-tar creosote, creosote oil

Steinkohlenmazeral *n* coal maceral (constituent)
Steinkohlenmühle *f* coal (coal-dust, coal-grinding, coal-pulverizing, pulverized-coal) mill, coal pulverizer
Steinkohlenöl *n* coal[-tar] oil
Steinkohlenpech *n* coal-tar pitch
Steinkohlenschwelteer *m* low-temperature coal tar
Steinkohlenschwelung *f* coal carbonization
Steinkohlenstaub *m* pulverized (powdered) coal
Steinkohlenstaubbrenner *m* pulverized-coal burner
Steinkohlenteer *m* coal tar
Steinkohlenteerdestillat *n* coal-tar distillate
Steinkohlenteerkreosot *n* coal-tar creosote, creosote oil
Steinkohlenteeröl *n* coal[-tar] oil
Steinkohlenteerpech *n* coal-tar pitch
Steinkohlen-Tieftemperaturteer *m*, **Steinkohlenurteer** *m* low-temperature coal tar
Steinkohlenverkokung *f* coal carbonization
Steinkohlenverschwelung *f* coal carbonization
Steinkohlenwäsche *f* coal washing (cleaning)
Steinkugel *f* pebble *(as used in a pebble mill or pebble heater)*
Steinmahlgang *m* burstone mill, buhrmill
Steinmeteorit *m* stone meteorite, meteoric stone, aerolite, aerolith
Steinnuß *f* vegetable ivory *(fruit from Phytelephas macrocarpa Ruiz et Pav.)*
Steinoberfläche *f (pap)* stone surface
Steinretorte *f* brick retort
Steinsalz *n (min)* rock salt, halite *(sodium chloride)*
Steinsalzgitter *n (cryst)* rock-salt (sodium chloride) lattice
Steinsalzkristall *m* rock-salt (sodium chloride) crystal
Steinsalzplättchen *n* rock-salt plate
Steinsalzprisma *n* rock-salt prism
Steinschicht *f (ceram)* course
Steinschmelzen *n (met)* matte smelting
Steinstich *m*, **Steinstichloch** *n (met)* taphole for matte
Steinumfangsgeschwindigkeit *f (pap)* stone speed
Steinwalze *f* stone roll
Steinwolle *f* rock wool
Steinzeug *n (ceram)* stoneware
 S. für die chemische Industrie *s.* chemisches S.
 braunes S. brown ware
 chemisches S. chemical stoneware
 Delfter S. delft[ware], delph[ware]
 säurefestes S. acid-proof stoneware
Steinzeugfüllkörper *m* stoneware packing
Steinzeugrohr *n* stoneware pipe
Steinzeugton *m* stoneware clay
stellen / alkalisch (basisch) *(dye)* to alkal[in]ize, to make alkaline, to basify
 größer s. *(lab)* to turn up *(a burner flame)*
 typkonform s. *(dye)* to reduce to standard (type strength)

Stellglied *n* controller *(control engineering)*
 S. für Massenstrom (Mengenstrom) flow controller
Stellhefe *f (ferm)* pitching (inoculating) yeast
Stellmotor *m* actuator, operator *(control engineering)*
Stellung *f* position
***m*-Stellung** *f* meta position ✦ **sich in** *m*-Stellung **befinden** to be [located, situated] meta
***o*-Stellung** *f* ortho position
***p*-Stellung** *f* para position
1,2-Stellung *f s. o*-Stellung
1,2,3-Stellung *f* vicinal (near-by, neighbouring, adjacent) position
1,3-Stellung *f s. m*-Stellung
1,4-Stellung *f s. p*-Stellung
Stellungsisomer[es] *n* position[al] isomer
Stellungsisomerie *f* position (place) isomerism
Stellungssymbol *n* symbol of position
Stellungsziffer *f* position number
 akzentuierte (apostrophierte) S. *s.* gestrichene S.
 gestrichene S. primed number
 ungestrichene S. unprimed number
Stellventil *n* control valve
Stellwerk *n* positioner
Stelzner-Methode *f (nomencl)* Stelzner method
Stempel *m* 1. *(tech)* plunger, piston; 2. *(glass)* needle, plunger *(of a feeder)*; plug, *(sometimes also)* plunger *(of a glass-blowing machine for hollow-ware)*; 3. *(plast)* force (moulding) plug, male form (mould); 4. *(ceram)* punch
Stempel[kissen]farbe *f* stamping ink
Stempelplatte *f (plast)* force plate (retainer)
Stempelprofil *n (plast)* force (moulding) plug, male form (mould)
Stengel *m (cryst)* column
 flacher S. blade
Stengel[bast]faser *f* stem (stalk) fibre
stengelig *(cryst)* columnar
Stengelkristallisation *f* columnar crystallization
Stengel-Verfahren *n* Stengel process *(for making ammonium nitrate fertilizer)*
Stephanit *m (min)* stephanite *(antimony(III) silver sulphide)*
Steppenraute *f* / **Syrische** harmal[a], harmel, Peganum harmala L.
Stereochemie *f* stereochemistry
stereochemisch stereochemical
stereoelektronisch stereoelectronic
Stereoformel *f* space formula
stereoisomer stereoisomeric
Stereoisomer[es] *n* stereo[iso]mer
Stereoisomerie *f* stereoisomerism, space isomerism
Stereomer[es] *n s.* Stereoisomeres
Stereometer *n* stereometer, volumenometer
Stereometrie *f* stereometry
stereoregulär, stereoreguliert stereoregular
stereoselektiv stereoselective
Stereoselektivität *f* stereoselectivity
Stereospektrogramm *n* stereospectrogram

stereospezifisch stereospecific
Stereospezifität *f* stereospecificity
steril sterile, *(if achieved by physical methods also)* aseptic; *(soil)* barren, infertile
Sterilfiltration *f* sterilization by filtration, filtration sterilization
Sterilisation *f* sterilization
Sterilisationsapparat *m*, **Sterilisator** *m* sterilizer
Sterilisierbüchse *f* Petri dish box
Sterilisierdrahtkorb *m* test-tube basket
sterilisieren to sterilize
Sterilisierung *f* sterilization
Sterilität *f* sterility, *(if achieved by physical methods also)* asepsis; *(soil)* barrenness, infertility
Sterilitätsprüfung *f* test for sterility
Sterilmilch *f* sterilized (sterile) milk
Sterilsahne *f* sterilized cream
Sterin *n (org chem)* sterol
  pflanzliches S. plant sterol, phytosterol
  tierisches S. animal sterol, zoosterol
sterisch steric
  s. gehindert sterically hindered
  s. möglich sterically feasible
  s. regelmäßig stereoregular
Sterkobilin *n (bioch)* stercobilin
Sterkuliagummi *n* sterculia gum, karaya [gum], Indian tragacanth *(chiefly from Sterculia urens Roxb.)*
Sternanis *m* star aniseed, badian *(a condiment from Illicium verum Hook.fil.)*
Sternanisöl *n* Japanese anise oil *(from Illicium verum Hook.fil.)*
Sternbergit *m (min)* sternbergite *(iron(II,III) silver sulphide)*
Sterndämpfer *m (text)* star steamer
Sternrahmen *m*, **Sternreifen** *m (text)* star frame
Sternutatorium *n (pharm)* sternutator[y], sternutative
Steroid *n* steroid [compound]
Steroidalkaloid *n* steroid alkaloid
Steroidchemie *f* steroid chemistry
Steroidhormon *n* steroid hormone
Sterol *n* sterol
Sterrettit *m (min)* sterrettite, eggonite *(scandium phosphate)*
stetig continuous *(operation)*; constant *(e.g. flow)*
Stetigmischer *m* continuous mixer
Stetigschleifer *m (pap)* continuous grinder
Steuerdruck *m* control pressure
Steuereinrichtung *f* control equipment
Steuergerät *n* controlling device, controller
steuern to control *(a process)*
Steuerung *f* 1. control; 2. *s.* Steuereinrichtung
  automatische S. automatic control
  manuelle (nichtselbsttätige) S. manual control
  selbsttätige S. *s.* automatische S.
Steuerungsrechner *m* control computer
Steuerventil *n* control valve
Stevens-Umlagerung *f* Stevens rearrangement *(of a benzyl group in a quaternary ammonium salt)*
STH *s.* Hormon / somatotropes
Stibikonit *m (min)* stibiconite *(an oxidic ore containing antimony)*
Stibin *n* 1. SbH_3 stibine, antimony trihydride, antimony(III) hydride; 2. SbR_3 stibine *(any of a class of organic antimony compounds)*
  organisches S. *s.* Stibin 2.
Stibinsäure *f* RR'SbOOH stibinic acid
Stibnit *m (min)* stibnite, grey (dark) antimony, antimony glance, antimonite *(antimony(III) sulphide)*
Stich *m* 1. *(dye)* hue, shade, tone, tinge, tint; 2. *(met)* tapping; 3. *s.* Stichloch
Stichausreißfestigkeit *f (text)* stitch tear strength
stichig sour *(wine)*
  s. werden to sour *(of wine)*
Stichigwerden *n* souring *(of wine)*
Stichkultur *f (bioch)* stab culture
Stichloch *n (met)* taphole, tapping hole
Stichprobe *f* random (spot) sample
Stickoxid *n* nitrogen oxide, *(specif)* NO nitrogen(II) oxide, nitrogen monooxide
Stickstoff *m* N nitrogen
  aktiver S. active nitrogen
  atmosphärischer S. atmospheric nitrogen
  fixierter S. fixed nitrogen
  flüssiger S. liquid nitrogen
  gebundener S. fixed nitrogen
  langsamwirkender S. *(agric)* slow-release nitrogen
  verflüssigter S. liquid nitrogen
stickstoffarm poor in nitrogen
Stickstoffarmut *f (agric)* poverty in nitrogen
Stickstoffaufnahme *f* nitrogen uptake
Stickstoffbad *n* / **flüssiges** liquid-nitrogen bath
Stickstoffbase *f* nitrogenous base
Stickstoffbestimmung *f* determination (estimation) of nitrogen, nitrogen determination (analysis)
  S. nach Dumas Dumas nitrogen analysis
  S. nach Kjeldahl Kjeldahl nitrogen analysis, Kjeldahl determination
  N. nach Neßler Nessler nitrogen analysis, Nessler test, nesslerization
Stickstoffbestimmungsapparat *m* nitrogen determination apparatus
  S. nach Kjeldahl Kjeldahl digestion apparatus
Stickstoffbindung *f (agric)* nitrogen fixation
  S. durch freilebende Bakterien *s.* nichtsymbiotische S.
  nichtsymbiotische S. non-symbiotic fixation, free fixation, azofication
Stickstoffbrücke *f* nitrogen bridge
Stickstoff(III)-chlorid *n* NCl_3 nitrogen(III) chloride, nitrogen trichloride
Stickstoffdioxid *n* NO_2 nitrogen dioxide
Stickstoffdon[at]or *m* nitrogen donor
Stickstoffdüngemittel *n* nitrogen[ous] fertilizer
Stickstoffdüngung *f* nitrogen fertilization
Stickstoffentzug *m* denitrogenation
Stickstoff-Fixierung *f s.* Stickstoffbindung
Stickstoff(III)-fluorid *n* NF_3 nitrogen(III) fluoride, nitrogen trifluoride

stickstofffrei nitrogen-free, non-nitrogenous
Stickstoffgehalt *m* nitrogen content
Stickstoffgruppe *f* nitrogen family (group)
stickstoffhaltig containing nitrogen, nitrogenous
Stickstoffhärtung *f (met)* nitrogen [case-]hardening, nitride hardening, nitriding, nitridation
Stickstoff(III)-jodid *n* NI_3 nitrogen(III) iodide, nitrogen triiodide
Stickstoffkreislauf *m* nitrogen cycle
Stickstofflost *n s.* Stickstoffyperit
Stickstoffmehrer *m s.* Stickstoffsammler
Stickstoffmodell *n* / **Dreidingsches** Dreiding nitrogen model
Stickstoffmonoxid *n s.* Stickstoff(II)-oxid
Stickstoffnachweis *m* detection of nitrogen
Stickstoffoxid *n* nitrogen oxide
Stickstoff(I)-oxid *n* N_2O nitrogen(I) oxide, dinitrogen monooxide
Stickstoff(II)-oxid *n* NO nitrogen(II) oxide, nitrogen monooxide
Stickstoff(III)-oxid *n* N_2O_3 nitrogen(III) oxide, dinitrogen trioxide, nitrogen trioxide
Stickstoff(IV)-oxid *n* nitrogen(IV) oxide, *(specif) s.* Stickstoffdioxid *(or)* Stickstofftetroxid
Stickstoff(V)-oxid *n* N_2O_5 nitrogen(V) oxide, dinitrogen pentaoxide, nitrogen pentaoxide
Stickstoffoxidchlorid *n* NOCl nitrosyl chloride
Stickstoffoxidfluorid *n* NOF nitrosyl fluoride
Stickstoffpentasulfid *n s.* Stickstoff(V)-sulfid
Stickstoff-Phosphor-Gruppe *f* nitrogen family (group)
Stickstoffprobe *f* **nach Lassaigne** Lassaigne test [for nitrogen]
Stickstoffquelle *f (agric)* nitrogen reservoir
stickstoffreich rich in nitrogen, high-nitrogen-content
Stickstoffreservoir *n (agric)* nitrogen reservoir
Stickstoffsammler *m (agric)* nitrogen-gathering (nitrogen-storing) plant
Stickstoff(V)-sulfid *n* N_2S_5 nitrogen(V) sulphide, dinitrogen pentasulphide, nitrogen pentasulphide
Stickstofftetroxid *n* N_2O_4 dinitrogen tetraoxide
Stickstoffträger *m (agric)* nitrogen carrier
Stickstofftri . . . *s.* Stickstoff(III)- . . .
Stickstoffwasserstoffsäure *f* HN_3 hydrazoic acid, hydrogen azide, azoimide
Stickstoffyperit *n* $RN(CH_2CH_2Cl)_2$ nitrogen mustard *(any of several halogen alkylamines)*
Stickwetter *pl (mine)* afterdamp
Stiefel *m* 1. *(glass)* hood, boot, potette; 2. *s.* Freiberger S.
  Freiberger S. *(text)* J box
Stiefelwanne *f (glass)* potette tank
Stiel *m* stem *(of a funnel)*
Stift *m* 1. *(tech)* pin, tack; 2. *(pap)* sliver, skim; 3. *(cosmet)* stick
  desodor[is]ierender S. deodorant stick
Stift[scheiben]mühle *f* pinned-disk disintegrator (mill), pin mill
Stilben *n* $C_6H_5CH{=}CHC_6H_5$ stilbene, trans-1,2-diphenylethylene
Stilbenfarbstoff *m* stilbene dye
Stilbit *m (min)* stilbite, desmine *(a tectosilicate)*
stillegen to shut [down]
Stillegung *f* shut[down]
Stillingia-Öl *n* stillingia (tallow-seed) oil *(from Sapium sebiferum (L.) Roxb.)*
Stillingia-Talg *m* Chinese vegetable tallow *(from Sapium sebiferum (L.) Roxb.)*
Still-Koksofen *m*, **Still-Regenerativofen** *m* Still oven
stillsetzen to shut [down]
Stillsetzregelung *f* automatic-start-and-stop control
Stillsetzung *f* shut[down]
Stillstand *m* standstill *(of a reaction or machine)*; shutdown *(of an industrial plant)*; ✦ **zum S. kommen** 1. *(relating to a reaction)* to die down; 2. *(relating to a machine)* to break down
Stillstandszeit *f* downtime, down period, outage time
Stillwein *m* still wine
Stilpnomelan *m (min)* stilpnomelane *(a phyllosilicate)*
Stilpnosiderit *m (min)* stilpnosiderite *(an iron ore)*
Stimulans *n* stimulant, stimulus, *(pharm also)* stimulatory drug
Stinkasant *m* devil's dung, food of the gods *(gum resin from Ferula specc.)*
Stinkäscher *m (tann)* dead (rotten) lime, mellow lime liquor
Stippen *fpl* 1. *(rubber)* filler specks; 2. *(text)* specks
Stobbe-Kondensation *f*, **Stobbe-Reaktion** *f* Stobbe condensation (reaction) *(with esters of succinic acid)*
Stocherloch *n*, **Stocheröffnung** *f* pokehole, poking hole
Stochervorrichtung *f* poker
Stöchiometrie *f* stoichiometry
stöchiometrisch stoichiometric
Stochloch *n*, **Stochöffnung** *f s.* Stocherloch
Stochvorrichtung *f s.* Stochervorrichtung
Stockblender *m (rubber)* stock blender
Stocklack *m* stick lac
Stockpunkt *m* setting (solidification, solidifying, solid) point, s.p. *(of oils)*
Stockthermometer *n* rod thermometer
Stoddard-Solvent *n* Stoddard solvent *(a refined petroleum product for use in dry cleaning)*
Stoff *m* 1. substance, matter, material *(for compounds s.a.* Substanz); 2. *(pap)* pulp, fibrous pulp (mass), pulp slurry (stock), slush [of] stock, fibre suspension (*s.a.* Halbstoff, Ganzstoff); 3. *(text)* fabric [cloth], cloth, tissue, textile
  absorbierender S. absorbing substance (agent), absorbent, absorber
  absorbierter S. absorbed substance, absorbate
  adsorbierender S. adsorbing substance (agent), adsorbent
  adsorbierter S. adsorbed substance, adsorbate, adsorptive

aggregierend wirkender S. *(soil)* aggregating agent
aktiver S. active substance (agent)
am schwersten siedender S. *(distil)* heavier-than-heavy component
amphoterer S. amphoteric (amphiprotic) substance (electrolyte), ampholyte
anionaktiver S. anionic surfactant
antibiotischer S. antibiotic agent (substance), antibiotic
antimikrobieller S. microbicide
antiseptischer S. antiseptic [agent]
arsenhaltiger S. arsenical
aufgelöster S. dissolved substance, solute
aufgenommener S. sorbate
aufnehmender (aufsaugender) S. sorbent
bituminöser S. bituminous substance
büttenfertiger S. *(pap)* accepted (screened) stock, accepts
diamagnetischer S. diamagnetic [substance]
färbender (farbgebender) S. colouring matter (substance)
ferroelektrischer S. ferroelectric [substance]
ferromagnetischer S. ferromagnetic [substance]
fester S. solid [matter, substance]
filmbildender S. film-forming substance, film former
flüchtiger S. volatile matter, v.m., VM
fremder S. foreign (extraneous) substance
gelöster S. dissolved substance, solute
gerbender S. tanning agent (substance), tan
glaskeramischer S. neo-ceramic glass, devitrified glass, glass ceramic, vitroceramic, vitrokeram
grenzflächenaktiver S. tension depressor
gummierter S. rubberized (rubber-coated, rubbered, proofed) fabric, proofing
halbleitender S. semiconducting material
halluzinogenisierender S. *(pharm)* hallucinogenic substance, hallucinogen
hautreizender S. skin irritant
indifferenter S. inert substance
ionogener grenzflächenaktiver S. ionic surfactant
kanzerogener (karzinogener) S. carcinogen
kationaktiver S. cationic surfactant
komplexbildender S. complexing agent
kurzröscher S. *(pap)* short free stock
kurzschmieriger S. *(pap)* short wet stock
langröscher S. *(pap)* long free stock
langschmieriger S. *(pap)* long wet stock
maskierender S. masking (sequestering) agent, sequestrant
mineralischer S. mineral matter (substance)
nichteiweißartiger S. non-protein
nichtflüchtiger S. non-volatile matter
nichtionischer (nichtionogener) grenzflächenaktiver S. non-ionic [surfactant]
nichtsystemischer S. non-systemic chemical *(crop protection)*
oberflächenaktiver S. surface-active agent (substance), surfactant
paramagnetischer S. paramagnetic [substance]
protogener S. *s.* Dysprotid
protophiler S. *s.* Emprotid
reagierender S. reacting substance, reactant
röscher S. *(pap)* fast stock, free (free-beaten, free-running, free-working) stock, free pulp
schmieriger S. *(pap)* wet (slow, shiny, soft, greasy) pulp
sorbierter S. sorbate
textiler S. *s.* Stoff 3.
therapeutisch wirksamer S. therapeutic agent
thixotroper S. thixotropic substance
Thixotropie erzeugender S. thixotroping (thixotropic) agent
tranquil[l]isierender S. *(pharm)* tranquillizer, tranquillizing drug
wachstumsfördernder S. growth-promoting (growth-stimulating, growth) substance
waschaktiver S. detergent surfactant
wasserlässiger S. *s.* röscher S.
wehenerregender S. oxytocic [agent]
wiedergewonnener S. recovered substance; *(pap)* recovered stock
wirksamer S. active substance (agent), agent
zu absorbierender S. material to be absorbed
zu adsorbierender S. material to be adsorbed
zu sublimierender S. sublimand
zurückgewonnener S. *s.* wiedergewonnener S.

Stoffang *m s.* Stoffänger
Stoffänger *m (pap)* pulp saver, save-all [tray], stuff catcher
nach dem Absetzprinzip (Sedimentationsprinzip) arbeitender S. gravity (sedimentation) save-all
Stoffaufbereitung *f (pap)* stock (fibre) preparation
Stoffauflauf *m (pap)* 1. stock flow to (onto) the wire, stock inlet; 2. headbox, stuff (breast, flow) box
geschlossener S. enclosed (closed-type) headbox
offener S. open-type (gravity-type) headbox
vakuumgesteuerter S. vacuum headbox
Stoffauflaufkasten *m s.* Stoffauflauf 2.
Stoffauflöser *m (pap)* [hydra]pulper, pulping engine
Stoffaufschläger *m (pap)* refiner, refining machine (engine), perfecting engine
Stoffaustritt *m (pap)* outgo of pulp
Stoffbahn *f* [paper] web, web of fibre[s], mat
Stoffbehälter *m (pap)* stock tank
stoffbespannt cloth-covered
Stoffbewegung *f (pap)* stock circulation *(in a Hollander beater)*
Stoffbrei *m (pap)* pulp, fibrous pulp (mass), pulp slurry (stock), slush [of] stock
Stoffbütte *f (pap)* pulp (stuff, supply, machine) chest, supply tank (vat)
Stoffdichte *f (pap)* stock (pulp) density (consistency)
Stoffdichteregler *m (pap)* consistency regulator
Stoffdruck *m* textile printing
Stoffeintrag *m (pap)* furnish[ing]
Stoffeintritt *m (pap)* inflow of pulp

Stoffentlüfter *m (pap)* stock deaerator, deculator
Stoffentlüftung *f (pap)* stock deaeration
Stofffluß *m (pap)* stock flow
stoffgeleimt *(pap)* engine-sized, E.S., pulp-sized, beater-sized
Stoffgeschwindigkeit *f (pap)* spouting (stock) velocity, speed of the stock
Stoffgrube *f (pap)* blow pit (tank, vat), wash (receiving) tank
stoffhaltig *(pap)* fibre-bearing *(water)*
Stoffilter *n* cloth (fabric, woven) filter
Stoffkasten *m s.* Stoffgrube
Stoffkette *f (phys chem)* material chain
Stoffklasse *f* family
Stoffkonsistenz *f*, **Stoffkonzentration** *f (pap)* stock (pulp) consistency (density)
Stofflauf *m (pap)* stock flow
Stoffleitung *f (pap)* stock line
stofflich material
Stofflöser *m (pap)* [hydra]pulper
  S. mit vier Auflösescheiben quatropulper
  S. mit zwei Auflösescheiben duopulper
Stoffmahlung *f (pap)* stock disintegration, beating of the stock
Stoffmenge *f* amount of substance
Stoffmengenregler *m (pap)* stock proportioner, stockmaker
Stoffmühle *f (pap)* beating engine, beater
Stoffpumpe *f (pap)* stock (stuff) pump
Stoffqualität *f (pap)* stock (pulp) quality
Stoffregulierkasten *m (pap)* regulating box
Stoffreinigung *f (pap)* stock cleaning (cleanup)
Stoffrinne *f (pap)* stock sewer (line) *(groundwood pulping)*
Stoffrückgewinnung *f* recovery, reclaim *(of useful material)*; *(pap)* fibre recovery
Stoffrückgewinnungsanlage *f* recovery plant; *s.* Stoffänger
Stoffsortierung *f (pap)* stock separation
Stoffstrom *m (pap)* stock flow
Stoffsuspension *f (pap)* pulp suspension (slurry), aqueous (watery) pulp (stuff), slush [of] stock
Stoffteilchen *n* particle of matter, material particle
Stofftemperatur *f (pap)* temperature of the stock
Stofftransport *m* mass transport
Stofftreiber *m (pap)* propeller-type agitator *(of a Hollander beater)*
Stofftrog *m* [cylinder] vat *(of a cylinder paper machine)*
Stofftrübe *f (min tech)* pulp of ore
Stoffturm *m (pap)* retention tower
Stoffübergang *m (phys chem)* mass transfer
Stoffübergangszahl *f* mass-transfer coefficient
Stofführung *f (pap)* delivery of the stock
Stoffumsetzer *m* reactor
Stoffumtrieb *m (pap)* stock circulation *(in a Hollander beater)*
Stoffumwandlung *f* transformation, conversion, *(esp nucl)* transmutation
Stoffverteiler *m (pap)* flow distributor
Stoffwanne *f (pap)* beater tub (vat, tank, pan)
Stoffwasser *n s.* Stoffsuspension
Stoffwechsel *m* metabolism
Stoffwechselenergie *f* metabolic energy
Stoffwechselpool *m (bioch)* metabolic pool
Stoffwechselprodukt *n* metabolic product, metabolite
Stoffwechselreaktion *f* metabolic reaction
Stoffwechselstörung *f* metabolic disturbance
Stoffwechselvorgang *m* metabolic process
Stoffweiße *f (pap)* pulp brightness
Stoffwiedergewinnung *f s.* Stoffrückgewinnung
Stoffzuführung *f (pap)* feed (delivery) of the stock
Stoffzuteiler *m s.* Stoffmengenregler
Stokes *n* stoke *(a unit of kinematic viscosity)*
Stokes-Fluoreszenz *f* Stokes fluorescence
Stokes-Gesetz *n* Stokes law
stollen *(tann)* to stake
Stollmaschine *f (tann)* staking machine, staker
Stolzit *m (min)* stolzite *(lead wolframate)*
Stomachikum *n (pharm)* stomachic, digestive stimulant
Stonitewalze *f (pap)* stonite roll
Stopfbuchse *f* stuffing (packing) box, packing (packed) gland
Stopfbuchsverbindung *f* packed-gland joint *(of pipes)*
Stopfen *m* stopper, plug, cork
  kalter S. *(plast)* cold slug
stopfend *(pharm)* styptic
Stopfwachs *n* propolis, bee glue
Stoppbad *n (phot)* stop bath
  saures S. acid stop bath
Stopper *m (plast)* chain stopper
Stöpsel *m* stopper, plug, cork
Störatom *n* foreign (impurity) atom
Storax *m s.* Styrax
Störgröße *f* disturbance [variable]
Stör[halb]leiter *m s.* Störstellenhalbleiter
Störleitung *f s.* Störstellenleitung
Störniveau *n* impurity level
Störstelle *f* point (lattice) imperfection (defect), crystal defect
  stöchiometrische S. stoichiometric lattice defect
Störstellenatom *n* impurity (foreign) atom
Störstellenband *n* impurity band
Störstellen[halb]leiter *m* impurity (extrinsic) semiconductor
Störstellenleitung *f* impurity [electric] conduction
Störstellenniveau *n*, **Stör[stellen]term** *m* impurity level
Störung *f* 1. disturbance, trouble; 2. *(cryst)* imperfection
störungsfrei undisturbed, trouble-free
Stoß *m* push, shock, impact, *(esp relating to particles)* impingement; *(phys chem)* collision
  S. erster Art collision of the first kind
  S. zweiter Art collision of the second kind
  aktivierender (anregender) S. *s.* S. erster Art
  desaktivierender S. *s.* S. zweiter Art
  unelastischer S. inelastic collision

Stoßabscheidekammer *f*, **Stoßabscheider** *m* impingement separator, inertia scrubber *(gas cleaning)*
Stoßanregung *f (phys chem)* collision excitation
Stoßappretur *f (tann)* friction finish
Stoßbelastung *f* shock load *(as of waste water)*
Stoßdämpfer *m* dash pot
Stoßdauer *f (phys chem)* duration of collision
Stoßdichte *f (phys chem)* collision density
Stoßdruck *m* impact pressure *(of a moving fluid)*
Stoßdurchmesser *m (phys chem)* collision diameter
Stoßeisen *n (tann)* slicker
Stößel *m* 1. *(tech)* plunger; 2. *(lab)* pestle *(of a mortar)*; 3. *(glass)* needle *(of a feeder)*
Stoßelastizität *f* [impact] resilience, rebound [resilience, elasticity]
Stoßelektron *n* impact electron
stoßempfindlich sensitive to impact (shock)
Stoßempfindlichkeit *f* sensitiveness to impact (shock)
stoßen 1. to crush, to chop *(ice)*; 2. to bump *(of boiling liquids)*
Stoßfaktor *m (phys chem)* collision factor
stoßfest impact-resistant, shockproof
Stoßfestigkeit *f* impact resistance (strength), resistance to shock
Stoßfluoreszenz *f* impact fluorescence
Stoßfrequenz *f (phys chem)* collision frequency
Stoßgalvanometer *n* ballistic galvanometer
Stoßhäufigkeit *f (phys chem)* collision frequency
Stoßherd *m (min tech)* shaking table
Stoßintegral *n (phys chem)* collision integral
Stoßkomplex *m* collision complex
Stoßkraftabscheider *m s.* Stoßabscheider
Stoßmischer *m* batch mixer
Stoßofen *m (ceram)* pusher-type kiln, pusher
Stoßquerschnitt *m (phys chem)* collision cross section
stoßsicher impact-resistant, shockproof
Stoßspektrum *n* impact spectrum
Stoßstange *f (coal)* [coke] pusher ram
Stoßstrahlung *f (phys chem)* collision radiation
Stoßtheorie *f (phys chem)* collision theory
Stoßverdampfung *f* flash evaporation
Stoßwahrscheinlichkeit *f (phys chem)* probability of collision
Stoßwelle *f* shock wave, *(relating to explosives also)* explosive (detonation) wave
Stoßwirkung *f* impact effect
Stoßzahl *f* **[je Zeiteinheit]** *(phys chem)* collision number
Stoßzahldichte *f (phys chem)* collision density
Stoßzündung *f* priming
Strahl *m* 1. ray *(of light or other radiation)*, *(if bundled)* beam; 2. jet *(of liquid or gas)*
  außerordentlicher S. extraordinary ray
  ordentlicher S. ordinary ray
  positiver S. positive (canal) ray
Strahlapparatmischer *m* jet mixer
strahlen to radiate, to emit rays
Strahlen ... *s.a.* Strahlungs ...
Strahlen *mpl* / **kosmische** cosmic rays
  weiche S. soft rays
α-Strahlen *mpl* alpha rays, α-rays
β-Strahlen *mpl* beta rays, β-rays
γ-Strahlen *mpl* gamma rays, γ-rays
Strahlenabschirmung *f* radiation shielding
Strahlenabsorption *f* radiation absorption
Strahlenbiologie *f* radiation biology, radiobiology
strahlenbiologisch radiobiologic[al]
Strahlenbündel *n* beam of rays
Strahlenchemie *f* radiation chemistry
strahlenchemisch radiation-chemical
strahlend radiant, radiative
Strahlendetektor *m* radiation detector
Strahlendosimetrie *f* radiation dosimetry
Strahlendosis *f* radiation dosage (dose)
strahlendurchlässig radiolucent
Strahlendurchlässigkeit *f* radiolucency
strahlenempfindlich radiosensitive
Strahlenempfindlichkeit *f* radiosensitivity
Strahlenfilter *n* radiation filter
Strahlengefahr *f* radiation hazard
Strahlengefährdung *f* radiation hazards
strahleninduziert radiation-induced
β-Strahlenionisationsdetektor *m* β-ray ionization detector
Strahlenkrankheit *f* radiation sickness (syndrome)
Strahlenkunde *f* radiology
Strahlennachweis *m* detection of radiation
Strahlenpasteurisierung *f* radiation pasteurization, radiopasteurization
Strahlenquelle *f* radiation source, emitter
Strahlenreaktion *f* radiation reaction
Strahlenschaden *m* radiation damage
Strahlenschutz *m* radiation protection
Strahlenschutzbeton *m* radiation [shielding] concrete, concrete for [atomic] radiation shielding
Strahlenspurverfolgung *f* ray tracing
Strahlensyndrom *n* radiation syndrome (sickness)
Strahlentherapie *f* radiation therapy, radiotherapy
Strahlenüberwachung *f* radiation monitoring (survey)
Strahlenüberwachungsgerät *n* radiation monitor
strahlenundurchlässig radiopaque
Strahlenundurchlässigkeit *f* radiopacity
Strahlenvernetzung *f (rubber)* vulcanization by high-energy radiation
Strahlenwarngerät *n* radiation monitor
Strahlenwirkung *f* radiation effect
Strahlenzähler *m*, **Strahlenzählrohr** *n* radiation counter
Strahler *m (nucl)* emitter, radiator; *(phys chem)* radiator; radiation (radiating) element *(for heating purposes)*
  grauer S. *(phys chem)* grey body [radiator]
  schwarzer S. black body [radiator]
α-Strahler *m (nucl)* alpha emitter, α-emitter

β-Strahler *m (nucl)* beta emitter, β-emitter
Strahlmischer *m* jet mixer
Strahlmühle *f* jet (fluid-energy) mill
Strahlprallmühle *f* flash (nozzle) pulverizer, pneumatic mill
Strahlpumpe *f* jet pump
Strahlstein *m (min)* actinolite *(an inosilicate)*
Strahlung *f* radiation
  energiereiche S. high-energy radiation
  fotochemische S. photochemical radiation
  harte S. high-energy radiation
  infrarote S. infrared radiation
  ionisierende S. ionizing radiation
  kosmische S. cosmic radiation
  kurzlebige S. short-lived radiation
  langlebige S. long-lived radiation
  schwarze S. black body radiation
  thermische S. thermal (calorific, caloric, heat) radiation
  ultrarote S. infrared radiation
Strahlungs ... *s. a.* Strahlen ...
strahlungsangeregt radiation-induced
strahlungsbeständig resistant to radiation
Strahlungsbeständigkeit *f* radiation resistance
Strahlungsdämpfung *f* radiation damping
Strahlungsdichte *f* radiation density
Strahlungsdruck *m* radiation pressure
Strahlungseinfang *m* radiative capture
Strahlungsemission *f* emission of radiation
Strahlungsenergie *f* radiant (radiation) energy
Strahlungsenergiedichte *f* radiant energy density
Strahlungsfeld *n* radiation field
Strahlungsfluß *m* radiant flux (power)
Strahlungsflußdichte *f* radiant flux density
Strahlungsformel *f* radiation formula
  Plancksche S. Planck radiation formula
strahlungsfrei radiationless, non-radiative
Strahlungsgesetz *n* radiation law
  Kirchhoffsches S. Kirchhoff radiation law
  Plancksches S. Planck radiation law
  Rayleigh-Jeanssches S. Rayleigh-Jeans radiation law
Strahlungsgleichgewicht *n* radiative equilibrium
Strahlungsheizung *f* radiation heating
strahlungsinitiiert initiated by irradiation *(e.g. a polymerization)*
Strahlungsintensität *f* radiant (radiation) intensity
Strahlungskatalyse *f* radiation catalysis
Strahlungskonstante *f* radiation constant
Strahlungslänge *f* radiation length
Strahlungsleistung *f* radiant power (flux)
strahlungslos radiationless, non-radiative
Strahlungsmeßgerät *n* radiation-measuring device
Strahlungsmessung *f* radiation measurement
Strahlungsofen *m* radiation oven
Strahlungspuffer *m* radiation buffer
Strahlungspyrometer *n* radiation pyrometer
Strahlungsquant *n* quantum of radiation, photon
Strahlungsspektrum *n* radiation spectrum
Strahlungstemperatur *f* radiation temperature
Strahlungsthermometer *n* pyrometer
Strahlungstrockner *m* radiation (radiant-heating, radiant heat) dryer
Strahlungstrocknung *f* radiation drying
Strahlungsübergang *m* radiative transition
Strahlungsverlust *m* radiation loss
Strahlungswärme *f* radiant heat
Strahlungswiderstand *m* radiation resistance
Strahlverlust *m* radiation loss
Strahn *m*, **Strähn** *m*, **Strähne** *f (text)* skein, hank
strähnig *(plast)* nervy *(extrudate)*
Straight-run-Benzin *n* straight-run gasoline (benzine), S.R.B., distillate gasoline
Straight-run-Destillation *f* straight[-run] distillation
Strainer *m (rubber)* strainer, straining machine, screen head extruder
Strainerkopf *m (rubber)* straining (screen) head
strainern *(rubber)* to strain
Straintest *m (rubber)* strain test
Straintester *m (rubber)* strain tester
stramm *(rubber)* stiff
Strammheit *f (rubber)* stiffness
Strang *m* 1. strand *(as of molecules)*; 2. *(met)* billet; 3. *(text)* rope; skein, hank *(of yarn)*; 4. run *(of an endless belt)*
Strangfärben *n (text)* skein (hank) dyeing *(of yarns)*; rope dyeing *(of cloth)*
Stranggarnfärbemaschine *f (text)* skein-dyeing (hank-dyeing) machine
Stranggarnmerzerisiermaschine *f (text)* hank-mercerizing machine
Stranggarnschlichtmaschine *f (text)* hank-sizing machine
Stranggarntrockenmaschine *f (text)* hank-drying machine
Stranggarnwaschmaschine *f (text)* hank-washing machine, hank washer
Stranghaspel *f(m) (text)* hank reel
Strangimprägniermaschine *f (text)* rope saturator
Strangmerzerisiermaschine *f (text)* hank-mercerizing machine
Strangöffner *m (text)* cloth opener
Strangpresse *f* 1. *(plast)* extruder, extruding (forcing) machine; 2. *(ceram)* extrusion press
strangpressen to extrude
Strangpressen *n* extrusion [moulding]
  S. ohne Lösungsmittel *s.* S. von Trockenmischung
  S. von Folien *(plast)* film extrusion
  S. von Trockenmischung *(plast)* dry[-blend] extrusion
Strangpressenkopf *m* extruder (extrusion, die) head
Strangpressenmundstück *n* extruder (extrusion) die
Strangpressenzylinder *m* extruder barrel
Strangpreßerzeugnis *n* extrudate
Strangpreßkopf *m s.* Strangpressenkopf
Strangpreßmischung *f* extrusion compound

Strangpreßziegel *m* wire-cut brick
Strangschlichtmaschine *f (text)* hank-sizing machine
Strangtrockenmaschine *f (text)* hank-drying machine
Strangwäsche *f (text)* rope scouring
Strangwaschmaschine *f (text)* rope-scouring machine, dolly [washer]; *(for yarn)* hank-washing machine, hank washer
Straß *m* strass *(a glass of high lead content)*
Straßenmarkierungsfarbe *f* road marking paint, roadline paint
Straßenoktanzahl *f* road octane number
Straßenöl *n* road oil *(for laying dust and for waterproofing)*
Straßenprüfung *f (rubber)* road test
Straßentankwagen *m* road tank wag[g]on, road tanker, tank truck
Straßenteer *m* road tar
streckbar stretchable, extensible, extendible, extensile, *(esp relating to metal)* ductile
Streckbarkeit *f* stretchability, extensibility, *(esp relating to metal)* ductility
strecken 1. to stretch, to extend *(mechanically)*; *(glass)* to flatten; *(tann)* to break *(after bleaching)*; 2. *(using additives)* to extend; to dilute *(liquids)*
mit Wasser s. to water [down] *(e.g. wine)*
Strecken *n* **auf dem Baum** *(tann)* beaming
zweiachsiges S. *(plast)* biaxial stretching
Strecker-Synthese *f* Strecker synthesis *(of α-amino acids)*
Streckformen *n (plast)* stretch forming; drape forming, draping *(of sheets in vacuo)*
S. mit pneumatischer Vorstreckung air-slip forming
S. mit Ring plug-and-ring forming
Streckgrenze *f (plast, met)* yield point
untere S. yield value
Streckmetall *n* expanded (protruded) metal
Streckmetallboden *m (distil)* expanded-metal tray
Streckmittel *n* extender, [extending] filler, cheapener, *(relating to liquids also)* diluting (thinning) agent, diluent, thinner
Streckofen *m (glass)* flattening kiln (oven)
Streckspannung *f* yield stress
Streckspinnen *n* stretch (reel) spinning
Streckstoff *m s.* Streckmittel
Streckung *f* 1. stretch[ing], extension *(mechanically)*; 2. *(using additives)* extension; dilution *(of liquids)*
Streckungsmittel *n s.* Streckmittel
Streckziehen *n (plast)* drape forming *(of sheets in vacuo)*
Streichanlage *f (pap)* coating plant (mill)
Streichappretur *f (text)* doctor finish
streichen 1. *(plast, pap)* to coat; 2. *(tann)* to scud; 3. *(rubber)* to spread; 4. *(ceram)* to mould *(e.g. tiles)*; 5. *(coat)* to paint; to brush *(the paint onto a surface)*
beidseitig (doppelseitig) s. *(pap)* to double-coat, to coat on both sides
einseitig s. *(pap)* to coat on one side
zweiseitig s. *s.* beidseitig s.
Streichen *n* **außerhalb der Maschine** *(pap)* conversion (off-machine, separate) coating
S. in der Maschine *(pap)* [on-]machine coating
S. mit Bürste *(plast)* brush coating (spreading)
S. mit Rakel *(plast, pap)* knife coating
beidseitiges (doppelseitiges) S. *(pap)* double[-sided] coating
einseitiges S. *(pap)* one-sided (single-sided) coating
Streicherei *f (pap)* coating plant (mill)
streichfähig *(food)* spreadable
Streichfähigkeit *f (food)* spreadability
Streichfarbe *f* 1. *(coat)* brushing paint; 2. *s.* Streichmasse
Streichgarngewebe *n* wool[l]en fabric
Streichgerät *n* spreading device, spreader *(chromatography)*
Streichgießverfahren *n (pap)* cast coating
Streichkarton *m* coated [card]board
Streichlösung *f (rubber)* dough, spreading mix[ture]
Streichmaschine *f* 1. *(plast, pap)* coating machine, coater; 2. *(rubber)* spreading machine, spreader
S. für beidseitigen (doppelseitigen) Strich *(pap)* double (duplex, two-side) coater
S. für einseitigen Strich *(pap)* single (one-side) coater
Streichmasse *f (pap)* coating slip (mixture, slurry, substance), coating
Streichmassentrog *m (pap)* coating pan
Streichmesser *n* doctor blade (knife), *(rubber also)* spreading (spreader) knife
Streichmischung *f (rubber)* spreading mix[ture], dough
Streichpapier *n* coated (surfaced) paper
Streichpresse *f (pap)* coating press
Streichrohpapier *n* coated base paper
Streichteig *m s.* Streichmischung
Streichtrog *m (pap)* coating pan
Streichvorrichtung *f (plast, pap)* coating (coater) unit; *(rubber)* spreading unit
Streichwalze *f (plast, rubber)* doctor roll; *(pap)* coating roll
Streifen *m* 1. tape, band; 2. *(phot)* streak *(a defect)*; *(phys chem)* band *(in a spectrum)*
Streifenart *f (coal)* type of band, microlithotype
Streifenbegiftung *f (agric)* band treatment [with pesticides]
Streifenbehandlung *f (agric)* band treatment
Streifendüngung *f (agric)* band placement [of fertilizer], band treatment [with fertilizer]
Streifenkohle *f* banded coal
Streifenschneidemaschine *f*, **Streifenschneider** *m (pap)* slitting machine (device), slitting and [re]winding machine, slitter
Streifenschreiber *m* strip chart recorder
Streifenstruktur *f (coal)* banded structure
streifig banded
Streifung *f (coal)* banded structure; *(geol)* striation

strengflüssig viscous, thick
Strengit *m (min)* strengite *(iron(III) orthophosphate)*
Strenglot *n* hard solder
Streptomyzin *n* streptomycin *(antibiotic)*
Streptose *f* streptose *(a monosaccharide)*
Stretch *m (text)* stretch[ing] *(of filaments)*
Stretchfaktor *m (text)* stretch factor
Stretchgarn *n (text)* stretch yarn
Streuamplitude *f* scattering amplitude
streubar *(agric)* drillable
Streubarkeit *f (agric)* drillability
Streuelektron *n* scattered electron
streuen 1. to sprinkle *(e.g. powder)*; to drill *(fertilizers or pesticides)*; 2. *(phys chem)* to scatter *(e.g. rays, electrons)*
streufähig *(agric)* drillable
Streufaktor *m* scattering factor
Streufrequenz *f* scattering frequency
Streugerät *n (agric)* distributor
Streukoeffizient *m* scattering coefficient
Streulicht *n* scattered light
Streulichtmeßgerät *n* nephelometer
Streulichtmessung *f* nephelometry
Streuneutron *n* scattered neutron
Streupuder *m* dusting powder
Streuquerschnitt *m* scattering cross section
Streusand *m (ceram)* placing sand
Streustrahlung *f* scattered radiation
Streustromkorrosion *f* electrocorrosion
Streuteller *m* whizzer *(classifying)*
Streuung *f (phys chem)* scatter[ing]
  S. der Röntgenstrahlen X-ray scattering
  S. des Lichts light scattering
  elastische S. *(nucl)* elastic scattering
  unelastische S. *(nucl)* inelastic scattering
Streuungs... *s.* Streu...
Streuvermögen *n* 1. *(phys chem)* scattering power; 2. throwing power *(electroplating)*
Streuwinkel *m* scattering angle
Strich *m* 1. *(pap)* coating *(act)*; coat[ing] *(substance)*; 2. *(min)* streak colour; 3. *(nomencl)* prime *(of a locant)*
  beidseitiger (doppelseitiger) S. *(pap)* double[-sided] coating
  einseitiger S. *(pap)* one-sided (single-sided) coating
Strichauftrag *m (pap)* coating application
Strichfarbe *f (min)* streak colour
Strichgitter *n* line grating *(spectroscopy)*
Strichmarke *f* mark *(as of a volumetric flask)*
Strichskale *f* division scale
Strichteilung *f* division *(of a gauge)*
Strippdampf *m* stripping steam
strippen *(distil)* to strip [off, out]
Strippen *n* **mit Luft** *(distil)* air stripping
Stripper *m (distil)* stripper
Stripperkolonne *f*, **Strippingkolonne** *f s.* Strippkolonne
Strippkolonne *f (distil)* stripping (stripper) column
strohähnlich straw-like
Strohaufschluß *m* pulping of cereal straw
Strohhäcksel *m* chopped straw
Strohhäckselmaschine *f* straw chopper (cutter)
Strohkarton *m* straw cardboard
Strohkochung *f (pap)* cooking of straw
Strohmischpappe *f* mixed strawboard
Strohpackpapier *n* straw wrapping paper
Strohpapier *n* straw paper
Strohpappe *f* straw cardboard, strawboard
Strohstoff *m* 1. straw pulp; 2. *s.* Strohzellstoff
  gelber S. coarse (yellow mechanical) straw pulp
  vollaufgeschlossener S. *s.* Strohzellstoff
Strohzellstoff *m* **[/ vollaufgeschlossener]** straw cellulose, fine straw pulp
Strom *m* 1. *(tech)* current, stream, flow; 2. [electric] current
  abgehender S. effluent
  abwärtsgerichteter S. downward (descending) current
  aufsteigender S. upward (ascending) current
  fotoelektrischer S. photocurrent
Stromapparat *m* hydraulic classifier
Stromausbeute *f* current efficiency
Stromdichte *f* current density, C.D.
stromdurchflossen current-carrying
Stromdurchgang *m* current passage
strömen to stream, to flow, to run, to pass, *(if swiftly)* to flush
Stromeyerit *m (min)* stromeyerite *(copper(I) silver sulphide)*
Stromfluß *m* current flow
stromführend current-carrying
Stromklassieren *n* hydraulic classification
  nasses S. wet classification
Stromklassierer *m* hydraulic (trough, launder) classifier
Stromkreis *m* electrical circuit
Stromleiter *m* conductor [of electricity]
Stromlinienfilter *n* streamline filter
stromlos dead
Strommesser *m* amperemeter, ammeter
Stromquelle *f* power (current) source
Stromrichtung *f* direction of flow
Stromrinne *f* trough washer
Stromrohr *n* drying duct *(of a dryer)*
Stromschlüssel *m* **[/ elektrolytischer]** salt bridge
Stromsichtermühle *f* air-swept mill
Strom-Spannungs-Kennlinie *f*, **Strom-Spannungs-Kurve** *f* current-voltage curve
Stromstärke *f* current strength (intensity)
Stromstärkeeinheit *f* unit of current
Stromstoßgalvanometer *n* ballistic galvanometer
Stromtrockner *m* pneumatic [conveying] dryer
Strom- und Spannungsmeßgerät *n* voltammeter
Strömung *f* current, stream, flow
  abwärtsgerichtete S. downward (descending) current
  aufwärtsgerichtete S. upward (ascending) current
  laminare S. laminar (streamlined) flow

turbulente S. turbulent flow
Strömungsbild *n* flow diagram (pattern, sheet)
Strömungsdoppelbrechung *f* streaming birefringence (double refraction), double refraction of flow
Strömungsgeschwindigkeit *f* flow rate
Strömungskalorimeter *n* continuous-flow calorimeter
Strömungsmesser *m* flowmeter
S. für Gase gas flowmeter
Strömungsmethode *f* continuous-flow method *(for determining the specific heat of gases)*
Strömungspotential *n* streaming potential
Strömungsquerschnitt *m* flow area
Strömungsrichtung *f* direction of flow
Strömungsumkehr *f* change of direction of flow
Strömungsverfahren *n* stream method *(underground gasification of coal)*
Strömungswiderstand *m* resistance to fluid flow
Stromzuführung *f* power connection
Strontian *n*, **Strontianerde** *f s.* Strontiumoxid
Strontianit *m (min)* strontianite *(strontium carbonate)*
Strontianverfahren *n* strontia process *(for desugaring molasses)*
Strontium *n* Sr strontium
radioaktives S. radioactive strontium, radiostrontium, *(specif)* ^{90}Sr strontium-90
Strontium-90 *n* ^{90}Sr strontium-90
Strontiumazetat *n* $Sr(CH_3COO)_2$ strontium acetate
Strontiumbromid *n* $SrBr_2$ strontium bromide
Strontiumchlorat *n* $Sr(ClO_3)_2$ strontium chlorate
Strontiumchlorid *n* $SrCl_2$ strontium chloride
Strontium-Einheit *f (biol)* sunshine unit, S.U. *(activity of 1 picocurie ^{90}Sr/g Ca)*
Strontiumfluorid *n* SrF_2 strontium fluoride
Strontiumhydrogenorthophosphat *n* $SrHPO_4$ strontium hydrogenorthophosphate, strontium hydrogenphosphate
Strontiumhydroxid *n* $Sr(OH)_2$ strontium hydroxide
Strontiumjodid *n* SrI_2 strontium iodide
Strontiumkarbonat *n* $SrCO_3$ strontium carbonate
Strontiumnitrat *n* $Sr(NO_3)_2$ strontium nitrate
Strontiumorthoarsenat(III) *n* $Sr_3(AsO_3)_2$ strontium orthoarsenite, strontium arsenite
Strontiumoxid *n* SrO strontium oxide
Strontiumperoxid *n* SrO_2 strontium peroxide
Strontiumsaccharat *n* strontium saccharate, strontium sucrate
Strontiumsulfat *n* $SrSO_4$ strontium sulphate
Strophanthin *n* strophanthin *(any of a group of glycosides)*
Strophanthin G *n* G-strophanthin, g-strophanthin, ouabain
Strophanthin K *n* K-strophanthin
Strudel *m* vortex, eddy
Struktur *f* structure, *(relating to chemical compounds also)* constitution
anisodesmische S. *(cryst)* anisodesmic structure
chemische S. chemical structure
dipolare S. dipolar structure
elektronische S. electronic structure
feinkörnige S. fine-grain structure
geradkettige S. straight-chain structure
isotaktische S. isotactic structure
körnige S. granularity
ringförmige S. annular structure
syndiotaktische S. syndiotactic structure
übermolekulare S. supermolecular structure
verzweigte S. branched-chain structure
wabenartige (wabenförmige) S. honeycomb structure
Widmannstättensche S. Widmannstätten structure *(metallography)*
zellartige (zellige) S. cellular structure
zybotaktische S. cybotaxis *(of liquids)*
Strukturamplitude *f (cryst)* structure amplitude
Strukturanalyse *f* structure analysis
S. mit Röntgenstrahlen X-ray structure (structural) analysis
Strukturaufklärung *f* structure elucidation
Strukturbestimmung *f* structure determination
Strukturbeweis *m* proof (evidence) of structure
Strukturchemie *f* structure (structural) chemistry
strukturchemisch structure-chemical
Struktureffekt *m* structural effect
Struktureinheit *f*, **Strukturelement** *n* structural element (entity, unit), base (repeating) unit
strukturell structural
Strukturfaktor *m (cryst)* structure factor
Strukturfarbe *f (biol)* structural colour
Strukturformel *f* structural (constitutional, graphic) formula
geometrische S. space formula
strukturiert with structure
wabenartig s. honeycombed
strukturisomer structurally isomeric
Strukturisomer[es] *n* structural isomer
Strukturisomerie *f* structural isomerism
strukturlos structureless, devoid of structure
Strukturresonanz *f* resonance, mesomerism
Strukturtyp *m (cryst)* lattice (structure) type
Strukturuntersuchung *f* investigation of structure, structural study
Strukturviskosität *f* structural viscosity
Struvit *m (min)* struvite *(ammonium magnesium orthophosphate)*
Strychnin *n* strychnine *(alkaloid)*
Strychninhydrochlorid *n* strychnine hydrochloride
Strychninnitrat *n* strychnine nitrate
Strychninsulfat *n* strychnine sulphate
Strychnosalkaloid *n* strychnos alkaloid
Stuart-Kalotte *f*, **Stuart-Modell** *n* Stuart model *(for illustrating molecular structures)*
Stückarsenik *m(n) (glass)* glassy (dense) arsenic
Stückbrennstoff *m* lump fuel
Stückbrikett *n* block briquette
Stückenharz *n* crushed resin
Stückerz *n* lump ore
Stückfärbemaschine *f (text)* piece-dyeing machine

Stückfärben *n (text)* piece dyeing
Stuckgips *m* stucco, estrich plaster
Stückgröße *f* particle size; *(rubber)* batch size
stückig lumpy
Stückigkeit *f* lumpiness
Stückigmachen *n* agglomeration
S. auf einem Rütteltisch agglomeration tabling
Stückkalk *m* lump lime
Stückkohle *f* lump coal
Stückkoks *m* lump coke
Stückmasse *f (rubber)* batch weight
Stückware *f (text)* broad cloth
Stückzucker *m* cut (lump) sugar
Stufe *f* 1. step, stage *(as of a reaction)*; 2. *(qualitatively)* grade; 3. effect *(of an evaporator)*
erste S. initiating step *(of a reaction)*
letzte S. completing step *(of a reaction)*
polarografische S. polarographic wave
Stufeneluierung *f* gradient elution
Stufenfolge *f* gradation, scale
Stufengitter *n* echelon grating *(any of various diffraction gratings)*
Stufenheizung *f (rubber)* step (step-up) cure, vulcanization in stages
Stufenhöhe *f* wave height *(polarography)*
Stufenkeil *m (phot)* step wedge
Stufenreaktion *f* step (stepwise, successive) reaction
Stufenregel *f* / **Ostwaldsche** Ostwald rule, law of intermediate reactions (stages), successive reactions law
Stufenrost *m* step grate
Stufenseparator *m (petrol)* multistage separator
Stufenversetzung *f (cryst)* edge dislocation
Stufenwäsche *f* fractional washing
stufenweise stepwise, step-by-step, gradual
Stufenzahl *f* number of stages
Stuhlzäpfchen *n (pharm)* suppository
Stulpdichtung *f*, **Stulpe** *f* cup [ring]
stumpf dull *(surface)*; dead, flat, dull *(colour)*
s. werden to dull
Stumpfheit *f* dullness *(of a surface or colour)*
Stumpfwerden *n* dulling *(of a surface or colour)*
Stupp *f* stupp *(a deposit obtained in distilling mercury ores)*
stürmisch vigorous
Sturtevant-Ring[rollen]mühle *f* Sturtevant mill
stürzen to drop *(e.g. coke)*
die Mischung über Kopf s. *(rubber)* to pass the stock endwise through the mill
Sturzfestigkeit *f* shatter resistance (strength) *(of coke)*
Sturzgießverfahren *n* slush moulding *(for producing hollow articles from polyvinyl chloride plastisols)*
Sturzmühle *f* tumbling mill
autogen arbeitende S. autogenous tumbling mill
Sturzprobe *f*, **Sturzversuch** *m* [drop] shatter test *(of coke)*
Stütze *f (ceram)* post, upright, prop
Stutzen *m* 1. *(tech)* port; fitting, pipe connection; 2. *(lab)* glass cylinder *(broad form)*
Stutzenflasche *f (lab)* aspirator
Stützfaden *m (text)* carrier thread
Stützgewebe *n (text)* back cloth, back grey [cloth]
Stützplatte *f* backup plate *(of a filter)*
Stützrolle *f* idler [roller] *(as of a conveyor belt)*
Stützschneide *f* central knife edge *(of a balance)*
Stützsieb *n* support screen *(of a filtering centrifuge)*; backup plate *(of a filter)*
Stützweite *f* support span *(in installing piping)*
St. W. *s.* Stärkewert
Styphninsäure *f* $C_6H(OH)_2(NO_2)_3$ styphnic acid, 2,4,6-trinitroresorcinol
Styptikum *n (pharm)* styptic
styptisch *(pharm)* styptic
Styrax *m* storax, styrax *(a balsam obtained from Liquidambar specc.)*
Amerikanischer S. American storax (styrax), sweet (red) gum, white amber *(from Liquidambar styraciflua L.)*
Asiatischer S. *s.* Levantiner S.
Levantiner S. Levant (oriental) storax, oriental sweet gum *(from Liquidambar orientalis Mill.)*
Styraxöl *n (cosmet)* storax (styrax) oil *(from Liquidambar specc.)*
Styrol *n* $C_6H_5CH{=}CH_2$ styrene, styrol, cinnamene, phenylethylene
Styrolalkydharz *n* styrenated alkyd
Styrol-Butadien-Kautschuk *m* styrene-butadiene rubber, SBR
Styrol-Chloropren-Kautschuk *m* styrene-chloroprene rubber, SCR
Styroldibromid *n* $C_6H_5CHBrCH_2Br$ styrene dibromide, α,β-dibromoethylbenzene
styrolisieren to styrenate
Styrol-Isopren-Kautschuk *m* styrene-isoprene rubber, SIR
Styrolkautschuk *m s.* Styrol-Butadien-Kautschuk
Styron *n* $C_6H_5CH{=}CHCH_2OH$ cinnamyl alcohol, styryl alcohol, 3-phenyl-2-propen-1-ol
Suakingummi *n* Suakin gum *(from Acacia stenocarpa Hochst.)*
subaquatisch subaqueous, subaquatic
subatomar subatomic
subazid *(med)* subacid
Subazidität *f (med)* subacidity
Subbild *n (phot)* sub-image
subbituminös subbituminous
Suberan *n* suberane, cycloheptane
Suberin *n (bioch)* suberin
Suberinsäure *f* $HOOC[CH_2]_6COOH$ suberic (suberinic) acid, octanedioic acid
Suberol *n s.* Suberylalkohol
Suberon *n* suberone, cycloheptanone
Suberylalkohol *m* suberyl alcohol, suberol, cycloheptanol
Subkeim *m (phot)* sub-image speck
Sublimand *m* sublimand
Sublimat *n* 1. sublimate *(a product obtained by sublimation)*; 2. $HgCl_2$ sublimate, mercury(II) chloride, mercury dichloride
Sublimation *f* sublimation

Sublimationsdruck *m* sublimation pressure
Sublimationsenergie *f* sublimation energy
Sublimationsenthalpie *f* latent heat of sublimation
Sublimationsgut *n* sublimand, material being (*or* to be) sublimated; sublimate *(product of sublimation)*
Sublimationskurve *f* sublimation curve
Sublimationspunkt *m* sublimation point
Sublimationstemperatur *f* sublimation temperature
Sublimationstrocknung *f* lyophilization, freeze drying
Sublimationsvorlage *f* condenser
Sublimationswärme *f* heat of sublimation
Sublimatpapier *n* mercury (mercuric) chloride paper *(a test paper)*
Sublimatprobe *f (med)* sublimate test
sublimierbar sublimable
Sublimierbarkeit *f* sublimability
Sublimierblase *f* sublimer
sublimieren to sublimate, to sublime
Sublimiergut *n s.* Sublimationsgut
submers submerged, submersed
Submikrogefüge *n* submicrostructure
Submikron *n* submicron *(a minute particle visible only with an ultramicroscope)*
submikroskopisch submicroscopic
Submikrostruktur *f* submicrostructure
Submolekül *n* submolecule
Suboxid *n* suboxide
Subsilikatschlacke *f* subsilicate slag
substantiv *(dye)* substantive, direct
Substantivfarbstoff *m* substantive (direct) dye (dyestuff)
Substantivität *f (dye)* substantivity
Substanz *f* substance, matter, material (*for compounds s.a.* Stoff)
  abgewogene S. weighed substance
  abzuwägende S. substance to be weighed
  adstringierende S. astringent, styptic
  bakterientötende (bakterizide) S. bactericide
  giftige S. toxic [substance], toxicant
  inkrustierende S. encrustant, incrustant, encrusting (incrusting) substance
  mineralische S. mineral matter (substance)
  oberflächenaktive S. surface-active agent (substance), surfactant
  paramagnetische S. paramagnetic [substance]
  pflanzliche S. vegetable matter
  standardisierte S. standard substance
  synthetisch gewonnene S. synthetic
  unbekannte S. unknown [substance]
  unlösliche S. insoluble
  vegetabilische S. vegetable matter
  wirksame S. active substance (agent)
Substanzbarren *m* ingot *(in zone melting)*
Substanzformel *f* stoichiometric (empirical, simplest) formula
Substanzmenge *f* amount of substance
Substanzpolymerisation *f* mass (bulk) polymerization
Substanzprobe *f* sample
Substanzschiffchen *n (lab)* boat
  S. aus Graphit graphite boat
  S. aus Quarz quartz boat
  S. aus Tonerde alumina boat
Substanztransport *m* matter transport
Substanzverlust *m* loss of material
Substituent *m* substituent
  S. erster Ordnung *s.* ortho-para-dirigierender S.
  S. mit aktivierender Wirkung activating group
  S. mit desaktivierender Wirkung deactivating group
  S. zweiter Ordnung *s.* meta-dirigierender S.
  äquatorialer S. equatorial substituent
  ausgewechselter S. substituted group
  axialer S. axial substituent
  meta-dirigierender S. meta-directing group, meta director
  ortho-para-dirigierender S. ortho-para-directing group, ortho-para director
  verdrängter S. substituted group
Substituenteneffekt *m* substituent effect
  mesomerer S. mesomeric (electromeric) effect
Substituentenkonstante *f* substituent constant
substituierbar substitutable, replaceable
Substituierbarkeit *f* substitutability, replaceability
substituieren to substitute, to replace
substituiert / dreifach trisubstituted
  einfach s. monosubstituted
  mehrfach s. polysubstituted
  zweifach s. disubstituted
1,2-substituiert *s. o*-substituiert
1,3-substituiert *s. m*-substituiert
1,4-substituiert *s. p*-substituiert
***m*-substituiert** *m*-substituted
***o*-substituiert** *o*-substituted
***p*-substituiert** *p*-substituted
Substitution *f* substitution, replacement
  anionoide S. *s.* nukleophile S.
  elektrophile S. electrophilic substitution
  kationoide S. *s.* elektrophile S.
  nukleophile S. nucleophilic substitution, S_N reaction
Substitutionsbereitschaft *f* susceptibility to substitution
Substitutionsgrad *m* degree of substitution, D.S.
Substitutionsisomerie *f* substitution[al] isomerism
Substitutionslegierung *f* substitutional alloy
Substitutionsmethode *f* substitution method
  S. nach Borda Borda's method [of substitution] *(in weighing)*
  fotometrische S. substitution method in photometry
Substitutionsmischkristall *m* substitutional solid solution
Substitutionsname *m* substitutive name
Substitutionsprodukt *n* substitution product
Substitutionsreaktion *f* substitution (replacement) reaction

Substitutionstautomerie *f* substitution tautomerism
Substitutionswägung *f* substitution weighing
Substrat *n* substrate, *(enzymology also)* reactant
Subtraktionsmischkristall *m* subtraction solid solution
Subtraktionsname *m*, **Subtraktivname** *m* subtractive name
Suchtbildung *f (pharm)* habit formation
suchterzeugend *(pharm)* habit-forming
Sucrochemie *f* sucrochemistry
Sud *m (ferm)* boiling, brewing *(act)*; brew[ing], gyle *(product)*; *(pharm)* decoction
Sudangummi *n* gum arabic, Arabian (acacia) gum *(from Acacia specc.)*
Sudhaus *n* brewhouse
  S. mit doppeltem Sudwerk (Sudzeug) double brewhouse
  S. mit einfachem Sudwerk (Sudzeug) simple brewhouse
  S. mit Mehrgerätesudwerk (Mehrgerätesudzeug) double brewhouse
  S. mit Zweigerätesudwerk (Zweigerätesudzeug) simple brewhouse
Sudhausausbeute *f (ferm)* copper-yield
Sukzinaldehyd *m s.* Sukzindialdehyd
Sukzinamidsäure *f* $NH_2OC-CH_2CH_2COOH$ succinamic acid, succinic acid monoamide
Sukzinat *n* succinate *(a salt or ester of succinic acid)*
Sukzinbromimid *n N*-bromosuccinimide, NBS
Sukzindialdehyd *m* $OHC-CH_2CH_2-CHO$ succindialdehyde, 1,4-butanedial
Sukzinimid *n* succinimide, 2,5-dioxopyrrolidine
Sukzinit *m* succinite, amber
Sukzinodehydrogenase *f* succinodehydrogenase, succinic [acid] dehydrogenase
Sukzinoxydase *f* succinoxidase, succinic [acid] oxidase
Sukzinsäure *f* $HOOC-CH_2CH_2-COOH$ succinic acid, butanedioic acid
Sukzinylchlorid *n* $ClCOCH_2CH_2COCl$ succinyl chloride
Sulfachinoxalin *n* sulphaquinoxaline
Sulfadiazin *n* sulphadiazine, sulphanilamidopyrimidine
Sulfaguanidin *n* sulphanilylguanidine, sulphaguanidine
Sulfamat *n s.* Sulfamidat
Sulfamerazin *n* sulphamerazine, 2-sulphanilamido-4-methylpyrimidine
Sulfamethazin *n* sulphamethazine, sulphadimethylpyrimidine
Sulfamid *n s.* Sulfonamid
Sulfamidat *n* $NH_2SO_2OM^I$ amidosulphate, sulphamate
Sulfamidsäure *f* sulphamic acid *(a compound H_2NSO_3H or any of its organic derivatives $RNHSO_3H$ or R_2NSO_3H)*
Sulfaminat *n s.* Sulfamidat
Sulfaminsäure *f s.* Sulfamidsäure
Sulfan *n* sulphane, hydrogen sulphide, *(specif)* H_2S monosulphane, hydrogen sulphide
Sulfanilamid *n* $NH_2C_6H_4SO_2NH_2$ sulphanilamide, *p*-aminobenzenesulphonamide
Sulfanilsäure *f* $NH_2C_6H_4SO_3H$ sulphanilic acid, aniline-*p*-sulphonic acid
Sulfapyridin *n* sulphapyridine, 2-sulphanilamidopyridine
Sulfat *n* $M_2^ISO_4$ sulphate
Sulfatangriff *m* sulphate attack, attack by sulphates
Sulfat-Anion *n s.* Sulfat-Ion
Sulfatase *f* sulphatase *(any of various esterases)*
Sulfataufschluß *m (pap)* sulphate pulping
Sulfatbeständigkeit *f* sulphate resistivity
Sulfatblase *f* [sulphate] scab, whitewash *(a defect in glass)*
Sulfathiazol *n* sulphathiazole, 2-sulphanilamidothiazole
Sulfathüttenzement *m* super-sulphated cement, slag sulphate cement
sulfatieren to sulphatize, to sulphate
Sulfatierung *f* sulphation, sulphatizing
Sulfat-Ion *n* sulphate ion
sulfatisieren 1. to sulphatize *(e.g. sulphide ores by roasting)*; 2. to sulphate *(the plates of a lead-acid accumulator)*
Sulfatisierung *f* 1. sulphatizing *(as of sulphide ores by roasting)*; 2. sulphation *(of the plates of a lead-acid accumulator)*
Sulfatkochlauge *f (pap)* sulphate (kraft) cooking liquor, sulphate [digestion] liquor
Sulfatkochung *f (pap)* sulphate cook
  S. für Kraftzellstoff kraft cook
Sulfatlauge *f s.* Sulfatkochlauge
Sulfatlignin *n (pap)* sulphate lignin
Sulfatpapier *n* sulphate paper
Sulfatreduktase *f* sulphate reductase
Sulfatseife *f (pap)* sulphate (tall-oil) soap
Sulfatterpentin *n (pap)* sulphate turpentine
Sulfatverfahren *n (pap)* sulphate process
Sulfatzellstoff *m* sulphate pulp
  ungewaschener S. brown stock (pulp)
Sulfatzellstoffabrik *f* sulphate (kraft) mill
Sulfatzellstoffkocher *m (pap)* sulphate (kraft) digester
Sulfatzellstoffwäsche *f (pap)* brown-stock washing
Sulfazetamid *n* sulphacetamide *(a sulphonamide)*
Sulfenamid *n* sulphenamide
Sulfenamidbeschleuniger *m (rubber)* sulphenamide accelerator
Sulfensäure *f (org chem)* RSOH sulphenic acid
Sulfhydrylgruppe *f* –SH sulphydryl (thiol, mercapto) group
Sulfid *n* M_2^IS sulphide
  saures S. *s.* Hydrogensulfid
Sulfidäscher *m (tann)* sulphide lime
Sulfiderz *n* sulphide ore
Sulfidgehalt *m* sulphide content, *(pap also)* sulphidity
sulfidieren to sulphidize, to sulphide, *(text also)* to xanthate, to churn

Sulfidiertrommel *f (text)* xanthator, [xanthating] churn, baratte
Sulfidierung *f* sulphidizing, *(text also)* xanthation, churning
sulfidisch sulphidic
Sulfidität *f* sulphidity
Sulfidoxidschale *f*, **Sulfidoxidzone** *f (geoch)* sulphide-oxide shell
Sulfidphosphor *m* sulphide phosphor
Sulfidschwefel *m* sulphide sulphur
Sulfieranlage *f* sulphonation plant (unit)
Sulfierapparat *m* sulphonator
sulfieren to sulphonate *(to treat an organic compound with sulphuric acid or a related agent)*
Sulfiergefäß *n*, **Sulfierkessel** *m* sulphonation (sulphonating) pan
Sulfierkolben *m* sulphonation flask
Sulfierung *f* sulphonation *(treatment of an organic compound with sulphuric acid or a related agent)*
Sulfierungs ... *s.* Sulfier ...
Sulfinfarbe *f* sulphide (sulphur) dye (dyestuff)
Sulfinsäure *f (org chem)* RSO_2H sulphinic acid
Sulfit *n* $M_2^ISO_3$ sulphite
Sulfitablauge *f (pap)* spent (waste) sulphite liquor, red liquor, sulphite lye
Sulfitablaugehefe *f* spent sulphite liquor yeast
Sulfitation *f (sugar)* sulphitation
Sulfitaufschluß *m (pap)* sulphite pulping
sulfitieren *(sugar)* to sulphite
Sulfitierung *f (sugar)* sulphitation
Sulfitkochsäure *f (pap)* sulphite (bisulphite) cooking liquor
Sulfitkochsäureherstellung *f (pap)* cooking-liquor manufacture, acid preparation (making)
Sulfitkochung *f (pap)* sulphite cook
Sulfitkraftpapier *n* sulphite kraft paper
Sulfitokobaltat(III) *n* $M_3^I[Co(SO_3)_3]$ sulphitocobaltate(III)
Sulfitomerkurat(II) *n* $M_2^I[Hg(SO_3)_2]$ sulphitomercurate(II), disulphitomercurate(II)
Sulfitpapier *n* sulphite paper
Sulfitsäure *f s.* Sulfitkochsäure
Sulfitstoff *m s.* Sulfitzellstoff
Sulfitverfahren *n (pap)* sulphite pulping [process]
Sulfitzellstoff *m* sulphite pulp
  S. in Bogenform (Pappenform) sulphite laps
Sulfitzellstoffabrik *f* sulphite [pulp] mill
Sulfitzellstoffblätter *npl* sulphite laps
Sulfitzellstoffindustrie *f* sulphite pulp industry
Sulfitzellstoffkocher *m* sulphite digester
Sulfitzellstoffwerk *n* sulphite [pulp] mill
Sulfochlorid *n* RSO_2Cl sulphonyl chloride
sulfochlorieren to sulphochlorinate
Sulfochlorierung *f* sulphochlorination
Sulfogruppe *f* $-SO_3H$ sulpho group
Sulfokarbamid *n s.* Thioharnstoff
Sulfon *n* sulphone
Sulfonal *n* $(CH_3)_2C(SO_2C_2H_5)$ sulphonal, propane-2,2-diethyldisulphone
Sulfonamid *n* sulphonamide *(any of a class of compounds characterized by the radical* $-SO_2NHR$*)*
Sulfonamidpräparat *n* sulpha drug
Sulfonat *n* sulphonate
Sulfonator *m* sulphonator
Sulfongruppe *f s.* Sulfogruppe
Sulfonieranlage *f* sulphonation plant (unit)
sulfonieren to sulphonate
Sulfonierer *m* sulphonator
Sulfonierung *f* sulphonation
Sulfonierungsagens *n* sulphonating [re]agent
Sulfonierungsanlage *f* sulphonation plant (unit)
Sulfonierungsgemisch *n* sulphonation mixture
Sulfonierungsgrad *m* degree of sulphonation
Sulfonierungskessel *m* sulphonation (sulphonating) pan
Sulfonierungsmittel *n* sulphonating [re]agent
Sulfoniumverbindung *f* sulphonium compound
Sulfonsäure *f* sulphonic acid, sulpho acid, sulphacid
Sulfonsäureamid *n s.* Sulfonamid
Sulfonsäurechlorid *n s.* Sulfochlorid
Sulfonsäuregruppe *f s.* Sulfogruppe
Sulfonylchlorid *n s.* Sulfochlorid
Sulfonylgruppe *f* $=SO_2$ sulphonyl group
Sulfosalizylsäure *f* $C_6H_3(OH)(SO_3H)COOH$ sulphosalicylic acid *(either of two hydroxy-sulphobenzoic acids)*
Sulfosäure *f s.* Sulfonsäure
Sulfoxid *n* R^1SOR^2 sulphoxide
Sulfoxylat *n* sulphoxylate
Sulfoxylsäure *f* HSO_2H sulphoxylic acid
Sulfurationsapparat *m* sulphonator
Sulfurationskessel *m* sulphonation (sulphonating) pan
sulfurieren to sulphonate
Sulfurierung *f* sulphonation *(broadly, regardless of the nature of the products)*
Sulfurikation *f (soil)* sulphofication *(microbial oxidation of organically bound sulphur to sulphate)*
Sulfuröl *n* sulphur [olive] oil *(olive oil of inferior grade)*
Sulfurylchlorid *n* SO_2Cl_2 sulphuryl chloride
Sumach *m (tann)* sumac *(leaves from Rhus specc., chiefly from Rhus coriaria L.)*
Sumatrabenzoe *f* Sumatra benzoin [gum] *(from Styrax benzoin Dryander)*
Sumatrakampfer *m* Sumatra (Borneo, Malayan, Baros) camphor *(from Dryobalanops aromatica Gaertn.f.)*
Sumbulöl *n* sumbul oil *(from Ferula sumbul Hook.)*
Summenformel *f* empirical (total molecular) formula
  einfachste S. simplest [possible] formula, stoichiometric formula
  wahre S. true (empirical molecular) formula
Summenregel *f* / **Burger-Dorgelo-Ornsteinsche** Burger-Dorgelo-Ornstein sum rule *(for atomic spectra)*
Sumpf *m (tech)* pond, pit, sump; *(distil)* bottom *(of a column)*; pool *(polarography)*
Sumpfaufgabe *f* top feed *(of a double-drum dryer)*

sumpfen *(ceram)* to soak, to mature
Sumpfgas *n* marsh gas
Sumpfgrube *f (ceram)* soak[ing] pit
Sumpfofen *m s.* Sumpfphasehydrierofen
Sumpfphase *f* liquid phase *(in coal hydrogenation)*
Sumpfphasebenzin *n* liquid-phase gasoline (petrol) *(obtained in coal hydrogenation)*
Sumpfphasehydrierofen *m* liquid-phase converter *(in coal hydrogenation)*
Sumpfphasehydrierung *f* liquid-phase hydrogenation *(of coal)*
Sumpfphasekammer *f* liquid-phase stall *(in coal hydrogenation)*
Sumpfphasekatalysator *m* liquid-phase catalyst *(in coal hydrogenation)*
Sumpfphaseofen *m s.* Sumpfphasehydrierofen
Sumpfprodukt *n (distil)* bottoms, bottom product
Sumpftiefe *f (tech)* pond depth
Sunn *m (text)* sunn, sun[n] hemp *(fibres from Crotalaria specc.)*
Super-Abrasion-Furnace-Ruß *m (rubber)* super abrasion furnace black, SAF black
Superadditivität *f (phot)* superadditivity
Superbenzin *n* premium gasoline (motor fuel, spirit)
superfluid superfluid, superliquid
Superfluidität *f* superfluidity, superfluid state
superflüssig *s.* superfluid
Superhelix *f* coiled-coil *(fine structure of polypeptide chains)*
Superkalander *m (pap)* supercalender
Superkraftstoff *m* premium fuel (gasoline, spirit), premium motor fuel
Superlegierung *f* superalloy
Supermultiplett *n* supermultiplet
Superphosphat *n* superphosphate
einfaches (normales) S. ordinary (normal) superphosphate
Supertemperatur *f* super temperature *(more than 5000°C)*
Superzentrifuge *f* supercentrifuge
Suppositorium *n (pharm)* suppository
suprafluid superfluid, superliquid
Suprafluidität *f* superfluidity, superfluid state
supraflüssig *s.* suprafluid
Supraflüssigkeit *f s.* Suprafluidität
supraleitend superconducting, superconductive
Supraleiter *m* superconductor
supraleitfähig *s.* supraleitend
Supraleitfähigkeit *f* superconductivity, supraconductivity, superconduction ✦ **S. zeigen** to superconduct
Supraleitung *f s.* Supraleitfähigkeit
Supraleitungselektron *n* superconducting electron
Surrogat *n* substitute
suspendierbar suspensible
Suspendierbarkeit *f* suspensibility
suspendieren to suspend
Suspendiermittel *n s.* Suspensionsmittel
Suspendiervermögen *n* suspending capacity
Suspension *f* suspension ✦ **in S. halten** to keep in suspension
S. mit Ölflockung *(agric)* oil-flocculated suspension
Suspensionsgrenze *f* sludge line
Suspensionskolloid *n* suspensoid [colloid]
Suspensionsmittel *n* suspending agent (medium)
Suspensionspolymerisation *f* suspension (bead) polymerization
Suspensionspolyvinylchlorid *n* suspension polyvinyl chloride
Suspensionsröstung *f* suspension roasting
Suspensionsspritzmittel *n* wettable powder *(crop protection)*
Suspensionszulauf *m (filtr)* feed inlet
Suspensoid *n* suspensoid [colloid]
Suspensoid-Kracken *n (petrol)* suspensoid [catalytic] cracking
Suspensoid-Krackverfahren *n (petrol)* suspensoid[-catalytic-cracking] process
süß sweet *(taste)*; sweet[ened] *(petroleum distillate)*
Süße *f* sweetness
süßen *(food, petrol)* to sweeten
Süßerde *f* beryllia, beryllium oxide
Sussexit *m (min)* sussexite *(manganese(II) hydrogenorthoborate)*
Süßholzsaft *m* [pure] licorice, liquorice *(from Glycyrrhiza glabra L.)*
süßlich sweetish
Süßmaische *f* sweet mash
Süßmolke *f* sweet whey
Süßmolkenpulver *n* sweet-whey powder
Süßmost *m* fruit juice
Süßrahm *m* sweet cream
Süßrahmbutter *f* sweet[-cream] butter
Süßrahmbuttermilch *f* sweet-cream buttermilk
Süßstoff *m* sweetening agent, sweetener
Süßung *f* sweetening *(of petroleum distillates)*
Süßungsmittel *n (petrol)* sweetening agent
Süßungsreaktion *f (petrol)* sweetening reaction
Süßungsverfahren *n (petrol)* sweetening process
oxydatives S. oxidation [sweetening] process
Süßwaren *fpl* confectionery
Süßwasser *n* sweet (fresh) water
Süßwassergewinnungsanlage *f* desalination (desalinating) plant
Süßwein *m* sweet wine
Suszeptibilität *f* susceptibility
diamagnetische S. diamagnetic susceptibility
ferromagnetische S. ferromagnetic susceptibility
molare S. molar susceptibility, susceptibility per gram mole
paramagnetische S. paramagnetic susceptibility
spezifische S. specific (mass) susceptibility, susceptibility per gram
Sweetland-Filter *n*, **Sweetland-Presse** *f* Sweetland filter, Sweetland [filter] press
Swellingindex *m (coal)* swelling index (number)

Swenson-Walker-Kristallisator *m* Swenson-Walker crystallizer
S-Wert *m (soil)* S value *(sum of exchangeable bases)*
swl. *s.* löslich / sehr schwer
Sydnon *n* sydnone *(any of a class of mesoionic compounds)*
Syenit *m* syenite *(an igneous rock composed essentially of alkali feldspar)*
Sylvan *n* sylvan, 2-methylfuran
Sylvanit *m (min)* sylvanite *(a gold silver telluride)*
Sylvestren *n* sylvestrene *(a cyclic terpene)*
Sylvin *m (min)* sylvite, sylvin[e] *(potassium chloride)*
Symbol *n* symbol, sign, *(for representing chemical elements also)* chemical symbol (sign)
Symbole *npl* / **Bravaissche** *(cryst)* Bravais-Miller indices
Symbolik *f* symbolism
Symmetrie *f* symmetry
äußere S. *(cryst)* external symmetry
Symmetrieachse *f (cryst)* axis of symmetry
sechszählige S. sixfold axis of symmetry
Symmetrieebene *f (cryst)* plane of symmetry
Symmetrieelement *n (cryst)* element of symmetry, symmetry element
Symmetriegrad *m* degree of symmetry
Symmetrieklasse *f* class of crystal symmetry, crystallographic class
Symmetrieordnung *f s.* Symmetriegrad
Symmetriezentrum *n (cryst)* centre of symmetry
symmetrisch symmetric[al]
Symons-Kegelbrecher *m* Symons cone crusher
Symons-Kegelgranulator *m* Symons short-head cone crusher
Symons-Scheibenbrecher *m*, **Symons-Tellerbrecher** *m* Symons disk crusher
Sympathikolytikum *n* sympatholytic [drug]
sympathikolytisch sympatholytic
Sympathikomimetikum *n* sympathomimetic [drug]
sympathikomimetisch sympathomimetic
sympatho ... *s.* sympathiko ...
Sympatho ... *s.* Sympathiko ...
Synaerese *f s.* Synärese
synantetisch *(geoch)* synantetic
syn-anti-Isomerie *f* syn-anti isomerism
Synärese *f (coll)* synaeresis
Synchrotron *n* synchrotron
Synchrozyklotron *n* synchrocyclotron *(frequency-modulated cyclotron)*
syndiotaktisch, syndyotaktisch syndiotactic, syn[dyo]tactic
Synergismus *m* synergism
Synergist *m* synergist
synergistisch synergistic
syn-Form *f* syn form, skew (gauche) form *(stereochemistry)*
Syngenit *m (min)* syngenite *(calcium potassium sulphate)*
Synionie *f* synionism *(mesomerism in the anion of tautomeric compounds)*
syn-Isomer[es] *n* syn isomer
Synovia[flüssigkeit] *f (med)* synovial fluid
±**syn-periplanar** syn-periplanar, eclipsed, opposed *(stereochemistry)*
syn-Stellung *f* syn position ✦ **in syn-Stellung [befindlich]** syn ✦ **in syn-Stellung stehen** to be syn
Syntan *n* syntan, synthetic tannin (tanning agent)
Syntangerbung *f* syntan tannage
Syntexis *f (geol)* syntexis
Synthese *f* synthesis
S. nach Müller-Rochow Müller-Rochow synthesis *(of organosilicon compounds)*
asymmetrische S. asymmetric synthesis
biologische S. biosynthesis, biogenesis
Conrad-Limpachsche S. Conrad-Limpach synthesis *(of 4-hydroxyquinolines)*
Doebnersche S. Doebner synthesis *(of substituted cinchoninic acids)*
Erlenmeyersche S. Erlenmeyer-Plöchl azlactone synthesis *(of α-amino acids)*
Fittigsche S. Fittig synthesis *(of aromatic hydrocarbons)*
Friedel-Craftssche S. Friedel-Crafts synthesis *(of aromatic hydrocarbons or ketones)*
Gattermann-Kohlsche S. Gattermann-Kohl synthesis *(of phenolic aldehydes)*
Kilianische S. Kiliani[-Fischer] synthesis *(for increasing the number of C atoms in the carbon chain of sugars)*
Knoevenagelsche S. Knoevenagel condensation (synthesis) *(of α,β-unsaturated acids or esters)*
Kolbesche S. Kolbe synthesis *(1. of hydrocarbons by electrolysis; 2. of phenolic acids)*
Kolbe-Schmittsche S. Kolbe-Schmitt synthesis *(of aromatic hydroxy acids)*
Perkinsche S. Perkin condensation (synthesis) *(of unsaturated carboxylic acids)*
Reformatskysche S. Reformatsky synthesis *(of 3-hydroxycarboxylic acid esters)*
Reimer-Tiemannsche S. Reimer-Tiemann synthesis *(of hydroxyaldehydes or hydroxyacids)*
Skraupsche S. Skraup [quinoline] synthesis
Streckersche S. Strecker synthesis *(of α-amino acids)*
technische (technisch brauchbare) S. commercial synthesis
Wurtzsche S. Wurtz synthesis *(of aliphatic hydrocarbons)*
Syntheseammoniak *n* synthetic ammonia
Synthesechemie *f* synthetic chemistry
Synthesefaser *f* synthetic [polymer] fibre
Synthesefaserstoff *m* synthetic [polymer] fibre
Synthesefett *n* synthetic (artificial) fat
Synthesegas *n* synthesis gas
Synthesekautschuk *m* synthetic (artificial, man-made, chemical) rubber
Synthesekautschukkleber *m* synthetic-rubber adhesive
Synthesekautschuklatex *m* synthetic[-rubber] latex
Synthesekautschukmischung *f* synthetic-rubber mix (stock, compound)
Syntheselatex *m s.* Synthesekautschuklatex

Syntheseprodukt *n* synthetic
synthetisch synthetic[al], artificial, man-made, manufactured, non-natural
synthetisieren to synthesize
S-Yperit *n s.* Schwefelyperit
Syringaaldehyd *m*, **Syringaldehyd** *m* $(CH_3O)_2C_6H_2(OH)CHO$ syringic aldehyde, syringaaldehyde, 4-hydroxy-3,5-dimethoxybenzaldehyde
Syringasäure *f* $(CH_3O)_2C_6H_2(OH)COOH$ syringic acid, 4-hydroxy-3,5-dimethoxybenzoic acid
Syringin *n* syringin, methoxyconiferin
System *n* system
  S. der Chemical Abstracts *(nomencl)* Chemical Abstracts system
  S. des Ring-Index *(nomencl)* Ring Index system
  S. mit konjugierten Doppelbindungen conjugated dienoid system
  abgeschlossenes S. isolated system
  Baeyersches S. Baeyer system *(of naming bridged hydrocarbons)*
  binäres S. binary (two-component) system
  dispergierendes S. dispersive system *(of a spectroscope)*
  disperses S. disperse system
  divariantes S. divariant system
  einphasiges S. *s.* homogenes S.
  endozyklisches S. endocyclic (caged ring) system
  geschlossenes S. *(phys chem)* closed system
  heterogenes S. heterogeneous system
  heterozyklisches S. heterocyclic system
  homogenes S. homogeneous (one-phase) system
  inhomogenes S. *s.* heterogenes S.
  isodisperses S. *(coll)* isodisperse system, isodispersion
  isoliertes S. isolated system
  kolloiddisperses (kolloides) S. colloidal system
  kondensiertes S. condensed system
  konjugiertes S. conjugated system
  kristallografisches S. *s.* Kristallsystem
  mehrphasiges S. multiphase system
  metastabiles S. metastable system
  monodisperses S. monodisperse system
  monovariantes S. monovariant system
  offenes S. *(phys chem)* open system
  offenkettiges S. *(org chem)* open-chain system
  periodisches S. *s.* Periodensystem [der Elemente]
  polydisperses S. *(coll)* polydisperse system, polydispersion
  quaternäres S. quaternary (four-component) system
  Stocksches S. Stock system *(of indicating the oxidation state)*
  ternäres S. ternary (tertiary, three-component) system
  unitäres S. one-component system
  zweiphasiges S. *s.* binäres S.
Systeminsektizid *n* systemic insecticide
systemisch systemic *(pesticide)*
Systemname *m* / **Genfer** *(nomencl)* Geneva name
SZ *s.* Säurezahl
Szent-Györgyi-Krebs-Zyklus *m* Krebs (citric-acid, tricarboxylic-acid) cycle, TCA cycle
Szilard-Chalmers-Detektor *m (nucl)* Szilard-Chalmers detector
Szilard-Chalmers-Effekt *m (nucl)* Szilard-Chalmers effect
Szilard-Chalmers-Methode *f (nucl)* Szilard-Chalmers method
Szilard-Chalmers-Reaktion *f (nucl)* Szilard-Chalmers reaction
Szintillation *f* scintillation
Szintillationsmethode *f* scintillation method
Szintillationsspektrometer *n* scintillation spectrometer
Szintillationszähler *m* scintillation counter
szintillieren to scintillate
SZT *s.* Selbstzündtemperatur

T

2,4,5-T *s.* 2,4,5-Trichlorphenoxyessigsäure
TA *s.* Triazetatfaserstoff
Tabak *m* tobacco
Tabakalkaloid *n* tobacco alkaloid
Tabaklauge *f* tobacco liquor (water)
Tabakpapier *n* tobacco paper
Tabaksamenöl *n* tobacco[seed] oil
Tablette *f* tablet; *(plast)* biscuit, bisque *(for manufacturing disk records)*
Tablettenpresse *f s.* Tablettiermaschine
tablettieren to tablet
Tablettiermaschine *f* tabletting (tablet-compressing) machine, tablet press
Tablettierung *f* tabletting
Tabun *n* $(CH_3)_2NP(O)(C_2H_5O)(CN)$ tabun *(ethyl ester of dimethylphosphoramidocyanidic acid)*
Tachyhydrit *m (min)* tachyhydrite, tachydrite *(calcium magnesium chloride)*
Tachysterin *n* tachysterol *(formed by ultraviolet irradiation of ergosterol)*
Taenit *m* taenite *(a nickel-iron alloy occurring in iron meteorites)*
Tafel *f* plate, *(if comparatively thick)* slab, *(if thin)* sheet; *(cryst)* plate
Tafel-Diagramm *n (el chem)* Tafel diagram (plot)
Tafelessig *m* table vinegar
tafelförmig tabular; *(cryst)* platy
Tafel-Gerade *f (el chem)* Tafel line
Tafelglas *n* sheet glass
Tafelglasziehverfahren *n (glass)* [flat] sheet drawing process
  Fourcaultsches T. Fourcault [sheet-drawing] process
Tafel-Gleichung *f (el chem)* Tafel equation
tafelig *s.* tafelförmig
Tafelmargarine *f* table margarine
Tafelöl *n* table oil
Tafelsalz *n* table salt
Tafelschmiere *f (tann)* dubbin[g], stuffing mixture
Tafelwasser *n* table (mineral) water
Tagatose *f* tagatose *(a ketose)*

Tagebaubetrieb *m*, **Tagebauförderung** *f* surface (open-cast, open-cut) mining, open-pit method, *(esp relating to ores)* [surface] quarrying

Tagescreme *f (cosmet)* vanishing cream

Tagesdosis *f (tox)* daily intake [dose]
  bedingt duldbare (zulässige) T. conditional acceptable daily intake
  duldbare (zulässige) T. acceptable daily intake, ADI

Tageslichtbeständigkeit *f* sunlight resistance (stability)

Tageslicht-Leuchtpapier *n* luminous paper

Tageslichtpapier *n (phot)* daylight paper

Tageswanne *f (glass)* day tank

Tagliabue-Prüfer *m* Tag[liabue] closed tester *(for determining the flash point)*

Taigusäure *f* lapachol, taiguic acid *(a naphthoquinone derivative)*

Taktizität *f* tacticity

Taktoid *n (coll)* tactoid

Taktosol *n (coll)* tactosol

Talalay-Treibverfahren *n* Talalay process *(for producing foamed rubber)*

Talbotypie *f (phot)* calotype process

Talca-Gummi *n s.* Talha-Gummi

Talg *m* tallow
  Chinesischer (vegetabilischer) T. Chinese vegetable tallow *(from Sapium sebiferum (L.) Roxb.)*

talgartig tallowy, sebaceous

Talgbaum *m* / **Westafrikanischer** butter tree, Pentadesma butyraceum Sabine

Talgdrüse *f* sebaceous (fat) gland

talgig tallowy, sebaceous

Talh[a]-Gummi *n* talha (talh, talca, suakin) gum *(a gum arabic, chiefly from Acacia stenocarpa Hochst.)*

Talk *m (min)* talc[um] *(magnesium dihydrogentetrasilicate)*; $Mg_3[Si_4O_{10}](OH)_2$ *(synthetic)* talc[um]; talcum [powder], talc

talkieren *(rubber)* to soapstone

talkig talcose, talcous, talcky

Talkpuder *m s.* Talkum

Talkschiefer *m* talc (talcose) schist (slate)

Talkum *n* talcum [powder], talc

talkumieren *(rubber)* to soapstone

Talkumpuder *m s.* Talkum

Tallöl *n* tall oil, tallo[e]l, liquid resin (rosin)

Tallölkolophonium *n* tall-oil (sulphate wood) rosin

Tallölseife *f (pap)* tall-oil soap

Talomethylose *f* talomethylose *(a 6-deoxyhexose)*

Talonsäure *f* $HOCH_2[CHOH]_4COOH$ talonic acid

Taloschleimsäure *f* talomucic acid

Talose *f* talose *(a hexose)*

Tamarindenmus *n (pharm)* tamarind pulp, tamarind[o] *(from Tamarindus indica L.)*

Tambour *m s.* 1. Tambourrolle; 2. Tambourwalze

Tambourrolle *f (pap)* reel (roll) of paper

Tambourwalze *f (pap)* reel-up (reeling-up) drum (cylinder), reel (reeling) drum (cylinder)

Tanazetketon *n* tanacetketone, thujaketone, 6-methyl-5-methylene-2-heptanone

Tanazetöl *n* tansy oil *(from Chrysanthemum vulgare (L.)Bernh.)*

Tanazetylalkohol *m* tanacetyl alcohol, thujyl alcohol, 3-thujanol

Tandemanlage *f (rubber)* tandem calender

Tangelrendzina *f (soil)* subalpine rendzina

Tangentialkammer *f* **nach Meyer** Meyer tangential chamber *(sulphuric-acid manufacture)*

Tank *m* tank, reservoir

Tank-Absetz-Verfahren *n (petrol)* cold-settling process

Tankanhänger *m* tank trailer *(for lorries)*

Tankauto *n* tank truck, [road] tanker, road tank wag[g]on

Tankbehandlung *f* bulk handling *(as of milk)*

Tankboden *m* tank bottom

Tankbodenparaffin *n (petrol)* tank-bottom wax

Tankbodenrückstände *mpl* tank bottoms

Tankbodenwachs *n (petrol)* tank-bottom wax

Tankdialysator *m* tank-type dialyzer

Tankentwickler *m (phot)* tank developer

Tankentwicklung *f (phot)* tank development

Tankgärverfahren *n* bulk champagnization, bulk (charmat) process

Tanklager *n* tank farm

Tankmischmethode *f* tank-mix method *(crop protection)*

Tankrückstandsparaffin *n*, **Tankrückstandswachs** *n (petrol)* tank-bottom wax

Tankschlamm *m* tank sludge

Tankwaage *f* weighing tank

Tannase *f* tannase

Tannat *n* tannate *(salt or ester of tannic acid)*

tanniert *(dye)* tannin-mordanted

Tannin *n* tannic acid, gallotannic acid, gallotannin, tannin; *(broadly)* vegetable tannin *(any of a large number of substances used in leather tanning)*
  eigentliches T. tannic acid proper

tanningebeizt *(dye)* tannin-mordanted

tanninhaltig tanniferous

Tanninlösung *f* tannic-acid solution, *(broadly)* tannin solution

Tannin-Nachbehandlung *f* back-tanning

Tanninreaktiv *n* / **Weingärtners** Weingärtner solution *(for precipitating basic coal-tar dyes)*

Tannin-Solutizerverfahren *n* tannin-solutizer process *(for desulphurizing petroleum distillates)*

Tantal *n* Ta tantalum

Tantal(III)-bromid *n* $TaBr_3$ tantalum(III) bromide, tantalum tribromide

Tantal(V)-bromid *n* $TaBr_5$ tantalum(V) bromide, tantalum pentabromide

Tantal(III)-chlorid *n* $TaCl_3$ tantalum(III) chloride, tantalum trichloride

Tantal(V)-chlorid *n* $TaCl_5$ tantalum(V) chloride, tantalum pentachloride

Tantal(V)-fluorid *n* TaF_5 tantalum(V) fluoride, tantalum pentafluoride

Tantal(V)-hydroxid *n* $Ta(OH)_5$ tantalum(V) hydroxide

Tantalit *m (min)* tantalite *(an oxide of iron, manganese, tantalum, and niobium)*

Tantalkarbid *n* tantalum carbide

Tantal(V)-oxid *n* Ta_2O_5 tantalum(V) oxide, ditan-

talum pentaoxide, tantalum pentaoxide
Tantalpent..., Tantalpenta... *s.* Tantal(V)-...
Tantalsäure *f* tantalic acid
Tantaltri... *s.* Tantal(III)-...
Tapetenpapier *n* wall paper, hanging [paper]
Tapetenrohpapier *n* hanging base paper
Tapioka[stärke] *f* tapioca [starch] *(from Manihot utilissima Pohl)*
Tapiolit *m (min)* tapiolite *(an oxide of iron, manganese, tantalum, and niobium)*
Tara *m (tann)* tara *(pods of Caesalpinia tinctoria Domb.)*
Tarapacait *m (min)* tarapacaite *(potassium chromate)*
Tarelaidinsäure *f* tarelaidic acid, *trans*-6-octadecenoic acid
tarieren to tare
Taririnsäure *f* $CH_3[CH_2]_{10}C{\equiv}C[CH_2]_4COOH$ tariric acid, 6-octadecynoic acid
Taroxylsäure *f* $CH_3[CH_2]_{10}COCO[CH_2]_4COOH$ taroxylic acid, 6,7-dioxo-octadecanoic acid
Tartramid *n* $NH_2COCH(OH)CH(OH)CONH_2$ tartramide
Tartramidsäure *f* $NH_2COCH(OH)CH(OH)COOH$ tartramidic acid, tartaric monoamide
Tartranilsäure *f* $C_6H_5NHCOCH(OH)CH(OH)COOH$ tartranilic acid, tartaric monoanilide
Tartrat *n* tartrate *(salt or ester of tartaric acid)*
Tartratokomplex *m* tartrato complex
Tartrazin *n* tartrazine, hydrazine yellow *(a pyrazole derivative)*
Tartronsäure *f* $HOCH(COOH)_2$ tartronic acid
Tartronylharnstoff *m* tartronylurea, dialuric acid, 5-hydroxybarbituric acid
Tasche *f (pap)* pocket *(of a pulpwood grinder)*
Taschenspektroskop *n* pocket spectroscope
Tassendrehmaschine *f (ceram)* cup jolley
Tassengarniermaschine *f (ceram)* cup-handling machine
Tastpolarografie *f* tast polarography
taub *(mine)* barren, dead
tauchaluminieren to dip-aluminize
Tauchanlage *f* dipping plant
Tauchapparat *m* dipping machine (apparatus)
Tauchartikel *mpl* dipped goods (articles)
Tauchauftrag *m* dip feed *(to a drum dryer)*
Tauchbad *n* dipping bath, *(agric also)* dip
Tauchbehälter *m* dip[ping] tank
Tauchbeschichtung *f* dip coating
Tauchbeschichtungseinrichtung *f* dip coater
Tauchblattfilter *n* open-tank leaf filter
Tauchbleiche *f (tann)* dip bleaching
Tauchbottich *m* dip[ping] tank
Tauchbrenner *m* submerged burner, submerged combustion burner (heater)
Tauchbütte *f (pap)* dipping (working) vat, vat of pulp
Tauchdauer *f* dipping time
Tauchelektrode *f* immersion (dipped) electrode
Tauchemaillieren *n* dip enamelling
tauchen to immerse, to immerge, to plunge, *(for a short time)* to dip, *(totally)* to submerge
Tauchen *n* immersion, plunge, *(for a short time)* dip[ping], *(totally)* submergence
T. in Lösungen *(plast)* solvent moulding
T. mit heißen Formen *(rubber)* hot former dipping
T. mit Koagulationsmitteln coagulant dipping
T. von Hand *(ceram)* hand dipping
Tauchfärbemaschine *f* dip-dyeing machine
tauchfärben to dip-dye
Tauchfärbung *f* dip dyeing
Tauchfilter *n* open-tank leaf filter
Tauchform *f* dipping form, *(rubber also)* former
graphit[is]ierte T. *(glass)* paste mould
Tauchgefäß *n* dip[ping] tank
Tauchgestell *n* dipping rack
Tauchglasieren *n* dip glazing
Tauchglasiermaschine *f* dip-glazing machine
Tauchgummiwaren *fpl* dipped goods (articles)
tauchhärten to dip-harden
Tauchhärtung *f* dip hardening
Tauchkolben *m* plunger, ram
Tauchkolbenpumpe *f* plunger (ram) pump
Tauchkolorimeter *n* immersion colorimeter
Tauchkühler *m* pond cooler
Tauchlack *m* dipping varnish
tauchlackieren to dip
Tauchlackierung *f* dipping
Tauchlösung *f* dipping solution
tauchlöten to dip-solder; *(using hard solder)* to dip-braze
Tauchmischung *f* dipping compound, dip mix
tauchpatentieren to dip-patent
Tauchpresse *f* steeping press
Tauchprüfung *f* total immersion test *(corrosion testing)*
Tauchpumpe *f* submerged (wet-pit) pump
Tauchrohrkondensator *m*, **Tauchrohrverflüssiger** *m* submerged-coil condenser
Tauchrüttler *m* immersion (poker, needle) vibrator
Tauchsieder *m* immersion heater
Tauchstreichmaschine *f (pap)* dip coater
Tauchtank *m* dip[ping] tank
Tauchtest *m* total immersion test *(corrosion testing)*
Tauchtiefe *f* submergence
Tauchtränkung *f* steeping
Tauchüberzugseinrichtung *f* dip coater
tauchveraluminieren to dip-aluminize
tauchverbleien to lead-dip
Tauchverbrennung *f* submerged combustion
Tauchverfahren *n* dipping method (process); *(agric)* immersion (pickling) methode
Tauchversuch *m* total immersion test *(corrosion testing)*
Tauchwalze *f* dipping (immersion) roll, *(pap also)* size (fountain) roll
Tauchwalzentrockner *m* dip-feed drum dryer
Tauchwanne *f* dip[ping] tank
Tauchwaren *fpl (rubber)* dipped goods (articles)
Tauchzeit *f* dipping time
tauen to thaw
Tauen[pack]papier *n* rope wrapping (brown)

Taukurve *f*, **Taulinie** *f (distil)* dew-point curve, condensation curve
Taumelscheibenzähler *m* disk (nutating-piston) meter
Taupunkt *m* dew point, saturation temperature
Taupunktmethode *f (phys chem)* dew-point method
Taurin *n* $NH_2CH_2CH_2SO_3H$ taurine, aminoethylsulphonic acid
Taurocholsäure *f* taurocholic acid *(a bile acid)*
Taurokarbamidsäure *f* $NH_2CONHCH_2CH_2SO_3H$ taurocarbamic acid, 2-ureidoethane-1-sulphonic acid
Tauröste *f*, **Taurotte** *f (text)* dew-ret[ting]
tautomer tautomeric
Tautomer *n* tautomer, dynamic isomer
Tautomerengleichgewicht *n* tautomeric equilibrium
Tautomerenkonstante *f* tautomerization constant
Tautomeres *n s.* Tautomer
Tautomerie *f* tautomerism, tautomery, dynamic isomerism
  zyklisch-offene T. ring-chain tautomerism
Tautomeriegleichgewicht *n* tautomeric equilibrium
Tautomeriekonstante *f* tautomerization constant
tautomerisierbar tautomerizable
tautomerisieren to tautomerize
Tautomerisierung *f* tautomerization
TBA *s.* Trichlorbenzoesäure
TCA *s.* Trichloräthansäure
TCC-Verfahren *n (petrol)* Thermofor [catalytic-cracking] process, TCC process *(with moving catalyst)*
TC-Kautschuk *m* technically classified rubber, T.C. rubber
TCP-Verfahren *n (petrol)* Thermofor continuous-percolation process
Technetat(VII) *n* M^ITcO_4 pertechnate
Technetium *n* Tc technetium
Technetium(VII)-säure *f* $HTcO_4$ pertechnetic acid
Technik *f* 1. *(science)* engineering; 2. *(applied to practical purposes)* technology; 3. *(manner of performing technical details, esp lab)* technique, procedure, method
  biochemische T. biochemical technology
  chemische T. chemical technology
  metallurgische T. metallurgical technology
  mikroskopische T. microscopic technique, microtechnique, micrology
technisch durchführbar feasible, practicable
  t. rein technical
Technologie *f* technology
  T. der Kunststoffe (Plaste) plastics technology
  chemische T. chemical technology
  petrolchemische T. petrochemical technology
Teegerbstoff *m* tea tannin
Teeöl *n* tea[seed] oil
Teer *m* tar
  destillierter (präparierter) T. distilled (prepared) tar
  schwedischer T. *s.* Stockholmer T.
  Stockholmer T. Stockholm tar *(a pine tar)*
Teerabscheider *m* tar separator (extractor), detarrer
  elektrostatischer T. electrostatic tar filter
Teerabscheidung *f* tar separation
teerartig tarry
Teerdachpappe *f* tarred (asphaltic) felt, tar[red] board
Teerdampf *m* tar vapour
Teer-Elektrofilter *n* electrostatic tar filter
teeren to tar
Teerentfernung *f* detarring, tar separation
Teerfarbe *f s.* Teerfarbstoff
Teerfarbstoff *m* coal-tar dye[stuff], *(Am also)* coal-tar color
Teerfilter *n* tar filter
Teerfraktion *f* tar fraction
teerfrei tarfree
teerig tarry
Teerkrebs *m* tar cancer
Teerkresol *n* coal-tar-derived cresylic acid *(a mixture of o-, m-, and p-cresol and other phenolic compounds)*
Teeröl *n* tar oil
Teerpapier *n* tarred [brown] paper, tar (asphalt, pitch) paper
Teerpappe *f* tar[red] board
Teerpech *n* tar pitch
Teersand *m* tar sand
Teersäure *f* tar acid
Teerscheider *m s.* Teerabscheider
Teerscheidung *f s.* Teerabscheidung
Teerwäscher *m* tar scrubber
Teesamenöl *n* tea[seed] oil
Teig *m* dough, paste; *(food)* dough
teigartig dough-like, doughy
Teigformmaschine *f (food)* dough forming (moulding) machine
teigig dough-like, doughy
Teigkneter *m*, **Teigknetmaschine** *f* dough kneader (mixer), dough kneading (mixing) machine
Teiglockerungsmittel *n s.* Teigtriebmittel
Teigmischer *m*, **Teigmischmaschine** *f s.* Teigkneter
Teigtriebmittel *n (food)* leaven, leavening [agent]
Teil *m* part; portion, moiety; section *(as of an industrial plant)*
  aliquoter T. aliquot [part]
  durchhängender T. *(phot)* region of underexposure, toe, foot *(of the characteristic curve)*
  geradliniger T. *(phot)* region of correct (normal) exposure, straight[-line] portion, straight line *(of the characteristic curve)*
Teil *n* part
  grünes T. green compact *(in powder metallurgy)*
Teilbindung *f* fractional bond *(of atoms)*
Teilchen *n* particle
  energiereiches T. high-energy particle
  materielles T. material particle, particle of matter
  seltsames T. *(nucl)* strange particle
α-Teilchen *n* alpha particle, α-particle
β-Teilchen *n* beta particle, β-particle
Teilchenaggregat *n* particle (particulate) aggregate
Teilchenbahn *f s.* Teilchenflugbahn
Teilchenbeschleuniger *m* particle accelerator

linearer T. linear accelerator
Teilchenbeschleunigung *f* particle acceleration
Teilchenenergie *f* particle energy
Teilchenerzeugung *f* particle production
Teilchenfeinheit *f* particle fineness
Teilchenflugbahn *f* particle trajectory, path of the particle
Teilchenform *f* particle shape
Teilchengeschwindigkeit *f* particle velocity
Teilchengröße *f* particle size, *(screen classification also)* screen size
Teilchengrößenbereich *m* range of particle sizes, *(screen classification also)* range of screen sizes
Teilchengrößenbestimmung *f* particle size determination
Teilchengrößenverteilung *f* distribution of particle size
Teilchenladung *f* particle charge
Teilchenspin *m* particle spin
Teilchenstrahlung *f* corpuscular (particle) radiation
Teilchenstruktur *f* particle structure
Teilchenverteilung *f* distribution of particles
Teilchenzähler *m* particle counter
Teildampfdruck *m* partial vapour pressure
Teildruck *m* partial pressure
teilen / in Grade to graduate
Teilentsalzung *f* undersoftening *(water treatment)*
Teilentstaubungsgrad *m* fractional[-weight collection] efficiency *(classifying)*
Teilfäserchen *n (text)* fibril[la]
Teilfuge *f* [mould-]parting line
Teilkondensation *f* partial condensation, dephlegmation
geschlossene (integrale) T. equilibrium partial condensation
Teilkondensator *m* [countercurrent] partial condenser, partial-condensation head
Teilkreis *m* divided circle *(as of a refractometer)*
Teilpipette *f* graduated (measuring) pipette
Teilschale *f (nucl)* subshell
Teilschritt *m* step *(as of a reaction)*
Teilsterilisation *f* partial sterilization *(as of soils)*
Teilstrahlungspyrometer *n* [partial-]radiation pyrometer
Teilstrich *m* graduation (division) mark (line), graduation, division
Teilstrichabstand *m* [scale] division
Teilungsbild *n* graduation *(as on a measuring vessel)*
Teilungsebene *f (cryst)* cleavage (parting) plane
Teilverbrennung *f* partial combustion
Tein *n* caffeine, theine, 1,3,7-trimethylxanthine
T-Einheit *f* trifunctional unit, T unit *(structural element)*
Teinochemie *f* teinochemistry
Tektochinon *n* tectoquinone, 2-methylanthraquinone
Tektosilikat *n (min)* tectosilicate *(any of a class of polymeric silicates)*
Telepathin *n s.* Harmin
TEL-Fluid *n* ethyl fluid *(an antiknock additive mainly consisting of tetraethyl lead)*
Telinit *m* tel[l]inite *(a coal maceral)*
Teller *m (tech)* disk, plate; disk *(of a valve)*
Telleraufgabegerät *n*, **Telleraufgeber** *m s.* Tellerbeschicker
Tellerbeschicker *m* [revolving-]disk feeder
Tellerbrecher *m* disk crusher
Tellerdrehmaschine *f (ceram)* plate-jiggering machine
Tellerkneter *m* [rotary] kneading table
Tellermesser *n (pap)* disk (circular slitting) knife, slitter
Tellermühle *f* disk [attrition] mill, disk grinder
Tellersatz *m* disk stack *(of a centrifuge)*
Tellerschleuder *f*, **Tellerseparator** *m s.* Tellerzentrifuge
Tellerspeiser *m s.* Tellerbeschicker
Tellertrockner *m* disk dryer
Tellerventil *n* disk valve
pilzförmiges T. mushroom-seated valve
Tellerwäscher *m* plate scrubber *(gas cleaning)*
T. nach Theisen Theisen disintegrator
Tellerzentrifuge *f* disk [bowl] centrifuge, *(food also)* disk separator
T. mit Düsenaustrag nozzle-discharge disk centrifuge
T. mit Schlitzaustrag annular solids-discharge disk centrifuge
Tellur *n* Te tellurium
Tellurat *n* tellurate
Tellurblei *n (min)* altaite *(lead telluride)*
Tellur(II)-bromid *n* $TeBr_2$ tellurium(II) bromide, tellurium dibromide
Tellur(IV)-bromid *n* $TeBr_4$ tellurium(IV) bromide, tellurium tetrabromide
Tellur(II)-chlorid *n* $TeCl_2$ tellurium(II) chloride, tellurium dichloride
Tellur(IV)-chlorid *n* $TeCl_4$ tellurium(IV) chloride, tellurium tetrachloride
Tellurdibromid *n s.* Tellur(II)-bromid
Tellurdichlorid *n s.* Tellur(II)-chlorid
Tellurdioxid *n s.* Tellur(IV)-oxid
Tellur(VI)-fluorid *n* TeF_6 tellurium(VI) fluoride, tellurium hexafluoride
Tellurhexafluorid *n s.* Tellur(VI)-fluorid
Tellurhydrid *n s.* Tellurwasserstoff
Tellurid *n* M_2^ITe telluride
Tellurit *m (min)* tellurite, telluric ochre *(tellurium(IV) oxide)*
Tellur(IV)-jodid *n* TeI_4 tellurium(IV) iodide, tellurium tetraiodide
Tellurmonoxid *n s.* Tellur(II)-oxid
Tellurnickel *n (min)* melonite *(nickel telluride)*
Tellurocker *m s.* Tellurit
Tellur(II)-oxid *n* TeO tellurium(II) oxide, tellurium monooxide
Tellur(IV)-oxid *n* TeO_2 tellurium(IV) oxide, tellurium dioxide
Tellur(VI)-oxid *n* TeO_3 tellurium(VI) oxide, tellurium trioxide
Tellursäure *f* H_6TeO_6 orthotelluric acid, telluric acid, hexaoxotelluric acid
Tellurtetra... *s.* Tellur(IV)-...

Tellurtrioxid *n s.* Tellur(VI)-oxid
Tellurwasserstoff *m* H_2Te hydrogen telluride, tellurium hydride
Telomer[es] *n,* **Telomerisat** *n* telomer
Telomerisation *f (org chem)* telomerization
Telomerisationsreaktion *f (org chem)* telomerization reaction
Telsmith-Kegelbrecher *m* Telsmith cone crusher
Telsmith-Kreiselbrecher *m* Telsmith gyratory crusher, Telsmith breaker
Temperafarbe *f* tempera paint
Temperatur *f* temperature
T. der Phasenumwandlung first-order transition temperature
absolute T. absolute temperature
charakteristische T. characteristic (Debye) temperature
erhöhte T. elevated temperature
eutektische T. eutectic temperature
gewöhnliche T. normal (ordinary) temperature
kritische T. critical temperature
potentielle T. potential temperature
Temperaturabfall *m* temperature drop
temperaturabhängig temperature-dependent
Temperaturabhängigkeit *f* temperature dependence
Temperaturänderung *f* temperature change, change in temperature
Temperaturanregung *f* thermal excitation
Temperaturanstieg *m* temperature rise, rise (increase) in temperature
Temperaturausgleich *m* temperature equalization (compensation)
Temperaturbeiwert *m* temperature coefficient
Temperaturbereich *m* temperature range
temperaturbeständig temperature-resistant, temperature-resisting, temperature-stable
Temperaturbeständigkeit *f* temperature resistance (stability)
Temperaturdifferenz *f* temperature difference
temperaturempfindlich temperature-sensitive
Temperaturempfindlichkeit *f* temperature sensitivity
Temperatur-Entropie-Diagramm *n* temperature-entropy chart, entropy chart (diagram)
Temperaturentwicklung *f (rubber)* temperature (heat) build-up *(by dynamic stress)*
Temperaturerhöhung *f* temperature raising; temperature rise, rise (increase) in temperature
Temperaturgefälle *n* temperature drop
Temperaturgleichgewicht *n* temperature equilibrium
Temperaturgrad *m* degree of temperature
Temperaturgradient *m* temperature gradient
Temperaturgrenze *f* temperature limit
Temperaturkoeffizient *m* temperature coefficient
T. der Viskosität (Zähigkeit) viscosity-temperature coefficient
Temperaturkompensation *f* temperature compensation
Temperaturleitfähigkeit *f,* **Temperaturleitvermögen** *n* thermal diffusivity
Temperaturleitzahl *f* thermal diffusivity
Temperaturmeßfarbe *f* temperature-indicating paint
Temperaturmeßfarbstift *m* temperature crayon
Temperaturmessung *f* temperature measurement, thermometry
Temperaturregelung *f* temperature control
Temperaturregler *m,* **Temperaturregulator** *m* temperature controller, thermoregulator, thermostat
temperaturreversibel thermally reversible
Temperaturschreiber *m* temperature recorder, thermograph
Temperaturskale *f* temperature scale
absolute (thermodynamische) T. absolute (thermodynamic) temperature scale
temperaturstabil temperature-resistant, temperature-resisting, temperature-stable
Temperatursteigerung *f s.* Temperaturerhöhung
temperaturunabhängig temperature-independent
Temperaturunterschied *m* temperature difference
temperaturwechselbeständig thermal-shock resistant
Temperaturwechselbeständigkeit *f* thermal-shock resistance, *(ceram also)* [thermal] spalling resistance, thermal stability (endurance)
Temperguß *m* malleable [cast] iron
temperieren to thermostat
Temperiergefäß *n* tempering vessel
Temperierkessel *m (food)* tempering tank (vat) *(as in margarine making)*
tempern to malleabl[e]ize, to anneal *(cast iron); (plast)* to anneal, to temper; *(glass)* to anneal
Tengerit *m (min)* tengerite *(a water-containing yttrium carbonate)*
Tennenmälzerei *f,* **Tennenvermälzung** *f (ferm)* floor malting, flooring
Tenorit *m (min)* tenorite *(copper(II) oxide)*
Tensid *n* surfactant, surface-active agent
Tensionsthermometer *n* vapour-pressure thermometer
Tensometer *n* extensometer *(materials testing)*
Tephroit *m (min)* tephroite *(a nesosilicate)*
TEPP *s.* Tetraäthylpyrophosphat
Teppichtyp *m* carpet grade (quality) *(of chemical fibres)*
Teratolith *m (min)* teratolite *(a blue bole)*
Terbinerde *f s.* Terbiumoxid
Terbium *n* Tb terbium
Terbiumchlorid *n* $TbCl_3$ terbium chloride
Terbiumfluorid *n* TbF_3 terbium fluoride
Terbiumnitrat *n* $Tb(NO_3)_3$ terbium nitrate
Terbiumoxid *n* Tb_2O_3 terbium oxide
Terbiumsulfat *n* $Tb_2(SO_4)_3$ terbium sulphate
Terebinsäure *f* terebic acid, 2,2-dimethyl-5-oxo-oxolan-3-carboxylic acid
Terephthalsäure *f* $C_6H_4(COOH)_2$ terephthalic acid, *p*-phthalic acid, benzene-1,4-dicarboxylic acid
Teri *pl (tann)* teri pods *(from Caesalpinia digyna Rottl.)*
Term *m* term, energy level
fester (konstanter) T. constant term

variabler T. variable (current) term
Termbezeichnung *f s.* Termsymbol
terminal terminal
Termschema *n* term scheme, scheme of energy levels
Termsymbol *n* term symbol (signature)
T. eines Atoms atomic term symbol
Termsystem *n* term system
ternär ternary
Terpen *n* terpene
Terpenkohlenwasserstoff *m* terpene (terpenoid) hydrocarbon
Terpentin *n* turpentine [oleoresin] *(balsam obtained from coniferous trees)*
kanadisches T. Canada turpentine *(from Abies balsamea (L.) Mill.)*
Terpentinersatz *m* turpentine substitute
Terpentinessenz *f* rosin (resin) spirit, pinolin[e], pinolene
Terpentingallen *fpl (dye)* turpentine (carob) galls *(from Pistacia terebinthus L.)*
Terpentinharz *n* rosin, pine resin (rosin)
Terpentinharzöl *n* rosin oil, rosinol *(by fractional distillation of rosin)*
Terpentinöl *n* oil (spirit) of turpentine
Terpentinölersatz *m*, **Terpentinölsurrogat** *n* turpentine substitute
Terpenylsäure *f* terpenylic acid, 2,2-dimethyl-5-oxo-oxolan-3-acetic acid
Terphenyl *n* $(C_6H_5)_2C_6H_4$ terphenyl, diphenylbenzene, *(specif)* 1,4-diphenylbenzene
Terpin *n* terpin *(a cyclohexanol derivative)*
Terpinen *n (org chem)* terpinene
Terpineol *n* terpineol *(any of 3 isomeric terpene alcohols)*
Terpinolen *n* terpinolene, 4-isopropylidene-1-methylcyclohexane
Terpolymer[es] *n* terpolymer
Terra rossa *f (ceram)* terra rossa
Terra sigillata *f (ceram)* terra sigillata
Terrakotta *f (ceram)* terra cotta
Terreinsäure *f* terreic acid *(a benzoquinone derivative)*
tert. *s.* tertiär
tertiär tertiary
Tertiärluft *f* tertiary air
Test *m* test
zerstörungsfreier T. non-destructive test
testen to test
Tester *m* tester
Testkohle *f* standard coal
Testmaterial *n* test material
Testmischung *f (rubber)* test compound (mix, mixture)
Testmuster *n* test piece (specimen)
Testorganismus *m* test organism
Testosteron *n* testosterone, 17β-hydroxyandrost-4-en-3-one *(a sex hormone)*
Testriol *n* $C_{16}H_{33}OCH_2CH(OH)CH_2OH$ testriol, chimyl alcohol, 2,3-dihydroxypropyl hexadecyl ether
Testtoxin *n* test toxin
Testverfahren *n* test[ing] method
Tetanusserum *n* antitetanus serum
Tetanustoxin *n* tetanus toxin
tetartoedrisch *(cryst)* tetartohedral
Tetra *s.* Tetrachlorkohlenstoff
Tetraarsentetrasulfid *n* As_4S_4 tetraarsenic tetrasulphide
Tetraäthylblei *n* $(C_2H_5)_4Pb$ tetraethyl lead, TEL
✦ **mit T. versetzen** to lead *(motor fuel)*
Tetraäthylpyrophosphat *n* $(C_2H_5)_4P_2O_7$ tetraethylpyrophosphate, TEPP
Tetraäthylrhodamin *n* tetraethyl rhodamine
Tetraäthylsilan *n* $Si(C_2H_5)_4$ tetraethylsilane
Tetraäthylthiuramdisulfid *n* tetraethylthiuram disulphide, disulphiram
Tetraazetat *n* tetraacetate
Tetraboran *n* B_4H_{10} tetraborane
Tetraborat *n* $M_2^IB_4O_7$ tetraborate, heptaoxotetraborate
Tetraborid *n* tetraboride
Tetraborsäure *f* $H_2B_4O_7$ tetraboric acid, heptaoxotetraboric acid
Tetrabromfluoreszein *n* tetrabromofluorescein, eosin
Tetrabromid *n* tetrabromide
Tetrabromindigo *m (n)* tetrabromoindigo
Tetrabromkohlenstoff *m*, **Tetrabrommethan** *n* CBr_4 carbon tetrabromide, tetrabromomethane, perbromomethane
Tetrabromoborat *n* tetrabromoborate
Tetrabromogold(III)-säure *f* $H[AuBr_4]$ tetrabromoauric(III) acid, bromoauric acid
Tetrabromoplatinat(II) *n* $M_2^I[PtBr_4]$ tetrabromoplatinate(II), bromoplatinate(II)
Tetrabromoplatin(II)-säure *f* $H_2[PtBr_4]$ tetrabromoplatinic(II) acid, bromoplatinic(II) acid
Tetrabromphenolsulfonphthalein *n* tetrabromophenolsulphonphthalein, bromophenol blue *(a pH indicator)*
Tetrabromsilan *n* $SiBr_4$ tetrabromosilane
Tetrachloranthrachinon *n* tetrachloroanthraquinone
1,1,2,2-Tetrachloräthan *n*, *symm.*-**Tetrachloräthan** *n* $CHCl_2CHCl_2$ 1,1,2,2-tetrachloroethane
Tetrachloräthen *n*, **Tetrachloräthylen** *n* $Cl_2C{=}CCl_2$ tetrachloroethylene, perchloroethylene
Tetrachlor-*p*-benzochinon *n* tetrachloro-*p*-benzoquinone, tetrachloroquinone, chloranil
Tetrachlorid *n* tetrachloride
Tetrachlorkohlenstoff *m*, **Tetrachlormethan** *n* CCl_4 carbon tetrachloride, tetrachloromethane, perchloromethane
Tetrachloroborat *n* tetrachloroborate
Tetrachlorodiamminplatin(IV) *n* $[Pt(NH_3)_2Cl_4]$ diamminetetrachloroplatinum(IV)
Tetrachlorogold(III)-säure *f* $H[AuCl_4]$ tetrachloroauric(III) acid, chloroauric(III) acid
Tetrachloropalladat(II) *n* $M_2^I[PdCl_4]$ tetrachloropalladate(II), chloropalladate(II)
Tetrachloroplatinat(II) *n* $M_2^I[PtCl_4]$ tetrachloroplatinate(II), chloroplatinate(II)
Tetrachlorsilan *n* $SiCl_4$ tetrachlorosilane
Tetradekanal *n* $CH_3[CH_2]_{12}CHO$ tetradecanal, tetradecylaldehyde, myristaldehyde

1-Tetradekanol *n* $CH_3[CH_2]_{12}CH_2OH$ 1-tetradecanol, myristyl alcohol
Tetradekansäure *f* $CH_3[CH_2]_{12}COOH$ tetradecanoic acid
Tetradezensäure *f* $C_{13}H_{25}COOH$ tetradecenoic acid
Tetradymit *m (min)* tetradymite *(bismuth tellurium sulphide)*
Tetraeder *n (cryst)* tetrahedron
Tetraederanordnung *f* tetrahedral arrangement
Tetraederformel *f* tetrahedral formula
Tetraedermodell *n* tetrahedral model
Tetraederorbital *n* tetrahedral orbital
Tetraedervalenz *f* tetrahedral valency
Tetraederwinkel *m* tetrahedral angle
tetraedrisch tetrahedral
Tetraedrit *m (min)* tetrahedrite, grey copper [ore], fahlerz, fahlore *(a sulphide of copper and antimony often containing zinc)*
Tetraen *n (org chem)* tetraene
Tetrafluoräthen *n*, **Tetrafluoräthylen** *n* $F_2C{=}CF_2$ tetrafluoroethylene, T.F.E., perfluoroethylene
Tetrafluorid *n* tetrafluoride
Tetrafluoroborsäure *f* $H[BF_4]$ tetrafluoroboric acid
Tetrafluorsilan *n* SiF_4 tetrafluorosilane
tetrafunktionell tetrafunctional
tetragonal *(cryst)* tetragonal
Tetrahalogenid *n* tetrahalogenide, tetrahalide
Tetrahydrat *n* tetrahydrate
Tetrahydrid *n* tetrahydride
Tetrahydridoborat *n* $M^I[BH_4]$ tetrahydridoborate, hydridoborate
Tetrahydrobenzol *n* tetrahydrobenzene, cyclohexene
Tetrahydrofuran *n*, **Tetrahydrofurfuran** *n* tetrahydrofuran, THF
Tetrahydroisochinolin *n* tetrahydroisoquinoline
Tetrahydroisochinolinbase *f* tetrahydroisoquinoline base
Tetrahydronaphthalin *n* tetrahydronaphthalene
Tetrahydrothiophen *n* tetrahydrothiophene
Tetrajodid *n* tetraiodide
Tetrajodoaurat(III) *n* $M^I[AuI_4]$ tetraiodoaurate(III), iodoaurate(III)
Tetrajodomerkurat(II) *n* $M_2^I[HgI_4]$ tetraiodomercurate(II), iodomercurate(II)
Tetrajodsilan *n* SiI_4 tetraiodosilane
Tetrakain *n* decicaine, tetracaine *(hydrochloride of 2-dimethylaminoethyl-p-butylaminobenzoate)*
Tetrakarboximidfarbstoff *m* tetracarboxyimide dye
Tetrakisazofarbstoff *m* tetrakisazo dye
Tetrakistri ... *s.* Dodeka ...
Tetrakosansäure *f* $C_{23}H_{47}COOH$ tetracosanoic acid
tetrakovalent tetracovalent, quadricovalent
Tetralakton *n* tetralactone
Tetramer[es] *n* tetramer
Tetramethylarsin *n s.* Tetramethyldiarsin
Tetramethyläthylenglykol *n* $(CH_3)_2C(OH)COH(CH_3)_2$ tetramethylethylene glycol, 2,3-dimethyl-2,3-butanediol, pinacol
Tetramethylbiarsyl *n s.* Tetramethyldiarsin
Tetramethylblei *n* $(CH_3)_4Pb$ tetramethyl lead, TML
Tetramethyldiarsin *n* $(CH_3)_2AsAs(CH_3)_2$ tetramethyldiarsine, cacodyl
Tetramethylen *n s.* Zyklobutan
Tetramethylendiamin *n* tetramethylenediamine, putrescine
Tetramethylenimin *n* pyrrolidine, tetramethyleneimine
Tetramethylenoxid *n s.* Tetrahydrofuran
Tetramethylensulfid *n s.* Tetrahydrothiophen
Tetramethylglukose *f* tetramethylglucose
Tetramethylierung *f* tetramethylation
Tetramethylmethan *n* $C(CH_3)_4$ 2,2-dimethylpropane, tetramethylmethane, neopentane
Tetramethylsilan *n* $Si(CH_3)_4$ tetramethylsilane
Tetrammin *n* tetraammine
Tetramminkupfer(II)-sulfat *n* $[Cu(NH_3)_4]SO_4$ tetraamminecopper(II) sulphate, cupric tetraammine sulphate
Tetramminnickel(II)-nitrat *n* $[Ni(NH_3)_4](NO_3)_2$ tetraamminenickel(II) nitrate
Tetramminpalladium(II)-chlorid *n* $[Pd(NH_3)_4]Cl_2$ tetraamminepalladium(II) chloride
Tetramminsalz *n* tetraammine salt
tetramolekular tetramolecular, quadrimolecular
Tetramolybdat *n* tetramolybdate
Tetranatriumsalz *n* tetrasodium salt
Tetranitrid *n* tetranitride
Tetraoxalat *n* tetraoxalate
Tetrapeptid *n* tetrapeptide
Tetraphosphor *m* tetraphosphorus, white phosphorus
Tetraphosphorheptasulfid *n* P_4S_7 tetraphosphorus heptasulphide, phosphorus heptasulphide
Tetraphosphormonoselenid *n* P_4Se tetraphosphorus monoselenide
Tetraphosphorpentasulfid *n* P_4S_5 tetraphosphorus pentasulphide
Tetraphosphortrisulfid *n* P_4S_3 tetraphosphorus trisulphide
Tetraquoeisen(II)-Ion *n* $[Fe(H_2O)_4]^{2+}$ tetraaquairon(II) ion
Tetrarhodanid *n s.* Tetrathiozyanat
Tetrarsentetrasulfid *n* As_4S_4 tetraarsenic tetrasulphide
Tetraschwefeltetranitrid *n* S_4N_4 tetrasulphur tetranitride
Tetraselentetranitrid *n* Se_4N_4 tetraselenium tetranitride, selenium nitride
Tetrasilan *n* Si_4H_{10} tetrasilane
Tetrasiloxan *n* $Si_4H_{10}O_3$ tetrasiloxane
tetrasubstituiert tetrasubstituted
Tetrasubstitution *f* tetrasubstitution
Tetrasubstitutionsprodukt *n* tetrasubstitution product
Tetrasulfan *n* H_2S_4 tetrasulphane, hydrogen tetrasulphide
Tetrasulfid *n* tetrasulphide
Tetrathionat *n* $M_2^IS_4O_6$ tetrathionate
Tetrathiozyanat *n* tetrathiocyanate, tetrarhodanide
tetravalent tetravalent, quadrivalent, *(of molecules also)* tetraatomic
Tetravalenz *f* tetravalence, quadrivalence
Tetrazen *n* 1. tetracene, 1-(5-tetrazolyl)-4-guanyltetrazene hydrate *(a primary explosive)*; 2. naph-

thacene, tetracene, 2,3-benzanthracene
Tetrazin *n* tetrazine
Tetrazol *n* tetrazole
tetrazotieren *(dye)* to tetraazotize
Tetrazotierung *f (dye)* tetraazotization
Tetrazoverbindung *f* tetraazo compound
Tetrazyanoaurat(III) *n* $M^I[Au(CN)_4]$ tetracyanoaurate(III), cyanoaurate(III)
Tetrazyanoplatinat(II) *n* $M_2^I[Pt(CN)_4]$ tetracyanoplatinate(II), cyanoplatinate(II)
Tetrazyklin *n* tetracycline *(a dioxanaphthacene derivative)*
Tetrit *m* tetritol, tetrahydroxy alcohol, tetrahydric alcohol
Tetrose *f* tetrose *(monosaccharide containing 4 carbon atoms per molecule)*
Tetroxalat *n s.* Tetraoxalat
Tetroxid *n* tetraoxide
Tetroxojodat(VII) *n* M^IIO_4 tetraoxoiodate(VII), periodate, metaperiodate
Tetroxojod(VII)-säure *f* HIO_4 tetraoxoiodic(VII) acid, periodic acid, metaperiodic acid
Tetroxokieselsäure *f* H_4SiO_4 tetraoxosilicic acid, orthosilicic acid
Tetroxoosmat(VI) *n* $M_2^IOsO_4$ tetraoxoosmate(VI), osmate(VI)
Tetroxoplumbat(IV) *n* $M_4^IPbO_4$ tetraoxoplumbate(IV), orthoplumbate
Tetroxorhenat(VII) *n* M^IReO_4 tetraoxorhenate(VII), perrhenate
Tetroxosilikat *n* $M_4^ISiO_4$ tetraoxosilicate, orthosilicate, silicate
Tetroxostannat(IV) *n* $M_4^ISnO_4$ tetraoxostannate(IV), orthostannate
Tetroxotellurat(VI) *n* $M_2^ITeO_4$ tetraoxotellurate(VI), metatellurate(VI)
Tetroxovanadat(V) *n* $M_3^IVO_4$ tetraoxovanadate(V), orthovanadate
Tetroxozinnsäure *f* H_4SnO_4 tetraoxostannic acid, orthostannic acid
Teufelsdreck *m* asafetida, devil's dung, food of the gods *(gum resin from Ferula specc.)*
Textilausrüstung *f* textile finishing
Textilchemie *f* textile chemistry
Textilchemiker *m* textile chemist
Textildruck *m* textile printing
Textileinlage *f* textile insertion (casing) *(as in rubber products)*
Textilerzeugnisse *npl* textiles
Textilfaser *f* textile fibre
Textilfaserstoff *m* textile fibre
Textilgewebe *n* fabric [cloth], tissue
Textilhilfsmittel *n* textile auxiliary
Textilien *pl* textiles
schaumstoffkaschierte T. foambacks
ungewebte T. non-wovens, non-woven fabrics, bonded fabrics
Textilöl *n* textile oil
Textilreinigungsmittel *n* textile cleanser
Textilschlichte *f* textile size
Textilverbundstoff *m* non-woven fabric, bonded fabric
Textilveredlung *f* textile finishing
Textilveredlungsmittel *n* textile auxiliary
Textilzellstoff *m* rayon (dissolving) pulp
Textur *f* texture; *(ceram)* lamination *(a defect)*
texturieren *(text)* to texture, to bulk
Texturseide *f* textured (bulked) yarn
Thalenit *m (min)* thalenite *(yttrium disilicate)*
Thallium *n* Tl thallium
Thalliumalaun *m* $TlAl(SO_4)_2 \cdot 12H_2O$ thallium alum
Thallium(I)-azetat *n* CH_3COOTl thallium(I) acetate, thallous acetate
Thallium(III)-azetat *n* $(CH_3COO)_3Tl$ thallium(III) acetate, thallic acetate
Thallium(I)-bromid *n* TlBr thallium(I) bromide, thallium monobromide, thallous bromide
Thallium(III)-bromid *n* $TlBr_3$ thallium(III) bromide, thallium tribromide, thallic bromide
Thallium(I)-chlorid *n* TlCl thallium(I) chloride, thallium monochloride, thallous chloride
Thallium(III)-chlorid *n* $TlCl_3$ thallium(III) chloride, thallium trichloride, thallic chloride
Thallium(I)-fluorid *n* TlF thallium(I) fluoride, thallium monofluoride, thallous fluoride
Thallium(III)-fluorid *n* TlF_3 thallium(III) fluoride, thallium trifluoride, thallic fluoride
Thalliumhexachloroplatinat(IV) *n* $Tl_2[PtCl_6]$ thallium hexachloroplatinate(IV)
Thallium(I)-hexafluorosilikat *n* $Tl_2[SiF_6]$ thallium(I) hexafluorosilicate
Thallium(I)-hydroxid *n* TlOH thallium(I) hydroxide, thallous hydroxide
Thalliummono ... *s.* Thallium(I)-...
Thallium(I)-nitrat *n* $TlNO_3$ thallium(I) nitrate, thallous nitrate
Thallium(III)-nitrat *n* $Tl(NO_3)_3$ thallium(III) nitrate, thallic nitrate
Thallium(I)-orthophosphat *n* Tl_3PO_4 thallium(I) orthophosphate, thallium(I) phosphate
Thallium(I)-oxid *n* Tl_2O thallium(I) oxide, thallium monooxide, thallous oxide
Thallium(III)-oxid *n* Tl_2O_3 thallium(III) oxide, thallium trioxide, thallic oxide
Thallium(I)-sulfat *n* Tl_2SO_4 thallium(I) sulphate, thallous sulphate
Thallium(III)-sulfat *n* $Tl_2(SO_4)_3$ thallium(III) sulphate, thallic sulphate
Thallium(I)-sulfid *n* Tl_2S thallium(I) sulphide, thallium monosulphide, thallous sulphide
Thallium(III)-sulfid *n* Tl_2S_3 thallium(III) sulphide, thallium trisulphide, thallic sulphide
Thalliumtri... *s.* Thallium(III)-...
Thallium(I)-zyanid *n* TlCN thallium(I) cyanide, thallous cyanide
Thebain *n* thebaine, paramorphine *(alkaloid)*
Thein *n s.* Tein
Theisen-Desintegrator *m*, **Theisen-Wäscher** *m* Theisen disintegrator *(gas cleaning)*
Thelephorsäure *f* thelephoric acid *(a lichen acid)*
Thenardit *m (min)* thenardite *(sodium sulphate)*
Theobromin *n* theobromine, 2,6-dihydroxy-3,7-dimethylpurine *(alkaloid)*
Theobrominnatrium *n* theobromine sodium

Theobrominnatriumazetat *n*, **Theobrominnatrium-Natriumazetat** *n* theobromine sodium [and sodium] acetate
Theobrominnatrium-Natriumsalizylat *n*, **Theobrominnatriumsalizylat** *n* theobromine sodium [and sodium] salicylate
Theophyllin *n* theophylline, 2,6-dihydroxy-1,3-dimethylpurine *(alkaloid)*
Theophyllinnatrium *n* theophylline sodium
Theophyllinnatriumazetat *n*, **Theophyllinnatrium-Natriumazetat** *n* theophylline sodium [and sodium] acetate
Theorem *n* / **Babinetsches** Babinet absorption rule
Theorie *f* theory ✦ **etwas mehr als der T. entspricht** in slight excess of theory
  T. der absoluten Reaktionsgeschwindigkeit theory of absolute reaction rate
  T. der Böden plate theory *(as in distillation and chromatography)*
  T. der Elektronenpaarbindungen *s.* T. der Valenzstrukturen
  T. der frei beweglichen Elektronen free-electron theory
  T. der Molekülorbitale molecular-orbital theory, Hund-Mulliken-Lennard-Jones-Hückel theory
  T. der Partialvalenzen partial-valence theory
  T. der übereinstimmenden Zustände *(phys chem)* theory of corresponding states
  T. der Valenzstrukturen valence-bond (electron-pair) theory, VB theory, Heitler-London-Slater-Pauling theory, HLSP theory
  T. des radioaktiven Zerfalls theory of radioactive disintegration (decay)
  Debyesche T. der spezifischen Wärme Debye theory of specific heat
  Diracsche T. [des Elektrons] Dirac electron theory
  Heitler-Londonsche T. Heitler-London theory, HL theory *(of valency)*
  Sachse-Mohrsche T. [spannungsfreier Ringe] Sachse-Mohr concept [of strainless rings]
  Wernersche T. Werner theory *(of coordination)*
Therapeutikum *n* therapeutic agent
therapeutisch therapeutic[al]
Therapie *f* / **medikamentöse** pharmacotherapy
Thermalruß *m* thermal [carbon] black, thermal-decomposition black, furnace thermal black
Thermalspaltprozeß *m* [furnace] thermal process *(carbon-black manufacture)*
Thermion *n* thermion
Thermionenemission *f* thermionic emission
thermionisch thermionic
thermisch thermal
Thermitverfahren *n* aluminothermic (thermite, Goldschmidt's) process, aluminothermics, aluminothermy
Thermoanalyse *f* thermal analysis
Thermochemie *f* thermochemistry, chemical thermodynamics
Thermochemiker *m* thermochemist
thermochemisch thermochemical
Thermochromie *f* thermochromism
Thermodiffusion *f* thermal diffusion
Thermodiffusionsverfahren *n* thermal-diffusion process
Thermodynamik *f* thermodynamics
  T. der Nichtgleichgewichtsprozesse *s.* T. irreversibler Prozesse
  T. irreversibler Prozesse thermodynamics of irreversible processes, non-equilibrium thermodynamics
  chemische T. chemical thermodynamics
  irreversible T. *s.* T. irreversibler Prozesse
  statistische T. statistical thermodynamics
  technische T. chemical engineering thermodynamics
thermodynamisch thermodynamic[al]
thermoelastisch thermoelastic
Thermoelastizität *f* thermoelasticity
thermoelektrisch thermoelectric
Thermoelektrizität *f* thermoelectricity
Thermoelektron *n* thermoelectron, thermionic electron
Thermoelement *n* thermocouple
thermofixieren to heat-set
Thermofixierung *f* heat setting
Thermofor-Continuous-Percolation-Verfahren *n* Thermofor continuous-percolation process
Thermoformmaschine *f* *(plast)* thermoforming machine
Thermoformung *f* *(plast)* thermoforming
Thermofor-Verfahren *n* Thermofor [catalytic-cracking] process, TCC process
Thermogalvanometer *n* thermogalvanometer
Thermograf *m* thermograph
Thermografie *f* thermography
thermografisch thermographic
Thermogramm *n* thermogram
Thermogravimetrie *f* thermogravimetry
  derivative T. derivative (differential) thermogravimetry
thermogravimetrisch thermogravimetric
Thermokompression *f* thermocompression
Thermokompressor *m* thermocompressor
Thermokonvektion *f* thermal convection (siphoning)
Thermokraft *f* thermoelectric power
thermolabil thermolabile, heat-labile
Thermolumineszenz *f* thermoluminescence, thermal luminescence
Thermolyse *f* thermolysis
thermolytisch thermolytic
thermomagnetisch thermomagnetic
thermomechanisch thermomechanical
Thermometer *n* thermometer
  Beckmannsches T. Beckmann thermometer
  feuchtes T. wet-bulb thermometer
Thermometerfehler *m* thermometer error
Thermometergefäß *n* thermometer bulb
Thermometerglas *n* thermometer glass
Thermometerrohr *n* thermometer tube (pipe)
Thermometerskale *f* thermometer (thermometric) scale
Thermometerstutzen *m* thermometer pocket
Thermometersubstanz *f* thermometer substance

Thermometrie *f* thermometry
thermometrisch thermometric[al]
Thermonatrit *m (min)* thermonatrite *(sodium carbonate-1-water)*
thermonuklear thermonuclear
thermooxydativ thermal-oxidative
Thermopaar *n* thermocouple
standardisiertes T. standard thermocouple
thermophil thermophilic, thermophilous
Thermoplast *m* thermoplastic [material]
verstärkter T. reinforced thermoplastic
thermoplastisch thermoplastic
Thermoplastizität *f* thermoplasticity
Thermoregulator *m* thermoregulator
thermoresistent *s.* thermostabil
Thermosäule *f* thermopile
Thermoskop *n* thermoscope
Thermosol-Klotz-Dämpfverfahren *n (dye)* thermosol pad-steam process
Thermosolverfahren *n (dye)* thermosol method
Thermospannung *f* thermoelectric power
thermostabil thermostable, thermoresistant, heat-resistant, *(relating to microorganisms also)* thermoduric
Thermostabilität *f* thermostability, thermal (heat) stability (resistance)
Thermostat *m* thermostat
thermostatieren to thermostat
Thermostatierung *f* thermostatting
Thermostrom *m* thermoelectric current
Thermoumformer *m* thermoelement
Thermovulkanisation *f* thermal vulcanization
Thermowaage *f* thermobalance
Thiamin *n* thiamin, aneurin *(vitamin B_1)*
Thiaminpyrophosphat *n* thiamin pyrophosphate, TPP
Thiaphen *n s.* Thiophen
Thiazinfarbstoff *m* thiazine dye
Thiazolbeschleuniger *m (rubber)* thiazole accelerator
Thiazolfarbstoff *m* thiazole dye
Thiazolgelb *n* thiazole (titan, Clayton) yellow
Thiazolidin *n* thiazolidine, tetrahydrothiazole
Thiele-Addition *f* Thiele addition *(of acetic anhydride to quinones)*
Thielepape-Aufsatz *m* Thielepape head *(for extracting)*
Thielepape-Extraktor *m* Thielepape extractor
Thioalkohol *m* thiol, thioalcohol, mercaptan
Thioantimonat(III) *n* $M_3{}^ISbS_3$ thioantimonite, trithioantimonite
Thioantimonat(V) *n* $M_3{}^ISbS_4$ thioantimonate, tetrathioantimonate
Thioarsenat(III) *n* $M_3{}^IAsS_3$ thioarsenite, trithioarsenite
Thioarsenat(V) *n* $M_3{}^IAsS_4$ thioarsenate, tetrathioarsenate
Thioäthanol *n s.* Äthanthiol
Thioäther *m* thioether, thiaalkane, alkyl sulphide
Thiobakterien *pl* sulphur bacteria
Thioderivat *n* thio derivative
Thioessigsäure *f* CH_3COSH thioacetic acid

Thiofuran *n s.* Thiophen
Thioglykolsäure *f* $HSCH_2COOH$ thioglycollic acid, mercaptoacetic acid
Thioglyzerin *n* $HOCH_2CH(OH)CH_2SH$ thioglycerol
Thioharnstoff *m* NH_2CSNH_2 thiourea, thiocarbamide
Thioharnstoff-Formaldehydharz *n* thiourea-formaldehyde resin
Thioharnstoffharz *n* thiourea resin, polythiourea
Thiohypophosphat *n* $M_4{}^IP_2S_6$ thiohypophosphate, hexathiohypophosphate
Thioindigo *m (n)* thioindigo
Thioindoxyl *n* thioindoxyl, 3-hydroxy-benzo[*b*]thiophene
Thiokarbamid *n s.* Thioharnstoff
Thiokarbanilid *n* $CS(NHC_6H_5)_2$ thiocarbanilide
Thiokarbonat *n* $M_2{}^ICS_3$ thiocarbonate, trithiocarbonate
Thiokarbonyldichlorid *n* $CSCl_2$ thiocarbonyl chloride, thiophosgene
Thiokarbonylselenid *n* CSSe thiocarbonyl selenide, carbon selenide sulphide
Thiokarbonyltellurid *n* CSTe thiocarbonyl telluride, carbon sulphide telluride
Thiokarbonyltetrachlorid *n* CCl_3SCl trichloromethanesulphenyl chloride, thiocarbonyl tetrachloride
Thiokohlensäure *f* H_2CS_3 thiocarbonic acid, trithiocarbonic acid
Thioktansäure *f s.* Thioktinsäure
Thioktinsäure *f* thioctic acid, α-lipoic acid, 3-(4-carboxybutyl)-1,2-dithiolane
Thioktsäure *f s.* Thioktinsäure
Thiolgruppe *f* –SH thiol (sulphydryl, mercapto) group
Thiolignin *n* thiolignin
Thiolthionkohlensäure *f* HOCSSH dithiocarbonic acid *(one form)*
Thionaphthol *n* $C_{10}H_7SH$ thionaphthol, naphthalenethiol
Thionylbromid *n* $SOBr_2$ thionyl bromide
Thionylchlorid *n* $SOCl_2$ thionyl chloride
Thionylfluorid *n* SOF_2 thionyl fluoride
Thiooxalat *n* $M^IO–CS–COOM^I$ thiooxalate
Thiophen *n* thiophene
thiophenfrei thiophene-free
Thiophenprobe *f* thiophene test
Thiophosgen *n s.* Thiokarbonyldichlorid
Thiophosphat *n* thiophosphate
Thiophosphorsäuretriamid *n s.* Thiophosphoryltriamid
Thiophosphorylbromid *n* $PSBr_3$ thiophosphoryl bromide
Thiophosphorylbromiddichlorid *n* $PSBrCl_2$ thiophosphoryl bromide dichloride
Thiophosphorylchlorid *n* $PSCl_3$ thiophosphoryl chloride
Thiophosphoryldibromidchlorid *n* $PSClBr_2$ thiophosphoryl dibromide chloride
Thiophosphorylfluorid *n* PSF_3 thiophosphoryl fluoride
Thiophosphoryltriamid *n* $PS(NH_2)_3$ thiophosphoryl amide, thiophosphoric triamide

Thioplast *m* thioplast, polysulphide rubber
Thiopropylalkohol *m* $CH_3CH_2CH_2SH$ 1-propanethiol, thiopropyl alcohol
Thiosalizylsäure *f* $C_6H_4(SH)COOH$ thiosalicylic acid, *o*-mercaptobenzoic acid
Thiosäure *f* thioacid
Thiosemikarbazid *n* $NH_2CSNHNH_2$ thiosemicarbazide, aminothiourea
Thiosulfat *n* $M_2^IS_2O_3$ thiosulphate
Thiosulfatentfernung *f (phot)* hypo elimination
Thiosulfit *n* $M_2^IS_2O_2$ thiosulphite
Thiotolen *n* thiotolene, methylthiophene
Thiozinn(II)-säure *f* H_2SnS_2 thiostannous acid, dithiostannous acid
Thiozinn(IV)-säure *f* H_2SnS_3 thiostannic acid, trithiostannic acid
Thiozyanat *n* M^ISCN thiocyanate, rhodanide
Thiozyanatoaurat *n* thiocyanatoaurate
Thiozyanatoferrat *n* thiocyanatoferrate
Thiozyanatowolframat *n* thiocyanatowolframate, thiocyanatotungstate
Thiozyansäure *f* HSCN thiocyanic acid, hydrogen thiocyanate, rhodanic acid
Thiuramdisulfidvernetzung *f (rubber)* thiuram disulphide cure
Thiuramvernetzung *f*, **Thiuramvulkanisation** *f* thiuram cure
thixotrop *(coll)* thixotropic
Thixotropie *f (coll)* thixotropy
Thixotropier[ungs]mittel *n (coll)* thixotroping (thixotropic) agent
Thomas-Birne *f s.* Thomas-Konverter
Thomas-Gasmesser *m* Thomas meter
Thomas-Konverter *m (met)* Thomas (basic Bessemer) converter
Thomas-Konverterstahl *m s.* Thomas-Stahl
Thomas-Konverterverfahren *n s.* Thomas-Verfahren
Thomas-Mehl *n*, **Thomas-Phosphat** *n* Thomas meal (phosphate)
Thomas-Schlacke *f* Thomas (basic, Belgian) slag
Thomas-Stahl *m* Thomas steel, basic [Bessemer, converter] steel
Thomas-Verfahren *n* Thomas[-Gilchrist] process, basic [Bessemer, converter] process
Thomsenolith *m (min)* thomsenolite *(calcium sodium hexafluoroaluminate)*
Thomson-Effekt *m* Thomson [thermoelectric] effect
Thomsonit *m (min)* thomsonite *(a tectosilicate)*
Thomson-Überfall *m* triangular notch *(flow measurement)*
Thorakalapplikation *f* topical application *(for testing the efficiency of an insecticide)*
Thorat *n* $M_2^IThO_3$ thorate
Thorerde *f s.* Thoriumoxid
thorieren to thoriate *(e.g. tungsten filaments)*
Thorit *m (min)* thorite *(thorium orthosilicate)*
Thorium *n* Th thorium ✦ **mit T. überziehen** to thoriate *(e.g. tungsten filaments)*
Thorium-228 *n* ^{228}Th thorium-228, radiothorium
Thoriumbromid *n* $ThBr_4$ thorium bromide
Thoriumchlorid *n* $ThCl_4$ thorium chloride
Thoriumdioxid *n s.* Thoriumoxid
Thorium-Emanation *f s.* Radon-220
Thoriumhydroxid *n* $Th(OH)_4$ thorium hydroxide
Thoriumjodid *n* ThI_4 thorium iodide
Thoriumkarbid *n* ThC_2 thorium carbide
Thoriumnitrat *n* $Th(NO_3)_4$ thorium nitrate
Thoriumoxid *n* ThO_2 thorium oxide, thorium dioxide
Thoriumreihe *f (nucl)* thorium [decay] series
Thoriumsulfat *n* $Th(SO_4)_2$ thorium sulphate
Thoriumsulfid *n* thorium sulphide
Thoriumzerfallsreihe *f (nucl)* thorium [decay] series
Thorne-Bleichturm *m (pap)* Thorne bleacher
Thorne-Entrindungsmaschine *f (pap)* Thorne (waterous) barker
Thoron *n s.* Radon-220
Thr *s.* Threonin
Threit *m* $CH_2OH[CHOH]_2CH_2OH$ threitol, *anti*-1,2,3,4-tetrahydroxybutane
Threonin *n* $CH_3CH(OH)CH(NH_2)COOH$ threonine, 2-amino-3-hydroxybutyric acid
Threose *f* threose *(a monosaccharide)*
Thresholdverfahren *n* threshold treatment *(for softening water)*
Thrombin *n* thrombin *(an enzyme promoting the clotting of blood)*
Thrombokinase *f*, **Thromboplastin** *n* thrombokinase, thromboplastin
Thujaöl *n* thuja (cedar-leaf) oil *(chiefly from Thuja occidentalis L.)*
Thujasäure *f* thujic acid, 4,4-dimethylcycloheptatriene-1-carboxylic acid
Thujopsen *n* thujopsene, widdrene *(a tricyclic sesquiterpene)*
Thujylalkohol *m*, **β-Thujylalkohol** *m* thujyl alcohol, 3-thujanol
Thulium *n* Tm thulium
Thuliumoxid *n* Tm_2O_3 thulium oxide
Thuringit *m (min)* thuringite *(a phyllosilicate)*
Thymiankampfer *m s.* Thymol
Thymianöl *n* thyme oil *(from Thymus vulgaris L. and Th. zygis L.)*
Thymiansäure *f s.* Thymol
Thyminose *f* thyminose, *D*-2-deoxyribose *(a monosaccharide)*
Thymochinon *n* thymoquinone, 2-isopropyl-5-methyl-1,4-benzoquinone
Thymol *n* $CH_3C_6H_3(C_3H_7)OH$ 6-isopropyl-*m*-cresol
Thymolblau *n* thymol blue, thymolsulphonphthalein *(a pH indicator)*
Thymolphthalein *n* thymolphthalein *(a pH indicator)*
Thymolsulfophthalein *n s.* Thymolblau
Thymonukleinsäure *f*, **Thymusnukleinsäure** *f s.* Desoxyribonukleinsäure
Thyreoglobulin *n (bioch)* thyroglobuline
thyreotrop thyrotropic
Thyreotropin *n* thyrotropin, thyrotropic hormone, thyroid-stimulating hormone
Thyssen-Gálocsy-Verfahren *n* Thyssen-Gáloczy process *(of coal gasification)*
Tiefätzung *f (glass)* deep etching

Tiefbaubetrieb *m*, **Tiefbauförderung** *f* deep (underground) mining, underground work[ing]
Tiefbrunnen *m* deep well
Tiefdruck *m* intaglio printing
Tiefdruckfarbe *f* intaglio [printing] ink
Tiefdruckpapier *n* intaglio [printing] paper
Tiefdruckverfahren *n* intaglio [printing] process
Tiefenentwickler *m (phot)* depth developer
Tiefenentwicklung *f (phot)* depth development
Tiefenfiltration *f* filter-medium filtration (operation)
Tiefengestein *n* plutonic (hypogene, deep-seated, irruptive) rock, plutonite
Tiefenmagma *n (geol)* hypomagma
Tiefenrüttler *m* immersion (poker) vibrator
tieffärbend deep dyeing
Tieffassung *f* low-position shoe *(of an electrode)*
tiefgefrieren to deep-freeze
Tiefgefriermilch *f* deep-frozen milk
Tiefkühlanlage *f* deep-cooling plant
Tiefkühlung *f* deep cooling
Tiefkühlvorlage *f (distil)* low-temperature receiver
Tiefkultur *f* submerged culture *(microbiology)*
Tiefkupferglanz *m (min)* low-chalcocite *(copper(I) sulphide)*
tiefmatt *(text)* very dull
Tiefofen *m (met)* soak[ing] pit
Tiefpumpe *f* subsurface pump
Tiefquarz-Modifikation *f* lowquartz modification *(of germanium)*
tiefschmelzend low-melting[-point], low-fusion
Tiefseeablagerung *f (geol)* deep-sea deposit
tiefsiedend low-boiling, light
Tieftankverfahren *n* submersion (submerged culture) process *(microbiology)*
Tieftemperaturabscheidung *f* low-temperature separation
tieftemperaturbeständig low-temperature-resistant, resistant to cold, cold-resistant
Tieftemperaturbeständigkeit *f* low-temperature resistance, resistance to cold, cold resistance
Tieftemperaturchlorierung *f* low-temperature chlorination
Tieftemperatureigenschaften *fpl* low-temperature properties (characteristics, behaviour)
Tieftemperaturentgasung *f s.* Tieftemperaturverkokung
Tieftemperaturerzeugung *f* production of low temperatures
Tieftemperaturflexibilität *f* low-temperature flexibility
Tieftemperaturform *f* low-temperature form
Tieftemperaturhydrierung *f* low-temperature hydrogenation
Tieftemperaturkautschuk *m* cold (cold polymerized, low-temperature) rubber
Tieftemperaturkoks *m* low-temperature coke
Tieftemperaturpolymer[es] *n*, **Tieftemperaturpolymerisat** *n* low-temperature polymer, cold polymer
Tieftemperaturpolymerisation *f* low-temperature (cold) polymerization
Tieftemperaturtechnik *f* cryogenic engineering
Tieftemperaturteer *m* low-temperature tar
Tieftemperaturverdampfer *m* low-temperature evaporator
Tieftemperaturverfahren *n* 1. cryogenic process; 2. *(coal)* low-temperature carbonization (carbonizing) process
Tieftemperaturverkokung *f* low-temperature carbonization
tiefziehen to deep-draw
Tiefziehen *n* deep drawing
  T. mit Gleitvorrichtung *(plast)* slip forming
  T. mit Ziehring *(plast)* slip ring forming
Tiefziehteil *n* deep-drawing part
Tiegel *m* 1. crucible; 2. [flash] cup *(of a flash-point tester)*
  geschlossener T. closed [flash] cup
  offener T. open [flash] cup
  offener T. nach Cleveland Cleveland open cup
Tiegelblähprobe *f (coal)* crucible swelling test
Tiegeldeckel *m* crucible lid
Tiegelkoks *m* crucible coke
Tiegelofen *m* crucible [melting] furnace
Tiegelofenverfahren *n (met)* crucible process
Tiegelring *m* crucible ring
Tiegelschmelzverfahren *n s.* Tiegelofenverfahren
Tiegelstahl *m* crucible [cast] steel
Tiegelstahlverfahren *n s.* Tiegelofenverfahren
Tiegeluntersatz *m* crucible stand
Tiegelverfahren *n s.* Tiegelofenverfahren
Tiegelzange *f* crucible tongs
Tiemannit *m (min)* tiemannite *(mercury(II) selenide)*
Tierexperiment *n s.* Tierversuch
Tierfaser *f* animal fibre
Tierfett *n* animal fat
Tierhaar *n* animal hair
tierisch animal
Tierkohle *f* animal char[coal]
Tierkörpermehl *n* [animal, garbage] tankage *(a fertilizer)*
Tierleim *m* animal glue (gelatine), *(pap also)* animal size
Tieröl *n* animal oil
  Dippels T. Dippel's oil, oil of hartshorn *(for denaturing ethanol)*
Tierstärke *f* animal starch
Tierversuch *m (pharm, tox)* bioassay, biological assay
Tierwachs *n* animal wax
Tiffeneau-Umlagerung *f* Tiffeneau rearrangement *(of amino alcohols forming carbonyl compounds)*
Tigerauge *n (min)* tiger's eye *(a variety of quartz)*
Tiglinsäure *f* $CH_3CH{=}C(CH_3)COOH$ tiglic acid, 2-methylcrotonic acid, *trans*-2-methyl-2-butenoic acid
Tinkal *m (min)* tincal, borax *(sodium tetraborate-10-water)*
Tinktur *f (pharm)* tincture
Tinte *f* [writing] ink
  sympathetische T. sympathetic (secret) ink
Tintenfarbstoff *m* ink dye

Tintenfestigkeit *f (pap)* ink resistance
Tintenschreiber *m* pen-and-ink recorder
Tintentablette *f* ink tablet
Tintometer *n* tintometer, colorimeter
T. nach Lovibond Lovibond tintometer
Tirolit *m (min)* tyrolite, copper froth, froth copper *(an arsenate containing calcium and copper)*
Tischrüttler *m* table vibrator
Tischverfahren *n (glass)* table [casting] process
Titan *n* Ti titanium
Titanat *n* titanate
Titan(II)-bromid *n* $TiBr_2$ titanium(II) bromide, titanium dibromide
Titan(IV)-bromid *n* $TiBr_4$ titanium(IV) bromide, titanium tetrabromide
Titan(II)-chlorid *n* $TiCl_2$ titanium(II) chloride, titanium dichloride
Titan(III)-chlorid *n* $TiCl_3$ titanium(III) chloride, titanium trichloride
Titan(IV)-chlorid *n* $TiCl_4$ titanium(IV) chloride, titanium tetrachloride
Titanchloridmethode *f* **von Edmund Knecht** titanous-chloride method of E. Knecht *(for identifying azo dyes)*
Titandi... *s.a.* Titan(II)-...
Titandioxid *n s.* Titan(IV)-oxid
Titandiphosphat *n* TiP_2O_7 titanium diphosphate, titanium pyrophosphate
Titandisulfid *n s.* Titan(IV)-sulfid
Titanerde *f s.* Titan(IV)-oxid
Titan(III)-fluorid *n* TiF_3 titanium(III) fluoride, titanium trifluoride
Titan(IV)-fluorid *n* TiF_4 titanium(IV) fluoride, titanium tetrafluoride
titanführend titaniferous
Titangelb *n (dye)* titan (Clayton, thiazole) yellow
titanhaltig titaniferous
Titanhydrid *n* TiH_2 titanium hydride
Titan(IV)-hydroxid *n* $Ti(OH)_4$ titanium(IV) hydroxide
Titanit *m (min)* titanite *(calcium titanium(IV) oxide orthosilicate)*
Titan(II)-jodid *n* TiI_2 titanium(II) iodide, titanium diiodide
Titan(IV)-jodid *n* TiI_4 titanium(IV) iodide, titanium tetraiodide
Titanmonokarbid *n* TiC titanium monocarbide
Titanmonosulfid *n s.* Titan(II)-sulfid
Titanmonoxid *n s.* Titan(II)-oxid
Titannitrid *n* TiN titanium nitride
Titanometrie *f* titanometry
Titan(II)-oxid *n* TiO titanium(II) oxide, titanium monooxide
Titan(III)-oxid *n* Ti_2O_3 titanium(III) oxide, dititanium trioxide
Titan(IV)-oxid *n* TiO_2 titanium(IV) oxide, titanium dioxide
Titanoxidsulfat *n* $TiOSO_4$ titanium oxide sulphate
Titanporzellan *n* titania porcelain
Titanpyrophosphat *n s.* Titandiphosphat
Titansäure *f* titanic acid *(any of various hydrates of titanium dioxide)*
Titan(IV)-sulfat *n* $Ti(SO_4)_2$ titanium(IV) sulphate
Titan(II)-sulfid *n* TiS titanium(II) sulphide, titanium monosulphide
Titan(III)-sulfid *n* Ti_2S_3 titanium(III) sulphide, dititanium trisulphide
Titan(IV)-sulfid *n* TiS_2 titanium(IV) sulphide, titanium disulphide
Titantetra... *s.* Titan(IV)-...
Titantri... *s.* Titan(III)-...
Titanweiß *n* titanium white *(a pigment consisting mainly of TiO_2)*
Titanweißware *f (ceram)* titania whiteware
Titer *m* 1. titre, titer *(strength of a solution)*; 2. titre, titer *(for defining the fineness of yarn)*; 3. titre [value] *(the solidifying point of fatty acids)*
Titerlösung *f* standard solution
Titerpumpe *f* spinning (metering) pump
Titersubstanz *f* standard reagent (titrant, titrimetric substance)
Titertest *m* titre test *(for determining the solidifying point of fatty acids)*
Titerwert *m s.* Titer 3.
Titrans *n* titrant
Titration *f* titration
T. in nichtwäßriger Lösung non-aqueous titration
amperometrische T. amperometric titration
chelatometrische T. chelatometric titration, chelatometry
coulometrische T. coulometric titration
elektrometrische T. *s.* potentiometrische T.
jodometrische T. iodometric titration
komplexometrische T. complexometric titration, complexometry
konduktometrische T. conductometric (conductance) titration
manganometrische T. permanganate titration
nephelometrische T. nephelometric titration
potentiometrische T. potentiometric (electrometric) titration
thermometrische T. thermometric titration
turbidimetrische T. turbidimetric (turbidity) titration
Titrationsapparat *m* titration (titrating) apparatus
Titrationsazidität *f (soil)* total acidity
Titrationscoulometer *n* titrating coulometer
Titrationsgefäß *n* titration vessel
Titrationskurve *f* titration curve
Titrationszelle *f* titration cell
Titrator *m* titrator
Titrieranalyse *f* titrimetric (volumetric, mensuration) analysis
Titrierapparat *m s.* Titrationsapparat
Titrierautomat *m* automatic titrator (titration apparatus)
titrierbar titr[at]able
Titrierbecher *m* titrating beaker
titrieren to titrate
Titrierfehler *m* titration error
Titriergerät *n* titrator
Titrierkolben *m* titration flask
Titrierung *f s.* Titration
Titriervorrichtung *f* titrating device

Titrimeter *n* titrimeter
Titrimetrie *f* titrimetry, volumetry
titrimetrisch titrimetric, volumetric
Tizerahextrakt *m (tann)* tizerah extract *(from Rhus pentaphylla Desf.)*
TNT *s.* Trinitrotoluol
TOA *s.* Alttuberkulin
Tobermorit *m (min)* tobermorite *(an inosilicate)*
Tobias-Säure *f* $C_{10}H_6(NH_2)(SO_3H)$ Tobias acid, 1-naphthylamine-1-sulphonic acid
Tödlichkeitsdosis *f (tox)* lethal (fatal) dose, LD
Tödlichkeitsindex *m (tox)* ct product *(product of concentration and survival time)*
Tödlichkeitsprodukt *n* / **Habersches** *s.* Tödlichkeitsindex
Toilettenpräparat *n* toilet preparation
Toilettenseidenpapier *n* toilet (sanitary) tissue
Toilettenseife *f* toilet soap
Toilettenwasser *n* toilet water
Tokopherol *n (bioch)* tocopherol
Tolan *n* $C_6H_5C{\equiv}CC_6H_5$ tolane, diphenylacetylene
Toleranz *f* tolerance, *(mechanically also)* allowance; *s.* Toleranzdosis
Toleranzdosis *f*, **Toleranzwert** *m (tox)* [maximum] tolerance
Tollens-Reagens *n* Tollen's reagent *(for detecting aldehydes)*
Tollkirsche *f* belladonna, deadly nightshade, Atropa belladonna L.
Tolubalsam *m* Tolu balsam *(from Myroxylon balsamum (L.) Harms var. balsamum)*
Toluidin *n* $CH_3C_6H_4NH_2$ toluidine, aminotoluene
Toluol *n* $C_6H_5CH_3$ toluene, methylbenzene, *(commercially also)* toluol
Toluoldiisozyanat *n* toluene diisocyanate, TDI
1'-Toluolkarbonsäure *f s.* α-Tolylsäure
***p*-Toluolsulfochlorid** *n* $CH_3C_6H_4SO_2Cl$ *p*-toluenesulphonyl chloride, tosyl chloride
***p*-Toluolsulfonat** *n* *p*-toluenesulphonate, tosylate
***p*-Toluolsulfonsäure** *f* $CH_3C_6H_4SO_3H$ toluene-*p*-sulphonic acid
Toluolsulfonsäurechloramidnatrium *n* sodium *p*-toluenesulphonchloramine, chloramine-T
***p*-Toluolsulfonsäurechlorid** *n s.* *p*-Toluolsulfochlorid
***p*-Toluolsulfonsäureester** *m s.* *p*-Toluolsulfonat
***p*-Toluolsulfonylchlorid** *n s.* *p*-Toluolsulfochlorid
***p*-Toluolsulfonylgruppe** *f* $CH_3C_6H_4SO_2-$ *p*-toluenesulphonyl group, tosyl group
Toluylendiamin *n* $CH_3C_6H_3(NH_2)_2$ toluylenediamine, diaminotoluene
Toluylenrot *n* toluylene (neutral) red *(an oxidation-reduction indicator)*
Toluylsäure *f* $CH_3C_6H_4COOH$ toluic acid, methylbenzoic acid
Tolylendiamin *n s.* Toluylendiamin
α-Tolylsäure *f*, **1'-Tolylsäure** *f* $C_6H_5CH_2COOH$ α-toluic acid, toluylic acid, phenylacetic acid
Tombak *m* tombac, tombak
Ton *m* 1. clay, argil[la]; 2. shade, hue *(of colour)* ✦ **einen T. treffen** *(dye)* to match a shade
aktivierter T. activated clay
aluminiumoxidreicher T. high-alumina clay
empfindlicher T. tender clay
fetter T. plastic clay
feuerfester T. refractory clay, fire-clay
gesumpfter T. soaked clay
kieselsäurereicher T. high-silica clay, siliceous clay
leichtschmelzender T. fusible clay
natürlicher T. natural (naturally occurring, non-activated) clay
plastischer T. plastic clay
primärer T. primary (residual) clay
reiner T. *s.* weißer T.
säureaktivierter T. acid clay *(for refining purposes)*
tonerdereicher T. high-alumina clay
weißbrennender T. white-firing clay
weißer T. white (china, porcelain) clay, kaolin[e], white bole, bolus alba
windgesichteter T. aeroclay
Tonabbau *m* clay mining
tonartig clayey, clayish, argillaceous
Tonaufbereitung *f* clay preparation
Tonaufbereitungsanlage *f* clay preparation plant
Tonboden *m* clay soil
Tondreieck *n* pipeclay triangle
Toneisenstein *m (min)* clay ironstone, ironstone clay, argillaceous haematite
tonen *(phot)* to tone
tönen to tint, to tinge, to tone, to shade
Toner *m (phot)* toning agent
Tonerde *f* 1. clay, argil; 2. *s.* Aluminiumoxid
[künstlich] aktivierte T. activated clay *(petroleum refining)*
naturaktive T. *s.* natürliche T.
natürliche T. natural (naturally occurring) clay, *(petrol also)* non-activated clay
säureaktivierte T. acid clay *(petroleum refining)*
Tonerdegel *n* alumina gel, gelatinous aluminium hydroxide
tonerdehaltig aluminiferous
Tonerdekatalysator *m*, **Tonerdekontakt** *m* clay catalyst (contact)
Tonerdeporzellan *n* alumina porcelain
Tonerdeschiffchen *n* alumina boat
Tonerdeschmelzzement *m s.* Tonerdezement
Tonerdesilikatglas *n* aluminosilicate glass
Tonerdeweißware *f (ceram)* alumina whiteware
Tonerdezement *m* aluminous (high-alumina) cement
Tonesse *f* burner guard *(for Bunsen burners)*
Tonfilter *n* earthenware filter
Tonfraktion *f (soil)* clay fraction
tonfrei free from clay, non-clay
Tongalle *f (geol)* clay gall
tongebunden *(ceram)* clay-bonded
Tongestein *n (geol)* claystone
Tongewinnung *f* clay mining
Tongrube *f* clay pit
Tongut *n s.* Tonware
tonhaltig containing clay, clayey, clayish, argilliferous, argillaceous

Tonhobel *m (ceram)* clay cutter
Ton-Humus-Komplex *m (soil)* clay-humus (organoclay) complex, colloidal complex
tonig clayey, clayish, argillaceous
Tonikum *n (pharm)* tonic
Ton-in-Ton-Färbung *f* tone-in-tone dyeing
tonisch *(pharm)* tonic
tonisieren *(pharm)* to tone
tonisierend *(pharm)* tonic
Tonkabohne *f* tonka (tonca, tonga, tonqua) bean *(from Dipteryx specc.)*
Tonkabohnenkampfer *m* tonka bean camphor, coumarin
Tonkalk *m* argillaceous limestone
Tonkneter *m*, **Tonknetmaschine** *f s.* Tonschneider
Tonlager *n*, **Tonlagerstätte** *f* clay deposit
Tonmasse *f (ceram)* clay body
Tonmergel *m (geol)* clay marl
Tonmineral *n* clay mineral
Tonne *f* cask, barrel, *(if large)* tun, *(if small)* keg
Tonraspler *m (ceram)* clay shredder
Tonrohr *n*, **Tonröhre** *f* clay (earthenware) pipe
Tonschicht *f* clay bank
Tonschiefer *m (geol)* argillite, [clay] slate
Tonschiffchen *n* clay combustion boat
Tonschlempe *f*, **Tonschlicker** *m* clay[-water] slurry
Tonschneider *m (ceram)* pug (clay) mill
Tonsilo *n(m) (ceram)* clay silo
Tonsubstanz *f* clay substance
Tontiegel *m* clay crucible
Tontopf *m* earthenware pot (vessel)
Tonumfang *m (phot)* tone range
Tonung *f (phot)* toning
Tönung *f* 1. tinting, tinging, toning, shading *(act)*; 2. tint, tinge, tone, shade, cast *(result)*
Tonvorkommen *n* clay deposit
Tonware *f* earthenware, clay ware
Ton[wert]wiedergabe *f (phot)* tone rendering (reproduction)
Tonzelle *f* **/ poröse** porous pot (cell, cup)
Tonziegel[stein] *m* clay [building] brick
Tonzylinder *m* **/ poröser** *s.* Tonzelle / poröse
Topas *m (min)* topaz *(an aluminium silicate)*
Topazolith *m (min)* topazolite *(a nesosilicate)*
Topf *m* 1. pot, *(lab also)* jar; 2. *(text)* [spinning] can; pot *(for pot spinning)*
  Wittscher T. Witt jar
Töpfer *m* potter
Töpferei *f* pottery
Töpferscheibe *f* potter's wheel
Töpferton *m* potter's clay
Töpferware *f* pottery, earthenware
Topffärben *n (text)* potting
topfglühen to pot-anneal
Topfglühofen *m* pot-annealing furnace
Topfglühung *f* pot annealing
Topfkurare *n* pot curare
Topfmanschette *f* cup [ring]
Topfspinnverfahren *n* [centrifugal] pot spinning
Topfzeit *f* pot life *(as of adhesives and organic coatings)*
Topf-Zentrifugenspinnverfahren *n* [centrifugal] pot spinning
Topochemie *f* topochemistry
topochemisch topochemical
Toppanlage *f (petrol)* topping (skimming) plant
Toppdestillation *f* topping
toppen *(petrol)* to top, to skim
Toppprodukt *n (petrol)* tops, overhead [product]
Topprückstand *m (petrol)* long residue (residuum) ✦ **Topprückstände aufarbeiten** to run resid
Toppung *f (petrol)* topping
Torbernit *m (min)* torbernite *(a hydrous uranium copper phosphate)*
Torf *m* peat
  handgestochener T. hand-cut peat
  maschinengeformter T. machine[-cut] peat
  terrestrischer T. terrestrial peat
  zerrissener T. macerated peat
torfbildend peat-forming
Torfbildung *f* peat formation
Torfbildungsprozeß *m* peat-forming process
Torfboden *m* peat soil
Torfbrikett *n* peat briquette
Torfdolomit *m (geol)* coal ball
Torffräsverfahren *n* milled-peat process
Torfgas *n* peat gas
Torfgewinnung *f* peat winning
Torfhumus *m* peat humus
torfig peaty
Torfkoks *m* peat coke
Torfmaschine *f* peat machine
Torfmasse *f* peat substance
Torfmoor *n* peat bog (moor)
Torfpappe *f* peat board
Torfstich *m* peat bank, peatery
Torfsubstanz *f* peat substance
Torfteer *m* peat tar
Torftrocknung *f* peat drying
Torkretbeton *m* gunite, gunned concrete
torkretieren to gunite
Törnebohmit *m (min)* törnebohmite *(a silicate containing cerium and lanthanum)*
torpedieren *(petrol)* to shoot
Torpedo *m (plast)* torpedo, *(in injection moulding also)* spreader
  rotierender T. rotating spreader *(of a plunger-type injection machine)*
Torsion *f* torsion, twist
Torsionsschwingung *f* torsional (twisting) vibration *(spectroscopy)*
Torsionssteifigkeit *f* stiffness in torsion
Torsionswaage *f* torsion balance
Torulahefe *f* torula yeast
Tosylat *n* tosylate, *p*-toluenesulphonate *(salt or ester of toluene-p-sulphonic acid)*
Tosylchlorid *n* $CH_3C_6H_4SO_2Cl$ tosyl chloride, *p*-toluenesulphonyl chloride
Tosylester *m* tosylate, *p*-toluenesulphonate
Tosylgruppe *f* $CH_3C_6H_4SO_2-$ tosyl group, *p*-toluenesulphonyl group
Tosylierung *f* tosylation *(introduction of the p-toluenesulphonyl group)*
tot *(rubber)* lifeless, dead

Totalherbizid *n* general herbicide, non-selective herbicide (weed-killer), soil sterilant
Totalkondensation *f (distil)* total condensation
Totalreflexion *f* total (complete) reflection
Totalsynthese *f* total synthesis
totbrennen to dead-burn *(e.g. gypsum, dolomite)*
totgerben *(tann)* to case-harden, to overtan
Totgerbung *f (tann)* case-hardening, overtannage
totmahlen to overgrind; *(pap)* to beat dead
totmastizieren *s.* totwalzen
Totraum *m* 1. dead spot *(in an extruder)*; 2. *s.* Totvolumen
totrösten *(met)* to dead-roast
Totröstung *f (met)* dead roasting
Totvolumen *n (phys chem)* dead volume (space)
totwalzen *(rubber)* to mill to death, to kill, to overmill, to overmasticate
Totwalzen *n (rubber)* dead milling, killing, overmilling, overmastication
Totweiche *f* oversteeping *(of malt)*
Totzeit *f* dead time *(of a counting tube)*
Tourill *n* tourill *(a kind of absorption vessel)*
Toxalbumin *n* toxalbumin
toxigen *(med)* toxi[co]genic
Toxikologe *m* toxicologist
  vereidigter T. official toxicologist
Toxikologie *f* toxicology
toxikologisch toxicologic[al]
Toxikum *n* toxic [substance], toxicant
Toxin *n (bioch)* toxin ✦ **T. erzeugend** toxigenic
toxisch toxic[al], poisonous
Toxizität *f* toxicity
  T. für Säugetiere mammalian toxicity
  akute T. acute toxicity
  chronische T. chronic toxicity
  orale T. oral toxicity
Toxoid *n (med)* toxoid
Tozer-Verfahren *n* Tozer process *(of low-temperature carbonization)*
TPN$^+$ *(abbr)* Triphosphopyridinnukleotid, *s.* Nikotinamid-adenin-dinukleotidphosphat
Tracer *m* / **radioaktiver** radioactive tracer (indicator), radiotracer
Tracerchemie *f* tracer chemistry
Tracermethode *f* tracer (indicator) method
Tracertechnik *f* tracer technique
Tracerversuch *m* tracer experiment
Trafoöl *n* transformer oil
Tragant *m* tragacanth gum, gum tragacanth *(any of various gums esp from Astragalus specc.)*
  Afrikanischer T. African tragacanth *(gum from Sterculia tragacantha Lindl.)*
  Indischer T. Indian tragacanth, sterculia gum, karaya [gum], gum karaya *(chiefly from Sterculia urens Roxb.)*
  Ostindischer T. *s.* Indischer T.
  Persischer T. Persian tragacanth *(from Astragalus specc.)*
Tragantgummi *n s.* Tragant
Tragantleim *m* tragacanth adhesive
Tragantschleim *m* tragacanth mucilage
Tragasol *n* gum tragasol *(a leather finish from seed shells of Ceratonia siliqua L.)*
träge inert, inactive, passive, indifferent; slow *(reaction)*
tragecht *(text)* fast (resistant) to wearing, wear-resistant, wearproof, tough
Tragechtheit *f (text)* fastness (resistance) to wearing, wear resistance
Trageeigenschaft *f (text)* wearing quality, wearability
Träger *m* 1. carrier, *(esp relating to heat transfer also)* medium; 2. support, supporting material *(for a layer containing active substances, as catalysts or photosensitive compounds)*; 3. substrate *(for pigments)*
  kolloidaler T. colloid carrier, protector *(corresponding to apoenzyme in recent terminology)*
Trägerdampfdestillation *f* steam distillation
Trägerelektrode *f* carrier electrode
Trägerelement *n (nucl)* carrier
Trägerfaden *m (text)* carrier thread
Trägerflüssigkeit *f* carrier liquid
trägerfrei carrier-free
Trägergas *n* carrier [gas] *(gas chromatography)*; entrainer, carrier *(sublimation)*
Trägergasstrom *m* carrier gas stream *(gas chromatography)*
Trägergassublimation *f* entrainer (carrier) sublimation
Trägergestein *n* carrier bed
Trägerkatalysator *m* supported catalyst
trägerlos unsupported *(plastic film)*
Trägerluft *f* transport air
Trägermaterial *n s.* Träger 2.
Trägerspeicherung *f* carrier storage
Trägerstoff *m s.* Träger 1.; 2.
Trägerstoffdestillation *f* carrier distillation
Trägersubstanz *f s.* Träger
Tragfähigkeit *f* [load-]bearing strength, load (bearing) capacity
tragfest *s.* tragecht
Traggestell *n (lab)* carrying frame
Trägheit *f* 1. *(chem)* inertness, inactivity, passivity, indifference; slowness *(of a reaction)*; 2. *(phys)* inertia
Trägheitsgesetz *n* law of inertia
trägheitslos inertia-free
Trägheitsmoment *n* moment of inertia, inertia effect
Tragkasten *m (lab)* carrying box
  T. für Flaschen bottle carrier
Tragkettenförderer *m* rigid arm elevator
Tragkraft *f* lifting capacity *(of cranes)*
Tragplatte *f* bearing plate
Tragrolle *f* idler [roll] *(as of a conveyor belt)*
  T. am Leertrum (Untertrum) return roll
Tragtrommel *f*, **Tragwalze** *f (pap)* carrying (support, supporting) roll, winder drum
Trajektorie *f* trajectory *(in diagrams)*
Traktorenkerosin *n s.* Traktorenpetroleum
Traktorenkraftstoff *m* tractor fuel
Traktorenöl *n s.* Traktorenpetroleum
Traktorenpetrol[eum] *n* tractor [vaporizing] oil
Trame[seide] *f* tram silk

Tran *m* fish (train) oil
Tranausharzung *f (tann)* fish-oil spew (spue)
Träne *f (coat)* tear *(in dipping)*
tränenerregend *s.* tränenreizend
Tränengas *n* tear gas, lachrymator
tränenreizend lachrymatory, lacrimatory, tear-producing
Tränenreizstoff *m* lachrymator, lacrimator
Tranfettung *f*,**Tranfüllung** *f (tann)* fish-oil stuffing
tränken to impregnate, to saturate, to imbibe, *(using an aqueous solution)* to steep, to soak, to water
  mit Harz t. *(plast)* to resin
Tränkflüssigkeit *f* steeping liquor
Tränkharz *n* impregnating resin
Tränklauge *f (pap)* impregnating liquor
Tränkmasse *f* impregnating material
Tränkstoff *m* impregnation material
Tränktrog *m* steeping pan
Tränkung *f* impregnation, saturation, imbibition, *(using an aqueous solution)* steep[age], soak[age], watering
Tranquilit *m* tranquilite *(a lunar mineral)*
trans-Addition *f* trans addition
Transaminase *f* transaminase
Transaminierung *f* transamination
transannular *(org chem)* transannular
Transferase *f*, **Transferenzym** *n* transferase, transferring enzyme
Transferformung *f (rubber)* transfer moulding
Transferpressen *n (plast)* transfer moulding
Transferpreßwerkzeug *n (plast)* transfer mould
Transfer-Ribonukleinsäure *f*, **Transfer-RNS** *f* transfer (soluble) ribonucleic acid, s-RNA
Transfer-Verfahren *n (rubber)* transfer moulding
trans-Form *f* trans form
Transformation *f (phys chem)* transformation
Transformationsintervall *n (phys chem)* transition interval, *(relating to glass)* transformation range
Transformationspunkt *m*, **Transformationstemperatur** *f (phys chem)* transition point, *(relating to glass)* transformation point
Transformatorenöl *n* transformer oil
Transfusion *f* diffusion *(of gases through a porous diaphragm)*
Transglykosidierung *f*, **Transglykosylierung** *f (bioch)* transglycosylation
trans-Isomer[es] *n* trans isomer
trans-Konformation *f s.* Konformation / ± antiperiplanare
Transkription *f (bioch)* transcription *(of genetical information from DNA on messenger-RNA)*
transkristallin transcrystalline, transgranular
trans-Lage *f* trans position
Translation *f (phys chem, cryst, bioch)* translation
Translationsbewegung *f* translational motion
Translationsebene *f (cryst)* translation (glide) plane
Translationsenergie *f* translational energy
  molare T. molar translational energy
Translationsentropie *f* translational entropy
Translationsfreiheitsgrad *m* translational degree of freedom, degree of translational freedom
Translationsgitter *n (cryst)* translational lattice
  Bravaissches T. Bravais lattice
Translationsverteilungsfunktion *f* translational partition function
translokal systemic *(pesticide)*
Translokation *f* translocation
Transmission *f (tech)* transmission; transmission, transmittance, transmittancy *(optics)*
Transmissionsgitter *n* transmission grating *(spectroscopy)*
Transmissionsgrad *m* transmission ratio, transmittance, transmittancy
Transmutation *f* transmutation
trans-orientiert trans-oriented
transparent transparent
  unvollkommen t. translucent, translucid
Transparentglasur *f (ceram)* transparent glaze
Transparentpapier *n* tracing paper
Transparentseife *f* transparent soap
Transparentzeichenpapier *n* tracing paper
Transparenz *f* transparency, light transmittance
  unvollständige T. translucence, translucency
transpirationshemmend *(cosmet)* antiperspirant
Transport *m (bioch)* transport; *(phys chem)* transport, transfer *(as of electrons or heat)*
  aktiver T. *(bioch)* active transport
transportabel [trans]portable
Transportband *n* conveyor (conveying) belt; *(pap)* delivery tape *(of a cross-cutter)*
Transportband-Konfektioniermaschine *f (rubber)* belt building machine
Transportbehälter *m* container
  T. für Flaschen *(lab)* bottle carrier
Transportbeton *m* ready-mixed concrete
Transportfilz *m (pap)* conveyor felt
transportieren *(bioch)* to transport; *(phys chem)* to transport, to transfer *(e.g. electrons or heat)*
Transportkübel *m* skip [car]
Transportreaktion *f* transport reaction
Transportschnecke *f* conveyor (conveying) screw (worm)
Transportwalze *f (pap)* support[ing] roll
  geriffelte T. fluted roll
trans-ständig trans
  t. [angeordnet] sein to be trans
trans-Stellung *f* trans position
trans-trans-Kohlenwasserstoff *m* trans-trans hydrocarbon
Transuran *n* transuranium (transuranic, transuranian) element
Traß *m (geol)* trass
Traßzement *m* trass cement
Trauben *fpl* clusters *(of fat in creaming milk)*
Traubenbildung *f* clustering, cluster formation *(of fat in creaming milk)*
traubenförmig botryoidal
Traubenkernöl *n* grape-seed (grape-stone) oil
Traubenmaische *f* grape pomace
Traubenmost *m* grape juice
Traubenpresse *f* grape (wine) press
Traubensaft *m* grape juice
Traubensäure *f* $HOOC[CHOH]_2COOH$ racemic acid, racemic (inactive) tartaric acid, (±)-tartaric acid

Traubentrester *pl* grape pomace
Traubenwein *m* grape wine
Traubenzucker *m* grape sugar, *D*-glucose, dextrose *(a monosaccharide)*
traubig botryoidal
träufeln to trickle
Traumatinsäure *f* HOOC[CH_2]$_8$CH=CHCOOH traumatic acid, 2-dodecenedioic acid
Trauzl-Block *m* [Trauzl] lead block *(for testing explosives)*
Trauzl-Blockausweitung *f* [Trauzl] lead-block expansion *(in testing explosives)*
Treber *pl* [brewer's] grains, spent grains
Treberschicht *f (ferm)* grains settling
Treffplatte *f* target *(as of an X-ray tube)*
Treibarbeit *f (met)* cupellation
Treibdampf *m* operating (motive) steam *(as in injector-type jet pumps)*
Treibdampfbrenner *m* steam-atomizing burner
Treibdruck *m* swelling pressure *(of coal)*
Treibdüsenbrenner *m* premix burner
treiben 1. *(met)* to cupel; 2. *(tann)* to paddle; 3. to expand *(of cement)*
  an die Oberfläche t. to buoy up
Treiben *n* 1. *(met)* cupellation; 2. *(tann)* paddling; 3. expansion *(of cement)*
Treiber *m (glass)* needle *(of a feeder)*
Treibgas *n* fuel gas; propellant [gas], propellent *(for liquids)*
  T. für Aerosole aerosol propellant (dispenser)
Treibladung *f* propellant charge
Treibladungspulver *n* propellant powder
  einbasiges T. single-base powder
  zweibasiges T. double-base powder
Treibmittel *n* 1. *(plast)* blowing (foaming, sponging) agent, foamer, *(for producing hollow articles)* blowing (inflating) agent; 2. *(food)* leavening [agent], leaven; 3. pumping (motivating) fluid *(of a jet pump)*; 4. expanding agent *(for cement)*; 5. *s.* Treibstoff 2.
Treibmittelpumpe *f* jet pump
Treibneigung *f* expansion *(of cement)*
Treibrolle *f* driving pulley
Treibspiritus *m* power (fuel) alcohol
Treibstoff *m* 1. fuel, *(esp for rockets)* propellant; 2. propellant [explosive], propellent, low explosive, deflagrating powder
  fester T. solid fuel, *(esp for rockets)* solid propellant
  homogener T. monofuel
  hypergoler T. hypergolic fuel (rocket propellant), hypergol
Treibtrommel *f* driving pulley
Treibversuch *m (coal)* swelling-pressure test
Tremolit *m (min)* tremolite *(an inosilicate)*
trennbar separable
Trennbarkeit *f* separability
Trenndiffundieren *n*, **Trenndiffusion** *f* separating diffusion
Trenneffekt *m* separation effect; *(plast)* release effect (action)
trennen 1. *(chem)* to separate, to segregate; to resolve *(racemates)*; 2. *(tech)* to disconnect *(e.g. the joints of an apparatus)*; to release *(mouldings from the mould)*
  in Schichten t. to delaminate
  nach Korn[größen]klassen t. to size, to grade
Trennentwässerung *f* separate [sewerage] system
Trennfaktor *m* separation factor; relative centrifugal force *(of a centrifuge)*
Trennfiltration *f* solids recovery filtration
Trennfläche *f (cryst)* cleavage (parting) plane
Trennflüssigkeit *f s.* Trennmedium
Trennfuge *f* 1. parting line *(of a mould)*; 2. joint (parting) line, match (mould) mark, mould seam *(on a moulding)*
Trenngrad *m* degree of separation, separation efficiency
Trenngüte *f*, **Trenngütegrad** *m (min tech)* efficiency of cut
Trennkolonne *f (distil)* rectifying (rectification) column
Trennkorngröße *f* size of separation, critical diameter, cut size (point)
Trennleistung *f* separation efficiency
Trennlinie *f* 1. interface line *(between two components)*; 2. *s.* Trennfuge
Trennmedium *n* separating fluid (liquid)
Trennmethode *f* separation method
Trennmittel *n* 1. release (parting, separating, antitack) agent *(for mouldings)*; 2. separating fluid (liquid)
  äußeres T. *(plast)* external lubricant
Trennplatte *f (ceram)* parting dish
Trennrohr *n* thermal-diffusion column *(for separating isotopes)*
Trennrohrverfahren *n* **[/ Clusiussches]** thermal-diffusion method *(for separating isotopes)*
Trennsäule *f* separation column; *(distil)* rectifying (rectification) column
Trennschärfe *f* sharpness (degree) of separation, separation efficiency, selectivity
Trennschicht *f* interlayer, interlining
Trennschleuder *f* centrifuge
Trennschnitt *m (distil)* cut
Trennstufe *f* separation stage, *(distil also)* distillation stage
  praktische T. *(distil)* actual (practice) plate
  theoretische T. *(distil)* theoretical separation stage, theoretical (ideal, perfect) plate
Trennstufenhöhe *f (distil)* height equivalent to a theoretical plate, HETP
Trennsystem *n* separate [sewerage] system *(waste-water treatment)*
Trenntank *m (petrol)* separator
Trenntechnik *f* separation technique
Trenntemperatur *f* separating temperature
Trenntrichter *m* separatory (separating) funnel
Trennung *f* 1. *(chem)* separation; segregation; resolution *(of racemates)*; 2. *(tech)* disconnection *(as of the joints of an apparatus)*; release *(of mouldings from the mould)*
  T. flüssig-fest solids-liquid separation
  T. in Gruppen group separation *(analytical chemistry)*

T. in Schweretrüben *(min tech)* heavy-medium (dense-medium) separation (cleaning)
T. nach der Dichte density separation (cut)
T. nach Gleichfälligkeit wet classification
T. nach Korn[größen]klassen grading [into size], size classification (grading, separation), sizing
elektrolytische T. electrolytic separation
flotative T. *(min tech)* floatation separation
säulenchromatografische T. column separation
Trennungs... *s. a.* Trenn...
Trennungsenergie *f* **[der Bindung]** bond dissociation energy
Trennungsfläche *f* surface of separation, boundary (bounding) surface, interface
Trennungsleuchten *n* triboluminescence
Trennverfahren *n* separation process
Trennvermögen *n* abhesiveness, release
Trennwand *f* partition; centre board *(of a plunger jig)*
halbdurchlässige (semipermeable) T. semipermeable membrane (partition)
Trennwirkung *f s.* Trenneffekt
Trennzentrifuge *f* separator
Treppenrost *m* step (cascade) grate
Treppenrostgenerator *m* step-grate producer
Trester *pl (ferm)* marc, rape, [grape] pomace, pummace
treten / in Wechselwirkung to interact
zutage t. *(geol)* to crop out, to outcrop
Tretgebläse *n* foot bellows
Triade *f* triad *(in the periodic table)*
Döbereinersche T. Döbereiner's triad
Triadenregel *f* **/ Döbereinersche** Döbereiner's law of triads
Triakontan *n* $CH_3[CH_2]_{28}CH_3$ triacontane
Triakontansäure *f* $CH_3[CH_2]_{28}COOH$ triacontanoic acid
Trialkylaluminium *n* trialkylaluminium
Triallylzyanurat *n* triallyl cyanurate
Triamidophosphorsäure *f* $OP(NH_2)_3$ triamidophosphoric acid, phosphoryl triamide, phosphoric triamide
Triaminchelat *n* triamine chelate
Triarylmethanfarbstoff *m* triarylmethane dye
Triäthanolamin *n* $N(CH_2CH_2OH)_3$ triethanolamine, tri(2-hydroxyethyl)amine
Triäthylaluminium *n* $(C_2H_5)_3Al$ triethylaluminium
Triäthylamin *n* $(C_2H_5)_3N$ triethylamine
Triäthylboran *n* $(C_2H_5)_3B$ triethylborane
Triäthylborin *n s.* Triäthylboran
Triäthylenglykol *n* triethylene glycol
Triäthylolamin *n s.* Triäthanolamin
Triäthylphosphat *n* $(C_2H_5)_3PO_4$ triethyl phosphate
Triäthylzellulose *f* ethylcellulose
Triäthylzitrat *n* triethyl citrate
Triazetat *n* triacetate; *(text)* $[C_6H_7O_2(OCOCH_3)_3]_x$ cellulose triacetate, primary [cellulose] acetate
Triazetatfaser *f* cellulose triacetate fibre
Triazetatfaserstoff *m* cellulose triacetate fibre
Triazin *n (org chem)* triazine
Triazol *n (org chem)* triazole
Tribolumineszenz *f* triboluminescence
triboluminesziérend triboluminescent
Triborid *n* triboride
Tribromanilin *n* $Br_3C_6H_2NH_2$ tribromoaniline
Tribromäthanal *n s.* Tribromazetaldehyd
Tribromäthanol *n,* **Tribromäthylalkohol** *m* CBr_3CH_2OH tribromoethanol, tribromoethyl alcohol
Tribromazetaldehyd *m* CBr_3CHO tribromoacetaldehyde, bromal
Tribromgerman *n* $GeHBr_3$ tribromogermane, germanium bromoform
Tribromid *n* tribromide
Tribrommethan *n* $CHBr_3$ tribromomethane, bromoform
Tribromsilan *n* $SiHBr_3$ tribromosilane
Trichit *m (cryst)* trichite
Trichloraldehyd *m s.* Trichlorazetaldehyd
1,1,1-Trichloräthan *n* CH_3CCl_3 1,1,1-trichloroethane, methylchloroform
1,1,2-Trichloräthan *n* $ClCH_2CHCl_2$ 1,1,2-trichloroethane
Trichloräthanal *n s.* Trichlorazetaldehyd
Trichloräthannitril *n* CCl_3CN trichloroacetonitrile
Trichloräthansäure *f* Cl_3CCOOH trichloroacetic acid, TCA
Trichloräthen *n,* **Trichloräthylen** *n* $CHCl{=}CCl_2$ trichloroethylene
Trichloräthylidenglykol *n s.* Trichlorazetaldehydhydrat
Trichlorazetaldehyd *m* CCl_3CHO trichloroacetaldehyde, chloral
Trichlorazetaldehydhydrat *n* $Cl_3CCH(OH)_2$ trichloroacetaldehyde hydrate, chloral hydrate
1,1,1-Trichlorazeton *n* Cl_3CCOCH_3 1,1,1-trichloroacetone
Trichlorbenzoesäure *f* $Cl_3C_6H_2COOH$ trichlorobenzoic acid
Trichlorbutylalkohol *m* $(CH_3)_2C(OH)CCl_3$ trichloro-*tert.*-butyl alcohol, chloretone
Trichlorderivat *n* trichloro derivative
Trichloressigsäure *f* Cl_3CCOOH trichloroacetic acid, TCA
Trichlorgerman *n* $GeHCl_3$ trichlorogermane, germanium chloroform
Trichlorid *n* trichloride
Trichlormethan *n* $CHCl_3$ trichloromethane, chloroform
Trichlormethansulfenylchlorid *n* CCl_3SCl trichloromethanesulphenyl chloride, perchloromethanethiol
α-Trichlormethylbenzol *n s.* α-Trichlortoluol
Trichlornitromethan *n* CCl_3NO_2 trichloronitromethane, chloropicrin
2,4,5-Trichlorphenoxyessigsäure *f* $Cl_3C_6H_2OCH_2COOH$ 2,4,5-trichlorophenoxyacetic acid, 2,4,5-T *(a herbicide)*
2-(2,4,5-Trichlorphenoxy)propionsäure *f* $CH_3CH(C_6H_2Cl_3)COOH$ 2-(2,4,5-trichlorophenoxy)-propionic acid, fenoprop, 2,4,5-TP *(a herbicide)*
2,3,6-Trichlorphenylessigsäure *f* $Cl_3C_6H_2COOH$ 2,3,6-trichlorophenylacetic acid, fenac *(a herbicide)*

Trichlorsilan *n* $SiHCl_3$ trichlorosilane
α-**Trichlortoluol** *n*, ω-**Trichlortoluol** *n* $C_6H_5CCl_3$ α,α,α-trichlorotoluene
Trichlor-*symm.*-triazin *n* 2,4,6-trichloro-1,3,5-triazine, cyanuric chloride
Trichroismus *m (cryst)* trichroism
Trichromat *n* $M_2^ICr_3O_{10}$ trichromate
Trichromdikarbid *n* Cr_3C_2 trichromium dicarbide
Trichromtetrasulfid *n* Cr_3S_4 trichromium tetrasulphide
Trichter *m* funnel; *(tech)* hopper; cup *(of a blast furnace)*
T. mit glatter Wandung plain glass funnel
T. mit kurzem Rohr (Stiel) short-stem[med] funnel
T. mit langem Rohr (Stiel) long-stem[med] funnel, Bunsen funnel
T. nach Hirsch Hirsch funnel
Trichtereinlage *f* **zum Filtrieren** filter cone
Trichterhalter *m* funnel holder
Trichterrohr *n*, **Trichterröhre** *f (lab)* thistle funnel (tube), funnel tube
T. mit Schleife und Kugel thistle funnel with safety bulb
Trichterstiel *m* funnel stem
Trichterstoffänger *m (pap)* cone save-all, settling cone
Trichtertrockner *m* hopper dryer
Trickle-[Phase-]Verfahren *n (petrol)* trickle [low] process *(hydrodesulphurization)*
Tridekan *n* $C_{13}H_{28}$ tridecane
Tridekandisäure *f* $HOOC[CH_2]_{11}COOH$ tridecanedioic acid, brassylic acid
Tridekansäure *f* $CH_3[CH_2]_{11}COOH$ tridecanoic acid
Tridymit *m* tridymite *(one form of silicon dioxide)*
Tridyne-Verfahren *n (plast)* Tridyne process *(transfer moulding)*
Triebkraft *f* 1. driving force (potential), [chemical] affinity *(of a reaction)*; 2. dough raising power *(as of yeast)*; 3. *(tech)* momentum
Triebmittel *n (food)* leaven, leavening [agent]
Trien *n* triene *(any of a class of hydrocarbons containing three carbon double bonds)*
Trifluoräthansäure *f s.* Trifluoressigsäure
Trifluorchloräthylen *n (incorrectly for)* Chlortrifluoräthylen
Trifluoressigsäure *f* CF_3COOH trifluoroacetic acid
Trifluorid *n* trifluoride
Trifluormethan *n* CHF_3 trifluoromethane, fluoroform
Trifluorsilan *n* $SiHF_3$ trifluorosilane
trifunktionell trifunctional
trig. *s.* trigonal
Trigerman *n* Ge_3H_8 trigermane, germanium octahydride
Trigermaniumdinitrid *n* Ge_3N_2 trigermanium dinitride
Trigermaniumtetranitrid *n* Ge_3N_4 trigermanium tetranitride
Triglykol *n s.* Triäthylenglykol
Triglyzerid *n* triglyceride
trigonal *(cryst)* trigonal, trig.
Trihalogenid *n* trihalide, trihalogenide
Trihydrat *n* trihydrate
Trihydroxid *n* trihydroxide
Trihydroxyanthrachinon *n* trihydroxyanthraquinone
Trihydroxybenzoesäure *f* $C_6H_2(OH)_3COOH$ trihydroxybenzoic acid
Trihydroxybenzol *n* $C_6H_3(OH)_3$ trihydroxybenzene
Trijodid *n* triiodide
Trijodmethan *n* CHI_3 triiodomethane, iodoform
Trijodsilan *n* $SiHI_3$ triiodosilane
Trikaliumorthophosphat *n*, **Trikaliumphosphat** *n* K_3PO_4 tripotassium orthophosphate, potassium phosphate
Trikalziumorthophosphat *n*, **Trikalziumphosphat** *n* $Ca_3(PO_4)_2$ tricalcium orthophosphate, calcium phosphate
Trikarballylsäure *f* tricarballylic acid, propane-1,2,3-tricarboxylic acid
Trikarbonsäure *f* tricarboxylic acid
Trikarbonsäurezyklus *m (bioch)* tricarboxylic-acid cycle, TCA cycle, citric-acid cycle, Krebs cycle
trikl. *s.* triklin
triklin *(cryst)* triclinic, tric., anorthic
Trikobalttetroxid *n* Co_3O_4 tricobalt tetraoxide, cobalt(II,III) oxide
Trikohlenstoffdioxid *n* C_3O_2 tricarbon dioxide
Trikohlenstoffdisulfid *n* C_3S_2 tricarbon disulphide
Trikosan *n* $C_{23}H_{48}$ tricosane
Trikresylphosphat *n* $[CH_3C_6H_4O]_3PO$ tritolyl phosphate, tricresyl phosphate, TCP
Trikupferphosphid *n* Cu_3P tricopper monophosphide, copper(I) phosphide, cuprous phosphide
Trillo *m (tann)* drillo *(from the cupulae of several oriental Quercus specc.)*
Trimellithsäure *f* $C_6H_3(COOH)_3$ trimellitic acid, benzene-1,2,4-tricarboxylic acid
trimer trimeric
Trimer[es] *n* trimer
Trimerisation *f*, **Trimerisierung** *f* trimerization
Trimesinsäure *f* $C_6H_3(COOH)_3$ trimesic acid, benzene-1,3,5-tricarboxylic acid
Trimethoxybenzoesäure *f* $(CH_3O)_3C_6H_2COOH$ trimethoxybenzoic acid
Trimethylaluminium *n* $(CH_3)_3Al$ trimethylaluminium
Trimethylamin *n* $(CH_3)_3N$ trimethylamine
Trimethylbenzoesäure *f* $(CH_3)_3C_6H_2COOH$ trimethylbenzoic acid, isodurylic acid
Trimethylbenzol *n* trimethylbenzene
Trimethylbor *n s.* Trimethylboran
Trimethylboran *n* $(CH_3)_3B$ trimethylborane
Trimethylbrommethan *n s.* 2-Brom-2-methylpropan
2,2,3-Trimethylbutan *n* 2,2,3-trimethyl-butane, triptane
Trimethylchinolin *n* trimethylquinoline
Trimethylen *n s.* Zyklopropan
Trimethylessigsäure *f (incorrectly for)* 2,2-Dimethylpropionsäure
Trimethylglykokoll *n* trimethylglycine, lycine, oxyneurine, betaine *(proper)*

Trimethylkarbinol *n s.* 2-Methylpropanol-(2)
Trimethylmethan *n s.* 2-Methylpropan
1,1,1-Trimethyloläthan *n* $CH_3C(CH_2OH)_3$ 1,1,1-trimethylolethane, pentaglycerol, pentaglycerine
trimolekular trimolecular, termolecular
Trimolybdat *n* $M_2^IMo_3O_{10}$ trimolybdate
Trimyristin *n* trimyristin, glycerol trimyristate
Trinatriumorthophosphat *n*, **Trinatriumphosphat** *n* Na_3PO_4 trisodium orthophosphate, sodium phosphate
Trinatriumsalz *n* trisodium salt
Trinickeltetrasulfid *n* Ni_3S_4 trinickel tetrasulphide, nickel(II,III) sulphide
Trinitrat *n* trinitrate
Trinitrid *n* trinitride
Trinitrierung *f* trinitration
2,4,6-Trinitrophenol *n* $C_6H_2(NO_2)_3OH$ 2,4,6-trinitrophenol, picric acid
2,4,6-Trinitroresorzin *n* $C_6H(OH)_2(NO_2)_3$ 2,4,6-trinitroresorcinol, styphnic acid
Trinitrotoluol *n* $C_6H_2(NO_2)_3CH_3$ trinitrotoluene, TNT
trinkbar potable
Trinkbarkeit *f* potability
Trinkbranntwein *m* potable spirit
trinkfertig ready-to-drink
Trinkwasser *n* drinking water
Trinkwasserenthärtung *f* drinking-water softening, municipal water softening
Trinkwasserversorgung *f* drinking-water supply
Triol *n* triol, trihydric alcohol
Triose *f* triose *(monosaccharide containing three carbon atoms per molecule)*
Triosephosphatdehydr[ogen]ase *f* triose-phosphodehydrogenase
Trioxalatochromat(III) *n* $M_3^I[Cr(C_2O_4)_3]$ trioxalatochromate(III), oxalatochromate(III)
Trioxalatokobaltat(III) *n* $M_3^I[Co(C_2O_4)_3]$ trioxalatocobaltate(III), oxalatocobaltate(III)
1,3,5-Trioxan *n* 1,3,5-trioxan, trioxymethylene, metaformaldehyde
Trioxid *n* trioxide
Trioxoborat *n* $M_3^IBO_3$ trioxoborate, orthoborate, borate
Trioxoborsäure *f* H_3BO_3 trioxoboric acid, orthoboric acid, boric acid
Trioxokieselsäure *f* H_2SiO_3 trioxosilicic acid, metasilicic acid
Trioxoplumbat(IV) *n* $M_2^IPbO_3$ trioxoplumbate(IV), metaplumbate(IV)
Trioxosilikat *n* $M_2^ISiO_3$ trioxosilicate, metasilicate
Trioxotitanat(IV) *n* $M_2^ITiO_3$ trioxotitanate(IV), metatitanate(IV)
Trioxovanadat(V) *n* M^IVO_3 trioxovanadate(V), metavanadate
Trioxozirkonat(IV) *n* $M_2^IZrO_3$ trioxozirconate(IV), metazirconate(IV)
Trioxymethylen *n* trioxymethylene, 1,3,5-trioxan, metaformaldehyde
Tripalmitin *n* tripalmitin, glycerol tripalmitate
Tripel *m (min)* trippel, tripoli *(schistose deposits of silica)*
Tripeleffekt *m* triple effect
Tripeleffektverdampfer *m* triple-effect evaporator (evaporating unit)
Tripelpunkt *m* triple point
Tripelsalz *n* triple salt
Tripeptid *n* tripeptide
Triphenylamin *n* $(C_6H_5)_3N$ triphenylamine
Triphenylaminfarbstoff *m* triphenylamine dye
Triphenylbor *n s.* Triphenylboran
Triphenylboran *n* $(C_6H_5)_3B$ triphenylborane
Triphenylen *n* triphenylene, 1,2,3,4-dibenznaphthalene
Triphenylkarbinol *n s.* Triphenylmethanol
Triphenylmethan *n* $(C_6H_5)_3CH$ triphenylmethane
Triphenylmethanfarbstoff *m* triphenylmethane (triarylmethane) dye
Triphenylmethanol *n* $(C_6H_5)_3COH$ triphenylmethanol
Triphenylphosphat *n* $(C_6H_5)_3PO_4$ triphenyl phosphate
Triphenylphosphin *n* $(C_6H_5)_3P$ triphenylphosphine
Triphenylstibin *n* $(C_6H_5)_3Sb$ triphenylstibine
Triphenylzinnchlorid *n* $(C_6H_5)_3SnCl$ triphenyltin chloride
Triphosphat *n* $M_5^IP_3O_{10}$ triphosphate
Triphosphopyridinnukleotid *n s.* Nikotinamid-adenin-dinukleotidphosphat
Triphosphorpentanitrid *n* P_3N_5 triphosphorus pentanitride, phosphorus nitride
Triphylin *m (min)* triphylite, triphyline *(a phosphate of lithium, iron, and manganese)*
Triplett *n*, **Triplettspektrallinie** *f* triplet
Triplettsystem *n* triplet system
Triplettzustand *m* triplet state
Triplexkarton *m* triplex board
Triplexpappe *f* triplex board
Triplexpumpe *f* triplex (three-throw) pump
Triplit *m (min)* triplite *(a phosphate of manganese and iron)*
Trippkeit *m (min)* trippkeite *(copper arsenite)*
Triptan *n s.* 2,2,3-Trimethylbutan
Trisaccharid *n* trisaccharide
Trisauerstoff *m* O_3 trioxygen, ozone
Trisazofarbstoff *m* trisazo dye
Trischwefelwasserstoff *m s.* Trisulfan
Trisilan *n* Si_3H_8 trisilane
Trisilazan *n* $H_3Si-NH-SiH_2-NH-SiH_3$ trisilazane
Trisilikat *n* trisilicate
Trisilikatschlacke *f* trisilicate slag
Trisiloxan *n* $H_3Si-O-SiH_2-O-SiH_3$ trisiloxane
Trisilthian *n* $H_3Si-S-SiH_2-S-SiH_3$ trisilthiane
Trispiro-Verbindung *f* trispiro compound (hydrocarbon)
Tristearin *n* tristearin, glyceryl tristearate
trisubstituiert trisubstituted
Trisulfan *n* H_2S_3 trisulphane, hydrogen trisulphide
Trisulfat *n* $M_2^IS_3O_{10}$ trisulphate
Trisulfid *n* trisulphide
Trisulfonsäure *f* trisulphonic acid
Tritan *n* $(C_6H_5)_3CH$ tritane, triphenylmethane
Triterpen *n (org chem)* triterpene
Trithioarsenat(V) *n* M^IAsS_3 trithioarsenate, metathioarsenate

Trithiokarbonat *n* $M_2^ICS_3$ trithiocarbonate, thiocarbonate
Trithiokohlensäure *f* H_2CS_3 trithiocarbonic acid, thiocarbonic acid
Trithionat *n* $M_2^IS_3O_6$ trithionate
Trithiostannat(IV) *n* $M_2^ISnS_3$ trithiostannate(IV), metathiostannate(IV)
Tritium *n* T, $_1^3H$ tritium
Tritol *n s.* Trinitrotoluol
Triton *n (nucl)* triton
Tritriakontan *n* $C_{33}H_{68}$ tritriacontane
Trituration *f (pharm)* trituration
Tritylchlorid *n* $(C_6H_5)_3CCl$ trityl chloride, α-chlorotriphenylmethane
Tritylfarbstoff *m* triarylmethane dye
Triuranoktoxid *n* U_3O_8 triuranium octaoxide, uranium(IV) uranate
trivalent trivalent, tervalent
Trivalenz *f* trivalency, tervalency
trivariant trivariant
Trivialname *m* trivial (common, unsystematic) name
verbotener T. abandoned [trivial] name
zugelassener (zulässiger) T. recognized [trivial] name
t-RNS *s.* Transfer-Ribonukleinsäure
trocken dry
t. und mineral[stoff]frei dry and mineral-matter-free, d.m.m.f., D.M.F.
absolut t. bone-dry, oven-dry, oven-dried, OD
Trockenabschnitt *m* beach section *(of a helical-conveyor centrifuge)*
Trockenanalyse *f* dry analysis
Trockenanlage *f* drying plant
Trockenapparat *m* drying apparatus (machine), dryer, dehydrator
Trockenaufbereitung *f* 1. *(min tech)* dry (pneumatic) cleaning; 2. *(ceram)* dry preparation (mixing, mix)
Trockenausschuß *m (pap)* dry broke
Trockenbatterie *f* dry battery
Trockenbeize *f (agric)* dry [seed] treatment, dust treatment
Trockenbestandteil *m* solid constituent
trockenblasen / mit Druckluft to air-blow
Trockenblech *n* [drying] tray
Trockenboden *m (ceram)* hot floor; *(pap)* drying loft ✦ **auf dem T. getrocknet** *(pap)* loft-dried
Trockenbrett *n (lab)* draining board
aufhängbares T. wall-mounting draining board
Trockendampf *m* dry [saturated] steam
Trockendämpfen *n*, **Trockendekatieren** *n*, **Trockendekatur** *f (text)* dry steaming (decatizing)
Trockendestillation *f* dry (pyrogenic) distillation
Trockenei *n* dried (desiccated, processed, dehydrated) eggs, egg powder
Trockeneigelb *n* dried egg yolk
Trockeneinfärben *n (plast)* dry colouring *(of moulding compounds)*
Trockeneis *n* dry ice, carbon dioxide ice (snow), solid carbon dioxide
Trockenelektroabscheider *m*, **Trockenelektrofilter** *n* dry precipitator
Trockenelement *n* dry cell
Trockenemulsion *f (phot)* dry emulsion
Trockenentschwefelung *f* dry desulphuration *(of gas)*
Trockenessig *m* dry vinegar
Trockenextrakt *m* dry extract
Trockenfeld *n* drying ground *(winning of peat)*
Trockenfestigkeit *f* dry strength; *(ceram)* green strength
Trockenfilter *n* dry filter
Trockenfilz *m (pap)* dry[er] felt
Trockenfilzleitwalze *f (pap)* dry-felt roll
Trockenfläche *f* drying area (surface)
Trockenflecken *mpl (phot)* drying marks
Trockenfließpapier *n* dry blotting paper
Trockenfurfural *n* dry furfural
Trockenfutterhefe *f* mineral yeast
Trockengas *n* 1. dry [natural] gas; 2. *s.* Trocknungsgas
Trockengasreinigung *f* dry gas cleaning
Trockengehalt *m (pap)* solid[s] content *(of sulphite waste liquor)*
Trockengel *n* dry gel
trockengepreßt *(ceram)* dry-pressed
Trockengerbung *f* dry tannage
Trockengeschwindigkeit *f* drying rate
Trockengestell *n (ceram)* drying rack
Trockengewicht *n* dry weight
Trockenglatt-Ausrüstung *f (text)* smooth-drying finish
Trockenglättwerk *n (pap)* calender, calender machine (section)
Trockengruppe *f (pap)* dryer group (section)
Trockenguß *m* dry-sand casting *(foundry)*
Trockengußform *f* dry-sand mould *(foundry)*
Trockengußformen *n* dry[-sand] moulding *(foundry)*
Trockenguß[form]sand *m* dry sand *(foundry)*
Trockengut *n* material being (*or* to be) dried; dry product
Trockenhänge *f (text)* festoon dryer
Trockenhaspel *f (text)* drying reel
Trockenhefe *f* dry yeast, yeast powder
Trockenheit *f* dryness
Trockenhitzebehandlung *f* baking
Trockenhorde *f* [drying] tray
Trockenimprägnieren *n* waterproofing
Trockenkammer *f* drying room (chamber), *(if small)* drying cabinet (box)
Trockenkarbonisation *f (text)* dry carbonizing
Trockenkautschukgehalt *m* dry rubber content
Trockenkleber *m* dry adhesive
Trockenköder *m (agric)* dry bait
Trockenkollergang *m* dry pan
Trockenkonzentrat *n (agric)* dry concentrate
Trockenlaufverdichter *m* non-lubricated compressor
Trockenlöscher *m* powder extinguisher, powder-type fire extinguisher, dry chemical fire extinguisher
Trockenmagermilch *f* skim-milk powder, dry skim milk, dried fat-free milk
Trockenmahlung *f* dry milling (grinding)

Trockenmaschine *f* drying machine, dryer
Trockenmasse *f* 1. solids; 2. [bone-]dry weight, moisture-free weight
T. des Holzes *(pap)* dry wood weight
fettfreie T. *(food)* non-fat[ty] solids, solids-not-fat, S.N.F.
Trockenmedium *n s.* Trockenmittel
Trockenmilch *f* milk powder, dried (dry, powdered, desiccated) milk
auf Milchkonzentration verdünnte T. reconstituted milk
Trockenmilchpulver *n s.* Trockenmilch
Trockenmischer *m* dry blender (mixer)
Trockenmischung *f* dry blend (mix)
Trockenmittel *n* desiccant, drying (dehydrating) agent (medium), drier, dehydrator, dehumidifier
Trockenmittelkolben *m* desiccant chamber *(of a drying pistol)*
Trockenmolke *f* whey powder, dried (dry, powdered) whey
Trockenofen *m* drying kiln; *(coat)* stove
Trockenoffsetdruck *m* dry offset printing
Trockenpartie *f* drying (dryer) part (section), dry part (end) *(of a paper-making machine)*
Trockenpatrone *f* balance desiccator
Trockenpektin *n* powdered pectin
Trockenpistole *f* drying pistol
Trockenplatte *f (phot)* dry plate
Trockenplatz *m* drying ground *(winning of peat)*
trockenpökeln to dry-cure, to dry-salt
Trockenpökelung *f* dry curing (salting), dry-salt cure
Trockenpräparat *n* dry preparation
Trockenpressen *n (ceram)* dry pressing, *(relating to tiles also)* dust pressing
Trockenpreßmasse *f (ceram)* dry pressing mix (body)
Trockenprobe *f (met)* dry assay
Trockenrahm *m* dry (dried) cream
Trockenraum *m* drying room (chamber)
Trockenreibechtheit *f (text)* fastness to dry rubbing
Trockenreiniger *m* dry cleaner (purifier)
Trockenreinigung *f* dry cleaning (*1. as of gases*; *2. of clothes*)
Trockenreinigungsechtheit *f (text)* fastness to dry cleaning
Trockenriß *m (ceram)* drying crack
Trockenrohr *n* drying tube
Trockensahne *f* dry (dried) cream
trockensalzen to dry-cure, to dry-salt
Trockensalzen *n* dry curing (cure, salting)
Trockensand *m* dry sand
Trockensandform *f* dry-sand mould *(foundry)*
Trockensandformen *n* dry[-sand] moulding *(foundry)*
Trockenschacht *m* drying shaft
Trockenschale *f* drying tray
Trockenschampun *n (cosmet)* dry shampoo
Trockenscheidung *f (sugar)* dry liming, defecation with dry lime
Trockenschleuder *f* hydroextractor
Trockenschmelze *f*, **Trockenschmelzen** *n (food)* dry rendering (fat melting)
Trockenschnecke *f* screw-conveyor dryer
Trockenschnitzel *npl (sugar)* dried [beet] pulp
Trockenschrank *m (lab)* drying cupboard (oven), [air] oven ✦ **im T. getrocknet** oven-dry, oven-dried, OD ✦ **im T. trocknen** to oven-dry
elektrischer T. electric drying oven
Trockenschrank-Heißluftsterilisator *m* hot-air sterilizer
Trockenschuppen *m* drying shed
Trockenschwindung *f (ceram)* drying shrinkage (contraction), air shrinkage
Trockenshampoo[n] *n s.* Trockenschampun
Trockensieben *n* dry screening
Trockenspeicher *m* drying loft
Trockenspiegel *m* plane of evaporation
Trockenspinnen *n* dry spinning
Trockenspritzen *n (coat)* dry spray
Trockenstoff *m s.* Trockenmittel
Trockenstoffmasse *f* [bone-]dry weight, moisture-free weight
Trockensubstanz *f* dry (solid) matter, solids, dry residue (substance)
T. der Schwarzlauge *(pap)* black liquor solids
Trockensubstanzmasse *f s.* Trockenstoffmasse
Trockentemperatur *f* drying temperature
Trockentorf *m* dry peat
Trockentreber *pl (ferm)* dry spent grains
Trockentrommel *f* 1. drying (dryer) cylinder (drum); 2. *s.* Trommeltrockner
Trockentunnel *m* drying tunnel
Trockenturm *m* 1. drying tower; 2. *(lab)* [gas] drying jar
Trockenverarbeitung *f* dry processing
Trockenverfahren *n* dry process
Trockenverlust *m* drying loss, loss on drying
Trockenvermahlung *f* dry milling (grinding)
Trockenverschluß *m* dry seal
Trockenvollmilch *f* dry whole milk
Trockenwalze *f* 1. drying (dryer) roll; 2. *s.* Walzentrockner
Trockenware *f (food)* dry product
Trockenwaschverfahren *n (text)* solvent scouring process *(for treating raw wool)*
Trockenzeit *f* drying period (time)
Trockenzentrifuge *f* centrifugal dryer, whizzer, hydroextractor
trockenzyanieren to dry-cyanide, to gas-cyanide, to carbonitride *(steel)*
Trockenzyanieren *n* dry (gas) cyaniding (cyanization), carbonitriding, ni-carbing, nitrocementation *(of steel)*
Trockenzylinder *m* 1. drying (dryer) cylinder (drum); 2. cylinder (drum) dryer (drying machine); *(pap)* can dryer
T. einer Selbstabnahmemaschine *(pap)* Yankee dryer
Trockenzylinderanordnung *f (pap)* dryer arrangement
Trockenzylindergruppe *f (pap)* dryer group (section)
Trockenzylinderoberfläche *f (pap)* dryer surface
Trockne *f* dryness ✦ **zur T. eindampfen** to evaporate to dryness

trocknen to dry, to dehydrate, to desiccate, to exsiccate; to dehumidify *(esp gases)*; to season *(wood)*
an der Luft t. to air-dry
an der Sonne t. to sun-dry
bis zur Gewichtskonstanz (Massekonstanz) t. to dry to constant weight
im Ofen t. *(coat)* to stove
im Sprühverfahren t. to spray-dry
im Trockenschrank t. to oven-dry
im Vakuum t. to vacuum-dry
im Zerstäubungsverfahren t. to spray-dry
lederhart t. *(ceram)* to dry to leather-hard
lyophil t. *(food)* to freeze-dry
schwach t. to dry soft
stark t. to dry hard
thermisch t. to stove, to bake
unter Abtropfen t. to drip-dry
unter Vakuum t. to vacuum-dry
unvollständig (unzureichend) t. to underdry
trocknend / chemisch (durch chemische Reaktion) convertible *(organic-coating material)*
langsam t. slow-drying
physikalisch t. non-convertible *(organic-coating material)*
schnell t. quick-drying
Trockner *m* 1. dryer, drying apparatus (machine), dehydrator; 2. *s.* Trockenmittel
T. mit bewegtem Trockengut moving-product dryer
T. mit Durchbelüftung air-through circulation dryer
T. mit unbewegtem Trockengut fixed dryer
atmosphärischer T. atmospheric (air) dryer
begehbarer T. walk-in dryer
kontinuierlich arbeitender T. continuous dryer
pneumatischer T. pneumatic [conveying] dryer
Trockneraufgabegut *n* material to be dried
Trocknerkammer *f (ceram)* dryer corridor
Trocknertrommel *f* dryer drum
Trocknerwalze *f* dryer roll
Trocknerzylinder *m* dryer cylinder
Trocknung *f* drying, dehydration, desiccation, exsiccation; dehumidification *(esp of gases)*; seasoning *(of wood)*
T. an der Luft *s.* atmosphärische T.
T. an der Sonne solar (sun) drying
T. bis zur Gewichtskonstanz (Massekonstanz) drying to constant weight
T. mittels Lösungsmitteln solvent drying
atmosphärische T. atmospheric (air, open-air) drying
dielektrische T. dielectric (radio-frequency) drying
künstliche T. artificial drying
lyophile T. lyophilic drying
natürliche T. *s.* atmosphärische T.
thermische T. thermal drying, stoving
übermäßige T. excessive drying
ungleichmäßige T. uneven drying
unvollständige (unzureichende) T. underdrying
Trocknungs... *s.a.* Trocken...
Trocknungsabschnitt *m* period of drying
Trocknungsdauer *f* drying period (time)
Trocknungsgas *n* drying gas
Trocknungsgruppe *f (pap)* dryer group (section)
Trocknungsgut *n* material being (*or* to be) dried
Trocknungskurve *f* drying curve
Trocknungsluft *f* drying air
Trocknungspotential *n* drying potential
Trocknungsrohr *n* heating chamber *(of a drying pistol)*
Trocknungstriebkraft *f* drying potential
Trocknungsverfahren *n* drying process
Trocknungsverlust *m s.* Trockenverlust
Trocknungsvorgang *m* drying process
Trog *m* trough, vat, tank, tub, pan
elektrolytischer T. electrolytic tank
Trogboden *m (pap)* floor of the pan (tub) *(of a Hollander beater)*
Trögerit *m (min)* trögerite *(a hydrous arsenate of uranium)*
trogförmig trough-shaped
Trogkettenförderer *m* skeleton flight (continuous-flow) conveyor, continuous (Redler, en masse) conveyor
Trogkneter *m*, **Trogmischer** *m* trough (open-pan) mixer
Trogpresse *f* pot press
Trogsohle *f s.* Trogboden
Trogtränkung *f* steeping *(of timber)*
T. mit Wärmestandsänderung hot-and-cold open tank treatment
Trommel *f* drum, cylinder; bowl, basket *(of a centrifuge)* ✦ **in der T. färben** *(text)* to drum-dye
Trommelausstoßmaschine *f (tann)* drum setting machine
Trommelentrindung *f (pap)* drum barking
Trommelerhitzer *m (food)* cylindrical batch pasteurizer
Trommelfallmühle *f* autogenous tumbling mill
Trommelfärbemaschine *f (text)* drum-dyeing machine
Trommelfärbung *f (text)* drum dyeing
Trommelfilter *n* drum (revolving) filter
zellenloses T. single-compartment drum filter
Trommelhülse *f (text)* winding head
Trommelkonverter *m* **nach Peirce-Smith** *(met)* Peirce-Smith converter
Trommelkühlung *f* drum cooling
trommellackieren *(coat)* to tumble
Trommellackierung *f (coat)* tumbling
Trommelmälzerei *f (ferm)* drum malting
Trommelmischer *m* drum mixer, blending drum, barrel blender (mixer)
Trommelmühle *f* drum mill
trommeln *(plast, coat)* to tumble
Trommelnaßmühle *f* wet-cylinder mill
Trommelneigung *f* drum slope
Trommelpolieren *n (plast)* tumbling, barrel polishing
Trommelprüfung *f (coal)* tumbler (trommel) test
T. nach Cochrane Cochrane test
Trommelreifen *m (rubber)* drum-built tyre
Trommelscheider *m* drum separator

Trommelschichtenfilter *n* drum layer filter
Trommelschieber *m* coal inlet valve *(of a gas retort)*
Trommelschlichtmaschine *f (text)* cylinder sizing machine
Trommelschneidmaschine *f (sugar)* drum beet slicer
Trommelschreiber *m* drum[-chart] recorder
Trommelsieb *n* drum (trommel, revolving) screen
Trommeltest *m s.* Trommelprüfung
Trommeltrockner *m* 1. rotary (rotatory) dryer; 2. *(sugar)* granulator
T. mit Kontaktheizung indirect rotary dryer
Trommelverfahren *n* dry [drum-]cooling method, chill-roll method *(of margarine making)*
Trommelversuch *m s.* Trommelprüfung
Trommelzellenfilter *n* multicompartment drum filter
Trommelzentrifuge *f* bowl (basket) centrifuge
T. mit Einsatztellern disk [bowl] centrifuge, disk separator
Trona *m(f) (min)* trona *(a hydrous acid sodium carbonate)*
Trona-Verfahren *n* Trona process *(potash industry)*
Troostit *m (min)* troostite *(a nesosilicate)*
Tropaalkaloid *n* tropane alkaloid
Tropakokain *n* tropacocaine *(alkaloid)*
Tropan *n (org chem)* tropane
Tropanalkaloid *n* tropane alkaloid
Tropäolin *n* tropaeolin *(any of several azo dyes)*
Tropäolin D *n* $(CH_3)_2NC_6H_4N{=}NC_6H_4SO_3Na$ tropaeolin D, methyl orange, *p*-(*p*-dimethylamino phenylazo)benzene sulphonate of sodium
Tropäolin O *n* $NaSO_3C_6H_4N{=}NC_6H_3(OH)_2$ tropaeolin O, resorcinol yellow, sodium azoresorcinolsulphanilate
Tropäolin OO *n* $C_6H_5NHC_6H_4N{=}NC_6H_4SO_3Na$ tropaeolin OO, orange IV, sodium *p*-diphenylaminoazobenzenesulphonate
Tropäolin R *n s.* Tropäolin O
Tropasäure *f* $HOCH_2CH(C_6H_5)COOH$ tropic acid, 3-hydroxy-2-phenylpropionic acid
Tropenas-Konverter *m (met)* Tropenas [side-blown] converter
tropenbeständig *s.* tropenfest
Tropenbeständigkeit *f s.* Tropenfestigkeit
tropenfest resistant to tropical conditions, stable under tropical conditions
Tropenfestigkeit *f* resistance to tropical conditions, stability under tropical conditions
Tropenfestmachen *n* tropicalization
tropfbar liquid
Tropfbenzoltank *m* drains tank *(of a benzole plant)*
Tröpfchen *n* droplet
Tröpfchenbildung *f* formation of droplets
Tröpfchengröße *f* droplet size
Tropfelektrode *f* drop[ping] electrode
tröpfeln to trickle
tropfen to drop, to drip
Tropfen *m* 1. drop; 2. *(coat)* tear, drip; 3. *(glass)* gob
Tropfenabgabe *f (glass)* gob delivery
Tropfenabziehen *n (coat)* detearing
elektrostatisches T. electrostatic detearing
Tropfenbildung *f* drop formation
Tropfenfänger *m (lab)* Kjeldahl connecting bulb
Tropfenform *f (glass)* gob shape
tropfenförmig drop-shaped
Tropfengewicht *n (glass)* gob weight
Tropfengewichtsmethode *f (phys chem)* drop-weight method
Tropfengröße *f* drop size
Tropfenkondensation *f* dropwise condensation
Tropfenspeiser *m (glass)* gob feeder
Tropfenverteilung *f* drop-size distribution
tropfenweise dropwise, drop by drop
Tropfenzähler *m (lab)* 1. drop counter; 2. *s.* Tropfflasche; 3. *s.* Tropfpipette
Tropfer *m s.* Tropfpipette
Tropfflasche *f*, **Tropfglas** *n (lab)* dropping bottle
Tropfkatode *f* dropping cathode
Tropfkörper *m* trickling (percolating) filter, bacteria bed *(waste-water treatment)*
Tropfpipette *f (lab)* dropping (teat) pipette, dropper
Tropfpunkt *m* drop[ping] point
T. nach Ubbelohde Ubbelohde drop[ping] point
Tropfrohr *n (lab)* drip tube (pipe)
Tropfspeisung *f (glass)* gob feeding
Tropftrichter *m (lab)* dropping (drop, tap) funnel
Tropfwanne *f (coat)* draining pan *(in flow coating)*
tropfwassergeschützt drip-proof
Tropfzeit *f* drop time *(as with a titration)*
Tropfzündpunkt *m* drop ignition temperature
Tropigenin *n* nortropine, tropigenin, 8-azabicyclo-[3,2,1]octan-3-ol
Tropinalkaloid *n* tropane alkaloid
Tropinmandelsäureester *m* mandelyltropine, homatropine, phenylglycollyltropine
Tropinsäure *f* tropinic acid *(a degradation product of atropine)*
Tropolon *n (org chem)* tropolone
Trotyl *n s.* Trinitrotoluol
Trub *m (ferm)* settling[s], lees, sediment, cloud
trüb[e] opaque, *(esp relating to liquids)* turbid, hazy, cloudy, feculent, thick *(esp wine)*; dull, dusky *(shade)*
Trübe *f* 1. slurry, pulp; *(filtr)* prefilt, prefilt slurry (feed), material being (*or* to be) filtered; feed slurry (pulp) *(centrifuging)*; fluid pulp *(classifying)*; *(min tech)* pulp [of ore]; 2. *s.* Trub; 3. *s.* Trübheit
schwere T. *(min tech)* dense (heavy, high-gravity) medium
zulaufende T. feed slurry (pulp)
Trübeaufgaberinne *f (min tech)* feed launder
Trübefeststoff *m* medium solid *(dense-medium separation)*
Trübeis *n* opaque (milky) ice
Trübekreislauf *m* medium circuit *(dense-medium separation)*
trüben / sich to opacify, *(esp of liquids)* to cloud, to become cloudy (turbid)
Trübeniveau *n* surface level of the medium *(dense-medium separation)*
Trübeumlauf *m* medium circuit *(dense-medium separation)*

Trübezulauf *m*, **Trübezuleitung** *f* feed (pulp) inlet
Trübglas *n* opaque (opal) glass
Trübglasur *f (ceram)* opaque glaze
Trübheit *f* opacity, *(esp of liquids)* turbidity, haziness, cloudiness, feculence, *(esp of wine)* thickness; dullness, duskiness *(of a colour)*
Trübheitsmessung *f* turbidimetry, turbidimetric (turbidity) measurement
Trüblauf *m (filtr)* bleeding
Trübung *f* 1. *s.* Trübheit; 2. haze *(in beer)*, casse *(in wine)*; 3. opacifying, *(esp of liquids)* clouding
  T. durch Reflexion *(coll)* reflection turbidity
  biologische T. biological haze *(in beer)*
  milchige T. milkiness
  nichtbiologische T. non-biological haze *(in beer)*
Trübungsgrad *m* degree of turbidity
Trübungsmesser *m* turbidimeter, turbidometer, nephelometer
Trübungsmessung *f* turbidimetry, turbidimetric (turbidity) measurement
Trübungsmittel *n* opacifier *(as for glass and plastics)*
Trübungspunkt *m* turbidity (cloud) point *(of solutions or oils)*; *(petrol)* cloud point; titre *(of a soap)*
Trübungstitration *f* turbidimetric (turbidity) titration
True-Vapour-Phase-Verfahren *n (petrol)* true vapour-phase [cracking] process, TVP process
Trum *m (n)* strand *(of a conveyor)*
Trümmergestein *n* fragmental rock
Truxillo-Koka *f* Truxillo coca *(from Erythroxylum novogranatense (Morris) Hieron.)*
Trypanblau *n* Trypan blue, direct blue 14
trypanozid [wirkend] *(pharm)* trypanocidal
Trypanrot *n* trypan red *(a biological stain)*
Trypsin *n* trypsin *(a proteolytic enzyme or a preparation containing proteolytic enzymes)*
Trypsin-Verfahren *n (rubber)* trypsin method *(heat sensitization of latex)*
Tryptophan *n* tryptophane, 2-amino-3,3′-indolylpropionic acid
Tschandu *n* chandu, chandoo *(a kind of prepared opium)*
Tscherenkow-Strahlung *f* Cherenkov radiation
Tschermigit *m (min)* tschermigite *(aluminium ammonium sulphate-12-water)*
Tschernosjom *m(n) (soil)* chernozem, black earth
Tschitschibabin-Reaktion *f* Chichibabin (Tschitschibabin) reaction *(for preparing amino derivatives of heterocyclic bases)*
Tschugajew-Reaktion *f* Chugaev (Tschugaeff) reaction *(for obtaining alkenes)*
T-s-Diagramm *n* temperature-entropy chart, entropy chart (diagram)
TSH *s.* Hormon / thyreotropes
T-Stück *n* T-shape connecting tube, T-type connector, T-piece, tee [connector]
  T. mit Reduzierung reducer tee
T-50-Test *m* T-50 test *(for determining the stage of vulcanization)*
TTT *s.* Tieftemperaturteer
Tubasäure *f* tubaic acid *(a benzofuran derivative)*
Tube-and-Tank-Verfahren *n (petrol)* tube-and-tank [cracking] process
Tuberkulin *n (med)* tuberculin
  T. Koch old tuberculin
Tuberkulosemittel *n*, **Tuberkulostatikum** *n* antitubercular (anti-tuberculosis) drug
Tubokurare *n* tubocurare, tube curare
Tuch *n* [woven] fabric, *(if of definite size)* cloth
tuchbespannt cloth-covered
Tuchfilter *n* cloth (fabric) filter, woven[-fabric] filter
Tuchherd *m (min tech)* blanket table
Tuchkleber *m* cloth adhesive
Tuchpapier *n* velour (flock) paper
Tuff *m* tufa, *(of volcanic origin also)* tuff
Tüllenbürste *f* test-tube brush
Tully-Anlage *f* Tully plant *(for making water gas)*
Tully-Verfahren *n* Tully process *(for making water gas)*
Tulpe *f (lab)* crucible adapter
Tünche *f* limewash, whitewash, distemper
Tungöl *n* tung oil *(from seeds of Aleurites fordii Hemsl.)*
  Japanisches T. Japanese tung oil *(from seeds of Aleurites cordata (Thunb.)R.Br. ex Steud.)*
Tungstenit *m (min)* tungstenite *(tungsten disulphide)*
Tungstit *m (min)* tungstite, tungstic ochre *(a hydrous tungsten trioxide)*
Tunkbad *n (tann)* dipping bath
Tunkfärbung *f (tann)* dip dyeing
Tunnelanguß *m (plast)* submarine gate
Tunnelglocke *f (distil)* tunnel cap
Tunnelofen *m* tunnel furnace (kiln)
Tunneltrockner *m* tunnel (canal) dryer (drying machine)
Tüpfelanalyse *f* spot (drop) analysis
tüpfeln to spot, to test by spotting
Tüpfelplatte *f* spot (spotting, cavity) plate
Tüpfelprobe *f* spot (drop) test
Tüpfelreaktion *f* spot (drop) reaction
tupfen to tip, to spot, to dot
Tupfen *m* spot
Türabheber *m* door extractor
Turanose *f* turanose *(a disaccharide)*
Turbidimeter *n* turbidimeter, turbidometer, nephelometer
Turbidimetrie *f* turbidimetry, turbidimetric (turbidity) measurement
turbidimetrisch turbidimetric
Turbine *f* turbine
  spülungsbetriebene T. *(petrol)* mud turbine
Turbinen[dreh]bohren *n (petrol)* turbine drilling, turbodrill[ing]
Turbinenöl *n* turbine oil
Turbinenpumpe *f* turbine (diffuser) pump
Turbinenrührer *m* turbine mixer (agitator)
Turbinentreibstoff *m* turbine fuel
  T. für Flugturbinen aviation (aircraft) turbine fuel
Turbinentrockner *m* turbo [shelf] dryer
Turbobohren *n s.* Turbinendrehbohren
Turbogebläse *n* turboblower, turbine blower
  dampfgetriebenes T. steam-driven turboblower

Turbogridboden *m* turbogrid tray
Turbokompressor *m* turbocompressor
Turbolöser *m* turbodissolver
Turbomischer *m*, **Turborührer** *m* turbine mixer (agitator)
turbostratisch turbostratic
turbulent turbulent
Turbulenz *f* turbulence
Turgeszenz *f (biol)* turgidity
Turgor[druck] *m (biol)* turgor [pressure]
Türheber *m* door extractor
Turille *f* tourill *(an absorption vessel)*
Türkis *m (min)* turquois[e], kalaite *(a hydrous basic aluminium copper phosphate)*
Türkischrot *n* Turkey red *(an alizarin lake)*
Türkischrotfärberei *f* Turkey-red dyeing
Türkischrotöl *n* Turkey-red oil, sulph[on]ated castor oil
Türkischrotölseife *f* Turkey-red oil soap
Turm *m* tower
Turmalin *m (min)* tourmaline *(a cyclosilicate)*
  blauer T. indicolite, indigolite
Turmalinisierung *f* tourmalinization
Turmbleiche *f (pap)* tower bleaching
Turmdämpfer *m (text)* tower steamer
Turmfertigbleiche *f (pap)* final tower bleaching
Turmkammer *f* **von Gaillard** Gaillard tower *(for concentrating sulphuric acid)*
Turmkühler *m* cooling tower
Turmreiniger *m* tower purifier
Turmrollenblock *m* crown block *(of a rotary-drilling installation)*
Turmsäure *f* 1. *(pap)* tower (storage, raw) acid, raw sulphite cooking acid; 2. Glover [tower] acid *(chamber process)*
Turmsäureherstellung *f (pap)* raw acid production, manufacture of raw acid
Turmsystem *n (pap)* tower system
Turmtrockner *m* drying tower
Turmwäscher *m* scrubbing tower, scrubber, gas washer
Tusche *f* **[/ chinesische]** India (China, Chinese) ink
  lithografische T. lithographic tusche
Tussaseide *f* tussah (tussur) silk
Tütenpapier *n* bag paper
TVP-Verfahren *n (petrol)* TVP process, true vapour-phase [cracking] process
Twaddell-Grad *m* degree Twaddell
T-Wert *m (soil)* total exchangeable bases
T-50-Wert *m (rubber)* T-50 value
Twistingpapier *n* twisting paper
Twitchell-Reagens *n* Twitchell reagent *(a sulphonated addition product of naphthalene and oleic acid)*
Twitchell-Verfahren *n* Twitchell process *(for hydrolyzing glycerides)*
Tyler-Normalsiebskala *f*, **Tyler-Siebreihe** *f* Tyler standard screen (sieve) scale, Tyler scale (series)
Tyndall-Effekt *m (coll)* Tyndall effect (phenomenon)
Tyndall-Kegel *m (coll)* Tyndall cone
Tyndall[o]meter *n (coll)* tyndallometer, Tyndall meter
Tyndallometrie *f (coll)* tyndallimetry
Tyndall-Phänomen *n s.* Tyndall-Effekt
Typfärbung *f (text)* standard dyeing
typkonform *(dye)* equal to type ✦ **t. stellen** to reduce to standard (type strength)
Tyr *s.* Tyrosin
Tyramin *n* tyramine, 2-*p*-hydroxyphenyl-ethylamine
Tyrosin *n* $HOC_6H_4CH_2CH(NH_2)COOH$ tyrosine, 2-amino-3-*p*-hydroxyphenyl-propionic acid
Tyrosinase *f* tyrosinase
Tysonit *m (min)* tysonite *(a cerium fluoride containing lanthanum)*

U

Ubbelohde-Viskosimeter *n* Ubbelohde viscometer
übel disgusting, disagreeable, objectionable *(smell)*
übelriechend malodorous, ill-smelling, foul-smelling, noxious-smelling, obnoxious
Überallzünder *m* strike-anywhere match
Überäscherung *f (tann)* excessive liming
überbelichten *(phot)* to overexpose
Überbelichtung *f (phot)* overexposure
Überbelüftung *f* cross [air] circulation *(in drying processes)*
Überbleiche *f* overbleaching
überbleichen to overbleach
überbrennen to overburn *(e.g. cement clinker)*; *(ceram)* to overfire; *(coat)* to overstove
überbrücken to bridge
überchloren to superchlorinate
Überchlorierung *f s.* Überchlorung
Überchlorsäure *f s.* Perchlorsäure
Überchlorung *f* superchlorination, excess (high) chlorination *(of water)*
überdecken to cover over; to mask *(a reaction)*
überdehnen to strain
überdestillieren 1. *(of human agent)* to distil over; 2. *(of a distillate)* to distil [over], to pass (come) over
überdosieren to overdose
Überdosierung *f* overdosage
Überdosis *f* overdose
überdrehen *(ceram)* to jigger
Überdrehmaschine *f (ceram)* jigger[ing machine]
Überdruck *m* 1. excess[ive] pressure, overpressure; 2. *(text) s.* Überdrucken
Überdrucken *n (text)* top printing, cross-printing
Überdruckfilter *n* pressure filter
Überdrucksicherung *f* pressure relief device
Überdruckventil *n* pressure relief valve
überdüngen to fertilize excessively, to overfertilize
Überdüngung *f* excessive fertilization, overfertilization
übereinanderlagern to superimpose *(stereochemistry)*
überempfindlich *(med, phot)* hypersensitive, supersensitive
Überempfindlichkeit *f (med, phot)* hypersensitiveness, hypersensitivity, supersensitivity

überentwickeln *(phot)* to overdevelop
Überentwicklung *f (phot)* overdevelopment
übereutektoid[isch] hypereutectic, hypereutectoid
Überfall *m* notch *(flow measurement)*
  dreieckiger Ü. triangular notch
  rechteckiger Ü. rectangular notch
Überfallwehr *n (distil)* overflow weir (lip)
überfangen *(glass)* to flash
Überfangglas *n* flashed glass
überfärbeecht fast to cross-dyeing
Überfärbeechtheit *f* fastness to cross-dyeing
überfärben *(text)* 1. *(unintentionally)* to overdye; 2. *(by the use of another dyestuff)* to dye over, to overdye; *(for dyeing a second component in fibre mixtures)* to cross-dye; 3. *(by applying further dye to achieve a desired final shade)* to top
Überfärbung *f (text)* 1. *(unintentionally)* overdyeing; 2. *(by the use of another dyestuff)* dyeing-over, overdyeing; cross-dyeing *(of fibre mixtures)*; 3. topping *(to achieve a desired final shade)*
Überfettung *f (tann)* excessive stuffing
Überfeuerung *f (ceram)* overfiring *(a defect)*
überfließen to overflow
überflüssig *s.* superfluid
überfluten to flood
Überflutungsgrenze *f (distil)* flooding point
überformen *(ceram)* to jigger
überfrischen *(met)* to overblow
überführen 1. *(chemically)* to convert; 2. *(mechanically)* to transfer
  in Dampf ü. to vaporize
  in den Fließbettzustand ü. to fluidize
  in Dextrin ü. to dextrinate, to dextrinize
  in ein Karbonat ü. to carbonate
  in ein Salz ü. to salify *(an acid or a base)*
  in einen Komplex ü. to complex
  in Gelee ü. to gelatinize, to gelatinate, to jellify, to jelly
  ineinander ü. to interconvert
Überführfilz *m (pap)* conveyor felt
Überführung *f* 1. *(chemically)* conversion; *(el chem)* transference; 2. *(mechanically)* transfer; 3. *s.* Überführungskanal
Überführungskanal *m* crossover flue *(of a coke oven)*
Überführungsventil *n* by-pass valve
Überführungszahl *f (el chem)* transference (transport) number
  anomale Ü. abnormal transference number
  Hittorfsche Ü. Hittorf number
  wahre Ü. true transference number
Übergang *m* 1. *(process)* transition, transformation, change; *(el chem)* transference; 2. *(position)* junction *(in semiconductors)*
  Ü. Normalleitung-Supraleitung superconducting transition
  Ü. vom flüssigen in den festen Zustand fluid-solid transition
  erlaubter Ü. allowed transition *(of electrons)*
  radioaktiver Ü. radioactive transformation
  strahlender Ü. radiative transition
  strahlungsfreier (strahlungsloser) Ü. radiationless transition
  verbotener Ü. forbidden transition *(of electrons)*
Übergangsbereich *m* transition region; transition interval *(of a colour indicator)*
Übergangseinheit *f (phys chem)* transfer unit
  Ü. in der Dampfphase vapour-phase transfer unit
  Ü. in der flüssigen Phase liquid-phase transfer unit
Übergangselement *n* transition element, transition[al] metal
Übergangsfließen *n* transient flow
Übergangsform *f* intermediate form
Übergangsfraktion *f (distil)* intermediate cut
Übergangsgebiet *n* transition region
Übergangsintervall *n* transition interval *(of a double salt)*
Übergangskriechen *n (cryst)* transient creep
Übergangsmetall *n* transition[al] metal, transition element, semimetal
Übergangsmetallhydrid *n* transition-metal hydride
  binäres Ü. transition-metal binary hydride
Übergangsmetallsalz *n* transition-metal salt
Übergangsmoment *n* transition moment *(for electrons)*
Übergangsreihe *f* transition series *(in the periodic table)*
Übergangsstück *n (distil)* adapter
  Ü. [Form] A reducing (reduction) adapter *(with small socket on larger cone)*
  Ü. [Form] B enlarging (expanding, expansion) adapter *(with large socket on smaller cone)*
  Ü. von Kegelschliff auf Kugelschliff conical-spherical adapter
  Ü. von Kugelschliff auf Kegelschliff (Normschliff) spherical-conical adapter
Übergangstemperatur *f* transition temperature
Übergangswahrscheinlichkeit *f (nucl)* transition[al] probability
  Einsteinsche Ü. Einstein transition probability
Übergangszone *f* transition region
Übergangszustand *m* transition state; *(reaction theory also)* activated complex
übergehen to pass, to be converted *(into another compound or modification)*; *(distil)* to come (pass) over, to distil [over]; to transfer *(as of electrons)*; to change *(as in colour)*
  in den Gelzustand ü. *(coll)* to gel
  in den Solzustand ü. *(coll)* to solate
Übergemengteile *mpl* accessory components (constituents, minerals) *(in a rock)*
Übergerbung *f* overtannage
Übergitter *n (cryst)* superlattice, superstructure
überhärten *(plast)* to overcure
Überhärtung *f (plast)* overcure, overcuring
überheizen to overheat
überhitzen to superheat, to overheat
Überhitzer *m* superheater
Überhitzung *f* superheating, overheating
  lokale (örtliche) Ü. local superheating
Überhitzungswärme *f* superheat
Überjodsäure *f s.* Metaperjodsäure

überkalken *(sugar, agric)* to overlime
Überkalkung *f (sugar, agric)* overliming
überkochen 1. to boil over; 2. *(pap)* to overcook
Überkochen *n* boilover
Überkochung *f (pap)* overcooking
überkohlen to supercarburize, to overcarburize *(steel)*
Überkohlung *f* supercarburization, overcarburization *(of steel)*
Überkopfprodukt *n (distil)* overhead [product]
Überkorn *n* oversize [material, product], tailings, tails *(classifying), (in screening process also)* screen oversize, plus material (mesh)
überkritisch *(nucl)* supercritical
überkrusten to encrust, to incrust
Überlademethode *f* F4 method *(for octane rating)*
überlagern to superimpose *(stereochemistry)*
Überlagerungsmethode *f* heterodyne beat method *(for measuring dielectric constants)*
überlappen to overlap *(chemical-bond theory)*
Überlappung *f* overlap[ping] *(chemical-bond theory)*
Überlappungsintegral *n* overlap integral *(chemical-bond theory)*
überlasten to overload
Überlastungsmelder *m* overload alarm
Überlauf *m* 1. overflow [weir, lip, port], weir; *(met)* skimmer *(for separating the slag)*; 2. overflow [fraction, product], overs, tailings, tails *(classifying)*; 3. overflow rate
Überlaufdamm *m s.* Überlauf 1.
überlaufen to run over, to overflow; *(with heat)* to boil over
Überlaufen *n* overflow; *(with heat)* boiling over
Überlauffraktion *f* overflow fraction
Überlaufgut *n s.* Überlauf 2.
Überlaufkasten *m* overflow box *(as of a thickener)*; *(pap)* weir box
Überlaufklasse *f* overflow fraction *(classifying)*
Überlauföffnung *f* overflow port
Überlaufprodukt *n s.* Überlauf 2.
Überlaufrinne *f* overflow launder
Überlaufrohr *n*, **Überlaufstutzen** *m* overflow pipe (tube)
Überlaufwehr *n* overflow weir (lip)
Überlebensdauer *f (tox)* survival time
überleiten to pass (carry) over *(e.g. vapour)*
übermahlen to overgrind
Übermaß *n* excess, surplus
übermäßig excess[ive]
übermastizieren *(rubber)* to overmasticate, to overmill, to kill
Übermastizieren *n (rubber)* overmastication, overmilling, killing, dead milling
Übermikroskop *n* electron microscope
Übermöllerung *f (met)* overburdening
überpolen to overpole *(copper)*
überprüfen to recheck
Überrest *m* remainder, remains, residue
Überröste *f (text)* excess retting
übersättigen to supersaturate, to oversaturate
Übersättigung *f* supersaturation, oversaturation
Übersättigungsgrad *m* degree of supersaturation
Übersättigungskurve *f* supersolubility curve
Übersaturation *f (sugar)* oversaturation
Übersäure *f* peracid
überschäumen to froth (foam) over
überschichten to cover with a layer *(of another liquid)*
Überschönung *f* overfining *(of wine)*
Überschuß *m* excess, surplus ✦ **mit einem geringen Ü. an** in slight excess of
Überschußchlor[ier]ung *f* excess (high) chlorination, superchlorination *(of water)*
Überschußelektron *n* excess electron
Überschußenergie *f* excess energy
Überschußgas *n* surplus gas
Überschußhalbleiter *m n*-type semiconductor
überschüssig excess[ive]
Überschußladungsträger *m (phys chem)* excess (unbalanced) carrier
Überschußluft *f* excess air
Überschußträger *m s.* Überschußladungsträger
Überschußträgerspeicherung *f (phys chem)* excess carrier storage
Überschußwasserstoff *m* excess hydrogen
überschwänzen *(ferm)* to sparge
Überschwänzwasser *n (ferm)* sparge water
Überseepostpapier *n* foreign note paper
Übersensibilisator *m (phot)* supersensitizer
übersensibilisieren to hypersensitize, *(phot also)* to supersensitize
Übersensibilisierung *f* hypersensitization, *(phot also)* supersensitization
übersetzen to top *(leather previously dyed in acid medium with basic dyestuff)*
Überspannung *f* / **[elektrochemische, elektrolytische]** overvoltage, overpotential, excess potential
katodische Ü. cathodic overvoltage
übersprühen to spray
Übersprühen *n* **mit Wasser** water spraying
überstehend supernatant *(liquid)*
übersteigen to exceed
überströmen to overflow
Überströmkanal *m* crossover flue *(as of a coke oven)*
Überströmrohr *n* overflow pipe (tube)
Überstruktur *f (cryst)* superstructure, superlattice
übertragen to transfer
Überträger *m* carrier
Übertragung *f* transfer[ence]
Übertragungsfunktion *f* transfer function
Übertragungsreaktion *f (plast)* transfer reaction
Übertragungsregler *m* chain-transfer agent *(polymerization)*
übertreiben *(distil)* to carry (distil, pass) over, to prime
langsam ü. to sweat off
Übertrieb *m s.* Übertriebsäure
Übertriebgas *n (pap)* [digester] relief gas, release, blow-off
Übertriebsäure *f (pap)* relief liquor, release, blow-off
übertrocknen to overdry
Übertrocknung *f* overdrying, excessive drying

übervernetzen *s.* übervulkanisieren
Übervernetzung *f s.* Übervulkanisation
Übervulkanisation *f* overcure, overcuring, overvulcanization
übervulkanisieren to overcure, to over-vulcanize
überwachen to supervise, to inspect, to control
Überwachsung *f (cryst)* overgrowth
Überwachung *f* supervision, inspection, control
Überwachungseinrichtung *f* 1. control device, monitor; 2. *(tox)* regulatory agency
Überweiche *f* oversteeping *(of malt)*
überziehen to coat, to cover; *(esp relating to metals)* to clad; *(relating to an inner surface)* to line; to engobe *(with slip)*
  mit Blei ü. to lead-coat
  mit einer Kruste ü. to encrust
  mit einer Rhodiumschicht ü. to rhodanize
  mit Gips ü. to plaster
  mit Graphit ü. to coat with graphite, to graphitize, *(Am also)* to graphite
  mit Gummi ü. to rubber-coat, to rubber-cover
  mit Platin ü. to platinize, to platinate
  mit Thorium ü. to thoriate
  mit Zuckerglasur ü. *(food)* to glaze
  sich ü. to become coated (covered)
Überzug *m* coat[ing], cover, *(esp if consisting of metal)* cladding; lining *(of an inner surface)*; *(phot)* supercoat *(for protecting the emulsion)*; cover coat *(for protecting a ground coat)*
  abstreifbarer Ü. strippable coating
  dünner Ü. film
  feuerfester Ü. refractory coating
  flüssig aufgetragener Ü. wash
  galvanischer Ü. electrodeposit, plating, plate
  gelartiger Ü. gel coating
  heißschmelzender Ü. hot-melt coating
  metallischer Ü. metallic coating
  oxidischer Ü. oxide coating
  schützender Ü. protective coating, protecting (protection) layer, pro-coating
Überzugsharz *n* coating resin
Überzugslack *m* coating varnish, finish
Überzugsmetall *n* coating metal
Überzugsmittel *n* surface coating
Überzugspapier *n* lining paper, pasting [paper], liners
Überzugstechnik *f* coating technology
Überzugswerkstoff *m* coating material, *(esp relating to metals)* cladding material
UDP *s.* Uridindiphosphat
UDPG, UDP-Glukose *f s.* Uridindiphosphoglukose
U-Eisen *n* channel iron *(for producing channel black)*
U-Feuerung *f* fantail firing *(a suspension firing)*
U-Flammen-Wanne *f (glass)* end-fired (end-port) furnace
U-förmig U-shaped
ug-Kern *m* odd-even nucleus
Uhde-Zelle *f* Uhde cell *(a mercury cell)*
Uhrenöl *n* watch oil
Uhrglas *n*, **Uhrglasschale** *f (lab)* watch (clock) glass
Uhrzeigersinn / entgegen dem counterclockwise
  im U. clockwise
U-I-Kennlinie *f* current-voltage curve
Ulexit *m (min)* ulexite *(a soroborate)*
Ullmannit *m (min)* ullmannite, nickel stibine *(sulphantimonide of nickel)*
Ullmann-Reaktion *f* Ullmann reaction *(for synthesizing diaryls)*
Ulmin *n (soil)* ulmin
Ulminstoff *m (soil)* ulmin material
Ultrabeschleuniger *m (rubber)* ultra-accelerator, super-accelerator, ultrafast (ultrarapid) accelerator
Ultrafilter *n* ultrafilter, millipore filter
Ultrafiltrat *n* ultrafiltrate
Ultrafiltration *f* ultrafiltration
ultrafiltrieren to ultrafilter
Ultraformen *n* ultraforming *(a variety of catalytic reforming)*
Ultrakurzwelle *f* ultrashort wave
Ultramarin *n* ultramarine blue
  gelbes U. 1. yellow ultramarine *(a mixture of zinc and calcium chromate)*; 2. *s.* Ultramaringelb
Ultramarinblau *n* ultramarine blue
Ultramaringelb *n* yellow ultramarine, lemon chrome, Steinbühl yellow, ultramarine (barium, baryta) yellow *(barium chromate)*
Ultramikroanalyse *f* ultramicroanalysis
Ultramikrobestimmung *f* ultramicrodetermination
Ultramikrochemie *f* ultramicrochemistry
Ultramikroskop *n* ultramicroscope
ultramikroskopisch ultramicroscopic
Ultrapasteurisation *f (food)* uperization
ultrarein ultrapure
Ultrarot *n s.* Infrarot
Ultraschallbehandlung *f* ultrasonic treatment
Ultraschallfeld *n* ultrasonic field
Ultraschallkoagulation *f* ultrasonic coagulation
Ultraschallprüfung *f* ultrasonic (supersonic) testing
Ultraschallschweißen *n (plast)* ultrasonic welding
Ultraschallsirene *f* ultrasonic agglomerator *(gas cleaning)*
Ultraschallwellen *fpl* ultrasonic waves
Ultraviolett *n* ultraviolet [radiation]
Ultraviolettabsorber *m* ultraviolet absorber
Ultraviolettabsorption *f* ultraviolet absorption
Ultraviolettanalyse *f* ultraviolet analysis
Ultraviolettbeständigkeit *f* ultraviolet resistance
Ultraviolettbestrahlung *f* ultraviolet irradiation
Ultraviolettfotografie *f* ultraviolet photography
Ultraviolettspektrograf *m* ultraviolet spectrograph
Ultraviolettspektrografie *f* ultraviolet spectrography
Ultraviolettspektrum *n* ultraviolet spectrum
Ultraviolettstrahlung *f* ultraviolet radiation
Ultrazentrifuge *f* ultracentrifuge, high-speed centrifuge
Ultrazentrifugieren *n* ultracentrifugation, high-speed centrifugation
ULV-Verfahren *n* ultralow (very low) volume method (spraying) *(spraying of pesticides without water in amounts of only 1 to 5 gallons liquid/acre)*
Umbelliferon *n* umbelliferone, 7-hydroxy-2H-chromen-2-one

Umber *m s.* Umbra
Umblattpapier *n* cigar wrapping paper
Umbra *f,* **Umbraun** *n* umber *(an earth pigment)*
umdestillieren to redistil, to rerun
Umdruckpapier *n* transfer paper
umestern to transesterify
Umesterung *f* transesterification, interesterification, ester interchange
gelenkte U. directed transesterification
umfällen to reprecipitate
Umfällung *f* reprecipitation
Umfangsgeschwindigkeit *f* peripheral speed
umfärben *(text)* to redye
umformen to change *(a structure)*; *(plast)* to form
sich u. to change *(of a structure)*
Umformung *f* change *(of a structure)*
umfüllen to transfer *(substances into another vessel)*
Umfüllen *n* transfer *(of substances)*
umgeben to pack, to invest
umgebend ambient
Umgebung *f* environment; vicinity
korrodierend wirkende U. corrosive environment
Umgebungsdruck *m* ambient pressure
umgebungsempfindlich environment-sensitive
Umgebungsfeuchtigkeit *f* ambient humidity
Umgebungstemperatur *f* ambient temperature
umgehen to by-pass
umgruppieren to rearrange
sich u. to rearrange
Umgruppierung *f* rearrangement
umhüllen to jacket, to coat, to cover, *(esp relating to metal)* to clad
Umhüllung *f* jacket, coat[ing], cover[ing], *(esp relating to metal)* cladding
Umkehr *f* reversal
umkehrbar reversible, invertible
nicht u. non-reversible, irreversible
Umkehrbarkeit *f* reversibility, invertibility
Umkehremulsion *f (phot)* reversal emulsion
umkehren to reverse, to invert *(a process)*
Umkehrentwickler *m (phot)* reversal developer
Umkehrentwicklung *f (phot)* reversal development (processing)
Umkehrfilm *m* reversal film
Umkehrkammer *f* / **frei bewegliche** floating head *(of a heat exchanger)*
Umkehrmaterial *n (phot)* reversal material
Umkehrphasenchromatografie *f* reversed-phase chromatography
Umkehrung *f* reversion, reversal, inversion
Waldensche U. Walden inversion
Umkehrverfahren *n (phot)* reversal process
Umklappen *n* turnover, flipping-over, flip *(of bonds as in the Walden inversion)*
U. des Ringes ring flip
Umklappprozeß *m (nucl)* umklapp process
umkleiden *s.* umhüllen
Umkristallisation *f* recrystallization
postkinematische (posttektonische) U. *(geoch)* posttectonic recrystallization
umkristallisieren to recrystallize
Umladung *f (phys chem)* charge reversal
umlagern to rearrange
sich u. to rearrange
Umlagerung *f* rearrangement
U. am Aromaten aromatic rearrangement
Beckmannsche U. Beckmann rearrangement, Beckmann molecular transformation *(of a ketoxime into an amide derivative)*
Friessche U. Fries rearrangement *(as for synthesizing phenolic ketones)*
innermolekulare (intramolekulare) U. intramolecular change
Lossensche U. Lossen rearrangement *(of aromatic hydroxamic acids into isocyanates)*
Stevenssche U. Stevens rearrangement *(of a benzyl group in quaternary ammonium salts)*
Wolffsche U. Wolff rearrangement *(of diazoketones)*
1,2-Umlagerung *f* 1,2-shift
Umlagerungsgeschwindigkeit *f* velocity of rearrangement
Umlagerungspolymerisation *f* rearrangement polymerization
Umlagerungsreaktion *f* rearrangement reaction
Umlauf *m* [re]circulation
natürlicher U. natural circulation
sauberer U. clean circulation *(in thermal cracking)*
Umlaufbahn *f* orbit
kreisförmige U. circular orbit
Umlaufbeladung *f* circulating load
umlaufen 1. *(of liquids)* to circulate; 2. *(as of rolls)* to rotate, *(if rapidly)* to spin
Umlauffilter *n* band filter, [linear] belt filter
Umlaufgas *n* recycle gas
Umlaufgut *n* circulating load
Umlaufkolbengebläse *n* positive-displacement blower, [positive] rotary blower
Umlaufkolbenpumpe *f* rotary pump
Umlaufkonche *f (food)* rotation conche
Umlauföl *n* cycle oil
Umlaufpumpe *f* recirculation (recycle, circulating) pump, circulator
Umlaufspannung *f* magnetomotive force
Umlaufsystem *n* circulation system
Umlaufverdampfer *m* forced-circulation reboiler
Umlaufwasser *n* circulating water
umlegen to reverse *(e.g. the direction of flow)*
Umlenkblech *n* baffle [plate] ✦ **mit Umlenkblechen [versehen]** baffled ✦ **ohne Umlenkbleche** unbaffled
Umlenkplatte *f s.* Umlenkblech
Umlenkrinne *f (glass)* deflector
ummanteln to jacket, to coat, to cover, *(esp relating to metal)* to clad
Ummantelung *f* jacket, coat[ing], cover[ing], *(esp relating to metal)* cladding
Umpherston-Holländer *m (pap)* Umpherston beater
Umpump *m* forced circulation
umpumpen to recirculate, to recycle

umrechnen to convert
Umrechnung *f* conversion
Umrechnungsfaktor *m* conversion factor
Umrechnungstabelle *f* conversion table
umrollen *(pap)* to rereel, to rewind
Umroller *m* *(pap)* reel-slitting (roll-slitting) machine, rereeling (rewinding, slitting) machine, rewinder, slitter
umrühren to agitate, *(esp relating to molten metal)* to puddle
Umrühren *n* agitation, *(esp relating to molten metal)* puddling
Umsatzgleichung *f* **[/ chemische]** reaction (chemical) equation
Umsatzgrad *m* degree of conversion
umschalten to reverse
Umschlag *m* break *(as in titration)*; change, transition *(of an indicator)*
umschlagen 1. to change *(of colours or reactions)*; 2. to reverse *(of emulsions)*
Umschlagen *n* 1. change *(of colours or reactions)*; 2. reversion *(of emulsions)*
Umschlagpapier *n* cover paper
Umschlagsbereich *m*, **Umschlagsgebiet** *n s.* Umschlagsintervall
Umschlagsintervall *n* transition (colour change) interval *(of a pH indicator)*
Umschlagspunkt *m* equivalent (final, end) point *(titration)*
Umschlingungswinkel *m* angle of wrap *(conveying)*
umschmelzen to remelt, *(met also)* to re-fuse
umschwenken to swirl
umsetzen to convert, to react
sich u. to react, to convert
Umsetzer *m* reactor
Umsetzung *f* conversion, reaction
doppelte U. metathesis, double decomposition, exchange
einfache U. simple decomposition
fotochemische U. photochemical reaction, photoreaction
Umsetzungsgeschwindigkeit *f* conversion rate (velocity)
Umsetzungsgrad *m* degree of conversion
umsieden to distil
Umsieden *n* distillation
umspülen to flow round
umstellen 1. to rearrange; 2. to reverse *(e.g. the direction of flow)*
Umstellung *f* 1. rearrangement; 2. reversal *(as of the direction of flow)*
Umstellventil *n* reversing valve
Umtauschkapazität *f* exchange capacity; *(soil quantitatively)* total exchangeable bases
Umtriebpropeller *m* *(pap)* propeller-type agitator, revolving paddles *(of a Hollander beater)*
umwälzen to circulate
Umwälzgeschwindigkeit *f* circulation rate
Umwälzheizeinrichtung *f* circulation heater
Umwälzpumpe *f* recirculation (recycle, circulating) pump, circulator
Umwälztrockner *m* convection dryer
Umwälzung *f* circulation
umwandelbar convertible, transformable
ineinander (wechselseitig) u. interconvertible; *(relating to modifications)* enantiotropic
Umwandelbarkeit *f* convertibility, transformability
wechselseitige U. interconvertibility; enantiotropy *(of modifications)*
umwandeln to transform, to convert, to change; to reform *(hydrocarbons)*; *(nucl)* to transform, to transmute; *(cryst)* to invert
in Dextrin u. to dextrinate, to dextrinize
in ein Karbonat u. to carbonatize, to carbonate
in Kohlenstoff u. to carbonize
in Zucker u. to saccharify, to saccharize
ineinander u. to interconvert
sich u. to convert, to transform, to undergo change, to be converted; *(nucl)* to transform, to transmute; *(cryst)* to invert
sich ineinander u. to interconvert
Umwandlung *f* transformation, conversion, change; reforming *(of hydrocarbons)*; *(nucl)* transformation, transmutation, *(by decay also)* devolution; transition *(into another phase)*; *(cryst)* transition, inversion
U. bei gleichbleibender Temperatur isothermal (constant-temperature) transformation
U. bei kontinuierlicher (stetiger) Abkühlung continuous-cooling transformation
U. erster Ordnung first-order transition
U. in ein Karbonat carbonatization, carbonation
U. in Kohlenstoff carbonization
U. zweiter Ordnung second-order transition, *(plast, rubber also)* vitrification, glass transition
gegenseitige U. interconversion
isotherm[isch]e U. isothermal (constant-temperature) transformation
radioaktive U. radioactive transformation (transmutation)
Umwandlungsbeginn *m* start of transformation
Umwandlungsende *n* finish of transformation
Umwandlungsgeschwindigkeit *f* velocity of transformation, conversion rate, rate of change
Umwandlungsgrad *m* degree of conversion
Umwandlungsintervall *n* transition interval *(of a double salt)*
Umwandlungsprodukt *n* transformation (conversion) product
Umwandlungspunkt *m* transition point (temperature), *(cryst also)* inversion point (temperature)
U. erster Ordnung first-order transition point (temperature)
U. zweiter Ordnung second-order transition point (temperature), *(plast, rubber also)* glass transition point, Tg [point]
Umwandlungsschaubild *n* transformation diagram
U. für gleichbleibende Temperatur isothermal transformation diagram
U. für kontinuierliche Abkühlung continuous-cooling transformation diagram
Umwandlungstemperatur *f s.* Umwandlungspunkt
Umwandlungswärme *f* heat of transition, latent heat *(of phases or crystals)*

Umweltbedingungen *fpl* environmental factors
Umwelteinfluß *m* environmental influence
Umweltfaktor *m* environmental factor
Umweltverschmutzung *f* environmental pollution
umwickeln *s.* umrollen
Umxanthogenierung *f* rexanthation
unabgesättigt unsaturated
Unabhängigkeitsverbundwirkung *f* *(tox)* independent joint action *(without reciprocal influence of the components)*
unabtrennbar inseparable
unangegriffen unchanged, unattacked, unaffected
unangenehm disagreeable, disgusting, unpleasant, objectionable *(e.g. smell)*
  u. riechend ill-smelling, foul-smelling, obnoxious
  u. schmeckend ill-tasting
unangreifbar / chemisch stable (resistant) to chemical attack, chemically stable (resistant)
Unangreifbarkeit *f* / **chemische** stability (resistance) to chemical attack, chemical stability (resistance)
unaufgeschlossen undigested; uncooked *(paper pulp)*
unbehandelt untreated, unprocessed, unfinished, raw, virgin
unbelichtet *(phot)* unexposed, raw
unbenetzbar non-wettable
unbesetzt vacant *(e.g. lattice position)*
unbeständig instable, unstable, transient
Unbeständigkeit *f* instability, transience
Unbestimmtheitsbeziehung *f* / **Heisenbergsche** Heisenberg indeterminacy (uncertainty) principle
Unbestimmtheitsrelation *f s.* Unbestimmtheitsbeziehung / Heisenbergsche
unbeweglich immobile, non-mobile
unbrennbar incombustible, non-combustible, fireproof ✦ **u. machen** to fireproof
Unbrennbarkeit *f* incombustibility
Unbrennbarmachen *n* fireproofing
Undekan *n* $CH_3[CH_2]_9CH_3$ undecane
Undekanal *n* $CH_3[CH_2]_9CHO$ undecanal, undecyl aldehyde
Undekandisäure *f* $HOOC[CH_2]_9COOH$ undecanedioic acid
Undekanol *n* undecanol *(any of several isomeric alcohols $C_{11}H_{23}OH$)*
Undekanon *n* undecanone *(any of several isomeric ketones $C_{11}H_{22}O$)*
Undekansäure *f* $CH_3[CH_2]_9COOH$ undecanoic acid
Undezen *n* undecene *(any of several isomeric alkenes $C_{11}H_{22}$)*
Undezenol *n* undecenol *(any of several isomeric alcohols $C_{11}H_{21}OH$)*
Undezensäure *f* undecenoic acid *(any of several isomeric alkenoic acids $C_{11}H_{20}O_2$)*
Undezen-(9)-säure *f* $CH_3CH{=}CH[CH_2]_7COOH$ 9-undecenoic acid
Undezin *n* undecyne *(any of several isomeric alkynes $C_{11}H_{20}$)*
Undezinsäure *f* undecynoic acid *(any of several isomeric alkynoic acids $C_{11}H_{18}O_2$)*
Undezylaldehyd *m s.* Undekanal
Undezylalkohol *m* $CH_3[CH_2]_9CH_2OH$ undecyl alcohol, 1-undecanol
Undezylamin *n* undecylamine, aminoundecane *(any of several isomeric amines $C_{11}H_{25}N$)*
Undezylen *n s.* Undezen
Undezylensäure *f s.* Undezensäure
Undezylsäure *f s.* Undekansäure
undicht leaky, pervious, porous, porose
  u. sein to leak, to run
  u. werden to spring a leak
Undichtheit *f*, **Undichtigkeit** *f* leakiness
undissoziiert undissociated
undurchdringlich *s.* undurchlässig
undurchgängig *s.* undurchlässig
undurchlässig impermeable, impenetrable, impervious, [leak]tight, [leak]proof; *(optically)* opaque ✦ **u. machen** to make impervious
  u. für Röntgenstrahlen radiopaque
Undurchlässigkeit *f* impermeability, impenetrability, imperviousness, tightness, proofness; *(optically)* opacity
  U. für Röntgenstrahlen radiopacity
Undurchlässigmachen *n* proofing
undurchsichtig opaque, non-transparent
Undurchsichtigkeit *f* opacity, non-transparency
unecht 1. artificial, factitious; 2. *(dye)* not fast, fugitive
unedel base *(metal)*
Unedelmetall *n* base metal
uneinheitlich non-uniform
unelastisch inelastic
unempfindlich insensitive ✦ **u. machen** *(phot)* to desensitize
Unempfindlichkeit *f* insensitiveness
unentbehrlich *(bioch, agric)* essential
Unentbehrlichkeit *f (bioch, agric)* essentiality
unentflammbar non-[in]flammable, non-flam, flameproof
Unentflammbarkeit *f* non-flammability, flameproofness
Unentflammbarmachen *n* flameproofing
unfruchtbar *(soil)* infertile, barren
Unfruchtbarkeit *f (soil)* infertility, barrenness
ungealtert *(rubber)* unaged
ungebeizt unmordanted
ungebleicht unbleached
ungebleit unleaded *(motor fuel)*
ungebrannt *(ceram)* unburned, unburnt, raw, green
ungebunden uncombined
ungefährlich harmless
Ungefährlichkeit *f* harmlessness
ungefüllt filler-free, unfilled, unloaded, *(rubber also)* non-pigmented
ungeglättet *(pap)* unfinished, unglazed
ungekocht uncooked, raw
ungeladen uncharged
ungeleimt *(pap)* unsized
ungemahlen unground; *(pap)* unbeaten
ungemälzt *(ferm)* unmalted
Ungenauigkeitsbeziehung *f*, **Ungenauigkeitsrelation** *f s.* Unbestimmtheitsbeziehung / Heisenbergsche

ungenießbar inedible
Ungenießbarkeit *f* inedibility
ungeordnet unordered, random
ungepaart unpaired *(e.g. electron)*
ungepreßt un[com]pressed
ungepuffert unbuffered
Ungerade-gerade-Kern *m* odd-even nucleus
Ungerade-ungerade-Kern *m* odd-odd nucleus
ungeradzahlig odd-numbered
ungereckt *(text)* undrawn *(man-made fibres)*
ungesalzen unsalted, *(tann also)* fresh
ungesättigt unsaturated *(solution or compound)*
  doppelt u. doubly unsaturated, diunsaturated
  dreifach u. triply unsaturated
  einfach u. monounsaturated
  mehrfach u. polyunsaturated
Ungesättigte *npl (petrol)* unsaturate[d]s
Ungesättigtheit *f* unsaturation
  geringe U. low unsaturation
Ungesättigtheitsgrad *m* degree of unsaturation
ungesäuert unsoured
ungeschält undecorticated; *(pap)* unbarked
ungeschlichtet *(text)* unsized
ungesintert unsintered
ungespannt unstrained
ungiftig non-toxic, atoxic, non-poisonous
Ungiftigkeit *f* non-toxicity, freedom from toxicity
unglasiert unglazed
ungleich[artig] dissimilar
  u. zusammengesetzt heterogeneous, inhomogeneous
Ungleichartigkeit *f* heterogeneity, inhomogeneity
ungleichförmig non-uniform
Ungleichgewicht *n* unbalance
ungleichmäßig non-uniform
unhaltig *(mine)* barren
unharmonisch anharmonic
unhybridisiert unhybridized *(chemical-bond theory)*
unhydrierbar unhydrogenable
Unifarbe *f* self-colour
Unifärben *n* union dyeing
Unifining-Verfahren *n (petrol)* unifining process
unimolekular unimolecular, monomolecular
Union *f* **für reine und angewandte Chemie / Internationale** International Union of Pure and Applied Chemistry, IUPAC
unionisiert unionized
unitarisch unitarian, unitary, homopolar, non-polar, covalent *(chemical-bond theory)*
univariant univariant, monovariant
Universaldoppelmuffe *f (lab)* swivel clamp holder
Universalechtheit *f (text)* all-round fastness
Universalemulsion *f (phot)* universal emulsion
Universalentwickler *m (phot)* universal developer
Universalindikator *m* universal indicator
Universalindikatorpapier *n* universal indicator paper
Universalinsektizid *n* general-purpose insecticide
Universalmittel *n (pharm)* panacea, cure-all
Universal-pH-Meter *n* general-purpose pH meter
Universalstativ *n (lab)* assembly
Universalstativklemme *f (lab)* universal stand-clamp
Universalstrom- und -spannungsmesser *m* voltammeter
Universalwaschmittel *n* all-purpose washing agent
unkoordiniert uncoordinated
Unkrautbekämpfung *f* weed control (eradication), weeding
Unkrautbekämpfungsmittel *n* herbicide, weed-killer, weed control agent
  U. auf Wuchsstoffbasis hormone weedkiller
Unkrautvertilgungsmittel *n s.* Unkrautbekämpfungsmittel
unl. *s.* unlöslich
unlegiert unalloyed
unlöslich in[dis]soluble, i.s., non-soluble ✦ **u. machen** to insolubilize
  u. in Alkali insoluble in alkali
  u. in Alkohol insoluble in alcohol
  u. in Wasser water-insoluble
Unlösliches *n* insoluble, i.; *(in tar)* free carbon
Unlöslichkeit *f* insolubility, insolubleness
unmagnetisch non-magnetic
unmarkiert unlabelled
unmischbar immiscible
Unmischbarkeit *f* immiscibility
unmodifiziert unmodified
unnatürlich unnatural
Unordnung *f (cryst)* disorder, randomness
unpaar[ig] unpaired
unpigmentiert unpigmented
unplastifiziert unplasticized
unplastisch non-plastic
unplastiziert unplasticized
unpolar non-polar, homopolar, covalent, unitarian, unitary *(chemical-bond theory)*
unpolarisierbar non-polarizable
unraffiniert unrefined
unreagiert unreacted
unregelmäßig anomalous
Unregelmäßigkeit *f* anomaly
unrein impure
Unreinheit *f* impurity, *(if localized also)* speck, spot
uns. *s.* unsymmetrisch
unschädlich harmless, innocuous
Unschädlichkeit *f* harmlessness, innocuousness, innocuity
Unschärfebeziehung *f* / **Heisenbergsche** Heisenberg indeterminacy (uncertainty) principle
Unschärferelation *f s.* Unschärfebeziehung / Heisenbergsche
unschmackhaft unpalatable, distasteful
unschmelzbar infusible
Unschmelzbarkeit *f* infusibility
Unsicherheitsbeziehung *f s.* Unschärfebeziehung/ Heisenbergsche
unstabil unstable, instable
unstabilisiert unstabilized
unstetig discontinuous, batch[wise]
  u. arbeitend batch *(apparatus)*
unstrukturiert structureless, devoid of structure
Unsymmetrie *f* asymmetry, dissymmetry

unsymmetrisch asymmetric[al], dissymmetric[al], unsymmetric[al]
Untenaustrag *m*, **Untenentleerung** *f* bottom discharge
Unteranstrich *m* undercoat
Unterbainit *m (met)* lower bainite
Unterbau *m* 1. *(tech)* substructure; 2. *(rubber)* case, casing, carcass, carcase *(of a pneumatic tyre)*
unterbelichten *(phot)* to underexpose
Unterbelichtung *f (phot)* underexposure
Unterbezugskraftstoff *m* secondary reference fuel
Unterbleiche *f (pap)* underbleaching
unterbleichen *(pap)* to underbleach
Unterboden *m (agric)* subsoil
Unterbrecherbad *n (phot)* stop bath
Unterbrenner *m* underjet
Unterbrennerkoksofen *m* underjet coke oven
Unterbrennerleitungen *fpl* underjet piping *(of a coke oven)*
unterbringen to locate *(e.g. atoms in interstitial lattice sites)*
Unterdruckdestillation *f* distillation under vacuum (reduced pressure)
unterdrücken to suppress *(e.g. a reaction)*
Unterdruckentgaser *m* vacuum degasifier
Unterdruckfilter *n* vacuum (suction) filter
Unterdruckfiltration *f* vacuum filtration, filtration under reduced pressure
Unterdruckkokillenguß *m* vacuum diecasting
Unterdruckpresse *f (plast)* bottom ram press
Unterdruckventil *n* vacuum relief valve, vacuum breaker
unterentwickeln *(phot)* to underdevelop
Unterentwicklung *f (phot)* underdevelopment
untereutektoid[isch] hypoeutectic, hypoeutectoid
Unterfilz *m (pap)* bottom (lower, mould) felt *(of a cylinder board machine)*
Unterflottenjigger *m (text)* immersion jigger
untergärig bottom-fermenting, bottom-fermented
Untergärung *f* bottom fermentation
untergetaucht submerged, submersed
Unterglasurdekor *m (ceram)* underglaze decoration
Unterglasurfarbe *f (ceram)* underglaze colour
Unterglasurmalerei *f (ceram)* underglaze painting
Untergrund *m* 1. *(coat)* ground, undersurface; 2. *(agric)* subsoil, C-horizon, bedrock; 3. *(nucl)* background
Untergrundanstrich *m* priming (primer, ground) coat, priming
Untergrunddüngung *f* fertilization of subsoil
Untergrundstrahlung *f* background radiation
Untergruppe *f* subgroup
Untergruppennummer *f*, **Untergruppenziffer** *f* subgroup number *(of the international coal classification system)*
unterhalten *(tech)* to maintain; to support *(e.g. combustion)*
Unterhaltung *f (tech)* maintenance, upkeep; support *(as of combustion)*
Unterhaltungskosten *pl* maintenance cost[s], cost of upkeep
Unterhärtung *f (plast)* undercure
Unterhaut *f (tann)* subcutis
Unterhautbindegewebe *n (tann)* adipose tissue
Unterhefe *f* bottom[-fermentation] yeast
unterkochen *(pap)* to undercook
Unterkochung *f (pap)* undercooking
Unterkolbenpresse *f (plast)* up-stroke press, bottom-ram press
Unterkorn *n* undersize [material], fines, minus material *(classifying)*
unterkritisch *(nucl)* subcritical
unterkühlen to supercool, to subcool, to overcool
Unterkühler *m* subcooler
Unterkühlung *f* supercooling, subcooling, overcooling
Unterlage *f* base, pad, *(esp phot)* support
Unterlagspapier *n* carpet felt
Unterlagspappe *f* carpet felt
Unterlagsplatte *f* bed plate
Unterlauf *m* underflow *(classifying)*
Unterlaufprodukt *n* underflow *(classifying)*
Unterluftzelle *f (min tech)* sub-aeration floatation cell
Untermesser *n (pap)* bed knife *(of a cross-cutter)*
untersättigt subsaturated, undersaturated
Untersättigung *f* subsaturation, undersaturation
Untersatzschale *f* **für Säureflaschen** *(lab)* bottle tray, acid dish
Unterschale *f (nucl)* subshell
unterschichten to add to form a lower layer
Unterschubfeuerung *f* underfeed firing, underfiring
Unterschubrost *m* underfeed stoker
Unterseite *f (pap)* wire side
untersinken to sink
Unterstempel *m (plast)* bottom ram; *(ceram)* bottom punch
Unterstruktur *f (cryst)* substructure
untersuchen to investigate, to assay, to examine, to analyze
Untersuchung *f* investigation, research, examination, assay, analysis; *(for detecting a specified substance)* test
mikroskopische U. microexamination
röntgenografische U. X-ray investigation
Untersuchungsbohrung *f (petrol)* exploration drilling, wildcat
Untersuchungsflasche *f* test bottle
Untersuchungsflüssigkeit *f* liquid for (*or* under) investigation (examination)
Untersuchungslösung *f* solution for (*or* under) investigation (examination), solution being (*or* to be) tested, test solution
Untersuchungsmaterial *n* material for (*or* under) investigation, experimental (test) material
Untersuchungsmethode *f* investigational (test, testing) method
standardisierte U. standard test[ing] method
Untersuchungssubstanz *f s.* Untersuchungsmaterial
Untertagebau *m* underground mining (working)
Untertagespeicher *m* underground reservoir (storage tank)

Untertagespeicherung *f* underground storage
Untertagevergasung *f* underground gasification
untertauchen to submerge, to immerse
Untertauchen *n* submersion, immersion
Unterteil *n (plast)* bottom plug
unterteilen to classify; to graduate
Unterteilung *f* classification; graduation
Untertrum *m(n)* return side *(conveying)*
untervernetzen *s.* untervulkanisieren
Untervernetzung *f s.* Untervulkanisation
Untervulkanisation *f* undercure, undercuring, undervulcanization
untervulkanisieren to undercure, to undervulcanize
Unterwalze *f* bottom roll
U. der Leimpresse *(pap)* bottom size-press roll
Unterwasserbohrung *f (petrol)* marine drilling
Unterwasserjigger *m (text)* immersion jigger
Unterwasserkorrosion *f* subaqueous corrosion
Unterwasserpumpe *f* submerged (wet-pit) pump
Unterwerkzeug *n (plast)* die
Unterwind *m* downdraught
Unterwindfeuerung *f* overfeed firing
Unterwindfrischkonverter *m (met)* bottom-blown converter
Unterwindfrischverfahren *n (met)* bottom-blown-converter process
Unterwindgebläse *n* forced-draught fan (blower)
untoxisch non-toxic, atoxic, non-poisonous
untrennbar inseparable
ununterbrochen continuous *(operation)*
u. arbeitend continuous *(apparatus)*
unveränderlich invariable
unverändert unchanged, unaltered
unverbleit unleaded *(fuel)*
unverbrannt unburned, unburnt
Unverbranntes *n* unburned combustible [matter]
unverbraucht unreacted
unverbrennbar incombustible, non-combustible
Unverbrennbares *n*, **Unverbrennliches** *n* incombustible [matter], non-combustible [matter]
unverdampft unevaporated
unverdaulich indigestible
Unverdaulichkeit *f* indigestibility
unverdickt *(coat)* unbodied
unverdünnt undiluted
unverfestigt unconsolidated
unvergärbar unfermentable
unvergast ungasified
unvermischbar immiscible
Unvermischbarkeit *f* immiscibility
unvermischt unmixed
unvernetzt *(rubber)* uncured, unvulcanized
unverpackt unpacked
unverrohrt *(petrol)* uncased
unverseifbar unsaponifiable, non-saponifiable
Unverseifbares *n* unsaponifiable matter (residue)
unverseift unsaponified
unverstärkt unreinforced *(e.g. plastic material)*
unverstreckt undrawn *(man-made fibres)*
unverträglich incompatible
Unverträglichkeit *f* incompatibility
unverwitterbar unweatherable
unverzweigt unbranched[-chain], non-branched
unvulkanisiert *(rubber)* uncured, unvulcanized
unwirksam ineffective ✦ **u. machen** to inactivate, to block
unzerbrechlich unbreakable
unzersetzt undecomposed
unzugänglich *(agric)* unavailable *(nutrients)*
unzulässig *(tox)* undue
Upas[gift] *n* upas, Malay poison *(any of several arrow poisons)*
Uperisation *f (food)* uperization
uperisieren *(food)* to uperize
Uralit *m* uralite *(any of a series of inosilicates)*
uralitisieren *(min)* to uralitize
Uralitisierung *f (min)* uralitization
Uramil *n* uramil, 5-aminobarbituric acid
Uran *n* U uranium
Uran-Aktinium-Zerfallsreihe *f* actinium series, uranium-actinium disintegration (decay) series
Uranat *n* uranate *(salt or ester of uranic acid)*
Uranblüte *f (min)* zippeite *(a uranyl sulphate)*
Urandioxid *n s.* Uran(IV)-oxid
Urangelb *n* uranium yellow *(sodium diuranate-6-water)*
Uranglas *n* uranium glass
uranhaltig uraniferous
Uranhydrid *n* UH_3 uranium hydride
Uranid *n s.* Uranoid
Uranin *n* uranin[e] *(disodium salt of fluorescein)*
Uraninit *m (min)* uraninite *(uranium(IV) oxide)*
Uranmineral *n* uranium mineral
Uranocker *m s.* Uranopilit
Uranoid *n* uranoid *(any of a series of transuranium elements)*
Uranophan *m (min)* uranophane, uranotil *(a hydrous calcium uranium silicate)*
Uranopilit *m (min)* uranopilite, uranic ochre *(a uranyl sulphate)*
Uranosphärit *m (min)* uranosphaerite *(a uranyl bismuth hydroxide)*
Uran(IV)-oxid *n* UO_2 uranium(IV) oxide, uranium dioxide
Uran(IV,VI)-oxid *n* U_3O_8 triuranium octaoxide, uranium(IV) uranate
Uran(VI)-oxid *n* UO_3 uranium(VI) oxide, uranium trioxide
Uranperoxid *n s.* Urantetroxid
Uranpyrochlor *m (min)* uranpyrochlore
Uran-Radium-Zerfallsreihe *f* uranium-radium disintegration (decay) series, uranium (radium) series
Uransäure *f* uranic acid
Uranspaltung *f* uranium fission
Uran(IV)-sulfat *n* $U(SO_4)_2$ uranium(IV) sulphate
Uran(III)-sulfid *n* U_2S_3 uranium(III) sulphide, diuranium trisulphide
Urantetroxid *n* UO_4 uranium tetraoxide
Urantrioxid *n s.* Uran(VI)-oxid
Uran(IV)-uranat *n* $U(UO_4)_2$ uranium(IV) uranate, triuranium octaoxide
Uranvitriol *m (min)* uranvitriol, johannite *(a uranyl sulphate)*
Uranylazetat *n* $UO_2(CH_3COO)_2$ uranyl acetate

Uranylbromid *n* UO_2Br_2 uranyl bromide
Uranylchlorid *n* UO_2Cl_2 uranyl chloride
Uranylhexazyanoferrat(II) *n* $(UO_2)_2[Fe(CN)_6]$ uranyl hexacyanoferrate(II)
Uranylhydroxid *n* $UO_2(OH)_2$ uranyl hydroxide
Uranylnitrat *n* $UO_2(NO_3)_2$ uranyl nitrate
Uranylsalz *n* uranyl salt
Uranylverbindung *f* uranyl compound
Urat *n* urate *(a salt of uric acid)*
Uratoxydase *f* uricase
Urazil *n* uracil, 2,4-pyrimidinedione
Urazil-4-karbonsäure *f* uracil-4-carboxylic acid, orotic acid
p-**Urazin** *n* *p*-urazine, hexahydro-*s*-tetrazine-3,6-dione
Urbarmachungskrankheit *f (agric)* reclamation disease *(caused by Cu shortage)*
Urbezugskraftstoff *m* primary reference fuel
Urease *f* urease
Ureid *n* ureide *(an acyl derivative of urea)*
  offenes U. acyclic ureide
  zyklisches U. cyclic ureide
2-Ureidoäthan-1-sulfonsäure *f* $NH_2CONHCH_2CH_2SO_3H$ 2-ureidoethane-1-sulphonic acid, taurocarbamic acid
Ureidoessigsäure *f* $NH_2CONHCH_2COOH$ ureidoacetic acid, hydantoic acid
Ureidosäure *f* ureido acid
Ureometer *n* ureometer, ureameter *(for determining the amount of urea)*
Urethan *n* urethane
  polymeres U. polyurethane
Urethanelastomer[es] *n* [poly]urethane elastomer
  auf Polyätherbasis aufgebautes U. polyether urethane
  auf Polyesteramid aufgebautes U. polyesteramide urethane
  auf Polyesterbasis aufgebautes U. polyester urethane
Urethankautschuk *m* [poly]urethane rubber, isocyanate rubber
Urethanöl *n* urethane oil
Urethanschaumstoff *m* urethane foam
Uridin *n (org chem)* uridine
Uridindiphosphat *n* uridine diphosphate, UDP
Uridindiphosphatglukose *f* uridinediphosphate-glucose, uridinediphosphoglucose, UDPG
Uridindiphosphoglukose *f s.* Uridindiphosphatglukose
Uridinphosphat *n* uridine phosphate, *(esp)* uridine monophosphate, uridine phosphoric acid, UMP
Uridinphosphorsäure *f* uridine phosphoric acid
Uridin-3'-phosphorsäure *f s.* Uridylsäure
Uridintriphosphat *n*, **Uridintriphosphorsäure** *f* uridine triphosphate, UTP
Uridylsäure *f* uridylic acid, uridine 3'-phosphate, UMP-3'
Urikase *f* uricase
Urikolyse *f* uricolysis *(the conversion of uric acid to urea)*
urikolytisch uricolytic
Urlauge *f* mother liquid (liquor); *(pap)* red (spent sulphite) liquor, sulphite lye (waste liquor)
Urobilin *n* urobilin *(a pigment of urine)*
Urochrom *n* urochrome *(a pigment of urine)*
Urochromogen *n* urochromogen *(the colourless precursor of urochrome)*
U-Rohr *n* U-tube
U-Rohr-Wärmeaustauscher *m*, **U-Rohr-Wärmeübertrager** *m* U-tube heat exchanger
Urokaninsäure *f*, **Urokansäure** *f* urocanic acid, urocaninic acid, 4-imidazoleacrylic acid
Urometer *n* urinometer *(a hydrometer for determining the specific gravity of urine)*
Uronsäure *f* $OHC[CHOH]_nCOOH$ uronic acid *(any of a series of aldehyde acids)*
Ursäure *f* ur-acid *(any of a class of hydrolytic products of ureides)*
Urspannung *f* **[/ elektrische]** electromotive force
Urspannungsnormal *n* standard of emf
Ursprung *m* origin, source ✦ **pflanzlichen Ursprungs** of vegetable (plant) origin, plant-derived ✦ **tierischen Ursprungs** of animal origin ✦ **vulkanischen Ursprungs** of volcanic origin, *(esp of gases)* plume-borne
Ursprungsgestein *n* parent (mother, source) rock
Ursprungssubstanz *f* parent substance
Ursubstanz *f s.* Urtitersubstanz
Urteer *m* low-temperature tar
Urtitersubstanz *f* primary standard *(in titrimetric analysis)*
Urundayextrakt *m(n) (tann)* urunday extract *(from Astronium balansae Engl.)*
Usnetininsäure *f* usnetinic acid *(a lichen acid)*
Usnetinsäure *f* lobaric acid, usnetic acid *(a lichen acid)*
Usnidinsäure *f s.* Usnetinsäure
Usninsäure *f* usnic acid *(a lichen acid)*
Uterustonikum *n (pharm)* uterotonic
uu-Kern *m* odd-odd nucleus
UV- ... *s.* Ultraviolett ...
Uvinsäure *f* uvinic acid, 2,5-dimethylfuran-3-carboxylic acid
Uwarowit *m (min)* uvarovite, uwarowite *(a calcium-chromium garnet)*

V

Vacanceine-Rot *n* Vacanceine red *(obtained by coupling β-naphthol with β-naphthylamine)*
vakant vacant
Vakuum *n* vacuum
  Torricellisches V. Torricellian vacuum
Vakuumanschluß *m* vacuum connection (intake, port)
Vakuumapparat *m* **mit Heizschlangen** *(sugar)* strike pan
Vakuumarbeitsraum *m* vacuum chamber
Vakuumaufdampfung *f* vacuum coating by evaporation, vacuum deposition
Vakuumbegasung *f* vacuum fumigation

Vakuumbrennen *n (ceram)* vacuum firing
Vakuumdestillation *f* vacuum distillation
Vakuumdestillationsanlage *f* vacuum-distillation plant
Vakuumdestillierapparat *m* vacuum still
vakuumdestilliert vacuum-distilled, distilled in vacuo
vakuumdicht vacuum-tight
Vakuumdrehfilter *n* rotary vacuum filter, vacuum drum filter
Vakuumdruckguß *m* vacuum diecasting
Vakuumentgaser *m* vacuum degasifier
Vakuumentgasung *f* vacuum degasification; vacuum deaeration *(as in water treatment)*
Vakuumexsikkator *m* vacuum desiccator
Vakuumfett *n* vacuum grease
Vakuumfilter *n* vacuum (suction) filter
Vakuumfilternutsche *f* vacuum nutsche
Vakuumfiltration *f* vacuum filtration
Vakuumformen *n* **mit [mechanischer] Vorstreckung** *(plast)* plug-assist [vacuum] forming
Vakuumformmaschine *f* vacuum-forming machine
Vakuumformung *f* vacuum forming
Vakuumfülltrichter *m* vacuum hopper
Vakuumgasöl *n* vacuum gas oil
Vakuumgummisackverfahren *n (plast)* vacuum-bag moulding
Vakuumheizplattentrockenschrank *m* vacuum shelf dryer
Vakuumisolierung *f* vacuum insulation
Vakuumkitt *m (plast)* vacuum cement
Vakuumkneter *m* vacuum kneader (blender)
Vakuumkolben *m* vacuum flask
Vakuumkolonne *f* vacuum column (still)
Vakuumkristallisation *f* vacuum crystallization
Vakuumkristallisator *m* vacuum crystallizer
Vakuumkühler *m* vacuum cooler
Vakuumkühlung *f* vacuum cooling
Vakuum-Lecksuchgerät *n* vacuum tester, leak detector
Vakuumleitung *f* vacuum line
Vakuummantel *m* vacuum jacket
Vakuummeßgerät *n* vacuum gauge, vacuometer
Vakuummetallisierung *f* vacuum metallizing (coating by evaporation), vapour (gas) plating
Vakuummeter *n* vacuum gauge, vacuometer
Vakuummischer *m* vacuum mixer
Vakuumnutsche *f* vacuum nutsche
Vakuumpfanne *f* vacuum pan
Vakuumplattentrockner *m* vacuum shelf dryer
Vakuumpolarisation *f* vacuum polarization
Vakuumpumpe *f* vacuum pump
Vakuumpumpenanschluß *m* vacuum-pump connection
Vakuumraum *m* vacuum vessel
Vakuumrotationsfilter *n* rotary vacuum filter
Vakuumrückstand *m (distil)* vacuum residue, short residue (residuum)
Vakuumsack *m (plast)* vacuum bag
Vakuumsaugverfahren *n (plast)* straight vacuum forming
Vakuumschaufeltrockner *m* rotary vacuum dryer
Vakuumschlauch *m* vacuum tubing (hose)
Vakuumschmelzen *n* vacuum melting
Vakuumspektrograf *m* vacuum spectrograph
Vakuumspektroskopie *f* vacuum spectroscopy
Vakuumstoffauflauf *m (pap)* vacuum headbox
Vakuumstrangpresse *f (ceram)* vacuum extrusion press, de-airing auger (machine, pug mill)
Vakuumtaumeltrockner *m* rotating vacuum dryer
Vakuumtechnik *f* vacuum technology
Vakuumtiefziehen *n (plast)* straight vacuum forming
Vakuumtonschneider *m s.* Vakuumstrangpresse
Vakuumtrockenanlage *f* vacuum-drying plant
Vakuumtrockenapparat *m* vacuum dryer
Vakuumtrockenpartie *f (pap)* vacuum dryer
Vakuumtrockenschrank *m* vacuum[-drying] oven
  V. nach Paßburg Passburg dryer
Vakuumtrockner *m* vacuum dryer
Vakuumtrocknung *f* vacuum drying
Vakuumtrommelfilter *n* vacuum drum filter, rotary vacuum filter
Vakuumverdampfapparat *m*, **Vakuumverdampfer** *m* vacuum evaporator
Vakuumverdampfung *f* vacuum evaporation
Vakuumverformbarkeit *f* vacuum formability
Vakuumverformung *f* vacuum forming
Vakuumwalzentrockner *m* vacuum drum dryer
Vakuumwäscher *m* vacuum washer
Vakuumzellenfilter *n (pap)* rotary vacuum drum-type save-all
Vakuumziehen *n (plast)* vacuum forming
  V. mit Vorstreckung plug-assist vacuum forming
Vakzensäure *f* $CH_3[CH_2]_5CH{=}CH[CH_2]_9COOH$ vaccenic acid, 11-octadecenoic acid *(specif trans form)*
Vakzine *f* vaccine
Val *n* val, gram equivalent
Valentinit *m (min)* valentinite, antimony bloom, flowers of antimony *(antimony(III) oxide)*
Valenz *f* 1. valency, *(Am)* valence *(unit of valence)*; 2. *s.* Wertigkeit; 3. *s.* Bindungskraft 1.
  gerichtete V. directed valency
Valenzabsättigung *f* valency saturation
Valenzband *n* valency band
Valenzbindung *f* valency (valence) bond
Valenzbindungsmethode *f* valence-bond method, VB method, electron-pair method, Heitler-London-Slater-Pauling (HLSP) method
Valenzbindungsresonanz *f* valency-bond resonance
Valenzbindungstheorie *f* valence-bond theory, VB theory, electron-pair theory, Heitler-London-Slater-Pauling (HLSP) theory
valenzchemisch valence-chemical
Valenzelektron *n* valency (outermost, optical) electron
Valenzelektronenpaar *n* valency electron pair
Valenzfrequenz *f* stretching frequency

Valenzisomerisierung *f* valence isomerization
Valenzkraft *f* valency force
Valenzlehre *f* theory of valency, valency theory
Valenzorbital *n* valency (bond) orbital
Valenzschale *f* valency shell
Valenzschwingung *f* stretching vibration
Valenzstrich *m* valency dash
Valenzstrichformel *f* valency-dash formula
Valenzstruktur *f* valency-bond structure
Valenzstrukturmethode *f s.* Valenzbindungsmethode
Valenzstrukturtheorie *f s.* Valenzbindungstheorie
Valenzstufe *f* valence state
Valenztautomerie *f* valence tautomerism
Valenztheorie *f* theory of valency, valency theory
Valenzwechsel *m* valency change
Valenzwinkel *m* valency angle
Valenzwinkelabweichung *f* valency[-angle] deviation, valency-bond deviation
Valenzzahl *f* valency number
Valeraldehyd *m* $CH_3[CH_2]_3CHO$ valeraldehyde, pentanal
Valerat *n s.* Valerianat
Valerianat *n* valerate *(salt or ester of a valeric acid)*
Valeriansäure *f* C_4H_9COOH valeric acid, pentanoic acid, *(specif)* $CH_3[CH_2]_3COOH$ *n*-valeric acid, *n*-pentanoic acid
Valeriansäureanhydrid *n* $(CH_3[CH_2]_3CO)_2O$ valeric anhydride
Valeriansäureäthylester *m* $CH_3[CH_2]_3CO-OC_2H_5$ ethyl valerate
Valeriansäureisoamylester *m* $(CH_3)_2CHCH_2CO-OC_5H_{11}$ isoamyl valerate
Valerolaktam *n* valerolactam, α-piperidone
Valin *n* $(CH_3)_2CHCH(NH_2)COOH$ valine, 2-amino-3-methylbutyric acid
Vallez-Filter *n* Vallez filter, horizontal-tank sluicing filter
Valonea *f* valonea, valonia *(acorn cups rich in tannin from several oriental Quercus specc.)*
Valoneasäure *f* valoneaic acid *(a depside)*
Vanadat *n* vanadate
Vanadin *n* V vanadium
Vanadin(III)-bromid *n* VBr_3 vanadium(III) bromide, vanadium tribromide
Vanadin(II)-chlorid *n* VCl_2 vanadium(II) chloride, vanadium dichloride
Vanadin(III)-chlorid *n* VCl_3 vanadium(III) chloride, vanadium trichloride
Vanadindichlorid *n s.* Vanadin(II)-chlorid
Vanadindijodid *n s.* Vanadin(II)-jodid
Vanadindioxid *n s.* Vanadin(IV)-oxid
Vanadin(III)-fluorid *n* VF_3 vanadium(III) fluoride, vanadium trifluoride
Vanadin(V)-fluorid *n* VF_5 vanadium(V) fluoride, vanadium pentafluoride
Vanadinglimmer *m (min)* vanadium mica, roscoelite
vanadinhaltig vanadiferous
Vanadinit *m (min)* vanadinite *(a lead vanadate and chloride)*
Vanadin(II)-jodid *n* VI_2 vanadium(II) iodide, vanadium diiodide
Vanadinmonosulfid *n s.* Vanadin(II)-sulfid
Vanadinmonoxid *n s.* Vanadin(II)-oxid
Vanadin(II)-oxid *n* VO vanadium(II) oxide, vanadium monooxide
Vanadin(III)-oxid *n* V_2O_3 vanadium(III) oxide, vanadium trioxide, divanadium trioxide
Vanadin(IV)-oxid *n* VO_2 vanadium(IV) oxide, vanadium dioxide
Vanadin(V)-oxid *n* V_2O_5 vanadium(V) oxide, vanadium pentaoxide, divanadium pentaoxide
Vanadin(III)-oxidchlorid *n* VOCl vanadium(III) monochloride oxide
Vanadin(IV)-oxidchlorid *n* $VOCl_2$ vanadium(IV) dichloride oxide
Vanadin(V)-oxidchlorid *n* $VOCl_3$ vanadium(V) trichloride oxide
Vanadinoxiddichlorid *n s.* Vanadin(IV)-oxidchlorid
Vanadinoxidmonochlorid *n s.* Vanadin(III)-oxidchlorid
Vanadin(IV)-oxidsulfat *n* $VOSO_4$ vanadium(IV) oxide sulphate
Vanadinoxidtrichlorid *n s.* Vanadin(V)-oxidchlorid
Vanadinpentafluorid *n s.* Vanadin(V)-fluorid
Vanadinpentasulfid *n s.* Vanadin(V)-sulfid
Vanadinpentoxid *n s.* Vanadin(V)-oxid
Vanadinsäure *f* vanadic acid
Vanadin(II)-sulfat *n* VSO_4 vanadium(II) sulphate
Vanadin(II)-sulfid *n* VS vanadium(II) sulphide, vanadium monosulphide
Vanadin(III)-sulfid *n* V_2S_3 vanadium(III) sulphide, vanadium trisulphide, divanadium trisulphide
Vanadin(V)-sulfid *n* V_2S_5 vanadium(V) sulphide, vanadium pentasulphide, divanadium pentasulphide
Vanadintri ... *s.* Vanadin(III)- ...
Vanadium *n s.* Vanadin
Vanadyl ... *s.* Vanadinoxid ...
Van-de-Graaff-Bandgenerator *m*, **Van-de-Graaff-Beschleuniger** *m s.* Van-de-Graaff-Generator
Van-de-Graaff-Generator *m* Van de Graaff generator (accelerator) *(an electrostatic machine)*
Van-der-Waals-Anziehung *f*, **Van-der-Waals-Attraktion** *f* von der Waals attraction
Van-der-Waals-Bindung *f* van der Waals bond
Van-der-Waals-Kräfte *fpl* van der Waals forces [of attraction], van der Waals attractive forces
Van-der-Waals-Kristall *m* van der Waals crystal
Van-der-Waals-Molekül *n* van der Waals molecule
Van-der-Waals-Radius *m* van der Waals radius
Van-Dyck-Braun *n* Vandyke brown *(a natural pigment)*
Van-Dyck-Rot *n* Vandyke red *(copper(II) hexacyanoferrate(II))*
Van-Dyke- ... *s.* Van-Dyck- ...
Vanillal *n* $(C_2H_5O)(OH)C_6H_3CHO$ vanillal, bourbonal, 3-ethoxy-4-hydroxybenzaldehyde

Vanillaldehyd *m s.* Vanillin
Vanilleschote *f* vanilla bean
Vaniliezucker *m* vanillin sugar
Vanillin *n* $(CH_3O)(OH)C_6H_3CHO$ vanillin, 4-hydroxy-3-methoxybenzaldehyde
Vanillinsäure *f* $(CH_3O)(OH)C_6H_3COOH$ vanillic acid, 4-hydroxy-3-methoxybenzoic acid
Vankomyzin *n* vancomycin *(an antibiotic)*
Van-Slyke-Methode *f* Van Slyke method *(for determining free amino groups)*
Van't-Hoff-Gleichung *f* van't Hoff equation
Va-Purge-Verfahren *n (pap)* Va-Purge process
Variabilitätskoeffizient *m* coefficient of variation *(statistics)*
Varianz *f* variance *(statistics)*
Varietät *f (min)* variety
Variole *f (geol)* variole *(a spherule of a variolite)*
Variszit *m (min)* variscite *(aluminium orthophosphate)*
Vasodilatans *n*, **Vasodilat[at]or** *m (pharm)* vasodilator
vasodilatatorisch *(pharm)* vasodilating
Vasokonstriktor *m (pharm)* vasoconstrictor
vasokonstriktorisch *(pharm)* vasoconstrictive
Vateriafett *n* Dhupa fat, piney tallow *(from Vateria indica L.)*
VB-Methode *f s.* Valenzbindungsmethode
VB-Näherung *f* valence-bond approximation (approach), VB approximation
VB-Theorie *f s.* Valenzbindungstheorie
vegetabilisch vegetable
Velinglasur *f (ceram)* vellum (satin, satin-vellum) glaze
Velinpapier *n* wove paper
Vello-Verfahren *n (glass)* Vello process
Velourleder *n* suede leather
Velourpapier *n* velour (flock) paper
Venetianischrot *n* Venetian red *(iron(III) oxide)*
Ventil *n* valve
  V. mit geteiltem Gehäuse split-body valve
  V. mit Hilfssteuerung pilot-controlled valve
  V. mit kugeligem Gehäuse globe valve
  doppelsitziges V. double-seat[ed] valve
  einsitziges V. single-seat[ed] valve
  membranbetätigtes V. diaphragm motor valve
  vorgesteuertes V. pilot-controlled valve
Ventilator *m* blower, fan
  V. mit geraden Schaufeln straight-blade fan
  V. mit rückwärtsgekrümmten Schaufeln backward-curved-blade fan
  V. mit vorwärtsgekrümmten Schaufeln forward-curved-blade fan
Ventilatorkühlturm *m* forced-draught (induced-draught, mechanical-draught) cooling tower
Ventilatoschwefel *m* winnowed (wind-blown) sulphur
Ventilauskleidung *f* valve trim
Ventilboden *m* valve tray
Ventilgehäuse *n* valve body
Ventilhaube *f* valve bonnet
Ventilkörper *m* valve body
Ventilspindel *f* valve stem
Ventiltellerzentrifuge *f* nozzle discharge disk centrifuge
Venturi-Abscheider *m* Venturi scrubber (washer)
Venturi-Düse *f* Venturi tube
Venturi-Kanal *m* Venturi flume *(for flow measurement)*
Venturi-Messer *m* Venturi meter
Venturi-Rohr *n* Venturi tube
Venturi-Skrubber *m s.* Venturi-Wäscher
Venturi-Wäscher *m* Venturi scrubber (washer), Venturi-type water jet scrubber
Venushaar *n (min)* love arrows, flèche d'amour, cupid's darts *(a fibrous variety of rutile)*
verabreichen to apply, *(pharm also)* to administer
Verabreichung *f* application, *(pharm also)* administration
veraluminieren to aluminize, to alumetize
verarbeitbar processable, processible, workable; *(pap)* runnable
Verarbeitbarkeit *f* processability, processibility, workability; *(pap)* runnability
verarbeiten to process, to work
  auf Papier v. to make into paper
  bis zur Fellbildung v. *(rubber)* to sheet [out]
  zu Krümeln v. *(rubber)* to pellet[ize]
  zu Pellets v. to pellet[ize]
Verarbeitung *f* processing, working; *(petrol)* refining
Verarbeitungsbereich *m* working range
Verarbeitungseigenschaften *fpl* processing characteristics (properties)
verarbeitungsfähig *s.* verarbeitbar
Verarbeitungshilfsmittel *n* processing agent (aid)
Verarbeitungsindustrie *f* processing industry
Verarbeitungsprozeß *m* manufacturing process
verarbeitungssicher safe for [factory] processing
Verarbeitungssicherheit *f* processing safety
Verarbeitungstechnik *f* processing technology *(branch of science or its application)*
Verarbeitungstechnologie *f* processing technology *(sequence of processing steps)*
Verarbeitungsverfahren *n* manufacturing process
verarmen to become poor; *(soil)* to impoverish
Verarmung *f* loss, depletion; *(soil)* impoverishing
veraschen to ash, to reduce to ashes, to incinerate
Verascher *m (pap)* incinerator
verascht / feucht wet-ashed
  trocken v. dry-ashed
Veraschung *f* ashing, incineration
Veraschungsschälchen *n* incineration dish
verästelt *(cryst)* dendritic[al]
veräthern to etherify
Verätherung *f* etherification
Veratraldehyd *m s.* Veratrumaldehyd
Veratrol *n* $C_6H_4(OCH_3)_2$ veratrol, 1,2-dimethoxybenzene
Veratrumaldehyd *m* $(CH_3O)_2C_6H_3CHO$ veratric aldehyde, 3,4-dimethoxybenzaldehyde
Veratrumsäure *f* $(CH_3O)_2C_6H_3COOH$ veratric acid, dimethoxybenzoic acid, *(specif)* 3,4-dimethoxybenzoic acid

verätzen to burn, *(agric also)* to scorch; *(med)* to cauterize, to burn *(e.g. a wound)*
Verätzung *f* 1. *(act)* burning, *(agric also)* scorching; *(med)* cauterization, cautery *(as of wounds)*; 2. *(result)* burn, *(agric also)* scorch
verbacken to bake *(amines for sulphonation)*; to set up *(as of hygroscopic material)*
Verbacken *n* baking *(of amines for sulphonation)*; set-up *(as of hygroscopic material)*
Verbandwatte *f* sanitary cotton
Verbenaöl *n* / **echtes** *(cosmet)* verbena oil *(from Lippia triphylla (L'Hérit.) O. Kuntze)*
Indisches V. East Indian verbena oil, lemon-grass oil *(from Cymbopogon specc.)*
verbessern to improve
die Oberflächenbeschaffenheit (Oberflächengüte) v. to finish
Verbesserung *f* improvement
V. der Oberflächenbeschaffenheit (Oberflächengüte) finish[ing]
Verbesserungsmittel *n (food)* improver
verbinden *(chemically)* to combine *(elements, compounds)*; to bond [together], to link [together] *(atoms)*; *(mechanically)* to connect, to join, to link [together], *(esp by means of an adhesive)* to bond
durch Rohrleitung v. to pipe
miteinander v. *s.* verbinden
sich [miteinander] v. *(of elements, compounds)* to combine; *(of atoms)* to bond, to link [together]
Verbinder *m* fitting *(for pipes and hoses)*
Verbindung *f* 1. *(act or process)* combination *(of elements or compounds)*; bonding, linking, linkage *(of atoms)*; *(mechanically)* connection, joining, linkage; 2. *(substance)* compound; *(piece)* connection, joint, link, union, *(esp for pipes and hoses)* fitting; 3. *(state)* connection, *(esp by an adhesive)* bond ✦ **eine chemische V. eingehen** to enter into chemical combination
V. mit großem Ring large-ring compound, macrocyclic (macroring) compound
V. mit kleinem Ring small-ring compound
V. mit mittelgroßem (mittlerem) Ring medium-ring (medium-size ring) compound
V. mit Überwurfmutter union joint
V. mit verdecktem Maximum *(cryst)* hidden maximum system
V. von nichtkonstanter Zusammensetzung *s.* berthollide V.
additive V. additive (addition) compound
aliphatische V. aliphatic compound
alizyklische V. alicyclic compound
analoge V. analogue
anorganische V. inorganic compound
asymmetrische V. unsymmetrical compound
ausgegossene V. poured joint *(of tubes)*
azyklische V. acyclic compound
berthollide V. berthollide (non-stoichiometric, non-Daltonian, non-daltonide) compound
berylliumorganische V. organoberyllium compound
binäre V. binary compound
bororganische V. organoboron compound
daltonide V. daltonide compound
diastereomere V. diastereo[iso]mer
druckgespannte V. *(tech)* pressure-seal joint
geometrisch isomere V. geometric[al] isomer
halbleitende V. semiconducting compound
Herzsche V. *(dye)* Herz compound
heteropolare V. heteropolar (polar, ionic) compound
heterozyklische V. heterocycle, heterocyclic [compound]
homologe V. homolog[ue]
homöopolare V. homopolar (non-polar, non-ionic) compound
homozyklische V. *s.* isozyklische V.
hydroaromatische V. hydroaromatic compound
inaktive V. inactive compound
innerkomplexe V. inner complex salt
intermediäre V. intermediate [compound]
intermetallische V. intermetallic (metal) compound, intermediate constituent (phase)
interstitielle V. interstitial compound
isomere V. isomeric compound
isostere V. isosteric compound, isostere
isozyklische V. isocyclic (homocyclic, carbocyclic) compound
kadmiumorganische V. organocadmium compound
karbozyklische V. *s.* isozyklische V.
kettenförmige V. chain compound
komplexe V. complex compound
lamellare V. lamellar (intercalation, intercalate) compound
lithiumorganische V. organolithium compound
makrozyklische V. macrocyclic (macroring, large-ring) compound
markierte V. labelled (tagged) compound
mehrkernige V. *(org chem)* multinuclear compound
mesoionische V. mesoionic compound
mesomere V. mesomeric (resonance) compound
metallische intermediäre V. *s.* intermetallische V.
metallorganische V. organometallic (metallo-organic) compound
nichtdaltonide (nichtdaltonische) V. *s.* berthollide V.
nichtradioaktive V. non-radioactive compound, inactive compound
oberflächenaktive V. surface-active compound, surfactant
offenkettige V. open-chain compound
optisch isomere V. optical isomer, antimer
organische V. organic compound
organometallische V. *s.* metallorganische V.
paramagnetische V. paramagnetic compound
phosphororganische V. organophosphorus compound
polare V. *s.* heteropolare V.

polyzyklische V. polycyclic compound, aggregate
quecksilberorganische V. organomercury compound
razemische V. racemic compound
ringförmige V. cyclic (ring) compound
selbstdichtende V. *(tech)* pressure-seal joint
siliziumorganische V. organosilicon compound
spirozyklische V. spiro compound, spiran[e]
stabile V. stable compound
stöchiometrisch zusammengesetzte V. *s.* daltonide V.
ternäre V. ternary compound
trimere V. trimer
unpolare V. non-polar compound
unsymmetrische V. unsymmetrical compound
zinnorganische V. organotin compound
zyklische V. cyclic (ring) compound
zykloaliphatische V. alicyclic compound
m-**Verbindung** *f* meta compound
o-**Verbindung** *f* ortho compound
p-**Verbindung** *f* para compound
Verbindungsbildung *f* compound formation
Verbindungsfähigkeit *f* combining ability (capacity)
Verbindungsgewicht *n* combining weight
Verbindungshalbleiter *m* compound semiconductor
Verbindungsklasse *f* family
Verbindungskraft *f* combining force (power)
Verbindungsneigung *f* tendency to combine
Verbindungsrohr *n* connecting tube
Verbindungsstück *n* connection, connector; *(for pipes and hoses)* fitting
Y-förmiges V. Y-shape connecting tube
Verbindungswärme *f* heat of combination
verblasen *(glass, met)* to blow
im Konverter v. *(met)* to convert, to bessemerize
Verblaserösten *n (met)* blast roasting
verblassen to fade, to discolour
Verblassen *n* fading, discoloration
verblassend / nicht *(dye)* fadeless
verbleichen *s.* verblassen
verbleien 1. to lead *(motor fuel)*; 2. to lead-coat *(e.g. metal piping)*
verbleit leaded *(motor fuel)*
schwach v. low-leaded
stark v. high-leaded
Verbrauch *m* consumption
verbrauchen to consume; to spend, to exhaust *(e.g. dye liquor)*
Verbrauchsstoff *m* consumption material
Verbrauchszucker *m* consumption (white) sugar
verbraucht spent, exhausted *(e.g. solution)*; *(tech)* worn out
verbrennbar combustible
Verbrennbarkeit *f* combustibility
verbrennen to burn; *(agric)* to burn, to scorch *(e.g. leaves by excess fertilization)*
verbrennlich combustible
Verbrennliches *n* combustible
Verbrennlichkeit *f* combustibility
Verbrennung *f* combustion; *(agric)* burn, scorch *(as of leaves by excess fertilization)*
V. unter vermindertem Sauerstoffzutritt restricted combustion
direkte V. direct combustion
stille V. slow combustion
unmittelbare V. direct combustion
unvollkommene (unvollständige) V. partial combustion
Verbrennungsabgas *n* flue (furnace, burner) gas
Verbrennungsanalyse *f* combustion analysis
Verbrennungsbombe *f* calorimeter (calorimetric, explosion) bomb, bomb calorimeter
Verbrennungsenthalpie *f* heat of combustion at constant pressure
Verbrennungsgeschwindigkeit *f* rate of combustion, burning rate
Verbrennungskammer *f* combustion (furnace) chamber
Verbrennungslöffel *m* combustion (deflagration) spoon
Verbrennungsluft *f* combustion air
Verbrennungsmarkierung *f (plast)* burned spot *(an injection defect)*
Verbrennungsofen *m* combustion furnace, burner
rotierender V. rotary burner
Verbrennungsprodukt *n* product of combustion
Verbrennungsprozeß *m* combustion process
Verbrennungsraum *m* combustion space
Verbrennungsrohr *n*, **Verbrennungsröhre** *f* combustion tube
Verbrennungsrückstand *m* residue of combustion
Verbrennungsschaden *m (agric)* burn, scorch *(as by excess fertilization)*
V. durch Schwefel *(agric)* sulphur scald
Verbrennungsschale *f* combustion barge
Verbrennungsschiffchen *n* combustion boat
V. aus Ton clay combustion boat
Verbrennungstemperatur *f* combustion temperature
Verbrennungsvorgang *m* combustion process
Verbrennungswärme *f* heat of combustion, *(specif)* gross heat of combustion, gross calorific value, higher heating value
molare V. heat of combustion per mole
Verbrennungszone *f* combustion (oxidation) zone
verbrühen *(tann)* to scald *(pelts in the bate)*
Verbundbauweise *f (plast)* sandwich construction
verbunden / über Wasserstoff hydrogen-bonded
Verbundfolie *f* laminated (composite) film
Verbundglas *n* laminated glass
Verbundkoksofen *m* combination oven
V. mit Unterbrennern combination underjet coke oven
Verbundmühle *f* compound (compartment, multicompartment) mill

Verbundname *m (nomencl)* conjunctive name
Verbundofen *m* combination oven
V. nach Still Still oven
Verbundplatte *f* sandwich panel
Verbundplattenbauweise *f* sandwich construction
Verbundpreßstoff *m* combined plastic
Verbundrohrmühle *f s.* Verbundmühle
Verbundsicherheitsglas *n* laminated safety [sheet] glass
Verbundstoff *m s.* Verbundwerkstoff
Verbundtrommeltrockner *m* direct-indirect rotary dryer
Verbund-Unterbrennerofen *m* combination underjet coke oven
Verbundwerkstoff *m* composite
V. mit Wabenkern honeycomb sandwich material
Verbundwirkung *f* joint action
verbuttern to churn [to butter]
verchromen *(el chem)* to chrome-plate; *(esp by thermal diffusion)* to chromize
Verchromung *f (el chem)* chromium plating; *(esp by thermal diffusion)* chromizing
verd. *s.* verdünnt
Verdampfanlage *f (sugar)* evaporating (evaporator) station
Verdampfapparat *m* evaporator (*for compounds s.* Verdampfer)
verdampfbar [e]vaporable, vaporizable
leicht v. volatile, volatilizable
Verdampfbarkeit *f* [e]vaporability
verdampfen 1. *(of human agent)* to vaporize, to boil away (down), to evaporate, to volat[il]ize, *(suddenly as by expansion)* to flash; 2. *(of a liquid)* to evaporate, to vaporize, to volat[il]ize, *(suddenly as by expansion)* to flash
Verdampfer *m* evaporator, vaporizer; *(in cryogenic processes)* expander
V. für Extraktlösung extract solvent evaporator
V. mit Brüdenverdichtung thermocompression evaporator
V. mit eingehängtem Rohrkorb basket evaporator
V. mit Heizschlange coiled-tube evaporator
V. mit natürlichem Flüssigkeitsumlauf (Umlauf) natural-circulation evaporator
V. mit Plattenheizkörpern flat-plate evaporator
V. mit Thermokompression thermocompression (vapour-compression) evaporator
mehrstufiger V. multistage evaporator
Verdampferapparat *m s.* Verdampfer
Verdampferdruck *m* evaporator pressure
Verdampfereinheit *f* evaporator unit (*s.a.* Verdampferkörper)
Verdampferfläche *f* evaporator area
Verdampferkörper *m* body, *(in a series of evaporators preferably)* effect
Verdampferreaktor *m* boiling [water] reactor
Verdampferschlange *f* evaporator (evaporation) coil; *(in cryogenic processes)* expansion coil
Verdampfkristallisator *m* evaporative (evaporator) crystallizer
Verdampfstation *f (sugar)* evaporating (evaporator) station
Verdampfung *f (esp process)* evaporation, *(esp act)* vaporization, *(esp if readily)* volatilization; *(rapidly as by expansion)* flash
geschlossene V. *(distil)* equilibrium [flash] vaporization
Verdampfungsentropie *f* entropy of evaporation (vaporization)
verdampfungsfähig *s.* verdampfbar
Verdampfungsfläche *f* evaporative (evaporating) surface (area)
Verdampfungsgeschwindigkeit *f* rate of evaporation (vaporization)
Verdampfungskammer *f* flash chamber *(of a flash evaporator)*
Verdampfungskristallisation *f* evaporative crystallization
Verdampfungskristallisator *m* evaporative crystallizer, crystallizing (salting-out) evaporator
Verdampfungskühlung *f* evaporative (evaporation) cooling
Verdampfungsleistung *f* evaporating efficiency
Verdampfungsofen *m (distil)* still pot (body), reboiler
Verdampfungspfanne *f* evaporating (boiling-down) pan
Verdampfungstemperatur *f* evaporation (evaporating) temperature
Verdampfungstrocknung *f* drying by evaporation, evaporation drying
V. im Vakuum puff drying
Verdampfungsverlust *m* evaporation (evaporative, volatile) loss
Verdampfungswärme *f* heat of evaporation (vaporization)
verdauen to digest
verdaulich digestible
Verdaulichkeit *f* digestibility
Verdauung *f* digestion
Verdauungsferment *n* digestive ferment
Verdauungskoeffizient *m*, **Verdauungsquotient** *m* digestibility (digestion) coefficient
Verdauungssaft *m* digestive juice
Verdauungswert *m s.* Verdauungskoeffizient
verdeckt opposed, eclipsed *(stereochemistry)*
Verderb *m* decay, deterioration, spoilage
verderben to decay, to deteriorate, to spoil, to perish
verderblich perishable
Verderblichkeit *f* perishability
Verdet-Konstante *f* Verdet constant *(of the magnetic rotatory power)*
verdichtbar compressible; condensable *(vapour)*
Verdichtbarkeit *f* compressibility; condensability *(of vapour)*
verdichten to densify, to compress, *(esp relating to bulk material)* to compact; to condense *(vapours)*; to thicken, to concentrate, to inspissate *(a solution)*
sich v. to condense *(of vapours)*
Verdichten *n* **von Hand** hand compaction *(as of concrete)*

Verdichter *m* compressor
einstufiger V. single-stage compressor
mehrstufiger V. multistage compressor
ölfreier V. non-lubricated compressor
rotierender V. rotary compressor
schmierloser V. *s.* ölfreier V.
Verdichtung *f* densification, compression, *(esp relating to bulk material)* compacting; condensation *(of vapours)*; concentration *(of a solution)*
beidseitige (doppelseitige) V. double-action compacting *(powder metallurgy)*
einseitige V. single-action compacting *(powder metallurgy)*
zweiseitige V. *s.* beidseitige V.
Verdichtungsarbeit *f* work of compression *(as of a pump)*
Verdichtungsdruck *m* compression pressure; *(relating to bulk material)* compacting pressure
Verdichtungsfähigkeit *f s.* Verdichtbarkeit
Verdichtungsgrad *m (plast)* bulk factor, *(Am)* compression ratio
Verdichtungsverhältnis *n* compression ratio *(in an internal-combustion engine)*
Verdichtungszone *f* compression zone
verdicken to thicken, to concentrate, to inspissate *(a liquid)*; *(coat)* to body; *(pap)* to decker
Verdicker *m s.* Verdickungsmittel
Verdickungsmasse *f* thickening paste
Verdickungsmittel *n* thickening agent, thickener
Verdickungspaste *f* thickening paste
verdrängbar *(chem)* displaceable
verdrängen *(chemically)* to displace; *(physically)* to expel, to push out, to eject *(e.g. gases)*
Verdränger *m* displacer *(chromatography)*
Verdrängerpumpe *f* positive-displacement pump
Verdrängung *f (chemically)* displacement; *(physically)* expulsion, ejection *(as of gases)*
Verdrängungsanalyse *f* displacement analysis
Verdrängungschromatografie *f* displacement chromatography
Verdrängungsentwicklung *f* displacement development *(chromatography)*
Verdrängungskörper *m* 1. displacer *(of a liquid-level gauge)*; 2. *(plast)* torpedo
Verdrängungsmethode *f* displacement method
Verdrängungsmittel *n* displacer *(chromatography)*
Verdrängungsname *m* replacement name
Verdrängungsreaktion *f* displacement (substitution) reaction
Verdrängungstechnik *f* displacement technique *(gas chromatography)*
Verdrängungsvolumenzähler *m* positive-displacement flowmeter, displacement meter
Verdrängungswaschung *f* displacement wash[ing]
Verdrängungszähler *m s.* Verdrängungsvolumenzähler
Verdrehungsschwingung *f* torsional (twisting) vibration *(spectroscopy)*
Verdrehungssteifigkeit *f* stiffness in torsion
Verdunklung *f (phot)* blackout
verdünnbar dilutable *(liquid)*
mit Wasser v. water-dilutable
Verdünnbarkeit *f* dilutability
verdünnen to dilute, to thin, to attenuate, *(esp relating to gases)* to rarefy; to potentiate *(homoeopathy)*
Verdünnen *n* dilution, thinning, attenuation, *(esp relating to gases)* rarefaction; potentiation *(homoeopathy)*
Verdünner *m s.* Verdünnungsmittel
verdünnt dilute
Verdünnung *f* 1. dilution *(diluted liquid)*; potency *(homoeopathy)*; 2. *s.* Verdünnungsmittel; 3. *s.* Verdünnen
V. in der Stoffgrube *(pap)* blow-pit (blow-tank) dilution
unendliche V. infinite dilution
Verdünnungsenthalpie *f* enthalpy of dilution
Verdünnungsfaktor *m* dilution factor
Verdünnungsgesetz *n* / **Ostwaldsches** Ostwald dilution law
Verdünnungsgrad *m* degree of dilution
Verdünnungsmittel *n* diluting agent, diluent; *(coat)* thinner, thinning agent
Verdünnungsverhältnis *n* dilution ratio
Verdünnungswärme *f* heat of dilution
differentiale (differentielle) V. differential (partial) heat of dilution
integrale V. integral heat of dilution
verdunstbar [e]vaporable, vaporizable
Verdunstbarkeit *f* [e]vaporability
verdunsten to evaporate, to vaporize, to volat[il]ize *(below normal boiling point)*
Verdunstung *f* evaporation, vaporization, volatilization *(below normal boiling point)*
Verdunstungsfläche *f* evaporative (evaporating) surface (area)
Verdunstungsgeschwindigkeit *f* rate of evaporation (vaporization)
Verdunstungshaube *f* hood
Verdunstungskühlung *f* evaporative (evaporation) cooling
Verdunstungsmesser *m*, **Verdunstungsmeßgerät** *n* atmometer, evaporimeter
Verdunstungstemperatur *f* evaporation (evaporating) temperature
Verdunstungstrocknung *f* drying by evaporation
Verdunstungsverlust *m* evaporation (evaporative) loss
Verdunstungswärme *f* heat of evaporation (vaporization)
Verdunstungszahl *f* evaporation number
verdüsen to atomize, to spray
Verdüsung *f* [nozzle] atomization, spraying
veredeln to refine, to improve; to finish *(a surface)*; *(text)* to finish, to process; *(pap)* to refine *(pulp)*
Vered[e]lung *f* refining [treatment], refinement, improvement; finish[ing] *(of a surface)*; *(text)* finish[ing], processing; *(pap)* [alkali] refining *(of pulp)*
vereinigen to combine *(elements)*; to mix *(filtrates)*;

to assemble *(parts of an apparatus)*
sich v. to combine *(of elements)*; to coalesce *(as of gas bubbles)*
sich paarweise v. to pair *(as of electrons)*
Vereinigung *f* combination *(of elements)*; mixing *(of filtrates)*; assembly *(of parts of an apparatus)*; coalescence *(as of gas bubbles)*
Vereinigungsbestreben *n* tendency to combine
vereisen to ice
Vereisung *f* icing
Vereisungsverhinderer *m* anti-icing additive (agent)
Verengung *f* throat, restriction *(as of a nozzle)*
V. am Venturi-Rohr Venturi throat
Vererzung *f (geoch)* metallization
verestern to esterify
Veresterung *f* esterification
extraktive V. extractive esterification
Veresterungsgrad *m* degree of esterification
Veresterungskatalysator *m* esterification catalyst
verfahren to proceed
Verfahren *n* process, method; *(manner of performing technical details, esp lab)* method, technique, procedure
V. der Dampfdichtebestimmung vapour-density method *(for determining molecular weight)*
V. mit bewegtem Katalysatorbett moving-bed process *(as for catalytic cracking)*
V. mit umlaufender Beschickung moving-burden process *(for the complete gasification of coal)*
V. von Pott-Broche Pott-Broche process *(extraction of coal)*
abgekürztes V. short-cut method
aluminothermisches V. aluminothermic process
basisches V. basic process *(of steelmaking)*
Béchampsches V. Béchamp method *(of reducing aromatic nitro compounds to amines)*
bodenblasendes V. *(met)* bottom-blown-converter process
direktes V. direct process *(as for producing ammonia or chlorosilanes)*
diskontinuierliches V. batch process
ebullioskopisches V. ebullioscopic method *(for determining molecular weight)*
großtechnisches V. large-scale process
halbdirektes V. semidirect process *(as for producing ammonia)*
Heumannsches V. Heumann method *(for synthesizing indigo)*
indirektes V. indirect process *(as for producing ammonia)*
Kassnersches V. Kassner process *(for producing oxygen)*
konduktometrisches V. conductometric (conductance) method *(titration)*
kontinuierliches V. continuous process
Linz-Donawitzer V. Linz-Donawitz process *(of steelmaking)*
maschinelles V. machine process
mechanisch-thermisches V. *(rubber)* thermomechanical process
Mondsches V. Mond process *(1. for producing town gas; 2. for producing nickel)*
Münchner V. *(pap)* Kesting electrolytic process *(for producing chlorine-dioxide bleaching liquor)*
nasses V. wet process
periodisches V. batch process (operation)
plastisches V. wire-cut process *(for producing bricks)*
potentiometrisches V. potentiometric method *(titration)*
pulvermetallurgisches V. powder-metallurgical (powder-metallurgy) process (technique)
röntgenografisches V. X-ray method *(as in crystallography)*
rotationsviskosimetrisches V. rotating-cylinder method *(viscometry)*
saures V. acid process *(of steelmaking)*
seismisches V. seismic method *(of prospecting for petroleum)*
sekundäres V. *(petrol)* secondary recovery method
selbstkonsistentes V. *(phys chem)* self-consistent method
silikothermisches V. *(met)* silicothermic process
standardisiertes V. standard method
technisches V. commercial process
thermomechanisches V. *(rubber)* thermomechanical process
trockenes V. dry process
turbidimetrisches V. turbidimetric method
weichplastisches V. *(ceram)* soft-mud process
zweistufiges V. two-stage (two-step) process
Verfahrenschemiker *m* process chemist
Verfahrensfehler *m* error of method
Verfahrensgröße *f* process variable
Verfahrensindustrie *f* process industry
Verfahrensingenieur *m* process engineer
Verfahrensregelung *f* process control
Verfahrenssteuerung *f* process control
Verfahrenstechnik *f* process engineering (technology)
chemische V. chemical engineering
Verfahrenstechniker *m* process engineer
Verfahrensveränderliche *f* process variable
Verfahrenszug *m* processing train
Verfall *m* decay *(as of organic matter)*
verfallen 1. to decay *(as of organic matter)*; 2. *(tann)* to fall *(of pelts)* ✦ **v. machen** *(tann)* to bring down, to deplete, to fall *(pelts)*
Verfallen *n (tann)* falling, depletion *(of pelts)*
verfälschen to adulterate
Verfälschung *f* adulteration
Verfaltung *f* fold, lap *(a defect in glass)*
verfärben to discolour *(esp if locally)* to stain
sich v. to discolour, *(esp if locally)* to stain; to fade *(to lose intensity of colour)*
Verfärbung *f* 1. *(process)* discoloration, *(esp if locally)* staining; fading *(loss of colour intensity)*; 2. *(state)* discolo[u]ration, stain

verfaulen to putrefy, to decompose
verfeinern to improve *(e.g. an analytical method)*
Verfeinerung *f* improvement *(e.g. of an analytical method)*
verfestigen 1. to compact *(bulk material)*; 2. to work-harden *(by cold forming)*
sich v. 1. to compact *(as of bulk material)*; 2. to harden *(as by cold forming)*; 3. to set *(of a gel)*; 4. *(geol)* to lithify; 5. to consolidate *(of coal)*
Verfestigung *f* 1. compaction *(as of bulk material)*; 2. [work-]hardening *(as by cold forming)*; 3. set[ting] *(of a gel)*; 4. *(geol)* lithification; 5. consolidation *(of coal)*
verfilzen / sich to felt [together], to mat [together] *(as of fibres)*
Verfilzungsvermögen *n* felting power *(of fibres)*
Verflocker *m* flocculation tank, flocculator *(water treatment)*
verflüchtigen to volat[il]ize
sich v. to volat[il]ize
verflüchtigend / leicht zu volatilizable
Verflüchtigung *f* volatilization
verflüssigbar liquefiable
verflüssigen 1. to liquefy *(a solid or gas)*; 2. *(ceram)* to deflocculate *(a glaze slip)*
sich v. to liquefy *(of a solid or gas)*; to run *(of metal)*
Verflüssiger *m* 1. liquefier *(in gas liquefaction)*; 2. condenser *(of a refrigerating machine)*
Verflüssigerdruck *m* condenser pressure *(refrigeration)*
Verflüssigung *f* 1. liquefaction *(of a solid or gas)*; 2. *(ceram)* deflocculation *(of a glaze slip)*
Verflüssigungsdruck *m* condensation (condensing) pressure *(refrigeration)*
Verflüssigungsmittel *n (ceram)* deflocculant, deflocculent, deflocculating agent
verformbar mouldable, plastic, *(ceram also)* workable, *(of metals also)* ductile
Verformbarkeit *f* mouldability, plasticity, *(ceram also)* workability, *(of metals also)* ductility
verformen 1. to deform, *(relating to elastic material also)* to strain; 2. to form, to shape, to mould
Verformung *f* 1. deformation, *(relating to elastic material also)* strain; 2. forming, shaping, moulding
bleibende V. permanent (residual, plastic) deformation, permanent set
bleibende V. nach Druckbelastung (Druckeinwirkung) [permanent] compression set
elastische V. elastic deformation
irreversible (plastische) V. *s.* bleibende V.
Verformungsarbeit *f* resilience, resiliency
Verformungsrest *m s.* Verformung / bleibende
verfügbar available
nicht v. unavailable
Verfügbarkeit *f* availability
vergällen to denature, to denaturate, to denaturize, *(ferm also)* to methylate
Vergällung *f* denaturation, denaturing, *(ferm also)* methylation
Vergällungsmittel *n* denaturant, denaturing agent
vergärbar fermentable
Vergärbarkeit *f* fermentability
vergären to ferment
Vergärung *f* fermentation
vergasen 1. to gasify *(e.g. coal)*; to carburet *(motor fuel)*; to volat[il]ize *(pesticides)*; 2. *(met)* to gas *(a melt)*
Vergaser *m* 1. gasifier *(of a gasification plant)*; 2. carburet[t]or, carburetter *(of an internal-combustion engine)*
Vergaserkraftstoff *m* carburetting (gasifiable motor) fuel, carburant
Vergasung *f* 1. gasification *(e.g. of coal)*; carburetting *(of motor fuel)*; volatilization *(of pesticides)*; 2. *(met)* gassing *(of a melt)*
V. im Flöz underground gasification *(of coal)*
V. in der Flugstaubwolke entrainment gasification *(of coal)*
restlose (rückstandslose) V. *s.* vollständige V.
unterirdische V. underground gasification
vollständige V. complete (total) gasification
Vergasungsanlage *f* gasification plant
Vergasungsapparat *m* gasifier
Vergasungskammer *f* gasification chamber
Vergasungsmedium *n* gasification medium *(for pulverized coal)*
Vergasungsmittel *n* 1. fumigant, fumigator *(crop protection)*; 2. *s.* Vergasungsmedium
Vergasungsraum *m* gasification chamber
Vergasungszone *f* gasification zone
Vergépapier *n* laid paper
vergesellschaften / sich *(min)* to associate
Vergesellschaftung *f (min)* association
vergießbar castable, pourable
Vergießbarkeit *f* castability, pourability
vergießen to cast, to pour *(a resin or a melt)*
durch Druck v. to [pressure-]diecast
in der Kokille v. to diecast
in metallischen Dauerformen v. to gravity-diecast
vergiften to poison, to toxify
mit Gas v. to gas
Vergiftung *f* poisoning ✦ **durch V. bedingt (entstanden)** toxi[co]genic
absichtliche V. deliberate poisoning *(of a catalyst)*
tödliche V. fatal poisoning
vergilben to [go] yellow, *(pap also)* to discolour, to age
Vergilbung *f* yellowing, *(pap also)* discolo[u]ration, ag[e]ing
vergilbungsbeständig non-yellowing
vergipsen to plaster
verglasen *(ceram, geoch)* to vitrify
Verglasung *f (ceram, geoch)* vitrification
Verglasungsbereich *m (ceram)* vitrification range
Vergleichselektrode *f* reference electrode
Vergleichslösung *f* standard (reference) solution
Vergleichsprobe *f* standard sample
Vergleichsspannung *f* reference voltage
Vergleichsspektrum *n* reference (comparison) spectrum

Vergleichsstandard *m* reference standard
Vergleichsstrahl *m* reference beam *(spectroscopy)*
Vergleichssubstanz *f* reference substance
Vergleichsverbindung *f* reference (comparison) compound
Verglühbrand *m (ceram)* biscuit firing, biscuitting; hardening-on *(of the decoration before glazing)*
verglühen *(ceram)* to biscuit-fire; to harden on *(the decoration before glazing)*
vergolden to gild; *(electrically)* to gold-plate
Vergoldung *f* gilding; *(electrically)* gold plating
vergrauen *(text)* to grey
Vergrößerungspapier *n (phot)* enlarging paper
Vergußharz *n* cast[ing] resin
vergüten *(met)* to heat-treat, to harden and temper; *(glass)* to coat
Vergütung *f (met)* heat-treatment, hardening and tempering; *(glass)* antireflection (antiflare) coating
Verhakung *f* entanglement *(of polymer chains)*
verhalten / sich to behave *(of a substance)*
Verhalten *n* properties, characteristics, behaviour *(of a substance)*
V. bei hohen Temperaturen high-temperature properties
V. bei Kälte low-temperature properties
dynamisches V. dynamic properties
fettabweisendes V. grease repellency
ölabweisendes V. oil repellency
rheologisches V. rheological properties
verkokungstechnisches V. coking properties
wasserabweisendes V. water repellency
Verhaltensresistenz *f (biol)* behaviouristic resistance *(as against pesticides)*
Verhältnis *n* relation; *(quantitatively)* ratio, proportion ✦ **in jedem V. mischbar** miscible in all proportions ✦ **in molekularem V.** in molecular proportions
V. Deckgebirge zu Kohle *(mine)* ratio of overburden to coal
V. von Harz zu Öl *(coat)* resin/oil ratio, resin-to-oil ratio
V. von Katalysator (Kontakt) zu Öl catalyst/oil ratio, catalyst-to-oil ratio
V. von Öl zu Harz *(coat)* oil length, oil/resin ratio, oil-to-resin ratio
gyromagnetisches V. [des Atomkerns] [nuclear] gyromagnetic ratio
Verhältnisformel *f* stoichiometric formula, simplest [possible] formula
Verhältnisse *npl* / **aerobe** aerobic conditions
anaerobe V. anaerobic conditions
verhängen *(text)* to expose to [the] air
Verhängen *n (text)* exposure to [the] air
verhärten to harden, to solidify
Verhärtung *f* hardening, solidification
verharzbar resinifiable
verharzen to resinify, *(specif petrol)* to gum
Verharzung *f* resinification, *(specif petrol)* gum formation, gumming
Verharzungsprodukte *npl* / **aktuelle** *(petrol)* existent (preformed) gum
potentielle V. potential gum
Verhieb *m (mine)* working
verhindern to prevent, to hinder, to inhibit
verholzen to lignify
Verholzung *f* lignification
verhornen *(biol)* to keratinize
Verhornung *f (biol)* keratinization
verhüten *s.* verhindern
verhütten to smelt *(ores)*
Verhüttung *f* smelting *(of ores)*
verjagen to drive off (out), to expel, to dispel *(volatile matter)*
verjüngen / sich to taper *(as of a ground-glass joint)*
verjüngend / sich taper[ing]
Verjüngung *f* taper *(as of a ground-glass joint)*
verketten to link *(molecules)*
Verkettung *f* linking, linkage *(of molecules)*
verkieseln *(geoch)* to silicify
Verkieselung *f (geoch)* silicification
verkirnen to churn *(margarine)*
verkitten to putty, to lute, to cement, to seal *(e.g. joints of tubes)*
Verkittung *f* puttying, luting, cementation, sealing
verkleben to cement, to glue, to bond, to agglutinate; to stick together, to agglutinate
Verklebung *f* 1. *(act)* cementation, gluing, bond, bonding, agglutination; 2. *(process)* agglutination
verkleiden to cover, to case, to jacket, *(esp relating to metal)* to clad; to line *(an inner surface)*
Verkleidung *f* cover[ing], casing, jacket, *(esp relating to metal)* cladding; lining *(of an inner surface)*
verklumpen to agglomerate, to clog, to clot, to lump, to agglutinate, to curd[le]
Verklumpung *f* agglomeration, clogging, clotting, lumping, agglutination, curdling
verknüpfen [/ miteinander] to link [together], to bond [together], to couple *(atoms)*
Verknüpfung *f* linking, linkage, bonding, coupling *(of atoms)* (*for compounds s.* Bindung)
Verknüpfungsstelle *f* point of linkage, position of attachment *(of atoms)*
verkochen to boil down, to concentrate; to boil away
auf Korn v. *(sugar)* to boil to grain, *(Am also)* to sugar off *(esp maple sap)*
Verkochen *n* boildown, boiling-down, concentration; boiling-away
verkohlen to char, to carbonize
Verkohlung *f* charring, carbonization
verkoken to coke, to carbonize
Verkokung *f* coking, carbonization, *(specif)* high-temperature coking (carbonization)
kontinuierliche katalytische V. *(petrol)* continuous contact coking
verzögerte V. *(petrol)* delayed coking
Verkokungsanlage *f* carbonizing (coking, coke) plant

Verkokungsbatterie *f* retort (coke-oven) battery, carbonizing bench
Verkokungsblase *f* coking still
Verkokungseigenschaften *fpl* coking properties
Verkokungsfähigkeit *f* coking power
Verkokungsgas *n* carbonization gas
Verkokungsgefäß *n (petrol)* coking furnace
V. für den Ramsbottom-Test Ramsbottom coking furnace
Verkokungskammer *f (petrol, coal)* coking (coke) chamber
Verkokungskohle *f* coking coal
Verkokungskrackverfahren *n* non-residue method *(of cracking)*
Verkokungsofen *m* coke (coking) oven
Verkokungsprobe *f s.* Verkokungstest
Verkokungsrückstand *m* carbon (coke) residue
V. nach Conradson Conradson carbon (coke) residue
V. nach Ramsbottom Ramsbottom carbon (coke) residue
Verkokungstest *m* carbon-residue test, coking test
V. nach Conradson Conradson [carbon-residue] test
V. nach Ramsbottom Ramsbottom [carbon-residue] test
Verkokungsverfahren *n* 1. carbonization (coking) process; 2. non-residue method *(of cracking)*
Verkokungsverhalten *n* coking properties
Verkokungsvermögen *n* coking power
Verkokungsvorgang *m* coking process
Verkokungswärme *f* coking heat
obere V. gross coking heat
untere V. net coking heat
Verkokungswert *m*, **Verkokungszahl** *f* carbon-residue value, coke value (number)
verkorken 1. to cork; 2. *(bot)* to suberize
Verkorkung *f* 1. corking; 2. *(bot)* suberization, suberification
verkrusten to encrust, to incrust, to scale
Verkrustung *f* encrustation, incrustation, scaling
verküpbar *(dye)* vattable
verküpen *(dye)* to vat
verkupfern to copper; *(electrically)* to copper-plate
Verkupferung *f* coppering; *(electrically)* copper plating
Verkupferungsbad *n* copper[-plating] bath
Verküpung *f (dye)* vatting
Verküpungsverfahren *n (dye)* vat process
verlaben to rennet *(milk)*
Verlabung *f* renneting *(of milk)*
verlagern to displace; to shift *(e.g. an equilibrium)*; to translocate *(e.g. nutrients in plants)*
sich v. to shift *(as of an equilibrium)*
Verlagerung *f* displacement; shift *(as of an equilibrium)*; translocation *(as of nutrients in plants)*
verlangsamen to slow down, to decelerate, to retard; *(nucl)* to moderate
sich v. to slow down, to decelerate
Verlangsamung *f* slowing-down, deceleration, retardation; *(nucl)* moderation
Verlauf *m* 1. course *(as of a reaction)*; 2. *(coat)* flow
schlechter V. *(coat)* bad flow
verlaufen 1. to proceed *(as of reactions)*; 2. *(coat)* to flow
Verlaufmittel *n (coat)* flow-control agent
verleimen to cement, to glue
verlöschen to go out, *(gradually)* to die out
Verlöschen *n* going-out, *(gradually)* dying-out
Verlust *m* loss, wastage
V. durch Mitreißen entrainment loss
V. durch Schäumen foaming loss
V. durch Unverbranntes unburned combustible (fuel) loss
dielektrischer V. dielectric loss
Verlustbeiwert *m* **der Düse** nozzle coefficient
Verlustfaktor *m* **[/ dielektrischer]** loss tangent, dissipation factor
Verlustwinkel *m* loss angle
Verlustziffer *f* **[/ dielektrische]** [dielectric] loss factor
vermahlbar grindable
Vermahlbarkeit *f* grindability
vermahlen to grind, to mill, *(relating to soft material also)* to disintegrate, *(relating to hard material also)* to pulverize
Vermahlen *n* grinding, milling, *(relating to soft material also)* disintegration, *(relating to hard material also)* pulverization
vermälzen to malt
vermehren *(nucl)* 1. to breed *(fissionable material)*; 2. to multiply *(neutrons)*
sich v. *(nucl)* to multiply *(of neutrons)*
Vermehrung *f (nucl)* 1. breeding *(of fissionable material)*; 2. multiplication *(of neutrons)*
Vermehrungsfaktor *m (nucl)* multiplication factor (constant)
vermengen to mix, to blend, to mingle
Vermikulit *m (min)* vermiculite *(a group of phyllosilicates)*
Vermillon[-Zinnober] *m* vermil[l]ion, Victoria red *(precipitated red mercury(II) sulphide)*
Verminderung *f* **der Überschußladungsträger** excess carrier resorption
vermischen to mix, to mingle, *(specif relating to solids)* to blend
sich v. to mix
Vermischen mix[ing], mixture, mingling, *(specif relating to solids)* blending
V. im Fertigtank *(petrol)* batch blending
V. in der Pumpleitung *(petrol)* in-line blending
V. von Feststoffkomponenten dry mixing (mix, blending)
Vermizid *n* vermicide
Vermoderungshorizont *m (soil)* fermentation layer, F-layer
vernachlässigbar negligible
vernachlässigen to neglect
vernachlässigen[d] / zu negligible
vernebeln to nebulize, to aerosolize
Verneb[e]lung *f* nebulization, aerosolization

vernetzbar cross-linkable, curable, *(rubber also)* vulcanizable
mit Schwefel v. sulphur-vulcanizable, sulphur-curable
nicht v. non-curing, *(rubber also)* non-vulcanizable
vernetzen to cross-link, to cure, *(rubber also)* to vulcanize
Vernetzer *m s.* Vernetzungsmittel
Vernetzerkombination *f s.* Vernetzungssystem
Vernetzung *f* cross-linkage, cure, *(rubber also)* vulcanization
V. durch Bestrahlung *(rubber)* radiation vulcanization
V. durch energiereiche Strahlen (Strahlung) *(rubber)* vulcanization by high-energy radiation
V. mit Metalloxiden *(rubber)* metallic-oxide vulcanization
V. mit Schwefel *(rubber)* [elemental] sulphur vulcanization
V. über C-C-Verknüpfung carbon-carbon cross-linking
peroxidische V. peroxide cross-linking
schwefelfreie V. *(rubber)* sulphurless (non-sulphur) vulcanization
Vernetzungsdichte *f* cross-linking density
vernetzungsfähig *s.* vernetzbar
Vernetzungsgrad *m* degree of cross-linking
Vernetzungsmittel *n* cross-linking agent, curing agent, *(rubber also)* vulcanizing agent
Vernetzungsreaktion *f* cross-linking reaction
Vernetzungsstelle *f* cross-link[age]
Vernetzungssystem *n* cross-linking system, curing (cure, curative) system, *(rubber also)* vulcanizing (vulcanization) system
vernichten *(phys chem)* to annihilate
Vernichtung *f (phys chem)* annihilation
strahlungslose V. radiationless annihilation
Vernichtungsrate *f (phys chem)* annihilation rate
Vernichtungsspektrum *f (phys chem)* annihilation spectrum
Vernichtungsstrahlung *f (phys chem)* annihilation radiation
vernickeln to nickel[ize]; *(electrically)* to plate with nickel, to nickel-plate
Vernickelung *f* nickelization; *(electrically)* nickel plating
Vernickelungsbad *n* nickel-plating bath
verolen to olate *(of hydroxo compounds)*
Verolung *f* olation *(of hydroxo compounds)*
verpacken to pack[age]; *(met)* to pack *(with a carburizing powder)*
in Dosen v. to tin
Verpacken *n* 1. pack[ag]ing, package; 2. *(met)* packing *(with a carburizing powder)*
Verpackung *f* 1. packing, package, pack[ag]ing material; 2. *s.* Verpacken 1.
Verpackungsfolie *f* packaging film; *(of metal)* packaging foil
Verpackungsmaterial *n* pack[ag]ing material, package, packing, *(relating to paper, plastic film or cloth also)* wrapping
Verpackungspapier *n* packing (wrapping, package) paper
verperlen to shot
verpreßbar *(plast)* mouldable
Verpreßbarkeit *f (plast)* mouldability
verpressen *(plast)* to mould
heiß v. to hot-press, *(relating to powders also)* to sinter under pressure
zu Briketts v. to briquette
Verpressen *n* **von Schnitzelpreßmasse** *(plast)* macerate moulding
verpuffen to deflagrate
Verpuffung *f* deflagration
Verpuffungstemperatur *f* deflagration temperature
verputzen to trim, to fettle, *(plast also)* to deburr, to deflash
verreiben to grind, to powder, to pulverize, *(esp pharm)* to triturate
gemeinsam v. to powder together
Verreiben *n* grinding, powdering, pulverization, *(esp pharm)* trituration
Verreibung *f (pharm)* trituration
verrohren *(petrol)* to case
Verrohrung *f* tubing; *(petrol)* casing *(act or material)*
verrosten to rust, to corrode
verrotten to rot
verrottungsbeständig *s.* verrottungsfest
verrottungsfest rot-resistant, rotproof
Verrottungsfestappretur *f* rot-resistant finish
Verrottungsfestigkeit *f* rot-resistance, rotproofness
Verrottungsschutzmittel *n* rotproofing agent
verrühren to stir up, to mix up
versagen to fail, to break down
Versagen *n* failure, breakdown
V. durch Ermüdung fatigue failure
versalzen *(soil)* to salinize
Versalzung *f (soil)* salinization
Versandvorschrift *f* shipping regulation
Versatz *m* 1. *(ceram)* batch [composition]; 2. *(tann)* layers, dusters, layer pits; 3. *(mine)* waste material
Versatzformel *f (ceram)* batch formula
Versatzgut *n*, **Versatzmaterial** *n (mine)* waste material
versauern *(agric)* to acidify *(a soil, as of fertilizers)*; to sour, to become acid *(of soils)*
Versauerung *f (agric)* acidification, souring *(of soils)*
verschäumbar *(plast)* foamable, expandable, expandible
verschäumen *(plast)* to foam, to expand
Verschäumung *f (plast)* foaming, expanding
verschiebbar displaceable *(atom or radical)*; shiftable *(electron or equilibrium)*
verschieben to displace *(atoms or radicals)*; to shift *(a frequency, an electron or an equilibrium)*
sich v. to shift *(of a frequency, an electron or an equilibrium)*
Verschiebestempel *m (plast)* sliding punch

Verschiebung *f* shift
V. nach Rot shift to the red
bathochrome V. bathochromic shift
chemische V. chemical shift
Friessche V. Fries reaction (rearrangement) *(of phenyl esters or phenyl ethers)*
Verschiebungsgesetz *n s.* Verschiebungssatz
Verschiebungspolarisation *f* induced (distortion) polarization
Verschiebungssatz *m (nucl)* [radioactive-]displacement law, group displacement law
Sommerfeld-Kosselscher V. spectroscopic displacement law of Kossel and Sommerfeld
Wienscher V. Wien displacement law
verschiedenartig heterogeneous, inhomogeneous, dissimilar
Verschiedenartigkeit *f* heterogeneity, inhomogeneity, dissimilarity
verschießen *(text)* to fade, to discolour
Verschießen *n (text)* fading, discolo[u]ration
V. in Abgasatmosphäre gas fading
verschießend / nicht fadeless
verschimmeln to mould
verschlacken to slag, to scorify *(impurities)*; to slag *(of impurities)*; *(coal)* to clinker
Verschlackung *f* slagging, scorification *(act, process, or result)*; *(coal)* clinkering
Verschlackungsbeständigkeit *f* resistance to slagging
verschlammen to clog with mud
verschlechtern to deteriorate
sich v. to deteriorate
Verschlechterung *f* deterioration
verschleiern *(phot)* to fog
verschleifen to grind
zu Holzschliff v. *(pap)* to pulp, to make into [a] pulp, to reduce to pulp
Verschleiß *m* wear
reibender V. abrasive wear, abrasion
verschleißbeständig *s.* verschleißfest
verschleißen to wear [out]
Verschleißfaktor *m* wear factor
verschleißfest wear-resistant, resistant to wear, *(tech also)* abrasion-resistant, resistant to abrasion
Verschleißfestigkeit *f* wear resistance, resistance to wear, *(tech also)* abrasive resistance, resistance to abrasion
Verschleißschutzschicht *f* antiabrasion layer
Verschleißwiderstand *m s.* Verschleißfestigkeit
verschleppen to carry over (through)
Verschleppen *n* carryover
verschließen to close; to stopper, to plug *(e.g. a flask)*; *(if gas-tight or water-tight)* to seal
verschlucken to absorb *(rays)*
Verschluß *m* 1. closure; *(gas-tight or water-tight)* seal; cover, cap; 2. closure *(act)*
hydraulischer V. water seal
Verschlußdüse *f (plast)* shut-off nozzle
Verschlußglocke *f* bell *(of a blast furnace)*
verschmelzen to fuse, to melt; to smelt *(ore)*; to seal *(as in working glass)*; to coalesce *(as of gas bubbles)*; *(nucl)* to fuse, to merge
auf Stein v. *(met)* to matte-smelt
Verschmelzen *n* fusion, melting; smelting *(of ore)*; sealing *(as in working glass)*
V. auf Stein *(met)* matte smelting
autogenes V. *(met)* autogenous smelting
direktes (unmittelbares) V. *(met)* direct smelting
Verschmelzung *f* coalescence *(as of gas bubbles)*
Verschmelzungsname *m (nomencl)* fusion name
verschmieren to lute *(e.g. a pipe joint)*; *(unintentionally)* to choke [up] *(e.g. screen apertures)*
verschmutzen to soil, *(esp relating to the environment)* to pollute, *(esp with poisonous matter)* to contaminate; to be soiled
Verschmutzung *f* 1. *(act or process)* soiling, *(esp relating to the environment)* pollution, *(esp with poisonous matter)* contamination; 2. *(state)* pollution, contamination, dirtiness
Verschmutzungsgrad *m (text)* degree of soiling; degree of pollution *(of air or water)*
verschneidbar *(coat)* dilutable
Verschneidbarkeit *f (coat)* dilutability; *(quantitatively)* dilution ratio, hydrocarbon tolerance
verschneiden *(food, rubber)* to blend; *(coat)* to dilute *(solvents)*; *(coat)* to extend *(pigments)*; to cut back, to flux *(high-boiling petroleum fractions)*
Verschneidmittel *n (coat)* diluent, diluting agent, indirect (latent) solvent; [paint] extender, [extending] filler *(for pigments)*
Verschneidwert *m (coat)* dilution value
Verschnitt *m (food, rubber)* 1. blending *(process)*; 2. blend *(product)*
Verschnittbitumen *n* cutback [bitumen], bitumen cutback
verschnittfähig *s.* verschneidbar
Verschnittmittel *n s.* Verschneidmittel
Verschnittöl *n (petrol)* flux [oil]
Verschnittpigment *n* extender pigment, extending filler
Verschnittprodukt *n (petrol)* cutback product
Verschönerungsmittel *n (cosmet)* makeup [preparation]
verschreiben to prescribe *(a medicine)*
verschrühen *(ceram)* to biscuit-fire; to harden on *(for fixing the decoration before firing)*
verschütten to spill
verschwammen *(ceram)* to sponge
verschwefeln to thionate *(organic compounds for producing dyestuffs)*
Verschwefelung *f* thionation *(of organic compounds for producing dyestuffs)*
verschweißen to weld, *(plast also)* to seal
Verschweißen *n* **durch Wärme** *(plast)* heat welding, *(esp relating to films)* thermal (heat) sealing
verschwelen 1. to carbonize *(e.g. coal at low temperature)*; to char *(organic matter by heat)*; 2. to char *(of organic matter under the influence of heat)*

Verschwelung *f (coal)* [low-temperature] carbonization, low-temperature coking
versehen / mit einem Deckanstrich (Schlußanstrich) *(coat)* to finish
  mit Rohren v. to pipe
verseifbar saponifiable
  nicht v. unsaponifiable, non-saponifiable
Verseifbares *n* saponifiable matter
verseifen to saponify
verseift / partiell partly saponified
Verseifung *f* saponification
  alkalische V. alkaline saponification
  oberflächliche V. *(text)* surface saponification, S finishing
Verseifungsgeschwindigkeit *f* rate of saponification
Verseifungskolben *m* saponification flask
Verseifungszahl *f* saponification value (number), sap. value, S.V.
versengen to scorch
Versenk *m(n) (tann)* handlers, lay-away pits (vats), floaters
versenken to sink, to submerge
Versenkung *f* sinking, submersion
  V. ins Meer sea disposal (burial) *(as of radioactive waste)*
versetzen 1. *(filtr)* to clog; 2. to add
  in Schwingungen v. to vibrate
  mit Borverbindungen v. to boronate *(fertilizers)*
  mit Hefe v. to yeast
  mit Hopfen v. to hop *(the wort)*
  mit Lab v. to rennet *(milk)*
  mit Luft v. to aerate
  mit Neßlers Reagens v. to nesslerize
  mit Tetraäthylblei v. to lead *(motor fuel)*
versetzt / gegeneinander *(cryst)* dislocated; staggered *(stereochemistry)*
Versetzung *f* [crystal] dislocation
Versetzungsenergie *f (cryst)* energy of dislocation
verseuchen [/ radioaktiv] to contaminate
Verseuchung *f* **[/ radioaktive]** [radioactive] contamination
versickern to seep [away], to soak [away], to percolate, *(esp if slowly)* to ooze [away]
Versickerung *f* 1. seepage, soaking, percolation, *(esp if slowly)* oozing; 2. *(act)* floor drain *(wastewater treatment)*
versiegeln to seal
versilbern to silver; *(electrically)* to silver-plate
Versilberung *f* silvering; *(electrically)* silver plating
Versilberungsbad *n* silver-plating bath
verspannen *(glass)* to temper
versperren to choke [up] *(e.g. filter pores)*
Verspinnbarkeit *f* spinnability
verspinnen to spin *(a dope)*
verspritzen to splash, to spatter, to spill
verspröden to embrittle
Versprödung *f* embrittlement
Versprödungstemperatur *f* brittle (brittleness, brittle-point) temperature
versprühen to atomize, to spray *(liquids)*
Versprüher *m* atomizer, sprayer
  V. mit Hilfsstoff auxiliary-fluid atomizer
  rotierender V. rotating (rotary) atomizer (sprayer)
  rotierender V. mit Beschleunigungsschaufeln vaned-disk atomizer, rotary-vane atomizer (sprayer)
Versprühung *f* atomization, spraying
  V. durch Düsen nozzle atomization
  V. ohne Hilfsstoff single-fluid atomization
verspunden to bung
verstärken 1. to fortify *(e.g. a construction)*; to reinforce *(e.g. plastics)*; *(text)* to splice; 2. to intensify, to potentiate *(e.g. an action)*; 3. to fortify, to strengthen *(solutions)*, *(esp by evaporation)* to graduate; 4. *(phot)* to intensify
Verstärker *m* 1. *(phot)* intensifier; 2. *(met)* energizer; 3. *s.* synergetischer V.; 4. *s.* Verstärkerfüllstoff
  proportionaler V. *(phot)* proportional intensifier
  superproportionaler V. *(phot)* superproportional intensifier
  synergetischer V. activator, promoter, activating (promoting) agent
  überproportionaler V. *s.* superproportionaler V.
Verstärkerfüllstoff *m (rubber)* reinforcing filler (ingredient, pigment), active filler
  heller V. white (non-black) reinforcing filler
Verstärkerwirkung *f* reinforcing (strengthening) action
Verstärkung *f* 1. fortification *(as of a construction)*; reinforcement *(as of plastics)*; 2. intensification, potentiation *(as of an action)*; 3. fortification, strengthening *(of solutions)*, *(esp by evaporation)* graduation; 4. *(phot)* intensification
  V. des latenten Bildes *(phot)* latent-image intensification, latensification
  chemische V. *(phot)* chemical intensification
Verstärkungsgerade *f*, **Verstärkungslinie** *f (distil)* enrichment (rectifying operating) line
Verstärkungsmaterial *n* reinforcing agent (material) *(as for plastics)*
Verstärkungsteil *m (distil)* enriching (enrichment, rectifying) section
Verstärkungsverhältnis *n* **/ mittleres** *(distil)* overall column (plate) efficiency *(ratio of number of theoretical plates to number of actual plates)*
Verstäubbarkeit *f* dustability
verstäuben to dust *(e.g. pesticides)*
Verstäuber *m (agric)* duster, dusting machine (appliance)
  tragbarer V. rotary hand duster
Verstäubungsgerät *n s.* Verstäuber
Verstaubungsverlust *m* dust loss
versteifen to stiffen
Versteifung *f* stiffening
versteinern *(geol)* to lithify, to petrify
Versteinerung *f (geol)* lithification, petrifaction
Verstellpumpe *f* variable-displacement pump

versticken *(met)* to nitride
Verstickung *f (met)* nitride hardening, nitriding, nitridation, nitrogen [case-]hardening
verstopfen to clog [up], to choke [up], to blind, to plug *(e.g. screen openings)*
sich v. to clog, to choke, to plug *(as of screen openings)*
verstöpseln to stopper, to cork
verstrammen *(rubber)* to stiffen
verstrecken *(text, plast)* to stretch
Verstrecken *n (text, plast)* stretch[ing]
Verstreckungsgrad *m (text, plast)* degree of stretching
Verstreckwiderstand *m (text)* stretch resistance
Versuch *m* experiment, test, trial, assay, run
V. mit Leitisotopen *(bioch)* tracer experiment
V. mit Molekularstrahlen molecular-beam experiment
halbtechnischer V. semicommercial-scale test
statischer V. *(rubber)* static test
Stern-Gerlachscher V. Stern-Gerlach experiment *(for determining magnetic moments of atoms)*
Versuchsanlage *f* experimental plant
halbtechnische V. pilot plant, semiworks
Versuchsbetrieb *m* 1. test run; 2. *s.* Versuchsanlage
Versuchsdauer *f* duration of experiment (test)
Versuchsdurchführung *f* performance of the experiment, test practice
Versuchselektrode *f* working electrode
Versuchsfärbung *f* trial dyeing
Versuchsfehler *m* experimental error
Versuchskammer *f* model oven *(for determining the swelling pressure of coal)*
Versuchskochung *f (pap)* experiment boil
Versuchslösung *f* experimental solution
Versuchsperson *f* / **freiwillige** *(tox)* volunteer
Versuchsreaktor *m* experimental (research) reactor
Versuchsstadium *n* experimental stage
Versuchsstation *f* experiment[al] station
Versuchssubstanz *f* experimental (test) substance (material), substance for (*or* under) investigation
Versuchstier *n* experimental (test) animal
Versuchswert *m* experimental value
Vertauschungswägung *f* [Gauss's method of] double weighing
verteilen to distribute, to partition; to spread [out] *(over a surface)*; to disperse *(e.g. solid particles in a liquid)*; to homogenize *(e.g. for forming an emulsion)*
Verteiler *m* 1. distributor, distributing device; 2. *s.* Verteilerplatte
Verteilerbürste *f (distil)* wiper
Verteilerplatte *f* deflector [plate], distributor plate
Verteilerrohr *n* distributing pipe
Verteilerwirksamkeit *f* spreading efficiency *(of surface-active agents)*
Verteilplatte *f s.* Verteilerplatte
verteilt / fein finely divided, disperse
normal v. normally distributed
regellos (statistisch, zufällig) v. randomly distributed
Verteilung *f* distribution, partition; spreading *(over a surface)*; dispersion *(as of solid particles in a liquid)*; homogenization *(as for forming an emulsion)* ✦ **in feiner V.** finely divided, disperse
V. der relativen Molekülmassen molecular-weight distribution
V. mit Phasenumkehr reversed-phase partition *(chromatography)*
Bose-Einsteinsche V. Bose-Einstein distribution *(of gas particles or photons)*
klassische V. *(phys chem)* classical distribution
Maxwell-Boltzmannsche V. Maxwell-Boltzmann distribution
Maxwellsche V. Maxwell distribution
regellose (statistische, zufällige) V. random distribution
Verteilungschromatografie *f* partition chromatography
V. mit Phasenumkehr reversed-phase partition chromatography
Verteilungseigenschaften *fpl* spreading properties *(of surface-active agents)*
Verteilungsfunktion *f* distribution function; *(phys chem)* partition function
V. der Rotationsenergie rotational partition function
V. der Schwingungsenergie vibrational partition function
V. der Translationsenergie translational partition function
elektronische V. electronic partition function
radiale V. radial distribution (probability) function *(for electrons)*
vollständige V. complete partition function *(for energy)*
Verteilungsgasleitung *f* distribution gas main
Verteilungsgefäß *n* partition vessel
Verteilungsgesetz *n* distribution (partition) law
Boltzmannsches V. Boltzmann distribution law *(of energy)*
klassisches V. *s.* Maxwell-Boltzmannsches V.
Maxwell-Boltzmannsches V. Maxwell-Boltzmann [velocity-]distribution law, classical distribution law
Nernstsches V. *s.* Verteilungssatz / Nernstscher
Rosin-Rammlersches V. Rosin-Rammler exponential law *(relating to particle size in grinding)*
Verteilungskoeffizient *m* distribution (partition) coefficient (ratio), *(in zone refining also)* segregation coefficient
effektiver V. effective distribution (segregation) coefficient *(in zone refining)*
idealer V. equilibrium distribution (segregation) coefficient *(in zone refining)*
Verteilungskonstante *f s.* Verteilungskoeffizient
Verteilungskurve *f* distribution curve
Gaußsche V. Gaussian curve *(statistics)*

Verteilungsrohr *n* distributing pipe
Verteilungssatz *m* / **Nernstscher** Nernst distribution (partition) law
Verteilungsverfahren *n* dispersion method *(of preparing colloidal solutions)*
Verteilungsvorrichtung *f* distributing device
vertiefen to intensify *(a colour)*
Vertiefung *f* intensification *(of a colour)*
Vertikalbeziehung *f* vertical relationship *(in the periodic system)*
Vertikalelektrofilter *n* vertical-flow electrical precipitator
Vertikalkammer *f* vertical chamber
Vertikalofen *m* vertical oven
Vertikalretorte *f* vertical retort
V. mit kontinuierlicher Beschickung continuous vertical retort
stetig betriebene V. continuous vertical retort
Vertikalrohrverdampfer *m* vertical [tube] evaporator
V. mit Innenheizkammer standard (calandria, Robert) evaporator
Vertikalziehverfahren *n (glass)* vertical sheet drawing process
Vertikalzug *m* vertical flue *(of an oven)*
Vertorfung *f* peat formation
verträglich compatible
Verträglichkeit *f* compatibility
Vertrauensbereich *m* confidence interval *(statistics)*
Vertrauensgrenze *f* confidence limit *(statistics)*
vertreiben to drive off (out), to expel, to dispel *(volatile matter)*; to repel *(e.g. insects by repellents)*
verunreinigen to soil, *(esp relating to the environment)* to pollute, *(esp with poisonous matter)* to contaminate
Verunreinigung *f* 1. *(act)* soiling, *(esp relating to the environment)* pollution, *(esp with poisonous matter)* contamination; 2. *(substance)* contaminant, impurity, admixture, foreign substance (material, matter), *(esp in environment)* pollutant, *(esp in zone melting)* solute
zufällige V. chance contaminant
Verunreinigungsniveau *n* impurity level *(of insulators and semiconductors)*
Verunreinigungsstoff *m*, **Verunreinigungssubstanz** *f s.* Verunreinigung 2.
Vervielfacherfotozelle *f s.* Sekundärelektronenvervielfacher
Vervielfältigungspapier *n* duplicating (duplicator) paper
verwachsen *(cryst, coal)* to intergrow
Verwachsenes *n (coal)* intergrown material
Verwachsung *f (cryst, coal)* intergrowth
Verwachsungszwillinge *mpl (cryst)* penetration twins
verwandeln *s.* umwandeln
Verwandlung *f s.* Umwandlung
verwandt allied *(chemical substances)*
Verwandtschaftsgruppe *f* family *(in the periodic system)*
verwaschen to fade *(of colours)*
Verwehung *f* **des Sprühmittels** *(agric)* spray drift
Verweilzeit *f* residence (hold-up, holding, detention, retention) time, *(bioch also)* turnover time
mittlere V. average retention time
verwendbar applicable
allgemein (universell) v. generally applicable, general-purpose
Verwendbarkeit *f* applicability
verwenden to apply, to use
lokal v. *(pharm)* to use topically (in topical applications)
Verwendungsmöglichkeit *f* applicability
verwerfen *(lab)* to discard, to reject
verwesen to decay, to decompose
Verwesung *f* decay, decomposition
Verwirbelung *f* vortexing *(of material in a cyclone)*
verwitterbar weatherable
verwittern to weather, to decay, to disintegrate
unter Kristallwasserverlust v. *(cryst)* to effloresce
verwitternd / nicht non-weathering
Verwitterung *f* weathering, decay, disintegration
V. unter Kristallwasserverlust *(cryst)* efflorescence
Verwitterungslehm *m* residual loam
Verwitterungsprodukt *n* weathered product
Verwitterungsschutt *m (geol)* detritus, detrital material, weathering (rock) waste
Verwitterungston *m* residual (primary) clay
verzerren to distort *(e.g. an electron cloud)*
Verzerrung *f* distortion *(as of an electron cloud)*
verziehen / sich to be distorted (deformed), *(esp of metal)* to buckle; *(of ceramics or wood)* to warp
verzinken to zinc, to galvanize
galvanisch v. to electrogalvanize
Verzinkung *f* zinc[k]ing, galvanization
galvanische V. zinc plating, electrogalvanizing, wet (cold) galvanizing
verzinnen to tin
galvanisch v. to tin-plate, to electrotin
Verzinnung *f* tinning
galvanische V. tin plating, electrotinning
Verzinnungsbad *n* tin bath
Verzinsungsgabe *f (agric)* interest dosage *(of fertilizers)*
Verzögerer *m* retarder, retarding agent, inhibitor, anticatalyst, negative catalyst; *(rubber)* retarder, antiscorcher, antiscorching agent; *(phot)* restrainer, restraining agent
verzögern to retard, to inhibit, to delay
sich v. to retard
verzögernd retardant, inhibitory
Verzögerung *f* retardation, inhibition, delay
Verzögerungsfaktor *m* retardation (retention) factor, Rf value *(chromatography)*
Verzögerungskolonne *f s.* Verzögerungssäule
Verzögerungsmittel *n s.* Verzögerer
Verzögerungssäule *f* retardation column *(chromatography)*

verzuckern to saccharify
Verzuckerung *f* saccharification
Verzuckerungsapparat *m*, **Verzuckerungsgerät** *n* saccharifier
Verzuckerungsrast *f*, **Verzuckerungszeit** *f* *(ferm)* saccharification period (rest, time)
Verzug *m s.* Verzögerung
verzundern *(met)* to scale
Verzunderung *f* *(met)* scaling
verzweigen / sich to branch
verzweigt branched; *(cryst)* dendritic[al]
schwach v. lightly branched
stark v. highly branched
verzweigtkettig branched-chain
Verzweigung *f* 1. branching *(process)*; branch *(state)*; 2. *s.* radioaktive V.
radioaktive V. *(nucl)* branched (multiple) disintegration (decay)
Verzweigungsanteil *m* branching fraction
Verzweigungsgrad *m* degree of branching
Verzweigungsstelle *f* branch point
Verzweigungsverhältnis *n* branching ratio
Verzwillingung *f* *(cryst)* twinning
polysynthetische V. polysynthetic twinning
Verzwitterung *f* hybridization
Vesuvian *m* *(min)* vesuvian[ite] *(a sorosilicate)*
Vesuvin *n* *(dye)* vesuvine brown, vesuvin
Vetiverol *n* *(cosmet)* vetiverol, vetivol *(a mixture of alcohols from Vetiveria zizanioides (L.) Nash)*
Vetiveröl *n* vetiver (cuscus) oil, vetivert *(from Vetiveria zizanioides (L.) Nash)*
Vetiverylazetat *n* *(cosmet)* vetivert acetate *(a mixture of esters)*
VFA-Zahl *f* *(rubber)* VFA number
VI *s.* 1. Viskosefaserstoff; 2. Viskositätsindex
Vibration *f* vibration, oscillation
Vibrationsbandtrockner *m* vibrating conveyor dryer
Vibrationsbeanspruchung *f* vibrating stress
Vibrationsbeton *m* vibrated concrete
Vibrationsdüse *f* vibrating nozzle
Vibrationsfilter *n* vibration filter
Vibrationsgalvanometer *n* vibration galvanometer
Vibrationsknotenfänger *m* *(pap)* vibrating screen
Vibrationsmischer *m* reciprocating-impeller agitator
Vibrationsplattentrockner *m* vibrating tray dryer
Vibrationsquantenzahl *f* vibrational quantum number
Vibrationsrührer *m* reciprocating-impeller agitator
Vibrationsschaber *m* *(pap)* vibrating (oscillating) doctor
Vibrationsschüttelsieb *n* *(petrol)* vibrating mudscreen *(of a rotary-drilling installation)*
Vibrationssieb *n*, **Vibrationssortierer** *m* vibrating (oscillating, reciprocating) screen
Vibrationsverdichtung *f* vibrational compaction
Vibrator *m* vibrator
Vibratorsieb *n s.* Vibrationssieb
vibrieren to vibrate, to oscillate
v. lassen to vibrate, to oscillate
Vibrieren *n* vibration, oscillation
Vibromischer *m* reciprocating impeller agitator
vic. *s.* vicinal
Vicat-Apparat *m*, **Vicat-Nadel** *f* Vicat apparatus (needle) *(materials testing)*
Vicat-Zahl *f* *(plast)* Vicat softening point (temperature), V.S.P., Vicat needle point
vicinal vicinal, vic
Vickers-Härte *f* Vickers hardness, diamond pyramid hardness, DPH
Vickers-Härteprüfer *m* Vickers tester
Vickery-Filzinstandhalter *m* *(pap)* Vickery felt conditioner
Vidal-Schwarz *n* *(dye)* Vidal black
Viehbademittel *n* stock dip *(for insect control)*
Viehbesprühung *f* cattle spraying *(for insect control)*
Viehsalz *n* cattle (fodder) salt, cattle lick
vielatomig polyatomic
Vielfachbeschleuniger *m* multiple accelerator
Vielfachbeschleunigung *f* multiple acceleration
Vielfachstreuung *f* multiple scattering
Vielfachzyklon *m* multiple[-unit] cyclone
Vielfarbeneffekt *m* *(text)* multicolour[ed] effect
vielflächig *(cryst)* polyhedral
Vielflächner *m* *(cryst)* polyhedron
vielgestaltig polymorphic, polymorphous
Vielgestaltigkeit *f* polymorphism
vielgliedrig multimembered, polymembered, many-membered *(ring)*
Vielmesserhackmaschine *f* *(pap)* multiknife chipper
Vielstoffgemisch *n* multicomponent mixture
Vielstufenbleiche *f* *(pap)* multistage bleaching
vielwertig multivalent, polyvalent, polyad
vielzähnig multidentate, polydentate *(ligand)*
Vielzellenabscheider *m*, **Vielzellenentstauber** *m* multicell dust collector (extractor)
Vielzellenverdichter *m* [sliding-]vane compressor
Vielzweckkalander *m* universal calender
Vielzweckkleber *m* all-purpose adhesive
vieratomig tetratomic
vierbasig tetrabasic, quadribasic *(acid)*
vierbindig tetracovalent, quadricovalent
Viererring *m s.* Vierring
vierfachpositiv [geladen] tetrapositive
Vierfarbendruck *m* four-colour[-process] printing
vierflächig *(cryst)* tetrahedral
Vierflächner *m* *(cryst)* tetrahedron
viergliedrig four-membered
Vierhalskolben *m* four-neck[ed] flask
Vierhalsrundkolben *m* round-bottom four-neck flask
Vierkomponentensystem *n* four-component system
Vierkörperverdampfer *m* quadruple-effect [evaporator]
Vierring *m* four-membered ring
viersäurig tetraacid *(base)*
Vierstoffsystem *n* four-component system

Vierstufenreaktion *f* four-step reaction
Vierstufenverdampfer *m* quadruple-effect [evaporator]
Vierstufenverfahren *n* four-step process
viertelgeleimt *(pap)* quarter-sized, $^1/_4$ sized
Vierwalzenbrustkalander *m* inverted L type of calender
Vierwalzenkalander *m* four-bowl (four-roll) calender
  V. mit oberer Brustwalze inverted L type of calender
  V. mit übereinanderliegenden Walzen four-bowl stack type of calender
  V. mit unten vorliegender Walze L type of calender
  V. mit Z-Anordnung Z type of calender
Vierwegehahn *m*, **Vierwegeventil** *n* four-way valve
Vierwegkreuzung *f* four-way juncture
vierwertig tetravalent, quadrivalent
Vierwertigkeit *f* tetravalence, quadrivalence
vierzählig 1. *(cryst)* tetrad; 2. *s.* vierzähnig
vierzähnig tetradentate, quadridentate *(ligand)*
Vigoureuxdruck *m (text)* vigoureux (top) printing
Vigreux-Kolonne *f (distil)* Vigreux column
Viktoriagrün *n* Victoria (malachite) green *(a triphenylmethane dye)*
Viktoriarot *n s.* Chromrot
Villard-Effekt *m (phot)* Villard effect
Vinakonsäure *f* vinaconic acid, cyclopropane-1,1-dicarboxylic acid
3-Vinylakrylsäure *f* $CH_2=CHCH=CHCOOH$ 3-vinylacrylic acid, 2,4-pentadienoic acid
Vinylalfaser *f* polyvinyl alcohol fibre
Vinylalfaserstoff *m* polyvinyl alcohol fibre
Vinylalkohol *m* $CH_2=CHOH$ vinyl alcohol, ethenol
Vinyläthen *n s.* Vinyläthylen
Vinyläther *m* $CH_2=CH-OR$ vinyl ether; *(incorrectly for)* Divinyläther
Vinyläthin *n s.* Vinylazetylen
Vinyläthylen *n* $CH_2=CHCH=CH_2$ 1,3-butadiene, vinylethylene
Vinylazetat *n* $CH_3COOCH=CH_2$ vinyl acetate
Vinylazetylen *n* $CH_2=CHC\equiv CH$ 1-buten-3-yne, vinylacetylene
Vinylbenzol *n* $C_6H_5CH=CH_2$ vinylbenzene, phenylethylene, styrene
Vinylbromid *n* $CH_2=CHBr$ bromoethylene, vinyl bromide
Vinylchlorid *n* $CH_2=CHCl$ chloroethylene, vinyl chloride
Vinylchlorid-Dichloräthen-Kopolymerisat *n* vinyl chloride dichloroethylene copolymer
Vinylgruppe *f* $CH_2=CH-$ vinyl group (residue)
Vinylhalogenid *n* vinyl halogenide (halide)
Vinylharz *n* vinyl resin
Vinylharzkunststoff *m* vinyl plastic
Vinylharzlack *m* vinyl lacquer
Vinylharzschaum *m* vinyl foam
vinylhomolog *s.* vinylog
Vinylhomolog[es] *n s.* Vinyloges
Vinylidenchlorid *n* $CH_2=CCl_2$ 1,1-dichloroethylene, vinylidene chloride
Vinylidenzyanid *n s.* 1,1-Dizyanäthen
Vinylierung *f* vinylation
vinylisch vinylic
Vinylkarbinol *n s.* Propen-(2)-ol-(1)
Vinylkarbonsäure *f* $CH_2=CHCOOH$ vinylformic acid, ethylenecarboxylic acid, acrylic acid, propenoic acid
Vinylkunststoff *m* vinyl plastic
vinylog vinylogous
Vinyloges *n* vinylog[ue], vinyl homologue
Vinylogie *f* vinylogy
Vinylradikal *n* vinyl radical, *(specif)* $-CH=CH_2$ free vinyl radical
Vinylrest *m s.* Vinylgruppe
Vinylsilikon *n* vinyl silicone
Vinylsulfonfarbstoff *m* vinyl sulphone dye
Vinylsulfongruppe *f* vinyl sulphone group
Vinyltrichlorid *n s.* 1,1,2-Trichloräthan
Vinylverbindung *f* vinyl compound
Vinylzyanid *n* $CH_2=CHCN$ vinyl cyanide, cyanoethylene, acrylonitrile
Violarit *m (min)* violarite *(iron nickel sulphide)*
Violett *n* / **Döbners** Döbner's violet *(a triphenylmethane dye)*
  Hofmanns V. Hofmann's violet *(an aniline dye)*
Violursäure *f* violuric acid, 5-isonitrosobarbituric acid
virentötend *s.* viruzid
Virialgleichung *f (phys chem)* virial equation
Virialkoeffizient *m (phys chem)* virial coefficient
Virialsatz *m (phys chem)* virial equation
Viridian *n s.* Smaragdgrün
Virologe *m* virologist
virozid *s.* viruzid
Viruseiweiß *n* viral protein
Virusforscher *m* virologist
Virusprotein *n* viral protein
virustötend *s.* viruzid
viruzid virucidal, viricidal, antiviral
viskoelastisch viscoelastic
Viskoelastizität *f* viscoelasticity, viscous elasticity
viskos viscous, viscose, viscid
Viskose *f* viscose
Viskoseerspinnlösung *f* viscose [spinning] solution, viscose dope
Viskosefaser *f* viscose [staple] fibre
Viskosefaserstoff *m* viscose fibre
Viskosefolie *f* viscose film
Viskosemodifikator *m*, **Viskosemodifizierungsmittel** *n* viscose modifier
Viskosereifung *f* viscose ripening
Viskoseschicht *f* viscose film
Viskoseseide *f* viscose rayon
Viskosespinnlösung *f s.* Viskoseerspinnlösung
Viskoseverfahren *n (text)* viscose [rayon] process
Viskosimeter *n* viscometer, viscosimeter
  Sayboltsches V. *(petrol)* Saybolt viscometer
Viskosimetrie *f* viscometry, viscosimetry
Viskosität *f* viscosity
  absolute V. *s.* dynamische V.
  dynamische V. dynamic (absolute) viscosity;

(quantitatively) viscosity coefficient
kinematische V. kinematic viscosity, viscosity/density ratio
relative V. relative viscosity
scheinbare V. apparent viscosity
spezifische V. specific viscosity
Viskositätsbeständigkeit *f* viscosity stability
Viskositätsbrechen *n (petrol)* viscosity breaking, visbreaking
Viskositäts-Dichte-Verhältnis *n* viscosity/density ratio, kinematic viscosity
Viskositätsindex *m* viscosity index, V.I. *(of lubricating oils)*
Viskositätsindexerhöher *m*, **Viskositätsindexverbesserer** *m* viscosity index improver
Viskositätskoeffizient *m*, **Viskositätskonstante** *f* viscosity coefficient
Viskositätskonstanz *f* viscosity stability
Viskositätsmessung *f* viscometry, viscosimetry
Viskositätsstabilisator *m* viscosity stabilizer
Viskositätsstabilität *f* viscosity stability
Viskositäts-Temperatur-Koeffizient *m* viscosity-temperature coefficient, temperature coefficient of viscosity
Viskositäts-Temperatur-Kurve *f* viscosity-temperature curve (slope)
Viskositätsverbesserer *m* viscosity modifier
Viskositätszahl *f* viscosity number
logarithmische V. logarithmic viscosity number
Vitalfarbstoff *m (biol)* vital stain
Vitalfärbung *f (biol)* vital staining
Vitamin *n* vitamin ✦ **mit Vitaminen anreichern** to vitaminize, to fortify (enrich) by vitamins
antihämorrhagisches V. antihaemorrhagic (coagulation) vitamin, phylloquinone, vitamin K_1
antirachitisches V. antirachitic vitamin, vitamin D
antiskorbutisches V. antiscorbutic vitamin, vitamin C
antixerophthalmisches V. antixerophthalmic vitamin, vitamin A
fettlösliches V. lipovitamin
Vitaminantagonist *m* antivitamin
vitaminarm poor in vitamins
Vitaminchemie *f* vitamin chemistry
vitaminfrei vitamin-free
Vitamingehalt *m* vitamin content
vitaminisieren to vitaminize, to fortify (enrich) by vitamins
Vitaminisierung *f* vitaminization
Vitaminkonzentrat *n* vitamin concentrate
Vitaminmangel *m* vitamin deficiency
vitaminreich rich in vitamins, vitamined
Vitaminwirksamkeit *f* vitamin potency
Vitellin *n* egg vitellin, ovovitellin *(a phosphoprotein)*
Vitellus *m* vitellus, [egg] yolk
Vitrain *m (coal)* vitrain
Vitrinertit *m (coal)* vitrinertite
Vitrinit *m (coal)* vitrinite
vitrinitisch *(coal)* vitrinitic
Vitriol *n* vitriol *(a hydrated sulphate of a bivalent metal)*
Vitriolschiefer *m* alum schist (shale, slate)
Vitrit *m (coal)* vitrain
vitritähnlich *(coal)* vitrainlike
Vitritlinse *f (coal)* vitrain lens
Vitrokeram *n* vitroceramic, vitrokeram, devitrified (neo-ceramic) glass, glass ceramic
vitro[por]phyrisch *(geol)* vitrophyric
Vivianit *m (min)* vivianite *(iron(II) orthophosphate)*
vizinal vicinal, adjacent, neighbouring *(substituent)*
Vizinalfunktion *f* vicinal function
VK *s.* 1. Vergaserkraftstoff; 2. Vulkanisationskoeffizient
Vlies *n (text)* fleece; *(glass)* mat
Vliesfolie *f* non-woven fabric
Vliesstoffe *mpl*, **Vlieswaren** *fpl* non-wovens
Vlieswolle *f* fleece wool
V-Lunker *m (met)* secondary pipe
V-Mischer *m* vee-type (twin-shell) blender (mixer)
Vogelfraß-Abwehrmittel *n* bird repellent
Vogelguano *m* bird guano
Vol.% *s.* Volumenprozent
volatil volatile
Volborthit *m (min)* volborthite *(copper(II) vanadate)*
Vollanalyse *f* complete analysis
Volldünger *m* complete fertilizer
Volldüngung *f* complete fertilization
Volleipulver *n* whole-egg powder
Vollentsalzung *f* complete softening *(of water)*
Vollfeuer *n (ceram)* full fire
vollflächig *(cryst)* holohedral
Vollflächnerkristall *m* holohedral crystal
Vollflußventil *n* inclined-seat valve
vollgeleimt *(pap)* hard-sized, H.S., strongly sized
Vollgerbstoffsyntan *n* replacement [syn]tan
Vollgummireifen *m* solid[-rubber] tyre, band tyre
Vollguß *m (ceram)* solid casting
Vollkasein *n* whole casein
Vollkegeldüse *f* solid-cone nozzle
Vollkornmehl *n* whole meal
vollkristallin[isch] holocrystalline
Vollküken *n* solid stopper *(of a glass stopcock)*
Vollmantelschleuder *f*, **Vollmantelzentrifuge** *f* solid-bowl (solid-wall) centrifuge
Vollmilch *f* whole (rich) milk, full[-cream] milk
eingedickte (kondensierte) V. condensed whole milk
Vollmilchpulver *n* dry whole milk
vollmundig palateful, rich in flavour *(e.g. beer)*
Vollmundigkeit *f* palatefulness, full[l]ness *(as of beer)*
Vollpipette *f* transfer (volumetric, bulb, ordinary) pipette
V. mit einer Marke one-mark pipette
vollpumpen to charge by pumping
Vollrahmmargarine *f* whipped margarine
Vollrohrmodell *n* full-scale model *(as of a pipe system)*

vollsaugen / sich to soak
Vollsprühkegel *m* solid spray cone
Vollsprühkegeldüse *f* solid-cone nozzle
Vollständigglühen *n (met)* full (true) annealing
Vollstromwechselofen *m* **nach Collin** Collin oven *(a coke oven)*
Volltränkverfahren *n* full-cell process *(wood preservation)*
Volltrum *m(n)* carrying side, top strand, *(relating to belt conveyors also)* drive belt
Vollverkokung *f* normal carbonization, coking
vollverseift completely saponified
Vollwaschmittel *n* heavy-duty detergent
Vollwelle *f* solid shaft
Vollzellstoff *m* chemical pulp
Voltait *m (min)* voltaite *(a hydrous sulphate of potassium and iron)*
Voltameter *n* voltameter, coulo[mb]meter
Voltammetrie *f* voltammetry
  spannungsgeregelte V. voltage-scan voltammetry
  stromgeregelte V. current-scan voltammetry
voltammetrisch voltammetric
Voltzin *m* voltzine *(a sulphidic zinc mineral containing arsenic)*
Volum ... *s.* Volumen ...
Volumen *n* volume ✦ **mit konstantem V.** constant-volume
  V. der Zwischenräume void volume
  kritisches V. critical volume
  partielles molares V. partial molar (molal) volume
  spezifisches V. specific volume
Volumenabnahme *f* decrease in volume
Volumenänderung *f* volume change
Volumenausdehnung *f* cubical expansion
volumenbeständig volume-stable
Volumenbeständigkeit *f* volume stability
Volumencoulometer *n* volumetric coulometer
Volumendosiervorrichtung *f* volumetric feeder (feeding device)
Volumenelement *n* volume element
Volumengesetz *n* law of volumes
Volumengetterung *f* dispersal gettering *(vacuum technology)*
Volumenkontraktion *f* volume contraction
Volumenkonzentration *f* volume concentration
  molare V. molar concentration, molarity
Volumenlebensdauer *f* volume lifetime *(of a semiconductor crystal)*
volumenmolar molar
Volumenmolarität *f* molarity, molar concentration
Volumenometer *n* volumenometer, stereometer
Volumenprozent *n* volume percentage, percentage (per cent) by volume
Volumenschwindung *f* volume (cubic) shrinkage
Volumenstrom *m* volumetric rate of flow
Volumensuszeptibilität *f* volume susceptibility *(magnetochemistry)*
Volumenteil *m* part by volume
Volumenvergrößerung *f* increase in volume
Volumenverhältnis *n* proportion by volume
Volumetrie *f* volumetry, titrimetry, volumetric (titrimetric, mensuration) analysis
volumetrisch volumetric, *(analytical chemistry also)* titrimetric
voluminös voluminous
Volumometer *n* volumenometer, stereometer
Vorappretur *f (text)* grey finish
Vorauflaufbehandlung *f (agric)* pre-emergence treatment
Vorauflaufherbizid *n* pre-emergence herbicide (weed-killer)
Vorausrüstung *f (text)* grey finish
Voraussaat ... *s.* Vorsaat ...
Vorauswahl *f* screening *(as of newly developed preparations)*
Vorbad *n (phot)* forebath
Vorbedingung *f* precondition
vorbehandeln to pretreat, to prepare, to precondition
Vorbehandlung *f* preparatory (preliminary) treatment, pretreatment, preparation, preconditioning
  V. der Trübe *(filtr)* slurry preparation
vorbeharzen *(plast)* to preimpregnate, to precompound
vorbeiführen to by-pass
vorbelichten *(phot)* to pre-expose
Vorbelichtung *f (phot)* pre-exposure
vorbelüften to preaerate *(waste water)*
Vorbelüftung *f* preaeration *(of waste water)*
vorbereiten to prepare, to prime
Vorbereitung *f* preparation, priming
Vorbereitungsabschnitt *m (glass)* conditioning section (zone) *(of the feeder channel)*
vorbestrahlen *(plast)* to preirradiate
Vorbestrahlung *f (plast)* preirradiation
vorblasen *(glass)* to puff, to blow back
Vorblasen *n (glass)* puffing, counter blow
Vorbleiche *f (pap)* prechlorination
vorbleichen *(pap)* to prechlorinate
Vorblütenspritzung *f (agric)* pre-bloom spray
Vorbrecher *m* prebreaker, precrusher, primary crusher
vorbrennen *(ceram)* to prefire
Vorce-Zelle *f* Vorce cell *(dialysis)*
vorchloren to prechlorinate *(water)*
vorchlorieren *(pap)* to prechlorinate
Vorchlorierung *f (pap)* prechlorination
Vorchlorung *f* prechlorination *(of water)*
vorchromieren to prechrome
vordämpfen *(pap)* to presteam *(chips)*; *(text)* to preset
Vorderwürze *f (ferm)* first (original) wort
Vordestillationskolonne *f* primary column
Vordruckwalze *f (pap)* watermarking dandy [roll], dandy [roll]
Voreindickung *f* preliminary concentration
Voremulsion *f* preliminary emulsion
Vorentgasung *f* preliminary degassing
vorerhitzen to preheat, to forewarm
Vorerhitzer *m* preheater, forewarmer

Vorerhitzung *f* preheating, forewarming
Vorexpansionseinrichtung *f (plast)* pre-expander
vorfertigen to prefabricate; to precast *(e.g. concrete slabs)*
Vorfeuer *n (ceram)* prefire
Vorfilter *n* roughing filter
vorfixieren *(text)* to preset; *(cosmet)* to prefix
Vorfixierung *f (text)* presetting
Vorform *f (glass)* blank [mould], parison mould
Vorformboden *m (glass)* baffle
Vorformbodennaht *f*, **Vorformbodennarbe** *f (glass)* baffle mark
vorformen to preform
Vorformkammer *f (plast)* plenum chamber
Vorformling *m (plast)* preform, parison, pill
extrudierter V. extruded parison *(in extrusion blowing)*
geschichteter V. laminated preform
schlauchförmiger V. tubular parison
spritzgegossener V. injection-moulded parison
Vorformmaschine *f* preform machine, preformer
Vorformpresse *f* preforming press
Vorformschirm *m* preform screen *(for glass-fibre reinforced plastics)*
Vorformverfahren *n* preform moulding (process)
Vorformwerkzeug *n* preforming tool
vorfraktionieren to prefractionate
Vorfraktionierturm *m (petrol)* prefractionator
vorfüllen to prime *(a pump)*
Vorgalvanisierbad *n* strike bath (solution)
Vorgang *m* process
irreversibler (nicht umkehrbarer) V. irreversible process
reversibler (umkehrbarer) V. reversible process
Vorgarn *n* 1. roving; 2. *(glass)* sliver
Vorgärung *f* pre-fermentation, primary fermentation
vorgerben to pretan
Vorgerbung *f* pretannage
vorgeschrumpft preshrunk
vorgesteuert pilot-controlled, pilot-operated *(hydraulic and pneumatic valves)*
Vorhaltwirkung *f* derivative (rate) action *(process control)*
Vorhang *m (coat)* curtain *(film fault)*
Vorhangbildung *f (coat)* curtaining, sagging
vorhärten *(plast)* to precure
Vorhärtung *f (plast)* precure, precuring
vorheizen to preheat; *(rubber)* to prevulcanize, to precure
Vorheizer *m*, **Vorheizofen** *m* preheater
Vorheizung *f* preheating; *(rubber)* prevulcanization, precure, set cure, semicure
Vorheizzone *f* preheating zone (compartment)
Vorherd *m (met, glass)* forehearth
Vorhydrolyse *f (pap)* preimpregnation, preliminary impregnation (penetration), presoaking, steeping, steepage *(of chips)*
vorimprägnieren to preimpregnate, *(plast also)* to precompound, *(pap also)* to presoak, to steep
Vorimprägnierung *f* 1. preimpregnation, preliminary impregnation (penetration), *(plast also)* precompounding; 2. *s.* Vorhydrolyse
Vorkalkung *f (sugar)* preliming, predefecation
Vorkammeranguß *m (plast)* tab gate
vorklären *(filtr)* to preclarify
Vorklärung *f (filtr)* preclarification
Vorklassierung *f* preliminary screening (sizing)
vorkochen *(pap)* to predigest
Vorkochung *f (pap)* predigestion
vorkommen to occur
Vorkommen *n* occurrence; *(geol)* deposit
natürliches V. occurrence in nature
vorkommend / natürlich naturally occurring, found in nature, from natural sources, native
Vorkondensat *n (plast)* precondensate
Vorkondensation *f (plast)* precondensation, precure, precuring
Vorkondensationsprodukt *n (plast)* precondensate
Vorkonzentrierung *f* preliminary concentration
Vorkristaller *m* precrystallizer
Vorkristallisation *f* precrystallization
Vorkristallisator *m* precrystallizer
vorkühlen to precool, to prechill
Vorkühler *m* precooler, primary cooler
Vorkühlung *f* precooling, preliminary cooling, prechill
Vorlage *f* [distillate, distillation, still] receiver, *(lab also)* receiving flask
Vorlauf *m (distil)* first runnings, forerun[ning]s, *(esp in distilling alcohol)* foreshot[s]; *(sugar)* high green syrup
Vorläufer *m* precursor, progenitor
Vorlaufherbizid *n s.* Vorauflaufherbizid
Vorleitschaufel *f* inlet vane
Vormaischapparat *m* premasher
vormaischen to premash
Vormastikation *f (rubber)* premastication
vormastizieren *(rubber)* to premasticate
vormetallisieren to premetallize
vormischen to premix, to preblend
Vormischgefäß *n* premixer
Vormischgerät *n* premixer
Vormischung *f* premix, preblend, initial mixture; *(rubber)* masterbatch, mother stock ✦ **Vormischungen herstellen** *(rubber)* to masterbatch, to mix into a masterbatch ✦ **mit Vormischungen mischen** *(rubber)* to masterbatch
V. aus Polymer und Ruß *(rubber)* [carbon] black masterbatch
Vorneutralisation *f* preliminary neutralization
vorneutralisieren to preneutralize
Vorperiode *f* pre-period, initial period *(as of calorimetric measurement)*
vorplastifizieren *s.* vorplastizieren
vorplastizieren *(plast)* to preplasticate, to preplasticize
Vorplastiziersystem *n (plast)* preplasticating system
Vorplastizierung *f (plast)* preplastication, preplasticating, preplasticizing
Vorplastizierzylinder *m (plast)* preplasticator cylinder

Vorpolymer[es] *n*, **Vorpolymerisat** *n* prepolymer
Vorpresse *f (pap)* baby (pony) press
Vorpreßling *m (plast)* preform
Vorpreßwalze *f (pap)* 1. baby press roll, pony roll; 2. watermarking dandy [roll], [water]marking roll, dandy roll
Vorprobe *f* preliminary (crude) test
Vorprodukt *n* initial product
Vorprüfung *f* preliminary examination
Vorpumpe *f* forepump, backing pump *(vacuum technology)*
Vorrang *m (nomencl)* seniority
vorrangig *(nomencl)* senior
Vorrat *m* stock, store
Vorratsbehälter *m* storage vessel, *(tech also)* storage tank (bin), stock bin; *(pap)* storage chest
V. für Aufschlußchemikalien *(pap)* chemical storage tank
V. für Hackschnitzel *(pap)* chip bin (storage bin, silo)
Vorratsbunker *m* storage (stock) bin
Vorratsbürette *f* dispensing burette
Vorratsbütte *f (pap)* storage chest
Vorratsdüngung *f* stock (reserve) fertilization
Vorratsflasche *f* storage bottle
Vorratsgefäß *n* storage jar
Vorratshaltung *f* stockpiling, storing
Vorratskasten *m (pap)* storage cell
Vorratslösung *f* stock solution
Vorratsraum *m* store (stock) room, storage, store
Vorratsschutz *m* protection of stored products
Vorratssilo *n(m)* storage (stock) bin
Vorratstank *m* storage (store) tank; *(pap)* storage chest
Vorratstrichter *m* storage hopper
Vorreiniger *m* precleaner
Vorreinigung *f* preliminary cleaning (purification); primary treatment *(of waste water)*
Vorrichtung *f* contrivance, device
vorrösten to preroast
Vorsaatanwendung *f* pre-sowing application *(of pesticides)*
Vorsaatbehandlung *f* pre-sowing treatment *(of soils with pesticides)*
Vorsaatherbizid *n* pre-sowing herbicide (weedkiller)
Vorsatzkuchen *m (glass)* shear cake
Vorsatzpapier *n* [book] end paper, book-lining paper
vorschärfen *(text)* to sharpen
vorschäumen *(plast)* to prefoam
Vorscheidung *f (sugar)* predefecation, preliming
heiße V. hot predefecation
kalte V. cold predefecation
progressive V. progressive predefecation
warme V. *s.* heiße V.
vorschmelzen to premelt, to prefuse
Vorschmelzkammer *f (pap)* sulphur melter *(of a sulphur burner)*
vorschreiben to order, to prescribe; to specify *(the qualities of a product)*; to formulate *(e.g. the mixture of a batch)*
Vorschrift *f* instruction, direction; specification *(relating to the qualities of a product)*; formula *(as for mixing a batch)*; *(pharm)* prescription
vorschrumpfen *(text)* to preshrink
Vorschrumpfung *f (text)* preshrinking
Vorschub *m* feed[ing]
Vorschubgeschwindigkeit *f* rate of feed[ing]
vorschwelen to precarbonize
Vorschwelung *f* precarbonization, *(in producing gas also)* predistillation
Vorsieb *n* preconditioning screen
Vorsilbe *f* / **vervielfachende** *(nomencl)* multiplying prefix, multiplicative numeral (numerical) prefix
vorsintern to presinter
Vorsinterung *f* presintering
vorsortieren *(pap)* to prescreen
Vorsortierer *m (pap)* preknotter
Vorsortierung *f (pap)* prescreening, [pre]knotting
vorspannen to prestress; *(glass)* to temper
Vorspannung *f* prestressing; *(glass)* tempering
vorspinnen *(text)* to rove
Vorspinnmaschine *f (text)* flyer
vorstabilisieren *(text)* to preset
Vorstabilisierung *f (text)* presetting
Vorsteuerleitung *f* pilot-supply line *(hydraulics and pneumatics)*
Vorsteuerventil *n* pilot valve *(hydraulics and pneumatics)*
Vorstoß *m* adapter; *(distil)* [condenser] adapter, receiver (delivery) tube
V. für Frittentiegel *(lab)* crucible adapter
Vorstoßpapier *n s.* Vorsatzpapier
Vorstufe *f* precursor, progenitor
vorstufenfrei precursor-free
Vortex *m* vortex
Vortrockenzylinder *m (pap)* predryer, baby (receiving) dryer
vortrocknen to predry
Vortrockner *m* predryer; *(ceram)* preheating dryer; *s.* Vortrockenzylinder
Vortrocknung *f* predrying, preliminary drying
Vorturm *m (pap)* strong[-acid] tower *(of a two-tower system)*
Voruntersuchung *f* preliminary examination
Vorvakuum *n* initial vacuum, forevacuum
Vorvakuumpumpe *f* forepump
Vorverdampfer *m* pre-evaporator
Vorversilberungsbad *n* strike solution for silver plating
Vorversuch *m* preliminary test
V. mit kleinen Substanzmengen small-scale [preliminary] test
Vorvulkanisation *f (rubber)* prevulcanization, precure, semicure, set cure
vorvulkanisieren *(rubber)* to prevulcanize, to precure
Vorwachs *n* propolis, bee glue
vorwärmen to preheat, to warm [up]; *(glass)* to warm in
Vorwärmer *m* preheater, feed heater
regenerativer V. [heat] regenerator

rekuperativer V. recuperator
Vorwärmgerät *n* preheater
Vorwärmkammer *f* regenerator chamber
Vorwärmung *f* preheating, forewarming, warming[-up]; *(glass)* warming-in
dielektrische V. dielectric preheating
Vorwärmwalze *f (rubber)* preheater
Vorwärmwalzwerk *n (rubber)* preheating (warming, warm-up) mill
Vorwärmzeit *f* preheat time
Vorwärmzone *f* preheating zone (compartment); non-boiling zone *(of an evaporator)*
Vorwäsche *f (text)* bottoming *(before bleaching)*
vorwaschen *(text)* to bottom *(before bleaching)*
Vorweiche *f (tann)* presoaking
vorweichen *(tann)* to presoak; *(pap)* to preimpregnate, to presoak, to steep
Vorweichen *n (pap)* preimpregnation, preliminary impregnation (penetration), presoaking, steeping, steepage *(of chips)*
Vorzugsbenennung *f (nomencl)* preferred name
Vorzugsmilch *f* certified milk
Vorzugsorientierung *f* preferred orientation
Vorzugsretention *f (bioch)* preferential retention
Vorzugsrichtung *f* preferred direction, directional preference
Votator *m*, **Votatoranlage** *f* votator [chilling unit], votator margarine plant, margarine votator
Votatormargarine *f* votator margarine
Votatorverfahren *n* votator process *(for continuous margarine making)*
VPI *s.* VPI-Stoff
VPI-Korrosionsschutzpapier *n*, **VPI-Papier** *n* vapour-phase inhibitor paper
VPI-Stoff *m* vapour-phase inhibitor, V.P.I.
VTC *s.* Viskositäts-Temperatur-Koeffizient
VT-Kurve *f* viscosity-temperature curve (slope)
Vulkanfiber *f* vulcanized fibre
Vulkanfiberersatz *m* semivulcanized board
Vulkanfiberrohpapier *n*, **Vulkanfiberrohstoff** *m* base paper for vulcanized fibre
Vulkanglas *n* volcanic glass
Vulkanisat *n (rubber)* vulcanizate, vulcanized product
γ-Vulkanisat *n* gamma vulcanizate
Vulkanisatabfälle *mpl* [vulcanized] rubber scrap, [vulcanized] waste rubber
Vulkanisateigenschaften *fpl* vulcanizate properties
Vulkanisation *f (rubber)* vulcanization, vulcanizing, cure, curing
V. bei Raumtemperatur room-temperature vulcanization, room cure
V. durch Bestrahlung radiation vulcanization (cure)
V. durch energiereiche Strahlung vulcanization (cure) by high-energy radiation
V. in Dampf steam vulcanization (cure)
V. in der Presse (Preßform) press vulcanization (cure)
V. in Formen mould vulcanization (cure), moulding
V. in Heißluft hot-air (dry-air) vulcanization (cure), HAV, heat vulcanization (cure), air cure
V. in offenem Dampf open-steam vulcanization (cure)
V. in Wasser hydraulic vulcanization (cure)
V. in zwei Stufen two-step vulcanization (cure)
V. mit Epoxidharzen epoxy vulcanization (cure)
V. mit Peroxid peroxide vulcanization (cure)
V. mit Schwefelspendern sulphur-donor vulcanization (cure), non-elemental sulphur cure, NES cure
V. mit Thiuramdisulfiden thiuram disulphide vulcanization (cure)
V. nach Peachey Peachey vulcanization (cure)
V. ohne freien Schwefel *s.* V. mit Schwefelspendern
V. unter Blei lead press vulcanization (cure)
absatzweise V. length-by-length vulcanization (cure)
kontinuierliche V. continuous vulcanization (cure), CV
kontinuierliche V. im Dampfrohr continuous steam vulcanization
optimale V. optimum vulcanization (cure)
schwefelfreie V. sulphurless (non-sulphur) vulcanization (cure)
stückweise V. *s.* absatzweise V.
Vulkanisations ... *s.a.* Vulkanisier ...
Vulkanisationsagens *n s.* Vulkanisationsmittel
Vulkanisationsbeschleuniger *m* vulcanization (cure, rubber) accelerator
Vulkanisationseigenschaften *fpl* vulcanization (curing) characteristics
Vulkanisationseinsatz *m* starting (beginning, onset) of vulcanisation, starting of cure
später (verzögerter) V. scorch delay
Vulkanisationsfreudigkeit *f* susceptibility to vulcanization (cure)
Vulkanisationsgrad *m s.* Vulkanisationskoeffizient
Vulkanisationskoeffizient *m* vulcanization coefficient, degree of vulcanization (cure)
Vulkanisationskurve *f* vulcanization (curing) curve
Vulkanisationsmittel *n* vulcanizing (curing) agent
Vulkanisationsmittel *npl* vulcanizing (curing) ingredients, curatives
Vulkanisationsoptimum *n* optimum of vulcanization (cure)
Vulkanisationsplateau *n* cure plateau, plateau effect ✦ **mit breitem V.** flat-curing ✦ **mit kurzem V.** peaky-curing
Vulkanisationsreaktion *f* vulcanization (curing) reaction
Vulkanisationssystem *n* vulcanization (vulcanizing, curing, curative) system
Vulkanisationsverfahren *n* vulcanization (curing) process
Vulkanisationsverhalten *n* vulcanization (curing) characteristics
Vulkanisationsverlauf *m* course of vulcanization (cure)

Vulkanisationsverzögerer *m* antiscorcher, antiscorching agent, retarder
Vulkanisationsverzögerung *f* retardation of vulcanization (cure)
Vulkanisationsvorgang *m* vulcanization (curing) process
Vulkanisationszeit *f* vulcanization (vulcanizing, curing) time
Vulkanisationszustand *m* state of vulcanization (cure)
Vulkanisator *m s.* Vulkanisierapparat
vulkanisch volcanic, vulcanic, igneous
Vulkanisier... *s.a.* Vulkanisations...
Vulkanisierapparat *m* vulcanizer, vulcanizing apparatus, heater
vulkanisierbar vulcanizable, curable
  mit Schwefel v. sulphur-vulcanizable, sulphur-curable, sulphur-curing
  nicht v. non-vulcanizable, non-vulcanizing, non-curing
vulkanisieren to vulcanize, to cure
  bei Raumtemperatur v., in der Kälte v. to vulcanize (cure) at room temperature
  in Formen v. to mould
vulkanisierend / bei Zimmertemperatur room-temperature-vulcanizing, RTV
  langsam v. slow-curing
  rasch (schnell) v. fast-curing, quick-curing
Vulkanisierform *f* vulcanizing (curing) mould
Vulkanisiergerät *n s.* Vulkanisierapparat
Vulkanisierkessel *m* vulcanizing autoclave (boiler), open-steam vulcanizer
  liegender V. horizontal vulcanizer
Vulkanisiermaschine *f* continuous vulcanizer (vulcanizing machine)
Vulkanisierofen *m* vulcanizing (curing) oven
Vulkanisierpresse *f* vulcanizing (curing, moulding) press
  hydraulische V. hydraulic vulcanizing press
vulkanisiert / mit Peroxid peroxide-cured
  mit Schwefel v. sulphur-cured
Vulkanisiertrommel *f* vulcanizing (curing) drum, cylinder
Vulkanisierung *f* vulcanization, cure
Vulkanit *m* volcanic rock
VZ *s.* Verseifungszahl

W

Waage *f* balance, *(of simple construction also)* scales
  W. mit Dämpfung[seinrichtung] damped balance
  analytische W. analytical balance
  chemische W. chemical (assay) balance
  gleicharmige W. equal-arm balance
  Mohr-Westphalsche W. Mohr-Westphal balance
  Westphalsche W. Westphal balance
Waagebalken *m* balance beam
Waagekasten *m* balance case
Waagengehäuse *n* balance case
Waagensäule *f* balance column
Waagerechtziehverfahren *n* **für Tafelglas** horizontal sheet drawing process
Waagschale *f* balance pan
  linke W. left-hand pan *(of an analytical balance)*
  rechte W. right-hand pan *(of an analytical balance)*
Wabe *f* honeycomb
wabenartig strukturiert honeycombed
Wabenbauweise *f (plast)* sandwich construction
Wabenhonig *m* comb honey
Wabenkern *m (plast)* honeycomb core *(of a sandwich construction)*
Wabenmittellage *f (plast)* honeycomb sandwich
Wabenstruktur *f* honeycomb structure ✦ **mit W.** honeycombed
Wacholder[beer]öl *n* juniper (Jupiter) oil, oil of juniper berries *(from Juniperus communis L.)*
Wacholderteer *m* gum juniper *(from Juniperus specc.)*
Wachs *n* wax
  amorphes W. *s.* mikrokristallines W.
  Chinesisches W. Chinese [tree] wax, insect wax, vegetable spermaceti *(secreted by scale lice)*
  gelbes W. yellow wax *(unbleached beeswax)*
  grünes W. *(incorrectly for)* Myrikatalg
  Japanisches W. *s.* vegetabilisches W.
  mikrokristallines W. micro[crystalline] wax, microparaffin
  mineralisches W. mineral (fossil) wax
  tierisches W. animal wax
  vegetabilisches W. vegetable wax, *(specif)* sumac wax, Japan tallow (wax) *(from Rhus succedanea L.)*
  weißes W. white wax, bleached beeswax
Wachsappretur *f* wax finish
wachsartig wax-like, waxy
Wachsausschmelzmodell *n (met)* lost-wax investment pattern, wax pattern
Wachsausschmelzverfahren *n (met)* lost-wax (cire-perdue) process
wachsen 1. to wax; 2. to grow *(as of crystals or polymer chains)*; to increase *(as of entropy)*
Wachsen *n* 1. waxing; 2. growth *(as of crystals or polymer chains)*; increase *(as of entropy)*
wächsern waxy
wachsfrei wax-free
Wachsglanz *m (min)* waxy lustre
wachshaltig wax-bearing
Wachskerze *f* wax candle
Wachsleim *m (pap)* wax size
Wachsmodell *n* **[/ verlorenes]** *(met)* lost-wax investment pattern, wax pattern
Wachspapier *n* wax[ed] paper
Wachstuch *n* oilcloth
Wachstum *n* growth *(as of crystals)*; growth, propagation *(of polymer chains)*; increase *(as of entropy)*
Wachstumsbedingung *f* condition of growth
Wachstumsfaktor *m* growth[-promoting] factor, growth-stimulating factor

wachstumsfördernd growth-promoting
Wachstumsgeschwindigkeit *f* rate of growth, *(relating to polymer chains also)* rate of propagation
Wachstumshemmung *f* inhibition of growth
Wachstumshormon *n* growth hormone
Wachstumsperiode *f* period of growth; *(relating to polymers)* period of chain growth (propagation)
Wachstumsregulator *m* growth-regulating substance, growth regulator (modifier)
Wachstumsstadium *n* growth (propagation) stage *(of polymers)*
Wächter *m (tech)* safeguard
Wad *n(m) (min)* wad, bod manganese, black ochre *(manganese(IV) oxide)*
waf *s.* wasser- und aschefrei
Wägebehälter *m* weighing tank
Wägebürette *f* weighing burette
Wägefehler *m* weighing error
Wägefläschchen *n* **[für Dichtemessungen]** density (specific gravity) bottle (flask), pycnometer
Wägeform *f* weighed form *(gravimetric analysis)*
Wägegefäß *n* weighing tank
Wägeglas *n*, **Wägegläschen** *n* weighing bottle
Wägeglaskappe *f* ground-glass lid
Wägegut *n* material being (*or* to be) weighed
wägen to weigh
Wägeraum *m* weighing (balance) room
Wägeröhrchen *n* weighing tube
Wägesatz *m* set of weights
Wägeschiffchen *n* weighing scoop (boat), balance dish
Wägestück *n* balance weight
Wägetisch *m* balance table
Wägezimmer *n* weighing (balance) room
wagging-Schwingung *f* wagging vibration *(spectroscopy)*
Waggonwaage *f* wagon balance
Wagnerit *m (min)* wagnerite *(a magnesium phosphate containing fluorine)*
Wagner-Meerwein-Umlagerung *f (org chem)* Wagner-Meerwein rearrangement
Wägung *f* weighing
  Bordasche W. Borda's method [of weighing]
  direkte (einfache) W. direct weighing
  Gaußsche W. Gauss's method of double weighing
Wägungs... *s.* Wäge...
Wahrscheinlichkeit *f* probability
  thermodynamische W. thermodynamic probability
Wahrscheinlichkeitsamplitude *f* probability amplitude
Wahrscheinlichkeitsdichte *f* probability density
Wahrscheinlichkeitsfaktor *m* probability (steric) factor *(reaction kinetics)*
Wahrscheinlichkeitsverteilung *f* probability distribution
Waid *m* 1. woad, Isatis tinctoria L.; 2. woad *(natural dye from 1.)*
Waldboden *m* / **brauner** brown forest soil
Waldtorf *m* forest (wood) peat
Walke *f* 1. *(text)* milling, fulling; 2. *s.* Walkmaschine
  alkalische W. alkali[ne] milling
  saure W. acid milling
Walkechtheit *f (text)* fastness to milling
walken 1. *(text, tann)* to mill, to full; 2. *(ceram)* to wedge
Walk[er]erde *f* fuller's earth
Walkfett *n (tann)* dressing grease
Walkmaschine *f* milling (fulling) machine, mill, beater
Wallace-Härtemesser *m (rubber)* Wallace pocket meter
Wallebertran *m* whale-liver oil
wallen to bubble, to boil up
Wallner-Linien *fpl (glass)* Wallner lines
Wallonen *fpl* valonea, valonia *(acorn cups rich in tannin from several oriental Quercus specc.)*
Walnußöl *n* walnut oil *(from Juglans regia L.)*
Walöl *n* 1. whale (train) oil; 2. *s.* Walratöl
Walrat *m(n)* spermaceti [wax]
Walratöl *n* sperm [whale] oil
  mineralisches W. mineral sperm [oil]
Walspeck *m* [whale] blubber
Waltran *m* whale (train, blubber) oil
Waltranhartfett *n* hydrogenated whale oil
Walzblock *m (met)* billet
Walzdruck *m* roll pressure
Walze *f* roll[er]; *(if hollow)* cylinder, drum ✦ **mit mehreren Walzen [versehen]** multiroll[er], multiple-roll
  W. der Greiferpresse (Transportpresse, Zugpresse) *(pap)* tag (squeeze, squeezing, drawing-in) roll
  elastische W. *(pap)* filled (resilient) roll *(of a supercalender)*
  geriffelte W. corrugated (fluted) roll
  gummierte W. rubber-covered roll
  parallel geführte W. parallel roll
  schwimmende W. *(text)* swimming roll
walzen to roll, to mill
Walzen *fpl* / **gegenläufige** contrarotating rolls
wälzen *(glass)* to marver *(a gather on a flat plate)*
Walzenabnahme *f* roll discharge
Walzenanpreßdruck *m* nip pressure
Walzenauftragmaschine *f* roll coater
Walzenaushebeverfahren *n (glass)* machine cylinder method
Walzenbelastung *f* roll loading
Walzenblasverfahren *n (glass)* hand cylinder method
Walzenbombage *f* roll crown
Walzenbrecher *m* roll[ing] crusher
  W. mit geriffelten Walzen corrugated-roll crusher
  W. mit glatten Walzen smooth-roll crusher
Walzenbrikett[ier]presse *f* roll-type briquette (briquetting) machine, Belgium roll machine
Walzendruck *m* 1. roll[er] pressure; 2. *(text)* roll[er] printing
Walzendruckmaschine *f (text)* roll[er] printing machine

Walzendünnschichttrockner *m* drum film dryer
Walzenegreniermaschine *f (text)* roller gin
Walzenentfleischmaschine *f (tann)* rubber-roll fleshing machine
Walzenglättwerk *n* calender [machine]
Walzenlager *n* roll[er] bearing
Walzenmesser *npl (pap)* [beater] roll bars, fly bars, roll blades *(of a Hollander beater)*
Walzenmilchpulver *n* roller[-dried] milk powder
Walzenmischer *m* mixing mill (rolls)
Walzenmischung *f (rubber)* roll compound
Walzenmühle *f* roller mill
Walzenoberfläche *f* roll surface, face of the roll
Walzenpaar *n* roll[er] pair
Walzenpresse *f* 1. roll[er] press; 2. *(lab)* cork roller
Walzenringmühle *f* ring-roll mill *(with vertical grinding ring)*
Walzensatz *m* set (stack) of rolls
Walzenscheider *m* rotor separator, induced-roll [magnetic] separator, magnetic drum [separator]
Walzenschiff *n (rubber)* mill pan *(in a mixing mill)*
Walzenschränkung *f* skew (cross-axis) mounting
Walzenschüsselmühle *f* [roller-and-]bowl mill
Walzensinter *m (met)* roll (mill) scale
Walzenspalt *m* roll nip, nip of the rolls, bite *(clearance between the rolls)*
Walzenspeiser *m* roll feeder
Walzenstreichmaschine *f* roll coater
Walzenstuhl *m* roller mill
Walzentrockner *m* drum dryer
Walzentrocknung *f* drum drying
Walzenzunder *m (met)* roll (mill) scale
Walzfell *n (plast)* rolled sheet
Walzgerüst *n* rolling-mill stand
Walzglas *n* rolled glass
Walzhaut *f (met)* skin
Wälzkolbengebläse *n* Roots (cycloidal) blower
Wälzkolbenverdichter *m* straight-lobe compressor
Wälzlager *n* rolling bearing
Wälzlagerfett *n* antifriction bearing grease
Walzmaschine *f (glass)* rolling (casting) machine
Wälzmühle *f* roller mill
Wälzplatte *f (glass)* marver [plate]
Walzpuppe *f (plast)* billet
Walzsinter *m (met)* roll (mill) scale
Walzwerk *n* rolling mill; *(rubber)* open roll mill
  ✦ **auf dem W. mischen** *(rubber)* to mill-mix
  W. zum Mastizieren *(rubber)* masticating mill
  enggestelltes W. tight mill
Walzzunder *m (met)* roll (mill) scale
Wand *f* wall
  Blochsche W. *(cryst)* Bloch wall
  halbdurchlässige (semipermeable) W. semipermeable membrane
Wanddicke *f* wall thickness
Wanddruck *m (biol)* turgor [pressure]
Wanderbett *n (tech)* moving bed
Wanderbettverfahren *n (tech)* moving-bed process
Wanderfläche *f (nucl)* migration area
Wanderlänge *f (nucl)* migration length
wandern 1. to travel *(as of bulk material being processed)*; 2. to migrate *(as of ions, groups, or pigments)*; 3. *(petrol)* to migrate; 4. to drift *(as of zero point)*
Wandern *n* **der Schmelzzone** zone travel[ling] *(in zone melting)*
Wandernutsche *f* travelling-pan filter, TP filter
Wanderrost *m* travelling (chain) grate
Wanderrostfeuerung *f* chain-grate stoker
Wanderung *f* 1. travel *(as of bulk material being processed)*; 2. migration *(as of ions, groups, or pigments)*; 3. *(petrol)* migration; 4. drift *(as of zero point)*
wanderungsbeständig *(coat)* non-migrating
Wanderungsgeschwindigkeit *f* 1. travel rate *(as of bulk material being processed)*; 2. migration speed (velocity) *(as of ions)*
  W. der Schmelzzone rate of zone travel, zone (zoning) speed *(in zone melting)*
Wanderungsrichtung *f* **der Schmelzzone** direction of zoning (zone travel) *(in zone melting)*
Wanderungsstrom *m (el chem)* migration current
Wandfliese *f* wall tile
Wandkatalyse *f* wall catalysis
Wandpappe *f* wall board
Wandreaktion *f* wall reaction
Wandreibung *f* wall friction
Wandung *f* wall
Wanne *f* 1. *(tech)* trough, tub, vat; *(glass)* tank; crucible *(of an arc furnace)*; 2. *s.* Wannenform
  kontinuierliche W. *(glass)* continuous tank
  periodische W. *(glass)* day tank
  pneumatische W. pneumatic trough
Wannenform *f* boat conformation (form), boat *(stereochemistry)*
Wannenglas *n* tank glass
Wannenofen *m (glass)* tank furnace
  W. mit Längsfeuerung end-fired (end-port) furnace
  W. mit Querbrennern (Querfeuerung, querziehender Flamme) *s.* querbeheizter W.
  querbeheizter W. side-fired (side-port) furnace
Wannenstein *m (glass)* tank block
Ward-Feuerung *f* Ward bagasse furnace
Ware *f (ceram)* ware
  Delfter W. delft[ware], delph[ware]
  einmal gebrannte W. once-fired ware
  geschrühte W. biscuit, bisque, biscuitted (biscuit-fired) ware
Warendichte *f (text)* gauge, gg.
Warendurchlauf *m (text)* cloth passage
Warenname *m* **/ freier** non-proprietary name
Warenprobe *f* sample
Warenspeicher *m (text)* J box *(as for bleaching)*
Warenzeichen *n* trademark
  eingetragenes (registriertes) W. registered trademark
Warenzeicheninhaber *m* registrant
warm halten to stove
warmbehandeln to heat-treat

Warmbehandlung *f* heat treatment
Warmbehandlungsofen *m* heat-treatment furnace
warmblasen to blast, to blow *(in producing water gas)*
Warmblasen *n* blasting, blow[ing] *(in producing water gas)*
Warmblaseperiode *f* blow period *(in producing water gas)*
Warmbleiche *f (pap)* warm bleach[ing]
Wärme *f (tech)* heat; *(sensation)* warmness, warmth
differentielle W. differential heat
freie W. uncombined heat
fühlbare W. sensible heat
gebundene W. latent heat
integrale W. integral heat
intermediäre W. *s.* differentielle W.
latente W. latent heat
mäßige W. moderate heat
radiogene W. radiogenic (radioactive) heat
spezifische W. specific heat [capacity], specific thermal capacity
ungebundene W. uncombined heat
wärmeabgebend exothermic, heat-giving
Wärmeableitung *f* dissipation of heat
Wärmealterung *f* heat ageing, thermosenescence
Wärmeänderung *f* heat change
Wärmeanstieg *m* heat rise
Wärmeanwendung *f* application of heat
Wärmeäquivalent *n* equivalent of heat
elektrisches W. electrical equivalent of heat
mechanisches W. mechanical equivalent of heat
wärmeaufnehmend endothermic
Wärmeausdehnung *f* thermal expansion
Wärmeausdehnungskoeffizient *m*, **Wärmeausdehnungszahl** *f* coefficient of thermal expansion
Wärmeausgleichgrube *f (met)* soak[ing] pit
Wärmeausstrahlung *f* radiation of heat, thermal (calorific, caloric) radiation
Wärmeaustausch *m* heat exchange (interchange)
Wärmeaustauschabteilung *f* / **regenerative** regeneration section *(of a plate pasteurizer)*
Wärmeaustauscher *m* heat exchanger (interchanger)
direkter W. direct-contact heat exchanger
Wärmeaustauschfläche *f* heat-exchanging (heat-exchange) surface
Wärmebad *n* heating bath
Wärmebearbeitung *f (chem)* heat processing
Wärmebedarf *m* heat requirement[s], *(quantitatively also)* amount of heat required
Wärmebehälter *m* heat reservoir
wärmebehandeln to heat-treat
Wärmebehandlung *f* heat treatment (processing)
Wärmebehandlungsofen *m* heat-treatment furnace
Wärmebelastung *f* heat load
wärmebeständig *(relating to deformation)* heat-resistant; *(relating to decomposition)* thermally stable
Wärmebeständigkeit *f (relating to deformation)* resistance to heat, *(plast also)* plastic yield; *(relating to decomposition)* thermal stability, thermostability
Wärmebewegung *f* thermal motion (agitation)
Wärmebilanz *f* thermal (heat) balance
wärmebindend endothermic
wärmedämmend heat-insulating
Wärmedämmstoff *m* heat-insulating material, thermal (heat) insulator
Wärmedämmung *f* thermal (heat) insulation
Wärmedehnung *f* thermal expansion
wärmedurchlässig diathermanous, diathermic
Wärmedurchlässigkeit *f* diatherma[n]cy
Wärmeeffekt *m* thermal (heat) effect
Wärmeeinheit *f* unit of heat, heat unit
wärmeelastisch thermoelastic
Wärmeelastizität *f* thermoelasticity
wärmeempfindlich heat-sensitive
Wärmeempfindlichkeit *f* heat sensitivity
Wärmeenergie *f* thermal (heat) energy
Wärmeentwicklung *f* 1. evolution of heat; 2. *(rubber)* heat (temperature) build-up *(with dynamic stress)*
wärmeerzeugend heat-generating, calorific, calorigenic
wärmefest *s.* wärmebeständig
Wärmefluß *m* heat flow
Wärmefreisetzung *f* heat release
Wärmefunktion *f* / **Gibbssche** Gibbs function, free enthalpy
Wärmegrube *f (met)* soak[ing] pit
wärmehärtbar thermosetting, thermoreactive
Wärmehaushalt *m* thermal (heat) balance
Wärmeimpulssiegeln *n (plast)* thermal-impulse heat sealing
Wärmeinhalt *m* heat content, *(per unit mass also)* enthalpy, heat function at constant pressure
Wärmeinhalts-Konzentrations-Diagramm *n* enthalpy/concentration diagram, heat-content/concentration diagram (chart)
Wärmeinhalts-Temperatur-Diagramm *n* enthalpy/temperature diagram, heat-content/temperature diagram (chart), It diagram
Wärmeisolation *f* thermal (heat) insulation
Wärmeisolator *m*, **Wärmeisolierstoff** *m* heat-insulating material, heat insulator
wärmeisoliert heat-insulated
Wärmeisolierung *f* thermal (heat) insulation
Wärmekammer *f* heated chamber
Wärmekapazität *f* heat (thermal) capacity
spezifische W. specific heat [capacity], specific thermal capacity
Wärmekonvektion *f* thermal convection (siphoning)
Wärmeleiter *m* heat conductor
Wärmeleitfähigkeit *f* 1. thermal (heat) conductivity; 2. *s.* Wärmeleitzahl
Wärmeleitfähigkeitszelle *f* thermal-conductivity cell
Wärmeleitung *f* heat (thermal) conduction
Wärmeleitvakuummeter *n* thermal conductivity [vacuum] gauge

Wärmeleitvermögen *n* 1. thermal (heat) conductivity; 2. *s.* Wärmeleitzahl
spezifisches W. *s.* Wärmeleitzahl
Wärmeleitzahl *f* coefficient of thermal conduction, specific thermal conductivity
wärmeliebend thermophilic, thermophilous
wärmeliefernd exothermic, heat-giving
Wärmemenge *f* quantity (amount) of heat
Wärmemitführung *f* thermal convection (siphoning)
wärmen to warm [up]
Wärmenachbehandlung *f* postheat treatment
Wärmeofen *m* heating furnace
wärmeoxydativ thermal-oxidative
Wärmeplastizität *f* thermoplasticity
Wärmepolymer[es] *n,* **Wärmepolymerisat** *n* heat (high-temperature, hot) polymer
Wärmepolymerisation *f* heat (high-temperature, hot) polymerization
Wärmepumpe *f* heat pump
Wärmequelle *f* heat source
Wärmeregler *m* thermoregulator
Wärmereservoir *n* heat reservoir
Wärmerückgewinnung *f* recovery of heat
Wärmesatz *m* / **Nernstscher** Nernst heat theorem, third law of thermodynamics
Wärmeschutz *m* thermal (heat) insulation
Wärmeschutzglas *n* heat-absorbing glass
Wärmeschutzstoff *m* heat-insulating material, heat insulator
Wärmeschwingung *f (cryst)* thermal vibration
wärmesensibel eingestellt *(rubber)* heat-sensitized
Wärmesensibilisierung *f* heat sensitization
Wärmesensibilisierungsmittel *n* heat sensitizer
Wärmespeicher *m* heat reservoir (accumulator)
Wärmespeicherung *f* heat storage
Wärmespritzen *n (plast)* flame spraying
wärmestabil *s.* wärmebeständig
Wärmestabilisator *m* heat stabilizer
Wärmestabilisierung *f* heat stabilization
Wärmestabilisierungsmittel *n* heat stabilizer
Wärmestein *m* pebble *(of a pebble heater)*
Wärmesteinerhitzer *m* pebble heater (stove)
Wärmesteinschüttung *f* pebble bed *(in a pebble heater)*
Wärmestrahler *m* heat lamp
Wärmestrahlung *f* radiation of heat, thermal (calorific, caloric) radiation
Wärmestrom *m* heat flow
Wärmetauscher *m* heat exchanger (interchanger)
Wärmetheorem *n* [/ **Nernstsches**] [Nernst] heat theorem
Wärmetönung *f* heat tonality
Wärmeträger *m* heat carrier, heat-exchanging (heat-transfer) medium, heating medium
flüssiger W. thermal liquid
Wärmetransmissionskoeffizient *m* coefficient of heat transfer
Wärmeübergang *m* heat (thermal) transmission (transfer), heat exchange (interchange)
Wärmeübergangszahl *f* heat-transfer coefficient
Wärmeübertrager *m,* **Wärmeüberträger** *m* heat exchanger (interchanger)
Wärmeübertragung *f* heat (thermal) transmission (transfer), heat exchange (interchange)
Wärmeübertragungsfläche *f* heat-exchanging (heat-exchange) surface
Wärmeübertragungsmasse *f* heat-transfer cement
Wärmeübertragungsmittel *n s.* Wärmeträger
Wärmeübertragungsöl *n* heat-transfer oil
Wärmeübertragungssalz *n* heat-transfer salt, HTS
wärmeunbeständig thermolabile, heat-labile
wärmeundurchlässig atherm[an]ous
wärmeverbrauchend endothermic
Wärmeverhalten *n* thermal (high-temperature) properties
Wärmeverlust *m* heat loss
Wärmeverschiebung *f* heat displacement
Wärmeverteilung *f* heat distribution
wärmevulkanisierbar *(rubber)* heat-curable
wärmevulkanisiert *(rubber)* heat-cured
Wärmewiderstand *m* thermal resistance
Wärmewirkungsgrad *m* thermal efficiency
Wärmezufuhr *f* heat input
Warmfestigkeit *f (plast)* hot strength
Warmfetten *n (tann)* hot[-air] stuffing
Warmformen *n* hot forming (working), *(plast also)* thermoforming
Warmformgebung *f* hot forming (working)
Warmformmaschine *f (plast)* thermoforming machine
Warmhalteofen *m* heating (holding) furnace
Warmhärte *f* hot hardness
Warmkammerdruckgießen *n,* **Warmkammerdruckguß** *m* hot-chamber die (pressure) casting
Warmkammer[druckguß]maschine *f* hot-chamber [die-casting] machine
Warmkleber *m* hot-setting (hot-melt) adhesive, hot glue
Warmluft *f* hot air
Warmlufteintritt *m* hot-air inlet
warmpressen to hot-press
Warmpressen *n* hot pressing
Warmräucherei *f* hot smoking *(at 25°C)*
warmräuchern to hot-smoke *(at 25°C)*
Warmspritzen *n (coat)* warm spray
Warmstreckgrenze *f (plast)* yield point at elevated temperatures
Warmtonentwickler *m (phot)* warm-tone developer
Warmumformung *f* hot forming (working)
warmverarbeiten to hot-work
Warmverarbeitung *f* hot working
Warmverformen *n s.* Warmformen
Warmverpressen *n (coal)* warm briquetting
Warmwassertrichter *m* hot-water funnel, funnel heater
Wärmzone *f* heating zone
Warnetikett *n* caution label

Warnstoff *m* warning agent
Warnvorrichtung *f* alarm
Wartefrist *f (agric)* preharvest interval *(after application of pesticides)*
Wartezeit *f* 1. downtime, down period; 2. *s.* Wartefrist
  geschlossene W. *(plast)* closed assembly time
  offene W. *(plast)* open assembly time
WAS *s.* Stoff / waschaktiver
Waschaggregat *n (text)* scouring train
Waschanlage *f* 1. cleaning plant; 2. *(text)* scouring train
Waschapparat *m* washing device (machine), washer
Waschbad *n (text)* scouring bath
Waschband *n (filtr)* washing blanket
waschbar *s.* waschecht
Waschbehandlung *f* washing treatment
Waschbenzin *n* cleaner's naphtha (solvent)
waschbeständig *s.* waschecht
Waschblau *n* laundry blue
Waschbottich *m* wash tub (tank)
Waschbrause *f* washing spray
Waschbrett *n (glass)* washboard *(a surface defect)*
Waschbütte *f* wash tub (tank)
Waschdauer *f* washing period (time)
Waschdüse *f (text)* scour nozzle
Wäsche *f* 1. *(tech)* wash[ing], wet cleaning, wash-up; 2. *s.* Waschanlage
  W. mit Wasser water wash[ing]; *(pap)* water-washing stage
  alkalische W. *(pap)* alkaline washing stage, alkali [extraction] stage, caustic extraction [stage]
  mechanische W. mechanical washing
waschecht fast to washing, wash-fast, washable
Waschechtheit *f* fastness to washing, wash[ing] fastness, washability
Wascheffekt *m* cleaning (detergent) effect
Wascheigenschaften *fpl* detergent properties
Wascheindicker *m* washing [tray] thickener *(as in countercurrent decantation)*
waschen to wash, to wet-clean; to scrub *(gases)*; to launder *(textiles)*
  durch Rückspülung w. to backwash
  im Gegenstrom w. to wash countercurrently
  in der [Naß-]Setzmaschine w. *(min tech)* to jig-wash
Waschen *n* **im Strang** *(text)* rope scouring
  W. mit Alkalien alkaline wash[ing]
  W. mit Säure acid wash[ing]
  W. mit Wasser water wash[ing]
Wascher *m s.* Wäscher
Wäscher *m* 1. washer; 2. scrubber, wet collector, gas washer
Wascherz *n* placer [deposits]
waschfest *s.* waschecht
Waschflasche *f (lab)* wash[ing] bottle, absorption bottle, bubbler
  W. nach Drechsel Drechsel bottle
Waschflaschenbatterie *f* scrubbing train
Waschflotte *f (text)* washing (scouring) liquor, washing bath (liquid), detergent solution
Waschflüssigkeit *f* 1. wash[ing] liquid, washing liquor (solvent), *(if spent also)* washings; 2. scrubbing liquid *(for gases)*; 3. *s.* Waschflotte
Waschflüssigkeitszulauf *m* wash inlet
Waschgold *n* placer (stream) gold
Waschhilfsmittel *n* washing aid
Waschholländer *m (pap)* washer beater, Hollander washer, washing (potching) engine
Waschkasten *m* washing tank
Waschkolonne *f* scrubber (wash) column
Waschkraft *f* washing (cleaning) efficiency, detergent power, detergency
Waschkufe *f* wash tub
Waschkühler *m* washer-cooler
Waschkurve *f (min tech)* washability curve
Waschlauge *f s.* Waschflotte
Waschleistung *f s.* Waschkraft
Waschmaschine *f* washing machine, washer
Waschmittel *n* washing agent, detergent *(in a larger sense)*
  alkalisches W. alkaline detergent
  enzymatisches W. enzyme detergent
  synthetisches W. [synthetic] detergent, syndet
Waschmittelherstellung *f* detergent manufacture
Waschmittelindustrie *f* detergent industry
Waschöl *n* wash oil; *(petrol)* scrubbing (stripping, absorption) oil
  armes W. *(petrol)* lean oil *(for an absorption column)*
  beladenes W. *s.* reiches W.
  frisches (regeneriertes) W. *s.* armes W.
  reiches W. *(petrol)* rich oil *(in an absorption column)*
Waschölabsorption *f* wash-oil absorption
Waschplatte *f (filtr)* washing plate
Waschprobe *f* wash test
  W. nach Barret *(dye)* Barret's [wash] test
Waschpulver *n* washing powder
Waschrinne *f (min tech)* sluice
Waschsieb *n* rinse screen
Waschsoda *f* washing soda, salt of soda, [sal] soda, natron *(sodium carbonate-10-water)*
Waschtrommel *f* washing cylinder (roll, drum)
Waschturm *m* washing (scrubbing) tower, gas washer, scrubber
Wasch-und-Trage-Erzeugnis *n (text)* wash and wear product
Waschung *f* wash[ing]
Waschvermögen *n s.* Waschkraft
Waschwalzwerk *n (rubber)* washing mill (machine), wash mill
Waschwanne *f* wash tub
Waschwasser *n* washing (wash, washery, rinsing, rinse) water, wash [liquid], *(if spent also)* washings
Waschwirkung *f* cleaning (scouring, detergent) effect
Waschzyklon *m* cyclone washer (scrubber)
Wash-Primer *m (coat)* wash (reaction) primer, self-etching primer, self-etch pretreatment primer
Wasser *n* 1. H_2O water; 2. *(tech)* water, liquor, liquid; 3. *(cosmet)* wash, waters; 4. *(pharm)* wa-

ters; 5. *(min)* water *(limpidity and lustre of gems)* ✦ **mit W. verdünnbar** water-dilutable
chemisch gebundenes W. combined water
destilliertes W. distilled water
fließendes W. running water
freies W. free water (moisture), *(coal also)* accidental moisture
gebundenes W. bound water (moisture)
geklärtes W. clarified water
Goulardsches W. Goulard's extract, vinegar of lead *(aqueous solution of basic lead acetates)*
hartes W. hard water
hygroskopisches W. hygroscopic[al] water (moisture), water of deliquescence
inneres W. inherent moisture
kohlensäurehaltiges W. carbonated water, soda [water]
kohlensaures W. naturally carbonated water
konstitutiv gebundenes W. constitutional water, water of constitution
mechanisch gebundenes W. mechanically-held water, mechanical water
salzhaltiges W. saline (salt) water
schweres W. D_2O heavy water, deuterium oxide
weiches W. soft water
zeolithisches (zeolithisch gebundenes) W. zeolitic water
Wasserabgabe *f (chem)* liberation of water
Wasserablaßhahn *m* pet cock
Wasserablösung *f (met)* water quenching
Wasserabscheider *m* water separator
Wasserabschluß *m* water seal
Wasserabschreckung *f (met)* water quenching
wasserabsorbierend water-absorbing
Wasserabsorption *f* water absorption
Wasserabspaltung *f* elimination of water, dehydration
wasserabstoßend, wasserabweisend water-repellent, hydrophobic, hydrophobe
Wasserabweisung *f* water repellency
Wasserabweisungsvermögen *n* water repellency
Wasseranalyse *f* water analysis
Wasseranlagerung *f* water addition
wasseranziehend water-attracting, hygroscopic[al]
wasserartig aqueous
Wasseraufbereitung *f* water conditioning
Wasseraufnahme *f* water absorption, uptake of water
wasseraufnehmend water-absorbing, hydrophilic, hydrophile
Wasseraufsichtsbehörde *f* water pollution control authority
Wasserbad *n* [hot-]water bath
kupfernes W. copper bath
Wasserbebrausung *f* water spraying
Wasserbedarf *m* water requirements
Wasserbehälter *m* water tank
Wasserbehandlung *f* water treatment
Wasserbeize *f (dye)* water stain
Wasserberegnung *f*, **Wasserberieselung** *f* water spraying
wasserbeständig water-resistant
Wasserbeständigkeit *f* water resistance
Wasserbeständigmachen *n* waterproofing
Wasserbestimmungsapparat *m* **nach Dean-Stark** Dean and Stark apparatus
Wasserbilanz *f* water balance
Wasserchemie *f* water chemistry, hydrochemistry
Wasserdampf *m* steam, water vapour ✦ **mit W. behandeln** to steam
Wasserdampfaufnahme *f* steam (water vapour) absorption
Wasserdampfdestillation *f* steam distillation
wasserdampfdicht *s.* wasserdampfundurchlässig
Wasserdampfdruck *m* steam (water vapour) pressure
wasserdampfdurchlässig permeable to water vapour
Wasserdampfdurchlässigkeit *f* water vapour permeability (transmission), WVT
wasserdampfflüchtig steam-volatile
Wasserdampfpartialdruck *m* partial pressure of water vapour
wasserdampfundurchlässig resistant to water vapour, moisture-resistant, moistureproof
Wasserdampfundurchlässigkeit *f* water vapour resistance, moisture resistance (proofness)
wasserdicht water-tight, waterproof ✦ **w. abschließen** to waterproof ✦ **w. machen** *(text)* to waterproof
Wasserdichtausrüstung *f (text)* waterproof finish
Wasserdichtmachen *n (text)* waterproofing
Wasserdichtmacher *m (text)* waterproofing agent
Wasserdruck *m* water pressure; *(tech)* hydraulic pressure; *(petrol)* water drive
wasserecht fast to water
Wasserechtheit *f* fastness to water
Wassereis *n* water ice
Wasserelektrolyse *f* electrolysis of water
wasserenthärtend water-softening
Wasserenthärter *m* water softener
Wasserenthärtung *f* water softening
Wasserenthärtungsanlage *f* water-softening plant
Wasserenthärtungsmittel *n* water softener
Wasserentsalzung *f* desalinization (demineralization) of water
Wasserentsalzungsapparat *m* desalter, demineralizer
Wasserentzug *m* abstraction (removal) of water, dehydration
Wasserfarbe *f* water[-base] paint, *(Am also)* watercolor; water colour *(art)*
Wasserfassungsvermögen *n (soil)* water-holding capacity, WHC
wasserfeindlich hydrophobic, hydrophobe
wasserfest water-resistant
Wasserfestigkeit *f* water resistance
Wasserfestmachen *n s.* Wasserdichtmachen

Wasserflecken *mpl (phot)* drying marks
Wasserfluten *n (petrol)* water flooding
wasserfrei free from water, moisture-free, mf, moistureless, anhydrous, anh.
wasserfreundlich hydrophilic, hydrophile
wasserführend water-bearing
Wassergas *n* water gas
blaues W. blue [water] gas
karburiertes W. carburetted (enriched) water gas
Wssergasanlage *f* water-gas plant
Wassergasgenerator *m* water-gas generator
Wassergasgleichgewicht *n* water-gas equilibrium
Wassergasmaschine *f* water-gas machine
Wassergasreaktion *f* water-gas (steam-carbon) reaction *(reduction of water by means of carbon)*
Wassergasteer *m* water-gas tar
Wassergasverfahren *n* water-gas process
wassergebremst *(nucl)* water-moderated
Wassergehalt *m* water content
wassergekühlt water-cooled
wassergesättigt water-saturated
wassergeschützt waterproof
Wassergesetz *n* water pollution regulation
Wasserglas *n* water (liquid, soluble) glass *(sodium or potassium silicate)*
Wasserglaskitt *m* water glass cement
Wasserglasleim *m* water glass glue
Wasserglaslösung *f* water glass solution
Wasserhaltevermögen *n (soil)* water-holding capacity, WHC
wasserhaltig hydrous, aqueous; *(min)* enhydritic, enhydrous
Wasserhärte *f* water hardness
wasserhärten *(met)* to water-harden
Wasserhärter *m*, **Wasserhärtestahl** *m* water-hardening steel
Wasserhärtung *f (met)* water hardening
Wasserhaushalt *m (biol)* water metabolism
wasserhell water-white
wässerig *s.* wäßrig
Wasser-in-Fett-Emulsion *f* water-in-fat emulsion
Wasser-in-Öl-Emulsion *f* water-in-oil emulsion, W/O emulsion, *(agric also)* mayonnaise *(as of pesticides)*
Wasser-in-Öl-Typ *m* water-in-oil type, W/O type *(of emulsions)*
Wasserkalander *m (text)* water mangle
Wasserkalk *m* hydraulic (water) lime *(a lime which will harden under water)*
Wasserkalorimeter *n* water calorimeter
Wasserkanal *m* water channel
Wasserkapazität *f (soil)* water-holding capacity, WHC
Wasserkesselreaktor *m*, **Wasserkocherreaktor** *m* water-boiler reactor
Wasserkreislauf *m* water [re]cycle
Wasserkühler *m* water-cooled condenser
Wasserkühlung *f* water cooling
Wasserkultur *f (agric)* water (hydroponic) culture, hydroponics
Wasserlack *m* water varnish
Wasserleistung *f* capacity of water evaporation *(of a dryer)*
Wasserleitung *f* 1. water main (piping), water-supply line; 2. *(biol)* water transport
Wasserlinien *fpl (pap)* watermarked lines, waterlines
Wasserlinienpapier *n* laid paper
wasserlöslich water-soluble, hydrosoluble
Wasserlöslichkeit *f* water (aqueous) solubility
Wassermantel *m* water jacket ✦ **mit W.** water-jacketed
Wassermargarine *f* water margarine
Wassermesser *m* water meter
wassermoderiert *(nucl)* water-moderated
Wassermolekül *n* water molecule
Wassermörtel *m* hydraulic mortar
wässern 1. *(of human agent)* to steep, to soak; *(in running water)* to rinse; *(phot)* to wash, to rinse, to soak; 2. *(of material)* to soak
Wassernachweis *m* test for water
Wasseroberfläche *f* water surface
Wasserphase *f* aqueous (water) phase
Wasserreaktor *m* water-moderated (water-boiler) reactor
wasserreich rich in water
Wasserreinigung *f* water purification
Wasserreinigungsanlage *f* water purification plant
Wasserrohr *n* water tube (pipe)
Wasserrohrdampferzeuger *m*, **Wasserrohrkessel** *m* water-tube boiler
Wasserröste *f (text)* water ret[ting] *(of flax)*
W. in stehenden Gewässern dam retting
Wasserrotte *f s.* Wasserröste
Wasserschlag *m* water hammer
Wasserschleier *m* water curtain
Wasserschlepper *m (distil)* water entrainer
Wasserschüssel *f* water-sealed trough
Wasserspiegel *m* water level
Wasserspülung *f (petrol)* water-base mud
Wasserstand *m* water level
Wasserstand[s]anzeiger *m* gauge glass
Wasserstand[s]regler *m* water level regulator (controller), constant-level device
Wasserstoff *m* H hydrogen ✦ **über W. verbunden** hydrogen-bonded
aktiver W. active hydrogen
atomarer (einatomiger) W. atomic hydrogen, monohydrogen
indizierter W. indicated hydrogen
leichter W. ^{1}H protium, light hydrogen
schwerer W. $^{2}_{1}H$, D deuterium, heavy hydrogen
überschwerer W. $^{3}_{1}H$, T tritium
Wasserstoffabscheidung *f* liberation (evolution) of hydrogen
Wasserstoffabspaltung *f* elimination of hydrogen, dehydrogen[iz]ation
wasserstoffähnlich hydrogen-like
Wasserstoffakzeptor *m* hydrogen acceptor
wasserstoffarm poor in hydrogen
Wasserstoffatmosphäre *f* hydrogen atmosphere

Wasserstoffatom *n* hydrogen atom
bewegliches W. mobile hydrogen atom
Wasserstoffatomspektrum *n* [atomic] spectrum of hydrogen, hydrogen spectrum
Wasserstoffaustausch *m* hydrogen[-cycle] cation exchange *(as in water treatment)*
Wasserstoffaustauscher *m* hydrogen cation exchanger
Wasserstoffbindung *f* hydrogen bond (linkage) *(state)*; hydrogen bonding *(process)*
Wasserstoffbrüchigkeit *f* hydrogen embrittlement
Wasserstoffbrücke *f* hydrogen bridge
Wasserstoffbrückenbindung *f* hydrogen bridge bond (linkage)
Wasserstoffdisulfid *n* H_2S_2 hydrogen disulphide, disulphane
Wasserstoffdon[at]or *m* hydrogen donor
Wasserstoffdruck *m* hydrogen pressure
Wasserstoffelektrode *f* hydrogen [gas] electrode
Wasserstoffentschwefelung *f* hydrodesulphurization, HDS
Wasserstoffentwicklung *f* evolution of hydrogen
Wasserstoffentzug *m* abstraction of hydrogen, dehydrogen[iz]ation
Wasserstoffexponent *m s.* Wasserstoffionenexponent
Wasserstoffgas *n* hydrogen gas
Wasserstoffhalbelement *n* hydrogen half-cell
wasserstoffhaltig containing hydrogen, hydrogenous
Wasserstoffion *n* hydrogen ion
Wasserstoffionenaktivität *f* hydrogen-ion activity
Wasserstoffionenexponent *m* hydrogen ion exponent, pH value (number)
Wasserstoffionenkonzentration *f* hydrogen-ion concentration
Wasserstoffionenmessung *f* hydrogen-ion measurement
Wasserstoffkern *m* hydrogen nucleus
Wasserstoffnachweis *m* detection of hydrogen
Wasserstoffnormalelektrode *f* normal hydrogen electrode
Wasserstoffoxid *n* H_2O hydrogen oxide, water
Wasserstoffpentasulfid *n* H_2S_5 hydrogen pentasulphide, pentasulphane
Wasserstoffperoxid *n* H_2O_2 hydrogen peroxide
Wasserstoffradius *m* / **Bohrscher** Bohr radius
Wasserstoffreduktion *f* hydrogen reduction
wasserstoffreich rich in hydrogen, hydrogen-rich
Wasserstoff-Sauerstoff-Schweißen *n* oxyhydrogen welding
Wasserstoffsäure *f* hydrogen acid, hydracid
Wasserstoffskale *f* hydrogen scale *(for electrode potentials)*
Wasserstoffspektrum *n* hydrogen spectrum
Wasserstoffsulfid *n* H_2S hydrogen sulphide, monosulphane
Wasserstofftetrasulfid *n* H_2S_4 hydrogen tetrasulphide, tetrasulphane
Wasserstofftrisulfid *n* H_2S_3 hydrogen trisulphide, trisulphane
Wasserstoffüberspannung *f (el chem)* hydrogen overvoltage
Wasserstoffübertragung *f* hydrogen transfer
Wasserstoffverbindung *f* hydrogen compound
Wasserstoffversprödung *f* hydrogen embrittlement
Wasserstoffwechsel *m (biol)* water metabolism
Wasserstrahl *m* water jet
scharfer W. high-pressure water jet
Wasserstrahlentrinder *m (pap)* hydraulic (stream) barker
Wasserstrahlentrindung *f (pap)* hydraulic (stream) barking
Wasserstrahlgebläse *n s.* Wasserstrahlverdichter
Wasserstrahlpumpe *f* water-jet [vacuum] pump, water pump (aspirator)
Wasserstrahlverdichter *m* water[-operated] ejector
Wasserstrom *m* stream (current) of water, water flow
Wassertank *m* water tank
Wassertechnologie *f* water technology
Wassertransport *m (biol)* water transport
Wassertrieb *m (petrol)* water drive
Wassertrieblagerstätte *f (petrol)* water-drive field (reservoir)
Wassertröpfchen *n* water droplet
Wassertropfenprobe *f* water drop test
Wasser-Tupelo *m* tupelo (cotton) gum, Nyssa aquatica L.; black (sour) gum, Nyssa sylvatica Marsh.
Wasserturbine *f (lab)* water turbine
Wasserturm *m* water tower
Wasserüberlauf *m* clear overflow *(of a thickener)*
Wasserüberschuß *m* excess water
Wasseruhr *f* water meter
Wasserumlauf *m* water [re]cycle
wasser- und aschefrei moisture-and-ash-free, maf
wasserundurchlässig impermeable to water, water-tight, waterproof ✦ **w. machen** to waterproof
Wässerung *f* steep[ing], steepage, soaking; *(in running water)* rinse, rinsing; *(phot)* [water] washing, wash, rinse, rinsing, soaking
Wässerungstank *m (phot)* washing tank, print washer
Wässerungszeit *f (phot)* washing period (time)
wasserunlöslich insoluble in water, water-insoluble
Wasseruntersuchung *f* water analysis
Wasserverbrauch *m* water consumption (use, usage)
wasserverdünnbar water-dilutable
wasservernetzt *(rubber)* water-cross-linked
Wasserverschluß *m* water seal ✦ **mit W.** water-sealed
Wasserverschmutzung *f* water pollution, *(esp relating to poisons or bacteria)* water contamination
Wasserversorgung *f* water supply
Wasserversorgungsanlage *f* water-supply plant
Wasserversorgungsleitung *f* water-supply line

Wasserversorgungspumpe *f* water-supply pump
Wasserverteilung *f* water distribution
Wasserverteilungssystem *n* water distribution system
Wasserverunreinigung *f* 1. *(substance)* water impurity; 2. *s.* Wasserverschmutzung
Wasservorhang *m* water curtain
Wasservorrat *m* water supply
Wasservorwärmer *m* boiler feed preheater
Wasservulkanisation *f* hydraulic cure
Wasserwäsche *f* water wash[ing]
Wasserwerk *n* water works
Wasserwerkstatt *f (tann)* beamhouse
Wasserwert *m* water equivalent *(of a calorimeter)*
Wasserzähler *m* water meter
Wasserzeichen *n (pap)* watermark, w/m. ✦ **W. einarbeiten** to watermark
W. mit Schattierungen shaded mark
echtes W. genuine watermark
Wasserzeichenherstellung *f (pap)* watermarking
Wasserzeichenlinien *fpl (pap)* watermarked lines, water-lines
Wasserzeichenpapier *n* watermarked paper
Wasserzeichenwalze *f (pap)* [water]marking roll, dandy roll, watermarking dandy [roll]
Wasserzement *m* hydraulic cement
Wasser-Zement-Faktor *m*, **Wasser-Zement-Wert** *m* water-cement ratio
Wasserzusatz *m* water addition
wäßrig aqueous, hydrous, watery
nicht w. non-aqueous, anhydrous
Watkin-Kennkörper *m (ceram)* Watkin [heat] recorder
Watkins-Faktor *m (phot)* Watkins [development] factor
Watte *f* wadding, *(esp med)* cotton wool
W. in Lagen batting
Wattebausch *m* wad of cotton [wool]
Wavellit *m (min)* wavellite *(a hydrous basic aluminium phosphate)*
Webkette *f (text)* warp
Wechsel *m* change, alteration
W. der Strömungsrichtung change of direction of flow
Wechselbeanspruchung *f* alternating stress
Wechselbeziehung *f* correlation
Wechselbiegeprüfung *f (plast)* alternating bending test
wechseln to change
Wechselpapier *n* bill paper
wechselseitig mutual
Wechselstrom *m* alternating current, a.c., AC
Wechselstrom[licht]bogen *m* alternating-current arc, a.c. arc
Wechselstrompolarografie *f* alternating-current polarography, a.c. polarography
Wechseltauchprüfung *f*, **Wechseltauchversuch** *m* alternate immersion test *(a corrosion test)*
Wechselventil *n* reversing valve
Wechselwirkung *f* interaction ✦ **in W. stehen (treten)** to interact
interionische W. interionic action, ionic interaction
zwischenmolekulare W. molecular interaction
Wechselwirkungskraft *f* force of interaction
interionische W. interionic force
Wechselzahl *f (bioch)* turnover number
Wechselzersetzung *f* metathesis, double decomposition, exchange
Wedge-Ofen *m (met)* Wedge furnace
Weg *m* path *(of a reaction)*; trajectory, path *(of a particle)*
wegbrennen to burn off
Weglänge *f (phys chem)* length of path
freie W. free path
mittlere freie W. mean free path
weglösen to disperse *(printing-ink adhesive from paper fibres)*
wegschmelzen to deliquesce
wegspülen to rinse off
wegwandern to migrate out *(as of ions)*
wegwaschen to wash away
Wegwerfpatrone *f (filtr)* throw-away cartridge
wehenerregend *(pharm)* oxytocic
Wehr *n* weir
Wehrhöhe *f* weir level *(of a bubble-cap plate)*
weich soft ✦ **w. machen** to soften *(e.g. water, plastics)*
Weichasphalt *m* soft asphalt
Weichblei *n* soft (chemical) lead *(of more than 99.9 per cent purity)*
Weichdichtung *f* soft packing
Weiche *f* 1. *(tann)* soaking; 2. *(ferm)* steep tank, steeper
weichen 1. to steep, to soak; 2. *(of material)* to soak
Weichen *n* 1. steep[ing], steepage, soaking; 2. *(of material)* soaking
Weichferrit *m (ceram)* soft ferrite
Weichglas *n* soft [sealing] glass
weichglühen to soft-anneal, to spheroidize
Weichglühen *n* soft annealing, softening anneal, spheroidizing [anneal], spheroidization
Weichgrad *m (ferm)* degree of steeping
Weichgriffigkeit *f (text)* soft handle
Weichgummi *m* soft rubber
Weichgummiwalze *f* soft rubber-covered roll
Weichharz *n* soft resin
Weichheitsgrad *m* degree of softness
Weichheitsprüfgerät *n* hardness tester (meter)
Weichheitszahl *f (plast)* softness index (number)
Weichholz *n* soft wood
weichkochen to cook soft *(cellulose)*
Weichkohle *f* soft coal
Weichlot *n* soft solder
weichlöten to soft-solder
Weichlöten *n* soft soldering
weichmachen 1. to soften, to plasticize, to plastify, *(plast also)* to flux, to flexibilize; 2. *(tann)* to dress *(chamois leather)*
Weichmachen *n* 1. softening, plasticization, plastification, *(plast also)* fluxing; 2. *(tann)* dressing *(of chamois leather)*
Weichmacher *m* softener, plasticizer, softening (plasticizing) agent, flexibilizer

innerer W. polymerizable plasticizer
weichmacherfrei unplasticized
Weichmacheröl *n (rubber)* process[ing] oil
Weichmacherwanderung *f* migration of plasticizer
Weichmasse *f (ceram)* soft paste, pâte tendre
Weichnitrieren *n* soft-nitriding *(of steel)*
Weichpackung *f s.* Weichdichtung
Weichparaffin *n* soft paraffin [wax]
Weichporzellan *n* soft[-paste] porcelain
Weich-PVC *n* flexible (plasticized) PVC
Weichseide *f* souple silk
Weichstahl *m* soft (mild) steel
Weichstock *m (ferm)* steep tank, steeper
Weich- und Spritzmaschine *f (food)* soaker-hydro (soaker-sprayer) bottle washing machine
Weichwachs *n* soft wax
Weichwasser *n* 1. *(ferm)* steep[ing] water; *(tann)* soak liquor; 2. soft water
Weichwerden *n* softening
Weichzerkleinerung *f* size reduction of soft materials
Weihnachtsbaum *m (petrol)* Christmas tree
Weihrauch *m* [frank]incense, olibanum *(a gum resin from Boswellia specc.)*
Weihrauchkiefer *f* loblolly pine, Pinus taeda L.
Weihrauchöl *n (pharm)* frankincense (olibanum) oil
Weihrauchpapier *n* incense paper
Wein *m* wine
alkoholisierter W. *s.* gespriteter W.
ausgebauter (geschulter) W. aged wine
gespriteter W. fortified wine
halbtrockener W. semidry wine
kahmiger W. ropy wine
trockener W. dry wine
zäher W. ropy wine
Weinbrand *m* brandy
Weinbrandverschnitt *m* blended brandy
Weinessig *m* grape (wine) vinegar
Weinfachkunde *f* oenology
weinfachkundlich oenological
Weingeist *m* spirit[s] of wine *(aqueous ethyl alcohol)*
Weinhefe *f* wine yeast
Weinhold-Gefäß *n* Dewar flask (jar), Dewar
Weinkunde *f* oenology
weinkundlich oenological
Weinsäure *f* HOOC[CHOH]$_2$COOH tartaric acid
gewöhnliche W. *s.* *L*(+)-Weinsäure
linksdrehende W. *s.* *D*(−)-Weinsäure
razemische W. *s.* *DL*-Weinsäure
rechtsdrehende W. *s.* *L*(+)-Weinsäure
anti-**Weinsäure** *f s.* *meso*-Weinsäure
d-**Weinsäure** *f s.* *L*(+)-Weinsäure
D(−)-**Weinsäure** *f* (−)-tartaric acid, *D*-tartaric acid, laevotartaric acid, laevorotatory tartaric acid
dl-**Weinsäure** *f s.* *DL*-Weinsäure
DL-**Weinsäure** *f* (±)-tartaric acid, racemic (inactive) tartaric acid
l-**Weinsäure** *f s.* *D*(−)-Weinsäure
L(+)-**Weinsäure** *f* (+)-tartaric acid, *L*-tartaric acid, dextrotartaric acid, dextrorotatory tartaric acid
meso-**Weinsäure** *f* mesotartaric acid
para-**Weinsäure** *f s.* *DL*-Weinsäure
Weinsäurediamid *n* tartramide, diamide of tartaric acid
Weinsäuremonoamid *n* tartaric monoamide, tartramidic acid
Weinsäuremonoanilid *n* tartaric monoanilide, tartranilic acid
Weinschönungsmittel *n* wine-fining agent
Weinspiritus *m s.* Weingeist
Weinstein *m* tartar
gereinigter W. KOOC(CHOH)$_2$COOH cream of tartar, potassium monotartrate
roher W. wine stone, crude cream of tartar, argol, argal
Weinsteinrahm *m s.* Weinstein / gereinigter
Weinsteinsalz *n* salt of tartar, potassium carbonate
Weinsteinsäure *f s.* *L*(+)-Weinsäure
Weintraubenkernöl *n* grape-seed (raisin-seed, grape-stone, wine-stones) oil
Weinwaage *f* wine gauge
Weisel[zellen]futtersaft *m (pharm)* royal jelly, queen-bee's nutrient jelly
weiß anlaufen to blush *(esp of nitrocellulose lacquer)*
w. werden 1. to whiten; 2. *s.* weiß anlaufen
Weiß *n* white
Pariser W. Paris white *(finely ground calcium carbonate)*
Weißablauf *m (sugar)* wash syrup, second molasses
Weißanlaufen *n* blushing *(esp of nitrocellulose lacquer)*
Weißarsenik *n* As_2O_3 white arsenic, arsenic trioxide
Weißätze *f (text)* white discharge
weißätzen *(text)* to white-discharge
Weißätzung *f (text)* white discharge
Weiß-Bezirk *m (cryst)* Weiss [molecular magnetic] field
Weißbier *n* weiss beer
Weißblech *n* tinned sheet iron, tin plate
Weißblechbüchse *f*, **Weißblechdose** *f* tin can, tin
Weißbleierz *n (min)* cerussite *(lead carbonate)*
weißbrennend white-burning
Weiße *f* degree of white[ness], *(pap also)* brightness [level]
W. vor der Bleiche *(pap)* initial brightness
Weißei *n* [egg] albumen
Weißenberg[-Böhm]-Verfahren *n (cryst)* Weissenberg method (technique)
Weißfäule *f* white rot
weißgar alum-tanned, alum-dressed, alumed
Weißgehalt *m (pap)* degree of white[ness], brightness [level]
W. des Zellstoffs pulp brightness
W. vor der Bleiche initial brightness
Weißgehaltserhöhung *f (pap)* increase in brightness
weißgerben to taw

Weißgerber-Degras *m (tann)* sod oil *(a by-product of the chamois tannage)*
Weißgerberei *f*, **Weißgerbung** *f* tawing, alum tannage
Weißglas *n* white flint
weißglühen to incandesce
weißglühend incandescent, white-hot
Weißglut *f* incandescence, incandescency, white heat, W.H. ✦ **auf (bis zur) W. erhitzen** to incandesce
Weißgrad *m s.* Weißgehalt
Weißguß *m* white cast iron
Weißkalk *m* white (fat, rich) lime; *(dye)* pyrolignite of lime
Weißkalkäscher *m (tann)* straight lime liquor, fresh lime
Weißlauge *f (pap)* white (fresh cooking) liquor
Weißlaugenbehälter *m (pap)* white-liquor storage tank
Weißleder *n* white leather
Weißmaterial *n* limestone *(in the manufacture of calcium carbide)*
Weißmetall *n* white metal *(1. any of several bearing metals; 2. copper matte having its iron removed)*
Weißnickelkies *m (min)* white nickel, chloanthite, nickel-skutterudite *(nickel arsenide)*
Weißöl *n* [petroleum] white oil, white mineral (petroleum) oil
technisches W. technical white oil
Weißprodukt *n (petrol)* white product
Weißruß *m* white carbon [black]
Weißsirup *m (sugar)* wash syrup, second molasses
Weißtöner *m (text)* optical brightener (bleaching agent, bleach, whitening agent), fluorescent brightener; *(plast)* whitening agent
Weißtünche *f* whitewash
Weißware *f (ceram)* whiteware
Weißwein *m* white wine
Weißwerden *n* 1. whitening; 2. *(esp of cellulose nitrate lacquer)* blushing
Weißzucker *m* white sugar
Weißzuckerfüllmasse *f (sugar)* first fillmass, white (high-grade) massecuite
Weißzuckervakuumapparat *m (sugar)* white pan
weiten to stretch
Weitenverteilung *f* size distribution *(as of pores)*
Weiternitrierung *f* further nitration
Weiterreißen *n (pap)* further tearing
Weiterreißfestigkeit *f* tear (tearing) resistance (strength), tear propagation strength, resistance to further tearing, *(tann also)* tongue tear strength
Weiterreiß[festigkeits]prüfung *f* tear[ing] test, tear propagation test
Weiterreißwiderstand *m s.* Weiterreißfestigkeit
weitertragen to propagate *(e.g. a reaction)*
Weitertragen *n* propagation *(as of a reaction)*
Weithalsartikel *mpl (glass)* wide-mouth ware
Weithalsflasche *f* wide-mouth (wide-neck) bottle
Weithalskolben *m* wide-mouth (wide-neck) flask
Weithalsrundkolben *m* round-bottom wide-mouth flask
Weitholz *n* spring wood
weitlumig coarse-grained *(wood)*
weitmaschig large-meshed, wide-meshed, macroporous
Weitwinkelaufnahme *f (cryst)* wide-angle X-ray pattern
Weizenkeimöl *n* wheat germ oil
Weizenkleie *f* wheat bran
feine W. shorts
Weizenmalz *n* wheat malt, malted wheat
Weizenmehl *n* wheat flour, *(if coarse)* wheat meal
Weizenstärke *f* wheat starch
Weldon-Verfahren *n* Weldon [chlorine] process
Wellasbest *m* corrugated asbestos
Welle *f* 1. wave; 2. *(tech)* shaft
außerordentliche W. *(cryst)* extraordinary wave
polarografische W. polarographic wave
zentrale W. *(tech)* central shaft
Wellenausbreitung *f* wave propagation
Wellenbewegung *f* wave motion
Wellenbildung *f* casing (outer layer) effect *(a defect in glass)*
Wellendichtring *m* oil seal ring
Welleneigenschaft *f* wave property
Wellenfunktion *f* wave function
molekulare W. molecular wave function
selbstkonsistente W. self-consistent wave function
Wellengleichung *f* wave equation
Schrödingersche W. Schrödinger [wave] equation
Wellengruppe *f* wave packet
Wellenkatode *f* angular cathode *(in a Billiter cell)*
Wellenlänge *f* wavelength
Wellenlängenbereich *m* wavelength range
Wellenlängenskale *f* wavelength scale
Wellenmechanik *f* wave mechanics
Wellennatur *f* wave nature
Wellenpaket *n* wave packet
Wellentheorie *f* wave theory
Wellenvektor *m* wave vector
Wellenzahl *f* wave number
Wellenzahlvektor *m* wave vector
Welligwerden *n (pap)* cockling
Wellkarton *m* corrugated cardboard
Wellman-Galusha-Generator *m* Wellman-Galusha producer
Wellpapier *n s.* Wellpappenpapier
Wellpappe *f* corrugated (cellular) board
Wellpappenpapier *n* corrugated (corrugating) paper
Wellplatte *f (plast)* corrugated sheet
Wellrohr *n* corrugated tube (pipe)
Wellrohr[dehnungs]ausgleicher *m*, **Wellrohrkompensator** *m* corrugated (bellows-type) expansion joint
Weltraumstrahlung *f* cosmic radiation
Wemco-Trommel[sink]scheider *m* Wemco separator

Wendefilz *m (pap)* reverse press felt
Wendel *f (distil)* helix
Wendelrutsche *f* spiral chute
Wendelscheider *m* spiral separator
Wendepflug *m* turnover plough *(in margarine making)*
Wendepresse *f (pap)* reverse press
Wendepunkt *m* point of inflection *(of a curve)*
Wenderschaufel *f (ferm)* malt shovel
Wendevorrichtung *f (ferm)* floor plough
Wendewalze *f (pap)* hitch roll
werfen / sich to warp
Werfen *n* warpage, warping
Werg *n* tow, oakum
Werkblei *n* crude (pig) lead
Werkdruckpapier *n* book[-printing] paper, *(Am also)* text paper
Werksabwasser *n s.* Industrieabwasser
Werkschemiker *m* industrial chemist
Werkstatt *f* / **galvanische** [electro]plating shop
Werkstoff *m* material, matter
W. auf Hartmetallbasis cemented carbide material
W. für Tieftemperaturtechnik low-temperature material
geschichteter W. laminated material (plastic), laminate
halbleitender W. semiconducting material
hochtemperaturbeständiger W. high-temperature material
keramischer W. ceramic material
keramometallischer W. *s.* metallkeramischer W.
metallkeramischer W. cermet, ceramal, ceramel, ceramet
mischkeramischer W. *s.* metallkeramischer W.
poriger (poröser) W. porous material
warmfester W. high-temperature material
Werkstoffkunde *f* materials science
Werkstoffprüfung *f* materials testing
Werkzeug *n* 1. *(plast)* mould, form; 2. tool
W. mit Heizkanälen cored mould
geschlossenes W. closed mould
offenes W. open mould
zusammengesetztes W. composite (split) mould
Werkzeugaufnahmegestell *n (plast)* spider *(in centrifugal casting)*
Werkzeugbau *m (plast)* mouldmaking
Werkzeugeinsätze *mpl (plast)* splits of mould
Werkzeughälfte *f (plast)* mould half
bewegliche W. moving mould half
feststehende W. stationary mould half
Werkzeugharz *n (plast)* tooling resin
Werkzeughohlraum *m (plast)* mould cavity
Werkzeugkonizität *f (plast)* taper
Werkzeugkonstrukteur *m (plast)* mould designer
Werkzeugkonstruktion *f (plast)* mould design
Werkzeugmacher *m (plast)* mould maker
Werkzeugneigung *f (plast)* taper
Werkzeugplatte *f* / **bewegliche** *(plast)* movable (moving) platen
feststehende W. *(plast)* stationary platen
Werkzeugschließdruck *m (plast)* mould-locking pressure
Werkzeugschließeinheit *f (plast)* mould clamp
Werkzeugschließhub *m (plast)* mould-clamping stroke
Werkzeugschließsystem *n (plast)* [mould-]clamping mechanism
Werkzeugschließweg *m (plast)* mould-clamping stroke
Werkzeugschließzeit *f (plast)* mould-closing time
Werkzeugschluß *m (plast)* mould closing
Werkzeugschwindmaß *n (plast)* mould shrinkage
Werkzeugstahl *m* tool steel
Werkzeugtemperatur *f (plast)* mould temperature
Werkzeugtrennmittel *n (plast)* release agent
Werkzeugzuhaltekraft *f (plast)* mould-clamping force
Werkzeugzuhaltung *f (plast)* hydraulic clamp
Wermut *m* absinth, wormwood, Artemisia absinthium L.
Wermutöl *n* absinth, wormwood oil
Werner-Komplex *m* Werner complex *(chemical-bond theory)*
Werner-Pfleiderer-Kneter *m* Werner-Pfleiderer mixer
Wert *m* value ✦ **einen konstanten W. aufweisen** to show a constant reading ✦ **etwas über dem theoretischen W.** in slight excess of theory, slightly over the theoretical ✦ **etwas unter dem theoretischen W.** slightly short of theory
absoluter W. absolute value
experimenteller W. experimental value
kalorischer W. *(food)* calorific value, c.v.
ν-Wert *m* Abbe number (value), ν-value, constringence *(reciprocal relative dispersion)*
wertarm low-grade
Wertbestimmung *f* / **biologische** biological assay, bioassay
Wertigkeit *f* valence, valency
elektrochemische W. electrochemical valence, electrovalence
koordinative W. coordination (covalence) number, ligancy
kovalente W. covalence, covalency
maximale W. maximum valence
stöchiometrische W. stoichiometric valence
Wertigkeitsänderung *f* valence change
Wertigkeitsbezeichnung *f* indication of valence
Stocksche W. Stock notation
Wertigkeitsstufe *f* valence state
Wertigkeitstheorie *f* theory of valence, valence theory
Wertigkeitswechsel *m* valence change
Wertminderung *f* deterioration
Wertpapierdruck *m* bond printing
Wertpapierdruckfarbe *f* bond ink
Wertsteigerung *f* upgrading
Wertstoff *m* useful material, valuable product, product of value
Wertstoffrückgewinnung *f* recovery (reclaim) of useful material

wesentlich essential
West-Kühler *m (distil)* West-type condenser
Weston-Element *n* Weston cell, Weston normal (standard) cell, Weston saturated cadmium cell
  ungesättigtes W. unsaturated Weston cell
Weston-Normalelement *n s.* Weston-Element
Weston-Standardelement *n* unsaturated Weston cell
Weston-Zahl *f (phot)* Weston figure
Wetherill-Ofen *m (met)* Wetherill furnace
Wetter *pl* / **matte** *(mine)* dead air
  schlagende W. firedamp
wetterbeständig weather-resistant, resistant to weathering; *(text)* weatherproof
Wetterbeständigkeit *f* weather[ing] resistance, resistance to weathering, weatherability, outdoor (exterior) durability; *(text)* weatherproofness
Wetterbeständigkeitsprüfgerät *n* weatherometer
wetterfest *s.* wetterbeständig
wettergeschützt weather-protected, weatherproof, weather-resistant
wettersicher *s.* 1. wetterbeständig; 2. wettergeschützt
Wettersprengstoff *m (mine)* safety explosive
wf *s.* wasserfrei
Whatman-Papier *n* Whatman paper *(a filter paper)*
Wheatstone-Brücke *f* Wheatstone bridge
Wheeler-Mühle *f* Wheeler mill *(a jet mill)*
Whisky *m* whisky, whiskey
Wichte *f* specific weight, *(sometimes also)* density *(the weight of a substance per unit volume)*
Wickel *m (text)* lap
Wickelhülse *f (pap)* core, centre
Wickelkörper *m (text)* package
Wickeltrommel *f (rubber)* casemaking (building, tyre-building) drum
Widdren *n* widdrene, thujopsene *(a tricyclic sesquiterpene)*
Widdrol *n* widdrol *(a sesquiterpene alcohol)*
widerlich disagreeable, obnoxious, repulsive *(smell)*
Widerstand *m* 1. resistance; 2. resistor
  W. gegen Einreißen tear initiation strength
  W. gegen Rißwachstum resistance to crack growth
  W. gegen Schnittwachstum resistance to cut growth
  W. gegen Temperaturwechsel thermal-shock resistance
  aerodynamischer W. drag force
  elektrischer W. 1. electrical resistance; 2. resistor
  fotoelektrischer W. photoconductive cell, photoresistor
  hydrodynamischer W. fluid friction, drag force
  lichtelektrischer W. *s.* fotoelektrischer W.
  spezifischer W. specific resistance, resistivity
  thermischer W. thermal resistance
Widerstandsabschwächer *m (tox, agric)* antiresistant
Widerstandsbeheizung *f* [/ **elektrische**] resistance heating
Widerstandsbrücke *f* [/ **elektrische**] resistance bridge
Widerstandsdraht *m* resistance wire
Widerstandserhitzung *f* [/ **elektrische**] resistance heating
widerstandsfähig resistant, stable, fast
Widerstandsfähigkeit *f* resistance, stability, fastness
  W. bei hohen Temperaturen stability at high temperatures
  W. gegen Abblättern (Abplatzen) spalling resistance
  W. gegen Abrieb resistance to abrasion (wear), abrasion (wear) resistance
  W. gegen Bruch resistance to breakage
  W. gegen Chemikalien resistance to chemicals
  W. gegen Gase resistance to gases
  W. gegen hohe Temperaturen resistance to high temperatures, high-temperature resistance (stability, durability)
  W. gegen Lösungsmittel solvent resistance
  W. gegen Ritzen resistance to scratching
  W. gegen Schimmel mildew resistance
  W. gegen Verschlackung resistance to slagging
  chemische W. resistance to chemical attack, chemical resistance (stability)
  thermische W. *(relating to decomposition)* thermal stability, thermostability; *(relating to deformation)* resistance to heat
Widerstandsfotozelle *f* photoconductive cell, photoresistor
Widerstandsheizelement *n*, **Widerstandsheizkörper** *m* resistance (resistive) heater
Widerstandsheizung *f* [/ **elektrische**] resistance heating
Widerstandskapazität *f* resistance capacity; cell constant *(of a conductance cell)*
Widerstandskurve *f* resistivity curve
Widerstandsmaterial *n* resistor material
Widerstandsofen *m* [/ **elektrischer**] resistance (resistance-heated, resistor) furnace
Widerstandsschweißen *n* resistance welding
Widerstandsthermometer *n* resistance thermometer
Widerstandswerkstoff *m* resistor material
Widerstandswert *m* resistance
Widerstandswicklung *f* resistance coil (winding)
Widerstandszelle *f s.* Widerstandsfotozelle
Widmer-Kolonne *f (distil)* Widmer spiral column
wiederanreichern to re-enrich
wiederaufarbeiten to reprocess, to rework, to reclaim
wiederaufbereiten to remanufacture *(e.g. waste paper)*
Wiederaufbereitung *f* remanufacture *(as of waste paper)*
wiederaufkohlen *(met)* to recarburize
Wiederaufkohlung *f (met)* recarburization
Wiederaufkohlungsmittel *n (met)* recarburizing agent, recarburizer

wiederauflösen to redissolve
wiederaufschwemmen to repulp, to reslurry
Wiederaufziehen *n (text)* redeposition *(of soil)*
wiederbeleben to revivify, to revivificate, to regenerate *(a sorbent)*
Wiederbelebung *f* revivification, regeneration *(of a sorbent)*
Wiederbelebungsmittel *n* regenerant *(for sorbents)*
wiederbenutzbar reusable
Wiederbenutzung *f* reuse
Wiedereinfangen *n (nucl)* recapture
Wiedergabetreue *f (phot)* fidelity of reproduction
wiedergeben to reproduce, to represent, *(phot also)* to render
Wiedergebrauch *m* reuse
wiedergewinnbar recoverable
wiedergewinnen to recover, to reclaim
Wiedergewinnung *f* recovery, reclaim[ing]
Wiedergewinnungsanlage *f* recovery plant
wiederherstellen to restore
wiederholbar repeatable *(experiment)*
Wiederholbarkeit *f* repeatability *(of an experiment)*
Wiederholstreubereich *m* repeatability *(statistics)*
Wiederholung *f* replicate, replication *(statistics)*
wiederverarbeiten to reprocess
Wiederverarbeitung *f* reprocessing
wiedervereinigen *(chem)* to recombine
  sich w. *(chem)* to recombine
Wiedervereinigung *f (chem)* recombination
Wiedervereinigungsgeschwindigkeit *f* recombination rate (velocity)
Wiederverfestigung *f* resolidification
Wiederverfestigungstemperatur *f* resolidification temperature
wiederverwendbar reusable
wiederverwenden to reuse
Wiederverwendung *f* reuse
wiegen to weigh
Wieland-Gumlich-Aldehyd *m* Wieland-Gumlich aldehyde, caracurine-VII
Wien-Effekt *m (el chem)* Wien effect
Wiesenkalk *m*, **Wiesenmergel** *m (soil)* bog lime
Wijs-Lösung *f* Wijs [iodine monochloride] solution *(for determining the iodine number of fats)*
Wildcat-Bohrung *f (petrol)* wildcat, exploration drilling
Wildhefe *f* wild yeast
Wildkautschuk *m* wild rubber
Wildleder *n* suede leather
Wildseide *f* wild silk *(e. g. tussah silk)*
Wildverbißschutzmittel *n* anti-game protective agent
Wilfley-Herd *m (min tech)* Wilfley table
Willemit *m (min)* willemite *(zinc orthosilicate)*
Williams-Abriebprüfer *m (rubber)* Williams abrader, Du Pont-Grasselli-Williams machine
Williams-Einheit *f (text)* Williams unit *(a roller vat)*
Williamson-Ofen *m* Williamson kiln *(a tunnel kiln)*
Williamson-Synthese *f* Williamson [ether] synthesis
Williams-Plastometer *n (rubber)* Williams plastometer
Williams-Prüfer *m s.* Williams-Abriebprüfer
Willstätter-Nagel *m* Willstätter nail *(a filter aid)*
Wilputte-Koksofen *m* Wilputte oven
Wilson-Aufnahme *f (nucl)* cloud-chamber photograph
Wilson-Kammer *f* cloud (Wilson) chamber, [Wilson] cloud-track apparatus
Wind *m* 1. wind; 2. *(met)* [air] blast
  elektrischer W. electric (ionic) wind
  heißer W. *(met)* hot[-air] blast
  kalter W. *(met)* cold[-air] blast
Winddüse *f s.* Windform
winden *(text)* to reel [up]
Winderhitzer *m (met)* hot-blast (air-blast, blast-furnace) stove, air heater
Windform *f (met)* air-blast tuyère
Windformebene *f (met)* tuyère level
Windformenzone *f (met)* tuyère zone
Windformkühlkasten *m (met)* tuyère cooler
windfrischen *(met)* to bessemerize, to air-refine, to convert
Windfrischen *n (met)* bessemerizing, air (blast) refining, air converting
Windfrischkonverter *m (met)* Bessemer converter
Windfrischstahl *m (met)* Bessemer (converter) steel
Windfrischverfahren *n (met)* Bessemer (converter) process
  basisches W. basic Bessemer (converter) process, Thomas[-Gilchrist] process
  saures W. acid Bessemer (converter) process
Windgebläse *n* air blower
windgesichtet air-floated
Windkasten *m* wind box
Windkessel *m* air vessel (receiver, chamber), surge chamber
Windleitung *f (met)* [air-]blast main
Windmantel *m (met)* wind belt *(of a cupola)*
Windmesser *m* anemometer
Windröstverfahren *n* **[nach Huntington und Heberlein]** *(met)* Huntington-Heberlein process
Windsichten *n* air classifying (separating, classification, separation, sweeping, sizing), pneumatic classification
Windsichter *m* air (air-swept, pneumatic) classifier, air separator
Windsichtermühle *f* air-swept mill
Windsichtung *f s.* Windsichten
Windtemperatur *f (met)* [air-]blast temperature
Windzufuhr *f* air supply
Winkel *m* / **Braggscher** *(cryst)* Bragg angle
  Brewsterscher W. Brewster (polarizing) angle
Winkelbeschleunigung *f* angular acceleration
Winkelgeschwindigkeit *f* angular velocity
Winkelkopf *m* [fixed-]angle head, [fixed-]angle rotor *(of a centrifuge)*
Winkelmesser *m (cryst)* goniometer
Winkelpresse *f* angle press
Winkelprobe *f (rubber)* angle test; angle test piece *(for determining the tear propagation strength)*
Winkelrotor *m s.* Winkelkopf
Winkelspektrum *n* angular spectrum
Winkelstück *n* angle connector, elbow [fitting], ell

Winkelthermometer *n* angle thermometer
Winkelverteilung *f* angular distribution
Winkler-Gaserzeuger *m s.* Winkler-Generator
Winkler-Gaserzeugung *f* fluidized-bed gasification
Winkler-Generator *m* fluidized-bed gasifier, Winkler generator
Winkler-Generatorverfahren *n s.* Winkler-Vergasungsverfahren
Winkler-Koch-Spaltverfahren *n (petrol)* Winkler-Koch process
Winkler-Vergasungsverfahren *n* Winkler (fluid) process *(for gasifying lignite)*
Wintergrünöl *n* oil of wintergreen *(from Gaultheria procumbens L.)*
  künstliches W. $C_6H_4(OH)CO\text{-}OCH_3$ artificial oil of wintergreen, methyl salicylate
Winterisation *f s.* Winterung
wintern 1. to winterize, to demargarinate, to destearinate, to destearinize *(oils)*; 2. *(ceram)* to winter, to weather
Winterrahm *m* winter cream
Winterspritzmittel *n (agric)* dormant spray, winter wash
  öliges W. dormant oil [spray]
Winterspritzung *f (agric)* dormant spray[ing], dormant winter spray[ing]
Winterung *f* winterization, demargarination, destearinization *(of oils)*
Wipprahmen *m (tann)* rocker [frame] *(for moving pelts in suspender pits)*
Wirbel *m* vortex, whirl, swirl, eddy
Wirbelabscheider *m* cyclone collector (separator), cyclone
Wirbelbett *n* fluid (fluidized, boiling) bed
Wirbelbettkatalysator *m* fluidized (fluid-bed, fluid) catalyst
Wirbelbetttrockner *m* fluid-bed dryer
Wirbelbrenner *m* vortex burner
Wirbelfeuerung *f* cyclone firing
Wirbelgeschwindigkeit *f* fluidizing velocity *(in a fluidized bed)*
wirbelig turbulent
Wirbelkammervergaser *m* vortex gasifier
Wirbelreiniger *m s.* Wirbelsichter
Wirbelscheider *m s.* Wirbelabscheider
Wirbelschicht *f* fluid (fluidized, boiling) bed
Wirbelschichtadsorption *f* fluidized (moving-bed) adsorption
Wirbelschichtgenerator *m* fluidized-bed gasifier, Winkler generator
Wirbelschichtkracken *n* / **katalytisches** fluid catalytic cracking
Wirbelschichtröstofen *m* fluid-bed roaster (roasting furnace)
Wirbelschichttechnik *f* fluidized-bed (fluid-bed, boiling-bed) technique
Wirbelschichttrockner *m* fluid-bed dryer
Wirbelschichtverbrennung *f* fluid-bed burning
Wirbelschichtverfahren *n* fluid (fluidized) process (operation)
Wirbelschleuder *f s.* Wirbelsichter
Wirbelsichter *m* centrifugal cleaner, *(Am)* centrifiner
Wirbelsintern *n (plast)* fluid-bed coating, dip coating in powder
Wirbelstraße *f* vortex street
Wirbelstrombrenner *m* vortex burner
Wirbelstromheizung *f* eddy current heating
Wirbelströmung *f* eddy (turbulent, sinuous) flow
Wirbelsucher *m* vortex finder
Wirbeltrockner *m* vortex dryer
Wirbelzellenwärmeaustauscher *m* plate-fin heat exchanger
Wirkdruck *m* differential pressure (head)
Wirkdruck-Durchfluß[mengen]messer *m*, **Wirkdruck-Mengenstrommesser** *m* head flowmeter
Wirkdruckmesser *m* differential-pressure meter
wirken to act
  als Kristallisationskeim w. to nucleate
  antifungal (fungizid) w. to have antifungal activity
  füllend w. *(coat)* to body
wirkend / schleichend *(tox)* slow-acting
Wirkgruppe *f* active group, coenzyme, prosthetic group, agon
Wirkkraft *f (pharm)* potency
Wirkquerschnitt *m s.* Wirkungsquerschnitt
wirksam active; efficient, effective, potent
Wirksamkeit *f* activity, efficiency, potency
  antibakterielle W. antibacterial activity
  antibiotische W. antibiotic activity
  fotochemische W. actinic activity
  fungizide W. fungicidal activity, fungitoxicity
  herbizide W. herbicidal activity
  insektizide W. insecticidal activity
  katalytische W. catalytic activity
  oxytozische W. *(pharm)* oxytocic activity
  relative biologische W. relative biological effectiveness, RBE *(radiation biology)*
  systemische W. systemic activity *(of a pesticide)*
  therapeutische W. therapeutical activity
Wirksamkeitsverlust *m* loss of activity
Wirkstoff *m* agent, active substance (agent, material), *(biol also)* ergone
  antibiotischer W. antibiotic
  chemischer W. chemical agent
  giftiger W. toxicant, toxic [substance]
  therapeutischer W. therapeutic agent
Wirkstoffgehalt *m* content of active component
Wirkstoffkonzentration *f* concentration of active constituents
Wirkstoffnebel *m* aerosol
Wirkstoffzubereitung *f* **mit aktivierendem Zusatz** activated formulation *(crop protection)*
Wirksubstanz *f s.* Wirkstoff
Wirkung *f* 1. action; influence; 2. *(result)* effect
  adstringierende W. *(pharm)* stypticity
  anästhe[ti]sierende W. anaesthetic action
  antagonistische W. antagonistic action
  betäubende W. anaesthetic action
  dirigierende W. directive influence *(as on substituents)*
  dispergierende W. dispersing (dispersant) action

dominierende W. overriding influence
erweichende W. *(rubber)* peptizing action
farberhöhende W. hypsochromic action
fungizide W. fungicidal action
geschwulsterregende W. tumorigenicity
heilende W. curative action
herbizide W. herbicidal action
insektizide W. insecticidal action
katalytische W. catalytic action
klopfhemmende W. antiknock action
korrodierende W. corrosiveness, corrosivity
krebserregende W. carcinogenicity
kumulative W. cumulative action
lösungsvermittelnde W. solubilizing action
netzende W. wetting action
nitrierende W. *(met)* nitriding action
oxydierende W. oxidizing action
reduzierende W. reducing action
reinigende W. detergent action
selektive W. selectivity
synergistische W. synergistic action
toxische W. toxic action
verstärkende W. strengthening action, *(rubber also)* reinforcing action
zementierende W. *(met)* carburizing action
Wirkungsbereich *m s.* Wirkungsbreite
Wirkungsbreite *f* spectrum of activity *(as of pesticides)* ✦ **mit großer W.** broad-spectrum, broad-scale
umfassende W. all-round efficiency
Wirkungsdosis *f* / **minimale** *(tox)* activation threshold
Wirkungsgrad *m* 1. *(chem)* strength *(of a reagent)*; 2. *(tech)* efficiency [factor] ✦ **mit hohem W.** *(tech)* efficient
elastischer W. *(rubber)* resilience, resiliency
elastischer W. bei Stoß-Druckbeanspruchung *(rubber)* rebound, rebound elasticity (resilience), impact resilience
hydraulischer W. hydraulic efficiency
katalytischer W. catalytic efficiency
maximaler W. maximum efficiency
thermischer W. thermal efficiency
volumetrischer W. volumetric efficiency *(of pumps)*
wirkungslos ineffective
Wirkungsmechanismus *m* mechanism of action
Wirkungsprodukt *n* / **Habersches** *(tox)* ct product *(product of concentration and survival time)*
Wirkungsquantum *n* / **Plancksches** Planck [action] constant, [Planck] quantum of action
Wirkungsquerschnitt *m (phys chem)* effective cross section; *(nucl)* cross-section target area, [nuclear] cross section
Wirkungsspektrum *n* spectrum of activity *(as of pesticides)* ✦ **mit breitem W.** broad-spectrum, broad-scale
Wirkungsstärke *f (pharm)* strength, potency
Wirkungssteigerung *f* fortification
Wirkungsweise *f* mode (mechanism) of action
Wirkungswert *m* action (effect) value *(as of fertilizers)*
Wirt *m* host [component] *(of an inclusion compound)*
Wirtselement *n (min, geoch)* host element
Wirtsgestein *n* host rock
Wirtsgitter *n (cryst)* host lattice
Wirtskomponente *f* host [component] *(of an inclusion compound)*
Wirtspflanze *f* host plant
Wismut *n* Bi bismuth
Wismut(III)-bromid *n* $BiBr_3$ bismuth(III) bromide, bismuth tribromide
Wismutbronze *f* bismuth bronze
Wismut(III)-chlorid *n* $BiCl_3$ bismuth(III) chloride, bismuth trichloride
Wismutdioxid *n* BiO_2 bismuth dioxide
Wismuterz *n* bismuth ore
Wismut(III)-fluorid *n* BiF_3 bismuth(III) fluoride, bismuth trifluoride
Wismutglanz *m (min)* bismuth glance, bismuthin[it]e *(bismuth(III) sulphide)*
Wismut-Glaskitt *m* bismuth putty *(an alloy consisting of Bi, Pb, and Sn)*
Wismutgold *n (min)* bismuth gold, maldonite *(a natural alloy)*
wismuthaltig bismuthiferous
Wismut(III)-hydroxid *n* $Bi(OH)_3$ bismuth(III) hydroxide, bismuth trihydroxide
Wismut(III)-jodat *n* $Bi(IO_3)_3$ bismuth(III) iodate
Wismut(III)-jodid *n* BiI_3 bismuth(III) iodide, bismuth triiodide
Wismutmonosulfid *n s.* Wismut(II)-sulfid
Wismut(III)-nitrat *n* $Bi(NO_3)_3$ bismuth(III) nitrate, bismuth trinitrate
Wismutocker *m (min)* bismuth ochre
Wismut(III)-oxalat *n* $Bi_2(C_2O_4)_3$ bismuth(III) oxalate
Wismut(III)-oxid *n* Bi_2O_3 bismuth(III) oxide, bismuth trioxide
Wismut(V)-oxid *n* Bi_2O_5 bismuth(V) oxide, bismuth pentaoxide
Wismutoxidbromid *n* BiOBr bismuth bromide oxide
Wismutoxidchlorid *n* BiOCl bismuth chloride oxide
Wismutoxidfluorid *n* BiOF bismuth fluoride oxide
Wismutoxidjodid *n* BiOI bismuth iodide oxide
Wismutoxidkarbonat *n* $(BiO)_2CO_3$ bismuth carbonate oxide
Wismutoxidnitrat *n* $BiO(NO_3)$ bismuth nitrate oxide
Wismutpentoxid *n s.* Wismut(V)-oxid
Wismut(III)-phosphat *n* $BiPO_4$ bismuth(III) phosphate, bismuth orthophosphate
Wismutsäure *f* bismuthic acid
Wismut(III)-selenid *n* Bi_2Se_3 bismuth(III) selenide, dibismuth triselenide
Wismut(III)-sulfat *n* $Bi_2(SO_4)_3$ bismuth(III) sulphate
Wismut(II)-sulfid *n* BiS bismuth(II) sulphide, bismuth monosulphide
Wismut(III)-sulfid *n* Bi_2S_3 bismuth(III) sulphide, bismuth trisulphide
Wismuttri... *s.* Wismut(III)-...
Wismutwasserstoff *m* BiH_3 bismuth hydride, bismuthine
Wismutweiß *n* bismuth (pearl, Spanish) white,

cosmetic bismuth *(consisting of bismuth nitrate oxide or bismuth chloride oxide)*
Wiswesser-Notationssystem *n (nomencl)* Wiswesser [notation] system
Witherit *m (min)* witherite *(barium carbonate)*
witterungsbeständig *s.* wetterbeständig
Wittig-Reaktion *f* Wittig reaction *(between alkylidene phosphoranes and carbonyl compounds)*
Wittig-Umlagerung *f* Wittig rearrangement *(of aryl-alkyl or dialkyl ethers to alcohols)*
Wohl-Abbau *m* Wohl degradation *(of sugars)*
Wöhlerit *m (min)* wöhlerite *(a sorosilicate)*
Wohlgeruch *m* fragrance, fragrancy, aroma, scent, odour
wohlriechend fragrant, aromatic
wohlschmeckend tasty, aromatic
Wolff-Kishner-Reaktion *f (org chem)* Wolff-Kishner reaction (reduction)
Wolfram *n* W tungsten, wolfram
Wolframat *n* wolframate, tungstate
normales W. $M_2^IWO_4$ normal wolframate, normal tungstate
Wolframatoarsenat *n* wolframoarsenate, tungstoarsenate
Wolframatoborat *n* wolframoborate, tungstoborate
Wolframatoborsäure *f* wolframoboric acid, tungstoboric acid
Wolframatokieselsäure *f* wolframosilicic acid, tungstosilicic acid
Wolframatophosphat *n* wolframophosphate, tungstophosphate
Wolframatophosphorsäure *f* wolframophosphoric acid, tungstophosphoric acid
Wolframatosilikat *n* wolframosilicate, tungstosilicate
Wolframatphosphor *m* tungstate phosphor
Wolframbogenlampe *f* tungsten point (arc) lamp
Wolfram(II)-bromid *n* WBr_2 tungsten(II) bromide, wolfram(II) bromide, tungsten dibromide, wolfram dibromide
Wolfram(V)-bromid *n* WBr_5 tungsten(V) bromide, wolfram(V) bromide, tungsten pentabromide, wolfram pentabromide
Wolfram(VI)-bromid *n* WBr_6 tungsten(VI) bromide, wolfram(VI) bromide, tungsten hexabromide, wolfram hexabromide
Wolframbronze *f* tungsten bronze
Wolfram(II)-chlorid *n* WCl_2 tungsten(II) chloride, wolfram(II) chloride, tungsten dichloride, wolfram dichloride
Wolfram(IV)-chlorid *n* WCl_4 tungsten(IV) chloride, wolfram(IV) chloride, tungsten tetrachloride, wolfram tetrachloride
Wolfram(V)-chlorid *n* WCl_5 tungsten(V) chloride, wolfram(V) chloride, tungsten pentachloride, wolfram pentachloride
Wolfram(VI)-chlorid *n* WCl_6 tungsten(VI) chloride, wolfram(VI) chloride, tungsten hexachloride, wolfram hexachloride
Wolframdi ... *s.a.* Wolfram(II)- ...
Wolframdioxid *n s.* Wolfram(IV)-oxid
Wolframdioxiddichlorid *n* WO_2Cl_2 tungsten dichloride dioxide, wolfram dichloride dioxide
Wolframdisulfid *n s.* Wolfram(IV)-sulfid
Wolfram(VI)-fluorid *n* WF_6 tungsten(VI) fluoride, wolfram(VI) fluoride, tungsten hexafluoride, wolfram hexafluoride
Wolframglühfaden *m* tungsten filament
Wolframhexa ... *s.* Wolfram(VI)- ...
Wolframit *m (min)* wolframite
Wolfram(II)-jodid *n* WI_2 tungsten(II) iodide, wolfram(II) iodide, tungsten diiodide, wolfram diiodide
Wolfram(IV)-jodid *n* WI_4 tungsten(IV) iodide, wolfram(IV) iodide, tungsten tetraiodide, wolfram tetraiodide
Wolframkarbid *n* tungsten carbide
gesintertes W. cemented tungsten carbide
Wolframkarbidhartmetall *n* cemented tungsten carbide
Wolframleuchtfaden *m* tungsten filament
Wolframocker *m (min)* tungstite, tungstic ochre
Wolfram(IV)-oxid *n* WO_2 tungsten(IV) oxide, wolfram(IV) oxide, tungsten dioxide, wolfram dioxide
Wolfram(VI)-oxid *n* WO_3 tungsten(VI) oxide, wolfram(VI) oxide, tungsten trioxide, wolfram trioxide
Wolframoxidtetrachlorid *n* $WOCl_4$ tungsten tetrachloride oxide, wolfram tetrachloride oxide
Wolframpenta ... *s.* Wolfram(V)- ...
Wolframpulver *n* tungsten powder
Wolframpunktlichtlampe *f* tungsten point (arc) lamp
Wolframsäure *f* wolframic acid, tungstic acid
Wolframsinterkarbid *n* cemented tungsten carbide
Wolframstahl *m* tungsten steel
Wolfram(IV)-sulfid *n* WS_2 tungsten(IV) sulphide, wolfram(IV) sulphide, tungsten disulphide, wolfram disulphide
Wolfram(VI)-sulfid *n* WS_3 tungsten(VI) sulphide, wolfram(VI) sulphide, tungsten trisulphide, wolfram trisulphide
Wolframtetra ... *s.* Wolfram(IV)- ...
Wolframtrioxid *n s.* Wolfram(VI)-oxid
Wolframtrisulfid *n s.* Wolfram(VI)-sulfid
Wolke *f* cloud
π-Wolke *f* π cloud *(chemical-bond theory)*
wolkig cloudy
Wolkigkeit *f (pap)* cloud effect
Wollastonit *m (min)* wollastonite *(calcium metasilicate)*
Wolle *f (from sheep)* wool; *(from other animals)* hair fibre
wollen wool[l]en
Wollfarbstoff *m* wool dye
Wollfett *n* wool fat (grease), *(text also)* yolk, suint
gereinigtes W. lanolin
rohes W. crude wool fat
Wollfettalkohol *m* wool alcohol
Wollfilz *m (pap)* wool[len] felt
Wollfilzpapier *n* wool-felt paper
Wollfilzpappe *f* wool-felt board, rag felt

Wollgewebe *n* wool[len] fabric
Wollschweiß *m s.* Wollfett
Wollstoff *m* wool[len] fabric
Wolltrockenfilz *m (pap)* wool dry[er] felt
Wollwachs *n* wool wax
Wollwäsche *f* wool scouring
Wollwaschmaschine *f* wool-scouring machine
Woodall-Duckham-Ofensystem *n (coal)* Woodall-Duckham system
Woodall-Duckham-Retorte *f (coal)* Woodall-Duckham continuous vertical retort
Wood-Glas *n* Wood's glass
wss., wssr. *s.* wäßrig
Wuchshormon *n* growth hormone
Wuchsstoff *m* growth substance
selektiver W. selective growth promoter
Wuchsstoffherbizid *n* hormone weedkiller, [plant-]growth regulator
Wuchsstoffmittel *n* growth-regulating substance, growth regulator (modifier)
Wuchsstoffpräparat *n s.* 1. Wuchsstoffmittel; 2. Wuchsstoffherbizid
wulchern *(glass)* to marver *(a gather in an ovoid mould)*
Wulfenit *m (min)* wulfenite, yellow lead ore *(lead molybdate)*
Wulff-Bock-Kristallisator *m* Wulff-Bock crystallizer
Wulff-Verfahren *n* Wulff process *(for manufacturing acetylene)*
Wulst *m(f)* bead, boss; *(rubber)* bead, *(in the mill-opening)* bank
Wulstmischung *f (rubber)* bead compound
Wulstschneidemaschine *f (rubber)* bead cutter, debeader, debeading machine
Wulstschutzstreifen *m (rubber)* chafer, [strip]
Wundererde *f* / **Sächsische** *(min)* teratolite *(a blue bole)*
Wunderkerze *f* sparkler
Wundverband *m* wound dressing *(crop protection)*
Würfelbruch *m (glass)* dice
Würfelfestigkeit *f* cube strength *(testing of concrete)*
würfelförmig cubic[al]
Würfelgitter *n (cryst)* cubic lattice (system, pattern, structure)
flächenzentriertes W. face-centred cubic lattice
raumzentriertes W. body-centred cubic lattice
Würfelmischer *m* cubical blender, cube mixer
würfeln *(plast)* to dice
Würfelschneider *m (plast)* dicing cutter (machine), dicer
Würfelzucker *m* cube (lump) sugar
Wurfförderer *m,* **Wurfförderrinne** *f* directional-throw conveyor
Wurfherd *m (min tech)* shaking table
Wurmfarnextrakt *m (pharm)* male fern extract
Wurmmittel *n (pharm)* vermifuge, anthelminthic, helminthagogue, *(if potent)* helminthicide
Wurmsamenöl *n (pharm)* chenopodium oil *(from Chenopodium ambrosioides L. var. anthelminthicum)*
wurmvertreibend, wurmwidrig *(pharm)* anthelminthic
Würth-Abstichgaserzeuger *m,* **Würth-Generator** *m* Würth producer
Wurtz-Fittig-Synthese *f* Wurtz-Fittig synthesis *(of hydrocarbons)*
Wurtzit *m (min)* wurtzite *(zinc sulphide)*
Wurtz-Synthese *f* Wurtz reaction, Wurtz synthesis *(of hydrocarbons)*
Würze *f* 1. *(ferm)* [beer] wort; 2. aroma, flavour *(of wine)*; 3. *s.* Gewürz
gehopfte W. *(ferm)* hopped wort
Würze[koch]kessel *m s.* Würzepfanne
Würzekühler *m (ferm)* wort cooler
Würzekühlung *f (ferm)* wort cooling
Wurzelausscheidung *f,* **Wurzeldiffusat** *n* root excretion (exudate, diffusate)
Wurzelharz *n,* **Wurzelkolophonium** *n* wood rosin
Würzepfanne *f (ferm)* wort copper
Würzepumpe *f (ferm)* wort pump
würzig aromatic, spicy
Würzigkeit *f* aromaticity, spiciness
Würzmittel *n* seasoning matter
Wüstenkruste *f,* **Wüstenlack** *m,* **Wüstenrinde** *f (geol)* desert varnish
Wz. *s.* Warenzeichen
WZ-Faktor *m* water-cement ratio

X

Xanthan *n s.* Xanthen
Xanthansäure *f s.* Xanthen-9-karbonsäure
Xanthanwasserstoff *m* xanthane hydride, isoperthiocyanic acid
Xanthat *n s.* Xanthogenat
Xanthatin *n* xanthatin *(an antibiotic)*
Xanthatkneter *m (text)* xanthator, xanthating churn, baratte
Xanthen *n* xanthene, *(specif)* dibenzo*[a,e]*pyran
Xanthenfarbstoff *m* xanthene dye[stuff]
Xanthen-9-karbonsäure *f* xanthene-9-carboxylic acid
Xanthenringsystem *n* xanthene ring system
Xanthenthion *n s.* Xanthion
xanthieren *s.* xanthogenieren
Xanthin *n* xanthine, *(specif)* 2,6-dihydroxypurine
Xanthinbase *f* xanthine base
Xanthinoxydase *f* xanthine oxidase
Xanthion *n* xanthion, xanthene-9-thione
Xanthogenamid *n* $NH_2CSOC_2H_5$ xanthogenamide *(one form of ethyl aminothioformate)*
Xanthogenat *n* xanthate, xanthogenate *(salt or ester of a xanthogenic acid)*
xanthogenieren *(text)* to xanthate, to sulphidize, to sulphide
Xanthogenierung *f (text)* xanthation, sulphidizing
Xanthogensäure *f* 1. $C_nH_{2n+1}O-CSSH$ xanthogenic acid, *(specif)* $C_2H_5O-CSSH$ dithiocarbonic *O*-ethyl ester, ethylxanthogenic acid; 2. *(recently proposed name for)* HOCSSH dithiocarbonic acid
Xanthogensäureester *m* xanthogenic-acid ester
Xanthon *n* xanthone, xanthen-9-one, dibenzopyrone

Xanthophyllit *m (min)* xanthophyllite *(a phyllosilicate)*
Xanthoprotein *n* xanthoprotein
Xanthoproteinreaktion *f* xanthoproteic (xanthoprotein) reaction
Xanthoproteinsäure *f* xanthoproteic acid
Xanthosin *n* xanthosine, xanthine 9-ribofuranoside *(a nucleoside)*
Xanthurensäure *f* xanthurenic acid, 4,8-dihydroxyquinoline-2-carboxylic acid
Xanthyliumsalz *n* xanthylium salt
XE *s.* X-Einheit
X-Einheit *f* X unit
Xenat *n* xenate *(a salt of xenic acid)*
Xenoblast *m (geol)* xenoblast
xenoblastisch *(geol)* xenoblastic
Xenolith *m (geol)* xenolith *(exogenous enclosure)*
xenomorph *(cryst)* xenomorphic, allotriomorphic, anhedral
Xenon *n* Xe xenon
Xenonsäure *f* H_6XeO_6 xenic acid
Xenotim *m (min)* xenotime *(yttrium phosphate)*
Xenylamin *n* $C_6H_5C_6H_4NH_2$ xenylamine, *p*-aminobiphenyl
Xerogel *n (coll)* xerogel
Xerografie *f* xerography
xerografisch xerographic
Xeroradiografie *f* xeroradiography
Ximeninsäure *f* ximenynic acid, santalbic acid, 11-octadecen-9-ynoic acid
Xylan *n* xylan *(a polysaccharide)*
Xylarsäure *f* $HOOC[CHOH]_3COOH$ xylaric acid, xylo-trihydroxyglutaric acid
Xylenol *n* $(CH_3)_2C_6H_3OH$ xylenol, hydroxyxylene, dimethylphenol
Xylenolharz *n* xylenol resin
Xylidin *n* $(CH_3)_2C_6H_3NH_2$ xylidine, aminodimethylbenzene
Xylit *m* 1. $CH_2OH[CHOH]_3CH_2OH$ xylitol *(a pentanepentol)*; 2. xylite, woody lignite (brown coal)
Xylobiose *f* xylobiose *(a disaccharide)*
Xylochinon *n* xyloquinone, dimethylbenzoquinone
Xylodesose *f* deoxyxylose, xylodesose *(a monosaccharide)*
Xylohexaose *f* xylohexaose *(an oligosaccharide)*
Xyloketose *f s.* Xylulose
Xylol *n* $C_6H_4(CH_3)_2$ xylene, dimethylbenzene, *(esp commercially for impure products)* xylol
Xylollichtgelb *n* xylene light yellow
Xylolmoschus *m* $(NO_2)_3C_6(CH_3)_2C(CH_3)_3$ musk xylol, musk xylene, xylene musk, 2,4,6-trinitro-1,3-dimethyl-5-*tert*-butylbenzene
Xylolsulfonsäure *f* xylenesulphonic acid
Xylonsäure *f* $HOCH_2[CHOH]_3COOH$ xylonic acid *(a tetrahydroxypentanoic acid)*
Xylopentaose *f* xylopentaose *(an oligosaccharide)*
Xylorzin *n* $(CH_3)_2C_6H_2(OH)_2$ xylorcinol, dihydroxyxylene
Xylose *f* xylose, wood sugar *(a monosaccharide)*
Xylotetraose *f* xylotetraose *(an oligosaccharide)*
Xylotrihydroxyglutarsäure *f s.* Xylarsäure
Xylotriose *f* xylotriose *(an oligosaccharide)*
Xyloylgruppe *f* $(CH_3)_2C_6H_3CO-$ xyloyl group
Xylulose *f* $HOCH_2[CHOH]_2COCH_2OH$ xylulose, xyloketose
Xylylbromid *n* $CH_3C_6H_4CH_2Br$ α-bromoxylene, xylyl bromide
Xylylchlorid *n* $CH_3C_6H_4CH_2Cl$ α-chloroxylene, xylyl chloride
Xylylenbromid *n* $C_6H_4(CH_2Br)_2$ α,α'-dibromoxylene, xylylene dibromide
Xylylenchlorid *n* $C_6H_4(CH_2Cl)_2$ α,α'-dichloroxylene, xylylene dichloride
Xylylsäure *f* $(CH_3)_2C_6H_3COOH$ xylylic acid, dimethylbenzoic acid
x-y-Schreiber *m* X-Y recorder

Y

Yagein *n s.* Harmin
Yankee-Maschine *f (pap)* Yankee (single-cylinder, lick-up) machine
Ylang-Ylang-Öl *n* ilang-ilang (ylang-ylang) oil *(from Cananga odorata (Lam.) Hook. fil. et Thomson)*
Ylid *n* ylide *(any of a class of internal salts)*
Ylidreaktion *f* ylide reaction
Yohimbealkaloid *n* yohimbé alkaloid
Yohimbin *n* yohimbine, quebrachine *(alkaloid)*
Yohimbinsäure *f* yohimbic acid
Yperit *n* $(ClCH_2CH_2)_2S$ yperite, di-2-chloroethyl sulphide
Ypsilonstück *n* Y-shape[d] connecting piece (tube)
Ysopöl *n* hyssop oil *(from Hyssopus officinalis L.)*
Y-Stück *n s.* Ypsilonstück
Ytterbinerde *f* 1. ytterbium earth *(class name)*; 2. *s.* Ytterbiumoxid
Ytterbium *n* Yb ytterbium
Ytterbium(II)-chlorid *n* $YbCl_2$ ytterbium(II) chloride, ytterbium dichloride
Ytterbium(III)-chlorid *n* $YbCl_3$ ytterbium(III) chloride, ytterbium trichloride
Ytterbiumdichlorid *n s.* Ytterbium(II)-chlorid
Ytterbiumoxid *n* Yb_2O_3 ytterbium oxide
Ytterbium(III)-sulfat *n* $Yb_2(SO_4)_3$ ytterbium(III) sulphate
Ytterbiumtrichlorid *n s.* Ytterbium(III)-chlorid
Yttererde *f* 1. yttrium earth *(class name)*; 2. *s.* Yttriumoxid
Yttrialith *m (min)* yttrialite *(a sorosilicate containing yttrium and thorium)*
Yttrium *n* Y yttrium
Yttriumazetat *n* $(CH_3COO)_3Y$ yttrium acetate
Yttriumbromid *n* YBr_3 yttrium bromide
Yttriumchlorid *n* YCl_3 yttrium chloride
Yttriumfluorid *n* YF_3 yttrium fluoride
Yttriumhydroxid *n* $Y(OH)_3$ yttrium hydroxide
Yttriumjodid *n* YI_3 yttrium iodide
Yttriumkarbid *n* YC_2 yttrium dicarbide
Yttriumkarbonat *n* $Y_2(CO_3)_3$ yttrium carbonate
Yttriumnitrat *n* $Y(NO_3)_3$ yttrium nitrate
Yttriumoxid *n* Y_2O_3 yttrium oxide
Yttriumsulfat *n* $Y_2(SO_4)_3$ yttrium sulphate
Yttrotantalit *m (min)* yttrotantalite

Yttrotitanit *m (min)* yttrotitanite, keilhauite *(a neso-subsilicate)*
Yukkafaser *f* yucca fibre *(leaf fibre from various Yucca specc.)*

Z

Z *s.* Ordnungszahl
Zacke *f,* **Zacken** *m* peak *(as of a chromatogram)*
Zaffer *m* zaffre, zaffar, zaffir *(a roasted mixture of cobalt ore and sand)*
zäh[e] 1. tough, tenaceous *(material)*; *(soil)* tenaceous; 2. *s.* zähflüssig
zähflüssig viscid, viscous, viscose, semiliquid, semifluid, ropy, thick ✦ **z. machen** to thicken
z. werden to inspissate, to thicken
Zähflüssigkeit *f* viscosity, ropiness, thickness
Zähigkeit *f* 1. toughness, tenacity; 2. *(phys chem)* viscosity
absolute (dynamische) Z. *(phys chem)* dynamic viscosity
kinematische Z. kinematic viscosity
Zähigkeitskraft *f* viscous force
Zähigkeitsstabilität *f* viscosity stability
Zahl *f* 1. *(for characterizing a property)* number, value; 2. *s.* Zahlenindex
Abbesche Z. Abbe number (value), ν-value, constringence *(reciprocal relative dispersion)*
Avogadrosche Z. 1. Avogadro number (constant) *(the number of molecules in a gram molecule of any substance)*; 2. *(up to now)* Loschmidt number *(the number of molecules in 1 cm^3 of an ideal gas at 0 °C and 1 atm)*
Loschmidtsche Z. *s.* Avogadrosche Z. 1.
Poissonsche Z. Poisson's ratio *(the ratio of transverse to longitudinal strain in a material under tension)*
Reichert-Meisslsche Z. Reichert-Meissl number (value), R-M number *(for fats)*
Reynoldssche Z. Reynolds number, Re. *(rheology)*
Zählapparat *m s.* Zählwerk
zählen to enumerate *(e.g. the atoms of a ring system)*
Zahlenbuna *m* numbered buna rubber, numeral grade of buna
Zahlenindex *m (nomencl)* numeral
hochgestellter Z. superscript [numeral], numeric superscript, upper index
tiefgestellter Z. subscript [numeral], numeric subscript, lower index
Zahlenpräfix *n* numerical prefix
Zahlenverhältnis *n* numerical ratio
Zahlenvorsatz *m* numerical prefix
Zahlenwert *m* numerical value
Zähler *m* 1. meter *(for fluids)*; 2. *(nucl)* counter, counting tube; 3. *s.* Zählwerk
Zählgerät *n s.* Zählwerk
Zähligkeit *f* ligancy, coordination (covalence, covalency) number
Zählrohr *n (nucl)* counting tube, counter
selbstlöschendes Z. self-quenched (self-quenching) counter
Zählung *f* enumeration *(as of the atoms of a ring system)*
Zählvorrichtung *f s.* Zählwerk
Zählwalze *f (pap)* counting roll
Zählweise *f (nomencl)* enumeration system
Zählwerk *n* counting apparatus (device), counter
Zahlwort *n* / **multiplikatives** *s.* Zahlwortpräfix
Zahlwortpräfix *n (nomencl)* numerical (multiplying) prefix
Zahnbein *n* dentine
Zahnpaste *f* tooth-paste
Zahnpflegemittel *n* dentifrice
Zahnporzellan *n* dental porcelain
Zahnputzmittel *n* dentifrice
Zahnradpumpe *f* gear pump
Zahnscheibenmühle *f* toothed-disk mill
Zahnschmelz *m* [dental] enamel
Zahnstein *m* scale
Zahnwalzenbrecher *m* toothed-roll crusher
Zahnzement *m* dental cement
zähpolen *(met)* to toughen by poling, to pole
Zange *f* [pair of] tongs, forceps, *(if small)* pincers
zapfen *(rubber)* to tap
Zapfen *m* 1. trunnion, pivot, journal, neck *(as of a shaft or cylinder)*; 2. bung, spigot, plug, plug cock (bib) *(on a cask)*; 3. tit *(a defect in glass)*
Zapfenlager *n* journal bearing
Zapfloch *n* bunghole
Zapfmesser *n (rubber)* tapping knife
Zapfschnitt *m (rubber)* tapping cut (incision)
Zapfung *f (rubber)* tapping
Zaratit *m (min)* zaratite *(a hydrous basic nickel carbonate)*
Zäsium *n* Cs caesium
Zäsiumalaun *m* $CsAl(SO_4)_2 \cdot 12H_2O$ caesium alum
Zäsiumaluminiumsulfat *n* $CsAl(SO_4)_2$ aluminium caesium sulphate
Zäsiumazetat *n* CH_3COOCs caesium acetate
Zäsiumbromat *n* $CsBrO_3$ caesium bromate
Zäsiumbromid *n* CsBr caesium bromide
Zäsiumchlorat *n* $CsClO_3$ caesium chlorate
Zäsiumchlorid *n* CsCl caesium chloride
Zäsiumchromat *n* Cs_2CrO_4 caesium chromate
Zäsiumdichromat *n* $Cs_2Cr_2O_7$ caesium dichromate
Zäsiumdisulfid *n* Cs_2S_2 cesium disulphide
Zäsiumfluorid *n* CsF caesium fluoride
Zäsiumhexafluorosilikat *n* $Cs_2[SiF_6]$ caesium hexafluorosilicate
Zäsiumhexasulfid *n* Cs_2S_6 caesium hexasulphide
Zäsiumhydrid *n* CsH caesium hydride
Zäsiumhydroxid *n* CsOH caesium hydroxide
Zäsiumjodid *n* CsI caesium iodide
Zäsiumkarbonat *n* Cs_2CO_3 caesium carbonate
Zäsiummetaperjodat *n* $CsIO_4$ caesium metaperiodate, caesium periodate, caesium tetraoxoiodate(VII)
Zäsiumnitrat *n* $CsNO_3$ caesium nitrate
Zäsiumoxid *n* Cs_2O caesium oxide
Zäsiumpentajodid *n* CsI_5 caesium pentaiodide

Zäsiumpermanganat *n* $CsMnO_4$ caesium permanganate
Zäsiumsulfat *n* Cs_2SO_4 caesium sulphate
Zäsiumsulfid *n* Cs_2S caesium sulphide
Zäsiumtetrasulfid *n* Cs_2S_4 caesium tetrasulphide
Zäsiumtetroxojodat(VII) *n s.* Zäsiummetaperjodat
Z-Drehung *f (rubber)* Z twist
Zeaxanthin *n* zeaxanthin *(a leaf pigment)*
Zedernblätteröl *n* cedar-leaf oil, thuja oil *(chiefly from Thuja occidentalis L.)*
Zedernholzöl *n* cedarwood oil *(from Cedrus atlantica Manetti or Juniperus specc.)*
Zedernkampfer *m s.* Zedrol
Zedren *n* cedrene *(a tricyclic sesquiterpene)*
Zedrol *n* cedrol, cedar (cedarwood) camphor *(a sesquiterpenoid alcohol)*
Zeeman-Effekt *m* Zeeman effect
Zehntelnormallösung *f* decinormal solution
Zeichen *n* / **chemisches** chemical symbol (sign)
Zeichenkarton *m* drawing (painting) cardboard (board), painters' (artists') cardboard
Zeichenkreide *f* crayon
Zeichenpapier *n* drawing paper
Zeichensprache *f* / **chemische** chemical shorthand
Zeiger *m* 1. pointer, indicator, index *(of an apparatus)* ; 2. *(biol)* indicator *(of soil conditions)*
Zein *n* zein *(a prolamin obtained from maize)*
Zeinfaser *f* zein staple [fibre], maize protein staple
Zeinkunststoff *m* zein plastic
Zeisel-Bestimmung *f* Zeisel determination *(of alkoxy groups)*
Zeiß-Abbe-Refraktometer *n* Abbe refractometer
Zeit *f* / **mittlere freie** *(phys chem)* mean free time *(between two collisions)*
Zeitentwicklung *f (phot)* development by time
Zeit-Gamma-Kurve *f (phot)* time-gamma curve
Zeitkonstante *f* time constant
Zeitreaktion *f* clock (time) reaction
Zeitschrift *f* / **referierende** abstract[ing] journal
Zeitschriftenpapier *n* magazine paper
Zeit-Temperatur-Umwandlungsdiagramm *n* time-temperature-transformation diagram, T.T.T. diagram
  Z. für kontinuierliche Abkühlung continuous-cooling transformation diagram
  isothermes Z. isothermal transformation diagram
Zeitungsdruck *m* newsprint
Zeitungsdruckfarbe *f* newsprint (news) ink
Zeitungsdruckpapier *n* newsprint paper, news [print]
Zeitungsdruckpapiermaschine *f* newsprint paper machine, news machine
Zellatmung *f* cellular respiration
Zelle *f* cell, *(tech also)* compartment, chamber
  elektrochemische Z. *s.* galvanische Z.
  elektrolytische Z. electrolytic (electrolysis) cell
  fotoelektrische Z. photoelectric cell, photocell, phototube, photovalve
  galvanische Z. voltaic (galvanic, electrical, chemical, electrochemical) cell
  lichtelektrische Z. *s.* fotoelektrische Z.
Zellenaggregat *n (el chem)* cell assembly
zellenartig cellular
Zellenbeton *m* cellular concrete
Zellendolomit *m* cellular dolomite
Zellenfilter *n* cellular (cell-type) filter
Zellenflüssigkeit *f (el chem)* cell liquor
Zellengefüge *n* cellular structure
Zellenkalk *m (geol)* cellular lime
Zellenofen *m* cell-type oven
Zellenrad *n*, **Zellenradzuteiler** *m* star wheel (valve), star (rotary-vane) feeder
Zellenreaktion *f (phys chem)* cell reaction
Zellenschleuse *f s.* Zellenrad
Zellenspannung *f (phys chem)* cell voltage
Zellenspeicher *m* bin, silo
Zellenstruktur *f* cellular structure
Zellenverdichter *m* [sliding-]vane compressor
Zellflüssigkeit *f (biol)* cell sap
Zellgefüge *n* cellular structure
Zellgewebe *n (text)* cellular tissue
Zellgift *n* cytotoxin
Zellglas *n* cellophane, cellulose film
Zellgummi *m* cellular (expanded) rubber
Zellhartgummi *m* cellular ebonite
Zellhaut *f s.* Zellglas
Zellhormon *n* cell hormone
Zellhorn *n* celluloid
zellig cellular
Zellinhalt *m (biol)* cell contents
Zellkern *m (biol)* nucleus
Zellmembran *f* cell membrane
Zellobiose *f* cellobiose *(a disaccharide)*
Zellotetraose *f* cellotetraose *(a tetrasaccharide)*
Zellotriose *f* cellotriose *(a trisaccharide)*
Zellphysiologie *f* cell physiology
Zellrad *n*, **Zellradschleuse** *f s.* Zellenrad
Zellsaft *m (biol)* cell sap
zellschädigend cytotoxic
Zellstoff *m* 1. *(chem)* $(C_6H_{10}O_5)_x$ cellulose; 2. *(tech)* [chemical] pulp, *(specif)* [chemical] wood pulp, CWP
  Z. für die Chemiefaserindustrie rayon (dissolving) pulp
  Z. für die Papierindustrie paper[making] pulp
  Z. in Bogenform (Pappenform) wood-pulp board, lap[ped] pulp, sheets (laps) of chemical wood pulp
  Z. mit hohem Weißgehalt high-brightness pulp
  alkalisch aufgeschlossener (erkochter, gekochter) Z. alkaline-cooked pulp, alkali pulp
  bleichbarer Z. bleaching pulp
  harter (hartgekochter) Z. hard (strong, low-boiled) pulp
  klassischer Z. chemical pulp
  leicht bleichbarer Z. easy-bleaching pulp
  schwer bleichbarer Z. hard-bleaching pulp
  sehr harter (roher) Z. *s.* unaufgeschlossener Z.
  unaufgeschlossener Z. high-strength pulp, prime strong pulp
  weicher (weichgekochter) Z. soft (well-cooked, high-boiled) pulp
  weit heruntergekochter Z. *s.* weicher Z.

wenig aufgeschlossener Z. *s.* harter Z.
Zellstoffabrik *f* pulp mill
Zellstoffabrikabwasser *n* chemical pulp factory wastes
Zellstoffaufschluß *m* chemical pulping
Zellstoffausbeute *f* pulp yield
Zellstoffbahn *f* web of pulp
Zellstoffblätter *npl s.* Zellstoff in Bogenform
Zellstoffbleiche *f* pulp bleaching
Zellstoffbogen *m* pulp sheet
Zellstoffbrei *m* pulp
Zellstoffentwässerungsmaschine *f* pulp[-drying] machine, wet machine (press), half-stuff machine, press-pâte
Zellstofferzeugung *f* pulp making (manufacture)
Zellstoffestigkeit *f* pulp strength
Zellstoffhersteller *m* pulpmaker
Zellstoffindustrie *f* pulp industry
Zellstoffkarton *m* pulp cardboard
Zellstoffkocher *m* digester
Zellstoffpapier *n* wood-free paper
Zellstoffpappe *f s.* Zellstoff in Bogenform
Zellstoffproduzent *m* pulpmaker
Zellstoffreinigung *f* pulp purification
Zellstoffsuspension *f (pap)* fibre suspension, fibrous pulp, pulp slurry (stock)
Zellstofftrocknung *f* pulp drying
Zellstoff- und Papierfabrik *f* integrated mill
Zellstoffveredelung *f* pulp refining (purification), [alkaline] refining of pulp
Zellstoffveredelungslauge *f* / **alkalische** alkali refining liquor
Zellstoffwäsche *f* pulp washing
Zellstoffwäscher *m* pulp washer
Zellstoffwatte *f* artificial cotton, *(Am)* cellulose (pulp) wadding
Zellstoffwerk *n* pulp mill
Zellstruktur *f* cellular structure
zellulär cellular
Zelluloid *n* celluloid
Zellulose *f* $(C_6H_{10}O_5)_x$ cellulose
native (natürliche) Z. natural cellulose
oxydierte Z. oxycellulose
regenerierte Z. regenerated cellulose, hydrate[d] cellulose, cellulose hydrate
α-Zellulose *f* alpha cellulose, α-cellulose
β-Zellulose *f* beta cellulose, β-cellulose
γ-Zellulose *f* gamma cellulose, γ-cellulose
Zelluloseabbau *m* cellulose decomposition
Zelluloseabkömmling *m* cellulose derivative, cellulosic
Zelluloseäther *m* cellulose ether
Zelluloseäthyläther *m* ethylcellulose
Zelluloseazetat *n* cellulose acetate, acetylated cellulose
Zelluloseazetatbutyrat *n* cellulose acetate butyrate
Zelluloseazetatfaser *f* cellulose acetate fibre
Zelluloseazetatfaserstoff *m* cellulose acetate fibre
Zelluloseazetatpropionat *n* cellulose acetate propionate
Zelluloseazetatseide *f* cellulose acetate rayon, acetate filament yarn
Zelluloseazetatspinnlösung *f* cellulose acetate dope
Zellulosechemiefaser *f* cellulosic fibre
Zellulosechemiefaserstoff *m* cellulosic fibre
Zellulosederivat *n* cellulose derivative, cellulosic
Zellulosediazetat *n* cellulose diacetate, *(text also)* secondary [cellulose] acetate
Zellulosedichtung *f* cellulose gasket
Zelluloseerzeugnis *n* cellulosic
Zelluloseester *m* cellulose ester
Zellulosefaser *f* cellulose fibre
Zellulosefaserstoff *m* cellulose fibre
Zellulosefibrille *f* cellulose fibril
Zelluloseformiat *n* cellulose formate
Zelluloseglykolsäure *f* carboxymethylcellulose, C.M.C.
Zellulosehydrat *n* cellulose hydrate
Zellulosekohle *f* cellulose coal
Zelluloselack *m* cellulose lacquer
Zellulosenitrat *n* cellulose nitrate
Zellulosenitratlack *m* cellulose nitrate lacquer
Zellulosepropionat *n* cellulose propionate
Zelluloseregeneratfaser *f* regenerated cellulose fibre
Zelluloseregeneratfaserstoff *m* regenerated cellulose fibre
Zellulosesalpetersäureester *m* cellulose nitrate
Zellulosetriazetat *n* cellulose triacetate, *(text also)* primary [cellulose] acetate
Zellulosetrinitrat *n* cellulose trinitrate
Zellulosexanthat *n s.* Zellulosexanthogenat
Zellulosexanthogenat *n* cellulose xanthate
Zellulosexanthogensäure *f* cellulose xanthic acid
Zellulosezersetzer *mpl (agric)* cellulose decomposers *(bacteria)*
Zellwand *f (biol)* cell wall
Zellwolle *f* viscose staple fibre
Zeltbegasung *f (agric)* tent fumigation
Zement *m* 1. *(build, rubber)* cement; 2. *(geoch)* cement, cementing agent, binder, agglutinant
hydraulischer Z. hydraulic cement
loser Z. bulk cement
sulfatbeständiger Z. sulphate-resisting cement
treibender Z. expanding (expansive) cement
wasserabstoßender (wasserabweisender) Z. hydrophobe (water-repellent) cement
wasserbindender Z. hydraulic cement
wasserdichter Z. *s.* wasserabstoßender Z.
weißer Z. white cement
zementartig cementitious
Zementation *f* 1. *(met, geol)* cementation *(precipitation of a metal from salt solutions by the action of a less noble one)*; 2. *(met)* carburization, cementation
Zementationsbad *n (met)* carburizing bath
Zementationsgas *n (met)* carburizing gas
Zementationsgemisch *n (met)* carburizing mixture
Zementationshärten *n (met)* carburization, cementation *(and subsequent hardening)*
Zementationskasten *m (met)* carburizing (cementing) box
Zementationsmittel *n (met)* carburizing (cement-

ing) medium (agent), [case-hardening] carburizer
Zementationspulver *n (met)* carburizing powder
Zementationsschicht *f (met)* carburized case
Zementationstemperatur *f (met)* carburizing temperature
Zementationstiefe *f (met)* carburizing (carburization) depth, depth of case
Zementationszeit *f (met)* carburizing time
Zementationszone *f (geoch)* zone of cementation
Zementbrennen *n* cement burning
Zementchemie *f* cement chemistry
Zementdreh[rohr]ofen *m* rotary cement kiln
Zementformsand *m* cement-bonded sand
Zementherstellung *f* manufacture (production) of cement
Zementier ... *s.a.* Zementations ...
zementieren 1. *(build)* to cement; 2. *(met, geol)* to cement *(to precipitate from salt solutions by the action of a less noble metal)*; 3. *(met)* to carburize, to cement
im Salzbad z. *(met)* to liquid-carburize, to bath-carburize
in der Randschicht (Randzone) z. *(met)* to case-carburize
in festem Einsatz z., in festen Mitteln z. *(met)* to pack-carburize
in flüssigen Mitteln z. *(met)* to liquid-carburize, to bath-carburize
in gasförmigen Mitteln z. *(met)* to gas-carburize
Zementieren *n* 1. *(build)* cementing; 2. *(met)* cementing *(precipitating of a metal from salt solutions by a less noble one)*; 3. *(met)* carburizing, cementing
Z. im Salzbad *(met)* liquid (liquid-salt, bath) carburizing
Z. in festem Einsatz, Z. in festen Mitteln *(met)* solid (pack, solid-pack) carburizing
Z. in flüssigen Mitteln *(met)* liquid (liquid-salt, bath) carburizing
Z. in gasförmigen Mitteln *(met)* gas carburizing
Zementierofen *m (met)* carburizing furnace (oven)
Zementierung *f s.* Zementation
Zementit *m (met)* Fe_3C cementite, cemented carbide
Zementkalk *m s.* Kalk / hydraulischer
Zementklinker *m* [cement] clinker
Zementkuchen *m* cement pat
Zementkupfer *n* cement (precipitated) copper
Zementleim *m* cement paste
Zementmilch *f* laitance
Zementmörtel *m* cement mortar
Zementofen *m s.* Zementschachtofen
Zementsand *m* cement-bonded sand
Zementschachtofen *m* cement kiln
Zementschlamm *m,* **Zementschlempe** *f* laitance
Zementschwarz *n* manganese black *(manganese dioxide)*
Zentigramm-Methode *f* centigramme method *(semimicroanalysis)*
Zentralanguß *m (plast)* centre gate
Zentralatom *n* central atom
Zentralgenerator *m* independent producer *(gas making)*
Zentralion *n* central ion
Zentralkohlenstoffatom *n* central carbon atom
Zentralmetall *n* central metal *(of a complex compound)*
Zentralrohr *n* central pipe (tube)
zentralsymmetrisch centrosymmetric
Zentralwelle *f* central shaft
zentrifugal centrifugal
Zentrifugalabscheider *m* centrifugal collector (separator)
Zentrifugalabsorber *m* centrifugal [gas] absorber
Zentrifugalbecherwerk *n* centrifugal-discharge bucket elevator
Zentrifugalbeschleunigung *f* centrifugal acceleration
Zentrifugalchromatografie *f* centrifugal chromatography
Zentrifugalextraktor *m* centrifugal extractor
Zentrifugalfeld *n* centrifugal field
Zentrifugalfilter *n* centrifugal filter
Zentrifugalgasabsorber *m* centrifugal gas absorber
Zentrifugalgaswäscher *m* disintegrator [washer]
Zentrifugalgießen *n,* **Zentrifugalguß** *m* centrifugal casting
Zentrifugalklassierer *m* centrifugal classifier
Zentrifugalkorb *m* rotary-cup atomizer
Zentrifugalkraft *f* centrifugal force
Zentrifugalkraftklassierer *m,* **Zentrifugalkraftsichter** *m* centrifugal classifier
Zentrifugalpumpe *f* centrifugal pump
Zentrifugalreiniger *m* centrifugal cleaner; *(pap)* centrifugal strainer
Zentrifugalscheibe *f* centrifugal disk
Zentrifugalscheider *m* centrifugal collector (separator)
Zentrifugalsichter *m* centrifugal classifier; *(pap)* centrifugal-type screen *(for cleaning mechanical wood pulp)*
Zentrifugalsortierer *m* centrifugal classifier; *(pap)* centrifugal strainer, erkensator *(for cleaning wet pulp)*
Zentrifugaltrockenmaschine *f* whizzer
Zentrifugalversprüher *m* centrifugal atomizer
Zentrifugalversprühung *f* centrifugal atomization
Zentrifugalwäscher *m* disintegrator [washer]
Zentrifugalzerstäuber *m* centrifugal atomizer
Zentrifugalzerstäubung *f* centrifugal atomization
Zentrifugat *n* centrifugate
Zentrifuge *f* centrifuge, *(tech esp for filtering)* centrifugal
Z. mit seitlicher Entleerung, Z. mit Umfangsaustrag peripheral discharge centrifuge
Z. mit Untenaustrag bottom discharge centrifuge
stehende Z. underdriven (bottom-driven, base-bearing) centrifuge
unten entleerende Z. *s.* Z. mit Untenaustrag
von unten angetriebene Z. *s.* stehende Z.
Zentrifugenglas *n* centrifuge tube

Zentrifugenrotor *m* centrifuge rotor (head)
Zentrifugenspinnen *n* [centrifugal] pot spinning
Zentrifugentrommel *f* centrifuge basket
Zentrifugenwickel *m* cake
Zentrifugenzahl *f* relative centrifugal force
zentrifugieren to centrifugate, to centrifuge
Zentrifugierung *f* centrifugation
zentripetal centripetal
Zentripetalkraft *f* centripetal force
Zentrireiniger *m (pap)* centricleaner
zentrisch centric, central, on-centre
Zentrosphäre *f (geoch)* centrosphere *(core of the earth)*
zentrosymmetrisch centrosymmetric
Zentrum *n* centre
  aktives Z. active centre (patch) *(as of a catalyst)*
Zeolith *m (min)* zeolite
  künstlicher Z. artificial (synthetic) zeolite, zeolite exchange resin *(ion exchange technology)*
Zeolithisation *f* zeolitization
zeolithisch zeolitic
Zeolithisierung *f* zeolitization
Zeolithwasser *n* zeolitic water
Zephaelin *n* cephaelin[e] *(alkaloid)*
Zephalosporansäure *f* cephalosporanic acid *(parent compound of an antibiotic)*
Zer *n* Ce cerium
Zerasinsäure *f s.* Lignozerinsäure
zerbrechen to break
zerbrechlich breakable, fragile, brittle
zerbröckeln 1. to crumble; 2. *(of material)* to crumble [away], to slake
Zer(III)-bromid *n* $CeBr_3$ cerium(III) bromide, cerium tribromide
Zer(III)-chlorid *n* $CeCl_3$ cerium(III) chloride, cerium trichloride
Zerdioxid *n s.* Zer(IV)-oxid
zerdrücken to crush, to mill
Zerebrosid *n* cerebroside *(a lipoid)*
Zererde *f s.* Zer(IV)-oxid
Zeresin *n* ceresin [wax], ceresine
Zerewitinow-Bestimmung *f* Zerewitinoff determination *(of active H atoms)*
Zerfall *m (tech)* disintegration, breakdown of size; *(chem)* decomposition, breakdown, *(into ions)* dissociation; *(nucl)* disintegration, fission, decay; *(biol)* decomposition, decay ✦ **durch radioaktiven Z. entstanden** radiogenic
  dualer Z. *(nucl)* branched (multiple) disintegration (decay), branching
  radioaktiver Z. radioactive disintegration (decay)
  thermischer Z. thermal decomposition (degradation)
  verzögerter Z. delayed disintegration
  verzweigter Z. *s.* dualer Z.
β-**Zerfall** *m (nucl)* beta (β-ray) disintegration (decay)
zerfallen *(mechanically)* to disintegrate, to break down, to crumble [away], *(esp of coal)* to slake; *(chemically)* to decompose, *(into ions)* to dissociate; *(nucl)* to disintegrate, to decay; *(biol)* to decompose, to decay
  in Schichten z. to delaminate
  in Stücke z. to crumble to pieces
Zerfallselektron *n* disintegration (decay) electron
Zerfallsenergie *f (nucl)* disintegration (decay) energy
Zerfallsgeschwindigkeit *f (nucl)* rate of disintegration (decay); *(chem)* decomposition rate
Zerfallsgesetz *n (nucl)* [radioactive-]decay law
Zerfallskonstante *f (nucl)* disintegration (decay, transformation) constant, decay coefficient (factor)
Zerfallskurve *f (nucl)* disintegration (decay) curve
Zerfallsleuchten *n (nucl)* decay luminescence
Zerfallsprodukt *n (nucl)* disintegration (decay) product; *(chem)* decomposition product
Zerfallsrate *f (nucl)* rate of disintegration (decay)
Zerfallsreaktion *f (chem)* decomposition reaction
Zerfallsreihe *f (nucl)* disintegration (decay) series, radioactive series (family, chain)
Zerfallsteilchen *n (nucl)* disintegration (decay) particle
Zerfallswahrscheinlichkeit *f (nucl)* probability of disintegration (decay)
Zerfallswärme *f (nucl)* heat of radioactivity
Zerfallszeit *f (nucl)* disintegration (decay) period (time)
Zerfaserer *m (pap)* shredding (kneading) machine, shredder, kneader *(recovery of waste paper)*
zerfasern *(pap)* to defibre, to defibrate, to reduce to fibres, to shred, *(Am also)* to [de]fiberize
Zerfasern *n (pap)* defib[e]ring, defibration, shredding, *(Am also)* [de]fiberization
Zerfaserungsmaschine *f s.* Zerfaserer
zerfließen to deliquesce, to liquefy
Zerfließen *n* deliquescence
zerfließlich deliquescent
Zerfließlichkeit *f* deliquescence
Zer(III)-fluorid *n* CeF_3 cerium(III) fluoride, cerium trifluoride
Zer(IV)-fluorid *n* CeF_4 cerium(IV) fluoride, cerium tetrafluoride
zerfressen to corrode, to eat [away]
Zerfressen *n* corrosion
zerfressend corrosive
zergehen *s.* zerfließen
zerhacken to chop, to chip
Zer(III)-hydroxid *n* $Ce(OH)_3$ cerium(III) hydroxide, cerium trihydroxide
Zer(IV)-hydroxid *n* $Ce(OH)_4$ cerium(IV) hydroxide, cerium tetrahydroxide
Zer(IV)-hydroxidnitrat *n* $Ce(OH)(NO_3)_3$ cerium(IV) hydroxide nitrate
Zerimetrie *f* ceri[o]metry
Zerit *m* cerite, cerine *(a siliceous cerium mineral)*
Zeriterde *f* cerite earth *(any of one group of rare-earth metal oxides)*
Zerkarbid *n* CeC_2 cerium carbide
Zer(III)-karbonat *n* $Ce_2(CO_3)_3$ cerium(III) carbonate
zerkleinern to comminute, to reduce; to mill, to grind; *(into chips)* to chip, to chop; to crush *(brittle material)*
Zerkleinerung *f* comminution, [size] reduction;

milling, grinding; chipping, chopping; crushing *(of brittle material)*
Zerkleinerungsanlage *f* comminution plant
Zerkleinerungsgesetz *n* **nach Bond** Bond's law
Z. nach Kick Kick's law
Zerkleinerungsgrad *m* degree of reduction, [size-]reduction ratio
Zerkleinerungsmaschine *f* comminuting machine
zerkrümeln to crumble
zerlegen *(physically)* to fractionate, to split [up], to separate; *(chemically)* to decompose, to fragment, to split [up]
elektrolytisch z. to electrolyze
in Einzelfasern z. *s.* zerfasern
in Schichten z. to delaminate
nach Korn[größen]klassen z. to size, to fractionate
Zerlegung *f (physically)* fractionation, splitting[-up], separation; *(chemically)* decomposition, fragmentation, splitting[-up]
zermahlen to grind, to mill, to triturate, to grate
zermalmen to bruise, to crush
Zer-Mischmetall *n* misch metal
Zer(III)-nitrat *n* $Ce(NO_3)_3$ cerium(III) nitrate, cerium trinitrate
Zer(IV)-nitrat *n* $Ce(NO_3)_4$ cerium(IV) nitrate, cerium tetranitrate
Zer(III)-orthophosphat *n* $CePO_4$ cerium(III) orthophosphate
Zerotinsäure *f* $CH_3[CH_2]_{24}COOH$ cerotic acid, cerinic acid, hexacosanoic acid
Zer(III)-oxid *n* Ce_2O_3 cerium(III) oxide, cerium trioxide
Zer(IV)-oxid *n* CeO_2 cerium(IV) oxide, cerium dioxide
Zer(III)-oxidchlorid *n* $CeOCl$ cerium(III) oxide chloride
zerplatzen to crack, *(violently)* to explode, to blow up
zerpulvern to pulverize, to powder, to triturate
zerquetschen to crush, to mash
zerreibbar *s.* zerreiblich
zerreiben to triturate, to grate, to grind, to mill
zerreiblich friable, triturable
Zerreiblichkeit *f* friability
Z. des Kokses coke friability
Zerreißdehnung *f* elongation (extension, strain) at break, elongation at rupture (failure), breaking (ultimate) elongation (extension)
zerreißen 1. to tear, to rupture, to break; to macerate *(fibrous material)*; 2. *(of material)* to tear, to break, to crack
Zerreißfestigkeit *f* resistance to tear[ing], tearing (tensile, bursting) strength
Zerreißgrenze *f* [ultimate] tensile strength, bursting strength
Zerreißmaschine *f* 1. tensile-strength tester (testing machine); 2. macerator *(for fibrous material)*
Zerreißprüfung *f* [ultimate] tensile-strength test
Zerreißpunkt *m* breaking point
Zerreißwerk *n* macerator *(for fibrous material)*
zerrieseln 1. to disintegrate *(of granulated fertilizers)*; 2. *(ceram)* to dust *(of materials with a high content of calcium orthosilicate)*
Zerrieseln *n* 1. disintegration *(of granulated fertilizers)*; 2. *(ceram)* dusting *(of materials with a high content of calcium orthosilicate)*
zerschlagen to crush, to smash, to break up
zerschleifen to grind *(wood)*
zerschmelzen to melt, to deliquesce
Zerschmelzen *n* melting, deliquescence
zerschmelzend deliquescent
zersetzbar decomposable
zersetzen *(chem, biol)* to decompose
durch Solvolyse z. to solvolyze
elektrolytisch z. to electrolyze
sich z. *(chem)* to decompose; *(biol)* to decay, to decompose
zersetzend destructive
Zersetzer *m* decomposer
Zersetzerzelle *f* decomposer
zersetzlich decomposable
leicht z. easily decomposable
Zersetzlichkeit *f* decomposability
Zersetzung *f (chem)* decomposition; *(biol)* decay, decomposition
Z. durch Licht photolysis, photodecomposition, photodegradation
elektrolytische Z. electrolytic dissociation
strahlenchemische Z. radiolysis
thermische Z. thermal decomposition (degradation)
Zersetzungsdestillation *f* destructive distillation
Zersetzungserscheinung *f* decomposition phenomenon
Zersetzungsgeschwindigkeit *f* decomposition rate
Zersetzungskatalysator *m* decomposition catalyst
Zersetzungspotential *n* decomposition potential
Zersetzungsprodukt *n* decomposition product
Zersetzungspunkt *m* decomposition point
Zersetzungsreaktion *f* decomposition reaction
Zersetzungsspannung *f* decomposition voltage
Zersetzungswärme *f* heat of decomposition
Zersetzungszelle *f* decomposer
Zersilizid *n* ceric silicide
zerspalten *s.* spalten
zersplittern to spall, to shatter; *(min)* to delaminate
zerspringen to crack, to shatter
zersprühen to atomize
Zersprühen *n* atomizing, atomization
zerstäuben 1. to atomize *(solids)*; 2. *(incorrectly for)* versprühen
Zerstäuber *m* 1. atomizer *(for solids)*; 2. *(incorrectly for)* Versprüher
Z. mit Hilfsstoff auxiliary-fluid atomizer
Zerstäuberdüse *f* atomizing nozzle
Zerstäuberscheibe *f (incorrectly for)* Sprühscheibe
Zerstäubung *f* atomization
Z. durch Düsen nozzle atomization
Z. ohne Hilfsstoff single-fluid atomization
elektrische Z. electrical dispersion, electrodispersion
Zerstäubungsbrenner *m* atomizing (spray) burner

Zerstäubungsmilchpulver *n* spray-dried milk [powder], spray powder milk
Zerstäubungsschwefelverbrennungsofen *m* *(pap)* spray-type sulphur burner
Zerstäubungstrockner *m* spray (flash) dryer
Zerstäubungstrocknung *f* spray (flash) drying
zerstören / oberflächlich to corrode
Zerstörung *f* **/ oberflächliche** corrosion
zerstoßen to crush, to triturate
Zerstoßen *n* crush[ing], trituration
zerstrahlen to undergo annihilation by radiation
Zerstrahlung *f* annihilation radiation
Zerstrahlungsspektrum *n* annihilation spectrum
Zer(III)-sulfat *n* $Ce_2(SO_4)_3$ cerium(III) sulphate
Zer(IV)-sulfat *n* $Ce(SO_4)_2$ cerium(IV) sulphate
Zer(III)-sulfid *n* Ce_2S_3 cerium(III) sulphide, cerium trisulphide
zerteilen to disperse; *(coll)* to deflocculate; *(relating to solids)* to comminute, to reduce
Zerteilung *f* dispersion, dispersal; *(coll)* deflocculation; *(relating to solids)* comminution, [size] reduction
Zerteilungsgrad *m* degree of dispersion
Zertetrafluorid *n s.* Zer(IV)-fluorid
Zertri ... *s.* Zer(III)- ...
zertrümmern to shatter, to destroy, to break down; *(esp for further processing)* to break up *(e. g. ore)*
Zertrümmerung *f* shatter[ing], destruction
Zerussit *m* *(min)* cerussite *(lead carbonate)*
Zerwellen *n* wavy sheet disintegration *(in atomizing liquids)*
Zerylalkohol *m* $CH_3[CH_2]_{24}CH_2OH$ ceryl alcohol, 1-hexacosanol
Zetan *n* $C_{16}H_{34}$ cetane, hexadecane
Zetanol *n s.* Zetylalkohol
Zetanzahl *f*, **Zetanziffer** *f* cetane number (rating)
Zeta-Potential *n* zeta (double-layer, electrokinetic) potential
Zetoleinsäure *f* cetoleic acid, 11-docosenoic acid
Zetylalkohol *m* $CH_3[CH_2]_{15}OH$ cetyl alcohol, 1-hexadecanol
Zetylessigsäure *f* $CH_3[CH_2]_{16}COOH$ cetylacetic acid, stearic acid, octadecanoic acid
Zetylmerkaptan *n s.* 1-Hexadekanthiol
Zetylpalmitat *n* $C_{15}H_{31}CO-OC_{16}H_{33}$ cetyl palmitate, cetin
Zetylsäure *f* $CH_3[CH_2]_{14}COOH$ cetylic acid, palmitic acid, hexadecanoic acid
Zeunerit *m* *(min)* zeunerite *(a hydrous copper uranium arsenate)*
Zibet *m* *(cosmet)* civet
Zickzackanordnung *f* zigzag arrangement
Zickzackkette *f* zigzag chain
Zickzackofen *m* *(ceram)* zigzag kiln
Ziegel *m* brick
  gebrannter Z. fired brick
  geschnittener Z. wire-cut brick
  glasierter Z. glazed brick
  hochfeuerfester Z. highly refractory brick
Ziegelauskleidung *f* brick lining
Ziegelei *f* brickworks
Ziegeleierzeugnis *n* brickware
Ziegelerde *f* brick earth
Ziegelformmaschine *f* *(ceram)* brick machine
Ziegelmehl *n* brick (tile) dust
Ziegelofen *m* brick kiln
Ziegelpresse *f* *(ceram)* brick machine
Ziegelsplitt *m* broken brick
Ziegelstein *m s.* Ziegel
Ziegeltee *m* brick tea
Ziegelton *m* brick (brick-making, tile) clay
Ziegenhaar *n* goat hair
Ziegenkäse *m* goat's milk cheese
Ziegenleder *n* goatskin leather
Ziegenmilch *f* goat milk
Ziegler-Katalysator *m*, **Ziegler-Kontakt** *m* Ziegler catalyst
Ziegler-Methode *f* Ziegler method *(for obtaining cyclic ketones)*
Ziegler-Verfahren *n* Ziegler process *(a polymerization process)*
Ziehbalken *m* *(glass)* draw bar
ziehbar ductile *(metal)*
Ziehbarkeit *f* ductility
Ziehdüse *f* *(glass)* debiteuse *(in the Fourcault sheet-drawing process)*
ziehen 1. to draw, to pull *(e. g. glass or plastics)*; to pull *(a reaction)*; 2. *(of semiliquid material)* to slide *(e. g. to the discharge point)*
  auf Flaschen z. to bottle [in, up]
  direkt auf Baumwolle z. *(dye)* to be direct to cotton
  Elektronen z. to withdraw electrons
  Platten z. *(rubber)* to sheet [out] *(on the calender)*
Ziehen *n* **von Bohrkernen** *(geol)* coring
  Z. von Platten *(rubber)* sheet calendering, sheeting[-out]
  Z. von Profilen *(rubber)* profiling
  Z. von Seitenkernen *(petrol)* sidewall coring
Ziehfett *n* pastry fat
Ziehglas *n* drawn glass
Ziehherd *m* *(glass)* drawing pot
Ziehkammer *f* *(glass)* drawing chamber
Ziehmargarine *f* puff (flaky) pastry margarine
Ziehmaschine *f* *(glass)* drawing machine
Ziehschacht *m* *(glass)* drawing shaft
Ziehstreifen *m* *(glass)* drawing mark *(a defect)*
Ziehverfahren *n* *(glass)* drawing process
Ziehvermögen *n* *(dye)* absorptive (absorbing) capacity (power)
Ziehwalze *f* *(glass)* forming roll *(in the sheet-drawing process)*
Ziehzwiebel *f* *(tech)* bulb
Zierfliese *f* decorative tile
Ziervogel-Verfahren *n* Ziervogel process *(for extracting silver from sulphide ores)*
Ziffer *f* number
  akzentuierte (apostrophierte) Z. *s.* gestrichene Z.
  gestrichene Z. *(nomencl)* primed number
  ungestrichene Z. *(nomencl)* unprimed number
Ziffernfolge *f* *(nomencl)* series of locants
Zigarettenpapier *n* cigarette paper, *(Am)* cigarette tissue

Zigarrendeckblattpapier *n* cigar wrapping paper
Zimmermann-Reaktion *f* Zimmermann reaction *(for determining ketonic steroids)*
Zimmertemperatur *f* room (normal, ordinary) temperature
Zimt *m* cinnamon
 Chinesischer Z. Chinese cinnamon *(from Cinnamomum aromaticum Nees)*
 Echter Z. Ceylon cinnamon *(from Cinnamomum zeylanicum Bl.)*
Zimtaldehyd *m* $C_6H_5CH{=}CHCHO$ cinnamaldehyde, cinnamic aldehyde, 2-phenylacrolein
Zimtalkohol *m* $C_6H_5CH{=}CHCH_2OH$ cinnamyl alcohol, 3-phenyl-2-propen-1-ol
Zimtblätteröl *n* cinnamon leaf oil *(from Cinnamomum zeylanicum Bl.)*
Zimtkassia *f s.* Zimt / Chinesischer
Zimtöl *n* cinnamon oil
 Chinesisches Z. cassia oil *(from Cinnamomum aromaticum Nees)*
Zimtrinde *f* cinnamon bark
Zimtsäure *f* $C_6H_5CH{=}CHCOOH$ cinnamic acid, 3-phenylacrylic acid
 gewöhnliche Z. *s. trans*-Zimtsäure
***trans*-Zimtsäure** *f trans*-cinnamic acid, ordinary cinnamic acid
Zimtsäureäthylester *m* $C_6H_5CH{=}CHCO{-}OC_2H_5$ ethyl cinnamate
Zimtsäurebenzylester *m* $C_6H_5CH{=}CHCO{-}OCH_2C_6H_5$ benzyl cinnamate, cinnamein
Zinchonidin *n* cinchonidine *(alkaloid)*
Zinchonin *n* cinchonine *(alkaloid)*
Zinchoninsäure *f* cinchoninic acid, quinoline-4-carboxylic acid
Zinckenit *m (min)* zinckenite *(antimony(III) lead sulphide)*
Zineb *n* zineb, zinc ethylenebisdithiocarbamate
Zineol *n* cineole, cineol-1,8, eucalyptol
Zingeron *n* zingerone, 4-hydroxy-3-methoxybenzylacetone
Zingiberen *n* zingiberene *(a monocyclic sesquiterpene)*
Zingiberol *n* zingiberol *(a monocyclic sesquiterpene alcohol)*
Zink *n* Zn zinc
Zinkalkyl *n s.* Dialkylzink
Zinkamalgam *n* zinc amalgam
Zinkat *n* zincate
Zinkäthid *n*, **Zinkäthyl** *n s.* Diäthylzink
Zinkblende *f (min)* zinc blende, sphalerite *(zinc sulphide)*
Zinkblendegitter *n (cryst)* zinc-blende lattice
Zinkblumen *fpl* flowers of zinc *(white woolly zinc oxide)*
Zinkblüte *f (min)* hydrozincite *(a basic zinc carbonate)*
Zinkbutter *f* butter of zinc *(zinc chloride)*
Zinkchlorid *n* $ZnCl_2$ zinc chloride
Zinkchromat *n* 1. $ZnCrO_4$ zinc chromate; 2. *s.* Zinkgelb
Zinkchromgelb *n s.* Zinkgelb
Zinkdampf *m* zinc vapour
Zinkdialkyl *n s.* Dialkylzink
Zinkdiäthyl *n s.* Diäthylzink
Zinkdichromat *n* $ZnCr_2O_7$ zinc dichromate
Zinkdimethyl *n s.* Dimethylzink
Zinkdithionit *n* ZnS_2O_4 zinc dithionite
Zinkelektrode *f* zinc electrode
Zinkgelb *n* zinc yellow, zinc[-potassium] chromate
Zinkgrau *n* zinc grey *(any of several preparations containing zinc dust or zinc oxide)*
Zinkhexafluorosilikat *n* $Zn[SiF_6]$ zinc hexafluorosilicate
Zinkhexazyanoferrat(II) *n* $Zn_2[Fe(CN)_6]$ zinc hexacyanoferrate(II)
Zinkhydrid *n* ZnH_2 zinc hydride
Zinkhydrosulfitbleiche *f (pap)* zinc-hydrosulphite bleaching
Zinkit *m (min)* zincite, red zinc ore *(zinc oxide)*
Zinkkaliumchromat *n s.* Zinkgelb
Zink-Kalk-Küpe *f (text)* zinc-lime vat
Zinkkronglas *n* zinc crown glass
Zinkmanganat(VII) *n s.* Zinkpermanganat
Zinkmetasilikat *n* $ZnSiO_3$ zinc metasilicate, zinc trioxosilicate
Zinkmethid *n*, **Zinkmethyl** *n s.* Dimethylzink
Zinkmonochromat *n s.* Zinkchromat
Zinkorthoarsenat(V) *n* $Zn_3(AsO_4)_2$ zinc orthoarsenate
Zinkorthophosphat *n* $Zn_3(PO_4)_2$ zinc orthophosphate
Zinkorthosilikat *n* Zn_2SiO_4 zinc orthosilicate, zinc tetraoxosilicate
Zinkoxid *n* ZnO zinc oxide
Zinkoxidkatalysator *m* zinc-oxide catalyst
Zinkpermanganat *n* $Zn(MnO_4)_2$ zinc permanganate
Zinkrhodanid *n s.* Zinkthiozyanat
Zinksalz *n* zinc salt
Zinkspat *m (min)* zinc spar, smithsonite, dry-bone ore *(zinc carbonate)*
Zinkspinell *m (min)* zinc spinel, gahnite *(zinc aluminate)*
Zinkstaub *m* zinc dust
Zinkstaubdestillation *f* zinc-dust distillation
Zinktetroxosilikat *n s.* Zinkorthosilikat
Zinkthiozyanat *n* $Zn(SCN)_2$ zinc thiocyanate, zinc rhodanide
Zinktrioxosilikat *n s.* Zinkmetasilikat
Zinkvitriol *n* $ZnSO_4 \cdot 7H_2O$ zinc vitriol, zinc sulphate-7-water
Zinkvitriol *m (min)* goslarite, zinc vitriol, white vitriol (copperas) *(zinc sulphate-7-water)*
Zinkwasserstoff *m* ZnH_2 zinc hydride
Zinkweiß *n* zinc white *(crude zinc oxide)*
Zinn *n* Sn tin
 graues Z. *s.* α-Zinn
α-Zinn *n* alpha (grey) tin, α tin
β-Zinn *n* beta (white) tin, β tin
γ-Zinn *n* gamma (brittle) tin, γ tin
Zinnabarit *m (min)* cinnabar, liver ore *(mercury(II) sulphide)*
Zinnamal *n*, **Zinnamaldehyd** *m s.* Zimtaldehyd
Zinnamat *n* cinnamate *(salt or ester of cinnamic acid)*

Zinnamein *n* $C_6H_5CH{=}CHCO{-}OCH_2C_6H_5$ cinnamein, benzyl cinnamate
Zinnamsäure *f s.* Zimtsäure
Zinnamylaldehyd *m s.* Zimtaldehyd
Zinnamylalkohol *m s.* Zimtalkohol
Zinnasche *f* SnO_2 flowers of tin, tin ash[es] *(tin(IV) oxide)*
Zinn(II)-azetat *n* $Sn(CH_3COO)_2$ tin(II) acetate, tin diacetate, stannous acetate
Zinnbad *n* tin bath
Zinn-Blei-Lot *n* tinman's solder
Zinn(II)-bromid *n* $SnBr_2$ tin(II) bromide, tin dibromide, stannous bromide
Zinn(IV)-bromid *n* $SnBr_4$ tin(IV) bromide, tin tetrabromide, stannic bromide
Zinnbronze *f* tin bronze
Zinnbutter *f* butter of tin *(tin(IV) chloride-5-water)*
Zinn(II)-chlorid *n* $SnCl_2$ tin(II) chloride, tin dichloride, stannous chloride
Zinn(IV)-chlorid *n* $SnCl_4$ tin(IV) chloride, tin tetrachloride, stannic chloride
Zinndi... *s.a.* Zinn(II)-...
Zinn(II)-dihydrogenorthophosphat *n* $Sn(H_2PO_4)_2$ tin(II) dihydrogenorthophosphate, tin dihydrogenphosphate, stannous dihydrogenorthophosphate
Zinndioxid *n s.* Zinn(IV)-oxid
Zinn(II)-diphosphat *n* $Sn_2P_2O_7$ tin(II) diphosphate, tin(II) pyrophosphate, stannous diphosphate
Zinndisulfid *n s.* Zinn(IV)-sulfid
Zinnerz *n* tin ore
Zinn(II)-fluorid *n* SnF_2 tin(II) fluoride, tin difluoride, stannous fluoride
Zinn(IV)-fluorid *n* SnF_4 tin(IV) fluoride, tin tetrafluoride, stannic fluoride
Zinnfolie *f* tinfoil
Zinnfolienpapier *n* tinfoil paper
Zinngeschrei *n* tin cry
Zinnglasur *f (ceram)* tin glaze
zinnhaltig tin-bearing
Zinnhydrid *n* SnH_4 tin hydride, stannic hydride, stannane
Zinn(II)-hydrogenorthophosphat *n* $SnHPO_4$ tin(II) hydrogenorthophosphate, tin(II) hydrogenphosphate, stannous hydrogenorthophosphate
Zinn(II)-hydroxid *n* $Sn(OH)_2$ tin(II) hydroxide, stannous hydroxide
Zinn(II)-jodid *n* SnI_2 tin(II) iodide, tin diiodide, stannous iodide
Zinn(IV)-jodid *n* SnI_4 tin(IV) iodide, tin tetraiodide, stannic iodide
Zinnkies *m (min)* tin pyrites, stannite
Zinnlegierung *f* tin alloy
Zinnmonosulfid *n s.* Zinn(II)-sulfid
Zinn(II)-nitrat *n* $Sn(NO_3)_2$ tin(II) nitrate, stannous nitrate
Zinn(IV)-nitrat *n* $Sn(NO_3)_4$ tin(IV) nitrate, stannic nitrate
Zinnober *m* cinnabar, red mercuric sulphide, *(min also)* liver ore *(mercury(II) sulphide)*
  Chinesischer Z. Chinese vermil[l]ion *(red mercury(II) sulphide with small amounts of antimony sulphide)*
  Grüner Z. green cinnabar, chrome [oxide] green, oil green *(chromium(III) oxide)*
Zinnöl *n (ceram)* tin oil *(a mixture of stannic and stannous chlorides and oils)*
Zinnolin *n* cinnoline, 1,2-benzodiazine
Zinn(II)-orthophosphat *n* $Sn_3(PO_4)_2$ tin(II) orthophosphate, tin(II) phosphate, stannous orthophosphate
Zinn(II)-oxalat *n* $Sn(COO)_2$ tin(II) oxalate, stannous oxalate
Zinn(IV)-oxid *n* SnO_2 tin(IV) oxide, tin dioxide, stannic oxide
Zinnpest *f* tin pest (plague, disease)
Zinn(II)-pyrophosphat *n s.* Zinn(II)-diphosphat
Zinnsalz *n* $SnCl_2 \cdot 2H_2O$ tin salt *(tin(II) chloride-2-water)*
Zinnsäure *f* stannic acid
  gewöhnliche Z. *s.* α-Zinnsäure
a-Zinnsäure *f s.* α-Zinnsäure
b-Zinnsäure *f s.* β-Zinnsäure
α-Zinnsäure *f* metastannic α acid
β-Zinnsäure *f* metastannic β acid
Zinnschrei *m* tin cry
Zinnstanniol *n* tin foil
Zinnstein *m (min)* tin stone, cassiterite *(tin(IV) oxide)*
Zinn(II)-sulfat *n* $SnSO_4$ tin(II) sulphate, stannous sulphate
Zinn(IV)-sulfat *n* $Sn(SO_4)_2$ tin(IV) sulphate, stannic sulphate
Zinn(II)-sulfid *n* SnS tin(II) sulphide, tin monosulphide, stannous sulphide
Zinn(IV)-sulfid *n* SnS_2 tin(IV) sulphide, tin disulphide, stannic sulphide
Zinntetra... *s.* Zinn(IV)-...
Zinnwaldit *m (min)* zinnwaldite *(a phyllosilicate)*
Zinnwasserstoff *m s.* Zinnhydrid
Zipfel *m* tit *(a defect in glass)*
Zippeit *m (min)* zippeite *(a basic uranyl sulphate)*
Ziram *n* ziram, zinc dimethyldithiocarbamate
Zirkon *m (min)* zircon *(zirconium orthosilicate)*
Zirkonat *n* zirconate
Zirkonerde *f s.* Zirkonium(IV)-oxid
Zirkongerbung *f* zirconium tannage
Zirkonium *n* Zr zirconium
Zirkonium(II)-bromid *n* $ZrBr_2$ zirconium(II) bromide, zirconium dibromide
Zirkonium(II)-chlorid *n* $ZrCl_2$ zirconium(II) chloride, zirconium dichloride
Zirkonium(III)-chlorid *n* $ZrCl_3$ zirconium(III) chloride, zirconium trichloride
Zirkonium(IV)-chlorid *n* $ZrCl_4$ zirconium(IV) chloride, zirconium tetrachloride
Zirkoniumdibromid *n s.* Zirkonium(II)-bromid
Zirkoniumdichlorid *n s.* Zirkonium(II)-chlorid
Zirkoniumdioxid *n s.* Zirkonium(IV)-oxid
Zirkoniumdioxidhydrat *n s.* Zirkonsäure
Zirkonium(IV)-fluorid *n* ZrF_4 zirconium(IV) fluoride
Zirkoniumhydrid *n* zirconium hydride
Zirkonium(IV)-jodid *n* ZrI_4 zirconium(IV) iodide, zirconium tetraiodide
Zirkoniumkarbid *n* ZrC zirconium carbide

Zirkonium(IV)-oxid *n* ZrO_2 zirconium(IV) oxide, zirconium dioxide
Zirkoniumoxidbromid *n* $ZrOBr_2$ zirconium dibromide oxide
Zirkoniumoxidchlorid *n* $ZrOCl_2$ zirconium dichloride oxide
Zirkoniumoxidhydroxid *n* $ZrO(OH)_2$ zirconium dihydroxide oxide
Zirkoniumoxidjodid *n* $ZrOI_2$ zirconium diiodide oxide
Zirkoniumsilikat *n* $ZrSiO_4$ zirconium silicate, zirconium orthosilicate
Zirkonium(IV)-sulfat *n* $Zr(SO_4)_2$ zirconium(IV) sulphate
Zirkonium(IV)-sulfid *n* ZrS_2 zirconium(IV) sulphide
Zirkoniumtetra ... *s.* Zirkonium(IV)- ...
Zirkoniumtrichlorid *n s.* Zirkonium(III)-chlorid
Zirkonmasse *f (ceram)* zircon body
Zirkonporzellan *n* zircon porcelain
Zirkonsand *m* zircon sand
Zirkonsäure *f* $ZrO_2 \cdot xH_2O$ zirconic acid
Zirkonweißware *f (ceram)* zircon whiteware
Zirkonylbromid *n s.* Zirkoniumoxidbromid
Zirkonylchlorid *n s.* Zirkoniumoxidchlorid
Zirkonylhydroxid *n s.* Zirkoniumoxidhydroxid
Zirkularchromatografie *f* circular [paper] chromatography
Zirkularkomponente *f* circular component *(of polarized light)*
Zirkularpolarisation *f* circular polarization
zirkularpolarisiert circularly polarized
Zirkulation *f* [re]circulation
Zirkulationsfärbeapparat *m* circulating-liquor dyeing machine
Zirkulationsgas *n* recycle gas
Zirkulationspumpe *f* circulating pump, circulator
Zirkulationssystem *n* circulation system
zirkulieren to circulate
z. lassen to circulate
Zitrakonsäure *f* $CH_3C(COOH){=}CHCOOH$ citraconic acid, *cis*-methylbutenedioic acid
Zitral A *n* citral a, geranial, *trans*-3,7-dimethyl-2,6-octadienal
Zitral B *n* citral b, neral, *cis*-3,7-dimethyl-2,6-octadienal
***cis*-Zitral** *n s.* Zitral B
***trans*-Zitral** *n s.* Zitral A
Zitrat *n* citrate
zitratlöslich *(soil)* citrate-soluble
Zitratlöslichkeit *f (soil)* citrate solubility
Zitratzyklus *m* citric-acid (tricarboxylic-acid) cycle, TCA cycle, Krebs cycle
Zitrazinsäure *f* citrazinic acid, 2,6-dihydroxypyridine-4-carboxylic acid
Zitrin *m (min)* citrine, false topaz *(a variety of quartz)*
Zitrinin *n* citrinin *(antibiotic)*
Zitronellal *n* citronellal, 3,7-dimethyl-6-octenal
Zitronellaldehyd *m s.* Zitronellal
Zitronellöl *n* citronella oil *(from Cymbopogon specc.)*
Zitronengelb *n* 1. lemon chrome, chrome yellow *(lead chromate)*; 2. *s.* Zinkgelb
Zitronengrasöl *n (cosmet)* lemon-grass oil, East Indian verbena oil *(from Cymbopogon specc.)*
Zitronenöl *n* citrus oil
Zitronensäure *f* $HOOCCH_2C(OH)(COOH)CH_2COOH$ citric acid, 2-hydroxypropane-1,2,3-tricarboxylic acid
Zitronensäuregärung *f* citric acid fermentation
Zitronensäuretriäthylester *m* triethyl citrate
Zitronensäurezyklus *m s.* Zitratzyklus
Zitronin A *n* citronin A *(sodium or potassium salt of 2,4-dinitro-1-naphthol-7-sulphonic acid)*
Zitrovorumfaktor *m* citrovorum factor, leucovorin, N^5-formyltetrahydrofolic acid, folinic acid
Zitrusöl *n* citrus oil
Zittersieb *n* reciprocating screen
Zitwer *m* / **Gelber** Cassumunar ginger, Zingiber cassumunar Roxb.
Z-Kalander *m* Z type of calender
Zoisit *m (min)* zoisite *(a basic aluminium calcium silicate)*
Zölestin *m (min)* coelestine *(strontium sulphate)*
Zone *f* zone, region
Z. freien Absetzens free-settling zone
aufgeschmolzene Z. molten zone *(in zone melting)*
freischwebende Z. floating zone *(in zone melting)*
glühende Z. incandescent zone
schwebende Z. floating zone *(in zone melting)*
verbotene Z. *(phys chem)* forbidden (unallowed) band, energy gap
Zonenachse *f (cryst)* zone axis
Zonenbreite *f* zone width *(in zone melting)*
Zonendurchgang *m* zone pass
Zonendurchgangszahl *f* number of zone passes *(in zone melting)*
Zonenebnen *n* zone levelling *(in zone melting)*
Zonenelektrophorese *f* zone electrophoresis
Zonenfällung *f* zone precipitation
Zonengeschwindigkeit *f* rate of zone travel, zone (zoning) speed *(in zone melting)*
Zonenkante *f (cryst)* zone axis
Zonenlänge *f* zone length *(in zone melting)*
Zonenlegieren *n s.* Zonennivellieren
Zonennivellieren *n* zone levelling *(in zone melting)*
Zonenplanieren *n s.* Zonennivellieren
zonenreinigen to zone-refine, to zone-purify
Zonenreinigung *f* zone refining (purification)
Z. nach dem Schwebezonenverfahren *s.* tiegelfreie Z.
tiegelfreie (tiegellose) Z. floating-zone refining
Zonenreinigungsanlage *f* zone-refining apparatus (unit), zone refiner
Zonenrichtung *f* direction of zoning (zone travel) *(in zone melting)*
Zonenschmelzanlage *f* zone-melting apparatus (unit)
Zonenschmelzdurchgang *m* zone pass
Zonenschmelze *f* zone melting
zonenschmelzen to zone-melt
Zonenschmelzen *n* zone melting
Z. nach dem Schwebezonenverfahren *s.* tiegelfreies Z.

tiegelfreies (tiegelloses) Z. floating-zone melting
Zonenschmelzgerät *n* zone-melting apparatus (unit)
Zonenschmelzofen *m* zone-melting furnace
Zonenschmelzverfahren *n* zone-melting process
Zonenwanderung *f* zone travel[ling] *(in zone melting)*
Zonenzahl *f* number of zone passes *(in zone melting)*
Zoomarinsäure *f* $CH_3[CH_2]_5CH=CH[CH_2]_7COOH$ zoomaric acid, palmitoleic acid
Zoosterin *n (bioch)* zoosterol, animal sterol
Zopfwinde *f (pap)* rag catcher, [de]ragger
Zschocke-Desintegrator *m* Zschocke disintegrator
ZT *s.* Zimmertemperatur
ZTU-Diagramm *n* time-temperature-transformation diagram, T.T.T. diagram
Z. für kontinuierliche Abkühlung continuous-cooling transformation diagram
isothermes Z. isothermal transformation diagram
Zuber *m* tub, vat
zubereiten to prepare
zubereitet / frisch freshly prepared
Zubereitung *f* preparation, formulation *(act or substance)*
galenische Z. *(pharm)* galenical
kosmetische Z. cosmetic preparation
zubessern *(tann)* to mend *(the lime liquor)*
Zubringevorrichtung *f* feed mechanism
Zubringewagen *m* transfer car
züchten to grow *(crystals)*
Zuchthefe *f* cultivated (culture) yeast
Zucker *m* sugar ✦ **in Z. umwandeln** to saccharify, to saccharize
affinierter Z. affinated (affination, washed raw) sugar
brauner Z. brown sugar
einfacher Z. simple sugar, monosaccharide
geblauter Z. blued sugar
gebrannter Z. *s.* karamelisierter Z.
karamelisierter Z. caramelized sugar, caramel
reduzierender Z. reducing sugar
seltener Z. unusual sugar
Zuckerabkömmling *m* sugar derivative
Zuckerahorn *m* sugar (rock) maple, Acer saccharum Marsh.
Zuckeralkohol *m* sugar alcohol
Zuckeranhydrid *n* sugar anhydride
zuckerartig sugar-like, saccharine, saccharoid
Zuckeräther *m* sugar ether
Zuckerbirke *f* cherry birch, Betula lenta L.
Zuckerbruchstück *n* sugar fragment
Zuckercouleur *f* sugar colouring (dye)
Zuckerderivat *n* sugar derivative
Zuckerersatzstoff *m* sugar substitute
Zuckerester *m* sugar ester
Zuckerfabrik *f* sugar refinery (factory)
Zuckerfabrikabwasser *n* sugar-factory waste
Zuckerfarbe *f* sugar colouring (dye)
Zucker-Fettsäureester *m* fatty acid ester of sugar
zuckerfremd non-sugar *(e.g. portion of a glycoside)*
Zuckergehalt *m* sugar content
zuckerhaltig sacchariferous, saccharine, sugary
Zuckerhaus *n* sugar house
Zuckerhut *m* sugar loaf
Zuckerindustrie *f* sugar industry
Zuckerkand[is] *m* candy [sugar], sugar candy, rock candy
Zuckerkohle *f* sugar charcoal
Zuckerkristall *m* sugar crystal
zuckerliefernd sacchariferous
Zuckerpalme *f* sugar palm, Arenga Labill.
Zuckerpflanze *f* sugar plant *(a plant not converting its assimilation products into starch)*
Zuckerraffination *f* sugar refining
Zuckerraffinerie *f* sugar refinery
Zuckerraffineur *m* sugar refiner
Zuckerreihe *f* sugar series
Zuckerrohr *n* sugar cane, Saccharum officinarum L.
Zuckerrohrbagasse *f* [sugar cane] bagasse, bagass[e], megass[e]
Zuckerrohrmelasse *f* [sugar] cane molasses
Zuckerrohrquetsche *f* sugar crusher (mill)
Zuckerrohrwachs *n* [sugar] cane wax
Zuckerrübe *f* sugar beet, Beta vulgaris L. var. altissima Doell
zuckerreiche Z. high-sugar beet
Zuckerrübenanbau *m* beet growing
Zuckerrübenmelasse *f* beet molasses
Zuckerrübenschnitzel *npl* sugar beet chips (slices), [beet] cossettes
Zuckersäure *f* sugar acid, *(specif)* $HOOC[CHOH]_4COOH$ saccharic acid, glucosaccharic acid *(one form of 2,3,4,5-tetrahydroxyhexanedioic acid)*
zuckerspeichernd *(biol)* sacchariferous
Zuckertechnologie *f* sugar technology
Zuckerveraschungsschale *f* sugar incinerating dish
Zuckerwerk *n* confectionery, sweetmeats, *(Am)* candy
Zuckerzerfall *m* sugar fragmentation
Zuckerzerfallsprodukt *n* sugar fragment
zudosieren to charge
Zufallsbeobachtung *f* chance observation
Zufallsfehler *m* random (accidental) error
Zufallslage *f (phys chem)* random (disordered) fashion
Zufallsorientierung *f (cryst)* random orientation
zufließen to flow in, to run in
z. lassen to run in
Zufluß *m s.* Zulauf
zufügen to add
Zufügung *f* addition
Zufuhr *f* feed[ing]
zuführen to feed *(a material)*
wieder z. to return
Zuführung *f* 1. feed[ing]; 2. [feed] inlet
Zuführungsleitung *f* supply (lead) line
Zuführungsrohr *n* feed pipe
Zuführungstrichter *m* [feed] hopper; cup *(of a blast furnace)*

Zug *m* 1. *(passageway)* flue *(as of an oven)*; 2. *(phenomenon)* draught, pull, suction *(as of a stack or fan)*; 3. *(tann)* stretch
künstlicher Z. forced draught
natürlicher Z. natural draught
senkrechter Z. vertical flue
waagerechter Z. horizontal flue
Zugabe *f* addition
zugänglich *(agric)* available *(nutrients)*
Zugänglichkeit *f (agric)* availability *(of nutrients)*
Zugbruch *m* tensile failure
Zugdehnung *f* tensile elongation
Zug-Dehnungs-Diagramm *n* stress-strain diagram *(expressed as tons or pounds per square inch)*; load-elongation (load-extension) diagram *(expressed as inches per inch)*
Zugdehnungseigenschaften *fpl* tensile [stress-strain] characteristics
Zug-Dehnungs-Kurve *f* stress-strain curve *(expressed as tons or pounds per square inch)*; load-elongation (load-extension) curve *(expressed as inches per inch)*
Zugdehnungsprüfgerät *n* tensile-strength tester (testing machine), tensile tester (testing machine)
Z. Bauart Schopper *(rubber)* Schopper machine
Zugdehnungsverhalten *n s.* Zugdehnungseigenschaften
zugeben to add
chargenweise (portionsweise) z. to charge
Zugeben *n* addition
chargenweises (portionsweises) Z. charging
tropfenweises Z. dropwise addition
zügeln to control *(e.g. a reaction)*
zugestellt / basisch *(met)* basic-lined
sauer z. *(met)* acid-lined
Zugfestigkeit *f* tensile strength, tenacity, *(quantitatively also)* ultimate tensile strength (stress); *(pap)* breaking strength
Zugfestigkeitsprüfgerät *n* tensile-strength tester (testing machine), tensile tester
zügig 1. tacky *(printing ink)*; 2. *(tann)* stretchy
Zügigkeit *f* 1. tack[iness] *(of printing ink)*; 2. *(tann)* stretch
Zugkraft *f* tensile force
Zugmesser *m* draught gauge *(as for chimneys)*
Zugpflaster *n (med)* blistering plaster
Zugspannung *f* tensile stress, tension
Zugstange *f* pitman *(of a jaw crusher)*
Zugverformungseigenschaften *fpl s.* Zugdehnungseigenschaften
Zugverformungsrest *m* permanent [set at] elongation, permanent (tensile) set, residual elongation
Zugversuch *m* tensile[-strength] test
zukorken to cork
Zulassung *f* permit, registration *(permission for manufacturing a pesticide)*
Zulauf *m* 1. *(liquid)* influent, affluent, afflux, inflow, *(esp distil)* feed [stream]; 2. feed inlet; 3. *(process)* inflow, afflux
Z. eines binären Gemisches *(distil)* binary feed
Z. eines Zweistoffgemisches *(distil)* binary feed
Zulaufbehälter *m* feed tank
Zulaufboden *m (distil)* feed tray (plate)
zulaufen to run in, to flow in
z. lassen to run in
konisch z. to taper
Zulaufgefäß *n* feed box
Zulaufrohr *n* feed (charging, influent) pipe, feed tube
Zulaufschieber *m* feed gate
Zulaufseite *f* feed end
Zulaufstutzen *m* charging pipe
Zulauftauchrohr *n* influent well
Zulaufventil *n* feed valve
Zulaufwehr *n (distil)* inlet weir
Zulaufzusammensetzung *f (distil)* feed composition
Zuleitung *f* feed, inlet
Zuluft *f* inlet air
zumessen to meter, to dose, to proportion
Zumeßpumpe *f* metering (dosing, proportioning, controlled-volume) pump
Zumessung *f* metering, dosing, dosage, proportioning
zumischen to admix
Zumischung *f* admixture
Zunahme *f* increase, growth, rise
Zündanlage *f* ignition system
Zündbeschleuniger *m* pro-ignition dope
Zünddraht *m* ignition (firing) wire
Zündeigenschaften *fpl* ignition performance (quality)
zünden to ignite, to spark, to prime, to fire
Zunder *m* scale, cinder *(thrown off in forging metal)*
Zünder *m* igniter, primer
zunderbeständig non-scaling
Zunderbildung *f* scale formation, formation of cinder *(on forging metal)*
zunderfrei non-scaling
zundern to scale
Zunderschicht *f* scale (oxide) layer
Zündhaube *f (met)* igniter *(of a sintering machine)*
Zündholzparaffin *n* match wax
Zündhütchen *n* ignition cap, primer
Zündkirsche *f* ignition pellet
Zündmaschine *f* / **Döbereinersche** Döbereiner's lamp
Zündmittel *n* igniting agent, igniter, priming medium
Zündofen *m (met)* ignition furnace, igniter *(of a sintering machine)*
Zündpol *m* firing terminal *(as of a calorimetric bomb)*
Zündpunkt *m* spontaneous-ignition temperature, S. I. T.
Zündsatz *m* priming composition
Zündschnur *f* [blasting] fuse
detonierende Z. detonating fuse
Zündspannung *f* sparking potential *(as in electrical precipitation)*
Zündsprengstoff *m* initiating (primary) explosive, initiator, primer, detonator

Zündstein *m* lighter flint [tip]
Zündstoff *m s.* 1. Zündmittel; 2. Zündsprengstoff
Zündsystem *n* ignition system
Zündtemperatur *f* ignition temperature
Zündung *f* ignition, inflammation, *(relating to explosives also)* firing
Zündverhalten *n* ignition performance (quality)
Zündverzögerung *f*, **Zündverzug** *m* ignition delay
Zündverzugszeit *f* ignition delay period
Zündvorrichtung *f* ignition device, igniter
Zündwilligkeit *f* ignition performance (quality)
Zündzeit *f* ignition time
zunehmen to increase, to grow, to rise
Zunge *f* index, pointer *(as of a measuring device)*
Zungendüse *f* fan nozzle
Zungenreißfestigkeit *f* *(tann)* tongue tear strength
zupfropfen to stopper, to cork
zurichten *(tann)* to dress, to finish, to curry
Zuricht[hilfs]mittel *n* *(tann)* dressing (finishing) auxiliary (agent)
zurückbilden to reconvert, to re-form
sich z. to reconvert, to re-form
zurückbleiben to remain
zurückfließen to flow back, to return
zurückführen to return
in den Kreislauf z. to return to the circuit, to recirculate, to recycle
zurückgewinnen to recover, to reclaim
zurückhalten to retain
Zurückhaltung *f* retention
zurücklaufen to run back
zurückoxydieren to reoxidize, to oxidize back
Zurückoxydieren *n* reoxidation
zurückprallen to rebound; *(coat)* to bounce back
Zurückprallen *n* rebound; *(coat)* bounce-back *(of pigment dust)*
zurücksaugen to suck back
zurückschlagen to strike (flash) back *(of flame in a burner)*
Zurückschlagen *n* striking-back, flashback, flare-back *(of flame in a burner)*
zurückspringen to rebound, to spring back; *(rubber also)* to recover
Zurückspringen *n* rebound, springing back, recoil, resilience, *(rubber also)* [elastic] recovery
zurückstoßen to repel
Zurückstoßung *f* repulsion
zurückstrahlen to reflect [back]
zurücktitrieren to back-titrate
Zurücktitrieren *n* back-titration
zurückverwandeln to reconvert
sich z. to reconvert
zurückwandern to shift back
zurückwerfen to reflect [back]
zusammenbacken to cake, to agglomerate, to set up, to stick together; *(plast)* to bridge *(on filling)*
Zusammenbacken *n* caking, agglomeration, set-up; *(plast)* bridging *(on filling)*
zusammenballen to agglomerate, to agglutinate, to ball [up]
sich z. to agglomerate, to aggregate, to agglutinate, to ball [up], to clog, to clot, to lump
Zusammenballen *n* 1. *(of human agent)* agglomeration; 2. *(of material)* agglomeration, aggregation, agglutination, balling, clogging, clotting, lumping
Zusammenbau *m* 1. assembly; 2. *(rubber)* assembly, building
zusammenbauen 1. to assemble; 2. *(rubber)* to assemble, to build
zusammenbrechen to collapse, to break down *(as of foam)*
Zusammenbruch *m* collapse, breakage *(as of foam)*
zusammendrückbar compressible
Zusammendrückbarkeit *f* compressibility
zusammendrücken to compress
Zusammendrücken *n* compression, compressing
Zusammendrückungsrest *m* [permanent] compression set *(materials science)*
zusammenfließen to flow together, to coalesce
Zusammenfließen *n* coalescence
zusammenflocken / sich to flocculate
zusammenfrieren to freeze *(as of coal)*
zusammenfritten to agglomerate *(under the influence of heat)*
Zusammenfritten *n* agglomeration *(under the influence of heat)*
zusammengautschen *(pap)* to couch together, to laminate, to line
Zusammenhalt *m* coherence, coherency
zusammenhalten to cohere, to hold together
zusammenhaltend coherent, cohesive
zusammenkitten to cement together
zusammenkleben to stick together
zusammenlaufen to coalesce
zusammenmischen to mix, *(esp relating to bulk material)* to blend
Zusammenprall *m* impact, impingement, collision
zusammenprallen to impact, to impinge, to collide
zusammenpressen to compact, to compress
Zusammenpressen *n* compaction, compression
zusammenrühren to stir together
zusammenschmelzen to melt (fuse) together; to melt down
zusammensetzen to compose, to compound, to combine
Zusammensetzung *f* composition, combination, *(act also)* compounding ✦ **von unklarer Z.** ill-defined
chemische Z. chemical composition
prozentuale Z. percent[age] composition
zusammensinken to break, to collapse *(of foam)*
Zusammensinken *n* breakage, collapse *(of foam)*
zusammenstellen to arrange; to compound, to prepare, to put up *(e.g. a prescription)*
Zusammenstellung *f* arrangement; compounding, composition, preparation *(as of a prescription)*
Zusammenstoß *m* impact, impingement, collision
zusammenstoßen to impact, to impinge, to collide
zusammentreten / paarweise to pair *(of electrons)*
zusammenwachsen *(cryst)* to coalesce, to grow together, to intergrow

Zusammenwachsen *n* *(cryst)* coalescence, intergrowth
zusammenziehen / sich to contract
zusammenziehend astringent, styptic
Zusatz *m* 1. addition, admixture; 2. additive [substance], admixture *(s. a.* Zusatzstoff)
aktivierender (belebender) Z. *(min tech)* activator
drückender Z. *(min tech)* depressant
geschmacksverbessernder Z. *(pharm)* corrigent, corrective
keimbildender Z. *(cryst)* nucleation (nucleating) agent
passivierender Z. *(min tech)* depressant
pH-regelnder Z. pH regulator
reinigender Z. detergent additive
staubverhindernder Z. dust-preventing additive, anti-dust
weichmachender Z. plasticizer, plasticizing agent
zündbeschleunigender Z. pro-ignition dope *(for motor fuel)*
Zusatzbeschleuniger *m* *(rubber)* secondary accelerator
aktivierender Z. activating accelerator, booster
Zusatzchemikalie *f* make-up [chemical] *(for compensating losses)*
Zusatzdünger *m* supplemental fertilizer, side-dressing
Zusatzdüngung *f* supplemental fertilizing, side-dressing
Zusatzkesselspeisewasser *n* make-up boiler [feed] water
Zusatzkomponente *f* *(distil)* codistillant
Zusatzmittel *n* *s.* Zusatzstoff
Zusatzstoff *m* additive [substance], accessory agent, *(esp for cheapening or weighting)* loading material, load; *(for improving physical properties)* conditioner; *(distil)* codistillant, separating agent
luftporenbildender Z. *(build)* air-entraining additive (admixture)
Zusatzwasser *n* make-up water, *(specif)* make-up boiler water, boiler make-up
Zuschlag *m* 1. loading *(as of filling materials)*; 2. loading material, load; *(met)* flux; *(build)* aggregate
feinkörniger Z. *(build)* fine aggregate
grober Z. *(build)* coarse aggregate
künstlicher Z. *(build)* manufactured aggregate
leichter Z. *(build)* lightweight aggregate
Zuschlagstoff *m* *s.* Zuschlag 2.
zuschmelzen to seal [up]
Zuschußwasser *n* *s.* Zusatzwasser
zusetzen 1. to add, to admix; 2. to clog [up], to blind *(e. g. screen openings)*
sich z. to clog, to plug *(as of screen openings)*
zuspeisen to feed
Zuspeisen *n* feed[ing]
zuspunden to stopper *(casks)*
Zustand *m* state, condition ✦ **in fein verteiltem Z.** in a fine state ✦ **in freiem Z.** in the free state
angeregter Z. excited state
atomarer Z. atomic state
fester Z. solid state
flüssiger Z. liquid state
freier Z. free state
gasförmiger Z. gaseous state
gebundener Z. combined state
glasartiger (glasiger) Z. vitreous (glass-like, glassy) state
grüner Z. *(ceram)* green state
liquokristalliner Z. *s.* mesomorpher Z.
mesomorpher Z. mesomorphic state
metallischer Z. metallic state
metastabiler Z. metastable state
plastischer Z. plastic state; *(coal)* stage of plasticity
roher Z. *(ceram)* green state
stabiler (stationärer) Z. steady (stationary) state
superfluider (superflüssiger) Z. superfluid state, superfluidity
suprafluider (supraflüssiger) Z. *s.* superfluider Z.
supraleitender Z. superconducting state
überflüssiger Z. *s.* superfluider Z.
zähgepolter Z. *(met)* tough-pitch condition
Zustände *mpl* / **übereinstimmende** *(phys chem)* corresponding states
Zustandsänderung *f* change in (of) state
adiabatische Z. adiabatic change
Zustandsdiagramm *n* constitution (equilibrium, phase) diagram
Zustandsenergie *f* potential energy
Zustandsgleichung *f* equation of state
Z. idealer Gase ideal gas equation
Beattie- und Bridgemansche Z. Beattie and Bridgeman equation
Berthelotsche Z. Berthelot equation
Clausiussche Z. Clausius equation
reduzierte Z. reduced equation of state
van-der-Waalssche Z. van der Waals equation [of state]
Zustandssumme *f* *s.* Verteilungsfunktion
zustellen *(met)* to line
Zustellung *f* *(met)* lining *(act or material)*
basische Z. basic lining
saure Z. acid lining
zustöpseln to stopper, to cork, to plug
Zutageliegen *n*, **Zutagetreten** *n* *(geol)* outcrop[ping]
Zuteileinrichtung *f* dosing (proportioning) apparatus, feeder
zuteilen to dose, to proportion, to charge
Zuteiler *m* *s.* Zuteileinrichtung
zutropfen to add dropwise, to drop in
Zutropfen *n* dropwise addition
Zuwachs *m* increase
zuwandern to migrate in *(as of ions)*
Z-Walzenkalander *m* Z type of calender
Zwangdurchlaufdampferzeuger *m*, **Zwangdurchlaufkessel** *m* once-through[-flow] boiler
Zwanglauf *m* forced circulation
Zwangsumlauf *m*, **Zwangszirkulation** *f* *s.* Zwangumlauf
Zwangumlauf *m* forced circulation

Zwangumlaufverdampfer *m* forced-circulation evaporator
Z. mit Brüdenverdichtung forced-circulation vapour compression plant
zweiachsig *(cryst)* biaxial
zweiatomig diatomic, biatomic
Zweibadchromgerbung *f* two-bath chrome tannage (tanning)
Zweibadentwicklung *f (phot)* two-bath development
Zweibadfixierung *f (phot)* two-bath fixation
Zweibadverfahren *n* two-bath method
zweibasig dibasic *(acid)*
Zweidecker-Siebmaschine *f* two-deck sifter
zweidimensional two-dimensional
Zweielektronenbindung *f* two-electron bond
Zweielektronenkonfiguration *f* two-electron configuration
Zweielektronenreduktion *f* two-electron reduction
Zweielektronensystem *n* two-electron system
Zweierschale *f* duplet shell
Zweierstoß *m (phys chem)* dual (binary, two-body) collision
Zweietagenofen *m (ceram)* double-deckle (two-tier) kiln
Zweifachbindung *f* double bond (link, linkage)
Zweifachfilter *n* dual media filter
zweifachfrei divariant, bivariant
zweifachnegativ [geladen] dinegative
zweifachpositiv [geladen] dipositive
Zweifachzucker *m* disaccharide
Zweifarbeneffekt *m (text)* bicolour effect
Zweifarbenspritzgießen *n (plast)* double-shot moulding
zweifarbig dichromatic
Zweifarbigkeit *f* dichromatism
Zweifilmtheorie *f* two-film theory *(of phase boundaries)*
Zweiflächner *m (cryst)* pinacoid
Zweig *m* 1. branch *(spectroscopy)*; 2. bridge *(of a bridge-ring compound)*
negativer Z. P-branch *(of a band spectrum)*
positiver Z. R-branch *(of a band spectrum)*
zweigliedrig binary
Zweigutapparat *m* two-product unit
zweihalsig two-neck[ed]
Zweihalskolben *m* two-neck[ed] flask
Zweihalsrundkolben *m* round-bottom two-neck flask
Zweihordendarre *f* two-floor (double-floor) kiln
zweikernig binuclear, binucleate
Zweikomponentenfarbe *f* two-can (two-pack) paint
Zweikomponentenkleber *m* two-component adhesive
Zweikomponentensystem *n (phys chem)* two-component system, binary system; *(coat)* two-can (two-pack) system
Zweikörperverdampfer *m* double-effect evaporator
Zweilösungsmittelverfahren *n* two-solvent process
Zweiphasenschaum *m* two-phase foam
Zweiphasensystem *n* two-phase system
Zweiplattenschieber *m* double-disk gate valve
Zweipressenschleifer *m (pap)* two-pocket grinder
Zweiproduktenapparat *m* two-product unit
Zweipunktregelung *f* two-position (on-off) control
zweisäurig diacid[ic], biacid *(base)*
Zweischalenentwicklung *f (phot)* two-bath development
Zweischicht[en]film *m* double-coated film
Zweischneckenextruder *m* twin-screw extruder
Zweiseitenstoff *m (text)* double-face[d] fabric
zweiseitig *(pap)* two-sided
nicht z. non-two-sided
Zweiseitigkeit *f (pap)* two-sidedness, two-sided effect
Z. in der Färbung colour two-sidedness
Zweisiebpapiermaschine *f* twin-wire paper machine
Zweistoffdüse *f* two-fluid (gas-atomizing) nozzle
Zweistoffgemisch *n* binary mixture
Zweistofflegierung *f* binary alloy
Zweistoffsystem *n* two-component system, binary system
Zweistoffversprüher *m* two-fluid (auxiliary-fluid) atomizer
Zweistoffversprühung *f* two-fluid [nozzle] atomization, pneumatic nozzle atomization
Zweistoffzerstäuber *m s.* Zweistoffversprüher
Zweistoffzerstäubung *f s.* Zweistoffversprühung
Zweistrahlspektralfotometer *n* double-beam spectrophotometer
Zweistrahlspektrometer *n* double-beam spectrometer
zweisträngig double-strand *(e. g. conveyor)*
Zweistrangkette *f* double-strand chain *(conveying)*
Zweistufenbleiche *f (pap)* two-stage bleaching
Zweistufenmechanismus *m* two-step reaction mechanism
Zweistufenreaktion *f* two-step reaction
Zweistufenspritzgießen *n (plast)* double-shot moulding
Zweistufenverdampfer *m* double-effect evaporator
Zweistufenverfahren *n* two-stage (two-step) process; *(met)* duplex process
zweistufig two-stage, two-step, double-stage
Zweitbeschleuniger *m (rubber)* secondary accelerator
aktivierender Z. activating accelerator, booster
Zweitdestillation *f* redistillation, rerun[ning]
Zweitemperaturverfahren *n (nucl)* dual temperature process
Zweitischmaschine *f (glass)* two-table machine
Zweitkomponente *f (dye)* secondary component
Zweitluft *f* secondary air
Zweitsubstituent *m* second substituent
Zweiturmsystem *n (pap)* two-tower [acid] system, Jensen two-tower [acid-making] system
Zweiwalzen[feucht]kalander *m (pap)* two-roll calender, nip (intermediate) rolls

Zweiwalzenmühle *f* two-roll mill
Zweiwalzentrockner *m* double-drum dryer
Zweiwalzen-Walzwerk *n* two-roll mill
Zweiweghahn *m* two-way stopcock
Zweiwegmundstück *n* two-way outlet
zweiwertig divalent, bivalent; dihydric *(alcohol)*
Zweiwertigkeit *f* divalence, bivalence
zweizählig *(cryst)* diadic; *s.* zweizähnig
zweizähnig bidentate *(ligand)*
Zweizentrenbindung *f* two-centre bond *(chemical-bond theory)*
Zweizentrenmolekülorbital *n* two-centre molecular orbital
Zwiebel *f* 1. *(tech)* bulb; 2. *(glass)* onion, meniscus *(in vertical sheet drawing)*
Zwiebelhautpapier *n*, **Zwiebelschalenpapier** *n* onion skin
Zwilling *m* twin [crystal]
Zwillingsbildung *f (cryst)* twinning, twin formation
 polysynthetische Z. polysynthetic twinning
Zwillingskalorimeter *n* differential calorimeter
Zwillingskristall *m* twin [crystal]
Zwillingspumpe *f* duplex pump
Zwillingstunnelofen *m (ceram)* twin-tunnel kiln
Zwillingstunneltrockner *m (ceram)* twin-tunnel dryer
Zwillingsverbundofen *m* **nach Otto** Otto oven
Zwinge *f* clamp, clip
Zwischenabzweigstück *n* / **T-förmiges** T-shape connecting tube, T-piece, T-type connector, tee [connector]
 Y-förmiges Z. Y-shape connecting tube, Y-piece, Y-type connector
Zwischenanstrich *m* undercoat
Zwischenanstrichfarbe *f* undercoat material, undercoater
Zwischenbehälter *m* intermediate container; *(petrol)* engaging chamber *(in the pebble-heater process)*
Zwischenbehandlung *f* intermediate treatment
 alkalische Z. *(pap)* in-between alkali stage *(bleaching step)*
Zwischenboden *m* false bottom
Zwischenchelat *n* intermediate chelate
Zwischenchelatform *f* intermediate chelate form
Zwischenerhitzer *m* intermediate heater
Zwischenerzeugnis *n* intermediate [product]
Zwischenfaserbindung *f (pap)* interfibre bonding
Zwischenform *f* intermediate form
Zwischenfraktion *f* 1. *(distil)* intermediate fraction (cut); 2. intermediate material, middlings product *(classifying)*
Zwischengitteratom *n* interstitial [atom]
Zwischengitterion *n* interstitial ion
Zwischengitterpaar *n* interstitial pair
Zwischengitterplatz *m* interstitial [lattice] site, interstitial (interlattice) position
Zwischengitterplatzdiffusion *f* interstitial diffusion
Zwischenglasurdekor *n (ceram)* inter-glaze (in-glaze) decoration
Zwischenglied *n (tech)* link
Zwischenglühen *n* process (commercial) annealing *(of steel)*
Zwischengut *n* intermediate [product]
Zwischenkern *m* intermediate nucleus
Zwischenkomplex *m* intermediate complex
Zwischenkornvolumen *n* void volume
zwischenkristallin intercrystalline
Zwischenkühler *m* intercooler
Zwischenlage *f* intermediate layer, interlayer, interlining, ply
Zwischenlagebogen *m (pap)* set-off sheet, *(Am)* slip sheet
Zwischenlagepapier *n* set-off paper, tympan paper, *(Am)* slip-sheet paper
Zwischenlauf *m (distil)* intermediate cut
Zwischenläufer *m (rubber)* wrapper, leader
Zwischenlegepapier *n s.* Zwischenlagepapier
Zwischenleinen *n (rubber)* wrapper, leader
zwischenmolekular intermolecular
Zwischenphase *f* intermediate phase
Zwischenplatte *f (plast)* backing plate
 bewegliche Z. *(plast)* floating plate
Zwischenprodukt *n* intermediate [product]
 chemisches Z. intermediate chemical
Zwischenraum *m* interstice, spacing
Zwischenreaktion *f* intermediate reaction
Zwischenschicht *f* intermediate layer, interlayer, interlining, ply
Zwischenstadium *n* intermediate stage
Zwischenstoff *m* intermediate [substance, product]
Zwischenstoffreaktion *f* intermediate-product reaction
Zwischenstoffwechsel *m (bioch)* intermediary metabolism
Zwischenstück *n* spacer; adapter, connector, *(if tubular)* connecting tube
Zwischenstufe *f* intermediate stage
Zwischenstufengefüge *n (met)* bainite [structure], bainitic structure
Zwischenverbindung *f* intermediate [compound]
 Herzsche Z. *(dye)* [intermediate] Herz compound
Zwischenwand *f* partition [wall], *(if thin)* diaphragm; *(pap)* midfeather, mid-wall, midriff, centre division *(of a Hollander beater)*
Zwischenzahl *f (rubber)* tensile modulus
Zwischenzustand *m* intermediate (transitory) state (condition)
Zwitterion *n* zwitterion, dual (dipolar, hybrid, amphoteric, ampholyte) ion, amphion
zwitterionisch zwitterionic
Zwölfflächner *m (cryst)* dodecahedron
Zyan *n* $N{\equiv}C{-}C{\equiv}N$ cyanogen, oxalonitrile
Zyanamid *n* $NC{-}NH_2$ cyanamide, carbodiimide
Zyanamidverfahren *n* cyanamide process *(for producing ammonia)*
Zyanat *n* $M^{I}OCN$ cyanate
Zyanäthylierung *f* cyanoethylation
Zyanbad *n* cyanide [salt] bath *(for hardening steel)*
Zyanbadhärten *n* cyanide [case-]hardening, cyaniding
Zyanbenzol *n* C_6H_5CN cyanobenzene, benzonitrile

Zyanchlorid *n* ClCN chlorine cyanide, cyanogen chloride
Zyangruppe *f* CN− cyano group
Zyanhydrin *n* R′−C(OH)(CN)−R″ cyanohydrin
Zyanid *n* M^ICN cyanide
Zyanidbad *n* cyanide [salt] bath *(for hardening steel)*
Zyanidlaugung *f (met)* cyanidation, cyaniding
Zyanidsalzbad *n s.* Zyanidbad
zyanieren to cyanide *(steel)*
Zyanieren *n* cyaniding, cyanide [case-]hardening *(of steel)*
Zyanin *n* 1. cyanin *(an anthocyanin)*; 2. *s.* Zyaninfarbstoff
Zyaninfarbstoff *m* cyanine dye
Zyanit *m (min)* cyanite, kyanite, disthene *(aluminium oxide orthosilicate)*
Zyanjodid *n* ICN iodine cyanide, cyanogen iodide
Zyankali *n s.* Kaliumzyanid
Zyanoargentat *n* cyanoargentate
Zyanoaurat(I) *n* $M^I[Au(CN)_2]$ cyanoaurate(I), dicyanoaurate(I)
Zyanoaurat(III) *n* $M^I[Au(CN)_4]$ cyanoaurate(III), tetracyanoaurate(III)
Zyanochromat(II) *n* $M_4^I[Cr(CN)_6]$ cyanochromate(II), hexacyanochromate(II)
Zyanochromat(III) *n* $M_3^I[Cr(CN)_6]$ cyanochromate(III), hexacyanochromate(III)
Zyanoferrat(II) *n* $M_4^I[Fe(CN)_6]$ cyanoferrate(II), hexacyanoferrate(II)
Zyanoferrat(III) *n* $M_3^I[Fe(CN)_6]$ cyanoferrate(III), hexacyanoferrate(III)
Zyanoferrat(III)-komplex *m* cyanoferrate(III) complex, hexacyanoferrate(III) complex
Zyanoiridat *n* cyanoiridate
Zyanokobaltat(II) *n* $M_4^I[Co(CN)_6]$ cyanocobaltate(II), hexacyanocobaltate(II)
Zyanokobaltat(III) *n* $M_3^I[Co(CN)_6]$ cyanocobaltate(III), hexacyanocobaltate(III)
Zyanokuprat *n* cyanocuprate
Zyanomanganat(II) *n* $M_4^I[Mn(CN)_6]$ cyanomanganate(II), hexacyanomanganate(II)
Zyanomanganat(III) *n* $M_3^I[Mn(CN)_6]$ cyanomanganate(III), hexacyanomanganate(III)
Zyanomerkurat *n* cyanomercurate
Zyanomolybdat *n* cyanomolybdate
Zyanoniccolat *n* cyanoniccolate, cyanonickelate
Zyanoosmat(II) *n* $M_4^I[Os(CN)_6]$ cyanoosmate(II), hexacyanoosmate(II)
Zyanopalladat *n* cyanopalladate
Zyanoplatinat(II) *n* $M_2^I[Pt(CN)_4]$ cyanoplatinate(II), tetracyanoplatinate(II)
Zyanoplatinat(IV) *n* $M_4^I[Pt(CN)_6]$ cyanoplatinate(IV), hexacyanoplatinate(IV)
Zyanorhodat *n* cyanorhodate
Zyanoruthenat *n* cyanoruthenate
Zyanose *f (med)* cyanosis
Zyanotypie *f* 1. blue-printing, blueprint, ferroprussiate process; 2. cyanotype, blue print *(product)* ✦ **Zyanotypien herstellen** to blueprint
Zyanovanadat *n* cyanovanadate
Zyanowolframat(IV) *n* $M_4^I[W(CN)_8]$ cyanowolframate(IV), cyanotungstate(IV), octacyanowolframate(IV), octacyanotungstate(IV)
Zyanozinkat *n* cyanozincate
Zyanradikal *n* cyano radical, *(specif)* free cyano radical
Zyanrest *m s.* Zyangruppe
Zyansalzbad *n*, **Zyansalzschmelze** *f s.* Zyanidbad
Zyansäure *f* HOCN cyanic acid
Zyanurchlorid *n* cyanuric chloride
Zyanursäure *f* cyanuric acid
Zyanursäureamid *n* cyanuramide, triaminotriazine, melamine
Zyanursäurechlorid *n s.* Zyanurchlorid
Zyanwasserstoff *m* HCN hydrogen cyanide, hydrocyanic acid
zybotaktisch *(phys chem)* cybotactic
zyklisch cyclic
zyklisieren *(org chem)* to cyclize; *(rubber)* to cyclize *(under the influence of oxygen)*
Zyklisierung *f (org chem)* cyclization; *(rubber)* cyclization, oxygen vulcanization
Zyklisierungsreaktion *f* cyclization reaction
Zyklit *m* cyclitol *(an isocyclic polyalcohol)*
zykloaliphatisch cycloaliphatic, alicyclic
Zykloalkan *n* cycloalkane, cyclane, cycloparaffin, naphthene
Zykloalken *n* cycloalkene
Zykloalkin *n* cycloalkyne
Zyklobutan *n* cyclobutane
Zyklodehydratisierung *f* cyclodehydration
Zyklodehydrierung *f* cyclodehydrogenation
Zyklodekan *n* cyclodecane
Zykloheptan *n* cycloheptane
Zykloheptanol *n* cycloheptanol, suberyl alcohol
Zyklohexadien *n* cyclohexadiene, dihydrobenzene
Zyklohexadiendion-(1,4) *n* 1,4-cyclohexadienedione, *p*-benzoquinone
Zyklohexan *n* cyclohexane, hexahydrobenzene
Zyklohexan-1,2-dikarbonsäure *f* cyclohexane-1,2-dicarboxylic acid, hexahydrophthalic acid
Zyklohexanol *n* cyclohexanol, hexahydrophenol
Zyklohexanring *m* cyclohexane ring
Zyklohexen *n* cyclohexene, tetrahydrobenzene
Zykloheximid *n* cycloheximide, actidione *(antibiotic)*
Zyklokautschuk *m* cyclorubber, cyclized rubber
Zyklokautschuklatex *m* cyclized latex
Zyklokohlenwasserstoff *m* cyclic hydrocarbon
Zyklon[abscheider] *m* [centrifugal] cyclone separator, centrifugal (cyclonic) separator (collector), cyclone
Zyklonbatterie *f* multiple[-unit] cyclone
Zyklonentstauber *m* cyclone dust collector
Zyklonfeuerung *f* cyclone firing
Zyklonieren *n* cyclonic (cyclone) separation
Zyklon-Oberflächenverdampfer *m* cyclone evaporator
Zyklononan *n* cyclononane
Zyklonscheider *m s.* Zyklonabscheider
Zyklonskrubber *m*, **Zyklonwäscher** *m* cyclonic (cyclone) scrubber (washer)
Zyklooktan *n* cyclooctane

Zykloolefin *n* cycloolefin
Zykloparaffin *n s.* Zykloalkan
Zyklopentan *n* cyclopentane
Zyklopeptid *n* cyclic peptide
Zyklopropan *n* cyclopropane
Zyklopropandikarbonsäure *f* cyclopropanedicarboxylic acid
Zyklopropanring *m* cyclopropane ring
Zyklosilikat *n (min)* cyclosilicate, ring-silicate *(any of a class of polymeric silicates)*
Zyklosiloxan *n* cyclosiloxane, cyclic siloxane
Zyklotron *n* cyclotron
frequenzmoduliertes Z. frequency-modulated cyclotron, synchrocyclotron
Zyklus *m* cycle
Zylinder *m* cylinder
rotierender Z. rotating cylinder
zylinderförmig cylindrical
Zylindergruppe *f (pap)* dryer group (section)
Zylindermantel *m (pap)* dryer shell
Zylindermethode *f s.* Zylinderplattenmethode
Zylinderöl *n* cylinder [lubricating] oil
Zylinderölstock *m* cylinder stock
Zylinderplattenmethode *f* Oxford (cup) method *(for evaluating penicillin)*
Zylinder[rohr]schlange *f* helical coil
Zylinderschlichtmaschine *f (text)* cylinder sizing machine
Zylindertrockner *m (pap, text)* cylinder (can) dryer
Zylinderverfahren *n (glass)* cylinder process
Zylinderwalke *f (text)* rotary milling machine
Zylinderzapfen *m (pap)* dryer journal
Zylinderzelle *f* round cell *(electrolysis)*
Zylinderzellenapparat *m (petrol)* vertical tube sweating stove
Z. von Henderson Henderson stove
zylindrisch cylindrical
Zymase *f* zymase *(a complex of enzymes isolated from yeast)*
zymogen zymogenic, zymogenous
Zymogen *n (bioch)* zymogen
Zymol *n* $CH_3C_6H_3CH(CH_3)_2$ cymene, isopropyltoluene, isopropylmethylbenzene
Zymologie *f* zymology
Zymophenol *n* $(CH_3)_2C_6H_3(OH)CH_3$ cymophenol, carvacrol, 2-hydroxy-4-isopropyl-1-methylbenzene
Zymotechnik *f* zymotechnics
zymotechnisch zymotechnic[al]
zymotisch *s.* zymogen
Zypressenkampfer *m* cedrol, cedar [wood] camphor *(a sesquiterpenoid alcohol)*
Zystein *n* cysteine, 2-amino-3-mercaptopropionic acid
Zysteinsäure *f* $HO_3S-CH_2CH(NH_2)COOH$ cysteic acid, 2-amino-3-sulphopropionic acid
Zystin *n* cystine *(an amino acid)*
Zystin-Bindeglied *n*, **Zystin-Brücke** *f* cystine link
Zytase *f* cytase
Zytisin *n* cytisine *(a lupine alkaloid)*
Zytochemie *f* cytochemistry
Zytochrom *n* cytochrome
Zytochromoxydase *f* cytochrome (indophenol) oxidase, Warburg's respiratory enzyme
Zytochromreduktase *f* cytochrome reductase
Zytohormon *n* cell hormone
Zytokinin *n* cytokinin *(any of various plant growth factors)*
Zytoplasma *n* cytoplasm
Zytotoxin *n* cytotoxin
zytotoxisch cytotoxic